2014 19th Asia and South Pacific Design Automation Conference

(ASP-DAC 2014)

Singapore
20-23 January 2014

IEEE Catalog Number: CFP14ASP-POD
ISBN: 978-1-4799-2817-0

Copyright © 2014 by the Institute of Electrical and Electronic Engineers, Inc
All Rights Reserved

Copyright and Reprint Permissions: Abstracting is permitted with credit to the source. Libraries are permitted to photocopy beyond the limit of U.S. copyright law for private use of patrons those articles in this volume that carry a code at the bottom of the first page, provided the per-copy fee indicated in the code is paid through Copyright Clearance Center, 222 Rosewood Drive, Danvers, MA 01923.

For other copying, reprint or republication permission, write to IEEE Copyrights Manager, IEEE Service Center, 445 Hoes Lane, Piscataway, NJ 08854. All rights reserved.

***This publication is a representation of what appears in the IEEE Digital Libraries. Some format issues inherent in the e-media version may also appear in this print version.

IEEE Catalog Number: CFP14ASP-POD
ISBN 13:   978-1-4799-2817-0

Additional Copies of This Publication Are Available From:

Curran Associates, Inc
57 Morehouse Lane
Red Hook, NY 12571 USA
Phone: (845) 758-0400
Fax: (845) 758-2633
E-mail: curran@proceedings.com
Web: www.proceedings.com

2014 19th Asia and South Pacific Design Automation Conference (ASP-DAC)
Date: Jan 20-23, 2014
Place: Suntec, Singapore

Table of Contents

Technical Papers

No	Title	Page No
1	Normally-Off Computing Project : Challenges and Opportunities	1
2	Novel Nonvolatile Memory Hierarchies to Realize "Normally-Off Mobile Processors"	6
3	Normally-Off MCU Architecture for Low-Power Sensor Node	12
4	Normally-Off Technologies for Healthcare Appliance	17
5	A Dual-Loop Injection-Locked PLL with All-Digital Background Calibration System for On-Chip Clock Generation	21
6	A 950µW 5.5-GHz Low Voltage PLL with Digitally-Calibrated ILFD and Linearized Varactor	23
7	A Swing-Enhanced Current-Reuse Class-C VCO with Dynamic Bias Control Circuits	25
8	Design of A High-Performance Millimeter-Wave Amplifier Using Specific Modeling	27
9	A Multi-Mode Reconfigurable Analog Baseband with I/Q Calibration for GNSS Receivers	29
10	An 8b Extremely Area Efficient Threshold Configuring SAR ADC with Source Voltage Shifting Technique	31
11	A Single-Inductor 8-Channel Output DC-DC Boost Converter with Time-Limited Power Distribution Control and Single Shared Hysteresis Comparator	33

12	A DC-DC Boost Converter with Variation Tolerant MPPT Technique and Efficient ZCS Circuit for Thermoelectric Energy Harvesting Applications	35
13	7.3 Gb/s Universal BCH Encoder and Decoder for SSD Controllers	37
14	A High-Speed and Low-Complexity Lens Distortion Correction Processor for Wide-Angle Cameras	39
15	Analytical Placement of Mixed-Size Circuits for Better Detailed-Routability	41
16	Lithographic Defect Aware Placement Using Compact Standard Cells Without Inter-Cell Margin	47
17	Structural Planning of 3D-IC Interconnects by Block Alignment	53
18	Comprehensive Die-Level Assessment of Design Rules and Layouts	61
19	Prefetching Techniques for STT-RAM Based Last-Level Cache in CMP Systems	67
20	CNPUF: A Carbon Nanotube-based Physically Unclonable Function for Secure Low-Energy Hardware Design	73
21	3DCoB: A New Design Approach for Monolithic 3D Integrated Circuits	79
22	Emulator-Oriented Tiny Processors for Unreliable Post-Silicon Devices: A Case Study	85
23	Applying VLSI EDA to Energy Distribution System Design	91
24	A Model-Based Design of Cyber-Physical Energy Systems	97
25	The Data Center as a Grid Load Stabilizer	105
26	Bounding Buffer Space Requirements for Real-Time Priority-Aware Networks	113
27	Task- and Network-Level Schedule Co-Synthesis of Ethernet-Based Time-Triggered Systems	119
28	Service Adaptions for Mixed-Criticality Systems	125
29	Efficient Feasibility Analysis of DAG Scheduling with Real-Time Constraints in the Presence of Faults	131
30	Flexible Packed Stencil Design with Multiple Shaping Apertures for E-Beam Lithography	137
31	Self-Aligned Double Patterning Layout Decomposition with Complementary E-Beam Lithography	143
32	Fixing Double Patterning Violations with Look-Ahead	149
33	EUV-CDA: Pattern Shift Aware Critical Density Analysis for EUV Mask Layouts	155
34	Statistical Analysis of Random Telegraph Noise in Digital Circuits	161
35	Semi-Analytical Current Source Modeling of FinFET Devices Operating in Near/Sub-Threshold Regime with Independent Gate Control and Considering Process Variation	167
36	2-SAT Based Linear Time Optimum Two-Domain Clock Skew Scheduling	173
37	Power Minimization of Pipeline Architecture through 1-Cycle Error Correction and Voltage Scaling	179
38	A Silicon Nanodisk Array Structure Realizing Synaptic Response of Spiking Neuron Models with Noise	185
39	Energy Efficient In-Memory Machine Learning for Data Intensive Image-Processing by Non-Volatile Domain-Wall Memory	191
40	Lessons from the Neurons Themselves	197

41	Leveraging the Error Resilience of Machine-Learning Applications for Designing Highly Energy Efficient Accelerators	201
42	ArISE: Aging-Aware Instruction Set Encoding for Lifetime Improvement	207
43	DRuiD: Designing Reconfigurable Architectures with Decision-Making Support	213
44	Edit Distance Based Instruction Merging Technique to Improve Flexibility of Custom Instructions Toward Flexible Accelerator Design	219
45	A Network-Flow-Based Optimal Sample Preparation Algorithm for Digital Microfluidic Biochips	225
46	Exploring Speed and Energy Tradeoffs in Droplet Transport for Digital Microfluidic Biochips	231
47	General Purpose Cross-Referencing Microfluidic Biochip with Reduced Pin-Count	238
48	Wash Optimization for Cross-Contamination Removal in Flow-Based Microfluidic Biochips	244
49	ABCD-NL: Approximating Continuous Non-Linear Dynamical Systems Using Purely Boolean Models for Analog/Mixed-Signal Verification	250
50	Toward Efficient Programming of Reconfigurable Radio Frequency (RF) Receivers	256
51	Efficient Matrix Exponential Method Based on Extended Krylov Subspace for Transient Simulation of Large-Scale Linear Circuits	262
52	SDG2KPN: System Dependency Graph to Function-Level KPN Generation of Legacy Code for MPSoCs	267
53	Low Power Design of the Next-Generation High Efficiency Video Coding	274
54	Mapping Complex Algorithm into FPGA with High Level Synthesis	282
55	Leveraging Parallelism in the Presence of Control Flow on CGRAs	285
56	Physical-Aware Task Migration Algorithm for Dynamic Thermal Management of SMT Multi-Core Processors	292
57	Agile Frequency Scaling for Adaptive Power Allocation in Many-Core Systems Powered by Renewable Energy Sources	298
58	Variation Aware Voltage Island Formation for Power Efficient Near-Threshold Manycore Architectures	304
59	An Evaluation of an Energy Efficient Many-Core SoC with Parallelized Face Detection	311
60	Energy Aware Real-Time Scheduling Policy with Guaranteed Security Protection	317
61	A Comprehensive and Accurate Latency Model for Network-on-Chip Performance Analysis	323
62	A Low-Latency Asynchronous Interconnection Network with Early Arbitration Resolution	329
63	A Vertically Integrated and Interoperable Multi-Vendor Synthesis Flow for Predictable NoC Design in Nanoscale Technologies	337
64	Fuzzy Flow Regulation for Network-on-Chip Based Chip Multiprocessors Systems	343
65	Adjustable Contiguity of Run-Time Task Allocation in Networked Many-Core Systems	349
66	STD-TLB: A STT-RAM-Based Dynamically-Configurable Translation Lookaside Buffer for GPU Architectures	355

67	Training Itself: Mixed-Signal Training Acceleration for Memristor-Based Neural Network	361
68	HDTV1080p HEVC Intra Encoder with Source Texture Based CU/PU Mode Pre-decision	367
69	Fast Large-Scale Optimal Power Flow Analysis for Smart Grid through Network Reduction	373
70	Storage-Less and Converter-Less Maximum Power Point Tracking of Photovoltaic Cells for a Nonvolatile Microprocessor	379
71	Soft Error Resiliency Characterization on IBM BlueGene/Q Processor	385
72	Resiliency for Many-Core System on a Chip	388
73	Rethinking Error Injection for Effective Resilience	390
74	Amphisbaena: Modeling Two Orthogonal Ways to Hunt on Heterogeneous Many-Cores	394
75	Co-Simulation Framework for Streamlining Microprocessor Development on Standard ASIC Design Flow	400
76	Annotation and Analysis Combined Cache Modeling for Native Simulation	406
77	A Scorchingly Fast FPGA-Based Precise L1 LRU Cache Simulator	412
78	Redundant-Via-Aware ECO Routing	418
79	A Fast and Provably Bounded Failure Analysis of Memory Circuits in High Dimensions	424
80	Predicting Circuit Aging Using Ring Oscillators	430
81	Statistical Analysis of Process Variation Based on Indirect Measurements for Electronic System Design	436
82	Symbolic Computation of SNR for Variational Analysis of Sigma-Delta Modulator	443
83	Sparse Statistical Model Inference for Analog Circuits under Process Variations	449
84	Time-Domain Performance Bound Analysis for Analog and Interconnect Circuits Considering Process Variations	455
85	A Robustness Optimization of SRAM Dynamic Stability by Sensitivity-Based Reachability Analysis	461
86	Accurate and Inexpensive Performance Monitoring for Variability-Aware Systems	467
87	Quantifying Workload Dependent Reliability in Embedded Processors	474
88	QED Post-Silicon Validation and Debug: Frequently Asked Questions	478
89	Efficient Synthesis of Quantum Circuits Implementing Clifford Group Operations	483
90	Optimal SWAP Gate Insertion for Nearest Neighbor Quantum Circuits	489
91	Qubit Placement to Minimize Communication Overhead in 2D Quantum Architectures	495
92	A Novel Wirelength-Driven Packing Algorithm for FPGAs with Adaptive Logic Modules	501
93	A Topology-Based ECO Routing Methodology for Mask Cost Minimization	507
94	BOB-Router: A New Buffering-Aware Global Router with Over-the-Block Routing Resources Optimization	513
95	Routability-Driven Bump Assignment for Chip-Package Co-Design	519

96	VFGR: A Very Fast Parallel Global Router with Accurate Congestion Modeling	525
97	Efficient Simulation-Based Optimization of Power Grid with On-Chip Voltage Regulator	531
98	Walking Pads: Fast Power-Supply Pad-Placement Optimization	537
99	Power Supply Noise-Aware Workload Assignments for Homogenous 3D MPSoCs with Thermal Consideration	544
100	SwimmingLane: A Composite Approach to Mitigate Voltage Droop Effects in 3D Power Delivery Network	550
101	Spiking Brain Models: Computation, Memory and Communication Constraints for Custom Hardware Implementation	556
102	Advanced Technologies for Brain-Inspired Computing	563
103	GPGPU Accelerated Simulation and Parameter Tuning for Neuromorphic Applications	570
104	A Scalable Custom Simulation Machine for the Bayesian Confidence Propagation Neural Network Model of the Brain	578
105	NoΔ :Leveraging Delta Compression for End-to-End Memory Access in NoC Based Multicores	586
106	DPA: A Data Pattern Aware Error Prevention Technique for NAND Flash Lifetime Extension	592
107	Scattered Refresh: An Alternative Refresh Mechanism to Reduce Refresh Cycle Time	598
108	A Read-Write Aware DRAM Scheduling for Power Reduction in Multi-Core Systems	604
109	A Coherent Hybrid SRAM and STT-RAM L1 Cache Architecture for Shared Memory Multicores	610
110	Allocation of FPGA DSP-Macros in Multi-Process High-Level Synthesis Systems	616
111	Array Scalarization in High Level Synthesis	622
112	Data Compression via Logic Synthesis	628
113	Synthesis of Power- and Area-Efficient Binary Machines for Incompletely Specified Sequences	634
114	Multi-Mode Trace Signal Selection for Post-Silicon Debug	640
115	Implicit Intermittent Fault Detection in Distributed Systems	646
116	A Segmentation-Based BISR Scheme	652
117	Fault-Tolerant TSV by Using Scan-Chain Test TSV	658
118	Suppressing Test Inflation in Shared-Memory Parallel Automatic Test Pattern Generation	664
119	A Volume Diagnosis Method for Identifying Systematic Faults in Lower-Yield Wafer Occurring during Mass Production	670
120	An Overview of Spin-Based Integrated Circuits	676
121	Advances in Spintronics Devices for Microelectronics - from Spin-Transfer Torque to Spin-Orbit Torque	684
122	Hybrid CMOS/Magnetic Process Design Kit and SOT-Based Non-Volatile Standard Cell Architectures	692
123	Architectural Aspects in Design and Analysis of SOT-Based Memories	700
124	Timing Anomalies in Multi-Core Architectures due to the Interference on the Shared Resources	708
125	A Unified Online Directed Acyclic Graph Flow Manager for Multicore Schedulers	714

126	Variation-Aware Statistical Energy Optimization on Voltage-Frequency Island Based MPSoCs under Performance Yield Constraints	720
127	QoS-Aware Dynamic Resource Allocation for Spatial-Multitasking GPUs	726
128	Automated Debugging of Missing Assumptions	732
129	Property Directed Reachability for QF_BV with Mixed Type Atomic Reasoning Units	738
130	Adaptive Interpolation-Based Model Checking	744
131	Efficient Parallel GPU Algorithms for BDD Manipulation	750
132	Efficient Techniques for the Capacitance Extraction of Chip-Scale VLSI Interconnects Using Floating Random Walk Algorithm	756
133	3DLAT: TSV-Based 3D ICs Crosstalk Minimization Utilizing Less Adjacent Transition Code	762
134	Tackling Close-to-Band Passivity Violations in Passive Macro-Modeling	768
135	HIE-Block Latency Insertion Method for Fast Transient Simulation of Nonuniform Multiconductor Transmission Lines	774
136	The Role of Photons in Cryptanalysis	780
137	SPADs for Quantum Random Number Generators and Beyond	788
138	Quantum Key Distribution with Integrated Optics	795
139	Constraint-Based Platform Variants Specification for Early System Verification	800
140	A Transaction-Oriented UVM-Based Library for Verification of Analog Behavior	806
141	Automata-Theoretic Modeling of Fixed-Priority Non-Preemptive Scheduling for Formal Timing Verification	812
142	PROCEED: A Pareto Optimization-Based Circuit-Level Evaluator for Emerging Devices	818
143	Modeling and Design Analysis of 3D Vertical Resistive Memory - A Low Cost Cross-Point Architecture	825
144	The Stochastic Modeling of TiO2 Memristor and Its Usage in Neuromorphic System Design	831
145	Through-Silicon-Via Inductor: Is It Real or Just A Fantasy?	837
146	Design and Control Methodology for Fine Grain Power Gating Based on Energy Characterization and Code Profiling of Microprocessors	843
147	A Hybrid Random Walk Algorithm for 3-D Thermal Analysis of Integrated Circuits	849
148	LightSim : A Leakage Aware Ultrafast Temperature Simulator	855
149	Fast Vectorless Power Grid Verification Using Maximum Voltage Drop Location Estimation	861

Call for Designs

University LSI Design Contest
ASP-DAC 2015
http://www.aspdac.com/aspdac2015/
January 19-22, 2015
Japan

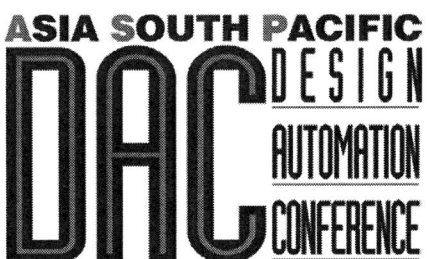

Aims of the Contest:

As a unique feature of ASP-DAC 2015, the University LSI Design Contest will be held. The aim of the Contest is to encourage education and research on VLSI design at universities and other educational organizations. We solicit designs that fit in one or more of the following categories:

(1) Designed, and actually implemented on chips in universities or other educational organizations during the last two years;
(2) Designs that report actual measurements from implementations;
(3) Innovative design prototypes.

Interesting or excellent designs selected will be honored by providing the opportunities for presentation in a special session at the conference. Award(s) will be given to a few numbers of outstanding designs, selected from those presented at the conference.

Areas of Design:

Application areas or types of circuits of the original LSI circuit designs include (but are not limited to):
    (1) Analog, RF and Mixed-Signal Circuits, (2) Digital Signal Processing, (3) Microprocessors, (4) Custom ASIC.
Methods or technology used for implementation include:
    (a) Full Custom and Cell-Based LSIs, (b) Gate Arrays, (c) FPGA/PLDs.

Submission of Design Descriptions:

A camera-ready summary is requested to be prepared within 2 pages including figures, tables, and references. It is strongly recommended that measured experimental results and a chip micrograph are included in the summary. Please do not submit the same paper as a regular paper.

Specification of the submission format will be available at **http://www.aspdac.com/aspdac2015/**

Deadline for summary:	5PM JST (UTC+9) July 11 (Fri.), 2014
Notification of acceptance:	Sep. 15 (Mon.), 2014
Deadline for camera-ready:	5PM JST (UTC+9) Nov. 10 (Mon.), 2014

Review:

Submitted designs will be reviewed by the Design Contest Committee in a process similar to the review process for the technical papers. The following criteria will be applied in the selection of designs:
    (1) Reliability of design and implementation, (2) Quality of implementation, (3) Performance of the design,
    (4) Novelty of application, algorithm, architecture, (5) Others.
Interesting or excellent designs selected will be presented at a special session of the conference.

Presentation:

An author of each selected design will be required to make a short presentation at a special session of ASP-DAC 2015. A digest of each design to be presented will be included in the conference proceedings.

Contact Email: aspdac2015-udc@mls.aspdac.com

ASP-DAC 2015 Chairs

General Chair:	**Kunio Uchiyama (Hitachi)**
Technical Program Chair:	**Naehyuck Chang (Seoul National University)**
Design Contest Co-Chairs:	**Hiroyuki Ito (Tokyo Institute of Technology)**
	Noriyuki Miura (Kobe University)

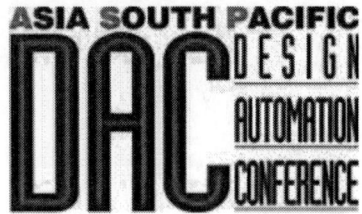

Call for Papers
ASP-DAC 2015
http://www.aspdac.com/aspdac2015/
January 19-22, 2015
Japan

ASP-DAC 2015 is the 20th annual international conference on VLSI design automation in Asia and South Pacific regions, one of the most active regions of design and fabrication of silicon chips in the world. The conference aims at providing the Asian and South Pacific CAD/DA and Design community with opportunities of presenting recent advances and with forums for future directions in technologies related to Electronic Design Automation (EDA). The format of the meeting intends to cultivate and promote an instructive and productive interchange of ideas among EDA researchers/developers and system/circuit/device designers. All scientists, engineers, and students who are interested in theoretical and practical aspects of VLSI design and design automation are welcomed to ASP-DAC.

Areas of Interest:
Original papers in, but not limited to, the following areas are invited.

Paper Submission Deadline:
July 11, 2014, 5:00 PM JST (UTC +09:00)

[1] **System-Level Modeling, Simulation and Verification:**
- System-level modeling, specification, language, etc.
- Performance analysis of architectures and systems
- System-level simulation and verification
- HW-SW co-simulation/co-verification
- Formal verification
- Transaction-level modeling and validation
- RTL/gate-level modeling and validation
- Logic and symbolic simulation

[2] **Interconnect/Device/Circuit/Gate Modeling, Simulation and Verification:**
- Clock and bus analysis
- Interconnect and substrate modeling/extraction
- Device modeling and simulation
- Circuit simulation

[3] **System-Level Architecture and Design Methodologies:**
- SoC and MPSoC design methodology
- HW-SW co-design (partitioning, synthesis and scheduling)
- IP/platform-based design
- Application-specific instruction-set processor (ASIP) system design
- Reconfigurable architectures and systems

[4] **Power and Thermal Modeling, Simulation and Optimization/Management:**
- Device- and circuit-level power and thermal modeling, simulation and estimation
- Device- and circuit-level power and thermal management
- Cross-layer power and thermal modeling, simulation and estimation
- Cross-layer power and thermal management
- Domain-specific power and thermal management
- Aging-aware design and reliability

[5] **On-chip and System-level Communication Design, I/O, Networks on Chip, and Memory Systems:**
- Communication-based architecture design
- Network-on-chip (NoC) design methodologies
- Interface and I/O design, synthesis and optimization
- System communication architecture
- Architecture and compiler techniques for memory systems

[6] **Embedded Systems:**
- Embedded software (OS, middleware and compilation)
- Real-time system design
- Advanced storage systems and applications
- Compilation techniques
- Human-computer interface design
- Safety-critical and secure systems

[7] **Logic/Behavioral/High-Level Synthesis and Optimization:**
- High-level/behavioral/RTL synthesis
- Technology-independent optimization
- Technology mapping

- Interaction between logic design and layout
- Sequential and asynchronous logic synthesis
- Resource scheduling, allocation, synthesis

[8] **Physical Design:**
- Floorplanning, partitioning and placement
- Buffer insertion and interconnect planning
- Post-place optimization and routing
- Clock network synthesis
- Post-routing optimization and layout verification
- High-level physical design and synthesis
- Package/PCB routing
- Gate sizing and cell library design

[9] **Timing and Signal/Power Integrity Analysis and Verification:**
- Deterministic timing and performance analysis and optimization
- Statistical timing and performance analysis and optimization
- Power/ground and package modeling, analysis and optimization
- Signal and power integrity

[10] **Design for Manufacturability, Yield and Statistical Design:**
- DFM, DFY, CAD support for OPC and RET
- Variability analysis, yield analysis and optimization
- Reliability analysis, design for resilience and robustness
- Cell library design

[11] **Test and Design for Testability:**
- Testable design and fault modeling
- ATPG, BIST and DFT
- Memory test and repair
- Core and system test
- Delay test
- Analog and mixed signal test

[12] **Analog, RF and Mixed Signal Design and CAD:**
- Analog/RF synthesis
- Analog layout, verification and simulation techniques
- Noise analysis
- High-frequency electromagnetic simulation of circuits
- Mixed-signal design considerations

[13] **CAD and Design Methodologies for Emerging Technologies:**
- Nanotechnology CAD and design methodologies
- Quantum computing and CAD
- Emerging memory and logic technologies
- 3D integration
- On-chip wireless and optic communication and applications

[14] **CAD for Emerging Applications and Cyber-Physical Systems:**
- Cyber-physical systems
- Biological, bioelectronics, and biomedical systems
- Automotive applications
- Energy system design and optimization

ACM, IEEE, and IEICE reserve the right to exclude a paper from distribution after the conference (e.g., removal from ACM Digital Library and IEEE Xplore) if the paper is not presented at the conference by the author of the paper. ASP-DAC does not allow double submissions and parallel submissions of similar work to any other conferences and symposia as well as journals and transactions in the field.

Submission of Papers:

Deadline for submission:	5 PM JST (UTC+9)	July 11 (Fri), 2014
Notification of acceptance:		Sep. 15 (Mon), 2014
Deadline for final version:	5 PM JST (UTC+9)	Nov. 10 (Mon), 2014

For detailed instructions for submission, please refer to the "Authors' Guide" at:
http://www.aspdac.com/aspdac2015/

Panels, Special Sessions, and Tutorials:
Suggestions and proposals are welcome and have to be addressed to the Conference Secretariat (aspdac2015-sec@mls.aspdac.com) no later than May 30 (Fri), 2014.

Prospective Sponsors:
ACM SIGDA, IEEE CASS, IEEE CEDA, IEICE ESS, IPSJ SIGSLDM

ASP-DAC2015 Chairs:
General Chair:
Kunio Uchiyama (Hitachi, Japan)
Technical Program Chair:
Naehyuck Chang (Seoul National University, Korea)
Technical Program Vice Chairs:
TingTing Hwang (National Tsing Hua University, Taiwan)
Yasuhiro Takashima (University of Kitakyushu, Japan)

Contact: Conference Secretariat: aspdac2015-sec@mls.aspdac.com   TPC Secretariat: aspdac2015-tpc@mls.aspdac.com

Welcome to ASP-DAC 2014

On behalf of the ASP-DAC 2014 Organizing Committee, we would like to invite all colleagues from academia and industry working on the LSI design and design automation areas to the 19th Asia and South Pacific Design Automation Conference (ASP-DAC 2014). ASP-DAC 2014 will be held from 20th January (Monday) to 23rd January (Thursday), 2014 at Suntec City Singapore.

Even for frequent visitors to Singapore, this island city country is keeping on transforming its landscape and attracting visitors with new wonders. The conference site is Suntec City, which is designed to be a 'city within a city'. It is the single largest integrated commercial development in Singapore with an international convention and exhibition centre, a shopping mall, five office towers, and a fountain all connected to each other by street level plazas, walkways and courtyards. The conference banquet will be in the nearby Flower Field Hall of Gardens by The Bay, a newly developed oasis where over 250,000 rare plants greet you in full bloom. We believe you will enjoy your stay in Singapore, a truly world-class modern city in Southeast Asia with over 5 million people and diverse culture like no other.

ASP-DAC 2014 attracted 343 submissions from 29 countries from our worldwide colleagues in academic, industry and government institutions. Under the leadership of Technical Program Co-Chairs, Nagisa Ishiura, Naehyuck Chang, and Tulika Mitra, the Technical Program Committee members conducted rigorous and thorough reviews and a full-day face-to-face meeting to select excellent papers for the technical program of ASP-DAC 2014. 108 papers have been accepted for regular presentation that cover key topics from system design to physical design. 9 Special Sessions have also been organized based on invited talks by the Technical Program Committee to discuss up-to-date topics.

We are happy to report that we have invited 4 distinguished keynote speakers from both academia and industry to discuss topics with programmable platform, analog, digital, and local focuses. The first keynote speaker Dr. Ivo Bolsens is Senior Vice President and CTO from Xilinx, USA. He will give a talk on how programmable platforms can contribute to big data applications. The second keynote speaker Prof. Georges Gielen is from Katholieke Universiteit Leuven, Belgium. He will give a talk on the building analog functions without analog transistors. The third Keynote speaker Prof. Kaushik Roy is from Purdue University. He will give a talk on the usage of spin instead of charge as state variable to achieve high density memory and ultra-low voltage/power logic. The fourth keynote speaker Mr. Ulf Schneider is Managing Director from Lantiq Asia Pacific. He is currently also serving as the President of Singapore Semiconductor Industry Association (SSIA). He will give us the stories of past, current and future Singapore Semiconductor industry.

Eight tutorials have been arranged on 20th January (Monday), 2014. This year, we have changed the tutorial style: instead of full-day, in-depth tutorials, participants can choose two 3-hour tutorials – one in the morning session and the other in the afternoon. For each session, four options are available – two in the physical-design (PD) domain and two in the system-design (SD) domain. In addition, as an important annual event of ASP-DAC, 10 designs were selected by the University Design Contest for presentation on 21st January (Tuesday), 2014.

ASP-DAC 2014 offers you an ideal opportunity to touch the recent technologies and the future directions on the LSI design and design automation areas. The success of ASP-DAC 2014 is indebted to the support of authors and the Organizing Committee members. Without authors' contributions, we would not have the opportunity to form the excellent technical program of ASP-DAC 2014. Every member of the Organizing Committee have devoted countless volunteering hours and days to ensure the success of conference.

Thank you very much for coming to Singapore for ASP-DAC 2014. We extend a warm welcome to all participants!

Yong Lian, Yajun Ha
General Co-Chairs, ASP-DAC 2014

Message from Technical Program Committee

On behalf of the Technical Program Committee of the Asia and South Pacific Design Automation Conference (ASP-DAC) 2014, we would like to welcome all of you to the conference scheduled from January 20 to 23, 2014 at Suntec City, Singapore.

This year, we received 343 submissions from 29 countries/regions, with the majority of them from Asia, North America, and Europe. Paper selection was really a challenge.

To support the selection process, we organized the Technical Program Committee consisting of 110 leading experts on EDA, IC design, and system design, who are from 15 countries/regions. The TPC was organized into 15 subcommittees. All committee members contributed to in-depth and thorough reviews, through a rigorous double-blind review process that involved vigorous discussions. Through a full day face-to-face discussion at the TPC Meeting held on September 2, 2013, at Kyoto Research Park in Japan, 108 high-quality papers were accepted, resulting in a very competitive acceptance rate of 31.5%.

Along with the selection of the regular papers, invitation of keynote speeches and special sessions were done. Then all the presentations were compiled into a three-day, four parallel-session program.

Each day, technical session starts with a keynote address. This year, we have also an extra keynote speech during the banquet. We have 9 special sessions on Track S (1S through 9S), which consists of invited talks on the state-of-the-art topics, including EDA and methodologies for ultra-large scale, ultra-low power, and high reliability design, emerging technologies and applications such as magnetoresistive memories, brain and neuron inspired computing, quantum devices, and EDA for energy. On the first day, we have a University LSI Design Contest session (1A). Regular papers are presented in 26 sessions on tracks A, B, and C.

Among the accepted regular papers, 13 were nominated for the Best Paper from each subcommittee. These Best Paper candidates went through a thorough evaluation process by the Best Paper Award Committee composed of 15 TPC members, and finally one of the candidate papers was selected for the ASP-DAC 2014 Best Paper Award.

The Technical Program of ASP-DAC 2014 is the fruit of the hard work of many people. We would like to thank all the people who have contributed to the technical program. In particular, we thank all the authors who submitted excellent papers that continue to make ASP-DAC a very vibrant conference and a premier forum for exchanging ideas and results. We would also like to thank TPC Secretaries and TPC members for their hard work. Finally, we also would like to thank the members of the Organizing Committee for their excellent services.

We hope that you will enjoy the ASP-DAC 2014 technical program.

Nagisa Ishiura
TPC Chair, ASP-DAC 2014

Naehyuck Chang
TPC Vice Chair, ASP-DAC 2014

Tulika Mitra
TPC Vice Chair, ASP-DAC 2014

University LSI Design Contest

The University LSI Design Contest has been conceived as a unique program at ASP-DAC. The purpose of the contest is to encourage research in LSI design at universities and its realization on a chip by providing opportunities to present and discuss the innovative and state-of-the-art design. The scope of the contest covers circuit techniques for (1) Analog / RF / Mixed-Signal Circuits, (2) Digital Signal Processer, (3) Microprocessors, (4) Custom Application Specific Circuits / Memories, and methodologies for (a) Full-Custom / Cell-Based LSIs, (b) Gate Arrays, (c) Field Programmable Devices.

This year, the University LSI Design Contest Committee received 19 designs from five countries/areas, and selected 10 designs out of them. The selected designs will be disclosed in Session 1A with four-minute presentations, followed by live discussions in front of their posters. For the two outstanding designs, The Best Design Award and The Special Feature Award will be presented in the banquet. We sincerely acknowledge the other contributions to the contest, too. It is our earnest belief to promote and enhance research and education in LSI design in academic organizations. Please come to the University LSI Design Contest and enjoy the stimulating discussions.

Last but not least, we would like to express our sincere gratitude to the Council on Electronic Design Automation (CEDA) for their generous sponsorship of this design contest. We would also like to thank all committee members within the UDC review panel for their efforts in reviewing and selecting the papers.

Date: Tuesday, January 21, 2014, 10:40 – 12:20
Location: Suntec City, 3rd level floor

Oral Presentation Room: Room 300
Poster Presentation Room: Room 304

University LSI Design Contest Committee Chair

Chun-Huat Heng
(National Univ. of Singapore, Singapore)

Keynote Addresses

Opening & Keynote I
Tuesday, January 21, 08:30 – 10:00

"All Programmable SOC FPGA for Networking and Computing in Big Data Infrastructure"

Dr. Ivo Bolsens
Senior VP and CTO, Xilinx, U.S.A.

Abstract: Today's FPGAs have become 'All Programmable SOC Platforms' that integrate in one single device multi-core CPU's, programmable DSP functions, programmable IO and programmable logic, all immersed in a rich and configurable interconnect network. These programmable platform FPGA's allow for the implementation of heterogeneous multi-core architectures that combine traditional CPU's with application-specific processing cores and dedicated data transfer and storage functions. This is enabled by tools that guide designers during the partitioning and mapping of high-level specifications onto a combination of software running on embedded processors and hardware implemented in programmable logic.

FPGAs are well placed to continue to benefit from Moore's law. Advances in process scaling will be augmented with new circuit and architectural improvements along with innovations in system-in-package technology to solve IO challenges and integrate heterogeneous technologies. These innovations will allow designers to build higher performance and lower power systems that optimally exploit the programmable FGPA architecture.

As FPGA platforms continue to deliver more performance at lower cost and lower power, they are becoming the heart of embedded applications such as complex packet processing for networks with line rates of 400+ Gbps; high performance digital signal processing in novel wireless baseband and radio functions; and high flexibility to enable programmable networking and data storage functions in cloud infrastructure.

Keynote II
Wednesday, January 22, 08:30 – 09:30

"Designing Analog Functions without Analog Transistors"

Prof. Georges Gielen
Katholieke Universiteit Leuven, Belgium

Abstract: Analog functions are indispensable for most electronic applications, ranging from telecom to biomedical or automotive applications. Yet, designing the analog circuits has become a large burden, especially in advanced CMOS technologies where reduced voltage headrooms and increased variability and reliability problems challenge the design of power-efficient analog circuits. Together with the lack of adequate EDA tools this also jeopardizes efficient analog circuit design. This keynote describes a possible way forward. The industry clearly has reached a bifurcation point. Many applications will leave the scaling race, and adopt older or nonstandard (e.g. flexible organic) technologies for the analog circuits, offering the increased functionality essentially through heterogeneous integration. Many other applications will stick to advanced CMOS, but will shift the analog design paradigm from analog-heavy to digital-heavy minimalistic-analog circuits. The presentation will discuss and illustrate the challenges and solutions in such approach to design analog functions without analog transistors.

Keynote III
Thursday, January 23, 08:30 – 09:30

"Beyond Charge-Based Computing"

Prof. Kaushik Roy
Purdue Univ., U.S.A.

Abstract: The trend towards ultra low power logic and low leakage embedded memories for System-On-Chips, has prompted researcher to consider the possibility of replacing charge as the state variable for computation. Recent experiments on spin devices like magnetic tunnel junctions (MTJ's), domain wall magnets (DWM) and spin valves have led to the possibility of using "spin" as state variable for computation, achieving very high density on-chip memories and ultra low voltage logic. High density of memories can be exploited to develop memory-centric reconfigurable computing fabrics that provide significant improvements in energy efficiency and reliability compared to conventional FPGAs. While the possibility of having on-chip spin transfer torque memories is close to reality, several questions still exist regarding the energy benefits of spin as the state variable for logic computation. Latest experiments on lateral spin valves (LSV) have shown switching of nano-magnets using spin-polarized current injection through a metallic channel such as Cu. Such lateral spin valves having multiple input magnets connected to an output magnet using metal channels can be used to mimic "neurons". The spin-based neurons can be integrated with CMOS and other devices like Phase change memories to realize ultra low-power data processing hardware based on neural networks, and are suitable for different classes of applications like, cognitive computing, programmable Boolean logic and analog and digital signal processing. Note, for some of these applications, CMOS technologies may not be suitable for ultra low power implementation. In this talk I will first discuss the advantages of using spin (as opposed to charge) as state variable for both memory and logic and then present how a cellular array of magneto-metallic devices, operating at terminal voltages ~20mV, can do efficient hybrid digital/analog computation for applications such as cognitive computing. Finally, I will consider recent advances in other non-charge based computing paradigm such as magnetic quantum cellular automata.

Banquet Keynote
Wednesday, January 22, 18:30 – 21:30

"The Art of Innovation - How Singapore Will Continue to Drive the Progress in Semiconductor Technologies"

Mr. Ulf Schneider
Managing Director, Lantiq Asia Pacific/President, SSIA, Singapore

Abstract: Since the mid 1960's Singapore has been an important pillar of the worldwide semiconductor industry, reinventing its portfolio, focus and strategy a few times to keep up with overall trends. Preparing for the next decade, Singapore's industry, research and academia has to put up again the right directions and strategy to keep up with the pace in a more and more competitive global environment. The talk will cover some of the really unique opportunities which Singapore has in this aspect.

Best Paper Award

Award Winner

2B-1: **"Flexible Packed Stencil Design with Multiple Shaping Apertures for E-Beam Lithography"**
Chris Chu (Iowa State Univ., U.S.A.), Wai-Kei Mak (National Tsing Hua Univ., Taiwan)

Candidates

1B-1: **"Analytical Placement of Mixed-Size Circuits for Better Detailed-Routability"**
Shuai Li, Cheng-Kok Koh (Purdue Univ., U.S.A.)

1C-1: **"Prefetching Techniques for STT-RAM Based Last-Level Cache in CMP Systems"**
Mengjie Mao (Univ. of Pittsburgh, U.S.A.), Guangyu Sun (Peking Univ., China), Yong Li, Alex K. Jones, Yiran Chen (Univ. of Pittsburgh, U.S.A.)

2A-1: **"Bounding Buffer Space Requirements for Real-Time Priority-Aware Networks"**
Hany Kashif, Hiren D. Patel (Univ. of Waterloo, Canada)

2C-1: **"Statistical Analysis of Random Telegraph Noise in Digital Circuits"**
Xiaoming Chen, Yu Wang (Tsinghua Univ., China), Yu Cao (Arizona State Univ., U.S.A.), Huazhong Yang (Tsinghua Univ., China)

3B-1: **"A Network-Flow-Based Optimal Sample Preparation Algorithm for Digital Microfluidic Biochips"**
Trung Anh Dinh, Shigeru Yamashita (Ritsumeikan Univ., Japan), Tsung-Yi Ho (National Cheng Kung Univ., Taiwan)

4A-1: **"Physical-Aware Task Migration Algorithm for Dynamic Thermal Management of SMT Multi-Core Processors"**
Bagher Salami (Ferdowsi Univ. of Mashhad, Iran), Mohammadreza Baharani (Univ. of Tehran, Iran), Hamid Noori (Ferdowsi Univ. of Mashhad, Iran), Farhad Mehdipour (Kyushu Univ., Japan)

5B-1: **"Redundant-Via-Aware ECO Routing"**
Hsi-An Chien, Ting-Chi Wang (National Tsing Hua Univ., Taiwan)

5C-1: **"Symbolic Computation of SNR for Variational Analysis of Sigma-Delta Modulator"**
Jiandong Cheng, Guoyong Shi (Shanghai Jiao Tong Univ., China)

6A-1: **"Efficient Synthesis of Quantum Circuits Implementing Clifford Group Operations"**
Philipp Niemann (Univ. of Bremen, Germany), Robert Wille (Univ. of Bremen/Cyber Physical Systems DFKI GmbH/Technical Univ. Dresden, Germany), Rolf Drechsler (Univ. of Bremen/Cyber Physical Systems DFKI GmbH, Germany)

6C-2: **"Walking Pads: Fast Power-Supply Pad-Placement Optimization"**
Ke Wang (Univ. of Virginia, U.S.A.), Brett Meyer (McGill Univ., Canada), Runjie Zhang, Kevin Skadron, Mircea Stan (Univ. of Virginia, U.S.A.)

7A-1: **"NoΔ: Leveraging Delta Compression for End-to-End Memory Access in NoC Based Multicores"**
Jia Zhan, Matt Poremba (Pennsylvania State Univ., U.S.A.), Yi Xu (AMD Research, China), Yuan Xie (AMD, China/Pennsylvania State Univ., U.S.A.)

7C-4: **"Suppressing Test Inflation in Shared-Memory Parallel Automatic Test Pattern Generation"**
Jerry C. Y. Ku, Ryan H.-M. Huang, Louis Y. -Z. Lin, Charles H.-P. Wen (National Chiao Tung Univ., Taiwan)

University LSI Design Contest Awards

Best Design Award

1A-1: **"A Dual-Loop Injection-Locked PLL with All-Digital Background Calibration System for On-chip Clock Generation"**

Wei Deng, Ahmed Musa, Teerachot Siriburanon, Masaya Miyahara, Kenichi Okada, Akira Matsuzawa (Tokyo Institute of Technology, Japan)

Special Feature Award

1A-7: **"A Single-Inductor 8-Channel Output DC-DC Boost Converter with Time-Limited Power Distribution Control and Single Shared Hysteresis Comparator"**

Jungmoon Kim, Chulwoo Kim (Korea Univ., Republic of Korea)

10-Year Retrospective Most Influential Paper Award

Award Winner

(ASP-DAC 2004)

2E-2: **"Design Diagnosis Using Boolean Satisfiability"**
Alexander Smith, Andreas Veneris, Anastasios Viglas (Univ. of Toronto, Canada)

Candidates

1D-1: **"Register Binding and Port Assignment for Multiplexer Optimization"**

Deming Chen, Jason Cong (Univ. of California, Los Angeles, USA)

1E-1: **"TranGen: A SAT-Based ATPG for Path-Oriented Transition Faults"**

Kai Yang, Kwang-Ting Cheng, Li-C. Wang (Univ. of California, Santa Barbara, U.S.A.)

3D-1: **"Efficient Translation of Boolean Formulas to CNF in Formal Verification of Microprocessors"**

Miroslav N. Velev (Carnegie Mellon Univ., U.S.A.)

Organizing Committee

General Co-Chairs	**Yong Lian** (National Univ. of Singapore, Singapore)
	Yajun Ha (National Univ. of Singapore, Singapore)
Past Chair	**Shinji Kimura** (Waseda Univ., Japan)
Technical Program Chair	**Nagisa Ishiura** (Kwansei Gakuin Univ., Japan)
Technical Program Vice Chairs	**Naehyuck Chang** (Seoul National Univ., Republic of Korea)
	Tulika Mitra (National Univ. of Singapore, Singapore)
Design Contest Chair	**Chun-Huat Heng** (National Univ. of Singapore, Singapore)
Finance Chair	**Lap-Pui Chau** (Nanyang Technological Univ., Singapore)
Tutorial Chair	**Hao Yu** (Nanyang Technological Univ., Singapore)
Publicity Chair	**Weng-Fai Wong** (National Univ. of Singapore, Singapore)
Publication Chair	**Kwen-Siong Chong** (Nanyang Technological Univ., Singapore)
Exhibition Chair	**Zhi Yang** (National Univ. of Singapore, Singapore)
Local Arrangement Chair	**Say-Wei Foo** (Nanyang Technological Univ., Singapore)
Industry Liaison	**Fan-Yung Ma** (Infineon, Singapore)
Registration Chair	**Akash Kumar** (National Univ. of Singapore, Singapore)
Logistic Chair	**Yajun Yu** (Nanyang Technological Univ., Singapore)
Web Chair	**Shaobo Luo** (National Univ. of Singapore, Singapore)
Conference Secretariat	**Stanley Teng** (A'Tenga C. E., Singapore)
	Mary Teng (A'Tenga C. E., Singapore)
IT Secretariat	**Joseph Lim** (ELITE, Singapore)

Steering Committee

Chair	**Hiroto Yasuura** (Kyushu Univ., Japan)
Vice Chair	**Hidetoshi Onodera** (Kyoto Univ., Japan)
Secretaries	**Atsushi Takahashi** (Tokyo Institute of Technology, Japan) **Yutaka Tamiya** (Fujitsu Laboratories) **Nozomu Togawa** (Waseda Univ., Japan)
ASP-DAC 2014 General Co-Chairs	**Yong Lian** (National Univ. of Singapore, Singapore) **Yajun Ha** (National Univ. of Singapore, Singapore)
ASP-DAC 2013 General Chair	**Shinji Kimura** (Waseda Univ., Japan)
ACM SIGDA Representative	**Naehyuck Chang** (Seoul National Univ., Republic of Korea)
IEEE CAS Representative	**Takao Onoye** (Osaka Univ., Japan)
IEEE CEDA Representative	**Yao-Wen Chang** (National Taiwan Univ., Taiwan)
DAC Representative	**Patrick Groeneveld** (Magma Design Automation)
DATE Representative	**Wolfgang Nebel** (Carl von Ossietzky Univ. Oldenburg)
ICCAD Representative	**Youngsoo Shin** (Korea Advanced Institute of Science &Technology, Republic of Korea)
International Members	**Kiyoung Choi** (Seoul National Univ., Republic of Korea) **Oliver C.S. Choy** (The Chinese Univ. of Hong Kong) **Yajun Ha** (National Univ. of Singapore, Singapore) **Yuchun Ma** (Tsinghua Univ., China) **Sri Parameswaran** (The Univ. of New South Wales, Australia) **Ren-Song Tsay** (National Tsing Hua Univ., Taiwan) **Xiaoyang Zeng** (Fudan Univ., China)
Advisory Members	**Kunihiro Asada** (Univ. of Tokyo, Japan) **Satoshi Goto** (Waseda Univ., Japan) **Fumiyasu Hirose** (Cadence Design Systems, Japan) **Masaharu Imai** (Osaka Univ., Japan) **Takashi Kambe** (Kinki Univ., Japan) **Tokinori Kozawa** **Chong-Min Kyung** (Korea Advanced Institute of Science & Technology, Republic of Korea) **Youn-Long Steve Lin** (National Tsing Hua Univ., Taiwan) **Isao Shirakawa** (Univ. of Hyogo) **TingAo Tang** (Fudan Univ., China) **Kazutoshi Wakabayashi** (NEC) **Kenji Yoshida** (D2S KK)

Technical Program Committee

Technical Program Chair
Nagisa Ishiura (Kwansei Gakuin Univ., Japan)

Technical Program Vice Chairs
Naehyuck Chang (Seoul National Univ., Republic of Korea)
Tulika Mitra (National Univ. of Singapore, Singapore)

Secretary
Hyung Gyu Lee (Daegu Univ., Republic of Korea)
Yasuhiro Takashima (Univ. of Kitakyushu, Japan)
Shigeru Yamashita (Ristumeikan Univ., Japan)

Subcommittee Chairs and Subcommittees (*: Subcommittee Chairs)

[01] System-Level Modeling and Simulation/Verification

* Derek Chiou (Univ. of Texas at Austin, U.S.A.)
Shih-Hao Hung (National Taiwan Univ., Taiwan)
Makoto Sugihara (Kyushu Univ., Japan)
Yosinori Watanabe (Cadence Design Systems, U.S.A.)
Dongrui Fan (Chinese Academy of Sciences, China)
Atushi Ike (Fujitsu Laboratories, Japan)
Lei Wang (Univ. of Connecticut, U.S.A.)

[02] System-Level Synthesis and Optimization

* Yun (Eric) Liang (Peking Univ., China)
Paolo Ienne (EPFL, Switzerland)

Farhad Mehdipour (Kyushu Univ., Japan)
Sri Parameswaran (Univ. of New South Wales, Australia)
Unmesh Bordoloi (Linkoping Univ., Sweden)
Sungchan Kim (Chonbuk National Univ., Republic of Korea)
Alexandros Papakonstantinou (Nvidia, U.S.A.)
Jiang Xu (Hong Kong Univ. of Science and Technology, Hong Kong)

[03] System-Level Memory/Communication Design and Networks on Chip

* TingTing Hwang (National Tsing-Hua Univ., Taiwan)
Li-Pin Chang (National Chiao Tung Univ., Taiwan)
Yinhe Han (Chinese Academy of Sciences, China)
Chung-Ta King (National Tsing Hua Univ., Taiwan)
Jin Ouyang (Nvidia, U.S.A.)

Hiroyuki Tomiyama (Ritsumeikan Univ., Japan)
Paul Bogdan (Univ. of Southern California, U.S.A.)
Masoud Daneshtalab (Univ. of Turku, Finland)
Koji Inoue (Kyushu Univ., Japan)
Hsien-Hsin Lee (Georgia Institute of Technology, U.S.A.)
Muhammad Shafique (Karlsruhe Institute of Technology, Germany)

[04] Embedded and Real-Time Systems

* Jian-Jia Chen (Karlsruhe Institute of Technology, Germany)
Sudipta Chattopadhyay (National Univ. of Singapore, Singapore)
Song Han (Univ. of Connecticut, U.S.A.)
Kyungsoo Lee (Kyoto Univ., Japan)
Hyunok Oh (Hanyang Univ., Republic of Korea)
Jason Xue (City Univ. of Hong Kong, Hong Kong)
Philip Brisk (Univ. of California, Riverside, U.S.A.)

Nan Guan (Northeastern Univ., China)

Shinpei Kato (Nagoya Univ., Japan)
Hiroki Matsutani (Keio Univ., Japan)
Sebastian Steinhorst (TUM CREATE, Singapore)

[05] High-Level/Behavioral/Logic Synthesis and Optimization

* Robert Wille (Univ. of Bremen, Germany)

Deming Chen (Univ. of Illinois, Urbana-Champaign, U.S.A.)

Kiyoung Choi (Seoul National Univ., Republic of Korea)

Yuko Hara-Azumi (Nara Institute of Science and Technology, Japan)

Christian Haubelt (Univ. of Rostock, Germany)

Yusuke Matsunaga (Kyushu Univ., Japan)

Zhiru Zhang (Cornell Univ., U.S.A.)

[06] Validation and Verification for Behavioral/Logic Design

* Miroslav Velev (Aries Design Automation, U.S.A.)

Charles H.-P. Wen (National Chiao Tung Univ., Taiwan)

Kiyoharu Hamaguchi (Shimane Univ., Japan)

Tsuyoshi Iwagaki (Hiroshima City Univ., Japan)

[7a] Physical Design (Placement)

* Ting-Chi Wang (National Tsing Hua Univ., Taiwan)

Hung-Ming Chen (National Chiao Tung Univ., Taiwan)

Guojie Luo (Peking Univ., China)

Shigetoshi Nakatake (Univ. of Kitakyushu, Japan)

Jia Wang (Illinois Institute of Technology, U.S.A.)

[7b] Physical Design (Routing)

* Evangeline Young (Chinese Univ. of Hong Kong, Hong Kong)

Mark Po-Hung Lin (National Chung Cheng Univ., Taiwan)

Wen-Hao Liu (National Tsing Hua Univ., Taiwan)

Gi-Joon Nam (IBM Research, U.S.A.)

Toshiyuki Shibuya (Fujitsu Laboratories, Japan)

[08] Timing, Power, Thermal Analysis and Optimization

* Masanori Hashimoto (Osaka Univ., Japan)

Mango Chia-Tso Chao (National Chiao Tung Univ., Taiwan)

Lih-Yih Chiou (National Cheng Kung Univ., Taiwan)

Mineo Kaneko (Japan Advanced Intistute of Science and Technology, Japan)

Bing Li (Technical Univ. of Munich, Germany)

Takashi Sato (Kyoto Univ., Japan)

Yiyu Shi (Missouri Univ. of Science and Technology, U.S.A.)

Youngsoo Shin (Korea Advanced Institute of Science and Technology, Republic of Korea)

[09] Signal/Power Integrity, Interconnect/Device/Circuit Modeling and Simulation

* Ram Achar (Carleton Univ., Canada)

Luca Daniel (Massachusetts Institute of Technology, U.S.A.)

Dipanjan Gope (Indian Institute of Science, India)

Rung-Bin Lin (Yuan Ze Univ., Taiwan)

Fan Yang (Fudan Univ., China)

Wenjian Yu (Tsinghua Univ., China)

[10] Design for Manufacturability/Yield and Statistical Design

* Xuan Zeng (Fudan Univ., China)

Steven (Chien-Wen) Chen (TSMC, Taiwan)

Puneet Gupta (Univ. of California, Los Angeles, U.S.A.)

Shigeki Nojima (Toshiba Corporation, Japan)

Martin Wong (Univ. of Illinois, Urbana-Champaign, U.S.A.)

Jae-seok Yang (Samsung, Republic of Korea)

Hai Zhou (Northwestern Univ., U.S.A.)

[11] Test and Design for Testability

* Tomokazu Yoneda (NAIST, Japan)
Jiun-Lang Huang (National Taiwan Univ., Taiwan)
Kohei Miyase (Kyushu Institute of Technology, Japan)
Yu Hu (Chinese Academy of Sciences, China)
Yu Huang (Mentor Graphics, U.S.A.)
Dong Xiang (Tsinghua Univ., China)

[12] Analog, RF and Mixed Signal Design and CAD

* Sheldon Tan (Univ. of California, Riverside, U.S.A.)
Shi Guoyong (Shanghai Jiaotong Univ., China)
Hai Wang (Univ. of Electronic Science and Technology of China, China)
Hideki Asai (Shizuoka Univ., Japan)
Wong Ngai (Univ. of Hong Kong, Hong Kong)

[13a] EDA and Design Methodologies for Emerging Technologies

* Hai (Helen) Li (Univ. of Pittsburgh, U.S.A.)
Jae-Joon Kim (Pohang Univ. of Science and Technology, Republic of Korea)
Guangyu Sun (Peking Univ., China)
Danghui Wang (Northwestern Polytechnical Univ., China)
Ik-Joon Chang (Kyunghee Univ., Republic of Korea)
Yongpan Liu (Tsinghua Univ., China)

Yvain Thonnart (CEA-LETI, France)

[13b] Emerging Applications

* Tsung-Yi Ho (National Cheng Kung Univ., Taiwan)
Shanq-Jang Ruan (National Taiwan Univ. of Science and Technology, Taiwan)
Yu Wang (Tsinghua Univ., China)
Dajiang Zhou (Waseda Univ., Japan)
Jongsun Park (Korea Univ., Republic of Korea)
Yasushi Sugama (Fujitsu Laboratories, Japan)

Xiaoyang Zeng (Fudan Univ., China)

University LSI Design Contest Committee

Chair   **Chun-Huat Heng** (National Univ. of Singapore, Singapore)

Members   **Pak Kwong Chan** (Nanyang Technological Univ., Singapore)
Bah Hwee Gwee (Nanyang Technological Univ., Singapore)
Hiroyuki Ito (Tokyo Institute of Technology, Japan)
Chulwoo Kim (Korea Univ., Republic of Korea)
Tony Kim (Nanyang Technological Univ., Singapore)
Tsung Hsien Lin (National Taiwan Univ., Taiwan)
Noriyuki Miura (Kobe Univ., Japan)
Sam Chun Geik Tan (Mediatek, Singapore)

List of Reviewers

Ram Achar
Junwhan Ahn
Sidharta Andalam
Hideki Asai
Yasmine Badr
Fang Bao
Andrew J. Becker
Paul Bogdan
Unmesh Bordoloi
Philip Brisk
Hua-Yu Chang
Li-Pin Chang
Ik-Joon Chang
Wanli Chang
Yuan-Ying Chang
Mango Chia-Tso Chao
Sudipta Chattopadhyay
Jian-Jia Chen
Deming Chen
Hung-Ming Chen
Steven (Chien-Wen) Chen
Yuankai Chen
Wu-Tung Cheng
Derek Chiou
Lih-Yih Chiou
Kiyoung Choi
Steve Dai
Masoud Daneshtalab
Luca Daniel
Masoumeh Ebrahimi
Yibo Fan
Dongrui Fan
Eric J.-W. Fang
Shao-Yun Fang
Ping Gao
Nithin George
Aaron Gonzales
Dipanjan Gope
Mark Gottscho
Nan Guan
Shi Guoyong
Puneet Gupta
Charles H.-P. Wen

Kiyoharu Hamaguchi
Kyuseung Han
Yinhe Han
Song Han
Kecheng Hao
Yanan Hao
Yuko Hara-Azumi
Masanori Hashimoto
Christian Haubelt
Tsung-Yi Ho
Chih-Hsien Hsia
Yu Hu
Yu Huang
Jiun-Lang Huang
Yoshi Shih-Chieh Huang
Shih-Hao Hung
TingTing Hwang
Paolo Ienne
Atushi Ike
Ryoichi Inanami
Koji Inoue
Tsuyoshi Iwagaki
Ramkumar Jayaseelan
Iris Hui-Ru Jiang
Ke Jiang
Manhwee Jo
Abde Ali Kagalwalla
Tsang-Chi Kan
Mineo Kaneko
Peng Kang
Shinpei Kato
Matthias Kauer
Martin Keim
Oliver Kesz$BC6(Bcze
Kyoung Hoon Kim
Myung-Chul Kim
Sungchan Kim
Jae-Joon Kim
Chung-Ta King
Chikaaki Kodama
Liangzhen Lai
Liyang Lai
Imyong Lee

Jinho Lee
Hsien-Hsin Lee
Kyungsoo Lee
Bing Li
Hai (Helen) Li
Li Li
Yun (Eric) Liang
Mark Po-Hung Lin
Rung-Bin Lin
Xijiang Lin
Chang Hong Lin
Gai Liu
Wen-Hao Liu
Yongpan Liu
Chunsheng Liu
Brandon Lu
Yinghai Lu
Martin Lukasiewycz
Yan Luo
Guojie Luo
Yusuke Matsunaga
Tetsuaki Matsunawa
Hiroki Matsutani
Farhad Mehdipour
Lars Middendorf
Kohei Miyase
Philipp Mundhenk
Fumiharu Nakajima
Shigetoshi Nakatake
Gi-Joon Nam
Swaminathan
Narayanaswamy
Wong Ngai
Philipp Niemann
Shigeki Nojima
Kenneth O'Neal
Hyunok Oh
Jin Ouyang
Alexandros
Papakonstantinou
Sri Parameswaran
Hanmin Park
Jongsun Park

Brendon Parker
Ke Peng
Judith Peters
Michele Petracca
Nils Przigoda
Zhiliang Qian
Shanq-Jang Ruan
Florian Sagstetter
Keishi Sakanushi
Soheil Samii
Takashi Sato
Eleonora Schoenborn
Julia Seiter
Muhammad Shafique
Chung-An Shen
Yiyu Shi
Toshiyuki Shibuya
Youngsoo Shin
Hyunjik Song
Sebastian Steinhorst
Jannis Stoppe
Yasushi Sugama
Makoto Sugihara
Guangyu Sun
Chi-Chia Sung
Stephen Sunter

Aditya Tammewar
Mingxing Tan
Sheldon Tan
Huaxing Tang
Yvain Thonnart
Hiroyuki Tomiyama
Kun-Lin Tsai
Tsun-Ming Tseng
Mihai Udrescu
Miroslav Velev
Shaodi Wang
Xiren Wang
Yuxin Wang
Lei Wang
Ting-Chi Wang
Jia Wang
Hai Wang
Danghui Wang
Yu Wang
Chen Wang
Peter Waszecki
Yoshi Watanabe
Yaoguang Wei
Robert Wille
Martin Wong
Chin-Hsien Wu

Dong Xiang
Bingjun Xiao
Kan Xiao
Yang Xu
Jiang Xu
Qiang Xu
Jason Xue
Yuta Yamato
Fan Yang
Jae-seok Yang
Zuochang Ye
Jing Ye
Yoko Yokoyama
Tomokazu Yoneda
Evangeline Young
Wenjian Yu
Christian Zebelein
Xuan Zeng
Xiaoyang Zeng
Jia Zhan
Zhiru Zhang
Jinjia Zhou
Hai Zhou
Dajiang Zhou
Yi Zou
Wei Zou

Normally-Off Computing Project : Challenges and Opportunities

Hiroshi Nakamura, Takashi Nakada, Shinobu Miwa
The University of Tokyo
Tokyo, 113-8656, JAPAN
e-mail: {nakamura, nakada, miwa}@hal.ipc.i.u-tokyo.ac.jp

Abstract— Normally-Off is a way of computing which aggressively powers off components of computer systems when they need not to operate. Simple power gating cannot fully take the chances of power reduction because volatile memories lose data when power is turned off. Recently, new non-volatile memories (NVMs) have appeared. High attention has been paid to normally-off computing using these NVMs. In this paper, its expectation and challenges are addressed with a brief introduction of our project started in 2011.

I. INTRODUCTION

A. Background

It is indispensable for advanced information society to reduce power consumption of computer systems. As power consumption of VLSIs dominates significant part of total power consumption, low-power techniques of VLSIs are highly required. The power consumption is classified into dynamic and static power. The former is caused by switching activities of transistors and essentially consumed by computing. On the other hand, the latter is caused by leakage current and always consumed whenever power is supplied.

As technology improves, static power increases more rapidly than dynamic power [1], and now it gets comparable to dynamic power consumption. As static power is consumed without any contribution to computing, its reduction is strongly required.

A wide variety of power reduction techniques has been proposed and realized, including clock gating, DVFS, power gating and so on. All these techniques trade performance and power and provide a knob that can adjust the trade-off point. If these exist components which are not bottlenecks in performance, they should be turned into low performance but low power mode by the knob.

B. Limitation of Power Gating

Power gating (PG) is a promising way to reduce static power and used mainly in embedded systems. In modern computer system, all the components need not work all the time during computation. Based on this observation, there exist many chances for PG. So far, power gating is applied in coarse manner. Especially its temporal granularity is quite coarse. Recently, however, fine-grain power gating receives much attention because finer granularity increases the chances of PG. For example, Geyser-2 [2] and Geyser-3 [3], which are MIPS compatible microprocessors, implement fine-grained run-time

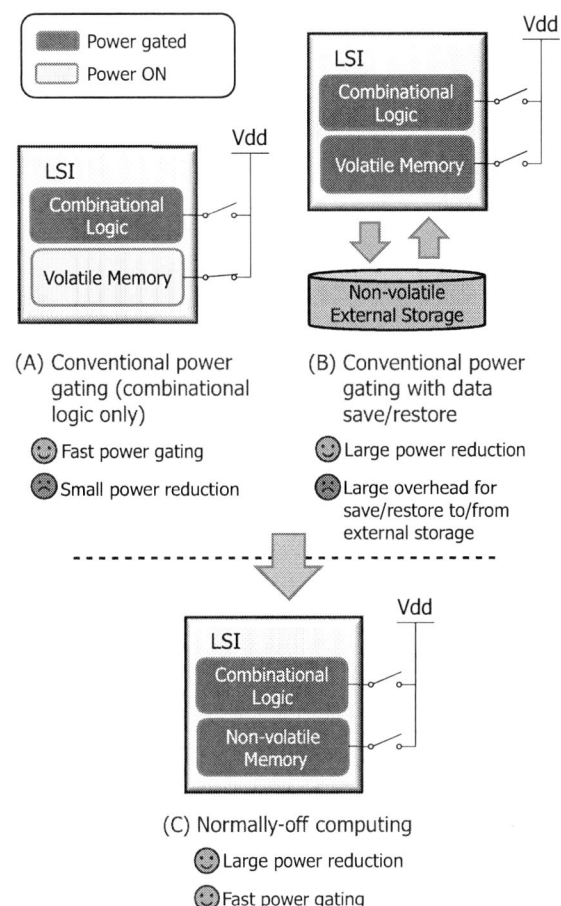

Fig. 1. Conventional power managements and Normally-off computing

PG. In these processors, PG is applied to function units (FUs). Each FU can be powered on and off instruction by instruction. In other words, instruction-level power gating is implemented in these processors. They confirmed that Geyser-2 successfully operates and 60% leakage power reduction at 25°C with 200MHz operation[2].

However, there remain problems to be solved for further power reduction by fine-grain power gating. Currently, volatile memories such as SRAM or DRAM are used in traditional VLSIs. When power is turned off, contents or values in the volatile memories are lost. Then, systems cannot resume computation. To avoid this problem, existing power management (1) doesn't power off the volatile memories (Fig. 1-(A)) or (2) save/restore

TABLE I
COMPARISON OF MEMORY PARAMETERS

	SRAM	DRAM	NAND FLASH	STT-MRAM	FeRAM	PCM	ReRAM
Non-volatility	-	-	√	√	√	√	√
Cell Size (F2)	50~120	6~10	2~5	6~20	15~23	6~12	<1
Read Lat. (ns)	1	30	50	2~20	<55	20~50	<50
Write Lat. (ns)	1	50	$10^5 \sim 10^6$	2~20	<55	50~120	<100
Endurance (Write cycles)	10^{16}	10^{16}	10^5	10^{15}	10^{14}	10^{10}	10^6
Read/Write Energy	Small	Small	Large	Small	Small	Small	Small
Other Energy	Large Leakage	Refresh required	-	-	-	-	-
Supply Voltage	Same as CMOS	2V	16-20V	Same as CMOS	1.5-3V	1.5-3V	1.6-2V

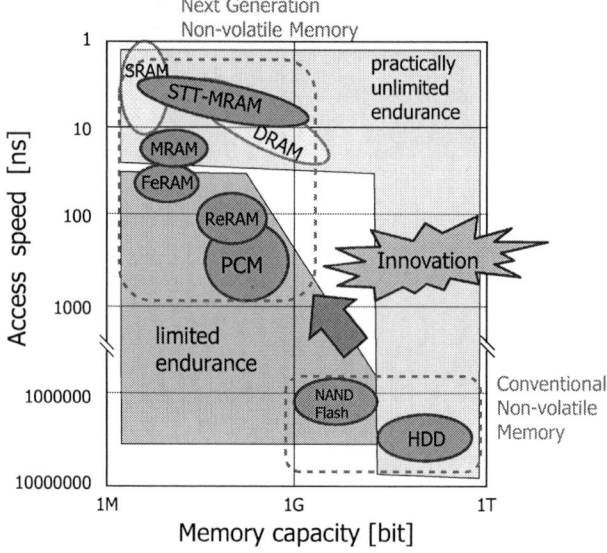

Fig. 2. Paradigm shift of non-volatile memory

the contents to/from an external non-volatile memory for every power gating cycle (Fig. 1-(B)). For the method (1), the power reduction is limited due to remaining static power of volatile memory. For the method (2), the power and the performance overheads for save and restore decrease its effectiveness.

C. Emerging Technology: Non-volatile Memory

Recently, new generation non-volatile memories with new materials get available. These non-volatile memories are more than 1,000x faster than conventional non-volatile memory such as NAND FLASH memory. Their access speed is comparable to that of SRAM or DRAM (Fig. 2).

Important characteristics of these memories are summarized in Tab. I. ReRAM is good for integration density and short write latency, and expected to replace FLASH memories as a new storage class memory. PCM is good for endurance and expected to replace DRAM. STT-MRAM realized fast ac-

cess and long endurance and is a good candidate to replace SRAM. FeRAM is already available and replacing EEPROM and FLASH memories in embedded systems.

In addition to these advantages, these new generation non-volatile memories have good compatibility with CMOS process. Therefore it is strongly expected to replace the on-chip SRAM and DRAM.

D. Objective of Normally-Off Computing

Normally-Off is a way of computing which aggressively powers off components of computer systems when they need not to operate. The new generation non-volatile memories are very fast and keep its contents without power supply. Thus, by making use of these memories, very fine-grain power control, that is normally-off computing, would become available (Fig. 1-(C)).

In this paper, we describe opportunities and challenges of the normally-off computing, Also we introduced our "Normally-Off Computing Project"[4] supported by NEDO (New Energy and Industrial Technology Development Organization)/METI (Ministry of Economy, Trade and Industry) in Japan. This is a 5 year project started in 2011. Toshiba Corporation, Renesas Electronics Corporation, ROHM Corporation, and the University of Tokyo join this project. All the industrial members have their own strong application fields and are responsible for developing world-leading normally-off computing in their fields. The University of Tokyo is working with them to put all their knowledge together and establish general normally-off computing design methodology for forthcoming comfortable and power efficient information society.

II. CHALLENGES OF NORMALLY-OFF COMPUTING

To realize the normally-off computing, the most important problems to be solved are (1) organization of memory hierarchy with the non-volatile memory and (2) optimization of the power gating dedicated for this memory hierarchy. These two issues are essentially related to their non-volatility.

978-1-4799-2817-0/14 $31.00 © 2014 IEEE

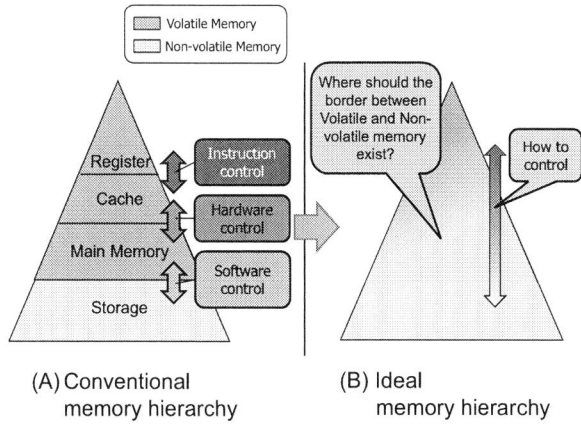

Fig. 3. Conventional and ideal memory hierarchies

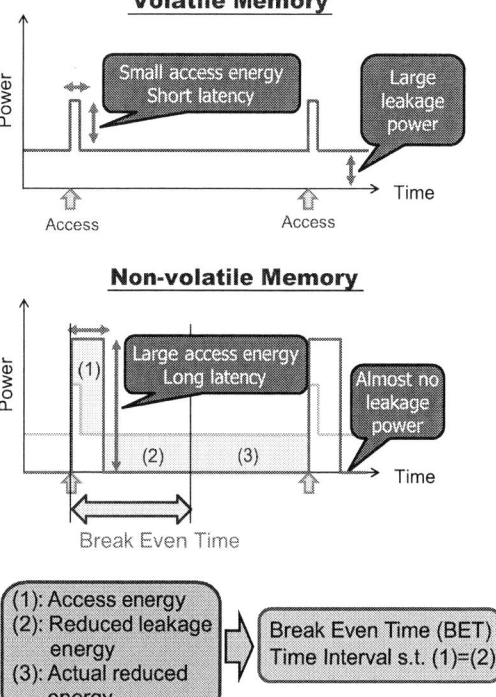

Fig. 4. Definition of Break Even Time (BET)

A. Memory Hierarchy

In general, both fast access and large capacity are preferable features for the memory system. However, since there is a trade-off between these characteristics, conventional computer systems employ memory hierarchy combining small and fast memory and large and slow memory to provide large capacity and low latency memory for processor core. To realize better trade-off, multi-level cache is commonly used especially for high performance systems.

Unfortunately, the next generation non-volatile memories are still not as fast as SRAM. Therefore, if we simply replace SRAM and DRAM with non-volatile memories, the performance of the memory system would be degraded. Then the execution time gets longer, and as a result, the total energy consumption increases. Thus straightforward introduction of non-volatile memories, which have smaller leakage power, does not always reduce the total energy consumption.

To improve energy efficiency, more sophisticated combination of the non-volatile and the volatile memories would be a better solution. Namely, it is worth considering to utilize the fast and small volatile memories for performance sensitive parts and the non-volatile memories for the rest parts. This solution includes integration of volatile and non-volatile memories within a layer (Fig. 3-(B)) unlike the traditional configuration which integrating with uniform memory in each memory layer (Fig. 3-(A)).

Moreover, high attention should be paid to data movement between memory layers. In conventional memory hierarchy, hardware manages data movement between cache memory and main memory, whereas system software manages between the main memory and the external memories. In contrast, in the new memory hierarchy (Fig. 3-(B)), memories with different speed are used within the same layer. Thus, more sophisticated memory management is required. For example, it would be worth considering that even in the same memory layer the volatile memory is managed by hardware and the non-volatile memory is managed by system software.

When PG is applied to the volatile memory, it is required to save/restore the contents. In conventional systems, since the external non-volatile memory is slow (Fig. 1-(B)), this manage-ment is usually handled by system software. The non-volatile memory, however, is quite fast and may be implemented on the same package as cores (Fig. 1-(C)). Then, management should be done by hardware. In this way, it should be carefully considered when and which contents should be saved/restored and who manages it.

B. Temporal Granularity

The principle of the next-generation non-volatile memory is not based on capacitive phenomena, which is completely different from conventional volatile memory. Generally speaking, their leakage power is drastically reduced, but required energy for access, especially for write access, is higher compared with capacitive volatile memory.

Therefore, to replace the volatile memory with the non-volatile memory, the Break Even Time (BET) must be considered as shown in Fig. 4. This figure illustrates power consumption of volatile and non-volatile memories. Volatile memory consumes fairly large leakage power at all the time but its access energy is small. Non-volatile memory, however, consumes large energy when accessed though its leakage power is almost zero. Therefore, there exists a certain time interval between two consecutive accesses where the increased access energy (1) gets equal to the reduced leakage energy (2). The length of this time interval is defined in Break Even Time or BET. If the interval between two consecutive accesses is longer than the BET, then the power consumption memory is successfully reduced by using non-volatile memory, and the save energy is represented by (3) in Fig. 4. On the other hand, if the interval

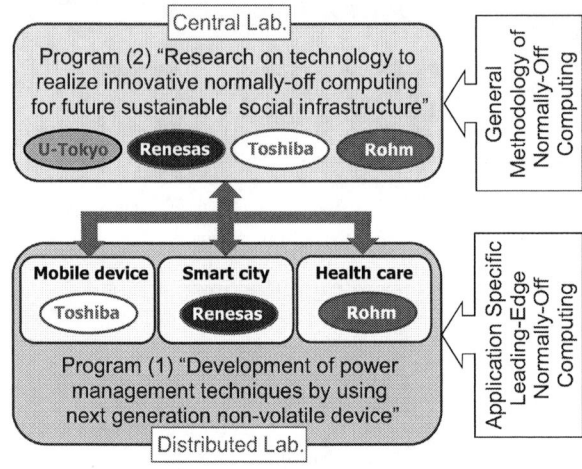

Fig. 5. Organization of Normally-Off Conputing Project

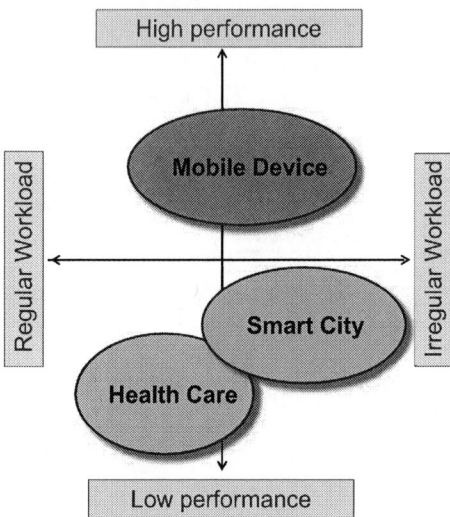

Fig. 6. Characteristics of target applications

is shorter than the BET, then use of non-volatile memory leads to the increase of the total power consumption.

Therefore, in order to reduce the total power consumption, it is indispensable to control the access frequency and keep the access interval longer than the BET. The BETs of non-volatile memories differ depending on their physical mechanisms, and are not the same for read access and write access. Thus, the access interval, or the temporal granularity of memory accesses, should be carefully controlled by memory access scheduling. This optimization on temporal granularity is a hard problem and cannot be solved without cooperation between algorithm, OS, compiler, architecture, circuit and device.

III. NORMALLY-OFF COMPUTING PROJECT

A. Organization

Our "Normally-Off Computing Project," started in 2011, aims to achieve higher performance per power for wide varieties of computing systems including the next generation sensor networks, mobile devices, servers and other equipment. Its strategy is to make full use of non-volatile memory by cooperative development of hardware and software.

The two challenges mentioned in section II depend on the characteristic of memory access required by applications. Thus, the solutions also depend on the target fields or applications. To cope with these problems, our project is performed under the organization of Fig. 5. This project has two research programs. The first program (1) corresponds to application specific development of normally-off computing and the second program (2) is for general purpose development.

The goal of program (1) is to realize drastic power reduction dedicated for each application field. This program is performed by three industrial companies, that is Toshiba Corporation, Renesas Electronics Corporation, and Rohm Corporation, independently. We call the laboratory engaged in program (1) distributed laboratory.

The goal of program (2) is to investigate and develop generalized normally-off computing, which can be applied widely to our future sustainable and sophisticated information society. Its goal is development of computer platform independent from specific applications rather than application specific appliances. This program is carried out in central laboratory located in the University of Tokyo. Researchers from the companies also join this laboratory.

The distributed and the central laboratories tightly collaborates each other through generalization and specialization feedback loops. We are trying to develop novel and efficient normally-off computing through these feedback loops.

B. Target Application

For the success of program (2), the development of generalized normally-off computing, it would be better to have experience and insight of research and development on wider varieties of applications. Then, the following question arises; how many applications are required? Before starting this project, we summarized what kinds of application characteristics significantly affect the development of normally-off computing, or the two challenges mentioned in section II.

Applications are categorized according to two characteristics, "peak performance" and "workload regularity" as shown in Fig. 6. We picked up these two characteristics due to the following reasons. First, peak performance largely affects system configuration and memory hierarchy. Secondly, temporal granularity of activity is significantly affected by workload regularity.

Then we selected three application fields, mobile device, smart city and health care, because these applications cover a wide area of the whole application field as shown in from Fig. 6.

The mobile devices require high peak performance to provide intelligent data processing. Some applications have steady workload such as encoding or decoding, but other applications have irregular workload such as web browsers or games. In the smart city, there are lots of smart sensors. In most cases, these

978-1-4799-2817-0/14 $31.00 © 2014 IEEE

sensors periodically collect data. However once unusual situation is detected, additional workload may be invoked. Thus we categorize it into irregular workload. Finally, the health care systems also consist of many sensors, but its application requires long term data collection and fixed statistical and periodical processing.

IV. OPPORTUNITIES OF NORMALLY-OFF COMPUTING

Low power technologies of computer systems are indispensable for the forthcoming sustainable and sophisticated information society. Normally-off computing is one of the promising ways to achieve this goal. In this paper, we described its expectation and introduced our normally-off computing project.

In this project, we attempt to find the best way to make full use of the new generation non-volatile memories. So far, from the view point of memory system optimization, only two properties of memory are focused on and discussed, that is latency and capacity. Now, non-volatility appears as the third property.

This property absolutely contributes further power reduction. However, it will not effectively lead to power reduction without careful considerations because of two major problems or challenges, that is, memory hierarchy and temporal granularity. In other words, once we overcome these problems, further power reduction would be achieved by using dreamy zero-leakage non-volatile memories. Cooperation and cooptimization of different design layers, including algorithm, OS, compiler, architecture, circuit and device, are definitely required.

Currently a lot of researches on new memory devices are performed. Other types of useful memory devices other than what introduced in this paper may become available in the future. However, it would not be fruitful if we discuss the device properties only. The viewpoint of "computing technology," that is the technology how to make full use of the attractive property is essentially important.

ACKNOWLEDGMENTS

This work is supported by Normally-Off Computing Project of the New Energy and Industrial Technology Development Organization (NEDO) in Japan.

REFERENCES

[1] R. Puri, L. Stok, and S. Bhattacharya, "Keeping hot chips cool," in Proceeding of the 42nd annual Design Automation Conference (DAC), pp.285–288, June 2005

[2] L. Zhao, D. Ikebuchi, T. Saito, M. Kamata, N. Seki, Y. Kojima, H. Amano, S. Koyama, T. Hashida, Y. Umahashi, D. Masuda, K. Usami, K. Kimura, M. Namiki, S. Takeda, H. Nakamura, M. Kondo, "Geyser-2: The second prototype CPU with fine-grained run-time power gating," in Proceeding of 16th Asia and South Pacific Design Automation Conference (ASP-DAC), pp.87–88, Jan. 2011

[3] K. Usami, M. Kudo, K. Matsunaga, T. Kosaka, Y. Tsurui, W. Wang, H. Amano, H. Kobayashi, R. Sakamoto, M. Namiki, M. Kondo, H. Nakamura, "Design and Control Methodology for Fine Grain Power Gating based on Energy Characterization and Code Profiling of Microprocessors," in Proceeding of 19th Asia and South Pacific Design Automation Conference (ASP-DAC), Jan. 2014

[4] "Normally-Off Computing Project": http://noff-pj.jp/en/

Novel Nonvolatile Memory Hierarchies to Realize "Normally-Off Mobile Processors"

Shinobu Fujita, Kumiko Nomura, Hiroki Noguchi, Susumu Takeda , Keiko Abe
Toshiba Corporation, R&D Center
Advanced LSI technology laboratory
1 Komukai-Toshiba, Kawasaki, Kanagawa, Japan 2128582
Tel : +81-44-549-2315
Fax : +81-44-549-2318
e-mail : shinobu.fujita@toshiba.co.jp

Abstract - **This paper presents novel processor architecture for HP-processor with nonvolatile/volatile hybrid cache memory. By simulations of high-performance (HP)-processor using MTJs, it has been clarified that total power of the HP-processor using perpendicular-(p-)STT-MRAM can be reduced by over 90 % with little degradation of processor performance. The presented architecture with nonvolatile memory hierarchy will realize the "normally-off computers".**

I Introduction

Static leakage power has been rapidly increasing in recent CMOS logics with CMOS scaling. Advanced microprocessors use the control of power supply to each circuit block, called "power gating (PG) technique" to decrease the static CMOS power. As a result, low-power and high-performance operation (500mW~ 2 W) has been achieved[1]. To decrease the static power further, PG using nonvolatile memories have been tried for power reduction of high-performance (HP)-processors. The most promising candidate in nonvolatile memories recently developed is STT-MRAM (spin torque transfer magnetic random access memory) especially using magnetic tunnel junction having perpendicular magnetic anisotropy(p-MTJ) [2,3]. This has high speed close to that of SRAM and unlimited memory endurance, as shown in Fig.1.

in 2001[4], where nonvolatile memory is used for every memory hierarchy in computer systems and power is always down whenever CPU core is in an idle state. According to this concept, an ideal processor, called as a "normally-off processor (Noff-P)" in this paper, should use nonvolatile memories in every memory hierarchy, such as flip flops, resistors, resistor files, cache memories etc. in the processor, as shown in Fig.2(a). This concept based on all nonvolatile memory is herein defined as N0ff-P version 0 (ver.0).

To realize the concept Noff-P ver.0, MTJ-based nonvolatile circuits such as nonvolatile flip flops, nonvolatile ALU and etc. were proposed[5-10]. When the processor is in an idle state, logic data are stored in these MTJ-based nonvolatile circuits, and then power supply is cut off to the logic circuits to stop leakage current. However, since both energy and time for storing data into MTJs is much larger than that in conventional SRAM, when the clock frequency is high, total power of nonvolatile circuits is not decreased but rather increased compared with that of volatile circuits. For upper level memory hierarchy, since active power is dominant, as shown in Fig.2, it is hard to reduce the power using nonvolatile circuits. Also, the processor performance is worsen at the same time. Noff-P ver.0 based on MTJ-based nonvolatile circuits as described in[10] can not, therefore, be applied to HP-processors. It should be used for low-frequency applications with ultra-low power such as sensor network devices considering its feature.

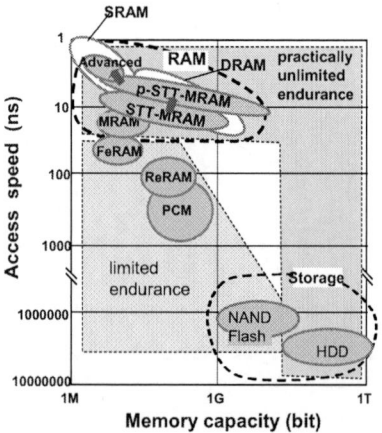

Fig.1. Memory capacity and speed for various nonvolatile memories recently developed.

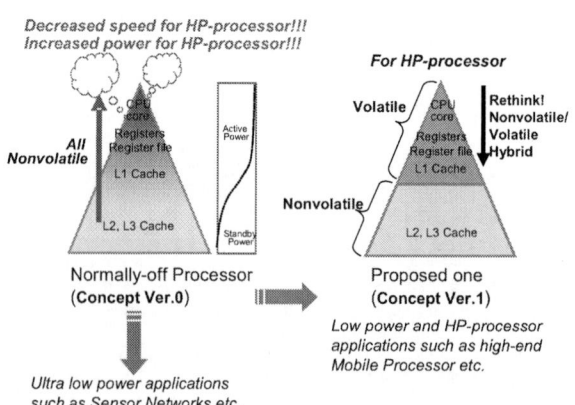

Fig.2. Original concept ver.0 of "Normally-off processor" and proposed one ver.1 based on nonvolatile/volatile (STT-MRAM/SRAM) hybrid cache systems.

A unique concept regarding ultimately low power computers named as "normally-off computer" was proposed

For a conventional processor architecture, volatile memories based on SRAM are used for high frequency access memories like resistor files, L1 cache memory, and L2 cache memory. Based on conventional PG technique[1], power supply is cut off to CPU cores but power supply is not cut off to L2 cache to keep data in it (transition to "CPU core sleep state") during relatively short standby state, where the average power is x2 to x3 lower than that in active state. When time of standby state is long, power supply is also cut off to L2 cache except for small SRAM blocks to retain resistor state. This is a critical transition to "deep power down state (**DPS**)" having more than x10 lower power than that in active state. Clearly, frequent state transition to DPS plays a vital role to reduce processor power effectively. However, since L2 cache cannot retain data during DPS, after "wake-up" the processor system must recover the lost data from main memory. From this reason, the "wake-up time", time needed for recovery from DPS, is so long, which can degrade performance largely. Therefore, there is severe trade off between power and performance based on conventional PG. Although speed and frequency for PG has been improved, ratio of leakage power for L2 cache in total consumed power for CPU is increased with increasing speed of PG, since it is fundamentally difficult to perform PG for volatile cache.

It is, hence, considered that average CPU power determining battery life depends strongly on L2 or L3 cache memory capacity. Cache memory capacity has been increased as shown in Fig. 3. This fact can be attributed to two kinds of processor architecture trend, as described as follows. Firstly, since it is currently difficult to increase clock frequency over 2~3 GHz from the limitation due to power consumption and heating for the processor, cache memory capacity has been increased to improve the performance. Secondly, for multi-core processors based on conventional architectures, shared cache memory capacity, L2 or L3, is increased with increasing number of core. As a result, power consumed by cache memory occupies major part in the recent processor chips.

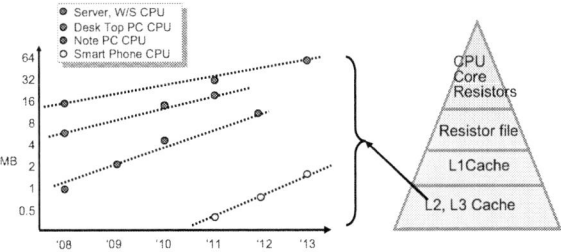

Fig.3. Increment of L2, L3 cache memory capacity.

Considering such situation of HP-mobile processors, the authors have proposed STT-MRAM/SRAM hybrid memory hierarchy, as shown in Fig.2 [11,12]. Here, SRAM-based cache memory is applied to level-1 (L1) cache, and STT-MRAM-based cache is applied to level-2 (L2) cache. This is because, whereas active power is dominant for L1 cache due to frequent access to L1, L2 cache infrequently operates only in the case of L1 cache miss. Further, L2 cache capacity is 10 times or more larger than that of L1. Leakage power in all cache memory is, therefore, dominated by L2 cache. To reduce the leakage power in CPU, STT-MRAM should be applied to L2 cache.

However, general STT-MRAM have large active power especially for write operation and/or large latency. Total energy consumption is, hence, largely increased, although leakage power can be largely decreased by nonvolatile cache memory for high-performance processors, as shown in Fig.4. In general, nonvolatile memories have little static power but large active power. This issue is "dilemma of nonvolatile memory"[11].

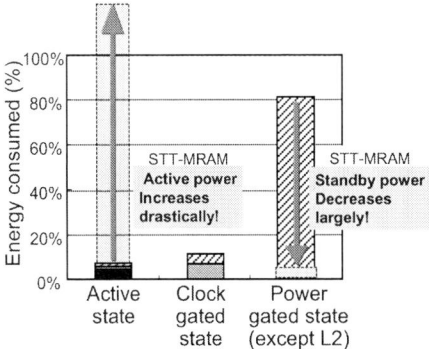

Fig.4. Consumed energy analysis of each state in a conventional low power mobile processor and its change by applying STT-MRAM to nonvolatile L2 cache. Issus of "Dilemma of Nonvolatile Memory" is highlighted by yellow color.

Reduction in write power and time for STT-MRAM is inevitable to solve this issue. Figures 5 shows how fast write is needed to reduce the total average power including active and static power compared with that of SRAM based on preliminary case studies. The results suggest that 5 ns is a target access time. Recently, a **p-STT-MRAM with *the most advanced* p-MTJ**[12,13] has demonstrated 3ns fast write with write current 50uA per bit. This advanced p-STT-MRAM has, thus, a strong potential to replace SRAM for the lower power cache memory considering results in Fig.5

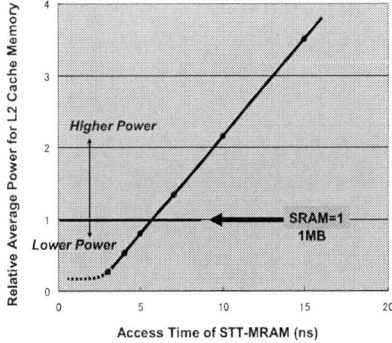

Fig.5. Relative average power for STT-MRAM based L2 cache memory (1MB) compared with SRAM as a function of access time of STT-MRAM. Write current is assumed to be 50uA.

Although write time of 3 ns per bit is short, it is still longer than that of SRAM. Influence of this fact on CPU performance has also been estimated based on a preliminary case study, and obtained results are shown in Fig.6. These results suggest that L2 cache latency does not degrade CPU performance much, if STT-MRAM access time is shorter than 5ns.

This paper describes the details of system designs for the proposed novel memory hierarchy having nonvolatile L2 or last level cache(LLC) using advanced p-STT-MRAM to realize Noff-P. In this work, p-STT-MRAM has a "**normally-off type memory cell design**" that enables almost-zero leakage power in the memory area *without PG* even for a very short standby state of L2 cache during a CPU active state. Nonvolatile cache memory also enables faster wake-up and faster shut-down compared with conventional SRAM-based cache.

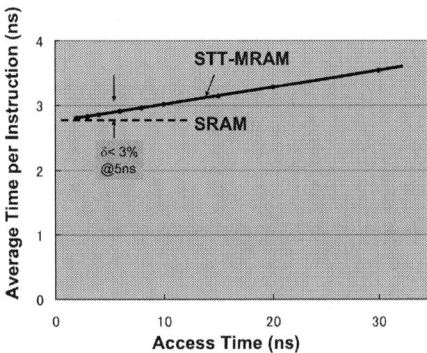

Fig.6. Average time per instruction of CPU with STT-MRAM based L2 cache as a function of access time of STT-MRAM. In this case study, L1, L2 and main memory access latency are 1ns, 10ns and 100ns, respectively. Respective cache miss rate of L1 and L2 are 90% and 80%.

II Decrease in power for CPU active state by applying Normally-off type memory cell design

Previously, some of the authors proposed a "nonvolatile SRAM" (NV-SRAM) for L2 cache[14,15] using NV-latches combined with two p-MTJs. However, these NV-SRAMs include leakage paths like SRAM, and we call "normally-on" type memory cells, as shown in Fig.7. Although other kinds of NV-SRAM combined with SRAM cells and MTJs were recently proposed[16,17], as shown in Fig.7, all of them have normally-on type memory cells. For normally-on type, highly-frequent PG is needed to reduce leakage power. The highly-frequent PG with NV-SRAMs, however, increases circuit area largely and degrade computing performance[18].

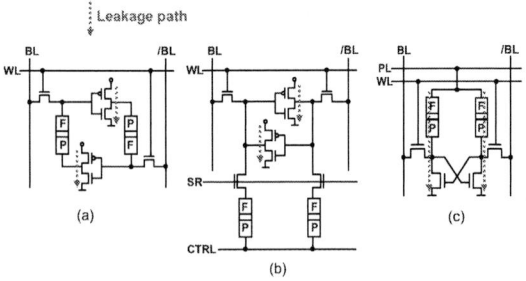

Fig.7. NV-SRAM with (a) 6T-2MTJ[14], (b) 8T-2MTJ [16] and (c)4T-2MTJ [17].

The authors have proposed to change the memory design concept from normally-on type with PG to "normally-off" type without PG that enable to reduce power effectively even

for just one clock-cycle standby state. Figures 8 show the novel normally-off type memory cells without leakage paths designed by some of the authors[18-21] based on the most advanced p-STT-MRAM. No leakage paths can decrease in leakage power even for a very short standby state during active state of CPU, as shown in Fig. 9.

Furthermore, the memory cell area of normally-off type shown in Figs. 8 is about x1.5 to x4 smaller than that of SRAM. On the contrary, the memory cell area of "normally-on type" shown in Figs. 7 is x1.2 to x2.5 "larger" than that of SRAM[18], which is severe disadvantage for L2 or LLC. Some of these memory cells were fabricated in in-house labs and high-performance was demonstrated.

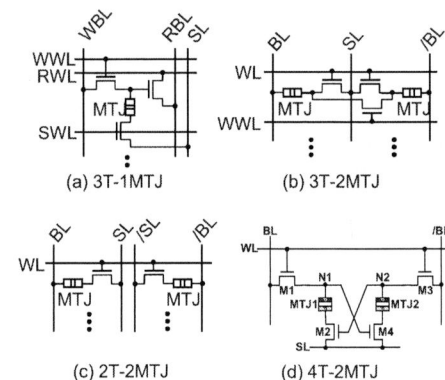

Fig.8 Four kinds of normally-off type memory cell designs using 3T-1MTJ[18], 3T-2MTJ[19], 2T-2MTJ[20] and 4T-2MTJ[21]. Note that there are no leakage path in these cells.

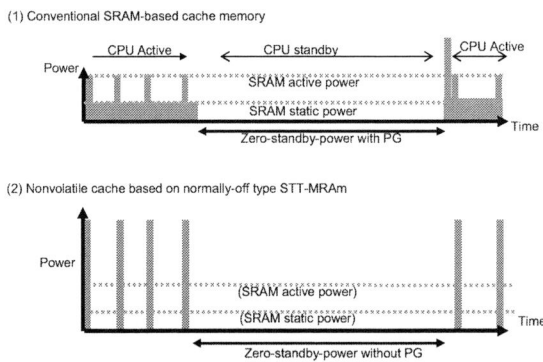

Fig.9 Decrease in leakage power for the short waiting state during active state of CPU by normally-off type STT-MRAM.

Power and performance in processor with 2T-2MTJ STT-MRAM based cache memory for LLC with was evaluated based on CPU simulator customized using gem5 [20]. The CPU core uses 3-way issue out-of-order ARMv7-a single core architecture. Processor simulations conducted with SPEC CPU2006 benchmarks [20]. Our proposed STT-MRAM, compared to the typical 0.8 V operating SRAM design, can reduce the energy per instruction (EPI) of the total cache memory reduced by 64%, while remaining the instruction per cycle (IPC) performance degradation within 6 % in spite of its latency overhead, as shown in Fig. 10. On the other hand, the EPI for other STT-MRAMs previously reported is largely increased. For our proposed STT-MRAM, I/O width can be

978-1-4799-2817-0/14 $31.00 © 2014 IEEE

expanded to 256 bit or 512 bit that needed for the cache-line width (e.g. 64 B) thanks to low power operation, whereas previous ones are less than 64 bit as listed in Table II. It should be noted that decrease in power for CPU active state is attributable to decrease in leakage power of L2 cache memory.

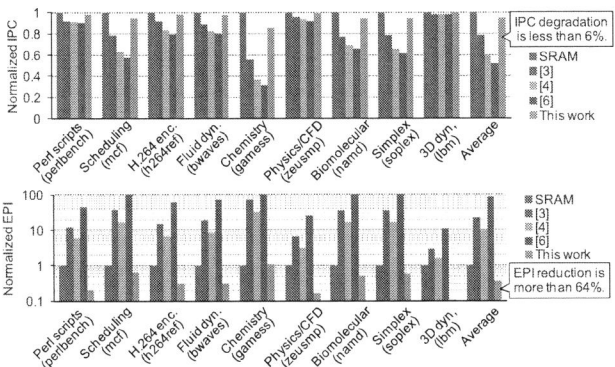

Fig. 10. Evaluated IPC and EPI performances of 1MB L2cache using 2T-2MTJ design[20].

TABLE I : Comparisons of features for various STT-MRAMs.

	Efficiency (%)	Cell size (F^2)	Latency (ns)	IO (bits)	Iw (uA)	Dynamic power (mW)	Energy per access (nJ)*
[22]	63	169.1	Read: 12, write: 12	32	600	Read: 60(66MHz), write: 91(66MHz)	Read: 14.5, write: 22.0
[23]	51	42.6	Read: 11, write: 30	16	N/A	Read: 7.8(33MHz), write: 9.3(33MHz)	Read: 7.56, write: 9.02
[24]	38	50.7	Read: 8, write: 10	32 /64	N/A	N/A	N/A
[25]	64.8	270.3	Read: 8, write: 40	1	100	Read: 10.7(100MHz), write: 4.3(25MHz)	Read: 54.8, write: 88.1
This work	75.1	106.5	Read: 4, write: 4	256	50	Read: 17.8(250MHz), write: 46.5(250MHz)	Read: 0.142, write: 0.372

*Converted to a cache line (=512 bits) access

III Power Decrease for long CPU standby state by Ultra-Fast-Power Gating

Next, power decrease for long standby state is investigated. When time of standby state is long for a conventional processor with PG, power supply is cut off to not only CPU core but also L2 cache (transition to DPS) except for small memory blocks, as shown in table II (a). Since L2 cache loses all data after DPS, the processor system must recall the lost data from main memory again, which is large overhead of delay and power. For example, the "wake-up time", time needed for recovery from the DPS, is so long, which can degrade performance largely.

When nonvolatile (NV-) cache is applied, wake-up time from DPS can be decreased to less than 100 ns based on our analysis, as cache data can be retained in nonvolatile memory without power supply. It can increase frequency for power gating, since frequency of standby state is increased with a decrease in standby state time, as shown in table II (b). This fact enables power to be further reduced, since power down is feasible even for short standby state less than 100 ns.

We evaluated decrease in processor power using a power model for ultra-high-speed power gating. Power down block for each power gated-states for a conventional PG and a ultra-fast power gating model using NV-cache was used as shown in table II.

Table II. Transition policy of power-gated states for (a) a conventional PG and (b) ultra-fast PG with nonvolatile cache for a case study.

(a)

<Conventional PG>	Clock	L1-$	L2-$	Others	Recovery time to active-state
Active state					
Clock gated state	OFF				~1ns
CPU core sleep (L2-Cache retention)	OFF	OFF			10µs
CPU core Sleep (L2-Cache decay)	OFF	OFF	decay		20µs
Deep power-down state (DPS)	OFF	OFF	OFF	OFF except State SRAM	100µs

(b)

<High-speed PG with NV-cache>	Clock	L1-$	L2-logic +prepheral	L2-memory	Recovery time to active-state
Active state				Normally OFF	
Clock gated state	OFF			Normally OFF	~1ns
CPU core sleep; Deep power-down state (DPS)	OFF		OFF	Normally OFF	~10ns
Deeper power-down state	OFF	OFF	OFF	Normally OFF	~100ns

It should be noted that as the memory core in the LLC is normally-off, the leakage power is always decreased even when CPU core is active. Also, a vital state for the NV-cache case is the "CPU core sleep" in Table II(b) where whole L2 cache is powered down, whereas power for L1 cache is supplied for retaining CPU state, is equivalent to the conventional DPS, as the power is almost the same as that for the conventional DPS. Since the wake-up time from DPS with NV-cache is decreased by three orders compared with that from conventional DPS, there is much more opportunity for highly-frequent shift to DPS, which can decrease average power largely.

A case study has been done for comparison between conventional PG and ultra-fast PG with NV-cache, based on a case study according to table II. Figure 11 shows relative power consumption of an advanced mobile processor based on conventional PG and that of the ultra-fast PG using NV-cache. Here, the state transition policy for the PG is that a state shifts to a lower power state when CPU idle time becomes two times of the wake-up time for each state listed in table II. For example, for a conventional PG, the CPU core shifts to the CPU core sleep state with L2 cache retention when CPU idle time is 20us, as its wake-up time is 10us.

Figure 12 shows three kinds of simulation results on total processor power consumption according to the transition policy shown in Fig.11. The case 1 is the CPU active state dominated, the case 3 is the CPU idle state dominated, and the case 2 is in the middle of these two cases. The result of case 1 indicates that the average power is reduced by 29%, which is attributed to the reduction of leakage power in memory core in L2 cache. It should be noted that normally-off memory core design can, thus, reduce even active power in CPU, which has been suggested from results shown in Fig. 10. In the case 2, ultra-fast PG enables the processor to shift to DDS quickly with the NV-L2 cache, whereas the state for the conventional one remains in active state or clock gated state. As a result, the average power is reduced by 66% using NV-L2 cache. In the case 3, whereas CPU state data is retained in NV-cache without power supply during long standby state, small capacitance SRAM with power supply has to be used during it. Consequently, the average power is reduced up to 1/10 using NV-L2 cache. These case studies suggest that ultra-fast PG with NV-cache is much effective method to reduce processor power in every situation. In our future work, statistics of CPU status in real usage cases will be analyzed, which enables to raise the precision of reduction in processor power.

ver.2 has been also studied to improve power and performance. In the ver.2, volatile/nonvolatile hybridization is also used in the same memory hierarchy such as L1 cache L2 cache, and LLC, as shown in Fig.13, as it is more effective to use both volatile and nonvolatile memory considering both high-speed operation and low-standby power operation. As an example, hybrid 3T-1MTJ based STT-MRAM has been studied[16]. 3T-1MTJ based STT-MRAM is a hybrid memory that can be used as DRAM mode or MRAM mode. When data in the cache is predicted to be used for a long term, the data should be stored in a MRAM mode. When data in the cache is predicted to be used for a short term, the data should be stored in a DRAM mode. The selection policy has been analyzed based on cache operation architectures[27]. It is also expected that this hybridization is effective for heterogeneous multi-core processors[28]. These works are in progress and various novel architectures will be presented according to applications in near future.

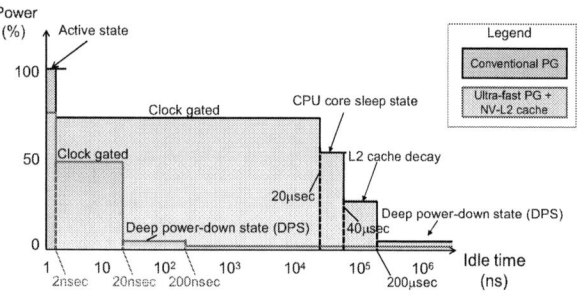

Fig.11. Timing for state shift with PG and relative power consumption for conventional PG and for ultra-fast PG using nonvolatile LLC.

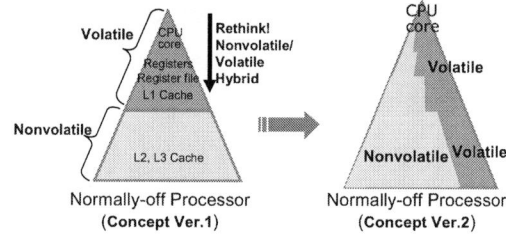

Fig.13. Evolution from Noff-P concept ver.1 to ver.2 to further improve power and performance.

IV. Conclusion

The proposed processor with normally-off type p-STT-MRAM based nonvolatile LLC is very close to the idealistic ultra-low power mobile processor, called as normally-off processor, considering its function and energy per performance. It has been also found that performance degradation is limited even using nonvolatile memory whose access speed is rather low compared with SRAM, and that considering advantage for operation speed of ultra-fast PG, there is large possibility that processor performance can become superior to that of conventional one.

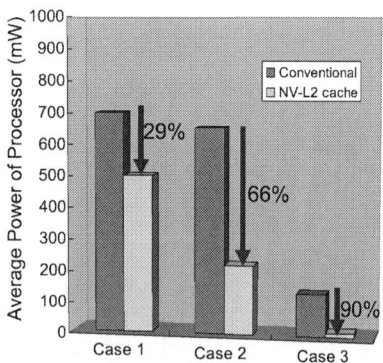

Fig.12. Comparison of processor power reduction by applying nonvolatile (NV)-L2 cache with ultra-fast PG for three cases.

Acknowledgments

This work was partly supported by the Normally-Off Computing Project of the New Energy and Industrial Technology Development Organization (NEDO) in Japan. Normally-off processor architecture has been studied based on collaboration with Prof. Hiroshi Nakamura and his group of University of Tokyo. Embedded STT-MRAM technologies and circuit designs have been studied with K. Ikegami, K. Ikegami, C. Kamata, M. Amano, E. Kitagawa, T. Ochiai, N. Shimomura, J. Ito, A. Kawasumi, K. Kushiba and H. Hara for Toshiba. The first author specially thanks Dr. K. Ando for the

IV Evolution Towards Normally-off Processor Ver.2

As described above, our proposed Noff-P ver.1 with normally-off type p-STT-MRAM based nonvolatile LLC, is very close to the idealistic Noff-P concept considering its function and energy per performance. Furthermore, Noff-P

Japan National Institute of Advanced Industrial Science and Technology (AIST) for discussion on concept of normally-off processors.

References

[1] G. Gerosa et al., "A sub-1W to 2W Low-Power IA Processor for Mobile Internet Devices and Ultra-Mobile PCs in 45nm High-k metal Gate CMOS", IEEE, ISSCC Technical Digest, 2008, p.p. 256-25

[2] H. Yoda et. al, "High efficient spin transfer torque writing on perpendicular magnetic tunnel junctions for high density MRAMs", Current Appl. Phys. 10, e87-89.

[3] T. Kishi et al., "Lower-current and fast switching of a perpendicular TMR for high speed and high density spin-transfer-torque MRAM", International Electron Devices Meeting (IEDM) Technical Digest 2008, p.p. 309-312.

[4] K. Ando, "A Normally-off Computer", FED Journal, 12, 89, 2001(in Japanese).

[5] K. Abe, S. Fujita and T. H. Lee, "Architecture of Three-Dimensional Circuit Using Nanoscale Memory Devices, " European Micro and Nano Systems (EMN04), pp. 225-229, Oct. 2004.

[6] K. Abe, S. Fujita and T. H. Lee, "Novel Nonvolatile Logic Circuits with Three-Dimensionally Stacked Nanoscale Memory Device," Proceedings of the 2005 NSTI Nanotechnology Conference, vol. 3, pp 203–206, May 2005.

[7] N. Sakimura et al., "Nonvolatile Magnetic Flip-Flop for Standby-power-free SoCs", P.P. 355-358, IEEE 2008 Custom Integrated Circuits Conference (CICC).

[8] R. Nebashi et al. "A Content Addressable Memory Using Magnetic Domain Wall Motion Cells" 28-3, Symposium on VLSI Circuits 2011, Kyoto, Japan

[9] S. Matsunaga et al., "Fabrication of a Nonvolatile Full Adder Based on Logic-in-Memory Architecture Using Magnetic Tunnel Junctions", Appl. Phys. Express, 2008, 1(9), 091301.

[10] T. Kawahara, "Nonvolatile Memory and Normally-off Computing" ASP-DAC, 2011.

[11] K. Ando, S. Ikegawa, K. Abe, S. Fujita, and H. Yoda, "Normally-Off Computer: new roles of nonvolatile devices in future computer systems", in *Sustainable Green Computing*, IGI press, ISBN 978-1-4666-1842-8, June, 2012.

[12] H. Yoda, "Progress of STT-MRAM Technology and the Effect on Normally-off Computing Systems", Session 11.3, Technical Digest of International Electron Devices Meeting (IEDM), 2012.

[13] E. Kitagawa, S. Fujita, K. Nomura, H. Noguchi, K. Abe, N. Shi-momura, J. Ito, H. Yoda, T. Daibou, Y. Kato, C. Kamata, and S. Ka-shiwada, "Impact of ultra low power and fast write operation of ad-vance perpendicular MTJ on power reduction for high-performance mobile CPU", Session 29.4, Technical Digest of International Electron Devices Meeting (IEDM), 2012.

[14]K. Abe, K. Nomura, S. Ikegawa, T. Kishi, H. Yoda and S. Fujita, "Hierarchical Nonvolatile Memory with Perpendicular Magnetic

Tunnel Junctions for Normally-Off Computing," the 2010 International Conference on Solid State Devices and Materials (SSDM), Tokyo, pp. 1144-1145, Sep. 2010.

[15] K. Nomura et. al, "Ultra low power processor using perpendicular-STT-MRAM/SRAM based hybrid cache toward next generation normally-off computers", in proceedings of MMM 2011, and 111, 7, J. of App. Phys., p.p. 07E330-333, 2012.

[16] S. Yamamoto and S. Sugahara, "Nonvolatile Static Random Access Memory Using Magnetic Tunnel Junctions with Current-Induced Magnetization Switching Architecture," Japn. J. Appl. Phys. 48 (2009) 043001, Apr. 2009.

[17] T. Ohsawa, H. Koike, S. Miura, et al., "1Mb 4T-2MTJ nonvolatile STT-RAM for embedded memories using 32b fine-grained power gating technique with 1.0ns/200ps wake-up/power-off times", Symp. VLSI Circuits, pp.46-47, July 2012.

[18] K. Abe et al., " Novel Hybrid DRAM/MRAM Design for Reducing Power of High Performance Mobile CPU", IEDM Tech. Dig., pp.243-246, 2012.

[19]A. Kawasumi et. al, "Circuit techniques in realizing voltage-generator-less STT MRAM suitable for normally-off-type non-volatile L2 cache memory", 5th IEEE International Memory Workshop (IMW), p.p. 76 - 79, 2013.

[20] H. Noguchi et. al, " A 250-MHz 256b-I/O 1-Mb STT-MRAM with advanced perpendicular MTJ based dual cell for nonvolatile magnetic caches to reduce active power of processors", VLSI Technology Symposium, C108 - C109, 2013.

[21] C. Tanaka et. al, " Normally-off type Nonvolatile SRAM with perpendicular STT-MRAM cells and smallest number of transistors" , International Conference on Solid State Devices and Materials (SSDM), p.p. 1092-1093, 2013.

[22] R. Nebashi et al., "A 90nm 12ns 32Mb 2T1MTJ MRAM", IEEE ISSCC Technical Digest, 2009, p.p. 462-463.

[23] K. Tsuchida et al., "A 64Mb MRAM with Clamped-Reference and adequate-Reference Schemes", 2010 IEEE International Solid-State Circuits Conference (ISSCC) Technical Digest, 258-260, 2010.

[24]J. P. Kim et al., Symposium on VLSI Circuits, pp.296-297, 2011.

[25] T. Ohsawa, H. Koike, S. Miura, et al., "1Mb 4T-2MTJ nonvolatile STT-RAM for embedded memories using 32b fine-grained power gating technique with 1.0ns/200ps wake-up/power-off times", Symp. VLSI Circuits, pp.46-47, July 2012.

[26] B. Cooper, "Optimizing Battery Life on Intel Architecture Based PCs with Windows 8", EBL S001_100, Intel Developer Forum, 2012.

[27] H. Noguchi, "D-MRAM Cache: Enhancing Energy Efficiency with 3T-1MTJ DRAM/MRAM Hybrid Memory" Design Automation and Test in Europe(DATE), p.p. 1813 - 1818, 2013.

[28] S. Fujita et al., " Progress of STT-MRAM and its Challenge towards Normally-off-Multi-core SoC" in proceedings of International Forum on Embedded MPSoC and Multicore.

978-1-4799-2817-0/14 $31.00 © 2014 IEEE

Normally-Off MCU Architecture for Low-power Sensor Node

Masanori Hayashikoshi, Yohei Sato, Hiroshi Ueki, Hiroyuki Kawai, Toru Shimizu

Renesas Electronics Corporation
Itami-shi, Hyogo, 664-0005, Japan

e-mail : {masanori.hayashikoshi.cj, yohei.sato.zc, hiroshi.ueki.jy, hiroyuki.kawai.vf,
toru.shimizu.xn}@renesas.com

Abstract - **Sensor nodes are used extensively in order to gather real-time information in the social environment and natural environment. And the production volume of sensor nodes is much increased with the development of cyber-physical systems. Therefore, it becomes important how to reduce the power consumption of huge sensor nodes. In this work, normally-off architecture of microcontroller for future low-power sensor node is proposed. To realize true low-power effects with normally-off computing technology, a co-design of hardware and software technology is much important. In this work, the power consumption of sensor nodes is possible to reduce of around 70% by using normally-off MCU architecture in sensor node.**

I. Introduction

As a challenge to reduce the power consumption, we describe the fusion of the normally-off computing technology and the non-volatile memory device technology.

The conventional embedded memories (SRAM, Flash memory) in the microcontroller have below problems.
- Leakage current of SRAM due to process miniaturization.
- Overhead time and power consumption to restore data from Flash memory at power-on period.

It is possible to overcome the above problems by the next generation non-volatile memory (NVRAM), and realize
- Frequent intermittent operation is possible (Ultra-low power consumption),
- Full power-off standby operation is possible (High reliability of peripheral circuits)
(Fig. 1).

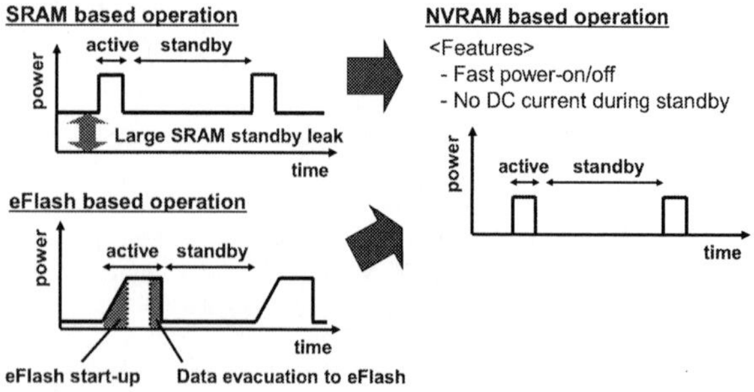

Fig. 1. The paradigm shift in the low-power consumption

However, it is concerned that rather increases power consumption with frequent intermittent operation. In order to improve the low-power effect, the power-off period needs to be as long as a breakeven time (BET). So, the task scheduling technique for activity localization on the time axis is important (Fig. 2).

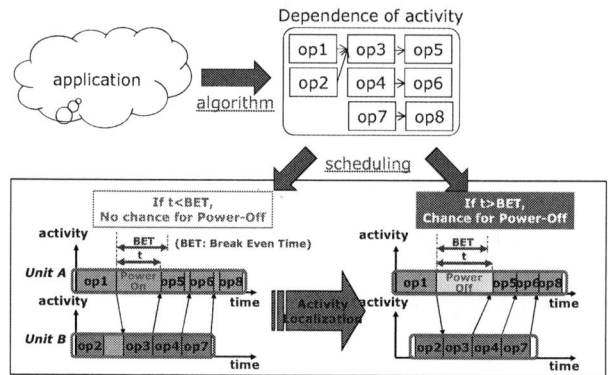

Fig. 2. Normally-off computing technology with activity localization

True normally-off computing technology enable to cut off the power, except for the component in work truly, even in operation as a system, by the synergy of nonvolatile memory devices (NVRAM) and power gating technology. Thus, it is expected to realize the ultra-low-power consumption.

The next generation non-volatile memory (NVRAM) such as STT-MRAM has the characterization of high speed writing compared to the Flash memory, and the data storing time at power-off period can be reduced, and the overhead time associated with the intermittent operation can be minimized. The start-up time at power-on is faster than Flash memory with no need to high voltage generator as Flash memory. As the results, the efficient intermittent operation of the system is possible.

In other words, the fine-grained power gating becomes possible by normally-off computing technology, and the advent of microcontrollers embedded with the next-generation non-volatile memory is expected that the requirements of the normally-off technology is increased, and the paradigm shift to the next generation non-volatile memory era is expected.

978-1-4799-2817-0/14 $31.00 © 2014 IEEE

As the technologies to realize normally-off system,

(1) Hardware technology
- Implementation of the fine-grained power gating
- Optimization of the memory architecture of NVRAM
- High-speed power-up and stabilization technology of analog circuit

(2) Software technology
- Optimization and scheduling of task

(3) Non-volatile memory device technology

And it is needed the co-design of hardware and software technology (Fig. 3).

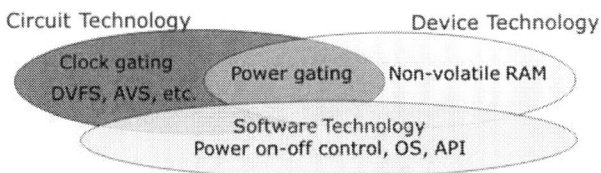

Combining hardware and software technologies.

Fig. 3. Cooperation of normally-off technology

Normally-off computing technology can be broadly applicable to the high-speed processor, and health care and sensor network to achieve smart society that do not require high speed.

II. Lower power requirements of sensor nodes in smart society

The so-called cyber-physical system [1], which combines the large-scale data processing by the server and cloud on the Internet and the sensor nodes controlling sensors and actuators by the network, is a figure of next generation embedded systems aim. Future, information equipment, human and things are connected in the network, and the large amount of data is handled.

The cyber physical system is a foundation technology of smart community, and aims the realization of advanced services by processing the data of the natural environment and social environment that is poured in real time to the internet from the sensor (Fig. 4).

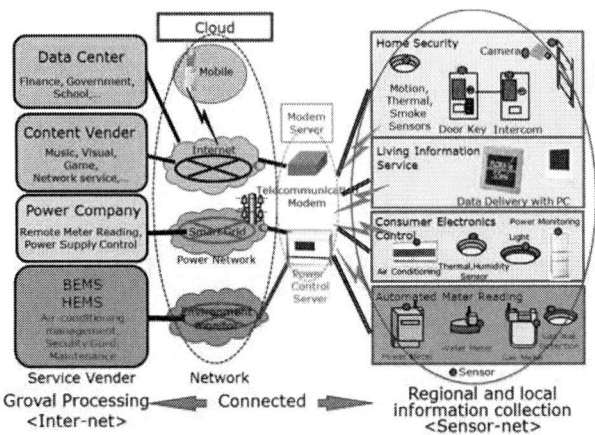

Fig. 4. The cyber-physical systems in smart community

Thus, sensor nodes are used extensively in order to gather real-time information in the social environment and natural environment. The production volume of sensor nodes is currently 10 billion units scale, and this volume will be increased with the development of cyber-physical systems. Therefore, it becomes important how to reduce the power consumption of huge sensor nodes.

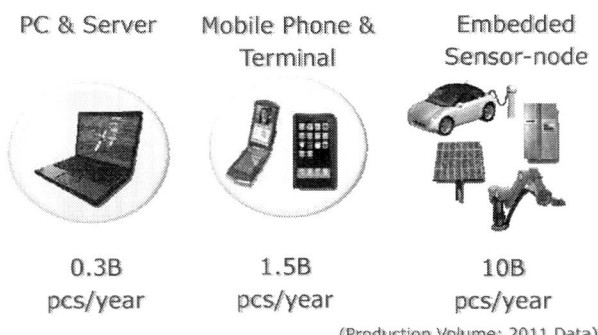

Fig. 5. Production scale of each application equipment

In this paper, we show the feasibility of reducing power consumption by normally-office computer computing control toward the low power consumption of sensor terminal, was taking advantage of environmental data acquisition of the sensor, as an application of these sensors such terminals I will explain about the possibility of low-power consumption at the system level and application study of bus stop on-demand bus.

III. Challenges for low-power sensor node

A. Architecture of current sensor node

The conventional sensor node is consists of sensor modules, and microcontroller (Fig. 6).

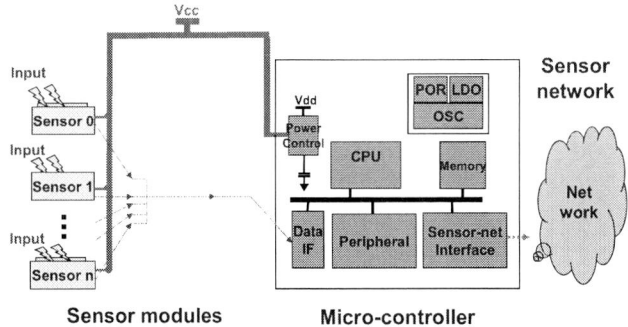

Fig. 6. Architecture for the conventional sensor node

Power is always supplied to the sensor modules and microcontroller. And the processing of sampling, average and etc., of sensor data is performed periodically at microcontroller according to the request from systems, and the processing data is sent out to sensor networks through the sensor-net interface.

978-1-4799-2817-0/14 $31.00 © 2014 IEEE   13

B. Possibility of power reduction for the sensor node

In the smart community, the sensors used in environment system (temperature, brightness, motion sensors and etc.) are performed at the data sampling intervals of 50ms~10s. Since power is always supplied to sensor modules and microcontroller, the unnecessary power is consumed during the period that has not been processed (Fig. 7).

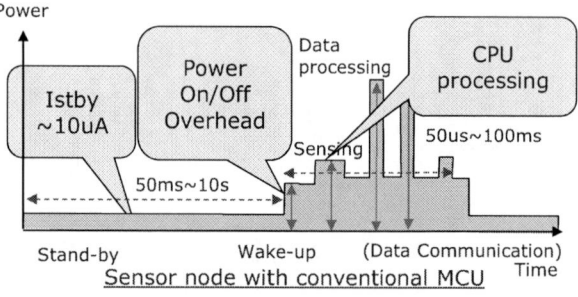

Fig. 7a. Transition of processing and power consumption in sensor system by microcontroller

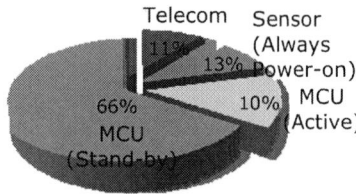

Fig. 7b. Distribution of power consumption of the sensor system

The challenges to realize the low power consumption of above sensor nodes,
- To supply power at sensor modules and microcontroller on when needed,
- To reduce the active currents of microcontroller.

Therefore, it is necessary to minimize the overhead power and time at power return, and reduce the processing load of the microcontroller and maximize the power-off time as long as possible, and also minimize the power-on area. (Fig. 8)

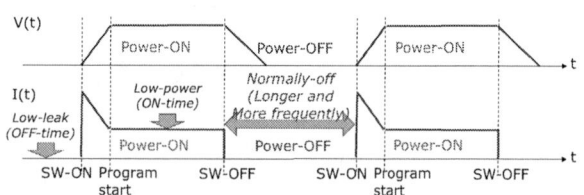

Fig. 8. Waveform of intermittent operation

In order to realize low power sensor node, we propose a normally-off architecture of microcontroller, aiming to achieve a normally-off computing system in future sensor nodes.

IV. Proposal for normally-off architecture for low-power sensor node

A. Normally-off architecture for low-power sensor node

Normally-off architecture for future low-power sensor node is consists of
- normally-off power manager (a),
- sensors,
- sensor controller (b),
- sensor data buffer (c),
- microcontroller
(Fig. 9).

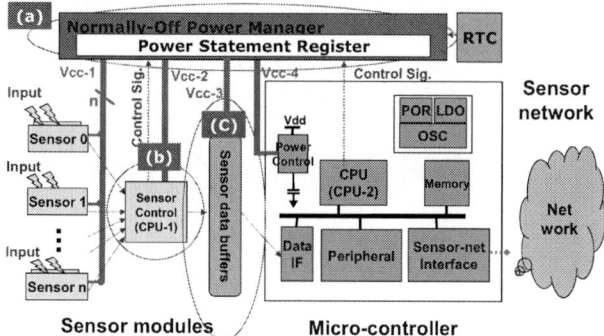

Fig. 9. Normally-off architecture for low-power sensor node

In this work, normally-off power manager (a) and sensor controller (b) and sensor data buffer (c) are applied to the conventional sensor node. And the roles and an example of the processing of these elements are described.

(a) Normally-off Power Manager

Normally-off power manager manages the task scheduling and power control of sensor, sensor controller, sensor data buffer and microcontroller.

(b) Sensor controller

The relatively simple processing of the data sampling, etc., which were carried out in the microcontroller so far, are performed in the sensor controller.

(c) Sensor data buffer

The data processed in sensor controller (b) are temporarily stored in sensor data buffer, and are processed in microcontroller only when needed for the averaging process , etc.. Therefore, it is possible to reduce the operating frequency of the microcontroller, and to maximize the power-off time of microcontroller.

In the normally-off architecture in this work, the co-design of hardware technology (optimization of power control) and software technology (maximization of power-off time by the task scheduling) is important in order to reduce power consumption effectively in sensor node.

B. Normally-off power management scheme

The target of normally-off power management in this work

is to achieve optimal power control in consideration of the breakeven time (BET) in sensor nodes and new ideas of two are proposed. One is an autonomous standby mode transition technology and another is a task scheduling technology. By these ideas, it becomes possible of energy optimization in a multi-sensor network system and of usability improvement by ease of application software development. The details of the two ideas are described below.

(1) Autonomous standby mode transition technology

Now, the development of low-power technology of microcontroller (MCU), various attempts have been made, and are classified as below (Fig. 10).
- Reduction of the CPU activating time by the low-power / high-performance CPU.
- Minimization of standby power by the standby mode setting of multiple.
- Minimization of CPU activating time and CPU start-up cycles by the improved peripheral circuit.
- Minimization of wasted operating time by the CPU start-up time reduction.

These are low power technologies for efficient use of standby mode. The purpose is to reduce the power consumption energy, by allowing transition period as long as possible in lower power standby mode and by reducing of the overhead time of the state transition of standby and active.

The software development is performed while checking the power consumption of each operation in real time by using the development tools in order to minimize the power consumption energy. However, it is difficult to select the optimum standby mode in consideration of the dynamic changes in condition and environment of use.

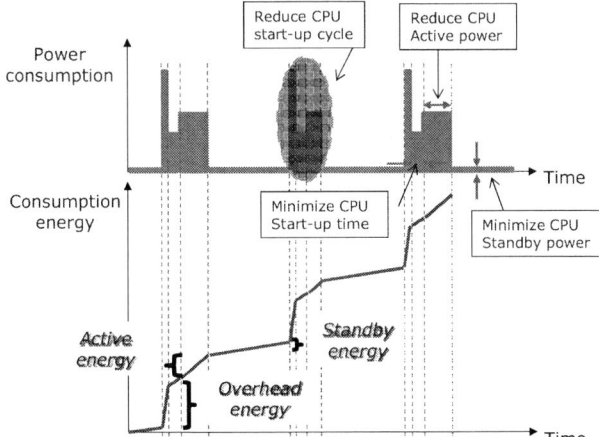

Fig. 10. Power consumption transition of CPU

In order to maximize the effect of low-power consumption technology of MCU, the autonomous standby mode transition technology is to make select and transition the optimum standby mode autonomously to minimize power consumption energy by quantifying the constraints of hardware and software.

Understanding of the constraint condition of hardware side and software side is needed in order to select the optimal standby mode. The constraint of the hardware side is the

breakeven time (BET), which is derived from standby power consumption specific to each device and the energy overhead of the state transition of the standby and active. The constraint of the software side is the waiting time depending on the application.

The power consumption energy in the standby state and the energy overhead of returning from the standby state are schematically described about supported standby mode in microcontroller (Fig. 11). Some standby modes are existed in microcontroller. One case is that energy consumption in standby mode is very low as power off, but the overhead energy at return from standby mode is very large. Other case is that the return from standby mode is fast, but power consumption energy in standby mode is not reduced as IDLE mode. Each standby mode has different breakeven time (BET). When the waiting time at the request of the software is more than BET, it is better to make a transition to a deeper standby mode. In this work, the optimal standby mode can be selected autonomously by normally-off power manager, which include the autonomous standby mode selection scheme.

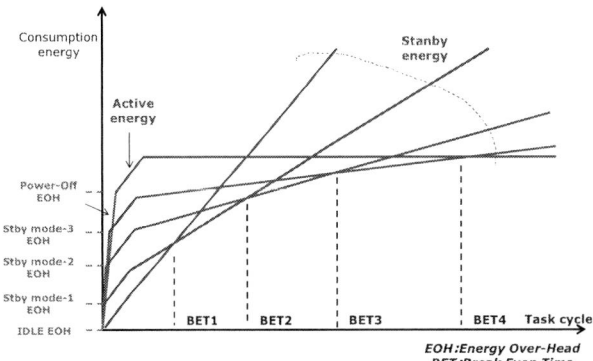

Fig. 11. Break-even time of each standby mode

(2) Activity localization technology with task scheduling method

The activity localization technology by using task scheduling method is described to be utilized for the low-power sensor application.

In the conventional power gating (PG) control, the microcontroller is activated when every data sampling process of sensors is performed (Fig. 12). Therefore, total of power consumption energy is large with increase of number of power-on/off cycles of microcontroller.

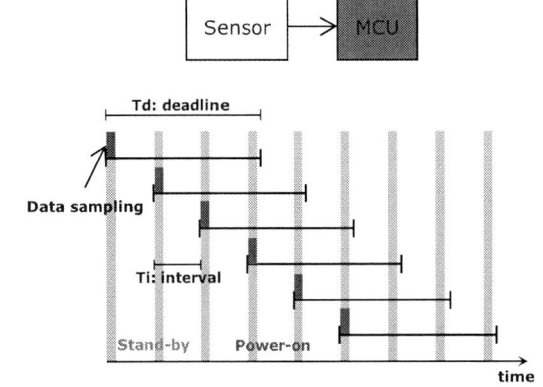

Fig. 12. Conventional power-gating (PG) control

978-1-4799-2817-0/14 $31.00 © 2014 IEEE   15

In this proposed hierarchical power gating (PG) control, sensing data is buffered in sensor data buffer once, and after then microcontroller is activated and performed the process together (Fig. 13). Therefore, it is possible to optimize the number of power on/off cycles and much decrease total of power consumption energy.

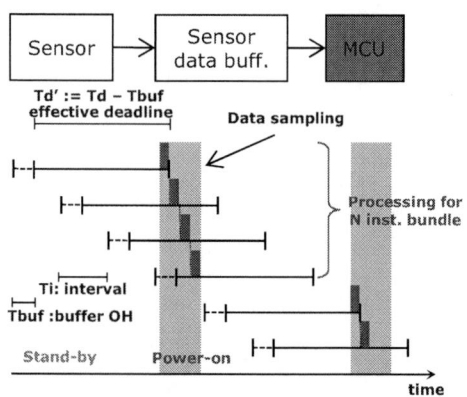

Fig. 13. Hierarchical power-gating (PG) control

In our proposed normally-off architecture for low-power sensor node, the activity localization can be realized with effects of sensor controller (b) and sensor data buffer (c).

C. Evaluation Results

We evaluated the low-power effects at sensor application in our proposed normally-off architecture by using the evaluation board (Fig.14).
The evaluation board is consist of,
- Sensor (Temperature, Brightness, Pyroelectric(Motion),
- Microcontroller,
- Other peripheral circuits.

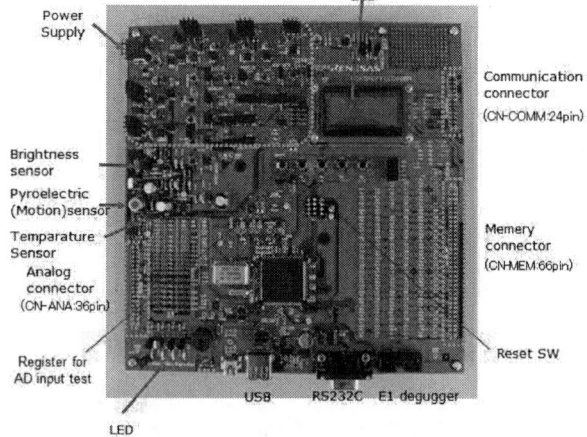

Fig. 14. Normally-off evaluation board

Evaluation results are below (Fig. 15). In conventional sensor node, sensors are always power on, and microcontroller (MCU) is applied with conventional power gating (PG) control. In this work, sensors are intermittent power on, and MCU is applied with the proposed hierarchical PG control.

As the results, it can be obtained the power consumption energy reduction of 67%.

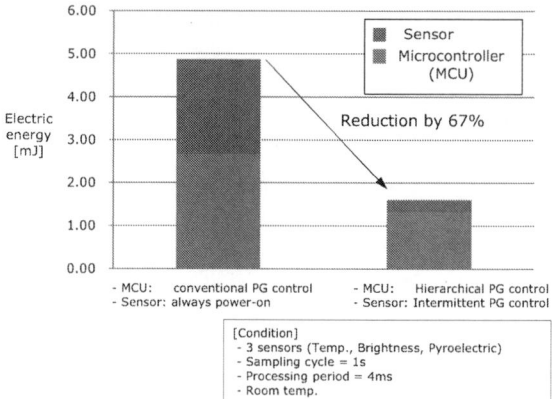

Fig. 15. Evaluation results

The lower consumption of active currents is possible by reducing the operating time of the microcomputer further, and a further power reduction in sensor nodes is expected by task optimization in normally-off architecture.

V. Summary and Conclusions

Sensor nodes are used extensively in order to gather real-time information in the social environment and natural environment. And the production volume of sensor nodes is much increased with the development of cyber-physical systems. Therefore, it becomes important how to reduce the power consumption of huge sensor nodes. In this work, normally-off architecture of microcontroller for future low-power sensor node has been proposed. To realize true low-power effects with normally-off computing technology, a co-design of hardware and software technology is much important. In this work, the power consumption of sensor nodes is possible to reduce of 67% by using normally-off MCU architecture in sensor node. The lower consumption of active currents is possible by reducing the operating time of the microcomputer further, and a further power reduction in sensor nodes is expected by task optimization in normally-off architecture.

Acknowledgments

This work is supported by Normally-Off Computing Project of the New Energy and Industrial Technology Development Organization (NEDO) in Japan.

References

[1] Cyber-Physical Systems Executive Summary, Prepared by the CPS Steering Group, March 6, 2008.

Normally-Off Technologies for Healthcare Appliance

Shintaro Izumi, Hiroshi Kawaguchi, and Masahiko Yoshimoto

Kobe University
Kobe, Japan, 6578501
Tel : +81-78-803-6629
e-mail : {shin, kawapy, yosimoto}@cs28.cs.kobe-u.ac.jp

Yoshikazu Fujimori

Rohm
Kyoto, Japan, 8891695
Tel : +81- 985-85-5111
yoshikazu.fujimori@lsi.rohm.co.jp

Abstract - Battery mass and power consumption of wearable system must be reduced because the key factors affecting wearable system usability are miniaturization and weight reduction. This report describes a wearable biosignal monitoring system using normally-off technologies to minimize the power consumption. Especially we focused on daily-life monitoring and electrocardiograph (ECG) processor. Our system employs Ferroelectric Random Access Memory (FeRAM) and Near Field Communication (NFC) for normally-off data logging and normally-off data communication. A robust heart rate monitor and Cortex M0 core are used to on-node processing for logging data reduction.

I Introduction

Because of the advent of an aging society in Japan, mobile health plays an ever more prominent role [1]. Daily-life monitoring is especially important in preventing lifestyle diseases, which have rapidly increased the number of patients and elderly people requiring nursing care. Our goal is the monitoring and display of vital signals and physical activity in daily life to improve users' quality of life and realize a smart society.

This report specifically describes a wearable biosignal monitoring system, which can acquire long-term instantaneous heart rate (IHR) data and an acceleration value. The physical activities in daily life (e.g., locomotive, household activities) are classifiable using a triaxial accelerometer [2]. The IHR, which is calculated from the interval of R-waves in electrocardiogram (ECG), is useful for heart disease detection, heart rate variation (HRV) analysis [3], and exercise intensity estimation [4].

The key factors affecting wearable system usability are miniaturization and weight reduction. Battery mass and power consumption must be reduced because battery mass dominates wearable systems. To reduce the power consumption, a wearable and wireless ECG telemetry system [5, 6] and single-chip ECG monitoring system LSIs [7-9] have been developed. However, strict limitations on power consumption and electrode distance of wearable ECG monitors render them sensitive to noise of various kinds. Especially, if a subject is not at rest (e.g. during exercise), the signal-to-noise ratio (SNR) of ECG signals will be significantly degraded.

To realize the low-power and noise-tolerant system, we proposed a ECG processor using normally-off architecture and robust IHR monitor [10].

II. Normally-off Wearable Biosignal Monitor Design

A. System Description and Architecture

Fig. 1 presents an overview of the wearable healthcare system, comprising the proposed ECG processor, Near Field Communication (NFC) tag IC, and accelerometer IC. The NFC is used for program loading, individual optimization, and data retrieval from the ECG processor.

Fig. 2 presents a block diagram showing the proposed ECG processor, which consists of an ECG sensing block, 64-Kbyte Ferroelectric Random Access Memory (FeRAM), 32-bit Coretex-M0 core, and extra interfaces.

The ECG sensing block has an analog front end (AFE), a 12-bit SAR ADC, and a robust IHR monitor. The AFE includes a 34-dB gain instrumental amplifier and a 20-dB gain amplifier. The ADC sampling rate can be set to 1 kSamples/s for ECG processing mode and 128 Samples/s for IHR monitoring mode.

The operating frequency of the Cortex M0 core, which is used for an on-node vital signal processing, is 24 MHz, whereas the operating frequency of other digital blocks is 32 kHz. The slow signals in the 32-kHz domain are synchronized at the low-speed bus to the 24-MHz domain. When the Cortex M0 core is in a deep sleep state, the on-chip 24-MHz oscillator is also stopped.

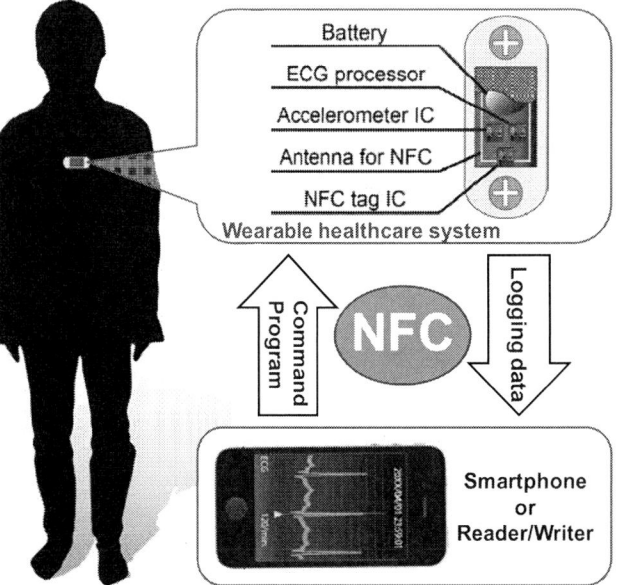

Fig. 1.   Wearable healthcare system overview.

978-1-4799-2817-0/14 $31.00 © 2014 IEEE

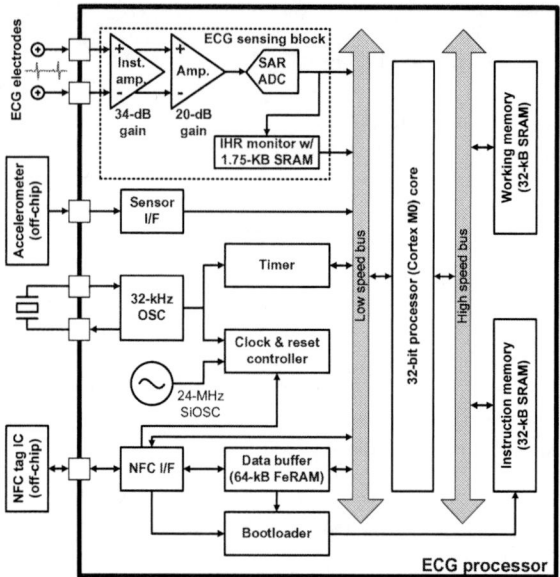

Fig. 2. Block diagram of ECG processor.

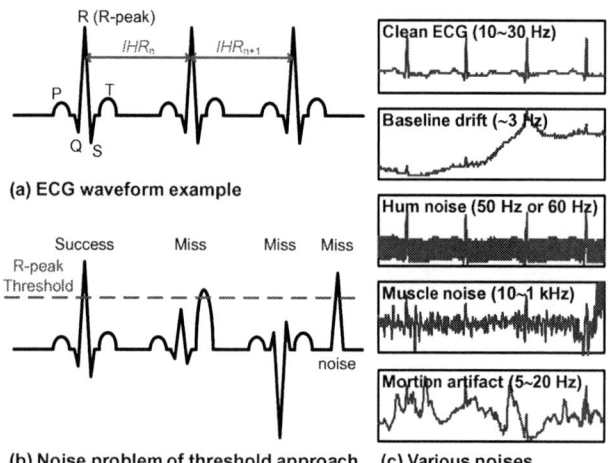

(a) ECG waveform example

(b) Noise problem of threshold approach (c) Various noises

Fig. 3. Threshold based R-wave detection and its noise problems in wearable healthcare systems.

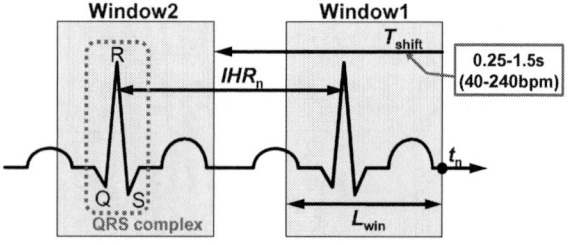

Fig. 4. IHR extraction with short-term autocorrelation (STAC).

B. Normally-off data logging using FeRAM and IHR monitor

Because the frequency range of vital signals is low (less than 1 kHz), both the standby power reduction and sleep time maximization are important to minimize the total power consumption. The 64-Kbyte FeRAM is integrated as a data buffer for daily life monitoring because the leakage current of the data buffer is dominant in the standby state. However, the

FeRAM has larger read/write power, lower speed and endurance compared with SRAM. To mitigate these issues of FeRAM, our system has on-node IHR monitoring system and CM0 core to reduce the logging data. The detail of robust IHR monitor is described in Sect. III.

C. Normally-off data communication using NFC

The power consumption of transceiver circuits is ten or hundred times larger than other blocks, although the active time can be suppressed by data compression and on-node processing. The stand-by power of receiver circuit is also a problem. Our proposed system employs NFC to cooperate with smartphone (or a reader/writer). Several smartphones and tablet devices now support NFC technology.

Generally, a wireless transceiver consumes the greatest amount of power in the biosignal monitoring system. However, compared with Bluetooth Low Energy or ZigBee, the standby power of NFC is extremely small. Moreover, the active communication energy is only consumed on the reader/writer side when using a passive mode, in which only the initiator generates a carrier during communications [11]. In other words, the transceiver of our wearable biosignal monitor can achieve normally-off communication. In addition, NFC enables us to initiate communication without manually configuring the communication link, unlike Bluetooth [12]. The NFC also has high security because it has a short communication range. Some secure payment services already use NFC [12].

In our system, the NFC controller can update the stored program to configure the logging parameters. The logging data can be retrieved directly from the data buffer. The user can also directly access the low speed data bus to communicate with CM0. The logging start, logging stop, and system reset command are issued from a smartphone to CM0.

III Robust IHR Monitor

A. IHR Extraction Algorithm

The wearable ECG monitor is sensitive to extraneous noise because its electrodes are close together. The SNR of ECG signals will be especially degraded if a user is not at rest. Consequently, a sophisticated and costly analog front end is usually required. However, the feature and purpose of our approach is digital signal processing to reduce the performance requirements of the analog portion and to minimize the overall system power consumption.

Extracting R-waves (see Fig. 3(a)) with threshold determination is a general approach. Recently, various statistical approaches have been proposed for noise-tolerant threshold calculation such as using root-mean-squares (RMS) [13], standard deviations (SD), and mean deviations (MD) [14]. However, as depicted in Fig. 3, both misdetection and false detection are increased in the wearable healthcare system by noise from various sources such as myoelectric signals from muscle and electrode movement because the power consumption and electrode distance of the wearable sensor are strictly limited to reduce its size and weight.

Autocorrelation [15, 16] and template matching [17] are

978-1-4799-2817-0/14 $31.00 © 2014 IEEE   18

more robust approaches to prevent incorrect detection because these algorithms use the similarity of QRS complex waveforms and have no threshold calculation process. Autocorrelation has been used in a non-invasive monitoring system [16]. However, the method necessitates numerous computations because it calculates the average heart-rate over a long duration (30 s). In our previous work, a short-term autocorrelation (STAC) technique was proposed for IHR detection [18].

Fig. 4 portrays IHR extraction using STAC. As depicted in Fig. 4 and (1–4), the IHR at time t_n (IHR_n) is obtained as a window shift length (T_{shift}) that maximizes the correlation coefficient between the template window and the search window (CC_n). The STAC method can improve the noise tolerance about 5.6 dB with a 95% success rate.

$$CC_n\left[T_{shift}\right] = w_1 \cdot \sum_{i=0}^{L_{win}-1} Q_w\left[t_n - i\right] \cdot Q_w\left[\left(t_n - T_{shift}\right) - i\right] \quad (1)$$

$$IHR_n = \arg_{T_{shift}} \max_{0.25 \times F_s \leq T_{shift} \leq 1.5 \times F_s} \left\{CC_n\left[T_{shift}\right]\right\} \quad (2)$$

$$L_{win} = 1.5 \times F_s \quad (3)$$

$$w_1 = \begin{cases} 1 & \left(T_{shift} \leq 0.546 \times F_s\right) \\ 0.75 & \left(0.546 \times F_s < T_{shift} \leq 0.983 \times F_s\right) \\ 0.5 & \left(0.983 \times F_s < T_{shift}\right) \end{cases} \quad (4)$$

In the equations presented above, F_s, L_{win}, and w_1 respectively denote the sampling rate (samples/s), the window length, and the weight coefficient. The value of T_{shift} is set as 0.25 s to 1.5 s because the heart rate of a healthy subject is 40 bpm to 240 bpm. The L_{win} is updated according to the estimated IHR to reduce the computational amount and to improve the IHR estimation accuracy. Then, the range of L_{win} and w_1 is determined by the maximum rate of the beat-to-beat variation, which is generally 20% in a healthy subject [19].

B. Hardware Implementation

We proposed the robust IHR monitor hardware, which employs two-step noise reduction technique. In the first stage, a quadratic spline wavelet transform (QSWT) [20] is used to mitigate the baseline drift and hum noise. The QSWT requires few calculations and low hardware cost because it can be implemented using only adders and shift operators. Fig. 5 presents a block diagram and frequency characteristics of the QSWT with 128-Hz sampling rate. The baseline drift and hum noise can be removed easily using QSWT. Unfortunately, it is difficult to remove the myoelectric noise and electrode motion artifacts only using QSWT because these frequency ranges are similar to the desired ECG signal.

Therefore, in the second stage, the IHR is extracted using the STAC method. The STAC is also implemented as dedicated hardware to minimize the power overhead. Fig. 6 presents the block diagram of the IHR monitor and STAC processing core. Each STAC core has CC buffer to store the intermediate value of $CC_n[T_{shift}]$ in (1). The CC buffer is updated in synchronization with ADC output (see Fig. 7). Since the L_{win} is 1.5 s and because IHR is updated every second, two STAC cores alternately calculate IHR with 0.5 s overlap.

The gate level simulation result shows the IHR monitor block, which contains QSWT, two STAC cores, and SRAMs, consumes 1.21 μA. The digital logic and SRAMs respectively consume 0.26 μA and 0.95 μA.

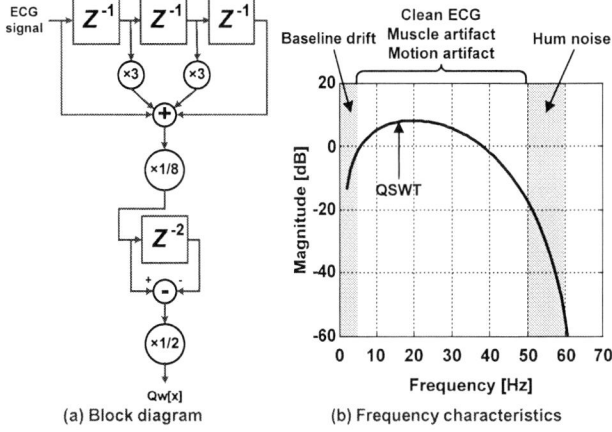

(a) Block diagram   (b) Frequency characteristics

Fig. 5. Block diagram and frequency characteristics of QSWT.

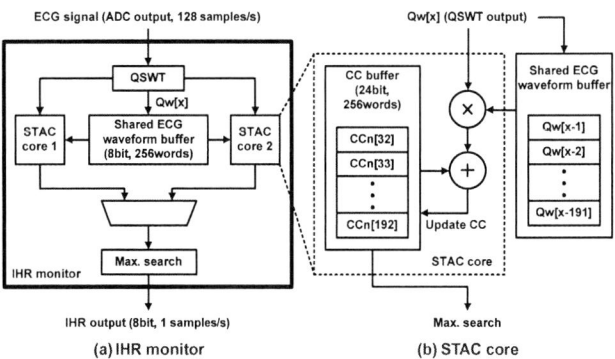

(a) IHR monitor   (b) STAC core

Fig. 6. Block diagram of robust IHR monitor.

Fig. 7. Timing chart of IHR extraction.

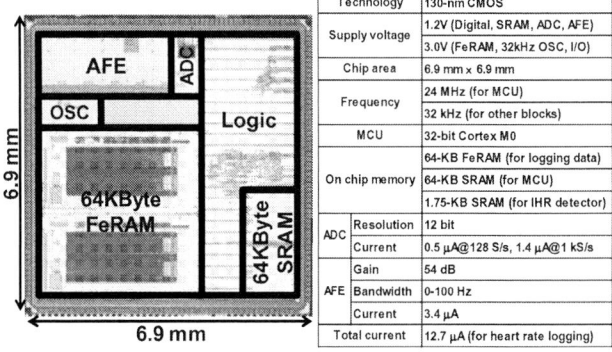

Fig. 8. Chip photograph and chip specifications.

Technology	130-nm CMOS
Supply voltage	1.2V (Digital, SRAM, ADC, AFE)
	3.0V (FeRAM, 32kHz OSC, I/O)
Chip area	6.9 mm × 6.9 mm
Frequency	24 MHz (for MCU)
	32 kHz (for other blocks)
MCU	32-bit Cortex M0
On chip memory	64-KB FeRAM (for logging data)
	64-KB SRAM (for MCU)
	1.75-KB SRAM (for IHR detector)
ADC	Resolution	12 bit
	Current	0.5 μA@128 S/s, 1.4 μA@1 kS/s
AFE	Gain	54 dB
	Bandwidth	0-100 Hz
	Current	3.4 μA
Total current	12.7 μA (for heart rate logging)

978-1-4799-2817-0/14 $31.00 © 2014 IEEE

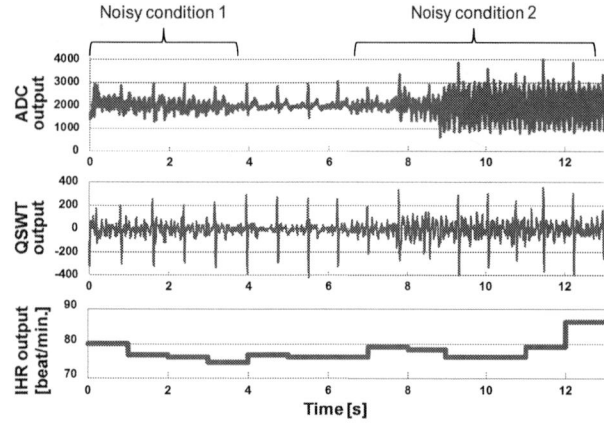

Noisy condition 1   Noisy condition 2

Fig. 9.   Measured waveform of IHR monitor in a noisy condition.

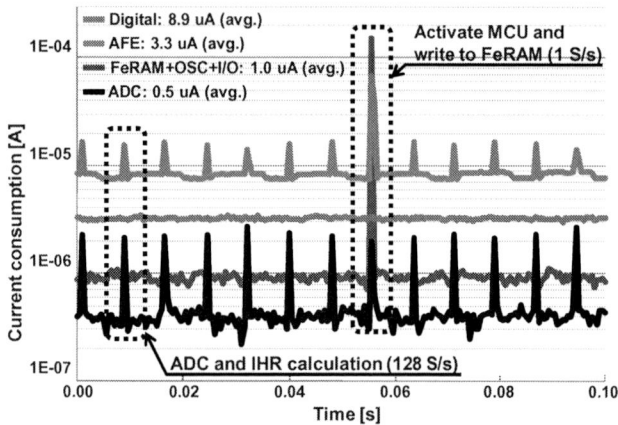

Fig. 10.   Measurement result of current consumption with a heart rate logging application.

IV. Implementation Result

The test chip is fabricated using 130-nm CMOS technology. Fig. 8 presents a chip photograph and a performance summary. The operating voltage is 1.2 V for AFE, ADC, SRAM, 24-MHz oscillator, and digital blocks. The FeRAM, 32-KHz oscillator, and IO circuits are operated with 3.0 V supply voltage.

To demonstrate the test chip performance, we implemented a heart rate logging application. As portrayed in Fig. 9, the IHR is extracted correctly in a noisy condition. Fig. 10 portrays the current consumption with a heart rate logging application. In this experiment, the ADC sampling rate and the logging interval of IHR are set respectively to 128 Samples/s and 1 Sample/s. Then the AFE, 32-kHz OSC, and Timer block are always activated. The measurement results show that the test chip consumes 13.7 μA on average for the heart rate logging application. The peak current, which is consumed when the Cortex and FeRAM are activated to store the logging IHR data every second, is less than 1 mA.

Acknowledgments

This research was supported by the Ministry of Economy, Trade and Industry (METI) and the New Energy an Industrial Technology Development Organization (NEDO).

References

[1] H. Nakajima, T. Shiga, and Y. Hara, "Systems Health Care," In Proc. of IEEE SMC, pp. 1167-1172, Oct. 2011.

[2] Y. Oshima, K. Kawaguchi, S. Tanaka, K. Ohkawara, Y. Hikihara, K. Ishikawa-Tanaka, and I. Tabata, "Classifying household and locomotive activities using a triaxial accelerometer," Gait and Posture, vol. 31, pp. 370-374, 2010.

[3] W. Roel and M. John, "Comparing Spectra of a series of Point Events Particularly for Heart Rate Variability Data," IEEE T-BME, BME-31, no.4, pp.384-387, Apr. 1984.

[4] S. Yazaki and T. Matsunaga, "Evaluation of activity level of daily life based on heart rate and acceleration," In Proc. of SICE, pp. 1002-1005, Aug. 2010.

[5] K. Itao, T. Umeda, G. Lopez, M. Kinjo, "Human Recorder System Development for Sensing the Autonomic Nervous System," In Proc. of IEEE Sensors, pp.423-426, Oct. 2008.

[6] K. Itao, T.Ito, "Integrated Sensing Systems for Health and Safety," In Proc. of DTIP Symposium, pp.212-216, May 2011.

[7] F. Zhang, Y. Zhang, J. Silver, et al., "A Battery-less 19μW MICS/ISM-Band Energy Harvesting Body Area Sensor Node SoC," ISSCC, pp. 298-299, Feb. 2012.

[8] H. Kim, R. F. Yazicioglu, S.Kim, et al., "A configurable and low-power mixed signal SoC for portable ECG monitoring applications," VLSI Symp., pp.142-143, Jun. 2011.

[9] S. Y. Hsu, Y. L. Chen, P. Y. Chang, et al., "A Micropower Biomedical Signal Processor for Mobile Healthcare Applications," In Proc. of IEEE ASSCC, pp.301-304, Nov. 2011.

[10] S. Izumi, K. Yamashita, M. Nakano, et al., " A 14 μA ECG Processor with Robust Heart Rate Monitor for a Wearable Healthcare System," In Proc. of IEEE ESSCIRC, pp.145-148, 2013.

[11] E. Strommer, J. Kaartinen, J. Parkka, A. Ylisaukko-oja, and I. Korhonen, "Application of Near Field Communication for Health Monitoring in Daily Life," In Proc. of IEEE EMBS, pp. 3246-3249, Aug. 2006.

[12] J. Morak, H. Kumpusch, D. Hayn, R. Modre-Osprian, and G. Schreier, "Design and Evaluation of a Telemonitoring Concept Based on NFC-Enabled Mobile Phones and Sensor Devices," IEEE T-ITB, vol. 16, no. 1, pp. 17-23, 2012.

[13] H. Kim, R.F. Yazicioglu, et al., "ECG Signal Compression and Classification Algorithm with Quad Level Vector for ECG Holter System," IEEE T-ITB., vol. 14, no. 1, pp. 93-100, Jan, 2010.

[14] J. P. Martinez, R. Almeida, S. Olmos, et al., "A wavelet-based ECG delineator: evaluation on standard databases," IEEE Trans. Biomed. Eng., vol. 51, no. 4, pp. 570-581, Apr. 2004.

[15] Y. Takeuchi, M. Hogaki, "An adaptive correlation rate meter: a new method for Doppler fetal heart rate measurements," Ultrasonics, pp. 127-137, May 1978.

[16] M. Sekine, K. Maeno, "Non-Contact Heart Rate Detection Using Periodic Variation in Doppler Frequency," In Proc. of IEEE SAS, pp. 318-322, Feb. 2011.

[17] H. L. Chan, G. U. Chen, M. A. Lin, and S. C. Fang, "Heartbeat Detection Using Energy Thresholding and Template Match," In Proc. of IEEE EMBC, pp. 6668-6670, Aug. 2005.

[18] M. Nakano, T. Konishi, et al., "Instantaneous Heart Rate detection using short-time autocorrelation for wearable healthcare systems, " In Proc. of EMBC, pp. 6703-6706, Aug. 2012.

[19] M. Malik, "Heart Rate Variability; standards of Measurement, Physiological Interpretation, and Clinical Use," Circulation, 1996; 93: 1043-1065.

[20] C. I. Ieong, M. I. Vai, P. U. Mak, P. I. Mak, "ECG Heart Beat Detection via Mathematical Morphology and Quadratic Spline Wavelet Transform," In Proc. of IEEE ICCE, pp. 609-610, Jan. 2011.

978-1-4799-2817-0/14 $31.00 © 2014 IEEE

A Dual-loop Injection-locked PLL with All-digital Background Calibration System for On-chip Clock Generation

Wei Deng, Ahmed Musa, Teerachot Siriburanon, Masaya Miyahara, Kenichi Okada, and
Akira Matsuzawa

Dept. Physical Electronics, Tokyo Institute of Technology
2-12-1-S3-27, Ookayama, Meguro-ku, Tokyo 152-8552 Japan.
Tel/Fax: +81-3-5734-3764, E-mail:deng@ssc.pe.titech.ac.jp

Abstract – This paper presents a compact, low power, and low jitter dual-loop injection-locked PLL with synthesizable all-digital background calibration system for clock generation. Implemented in a 65nm CMOS process, this work demonstrates a 0.7-ps RMS jitter at 1.2 GHz while having 0.97-mW power consumption resulting in an FOM of -243dB. It also consumes an area of only 0.022mm² resulting in the best performance-area trade-off system presented up-to-date.

I. Introduction

For modern SoC systems, stringent requirements on on-chip clock generators include small-area, low-power consumption, PVT-insensitive, and the lowest possible jitter performance [1-5]. Sub-harmonically injection-locked technique can significantly improve random jitter characteristic of a clock generator. To compensate the free-running frequency shift caused by temperature and voltage variations, this paper [4] proposes the use of a dual-loop topology with one free-running VCO (Replica VCO) placed inside a FLL for tracking temperature and voltage drift. The other VCO (Main VCO) is injection-locked for producing a low-jitter clock, while the free-running frequency shift can be compensated by the replica loop.

II. Proposed IL-PLL with all-digital calibration

Block diagram of the proposed system is given in Fig.1. Aside from the VCO and the 10-bit DAC, all circuits that makeup the FLL are implemented using digital standard cells to have a synthesizable all-digital FLL (AD-FLL) design. Therefore, the proposed work can significantly scale down in area due to the absence of a passive loop filter. The oscillation frequency of replica VCO is measured by a digital counter. This counted number is compared to a predefined frequency control word that maps to the expected number of pulses during one cycle for a certain division ratio. Depending on the sign of the output, the up/down counter adjusts both the main and replica VCO frequency.

Fig.2 shows the detailed calibration algorithm. After the initial frequency calibration of the replica VCO is carried out, the counter connected to the replica VCO is disabled and the multiplexer control lines are set to choose the output of the main VCO counter for calibrating frequency offset. In phase II, an offset is added or subtracted from the main VCO counter to compensate frequency difference between it and the replica VCO that might arise from process variations. By comparing the number of pulse of main VCO to the predefined value and incrementing or decrementing the main VCO counter, the required offset to calibrate any difference in frequency will be added or subtracted. Finally, as both VCOs are having the same frequency, the loop is returned to its initial settings to maintain both VCOs frequency over temperature and voltage variations for robust operation. Schematic of the injection-locked VCO is given in Fig. 3, and both main and replica VCOs share the same topology

and layout to minimize mismatches.

III. Measurement Results

The proposed circuit is fabricated in a 65 nm CMOS process. The phase noise before and after performing injection-locking is measured by using a signal source analyzer and is given in Fig.4. This phase noise maps to a 0.7 ps jitter when integrated from 10 kHz to 40 MHz. The proposed dual-loop IL-PLL has an operating range of 0.5-to-1.6 GHz, and it consumes a total power consumption of 0.97 mW excluding output buffer, from a 1 V power supply. The measured reference spur is -57 dBc. The reference clock can be varied from 40 to 300 MHz. All the above-mentioned measurements are performed at the room temperature.

Fig. 5 shows the effect of sweeping the temperature on the measured peak-to-peak and RMS jitter for IL-PLL with and without replica loop, at a carrier of 1.2 GHz. Configured as a conventional IL-PLL without the replica loop, it is observed that the locking point shifts to the edge of the locking range as temperature gradually increases. This causes a significantly deterioration of the peak-to-peak and RMS jitter. On the other hand, it is clearly shown that the proposed IL-PLL with replica loop manages to maintain the locked state and correct frequency over the range from 0 °C to 80 °C, which validates the effectiveness of the IL-PLL with the dual-loop PVT calibration.

Fig .6 gives a comparison between this work and previously published work. The proposed dual-loop IL-PLL achieves comparable performance with the-state-of-the-art, while only occupies 0.022 mm² chip area. The figure of merit (FOM) is -243dB at a 1.2 GHz carrier. The proposed circuit realizes a low-jitter and small-area clock generation with a robust operation over different operating conditions. The die micrograph is shown in Fig. 7.

IV. Conclusion

This paper [4] proposes a dual-loop injection-locked PLL with synthesizable all-digital PVT calibration circuits. With carful design, the proposed PLL can be suited for clock generation in wireline and wireless systems.

Acknowledgements

This work was partially supported by STARC, MIC, SCOPE, MEXT, Canon Foundation and VDEC in collaboration with Cadence Design Systems, Inc., and Agilent Technologies Japan, Ltd.

References

[1] A. Elshazly, et al., "A 1.5GHz 890μW digital MDLL with 400fsrms integrated jitter, -55.6dBc reference spur and 20fs/mV supply-noise sensitivity using 1b TDC," *ISSCC Dig. Tech. Papers*, pp. 242-243, Feb. 2012.

[2] B. Helal, et al., "A highly digital MDLL-based clock multiplier that leverages a self-scrambling time-to-digital converter to achieve subpicosecond jitter performance," *IEEE J. Solid-State Circuits*, vol.43, no.4, pp. 855-863, Apr. 2008.

978-1-4799-2817-0/14 $31.00 © 2014 IEEE

[3] G. Xiang, *et al.*, "A 2.2GHz 7.6mW sub-sampling PLL with −126dBc/Hz in-band phase noise and 0.15psrms jitter in 0.18μm CMOS," *ISSCC Dig. Tech. Papers*, pp.392-393, Feb. 2009.

[4] Wei Deng, Ahmed Musa, Teerachot Siriburanon, Masaya Miyahara, Kenichi Okada, and Akira Matsuzawa, "A 0.022mm² 970μW dual-Loop injection-locked PLL with -243 dB FOM using synthesizable all-digital PVT calibration circuits," *ISSCC Dig. Tech. Papers*, pp. 248-249, Feb. 2013.

[5] C. Liang and K. Hsiao, "An Injection-Locked Ring PLL with Self-Aligned Injection Window," *ISSCC Dig. Tech. Papers*, pp. 90-91, Feb. 2011.

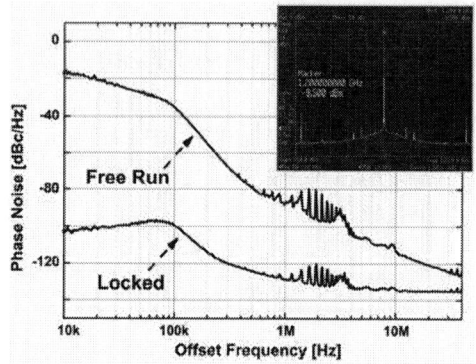

Fig.4. Measured phase noise and spectrum at a carrier of 1.2GHz.

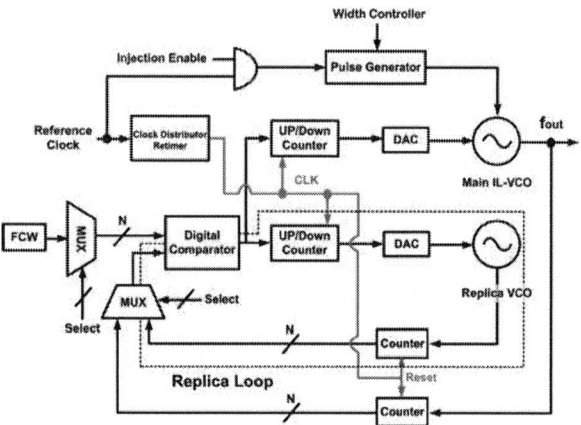

Fig.1. Block diagram of the proposed dual-loop IL-PLL.

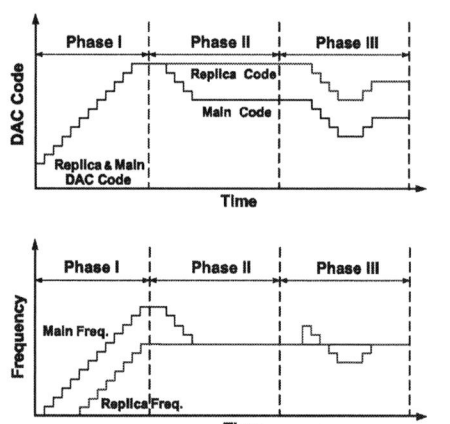

Fig.2. Main and replica DACs output code and output frequency at different calibrations phases.

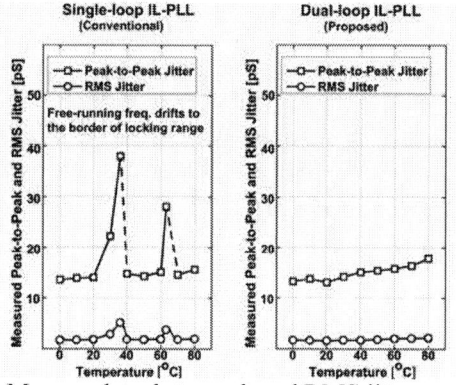

Fig.5. Measured peak-to-peak and RMS jitter against temperature variations.

	This work	[1]		[2]	[5]
	IL-PLL	DMDLL	DPLL	MDLL	IL-PLL
Freq. [GHz]	0.5-1.6	0.8-1.8	0.8-1.8	1.6	0.216
Ref. [MHz]	40-300	375	375	50	27
Power [mW]	0.97	0.89	1.35	12	6.9
Area [mm²]	0.022	0.25	0.25	0.058	0.03
Integ. Jitter [ps]	0.7	0.4	3.2	0.68	2.4
Jitter RMS/PP [ps]	1.81/19.4 10M hits	0.92/9.2 5M hits	4.2/33 5M hits	0.93/11.1 30M hits	N.A.
Ref. Spur [dBc]	-57	-55.6	-40.5	-58.3	-70.7
FOM [dB]	-243	-248.46	-228.59	-233.76	-225
CMOS Technology	65nm	130nm	130nm	130nm	55nm

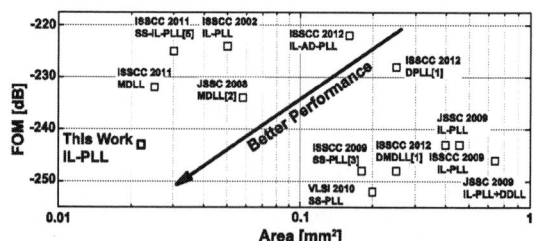

Fig.6. Performance summary and comparison with state-of-the-art prior work.

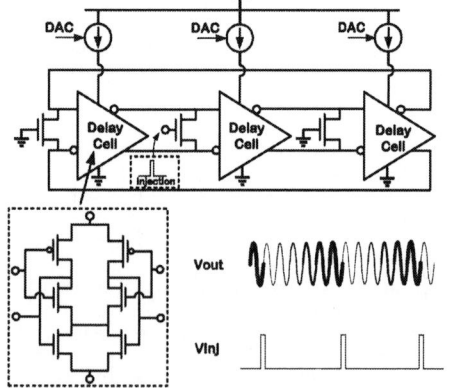

Fig.3. Schematics of the injection-locked ring oscillator.

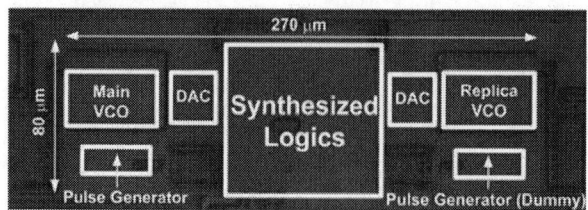

Fig.7. Chip micrograph.

978-1-4799-2817-0/14 $31.00 © 2014 IEEE   22

A 950μW 5.5-GHz Low Voltage PLL with Digitally-Calibrated ILFD and Linearized Varactor

Sho Ikeda, Tatsuya Kamimura, Sangyeop Lee, Hiroyuki Ito, Noboru Ishihara, and Kazuya Masu

Solutions Research Laboratory, Tokyo Institute of Technology
4259-S2-14 Nagatsuta, Midori-ku, Yokohama 226-8503, Japan
E-mail: paper@lsi.pi.titech.ac.jp

Abstract— This paper proposes an ultra-low-power 5.5-GHz PLL which employs a divide-by-4 injection-locked frequency divider (ILFD), which is calibrated by digital circuits, and linearity-compensated varactors for low supply-voltage operation. The proposed PLL was fabricated in 65 nm CMOS. It shows a 1-MHz-offset phase noise of −106 dBc/Hz and the total power consumption of 950 μW at 5.5 GHz.

I. INTRODUCTION

A phase locked loop (PLL) is one of the key components to save power of an analog/RF front-end. The use of low-voltage supply is a common approach to reduce power of PLLs, especially voltage-controlled oscillators (VCO) and frequency dividers (FD) that consume larger power than other PLL components [1]. However, it is quite difficult to achieve low voltage operation in PLL because of the limited threshold voltage of MOS transistors.

This paper proposes a PLL that employs an ultra-low-power divide-by-4 injection-locked frequency divider (ILFD), whose output frequency is controlled by the digital calibration circuit automatically. Furthermore, the linear tuning characteristics with a forward-body-biased (FBB) varactor make the PLL operation stable even under the supply of 0.5 V.

II. PROPOSED PLL

Fig. 1 shows proposed PLL architecture. The PLL contains a Class-C VCO and a divide-by-4 ILFD with its digital calibration circuit.

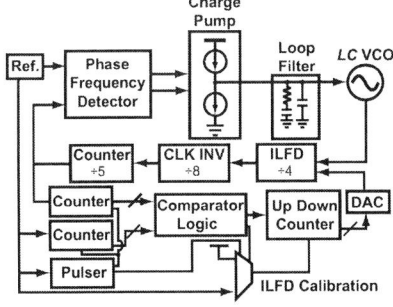

Fig. 1. Proposed PLL.

A. Proposed ILFD

The first stage of divider chain usually requires much power consumption in high-frequency-operation PLL. To achieve lower power consumption, a divide-by-4 ILFD, which can reduce the number of divider stages, is employed as shown in Fig. 2 (a). To achieve divide-by-4 characteristics, the ILFD employs the *double-switch injection* technique. By applying

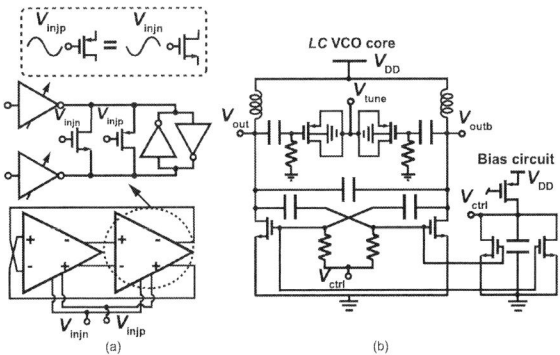

Fig. 2. Proposed (a) divide-by-4 ILFD, (b) Class-C VCO.

the injection switches for the I/Q output of the ILFD simultaneously, the ILFD has wider lock range in the case of dividng-by-4 [2]. To achieve further wide lock range, the proposed ILFD has differential input nodes. These two differential inputs, that are injected to the nMOS and pMOS respectively, have the same function. Therefore, the balance between differential VCO outputs can be improved and the lock range of the ILFD is also widened.

To adjust the free-running frequency of the ILFD, circuits for the digital calibration is applied as shown in Fig. 1. The digital circuit controll the DAC and converts the digital code into analog control voltage for frequency tuning of ILFD. In steady state, the MUX stops to supply the clock for the up-down counter and the DAC keeps the control voltage constant. The pulser makes pulses every eighteen periods of the reference signal, and resets the counter to prevent overflowing of counters.

B. VCO

Fig. 2 (b) shows the proposed Class-C VCO [4], which can achieve low power and low phase noise, with bias circuit [3] to control the gate bias for Class-C operation. To achieve the stable operation of PLL under low supply voltage, a linear-region-compensated varactor is proposed.

Using the low supply voltage causes non linearity of the VCO because the active control voltage range of the VCO would be narrower. In this case, the active voltage range cannot match the linear region of the varactor which is fundamentally used in the normal supply voltage operation. The non-linearity of the VCO may deteriorate the PLL stability and phase noise characteristics.

To solve this problem, a linearized varactor for the low voltage operation is proposed as shown in Fig. 2 (b). To shift the linear region of the varactor to lower control voltage, the FBB is applied to the pMOS varactor to lower the threshold voltage.

978-1-4799-2817-0/14 $31.00 © 2014 IEEE

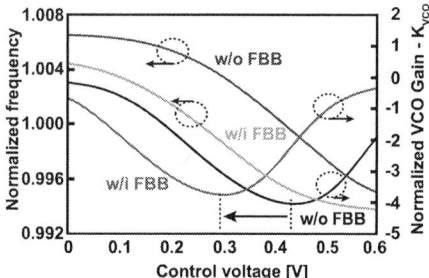

Fig. 3. VCO tuning characteristic and VCO Gain w/o and w/i FBB.

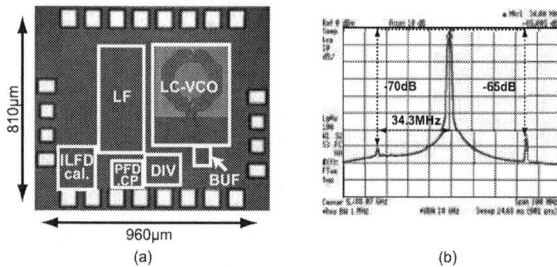

Fig. 4. (a) A chip micrograph of the proposed PLL. (b) Measured spectrum at f_{out} = 5.49 GHz.

With this technique, the linear region of the varactor is shifted to about half V_{DD} when the supply voltage is 0.5 V.

Fig. 3 shows the simulation results of VCO tuning characteristics and its VCO gain (K_{VCO}) with and without FBB for the varactor. The linear region without FBB is located at higher voltage than 0.4 V. On the other hand, Fig. 3 also depicts that the linear region can be shifted by about 0.1 V and be around $V_{DD}/2$ by means of the threshold voltage reduction caused by the FBB.

III. MEASUREMENT RESULTS

Fig. 4 (a) shows a chip micrograph of the proposed PLL. The area is 960 μm × 810 μm. The PLL is fabricated in a 65nm Si CMOS process and measured with a 0.54 V supply voltage for divider and 0.5 V for other components at the output frequency of 5.49 GHz. The total power consumption was 950 μW. As a reference signal, 34.3-MHz sine wave (f_{ref}) was input during the measurement. Fig. 4 (b) shows a frequency spectrum of the PLL output, and the reference spurious level of -65 and -70 dBc were observed. Measured phase noise at a 1-MHz offset frequency was -106 dBc/Hz as shown in Fig. 5.

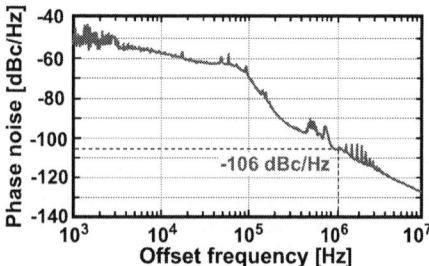

Fig. 5. Measured phase noise at f_{out} = 5.49 GHz.

Performance summary of the proposed PLL and its comparison with other low-power-supply PLLs are given in Ta-

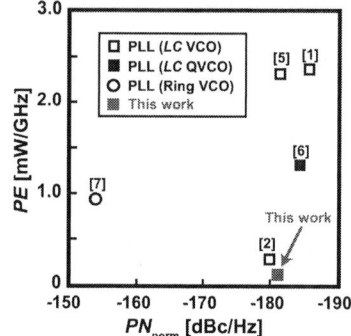

Fig. 6. Performance comparison among low-power-supply PLLs.

TABLE I
PERFORMANCE SUMMARY AND COMPARISON OF LOW-POWER-SUPPLY PLLs.

Ref.	V_{DD} [V]	f_{out} [GHz]	PN@1 MHz [dBc/Hz]	Power [mW]
This work	0.5/0.54*	5.49	−106	0.95
[1]	0.5	1.9	−120	4.5
[2]	0.5	5.54	−105	1.6
[5]	0.5/0.65**	2.59	−113	6.0
[6]	0.5/0.8***	9.12	−105	12
[7]	0.5	2.24	−87	2.1

*0.54 V: Divider, 0.5 V: Others supply, **0.5 V: Analog, 0.65 V: Digital supply, ***0.5 V: Analog, 0.8 V: Digital supply.

ble I. In addition, to make a fair power consumption and phase noise comparison, the power efficiency (PE) and normalized phase noise (PN_{norm}) are used in Fig. 6. These factors can be expressed as PE =(power consumption)/f_{out} and $PN_{norm} = PN - 20\log(f_{out}/\Delta f)$, respectively. As shown in Fig. 6, the proposed PLL achieves the finest PE with relatively good PN_{norm}.

IV. CONCLUSION

We proposed the ultra-low-power PLL with an ILFD and a linearized varactor under low supply voltage operation. The FBB is applied in the MOS varactor to compensate the linearity under low voltage operation, which can achieve the stability of PLL operation. To adjust the control voltage of ILFD automatically, the digital calibration circuit is applied.

The concept was verified with a fabricated chip in a 65 nm Si CMOS process. The proposed PLL shows the great power efficiency with acceptable phase noise characteristics under low supply voltage, at the output frequency of 5.49 GHz.

REFERENCES

[1] H.-H. Hsieh, et al., "A 0.5-V 1.9-GHz low-power phase-locked loop in 0.18-μm CMOS," *VLSI Circuits*, pp. 164–165, Jun. 2007.

[2] S. Ikeda, et al., "A 0.5-V 5.5-GHz Class-C-VCO-Based PLL with Ultra-Low-Power ILFD in 65 nm CMOS," *ASSCC*, pp. 357–360, Nov. 2012.

[3] M. Tohidian, et al., "High-swing class-C VCO," *ESSCIRC*, pp. 495–498, Sep. 2011.

[4] A. Mazzanti, et al., "Class-C harmonic CMOS VCOs, with a general result on phase noise," *JSSC*, vol. 43, no. 12, pp. 2716–2729, Dec. 2008.

[5] S.-A. Yu, et al., "A 0.65-V 2.5-GHz fractional-N synthesizer with two-point 2-Mb/s GFSK data modulation," *JSSC*, vol. 44, no. 9, pp. 2411–2425, Sep. 2009.

[6] C.-Y. Yang, et al., "A 0.5/0.8-V 9-GHz frequency synthesizer with doubling generation in 0.13-μm CMOS," *TCASII*, vol. 58, no. 2, pp. 65–69, Feb. 2001.

[7] K.-H. Cheng, et al., "A 0.5-V 0.4–2.24-GHz inductorless phase-locked loop in a system-on-chip," *TCASI*, vol. 58, no. 5, pp. 849–859, May. 2011.

A Swing-Enhanced Current-Reuse Class-C VCO with Dynamic Bias Control Circuits

Teerachot Siriburanon, Wei Deng, Kenichi Okada, and Akira Matsuzawa

Dept. Physical Electronics, Tokyo Institute of Technology

2-12-1-S3-27, Ookayama, Meguro-ku, Tokyo 152-8552 Japan.

Tel/Fax: +81-3-5734-3764, E-mail: tee@ssc.pe.titech.ac.jp

Abstract – A swing-enhanced current-reuse class-C VCO which can theoretically achieve same phase noise figure-of-merit (FoM) as other class-C VCOs at the lowest power consumption is presented. A swing enhancement in class-C operation and an oscillation robustness are achieved through dynamic bias control circuits for both NMOS and PMOS transistors. The proposed VCO has been fabricated in 180nm CMOS process while oscillating at 4.6 GHz. The measured phase noise is -119 dBc/Hz at 1 MHz offset while consuming 1.6 mA from 1.5 V supply. An FoM of -189 dBc/Hz is achieved.

I. Introduction

Recently, a rapid growth of personal wireless communication has driven the demands for low-cost low-power wireless transceivers. To perform a reliable communication of a low-power system, a design of low-power and low-phase-noise VCOs has attracted great attentions from both industry and academia.

Due to phase noise performance superiority, LC-based VCOs are usually chosen over ring-based oscillators. For a classical LC-tank oscillators, NMOS and CMOS VCOs are standard choices for RF circuit design. As shown in Fig. 1, even though CMOS VCO shares the same maximum phase noise FoM comparing to an NMOS counterpart, it can achieve 6 dB better FoM with the same power budget and LC tank while operating in current-limited regime. Moreover, its oscillation swing is within the supply rail which help alleviates reliability issue. To achieve similar performance at lower power consumption, a current-reuse VCO consumes only half power that of CMOS VCO [1]. However, in order to avoid asymmetrical waveforms, additional components are required to operate both transistors in current-limited region at expense of a swing headroom. To further improve performance, Class-C VCOs which have higher DC-RF current conversion efficiency can provide a theoretical 3.9dB lower FoM comparing to conventional Class-B VCOs [2]. Similar to conventional class of VCOs, a class-C CMOS VCO can provide 6dB better phase noise compared to class-C NMOS VCO when operating in the current-limited region. In this work, a current-reuse class-C VCO which has a same theoretical FoM comparing to NMOS and CMOS class-C VCOs [2]-[3] while consuming the lowest power consumption is proposed as shown in Fig.1.

II. Circuit Design and Implementation

For a conventional Class-C VCO based on NMOS topology, there exists a tradeoff for larger oscillation swing and robustness in oscillation startup. An approach to overcome this issue is to provide an adaptive gate bias for NMOS transistors to be higher at the initial state and lower at the steady state [4]-[5]. In order to further improve power efficiency, a lower-power current-reuse class-C VCO [6] is proposed with a dynamic biasing for both NMOS and PMOS transistors to provide a robustness in oscillation at the initial state and provide high oscillation swing at the steady state.

The detailed schematic of the proposed swing-enhanced current-reuse class-C VCO using dynamic bias control circuits is depicted in Fig.2. It is composed of three main components, *i.e.*, a core VCO and dynamic bias circuits for NMOS and PMOS transistors. The core VCO is composed of only one pair of cross-coupled PMOS and NMOS transistors. The dynamic bias control circuits are composed of two modified current mirrors as shown in Fig.2.

If the current-reuse VCO enters voltage-limited region, the outputs will have asymmetric waveforms. Moreover, the conduction angle of devices increases and the drain current shapes are widened which loses its high DC-to-RF current conversion efficiency [2]. To avoid such issue, both NMOS and PMOS transistors should remain in active region. By assuming the same common voltage, the maximum oscillation amplitude is limited by the difference between steady state bias of gate bias and threshold voltage of PMOS and NMOS transistors [6]. To ensure its oscillation start-up, higher and lower gate biases for NMOS and PMOS, respectively, are necessary. Once the steady oscillation has been built, $V_{g,p}$ and $V_{g,n}$ can adaptively change its value to become higher and lower, respectively, which in turns, maximizing oscillation swing at the steady state. The circuit operation can be described as follows. From Fig. 3 and 4, before VCO starts to oscillate, initial gate biases of NMOS and PMOS transistors are determined by $I_{REF,n}$ and $I_{REF,p}$. Once the current in the core oscillator provides enough transconductance to meet the oscillation conditions, output voltage starts to swing across V_{CM}. Then, the adaptive bias scheme acts like a negative feedback which senses an oscillation swing and adaptively changes gate biases of both transistors to enhance its maximum swing in a current-limited regime at the steady state.

From simulation, under a 1.5-V supply, at the initial state, gate biases of NMOS and PMOS are 0.6V and 0.8V, respectively, from 68μA of reference current. At the steady state, the gate biases of NMOS and PMOS adaptively change to 0.25V and 1.05V, respectively, as shown in Fig.4. This results to a swing enhancement at steady state. Moreover, the proposed VCO can maintain balanced waveform as shown in Fig.5 and achieve better phase noise performance since both transistors never enter deep triode region.

III. Measurement Results

The proposed circuit is fabricated in a 180-nm CMOS process. The die micrograph is shown in Fig.6. The measured phase noise is given in Fig.7. The measured tuning range is 4.5GHz to 4.6GHz. For 1.5V supply, a phase noise of -119dBc/Hz at 1MHz offset can be achieved while consuming 1.6mA resulting in a figure of merit of -189 dBc/Hz. Table I summarizes the comparison to the state-of-the-art VCOs in current-reuse topologies. The proposed VCO achieves a few dB better than the state-of-the-art CMOS VCO and similar performance comparing to the state-of-the-art current-reuse VCOs.

IV. Conclusion

In this paper, a swing-enhanced current-reuse class-C VCO [6] using dynamic bias control circuits is proposed.

The proposed dynamic bias scheme keeps both transistors in a proper operation for high current efficiency and a balanced tank waveform.

Acknowledgements

This work was partially supported by MIC, SCOPE, MEXT, STARC, Canon Foundation, and VDEC in collaboration with Cadence Design Systems, Inc., and Agilent Technologies Japan, Ltd.

References

[1] S-J. Yun, *et. al*, "A 1 mW Current-Reuse CMOS Diffential LC-VCO with Low Phase Noise," *IEEE International Solid-State Circuits Conference (ISSCC), Digest of Technical Papers*, pp. 540-541, 2005

[2] A. Mazzanti and P. Andreani, "Class-C harmonic CMOS VCOs, with a general result on phase noise," *IEEE Journal of Solid-State Circuits*, vol. 43, no. 12, pp. 2716-2729, 2008.

[3] A. Mazzanti and P. Andreani, "A Push-Pull Class-C CMOS VCO," *IEEE Journal of Solid-State Circuits*, vol. 48, no. 3, pp. 724-732, 2013.

[4] Wei Deng, Kenichi Okada, and Akira Matsuzawa, "A Feedback Class-C VCO With Robust Startup Condition over PVT Variations and Enhanced Oscillation Swing," *IEEE European Solid-State Circuits Conference (ESSCIRC)*, pp. 499-502, Sep., 2011.

[5] Wei Deng, Kenichi Okada, and Akira Matsuzawa, "Class-C VCO With Amplitude Feedback Loop for Robust Start-Up and Enhanced Oscillation Swing," *IEEE Journal of Solid-State Circuits*, vol. 48, no. 2, pp. 429-440, 2013.

[6] Teerachot Siriburanon, Wei Deng, Kenichi Okada, and Akira

Matsuzawa, "A Current-Reuse Class-C LC-VCO with an Adaptive Bias Scheme," *IEEE Radio Frequency Integrated Circuits Symposium (RFIC)*, pp. 35-38, June, 2013.

[7] S. Levantino, M. Zanuso, C. Samori, and A. Lacaita, "Suppression of Flicker Noise Upconversion in a 65nm CMOS VCO in the 3.0-to-3.6GHz Band," *IEEE International Solid-State Circuits Conference (ISSCC), Digest of Technical Papers*, pp. 50-51, 2010.

[8] M. Taghivand, M. Ghahramani, and M. P. Flynn, "A Low Voltage Sub 300uW 2.5GHz Current Reuse VCO," in *Proceedings of IEEE Asian Solid-State Circuits Conference Digest of Technical Papers*, Nov. 2012.

[9] C. Yang and Y. Chiang, "Low Phase-Noise and Low-Power CMOS VCO Constructed in Current-Reused Configuration," *IEEE Wireless and Microwave Components Letters*, Vol. 18, pp. 136-138, Feb. 2008.

[10] M. Wei, S. Chang, and S. Huang, "An amplitude-balanced current-reused CMOS VCO using spontaneuos transconductance match technique," *IEEE Wireless and Microwave Components Letters*, Vol.19, pp.395-397, 2009.

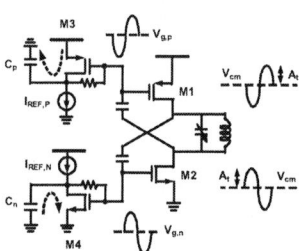

Fig.3. Simplified diagram of the proposed swing-enhanced class-c current reuse VCO to describe its operation.

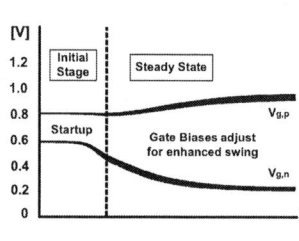

Fig.4. Simulated transient waveforms of dynamic gate biases of PMOS, NMOS.

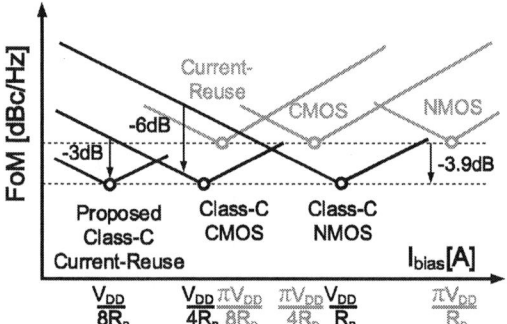

Fig.1. Theoretical FoM limit of conventional class of VCOs and its corresponding Class-C VCOs.

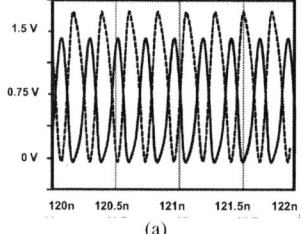

(a)

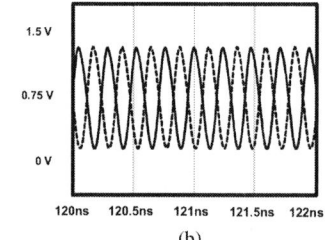

(b)

Fig.5 Output amplitude of current-reuse VCO (a) without and (b) with dynamic bias control circuits.

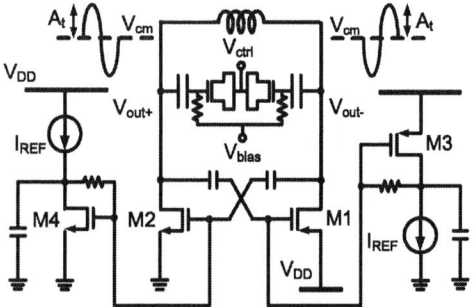

Fig.2. Proposed swing-enhanced current-reuse class-C VCO using dynamic bias control circuits

Fig.6. Chip Micrograph

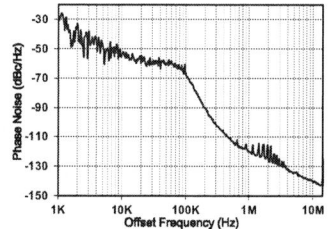

Fig.7. Measured phase noise plot at 4.63GHz

TABLE I COMPARISON WITH THE-STATE-OF-THE-ART VCOs IN A CURRENT-REUSE TOPOLOGY

Ref.	Tech.	VCO Topology	Frequency (GHz)	Phase Noise (dBc/Hz)	Power (mW)	FoM (dBc/Hz)	Additional components for Balanced Amplitude
[7]	65nm	CMOS	3	-114@1MHz offset	0.7	-187	-
[8]	65nm	Current-Reuse	2.1	-126.1@3MHz offset	0.28	-190	-
[9]	180nm	Current-Reuse	16	-111@1MHz offset	8.1	-187	Required
[10]	180nm	Current-Reuse	2.9	-122@1MHz offset	1.7	-188	Required
[1]	180nm	Current-Reuse	2.0	-123@1MHz offset	1	-189	Required
This	180nm	**Class-C Current-Reuse**	4.6	-119 @1MHz offset	2.4	-189	**Not required**

Design of A High-Performance Millimeter-wave Amplifier Using Specific Modeling

X. J. Bi[1,2], Y. X. Guo[1,3], M. Annamalai. Arasu[2], M. S. Zhang[1], Y. Z. Xiong[2] and M. K. Je[2]

[1]National University of Singapore, 117576, Singapore
[2]Institute of Micro-electronics, Agency for Science, Technology and Research, 117685, Singapore
[3]National University of Singapore (Suzhou) Research Institute, 215123, Suzhou, China

Abstract - In this design contest, the design methodology leading to a high performance Millimeter-wave amplifier in 0.13 μm SiGe BiCMOS is elaborated. Equivalent circuit models of the utilized cascode shielding structure are developed to assist the amplifier design. Meanwhile, final layouts of the passive connections are verified by 3D electromagnetic simulation in ANSYS HFSS. The implemented amplifier obtained a gain more than 45 dB in band, which is the gain record of silicon-based amplifiers in W-band.

I Introduction

The current millimeter-wave radiometers reply on multi-LNAs in cascade to obtain above 40 dB gain with a 20 GHz bandwidth [1]. However, the current gain record of a single LNA is still below 35 dB. Neutralized differential pairs can improve the power amplifiers' performance through enhancing the Q factor of a single stage, nonetheless, they are more suitable for low voltage supply applications, and they still demonstrate lower gain than cascode stages. In this design, Q-enhanced cascode stages together with cascode shielding structures are used to enhance the gain and stability of the amplifier.

Moreover, specific models, including equivalent circuit models for shielding structure optimization and 3D electromagnetic models in ANSYS HFSS for final layout verification, are built to guarantee the success of the design.

II. Design

A. Amplifier Architecture

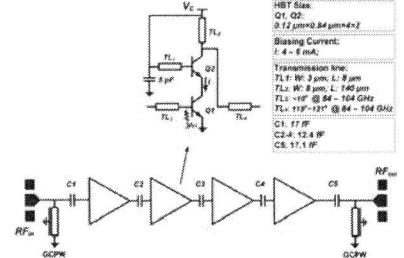

Fig. 1. Schematic of the proposed four-stage cascode amplifier.

Fig. 1 illustrates the schematic of our proposed four-stage cascode LNA [2]. The input matching network uses the L-type LC matching network, which aims to minimize noise figure (NF). The inter-stage matching networks are constructed from simple shunt-series matching networks. Q1 and Q2 in the four cascode stages are both constructed by two parallel 0.12 μm×0.84 μm×4 transistors. As shown in Fig. 2 (a), the top 3 μm thick metal layer (TM2) is employed

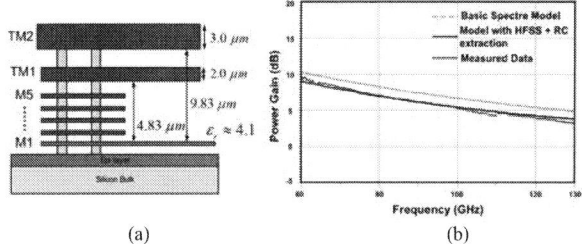

Fig.2 The model verification: (a) Metal layers information for passive modeling; (b) Power gain verification of the used HBT with size of 0.12 μm × 0.84 μm × 4 under bias condition: I_b=25 μA, V_c=1.2 V.

for signal lines and the 0.4 μm thick bottom metal layer (M1) is used for the ground plane. The supply voltage is 1.2 V while the base voltage (V_{b1}) of Q1 is around 0.89 V which leads to a collector current of 4 mA. The first two stages are biased at 4 mA for a minimal noise figure, while the cascode structures at the 3rd and 4th stage facilitate a different biasing V_{b1} for Q1 (I=4-6 mA) in order to assist the optimization of the gain and dynamic range.

B. HBT model verification

The HBTs of T1 and T2 in the four cascode stages are all in size of 0.12 μm × 0.84 μm × 4 × 2. Meanwhile, the used HBT model was verified using both the ANSYS HFSS and Assura RC extraction. Power gain verification of the used HBT with size of 0.12 μm × 0.84 μm × 4 under a bias condition: I_b=25 μA, V_c=1.2 V is performed. As shown in Fig. 2(b), the variation caused by parasitic coupling is underestimated by the basic Spectre model. With the assistance of the above two tools, the simulation result fit well with the measured power gain from 70 GHz to 120 GHz. Since 0.12 μm × 0.84 μm × 4 model demonstrated the best fit with our device characterization results among all provided devices, to guarantee the design precision, the cascode stage are not optimized through sizing M1 and M2.

C. Model of the proposed shielding structure

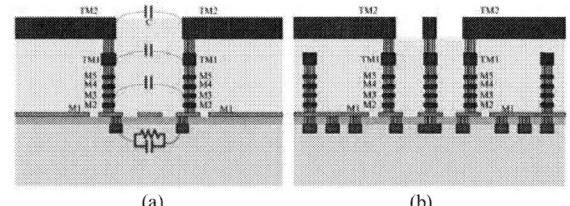

Fig. 3.Cascode shielding structure:(a) Conventional transistor connection (b) Cross view of isolation design

978-1-4799-2817-0/14 $31.00 © 2014 IEEE

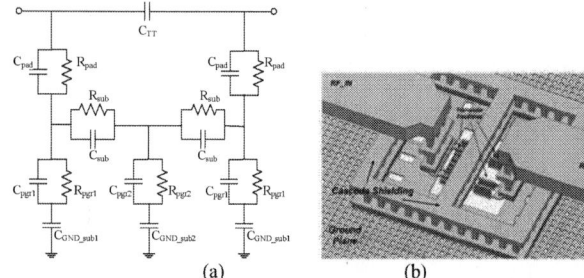

Fig. 4. (a) Simplified circuit model of the shielding structure; (b) 3-D model of the shielding structure in ANSYS HFSS.

Significant coupling, existing between the input and output of each cascode stage, deteriorates the gain and stability. As shown in Fig. 3 (a), there are capacitive coupling between two vias in both the oxide layers and the silicon substrate. As such, considerable reverse coupling is caused. Rectangular grounded shielding structure is built for each cascode stage to increase their reverse isolation. Metal stack arrays, from Top metal down to the P+ layer, are employed for substrate noise suppression between Q1 and Q2. The case using the shielding structure results in a 7 dB reduction of coupling in the oxide layers, compared to the conventional case. Meanwhile, adding metal stack arrays from M1 to P+ improves the isolation by about 3 dB further. The total coupling is reduced from -30 dB to -40 dB for each stage. The above parasitic coupling in dielectric and substrate, and passive loss of the metal stack can be simplified as the circuit model presented in Fig. 4 (a). The capacitive coupling between the two feedlines can be approximated by a capacitor C_{TT}. Substrate coupling between the two transistors in close area is modeled by the rest network excluding C_{TT}. The detailed component value and equations cannot be covered here due to the limited content. With the assistance of the above mentioned circuit model, the optimum shielding structure size can be determined. The final layout of the shielding structure is verified in ANSYS HFSS as shown in Fig. 4(b).

D. Q-enhancement

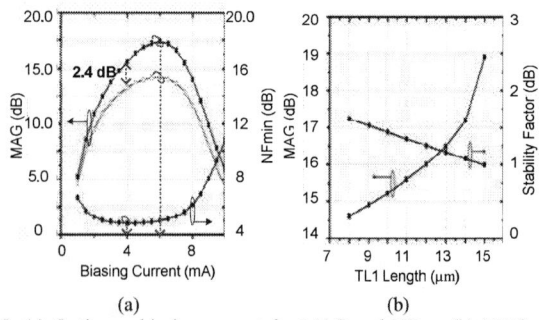

Fig. 5. (a) Optimum biasing current for MAG and NF_{min}; (b) MAG and stability factor versus TL1 length.

The high-Q amplifying stages are realized by connecting a transmission line, TL1 to the base of Q2 as shown in Fig. 1; As shown in Fig. 5 (a), the biasing current is 4 mA per stage, the MAG can be boosted for 2.4 dB. As shown in Fig. 5 (b), the stability approaches 1 if TL1 increases to 15 μm. Therefore, the gain performance of this cascode stage is

eventually comprised with stability. On the other hand, due to the aforementioned shielding structure, the stability has been compensated due to the reduced reverse isolation. From the layout optimization work, an optimal TL1 length of 8 μm is finally chosen to maintain a K>1.5 for each stage.

III. EXPERIMENTAL RESULTS

The amplifier was implemented in 0.13-μm SiGe BiCMOS technology. With 4 mA collector biasing current of the used HBT, the f_T is 228 GHz. For measurement of the gain and noise of this amplifier in W-band, the Y-factor technique was used. The noise characterization system contains a W-band noise source, WR-10 waveguides, WR-10 to coaxial adaptors, 1.85 mm coaxial cables and on-wafer probing. Furthermore, the measuring frequency range is extended to 110 GHz by a 67-GHz spectrum analyzer (R&S FSU) through an external harmonic mixer.

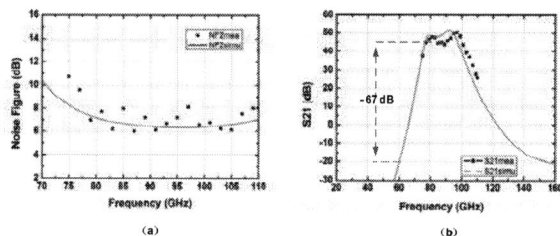

Fig. 6. Simulated and measured (a) Noise Figure; (b) Gain.

As shown in Fig. 6 (a) and (b), the implemented amplifier has obtained a noise figure of 6.0 to 8.1 dB, a gain of > 45 dB in band and power consumption of 19.2 mW. The simulated OP1dB of this amplifier is around +0.2 dBm and the corresponding IP1dB is around -43.8 dBm at 83 GHz. The IP1dB can be optimized to -29 dBm with 29 dB gain and 12 mW power consumption, according to simulation.

IV. CONCLUSION

In this design contest, a high gain W-band amplifier in 0.13 μm BiCMOS technology is presented. Shielding structures for each stage and Q-enhanced cascode topology are deployed to improve the gain and stability of the amplifier. Using specific modeling, the implemented amplifier has obtained a gain of > 45 dB which is the gain record in W-band.

Acknowledgement

This work was supported by NUSRI under the grant number NUSRI-R-2012-N-010 and IME, A*Star, Singapore under grant number IME-10-220003.

References

[1] C. D. Dietlein, A. Luukanen, F. Meyer, Z. Popovic, and E. N. Grossman, "Phenomenology of passive broadband terahertz images," presented at the 4th ESA Millimeter Wave Technol. Applicat. Workshop, Espoo, Finland, 2006.

[2] X. J. Bi, Y. X. Guo, Y. Z. Xiong, M. A. Arasu, and M. K. Je, "A 19.2 mW, >45 dB Gain and High-Selectivity 94 GHz LNA in 0.13 μm SiGe BiCMOS," *IEEE Microw. Wireless Compon. Lett.*, vol.23, no.5, pp.261-263, May 2013

A Multi-Mode Reconfigurable Analog Baseband with I/Q Calibration for GNSS Receivers

Zheng Song, Nan Qi, Baoyong Chi, and Zhihua Wang

Institude of Microelectronics, Tsinghua University

Beijing 100084, China

Abstract-A multi-mode reconfigurable analog baseband with I/Q calibration for GNSS receivers is presented. The baseband circuit consists of an I/Q calibration circuit, a reconfigurable complex band-pass filter (C-BPF) and an AGC loop. It provides I/Q mismatch auto-calibration with the aid of a FPGA. The 3rd/ 5th-order reconfigurable C-BPF supports various bandwidths from 2.2 to 10MHz and with different center frequencies from 3.996 to 16MHz. The AGC loop features 5-50dB gain range and 1dB step, and digital AGC algorithms. The auto DC-offset cancellation is also integrated on-chip. The analog baseband circuit has been implemented in 180nm CMOS and consumes 6.5-13mA variable current thanks to the power scaling technique. The measured image-rejection ratio is 45-55dB, improved by 22dB after I/Q calibration.

I Introduction

With the development of the new generation Global Navigation Satellite System (GNSS), reconfigurable multi-system monolithic receivers are now demanding [1]. The analog baseband providing bass-band filtering and programmable gain amplification is one of the most essential parts, which should be reconfigurable to cover different operating modes. Low-IF receiver suffers from the limited image rejection, which is caused by I/Q path imbalances. In typical GNSS receivers, low-resolution ADCs (2-4 bit outputs) make it impossible to achieve high IRR with the aid of digital-only calibration.

This paper presents the analog baseband which consist of I/Q calibration, reconfigurable C-BPF and AGC. In order to improve the image rejection ratio (IRR), the I/Q calibration are implemented before the filter. The C-BPF can be reconfigurable to work in either 3rd or 5th order mode, and the center frequency (3.996, 7.161, 10.324, 13.29 and 16 MHz) and the bandwidth (2.2, 4.2, 6 and 10MHz) are reconfigurable.

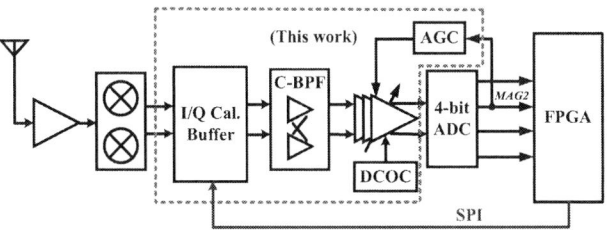

Fig1. The analog baseband in the whole receiver

II. Implementation of analog baseband circuits

A. Complex band-pass filter (C-BPF)

A reconfigurable complex band-pass filter is employed in the analog baseband to provide the out-of-band suppression as well as the image rejection. Switchable resistors R_C are cross-coupled between the I/Q path, adjusting the C-BPF's

center frequency (Fig. 2c). The filter's bandwidth can be reconfigured by adjusting R_F, and the process-induced variation can be compensated by the fine tuning of C_F with the help of on-chip frequency tuner. Besides, the filter's order can be switched from 5th to 3rd by shorting two integrator stages to save power. In addition, each of the OTA served in the filter is realized in arrays, and is power scalable along with bandwidth requirements. More sub-arrays would be enabled for high frequency modes [2].

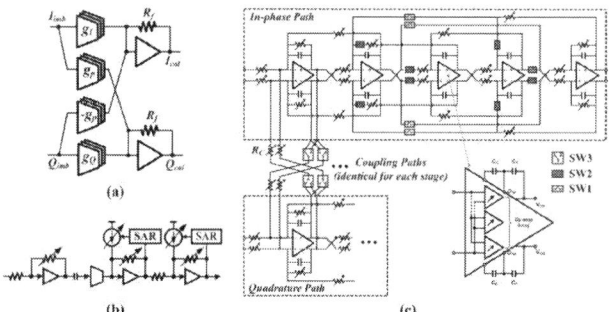

Fig2. Schematic of the analog baseband: (a) I/Q calibration, (b) PGA, (c) C-BPF.

B. I/Q calibration

In general, the in-phase and quadrature signals have the same amplitude and 90 degree phase difference. I/Q amplitude and phase imbalance would leads to the leakage of image interference.

Many digital I/Q calibration techniques are proposed, and the I/Q calibration is implemented in the digital baseband after high-resolution ADC. However, only with the high-resolution ADC, the high IRR can be achieved. In our work, a self-calibrated technique with low resolution ADC is implemented (Fig.2a), in which the I/Q calibration is placed before the filter and the digital control module is implemented by FPGA after low-resolution ADC. A test tone signal should be pumped into the LNA, the image interferences will be sensed by the power detector after ADC. The FPGA will continuously change the control words dependent on the last received power until the detected power reaches its minimum value.

C. PGA and AGC Loop

A programmable-gain amplifier with auto gain-control is realized in this work. The PGA contains three stages providing 5-50dB dynamic gain range with 1dB step. Traditionally, DC-offset would be removed with high-pass passive filter in high voltage gain mode which consumes a lot of die area. In our work, a DC-offset auto-cancellation is proposed to save chip area. The SAR logic adjusts the current value injecting in the amplifier input of PGA, dependent on the detecting result of the comparator following the PGA, until the DC-offset is negligible(Fig.2b).

An AGC loop is employed to regulate the PGA's output signal strength, which monitors the highest magnitude output of the flash ADC, accumulates its value within a long time, and compares it to the predetermined threshold.

III. Measured Results

The proposed analog baseband has been implemented in 180nm CMOS process, Fig3 shows its die photo.

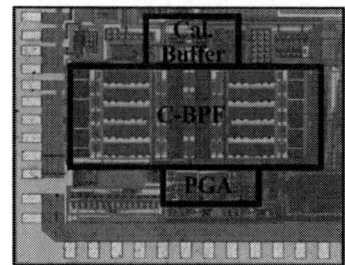

Fig.3 Die photo.

A. I/Q calibration

The IRR is tested by injecting RF test-tones at the image and desired band. Just depicted in the Fig.4, the IRR is improved from 28.2dB to 49.9dB after I/Q calibration.

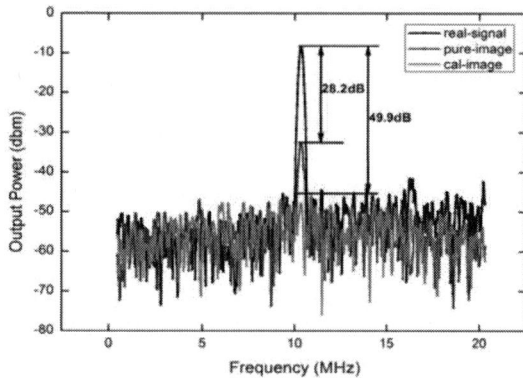

Fig.4 Tested IRR results, before and after I/Q calibration

B. C-BPF

The frequency response is measured by sweeping the RF input signal and observing the output of the PGA on a spectrum analyzer. As shown in Fig.5, the bandwidths vary from 2.2MHz to 16MHz.

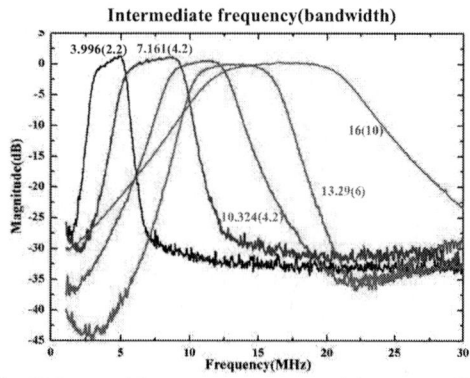

Fig.5 Measured Frequency Response of the C-BPF filter

C. PGA

The PGA is tested on a spectrum analyzer injecting a RF signal and observed in the PGA output. The measured result indicates that the PGA has a gain range from 5dB to 50dB with 1dB step.

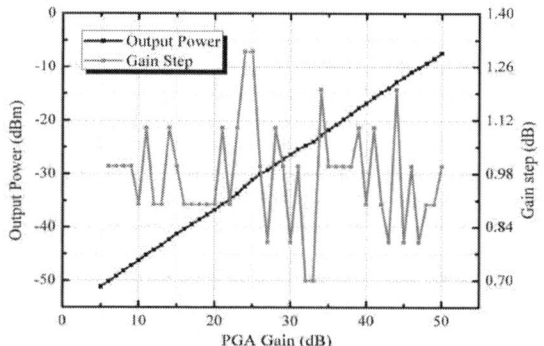

Fig.6 Measured PGA gain and gain step

TABLE I   Comparison of Analog Baseband Performance

Index	This work	[4]	[5]
Technology	180nm	90nm	90nm
VDD(V)	1.7	0.6	1
IF(MHz)	4-16	0	0
BW(MHz)	2.2-10	0.9-2.4	0.4-30
Power(mA)	6.1-13	10	13
IRR(dB)	45~55	~33	-

IV. Conclusions

A power-scalable analog baseband circuit for GNSS receiver is implemented in 180nm CMOS process. The C-BPF can be reconfigured to work in fifth-order or third-order mode with the highest IF to 21MHz. The PGA have a gain range of 5-50dB with 1 dB step. Moreover, closed-loop I/Q calibration can achieve 50dB IRR with the aid of FPGA.

References

[1] Nan Qi, et al., "A Dual-Channel Compass/GPS/Glonass/Galileo Reconfigurable GNSS Receiver in 65nm CMOS With On-Chip I/Q Calibration," *IEEE Trans. Circuits and Systems-I*, vol. 59, no. 9, pp. 1720-1732, Aug. 2012.

[2] V. Giannini, et al., "Flexible baseband analog circuits for software-defined radio front-ends," *IEEE J. Solid-State Circuits*, vol. 42, no. 7, pp. 1501–1512, Jul. 2007.

[3] S. Lerstaveesin and B.S. Song, "A complex image rejection circuit with sign detection only," *IEEE J. Solid-State Circuits*, vol. 41, no. 12, pp. 2693–2702, Dec. 2006.

[4] A. Balankutty, et al., "A 0.6-V zero-IF/low-IF receiver with integrated fractional-N synthesizer for 2.4-GHz ISM-band applications," *IEEE J. Solid-State Circuits*, vol. 45, no. 3, pp. 538–553, Mar. 2010.

[5] M. Kitsunezuka, et al., "A Widely-Tunable, Reconfigurable CMOS Analog Baseband IC for Software-Defined Radio," *IEEE J. Solid-State Circuits*, vol. 44, no. 9, pp. 2496-2502, Sept. 2009.

An 8b Extremely Area Efficient Threshold Configuring SAR ADC with Source Voltage Shifting Technique

Kentaro Yoshioka, Akira Shikata, Ryota Sekimoto, Tadahiro Kuroda, and Hiroki Ishikuro

Department of Electronics and Electrical Engineering, Keio University, Yokohama, Japan

Email:yoshioka@iskr.elec.keio.ac.jp

Abstract

An extremely low power and area efficient threshold configuring ADC (TC-ADC) for time interleaved ADC is proposed. The threshold configuring comparator (TCC) performs a binary search. 5b conversion is carried out by TCC with source voltage shifting technique. Additional 2b resolution is achieved by the proposed threshold interpolation (TI) technique with only 15% power overhead. Prototype ADC in 40nm CMOS occupies a core area of only 0.0038mm^2 and when calibration circuit included, 0.0058 mm^2. With a supply voltage of 0.7V, the ADC achieves 7.0 ENOB with 24MS/s. Peak FoM of 9.8fJ/conv. is obtained at 0.5V supply, which is over 15x improvement compared with conventional TC-ADC.

Introduction

ADC which operates at few GS/s with low power and small area is required for high speed wireline and UWB applications. By time interleaving multi-channel ADCs, excellent power efficiency can be achieved even operated at few GS/s[1]. SAR ADCs are often used for the channel ADC because of superior power efficiency compared with other architectures. However, its area is limited by C-DAC process mismatch[2] .

On the other hand, TC-ADCs which perform a binary search by directly configuring the comparator threshold voltage (V_{THcomp}) can aggressively reduce the ADC area[3]. However, to control V_{THcomp}, ref.[3] places large number of capacitors at the output node of the comparator. This increases power consumption and decision time significantly. Furthermore, complexity of V_{THcomp} control increases exponentially as target resolution rise (TABLE I) and ENOB is limited to 5b so far.

In this work[6], we present an 8b TC-ADC which requires only 1/8 of control circuit of V_{THcomp} compared with the conventional technique. The TCC is a hybrid of V_{THcomp} configuration techniques with wide range V_{THcomp} configuring by source voltage shifting and 2b interpolation performed by simple capacitor load switching. The proposed method greatly reduces the digital controlling circuit and power efficiency is improved simultaneously.

Proposed Threshold Configuring ADC

A. Architecture

The block diagram of the proposed circuit is shown in Fig.1. As soon as the input is sampled, MSB (D[7]) decision is made. The 1b-DAC shifts the sampled input reflecting the results[4]. TCC performs 7b (D[6:0]) conversion, where the first 5b is resolved by binary search by threshold configuring and 2b LSB by threshold interpolation. As a result, 8b output (D[7:0]) is obtained.

B. Threshold Configuring Comparator

Variable current sources (VCS)[5] can configure V_{THcomp} and consume less power while keeping high speed operation. However offset current by VCS must be widely changed, especially at sub-threshold region, because the drain current through input transistor changes several order of magnitude depending on the input level. To solve this issue, TCC

TABLE I Required V_{THcomp} control signals for differential implement

	SAR	TC-ADC [3]	TC-ADC (Proposed)
6bit	12	63	31
7bit	14	127	32 (w/1b interpolation)
8bit	16	255	34 (w/2b interpolation)

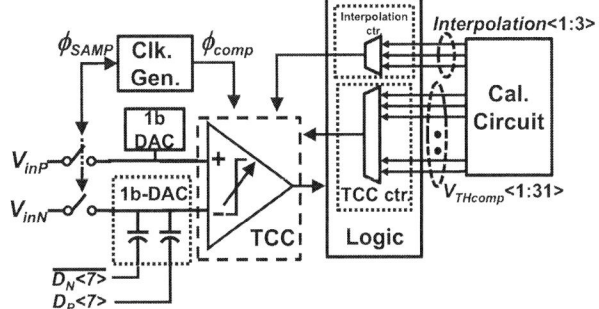

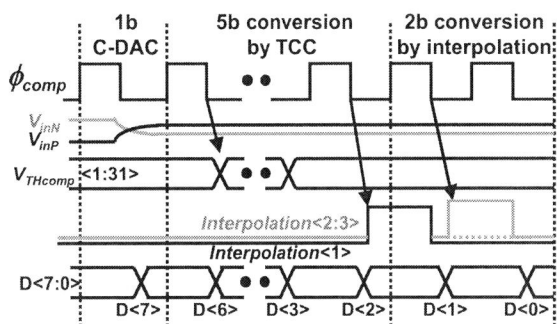

Fig. 1 Block Diagram and timing chart of the proposed TC-ADC

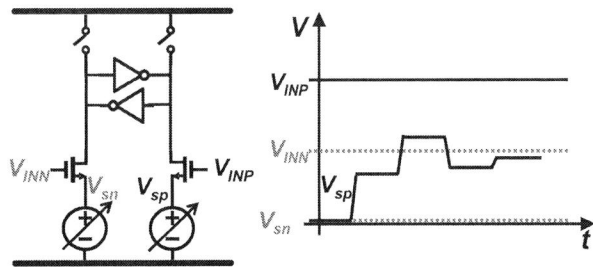

Fig. 2 Operation of TCC w/Source Voltage Shift

configuring the source voltage of input transistor is proposed. The binary search by TCC with source voltage shifting technique is described in Fig.2. For simplicity, the source voltage shift is explained with a configurable bias generator attached to the source node of the input transistors. By configuring the source voltage, the gate-source voltage (V_{GS}) of the input transistor is configured simultaneously and realizes wide-range threshold configuring.

The first comparison is done without threshold configuring. If V_{INP} is higher than V_{INN}, the source voltage V_{sp} is configured. When V_{sp} is controlled with enough accuracy, 5b binary search can be realized.

978-1-4799-2817-0/14 $31.00 © 2014 IEEE

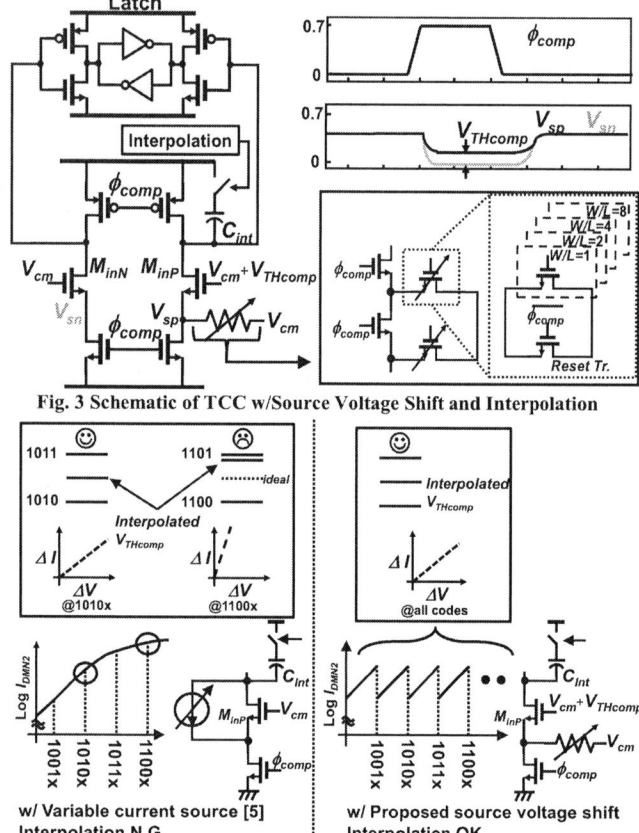

Fig. 3 Schematic of TCC w/Source Voltage Shift and Interpolation

Fig. 4 Interpolation by offset current versus source voltage shift

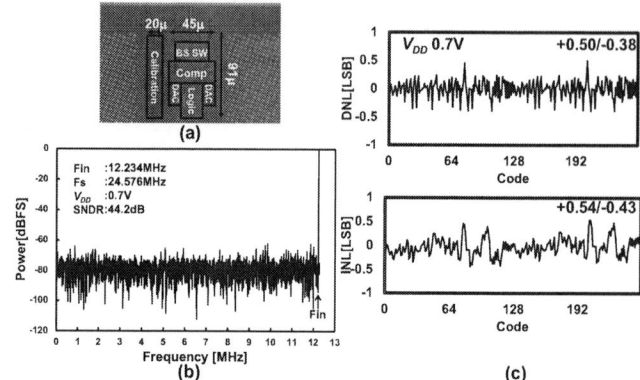

Fig. 5 (a)Chip microphotograph (b) 4096 FFT spectrum (c) DNL/INL

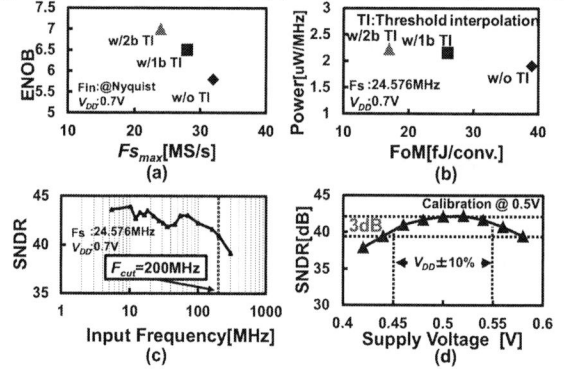

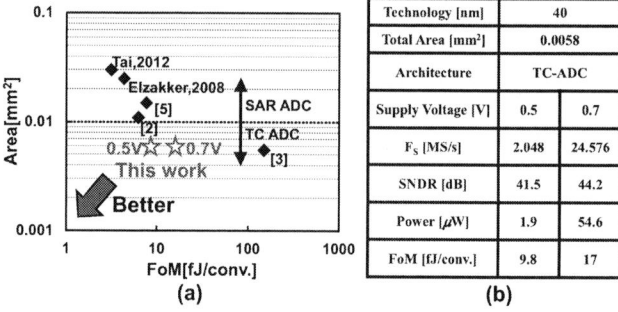

Fig. 6 Dynamic performance of ADC

Fig.3 show the schematic of TCC and operation when the desired threshold is V_{THcomp}. Source voltage shifting is realized by digitally controlled variable resistor connected to V_{sp}, which push up the source voltage by V_{THcomp}. As a result, the threshold of the comparator is set to V_{THcomp} and the current of M_{inP} can be kept constant.

The specific schematic of variable resistor, which is fully constructed by NMOS transistors, is shown in Fig.3 as well. The cascading of 4b binary weighed transistor array enables fine tuning of V_{THcomp} with small area. To avoid hysteresis effects, the source nodes pre-charge to V_{cm} during the comparator reset phase. The mapping between the V_{THcomp} and setting code are stored beforehand by foreground calibration. By V_{cm} biasing technique proposed in ref.0, the TCC is free from power supply variation.

A simple threshold interpolation can be realized by connecting C_{Int} at the output node of the comparator (Fig.4). Threshold interpolation by capacitor strongly depends on I_{DMinP}. In case of threshold configuring by VCS [5], since I_{DMinP} is strongly non-linear, optimum C_{Int} must be set depending on V_{THcomp}. On the other hand, threshold configuring by source voltage shifting "folds" the drain current of M_{inP}. As a result, interpolation by simple switching of C_{Int} is realized.

Measurement Results

The prototype ADC was designed and fabricated in 40nm standard CMOS process. Fig.5(a) shows the microphotograph. The foreground calibration was done automatically by serial input registers controlled via Matlab. Fig.5(b) shows the measured FFT spectrum with 24.576 MS/s sampling frequency and nyquist input. The DNL and INL in Fig.5(c) were +0.50/-0.38 and +0.54/-0.43 respectively.

With 2b interpolation, the resolution improves 1.2b as in Fig.6(a). Furthermore, the power increase is only 15% which greatly improves FoM (Fig.6(b)). Since the input capacitance is very small, the input frequency of 3dB cutoff (F_{CUT}) is 200MHz (Fig.6(c)). Fig.6(d) shows the SNDR dependence on the power supply voltage variation.

ADC performance comparison in Fig.7 show TC-ADC with TI technique can achieve FoM competing with state-of-the-art SAR ADCs with area smaller than 50%.

Fig. 7 (a) Comparison with state-of-the-art ADC (b) Performance summary

Technology [nm]	40	
Total Area [mm²]	0.0058	
Architecture	TC-ADC	
Supply Voltage [V]	0.5	0.7
F_S [MS/s]	2.048	24.576
SNDR [dB]	41.5	44.2
Power [μW]	1.9	54.6
FoM [fJ/conv.]	9.8	17

(b)

Acknowledgement

This work was carried out as a part of ELP project supported by METI and NEDO.

References

[1] E. Janssen et al, "An 11b 3.6 GS/s time-interleaved SAR ADC in 65nm CMOS," *IEEE ISSCC*, pp.464-465,2013.

[2] A. Shikata et al, "A 0.5 V 1.1 MS/sec 6.3 fJ/conversion-step SAR-ADC with tri-level comparator in 40 nm CMOS," *IEEE JSSC*, pp.1022-1030, 2012.

[3] P. Nuzzo et al, "A 6-bit 50-MS/s threshold configuring SAR ADC in 90-nm digital CMOS," *IEEE VLSI Symp.*, pp.238-239, 2009.

[4] G. Van der Plas et al, "A 150 MS/s 133uW 7 bit ADC in 90 nm Digital CMOS," *IEEE JSSC*, pp.2631-2640, 2008.

[5] K. Yoshioka et al, "An 8bit 0.35–0.8 V 0.5–30MS/s 2bit/step SAR ADC with wide range threshold configuring comparator," *IEEE ESSCIRC*, pp.381-384, 2012.

[6] K. Yoshioka et al, "A 0.0058 mm2 7.0 ENOB 24MS/s 17fJ/conv. threshold configuring SAR ADC with source voltage shifting and interpolation technique," *IEEE VLSI Symp.*, pp.266-267, 2013.

978-1-4799-2817-0/14 $31.00 © 2014 IEEE

A Single-Inductor 8-Channel Output DC-DC Boost Converter with Time-limited Power Distribution Control and Single Shared Hysteresis Comparator

Jungmoon Kim and Chulwoo Kim
Korea University, Seoul, Korea
Email : kjm@kilby.korea.ac.kr

Abstract

This paper describes a time-limited power distribution control (TPDC) technique that can be used for single-inductor multiple-output (SIMO) DC-DC converter with many unbalanced loads. Furthermore, the true all-comparator control technique that raises no stability or complexity issues is proposed. This all-comparator technique for SIMO converters is realized only with a single shared hysteresis comparator at a constant switching frequency of 800 kHz. The maximum efficiency reaches 92%. The fabricated chip with 8-channel outputs occupies 2.4×2.1 mm^2 in a 0.35-μm CMOS process.

Introduction

Future PMIC with a huge number of outputs needs an alternative to replace the combination of LDOs and DC-DC converters. Although the SIMO DC-DC converters can be a good alternative for mobile equipment due to their small area and low cost, the conventional architectures [1-6] could not address issues regarding supporting unbalanced loads with high stability and extending the number of outputs without limit simultaneously. Therefore, TPDC technique and true all-comparator control using a constant switching frequency is presented in this paper. In addition, this leads to the development of a new inductor peak-current control method.

Techniques for unlimited outputs with unbalanced loads

Fig. 1 shows the architecture of the proposed 8-channel SIMO DC-DC boost converter. Scaled feedback voltages and reference voltages are applied to the shared hysteresis comparator via two MUXs. True all-comparator control with a fixed switching frequency adopts a new load-dependent inductor peak-current control (IPCC) method to limit the peak level of the inductor current sensed by the I_L sensor. P-Switch controller distributes a portion of the inductor current to each output in accordance with its demand, and the amount of inductor current released to each individual output is controlled by an on-time manager (OTM). It is easy and simple to design this converter, because all of the blocks are digitally controlled, apart from the I_L sensor and the hysteresis comparator.

A. Time-limited Power Distribution Control

Fig. 2 illustrates the timing diagram of the single-inductor DC-DC converter with a large number of outputs to describe the proposed TPDC scheme. As shown in Fig. 2(a), if some of the heavy loads occupy inductor currents over a wide range of time periods, the output voltages of the lighter loads may be out of the expected range, due to the delayed delivery of the inductor current. Thus, it is desirable that the time slot that each output can take up with the inductor current be time-limited. The time-limited delivery of the inductor current to each output permits all of the outputs to be frequently supplied with energy, even under unbalanced loads, as shown in Fig. 2(b). Thanks to the limited time-slot for every output, heavy loads do not monopolize the energy from the inductor when other outputs demand a large amount of energy simultaneously. Therefore, the TPDC technique allows the SIMO converter to reliably support unbalanced loads and extend the number of outputs without limit.

B. Single Shared Hysteresis Comparator Control

Fig. 3 shows the block diagram for shaping the inductor current described in Fig. 2(b). The inductor current is distributed to each output by observing output voltages to regulate output voltages to reference voltages. We can regulate multiple output voltages in order with only one hysteresis comparator, because the inductor current can be supplied to just one output at any time. If $V_{H,i}$ is low, the P-Switch controller turns on the load switch, $M_{P,i}$, with discrete duties based on $OS_{PW,i}$ to increase its output voltage. If $V_{H,i}$ is high, $M_{P,i}$ can be skipped quickly, so that energy from the inductor can be instantly supplied to each of the other outputs requiring energy. In this regard, the proposed hysteresis scheme is suitable for a SIMO converter with many outputs. Furthermore, the all-comparator control technique makes the SIMO converter completely free from the stability problem, because it does not require any compensation.

Circuit Implementation

A. On-Time Manager and New Inductor Peak-Current Control

OTM in Fig. 3 controls the period of the inductor current delivered to each output. This can be done just by observing the output voltages. OTM counts the number of times (NR_i) $M_{P,i}$ is turned on during the rise of each $V_{OUT,i}$. If NR_i exceeds the upper bound ($NR_{U,i}$), OTM extends the on-time ($T_{OS,i}$) of $M_{P,i}$ via the signal, OS_{UP}. Conversely, OS_{DN} is triggered when NR_i is less than the lower bound ($NR_{D,i}$). The lower and upper bounds of NR_i are adaptively tuned

according to the inductor current level, as shown in Fig. 4(a). In addition, OTM controls the inductor peak-current. As shown in Fig. 4(b), if the increased $T_{OS,i}$ cannot handle the present load condition alone, the inductor peak-current will increase. Conversely, if the decreased $T_{OS,i}$ can no longer be shorter, the inductor peak-current will decrease. That is, the inductor peak-current, $I_{L,pk}$, is coarsely tuned for non-frequent changes of the total output current. $I_{L,pk}$ is also finely tuned by the number (Fall_N) of falling outputs not requiring the delivery of the inductor current. IPCC generates reference voltage for inductor peak-current control and the reference voltage is controlled by $VP_{U/D}$ from OTM. Discretely-controlled inductor peak-current does not degrade the efficiency, because all outputs are supplied from the output capacitors, while all of the output voltages fall down to the lower boundary of the hysteresis window, $V_{RL,i}$.

B. P-Switch Controller

Fig. 5 shows a schematic of P-Switch controller. $OS_{PW,i}$ controls the on-time of the load switch, $M_{P,i}$. The on-time of $M_{P,i}$ is not constant and the number of successive skipped pulses for $M_{P,i}$ is also variable. The additional switch of the proposed selective pulse generator, as shown in Fig. 6, can make V_{CH2} discharge fully and the rising-edge detector does not allow two successive inputs to be affected by each other. The capacitor is charged by the correct number of multiples of a unit current source in cascode mirrors to improve the matching property between various on-times.

Measurement Results

The circuit is implemented in a 0.35-μm CMOS process. Fig. 7 shows the chip photo of the proposed converter with an active area of 5.04 mm^2. Fig. 8(a) shows the measured steady-state waveforms. The inductor current is supplied to each load with the limited maximum period. The proposed TPDC technique reliably supports the unbalanced loads, as shown in Fig. 8(b). Fig. 8(c) shows that the inductor peak-current is discretely controlled by the OTM and the IPCC. Fig. 8(d) shows the measured waveforms for skip operation. Fig. 8(e) shows the ripple voltage of the sixth output that is observed over a long period of time with a resolution of 400 μs. Fig. 8(f) shows the measured waveforms with the load transient at V_{OUT6} to confirm that the proposed techniques minimize the cross-regulation effect. One output (V_{OUT7}) has a load of 100 mA, whereas the other six outputs ($V_{OUT,i}$, i = 1–5, 8) have loads of 5 mA. The converter operates at a fixed switching frequency of 800-kHz when a single shared hysteresis comparator is used. As shown in Fig. 8(g), the output power spectrum in the proposed SIMO converter does not have a single fixed dominant component because the switching frequencies of individual outputs vary. However, these high switching frequency noises from outputs of the SIMO converter do not affect other switching converters on the same input line because the inductor current ripple of the boost converter is continuous and its frequency is the same as that of the switching frequency. Fig. 8(h) shows the output power spectrum of the converter with four light loads ($R_L=3k\Omega$) and four unused loads. The skip operation for load switches causes much higher frequency peaks than the main switching frequency. The maximum efficiency reaches 92% at a total output power of 320 mW with V_{IN} of 2 V and all $V_{OUT,i}$ of 3 V, as shown in Fig. 9(a). The efficiency strongly depends on the total output power rather than individual loads. Nevertheless, because the sizes of load switches are optimized for medium load of individual outputs, the efficiency curve of the same output power displays a curvature, as shown in Fig. 9(b). Table I summarizes the performance of the proposed SIMO converter.

Conclusion

We demonstrate the 8-channel SIMO boost converter with the TPDC technique for a large number of outputs and unbalanced loads. This converter operates at a constant frequency using comparator-sharing technique and a new inductor peak-current control scheme. Table II compares the proposed architecture with the conventional works. To the best of the author's knowledge, the number of outputs of the proposed SIMO converter with the TPDC scheme exceeds that of any known SIMO DC-DC converters.

References

[1] D. Ma *et al.*, *IEEE JSSC*, vol. 38, pp. 89-100, Jan. 2003.
[2] D. Ma *et al.*, *IEEE JSSC*, vol. 38, pp. 1007-1014, Jun. 2003.
[3] H.-P. Le *et al.*, *IEEE JSSC*, vol. 42, pp. 2706-2714, Dec. 2007.
[4] K.-C. Lee *et al.*, in *IEEE ISSCC*, Feb. 2010, pp. 200-201.
[5] M.-H. Huang *et al.*, *IEEE JSSC*, vol. 44, pp. 1099-1111, Apr. 2009.
[6] X. Jing *et al.*, *IEEE JSSC*, vol. 46, pp. 2350-2362, Oct. 2011.

978-1-4799-2817-0/14 $31.00 © 2014 IEEE

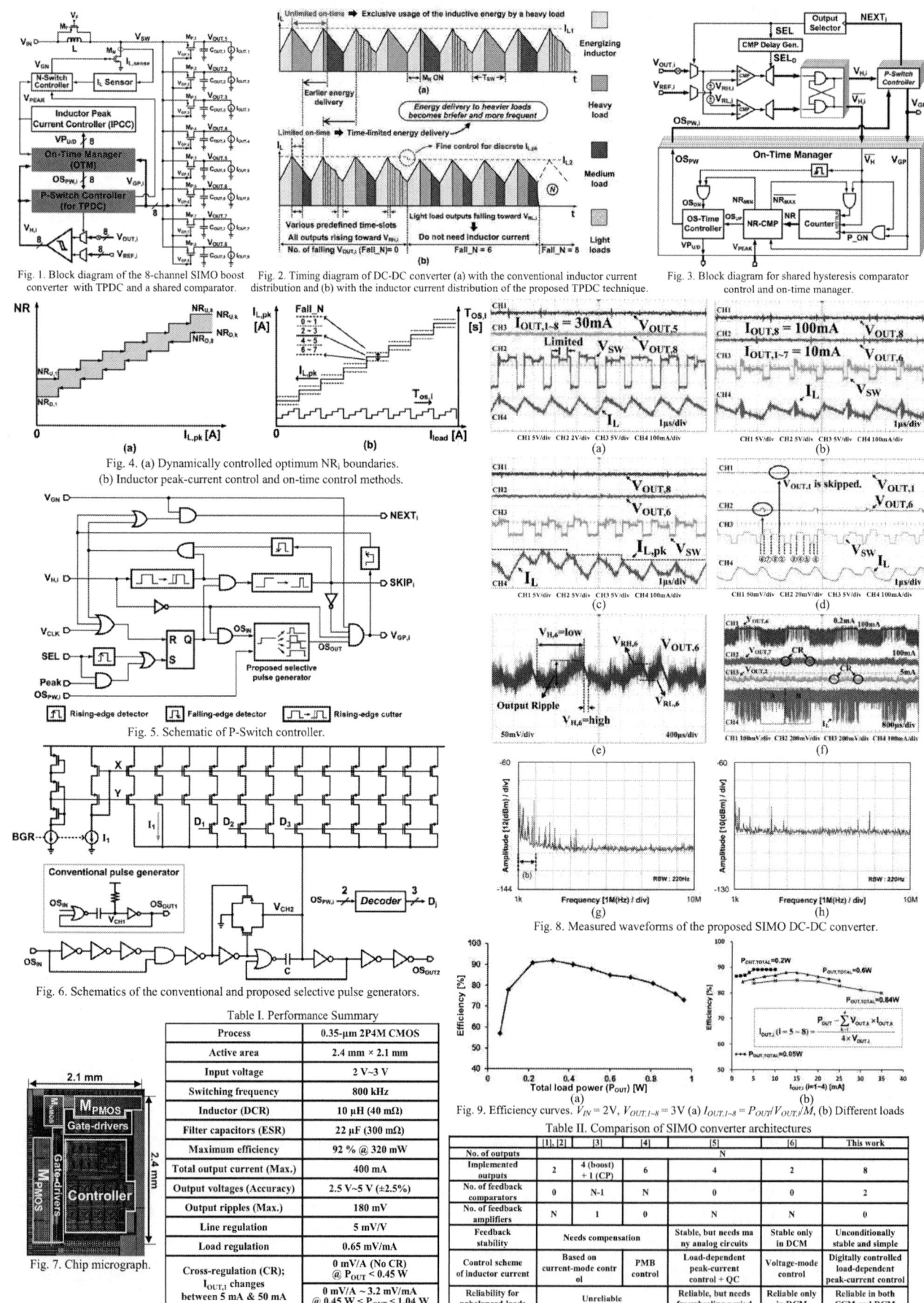

Fig. 1. Block diagram of the 8-channel SIMO boost converter with TPDC and a shared comparator.

Fig. 2. Timing diagram of DC-DC converter (a) with the conventional inductor current distribution and (b) with the inductor current distribution of the proposed TPDC technique.

Fig. 3. Block diagram for shared hysteresis comparator control and on-time manager.

Fig. 4. (a) Dynamically controlled optimum NR_i boundaries. (b) Inductor peak-current control and on-time control methods.

Fig. 5. Schematic of P-Switch controller.

⊓ Rising-edge detector   ⊔ Falling-edge detector   ⊓→⊓ Rising-edge cutter

Fig. 6. Schematics of the conventional and proposed selective pulse generators.

Fig. 7. Chip micrograph.

Fig. 8. Measured waveforms of the proposed SIMO DC-DC converter.

Fig. 9. Efficiency curves. $V_{IN} = 2V$, $V_{OUT,1-8} = 3V$ (a) $I_{OUT,1-8} = P_{OUT}/V_{OUT}/M$, (b) Different loads

Table I. Performance Summary

Process	0.35-μm 2P4M CMOS
Active area	2.4 mm × 2.1 mm
Input voltage	2 V~3 V
Switching frequency	800 kHz
Inductor (DCR)	10 μH (40 mΩ)
Filter capacitors (ESR)	22 μF (300 mΩ)
Maximum efficiency	92 % @ 320 mW
Total output current (Max.)	400 mA
Output voltages (Accuracy)	2.5 V~5 V (±2.5%)
Output ripples (Max.)	180 mV
Line regulation	5 mV/V
Load regulation	0.65 mV/mA
Cross-regulation (CR); $I_{OUT,1}$ changes between 5 mA & 50 mA	0 mV/A (No CR) @ P_{OUT} < 0.45 W
	0 mV/A ~ 3.2 mV/mA @ 0.45 W < P_{OUT} < 1.04 W

Table II. Comparison of SIMO converter architectures

	[1], [2]	[3]	[4]	[5]	[6]	This work
No. of outputs				N		
Implemented outputs	2	4 (boost) + 1 (CP)	6	4	2	8
No. of feedback comparators	0	N-1	N	0	0	2
No. of feedback amplifiers	N	0	0	N	N	0
Feedback stability	Needs compensation			Stable, but needs many analog circuits	Stable only in DCM	Unconditionally stable and simple
Control scheme of inductor current	Based on current-mode control		PMB control	Load-dependent peak-current control + QC	Voltage-mode control	Digitally controlled load-dependent peak-current control
Reliability for unbalanced loads	Unreliable			Reliable, but needs freewheeling period	Reliable only in DCM	Reliable in both CCM and DCM

978-1-4799-2817-0/14 $31.00 © 2014 IEEE

A DC-DC Boost Converter with Variation Tolerant MPPT Technique and Efficient ZCS Circuit for Thermoelectric Energy Harvesting Applications

Jungmoon Kim, Minseob Shim, Junwon Jung, Heejun Kim, and Chulwoo Kim

Korea University, Seoul, Korea

Email : kjm@kilby.korea.ac.kr

Abstract-This paper presents a boost converter with the maximum power point tracking (MPPT) technique for thermoelectric energy harvesting (EH) applications. The technique realizes variation tolerance by adjusting the switching frequency f_{SW} of the converter. A finely controlled zero-current switching (ZCS) scheme together with the accurate MPPT technique enhances the overall efficiency (η) of the converter because of an optimal turn-on time generated by a one-shot pulse generator that is proposed. Moreover, the ZCS technique can deal with low and high temperature differences applied to the thermoelectric generator. Experimentally, the converter implemented in a 0.35 μm BCDMOS process had a peak η of 72% at the input voltage V_{IN} of 500mV while supplying a 5.62V output.

I. Introduction

Emerging EH devices are being adopted by various applications such as medical sensors, automotive sensors, and smart buildings [1-5]. Recent works on the thermal EH convert even body heat into electricity to supply power to electronics devices. However, some challenging problems in thermal EH need to be overcome in mobile applications. Generally, the output voltage of a thermoelectric generator (TEG) is proportional to the temperature difference ΔT between both sides of the TEG. For example, the temperatures of components, such as the baseband chips, in cellular phones, may sometimes be increased to more than 70°C, which will generate a few hundred mV at the output of the TEG. In contrast, the normal temperature difference between the two sides of a TEG body and chips, is only 1–2°C, which generates just around 50 mV. Therefore, the thermal EH circuit in some applications such as portable devices should be able to deal with a wide range of TEG output voltages. In this paper, a boost converter with the high accuracy ZCS control technique for a wide range of conversion ratios is presented. This ZCS technique is realized by a proposed one-shot pulse generator. The MPPT technique that adjusts f_{SW} of the boost converter according to system variations is also demonstrated.

II. Proposed Thermal EH Circuit

Fig. 1 shows the overall architecture of a power delivery system from a TEG to a power management unit (PMU). Because the voltage generated by the TEG is very small, the start-up of the converter must be supported by an instantaneous high voltage from a battery. Fig. 2 shows the timing diagram of the main signals during the start-up operation. While the switch *S1* is closed during the start-up process, the charge pump *CP1* supplies power to the controller of the boost converter. When the oscillator starts to operate, the boost converter starts to convert the small TEG voltage to a high voltage output for the PMU as well. *S1* is then open to cut the power supply from the battery. Moreover, by closing switch *S2*, the capacitor C_{DD} is utilized, and the controller of the converter is powered by its own output. The input of the converter is periodically disconnected from the TEG by the switch *S3*. To implement the system variation tolerant MPPT, the open circuit voltage V_T of the TEG is measured periodically during the short disconnection period.

III. Circuit Implementation

A. Thermoelectric Energy Harvesters

The TEG can be modeled as a voltage source V_T with an internal resistance R_T, where V_T is an open circuit voltage proportional to the temperature difference between both sides of the TEG as shown in Fig. 3(a). Fig. 3(b) shows the output characteristics of a TEG according to two different temperature gradients. If the load impedance matches R_T, MPPT can be achieved. In [4], f_{SW} of the converter is determined by the R_T of the TEG and the inductor value L. This frequency, $f_{SW}=R_T/8L$ (Eq. 1), is used to realize the MPPT for extracting maximum power from the TEG. Thus, just setting the frequency to a constant value activates the MPPT. The first problem of this conventional approach is that the frequency of general on-chip clock oscillators varies over a wide range depending on process, voltage, and temperature conditions. Secondly, because R_T of the TEG depends on ΔT, it can deviate by 12% from the design value, as shown in Fig. 3(c). Finally, in addition to the variation of R_T, a general inductor has an inductance tolerance of at least ±20%. Therefore, unless a tracking method is used to deal with these variations, maximum power cannot be obtained from the TEG.

B. Variation Tolerant MPPT Technique

The conventional approach for MPPT in thermal EH uses f_{SW} that is calculated with R_T [4]. However, because this simple technique is affected by variations of many parameters, the optimal f_{SW} needs to be adaptively controlled. Fig. 4 shows the implementation of the proposed converter. To obtain the maximum power from the thermoelectric harvester, the load impedance should be equal to R_T. In other words, V_{TH} of the TEG needs to be regulated to half of V_T. Based on this MPPT theory, the half value V_{TH_HF} of V_{TH2}, which is sensed whenever switch *S4* is momentarily closed, is generated by the converter with a conversion ratio of 1/2. A comparator (CMP) *CMP1* then compares V_{TH_HF} to V_{TH_AVG}. This comparison for V_{IN} regulation achieves the optimal f_{SW} by controlling the input impedance. Finally, the ZCS control circuit allows the converter to deliver maximum power to the load at the adjusted optimal frequency.

C. ZCS for a Wide Range of V_{IN} and the Optimal Turn-On Time

The DCM operation which prevents the inductor current from flowing negative is adopted in this design for low-power applications. While the on-time τ_N of the low-side switch *LS* in the power stage is set to half of the switching period T_{SW}, the on-time τ_P of the high-side switch *HS* is finely tuned for the accurate ZCS.

Because the output voltage of the TEG in applications utilizing body heat is very small, $\tau_{P,MIN}$ is very short, as shown in Fig. 5. Thus, the on-time can be controlled just by using the delay of the digital logic gates in the conventional ZCS control circuit. However, this conventional ZCS control technique has problems associated with circuit complexity, area, and power consumption in terms of a wide V_{IN} range and η of the boost converter.

Therefore, the logic gates for generating the control signal *PG* of the switch *HS* should be replaced with a new one-shot pulse generator. Fig. 6(a) shows the conventional one-shot pulse generator using an RC delay. This conventional one-shot pulse generator has two limitations; first, the values of both *R* and *C* are fixed so the output pulse width cannot be controlled and second, the rate at which the output one-shot pulse is retriggered is limited by the time it takes for V_{M1} to decay back down to V_{DD}. As shown in Fig. 6(b), if the input pulses are close to each other, the output pulses have different widths. Therefore, a one-shot pulse generator is proposed, as shown in Fig. 6(c). In order to control the output pulse width, the resistor is replaced with a controllable current source. Controlling the current source allows various pulse widths. When *IN* is triggered, the short output pulse of the rising-edge detector closes the switch *S6*, which causes V_{M2} to discharge to ground. Because the initial potential of V_{M2} at which the current source starts to charge the capacitor always becomes zero, the retriggered one-shot can be independent of inputs.

Fig. 7 shows the circuit implementation to realize the ZCS control of the inductor current for a wide range of conversion ratios and high resolution optimal turn-on times. When *CMP3* senses that V_{OUT_FB} is less than V_{REF}, the inductor starts to be charged for one half of T_{SW}. Then, the inductive energy is transferred to the output through *HS*. The on-time of this switch should be finely controlled to obtain high converter η. *CMP2* determines whether *HS* has been turned off too early, generating the signal *UP*, or has been turned off too late, generating the signal *DN*. The proposed one-shot pulse generator is used in the ZCS control circuit. Therefore, the output signal *PG* can be defined accurately. This means that the converter η degradation due to the quantization error from the finite on-time resolution can be minimized.

Fig. 8 shows the timing diagram of the tracking operation for optimum ZCS control. I_{Lf} is the falling inductor current flowing through the high side switch *HS* of the converter. The gate control signal *PG* for *HS* is initially set to zero, and the clock signal *PS* for *CMP2* is a very short pulse triggered a little after *PG* falls. At this time, if V_X is higher than V_{OUT}, *CMP2* outputs *UP*. Until V_{OUT} increases enough, only *UP* is continuously generated and the V_{CONT} for the current source in the ZCS control circuit is increased. Finally, once V_X is lower than V_{OUT}, *DN* at the output of *CMP2* is triggered and the inductor reverse current is sensed. To remove this inverse inductor current, the on-time of *PG* is decreased. Then, *UP* is triggered again, and both *UP* and *DN* alternately occur. Because *PS* is very short, V_{CONT} can be finely controlled to minimize the body conduction loss.

Fig. 9 depicts the simulated waveforms to show the tracking of the optimal duty of *PG* and the frequency locking for MPPT. The initial f_{SW} is fixed as *S1* closes. In the meantime, the duty of *HS* increases toward the optimal value, and V_{OUT} also increases. *S1* is then opened when V_{OUT} increases sufficiently, and then, the frequency tracking starts. Finally, f_{SW} oscillates around the optimal frequency.

IV. Measurement Results

The circuit is implemented in a 0.35-μm BCDMOS process. Fig. 10 shows the photograph of the chip. The fabricated converter occupies an active area of 1.9 mm². The TEG used for the measurement can output 25 mV/K with an internal 8 Ω resistor. The inductor is 22 μH with a parasitic resistance of 120 mΩ. The input capacitor is 10 μF, and the output capacitor is 1 μF.

Fig. 11 shows the measured waveforms that describe the start-up of the proposed converter. By adjusting f_{SW}, V_{TH} is regulated to 50 mV, or one half of V_T. This corresponds to a 4 K temperature difference across the TEG.

The on-time of HS should be the same as the optimal time required to minimize a loss in the converter. Fig. 12 shows the tracking process that was carried out for an accurate ZCS of the converter. The body conduction loss of HS nearly disappears at the steady state, which is the main contribution to the improvement of the overall η.

Fig. 13 shows the magnified measured waveform of the TEG open circuit voltage (V_T) sensed during the short period. V_T (100 mV) of the TEG is sensed for 4 μs, and V_{IN} is regulated to 50 mV, one half of V_T.

Fig. 14 shows the measured power η curves of state-of-the-art thermoelectric EH converters as V_T changes. The input voltage of the

PMU Block is specified from 3V to 5.8V in this design. As the load resistance is varied, the output power is measured. Fig. 14(a) shows that η of the proposed circuit at 0.1 V V_{IN} is 8% greater than that of [4] and 35% greater than that of [5] in the low V_T domain. Fig. 14(b) shows η of the converter in the high V_{IN} domain, which cannot be covered by the ZCS control circuit of the conventional converter [4] for thermal EH. Compared to [5], the proposed techniques also contribute to η improvement in the high V_T domain as well as in the low V_T domain. The η of the obtained output power in the proposed converter reaches 72.2% of the theoretical maximum power available from the TEG with V_{OUT} of 5.62V and R_L of 5.6kΩ. At this time, V_T of 500mV exceeds the input range that can be supported by the converter in [5].

Table I shows the comparison of the performance of the presented converter with those of other switching converters for thermoelectric EH. The proposed ZCS circuit supports a wide range of V_{IN} from the TEG. The finely-controlled and accurate ZCS circuit and the variation tracking technique result in the best peak η among converters with the MPPT scheme.

References

[1] ECT 310 Perpetuum, EnOcean Inc., 2010: http:// www.enocean.com
[2] H. Lhermet *et al.*, *IEEE JSSC*, vol. 43, no. 1, pp. 246-255, Jan. 2008.
[3] E. J. Carlson *et al.*, *IEEE JSSC*, vol. 45, no. 4, pp. 741-750, Apr. 2010.
[4] Y. K. Ramadass *et al.*, *IEEE JSSC*, vol. 46, no. 1, pp. 333-341, Jan. 2011.
[5] J.-P. Im *et al.*, *IEEE ISSCC*, Feb. 2012, pp. 104–105.

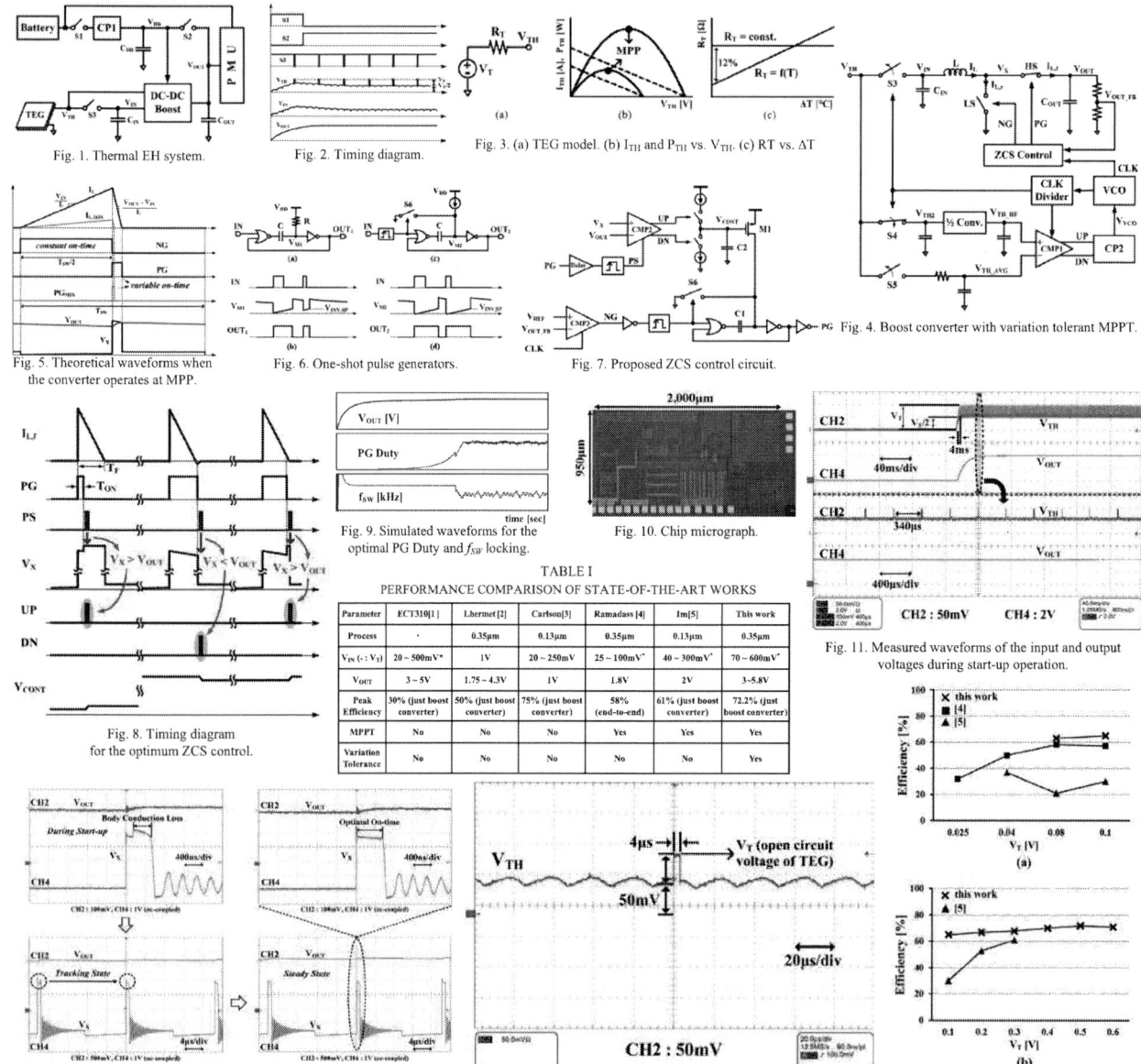

Fig. 1. Thermal EH system.

Fig. 2. Timing diagram.

Fig. 3. (a) TEG model. (b) I_{TH} and P_{TH} vs. V_{TH}. (c) RT vs. ΔT

Fig. 4. Boost converter with variation tolerant MPPT.

Fig. 5. Theoretical waveforms when the converter operates at MPP.

Fig. 6. One-shot pulse generators.

Fig. 7. Proposed ZCS control circuit.

Fig. 8. Timing diagram for the optimum ZCS control.

Fig. 9. Simulated waveforms for the optimal PG Duty and f_{SW} locking.

Fig. 10. Chip micrograph.

Fig. 11. Measured waveforms of the input and output voltages during start-up operation.

TABLE I
PERFORMANCE COMPARISON OF STATE-OF-THE-ART WORKS

Parameter	ECT310[1]	Lhermet [2]	Carlson[3]	Ramadass [4]	Im[5]	This work
Process	0.35μm	0.13μm	0.35μm	0.13μm	0.35μm
V_{IN} ($\cdot : V_T$)	20 ~ 500mV*		20 ~ 250mV	25 ~ 100mV*	40 ~ 300mV*	70 ~ 600mV
V_{OUT}	3 ~ 5V	1.75 ~ 4.3V	1V	1.8V	2V	3 ~ 5.8V
Peak Efficiency	30% (Just boost converter)	50% (Just boost converter)	75% (Just boost converter)	58% (end-to-end)	61% (Just boost converter)	72.2% (just boost converter)
MPPT	No	No	No	Yes	Yes	Yes
Variation Tolerance	No	No	No	No	No	Yes

Fig. 12. Measured waveforms of tracking of the optimal turn-on time for HS.

Fig. 13. Magnified waveforms of the TEG open circuit voltage .

Fig. 14. Efficiency comparison.

978-1-4799-2817-0/14 $31.00 © 2014 IEEE

7.3 Gb/s Universal BCH Encoder and Decoder for SSD Controllers

Hoyoung Yoo, Youngjoo Lee, and In-Cheol Park

Dept. of Electrical Engineering
Korea Advanced Institute of Science and Technology (KAIST)
Daejeon 305-701, Republic of Korea
Tel : +82-42-351-9884
Fax : +82-42-350-8061
e-mail : {hyyoo, yjlee}.ics@gmail.com, icpark@kaist.edu

Abstract - This paper presents a universal BCH encoder and decoder that can support multiple error-correction capabilities. A novel encoding architecture and on-demand syndrome calculation technique is proposed to reduce both hardware complexity and power consumption. Based on the proposed methods, 32-parallel universal encoder and decoder are designed for BCH (8192+14t, 8192, t) codes, where the error-correction capability t is configurable to 8, 11, 16, 24, 32, and 64. The prototype chip achieves a throughput of 7.3 Gb/s and occupies 2.24 mm^2 in 0.13μm CMOS technology.

I Introduction

The binary BCH code that can recover multiple erroneous bits has widely employed for error control in solid-state drives (SSDs) due to its superior error-correction performance and affordable hardware complexity [1]-[3]. The binary BCH (n, k, t) code is generally characterized by code length n, message length k, and error-correctional capability t [4]. In SSD controllers, the error-correction capability of the BCH code is mainly determined by the size of the spare region in flash memory. As the error-correction capability in commercial NAND flash memories is varying respective to vendors and technologies, it is necessary to develop a universal error-correction unit that can support multiple error-correction capabilities. Otherwise, the error-correction unit requires several circuitries each dedicated to a specific error-correction capability, leading to a tremendous hardware complexity.

In general, a SSD controller accesses the NAND flash memories in accordance with the request of the host system as depicted in Fig. 1. When the host requests a write operation, the message is encoded to generate a BCH codeword, and the codeword including both the message and the parity information is stored in the NAND flash memory. On the other hand, when the host requests a read operation, the codeword read from the NAND flash memory is decoded by passing through three stages: syndrome calculation (SC), key-equation solving (KES), and Chien search (CS). Recently, a SSD controller is connected to the host through advanced high-speed serial interfaces such as SATA III with a maximum transfer rate of 6Gb/s [3]. Since the performance of a SSD is normally determined by the throughput of the error-correction unit, developing an efficient universal BCH encoder and decoder are crucial in achieving a high-performance SSD controller.

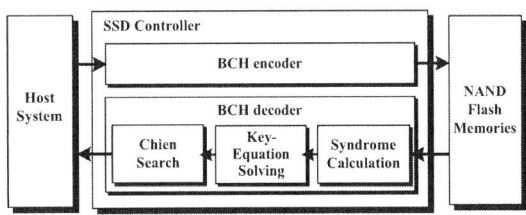

Fig. 1. The error-correction unit in a typical SSD controller.

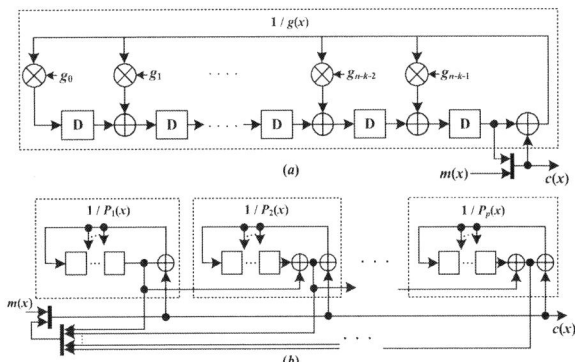

Fig. 2. BCH encoding architectures, (a) conventional single-mode encoder, (b) proposed multi-mode encoder.

II. Encoding Architecture

Based on the generator polynomial $g(x)$, a typical BCH encoder can be implemented by using the LFSR architecture illustrated in Fig. 2(a) [4]. To realize a universal encoder that can support multiple error-correction capabilities, the generator polynomial $g(x)$ is decomposed into several short polynomials $P(x)$. Note that the decomposition of the generator polynomial is always guaranteed, as the generator polynomial $g(x)$ is the least common multiple of $2t$ minimal polynomials $M_i(x)$, $1 \leq i \leq 2t$ in a binary BCH code. As shown in Fig. 2(b), the proposed universal encoder consists of several short LFSRs cascaded in serial to achieve the same functionality as the long LFSR corresponding to $g(x)$, and the output of each short LFSR is fed back to the input side to support multiple error-correction capabilities. While the conventional encoder needs multiple encoders as many as the number of configurable error-correction capabilities, the proposed encoder facilitates the multi-mode encoder to preserve the hardware complexity similar to that of the

978-1-4799-2817-0/14 $31.00 © 2014 IEEE

single-mode encoder. The proposed universal encoder can support multiple error-correction capabilities of 8, 11, 16, 24, 32, and 64, and saves 56.4% equivalent gates compared to the conventional approach.

III. Decoding Architecture

Since the decoding architecture with error-correction capability t can easily cover error-correction capabilities less than t, the *de facto* method that activates only necessary parts of the configurable error-correction circuit is commonly used in realizing multi-mode decoder [1]. Taking the *de facto* method to support a universal decoding property, we propose a new on-demand syndrome calculation technique to reduce the hardware complexity further. Although all the previous decoders generate both odd- and even-indexed syndromes in parallel in their SC, all the $2t$ syndromes are not necessary in the beginning of KES. The main concept of the proposed architecture is to schedule the syndrome calculation to reduce hardware; odd-indexed syndromes are computed in the SC, while even-indexed syndromes are computed when they are needed in the KES using the property that $S_{2i} = (S_i)^2$ where S_i represents the i-th syndrome. Fig. 3 illustrates the hardware architecture supporting the proposed on-demand syndrome calculation. As a result, the hardware complexity of the conventional SC is halved at the cost of a small hardware increase in the KES. The proposed architecture saves 48.7% of hardware resources in the SC compared to the conventional architecture. To achieve a further hardware reduction, the folding technique and the sharing of common sub-expressions that have been applied in the previous single-mode BCH decoder [3] are also employed in the universal decoder.

IV. Implementation Results and Conclusion

To support BCH $(8192+14t, 8192, t)$ codes, where t is configurable to 8, 11, 16, 24, 32, and 64, a 32-parallel universal encoder and decoder are designed in a 0.13 um technology. In virtue of the proposed methods, the universal encoder and decoder demand only 77k and 180k equivalent gates, respectively, occupying overall 2.24 mm². Both the proposed encoder and decoder operate at 230 MHz, leading to 7.3 Gb/s. Fig. 4 shows the micrograph of the proposed architecture and how much the proposed methods reduce the hardware. Moreover, Table I compares the proposed universal error-correction unit and the recent works. For a fair comparison, the normalized gate counts, taking the error-correction capability and the parallel factor into account, is compared. Considering that the proposed encoder and decoder are associated with a high parallel factor of 32 and configurable strong error-correction capabilities up to 64, the proposed architecture is superior to the previous works in many aspects of hardware complexity and energy efficiency.

Acknowledgments

This work was supported in part by Korea IT R&D program of MOTIE/KEIT No.KI10035202 and by the IC Design Education Center (IDEC).

References

[1] T. H. Chen *et al.*, "An Adaptive-Rate Error Correction Scheme for NAND Flash Memory," in *Proc. IEEE VLSI Test Symp. (VLSI-TST)*, 2009, pp. 53-58.

[2] K. Lee, S. Lim and J. Kim, "Low-cost, Low-power and High-throughput BCH decoder for NAND Flash Memory," in *Proc. Symp. Circuits Syst. (ISCAS)*, May 2012, pp. 413-415.

[3] Y. Lee, H. Yoo, I. Yoo, and I.-C. Park "6.4Gb/s Multi-Threaded BCH Encoder and Decoder for Multi-Channel SSD Controllers," *ISSCC Dig. Tech. Papers*, Feb. 2012, pp. 426-427.

[4] S. Lin and D. J. Costello, *Error control coding*, 2nd ed. Englewood Cliffs, NJ: Prentice-Hall Inc., 2004.

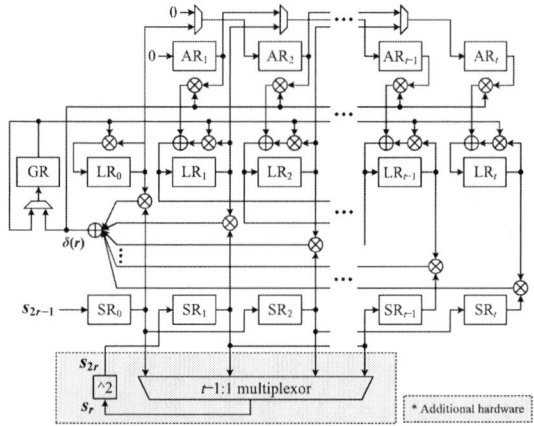

Fig. 3. A new KES block supporting the proposed on-demand syndrome calculation.

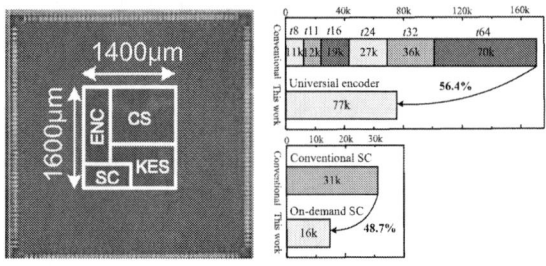

Fig. 4. The micrograph of the proposed universal error-correction unit and complexity comparison to the conventional structure.

TABLE I
COMPARISON OF BCH ERROR-CORRECTION UNIT

Architecture	This work	[1]	[2]	[3]
Functionality	Enc./Dec.	Enc./Dec.	Dec.	Dec.
Multi-mode	○	○	×	×
Technology	130nm	130nm	45nm	130nm
Operating speed	230MHz	125MHz	400MHz	200MHz
User data length	8192	4096	8192	8192
Correctable errors	8,11,16, 24,32,64	9,14, 19,24	64	32
Parallel factor	32	8	16	32
Throughput	7.3Gb/s	1.0Gb/s	6.4Gb/s	6.4Gb/s
Gate counts (enc./dec.)	77k/180k	11k/147k	−/230k	−/110k
Normalized gate counts[a]	1.2k/2.8k	1.8k/24.5k	−/7.2k	−/3.4k
Chip area	2.24mm²	0.81mm²	N.A.	0.85mm²
Energy efficiency[b]	8.62pJ/bit	N.A.	52.07pJ/bit	4.92pJ/bit

[a] Gate counts / Max. error-correction capability × (32/Parallel factor)
[b] Energy efficiency = Power / Throughput

978-1-4799-2817-0/14 $31.00 © 2014 IEEE

A High-Speed and Low-Complexity Lens Distortion Correction Processor for Wide-Angle Cameras

Won-Tae Kim, Hui-Sung Jeong, Gwang-Ho Lee and Tae-Hwan Kim

School of electronics, telecommunication and computer engineering, Korea Aerospace University
76, Hanggongdaehang-ro, Deogyang-gu, Goyang-si, Gyeonggi-do, 412-791, Republic of Korea
e-mail : taehwan.kim@kau.ac.kr, {wonatekim731, heesung.jeong.kr, yikwangho1}@gmail.com

Abstract- This paper presents a high-speed and low-complexity lens distortion correction processor for wide-angle cameras. In the proposed processor, the conventional correction process is modified to be performed incrementally so as to reduce the hardware complexity. In addition, an efficient memory interface is proposed by utilizing the locality of the memory access in the correction process. The proposed processor is implemented with 17.2K logic gates in a 0.11 μm CMOS process and its correction speed is 205 Mpixels/s.

I. Introduction

The wide-angle cameras are extensively used in many applications such as endoscopes and surveillance systems; however, they usually have a radial defect called the lens distortion. Because the lens distortion can make serious problems in the applications above, its correction is crucial.

The lens distortion correction (LDC) is usually performed in two steps [1]–[3]: the backward mapping and the linear interpolation, which is illustrated in Fig. 1. In the backward mapping, we have to find a distorted image space (DIS) pixel location corresponding to a corrected image space (CIS) pixel location. Since the DIS pixel location may not be an integer, the intensity of the CIS pixel has to be calculated by using those of neighboring pixels, for which the bilinear interpolation is usually employed [1]-[3]. By performing the LDC process for every pixel, CIS can be reconstructed with given DIS.

II. Proposed Lens Distortion Correction Processor

The proposed processor performs the LDC in a seven-stage pipeline. Fig. 2 shows the pipeline structure, where Stage1-4 and Stage 5-7 perform the backward mapping and the bilinear interpolation, respectively.

A. Low-Complexity Backward Mapping Architecture

In the backward mapping, every CIS pixel is mapped into a DIS pixel, which can be explained by introducing a scaling factor, as shown in Fig. 1. The scaling factor for each pixel is usually obtained by calculating a high-order polynomial of r [1]-[3]; and thus, the complexity of the hardware units required for the backward mapping is very high.

The proposed processor runs the backward mapping in the raster scanning order. In other words, the backward mapping is performed sequentially for pixels, horizontally left-to-right, vertically top-to-bottom. Based on such sequential operation, the proposed processor calculates the scaling factor incrementally; whereas in the conventional processor, it is calculated straightforwardly. As shown in Fig. 2, the

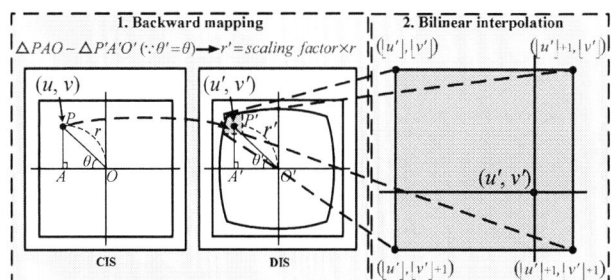

Fig. 1. LDC process, where P and P' are a CIS pixel and a DIS pixel, respectively. The optical centers of CIS and DIS are denoted by O and O', respectively.

proposed processor calculates s_n incrementally, where s_n is the n-th scaling factor and r_n is the radial distance between the n-th pixel and the optical center. In addition, r_n^2 is calculated incrementally in the proposed processor. Accordingly, the calculation of powers of r_n is eliminated effectively in the proposed processor. Table I shows that the number of required arithmetic units for the proposed processor is quite smaller than that for the conventional one. In Fig. 2, Stage 1-4 perform the proposed incremental backward mapping, where the multiply-accumulate (MAC) units based on Wallace tree reduction are employed for efficiently fused arithmetic operations.

B. Efficient Memory Interface

In the bilinear interpolation for the LDC process, four DIS pixel intensities are required for calculating one CIS pixel intensity. If the bilinear interpolation could be performed in a single cycle, four copies of a DIS memory would be necessitated in order to read four pixel intensities simultaneously. Because it is not efficient in terms of the cost, an LDC processor with a single DIS memory was proposed in [3]; however, it suffers from a low correction speed by the limited memory bandwidth. Considering the raster scanning order of the LDC process, we can expect that the read operations for the bilinear interpolation have a locality. To exploit the locality, an efficient memory interface was designed by employing a fully associative buffer, where its detailed structure is shown in Fig. 3. The pixel intensity accessed from the DIS memory is stored into the buffer temporarily, since it is probable to be reused for the subsequent correction process. As this can reduce the traffic to the memory, the proposed processor achieves a high correction speed even with a single memory. By performing simulations for a specific wide angle lens whose angle of view is 170°, the size of the buffer is determined to 8 at which the memory traffic is reduced by 55% as shown in Fig.

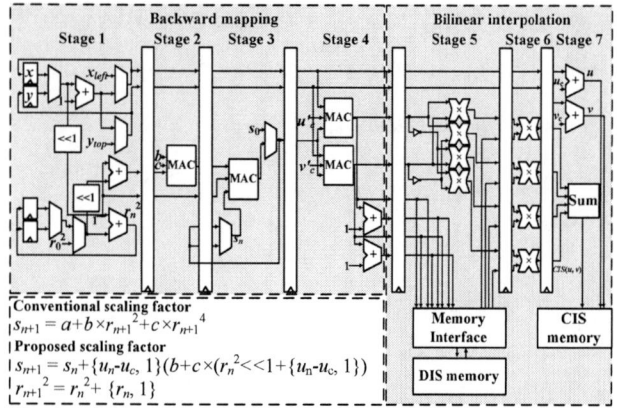

Fig. 2. Pipeline structure for the proposed LDC processor, where $\{x, y\}$ means the concatenation of x and y and the mapping coefficients are denoted by a, b, and c.

4. Consequently, the proposed processor achieves a high correction speed despite the limited bandwidth of a single DIS memory.

III. Implementation Results

The proposed processor is implemented with 17.2K logic gates in a 0.11 μm CMOS process and its layout is shown in Fig. 5. Fig. 6 shows the demonstration system of the proposed processor. Table II compares the implementation results of the proposed processor with those of conventional ones. The correction speed of the proposed processor is higher than other LDC processors in previous work, and its gate count is very small. When compared to the results presented in [2], the correction speed of the proposed processor is increased by 46.42 %, while the gate count is reduced by 61.72 %. It should also be noted that the correction speed of the proposed processor is very high even with a single DIS memory.

IV. Conclusion

This paper presented a high-speed and low-complexity LDC processor for wide-angle cameras. The proposed processor is based on the low-complexity backward mapping by the incremental calculation and the efficient memory interface utilizing the locality of the memory access. The correction speed of the proposed processor is 205 Mpixels/s and its gate count is 17.2K.

Acknowledgement

This work was supported by the Gyunggi Regional Research Center (GRRC) support program and the IC Design Education Center (IDEC).

References

[1] H. T. Ngo and V. K. Asari, "A pipelined architecture for real-time correction of barrel distortion in wide-angle camera images," *IEEE Trans. Circuits & Systems for Video Technology.*, vol. 15, no. 3, pp. 436-444, Mar. 2005.
[2] P. Y. Chen, C. C. Huang, Y. H. Shiau, and Y. T. Chen, "A VLSI implementation of barrel distortion correction for wide-angle

camera images," *IEEE Trans. Circuits & Systems II: Express Briefs.*, vol. 56, no. 1, pp. 51-55, Jan. 2009.
[3] S. Chen, H. Huang, and C. Luo, "Time multiplexed VLSI architecture for real-time barrel distortion correction in video-endoscopic images," *IEEE Trans. Circuits & Systems for Video Technology.*, vol. 21, no. 11, pp. 1612-1621, Nov. 2011.

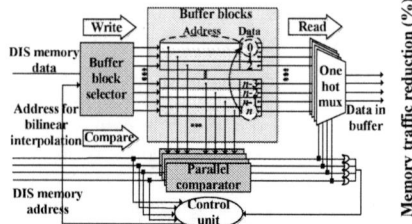

Fig. 3. Proposed memory interface.

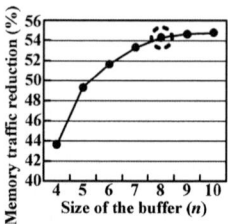

Fig.4. Memory traffic reduction by the proposed memory interface scheme.

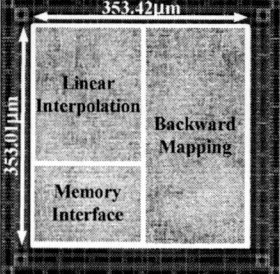

Fig.5. Layout of the proposed LDC processor.

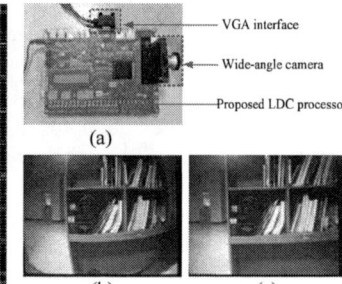

Fig. 6. Demonstration of the proposed LDC processor: (a) prototype system, (b) distorted image and (c) corrected result.

TABLE I
Number of Required Arithmetic Units

Architecture	This work	[2]
24 bit × 24 bit multiplier	0	7
24 bit × 16 bit multiplier	0	2
19 bit × 12 bit multiplier	1	0
16 bit × 16 bit multiplier	0	2
12 bit × 11 bit multiplier	2	0
11 bit × 7 bit multiplier	1	0

TABLE II
Implementation Results

	This work	[2]	[3]
CMOS technology (μm)	0.11	0.18	0.18
Equivalent gate count [a]	17,223	44,992	13,917
Max. frequency (MHz)	370	200	200
DIS memory	1	4	1
Correction speed (Mpixels/s)	205	140	40
Supported frame size	2048 × 2048	1024 × 1024	1024 × 1024

[a] The smallest 2-input NAND cell is counted as one.

Analytical Placement of Mixed-size Circuits for Better Detailed-Routability

Shuai Li and Cheng-Kok Koh

School of Electrical and Computer Engineering, Purdue University

West Lafayette, IN, 47907-2035

{li263, chengkok}@purdue.edu

Abstract— We propose an analytical placer for generating placement results that can be detailed-routed faster and have fewer violation of design rules. By including a group of pin density constraints in its mathematical formulation, the placer manages to alleviate pin congestion when distributing cells. Moreover, for mixed-size circuits, we adopt a scaled smoothing method to minimize the possible negative influence of fixed macro blocks in placement. As a result, we have few cells overlapping with fixed blocks after global placement, implying that the global placement solution resembles a legal solution more and that legalization has less perturbance to the placement quality. Also, in the final placement result, fewer cells are around macro blocks, whose negative effect in the future routing stage can thus be reduced. Routing solutions obtained by a commercial router show that for most benchmark circuits, detailed routing solutions with fewer violations can be achieved on the placement results generated by our analytical placer.

I. Introduction

Routability is one of the most critical issues in modern circuit design [1]. Placement, the immediate step before routing, should play an important role in alleviating routing congestion. With the promotion of the ISPD11, DAC12 and ICCAD12 routability-driven placement contests [2][3][4] (and earlier ISPD contests), a number of routability-driven placers, including NTUplace4 [5], Ripple [6], SimPLR [7], etc., have been proposed in recent years. Most existing routability-driven placers resort to placement refinement for the alleviation of possible routing congestions. After an initial solution is generated by a traditional placement algorithm, congested regions are identified based on global routing overflows and/or pin density. Then the placement result is refined accordingly by means of techniques such as white space allocation [8], cell bloating [7][9], net-based movement [6], etc. This process is usually iterative and continues until no significant improvement can be made. Some exceptions are analytical placers proposed in [5][10], which alleviate routing congestions by integrating wire density, pin density, and/or overflow refinement constraints into analytical placement formulations.

Most existing routability-driven placers use global routing over-flow as the main congestion estimation factor. In global routing, the chip is typically partitioned into uniform bins, and the bin-to-bin routes are determined for global nets (i.e., a net connecting pins in different bins), whereas all the local nets (i.e., a net connecting only pins inside a bin) are ignored, which, however, may cause detrimental local congestions, too. As a result, the congestion information provided by global routing may not be exact, guided by which placers are likely to generate placement results that have fewer global routing overflows but poor routability in reality. Such a mismatch between the quality of a global routing solution and the quality of a detailed routing solution has been highlighted in [11][12][13].

However, it is infeasible to include in a placer a detailed routing step to guide the placement optimization process. In other words, the question of how to generate placement results that can be (detailed) routed easily remains. In this paper, we present a new analytical placer for routability-driven placement of mixed-size circuits. The goal is to achieve good detailed routability without invoking a global router or a detailed router. Two main contributions we make are:

(1) A new analytical placement formulation with pin density (or more precisely, pin count) constraints.

Each pin on a cell is involved in some net. The pin count within a region thus is related to the routing demand within that region. Local peaks of pin count usually lead to routing congestions. Therefore, in [9][14], pin density is used in congestion estimation and guiding placement refinement. The placement refiner in [14] is shown to be effective in improving the detailed-routability of placement results. Alternatively, in order to achieve even pin allocation with fewer local pin congestion, NTUplace4 [5] incorporates a pin density penalty term in its analytical placement formulation.

In our work, however, we adopt a new analytical placement formulation with pin density constraints to accomplish even pin distribution. Typically, the formulation of many existing analytical placers, including NTUplace3 [15], Aplace [16], use cell density constraints, with which the potential of cell occupation in a region is forced to be less than the area unoccupied by fixed blocks in the region. This formulation has been shown to be effective in distributing cells evenly. However, pin congestion may still exist in the generated placement results, because the pin density on cells may vary.

Therefore, we replace cell density constraints with pin density constraints in our formulation, such that the potential of pin count in a region is constrained to be less than an upper bound. The upper bound is defined as the product of the unoccupied area in the region and the average pin density of all cells. In this way, our placer can generally distribute cells as other analytical placers, while the pin count in each region is also under control.

(2) Scaled smoothing method to cope with fixed macro blocks.

Fixed blocks may have many adverse effects on placement and routing. They are obstacles that affect the distribution of cells in global placement. In addition, if there are many overlaps between cells and fixed blocks in a global placement result, legalizing these cells will greatly perturb the quality of placement results.

To solve this problem, we adopt a scaled smoothing method. The allocation of fixed blocks on the chip is smoothed with Gaussian smoothing, and the smoothed allocation of fixed blocks is properly scaled up. In this way, cells are more likely to be kept away from macro blocks in global placement, and cell displacement in legalization is reduced. Moreover, since macro blocks are blockages in the routing stage, keeping cells away reduces the negative effect of macro blocks in routing, too.

Experimental results show that both proposed methods are ef-fective in helping our placer generate placement results with good routability. Evaluated with the same global routers used in the routability-driven placement contests, many of our placement results have metrics as good as those of the top placers in the contests.

978-1-4799-2817-0/14 $31.00 © 2014 IEEE

Furthermore, by using a commercial router, we also obtained detailed routing solutions for placement results on ISPD11 and DAC12 benchmark circuits. Detailed routing runtime usually is long since many violations of spacing rules may occur, and repairing as many violations as possible is time-consuming. It turns out that for most circuits, our placer's results can be routed in a shorter time and the final routing solutions have fewer routing violations, when compared with the results generated by the top placers in the contests.

The rest of the paper is organized as follows. Section II introduces the formulation of our analytical placer. Section III discusses the scaled smoothing method, as well as the implementation details of our placer. Section IV briefly presents the legalization and detailed placement techniques used in our placer. Section V demonstrates experimental results. Section VI concludes the paper.

II. FORMULATION

The problem of global placement is formulated as a non-linear constrained optimization problem in our analytical placer. All the macro blocks are considered to be fixed and only cells are movable in circuits as in the routability-driven placement contests. The whole circuit is partitioned into uniform nonoverlapping rectangle *bins* for the control of routing congestions.

The definitions used in formulations are as follows:

$\mathcal{N}$	set of all nets;
$\mathcal{C}$	set of all cells;
$\mathcal{B}$	set of all uniform rectangle bins;
F_b	area occupied by fixed macro blocks in bin $b \in \mathcal{B}$;
S_b	available area for placing cells in bin $b \in \mathcal{B}$;
P_{bc}	overlapping portion of cell $c \in \mathcal{C}$ in bin $b \in \mathcal{B}$;
d_c	number of pins on cell $c \in \mathcal{C}$;
r_c	normalization factor for cell $c \in \mathcal{C}$;
w_c, h_c	width, height of cell $c \in \mathcal{C}$;
w_b, h_b	width, height of bin $b \in \mathcal{B}$;
(x_c, y_c)	coordinates of the center of cell $c \in \mathcal{C}$;
(x_b, y_b)	coordinates of the center of bin $b \in \mathcal{B}$;
t_{den}	target placement density;
avg_{pd}	average pin density of all cells.

A. Pin Density Oriented Formulation

The formulation for global placement is:

$$\text{min} \quad \text{HPWL} \tag{1}$$

$$\text{s.t.} \quad \sum_{c \in \mathcal{C}} r_c P_{bc}(x_c, y_c) \leqslant avg_{pd} S_b \quad \forall b \in \mathcal{B}. \tag{2}$$

The variables in the formulation are the locations of cells, (x_c, y_c). The objective is to minimize the total wirelength estimated with the summation of half perimeter wirelength (HPWL) for all the nets. Meanwhile, a *pin density constraint* is defined for each bin to avoid pin congestion in it. The right hand side of (2) denotes the maximum number of pins that bin b is supposed to accommodate. It is defined as the product of the average pin density of all the cells, avg_{pd}, and the available area unoccupied by fixed blocks in bin b, S_b:

$$avg_{pd} = \sum_{c \in \mathcal{C}} d_c \Big/ \sum_{c \in \mathcal{C}} (w_c h_c), \quad S_b = t_{den}(w_b h_b - F_b) \quad \forall b \in \mathcal{B}. \tag{3}$$

The left hand side of (2) gives the potential of pin count in bin b in current placement solution. The pins on each cell c are distributed to

bins proportionally based on P_{bc}, the overlapping portion between the cell and each bin b. And r_c here is a normalization factor to guarantee that $\sum_b (r_c P_{bc}) = d_c$, i.e, cell c contributes a total potential that is equal to the number of pins on it.

The non-differentiable functions in the formulation, HPWL and P_{bc}, must be estimated with differentiable continuous functions so that the minimization problem could be solved with Newton-like methods. The definition of HPWL is as follows,

$$\text{HPWL} = \sum_{n \in \mathcal{N}} (\max_{c \in n}\{x_c\} - \min_{c \in n}\{x_c\} + \max_{c \in n}\{y_c\} - \min_{c \in n}\{y_c\}).$$

where max and min functions are non-differentiable. As in [17], we use the function below to smooth HPWL function:

$$CHKS(x_1, x_2) = (\sqrt{(x_1 - x_2)^2 + \alpha^2} + x_1 + x_2)/2.$$

The two-variable CHKS function can be used as a smooth function for $max(x_1, x_2)$, and its smoothness can be tuned with factor α. In a nested way, multiple-variable max, min functions can also be smoothed with it [17].

The overlapping portion function, P_{bc}, can be defined as the product of two non-differentiable functions:

$$P_{bc}(x_c, y_c) = h_{bc}(x_c) v_{bc}(y_c).$$

where h_{bc} and v_{bc} are the overlapping portion of cell c and bin b in horizontal and vertical dimensions, respectively. $h_{bc}(x_c)$ can be smoothed with $p(x_c - x_b)$, where $p(x)$ is a continuous differentiable "bell-shape" function [16]:

$$p(x) = \begin{cases} 1 - x^2/(2w_b^2), & 0 \leq |x| \leq w_b, \\ (|x| - 2w_b)^2/(2w_b^2), & w_b \leq |x| \leq 2w_b, \\ 0, & |x| \geq 2w_b. \end{cases}$$

v_{bc} can be smoothed in the same way, too.

With the smoothed functions, the constrained optimization problem (1)-(2) is "solved" as a sequence of unconstrained optimization problems:

$$\text{min} \quad \text{HPWL} + \lambda \sum_{b \in \mathcal{B}} [max(avg_{pd}S_b - \sum_{c \in \mathcal{C}} (r_c P_{bc}), 0)]^2, \tag{4}$$

i.e., pin density constraints are incorporated in the objective function as a weighted penalty term. The weight factor λ is doubled iteratively in the solution sequence. With each λ, the unconstrained optimization problem is solved with *L-BFGS-G*, a quasi-Newton solver with boundary constraints [18]. Note that the two-variable max function in (4) is also smoothed with the CHKS function.

B. Comparison with Cell Density Oriented Formulation

Instead of (2), the formulation of many existing analytical placers make use of *cell density constraints*, as shown below:

$$\sum_{c \in \mathcal{C}} k_c P_{bc} \leqslant S_b \quad \forall b \in \mathcal{B}. \tag{5}$$

The potential of area occupied by cells in each bin b is constrained to be no larger than the available area in it. The area of each cell c is allocated to bins proportionally, too, and k_c is the normalization factor such that cell c contributes a total potential equal to its area. In NTUplace4 [5], a penalty term is also added to the right hand side of (5), such that the available space in bins are adjusted by pin density, and fewer cells will be placed in bins with higher pin density.

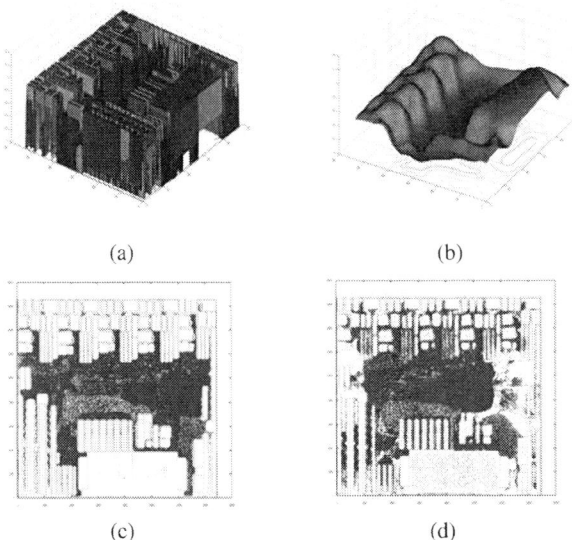

(a)   (b)

(c)   (d)

Fig. 1: ISPD11 benchmark circuit superblue4: (a) Contour of $F_b/(w_b h_b)$; (b) Contour of $F_b'/(w_b h_b)$; (c) Plot of placement result with the proposed scaled smoothing technique; (d) Plot of placement result with the smoothing technique but without any scale-up.

In the special case when the pin density on all cells are the same, constraints (5) are equivalent to constraints (2). However, in reality, pin densities can vary. For instance, on ISPD11 benchmark, superblue4, the pin density of a cell, calculated as the number of pins on it divided by its area, varies from 0.0024 to 0.1111, whereas avg_{pd} of all cells equals 0.0299. Consequently, it is possible that a placement solution that is optimized to meet (5) may have even cell distribution, but have uneven pin distribution, which may have adverse effect in routability.

Our experimental results in Section V-A show that a simple replacement of (5) by (2) can have an obvious impact on the routability of placement results.

III. SCALED SMOOTHING METHOD

One main feature of modern circuit designs is the increasing number of fixed blocks, such as the analog blocks, memory blocks, etc., on the die. Fixed blocks have at least two negative effects in placement. First, in global placement, they are obstacles preventing cells from spreading. Second, if many cells end up being placed on macro blocks, legalizing these cells may result in great perturbance to the quality of placement solutions.

Moreover, macro blocks are blockages in routing. There are relatively fewer routing resources over macro blocks. Routing wires passing through them usually have to "climb" to higher routing layers. Pins placed too close to blocks are hard to access because of spacing rules. Routing wires connecting to these pins are also more likely to pass through macro blocks, resulting in more vertical vias. As most state-of-the-art academic placers do not consider much of macro blocks' negative influence on routability, it is shown in [13] that in the detailed routing solutions of their placement results, many violation of spacing rules may occur near macro blocks.

A. The Smoothing Technique

To overcome the challenge imposed by fixed macro blocks on cell spreading, we can adopt the Gaussian smoothing technique proposed in [15]. F_b in (3) is replaced with the smoothed allocation of fixed blocks, F_b', which is calculated with the 2-D Gaussian function:

$$G(x,y) = \frac{1}{2\pi\sigma^2} e^{-\frac{x^2+y^2}{2\sigma^2}}.$$

F_b's are normalized such that $\sum_b F_b' = \sum_b F_b$. Normalized contours of F_b and F_b' of superblue4 are illustrated in Fig.1. The smoothness of F_b' can be tuned with σ. Larger σ leads to smoother F_b'. Conversely, when σ is sufficiently small, F_b' is nearly equivalent with F_b. At the beginning of global placement, we can calculate F_b' with a larger σ, and then, we recalculate F_b' with gradually decreased σ's. In this way, the movement of cells are less likely to be prevented by those steep "mountains" as illustrated in Fig.1(a).

However, even with the smoothing technique, many cells may still end up overlapping with blocks after global placement. One such example is shown in Fig.1(d). One of the reasons is that after smoothing, many fully occupied bins (i.e., bins with $F_b = w_b h_b$) may have empty spaces temporarily because $F_b' < w_b h_b$. As a result, many cells may be placed in these bins, or more exactly, on top of blocks. Then, after σ is decreased, they may be located in the middle of a relatively "flat" mountain, and thus be trapped in local optimality and never get off, because no matter to which direction they move, the contribution to the penalty term in (4) stays about the same.

B. The Scaled Smoothing Technique

To solve this problem, we choose to properly scale up F_b' so that most fully occupied bins have no temporarily empty space available for cells. To maintain the smoothness of fixed-block allocation, we scale up F_b' of all bins with the same factor, α,

$$F_b'' = \alpha F_b' \quad \forall b \in \mathcal{B}.$$

After Gaussian smoothing, empty bins (i.e., bins with $F_b=0$) far from fixed blocks have F_b' close to 0. Thus, as long as α is not extremely large, the scale-up has little influence on these empty bins, and the placement in block-free areas will not be affected significantly. Meanwhile, F_b' of empty bins around macro blocks may be scaled up a lot, but this is positive in helping keep cells away from blockages in the future routing stage.

On the other hand, the scale-up does no good to cell spreading. Therefore, our global placer is implemented as a two-stage work.

Stage 1: Spreading cells. In this stage, we set σ to be a quarter of the chip width, and we do not scale up F_b's. The optimization in this stage continues until cells are well spread with even pin distribution, which is evaluated with the pin density deviation defined below:

$$\text{den\_dev} = \sqrt{\frac{\sum_b [max(avg_{pd}S_b - \sum_c(r_c P_{bc}), 0)]^2}{|\mathcal{B}|}}.$$

Stage 2: Relocating cells overlapping with blocks. This stage is composed with multiple iterations. In the first iteration, σ is set to be large enough so that no "flat" top would form on large blocks. Given the width and height of the largest macro block on the chip, w_m and h_m, we obtain a satisfying σ by calculating the equation below:

$$\frac{G(w_m/2, h_m/2)}{G(0,0)} = tol$$

where $tol \in (0, 1)$ is an user-set constant, e.g., 0.01. Then, all F_b''s are scaled up to get F_b'', with which the problem (1)-(2) is optimized.

Next, σ is halved and the second iteration starts. With the new σ, scaled-up F_b'' are recalculated, and the problem (1)-(2) is optimized again. This iterative process continues until σ is small enough, when F_b'' equals F_b in most empty bins.

The setting of the scale-up factor, α, is critical in each iteration. With too small a factor, many cells may still end up overlapping with macro blocks, whereas the placement in empty bins may be influenced a lot with too large a factor. In our implementation, α is set such that the number of bins with $F_b'' \geq w_b h_b$ equals the number of originally fully occupied bins with $F_b = w_b h_b$. In general, factors determined by this method are between 1.3 and 2.0.

In addition, in order to improve the scalability of the two-stage placer, we adopt the multi-level placement technique [15][19] in both stages of the placer. The first choice (FC) clustering algorithm [19] is used in building the hierarchy of clusters.

The effectiveness of scaled smoothing is shown in Fig.1. It contains plots of the placement results generated by our placer. Cells are shown blue in the plots. (c) illustrates the result when the scaled smoothing technique is applied, whereas (d) illustrates the result when no scale-up is performed, i.e., the scale-up factor is always 1.

Only 0.08% of cells still overlap with blocks in (c), whereas 7.71% of cells overlap with blocks in (d). Many cells in (d) are trapped in the center of large blocks. As a consequence, for the result in (d), the average displacement of cells in legalization is 45.6, whereas it is only 22.9 for the result in (c). Moreover, in (c), fewer cells are placed in channels between blocks.

In our experiments on DAC12 benchmark circuits, with the scaled smoothing technique, the number of cells overlapping with macro blocks after global placement is also reduced by over 10 times on average, and the displacement in legalization is decreased accordingly.

IV. LEGALIZATION & DETAILED PLACEMENT

Our legalizer is similar to the Abacus algorithm [20]. Cells are sorted based on their x-coordinates. From left to right, cells are placed into legal positions one by one. Each cell is placed in a row such that inserting the cell in the row leads to the smallest displacement. Instead of packing cells tightly from left to right in each row, in our implementation, we place cells as close to their original position as possible, so that the perturbance to pin distribution is reduced. We apply the sliding window technique [8] for detailed placement to further optimize the placement.

V. EXPERIMENTAL RESULTS

By applying the two-stage global placer discussed in Section III, followed by the legalizer & detailed placer introduced in Section IV, we have generated placement results for different benchmark circuits. Target density t_{den} is set to be 0.80 for all benchmark circuits. The routability of the generated placement results have been evaluated by the same global routers as in the routability-driven placement contests.

Moreover, the routability quality of placement results on ISPD11 and DAC12 benchmark circuits [2][3] have also been evaluated by the commercial router Wroute [21] (version 3.1.61), with the application of the LEF/DEF file translator developed under some

TABLE I: Routing metrics and placement runtime of our placer(SSmth&PinDen)'s results on DAC12 benchmark circuits.

	VIO	WL(e7)	VIA(e6)	TR(m)	NUN	OC(%)	TP(m)
s2	681	7.00	11.90	312	253	3.81	150
s3	193	4.09	10.67	196	68	6.12	181
s6	202	4.08	11.31	217	295	6.61	207
s7	418	4.89	16.45	280	127	8.85	391
s9	33	2.94	9.40	156	13	6.76	229
s11	425	4.00	10.12	214	24	3.01	152
s12	104	4.34	16.38	286	0	15.87	497
s14	621	2.77	7.00	178	378	6.51	123
s16	36	3.01	7.38	134	0	7.1	133
s19	106	1.80	5.76	106	93	6.56	130
Norm	1.00	1.000	1.000	1.000	1.00	1.000	–
Ropt	1.00	1.000	1.000	1.000	1.00	1.000	–

28nm design rules in [13]. The translator cannot be applied on ICCAD12 benchmark circuits [4] so far.

The runtime of Wroute is mainly spent on detailed routing. Repairing violations of spacing rules usually takes Wroute a long time during detailed routing. Thus, we use two metrics to evaluate the detailed routability of a placement result: (1) The runtime of Wroute; (2) The number of routing violations left in the final routing solution.

As in [13], we run Wroute for two iterations. First, Wroute is run in the default routing mode. Wroute may report some nets as unroutable after global routing, and ignore these nets. Second, Wroute is run in the post routing mode. It tries to route all nets and repairs as many routing violations as possible. In each iteration, a time limit of 24 hours is set.

All our experiments on placement and routing are performed on the same PC workstation with a quad-core 3.10 GHz CPU (only one core is used) and a 8GB memory.

A. Effectiveness of the two proposed methods

For our placer's results on DAC12 benchmark circuits, Table I gives the runtime of our placer, TP in minutes, and the metrics of the routing solutions obtained by Wroute. VIO, WL(e7), VIA(e6) are detailed routing metrics, denoting the number of routing violations, wirelength in nanometers and the number of vertical vias, respectively. TR in minutes is the CPU time taken by Wroute in the two-iteration routing process. The other two are global routing metrics: NUN gives the number of unroutable nets reported after the global routing in the first iteration; OC(%) gives the percentage of overcapacity gcells after the global routing in the second iteration.

By default, both the scaled smoothing technique and pin density constraints are used in our placer (SSmth&PinDen). However, the placement results generated by our placer with the smoothing technique and cell density constraints (Smth&CellDen), with the smoothing technique and pin density constraints (Smth&PinDen), with the scaled smoothing technique and cell density constraints (SSmth&CellDen) are evaluated by Wroute, too, as shown in Table II. For different placers, the summation of the same metric are normalized, and shown in the "Norm" row in the tables.

Smth&CellDen is an analytical placer without considering routability, which is similar to NTUplace3 [15]. The number of violations is high for seven out of ten circuits. Then, in SSmth&CellDen, with the application of scaled smoothing, both the number of violations and routing runtime are reduced for all circuits.

TABLE II: Routing metrics of the placement results generated by three restricted versions of our placer on DAC12 benchmark circuits.

	Smth&CellDen						Smth&PinDen						SSmth&CellDen					
	VIO	WL	VIA	TR	NUN	OC	VIO	WL	VIA	TR	NUN	OC	VIO	WL	VIA	TR	NUN	OC
s2	637984	7.36	14.10	3326	9197	5.43	450587	7.07	13.68	2945	4679	5.25	79434	7.32	13.31	1857	3042	4.97
s3	24415	4.06	11.81	1337	6656	8.08	1482	4.00	11.53	345	3115	8.05	418	4.23	11.34	285	1774	7.20
s6	60951	4.04	12.03	1604	4415	7.63	14253	4.10	12.06	587	2917	8.44	1380	4.13	11.65	527	1282	6.87
s7	248	4.76	16.97	317	1763	9.57	194	4.95	16.60	286	744	9.44	102	5.00	16.66	281	114	8.61
s9	43468	2.92	9.92	549	2266	8.02	5562	2.88	9.86	294	2466	8.07	58	2.93	9.59	167	56	7.00
s11	431	3.88	10.66	264	1404	3.65	12164	3.92	10.53	372	1560	3.79	425	3.85	10.43	257	581	3.18
s12	11382082	4.66	19.64	3115	8296	18.87	6584968	4.54	19.82	2988	2599	21.06	5210636	4.80	19.10	3026	2682	18.51
s14	139958	2.69	7.69	1007	3340	7.90	20032	2.73	7.24	276	1174	7.66	7397	2.79	7.41	276	503	7.22
s16	118225	2.84	7.92	818	651	7.26	140017	2.86	7.63	869	757	7.63	54	2.86	7.70	144	62	6.68
s19	870	1.86	6.20	236	1389	7.69	111	1.78	5.93	109	403	7.38	362	1.87	6.07	150	755	6.99
Norm	–	1.005	1.099	6.048	31.48	1.181	–	0.998	1.080	4.363	16.32	1.219	–	1.023	1.065	3.353	8.67	1.085

TABLE III: Routing metrics for the placement results generated by NTUplace4, Ripple and SimPLR in the DAC12 contest [3].

	NTUplace4						Ripple						SimPLR					
	VIO	WL	VIA	TR	NUN	OC	VIO	WL	VIA	TR	NUN	OC	VIO	WL	VIA	TR	NUN	OC
s2	1015942	6.81	12.01	2086	2311	3.57	52155	7.32	12.40	1187	1973	3.83	692	7.01	12.08	301	599	4.02
s3	1099	4.01	10.80	307	1853	6.2	205	4.39	11.39	260	1491	7.17	162	4.58	11.31	225	601	7.71
s6	485	3.94	11.31	247	1100	6.4	223	4.14	11.62	230	759	6.98	265	4.17	11.49	220	589	7.19
s7	181	4.85	16.50	272	599	8.75	257	5.43	17.38	300	895	10.24	8181	5.57	17.59	426	634	11.25
s9	65	2.94	9.41	190	1164	6.61	98	3.30	9.86	169	141	7.97	48	3.13	9.69	173	834	7.67
s11	583	3.87	9.95	236	348	2.94	441	4.04	10.28	229	303	3.35	808	3.94	10.25	330	1020	3.46
s12	106	4.31	16.73	442	5104	13.03	217	4.82	17.73	481	5966	15.71	166	4.74	17.27	296	1195	16.17
s14	17476	2.69	7.30	632	2518	6.69	28250	2.80	7.45	290	1387	7.45	262930	2.88	7.48	541	1555	7.66
s16	24	2.93	7.58	136	107	6.55	36	2.93	7.81	144	119	7.17	56	3.05	7.85	154	470	7.44
s19	12574	1.82	5.86	404	1158	6.36	105	1.96	6.10	135	668	7.38	74199	1.88	5.97	449	534	6.96
Norm	–	0.981	1.010	2.382	13.00	0.942	–	1.058	1.053	1.647	10.95	1.085	–	1.053	1.043	1.498	6.42	1.117
Norm*	2.38	0.980	1.008	1.295	24.19	0.929	1.23	1.052	1.053	1.258	21.95	1.063	1.52	1.052	1.040	1.162	11.77	1.092
Ropt	8.91	0.985	1.017	1.360	8.34	0.963	7.40	1.045	1.054	1.430	7.44	1.081	41.03	1.039	1.031	1.246	5.40	1.070

*normalization excluding s2, s7, s14 and s19

Furthermore, by replacing cell density constraints with pin density constraints, SSmth&PinDen generates placement results with even better routability. As shown in Table I, for all the ten circuits, the number of violations is reduced to be less than 1000. NUN and OC columns show that the quality of global routing is improved, too.

Moreover, compared with Smth&CellDen, HPWL on average is increased by around 9% in SSmth&PinDen (not shown in the tables for lack of space), but the average routed wirelength is not increased and the average via count is decreased by nearly 10%.

B. Comparison with other routability-driven placers' results

Table III contains the routing solutions for the placement results generated by the top three placers in the DAC12 contest [3], NTU-place4 [5], Ripple [6] and SimPLR [7]. The metrics in the table are also normalized, as shown in the "Norm" row.

However, it is important to note that these placers were developed to reduce global routing congestions. It is therefore not fair to compare their detailed routing solutions because the quality of their detailed routing solutions can definitely be improved if that has been the objective. Any future reference of the detailed routing solutions presented here should always include this caveat.

For seven out of the ten circuits, the number of violations in our placer's results is the smallest. For some circuits, the routing solutions of other placers' results end up with over 5000 violations even after a long time of violation repairment, because of which the skewed normalization of the VIO metric are not listed in the tables.

However, even excluding these "abnormal" circuits, our placement results still have shorter routing runtime and fewer violations on average, as shown in the "Norm*" row in Table III.

The average wirelength and via count in routing solutions of our placement results are over 4% smaller than those of Ripple and SimPLR. Compared with the results of NTUplace4, the average wirelength of our results is 1.9% larger, whereas the average via count is 1.0% smaller. The NUN and OC columns in Table II and Table III also reveal that for most of our placement results, the global routing solutions are also reasonable.

As mentioned in Section I, a placement refiner called the routability optimizer (Ropt) is proposed in [14] and has been shown to be effective in improving the detailed-routability of NTUplace4's, Ripple's and SimPLR's results such that their routing solutions have fewer violations. Ropt is effective in refining our placer's results, too, as the routing solutions of refined results have fewer violations for eight out of ten DAC12 benchmark circuits. The normalized routing metrics for all placers' refined results are shown in the "Ropt" rows in Table I and Table III. On average, our placer's refined results still have the fewest violations and the shortest routing runtime.

Table IV gives the DAC12 metrics (without runtime penalty) of our placer's results. As in the DAC12 contest, two academic global routers, NCTU-GR 2.0 [22] and BFG-R 2.0 [23], with the given parameter settings, are performed on the placement results. In the generated global routing solutions, routing congestion (RC) are evaluated, and based on HPWL and RC, the scaled wirelength w.r.t. two global placers (SW_NCTU&SW_BFG) are calculated. The same

TABLE IV: DAC12 metrics of our placer's results, and the evaluation of the results by NCTU-GR with a different parameter setting.

	DAC12 metric(e8)		NCTU-GR evaluation		
	SW_NCTU(RC)	SW_BFG(RC)	TOF	MOF	TGR(s)
s2	11.29(126.5)	11.95(130.0)	75310	24	25507
s3	4.96(116.8)	3.97(106.8)	0	0	294
s6	4.30(108.3)	4.68(111.9)	0	0	124
s7	4.30(102.5)	4.27(102.2)	0	0	92
s9	2.42(100.4)	2.52(101.9)	0	0	49
s11	3.96(103.8)	3.65(100.9)	0	0	71
s12	5.00(123.5)	5.26(126.5)	0	0	425
s14	2.39(100.6)	2.52(102.2)	0	0	67
s16	2.96(102.8)	2.96(102.8)	0	0	87
s19	1.85(107.4)	1.66(103.1)	0	0	48
NTUplace4	0.759(0.93)	0.801(0.95)	–	–	–
Ripple	0.838(0.94)	0.903(0.97)	–	–	–
SimPLR	1.174(1.04)	1.035(1.00)	–	–	–

TABLE V: Routing metrics and placement runtime of our placer's placement results on ISPD11 benchmark circuits.

	VIO	WL(e7)	VIA(e6)	TR(m)	NUN	OC(%)	TP(m)
s1	64	3.35	9.86	173	6	5.2	187
s2	686	6.93	12.22	288	243	3.43	150
s4	221	2.57	6.41	122	0	5.61	93
s5	5062	4.15	8.96	239	578	3.44	114
s10	115	6.48	13.16	318	381	4.41	158
s12	104	4.34	16.38	262	0	15.87	497
s15	24	3.69	12.38	184	24	12.26	192
s18	57	2.09	6.51	127	506	8.04	128
NTUplace	–	1.097	1.178	6.528	20.05	1.373	–
Ripple	–	1.142	1.067	1.490	7.56	1.149	–
SimPLR	–	1.125	1.083	2.860	8.22	1.283	–

metrics for other placers' results in the contest are available at [3]. Using our placer's results as baseline, we normalized the metrics of other placers' results, as shown in the bottom rows in Table IV. The runtime penalty is not considered because our placer's results were not generated on the same machine as in the contest. Table I shows that the runtime of our placer is comparable with that of NTUplace4.

Although our placer does not invoke global routing, for some circuits, the generated results has RC close to 100, implying a positive evaluation of the global routing solution. Even though the metric for other results are high, the data in Table I show that these may not reflect the true routability quality of the placement results. The routing solution of circuit superblue2 is one such example.

Moreover, we have also tried running global router NCTU-GR 2.0 [22] with a different parameter setting, optimizing global routing until all overflows are eliminated or no significant improvement can be made. Table IV lists the number of total overflow, TOF, and maximum overflow, MOF, in the global routing solutions, as well as the global routing runtime in seconds, TGR. For most of our results, overflow-free solutions are generated. Again, the supposedly poor performance of superblue2 in the context of global routing does not reflect its better detailed routability. With the new parameter setting, overflow-free global routing solutions can be generated for nearly all other placers' results, too.

Besides the results on DAC12 benchmark circuits, the routing metrics for our placer's results on ISPD11 benchmark circuits are given in Table V. Using metrics of our placement results as baseline, we normalized the same metrics for NTUplace's, Ripple's, SimPLR's results in the ISPD11 contest [2], which are shown in the bottom three

rows, except the skewed normalization in the VIO metric. In general, for all the circuits except superblue5, the routing solutions of our placer's results have the smallest number of violations among all.

VI. CONCLUSION

In order to alleviate possible routing congestions, our analytical placer makes use of pin density constraints in its mathematical formulation. Also, our placer adopts a scaled smoothing technique to avoid cells overlapping with macro blocks in global placement results, and keep more cells away from routing blockages in the future routing stage. Experimental results in Section V show that both techniques are effective in improving the routability of placement results. Compared with other routability-driven placers' placement results, in most cases, detailed routing solutions with fewer violations can be generated in a shorter time for our placer's placement results.

REFERENCES

[1] N. Viswanathan et al. The ISPD-2011 routability-driven placement contest and benchmark suite. In Proc. ISPD, pages 141–146, 2011.

[2] http://www.ispd.cc/contests/11/ispd2011_contest.html.

[3] http://archive.sigda.org/dac2012/contest/dac2012_contest.html.

[4] http://cad_contest.cs.nctu.edu.tw/CAD-contest-at-ICCAD2012 /problems/p2/p2.html.

[5] M.-K. Hsu et al. Routability-driven analytical placement for mixed-size circuit designs. In Proc. ICCAD, pages 80–84, 2011.

[6] X. He et al. Ripple: an effective routability-driven placer by iterative cell movement. In Proc. ICCAD, pages 74–79, 2011.

[7] M.-C. Kim et al. A SimPLR method for routability-driven placement. In Proc. ICCAD, pages 67–73, 2011.

[8] C. Li et al. Routability-driven placement and white space allocation. IEEE TCAD, 26(5):858–871, May 2007.

[9] J. A. Roy et al. CRISP: congestion reduction by iterated spreading during placement. In Proc. ICCAD, pages 357–362, 2009.

[10] K. Tsota, C.-K. Koh, and V. Balakrishnan. Guiding global placement with wire density. In Proc. ICCAD, pages 212–217, 2008.

[11] T. Taghavi et al. New placement prediction and mitigation techniques for local routing congestion. In Proc. ICCAD, pages 621–624, 2010.

[12] C. J. Alpert et al. What makes a design difficult to route. In Proc. ISPD, pages 7–12, 2010.

[13] W.-H. Liu et al. Case study for placement solutions in ISPD11 and DAC12 routability-driven placement contests. In Proc. ISPD, pages 114–119, 2013.

[14] W.-H. Liu, C.-K. Koh, and Y.-L. Li. Optimization of placement solutions for routability. In Proc. DAC, pages 153:1–153:9, 2013.

[15] T.-C. Chen et al. NTUplace3: An analytical placer for large-scale mixed-size designs with preplaced blocks and density constraints. IEEE TCAD, 27(7):1228–1240, July 2008.

[16] A. B. Kahng et al. APlace: a general analytic placement framework. In Proc. ISPD, pages 233–235, 2005.

[17] C. Li et al. Recursive function smoothing of half-perimeter wirelength for analytical placement. In Proc. ISQED, pages 829–834, 2007.

[18] C. Zhu et al. Algorithm 778: L-BFGS-B: Fortran subroutines for large-scale bound-constrained optimization. ACM Trans. Math. Softw., 23(4):550–560, December 1997.

[19] T. F. Chan et al. An enhanced multilevel algorithm for circuit placement. In Proc. ICCAD, pages 299–, 2003.

[20] P. Spindler et al. Abacus: fast legalization of standard cell circuits with minimal movement. In Proc. ISPD, pages 47–53, 2008.

[21] http://www.cadence.com/products/di/soc_encounter/.

[22] W.-H. Liu et al. Multi-threaded collision-aware global routing with bounded-length maze routing. In Proc. DAC, pages 200–205, 2010.

[23] J. Hu, J. A. Roy, and I. L. Markov. Completing high-quality global routes. In Proc. ISPD, pages 35–41, 2010.

Lithographic Defect Aware Placement Using Compact Standard Cells Without Inter-Cell Margin

Seongbo Shim, Yoojong Lee, and Youngsoo Shin
Department of Electrical Engineering, KAIST
Daejeon 305-701, Korea
Email: sbshim@dtlab.kaist.ac.kr

Abstract— Conventional standard cells contain extra space, called inter-cell margin, to prevent potential defects caused by lithography process. Margin is indeed necessary between some cell pairs, but there are also lots of cell pairs that do not yield any defects (or have very low probability of defects) when they are placed without margin. We address a new placement problem using standard cells without inter-cell margin. Placement should be done such that defect probability is made as small as possible while standard objectives such as wirelength is also pursued. The key in this approach is efficient computation of defect probabilities of all cell pairs and arranging them as a table that is referred to by a placer. We study how the cell pairs can be grouped by examining similar patterns along cell boundary, which greatly reduces the number of defect probability computation. The proposed placement method was evaluated on a few test circuits using 28-nm technology. Chip area was reduced by 10.8% on average with average and maximum defect probability kept below 0.4% and 4.1%, respectively.

I. INTRODUCTION

Any kind of pattern failure, e.g. contact bridge and metal short, originated from lithography process is called lithographic defect. It has been become more important as patterns are scaled down to nano-meter geometry but lithography process does not fully support that fine feature.

A modern standard cell is designed in such a way that lithographic defect never arises within the region of a cell. This is done by repeated layout, retargeting and OPC, and verification through lithography simulation [1], [2]. Potential defects when cells are abutted together are avoided by reserving extra space, called inter-cell margin. Fig. 1(a) compares a cell before and after margin is added. A margin is typically the width of a single poly pitch and a dummy poly is inserted in the margin so that polys can be regularly placed for better lithography, as shown in Fig. 1(b).

Inter-cell margin, however, is not always necessary. In 28-nm technology we use in this paper, about 20% of cell pairs can be abutted together without margin but with zero defects, and about 30% of cell pairs with defect probability less than 10%. This suggests a possibility of new library consisting of compact cells without inter-cell margin. Some cells may be safely abutted for the benefit of reduction in area and wirelength; some others may be placed with extra whitespace in-between as shown in Fig. 1(c) if defect probability is expected to be high.

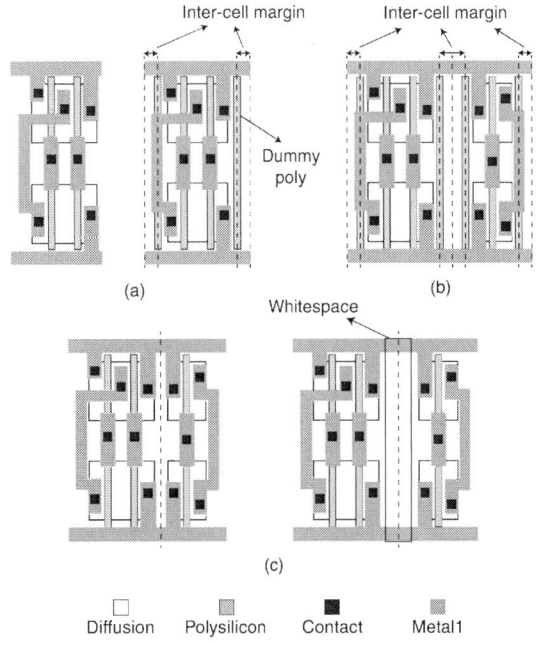

Fig. 1. (a) A cell without vs with inter-cell margin, (b) abutted cells with inter-cell margin, and (c) abutted cells without inter-cell margin vs abutment with extra whitespace in-between.

A. Contributions

We address a new placement problem, in which average (and maximum) defect probability is kept as low as possible; total wirelength, which is a popular objective, is also minimized. A key component in this problem is a table of defect probability. A challenge lies in probability computation time because it takes more than a minute for one cell pair and a practical library has a million number of cell pairs. This is addressed by two approaches: we experimentally show that the computation can be confined within 2-pitch range from cell boundary with tolerable error; the range of 2-pitch, called extent, can be classified into a few groups by identifying geometrically similar extents, which effectively reduces the number of cell pairs in great amount.

The remainder of this paper is organized as follows. Defect probability is defined in Section II, which also studies how probability computation can be approximated. Section III

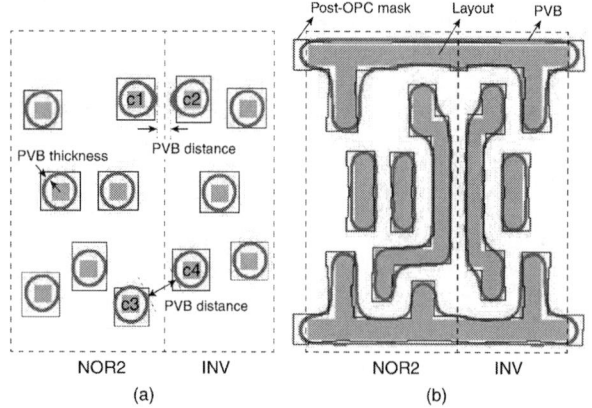

Fig. 2. PVBs in (a) contact layer and (b) metal 1 layer of two abutted cells.

addresses extent grouping. A prototype placement tool is described in Section IV, together with its experimental assessment. We draw conclusions in Section V.

II. DEFECT PROBABILITY COMPUTATION

A. Defect Probability

A pattern failure or defect caused by lithography process is usually modeled by using process variation band (PVB) [1], [3]. Fig. 2(a) illustrates PVBs of contacts in two abutted cells; PVBs of metal 1 layer is shown in Fig. 2(b). PVBs are obtained, after retargeting and OPC are applied to layout, by repeated lithography simulation at various lithography settings, which are combinations of extreme (and nominal) conditions of scanner focus, exposure energy, and mask error [3]. Thus, a PVB of each contact (or metal 1) is made of multiple contours, in which one contour corresponds to an exposed image on a wafer in particular lithography setting [1].

Note that contacts c1 and c2 are relatively close. Their PVBs along cell boundary look thicker, which implies more variation in the exposed image. This is due to the increased light interference as the two contacts are located closer. PVBs of c3 and c4, on the other hand, are thinner.

The probability that lithographic defect occurs, which we call defect probability, can be modeled by the minimum distance between PVB pairs, called PVB distance. Conventional cell is designed in a way that PVB distance within a cell is kept large enough so that defect never arises; it also includes inter-cell margin on both sides, which guarantees no defect when it is abutted to arbitrary cells. In our approach, we aim to remove all inter-cell margin, so some cell pairs may have non-zero defect probability. The defect probability when cells i and j are abutted together is modeled in a linear fashion and is given by

$$D(i,j) = \frac{M_{\mathrm{pvb}} - PD(i,j)}{M_{\mathrm{pvb}} - m_{\mathrm{pvb}}} \times 100\%, \quad (1)$$

where M_{pvb} is PVB distance beyond which defect probability is 0%, m_{pvb} is PVB distance below which defect occurs in

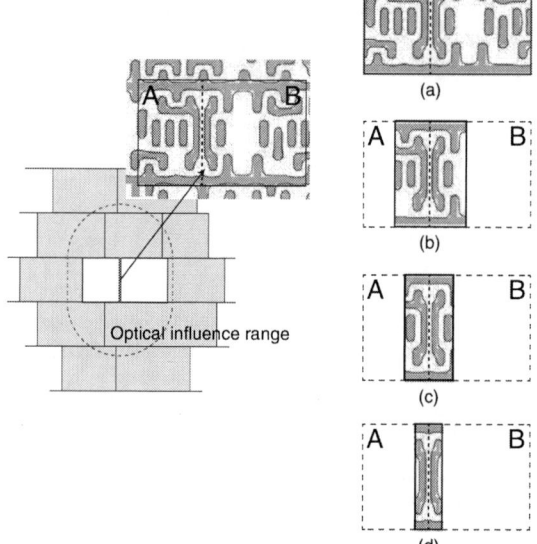

Fig. 3. Different settings for defect probability computation; lithography simulation in (a) full cell width, (b) within 3-pitch from boundary, (c) within 2-pitch from boundary, and (d) within 1-pitch from boundary.

TABLE I

MAXIMUM ERROR OF DEFECT PROBABILITY IN DIFFERENT SETTINGS OF LITHOGRAPHY SIMULATION

Layer	Maximum error of defect probability (%)			
	Full cell width	3-pitch from boundary	2-pitch from boundary	1-pitch from boundary
Contact	1.0	2.3	4.3	15.7
Metal1	1.0	3.7	5.3	37.0

100%, and $PD(i,j)$[1] is PVB distance between i and j. The values of M_{pvb} and m_{pvb} are typically available from foundry fab [3], [4]. Contact and metal 1 layers are the most critical layers for lithographic defect, so we measure $D(i,j)$ on two layers and consider the larger one as its value.

B. Approximate Computation of Defect Probability

We assume 193nm ArF as an illumination source with immersion lithography in this paper. Optical influence range in this setting is about 1 μm, which reaches a few cells beyond cells A and B as shown in Fig. 3, whose defect probability we want to compute. Lithography simulation, which is performed at two layers and is repeated at various lithography settings as we discussed in Section II-A, turns out to take more than a minute. This is prohibitive because there are a million cell pairs for defect probability computation if a library contains 1000 cells.

Considering only abutted cells as shown in Fig. 3(a) fortunately yields negligible amount of error in defect probabil-

---

[1]$PD(i,j)$ and thus $D(i,j)$, in fact, have four different values because cells can be flipped along y-axis and the two cells can be abutted in two different orders.

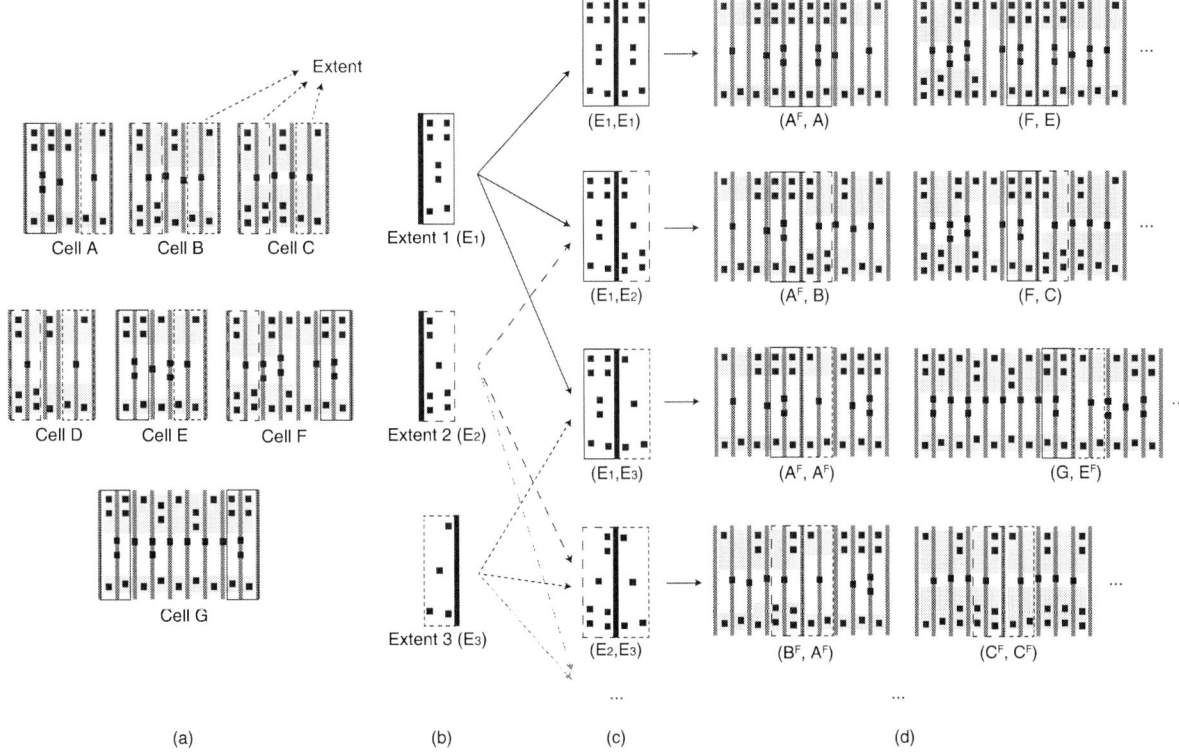

Fig. 4. (a) Library cells and their extents, (b) extent grouping, (c) extent abutment, and (d) mapping to cell abutment.

ity [5], [6]. The second column of Table I reports maximum error of defect probability for 1000 cell pairs, in which lithography simulation within full optical influence range serves as a reference; errors are 1% in both layers. We further decrease the range of defect probability computation to 3-pitch (3 times of minimum pitch of each layer) from cell boundary as shown in Fig. 3(b), 2-pitch in Fig. 3(c), and 1-pitch in Fig. 3(d), and assess the maximum error. As shown in Table I, 2-pitch from boundary seems to be a reasonable choice, which we assume for fast computation of defect probability in this paper.

III. GENERATION OF DEFECT PROBABILITY TABLE

Even though we limit defect probability computation within 2-pitch range from cell boundary, a huge number of cell pairs still makes the computation intractable. We approach the problem by identifying the patterns along cell boundary and grouping them.

A. Extent Grouping

Fig. 4 illustrates the process. Let a library contain 7 cells. Contact patterns within 2-pitch range from cell boundary, called extents, are identified as shown in Fig. 4(a). They are compared and the same patterns are grouped as shown in Fig. 4(b). We then figure out how extents can be abutted as shown in Fig. 4(c). Note that extent, by its definition, can be abutted only along one side, which is denoted by thick line. Therefore, if there are n extent groups, there are $\left(\binom{n}{2}\right) =$

$\frac{n(n+1)}{2}$ different ways extents can be abutted, which now serves as the number of defect probability computation; this is usually much smaller than the number of ways cells can be abutted (see Fig. 4(a)), $(2N)^2$, where N is the number of cells and $N > n$. Finally, cell pairs are mapped to corresponding extent pairs as shown in Fig. 4(d), in which A^F indicates the cell A with its orientation along y-axis flipped.

Our 28-nm library consists of 1043 cells thus 2086 extents. They are grouped into 944 in contact layer, and 1117 in metal 1 layer. Defect probability computation in this case is expected to take 198.0 and 346.6 hours, respectively, which are much less than 967.0 and 1208.7 hours when defect probability is computed for all cell pairs without any extent grouping.

B. Further Extent Grouping by Pattern Similarity Check

Defect probability computation still takes too much of time even after extent grouping. We try to reduce the number of extent groups by grouping similar (as well as exact) patterns together. We have tried three different criterion of similarity, 90%, 80%, and 70%. In 90% of similarity, for example, two extents are grouped together if more than 90% area is filled with 0 once we form geometric XOR of the extents.

The result of this new grouping is presented in Table II; the method of 100% similarity corresponds to the method of Section III-A. The number of extent groups is greatly reduced if we choose 90% as similarity criterion, e.g. from 944 to 388 in contact layer. The reduction in computation time of defect probability is roughly proportional to the square of

TABLE II
VARIOUS EXTENT GROUPING METHODS

Grouping method		# Extent groups	Defect probability computation	
			Time (hours)	Max error (%)
Contact	Similarity (100%)	944	198.0	5.7
	(90%)	388	18.6	6.6
	(80%)	236	5.3	19.7
	(70%)	116	1.6	34.1
Metal 1	Similarity (100%)	1117	346.6	4.8
	(90%)	416	23.7	6.1
	(80%)	299	19.9	24.7
	(70%)	185	2.3	36.2

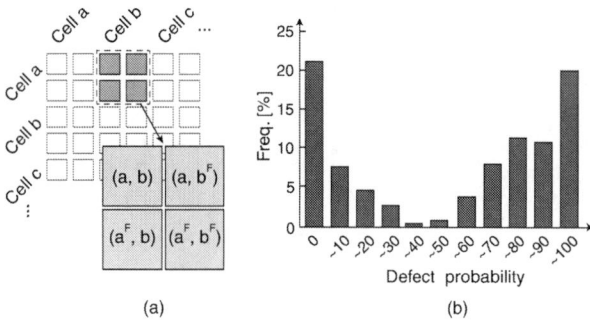

Fig. 5. (a) Defect probability table, and (b) defect probability histogram.

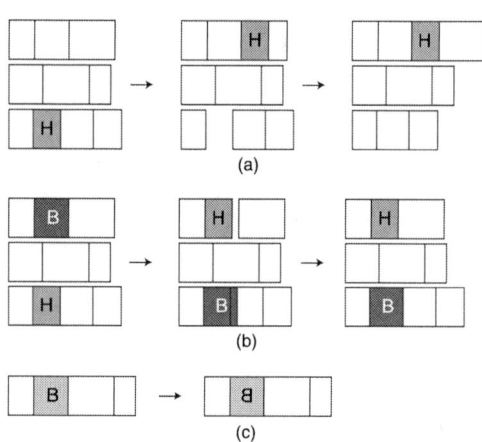

Fig. 6. Operations to generate a new placement.

the reduction in extent groups, which explains even greater reduction of time from 198.0 to 18.6 hours. Fortunately maximum error increases only marginally, from 5.7% to 6.6%. In similarity criterion of 80% and 70%, maximum error is now substantial, even though the number of extent groups and time for defect probability computation are further reduced. This is because a single extent group usually originates from a similar gate family (e.g. INV1X, INV2X, etc) in 90% similarity, while more than one gate family may constitute a single extent group in 80% and 70% similarity. We thus choose 90% similarity in our defect probability computation.

Fig. 5(a) pictorially shows defect probability table, which has 2086 rows and 2086 columns. Fig. 5(b) summarizes defect probabilities as a histogram; it indicates that defect probability is widely distributed, which should be carefully taken into account when cells are placed.

IV. DEFECT PROBABILITY AWARE PLACEMENT

As shown in Fig. 5(b), many cell pairs have non-negligible defect probability when they are abutted. This fact lends itself to automatic placement problem, in which average (and maximum) defect probability is kept as low as possible; other usual objectives such as wirelength should also be taken into account. A prototype tool has been developed based on simulated annealing to solve the problem [7], [8].

A. Cost Function

Each placement during simulated annealing is evaluated by using a cost function:

$$\Gamma = \alpha C_w + \beta C_r + \gamma C_d, \qquad (2)$$

where C_w is total wirelength measured by using half-perimeter of bounding box. C_r is introduced to reserve enough amount of whitespace in each circuit row, so that cell pairs with larger defect probability can be separated apart; it is given by

$$C_r = \sum_{\forall \text{row } i} \left(WS_{req}(i) + RL(i) - RL_0 \right)^2, \qquad (3)$$

where $WS_{req}(i)$ is the whitespace that is required to have defect probability of all cell pairs in row i below some threshold (which is a parameter to control the maximum defect probability), $RL(i)$ denotes the sum cell width of row i, and RL_0 is the minimum $RL(i)$ in the initial placement. C_d corresponds to the average defect probability of all cell pairs.

Coefficients α, β, and γ are employed to balance the quantity of the three terms. Since C_w and C_r tend to increase with the number of cells while C_d does not, we empirically set α and β to 1 and γ to the number of cells divided by 100.

B. Implementation of Placement

The input to our placement tool includes a target aspect ratio of placement region and the amount of whitespace as

TABLE III
COMPARISON OF THREE PLACEMENT METHODS

Circuit			Standard placement with conventional cells (A)		Standard placement with compact cells (B)			Proposed placement with compact cells (C)			
Name	# Cells	# Nets	C_w (mm)	Area (um^2)	C_w (%)	C_d	Area (%)	C_w (%)	C_d	Max defect prob	Area (%)
sasc	349	365	2	551	93.3	71.2	90.3	101.7	0.9	4.2	90.3
i2c	515	550	3	682	91.9	66.1	88.9	100.2	0.4	4.1	88.9
spi	1249	1330	10	1637	99.0	66.3	88.8	98.3	0.3	4.2	88.8
wb_dma	2160	2423	21	2728	92.2	67.5	88.5	93.8	0.4	4.1	88.5
tv80	3932	3980	44	4356	94.8	64.9	86.8	96.8	0.2	4.2	86.8
mem_ctrl	4511	4663	56	5937	95.0	44.0	88.9	101.7	0.3	4.1	88.9
ac97	6310	6382	104	10299	90.5	66.6	91.0	97.6	0.5	4.1	91.0
usbf	7492	7991	132	9452	93.2	59.8	88.5	98.2	0.4	4.0	88.5
pci	9996	10297	247	16664	91.8	63.4	91.2	93.7	0.5	4.1	91.2
aes_cipher	11112	11371	152	9974	91.3	80.9	83.8	97.1	0.2	4.0	83.8
ethernet	31034	31286	469	54756	92.0	70.6	91.7	98.1	0.4	4.0	91.7
vga_lcd	46203	46339	990	86162	93.5	69.6	92.2	97.2	0.3	3.9	92.2
Average			—	—	93.2	65.9	89.2	97.3	0.4	4.1	89.2

percentage of that region. The number of circuit rows is then identified from the input and the sum area of all cells that will be placed.

The placement consists of two steps: repeated placement using simulated annealing and whitespace distribution to further reduce defect probability. Three operations are performed to generate a new placement for the first step. The first operation shown in Fig. 6(a) displaces a randomly picked cell (cell H) to a randomly chosen place; cell overlap and inter-cell whitespace are removed accordingly. The second operation shown in Fig. 6(b) switches the location of randomly picked cell pairs (B and H). In the final operation shown in Fig. 6(c), a cell (B) is picked, and its orientation along y-axis is flipped. The operations are applied with different probability, which is heuristically determined.

While the three operations are applied to generate a new placement, whitespace is allowed only at the right end of each circuit row. The second step of placement starts once the simulated annealing completes, and distributes whitespace so that defect probability can further be reduced. All cell pairs with the defect probability being larger than some threshold (the same threshold to control C_r) become the candidates, and whitespace of 1-poly pitch (to keep pitch regularity of polysilicon for stable patterning result [9]) is distributed in greedy fashion.

C. Experimental Results

The proposed placement method was evaluated using 12 test circuits from open cores [10]; circuit information is listed in the first three columns of Table III. A square placement region was assumed and 20% of total region was dedicated to whitespace. The experiment was based on commercial 28-nm technology.

Three placement methods were implemented and compared. The first method (placement A) is a standard placement, which places conventional cells having inter-cell margin and minimizes total wirelength; in this method, γ is set to 0 in (2) and $WS_{req}(i)$ is dropped from (3). Resulting C_w and chip

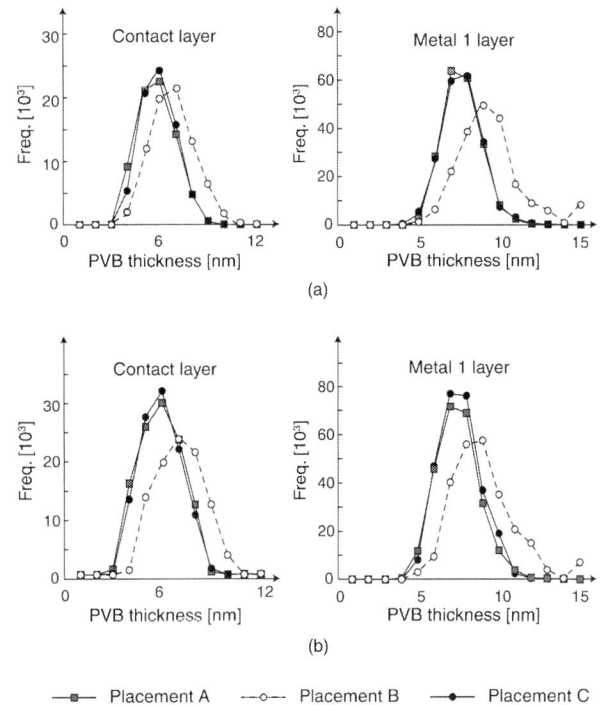

Fig. 7. Histogram of PVB thickness of contact and metal 1 layer: (a) mem_ctrl and (b) ac97.

area are reported in columns 4–5; defect probability is always 0 thanks to inter-cell margin. The second method (placement B) is the same as placement A but cells are now without inter-cell margin. Area is shown in percentage of the area from placement A; it is reduced by 10.8% on average, which is the main advantage of deploying cells without inter-cell margin. Total wirelength, which is also shown in percentage number, is reduced too as a result of reduced chip area; but, defect probability is necessarily very high.

The proposed method is named placement C. It achieves the same area reduction as placement B. But, average defect

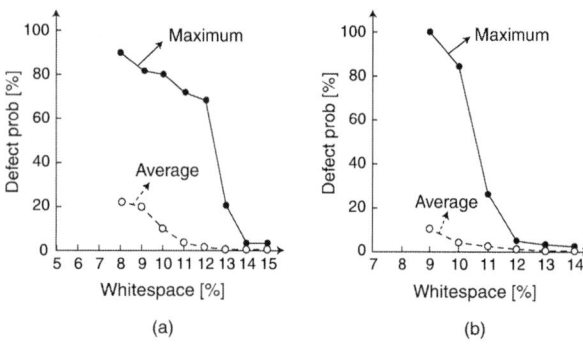

Fig. 8. Defect probability with varying whitespace: (a) spi and (b) sasc.

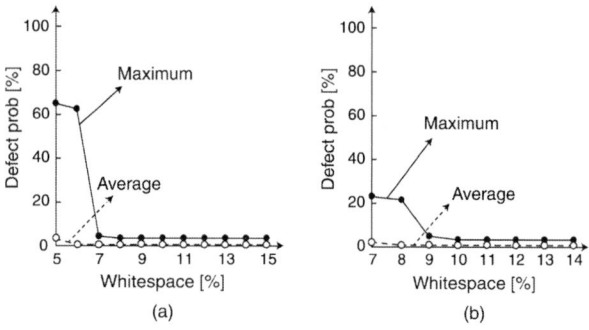

Fig. 9. Defect probability when circuits are synthesized using defect–friendly cells: (a) spi and (b) sasc.

probability C_d is kept very low, less than 1% in all test circuits. This comes at the cost of increase in the total wirelength if C_w from placement B and C is compared. We also report maximum defect probability from placement C; it is affected by the threshold used to control the quantity of (3) and to perform whitespace distribution; we set that value to 5.0%, which was honored by all test circuits. PVB thickness is also the measure of lithographic defect [11], so we extracted its distribution as a histogram from three placement methods and compared. As shown in Fig. 7 placement C achieves almost identical histogram as placement A, which is understandable consequence of very low defect probability from placement C.

The amount of whitespace clearly affects average and maximum defect probability from our placement method. We repeated the placement while we vary whitespace percentage for a few test circuits. The result for two sample circuits is shown in Fig. 8; 15% whitespace, which represents very tight placement [12], was the point beyond which maximum defect probability is kept below 5%.

Fig. 5(b) indicates that there are many good cells in terms of defect probability but there are also many bad cells. We may try to avoid those bad cells right from logic synthesis, which will make placement easier. We excluded cells whose average defect probability (over all cell pair combinations that include those cells) exceeds 50% from a library; 617 out of 1043 cells fell in that category, in which AND2X and BUF11X were examples. We took the same circuits of Fig. 8, re–performed logic synthesis, and repeated the same experiment of Fig. 8.

The result is shown in Fig. 9. Both average and maximum defect probability can now be kept low using smaller amount of whitespace, which can be expected. This comes at the cost of increased circuit area due to less freedom of logic synthesis; sum of cell area increases from 1277 to 1459 in spi and from 574 to 609 in sasc.

V. Conclusion

Conventional standard cells contain inter-cell margin to prevent potential lithographic defects. We advocate that the margin can be removed for the benefit of chip area and wirelength, while defect probability is affected very little when cells are carefully places. A prototype placement tool has been developed and experiments have been performed to prove our argument.

A key component in this approach is the table of defect probability of all cell pairs, which placement refers to for calculating its cost function. A straightforward computation of defect probability is intractable because of large lithography simulation time and large number of cell pairs. We have shown that the computation can be confined within 2-pitch range from cell boundary with tolerance amount of error; the range of 2-pitch, called extent, has been shown to be classified into a few groups by grouping geometrically similar extents, which greatly reduces the number of computation.

References

[1] K. Peter, R. Marz, S. Grondahl, and W. Maurer, "Litho-friendly design (LfD) methodologies applied to library cells," in *Proc. SPIE*, Mar. 2013, pp. 8 684 031–8 684 039.
[2] S. Paek, J. Kang, N. Ha, and B. Kim, "Yield enhancement with DFM," in *Proc. SPIE*, Oct. 2012, pp. 128–138.
[3] L. Liebmann, S. Mansfield, G. Han, J. Culp, J. Hibbeler, and R. Tsai, "Reducing DfM to practice: the lithography manufacturability assessor," in *Proc. SPIE*, Feb. 2006, pp. 786–98.
[4] J. P. Cain, "Design for manufacturability: a fabless perspective," in *Proc. SPIE*, Mar. 2013, pp. 8 684 031–8 684 039.
[5] A. Wong, *Resolution Enhancement Techniques in Optical Lithography*. SPIE Press, 2001.
[6] S. Hu and J. Hu, "Pattern sensitive placement for manufacturability," in *Proc. Int. Symp. Phys. Des.*, Mar. 2007, pp. 28–34.
[7] S. Kirkpatric, C. D. Gelatt, and M. P. Vecchi, "Optimization by simulated annealing," *Science*, vol. 220, no. 4598, pp. 671–680, May 1983.
[8] C. Sechen and A. Sangiovanni-Vincentelli, "The TimberWolf placement and routing package," *IEEE J. Solid-State Circuits*, vol. 20, no. 2, pp. 510–522, Apr. 1985.
[9] J. Wang, A. K. K. Wong, and E. Y. Lam, "Standard cell design with regularly placed contacts and gates," in *Proc. SPIE*, vol. 5379, Mar. 2004, pp. 55–66.
[10] OpenCores. [Online]. Available: http://www.opencores.org/
[11] J. Andres and T. Robles, "Integrated circuit layout design methodology with process variation bands," U.S. Patent 0 251 771 A1, Nov. 10, 2005.
[12] J. Hu, Y. Shin, N. Dhanwada, and R. Marculescu, "Architecting voltage islands in core–based system–on–a–chip design," in *Proc. Int. Symp. Low Power Electron. Des.*, Aug. 2004, pp. 180–185.

Structural Planning of 3D-IC Interconnects by Block Alignment

Johann Knechtel*, Evangeline F. Y. Young**, and Jens Lienig*

* Institute of Electromechanical and Electronic Design, Dresden University of Technology, Germany
** Department of Computer Science and Engineering, The Chinese University of Hong Kong, Hong Kong
johann.knechtel@ifte.de, fyyoung@cse.cuhk.edu.hk, jens@ieee.org

Abstract—Three-dimensional integrated circuits rely on optimized interconnect structures for blocks which are spread among one or multiple dies. We demonstrate how 2D and 3D block alignment can be efficiently utilized for structural planning of different interconnects. To realize this, we extend the corner block list and provide effective techniques for 3D layout generation, i.e., block placement and alignment. Our techniques are made available in an open-source, simulated-annealing-based tool. Besides block alignment, it accounts for key objectives in 3D design like fast thermal management and fixed-outline floorplanning. Experimental results on GSRC and IBM-HB+ circuits demonstrate the capabilities of our tool for both planning 3D-IC interconnects by block alignment and for 3D floorplanning in general.

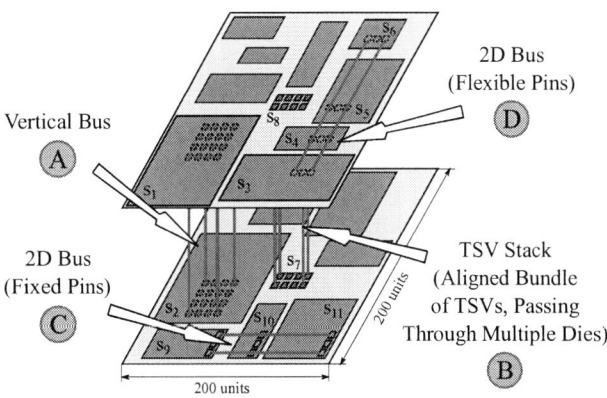

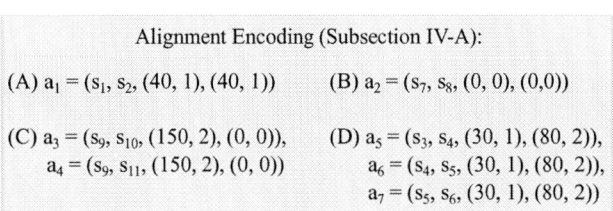

Fig. 1. Interconnect structures in a 3D IC and related block-alignment configurations. Vertical buses (A) are essential to connect (split-up) blocks among adjacent dies. TSV stacks (B) comprise aligned bundles of TSVs, are passing two or more dies, and are for example used in 3D NoCs. Both interconnect structures rely on *inter-die alignment*, i.e., blocks spread among several dies are to be aligned. Regular 2D buses with fixed or flexible pins (C, D) are traditionally considered to optimize datapaths or similar structures; they require blocks to be aligned within one die, i.e., rely on *intra-die alignment*.

I. INTRODUCTION

Three-dimensional (3D) stacking of active dies is recognized as a promising approach to meet demands on todays' and future chips regarding their performance, functionality and power consumption. Vertical plugs connecting through separate dies, mainly the through-silicon vias (TSVs), facilitate short and low-power interconnects and thus enable high-performance 3D integrated circuits (3D ICs). 3D network on chip (NoC) architectures have been proposed to increase communication capabilities for logic integration [1] or memory integration [2]. Complementing such approaches, the well-known concept of *bus planning*, i.e., grouping multiple signals into adjacent wires, remains also relevant for 3D-IC design.

Although the concept of *block alignment* has been successfully applied in 2D layout representations for bus planning [3, 4], it has been every so often neglected in 3D representations. Some studies, e.g., [5–7], enable *fixed alignment*, i.e., blocks are to be aligned (possibly across several dies) such that their relative positions fulfill fixed distances. However, an application to vertical-bus planning is only indicated in [6]. To the best of our knowledge, none of the existing studies considers *alignment ranges*, i.e., blocks are to be aligned such that their relative positions fulfill upper and/or lower distance limits. Thus, "flexible" block alignment is not supported so far. We observe that utilizing these different alignment approaches enables structural planning of interconnects for 3D ICs—as illustrated in Figs. 1 and 2, processing block alignment allows one to design dedicated, straight interconnect structures.

To address previously inadequate support for such interconnect structures during 3D floorplanning, we present a methodology based on orchestrated block placement and alignment. In our study, we consider interconnects for different practical scenarios in 3D ICs, as further motivated in Section II.

Our contributions can be summarized as follows.

1. We propose an extension of the corner block list (CBL) (Section IV). Our extension can (*i*) encode both fixed alignment and alignment ranges, as well as (*ii*) handle inter- and intra-die alignment in a unified manner. (See Section II for these terms and related background.)

2. We develop effective techniques for 3D layout generation, i.e., block placement and alignment as well as layout packing (Subsection IV-B).

3. We provide an open-source 3D-floorplanning tool based on our CBL extension and simulated annealing (SA) (Section V). Besides block alignment, our tool considers these key objectives: fixed outlines, fast thermal management, layout packing, and wirelength optimization.

II. BUS AND VIA STRUCTURES IN 3D ICs AND RELATED BLOCK ALIGNMENT

As for the 3D design style, we consider block-level integration of 2D blocks. This style is acknowledged as a reliable and efficient approach, especially for first commercial 3D-IC applications [8, 9]. In such 3D ICs, routing paths for (massively parallel) interconnect structures can be enabled by means of block alignment (Figs. 1, 2). Block alignment in 3D ICs can be generally classified into *inter-die alignment*, i.e., blocks spread

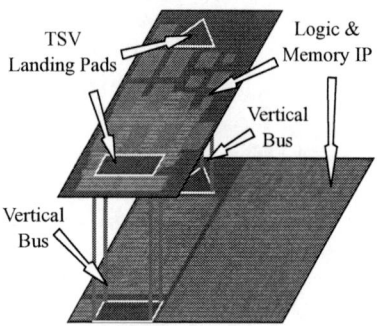

TSV Landing Pads

Logic & Memory IP

Vertical Bus

Vertical Bus

Fig. 2. Die shots of two macroblocks, partitioned across adjacent dies for delay and power optimization [14]. Embedded vertical buses require the macros to be aligned such that TSVs and related landing pads can be included.

among several dies are to be aligned, and *intra-die alignment*, i.e., blocks are aligned within one die. The variety of alignment specifics arise from different scenarios for 3D integration and interconnects, which are reviewed next. Note that we focus on signal interconnects in this work. For optimized planning of other interconnect types refer to, e.g., [10].

Monolithic integration has recently gained more interest due to advances in manufacturing processes; for block-level integration, this technology is advantageous in terms of improved interconnectivity [11]. In the general context of massively-interconnected dies, planning *vertical buses* which connect particular (split-up) blocks spread on separate dies is critical and should thus be considered from early design phases on. It is important to note that TSV-based integration can also exploit such buses, assuming that blocks can be adapted to include TSVs. For example, consider the two macroblocks in Fig. 2: this arrangement of tightly interconnected (for delay and power consumption optimized) modules relies on vertical buses, which are implemented by groups of TSVs. Accounting for such vertical buses during floorplanning requires capabilities for inter-die alignment. That is, in order to include a large number of vertical interconnects, the related blocks have to exhibit some intersecting regions.

A special case of vertical buses are *aligned TSV stacks*, i.e., TSVs are grouped and placed such that straight interconnects are passing through multiple dies. TSV stacks are relevant for different applications, e.g., to realize regular 3D NoCs, or to limit power-supply noise and to improve thermal distribution [12,13]. Consideration of aligned TSV stacks during floorplanning requires inter-die alignment with fixed offsets.

Regular (2D) bus structures connecting blocks within dies are independent of the 3D-integration technology. These buses are traditionally considered for several scenarios, e.g., to optimize datapath interconnects. Note that such structures rely on intra-die alignment of related blocks. Depending on fixed / flexible block pins, the planning of 2D buses requires support for fixed alignment / alignment ranges.

III. BASIC PRINCIPLES OF CORNER BLOCK LIST

The corner block list (CBL) [15] is a topological 2D layout representation. In our work, we utilize it mainly for its efficiency (layout generation has a $\mathcal{O}(n)$ complexity) and feasible expandability towards a 3D representation (Section IV).

The CBL encodes a floorplan solution as tuple (S, L, T) where S is the *block-insertion sequence*, L the *insertion-direction sequence*, and T the *sequence of covered T-junctions* (see next paragraph). Note that conceptual *rooms*, i.e., dimen-

sionless entities, are encoded in S. Each block is associated with a room; to obtain the physical layout, a transformation from the room topology to block coordinates is required.

During sequential layout generation, two criteria are to be considered for each block (within a room) $s_i \in S$: first, the *insertion direction* where $l_i = 0$ encodes vertical placement and $l_i = 1$ horizontal placement, respectively; second, the number t_i of *T-junctions* to be covered. The notion of T-junctions is a verbatim encoding; for example, $t_i = 1$ requires to (perpendicularly) cover the common boundary of two adjacent blocks.

IV. CORBLIVAR: CORNER BLOCK LIST FOR VARIED ALIGNMENT REQUESTS

To enable interconnect structures during 3D floorplanning, we propose an extension of the classical 2D CBL. Our extension is named *corner block list for varied alignment requests (Corblivar)*. It encodes a 3D-IC design integrated on n dies using an ordered sequence $\{CBL_1, \ldots, CBL_n\}$ of CBL tuples and one global *alignment sequence* A. (Thus, Corblivar is a so-called 2.5D layout representation. Refer to [16] for an investigation of several previous 2.5D and 3D representations.) The *alignment tuples* $a_k \in A = \{a_1, \ldots, a_n\}$ are designed to encode different types of alignment requests as defined below and illustrated in Fig. 1. Like any layout representation, we need to embed Corblivar in a floorplanning tool; core parts and main features are outlined in Fig. 3.

A. Alignment Tuples

Definition of alignment tuples – Assume the placement of block s_j has to consider some alignment request with regard to (w.r.t.) s_i. The request is then defined as tuple $a_k = (s_i, s_j, (AR_x, ART_x), (AR_y, ART_y))$ where (AR_x, ART_x) and (AR_y, ART_y) denote the partial requests with respect to the x- and y-coordinate. These requests can be *independently* defined as *fixed offset* $(ART = 0)$, as *minimal overlap* $(ART = 1)$, as *maximal distance* $(ART = 2)$ or as *don't care* $(ART = -1)$; the meaning of these types is explained next.

Alignment types – Given a *fixed offset*, s_j is to be placed AR_x/AR_y units to the right/top $(AR_x/AR_y \geq 0)$ or to the left/bottom $(AR_x/AR_y < 0)$ of s_i, respectively, w.r.t. the blocks' lower-left corners. Fixed-offset alignment is required for restricted placement, e.g., of blocks with fixed pins.

For a (positive) *minimal overlap*, the projected intersection of blocks s_i and s_j must be *at least* AR_x units wide and/or AR_y units high. The intention of such alignment is to ensure straight but locally flexible paths for subsequent bus routing / placement of vertical interconnects.

An alignment request defining a *maximal distance* requires that the center points of blocks s_i and s_j are *at most* AR_x/AR_y units apart. This way, interconnects structures can be easily limited in their length and/or width.

It may not be necessary to define a request for both x- and y-coordinates; we label the unrestricted coordinate's request simply as *don't care*.

Dynamic interpretation of requests – Note that the introduced tuples can be easily utilized for both intra- or inter-die alignment by assigning related blocks to one common or to separate CBLs (dies). In other words, the proposed encoding does not restrict blocks to particular dies. For requests spanning multiple blocks (discussed next), it is also possible to combine intra- and inter-die alignment for several blocks.

Definition of tuples to align multiple blocks – For 3D-IC interconnects, implementing links among multiple blocks

978-1-4799-2817-0/14 $31.00 © 2014 IEEE

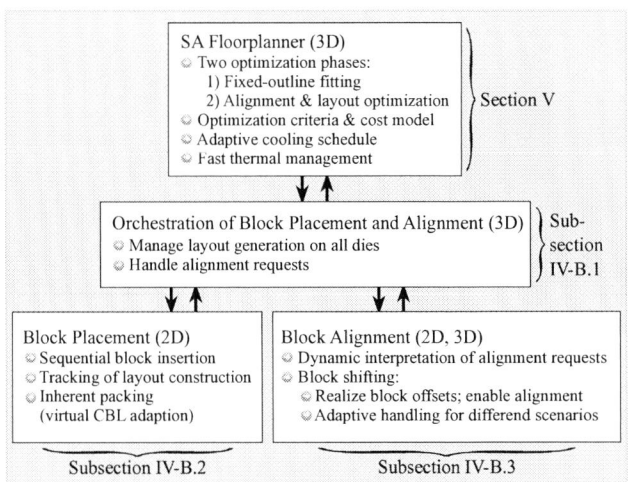

Fig. 3. Corblivar's components, embedded in a SA-based floorplanning tool. *Orchestration of Block Placement and Alignment* interacts with the SA heuristic for layout optimization, monitors the overall layout process, and delegates to *Block Placement* and *Block Alignment* in a synchronized manner.

is an essential scenario. Thus, let us assume the placement of blocks $s_1 \ldots s_n$ has to consider several, combined alignment requests for interconnects planning. The required set of tuples can be derived in any desired fashion. For example, for requests requiring *one reference block* s_1 (e.g., to represent one specific end of a bus), the tuples would be defined as $(s_1, s_2, (AR_{x_2}, ART_{x_2}), (AR_{y_2}, ART_{y_2}))$, $\ldots\ldots\ldots\ldots\ldots, (s_1, s_n, (AR_{x_n}, ART_{x_n}), (AR_{y_n}, ART_{y_n}))$. To give another example, we can encode alignments in a chain-like fashion to enable flexible interconnect structures (i.e., allowing local deviations from a straight, global path): $(s_1, s_2, (AR_{x_2}, ART_{x_2}), (AR_{y_2}, ART_{y_2}))$, $(s_2, s_3, (AR_{x_3}, ART_{x_3}), (AR_{y_3}, ART_{y_3})), \ldots\ldots\ldots\ldots$, $(s_{n-1}, s_n, (AR_{x_n}, ART_{x_n}), (AR_{y_n}, ART_{y_n}))$.

B. Layout Generation

We extend the CBL technique [15] in order to (i) handle inter- and intra-die alignment simultaneously, (ii) consider fixed offsets as well as alignment ranges, and (iii) perform effective layout packing. In the following subsections, we first discuss the orchestration of block placement and alignment and then provide techniques for these steps themselves.

B.1 Orchestration of Block Placement and Alignment

We next discuss the overall process of 3D layout generation. As illustrated in Fig. 3, this requires to (i) manage the layout-generation progress on all dies, (ii) handle the alignment requests, and (iii) interact with block placement and alignment (Subsections B.2 and B.3). In the following, we label "calls" to latter techniques as PLACE and ALIGN, respectively.

Auxiliary data structures – We memorize alignment requests in progress using the *alignment stack AS*. *Progress pointers* $p_i = s_j$ denote the currently processed block s_j for each die d_i. A *die pointer* $p = d_i$ is used to keep track of the currently processed die.

Process flow (Algorithm 1) – We perform the following steps for each block s_i. Initially, we check whether the associated die d is currently marked as *stalled* (line 5), i.e., layout generation is halted due to another alignment request in progress—this occurs for *intersecting requests*, i.e., related

blocks are arranged in the CBL sequences such that their placement is interfering. To resolve this, we need to unlock die d— we PLACE the current block s_i, mark related changes, and proceed with the next block (lines 6–9). Otherwise (for non-stalled dies), we check if some alignment requests a_k are applying to s_i (line 11). If no a_k are found, we directly PLACE s_i and proceed with the next block (lines 28–29). If some request(s) a_k are defined, we need to handle them appropriately (lines 12–23), as described next. For any given a_k, we search the stack AS for it and continue accordingly. *Case a:* if a_k is found, it was previously handled while processing s_j, that is the block to be aligned with s_i. Thus, it is assured that preceding blocks on both related dies are placed at this point. We can now safely ALIGN both s_i and s_j, mark them as placed, and drop the request a_k (lines 14–17). Note that only in cases where *all* requests for s_i are handled, we proceed on the current die d (line 25). Otherwise, we continue layout generation without loss of generality (w.l.o.g.) on s_j's die d' (line 21). *Case b:* if a_k is not found in AS, s_j was not processed yet. We then memorize a_k as in progress, halt layout generation on d, and continue on d' (lines 19–21). Finally, if layout generation is done on d, we proceed on yet unfinished dies until the whole 3D layout is generated (lines 32–38).

Be aware that *deadlock situations*, i.e., layout generation on different dies is waiting for each other until particular blocks can be aligned, *cannot* occur due to resolving of stalled dies. This is true for any alignment request; see also Subsection B.3 for implications on block alignment.

B.2 Block Placement

To maintain a valid layout during placement, it is necessary to consider previously placed blocks. We propose a technique which allows us to (i) efficiently keep track of relevant blocks, (ii) fix CBL tuples w.r.t. exceeding T-junctions, and (iii) virtu-

---

Algorithm 1 Orchestration of Block Placement and Alignment

```
1:  p ← d₁                                    ▷ start without loss of generality on bottom die
2:  pᵢ ← s₁
3:  loop
4:      sᵢ ← pᵢ ← p
5:      if die d ← p is stalled then
6:          PLACE(sᵢ)
7:          mark sᵢ in any aᵢ′ ∈ A as placed
8:          mark d as not stalled
9:          pᵢ ← pᵢ₊₁
10:     else                                                      ▷ d is not stalled
11:         if some aₖ are defined for sᵢ then
12:             for all aₖ do                        ▷ consider aₖ w/ placed blocks first
13:                 if aₖ in AS then
14:                     ALIGN(aₖ)
15:                     remove aₖ from AS
16:                     mark sᵢ, sⱼ in any aᵢ′ ∈ A as placed
17:                     mark dⱼ = die(sⱼ) as not stalled
18:                 else
19:                     add aₖ to AS
20:                     mark d as stalled
21:                     p ← die(sⱼ)
22:                 end if
23:             end for                                          ▷ all aₖ considered
24:             if d is not stalled then
25:                 pᵢ ← pᵢ₊₁
26:             end if
27:         else                                           ▷ no aₖ defined for sᵢ
28:             PLACE(sᵢ)
29:             pᵢ ← pᵢ₊₁
30:         end if
31:     end if                                                  ▷ sᵢ processed
32:     if pᵢ = end then
33:         if some pⱼ ≠ end then
34:             p ← pⱼ
35:         else
36:             return done
37:         end if
38:     end if
39: end loop
```

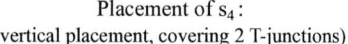

Placement of s_4:
(vertical placement, covering 2 T-junctions)

0) Previous state of placement stacks:
$H_j = \{s_3\}$, $V_j = \{s_3, s_2, s_1\}$
1) Pop relevant blocks: $\{s_3, s_2, s_1\}$
2) Determine (lower) x-coordinate
 - from relevant blocks' left front;
 - $x(s_4) = \min\{x(s_3), x(s_2), x(s_1)\} = 0$
3) Determine (left) y-coordinate
 - consider in x-dimension intersecting blocks' upper front;
 - $y(s_4) = \max\{y(s_1)\} = 45$
4) Update placement stacks: $H_j = \{s_3\}$, $V_j = \{s_3, s_2, s_4\}$

Fig. 4. Placement steps, to be performed for exemplary vertical insertion of block s_4 while covering 2 T-junctions.

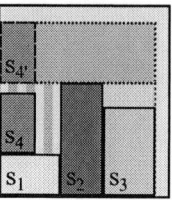

CBL Encoding:	Equivalent Encoding Example:
$S = \{s_1, s_2, s_3, s_4\}$	$S = \{s_1, s_4, s_2, s_3\}$
$L = \{1, 1, 1, 0\}$	$L = \{1, 0, 1, 1\}$
$T = \{0, 0, 0, 2\}$	$T = \{0, 0, 1, 0\}$

Fig. 5. Implications of virtual CBL adaption. Rooms and their assigned blocks are similarly colored. Block s_4 is placed "into the room" of s_1, thereby enabling a more compact layout. Without applying virtual CBL adaption, s_4 would be placed as $s_{4'}$. Virtual CBL adaption can result in different, equivalent encodings for the same, compact layout; hence, it supports an efficient solution-space exploration towards compact layouts.

ally adapt CBL tuples for implicit layout compaction. Our approach differs from [15] in these features but follows the same principle of sequential block placement into dissected rooms.

Auxiliary data structures – We keep track of placed blocks using two stacks H_j / V_j for each CBL_j. More precisely, these stacks are governed to contain CBL_j's blocks currently covering the vertical right / horizontal upper front of die d_j; these are considered as the *boundary fronts* for further placement.

Placement flow – We determine each blocks' $s_i \in S_j$ lower-left coordinates (x_i, y_i) as follows. (See Fig. 4 for an example.) *First*, we retrieve $t_i + 1$ previously placed blocks from the respective stacks H_j / V_j. These blocks are referred to as *relevant blocks* in the following. Note that in cases where only $t_{max} < t_i + 1$ blocks are available, the related CBL tuple is technically infeasible [15]. To fix such invalid tuples, we simply consider all t_{max} blocks in order to fulfill the desired covering of T-junctions as best as possible. *Second*, we determine s_i's y / x-coordinates (orthogonal to the horizontal / vertical insertion direction) by considering the structural change of CBL_j's room dissection. In case a new column / row is implicitly defined due to covering all relevant blocks during placement of s_i, we set the respective y / x-coordinate to 0. In the remaining cases, we derive the coordinate from the relevant blocks' lower / left front. This can be also thought of as placing a new column / row into the existing room dissection. *Third*, we determine s_i's x / y-coordinates (along the insertion direction) by considering the right / upper front of *previously placed blocks* which are intersecting with s_i in its orthogonal, recently determined y / x-coordinates. *Fourth*, we update the placement stacks to follow the changed layout's fronts as follows. We push s_i onto the insertion-direction-related stack H_j / V_j. In case s_i is not covered by some relevant block to its top / right front, we also push s_i to the (unrelated) stack V_j / H_j. Finally, we push relevant blocks not covered by s_i back to H_j / V_j; these blocks remain part of the layout's boundary front and are thus to be furthermore considered.

Virtual CBL adaption – For any block smaller than the room it is supposed to cover, the next, adjacent block(s) will be packed "into the room" of this smaller block (Fig. 5). We refer to this feature as virtual CBL adaption since it results in practice in different CBL tuples encoding the same (compact) layout. Note that virtual CBL adaption is generally applied during block placement.

B.3 Block Alignment: Inter- and Intra-Die Alignment

Recall that our alignment tuples support different alignment types and can be interpreted as inter- or intra-die requests.

We observe that such diverse requests all depend on their assigned blocks' planar offsets, i.e., relative distances considering their projections onto a plane. This implies that we can rely on adjusting the blocks' offsets in order to handle alignment requests. Such adjustments are practical since our layout-generation process is synchronized across the whole 3D IC, i.e., blocks to be aligned "wait for each other's die to be ready", that means until preceding blocks are placed. This "waiting" might result in circular dependencies; layout generation handles such cases by resolving stalled dies (Subsection B.1).

It is also important to note that, depending on particular alignment and CBL configurations, it may be infeasible to fulfill all requests.[1] One resolution (exclusively applying) for failing *intra-die* alignments includes preprocessing CBL tuples and adjusting topologically infeasible configurations [17]. Yet, such preprocessing is not warranted in the presence of different alignment requests—the applicability of *inter-die* alignments depends on the layout of all dies, that is on the entire layout-generation process. The flow described below includes layout-aware techniques, i.e., enables alignment in some cases of previously placed blocks.

Alignment flow – Remember that alignment tuples always cover two blocks; requests spanning more blocks (and tuples) are thus (implicitly) handled stepwise. Initially, we need to check whether one or both blocks have been previously placed. Depending on preceding placement, three different scenarios are to be distinguished for handling request a_k.

Scenario I: both blocks are placed – In this case, we cannot fulfill a_k since we omit post-placement shifting.[2]

Scenario II: one block is yet unplaced – Here, we assume w.l.o.g. that $s_i \in a_k$ is yet unplaced and $s_j \in a_k$ was previously placed. Depending on both the coordinates of (placed) s_j and the properties of a_k, we may be able to fulfill a_k as follows. First, we determine s_i's y / x-coordinates orthogonal to its horizontal / vertical insertion direction ("Second", Subsection B.2). Next, based on both the inherent offset between s_i and s_j and the defined alignment of a_k, we derive the *required shifting range* $rs_y(s_i, s_j)$ / $rs_x(s_i, s_j)$, i.e., the remaining offset of s_i w.r.t. s_j in order to fulfill a_k. Note that $rs(s_i, s_j) = -rs(s_j, s_i)$, i.e., the shifting range is directed and invertible. In cases $rs(s_i, s_j) = 0$, a_k is already fulfilled. In cases $rs(s_i, s_j) < 0$, we would need to shift s_i downward / leftward which is trivially prohibited while maintaining a valid layout. Alternatively, we could shift placed s_j upward / rightward;

---

[1] We would like to stress that this limitation only applies to particular configurations. That means, adapting the CBL configurations during alignment-aware 3D floorplanning (as proposed in Section V) can resolve this issue.

[2] Based on our observations, shifting placed blocks most likely requires adjacent (or nearby) blocks to be shifted as well in order to maintain a valid layout. This is impractical in the presence of different alignment requests—such shifting can then undermine handling of remaining requests and even invalidate previously processed ones.

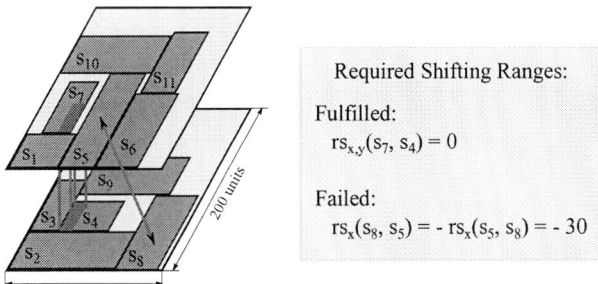

Required Shifting Ranges:

Fulfilled:

$$rs_{x,y}(s_7, s_4) = 0$$

Failed:

$$rs_x(s_8, s_5) = -rs_x(s_5, s_8) = -30$$

Fig. 6. Examples for required shifting ranges. During alignment of block s_7 with previously placed s_4, shifting s_7 to the right was applicable such that $rs(s_7, s_4)$ is resolved. In contrast, we cannot shift previously placed s_5 (in order to align it with s_8) since we omit post-placement shifting. Also, the inverted shifting range $rs(s_8, s_5) = -rs(s_5, s_8)$ cannot be resolved since this would require a shift of s_8 to the left, which is hindered by s_2.

however, this is not applicable, as mentioned before. (See also Fig. 6 for illustration.) Third, if and only if $rs_{y/x}(s_i, s_j) \geq 0$, we can perform a corresponding forwards shift of s_i in y / x-direction and thus satisfy a_k's partial request. Next, we perform above steps similarly for s_i's x / y-coordinates along its insertion direction. Finally, we handle the placement stacks. In case block shifting was conducted, we need to *rebuild* them, i.e., the stacks' current blocks along with s_i are sorted by their coordinates in descending order; afterwards uncovered (*relevant*) blocks redefine the stack. In case block shifting was not required, we simply update the stacks (Subsection B.2).

Scenario III: both blocks are yet unplaced – We are free to shift both blocks, thus we can satisfy a_k. Depending on the blocks' insertion direction and coordinates, we process block shifting in parallel or sequentially (similar as described above).

V. 3D-FLOORPLANNING TOOL

We provide Corblivar along with our C++ implementation of a SA-based 3D-floorplanning tool [18]. Our holistic concept of (orchestrated) block placement and alignment differs notably from previous works. Applying SA-based floorplanning during initial experiments, we observe limitations of existing techniques w.r.t. solution-space exploration for block alignment as well as for "classical" 3D floorplanning. Thus, some effective extensions are needed; notable features of our tool are (i) a SA framework including two optimization phases and specific cost models, (ii) an adaptive SA cooling schedule, and (iii) a fast yet sufficiently accurate thermal analysis.

A. Optimization Criteria & Cost Models

We next discuss applied optimization criteria along with their cost functions/models. Note that cost functions are formulated for SA's classical cost-minimization approach, i.e., lower cost correspond to more optimized layouts.

Outline – This criteria unifies evaluation of the layout's bounding boxes, i.e., packing density, as well as fixed-outline fitting. This is achieved by extending Chen and Chang's aspect-ratio-based cost model [19]. Our model is defined as

$$c_{OL} = c_{PD} + c_{ARV}$$
$$c_{PD} = \frac{1}{2}\alpha(1 + \frac{n_{feasible}}{n}) \times \max_{d_i}\left(\frac{A_{outline}(b \in d_i)}{A_{outline}(d_i)}\right)$$
$$c_{ARV} = \frac{1}{2}\alpha(1 - \frac{n_{feasible}}{n}) \times \max_{d_i}\left(\Delta_{AR}(d_i)^2\right)$$
$$\Delta_{AR}(d_i) = AR_{outline}(b \in d_i) - AR_{outline}(d_i)$$

where c_{PD} and c_{ARV} denote the respective cost terms for *packing density* and *aspect-ratio violation*. Functions $A_{outline}$ and $AR_{outline}$ determine the outline's area and aspect ratio w.r.t. a set of blocks $b \in d_i$ / a die d_i, respectively. Note that we perform cost calculation for previous n layout operations where $n_{feasible} \leq n$ operations resulted in a valid layout, i.e., blocks on all dies are fitting into the fixed outline.

Wirelength and TSV count – We assume that the lowermost die d_0 is connected to the package board. For each net n, we determine the half-perimeter wirelength (HPWL) on each related die d_i separately, denoted as $HPWL(n, d_i)$. To do so, we construct the bounding box by encircling connected terminal pins (only for d_0) and assigned blocks on d_i and on the upper die $d_j, j > i$ as well. The latter is required to model wires for connecting blocks with TSV landing pads in upper dies. Overall cost terms are defined as

$$c_{WL} = \sum_n \left(l_{TSV} \times TSVs(n) + \sum_{d_i \in n} HPWL(n, d_i)\right)$$
$$c_{TSVs} = \sum_n TSVs(n)$$

where $TSVs(n)$ denotes the required TSV count for net n. Note that we also account for "TSV wirelength" l_{TSV} in c_{WL}.

Thermal management – Cost c_T is the *estimated maximal temperature* of the critical die furthest away from the heatsink. Details on thermal modelling are given in Subsection E.

Alignment mismatch – For an alignment tuple a_k, we describe the spatial mismatch between desired alignment and actual layout as cost $c_{AMM}(a_k) = |rs(s_i, s_j)|$ (Subsection IV-B.3). Overall cost is then calculated as $c_{AMM} = \sum_{a_k \in A} c_{AMM}(a_k)$.

B. Optimization Phases

We consider two different phases for SA optimization; these phases support efficient solution-space exploration and layout optimization for 3D floorplanning with block alignment.

Phase I, "Fixed-Outline Fitting" – The cost function is defined as $c_{FOF} = c_{OL}$. Note that we do not perform alignment (Subsection IV-B.3) in this phase. The reason for initially focusing SA's search solely on the fixed-outline is simply that non-fitting layouts are a "knock-out", regardless of any achieved block alignment and layout optimization. The transition to phase II is made when the SA search triggers the first, fixed-outline fitting layout.

Phase II, "Alignment and Layout Optimization"– We compose the cost function as

$$c_{ALO} = c_{OL} + (1 - \alpha) \times \sum_{c' \in C'} c'$$
$$C' : \{\beta(c_{WL}/c_{WL_{init}}), \gamma(c_{TSVs}/c_{TSVs_{init}}),$$
$$\delta(c_T/c_{T_{init}}), \epsilon(c_{AMM}/c_{AMM_{init}})\}$$

with $\beta + \gamma + \delta + \epsilon \leq 1$. Note that we memorize *initial cost terms* like $c_{WL_{init}}$ during transition to phase II, i.e., we derive them from the first valid solution. Furthermore, note that we consider c_{OL} as essential term in this phase as well; based on our experiments, the SA search for comprehensively optimized layouts still depends on outline fitting/optimization.

C. Layout Operations

We consider the following set of layout operations to support the SA heuristic in effective exploration of Corblivar's solution space: swapping blocks within or across dies (CBL sequences), swapping or moving whole CBL tuples within or across CBL sequences, switching a block's insertion direction, switching a block's T-junctions, rotating hard blocks, and guided shaping of soft blocks as proposed in [19].

For optimization phase I, operations and blocks / CBL tuples are selected randomly. In phase II, blocks related with failed alignment requests are particularly selected. These blocks are swapped with adjacent blocks such that $|rs(s_i, s_j)|$ is reduced, i.e., such that the alignment is more likely to be fulfilled.

D. Cooling Schedule

As indicated earlier, we require an adaptive cooling schedule for improved efficiency of solution-space exploration. Our schedule is capable of (i) guiding the SA search within the global phases and (ii) escaping local minima. The schedule is composed of three different phases, explained below. Note that i labels the current step of i_{max} total temperature steps.

Phase "Adaptive Cooling" – We apply this cooling phase during SA phase I, which is aiming for fixed-outline fitting.

$$T_{i+1} = \left(cf_1 + \frac{i-1}{i_{max}-1} \times (cf_2 - cf_1) \right) \times T_i$$

The cooling rate slows down (given that $cf_1 < cf_2 < 1.0$); our intention here is to achieve initially fast cooling for the global scope, followed by slower cooling in a confined, "local" solution space.

Phase "Reheating and Freezing" – This is applied for SA phase II, i.e., after a fitting layout was found in step i_{first}.

$$T_{i+1} = \left(1 - \frac{i - i_{first}}{i_{max} - i_{first}} \right) \times cf_3 \times T_i$$

The cooling rate increases steadily; however, setting $cf_3 > 1.0$ results in an initial reheating for $i \geq i_{first}$. This way, the SA search has an increased flexibility for accepting high-cost solutions in this "interesting solution-space region" covering the first fitting layout. According to experiments, this limits the risk for being subsequently trapped in solution-space minima.

Phase "Brief Reheating" – This phase enables a somewhat "autonomous" and robust cooling schedule.

$$T_{i+1} = cf_4 \times T_i, cf_4 > 1.0$$

It is applied in alternation with the phase "reheating and freezing" for *individual* temperature steps during SA phase II. Such brief reheating helps the SA search to escape local minima; it is applied when we observe $\sigma(c_{ALO}) \sim 0$ during previous k steps, that is when the search reached a local "cost plateau". This technique is inspired by Chen and Chang's study [19]; their approach, however, proposes reheating solely at one particular temperature step, which we believe is not as effective as our cost-controlled reheating.

E. Fast Evaluation of Thermal Distribution

For fast yet accurate (steady-state) temperature analysis, we extend the work of Park et al. on *power blurring* [20]. Instead of using computationally intensive finite-differences or finite-elements analysis (FEA), power blurring is based on simple matrix convolution of thermal impulse responses and power-density distributions. (Park et al. reveal promising results when comparing to *ANSYS FEA* runs; they achieve maximal errors of less than 2% with computation speedups of $\sim 60\times$.)

For improved efficiency and to provide an integrated floorplanning tool [18], we refrain from time-consuming FEA runs for retrieving thermal masks [20]. Instead, we model the masks' underlying thermal impulse responses as *2D gauss functions* $g(x, y, w, s) = w \exp\left(-\frac{1}{s}x^2\right) \exp\left(-\frac{1}{s}y^2\right)$ with w as amplitude-scaling factor and s as lateral-spreading factor. To obtain the whole set of required masks [20], we need some scaling measure for g. We thus adapt w for each die d_i's mask such that $w_i = w/(i^{w_s})$, where $\max(i)$ represents the uppermost die next to the heatsink and w_s denotes a scaling parameter. For actual parametrization of w, w_s and s, we determine

for each different 3D-IC setup (w.r.t. die count and dimensions) (i) an exemplary thermal distribution using a 3D-IC extension of *HotSpot* [21], a state-of-the-art academic thermal analyzer, and (ii) a best fit for above parameters based on a local search using *HotSpot*'s solution as reference model.

VI. EXPERIMENTAL RESULTS

We conduct several experiments described below to validate Corblivar's capabilities. Relevant configuration details are given in Subsection A; results are discussed in Subsection B.

Structural planning of interconnects – We consider a set of several interconnects running both within and across dies; the (arbitrarily defined) set contains 10 width- and length-limited buses, each covering up to 5 blocks, along with 3 block pairs to be vertically aligned. We assume that each interconnect structure bundles 64 signals. For structural planning of these interconnects, in total 18 blocks have to be aligned simultaneously; related alignment tuples can be retrieved from [18]. Such scenario has not been considered in previous studies, thus we cannot meaningfully compare to other work.[3]

Regular and large-scale 3D floorplanning – To evaluate Corblivar's efficiency w.r.t. key 3D floorplanning objectives, we look into layout packing, wirelength and thermal optimization. (Note that we refrain from deriving alignment tuples for the considered benchmarks' nets. In other words, here we do not apply block alignment for interconnects planning and/or wirelength optimization.) We compare our work to relevant previous studies [22, 23]. Furthermore, we demonstrate Corblivar's scalability by utilizing the *IBM-HB+* benchmark suite [24]. To the best of our knowledge, this is the first time that these large-scale circuits are considered for 3D floorplanning.

A. Configuration

3D-IC configuration and benchmarks – We assume face-to-back stacking of two or three dies. Dies are $100\mu m$ thick; further properties are given in [18]. Terminal pins are only available on the lowermost die which is assumed to be connected to the package board. Practical (i.e., stackable) fixed outlines are ranging from $10mm \times 10mm$ up to $15mm \times 15mm$. We consider *GSRC* [25] and *IBM-HB+* [24] circuits. For reasonable utilization of die outlines, benchmarks are enlarged. In this context, power-density values are scaled down by factor 10. Also, results are referring to packed layouts where feasible, i.e., reduced outlines are reported. Deadspace utilization by $10\mu m \times 10\mu m$-sized TSVs was negligible in most cases, we thus refrain from optimizing and reporting TSV counts.

Setup – We conduct all experiments on a *Intel Core 2* system; reported runtimes are thus comparable. Corblivar and [23] are embedded in SA-based tools; best results are chosen from 5 up to 25 runs. Applied Corblivar parameters are retrievable from [18]. For *HotSpot*, default settings are applied [21].

---

[3]Previous studies on block alignment for 3D ICs have looked into differing scenarios. Nain and Chrzanowska-Jeske [5] propose techniques to split up and align (sub-)modules among adjacent dies with fixed (zero) offsets. They neglect to provide derived benchmarks containing split-up blocks, thus a comparison is hindered. Law et al. [6] consider a more flexible problem formulation; for vertical bus planning, they define sets of blocks for each die separately and require (at least) one block from each set to be vertically aligned with one block from the other sets. This simplified alignment problem is not compatible with our approach where we require all specified blocks to be aligned. Li et al. [7] indicate capabilities for block alignment but refrain from providing further details and related experimental results. Finally note that all aforementioned studies exclusively consider vertical alignment with fixed offsets.

B. Results

Structural planning of interconnects – We observe that the entire set of interconnects is successfully integrated, i.e., all related blocks can be simultaneously aligned (upper part of Table I). Compared to experiments where planning of interconnects is ignored (lower part of Table I), we expect and observe an increase of die outlines and deadspace—block alignment limits the flexibility of layout packing. More importantly, however, we observe notable wirelength increases in case of neglected interconnects planning; these overheads arise from routing detours for interconnects embedded in unaligned blocks. Finally, fixed die outlines were fulfilled in any case, i.e., the proposed SA optimization phases are effective.

An example for successful interconnects planning with corresponding block alignment is illustrated in Fig. 7.

Regular 3D floorplanning – Next, we discuss results on conducting floorplanning with applied layout packing and (equal) consideration of thermal and wirelength optimization (Table III). We observe that Corblivar is competitive with a force-directed tool [22] and superior to a SA-based tool [23]; both represent state-of-the-art academic works. In particular, we achieve comparable wirelengths and temperatures as [22] but with reduced die outlines and deadspace ratios. This indicates the efficiency of layout packing, which is most likely achieved by the proposed virtual CBL adaption. Comparing to [23], however, we note that Corblivar's layouts exhibit larger deadspace ratios and thus reduced packing densities. Nonetheless, we achieve reduced wirelengths in most cases. Also, the high packing density of [23] comes at a price; maximal temperatures are notably increased by tens of Kelvins compared to Corblivar. Thus, our tool effectively addresses the trade-off between packing density and maximal temperature. Furthermore, fixed outlines were fulfilled in these experiments as well.

As for our temperature analysis, we observe that it shows some local deviations compared to *HotSpot*-verification runs (Fig. 8). As indicated in [20], convolution-based thermal analysis particularly induces estimation errors at die boundaries. Thanks to our proposed mask parametrization, the actual thermal-distribution scale (i.e., the scale w.r.t. *HotSpot* runs) is

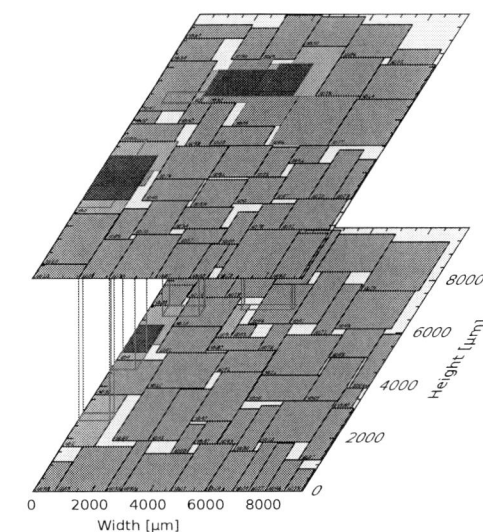

Fig. 7. Planned interconnect structures with corresponding block alignment, for enlarged benchmark *n100*. Vertical-bus sites are indicated by green, vertically extended rectangles; sites for 2D buses are colored as dark-brown. For interconnects planning aligned blocks are colored orange. For illustration purposes, we consider a reduced set of buses, covering blocks sb_1 to sb_9.

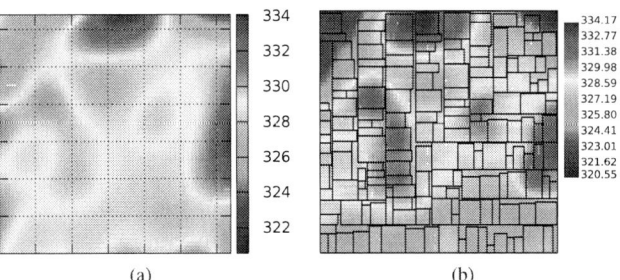

Fig. 8. Thermal maps of the critical die (furthest away from the heatsink), benchmark *n300*. The temperature scale of our fast analysis (a) matches with the scale obtained by running *HotSpot* (b); yet, local deviations are visible.

matched nevertheless. For analysis during layout optimization, i.e., maximal-temperature estimation, our approach is thus applicable. It is also efficient due to fast computation; one run can be conducted in ~20ms. (For comparison, one *HotSpot* run can take tens of seconds up to a few minutes.)

Large-scale 3D floorplanning – The IBM-HB+ suite does not include power information; we thus configured Corblivar only for wirelength and packing optimization (including successful consideration of fixed-outline constraints). Results on arbitrarily selected circuits are provided in Table II. We observe that total deadspace for these experiments is on average larger than for experiments on some GSRC circuits. This is expected and likely due to the fact that IBM-HB+ circuits contain up to ~1,500 blocks where largest blocks are ~33,000 times bigger than smallest ones; such designs are difficult to floorplan [26].

VII. SUMMARY

In this work, we extend 3D floorplanning towards structural planning of interconnects—an important yet inadequately addressed scenario for (future) massively interconnected 3D ICs. To tackle this omission of previous works, we promote block alignment. We initially discuss how 3D (inter-die) and 2D (intra-die) alignment can be applied for planning of diverse interconnects like vertical buses connecting (split-up) blocks on separate dies or classical 2D buses. We then introduce *Corb-*

TABLE I
RESULTS ON ENLARGED GSRC BENCHMARKS FOR APPLIED INTERCONNECTS PLANNING, I.E., BLOCK ALIGNMENT (UPPER PART), COMPARED TO RESULTS FOR FLOORPLANNING WITHOUT INTERCONNECTS PLANNING (LOWER PART)

Metric	2 Dies			3 Dies		
	n100	n200	n300	n100	n200	n300
Wirelength ($cm \times 10^3$)	1.18	1.81	1.97	1.10	1.93	2.07
Die Outlines (cm^2)	1.14	1.14	1.14	0.73	0.84	0.91
Total Deadspace (%)	29.21	30.39	31.14	26.81	37.20	42.06
Runtime (s)	80	359	891	81	360	858
Wirelength ($cm \times 10^3$)*	1.83	2.60	2.53	1.34	2.59	2.76
Die Outlines (cm^2)	1.00	1.08	1.07	0.77	0.82	0.35
Total Deadspace (%)	18.82	27.04	26.39	29.81	36.00	32.19
Runtime (s)	59	304	726	59	304	734

*Estimated routing detours for (unaligned) interconnect structures are included.

TABLE II
RESULTS ON ENLARGED IBM-HB+ BENCHMARKS FOR LAYOUT PACKING AND WIRELENGTH OPTIMIZATION

Metric	2 Dies			3 Dies		
	ibm01	ibm03	ibm07	ibm01	ibm03	ibm07
Wirelength ($cm \times 10^3$)	4.77	7.29	1.77	4.67	7.39	1.67
Die Outlines (cm^2)	0.64	0.65	0.79	0.46	0.48	0.57
Total Deadspace (%)	17.24	19.26	18.59	24.97	27.86	24.88
Runtime (s)*	1195	3611	3081	1285	3792	3895

TABLE III
COMPARATIVE RESULTS ON GSRC BENCHMARKS FOR LAYOUT PACKING WITH THERMAL AND WIRELENGTH OPTIMIZATION – BENCHMARKS ARE
NOT ENLARGED FOR FAIR COMPARISON

Metric	Corblivar, 2 Dies				Corblivar, 2 Dies			Corblivar, 3 Dies				Corblivar, 3 Dies		
	n100	n200	n300	Avg	ami33	xerox	Avg	n100	n200	n300	Avg	ami33	xerox	Avg
Wirelength ($\mu m \times 10^5$)	3.70	6.57	9.07	6.45	2.02	13.89	7.96	4.24	7.19	10.28	7.24	2.06	16.36	9.21
Die Outlines ($\mu m^2 \times 10^5$)	1.01	0.99	1.61	1.20	10.38	143.15	76.77	0.75	0.68	1.08	0.84	8.98	117.48	63.23
Total Deadspace (%)	11.98	12.01	15.65	13.21	44.31	32.41	38.36	20.53	14.62	16.27	17.14	57.08	45.09	51.09
Max Temp [21] ($^\circ$K)	313.81	314.53	315.95	314.76	309.14	353.85	331.49	355.62	363.94	363.35	360.97	333.17	416.61	374.89
Runtime (s)	108	286	548	314	50	14	32	154	380	704	413	71	22	47

Metric	[22], 2 Dies				[23], 2 Dies			[22], 3 Dies				[23], 3 Dies		
	n100	n200	n300	Avg	ami33	xerox	Avg	n100	n200	n300	Avg	ami33	xerox	Avg
Wirelength ($\mu m \times 10^5$)	3.65	6.18	9.53	6.45	1.81	17.14	9.48	4.59	7.17	10.61	7.46	2.22	21.86	12.04
Die Outlines ($\mu m^2 \times 10^5$)	1.19	1.21	2.15	1.52	9.65	125.17	67.41	0.97	0.82	1.48	1.09	7.54	88.93	48.24
Total Deadspace (%)	25.02	27.87	36.69	29.86	40.09	22.70	31.40	38.89	28.64	38.65	35.39	48.87	27.47	38.17
Max Temp [21] ($^\circ$K)	313.31	313.74	314.63	313.89	336.36	366.48	351.42	348.55	360.60	361.35	356.83	384.39	482.16	433.28
Runtime (s)	439	446	526	470	193	47	120	266	497	574	446	193	48	121

livar, a 3D layout representation based on an extended corner block list with novel alignment tuples. To this end, we also develop effective techniques for block placement and alignment. We note that it is essential to synchronize alignment across the whole 3D IC—in particular, inter-die alignment requires to consider each die's layout in progress. Our techniques handle this appropriately for different scenarios of blocks to be aligned and/or to be placed. We embed Corblivar into an open-source, SA-based floorplanning tool; we also develop necessitated extensions like adaptive SA cooling and convolution-based fast thermal analysis. Experimental results on GSRC and large-scale IBM-HB+ benchmarks demonstrate Corblivar's applicability for structural planning of interconnects, i.e., block alignment, as well as its competitive performance for "classical" 3D floorplanning while considering fixed outlines, layout packing, thermal and wirelength optimization.

ACKNOWLEDGMENTS

The work of J. Knechtel was supported by the German Research Foundation (Project 1401/1). For the parametrization of the thermal model (Subsection V-E), Timm Amstein provided *Octave* scripts which are included in [18]. The authors thank Igor L. Markov for comments on drafts of this paper.

REFERENCES

[1] I. Loi et al., "Characterization and implementation of fault-tolerant vertical links for 3-D networks-on-chip," *Trans. Comput.-Aided Des. Integr. Circuits Sys.*, vol. 30, no. 1, pp. 124–134, 2011.

[2] F. Li et al., "Design and management of 3D chip multiprocessors using network-in-memory," in *Proc. Int. Symp. Comput. Archit.*, 2006, pp. 130–141.

[3] H. Xiang et al., "Bus-driven floorplanning," in *Proc. Int. Conf. Comput.-Aided Des.*, 2003, pp. 66–73.

[4] J. H. Law and E. F. Young, "Multi-bend bus driven floorplanning," *Integration*, vol. 41, no. 2, pp. 306–316, 2008.

[5] R. Nain and M. Chrzanowska-Jeske, "Fast placement-aware 3-D floorplanning using vertical constraints on sequence pairs," *Trans. VLSI Syst.*, vol. 19, no. 9, pp. 1667–1680, 2011.

[6] J. H. Law et al., "Block alignment in 3D floorplan using layered TCG," in *Proc. Great Lakes Symp. VLSI*, 2006, pp. 376–380.

[7] X. Li et al., "A novel thermal optimization flow using incremental floorplanning for 3D ICs," in *Proc. Asia South Pacific Des. Autom. Conf.*, 2009, pp. 347–352.

[8] J. Knechtel et al., "Assembling 2-D blocks into 3-D chips," *Trans. Comput.-Aided Des. Integr. Circuits Sys.*, vol. 31, no. 2, pp. 228–241, 2012.

[9] D. H. Kim et al., "Block-level 3D IC design with through-silicon-via planning," in *Proc. Asia South Pacific Des. Autom. Conf.*, 2012, pp. 335–340.

[10] J. Knechtel et al., "Multiobjective optimization of deadspace, a critical resource for 3D-IC integration," in *Proc. Int. Conf. Comput.-Aided Des.*, 2012, pp. 705–712.

[11] S. Panth et al., "High-density integration of functional modules using monolithic 3D-IC technology," in *Proc. Asia South Pacific Des. Autom. Conf.*, 2013, pp. 681–686.

[12] H.-T. Chen et al., "A new architecture for power network in 3D IC," in *Proc. Des. Autom. Test Europe*, 2011, pp. 1–6.

[13] K. Athikulwongse et al., "Exploiting die-to-die thermal coupling in 3D IC placement," in *Proc. Des. Autom. Conf.*, 2012, pp. 741–746.

[14] S. K. Lim, "Folded 2-die block," Personal communication, March 2013.

[15] X. Hong et al., "Corner block list: an effective and efficient topological representation of non-slicing floorplan," in *Proc. Int. Conf. Comput.-Aided Des.*, 2000, pp. 8–12.

[16] R. Fischbach et al., "Investigating modern layout representations for improved 3D design automation," in *Proc. Great Lakes Symp. VLSI*, 2011, pp. 337–342.

[17] S. Chen et al., "VLSI block placement with alignment constraints based on corner block list," in *Proc. Int. Symp. Circ. Sys.*, vol. 6, 2005, pp. 6222–6225.

[18] J. Knechtel. (2013) Corblivar floorplanning suite. [Online]: http://www.ifte.de/english/research/3d-design/index.html

[19] T.-C. Chen and Y.-W. Chang, "Modern floorplanning based on B*-Tree and fast simulated annealing," *Trans. Comput.-Aided Des. Integr. Circuits Sys.*, vol. 25, no. 4, pp. 637–650, 2006.

[20] J.-H. Park et al., "Fast thermal analysis of vertically integrated circuits (3-D ICs) using power blurring method," in *Proc. ASME InterPACK*, 2009, pp. 701–707.

[21] A. Coskun et al. (2011, November) Hotspot 3D extension. [Online]: http://lava.cs.virginia.edu/HotSpot/links.htm

[22] P. Zhou et al., "3D-STAF: scalable temperature and leakage aware floorplanning for three-dimensional integrated circuits," in *Proc. Int. Conf. Comput.-Aided Des.*, 2007, pp. 590–597.

[23] Y. Chen. (2010) 3DFP – thermal-aware floorplanner for three-dimensional ICs. [27]. [Online]: http://www.cse.psu.edu/~yxc236/3dfp/index.html

[24] A. N. Ng et al. (2006) IBM-HB+ benchmarks. [26]. [Online]: http://vlsicad.eecs.umich.edu/BK/ISPD06bench/

[25] (2000) GSRC benchmarks. [Online]: http://vlsicad.eecs.umich.edu/BK/GSRCbench/

[26] J. A. Roy et al., "Solving modern mixed-size placement instances," *Integration, the VLSI Journal*, vol. 42, no. 2, pp. 262–275, 2009.

[27] W.-L. Hung et al., "Interconnect and thermal-aware floorplanning for 3D microprocessors," in *Proc. Int. Symp. Quality Elec. Des.*, 2006, pp. 98–104.

Comprehensive Die-Level Assessment of Design Rules and Layouts

Rani S. Ghaida[*], Yasmine Badr[†], Mukul Gupta[‡], Ning Jin[*], and Puneet Gupta[†]

[*] GLOBALFOUNDRIES, Inc.
[†] EE Department., Univ. of California, Los Angeles
[‡] QUALCOMM, Inc.
rani.ghaida@globalfoundries.com, ybadr@ucla.edu, puneet@ee.ucla.edu

Abstract— Co-development of design rules and layout methodologies is the key to successful adoption of a technology. In this work, we propose Chip-level Design Rule Evaluator (ChipDRE), the first framework for systematic evaluation of design rules and their interaction with layouts, performance, margins and yield at the chip scale (as opposed to standard cell-level). A "good chips per wafer" metric is used to unify area, performance, variability and yield. The framework uses a generated virtual standard-cell library coupled with a mix of physical design, semi-empirical, and machine-learning-based models to estimate area and delay at the chip level. The result is a unified design-quality estimate that can be computed fast enough to allow using ChipDRE to optimize a large number of complex design rules. For instance, a study of well-to-active spacing rule reveals a non-monotone dependence of rule value to chip area (although the dependence to cell area is monotone) due to delay changes coming from well-proximity effect.

I. INTRODUCTION

Semiconductors have fueled wealth creation, making new applications (cost-) feasible with each successive technology generation. Keeping Moore's law alive would require rapid technology changes over the next decade and beyond. Accurate projection of the design impact of device and technology changes is key for making informed technology/design decisions, thereby, ensuring timely and cost-effective development of technology and design flows.

The evaluation of technology impact on design is traditionally inferred from the evaluation of Design Rules (DRs), which are the biggest design-relevant quality metric for a technology. Unfortunately, even after decades of existence, DR evaluation is largely unsystematic and empirical in nature; it relies on limited and small-scale experiments and manufacturing tests and much on speculations based on technologists/designers experience with previous technology generations [1]–[4]. The work in [5] presents a flow for the optimization of double-patterning design rules. The method consists of an optimization loop in which rules are modified, standard-cell layouts are generated, and printability is analyzed. Although this approach, like [3,4], may be suited for exploring rules from a pure printability perspective, it does not examine the electrical effects of rules. Moreover, because actual layout generation and printability analysis are excessively time-consuming, exploring a wide range of rules and rule combinations is impractical with these approaches.

More recently, the work of [6] offers a framework for evaluating design rules, at early stages of technology development, through fast layout-topology generation of standard-cell layouts and estimation of variability and manufacturability using first-order models. This work has two major limitations. First, the evaluation was performed at the cell-level, which may lead to false conclusions because most designs are routing-limited and, hence, *not every change in cell area results in a corresponding change in chip area*. Second, delay was not evaluated but it is well-known that *delay-change can affect chip area* due to techniques like buffering and gate sizing required to meet timing requirements.

In this work, we propose Chip-level Design Rule Evaluator (ChipDRE), the first framework for systematic evaluation of design rules and their interaction with layouts, performance, margins and yield at the chip scale.

ChipDRE uses a "good chips per wafer" (GCPW) metric to unify area, performance, variability and functional yield. It uses a generated virtual standard-cell library coupled with a mix of physical design and semi-empirical models to estimate area, delay and yield at the chip level. To predict the design-rule/layout impact on delay and delay variability, ChipDRE employs a Static Timing Analysis model to estimate cell-delay and a neural network-based model to predict delay-margin dependent area penalty. Chip-level area is estimated from cell area – including the delay-margin area penalty – and a cell-area to chip-area model that is calibrated using actual Synthesis, Place and Route (SPR) data. Finally, GCPW is calculated taking into consideration a chip-level functional yield estimate. The result is a unified design-quality estimate that can be computed fast enough to allow using ChipDRE to optimize a large number of complex design rules and achieve "true" design/technology co-optimization.

We make the following contributions.

- We offer ChipDRE, the first framework for collective evaluation of design rules, layout styles, and library architectures *at the chip scale*. ChipDRE is designed to be *used for design/technology co-optimization* and supports state-of-the-art technologies including FinFETs and Local Interconnects (LI). It aims at making rule generation and optimization easier and much faster. Rather than exploring the entire search space of design rules manually or with conventional compute-expensive methods, the framework can be used to quickly eliminate poor rule choices.
- We develop a cell-delay estimator and a neural network-based model to project the impact of cell-delay change on the overall chip area.
- We propose a cell-area to chip-area model to project how cell area translates into chip area.
- We evaluate the rule impact on delay and report the evaluation in terms of GCPW unifying area, performance, variability and functional yield metrics. This comprehensive evaluation allows studying interesting trade-offs that occur at the chip level like the one between variability, performance and area.
- We perform evaluation studies of major design rules at advanced nodes (some FinFET-specific) including: gate to local-interconnect spacing, gate-to-well edge spacing and fin pitch.

The remaining paper is organized as follows. Section II gives an overview of our approach. Sections III elaborates the cell-delay estimation and the virtual standard-cell layout generation including I/O pin-access estimation and supporting FinFET and local-interconnect technologies. The cell-area to chip-area model is described in Section V, while the model to predict delay-margin dependent area penalty is described in Section IV. Section VI presents the results of a number of evaluation studies at 45nm technology node using ChipDRE. Finally, Section VII gives a brief summary of the paper and some directions for future research.

II. OVERVIEW AND STANDARD-CELL LAYOUT ESTIMATION

In this section, we give an overview of ChipDRE and briefly describe its components. We also present our approach for cell-layout estimation.

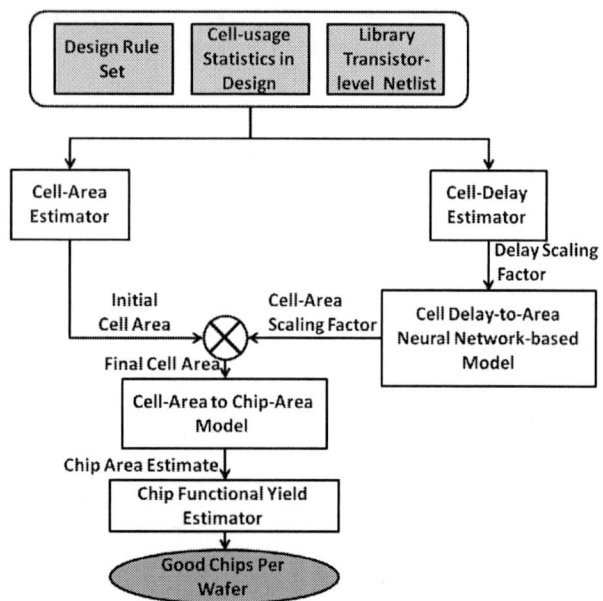

Figure 1. Overview of ChipDRE and its main components.

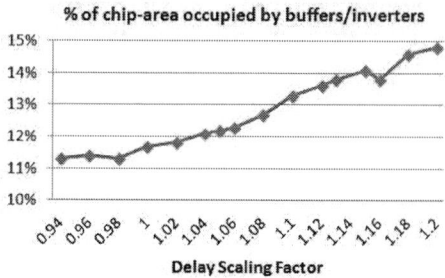

Figure 2. Empirical data from placement-aware synthesis commercial tool manifesting the impact of cell delay on the percentage of chip area that is occupied by buffers/inverters.

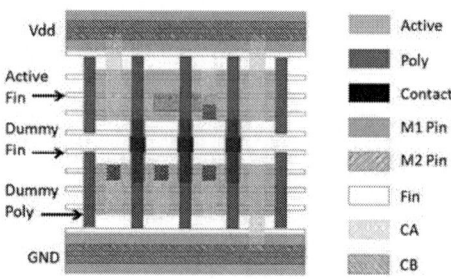

Figure 3. Example layout for OAI21_X1 cell generated by ChipDRE with FinFETs, local interconnects (i.e., CA and CB layers), and DR violation-free I/O pin segments.

A. Overview

An overview of ChipDRE is depicted in Figure 1. The framework takes the following inputs: transistor-level netlists (SPICE) of cells, rules and their values, estimates of process control (e.g., overlay error distribution), and cell-usage statistics of the design to evaluate the rules on. In ChipDRE, only the values of rules under evaluation are modified while all others remain unchanged. This modified set of rules is then used to estimate the cell-layout and perform the design-level evaluation.

Concisely, the first stage of ChipDRE is to estimate the cell layout/area and cell delay for a given set of rules. If the cell delay changes in comparison with the delay obtained using a base set of rules, the cell-delay change is converted into a delay-scaling factor which is used to scale the timing characteristics of the standard-cell library (in Liberty file format). A neural network-based model is then used to estimate the impact of cell-delay change on the design overall cell area (Figure 2 manifests the significance of this impact). The model essentially predicts gate-sizing and buffer-insertion to meet the timing requirements with the new cell-delay characteristics. In the second stage of ChipDRE, another semi-empirical model – fitted to SPR data – is used to predict how the cell area translates into chip area. The final stage of ChipDRE, chip-level functional yield is estimated and a unified design-quality metric, number of "good chips per wafer" (GCPW), is calculated.

B. Cell-area Estimation

The cell-area estimator is based on the virtual-cell generator from [6]. This generator[1] accurately estimates cell area ($< 1\%$ error [6]) through fast generation of front-end-of-line (FEOL) layers and congestion-based estimation of wiring area. In this work, we extend the cell-layout estimator of [6] to enable its application at the chip level and using state-of-the-art technologies (e.g., FinFETs).

For chip-level evaluation, we generate I/O pin segments and the physical specifications of the technology and standard-cells (in Library Exchange Format or LEF). In studies presented in this work, pin segments are kept at minimum possible dimensions while meeting the minimum area design rule. We first sort vertical pins from left to right and horizontal pins from bottom to top. We then assign pins sequentially to the closest available track without creating DR violations. It is worth noting that we allow three pin configurations: (1) all pins on M1, (2) all pins on M2, and (3) pins on either M1 or M2 layers. In case of (3), a pin will be assigned to M1 by default and moved to M2 if doing so helps

[1]Publicly available at nanocad.ee.ucla.edu.

resolving M1 congestion in the cell (see Figure 3 for an example). In all our experiments, we use pin configuration (3).

FinFET technology with local-interconnect layers will be standard across the industry at advanced nodes (22nm and below [7]). Hence, to enable rule-evaluation at advanced nodes, we extend the layout-generation of front-end-of-line layers to include additional local-interconnect and FinFET-specific layers. The additional layers are: CA, CB, and fin-layer. CA is the vertical local-interconnect layer and is used to connect the fins of the same FET together[2], primarily to make contact from the contact-layer to the fins. CA can also be optionally used to make power/ground connection to the FETs (when a local-interconnect power rail exists). CB is the horizontal local-interconnect layer and is used to make contact from the contact-layer to Poly and to make short Poly-to-Poly connections when possible. The fin layer constitutes the actual FinFETs, referred to as active fins, and dummy fins, which are necessary to conform the fin layer to a grid and ensure printability. The fin grid needs to be in accordance with the cell-height so that it is maintained after cell-placement in the design. This constraint makes finding a valid configuration of fin count and pitch in active regions (P/N networks) as well as top, bottom, and center overhead regions complex. Given a range of allowed fin pitches, we run an exhaustive search to find a working configuration with maximum number of total active fins in one column and the smallest active fin pitch. To improve the chances to reach a better solution, we optionally allow the dummy fin pitch in top/bottom/center overhead regions to differ and allow the cell top/bottom edges to coincide either with the center of the fin (as in Figure 3) or with the center of the dummy fin-to-fin spacing.

To migrate a planar FET-based netlist to a FinFET-based netlist, we employ the following model to determine the number of fins for every transistor:

$$n = \lceil \frac{W}{\alpha \times F_H} \rceil, \tag{1}$$

where W is the transistor width specified in the planar-based netlist, F_H is the fin-height, and α is a planar-to-finFET width translation parameter[3].

[2]Note this is optional when the source/drain is not contacted

[3]We use $\alpha = 2$ in our fin-pitch experiment like [8]. A higher value of α can be used to take into account the contribution of the top gate as well as the triangular profile of FinFETs.

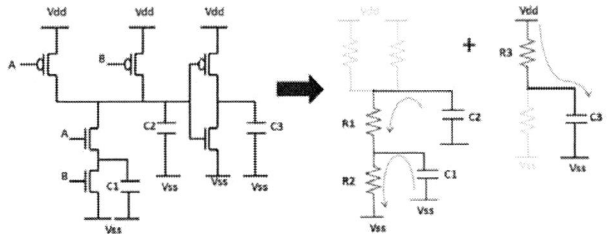

Figure 4. Estimation of low-to-high propagation delay for AND gate, equivalent RC tree and charge/discharge paths. It consists of two stages, the pull-down network for the NAND gate followed by the pull-up network of the inverter. Using Elmore delay and adding up delay stages, the propagation delay for the cell rise is estimated as: $t_{pLH} = R_1\, C_1 + (R_1 + R_2)\, C_2 + R_3\, C_3$. C_1, C_2 and C_3 include all the gate and diffusions capacitances connected at each of the 3 nets.

The rounding up of number of fins in Equation 1 is done to ensure the minimum transistor performance is preserved after the migration.

III. VARIABILITY-AWARE CELL-DELAY ESTIMATION

A crucial aspect of Design Rule Evaluation is the assessment of the impact of the DRs on performance. To characterize a digital chip-level delay, it is required to model the delay for each standard cell. First-order delay models are employed in order to have a fast and approximate delay estimation.

A. Cell Delay Model

To characterize the cell rise or fall delay, the cell is considered as a sequence of stages and the delays of these stages are then added up. For each stage, all paths connecting the output to the power supply (Vdd or ground) are enumerated. An RC tree is constructed for each path and Elmore delay [9] is applied to compute the path delay [10]. The worst case pull up and pull down delays are determined for each stage. Identical paths (paths that switch simultaneously) are considered as parallel resistances and their capacitances are added up.

B. Transistor Model

We apply an RC approximation for each transistor where the capacitance model [10] considers the gate capacitance (including channel and overlap capacitances) as well as diffusion capacitances, and accounts for Miller effect. The MOS switch model in [10] is used to estimate the equivalent resistance R_{on} of the transistor.

To model delay variability and consider the worst case delay, we use the current variability estimates from [6] which primarily models layout-dependent, lithography-induced variations in drive current. Variability is computed as 3σ change in current which is subtracted from the nominal current value before calculating resistance. As an example, we illustrate the pull-up of an AND gate in Figure 4.

C. Verification and Results

For verification, we used NCX [11] with HSPICE [12] to generate the liberty file for some standard cells from Nangate Standard Cell Library [13]. The worst cases for cell rise and cell fall were compared to the values reported by ChipDRE delay estimator, using the same load capacitance.

Gate Length Scaling Experiments. For these experiments, the gate length rule was scaled by 10%, and the scaling factor of the ChipDRE-estimated delay (i.e. the ratio between delay at the scaled gate length to the delay at the original length) was compared to the scaling factor obtained by our spice simulation setup. Table I lists the scaling factors obtained from ChipDRE and spice, as well as the magnitude of the error which does not exceed 3%.

Well-Proximity Effect (WPE) Experiment. To model the Well-Proximity effect, BSIM [14] model for WPE impact on threshold voltage and mobility was used. Values of the model's parameters were computed as in [15]. The gate-to-well distance value in the BSIM model was scaled down by 10%, and the corresponding delay values were computed.

TABLE I
VERIFICATION OF DELAY MODEL USING GATE-LENGTH SCALING EXPERIMENTS BY COMPARING THE CHIPDRE-ESTIMATED DELAY SCALING FACTOR TO THE SCALING FACTOR FROM SPICE

Cell	Pull-up			Pull-down		
	ChipDRE	Spice	Abs Error (%)	ChipDRE	Spice	Abs Error (%)
INV_X32	1.10	1.09	0.9	1.10	1.07	3
NAND2_X1	1.10	1.10	0.3	1.10	1.07	3
INV_X1	1.08	1.09	1.1	1.08	1.07	1.1
AND2_X4	1.10	1.09	0.8	1.10	1.09	1.2
OAI21_X2	1.10	1.09	0.7	1.10	1.07	2.6
AOI211_X1	1.10	1.09	0.5	1.10	1.08	2.2
OAI33_X1	1.10	1.10	0.3	1.10	1.08	2.1
AND2_X2	1.10	1.09	0.8	1.10	1.08	1.6
Average	1.1	1.09	0.7	1.1	1.08	2

TABLE II
VERIFICATION OF DELAY MODEL USING WELL-PROXIMITY EFFECT (WPE) EXPERIMENT BY COMPARING THE CHIPDRE-ESTIMATED DELAY SCALING FACTOR TO THE SCALING FACTOR FROM SPICE

Cell	Pull-up			Pull-down		
	ChipDRE	Spice	Abs Error(%)	ChipDRE	Spice	Abs Error(%)
INV_X32	0.96	0.96	0.6	0.96	0.97	0.8
NAND2_X1	0.76	0.78	2.4	0.85	0.88	2.8
INV_X1	0.76	0.78	2.1	0.79	0.84	5.4
AND2_X4	0.93	0.92	1.5	0.89	0.84	6.6
OAI21_X2	0.93	0.93	0.1	0.93	0.94	1.0
AOI211_X1	0.89	0.89	0.3	0.85	0.85	0.5
OAI33_X1	0.89	0.89	0.8	0.85	0.88	3.7
AND2_X2	0.89	1.00	11.0	0.87	0.94	7.3
OR2_X2	0.87	0.88	1.4	0.88	0.88	0.2
Average	0.88	0.89	2.4	0.88	0.89	3.5

The ratios of cell delay with scaled gate-to-well distance to original cell delay were compared to the equivalent ratios obtained using Spice [12] simulation. Table II shows the comparison between the ratios obtained by ChipDRE to those obtained by Spice and the corresponding error which is below 7.3%.

D. Liberty Delay File Generation

For the baseline set of design rules, we assume a Liberty file [4]. To generate the liberty file for virtual standard-cell library corresponding to the set of rules under evaluation, the worst-case pull-up and pull-down delays for the gates are computed as explained in section III-A. This is also done for the baseline set of design rules to create a reference gate delay (computed by ChipDRE). The ratios between the gate delays in the case of design rules under evaluation and those of the baseline design rules are used to scale the baseline liberty file to obtain an estimated Liberty file for the design rules under evaluation. For sequential elements, their hold and setup times are left unchanged (same as baseline liberty file), and their clock to output delay is scaled by the same scaling factor as inverter. The entire flow of generating layouts, estimating delays and generating the Liberty file within ChipDRE takes less than 49 minutes for a 100 cell library as opposed to commercial library characterization tools which take several CPU days.

IV. DELAY-TO-AREA MODELING

One of the major issues ChipDRE addresses which typical cell-based design rule optimization approaches suffer from is the effect of timing optimization - during physical synthesis - on area. Physical synthesis tools use several optimization techniques to meet timing constraints at the minimum possible area, like gate sizing, buffer insertion and logic restructuring. Thus, as delay of standard cell increases, we can expect an increase in the resultant chip area. Previous work [16] has experimentally characterized the impact of timing guardband reduction on some metrics of the circuit by running synthesis, place and route for several scaled

[4]This could be a characterized or scaled version from a previous technology node. The absolute values of delays in the Liberty file are not very important for ChipDRE as we are more interested in relative delay changes with rule changes.

978-1-4799-2817-0/14 $31.00 © 2014 IEEE

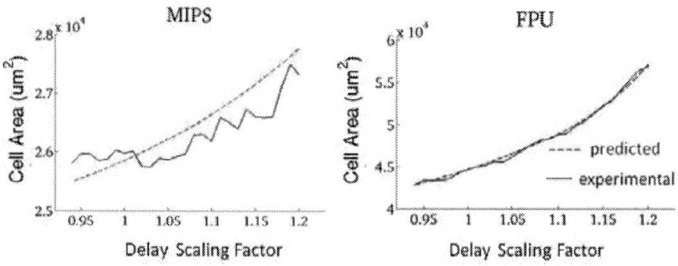

Figure 5. NN testing on MIPS design (a blind test case) and on FPU (used in training). This network has been trained resulting in a training mean square error of 8.16x10⁴

libraries. However, this is impractically slow to explore design rule choices. Moreover, the work of [17] has demonstrated that little noise can have huge effect on place-and-route solution quality; this makes using a model-based estimate even more attractive.

Modeling these optimization techniques analytically is complicated with a tremendous number of degrees of freedom. Thus we use a machine learning technique to predict the cell-area scaling factor (ratio between the cell area of the design at some delay scaling factor to the baseline cell area of the same design) as the standard cells delay scales (due to a change in DRs).

A neural network has been trained using data from physically-aware synthesis performed using [18][5]. To train the neural network (one hidden layer with 6 nodes), the following features have been used: number of instances on critical path, average fanout, average interconnect length, average delay and area of gates on critical path, utilization, timing constraint, ratio between area of critical paths to the total cell area and the delay scaling factor. Those features have been selected because they affect the amount of buffering and gate sizing performed by the tool to meet the timing constraints. We assumed that there is no change to the back end rules and only the front end rules are undergoing change and evaluation. Otherwise, other features need to be added like capacitance and resistance of the used metal layers per unit length.

The network was trained – using Matlab Neural Network Toolbox – on 27 delay scaling factors (each time the liberty file being scaled) from 9 test cases; 3 from [19] and 6 from ISCAS85 benchmarks. Upon testing the network on MIPS design from [19] (not used in training), the neural network was able to predict the cell-area scaling factor – used to calculate cell area – and rule out tool noise as shown in Figure 5. The figure also shows the performance of the neural network on one of the training test cases, the FPU design (from [19]).

V. Chip Area and Yield Modeling

A. Minimum Routable Area

Minimum Routable Area (MRA) of a design requires the estimation of maximum utilization at which the number of DR violations cease to be zero. This implies that for finding MRA multiple Place & Route (P&R) runs are required, making the whole process time consuming (detailed routing being the main culprit). For instance, an experiment to estimate MRA of AES (~10K gate design) using binary search took 14 hours (as shown in column 6 of Table III). Such excessive runtime makes chip-level evaluation of multiple design rules impractical.

Thus, we propose a new methodology, Area Estimation using Global Routing (AEGR) that estimates MRA using global routing congestion estimates. Global routing congestion estimates require the estimation of wiring demand and wiring supply on each of the global routing cell – called G-cell – which represents a fixed number of available routing tracks in each layer. If wiring demand exceeds supply, the detailed routing is

[5]Physically-aware synthesis, which performs placement to estimate interconnect delay, has been used since it takes less time than the complete time-consuming place and route and yet produces estimates that are accurate enough for our purpose.

unlikely to implement a design rule correct wire pattern. Congestion in an arbitrary G-cell is given by

$$C = \frac{\text{routing demand (d)}}{\text{routing supply (s)}}. \tag{2}$$

SPR tools cannot resolve all instances of congestion and for very high congestion values, the tool might not find enough unused G-cells to successfully route the design. Hence we propose that there exists a threshold on congestion beyond which tool cannot successfully route the design. Based on this we define a metric, $m(u)$, in the following manner

$$m(u) = \alpha \times C_{peak}(u) + \beta \times C_{avg}(u), \tag{3}$$

where C_{avg} is the average congestion over all G-cells and C_{peak} is the maximum congestion over all G-cells , and α and β are the tool dependent parameters. The utilization u_{max} for which m(u_{max}) is 1 is classified as the maximum utilization of the design.

To further refine the estimation of maximum utilization, we run detailed routing in the range $[0.9u_{max}, 1.1u_{max}]$ to get two utilization values where number of DR violations is greater than zero. Then linear extrapolation is done using these two points to estimate the utilization value where number of DR violations is equal to zero. This estimated utilization value is termed as the maximum utilization value. Using this methodology substantial runtime improvement was achieved as we show later in this section.

B. Model Formulation

Although AEGR gives substantial improvement in runtime, it still requires running Place & Route (P&R) for all the designs and large number of FEOL design rules (increasing with every new technology node). Also, tool noise leads to problems in optimization. To overcome these problems, we model chip area as a function of total cell area thereby skipping P&R to the maximum possible extent. Our proposed model in differential form is given in Equation (4). Here y is the chip area and x is the total cell area. $\frac{x}{y}$ is the utilization of the design. In the proposed model, as the utilization increases or equivalently white space decreases, change in chip area is more sensitive to any change in cell area. The final analytical equation is given in Equation (5).

$$\frac{dy}{dx} = k1 - k2 \times (y/x). \tag{4}$$

After solving Equation (4), we get

$$y = \frac{k1}{k2+1} \times x + \left(y0 - (\frac{k1}{k2+1}) \times x0 \right) \times (\frac{x}{x0})^{-k2}. \tag{5}$$

There are four unknowns in the model viz. $k1$, $k2$, $x0$ and $y0$. $y0$ can be thought of as the routing limited chip area. $x0$ can be thought of as any unutilized whitespace area[6] when the chip area is $y0$. $x0$ depends on the cell routability which in turn is dependent on the pin access and congestion within the cell [20]. Larger congestion implies router needs to drop more vias outside the cells to make connections with the cell instance pins, effectively decreasing any unutilized whitespace and hence decreasing $x0$.

To find $k1$ and $k2$ we apply the following boundary conditions

$$k1 - k2 = 1, \tag{6}$$
$$k1 - k2 \times \frac{y0}{x0} = 0. \tag{7}$$

Equation (6) is based on the fact that for very high utilization values, change in chip area is roughly equal to change in total cell area. This implies that as $u \to 1$, $\frac{dy}{dx} \to 1$. Hence the boundary condition follows from Equation (4). Similarly from the other extreme, for any total cell area less than $x0$, chip area is routing limited and is equal to $y0$. Hence, Equation (7) follows from Equation (4). Based on these

[6]chip area minus the area required by the router to make connection with the cell instance pins using M1 layer.

Table III
RUNTIME COMPARISON BETWEEN AREA ESTIMATION USING GLOBAL
ROUTING (AEGR) METHOD AND ACTUAL P&R FOR FINDING THE MINIMUM
ROUTABLE AREA.

Design	Routing Layers	AEGR Util.	P&R Util.	Runtime in mins (AEGR)	Runtime in mins (P&R)	Runtime Reduction
MIPS	3	0.83	0.83	97	322	3.3x
MIPS	4	0.97	0.97	23	145	6.3x
JPEG	3	0.93	0.93	345	892	2.6x
AES	3	0.44	0.47	57	1267	22x
AES	4	0.76	0.76	110	842	7.6x
AES	5	0.85	0.84	52.4	141	2.7x
FPU	3	0.91	0.90	52	261	5x
NOVA	4	0.88	0.88	296	519	1.8x

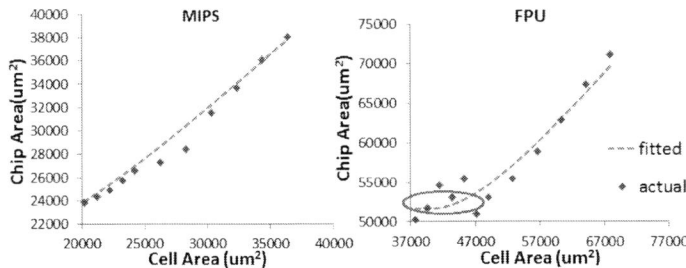

Figure 6. Plots showing MIPS and FPU chip-area vs. cell-area results obtained from actual P&R runs and those estimated using our analytical predictive model. Notice the circled region on FPU which exhibits a flat relationship between cell area and chip area. FPU is more routing-limited than MIPS.

boundary conditions, model coefficients and final analytical equation are given by

$$k1 = \frac{y0}{y0 - x0}, \tag{8}$$

$$k2 = \frac{x0}{y0 - x0}, \tag{9}$$

$$y = x + (y0 - x0) \times \left(\frac{x0}{x}\right)^{\frac{x0}{y0 - x0}} \text{ for } x > x0, \tag{10}$$

$$y = y0 \text{ for } x <= x0. \tag{11}$$

Since $y0$ and $x0$ are design dependent parameters, we estimate them by actual P&R runs for each design under consideration. $x0$ and $y0$ need to be estimated only once for a given back-end interconnect stack and library architecture. This gives substantial improvement in runtime making it possible to simultaneously evaluate large number of design rules.

Our experiments to validate our methodology were performed on 5 designs from [19], synthesized using Nangate Open Cell-Library [13], and FreePDK open-source process [21]. First, data for actual P&R were created for all the designs using cadence encounter, with router objective function as "minimize congestion", and for varying number of routing layers. Based on these runs α and β (in Equation (3)) were estimated to be $\frac{1}{3}$, i.e. the coefficients were estimated such that the metric agrees with the routability of designs confirmed by P&R runs. Runtime comparison between AEGR and actual P&R methods for MRA estimation is given in Table III. For actual P&R, maximum utilization was found using binary search algorithm.

To evaluate the area model, area of various cells was increased in the LEF file to closely imitate cell-area change due to FEOL design rule changes. However, the pin shapes and pin positions were not modified. Chip area was then estimated using AEGR for every increase in total cell area and the proposed model was fitted on the resulting data. The plots are shown in Figure 6 and values of $x0$ and $y0$ are shown in Table IV.

C. Functional Yield Modeling and GCPW Calculation

Functional yield at the cell-level is computed similarly to [6]. It includes three yield-loss sources: overlay error (i.e. misalignment between

Table IV
VALUES OF $x0$ AND $y0$ FOR VARIOUS DESIGNS (SEE PLOTS OF FIGURE 6).

Design Name	x0 (um^2)	y0 (um^2)
MIPS	12526	20437
FPU	30950	36760

Table V
CHIP AREA COMPARISON BETWEEN GOLDEN SPR AND MODEL BASED
PREDICTION ON MIPS. THE RUNTIME FOR CHIPDRE IS JUST THE CELL
ESTIMATION TIME: 49 MINUTES FOR A 100 CELL LIBRARY. GOLDEN FLOW
USES CHIPDRE-GENERATED LIBRARIES WITH COMMERCIAL TOOLS FOR
PHYSICAL DESIGN WITH THE AEGR METHOD PROPOSED IN THIS PAPER.
"EST" IS THE VALUE ESTIMATED BY CHIPDRE.

Well-to-active spacing [nm]	Run-time (SPR) [mins]	Cell Area (est.) [um^2]	Chip Area (est.) [um^2]	Chip Area (SPR) [um^2]	Error in %	GCPW (est.)
140	118	28171	30364	30130	0.8	667
185	356	28171	29709	29460	0.8	681
200	240	32527	33008	33913	-2.7	612
210	207	32554	32787	33554	-2.3	616

layers) coupled with lithographic line-end shortening (a.k.a. pull-back), contact-hole failure, and random particle defects. The yield at the cell level is extended to the chip level using the well-known negative binomial model [7]. GCPW can then be calculated as the ratio of $\frac{wafer\_area}{chip\_area} \times yield$.

VI. EXPERIMENTAL RESULTS

As examples, we study three interesting rules in ChipDRE: (1) well-to-active spacing rule which affects number of transistor folds (hence area and delay variability) as well as threshold voltage and mobility of transistors (hence delay); (2) local-interconnect to gate spacing rule which affects capacitances as well as area; and (3) fin-pitch rule for a candidate FinFET technology. We observe that simple cell-based estimates (as is the state-of-the-art) to assess rule quality can be misleading highlighting the importance of the ChipDRE framework [8].

A. Well-to-active Spacing Rule Exploration

ChipDRE was used to perform a study of the well-to-active spacing rule, which impacts cell delay as well as cell area. The rule values that were chosen are 140nm, 185nm, 200nm and 210nm with 140nm as the baseline value. SPR data were generated for MIPS design using the ChipDRE-generated LEF and LIB files for each spacing rule with timing optimization done at both placement and post-routing stages while keeping the congestion effort "high". The clock period was chosen such that minimum positive slack was achieved for the baseline case. The maximum possible cell-utilization with no DR violations and a positive timing slack is used to compute the chip area. Chip-area comparison between actual results from SPR and estimation from the proposed ChipDRE flow is given in Table V. The table also shows the GCPW metric for the design rule[9]. This study shows that a well-to-active spacing rule of value of 185nm results in the best number of GCPW even though it does not achieve the minimum cell area.

Table V results show that ChipDRE predictions are in strong agreement with the full SPR based flow and match the trends well. Interestingly, the dependence of GCPW and chip area on the rule value are *non-monotone*. This is primarily due to improved delay when well-to-active spacing is increased and despite the fact that the cell area monotonically increases as the rule value increases.

---

[7] Yield loss in routing-layers will be addressed in future work.

[8] We use 45nm rules from a publicly available pdk [21] to perform example studies which could be performed for future technology nodes.

[9] Note that for calculation of yield and GCPW, we assume the final design area is actually n copies of the indicated area (analogous to multiple cores), where n was selected to make the final design area roughly 100 mm^2 at the baseline design rule value.

978-1-4799-2817-0/14 $31.00 © 2014 IEEE

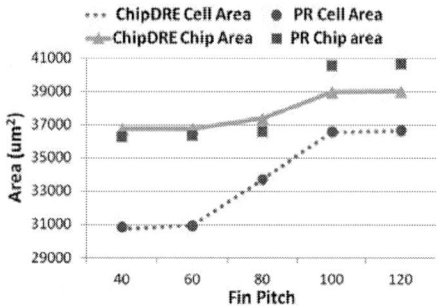

Figure 7. Plots for cell/chip area of FPU design as a function of fin pitch.

B. FinFET Fin-Pitch Study

Fin pitch value is a technology parameter that has a strong impact on the layout density. Although fin pitch is usually defined by process and technology constraints, exploring the design implications of this rule can help process developers decide which patterning technology to adopt (e.g. Self Aligned Double Patterning vs Directed Self Assembly). We use this fin pitch exploration as an example study to highlight the difference between chip-level and cell-level assessment of DRs. Hence, we use our framework to evaluate the impact of fin pitch on cell/chip area [10]. The impact of fin pitch on delay was ignored in this experiment since its impact on parasitic capacitances was not modeled in this work. Fin pitch was varied from 60nm to 120nm in steps of 20nm and for each value standard cell layouts were generated. Based on the standard cell usage of FPU design, total cell area was computed. The cell area was then plugged into cell-area to chip-area model' and chip area was computed. This has been verified against PR runs, and the maximum error in the model predictions was found to be 5%. Figure 7 shows the chip area and cell area variations as the fin pitch is varied, both from ChipDRE and PR experiments. The figure shows that for a fin pitch of 60nm through 80nm, the cell area is steeply increasing with a very slight change in chip area, which emphasizes the importance of chip-level evaluation as opposed to cell-level evaluation. It is also observed that the fin pitch can be increased from 40nm to 60nm with a negligible impact on cell area. GCPW trends are similar to chip-area trends in this case.

C. LI-to-gate spacing

Local interconnect is used in modern technologies to relieve congestion on local metal layers. One of the primary purposes is to make the power and ground rail connections from corresponding active areas in the devices. These connections replace contacts and metal. Unfortunately, these long contacts also increase capacitive coupling between gate and the local interconnect resulting in increased C_{gs}. To complicate matters further, increased spacing between gate and local interconnect can cause increase in the active area resulting in increased diffusion capacitance as well. We model both these effects in ChipDRE for the planar process and explore this spacing rule. Figure 8 shows the effect of changing the LI-to-gate spacing on the chip area (with GCPW trends being similar). In this case, the cell-area increase due to rule-value increase dominates the potential area reduction coming from delay improvement brought by a reduced gate-to-LI coupling capacitance (unlike the well-to-active rule experiment which showed a stronger delay impact).

VII. CONCLUSIONS

We presented ChipDRE, the first framework for *fast*, *early* and *systematic* collective evaluation of design rules, layout styles, and library architectures at the chip-scale. The framework makes rule definition and optimization easier, efficient, and much more systematic. Rather than exploring the entire search space of design rules manually or with conventional compute-expensive methods, the framework can be used to

---

[10]We realize there is no finfets in a 45nm process, but the study is performed for demonstration purposes.

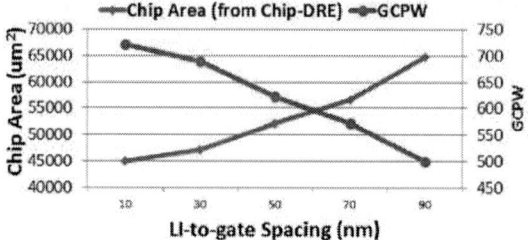

Figure 8. LI-to-gate design rule evaluation and effect on chip area for FPU.

quickly eliminate poor rule and technology choices. By using fast layout-estimation methods coupled with semi-empirical and neural network-based models for cell-area/cell-delay impact and trade-offs at the chip-level, the ChipDRE framework unifies area, performance, variability, and yield a "good chips per wafer" metric. To show potential applications of ChipDRE, we use it to perform evaluation studies of debatable rules for state-of-the-art technologies, including FinFETs and local-interconnects, at the chip-scale. For instance *a study of well-to-active spacing rule reveals a non-monotone dependence of rule value to chip area* (although the cell-area relationship is monotone) due to delay changes coming from well-proximity effect.

VIII. ACKNOWLEDGEMENTS

This work was partly supported by IMPACT+ center (http://impact.ee.ucla.edu), SRC and NSF CAREER award.

REFERENCES

[1] L. Capodieci, P. Gupta, A. B. Kahng, D. Sylvester, and J. Yang, "Toward a methodology for manufacturability-driven design rule exploration," in *Proc. DAC*, 2004, pp. 311–316.

[2] Y. Zhang, J. Cobb, A. Yang, J. Li, K. Lucas, and S. Sethi, "32nm design rule and process exploration flow," in *Proc. SPIE*, vol. 7122, 2008, p. 71223Z.

[3] V. Dai, L. Capodieci, J. Yang, and N. Rodriguez, "Developing DRC Plus rules through 2D pattern extraction and clustering techniques," in *Proc. SPIE*, vol. 7275, 2009, p. 727517.

[4] S. Chang, J. Blatchford, S. Prins, S. Jessen, T. Dam, G. Xiao, L. Pang, and B. Gleason, "Exploration of complex metal 2D design rules using inverse lithography," in *Proc. SPIE*, vol. 7275, 2009, p. 72750D.

[5] Y. Deng, Y. Ma, H. Yoshida, J. Kye, H. J. Levinson, J. Sweis, T. H. Coskun, and V. Kamat, "Dpt restricted design rules for advanced logic applications," in *Proc. SPIE*. International Society for Optics and Photonics, 2011, pp. 79 730H–79 730H.

[6] R. S. Ghaida and P. Gupta, "DRE: A Framework for Early Co-Evaluation of Design Rules, Technology Choices, and Layout Methodologies," *IEEE TCAD*, vol. 31, no. 9, Sept 2012, pp. 1379–1392.

[7] D. McGrath. Globalfoundries looks to leapfrog fab rivals. Http://www.eetimes.com/.

[8] P. Mishra, A. Muttreja, and N. K. Jha, "Finfet circuit design," in *Nanoelectronic Circuit Design*. Springer, 2011, pp. 23–54.

[9] W. C. Elmore, "The Transient Response of Damped Linear Networks with Particular Regard to Wideband Amplifiers," *Journal of Applied Physics*, vol. 19, no. 1, 1948, pp. 55–63.

[10] J. M. Rabaey, A. Chandrakasan, and B. Nikolic, *Digital integrated circuits- A design perspective*, 2nd ed. Prentice Hall, 2004.

[11] Synopsys Liberty NCX.

[12] (2012) Synopsys HSPICE.

[13] Nangate Open Cell Library v1.3. 2009. Http://www.si2.org/openeda.si2.org/projects/nangatelib.

[14] BSIM4.6.2 User Manual.

[15] C. M. Council. Guidelines for Extracting Well Proximity Effect Instance Parameters.

[16] K. Jeong, A. Kahng, and K. Samadi, "Impact of Guardband Reduction On Design Outcomes: A Quantitative Approach," *IEEE TSM*, vol. 22, no. 4, Nov. 2009, pp. 552 –565.

[17] A. Kahng and S. Mantik, "Measurement of inherent noise in EDA tools," in *Proc. ISQED*, 2002, pp. 206–211.

[18] Cadence RTL Compiler Advanced Physical Option.

[19] Open Cores. Http://www.opencores.org/.

[20] T. Taghavi, C. Alpert, A. Huber, Z. Li, G.-J. Nam, and S. Ramji, "New placement prediction and mitigation techniques for local routing congestion," in *Proc. ICCAD*, 2010, pp. 621–624.

[21] FreePDK. Http://www.eda.ncsu.edu/wiki/FreePDK.

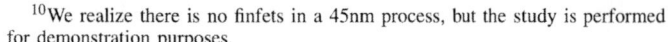

Prefetching Techniques for STT-RAM based Last-level Cache in CMP Systems*

Mengjie Mao, Guangyu Sun[†], Yong Li, Alex K. Jones, Yiran Chen

University of Pittsburgh, Pittsburgh, PA 15261, USA

[†]CECA, Peking University

{mem231, yol26, akjones, yic52}@pitt.edu, [†]gsun@pku.edu.cn

Abstract—Prefetching is widely used in modern computer systems to mitigate the impact of long memory access latency by paying extra cost in memory and cache accesses. However, the efficacy of prefetching significantly degrades in the memory hierarchy using the emerging *spin-transfer torque random access memory* (STT-RAM) as last-level cache (LLC) due to the long write access latency. In this work, we propose two orthogonal but complimentary techniques to improve the prefetching efficacy of STT-RAM based LLC in chip multi-processor (CMP) systems, namely, request prioritization (RP) and hybrid local-global prefetch control (HLGPC). Simulation results show that by combining these two techniques, we can achieve 6.5%~11% system performance improvement and 4.8%~7.3% LLC energy saving in a quadcore system with a 2MB~8MB STT-RAM based LLC, compared to the system with only basic prefetching.

I. INTRODUCTION

Prefetching technique [1], [6] is widely adopted in modern computer systems to alleviate the impact of long memory access latency. Data is pre-loaded into a cache based on run-time prediction before it is actually requested. The memory access latency is hidden though extra burdens are imposed on the traffic from the memory to the cache. Potential access conflicts, e.g., bank access contention and cache pollution, may be introduced.

Very recently, *spin-transfer torque random access memory* (STT-RAM) gains increasing attentions in on-chip cache implementation [16]. Compared to SRAM, STT-RAM has nearly zero leakage power consumption, much higher cell density, and similar read performance. However, the long write latency and high write energy primarily impede the adoption of STT-RAM as last-level cache (LLC). Although write performance can be improved by raising write current, the storage density must be sacrificed because a large-size driving transistor is required.

In a chip multi-processor (CMP) system with prefetching, LLC commonly serves prefetch requests (writes) besides common accesses. The introduction of STT-RAM based LLC results in some potential adverse impacts on the system performance and energy: For a LLC built with high-density/small-size STT-RAM cells, the long write access latency increases the possibility of bank access conflicts; for a LLC built with low-density/large-size STT-RAM cells, the limited LLC capacity makes the system more sensitive to the prefetching-incurred cache pollution, causing the increases in cache miss rate and the number of write accesses.

In this work, we propose two orthogonal but complimentary techniques to mitigate the impacts of long write latency of STT-RAM based LLC on the performance of the CMP systems with aggressive prefetching. The first technique is named as *request prioritization* (RP) where different access types of LLC are prioritized based on their criticality to the system performance. Requests with low priority may be preempted by the ones with higher priority during executions. A *hybrid local-global prefetch control* (HLGPC) mechanism is then introduced to dynamically tune the aggressiveness of prefecher for alleviating the LLC access contention.

Compared to the existing works on the memory hierarchy built with STT-RAM, our two major contributions are:

- We quantitatively analyze the degradation of prefetching efficacy in the CMP systems with STT-RAM based LLC.
- Two complimentary techniques – RP and $HLGPC$ are innovated to mitigate the adverse impacts of aggressive prefetching on CMP systems, achieving substantial performance speedup and energy saving under different cache design configurations.

To the best of our knowledge, this is the first work quantitatively analyzing the impact of adopting STT-RAM as LLC on the prefetching efficacy of CMP systems. Simulation results show that the combination of RP and $HLGPC$ techniques can achieve 6.5%~11% system performance improvement (geometric mean) as well as 4.8%~7.3% LLC energy saving for a quadcore system with 2MB~8MB STT-RAM based LLC.

The rest of our paper is organized as follows: Section II gives a review of STT-RAM based LLC design and the recent prefetching efficacy enhancement techniques; Section III depicts the motivations of our work; Section IV shows the implementation details on the RP and $HLGPC$; Section V presents our experiment setup; Section VI shows the simulation results; Section VII concludes our work.

II. BACKGROUND

A. STT-RAM based LLC

The most popular STT-RAM cell design is "1T1J", which contains one magnetic tunneling junction (MTJ) device connected with an access transistor. The resistance of the MTJ can be switched between high- and low-state when a current with different polarization is applied. The MTJ switching can be speeded up by raising the switching current. However, it needs to enlarge the NMOS transistor size and results in the increase of the cell area. Table 1 shows the parameters of three STT-RAM cache designs built with the access transistors with different driving abilities; all parameters are extracted from NVSim [15]. When the MTJ switching time is 3ns, 10ns, and 30ns, the corresponding STT-RAM cell area is about $71F^2$, $25F^2$, and $15F^2$, respectively. Here F is the feature size fabrication technology, say, 45nm here.

TABLE I: The timing and energy parameters of 2/4/8MB STT-RAM LLC at 45nm technology

		2MB		4MB		8MB	
		Write	Read	Write	Read	Write	Read
Timing(ns)	Pulse	3	3.82	10	3.06	30	4.04
	Peripheral	1.69		1.75		2.07	
Cell area(F^2)		71		25		15	
Dynamic energy(nJ)		1.268	0.77	1.08	0.877	1.310	1.197
Memory area(mm^2)		10.429		9.733		10.163	
Leakage power(mW)		924.7		800.4		859.0	

Dong *et al.* first proposed using STT-RAM to implement LLC in microprocessors and discovered its advantages on leakage power reduction [14]. Sun *et al.* extended STT-RAM based LLC design

---

*This work is supported in part by NSF awards CNS-1116171, NSF of China (No. 61202072), and National High-tech R&D Program of China (No. 2013AA013201).

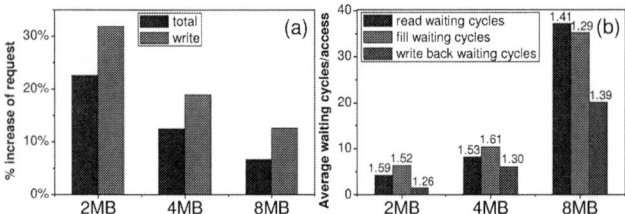

Fig. 1: (a) The increase of total LLC access requests and write requests after applying prefetching; (b) The average waiting cycles of different types of LLC accesses (values atop each bar are normalized to the results without prefetching).

to CMP systems and identified the negative impacts of long write latency of STT-RAM on the system performance [5]. A read-preemptive write buffer, which allows the read request to preempt the write back request that is accessing the same cache bank, was also proposed to alleviate the blocking effect of the write access to STT-RAM based cache. The reliability of STT-RAM read/write operation has been well studied [18], [13], [12] promising the installation of STT-RAM on modern microprocessor design.

B. Prefetching Efficacy Enhancement

The inaccuracy of data prefetcher introduces significant memory resource wasting and degrades system performance. For example, inaccurate prefetch requests incur excessive write accesses to the LLC, which compete with normal LLC requests. As a result, the induced cache pollution and the increased cache miss rate trigger more LLC requests and further aggravate the competition among the requests. In [11], [17], filter techniques are proposed to dynamically tune the prefetch requests based on the prefetech accuracy while a hardware-based cache pollution filter is constructed for the prefetcher. If the predicted accuracy of a particular prefetch request is lower than a preset threshold, the corresponding request will be prevented from being issued. In [2], [9], the aggressiveness of prefetchers, i.e., the prefetch degree[1] and prefetch distance[2], is dynamically adjusted based on the real-time system feedback information like temporal accuracy, prefetch timeliness, cache pollution, and memory bandwidth utilization. These techniques are traditionally applied to shared main memory in the presence of multiple prefetchers.

III. TECHNICAL MOTIVATION

Figure 1(a) shows the increases in the total LLC access requests and the write requests after applying prefetching to a STT-RAM based LLC with a capacity of 2MB, 4MB, and 8MB, respectively, compared to the case that no prefetching is applied. The data is averaged over six workloads, each of which consists of 4 different applications running simultaneously on a quadcore CMP (more simulation setup details will be given in Section V). As shown in Table I, all LLC configurations have the similar areas ($\sim 10mm^2$) but different performances: large cache capacity is associated with long write access latency due to the small transistor size. The shortest read latency, which is the sum of delays on memory array and peripheral circuits, happens in the 4MB LLC design. Figure 1(b) depicts the average waiting cycles of each types of LLC access requests caused by cache bank access conflicts after prefetching is applied. Based on Figure 1(a) and (b), we made the following observations:

- With prefetching, the numbers of total access requests as well as write requests climb up, followed by the increase in the waiting

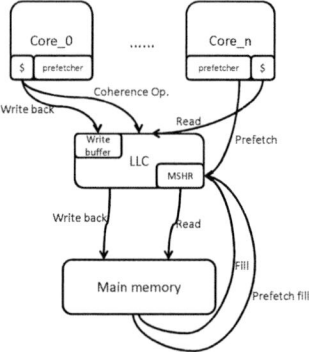

Fig. 2: Access types of the shared LLC in CMP systems.

time for all types of access requests. As the capacity of LLC increases, the average waiting time of all types of access requests keeps raising due to the prolonged write access latency. Figure 1 (b) also show that read requests and fill requests, which are commonly located on the critical path of execution, experience much longer waiting time compared to the less critical requests, e.g., write back. Therefore, the system performance may be potentially improved by assigning higher priority to the critical LLC access requests during execution to avoid long waiting time.

- The prefetching efficiency is affected by the capacity (cell size) of STT-RAM based LLC. A large cell size helps to shorten the average waiting time of the LLC access requests by reducing the blocking time of write operations. However, the cache pollution incurred by prefetching also becomes severer due to the reduced total capacity. On the contrary, the average waiting time of the LLC access requests rises when the LLC capacity increases. The system performance is determined by the tradeoffs among the write request frequency, the average waiting time (or the LLC access conflicts), and the prefetching policy, all of which vary during the execution of a program. Hence, we may dynamically throttle the aggressiveness of prefetchers for system performance improvement by taking into account all these factors.

IV. PROPOSED TECHNIQUES
A. Request Prioritization (RP)

Figure 2 summarizes the common LLC access requests in a CMP system: read requests, coherence requests[3] and write back requests are initiated by the upper-level cache of each CPU core; prefetcher sends the prefetch requests to the main memroy and fills the corresponding data into the LLC. If the read request to the LLC encounters a cache miss, a fill request will be generated to fetch the data from the main memory to the LLC. Consequently, the following accesses to the corresponding cache bank of the STT-RAM based LLC may be blocked due to long write latency of memory cells, leading to a bank access conflict.

Notice that these LLC access requests shown in Figure 2 demonstrate different criticality to the system performance. For example, read requests must be handled immediately as the cache miss at upper-level cache significantly affects the system performance. Fill requests are also critical since they are triggered by LLC misses, especially if the requested data is loaded into LLC first before it is sent to upper-level cache for consistency. Prefetch fill requests, however, have lower criticality because the prefetched data is unlikely to be used immediately. Write back requests apparently have the lowest criticality because the data written back to the LLC may not be used in the near future.

---

[1] # of prefetch requests that can be issued in one access.

[2] The range that the prefetch requests locate into.

[3] Most cache coherence protocols only modify the states stored in the tag array/directory that is highly banked, without accessing the data array. Thus, coherence request is not included in our RP technique.

978-1-4799-2817-0/14 $31.00 © 2014 IEEE   68

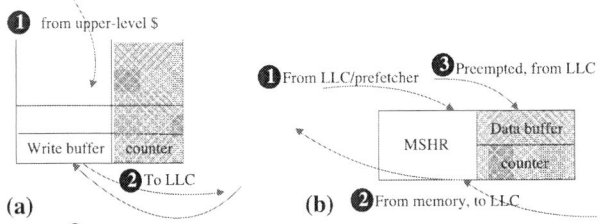

Fig. 3: Implementation of RP with augmented components: (a) Write back requests. (b) Fill and prefetch fill requests.

We proposed to prioritize the different types of LLC access requests based on their criticality to the system performance, and process the requests based on their priorities. For example, if multiple requests compete the access to the same cache bank, the one with the highest priority will be granted the right of access. However, a request with a high priority may also be blocked by a low priority request which is being processed. To address this issue, we allow the high-priority request to preempt the low-priority one if the elapse time of processing the low-priority request is below a threshold, say, the retirement accomplishment degree (RAD) [5]. For example, a 30% RAD of write back request means that the write back request can be preempted by a request with higher priority if the elapse time of the LLC write back access is below 30% of the STT-RAM write latency. We refer to this method as *request prioritization* (RP).

To record the elapse time of the LLC access request, each *miss state handling register* (MSHR) or write buffer entry must be augmented with a N-bit counter (e.g., 7-bit to record up to 128-cycle operation time for an 8MB LLC). The counter is initialized once the associated request grasps the access right of the cache bank, and reset after the request is preempted or successfully processed. As shown in Figure 3(a), when a write back request is preempted, the data being written is still retained at the head of the write buffer and waiting for the next try; otherwise, the buffer entry will be recycled once the write back completes.

However, if a fill or prefetch fill request is preempted, the request must be buffered for the future retry. As shown in Figure 3(b), besides the elapse time counter, a data buffer whose length is the same as one cache block is also augmented to the MSHR to buffer the data associated with the fill/prefetech fill request. The procedure which is similar to LLC miss handling can be applied to deal with the write preemption: Once the request is issued, a MSHR is assigned to it by LLC controller; the MSHR holds the data fetched from the main memory until it is successfully written into the LLC; if the request is preempted before the writing is completed, the request is buffered in the MSHR and waiting for the next retry; otherwise, the MSHR will be recycled after the writing is completed. Read requests will never be preempted.

Preempting write request may cause false LLC states since the tag array would be updated before the data array is updated. For instance, a write request to the data array is preempted after modifying the valid/dirty bits of the corresponding tag entry. If a read request accesses the same data entry before the retry of the write request, it may actually access the old or invalid data. To avoid this inconsistency, we define the policy that a write request cannot update the valid bit in the tag entry until it successfully updates the data array. The read request in the above example will trigger a cache miss which will be filtered by the MSHR entry occupied by the preempted write request.

Similar to the scheme in [5], a 20-entry write buffer is adopted in our design. To suppress the competition between the prefetch fill requests and fill requests and avoid overflow, the number of MSHRs must be sufficiently large, e.g., 128 in our simulated quadcare system. The augmented storage is built with SRAM.

TABLE II: GPC decision guideline

Case	$core_i$'s info		LLC's info	Decision
	Pref. Acc.	Pref. Freq.	Acc. Freq.	
1	High	Low	High	Allow local
2	High	High	Low	Allow local
3	High	High	High	Disable scale up
4	High	Low	Low	Allow local
5	Low	Low	High	Disable scale up
6	Low	High	Low	Allow local
7	Low	High	High	Force scale down
8	Low	Low	Low	Allow local

B. Hybrid Local-global Prefetch Ctrl (HLGPC)

As aforementioned in Section III, overaggressive prefetching in a CMP system may cause cache pollution and aggravate the access conflict of the shared memory resource [4]. This issue becomes severer in the system with small LLC because prefetch fill requests will have a higher probability to evict the useful cache blocks. The increased LLC miss rate will generate more prefetches and eventually put the system into a negative feedback loop. The general solutions of this issue includes *local (per core) prefetch control* (LPC) [3], [9] and prefetcher coordination by considering main memory access contention [2], [4], which dynamically control the aggressiveness of the prefetchers.

However, conventional LPC technique focuses on maximizing the performance of each CPU core, which does not necessarily optimize the overall system performance due to the LLC access conflicts among the different prefetchers. Also, the significant access conflicts in STT-RAM based LLC become another crucial factor affecting the prefetching efficacy besides main memory access contention. Based on the above observations, we propose *hybrid local-global prefetch control* (HLGPC) technique to achieve a balanced dynamic aggressiveness control across all prefetchers in a CMP system. The two integrated components of HLGPC are:

Local prefetch control (LPC): At core-level, we choose *feed-back directed prefetching* (FDP) [9] as the LPC scheme to manage the aggressiveness of the prefetcher. FDP periodically samples *the prefetcher's accuracy, prefetch timeliness* and *per-core-induced cache pollution* over every time interval and determines the aggressiveness of the prefetcher in the next interval accordingly.

Global prefetch control (GPC): At chip-level, GPC may retain or override the decision of LPC based on the runtime information of LLC. GPC periodically samples the *prefetch frequency* of each core and the *global access frequency* of LLC. Based on the values of these two metrics and the prefetcher's accuracy over the current interval, GPC applies the following three rules on the prefetcher of each core in the next interval, as shown in Table II.

- Disable scale up: In case 3, although the prefetch accuracy and frequency are high for $core_i$, the high global access frequency of the LLC indicates the severe access conflicts. Therefore, the scaling up of the prefetcher's aggressiveness is disabled even the LPC decides to do so. In case 5, continue scaling up the prefetcher's aggressiveness in $core_i$ will likely deteriorate access conflicts due to high LLC access frequency and low prefetch accuracy. Hence, the scaling up of the prefetcher's aggressiveness is also prevented;

- Force scale down: In case 7, low prefetch accuracy couples with high prefetch frequency, indicating high volume of useless prefetches. Considering the high global access frequency of the LLC (or say, the severe access conflicts), GPC will force $core_i$ to scale down the prefetcher's aggressiveness regardless the decision of LPC;

- Allow local decision: For the rest cases in Table II, GPC will stick to the decision of LPC which ensures the best performance

TABLE III: The aggressiveness levels of prefetchers

Aggressiveness	Prefetch Distance	Prefetch Degree
Very conservative	4	1
Conservative	8	2
Medium	16	4
Aggressive	32	4
Very aggressive	64	8

for each core as well as the overall system.

In our implementation of HLGPC, the design metrics like prefetch frequency of $core_i$ and the global access frequency over interval i are updated as [9]:

$$UpdatedCount_i = 0.5 \cdot UpdatedCount_{i-1} + 0.5 \cdot CurrentCount_i. \quad (1)$$

Here $UpdatedCount_i$ is the metric over the current interval i while $UpdatedCount_{i-1}$ is the metric over the previous interval $i-1$. $CurrentCount_i$ denotes the number of the prefetch requests issued by $core_i$ or the LLC accesses over the current interval i. Eq. (1) gives the credits to the metrics over both the current and all previous intervals and reflects the temporal correlation of the execution of the workload. The threshold value of each GPC metric adopted in our simulation will be shown in Section V-A.

V. EXPERIMENT SETUP

A. Simulation Platform

Without loss of generality, we adopt the stream prefetcher in IBM POWER4 processor [6] in our simulation among the exisiting prefetcher designs. The aggressiveness of stream prefetcher can be dynamically adjusted among five configurations, as shown in Table III. We also assume there is only one prefetcher per CPU core in the simulated CMP system.

The MacSim [19] simulator with Intel's Sandy Bridge configuration is used in our evaluation. The design parameters of STT-RAM based LLC at 45nm technology, which are summarized in Table I, are obtained from NVsim [15] with the appropriate device parameters in [7]. We also consider the energy overhead of the augmented write buffer and MSHR while the energy parameters of these components are extracted from CACTI [10] and Synopsys using VHDL. The configuration of the simulated quadcore system is summarized in Table IV. The length of an interval in the HLGPC scheme is set to 8192 evictions from LLC [9]. The threshold values for the high/low prefetch accuracy, the prefetch frequency, and the global access frequency are empirically set to 0.5, 246 and 8192, respectively. Total four 13-bit counters are needed in each core to measure the prefetch accuracy, the prefetch frequency, and the global access frequency of LLC (it needs two counters).

B. Selection of Benchmarks

We select 18 benchmarks from SPEC CPU 2000/2006 suite. The execution trace of each benchmark is collected by using a modified pin-based trace extraction tool packed with MacSim tool set. Each trace is composed of 400 million instructions after bypassing the initialization phase ranging from 5 billion to 100 billion instructions.

We construct 6 multi-programmed workloads which generally suffer from LLC conflicts with prefetching. Each workload consists of 4 benchmarks, among which at least two benchmarks are *LLC access intensive* (LLCI), one benchmark is *LLC non-intensive* (LLCNI), and one benchmark is *prefetch intensive* (PREFI). Here, LLCI is defined as the *accesses per 1 kilo instructions* (APKI) of LLC > 8; LLCNI is defined as APKI ≤ 8; and PREFI is defined as the *prefetch requests per 1 kilo instructions* (PPKI) of LLC > 1. The workloads that every benchmark belongs to and their properties are summarized in Table V. These properties are profiled by running each benchmark on only one core of a quadcore system with prefetching.

TABLE V: Properties of 18 SPEC benchmarks

benchmark	LLCI	PREFI	Workload
art	Y	Y	1, 4
ammp	Y	N	3
bzip2	N	N	1
gcc	N	N	6
gobmk	N	N	6
milc	Y	N	1
zeusmp	Y	Y	4, 6
gromacs	N	N	5
cactusADM	N	N	2
leslie3d	Y	Y	1, 2
dealII	N	N	2
hmmer	N	N	3
GemsFDTD	Y	Y	3
h264ref	N	N	5
lbm	Y	Y	2, 4
omnetpp	Y	Y	3, 5
astar	N	N	4
xalancbmk	Y	Y	5, 6

VI. RESULTS AND DISCUSSIONS

A. The Efficacy of RP

Table VI depicts three priority assignment schemes evaluated in our simulations, namely, P1, P2, and P3. The priorities among different requests under different priority assignments are represented by RAD. For example, read has a higher priority than fill (25% RAD under P2), which means that fill will be preempted by read if its elapse time is below 25% of the write access latency of the LLC. We conducted extensive experiments to explore the performance/energy impact of different RAD settings and extracted three representative priority assignments, as summarized in Table VI.

Figure 4 depicts the *Harmonic mean speedups* (HS) [8] of every workloads under different RP schemes (P1, P2, P3), respectively. The results are normalized to the HS of the same system without prefetching. For comparison purpose, we also provide the results of conventional prefetching technique (*pref. only*) and the geometric mean of the HS of the read-preemptive write buffer technique [5] (RAD_{org}). The results show that prioritizing LLC access requests always achieve system performance improvement w.r.t. conventional prefetching technique though the results of different RP schemes vary at different workloads and LLC configurations. The highest performance is achieved by P1 with a HS improvement of 4.6%/4.8%/8.3% for 2/4/8MB LLC compared to *pref. only*. RAD_{org}, however, receives very marginal performance improvement across all scenarios as it only differentiates the priorities between read and write back requests.

Figure 4 also shows that compared to no prefetching, conventional prefetching technique achieves marginal or even negative performance improvement (i.e., wl2) in all workloads. This fact clearly reflects the adverse impacts induced by the LLC access conflicts due to the long write access latency of STT-RAM. The detailed analysis on the efficacy of all RP schemes are:

- Among all RP schemes, P1 achieves the best performance because the critical requests like read and fill are always served in a timely manner;
- Comparing P1 and P2, lifting up the priority of fill request (P1) results in performance improvement, indicating that the criticality of read and fill requests are equivalent or at least close to each other in the simulated workloads;

TABLE VI: Priority assignment schemes

	read			fill			prefetch fill			write back		
	P1	P2	P3	P1	P2	P3	P1	P2	P3	P1	P2	P3
read	-	-	-	-	25%	100%	60%	60%	100%	60%	60%	100%
fill	-	-	-	-	-	-	60%	60%	60%	60%	60%	60%
prefetch fill	-	-	-	-	-	-	-	-	-	60%	60%	60%
write back	-	-	-	-	-	-	-	-	-	-	-	-

TABLE IV: System Configuration

Execution	4GHz, OOO 4 issues(up to 2 mem and 2 FP), 20 stages pipeline, gshare branch predictor, 1024-entry BTB, global history length 14, 20 cycles branch misprediction penalty, 256-enrty ROB
Upper-level cache	32KB L1IC, 4-way, 4 banks, 4 ports, 1-cycle latency, 64B line
	32 KB L1DC, 4-way, 4 banks, 4 ports, 2-cycle latency, write back, 64B line
	256 KB L2, 8-way, 4 banks, 1 r/w port, 8-cycle latency, write back, 64B line, 64 MSHRs
Prefetcher	Stream prefetcher with 64 streams, max prefetch degree 8, max prefetch distance 64
LLC	2/4/8 MB STT-RAM shared L3, 8/16/16-way, 8/8/8 banks, 1 r/w port, 16/12/16-cycle read, 20/48/128-cycle write, 64B line, write back, 128 MSHRs
Main memory	Fixed 200-cycle latency

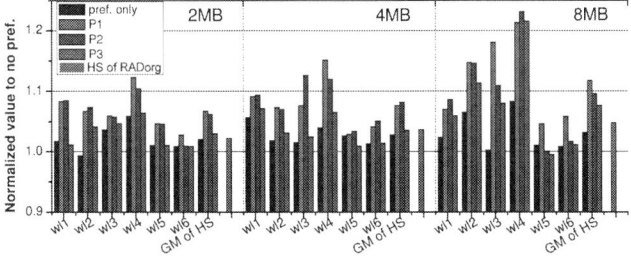

Fig. 4: HS of *perf. only* and different RP schemes. Geometric means of the HS of RAD_{org} are also included.

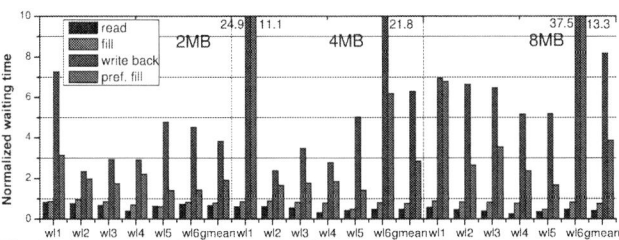

Fig. 5: Average waiting times of different types of LLC access requests of P1.

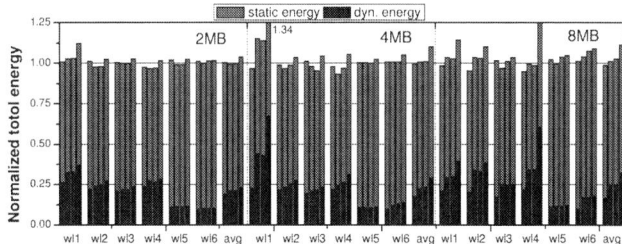

Fig. 6: LLC energy consumption of different prefetching schemes. From left to right for each workload: *pref. only*, P1, P2 and P3.

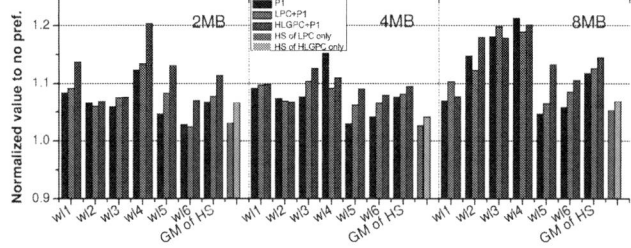

Fig. 7: HS of P1, LPC+P1 and HLGPC+P1. Geometric means of the HS of *LPC only* and *HLGPC only* are also included.

- Comparing P2 and P3, the aggressive priority assignment to read request (P3) introduces very little performance improvement. It is because of the overflowing of the write buffer incurred by the frequent preemption from the read requests as well as the equivalent criticality of read and fill requests.

- In some cases (e.g., wl3 under 4MB, wl1 and wl4 under 8MB), system performance is more sensitive to the priority assignment of read requests; thus, P2 noticeably outperforms P1 for those cases.

- RP scheme achieves more substantial performance improvement for large LLC with long write access latency than small LLC because of the severer LLC bank conflict. For an 8MB LLC, the geometric mean of HS for P1 can be as high as 8.3%.

To gain the insight into the efficacy of P1, Figure 5 lists the average waiting times of different types of LLC access requests of P1, which are normalized to the result of *pref. only*. The results show that most of the waiting time has been allocated from read and fill requests to prefetch fill and write back requests. Less waiting time of the critical requests is the main reason for system performance improvement.

Figure 6 summarizes the LLC energy consumptions of the different prefetching schemes that are normalized to no prefetching. The extra write operations induced by the preemptions in RP schemes raise the dynamic energy of LLC. Since P1 does not preempt fill requests, it consumes the lowest dynamic energy among all RP schemes. Because all types of write accesses can be preempted by read requests (i.e., RAD = 100%), the dynamic energy consumption of P3 is significantly higher than any other schemes. Following the increase of LLC capacity, the probability and the number of the low-priority requests being preempted climb up due to the prolonged write access latency, resulting in dynamic energy increase. Luckily, the leakage energy of the LLC is substantially reduced as the execution time

decreases. The average LLC energy consumptions increase less than 1% across all LLC configurations. Since P1 achieves well-balanced tradeoff between system performance and energy consumption, we select it as our baseline RP scheme in the evaluation of HLGPC.

B. The Efficacy of HLGPC

Figure 7 depicts the HS for every workload before and after applying LPC+P1 and HLGPC+P1, respectively. The selected RP scheme is P1 and the selected LPC scheme is FDP [9]. The geometry means of the HS of *LPC only* and *HLGPC only* are also included to illustrate the result of the prefetch control without RP scheme. All results are normalized to no prefetching. The performance of *LPC only* is visibly worse than that of *HLGPC only*, indicating the necessity of combining both global and local prefetch control. Compared to applying only P1, both LPC+P1 and HLGPC+P1 improve the average system performance while HLGPC+P1 achieves the highest improvement (4.3%/1.7%/2.5% HS for 2/4/8MB LLC compared to P1 or 9.6%/6.5%/11% HS for 2/4/8MB LLC compared to *pref. only*). Small LLC benefits the most from HLGPC+P1 as the corresponding LLC access conflicts are effectively alleviated.

In HLGPC+P1, the rules applied at GPC level may mitigate the prefetcher's aggressiveness and cause the increase in LLC miss rate. If system performance is more sensitive to LLC miss rate than access conflicts, LPC+P1 could outperform HLPGC+P1, as shown in "wl1" and "wl3" under 8MB LLC. In "wl2" under 4MB LLC, the LLC access conflicts induced by prefetching is very insignificant. Applying extra control on prefetch in LPC+P1 or HLGPC+P1 indeed degrades the system performance compared to applying no prefetch control (P1).

Figure 8 presents the profile of the number of fill and prefetch fill requests and the average waiting times of read and fill requests before

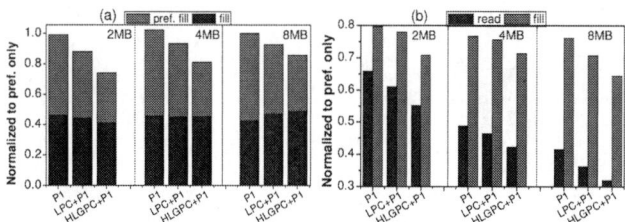

Fig. 8: The normalized (a) Number of fill and prefetch fill requests of P1, LPC+P1 and HLGPC+P1; (b) Average waiting times of critical requests of P1, LPC+P1 and HLGPC+P1.

and after applying different prefetch controls. All data are normalized to *pref. only*. As shown in Figure 8(a), HLGPC+P1 substantially suppresses prefetch fill requests without significantly increasing LLC miss rate denoted by the number of fill requests. It also implies that the miss rate may decrease with less aggressive prefetching because of the alleviation of cache pollution. As shown in Figure 8(b), the average waiting time of read and fill requests for each scheme is also significantly reduced due to the alleviated LLC access conflicts.

Figure 9 summarizes the LLC energy consumptions of different prefetch controls that are normalized to no prefetching. By throttling the prefetch requests, the dynamic energies of both LPC+P1 and HLGPC+P1 are all below that of P1 while HLGPC+P1 achieves the lowest dynamic energy consumption. Nonetheless, as the system performance improves, the total energy efficiency of the LLC with HLGPC+P1 is enhanced by 7.3%/4.8%/5.6% for 2MB/4MB/8MB LLC compared to *pref. only*.

To gain the insight into how HLGPC+P1 globally balances the aggressiveness of different prefetchers in the CMP system, we conducted a case study on the execution of four applications of "wl4" under a 2MB LLC, as shown in Figure 10. Figure 10(a) shows the individual application performance, which is normalized to that when only one application running exclusively on only one core of the quadcore system. After applying P1, the performances of `art` and `lbm` are substantially improved while that of the other two benchmarks are only slightly improved. After introducing LPC+P1, the performance of `art` and `astar` are further improved but those of `lbm` and `zeusmp` slow down a little bit. The microscope analysis in Figure 10(b) shows that after applying LPC+P1, `art` and `zeusmp` experience noticeable prefetch de-aggressiveness that relieve the LLC access conflicts. Simulations also show that the prefetch accuracy in these two applications are low, i.e., 34% for `art` and 46% for `zeusmp` in LPC+P1, while the volumes of prefetch requests are high. Therefore, HLGPC+P1 is able to improve the system performance by further throttling the prefetch aggressiveness and mitigating the LLC access conflicts. As shown in Figure 10(b), HLGPC+P1 significantly throttles the prefetch aggressiveness of `art` and `zeusmp` while moderately throttles the prefetch agggressiveness of `lbm`. However, the performance improvement of HLGPC+P1 in `lbm` is limited as the prefetch accuracy is high (e.g., 91% in LPC+P1). `astar` is a memory non-intensive application with extremely low LLC access frequency and contention. Hence, HLGPC+P1 actually degrades the performance of `astar` because of the conservative prefetch

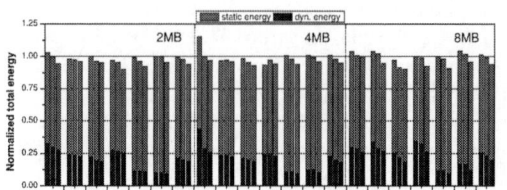

Fig. 9: Total energy consumptions of different prefetch control schemes. From left to right for each workload: P1, LPC+P1, HLGPC+P1.

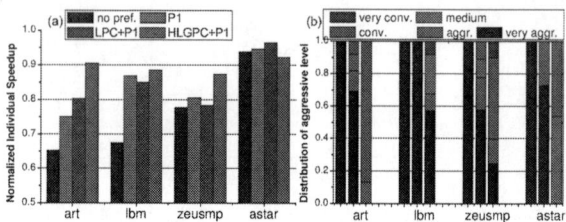

Fig. 10: A case study of workload 4: (a) Individual speedup; (b) The distribution of prefetch aggressiveness, from left to right is P1, LPC+P1 and HLGPC+P1, respectively.

aggressiveness.

VII. CONCLUSION

In this work, we identify the adverse impact of long write access latency of STT-RAM LLC on the prefetching efficacy in CMP systems. Two orthogonal and complimentary techniques, *request prioritization* (RP) and *hybrid local-global prefetch control* (HLGPC), are proposed to alleviate the prefetch-induced LLC access conflicts and improve the processing efficiency for LLC access requests. Our simulation results on a quadcore system show that combining RP and HLGPC, the system performance can be improved by 9.1%, 6.5%, and 11.0% for 2MB, 4MB, and 8MB STT-RAM LLCs, respectively, compared to the design without any LLC request prioritization or prefetch control. The corresponding LLC energy consumption is also saved by 7.3%, 4.8%, and 5.6%, respectively.

REFERENCES

[1] J. Doweck. Inside intel core microarchitecture and smart memory access. *Intel White Paper*, 2006.

[2] E. Ebrahimi et al. Coordinated control of multiple prefetchers in multi-core systems. In *MICRO*, pages 316–326, 2009.

[3] E. Ebrahimi et al. Techniques for bandwidth-efficient prefetching of linked data structures in hybrid prefetching systems. In *HPCA*, pages 7–17, 2009.

[4] E. Ebrahimi et al. Prefetch-aware shared resource management for multi-core systems. In *ISCA*, pages 141–152, 2011.

[5] G. Sun et al. A novel architecture of the 3d stacked mram l2 cache for cmps. In *HPCA*, pages 239–249, 2009.

[6] J. Tendler et al. Power4 system microarchitecture. *IBM Technical White Paper*, Oct. 2001.

[7] M. Hosomi et al. A novel non-volatile memory with spin torque transfer magnetization switching: Spin-ram. In *IEDM*, pages 459–462, 2005.

[8] S. Eyerman et al. System-level performance metrics for multiprogram workloads. *IEEE Micro*, 28(3):42–53, 2008.

[9] S. Srinath et al. Feedback directed prefetching: Improving the performance and bandwidth-efficiency of hardware prefetchers. In *HPCA*, pages 63–74, 2007.

[10] S. Thoziyoor et al. A comprehensive memory modeling tool and its application to the design and analysis of future memory hierarchies. In *ISCA*, pages 51–62, 2008.

[11] W.-F. Lin et al. Filtering superfluous prefetches using density vectors. In *ICCD*, 2001.

[12] W. Wen et al. PS3-RAM: A fast portable and scalable statistical STT-RAM reliability analysis method. In *49th DAC*, pages 1187–1192, June 2012.

[13] W. Wen et al. Loadsa: A yield-driven top-down design method for STT-RAM array. In *ASP-DAC 2013*, pages 291–296, 2013.

[14] X. Dong et al. Circuit and microarchitecture evaluation of 3d stacking magnetic ram (mram) as a universal memory replacement. In *DAC*, pages 554–559, 2008.

[15] X. Dong et al. Nvsim: A circuit-level performance, energy, and area model for emerging nonvolatile memory. *IEEE TCAD of ICS*, 31(7):994–1007, 2012.

[16] X. Wu et al. Hybrid cache architecture with disparate memory technologies. In *ISCA*, pages 34–45, 2009.

[17] X. Zhuang et al. A hardware-based cache pollution filtering mechanism for aggressive prefetchers. In *ICPP*, 2003.

[18] Y. Chen et al. On-chip caches built on multilevel spin-transfer torque RAM cells and its optimizations. *J. Emerg. Technol. Comput. Syst.*, 9(2):16:1–16:22, May 2013.

[19] MacSim. http://code.google.com/p/macsim/.

978-1-4799-2817-0/14 $31.00 © 2014 IEEE

CNPUF: A Carbon Nanotube-based Physically Unclonable Function for Secure Low-Energy Hardware Design

S. T. Choden Konigsmark, Leslie K. Hwang, Deming Chen, Martin D. F. Wong

Department of Electrical and Computer Engineering, University of Illinois at Urbana-Champaign
{konigsm2, lkhwang, dchen, mdfwong}@illinois.edu

Abstract – Physically Unclonable Functions (PUFs) are used to provide identification, authentication and secret key generation based on unique and unpredictable physical characteristics. Carbon Nanotube Field Effect Transistors (CNFETs) were shown to have excellent electrical and unique physical characteristics and are promising candidates to replace silicon transistors in future Very Large Scale Integration (VLSI) designs. We present Carbon Nanotube PUF (CNPUF), the first PUF design that takes advantage of unique CNFET characteristics. CNPUF achieves higher reliability against environmental variations and increased resistance against modeling attacks. Furthermore, CNPUF has a considerable power and energy reduction in comparison to previous ultra-low power PUF designs of 89.6% and 98%, respectively. Additionally, CNPUF allows power-security tradeoff.

I. INTRODUCTION

Modern life is heavily surrounded by electronics. Not only companies' valuable and confidential assets are stored and managed by technology but also our daily lives are connected with technology. Therefore, our privacy and confidential assets are vulnerable to attacks against the technologies we use. This trend leads to an increased interest in security. In addition to these common security concerns, wireless sensor networks and wearable technology have emerged as trends in new devices and can pose significant security risks for our society. Nodes in a sensor network are typically exposed to the public and can contain or handle sensitive data, e.g. power grid information or military defense mechanisms [1]. Wearable technology is emerging as part of ubiquitous computing and may accumulate as much information as the actual wearer, which represents a threat against privacy. Due to the nature of these new devices, they do not only require higher security and privacy, but are also critically limited in circuit area, power, and energy budgets. This trend was already observed in current mobile devices, such as smartphones or tablets, but wearable technology strengthens these constraints [2].

Software security typically assumes correctness and security of hardware and can only discover hardware based intrusions on a very limited scale [3]. Hardware security provides the building block for secure devices and aims to reduce hardware vulnerability to imaging, probing and intrusion. It is generally designed to take advantage of each chip's unique physical aspects.

Gassend et al. introduced the concept of silicon based physically unclonable functions (PUFs)[4], which has gained attention as an emerging hardware security technology. It maps a digital input, considered to be a challenge, to a digital output, defined as the response, based on intrinsic physical parameters of the circuit. Therefore, the mapping between the input and output of a PUF is called the challenge-response behavior.

A major advantage of PUF is the fast and simple response generation, but extremely difficult challenge-response prediction and duplication. Although manufactured instances of PUF share an identical design, the manufacturing process introduces unpredictable variations to the intrinsic physical parameters of the chip. Along with the large size of the challenge-response space, this leads to nearly impossible challenge-response behavior replication [5].

Various PUF designs were studied using different physical parameters of the device; silicon PUF [4], arbiter PUF [6], ring-oscillator PUF (RO-PUF) [7], butterfly PUF [8], clock PUF [9], and low-power current-based PUF [10] are examples of PUF designs taking advantage of very different circuit characteristics. The first silicon PUF, presented by Gassend et al., uses the delay of wires and digital logic devices within one circuit. RO-PUF is also designed to evaluate and compare inherent delay characteristics of wires and transistors, but compares distinct circuits. It is based on ring oscillators and uses the unique oscillation frequency for response generation. Butterfly PUF is based on FPGA-specific physical variations. ClockPUF is designed using the clock skew at the sink of the clock network. Ultra-low power current-based PUF converts analog current variations to unique digital quantities.

Critical characteristics of PUF are reliability and uniqueness. The former is measured as the reverse of the average hamming distance of a single chip under varying environment conditions and the same challenge. The latter is the average hamming distance between multiple manufactured instances of the same design, and is desired to be 50%.

Most of the existing PUFs focus on conventional silicon devices, and several of them are not geared towards low power operation. Furthermore, as technology moves forward, silicon devices for low-power, high-speed applications are facing a miniaturization bottleneck. Carbon-based structures, such as carbon nanotubes (CNT), are one of the promising emerging technologies that are considered as possible replacements for current silicon technology. Moreover, CNTs have great potential in flexible or wearable electronics [11].

In this work, we present a novel carbon-nanotube based PUF (CNPUF), which uses carbon nanotube field-effect transistors (CNFET). Our contributions and the advantages of CNPUF are as follows:

- A PUF design that takes advantage of its unique CNFET characteristics and actively uses metallic CNTs, which are currently inevitable but typically considered a major issue for digital designs.
- Considerable reduction in footprint by using CNFET-unique properties to reduce the transistor count.
- Extremely low power and energy consumption that is 89.6% and 98% lower than ultra-low power current based PUF [10] at 90 nm.

978-1-4799-2817-0/14 $31.00 © 2014 IEEE

- Very high reliability against environment variations and SPICE-accurate experimental evaluation in two different settings.
- An extended design that enables a power-security tradeoff, highly relevant for practical usage scenarios.
- Evaluation of PUF behavior with regard to different CNT technology parameters

The rest of this paper is structured as follows. In section II, we provide background knowledge and explain CNT behavior and characteristics. In section III, we propose CNPUF and theoretically evaluate it. An extension of CNPUF, which can be used for high security applications, is presented in section IV. A SPICE accurate experimental evaluation and comparison is provided in section V. Finally, we summarize our findings and give an outlook in section VI.

II. BACKGROUND

Carbon nanotubes (CNT) are cylindrical carbon molecules that have superior electrical, mechanical and thermal properties [12]. Thus, CNT technology is considered as one of the potential candidates for future electronics [12]. CNFETs, first introduced by S. Tans el al. [13], are transistors with channels consisting of CNTs instead of bulk silicon. Conventional methods in technology scaling will likely encounter physical limitations and these molecular electronics have attracted much interest as they can lead to further technology scaling.

Regardless of the superior properties, there exist fundamental limitations and obstacles in fabrication of CNTs. Due to intense research studies, yield and performance of CNFETs are fast improving and they will be realized as digital circuits in the near future [14]. Most recently, a first fully CNT based subsystem was presented [15], and a first digital Carbon Nanotube Computer was created [16]. Nevertheless, it is still impossible to guarantee perfect alignment, semiconducting property, and uniform distribution, which lead to performance variations [17]. The major CNT variations [18] are: (i) chirality, which defines the type to metallic or semiconducting; (ii) diameter; (iii) growth density (iv) alignment; (v) doping concentration.

Some of these variations correlate with one another and all of them result in electrical property variation and can lead to malfunction of digital circuits in the worst case. Particularly the lack of chirality (and thus type) control is a major issue for CNT usage in digital circuits, as metallic CNTs in transistors lead to direct drain-to-source shorting. Due to the lack of precise chirality control, removal of metallic CNTs became necessary [19]. However, removal of undesired CNTs can lead to further circuit variations and even malfunctioning. For CNPUF design, we inherently consider this CNT specific property and take advantage of the type variation for extremely high efficiency digital secret generation.

The CNT chirality is a pair of indices (n, m) and represents the 2-dimensional wrapping of graphene and determines the type to either metallic or semiconducting [20]. Based on the chirality, CNTs are categorized into three categories. If $n = m$, it is called an *armchair* nanotube. Another structure is *zigzag* nanotube with $m = 0$. Otherwise, it is a *chiral* nanotube. For a given chirality (n, m), a CNT is metallic when it satisfies either of following two cases.
1) $n = m$
2) $n - m = 3N$ for any $N \in \mathbb{N}$

Metallic and semiconducting CNTs impact differently on different operating modes. When the CNFET is turned off, the metallic CNTs are still conducting, as they lead to a direct gate-source shorting. However, both metallic and semiconducting CNTs contribute when the transistor is on. Hence, the ratio of the semiconducting to metallic CNT is closely related to the on and off current (I_{on}, I_{off}) ratio. By utilizing this CNT specific characteristic, our CNPUF design provides a simple, but unconventional and extremely energy efficient security solution.

III. CARBON-NANOTUBE PUF

In this section, we will explain the CNFET based PUF design. Subsection A gives an overview of the basic design and explains the challenge-response behavior. Then, the main internal block of the CNPUF design, the CNPUF Parallel-Element (CNPUF-PE), will be discussed in subsection B. An analysis of area cost is provided in subsection C. In subsection D, we will elaborate on the design characteristics responsible for providing high reliability, which is experimentally shown in section V. Subsection E shows the complexity of the PUF design and the dimensions that affect the challenge-response behavior, which leads to resistance against modeling attacks.

A. Basic Design

The design of the CNPUF is shown in Fig. 1. We define a CNPUF Parallel-Element (CNPUF-PE) as a pair of CNFETs that share the same gate voltage, which is an input to the CNPUF-PE. These inputs form the challenge. Each bit of the challenge is associated with a single CNPUF-PE, by directly providing the input challenge as the gate voltage. A high gate voltage corresponds to logic 1, and a low gate voltage corresponds to logic 0.

Each of the CNPUF-PE has two distinct states, one for a high input gate voltage and one for a low voltage. These two states differ among all CNPUF-PEs because of the static variations caused by the manufacturing process. A detailed explanation of the variation sources is provided in subsection E. The output bit of CNPUF is generated by comparing two currents through a series connection of CNPUF-PEs. In Fig. 1, when $I_1 > I_2$, the output is 1; otherwise, it is 0.

CNPUF can be used in different configurations to achieve multi-bit responses: The simplest approach is to replicate and parallelize a one-output-bit CNPUF to achieve multiple output bits. While this appears costly, each output bit is truly generated by different physical circuits and thus they are fully independent of each other. An area saving alternative to this approach can be achieved by reusing a one-bit CNPUF with a Pseudo Random Number Generator (PRNG) as a challenge translator. However, the output bits are no longer independent and the area saving can be very limited, as the CNPUF can be implemented with a small number of transistors, whereas a multi-bit PRNG can introduce high area overhead.

B. CNPUF Parallel-Element

The CNPUF mainly consists of a serial connection of CNPUF-PEs. Internally, the CNPUF-PE consists of two parallel CNFETs that share the same gate voltage. Each CNFET consists of a large number of semiconducting CNTs and few metallic CNTs, typically with a metallic-to-semiconducting ratio between 10% and 33% [3]. Due to this difference, a CNPUF-PE has two distinct and nearly independent states for a high gate voltage and a low gate voltage. The current characteristics for a low gate voltage are

978-1-4799-2817-0/14 $31.00 © 2014 IEEE

dominated by the metallic CNTs, as the off-current for semiconducting CNTs is considerably lower than the current for metallic CNTs, even when the number of semiconducting CNTs is large. For a high gate voltage, the semiconducting CNTs dominate the current characteristics due to their much larger number. As CNT technology improves, we expect the ratio of semiconducting to metallic CNTs to increase, and thus the correlation between both states to further reduce.

C. Area comparison

In practice, the number of challenge bits is larger than 128 Bit. In this case, the area cost per bit of a one-output-bit CNPUF, $A_{CNPUF,bit}$, is approximately the area of a CNPUF-PE and thus two times the area of a transistor, $A_{Transistor}$:

$$A_{CNPUF,bit} \cong 2 * A_{Transistor}.$$

This compares favorably to the basic implementation of the Arbiter PUF [1], which requires two multiplexers per input bit (assuming that the challenge bit length is large enough to neglect the Arbiter). Using a transmission gate implementation, the area cost per bit of a basic Arbiter PUF, $A_{A-PUF,bit}$, is

$$A_{A-PUF,bit} \cong 2 * A_{MUX} \cong 8 * A_{Transistor}.$$

Here, A_{MUX} is the area of a 2:1 multiplexer (or MUX).

The area advantage of CNPUF is even larger when compared with a Ring Oscillator PUF (RO-PUF) [2]: A RO-PUF requires at least $2^N - 1$ Ring Oscillators (ROs) for N input bits. Considering the usage of very small ROs consisting of only three inverters, the cost of a RO-PUF per bit $A_{RO-PUF,bit}$ depends on the challenge length N and is

$$A_{RO-PUF,bit} \cong \frac{(2^N - 1)}{N} * 6 * A_{Transistor}.$$

D. High reliability

The reliability of PUF designs is typically evaluated with regard to variations in environmental parameters, such as the temperature, and variations in the operating conditions, e.g. the supply voltage. As explained in section II, CNTs have exceptional electrical characteristics and are known to have high stability under environment variations. The reliability of our proposed CNPUF builds on the highly stable characteristics of CNTs, but has two additional features that support reliability:

- Strong impact of physical variations
- Regular design

The physical variations are large and have a strong impact on the PUF challenge-response behavior. Particularly the ratio variations between semiconducting and metallic CNTs have a very strong impact on the circuit behavior. Our design takes advantage of the difficulty to control CNT chirality to create a secret by comparing the current through a series connection of CNPUF-PE. Since metallic and semiconducting CNTs are randomly mixed during the fabrication process, the exact ratio of metallic and semiconducting CNTs for each FET is not predictable and can be treated as a random number.

Except for the comparator, our design is very regular and therefore, the dynamic effects in the upper and lower path can average out. Only the comparator has to be specifically designed to be resilient to dynamic variations such as temperature changes or voltage peaks. In comparison, a Ring Oscillator PUF has MUXes, counters and also a comparator

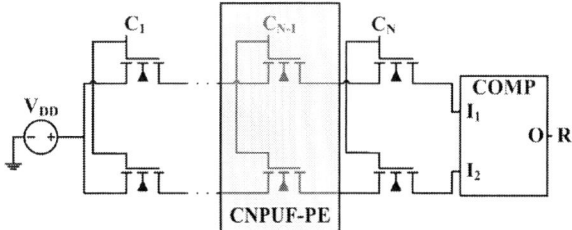

Fig. 1. Row of the CNPUF design. A series of CNPUF-PE are evaluated by a comparator (COMP) to generate the output bit.

that have to be tuned, because they affect both paths and therefore have to treat these paths equally. This can be difficult in practice, as the variations on different metal layers and in different chip regions can be different, so that the number of circuit elements that require fine tuning has to be minimized.

E. Resistance against Modeling Attacks

The proposed CNPUF design takes advantage of inherent CNT properties and variations to represent a complex challenge-response behavior to modeling attacks. Some of this is due to the many varying physical factors as described in Section II. Even more important is that these varying physical properties can have a different level of impact on the challenge-response behavior. Therefore, a model based attack has to accurately identify and mirror different physical characteristics. Due to an increase in this dimension, CNPUF has a high resistance against such attacks. However, the basic implementation of CNPUF has independent output bits and thus does not take full advantage of the existing circuitry. Furthermore, some applications, such as secret key generation, require very high randomness and must have an even more complex challenge-response-behavior. Therefore, we also present an extended version of CNPUF that introduces a power vs. security tradeoff and has area overhead to create a feedback based design with even higher modeling complexity.

IV. EXTENDED CNPUF

A. Extended Design

While the basic CNPUF has many desirable properties, such as low power consumption and minimal area requirements, it is of static nature and doesn't allow fine-tuning by the designer to specific application needs. Therefore, we also present the extended CNPUF (ex-CNPUF), which is shown in Fig. 2. As a basic element, the ex-CNPUF contains one basic CNPUF per output bit, shown in Fig. 1. However, ex-CNPUF buffers each bit response and feeds it back into the CNPUF-PE elements through an XOR-element with the original challenge. This may be repeated a specific number of times for each challenge, depending on the application specific requirements. As a result of this response feedback, the complexity at the bit level increases, which further improves the resistance against modeling attacks. Furthermore, the output bits will no longer be independent of each other, a feature that drastically increases the design complexity a modeling attack has to consider.

The additional area overhead compared to the basic CNPUF is one XOR gate per CNPUF-PE element. Considering that the CNPUF-PE was designed for minimal area and consists of only two transistors, the area overhead of the extended design is approximately 300%. The area of ex-CNPUF is:

$$A_{ex-CNPUF} \approx 8A(T)$$

Although the area overhead over CNPUF is not negligible, ex-CNPUF still compares favorably to several PUF implementations. It has approximately the same area as a simple Arbiter PUF and less area than an RO-PUF. However, the greatest advantage is the power-security tradeoff, which is explained in subsection B.

B. Power-Security tradeoff

As shown in subsection A, ex-CNPUF brings several advantages in flexibility and security, but introduces area overhead relative to the basic CNPUF design. The design of ex-CNPUF is very feasible for high-security applications that require more modeling resistance or higher randomness than what CNPUF can provide. By increasing the number of ex-CNPUF iterations, the complexity and randomness of the response increases. This comes at the cost of higher energy consumption and reduced reliability. The reduced reliability is a result of error propagation within the PUF and the fact that even a single bit-error in the challenge can profoundly change the response, if no error correction schemes are used.

Note that the area tradeoff is static and has to be made at design time, but the power tradeoff can be dynamically adjusted. Thus, ex-CNPUF can power a security interface that provides different degrees of resistance for different domains.

V. EXPERIMENTAL EVALUATION

A. Simulation Setup

For the experimental evaluation of CNPUF, we simulated the design in HSPICE[1] in a Linux environment. To simulate the CNFETs contained in CNPUF, we employed the Stanford CNFET HSPICE model [21], [22]. For standard logic and comparison purposes, we employed the Predictive Technology Model (PTM)[23]. Our CNT simulation is based on zigzag structure with a nominal chirality of metallic nanotubes of $(n, m) = (10, 0)$ and a nominal chirality of semiconducting nanotubes of $(9, 0)$.

As a solution to the lack of support for metallic CNTs in the HSPICE model, we simulate real CNFETs by splitting them into one ideal metallic and one ideal semiconducting CNFET, as shown in Fig. 3. The metallic CNFET is modeled by assigning the appropriate chirality and setting the gate voltage to always-on, independent of the challenge. Note that this separation is solely for the purpose of experimental evaluation through simulations; one metallic CNFET and one semiconducting CNFET in the simulation model represent a single transistor. In this regard, our design does not require ideal or pure-semiconducting CNFETs, but instead takes advantage of metallic CNTs.

For this experimental evaluation, we implemented an 8-Bit input/output CNPUF and analyzed reliability, inter-chip variability, and power consumption. A small design was chosen, such that a large number of SPICE-accurate simulations can be performed to provide detailed insight into the reliability under environment variations. As a proof of concept, we have also evaluated a larger design of 128-bits and observed in a smaller number of simulations that the behavior is very similar to the 8-bit implementation.

---
[1] HSPICE Version E-2010.12-SP2 32-BIT

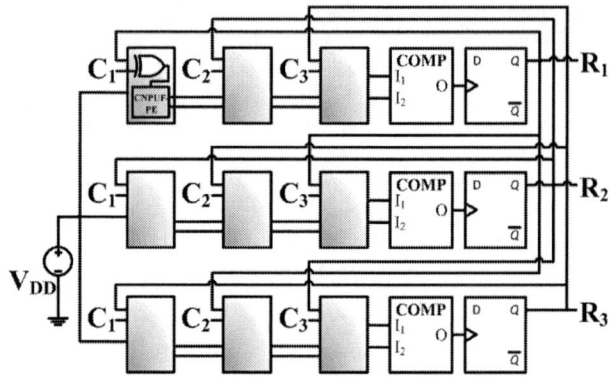

Fig. 2. Design of the extended CNPUF (ex-CNPUF) to allow a power vs. security tradeoff and more complex challenge-response behavior.

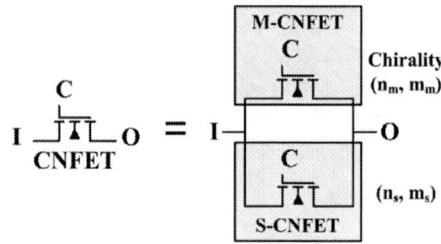

Fig. 3. Simulation model for the metallic CNTs in the CNFETs used by CNPUF.

B. Reliability

Reliability is one of the main criteria for PUF quality and quantifies the capability to repeatedly and consistently produce the same challenge-response behavior. As explained in section II, CNTs generally have a high reliability that is further improved by specific design measures described in section III.D. To experimentally validate the reliability, CNPUF is evaluated under a standard simulation environment that is used in most literature, and a simulation environment with more dynamic variations to simulate real circuit performance.

The standard simulation environment is similar to that employed in several publications [10], [24] and has the following characteristics:

- *Temperature variations*: The temperature of the whole circuit is evaluated at specific temperature points or at random temperature values.
- *Supply voltage variations*: The supply voltage of the whole circuit is varied. This is implemented by having a common random voltage variation with $\sigma_{V,supply}$ at every voltage source (including challenges).

To provide a stronger comparison with implemented PUF circuits, we also propose an extended simulation environment with the following parameters:

- *Dynamic temperature*: In addition to the static temperature variation, we model dynamic temperature variation by adding a different random temperature T_{rand} to the static temperature for each simulation. Therefore, we compare challenges that were acquired at different temperatures.
- *Local voltage variation*: Simulation of dynamic local voltage variation by introducing a local variation

$\sigma_{V,local}$ at each voltage source and at each gate in addition to the common voltage variation.

This is particularly relevant to real physical devices, as there is additional circuitry on chip besides the PUF that can influence the power and temperature with different levels of activity. The experimental data for both simulation environments is shown in Fig. 4. The detailed averages for each ratio are provided in Table I. The range for these parameters is shown in Table II. For the standard environment, we conducted over 3500 HSPICE simulations and evaluated the intra-chip hamming distance to:

$$HD_{Intra,std} = 1.9\%$$

In the extended simulation environment for accurate comparison against actual circuit implementations, the intra-chip hamming distance was determined from more than 6000 HSPICE simulations:

$$HD_{Intra,ext} = 3.5\%$$

The reported data was gathered for three different ratios of metallic CNTs to semiconducting CNTs, as the ratio between metallic and semiconducting CNTs can have different nominal values. This ratio is typically between 10% and 33%, therefore we provide evaluations with 10%, 20%, and 30% ratios and show that CNPUF is viable in the whole range. The graph further shows that a higher metallic ratio actually leads to a slightly higher reliability. This is an effect of increasing dominance of the metallic CNTs, as their share becomes larger. To achieve a well-balanced design, we therefore propose the usage of CNPUF with a nominal metallic ratio of around 20%, which can be achieved without any breakdown in current technology.

Due to simulation complexity, ex-CNPUF was only evaluated at a 20% metallic ratio. As expected, the reliability slightly decreases as the complexity of the design increases. However, the intra-chip hamming distance is still well below the 10%, typically considered a limit for error correction [9], and competitive with other PUF designs

C. Inter-chip variability

For security applications, different PUF instantiations require sufficiently different challenge-response behavior, so that the behavior of one PUF instance may not be inferred through ownership of another. Ideally, all pairs of responses from different instances would share 50% of their output bits on average. As a metric for this variability between physical instances, the inter-chip hamming distance, HD_{inter}, is used.

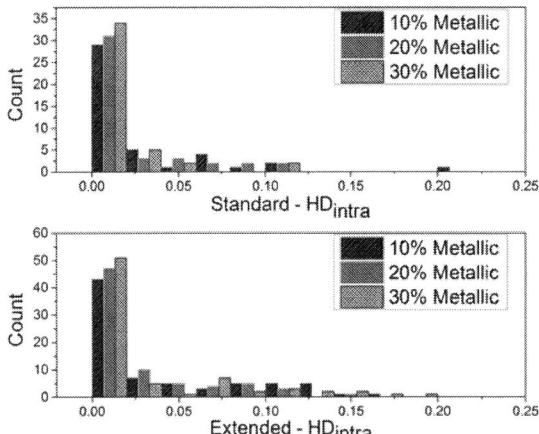

Fig. 4. Robustness of CNPUF in a standard simulation environment (top) and in an extended simulation environment (bottom).

TABLE II
SIMULATION PARAMETERS FOR CNPUF

Parameter	Range
Temperature T	$-20°$ to $80°C$
Dyn. Temp. T_{rand}	$0°$ to $20°C$
Voltage variation	$\mu = 0.8V$ $3\sigma_{supply} = 22.5\%$ $3\sigma_{dynamic} = 7.5\%$
CNT ratio variation	$\mu_{ratio} = \{0.1, 0.2, 0.3\}$ $3\sigma_{phy} = 22.5\%$
Channel length variation	$\mu_{channel} = 14nm$ $3\sigma_{channel} = 7.5\%$

For the generation of HD_{inter}, we created 10 groups of 10 PUF instances. To each group of PUF, we issued 100 randomly generated challenges. The average HD_{inter} of CNPUF is 49.67% and therefore very close to perfect (50%).

D. Power consumption

By using intrinsic CNFET-unique properties, such as the metallic-semiconducting ratio, CNPUF allows secret-key generation at a very low cost. An area comparison was provided in section III.C and showed that CNPUF requires less logic than other PUF designs. These advantages combine to greatly reduce the power and energy consumption, as shown in Table V. Based on SPICE-accurate simulations, we report that CNPUF achieves the highest power and energy efficiency to the best of our knowledge and reduces the power consumption per bit to $1.26\mu W$ and energy consumption to $0.0334fJ/bit$. We compare CNPUF with ultra-low power current based PUF [10] at 90nm and at 14nm. According to data provided by the authors, CNPUF reduces power by 89.6% and energy by 98% when implemented in 90nm technology. For comparison purposes, we reimplemented [10] into 14nm technology and conducted power and energy measurement under ideal conditions. At this technology node, CNPUF reduces the power by 94.75% and energy by 72.16%. However, it is very likely that the quality of PUF designs that require sub-threshold operation, e.g. [10], will degrade at smaller technology nodes, as susceptibility to environment variations greatly increases.

TABLE I
EXPERIMENTAL RESULTS FOR RELIABILITY

Nominal Metal Ratio	Intra Chip Hamming Distance HD_{intra}			
	Basic CNPUF		Ex-CNPUF	
	Std Env	Ext Env	Std Env	Ext Env
10%	0.025	0.040	-	-
20%	0.019	0.034	0.045	0.05
30%	0.012	0.030	-	-
Total Average	0.019	0.035	0.045	0.05

TABLE III
COMPARISON OF HD_{intra} IN DIFFERENT SIMULATED PUF DESIGNS. LOWER PERCENTAGES MEAN HIGHER ROBUSTNESS.

CNPUF	SCANPUF[24]	ROPUF[9]	CLOCKPUF[9]	CURRENT PUF[10]
1.9%	5%	9.51%	5.07%	~3%

TABLE IV
COMPARISON OF HD_{intra} BETWEEN REAL PUF CIRCUITS AND CNPUF UNDER EXTENDED ENVIRONMENT SIMULATION.

CNPUF	BUTTERFLY PUF [8]	SRAM-PUF [25]	SCANPUF [24]
3.5%	6%	~8%-18%	3.2% (*)

TABLE V
POWER AND ENERGY COMPARISON BETWEEN CNPUF AND ULTRA -LOW POWER CURRENT-BASED PUF [10] AT 14NM AND 90NM.

Designs	CNPUF		Current based PUF [10]	
Technology	90nm, 1.2V	14nm, 0.8V	90nm	14nm, 0.8V
Power	15.6µW/bit	1.26µW/bit	150µW/bit	24µW/bit
Delay	43ps	26.5ps	250ps	~5ps
Energy	0.67fJ/bit	0.0334fJ/bit	37.5fJ/bit	0.12fJ/bit

E. Overall comparison with other PUF designs

In Table III, we compare the reliability of CNPUF against other PUF designs with simulated results and show that CNPUF can outperform them. In Table IV, the evaluation under extended environment conditions of CNPUF is compared against physical implementations of other PUF designs. Note that the authors only evaluated ScanPUF at a single temperature, which reduces the comparability, as all other designs were evaluated at a wide range of temperatures. The numbers show that in addition to a considerable reduction in area and power consumption that we previously showed, CNPUF can also achieve higher reliability. The inter-chip distance of all PUF designs, including ours, is very comparable and close to the desired 50%.

VI. CONCLUSION AND OUTLOOK

We presented a PUF design based on intrinsic physical variations of CNTs. It takes advantage of the metallic to semiconducting CNT ratio in CNFETs to increase reliability, while strongly reducing the average power consumption and energy usage per bit. CNPUF was experimentally evaluated with SPICE-accurate simulations and showed strong results for security relevant properties such as reliability and inter-chip distance. Furthermore, we presented and evaluated an extension of CNPUF that allows a power vs. security tradeoff for dynamic usage in high security circuits.

CNPUF and ex-CNPUF provide the future basis for authentication and secret key generation by offering security at a very low area and power cost. This can open the field of PUF for a variety of new applications and is especially relevant for current research areas such as wireless sensor networks or ubiquitous computing.

REFERENCES

[1] I. F. Akyildiz, W. Su, Y. Sankarasubramaniam, and E. Cayirci, "Wireless sensor networks: a survey," *Comput. networks*, vol. 38, no. 4, pp. 393–422, 2002.

[2] T. Starner, "The challenges of wearable computing: Part 2," *Micro*, vol. 21.4, no. IEEE, pp. 54–67, 2001.

[3] R. Lee, S. Sethumadhavan, and G. E. Suh, "Hardware enhanced security," *Proc. 2012 ACM Conf. Comput. Commun. Secur. - CCS '12*, p. 1052, 2012.

[4] B. Gassend, D. Clarke, M. van Dijk, and S. Devadas, "Silicon Physical Random Functions," *Proc. 9th ACM Conf. Comput. Commun. Secur.*, 2002.

[5] R. Nithyanand and J. Solis, "A Theoretical Analysis: Physical Unclonable Functions and the Software Protection Problem," *2012 IEEE Symp. Secur. Priv. Work.*, pp. 1–11, May 2012.

[6] D. Lim, J. W. Lee, B. Gassend, G. E. Suh, M. Van Dijk, and S. Devadas, "Extracting Secret Keys From Integrated Circuits," *IEEE Trans. Very Large Scale Integr. Syst.*, vol. 13, no. 10, pp. 1200–1205, 2005.

[7] G. E. G. Suh and S. Devadas, "Physical Unclonable Functions for Device Authentication and Secret Key Generation," in *44th ACM/IEEE Design Automation Conference*, 2007.

[8] S. S. Kumar, J. Guajardo, R. Maes, G. Schrijen, and P. Tuyls, "Extended Abstract: The Butterfly PUF Protecting IP on every FPGA," IEEE International Workshop on Hardware-Oriented Security and Trust, 2008.

[9] Y. Yao, M. Kim, J. Li, I. Markov, and F. Koushanfar, "ClockPUF: Physical Unclonable Functions based on Clock Networks," in *Design, Automation & Test in Europe*, 2013.

[10] M. Majzoobi, G. Ghiaasi, F. Koushanfar, and S. R. Nassif, "Ultra-low power current-based PUF," *2011 IEEE Int. Symp. Circuits Syst.*, pp. 2071–2074, May 2011.

[11] S. Selvarasah, "Ultrathin and highly flexible parylene-c packaged carbon nanotube field effect transistors," 2010.

[12] M. Zhang and J. Li, "Carbon nanotube in different shapes," *Mater. Today*, vol. 12, no. 6, pp. 12–18, Jun. 2009.

[13] S. J. Tans, A. R. M. Verschueren, and C. Dekker, "Room-temperature transistor based on a single carbon nanotube," *Nature*, vol. 672, no. 1989, pp. 669–672, 1998.

[14] H. Park, A. Afzali, S.-J. Han, G. S. Tulevski, A. D. Franklin, J. Tersoff, J. B. Hannon, and W. Haensch, "High-density integration of carbon nanotubes via chemical self-assembly.," *Nature. Nanotechnology.*, vol. 7, no. 12, pp. 787–91, Dec. 2012.

[15] M. Shulaker, J. Van Rethy, G. Hills, H. Chen, G. Gielen, H. Philip, and S. Mitra, "Sacha: the Stanford Carbon Nanotube Controlled Handshaking Robot," in *Proceedings of the 50th Annual Design Automation Conference*, 2013.

[16] M. M. Shulaker, G. Hills, N. Patil, H. Wei, H.-Y. Chen, H.-S. P. Wong, and S. Mitra, "Carbon nanotube computer," *Nature*, vol. 501, no. 7468, pp. 526–30, Sep. 2013.

[17] J. Zhang, A. Lin, N. Patil, H. Wei, L. Wei, H. S. Wong, S. Mitra, "Robust Digital VLSI using Carbon Nanotubes," IEEE Transactions on Computer-aided Design of integrated circuits and systems, 31.4, 2012.

[18] J. Zhang, "Variation-Aware Design of Carbon Nanotube Digital VLSI Circuits," *Dissertation*, August, 2011.

[19] H. Wei, N. Patil, J. Zhang, A. Lin, H.-Y. Chen, H.-S. Wong, and S. Mitra, "Efficient metallic carbon nanotube removal readily scalable to wafer-level VLSI CNFET circuits," in *VLSI Technology (VLSIT), 2010 Symposium on*, 2010.

[20] J.W.G. Wildoer, L.C. Venema, A.G. Rinzler, R. E. Smalley, C. Dekker "Electronic structure of atomically resolved carbon nanotubes ¨.," *Nature*, vol. 584, no. 10, pp. 1996–1999, 1998.

[21] J. Deng and H.-S. Wong, "A compact SPICE model for carbon-nanotube field-effect transistors including nonidealities and its application—Part I: Model of the intrinsic channel region," *Electron Devices, IEEE Trans.*, vol. 54, no. 12, pp. 3186–3194, 2007.

[22] J. Deng and H.-S. Wong, "A compact SPICE model for carbon-nanotube field-effect transistors including nonidealities and its application—Part II: Full device model and circuit performance benchmarking," *Electron Devices, IEEE Trans.*, vol. 54, no. 12, pp. 3195–3205, 2007.

[23] S. Sinha, G. Yeric, V. Chandra, B. Cline, and Y. Cao, "Exploring sub-20nm FinFET design with predictive technology models," in *Proceedings of the 49th Annual Design Automation Conference*, 2012.

[24] Y. Zheng, A. R. Krishna, S. Bhunia, "ScanPUF: Robust ultralow-overhead PUF using scan chain," *ASP-DAC*. 2013.

[25] C. Bohm, M. Hofer, and W. Pribyl, "A microcontroller SRAM-PUF," *Network and System Security (NSS), 2011 5th International Conference on. IEEE*, 2011.

978-1-4799-2817-0/14 $31.00 © 2014 IEEE

3DCoB: A new design approach for Monolithic 3D Integrated Circuits

Hossam Sarhan, Sebastien Thuries, Olivier Billoint and Fabien Clermidy

CEA-LETI, Minatec Campus, Grenoble, 38054, France.
{hossam.sarhan; sebastien.thuries; olivier.billoint; fabien.clermidy}@cea.fr

Abstract – 3D Monolithic Integration (3DMI) technology provides very high dense vertical interconnects with low parasitics. Previous 3DMI design approaches provide either cell-on-cell or transistor-on-transistor integration. In this paper we present 3D Cell-on-Buffer (3DCoB) as a novel design approach for 3DMI. Our approach provides a fully compatible sign-off physical implementation flow with the conventional 2D tools. We implement our approach on a set of benchmark circuits using 28nm-FDSOI technology. The sign-off performance results show 35% improvement compared to the same 2D design.

I. Introduction

3D integration becomes an emerging technology to overcome the limitations of scaling of the advanced technology nodes [1, 2]. 3D integrated circuits can be fabricated by two main process techniques; parallel integration and sequential integration.

Parallel integration is achieved by fabricating two dies separately then vertically stacking using 3D integration technologies such as 3D Through-Silicon-Vias (TSVs) or 3D interposers [3].

On the other hand, sequential integration (3D monolithic integration) is a methodology to fabricate the 3D IC where a second transistor layer is directly fabricated on top of the first one [4]. The two transistors layers are connected through high-density inter-tier vias (about 100 times smaller compared to the TSVs [2, 4]).

Stacking granularity varies according to the technology used and the performance targeted [5]. Coarse grain integration can be achieved using parallel integration technologies, such as cores stacking [6], memory-on-logic [7] or heterogeneous integration [8]. However fine grain integration (gate-level or transistor-level) requires small vertical interconnects with low RC parasitics which can be achieved using monolithic integration technology [9].

Several previous works have studied design techniques using 3D monolithic technology [10-15]. They have investigated two design approaches using 3D monolithic integration: gate-level integration (cell-on-cell), and transistor-level integration. Cell-on-cell approach is done by assigning the standard cells either in the top tier or the bottom tier. This approach uses the conventional 2D standard cells however it suffers from the incompatibility of

the 2D design tools [13]. On the other hand, the transistor-level integration is achieved by splitting each standard cell between two tiers (NMOS transistors are placed in one tier and PMOS transistors are placed in another tier). Transistor-level integration provides compatible approach with the 2D conventional physical implementation tools but it requires re-designing of the standard cells layout [14].

In this paper, we introduce a new design approach for 3D monolithic integration providing a full compatibility with the conventional 2D sign-off physical implementation flow, in addition to improve the performance compared to the 2D design.

The proposed idea is to split the non-minimum drive standard cell into a logic stage and a driving stage. The logic stage is implemented by the equivalent minimum-drive cell. The driving stage is implemented by a buffer with the same drive as the original cell. The min-drive cell and the driving buffer are vertically stacked which we call 3D Cell-on-Buffer (3DCoB). Using this approach, the minimum-drive logic cell provides the same logical function as the original cell, while the driving buffer guarantees the same driving capability of the cell. We implement our approach on different benchmark circuits using 28nm FDSOI technology.

The paper is organized as following. Section II presents brief introduction monolithic 3D integrated circuits as well as the previously developed design techniques. The 3DCoB approach is introduced in section III. In section IV, the whole design flow is presented. Performance results and conclusion are shown in section V and section VI, respectively.

II. Monolithic 3D ICs

3D Monolithic Integration (3DMI) is the stacking process where a top transistor layer is fabricated sequentially with the bottom one. In [4], it is demonstrated that monolithic 3D integration can be achieved at low temperature process for the top tier (<600°C), providing inter-tier via diameter ~50 nm. The main advantages of 3DMI technology are (i) fine-grain partitioning because of smaller inter-tier vias compared to TSVs, (ii) less parasitics of the vertical interconnects compared to TSVs.

978-1-4799-2817-0/14 $31.00 © 2014 IEEE

Few previous works have studied the design techniques using 3DMI with two design approaches: transistor-level integration (N/P) and gate-level (cell-on-cell) integration. The N/P approach is implemented by splitting the standard cell (digital gate) so that the NMOS transistors are placed on one tier and the PMOS transistors on another tier [12, 14]. The monolithic 3D inter-tier vias are then used as connections inside the cell as shown in Fig.1(a). On the other hand, the cell-on-cell approach is achieved by implementing the whole cell in one tier, either in the top or the bottom tier [13]. In this case, the inter-tier vias are used to connect top and bottom cells if needed, as shown in Fig.1(b). In 3DMI technology, limited number of inter-tier tungsten metal layers can be used for routing the connectivity of the bottom tier [9]. These inter-tier metal layers can be used to reduce the global wiring congestion for the cell-on-cell approach.

The design trade-offs for the 3DMI have been shown in [13]. It compares between the transistor-level design (N/P) and gate-level design (cell-on-cell). The N/P design approach has the advantages of (i) compatibility with 2D design flow (but it requires designing of new 3D transistor-stacked standard cells) (ii) no inter-tier metal layers are needed for routing. The chip area reduction reported is about 30% in 45nm technology node. On the other hand, cell-on-cell design approach has the advantage of using existing 2D standard cells, with the disadvantage of the incompatibility with the conventional physical implementation tools. Also, the cell-on-cell gate-level approach has other design issues such as the need of routing

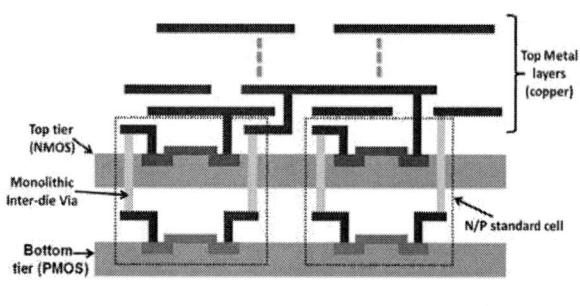

(a)

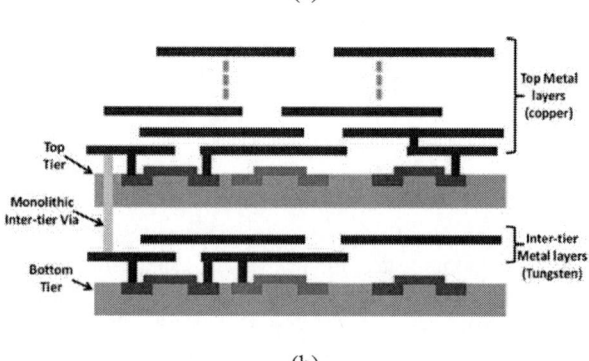

(b)

Fig 1. Monolithic 3D integration approaches, (a) transistor-level (N/P), (b) gate-level (cell-on-cell)

TABLE I
Comparison for different 3DMI approaches

	N/P [14]	Cell-on-Cell [13]	3DCoB (This work)
2D Design flow compatibility	Yes	No	Yes
Using 2D standard cells	No	Yes	Yes
Using inter-tier routing layers	No	Yes	No
Usage of inter-tier Vias	In every cell	Between cells (if req.)	Only in the 3D cells

the bottom tier with the inter-tier metal layers to decrease the global routing congestion which will increase the fabrication costs. In addition, the clock tree balancing between the two tiers in the gate-level approach requires some additional design add-ons [16].

CELONCEL has been introduced to solve the design tool issue for the gate-level approach. It is a 3D placement add-on for the monolithic 3D gate-level integration [15]. The main idea is to generate a virtual library by transforming the initial cells to half-sized cells (DEFLATE), then use these libraries with conventional 2D design flow. After full synthesis, the cells are re-transformed to regain the original size (INFLATE) and then remove the overlapping between cells by assigning different tiers to each overlapped pair of cells.

Table I shows a comparison between the different 3DMI design methodologies. To avoid design issues and tools incompatibility for the gate-level cell-on-cell, or redesigning the standard cells for the transistor-level N/P, we are proposing the 3DCoB to be compatible with the physical implementation tools without the need to redesign the standard cells.

III. 3D Cell-on-Buffer (3DCoB)

As we showed above, the previous monolithic 3D design approaches suffer either from the incompatibility of the conventional design flow or the need of re-designing the standard cell libraries, which affects the time-to-market needed to deliver the 3D blocks gaining from such advanced technology.

The main idea of the 3DCoB is to split the non-minimum drive 2D standard cells into a logical function stage and a driving stage. As a result, one tier contains the equivalent minimum-drive gate to perform the same functionality while the other tier contains a buffer to maintain the same driving capabilities of the cell, as shown in Fig. 2.

The external inputs of the 3DCoB cell are connected to the inputs of the minimum-drive gate, while the cell output is taken from the output of the driving buffer. An internal connection is needed to connect the output of the minimum-drive gate and the input of the driving buffer. The

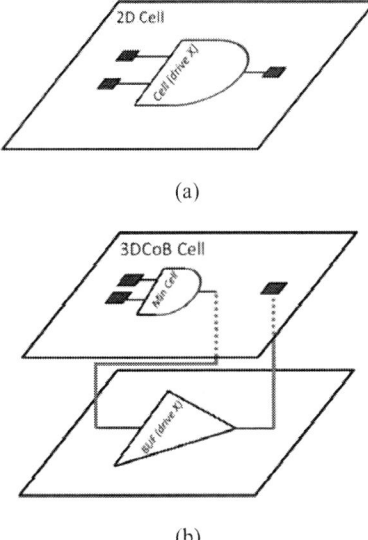

(a)

(b)

Fig 2. Cell-on-Buffer concept, (a) conventional 2D cell, (b) its equivalent 3DCoB cell (min-size cell on equivalent-size buffer)

3DCoB cell has internally two inter-tier 3DMI vias. The first via is to connect the output of the functioning minimum-drive gate to the input of the driving-buffer. The second via is to connect the output of driving-buffer to the output pin of the 3DCoB cell to be routed on the same tier as the input pin. Consequently, 3DCoB cells can be used by the conventional 2D place and route tools.

Another advantage of the 3DCoB approach is decreasing the input gate capacitance of the standard cells. As in the 3DCoB cells the inputs are connected to the minimum-drive gate instead of the original-drive gate, the input gate capacitance in the 3DCoB cells is less than that of the 2D cells. Fig. 3(a) shows the difference in input gate capacitances between 2D and 3DCoB cells for a 2-input AND gate cell at different gate drives in 28nm FDSOI technology. We can notice that the input gate capacitance of the 2D cell is 2.5 times that of 3DCoB cell for the AND2x33 cell. Fig 3(b) shows the input gate capacitances for different cells at 2D and 3DCoB implementations. This reduction of the input gate capacitance will decrease the signal path delay which will increase the overall performance.

Consequently the 3DCoB approach, compared to the previous approaches, can provide, for the design methodology:
1. Full compatibility with the conventional sign-off physical implementation flow.
2. No need for inter-tier routing metal layers between cells.
3. No clock synchronization issue between the two tiers.
And for the performance:
4. Improve the overall performance by decreasing the input gate capacitance of the cells.
5. Increase the cell-driving capability (increase the bottom-tier driving-buffer), with limited increase in the cell area.

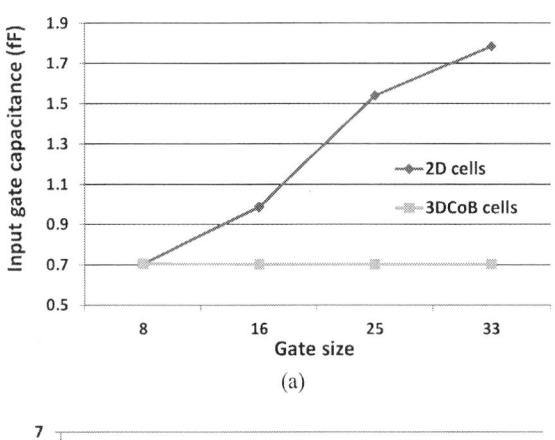

(a)

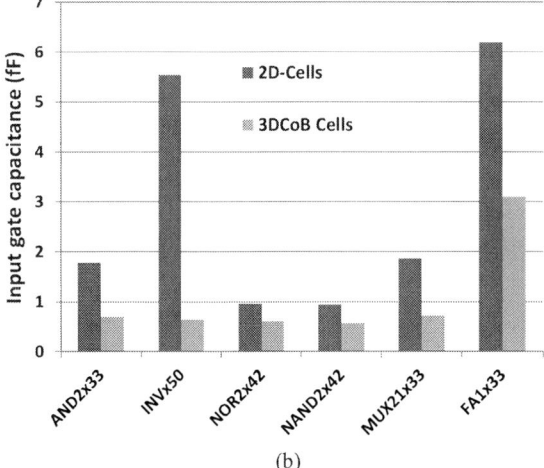

(b)

Fig 3. 2D and 3DCoB input gate capacitances in fF (y-axis) for (a) 2-input AND gate at different gate size and (b) different cell types.

IV. Design Flow

This section introduces the design flow methodology for the 3DCoB approach. First the 3DCoB cell generation is presented, and then area and pin modifications of the new cells are explained. Finally the full design flow is shown. In this work we used commercial tools for the whole design and physical implementation flow.

To generate the 3DCoB cells, we developed an algorithm to create spice netlist of the 3DCoB standard cells (.spi) by connecting the suitable 2D gate cell and buffer, and then we use an existing tool (Cadence Virtuoso Liberate) to generate the 3DCoB libraries (.LIB) from the spice netlist (.spi). To validate the generation of the libraries, we used the original 2D spice netlists (.spi), from the process design kit, and generate a new 2D libraries (.LIB) with the same methodology. The new generated libraries give similar performance as the original foundry libraries.

Algorithm 1 describes the 3DCoB generation process. To create the 3DCoB cells, first we replace the original-drive

978-1-4799-2817-0/14 $31.00 © 2014 IEEE   81

Algorithm 1: 3DCoB cell generation
1   **foreach** cell of the library **do**
2   get current_cell_drive;
3   get current_cell_type;
4   **if** (*current_cell is the min_drive cell*)
5   skip this current_cell;
6   **else**
7   Cellx = min_drive_cell (current_cell_type);
8   BUFx = select_buffer (current_cell_drive);
9   3DCoB_cell = Cellx and BUFx in series;
10   add inter-tier via parasitics to 3DCoB_cell;
11   **end**
12 **end**

gate cell with its equivalent min-drive gate to perform the same functionality, and then pick from the library the buffer with a similar drive as the original gate. For example, if we have 2-input AND gate (AND2) with drive 42 (AND2x42), and the minimum drive for all the 2-input AND gates is 4 (AND2x4). So the 3DCoB cell will be the min-drive AND2 cell (AND2x4) connected to the equivalent drive buffer (BUFx42). Then the new 3DCoB spice netlists are used to be characterized and generate the new 3DCoB library (.LIB).

Also, the 3DCoB approach needs to modify the Library Exchange Format (LEF) file which contains an abstract of the cells and design rules for the cell placement and route.

The area of the new 3DCoB cell depends on the difference between the areas of top min-drive cell and the bottom buffer. The min-drive cell area depends on the cell type.

Fig. 4 shows 3DCoB cell area in different cases. Adding inter-tier via directly between the top and bottom active regions is an invalid assumption because of the technology process constraints [4]. Consequently, in case of the top cell and the buffer have equal areas or the buffer is smaller than the top cell, an additional area is needed for the vertical inter-tier vias. However, in case of the buffer is larger than

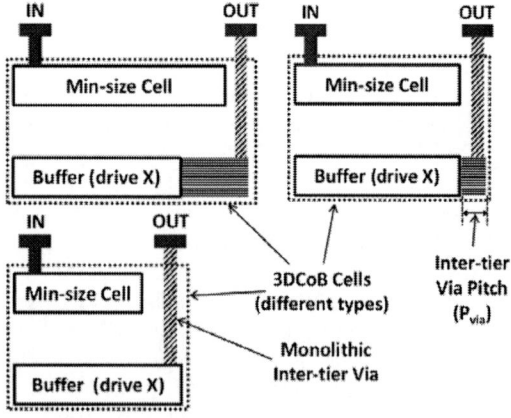

Fig 4. Different 3DCoB cells of the same drive X, showing the effective cell area and pin locations depending on cell type.

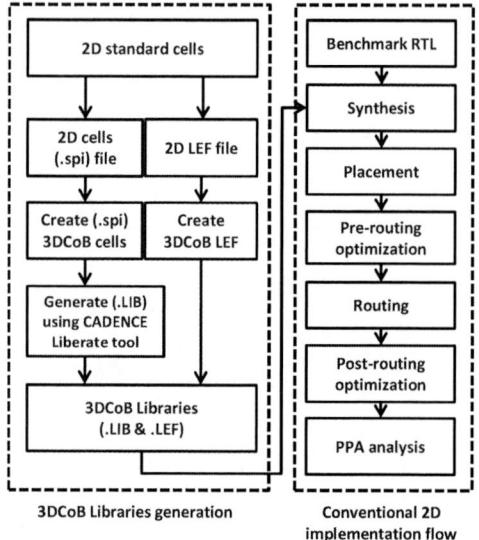

Fig 5. Full design flow for 3DCoB approach.

the top cell by equal or more than the inter-tier via pitch, the new 3DCoB cell area will be the same as the buffer area.

The area of the new 3DCoB cell can be calculated as shown in Eq. 1.

$$Area_{3DCoB}$$
$$= \begin{cases} A_{BUF}, & (A_{Min-Cell} + P_{via}) < A_{BUF} \\ \max(A_{Min-Cell}, A_{BUF}) + P_{via}, & (A_{Min-Cell} + P_{via}) \geq A_{BUF} \end{cases}$$
$$(1)$$

Where, $Area_{3DCoB}$ is the area of the new 3DCoB cell, $A_{Min-Cell}$ is the area of the top cell, A_{BUF} is the area of the bottom buffer cell, and P_{via} is the pitch needed to place the monolithic inter-tier via.

Using this methodology, the new 3DCoB libraries (.LIB and .LEF) can be generated and then used as normal libraries in a conventional 2D physical implementation design tools.

Fig. 5 shows the whole design flow as the following; RTL synthesis, placement, pre-routing optimization, routing, post-routing optimization, and finally get sign-off power, performance, and area (PPA) reports.

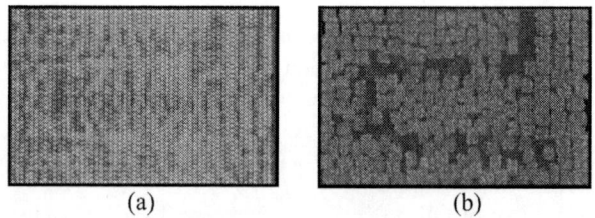

Fig 6. Place and route snapshots for the AES block using 3DCoB cells at 90% cell density; (a) with, and (b) without, top metal layers and cell fillers.

978-1-4799-2817-0/14 $31.00 © 2014 IEEE

TABLE II
Results comparison between 2D and 3DCoB implementations for FFT and AES blocks

		# Std Cells	Cell Density	Setup Slack reg2reg (ns)	Perf (GHz)	Performance Gain (%)
FFT Target clk=0.8ns Area=0.027mm²	2D	27362	91%	-0.009	**1.24**	--
	3DCoB	30160	86%	0.001	**1.25**	**0.8%**
FFT Target clk=0.6ns Area=0.027mm²	2D	24771	95%	-0.150	**1.33**	--
	3DCoB	26210	93%	-0.064	**1.51**	**13.53%**
AES Target clk=0.6ns Area= 0.119mm²	2D	184098	81%	-0.020	**1.61**	--
	3DCoB	187012	83%	-0.013	**1.63**	**1.24%**
AES Target clk=0.4ns Area= 0.119mm²	2D	164174	89%	-0.100	**2.00**	--
	3DCoB	166749	90%	-0.015	**2.41**	**20.5%**

V. Results

The 3DCoB design approach was implemented in 28nm FDSOI technology and using 3D Monolithic Integration technology demonstrated in [4]. We used a sign-off physical implementation flow using commercial tools as presented in Fig 5. CADENCE Liberate tool is used for the library (LIB) generation from the spice (.spi) files of the standard cells. Also, the design blocks are implemented using CADENCE Encounter. We used two benchmark circuits; Fast Fourier Transform (FFT) and 128-bit Advanced Encryption System (AES).

Fig.6 shows the place and route snapshots for the AES block using 3DCoB cells with 90% cell density, where the cell density is the percentage of total area of the cells compared to the block area.

Table II summarizes the physical implementation results for the FFT and AES blocks at different targeted clock period in the place and route tool. To show the difference in implementation, we present two points for each block, one is at low clock frequency and the other one is at high clock frequency. By comparing 2D and 3DCoB, the gain in

performance can be observed. The performance is measured as the effective clock frequency (the targeted clock period without the slack time). The gain in performance increases as the targeted clock period decreases. This is because as the targeted clock period decreases, the EDA tools tends to use more high drive cell, which increases the percentage of stacked cells used (non-min cells). By comparing both 2D and 3DCoB designs at the same targeted clock period, the performance gain reaches 13.5% for the FFT and 20.5% for the AES.

For a certain area, the design implementaion has a specific max performance limit that can be reached (max clock frequency). However, the 3DCoB design approach can achieve higher performance limit than that of 2D designs at the same area. Figure 7 shows the AES block performance at different targeted clock frequency. As the targeted clock frequency increases, the setup slack increases which decreases the block performance (the effective clock frequency) till reaching the performance limit. As a result, for the same AES block area, the 3DCoB approach provides 35% gain in the performance compared to the 2D. The performance limit for the 2D implementation is 2 GHz while

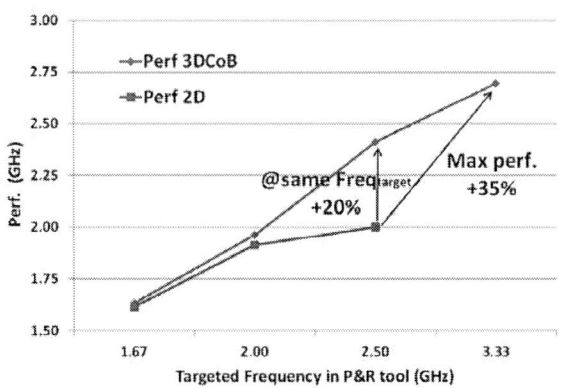

Fig 7. Comparison of max Performances between 2D and 3DCoB for AES 128 block at Area= 0.119 mm²

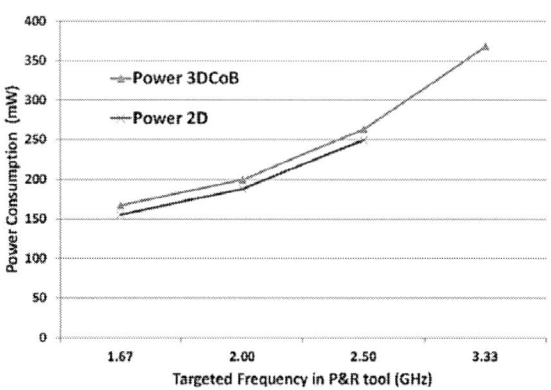

Fig 8. Power Consumption for 2D and 3DCoB for AES 128 block at Area= 0.119 mm²

the 3DCoB can reach performance of 2.7 GHz.

For the power analysis, we used power reports from Encounter tool with an activity factor of 15%. Figure 8 shows the power results of the AES block for both 2D and 3DCoB implementations for different targeted clock frequency at the same area. As shown, the two power curves have the same power trend without a significant power increase. The physical implementation of 2D block has a limit of target clock frequency equals to 2.5GHz, consequently there are no performance and power results presented for higher targeted clock frequency.

VI. Summary and Conclusions

3D monolithic integration (3DMI) technology offers a fine grain integration level thanks to high density and low-parasitics inter-tier vias. The previous design approaches to implement 3DMI suffers from two main drawbacks; either the incompatibility with the conventional 2D design tools, or the need to redesign the standard cell libraries.

In this work we introduced 3D Cell-on-Buffer as a new design approach for 3DMI which is fully compatible with the conventional 2D implementation design flow tools.

The main idea for 3DCoB is to split the conventional 2D cells into two stages; logical stage (min-drive cells) and driving stage (same-drive buffer). The logical cell is stacked over the driving buffer using the advantage of the 3DMI technology. By creating the 3DCoB cells and generating the 3DCoB libraries, the design can use the conventional 2D sign-off physical implementation flow. The implementation results using the 3DCoB cells show 35% performance improvement compared to the same 2D design.

References

[1] K. Banerjee, S. Souri, P. Kapur, and K. Saraswat, "3-D ICs: A novel chip design for improving deep submicron interconnect performance and systems-on-chip integration", Proceedings of IEEE 89.5, 2001.
[2] Semiconductor Industry Association, "The International Technology Roadmap for Semiconductors (ITRS)", 2011 Edition.
[3] F. Clermidy, F. Darve, D. Dutoit, W. Lafi, and P. Vivet, "3D Embedded Multi-core: Some Perspectives", Design, Automation & Test in Europe Conference & Exhibition (DATE), IEEE, 2011.
[4] P. Batude et al., "Demonstration of low-temprature 3D sequential FDSOI integration down to 50 nm gate length" VLSI Technology (VLSIT), symposium on, IEEE, 2011.
[5] G. Loh, X. Yie, and B. Bryan, "Processor design in 3D die-stacking technologies", Micro, IEEE, 27(3), 2007.
[6] V. Pavlidis and E. Friedman, "3-D Topologies for Networks-on-Chip," Very Large Scale Integration (VLSI) Systems, IEEE Transactions on , vol.15, no.10, pp.1081,1090, Oct. 2007.
[7] D. Dutoit, E. Guthmuller, and I. Miro-Panades, "3D integration for power-efficient computing," Design, Automation & Test in Europe Conference & Exhibition (DATE), vol., no., pp.779,784, 18-22 March 2013.
[8] H. Kim, J. Yoon, K. Hwang, Y. Kim, J. Park, and L. Kim, "A 275mW heterogeneous multimedia processor for IC-stacking on Si-interposer," Solid-State Circuits Conference Digest of Technical Papers (ISSCC), 2011 IEEE International , vol., no., pp.128,130, 20-24 Feb. 2011.
[9] P. Batude et al., "GeOI and SOI 3D Monolithic Cell integrations for High Density Applications" VLSI Technology (VLSIT), symposium on, IEEE, 2009.
[10] S. Wong, A. El-Gamal, P. Griffin, Y. Nishi, F. Pease, and J. Plummer, "Monolithic 3D Integrated Circuits," VLSI Technology, Systems and Applications, (VLSI-TSA), International Symposium on , pp.1,4, 23-25 April 2007.
[11] T. Naito et al., "World's first monolithic 3D-FPGA with TFT SRAM over 90nm 9 layer Cu CMOS." VLSI Technology (VLSIT), 2010 Symposium on, IEEE, 2010.
[12] O. Thomas, M. Vinet, O. Rozeau, P. Batude, and A. Valentian, "Compact 6T SRAM cell with robust read/write stabilizing design in 45nm Monolithic 3D IC technology", IC Design and Technology, ICICDT'09, IEEE International Conference on, IEEE, 2009.
[13] C. Liu and S. Lim, "A Design Tradeoff with Monolithic 3D Integration", Quality Electronic Design (ISQED), 13th International Symposium on, IEEE, 2012.
[14] Y. Lee, D. Limbrick, and S. Lim, "Power Benefit Study for Ultra-High Density Transistor-Level Monolithic 3D ICs", Proceedings of the 50th Annual Design Automation Conference (DAC), ACM, 2013.
[15] B. Shashikanth et al., "CELONCEL: Effective Design Technique for 3-D Monolithic Integration targeting High Performance Integrated Circuits", proceedings of the 16th Asia and South Pacific Design Automation Conference, IEEE Press, 2011.
[16] L.Pang et al., "A Shorted Global Clock design for Multi-GHz 3D Stacked Chips", *VLSI Circuits (VLSIC),Symposium on,* IEEE, 2012.

Emulator-Oriented Tiny Processors for Unreliable Post-Silicon Devices: A Case Study

Yuko Hara-Azumi[†‡]   Masaya Kunimoto[†]   Yasuhiko Nakashima[†]

†Graduate School of Information Science, Nara Institute of Science and Technology,   ‡JST, PRESTO

e-mail: †{yuko-ha, kunimoto, nakashim}@is.naist.jp

Abstract— Although various post-silicon devices have been invested in recent years, they still have a major issue of reliability. Because circuit area is an essential factor of reliability, especially for such unreliable post-silicon devices, it is desired to build small circuits which can reuse as many today's application programs as possible even if the performance is not very high. This paper presents the very first work to study novel, efficient techniques of emulating wider-bit guest processors (e.g., 32-bit) on a narrower-bit host processor (e.g., 8-bit) with very limited hardware resources while mitigating performance degradation. We propose three types of emulator-oriented tiny processors varying in available hardware resources and reliability enhancement approaches. Quantitative evaluation and discussions are done for comparing those three processors. We believe that this work will will be a good help of making breakthrough for further development of new device technologies and computers based on them.

I. Introduction

In recent years, various kinds of new materials have been invented and developed to build digital circuits, which have been attracting attentions as post-silicon devices [1, 2, 3]. One of their major advantages over silicon processes is to be environment-friendly in resources and energy. Some products such as displays and sensors [4] have been already made and available in the market. However, the infancy of their manufacturing technology and quality still makes it difficult to realize "computers" based on them. It is predicted that it will take a decade or longer for them to become mature enough to build such performance- and/or reliability-centric circuits. However, looking at it from the opposite perspective, in order to boost the development of those technologies, it is essential to establish computer-design techniques based on those devices even if such computers may work very slowly at present, e.g., only at a 1/100 speed of conventional silicon-computers.

Post-silicon devices have several-order higher defect rates in manufacturing/operation than silicon ones. Their integration rate is not very high, either [1]. Thus, it is crucially important to design as small circuits as possible for lessening these impacts. On the other hand, most of today's practical applications use 32 or 64 bits, that is, their underlying processors, even with 32-bit wide, are on thousands or millions transistors. Taking these into account, design technologies of tiny processors where the current applications are usable *as-is* should be developed – not going back to decades-ago processors – in order to accelerate the practical use of post-silicon-based processors.

One possible means is to emulate those applications. There have been a number of researches on emulation techniques, such as just-in-time compiler (JIT) [5, 6, 7], object code translation (OCT) [8], and dynamic OCT (DOCT) [9, 10]. They all target emulation of a guest processor with the same as or smaller bitwidth than the host processor, for providing good speedup by efficiently using large address space and/or a rich instruction set of the host processor, at the sacrifice of large circuit area. On the other hand, to the best of our knowledge, there have been no works enabling efficient emulation of a guest processor with wider bitwidth (e.g., 32-bit) on a host processor with narrower bitwidth (e.g., 8-bit). Because no chance of speedup is expected due to less address space on the host processors, this approach has not been employed when more speedup is quested exploiting large address space, which has been available along with higher integration thanks to the CMOS scaling. However, actually no works have yet to evaluate how much performance degradation and area reduction would be led by this emulation approach.

To address the above issue, in this paper, we present 8-bit tiny processors which efficiently emulate today's 32-bit applications. As aforementioned, this is the very first work which enables efficient emulation of a wider-bit guest processor on a narrower-bit host processor. For this, we propose a self-defined instruction set which well-balances the performance and circuit area of the host processor, and conduct a case study by implementing three types of processors which are all tiny but have some differences in the subset of instruction set they use and emulation mechanisms. Through evaluation on 24 MiBench benchmarks, we quantitatively evaluate performance and circuit area (which is also a metric of yield or reliability) and reveal our useful findings from observations.

To sum up, this paper has the following four contributions:

- The first attempt of developing an emulation technique to execute 32-bit practical applications and an OS on an 8-bit host processor (Section II.A)
- A self-defined instruction set which well-balances the performance and circuit area of the host processor (Sections II.B and II.C)
- A case study to implement three types of processors which trade-off the performance and circuit area (Sections II.B, II.C, and II.D)
- Quantitative evaluation and discussions of the three processors in terms of performance overhead and area reduction on 24 32-bit ARM applications and an OS (μCLinux) (Section IV)

We believe that this work will lead to not only acceleration of the development of post-silicon technologies themselves but also smooth migration to such new technologies without waiting for the maturity of those device technologies.

978-1-4799-2817-0/14 $31.00 © 2014 IEEE

The reminder of this paper is organized as follows: Section II describes our proposed instruction sets and emulator-oriented processors. Next, Section III explains metrics used for our evaluation. Section IV shows a case study where the proposed processors are quantitatively evaluated using a number of practical benchmark programs, through which we reveal our findings. Finally, Section V concludes this paper.

II. EMULATION-ORIENTED TINY PROCESSORS

As a case study, we propose three types of 8-bit emulator-oriented processors targeting 32-bit ARM (ARMv4T) as a guest processor. They have similar emulation mechanisms but vary in some structural/functional differences. In the following, we first describe the overview of those three processors and then explain their features.

A. Overview

It can be expected that if the guest processor is a simple one like MIPS, the host processor may be also made relatively simple even if the guest has the wider bitwidth than the host. On the other hand, widely-used processors like ARM, which is actually de-facto standard as being used in various applications including mobile devices, have much complex structures: although every field of most instructions can be aligned by 8 bits, some instructions have an irregular field, such as 5-bit immediate – ARM adopts such complex instruction formats to let a single complex instruction conduct multiple functions [11]. Thus, the 8-bit host processor itself may also need complex structures for supporting efficient emulation, i.e., mitigating performance degradation, for handling 32-bit ARM as a guest.

In this paper, we propose to emulate 32-bit ARM on a 8-bit emulator-oriented processor which consists of a small CPU and a moderate-sized emulator implemented in a RAM. The complexity of the CPU (i.e., the circuit area) and the size of the emulator (i.e., performance[1]) are in a trade-off relationship. Furthermore, the circuit area, which largely affects the manufacturing yield and reliability, causes another trade-off in reliability, i.e., it is harder to protect logic (CPU) than memories (RAM), whose reliability can be ensured relatively easily by ECC [12] and etc. Therefore, we present three types of 8-bit emulator-oriented processors, which will be explained in the following subsections, for well trading off the circuit area and performance, i.e., as small a CPU as possible while mitigating the emulator size. Those three processors have in common in that they allocate all registers (32-bit ARM registers including program counter (PC), current program status register (CPSR), etc. and registers for OS interrupts – hereafter, guest registers (GRs)) in the RAM, other than 8-bit on-chip host registers (HRs) used by the emulation program, but vary in the emulation processes and internal structures of CPUs. Caches or floating-point units (FPUs), which are to improve performance at the cost of the circuit area, are not contained.

---
[1] The more the emulator grows (i.e., the more functions the emulator holds), the less performance.

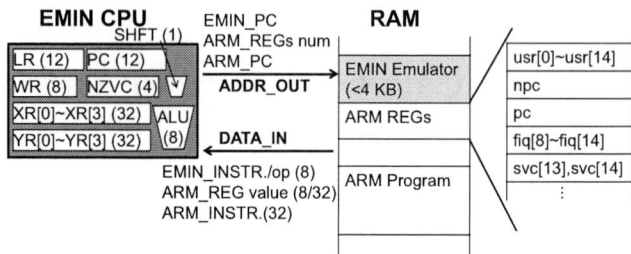

Fig. 1. The block diagram of EMIN.

Note that although our work is based on 32-bit ARM as a guest processor in this paper, our emulation techniques themselves are also applicable to other CPUs (MIPS, SH, PowerPC, etc).

B. EMIN: Emulator-oriented Minimal INstruction set computer

First, we present EMIN, which mitigates the CPU complexity while suppressing performance degradation by using performance-efficient instructions. It employs 1-byte simple host instructions and 2/3-byte complex ones (e.g., table jump), which internally express 4-byte guest instructions for better performance-efficiency. It basically uses 1-byte ones for emulation, but also uses 4-byte ones when it will take a large number of steps if done only with 1-byte ones. Employing variable-length instructions can relax the CPU complexity and the emulator size at the same time.

The block diagram of EMIN is drawn in Fig. 1, where numbers in parentheses represent the bit-width of each register. EMIN contains a 12-bit link register (LR), a 12-bit PC, a 8-bit work register (WR), a 4-bit register (NZVC) to specify condition flags, in total 32-bit X/Y registers (XR/YR)[2], an 8-bit ALU, and a 1-bit shifter (SHFT) in the CPU with the less-than-4Kbyte emulator in the RAM. Table I shows all instructions used in our proposed processors and their supplemental information. The EMIN instructions are listed in the third column.

Given an assembly code of an ARM program, EMIN emulates it using the hardware components and the emulator shown in Fig. 1. Three representative emulation processes are briefly explained below.

- **Decode of a guest instruction:** Irrespective of the guest processor, the sub-opcode based on the main-opcode needs to be found and extracted. For efficient emulation, EMIN first stores a 4-byte guest instruction to HRs temporarily, extracts its fields by the 4 bits, and then performs table jump depending on the specified value. More specifically, as shown in an example of Fig. 2(a), decode is performed in two steps whose processes are: (1) the value in NPC is loaded to YR, (2) the RAM is referred to using YR in order to load the corresponding guest instruction to XR, (3) the next NPC is calculated using YR, (4) the result in YR is updated to NPC, (5) the guest instruction in XR is decoded through table jump, and (6) the decoded result is loaded to XR. This two-step decode uses less registers while effectively reducing the emulator size.

---
[2] Each of XR and YR has four 8-bit entries.

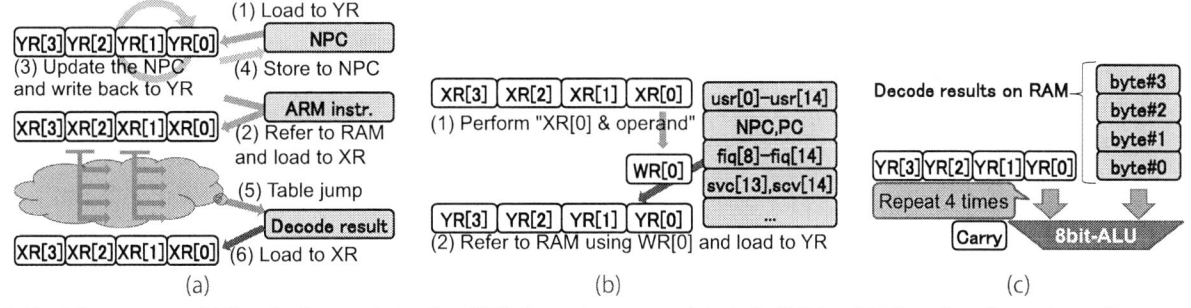

Fig. 2. Emulation processes: (a) Decode of a guest instruction, (b) Reference to a guest register in the RAM, and (c) Execution of a guest operation.

TABLE I HOST INSTRUCTION SETS

Group	Bytes	Types E/R	EE	mR	Description
Direct reference to GR/WR	2	12		8	Refer to the RAM using an operand as a GR/WR
Load & merge	2	8		8	Refer to the RAM using an operand as a GR/WR
8-bit load/store	2		6		Take an operand
	1		9		
32-bit load/store	1		9		
Indirect reference to GR/WR	1	9			Between GR & WR and the RAM & a HR
Data transfer between HRs	2	7			Take an operand
	1	9		1	
Table jump	2	9	2	1	Specify a reduction condition by an operand
8-bit arith./logic	2	9		14	Take an immediate value
32-bit arith./logic	2	6	6		Refer to the RAM using an operand as a GR/WR
1-bit shift	1	7			
Branch	3	2		2	Take a 2-byte offset
	2	9		5	Take a 1-byte offset
	1	1		1	
Indirect load/store of WR	1			2	Take a value in the ADDR space in the RAM and use it as the address for load/store of a WR

- **Reference to a guest register in the RAM:** EMIN reduces the address information by using a GR index, not the RAM address directly. An example is shown in Fig. 2(b): (1) The address is calculated from the ANDed a GR index (XR[0]) with an operand, and the result is stored in WR, and then (2) The RAM is referred to using the result in WR.

- **Execution of a guest operation:** Irrespective of the guest processor, generally, 32-bit CPUs adopt 32-bit arithmetic, comparison, and logic, and shift operations. As shown in Fig. 2(c), EMIN operates such 32-bit arithmetic/comparison/logic operations by repeating 8-bit corresponding operations four times on the 8-bit ALU through the combination of several host instructions and a condition code. Similarly, EMIN performs 32-bit shift operations by repeating the 1-bit shift operation on the 1-bit shifter (SHFT) for the specified number of times. Although this indicates that the large number of steps may be concerned by repeating the 1-bit shift (i.e., 32 times

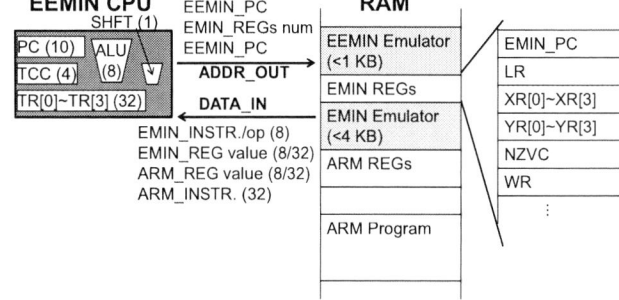

Fig. 3. The block diagram of EEMIN.

at the worst case), we observed that its incidence is negligibly low through evaluations on a number of MiBench applications (in Section IV), from which we conclude that a 1-bit SHFT is sufficient for most applications.

C. EEMIN: Emulator of EMIN

Secondly, for further reducing the circuit area, we propose EEMIN for emulating EMIN on a smaller CPU. The emulator size may become larger by directly emulating 32-bit ARM programs on fewer 8-bit hardware resources. Thus, given an assembly code of an ARM program and its corresponding emulation code of EMIN, whose registers are now allocated in the RAM, EEMIN indirectly emulates the ARM program through the EMIN code. Because the bitwidth of EEMIN is the same as that of EMIN (unlike 8-bit EMIN vs. 32-bit ARM), it is possible to mitigate the size of the EEMIN emulator even when using the further limited number of hardware resources.

The block diagram of EEMIN is shown in Fig. 3. EEMIN contains a 10-bit PC, a 4-bit register (TCC) to specify condition flags (corresponding to the 4-bit NZVC register in EMIN), in total 32-bit work registers (TR) (four 8-bit entries), an 8-bit ALU, and a 1-bit shifter (SHFT). As shown in the figure, the size of the EEMIN emulator itself is less than 1Kbyte, which led to the reduction in the PC bitwidth of the EEMIN CPU by 2 bits compared with that of the EMIN CPU. Thus, in EEMIN, in total less than 5Kbyte-address space (EMIN+EEMIN) in the RAM is necessary. Since the guest processor and its bitwidth are both different from EMIN, EEMIN uses a different set of instructions from EMIN. The newly defined instruction set for EEMIN is listed in the fourth column of Table I.

As handling the same bitwidth in the guest and host proces-

978-1-4799-2817-0/14 $31.00 © 2014 IEEE

sors, emulation processes of EEMIN themselves are simpler than those of EMIN. EEMIN, however, needs to emulate ARM programs in two steps, meaning that it takes much more steps to complete. Consequently, larger performance degradation is expected in exchange of achieving the smaller circuit area. The effects on the circuit area and performance will be evaluated in Section IV.

D. REMIN: Reliablity-aware EMIN

Although improvement of the manufacturing yield and reliability can be expected by making CPUs small, EMIN/EEMIN may not work anymore when a single gate in their CPUs gets defected. EEMIN tries to reduce such a risk by decreasing its circuit area, but does not have a direct approach for improving higher yield or reliability. Particularly EMIN has a higher risk since it employs performance-efficient, complex instructions passing a lot of gates, which makes such instructions more vulnerability to defects. On the other hand, simple instructions would have less possibility to be affected by defects. In the case that the former instructions cannot work correctly while the latter still work, the CPUs of EMIN/EEMIN should be still available by bypassing such defected sub-circuitries, but they have no such mechanisms of bypassing.

Our third proposal is REMIN, which adds a reliability-improvement mechanism to EMIN. Assume that hardware components are regularly checked either on-line or off-line by running test programs and so forth, and instructions using defected circuitries (hereafter, faulty instructions) are identified, REMIN replaces such a faulty instruction with a sequence of functionally-equivalent fault-free instructions so that it can keep emulating ARM programs correctly while bypassing defected circuitries. From our observation that XR/YR and their peripheral circuitries are complex and large, the replacement is applied to the instructions using XR and/or YR. For that, the EMIN instruction set was extended so that such replacement can be fully supported. It should be noted that this approach is to *increase the expectation of the lifetime extension*, not to ensure it – the lifetime may be extended depending on which parts of the circuit are still healthy, i.e., which instructions are fault-free. Also, note that the instruction extension is to complete EMIN's atom instructions, which were originally supposed to be included in the EMIN instruction set but removed because of employing some performance-efficient instructions (i.e., the ones which would be replaced), in order to exploit functional redundancy, not to add spatial redundancy. By the instruction replacement technique, REMIN can cover about 60% of the instructions in EMIN, meaning that a chance of longer lifetime can be expected. REMIN starts working as fast as EMIN and gradually slows down as replacing more performance-efficient, complex instructions with simple-but-more-robust ones.

The block diagram of REMIN is depicted in Fig. 4. The hardware components contained are basically the same as EMIN, but the bitwidth of LR and PC are extended by 2 bits, compared with EMIN, for supporting instruction replacement, which will expand the emulator size. As explained above,

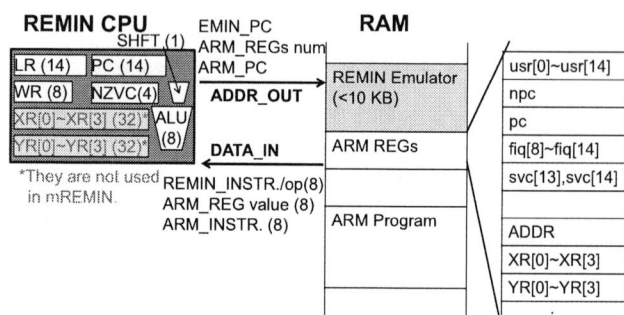

Fig. 4. The block diagram of REMIN.

REMIN first also uses XR/YR and their peripheral circuitries and works as fast as EMIN, but would slow down gradually by not using those circuitries as more parts get defected and efficient instructions cannot be used anymore. Table I shows the REMIN instruction sets when it is completely fresh (same as EMIN) and when it gets ages the most (i.e., any efficient instructions are not available) in the third and fifth columns, respectively. REMIN can work as long as the latter instructions are all available – hereafter, we call this minimum set of REMIN "mREMIN." One may think that it would be better to use only mREMIN, without supporting such mode-switching mechanisms. However, using only mREMIN would be quite slow, as will be evaluated in Section IV. Moreover, at the time of designing processors, the designers cannot expect how much performance would be required. Motivated by this, REMIN is designed for well trading off the performance and reliability depending on the applications and the device aging.

III. METRICS

The three processors proposed in the previous section are all supposed to emulate 32-bit ARM application programs using their self-defined instruction set. As have been explained, they have very limited resources in their CPU and the other things (the assembly code of an ARM program to be emulated and the emulator itself) in an external memory (RAM). Thus, the processors will be evaluated using the following two metrics:

1. **Circuit area** – because circuit area largely affects the manufacturing yield and chip lifetime, in our evaluation, the circuit area will be also used as a metric of reliability.
2. **Performance (dynamic steps)** – access time to the RAM is not included in order to evaluate the performance of CPUs solely[3].

One may think that the size of emulator (i.e., the number of static steps) may be large since those processors emulate 32-bit ARM applications on an 8-bit CPU. However, actually it is small enough (up to 10Kbyte even for executing μCLinux) compared with the size of applications (about at least a few Mbyte). The emulator is contained in the RAM together with

---

[3]EEMIN and mREMIN, which originally have more dynamic steps, have more accesses to the RAM, and thus, the relative evaluation results will not change.

the applications, and thus, its overhead is negligible. Furthermore, memories can be protected using ECC [12] and so on, at relatively low cost. Consequently, we will focus only on the circuit area for evaluating hardware cost and reliability issues.

IV. EVALUATION

This section quantitatively evaluates the three processors described in Section II, over a 32-bit native ARM processor.

A. Setup

To assess the effects of our proposed processors on performance and circuit area (which leads to manufacturing yield and reliability as aforementioned), we conducted experiments using 24 practical benchmark programs from MiBench and μCLinux[4]. For fair comparison, ARM7TDMI, which employs ARMv4T as its architectural type and contains no caches or FPUs, was chosen as the baseline native-ARM processor.

For executing MiBench benchmarks, first, their ARM assembly codes were generated through the GNU assembler. Then, they were fed to our in-house emulator and simulator for each type of our proposed processors, in order to evaluate performance (i.e., the number of dynamic steps). Also, their RTL circuits were implemented in Verilog-HDL, for which, in order to evaluate the circuit area, logic synthesis was performed by Design Compiler (F-2011.09-SP2) using Rohm180nm library. To be independent from the underlying device technology, we evaluated the circuit area in terms of the number of 2-input-NAND cells and that of transistors.

B. Circuit area

First, let's look at Table II, where the results of circuit area of the four processors, including the native ARM processor (ARM7TDMI)[5], are described. The results in parentheses of REMIN are for mREMIN.

As can be seen from Table II, the area of our proposed processors are about 1/11-1/6 of that of the native ARM. Actually, due to its simple set of instructions and hardware supports (i.e., no caches/FPUs), ARM7TDMI itself is already relatively small among various ARM processors available in the market. Thus, it can be concluded that our processors are all tiny and useful by significant reduction of hardware cost, leading to the improvement of manufacturing yield and reliability, which is crucial especially for unreliable post-silicon devices.

Comparing our proposed processors, EMIN vs. REMIN and EEMIN vs. mREMIN are comparable, respectively. Among all, EEMIN is the smallest thanks to its nested emulation mechanism (i.e., emulation of an emulator (EMIN)). Actually, we have expected that mREMIN would be the smallest since it excluded more registers than EEMIN. Contrary to our expectation, REMIN became slightly bigger than EEMIN because of

[4]μCLinux was used for confirming the usefulness of our proposed processors, not for evaluating their performance.

[5]ARM manual was referred to for the approximate transistor counts of ARM7TDMI, from which the cell counts were calculated

TABLE II AREA

Design	Native (ARM7TDMI)	EMIN	EEMIN	REMIN (mREMIN)
Cells (2in-NAND)	≈ 18,500	2,644	1,748	2,844 (1,892)
Transistors	≈ 74,000	10,576	6,992	11,376 (7,568)

an increase in the complexity of the REMIN's controller along with the switching mechanism of replacing instructions. From this, we can say that excluding more components does not always reduce the area.

C. Performance

We evaluated the number of dynamic steps for executing 24 MiBench benchmark programs on three processors. Their results are normalized by those of the native ARM and described in Fig. 5(a) ("1" represents the results of the native ARM). For each processor, the average results are also depicted. Note that REMIN runs as fast as EMIN when the entire circuit is healthy (i.e., "fresh-REMIN"), but later slows down as more parts of REMIN become faulty and more performance-efficient instructions are replaced with simple ones. In Fig. 5(a), "aged-REMIN" represents the case that REMIN runs using the instruction set of mREMIN.

The overhead of EMIN (and fresh-REMIN), EEMIN, and aged-REMIN is approximately 50x-80x (on average 65x), 800x-1,200x (on average 1,030x), and 430x-850x (on average 610x), respectively. The superiority of EMIN and fresh-REMIN over aged-REMIN demonstrates the performance-efficiency of 2/3-byte instructions used in EMIN and fresh-REMIN. Although EEMIN and REMIN have in common with not using such performance-efficient instructions, EEMIN has quite large overhead (i.e., further larger even than aged-EMIN) because, as discussed in Section IV.B, EEMIN is a nested emulator, and thus takes further more steps for emulating a single ARM instruction.

D. Discussion

Finally, the area-performance products of the processors for each application are compared. Again, the results are normalized by those of the native ARM, as shown in Fig. 5(b). For fresh- and aged-REMINs, the area of REMIN and mREMIN, respectively, were used to obtain the results.

Although the area is significantly small in EEMIN and mREMIN, their area-performance products are quite large (i.e., on average about 98x and 63x, respectively). This comes from their emulation inefficiency by using the very limited instructions only. On the other hand, the results for EMIN and fresh-REMIN are suppressed to about 9x and 10x, respectively, thanks to less performance degradation by 2/3-byte efficient instructions. These results indicate that aggressive area reduction may not well-tradeoff performance overhead.

Here, taking into account the reliability issue, EMIN may have a higher risk of early breakdown because of using more

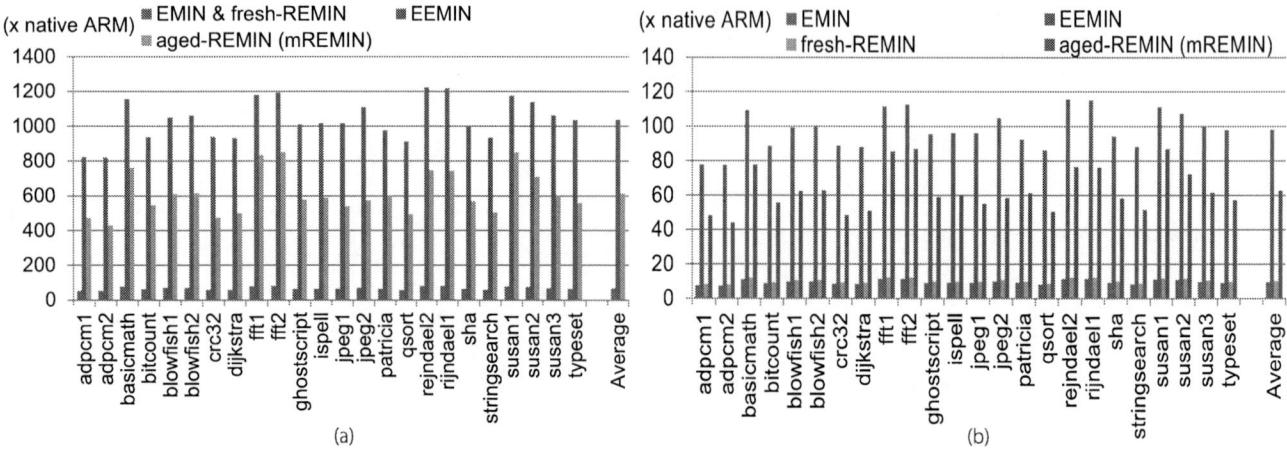

Fig. 5. Comparison against the native ARM processor: (a) Performance and (b) Area-performance product ("1" represents the native ARM in both graphs).

transistors (1.5x) than EEMIN and mREMIN. Comprehensively considering area-performance trade-offs and reliability concerns, REMIN is the most beneficial in that it not only runs fast using performance-efficient instructions at the early time, but also can expect a chance of extending its lifetime by detouring faulty parts of the circuit through instruction replacement.

From the above results, we would like to suggest the following three-hold scenarios along with the maturity of the underlying device technology; (1) When the device technology is in very early infancy, the circuits are preferred to be made as small as possible. At such time, reliability is the first-class priority. Then, EEMIN will be a help of anyway enabling executing applications although it is quite slow; (2) As the bigger circuits can be made but their reliability is unsure, REMIN or designs with such switching mechanisms may be employed so that good performance at the early time and slow-yet-longer-lifetime execution later are both available; and (3) When affording to quest performance efficiency, EMIN will be useful.

These three emulator-oriented processors are just an example of realizing computers on unreliable post-silicon devices. A variety of new types of reliability-aware processors may be studied. We believe our proposed processors will be a good help of making breakthrough for further development of new device technologies and architectural designs.

V. CONCLUSION

This paper addressed a case study in presenting three types of tiny processors for emulating 32-bit ARM on 8-bit CPUs. Those three processors use very limited hardware resources, for the purpose of improving reliability, while mitigating performance degradation. Through quantitative evaluation using 24 MiBench benchmarks and μCLinux , we compared the area-performance trade-offs of the three processors and concluded that they all have significant usefulness, depending on the maturity level of the underlying device technology. We believe that this work will open up new possibilities of exploiting emulation techniques, especially for boosting the development and investment of various post-silicon devices.

Not only the instruction sets but also the bitwidth of the

host processors affect the trade-offs of the circuit area and performance. Although, in this paper, we employed 8-bit host processors empirically and intuitively, the effects of the CPU bitwidth should be also quantitatively evaluated, which will be a subject of our future work.

ACKNOWLEDGEMENT

This work is in part supported by KAKENHI 24240005 and 24650020.

REFERENCES

[1] T.-C. Huang, J.-L. Huang, and K.-T. T. Cheng, "Robust circuit design for flexible electronics," *IEEE on Design & Test of Computers*, vol. 28, no. 6, pp. 8–15, 2011.

[2] T. Sekitani *et al.*, "Electrical artificial skin using ultra-flexible organic transistor," in *Proc. DAC*, no.125, 2013.

[3] M. Shulaker *et al.*, "Sacha: The Stanford carbon nanotube controlled handshaking robot," in *Proc. DAC*, no.124, 2013.

[4] K.-T. T. Cheng and T.-C. Huang, "What is flexible electronics?" *ACM SIGDA Newsletter*, vol. 39, no. 4, p. 1, 2009.

[5] T. Suganuma *et al.*, "A dynamic optimization framework for a Java just-in-time compiler," in *Proc. OOPSLA*, 2001, pp. 180–195.

[6] M. Kawahito *et al.*, "A new idiom recognition framework for exploiting hardware-assist instructions," in *Proc. AS-PLOS*, 2006, pp. 382–393.

[7] H. Inoue *et al.*, "A trace-based Java JIT compiler retrofitted from a method-based compiler," in *Proc. Code Generation and Optimization*, 2011, pp. 246–256.

[8] K. Andrews and D. Sand, "Migrating a CISC computer family onto risc via object code translation," in *Proc. AS-PLOS*, 1992, pp. 213–222.

[9] R. A. Lethin *et al.*, "Dynamic optimizing object code translator for architecture emulation and dynamic optimizing object code translation method," in *US 6463582 B1*, 2002.

[10] G. A. Mann, B. A. Noyes, and R.-J. Chevance, "Method and apparatus for dynamic management of translated code blocks in dynamic object code translation," in *US 6529862 B1*, 2003.

[11] ARM Limited, *ARM Architecture Reference Manual*. ARM DDI 0100E, 2000.

[12] A. N. Udipi *et al.*, "LOT-ECC: localized and tiered reliability mechanisms for commodity memory systems," in *Proc. ISCA*, 2012, pp. 285–296.

Applying VLSI EDA to Energy Distribution System Design

Sani Nassif, Gi-Joon Nam, Jerry Hayes
IBM Research - Austin
e-mail: nassif@us.ibm.com, gnam@us.ibm.com jerryh@us.ibm.com

Sani Fakhouri
University of California, Irvine
e-mail sani@uci.edu

Abstract— Energy distribution networks refer to that part of the electricity network that delivers power to homes and business. It is reported that significant amounts of energy are being wasted simply due to inefficiencies in this network. Further, this domain is rapidly changing with new types of loads such as electric vehicles or the spread of new types of energy sources such as photo-voltaic and wind. In this paper, we demonstrate a comprehensive design automation capability for energy distribution networks leading to much more flexible yet effective system. The new system's capabilities include power load distribution and transfers, equipment upgrading, geospatial-aware network optimization, outage identification, contingency planning and loss analysis/reduction. These features are enabled by advanced simulation, analysis and optimization engines that are adapted from those available in the traditional VLSI design automation area. The paper will conclude with potential future research directions that require further innovations in energy distribution networks.

I. INTRODUCTION

Ever since its introduction in the late 1800's, electricity has been an integral part of human life. While the reliability of electric energy supply varies significantly across the world and consumption per person varies across about 4 orders of magnitude [1], for many it is completely in the background. In fact it is depended on for so many aspects of our lives like lighting, refrigeration, entertainment, and transportation that we largely take it for granted.

In much of the world, electric energy generation and delivery is so important that it is controlled by the government. In some areas, however, *market de-regulation* has been instituted leading to a more open market where many providers can compete to provide the needed services. This is the case in many states in the USA, including Texas where this research was performed. De-regulation creates an open market for energy distribution where competition is intense and profit margins are relatively small. In fact, some of the local electric energy cooperatives in the Texas area are *non profits* and return any excess profits to their members! The result of this situation is that US industrial R&D investment in the area of energy distribution is quite small compared to -say- the Electronics industry.

While there has been no *Moore*[1] in the energy distribution area, recent developments are definitely changing the area. These include:

- Development of new *green* sources of energy such as wind, tide, and large-scale solar that are not *constant* and

therefore significantly change the patterns of energy flow in the network.
- Large scale deployment of small distributed sources of energy via small-scale solar installations which act to amplify local variability in loading over time.
- The introduction of new types of loads, specifically *electric vehicles* which are expected to create significant disruption to the grid due to large demand and temporal characteristics [2].

These changes are expected to make the energy distribution market significantly more *dynamic* and can be viewed as an opportunity to do constructive research in this area. In this paper we will assert that this is an excellent opportunity for the application of the large body of work in Electronic Design Automation (EDA) to the emerging and important area of energy distribution network design, analysis, and optimization.

II. RATIONALE: WHY EDA FOR ENERGY

The Integrated Circuit industry is unique in having been driven by *Moore's law* for more than 40 years. Over that time period constant and aggressive scaling of device dimensions has resulted in:

- Device Dimensions have scaled by a factor of 10^3.
- Device Density has scaled by a factor of 10^6.
- CPU Frequency has scaled by a factor of 5×10^4.
- CPU Performance has improved by a factor of 3×10^6.

These dramatic improvements came as a direct result of large amounts of R&D investments by key industry players such as Intel, IBM, TSMC and many others. Assuming a conservative 18% of the revenues of the top 20 semiconductor firms went into R&D for the last 20 years, the total is in excess of $1T$!

An important corollary of Moore's scaling trend is that integrated circuits in general, and CPUs in particular have become much more complicated. A modern server-class processor has more than a million times as many devices as the Intel 4004 CPU. This vastly increased complexity, however, is achieved without substantial changes to design time or the size of design team. Furthermore, this is accomplished in spite of the fact that ensuring the correctness of such complex designs is far more difficult.

The key enabler to this rapid pace of progress has been *design automation*, whereby sophisticated tools and algorithms were developed in order to reduce the burden on designers and to enable ever larger design sizes by improving designer *productivity*. Contrast this situation with that observed in the energy distribution market area, where tools are available for *simulation* but the design process is otherwise largely manually. Such manual processes have two significant impacts:

---

[1]We refer, of course, to the well known law in semiconductor integrated circuit scaling.

1. Results end up being highly dependent on the experience and expertise of the particular engineer doing the work.
2. There is no guarantee that any given result is *optimal* in any sense. The authors are aware of cases where multiple commercially-produced solutions to grid design problems are dramatically different from each other in both implementation cost as well as solution quality.

In contrast, EDA tools and algorithms are based on rigorous analysis and optimization algorithms that are the results of many decades of work, and that are frequently compared based on published benchmarks. We believe that the investment made in EDA tools is largely portable to the area of distribution network design and analysis, and this paper will highlight some areas of clear potential in this regard.

III. REPRESENTATIONS OF ENERGY NETWORKS

It is common in the energy utility industry to use Geographic Information Systems(GIS) to represent distribution networks due to it's ability to manage, visualize, and analyze location based data. The degree to which utility companies leverage these capabilities can vary greatly and as a general rule the larger the utility company the higher the level of GIS integration achieved within their business flows. However, even large utility companies fail to integrate the full potential of GIS. Some of the primary uses of GIS today in the energy distribution industry include asset management, outage management, and risk management during the planning and design stages of grid development. One example of using GIS in risk management is to insure that no natural gas pipelines are within a certain distance of the electric grid that would increase the risk of a disaster due to component failure. While these issues are extremely important to consider from a design planning perspective, they are manually performed as a final verification step and not integrated into an automated search for an optimal grid design solution.

The lack of integration is most apparent in the tools used for evaluating the electrical integrity and cost viability of potential grid designs. During the design phase, grid components are manually placed based upon rule-of-thumb spatial constraints that historically produce reasonable electrical performance. The placement is followed by additional manual optimization to tune the electrical parameters of each component to meet electrical constraints. However, interactions between the environment and the electric grid on various time scales ranging from land use classifications (*long*) to weather events (*short*) can impact grid performance. Another example of interactions between the grid and the environment is growth planning. Often such growth plans consider the temporal growth rate in power demand with population increase, but apply this increase uniformly across the service area, ignoring spatial variations in demand that can dramatically alter the optimal design of the grid. To adequately address the predictive modeling requirements of growth planning, spatial-temporal representations of future power demand can -for example- be derived from U.S census data that depict the historical spatial distribution of population growth.

A platform for integrating energy utility GIS within an EDA design flow is shown in Figure 1. Context awareness and pre-

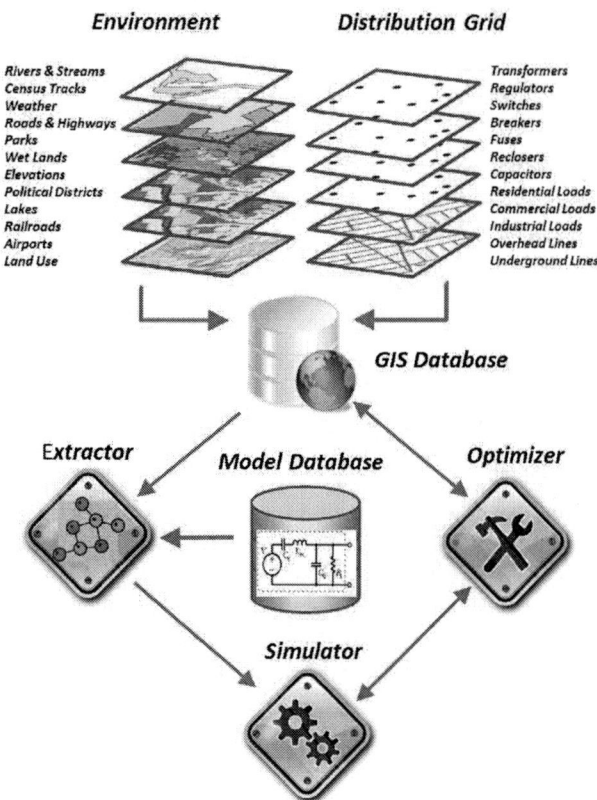

Fig. 1. Platform used to integrate GIS data into an EDA design flow.

dictive modeling of the environment are achieved by integrating *layers* of geographic information such as roads, river networks, lakes, parks, and cemeteries -all of which may represent *blockage areas* to avoid when routing power grid wires. Electric grid components such as transformers, regulators, switches, etc., are represented as point objects in the GIS database, while the power distribution is represented as lines. The spatial database is accessed by the *Extractor* to build an electrical description of the distribution network. Each object instance in the database has a model type in the Model Database that defines an equivalent electrical model topology and corresponding electrical parameters to use.

To give the reader a sense of scale, the total number of grid objects for a service area of roughly 100 square miles used in this study exceeded 30,000. The Extractor interfaces directly to a highly efficient *Simulator* discussed in section IV was developed that could perform an AC analysis on the entire electrical grid, i.e. all 30,000 objects, in a few seconds. The results of the analysis are then used by an *Optimizer* discussed in sections A and B for evaluating the electrical viability of potential solutions.

IV. EFFICIENT SIMULATION OF ENERGY NETWORKS

The simulation of energy networks has been a rich subject with many dedicated journals, conferences, and established simulation packages for a long time [3]. In this work we had some specific needs:

1. We did not require *dynamic* simulation of the distribution network, so we limited our attention to steady state or *static* analysis only.
2. Since the simulator was often in the *inner loop* of complex optimization algorithms we needed the simulator to be as efficient as possible.
3. To enable the most sophisticated type of optimization algorithms (namely gradient-based quadratic programming types of algorithms) we required the calculation of the *sensitivities* of various performance metrics to component values.

One of us (Nassif) had done work previously on the area of Integrated Circuit power grid simulation, and we re-used some infrastructure components we had to develop a static simulator for distribution networks. We write the system of complex linear equations as:

$$Gv = \frac{P}{v} \qquad (1)$$

where G is a matrix of conductances, v is a vector of node voltages, and P is a vector of power loads. This system is non-linear and would normally be solved via Newton-Raphson or some other similar method. Since we expect that the system is well behaved, we postulate that *nominal* value of v is known. This allows us to solve Eq. 1 via an iteration[2] of the form:

$$Gv^{i+1} = \alpha\frac{P}{v^i} + (1-\alpha)\frac{P}{v_{nom}} \qquad (2)$$

where v_{nom} is the expected nominal value of v and the system in Eq. 2 is solved from the parameter α from zero to one. Since the matrix G is not a function of α, the cost of N iterations of Eq. 2 is one factorization of G (the expensive part) and N solutions (relatively inexpensive). In our implementation, which was done using the Python programming language [5], we find that for a network containing 15 thousand nodes, and using $N \approx 10$ the solution can be found in just 3 seconds on a typical modern workstation.

The calculation of sensitivities in linear-time-invariant networks is a well studied area [6]. For our applications, we elected to implement a simple approximation to compute the sensitivity of voltages v to parameter values which requires just another solve of the the converged Eq. 2. The method is similar to [7], but adapted to handle the non-linearity of the system by using a small step size. When converged, with $\alpha = 1$, the solution of Eq. 2 would be:

$$v = G^{-1}q \qquad (3)$$

with $q = P/v$. Consider the case where we need the sensitivity of v to the value of a single conductance element g in the system, we can write the finite difference approximation to that sensitivity as:

$$\frac{dv}{dg} \approx \frac{V'-V}{\Delta g} = \frac{((G+\Delta g)^{-1} - G^{-1})q}{\Delta g} \qquad (4)$$

The key idea is to determine a perturbation of the right hand side q, Δq which allows us to approximate Eq. 4 as:

$$\frac{dv}{dg} \approx \frac{G^{-1}\Delta q}{\Delta g} \qquad (5)$$

[2]Also referred to as a *homotopy* [4].

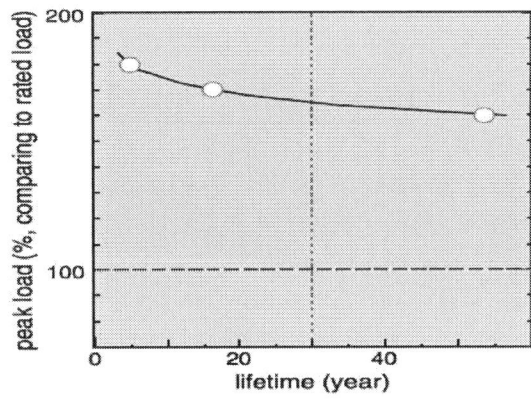

Fig. 2. Lifetime evaluation example of over loaded pole transformer. [10]

The extra term Δq is the amount of extra current that would flow in g if its value was perturbed by Δg and the solution v remained the same. This last point is important since Eq. 2 is non-linear.

The simulator was implemented in the Python language, and rigorously checked against other tools to ensure that it is correct. In the end, the most difficult part of the implementation was not the simulation algorithms, but rather the component models required to model entities such as switches, transformers, regulators, and so on, some of which (e.g. the regulator) are not linear. But a full discussion of those models, and how they are implemented in the context of the homotopy in Eq. 2 is beyond the scope of this paper and will have to wait for a dedicated paper on that topic.

V. OPTIMIZATION OF ENERGY NETWORKS

A. Load Balancing

The power distribution network consists of various types of components such as transformers that provide split-phase power for residential and light commercial services, voltage regulators that maintain a constant voltage level, electrical switches that can be configured to connect or break the network, and conductor wires, just to name a few. Among these components, transformers -which are costly- are the source of power and the useful life of a transformer depends on its usage level relative to its rating (Figure 2). In order to maximize the lifetime of a transformer, it should operate at the minimum level possible, preferably under its rating (dotted horizontal line in Figure 2). A typical distribution network has multiple transformers and ideally the relative usage levels of all transformer need to be evenly distributed to maximize lifetime. In reality, however, the system can be unbalanced and overloaded transformers can become critical failure points. This is the essence of the load balancing problem.

As discussed in section III, a power distribution network can be represented as a graph data structure for further optimizations. We annotate each vertex by the power consumption at that location, as well as total downstream power consumption. For a given vertex v, we define a function $DOWN(v)$ that returns the downstream loads[3]. For each edge e that represents a physical wire a slack can be calculated as follows:

$$SLACK(e) = RATE(e) - CURRENT(e) \qquad (6)$$

where $RATE()$ is a the wire's maximum allowable current rating, and $CURRENT()$ returns the current flowing on the given edge (wire). The slack of a wire indicates the amount of extra current allowed on it without causing thermal overload. Then, the load balancing problem is defined as:

Load Balancing :   In a power distribution network with multiple transformers, redistribute the loads such that 1) each transformer's usage level is below its rating and 2) the downstream loads among them are evenly distributed.

A typical power distribution network consists of multiple sub-networks, connected with switches and extra wires in order to provide redundancy in case of outages. The load balancing algorithm uses this structure by employing the following techniques:

- *Tieline addition* : By adding a new wire that connects a sub-tree of an overloaded network to less loaded network, loads can be shifted.
- *Switch configuration* : If there are turned-off switches that connect multiple networks driven by different transformers, by turning on these switches, some of loads can be shifted from one network to others.

Algorithms 1 and 2 represent heuristic methods to apply tieline addition and switch configuration to solve the load balancing. First the most/least loaded transformers and the amount of load that needs to be shifted are determined (line 2 - 4). Next, in the most loaded transformer network, a set of candidate sub-trees (cut-off points) are identified whose downstream loads are sufficiently large to improve the load distribution (line 6 - 8). Similarly, in the least loaded transformer network, a set of candidate tap points whose slack is large enough to accommodate additional loads from line 4, are generated (line 9 - 12). From these two candidate sets, the best pair is calculated based on the target objective function defined. In the tieline algorithm, the distance between the source and target sub-trees is the target because it is directly related to the cost of a solution. This process is repeated until no further improvement is achieved.

Figure 3 (a) and (b) show the before/after example where the overloaded power distribution network (colored in blue) is relieved by shifting parts of its sub-tree to another network (colored in green). The load shifting was implemented by adding a tieline (pink line) in figure (b).

Algorithm 2 solves the load balancing problem via switch configuration. First, the algorithm tries to find a pair of closed and open switches in the most heavily loaded transformer network (line 6 - 11). The sub-tree of a closed switch sw_j in the most heavily loaded transformer X_s will be shifted to a less loaded transformer after switch configuration. The associated implementation cost of switch reconfiguration is 0 since no additional equipment is required. Unlike the tieline addition algorithm, the criteria for selecting the *best* candidate is rather arbitrary (line 16). One potential criteria is to minimize the difference between the amount of load being shifted (δ_i) and the

---

[3]This is a similar concept to the downstream capacitance of Elmore delay in EDA

Algorithm 1 Load Balancing via Tieline

   ▷ Input: Power Graph PG with Transformer $X_i \in X$
1 **repeat**
2   Let X_s be X_i with max $DOWN(X_i)$
3   Let X_t be X_i with min $DOWN(X_i)$
4     $\delta = DOWN(X_s) - DOWN(X_t)$
5     $S = \varnothing; T = \varnothing$
6   **for each** vertex $v \in X_s$ sub-tree
7   **if**($DOWN(v) \geqslant \delta$)
8   **then** Add v to S
9   **for each** vertex $v \in X_t$ sub-tree
10   **if**($SLACK(v) \geqslant \delta$)
11   **then** Add v to T
12   Find a pair ($s \in S$, $t \in T$) with the shortest distance
13   Add a tieline between (s, t)
14 **until** No further improvement

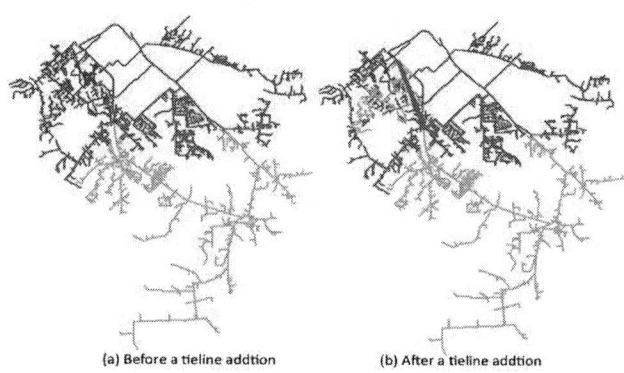

(a) Before a tieline addition   (b) After a tieline addition

Fig. 3. Tieline based load balancing example.

---

slack of the sub-tree from the target transformer $SLACK(p)$ (line 14). Note that since the algorithm relies only on existing switches in the network, it may be more difficult to achieve a precisely balanced network.

The two algorithms are different in terms of cost and granularity and a single application of either method does not necessarily provide a complete solution. We are currently working on more advanced formulations that allow mixing of the two methods.

B. Contingency Planning

Natural disasters are unavoidable, as is concern about possible terrorist attacks on critical infrastructure. Whenever an outage happens, power needs to be re-routed via alternative wires and switches. Reacting to such potential outages is the main subject of *Contingency Planning* research. There are two different angles to the contingency planning. *Reaction plan to outages* is an operational management perspective and concerns how to recover from the power failure. The techniques reviewed in section A can be employed to recover from potential failure of any one device. *Robust planning* is a design planning perspective and is concerned with how to add redun-

Algorithm 2 Load Balancing via Switch reconfiguration

▷ Input: Power Graph PG with Transformer $X_i \in X$
1 **repeat**
2   Let X_s be X_i with $max\ DOWN(X_i)$
3    $\delta = DOWN(X_s) - DOWN(X_t)$
4    $C = \varnothing$
5   **for each** vertex $v \in X_s$ sub-tree
6   **if**($DOWN(v) \geqslant \delta$
7   and there is an open switch sw_i connected to $X_i \neq X_s$
8   and there is a closed switch sw_j between v and X_s)
9   **then**$\delta_i = DOWN(sw_i)$
10   Add (sw_i, sw_j, δ_i) to C
11   **for each** $(sw_i, sw_j, \delta_i) \in C$
12   Let p be a parent of sw_j
13   **if**($SLACK(p) \geqslant \delta_i$)
14   **then** Add v to T
15   Let c be the best candidate $\in C$
16   From $c = (sw_i, sw_j, \delta_i)$, turn on sw_j, turn off sw_i
17 **until** No further improvement

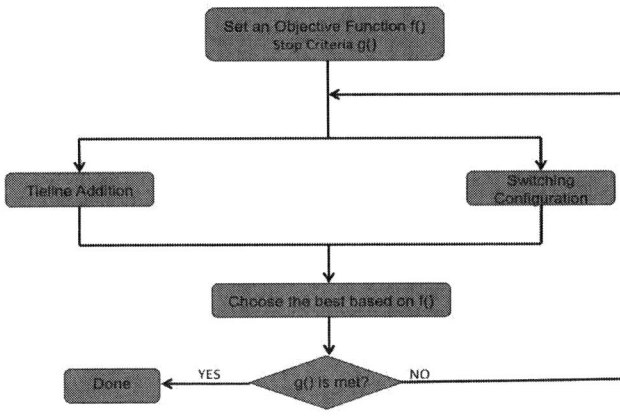

Fig. 4. Flow of load balancing with multiple optimization techniques.

dant resources *a priori* to the network in order to increase fault tolerance. The goal of contingency planning is to determine the optimal set of alternative resources which can achieve a specific recovery level.

With contingency planning formulation, the following aspects can be integrated:

- Geographical model: A certain region is more prone to an outage and cost more to recover. For example, a probability of failure can be calculated based on geographical location.
- Weather model: With advanced weather forecast model, accurate prediction is possible when and where a natural disaster might occur.
- Societal model: A certain class of customers such as a hospital or government buildings may be deemed more important than normal end users and requires the higher reliability standards.

TABLE I
WIRE PARAMETER MATRIX

Wire Type	Thermal Limit (A)	Impedance (Ohms/mile)	Cost ($/mile)
#4 ACSR	140	0.399	10,000
#1/0 ACSR	240	0.257	23,000
#4/0 ACSR	360	0.200	41,500

C. System Upgrades

Optimization of energy networks in steady-state operation requires meeting multiple parameters for the purpose of both power quality and electrical device safety. These parameters are described by ANSI standards from the power systems industry [8]. Typical metrics to monitor and optimize include voltage drop, load current with respect to capacity, and the power factor at end-user loads. The criteria to follow for these parameters is specified by the owner of the energy network utility and can be based on standards by regulatory bodies (such as the Federal Electric Reliability Commission (FERC) in the USA), though some utilities may choose to follow more stringent power quality standards.

As usage increases or equipment degrades, it is common to *upgrade* outmoded energy network components in order to meet system quality criteria. Common components to upgrade include: overhead wires, underground wires, automatic voltage regulators, substation transformers, power-factor correction capacitors, and even protective devices such as fuses and reclosers (when the criteria pertains to reliability in non-steady state conditions).

Performing a system upgrade on a certain component that is currently not meeting some criteria requires some information about the component. As an example, suppose that there are some overhead wires that are loaded beyond their rated thermal loading. Algorithm 3 shows one possible methodology to find and replace such wires.

Algorithm 3 Wire Upgrade

▷ Input: Power Graph PG with wires $w_i \in W$
1 **repeat**
2   **for each** $w_i \in W$
3   **if** $(SLACK(w_i) \leqslant 0)$
4   **then**
5   Find cheapest T with $RATE(T) \geqslant CURRENT(w_i)$
6   Upgrade w_i to a new wire type T
7 **until** No further improvement

The algorithm identifies wires with negative or zero slack, and replaces each such wire with the lease expensive alternative that meets the slack requirements. The available wire types are assumed to known, and Table I shows a small example.

However, such a brute force method of upgrading every single overloaded wire is inefficient. Wire types with higher thermal limits will typically have lower impedance which will in turn reduce loss and result in lower overall currents in the system. While this is a second order effect, it shows that the situation can be a little more complex. Upgrades can also occur in response to excessive voltage drop or other performance problems. We are working on more sophisticated multi-objective upgrade algorithms that can account for such interactions.

D. Other Optimization Problems

In the previous sections we described a number of optimization problems in the energy distribution area. This is by no means an exhaustive list, and there are a number of other important and practical problems that require solutions.

One problem that we will defer to a follow-on work is *loss minimization*. Approximately 6.7% of the electricity generated in the United States is lost due to inefficiencies in the transmission and distributions system. This amounts to about $25 billion dollars per year [9]. Preliminary work has shown that there is wide divergence between network efficiency across various operators and systems in the United States, with loss varying from 3% to as much as 15%. Systematic differences such as urban vs. rural do not explain the magnitude of this wide range of losses. We believe that applying EDA algorithms in this area is both interesting and societally important.

VI. FUTURE DIRECTIONS

There is no doubt that energy grid optimization is an important problem that can generate significant benefits for society. However, most current research appears to be focused on *operational management* addressing problems such as the reaction to outages, monitoring/adjusting power quality, and equipment management. At the other end of spectrum, a whole new research opportunity crops up in the *design planning* realm. Planning for outages, scheduling equipment upgrades, planning for power demand growth, building a fault tolerant distribution network are just a few examples of promising research topics which are heavily under researched in our view. With the tight integration of geographical, societal and weather models into energy distribution system, we believe that there are significant new research opportunities arising in this area.

ACKNOWLEDGMENTS

Authors would like to thank Hala Ballouz and Billy Yancey from Electric Power Engineers, Inc. in Texas for their valuable discussions.

REFERENCES

[1] Wikipedia Online Encyclopedia, http://en.wikipedia.org/wiki/List_of_countries_by_electricity_consumption

[2] Hadley, S.W., "Evaluating the impact of Plug-in Hybrid Electric Vehicles on regional electricity supplies," *iREP Symposium*, Aug. 2007 http://ieeexplore.ieee.org/stamp/stamp.jsp?tp=&arnumber=4410538&isnumber=4410507

[3] Mo-Shing Chen and Dillon, W.E., "Power system modeling," *Proceedings of the IEEE*, July 1974 http://ieeexplore.ieee.org/stamp/stamp.jsp?tp=&arnumber=1451473&isnumber=31182

[4] Melville, R.C. and Trajkovic, L. and Fang, S.-C. and Watson, L.T., "Artificial parameter homotopy methods for the DC operating point problem," *IEEE Transactions on Computer-Aided Design of Integrated Circuits and Systems*, Jun 1993 http://ieeexplore.ieee.org/stamp/stamp.jsp?tp=&arnumber=229761&isnumber=5930

[5] Python Official Website, http://www.python.org/

[6] Director, S., "LU Factorization in Network Sensitivity Computations," *IEEE Transactions on Circuit Theory*, Jan 1971 http://ieeexplore.ieee.org/stamp/stamp.jsp?tp=&arnumber=1083206&isnumber=23409

[7] Swamy, M.N.S. and Roytman, L.M., "An alternate approach to deriving sensitivity formulas for linear time invariant networks," *Proceedings of the IEEE*, Aug. 1980 http://ieeexplore.ieee.org/stamp/stamp.jsp?tp=&arnumber=1456049&isnumber=31292

[8] "IEEE Recommended Practice for Electric Power Distribution for Industrial Plants (IEEE Red Book)," *ANSI/IEEE Std 141-1986*, pp.1,609, Oct. 10 1986 http://ieeexplore.ieee.org/stamp/stamp.jsp?tp=&arnumber=35071&isnumber=1456

[9] "State Electricity Profiles," U.S. Energy Information Administration http://www.eia.gov/electricity/state/unitedstates/index.cfm/

[10] "Power Distribution Planning Reference Book, Second Edition" *CRC Press*, Dec. 2010

A Model-Based Design of Cyber-Physical Energy Systems

Mohammad Abdullah Al Faruque, Fereidoun Ahourai
Department of Electrical Engineering and Computer Science
University of California, Irvine
Irvine, CA, USA
Email: {alfaruqu, fahourai}@uci.edu

Abstract— *Cyber-Physical Energy Systems* (CPES) are an amalgamation of both power gird technology, and the intelligent communication and co-ordination between the supply and the demand side through distributed embedded computing. Through this combination, CPES are intended to deliver power efficiently, reliably, and economically. The design and development work needed to either implement a new power grid network or upgrade a traditional power grid to a CPES-compliant one is both challenging and time consuming due to the heterogeneous nature of the associated components/subsystems. The *Model Based Design* (MBD) methodology has been widely seen as a promising solution to address the associated design challenges of creating a CPES. In this paper, we demonstrate a MBD method and its associated tool for the purpose of designing and validating various control algorithms for a residential microgrid. Our presented co-simulation engine GridMat is a MATLAB/Simulink toolbox; the purpose of it is to co-simulate the power systems modeled in GridLAB-D as well as the control algorithms that are modeled in Simulink. We have presented various use cases to demonstrate how different levels of control algorithms may be developed, simulated, debugged, and analyzed by using our GridMat toolbox for a residential microgrid.

I. INTRODUCTION AND RELATED WORK

Currently in the power systems industry, there is a paradigm shift from the traditional, non-interactive, manually-controlled, power grid to the tight integration of both cyber information (computation, communications, and control - discrete dynamics) and proper physical representations (the flow of electricity governed by the laws of physics - continuous dynamics) at all scales and levels of the power grid network. This new grid which features this cyber and physical combination is termed as *Cyber-Physical Energy Systems* (CPES) [1], and it is expected to improve the reliability, flexibility, efficiency, cost-effectiveness, and security of the future electric grid [2, 3]. However, the introduction of *Distributed Energy Resources* (DERs) including renewable sources, and new types of loads (specifically *Electric Vehicles* (EVs)) in the residential distribution grid presents the challenge of multi-level monitoring and control for supply and demand management to an already complex and heterogeneous grid. A consequence of the rapid addition of these DERs would be that traditional power system design methodologies become more time-consuming to perform and that the ability to preempt grid problems would be more difficult. Authors in [4] demonstrate a potential model of future energy systems: a cyber-physical network which consists of many diverse energy components, each of which is equipped

with a local embedded system. Developing an optimal grid is a complex task, due in part to the need for modeling and analyzing the system at both different scales and across different information domains. To solve this issue, *Model-Based Design* (MBD), which has been proposed in recent publications [4, 5, 6], allows for modeling both the physical and cyber components concurrently. Moreover, it allows cyber-physical co-simulation to explore for various design alternatives required for designing, validating, and testing. With such an implementation, power systems may be virtually analyzed and advanced control algorithms may be developed without the need for physical prototypes because software would be able to estimate the dynamic behavior of the system under a large variety of conditions.

In [5], a generic MBD methodology for a cyber-physical system design has been discussed. In [4], a novel cyber-based dynamic model is proposed where the mathematical model is primarily based on the cyber world which is helpful for distribution decision making. While various domain-specific power system simulation tools currently exist, there is a critical gap of advanced system-level design methodology and tools in the modeling and simulating (of both discrete and continuous dynamics) of the cyber and physicals portions of CPES concurrently. State-of-the-art domain-specific power system modeling tools lack the capability to capture the cyber-components of CPES during modeling and simulation. On the other hand, tools capable of describing the discrete event dynamics of the cyber components are not well equipped with the models needed to best represent the physical dynamics of power systems. Therefore, cyber-physical co-simulation of different domain specific tools has been seen as the possible enabling technology [7, 8, 9].

In [6], a MBD framework that uses SystemC for the purpose of designing embedded systems for energy management is proposed. A residential electrical energy simulation platform (HomeSim) is proposed in [10] to evaluate the impact of technologies such as renewable energy and improved battery storage through centralized and distributed energy managements as well as smart appliances. A co-simulation tool MATLAB & EnergyPlus [11] for building energy automation solution (MLE+) is proposed in [7]. In [9], a Ptolemy II-based [8] co-simulation software environment, *Building Controls Virtual Test Bed* (BCVTB), for coupling different simulation programs is presented (co-simulation including EnergyPlus [11] and Matlab). In [12], a co-simulation of GridLAB-D [13] and MATLAB is used to model integrated renewable energy and

---

978-1-4799-2817-0/14 $31.00 © 2014 IEEE

Demand Response[1] (DR). In this co-simulation, the control developed in MATLAB (which acts as slave of the simulation) is executed as a co-process of GridLAB-D and cannot support a debugging utility function for the control engineers. To achieve cyber-physical co-simulation capability and to enhance the physical library of GridLAB-D, in [14], authors have presented a technique of using a functional mockup interface to integrate Modelica-based components with GridLAB-D. In [15], a co-simulation platform that consists of communication and power systems (ns-3 [16] and GridLAB-D) is presented.

CPES covers various scales and levels including residential/commercial/industrial buildings [17, 18], microgrid (see Section III for details), distribution (substation and feeder levels) and transmission levels [19], and generation side [20]. In this paper, we consider a residential microgrid as an example of a CPES and demonstrate the capability of the MBD methodology. Moreover, we present our cyber-physical co-simulation tool GridMat (see Section IV.C) for the purpose of residential microgrid modeling and simulation.

The rest of the paper is organized as follows. Section II presents the itemized content of the paper. After describing the residential microgrid in Section III, we present the MBD of the residential microgrid in Section IV. Experimental results are discussed in Section V before concluding the paper in Section VI.

II. CONTENT OF THIS PAPER

- We present a model-based design methodology for a residential microgrid which is an example of a CPES. We also present a MATLAB/Simulink toolbox (GridMat) where structural and behavioral aspects of a residential microgrid may be modeled using GridLAB-D from the *Graphical User Interface* (GUI) of the GridMat tool, and various control algorithms may be modeled using the graphical language of the Simulink. Moreover, using this GridMat tool, the developed cyber-physical model may be simulated, debugged, and analyzed.

- In the MBD methodology, MATLAB/Simulink utilities such as the embedded code generator are used for implementable software generation in order to support a high fidelity *Hardware-In-Loop Simulation* (HILS).

- To demonstrate the capability of the presented MBD methodology for a residential microgrid, we have modeled a large residential microgrid (using IEEE 13 node test feeder and 1000 houses with various appliances, etc.) and have studied various use cases of developing control algorithms that would be able to address some of the challenges that might occur in a residential microgrid such as demand response, peak load reduction, and reliability (voltage drop control at the demand side).

III. RESIDENTIAL MICROGRID

The traditional power grid where electricity typically flows only from large scale generation sites (mostly fossil-fuel-powered and nuclear powered) to a distribution grid through the transmission lines has been transformed to a more (bi-/)multi-

directional electricity flow system due to integration of various DER including renewable sources. This shift in electricity flow, rising costs of electricity production, and a better understanding of environmental impacts require balancing mechanisms between the supply and the demand of electricity [21]. As a solution to this challenge, microgrid solutions have been proposed both in academia and industry. A microgrid is a localized and semi-autonomous group of electrical energy resources (which consists of different storage and generator technologies such as photovoltaics, wind turbines, fuel cells, and microturbines) and electrical loads (industrial, commercial, and residential consumers) that connects to the traditional power grid (macrogrid) [21]. A microgrid can operate in two modes: (1) grid-connected mode where the microgrid is connected directly to the macrogrid, and (2) island mode where the microgrid can disconnect itself from the macrogrid so as to operate autonomously in physical and economic conditions. This capability to operate in these modes allows a microgrid to provide additional reliability to the demand side.

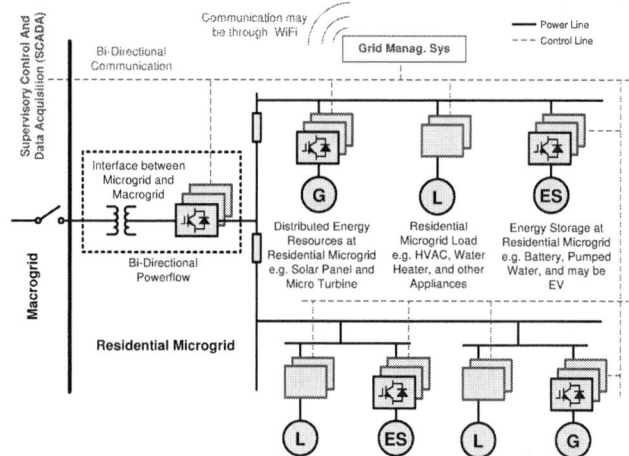

Fig. 1. A conceptual view of a residential microgrid

Besides typical deployment of microgrid in military installations, universities, and remote locations, a microgrid may be deployed in a typical residential area[2]. Fig. 1 illustrates a high level overview of a residential microgrid which consists of residential loads as well as DERs such as an energy storage.

Safety, security, protection, demand-side energy management, power quality, imbalance/asymmetry, plug and play operation of DER systems, distributed voltage/frequency profile control, and non-autonomous/autonomous operation are some of the challenges in designing and operating a microgrid in the residential area [22]. To solve these challenges, various local distributed controllers will need to be designed typically at the distribution level.

The design and validation of such a residential microgrid is a complex and demanding task because of the heterogeneous multi-domain components of the system. MBD method enables the designers to develop and validate the CPES by abstracting the details of a large system at a high level so as to perform the needed design space exploration in a short time; moreover, it is useful to validate and evaluate the design quickly without using real physical plant.

---

[1] Demand response in power grids is a dynamic demand mechanism to manage customer power consumption in response to supply conditions due to the balancing of supply and demand of power.

[2] In the case study of this paper, we have considered a microgrid that is deployed in a residential area, and name it as a residential microgrid.

978-1-4799-2817-0/14 $31.00 © 2014 IEEE

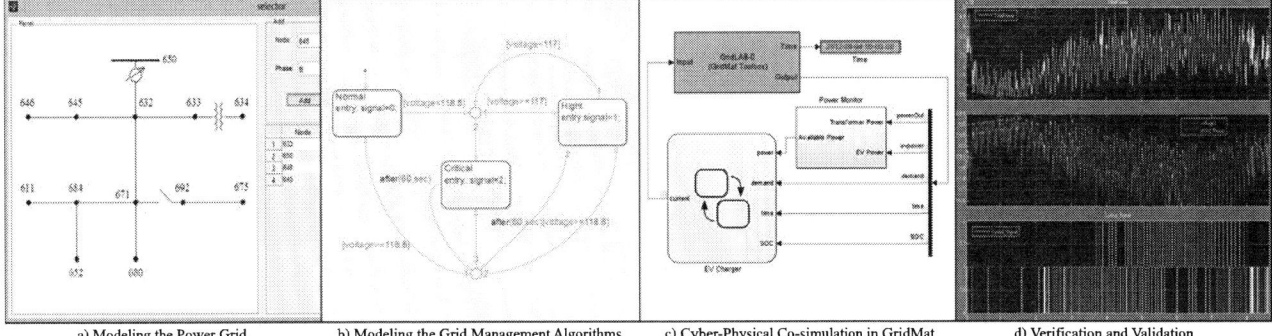

| a) Modeling the Power Grid | b) Modeling the Grid Management Algorithms | c) Cyber-Physical Co-simulation in GridMat | d) Verification and Validation |

Fig. 2. Using GridMat tool for MBD of the residential microgrid

IV. MODEL-BASED DESIGN

In the scope of the presented MBD methodology, we develop a *model-in-the-loop simulation* (MILS) environment using multi-tool co-simulation capabilities. Our methodology is capable of generating control software which will allow us to validate our model and increase the fidelity of the simulation by using both *Software-In-the-Loop Simulation* (SILS) and HILS environments.

Before starting to model a residential microgrid, the requirements, components/subsystems, structure, location, and problems should be defined. For example, we want to design a residential microgrid which consists of 1000 single family houses with multiple instances of level II EV chargers and low-rate step-down transformers in a mid-size distribution power grid. This residential microgrid should be able to manage and orchestrate the operation of all the individual chargers and transformers with respect to EV charging, demand response, and reliability of power system using residential load control. By using this information, the CPES could be modeled, simulated, and validated according to the MBD methodology. Fig. 2 shows the main steps of the MBD methodology of a residential microgrid. We explain these steps in detail as follow:

A. Modeling the Power Grid for the Residential Microgrid

To develop and analyze a CPES, the model of the physical plant which consists of both the structural and the behavioral models should be developed according to the requirements. The structural model of a residential microgrid includes the building-architectural model of houses, end-use appliances, distributed energy resources, transformers, and distribution power grid. On the other hand, the dynamic parts of a residential microgrid such as the appliance tasks, energy demand, weather, etc. are captured by behavioral modeling.

GridLAB-D is the most promising state-of-the-art tool for modeling a distribution power grid, especially a residential microgrid. GridLAB-D is an open-source, agent-based, multi-domain (power domain, market domain, weather domain, and built-environment domain such as building-related structural parameters) modeling and simulation tool for distribution power systems which is developed by PNNL [13]. *Discrete Event* (DE) model of computation, and agent-based simulation enable GridLAB-D to model and simulate a large scale distribution power grid at different levels of granularity which allows for the modeling of the end-use appliances necessary for a residential microgrid. However, GridLAB-D is limited at modeling and designing the discrete dynamics required for the microgrid management, e.g., the embedded systems and the control algorithms in CPES.

B. Modeling the Grid Management Algorithms

We have used Simulink the data flow graphical programming language tool to design, model, and simulate various controllers for a residential microgrid. MATLAB/Simulink has a rich library to implement various types of control algorithms, e.g. *Model Predictive Control* (MPC).

C. Cyber-Physical Co-Simulation

After modeling both the continuous and discrete dynamics of the residential microgrid, the integrated model needs to be simulated on a desktop-based simulator to analyze, verify, and validate according to design requirements. By adjusting the parameters of the model, different behaviors of the model may be captured through cyber-physical co-simulation which allows designers to validate the design precisely. For a residential microgrid, we have developed GridMat [23] which is a multi-tool/multi-domain co-simulation platform to overcome the limitation of modeling and simulation of a cyber-physical distribution power system.

GridMat: is a cyber-physical MATLAB toolbox that supports modeling, simulation, analysis, and validation of a distribution power system such as a residential microgrid. The major features of GridMat are: (1) access to all MATLAB toolboxes as well as the Simulink graphical programming language to design, develop, and debug advanced, hierarchical, and distributed control algorithms. Therefore, device-level and various *Supervisory Control And Data Acquisition* (SCADA)-level control algorithms may be developed for the residential microgrid very effectively; (2) a user friendly GUI for the state-of-the-art GridLAB-D tool; (3) a model creator that helps designers create the structural and behavioral models of a residential microgrid; (4) a data analysis utility that enables the designers to analyze the simulation results and the grid impact under various scenarios and allows modification of the controllers; (5) an embedded C code generator allows the designer to conduct a HILS for the purpose of validating the fidelity of the design.

In GridMat, GridLAB-D and MATLAB/Simulink communicate together through the HTTP protocol over the TCP/IP stack. A HTTP client wrapper is developed in GridMat in order to handle the co-simulation data exchange through HTTP between GridLAB-D and MATLAB/Simulink. Fig. 3 illustrates

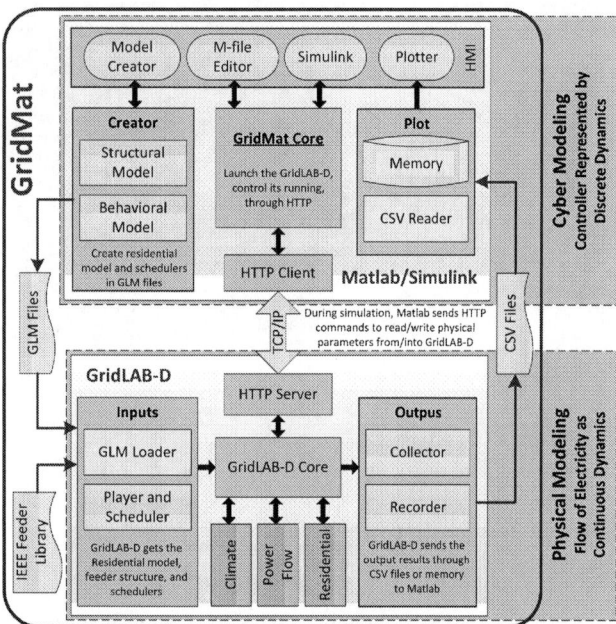

Fig. 3. Our GridMat architecture

TABLE I
HOUSE AND APPLIANCE SPECIFICATIONS

Object	Type 1	Type 2
Number of stories	1	2
Floor area	2100 sq.ft	2500 sq.ft
Heating system	GAS	GAS
Cooling system	Electric	Electric
Thermal integrity	NORMAL	ABOVE
Motor efficiency	AVERAGE	AVERAGE
Number of occupants	3	5
Heating set point	68° F	68° F
Cooling set point	72° F	72° F
Light power	1.2 kW	1.5 kW
Dishwasher power	1 kW	1.5 kW
Water tank volume	40 gal	50 gal
Water heater power	3 kW	4 kW
Clothes washer power	0.8 kW	1 kW
Miscellaneous	0.7 kW	0.8 kW
Compressor power	0.5 kW	0.6 kW
Oven	2.4 kW	3 kW
Oven set point	500° F	500° F
Dryer power	2 kW	3 kW

the structure of GridMat in detail. The GridMat core is the master of the co-simulation that runs an instance of GridLAB-D, makes a TCP/IP connection to the GridLAB-D instance, controls the co-simulation run time, and coordinates the writing and reading of parameters from and to GridLAB-D. Also, the core runs controllers which are implemented in the m-file editor or Simulink. The creator block in MATLAB creates GLM files (the GridLAB-D model files) based on the structural and behavior models which the user defines through the *Human-Machine Interface* (HMI). The GridLAB-D core loads the input GLM files and the IEEE test feeder library to simulate the physical plant of the microgrid by using different modules such as climate, power flow, and residential. It also generates some output files (CSV files) which may be used by the GridMat plotter in MATLAB to show simulation results to the user for validation and verification purposes.

D. Verification and Validation

The behavior of the control algorithms may be captured by formal verification and validation. In this step, the simulation result is analyzed and verified to ensure that the design requirements have been satisfied. An iterative approach may be applied here for debugging and validating the design by revisiting previous steps to modify the models of the physical plant, embedded computational system, control algorithm, and parameters of the models. The embedded software part may be tested without the real physical plant by using the HILS capabilities of GridMat. In HILS, a hardware platform should be selected to interact with the physical plant and run the control algorithm while it is able to persist in the environment. Then, the modeled controllers should be synthesized to generate an embedded software that may be executed on the real hardware. To perform HILS, the physical plant runs on the simulation tool (GridLAB-D) while the controller runs on the real embedded hardware and communicates with the simulated physical plant through our GridMat.

V. CASE STUDY EVALUATION

We demonstrate our tool and methodology using various use cases of a residential microgrid. The detailed residential microgrid model is present in [24]. In this model, 1000 residential single family houses are randomly distributed in the IEEE 13 node test feeder [25]. Each step-down transformer is connected to a node of the IEEE 13 node test feeder, and hosts a range of 3 to 7 houses. These step-down transformers range in rating from $15KVA$ to $35KVA$, depending on the number of connected houses. The houses are randomly selected from two types of houses (Type1 and Type2) which have a variety of end-use appliances such as dishwasher, lights, water heater, plug load (miscellaneous), refrigerator, clothes washer, dryer, and oven. Table I describes the specifications of our model.

We have chosen Newark, NJ, USA as the location for our residential microgrid and have simulated the model for both Summer and Winter seasons. All the structural and behavioral modeling of the power system (the physical plant) has been performed through our developed GridMat tool. Fig.4 illustrates the total average power consumption of a typical Summer and Winter day for two types of houses in our developed residential microgrid model. We have validated the fidelity of our model and the simulation results of various end-use loads and/or appliances as well as the average total power consumption of the houses with different sources [24, 26].

As we discussed in section III, demand response for peak-load reduction, power quality, and voltage drop control are some of the challenges for current residential microgrids. We propose implementing distributed control mechanisms where the controls of such are developed by the MBD methodology in GridMat by using different MATLAB/Simulink toolboxes such as MPC and Stateflow.

A. Residential Collaborative EV Charging

Higher rates of EV penetration will have a negative impact on the electrical power grid because uncoordinated EV charging on a mass scale at the secondary distribution grid would negatively affect both the total load and peak load power [27]. The results in [27] show that 30% of peak load power usage can be attributed to EV charging in the distribution grid. To address this negative effect, one solution could be to upgrade the power

978-1-4799-2817-0/14 $31.00 © 2014 IEEE

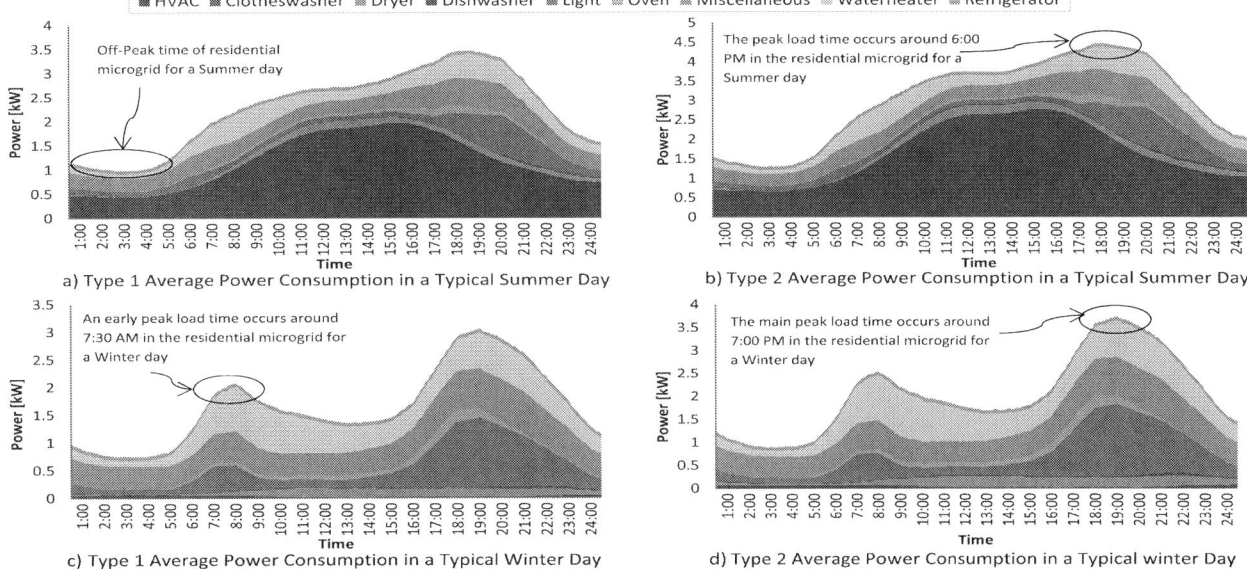

Fig. 4. Average power consumption of a house in our residential microgrid

system infrastructure by changing the transformers and adding more power plants to provide more energy to the residential grid [28, 29]. Unfortunately, this solution would undermine the economic and environmental benefits of EVs. Another solution, which would not have such a drastic downside, could be to control and coordinate the EV charging locally or at the substation level to mitigate the impact of EV charging from generating a peak [30]. In order to implement this solution and to balance the supply and demand, we need new control algorithms which may be tested, verified, and validated through MILS. Our presented MBD method and the GridMat tool may help in exploring the possible alternatives.

We have demonstrated the "fair-demand" control algorithm to coordinate the EV charging by adjusting the charging current of each EV according to its demand and the available power at the transformer level. EV demand is defined by the expected amount of energy that the EV needs to drive for the next trip divided by the duration that the EV stays connected to the charger. Equation 1 shows the demand definition in which D_i, E_i, and t_i are the demand, the expected needed energy for the next trip, and the time to leave for the next trip, respectively.

$$Di = Ei/ti \qquad (1)$$

The "fair-demand" algorithm uses weights based on the demand parameters to adjust the charging current. According to Equation 2, W_i is the weighted demand for an EV.

$$Wi = \frac{Di}{\sum Dj} \qquad (2)$$

We have assumed that the level II EV chargers with the maximum rate of $7.2KWh$ are used in the residential microgrid model. Since the level II EV charger works at $240V$, the maximum current for charging, per Ohm's Law, is $30A$. Equation 3 shows the C (current of charging) according to the weighted demand. P is the available power at the transformer level.

$$Ci = Min(30, \frac{Wi * P}{240}) \qquad (3)$$

The "fair-demand" algorithm will equally share the rest of the available power to charge the battery full after all EVs receive their demanded energy. Fig. 5 illustrates the state machine of the "fair-demand" algorithm implemented in Simulink.

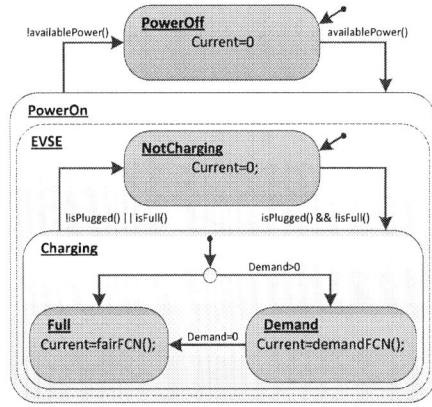

Fig. 5. Stateflow model of the "fair-demand" EV charging algorithm

We have modeled the behavioral aspects of the EV and the *Electric Vehicle Supply Equipment* (EVSE)[3] as per [24] where EV follows Gaussian distributions for the arrival and the departure time. The mean time of arriving and leaving are 5:30 PM and 7:30AM, respectively. The EVSE controls the charging current and receives some information from the EV such as the time to leave, the expected next trip energy demand, and the specifications of the EV. Besides the "fair-demand" control algorithm, we also demonstrate how to develop the "fair-shared" control algorithm [32] that may equally divide the available power among all the EVs connected to one transformer in our GridMat tool. Moreover, we implement a EV charging control policy that supports "deferral-based" EV charging (load-

---

[3]Our proposed model and algorithm are for advanced EVSEs which are able to provide variable rate current supply (multiple amperage adjustment capability) on-demand [31]

978-1-4799-2817-0/14 $31.00 © 2014 IEEE

shifting) where all householders defer the EVs charging to 9:00PM due to cost of electricity and peak load time of residential microgrid. In this mode, all EVs are charged at a rate of 3.2KWh to prevent the transformer from overloading. These control algorithms were experimented on, and the simulation results of them are shown in Fig. 6.

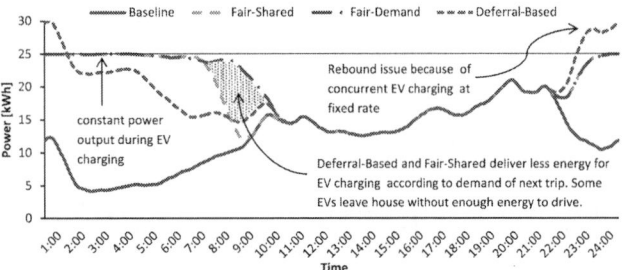

Fig. 6. EV charging simulation result at transformer level

While analyzing for a suitable control algorithm, we see that the result of the "deferral-based" algorithm shows a rebound effect due to all the EVs starting to charge after 9:00PM. However, the "fair-shared" and the "fair-demand" do not show such a rebound effect and actually keeps the power output of the transformer at its nominal rate (25KW with 5 connected houses). In general, the "fair-demand" delivers more energy to the EV compared to the "fair-shared" and the "deferral-based" control algorithms, because the "fair-demand" algorithm charges EVs according to their demand and tries to deliver all necessary energy to all the EVs before they leave the houses. However, in the "deferral-based" and the "fair-shared" algorithms all EVs receive equal energy, though at the cost of some of the EVs leaving their houses without sufficient energy to complete the expected trip. Such analysis at a high abstraction level in the MBD allows for development of control algorithms in a rapid and efficient manner, while also assessing the benefits and consequences of implementing them.

B. Residential Demand Response

DR mechanisms try to reduce the customer power consumption during peak load times or in response to market prices. Since the majority of energy demand, which accounts for 38% of the total energy usage, comes from residential buildings due to a large portion (on average 33%) being used in *Heating, Ventilation and Air Conditioning* (HVAC) systems, it is an attractive target for demand side energy management, especially during the peak load time period [33, 34]. Also, the water heater is another flexible load in most homes which significantly helps to reduce household power consumption during peak load time. Given that peak load time is generally a critical time period in the energy market, many utilities have a vested interest in encouraging their consumers to optimize their power consumption. If the demand side consumption could be appropriately curtailed, utilities would not need to add more power plants to compensate for the extra demand of energy. Some examples of demand control policies that some utilities might implement include *Time-Of-Use* (TOU) rate and *Direct Load Control* (DLC) [35] to encourage customers to reduce their demand during the peak load time. In DLC, customers make a contract with utilities, granting them the ability to control some of their appliances during the peak load time in return for some savings on

their electricity bills [35]. Examples of DLC that utilities commonly use include reducing the heating set point of the water heater and/or increasing the cooling set point of the HVAC system based on the weather. We have used MPC to control power consumption of the houses (see Fig. 7) according to a DLC signal while the inside temperature and water temperature are within the ANSI/ASHRAE specified [36] comfort zone[4].

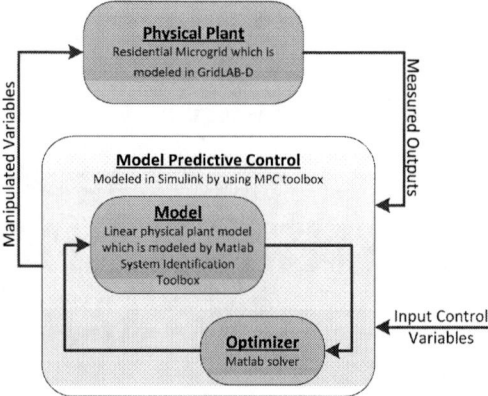

Fig. 7. Model-based design of MPC in residential microgrid

Equation 4 shows the cost function of the MPC controller that we used to control the power consumption of a house during DLC. P_{pred}, P_{ref}, u_i, and u_{ref} are predicted power output, desired power output, manipulated variable, and optimal resting value, respectively. In this cost function, w is the weight for each of the variables.

$$ I = w_1(P_{pred} - P_{ref})^2 + \sum_{MV} w_{2,i}(\Delta u_i)^2 + \sum_{MV} w_{3,i}(u_i - u_{rest})^2 \quad (4) $$

In such a DR scenario, the utility sends a *Demand Response Signal* (DRS) to the *Home Energy Manager* (HEM) which changes the cooling and heating set points of the HVAC and the water heater, respectively. During the experiment, the DRS is sent between 4:00 PM to 8:00 PM to reduce the power consumption to 3KW. We have validated the model for two different algorithms, MPC and DLC. In both the algorithms the cooling and heating set points change between $70°F$-$78°F$ and $115°F$-$130°F$ (comfort zone). In the DLC algorithm, the HEM increases the cooling set point to a maximum value of $78°F$ and decreases the heating set point to a minimum value of $115°F$ when the DRS is issued. However, these manipulated variables are changed by MPC according to a future prediction of the power output, weather, etc. Fig. 8 shows the results of MPC and DLC as well as the baseline which has no controller.

Our MBD methodology allows us to quickly validate various control algorithms for implementing a DR program by the utilities. During validation, we have observed that MPC reduces the power consumption to 3KW when DRS is generated; however, DLC is not able to keep power consumption at 3KW in this period. Moreover, DLC shows rebound effect after the DRS because of decreasing cooling set point and increasing heating set point.

---

[4]Comfort zone is defined as a term of the *predicted mean vote* (PMV) which is a function of the average temperature of the air surrounding the occupant, air speed, humidity, time, location, human activity, expected human clothing, etc.

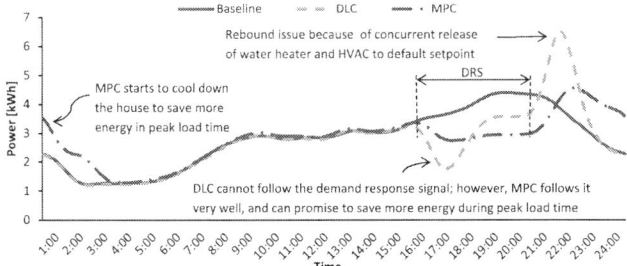

Fig. 8. Average power consumption of a house using demand response signal

C. Grid Reliability

Controlling the voltage drop seen at the extremities is one of the challenges in designing and developing a reliable residential microgrid. One of the ancillary services[5] in the power system is to maintain the voltage level at a particular range to maintain reliable operations of the power grid and prevent adverse effects on the operations of equipment. Typically ancillary services are provided by the generation side; however, recent researches show that ancillary services may be provided through demand side energy management in a residential microgrid [37]. The voltage is not constant through the power system because the conductors and power system components exhibit an impedance to the flow of current and the voltage tends to decrease as we move closer to the load. Therefore, voltage drop may be represented as:

$$\Delta V_{drop} = I_{load} \times Z_{cond} \qquad (5)$$

The power consumption of a load, and the impedance of conductors may be represented as:

$$P_{load} = I_{load}^2 \times Z_{load} \qquad (6)$$

$$Z_{cond} = Z_{line} + Z_{trans} \qquad (7)$$

In Equation 7, Z_{line} and Z_{trans} are the transmission line and transformer impedance. Finally, voltage drop is represented in Equation 8. This equation shows how power consumption, load impedance, line impedance, and transformer impedance impact the voltage drop.

$$\Delta V_{drop} = \sqrt{\frac{P_{load}}{Z_{load}}} \times (Z_{line} + Z_{trans}) \qquad (8)$$

In a typical power grid, the voltage drop can be controlled by using voltage regulators and transformer taps. However, in microgrid, these solutions are not very effective because the output voltage of the renewable distributed energy resources such as a solar photovoltaic may be reduced by cloudy weather [37]. In the scope of this paper, we demonstrate through our GridMat tool how to develop an algorithm to reduce the voltage drop for a residential microgrid by reducing demand during lower system voltage. In this case, we have used smart appliances to control local voltage drop by reducing its power consumption. A dryer consists of a coil to heat up the dryer and a motor to turn the drum. Since the coil of dryer consumes a large amount of power and that most consumers usually do not care (excessively) about the duration it takes to complete a drying task, we have developed a control algorithm to reduce the voltage drop by operating the dryer in an energy saving mode.

We have defined three modes of operation for a dryer: (1) normal; (2) high; and (3) critical. In normal mode, the dryer works with full capacity of its coil. The dryer reduces the coil capacity to 70% during high mode and turns off the coil in a critical mode. The motor of the dryer works normally during all of the above mentioned operation modes. The users can switch from the high or critical modes to the normal mode if they want to speed up the drying task at any time. The controller senses the local voltage and switches from normal to high mode when the voltage is less than $118V$ and from high to critical when the voltage is less than $116V$. In reverse direction, we add a $0.4V$ dead band; therefore, it is $118.4V$ to switch from high to normal and $116.4V$ to switch from critical to high.

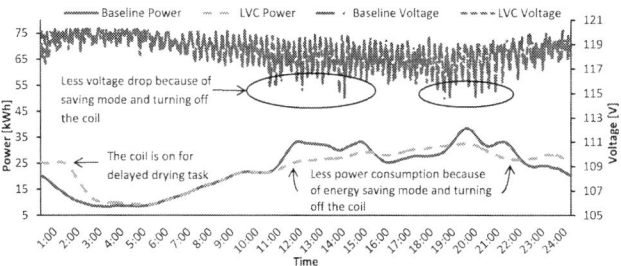

Fig. 9. Local voltage control by using smart dryer

Fig. 9 illustrates the results of the local voltage control that come from using a smart dryer. The *Local Voltage Control* (LVC) algorithm shows that there are smaller voltage drops compared to the baseline. LVC reduces the voltage drop by reducing the power consumption of the dryer when the voltage drop is high. The power consumption of the LVC algorithm is less than the baseline when the voltage goes below the thresholds. Our GridMat tool helps to quickly compare and validate different control algorithms to improve the power grid reliability before deployment into a real physical plant and the corresponding hardware.

D. Hardware-In-Loop-Simulation

After exploring the residential microgrid in MILS, we have generated the software as a C application by using MATLAB Embedded code generation for the purpose of running on real hardware. For this purpose, we have conducted a HILS by using GridMat and an open-source NETGEAR N300 wireless router which are both connected together through Ethernet. GridMat runs the model of the physical plant (residential microgrid) in GridLAB-D and handles the communication between controller and physical plant by using a HTTP wrapper. During HILS, GridLAB-D runs in real-time mode to emulate the physical plant. Fig. 10 shows the structure of the HILS which we have conducted to validate and test the controller on real hardware.

VI. Conclusions

In this paper, a model-based design methodology has been used to design, develop, and analyze a residential microgrid. Moreover, a cyber-physical co-simulation tool (GridMat) that has been developed by the authors for distributed power systems such as a residential microgrid is presented. Various use cases have been studied to demonstrate the capability of experimental MBD method for developing a heterogeneous multi-domain CPES. The results show that MBD is able to capture

---

[5]Ancillary services support the reliable operation of the transmission of power from generations to retail customers. Ancillary services include scheduling/dispatch, voltage/frequency control, load following, system protection, and energy imbalance.

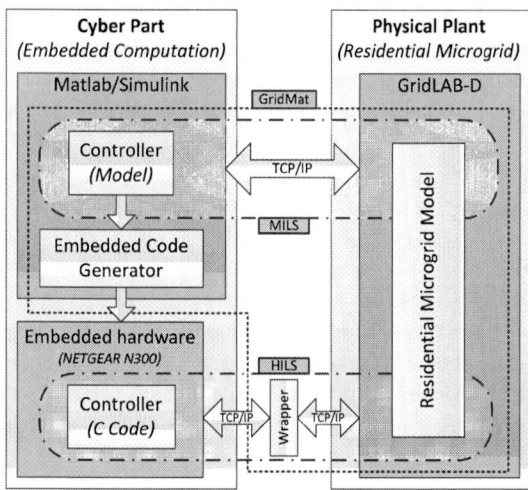

Fig. 10. Structure of MILS and HILS in GridMat

all aspects of CPES by modeling, simulating, and validating the design.

REFERENCES

[1] H. Gill, "Cyber-Physical Systems:Beyond ES, SNs, SCADA," *Trusted Computing in Embedded Systems (TCES) Workshop*, 2010.

[2] S. Karnouskos, "Cyber-Physical Systems in the SmartGrid," in *IEEE International Conference on Industrial Informatics (INDIN)*, pp. 20–23, 2011.

[3] T. Morris, A. Srivastava, B. Reaves, K. Pavurapu, S. Abdelwahed, R. Vaughn, W. McGrew, and Y. Dandass, "Engineering future cyber-physical energy systems: Challenges, research needs, and roadmap," in *North American Power Symposium (NAPS)*, pp. 1–6, 2009.

[4] M. Ilic, L. Xie, U. Khan, and J. Moura, "Modeling of Future Cyber-Physical Energy Systems for Distributed Sensing and Control," *IEEE Transactions on Systems, Man and Cybernetics, Part A: Systems and Humans*, vol. 40, no. 4, pp. 825–838, 2010.

[5] J. Jensen, D. Chang, and E. Lee, "A model-based design methodology for cyber-physical systems," in *Wireless Communications and Mobile Computing Conference (IWCMC)*, pp. 1666–1671, 2011.

[6] J. M. Molina, X. Pan, C. Grimm, and M. Damm, "A framework for model-based design of embedded systems for energy management," in *Modeling and Simulation of Cyber-Physical Energy Systems (MSCPES)*, pp. 1–6, 2013.

[7] W. Bernal, M. Behl, T. Nghiem, and R. Mangharam, "MLE+: a tool for integrated design and deployment of energy efficient building controls," *SIGBED Rev.*, vol. 10, pp. 34–34, July 2013.

[8] Center for Hybrid and Embedded Software Systems, "Ptolemy," *http://ptolemy.eecs.berkeley.edu/*, 2013.

[9] M. Wetter, "Co-simulation of building energy and control systems with the building controls virtual test bed," *Journal of Building Performance Simulation*, vol. 4, no. 3, p. 185203, 2011.

[10] J. Venkatesh, B. Aksanli, J.-C. Junqua, P. Morin, and T. Rosing, "Home-Sim: Comprehensive, smart, residential electrical energy simulation and scheduling," in *International Green Computing Conference (IGCC)*, pp. 1–8, 2013.

[11] U.S. Department of Energy, "EnergyPlus," *http://apps1.eere.energy.gov/buildings/energyplus/*, 2013.

[12] D. Wang, B. de Wit, S. Parkinson, J. Fuller, D. Chassin, C. Crawford, and N. Djilali, "A test bed for self-regulating distribution systems: Modeling integrated renewable energy and demand response in the GridLAB-D/MATLAB environment," in *Innovative Smart Grid Technologies (ISGT)*, pp. 1–7, 2012.

[13] D. Chassin, K. Schneider, and C. Gerkensmeyer, "GridLAB-D: An open-source power systems modeling and simulation environment," in *Transmission and Distribution Conference and Exposition*, pp. 1–5, 2008.

[14] A. Elsheikh, M. U. Awais, E. Widl, and P. Palensky, "Modelica-enabled rapid prototyping of cyber-physical energy systems via the functional mockup interface," in *Modeling and Simulation of Cyber-Physical Energy Systems (MSCPES)*, pp. 1–6, 2013.

[15] J. Fuller, S. Ciraci, J. Daily, A. Fisher, and M. Hauer, "Communication simulations for power system applications," in *Modeling and Simulation of Cyber-Physical Energy Systems (MSCPES)*, pp. 1–6, 2013.

[16] "ns-3," *http://www.nsnam.org/*, 2013.

[17] T. Weng and Y. Agarwal, "From Buildings to Smart Buildings-Sensing and Actuation to Improve Energy Efficiency," *Design Test of Computers*, vol. 29, no. 4, pp. 36–44, 2012.

[18] J. Kleissl and Y. Agarwal, "Cyber-physical energy systems: Focus on smart buildings," in *Design Automation Conference (DAC)*, pp. 749–754, 2010.

[19] I. Akkaya, E. A. Lee, and P. Derler, "Model-based evaluation of GPS spoofing attacks on power grid sensors," in *Modeling and Simulation of Cyber-Physical Energy Systems (MSCPES)*, pp. 1–6, 2013.

[20] K. W. Cheung, "Challenges of Generation Dispatch for Smart Grid," *IEEE Smart Grid Newsletter:http://smartgrid.ieee.org/may-2012/578-challenges-of-generation-dispatch-for-smart-grid*, 2012.

[21] Lawrence Berkeley National Laboratory, "Microgrid Concept," *http://der.lbl.gov/microgrid-concept*, 2013.

[22] A. Meliopoulos, "Challenges in simulation and design of uGrids," in *Power Engineering Society Winter Meeting*, vol. 1, pp. 309–314 vol.1, 2002.

[23] M. A. Faruque, F. Ahourai, "GridMat: Matlab Toolbox for GridLAB-D to Analyse Grid Impact and Validate Residential Microgrid Level Energy Management Algorithms," in *5th Innovative Smart Grid Technologies Conference (ISGT)*, 2014.

[24] F. Ahourai, M. A. Faruque, "Grid Impact analysis of a Residential Microgrid under Various EV Penetration Rates in GridLAB-D," Tech. Rep. TR 13-08, July 2013.

[25] IEEE Power and Energy Society, "IEEE 13 Node Test Feeders," *http://ewh.ieee.org/soc/pes/dsacom/testfeeders.html*.

[26] National Renewable Energy Laboratory, "Commercial and Residential Hourly Load Profiles for all TMY3 Locations in the United States," *http://en.openei.org/datasets/node/961*.

[27] J. Zhu, M. Jafari, and Y. Lu, "Optimal energy management in community micro-grids," in *Innovative Smart Grid Technologies - Asia (ISGT Asia)*, pp. 1–6, 2012.

[28] S. W. Hadley., "Impact of Plug-in Hybrid Vehicles on the Electric Grid," *ORNL Report*, 2006.

[29] P. D. K. Parks and T. Markel, "Costs and emissions associated with plug-in hybrid electric vehicle charging in the Xcel Energy Colorado Service Territory," *Technical Report, NREL/TP-640-1410*, May 2007.

[30] M. Faruque, L. Dalloro, S. Zhou, H. Ludwig, and G. Lo, "Managing residential-level EV charging using network-as-automation platform (NAP) technology," in *Electric Vehicle Conference (IEVC)*, pp. 1–6, 2012.

[31] SIEMENS, "VersiCharge," *http://www.usa.siemens.com/infrastructure-cities/us/en/*, 2013.

[32] F. Ahourai, I. Huang, M. A. Faruque, "Modeling and Simulation of the EV Charging in a Residential Distribution Power Grid," in *IEEE Green Energy and Systems Conference (IGESC)*, 2013.

[33] M. Vasak, A. Starcic, and A. Martincevic, "Model predictive control of heating and cooling in a family house," in *MIPRO*, pp. 739–743, 2011.

[34] A. Aswani, N. Master, J. Taneja, D. Culler, and C. Tomlin, "Reducing Transient and Steady State Electricity Consumption in HVAC Using Learning-Based Model-Predictive Control," *Proceedings of the IEEE*, vol. 100, no. 1, pp. 240–253, 2012.

[35] Federal Energy Regulatory Commission, "Assessment of Demand Response and Advanced Metering," *http://www.ferc.gov/legal/staff-reports/12-20-12-demand-response.pdf*, 2012.

[36] ANSI/ASHRAE, "Thermal environmental conditions for human occupancy, standard 55," 2008.

[37] B. Vyakaranam, G. B. Parker, J. C. Fuller, S. M. Leistritz, "Modeling of GE Appliances: Final Presentation," in *Pacific Northwest National Lab.*, 2013.

978-1-4799-2817-0/14 $31.00 © 2014 IEEE

The Data Center as a Grid Load Stabilizer

Hao Chen

Electrical and Computer Engineering
Boston University
Boston, MA 02215
haoc@bu.edu

Michael C. Caramanis

Systems Engineering
Boston University
Boston, MA 02215
mcaraman@bu.edu

Ayse K. Coskun

Electrical and Computer Engineering
Boston University
Boston, MA 02215
acoskun@bu.edu

Abstract — To accommodate the increasing presence of volatile and intermittent renewable energy sources in power generation, independent system operators (ISO) offer opportunities for *demand side regulation service (RS)* so as to stabilize the grid load. These power market features allow the demand side to earn monetary credits by modulating its power consumption dynamically following an RS signal broadcast by ISO. This paper studies the capacities and benefits of a major potential demand side, *the data center*, to provide RS. We propose a dynamic control policy that modulates the data center power consumption in response to ISO requests by leveraging server power capping techniques and various server power states. Results demonstrate that using our policy, data centers can provide fast reserves in quantities that are substantial proportions (around 50%) of their average energy consumption, with no major deterioration in quality of service (QoS). By doing so, data centers decrease their energy costs around 50%, while providing the ISOs and the society in general with cost effective demand side reserves that render massive renewable generation adoption affordable.

I. INTRODUCTION

Unlike traditional electric grids, today's smart grids incorporate a larger percentage of intermittent renewable energy sources in power generation. These new volatile energy sources create challenges for grid operators to stabilize the grid load and match the power supply with demand in real time. Therefore, ISOs adopt novel mechanisms in modern power markets to ensure stability. Demand side regulation service (RS) is one such mechanism, where the participant receives monetary benefits upon regulating its power consumption based on ISO requests.

This paper focuses on evaluating the benefits of data center participation in the power market for providing demand side RS reserves. We focus on demand side RS reserves because RS reserve market clearing prices are, on average, as valuable in today's markets as energy clearing prices [1, 2]. More importantly, we focus on RS reserves because on one hand their requirements are expected to increase rapidly with increasing renewable energy integration in the grid [19], while on the other hand data centers have a comparative advantage in offering RS reserves relative to other demand side reserve providers. The ability of data centers to offer RS is indeed significant due to their degrees of freedom in modulating their power consump-

tion and the diversity of jobs that they process ranging from high priority transactional jobs to less sensitive jobs that require a reasonable processing rate on average rather than an immediate response. The investigation of the ability of data centers to offer reserves is quite opportune given their increasing share in power consumption, which is 3% of total US electricity [16].

A considerable body of prior research has introduced techniques to reduce energy consumption of processors, servers, and of entire data centers (e.g., [17, 15, 20]). Rather than minimizing the energy consumption, this paper focuses on optimizing the design and real-time operation of data center power consumption in a way that offers RS reserves to ISOs in advanced power markets while, at the same time, maintaining appropriate levels of QoS to data center loads.

The main contributions of this paper are as follows: (1) We design a dynamic power control policy that enables data center to accurately track the RS signal power cap with no major deterioration in QoS, by leveraging server power capping techniques and various server states. (2) We introduce a method to estimate the average power consumption and regulation reserve amounts for data centers to bid in hour-ahead power markets. (3) We demonstrate the capabilities of data centers to participate in RS provision, and through this participation, the energy costs can be dramatically reduced by around 50%.

The rest of this paper starts with an overview of power markets and RS provision. Section III discusses our dynamic power control policy and how to estimate the average power consumption and the regulation reserves. Section IV describes the simulation methodology and provides the experimental results of data center RS provision. Section V discusses the related work and Section VI concludes the paper.

II. POWER MARKET AND REGULATION SERVICE

Power markets, introduced in the US in 1997 [23], have been widely adopted. Today they serve the majority of high-voltage-connected generators and large consumers. Soon after their introduction, power markets evolved to co-optimize or co-clear energy and capacity reserves (primary for frequency control, secondary for RS, tertiary, etc.), whose system-level requirements reflect contingency planning for uncertainty in energy balance, transmission, and generating capacity availability. Social-welfare contributions of competitive power markets are arguably due to the fact that they enable distributed, yet collaborative, decisions which (i) take advantage of lo-

978-1-4799-2817-0/14 $31.00 © 2014 IEEE

cally known uncertainty and dynamical-response-capability information, and (ii) can respond efficiently to price or other system-wide state sufficient statistics, such as frequency and Area Control Error (ACE) and associated reserve requirement signals. These sufficient statistics enable local decisions to be made efficiently and in a manner that is adaptive to power system requirements.

Synchronized power systems may become unstable when generation and consumption are not carefully balanced in practically real-time. To this end, ISOs solicit and secure sufficient quantities of a mix of reserves with different dynamic delivery properties. Bi-directional reserve contracts are secured at least an hour in advance and promise to respond in real-time to ISO-broadcasted fast changing system requirements. Fast reserves include primary (or frequency control) and secondary (or RS) reserves, and are more valuable than slower reserves such as spinning reserves. Each type of reserves is characterized by (i) the frequency with which the system-wide delivery request signal is updated, and (ii) a response time or speed at which that request must be met by each reserve provider. For example [1, 2], an RS reserve contract agreed upon an hour in advance promising to offer during the hour up to R MW of reserves, is obligated to respond to an ISO signal $z(t)$ re-broadcasted every 4 seconds, i.e., broadcasted at time $t = 0, 4, 8, ..., 3596$ sec and follow a response time of 300 seconds, namely a speed of $1/300$ percent per second. More precisely, $z(t)$ takes values in the interval $[-1, 1]$ setting a consumption target at time t of $P(t + 4) = \bar{P} + z(t)R$ where $\bar{P}$ is an average consumption set at the same time that R is offered in the hour ahead market. The demand side provider must change its consumption from $P(t)$ to $P(t + 4)$ at a positive or negative rate of change equal to $R/300$ MW/sec. In fact, the values of $z(t)$ are the outputs of an ISO specified integral proportional filter of system frequency and balancing ACE, and, as such, they are unpredictable and unaffected by a single individual market participant's behavior. The statistical behavior of $z(t)$, however, is well known at the beginning of the hour. Its average value over an hour is zero; i.e., the RS signal trajectory encourages energy neutral consumption modulation trajectory.

The objective of a data center decision support framework considered in this paper is to optimize the following hybrid discrete event system: Given discrete probabilistic arrivals of processing requests (jobs) and a well-defined stochastic process describing $z(t)$, determine a dynamic optimal control policy that maps the system state $x(t)$ to action $u(t)$ where:

- $x(t)$ contains (i) the state of servers in the data center: active, idle, asleep, off, in transition, (ii) the jobs waiting in buffer queues or being processed, (iii) the current value of $z(t)$, and (iv) the QoS achieved so far.
- $u(t)$ is a member of the allowable control set containing (i) initiation of server state transitions, (ii) assignment of jobs to servers and to virtual machines (VMs), (iii) rerouting jobs to other data centers, and (iv) taking power and performance management actions (e.g., CPU resource limits control in VMs, dynamic voltage and frequency scaling -DVFS-, etc.) at individual servers.

State dynamics depend on $u(t)$ and evolve with multiple time scale hybrid dynamics responding to discrete control actions (e.g., a server state transition), discrete events (e.g., a job

arrival), and continuous retired instruction rates in response to server power management settings.

Recent work [5, 6] on the decision support framework indicates that substantial cost reduction opportunities exist when we regulate the computational power in accordance with ISO RS requests. In this paper we show that after determining $\bar{P}$ and R levels based on workload estimates, physical limits of the servers, and constraints on performance and tracking error, it is possible to design a dynamic policy that maintains the desired QoS level while tracking the ISO signal with a small error. This ability translates into cost savings. Components of the optimization problem (workload estimation, determining $\bar{P}$ and R, dynamic policy, etc.) individually and their interactions introduce interesting yet complex challenges.

Before proceeding with a concrete proposal of the decision support framework used to investigate data center RS reserve offering, We describe the relevant features and pricing rules of RS transactions in the PJM and NYISO Power Markets [1, 2].

Consider a data center that purchases in the hour-ahead market $\bar{P}$ MWh of energy at the clearing price Π^E, and sells RS reserves R MW at the RS reserve clearing price or Π^R per MW traversed by the RS signal $z(t)$. The net cost of this transaction incurred by the data center when the hour ahead market clears is $\Pi^E \bar{P} - \Pi^R R$, provided that the data center tracks the RS signal $z(t)$ perfectly, modulating its power consumption to track the implied obligation $P(t + 4) = \bar{P} + z(t)R$ perfectly. However, perfect tracking is practically not possible. Hence at the end of the hour, RS reserve providers are charged an amount that reflects their relative tracking error (RTE). Moreover, if the RTE exceeds a certain threshold[1], the participant loses its qualification to participate in RS reserve transactions and has to repeat a rigorously defined qualification process to re-qualify [1]. More precisely, the RTE is the ratio of the sum, or tracking error per MW or R, over the length of the trajectory traversed by $z(t)$, namely:

$$RTE = \frac{\sum_{t=0,4,8...,3596}|P(t+4) - (\bar{P} + z(t)R)|}{\sum_{t=0,4,8...,3596}|z(t+4) - z(t)|R} \quad (1)$$

At the end of the hour an RS provider is charged an additional cost equal to $\Pi^R R * RTE$. Note that if during the hour the data center observes the RS signal $z(t)$ perfectly and modulates its power consumption to track the implied obligation $P(t + 4) = \bar{P} + z(t)R$ perfectly, then $RTE = 0$.

Note also that independent of power market transaction charges discussed above, a data center that provides RS reserves is bound to incur intrinsic costs from the operational level obligation to consume at or close to $P(t+4) = \bar{P}+z(t)R$. These consist of energy consumption efficiency losses associated with power consumption modulation as well as the value of possible reductions in the QoS provided to data center clients during, for instance, low $z(t)$ regulation signal values. Desirable operational level policies discussed in the next section must provide a reasonable tradeoff between market related charges and the above intrinsic costs. Moreover, they should ensure that the probability of exceeding maximum allowed RTE levels as well as client QoS guarantees is within pre-specified confidence intervals.

---

[1]Usually 30%.

III. DYNAMIC POWER CONTROL POLICY AND REGULATION RESERVES BIDDING

In this section, we first discuss a general data center model and various server states that are useful in power regulation. Then we propose a power control policy that dynamically modulates the data center power consumption $P(t+4)$ at each time interval t to track the RS signal related power capping, $\bar{P} + z(t)R$. We then introduce our estimates of the average power consumption, $\bar{P}$, and the regulation reserves, R, which the data center needs to bid in the hour-ahead power markets.

A. Data Center Model and Server States

A data center consists of computational nodes (servers) and cooling elements (computer room air conditioners, etc.). In this paper, we specifically focus on regulating the computational power. Our technique, however, can be combined with power budgeting techniques [31] that distribute a given total power cap into power caps of the sub-components of the data center.

Each server in the data center is in one of three states: active, idle and sleep [15]. When a server is running a job, it is "active", and its power consumption is P_{active}. P_{active} is composed of the dynamic power, P_{dyn}, and the static power, P_{static}. The dynamic power changes based on the characteristics of the running job, and can be modulated by power management techniques, such as DVFS [17], CPU resource limits [13], etc. The static power is a constant[2], and exists as long as the server is turned on. In our work, we use the CPU resource limits knob to modulate the server dynamic power. CPU resource limits change the resources allocated to a VM on the server, and as a result, adjust the server dynamic power and the throughput. It is a desirable control knob as it can be quickly changed at a very fine granularity [13, 6].

Regulating the dynamic power affects the server throughput. Previous work has shown a linear relation between P_{dyn} and the server throughput, represented by the retired instructions per second (RIPS), as $P_{dyn} = k * RIPS$ [6]. The server reaches its maximal throughput capacity by running at the peak power consumption rate, P_{peak}, with $P_{peak} = P_{dyn,max} + P_{static}$, where $P_{dyn,max}$ is the maximal dynamic power consumption that the server can achieve when the CPU resource limit is set to maximum.

A server is "idle" if it is turned on but is not running any jobs. An idle server consumes power at a constant rate, P_{idle}, which is equal to P_{static}. In the "sleep" state, the server consumes a very low constant power, P_{sleep}. In general, there are some time delays and energy costs of resuming a server from or suspending it to a "sleep" state. The suspending time delay, t_{susp}, usually is small and can be ignored, while the resuming time delay, t_{res}, is large [11, 15]. During both the suspending and resuming periods, the power consumption are similar, denoted as P_{tran}, which is often close to the peak power, P_{peak} [15]. The energy cost of the resuming period is $E_{loss} = t_{res} * P_{tran}$ and of the suspending period is ignored.

In fact, some servers in data centers can be completely turned off, which indicates a fourth state, "off", with no power consumption. However, the "off" state does not frequently appear due to the very large time delays and energy costs of resuming and suspending process. We do not consider the "off" state in this paper.

We assume there is a FIFO (first in first out) queue for holding the incoming jobs in the data center. Once a job arrives, it is first put into the queue. The job at the front of the queue is scheduled to a server using the policy introduced in Section III-B. In our model, each server can only serve one job a time; thus, we do not consider server consolidation.

We define the data center utilization U as the active time of the whole data center, or the number of active servers at each time interval. For example, $U = 50\%$ means each server is active for half of the whole period, and is in idle or sleep state for the rest of the time. We can also comprehend this as, at each moment, half of total servers in the data center are serving jobs. U is related to the arrival frequency of the workloads.

In this work, we study homogeneous data centers only, where all servers and jobs are of the same type. In fact, a heterogeneous data center with different types of servers and workloads can be split into homogeneous clusters. Also, many high performance computing (HPC) clusters include dedicated, optimized set of servers assigned to specific jobs.

B. Dynamic Power Control Policy

For real-time dynamic power tracking, we need to modulate the data center power consumption rate, $P(t+4)$, to match the dynamic RS signal related power value, $\bar{P} + z(t)R$. At the same time, workload QoS and overall energy waste also need to be considered. Our goals during the tracking process are as follows:

- Reduce the tracking error $|P(t+4) - (\bar{P} + z(t)R)|$;
- Improve the energy efficiency, including reducing the energy waste during the server state transition period, and reducing the static energy waste related to P_{static};
- Reduce the workload QoS performance degradation.

Apparently, there are tradeoffs among these goals. For example, reducing the tracking error prevents the servers from always running at their maximal capacity, which leads to performance degradation. Also, reducing the energy waste during the server state transition period requires reducing the number of server transitions, and reducing the static energy waste requires setting a fewer number of servers in idle and a larger number of servers in sleep mode. Both of these actions might violate the power tracking goal. Hence, our policy aims to optimize among these goals at each time interval by solving the following optimization problem:

$$\min_{u(t) \in U(x(t))} J(x(t), u(t)) = \alpha_1 |P(t+4) - (\bar{P} + z(t)R)|$$
$$+ \alpha_2 N_{tran}(t) - \alpha_3 N_{sleep}(t) - \alpha_4 N_{peak}(t) \quad (2)$$

where u(t) is our policy control, $x(t)$ is the dynamic state at t. $x(t) = (z(t), Q(t), S_i(t), P_i(t), J_i(t), R_i(t), t_{idle,i}(t), i = 1, 2..., N_{dc})$. $z(t)$ is the RS signal at time t, $Q(t)$ is the number of jobs waiting in the queue for scheduling. $S_i(t), P_i(t), J_i(t), R_i(t), t_{idle,i}(t)$ are the server state (active, idle, sleep), power consumption, the job in the server ($J_i(t) = 0$: has no job in the server i; $J_i(t) = 1$: has a job), the remaining number of instructions of the job in the server, and the time

---

[2]The static power, in fact, is temperature dependent. In this work, we assume there is no temperature change.

978-1-4799-2817-0/14 $31.00 © 2014 IEEE

of being in the idle state, of the server i, respectively. N_{dc} is the number of the servers in the data center, α_1, α_2, α_3 and α_4 are penalty coefficients. $N_{tran}(t)$ is the number of servers at time t that are suspending to or resuming from the sleep state, $N_{sleep}(t)$ is the number of servers in sleep, and $N_{peak}(t)$ is the number of servers running at their peak capacities. We have:

$$P(t) = \sum_{i=1}^{N_{dc}} P_i(t) \tag{3}$$

$$N_{sleep}(t) = \sum_{i=1}^{N_{dc}} \{S_i(t) == \text{``sleep''}\} \tag{4}$$

$$N_{tran}(t) = |N_{sleep}(t) - N_{sleep}(t-1)| \tag{5}$$

$$N_{peak}(t) = \sum_{i=1}^{N_{dc}} \{P_i(t) == P_{peak}\} \tag{6}$$

$$N_{idle}(t) = \sum_{i=1}^{N_{dc}} \{S_i(t) == \text{``idle''}\} \tag{7}$$

We include $N_{tran}(t)$ in Eq. (2) in order to reduce the transition energy waste. To reduce the static energy waste, we need to set a fewer number of servers in idle or non-peak active state; i.e., we need to increase the number of servers running at their peak capacity and put more servers in sleep state. Therefore, we include $N_{sleep}(t)$ and $N_{peak}(t)$ in Eq. (2).

We set different penalty weights for each goal by changing α_1, α_2, α_3 and α_4. For example, if the power tracking is the most important goal, we can simply assign α_1 much larger than α_2, α_3 and α_4. Then the optimal solution first aims at satisfying the power tracking constraint as accurate as possible at each t. We design some additional constraints and rules as follows:

- If $J_i(t) = 1$, i.e., the server is running a job, then the server must keep active, i.e., $S_i(t) = 1$, until the job is finished. In this way we provide guarantees for workload QoS. Furthermore, we set a lower bound of the minimal power rate P_{min} when serving a job, i.e., $P_i(t) \geq P_{min}$ if $J_i(t) = 1$, which forces the job to be served at a throughput with a lower bound and avoids the job being stalled in the server. P_{min} can be determined by the QoS requirements.

- Once a job is finished, i.e., $R_i(t) = 0$, the server immediately becomes idle.

- When $Q(t) = 0$, i.e., no jobs are waiting in the queue, then no idle server is allowed to be activated[3].

- Transition mechanism: if a server has been in idle longer than a timeout period, t_{tout}, then it automatically sleeps. This timeout mechanism is designed to avoid frequent transitions. We use the timeout value proposed by Gandhi et al. [11]: $t_{tout} = t_{res} * P_{peak}/P_{idle}$. In addition, in order to maximize the number of sleeping servers, we always select the server with smallest current $t_{idle}(t)$ to activate if a job is waiting to be served. Similarly, if we need to force some servers to sleep, we select the servers with the largest $t_{idle}(t)$.

Having these rules, our available control $u(t)$ can be (1) increase/decrease power consumption of active servers by using CPU resource limits; (2) resume sleeping servers; (3) put idle servers to sleep; (4) activate idle servers to run new jobs.

In our work, we assign a large value to α_1 to put power tracking as the high-priority goal. In addition, we set α_2 larger than α_3 and α_4, i.e., we are more reluctant to do the server state transition. The resulting policy from Eq.(2) is:

Case 1- If $P(t) < \bar{P} + z(t)R$, i.e., the power consumption needs to be increased:

- Increase power consumption of some active servers that are not running at maximal capacity to P_{peak};
- If $Q(t) > 0$ and $N_{idle}(t) > 0$, then activate some idle servers and run them at maximal capacity with power consumption at P_{peak};
- Resume sleeping servers following the transition mechanism.

We do the above three steps in order until $P(t+4)$ meets the power cap, $\bar{P} + z(t)R$.

Case 2- If $P(t) > \bar{P} + z(t)R$, i.e., the power consumption needs to be decreased:

- Decrease power consumption of some active servers that are not running at maximal capacity to P_{min};
- Decrease power consumption of some active servers that are working at maximal capacity to P_{min};
- Suspend idle servers following the transition mechanism.

We do the above three steps in order until $P(t+4)$ meets the power cap, $\bar{P} + z(t)R$.

C. Regulation Reserves

Now we estimate the average power consumption $\bar{P}$ and the regulation reserve R that the data center should bid in the power market for the next hour. We assume the arrival of the workloads is a Poisson process with an arrival rate λ (per hour). The value of λ can be controlled by allocating overall load among geographically dispersed data centers to exploit spatiotemporal variations in energy prices [29]. The λ considered here is the one after such allocation. Each job j is composed of a number of instructions, namely, I_j. Since we have homogeneous workloads, all $I_j, j = 1, 2...$ are equal and denoted as I. Finishing a job is equivalent to executing all the instructions.

Having λ, I and the coefficient k between P_{dyn} and RIPS, we are able to estimate the total dynamic energy needed during the hour, E_{dyn}, for finishing all workloads, which is $E_{dyn} = \lambda * kI$. Only active servers can consume dynamic power. As mentioned before, each server has the dynamic power range in $(0, P_{dyn,max}]$. However, our designed policy always tries to force active servers to run at peak capacity, and as a result, most of the active servers consume at the maximal dynamic power. Hence in order to provide sufficient dynamic energy E_{dyn} for serving all workloads in the hour, the average number of servers that should be active, $\bar{N}_{active}$, is:

$$\bar{N}_{active} = \frac{\int_0^{1h} N_{active}(t)dt}{1h}$$
$$= \frac{E_{dyn}}{P_{dyn,max} * 1h} = \frac{\lambda * kI}{P_{dyn,max} * 1h} \tag{8}$$

---

[3]In fact, it is possible to run synthetic workloads to help improve the power tracking performance. In this work we do not consider such loads.

While estimating the average power consumption $\bar{P}$ during the hour, the energy waste during transition periods needs to be considered. As introduced before, each resuming process has an energy loss as E_{loss}. We assume the total number of times the servers are resumed in the hour across the data center is N_{res}. Then, the total energy waste during the hour is: $E_{loss,1h} = E_{loss} * N_{res}$.

Next, we estimate the number of times the servers are resumed, N_{res}. As introduced before, the dynamic range of RS signal $z(t)$ is [-1,1]. At $z(t) = -1$, the data center is at the lowest power consumption, P_{low}, and at $z(t) = 1$, the data center is at the highest power consumption, P_{high}. In order to increase the regulation reserve R to gain more monetary savings, we should minimize P_{low} and maximize P_{high}. The minimal P_{low} we can achieve is by setting all servers in sleep, and the maximal P_{high} is by setting all servers actively running at the peak power. Thus, every time RS signal increases from -1 to 1, almost all the servers need to be resumed. On the other hand, our designed policy avoids the situation that a server is resumed and suspended back and forth when tracking minor changes in the RS signal. Therefore, in our policy, resuming servers only happens during the periods of the RS signal with large increases. We denote the number of RS signal periods with large increases during the hour as p_b. Then we have $N_{res} = p_b * N_{dc}$. A good estimate[4] of p_b is $p_b = 2$.

Now we can estimate the average power consumption $\bar{P}$ by the following equations:

$$\bar{P} = \frac{\int_0^{1h}(\bar{P} + Rz(t))dt}{1h} = \bar{N}_{active} * P_{active}$$
$$+ \bar{N}_{idle} * P_{idle} + \bar{N}_{sleep} * P_{sleep} + \frac{E_{loss,1h}}{1h} \quad (9)$$

and
$$N_{dc} = \bar{N}_{active} + \bar{N}_{idle} + \bar{N}_{sleep} \quad (10)$$

We have solved $\bar{N}_{active}$ and $E_{loss,1h}$ before. P_{idle} and P_{sleep} are constants and known. In our policy, most active servers run at peak capacity; hence, in the equation we can simply replace P_{active} with P_{peak}.

Due to the transition mechanism introduced before, $\bar{N}_{idle}$ is not 0. Moreover, for reducing the performance degradation caused by time delays for resuming the sleep servers, we can manually set aside some idle servers as the "performance guarantee slack". These idle servers are prepared for immediately serving coming jobs. Hence we are able to manually tune the $\bar{N}_{idle}$, and as a result, $\bar{P}$ is changed, i.e., $\bar{P}$ is a function of $\bar{N}_{idle}$. In our work, each sleeping server is coupled with an idle server for providing QoS guarantees; i.e., we use $\bar{N}_{idle} = \bar{N}_{sleep}$ in calculating $\bar{P}$.

Next, we estimate the regulation reserve R that we should bid. First, we have the constraints as:

$$\bar{P} - Rz(t) \geq N_{dc}P_{sleep},$$
$$\bar{P} + Rz(t) \leq N_{dc}P_{peak}, \; \forall t \quad (11)$$

As we know $z(t) \in [-1, 1]$, we have:

$$R \leq min\{N_{dc}P_{peak} - \bar{P}, \; \bar{P} - N_{dc}P_{sleep}\} \quad (12)$$

---

[4]This estimate is based on our observations. In fact, the experiments indicate that accuracy of p_b estimation is not critical.

Prior results on single server regulation show that the value of R does not notably affect the tracking performance or the QoS degradation [5]. Moreover, the results of the single server experiments show that the optimal R is indeed almost equal to $min\{\bar{P} - P_{idle}, \; P_{max} - \bar{P}\}$. Considering data centers provide even more flexibilities in providing RS compared to a single server, we assume a similar result for the data center; i.e., the RS reserve value R reaches the bound in Eq. (12).

IV. EXPERIMENTAL RESULTS

In this section, we first introduce the system simulation methodology, and then evaluate the bidding estimation and power control policy proposed in Section III under various data center scenarios.

A. Methodology

To determine the relationship between P_{dyn} and RIPS, we run each application from the PARSEC-2.1 [7] benchmark suite on a 1U server, which has an AMD Magny Cours (Opteron 6172) processor, with 12 cores on a single chip. The server is virtualized by VMware vSphere 5.1 ESXi hypervisor; hence, we are able to use the CPU resource limits knob to control the power-performance settings. Detailed results on P_{dyn} and RIPS are shown in our prior work [5].

We assume jobs arrive at the data center following a Poisson process. We generate the workload sequences using Monte Carlo simulation [5]. Without loss of generality, we assume a data center server cluster with $N_{dc} = 100$ servers. By default the data center utilization is 50%. We simulate a 1-hour period experiment 10 times and evaluate the power tracking, QoS performance, and the monetary cost. Regarding the sleep state, we assume $t_{res} = 10$s and $P_{sleep} = 10\%P_{peak}$.

B. Data Center RS

We next evaluate the data center level RS tracking performance and QoS degradation, as well as the monetary savings. Then we compare the results to the single server RS proposed in previous work [5, 6].

Fig. 1(a) shows the statistical distribution of the power tracking error, $\epsilon(t) = (P(t + 4) - (\bar{P} + z(t)R))/R$, over time t. The result shows that in most of the time, the tracking errors are close to 0 for the data center level RS, while for the single server the tracking errors are mostly around 0.1-0.2. Moreover, the data center RS has smaller deviation in the tracking error. The maximal tracking error of the data center is less than 1 while that of the single server case reaches close to 2.5. Some ISOs have strict limitations on the peak tracking error, thus the data center can perform much better than a single server.

Fig. 1(d) shows the statistical distribution of job servicing time degradation, D_i, for each job i. This degradation is the ratio of the job servicing time T_i when providing RS to the shortest processing time for the job, $T_{i,min}$, which refers to running the job without any power capping restrictions and without any waiting time in the queue. Thus, $D_i = T_i/T_{i,min} - 1$, and $D_i = 0$ means that there is no degradation. Our result shows that most jobs get almost no degradation in data center RS, while for the single server, the jobs suffer degradation with higher probabilities.

978-1-4799-2817-0/14 $31.00 © 2014 IEEE

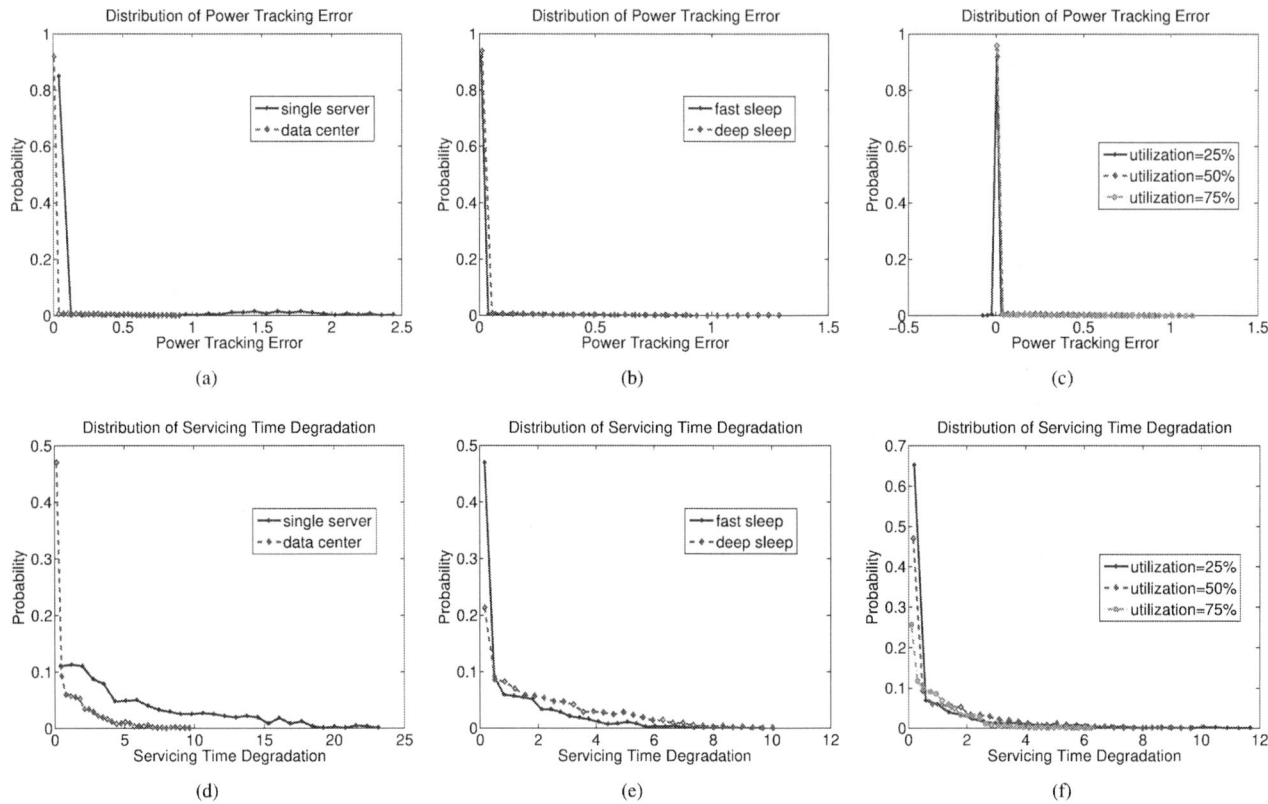

Fig. 1. Statistical distribution of tracking error (a, b, c) and job servicing time degradation (d, e, f). (a) and (d) are results of a single server and the data center both at 50% utilization. (b) and (e) are results of the data center at 50% utilization with a "fast sleep" server state and a "deep sleep" server state. (c) and (f) are results of the data center at various utilizations. All cases are tested with a (homogeneous) set of Blackscholes jobs.

Then, we check the monetary savings in both cases. The net cost of the power for providing RS is $\Pi_E \bar{P} - \Pi_R R$, with $\Pi_E \approx \Pi_R$ [5]. Therefore, $R/\bar{P}$ represents the percentage of the monetary savings. For the single server, we have optimal $R/\bar{P} = 29.7\%$, while in the data center it is 56.8%, which is around 2X improvement.

Thus, providing RS brings dramatic monetary savings (56.8%) to data centers or multi-server clusters, with zero power tracking error for most of the time, and no QoS degradation for most of the jobs. Compared to the single server RS, both QoS and the monetary savings are significantly improved. These results are expected as the multi-server clusters can provide more flexibility and control opportunities to perform power regulation.

C. Fast Sleep and Deep Sleep

Unlike desktops and laptops, in today's data centers, many servers do not have sleep states with fast transitions [15]. These servers are usually put in a deeper sleep mode and rebooted if needed. The time delay and the energy loss in the rebooting process are larger than those of resuming process from the sleep state, whereas servers can save more energy in the deep sleep state compared to the fast sleep state. Based on recent work [15], we assume the time delay of the rebooting process is $t_{reb} = 200$ seconds, and the power consumption in the deep sleep state is $P_{deepS} = 5\% P_{peak}$. The power consumption in the transition process, P_{tran}, is close to P_{peak}. We conduct an experiment with the assumption that the servers in the data cen-

ter have deep sleep and rebooting instead of fast sleep states.

Fig. 1(b) shows the statistical distribution of power tracking error of the two cases (fast sleep and deep sleep). The power tracking is accurate for both cases. This is because our policy puts power tracking as the highest priority, and unlike the server state, the power consumption rate during the transition process can be immediately changed without any delays, from P_{sleep} or P_{deepS} to P_{tran}, or vice versa. Hence the power tracking accuracy is not sensitive to whether the servers have fast sleep states or not. In addition, the figure shows that using slower deep sleep states leads to larger peak tracking errors.

Fig. 1(e) shows the statistical distribution of job servicing time degradation of two cases. The result shows that data center servers with fast sleep states have smaller degradation and better QoS. This is because the time delay of rebooting a server is very large, which strongly affects the job servicing performance. Overall, even though data centers without fast sleep states have more degradation, most of the degradation is still small, and the QoS is high.

For a data center with faster sleep states, we have monetary savings around $R/\bar{P} = 56.8\%$, while for the slower deep sleep states, the savings are only 36.9%, for the reason that the faster sleep state provides the ability to react more rapidly to ISO requests compared to rebooting. Thus, power RS in both cases can bring significant monetary savings with close to zero power tracking error for most of the time and small QoS degradation for most of the jobs, while having a fast sleep state further improves the monetary savings and QoS.

978-1-4799-2817-0/14 $31.00 © 2014 IEEE

D. Impact of Cluster Utilization

By default we assume the utilization of the data center is 50%. In real life, different clusters, or same clusters at different time, have different utilizations. Next, we evaluate the impact of cluster utilization in providing RS. Fig. 1(c) shows the statistical distribution of power tracking error under various utilizations. Tracking performance in various utilizations is similar, and most of the tracking errors are close to zero. This is because in all cases our policy gives the highest priority to power tracking.

Fig. 1(f) shows the statistical distribution of job servicing time degradation under various utilizations. The result shows that when utilization increases, the performance degradation increases. This is expected, as more jobs need to be processed under higher utilization and more servers are busy, which increases the performance degradation. Overall, the degradation of all three cases is mostly close to 0; hence, the data center can provide RS reserve without significantly influencing the job QoS in various utilization levels.

We compare the monetary savings of different utilization cases. For $U = 25\%, 50\%, 75\%$, we have savings $R/\bar{P} = 78.0\%, 56.8\%, 21.8\%$ correspondingly, which shows that the savings decrease when the utilization increases. This is due to the reason that higher utilization leads to higher average power consumption $\bar{P}$, which limits R. However, even with 75% utilization, we still can have around 22% monetary savings, which shows that providing RS on the data center has cost advantages regardless of the utilization.

E. Impact of Different Workloads

All previous experiments are conducted by using workloads made out of homogeneous Blackscholes jobs. In this part we study the data center RS problem with different types of workloads. Table I shows the experimental results on four different workloads. We list their power tracking statistics, QoS degradation statistics, and monetary savings. $\bar{D}$ and σ_D are the mean and standard deviation of performance degradation, $\bar{\epsilon}$ and σ_ϵ are the mean and standard deviation of the tracking error. The results show that the power tracking performance is not influenced by the workload type, while the performance degradation is. From the table, workloads with longer shortest processing time, i.e., $T_{i,min}$, such as Streamcluster and Facesim (whose shortest processing time is larger than 100 seconds, while Blackscholes and Canneal only have 20-40 seconds), have less QoS performance degradation. This is expected as the waiting time is relatively shorter (compared to the processing time) for longer processing time jobs. As our policy has rules (e.g., P_{min}) to guarantee the job processing time, waiting time becomes the main reason for degradation. Overall, both the performance degradation and the tracking error are quite small. In addition, in all cases, data centers can achieve more than 50% monetary savings. Hence data center level RS is expected to provide small tracking errors and QoS degradation along with dramatic monetary savings for a broad range of workloads.

V. RELATED WORK

Some previous work has investigated demand side RS in power market. Caramanis et al. [4] study the RS bidding prob-

TABLE I
CLUSTER LEVEL POWER REGULATION ON DIFFERENT WORKLOADS

	Blackscholes	Canneal	Streamcluster	Facesim
$\bar{P}$	$9.75 * 10^3$	$9.71 * 10^3$	$9.84 * 10^3$	$9.84 * 10^3$
R	$5.54 * 10^3$	$4.98 * 10^3$	$5.46 * 10^3$	$5.11 * 10^3$
$\bar{D}$	1.13	1.13	0.21	0.22
σ_D	1.54	0.69	0.26	0.27
$\bar{\epsilon}$	0.03	0.03	0.03	0.03
σ_ϵ	0.10	0.09	0.09	0.09
$R/\bar{P}$	56.8%	51.3%	55.5%	52.0%

[a] $\bar{D}$ and σ_D are mean and standard deviation of performance degradation; $\bar{\epsilon}$ and σ_ϵ are mean and standard deviation of tracking error.

lem by using optimal dynamic pricing policies. Paschalidis et al. [24] propose a market-based mechanism to enable the smart building to provide RS.

Data center level power management techniques have advanced significantly in the recent years. For the server level power management, DVFS, power gating and multi-core processor workload scheduling and allocation have been investigated [17, 27, 28]. Different power capping techniques, which are used for meeting the peak or average power constraints have also been widely studied [9, 8, 26, 18]. In virtualized servers, some CPU resource management and consolidation techniques have been applied to manage the power consumption [22, 14, 13]. On the data center level, application and server aware power budgeting have been researched [25, 21]. Zhan et al. [31] propose a system profile based energy-efficient data center power budgeting technique. Gandhi et al. [10] investigate the optimal power allocation in server farm by considering different complex situations. Server commitment is another hot topic in the data center level power management. Meisner et al. [20] propose PowerNap technique to eliminate the server idle power. Isci et al. [15] explore the feasibility of low-latency power states and demonstrate a power-aware virtualization management solution leveraging these states. Gandhi et al. [11] study the regime of sleep states that would be advantageous in data centers and propose some dynamic power management policy based on server commitment.

There are a few recent studies that investigated on data center participation in advanced power market. Ghamkhari et al. [12] build an analytical profit model to determine whether participation in an ancillary service market can be beneficial to data centers. Aikema et al. [3] analyze a number of different advanced power market for data centers to participate in potential. Wang et al. [30] propose to migrate the workload between geographically distributed and virtualized data centers situated in multiple regional electrical markets, to maximize the expected payoff. However, none of these work closely consider using data center power management techniques and designing servicing policy for providing RS. Chen et al. [6, 5] propose a data center level power management framework to provide RS, but then the work only focuses on a single server level power management.

Our work is the first to closely investigate the data center level power budgeting and management, the server commitment, as well as the workload scheduling and allocation, to enable the data center to participate in the advance power market. We propose a dynamic power control policy for the data center to provide RS, to achieve dramatic monetary savings while also guarantee no major deterioration in QoS.

VI. CONCLUSIONS AND FUTURE WORK

In this paper we have proposed a power control policy for data centers to dynamically track RS signal related power capping. We have also introduced an estimation method to calculate the RS provision bidding value. Experimental results show that data centers with our policy and estimation can accurately track the RS signal and achieve more than 50% energy monetary savings, with no major QoS performance degradation, regardless of types of workloads. The results also demonstrate the strong capacity and substantial monetary savings of data centers to provide RS in various scenarios, e.g., under different utilizations and with different server states. In addition, the significant improvement in both monetary savings and QoS of data center level RS provision has been investigated and compared to prior single server results, indicating that data-center-wide control is not only feasible but also more beneficial. Our ongoing work focuses on (i) leveraging heterogeneous workload and server RS provision by advanced power budgeting and job scheduling, and (ii) considering synergies with cooling power consumption.

ACKNOWLEDGMENTS

This research has been supported by NSF grant 1038230, Sandia National Labs, Oracle, and Decision Detective Corporation (SBIR).

REFERENCES

[1] PJM (2013). market-based regulation [online]. *http://pjm.com/markets-and-operations/ancillary-services/mkt -basedregulation.aspx.*

[2] Manual 2: Ancillary services manual, v3.26. *NYISO*, 2013.

[3] D. Aikema, R. Simmonds, and H. Zareipour. Data centres in the ancillary services market. In *Green Computing Conference (IGCC), 2012 International*, pages 1–10. IEEE, 2012.

[4] M. C. Caramanis, I. C. Paschalidis, C. G. Cassandras, E. Bilgin, and E. Ntakou. Provision of regulation service reserves by flexible distributed loads. In *Decision and Control (CDC), 2012 IEEE 51st Annual Conference on*, pages 3694–3700. IEEE, 2012.

[5] H. Chen, A. K. Coskun, and M. C. Caramanis. Real-time power control of data centers for providing regulation service. In *Decision and Control (CDC), 2013 IEEE 52nd Annual Conference on*, 2013.

[6] H. Chen, C. Hankendi, M. C. Caramanis, and A. K. Coskun. Dynamic server power capping for enabling data center participation in power markets. In *Intl. Conf. on Computer-Aided Design*, 2013.

[7] B. Christian. Benchmarking modern multiprocessors. *Ph.D.Thesis. Princeton University*, 2011.

[8] R. Cochran, C. Hankendi, A. K. Coskun, and S. Reda. Pack & cap: adaptive dvfs and thread packing under power caps. In *Proceedings of the 44th annual IEEE/ACM international symposium on microarchitecture*, pages 175–185. ACM, 2011.

[9] H. David, E. Gorbatov, U. R. Hanebutte, R. Khanna, and C. Le. Rapl: memory power estimation and capping. In *Low-Power Electronics and Design (ISLPED), 2010 ACM/IEEE International Symposium on*, pages 189–194. IEEE, 2010.

[10] A. Gandhi, M. Harchol-Balter, R. Das, and C. Lefurgy. Optimal power allocation in server farms. In *Proceedings of the eleventh international joint conference on Measurement and modeling of computer systems*, pages 157–168. ACM, 2009.

[11] A. Gandhi, M. Harchol-Balter, and M. A. Kozuch. Are sleep states effective in data centers? In *Green Computing Conference (IGCC), 2012 International*, pages 1–10. IEEE, 2012.

[12] M. Ghamkhari and H. Mohsenian-Rad. Data centers to offer ancillary services. In *Smart Grid Communications (SmartGridComm), 2012 IEEE Third International Conference on*, pages 436–441. IEEE, 2012.

[13] C. Hankendi, S. Reda, and A. K. Coskun. vcap: Adaptive power capping for virtualized servers. In *Low Power Electronics and Design (ISLPED), 2013 IEEE International Symposium on*, pages 415–420. IEEE, 2013.

[14] I. Hwang, T. Kam, and M. Pedram. A study of the effectiveness of cpu consolidation in a virtualized multi-core server system. In *Proceedings of the 2012 ACM/IEEE international symposium on Low power electronics and design*, pages 339–344. ACM, 2012.

[15] C. Isci, S. McIntosh, J. Kephart, et al. Agile, efficient virtualization power management with low-latency server power states. In *Proceedings of the 40th Annual International Symposium on Computer Architecture*, pages 96–107. ACM, 2013.

[16] J. Koomey. Growth in data center electricity use 2005 to 2010. *Oakland. CA: Analytics Press*, August, 1, 2010.

[17] J. Li and J. F. Martinez. Dynamic power-performance adaptation of parallel computation on chip multiprocessors. In *High-Performance Computer Architecture, 2006. The Twelfth International Symposium on*, pages 77–87. IEEE, 2006.

[18] K. Ma and X. Wang. Pgcapping: exploiting power gating for power capping and core lifetime balancing in cmps. In *Proceedings of the 21st international conference on Parallel architectures and compilation techniques*, pages 13–22. ACM, 2012.

[19] Y. V. Makarov, C. Loutan, J. Ma, and P. de Mello. Operational impacts of wind generation on california power systems. *Power Systems, IEEE Transactions on*, 24(2):1039–1050, 2009.

[20] D. Meisner, B. T. Gold, and T. F. Wenisch. Powernap: eliminating server idle power. In *Sigplan Notices*, volume 44, pages 205–216. ACM, 2009.

[21] R. Nathuji, C. Isci, E. Gorbatov, and K. Schwan. Providing platform heterogeneity-awareness for data center power management. *Cluster Computing*, 11(3):259–271, 2008.

[22] R. Nathuji, K. Schwan, A. Somani, and Y. Joshi. Vpm tokens: virtual machine-aware power budgeting in datacenters. *Cluster computing*, 12(2):189–203, 2009.

[23] A. L. Ott. Experience with PJM market operation, system design, and implementation. *Power Systems, IEEE Trans. on*, 18(2):528–534, 2003.

[24] I. C. Paschalidis, B. Li, and M. C. Caramanis. A market-based mechanism for providing demand-side regulation service reserves. In *Decision and Control and European Control Conference (CDC-ECC), 2011 50th IEEE Conference on*, pages 21–26. IEEE, 2011.

[25] K. Rajamani, H. Hanson, J. Rubio, S. Ghiasi, and F. Rawson. Application-aware power management. In *Workload Characterization, 2006 IEEE International Symposium on*, pages 39–48. IEEE, 2006.

[26] K. K. Rangan, G.-Y. Wei, and D. Brooks. Thread motion: fine-grained power management for multi-core systems. In *ACM SIGARCH Computer Architecture News*, volume 37, pages 302–313. ACM, 2009.

[27] K. Singh, M. Bhadauria, and S. A. McKee. Real time power estimation and thread scheduling via performance counters. *ACM SIGARCH Computer Architecture News*, 37(2):46–55, 2009.

[28] R. Teodorescu and J. Torrellas. Variation-aware application scheduling and power management for chip multiprocessors. *ACM SIGARCH Computer Architecture News*, 36(3):363–374, 2008.

[29] H. Wang, J. Huang, X. Lin, and H. Mohsenian-Rad. Exploring smart grid and data center interactions for electric power load balancing. In *SIGMETRICS, 2013 ACM Conference on*, 2013.

[30] R. Wang, N. Kandasamy, C. Nwankpa, and D. R. Kaeli. Data centers as controllable load resources in the electricity market. In *Intl. Conf. on Distributed Computing Systems*, 2013.

[31] X. Zhan and S. Reda. Techniques for energy-efficient power budgeting in data centers. In *Proceedings of the 50th Annual Design Automation Conference*, page 176. ACM, 2013.

Bounding Buffer Space Requirements for Real-Time Priority-Aware Networks

Hany Kashif and Hiren Patel

Electrical and Computer Engineering
University of Waterloo, Waterloo, Canada
e-mail: {hkashif, hiren.patel}@uwaterloo.ca

Abstract— One implementation alternative for network interconnects in modern chip-multiprocessor systems is priority-aware arbitration networks. To enable the deployment of real-time applications to priority-aware networks, recent research proposes worst-case latency (WCL) analyses for such networks. Buffer space requirements in priority-aware networks, however, are seldom addressed. In this work, we bound the buffer space required for valid WCL analyses and consequently optimize router design for application specifications by computing the required buffer space at each virtual channel in priority-aware routers. In addition to the obvious advantage of bounding buffer space while providing valid WCL bounds, buffer space reduction decreases chip area and saves energy in priority-aware networks. Our experiments show that the proposed buffer space computation reduces the number of unfeasible implementations by 42% compared to an existing buffer space analysis technique. It also reduces the required buffer space in priority-aware routers by up to 79%.

I. Introduction

Real-time embedded applications, and software in general, continue to evolve with increasing complexity. This translates into a need for higher computational power to satisfy their real-time requirements. However, increasing the performance of uniprocessors is not a viable solution anymore due to limits on power consumption and heat dissipation. Though, chip-multiprocessors (CMPs) present a solution to this performance bottleneck, they have only been cautiously adopted as a platform for real-time applications. The reason is that CMPs are designed to optimize average case performance, hence, complicating the provision of real-time guarantees to such applications. As a remedy, recent research focuses on developing custom network-on-chip (NoC) interconnects and their corresponding worst-case latency (WCL) analyses to facilitate their adoption as a platform for real-time applications.

Common NoC implementations include resource reservation (e.g., time-division multiplexing (TDM)) and priority-aware networks. Resource reservation networks statically allocate resources to prevent contention between communication flows in the network at runtime. Æthereal [1] is a NoC that implements TDM. Priority-aware networks, on the other hand, allow contention between communication flows. Contention is resolved at runtime using priority-aware arbitration routers (PARs). TDM networks guarantee schedulability of flows as part of their static allocation of resources and have minimal buffer requirements (one flit for guaranteed services and one packet for best-effort services) [1]. Priority-aware networks have better resource usage compared to TDM but require a worst-case latency (WCL) analysis to guarantee schedulability

of the network flows [2, 3].

WCL analysis for real-time applications deployed on priority-aware networks is essential to guarantee honoring their real-time constraints. WCL analysis computes the WCL of communication flows including worst-case interference from other flows. Recent WCL analyses [2, 3, 4, 5] have been developed for priority-aware networks with flit-level preemption. These priority-aware networks employ wormhole switching [6] and virtual channel (VC) resource allocation [7]. While these techniques reduce the required buffer space (flit level) and allow multiple flit buffers (VCs) to access the same physical channel, priority-aware networks are susceptible to chain-blocking (blocked flits spanning multiple routers). Chain-blocking creates back-pressure in the priority-aware network which eventually leads to blocking of the computation and communication tasks. Even though, it is clear that the blocking of tasks due to back-pressure affects the WCL of the communications flows, recent analyses ignore blocking due to back-pressure and, hence, assume infinite buffer space in the network. This assumption is a serious impediment to implementing priority-aware networks because the buffer space required to guarantee the validity of the computed WCLs is unknown.

Multiple works have focused on developing WCL guarantees for priority-aware networks. Shi and Burns [2, 3] present a flow-level analysis (FLA) to compute WCL estimates for the traffic flows. Kashif et al. introduce a link-level analysis (LLA) that provides tighter WCL bounds compared to FLA at the cost of a more detailed analysis. They use LLA for path selection of flows in a priority-aware network [4] and for the computation of end-to-end application WCLs by accounting for offsets between computation and communication tasks in the network [5]. Despite the usefulness of these works in providing WCL guarantees, they never discussed buffer space bounds or inherently assumed the existence of sufficient buffer space. This work, on the other hand, computes the necessary buffer space bounds to ensure the validity of WCLs and the compositionality of holistic WCL estimates.

The main contribution of this work is determining bounds on the buffer space required in the VCs of PARs in the priority-aware network. NoC designs usually target specific applications, and, hence, customizing the design to limit cost, such as buffer space, is applicable [8, 9]. This work computes buffer space bounds to provide an ability to implement priority-aware networks and enable existing WCL analyses to provide timing guarantees for real-time applications. While this paper focuses on computing buffer space bounds, these bounds are usually determined in the chip design phase. However, we can leverage this work for distributing buffer space between the VCs of reconfigurable routers and for guiding mapping and path-selection algorithms to guarantee a valid WCL analysis. First, we overview priority-aware networks and their respective WCL analyses: FLA and LLA. Then, we present a detailed buffer space analysis using each WCL analysis: namely link-level buffer-space analysis (LLBA) and flow-level buffer-space analysis (FLBA). Our experiments show that, compared to existing buffer space analysis [10], LLBA and FLBA reduce the number of unfeasible implementations by 42% and 27%, respectively. The results also show a reduction in the required buffer space by 79% and 67% using LLBA and FLBA, respectively.

II. RELATED WORK

Since NoC designs usually target specific applications (or application classes), multiple works investigate customizing NoC designs to optimize performance while limiting cost [8, 9]. Some of these works focus on limiting buffer space in NoCs. Hu and Marculescu [11] propose an algorithm for customizing buffer space in NoC routers at a system-level. Given a buffer space budget, they assign buffers to input channels to maximize performance. The proposed work, however, does not consider real-time requirements of flows and does not ensure timeliness. Manolache et al. [10] propose a technique for changing the mapping of data packets to network links and the release timing of packets to reduce buffer sizes and ensure timeliness. While the work does not consider wormhole switching (flit-level preemption) or VC resource allocation, it can be extended to apply to our work. Their proposed approach, however, is orthogonal to the approach presented in this paper. The authors use the worst-case response time analysis (WCRTA) proposed by Palencia and Gonzalez [12] to compute the buffer space requirements. This WCRTA, however, assumes all flows sharing resources (including indirectly interfering flows) to be directly interfering with each other, this produces higher buffer space upper bounds compared to FLBA and LLBA. A similar result applies to WCL bounds as shown in [5]. Kumar et al. [13] present a simulation based algorithm for reducing buffer sizes while considering latency requirements. The authors use simulation to capture the contention between flows on network resources which does not provide worst-case guarantees for real-time flows. Al Faruque and Henkel [14] propose an approach for reducing virtual channel buffers which also does not provide real-time guarantees.

Goosens et al. [1] proposed the Æthereal TDMA-based NoC. They provide a full implementation with buffers for handling best-effort and guaranteed services. Coenen et al. [15] present an algorithm to find the minimal buffer sizes required to decouple computation and communication in the TDMA NoC using credit-based flow control while maintaining real-time guarantees to flows. Similarly, in this work, we attempt to bound the buffer space requirements for priority-aware networks.

III. BACKGROUND

A. Network Model

In this work, we use the network model proposed in [4, 5] with slight extensions. We assume the deployment of a set of traffic flows to the priority-aware network. Each flow has a source and destination task pair that reside on nodes of the network. The communication between the task pair traverses a set of links in the network which is known as the path of the traffic

flow. The priority-aware network employs wormhole switching and a flow-control mechanism to prevent buffer overflow such as the credit flow-control used in the Æthereal NoC [1] and QNoC [9]. Each node in the network consists of a processing element and a PAR.

The set of traffic flows, deployed on the priority-aware network, $\Gamma := \{\tau_i : \forall i \in [1, n]\}$ has n traffic flows. A traffic flow τ_i is a 6-tuple $\langle P_i, T_i, D_i, J_i^R, L_i, F_i \rangle$ with priority P_i, a period T_i, a deadline D_i, and a release jitter J_i^R. We assume that each flow has a distinct priority level. While the proposed approach can be extended to consider priority sharing, we exclude it from this work for brevity and clarity. The flow sends packets either periodically with a period T_i or sporadically with a minimum interarrival time T_i. The basic link latency L_i of a flow is its WCL on one network link when it does not suffer any interferences on that link. The number of flits in one packet of the flow is F_i. The basic link latency can be computed as $\frac{flit\_size * F_i}{bandwidth}$, which is the total packet size divided by the link bandwidth. For clarity of presentation and without loss of generality, we assume that $\frac{flit\_size}{bandwidth} = 1$ to simplify the conversion between latency and buffer space. Note that the proportionality between transmission time and buffer space enables us to cleanly leverage WCL analyses for the buffer space computation.

The path δ_i of a traffic flow τ_i is a sequence of network links $(l_1, \ldots, l_{|\delta_i|})$. The basic latency of flow τ_i over its path is equal to $C_i = L_i + (|\delta_i| - 1)$ where the routing delay is one time unit at each router in the network. A flow is schedulable if its WCL R_i including interferences is less than or equal to its deadline D_i.

B. Router Model

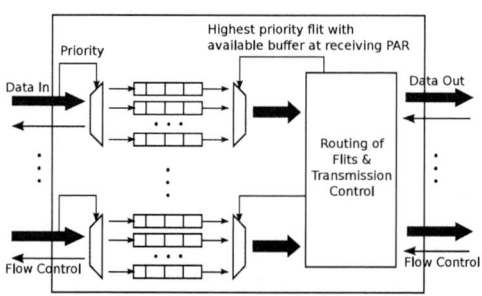

Fig. 1: Priority-aware router architecture [4].

We briefly overview the architecture of the PAR proposed in [2, 3, 4]. Figure 1 shows the internals of the PAR. Each input port has multiple VCs associated with it. A VC is a FIFO buffer that stores flits of a specific priority level. If each flow has a distinct priority level, then a VC will exist for each traffic flow. Otherwise, flows that share the same priority will utilize the same VC. According to a flit's destination and the routing algorithm, the router selects an appropriate output port for the flit. If multiple flits contend for the same output port, the router will forward the flit from the VC with the highest priority. Within the same VC, flits are forwarded based on a FIFO strategy. The flow-control scheme guarantees that a flit is forwarded from one router to another, only when there is enough buffer space at the destination router. If a flit from a certain priority level cannot access the output channel due to insufficient buffer space at a destination router, then a flit from a lower priority level can access the output channel.

C. Flow-Level Analysis

This section provides an overview of FLA. Additional details on the analysis and the proofs are available in [2, 3]. FLA computes the interference that a flow under analysis τ_i suffers along its path δ_i by considering the whole path as a single shared resource. There are two sources of interference for a packet of the flow under analysis τ_i: (1) direct and indirect interference from higher priority flows and (2) self-blocking from other packets of flow τ_i. Direct interference occurs when a flow τ_j preempts τ_i along its path δ_i. Flow τ_i suffers indirect interference from flow τ_k when flow τ_i has direct interference with flow τ_j which has direct interference with flow τ_k; however, flows τ_i and τ_k do not directly interfere with each other. The set of higher priority flows preempting flow τ_i is denoted by the symbol S_i^D. The set of higher priority flows indirectly interfering with flow τ_i is denoted by the symbol S_i^I.

FLA computes the interference on the path of τ_i to find its worst-case latency. It incorporates indirect interference from flows in the set S_i^I as interference jitter that directly interfering flows suffer. A busy period B_i is the contiguous time interval during which the flow τ_i suffers interference from flows of higher priority and from packets of the same flow. It is given by:

$$B_i = \left\lceil \frac{B_i + J_i^R}{T_i} \right\rceil * C_i + \sum_{\forall \tau_j \in S_i^D} \left\lceil \frac{B_i + J_j^R + J_j^I}{T_j} \right\rceil * C_j$$

The first term computes the latency of the packets of flow τ_i that are released in the busy period B_i. The second term accounts for interference from higher priority flows. The time interval during which interference occurs is $B_i + J_j^R + J_j^I$ where J_j^R and J_j^I are the release and interference jitters of the flow τ_j, respectively. The interference jitter J_j^I is the interference that τ_j suffers from flows in the indirect interference set of τ_i, S_i^I. The number of packets of flow τ_i in the busy period is $p_{B,i}$ which is equal to $\left\lceil \frac{B_i + J_i^R}{T_i} \right\rceil$. The worst-case latency R_i of τ_i is then the maximum response time of all packets and is given by:

$$R_i = \max_{p=1...p_{B,i}} \left(w_i(p) - (p-1)T_i + J_i^R \right)$$

where p is the index of the packet of τ_i. This response time of a packet is equal to its worst-case completion time measured from its activation time, $(p-1)T_i$, plus the packet's release jitter. The completion time of job p, $w_i(p)$, in the busy period B_i is given by:

$$w_i(p) = p * C_i + \sum_{\forall \tau_j \in S_i^D} \left\lceil \frac{B_i + J_j^R + J_j^I}{T_j} \right\rceil * C_j$$

D. Link-Level Analysis

LLA performs the WCL analysis at a finer granularity compared to FLA. As a result, LLA computes tighter WCL bounds at the expense of increased computation time. So far, LLA supports both direct and indirect interferences from higher priority

flows under the assumption that the deadline of a flow is less or equal to its period [4].

LLA computes the interference on each link along the path of the flow under analysis. It progressively adds new interferences on the flow's path while accounting only once for interferences that are common on consecutive links. If a higher priority flow τ_j interferes with the flow under analysis τ_i on a contiguous set of links, then the interference that τ_j causes should be accounted for only once [5]. The WCL of flow τ_i on any link l along its path is given by:

$$\underset{l}{R_i} = \underset{l'}{R_i} + \sum_{\forall \tau_j \in \underset{l}{S_i^D}} \left\lceil \frac{\underset{l}{R_i} + J_j^R + J_j^I}{T_j} \right\rceil * L_j$$
$$- \sum_{\forall \tau_k \in \underset{l}{S_i^D} \cap \underset{l'}{S_i^D}} \left\lceil \frac{\underset{l'}{R_i} + J_k^R + J_k^I}{T_k} \right\rceil * L_k \qquad (1)$$

where $\underset{l_0}{R_i} = L_i$. The first term represents the WCL on the previous link l'. The second term adds interferences from higher priority flows in the set $\underset{l}{S_i^D}$. The third term subtracts interferences from flows that are common on links l and l', i.e., in the set $\underset{l}{S_i^D} \cap \underset{l'}{S_i^D}$. The WCL on the first link of the path δ_i of flow τ_i is equal to the basic link latency L_i in addition to any interference suffered from higher priority flows on that link. The analysis considers both release and interference jitters as in FLA. For subsequent links, the WCL $\underset{l}{R_i}$ on a link l is equal to the WCL on the previous link l' plus new interferences on link l and subtracting common interferences on links l and l'. The overall WCL of flow τ_i is equal to the WCL on the last link of its path, plus its release jitter, plus the total routing delay along its path. Therefore, the overall WCL R_i is given by:

$$R_i = \underset{l_{|\delta_i|}}{R}_i + J_i^R + (|\delta_i| - 1) \qquad (2)$$

IV. BUFFER SPACE REQUIREMENTS

In this section, we introduce LLBA and FLBA to compute the buffer space required in the PARs of the priority-aware NoC to guarantee a valid WCL analysis. PARs implement a flow-control mechanism to prevent buffer overflow. If a VC in a receiving PAR is full, flits from the corresponding VC in the sending PAR will be blocked until there is space at the receiving PAR buffer. This might lead to chain-blocking and the creation of back-pressure in the network which might invalidate the WCL analysis. Therefore, our goal is to compute the buffer space required for each VC in each PAR to prevent back-pressure in the NoC. For each flow τ_i, we compute the buffer space X_i (measured in flits) required at the VCs used by τ_i along its path δ_i to avoid back-pressure in the network. First, we consider the simplest case:

Lemma 1. *Given a flow τ_i that suffers no interference from higher priority flows ($S_i^D = \emptyset$) and under the condition $D_i \leq T_i - J_i^R$ (no self-blocking), the buffer space required at each VC along δ_i is $X_i = 1$.*

Proof. Since the routers implement the wormhole switching protocol, then each router will directly forward received flits to the next router in the path δ_i of τ_i. For example, if one PAR forwards a flit to another PAR at time t_1, the receiving PAR will forward it in the next cycle $t_1 + 1$. And since the VCs used by τ_i are used exclusively by τ_i along its path δ_i and no contention occurs for the network links ($S_i^D = \emptyset$), then flits of τ_i are never blocked and are directly forwarded from one router to the other. Therefore, a buffer of one flit size suffices for the VCs of flow τ_i in that case. $\qquad\square$

Next, we consider two different interference scenarios:

S 1. *Interference from higher priority flows with no self-blocking from packets of the same flow under analysis.*

S 2. *Interference from higher priority flows with self-blocking from packets of the same flow under analysis.*

We introduce LLBA and FLBA to compute buffer space requirements for each interference scenario along the path of each flow. If enough buffer space exists at each router to accommodate the blocked flits, there will be no back-pressure in the network.

A. Link-Level Buffer-Space Analysis (LLBA)

LLBA computes the buffer space at each PAR on the path of the flow under analysis. Using the interference on a specific link on the path of flow τ_i, we can compute the buffer space required in the VC sending flits of τ_i on that specific link. Theorems 1 and 2 compute the buffer space at each VC for the interference scenarios S 1 and S 2, respectively.

Lemma 2. *Given a flow under analysis τ_i suffering a worst-case interference $\underset{l}{I_i}$ from higher priority flows in the set S_i^D on link l on its path δ_i, an upper bound to the buffer space required at the corresponding VC is given by $\underset{l}{X_i} = \underset{l}{I_i} + 1$.*

Proof. From Lemma 1, the buffer space required at each router along δ_i when no interference exists is one flit. Hence, if $\underset{l}{I_i} = 0$ on link l, a buffer space of one flit will be needed at the corresponding VC, i.e., $\underset{l}{X_i} = 1$. However, if $\underset{l}{I_i} > 0$, more buffer space will be required to prevent back-pressure.

A PAR forwards a flit to each of its output channels every cycle. Consider a flit of τ_i attempting to access link l in a certain cycle. This flit can only be blocked by higher priority flits in the set S_i^D. In the worst-case, this flit of τ_i will be blocked for a number of cycles equal to $\underset{l}{I_i}$. This will lead to other flits of τ_i accumulating in a FIFO order. To prevent back-pressure, enough buffer space is needed to buffer these flits of τ_i in the same VC. Since a PAR forwards a flit every cycle, then each cycle for which τ_i is blocked causes the accumulation of one more flit in the same VC. Hence, in the worst-case, an extra buffer space of $\underset{l}{I_i}$ flits is need to buffer flits accumulating from a maximum blocking time of $\underset{l}{I_i}$ cycles. Therefore, an upper bound to the buffer space required at the VC buffering flits of flow τ_i to access link l is equal to $\underset{l}{I_i} + 1$. Note that, by definition, the worst-case latency of τ_i is a contiguous time interval during which flits of τ_i and higher priority flits access link l, i.e., including the transmission of flits of flows τ_i. Hence, no interference other than $\underset{l}{I_i}$ can occur before all blocked flits of τ_i can access link l and the VC becomes empty. $\qquad\square$

We use Lemma 2 to derive the buffer space requirements for each of the different interference scenarios.

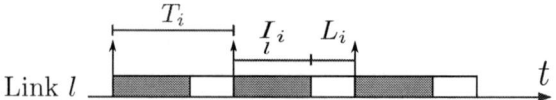

Fig. 2: Interference scenario S 1 on link l.

Theorem 1. *Given a flow under analysis τ_i with WCL $\underset{l}{R_i}$ on link l of path δ_i suffering interference only from higher priority flows in the set S_i^D under the condition $D_i \leq T_i - J_i^R$, the buffer space required at the corresponding VC is given by:*

$$\underset{l}{X_i} = \min\left(L_i, \sum_{\forall \tau_j \in S_i^D} \left\lceil \frac{\underset{l}{R_i} + J_j^R + J_j^I}{T_j} \right\rceil * L_j + 1\right)$$

Proof. Since $\underset{l}{R_i}$ is the WCL of flow τ_i on link l, then the time window during which a higher priority flow τ_j can interrupt τ_i is $\underset{l}{R_i} + J_j^R + J_j^I$. The number of times that flow τ_j interrupts τ_i is equal to $\left\lceil \frac{\underset{l}{R_i} + J_j^R + J_j^I}{T_j} \right\rceil$. And the interference that τ_i suffers from the flow τ_j is then equal to $\left\lceil \frac{\underset{l}{R_i} + J_j^R + J_j^I}{T_j} \right\rceil * C_j$. The maximum blocking $\underset{l}{I_i}$ that τ_i can suffer from all higher priority flows is equal to the sum of the interference suffered from all higher priority flows. Therefore, using Lemma 2, an upper bound to the buffer space $\underset{l}{X_i}$ is equal to $\sum_{\forall \tau_j \in S_i^D} \left\lceil \frac{\underset{l}{R_i} + J_j^R + J_j^I}{T_j} \right\rceil * L_j + 1$.

The required buffer size is, however, also bounded by the number of flits in one packet of the flow, i.e., $F_i = L_i$. We use Figure 2 to prove that both bounds prevent back-pressure in the network. The figure shows flits of flow τ_i (white) accessing link l while suffering interference from higher priority flits (grey). Assume that the first flit of τ_i attempts to access link l at a time t_1 (upward arrow in the figure). This flit will be blocked for an amount of time $\underset{l}{I_i} = \underset{l}{R_i} - L_i$.

Case $\underset{l}{I_i} + 1 < L_i$: At time $t_2 = t_1 + \underset{l}{I_i}$, the buffer will have $\underset{l}{I_i} + 1$ flits and the first flit will be accessing link l, creating an empty slot for a new incoming flit. The last flit in the packet of flow τ_i accesses the link l at time $t_3 = t_1 + \underset{l}{I_i} + L_i$.

Case $L_i < \underset{l}{I_i} + 1$: At time $t_2 = t_1 + L_i$, the buffer will have all flits of the packet under analysis of flow τ_i while the first flit is still blocked. After a blocking time $\underset{l}{I_i}$, the last flit in the buffered packet of flow τ_i will access the link l at time $t_3 = t_1 + \underset{l}{I_i} + L_i$.

In both cases, buffers sizes of $\underset{l}{I_i} + 1$ and L_i flits, respectively, will be sufficient under two conditions: (1) no new interference occurs before time t_3 and (2) no flits of another packet of τ_i arrive to the buffer before time t_3. By definition, the WCL $\underset{l}{R_i}$ is measured from the time the first flit of a packet of τ_i attempts accessing link l (at time t_1) until the time when the last flit of the packet accesses the link (at time t_3). Hence, no interference other than $\underset{l}{I_i}$ (used to compute $\underset{l}{R_i}$) can occur before t_3 or else it would have been part of I_i. Thus, satisfying the first condition. The earliest time at which the first flit of a new packet attempts to access link l is $t_4 = t_1 + T_i - J_i^R$. The flit of the new packet will be blocked if $t_4 < t_3$, i.e., $t_1 + T_i - J_i^R < t_1 + L_i + I_i$. Since $I_i = R_i - L_i$, then the blocking occurs when $T_i - J_i^R < R_i$. And for schedulability to be satisfied $R_i \leq D_i$. So the blocking occurs when $T_i - J_i^R < D_i$ which contradicts the no self-blocking condition $D_i \leq T_i - J_i^R$. Hence, the second condition is satisfied as well. Therefore, $\min(L_i, I_i + 1)$ is a safe upper bound to the buffer space X_i to prevent back-pressure. $\square$

To compute the buffer space in the presence of self-blocking, we consider the busy period of flow τ_i on link l. The busy period is the longest contiguous time interval of flits of priority equal to or higher than that of τ_i accessing link l before the link becomes idle.

Theorem 2. *Given a flow under analysis τ_i with a busy period B_i suffering interference from higher priority flows in the set S_i^D on link l along the path δ_i, the buffer space required at the corresponding VC is given by:*

$$X_i = \min(p_{B,i} * L_i, \sum_{\forall \tau_j \in S_i^D} \left\lceil \frac{B_i + J_j^R + J_j^I}{T_j} \right\rceil * L_j + 1)$$

Proof. The proof is similar to the proof of Theorem 1. The busy period B_i is the time window during which interference from higher priority flows can occur. Using Lemma 2, an upper bound to the buffer space X_i is equal to $\sum_{\forall \tau_j \in S_i^D} \lceil \frac{B_i + J_j^R + J_j^I}{T_j} \rceil * L_j + 1$. The buffer space is also bounded by the number of flits of τ_i in the busy period which is equal to the number of packets $p_{B,i}$ multiplied by the latency (number of flits) in one packet L_i.

If time t_1 is the time at which the first flit of τ_i attempts transmission in the busy period, then the time at which the last flit of τ_i accesses link l is equal to $t_3 = t_1 + I_i + p_{B,i} * L_i$. We need to prove that the buffer space bound is sufficient under the two conditions mentioned in the proof of Theorem 1. The first condition (no further interference) is satisfied by the definition of a busy period. If further interference occurs before t_3, it would have already been accounted for by I_i. The first flit of a new packet (beyond the busy period) can attempt to access link l at time $t_4 = t_1 + p_{B,i} * T_i - J_i^R$. This new flit can only be blocked if $t_4 < t_3$, i.e., when $p_{B,i} * T_i - J_i^R < I_i + p_{B,i} * L_i$. Since $I_i = B_i - p_{B,i} * L_i$, then blocking happens when $p_{B,i} * T_i - J_i^R < B_i$. This contradicts the definition of a busy period because if the length of the busy period exceeded the activation time of the new packet, it would have been part of the busy period. Therefore, $\min(p_{B,i} * L_i, I_i + 1)$ is a safe upper bound to the buffer space X_i to prevent back-pressure. $\square$

B. Flow-Level Buffer-Space Analysis (FLBA)

FLBA is applied for each flow in the network while considering the whole path of the flow as an indivisible unit during analysis. In this case, the buffer space computed using FLA will apply to each router along the path of the flow under analysis. For space constraints, we omit the proofs of the theorems. The theorems are similar to those used by LLBA but applied to the whole path of the flow instead of each link on its path.

Theorem 3. *Given a flow under analysis τ_i with WCL R_i suffering interference only from higher priority flows in the set S_i^D along its path δ_i under the condition $D_i \leq T_i - J_i^R$, the buffer space required at each VC used by τ_i along δ_i is given by:*

$$X_i = \min(L_i, \sum_{\forall \tau_j \in S_i^D} \left\lceil \frac{R_i + J_j^R + J_j^I}{T_j} \right\rceil * C_j + 1)$$

Theorem 4. *Given a flow under analysis τ_i with a busy period B_i suffering interference from higher priority flows in the set S_i^D along its path δ_i, the buffer space required at each VC used by τ_i along δ_i is given by:*

$$X_i = \min(p_{B,i} * L_i, \sum_{\forall \tau_j \in S_i^D} \left\lceil \frac{B_i + J_j^R + J_j^I}{T_j} \right\rceil * C_j + 1)$$

V. EXPERIMENTATION

We quantitatively evaluate the proposed buffer space computation techniques: LLBA and FLBA. We compare the proposed techniques to the analysis proposed by Palencia and Gonzalez [12] (PAL) which was used for buffer space computation in [10]. We perform the evaluation on a set of synthetic benchmarks as in [10, 4, 5]. Synthetic benchmarks allow us to assess the effect of different factors (as well as their extreme values) on the buffer space computation. We perform our experiments on 4×4 and 8×8 instances of the priority-aware NoC. Our goals from these experiments are: demonstrating the feasibility of computing buffer spaces for priority-aware networks, quantifying the reduction in the number of unfeasible implementations, quantifying the reduction in buffer space bounds computed by the proposed analyses, and comparing the computation times of buffer spaces using PAL, LLBA, and FLBA.

Our experiments involve changing multiple factors to evaluate their affect on buffer space computation. The number of communications flows varies from 1 flow to 100 flows (in steps of 1). The source and destination pairs of the flows are randomly mapped to the NoC. The paths for the flows are computed using a shortest-path algorithm. Packet sizes are uniformly distributed in the range (10,1000) flits. A uniform random distribution is used to assign periods (or minimum interarrival time for sporadic tasks) T_i to communication flows in the range (1000,1000000). The flow's deadline D_i is an integer multiple of its period. An arbitrary priority assignment scheme is used for selecting task priorities. The utilization of the NoC's communication resources varies from 10% to 6000% (in steps of 60%). Hence, we have 40000 possible configurations and we generate 100 different test cases for each configuration.

Figure 3a shows percentage of infeasible implementations against the utilization of the communication resources in the NoC. Increasing the utilization of the network links results in

978-1-4799-2817-0/14 $31.00 © 2014 IEEE

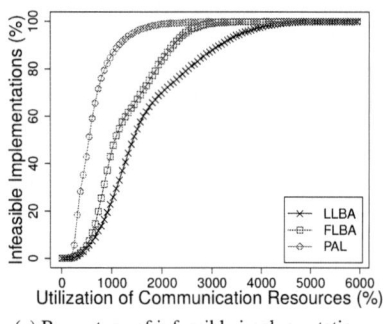 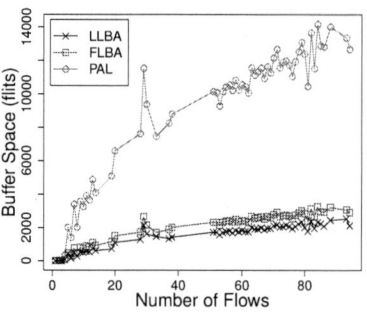 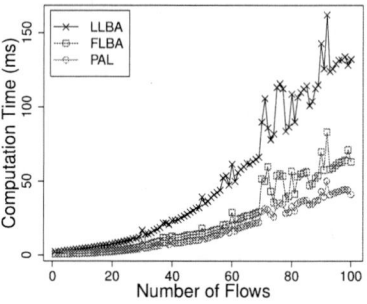

(a) Percentage of infeasible implementations.　　(b) Buffer space against number of flows.　　(c) Computation time against number of flows.

Fig. 3: Experimental results.

an increase in the interference suffered by the traffic flows in the network. As the interference increases, some of the flows will require an unbounded (infinite) buffer space. If one flow requires unbounded buffer space, the test case will be considered as an infeasible implementation. The graph shows that for any utilization, LLBA has the least percentage followed by FLBA then PAL. LLBA and FLBA reduce infeasible implementations by 42% and 27%, respectively compared to PAL.

Figure 3b shows the required buffer space against the number of flows at a network utilization of 900%. As the number of flows increases, the required buffer space increases due to the increase of VCs and increase of interference in the network. LLBA has the least buffer space requirements followed by FLBA then PAL On average, LLBA and FLBA reduce the buffer space by 79% and 67%, respectively compared to PAL.

Figure 3c shows the computation time of the buffer space analysis techniques against the number of flows. To guarantee fairness between the different techniques, the computation time includes the time required to run the corresponding WCL analyses needed to apply the buffer space analysis techniques. PAL has the least computation time followed by FLBA then LLBA. On average, LLBA, FLBA, and PAL have computation times of 49 ms, 23 ms, and 16 ms, respectively.

VI. CONCLUSION

The increase in computational requirements of real-time software and the performance limitations of uniprocessors make CMPs with NoC interconnects a viable platform for real-time software. Although recent research focuses on WCL analysis techniques for priority-aware networks, buffer space requirements were not investigated. Limiting buffer space is also important to reduce silicon area and energy. Typically, NoCs are designed for a specific application or application-classes, hence, designers can customize buffer space based on application requirements. In this work, we extend the two most-recent analyses for WCL computation in priority-aware networks to compute buffer space requirements in PARs which guarantee the validity of the WCL analyses. Our experiments show that LLBA and FLBA reduce the number of unfeasible implementations by 42% and 27% compared to PAL, respectively. LLBA and FLBA also reduce the required buffer space by 79% and 67%, respectively at the expense of an acceptable increase in computation time. Our future work includes developing a complete mapping algorithm for priority-aware networks, using both WCL and buffer space analysis, that guaran-

tees latency bounds and reduces buffer space in the network.

REFERENCES

[1] K. Goossens, J. Dielissen, and A. Radulescu, "Æthereal Network on Chip: Concepts, Architectures, and Implementations," *IEEE Design and Test*, vol. 22(5), 2005.

[2] Z. Shi and A. Burns, "Real-Time Communication Analysis for On-Chip Networks with Wormhole Switching," in *Proceedings of the Second ACM/IEEE International Symposium on Networks-on-Chip*, 2008.

[3] Z. Shi, "Real-Time Communication Services for Networks on Chip," Ph.D. dissertation, The University of York, UK, 2009.

[4] H. Kashif, H. D. Patel, and S. Fischmeister, "Using link-level latency analysis for path selection for real-time communication on nocs," in *Proceedings of the Asia South Pacific Design Automation Conference*, February 2012, pp. 499–504.

[5] H. Kashif, S. Gholamian, R. Pellizzoni, H. D. Patel, and S. Fischmeister, "ORTAP:An Offset-based Response Time Analysis for a Pipelined Communication Resource Model," in *IEEE Real-Time and Embedded Technology and Applications Symposium*, April 2013.

[6] L. Ni and P. McKinley, "A survey of wormhole routing techniques in direct networks," *Computer*, vol. 26, no. 2, 1993.

[7] W. Dally, "Virtual-channel flow control," *IEEE Transactions on Parallel and Distributed Systems*, vol. 3, no. 2, 1992.

[8] L. Benini and G. De Micheli, "Networks on chip: a new paradigm for systems on chip design," in *Proceedings of the conference on Design, Automation and Test in Europe*, 2002.

[9] E. Bolotin, I. Cidon, R. Ginosar, and A. Kolodny, "QNoC: QoS architecture and design process for network on chip," *JOURNAL OF SYSTEMS ARCHITECTURE*, vol. 50, 2004.

[10] S. Manolache, P. Eles, and Z. Peng, "Buffer space optimisation with communication synthesis and traffic shaping for nocs," in *Proceedings of the conference on Design, automation and test in Europe*, ser. DATE '06. European Design and Automation Association, 2006.

[11] J. Hu and R. Marculescu, "Application-specific buffer space allocation for networks-on-chip router design," in *IEEE/ACM International Conference on Computer Aided Design (ICCAD)*, 2004.

[12] J. C. Palencia and M. González Harbour, "Schedulability analysis for tasks with static and dynamic offsets," in *Proceedings of the IEEE Real-Time Systems Symposium*, ser. RTSS '98. IEEE Computer Society, 1998.

[13] A. Kumar, M. Kumar, S. Murali, V. Kamakoti, L. Benini, and G. De Micheli, "A simulation based buffer sizing algorithm for network on chips," in *IEEE Computer Society Annual Symposium on VLSI (ISVLSI)*, 2011.

[14] M. A. Al Faruque and J. Henkel, "Minimizing virtual channel buffer for routers in on-chip communication architectures," in *Proceedings of the conference on Design, automation and test in Europe*, ser. DATE '08. ACM, 2008.

[15] M. Coenen, S. Murali, A. Ruadulescu, K. Goossens, and G. De Micheli, "A buffer-sizing algorithm for networks on chip using TDMA and credit-based end-to-end flow control," in *Proceedings of the 4th international conference on Hardware/software codesign and system synthesis*, ser. CODES+ISSS '06. ACM, 2006.

978-1-4799-2817-0/14 $31.00 © 2014 IEEE

Task- and Network-level Schedule Co-Synthesis of Ethernet-based Time-triggered Systems

Licong Zhang, Dip Goswami, Reinhard Schneider, Samarjit Chakraborty
Institute for Real-Time Computer Systems, TU Munich, Germany
licong.zhang@rcs.ei.tum.de, dip.goswami@tum.de, reinhard.schneider@rcs.ei.tum.de, samarjit@tum.de

Abstract—In this paper, we study time-triggered distributed systems where periodic application tasks are mapped onto different end stations (processing units) communicating over a switched Ethernet network. We address the problem of application level (i.e., both task- and network-level) schedule synthesis and optimization. In this context, most of the recent works [10], [11] either focus on communication schedule or consider a simplified task model. In this work, we formulate the co-synthesis problem of task and communication schedules as a Mixed Integer Programming (MIP) model taking into account a number of Ethernet-specific timing parameters such as interframe gap, precision and synchronization error. Our formulation is able to handle one or multiple timing objectives such as application response time, end-to-end delay and their combinations. We show the applicability of our formulation considering an industrial size case study using a number of different sets of objectives. Further, we show that our formulation scales to systems with reasonably large size.

I. INTRODUCTION

Recently, several Ethernet based protocols are developed and adopted in real-time safety-critical domains. The examples are EtherCAT and Profinet [1] in industrial automation, AFDX [2] and TTEthernet [3] in avionics and the AVB networks [4]. The majority of these protocols are based on the original Ethernet standard IEEE802.3 and employ a *switched* network paradigm. That is, switches are used to connect the *end stations* via *full-duplex* links. A full-duplex link consists of two independent directed links and allows simultaneous transmission in both directions. Fig. 1 shows an example of such a network with four end stations and two switches.

In an Ethernet-based network, the end stations communicate by exchanging messages as Ethernet *frames*. The frames are forwarded link by link via the switches from the sender to receiver end station. In general, when the frames are forwarded to the same link simultaneously, they are stored in a queue and sent according to an output scheduler (e.g., strict priority). Thus, the queueing in switches results in a queueing delay which further depends on many factors such as the topology, traffic load and often introduces a considerably large delay. Such non-deterministic temporal behavior makes the traditional Ethernet-based networks unsuitable for safety-critical applications with stringent timing requirements. To address this problem, most of such protocols offer mixed-criticality services by dividing the traffic into different classes, applied to communication with different timing requirements. For example, Profinet has IRT, RT and NRT services in decreasing timing guarantee. Similarly, the traffic in TTEthernet is divided into TT, RC and BE traffic and in AVB networks Class A/Class B are differentiated with normal prioritized frames.

In the above context, the traffic class designed to provide most strict timing guarantee is the time-triggered traffic. Examples of this traffic class include the IRT in Profinet, the TT traffic in TTEthernet and the time-triggered traffic in the IEEE802.1Qbv standard still in development. The actual implementation of the time-triggered traffic class may vary from one protocol to another, but the basic idea is to schedule the

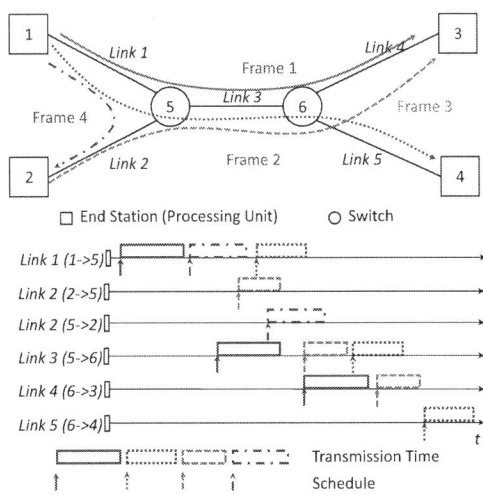

Fig. 1. An example of time-triggered traffic in switched Ethernet network.

frames so that the delays introduced by queueing in switches can be minimized or eliminated, thus achieving low network latency and jitter. Certain mechanisms are used to remove the influence from other traffic classes so that the frame transmission follows the pre-defined schedule. For example, Profinet uses a dedicated time period in a cycle to transmit IRT frames. All the frames in this traffic class are sent and forwarded in the network according to a pre-configured schedule. A schedule defines the exact time points when the end stations and switches send the frames on the Ethernet links. As shown in Fig. 1, the schedules should be chosen such that the frame transmission times do not collide with each other and thus ensure deterministic temporal behavior with low latency and jitter. In this paper, we consider the time-triggered traffic. Communication of other traffic classes can be added to the system without influencing the timing properties of time-triggered communication. We present our method considering the time-triggered traffic in a general switched Ethernet network. Our method can be tailored to fit into specific protocols such as Profinet and TTEthernet.

In a distributed system, a number of functions (or applications) are performed by a combination of *tasks* running on the end stations and data exchange by the communication over the network. For example, a distributed control system might have its *sensing*, *computing* and *actuating* tasks mapped onto different end stations and sensing/actuation data is sent via the network. In such cases, the performance of an application depends on schedules of both tasks and communication. Since optimal communication schedules may not necessarily result in optimal schedules at the application level, the above problem can not be simplified to that of the communication schedules alone. Moreover, the representation of design objectives is often itself a problem. For example, the design objective can be to minimize the overall latency of all applications while placing emphasis on certain ones. Such design requirements need multi-objective formulation.

978-1-4799-2817-0/14 $31.00 © 2014 IEEE

In this work, we aim to optimize the schedules for the tasks running on the end stations and for communication over the network considering one or multiple optimization objectives.

Related work: The related work in the above direction can be classified into three groups: (i) general time-triggered architecture [6] (ii) application level schedule synthesis and optimization of time-triggered (static) segment of FlexRay bus [7], [8], [9] (iii) optimization of time-trigged communication schedules in Ethernet-based network [10], [11], [12], [13]. Clearly, the works in direction (i) present general techniques [6] to deal with time-triggered systems without taking into account the Ethernet-based details which is the focus of our work. The recent works [7], [8], [9] in direction (ii) deal with both the network- [7] and application-level [8], [9] schedules taking into account the FlexRay specific details. However, FlexRay as a communication protocol differs considerably from a switched Ethernet network and works in [7], [8], [9] can not be applied in a straightforward manner. In direction (iii), [12] proposed a model formulation for the schedule synthesis for the Profinet IRT which is the time-triggered traffic pattern in Profinet. Similarly, [10] formulated the complete model of the time-triggered traffic in TTEthernet in SMT, which is later complemented by [11] for the bandwidth reservation of RC traffic. Further, [13] proposed an alternative schedule synthesis method of TTEthernet based on Tabu search, such that the deadlines of TT and RC messages are satisfied and end-to-end delays of RC messages are minimized.

Contributions: We propose a method for application-level schedule synthesis of distributed systems using time-triggered traffic in Ethernet communication. In this method, we formulate the scheduling problem as a Mixed Integer Programming (MIP) model. We further consider different optimization objectives and formulate a multi-objective optimization problem. We use Gurobi [16] to obtain a solution for the MIP formulation. Although there are a few recent works [10], [11] towards optimizing communication schedules in this context, our work focuses on application level optimization. Using an industrial size case-study, we illustrate various design aspects of such Ethernet-based time-triggered systems by considering different optimization objectives. To the best of our knowledge, this is the first work to consider the interplay between the task and communication schedules of a time-triggered Ethernet-based distributed system and optimize the schedules according to the (multi-)objectives at the application level.

II. ETHERNET-BASED TIME-TRIGGERED SYSTEMS

We consider a distributed system whose physical topology consists of a number of *end stations* connected by a switched Ethernet network. We denote the system as a graph $G(\mathcal{V}, \mathcal{E})$, where vertices $\mathcal{V}$ denote the set of network *nodes* (end stations and switches) and edges $\mathcal{E}$ denote the *full-duplex* Ethernet links. To differentiate between different types of nodes, we denote an end station as $v_i^e \in \mathcal{V}$ and a switch as $v_i^s \in \mathcal{V}$. A full-duplex link connecting two nodes v_m and v_n is denoted by $l_{m,n} \in \mathcal{E}$ and $l_{n,m} \in \mathcal{E}$, each representing a *directed* link from v_m to v_n and from v_n to v_m. In the distributed setup under consideration, we consider the following components which are described using an example shown in Fig. 2.

Application tasks: We define a task running on an end station as an application task. We consider periodic application tasks τ_i which are characterized by a tuple $\tau_i = \{\tau_i.p,\ \tau_i.o,\ \tau_i.e\}$ consisting of the period, offset and the WCET respectively. In the example shown in Fig. 2, there are five applications tasks τ_1 to τ_5 which are running in five different end stations.

Communication tasks: A communication task c_i can be defined as $c_i = \{f_i,\ c_i.tr,\ c_i.o,\ c_i.p\}$. Each communication

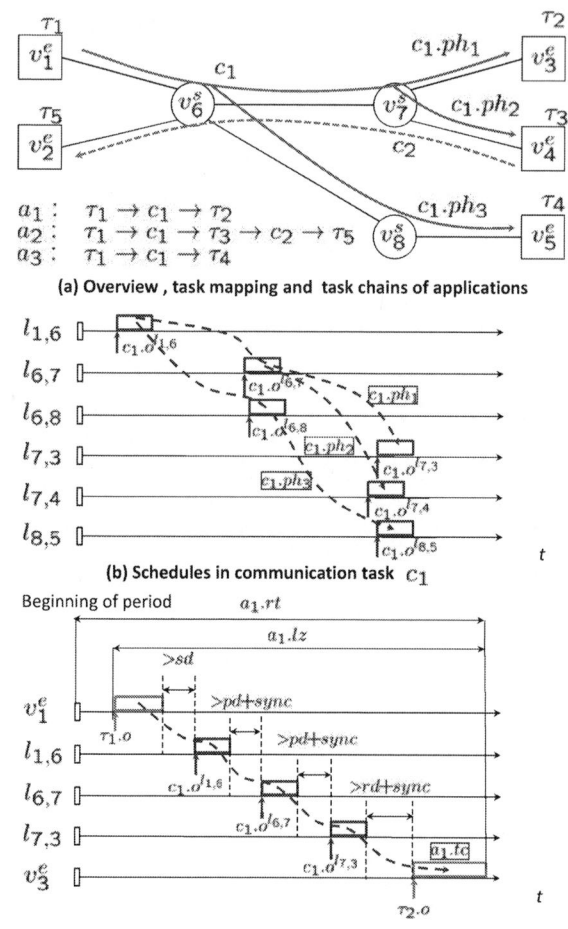

(a) Overview , task mapping and task chains of applications

(b) Schedules in communication task c_1

(c) Task chain and schedules of a_1

Fig. 2. Example of a distributed system of applications a_1 to a_3, application tasks τ_1 to τ_5 and communication tasks c_1 and c_2.

task is associated with an unique Ethernet frame with frame length $f_i.fl$. We define a *path* as a route from the sender to one receiver. For example, $c_1.ph_1$ is one path from v_1^e to v_3^e. Further, $c_i.tr$ denotes the *path tree* which consists of all the *paths* from a particular sender station to the receiver stations. A path tree may contain one path $c_i.tr = c_i.ph_1$ or multiple paths $c_i.tr = \{c_i.ph_1,\ c_i.ph_2,\ \ldots,\ c_i.ph_{\alpha_i}\}$, depending on whether the uni-cast or multi-cast operation is applied. For example, in Fig. 2 (a) and (b), path tree of c_1 contains three paths $c_1.ph_1$, $c_1.ph_2$ and $c_1.ph_3$. Furthermore, $c_i.o$ denotes the set of sending schedules of frame f_i on all the Ethernet links in the path tree. A single schedule on link $l_{m,n}$ is represented as $c_i.o^{l_{m,n}}$. In Fig. 2, the communication task c_1 involves six Ethernet links and there is one schedule for each link as indicated in Fig. 2 (b). Finally, $c_i.p$ denotes the period of the communication task. Note that as oppose to the traditional task, the communication task here represents the whole process of sending, forwarding and receiving a frame over the network.

Applications: An *application* a_i is a collection of certain application tasks and communication tasks that perform an independent function. An application is characterized by the tuple $a_i = \{a_i.tc,\ a_i.p,\ a_i.rt,\ a_i.lz\}$. Here, $a_i.tc$ is the *task chain*, containing the application and communication tasks that constitute the application in the temporal order. Three applications a_1, a_2 and a_3 are shown in Fig. 2 (a). We further denote the tasks contained in the task chain as $a_i.t_j$, where $1 \le j \le \beta_i$ with $a_i.t_1$ and $a_i.t_{\beta_i}$ denoting the first and last

978-1-4799-2817-0/14 $31.00 © 2014 IEEE   120

task respectively. Each task in the task chain either represents an application or a communication task. Fig. 2 (c) shows the details of task chain $a_1.tc$. Since there are three tasks in a_1, $\beta_1 = 3$ with $a_1.t_1 = \tau_1$, $a_1.t_2 = c_1$ and $a_1.t_3 = \tau_2$. $a_i.p$ denotes the period of the application and all the application and communication tasks share the same period with the application. That is, $\tau_1.p = c_1.p = \tau_2.p = a_1.p$ in the above example. Finally, $a_i.rt$, $a_i.lz$ are the *response time* and the *end-to-end latency*, which represent the time between the start of the period and the end of last task and the time between the start of the first task and end of last task. The significance of these two parameters will be discussed in details in the following sections.

Additional timing restrictions: In this paper, we consider all the end stations as single processors running under a time-triggered non-preemptive scheduling scheme. Such scheduling scheme is common in safety-critical domains in particular. If an application task finishes its execution, it needs some time before the data can be packed in frames and sent on the network. This time interval has an upper bound which is denoted by sd. Similarly, once a frame arrives at an end station, it needs some time to get unpacked and processed before the data can be utilized by the corresponding application task. We assume that this time has an upper bound denoted as rd. On the network, we use the general case where each Ethernet switch possesses a *dispatcher* and each frame arriving at a switch is *forwarded* according to a schedule. The maximum processing delay of a frame in a switch is bounded by pd. That is, pd is the time between reception of the last bit on the *input port* and the earliest possible transmission of the first bit on the *output port*. It should be noted that *store-and-forward* and *cut-through* mechanism [14] can be modeled by setting pd an appropriate value. We further denote the *bandwidth* as bw – time to send one bit on the link and *interframe gap* as ifg – minimal link idle time between the transmission of two consecutive frames. The *precision*, i.e., the maximum difference between the any two clocks in the system is denoted as $sync$. In this paper, all the schedules are referenced to the local time of the network nodes and we pad a $sync$ in the constraints with schedules from different nodes.

III. CONSTRAINTS FORMULATION

In this work, we present a framework to co-synthesize schedules for all the application tasks τ_i and communication tasks c_i such that some high-level objectives are optimized. We characterize the distributed system described in the previous section by the following set of constraints.

(C1) Collision free application tasks: This condition ensures that on every single processor, one application task is triggered only when the processor is idle, i.e., after the last task is finished. We define the set of all application tasks as $\mathcal{T}$ and that of those mapped on an end station v_m^e as $\mathcal{T}(v_m^e)$. This condition can be formulated as:

$$\forall m, v_m^e \in \mathcal{V}, \forall i, j, i \neq j, \tau_i, \tau_j \in \mathcal{T}(v_m^e) \tag{1}$$
$$\tau_i.p \times k_i + \tau_i.o + \tau_i.e < \tau_j.p \times k_j + \tau_j.o$$
or
$$\tau_j.p \times k_j + \tau_j.o + \tau_j.e < \tau_i.p \times k_i + \tau_i.o$$

where

$$\forall k_i \in \left[0, \frac{LCM(\tau_i.p, \tau_j.p)}{\tau_i.p} - 1\right]$$
$$\forall k_j \in \left[0, \frac{LCM(\tau_i.p, \tau_j.p)}{\tau_j.p} - 1\right].$$

and $LCM(\tau_i.p, \tau_j.p)$ denotes the least common multiple of periods $\tau_i.p$ and $\tau_j.p$. In Fig. 2, $\mathcal{T} = \{\tau_1, \tau_2, \tau_3, \tau_4, \tau_5\}$

and $\mathcal{T}(v_1^e) = \tau_1$ for example. Although there is only one application task mapped on each end station in this example, there can be multiple applications tasks running on an end station as we will see in the experimental section. The constraint (1) is applicable to those scenarios.

(C2) Collision free communication tasks: To ensure there is no collision of frames being sent on a single directed Ethernet link, a frame can only start its transmission ifg time units after the last frame is finished. We define $\mathcal{C}$ as the set of all communication tasks and $\mathcal{C}(l_{m,n})$ as the set whose path trees contain link $l_{m,n}$. This condition can be formulated as:

$$\forall m, n, l_{m,n} \in \mathcal{E}, \forall i, j, c_i, c_j \in \mathcal{C}(l_{m,n}) \tag{2}$$
$$c_i.p \times k_i + c_i.o^{l_{m,n}} + f_i.fl/bw + ifg < c_j.p \times k_j + c_j.o^{l_{m,n}}$$
or
$$c_j.p \times k_j + c_j.o^{l_{m,n}} + f_j.fl/bw + ifg < c_i.p \times k_i + c_i.o^{l_{m,n}}$$

where

$$\forall k_i \in \left[0, \frac{LCM(c_i.p, c_j.p)}{c_i.p} - 1\right]$$
$$\forall k_j \in \left[0, \frac{LCM(c_i.p, c_j.p)}{c_j.p} - 1\right].$$

In Fig. 2, $\mathcal{C} = \{c_1, c_2\}$, $\mathcal{C}(l_{6,8}) = \{c_1\}$ and similarly, $\mathcal{C}(l_{7,6}) = \{c_2\}$. Although there is only one communication task mapped on each link $l_{m,n}$ in this example, there can be multiple communication tasks sharing the same link. The constraint (2) is applicable to those scenarios.

(C3) Path dependency in communication: In a communication task, a frame can only be forwarded along the paths in the correct temporal order. We further represent the schedule $c_i.o^{l_{m,n}}$ on the link $l_{m,n}$ as $c_i.o^{l_{m,n}}[ph_j, q]$ or simply $c_i.o[ph_j, q]$, i.e., the qth schedule in path ph_j. Here, $q = 1$ and $q = \gamma_{i,j}$ represent the first and last schedule. The data dependency of the communication can be formulated as:

$$\forall i, c_i \in C, \forall j \in [1, \alpha_i], \forall q \in [2, \gamma_{i,j}] \tag{3}$$
$$c_i.o[ph_j, q - 1] + f_i.fl/bw + pd + sync < c_i.o[ph_j, q].$$

It should be noted that $f_i.fl/bw$ represents the transmission time of the frame f_i for a given bandwidth bw. In Fig. 2, let us consider the path $c_1.ph_1$. Here, $\gamma_{1,1} = 3$ and the three schedules $c_1.o^{l_{1,6}}[ph_1, 1]$, $c_1.o^{l_{6,7}}[ph_1, 2]$ and $c_1.o^{l_{7,3}}[ph_1, 3]$ should be in the correct temporal order satisfying the constraint (3). Note that $c_1.o^{l_{1,6}}[ph_1, 1]$, $c_1.o^{l_{1,6}}[ph_2, 1]$ and $c_1.o^{l_{1,6}}[ph_3, 1]$ represent the same schedule since all three paths share the same link $l_{1,6}$.

(C4) Data dependency in applications: Due to data dependency, the application and communication tasks in the task chain of an application have to be executed in the correct temporal order. In a task chain, two consecutive tasks must be one of the following cases: (i) both application tasks, (ii) an application task followed by a communication task (iii) a communication task followed by an application task. We define the set of all applications as $\mathcal{A}$. This condition can be formulated as follows:

$$\forall i, a_i \in \mathcal{A}, \forall j \in [1, \beta_i - 1]$$
if $a_i.t_j = \tau_h \in \mathcal{T} \wedge a_i.t_{j+1} = \tau_g \in \mathcal{T}$ **then**
$$\tau_h.o + \tau_h.e < \tau_g.o \tag{4}$$
if $a_i.t_j = \tau_h \in \mathcal{T} \wedge a_i.t_{j+1} = c_g \in \mathcal{C}$ **then**
$$\tau_h.o + \tau_h.e + sd < c_g.o[ph_u, 1] \tag{5}$$
if $a_i.t_j = c_h \in \mathcal{C} \wedge a_i.t_{j+1} = \tau_g \in \mathcal{T}$ **then**
$$c_h.o[ph_v, \gamma_{h,v}] + f_h.fl/bw + sync + rd < \tau_g.o. \tag{6}$$

where ph_u is any path within the path tree and ph_v represents the path which is utilized by the corresponding application.

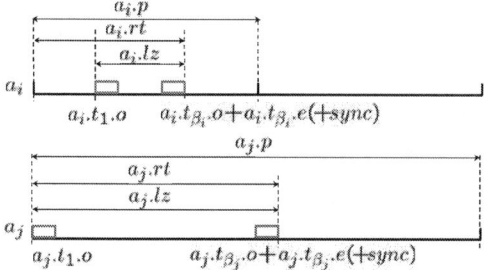

Fig. 3. Response time and end-to-end latency.

For illustration, let us consider the example in Fig. 2. For application a_1 in Fig. 2 (c), the application task τ_1 is followed by c_1 and the schedules $\tau_1.o$ and $c_1.o^{l_{1,6}}$ are constrained by (5). Similarly, c_1 is followed by τ_2 and the schedules $c_1.o^{l_{7,3}}$ and $\tau_2.o$ are constrained by (6).

(C5) Application response time constraint: The response time of an application a_i is given by:

$$a_i.rt = a_i.t_{\beta_i}.o + a_i.t_{\beta_i}.e(+sync) . \tag{7}$$

We denote the response time of an application as the time when the last task in the corresponding task chain is finished from the beginning of period. If observed from the local time of the sender station, the *sync* can be omitted. Fig. 3 shows an example of response time of two tasks a_i and a_j. In a_i, $a_i.t_{\beta_1}.o \neq$ *beginning of period* possibly because of unavailability of (computation or network) resources. In many real-life time-triggered systems, the platform status information needs to be updated within a maximum tolerable time bound which is reflected by response time of an application. A hard constraint on maximum tolerable response time $a_i.rt_{max}$ can be enforced by adding the condition

$$a_i.rt < a_i.rt_{max}, \forall i, a_i \in \mathcal{A} . \tag{8}$$

(C6) Application end-to-end latency constraint: End-to-end latency of an application can be formulated as:

$$a_i.lz = a_i.t_{\beta_i}.o + a_i.t_{\beta_i}.e - a_i.t_1.o(+sync) . \tag{9}$$

Clearly, the end-to-end latency of an application is the time between the start of the first task and the finishing of the last task of its task chain. Similar to the response time, a maximum tolerable end-to-end latency can be enforced by the condition

$$a_i.lz < a_i.lz_{max}, \forall i, a_i \in \mathcal{A} . \tag{10}$$

Fig. 3 illustrates the difference between the response time and the end-to-end latency. When $a_i.t_{\beta_1}.o =$ *beginning of period*, $a_i.rt = a_i.lz$. The application a_2 in Fig. 3 shows such an example. On the other hand, when $a_i.t_{\beta_1}.o \neq$ *beginning of period*,

$$a_i.rt = a_i.lz + a_i.t_1.o .$$

One such example is the application a_1 in Fig. 3.

A. Constraint formulation as Mixed Integer Programming

Here, we discuss how the constraints and optimization objectives described in the previous section can be converted into a MIP problem. Towards this, each constraint should be represented as a single inequity or equation. The constraints (C3) to (C6) are already in the form of a single inequity. Since the constraints (C1) and (C2) are *either-or* conditions, they have to be converted by introducing a binary decision variable [15]. Thus, (C1) can be represented as

$$\forall n, v_n^e \in V, \forall i, j, i \neq j, \tau_i, \tau_j \in T(v_m^e), \forall q \in [1, \ \lambda] \tag{11}$$
$$\tau_i.p \times k_i + \tau_i.o + \tau_i.e < \tau_j.p \times k_j + \tau_j.o + y_q \times M_q$$
$$\tau_j.p \times k_j + \tau_j.o + \tau_j.e < \tau_i.p \times k_i + \tau_i.o + (1 - y_q) \times M_q .$$

where y_q denotes the introduced binary variable, λ represents the total number of possible collisions between application tasks and M_q is a sufficiently large constant. Similar conversion can be applied to (C2).

IV. MULTI-OBJECTIVE OPTIMIZATION

Given the above representation of a time-triggered system, we consider three different classes of objectives. Let us consider a set of N applications $\mathcal{A}(obj) \in \mathcal{A}$.

(O1) Max/avg response time:

$$\forall i, a_i \in \mathcal{A}(obj) \tag{12}$$
$$\mathcal{A}(obj).rt_{max} = max(a_i.rt)$$
$$\mathcal{A}(obj).rt_{avg} = \sum a_i.rt/N$$

(O2) Max/avg end-to-end latency

$$\forall i, a_i \in \mathcal{A}(obj) \tag{13}$$
$$\mathcal{A}(obj).lz_{max} = max(a_i.lz)$$
$$\mathcal{A}(obj).lz_{avg} = \sum a_i.lz/N$$

(O3) Multi-objective optimization

$$\forall i, obj_i \in \mathcal{OBJ} \tag{14}$$
$$obj_M = \sum obj_i \times \omega_i$$

where $\mathcal{OBJ}$ denotes the set of objectives in the form of O1 or O2 and obj_M denotes the weighted multi-objective function.

A. Objective formulation as Mixed Integer Programming

The implementation of objective function as the response time or end-to-end latency of a single application or the average value of multiple applications is straight forward. When the objective is the maximum value of several applications, the *minimax* problem can be formulated in MIP by introducing a continuous variable z [15]. Thus the minimax objective functions can be converted to

$$obj = z, \forall i, a_i \in \mathcal{A}(obj), a_i.rt \leq z, a_i.lz \leq z . \tag{15}$$

Such an objective will introduce an extra continuous variable z and N inequities, which must be added to the constraints. Further, a multi-objective function can be formulated as (14).

V. EXPERIMENTAL RESULTS

In this section, we present the experimental results to show the applicability of our approach. We present an industrial size case study of a distributed system to illustrate the various design aspects considering different optimization objectives. Further, we present a scalability analysis and show that our formulation scales to systems with reasonably large size. We used Gurobi 5.10 [16] for solving the MIP model and all experiments are carried out on a computer with 1.87GHz CPU and 4GB memory.

A. Case Study

In the case study, we consider a distributed system consisting of 12 end stations connected via a switched Ethernet network. We explore four different network topologies as shown in Fig. 4. Further, 53 application tasks are mapped on the processors and 23 frames are sent on the network amongst which 7 are multi-cast. The application tasks and communication tasks constitute a total of 30 applications, with 5 of them having a task chain length of 5. The system parameters are $bw = 100$Mbps, $ifg = 0.96$us (12byte), $sd = pd = rd = 10$us, $sync = 5$us. The details of the system configuration is shown in Table I to Table III.

978-1-4799-2817-0/14 $31.00 © 2014 IEEE

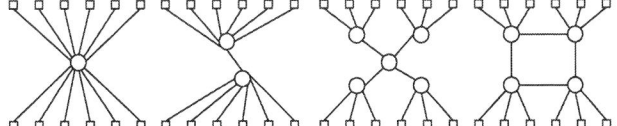

Fig. 4. Network topologies explored (from left to right): (a) Star, (b) Twinstar, (c) Tree, (d) Ring.

c_i	$f_i.fl$[B]	c_i	$f_i.fl$[B]	c_i	$f_i.fl$[B]	c_i	$f_i.fl$[B]
c_1	64	c_7	100	c_{13}	100	c_{19}	100
c_2	80	c_8	100	c_{14}	150	c_{20}	64
c_3	64	c_9	64	c_{15}	150	c_{21}	64
c_4	80	c_{10}	64	c_{16}	100	c_{22}	64
c_5	80	c_{11}	64	c_{17}	100	c_{23}	64
c_6	100	c_{12}	64	c_{18}	100		

TABLE III.   FRAME LENGTH OF COMMUNICATION TASKS.

v_i^e	τ_i	$\tau_i.e$[us]	send c_i	receive c_i
v_1^e	$\tau_1,\tau_2,\tau_3,\tau_4,\tau_5$	200	$c_1,c_2,-,-,-$	$-,-,c_3,c_5,c_7$
v_2^e	$\tau_6,\tau_7,\tau_8,\tau_9,\tau_{10}$	500	$c_3,c_4,-,-,-$	$-,-,c_{10},c_{17},c_{21}$
v_3^e	$\tau_{11},\tau_{12},\tau_{13},\tau_{14}$	550	$c_5,c_6,-,-$	$-,-,c_2,c_4$
v_4^e	τ_{15},τ_{16}	350	c_7,c_8	c_1,c_{13}
v_5^e	$\tau_{17},\tau_{18},\tau_{19},\tau_{20},\tau_{21}$	500	$c_9,-,-,-,-$	$-,c_6,c_{11},c_{17},c_{18}$
v_6^e	$\tau_{22},\tau_{23},\tau_{24},\tau_{25},\tau_{26},\tau_{27}$	400	$c_{10},c_{11},c_{12},-,-,-$	$-,-,c_{14},c_{15},c_{18}$
v_7^e	$\tau_{28},\tau_{29},\tau_{30},\tau_{31},\tau_{32}$	300	$c_{13},c_{14},-,-,-$	$-,-,c_8,c_{12},c_{19}$
v_8^e	$\tau_{33},\tau_{34},\tau_{35},\tau_{36},\tau_{37}$	500	$c_{15},c_{16},-,-,-$	$-,-,c_9,c_{23},c_3$
v_9^e	$\tau_{38},\tau_{39},\tau_{40},\tau_{41}$	500	$c_{17},-,-,-$	$-,c_{16},c_{21},c_9$
v_{10}^e	$\tau_{42},\tau_{43},\tau_{44},\tau_{45}$	300	$c_{18},c_{19},-,-$	$-,-,c_{17},c_{21}$
v_{11}^e	$\tau_{46},\tau_{47},\tau_{48},\tau_{49},\tau_{50}$	500	$c_{20},-,-,-,-$	$-,c_3,c_{19},c_{22},c_{18}$
v_{12}^e	$\tau_{51},\tau_{52},\tau_{53}$	600	c_{21},c_{22},c_{23}	c_3,c_{10},c_{20}

TABLE I.   CONFIGURATION OF APPLICATION TASKS.

a_i	$a_i.p$[ms]	$a_i.tc$	a_i	$a_i.p$[ms]	$a_i.tc$
a_1	5	$\tau_1,c_1,\tau_{15},c_7,\tau_5$	a_{16}	4	$\tau_{28},c_{13},\tau_{16},c_8,\tau_{30}$
a_2	10	τ_2,c_2,τ_{13}	a_{17}	4	$\tau_{29},c_{14},\tau_{25}$
a_3	5	τ_6,c_3,τ_3	a_{18}	20	$\tau_{33},c_{15},\tau_{26}$
a_4	5	τ_6,c_3,τ_{37}	a_{19}	20	$\tau_{34},c_{16},\tau_{39}$
a_5	5	τ_6,c_3,τ_{47}	a_{20}	10	τ_{38},c_{17},τ_9
a_6	5	$\tau_6,c_3,\tau_{51},c_{21},\tau_{45}$	a_{21}	10	$\tau_{38},c_{17},\tau_{20}$
a_7	10	τ_7,c_4,τ_{14}	a_{22}	10	$\tau_{38},c_{17},\tau_{44}$
a_8	10	τ_{11},c_5,τ_4	a_{23}	10	$\tau_{42},c_{18},\tau_{21}$
a_9	10	τ_{12},c_6,τ_{18}	a_{24}	10	$\tau_{42},c_{18},\tau_{27}$
a_{10}	10	τ_{17},c_9,τ_{35}	a_{25}	10	$\tau_{42},c_{18},\tau_{50}$
a_{11}	10	τ_{17},c_9,τ_{41}	a_{26}	10	$\tau_{43},c_{19},\tau_{32}$
a_{12}	5	τ_{22},c_{10},τ_8	a_{27}	10	$\tau_{43},c_{19},\tau_{48}$
a_{13}	5	$\tau_{22},c_{10},\tau_{52},c_{22},\tau_{49}$	a_{28}	5	$\tau_{46},c_{20},\tau_{53},c_{23},\tau_{36}$
a_{14}	10	$\tau_{23},c_{11},\tau_{19}$	a_{29}	5	$\tau_{51},c_{21},\tau_{10}$
a_{15}	10	$\tau_{24},c_{12},\tau_{31}$	a_{30}	5	$\tau_{51},c_{21},\tau_{40}$

TABLE II.   CONFIGURATION OF THE APPLICATIONS.

In the experiment, we define the following objectives to be minimized (response time and latency are observed from local time of the first sender station):

- obj_1: Maximal response time of all applications $\mathcal{A}$.
- obj_2: Maximal response time of applications a_1 to a_5.
- obj_3: Maximal response time of applications a_1 to a_{10}.
- obj_4: Average response time of all applications $\mathcal{A}$.
- obj_5: Maximal end-to-end latency of all applications $\mathcal{A}$.

Depending on the higher-level goals one or more of the objectives above can be optimized. For platform/system status applications, the response time is an important parameter since they need to be completed as early as possible. Towards this, obj_1 and obj_4 are useful objectives. Often, a subset of applications are system status applications while the rest are usual applications. In such cases, obj_2 and obj_3 allows to minimize the response time of the relevant applications. Moreover, for many applications such as feedback control loops, the maximum end-to-end latency plays an important role. Therefore, obj_5 also have application level relevance. In general, it is possible to form various combinations of response time and end-to-end latency as an objective depending on the exact design requirements. Further, in many real-life scenarios, a single objective might not suffice. In such cases, our approach offers a multi-objective formulation. In the case-study, we explore the results considering each single objective and several multi-objective cases. In the case of multi-objective optimization, we assign equal weights ω_i. The optimization results are shown in Table IV for four different topologies as shown in Fig. 4.

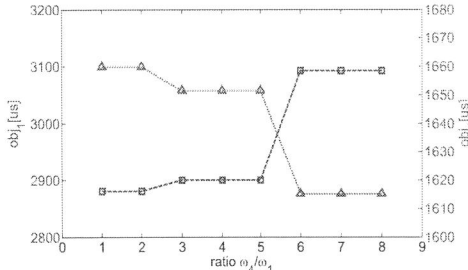

Fig. 5. Results for obj_1 and obj_4 in a multi-objective optimization with different weight ratio $\frac{\omega_4}{\omega_1}$.

Discussions: (i) Using our approach, it is possible to co-synthesize task and communication schedules in Ethernet-based time-triggered systems considering different application level objectives. For example, the definitions of obj_2 and obj_3 imply that $obj_2 \leq obj_3 \leq obj_1$, which is also reflected in our results. Similarly, the definition of obj_4 implies $obj_4 \leq obj_1$. (ii) In the case of single objective optimization, the first five rows of Table IV show the results for four different topologies. Clearly, the optimality was achieved for each individual objective (in bold font). For example, the result for obj_1 for tree topology is $obj_1 = 2880.96$us, which is the minimum among all the cases in first column. Naturally, such single objective cases often lead to non-optimal results for the others. E.g., in the case of tree topology, single objective optimization according to obj_2 and obj_3 leads to the minimal value of $obj_2 = 1342.48$us and $obj_3 = 2200$us respectively, but results in the undesirable results for obj_1 as 12395us and 18995us respectively. (iii) In the case of multi-objective optimization, it is often possible to achieve results close to the optimal one for all the objectives under consideration. The example of such optimization is illustrated using the combination of obj_1, obj_2 and obj_3 as shown in the 8th row of Table IV. For the tree topology, we obtain $obj_1 = 3092.72$us, $obj_2 = 1342.48$us, $obj_3 = 2252$us. Clearly, all three obj_i are close or equal to the results in the single objective cases. It should be noted that the results of the objectives that are not optimized (not underlined) can vary under different solver configurations. This is because it is possible that there are multiple solutions to such an optimization problem and the results in the table show only one of them. However, the results of objectives that are considered for optimization are independent of the solver configuration.

Further, we investigate the influence of the weights on the optimization results. Fig. 5 shows the result in the case of multi-objective optimization of obj_1 and obj_4 for different weight ratios ω_4/ω_1. As expected, the weight ratio of the objective can influence the optimization result. A higher ω_4/ω_1 should imply a relatively lower obj_4 which is also reflected in Fig. 5.

From the experimental results we can see that our formulation allows the designer to generate schedules according to various application level objectives. It also offers the possibility to combine single objectives to form multi-objective optimization problems that could meet more complicated higher-level requirements. Moreover, our approach is applicable

obj	STAR					TWINSTAR				
	1 [us]	2 [us]	3 [us]	4 [us]	5 [us]	1 [us]	2 [us]	3 [us]	4 [us]	5 [us]
1	**2800.48**	2800.48	2800.48	2164.52	2800.48	**2820.60**	2820.60	2820.60	1920.03	2820.60
2	18995.00	**1256.24**	9995.00	5445.84	8944.76	11126.36	**1256.24**	9995.00	5147.35	6027.00
3	12800.00	2200.00	**2200.00**	4601.62	6795.00	11200.00	1650.00	**2200.00**	4600.77	8188.52
4	3006.48	**1256.24**	2206.00	**1590.88**	2550.24	3006.48	1570.36	2206.00	**1597.34**	2570.36
5	10236.48	7785.76	7785.76	4485.34	**1700.48**	12504.96	9166.24	9166.24	6295.11	**1700.48**
1+2	3006.48	**1256.24**	3006.48	2157.80	3006.48	3006.48	**1256.24**	2906.48	1801.53	3006.48.
1+3	**2800.48**	2200.00	**2200.00**	2141.23	2800.48	**2820.60**	2200.00	**2200.00**	1979.30	2820.60
1+2+3	3006.48	**1256.24**	2206.00	1996.64	3006.48	3006.48	**1256.24**	2260.00	2009.12	3006.48
1+4	**2800.48**	1650.24	2602.80	1630.01	2800.48	**2820.60**	1670.36	2602.80	1638.19	2820.60
1+5	2900.48	2900.48	2900.48	2037.11	**1700.48**	2900.48	2900.48	2900.48	1961.06	**1700.48**
1+2+3+4	3006.48	**1256.24**	2206.00	**1590.88**	2550.24	3006.48	**1256.24**	2206.00	**1597.34**	2467.56
1+2+3+4+5	3056.48	**1356.00**	2206.00	1656.46	**1700.48**	3056.48	**1356.00**	2206.00	1689.81	**1700.48**

obj	TREE					RING				
	1 [us]	2 [us]	3 [us]	4 [us]	5 [us]	1 [us]	2 [us]	3 [us]	4 [us]	5 [us]
1	**2880.96**	2880.96	2880.96	1954.47	2880.96	**2840.72**	2840.72	2840.72	2093.28	2840.72
2	12395.00	**1342.48**	9995.00	6087.15	9995.00	12395.00	**1299.36**	9995.00	5915.79	8186.04
3	18995.00	2200.00	**2200.00**	5104.39	10601.48	18995.00	2200.00	**2200.00**	5569.66	11904.52
4	3092.72	**1342.48**	2252.00	**1615.28**	2590.48	3049.60	**1299.36**	2229.00	**1604.08**	2570.36
5	12962.20	8624.68	8624.68	6010.64	**1740.72**	14464.28	6170.60	6170.60	4703.60	**1720.60**
1+2	3092.72	**1342.48**	2902.00	2062.89	3092.72	3049.60	**1299.36**	3049.60	2290.29	3049.60
1+3	**2880.96**	2200.00	**2200.00**	2027.25	2880.96	**2840.72**	2200.00	**2200.00**	2009.24	2840.72
1+2+3	3092.72	**1342.48**	2252.00	2077.51	3092.72	3049.60	**1299.36**	2229.00	1915.68	3049.60
1+4	**2880.96**	1690.48	2602.80	1659.99	2880.96	**2840.72**	1670.36	2602.80	1651.58	2840.72
1+5	2902.00	2703.28	2902.00	2021.08	**1740.72**	2900.48	2683.16	2900.48	1972.93	**1720.60**
1+2+3+4	3092.72	**1342.48**	2252.00	**1615.28**	2487.68	3049.60	**1299.36**	2229.00	**1604.08**	2570.36
1+2+3+4+5	3132.96	**1356.00**	2252.00	1684.95	1750.00	3076.60	**1256.00**	2229.00	1682.22	**1720.60**

TABLE IV. OPTIMIZATION RESULTS ACCORDING TO DIFFERENT OBJECTIVES.

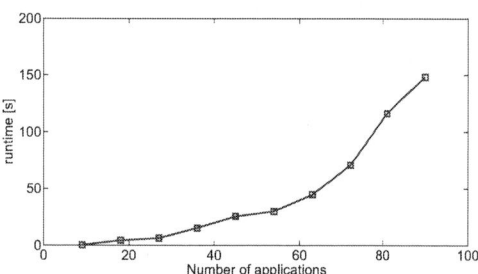

Fig. 6. Number of applications vs. runtime.

to any network topology and configuration of applications, application tasks and communication tasks.

B. Scalability analysis

To show the scalability of the proposed approach, we use a series of synthetic test systems in increasing size and measure the time required to solve the model. We construct 20 synthetic systems with different configurations for each system size from 9 to 90 applications. Initially, the systems employ a tree topology with three end stations and two switches. Subsequently, we add new applications and for each additional 9 applications, we add three end stations and one switch. All the applications consist of two application tasks and one uni-cast communication task. The periods of applications $a_i.p$ are chosen randomly from 4, 5, 10 and 20ms and the WCET $\tau_i.e$ are between 100 and 300us. Further, the frame lengths range from 80 to 200 bytes. Fig. 6 shows the average runtime of the overall system. Clearly, our approach scales up to 90 applications in a reasonable amount of time. In practice, the presented values are enough for many application domains such as automotive.

VI. CONCLUDING REMARKS

In this work, we presented a MIP formulation for the application-level schedule optimization problem for Ethernet-based time-triggered systems. Although the presented work mainly deals the application-level timing properties, our approach can be utilized to optimize a number of functional properties such as extensibility and sustainability. We plan to investigate in this direction in our future works.

REFERENCES

[1] R. Pigan, M. Metter, *Automating with PROFINET, 2nd edition*, Publicis Publishing, 2008.

[2] "664P7-1 Aircraft Data Network, Part 7, Avionics Full-Duplex Switched Ethernet Network", ARINC, 2009.

[3] "AS6802: Time-triggered Ethernet," SAE International, 2011.

[4] "Media access control (MAC) bridges and virtual bridge local area networks," IEEE Computer Society, 2011.

[5] "http://www.ieee802.org/1/pages/802.1bv.html"

[6] H. Kopetz, G. Bauer, "The time-triggered architecture," *Proceedings of the IEEE*, vol. 91, iss. 1, 112–126, 2003.

[7] M. Lukasiewycz, M. Glass, J. Teich, P. Milbredt, "FlexRay schedule optimization of the static segment," *CODES+ISSS*, 2009.

[8] H. Zeng, M. Di Natale, A. Ghosal, A. Sangiovanni-Vincentelli, "Schedule optimization of time-triggered systems communicating over the FlexRay static segment," *Industrial Informatics, IEEE Transactions on*, vol. 7, iss. 1, 1–17, 2011.

[9] D. Goswami, M. Lukasiewycz, R. Schneider, S. Chakraborty, "Time-triggered implementations of mixed-criticality automotive software," *DATE*, 2012.

[10] W. Steiner, "An evaluation of SMT-based schedule synthesis for time-triggered multi-hop networks," *RTSS*, 2010.

[11] W. Steiner, "Synthesis of static communication schedules for mixed-criticality systems," *ISORCW*, 2011.

[12] Z. Hanzalek, P. Burget, P. Sucha, "Profinet IO IRT message scheduling with temporal constraints," *Industrial Informatics, IEEE Transactions on*, vol. 6, iss. 3, 369–380, 2010.

[13] D. Tamas-Selicean, P. Pop W. Steiner, "Synthesis of communication schedules for TTEthernet-based mixed-criticality systems," *CODES+ISSS*, 2012.

[14] R. Seifert, J. Edwards *The All-New Switch Book: The Complete Guide to LAN Switching Technology*, Wiley Publishing, Inc., 2008.

[15] H. P. Williams, *Model Building in Mathematical Programming 4th edition*, John Wiley and Sons, LTD, 1999.

[16] "www.gurobi.com"

Service Adaptions for Mixed-Criticality Systems

Pengcheng Huang, Georgia Giannopoulou, Nikolay Stoimenov, Lothar Thiele
Computer Engineering and Networks Laboratory, ETH Zurich, 8092 Zurich, Switzerland
firstname.lastname@tik.ee.ethz.ch

Abstract— Complex embedded systems are typically mixed-critical, where heterogeneous guarantees must be provided for functionalities of different criticalities. We study in this paper the reconfiguration of services provided to low criticality tasks in reaction to the overruns of high criticality tasks. We further investigate the quantification of the resetting time of the system services. For both service reconfiguration and resetting, we derive tight analysis results under Earliest Deadline First (EDF) scheduling.

I. INTRODUCTION

Complex embedded systems are typically mixed-critical – functionalities of different criticalities (importances) coexist on a common computing platform [13]. Examples can be seen in automotive applications (e.g. smartcar systems) and in avionics applications (e.g. flight control and management systems). For such systems, it is crucial to meet requirements of different degrees of rigorousness for different criticality levels.

A large body of research already exists on specifying and scheduling of Mixed-Criticality (MC) systems, see e.g. [15, 5, 10, 3, 7, 4]. The common practice is to model a single task on all different criticality levels, with possible different worst-case execution times depending on the criticality levels. Since the rigorousness and conservatism considered on a higher criticality is higher, the worst-case execution times are non-decreasing from high to low criticality levels. The online scheduler then decides which tasks to guarantee according to the actual run-times of tasks: if any task exceeds its χ level worst-case execution time, then tasks of criticality levels no higher than χ are stopped after that. Various classical scheduling techniques are extended to support such dynamic behavior: Fixed Priority scheduling [15, 5], Earliest Deadline First (EDF) scheduling [10, 3, 7], Time-triggered scheduling [4].

On the one hand, it is often too pessimistic to reject all less critical tasks whenever a single high criticality task exceeds at runtime a certain execution time threshold. Indeed, this has already been confirmed in [12, 14]. Here, instead of complete killing of less critical tasks, either partial killing or best-effort scheduling are adopted for the less critical tasks. However, the service that the system can still guarantee to the less critical tasks when a critical one overruns is not known. It remains *an open problem to compute bounds on the service that can be provided for the less critical tasks.* Such information could be used to guide the online reconfiguration of the provided system services.

On the other hand, the guarantees made by the above scheduling techniques are rarely acceptable in practice. Consider a typical avionic MC usecase – the Flight Management System (FMS) application. If a localization task (certified at level B, with A being the highest criticality level and E being the lowest according to the DO-178B standard [1]) exceeds a

certain execution time threshold, we cannot simply reject the level C tasks (among them be the flightplan task) or perform best-effort scheduling for them, since the airplane constantly requires the flightplan information. Instead, a *degraded service* for the level C tasks should be guaranteed. Furthermore, for FMS, when the system can be *reset* to provide the original service to all tasks is an important measure to certify the runtime system. Therefore, offline analysis techniques need to be developed to bound the resetting time.

Contributions of this paper:

- We extend the EDF-VD [3] scheduling technique to guarantee a degraded service for the low criticality tasks when the high criticality tasks exceed their low criticality worst-case execution times.

- We extend the demand bound analysis for MC systems, which are scheduled by mode-switched EDF, see [7]. We present results on approximations of demand bounds for all tasks in different runtime modes.

- We propose an algorithm to compute the bounds on the degraded service provided to the low criticality tasks, as well as the configurations of EDF-VD to achieve such bounds.

- We present an analysis technique to bound offline the resetting time of the services provided to the low critical tasks. We demonstrate that a trade off between the degraded service and the resetting time exists.

- We show the applicability of our approach in a case study of an industrial application. For the considered Flight Management System application,the extended EDF-VD scheduling technique can schedule the application with a degraded service requirement for the low criticality tasks. Furthermore, service resetting time can be quantified, and traded for the degraded services of the LO criticality tasks.

Related Work The current research on mixed-criticality scheduling is initialized by the seminal work of Vestal et al. [15], where the common model for mixed-criticality systems is proposed. Various conventional scheduling techniques are subsequently extended to the mixed-criticality domain, see [6] for a comprehensive survey. Motivated with a realistic avionics use-case, the work in this paper considers service degradation for low criticality tasks instead of rejecting them. In addition, the quantification of service resetting time is investigated. The closest work to ours is that presented in [7, 8], where a general mixed-criticality mode switch protocol is introduced and analyzed. Our work is different from [7, 8] in that an elaborated analysis is proposed to *compute* the degraded service for the low criticality tasks. To the best of our knowledge, the quantifications of the degraded service and the service resetting time have not been investigated in existing results on mixed-criticality scheduling.

The work in this paper is conducted in the mixed-criticality context, where real-time services for all tasks must be provided even under abrupt scenario changes. During such scenario changes, the service adaptions of different functionalities depend on their criticalities. Existing results on conventional adaptive real-time scheduling either assume delayed system operation or task rejections, (see e.g. [11] for more references therein), and thus cannot directly apply in the mixed-criticality context.

II. SYSTEM MODEL

We consider a generalized mixed-criticality sporadic task model as proposed in [8]. Given is a dual-criticality task set $\tau = \{\tau_1, \tau_2, ..., \tau_n\}$ ($n \in \mathbb{N}$) scheduled on a uni-processor. Each task τ_i has a minimum inter-arrival time T_i and a relative deadline $D_i (D_i \leq T_i)$. The worst-case execution times (WCETs) of all tasks are modeled both on the HI (high) and on the LO (low) criticality levels. Only the HI criticality tasks are allowed to exceed their LO criticality WCETs at runtime. The system has *two operating modes*: whenever a HI criticality task exceeds its LO criticality WCET, the system transits *immediately* from the LO criticality mode to the HI criticality mode, during which all HI criticality tasks must finish their HI criticality level WCETs before their deadlines, and the LO criticality tasks are possibly guaranteed with *degraded service parameters* (e.g. longer periods and/or deferred deadlines).

We can then abstract each task τ_i by a tuple, $\{\{T_i(\text{HI}), C_i(\text{HI}), D_i(\text{HI})\}, \{T_i(\text{LO}), C_i(\text{LO}), D_i(\text{LO})\}, \chi_i\}$:

- $T_i(\chi) \in \mathbb{R}^+$ is the minimum inter-arrival time of τ_i in mode χ.
- $C_i(\chi) \in \mathbb{R}^+$ is the WCET of τ_i in mode χ.
- $D_i(\chi) \in \mathbb{R}^+$ is the relative deadline of τ_i in mode χ.
- $\chi = \{\text{HI}, \text{LO}\}$ is the set of criticality levels / operating modes.
- $\chi_i \in \chi$ is the criticality level of task τ_i.

$\{T_i(\chi), C_i(\chi), D_i(\chi)\}$ characterizes the service to be guaranteed for τ_i in mode χ. A dual-criticality task set is said *schedulable* if there exists a scheduling technique, such that the service requirements for all tasks in both operating modes can be satisfied. For a HI criticality task τ_i, its WCET is adjusted from $C_i(\text{LO})$ to $C_i(\text{HI})$ when transiting to the HI criticality mode. This can lead to a unschedulable system: consider a scenario when a job of this task just finishes its low criticality WCET at its deadline, and at the same time a mode transition happens, this implies that this task must finish $C_i(\text{HI}) - C_i(\text{LO})$ units of execution in zero time. In order to schedule the HI criticality tasks, their deadlines need to be tuned in the LO criticality mode ($D_i(\text{LO}) \leq D_i(\text{HI})$), such that they can still meet their deadlines when transiting to the HI criticality mode. This concept has been explored in [3, 2, 7].

In summary, the following assumptions are made: if $\chi_i = \text{HI}$, then

$$T_i(\text{LO}) = T_i(\text{HI}) = T_i, C_i(\text{HI}) \geq C_i(\text{LO}), D_i(\text{LO}) \leq D_i(\text{HI}), \quad (1)$$

if $\chi_i = \text{LO}$, then

$$T_i(\text{LO}) = T_i, T_i(\text{LO}) \leq T_i(\text{HI}), C_i(\text{HI}) = C_i(\text{LO}), D_i(\text{LO}) \leq D_i(\text{HI}). \quad (2)$$

For the original EDF-VD scheduling technique [3, 2], all tasks are scheduled by Earliest Deadline First (EDF) in both HI and LO criticality modes (i.e. *mode-switched EDF*). All LO criticality tasks are immediately dropped when the system transits to the HI criticality mode. This can be viewed as a special case by setting $T_i(\text{HI}) := D_i(\text{HI}) := \infty$ for all LO criticality tasks. Furthermore, for notational convenience, we denote by τ_{HI} (τ_{LO}) the set of HI (LO) criticality tasks.

III. MC SERVICE RECONFIGURATION

We present in this section the analysis of dual-criticality task sets scheduled by mode-switched EDF. We quantify the demand bounds of *all* tasks in the LO and HI criticality modes, and present the corresponding schedulability test, both in Section A. We present results on approximations of the analyzed demand bounds in Section B. We view the service degradation of the LO criticality tasks by a scaling factor (≥ 1) of their periods and deadlines. For implicit-deadline task sets ($T_i = D_i$), an algorithm is proposed in Section C to adjust the services of the LO criticality tasks in the HI mode. The results presented extend existing works in [3, 7, 8].

A. MC Demand Bound Analysis

The demand bound of a task in a given interval is defined as the sum of execution times of all task instances, which have arrival times and deadlines both in this interval. During the LO criticality mode, the demand bound dbf_{LO} of any task in an interval of length Δ can be bounded according to known results in [7]:

$$\text{dbf}_{\text{LO}}(\tau_i, \Delta) = \max\{\left\lfloor \frac{\Delta - D_i(\text{LO})}{T_i(\text{LO})} \right\rfloor + 1, 0\} \cdot C_i(\text{LO}). \quad (3)$$

For the HI criticality mode, we need to consider the impacts of unfinished jobs at the transition point. Those are the jobs for which the schedules may be changed (i.e. tasks are scheduled by EDF and their deadlines may be adjusted at the time of mode switch).

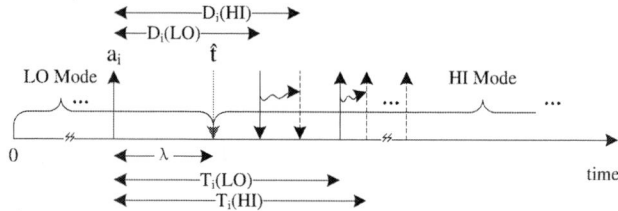

Fig. 1. A mode transition triggered by task overrun

We depict in Fig. 1 a system undergoing a mode transition at time $\hat{t}$. A job of task τ_i arrives in the LO mode at time a_i. A mode transition is signaled λ units after that. From this time on, all jobs of this task are guaranteed with the HI criticality parameters.

In [7], only the demand bounds of the HI criticality tasks in the HI criticality mode are derived. This is done by identifying a worst-case demand bound including that of the unfinished job at the transition time (4). We proceed to present a general approach to analyze the demand bounds of *all* tasks in the HI criticality mode. Let us further define a set of functions (5)-(8):

$$\text{RM}(\tau_i, \lambda) = C_i(\text{HI}) - C_i(\text{LO}) + \min\{D_i(\text{LO}) - \lambda, C_i(\text{LO})\}, \quad (4)$$

$$\text{dbf}_{\text{HI}}^1(\tau_i, \Delta) = \max\{\left\lfloor \frac{\Delta - D_i(\text{HI})}{T_i(\text{HI})} \right\rfloor + 1, 0\} \cdot C_i(\text{HI}), \quad (5)$$

$$\text{dbf}_{\text{RM}}(\tau_i, \lambda, \Delta) = \begin{cases} \text{RM}(\tau_i, \lambda) & \text{if } \Delta \geq D_i(\text{HI}) - \lambda, \quad (6a) \\ 0 & \text{if } \Delta < D_i(\text{HI}) - \lambda. \quad (6b) \end{cases}$$

$$\text{dbf}_{\text{HI}}^2(\tau_i, \lambda, \Delta) = \text{dbf}_{\text{RM}}(\tau_i, \lambda, \Delta) \quad (7)$$
$$+ \max\{\left\lfloor \frac{\Delta - D_i(\text{HI}) - (T_i(\text{HI}) - \lambda)}{T_i(\text{HI})} \right\rfloor + 1, 0\} \cdot C_i(\text{HI}).$$

$$\text{dbf}_{\text{HI}}(\tau_i, \Delta) = \sup\{\text{dbf}_{\text{HI}}^1(\tau_i, \Delta), \sup_{0 \leq \lambda \leq D_i(\text{LO})} \{\text{dbf}_{\text{HI}}^2(\tau_i, \lambda, \Delta)\}\}. \quad (8)$$

Formally, we have the following result.

978-1-4799-2817-0/14 $31.00 © 2014 IEEE

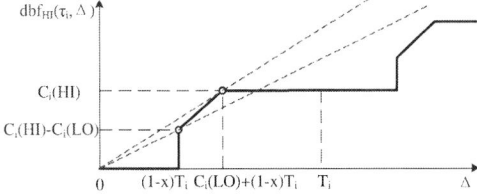

Fig. 2. Approximation of $\mathrm{dbf}_{\mathrm{HI}}(\tau_i, \Delta), \chi_i = \mathrm{HI}$

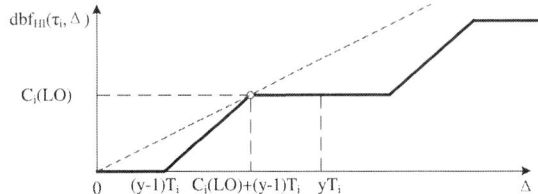

Fig. 3. Approximation of $\mathrm{dbf}_{\mathrm{HI}}(\tau_i, \Delta), \chi_i = \mathrm{LO}$

Lemma III.1. The demand bound function of any task in the HI criticality mode can be calculated by (8).

Proof. All proofs are presented in the Appendix. □

Based on the above computed demand bounds, schedulability of a task set can be tested using existing results.

Theorem III.1. [7] A dual-criticality task set τ is schedulable on an unit-speed processor, if $\forall \Delta \geq 0$:

$$\max\{\sum_\tau \mathrm{dbf}_{\mathrm{LO}}(\tau_i, \Delta), \sum_\tau \mathrm{dbf}_{\mathrm{HI}}(\tau_i, \Delta)\} \leq \Delta. \qquad (9)$$

B. MC Demand Bound Approximation

For the original EDF-VD scheduling technique, all LO criticality tasks are rejected in the HI criticality mode. The problem we are addressing in this paper is different, as we aim at computing the bounds on the service for the LO criticality tasks in the HI criticality mode. This is not a trivial problem as both HI and LO criticality tasks change their schedules immediately when transiting to the HI criticality mode, which makes the analysis difficult. In order to simplify the problem, we restrict ourselves to consider implicit-deadline task sets ($D_i = T_i$). According to the discussions in Section II, to ensure the schedulability of the HI criticality tasks in the HI criticality mode, their deadlines need to be tuned in the LO criticality mode. Similar to the original EDF-VD scheduling technique, we assume that the deadline tuning for the HI criticality tasks is characterized by a scaling factor x ($x \leq 1$):

$$T_i(\mathrm{LO}) = T_i(\mathrm{HI}) = D_i(\mathrm{HI}), D_i(\mathrm{LO}) = xD_i(\mathrm{HI}). \qquad (10)$$

In addition, for our problem, we assume the service degradation of the LO criticality tasks is characterized by another scaling factor y ($y \geq 1$):

$$T_i(\mathrm{LO}) = D_i(\mathrm{LO}), T_i(\mathrm{HI}) = D_i(\mathrm{HI}) = yT_i(\mathrm{LO}). \qquad (11)$$

We proceed to show that the demand bounds of all tasks in the HI criticality mode (Lemma III.1) can be tightly approximated based on (8),(9). The essential idea is to bound for both HI criticality and LO criticality tasks, the nonlinear function $\mathrm{dbf}_{\mathrm{HI}}(\tau_i, \Delta)$ by certain slopes. The following results are presented.

Lemma III.2. For a HI criticality task τ_i,

$$\mathrm{dbf}_{\mathrm{HI}}(\tau_i, \Delta) \leq \max\{\frac{C_i(\mathrm{HI}) - C_i(\mathrm{LO})}{(1-x)T_i}, \frac{C_i(\mathrm{HI})}{C_i(\mathrm{LO}) + (1-x)T_i}\} \cdot \Delta. \qquad (12)$$

Lemma III.3. For a LO criticality task τ_i,

$$\mathrm{dbf}_{\mathrm{HI}}(\tau_i, \Delta) \leq \frac{C_i(\mathrm{LO})}{C_i(\mathrm{LO}) + (y-1)T_i} \cdot \Delta. \qquad (13)$$

We plot now in Fig. 2 and Fig. 3 the approximations of demand bounds in the HI criticality mode for both HI and LO criticality tasks (dashed lines).

C. MC Service Reconfiguration

We now show how to reconfigure the system such that a maximum degraded service for the LO criticality tasks can be guaranteed. Our analysis is based on the demand bound approximations shown in the previous section. For notational convenience, we define two functions as follows:

$$\begin{aligned} h(x) &= \sum_{\tau_{\mathrm{HI}}} \frac{U_i(\mathrm{HI})}{U_i(\mathrm{LO}) + (1-x)}, \\ l(y) &= \sum_{\tau_{\mathrm{LO}}} \frac{U_i(\mathrm{LO})}{U_i(\mathrm{LO}) + (y-1)}, \end{aligned} \qquad (14)$$

where $U_i(\chi) = C_i(\chi)/T_i$. $h(x)$ ($l(y)$) represents the summed slopes of the HI (LO) criticality tasks in the HI criticality mode. We further use $U_{\chi_1}^{\chi_2}$ to denote $\sum_{\tau_i : \chi_i = \chi_1} U_i(\chi_2)$. We now explain the service reconfiguration in Algorithm 1. For the case when $U_{\mathrm{HI}}^{\mathrm{HI}} + U_{\mathrm{LO}}^{\mathrm{LO}} \leq 1$, the LO criticality tasks need not be reconfigured as the system can guarantee all tasks with their worst-case service requirements. For the case when $U_{\mathrm{HI}}^{\mathrm{LO}} + U_{\mathrm{LO}}^{\mathrm{LO}} > 1$, the system is even not schedulable during the LO criticality mode. Excluding the above conditions, the system needs to be further tested to see whether the reconfiguration of services in the HI criticality mode is feasible. This is ensured by enforcing that the total slopes of tasks in the HI criticality mode is ≤ 1 (line 8). A minimal service degradation factor y, that satisfies this condition, is computed based on (14).

Algorithm 1: Service reconfiguration

Input: τ

1 **if** $U_{\mathrm{HI}}^{\mathrm{HI}} + U_{\mathrm{LO}}^{\mathrm{LO}} \leq 1$ **then**
2     $x \leftarrow 1$;
3     $y \leftarrow 1$;
4 **else**
5   **if** $U_{\mathrm{HI}}^{\mathrm{LO}} + U_{\mathrm{LO}}^{\mathrm{LO}} \leq 1$ **then**
6         $x \leftarrow \frac{U_{\mathrm{HI}}^{\mathrm{LO}}}{1 - U_{\mathrm{LO}}^{\mathrm{LO}}}$;
7   **if** $h(x) \leq 1$ **then**
8             $y \leftarrow \inf\{y \geq 1 : h(x) + l(y) \leq 1\}$;
9   **else**
10   **return** false;
11   **end**
12   **else**
13   **return** false;
14   **end**
15 **end**
16 **return** true;

Formally, we have the following result.

Theorem III.2. Given a dual-criticality task set, Algorithm 1 can compute a minimized service degradation factor y for the LO criticality tasks, and a corresponding deadline tuning factor x for the HI criticality tasks, such that the task set is schedulable in all operating modes.

Example III.1. Consider the set of dual-criticality task set as shown in Table I.

TABLE I
EXAMPLE TASK SET

τ	τ_1	τ_2	τ_3	τ_4	τ_5
χ	HI	LO	LO	LO	LO
T/D	60	8	30	90	15
$C(\text{HI})$	18	4	4	6	3
$C(\text{LO})$	3	4	4	6	3

According to Algorithm 1, we can derive $x = 0.5$, $y = 2.6488$. In practice, one can ceil the value of y and multiply the periods of the LO criticality tasks by 3 in the HI criticality mode. This can be done by the scheduler skipping 2 task instances in every 3 arrivals of a task. Furthermore, according to Theorem III.2, there is a trade off between x and y. The more we shorten the deadlines of the HI criticality tasks (providing that all tasks are still schedulable in the LO criticality mode), the better degraded service we can get for the LO criticality tasks. We plot in Fig. 4 the *ceiling* of y as a function of x. Notice that for $x < 0.5$ the system will not be schedulable in the LO criticality mode and for $x > 0.75$ the system will not be schedulable in the HI criticality mode even assuming all LO criticality tasks are rejected.

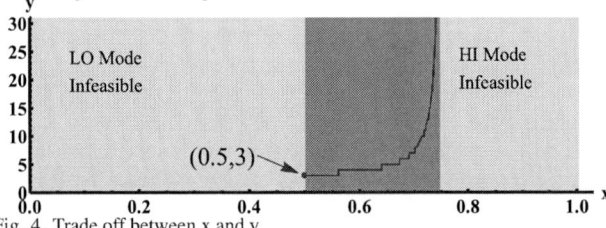

Fig. 4. Trade off between x and y

IV. MC SERVICE RESETTING

We present in this section how to quantify the resetting time of the system services (i.e. the elapsed time between when the system transits from the LO criticality mode to the HI criticality mode and when it can safely transit back). This is not a trivial problem since we do not statically know how many HI criticality tasks and for how long they will overrun. Our analysis in this section will not assume any such information. First, a sufficient condition for resetting the system service is given as follows.

Lemma IV.1. [12] A dual-criticality system can be safely reset to the LO criticality mode if the processor is idle.

Based on Lemma IV.1, a simple runtime mechanism can be implemented to reset the system to the LO criticality mode. However, it would be particularly interesting to quantify statically the worst-case resetting time, as an important measure of the guarantees that a mixed-criticality scheduling algorithm can provide.

We use the same notion as shown in Fig. 1 (a system transits to the HI criticality mode at time $\hat{t}$). Furthermore, we define the arrival demand function of a task (adf) in the interval $[\hat{t}, \hat{t} + \Delta]$ as the cumulative execution times of all task instances *issued* within this interval. We define a list of functions as follows:

$$\text{adf}_{\text{HI}}^1(\tau_i, \lambda, \Delta) = \text{RM}(\tau_i, \lambda) \\ + \max\left\{\left\lceil \frac{\Delta - (T_i(\text{HI}) - \lambda)}{T_i(\text{HI})} \right\rceil, 0\right\} \cdot C_i(\text{HI}), \quad (15)$$

$$\text{adf}_{\text{HI}}^2(\tau_i, \Delta) = \left\lceil \frac{\Delta}{T_i(\text{HI})} \right\rceil \cdot C_i(\text{HI}), \quad (16)$$

$$\text{adf}_{\text{HI}}(\tau_i, \Delta) = \sup\{\text{adf}_{\text{HI}}^2(\tau_i, \Delta), \sup_{0 \le \lambda \le D_i(\text{LO})}\{\text{adf}_{\text{HI}}^1(\tau_i, \lambda, \Delta)\}\}. \quad (17)$$

Formally, we have the following results.

Lemma IV.2. The arrival demand function of any task in the interval $[\hat{t}, \hat{t} + \Delta]$ can be calculated by (17).

Theorem IV.1. The service of the system can be reset at time $\hat{t} + \Delta_R$, where $\sum_\tau \text{adf}_{\text{HI}}(\tau_i, \Delta_R) \le \Delta_R$. And Δ_R can be lower bounded by $\frac{\sum_\tau C_i(\chi_i)}{1 - h(x) - l(y)}$.

Notice that according to Theorem IV.1, if we decrease x, then the resetting time will be reduced. Hence, one can simply set x as $\frac{U_{\text{HI}}^{\text{LO}}}{1 - U_{\text{LO}}^{\text{LO}}}$, which is the minimum to guarantee the schedulability of tasks in the LO criticality mode. For setting y, there is a trade off between the resetting time and the degraded service in the HI criticality mode: if we increase the degraded service for the LO criticality tasks (i.e. decrease y), then the resetting time will be increased.

Example IV.1. Consider the same task set as shown in Example III.1. We can now plot the service resetting time as a function of y.

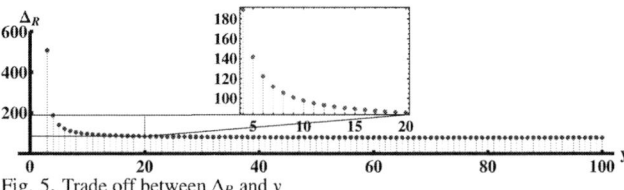

Fig. 5. Trade off between Δ_R and y

As one can see, the resetting time decreases with increasing y. This gives us the flexibility to trade the degraded service of the LO criticality tasks for the resetting time. Furthermore, as y increases, the gain of saving in resetting time will also decrease.

V. CASE STUDY

TABLE II
TASK PARAMETERS FOR THE FMS APPLICATION

τ	τ_1	τ_2	τ_3	τ_4	τ_5	τ_6
T/D	5000	200	1000	1600	100	1000
$C(\text{LO})$	$\{0,20\}$	$\{0,20\}$	$\{0,20\}$	$\{0,20\}$	$\{0,20\}$	$\{0,20\}$
χ	B	B	B	B	B	B
τ	τ_7	τ_8	τ_9	τ_{10}	τ_{11}	
T/D	1000	1000	1000	1000	1000	
$C(\text{LO})$	$\{0,20\}$	$\{0,200\}$	$\{0,200\}$	$\{0,200\}$	$\{0,200\}$	
χ	B	C	C	C	C	

We validate in this section our approach with an avionic use-case – the Flight Management System application (FMS). We show the applicability of our approach, such that service requirements for different runtime modes can be guaranteed. In addition, the service resetting time can be bounded in order to certify the scheduling of FMS.

We consider a subset of the original FMS, which contains the localization and the flightplan tasks (DO-178B level B and level C, where B corresponds to the HI criticality and C corresponds to the LO criticality). All tasks are abstracted as implicit-deadline sporadic tasks. The worst-case execution times (WCETs) of all tasks on level C are analyzed by an industrial partner and not available yet. However typical ranges

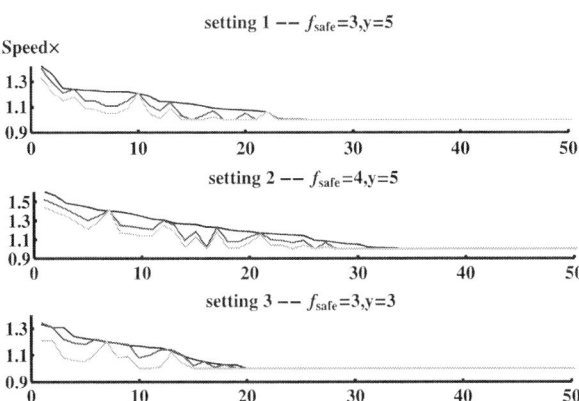

Fig. 6. Processor speedup factors comparison

TABLE III
RESETTING TIME

$f_{safe} = 3$			$f_{safe} = 4$			$f_{safe} = 5$		
y	x	Δ_R	y	x	Δ_R	y	x	Δ_R
1	1	0	3	0.25	21.6	22	0.25	2.1×10^3
-	-	-	4	0.25	7.8	23	0.25	661.8
-	-	-	5	0.25	5.92	24	0.25	406.1

can be assumed. We show the task parameters in Table II with timing information in units of *ms*. According to the discussions with our industrial partners, for avionics applications, a safety margin factor f_{safe} can be used to scale the WCETs on level C in order to get the WCETs on level B. We generate a set of FMS instances with random level C WCETs conforming to Table II.

We compare three different approaches: EDF with worst-case reservation (i.e. all tasks are guaranteed with WCETs on the HI criticality level), EDF-VD with degraded service guarantee (i.e. a degraded service requirement for the LO criticality tasks is guaranteed in the HI criticality mode), and the original EDF-VD scheduling technique (all LO criticality tasks are rejected in the HI criticality mode). In all three cases, we calculate the minimum speedup factors of the processor (by linear search) such that guarantees as aimed by those approaches are provided for FMS.

We show in Fig. 6 the speedup factors of all 3 approaches for 50 randomly generated FMS instances. The results are shown for 3 settings as indicated in Fig. 6. For presentation purpose, the computed results are sorted according to the speedup factors under worst-case reservation.

Based on the results, the original EDF-VD scheduling technique always has the minimum speedup factors among all three approaches. This is intuitive since schedulability of the HI criticality tasks is guaranteed by complete rejection of the LO criticality tasks. However, EDF-VD is not applicable to FMS, as a degraded service requirement for the LO criticality tasks must be provided. One solution to achieve this is by EDF scheduling with worst-case reservation. However, as one can expect, this approach always has the maximum processor speedup factors among all three approaches. The extension of EDF-VD with degraded service guarantee has a processor speedup factor in between of the other two approaches. Based on those results, we conclude that resource efficiency can be achieved by our proposed method in comparison to worst-case reservation, while degraded service for the LO criticality tasks can be guaranteed in comparison to the original EDF-VD.

Furthermore, as shown in Fig. 6, if we increase f_{safe} (setting 2), then the processor speedup factor will also be increased (the maximum speedup factors of all three approaches in this case come close to 1.5, while the maximum speedup factors come close to 1.35 in setting 1). In addition, if we increase the degraded service for the LO criticality tasks (i.e. decrease y in setting 3), the extension of EDF-VD with degraded service guarantee always has close/equal processor speedup factors to

the worst-case reservation approach. This implies that the gain in resource saving for our extension of EDF-VD decreases with decreasing y.

We continue to quantify the service resetting time for the FMS usecase. For this purpose, we pick one randomly generated FMS instance. We evaluate the impact of f_{safe} and y on the resetting time. The results are shown in Table III with resetting times given in units of second. For $f_{safe} = 3$, according to Algorithm 1, both x and y equal to 1. The FMS needs not be reconfigured in this case as the HI criticality WCETs of all tasks can be guaranteed. For $f_{safe} = 4$, the minimal y we calculate is 3, with a corresponding service resetting time of 21.6s. If we increase y (i.e. decrease the degraded service for the LO criticality tasks), we can reduce the resetting time ($y = 4, \Delta_R = 7.8s$). If we continue to increase f_{safe} to 5, the minimum y becomes 22 in this case, which has an associated service resetting time of $2.1 \times 10^3 s$.

VI. CONCLUSION

We present in this paper the reconfiguration of the services provided to the low criticality tasks when the high criticality tasks overrun. Our scheduling algorithm is an extension of the well known EDF-VD scheduling technique. We extend the demand bound analysis for testing the schedulability of the system under this setting. Furthermore, we provide an analytical method to bound offline the resetting time of the services provided to the low criticality tasks. The presented techniques are validated with a realistic avionic application.

ACKNOWLEDGMENT

The research leading to these results has received funding from the European Union Seventh Framework Programme (FP7/2007-2013) under grant agreement number 288175 (CERTAINTY project).

REFERENCES

[1] RTCA/DO-178B, Software Considerations in Airborne Systems and Equipment Certification, 1992.
[2] S. Baruah, V. Bonifaci, G. D'Angelo, H. Li, A. Marchetti-Spaccamela, S. van der Ster, and L. Stougie. The preemptive uniprocessor scheduling of mixed-criticality implicit-deadline sporadic task systems. In *ECRTS*, pages 145–154, 2012.
[3] S. Baruah, V. Bonifaci, G. D'Angelo, A. Marchetti-Spaccamela, S. Ster, and L. Stougie. Mixed-criticality scheduling of sporadic task systems. In *Algorithms - ESA*. 2011.
[4] S. Baruah and G. Fohler. Certification-cognizant time-triggered scheduling of mixed-criticality systems. In *RTSS*, pages 3–12, 2011.
[5] S. K. Baruah, A. Burns, and R. I. Davis. Response-time analysis for mixed criticality systems. In *RTSS*, pages 34–43, 2011.
[6] A. Burns and R. Davis. Mixed criticality systems-a review. 2013.
[7] P. Ekberg and W. Yi. Bounding and shaping the demand of mixed-criticality sporadic tasks. In *ECRTS*, pages 135–144, 2012.
[8] P. Ekberg and W. Yi. Bounding and shaping the demand of generalized mixed-criticality sporadic task systems. *Real-Time Systems*, pages 1–39, 2013.
[9] P. Huang, G. Giannopoulou, N. Stoimenov, and L. Thiele. Service adaptions for mixed-criticality systems. Technical Report 350, ETH Zurich, Laboratory TIK, 2013.
[10] T. Park and S. Kim. Dynamic scheduling algorithm and its schedulability analysis for certifiable dual-criticality systems. In *EMSOFT*, pages 253–262, 2011.

978-1-4799-2817-0/14 $31.00 © 2014 IEEE

[11] J. Real and A. Crespo. Mode change protocols for real-time systems: A survey and a new proposal. *Real-time systems*, 26(2):161–197, 2004.
[12] F. Santy, L. George, P. Thierry, and J. Goossens. Relaxing mixed-criticality scheduling strictness for task sets scheduled with fp. In *ECRTS*, pages 155–165, 2012.
[13] L. Sha. Resilient mixed-criticality systems. *The Journal of Defence Software Engineering*, 2009.
[14] H. Su and D. Zhu. An elastic mixed-criticality task model and its scheduling algorithm. In *DATE*, pages 147–152, 2013.
[15] S. Vestal. Preemptive scheduling of multi-criticality systems with varying degrees of execution time assurance. In *RTSS*, pages 239–243, 2007.

VII. APPENDIX

Proof. Lemma III.1

Let us consider two different cases.

1- $\lambda \in [0, D_i(\text{LO})]$: In this case, the current job of τ_i may have not finished its LO criticality WCET, the worst-case left-overs of the current job can be bounded by $\min\{D_i(\text{LO}) - \lambda, C_i(\text{LO})\}$. Since starting from $\hat{t}$ the current and future jobs of τ_i execute according to the HI criticality mode parameters, the total left-overs of the current job of τ_i can be bounded by (4). We have to further identify a time interval of length Δ within $[\hat{t}, +\infty)$, which has the worst-case demand bound. Suppose that such an interval is represented as $[\hat{t} + \eta, \hat{t} + \eta + \Delta]$ ($\eta \geq 0$), i.e. the interval starts η units of time after $\hat{t}$. Let us consider further two different cases:

- $\eta > 0$: In this case, the left-over job does *not* belong to the interval and its demand is not considered. Let γ represent the value of $((\lambda + \eta) \bmod T_i(\text{HI}))$, then the number of maximum complete arrivals of τ_i in $[\hat{t} + \eta, \hat{t} + \eta + \Delta]$ is bounded by:

$$
\begin{cases}
\max\left\{ \left\lfloor \dfrac{\Delta - D_i(\text{HI}) - (T_i(\text{HI}) - \gamma)}{T_i(\text{HI})} \right\rfloor + 1, 0 \right\} & \text{if } 0 < \gamma, \quad (18a) \\[2ex]
\max\left\{ \left\lfloor \dfrac{\Delta - D_i(\text{HI})}{T_i(\text{HI})} \right\rfloor + 1, 0 \right\} & \text{if } \gamma = 0. \quad (18b)
\end{cases}
$$

Hence, when $\gamma = 0$ we can get the maximum complete arrivals of τ_i within such an interval. And the demand bound of τ_i in $[\hat{t} + \eta, \hat{t} + \eta + \Delta]$ is upper bounded by (5).

- $\eta = 0$: In this case the left-over job needs to be considered for the demands in interval $[\hat{t}, \hat{t} + \Delta]$. Notice further that the left-over job is only considered when $\Delta \geq D_i(\text{HI}) - \lambda$, hence the demand of this job is given by (4).

The maximum number of future complete arrivals of τ_i within $[\hat{t}, \hat{t} + \Delta]$ is bounded by:

$$
\max\left\{ \left\lfloor \frac{\Delta - D_i(\text{HI}) - (T_i(\text{HI}) - \lambda)}{T_i(\text{HI})} \right\rfloor + 1, 0 \right\}. \quad (19)
$$

Hence the demand bound function of τ_i can be given by (7).

2- $\lambda \in (D_i(\text{LO}), T_i(\text{LO})]$: In this case τ_i has no active instance running in the system, we can derive similarly that the maximum number of complete arrivals of τ_i within an interval of length Δ happens when this interval starts with an arrival of τ_i. This number is again bounded by (18b). Hence the demand bound function in this case can be represented by (5).

Now, the demand bound function for any task τ_i in the HI criticality mode can be represented by (8). $\qquad\square$

Proof. Lemma III.2

Let us consider two different cases.

1- $0 \leq \Delta \leq T_i$: In this case, according to (5), (7), (8), we can get:

- $0 \leq \Delta < (1-x)T_i$: In this case, both $\text{dbf}_{\text{HI}}^1(\tau_i, \Delta)$ and $\text{dbf}_{\text{HI}}^2(\tau_i, \lambda, \Delta)$ (independent of λ) evaluate to zero. Hence, the demand bound is constantly zero.
- $(1-x)T_i \leq \Delta \leq C_i(\text{LO}) + (1-x)T_i$: In this case, $\text{dbf}_{\text{HI}}^1(\tau_i, \Delta)$ evaluates to zero. $\text{dbf}_{\text{HI}}^2(\tau_i, \lambda, \Delta)$ has the maximum value $C_i(\text{HI}) - C_i(\text{LO}) + \Delta - (1-x)T_i$ when $\lambda = T_i - \Delta$. Hence, the worst-case demand bound is $C_i(\text{HI}) - C_i(\text{LO}) + \Delta - (1-x)T_i$.

- $C_i(\text{LO}) + (1-x)T_i \leq \Delta \leq T_i$: In this case, $\text{dbf}_{\text{HI}}^1(\tau_i, \Delta)$ evaluates to $C_i(\text{HI})$. The worst-case of $\text{dbf}_{\text{HI}}^2(\tau_i, \lambda, \Delta)$ also evaluates to $C_i(\text{HI})$ (when $\lambda \leq xT_i - C_i(\text{LO})$). Hence, the maximum demand bound is $C_i(\text{HI})$.

2- $(k-1)T_i \leq \Delta \leq kT_i$ ($k \in \mathbb{N}^+$): Assume that $\Delta = (k-1)T_i + \delta$ ($0 \leq \delta \leq T_i$). First, we can derive that:

$$
\text{dbf}_{\text{HI}}^1(\tau_i, \Delta) = (k-1)C_i(\text{HI}) + \text{dbf}_{\text{HI}}^1(\tau_i, \delta). \quad (20)
$$

Second, $\text{dbf}_{\text{HI}}^2(\tau_i, \lambda, \Delta)\}$ achieves the maximum value when $\lambda = T_i - \delta$, then:

$$
\begin{aligned}
\sup_{0 \leq \lambda \leq xT_i} & \{\text{dbf}_{\text{HI}}^2(\tau_i, \lambda, (k-1)T_i + \delta)\} \\
&= \text{dbf}_{\text{HI}}^2(\tau_i, T_i - \delta, (k-1)T_i + \delta) \\
&= (k-1)C_i + \text{dbf}_{\text{HI}}^2(\tau_i, T_i - \delta, \delta) \\
&= (k-1)C_i + \sup_{0 \leq \lambda \leq xT_i} \{\text{dbf}_{\text{HI}}^2(\tau_i, \lambda, \delta)\}.
\end{aligned} \quad (21)
$$

Hence,

$$
\begin{aligned}
\text{dbf}_{\text{HI}}(\tau_i, \Delta) &= \sup\{\text{dbf}_{\text{HI}}^1(\tau_i, \Delta), \sup_{0 \leq \lambda \leq xT_i} \{\text{dbf}_{\text{HI}}^2(\tau_i, \lambda, \Delta)\}\} \\
&= (k-1)C_i(\text{HI}) + \text{dbf}_{\text{HI}}(\tau_i, \delta).
\end{aligned} \quad (22)
$$

Using the above results, we can draw the HI mode demand bound curve as shown in Fig. 2, and approximate it by two linear functions (dotted lines in Fig. 2). $\qquad\square$

Lemma III.3. This can be similarly derived as shown in Lemma III.2. $\qquad\square$

Proof. Theorem III.2

The schedulability of a task set in the LO criticality mode is tested by line 5 in Algorithm 1. We have to enforce further the schedulability in the HI mode according to Theorem III.1. According to Lemma III.2 and Lemma III.3:

$$
\begin{aligned}
& \sum_\tau \text{dbf}_{\text{HI}}(\tau_i, \Delta) \leq \Delta \\
\Leftarrow & \sum_{\tau_{\text{HI}}} \max\left\{ \frac{C_i(\text{HI}) - C_i(\text{LO})}{(1-x)T_i}, \frac{C_i(\text{HI})}{C_i(\text{LO}) + (1-x)T_i} \right\} \\
& + \sum_{\tau_{\text{LO}}} \frac{C_i(\text{LO})}{C_i(\text{LO}) + (y-1)T_i} \leq 1 \\
\Leftrightarrow & h(x) + l(y) \leq 1.
\end{aligned} \quad (23)
$$

If all LO criticality tasks are rejected in the HI criticality mode, the inequality becomes $h(x) \leq 1$, which is a sufficient condition for all HI criticality tasks to be schedulable in the HI criticality mode by rejecting all LO criticality tasks. Furthermore, we observe that the minimal value of y increases with increasing x. Hence, we can set x to $\frac{U_{\text{HI}}^{\text{LO}}}{1 - U_{\text{LO}}^{\text{LO}}}$ (the minimum x to guarantee schedulability in the LO criticality mode) in order to get the maximum degraded service for the LO criticality tasks (minimal y): $y = \inf\{y \geq 1 : h(x) + l(y) \leq 1\}$. $\qquad\square$

Proof Sketch. Lemma IV.2

This lemma can be similarly derived as shown in the proof of Lemma III.1.

$\qquad\square$

Proof. Theorem IV.1

Due to space reasons, we omit here the detailed proof, which can be found in [9]. $\qquad\square$

Efficient Feasibility Analysis of DAG Scheduling with Real-Time Constraints in the Presence of Faults

Xiaotong Cui

College of Computer Science
Chongqing University
Chongqing, China 400044
tel: +86 18716284896
e-mail:tywwdz@gmail.com

Jun Zhang

College of Computer Science
Chongqing University
Chongqing, China 400044
tel: +86 13193099498
e-mail:jeffjunzhang@gmail.com

Kaijie Wu

College of Computer Science
Chongqing University
Chongqing, China 400044
tel : +86 18523397246
e-mail : kaijie@gmail.com

Edwin Sha

College of Computer Science
Chongqing University
Chongqing, China 400044
tel: +86 15923005866
e-mail:edwinsha@gmail.com

Abstract – Tasks in hard real-time systems are required to meet deadlines in the presence of faults. We conclude that a sufficient condition of a task set experiencing its worst-case finish time (WCFT) is that its *critical task* (CT) incurs all faults. An algorithm is presented to identify the CT and the WCFT in $O(N^2)$ with N being the task number. A common practice that bet the WCFT using the task with the longest re-execution time could under estimate by up-to 35%!

Index Terms—Fault tolerance, feasibility test, frame-based real-time system, worst-case analysis, critical task

I. Introduction

Real-time embedded systems used in mission-critical applications such as navigation systems, process control, and surveillance systems demand high level of fault tolerance. Typically, real-time systems are designed to tolerate up to a specified number of faults in a given time interval by reserving enough redundancy in hardware and/or time to recover from fault(s). The upper bound on the expected number of faults is derived from the rate of faults under the designated operating conditions, i.e. the mean time to failure (MTTF) of hardware components and the rate of Single Event Upsets (SEU). Due to the severe resource constraints of embedded systems, optimizing redundancy overheads is of great importance. Among all the fault tolerance techniques, re-execution-based fault tolerance schemes have received broad attention due to their efficiency of resource usage and easiness in system design. Re-execution-based schemes usually assume that the faults that occur during the execution of a task are detected by the concurrent error detection mechanisms (c.g. watchdogs) supported by the processor. Therefore the focus is on determining and scheduling the re-executions of the (faulty) tasks. The schemes can be broadly classified into two categories – Passive and Active. An active scheme replicates tasks on multiple processing units such that all copies are executed (almost) simultaneously. If one of the replicas is found faulty, its result is replaced by the result of another replica. A passive scheme, on the other hand, executes only the primary replica of a task. Secondary replicas are executed only if the primary replica is found faulty. Feasibility test of both strategies need to estimate the amount of slack that can be devoted to tolerating the worst-case fault occurrences. Several strategies have been developed for a variety of applications [1][2][3][4][5][6][7].

Though feasibility analysis of fault-prone real-time systems has been studied for several years, majority of the attention has been paid to the task sets that comprise only independent tasks, such as [8][9][10][11][12][13]. None of these approaches, however, handles the task sets that have inter-task dependencies and are scheduled in multi-core systems. For such task sets, the classical processor demand analysis does not work anymore. This is because processors could be idle before all tasks are executed due to inter-task dependencies. In a quite recent work this is simply done by comparing all possible cases and choosing the worst one [6]. Such straightforward approach does not scale well as the number of cases grows rapidly [12]. In this paper we will present an algorithm that tests the feasibility in $O(N^2)$ where N is the number of tasks.

II. The System Models

A. System Model

Consider a closely-coupled M-processor real-time system where $M \geq 1$. Although processors used in examples are assumed to be identical, the analysis and conclusions directly apply to heterogeneous systems comprising of processors with different characteristics. Communications in closely-coupled real-time systems are carefully designed to guarantee invariable, minimum, and known overheads. This is often achieved by using special communication protocols, such as ultra-fast serial links. In a multi-processor system, inter-task communication overhead is determined (1) by the memory access latency and bandwidth in the case of shared-memory architecture, or (2) by the end-to-end link latency and bandwidth in the case of a networked architecture. In this work a common communication model described below is used for the two architectures:

• Communication between a pair of tasks scheduled on the same processor incurs constant worst-case delay. For simplicity, we assume the delay is 0.

• Communication between a pair of tasks scheduled on different processors incurs constant worst-case delay.

• Communication between a task and the tasks scheduled on different processors may experience different worst-case delay.

We believe the proposed communication delay model is practical and accurate in the context of closely-coupled systems. The underlying assumption is that the output generated by primary and secondary replicas of a task is identical (in terms of memory space or communication traffic) and hence the inter-task communication delay between a given pair of tasks remains independent of fault occurrences of tasks. It is assumed that system is rich in resources such that CPUs are the only resources tasks need to contend for doing their executions.

B. Frame-Based Systems

Presently there are two popular control paradigms of task sets in real-time systems. One is frame-based system and the other is event-triggered system. In a frame-based system, all tasks are released at time 0 and should be finished by the end of the frame. In an event-triggered system, each task is released according to its own period and is executed based on its order of priority. In general, while event-triggered systems are preferred by non-safety-critical applications due to its higher flexibility in resource allocations and higher efficiency in resource sharing, frame-based systems are more preferred in safety-critical applications due to its better hard real-time performance and easier incorporation of fault tolerance [14][15][16][17][17][19]. This work investigates frame-based systems.

978-1-4799-2817-0/14 $31.00 © 2014 IEEE

A scheduler first allocates the tasks to the processors according to some partitioning heuristics, and then derives the order between tasks in each processor. Hence, **a task set is "scheduled" if the task-to-processor allocation, and task-to-task order in each processor are determined**. Each task is defined by its Worst-case Execution Times C that denotes the maximum CPU time required by the fault-free execution of the task. C is the product of the worst-case CPU cycles of the task and the clock period of its host processor. Since the proposed analysis technique is applied after the schedule of tasks is determined, C of a task is a constant. The duration of a frame is denoted by D by which all tasks must finish their execution even in the presence of faults [20]. A dependent task cannot begin execution until all the tasks from which it expects inputs complete their executions. Due to non-negligible inter-processor delays, starting time of a dependent task may be delayed further if input data are communicated from tasks on other processor(s). While this analysis does not explicitly consider control dependencies where workload may change depending on the run-time resolutions, the proposed technique can still be applied if the control dependencies are cleared, e.g. by either taking one of possible outcomes or considering the worst case of all outcomes.

C. Fault Model and Re-execution

In hardware systems faults can be permanent, transient, or intermittent. In deep sub-micron and nano regimes, transient faults are the most common faults due to the increasing susceptibility to radiations – a result of continuously rising level of integration in semiconductor devices. The number of transient faults that may occur in a frame depends on fault susceptibilities of the underlying systems. For a scheduled task set, the maximum number of transient faults X is defined in the way that the probability of incurring more than X faults can be safely ignored. Let x denote the number of fault occurrences in a task , then the sum of fault occurrences of all tasks in a task set is no larger than X.

Fault occurrences during a task execution are manifested as erroneous outputs. It is assumed that faults are detected by concurrent error detection techniques [23]. Depending on the capabilities of deployed error detection mechanisms, a fault may be detected immediately after its occurrence or by the end of task execution. Immediate fault detection enables earlier re-executions thereby saving both energy and time. However, since this study focuses on the worst-case finish time, it is assumed that a fault is detected at the end of task execution. Since re-executions of a task can also incur faults, x fault occurrences in a row (i.e. burst faults) may affect the primary execution and up-to the $(x-1)^{\text{th}}$ re-executions, and the xth re-execution finishes fault-free. The worst-case execution time of executing and re-executing a task T is denoted by C_T and Cr_T respectively. We assume each task has only one version of re-execution, which is the case of many fault-tolerant systems [5][6][7][20]. Therefore, if T incurs total x faults, its actual execution time is $C_T + x \cdot Cr_T$. Many re-execution-based fault recovery techniques assume that a task's primary execution and re-executions use the same version of binary code, and therefore take the same amount of time, i.e. $C_T = Cr_T$. Our investigation relaxes the constraint to include the case where $C_T \neq Cr_T$ for the tasks using different (e.g. lighter) versions of binary code for re-executions.

Communication links between tasks are subject to faults as well. It is assumed that communication faults are detected and corrected at the communication level. In the case where memory-share architecture is used, Error-Correcting Codes (ECC) can be employed to protect system memory from transient faults [23]. In the case where end-to-end link is used, ECC are suggested as well [24][25]. Many state-of-art concurrent error detection techniques for processors make the same assumptions [26][27][28], which is in line with the proposed communication model.

D. The Problem to Be Addressed

Formally, given a scheduled task set, and the maximal number of faults that could occur during a frame, one wants to find the feasibility of the schedule by finding its worst-case finish time. A task set is scheduled if the task-to-processor allocation, and task-to-task order in each processor are determined. Without the schedule information, one cannot determine the feasibility of a task set since its worst-case finish time varies significantly with different schedules. Also because of this, the proposed technique works on both homogeneous and heterogeneous systems.

The problem is trivial if there are no data dependences among tasks. In this case, the worst-case finish time of the tasks allocated on a processor can be independently determined by considering only one case: the task with the longest re-execution time of this processor incurs all the faults. The problem, however, becomes more difficult if the otherwise. This is because that if a task's data dependences are not cleared, it may not be executed even if its processor is idle. From our work, it can be concluded that the worst-case finish time could be **under-estimated** by simply assuming the task with the longest re-execution time of the task set incur all the faults.

A motivational example is shown in Fig. 1. 6 tasks are scheduled on three processors. The arrows in the figure show the data dependences between tasks. The communication delay between tasks scheduled on the same processor or different processors is assumed to be 0 and 1, respectively. The schedule is shown in (a). Fig. 1(b) shows the case where T_4, the task with the longest re-execution time, incurs all two faults, while (c) shows the case where T_2, a task with shorter re-execution time, incurs all two faults. The shaded task bars are the faulty tasks. It is obvious that the second case is worse, 1 cycle longer than the first case (19 vs. 20). Hence the common practice that betting the worst-case finishing time of the whole task set using the first case could under-estimate the real worst case.

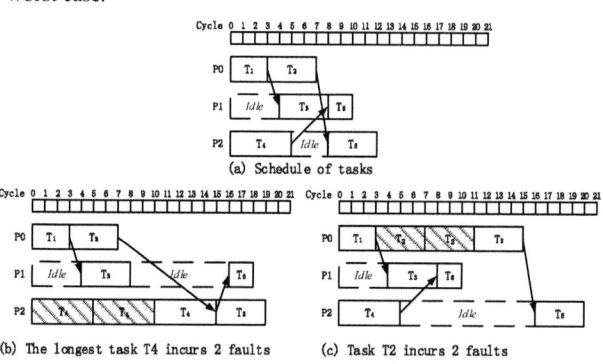

Fig. 1. A motivational example

III. Definitions and properties of a task set

In a multi-processor frame-based system, the worst-case finish time (*WCFT*) of a task depends upon itself, the tasks from which it receives inputs, and the task scheduled immediately before it on the same processor – its schedule predecessor. Fig. 2 (a) shows a task set with data dependencies among tasks (shown using arrows) and the C of each task. For simplicity the example assumes that the $C_T = Cr_T$ for all tasks, but this assumption is not required for the analysis to hold. Fig. 2 (b) shows a schedule that preserves the data dependencies and assumes the intra- and inter- processor communication delays between any tasks are 0 and 1 cycle, respectively. Schedule are introduced after the schedule is determined, and are shown

978-1-4799-2817-0/14 $31.00 © 2014 IEEE

using dotted lines in Fig. 2 (c). For example, T_3 is T_6's schedule predecessor, and T_6 cannot begin its execution until T_3 finishes. Since the given task-to-processor allocation and task-to-task execution order

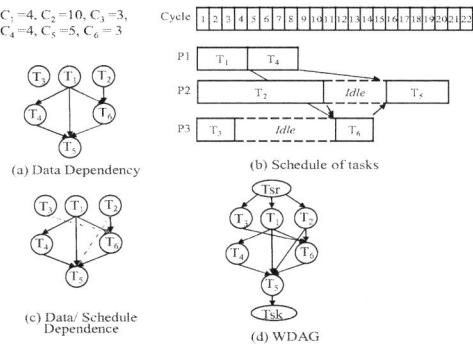

Fig. 2. An example task set and its schedule

are preserved throughout the analysis, it is not necessary to distinguish data and schedule dependencies afterwards.

In our analysis, the task graph is converted to a Weighted Directed Acyclic Graph (WDAG), $G = \{V, E\}$, where V is a set of vertices (tasks) and E is a set of directed edges (dependencies), as shown in Fig. 2 (d). In the following text, task and dependency are used interchangeably with vertex and edge respectively. The edge from T_i to T_j has a weight equal to the communication delay between T_i and T_j, and is denoted as $W_e(T_i, T_j)$. A vertex T also has a weight, denoted as $W_v(T)$, equal to the C_T of the task if it incurs zero fault, or $C_T + x \cdot Cr_T$ if it incurs x faults. All weights are non-negative. Two dummy vertices, *Source* and *Sink*, are added into G. The *Source* vertex, denoted as T_{sr} in Fig. 2 (d), has an edge to each vertex that has no incoming edge in the original G; and the *Sink*, denoted as T_{sk}, has an edge from each vertex that has no outgoing edge in the original G. The weights of both *Source* and *Sink* are zero and the weights of the edges from the *Source* and to the *Sink* are zero. A *parent* of vertex T_i is a vertex with an edge to T_i. Vertex T_i is then a *child* of its parents.

A simple path from vertex T_i to vertex T_j, denoted as $Path(T_i, T_j)$ is a sequence of no repeated vertices and edges by which the start vertex T_i and the end vertex T_j are connected together. Note that there may be no path, one path, or more than one path between a pair of vertices. Since the graph is directed and acyclic, paths are directed. If $Path(T_i, T_j)$ exists, $Path(T_j, T_i)$ must not exist. An *ancestor* of vertex T_i is a vertex that has at least one path to T_i. Vertex T_i is then a *descendant* of its ancestors. A parent (child) of a vertex is one of the ancestors (descendants) of the vertex. The *length* of a path, denoted as $|Path(T_i, T_j)|$, is the summation of the weights of the vertices and edges along the path. The *longest path* from vertex T_i to vertex T_j, denoted as $LPath(T_i, T_j)$, is the path that has the longest length among all paths from vertex T_i to vertex T_j, and its length is denoted as $|LPath(T_i, T_j)|$. There may be more than one longest path between a pair of vertices, and all of them have the same length. A *critical path* of vertex T, denoted as $CPath(T)$, is one of the longest paths from the *Source* T_{sr} to vertex T. A vertex may have multiple critical paths, and all of them have the same length, which is denoted as $|CPath(T)|$ and is equal to $|LPath(T_{sr}, T)|$. It is easy to see that the finish time of task T is equal to $|CPath(T)|$, and the start time of the task is equal to $|CPath(T)| - W_v(T)$, where $W_v(T) = C_T$ if T incurs zero fault, or $C_T + x \cdot Cr_T$ if T incurs x faults. It is important to note that the length of a path, the longest paths between two tasks, and the critical paths of a vertex VARY with the fault occurrences of involved vertices. Hence, a different case of fault occurrences will result in a different WDAG

because the weights of involved vertices have changed.

Let Tc be the current task under investigation, $BCFT_{Tc}$ be its best-case finish time when no fault occurs, and $WCFT_{Tc}$ be its worst-case finish time when all X faults have occurred during or before its execution. In order to derive $WCFT_{Tc}$, two useful task subsets are introduced.

- Parents Set of Tc (PS_{Tc}): Task subset comprising the parents of Tc.
- Ancestors Set of Tc (AS_{Tc}): Task subset comprising all ancestors of the current task Tc.

It is easy to see that only the fault occurrences of the tasks in AS_{Tc} could affect the start time of Tc. Obviously, $PS_{Tc} \subseteq AS_{Tc}$.

IV. The Critical Task

In this section we will prove that for any task Tc, there exists at least one critical task in $AS_{Tc} \cup \{Tc\}$, such that when this critical task incurs all the faults, Tc experiences the worst-case finish time. It is important to note that the critical task of Tc may not be the task that has the longest re-execution time in AS_{Tc}. We will start by proving Lemma 1, 2, and 3.

Lemma 1: Given Tc and a faulty task T, and T is not in any of the critical paths of Tc. All the critical paths of Tc when T incurs x faults are still critical paths when T incurs x' faults, and the finish time of Tc doesn't change, as long as $x > x' \geq 0$.

Explanation: This lemma has two implications. First, the critical paths of Tc when T incurs x faults will remain as the critical paths of Tc when T incurs x' faults, as long as $x > x'$. Second, if a path from the *Source* to Tc was not a critical path when T incurred x faults, it will not become a critical path when T incurs x' faults, as long as $x > x'$.

Proof: On one hand, because $x' < x$, the weight of vertex T is reduced, which further reduces the lengths of all the paths from the *Source* to Tc that include T. On the other hand, since the length of a path depends only on the weights of the vertices and edges along the path, the change of T's fault occurrences will not change the length of the paths that do not include T, including those critical paths. And since no path will get a longer length, those critical paths remain as critical paths. Therefore the finish time of Tc doesn't change. This proves Lemma 1.

Lemma 2: Assume there are two tasks, T_1 and T_2, that are in AS_{Tc} and incur x_1 and x_2 faults respectively. A worse or status-quo finish time of Tc can always be found by letting a task $T \in AS_{Tc}$ incur all the $x_1 + x_2$ faults. T could be either T_1 or T_2, or another task in AS_{Tc}.

Proof: If one of the two tasks, say T_1, is not in any critical path of Tc, a worse finish time of Tc can be easily obtained by letting T_1 incur 0 faults, and letting any task that is in a critical path of Tc incur x_1 more faults. This is because, from Lemma 1, as the number of faults occurring in T_1 decreases, all critical paths of Tc will not be affected and the finish time of Tc doesn't change since T_1 is not in any critical path of Tc. On the other hand, if there is a task T from a critical path of Tc incurs x_1 more faults, the length of this critical path increases, so does the $WCFT$ of Tc. This proves Lemma 2 if either T_1 or T_2 or neither are in critical paths of Tc.

If both T_1 and T_2 are in the critical paths of Tc, there could be two cases. The first case is that there exists one critical path that includes T_1 but not T_2. The critical path is thus denoted as $CPath(Tc)_{T1}$. The second case is that critical paths that include T_1 always include T_2. For both cases we have the following rearrangements of faults:

- Rearrangement 1: Let T_1 incur $x_1 + 1$ faults and let T_2 incur $x_2 - 1$ faults.
- Rearrangement 2: Let T_1 incur $x_1 - 1$ faults and let T_2 incur

x_2+1 faults

For the first case, Rearrangement 1 reallocates one fault from T_2 to T_1 and the total number of faults is not changed. The $|CPath(Tc)_{T1}|$ will be increased by the amount of Cr_{T1}, which will result in a worse finish time of Tc. Repeating this rearrangement will further worsen the finish time of Tc until T_1 incurs x_1+x_2 faults and T_2 incurs zero faults. This proves the Lemma 2 under this situation.

For the second case, Rearrangement 1 will change the length of the critical paths that include both T_1 and T_2 by the amount of $Cr_{T1} - Cr_{T2}$. If $Cr_{T1} > Cr_{T2}$, the length of the path is increased thereby resulting in a worse finish time of Tc. Repeating this rearrangement will further worsen the finish time of Tc until T_1 incurs x_1+x_2 faults and T_2 incurs zero faults. This proves Lemma 2. If $Cr_{T1} < Cr_{T2}$, Rearrangement 2 will increase the length of the critical path thereby resulting in a worse $WCFT$ of Tc. Repeating this rearrangement will further worsen the $WCFT$ of Tc until T_2 incurs x_1+x_2 faults and T_1 incurs zero faults. This proves Lemma 2. There is a special third case where $Cr_{T1} = Cr_{T2}$. Either letting T_1 or T_2 incur all the x_1+x_2 faults will reach a status-quo finish time of Tc. This also proves Lemma 2.

Lemma 3: If more than one task in AS_{Tc} incur faults, a worse or status-quo finish time of Tc can always be found by letting one task $T \in AS_{Tc}$ incur all the X faults.

Proof: Lemma 3 is a corollary of Lemma 2. Let's denote T_1, T_2, ..., T_k as these faulty tasks and x_1, x_2, ..., x_k as their fault occurrences respectively, $\sum_{i=1}^{k} x_i = X$. Lemma 3 can be proved by repeating Lemma 2. Every time two random faulty tasks are picked, a worse-case finish time of Tc can be obtained by rearranging the faults occurring in the two faulty tasks to the one in a critical path of Tc, or to a third task if neither of the two tasks is in a critical path. This procedure continues until there is only one faulty task T that incurs total X faults. Since the finish time of Tc is not advanced during all these rearrangements, Lemma 3 is proved.

Now it is time to prove the critical task theory as stated in Theorem 1, which gives a sufficient condition for Tc to experience the worst-case finish time.

Theorem 1: There exists at least one task for any task Tc such that if this task incurs all the expected X faults, task Tc experiences its worst-case finish time. This task, which could be Tc itself or a task in AS_{Tc}, is thus referred to as the Critical Task (Tct) of Tc.

Proof: Theorem 1 can be proved by contradiction. The counter statement is that Tc will only experience the worst-case finish time when more than one task incur faults. This statement directly contradicts to Lemma 3. Therefore Theorem 1 is proved.

Theorem 1 can be easily extended to identify the critical task of a processor, and then the critical task of the task set. The critical task of a processor is the critical task of the last task scheduled on this processor, and the critical task of the task set is the critical task of the *Sink*. Unfortunately, the critical task of a task set may not be the longest task in the set, and hence cannot be identified easily. We hence propose a recursive algorithm that identifies the critical task and the $WCFT$ of a task set in $O(N^2)$ where N is the number of tasks.

V. Algorithm to Identify the Critical Task and the Worst-Case Finish Time

Consider a task set that consists of N tasks and is subject to X faults. There are total $\binom{N+X-1}{X}$ distinct cases of fault occurrences [12]. Without the knowledge from the previous section, one needs to compute the $WCFT$ for each of these cases, and then chooses the worst of them. To compute the $WCFT$ for a given fault occurrence,

one needs to determine the longest path from the *Source* to the *Sink* which takes $O(N^2)$ [29]. Therefore the overall time complexity could be

$$O\left(\frac{(N+X-1)!}{X!(N-1)!}N^2\right)$$

We here propose a more efficient algorithm that identifies the critical task and calculates the $WCFT$ of a task in $O(N^2)$ time. The supporting lemma and theorem are given below.

Lemma 4: If a task is not a critical task of any parents of Tc, it will not be a critical task of Tc.

Proof: According to Theorem 1, Tc will experience its worst-case finish time when either Tc or one of the tasks in AS_{Tc} incurs all the faults. Apparently, Lemma 4 is applicable only when Tc is not a critical task of itself. The following proof thus assumes Tc is fault-free.

Tc could have more than one parent and each parent has its own critical tasks. A parent will experience the $WCFT$ when one of its critical tasks incurs all the faults. Let's denote the random task as $Tnct$. According to the condition of the lemma, $Tnct$ is not a critical task of any of these parents. If $Tnct$ incurs all the faults, none of the parents of Tc will experience their $WCFT$. As a result, Tc will not experience the $WCFT$. This indicates that $Tnct$ is not a critical task of Tc, thus proving Lemma 4.

Essentially Lemma 4 tells that the critical tasks of Tc can only be Tc itself or the critical tasks of Tc's parents. To determine the $WCFT$ of task Tc, we have the following two cases:

Case 1: Tc is its own critical task and it incurs all X faults while all tasks in AS_{Tc} finish fault-free. All parents of Tc thus experience the best-case finish time ($BCFT$). The $BCFT$ of each task can be calculated as long as the schedule is determined (i.e. no need to know the actual fault occurrences). Since Tc incurs all X faults, the re-execution time of Tc for tolerating X faults is $X \cdot Cr_{Tc}$. The $WCFT$ of Tc thus equals to:

$$BCFT_{Tc} + X \cdot Cr_{Tc}$$

Case 2: The critical task of Tc is a critical task of one of Tc's parents, and Tc finishes fault-free. Tc could have more than one parent. The $WCFT$ of Tc under this case equals to:

$$\text{Max}\{WCFT_{Tp} + W_E(Tp, Tc) + C_{Tc}, \forall Tp \in PS_{Tc}\}$$

where $W_E(Tp,Tc)$ is the communication delay from Tp and Tc. Therefore, to determine the worst-case finish time of Tc we have Theorem 2.

Theorem 2: The worst-case finish time of Tc is either $BCFT_{Tc}$ + $X \cdot Cr_{Tc}$, or Max$\{WCFT_{Tp} + W_E(Tp, Tc) + C_{Tc}, \forall Tp \in PS_{Tc}\}$, whichever has the larger value. If the former is larger, the critical task of Tc is itself. Otherwise, Tc inherits its critical task from its parent who has the maximum value of $WCFT_{Tp} + W_E(Tp, Tc)$.

Apparently, the critical task and the $WCFT$ of the *Sink* are the critical task and the $WCFT$ of the task set, respectively. Fig. 3 shows the pseudo code to compute the $WCFT$ and identify the critical task for a task Tc in a scheduled task set. It accepts Tc and X as inputs and outputs the $WCFT$ and one of the critical tasks of Tc. It also uses three arrays, $BCFT(T)$, $C(T)$ and $Cr(T)$, to store the best-case finish time, worst-case execution time, and worst-case re-execution time of all the tasks in a task set respectively. $BCFT(T)$ can be calculated from the schedule of the task set assuming zero fault occurrences. The communication delay between any two tasks T and T' is recorded in a two-dimension array $W_E(T, T')$. The algorithm updates two arrays, $WCFT(T)$ and $Tct(T)$ which are initialized to NULL and will be populated by the WCFT and critical tasks of all tasks respectively.

When the procedure is called the first time with the Tc, it begins by checking if $WCFT$ and Tct of Tc are already available. If yes, the procedure returns the recorded values and exits (line 1-2). Otherwise, it continues to determine if PS_{Tc} contains the *Source*, which indicates that this is the first task on a processor and does not depend on any other task. The *WCFT* of this task is computed as $BCFT(Tc) + X \cdot Cr(Tc)$, that is, the task is its own critical task and will experience its *WCFT* when itself incurs all X faults (line 3–7).

If PS_{Tc} does not contain the *Source*, which indicates that Tc has some parents, the procedure recursively calls itself to compute the *WCFT* of all tasks in PS_{Tc}. Since a task could be in the Parents Set of several tasks, it may have been visited before and its *WCFT* and critical task are available in $WCFT(T)$ and $Tct(T)$. If that is the case, the recursive call simply reads the values and returns (line 1-2). This ensures that the *WCFT* and *Tct* of any task will be computed only once. After visiting all tasks in PS_{Tc}, the task with the maximum $WCFT_{Tp}+W_E(T_p, Tc)$ is chosen to be the T_m (line 8 – 18).

Initialize *WCFT* array and *Tct* array to **NULL**;
Procedure FindWorstCase(Tc, X)
Input: (Tc, X); **Output**: ($WCFT(Tc)$, $Tct(Tc)$) {
1.   **if** ($WCFT(Tc)$!= **NULL**)
2.   **return** ($WCFT(Tc)$, $Tct(Tc)$);
3.   **if** (PS_{Tc} contains the *Source*) {
4. $WCFT(Tc) = BCFT(Tc) + X \cdot CR(Tc)$;
5. $Tct(Tc) = Tc$;
6.   **return** ($WCFT(Tc)$, $Tct(Tc)$);
7.   }
8. $WCFT_{Tm}$=0; W_{ETm}=0;
9.   **do** {
10.   Let Tp be a non-visited task in PS_{Tc};
11.   ($WCFT(Tp)$,$Tct(Tp)$) = FindWorstCase(Tp, X);
12.   **if** ($WCFT(Tp)$+$W_E(Tp, Tc)$ > $WCFT_{Tm}$+W_{ETm}) {
13. $WCFT_{Tm} = WCFT(Tp)$;
14. $W_{ETm} = W_E(Tp, Tc)$;
15. Tct_{Tm} = $Tct(Tp)$;
16.   }
17. Tp is marked as a visited task;
18.   } **until** all tasks in PS_{Tc} are visited.
19.   **if** $WCFT_{Tm}$+W_{ETm}+$C(Tc)$ > $BCFT(Tc)$+$X \cdot CR(Tc)$ {
20. $WCFT(Tc)$=$WCFT_{Tm}$+W_{ETm}+$C(Tc)$;
21. $Tct(Tc)$ =Tct_{Tm};
22.   }
23.   **else** {
24. $WCFT(Tc)$=$BCFT(Tc)$+$X \cdot CR(Tc)$;
25. $Tct(Tc) = Tc$;
26.   }
27.   **return** ($WCFT(Tc)$, $Tct(Tc)$);
28.}

Fig. 3 Identify the *WCFT* and a critical task of task Tc

Next, the algorithm determines which of the two cases results in a worse finish time of Tc: If $WCFT_{Tm}+W_{ETm}+ C(Tc) > BCFT(Tc)+X \cdot Cr(Tc)$, Tc inherits the critical task from T_m (line 19 – 22), otherwise Tc is its own critical task (line 23 – 26).

The *WCFT* and critical task of a processor can be obtained by calling the procedure with the last task scheduled on the processor as Tc. Similarly, the *WCFT* and critical task of the whole task set can be obtained by calling the procedure with the *Sink* as Tc. Since each task will be visited exactly once, the procedure will run only N times where N is the number of tasks in the task set. The loop from line 9 to line 18 will run no more than N times since a task can have at most N tasks in its Parents Set.

VI. Experimental Results

To estimate the *WCFT* of a task set, a common practice is to assume the task with the longest re-execution time incurs all the faults. In this section we want to show that this practice could significantly under-estimate the real worst case. We have developed a simulator for this purpose. The simulation is based on standard DAG benchmarks from DSPStone [30]. The experiments are performed on a Dell PC with Intel Core 2 Duo i5-2400 3.1G processor

and 4.00GB memory.

The simulator first uses list-scheduling to get a scheduled task set for the benchmarks. Then it identifies the critical task of the whole task set and calculates the *WCFT* in the presence of faults according to the proposed algorithm. At last, we compare the *WCFT* obtained by our algorithm to the finish time when all faults occur on the task with the longest re-execution time.

The simulation results are shown in Fig.4. The difference ratio reflects the difference of finish time between all faults incurring at critical task and longest task. The surface graph contains the results of 6 benchmarks from DSPStone with the number of processors varying from 2 to 8 and the number of faults varying from 1 to 12, respectively.

During the experiment, it is observed that: 1) the critical task may not be the longest task. This leads to different finish time. That is to say, in Fig. 4, the higher the curve is, the more significant the difference is. And the most significant difference ratio can be close to 35% in some cases; 2) for the same benchmark, the critical task may change with the number of processors and faults changing, while the task with the longest re-execution time remains the same. Since the critical task of a benchmark may not be the task with the longest re-execution time, if one simply bets the worst case by letting the latter one incur all the faults, the real worst case could be significantly under-estimated.

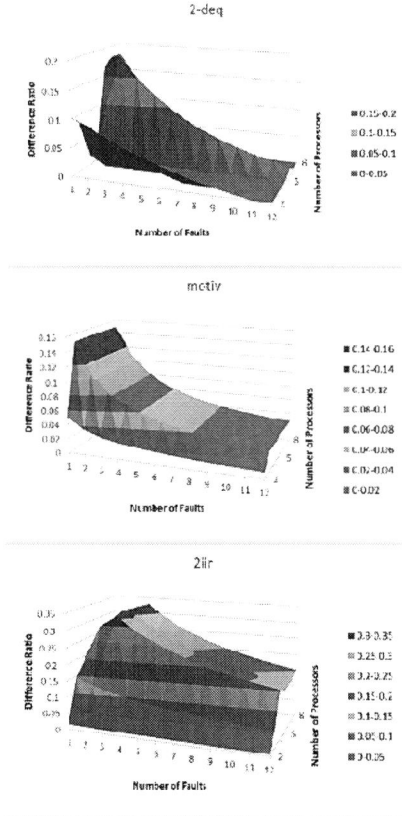

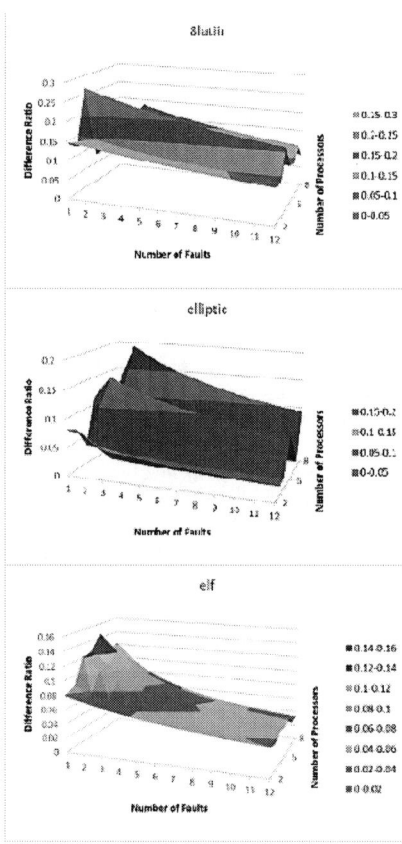

Fig. 4 Simulation Results for Benchmark

VII. Conclusions

In this paper, we have investigated the impact of the fault occurrences on the worst-case finish time of a task set scheduled in a frame-based multi-processor system. It is assumed that tasks could have inter-task dependency, and the number of fault occurrences in a frame is upper-limited by X. We conclude that there exists at least one critical task for each task. A task undergoes its worst-case finish time when one of its critical tasks incurs all X faults assuming one re-execution is dedicated to one fault. Based on the critical task theory, a recursive algorithm is then designed to identify the critical task and the worst-case finish time of a task set. For a task set with N tasks, the proposed algorithm takes only $O(N^2)$, while the state-of-the-art takes $O\left(\frac{(N+X-1)!}{X!(N-1)!}\right)N^2$.

References

[1] BERTOSSI, A. AND MANCINI, L., Scheduling algorithms for fault-tolerance in hard real-time systems. Real-Time Systems, 7, 3, 229–245, 1994.

[2] CHANDY, K. M., BROWNE, J. C., DISSLY, C. W., AND UHRIG, W. R., "Analytic models for rollback and recovery strategies in data base systems," *Software Engineering, IEEE Transactions on*, vol.SE-1, no.1, pp.100,110, March 1975.

[3] CHEVOCHOT, P. PUAUT, I., "Scheduling fault-tolerant distributed hard real-time tasks independently of the replication strategies," *Real-Time Computing Systems and Applications, 1999. RTCSA '99. Sixth International Conference on*, vol., no., pp.356,363, 1999.

[4] DIMA, P., GIRAULT, A., AND SOREL, Y., "Off-line real-time fault-tolerant scheduling," *Parallel and Distributed Processing, 2001. Proceedings. Ninth Euromicro Workshop on*, vol., no., pp.410,417, 2001.

[5] GIRAULT, A., KALLA, H., AND SIGHIREANU, M., "An algorithm for automatically obtaining distributed and fault-tolerant static schedules," *Dependable Systems and Networks, 2003. Proceedings. 2003 In-*

ternational Conference on, vol., no., pp.159,168, 22-25 June 2003.

[6] POP, P., IZOSIMOV, V., ELES, P., PENG, Z., "Design Optimization of Time- and Cost-Constrained Fault-Tolerant Embedded Systems With Checkpointing and Replication," *Very Large Scale Integration (VLSI) Systems, IEEE Transactions on*, vol.17, no.3, pp.389,402, March 2009.

[7] KANDASAMY, N., HAYES, J., AND MURRAY, B., "Transparent recovery from intermittent faults in time-triggered distributed systems," *Computers, IEEE Transactions on*, vol.52, no.2, pp.113,125, Feb. 2003.

[8] S. GHOSH, R. MELHEM, and D. MOSSE., "Enhancing real-time schedules to tolerate transient faults," *Real-Time Systems Symposium, 1995. Proceedings., 16th IEEE*, vol., no., pp.120,129, 5-7 Dec 1995.

[9] BURNS, A., DAVIS, R., AND PUNNEKKAT, S., "Feasibility analysis of fault-tolerant real-time task sets," *Real-Time Systems, 1996., Proceedings of the Eighth Euromicro Workshop on*, vol., no., pp.29,33, 12-14 Jun 1996.

[10] AUDSLEY, N. C., BURNS, A., RICHARDSON, M., "Applying new scheduling theory to static priority pre-emptive scheduling," *Software Engineering Journal*, vol.8, no.5, pp.284,292, Sep 1993.

[11] LIBERATO, F., MELHEM, R., AND MOSSE, D. "Tolerance to multiple transient faults for aperiodic tasks in hard real-time systems," *Computers, IEEE Transactions on*, vol.49, no.9, pp.906,914, Sep 2000.

[12] AYDIN, H., "Exact Fault-Sensitive Feasibility Analysis of Real-Time Tasks," *Computers, IEEE Transactions on*, vol.56, no.10, pp.1372,1386, Oct. 2007.

[13] M. CHROBAK, M. HURAND, J. SGALL, "Fast Algorithms for Testing Fault-Tolerance of Sequenced Jobs with Deadlines," in Proc. of 28th IEEE International Real-Time Systems Symposium, 2007

[14] BAUER, G. AND KOPETZ, H., "Transparent redundancy in the time-triggered architecture," *Dependable Systems and Networks, 2000. DSN 2000. Proceedings International Conference on*, vol., no., pp.5,13, 2000.

[15] BRETZ, A., "By-wire cars turn the corner," *Spectrum, IEEE*, vol.38, no.4, pp.68,73, Apr 2001.

[16] KOPETZ, H., "Why time-triggered architectures will succeed in large hard real-time systems," *Distributed Computing Systems, 1995., Proceedings of the Fifth IEEE Computer Society Workshop on Future Trends of*, vol., no., pp.2,9, 28-30 Aug 1995.

[17] OBERMAISSER, R., Event-Triggered and Time-Triggered Control Paradigms. Springer, ISBN 0387230432.

[18] POLEDNA, S., Fault-Tolerant Real-Time Systems: The Problem of Replica Determinism. Kluwer Academic Publishers, ISBN: 079239657X.

[19] SURI. N., Advances in ULTRA-Dependable Distributed Systems. IEEE Computer Society Press, ISBN 0818662875.

[20] POP, P., ELES, P. AND PENG, Z., Schedulability analysis for systems with data and control dependencies. In Proceedings of the 12th Euromicro Conference on Real-Time Systems, 201–208, 2000.

[21] ZHU, D., MELHEM, R., AND MOSSE, D. The effects of energy management on reliability in real-time embedded systems. In Proceedings of the 2004 IEEE ICCAD, 35–40.

[22] Y. LIU, H. LIANG and K. WU, Scheduling for Energy Efficiency and Fault Tolerance in Hard Real-Time Systems, In Proceedings of 2010 DATE, 1444-1449

[23] P.A. LAPLANTE, Real-Time Systems Design and Analysis. Wiley Press, 3rd ed, ISBN 9780471228554

[24] BALAKIRSKY, V.B., VINCK, A.J.H., Coding Schemes for Data Transmission over Bus Systems, IEEE International Symposium on Information Theory 2006.

[25] PIRIOU, E., AND JÉGO, C., ADDE, P., BIDAN, R.L., JÉZÉQUEL, M., Efficient architecture for Reed Solomon block turbo code. In Proc. IEEE International Symposium on Circuits and Systems, 2006

[26] GOMMA, M. A. AND VIJAYKUMAR, T. N., 2005. Oppotunistic transient-fault detection. In Proceedings of 2005 International Symposium on Computer Architecture (ISCA'05), 172–183.

[27] RAY, J., HOE, J. C., And Falsafi, B. Dual use of superscalar datapath for transient-fault detection and recovery. In Proceedings of 2001 IEEE International Symposium on Microarchitecture (MICRO'01), 214–224.

[28] WEAVER, C., EMER, J., MUKHERJEE, S. S., AND REINHARDT, S. K. Techniques to reduce the soft error rate of a high-performance microprocessor. In Proceedings of 2004 ISCA, 264–275.

[29] MANBER, U. 1989. Introduction to algorithms: a creative approach. Addson-Wesley Press, ISBN 0201120372.

[30] ZIVOJNOVIC V., VELARDE J. M., SCHLAGER C., And Mryr H., DSPstone: A DSP-oriented benchmarking methodology. In Proceedings of ICSPAT. 1994.

978-1-4799-2817-0/14 $31.00 © 2014 IEEE

Flexible Packed Stencil Design with Multiple Shaping Apertures for E-Beam Lithography[*]

Chris Chu
Department of Electrical and Computer Engineering
Iowa State University
Ames, Iowa 50011
e-mail: cnchu@iastate.edu

Wai-Kei Mak
Department of Computer Science
National Tsing Hua University
Hsinchu, Taiwan 300 R.O.C.
e-mail: wkmak@cs.nthu.edu.tw

Abstract— Electron-beam direct write (EBDW) lithography is a promising solution for chip production in the sub-22nm regime. To improve the throughput of EBDW lithography, character projection method is commonly employed and a critical problem is to pack as many characters as possible onto the stencil. In this paper, we consider two enhancements in packed stencil design over previous works. First, the use of multiple shaping apertures with different sizes is explored. Second, the fact that the pattern of a character can be located anywhere within its enclosing projection region is exploited to facilitate flexible blank space sharing. For this packed stencil design problem with multiple shaping apertures and flexible blank space sharing, a dynamic programming based algorithm is proposed. Experimental results show that the proposed enhancement and the associated algorithm can significantly reduce the total shot count and hence improve the throughput of EBDW lithography.

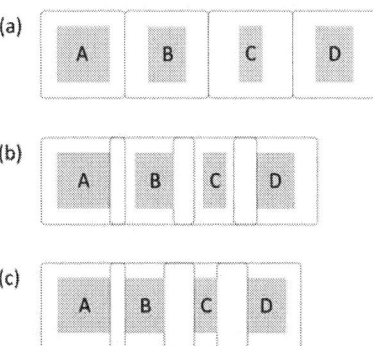

Fig. 1. (a) Four adjacent characters placed without overlapping their blank spaces. (b) The same characters are placed with their blank spaces overlapping to reduce the overall area as considered in [9, 10]. (c) Flexible blank space sharing by relocating the character patterns within their projection regions to further reduce the overall area as considered in this paper.

I. Introduction

With the continued delay of the introduction of extreme ultraviolet (EUV) lithography, the semiconductor industry is exploring other alternatives for manufacturing chips at ever smaller process nodes. E-beam direct write (EBDW)is a promising solution [1–3] and is being pursued by companies like TSMC. With the expected move to the next generation silicon wafers increasing the wafer size from 300 to 450mm, it will make massively parallel EBDW even more appealing [3].

E-beam direct write technology commonly employs the character projection method in which complex patterns, called characters, are printed onto a wafer [4]. An e-beam writing system has a stencil which can hold a set of characters. Patterns in a circuit that correspond to a character in the stencil can be projected in one shot. However, other patterns that do not match any character need to be fractured into constituent rectangles. Then each constituent rectangle requires its own shot and has to be printed in the variable shaped beam (VSB) mode [5].

Traditionally, a standard stencil adopts a grid-based layout with pre-designated spots for characters [6, 7]. Each character pattern can be of any size or shape up to the maximum allowed size, which is dictated by the shaping aperture. Since there is a fixed number of N pre-designated character spots on the grid,

one just needs to pick the N most beneficial characters and put them into these spots. While this arrangement is simple, it is too restrictive. It was pointed out in [7, 8] that by carefully arranging and packing the different sized characters on a stencil, the final number of characters that can be put on a stencil can be greatly increased. It is because two adjacent characters can be placed with their blank spaces overlapping as illustrated in Fig. 1(b). This will allow more patterns on the wafer to be shot as characters leading to reduced write time and cost. Packed stencil design was studied in [9] and [10].

For the optimal use of the stencil, we consider two enhancements in packed stencil design over previous works [9, 10]. Firstly, we note that previous works are limited to the case that there is only a single shaping aperture, but the use of multiple shaping apertures with different sizes is possible [4, 11] as shown in Fig. 2. Since the enclosing projection region size of a character is determined by the shaping aperture size, using a smaller shaping aperture for smaller character patterns is helpful for packing more characters into the stencil. Secondly, the existing packing algorithms did not exploit the fact that the pattern of a character can be located anywhere within its enclosing projection region (as long as it is not too close to the region boundary) to maximize blank space sharing of adjacent characters. For example, consider the characters in Fig. 1(a), we can reduce the total area occupied by the three characters even further compared to Fig. 1(b) through *flexible blank space*

---

[*]This work was supported in part by the National Science Council, under Grant NSC 102-2220-E-007-013.

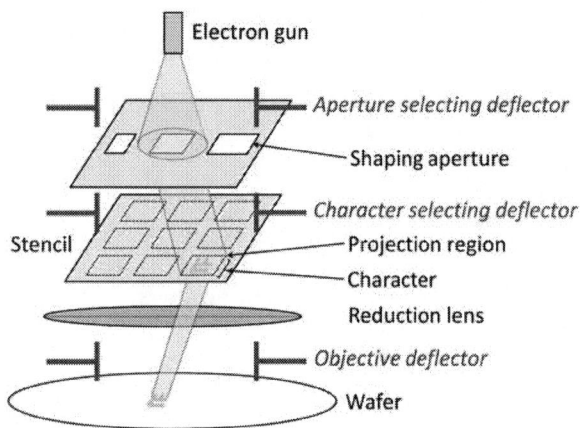

Fig. 2. A e-beam writing system with three shaping apertures.

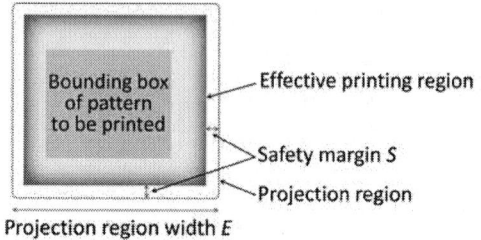

Fig. 3. The projection region and effective printing region for a character.

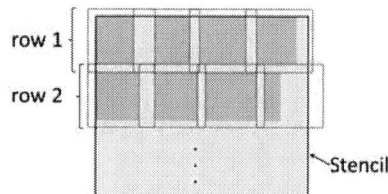

Fig. 4. Row-based stencil design.

sharing as illustrated in Fig. 1(c). Flexible blank space sharing can maximize total blank space sharing by relocating the character patterns within their projection regions.

In this paper, we investigate the packed stencil design problem with multiple shaping apertures and flexible blank space sharing. To the best of our knowledge, it has not been addressed by any existing work in the literature. We present the first algorithm targeting this problem and show that it can effectively increase the number of packed characters on a stencil and significantly reduce the total shot count.

The rest of the paper is organized as follows. In Section II, we introduce some preliminary information and the problem formulation. In Section III.C, we present a dynamic programming based approach for character selection and aperture size (or equivalently, projection region size) determination under flexible blank space sharing. Then we show how to actually pack the characters into the stencil while optimizing blank space sharing in Section III.D. Finally, we show some experimental comparisons with [9] and [10] in Section IV.

II. PROBLEM FORMULATION

Refer to the general e-beam machine in Fig. 2, it has multiple differently sized shaping apertures. The aperture selecting deflector can control the direction of the e-beam to shoot it through one of the shaping apertures. Subsequently, the character selecting deflector can direct the e-beam to shoot through any desired character on the stencil. Note that the dimensions of the projection region on the stencil when an e-beam is shot through a particular shaping aperture are dependent on the dimensions of the corresponding shaping aperture.

Due to scattering of electrons, features that are too close to the boundary of projection region may not be printed accurately. Hence, we define an effective printing region as the part of projection region that is at least a safety margin S away from the boundary as shown in Fig. 3. In order to print characters on the wafer with no loss of accuracy, the pattern of each character must lie within the effective printing region. In addition, the pattern of any character A cannot overlap with the e-beam projection region of another character B. Otherwise, part of character A would be printed erroneously when printing character B. However, the blank space of two neighboring

characters may overlap and is desirable to overlap in order to reduce the total area occupied by the characters.

We assume that standard cells are implemented by the characters. To minimize wastage of stencil area, the heights of different character projection regions should all be set to $2S$ plus the standard cell image height. In addition, characters on the stencil should be arranged in a row-based manner and each character pattern is always placed within its projection region such that the top blank space and the bottom blank space are equal to S. In this way, the bottom blank space of a row of characters can completely overlap with the top blank space of the row of characters below as in Fig. 4 and we can pack the maximum possible number of rows of characters into the stencil. Finally, we assume that the blank space of a character can lie outside of the available character area of the stencil [8] as in Fig. 4.

Next, we introduce the notations used in the rest of the paper.

- K denotes the number of choices for character projection region widths which is equal to the number of available shaping apertures.
- S denotes the safety margin from the effective printing region to the projection region boundary for an e-beam shot.
- W denotes the width of the available character area of the stencil.
- R denotes the number of character rows that can fit in the stencil.
- n denotes the number of different characters extracted from a design.
- w_c denotes the width of the pattern of character c. For simplicity, we will refer to the width of the pattern of a character as the character width in the rest of the paper.
- r_c denotes the number of occurrence of character c in a die.
- n_{VSB_c} denotes the number of e-beam shots required to print character c by the VSB method.

The problem of flexible packed stencil design with multiple shaping apertures is formally stated below.

978-1-4799-2817-0/14 $31.00 © 2014 IEEE 138

Def 1. *Suppose the values of K, S, W, and R for an e-beam machine with multiple shaping apertures are given. A set of n characters extracted from a design and the values of w_c, r_c, and n_{VSB_c} for each character c are also given. (i) Determine K different character projection region widths, (ii) select the characters to be put on the stencil, and (iii) pack the characters on the stencil in a row-based manner with flexible blank space sharing to minimize the total shot count for the design under the following constraints.*
(C1) The pattern of each character must lie within the effective printing region of the character.
(C2) The pattern of each character cannot lie within the projection region of any other character.
(C3) The patterns of all characters must lie within the available character area of the stencil.

Lemma 1. *The flexible packed stencil design problem is NP-complete.*

It can be shown that the 0-1 knapsack problem, which is a well-known NP-complete problem [12], can be reduced to the subproblem of selecting and packing of n characters on a stencil in which n character projection region widths are allowed. So the flexible packed stencil design problem is NP-complete.

III. Algorithm

III.A. Algorithm Outline

As the problem is very complicated, we break it down into three steps. First, we propose a dynamic programming algorithm to determine K projection region widths and to select a set of most beneficial characters that roughly can be packed on the stencil. Then we distribute the selected characters to different rows and construct tight linear packing for each row. At last, we greedily refine the solution by checking if any of the unselected characters can be added to the end of the tight linear packing of some row. The outline of the algorithm is given in Algorithm 1.

Algorithm 1 Flexible Packed Stencil Design

1: Determine K projection region widths and select a set of characters to be put on the stencil by dynamic programming.
2: Assign the characters selected in Step 1 to rows on the stencil and construct tight linear packing for each row.
3: Greedily pack some of the unselected characters at the end of each row, if possible.

Our algorithm is based on the key concepts of tight linear packing and effective character width. So, we will first introduce them in Section III.B. Then, we will describe the details of Step 1 of Algorithm 1 in Section III.C and the details of Steps 2 and 3 in Section III.D.

III.B. Tight Linear Packing and Effective Character Width

We introduce some definitions related to our problem.

Def 2. *A linear packing of a group of characters is a packing of all the given characters in a row with flexible blank space sharing satisfying constraints (C1) and (C2).*

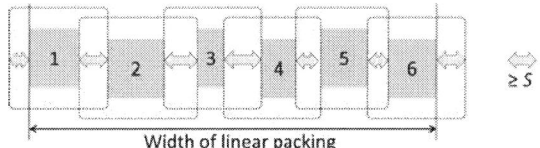

Fig. 5. A tight linear packing. Characters with even indexes are shifted down a bit to show the projection regions of all characters more clearly.

Def 3. *The width of a linear packing is the total span of the packing excluding the left blank space of the first character and the right blank space of the last character. (See Fig. 5.)*

In the following discussion, we assume that the characters in a linear packing are indexed from left to right by $1, 2, 3, \ldots$. And the width of the projection region for character i is denoted by E_i.

Def 4. *A tight linear packing is a linear packing such that for any two neighboring characters i and $i + 1$, the right blank space of character i and the left blank space of character $i + 1$ are exactly equal and completely overlap.*

For example, Fig. 5 shows a tight linear packing of six characters.

Def 5. *The effective width of a character c is defined as $(w_c + E_c)/2$.*

Below we show that the width of any tight linear packing is approximately equal to the total effective width of its characters.

Lemma 2. *For a linear packing P of k characters, let $W(P)$ denote the width of P, $w_i(P)$ denote the width of the i-th character $(i = 1, 2, \ldots, k)$, $s_0(P)$ denote the width of the left blank space of the first character, and $s_k(P)$ denote width of the right blank space of the last character in P.*
For a tight linear packing P_t,

$$W(P_t) = \sum_{i=1}^{k}(w_i(P_t) + E_i)/2 - s_0(P_t)/2 - s_k(P_t)/2$$

For an arbitrary linear packing P_a,

$$W(P_a) \geq \sum_{i=1}^{k}(w_i(P_a) + E_i)/2 - s_0(P_a)/2 - s_k(P_a)/2$$

Proof. For a tight linear packing P_t, the right blank space of the i-th character and the left blank space of the $(i+1)$-th character are exactly equal and completely overlap for $i = 1, 2, \ldots, k - 1$. So, its width can be expressed as $W(P_t) = \sum_{i=1}^{k} w_i(P_t) + \sum_{i=1}^{k-1} s_i(P_t)$ where $s_i(P_t)$ denote the width of the right blank space of the i-th character in P_t.

For each character i in P_t, we have $s_{i-1}(P_t) + w_i(P_t) + s_i(P_t) = E_i$ since its left and right blank spaces are equal to $s_{i-1}(P_t)$ and $s_i(P_t)$, respectively. Hence, $\sum_{i=1}^{k}(s_{i-1}(P_t) + w_i(P_t) + s_i(P_t)) = \sum_{i=1}^{k} E_i$. It implies that $2 \times \sum_{i=1}^{k-1} s_i(P_t) = \sum_{i=1}^{k} E_i - \sum_{i=1}^{k} w_i(P_t) - s_0(P_t) - s_k(P_t)$.

So, the width of a tight linear packing P_t can be re-written as

$$W(P_t) = \sum_{i=1}^{k} w_i(P_t)/2 + (\sum_{i=1}^{k} E_i - s_0(P_t) - s_k(P_t))/2$$
$$= \sum_{i=1}^{k} (w_i(P_t) + E_i)/2 - s_0(P_t)/2 - s_k(P_t)/2.$$

For an arbitrary linear packing P_a, we let $g_i(P_a)$ be the width of the gap between the i-th and $(i+1)$-th character patterns in P_a for $i = 1, 2, \ldots, k-1$, and let $g_0(P_a)$ be $s_0(P_a)$ and $g_k(P_a)$ be $s_k(P_a)$. Then P_a's width can be expressed as $W(P_a) = \sum_{i=1}^{k} w_i(P_a) + \sum_{i=1}^{k-1} g_i(P_a)$. Note that $g_{i-1}(P_a) + w_i(P_a) + g_i(P_a) \geq E_i$ for each character i. It implies that $2 \times \sum_{i=1}^{k-1} g_i(P_a) \geq \sum_{i=1}^{k} E_i - \sum_{i=1}^{k} w_i(P_a) - s_0(P_a) - s_k(P_a)$. So,

$$W(P_a) \geq \sum_{i=1}^{k} (w_i(P_a) + E_i)/2 - s_0(P_a)/2 - s_k(P_a)/2.$$

□

By applying Lemma 2, we can derive an upper bound of the difference between the width of a tight linear packing and the width of the minimum width linear packing and get Lemma 3.

Lemma 3. *Given a group of k characters, the width of any tight linear packing is less than $(E_1 + E_k)/2 - 2S$ away from the minimum width linear packing.*

It is apparent that if we can construct a tight linear packing, it will be a near optimal linear packing. As we will show in Section III.D, there is an efficient way to compute tight linear packing.

III.C. Projection Region Width and Character Selection by DP

In Step 1 of Algorithm 1, in order to determine the projection region widths and characters to be put into the stencil, we merge all rows of the stencil into a single row and use dynamic programming to maximize the overall shot saving subject to a total effective character width constraint.

Assume the characters are sorted in increasing order of width. We define $S[e, i, k, w]$ as the maximum shot saving using at most k different projection region widths for printing a subset of the first i characters such that the largest projection region width is e and the total effective width of the subset is at most w. The ranges of the parameters are $w_1 + 2S \leq e \leq w_n + 2S$, $0 \leq i \leq n$, $1 \leq k \leq K$, and $0 \leq w \leq RW$.

$S[e, i, k, w]$ can be expressed recursively as follow:

$$S[e, 0, k, w] = 0 \quad \text{for all } e, k, w$$
$$S[e, i, k, 0] = 0 \quad \text{for all } e, i, k$$
$$S[e, i, k, w]$$
$$= \max \begin{cases} S[e, i-1, k, w] \\ \begin{cases} r_i(n_{VSB_i} - 1) + S[e, i-1, k, w - \frac{w_i + e}{2}] \\ \quad \text{if } (w_i + 2S \leq e \text{ and } \frac{w_i + e}{2} \leq w) \\ 0 \quad \text{otherwise} \end{cases} \\ \begin{cases} S[w_i + 2S, i, k-1, w] \quad \text{if } k > 1 \\ 0 \quad \text{otherwise} \end{cases} \\ \quad \text{for all } e, i \neq 0, k, w \neq 0 \end{cases}$$

In the recursive expression above, $S[e, i, k, w]$ is the maximum over three cases. In the first case, character i is skipped. In the second case, character i is selected. It results in a shot saving of $r_i(n_{VSB_i} - 1)$ and a reduction of remaining effective width by $\frac{w_i + e}{2}$. In the third case, another projection region width of $w_i + 2S$ is used.

Then the maximum shot saving is given by $\max_i \{S[w_i + 2S, i, K, RW]\}$. It is clear that the above recursion to compute the values of $S[e, i, k, w]$ for all e, i, k, w can be implemented as a dynamic program. However, as we will see in Section 4, the memory requirement for typical problem instances can be up to tens of gigabytes. In other words, this dynamic programming formulation is not practical.

In order to reduce the memory requirement, we take advantage of the fact that many characters have the same width. Instead of considering each character separately, we group characters of the same width together. Let G_j be the group of characters of width w_j. Assume that the characters within each group are sorted in decreasing order of shot saving.

We define $S'[e, j, k, w]$ as the maximum shot saving using at most k different projection region widths for printing some subset of characters in each of the first j groups such that the largest projection region width is e and the total effective width of the subset is at most w.

$S'[e, j, k, w]$ can be expressed recursively as follow:

$$S'[e, 0, k, w] = 0 \quad \text{for all } e, k, w$$
$$S'[e, j, k, 0] = 0 \quad \text{for all } e, j, k$$
$$S'[e, j, k, w]$$
$$= \max \begin{cases} S'[e, j-1, k, w] \\ \max_{i \in \{1, 2, \ldots, |G_j|\}} \mathcal{R}(e, j, k, w, i) \\ \begin{cases} S'[w_j + 2S, j, k-1, w] \quad \text{if } k > 1 \\ 0 \quad \text{otherwise} \end{cases} \\ \quad \text{for all } e, j \neq 0, k, w \neq 0 \end{cases}$$

where $\mathcal{R}(e, j, k, w, i)$ is the shot saving if the first i characters in G_j (i.e., the i highest shot saving characters in G_j) are included in the stencil. Let $G_j[i]$ be the set of first i characters in G_j. Then $\mathcal{R}(e, j, k, w, i)$ is given by the following expression:

$$\mathcal{R}(e, j, k, w, i)$$
$$= \begin{cases} \sum_{c \in G_j[i]} r_c(n_{VSB_c} - 1) + S'[e, j-1, k, w - i \times \frac{w_j + e}{2}] \\ \quad \text{if } w_j + 2S \leq e \text{ and } i \times \frac{w_j + e}{2} \leq w \\ 0 \quad \text{otherwise} \end{cases}$$

The three cases for $S'[e, j, k, w]$ are similar to those for $S[e, i, k, w]$ except that a set of i characters in G_j is selected instead of a single character in the second case. As the number of groups are typically at least tens of times less than the number of characters, the memory requirement to compute $S'[e, j, k, w]$ would be reduced to less than 1 gigabyte in practice as shown in Section 4. Note that this group-based dynamic program and the character-based dynamic program above generate identical solutions and have the same runtime complexity. The only difference is that the group-based approach uses much less memory.

III.D. Tight Packing Construction

Let $\mathcal{E}$ be the set of K projection region widths selected in Step 1 of Algorithm 1. The selected characters are tightly packed into the stencils in Step 2 of Algorithm 1 by Procedure 1 below:

Procedure 1 Character Packing

1: For each character i, set E_i to be the smallest value in $\mathcal{E}$ such that $E_i \geq w_i + 2S$.
2: Sort the selected characters in increasing order of $E_i - w_i$.
3: **for** each character i in sorted order **do**
4:     Starting from the next row after the last packed character, find a row which i can be tightly packed at the end of it. Ignore character i if it cannot be packed to any row.
5: **end for**

In general, the projection region width E_i for each character i can be set to any value as long as it is at least $w_i + 2S$. However, E_i should be as small as possible to facilitate the packing of more characters into a row. Hence, E_i is set as in Step 1 above. According to Lemma 4 below, tight packings can be constructed by arranging the characters in increasing order of $E_i - w_i$. Hence, we can simply sort the characters in Step 2, then distribute the characters and tightly pack them to the rows in Steps 3-5.

Lemma 4. *Given a group of k characters, a tight linear packing can be constructed if the characters are ordered such that $E_i - w_i \leq E_{i+1} - w_{i+1}$ for $i = 1, \ldots, k - 1$.*

Proof. We can prove the lemma by induction on k.
1. The base case that $k = 1$ is trivially true.
2. We show that if the statement is true for $k = k' - 1$, then it is also true for $k = k'$. Suppose the statement is true for $k = k' - 1$ and we have a group of k' characters. We can construct a tight linear packing for the k' characters as follows. First, we sort the k' characters such that $E_i - w_i \leq E_{i+1} - w_{i+1}$ for $i = 1, \ldots, k' - 1$. By the assumption, a tight linear packing $P_{1..k'-1}$ can be formed for the first $k' - 1$ characters in the sorted order. We can append the k'-th character to the right end of $P_{1..k'-1}$ in such a way that the left blank space of the k'-th character completely overlap with the right blank space of the $(k' - 1)$-th character as shown in Fig. 6. It follows that the left blank space of the k'-th character must be no smaller than the safety margin S since the right blank space of the $k'-1$ character is at least S. But it remains to show that the resultant right blank space of the k'-th character is also at least S for the constructed packing to be a feasible tight linear packing. Let δ denote the left blank space of the k'-th character (equivalently, the right blank space of the $(k'-1)$-th character). The resultant right blank space of the k'-th character is equal to $E_{k'} - w_{k'} - \delta$. Since $E_{k'} - w_{k'} \geq E_{k'-1} - w_{k'-1}$, so $E_{k'} - w_{k'} - \delta \geq E_{k'-1} - w_{k'-1} - \delta \geq S$. Hence, the constructed packing is a tight linear packing. $\square$

After trying to pack all selected characters into the stencil, we further improve the shot saving in Step 3 of Algorithm 1 by trying to pack all unselected characters using the same packing procedure in Procedure 1.

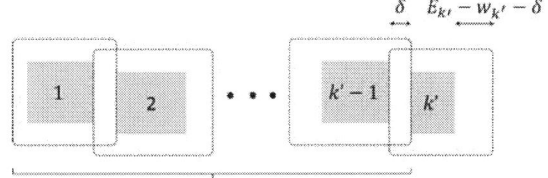

Fig. 6. Tight linear packing construction.

IV. EXPERIMENTAL RESULTS AND CONCLUSIONS

We implemented our approach in C and obtained the executable codes of [9] and [10] for comparison. All experiments were done on a Linux server powered by a 2.67 GHz Intel processor with 47 GB of memory.

In the first experiment, benchmarks 1D-1 to 1D-4 from [10] were used. The available character area of the stencil is $1000\mu m \times 1000\mu m$, and the number of character candidate in each benchmark is 1000. We set S to the minimum left/right blank space of the original characters in each benchmark. Recall that [9] and [10] assume that there is only a single shaping aperture and hence a single projection region width (i.e., $K = 1$). For our approach, we tried $K = 1$ and $K = 2$. Table I reports the comparison on total shot count, number of characters put on the stencil, and runtime. It also reports the shot count when using VSB only for reference.

Refer to columns 2, 3, 6, and 9 of Table I, it can be seen that even for $K = 1$ (i.e., all characters must use the same projection region width), our approach can reduce the shot count by $27.472\times$, $3.09\times$ and $1.65\times$ over VSB only, [9] and [10], respectively. The significant improvement over [9] and [10] is partially because they perform only simple blank space sharing while our approach performs flexible blank space sharing. Besides, their algorithms do not attempt to find the optimal projection region width while ours does.

Next, if we use two different sized shaping apertures (i.e., $K = 2$), a huge shot count reduction over using single shaping aperture can be obtained and we already can put all character candidates into the stencil for benchmarks 1D-1 and 1D-2. Our approach with $K = 2$ results in as much as $8.97\times$ and $4.28\times$ shot count reduction compared to [9] and [10], respectively.

For more testing, we generated some harder benchmarks (1D-1h to 1D-4h). We generated 200 extra character candidates into each of the original benchmarks while keeping the same stencil size, so it became impossible to put all characters on a stencil. Table II reports the results of our algorithm with $K = 1$, $K = 2$, and $K = 3$. It also reports the shot count when using VSB only for reference. As expected, the shot count reduction and the number of characters that can be put on the stencil increase with the value of K. And the greatest reduction occurs when K switches from 1 to 2.

We show in Table III the memory requirement of our program. The proposed character grouping technique in Section III.C can reduce the memory requirement of the dynamic program by roughly $40\times$. The resultant memory requirement after adopting the technique is less than 1 GB in each case.

Finally, we note that our algorithm also works for multibeam direct write system [1], where multiple beam columns

TABLE I
COMPARISON WITH [9] AND [10].

	VSB only	[9]			[10]			Ours ($K = 1$)			Ours ($K = 2$)		
	#shots	#shots	#ch	CPU(s)	#shots	#ch	CPU(s)	#shots	#ch	CPU(s)	#shots	#ch	CPU(s)
1D-1	770543	50809	926	12.13	29536	934	1.91	12972	980	6.90	10418	1000	22.21
1D-2	770543	93465	854	10.24	44544	863	1.74	28594	895	6.41	10418	1000	20.73
1D-3	770543	152376	749	7.85	78704	758	2.35	55761	797	6.59	30785	902	21.13
1D-4	770543	193494	687	6.52	107460	699	2.96	79275	734	7.47	44468	837	29.07
Normalized	27.472	3.090	0.944	1.355	1.650	0.955	0.325	1.000	1.000	1.000	0.570	1.102	3.388

TABLE II
RESULTS ON HARDER BENCHMARKS BY OUR ALGORITHMS FOR DIFFERENT VALUES OF K.

	VSB only	Ours ($K = 1$)			Ours ($K = 2$)			Ours ($K = 3$)		
	#shots	#shots	#ch	CPU(s)	#shots	#ch	CPU(s)	#shots	#ch	CPU(s)
1D-1h	922770	58648	980	8.82	26467	1114	27.22	17534	1163	43.33
1D-2h	922770	86176	905	8.43	48891	1018	25.58	39630	1068	40.67
1D-3h	922770	135332	800	8.63	93109	916	25.69	75709	948	40.89
1D-4h	922770	169105	739	9.76	116219	855	27.02	98204	886	42.91
Normalized	9.680	1.000	1.000	1.000	0.598	1.141	2.966	0.475	1.188	4.718

TABLE III
MEMORY USAGE (GB) WITH AND WITHOUT USING THE MEMORY SAVING TECHNIQUE.

	With memory saving			Without memory saving		
	$K = 1$	$K = 2$	$K = 3$	$K = 1$	$K = 2$	$K = 3$
1D-1	0.136	0.271	0.407	5.236	10.436	15.670
1D-2	0.131	0.262	0.392	5.044	10.087	15.092
1D-3	0.141	0.282	0.422	5.041	10.082	15.087
1D-4	0.156	0.311	0.467	5.205	10.377	15.582
1D-1h	0.136	0.271	0.407	6.282	12.564	18.800
1D-2h	0.131	0.262	0.392	6.051	12.102	18.154
1D-3h	0.141	0.282	0.422	6.048	12.096	18.144
1D-4h	0.156	0.311	0.467	6.245	12.490	18.696

furnished with their own stencils write different regions of a wafer in parallel. As identical dies are typically manufactured on a wafer, the stencil design for all beam columns in a multi-beam system should be the same and is not different from a single beam system.

REFERENCES

[1] T. Maruyama, Y. Machida, and S. Sugatani. CP based EBDW throughput enhancement for 22nm high volume manufacturing. In *Proceedings of SPIE 7637*, February 2010. 7637-1S.

[2] T. Maruyama et al. CP element based design for 14nm node EBDW high volume manufacturing. In *Proceedings of SPIE 8323*, April 2012. 8323-14.

[3] B. J. Lin. Future of multiple-E-beam direct-write systems. In *Proceedings of SPIE 8323*, March 2012.

[4] R. Inanami et al. Maskless lithography: Estimation of the number of shots for each layer in a logic device with character projection-type low-energy electron-beam direct writing system. In *Proceedings of SPIE 5037*, pages 1043–1050, 2003.

[5] H. Pfeiffer. Variable spot shaping for electron-beam lithography. *Journal of Vaccum Sci. and Tech.*, 15(3):887–890, May 1978.

[6] M. Sugihara et al. A character size optimization technique for throughput enhancement of character projection lithography. In *Proc. of ISCAS*, pages 2561–2564, 2006.

[7] A. Fujimura. Design for E-beam: Design insights for direct-write maskless lithography. In *Proceedings of SPIE 7823*, pages 137–140, Sep. 2010.

[8] K. Yoshida et al. Stencil design and method for improving character density for cell projection charged particle beam lithography. US Patent Application No. 2009/0325085 A1, December 2009.

[9] K. Yuan, B. Yu, and D. Z. Pan. E-beam lithography stencil planning and optimization with overlapped characters. *IEEE Trans. on Computer-Aided Design of Integrated Circuits and Systems*, 31(2):167–179, Feb 2012.

[10] B. Yu, K. Yuan, J.-R. Gao, and D.Z. Pan. E-BLOW: e-beam lithography overlapping aware stencil planning for MCC system. In *Proc. of DAC*, 2013.

[11] R. Inanami. Electron beam exposure apparatus, electron beam exposure method and method of manufacturing semiconductor device. US Patent No. 7449700, November 2008.

[12] M. R. Garey and D. S. Johnson. *Computers and Intractability: A Guide to the Theory of NP-Completeness*. Freeman, NY, 1979.

Self-aligned Double Patterning Layout Decomposition with Complementary E-Beam Lithography

Jhih-Rong Gao, Bei Yu, and David Z. Pan
Dept. of ECE, The University of Texas at Austin, Austin, TX 78712
Email: {jrgao, bei, dpan}@cerc.utexas.edu

Abstract— Advanced lithography techniques enable higher pattern resolution; however, techniques such as extreme ultraviolet lithography and e-beam lithography (EBL) are not yet ready for high volume production. Recently, complementary lithography has become promising, which allows two different lithography processes work together to achieve high quality layout patterns while not increasing much manufacturing cost. In this paper, we present a new layout decomposition framework for self-aligned double patterning and complementary EBL, which considers overlay minimization and EBL throughput optimization simultaneously. We perform conflict elimination by merge-and-cut technique and formulate it as a matching-based problem. The results show that our approach is fast and effective, where all conflicts are solved with minimal overlay error and e-beam utilization.

I. INTRODUCTION

The semiconductor industry demands advanced lithography to enable small feature size and high density designs. Currently, 193nm immersion lithography (193i) is still the mainstream for 32nm and 22nm technology nodes. Incorporating double patterning lithography (DPL) and multiple patterning lithography (MPL) can keep pushing the resolution limit [1]. However, DPL/MPL requires strict layout compliance, and to fabricate a layout with multiple masks increases the design complexity. Although some leading lithography technologies, such as extreme ultraviolet lithography (EUVL) [2], e-beam lithography (EBL) [3], direct self-assembly [4], etc., achieve good pattern quality, they are not yet ready for high volume production.

Complementary lithography is proposed to allow 193nm optical lithography work hand-in-hand with high-resolution lithography to enable advanced designs [5]. In the first step, base features are created by cheaper optical lithography or self-aligned double patterning (SADP); in the second step, high-resolution lithography techniques are applied to cut unnecessary lines. Such line cutting can be accomplished by costly quadruple patterning, EUVL, or EBL. By carefully arranging how features are generated with the combined lithography techniques, we can achieve good pattern quality with a reasonable manufacturing cost. The advantages of adopting complementary lithography include: (1) high throughput by generating base patterns with mature optical lithography; and (2) improved mask yield by partial EUVL or EBL patterning, while no heavy manufacturing cost introduced.

Recently, the technique to combine optical and complementary EBL becomes promising. Lam et al. [6,7] proposed using EBL to complement 193nm immersion lithography for 1D layout. The choice of SADP enables better overlay control compared with conventional Litho-Etch-Litho-Etch type double patterning. EBL is a maskless lithography which directly writes high-resolution patterns into the silicon wafer with e-beams. Although having their own advantage, standalone SADP and EBL are limited by low manufacturing flexibility and low throughput, respectively. By combining SADP with EBL together, we can improve manufacturability because EBL provides higher resolution; in the mean time, we need to reduce EBL utilization to ensure high productivity. This requires an overall optimization to consider complementary technique in the layout decomposition flow.

Several studies [6–8] have presented the effectiveness of applying SADP with line cutting technique on 1D layout designs. In order to optimize the overall throughput with hybrid SADP and EBL, an integer linear programming (ILP) -based approach [9] was proposed by properly distributing cutting patterns to the optical mask and e-beams. However, this work only targets at 1D gridded designs and allows wire-end extension that is not always permitted for general designs. There have been some studies [10–13] presented for pure SADP layout decomposition of 2D random patterns, where layout decomposition is the process to assign layout features into two different fabrication steps. These approaches impose strict SADP process rules to ensure the decomposed layout is SADP-manufacturable. However, the design flexibility is restricted and the layout may still not be decomposable given a complex layout.

In this paper, we solve 2D layout decomposition problem that enables SADP with complementary EBL. To the best of our knowledge, there is no existing study considering SADP with complementary lithography on 2D designs. In addition, we provide a systematic approach that allows conflict minimization during SADP layout decomposition. Our main contributions include:

- We present a new layout decomposition framework for SADP and complementary EBL, which considers overlay minimization and EBL throughput optimization simultaneously.

- We propose a new graph formulation and a matching-based approach that allows eliminating conflicts by the merge-and-cut technique.

- We show that for pure SADP layout decomposition problem, our approach can be adapted to minimize conflicts with overlay consideration.

- The results show that our approach is very efficient, and that all conflicts can be eliminated with minimal overlay error and e-beam utilization.

In the rest of this paper, we will introduce SADP process

978-1-4799-2817-0/14 $31.00 © 2014 IEEE

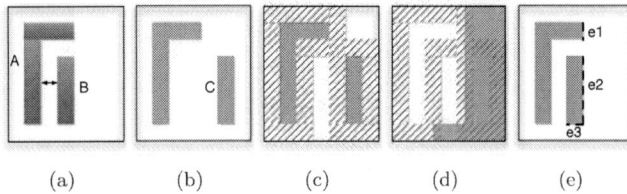

(a)   (b)   (c)   (d)   (e)

Fig. 1. SADP process. (a) Target layout patterns. (b) The mandrel mask. (c) Spacer deposition. (d) Mandrel removal and the trim mask. (e) Final patterns.

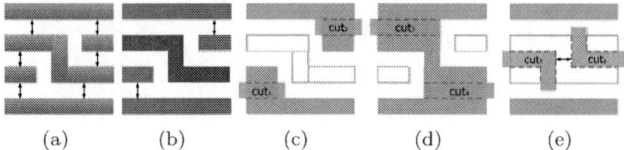

(a)   (b)   (c)   (d)   (e)

Fig. 2. Merge-and-cut example. (a) Target layout. (b) Two-coloring result. (c)~(e) Layout decomposition with merge-and-cut. Red lines show boundaries with overlay error risk.

and the layout decomposition problem in Sec. II. In Sec. III, we present our *face graph* formulation that embeds SADP constraints as well as the solution candidates for solving conflicts. Our layout decomposition that performs simultaneous overlay and EBL throughput optimization is presented in Sec. IV. We then explain an adapted conflict minimization approach that can be used for pure SADP in Sec. V. Finally, we will show experimental results in Sec. VI, followed by the conclusion in Sec. VII.

II. SADP AND CONFLICT SOLVING TECHNIQUE

A. SADP Process Overview

In SADP, the double patterning spacing S_{dp} restricts the minimum spacing between two patterns on the same mask. Any two patterns with distance less than S_{dp} must not be fabricated on the same masks, otherwise, it is called a *conflict*. In general, the layout decomposition process involves decomposing layout patterns into two sets; one is defined by the mandrel mask, and the other is co-defined by spacers and the trim mask.

Fig. 1 shows the SADP process, where the arrow indicates a conflict between the two target patterns, meaning they cannot be fabricated on the same mask. Part of the target layout is first defined by the mandrel mask as shown in Fig. 1(b). Pattern C is called an *assist mandrel*, which helps to define target patterns but will not appear on the final layout. Next, a spacer material is deposited around the boundary of the mandrels as shown in the slashed area in (c). The mandrels will then be removed as shown in (d). After that, the second mask, trim mask shown in the green area, will be applied to block the undesired layout region. A metal filling process will then fill the white area so that the final layout in (e) is obtained. We call pattern A a *mandrel pattern* since it is defined by the mandrel mask, and pattern B a *non-mandrel pattern*.

To achieve a valid layout decomposition, all patterns on the mandrel mask and the trim mask must satisfy the minimum spacing S_{dp}. One issue with SADP process is that the trim mask may not perfectly aligned to the mandrel mask. Consequently, the overlay error occurs and it may cause line-end or CD variation.

B. Merge-and-cut Technique

The layout decomposition problem is usually formulated as a two-coloring problem, where conflicting patterns must be assigned different colors. One color will be defined by mandrel pattens, as A in Fig. 1, while the other will be defined by non-mandrel patterns, as B. A two-coloring result of Fig. 2(a) is shown in 2(b).

The challenge of SADP layout decomposition is that two-coloring method may not necessarily avoid all conflicts. To further eliminate conflicts, *merge-and-cut* technique is utilized

[11, 14] to merge two conflicting patterns and then trim out the unwanted part by the trim mask. Fig. 2(b) shows two conflicts remaining after two-coloring, and thus we cannot generate those patterns by the mandrel and trim mask directly. Fig. 2(c)~(e) show possible merge-and-cut solutions by merging two conflicting patterns and then cutting the unwanted area by the *cutting patterns* cut_1~cut_6 defined by the trim mask. The pattern boundaries that directly touch the trim patterns would have potential overlay error; meanwhile, cutting patterns cannot violate the minimum spacing S_{dp} such as (e). Therefore, merge-and-cut solutions should be selected appropriately such that the cutting boundaries/overlay are as small as possible.

III. GRAPH FORMULATION WITH EMBEDDED SADP CONSTRAINTS AND CONFLICT SOLVING SOLUTIONS

Our graph formulation, *face graph*, is constructed by the flow shown in Fig. 3, and is explained in the following.

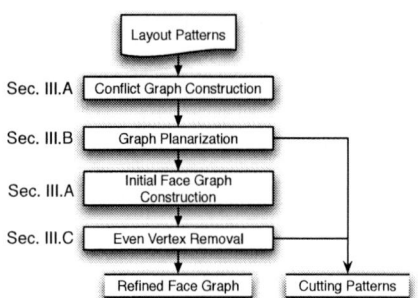

Fig. 3. Face graph construction flow.

A. Conflict Graph and Face Graph Construction

Given a 2D layout, we construct a conflict graph $G_c = (V_c, E_c)$ to express the relationship among layout patterns. Each vertex $v_c \in V_c$ represents a pattern, and each edge $e_c \in E_c$ is constructed when the distance between two patterns is less than S_{dp}. Fig. 5(a) show the conflict graph of the layout in Fig. 4(a). It has been shown that in two-coloring problem, a conflict occurs only when there is an odd cycle in the conflict graph [15]. To achieve a valid SADP layout decomposition, we have to eliminate all odd cycles in G_c.

Based on G_c, we define its dual graph, face graph $G_f = (V_f, E_f)$ where $V_f = V_{face} \cup V_{dummy}$. A vertex $v_{face} \in V_{face}$ corresponds to a face in G_c except the exterior face, and a *dummy vertex* v_{dummy} is created for each merge-and-cut candidate of a conflict. An edge $e_f \in E_f$ connects v_{face} and v_{dummy} if v_{dummy} is the solution candidate to solve the corresponding conflict of v_{face}. We call v_{face} as an *even (odd) vertex* if it corresponds to an even (odd) face in G_c. The initial face graph of the layout in Fig. 4(a) is shown in red in Fig. 5(b).

978-1-4799-2817-0/14 $31.00 © 2014 IEEE   144

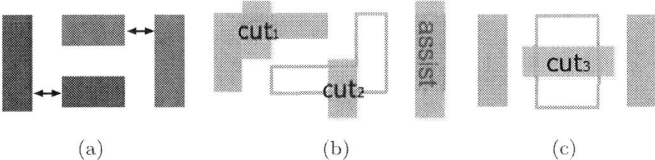

Fig. 4. (a) Target layout. (b)(c) Merge-and-cut solutions.

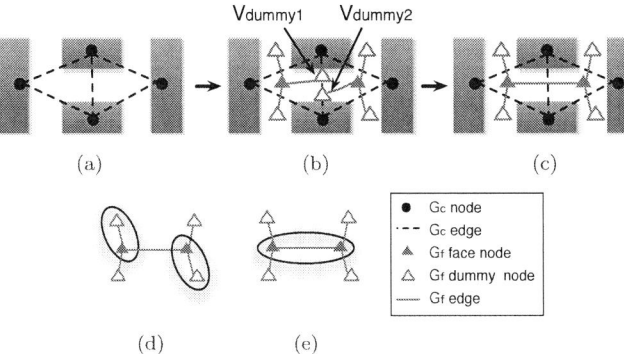

Fig. 5. (a)(b)(c) Conflict graph (black) and face graph (red) example. (d)(e) Matching results.

For adjacent odd vertices v_{f1} and v_{f2}, they may share the same merge-and-cut candidates. For example, v_{dummy1} and v_{dummy2} in Fig. 5(b) both refer to cut_3 in Fig. 4(c). In the case where $e_{f1} = (v_{f1}, v_{dummy1})$ and $e_{f2} = (v_{f2}, v_{dummy2})$ refer to the same merge-and-cut candidate, we combine e_{f1} and e_{f2}, and remove v_{dummy1} and v_{dummy2}, as shown in Fig. 5(c).

B. Conflict Graph Planarization

It has been shown in [16,17] that the planarity of the conflict graph is based on the setting of S_{dp}. The conflict graph is planar only if Eq. (1) is satisfied:

$$\left\{ \begin{array}{ll} S_{dp} < 2 \times S_{min} & \text{in the Manhattan distance} \\ S_{dp} < \sqrt{2} \times S_{min} & \text{in the Euclidean distance} \end{array} \right. \quad (1)$$

, where S_{min} is the minimum spacing between patterns on the layout. In the case that Eq. (1) is violated, we need to planarize G_c since G_f cannot be constructed based on a non-planar graph.

If a conflict graph G_c is highly non-planar, it implies that several patterns conflict with multiple patterns. Therefore it is less possible to find a valid DPL decomposition. Be assuming that the non-planar cases in the give layout is limited, we apply the following heuristic to solve the non-planar subgraph. In a non-planar graph G_c, assume $e_1 \in E_c$ and $e_2 \in E_c$ cross each other, we eliminate one of the two edges by merging their connected vertices. Conceptually, this planarization means we force two patterns to be merged to prevent a non-planar case. In order to minimize the overlay error and EBL cost of merging two patterns, the edge with smaller merging cost (defined in Sec. IV-B) will be eliminated.

C. Even Vertex Removal for Face Graph

Given an edge $e_f = (v_{face}, v_{dummy}) \in G_f$, it implies that we can use the merge-and-cut candidate e_f to reduce the degree of the corresponding face of v_{face} by one. Since V_f contains vertices from all faces, our merge-and-cut candidate may either make an odd face become an even face (meaning the conflict

---

Algorithm 1 RemoveEvenVertex

1: $F_{odd} \leftarrow$ all odd faces in G_c
2: **for all** $f \in F_{odd}$ **do**
3:      $\Gamma(f)_{even} \leftarrow$ even adjacent faces of f
4:      $\Gamma(f)_{odd} \leftarrow$ odd adjacent faces of f
5:      **if** $\Gamma(f)_{even} \neq \phi$ **then**
6:          Remove $v_i \in G_f, \forall f_i \in \Gamma(f)_{even}$
         where v_i is the corresponding face vertex of f_i
7:      **else**
8:          Solve f by the min-cost merge-and-cut candidate as Sec. IV-B
9:      **end if**
10: **end for**

---

is solved), or make an even face become an odd face (meaning a new conflict is introduced). Because applying merge-and-cut increases the risk of overlay error on the cutting boundaries, we would like to minimize new conflicts introduced by merge-and-cut. With this motivation, we apply a vertex removal heuristic in Algorithm 1 to greedily remove even vertices in G_f.

IV. LAYOUT DECOMPOSITION WITH SADP AND COMPLEMENTARY EBL

A. SADP with Complementary E-beam Lithography

A limitation of applying the merge-and-cut technique with the trim mask is that the distance between cutting patterns may violate the minimum DPL spacing S_{dp}. For example, the solution in Fig. 2(e) requires two cutting patterns cut_5 and cut_6, which actually conflict with each other.

EBL enables smaller feature width and spacing, and thus it achieve better pattern quality and design flexibility than SADP. However, EBL throughput is its biggest bottleneck as the write time is determined by the number of e-beam shots. Therefore, extensive use of e-beam cutting is not practical for manufacturing.

We adopt the conventional e-beam system where e-beam shots are variable-shaped (rectangular) beams (VSB). Cutting patterns formed by VSB require layout fracturing, meaning the patterns are decomposed into non-overlapping e-beam shots/rectangulars.

B. Min-Cost Matching based Conflict Elimination

In SADP layout decomposition problem, our objective is to eliminate conflicts with minimal overlay error and EBL utilization. We first explain our min-cost matching based conflict elimination algorithm on the *face graph*. Then we discuss how to utilize this algorithm for the overlay and EBL co-optimization, including a post processing based approach (Sec. IV-B.1) and a simultaneous optimization (Sec. IV-B.2).

The face graph defined in Sec. III has the following property:

PROPERTY A. An edge $e_f = (v_{odd}, v_{dummy}) \in E_f$ maps to a merge-and-cut candidate of the corresponding conflict of v_{odd}, where merge-and-cut reduces the degree of v_{odd} by one.

According to Property A, we can solve a conflict corresponding to an odd vertex by selecting one of its connecting edges $e_f \in E_f$. It is obvious that selecting more than one e_f for an odd vertex is unnecessary because a new conflict would be introduced. Therefore, we seek to find one merge-and-cut

Algorithm 2 Hybrid-Post

Input: $G_f = (V_f, E_f)$ // Sec. III
Output: $P_{cut} = P_{optical} \cup P_{ebl}$, with the objective of
    minimizing $\sum_{e_i \in P_{optical}} cost_{e_i}$ and $\sum_{e_j \in P_{ebl}} N_{shot}(e_j)$
1: AssignCost_SADP(E_f)
2: $P_{allcuts} \leftarrow$ RunMatching(G_f)
3: $G_{mc} \leftarrow$ ConstructConflictGraph($P_{allcuts}$)
4: $P_{optical} \leftarrow MIS(G_{mc})$
5: $P_{ebl} \leftarrow P_{allcuts} - P_{optical}$

Algorithm 3 Hybrid-Sim

Input: $G_f = (V_f, E_f)$ // Sec. III
Output: $P_{cut} = P_{optical} \cup P_{ebl}$, with the objective of
    minimizing $\sum_{e_i \in P_{optical}} cost_{e_i}$ and $\sum_{e_j \in P_{ebl}} N_{shot}(e_j)$
1: AssignCost_SADP(E_f)
2: $P_{conf} = \Phi$
3: **repeat**
4:    $P_{optical}, P_{ebl} \leftarrow$ RunMatching(G_f)
5:    $P_{conf} \leftarrow$ ValidateCut($P_{optical}$)
6:    SubstituteEBL(P_{conf})
7: **until** $P_{conf} == \Phi$

---

solution for each conflict, and we formulate the conflict elimination problem as the matching problem, where each match corresponds to solving a conflict. For example, Fig. 5(d) and (e) show two different matching results which corresponds to the final masks shown in Fig. 4(b) and (c), respectively.

B.1 Post Processing Based Conflict Elimination

Since the trim mask itself can be used for the merge-and-cut technique to solve conflicts, we can view EBL cutting as a back-up solution during conflict elimination. We propose a two-stage approach for overlay error and e-beam optimization. First, we solve all conflicts by applying the min-cost matching algorithm on G_f, then we assign the obtained cutting patterns to the trim mask and e-beam shots according to SADP constraint S_{dp}. The approach flow is shown in Algorithm 2, where $N_{shot}(e)$ represent the required number of e-beam shots for the merge-and-cut candidate corresponds to e.

In the beginning of Algorithm 2, G_f is constructed based on Sec. III. Since EBL is not considered in the first stage, we model the edge cost simply by the SADP overlay error according to Eq. (2) in Line 1.

$$cost_e = L_{boundary}(e) \quad \forall e \in E_f \tag{2}$$

where $L_{boundary}$ is the boundary length of the corresponding cutting pattern of e.

We then solve all conflicts by merge-and cut technique (Line 2). The cutting patterns $P_{allcuts}$ with the minimum overlay can be obtained by performing the min-cost matching algorithm in Line 2. However, there may exist conflicts among the cutting patterns because of the S_{dp} constraint. With e-beam available, we can carefully select a subset of cutting patterns $P_{optical}$ that do not violate S_{dp}, and let the rest of the cutting patterns P_{ebl} formed by EBL. We first construct a conflict graph G_{mc} for $P_{allcuts}$ in Line 3 to check if there is any conflict among $P_{allcuts}$. In order to minimize e-beam shot utilization, we apply maximal independent set (MIS) algorithm on G_{mc} to obtain the maximal number of valid patterns for $P_{optical}$ (Line 4), and assign the rest of the cutting patterns to be done by EBL (Line 5). Based on the property of MIS, the S_{dp} constraint is guaranteed to be satisfied for $P_{optical}$.

B.2 Conflict Elimination with Simultaneous Overlay and EBL Throughput Optimization

Although the approach in Sec. IV-B.1 can successfully solve conflicts with hybrid SADP and EBL, it only considers EBL in the last stage and does not include e-beam optimization when finding the min-cost matching. To further improve the decomposition result, we propose a simultaneous overlay error and EBL throughput optimization as shown in Algorithm 3. The main idea of the algorithm is to start from a restricted solution

space and gradually increase the solution space with more EBL merge-and-cut candidates until we find a valid solution. Based on the algorithm, only necessary e-beam candidates are considered, and the matching algorithm can simultaneously optimize SADP overlay error and e-beam utilization.

In the beginning of Algorithm 3, G_f is constructed based on Sec. III. All edges are initialized as optical cuts with the cost defined by Eq. (2). We then iteratively perform min-cost matching algorithm in Line 4 to find the cutting patterns. Since we may obtain conflicts among some optical cutting patterns in Line 5, we substitute those conflicting optical cuts by EBL cuts in Line 6 with the cost function in Eq. (3).

$$cost_e = C_{ebl} \times N_{shot}(e) \quad \forall e \in P_{conf} \tag{3}$$

where C_{ebl} is a user-defined parameter to control the cost of a e-beam shot.

In our implementation, we set C_{ebl} sufficiently large than the cost of any optical cut, such that optical cuts are always preferred than EBL cuts. By including both overlay and e-beam cost with Eq. (2) and Eq. (3), our min-cost matching solution can minimize the overlay error and e-beam utilization simultaneously. Note that we assume rectangular beam shape is applied. For example, an L-shaped pattern requires at least two beam shots. In addition, there are minimum size and maximum size limitation for beam shape, which would also affect the number of shots N_{shot} of each merge-and-cut candidate. Because a merge-and-cut candidate is formed between patterns with half-pitch width, generally the minimum shape constraint would not be violated; and the maximum shape constraint would apply for long cutting patterns.

V. OVERLAY-AWARE CONFLICT MINIMIZATION FOR PURE SADP

The approaches discussed so far are targeted at the layout decomposition for hybrid SADP and EBL. We find that our approach can be adapted for conflict minimization in pure SADP layout decomposition. Since there is no previous study that optimize cutting patterns in SADP layout decomposition, this approach can be very useful when the layout is highly complicated and complementary lithography is not available.

A. Adapted Face Graph for Conflicting Cuts

In the face graph defined in Sec. III, the merge-and-cut candidates may conflict each other. Therefore, when applying the min-cost matching algorithm to solve conflicts as explained in Sec. IV-B, we may obtain conflicting cutting patterns. These

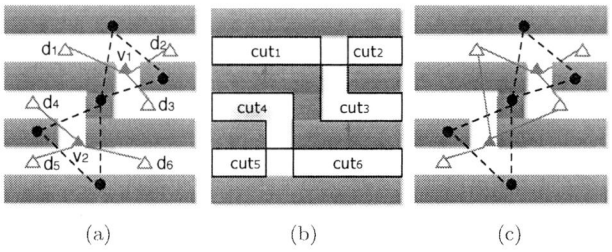

Fig. 6. Face graph adapted for conflicting cuts. (a) Conflict graph (black) and Face graph (red). (b) Cutting patterns that cannot co-exist are indicated by arrows. (c) Adapted face graph without conflicting cuts.

conflicts must be prevented in pure SADP layout decomposition.

Fig. 6(a) shows the layout patterns and its corresponding conflict and face graph. Two conflicts v_1 and v_2 are discovered because they forms odd cycles in the conflict graph. The merge-and-cut candidates of v_1 and v_2 are shown by $cut_1 \sim cut_3$ and $cut_4 \sim cut_6$ in Fig. 6(b), respectively. It can be seen that if we select cut_1 and cut_4 to solve the two conflicts, the solution would not be valid because the two cuts are too close to be fabricated on the same mask. Similarly, cut_3 and cut_6 cannot co-exist.

To add the conflicting cuts information into our face graph, we check all merge-and-cut candidates by traversing all edges, and make conflicting edges connect to the same dummy vertex. Finally, we can obtain an adapted face graph G'_f with the conflicting cuts information embedded. As shown in Fig. 6(c), after this graph adaption, d_1 and d_4 is merged, and so is d_3 and d_6.

B. Matching based Conflict Minimization

The adapted face graph G'_f has the following property:

PROPERTY B. If two merge-and-cut candidates cannot co-exist because their spacing is less then S_{dp}, the corresponding edges e_{f1} and e_{f2} connect to the same $v_{dummy} \in V_{dummy}$.

Property B guarantees that the matching algorithm would only select one edge that covers the same dummy vertex. This ensures that S_{dp} rule is satisfied for the cutting patterns. Besides, Property A in Sec. IV-B still holds for the adapted face graph. Consequently, by modeling the edge cost of G'_f with the desired SADP cost, we can minimize conflicts with the min-cost matching algorithm presented in Sec. IV-B. Here our objective is to minimize overlay error introduced by cutting patterns, therefore, Eq. (2) is adopted in the matching problem. This approach allows simultaneous optimization for overlay minimization and conflict minimization for pure SADP layout decomposition.

VI. EXPERIMENTAL RESULTS

The proposed algorithms are implemented in C++ and tested on Intel platform with 2.66 GHz CPU and 4G memory. We synthesize OpenSPARC T1 designs with Nangate 45nm standard cell library [18], and perform placement and routing with Cadence SOC Encounter [19] to generate the layouts. These layouts are then scaled down for 22nm technology node. For simplicity, we assume the sizes of the minimum pattern width, spacing, and spacer width are 50nm, and make the

corresponding adjustment for the benchmark. The double patterning spacing S_{dp} is set large enough to introduce conflicts to evaluate the performance of our algorithms. Table I shows the statistics of five designs with the number of 2D patterns (#Polygon) and the initial coloring conflicts (#Conf) before applying our approaches, where #Conf is obtained based on the two-coloring result.

A. Overlay-aware Layout Decomposition for SADP

We first apply the proposed approach in Sec. V to solve conflicts with the trim mask for conflict and overlay error minimization. Because the existing approaches for SADP layout decompositions [10–13] are performed for two-colorable cases, no solution would be generated for comparison on designs with conflicts. Although layout perturbation [20] can be applied to solve native conflicts, we do not allow layout change in our problem. An alternative to solve this problem is to minimize the total number of cutting patterns, by which we expect less overlay error and unsolved conflicts because less cutting patterns compete for the mask resource. As a baseline, we implement this alternative (SADP-Cut) by replacing Eq. (2) with Eq. (4) and compare it with the proposed overlay-aware layout decomposition (SADP-OV).

$$cost_e = 1 \quad \forall e \in E_f \tag{4}$$

Table I shows the results after layout decomposition in terms of the remaining conflicts (#Conf$_{rem}$), the total length of the overlay-risky boundaries touched by the trim mask ($Bndy_{ov}$), and the CPU time. It can be seen that the merge-and-cut technique is not sufficient to solve all conflicts because of the resolution limit of the trim mask. Compared with the baseline, SADP-OV successfully reduces the overlay error, whose effect can be represented by the length of the cutting boundaries. On average, the overlay error can be reduced by 28.36% with SADP-OV. Although there are still outstanding conflicts that cannot be resolved, our approach successfully resolve more than 95% of the initial conflicts, showing that merge-and-cut is promising for SADP layout decomposition. Note that our benchmarks are not targeting at any specific lithography process and thus are not SADP-friendly designs. By properly designing the layout for SADP or specify lithography-aware rules in early design stages, it would be easier to solve conflicts by our approach.

B. Overlay and EBL Throughput Co-optimization for SADP with Complementary EBL

By adopting complementary EBL, the conflicts that cannot be handled in Sec. VI-A can be solved. We compare the layout decomposition results when applying the two conflict elimination approaches, Hybrid-Post and Hybrid-Sim. Because Hybrid-Post is a two-stage approach, we would like to study how the result in the first stage affects the final result. Therefore, two versions of Hybrid-Post are implemented, one perform the min-cost matching algorithm based on Eq. (2) (Hybrid-Post-OV), while the other perform the min-cost matching algorithm based on Eq. (4) (Hybrid-Post-Cut).

The results are shown in Table II, where #VSB refers to the total number of variable shaped beams. Note that the cutting patterns are 2-dimensional and thus a conflict may require more than one VSB to solve. With complementary EBL, all conflicts in our benchmark are solved. It can be seen that Hybrid-Post-Cut requires a large number of VSB because it

978-1-4799-2817-0/14 $31.00 © 2014 IEEE

TABLE I
Overlay-aware layout decomposition for SADP.

Design	#Polygon	#Conf	SADP-Cut			SADP-OV			
			$\#Conf_{rem}$	$Bndy_{ov}$ (um)	CPU (s)	$\#Conf_{rem}$	$Bndy_{ov}$ (um)	CPU (s)	OV Imp%
alu	9792	1992	46	541.98	1.57	60	311.79	1.64	42.47%
byp	27675	5015	274	1413.04	2.56	25	1129.90	2.83	20.04%
div	20501	3914	222	901.16	3.32	39	680.20	3.36	24.52%
ecc	7922	2282	104	575.29	1.36	24	430.48	1.42	25.17%
efc	7173	1988	91	503.44	1.15	96	354.47	1.14	29.59%
Average									28.36%

TABLE II
Layout decomposition for Overlay and EBL throughput co-optimization

Design	Hybrid-Post-Cut			Hybrid-Post-OV			Hybrid-Sim		
	#VSB	$Bndy_{ov}$ (um)	CPU (s)	#VSB	$Bndy_{ov}$ (um)	CPU (s)	#VSB	$Bndy_{ov}$ (um)	CPU (s)
alu	240	541.98	1.59	219	311.79	1.66	24	329.00	1.85
byp	1347	1413.04	2.63	60	1129.90	2.91	11	1133.85	3.36
div	1249	901.16	3.37	119	680.20	3.41	72	685.81	3.77
ecc	479	575.29	1.38	72	430.48	1.45	37	433.41	1.52
efc	561	503.44	1.16	335	354.47	1.16	54	378.88	1.36
Avg Ratio	8.47	1.41	0.96	1.00	1.00	1.00	0.31	1.03	1.12

leaves more unsolved conflicts before applying e-beams. The simultaneous optimization Hybrid-Sim outperforms the two post-processing based approaches, which reduces VSB utilization by 69% while achieving comparable overlay error minimization with Hybrid-Post-OV.

Applying Hybrid-Post does not increase much computational time compared to Table I. Although Hybrid-Sim iteratively performs matching algorithm, the iterations converge quite fast and thus does not cause much runtime overhead. In our experiment, at most 4 iterations are needed to obtain a valid layout decomposition solution.

VII. Conclusion

We present a new layout decomposition framework for SADP and complementary EBL, which considers overlay minimization and EBL throughput optimization simultaneously. We show that conflict elimination by merge-and-cut can be formulated as a matching-based algorithm based on our graph formulation. Our approach is flexible to be applied for different lithography resources, including SADP with complementary EBL and pure SADP. The results show that applying merge-and-cut technique in hybrid SADP and EBL layout decomposition is promising, and that our approaches is efficient and effective in minimizing overlay error and e-beam utilization simultaneously.

Acknowledgment

This work is supported in part by NSF, SRC, NSFC, IBM and Intel.

References

[1] David Z. Pan, Bei Yu, and Jhih-Rong Gao. Design for manufacturing with emerging nanolithography. *IEEE Trans. on Computer-Aided Design of Integrated Circuits and Systems*, 2013.

[2] Yukiyasu Arisawa, Hajime Aoyama, Taiga Uno, and Toshihiko Tanaka. EUV flare correction for the half-pitch 22nm node. In *Proc. of SPIE*, 2010.

[3] Aki Fujimur. Beyond light: The growing importance of E-Beam. In *Proc. Int. Conf. on Computer Aided Design*, 2009.

[4] Chris Bencher, Jeffrey Smith, Liyan Miao, Cathy Cai, Yongmei Chen, Joy Y Cheng, Daniel P Sanders, Melia Tjio, Hoa D Truong, Steven Holmes, and William D Hinsberg. Self-assembly patterning

for sub-15nm half-pitch: a transition from lab to fab. In *Proc. of SPIE*, March 2011.

[5] Yan Borodovsky. MPProcessing for MPProcessors. In *Maskless Lithography and Multibeam Mask Writer Workshop*, 2010.

[6] David Lam, Dave Liu, and Ted Prescop. E-beam direct write (EBDW) as complementary lithography. In *Proc. of SPIE*, 2010.

[7] David Lam, Enden D. Liu, Michael C. Smayling, and Ted Prescop. E-beam to complement optical lithography for 1D layouts. In *Proc. of SPIE*, 2011.

[8] Kenichi Oyama, Eiichi Nishimura, Masato Kushibiki, Kazuhide Hasebe, Shigeru Nakajima, and et al. The important challenge to extend spacer DP process towards 22nm and beyond. In *Proc. of SPIE*.

[9] Yuelin Du, Hongbo Zhang, M.D.F. Wong, and Kai-Yuan Chao. Hybrid lithography optimization with E-Beam and immersion processes for 16nm 1D gridded design. In *Proc. Asia and South Pacific Design Automation Conf.*, 2012.

[10] Minoo Mirsaeedi, J. Andres Torres, and Mohab Anis. Self-aligned double patterning (SADP) layout decomposition. In *Proc. Int. Symp. on Quality Electronic Design*, 2011.

[11] Yongchan Ban, K. Lucas, and D. Z. Pan. Flexible 2D layout decomposition framework for spacer-type double pattering lithography. In *Proc. Design Automation Conf.*, June 2011.

[12] Hongbo Zhang, Yuelin Du, Martin D. F. Wong, and Rasit Topaloglu. Self-aligned double patterning decomposition for overlay minimization and hot spot detection. In *Proc. Design Automation Conf.*, 2011.

[13] Zigang Xiao, Hongbo Zhang, Yuelin Du, and Martin D. F. Wong. A polynomial time exact algorithm for overlay-resist self-aligned double patterning (SADP) layout decomposition. In *IEEE Trans. on Computer-Aided Design of Integrated Circuits and Systems*, 2013.

[14] Yuansheng Ma, Jason Sweis, Hidekazu Yoshida, Yan Wang, Jongwook Kye, and Harry J. Levinson. Self-aligned double patterning (SADP) compliant design flow. In *Proc. of SPIE*, 2012.

[15] Andrew B. Kahng, Chul-Hong Park, Xu Xu, and Hailong Yao. Layout decomposition for double patterning lithography. In *Proc. Int. Conf. on Computer Aided Design*, November 2008.

[16] Yue Xu and Chris Chu. A matching based decomposer for double patterning lithography. In *Proc. Int. Symp. on Physical Design*, 2010.

[17] Shao-Yun Fang, Szu-Yu Chen, and Yao-Wen Chang. Native-conflict and stitch-aware wire perturbation for double patterning technology. *IEEE Trans. on Computer-Aided Design of Integrated Circuits and Systems*, 2012.

[18] NanGate FreePDK45 Generic Open Cell Library [http://www.si2.org/openeda.si2.org/projects/nangatelib].

[19] Cadence SOC Encounter [http://www.cadence.com/].

[20] R.S. Ghaida, K.B. Agarwal, S.R. Nassif, Xin Yuan, L.W. Liebmann, and P. Gupta. Layout decomposition and legalization for double-patterning technology. *IEEE Trans. on Computer-Aided Design of Integrated Circuits and Systems*, 2013.

Fixing Double Patterning Violations With Look-Ahead

Sambuddha Bhattacharya   Subramanian Rajagopalan   Shabbir H. Batterywala

Synopsys India Pvt. Ltd.
RMZ Infinity, Benniganahallli
Bangalore, India - 560016
Email: {sbb, rsubbu, battery} @synopsys.com

Abstract: Double Patterning Technology (DPT) conflicts express themselves as odd cycles of spacing between layout shapes. One way of resolving these is by imposing a large spacing constraint between a pair of shapes participant in an odd cycle. However, this may shrink spacing in other parts of the layout and introduce DRC violations or new DPT conflicts. In this work, we model DPT conflict resolution as a constrained linear optimization problem, look ahead to upfront estimate potential violations and preclude them with additional constraints. We borrow the approach of Satisfiability Modulo Theory (SMT) solvers to simultaneously check satisfiability of linear constraint set and resolution of DPT conflicts. These two are interleaved and feed information to each other to churn out a feasible set of constraints that fixes DPT and DRC violations. We demonstrate the efficacy of the method on layouts at advanced nodes.

I. INTRODUCTION

Limitations in lithography at 20nm and below necessitate double patterning (DP) or splitting of a mask into two sub-mask patterns. Layout shapes need to be assigned to appropriate masks. This problem of mask assignment is a bi-colorability problem [1]. A layout is represented as a graph, where every layout shape represents a node and every pair of neighboring shapes on same layer contribute an edge. An edge denotes a manufacturing constraint between the associated layout shapes. A graph is free of DPT conflicts if it is bi-colorable. In other words, it must not have cycles with odd number of edges.

DPT color conflicts can be resolved by breaking such odd cycles. This is achieved in two ways: (a) splitting shapes into different masks, also known as the layout decomposition approach [2, 3], (b) increasing the spacing between two shapes participant in a conflict [4], also called the wire spreading approach. The goal for the first method is to reduce the number of splits as these can potentially lead to alignment errors during mask overlay. The second method attempts to minimize the layout perturbation as shapes are spread to break odd cycles.

Some recent work proposed combining the two methods so as to minimize the impact of both shape splitting and wire spreading [5, 6, 7]. The work in [7] proposes simultaneous optimization of splitting and spreading. The technique in [6] supplements splitting with subsequent layout legalization. The work in [4] also proposes layout legalization to fix 'native' conflicts that can not be resolved through decomposition. However, none of these model the non-DPT edges in the input lay-

out, i.e., cases where spacing is greater than DPT rule value in the input layout. As we shall see in Section VB, increasing the spacing to fix a DPT cycle may lead to shrinking of separation in other parts. In other words, a non-DPT 'edge' can get shrunk thereby introducing a new DPT edge. This in turn may create new conflict cycles.

In this paper, we focus on the legalization based approach to fix DPT violations. The legalization approach models the design rules within and across layers and naturally retains layout connectivity. Thus it can concurrently fix DPT conflicts and DRC violations. It is worth noting here that our approach can be easily combined with a layout decomposition approach similar to [6].

The main objective of this work is to address the issue of new DPT conflicts that may crop up as a result of spacing out shapes while fixing existing conflicts. For this, we present a systematic method by embedding a combination of constraint satisfiability checking and DPT conflict prediction as an intermediate step in legalization. The above step generates a set of constraints beyond what is usually extracted from the input layout. This precludes creation of new DPT conflicts in the subsequent steps of layout legalization. Specifically, our contributions are: (a) proposing a look-ahead method that predicts new DPT edges (b) setting up a tightly coupled interaction between a constraint feasibility checker and DPT conflict verifier similar to the approach of Satisfiability Modulo Theory (SMT) solvers (c) and using these to construct a rubust DPT fixing algorithm that can fix existing violations without introducing new ones.

The paper is organized as follows. In Section II, we outline the method of fixing DPT conflicts through layout legalization. In Section III, we describe the iterative resolution of constraints to simultaneously fix DPT conflicts and DRC violations. In Section IV, we discuss the details of resolving infeasibilities and generating a feasible set of constraints. In Section V, we present the key component of upfront predicting potential new DPT edges. We discuss the experimental results in Section VI and draw the conclusions in VII.

II. DPT FIXING AS LEGALIZATION

Here, we extend constraint solution based layout legalization methods to address DPT conflicts. Layout legalization refers to automatic fixing of DRC violations [8]. Legalization achieves DRC clean layouts by modifying top-shapes and re-positioning cells and macros. This is extremely useful in generating clean layout for complex cells, and large blocks in custom and analog design.

A legalization engine reads in an input layout and fractures its mask layer geometry into rectangles. Each rectangle edge is considered as a positional variable. Each rectangle, thus, has four positional variables associated with it, two each in horizontal and vertical directions. All applicable design rules are modeled as linear(-ized) *difference* constraints. These constraints are of the form $x_i - x_j \geq w_{ij}$, where x_i and x_j are the positional variables corresponding to two rectangle edges, and w_{ij} corresponds to the required distance between the variables to obey the particular design rule. This results into the following linear program.

$$\text{minimize} \sum_{i=1}^{N} |x_i - x_i^0| \quad \text{subject to}: \quad \mathbf{A}\mathbf{x} \geq \mathbf{b}, \quad (1)$$

where $\mathbf{x}$ is a column vector consisting of the variables corresponding to rectangle edges, $\mathbf{x}^0$ are the rectangle edge positions in input layout. The matrix $\mathbf{A}$ is the coefficient matrix for the constraints, where each row $\mathbf{A}_i$ represents a single constraint. The vector $\mathbf{b}$ represents the rule values. The term $|x_i - x_i^0|$ models minimum perturbation on the input layout. The objective function minimizes total perturbation across all layout edges. Heng et al [8] introduced a technique to linearize the minimum perturbation objective and solve it as a standard Linear Program (LP) [9].

Generally, the constraint set obtained from a realistic layout is often infeasible due to conflicts between complex rules and design intents, lack of leeway in moving shapes due to placed macros, locked shapes, and sometimes due to incomplete modeling. Therefore, it is required to detect and resolve such infeasibilities and obtain a consistent set of constraints. Since all constraints are of *difference* type, these can be expressed as a directed constraint graph G_C. An infeasibility in the corresponding linear program can be detected as positive cycle in the constraint graph. These positive cycles can be identified through longest path runs [10]. The constraints forming the positive cycles are modified to reduce the cycle weight to zero. Once all positive cycles are resolved the corresponding LP becomes feasible and can be optimized for minimum perturbation objective.

A. Modeling DPT Constraints

DPT conflicts express themselves as odd cycles of spacing between shapes. We break these cycles by 'spreading' or increasing the spacing between a pair of shapes in an odd cycle. Consider the layout in Figure 1(a). The shapes s_1, s_2 and s_3 have a dpt conflict, which is modeled through the odd cycle in the undirected dpt graph G_D in Figure 1(b). A large DPT spacing must be applied in one of c_1 and c_2, or c_3 or $c4$ to resolve this. In other words, a conflict cycle is a set $C = \{c_k : \bigvee_{1...i}(\bigwedge_{k=1...j} c_k)\}$.

Like DRC constraints, each DPT constraint c_k is also modeled in the *difference* form. Therefore, these can also be expressed into the directed constraint graph G_C. Figure 1(c,d) shows portions of the horizontal and vertical constraint graphs. In this example, the horizontal and vertical variables corresponding to the shapes are marked as x_i and y_i respectively. The associated spacing constraints are shown with arrows. Note that G_C edges $e_{2,3}^C$ and $e_{5,6}^C$ correspond to the G_D edge e_1^D. In other words, if the cycle in G_D has to be broken by removing the edge e_1^D, both constraints $e_{2,3}^C$ and $e_{5,6}^C$ must apply DPT rule value.

Generally, the constraints given in Equation 1 are solved together, i.e., in conjunction. Each constraint represented by

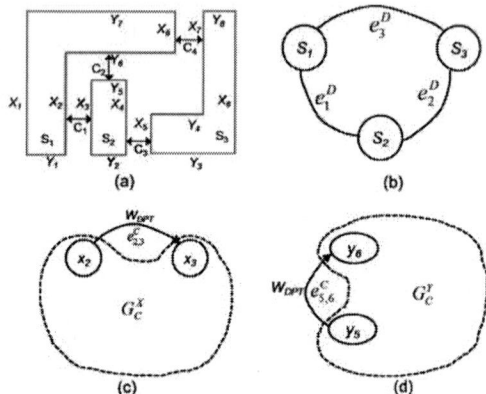

Fig. 1. A layout with dpt conflict and the relation between corresponding G_C and G_D

individual rows of $\mathbf{A}$ is met in the optimized solution vector $\mathbf{x}$. However, note that the DPT constraints are naturally disjunctive, i.e., only one set have to be satisfied to fix a conflict. If all DPT constraints are directly modeled as the rows in $\mathbf{A}$, then we get a sub optimal layout. For example, all shape pairs in an odd cycle would be spaced apart instead of only one to break the odd cycle. Moreover, such over-constraining can lead to infeasible LP instances.

In our formulation we allow disjunctive (aka OR) constraints, which is a set of difference constraints out of which only one constraint is kept active. The matrix $\mathbf{A}$ has only one representative difference constraint for each disjunctive set. We check the feasibility of the set of constraints in $\mathbf{A}$. If constraint set is infeasible, we replace the representative constraint of a disjunctive set with an alternative and repeat the feasibility check.

B. Modeling Complex Design Rules

It is easy to note that simple design rules like width, spacing, extension and overlap constraints are difference of two layout edges [11]. Complex design rules like width-length based spacing (table based spacing) can also be modeled as difference constraints across the two layout edges. In these cases, the existing widths and common run lengths need to be computed, table-based value looked up, and then applied as a difference constraint. Similarly rules like dual metal extension rules, complex via rules, end of line spacing rules, etc., can also be modeled by detecting the geometry context and applying the correct rule value as a difference constraint. In general, it is possible to model most of the intra and inter layer design rules (except rules like density rules, which are handled in other design creation steps) as difference constraints.

Design rules violations often have alternative fixes. For example, a minimum area rule can be imposed by expanding in horizontal, or vertical or both directions. Similarly, complex line-end rules can be satisfied by applying specified spacing at line ends or by increasing spacing between neighboring parallel lines. Likewise, dual extension rules require metal shapes to extend via-cut shapes by different distances in horizontal and vertical. Similar to DPT, the alternatives for fixing complex DRCs are also modeled through disjunctive constraints. As each individual alternative is still a difference constraint, these are expressed in the constraint graphs. These alternatives are explored to obtain a feasible set using longest path runs [10] on constraint graph. This is detailed in Section A.

978-1-4799-2817-0/14 $31.00 © 2014 IEEE   150

C. Methodology Overview

Figure 2(a) depicts the steps for straightforward DPT fixing in a legalization framework. First, the DPT conflicts are identified by detecting all DPT spacing violations in the input layout and identifying conflict cycles. The details are in Section IIIB. Then all design rules and intents are captured during constraint generation. Note that both methods employ standard line sweeps on layout geometry [11]. These two steps generate a combined set of constraints. Theses constraints are checked for feasibility which requires an exploration of disjunctive constraints. And finally the feasible set of constraints are solved by an LP solver with a minimum perturbation objective.

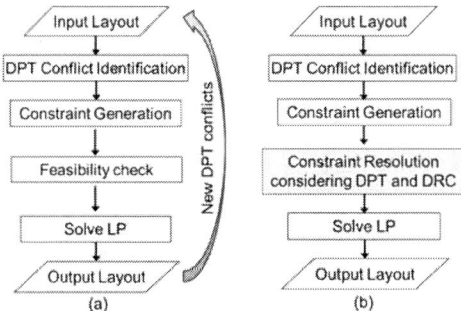

Fig. 2. Flow for DPT fixing through legalization: (a) straight-forward method, (b) proposed method

Unfortunately, in the straightforward approach, layout modifications could lead to new DPT conflicts and may require expensive iterations. This is shown in Figure 2(a). In the proposed approach of Figure 2(b), the feasibility check is replaced with a co-resolution step (shaded). In this, we process the constraint graph G_C and the DPT graph G_D simultaneously. We also use a look-ahead step to predict and thereby avoid new DPT conflicts. This is detailed below.

III. CO-RESOLUTION OF DPT AND DRC

As discussed in Section A, there is a strong relationship between G_D and G_C. Each edge in G_D has its counterpart in G_C. Intuitively, G_C should be made feasible and G_D should be freed of odd cycles. However, feasibility of G_C usually requires exploration of the disjunctive constraints due to DPT. Furthermore, as certain constraints get activated in G_C, these may absorb the slack in some paths in G_C. This may introduce new edges in G_D leading to new odd cycles. Therefore, G_C and G_D must interact to generate the desired feasible set. We discuss the ideas from Satisfiability Modulo Theory (SMT) which is an abstraction to address such interactions.

A. SMT-like Abstraction

Satisfiability Modulo Theory (SMT) [12] solvers extend the SAT [13] techniques beyond regular propositional satisfiability. This enables a much richer modeling of a variety of decision problems and have been incorporated into verification of datapath operations in microprocessors, bounded model checking, scheduling etc. An SMT problem instance is a generalization of a Boolean SAT instance in which regular Boolean variables are replaced by predicates over a set of non-Boolean variables. A predicate is a Boolean-valued function of non-Boolean variables including real, integer, data-types etc. These predicates are classified into appropriate 'theories'. Fundamentally, an SMT solver involves a SAT solver that interacts with a 'theory' solver that understands constraints defined on the non-Boolean variables.

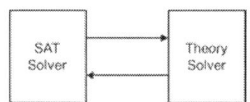

Fig. 3. SMT solver structure

Figure 3 shows an SMT solver structure where a SAT solver is tightly coupled with a Theory solver. The SAT solver drives the search and derives a Boolean-feasible assignment and also queries the auxiliary Theory solver to check consistency according to the underlying theory. The Theory solver detects conflicts involving theory reasoning and provides succinct explanations of inconsistency that the SAT solver further learns from.

In this work, we borrow the spirit of the SMT solvers to generate a satisfiable set of constraints that resolve DPT cycles. Similar to SMT solvers that employ a SAT solver for Boolean proposition, we detect infeasibilities in G_C using Positive Cycle Checker (PCC). Parallel to the Theory Solver in SMT, we detect DPT conflicts using a Bi-Colorability Checker (BCC). Like the SAT solver that creates feasible assignment, the PCC engine generates a feasible set of constraints. This feasible constraint set is then passed to the BCC engine to verify if it fixes or creates DPT conflicts. Akin to the learnings in SAT from inconsistencies reported by Theory solver, the BCC feeds additional sets of difference constraints to PCC. Figure 4 illustrates the process.

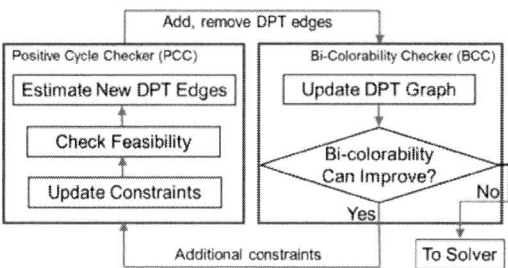

Fig. 4. Flowchart of DPT and DRC co-resolution

The PCC engine performs its operations on the constraint graph G_C. First, it checks for the feasibility of the combined set of DPT and DRC constraints through positive cycle detection on G_C (Section IV). A set of satisfiable DPT constraints result in deletion of the corresponding edges in G_D. Furthermore, we employ a look-ahead method to predict if new DPT edges get introduced while applying the set of constraints (Section V). These are fed into BCC from the PCC. The DPT graph G_D in BCC gets updated due to the deleted and added edges. The BCC engine then checks if the added edges create any new DPT cycle (Section IIIB). If new cycles get created, corresponding disjunctive constraints are fed back into the constraint graph G_C in PCC.

The interactions between the PCC engine and the BCC engine go on iteratively. Unlike prior approaches to legalization of DPT conflicts [4, 5, 6], the feedback loop is deep inside the flow. All refinement towards DPT fixing happens using fast graph algorithms, and prior to LP solution.

The co-resolution terminates with a set of feasible constraints that fix DPT conflicts and DRC violations. If BCC detects no conflict, iterations converge and the feasible constraint set is passed to the LP solver. The co-resolution also termi-

978-1-4799-2817-0/14 $31.00 © 2014 IEEE

nates if BCC detects no improvement in number of conflicts in consecutive iterations. This indicates that the remaining conflicts are un-resolvable. Such cases are seen in real layout due to presence of macros and 'locked' objects.

B. Detecting DPT Conflicts in BCC

As explained above, during the iterations, the BCC engine modifies the DPT graph G_D based on edge insertion and deletion information obtained from PCC, the constraint satisfiability checker. Then it identifies the odd cycles that remain in updated G_D. The sets of layout constraints corresponding to these odd cycles are fed back to PCC.

In this context, the feedback set should be devoid of redundancies. The total number of odd cycles in a graph can be exponential. Therefore, we avoid finding all odd cycles and reporting back corresponding constraints. Instead, we identify the odd-cyles in a cycle basis [4] from G_D. Similar to the concept of basis in vector space, a cycle basis forms a linearly independent spanning set of all cycles. Breaking the odd cycles in the cycle basis frees the DPT graph G_D from conflicts[4]. We adopt the depth first search based [10] method to obtain the cycle basis as suggested in [4].

IV. CONSTRAINT FEASIBILITY CHECK

This section describes the joint infeasibility resolution of DRC and DPT constraints. Each edge e_{ij}^C in the constraint graph G_C models corresponding constraints of the form $x_i - x_j \geq w_{ij}$. Generally, a pair of rectangles can have multiple constraints between them. Thus, an edge in the constraint graph needs to model these interactions. Figure 5 illustrates this where a DPT constraint of rule value of w_{DPT} and a DRC constraint of value w_{DRC} are bundled together on the edge. Of these constraints, only one is active on the edge at a given time. The weight on the edge corresponds to the RHS of the associated constraint. We also model the distance $|x_i^0 - x_j^0|$ between the nodes in the input layout on the edge if $|x_i^0 - x_j^0|$ is less than w_{DPT} and w_{DRC}. This is useful for infeasibility resolution.

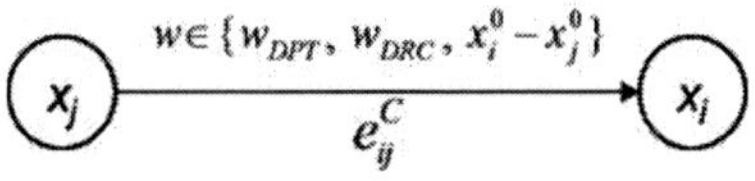

Fig. 5. Modeling DPT and DRC constraint in CG

A. Infeasibility Resolution

We first describe infeasibility resolution without considering DPT constraints. After that, we discuss how a suitable constraint from a DPT odd cycle is selected for co-resolution.

Inconsistencies in the constraints express themselves as positively weighted cycles in G_C. These *positive cycles* are detected by invoking a single source longest path algorithm (Bellman-Ford [10]). The algorithm is applied iteratively until all positive cycles are detected and resolved. In this work, we implemented the Goldberg-Radzik [14] variant of the longest path algorithm which is very efficient in practice.

For feasibility, the total cycle weight needs to be made non-positive. This is achieved by reducing the weights of certain edges participant to the cycle. In other words, if the corresponding constraint value is larger than the associated layout value, it is diluted to the layout value. Consider the model in

Figure 5. Assume that the DPT constraint were absent, and the active constraint is w_{DRC}, and $w_{DRC} > x_i^0 - x_j^0$. Then the edge weight gets reduced to $x_i^0 - x_j^0$. This dilution of constraint to input layout value is continued on other edges until the given cycle's weight becomes non-positive.

Resolving positive cycles in this way have two interesting properties. First, if a constraint is not satisfiable, the input layout value is retained i.e., it is made no worse than how it started. Second, the input layout positions represent a feasible solution (albeit with diluted constraints). Therefore, diluting infeasible constraints to input layout value ensures termination of the infeasibility resolution step.

Once the infeasibilities are resolved in G_C the graph is augmented with the DPT constraints. Recall that each DPT conflict corresponds to an ordered set of difference constraints of which only one needs to be satisfied. Initially, only the first difference constraint in the set is marked active. The difference constraints due to DPT are added to the corresponding edges. If no corresponding edge exists in G_C, a new one is created. Otherwise, the difference constraint due to DPT gets modeled in the edge as shown in Figure 5. If an edge was already relaxed to current layout value, the corresponding DPT difference constraint is deactivated and swapped with the next constraint in the ordered set.

Generally, DPT constraints are larger than DRC constraints. Therefore, imposing larger weight on the edge can introduce new positive cycles. We use the single source longest path method with some modifications to resolve infeasibilities in presence of DPT constraints. If a positive cycle involves an edge e_{ij}^C with DPT constraint, we look for an alternative difference constraint from the disjunctions.

The DPT constraint on the edge e_{ij}^C is deactivated, and the edge weight is restored to its prior DRC value. Since the graph G_C was feasible prior to augmenting it with DPT constraints, a cycle with only DRC constraints remains non-positive. Also, it implies that new DRC violations are not introduced in the layout. During this process, if none of the disjunctions for a DPT conflict are satisfiable, then the corresponding DPT conflict can not be resolved.

V. PREVENTING NEW DPT VIOLATIONS

Here we discuss possible introduction of new DPT edges due to spacing shrinkage while fixing existing DPT conflicts. As DPT conflict cycles are broken by increasing spacing, this can result in reduction of spacing in the neighborhood to a value between w_{DRC} and w_{DPT}. Such shrunk edges, when added to G_D, can 'complete' new odd cycles. We employ a look-ahead step to identify such shrunk edges.

A. Detecting New DPT Edges

Consider the example in Figure 6. The edge between x_2 and x_3 does not correspond to a DPT edge in the input layout because $x_3^0 - x_2^0 \geq 150 = w_{DPT}$. After the DPT constraint gets applied on the edge $e_{1,2}^C$, it forms part of a 0-weight cycle. The spacing between $x_3 - x_2$ gets reduced to $100(< w_{DPT})$ and cannot be further expanded. Therefore a new un-expandable DPT edge gets introduced into G_D.

In general, when a spacing is reduced below w_{DPT} new DPT edge gets added in G_D and this may create new odd cycles. We want to predict and prevent such new odd cycles.

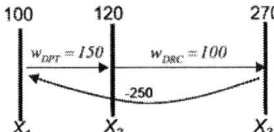

Fig. 6. Example of guaranteed new DPT edge

B. Predicting Potential DPT Edges

First, consider a node example depicted in Figure 7. The nodes x_1, x_2 and x_3 have input layout locations 100, 200 and 380 respectively. The constraints between the variables be the following: $x_2 - x_1 \geq 150$ and $x_3 - x_2 \geq 100$. Here, the regular spacing rule is $w_{DRC} = 100$ and the DPT rule value is $w_{DPT} = 150$. To apply the DPT rule value, the node x_2 can be minimally perturbed to location 250. This is exactly what an LP solver would do. However, this also reduces the spacing between x_2 and x_3 to 130 ($< w_{DPT}$) thereby introducing a new DPT edge. We want to simulate the behavior of the downstream LP solver by graph operations.

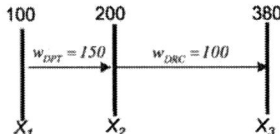

Fig. 7. Example where new DPT edge crops up

TABLE I
RANGES GENERATED USING BF AND LABF

Var	Lyt loc	BF LB	BF UB	LABF LB	LABF UB
x_1	100	0	130	50	100
x_2	200	150	280	200	250
x_3	380	250	380	380	380

As mentioned in Section A, the infeasibility checker in PCC uses Bellman-Ford algorithm (BF) to determine feasible constraint set. For the example in Figure 7 if we run BF from the left, we get the node positions as shown in column 3 of Table I. This is because all nodes get pulled towards the source node which is at 0. Similarly, running BF on the reversed graph from the right yields the node positions in column 4. These results indicate a compaction to the left and right respectively. In general, BF ranges may not include the minimum perturbation points achieved by LP.

To address this problem, we employ the Layout-aware Bellman-Ford Algorithm (LABF) recently proposed in [15]. Unlike regular BF which brings all nodes closest to the source (layout compaction [11]), the essence of this algorithm is to perturb a variable only if dictated by a constraint. This enables us to upfront simulate the behavior of the downstream LP-solver run. Therefore, we can predict which separations are reduced by LP.

For completeness, we briefly describe the technique. Unlike BF, the variables are initialized with their initial layout locations. In a forward run of LABF, a variable is updated only when a constraint pushes it to the right. Unlike BF, forward LABF run generates the upper bound. The reverse LABF (with graph edges and weight signs inverted) generates the lower bound. The ranges generated by LABF includes all points in the solution space where minimum layout perturba-

tion can be achieved [15]. The complexity of LABF is same as that of regular BF [10].

The bounds generated by the forward and reverse LABF runs are labeled LABF LB and LABF UB and listed in columns 5 and 6 of Table I. Note that the lower and upper bounds include the minimum perturbation points for each variable.

Let $L(x_i)$ and $U(x_i)$ represent the lower and upper bounds on the position of variable x_i as detected by LABF. We consider the case where a spacing is initially larger than w_{DPT}, i.e., $x_i^0 - x_j^0 \geq w_{DPT}$. We make the following observations:

- If $L(x_i) - L(x_j) < w_{DPT}$ or $U(x_i) - U(x_j) < w_{DPT}$, then the edge e_{ij}^C in G_C can lead to an edge in G_D.
- If $U(x_i) - L(x_j) < w_{DPT}$, then the edge e_{ij}^C in G_C definitely adds a DPT edge in G_D.
- If $L(x_i) - U(x_j) \geq w_{DPT}$, then the edge e_{ij}^C in G_C does not lead to a DPT edge in G_D.
- If $L(x_i) - L(x_j) = U(x_i) - U(x_j) = w_{DRC} < w_{DPT}$, then the edge e_{ij}^C in G_C participates in a tight cycle; therefore it leads to an unresolvable DPT edge in G_D.

The containment of the minimum perturbation point inside the range detected by LABF [15] allows reasonable prediction of new DPT edges. If addition of such edges to G_D creates new odd cycles, then these are fed back as additional constraints to G_C in next iteration (Figure 4).

VI. EXPERIMENTAL RESULTS

We now present experimental results using designs in 40nm and 20nm technology with appropriate representative DPT rules. These rules involved different spacing values for line side-side, tip-side, tip-tip and corner interactions. The design rules involved complex spacing tables, end-of-line and other context dependent rules across all layers. The experiments were conducted on a 2.8GHz, 4 core linux machine with 8 GB RAM.

Table II lists the results of our approach. Column 2 reports the number of layout objects; columns 3 and 4 list the number of nodes and edges in the corresponding constraint graph G_C. Columns 5 and 6 report the number of DPT conflicts and DRC violations in the input layout. The DPT conflicts include odd cyles of varying lengths 3,5,7,9 and up to 13. These also include conflicts due to 'pre-colored' shapes.

To demonstrate the efficacy of our approach, we report the results obtained through simple legalization under the heading "Simple Lgl". This is our implementation of the legalization methods proposed in [4, 5, 6]. In this, constraints due to DPT conflicts are extracted once from the input layout. No DPT edge prediction and iterative checking of DPT conflict is employed in the constraint resolution. Column 7 reports the number of DPT conflicts in the output layout generated by this "Simple Lgl" approach. Column 8 lists the corresponding DRC violations.

We present the results of our approach under the heading "Look Ahead". Column 9 reports the number of DPT conflicts in the output layout. Column 10 lists the number of DRC violations in the output layout. Note that the remaining violations are much less compared to that in "Simple Lgl". In some cases, the output layout is completely free of DPT conflicts and DRC violations. Column 11 reports the improvement observed using our approach; this is the ratio of number of DPT conflicts in the output layouts for "Simple Lgl" [4, 5, 6] to that of the look-ahead method of this work. Column 12 lists

TABLE II

NUMBER OF DPT VIOLATIONS REMAINING AFTER SIMPLE LEGALIZATION AND AFTER LOOK AHEAD METHODS.

Design	#lyt obj	#nodes	#edges	Input Violation		Simple Lgl [4, 6]		Look Ahead		improv	#iter	runtime(s)
				#dpt	#drc	#dpt	#drc	#dpt	#drc			
d_1	239	343	2255	389	452	248	389	198	380	1.7x	3	0.12
d_2	300	575	4718	1100	2835	415	394	251	324	1.6x	5	0.68
d_3	634	2513	14840	150	182	30	0	0	0	-	1	0.45
d_4	1354	5393	32257	330	370	70	0	0	0	-	1	1.49
d_5	1854	7173	41206	459	529	115	4	1	4	115x	5	4.86
d_6	2654	9953	59594	635	750	170	100	40	100	4.2x	3	5.87
d_7	3946	13749	80391	854	1066	234	19	40	19	5.8x	5	12.83

the number of iterations of co-resolution, DPT prediction and conflict checking before employing the LP solver. For these experiments, there was no perceptible area increase beyond that for "Simple Lgl" [4, 5, 6].

Not all DPT and DRC violations have been fixed for all designs. Design d_1 and d_2 include tricky scenarios with many pre-colored shapes, locked objects and small macros that have embedded DPT conflicts and DRC violations. Most of the violations are unfixable without editing the macros. In designs d_3 and d_4, the DPT conflicts and DRC violations are completely fixed using our approach. Designs d_6 and d_7 had some conflicts and violations inside macros that were left un-modified. In general, our approach shows considerable improvement (5.8x in design d_7) over "Simple Lgl" [4, 6] due to prediction and iterative co-resolution.

Column 13 reports the runtime for the main component of this work, i.e., constraint resolution with prediction and DPT cycle check. The runtime ranges from less than a second for d_1 to about 13s for total constraint resolution in d_7. Generally, the constraint resolution runtimes presented in this work are within 10% of total legalization runtime.

VII. CONCLUSIONS

We presented a DPT fixing approach based on wire spreading. The underlying framework of legalization ensures that DRC violations are modeled and fixed simultaneously, and layout structure and connectivity are retained naturally. We proposed an iterative method that tightly couples feasible constraint selection, new DPT edge prediction and DPT conflict cycle verification. This is added as an intermediate step in legalization. The proposed method addresses the problem of new DPT conflict creation during legalization. The experimental results demonstrate the efficacy of our approach on layouts at advanced technology nodes.

REFERENCES

[1] M. Drapeau, V. Wiaux, E. Hendrickx, S. Verhaegen, and T. Machida, "Double patterning design split implementation and validation for the 32 nm node," *Proc. SPIE*, vol. 6521, 2007.

[2] A. B. Kahng, C.-H. Park, X. Xu, and H. Yao, "Layout decomposition approaches for double patterning lithography," *IEEE Trans. Computer-Aided Design of Integrated Circuits and Systems*, vol. 29, pp. 939–952, Jun. 2010.

[3] X. Tang and M. Cho, "Optimal layout decomposition for double patterning technology," *Proc. Int. Conf. Computer Aided Design*, pp. 9–13, Nov. 2011.

[4] S.-Y. Fang, S. Y. Chen, and Y. W. Chang, "Native-conflict and stitch-aware wire perturbation for double patterning technology," *IEEE Trans. Computer-Aided Design of Integrated Circuits and Systems*, vol. 31, pp. 703–717, May 2012.

[5] C.-H. Hsu, Y. W. Chang, and S. R. Nassif, "Simultaneous layout migration and decomposition for double patterning technology," *IEEE Trans. Computer-Aided Design of Integrated Circuits and Systems*, vol. 30, pp. 284–294, Feb 2011.

[6] R. S. Ghaida, K. B. Agarwal, S. R. Nassif, X. Yuan, L. W. Leibmann, and P. Gupta, "Layout decomposition and legalization for double-patterning technology," *IEEE Trans. Computer-Aided Design of Integrated Circuits and Systems*, vol. 32, pp. 202–215, Feb. 2013.

[7] K. Yuan and D. Z. Pan, "Wisdom: Wire-spreading enhanced decomposition of masks in double patterning lithography," *Proc. Int. Conf. Computer Aided Design*, pp. 32–38, Nov. 2010.

[8] F.-L. Heng, Z. Chen, and G. Tellez, "A VLSI artwork legalization technique based on a new criterion of minimum layout perturbation," *Proc. Int. Symp. Physical Design*, pp. 116–121, Apr. 1997.

[9] L. Luenberger, *Linear and Nonlinear Programming*. Reading, MA, USA: Addison-Wesley, second ed., 1984.

[10] T. H. Cormen, C. E. Leiserson, and R. L. Rivest, *Introduction to Algorithms*. Cambridge, MA, USA: MIT Press, 1990.

[11] D. Boyer, "Symbolic layout compaction review," *Proc. IEEE/ACM Design Automation Conf.*, pp. 383–389, Jun. 1988.

[12] C. Barrett, R. Sebastiani, S. A. Seshia, and C. Tinelli, *Satisfiability Modulo Theories*. Handbook of Satisfiability, Ed: A. Biere, M. Heule, H. van Maaren and T. Walsch, IOS Press, 2008.

[13] C. P. Gomes, H. Kautz, A. Sabharwal, and B. Selman, *Satisfiability Solvers*. Handbook of Knowledge Representation Ed: F. van Harmelen, V. Lifschitz and B. Porter, Elsevier, 2008.

[14] A. V. Goldberg and T. Radzik, "A heuristic improvement of the bellman-ford algorithm," *Applied Math. Letters*, vol. 6, pp. 3–6, 1993.

[15] N. Salodkar, R. Subramanian, S. Bhattacharya, and S. H. Batterywala, "Automatic design rule correction in presence of multiple grids and track patterns," *Proc. IEEE/ACM Design Automation Conf.*, pp. 26–31, Jun. 2013.

978-1-4799-2817-0/14 $31.00 © 2014 IEEE

EUV-CDA: Pattern Shift Aware Critical Density Analysis for EUV Mask Layouts

Abde Ali Kagalwalla*, Michale Lam†, Kostas Adam† and Puneet Gupta*
(abdeali, puneet)@ee.ucla.edu
*Department of Electrical Engineering, University of California, Los Angeles
†Mentor Graphics Corp.

Abstract—Despite the use of mask defect avoidance and mitigation techniques, finding a usable defective mask blank remains a challenge for Extreme Ultraviolet Lithography (EUVL) at sub-10nm node due to dense layouts and low CD tolerance. In this work, we propose a pattern shift-aware metric called critical density, which can quickly evaluate the robustness of EUV layouts to mask defects ($300 - 1300\times$ faster than Monte Carlo, with average mask yield root mean square error (RMSE) ranging from $0.08\% - 6.44\%$), thereby enabling design-level mask defect mitigation techniques. Our experimental results indicate that reducing layout regularity improves the ability of layouts to tolerate mask defects via pattern shift.

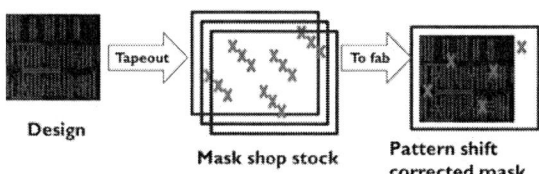

Fig. 1. Set of steps involved in a EUV mask shop involving pattern shift based mask defect mitigation.

I. INTRODUCTION

A. Background and Motivation

Extreme ultraviolet lithography is considered one of the most promising next generation lithography solutions to replace the current deep ultraviolet lithography [1]. Reflective EUV mask blanks suffer from hard-to-repair defects that can significantly alter the printed pattern on the wafer [2]. Mask blank defectivity is a key concern that could prevent the insertion of EUVL into volume manufacturing [1], [3].

Defect avoidance based techniques have emerged as an effective means to tolerate mask defects. These techniques rely on inspection of mask blanks to first determine defect locations. The position of the design pattern, which needs to be written on the mask, can be shifted relative to the mask to avoid the defects. Several approaches and results have been shown for such pattern shift based defect avoidance [4]–[7]. A similar, but more general mask floorplanning based defect avoidance has been proposed as well [8], [9]. Rotation of the mask pattern has also been explored, either small-angle [10] or 180°/flips [8]. Recent techniques have also looked at methods that can tolerate defect position inaccuracy [8], [11]. Elayat et. al. [12] and Jeong et. al. [13] provide a cost-benefit assesment of different defect avoidance and reticle planning strategies, respectively.

A likely design to fabrication flow for EUV masks is illustrated in Figure 1. A mask shop will typically have a collection of inspected mask blanks with known defect locations. Since each critical layer of the taped out design must be patterned on a defective mask blank, the mask shop must apply defect avoidance to find a defective mask blank that works for each layer. *Given a defect density and size distribution, the probability of finding a mask blank from a large set of blanks on which the given design layout can be patterned without causing any yield loss is referred to as mask yield.*

Certain layout topologies may be more capable of tolerating mask defects, and exploiting the benefits of defect avoidance strategies. An understanding of what characteristics of a layout can make it more robust to EUV mask defects can significantly aid EUV layout design and even the formulation of design rules. In order to develop any layout or design level techniques to create robust layouts, a quantifiable metric that characterizes the robustness of layouts to such

mask defects is needed. In addition to layout optimization, such a layout robustness metric can be used for mask blank assignment as well. Layouts with low mask yield can be assigned to a mask blank with lower defect density. This would save the computational effort of performing pattern shift for each layout-blank pair, as done in [14].

B. Key Contributions of this Work

In this work, we propose a new metric, *critical density*, that evaluates the robustness of EUV layouts to mask defects. To the best of our knowledge, this work is the first attempt towards developing such a metric. This metric allows us to estimate pattern shift aware mask yield for any defect density using a simple analytical expression, and enables us to distinguish between layouts that have different mask yield for any given defect density.

The need for a metric that quantifies robustness of layouts to mask defects bears resemblance to conventional critical area analysis (CAA) [15]. CAA is commonly used to check robustness of layouts to random *wafer* defects through the use of statistical metrics that estimate chip yield for some random distribution of defects. The key features of this work that also distinguish EUV-CDA and conventional CAA are the following:

- Unlike wafer defects, a single mask defect will print on every copy of the design on the wafer. Hence, the goal of EUV-CDA is to predict mask yield, not chip yield.
- The impact of defect avoidance techniques on mask yield must be probabilistically modeled as a part of EUV-CDA since actual defect locations are not known at the design stage. Pattern shift based defect avoidance is modeled in this work since it is the most popular defect avoidance strategy [12].
- The need to account for pattern shift means that mask failure depends on the simultaneous location and size of several defects on the mask. This is in contrast to conventional CAA, where every defect can cause failure independently. This dependence complicates the analysis significantly, requiring much more computational effort and modeling.

The remainder of this paper is organized as follows. Prohibited region of a layout is described in Section II. In Section III, we propose analytical methods to estimate mask yield for two limited scenarios.

978-1-4799-2817-0/14 $31.00 © 2014 IEEE

TABLE I
GLOSSARY OF TERMINOLOGY

Term	Description
s	Defect size (Height, full width half maximum pair)
PB^s	Prohibited Region for defect size s
$P(s)$	Probability of occurrence of defect size s
A_M	Mask area
D_P^s	Prohibited region density ($\frac{Area(PB^s)}{A_M}$)
D_P	Expected prohibited region density ($\sum_s P(s)D_P^s$)
K	Number of different defect sizes considered
N_d	Number of defects on mask
Δ_X	Available pattern shift in X direction
Δ_Y	Available pattern shift in Y direction
A_Δ	Total pattern shift area ($\Delta_X \times \Delta_Y$)
L_X	Design layout width
L_Y	Design layout height
A_L	Total design area ($L_X \times L_Y$)
ρ	Number of prohibited region shapes per unit area
N_X	Number of discrete X direction shifts
N_Y	Number of discrete Y direction shifts
(X_i, Y_j)	Potential pattern shift solution
E_{ij}^Δ	Event that pattern shift solution (X_i, Y_j) works
$P(E_{ij}^\Delta)$	Probability of event E_{ij}^Δ
PB_{ij}^s	Prohibited region for defect size s, shifted by (X_i, Y_j)
$Ar(PB_{ij}^s)$	Total area of all the polygons in the prohibited region
W_{pl}	Width of periodic parallel line structure
P_{pl}	Pitch of periodic parallel line structure
Y_M^{ct}	Mask Yield of periodic contact array layout
D_{cric}	Critical density
Y_M^{true}	Accurate mask yield (Monte Carlo method)
N_d^{min}	Minimum defect count considered
N_d^{max}	Maximum defect count considered
AC	Autocorrelation matrix
ACF	FFT of AC
pix	Pixel size (Sampling size) for computing AC
N_F	Number of terms from ACF used for predicting D_{cric}

We describe our critical density method in Section IV. Experimental results are then presented in Section V. We conclude this work in Section VI. All notation used in this paper is described in Table I.

II. PROHIBITED REGION

We define the prohibited region of a given layout, for a particular defect size s, as the set of polygons PB^s such that if the center of a mask defect lies inside any polygon $p \in PB^s$, the given mask layout pattern will not yield.

The method for constructing prohibited region is the same as proposed by Zhang et. al. [5] with the additional step of merging the constructed rectangles. It is similar to the process of using simple Boolean operations to compute critical area in conventional CAA [15]. *But the criteria for determining prohibited region is CD tolerance, in constrast to opens/shorts in conventional CAA.* We chose this pessimistic approach since we are dealing with mask defects. Assignment of this CD tolerance to layout shapes can be done by either setting a single pessimistic value for all the patterns (10% of the technology node, in our case), or by using some design information (timing slack, redundant/dummy patterns) to assign an appropriate CD tolerance, as done in [16]. Although EUV-CDA can be easily applied for such smart CD tolerance, for the sake of brevity we assign a single CD tolerance to all shapes in this work.

If pattern shift was not a part of EUV mask manufacturing, estimating mask yield from prohibited region would be fairly straight-

forward. Assuming a uniform spatial distribution of defects on the mask, mask yield could be estimated as $(1 - D_P^s)^{N_d}$ since every defect must lie outside the prohibited region[1]. This simple approach would imply that any two design layouts with the same prohibited region density will have the same mask yield. But if pattern shift is used to avoid mask defects, layout topology may also affect mask yield. To confirm this suspicion, we created four $20\mu m \times 20\mu m$ layouts such that their prohibited region is a set of parallel lines with pitch $80nm$. The width of the lines was treated as a Gaussian random variable with mean $20nm$. Different values of variance (σ) were used to construct the four layouts. We compared the mask yield of these four layouts, which is estimated using rigorous Monte Carlo simulation (described in Section III). The results, shown in Figure 2, highlight the huge difference in post pattern shift mask yield between the layouts which have very similar prohibited region density (confirmed by the pre-pattern shift mask yield of the four layouts, which are almost same).

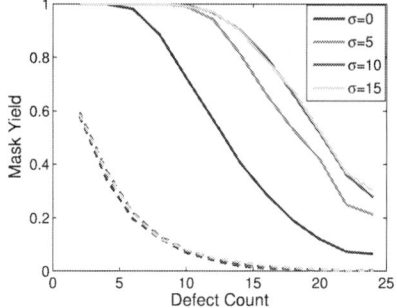

Fig. 2. Comparison of pre-pattern shift (dashed lines) and post-pattern shift (solid lines) mask yield of four parallel line layouts with same prohibited region density but different σ of Gaussian width.

III. APPROXIMATE ANALYTICAL METHODS

Monte Carlo based mask yield estimation is a simple, but computationally expensive strategy of generating random defect maps and performing pattern shift for each defect map. Mask yield can then be computing as the ratio of samples for which the final mask works, i.e. every defect on the mask is avoided. The Monte Carlo method starts off by constructing the prohibited region for different defect sizes. We then generate a defect map with N_d defects, assigning a size to each defect based on the given defect size distribution $P(s)$. Pattern shift is then applied for this defect map to determine if a feasible solution exists. Note that for each Monte Carlo iteration, N_d defects need to be generated since pattern shift makes mask failure dependent on location of several defects on the mask. This dependence necessitates a large number of Monte Carlo iterations to achieve convergence. The methodology we used for pattern shift is the same as the approach proposed by Wagner [6], which is optimal with respect to mask yield.

This naive method of estimating mask yield, although accurate, is cumbersome and slow, requiring many Monte Carlo iterations to give accurate results. Therefore this method of estimating mask yield is impractical for realistic layouts. Moreover, the method does not provide any design insights that could help improve the mask yield of a given layout. Despite these limitations, the accuracy of this method makes it appropriate for validating the faster, approximate method that we shall propose in this paper.

---

[1]All defects are assumed to be of size s in this example.

978-1-4799-2817-0/14 $31.00 © 2014 IEEE

A. Inclusion-Exclusion Method

In this section, we propose a method to estimate the mask yield with one simplifying assumption: pattern shift picks a feasible solution from a finite set of alternatives.

Let us first discretize all the potential defect sizes, into K discrete defect sizes, $s_1, s_2, ..., s_K$ with respective probabilities of occurrence, $P(s_1), P(s_2), ..., P(s_K)$. For a uniform spatial distribution of defects, we can then calculate the probability that a particular pattern shift solution (X_i, Y_j) works using Equation 1. Note that $\frac{Ar(PB^{s_k}_{ij})}{A_M}$ is equal to the prohibited region density for defect size s_k since shifting the polygons does not change area.

$$P(E^{\Delta}_{ij}) = \left(\sum_k P(s_k) \left(1 - \frac{Ar(PB^{s_k}_{ij})}{A_M} \right) \right)^{N_d} \quad (1)$$

Pattern shift aware mask yield can be estimated as the union of all the events E^{Δ}_{ij} since the mask will yield if any of the potential solutions work (Note that a solution is picked once the defect locations are known). Calculating this union of events can be done using inclusion-exclusion principle as shown in Equation 2.

$$P(\cup_{i,j} E^{\Delta}_{ij}) = \sum P(E^{\Delta}_{mn}) - \sum P(E^{\Delta}_{pq} \cap F^{\Delta}_{mn}) \ldots \quad (2)$$

The second order intersection term $P(E^{\Delta}_{pq} \cap E^{\Delta}_{mn})$ corresponds to the event that all defects lie in the non-prohibited region of both solution E^{Δ}_{pq} and E^{Δ}_{mn}. Hence, it can be computed using a polygon Boolean OR operation as shown in Equation 3.

$$P(E^{\Delta}_{pq} \cap E^{\Delta}_{mn}) = \left(\sum_k P(s_k) \left(1 - \frac{Ar(PB^{s_k}_{pq} \cup PB^{s_k}_{mn})}{A_M} \right) \right)^{N_d} \quad (3)$$

Computation of all the inclusion exclusion terms comprising $N_X \times N_Y$ orders (only first two order are shown above), is a #P-complete combinatorial enumeration problem since it requires the computation of $2^{N_X \times N_Y}$ terms [17]. This computational limitation, along with the quantization error incurred due to the assumption that the pattern shift solution space is discrete, make this method unsuitable for estimating mask yield for any realistic layouts.

Despite the impracticality of the inclusion-exclusion method, it does provide one interesting insight: *in addition to prohibited region density, mask yield depends on autocorrelation of the prohibited region of the input layout.* The second order terms in Equation 3, $Ar(PB^{s_k}_{pq} \cup PB^{s_k}_{mn})$, are linearly related to $Ar(PB^{s_k}_{pq} \cap PB^{s_k}_{mn})$, which measures the degree of overlap between the prohibited region PB^{s_k} with a shifted transform of PB^{s_k} (shifted by $(X_p - X_m, Y_q - Y_n)$). Each such overlapping area corresponds to one entry of the autocorrelation matrix of the 2D binary prohibited region signal, PB^{s_k}. Moreover, mask yield depends on the weighted sum of the prohibited region autocorrelation for different defect sizes. We will later leverage this dependence of mask yield on the weighted autocorrelation of prohibited region for critical density computation.

B. Spacings Method

In this sub-section, we will show that *if the prohibited region of a given layout is regular, the problem of finding the post-pattern shift mask yield can be mapped to the maximal spacing distribution problem, a classic geometric probability problem.* Note that pattern shift is assumed to be continuous here, unlike the previous subsection. Also, we consider only one defect size s here, and the assumption on regularity is for the prohibited region, PB^s.

First, suppose the prohibited region of the entire layout is a periodic parallel line structure. Assuming the lines are infinitely long and parallel to Y axis, any pattern shift in Y direction will not improve mask yield. Hence, only the X coordinates of the defects is relevant, and we can map all the defects to a single line. The periodicity assumption implies that the X coordinates of all the defects can be mapped to a modulo P_{pl} space with a single line of width W_{pl}. An optimal pattern shift based defect avoidance technique can successfully avoid all the defects, if and only if there exists a gap or spacing of size W_{pl} with no defect inside it. This mapping is illustrated in Figure 3.

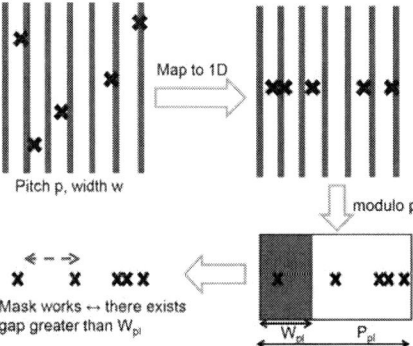

Fig. 3. Mapping mask yield estimation of parallel line to maximal spacing distribution.

This problem is equivalent to finding the probability of existence of a gap larger than $\frac{W_{pl}}{P_{pl}}$ on a unit circle with a uniform distribution of points [2]. This geometric probability, also referred to as the one dimensional maximal spacing problem [18], was first computed exactly by Stevens [19], which allows us to estimate pattern shift aware mask yield of a parallel line layout.

Similar to the parallel line case above, we can show that the the mask yield for a regular, square contact array pattern is equivalent to the two-dimensional maximal spacing distribution. Janson derived an asymptotic analytical expression for the multi-dimensional maximal spacing problem [20], that holds true as the number of random points (defects, in our case) tend to infinity. Using his expression, we can estimate pattern shift aware mask yield for an infinite contact array layout as shown in Equation 4. An additional condition is included for $N_d \leq \frac{2}{D_P}$ to correct for the anomaly that mask yield increases with increase in the number of defects. This analytical expression will be referred to as Janson's formula in the rest of this paper.

$$\begin{aligned} Y^{ct}_M &= 1 - e^{-N_d^2 D_P e^{-N_d D_P}} \quad && if \quad N_d \geq \frac{2}{D_P} \\ &= 1 && otherwise \quad (4) \end{aligned}$$

IV. CRITICAL DENSITY METHOD

Although the periodicity assumption of parallel line and contact arrays enables us to map the yield estimation problem to a maximal spacing distribution problem, deriving such analytical expressions for random layouts is not straight-forward. In order to address the issue of estimating the yield of realisitic layout patterns, we propose a two-step model that applies principles from Section III-A and III-B.

We first define critical density of a layout as follows. *For any given input layout, critical density is the value of D_P such that Equation*

[2]Since there is no mask defect distribution data available, we assume uniform spatial distribution of defects as it usually gives more pessimistic yield estimates compared to clustering [6], [13]

4 can most accurately predict the actual mask yield of the layout for any number of defects. Mathematically, we can use the ubiquitous least squares as the criteria for accuracy and thereby define critical density as given in Equation 5.

$$D_{crit} = \underset{0 \leq D_P \leq 1}{\operatorname{argmin}} \sum_{N_d = N_d^{min}}^{N_d = N_d^{max}} (Y_M^{ct}(N_d, D_P) - Y_M^{true}(N_d))^2 \quad (5)$$

With this definition of critical density, our two-step model first estimates critical density of an input layout using the weighted autocorrelation matrix of the prohibited region as the predictor variables (motivated by the derivation in Section III-A). Equation 4 is then used to estimate mask yield.

The use of critical density as a part of a two-step model to estimate mask yield abstracts out defect density thereby providing a defect density-independent metric that depends solely on the prohibited region of a given layout. This metric can be used to compare the robustness of different layouts to EUV mask defects. Additionally, critical density is similar to probability of failure (ratio of critical area to chip area) in conventional CAA.

For a realistic full chip layout, using the entire autocorrelation matrix of a layout as a feature set to predict critical density is infeasible, both from the perspective of computing the autocorrelation, and fitting a model ("curse of dimensionality"). We propose the following set of steps to reduce the dimension of the autocorrelation matrix which is then used to predict the critical density:

- *Limited autocorrelation:* Based on the derivation in Section III-A, only the first $\frac{\Delta_X}{pix} \times \frac{\Delta_Y}{pix}$ entries of the autocorrelation matrix of size $\frac{L_X}{pix} \times \frac{L_Y}{pix}$ need to be considered as a part of the feature set. Moreover, we scale all the entries of the autocorrelation matrix by the reticle area, to make it independent of design size.
- *Compression:* Only the low-frequency Fourier components of the limited autocorrelation matrix are used as features for predicting critical density. This is reasonable since layouts are dominated by lower frequency components due to design rule constraints.

With this reduced autocorrelation based feature set, we apply a simple multivariate linear regression model to predict critical density of any given layout. Since the autocorrelation matrix size depends on the maximum available pattern shift and is scaled by reticle size, it is independent of the layout size. Therefore the linear regression model can be trained using small layout clips, and the trained model can then be applied to large realistic designs. This makes the training of the model manageable.

The operations involved in computing critical density, and then mask yield, of a given random layout are specified in Algorithm 1. *Note that the two-step critical density model does not assume discrete pattern shift solutions.* It actually accounts for the optimal continuous pattern shift since the linear model that estimates critical density is fitted using the Monte Carlo method that uses the optimal continuous pattern shift.

The runtime for estimating critical density is dominated by the polygon Boolean operations to compute the autocorrelation matrix. Each polygon Boolean operation takes $O(\rho A_L \log \rho A_L)$ runtime. With a sampling pixel size of pix, fast fourier transform can be performed in $O(\frac{A_\Delta}{pix^2} \log \frac{A_\Delta}{pix^2})$. Hence, the runtime order complexity to compute the critical density is $O(K \times \frac{A_\Delta}{pix^2} \times \rho A_L \log \rho A_L + \frac{A_\Delta}{pix^2} \log \frac{A_\Delta}{pix^2})$. In constrast, the order complexity of Monte Carlo is $O(N_d \times (\rho A_\Delta \log \rho A_\Delta + (\log \rho A_L)^3)))$ per iteration[3]. This method can also be easily parallelized by computing each entry of the autocorrelation matrix independently.

Algorithm 1 Steps for estimating critical density

Input: Design layout of size $L_X \times L_Y$ and total shift size permitted $\Delta_X \times \Delta_Y$. Tunable parameters: sample size for autocorrelation pix, and number of fourier order to pick N_F
Output: Critical density of layout.
1: Construct prohibited region of layout PB^s for $s \in s_1, s_2, \dots s_K$
2: Define matrix AC of size $\frac{\Delta_X}{pix} \times \frac{\Delta_Y}{pix}$
3: reticleArea $= (L_X + \Delta_X) \times (L_Y + \Delta_Y)$
4: **for all** $X_i \in \{0, p, 2p, \dots \Delta_X\}$ **do**
5:    **for all** $Y_j \in \{0, p, 2p, \dots \Delta_Y\}$ **do**
6:       $AC(\frac{X_i}{pix}, \frac{Y_j}{pix}) = \sum_k P(s_k) \frac{Area(PB^{s_k} \cup PB_{ij}^{s_k})}{reticleArea}$
7:    **end for**
8: **end for**
9: $ACF = fft(AC)$
10: Pick all terms of ACF with Fourier order less than or equal to N_F
11: Apply fitted linear model to get critical density

V. EXPERIMENTAL RESULTS

Both the Monte Carlo method, and our proposed critical density method are implemented in C++. OpenAccess API [21] is used to read and query layouts. The polygon Boolean operations are performed using Boost Polygon Library [22]. Fourier transform of the autocorrelation matrix is done using FFTW library [23], and matrix operations are done using Eigen [24]. OpenMP is used to parallelize both the Monte Carlo Method[4] and the autocorrelation matrix construction step of our critical density method, with eight threads for execution. All our computation has been done on a high performance compute cluster. The reported runtime for the various testcases is the wall time on the compute nodes of the cluster.

The number of Monte Carlo iterations is kept fixed at $20,000$ for all the training clips and test layouts. For all our testcases, the Monte Carlo method is run 15 times, with defect density ranging from 10 defects to 150 defects. All of our reported average, maximum and average root mean square error (RMSE) values are across this range of defects.

All our analysis in this section is done on designs created using Synopsys $32nm$ standard cell library [26], and scaled down to $8nm$. The designs were synthesized, placed and routed using Cadence Encounter [27] with 90% cell utilization, unless otherwise stated.

All mask defects are taken as 3D Gaussian-shaped, with three discrete height values ($0.5nm, 1.0nm, 2nm$) and three discrete full width half maximum values ($25nm, 50nm, 75nm$). The corresponding nine discrete defect sizes are assigned probability of occurrence inversely proportional to their respective volume. The CD tolerance of all the shapes was set at $0.8nm$.

The available pattern shift ($\Delta_X \times \Delta_Y$) is taken as $0.5\mu m \times 0.5\mu m$. Current literature on pattern shift suggest that the total available shift is around $200\mu m \times 200\mu m$ [7]. A smaller shift value is chosen to demonstrate our methodology since the runtime of the validation Monte Carlo method becomes too slow with large shift area ($\geq 5,000$ hours based on the runtime of [5]). Moreover, the shift area we have

---

[3]Number of Monte Carlo iterations also depends on A_L and A_Δ
[4]Thread-safe random number generation is done here [25]

chosen is sufficient for comparing different layouts since it is large enough to cover several pitches at $8nm$ node. pix is set to $20nm$ [5], and the size of the autocorrelation based feature vector for each layout is 17. It comprises all the entries from the FFT matrix of the limited autocorrelation matrix with both row and column indices less than 4, and a constant.

The linear regression model to estimate critical density of a layout is trained using $5\mu m \times 5\mu m$ layout clips obtained from a large $8nm$ layout. We used a total of 400 layout clips from each of the four critical layers of a design as the training set: polysilicon, metal 1, contact and active. For these small layout clips, we used the Monte Carlo method to estimate the true mask yield for different number of defects. Using this data, we computed the critical density of the layout clips by solving Equation 5 in MATLAB with the interior point method. The autocorrelation based features are computed for each clip, and used to train the linear regression model for critical density in MATLAB.

A. Model Validation

The trained linear regression model is then applied on different layers of four benchmark layouts from ISCAS'89 [28], and one RISC processor layout. All these layouts are different from the training clips used to fit the model. The results are summarized in Table II [6].

Compared to the rigorous Monte Carlo method, our critical density method is able to predict critical density fairly accurately for all the test layouts with a runtime improvement ranging from $300\times$ to $1300\times$ The average RMSE in estimating mask yield ranges from $0, .08\% - 6.44\%$. Moreover, the method is able to track the general trend of how mask yield changes with defect density fairly accurately. This is illustrated in Figure 4, which plots the mask yield versus defect density of the Monte Carlo and the critical density method for the polysilicon and metal 1 layers of $s1423$.

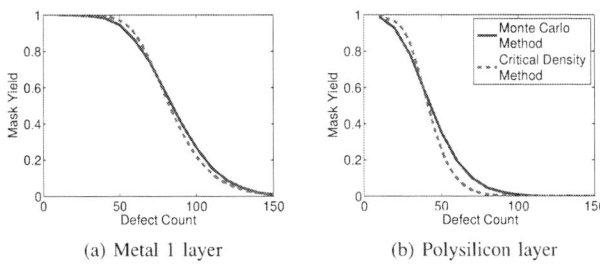

(a) Metal 1 layer   (b) Polysilicon layer

Fig. 4.   Mask yield versus defect density for two layers of s1423 design

B. Impact of Layout Density and Regularity

The critical density of a layout is strongly influenced by the layout density. This is confirmed by comparing two version of s1196 in Table II with cell utilization of 90% (default) and 70%. Reducing the cell utilization reduces the layout density of all the layers, thereby reducing the critical density as well. Moreover, note that different designs constructed with the same utilization have almost equal critical density for each respective layer. This suggests that critical density depends on global layout characteristics instead of local hotspot-like regions. Hence, improving this metric may necessitate changes in design rules or physical design techniques.

[5]Pixel size only dictates the shift values for which the Boolean AND operations are performed.

[6]The validation Monte Carlo method is too slow to be run for the larger processor layout, hence the missing entries in the table.

Although layout density plays an important role in determining critical density, it is not the only factor that affects critical density. Polysilicon layer (which is a fixed pitch regular grating) has higher critical density compared to irregular metal 1 layer of each design, despite lower layout density. This indicates that random layout patterns are better suited to exploit the benefits of pattern shift due to better autocorrelation properties. To further highlight this impact of regularity, we constructed two regular layouts, parallel line and contact array, which have the same layout density as the polysilicon layer of $s1423$ and metal 1 layer of $s1196 - u70$. The results, shown in Figure 5, highlight two key aspects of layouts that affect mask yield:

- 1D layout topology (parallel line and polysilicon), are significantly worse than 2D topology (contact array and metal 1) because 2D layouts can benefit from pattern shift in both X and Y directions, whereas 1D layouts benefit only in the direction perpendicular to the parallel lines.
- An irregular 2D layout like metal 1 is much better suited to derive the benefit of pattern shift compared to a regular 2D contact array.

Most manufacturing processes, especially lithography, favor 1D regular layouts. This has led to increasing layout regularity with each technology node. But our results show that the reduced mask yield of such regular layouts can significantly increase mask cost for EUV lithography. Hence, a systematic co-optimization to balance these two competing requirements may be required for EUV layouts.

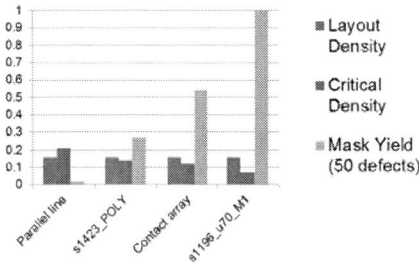

Fig. 5.   Illustration of impact of regularity on critical density (and consequently, mask yield) for layouts with same density.

VI. CONCLUSION AND FUTURE WORK

In this work, we proposed a new metric for evaluating the robustness of EUV layouts with respect to mask defects, critical density. Using critical density, layout designers and mask makers can quickly estimate the probability of finding a defective EUV mask blank on which the layout can be safely patterned for any defect density(mask yield). Our method accounts for the impact of pattern shift based defect avoidance technique on mask yield, which is the most challenging part of this methodology. We first solved the problem assuming discrete pattern shift solutions using inclusion-exclusion. Then we mapped the problem to a classic geometric probability problem, maximal spacings, for the limited case of parallel line and contact array patterns. Using principles from both these approaches, we defined our critical density metric and proposed a novel method to estimate critical density, which can then be used to estimate the pattern shift aware mask yield for any arbitrary layout. Our method was shown to be $300 - 1300\times$ faster than the rigorous Monte Carlo method for estimating mask yield and was able to predict mask yield with $0.08\% - 6.44\%$ average RMSE across a range of defect density for four critical layers (polysilicon, active, contact and metal1).

TABLE II

VALIDATION OF CRITICAL DENSITY METHOD. MASK YIELD RMSE IS AVG.RMSE BETWEEN THE MASK YIELD ESTIMATE OF MONTE CARLO AND PROPOSED METHOD ACROSS THE DEFECT RANGE. RUNTIME IS TOTAL WALL TIME REQUIRED TO COMPUTE MASK YIELD FOR THE DEFECT RANGE.

Design	Layer	Number of shape edges	Layout density	Monte Carlo method Runtime (sec.)	Critical density method		
					Critical density	Mask yield RMSE	Runtime (sec.)
s349-syn32nm	POLY	4084	0.16	10742	0.14	4.15%	19
	ACT	2384	0.32	14725	0.04	4.67%	16
Utilization 90%	CO	20132	0.03	57432	0.03	0.32%	62
	M1	9432	0.22	48382	0.08	2.84%	84
s1423-syn32nm	POLY	20672	0.16	66651	0.14	4.24%	86
	ACT	11348	0.33	24895	0.04	4.54%	79
Utilization 90%	CO	101856	0.03	239572	0.03	0.50%	349
	M1	47554	0.22	261816	0.08	1.98%	465
s1196-syn32nm	POLY	11788	0.16	28201	0.14	4.28%	47
	ACT	5800	0.29	23522	0.03	3.40%	39
Utilization 90%	CO	60448	0.03	253287	0.03	0.54%	192
	M1	26128	0.19	198613	0.07	2.91%	222
s1196-syn32nm-u70	POLY	11788	0.12	34597	0.10	6.44%	47
	ACT	6176	0.23	12797	0.03	0.94%	41
Utilization 70%	CO	62336	0.03	255483	0.03	0.08%	233
	M1	26502	0.16	208513	0.06	4.15%	227
Cortex M0	POLY	333748	0.15	NA	0.13	NA	858
	ACT	178396	0.29	NA	0.03	NA	999
Utilization 90%	CO	1746968	0.03	NA	0.03	NA	4743
	M1	767380	0.19	NA	0.07	NA	5852

The methodology for estimating critical density shows that mask yield is a strong function of the autocorrelation of the layout. Our analysis indicates that irregular 2D layouts have better mask yield for the same layout density, which is contrary to most manufacturing processes that demand layout regularity. By using dummy features to make irregular layouts regular with respect to printability, this problem can be addressed since defects on dummy features do not matter.

Fast estimation of critical density enabled by our method can allow us to develop techniques to improve the mask yield of EUV layouts. Our preliminary experiments to improve critical density by iteratively adjusting the whitespace after placement suggests that $3\% - 7\%$ improvement in mask yield is possible without any area penalty. Since we used a random search based whitespace optimization for each row independently. the method is too slow to be practically useful. In the future, we plan to develop a scalable layout optimization methodology to improve the robustness of EUV layouts. Our future work also includes further enhancing the critical density model to account for other defect avoidance strategies such as pattern rotation or floorplanning, and dealing with non-uniform spatial distribution of defects (if defect data necessitate so).

ACKNOWLEDGEMENT

The authors would like to acknowledge the generous support of IMPACT+ center (http://impact.ee.ucla.edu) and NSF CAREER award CCF-0846196, and valuable suggestions from Dr. Luigi Capodieci (Globalfoundries Inc.) and Mr. Pradiptya Ghosh (Mentor Graphics Corp.).

REFERENCES

[1] "International Technology Roadmap for Semiconductors(ITRS)," http://www.itrs.net/, 2009.
[2] C. H. Clifford, "Simulation and Compensation Methods for EUV Lithography Masks with Buried Defects," Ph.D. dissertation, EECS Department, University of California, Berkeley, 2010.
[3] H. J. Levinson, "Extreme ultraviolet lithography's path to manufacturing," SPIE JM3, 2009.
[4] J. Burns and M. Abbas, "EUV mask defect mitigation through pattern placement." Proc. SPIE/BACUS, 2010.
[5] H. Zhang, Y. Du, M. D. F. Wong, and R. Topalaglu, "Efficient pattern relocation for EUV blank defect mitigation," in Proc. ASP-DAC, 2012.
[6] A. Wagner, M. Burkhardt, , A. B. Clay, and J. P. Levin, "Mitigation of extreme ultraviolet mask defects by pattern shifting: Method and statistics," J. Vac. Sci. & Technol., B, 2012.

[7] P.-Y. Yan, Y. Liu, M. Kamna, G. Zhang, R. Chen, and F. Martinez, "EUVL multilayer mask blank defect mitigation for defect-free EUVL mask fabrication." Proc. SPIE, 2012.
[8] A. Kagalwalla and P. Gupta, "Design-Aware Defect-Avoidance Floor-planning of EUV Masks," IEEE TSM, 2013.
[9] Y. Du, H. Zhang, M. D. F. Wong, Y. Deng, and R. O. Topaloglu, "Efficient multi-die placement for blank defect mitigation in EUV lithography." Proc. SPIE, 2012.
[10] H. Zhang, Y. Du, M. D. F. Wong, Y. Deng, and P. Mangat, "Layout small-angle rotation and shift for EUV defect mitigation," in Proc. ICCAD, 2012.
[11] Y. Du, H. Zhang, and M. D. F. Wong, "Linear time EUV blank defect mitigation algorithm considering tolerance to inspection inaccuracy." Proc. SPIE Photomask, 2012.
[12] A. Elayat, P. Thwaite, and S. Schulze, "EUV mask-blank defect avoidance solutions assessment." Proc. SPIE Photomask, 2012.
[13] K. Jeong, A. B. Kahng, and C. J. Progler, "Cost-driven mask strategies considering parametric yield, defectivity, and production volume," SPIE JM3, 2011.
[14] Y. Du, H. Zhang, , M. D. F. Wong, and R. O. Topaloglu, "EUV mask preparation considering blank defects mitigation." Proc. SPIE/BACUS, 2011.
[15] P. Gupta and E. Papadopoulou, "Yield Analysis and Optimization," in The Handbook of Algorithms for VLSI Physical Design Automation. CRC Press, 2010.
[16] A. Kagalwalla, P. Gupta, C. Progler, and S. McDonald, "Design-aware mask inspection," IEEE TCAD, 2012.
[17] J. Kahn, N. Linial, and A. Samorodnitsky, "Inclusion-exclusion: Exact and approximate," Combinatorica, 1996.
[18] R. Pyke, "Spacings," Journal of the Royal Statistical Society. Series B (Methodological), 1965.
[19] W. L. Stevens, "Solution to a Geometrical Problem in Probability," Annals of Eugenics, vol. 9, no. 4, pp. 315–320, 1939.
[20] S. Janson, "Maximal spacings in several dimensions," The Annals of Probability, 1987.
[21] "Openaccess API," http://www.si2.org/.
[22] "Boost Polygon Library," http://www.boost.org/doc/libs/1_52_0/libs/polygon/doc/index.htm.
[23] "FFTW," http://www.fftw.org/.
[24] G. Guennebaud, B. Jacob et al., "Eigen v3," http://eigen.tuxfamily.org, 2010.
[25] M. Hoemmen, "Generating random numbers in parallel," 2007.
[26] "Synopsys 32nm library," 2010.
[27] "Cadence SOC Encounter," http://www.cadence.com/, 2011.
[28] "Iscas-89 Benchmark Circuits Verilog Files," http://www.pld.ttu.ee/~maksim/benchmarks/iscas89/verilog/.

Statistical Analysis of Random Telegraph Noise in Digital Circuits

Xiaoming Chen[†], Yu Wang[†], Yu Cao[§], Huazhong Yang[†]

[†]Department of Electronic Engineering, Tsinghua National Laboratory for Information Science and Technology,
Tsinghua University, Beijing 100084, China
[§]Department of ECEE, Arizona State University, Tempe, AZ, USA
Email: chenxm05@mails.tsinghua.edu.cn, yu-wang@tsinghua.edu.cn, yu.cao@asu.edu, yanghz@tsinghua.edu.cn

Abstract—Random telegraph noise (RTN) has become an important reliability issue at the sub-65nm technology node. Existing RTN simulation approaches mainly focus on single trap induced RTN and transient response of RTN, which are usually time-consuming for circuit-level simulation. This paper proposes a statistical algorithm to study multiple traps induced RTN in digital circuits, to show the temporal distribution of circuit delay under RTN. Based on the simulation results we show how to protect circuit from RTN. Bias dependence of RTN is also discussed.

Keywords—Random telegraph noise; Statistical analysis; Reliability

I. INTRODUCTION

As CMOS technology scales, many reliability mechanisms that affect circuits' reliability are becoming more serious. These issues must be evaluated and well addressed during the design time. In recent years, random telegraph noise (RTN) has attracted researchers' attention. RTN can cause random fluctuations in electrical parameters (such as V_{th} and I_d) [1]. RTN-induced I_d variation can be up to 40% in 30×30nm devices [2], and the V_{th} variation can be larger than 70mV for the smallest devices at 22nm technology node [3]. The RTN effect increases superlinearly with the scaling down of the device's size [4]. RTN is also a serious concern in CMOS logic circuits [5].

The physical mechanism of RTN has been studied for many years [1], [6]–[8]. The impact of single trap induced RTN on memories was widely studied [4], [9]–[15], some circuit-level simulation approaches were also proposed [16]–[19]. Most of them evaluate single trap induced RTN. Multiple traps induced RTN has been rarely studied [11]. However, there should be 2~3 detectable traps in each device, and the number of traps follows Poisson distribution [20]. The multi-trap problem is actually a statistical problem, which is more complex than the single-trap problem. This paper will evaluate circuit performance under multiple traps induced RTN.

On the other hand, many existing researches used time-consuming SPICE simulation to obtain the transient response under RTN [14]–[18]. This method can be used for predicting read/write failures in memories, but it is useless for evaluating the timing of CMOS logic circuits. To actually understand the impact of RTN on logic circuits, a temporal distribution of circuit delay should be used to obtain the statistical information. In other words, RTN should be integrated into timing analysis to see the real impact of RTN on logic circuits.

This work was supported by National Natural Science Foundation of China (No. 61373026, No.61261160501, No. 61028006), 973 project (2013CB329000), National Science and Technology Major Project (2011ZX01035-001-001-002), and the youth talent development plan of Beijing (YETP0099).

The contributions of this paper are summarised as follows.

- This paper integrates RTN into timing analysis to estimate the impact of RTN on logic circuits. A fast estimation framework which is based on a statistical algorithm is proposed to obtain the temporal distribution of circuit delay without SPICE or Monte-Carlo.

- This is the first time to analyze the impact of multiple traps induced RTN on logic circuits.

- We find that it is not practical to protect circuit from the maximum possible delay under RTN, but we should only ensure that circuit functions correctly with a certain probability during the whole lifetime, such that timing violation can hardly happen. By ensuring that circuit functions correctly with $1 - 10^{-9}$ probability, circuit should be protected from the degraded delay which is about 7% to 40% larger than the intrinsic delay.

- The impact of gate voltage on RTN effect is investigated. We show that the impact of V_{gs} on RTN has two trends: increasing and saturation. A simple V_{dd} tuning approach can effectively protect circuit from the effect of RTN.

The rest of the paper is organized as follows. Section II gives backgrounds and modeling of RTN. We introduce the proposed algorithm in Section III. The experimental results are shown in Section IV. Finally Section V concludes the paper.

II. MODELING RANDOM TELEGRAPH NOISE

A. Physics of RTN

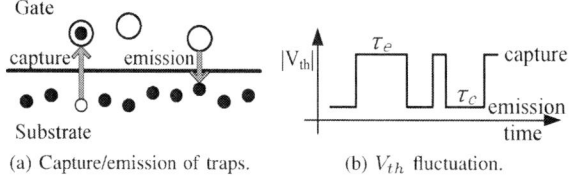

(a) Capture/emission of traps.   (b) V_{th} fluctuation.

Fig. 1: Physics of RTN and its impact on V_{th}.

As shown in Fig. 1a, RTN is caused by the capture/emission process of charge carriers by the oxide traps (defects) [1], [8]. A carrier in the channel is occasionally captured by a trap in the oxide, and the carrier will be emitted back after a period of time. The capture/emission process of a given trap can be described by a two-state Markov chain [16]. With reference to Fig. 1b, the high V_{th} state occurs when the carrier is captured by the trap (i.e. the trap is filled), and the low V_{th} state occurs when the carrier is emitted back (i.e. the trap is empty). The time spent in the high V_{th} state is emission time τ_e, which means "time to be emitted back", and the time spent in the low V_{th} state is capture time τ_c, which means "time to be captured". τ_e and τ_c strongly depend on gate overdrive ($\tau_c = 10^{-3}$s to 10^{-2}s and $\tau_e = 10^{-1}$s to 10^{+1}s [1]).

B. Dependence of Switched Bias Conditions

The capture and emission time constants also depend on the switched bias conditions. When a transistor is on, the time constants are $\tau_c^{(on)}$ and $\tau_e^{(on)}$; when the transistor is off, the time constants can be expressed as $\tau_c^{(off)} = \tau_c^{(on)} \times m_c$ and $\tau_e^{(off)} = \tau_e^{(on)}/m_e$ [21]. Consider a transistor with duty cycle SP (the probability of on is SP), its average time constants are

$$\begin{aligned} \tau_c &= SP \times \tau_c^{(on)} + (1 - SP) \times \tau_c^{(off)} \\ \tau_e &= SP \times \tau_e^{(on)} + (1 - SP) \times \tau_e^{(off)} \end{aligned} \quad (1)$$

A logic simulator is used to calculate the signal probability of all internal nodes in a circuit.

C. RTN-induced V_{th} Shift

Single trap induced V_{th} shift is given by [10]

$$\Delta V_{th} = \frac{q}{C_{ox}WL} \quad (2)$$

where q is the elementary charge, C_{ox} is the unit area capacitance, W and L are channel width and length respectively.

It is shown that the number of traps in each device follows Poisson distribution: $N_{tr} \sim Pois(\lambda)$ (N_{tr} is not the number of existing traps, which is constant, but that of detectable traps) [20], in which λ is the average number of traps. Based on the measured data plotted in [10], [11], [20], multiple traps induced maximum V_{th} shift can be approximately expressed as the sum of each individual trap induced V_{th} shift, so V_{th} shift caused by N_{fil} filled traps is

$$\Delta V_{th} = \frac{qN_{fil}}{C_{ox}WL} \quad (3)$$

D. RTN-induced Gate Delay Modeling

The propagation delay of a logic gate i is given by

$$D(i) = \frac{K_i C_{L,i} V_{dd}}{(V_{gs} - V_{th,i})^\alpha} \quad (4)$$

where K_i is a coefficient related with device physical parameters, $C_{L,i}$ is the load capacitance, and α is the velocity saturation index. The delay shift caused by V_{th} shift is

$$\Delta D(i) \approx \frac{\alpha \Delta V_{th,i}}{V_{gs} - V_{th0}} \times D(i) \quad (5)$$

For a given trap, its state can be described by a two-valued random variable X: 0 corresponding to empty state and 1 corresponding to filled state. When the capture/emission process is stationary, the two states have a stationary distribution, which is given by

$$P(X = 0) = \frac{\tau_c}{\tau_e + \tau_c}, P(X = 1) = \frac{\tau_e}{\tau_e + \tau_c} \quad (6)$$

Note that our target is statistical analysis so only stationary state is considered, the detailed capture/emission sequence is not considered. Combining Eq. (3), (5) and (6), delay shift of

gate i caused by multiple traps is described by a compound Poisson distribution:

$$\Delta D(i) = \sum_{i=1}^{N_{eff}} \Delta D_O(i), N_{eff} \sim Pois(r\lambda) \quad (7)$$

where $r = \frac{\tau_e}{\tau_e + \tau_c}$, and $\Delta D_O(i)$ is the delay shift caused by a single filled trap, which is calculated by Eq. (2) and (5).

III. STATISTICAL ALGORITHM FOR RTN SIMULATION

This section proposes an RTN simulation framework based on a statistical algorithm. The proposed algorithm can fast obtain the temporal distribution of circuit delay without Monte-Carlo or SPICE. The algorithm is similar to the idea of the well-known statistical static timing analysis (SSTA) algorithm [22], but they are different. SSTA is used to evaluate the impact of process variations on the yield. For each individual chip they are assumed to be constant, and SSTA predicts that certain percent of manufactured circuits functions correctly. Unlike that RTN affects gate delays differently in different time moments. Therefore, delay value varies in time randomly. Because of that if chip operates long enough path delay can always get its worst value. So we are more interested in the maximum circuit delay. In addition, SSTA assumes that any delay can be described by a canonical form, but according to Eq. (7), MAX of two compound Poisson distributions is not a compound Poisson distribution, so canonical forms cannot be used in RTN analysis.

A. Terms and Definitions

Several terms are defined as follows. The *arrival/leaving time* of gate i ($AT(i)/LT(i)$) are the maximum propagation delay from circuit primary inputs (PI) to the inputs/output of gate i. $LT(i) = AT(i) + D(i)$. The propagation delay from gate i's output to gate j's output is $PD(i,j)$. An example of AT, LT and PD are illustrated in Fig. 2. An *input path (IP)* of gate i is a path from PIs to gate i, and $IP(i)$ is the set of all input paths of gate i. The *critical input paths (CIP)* are the paths that have longest delay in $IP(i)$, denoted as $CIP(i)$. In the following contents, we use f_X and F_X to represent the probability density function (PDF) and cumulative distribution function (CDF) of random variable X. Φ and ϕ are the CDF and PDF of the standard normal distribution. Φ_2 and ϕ_2 are the CDF and PDF of the standard bivariate normal distribution. μ_X and σ_X are mean and standard deviation of random variable X.

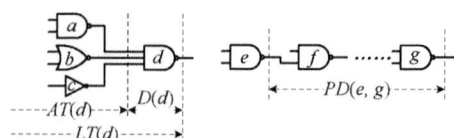

Fig. 2: Example to illustrate $AT(d)$, $LT(d)$ and $PD(e,g)$.

B. Simulation Framework

The RTN simulation framework is shown in Fig. 3. HSPICE is used to create a gate library which includes gate intrinsic delay and oxide capacitance of each gate type (i.e. INVX1, INVX4, NAND2X1, etc), based on the predictive technology model (PTM) [23]. STA tools are used to obtain the critical path information, which is used for correlation calculation.

A logic simulator is used to calculate the probability of all internal nodes. In STA, if $RT(i) - LT(i) <= \varepsilon$ (RT means "required time"), gate i is critical. $\varepsilon > 0$ to ensure that the criticality obtained in STA is also accurate in statistical analysis. Finally, the delay distribution of the circuit is calculated by the proposed statistical algorithm.

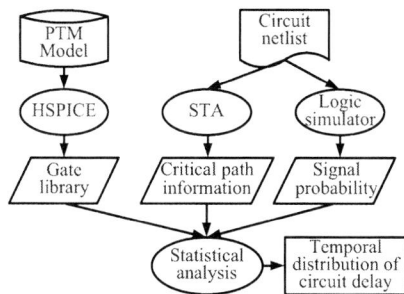

Fig. 3: RTN simulation framework.

C. Correlation of Gates

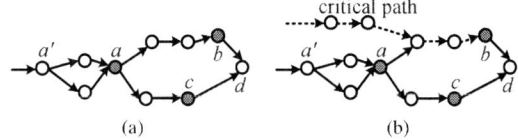

Fig. 4: Example to illustrate the correlation of gates.

In a traditional SSTA, the correlation is spatial correlation, i.e. global/local sources of variation. However, based on the delay model in Section II, there is no spatial correlation in RTN simulation. There is still correlation among gates, but the correlation is caused by path delay. The correlation comes from the case that the IPs of two gates have common gates, as illustrated in Fig. 4a, where gate a and a' are the common gates of $IP(b)$ and $IP(c)$. We are interested in the correlation of gate b and c, and we have

$$\rho_{LT(b),LT(c)} = \frac{E(LT(b)LT(c)) - \mu_{LT(b)}\mu_{LT(c)}}{\sigma_{LT(b)}\sigma_{LT(c)}} \quad (8)$$

Since $LT(b) = LT(a) + PD(a,b)$ and $LT(c) = LT(a) + PD(a,c)$, Eq. (8) can be written as

$$
\begin{aligned}
&\rho_{LT(b),LT(c)} \\
&= \{E[(LT(a) + PD(a,b)) \cdot (LT(a) + PD(a,c))] \\
&\quad -(\mu_{LT(a)} + \mu_{PD(a,b)}) \cdot (\mu_{LT(a)} + \mu_{PD(a,c)})\} \\
&\quad /(\sigma_{LT(b)}\sigma_{LT(c)}) \\
&= \frac{E(LT(a)^2) - (\mu_{LT(a)})^2}{\sigma_{LT(b)}\sigma_{LT(c)}} = \frac{(\sigma_{LT(a)})^2}{\sigma_{LT(b)}\sigma_{LT(c)}}
\end{aligned} \quad (9)
$$

This indicates that the correlation of gate b and c is only determined by the variance of the leaving time of gate b, c and a. Actually gate a' is also a common gate of $IP(b)$ and $IP(c)$; however, $\rho_{LT(b),LT(c)}$ does not depend on gate a'. So the "last" common gate should be used to calculate the correlation coefficient of two gates. Obviously the "last" common gate must have largest delay among all the candidate common gates. Another essential condition is that gate a should be in both $CIP(b)$ and $CIP(c)$; otherwise as shown in Fig. 4b, the dotted line is the CIP of gate b, so $LT(a)$ has

little impact on $LT(b)$, and $LT(b)$ is not correlated to $LT(c)$. The algorithm for calculating the correlation of two gates are shown in Algorithm 1.

Algorithm 1 Calculating the correlation of gate i and j.

1: $S = CIP(i) \cap CIP(j)$;
2: **if** $S == \emptyset$ **then**
3: $\quad \rho_{LT(i),LT(j)} = 0$;
4: **else**
5: $\quad$ Find gate $u \in S$ such that $\mu_{LT(u)} = \underset{v \in S}{MAX}\{\mu_{LT(v)}\}$;
6: $\quad \rho_{LT(i),LT(j)} = \sigma_{LT(u)}^2/(\sigma_{LT(i)}\sigma_{LT(j)})$;
7: **end if**

D. Proposed Statistical Algorithm

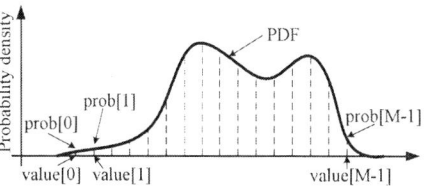

Fig. 5: Representing an arbitrary distribution by sampling the PDF.

The key idea of the proposed algorithm is that an arbitrary PDF that has no analytical form can be represented by sampling [24], as shown in Fig. 5. The number of sampled nodes is M, and larger M leads to better approximation. The sampled intervals can be uniform or non-uniform. Two vectors, $value$ and $prob$ with length M, are used to express the sampled value and the corresponding probability density, so the PDF becomes a discrete probability mass function (PMF). In this paper, $M = 100$ is used.

1) ADD operation: This is used to calculate the leaving time $LT(i) = AT(i) + D(i)$. Since $AT(i)$ and $D(i)$ are independent, their sum is the convolution of their PMFs. The convolution of two PMFs of sampled length M_1 and M_2 will have a length of $M_1 M_2$, which may exceeds M, so a grouping method is proposed to reconstruct the sum to be M-length. The pseudo code of ADD operation is shown in Algorithm 2.

Algorithm 2 Calculating $Z = X + Y$. X and Y are of sampled length M_X and M_Y.

1: Alloc two vectors, $value$ and $prob$, both of length $M_X M_Y$;
2: //convolution
3: **for** $i = 0$ to $M_Y - 1$ **do**
4: $\quad$ **for** $j = 0$ to $M_X - 1$ **do**
5: $\quad\quad value[i * M_X + j] = Y.value[i] + X.value[j]$;
6: $\quad\quad prob[i * M_X + j] = Y.prob[i] \times X.prob[j]$;
7: $\quad$ **end for**
8: **end for**
9: //grouping
10: $min = \underset{i < M_X M_Y}{MIN}\{value[i]\}$; $max = \underset{i < M_X M_Y}{MAX}\{value[i]\}$;
11: $step = (max - min)/M$;
12: Clear Z;
13: **for** $i = 0$ to $M_X M_Y - 1$ **do**
14: $\quad Z.value[(value[i] - min)/step] += prob[i] \times value[i]$;
15: $\quad Z.prob[(value[i] - min)/step] += prob[i]$;
16: **end for**
17: **for** $i = 0$ to $M - 1$ **do**
18: $\quad Z.value[i] /= Z.prob[i]$;
19: **end for**

978-1-4799-2817-0/14 $31.00 © 2014 IEEE

2) MAX operation: This is used to calculate the arrival time $AT(i) = MAX\{LT(j_1), LT(j_2), \cdots, LT(j_K)\}$, where gate $j_1, j_2, \cdots, j_K$ are the fan-ins of gate i. Consider a simple case $W = MAX(X, Y)$, X and Y are with correlation coefficient ρ_{XY}. The core operation is to calculate the integral

$$P(W \leq w) = F_{XY}(w, w; \rho_{XY}) = \int_{-\infty}^{w} \int_{-\infty}^{w} f_{XY}(u, v; \rho_{XY}) du dv \quad (10)$$

where f_{XY} is the joint PDF of X and Y and its analytical form is nonexistent. The integral is difficult to calculate. In this paper, we use an approximate method to fast obtain the result, based on the standard bivariate normal distribution.

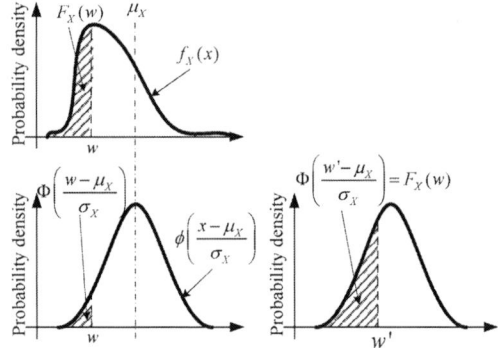

Fig. 6: Using normal distribution to calculate the approximate integral.

If the CDF of the corresponding bivariate normal distribution $(\Phi_2(\frac{w-\mu_X}{\sigma_X}, \frac{w-\mu_Y}{\sigma_Y}; \rho_{XY}))$ directly replaces $F_{XY}(w, w; \rho_{XY})$, it will generate large error when $f_{XY}(u, v; \rho_{XY})$ is far away from the PDF of the bivariate normal distribution $(\phi_2(\frac{u-\mu_X}{\sigma_X}, \frac{v-\mu_Y}{\sigma_Y}; \rho_{XY}))$. A univariate example is shown in Fig. 6, in which $\Phi\left(\frac{w-\mu_X}{\sigma_X}\right) << F_X(w)$. The reason of the error is that both F_X and Φ use the same integral end point w. If the integral end point of Φ is changed to $w' = \Phi^{-1}(F_X(w))\sigma_X + \mu_X$, we get $\Phi\left(\frac{w'-\mu_X}{\sigma_X}\right) = F_X(w)$. So the CDF of normal distribution can be used to replace the integral. Inspired by this result, in the bivariate case, the same method is used, which is given by

$$w_1 = \Phi^{-1}(F_X(w)), w_2 = \Phi^{-1}(F_Y(w))$$
$$F_{XY}(w, w; \rho_{XY}) \approx \Phi_2(w_1, w_2; \rho_{XY}) \quad (11)$$

Φ^{-1} is calculated by a lookup table. Though $\Phi_2(w_1, w_2; \rho)$ is still difficult to calculate, a fast and simple approximation is proposed in [25]. The pseudo code of MAX operation of X and Y is shown in Algorithm 3, in which X and Y are described by sampled PMFs.

We have validated this method using Monte-Carlo, and the accuracy is larger than 95% for some most common distributions. This accuracy is sufficient for RTN evaluation.

E. Complexity Analysis

The computational complexity of the proposed statistical algorithm is no more than $MAX\{O(M^2 n), O(KMn)\}$, where n is the number of gates in a given circuit, and K is the maximum number of gate fan-ins. Since M and K are both constant, the complexity is $O(n)$, which is linear to the size of circuit.

Algorithm 3 Calculating $W = MAX(X, Y)$, X and Y are of sampled length M_X and M_Y.

1: $min = MAX\{\underset{i<M_X}{MIN}\{X.value[i]\}, \underset{i<M_Y}{MIN}\{Y.value[i]\}\};$
2: $max = MAX\{\underset{i<M_X}{MAX}\{X.value[i]\}, \underset{i<M_Y}{MAX}\{Y.value[i]\}\};$
3: $step = (max - min)/M;$
4: **for** $i = 0$ to $M - 1$ **do**
5: $\quad w = min + step * (i + 1);$
6: $\quad p_1 = F_X(w); p_2 = F_Y(w);$
7: $\quad W.prob[i] = \Phi_2(\Phi^{-1}(p_1), \Phi^{-1}(p_2); \rho_{XY});$
8: $\quad W.value[i] = w;$
9: **end for**
10: **for** $i = M - 1$ to 1 **do**
11: $\quad W.prob[i] -= W.prob[i - 1];$
12: **end for**

IV. Experimental Results

A. Experiment Setup

The experiments are implemented on a PC with an Intel Q9550 CPU. The STA tool, the logic simulator, and the statistical algorithm in Fig. 3 are written in C++. Twenty-four benchmarks including ISCAS85 and some ALU circuits are used to evaluate the proposed algorithm for RTN simulation. The 16nm high-performance PTM [23] is used, with nominal $V_{dd} = 0.9V$ and $|V_{th0}| = 0.4V$. Some key parameters are listed: $\alpha = 1.5$ (in Eq. (4)), single trap induced $|\Delta V_{th}| = 30mV$ for smallest devices according to Eq. (2).

B. Results of RTN Evaluation

1) RTN-induced circuit delay degradation: In this experiment, the average number of detectable traps in each device (λ) is set to 2, $\tau_c^{(on)} = 0.01s$ and $\tau_e^{(off)} = 0.1s$, $m_c = m_e = 12$. The temporal distribution of circuit delay of four circuits obtained by the proposed algorithm and Monte-Carlo (MC) simulation are shown in Fig. 7. MC is implemented by 10000 times. The proposed algorithm can generally obtain accurate results compared with MC. We find that the RTN-induced temporal distribution of circuit delay has a long tail, as shown in Fig. 8. As mentioned above, RTN-induced circuit delay varies in time randomly, so if chip operates long enough circuit delay can always get its worst value. Consequently, for a reliable design, we should ensure that the maximum possible delay does not violates the design specification, leading to large design redundancy. However, as can be seen from Fig. 8, though the tail is very long, the large delay values can hardly appear during the whole lifetime. Ensuring that circuit functions correctly with $1 - 10^{-10}$ probability or $1 - 10^{-50}$ probability have no difference in essence. Therefor, it is not practical to consider the maximum possible delay which will lead to large design overheads, but we should only ensure that circuit functions correctly with a certain probability P during the whole lifetime, such that timing violation can hardly happen. Consider a 10-year circuit lifetime and the typical values of τ_e and τ_c, we choose $P = 1 - 10^{-9}$. So our algorithm predicts that circuit functions correctly with $1 - 10^{-9}$ probability. The corresponding delay value is called "RTN-induced maximum delay".

The results of all the benchmarks are shown in Table I. We show the circuit intrinsic delay and degradation ratio of the RTN-induced maximum delay. To ensure $1 - 10^{-9}$ correctness during the whole lifetime, circuit should be protected from the degraded delay rather than the intrinsic delay. The average degradation ratio is 20%. Our statistical algorithm is on

978-1-4799-2817-0/14 $31.00 © 2014 IEEE

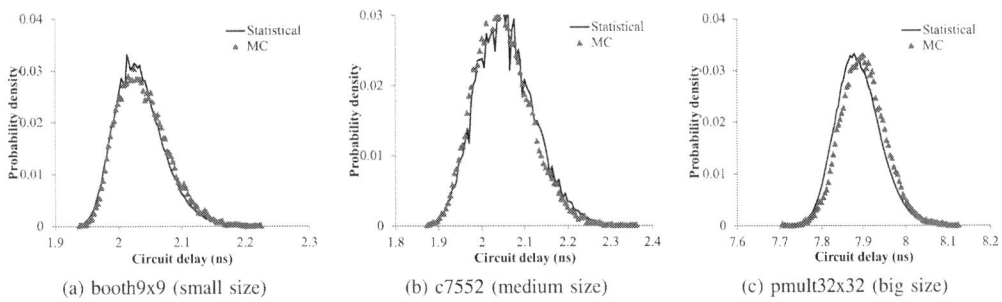

(a) booth9x9 (small size)   (b) c7552 (medium size)   (c) pmult32x32 (big size)

Fig. 7: Comparison with MC on circuit delay distribution.

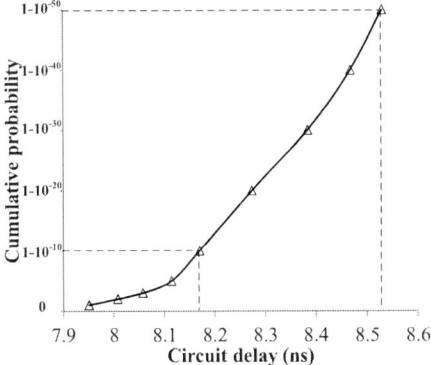

Fig. 8: Long tail of the RTN-induced temporal distribution of circuit delay, for pmult32x32.

average $41\times$ faster than MC (10000 times). The simulation time is at millisecond magnitude, which is also expected to be much faster than SPICE-based approaches. We also compare the accuracy between MC and our method in the error of the mean value of delay distribution. The average error is only 0.53%.

2) Impact of gate voltage: Several publications show that RTN strongly depends on gate voltage at device level. It is shown that the gate voltage V_{gs} has large impact on $\tau_e^{(on)}$ and $\tau_c^{(on)}$, and the dependence is approximately exponential [20], [26]. So $\tau_e^{(on)}$ and $\tau_c^{(on)}$ can be described as

$$\tau_c^{(on)} = \gamma_c e^{-\theta_c V_{gs}}, \tau_e^{(on)} = \gamma_e e^{\theta_e V_{gs}} \qquad (12)$$

Based on the data plotted in [26], we choose $\gamma_c = 32$, $\gamma_e = 3.2 \times 10^{-8}$, and $\theta_c = \theta_e = 11.5$. $m_c = m_e = 12$. λ also depends on V_{gs}, and the dependence is approximately linear [20]. When V_{gs} increases, λ also increases. The data plotted in [20] show that $\frac{d\lambda}{dV_{gs}} \approx 1.25$ and $\lambda \approx 2$ under nominal condition, so we choose

$$\lambda = 1.25 \times (V_{gs} - 0.9) + 2 \qquad (13)$$

Take pmult32x32 as an example, the impact of V_{gs} on RTN is shown in Fig. 9 and Fig. 10. We get two important observations from the results.

- RTN-induced maximum delay degradation $(d_{max} - d_0)$ has two different trends under different V_{gs} (Fig. 10). When $V_{gs} < 1$V, with V_{gs} increasing, τ_e increases, τ_c decreases, and the average number

TABLE I: Circuit delay degradation caused by RTN.

benchmark	#gate	d_0(ns)	MC	Statistical algorithm			error(%)
			T(s)	T(s)	Δ(%)	speedup	
c432	169	2.81	0.256	0.008	16.6	31	0.21
c499	204	2.23	0.316	0.007	39.9	46	2.03
c880	383	1.13	0.564	0.018	13.8	32	0.43
c1355	548	1.91	0.782	0.031	21.5	25	0.49
c1908	911	2.77	1.364	0.039	27.0	35	0.94
c2670	1279	1.38	2.044	0.061	27.5	34	0.73
c3540	1699	2.14	2.492	0.058	32.5	43	2.01
c5315	2329	1.87	3.667	0.110	28.6	33	0.13
c6288	2447	6.36	3.411	0.091	12.0	37	0.37
c7552	3566	1.80	5.541	0.193	30.8	29	0.33
array4x4	69	0.84	0.101	0.004	18.2	24	0.00
array8x8	375	2.86	0.541	0.025	17.5	22	2.93
bkung16	81	1.00	0.124	0.002	9.1	57	0.27
bkung32	165	1.94	0.257	0.004	7.1	63	0.13
booth9x9	412	1.90	0.593	0.014	22.4	42	0.23
kogge16	81	1.00	0.124	0.002	9.1	58	0.26
kogge32	164	1.97	0.259	0.004	7.0	64	0.13
log16	140	0.54	0.208	0.006	19.7	32	0.22
log32	371	0.85	0.556	0.020	31.0	28	0.34
log64	862	1.52	1.318	0.038	31.5	35	0.13
pmult4x4	72	0.93	0.107	0.004	18.2	25	0.01
pmult8x8	356	1.93	0.515	0.011	17.1	47	0.05
pmult16x16	1672	3.89	2.486	0.058	11.6	43	0.12
pmult32x32	6814	7.44	22.442	0.235	9.5	96	0.15
average					20.0	41	0.53

d_0 = circuit delay without RTN (intrinsic delay)
T = simulation time
$\Delta = \frac{d_{max} - d_0}{d_0}$, d_{max} = RTN-induced maximum delay
error = error of the mean value of delay distribution

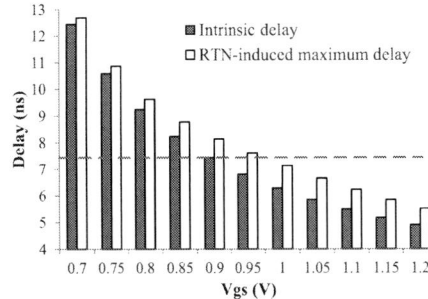

Fig. 9: Intrinsic delay (d_0) and RTN-induced maximum delay (d_{max}), for pmult32x32.

of filled traps $r\lambda = \frac{\lambda \tau_e}{\tau_e + \tau_c}$ also increases, so the maximum delay degradation increases. On the other hand, when $V_{gs} > 1$V, since $\tau_e \gg \tau_c$, the average number of filled traps is almost constant ($\frac{\lambda \tau_e}{\tau_e + \tau_c} \approx \lambda$), which means all the detectable traps are almost in

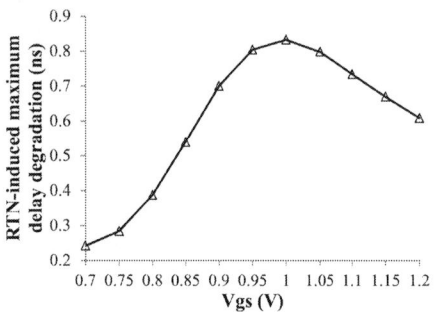

Fig. 10: RTN-induced maximum delay degradation ($d_{max} - d_0$), for pmult32x32.

filled state (although λ also depends on V_{gs}, the dependence is much weaker than that of the time constants); in the mean time, higher V_{gs} leads to lower intrinsic delay, so RTN-induced maximum delay degradation decreases. This phenomenon can be called "saturation" of traps.

• Although the RTN-induced maximum delay degradation ($d_{max} - d_0$) increases when $V_{gs} < 1$V, the maximum delay (d_{max}) still keeps decreasing with V_{gs} increasing. This indicates that a simple guardbanding approach can protect circuit from the effect of RTN. As shown in Fig. 9, if we choose the intrinsic delay when $V_{gs} = 0.9$V (nominal voltage) as the design constraint, the RTN-induced maximum delay satisfies the constraint when $V_{gs} \geq 1$V.

V. CONCLUSIONS

This paper integrates RTN into timing analysis and analyzes the impact of multiple traps induced RTN on the temporal performance of digital circuits. We show the temporal distribution of circuit delay under RTN, and circuit should be protected from a degraded delay which is on average 20% larger than intrinsic delay to ensure $1 - 10^{-9}$ correctness during the whole lifetime. The impact of gate voltage on RTN is investigated. Our results show that a simple guard-banding approach can effectively protect circuit from the effect of RTN.

REFERENCES

[1] J. Campbell, J. Qin, K. Cheung, L. Yu, J. Suehle, A. Oates, and K. Sheng, "The origins of random telegraph noise in highly scaled SiON nMOSFETs," in *IRW*, oct. 2008, pp. 1–16.

[2] A. Lee, A. R. Brown, A. Asenov, and S. Roy, "Random telegraph signal noise simulation of decanano MOSFETs subject to atomic scale structure variation," *Superlattices and Microstructures*, vol. 34, no. 3C6, pp. 293–300, 2003.

[3] N. Tega, H. Miki, F. Pagette, D. Frank, A. Ray, M. Rooks, W. Haensch, and K. Torii, "Increasing threshold voltage variation due to random telegraph noise in fets as gate lengths scale to 20 nm," in *VLSIT*, june 2009, pp. 50–51.

[4] A. Ghetti, C. Compagnoni, F. Biancardi, A. Lacaita, S. Beltrami, L. Chiavarone, A. Spinelli, and A. Visconti, "Scaling trends for random telegraph noise in deca-nanometer flash memories," in *IEDM*, dec. 2008, pp. 1–4.

[5] T. Matsumoto, K. Kobayashi, and H. Onodera, "Impact of random telegraph noise on cmos logic delay uncertainty under low voltage operation," in *IEDM*, 2012, pp. 25.6.1–25.6.4.

[6] M. J. Uren, D. J. Day, and M. J. Kirton, "1/f and random telegraph noise in silicon metal-oxide-semiconductor field-effect transistors," *Applied Physics Letters*, vol. 47, no. 11, pp. 1195–1197, dec 1985.

[7] K. Hung, P. Ko, C. Hu, and Y. Cheng, "A unified model for the flicker noise in metal-oxide-semiconductor field-effect transistors," *TED*, vol. 37, no. 3, pp. 654–665, mar 1990.

[8] T. Grasser, H. Reisinger, W. Goes, T. Aichinger, P. Hehenberger, P.-J. Wagner, M. Nelhiebel, J. Franco, and B. Kaczer, "Switching oxide traps as the missing link between negative bias temperature instability and random telegraph noise," in *IEDM*, dec. 2009, pp. 1–4.

[9] S.-M. Joe, J.-H. Yi, S.-K. Park, H. Shin, B.-G. Park, Y. J. Park, and J.-H. Lee, "Threshold voltage fluctuation by random telegraph noise in floating gate nand flash memory string," *TED*, vol. 58, no. 1, pp. 67–73, jan. 2011.

[10] N. Tega, H. Miki, M. Yamaoka, H. Kume, T. Mine, T. Ishida, Y. Mori, R. Yamada, and K. Torii, "Impact of threshold voltage fluctuation due to random telegraph noise on scaled-down SRAM," in *IRPS*, 27 2008-may 1 2008, pp. 541–546.

[11] N. Tega, H. Miki, T. Osabe, A. Kotabe, K. Otsuga, H. Kurata, S. Kamohara, K. Tokami, Y. Ikeda, and R. Yamada, "Anomalously large threshold voltage fluctuation by complex random telegraph signal in floating gate flash memory," in *IEDM*, dec. 2006, pp. 1–4.

[12] M. Tanizawa, S. Ohbayashi, T. Okagaki, K. Sonoda, K. Eikyu, Y. Hirano, K. Ishikawa, O. Tsuchiya, and Y. Inoue, "Application of a statistical compact model for random telegraph noise to scaled-SRAM Vmin analysis," in *VLSIT*, june 2010, pp. 95–96.

[13] S. O. Toh, Y. Tsukamoto, Z. Guo, L. Jones, T.-J. K. Liu, and B. Nikolic, "Impact of random telegraph signals on Vmin in 45nm SRAM," in *IEDM*, dec. 2009, pp. 1–4.

[14] K. Aadithya, A. Demir, S. Venugopalan, and J. Roychowdhury, "SAMURAI: An accurate method for modelling and simulating non-stationary random telegraph noise in SRAMs," in *DATE*, march 2011, pp. 1–6.

[15] K. Aadithya, S. Venogopalan, A. Demir, and J. Roychowdhury, "MUSTARD: A coupled, stochastic/deterministic, discrete/continuous technique for predicting the impact of random telegraph noise on SRAMs and DRAMs," in *DAC*, june 2011, pp. 292–297.

[16] K. Ito, T. Matsumoto, S. Nishizawa, H. Sunagawa, K. Kobayashi, and H. Onodera, "Modeling of random telegraph noise under circuit operation - simulation and measurement of RTN-induced delay fluctuation," in *ISQED*, march 2011, pp. 1–6.

[17] Y. Ye, C.-C. Wang, and Y. Cao, "Simulation of random telegraph noise with 2-stage equivalent circuit," in *ICCAD*, nov. 2010, pp. 709–713.

[18] T. B. Tang, A. Murray, and S. Roy, "Methodology of statistical RTS noise analysis with charge-carrier trapping models," *TCAS-I*, vol. 57, no. 5, pp. 1062–1070, may 2010.

[19] H. Luo, Y. Wang, Y. Cao, Y. Xie, Y. Ma, and H. Yang, "Temporal performance degradation under RTN: Evaluation and mitigation for nanoscale circuits," in *ISVLSI*, aug. 2012, pp. 183–188.

[20] T. Nagumo, K. Takeuchi, S. Yokogawa, K. Imai, and Y. Hayashi, "New analysis methods for comprehensive understanding of random telegraph noise," in *IEDM*, dec. 2009, pp. 1–4.

[21] A. Van Der Wel, E. Klumperink, L. K. J. Vandamme, and B. Nauta, "Modeling random telegraph noise under switched bias conditions using cyclostationary rts noise," *TED*, vol. 50, no. 5, pp. 1378–1384, 2003.

[22] C. Amin, N. Menezes, K. Killpack, F. Dartu, U. Choudhury, N. Hakim, and Y. Ismail, "Statistical static timing analysis: how simple can we get?" in *Design Automation Conference, 2005. Proceedings. 42nd*, 2005, pp. 652–657.

[23] Nanoscale Integration and Modeling (NIMO) Group, ASU, "Predictive Technology Model (PTM)." [Online]. Available: http://ptm.asu.edu/

[24] J.-J. Liou, K.-T. Cheng, S. Kundu, and A. Krstic, "Fast statistical timing analysis by probabilistic event propagation," in *DAC*, 2001, pp. 661–666.

[25] W.-J. Tsay and P.-H. Ke, "A simple approximation for bivariate normal integral based on error function and its application on probit model with binary endogenous regressor," *IEAS Working Paper : academic research 09-A011*, Nov. 2009.

[26] T. Nagumo, K. Takeuchi, T. Hase, and Y. Hayashi, "Statistical characterization of trap position, energy, amplitude and time constants by RTN measurement of multiple individual traps," in *IEDM*, dec. 2010, pp. 28.3.1–28.3.4.

Semi-Analytical Current Source Modeling of FinFET Devices Operating in Near/Sub-Threshold Regime with Independent Gate Control and Considering Process Variation

Tiansong Cui, Yanzhi Wang, Xue Lin, Shahin Nazarian, and Massoud Pedram
Department of Electrical Engineering
University of Southern California
Los Angeles, California, United States, 90089
{tcui, yanzhiwa, xuelin, shahin, pedram}@usc.edu

Abstract—Operating circuits in the near/sub-threshold regime can lower the circuit energy consumption at the expense of lowering the circuit speed. In addition near/sub-threshold can result in higher sensitivity to process-induced variations and transient noise. FinFETs have been proposed as an alternative to planar CMOS devices in sub-20nm CMOS technology nodes due to their more effective channel control, steep sub-threshold slope, high ON/OFF current ratio, low power consumption, and so on. Characteristics of FinFETs operating in the near/sub-threshold regime make it difficult to verify the timing of a circuit using conventional statistical static timing analysis (SSTA) techniques. Current source modeling (CSM) methods, which have been proposed to increase the accuracy of timing analysis in dealing with arbitrary shapes of the input signal waveforms, are the appropriate solution for performing SSTA on FinFET-based circuits. This paper thus extends the CSM to such circuits, operating in the near/sub-threshold voltage regime. In particular, FinFET devices with independent gate control and subject to process variations are modelled. The key idea of the proposed CSM approach is to combine non-linear analytical models and low-dimensional CSM lookup tables to simultaneously achieve high modeling accuracy and low time/space complexity.

I. INTRODUCTION

Power efficiency has gained growing attention in VLSI design with the increasing demand for extending the battery life of portable devices as well as lowering the electrical energy cost of high-performance computing nodes. Aggressive voltage scaling from the traditional super-threshold regime to the near/sub-threshold regime has shown effectiveness in reducing the power consumption of digital circuits [1][2][3]. It is especially beneficial for applications with relaxed performance requirements, such as wireless sensor processing and medical monitoring, because the operating frequency of near/sub-threshold logic is much lower than that of traditional strong-inversion circuits ($V_{DD} > V_{th}$) due to the smaller transistor ON-current in the near/sub-threshold regime. For example, according to [2], voltage scaling from super-threshold regime (e.g., 1.1V) down to the near-threshold regime (e.g., 0.5V) yields an energy reduction on the order of 10X at the expense of approximately 10X performance degradation.

With the dramatic downscaling of layout geometries, the traditional bulk CMOS technology is facing significant challenges due to several reasons such as the increasing leakage and short-channel effects (SCEs) [4]. FinFET devices, a special kind of quasi-planar double gate (DG) devices, have been proposed as an alternative for the bulk CMOS when technology scales beyond the 32nm technology node [5][6]. It has been proved in [22] that FinFET devices outperform bulk CMOS devices in ultra-low power designs by allowing for higher voltage scalability. Another unique feature of FinFET devices is the *independent gate control*, i.e., the front gate and the back gate can be controlled by separate signals, which enables more flexible circuit designs [8]. Due to the capacitor coupling of the front gate and the back gate, the threshold voltage of the front-gate-controlled FET varies in response to the back

gate biasing, and vice versa. Previous work [7] utilized the independent gate control for FinFETs in the pull-down network of an SRAM cell to keep the ~20 pA/μm standby power budget, whereas the authors of [8][9] studied joint gate sizing and negative biasing on the back gate of FinFET devices and demonstrated significant power reduction.

Our main goal in this work is to design an accurate delay model that accounts for variability while considering the afore-mentioned features of FinFET devices. *Statistical static timing analysis* (SSTA) is a well-known method to verify the circuit timing subject various sources of variation and considerable efforts have been invested in developing voltage-based statistical gate delay models [11]. However, their accuracy is limited as the input and output voltage dependencies are approximated with input slew and output load. This would be more severe for FinFET devices working in the near/sub-threshold regime as crosstalk noise more severely impacts the signal integrity [10]for such devices [10]. Also process variation is typically handled by applying a first-order correction to the quantities of interest using the variations of the process parameter [11]. Due to the exponential relationship between transistor's ON-current and threshold voltage for FinFET devices operating in the near/sub-threshold regime, process variations of important parameters such as the effective gate length L_{eff} and their impacts should be more carefully taken into account.

Current source-based logic cell modeling (CSM) has been introduced as an alternative approach for timing calculation and verification [12]~[16] in order to address key shortcomings of conventional voltage-based timing analysis methods. Instead of recording the delays and output slews in LUTs, the CSM method builds an equivalent circuit model for each logic gate using independent current sources and several equivalent capacitances. The values of current sources and capacitances are pre-characterized and recorded in *standard CSM LUTs*, where the terminal voltages are used as the index keys. The output waveforms are calculated in a discrete-time manner using pre-characterized LUTs based on given input waveforms. In presence of the input noise, the CSM method achieves very high accuracy in producing output waveforms and calculating delays, because the current and capacitances at various combinations of input and output voltages are all pre-characterized. In addition, the CSM method is much faster compared to a circuit simulator such as SPICE because the former indexes pre-characterized LUTs to obtain values of currents and capacitances. Finally, the LUT-based approach in CSM can be easily applied to different supply voltage regimes, and thus is very suitable for simulating and analyzing circuits that support burst-mode applications and operate in multiple supply voltage regimes. Thanks to these capabilities, CSMs are used in timing analysis and can effectively reduce errors in delay calculation.

To extend the CSM-based method to FinFET devices operating in the sub/near-threshold voltage regime, there are two key requirements as follows: (i) The model should appropriately capture the variations of the component values in the equivalent circuit model due to the variations of process parameters; (ii) It should account for the fact that the threshold voltage of the front-gate-controlled FET is affected by the back-gate voltage, and vice versa. In this work, two major sources of process variations are considered: *Line-Edge Roughness* (LER), which causes variations of the effective channel length, denoted by ΔL, and *Gate Work-Function Variation* (WFV), which causes variations of the intrinsic threshold voltage, denoted by ΔV_{th0} [19]. Considering these effects, the CSM method utilizes a general equivalent circuit model for FinFET devices. We need to determine all the driving current and equivalent capacitance values in this equivalent model given the applied voltages on the front-gate-controlled and back-gate-controlled fins, the output voltage, as well as process variation parameters ΔL and ΔV_{th0} for N-type and P-type FETs.

In order to reduce the memory storage requirements, previous work such as [14] and [16] store the LUTs for CMOS logic cells in the superthreshold regime under the nominal process conditions, and apply polynomial corrections for process variations. However, for FinFET devices operating in the sub/near-threshold regime, the relationship between the driving current and the gate voltages (and the threshold voltage) becomes exponential. Hence, the polynomial correction method is not sufficient to capture the effects of the back-gate (and front-gate) voltage and process variations for FinFET devices operating in the sub/near-threshold regime.

In this paper, we develop a semi-analytical approach for FinFET CSM operating in the sub/near-threshold regime, accounting for the unit feature of independent gate control as well as process variations. The proposed technique determines all the component values in this equivalent circuit model given the applied voltages on the front-gate-controlled and back-gate-controlled fins, the output voltage, as well as process variation parameters ΔL and ΔV_{th0} for N-type and P-type FETs. We use a simple example to illustrate the meaning of the term "semi-analytical". For one component value of interest, e.g., driving current I, we derive an analytical equation relating it to the terminal voltages x and y, and variations Δu and Δv in process parameters. The functional form of this equation is the same for all combinations of terminal voltages and process parameters. However, the equation also depends on a set of pre-characterized regression coefficients stored in LUTs. Suppose $I(x, y, \Delta u, \Delta v) = \mathbf{A}(x, y) \cdot g\left(\frac{x}{\mathbf{B}(x,y)}, y, \Delta u\right) \cdot h\left(y, \frac{\mathbf{C}(x,y)}{\Delta v}\right)$, and coefficients $\mathbf{A}(x, y)$, $\mathbf{B}(x, y)$, and $\mathbf{C}(x, y)$ are stored in LUTs corresponding to the (x, y) pair. The effect of $(\Delta u, \Delta v)$ on I has been captured by the function itself. This example captures the basic principle of the semi-analytical approach although the actual FinFET CSM is much more sophisticated. Notice that we only use 2D LUTs in our semi-analytical method to reduce the storage space requirement. Although the characterization process is extensive, it is done only once and the results are stored into compact low-dimensional LUTs.

II. Characteristics of FinFET Devices in Near/sub-Threshold Regime

A. Independent Gate Control for FinFET Devices

FinFET devices show better suppression of the short channel effect, lower energy consumption, higher supply voltage scaling capability, and higher ON/OFF current ratio compared with the bulk CMOS counterparts [6][7]. In addition to better control over the channel by using double gates, the FinFET structure allows for fabrication of separate front and back gates. In this structure, each fin is essentially the parallel connection of the *front-gate-controlled FET* and the *back-gate-controlled FET*, both with width H equal to the height of the fin. A unique feature of FinFET devices is the *independent gate control*, where the front and back gates are tied to different control signals.

Independent gate control makes it possible to apply different voltages to the front and back gates of a single fin, and thereby, allowing for more flexible circuit designs. Due to capacitor coupling of the front gate and the back gate of a FinFET transistor, the threshold voltage of the front-gate-controlled FET varies in response to the back-gate voltage, and vice versa. Under a relatively small back-gate voltage, a linear relationship between the change of the threshold voltage and the back-gate voltage is observed (suppose we consider N-type FETs):

$$\frac{dV_{th}}{dV_{BN}} = -\frac{C_{oxb} \cdot C_{si}}{C_{oxf} \cdot (C_{oxb} + C_{si})} \qquad (1)$$

where C_{si}, C_{oxf}, and C_{oxb} are the body capacitance, front-gate capacitance, and back-gate capacitance, respectively; V_{BN} is the voltage level applied to the back gate of the N-type fin. Eqn. (1) shows that decreasing the back-gate voltage of the N-type fin results in the increase of V_{th} of the front-gate-controlled N-type FET and therefore an exponential decrease of the leakage current.

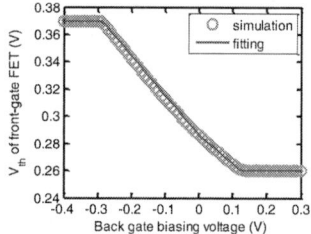

Figure 1. V_{th} of front-gate-controlled N-type FET v.s. back-gate voltage.

Figure 1 shows the relationship between the threshold voltage of the front-gate-controlled FET and back-gate voltage from the Hspice simulation. Please note that the threshold voltage will not further decrease (or increase) when we increase back-gate voltage larger than a special value $V_{BN,max}$ or smaller than a special voltage level $V_{BN,min}$.

Authors in [7][8][9] proposed and applied different implementation modes of FinFET logic gates to exploit the unique feature of independent gate control. For the N-type or P-type fin, there are two different connection modes: (i) the double gate (DG) mode, where the front gate and the back gate of the fin are tied together to the input signal, and (ii) the independent gate (IG) mode, where one of the gate is driven by the input signal and the other is connected to a biasing voltage or to the ground. These different modes achieve a trade-off between power consumption and rise/fall delay. We illustrate in Figure 2 two examples of implementations of an inverter with approximately the same rise and fall delays. In Figure 2 (a), we use the double gate mode for both the N-type and P-type fins. In Figure 2 (b), we use the independent gate mode for the N-type fin and double gate mode for the P-type fin. We may also use the independent gate mode for the P-type fin, which is however not considered due to space limitation.

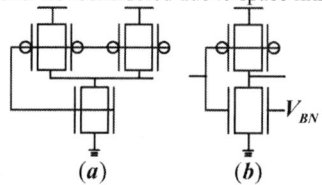

Figure 2. Different FinFET inverters in different modes.

B. Driving Currents for FinFET Devices in the Near/Sub-Threshold Regime

The driving current is a critical parameter in FinFET modeling. In the subthreshold regime, the driving current I_{FN} for the front-gate-controlled FET in an N-type fin follows an exponential relationship with the gate drive voltage V_{FN} and the drain-to-source voltage V_{ds}. The equation is given by:

$$I_{FN} = I_0 \frac{W}{L} \cdot e^{\frac{V_{FN} + \lambda V_{ds} - V_{th}(V_{BN})}{n \cdot v_T}} \left(1 - e^{\frac{-V_{ds}}{v_T}}\right) \qquad (2)$$

where I_0 is a technology-dependent parameter, λ is the drain voltage dependence coefficient (similar to but much smaller than the DIBL coefficient for bulk CMOS devices), n is the subthreshold slope factor, and v_T is the thermal voltage $\frac{kT}{q}$. Its threshold voltage V_{th} is affected by the voltage V_{BN} applied on the back gate of the same fin.

978-1-4799-2817-0/14 $31.00 © 2014 IEEE

In order to consider the near-threshold regime as well, we extend the method of [17] for FinFETs to provide a unified transregional model covering both subthreshold and near-threshold regimes. In this transregional model, the drain current I_{FN} is given as:

$$I_{FN} = I_0 \frac{W}{L} \cdot$$
$$e^{\frac{(V_{FN} + \lambda V_{ds} - V_{th}(V_{BN})) - a \cdot (V_{FN} + \lambda V_{ds} - V_{th}(V_{BN}))^2}{n \cdot v_T}} \left(1 - e^{\frac{-V_{ds}}{v_T}}\right) \quad (3)$$

where a is an empirical fitting parameter. We can extract the values of parameters I_0, a, and n from HSpice simulations. The transregional model provides an accurate FinFET modeling in both subthreshold and near-threshold regimes when V_{FN} (or V_{BN}) and V_{ds} are equal. Based on our experiments, the average and maximal errors of the proposed transregional model are 4.27% and 8.83%, respectively, compared with HSpice simulations.

III. CONVENTIONAL CURRENT SOURCE-BASED MODELING

The idea of current source-based modeling of logic cells was introduced about a decade ago with the goal of more accurately capturing the dependency of logic cell's timing behaviors on its input and output voltages. The CSM method builds an equivalent circuit model for each logic gate using independent current sources and several equivalent capacitors. An example of CSM for an inverter is presented in Figure 3.

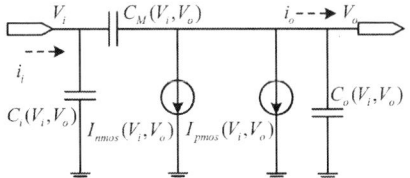

Figure 3. Equivalent CSM for a CMOS inverter [14].

In general, the CSM-based timing analysis is comprised of two phases: the characterization phase and the evaluation phase. In the characterization phase, an equivalent circuit model for each logic cell in the standard cell library is proposed and accurate circuit simulators (e.g., SPICE) are used to obtain the component values in the equivalent circuit model at different input and output voltages. In Figure 3, V_i is the input voltage level and V_o is the output voltage level of the inverter. Each component in Figure 3 is expressed as a function of V_i and V_o. $C_i(V_i, V_o)$ and $C_o(V_i, V_o)$ denote the equivalent capacitances at the input and output nodes of the inverter, respectively, and $C_M(V_i, V_o)$ denotes the Miller capacitance. $I_o(V_i, V_o)$ denotes the driving current, which is the sum of $I_{nmos}(V_i, V_o)$ and $I_{pmos}(V_i, V_o)$. Multiple 2D LUTs are generated to store the above component values at different V_i and V_o levels, as shown in Figure 4.

In the evaluation phase, we calculate the output waveform of a logic cell using pre-characterized driving currents and equivalent capacitances, as well as sample values of the input voltage. The accuracy of the CSM method depends on both the LUT precision, i.e., N and M values in Figure 4, and the sampling precision, i.e., the number of sampling points in a certain time period, of the input waveform.

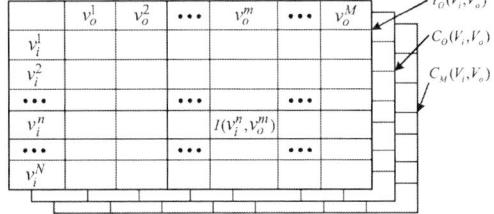

Figure 4. Conventional 2D look-up tables for the CSM [14].

CSMs for multiple-input logic cells such as NAND gates and NOR gates are more complex due to the existence of internal nodes, and will result in 3-D or 4-D LUTs. In order to reduce the storage overhead, several previous papers have focused on reducing the dimension of LUTs using approximation methods [24].

Both the driving current and parasitic capacitances can be significantly affected by process variations. Thus it is essential to account for the effects of variations of physical parameters in the CSM equivalent circuit model. A mathematical method is presented in [14] whereby the sensitivities of cell model elements are characterized with respect to the sources of variations. The nominal values of the LUTs are generated and a first-order (FO) correction is utilized to relate any component value in the logic cell with respect to the physical variation parameters. This process-variation induced CSM can be used to perform a statistical delay analysis for logic cells [14] or act as a fast Monte Carlo simulator.

IV. CURRENT SOURCE MODEL FOR FinFET DEVICES

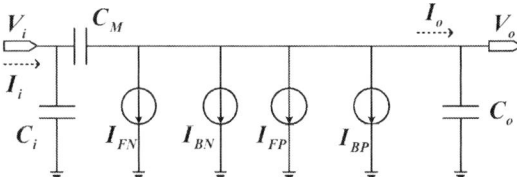

Figure 5. Equivalent CSM for a FinFET inverter.

To extend the CSM-based method to FinFET devices operating in near/sub-threshold voltage regimes, we should develop a model that can not only appropriately capture the variations of the component values in the equivalent circuit model due to the variations of process parameters, but also account for the effect that the threshold voltage of the front-gate-controlled FET is affected by the back-gate voltage, and vice versa. In this work, two major sources of process variations are considered in a FinFET transistor: *Line-Edge Roughness* (LER), which causes variations of the effective channel length, denoted by ΔL, and *Gate Work-Function Variation* (WFV), which causes variations of the intrinsic threshold voltage, denoted by ΔV_{th0} [19]. Considering these effects, the CSM method utilizes an equivalent circuit model for the FinFET inverter as shown in Figure 5. Figure 5 is a general model in that it can represent the two implementations of FinFET inverters shown in Figure 2 (although the parameters may be different.) For the FinFET inverter shown in Figure 2 (a), we have $V_{FN} = V_{BN} = V_i$, whereas $V_{FN} = V_i$ and V_{BN} is an additional biasing voltage for the FinFET inverter shown in Figure 2 (b). Given the values V_{FN} (V_{FP}), V_{BN} (V_{BP}), V_{out}, ΔL and ΔV_{th0} for both N-type and P-type FETs, we need to determine all the driving current and equivalent capacitance values shown in Figure 5.

One may extend the standard CSM LUTs to high dimensions to achieve this goal, e.g., storing different driving current values at various process parameter values and voltage levels. However, this simple treatment will result in an unacceptable memory space requirement. An alternative solution will be storing the LUTs when $\Delta V_{th0} = 0$ and $\Delta L = 0$, and apply polynomial corrections for process variations. However, this method turns out to be both inaccurate and cost-ineffective due to the following two reasons: (i) For FinFET devices operating in the sub/near-threshold regime, the relationship between the driving current and the gate voltages becomes exponential. (ii) The applied back-gate voltage exerts an influence on the threshold voltage of the front gate, and vice versa, which also significantly affects the timing behavior of the gate because the driving current is exponentially dependent on the threshold voltage. Hence, the polynomial correction method is not sufficient to capture the effects of the back-gate (and front-gate) voltage and process variations for FinFET devices operating in sub/near-threshold regime.

978-1-4799-2817-0/14 $31.00 © 2014 IEEE

169

In order to achieve higher modeling accuracy while maintaining time/space efficiency, we propose an efficient way to construct a semi-analytical CSM for the FinFET logic cells in this section based on the physical relations of the current, gate voltages and process parameters. Notice that we only use 2D LUTs in the proposed method to reduce the storage space requirement. In Section IV.A, we capture the change of threshold voltage of the front(back)-gate-controlled FET as a function of the back(front)-gate voltage and process variations. In Section IV.B, we analyze the modeling of the driving current of a fin with respect to the threshold voltages of both the front-gate-controlled and back-gate-controlled FETs and the variation of the effective channel length. Section IV.C focuses on the modeling of parasitic capacitances, and finally, the CSM LUT construction process is summarized in Section IV.D.

Please note that although we mainly describe FinFET inverters in this paper, the semi-analytical modeling framework can also be generated to the other types of FinFET gates, such as NAND and NOR gates. The generalization process will be similar to [24], and is not discussed in detail in this paper due to space limitation.

A. The Impact on the Threshold Voltage

We use a piecewise linear function to represent the impact of the back-gate voltage V_{BN} on the change of the threshold voltage $\Delta V_{th}(V_{BN})$:

$$\Delta V_{th}(V_{BN}) = \begin{cases} \Delta V_{th,max}, & V_{BN} < V_{BN,min} \\ k_1 V_{BN}, & V_{BN,min} \leq V_{BN} < 0 \\ k_2 V_{BN}, & 0 \leq V_{BN} < V_{BN,max} \\ \Delta V_{th,min}, & V_{BN} \geq V_{BN,max} \end{cases} \quad (4)$$

There are four fitting parameters in the above equation: k_1, k_2, $V_{BN,min}$, $V_{BN,max}$. k_1 and k_2 represent the $\frac{dV_{th}}{dV_{BN}}$ values in Eqn. (1) when $V_{BN} < 0$ and $V_{BN} > 0$, respectively, and are both less than 0. Notice that in general k_1 and k_2 are not equal, which means that the capacitances C_{si}, C_{oxf}, and C_{oxb} are not exactly the same when $V_{BN} < 0$ (i.e., reverse back-gate biasing) and $V_{BN} > 0$ (i.e., forward back-gate biasing.) Similarly, $\Delta V_{th}(V_{FN})$ is the threshold voltage change of the back-gate-controlled FET as a function of the front-gate voltage V_{FN}, which also satisfies Eqn. (4). In our experiment, we have the fitting results $(k_1, k_2, V_{BN,min}, V_{BN,max}) = (-0.2897, -0.2098, -0.29V, 0.12V)$.

Gate Work-Function Variation (WFV) is another important type of variation source which causes the variability of the threshold voltage. We denote the threshold voltage variation caused by WFV as ΔV_{th0}. Experimental results show that the fitting parameters in Eqn. (4) are independent of ΔV_{th0}.

In bulk CMOS, the threshold voltage V_{th} also depends on the variation of channel length due to the *drain-induced barrier lowering* (DIBL) effect. This effect is more pronounced in short-channel devices in the state-of-the-art CMOS technology [20], and *Halo doping* is used to compensate the DIBL effect, which will result in an opposite *reverse short channel effect* [21]. One the other hand, FinFET devices has a significantly reduced DIBL effect due to the more effective channel control by the gate voltages, which is one of the primary advantages [22] of the FinFET technology. Based on our experimental result on the 32nm PTM, ΔL has almost no effect on the threshold voltage for FinFET devices.

In summary, the threshold voltage of the front-gate-controlled FET, considering both the back-gate voltage and process variations, is given by

$$V_{th,F} = V_{th0} + \Delta V_{th}(V_{BN}) + \Delta V_{th0} \quad (5)$$

Similarly, the threshold voltage of the back-gate-controlled FET is given by $V_{th,B} = V_{th0} + \Delta V_{th}(V_{FN}) + \Delta V_{th0}$. Please note that the V_{th0} and ΔV_{th0} values are the same in both equations because both the front and back gates share the same fin. Experimental results show that the average and maximal fitting errors are 0.3% and 0.94%, respectively.

B. Modeling of the Driving Current

In the standard CSM shown in Figure 3, the driving current at the output node is a combination of the NMOS current, $I_N(V_i, V_o)$, and the PMOS current, $I_P(V_i, V_o)$. Thus, we characterize $I_N(V_i, V_o)$ and $I_P(V_i, V_o)$ for all possible combinations of V_i and V_o, which results in a set of 2-D lookup tables. For FinFET devices with the independent gate control and process variations, our goal is to use no larger than 2-D lookup tables to determine the driving current I_N (and I_P) under the applied voltage levels, i.e., V_{FN}, V_{BN} and V_{ds}, and specific process variation parameters, i.e., ΔL and ΔV_{th0}. Considering that each fin is essentially a parallel connection of the front-gate-controlled FET and the back-gate-controlled FET, I_N (or I_P) is the sum of the driving current of the front gate and that of the back gate:

$$I_N = I_{FN} + I_{BN} \quad (6)$$

Take the front gate of the N-type fin as an example. We fit I_{FN} with respect to ΔL and ΔV_{th0} based on Eqn. (3) in the near/sub-threshold regime using the following form:

$$I_{FN}(V_{FN}, V_{BN}, V_{ds}, \Delta L, \Delta V_{th0}) = \frac{\mathbf{C}(V_{FN}, V_{ds})}{L} \cdot e^{\mathbf{A}(V_{FN}, V_{ds}) \cdot V_{th,F}^2 + \mathbf{B}(V_{FN}, V_{ds}) \cdot V_{th,F}} \quad (7)$$

where $\mathbf{A}(V_{FN}, V_{ds})$, $\mathbf{B}(V_{FN}, V_{ds})$, and $\mathbf{C}(V_{FN}, V_{ds})$ are fitting parameters. The dependencies of the driving current on V_{FN} and V_{ds} in Eqn. (3) are absorbed into these fitting parameters. The value $L = L_0 + \Delta L$ where L_0 is the nominal effective length. The value $V_{th,F}$ depends on $\Delta V_{th}(V_{BN})$ as well as the WFV parameter ΔV_{th0}, as is shown in Eqn. (5). Notice that the front gate and the back gate have the symmetric structure. Hence, the same fitting parameters can be used to calculate the current of the back gate:

$$I_{BN}(V_{FN}, V_{BN}, V_{ds}, \Delta L, \Delta V_{th0}) = \frac{\mathbf{C}(V_{BN}, V_{ds})}{L} \cdot e^{\mathbf{A}(V_{BN}, V_{ds}) \cdot V_{th,B}^2 + \mathbf{B}(V_{BN}, V_{ds}) \cdot V_{th,B}} \quad (8)$$

The above method combines the non-linear analytical models and small-size CSM lookup tables, and simultaneously achieves high modeling accuracy and space/time efficiency. Compared with Eqn. (3), the lookup table-based model Eqns. (7) and (8) results in much higher accuracy (please see the experimental results) because some parameters in Eqn. (3), such as the subthreshold slope n, depend on V_{FN} and V_{ds} [25]. Moreover, the effect of V_{BN} on $V_{th,F}$ (and V_{FN} on $V_{th,B}$) and that of process variations are carefully accounted for without increasing the model complexity, since we are still using 2-D LUTs to store all the fitting parameters, like what is used in the standard CSM for bulk CMOS devices. Similarly, the proposed method can be applied to the P-type fin to determine the corresponding driving current I_P.

C. Modeling of Parasitic Capacitances

In this section, we analyze the impact of process variations on the values of equivalent capacitance in the CSM equivalent circuit model as shown in Figure 5. This equivalent circuit model is comprised of three non-linear voltage-dependent capacitances. Among then, C_i and C_o model the parasitic effects at the input and output nodes of the logic cell, whereas the Miller capacitance C_M models the Miller effect between these two nodes. Notice that for FinFET devices, these equivalent parasitic capacitances are contributed by the physical capacitances (e.g., C_{gs}, C_{gd}, C_{db} and so on) in both the front-gate-controlled and back-gate-controlled FETs in both the N-type and P-type fins.

We know that the equivalent capacitance values will be very different for the FinFET inverter shown in Figure 2 (a) and that in Figure 2 (b) due to different connection modes. Hence, we need to consider these two modes separately in the parasitic capacitance modeling. Take the inverter shown in Figure 2 (b) as an example. Both the process variations as well as the biasing voltage level V_{BN} will affect the equivalent capacitance values. For example, the LER effect affects the capacitance values as these capacitances are functions of the

978-1-4799-2817-0/14 $31.00 © 2014 IEEE   170

dimension of the transistors. In addition, the bias voltage also exerts an effect on the parasitic capacitances. Therefore, we perform curve fitting to relate the equivalent capacitances to the process parameters as well as the bias voltage V_{BN} for each (V_i, V_o) pair. As the effects of process variations and bias voltage on parasitic capacitances are relatively small compared with those on the driving current, experimental results show that the linear curve fitting is able to capture the dependency of the equivalent capacitances on the above-mentioned parameters. The variations of process parameters originate from both N-type and P-type fins, and thus the equivalent capacitance C_i is fitted as follow,

$$
\begin{aligned}
C_i&(V_i, V_o, \Delta L_P, \Delta L_N, \Delta V_{th0P}, \Delta V_{th0N}, V_{BN}) \\
&= \mathbf{C}_{i0}(V_i, V_o) + \boldsymbol{\alpha}_P(V_i, V_o) \cdot \Delta L_P + \boldsymbol{\alpha}_N(V_i, V_o) \cdot \Delta L_N \quad (9) \\
&+ \boldsymbol{\gamma}_P(V_i, V_o) \cdot \Delta V_{th0P} + \boldsymbol{\gamma}_N(V_i, V_o) \cdot \Delta V_{th0N} + \boldsymbol{\delta}_N(V_i, V_o) \cdot V_{BN}
\end{aligned}
$$

where $\mathbf{C}_{i0}$ is the LUT of the nominal input capacitance values in CSM, ΔL_P and ΔL_N are the variations of the channel length in the N-type fin and P-type fin, respectively. ΔV_{th0P} and ΔV_{th0N} are the WFV-induced threshold voltage variations in the N-type fin and P-type fin, respectively. Similar fittings are also performed for C_o and C_M.

The same fitting method can be used for the FinFET inverter shown in Figure 2 (a), except that no separate bias voltage exists in this inverter. Hence, we do not need the parameter $\boldsymbol{\delta}_N(V_i, V_o)$ in Eqn. (9).

D. CSM LUT Construction

After studying the modeling of driving current and the parasitic capacitances, the CSM LUT construction process can be concluded as follows: in the characterization phase, we perform characterization as well as curve fitting as mentioned earlier and record the coefficient into the LUTs with index of interested voltage levels. In the evaluation phase, we use the coefficient LUTs to construct the customized CSM LUTs including I_o, C_M, C_i and C_o under every voltage pair (V_i, V_o), under different corners of process variation parameters and bias voltage levels, e.g., ΔL, ΔV_{th0} and V_{BN}(or V_{FN}). The constructed CSM LUTs can be used to calculate the exact output waveform given the waveform of the input voltage.

The proposed semi-analytical CSM method enables accurate current-based timing analysis at various process corners and bias voltage levels without increasing the dimension (and thus the time/space complexity) of the conventional LUTs. In practice, although it is generally hard to know the detailed amounts of the variation (as the variations of the process parameters are random and usually normally described), the statistical method can be utilized to obtain the accurate joint distributions of the process parameters and reflect their variations statistically into the flow of constructing the standard CSM LUTs. With these distributions, we can perform the current-based statistical timing calculation using the proposed semi-analytical CSM method to study the distribution of the parameters of the output waveform. e.g., the delay and the transition time.

V. EXPERIMANTAL RESULTS

In this section, we evaluate the accuracy of the proposed semi-analytical CSM for FinFET devices in calculating the output waveform and delay. We adopt 32nm Predictive Technology Model for FinFET devices, in which the typical threshold voltages of the transistors are around ±0.3V. We set the supply voltage to 0.3V so that the circuits are operated in the near/sub-threshold regime. To ensure that voltage characterization covers the range of the noise, we sweep the input and output voltage from -100 mV to +400 mV with the interval of 5 mV. We consider 10% variation on the process parameters ΔL and ΔV_{th0}, and apply different bias voltage levels from -0.4V to 0.3V. The characterization is based on HSPICE, and the entire process for FinFET modeling and output waveform calculating takes less than an hour on a Debian 7 machine with 16 Intel E7-8837 2.66 GHz CPUs and 64 GB memory.

We compare our work with the CSM with first order correction like [14] in handing bias voltage and process variations. The proposed

method and baseline method are compared to the golden results generated using the HSPICE considering input noises. The fitting result of the impact of the back-gate voltage V_{BN} on the change of the threshold voltage $\Delta V_{th}(V_{BN})$ is already shown in the previous sections. In this section, we first verify the accuracy of the proposed semi-analytical CSM method in capturing the driving currents at different corners of process variations. After that, we demonstrate the accuracy of proposed CSM in calculating the real output waveforms under a noise input.

A. Modeling of the FinFET driving current

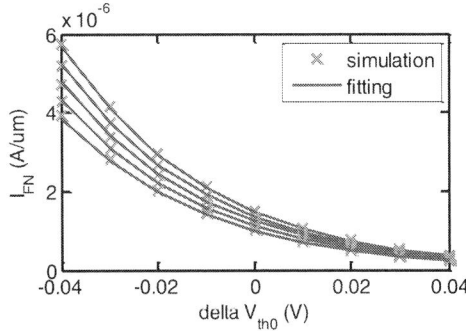

Figure 6. Curve fitting of N-type FET driving current under different corners of process variation.

Due to space limitation, we only show the fitting result of the driving current of N-type FET in Figure 6 under different corners of process variation parameters ΔL and ΔV_{th0} at the voltage pair $(V_{FN}, V_{ds}) = (0.1V, 0.2V)$. Our proposed method achieves very good fitting quality with an average error of 0.81%. The fittings of the driving currents are performed for every point (V_{FN}, V_{ds}).

B. Output waveform under noisy input

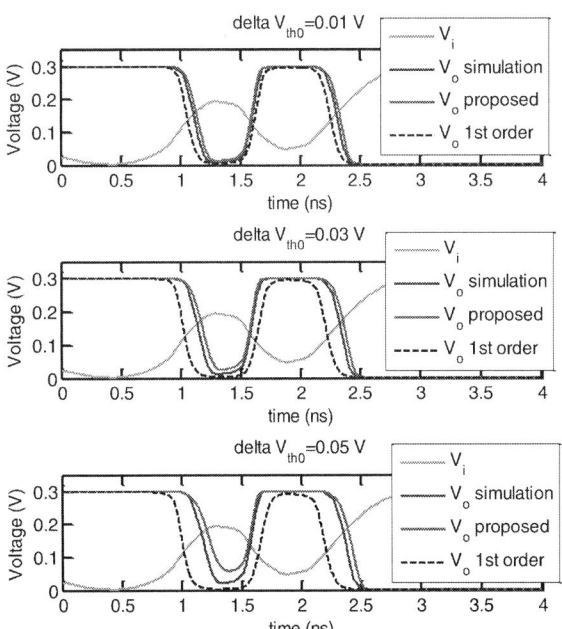

Figure 7. Output waveforms for different CSM variation handing techniques under a noise input at different threshold voltage variation levels.

We generate the customized CSM LUTs to calculate the output waveforms based on the pre-characterized LUTs with respect to different process variation parameters. We show the DG-mode inverter as an example and calculate output waveforms and compare them with the waveforms obtained using HSPICE simulation. Figure 7 shows three cases of ΔV_{th0} variation levels: 0.01V, 0.03V, and 0.05V, while ΔL is set to 1nm. The proposed CSM method consistently outperforms the first-order method. It reproduces the output waveform with a very high accuracy and achieves an average delay error reduction of more than 60% compared with the baseline first-order method. Among these two methods, the detailed comparison of the 50%-50% delay calculation errors at each transition point for different output waveforms is shown in Table 1. The high accuracy and low space requirement ensure the capability of performing statistical timing calculation and analysis based on the proposed CSM method, with the combination of some statistical tools.

Table 1. The delay error comparison for different output waveforms.

$\Delta L = 1nm, \Delta V_{th0} = 0.01V$			
	Point 1	Point 2	Point 3
Proposed method	2.33%	0.63%	1.03%
First-order method	5.21%	1.63%	2.42%
$\Delta L = 1nm, \Delta V_{th0} = 0.03V$			
	Point 1	Point 2	Point 3
Proposed method	3.31%	0.76%	1.37%
First-order method	10.82%	3.66%	5.38%
$\Delta L = 1nm, \Delta V_{th0} = 0.05V$			
	Point 1	Point 2	Point 3
Proposed method	4.57%	0.38%	1.94%
First-order method	15.03%	5.34%	7.43%

VI. CONCLUSION

In this paper, we present a semi-analynical current source modeling (CSM) method for FinFET devices operating in the near/sub-threshold voltage regime, accounting for the independent gate control and process variations. The driving currents and parasitic capacitances are analyzed under different bias voltages and process varations in the near/sub-threshold regime. A curve fitting step is performed to relate the driving currents and parasitic capacitances to the bias voltage levels as well as the sources of process variations, and fitting parameters are stored in low-dimensional look-up tables (LUTs). In circuit timing simulation, we derive the driving current and equivalent capacitances under the specific back-gate(or front-gate) bias voltage and process variation situation in practice. Experimental results demostrate the effectiveness of the proposed framework in both modeling accuracy and efficiency.

ACKNOWLEDGEMENT

This research is supported by grants from the PERFECT program of the Defense Advanced Research Projects Agency and the Software and Hardware Foundations of the National Science Foundation.

REFERENCES

[1] R. Dreslinski, M. Wiekowski, D. Blaauw, D. Sylvester, and T. Mudge, "Near-threshold computing: reclaiming Moore's law through energy efficient integrated circuits," *Proc. of IEEE*, 2010.

[2] D. Markovic, C. Wang, L. Alarcon, T. Liu, and J. Rabaey, "Ultralow-power design in near-threshold region," *Proc. of IEEE*, vol. 98, no. 2, pp. 237 – 252, Feb. 2010.

[3] A. Wang and A. Chandrakasan, "A 180 mV FFT processor using circuit techniques," *ISSCC*, pp. 292 – 293, Feb. 2004.

[4] L. Chang, Y. Choi, D. Ha, P. Ranade, S. Xiong, J. Bokor, C. Hu, T. King, "Extremely scaled silicon nano-CMOS devices," *Proc. of the IEEE*, vol. 91, no. 11, pp. 1860-1873, Nov. 2003.

[5] E. J. Nowak, I. Aller, T. Ludwig, K. Kim, R. V. Joshi, C.-T. Chuang, K. Bernstein, and R. Puri, "Turning silicon on its edge," *IEEE Circuits and Devices Magazine*, 2004, 20 – 31.

[6] T. Sairam, W. Zhao, and Y. Cao, "Optimizing FinFET technology for high-speed and low-power design", in *GLSVLSI*, 2007.

[7] T. Cakici, K. Kim, and K. Roy, "FinFET based SRAM design for low standby power applications," in *ISQED*, 2007.

[8] A. Muttreja, N. Agarwal, and N. K. Jha, "CMOS logic design with independent gate FinFETs," in *ICCD*, vol. 20, pp. 560 – 567, 2007.

[9] J. Ouyang and Y. Xie, "Power optimization for FinFET-based circuits using genetic algorithms," in *ISOCC*, 2008.

[10] D. Blaauw, V. Zolotov, and S. Sundareswaran, "Slope propagation in static timing analysis," *IEEE Trans. on Computer Aided-Design of Integrated Circuits & Systems*, pp. 1180 - 1195, 2002.

[11] C. Visweswariah, K. Ravindran, K. Kalafala, S. G. Walker, S. Narayan, D. K. Beece, J. Piaget, N. Venkateswaran, and J. G. Hemmett, "First-Order Incremental Block-Based Statistical Timing Analysis," *IEEE Transactions on CAD*, Oct. 2006.

[12] J. F. Croix, and D. F. Wong, "Blade and razor: cell and interconnect delay analysis using current-based models," *Design Automation Conference* (DAC), pp. 386-389, 2003.

[13] I. Keller, K. Tseng, and N. Verghese, "A robust cell-level crosstalk delay change analysis," *Proc. of Int'l Conf. on Computer Aided Design* (ICCAD), pp. 147-154, 2004.

[14] H. Fatemi, S. Nazarian, and M. Pedram, "Statistical logic cell delay analysis using a current-based model," *Design Automation Conference* (DAC), pp. 253-256, 2006.

[15] V. Veetil, D. Sylvester, and D. Blaauw, "Fast and Accurate Waveform Analysis with Current Source Models," *Proc. of Int'l Symp. on Quality Electronic Design*, 2008.

[16] A. Goel, and S. Vrudhula, "Statistical waveform and current source based standard cell models for accurate timing analysis," *Design Automation Conference* (DAC), 2008.

[17] D. M. Harris, B. Keller, J. Karl, and S. Keller, "A transregional model for near-threshold circuits with application to minimum-energy operation," in *ICM*, 2010.

[18] C. Kashyap, C. Amin, N. Menezes, and E. Chiprout, "A Nonlinear Cell Macromodel for Digital Applications", in *ICCAD*, 2007.

[19] T. Matsukawa, S. O'uchi, K. Endo, Y. Ishikawa, H. Yamauchi, Y.X. Liu, J. Tsukada, K. Sakamoto, M. Masahara, "Comprehensive analysis of variability sources of FinFET characteristics," 2009 *Symposium on VLSI Technology*, pp. 118¬119, 2009.

[20] K. Roy, S. Mukhopadhyay, and H. Mahmoodi-Meimand, "Leakage current mechanisms and leakage reduction techniques in deep-submicrometer CMOS circuits," *Proceedings of the IEEE*, vol.91, no.2, pp. 305- 327, Feb 2003 .

[21] T. H. Kim, J. Keane, H. Eom and C. H. Kim, "Utilizing reverse short-channel effect for optimal subthreshold circuit design," *IEEE Transactions on Very Large Scale Integration (VLSI) System*, vol. 15, no. 7, pp. 821-829, 2007.

[22] F. Crupi, M. Alioto, J. Franco, P. Magnone, M. Togo, N. Horiguchi and G. Groeseneken, "Understanding the Basic Advantages of Bulk FinFETs for Sub- and Near-Threshold Logic Circuits From Device Measurements," *IEEE Transactions on Circuits and Systems II: Express Briefs*, vol.59, no.7, pp.439,442, July 2012.

[23] W. Zhao, and Y. Cao, "New generation of predictive technology model for sub-45nm early design exploration," *IEEE Trans. on Electronic Devices*, vol. 53, no. 11, Nov 2006.

[24] B. Amelifard, H. Fatemi, S. Hatami, and M. Pedram, "A current source model for CMOS logic cells considering multiple input switching and stack effect," in *DATE*, 2008.

[25] B. Zhai, S. Hanson, D. Blaauw, and D. Sylvester, "Analysis and mitigation of variability in subthreshold design," in *ISLPED*, 2005.

2-SAT based Linear Time Optimum Two-Domain Clock Skew Scheduling

Yukihide Kohira
School of Computer Science and Engineering
The University of Aizu, Japan
kohira@u-aizu.ac.jp

Atsushi Takahashi
Dept. of Communications and Computer Engineering
Tokyo Institute of Technology, Japan
atsushi@eda.ce.titech.ac.jp

Abstract— **Multi-domain clock skew scheduling is an effective technique to improve the performance of sequential circuits by using practical clock distribution network. Although the upper bound of performance of a circuit increases as the number of clock domains increases in multi-domain clock skew scheduling, the improvement of the performance becomes smaller while the cost of clock distribution network increases much. In this paper, a linear time algorithm that finds an optimum two-domain clock skew schedule is proposed. Experimental results on ISCAS89 benchmark circuits and artificial data show that optimum circuits are efficiently obtained by our method in short time.**

I. Introduction

The semiconductor manufacturing process technology has improved the scale, speed, and power consumption of LSI circuits. However, increasing the ratio of the routing delay in the propagation delay bounds the amount of improvements in the conventional clock synchronous framework in which the simultaneous clock distribution to every register is assumed. The increases of the size and power consumption of a clock distribution circuit have become serious issues in the conventional clock synchronous framework. While, clock skew scheduling (CSS) [1–3], in which the clock is assumed to be distributed periodically to each register though not necessarily to all the registers simultaneously, is expected to give an essential solution. By CSS, the quality of circuit such as the clock frequency, area, power consumption, and peak power consumption are expected to be improved.

However, an unconstrained clock skew schedule with a large number of arbitrary clock delays cannot be realized reliably. Due to process variations, it is difficult to implement clock skew schedule with a large number of arbitrary clock delays. In practice, it is desirable that CSS is constrained to a limited number of clock delays. Multi-domain clock skew scheduling (MDCSS) was proposed to meet this practical design requirement in [4]. Instead of assigning arbitrary clock delays, MDCSS restricts the feasible clock delays to a small number, called clock domains.

Many methods were proposed to solve MDCSS problem. An algorithm based on simulated annealing was proposed in [5, 6]. In [4], the authors formulated MDCSS problem as a mixed integer programming problem and solved it by a SAT-based algorithm. Although this method guarantees the optimality, its computational time is long because it takes long time to solve a SAT problem. In [7], a multi-level clustering algorithm was proposed. The algorithm recursively merges half of the registers at each level until the total number of clusters is

small enough. Compared to the work of [4], this algorithm is much faster. However, the algorithm is heuristic and it does not guarantee the optimality. An exact algorithm based on branch-and-bound search framework with greedy speeding up heuristics was proposed in [8]. In [9], it is proved that MDCSS problem is NP-complete when the number of clock domains is $|V|/2$, where $|V|$ is the number of registers. Moreover, in [9], the algorithm was proposed to obtain an optimum clock skew schedule. The time complexity of the algorithm proposed in [9] is $O((k-1)!|V||E|^k)$, where k is the number of clock domains, $|V|$ is the number of registers, and $|E|$ is the number of register pairs with signal propagations. It means that MDCSS problem can be solved in polynomial time if the number of clock domains is restricted to a small constant. Of course, the amount of improvement becomes smaller if the number of clock domains is restricted. However, the performance of a circuit is usually improved much compared to a conventional single clock domain circuit even if the number of clock domains is two. Moreover, the most of performance improvement that can be achieved by multi-domain clock skew scheduling is achieved by two-domain clock skew scheduling. Therefore, the fast two-domain clock skew scheduling algorithm is desired in practical circuit design to improve the circuit performance by a practical clock distribution network. However, the algorithm in [9] takes to much time since the time complexity of the algorithm in [9] is $O(|V||E|^2)$ when $k = 2$.

In this paper, we propose an optimum linear time algorithm that maximizes the performance under the constraint that the number of clock domains is restricted to two. In our method, MDCSS problem is translated into 2-SAT problem. In our 2-SAT problem formulation, each variable corresponds to clock timing of each register and each set of four clauses corresponds to timing constraints for a register pair. Since 2-SAT problem formulation is obtained from MDCSS problem in $O(|V| + |E|)$ time in our proposed method and 2-SAT problem can be solved in $O(|V| + |E|)$ time [10], our proposed method is a liner time algorithm.

The contribution of this paper includes:

- We improve the time complexity of the algorithm for MDCSS problem where the number of clock domains is restricted to two. The time complexity of our proposed method is $O(|V| + |E|)$ and it is faster than the existing method proposed in [9] whose time complexity is $O(|V||E|^2)$.

- Experimental results on ISCAS89 benchmarks and artificial data show the optimality and efficiency of our method.

978-1-4799-2817-0/14 $31.00 © 2014 IEEE

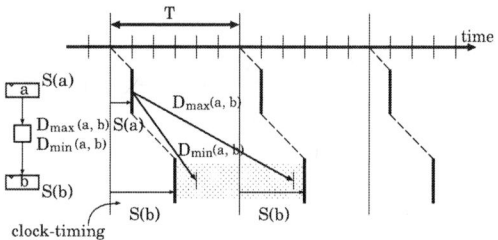

Fig. 1. Timing chart.

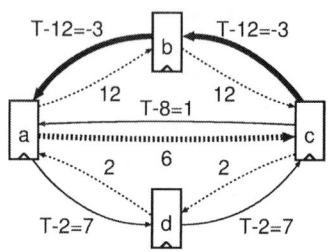

Fig. 2. Constraint graph $H(G, 9)$.

The rest of the paper is organized as follows: In section II, we discuss CSS and the formulation of MDCSS problem. We propose a linear time algorithm and improve the existing method [9] for MDCSS problem where the number of clock domains is restricted to two in section III. In section IV, an improvement of the algorithm proposed in [9] is discussed to compare with the proposed method fairly in experiments. Experimental results are presented and discussed in section V. The paper is concluded in section VI.

II. Preliminaries

A. Clock Skew Scheduling

A circuit in a clock synchronous framework works correctly with a clock period T if the following two types of constraints are satisfied for every register pair with signal propagations [1].

Setup (No-Zero-Clocking) Constraints

$$S(a) - S(b) \leq T - D_{\max}(a, b)$$

Hold (No-Double-Clocking) Constraints

$$S(b) - S(a) \leq D_{\min}(a, b),$$

where $S(a)$ ($S(b)$) is *clock timing* of a register a (b), which is defined as the difference in clock arrival time between a (b) and an arbitrary chosen reference register, $D_{\max}(a, b)$ ($D_{\min}(a, b)$) is the maximum (minimum) delay from a to b (Fig. 1).

Since a clock ticks all registers simultaneously in the conventional clock synchronous framework, the clock period must be larger than or equal to the maximum of delays between register pairs. Let $T_C(G)$ be the minimum clock period of a circuit G in the conventional clock synchronous framework which is equal to the maximum of delays between register pairs. On the other hand, in CSS, circuits can work correctly with the clock period which is smaller than the maximum of delays between register pairs, if all register pairs with signal propagations satisfy two types of constraints.

Let $T_S(G)$ be the minimum clock period of a circuit G in CSS under the assumption that the clock can be inputted to each register at an arbitrary designated timing. Hereafter, we simply call $T_S(G)$ the minimum clock period of G in CSS. Note that $T_S(G) \leq T_C(G)$ since the clock timing of each register can also be set to the same in CSS. $T_S(G)$ is determined by the

constraint graph $H(G) = (V_r, E_r)$ for G, where vertex set V_r corresponds to registers in G and directed edge set E_r corresponds to two types of constraints [2, 3]. An edge in E_r from a register a to a register b with weight $D_{\min}(a, b)$, called the D-edge, corresponds to the hold constraint, and an edge from a register b to a register a with weight $T - D_{\max}(a, b)$, called the Z-edge, corresponds to the setup constraint. Let $H(G, t)$ be the constraint graph in which the clock period T of Z-edges in $H(G)$ is set to t. Let the weight of a directed cycle in $H(G, t)$ be the sum of edge weights on the directed cycle. It is known that the minimum clock period $T_S(G)$ is the minimum t such that there is no cycle with negative weight in the constraint graph $H(G, t)$.

For example, the constraint graph $H(G, 9)$ shown in Fig. 2 has no cycle with negative weight and the weight of cycle (a, c, b, a) is negative when $T < 9$. Therefore, the minimum clock period $T_S(G) = 9$.

B. Problem Definition

Given the number k of clock domains, the objective of MDCSS is to decide the k domain values $\{s_1, s_2, \ldots, s_k\}$ as well as to assign each register one domain value such that the clock period T is minimized while the setup and hold constraints are satisfied. Since the clock timing is defined by the difference from that of an arbitrary chosen reference register, without loss of generality, we assume that $s_1 = 0$ and $s_i \leq s_{i+1} (1 \leq i \leq k-1)$ in the rest of the paper. MDCSS problem is formally formulated as follows:

MDCSS	
Input:	maximum and minimum delays between register pairs $D_{\max}(a, b)$ and $D_{\min}(a, b)$ for $\forall(a, b) \in E_r$
Output:	minimum clock period T_k, and a clock skew schedule $S : \forall a \rightarrow \{s_1(= 0), s_2, \ldots, s_k\}$
Constraint:	satisfy hold and setup constraints.

We focus on 2DCSS problem, in which the number of clock domains is restricted to two. 2DCSS problem is defined as follows:

978-1-4799-2817-0/14 $31.00 © 2014 IEEE

2DCSS

Input: maximum and minimum delays between register pairs $D_{\max}(a,b)$ and $D_{\min}(a,b)$ for $\forall (a,b) \in E_r$

Output: minimum clock period T_2, and a clock skew schedule $S : \forall a \to \{s_1(=0), s_2(\geq 0)\}$

Constraint: satisfy hold and setup constraints.

Moreover, the decision version of 2DCSS problem is defined as follows:

Decision problem of 2DCSS

Input: maximum and minimum delays between register pairs $D_{\max}(a,b)$ and $D_{\min}(a,b)$ for $\forall (a,b) \in E_r$, clock period T, and clock timing $s_2(\geq 0)$

Question: Does a clock skew schedule $S : \forall a \to \{s_1(=0), s_2(\geq 0)\}$ exist?

Constraint: satisfy hold and setup constraints.

C. 2-SAT

In our method, the decision problem of 2DCSS problem is translated into 2-SAT problem. 2-SAT problem is the problem of determining whether a collection of clauses with two-valued variables can be assigned values satisfying all the clauses. 2-SAT problem is often described using a Boolean expression with a conjunction of disjunctions, where each disjunction has two literals that are either variables or the negations of variables. Hereinafter, we also use Boolean expressions. 2-SAT problem and value assignment of 2-SAT problem are defined as follows:

2-SAT

Input: set of Boolean variables $X = \{x_1, x_2, \ldots, x_n\}$, and a collection of clauses $C(X) = \bigwedge_{i=1}^{m} c_i(X)$ (Each clause $c_i(X)$ has two literals.)

Question: Does a satisfying value assignment $t : X \to \{0, 1\}$ exist?

Value assignment of 2-SAT

Input: set of Boolean variables $X = \{x_1, x_2, \ldots, x_n\}$, and a collection of clauses $C(X) = \bigwedge_{i=1}^{m} c_i(X)$ (Each clause $c_i(X)$ has two literals.)

Output: if exists, a satisfying value assignment $t : X \to \{0, 1\}$.

It is known that 2-SAT problem and value assignment of 2-SAT problem can be solved in $O(n + m)$ time, where n is the number of variables and m is the number of clauses [10].

III. PROPOSED METHOD

A. Outline of Our Method

The outline of the proposed method is shown in Fig. 3.

Step1 $T_U = T_C, T_L = T_S$.

Step2 **while** $(T_U - T_L > \Delta T)$ **do**

Solve the decision problem of 2DCSS problem with $T = \frac{T_U + T_L}{2}$ and $s_2 = \max\{0, D_{\min}^-, T_C - T\}$ by corresponding 2-SAT problem.

if "yes" **then** $T_U = \frac{T_U + T_L}{2}$.

else $T_L = \frac{T_U + T_L}{2}$.

end while

Step3 Determine the clock timing of each register with $T = T_U$ and $s_2 = \max\{0, D_{\min}^-, T - T_C\}$ by corresponding value assignment of 2-SAT problem.

Fig. 3. Outline of the proposed method.

The proposed method determines the minimum clock period by a binary search on the clock period. In Fig. 3, the precision for the clock period is denoted by ΔT. If the decision problem of 2DCSS problem can be solved in $O(X)$ time, 2DCSS problem can be solved in $O(X \cdot \log T)$ by a binary search on the clock period T. As mentioned in [9], the upper bound of minimum clock period in 2DCSS problem is the minimum clock period T_C in the conventional clock synchronous framework and the lower bound of minimum clock period in 2DCSS problem is the minimum clock period T_S in CSS. Since the range of clock period can be assumed to a constant, 2DCSS problem can be solved in $O(X)$. In our method, the decision problem of 2DCSS problem is translated into 2-SAT problem. The decision problem of 2DCSS problem can be translated into 2-SAT problem in $O(|V_r| + |E_r|)$ time and 2-SAT problem can be solved in $O(|V_r| + |E_r|)$ time. Therefore, our method solves 2DCSS problem in $O(|V_r| + |E_r|)$ time.

In the proposed method, s_2 is set to $\max\{0, D_{\min}^-, T_C - T\}$, where $D_{\min}^- = -\min_{(a,b) \in E_r}\{0, D_{\min}(a,b)\}$. If there are D-edges with negative weights, $D_{\min}^- > 0$. Otherwise, $D_{\min}^- = 0$. Domain values $\{s_1, s_2, \ldots, s_k\}$ in MDCSS problem can be determined by the shortest path in the constraint graph for k clock domains [9]. In this paper, the number of clock domains is restricted to two. In this case, the shortest path in constraint graph for 2 clock domains depends on the minimum negative edge in the constraint graph. In the constraint graph, weights of D-edges with negative weights and Z-edges whose maximum delays $D_{\max}$ are larger than T are negative. Therefore, the shortest path in the constraint graph for 2 clock domains is determined by $\min\{0, -D_{min}^-, T - T_C\}$. Since we assume that $s_2 \geq 0$, we have $s_2 = \max\{0, D_{\min}^-, T_C - T\}$.

Theorem 1 *If a feasible clock skew schedule exists at the clock period T, there is a feasible clock skew schedule with $s_2 = \max\{0, D_{\min}^-, T_C - T\}$, where $D_{\min}^- = -\min_{(a,b) \in E_r}\{0, D_{\min}(a,b)\}$.*

978-1-4799-2817-0/14 $31.00 © 2014 IEEE

TABLE I

DEFINITION OF CLAUSE.

x_a	x_b	$S(a)$	$S(b)$	timing constraint for (a,b) satisfied	violated
0	0	0	0	$c_0^{(a,b)} = 1$	$c_0^{(a,b)} = x_a \vee x_b$
0	1	0	s_2	$c_1^{(a,b)} = 1$	$c_1^{(a,b)} = x_a \vee \overline{x_b}$
1	0	s_2	0	$c_2^{(a,b)} = 1$	$c_2^{(a,b)} = \overline{x_a} \vee x_b$
1	1	s_2	s_2	$c_3^{(a,b)} = 1$	$c_3^{(a,b)} = \overline{x_a} \vee \overline{x_b}$

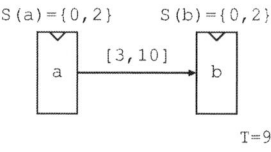

Fig. 4. An example which has a feasible clock schedule $(S(a), S(b)) = (0, 2)$.

B. Translation from 2DCSS to 2-SAT

In this sub-section, we discuss the translation from the decision problem of 2DCSS problem to 2-SAT problem.

A Boolean variable x_a in 2-SAT problem corresponds to clock timing $S(a)$ of a register a. $x_a = 0$ ($x_a = 1$), if and only if $S(a) = 0$ ($S(a) = s_2$).

Each set of four clauses corresponds to timing constraints of a register pair. A collection of clauses $C(X)$ is defined as follows:

$$C(X) = \bigwedge_{(a,b) \in E_r} (c_0^{(a,b)} \wedge c_1^{(a,b)} \wedge c_2^{(a,b)} \wedge c_3^{(a,b)})$$

The definition of each clause is shown in Table I. Each clause is defined by the timing constraints of a register pair. For example, if hold and setup constraints of a register pair (a,b) are satisfied when $(S(a), S(b)) = (0,0)$, $c_0^{(a,b)} = 1$. Otherwise, $c_0^{(a,b)} = x_a \vee x_b$.

If $(x_a, x_b) = (0,0)$ and if hold and/or setup constraint of a register pair (a,b) are violated when $(S(a), S(b)) = (0,0)$, $c_0^{(a,b)} = x_a \vee x_b = 0$. It means that $(x_a, x_b) = (0,0)$ is not a solution of value assignment of 2-SAT problem when hold and/or setup constraint of a register pair (a,b) are violated and when $(S(a), S(b)) = (0,0)$. On the other hand, $c_0^{(a,b)} = x_a \vee x_b = 1$ when $(x_a, x_b) = (0,1)$, $(1,0)$, and $(1,1)$. It means that even if hold and/or setup constraint of a register pair (a,b) are violated when $(S(a), S(b)) = (0,0)$, $(x_a, x_b) = (0,1)$, $(1,0)$, and $(1,1)$ are allowed as solutions of value assignment of 2-SAT problem.

If a satisfying value assignment exists, a clock skew schedule can be obtained by the satisfying value assignment. From the definition of 2-SAT problem, if the value assignment satisfies all clauses, the corresponding clock skew schedule satisfies the timing constraints.

For example, suppose the circuit such that $D_{\min}(a, b) = 3$, $D_{\max}(a, b) = 10$, $s_2 = 2$, and clock period $T = 9$ as shown in Fig. 4. In this case, setup constraints are violated when

$(S(a), S(b)) = (0, 0)$, $(2, 0)$, and $(2, 2)$. Consequently, we have

$$\begin{aligned} C(X) &= c_0^{(a,b)} \wedge c_1^{(a,b)} \wedge c_2^{(a,b)} \wedge c_3^{(a,b)} \\ &= (x_a \vee x_b) \wedge (\overline{x_a} \vee x_b) \wedge (\overline{x_a} \vee \overline{x_b}). \end{aligned}$$

Since this collection is satisfied only if $(x_a, x_b) = (0, 1)$, the feasible clock schedule $(S(a), S(b)) = (0, 2)$ is obtained by solving the value assignment of 2-SAT.

The number of Boolean variables $|X|$ in 2-SAT problem is equal to the number of registers $|V_r|$. The number of clauses in 2-SAT problem is at most $4 \cdot |E_r|$. Moreover, since the time complexity of check of the timing constraint for a register pair is constant, 2DCSS problem can be translated into 2-SAT problem in $O(|V_r| + |E_r|)$ time and 2-SAT problem can be solved in $O(|V_r| + |E_r|)$ time. Therefore, 2DCSS problem can be solved in $O(|V_r| + |E_r|)$ time.

The algorithm for 2-SAT problem proposed in [10] and examples are described in Appendix.

C. Enhancement of Our Method

The proposed method can be enhanced for the following generalized 2DCSS problem easily. The proposed method can be applied to various problems such as a 2DCSS problem with large delay variations and a post-silicon skew tuning problem with two values.

Generalized 2DCSS

Input: maximum and minimum delays between register pairs $D_{\max}(a, b)$ and $D_{\min}(a, b)$ for $\forall (a, b) \in E$, and clock period T

Output: a clock skew schedule $S : \forall a \to \{s_1^a, s_2^a\}$

Constraint: satisfy hold and setup constraints.

IV. IMPROVED METHOD OF EXISTING METHOD

Since the algorithm proposed in [9] focuses on MDCSS problem, it is not efficient for 2DCSS. To evaluate the proposed method fairly, an enhancement of the algorithm proposed in [9] for 2DCSS is discussed.

The time complexity of the existing method is $O((k-1)!|V_r||E_r|^k)$. Since there are at most $O((k-1)!|E_r|^{k-1})$ candidates of domain values $\{s_1, s_2, \ldots, s_k\}$ with k clock domains and the feasibility for each candidate can be checked by the modified Bellman-Ford algorithm [11] in $O(|V_r||E_r|)$ time, the time complexity of the existing method is $O((k-1)!|V_r||E_r|^k)$. Although the number of candidates whose feasibilities are checked is restricted by two pruning techniques in [9], its time complexity is still $O((k-1)!|V_r||E_r|^k)$. When $k = 2$, the time complexity of the existing method is $O(|V_r||E_r|^2)$.

As mentioned in Theorem 1, s_2 can be determined uniquely in 2DCSS problem. Consequently, the existing method can omit the enumeration of clock schedule candidates. We implemented the existing method omitting the enumeration and refer to this method as *improved BF method* hereafter. Note that the time complexity of the improved BF method is $O(|V_r||E_r|)$ time due to the modified Bellman-Ford algorithm.

978-1-4799-2817-0/14 $31.00 © 2014 IEEE

TABLE II
RESULT ON ISCAS89 BENCHMARKS.

Design	$|V_r|$	$|E_r|/2$	T_S	Existing [9] T_2/T_S	Existing [9] time[s]	Improved BF T_2/T_S	Improved BF time[s]	Proposed T_2/T_S	Proposed time[s]
s27	5	21	5.06	1.14	<0.1	1.14	<0.01	1.14	<0.01
s208	10	70	9.91	1.00	<0.1	1.00	<0.01	1.00	<0.01
s298	16	86	10.79	1.05	<0.1	1.05	<0.01	1.05	<0.01
s344	17	121	13.15	1.09	<0.1	1.09	<0.01	1.09	<0.01
s349	17	121	13.51	1.09	<0.1	1.09	<0.01	1.09	<0.01
s382	23	175	9.63	1.20	<0.1	1.20	<0.01	1.20	<0.01
s386	8	129	9.61	1.04	<0.1	1.04	<0.01	1.04	<0.01
s400	23	175	9.89	1.17	<0.1	1.17	<0.01	1.17	<0.01
s420	18	146	21.13	1.00	<0.1	1.00	<0.01	1.00	<0.01
s444	23	175	8.10	1.34	<0.1	1.34	<0.01	1.34	<0.01
s510	8	103	14.29	1.00	<0.1	1.00	<0.01	1.00	<0.01
s526	23	167	11.22	1.05	<0.1	1.05	<0.01	1.05	<0.01
s526n	23	167	11.31	1.05	<0.1	1.05	<0.01	1.05	<0.01
s641	21	486	29.51	1.00	<0.1	1.00	<0.01	1.00	<0.01
s713	21	486	30.58	1.00	<0.1	1.00	<0.01	1.00	<0.01
s820	7	213	16.74	1.00	<0.1	1.00	<0.01	1.00	<0.01
s832	7	213	16.22	1.00	<0.1	1.00	<0.01	1.00	<0.01
s838	34	298	44.66	1.00	<0.1	1.00	<0.01	1.00	<0.01
s1196	19	365	22.28	1.00	<0.1	1.00	<0.01	1.00	<0.01
s1238	19	365	24.33	1.00	<0.1	1.00	<0.01	1.00	<0.01
s1423	76	2235	73.13	1.03	<0.1	1.03	<0.01	1.03	<0.01
s1488	8	266	23.18	1.00	<0.1	1.00	<0.01	1.00	<0.01
s1494	8	266	23.85	1.00	<0.1	1.00	<0.01	1.00	<0.01
s5378	165	2180	22.88	1.13	<0.1	1.13	<0.01	1.13	<0.01
s9234	140	2226	33.76	1.01	<0.1	1.01	<0.01	1.01	<0.01
s13207	471	3885	53.36	1.03	<0.1	1.03	0.01	1.03	0.01
s15850	565	16375	85.27	1.08	<0.1	1.08	0.07	1.08	0.02
s38417	1465	31980	86.19	1.00	<0.1	1.00	0.01	1.00	0.01
s38584	1451	17900	286.62	1.00	<0.1	1.00	0.03	1.00	0.02

$|V_r|$ the number of registars
$|E_r|/2$ the number of register pairs with signal propagations
T_S the minimum clock period in CSS
T_2 the minimum clock period in 2DCSS

TABLE III
RESULT ON ARTIFICIAL DATA.

Type	$|V_r|$	$|E_r|/2$	Improved BF time[s]	Proposed time[s]
random	10000	100000	0.08	0.07
random	20000	200000	0.17	0.13
random	40000	400000	0.35	0.29
worst	10000	100000	20.02	0.03
worst	20000	200000	90.17	0.06
worst	40000	400000	376.76	0.15

V. EXPERIMENTAL RESULTS

We implemented the proposed algorithm and the improved BF method in C++, which were compiled by gcc4.3.2, and executed on a PC with 6 GB memory by using Intel core i7-940 of 2.93 GHz. Note that only one core is used for our experiments. Two methods are applied to ISCAS89 benchmarks obtained from authors of [7], which are the same as [9], and artificial data. However, since the number of registers and the number of register pairs with signal propagations in obtained s953 and s35932 are different from those shown in [7] and in [9], we ignore the results of these two circuits.

The results on remained 29 circuits in ISCAS89 benchmarks are shown in Table II. The number of registers, the number of register pairs with signal propagations, the minimum clock period in CSS, and the minimum clock period in 2DCSS are represented by $|V_r|$, $|E_r|/2$, T_S and T_2, respectively. The results of Existing [9] are directly copied from [9]. The results show that the minimum clock period obtained by our method is always the same as that obtained by the existing method and the improved BF method. Therefore, our method obtains an optimum solution as well as the existing method and the improved BF method. Although we cannot compare the execution time because the precisions of execution time shown in [9] are not

TABLE IV
THE RELATION OF A CLAUSE FOR TIMING VIOLATION AND DIRECTED EDGES ADDED INTO THE GRAPH.

x_a	x_b	clause	directed edges
0	0	$c_0^{(a,b)} = x_a \vee x_b$	$(\overline{x_a}, x_b), (\overline{x_b}, x_a)$
0	1	$c_1^{(a,b)} = x_a \vee \overline{x_b}$	$(\overline{x_a}, \overline{x_b}), (x_b, x_a)$
1	0	$c_2^{(a,b)} = \overline{x_a} \vee x_b$	$(x_a, x_b), (\overline{x_b}, \overline{x_a})$
1	1	$c_3^{(a,b)} = \overline{x_a} \vee \overline{x_b}$	$(x_a, \overline{x_b}), (x_b, \overline{x_a})$

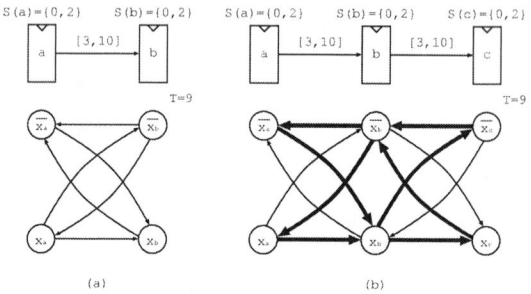

Fig. 5. Examples: (a) Feasible clock schedule exists. (b) Feasible clock schedule does not exist.

clear, the execution time of the proposed method and the improved BF method is almost same.

The results on artificial data are shown in Table III. We made two types: one denoted by random is made randomly, and another denoted by worst is made so that the shortest path has $|V_r|$ vertices. In each type, each FF has ten signal propagations. If the shortest path has $|V_r|$ vertices, the execution time of the improved BF method is expected to long since the number of iterations in the modified Bellman-Ford algorithm [11] is $|V_r|$. The results show that the execution time of the proposed method and the improved BF method is almost same in random since the number of iterations in the modified Bellman-Ford algorithm is small. However, the proposed method is much faster than the improved BF method in worst. The results show the effectiveness of the proposed method.

VI. CONCLUSION

In this paper, we proposed an optimum linear time algorithm that maximizes the performance under the constraint that the number of clock domains is restricted to two. In our method, the problem is translated into 2-SAT problem. Experiment results on ISCAS89 benchmarks and artificial data confirmed the optimality and efficiency of our method.

APPENDIX

A. Algorithm in [10]

2-SAT problem can be solved by using the concept of the strongly connected components in graph theory. For each variable x_i, two vertices named x_i and $\overline{x_i}$ are added in the graph. For each clause $(u \vee v)$, two directed edge $(\overline{u}, v)$ and $(\overline{v}, u)$ are added. 2-SAT problem is satisfiable if and only if no vertices x_i and $\overline{x_i}$ belong to the same strongly connected component in the graph.

B. Examples of 2DCSS

The relation between a clause for timing violation and directed edges added into the graph is shown in Table IV.

At first, suppose the circuit such that $D_{\min}(a, b) = 3$, $D_{\max}(a, b) = 10$, $s_2 = 2$, and clock period $T = 9$, which is shown in shown in Fig. 4 and shown in the top of Fig. 5 (a) again. In this case, timing constraints are violated when $(S(a), S(b)) = (0, 0), (2, 0)$, and $(2, 2)$. Therefore, edges $(\overline{x_a}, x_b), (\overline{x_b}, x_a), (x_a, x_b), (\overline{x_b}, \overline{x_a}), (x_a, \overline{x_b})$, and $(x_b, \overline{x_a})$ are

added into the graph (see the figure shown in the bottom of Fig. 5 (a)). In this case, both x_a and x_b belong to the different strongly connected components from $\overline{x_a}$ and $\overline{x_b}$ in the graph. Therefore, we can assign a satisfying value assignment of 2-SAT problem, and we obtain a feasible clock schedule $(S(a), S(b)) = (0, 2)$.

Next, we consider the circuit shown in the top of Fig. 5 (b). In this case, all vertices belong to the same strongly connected component in the graph shown in the bottom of Fig. 5 (b). Therefore, no satisfying value assignment of 2-SAT problem exists.

ACKNOWLEDGEMENTS

This work was partly supported by the Nakajima Foundation.

REFERENCES

[1] J. P. Fishburn, "Clock skew optimization," *IEEE Tranc. on Computers*, vol. 39, no. 7, pp. 945–951, 1990.

[2] R. B. Deoker and S. S. Sapatneker, "A Graph-Theoretic Approach to Clock Skew Optimization," in *ISCAS*, pp. 407–410, 1994.

[3] A. Takahashi and Y. Kajitani, "Performance and reliability driven clock scheduling of sequential logic circuits," in *ASP-DAC*, pp. 37–43, 1997.

[4] K. Ravindran, A. Kuehlmann, and E. Sentovich, "Multi-domain clock skew scheduling," in *ICCAD*, pp. 801–808, 2003.

[5] M. Toyonaga, K. Kurokawa, T. Yasui, and A. Takahashi, "A practical clock tree synthesis for semi-synchronous circuits," in *ISPD*, pp. 159–164, 2000.

[6] K. Kurokawa, T. Yasui, M. Toyonaga, and A. Takahashi, "A practical clock tree synthesis for semi-synchronous circuits," *IEICE Transactions on Fund.*, no. 11, pp. 2705–2713, 2001.

[7] J. Casanova and J. Cortadella, "Multi-level clustering for clock skew optimization," in *ICCAD*, pp. 547–554, 2009.

[8] Y. Zhi, H. Zhou, and X. Zeng, "A practical method for multi-domain clock skew optimization," in *ASP-DAC*, pp. 521–526, 2011.

[9] L. Li, Y. Lu, and H. Zhou, "Optimal multi-domain clock skew scheduling," in *DAC*, pp. 152–157, 2011.

[10] B. Aspvall, M. F. Plass, and R. E. Tarjan, "A Linear-Time Algorithm for Testing the Truth of Certain Quantified Boolean Formulas," *Information Processing Letters*, vol. 8, no. 3, pp. 121–123, 1979.

[11] J. P. Fishburn, "Solving a system of difference constraints with variables restricted to a finite set," *Inf. Process. Lett.*, vol. 82, no. 3, pp. 143–144, 2002.

Power Minimization of Pipeline Architecture through 1-Cycle Error Correction and Voltage Scaling

Insup Shin[1], Jae-Joon Kim[2], and Youngsoo Shin[1]

[1]Department of Electrical Engineering, KAIST, Daejeon 305-701, Korea

[2]Department of Creative IT Engineering, POSTECH, Pohang 790-784, Korea

Abstract—We present a new 1-cycle timing error correction method, which enables aggressive voltage scaling in a pipelined architecture. The proposed method differs from the state-of-the-art in that the pipeline stage where the timing error occurs can continue to receive input data without halting to avoid data collision. The feature allows the pipeline to avoid recurring clock gating when timing errors happen at multiple stages or timing errors continue to occur at a certain stage. Compared to a state-of-the-art method, the proposed method shows 2-6% energy reduction for a 5-stage pipeline and 7-11% reduction for a 10-stage pipeline. In addition, the proposed logic to propagate clock gating signal is much simpler than that of the previous method [1] by eliminating reverse propagation path of clock gating signal.

I. INTRODUCTION

Adding timing guardband has been an effective solution to address delay deviations due to process variations. Designers need to add suitable amount of timing guardband to critical path delay only to make the circuit operate correctly even in the worst case scenario. However, as the variation has become a major concern in nanoscale circuits, the amount of timing guardband increases significantly. This has led to growing interest in the design methodology to eliminate or reduce the timing guardband.

Timing speculation [2]–[6] is one of circuit techniques to eliminate the timing guardband. The key idea is to detect and correct timing errors; Razor [2] is a representative example. In this method, the output of a logic stage is sampled by a main flip-flop with nominal clock as well as by a shadow latch with the delayed clock. A timing error is checked by comparing the data from flip-flop and latch. The error is corrected by restoring the data from shadow latch into main flip-flop.

Another benefit of timing speculation is significant power saving when aggressive voltage scaling is employed. Since errors can be detected and corrected, supply voltage can be lowered beyond the minimum voltage that leads to zero error. The effectiveness of this voltage scaling depends on timing error rate below the minimum voltage. Various techniques [7]–[11] have been proposed to reduce minimum allowable voltage level given target throughput by reshaping distribution of path delays. Methods for adding slack to the frequently violated paths have been proposed in [7] (Blueshift). A cell sizing method was proposed in [8]; cell upsizing on critical and frequently exercised paths to extend voltage scaling and down sizing in noncritical and infrequently exercised paths to reduce leakage power. In [9], a dual-Vt assignment algorithm (Dynatune) was proposed to increase throughput based on the behavior curve of a circuit. The technique proposed in [10] changes the circuit structure by re-ordering fan-ins of some gates during logic optimization. The technique proposed in [11] uses new cost function which takes timing error probability into account during logic synthesis.

However, these techniques have limited impact due to the existence of critical operating point [12]. Since timing slacks are spent to optimize such objectives as area and power, many paths have almost same path delay as the critical one, which is called slack wall. Thus, there exists a critical voltage V_c for a fixed ambient temperature such that any voltage below V_c causes massive errors [12]. The techniques in [7]–[11] cannot handle such a large number of errors. Hence, critical operating point is the major challenge for the timing speculation based voltage overscaling.

Massive errors require very large timing penalty for error correction. Therefore, a challenge of timing speculation based voltage overscaling lies in reducing overall timing penalty for error corrections. There have been several error correction methods [1], [2], [5], [6], [13], [14]. Among them, instruction replay [5] has the smallest design overhead. When the instruction for which an error has been detected reaches the last stage, the processor flushes all stages and issues the failing instruction again. During re-execution of the failing instruction, clock frequency is reduced by half. Thus it has large timing penalty per error correction. Hence, the instruction replay has limitation for timing speculation based voltage overscaling technique.

Micro-rollback [13], [14] saves previous data of each pipeline stage to backup storage in each cycle. When the error signal reaches the last stage, the backup storage injects the last known correct data to each pipeline stage. However, since this method increases flip-flop energy consumption by 15% [13], it is also not suitable for voltage overscaling design.

In counterflow pipelining method [2], error is corrected by restoring shadow latch data into main flip-flop in the next cycle of the error detection. Then, to maintain proper operation, the stage which detected the error sends a bubble to the last stage and a flush signal to the first stage. When the flush signal

978-1-4799-2817-0/14 $31.00 © 2014 IEEE

reaches the first stage, the next-cycle instruction to the failed instruction is issued again. If an error is detected at the k-th stage, the completion time of the next-cycle instruction is delayed by $2k$. It has smaller timing penalty than that of instruction replay but the timing penalty is still too large to use for voltage overscaling scenario.

Recently, two 1-cycle error correction methods have been proposed [1], [6]. The first method [6] targets two-phase transparent latch based designs and the second method [1] is toward flip-flop or pulsed-latch based design. However, they also have limitation in handling massive errors. Both methods can correct multiple errors by one cycle penalty but only limited in scope.

In this paper, we propose new error correction method which has 1-cycle penalty with simpler control logic. More importanlty, the proposed method can handle massive errors (and even permanent error) with only a small number of timing penalties.

The remainder of this paper is organized as follows. Our proposed method is presented in Section II. Its extension to handle complex pipeline structure is addressed in Section III. Experimental results are provided in Section IV and Section V concludes the paper.

II. PROPOSED ARCHITECTURE

The main difficulty in error correction is loss of new input data while shadow latch data is restored to main flip-flop. Counterflow pipelining solves this problem by flushing all the subsequent instructions from the one that caused error and re-executing them after the error has been corrected. Instruction replay [5] avoids this problem by re-executing the failed instruction. This method mitigates the circuit complexity by using simple recovery signal so that it requires very small overhead in control logic. However many clock cycles are needed to flush instructions and re-executes them. A recently proposed error correction method [1] overcomes this problem by stalling each input stage in a wave-like fashion. However this method requires relatively complex control logic.

We propose a simpler method to overcome the problem. Shadow latch receives wider clock pulse than that main latch receives. Thus, after sending its data to main latch for error correction, shadow latch can still capture new input data. Our key idea is to alter the clock toward shadow latch in such a way that shadow latch opens after main latch closes. Then, shadow latch can send previous correct data to main latch for error correction and also capture new input data in the same cycle. Thus, the data conflict problem can be avoided.

Fig. 1(a) illustrates conventional Razor latch. We denote the clock for main and shadow latch as clk_m and clk_s, respectively. Pulse width of clk_s, denoted by W_s, is larger than that of clk_m, denoted by W_m. When an error occurs, restore signal becomes 1 so that shadow latch data SQ is restored into main latch. Fig. 1(b) illustrates modified Razor latch for the proposed method. It also uses shadow latch to capture the data when the error occurs, and send the stored data to main latch in the next cycle when the restore signal

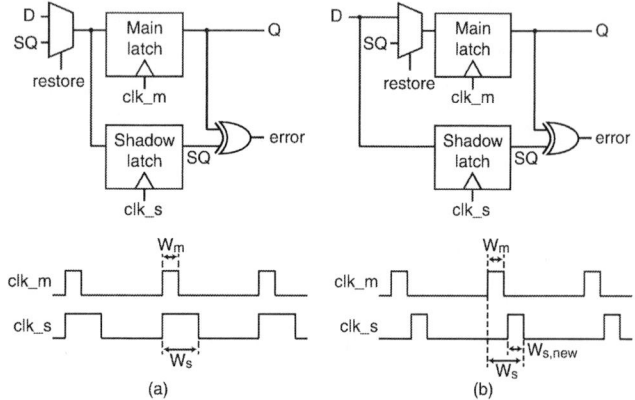

Fig. 1. Conceptual schematics of (a) conventional Razor latch and (b) modified Razor latch for our method.

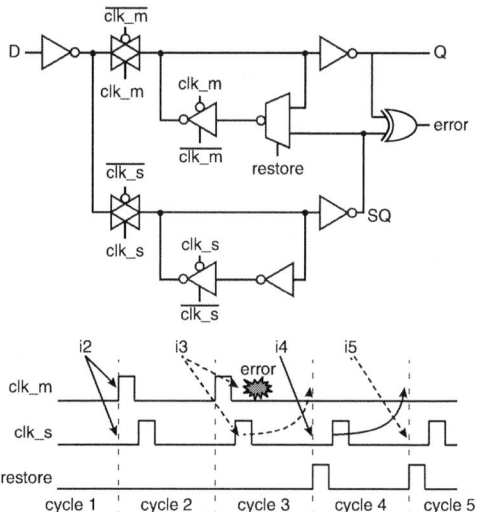

Fig. 2. Circuit level schematic of modified Razor latch.

becomes 1. The main difference of the modified Razor latch is to use non-overlapping clock signals to drive main and shadow latches. As shown 1(b), clk_s has been altered from conventional Razor latch such that shadow latch will start capturing input data after main latch stops capturing the data. Note that the window of timing speculation does not change from that of conventional Razor latch although clk_s has more narrow width. Also the input data is directly fed into the shadow latch so that it can capture the data even when main latch is gated; we will explain in the later section how this can help reduce control logic during error correction.

Fig. 2 shows circuit level schematic of modified Razor latch. The multiplexer at the input of Razor latch is moved to the feedback path of the main latch to reduce delay and power overhead. Assume that instruction $i3$ fails at cycle 3. After the error detection, clk_m is gated at each following cycle. In cycle 4, restore signal becomes 1 for the duration of W_m so that correct $i3$ is transmitted from shadow latch to main latch. Instruction $i4$ is captured by shadow latch after the restoration so that no data is lost.

978-1-4799-2817-0/14 $31.00 © 2014 IEEE

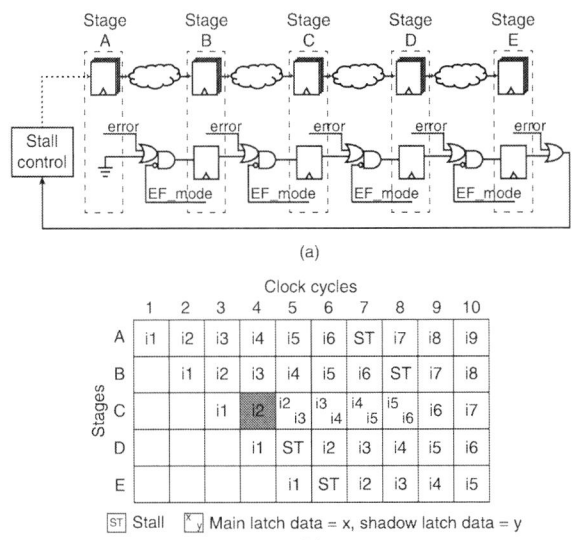

(a)

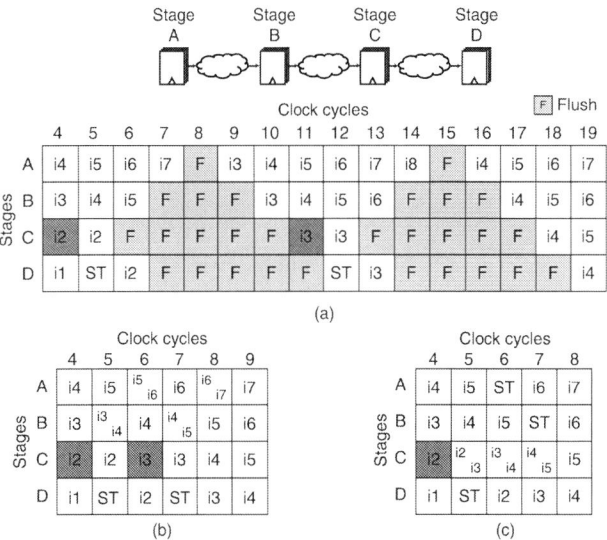

Fig. 3. (a) Modified pipeline organization for propagation of CG signal and (b) error correction example for our method when an error occurs at stage C in cycle 4.

Fig. 4. Error correction examples for (a) counterflow pipelining, (b) previous 1-cycle error correction [1], and (c) our method when instructions $i2$ and $i3$ will be failed at stage C.

From cycle 4 on, input data is stored into shadow latch and then is transmitted to main latch in the next cycle. This state will be retained until a stall is propagated to this stage. In order to send a stall to the stage where an error occurred, we add stall control logic as shown in Fig. 3.

Similar to the previous error correction methods, our method also uses clock gating control signal, called CG signal, to prevent incorrect data propagation through the pipeline due to the timing error at a stage. When an error occurs, a CG signal is propagated to output stages from the stage where the error occurred. The CG signal gates both clk_m and clk_s for one cycle. CG signal is propagated in a wave-like fashion such that it is transmitted to the next stage at each cycle. When the CG signal reaches the last stage, the stage sends stall signal to the stall control so that a stall is issued to the pipeline.

A. Error-Free Mode

Once an error occurs at a particular stage, input data is received by shadow latch alone and the data is transmitted back to main latch in the next cycle as mentioned before. No late timing error occurs in this operation mode, which we call error-free mode. Fig. 3 shows a conceptual pipeline organization for our method and error correction example. Assume that instruction $i2$ fails at stage C in cycle 4 as shown in Fig. 3(b). Stage C enters error mode and CG signal is propagated to last stage, stage E. Since stage E receives CG signal in cycle 5, it sends the CG signal to stall control in the next cycle. Thus, a stall is issued to the pipeline in cycle 7. When the stall is propagated to stage C, stage C gets out the error-free mode as shown in Fig. 3(b). Timing penalty per error correction is only one cycle.

One advantage of error-free mode is to offer the opportunity that multiple errors at the same stage are corrected with only one cycle. Consider a pipeline architecture that has four stages

as shown in Fig. 4 and assume that instructions $i2$ and $i3$ will fail at stage C. In counterflow pipelining, the completion time of the next-cycle instruction is delayed by $2k$ when an error occurs at k-th stage. Therefore, the completion time of instruction $i4$ is delayed by $2(3) + 2(3) = 12$ cycles as shown in Fig. 4(a). 1-cycle error correction method [1] requires two cycles to correct errors as shown in Fig. 4(b).

Compared to the previous error correction methods, our method can handle multiple errors at the same stage with less cycle penalty. Once an error occurs, the stage enters into error-free mode until a stall signal is propagated back to the stage. Thus, our method requires only one cycle delay to handle the case shown in Fig. 4(c). Furthermore, our method can handle a permanent error in the piepline more effectively. If a pipeline stage has a permanent timing error, that stage will simply stay in error-free mode; the pipeline stage will re-enter into error-free mode in the next cycle when the stall signal is propagated back to the stage and this will repeat as long as an error persists.

B. Multiple Timing Errors

CG signal is propagated from the stage where an error occurred to its output stages to prevent the propagation of incorrect data and then returns to the stage to bring it out of error-free mode. However, CG signal does not necessarily need to propagate back to the stage where is was issued since the propagation of CG signal will stop when it arrives at any stage already in error-free mode. Therefore, multiple errors occurring at different stages can be corrected simultaneously. In such cases, the number of stalled cycles is smaller than the number of errors.

In Fig. 5, timing errors occur at stage D in cycle 5 and at stage B in cycle 6. Propagation of CG signal which started from stage B stops at stage D in cycle 8 because stage D is already in error-free mode. Propagation of CG signal which

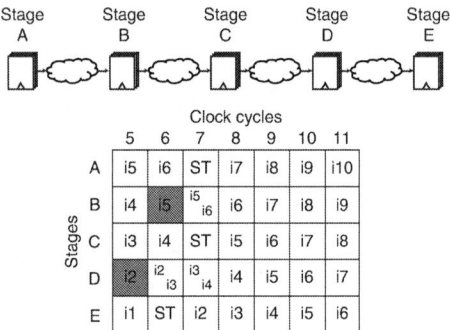

Fig. 5. Propagation of CG is stopped when it is propagated to the stage which has already entered into error-free mode.

started from stage D also stops at stage B in cycle 8 for the same reason. In this case, only one cycle is needed to correct both errors.

Previous 1-cycle error correction methods [1], [6] can also correct multiple errors by one cycle penalty but it is possible only when clock gating signals are met. Thus, our method has more opportunity to correct multiple error with one cycle penalty.

III. EXTENSION TO GENERAL PIPELINE ARCHITECTURE

A practical pipeline may have multiple fan-out and multiple fan-in stages or a loop. In order to handle more general pipeline architecture, additional control signal is needed. The main problem with a multiple fan-in stage lies in the fact that the pipeline fails after the multiple fan-in stage receives CG signals from part of its input stages alone. Assume that stage D has multiple fan-in as shown in Fig. 6. When an error occurs at stage A in cycle 4, CG signals are propagated to stage B and stage D in cycle 5. Stage D is therefore stalled in cycle 5 but stage C sends instruction $i2$ to stage D in the same cycle. Since instruction $i2$ cannot be captured in stage D, the pipeline fails after cycle 5.

The basic rule to maintain proper operation is to forbid all of its input stages from sending their data to the stage when a stage is stalled. Thus, the problem can be resolved by generating virtual errors at all the stages that did not send CG signals to a multiple fan-in stage. In Fig. 6(b), we generate a virtual error at stage C in cycle 4.

A new control signal VE is employed to realize the concept. When a stage receives a CG signal from any of its input stage, the stage sends VE signals back to all of its input stages to check whether each stage sends CG signal. If a stage which receives VE signal is already clock-gated or has generated an error in early part of the same cycle, the VE signal does not affect the stage. Otherwise, the stage gets into error-free mode. Fig. 6(c) shows a timing diagram for CG signal and VE signal. Since stage A sends CG signals to stage B and stage D in cycle 4, stage B sends VE signal to stage A, and stage D sends VE signals to its input stages (stage A and C) in the same cycle. The VE signal does not affect stage A because stage A already experiences a timing error in the earlier part of the cycle 4. Stage C gets into error-free mode immediately.

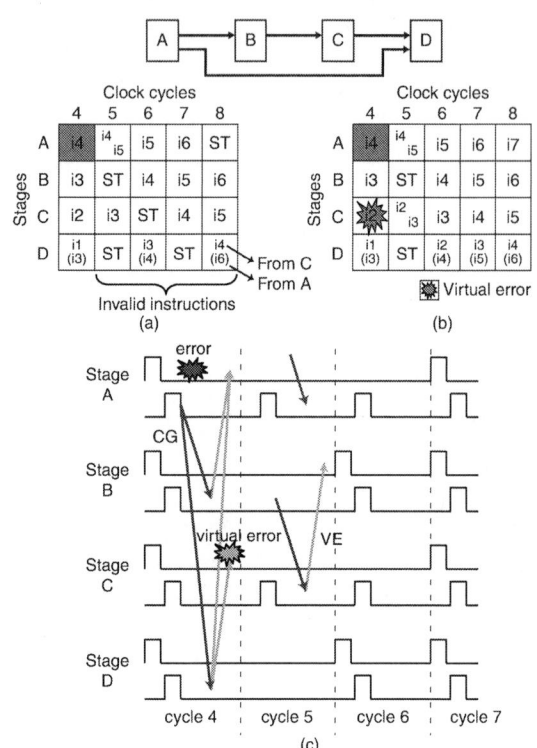

Fig. 6. A pipeline circuit which has multiple fan-out stage and multiple fan-in stage. (a) When VE control signal is not used and (b) when VE control signal is used and (c) its timing diagram.

In cycle 5, stage B sends CG signal to stage C and then stage C sends VE signal back to stage B. Due to the CG signal being propagated to stage C, the stage C gets out of error-free mode so that clk_m is not gated after cycle 7. Since stage B sends CG signal to stage C, the VE signal received by stage B is nullified.

Our method also correctly works in the loop condition by using VE control signal. The main challenge in such a condition is to prevent infinite looping. The infinite looping of CG propagation does not occur regardless of the location where the error happens. Note that the error can occur before the loop, in the loop, or after the loop. Consider a pipeline architecture that has a loop and assume that three errors occur at stage A, stage C, and stage E, respectively, as shown in Fig. 7. When an error occurs before the loop (stage A in cycle 5), the CG signal is inserted into the loop and generates a virtual error at stage D due to VE signal. Due to this virtual error, CG signals are propagated from stage D to its output stages (stage B and E) in cycle 6. The CG signal which was propagated to stage E is sent to stage A through the stall control so that stage A gets out of error-free mode in cycle 7. The CG signal which was sent to stage B is propagated to stage D and stage D also gets out of error-free mode in cycle 8. Timing penalty for the error correction is one cycle even in this case.

When an error occurs within the loop (stage C in cycle 9), CG signal is propagated from stage C to stage D in cycle 10. In cycle 11, stage D sends CG signal to its output stages (stage

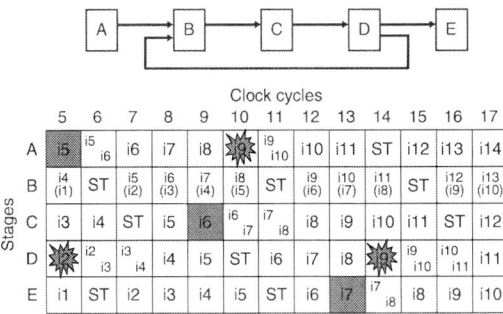

Fig. 7. Error correction example for loop case when errors occur before the loop, in the loop, and after the loop.

B and E). When stage B receives CG signal, VE signals are sent to stage A and stage D. Since stage D is already gated in cycle 10, the VE signal which was sent to stage D is nullified. Stage A is not gated nor detects an error in cycle 10. Thus, a virtual error is generated at stage A as shown in Fig. 7. Due to this virtual error, CG signal is propagated to stage C and thus stage C gets out of error-free mode in cycle 12. Stage A gets out of error-free mode by the CG signal which was propagated from stage C.

When an error occurs after the loop (stage E in cycle 13), CG signal is propagated to stage A through the stall control in cycle 14. When stage B receives CG signal, it sends VE signals to stage A and stage D. Due to this VE signal, a virtual error is generated at stage D in cycle 14. Stage E gets out of error-free mode in cycle 15 due to CG signal generated from stage D and stage D also gets out of error free mode in cycle 17 due to CG signal generated from stage E.

The control logic for proposed error correction method is illustrated in Fig. 8. All sequential elements in the figure are transparent latches. Once an error occurs, EF node becomes 1; CG_out node is also asserted to 1. EF node remains at 1 until CG signal arrives (CG_in node becomes 1). When CG signal arrives from any input stage, it is transmitted to all the output stages through CG_out node and VE signal is sent to all the input stages through VE_out node. However, the propagation of CG signal works only during normal operation mode (when both EF and EF_pre nodes are 0). When a stage receives VE signal, it generates a virtual error and then EF_node becomes 1 if CG_pre node is 0.

IV. EXPERIMENTAL RESULTS

To assess the effectiveness of our error correction method, we compiled nine pipelined circuits with 45-nm Open Cell Library [15]. Three different number of pipeline stages (5, 8, and 10) was tried; three different circuits (c1908, c3540, and c6288 from ISCAS benchmark) were assumed for each pipeline stage. The combinational circuits are synthesized using Design Compiler [16] with timing constraint set to 90% of critical path delay of the circuit synthesized without any contstraint. Clock periods of resulting c1908, c3540, and c6288 were 1.2 ns, 1.6 ns, and 2.4 ns, respectively. All pipelined circuits are pulsed-latch based Razor circuit. The pulse width of main latch is 105 ps, which is the minimum

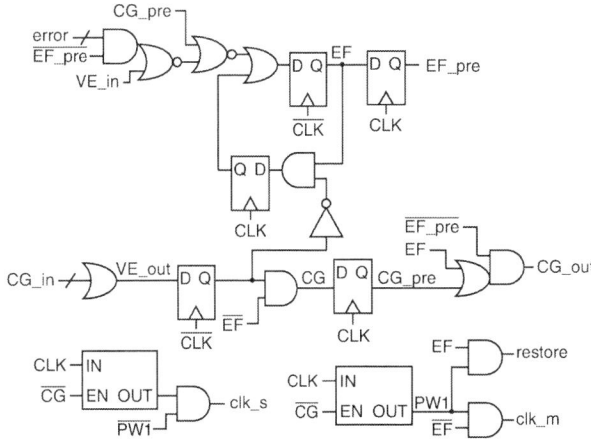

Fig. 8. Schematic of control logic.

width at 0.75 V under nominal process corner, and that of shadow latch is 400 ps. In all the circuits, extra delay buffers are inserted to fix hold violations.

We applied 100 random vectors to each pipelined circuit and determined its throughput via fast SPICE simulation [17]. The simulation was performed at the initial voltage (1.1 V) and the voltage was gradually reduced by an increment of 0.02 V until the throughput met a target; total energy dissipation was then measured. We performed this experiment for two target throughputs, 0.9 and 0.7, as shown in Table I and Table II. We used the same experimental setting to collect data for counterflow pipelining [2] and 1-cycle timing error correction (1-CTEC) [1] for comparison.

Since counterflow pipelining has the largest timing penalty per error correction ($2k$ cycles, where k is the order of the stage that an error was detected), its target voltage has the highest value as expected. Moreover, it requires the largest design overhead to implement flushing logic; this adds extra overhead to the total energy dissipation. Note that the normalized total energy dissipation of counterflow pipelining increases with more pipeline stages. This is because of two shortcomings compared to both 1-CTEC and our method: 1) its timing penalty per error correction depends on the number of pipeline stages and 2) it cannot correct multiple errors simultaneously.

Both our method and 1-CTEC have 1 cycle penalty per error correction. However, thanks to error-free mode, our method can handle multiple errors with less cycle penalties, which is why the target voltage level of our method is lower than that of 1-CTEC. Compared to 1-CTEC, our method reduces the total energy dissipation by 2% and 6% when target throughputs are 0.9 and 0.7, respectively, for 5-stage pipeline. Note that the benefit is more as the number of pipeline stage increases; for 10-stage pipeline, our method reduces the total energy dissipation by 7% and 11% when target throughputs are 0.9 and 0.7, respectively. This can be understood from the fact that the number of cycles that each stage runs in error-free mode increases with the number of pipeline stages; this increases the opportunity for correcting multiple errors occurred at different

978-1-4799-2817-0/14 $31.00 © 2014 IEEE   183

TABLE I

EXPERIMENTAL RESULTS WHEN TARGET THROUGHPUT IS 0.9

#Stages	Base circuit	Counterflow		1-CTEC		Ours	
		Voltage [V]	Energy [pJ]	Voltage [V]	Energy [pJ]	Voltage [V]	Energy [pJ]
5	c1908	0.92	783	0.84	707	0.84	716
	c3540	0.94	2107	0.88	1816	0.86	1751
	c6288	0.98	5108	0.90	4307	0.90	4221
	Avg.		1.16		1.00		0.98
8	c1908	0.92	1279	0.86	1156	0.84	1108
	c3540	0.94	3598	0.88	3056	0.86	2868
	c6288	0.98	8267	0.90	6774	0.90	6803
	Avg.		1.17		1.00		0.97
10	c1908	0.94	1591	0.88	1362	0.86	1316
	c3540	0.96	4931	0.90	3991	0.88	3692
	c6288	0.98	10449	0.90	8489	0.88	7596
	Avg.		1.21		1.00		0.93

TABLE II

EXPERIMENTAL RESULTS WHEN TARGET THROUGHPUT IS 0.7

#Stages	Base circuit	Counterflow		1-CTEC		Ours	
		Voltage [V]	Energy [pJ]	Voltage [V]	Energy [pJ]	Voltage [V]	Energy [pJ]
5	c1908	0.90	751	0.78	614	0.76	576
	c3540	0.92	1912	0.82	1594	0.80	1515
	c6288	0.98	5108	0.86	3994	0.84	3706
	Avg.		1.23		1.00		0.94
8	c1908	0.92	1279	0.80	967	0.76	914
	c3540	0.92	3435	0.82	2325	0.80	2045
	c6288	0.98	8267	0.86	6227	0.84	5696
	Avg.		1.38		1.00		0.91
10	c1908	0.92	1540	0.84	1254	0.80	1153
	c3540	0.96	4931	0.84	3411	0.80	2906
	c6288	0.98	10449	0.86	7457	0.84	6625
	Avg.		1.36		1.00		0.89

stages with less cycle penalties.

V. CONCLUSION

Fast error correction method is vital to timing error detection/correction mechanism especially when targeting for low voltage operation. Our main contributions in this paper include a simplified logic to propagate clock gating signal and the reduction of timing penalty when large number of errors occur at critical operation point voltage. Compared to previous fast (1-cycle) error correction methods, the number of stalled clock cycles can be reduced significantly by allowing input data to continuously flow into the pipeline stage where timing error has occurred.

ACKNOWLEDGMENT

The work of Jae-Joon Kim was supported by the MSIP (Ministry of Science, ICT, and Planning), Korea under the "IT Consilience Creative Program"(NIPA-2013-H0203-13-1001) supervised by NIPA (National IT Industry Promotion Agency).

REFERENCES

[1] I. Shin et al., "A pipeline architecture with 1-cycle timing error correction for low voltage operations," in Proc. Int. Symp. on Low Power Electronics and Design, Sept. 2013, pp. 199–204.

[2] D. Ernst et al., "Razor: a low-power pipeline based on circuit-level timing speculation," in Proc. Int. Symp. on Microarchitecture, Dec. 2003, pp. 7–18.

[3] S. Das et al., "Razor II: in situ error detection and correction for PVT and SER tolerance," IEEE Journal of Solid-State Circuits, vol. 44, no. 1, pp. 32–48, Jan. 2009.

[4] K. A. Bowman et al., "Energy-efficient and metastability-immune resilient circuits for dynamic variation tolerance," IEEE Journal of Solid-State Circuits, vol. 44, no. 1, pp. 49–63, Jan. 2009.

[5] K. A. Bowman et al., "A 45 nm resilient microprocessor core for dynamic variation tolerance," IEEE Journal of Solid-State Circuits, vol. 46, no. 1, pp. 194–208, Jan. 2011.

[6] M. Fojtik et al., "Bubble Razor: an architecture-independent approach to timing-error detection and correction," in Proc. Int. Solid-State Circuits Conf., Feb. 2012, pp. 488–490.

[7] B. Greskamp et al., "Blueshift: designing processors for timing speculation from the ground up," in Proc. Int. Symp. High Performance Computer Architecture, Feb. 2009, pp. 213–224.

[8] A. B. Kahng et al., "Slack redistribution for graceful degradation under voltage overscaling," in Proc. Asia South Pacific Design Automation Conf., Jan. 2010, pp. 825–831.

[9] L. Wan and D. Chen, "Dynatune: circuit level optimization for timing speculation considering dynamic path behavior," in Proc. Int. Conf. on Computer Aided Design, Nov. 2009, pp. 172–179.

[10] Y. Liu et al., "On logic synthesis for timing speculation," in Proc. Int. Conf. on Computer Aided Design, Nov. 2012, pp. 591–596.

[11] J. Cong and K. Minkovich, "Logic synthesis for better than worst-case designs," in Proc. Int. Symp. on VLSI Design, Automation & Test, Apr. 2009, pp. 166–169.

[12] J. Patel, "CMOS process variations: a critical operation point hypothesis," Online Presentation, 2008.

[13] D. Ernst et al., "Razor: circuit-level correction of timing errors for low-power operation," IEEE Micro, vol. 24, no. 6, pp. 10–20, Nov. 2004.

[14] J. Crop et al., "Error detection and recovery techniques for variation-aware CMOS computing: a comprehensive review," Journal of Low Power Electronic and Applications, vol. 1, no. 3, pp. 334–356, 2011.

[15] "Nangate 45nm open cell library," Available: http://www.nangate.com/.

[16] Design Compiler User Guide, Synopsys, Sept. 2011.

[17] NanoSim User Guide, Synopsys, Sept. 2011.

978-1-4799-2817-0/14 $31.00 © 2014 IEEE

A Silicon Nanodisk Array Structure Realizing Synaptic Response of Spiking Neuron Models with Noise

Takashi Morie, Haichao Liang, Yilai Sun, Takashi Tohara

Graduate School of Life Science and Systems Engineering
Kyushu Institute of Technology
Kitakyushu 808-0196 Japan
e-mail: morie@brain.kyutech.ac.jp

Makoto Igarashi, Seiji Samukawa

Institute of Fluid Science
Tohoku University
Sendai 980-8577 Japan

Abstract— In the implementation of spiking neuron models, which can achieve realistic neuron operation, generation of post-synaptic potentials (PSPs) is an essential function. We have already proposed a new nanodisk array structure for generating PSPs using delay in electron hopping among nanodisks. Generated PSPs have fluctuation caused by stochastic electron movement. Noise or fluctuation is effectively used in neural processing. In this paper, we review our proposed structure and show fluctuation controllability based on single-electron circuit simulation.

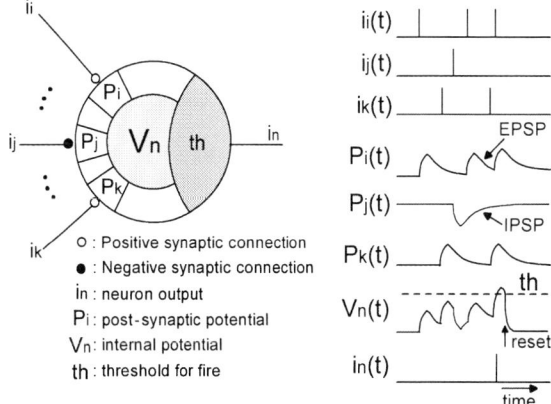

Fig. 1. Simple spiking neuron model: integrate-and-fire neuron.

I. INTRODUCTION

Mimicking brain functions and structures using nanotechnology is a big challenge to realize highly intelligent information processing such as association, perception and recognition. The brain consists of 10^{10} neurons, and each neuron receives many electric impulses called spikes via a few thousand synapses, and it outputs spike pulses. To achieve such massive parallelism in VLSI circuits, extremely small circuit footprint and ultimately low power consumption are essential. *Bottom-up*-type nanotechnology developed recently is useful to realize such massively parallel VLSI devices.

Computational neuroscientists have focused on *spiking neuron* models, which treat spike pulses directly [1]. In such models, generation of post-synaptic potentials (PSPs) is an essential function. However, in conventional CMOS LSI technology, a large footprint is required to implement this function [2]. Applying nanostructures to this implementation will ultimately achieve high packing density and low power dissipation.

One of the authors developed a fabrication technology of nanodisk array (NDA) structures using bio-nano-process and neutral beam etching techniques [3, 4]. Using this nanostructure, we have already proposed a spiking neuron device that generates PSPs by taking advantage of the delay in electron hopping movement among nanodisks [5]. A real NDA device was fabricated, and PSP generation operation was successfully verified [6].

In neural processing, noise or fluctuation is effectively used. It can improve the generalization ability of neural networks. The rising timing of PSPs can be used for calculating multiply-and-accumulation (MAC) [1]. Using MAC calculation, similarity measures can be evaluated. Similarity evaluation including noise can be used for efficient association and clustering algorithms, and they lead to intelligent information process-

ing such as recognition and concept creation [7]. When PSPs are generated in our nanostructures, fluctuation of PSPs is inevitably caused by random electron hopping movement among nanodisks. This fluctuation is included in MAC calculation.

Since one real biological neuron has a few thousand synapses, the key for very large scale integration of artificial neural networks is the design of a small synaptic circuit. Thus, we focus on realization of the synapse functions. The synapse part has to realize generation of controllable PSPs as well as synaptic weighting.

In this paper, we briefly review our proposed nanostructure and explain a design methodology using an electromagnetic simulator for analyzing nanostructures and a single-electron circuit simulator for analyzing electron movement. We also show simulation results for fluctuation controllability in the proposed nanostructure.

II. SPIKING NEURON MODELS AND INTELLIGENT INFORMATION PROCESSING WITH NOISE

In spiking neuron models, information is represented by spatiotemporal patterns in spike pulse trains. A simple spiking neuron model is shown in Fig. 1. A spike pulse inputted to a neuron via a synapse generates a PSP. There are excitatory and inhibitory synapses; these have positive and negative synaptic weights and generate an excitatory PSP (EPSP) and an in-

978-1-4799-2817-0/14 $31.00 © 2014 IEEE

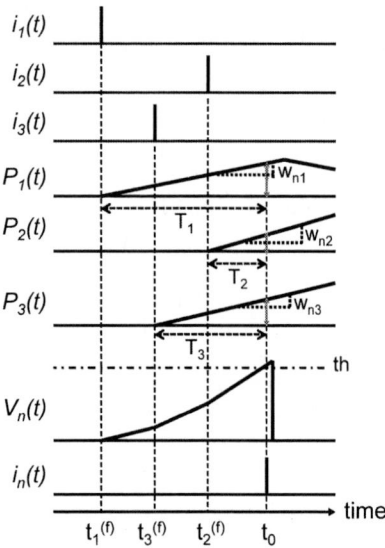

Fig. 2. Multiply-and-accumulation calculation using PSPs generated at spike timing.

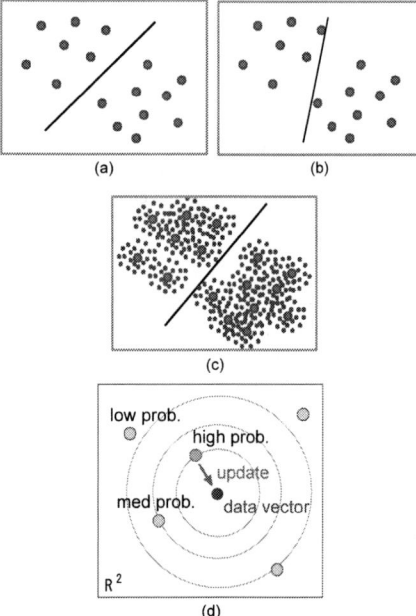

Fig. 3. Noise effect in discrimination learning problems: (a) ideal learning result, (b) simple supervised learning result with small number of inputs, and (c) simple learning result with many inputs added by noise. (d) Stochastic associative learning in clustering problems.

hibitory PSP (IPSP), respectively. A PSP temporarily increases or decreases according to whether the synaptic connection is excitatory or inhibitory.

The neuron's internal potential $V_n(t)$ is equal to the spatiotemporal summation of all PSPs generated by the input spike pulses. If $V_n(t)$ exceeds a certain threshold (th), the neuron emits a spike pulse and $V_n(t)$ is reset to the resting level. There

exists a refractory period after firing, in which the neuron cannot fire even if it receives many spikes.

Many neural network models use noise effectively, and the dynamics of a neuron state can be expressed as follows:

$$dV_n(t) = \{-gV_n(t) + \sum_{j,\, t_j^{(f)}} w_{nj}P(t - t_j^{(f)})\}dt + \sigma N(t), \quad (1)$$

where g is the decay coefficient, $t_j^{(f)}$ the firing timing of neuron j, w_{nj} the synaptic weight from neuron j to n, $P(\cdot)$ a PSP function, and $N(t)$ a noise term with influence coefficient σ. As can be seen in Eq. (1), synaptic weights and the noise coefficient should be controlled independently.

We can calculate multiply-and-accumulation (MAC) by only using rising timing of PSPs [1]. If we can assume that a PSP increases linearly when a spike inputs, as shown in Fig. 2, the internal state is given by the following equation:

$$V_n(t_0) = \sum_j w_{nj}T_j, \quad (2)$$

$$T_j = t_0 - t_j^{(f)}. \quad (3)$$

Thus, output spike timing t_0 represents the MAC calculation results.

Using MAC calculation, similarity measures can be evaluated. Similarity evaluation including noise can be used for efficient association and clustering algorithms. Figure 3 shows examples illustrating the effectiveness of noise in neural networks. For a discrimination learning problem, if the number of learning samples is small, the discrimination result obtained by a simple learning rule is far from ideal one, as shown in Fig. 3(b). However, if noise is added to the input data, and a sufficient number of learning trials are performed, the learning result becomes much better and can approach to the ideal one, as shown in Fig. 3(c). By using this principle, stochastic associative learning can be achieved in clustering problems, as shown in Fig. 3(d), and the clustering performance is much improved [7].

III. FABRICATION METHOD OF NANOSTRUCTURES

In order to develop a nanodevice for spiking neuron models with noise, we use in this work a newly-developed silicon nanodisk fabrication technology using a self-assembly bio-nanoprocess with ferritin supramolecules and neutral beam etching, as shown in Fig. 4 [3, 4].

Ferritin is a supramolecule that has an iron core inside of a protein shell. A silicon substrate, on which a polysilicon film is formed, is spin coated with solution including ferritin supramolecules, and a self-assembled ferritin array is formed on the substrate. After removing the protein shells by a heat treatment, a regular arrangement of iron cores remains on the substrate. By using this iron core array as an etching mask, the polysilicon film is etched with defect-free neutral beams. As a result, an array of polysilicon nanodisks is formed, and each nanodisk has the same diameter and gap between disks as the iron core. We measured temporal response signals through a 2D nanodisk array, and have verified that this 2D nanodisk array can be applied to generation of PSPs in spiking neuron devices. [6].

978-1-4799-2817-0/14 $31.00 © 2014 IEEE

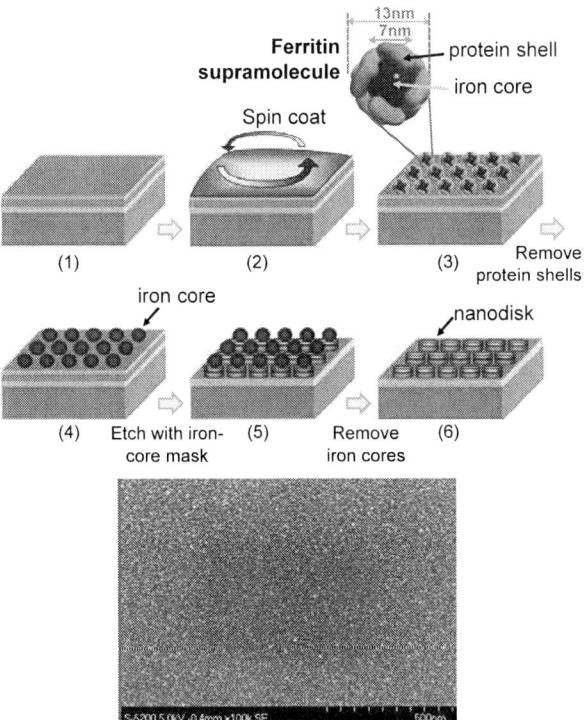

Fig. 4. Fabrication process flow of nanodisk array structures and microphotograph of ferritin supramolecule array.

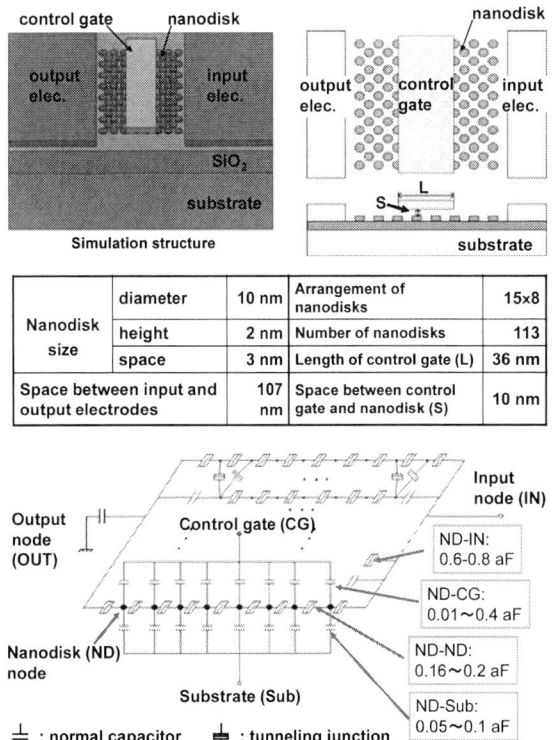

Nanodisk size	diameter	10 nm	Arrangement of nanodisks	15×8
	height	2 nm	Number of nanodisks	113
	space	3 nm	Length of control gate (L)	36 nm
Space between input and output electrodes		107 nm	Space between control gate and nanodisk (S)	10 nm

Fig. 6. 2D NDA structure used for analysis, its dimensions and equivalent circuit.

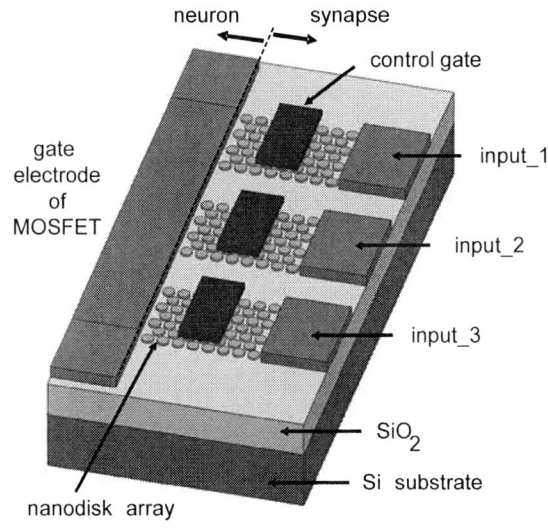

Fig. 5. Neuron device consisting of a MOSFET with NDA synapses.

IV. DESIGN OF SPIKING NEURON DEVICE WITH NANOSTRUCTURE

A. Spiking neuron device consisting of a MOSFET with nanodisk arrays

Using the nanostructure fabrication technology described above, in order to realize PSP generation in spiking neu-

ron models, we proposed a nanodevice structure [5] in which nanodisk-array synapses connect between the corresponding input electrode and the gate electrode of a MOSFET that acts as a neuron part by detecting charges, as shown in Fig. 5. When a spike pulse is fed into the input electrode, the voltage change is transferred to the gate electrode with large resistance realized by electron hopping in the NDA, which generates PSPs. The MOSFET collects all PSPs, and if the gate voltage exceeds the threshold, the transistor turns on, which means neuron firing. In order to control a synaptic weight, an electrode is placed over each nanodisk array. The voltage of this electrode can control the electron hopping behavior between nanodisks.

B. Design and analysis method

The procedure we adopted for designing a nanodevice is as follows. A sketch of the nanodevice is made, and then based on the structure, its electrical characteristics are analyzed using simulators, as shown in Fig. 6.

1. A nanostructure proposed using the parameters obtained through physical experiments is designed in a 3D simulator for electromagnetic analysis.
2. The equivalent electrical circuit consisting of capacitors between nodes is extracted.
3. In the extracted equivalent circuit, capacitance elements whose space between nodes is more than 3 nm are considered as capacitors without electron tunneling, and those less than 3 nm are considered as tunneling junctions.

4. Using the equivalent electrical circuit obtained, single-electron circuit simulation is conducted assuming an appropriate tunneling resistance.
5. If a single-electron/CMOS hybrid circuit simulator is available, such circuit simulation is conducted.
6. The above procedure is repeated to refine the device design and to improve the electrical characteristics.

In this work, we used ANSYS *Maxwell 3D* as a 3D electromagnetic simulator and a Monte Carlo based single-electron circuit simulator based on the following principle [8, 9]. A MOSFET connected with NDA was replaced by an output capacitance. On the basis of this equivalent circuit, we propose a control method for synaptic weighting and adding noise.

In the Monte Carlo based electron circuit simulator, an electron movement is determined by the energy difference ΔE_i between the current state and possible subsequent states i. The waiting time τ_i for tunneling process corresponding to each subsequent state is given by

$$\tau_i = \frac{1}{\Gamma_i} \ln \frac{1}{r}, \tag{4}$$

$$\Gamma_i = \frac{\Delta E_i}{e^2 R_T [1 - \exp(-\Delta E_i / k_B T)]}, \tag{5}$$

where r is a uniform random number ($0 < r < 1$), Γ_i the mean tunneling rate, R_T tunneling resistance, k_B the Boltzmann constant, e the elementary charge, and T temperature.

In our simulation, we set $R_T = 500 \text{M}\Omega$ and $T = 300$ K.

V. ANALYSIS OF SPIKING NEURON DEVICE WITH NANOSTRUCTURE

A. Simulation results

In our simulation, the MOSFET gate was replaced by an output capacitance of 100 aF. Figure 7 shows output voltage waveforms, which are considered as a PSP [5]. By changing control electrode voltage V_c, the peak voltage of PSP changes, which means that the synaptic weight can be controlled by V_c.

In order to analyze fluctuation in PSP rising operation, we conducted simulations of 1D NDA, as shown in Fig. 8(a). Figures 8(c)-(e) show simulation results with different input spike pulse amplitude voltages V_{in} and control gate voltages V_c, in each of which 100 trials of single-electron simulation were performed. From Fig. 8(c), the output voltage can be changed by V_{in} with a constant V_c, while the standard deviations of the output voltage are almost constant. Since the input information is expressed only by spike timing, not by spike amplitude V_{in}, as shown in Eq. (1), we can control the relative noise level by changing V_{in}. On the other hand, from Fig. 8(d), the output voltage can also be changed by V_c with a constant V_{in}, while the standard deviations of the output voltage are almost constant. Thus, we can control synaptic weighting by V_c with a nearly constant noise level.

The energy profiles corresponding to the conditions shown in Fig. 8(d) are also shown in Fig. 8(e), where the energy profile means the total electrostatic energy as a function of electron position in the nanodisk nodes. It is verified that V_c changes the energy profile, so that it changes the probability of electron movement, which determines the temporal output voltage.

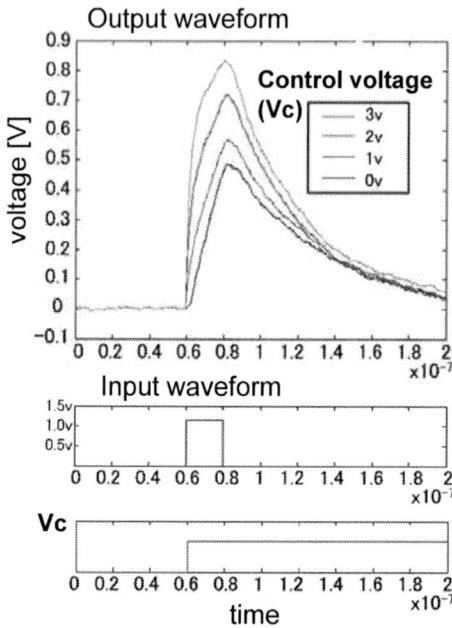

Fig. 7. Single-electron circuit simulation result: output voltage waveforms depending on control voltage V_c as a temporal response of an input spike pulse [5].

Next, we changed the nanodisk array pattern from the above 1D shape to various 2D shapes, as shown in Fig. 9. The simulation results showed that the 1D array generated a larger fluctuation rate than the 2D rectangle array, and that the sandglass shapes of nanodisk array had intermediate characteristics between 1D and 2D shapes. The reasons for these results are that random electron hopping movements in the NDA are averaged in 2D arrays, and that sandglass shapes have partial 1D property at the center part. Therefore, in order to generate larger fluctuation, it is important to reduce the dimension of the shape of nanodisk array.

VI. CONCLUSION

We analyzed the operation of the neuron device with nanodisk array structures, which can be formed by using self-assembly bio-nano-process and defect-free neutral-beam etching. We analyzed fluctuation in PSP generation by using an electromagnetic simulator and a single-electron circuit simulator. It was verified from the simulation results that fluctuation can be controlled by the voltage height of the input spike pulses and the control electrode voltage.

In future work, we will observe and analyze fluctuation in PSP generation using real fabricated nanodevices.

ACKNOWLEDGMENT

This work was supported by JSPS KAKENHI Grant Number 22240022.

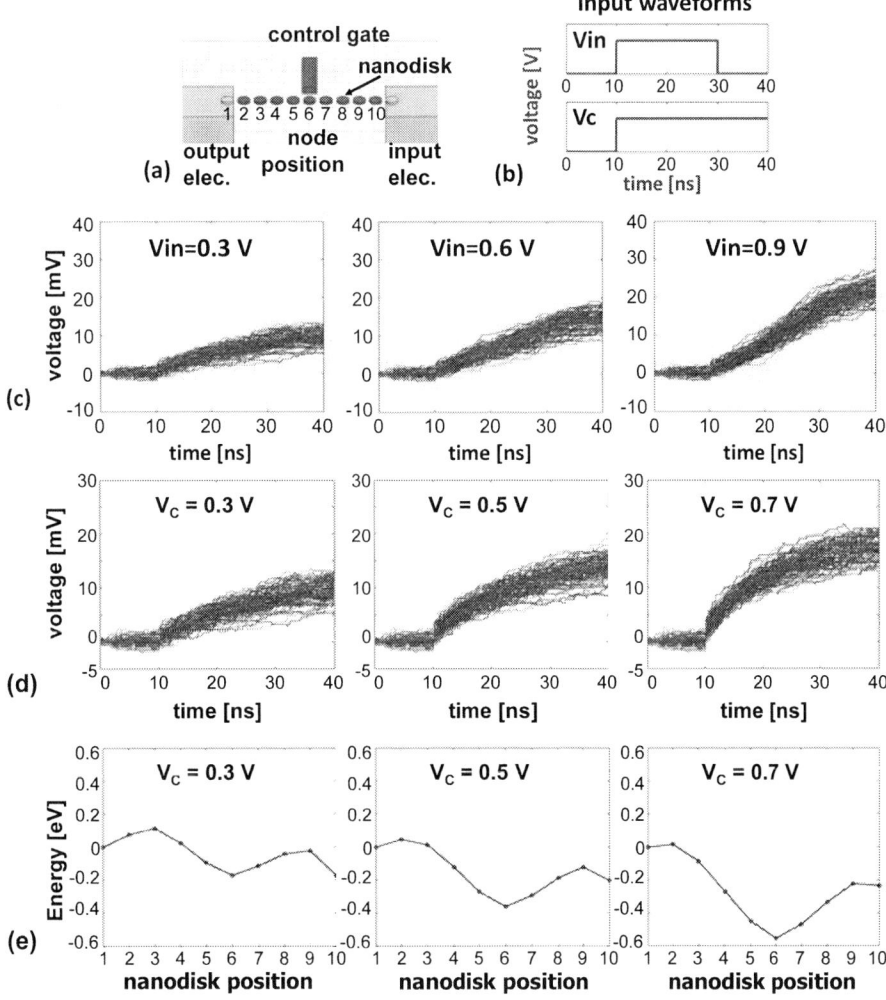

Fig. 8. Single-electron circuit simulation results: (a) 1D nanodisk array structure assumed in the simulation, (b) input waveforms, (c) simulation results when $V_C = 0.3\ V$, (d) simulation results when $V_{in} = 0.3\ V$, and (e) energy profiles corresponding to (d).

REFERENCES

[1] W. Maass and C. M. Bishop, editors *Pulsed Neural Networks*, MIT Press, Cambridge, MA, 1999.

[2] H. Tanaka, T. Morie, and K. Aihara, "A CMOS Spiking Neural Network Circuit with Symmetric/Asymmetric STDP Function," *IEICE Trans. Fundamentals*, vol. E92-A, no. 7, pp. 1690–1698, 2009.

[3] S. Samukawa, K. Sakamoto, and K. Ichiki, "Generating High-efficiency Neutral Beams by Using Negative Ions in an Inductively Coupled Plasma Source," *J. Vac. Sci. Tech.*, vol. A20, pp. 1566–1573, 2002.

[4] C. H. Huang, M. Igarashi, M. Wone, Y. Uraoka, T. Fuyuki, M. Takeguchi, I. Yamashita, and S. Samukawa, "Two-Dimensional Si-Nanodisk Array Fabricated Using Bio-Nano-Process and Neutral Beam Etching for Realistic Quantum Effect Devices," *Jpn. J. Appl. Phys.*, vol. 48, pp. 04C187–1–6, 2009.

[5] T. Morie, Y. Sun, H. Liang, M. Igarashi, C. Huang, and S. Samukawa, "A 2-Dimensional Si Nanodisk Array Structure for Spiking Neuron Models," in *IEEE Proc. of Int. Symp. Circuits and Systems (ISCAS)*, pp. 781–784, 2010.

[6] M. Igarashi, C. H. Huang, T. Morie, and S. Samukawa, "Control of Electron Transport in Two-Dimensional Array of Si Nanodisks for Spiking Neuron Device," *Appl. Phys. Express*, vol. 3, pp. 085202–1–3, 2010.

[7] T. Morie, T. Matsuura, M. Nagata, and A. Iwata, "An Efficient Clustering Algorithm Using Stochastic Association Model and Its Implementation Using Nanostructures," in T. G. Dietterich, S. Becker, and Z. Ghahramani, editors, *Advances in Neural Information Processing Systems*, volume 14, pp. 1115–1122. MIT Press, Cambridge, MA, 2002.

[8] N. Kuwamura, K. Taniguchi, and C. Hamaguchi, "Simulation of Single-Electron Logic Circuit," *IEICE Trans. Electron.*, vol. J77-C-II, pp. 221–228, 1994.

978-1-4799-2817-0/14 $31.00 © 2014 IEEE

pattern \ Vout	(a) 1D	(b) sandglass-1	(c) sandglass-2	(d) rectangle
NDA pattern				
standard deviation [mV]	1.88	4.33	4.32	4.93
average [mV]	14.1	71.6	85.5	106
fluctuation rate [%]	13.3	6.05	5.06	4.65

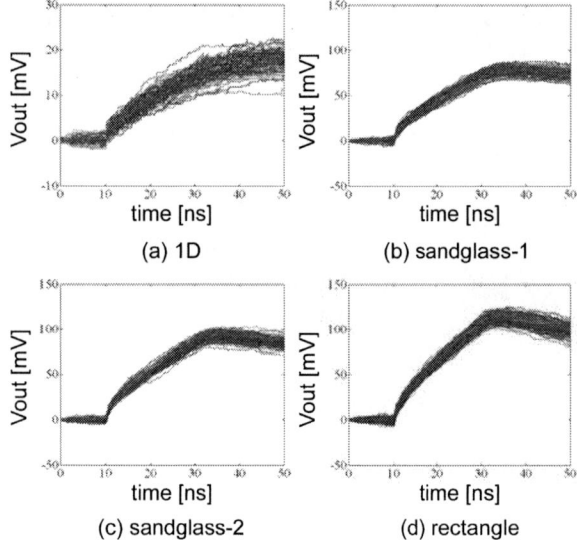

(a) 1D

(b) sandglass-1

(c) sandglass-2

(d) rectangle

Fig. 9. Single-electron circuit simulation results with various NDA patterns: (a) 1D pattern, (b) sandglass type-1, (c) sandglass type-2, and (d) rectangle.

[9] T. Yamanaka, T. Morie, M. Nagata, and A. Iwata, "A Single-Electron Stochastic Associative Processing Circuit Robust to Random Background-Charge Effects and Its Structure Using Nanocrystal Floating-Gate Transistors," *Nanotechnology*, vol. 11, no. 3, pp. 154–160, 2000.

Energy Efficient In-Memory Machine Learning for Data Intensive Image-Processing by Non-volatile Domain-Wall Memory

Hao Yu[1]*, Yuhao Wang[1], Shuai Chen[1], Wei Fei[1], Chuliang Weng[2], Junfeng Zhao[2] and Zhulin Wei[2]

[1]School of Electrical and Electronic Engineering, Nanyang Technological University, Singapore
[2]Huawei Shannon Laboratory, China
*Correspondent author: haoyu@ntu.edu.sg; Tel: +65-67904509, Fax: +65-6793 3318

Abstract - Image processing in conventional logic-memory I/O-integrated systems will incur significant communication congestion at memory I/Os for excessive big image data at exa-scale. This paper explores an in-memory machine learning on neural network architecture by utilizing the newly introduced domain-wall nanowire, called DW-NN. We show that all operations involved in machine learning on neural network can be mapped to a logic-in-memory architecture by non-volatile domain-wall nanowire. Domain-wall nanowire based logic is customized for in machine learning within image data storage. As such, both neural network training and processing can be performed locally within the memory. The experimental results show that system throughput in DW-NN is improved by 11.6x and the energy efficiency is improved by 92x when compared to conventional image processing system.

I Introduction

One exciting feature of future big-data storage system is to find implicit pattern of data and excavate valued behavior behind by big-data analytics such as image feature extraction during image search. Instead of performing the image search by calculating pixel similarity, image search by machine learning is a similar process as human brains. For example, each image feature extraction is performed to obtain the characteristics first, and then is matched by key words. As such, the image search becomes a traditional string matching problem to solve.

However, to handle big image data at exa-scale, there is memory-wall that has long memory access latency as well as limited memory bandwidth. For the example of the image search in one big-data storage system, there may be billions of images. In order to perform feature extraction for one of images, it will lead to significant congestion at I/Os when migrating data between memory and processor. Note that in-memory computing system [1,2,3,4] is promising as one future big-data solution to relieve the memory-wall issue. For example, domain specific accelerators can be developed within memory for big-data processing such that the data will be pre-processed before they are readout with the minimum number of data migrations.

In this paper, the big image data processing algorithm by machine learning is examined within the in-memory computing system architecture. Among numerous machine learning algorithms [5,6,7], neural-network based algorithm has shown low complexity with genetic adaptability. In particular, the extreme learning machine (ELM) [7,8] has one input layer, one hidden layer and one output layer, and hence it has tuning-free feature without expensive iterative training process, which makes it suitable for the low-cost hardware implementation. As such, the in-memory hardware

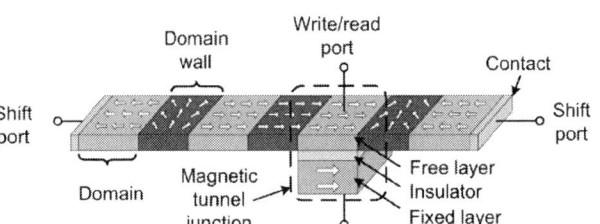

Fig. 1. The diagram of domain-wall nanowire device.

accelerators of ELM is studied here for the big image data processing.

The proposed in-memory ELM computing system is examined by the nano-scale non-volatile memory devices [9,10,11,12,13]. Domain-wall nanowire or racetrack memory [12,13], is one newly introduced spintronic NVM device that has not only potential for high density and high performance memory storage, but also feasible in-memory computing capability. In this paper, we show the feasibility of mapping the ELM to a full domain-wall nanowire based in-memory neural network computing system, called DW-NN. Compared to the scenario that ELM is executed in CMOS based general purpose processor, the proposed DW-NN improves the system throughput by 11.6x and energy efficiency by 92x.

The rest of paper is organized as following. Section II reviews the fundamental of domain-wall memory and domain-wall memory based in-memory computing architecture. Section III demonstrates algorithm and implementation of in-memory machine learning by domain-wall nanowire devices. The experiments are carried out in Section IV with conclusions in Section V.

II. In-memory Architecture on Domain-wall Devices

A. Domain-wall Nanowire Devices

Domain-wall nanowire, also known as racetrack memory [12,13], is the third-generation of spin-based NVM. As shown in Fig. 1, multiple bits of information are stored in a single ferromagnetic nanowire. Separated by domain-walls, each bit is represented by the magnetization direction. The domain-walls can be moved left or right by applying a current through the shift port at the two ends of the nanowire. The domain width of each bit remains unchanged, thus the stored information is preserved. All bits are shifted with a tape-like operation similarly as in a shift register.

In a domain-wall nanowire device, a strongly magnetized ferromagnetic layer is placed along the ferromagnetic nanowire at a desired position and separated by an insulator

978-1-4799-2817-0/14 $31.00 © 2014 IEEE

layer. Such a sandwich-like structure forms a magnetic-tunnel-junction (MTJ) through which the stored information can be accessed.

The formed MTJ exhibits different states depending on the alignment of the fixed layer and free layer. This is called the giant magnetoresistance (GMR) effect. MTJ shows different resistance with parallel or anti-parallel alignments. By detecting the resistance of MTJ, the stored bit can be read out. Moreover, write operation is achieved by altering the free layer magnetization with an injected current. The electron spin will force the free layer to be parallel or anti-parallel to fixed layer based on the current direction [14]. Note the write and read operation can only occur in the MTJ. Therefore, the bit to be operated needs to be shifted and aligned with the fixed layer, while the shift direction and velocity is controlled by the current direction and amplitude [15].

B. Domain-wall Memory based In-Memory Computing

Conventionally, all the data is maintained within memory that is separated from the processor but connected with I/Os. Therefore, during the execution, all data needs to be migrated to processor and written back afterwards. In the data-oriented applications, however, this will incur significant I/O congestions and hence greatly degrade the overall performance. In addition, significant standby power will be consumed in order to hold the large volume of data.

To overcome the above two issues, the in-memory non-volatile computing architecture is introduced. The overall architecture of domain-wall memory based in-memory computing platform is illustrated in Fig. 2(a). In particular, domain specific in-memory accelerators are integrated locally together with the stored data in distributed manner such that frequent involved operations can be performed without much communication with external processor. In addition, the distributed local accelerators can also provide great thread-level parallelism thus the throughput can be improved.

Fig. 2(b) shows how the in-memory distributed Map-Reduce [16] data processing is performed locally between one data array and local logic pair. Firstly, the external processor will issue commands to specific pair to perform in-memory logic computing. The commands will be received and interpreted by a controller in accelerator. Secondly, the controller will request related data to the data array with a read operation. As a result, the neural network based processing in ELM, mainly including weighted sum and sigmoid function, can be performed in a Map-Reduce fashion. Lastly, the results are written back to the data array.

In this platform, the domain-wall nanowire is intensively utilized towards ultra-low power big-data processing in both memory and logic, with significantly reduced leakage and operating power. What is more, domain-wall nanowire based adder, multiplier and look-up table for sigmoid function are adopted within non-volatile memory that further improves energy efficiency with small I/O overheads.

III. Domain-wall Nanowire based Extreme Learning Machine

In-memory ELM neural network offers two major advantages. Firstly, all domain-wall based arithmetic

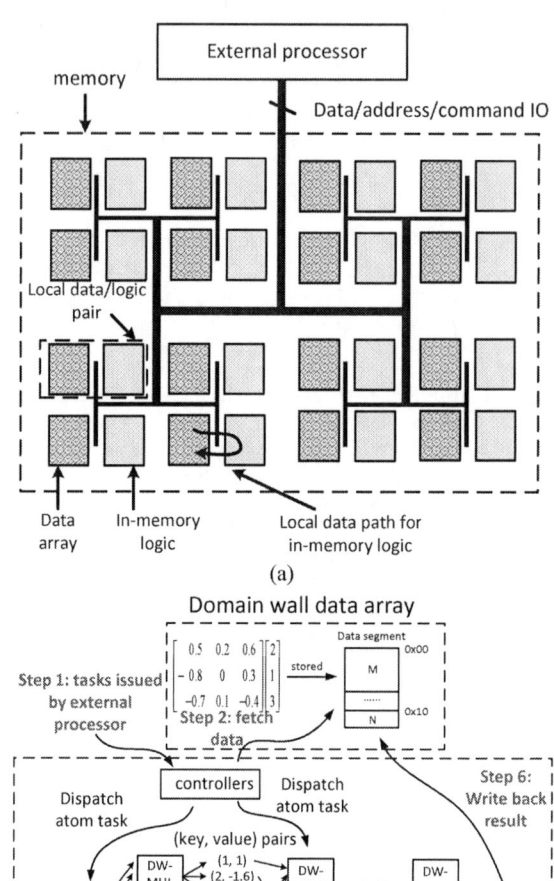

(a)

(b)

Fig. 2. (a) The overview of the in-memory computing architecture; (b) detailed domain-wall nanowire based machine learning platform in Map-Reduce fashion.

operations in ELM can be integrated into the memory, and these operations are performed directly on operands stored in non-volatile domain-wall memory. That is significantly different from the conventional memory-logic architecture where data need to be transferred from the memory to data path and then written back to the memory after being processed. Secondly, all the computing-intensive operations in ELM are implemented by domain-wall nanowire devices, which can also be used as storage units. This provides integration compatibility between data path and memory used in ELM, as well as the ability to reuse peripheral circuits like decoders and sense amplifiers.

A. Extreme Learning Machine

We first review the basic of the neural-network based ELM algorithm. Among numerous machine learning algorithm [3, 4], support vector machine (SVM) and neural network (NN) are widely discussed. However, both two algorithms have major challenging issues in terms of slow learning speed,

978-1-4799-2817-0/14 $31.00 © 2014 IEEE

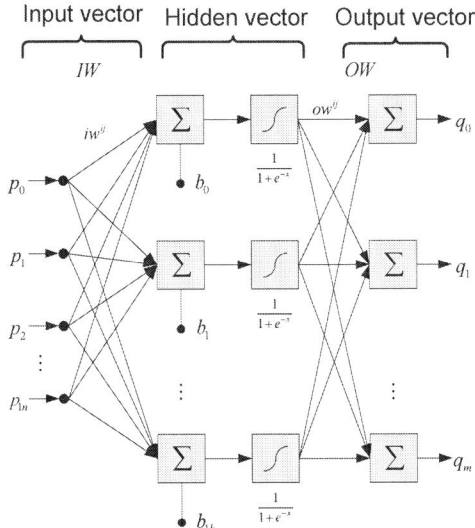

Fig. 3. Computation flow of extreme learning machine (ELM).

trivial human intervene (parameter tuning) and poor computational scalability [5].

Extreme Learning Machine (ELM) was initially proposed [7] for the single-hidden-layer feed-forward neural networks (SLFNs). Compared with traditional neural networks, ELM eliminates the need of parameter tuning in the training stage and hence reduces the training time significantly. The output function of ELM is formulated as (only one output node is considered)

$$f_L = \sum_{i=1}^{L} \beta_i h_i(X) = h(X)\beta, \qquad (1)$$

where $\beta = [\beta_1, \beta_2, \cdots, \beta_L]^T$ is the output weight vector storing the output weights between the hidden layer and output node. $h(X) = [h_1(X), h_2(X), \cdots, h_L(X)]^T$ is the hidden layer output matrix given input vector X and performs the transformation of input vector into L-dimensional feature space. The training process of ELM aims to obtain output weight vector β and to enhance the generalization ELM minimizes the training error as well as the norm of output weight

$$Minimize: ||H\beta\text{-}T|| \ and \ ||\beta|| \qquad (2)$$

where T is the target vector in the training process and β can be solved by minimal norm least square method as follows

$$\beta = H^\dagger T, \qquad (3)$$

where $H^\dagger$ is the Moore-Penrose generalized inverse of matrix H.

The application of ELM for image processing in this paper is an ELM based image super-resolution (SR) algorithm [8], which learns the image features of a specific category of images and improves low-resolution figures by applying learned knowledge. Note that ELM-SR is commonly used as pre-processing stage to improve image quality before applying other image algorithms. It involves intensive matrix operation, such as matrix addition, matrix multiplication as well as exponentiation on each element of a matrix. Fig. 3 illustrates

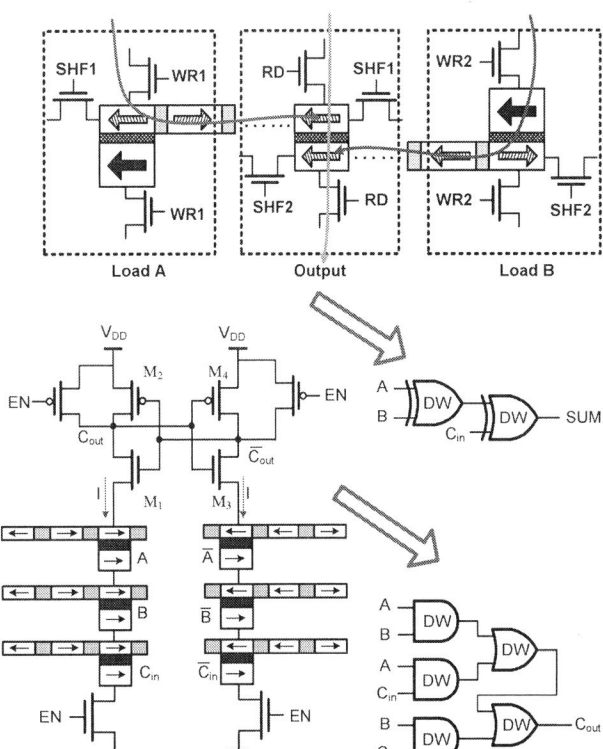

Fig. 4. Domain-wall nanowire based full adder with SUM operation by DW-XOR logic and CARRY operation by resistor comparator.

the computation flow for ELM-SR, where input vector obtained from input image is multiplied by input weight matrix. The result is then added with bias vector b to generate input of sigmoid function. Lastly sigmoid function outputs are multiplied with output weight matrix to produce final results. In the following, we will demonstrate how to map the fundamental addition, multiplication, and sigmoid function to domain-wall nanowires.

B. Weighted Sum by Domain-wall Adder and Multiplier

The GMR-effect can be interpreted as the bit-wise XOR operation of the magnetization direction of two thin magnetic layers, where the output is denoted by high or low resistance. In a GMR-based MTJ structure, however, the XOR-logic will fail as there is only one operand as variable since the magnetization in fixed layer is constant. Nevertheless, this problem can be overcome by the unique domain-wall shift-operation in the domain-wall nanowire device, which enables the possibility of DWL-based XOR logic for computing.

A bitwise-XOR logic implemented by two domain-wall nanowires is shown in Fig. 4. The bitwise-XOR logic is performed by constructing a new read-only-port, where two free layers and one insulator layer are stacked. The two free layers are in the size of one magnetization domain and are from two respective nanowires. Thus, the two operands, denoted as the magnetization direction in free layer, can both be variables with values assigned through the MTJ of the according nanowire. As such, it can be shifted to the operating port such that the XOR-logic is performed.

For example, the A xor B can be executed in the following steps

978-1-4799-2817-0/14 $31.00 © 2014 IEEE

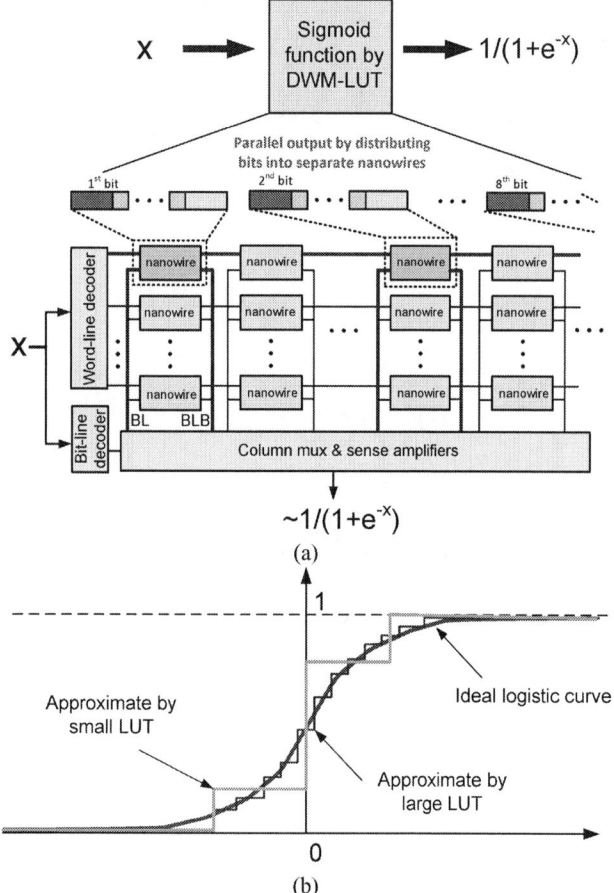

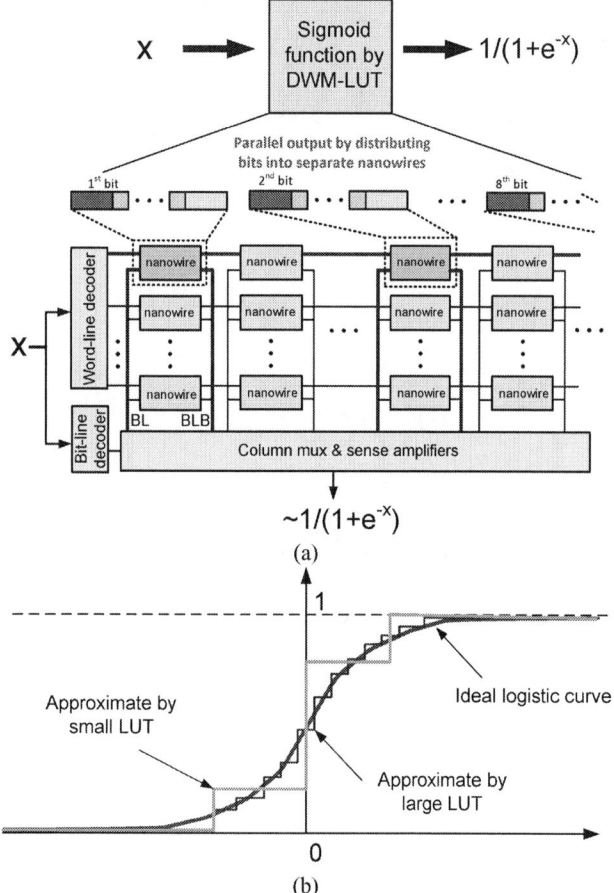

TABLE I
Domain-wall Operations and Logic Performance

Domain wall nanowire device		
Operation	Speed (cycles)	Energy (pJ)
read	1	0.5
write	1	0.1
shift	1	0.3
Domain wall nanowire logic		
Logic	Speed (cycles)	Energy (pJ)	Area (um^2)
8-bit full adder	54	40	2.6
8-bit multiplier	163	308	18.9
8-bit sigmoid (LUT)	2	116	31.8

TABLE II
System Area, Power, Throughput and Energy Efficiency
Comparison between In-Memory Architecture and
Conventional Architecture

Platform	DW-NN	GPP (with on-chip memory)	GPP (with off-chip memory)
Computational resources utilized	1×Processor 7714×DW-ADDER 7714×DW-MUL 551×DW-LUT 1×controller	1×Processor	1×Processor
Area of computational units	18 mm^2 (processor) + 0.5 mm^2 (accelerators)	18 mm^2	18 mm^2
Power (Watt)	10.1	12.5	12.5
Throughput (GBytes/s)	108MBytes/s	9.3MBytes/s	9.3MBytes/s
Energy efficiency (nJ/bit)	7	389	642

Fig. 5. (a) Sigmoid function implemented by domain wall nanowire based look-up table (DW-LUT); (b) DW-LUT size effect on the precision of the sigmoid function. The larger the LUT, the smoother and more precise the curve is.

- The operands A and B are loaded into two nanowires by enabling WL1 and WL2 respectively.
- A and B are shifted from their access-ports to the read-only ports by enabling SHF1 and SHF2 respectively;
- By enabling RD, the bit-wise XOR result can be obtained through the GMR-effect.

By deploying two DW-XOR logic units, the SUM operation of full adder can be achieved in by domain-wall nanowire devices with low power consumption.

In addition, to realize a full adder, the CARRY operation is also needed. Spintronics based CARRY operation is proposed in [17], where a pre-charge sensing amplifier (PCSA) is used for resistance comparison. The CARRY logic by PCSA and two branches of domain-wall nanowires is shown in Fig. 4. The three operands for CARRY operation are denoted by resistance of MTJ (low for 0 and high for 1), and belong to respective domain-wall nanowires in the left branch. The right branch is made complementary to the left one. Note that the C_{out} and $\overline{C_{out}}$ will be pre-charged high at first when PCSA EN signal is low. When the circuit is enabled, the branch with lower resistance will discharge its output to "0". For example, when left branch has no or only one MTJ in high resistance, i.e. no carry out, the right branch will have three or two MTJs in high resistance, such that the C_{out} will be 0. The complete truth table will confirm CARRY logic by this circuit. The

domain-wall nanowire works as the writing circuit for the operands by writing values at one end and shift it to PCSA.

Note that with the full adder implemented by domain-wall nanowires and intrinsic shift ability of domain-wall nanowire, a shift/add multiplier can be readily achieved purely by domain-wall nanowires.

C. Sigmoid Function by Domain-Wall Lookup Table

Sigmoid function includes exponentiation, division, and addition, which is a computing-intensive operation in ELM application. In particular, the exponentiation will take many cycles to execute in the conventional processor due to the lack of corresponding accelerator. Therefore, it is extremely economic to perform exponentiation by look-up table. Look-up table (LUT), essentially a pre-configured memory array, takes a binary address as input, finds target cells that contain result through decoders, and finally outputs correspondingly by sense amplifiers.

A domain-wall nanowire based LUT (DW-LUT) is illustrated in Fig. 5(a). Compared with the conventional SRAM or DRAM by CMOS, the DW-LUT can demonstrate two major advantages. Firstly, extremely high integration density can be achieved since multiple bits can be packed in one nanowire. Secondly, zero standby power can be expected as a non-volatile device does not require to be powered to

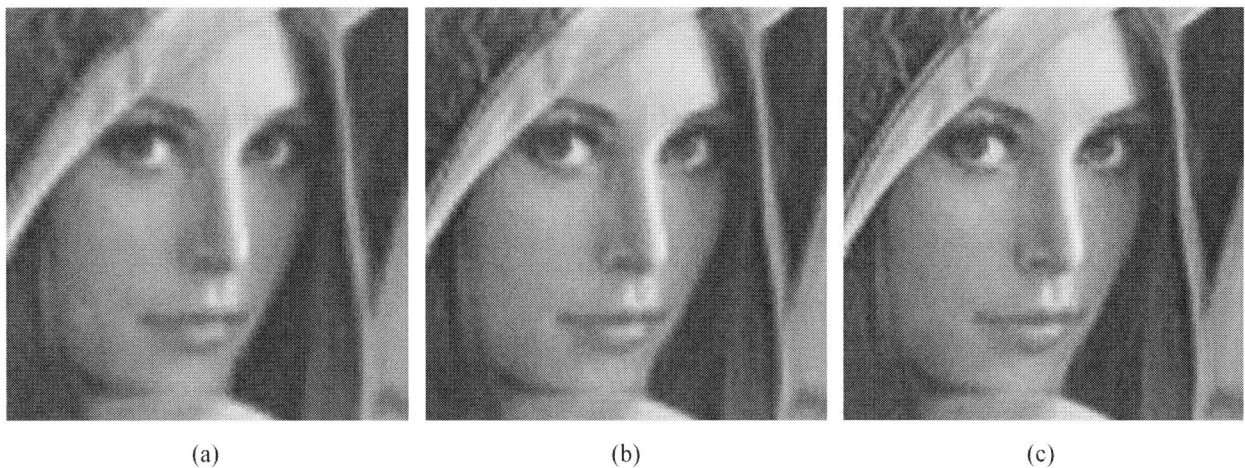

(a)   (b)   (c)

Fig. 6. (a) Original image before ELM-SR algorithm (SSIM value is 0.91); (b) Image quality improved after ELM-SR algorithm by DW-NN hardware implementation (SSIM value is 0.94); (c) Image quality improved by GPP platform (SSIM value is 0.97).

retain the stored data. By distributing the multiple bits of results in separated nanowires, the serial operation of nanowire can be avoided and the function can be done fast.

Note that the LUT size is determined by the input domain, the output range, and the required precision for the floating point numbers. Fig. 5(b) shows the ideal logistic curve and approximated curves by LUTs. It can be observed that the output range is bounded between 0 and 1, and although the input domain is infinite, it is only informative in the center around 0. The LUT visually is the digitalized logistic curve and the granularity, i.e. precision, depends on the LUT size. For machine learning application, the precision is not as sensitive as scientific computations. As a result, the LUT size for sigmoid function can be greatly optimized and leads to high energy efficiency for sigmoid function execution.

IV. Experiment Results

A. Domain-Wall Logic Performance

To evaluate domain-wall nanowire based in-memory ELM computing system, the following design platform has been set up. Firstly at device level, the transient simulation of MTJ read and write operations is performed within NVM-SPICE [18,19,20,21] to obtain accurate operation energy and timing for domain-wall nanowire. The shift-operation energy is modeled as the Joule heat dissipated on the nanowire when shift-current is applied. The shift-current density and shift-velocity relationship are based on [15]. The area of one domain-wall nanowire is calculated by its dimension parameters. Specifically, the technology node of 32nm is assumed with width of 32nm, length of 64nm per domain, and thickness of 2.2nm for one domain-wall nanowire; the R_{off} is set at 2600Ω, the R_{on} at 1000Ω, the writing current at $100\mu A$, and the current density at $6\times10^8 A/cm^2$ for shift-operation. Secondly at circuit level, the memory modeling tool CACTI [22] is modified with name as DW-CACTI. It can provide accurate power and area information for domain-wall nanowire memory peripheral circuits such as decoders and sense amplifiers (SAs). Together with the device level performance data, the DW-ADDER as well as the DW-LUT can be evaluated at circuit level.

Table I shows the energy cost and speed of basic domain-wall nanowire operations as well as logic circuits. Different from CMOS based logic, domain-wall nanowire based logic circuits need multiple cycles to operate, and thus the latency is expected to be much longer than its CMOS counterpart. However, computation throughput outweighs latency in the context of big-data applications where the data parallelism is high; benefited from the leakage free and high density property, more domain-wall logic resources are able to be allocated to significantly increase the system performance.

B. In-memory Throughput and Energy Efficiency

To compare proposed in-memory DW-NN platform and conventional general purpose processor (GPP) based platform, ELM based super resolution (ELM-SR) application is executed as the workload. The evaluation of ELM-SR in GPP platform is based on gem5 [23] and McPAT [24] for core power and area model. DW-NN is evaluated in our developed self-consistent simulation platform based on NVMSPICE, DW-CACTI, and DW-NN behavioral simulator. The processor runs at 3GHz while the accelerators run at 500MHz. System memory capacity is set as 1GB, and bus width is set as 128 bits. Based on [25], 3.7nJ and 6.3nJ per access are used for on-chip and off-chip I/O overhead respectively.

Table II compares ELM-SR in both DW-NN and GPP platforms. Due to the deployment of in-memory accelerators and high data parallelism, the throughput of DW-NN improves by 11.6x compared to GPP platform. In terms of area used by computational resources, DW-NN is 2.7% higher than that of GPP platform. Additional 0.5 mm^2 is used to deploy the domain-wall nanowire based accelerators. Thanks to the high integration density of domain-wall nanowires, the numerous accelerators are brought with only slight area overhead. In DW-NN, the additional power consumed by accelerators is compensated by the saved dynamic power of processor, since the computation is mostly performed by the in-memory logic. Overall, DW-NN achieves a power reduction of 19%. The most noticeable advantage of DW-NN is its much higher energy efficiency compared to GPP. Specifically, it is 56x and 92x better than that of GPP with on-chip and off-chip memory respectively. The advantage comes from three aspects: (a)

978-1-4799-2817-0/14 $31.00 © 2014 IEEE   195

in-memory computing architecture that saves I/O overhead; (b) non-volatile domain-wall nanowire devices that are leakage free; and (c) application specific accelerators.

Fig. 6 shows the image quality comparison between the proposed in-memory DW-NN hardware implementation and the conventional GPP software implementation. To measure the performance quantitatively, structural similarity (SSIM) [26] is used to measure image quality after ELM-SR algorithm. It can be observed that the images after ELM-SR algorithm in both platforms have higher image quality than the original low-resolution image. However, due to the use of LUT, which trades off precision against the hardware complexity, the image quality in DW-NN is slightly lower than that in GPP. Specifically, the SSIM is 0.94 for DW-NN, 3% lower than 0.97 for GPP.

V. Conclusion

With the use of the newly introduced domain-wall nanowire, this paper explores an in-memory architecture of machine learning on neural network, called DW-NN. In the proposed DW-NN, domain-wall nanowire based logic customized for machine learning is integrated within the image storage data such that machine-learning based image processing can be performed locally within the memory. We show that all operations involved in machine learning on neural network can be mapped to a logic-in-memory architecture by non-volatile domain-wall nanowire. The experimental results show that the I/O load in the proposed DW-NN is greatly alleviated with an energy efficiency improvement by 92x and throughput improvement by 11.6x compared to the conventional image processing system by general purpose processor.

Acknowledgments

This work is sponsored by Singapore MOE TIER-2 fund MOE2010-T2-2-037 (ARC 5/11) and NRF-CRP fund NRFCRP9-2011-01.

References

[1] S. Matsunaga and et al.,"Fabrication of a nonvolatile full adder based on logic-in-memory architecture using magnetic tunnel junctions," *Applied Physics Express,* vol. 1, p. 1301, 2008.

[2] H. Kimura and et al., "Complementary ferroelectric-capacitor logic for low-power logic-in-memory VLSI," *IEEE JSSC,* vol. 39, pp. 919-926, 2004.

[3] Y. Wang, H. Yu and D. Sylvester, "Energy efficient in-memory aes encryption based on nonvolatile domain-wall nanowire", *ACM/IEEE Design Automation and Test Conference in Europe,* March 2014.

[4] Y. Wang, P. Kong, and H. Yu, "Logic-in-memory based map-reduced computing by nonvolatile domain-wall nanowire devices", *IEEE Non-Volatile Memory Technology Symposium,* August, 2013.

[5] D. E. Goldberg and J. H. Holland, "Genetic algorithms and machine learning," *Machine learning,* vol. 3, pp. 95-99, 1988.

[6] S. Tong and E. Chang, "Support vector machine active learning for image retrieval," *Proceedings of the ninth ACM international conference on Multimedia,* pp. 107-118, Oct. 2001.

[7] G.-B. Huang, Q.-Y. Zhu, and C.-K. Siew, "Extreme learning machine: theory and applications," *Neurocomputing,* vol. 70, pp. 489-501, 2006.

[8] L. An and B. Bhanu, "Image super-resolution by extreme learning machine," *19th IEEE International Conference on Image Processing,* pp. 2209-2212, Sept. 2012.

[9] D. B. Strukov, and et al., "The missing memristor found," *Nature,* vol. 453, pp. 80-83, 2008.

[10] F. Bedeschi, and et al., "4-Mb MOSFET-selected phase-change memory experimental chip," *Proceeding of the 30th European Solid-State Circuits Conference,* pp. 207-210, Sept. 2004.

[11] M. Hosomi, and et al., "A novel nonvolatile memory with spin torque transfer magnetization switching: Spin-RAM," *IEEE International Electron Devices Meeting Technical Digest,* pp. 459-462, Dec. 2005.

[12] S. S. Parkin, M. Hayashi, and L. Thomas, "Magnetic domain-wall racetrack memory," *Science,* vol. 320, pp. 190-194, 2008.

[13] R. Venkatesan and et al., "TapeCache: a high density, energy efficient cache based on domain wall memory," *ACM/IEEE international symposium on Low power electronics and design,* July 2012.

[14] X. Wang, and et al., "Spintronic memristor through spin-torque-induced magnetization motion," *IEEEE lectron Device Letters,* vol. 30, pp. 294-297, 2009.

[15] C. Augustine and et al., "Numerical analysis of domain wall propagation for dense memory arrays," *IEEE International Electron Devices Meeting (IEDM),* pp. 17.6. 1-17.6. 4, Dec. 2011.

[16] J. Dean and S. Ghemawat, "MapReduce: simplified data processing on large clusters," *Communications of the ACM,* vol. 51, pp. 107-113, 2008.

[17] H.-P. Trinh and et al., D. Ravelsona, and C. Chappert, "Domain wall motion based magnetic adder," *Electronics Letters,* vol. 48, pp. 1049-1051, 2012.

[18] W. Fei and et al., "Design exploration of hybrid cmos and memristor circuit by new modified nodal analysis", *IEEE Transactions on Very Large Scale Integration Systems,* vol.20, no.6, pp.1012-1025, June 2012.

[19] Y. Shang, W. Fei, and H. Yu, "Analysis and modeling of internal state variables for dynamic effects of nonvolatile memory devices", *IEEE TCAS-I,* vol.59, no.9, pp.1906-1918, September 2012.

[20] Y. Wang, H. Yu, and W. Zhang, "Nonvolatile cbram crossbar based 3d integrated hybrid memory for data retention", IEEE Transactions on Very Large Scale Integration Systems, 2013.

[21] Y. Wang, and H. Yu, "An ultralow-power memory-based big-data computing platform by nonvolatile domain-wall nanowire devices", *ACM/IEEE International Symposium on Low Power Electronics and Design,* September 2013.

[22] S. J. Wilton and N. P. Jouppi, "CACTI: An enhanced cache access and cycle time model," *IEEE Journal of Solid-State Circuits,* vol. 31, pp. 677-688, 1996.

[23] N. Binkert and et al., "The gem5 simulator," *ACM SIGARCH Computer Architecture News,* vol. 39, no. 2, pp. 1–7, 2011.

[24] S. Li and et al., "Mcpat: an integrated power, area, and timing modeling framework for multicore and manycore architectures," in *IEEE/ACM International Symposium on Microarchitecture,* 2009, pp. 469–480.

[25] J.-K. Kim and et al., "A 3.6 Gb/s/pin simultaneous bidirectional (SBD) I/O interface for high-speed DRAM," *ISSCC Dig. Tech. Papers,* pp. 414-415, Feb. 2004.

[26] Z. Wang and et al., "Image quality assessment: From error visibility to structural similarity," *IEEE Transactions on Image Processing* vol. 13, pp. 600-612, 2004.

Lessons from the Neurons Themselves

Louis K. Scheffer
Howard Hughes Medical Institute
Ashburn, Virginia, 20147 USA
email: schefferl@janelia.hhmi.org

Abstract— Natural neural circuits, optimized by millions of years of evolution, are fast, low power, robust, and adapt in response to experience, all characteristics we would love to have in systems we ourselves design. Recently there have been enormous advances in understanding how neurons implement computations within the brain of living creatures. Can we use this new-found knowledge to create better artificial system? What lessons can we learn from the neurons themselves, that can help us create better neuromorphic circuits?

I. Introduction

Natural neural circuits, optimized by millions of years of evolution, are fast, low power, robust, and adapt in response to experience, all characteristics we would love to have in systems we ourselves design. A natural way to attempt to get these features is to copy the models (and perhaps mechanisms) that biological systems use for computation. Such circuits are called *neuromorphic* and many efforts have been made in this direction[1]. However, there are two main stumbling blocks to this strategy - first, we don't fully understand how many neural systems work, and second, of those we understand we don't know which features are vital and which just side effects of the biological implementation. Our existing neuromorphic circuits, such as neural nets and integrate-and-fire models, were inspired by biology and implemented features thought to be important. These systems, while producing some excellent results, certainly have not mimicked all the favorable properties of biology. Even software models, free from the inevitable constraints of hardware, cannot yet reproduce many of the aspects of biological computation. This indicates the more fundamental problem is the lack of understanding of how biological systems work, as opposed to the (in)availability of suitable components.

Recently there have been enormous advances in understanding how neurons implement computations within the brain of living creatures. Although there is still much that is not understood, it is clear that existing models do not capture all the tricks of biological computation. A few examples include:

- Neurons contain considerable internal state. This is not just a side effect but instead is often used in the computational process.

- At least some networks of neurons contain massively parallel connections and very large amounts of local, internal feedback.

- The supporting structure is not passive, but active in the computation.

- Neuromodulators provide intermediate range connections.

- Biological connectivity is not constant. Synapses disappear and new synapses are created as biological circuits change in response to experience.

Can we use this new-found knowledge to create better artificial systems? In Section II, we look at previous work and how well existing neuromorphic systems imitate real biology. In sections III-VII, we look at specific biological neural computation mechanisms, and see if they hold any lessons for how we might build neuromorphic equivalents. Section VIII contains conclusions and and potential directions for future research.

II. Previous work

Quantitative modeling of neurons began with the Hodgkin-Huxley model[2], a set of coupled differential equations describing the ion concentrations and resulting voltages in a nerve cell. This Nobel prize winning[3] model showed how non-linear interactions generated the spiking behavior seen in almost all neurons, and is the starting point for all attempts to realistically simulate biological neurons. However, for neuromorphic computation, this model is simultaneously too simple and too complex. It is too simple to describe the behavior of real neurons, and too complicated to construct analytically tractable networks[4].

Analytical tractabilty, with provable properties, demands even simpler models. For example, artificial neural networks keep the non-linearity, but give up the spiking behavior. Such networks are commonly built with many layers, each a linear summation followed by a monotonic but non-linear function. See Fig. 2, (a) and (b). These have excellent performance on specific tasks, but by now are largely divorced from their biological roots. These networks are typically trained through back propagation[5], basically the repeated implementation of the chain rule. This works well, but is very compute intensive, and not closely related to how biological learning is thought to take place[6].

Alternatively, integrate-and-fire models keep the spiking behavior but give up the non-linearity. Like neural nets, these have had some success, but duplicate only limited portions of real neural behavior.

Proceeding in the other direction, for modelling real neurons, the Hodgkin-Huxley model is too simple. Real neurons are very large in electrical terms, with many time constants separating distant branches. See, for example, Fig. 1 and compare it to the isopotential lengths computed in Table 1, lines 1 and 2. The scale bar in the figure is roughly equal to the length of an isopotential region, so a model with many compartments would be needed for an accurate electrical representation of this neuron. Even Hodgkin and Huxley themselves inserted thin silver wires down the length of each neuron to force them to be isopotential, since otherwise their limited computational tools could not handle even a single neuron.

It is now standard dogma that in normal operation, the different parts of a neuron often have very different potentials. This property is used heavily used in their operation, and is carefully controlled during the development process. As each neuron is growing and differentiating, some synapses will target the distal (far from the nucleus) branches of a cell, while others connections will explicitly target the

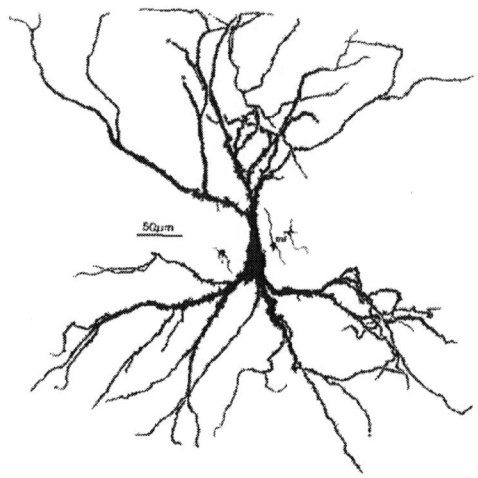

Fig. 1. Pyramidal cell from the hippocampus. The scale bar is 50 microns long and shows roughly the size of an equipotential region within the cell. From[7].

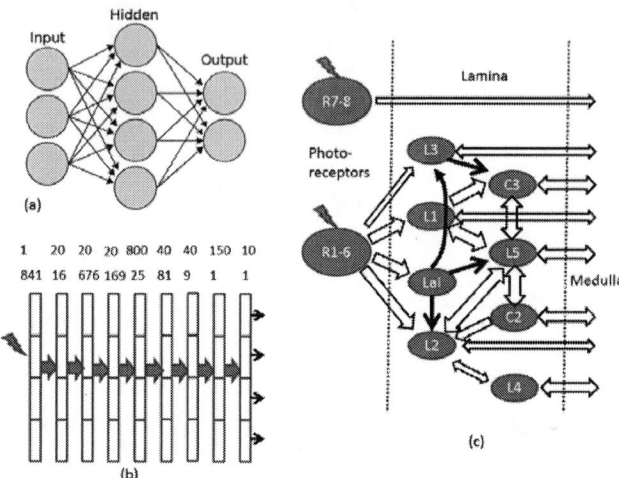

Fig. 2. Differences between neural nets and real neurons. (a) shows the prototypical neural net structure[19]. (b) shows a neural net with high performance on a task of recognizing handwritten digits[20]. The top line shows the number of neurons in each layer, and the bottom the number of variables in each neuron. (c) a biological neural network[21]. The widths of the arrows indicate the strength of the connection. The network is not nearly as neatly organized as (a) or (b) and has strong local feedback.

soma, near the nucleus. [8][9][10]. Example applications using different potentials in each part of the cell include:

- Amplification of distal inputs
- Coincidence detection
- Compressive dendritic nonlinearities
- Neurons within neurons

These neural behaviors are believed to be central to many animal behaviors.

Lesson: Slow interconnect may not be a problem. Designs may be able to take advantage of this characteristic in their calculations. On the other hand, trying to duplicate the neural results with a compartment model will require many compartments per neuron. Programs or hardware limited to a few compartments per neuron will not be able to simulate real neuron behavior. (This does not mean they cannot perform the desired calculations, only that they cannot do it exactly as neurons do.)

Interestingly, at least as far as analysis, EDA faced a very similar problem. At first, wiring within a chip could be treated as isopotential. But as chips got faster, and larger, this was no longer true, and in deep submicron technology, wires could be much slower than gates. (See Table 1, lines 3 and 4, for relative speeds of interconnect and gates in a 20nm technology). Accuracy could be recovered by going to the equivalent of a compartment model, as simulated by circuit simulators such as SPICE[15]. But this required considerable compute time.

The first step was to stick with a lumped model, but reduce the number of nodes by combining adjacent values subject to an error bound[16]. This also helps by removing very disparate time constants, which make simulation inefficient.

Alternatively, moment matching methods[17] were developed. These expand the response as a function of time in terms of moments. Each additional DC solution adds one additional moment to the series, which converges to the real answer in all physically realistic cases. In the case of a tree structured network, each individual DC solution can be computed in linear time. Since trees are the most common configuration of wires on chips, moments can be quickly calculated for all points on a wire, given one or more inputs. The moments themselves can give rough delays, and detailed voltages can be computed

at specific locations without computing them everywhere. This enabled detailed delay calculations on every wire within a chip[18].

Since neurons are almost always tree structured, moment evaluation should be fast for them as well. This could be one possible approach to calculating a neuron's response without a detailed, time domain, multi-compartment simulation.

III. LOCAL FEEDBACK

Neural nets, especially high performance ones, are commonly modelled as a deep collection of layers, each with inputs only from the previous layer, and outputs onto the next layer[20][22]. However, recent efforts in connectomics have revealed that biological networks often have considerable local feedback, many synapses in parallel[23],

and connections that skip many layers forward in the neural network. See Fig 2.(c) for a portion of the real network in the lamina (a part of the visual system) in the fruit fly.

Lesson: One lesson is that the individual synapse can have fairly poor properties, yet can still be used to construct solid systems. A biological synapse has considerable statistical variation in every step of its operation - the number of vesicles released, the molecule count of each vesicle, the diffusion across the cleft, and the uptake and ion channel operation on the post-synaptic side. Furthermore, they may come only in specific sizes and in any case cannot achieve purely arbitrary weights. They are also subject to considerable variation with temperature and their local chemical environment. Never the less, as shown by the fly visual system, the resulting networks can still achieve excellent performance over a wide dynamic range.

Another lesson is that we need new forms of analysis. Traditionally, circuits with both loops and non-linearities have been extremely hard to analyze. This is because the general case of such systems can give rise to oscillations, chaos, limit cycles, turbulence, and other hard-to-predict phenomena. Biology appears to use non-linearity in a more controlled way, beyond that of linear models but not needing the full power of non-linear analysis. Few mathematical tools currently exist for this regime.

978-1-4799-2817-0/14 $31.00 © 2014 IEEE

Tech	Size	R/mm	C/mm	T (100 μm)	Relative	T(1mm)	Relative	Breakeven
Bio	100 nm	130G	3.1pF	2ms	2	200ms	200	71 μm
Bio	1 um	1.3G	31pF	0.2ms	0.2	20ms	20	225 μm
Semi	20 nm	54K	200fF	54ps	6	5.4ns	600	43 μm
Semi	200 nm	5.4K	200fF	0.5ps	0.06	54ps	6	405 μm

TABLE I

DELAY AS A FUNCTION OF LENGTH, ASSUMING $\rho = 1$ OHM-M FOR CYTOPLASM[11] AND $1.68 \cdot 10^{-8}$ OHM-M FOR COPPER. MEMBRANE CAPACITANCE $= 0.01$ F/M^2[12], AND C COPPER INTERCONNECT $= 200$ FF/MM INDEPENDENT OF DIAMETER (LARGE WIRES HAVE LARGER SPACING, SEE ITRS[13].). RELATIVE DELAY IS RELATIVE TO A FAST OPERATION IN EACH SYSTEM, ASSUMED TO BE 1MS FOR BIO[14] AND 10PS FOR 20NM CMOS[13]. BREAKEVEN LENGTH IS WHERE A WIRE DELAY EQUALS AN OPERATION DELAY.

IV. THE SUPPORTING STRUCTURE IS ACTIVE

The conventional view is that the majority of computation is a consequence of the chemical synapses between cells. Cells with relatively little branching structure and no synapses are called glia, and are normally thought to be the scaffolding that holds and nourishes neurons[14]. But electron microscope studies of the glia in the lamina have shown they have synapses that terminate onto them. The T1 neuron in the lamina and medulla of the fruit fly is also strangely non-canonical - it looks like a typical neuron, but seems to have no outputs, just inputs of many different types. It receives histamine from cell types R2-R4, glutamate from the lamina amacrine cell[21], GABA from C3, and actylcholine from L2[23]. Despite the lack of obvious outputs, activation of this cell causes all sorts of visual problems in flies[24]. So what is it doing and how does it work?

Lesson: This is a case where more biological understanding is needed. Perhaps these cells are communicating only by gap junctions or neuromodulators. Another possibility is that the neurons are communicating specific needs for supplies or waste removal to the supporting structure. This could be a very local power optimization technique - if a neuron is not active, then it needs less resources. Or perhaps the read-only neuron is simply sucking neurotransmitters out of the clefts between cells, to make the response faster without affecting its peak strength.

V. NEUROMODULATORS

Normal operation of a nervous system is thought to be operated by synapses, where a cell excretes a *neurotransmitter*, which is then taken up by the directly adjacent cell and causes an excitory or inhibitory effect. However, since the operation is chemical, there is also the possibility that a chemical can diffuse, or spread through another system such as the bloodstream, and have effects far from where it is produced. Such effectors are called *neuromodulators* as opposed to the more local neurotransmitters. For a review of neuromodulator operation see Bargmann[25]. Also, is some systems it is now known that individual neurons produce both neurotransmitters, for local operation, and neuromodulators for longer range effects[26].

The range of a neuromodulator depends on how much is produced, the rate of diffusion, and where (and how fast) it is removed from the system. This can be quantified, with differential equations showing the strength and speed of the effect as a function of the distance from the source. One such paper, for example, shows that in the striatum, dopamine has a radius of action of about 10 microns[27].

If the range is long enough, a chemical can serve as a global variable to control some system property, as alcohol affects inhibition in humans, and seratonin affects mood. In this case the chemical is conventionally called a hormone.

Lesson: Computational models may need an intermediate range interaction. This can afford to be comparatively slow, if the analogy to biology holds.

Providing this interaction may require new architectures. One obvious problem is the definition of distance, on which the strength of the interaction depends. In biological networks this is defined (at least in the first approximation) as the 3-D distance between the release site and the measurement site. In an artificial network, which could be represented in planar, graph, or even more abstract form, computing the distance between neurons would need to be done differently. It is not at all clear how closely any solution would need to model biological operation in order to be useful.

If the artificial system was also in 3-D, for example, then a given neuromodulator could be represented by a high resistance grid, with sources and sinks defined on this grid. Then the local potential could correspond to the local concentration of a neuromodulator. Alternatively, since the system of neuromodulators is slow, a multiplexed circuit could be used. If each release event was sent on this bus, each receiver could compute its own response, taking into account time since release, release strength, and distance from the release site. This would be capable of coping with much more abstract definitions of distance.

VI. VARYING CIRCUIT CONNECTIVITY

Circuit connectivity in the adult brain changes as a result of experience[28].

Lesson: Current approaches to learning in neural networks can account for synapses disappearing, by setting their weight to zero. Accounting for newly created synapses seems harder. In the biological system referenced above, the new synapses appear to connect neurons that are already quite close; in the terminology of the paper, the larger connections, the axons and dendrites, are relatively static. The smallest connections, the spines, that bridge the last micron or so, were the portions that appear and disappear. Perhaps this could be modelled by adding connections to all nearby neurons, even if the initial training did not appear to need these connections. They would then be available for later network reconfiguration. This has the same problem, as mentioned above, over which neurons are "close".

VII. CONCLUSIONS AND FUTURE RESEARCH

Biological systems provide great examples showing certain computational tasks are possible. Inspired by these examples, researchers have built many forms of computation. Some, such as modern neural networks, are very useful, but have diverged far from their biological roots. Others, such as compartment models, are probably too biologically inspired, and are good for understanding but not efficient for computation. We do not yet have artificial systems with anywhere near the flexibility and high performance of living creatures.

There are three major problems in neuromorphic computing. The first is understanding how biological computations are performed. There is good progress in this area, but a long way to go.

The second is understanding which properties of biological systems are essential, and which are just kludges forced by the biological implementation.

Third is technologies with the required characteristics to imitate the functions that are essential. There are any number of technologies that can imitate portions of the behavior of a synapse. Each of these has weaknesses, but natural synapses too have weaknesses, and living creatures work around them. This is why the second goal, mentioned above, is essential.

If all these can be accomplished, the result should be a compact model that captures the essential computational aspects, but does not slavishly model every detail. Currently, the best form for such a model is still unclear.

REFERENCES

[1] Giacomo Indiveri, Bernabé Linares-Barranco, Tara Julia Hamilton, André Van Schaik, Ralph Etienne-Cummings, Tobi Delbruck, Shih-Chii Liu, Piotr Dudek, Philipp Häfliger, Sylvie Renaud, et al. Neuromorphic silicon neuron circuits. *Frontiers in neuroscience*, 5, 2011.

[2] Alan L Hodgkin and Andrew F Huxley. A quantitative description of membrane current and its application to conduction and excitation in nerve. *The Journal of physiology*, 117(4):500, 1952.

[3] Christof J Schwiening. A brief historical perspective: Hodgkin and Huxley. *The Journal of Physiology*, 590(11):2571–2575, 2012.

[4] LF Abbott and Thomas B Kepler. Model neurons: From Hodgkin-Huxley to Hopfield. In *Statistical mechanics of neural networks*, pages 5–18. Springer, 1990.

[5] David E Rumelhart, Geoffrey E Hintont, and Ronald J Williams. Learning representations by back-propagating errors. *Nature*, 323(6088):533–536, 1986.

[6] Donald Olding Hebb. *The organization of behavior: A neuropsychological theory*. Wiley, 1949.

[7] D. Beeman. Introduction to Realistic Neuron Modeling, 2005. https://en.wikipedia.org/wiki/File:Hippocampal-pyramidal-cell.png.

[8] Michael London and Michael Häusser. Dendritic computation. *Annu. Rev. Neurosci.*, 28:503–532, 2005.

[9] Sonia Gasparini and Jeffrey C Magee. State-dependent dendritic computation in hippocampal ca1 pyramidal neurons. *The Journal of neuroscience*, 26(7):2088–2100, 2006.

[10] Michael Häusser, Nelson Spruston, and Greg J Stuart. Diversity and dynamics of dendritic signaling. *Science*, 290(5492):739–744, 2000.

[11] Wilfrid Rall. Core conductor theory and cable properties of neurons. *Comprehensive Physiology*, 1977.

[12] Luc J Gentet, Greg J Stuart, and John D Clements. Direct measurement of specific membrane capacitance in neurons. *Biophysical Journal*, 79(1):314–320, 2000.

[13] ITRS. International Technology Roadmap for Semiconductors. http://www.itrs.net/Links/2012ITRS/Home2012.htm.

[14] Eric R Kandel, James H Schwartz, Thomas M Jessell, et al. *Principles of neural science*, volume 4. McGraw-Hill New York, 2000.

[15] Laurence William Nagel and Donald O Pederson. *SPICE: Simulation program with integrated circuit emphasis*. Electronics Research Laboratory, College of Engineering, University of California, 1973.

[16] Bernard N Sheehan. TICER: Realizable reduction of extracted RC circuits. In *Proceedings of the 1999 IEEE/ACM international conference on Computer-aided design*, pages 200–203. IEEE Press, 1999.

[17] Lawrence T Pillage and Ronald A Rohrer. Asymptotic waveform evaluation for timing analysis. *Computer-Aided Design of Integrated Circuits and Systems, IEEE Transactions on*, 9(4):352–366, 1990.

[18] Curtis L Ratzlaff, Nanda Gopal, and Lawrence T Pillage. RICE: Rapid interconnect circuit evaluator. In *Proceedings of the 28th ACM/IEEE Design Automation Conference*, pages 555–560. ACM, 1991.

[19] C. Burnett. Artificial neural network, 2011. https://commons.wikimedia.org/wiki/File:Artificial_-neural_network.svg.

[20] Dan Claudiu Cireşan, Ueli Meier, Luca Maria Gambardella, and Jürgen Schmidhuber. Deep, big, simple neural nets for handwritten digit recognition. *Neural computation*, 22(12):3207–3220, 2010.

[21] Marta Rivera-Alba, Shiv N Vitaladevuni, Yuriy Mishchenko, Zhiyuan Lu, Shin-ya Takemura, Lou Scheffer, Ian A Meinertzhagen, Dmitri B Chklovskii, and Gonzalo G de Polavieja. Wiring economy and volume exclusion determine neuronal placement in the *Drosophila* brain. *Current Biology*, 21(23):2000–2005, 2011.

[22] Dan Cireşan, Ueli Meier, and Jürgen Schmidhuber. Multi-column deep neural networks for image classification. In *Computer Vision and Pattern Recognition (CVPR), 2012 IEEE Conference on*, pages 3642–3649. IEEE, 2012.

[23] Shin-ya Takemura, Arjun Bharioke, Zhiyuan Lu, Aljoscha Nern, Shiv Vitaladevuni, Patricia K Rivlin, William T Katz, Donald J Olbris, Stephen M Plaza, Philip Winston, et al. A visual motion detection circuit suggested by *Drosophila* connectomics. *Nature*, 500(7461):175–181, 2013.

[24] John C Tuthill, Aljoscha Nern, Stephen L Holtz, Gerald M Rubin, and Michael B Reiser. Contributions of the 12 neuron classes in the fly lamina to motion vision. *Neuron*, 79(1):128–140, 2013.

[25] Cornelia I Bargmann. Beyond the connectome: How neuromodulators shape neural circuits. *Bioessays*, 34(6):458–465, 2012.

[26] Nicolas X Tritsch, Jun B Ding, and Bernardo L Sabatini. Dopaminergic neurons inhibit striatal output through non-canonical release of GABA. *Nature*, 490(7419):262–266, 2012.

[27] Margaret E Rice, Jyoti C Patel, and Stephanie J Cragg. Dopamine release in the basal ganglia. *Neuroscience*, 198:112–137, 2011.

[28] Joshua T Trachtenberg, Brian E Chen, Graham W Knott, Guoping Feng, Joshua R Sanes, Egbert Welker, and Karel Svoboda. Long-term *in vivo* imaging of experience-dependent synaptic plasticity in adult cortex. *Nature*, 420(6917):788–794, 2002.

978-1-4799-2817-0/14 $31.00 © 2014 IEEE

Leveraging the Error Resilience of Machine-Learning Applications for Designing Highly Energy Efficient Accelerators

Zidong Du[†]
duzidong@ict.ac.cn

Avinash Lingamneni[‡]
avinash.l@rice.edu

Yunji Chen[†]
cyj@ict.ac.cn

Krishna Palem[‡]
palem@rice.edu

Olivier Temam[§]
olivier.temam@inria.fr

Chengyong Wu[†]
cwu@ict.ac.cn

[†]CARCH[*], ICT, CAS, China

[‡]Rice University, USA

[§]INRIA, France

Abstract— In recent years, *inexact computing* has been increasingly regarded as one of the most promising approaches for reducing energy consumption in many applications that can tolerate a degree of inaccuracy. Driven by the principle of trading *tolerable* amounts of application accuracy in return for significant resource savings—the energy consumed, the (critical path) delay and the (silicon) area being the resources— this approach has been limited to certain application domains. In this paper, we propose to expand the application scope, error tolerance as well as the energy savings of inexact computing systems through neural network architectures. Such neural networks are fast emerging as popular candidate accelerators for future heterogeneous multi-core platforms, and have flexible error tolerance limits owing to their ability to be *trained*. Our results based on simulated 65nm technology designs demonstrate that the proposed inexact neural network accelerator could achieve 43.91%-62.49% savings in energy consumption (with corresponding delay and area savings being 18.79% and 31.44% respectively) when compared to existing baseline neural network implementation, at the cost of an accuracy loss (quantified as the Mean Square Error (MSE) which increases from 0.14 to 0.20 on average).

I. INTRODUCTION

Due to stringent energy constraints, researchers have started to accept that holistic approaches for saving energy are required, which span all computing system aspects, from circuits to applications. One of the most promising approaches for saving energy is to trade application accuracy for energy savings, popularly referred to as *inexact* (see [13] for additional references and [22] for early examples) or *approximate* [8] computing. The driving philosophy behind this approach is that some if not many applications do not require the degree of accuracy imposed by current circuit or programming methods; for instance, video decoding can tolerate a fair degree of inaccuracy because the image variations may not be perceived by the human eye. Some have started to take advantage of that observation by reducing the amount of logic, i.e., the logic operators' accuracy, used for the computations. These techniques advocated the reduction of logic density by either pruning/deletion of components that are not so significant [17], or by transformations to lesser power consuming yet similar logic that is "close" to the original design [6, 18].

While this concept is attractive, in practice, when applied to ASICs taking advantage of the specific interplay between logic operators within a given circuit datapath for the targeted application error [13], these ASICs (or programs run on them) have limited error tolerance to inexact computations as currently designed. Furthermore, the concept of approximate computing has had limited application to general-purpose processors because the control logic is highly susceptible to faults [8].

We seek to overcome these limitations by leveraging two trends: one in architecture and the other in applications. The trend in architecture is driven by increasing energy constraints: it is projected that multi-core architectures (embedded and high-performance) will increasingly rely on *accelerators* for most computationally intensive tasks (heterogeneous multi-cores) owing to the lack of voltage scaling leading to an effect known as Dark Silicon [20]. Beyond traditional accelerator solutions such as GPUs, FPGAs and ASICs, researchers are increasingly investigating accelerators or co-processors with an intermediate scope in terms of generality, covering more applications than ASICs but less than general-purpose processors, in order to reap significantly higher energy benefits than GPUs or FPGAs [23, 10]. Considering multiple such accelerators can be implemented within a chip today, the union of the application scopes of these accelerators can be very broad indeed.

The second trend is related to the nature of emerging computing-intensive applications. A few years ago, Intel has called the attention of the architecture community to the fact that its benchmarks were ill-designed, and encouraged the consideration of Recognition, Mining and Synthesis (RMS) applications [7] as more representative of emerging high-performance applications. As we know, this push later led to the PARSEC benchmarks [4], a collaboration between Intel and Princeton University.

Recently, it has been highlighted [24] that the combination of these two trends can lead to an atypical but highly attractive accelerator option based on *a hardware neural network*. Chen et al. [5] have shown that for a significant portion of the PARSEC benchmarks, 90% or more of their execution time could potentially be offloaded to a hardware neural network accelerator. Therefore, neural networks can be regarded as accelerators based on which a broad set of applications can be implemented. While several other accelerator options exist for supporting the PARSEC benchmarks, from our perspective, neural networks are particularly attractive because of their ability to inherently

---

[*]State Key Laboratory of Computer Architecture

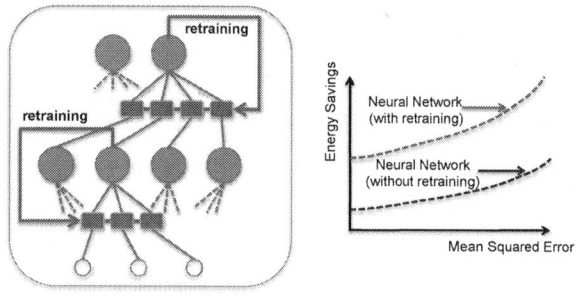

Fig. 1. Illustrating the benefit of retraining for neural network error tolerance.

tolerate errors. For instance, it was recently shown that a hardware neural network accelerator would be capable of tolerating transistor-level faults [24].

With this as a backdrop, we start from the following observations: (1) contrary to common wisdom, the application scope of a hardware neural network accelerator is very broad in light of emerging computing-intensive applications, (2) in spite of that broad application scope, a hardware neural network is just one large datapath circuit, with no complex control as in a core, and as a result, it can directly benefit from inexact arithmetic circuit optimizations, and (3) it implements an algorithm which has an unique capability of tolerating errors due to retraining.

Specifically, we take advantage of these three observations to extend the reach of inexact computing. First, by applying inexact computing to a hardware neural network accelerator, we demonstrate the energy savings for applications which can take advantage of that accelerator explicitly. Second, we take advantage of the error tolerance capability of neural networks to improve the degree of inaccuracy that can be tolerated and thus, resulting in greater energy savings. This capability derives from the learning/training algorithm used for neural networks (qualitatively shown in Figure 1). Using a hardware neural network accelerator with inexact logic operators, and a set of machine-learning tasks from the UCI Machine-Learning repository, we show that it is possible to achieve energy savings of 53.75% on average, and an area reduction of 31%, compared to a hardware neural network accelerator with standard logic operators, at a loss of only 0.06 in Mean Squared Error (MSE).

In Section II, we outline the main concepts of inexact computing and hardware neural networks, and we show how both of them can be combined, which represent the goal and the results of this paper. In Section III, we explore the design space formed by inexact hardware neural networks accelerators, the potential energy, delay and area savings, and the concomitant impact on task accuracy. We describe some of the related work that is most relevant in Section IV, and conclude in Section V.

II. *Inexact* HARDWARE NEURAL NETWORKS

In this section, we first present a brief primer on inexact computing and neural networks before proceeding to demonstrate how they can be combined to improve the energy efficiency of hardware neural networks without substantially degrading their accuracy. These techniques and concepts are based on previously known work and are thus, not claimed to be novel. The novel contribution of this paper is the characterization of the benefits and associated costs represented by increased error,

derived by combining inexact computing with Neural Network (hardware) accelerators as summarized in Section III.B.

A. A Primer on Inexact Computing

Let us consider a circuit that computes a completely-specified boolean function $\mathcal{F} : B^n \rightarrow B^m$ that maps n-input boolean vector of the form $\mathbf{x} = [x_1, x_2, \ldots x_n]$ to a vector $\mathbf{y} = [y_1, y_2, \ldots y_m]$ with an associated hardware cost function $\mathbf{C}_\mathcal{F}$. The goal of (heuristic) inexact logic minimization (paraphrased from [1]) is to find a Boolean function $\mathcal{F}' : B^{n'} \rightarrow B^{m'}$ where $n' \leq n$ and $m' \leq m$, such that its cost $\mathbf{C}_{\mathcal{F}'}$ is heuristically minimized subject to the following constraint:

$$\sum_{\forall \vec{i} \in \mathbf{I}} \frac{|\mathcal{F}(\vec{i}) - \mathcal{F}'(\vec{i})|}{L} \leq Er_{th} \qquad (1)$$

where $\mathbf{I}$ is a set of L-testbench vectors to this circuit and Er_{th} represents the average error (see [1] for a formulation).

Such inexact logic minimization (ILM) can be achieved by logic-level alteration of boolean function using the notion of intentional *bit flips*, which implies an intentional forcing of the output from $0 \rightarrow 1$ or $1 \rightarrow 0$ for certain input vectors. Some application-specific contexts where this idea has been explored can be found in [6, 1, 2], among others. One example of such a hardware building block synthesized and studied is a 3-input 2-output full adder cell and is shown in Figure 2 (from [2]).

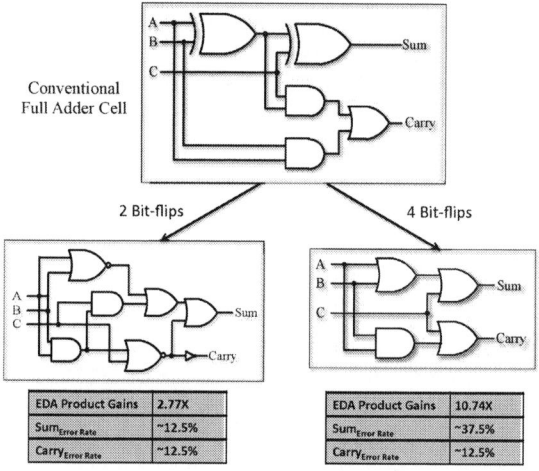

Fig. 2. Inexact design through inexact logic minimization: few examples from [2].

In this paper, we will use a portion of the inexact logic minimization approach from [1]. Specifically, to tackle the combinatorial explosion of the solution space for a given accuracy constraint from Equation (1) above, we use a *significance*-guided greedy heuristic combined with stochastic exploration to narrow down the search space. Given the richness of the solution space, even this heuristic-based exploration resulted in good design points. In our approach, as we are targeting datapath circuit design, we use an output-significance driven ranking of the nodes similar to previous works [18], i.e., nodes feeding outputs with higher (binary) significance are assigned higher rank and correspondingly are allotted lesser inexact configurations. As in the previous work [1], this significance or rank

978-1-4799-2817-0/14 $31.00 © 2014 IEEE

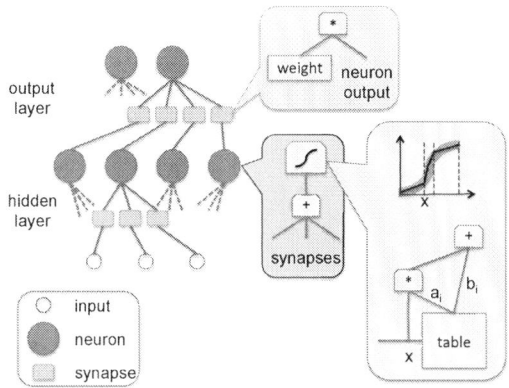

Fig. 3. Structure and logic operators of a 2-layer MLP, including the sigmoid interpolation.

is used to guide the decision on determining as to which elements of the circuit can accept more error in a relative sense. We then use this significance array as the template for guiding the amount of error introduced at each node.

We have performed such a heuristic-based exploration for the operators found to be most critical in a neural network, and built a library of these operators. The nature of the operators and the libary are described in Section II.D and in Section III.A.

B. A Primer on Hardware Neural Networks

Several types of artificial neural networks are used in machine-learning. A broadly used type of neural networks are Multi-Layer Perceptrons (MLPs) [12]. Convolutional Neural Networks (CNNs) [16] are also popular, and recently, a form of artificial neural networks, called Deep Belief Networks (DBNs), have been shown to be very powerful techniques again, outperforming Support Vector Machines (SVMs) and other techniques [15]. DBNs have a structure similar to MLPs or CNNs with broader and more layers. Here, we focus on MLPs, and the techniques proposed in this paper are compatible with any width or number of layers.

MLP. A typical MLP is a 2-layer feed-forward network: information flows come from an *input layer* (which contains no neurons) through a *hidden layer*, to an *output layer*. Following [12], each neuron is fully connected to the neurons of the previous layer, and each connection carries a synaptic weight. Each evaluation of one neuron j implements $\sum_{i=0}^{N} f(w_{ji} \times y_i)$ where y_i the output of neuron i of previous layer, w_{ji} is the synaptic weight between neurons i and j, and f is an activation function (sigmoid), typically $f(x) = \frac{1}{1+exp^{-x}}$. Following conventional approaches, in hardware, the activation function can be efficiently implemented using the piecewise linear approximation ($f(x) = a_i \times x + b_i$), through a small look-up table to store coefficients (a_i, b_i), an adder and a multiplier (Figure 3). Using a 16-segment approximation, we observed that the maximum sigmoid accuracy loss was 2.3% and this had no noticeable degradation of the MSE of the neural network.

Feed-Forward hardware. We use back-propagation [12], the most popular training algorithm for neural networks. However, we do not implement back-propagation in hardware; the hardware neural network only contains the feed-forward path.

There is a frequent and important misconception that *on-line* training is necessary for many applications. On the contrary, for many industrial applications *off-line* training is sufficient; for example, trained on handwritten digits, license plate numbers, a number of faces or objects to recognize, etc. The network can be taken offline periodically and retrained. We advocate this offline training approach in this paper.

C. Melding Inexact Computing and Neural Networks

In spite of its atypical nature, a hardware neural network is just one large datapath circuit, predominantly composed of multipliers (to compute the synaptic weight × neuron output, and activation function), and adders (sum of neuron inputs and the approximation to the activation function).

However, a neural network's desirable feature over other application contexts such as JPEG decompression, is its ability to lower the effects of faulty logic gates by retraining the circuit. As noted earlier, even a large number of faults due to the fabrication process can be compensated for by retraining [24]. In contrast to [24], our approach here is to *voluntarily* introduce introduce errors through inexactness in order to save energy, and to use the retraining capability of the neural network to lower the effects of the resulting functional errors. Retraining is done automatically re-balancing the synaptic weights so that the neurons producing the most errors are suppressed, or at least, their impact is diminished.

D. Sensitivity and Complexity of Exploration

The total exploration space of inexact neural networks is huge. An exhaustive consideration of all the parameters would result in 1.07×10^{68} different network configurations. Considering we need to train each network configuration in order to evaluate its quality in our experience, the evaluation of each design point is about 100 seconds. So, not only is an exhaustive exploration impossible, but in one year it would only be possible to evaluate a minuscule fraction of about $2.95 \times 10^{-61}\%$ of all possible configurations using a single workstation. At the same time, arbitrarily limiting the exploration to a tiny subset of the design space risks resulting in far too pessimistic solutions.

We address this issue by taking the specific nature of neural networks into consideration. We note that the adders accumulate the weight×output products in each neuron (see Figure 3), are potentially more susceptible to errors than each of the many synaptic weight multipliers which feed these adders. At the same time, the hardware cost of the adders is usually much smaller than the cost of multipliers, and as a result, the area/energy gains that can be expected from making adders inexact is not likely to be large even though their impact on the neuron output is significant. Similarly, the role of the multiplier and the adder used to approximate the activation function is likely to be significant from the standpoint of error, while providing savings equivalent to one multiplier at most. Based on this reasoning, we conclude that the most fruitful portions of the architecture for introducing inexactness are the synaptic weight multipliers. Restricting our attention to these multipliers reduces the total number of configurations to be explored as well.

978-1-4799-2817-0/14 $31.00 © 2014 IEEE

Exploration can be further reduced by focusing more on the hidden layers of the neural network. Since there is no synaptic weight after the neurons of the output layer, it is more difficult for the training algorithm to lower the error of these neurons, though training can reshuffle the role of each output and thus, still provide some form of error tolerance. Fortunately, most neural networks require large hidden layers, while the number of neurons in the output layer is usually small since the number of classes in a classification problem being solved for example is related to it. So, we bias the design space exploration towards the hidden layer.

To recapitulate and using common knowledge about the error sensitivity of the elements of the neural network architecture and focusing inexactness on the value of information, it is opportune that the significant gains can be realized by restricting the design space exploration to the less error sensitive elements such as the synaptic weight multipliers of the hidden layers while not considering the more error vulnerable adders or synaptic weight multipliers of the output layer.

III. PERFORMANCE EVALUATION

In this section, we describe the methodology, the design space which is explored, and the energy/accuracy tradeoffs achieved using the inexact hardware based neural networks. The material in Section III.B constitute the original contribution of this paper.

A. Methodology

Tool flow. We have explored a set of 7 inexact multipliers using the principles described in Section II.A, using a baseline of standard n-bit truncated multipliers with correction constant from [14]; an n-bit truncated multiplier has n-bit inputs and n-bit outputs, and as a result, it induces an error due to the truncation of output (as well as the corresponding logic generating these truncated outputs) from the full $2n$ bits down to n bits. The characteristics of these 7 inexact multipliers are shown in Table I and correspond to various tradeoffs in accuracy and efficiency.

The functional models of these inexact multipliers has been implemented as C++ subroutines (with corresponding area savings estimates) and are plugged into a C++ based neural network simulator. Specifically, we modified the implementation of the multi-layer perceptron in order to replace the calls to the standard multiplication with calls to the inexact multipliers, for all multiplications between synaptic weights and neuron inputs. This modified software neural network model is used to assess the impact of the inexact multiplier on the neural network accuracy before and after retraining.

The inexact multipliers have also been implemented as configurable Verilog HDL, along with the full neural network, in order to assess the energy, delay and area values. We synthesized the neural network using the Synopsys Design Compiler and the TSMC 65nm standard-V_{th} library, place and route using Synopsys IC compiler, and simulated using the Synopsys VCS and estimated the power using PrimeTime PX tool. The hardware neural network is a 2-stage pipelined architecture with one stage each for the hidden and output layers respectively.

TABLE I
INEXACT 16-BIT MULTIPLIERS.

Type	Rel.Err (%)	Area (um^2)	Power (uW)	Delay (ns)
#0 (truncated)	0.09	2068.20	931.41	11.47
#1 (inexact)	0.99	1883.52	725.50	10.46
#2 (inexact)	5.23	1642.32	558.70	8.48
#3 (inexact)	10.70	1537.92	494.59	7.99
#4 (inexact)	17.16	1520.64	468.02	8.84
#5 (inexact)	27.51	1384.92	380.86	7.03
#6 (inexact)	41.65	1298.88	370.81	7.03
#7 (inexact)	98.94	1249.92	364.10	7.03

Benchmarks and the neural network dimension. We evaluated the accuracy of the neural networks on tasks obtained from the UCI Machine-Learning repository [3]. The example cases in that repository are contributed by researchers and engineers from various domains in order to stimulate machine-learning research and thus, they are both diverse in nature and correspond to actual applications. In [24], it was shown that a neural network of 90 inputs, 10 hidden neurons, and 10 outputs could compute 90% of the tasks of the UCI repository and provide good classification results. We implemented a neural network with the same configuration for the 7 benchmarks mentioned in [24]. The benchmarks we used include: *glass, ionosphere, iris, robot, sonar, vehicle* and *wine*. Note that the *accuracy* of a neural network is typically reported as the mean squared classification error (MSE).

B. Design Space Exploration

The configurations we explored were determined by forcing the exploration on the synaptic multipliers of the hidden layers (Figure 3), due to the analysis of Section II.D. While the baseline configuration uses 16-bit operators, we considered bit widths of 8, 10, 12, 14 and 16; bit widths smaller than 8 generally yielded poor MSE. Any of the multipliers in the network can be replaced by any of the 8 multipliers (7 inexact and 1 exact). Overall, we explored 24500 configurations selected randomly out of the total possible configurations.

For each configuration, we estimate the MSE using the C++ neural network simulator on all 7 benchmarks. We separately train the neural network on each benchmark using 10-fold cross-validation, and repeat the experiments 5 times to account for statistical variations where the initial weights of the neural network are randomly set with an uniform distribution. We only retain the configurations for which the MSE is less than 2 times the MSE of the exact neural network, and estimate the area, delay and power cost of a configuration by simply extrapolating the resource cost of individual hardware objects such as multipliers, adders, registers, etc.

TABLE II
HARDWARE CHARACTERISTICS FOR BASELINE, BEST, BIT-WIDTH.

-	Best	Bit-Width	Baseline
Area (mm^2)	1.73	2.24	2.61
Delay (ns)	24.97	26.99	29.89
Power (W)	1.02	1.38	1.48
Energy (nJ)	25.40	37.15	44.34
Average MSE	0.20	0.15	0.14

978-1-4799-2817-0/14 $31.00 © 2014 IEEE

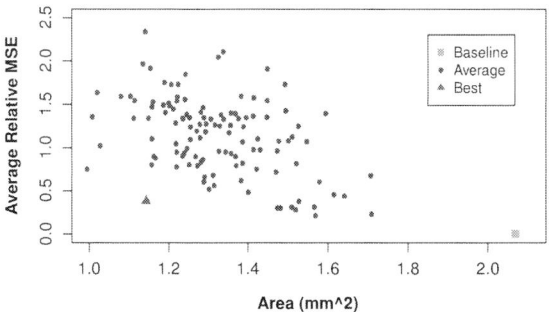

Fig. 4. Solutions of exploration for all 7 benchmarks.

In Figure 4, the relationship between the estimated area (from our C++ framework) and the accuracy across benchmarks is shown. In Table II, we selected a *best* configuration as the one on the area/MSE pareto-optimal front closest to the median between MSE and area (other area/MSE tradeoffs are naturally possible).

We observe that in this particular *best* configuration, there are no 8-bit nor 10-bit synaptic weight or multipliers operands, 44.56%, 44.56% of synaptic weights use 12-bit and 14-bit operands respectively, while the remaining are 16-bit synaptic weights. Among the multipliers in the hidden layer, 41.30% use 14-bit operands, 28.70% use 12-bit operands and the rest 30% use 16-bit operands. Also, we observed that all the types of inexact multipliers expect for the type 0 (standard truncated multiplier) listed in I are used in this *best* configuration (multiplier types 2, 4 and 6 are more dominant than other types). Another interesting observation in this *best* configuration is that all the multipliers in the output layer use 16-bit operands, suggesting that they were indeed more susceptible to errors, which was in line with the analysis of Section II.D.

Inexact Logic Minimization vs. Truncation. We also experimentally determined the impact of inexact multipliers over the more standard bit width truncation approach. In our exploration, the bit width truncation is applied by reducing the size of the operators, from 16 bits down to 8 bits.

In Table II, we showed the results of the *bit-width* configuration where the bit width of operators of the exact multipliers are being varied as freely as in the standard exploration. We can observe that varying the bit width of the operators provides only a small improvement over the baseline in terms of energy and area. On the other hand, introducing inexact multipliers provides more significant improvements, with only a small impact on accuracy.

C. Energy, Area and MSE of Inexact Hardware Neural Networks

In Figure 5, we compare the MSE of the baseline configuration against the MSE of the best configuration before and after training for that configuration; "before training" means that the synaptic weights of the baseline configuration are used. The MSE before training shows that inexact operators introduce a very significant loss of accuracy. This magnitude of accuracy loss would be unacceptable, but the training algorithm of neural networks can compensate for it. As a result, the MSE after training (0.20) is closer to the MSE of the baseline (0.14). This implies that the impact on accuracy of the inexact operators can be mostly overcome.

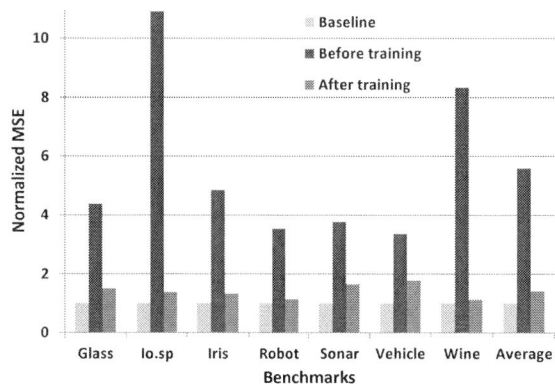

Fig. 5. Impact of inexact logic operators on error.

	Exact NN	Inexact NN
Area (mm^2)	4.23	2.90 (31.44%)
Delay (ns)	11.82	9.60 (18.79%)

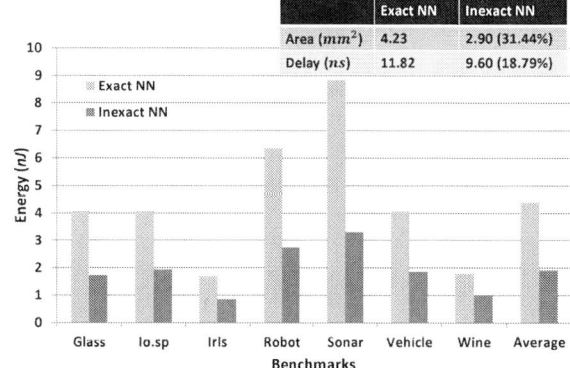

Fig. 6. Energy, delay and area benefits of inexact NN vs. exact NN.

At the same time, the energy and area savings results achieved by using inexact architectures are very significant, as shown in Figure 6. For the 7 benchmarks, the energy savings is at least 43.91%, and up to 62.49%. Clock and registers correspond to 7%-24% of the total energy cost, while most of the energy is consumed by the combinational logic.

The area cost also drops from 4.23 mm^2 to 2.90 mm^2 implying an improvement of 31.44%. Finally, thanks to ILM, the critical path delay has been reduced from 11.82ns to 9.60ns. We note that 900 multiplications are performed in the hidden layer with 90 inputs and 10 neurons within these 9.60ns.

IV. RELATED WORK

The early foundations of the principle of trading the hardware accuracy of the circuit for energy gains can be found in in [22] and since then, its application in many error-tolerant applications, in particular those whose output quality is judged by human sensory perception including vision or audition, has been exploited in a plethora of papers (for a partial summary, see [13]). Moving beyond datapaths to processors with control, the fact that the control in processors cannot be susceptible to errors proves to be a stringent limitation [8]. Mahdiani et al. study [19] considered the implementation of fuzzy logic and neural networks using imprecise adders and multipliers, but they sequentially execute all neurons on one single hardware neuron. Not only does this miss out on many of the performance and energy benefits of hardware neural networks in highly parallel implementations, but it creates the same susceptibility to control logic as in processors.

In this paper, we limit ourselves to previous work that is related only to inexact logic minimization that has been used here. These build on innovations at the logic- and architectural-level [6, 17, 2], which have emerged as a promising option providing *zero* (hardware) overhead. The readers are referred to earlier references for a discussion of other techniques with greater than zero overheads relying on voltage overscaling, for example. Also, we refer to some of the early examples where the notion of "significance" or value driven computation in conjunction with other techniques for energy minimization was done to realize inexact designs [11, 25].

Architectural approaches based on hardware neural networks are becoming increasingly popular [24] including considering concepts of using approximations at the computational level (in contrast to the datapath hardware level presented in this paper) and concerns of energy efficiency [9]. In the context of heterogeneous multi-cores, hardware neural networks are at the crossroad of multiple converging trends: the quest for both fault-tolerant and energy efficient accelerators, the need for accelerators with a broad application scope [5], and recent progress in machine-learning [15]. This convergence of trends has been acknowledged by companies such as IBM, whose hardware neural network accelerators [21] are now being developed.

V. CONCLUSIONS

In this article, we have proposed an approach for applying inexact computing to a broad set of applications by taking advantage of the inherent feature of hardware neural networks. Specifically, we take advantage of three distinct aspects of these neural networks: (a) many emerging high-performance applications can be competitively implemented using hardware neural networks, (b) these neural networks have much higher error tolerance owing to their learning/training ability, and (c) they can be realized as a circuit datapath using inexact logic operators. Using machine-learning benchmarks, we demonstrate that the proposed inexact neural network accelerator, implemented in 65nm technology, could achieve 43.91%-62.49% savings in energy consumption (with accompanying delay and area savings of 18.79% and 31.44% respectively) when compared to an existing baseline neural network implementation. These savings were achieved at only a slight cost in accuracy of output quantified through the MSE metric. Additionally, we also propose approaches to narrow down the huge search space of the inexact neural networks through a selective assignment of inexact logic operators and by identifying operators which yield higher energy savings and introducing relatively low error.

VI. ACKNOWLEDGMENT

This work is supported by the Intel ICRI-CI lab, a Google Faculty Research Award, ANR projects MHANN and NEME-SIS, and the Inria joint team YOUHUA. And this work is supported by the Strategic Priority Research Program of the Chinese Academy of Sciences (under Grant XDA06010403), the National Natural Science Foundation of China (under Grants 61003064, 61100163, 61222204, 61303158, 61050002,

61173006, and 61133004), the National 973 Program of China (under Grant 20118034), and the National 863 Program of China (under Grants 2012AA012202).

REFERENCES

[1] A. Lingamneni et al. Improving energy gains of inexact DSP hardware through reciprocative error compensation. *in the 50th Design Automation Conference*, pages 20:1–20:8, June 2013.

[2] A. Lingamneni et al. Synthesizing parsimonious inexact circuits through probabilistic design techniques. *in the ACM Transactions on Embedded Computing Systems*, 12(2s):93:1–93:26, May 2013.

[3] A. Asuncion and D. J. Newman. {UCI} Machine Learning Repository, 2007.

[4] C. Bienia, S. Kumar, J. P. Singh, and K. Li. The PARSEC benchmark suite: Characterization and architectural implications. In *International Conference on Parallel Architectures and Compilation Techniques*, New York, New York, USA, 2008. ACM Press.

[5] T. Chen, Y. Chen, M. Duranton, Q. Guo, A. Hashmi, M. Lipasti, A. Nere, S. Qiu, M. Sebag, and O. Temam. BenchNN: On the Broad Potential Application Scope of Hardware Neural Network Accelerators. In *International Symposium on Workload Characterization*, 2012.

[6] D Shin et al. Approximate logic synthesis for error tolerant applications. *in the proc. of DATE*, pages 957 – 960, 2010.

[7] P. Dubey. Recognition, Mining and Synthesis Moves Computers to the Era of Tera. *Technology@Intel Magazine*, 9(2):1–10, 2005.

[8] H. Esmaeilzadeh, A. Sampson, L. Ceze, and D. Burger. Architecture support for disciplined approximate programming. In T. Harris and M. L. Scott, editors, *ASPLOS*, pages 301–312. ACM, 2012.

[9] H. Esmaeilzadeh, A. Sampson, L. Ceze, and D. Burger. Neural Acceleration for General-Purpose Approximate Programs. In *International Symposium on Microarchitecture*, 2012.

[10] G Venkatesh et al. Conservation Cores: Reducing the Energy of Mature Computations. *in proc. of the Architectural Support for Programming Languages and Operating Systems*, pages 205–218, March 2010.

[11] J. George, B. Marr, B. E. S. Akgul, and K. V. Palem. Probabilistic arithmetic and energy efficient embedded signal processing. In *proc. of IEEE/ACM CASES*, pages 158 – 168, 2006.

[12] S. Haykin. *Neural Networks*. Prentice Hall Intl, London, UK, 2nd edition, 1999.

[13] K Palem et al. What to do about the end of moore's law, probably! *in the 49th Annual Design Automation Conference*, pages 924–929, 2012.

[14] E. J. King and E. E. Swartzlander Jr. Data-dependent truncation scheme for parallel multipliers. In *Signals, Systems & Computers, 1997. Conference Record of the Thirty-First Asilomar Conference on*, volume 2, pages 1178–1182. IEEE, 1997.

[15] H. Larochelle, D. Erhan, A. Courville, J. Bergstra, and Y. Bengio. An empirical evaluation of deep architectures on problems with many factors of variation. In *International Conference on Machine Learning*, pages 473–480, New York, New York, USA, 2007. ACM Press.

[16] Y. LeCun, L. Bottou, Y. Bengio, and P. Haffner. Gradient-Based Learning Applied to Document Recognition. *Proceedings of the IEEE*, 11(86):2278–2324, 1998.

[17] Lingamneni et al. Energy parsimonious circuit design through probabilistic pruning. *in proc. of DATE*, pages 764–769, Mar 2011.

[18] Lingamneni et al. Parsimonious circuit design for error-tolerant applications through probabilistic logic minimization. *in the proc. of the PATMOS*, pages 204–213, 2011.

[19] H. R. Mahdiani, A. Ahmadi, S. M. Fakhraie, and C. Lucas. Bio-inspired imprecise computational blocks for efficient VLSI implementation of soft-computing applications. *IEEE Transactions on Circuits and Systems I: Regular Papers*, 57(4):850–862, Apr. 2010.

[20] M. Muller. Dark Silicon and the Internet. In *EE Times "Designing with ARM" virtual conference*, 2010.

[21] P Merolla et al. A digital neurosynaptic core using embedded crossbar memory with 45pJ per spike in 45nm. In *IEEE Custom Integrated Circuits Conference*, pages 1–4. IEEE, Sept. 2011.

[22] K. V. Palem. Energy aware algorithm design via probabilistic computing: From algorithms and models to Moore's law and novel (semiconductor) devices. In *proc. of CASES*, pages 113 – 116, 2003.

[23] M. Stojilovic, D. Novo, L. Saranovac, P. Brisk, and P. Ienne. Selective flexibility: Breaking the rigidity of datapath merging. *in proc. of DATE*, pages 1543 –1548, march 2012.

[24] O. Temam. A Defect-Tolerant Accelerator for Emerging High-Performance Applications. In *International Symposium on Computer Architecture*, Portland, Oregon, 2012.

[25] Z. M. Kedem et al. Optimizing energy to minimize errors in dataflow graphs using approximate adders. In *in proc. of CASES*, pages 177–186, 2010.

978-1-4799-2817-0/14 $31.00 © 2014 IEEE

ArISE: Aging-aware Instruction Set Encoding for Lifetime Improvement

Fabian Oboril   Mehdi Tahoori

Karlsruhe Institute of Technology (KIT)
Chair of Dependable Nano Computing (CDNC)
Karlsruhe, Germany
e-mails: {Fabian.Oboril,Mehdi.Tahoori}@kit.edu

Abstract—Microprocessors fabricated at nanoscale nodes are exposed to accelerated transistor aging due to Bias Temperature Instability and Hot Carrier Injection. As a result, device delays increase over time reducing the Mean Time To Failure (MTTF) of the processor. To address this challenge, many (micro)-architectural techniques target the execution stage of the instruction pipeline, as this one is typically most critical. However, also the decoding stages can become aging-critical and limit the microprocessor lifetime, as we will show in this work. In this paper, we propose a novel aging-aware instruction set encoding methodology (ArISE), that improves the instruction encoding iteratively using a heuristic algorithm. Our experimental results show that MTTF of the decoding stages can be improved by 1.93x with negligible implementation costs.

I. INTRODUCTION

With the continuous downscaling of CMOS technology into nanoscale dimensions, reliability has emerged as an important design constraint besides performance, power and cost [3]. Among various reliability issues, accelerated transistor aging mainly caused by *Bias Temperature Instability (BTI)* [17, 22] and *Hot Carrier Injection (HCI)* [19] is a major challenge. Both phenomena manifest themselves in increasing switching and path delays and eventually cause faster wearout of the system. As a result *Mean Time To Failure (MTTF)* is reduced and timing violations can occur in the field. To avoid these, designers add safety margins to their designs to ensure a certain operational lifetime. However, such an overdesign is very expensive [13]. Hence, new aging-aware design techniques are necessary to take further advantage of scaled technology nodes in terms of performance, power, area and reliability.

As the microprocessor lifetime is mainly determined by the execution units [8], various aging-aware solutions have been proposed to extend the lifetime of these units. However, many of these techniques such as aging-aware instruction scheduling [15] or aging-aware NOP instructions [10] are hardly applicable to other stages of the instruction pipeline. Hence, by using such techniques and extending the lifetime of execution stages, the impact of other near-aging-critical stages on the overall microprocessor lifetime becomes more pronounced. As a result, MTTF improvements for the entire microprocessor become much smaller than the individual benefits for the execution units. Therefore, additional approaches are necessary to also improve MTTF of other pipeline stages. In particular the *decoding stages* have to be targeted, since these suffer from high wearout rates and are almost as critical as or even more critical than the execution stages, as we will show in this work. However, such approaches to improve lifetime of

pipeline stages in the frontend are still very rare.

To close this gap, we propose a novel aging-aware instruction set encoding technique[1] *called ArISE.* This technique exploits the fact that the instruction set encoding (ISE) has a considerable impact on wearout of the decoding stages, since it affects the input patterns (via the applied opcodes) and the gate-level implementation of these stages, which both influence wearout. To find a good instruction set encoding in terms of MTTF, our approach uses a hierarchical optimization algorithm based on simulated annealing to obtain an aging-aware opcode for each instruction in such a way, that the overall lifetime of the decoding stages is improved. Thereby, only the representing bit patterns are modified, while the opcode length remains unchanged. Together with existing aging mitigation techniques for other stages, this approach can help to significantly improve the overall microprocessor lifetime.

The results show that the proposed technique can improve MTTF[2] of the decoding stages of the FabScalar microprocessor [7] by 1.93x, while other pipeline stages MTTF are not (negatively) affected. In addition, performance is not impaired and the area costs are negligible.

In summary, the key contributions of this work are:

- We propose and evaluate an aging-aware ISE to increase MTTF of the decoding stages.
- We present a generic flow to obtain an aging-aware ISE

The rest of this paper is organized as follows. In Section II the considered transistor aging phenomena, BTI and HCI, are introduced followed by a discussion of related work. The novel ArISE approach is motivated in Section III and the methodology to obtain such an encoding is detailed in Section IV. Afterwards, we present in Section V our experimental results. Finally, Section VI concludes the paper.

II. PRELIMINARIES

BTI and HCI impair the threshold voltage of transistors, which results in longer path delays. This effect is explained in this section, followed by a discussion of related work.

A. Transistor Aging

A.1 Hot Carrier Injection (HCI)

HCI is a wearout mechanism of NMOS transistors. When accelerated channel-electrons collide with the gate oxide interface electron-hole pairs are created and free electrons get trapped inside the oxide layer. As a result, the threshold volt-

---

[1]The mapping of instructions (e.g. add) to their binary representation (e.g. 1101), i.e. opcode, is called Instruction Set Encoding.

[2]In this work MTTF is the time until first timing violation occurs

978-1-4799-2817-0/14 $31.00 © 2014 IEEE

age V_{th} increases irreversibly [22]. The magnitude of the V_{th}-shift has an exponential relation with temperature [5] and depends also on the number of transitions [19] (i.e. clock frequency, runtime and switching activity), as energetic electrons are only generated during transitions. To estimate the resulting V_{th}-shift, the model from [19] is used.

A.2 Bias Temperature Instability (BTI)

BTI consists of *positive* and *negative* BTI, affecting NMOS and PMOS transistors, respectively [17]. When the gate-source of a PMOS (NMOS) transistor is negatively (positively) biased, $|V_{th}|$ increases due to the creation of traps at the interface between gate oxide and channel as well as inside the gate oxide (*stress* phase). When the gate bias is removed, some of the previously generated traps are filled, which results in a decreasing $|V_{th}|$ (*recovery* phase). However, the initial shift cannot be entirely compensated leading to a gradual increase of $|V_{th}|$ over time. The wearout rate depends on several aspects, such as temperature and duty cycle, i.e the ratio of stress to total time. To estimate the overall V_{th}-shift, the model from [22] is used.

B. Related Work

B.1 Aging Mitigation

To alleviate transistor aging several design techniques have been already proposed of which we name just a few ones. Device and circuit-level techniques such as gate sizing [24], V_{th}-tuning [20], input vector control [23], path balancing [9] and stacking-based pin reordering [20] are orthogonal to our work and can be used together with our method.

At (micro)-architecture-level most techniques address transistor aging for the execution units of a microprocessor, as these are typically the lifetime-limiting factors [8]. In [15,18] aging-aware instruction scheduling techniques are evaluated and in [10] a special NOP-instruction is used to alleviate the impact of NBTI on an ALU. Also some techniques address wearout in memories. [14] uses cell-flipping to make duty cycles close to 0.5 and in [21] BTI-induced aging in a register-file is mitigated by flipping the leading bits of narrow-width values periodically.

In summary, although these (micro)-architecture-level techniques can increase MTTF of memory elements and execution units considerably, since other stages of the instruction pipeline do not benefit from these approaches, the lifetime of the entire microprocessor will be limited. Hence, aging mitigation techniques targeting other stages, particularly the decoding stages, are necessary. Besides this work, this issue is also targeted in [11]. The authors proposed to periodically invert the instruction opcode to make the signal duty cycles close to 0.5, which should mitigate transistor aging due to BTI. However, HCI is not addressed and the overall MTTF improvements are much lower than those of our proposed approach (see Section V).

B.2 Instruction Set Encoding (ISE)

Special ISEs are often applied to reduce the dynamic power consumption of instruction buffers and registers [1,6] by minimizing the input switching activity. Although we also propose ISE modifications, our approach is orthogonal to these techniques, since we target logic circuits instead of memory, and reliability (lifetime) instead of power. As a consequence,

power can be impacted. For example we observed a 14 % higher switching activity at the inputs of the instruction buffer (SRAM-based) for the aging-aware ISE (see Sec. V). This shows that, although power and wearout are coupled (via temperature), they require different optimization strategies. Hence, both approaches can be combined to achieve a good trade-off between power and MTTF.

III. MOTIVATION AND MAIN IDEA

As mentioned earlier pipeline stage delays increase considerably during runtime due to BTI and HCI. Thereby, the input stream of each stage has a significant influence on the aging rates, as it affects duty cycle and switching activity of internal signals. Hence, also the instruction opcodes as part of this stream contribute to the wearout of the affected stages. To underline this fact and motivate this work, we use the FabScalar[3] microprocessor as our case study [7].

To evaluate wearout of different stages for different ISEs, we extracted the delay at design time and after 3 years for each stage using the accurate gate-level aging-estimation flow detailed in Section IV.E in combination with the setup given in Section V and SPEC2000 benchmarks.

The results of this analysis are depicted in Fig. 1. Obviously ISE has a significant influence on the wearout of frontend stages, especially the decoding stages (Predecode and Decode). The effect on other stages is in contrast negligible. This is because the instruction opcode forms only a small part of the inputs of the other stages (or is no input at all). This difference in delay degradation rates for the Predecode stage translates to more than 2x difference in MTTF. Moreover, the results also show that the decoding stages can even become lifetime-limiting, if the ISE is not designed aging-aware, as shown in Fig. 1 (see Encoding 3). This underlines the need for using an aging-aware instruction encoding.

Improving the ISE is a challenging task, since many instruction encodings have to be modified to become aging-aware. For example, in case of the FabScalar microprocessor the instruction set architecture contains 131 instructions that are encoded using 8 bits. This means that there are $(2^8)!/(2^8 - 131)! \approx 10^{297}$ encoding possibilities. Since most encodings infer modifications in the gate-level implementation of the stages and affect signal duty cycles as well as switching activities, each encoding requires a new synthesis and simulation flow for aging analysis. Moreover, instructions cannot be considered in-

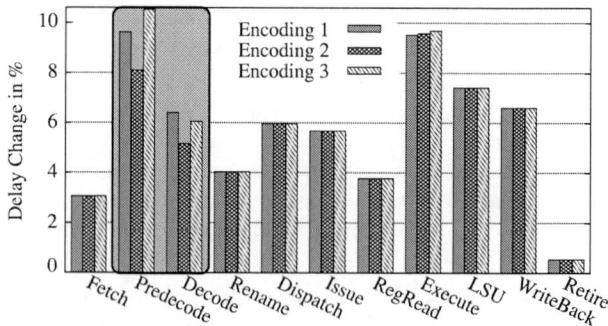

Fig. 1. Worst-case delay change after 3 years for different pipeline stages for 3 different ISEs (obtained with setup detailed in Section V)

---

[3]FabScalar is an out-of-order, 11-stage, superscalar processor

dependently, since duty cycles and switching activities depend not only on the actual instruction, but also on the preceding and subsequent instructions. Therefore, to see the effect of a particular encoding, or even optimizing the encoding for just one instruction, one has to simulate instruction streams. Due to this complexity an exhaustive method is infeasible. Hence, we use a hierarchical heuristic methodology based on simulated annealing to find an aging-aware ISE as described next.

IV. ArISE: Aging-aware Instruction Set Encoding

In this section the flow to create an aging-aware ISE considering the overall MTTF of the decoding stages is explained (IV.A) and a hierarchical approach to improve its efficiency is discussed (IV.B). Furthermore, a runtime analysis and further improvements are presented in part IV.C. The application of a modified ISE to a real system is described in IV.D and the aging estimation flow is detailed in Section IV.E.

A. Optimization Algorithm

The starting point of the optimization algorithm, which is shown in Fig. 2, is a random ISE (here: FabScalar's standard ISE). For this ISE all pipeline stages have to be synthesized and their MTTF values need to be extracted according to the flow presented in Section IV.E (Steps 1+2). Afterwards, a simulated annealing (SA) algorithm is invoked (Step 3). As a first step it generates a "neighbor" ISE for the current one (Step 3.1). For this new ISE the overall MTTF considering the decoding stages is estimated (Step 3.2). As neighbor definition we use in this work the following principal. Two ISEs are neighbors if and only if a) both ISEs differ only in one instruction opcode (only possible if there are more binary opcodes than instructions), or b) the second ISE can be derived from the first one by exchanging the opcodes of two instructions. Using this definition it is guaranteed that every possible ISE can be evaluated and that the difference between the new and the old ISE is not that huge that the optimization process is too random.

The next step is to evaluate if the new ISE will replace the old one (Step 3.3). Therefore we applied the exponential function detailed in Equation (1) as cost function:

$$P(\text{accept ISE}) = \exp\left(-(D_{new} - D_{old})/T\right) \overset{?}{>} P_{reject}. \quad (1)$$

The inputs for the cost function are the worst case delay (which can be used to represent MTTF) after 3 years for the new and old ISE. P_{reject} is the reject probability and T the annealing temperature, which can be iteratively reduced (Step 3.0) according to the SA principal. Since SA tries to find the global optimum, it does not only accept solutions with lower costs (i.e. better MTTF) but also intermediate solutions with higher costs to avoid local optima. In both cases a neighbor for the new ISE is created (Step 3.3.2) and the evaluation continues with this new neighbor. Otherwise, if the costs for the new ISE are too high, a new neighbor for the old ISE is generated (Step 3.3.1).

B. Hierarchical Optimization

To improve the efficiency of the optimization process, we try to minimize the number of steps until an acceptable solution is found. Therefore, we applied the *hierarchical optimization approach* shown in Fig. 3. First the instructions are

1. Select a starting instruction set encoding (ISE): ISE_old
2. Get overall MTTF and worst-case delay after X years
   for decode stages (see IV.E): MTTF_old & D_old
3. **While** solution is not good enough or number of steps < limit **do**
   3.0. Adjust temperature T
   3.1. Generate "neighbor" ISE of ISE_old: ISE_new
   3.2. Get overall MTTF and worst-case delay after X years
        for decode stages (see IV.E): MTTF_new & D_new
   3.3. **If** ISE_new is not acceptable according to Equation (1)
        3.3.1. **then** GoTo 3.1.
        3.3.2. **else** ISE_old = ISE_new
               ISE_best = ISE_old /* *memorize best ISE* */
               GoTo 3.1.
        EndIf
End

Fig. 2. Algorithm to generate an aging-aware ISE

classified into different (sub)groups based on their characteristics, e.g. all ALU-instructions are put into the same group (Step 1). Afterwards, at each hierarchy-level, the (sub)groups are ranked according to their aging impact (Step 2). However, as the aging rate strongly depends on the encoding, the real aging rates cannot be used for this ranking. As a matter of fact, instructions affecting the hardware-implementation of the decoder have the biggest aging impact. Therefore, the impact on hardware modifications due to a particular coding scheme is used to form the ranking. If there are several (sub)groups affecting the hardware-implementation, they are ranked depending on their occurrence frequency. The next step is to find the best encoding for each (sub)group, for all hierarchy-levels starting at the coarsest hierarchy-level (Step 3.1). Thereby, at each level, the group ranking determines the optimization order. As a result, an exhaustive technique can be applied if there are only a few (sub)groups at the same hierarchy level with considerable aging impact. Otherwise, the SA process of Section IV.A can be applied for the (sub)groups.

Using this approach, at each hierarchy-level, only the opcode bits corresponding to that level are modified. For example, if there are 16 groups at the coarsest level, only the four most significant opcode bits are modified (i.e. 16! configurations). As we used just 16 instruction groups, each of which contained at most 16 instructions the search space for the entire optimization process became much smaller than the original space ($16!^2 \approx 4 \cdot 10^{26}$ vs. 10^{297}), leading to fewer optimization steps.

We compared this approach with the optimization method described in the previous subsection. To limit the runtime for the latter, we enforced that one of the modified instructions in each iteration had to be one of the 10 most frequent instructions

1. Partition instructions into groups and subgroups
   /* *Instruction groups, subgroups inside groups,* */
   /* *instructions insides subgroups, etc.* */
2. Rank each group (and subgroups subsequently)
   2.1 Based on their hardware-impact
   2.2 **If** there are groups/subgroups with same ranking
       then use occurrence frequency to rank these
3. **For** the coarsest down to the finest hierarchy-level **do**
   For the highest ranked group down to the lowest **do**
      3.1 Find the best encoding for the elements within that group
          /* *Either exhaustive or with simulated annealing* */
      3.2 Stop as soon as MTTF is satisfactory
   Endfor
Endfor

Fig. 3. Hierarchical approach to obtain an aging-aware ISE

978-1-4799-2817-0/14 $31.00 © 2014 IEEE   209

to avoid modifying only the encoding of infrequent instructions, which have negligible effect on the overall aging rate. While this version usually finds a feasible ISE within the first 80 to 90 iterations, the hierarchical approach needs usually at most 30 iterations (i.e. at most 2 hours) to obtain an acceptable ISE, i.e 3x improvement in runtime.

Please note that this hierarchical approach can be divided into fewer or more optimization levels, depending on the size of the instruction set and maximum runtime. Fewer levels allows to investigate more ISEs and hence could find a better solution than an approach with more levels. However, more hierarchy depth results in a much shorter runtime.

Furthermore, please note that also other grouping schemes can be used for this purpose. In this work, we considered a two-level categorization and grouped the instructions based on their type, e.g. all branch, arithmetic, load, store and logic instructions had their own group, which is also the highest level of the hierarchy. The next hierarchy-level is the lowest level, i.e. all instructions inside a group are optimized independently. Other schemes, for example based on the occurrence frequency, are also possible and can affect the runtime as well as MTTF.

C. Runtime Analysis and Further Improvements

The runtime for a single simulated annealing step depends mainly on the time required for synthesizing the decoding stages, as the time for simulation and aging-estimation can be reduced to a negligible fraction (see Sec. IV.E). Doing so, a single step takes in our case less than 4 minutes, which means that 100 steps can be performed within 6 hours.

In addition, we avoid re-evaluating the same ISE in multiple simulated annealing steps, to reduce the runtime further. Also, the best solution found during simulated annealing steps is memorized, such that we can always apply the best ISE in terms of MTTF that was found and not just the last accepted one, which may not necessarily be the best one.

D. Applying Modified Instruction Encoding

An ISE modification affects the usability of existing software binaries since software compiled for the original ISE can

no longer be executed. To address this issue a software- or a hardware-based technique can be used.

The software-based solution re-compiles applications using a compiler based on the modified ISE either on-the-fly (run-time compilation) or when the application is started for the first time. In case of application-specific processors re-compilation is not necessary, since hardware and compiler are typically designed in close interaction and backwards compatibility is seldom required. In general, the advantage of this approach is that it infers no additional hardware costs and is easy to implement.

The hardware-based approach is intended for processors that have to be backward compatible to old software, or when re-compilation is not a feasible solution. Therefore, a mapper is used, which translates the standard ISE into the aging-aware ISE during runtime by using a lookup-table or logic-statements (if-else). This mapping can be done while the instructions are written to or read from the instruction cache. Our analysis shows that, the overhead of such a solution is negligible (less than 0.2 % area overhead for the FabScalar microprocessor). Besides area overhead this mapper can potentially impact performance, since it could increase the critical path length. However, in case of FabScalar using a mapper did not negatively affect the critical path, i.e there was no performance penalty.

E. Aging-Estimation Flow

The flow to calculate the aging rate (MTTF) of a pipeline stage is based on the one presented in [16]. In this flow, to accurately estimate the aging rate (MTTF) of a pipeline stage, the properties (duty cycle, switching activity) for all internal signals of this stage, its gate-level implementation and temperature behavior are analyzed. Hence, the first step is to generate a gate-level netlist of this stage. Afterwards gate-level simulations using real-world applications are performed to obtain the properties of all internal signals for the evaluated stage under realistic scenarios. This data is used to extract power and then temperature based on the floorplan and layout. Please note that since two neighboring stages affect the temperature behavior of each other, all stages have to be considered during the temper-

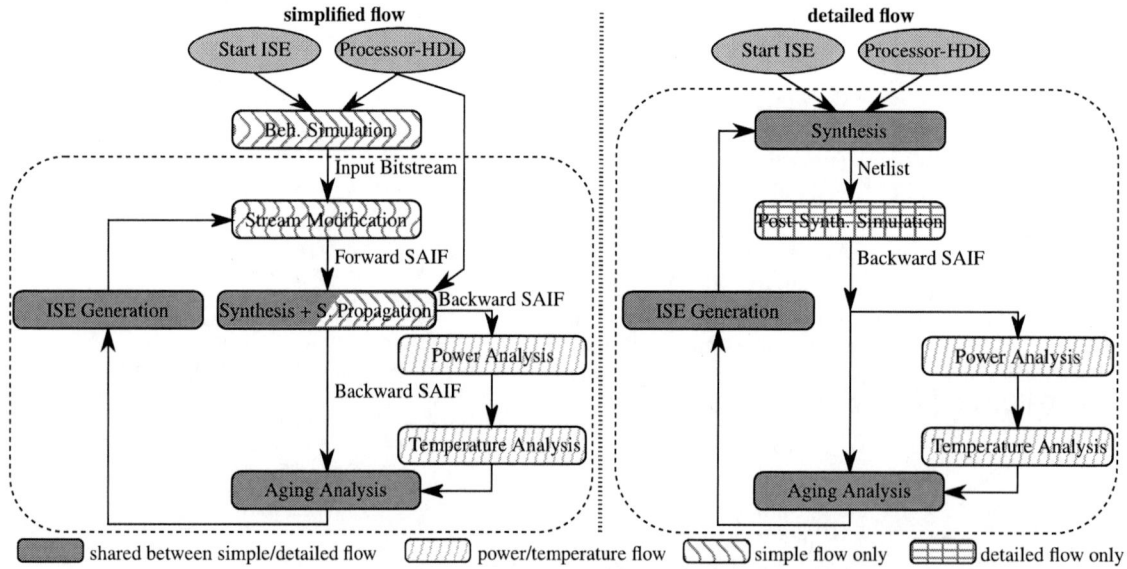

Fig. 4. ISE Optimization flows including aging analysis: Left: fast version without post-synthesis simulation, Right: slow and accurate version

Synthesis+Path Extraction	Synopsys Design Compiler D-2010.03
Simulation+VCD-Generation	Cadence NCSim 12.10 + SPEC2000
Power Extraction	Synopsys PrimeTime F-2011.06
Temperature Extraction	HotSpot 5.02 [12]
Aging Analysis	Inhouse C++-Tool

TABLE I
TOOLS USED FOR RESULT EXTRACTION

ature analysis. Then, the signal properties, temperature information and the netlist are used to estimate the V_{th}-shift of each transistor using accurate aging models based on [22] for BTI and [19] for HCI. Afterwards, the delay degradation of each gate is estimated based on an alpha-power delay model [4]. Finally, this information is given to the synthesis tool (in form of a Standard Delay Format, SDF, file) together with the netlist. The synthesis tool annotates the circuit with the degraded gate delays and afterwards the aged stage delay is extracted. As the synthesis tool considers all possible paths during the critical path extraction process, it is guaranteed that no path is excluded. Having the delay knowledge the stage's MTTF can be calculated. The tools used for these steps are given in Table I.

While the runtime of the aging estimation itself is negligible, the runtime of the post-synthesis simulations significantly affects the runtime of the optimization process. For this reason we propose to replace the post-synthesis simulations during the optimization phase as shown in Fig. 4. Instead prior to the optimization process a behavioral simulation is performed and the input-stream for the decoding stages is stored in a file. Then, during the optimization process, this input-stream is modified according to the ISE changes, i.e. old opcodes are replaced with modified ones. The resulting input signal properties are then given to the synthesis tool (in form of a Switching Activity Interchange Format, SAIF, file). The synthesis tool propagates these properties through the entire design and calculates the signal properties for all (internal) signals. By this means extracting the internal signal properties takes a negligible fraction of time compared to post-synthesis simulation (a few seconds vs. $\approx$ 30 minutes for 10^6 clock cycles). However, the aging estimation accuracy will be impacted, as signal correlations are not taken into account. Nevertheless, we observed that the inaccuracy in terms of delta delay is less than 0.5 % and hence accurate enough to be used in the optimization process.

Please note that for the final results again gate-level simulations are used to extract the signal properties. This way it is ensured that the presented results are accurate and that the optimization process was successful.

V. EXPERIMENTAL RESULTS

In this section the impact of an aging-aware ISE on the FabScalar microprocessor [7] is presented (Section V.A). The aging-aware ISE was generated using the hierarchical flow presented in Section IV. For the evaluation we used six SPEC2000 benchmarks (bzip, gap, gzip, mcf, parser, vortex) and simulated 10^6 cycles after a warmup. Furthermore, in Section V.B we compare our technique to the inversion method from [11].

A. Evaluation of Aging-aware Instruction Set Encoding

In case of FabScalar, the Predecode stage is more critical than the Decode stage, which means that first the instructions with considerable aging impact on the Predecode stage have

Stage	Standard Encoding			Best Encoding		
	Delay [ns] (0y)	Delay [ns] (3y)	MTTF [years]	Delay [ns] (0y)	Delay [ns] (3y)	MTTF [years]
Predecode	1.35	1.48	3.0	1.35	1.46	5.8
Decode	1.34	1.43	15.9	1.34	1.42	19.1
Overall	1.35	1.48	3.0	1.35	1.46	5.8 **+1.93x**

TABLE II
IMPROVEMENTS OF THE BEST ISE IN TERMS OF MTTF AND DELAY (BOTH WORST-CASE OVER THE USED SPEC2000 BENCHMARKS) FOR THE EVALUATED PREDECODE AND DECODE STAGE

to be optimized to extend the overall MTTF. As wearout of the Predecode stage is mainly affected by the branch instruction group, the best encoding for the corresponding instruction group was exhaustively determined using 16 iterations. Afterwards 100 simulated annealing iterations were performed to optimize the encoding for each branch instruction inside the group. Already after 25 iterations, i.e. after not even 100 minutes, the best ISE was obtained for which the delay degradation within the first 3 years dropped from 9.5 % to 8.1 % for the Predecode stage and from 6.4 % to 6.1 % for the Decode stage. Hence, the overall MTTF of the decoding stages can be improved by 1.93x from 3 years to 5.8 years as shown in Table II. This considerable improvement comes from the fact that the relation between runtime and delay degradation follows a root-like function. For example, in case of BTI the following relation can be used to estimate the delay degradation [2]:

$$\Delta d(t) \sim \delta^n \cdot t^n, \qquad (2)$$

where, d is the delay, δ the transistor's duty cycle, t the runtime and n is a technology constant equal to 0.25. Hence, a delta delay reduction from 9.5 %, to 8.1 % corresponds to a duty cycle, which is roughly 1.9x smaller than the original one. As a result, the lifetime improves by 1.9x. Since the behavior for HCI is very similar, however with $n = 0.5$, the real results shown in Table II slightly differ from this estimation.

Please note that the best ISE also improves power and area of the decoding stages. However, these are just positive side-effects and are negligible considering the entire processor. Furthermore, the switching activity at the inputs of the instruction buffer, that stores decoded instructions, increased in average by 14 %. Hence, depending on the memory technology also the power consumption of the memory elements can increase.

Stage	Standard Encoding			Best Encoding			
	Delay [ns]	Power [mW]	Area [μm^2]	Delay [ns]	Power [mW]	Area [μm^2]	Changes
Fetch	1.35	8.18	23262	1.35	8.18	23262	no
Predecode	1.35	11.2	35030	1.35	10.8	33459	yes
Decode	1.34	4.15	23616	1.34	3.60	23625	yes
Rename	1.33	0.91	4050	1.33	0.91	4050	no
Dispatch	1.33	0.12	1867	1.33	0.12	1867	no
Issue	1.35	9.59	30719	1.35	9.59	30719	no
RegRead	1.33	1.44	12061	1.33	1.44	12061	no
Execute	1.35	4.28	27529	1.35	4.06	27341	yes
LSU	1.35	34.1	107664	1.35	34.1	107664	yes
WriteBack	1.34	2.69	3183	1.34	2.69	3183	no
Retire	1.35	1.11	3201	1.35	1.11	3201	no
Overall	1.35	77.71	273133	1.35	76.54 -1.5 %	271383 -0.6 %	

TABLE III
COMPARISON OF THE FABSCALAR'S STANDARD ISE AND THE BEST OBTAINED ONE IN TERMS OF DESIGN-TIME DELAY, AVG. POWER (W/O MEMORY) AND AREA (W/O MEMORY) CONSIDERING ALL BENCHMARKS

978-1-4799-2817-0/14 $31.00 © 2014 IEEE

	Our Technique	Periodical Inversion [11]		
		never	always	every 10^3 cyc
Δ-Delay @ 3y	8.1 %	9.1 %	9.0 %	9.1 %
MTTF	5.8 years	4.0 years	4.1 years	4 years

TABLE IV

COMPARISON BETWEEN OUR PROPOSED TECHNIQUE AND PERIODICAL
ISE INVERSION [11] IN TERMS OF DELAY DEGRADATION AND MTTF
(NEVER = ISE IS NEVER INVERTED, ALWAYS = ISE IS ALWAYS INVERTED,
EVERY 10^3 CYC = INVERSION PERIOD =10^3 CYCLES)

Therefore, if power is another optimization objective our technique has to be combined with power reduction techniques.

In Table III the effect of ISE modification on the entire microprocessor is shown. As it can be seen, ISE changes also result in modifications of the Execution stage as well as Load-Store-Unit (LSU). However, in terms of delay, wearout or MTTF these changes are negligible. This is because the instruction opcode forms only a small part of the inputs.

B. Aging-aware ISE vs. Periodical Inversion

As mentioned in Section II.B.1, ISE can be periodically inverted to reduce BTI-induced wearout [11]. To compare this approach with our aging-aware ISE, we implemented that technique in the Predecode stage of FabScalar.

However, the improvements in terms of MTTF are much smaller than that of our proposed technique as summarized in Table IV. ISE inversion every 1000 cycles increases MTTF by just 1 year. Moreover, if ISE is inverted permanently, the improvements become even slightly better. Hence, our proposed technique significantly outperforms the periodic inversion technique of [11]. This is due to two reasons. First, HCI-induced wearout is not considered in [11]. Second, the periodical inversion is intended to balance wearout (duty cycles ≈ 0.5). However, this does not necessarily yield the best MTTF. The reason is that it is often better to reduce wearout of the most critical paths as much as possible (duty cycles $<< 0.5$) at the expense of faster wearout of non-critical paths (duty cycles $>> 0.5$). Overall this way a much better MTTF can be achieved.

Please note that the delay degradation with periodical inversion is not directly comparable with the degradation of the standard design, as some additional circuitry is necessary to re-invert the opcode everywhere it is used. Hence, also the overall circuit wearout is slightly different.

VI. CONCLUSION

Nanoscale microprocessors are exposed to accelerated transistor aging due to Bias Temperature Instability and Hot Carrier Injection. While various techniques exist to mitigate transistor aging in the execution units, only very few approaches target the frontend of the instruction pipeline. However, as we have shown, the decoding stages can become aging-critical and limit the microprocessor lifetime. To alleviate this problem we proposed a novel aging-aware instruction set encoding methodology (ArISE), which increases MTTF of the decoding stages. The ArISE technique exploits the fact that the instruction set encoding significantly affects the gate-level implementation as well as the signal behavior (switching activity, duty cycle) which all affect aging. To find an aging-aware instruction set encoding we presented a hierarchical heuristic algorithm that improves the binary instruction opcode with respect to MTTF. Our experimental results show that ArISE can improve MTTF of the decoding stages of the FabScalar microprocessor by 1.93x with negligible implementation costs.

VII. ACKNOWLEDGEMENT

This work was partly supported by the German Research Foundation (DFG) as part of the national focal program "Dependable Embedded Systems" (SPP-1500, `http://spp1500.ira.uka.de`).

REFERENCES

[1] L. Benini *et al.*, "Reducing Power Consumption of Dedicated Processors Through Instruction Set Encoding," in *GLSVLSI*, Feb. 1998, pp. 8–12.

[2] S. Bhardwaj *et al.*, "Predictive modeling of the nbti effect for reliable design," in *CICC*, Sep. 2006, pp. 189–192.

[3] S. Borkar, "Designing Reliable Systems from Unreliable Components: The Challenges of Transistor Variability and Degradation," *IEEE Micro*, pp. 10–16, Nov. 2005.

[4] K. A. Bowman *et al.*, "A physical alpha-power law MOSFET model," in *ISLPED*, 1999, pp. 218–222.

[5] A. Bravaix *et al.*, "Hot-Carrier Acceleration Factors for Low Power Management in DC-AC stressed 40nm NMOS node at High Temperature," in *IRPS*, Apr. 2009, pp. 531–548.

[6] A. Chattopadhyay *et al.*, "Power-efficient Instruction Encoding Optimization for Embedded Processors," in *VLSID*, Jan. 2007, pp. 595–600.

[7] N. Choudhary *et al.*, "FabScalar: Automating Superscalar Core Design," *IEEE Micro*, pp. 48–59, May 2012.

[8] M. DeBole *et al.*, "New-Age: A Negative Bias Temperature Instability-Estimation Framework for Microarchitectural Components," *Int'l J. of Parallel Prog.*, pp. 417–431, Aug. 2009.

[9] M. Ebrahimi *et al.*, "Aging-aware Logic Synthesis," in *Proc. of the Int'l Conf. on Computer-Aided Design*, Nov. 2013.

[10] F. Firouzi *et al.*, "NBTI Mitigation by NOP Assignment and Insertion," in *DATE*, Mar. 2012, pp. 218–223.

[11] E. Gunadi *et al.*, "Combating Aging with the Colt Duty Cycle Equalizer," in *MICRO*, Dec. 2010, pp. 103–114.

[12] W. Huang *et al.*, "HotSpot: A Compact Thermal Modeling Methodology for Early-Stage VLSI Design," *IEEE Trans. on VLSI Systems*, pp. 501–513, May 2006.

[13] K. Kang *et al.*, "Estimation of Statistical Variation in Temporal NBTI Degradation and its Impact on Lifetime Circuit Performance," in *ICCAD*, Nov. 2007, pp. 730–734.

[14] Y. Kunitake *et al.*, "Signal Probability Control for Relieving NBTI in SRAM Cells," in *ISQED*, Mar. 2010, pp. 660–666.

[15] F. Oboril *et al.*, "Reducing NBTI-induced Processor Wearout by Exploiting the Timing Slack of Instructions," in *CODES+ISSS*, Oct. 2012, pp. 443–452.

[16] F. Oboril and M. B. Tahoori, "MTTF-Balanced Pipeline Design," in *DATE*, Mar. 2013, pp. 270–275.

[17] S. Pae *et al.*, "BTI Reliability of 45 nm High-K + Metal-Gate Process Technology," in *IRPS*, May 2008, pp. 352–357.

[18] T. Siddiqua and S. Gurumurthi, "A Multi-Level Approach to Reduce the Impact of NBTI on Processor Functional Units," in *GLSVLSI*, May 2010, pp. 67–72.

[19] E. Takeda *et al.*, "New hot-carrier injection and device degradation in submicron MOSFETs," *IEEE Proc. I, Solid-State and Electron Devices*, pp. 144–150, Jun. 1983.

[20] R. Vattikonda *et al.*, "Modeling and Minimization of PMOS NBTI effect for Robust Nanometer Design," in *DAC*, Jun. 2006, pp. 1047–1052.

[21] S. Wang *et al.*, "Low Power Aging-Aware Register File Design by Duty Cycle Balancing," in *DATE*, Mar. 2012, pp. 546–549.

[22] W. Wang *et al.*, "Compact Modeling and Simulation of Circuit Reliability for 65-nm CMOS Technology," *IEEE Trans. on Device and Materials Reliability*, pp. 509–517, Dec. 2007.

[23] Y. Wang *et al.*, "On the efficiancy of Input Vector Control to mitigate NBTI effects and leakage power," in *ISQED*, Mar. 2009, pp. 19–26.

[24] X. Yang and K. Saluja, "Combating NBTI Degradation via Gate Sizing," in *ISQED*, 2007, pp. 47–52.

DRuiD: Designing Reconfigurable Architectures with Decision-making Support

Giovanni Mariani[*,†], Gianluca Palermo[†], Roel Meeuws[‡], Vlad-Mihai Sima[‡], Cristina Silvano[†], Koen Bertels[‡]

[*] Università della Svizzera Italiana - ALaRI. Email: marianig@alari.ch
[†] Politecnico di Milano. Email: {gpalermo, silvano}@elet.polimi.it
[‡] Delft University of Technology. Email: {v.m.sima, k.l.m.bertels}@tudelft.nl

Abstract—**Application development for heterogeneous platforms requires to code and map functionalities on a set of different computing elements. As a consequence, the development process needs a clear understanding of both, application requirements and heterogeneous computing technologies. To support the development process, we propose a framework called *DRuiD* capable of learning application characteristics that make them suitable for certain computing elements. The framework is composed of an expert system that supports the designer in the mapping decision and gives hints on possible code modifications to be applied to make the functionality more suitable for a computing element. The experimental results are tailored for a heterogeneous and reconfigurable platform (the Xilinx-ml510) including two computational elements, i.e. a Virtex5 FPGA and a PowerPC. The expert system identifies 88.9% of the times what are the functionalities that are accelerated efficiently by using the FPGA, without requiring the kernel porting. Additionally, we present two case studies demonstrating the potentialities of the framework to give hints on high level code modifications for an efficient kernel mapping on the FPGA.**

I. INTRODUCTION

Heterogeneous computing architectures are becoming more and more widespread solutions in both general-purpose and application-specific domains. Generally, heterogeneous architectures are coupling programmer-friendly general-purpose processing elements with specialized high-performance computing elements (such as FPGAs, GPUs and DSPs) increasing the degree of freedom for application development. To effectively utilize those architectures, it is important that the developers understand how to exploit the heterogeneous processing elements. Currently, existing tools are lacking of supporting the high level decision making process related to the design of heterogeneous systems [1]. Today solutions mainly rely on engineer experience, however this problem becomes even more evident once automatic tools are used as intermediate translator to port the code on specialized computing elements, such as High Level Synthesis (HLS) tools [2], [3] or parallelizing compilers for GPUs [4]. Given an application description, it is very hard to say whether or what application functionalities can be accelerated by using heterogeneous processing elements (e.g. FPGAs or GPUs). Moreover, exploring different application implementations to pick up the best one increases significantly the development costs.

To provide support during the development of heterogeneous systems, we propose to integrate the knowledge gathered during the development phase by different expert designers into an *expert system*. More specifically, we propose a framework called *DRuiD* based on machine learning approach combining *random forests* [5] and *genetic algorithms*. The underlying idea is to let the *DRuiD* framework to extract and learn what characteristics make application functionalities (or kernels) more suitable for a certain computing technology.

In this work, we focus on FPGA-based heterogeneous architectures. The methodology leverages a high-level synthesis tool to translate C code into a synthesizable hardware description language (such as VHDL). The availability of this tool allows us to describe both HW and SW by using a unique language, i.e. the ANSI C. Starting from the C implementation of the whole application, *DRuiD* selects the best computational element to be used for each application kernel and, if not accelerated by the computational elements, gives hints on code transformations to be applied for an efficient porting of the kernel. To demonstrate the effectiveness of the approach, we used a Xilinx-ml510 heterogeneous platform including the PowerPC and a Virtex-5 FPGA as computational elements.

The remainder of this paper is organized as follows. Section II reports the background work in the specific field. Section III presents the proposed *DRuiD* framework, the training methodology and the support provided at design-time. Section IV provides insight about the prediction accuracy and presents two case studies to encompass the potentialities of the proposed approach, while Section V concludes the paper.

II. BACKGROUND

So far, several *data mining* and *machine learning* techniques have been proposed to model performance of computer architectures at different abstraction level [6], [7], [8]. These works mainly differ in terms of the proposed performance model, target computing technology and/or performance metrics under study. While these models are precise enough to drive the design space exploration process of homogeneous computing architectures, they are not suitable for analyzing performance for heterogeneous systems.

Today solutions for the design of heterogeneous architectures either assume that costs related to mapping application kernels over different computational elements are available [9], [10] or they rely on virtual platform simulators (or models) where designer can plug-in different computational elements and evaluate the proposed solution [11], [12]. In both approaches, first the kernels of interest should be ported for execution on the target computational elements and then evaluated either by real execution or via simulation, making the application mapping a relatively simple problem.

The problem is that, porting different kernels on different components is a time consuming and error prone procedure.

978-1-4799-2817-0/14 $31.00 © 2014 IEEE

When considering FPGA-based technology, application kernels can be ported by using high-level synthesis tools such as *dwarv* or *Vivado* [2], [3]. These tools generally accept a subset of the input programming language. To be compliant with tool specific constraints one might need to rewrite significant parts of the kernels' code. Ideally, one would like to know the kernels to be mapped on the different computing elements before actually porting them, eventually employing a cross-component prediction method. Literature on this topic is very limited [13], [14], [15], [16], [17], [18]. The idea in these works is to predict performance for a target component given performance indices collected for a source component. The source component is either a host machine [13], [14], an intermediate code representation [15], [16], or simply a different configuration of the target architecture [17], [18]. In all cases, the computational element destination of the prediction belongs to a specific architecture family.

Performance prediction models suitable for truly heterogeneous architectures are today not available in literature. When dealing with the problem at hand, analytic performance prediction techniques are too much inaccurate. Thus, we propose a methodology that simply suggests the optimal kernel mapping while avoiding performance prediction.

III. The Proposed *DRuiD* Methodology

In this section we present the proposed *DRuiD* methodology. The implemented framework is based on an *expert system* that, given a parametric characterization of a specific kernel, classifies the kernel in terms of the processing element providing the best performance. *Decision trees* have been used in the past in the same area [14] as a flexible data-mining tool used to classify application performance depending on the code and machine characteristics. Additionally, it is rather easy to inspect such models to understand why a certain kernel is associated to a given class. Taking *decision trees* as baseline solution, we propose a methodology based on one extension known as *random forests* [5]. The random forest can be seen as a collection of trees where the classification is done by using a majority voter among the trees. This choice is driven by the evidence of a better classification accuracy (as it will be demonstrated in the experimental results section).

The basic assumption for the application of our methodology is that the translation process between the original software code and the code used for the computing elements is done by an automatic tool. This requirement is needed since the correlation between the original code and the generated code is higher with respect to a manual translation, making the results more predictable. In our work, we focused on the design of reconfigurable architectures (FPGA-based) and we employed as automatic translator for the kernels an high level synthesis tool starting from ANSI C code.

The proposed methodology requires to characterize the original version of the kernel by means of a vector v_k of SW parameters (or metrics). These metrics are generated by using two different technologies: (i) a *Static Code Analysis* tool (SCA) [19] to extract information related to the source code itself (such as average/maximum loop nesting, control flow complexity), and (ii) a Microarchitectural Independent Characterization (MICA) [13] to extract dynamic profiling

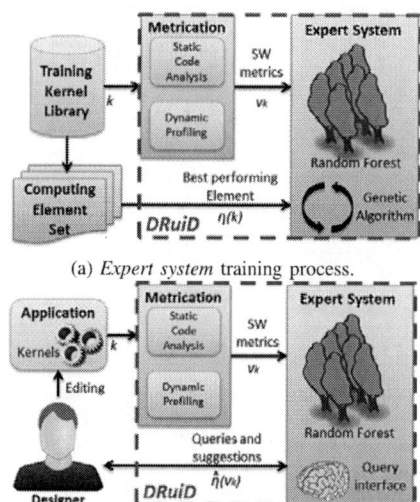

(a) *Expert system* training process.

(b) Application development using the *expert system*.

Fig. 1: The *DRuiD* framework.

information on a host machine (such as memory usage, branch predictability, instruction types). The *expert system* takes as input the SW metrics characterizing the kernel and returns as output the decision on the computing element to employ, that for reconfigurable architecture is the accelerator implementation on the FPGA or the SW execution on a core.

Before being used, the *expert system* needs to be trained (Figure 1a). During the training process, kernels belonging to the *training set* K are executed and profiled on the target *computing element set* C, and each kernel $k \in K$ is classified with a class $\eta(k) \in C$ that represents its best computational element. Formally:

$$\eta(k) = \operatorname*{argmin}_{c \in C} \delta(k, c) \qquad (1)$$

where $\delta(k, c)$ is the profiled execution time of kernel k on the computational element c.

Once the expert system is trained, it can be used to decide the best computational element $\hat{\eta}(v_k) \in C$ to be used for kernels coming from new, previously unobserved, applications (Figure 1b). Additionally, the *expert system* is provided with a query interface that generates a report about each of its decision. A report includes the relevance that each SW metric had during the selection of $\hat{\eta}(v_k)$, and suggests possible kernel code modifications to make it more suitable for the desired computational element.

Hereafter, we describe more in detail for the proposed *DRuiD* methodology the (i) kernel characterization, (ii) the training methodology and the (iii) expert system usage.

(i) Kernel Characterization. The proposed methodology requires that each kernel k is characterized with a vector v_k composed of 92 SCA metrics and 97 MICA metrics. For conciseness, we do not list here all 189 SW metrics, the reader can refer to [19], [13] for a more detailed analysis. Table I lists the groups of SW metrics and highlights the number of metrics in each group.

SCA metrics [19] report static code characteristics as follows. *Instruction types* present in the code, (e.g. float or integer

TABLE I: Groups of software metrics.

Source	Group name and acronym	Number of metrics
SCA	Instruction type (IT)	45
	Control flow (CF)	18
	Data flow (DF)	15
	Variables scope and location (VSL)	14
MICA	Instruction type (IT)	12
	Instruction level parallelism (ILP)	4
	Memory and register access pattern (MRAP)	66
	Branch predictability (BP)	15
Total		189

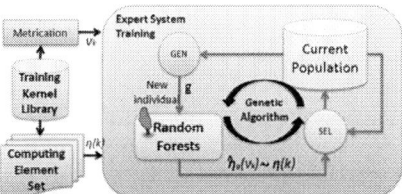

Fig. 2: Training of the expert system.

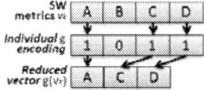

Fig. 3: Reducing the SW metric vector v_k using the individual g of the genetic algorithm.

addition/multiplication, bitwise and logical operations, etc.); *control flow* information (e.g. number and length of basic blocks, cyclomatic complexity, etc.); *data flow* information (e.g. number of operands and operators, the Elshof complexity metrics, etc.); *variable scope and location* (e.g. use of global/local variables, use of stack or heap memory, etc.).

MICA metrics [13] include information related to dynamic execution (on a host machine) as follows. *Instruction types* of executed code (e.g. arithmetic, control flow, floating point operations etc.); *instruction level parallelism* information about potential code parallelism on an ideal machine with infinite computing resources; *memory and register access pattern* information refers to read and write requests, access stride information, memory and register reuse distance; *branch predictability* gives insight on the efficiency of different branch prediction schemes.

SCA and MICA metrics represent complementary information (we will provide empirical evidence to validate this statement in the experimental results section). However, analyzing 189 SW metrics generates some complexity to be managed during the construction and comprehension of the model. Moreover, many of the 189 metrics are correlated to each other and some do not carry useful data for the tackled problem. The decision on what metrics to include in the model is not trivial and for this reason we included all the metrics in our analysis leaving to the training phase to select which ones to use.

(ii) Training the expert system. The training of the *expert system* is done through a combination of *genetic algorithms* (GA) and *random forests* (RF) (Figure 2). The GA is responsible for selecting the most suitable SW metrics to build an accurate decision-making system based on RF.

The GA processes a population of individuals using the traditional crossover and mutation operators (GEN block) and tournament based fitness selection (SEL block). Each individual g represents a function that reduces the vector v_k into a shorter vector $g(v_k)$. A bit vector of the same length of v_k is used to encode GA individuals g. A '1'(/'0') in the encoding of g means that the corresponding SW metric is reported(/not-reported) in $g(v_k)$. Figure 3 shows an example for 4 SW metrics. For each individual g, a *random forest* model $\hat{\eta}_g$ is trained to fit the kernel classes $\eta(k)$ for the kernels in the training library $k \in K$: $\hat{\eta}_g(v_k) \sim \eta(k)$. Each random forest model $\hat{\eta}_g$ differs in terms of the SW metrics it includes, i.e. $g(v_k)$. The GA fitness function is defined in terms of prediction accuracy of the trained RF and, in case of the same value, the configuration using the lower number of SW metrics is preferred. The GA population G is initialized by a random

sampling technique.

While the GA is in charge of the metric selection, the RF is the entity that is in charge of the mapping decision. The *random forest* (RF) is trained by using R [20]. In particular, each tree $\tau \in RF$ is trained separately by using a kernel set K_τ such that $K_\tau \subset K$ by using an *out of the bag* approach, i.e. removing randomly some of the kernels. The training of each tree τ is an iterative process. At each iteration, some variables in the vector $g(v_k)$ are randomly selected as next decision candidates. Among these ones, the one being the best discriminant (in terms of the Gini index [5]) for the identification of the class $\eta(k)$ is introduced in the tree. The introduction of randomness in the tree construction and the presence of multiple trees let RF be a more accurate and general model than a single *decision tree*. The output of the entire training process is the RF providing the best prediction accuracy and whose classification model is referred as $\hat{\eta}(v_k)$.

(iii) Querying the expert system. Once the RF is trained, it can be used for predicting the best computing element where to map a new kernel k. Each tree $\tau \in RF$ contributes with a vote $\tau(v_k) \in C$ for the computing element allocation $\hat{\eta}(v_k)$. A majority voter scheme is considered:

$$\hat{\eta}(v_k) = \arg\max_{c \in C} \left| \{\tau(v_k) \in RF : \tau(v_k) = c\} \right| \quad (2)$$

Additionally, the application developer willing to port a kernel k to a target component $c \in C$ can query the *expert system* to generate a report about the SW metrics that influence the decision of mapping k on c, with the goal of receiving hints on possible code modifications to apply. In fact, the presence of multiple trees $\tau \in RF$ lets the expert system be fuzzy, i.e. it is generally the case that some trees have a vote $\tau(v_k)$ that differs from the majority $\hat{\eta}(v_k)$. Let us call $\rho_{RF}(c, v_k)$ the number of trees $\tau \in RF$ such that $\tau(v_k) = c$. Let us define the SW metric vectors v_k^{+m} and v_k^{-m} such as the SW metric vector v_k where the SW metric m has been set respectively to its maximum and minimum values. We define the influence ι^{+m} and ι^{-m} as the variation in the votes with respect to a positive or negative variation of the metric m:

$$\rho_{RF}(c, v_k) = \left| \{\tau(v_k) \in RF : \tau(v_k) = c\} \right| \quad (3)$$

$$\iota^{+m}(c, v_k) = \rho_{RF}(c, v_k^{+m}) - \rho_{RF}(c, v_k) \quad (4)$$

$$\iota^{-m}(c, v_k) = \rho_{RF}(c, v_k^{-m}) - \rho_{RF}(c, v_k) \quad (5)$$

Given the target computational element c and the SW metric vector v_k, the expert system generates the following metrics: (i) Number of trees voting for the suggested component $\hat{\eta}(v_k)$ (the target component c might differ from $\hat{\eta}(v_k)$); (ii) Number of trees voting for the target component c; (iii) Influences $\iota^{+m}(c, v_k)$ and $\iota^{-m}(c, v_k)$ for each SW metric m considered in the RF model.

Usually, porting a kernel to a given component (e.g. GPU or FPGA) requires to edit the kernel code to be compliant with a subset of the C standard and/or to insert pragmas to specify where computations and variables shall be mapped to. This process might be long and tedious, depending on the target computational element and the tool-chain in use. The usage of *DRuiD* does not require to actually port the kernel on the target processing element, thus saving significant development time.

IV. Experimental Results

To validate and assess the proposed methodology, we applied it to a *computing element set* composed of the architectural elements available on an ml510 machine [21]: a Virtex5 FPGA and a PowerPC (PPC). The experimental campaign has been carried out by using a collection of 97 kernels/applications, some of them taken from *MediaBench* [22] and *MiBench* [23] originally written in C and coming from different application domains such as multimedia, cryptography, digital signal processing, automotive, mathematics, physics. The C to VHDL translation of the set of benchmarks is done by using *dwarv* [2] while the static code analysis (SCA) is implemented by using CoSy [24]. The dynamic profiling (MICA-based) and the subsequent evaluation of the kernels/applications set on the *computing element set*, (i.e. PPC and FPGA) has been done by using at least 5 different datasets from real scenarios. The SW metric values v_k and the execution time $\delta(k, c)$ used in this work refer to the median values over the different datasets. Finally, while the PPC operating frequency is 400 MHz and the maximum frequency supported for the FPGA is 325 MHz, the actual operating frequency of the kernels, once mapped on the FPGA, can be lower than 325 MHz, depending on the synthesis process.

Three sets of experimental results will be shown hereafter: (i) the prediction capability of the proposed approach; (ii) the analysis on SW metrics used to build the predictive model; (iii) two case studies where the predictive model has been used to give hints about how to modify the original C code to have benefits of using the FPGA instead of the PPC.

(i) Model Accuracy. To assess the prediction capability of the proposed *random forest* (RF) technique, we compared it with respect to a *decision-tree*-based methodology (*Tree*) that implements the state of the art technique proposed in [14] for heterogeneous systems. Additionally, to demonstrate that SCA and MICA metrics provide complementary information, the proposed *expert system* has been evaluated in three versions: *RF-SCA*, *RF-MICA* and *RF-all* where the *random forest* has been trained respectively by using SCA metrics only, MICA metrics only or, as in *DRuiD*, both set of metrics. Finally, to complete the validation we add a *random selection process* (*Random*), where the best computational element $\hat{\eta}(v_k)$ is randomly selected among the computing elements.

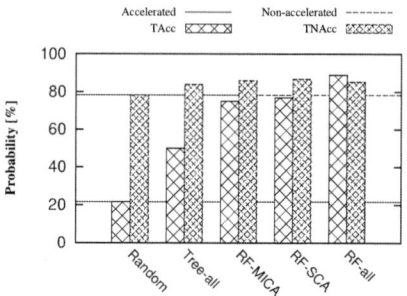

Fig. 4: Model accuracy comparison in terms of *truly accelerated* (*TAcc*) and *truly non-accelerated* (*TNAcc*) probabilities.

To evaluate the model accuracy in terms of prediction capability, we use a 30-fold cross-validation method. This method consists of first partitioning the set K of 97 kernels in 30 subsets and then, for each one of the 30 subsets, removing that subset, training the models with the remaining subsets, and perform the evaluation of the predictions $\hat{\eta}(v_k) \sim \eta(k)$ for the kernels k within the removed subset (not involved in the training phase).

In this work, we are interested in verifying if the model is capable to predict when the FPGA efficiently accelerate a kernel k with respect to its software execution. Let us define as *accelerated set Acc* the set of kernels that are successfully accelerated by the FPGA, $Acc = \{k \in K : \eta(k) = FPGA\}$, and *non-accelerated set* the set of kernels which FPGA implementation leads to a slowdown, $NAcc = \{k \in K : \eta(k) = PPC\}$. Similarly, we define the *accelerated* and *non-accelerated* kernel sets referring to model predictions; respectively $\widehat{Acc} = \{k \in K : \hat{\eta}(v_k) = FPGA\}$ and $\widehat{NAcc} = \{k \in K : \hat{\eta}(v_k) = PPC\}$.

Let us then define the *truly accelerated probability TAcc* as the probability that a kernel k classified by the model as accelerated is actually *accelerated*, $TAcc = |\{k \in \widehat{Acc} : k \in Acc\}|/|\widehat{Acc}|$, and the *truly non-accelerated probability TNAcc* is the probability for a kernel classified as non-accelerated to actually be non-accelerated: $TNAcc = |\{k \in \widehat{NAcc} : k \in NAcc\}|/|\widehat{NAcc}|$. Those probabilities represent the capability of the models to discriminate between accelerated and non-accelerated kernels and can be used as accuracy metrics for our goal. In particular, higher are the values better is the prediction accuracy.

Figure 4 shows the comparison results in terms of accuracy of all the methods involved in the validation of the proposed approach. Due to its nature the *Random* method has the $TAcc(/TNAcc)$ probability equal to the ratio of kernels that are accelerated(/non-accelerated) in the entire set (horizontal lines in the Figure 4). Regarding the other methods, both probabilities are always higher than those for the *Random* method, but while for the $TAcc$ it is evident, only small improvements occur for the $TNAcc$. The phenomenon on the $TNAcc$ is due to the large fraction of kernels that do not present advantages in being mapped on FPGA (79%) that makes less evident the learning capability of both *Tree* and *RF* methods. On the other hand, the $TAcc$ of the RF-based methods (regardless the type of used SW metrics) is higher

978-1-4799-2817-0/14 $31.00 © 2014 IEEE   216

TABLE II: Most relevant SW metrics used by *RF-all*.

Name	Source	Group	SW metric description
Stride32K	MICA	MRAP	% of memory accesses with stride lower than 32K
Operands	SCA	DF	Number of operands (variables or constants)
ArithEx	MICA	IT	% of arithmetic operations executed
Heap	SCA	VSL	% of code statements accessing heap memory
RegAge	MICA	MRAP	Probability that register age is lower than 16
FloatPre	SCA	IT	Number of floating point operators weighted by their precision
FloatExe	MICA	IT	% of floating point operations executed
MemReuse	MICA	MRAP	Memory reuse distance 32-64 [13]
BranchPre	MICA	BP	% of branch misprediction (using a Partial Match with global history table of length 12)
FloatALUs	SCA	IT	Required floating point ALUs

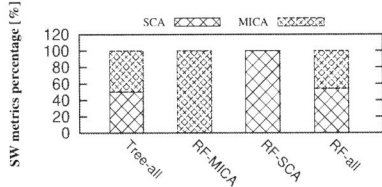

Fig. 5: Percentage of SCA and MICA metrics used.

(around 80% on average) with respect to the *Tree* method (50%). The proposed method (*RF-all*) outperforms all the others showing a $TAcc$ value of 89%.

(ii) SW metrics analysis. In this section, we analyze more in detail the usage of the SW metrics within the proposed methodology. Figure 5 shows, for each one of the methodologies considered in the validation, the percentage of SCA and MICA metrics used to build the models. The *Random* methodology has not been shown since it does not use any SW metric. It is possible to note that while the *RF-SCA* and *RF-MICA* use respectively only SCA and MICA metrics, both *RF-all* and *Tree-all* models have a good balance between the two sets of metrics, coming from both static code analysis (SCA) and dynamic profiling (MICA).

Analyzing more in detail the proposed *expert system (RF-all)*, it includes a total of 24 metrics, 11 out of them coming from SCA and 13 from MICA. Additionally, considering the metrics' relevance measured in terms of the mean Gini index [5] reduction obtained by introducing the metric within the trees of the *random forest* [20], Table II shows the 10 most relevant metrics out of the 24 used for the model. Table II reports the name of each metric together with a brief description, the source set (MICA or SCA) and the metric group type (see Table I). Almost all the groups of Table I are represented (except for SCA-CF) with predominance of the instruction type (MICA/SCA-IT) related metrics.

(iii) Querying the expert system. The proposed *expert system* can even be used to get hints on how to modify the code to move kernels from the *non-accelerated* set to the *accelerated*-one. In fact, the final user can query the *expert system* to generate a report about the SW metrics that influence the decision of mapping a kernel k on the component c. The report consists of the number of trees $\rho_{RF}(c, v_k)$ in the *random forest* that voted in favor of the computational element c and the possible variations in these votes that can be obtained by manipulating each metric m (i.e. ι^{+m} and ι^{-m}

from Equations 4 and 5). The generated reports are sorted by means of the maximum between ι^{+m} and ι^{-m}, giving the possibility to the user to better understand how to modify the code to accelerate it. For example, a pairs of metric-suggestions/modifications among the metrics of Table II are:

- *Operands* - Using different variables over different code statements reduces operation scheduling constraints improving parallelism opportunities. Additionally, some loop optimizations can implicitly increase the number of variables in use (e.g. unrolling).

- *FloatPre* - Despite of floating point operations are faster on FPGA since the PPC does not have a FP-unit, large floating point operators impact negatively on the FPGA operating frequency. Using *float* type rather than *double* type when possible mitigates this issue.

Hereafter two case studies are shown to clarify this concept.

- Susan edge case study. The *Susan edge* application is an image processing algorithm for edge detection taken from the *MiBench* benchmark suite [23]. This application was excluded from the training kernel set. We focus on porting on the FPGA a kernel named *susan_principle* (Listing 6). The kernel takes three arrays that are: *a)* the input image in gray scale (*in*), *b)* the gray scale output image where to store edge shapes (*r*) and *c)* an array of 516 gray scale levels to use (*bp*). The *expert system* classifies the kernel as a *non-accelerated* kernels meaning that the preferred computational element to use is the PPC. However, by querying the *expert system* about more details on the decision, it comes out that the first concern is related to memory usage, suggesting to reduce the memory access stride and the use of heap memory (the arrays passed as parameters). In the inner loop of Listing 6, a pixel is compared to its eight surrounding pixels using 9 memory reads (the array *in* is read through the pointer *p*). By caching the data used by the inner loop into small local arrays (Listing 7) implemented in HW as registers, we reduce the access to the shared memory and improve data locality. When mapped on the FPGA, while the original version of the kernel has a speedup of 0.38 (i.e. a slow-down) w.r.t. the PPC, the second version has a speedup of 3.96. The proposed optimization is technology dependent since the impact of the same kernel modification on the PPC brings to a performance slow-down.

- Bitreversal case study. As second case study, we synthesized a simple kernel that does not access memory arrays. We write a simple kernel (Listing 8) that takes a 32 bits unsigned integer and reverse them. Also in this case the expert system classifies the kernel as a *non-accelerated* kernel. Querying on the *expert system*, it mainly outlines two metrics: the first one was related to the number of floating point operations (*FloatExe*), while the second one is related to the total number of *Operands*, that if (in both cases) are increased could bring to a better implementation. We modified the code to improve the second metric since the improvement of the first metric was impossible being the kernel without any floating point operation. Summarizing, the original HW and the optimized HW versions of the kernel present w.r.t. the original PPC implementation a speed-up of 0.7 and 4.4 respectively.

V. CONCLUSION

In this work we propose the *DRuiD* framework for automatically identifying the most suitable computing element for

```
susan_principle(unsigned char *in,int *r,char *bp,int max_no,int x_size,int y_size){
  int   i, j, n;
  unsigned char *p,*cp;
  for (i=1;i<y_size-1;i++){
    for (j=1;j<x_size-1;j++){
      n=100;
      p=in + (i-1)*x_size + j - 1;// p <- up left
      cp=bp + in[i*x_size+j]; // Memory read center

      n+=*(cp--*p++);// Memory read up left
      n+=*(cp--*p++);// Memory read up
      n+=*(cp--*p); // Memory read up right
      p+=x_size-2;

      n+=*(cp--*p); // Memory read left
      p+=2;
      n+=*(cp--*p); // Memory read right
      p+=x_size-2;

      n+=*(cp--*p++);// Memory read down left
      n+=*(cp--*p++);// Memory read down
      n+=*(cp--*p); // Memory read down right

      if (n<=max_no)
        r[i*x_size+j] = 730 - n;
    }
  }
}
```

Fig. 6: Initial version of *susan_principle*.

```
susan_principle(unsigned char *in,int *r,char *bp,int max_no,int x_size,int y_size){
  int   i, j, n,x;
  unsigned char *cp;
  unsigned char cU[3],cC[3],cD[3];// Cache registers
  for (i=1;i<y_size-1;i++){
    for(x=0;x<3;x++){// Cache data into registers
      cU[x] = in[(i-1) * x_size + x];
      cC[x] = in[(i)   * x_size + x];
      cD[x] = in[(i+1) * x_size + x];
    }
    for (j=1;j<x_size-1;j++){
      n=100;
      cp=bp + cC[1];   // Register read center

      n+=*(cp--cU[0]); // Register read left
      n+=*(cp--cU[1]); // Register read up
      n+=*(cp--cU[2]); // Register read up right

      n+=*(cp--cC[0]); // Register read left
      n+=*(cp--cC[2]); // Register read right

      n+=*(cp--cD[0]); // Register read down left
      n+=*(cp--cD[1]); // Register read down
      n+=*(cp--cD[2]); // Register read down right

      // Shift registers
      cU[0] = cU[1];cC[0] = cC[1];cD[0] = cD[1];
      cU[1] = cU[2];cC[1] = cC[2];cD[1] = cD[2];
      cU[2] = in[(i-1) * x_size + j+2];// Mem read next up
      cC[2] = in[(i)   * x_size + j+2];// Mem read next
      cD[2] = in[(i+1) * x_size + j+2];// Mem read next down

      if (n<=max_no)
        r[i*x_size+j] = 730 - n;
    }
  }
}
```

Fig. 7: Optimized version of *susan_principle*.

```
int bitreversal(unsigned a)
{
    unsigned rev = 0;
    int  i=0;

    for(i=0; i < 32; i++)
    {
        rev = rev << 1;
        rev = rev | (a & 1);
        a = a >> 1;
    }
    return rev;
}
```

Fig. 8: The *bitreversal* kernel implementation.

mapping application functionalities in a context of heterogeneous platforms. The framework learns how to discriminate between the heterogeneous elements on the base of a kernel characterization based on both static code analysis and dynamic profiling that can be performed before porting the kernel onto the different computing elements. The learning phase is carried out by using *data mining* techniques such as *random forests* and *genetic algorithms*. The proposed methodology has been validated on a Xilinx-ml510 board that includes an FPGA and a PPC as computing elements.

Acknowledgment

This work was supported in part by the EC under the grants HARPA FP7-612069 and CONTREX FP7-611146, and by the Swiss National Science Foundation under the grant PBTIP2-142333 *"Run-time Resource Management for High Performance Embedded Heterogeneous Architectures"*.

References

[1] E.S. Chung, P.A. Milder, J.C. Hoe, and Ken Mai. Single-chip heterogeneous computing: Does the future include custom logic, FPGAs, and GPGPUs? In *Proceedings of MICRO*, pages 225–236, 2010.
[2] Y. Yankova, G. Kuzmanov, K. Bertels, G. Gaydadjiev, Yi Lu, and S. Vassiliadis. Dwarv: Delftworkbench automated reconfigurable vhdl generator. In *Proceedings of FPL*, pages 697–701, 2007.
[3] Stephen Neuendorffer and Fernando Martinez-Vallina. Building zynq accelerators with vivado high level synthesis. In *Proceedings of FPGA*, pages 1–2, New York, NY, USA, 2013. ACM.
[4] S. Lee, S.-J. Min, and R. Eigenmann. OpenMP to GPGPU: a compiler framework for automatic translation and optimization. In *Proceedings of PPoPP*, pages 101–110, 2009.
[5] Leo Breiman. Random forests. *Mach. Learn.*, 45(1):5–32, October 2001.
[6] G. Mariani, A. Brankovic, G. Palermo, J. Jovic, V. Zaccaria, and C. Silvano. A correlation-based design space exploration methodology for multi-processor systems-on-chip. In *Proceedings of DAC*, pages 120–125, 2010.
[7] H. Cook and K. Skadron. Predictive design space exploration using genetically programmed response surfaces. In *Proceedings of DAC*, pages 960–965, 2008.
[8] H.-Y. Liu and L. P. Carloni. On learning-based methods for design-space exploration with high-level synthesis. In *DAC*, pages 50–56, 2013.
[9] A. Stulova, R. Leupers, and G. Ascheid. Throughput driven transformations of synchronous data flows for mapping to heterogeneous mpsocs. In *Proceedings of SAMOS*, pages 144–151, 2012.
[10] G. Mariani, V. Sima, G. Palermo, V. Zaccaria, C. Silvano, and K. Bertels. Using multi-objective design space exploration to enable run-time resource management for reconfigurable architectures. In *Proceedings of DATE*, pages 1379 –1384, march 2012.
[11] A.D. Pimentel, C. Erbas, and S. Polstra. A systematic approach to exploring embedded system architectures at multiple abstraction levels. *Computers, IEEE Transactions on*, 55(2):99–112, 2006.
[12] F. Ferrandi, P. L. Lanzi, C. Pilato, D. Sciuto, and A. Tumeo. Ant colony heuristic for mapping and scheduling tasks and communications on heterogeneous embedded systems. 29(6):911–924, 2010.
[13] K. Hoste and L. Eeckhout. Microarchitecture-independent workload characterization. *Micro, IEEE*, 27(3):63 –72, may-june 2007.
[14] Qi Guo, Tianshi Chen, Yunji Chen, Ling Li, and Weiwu Hu. Microarchitectural design space exploration made fast. *Microprocessors and Microsystems*, 37(1):41 – 51, 2013.
[15] P. Giusto, G. Martin, and E. Harcourt. Reliable estimation of execution time of embedded software. In *Proc. of DATE*, pages 580–589, 2001.
[16] G. R. Gupta, M. Gupta, and P. R. Panda. Rapid estimation of control delay from high-level specifications. In *Proceedings of DAC*, pages 455–458, 2006.
[17] S. Khan, P. Xekalakis, J. Cavazos, and M. Cintra. Using predictive modeling for cross-program design space exploration in multicore systems. In *Proceedings of PACT*, pages 327 –338, sept. 2007.
[18] C. Zhang, A. Ravindran, K. Datta, A. Mukherjee, and B. Joshi. A machine learning approach to modeling power and performance of chip multiprocessors. In *Proceedings of ICCD*, pages 45 –50, oct. 2011.
[19] K.L.M. Bertels, S. A. Ostadzadeh, and R. J. Meeuws. Advanced profiling of applications for heterogeneous multi-core platforms. In *Proceedings of ERSA*, page 13, 2011.
[20] Julian J. Faraway. *Extending the Linear Model with R (Texts in Statistical Science)*. Chapman & Hall/CRC, 2005.
[21] XILINX. ml510 documentation, 2008. http://www.xilinx.com/support/documentation/ml510.htm.
[22] Jason E. Fritts, Frederick W. Steiling, Joseph A. Tucek, and Wayne Wolf. Mediabench ii video: Expediting the next generation of video systems research. *Microprocess. Microsyst.*, 33(4):301–318, June 2009.
[23] M. R. Guthaus, J. S. Ringenberg, D. Ernst, T. M. Austin, T. Mudge, and R. B. Brown. Mibench: A free, commercially representative embedded benchmark suite. In *Proceedings of WWC*, pages 3–14, 2001.
[24] A. A. C. Experts. Cosy: Compiler system.

978-1-4799-2817-0/14 $31.00 © 2014 IEEE

Edit Distance Based Instruction Merging Technique
to Improve Flexibility of Custom Instructions Toward Flexible Accelerator Design

Hui Huang

Department of Computer Science
University of California, Los Angeles
Los Angeles, CA 90095, USA
e-mail: huihuang@cs.ucla.edu

Taemin Kim, Yatin Hoskote

Intel Labs
Intel Corporation
Hillsboro, OR 97124, USA
(taemin.kim,yatin.hoskote)@intel.com

Abstract— Due to ever shortening time-to-market of a system-on-a-chip (SoC) and increasing NRE cost of designing accelerators in the SoC, a design methodology for a flexible accelerator is desirable. We propose a novel technique to make custom instructions (CIs) of an application specific instruction-set processor (ASIP) flexible. By doing so, CIs can support applications that were not considered at design time of the ASIP, which is difficult to do with a conventional CI design method. We have shown that custom instructions generated by our technique can support future applications by up to 7X better than those from a conventional method.

I INTRODUCTION

Modern system-on-a-chips (SoCs) integrate multiple fixed function accelerators to achieve better performance and lower power/energy than a general purpose processor. However, the performance and power/energy comes at the cost of losing flexibility. In other words, whenever new application needs to be accelerated, we need to design a new accelerator, which incurs significant non-recurring engineering (NRE) cost. Traditional flow of designing an accelerator is composed of application analysis, architecture definition, architecture implementation including register transfer level (RTL) description, logic synthesis and placement and routing. In general, new accelerator project has to go through all of the procedures aforementioned, which requires significant amount of NRE cost including design cost as well as monetary cost.

However, if a single accelerator[1] can support multiple applications that were not conceived at its design time, the NRE cost can be reduced significantly. In addition, since it is highly possible that the lifetime of a SoC including the accelerator is extended thanks to the flexible accelerator, a SoC company can respond market demand of new applications quickly. Moreover, it can respond the demand of ever decreasing time-to-market. Thus, a flexible accelerator makes it possible to significantly reduce time-to-market as well as NRE cost.

An application specific instruction-set processor (ASIP) is a potential solution toward such a flexible accelerator. It can support both flexibility and efficiency. Flexibility can be achieved by using a base instruction set whose granularity is small enough to support virtually all possible applications, and efficiency can be achieved through introducing custom instructions. However, if future applications are mostly supported by instructions in the base instruction set, we would not be able to achieve the efficiency due to the inherent inefficiency in a processor and base instructions. Thus, unless custom instruc-

tions have enough flexibility to support future applications, then the ASIP may not be a solution toward a *true* flexible accelerator. Thus, we need a way to improve flexibility of custom instructions.

In this paper, we propose a technique to achieve the objective of flexibility improvement. Before we discuss how flexibility of custom instructions can be achieved, we first discuss the similarity of applications in the same domain. Note that we do not target flexibility that supports arbitrary applications. We would like to achieve the flexibility with which an accelerator supports applications in the same domain. Our technique is based on three hypotheses that explain the nature of applications in the same domain in terms of their similarity. Although they are discussed in greater details in later section, main idea of the hypotheses is that applications in the same domain have significant similarity in terms of computational patterns. Thus, when we generate custom instructions for a given application, if they cover as many computational patterns in the application as possible, we can also improve the possibility that the custom instructions are used for future applications in the same domain that were not considered at design time of the custom instructions. Moreover, we empirically verify the hypotheses.

Based on the hypotheses, our technique constructs a custom instruction in such a way that it accommodates multiple computational patterns by merging them together. In order to limit the overhead of area and latency due to merging multiple computational patterns into a single custom instruction, we use *Edit Distance*[7] that confines the number of moves to transform one graph to another. It allows us to limit the difference between two computational patterns to be merged, thereby limiting additional hardware overhead. In conventional custom instruction generation flow, a single custom instruction can support only one computational pattern. By performing our technique, we can support larger number of computational patterns, which improves the possibility to support computational patterns in future applications according to the hypotheses discussed below. Our contributions can be summarized as follows.

- We constructed three hypotheses on the similarity of applications in the same domain in terms of their computational patterns, and verified them empirically.

- We proposed a new algorithm that improves the flexibility of custom instructions in such a way that they support as many computational patterns of future applications as possible.

- We have demonstrated effectiveness of the algorithm

---

[1] we call such an accelerator as a flexible accelerator hereafter

through various experiments.

The rest of this paper is organized as follows. In Section II, we briefly go over related works. Then, Section III defines terminologies to simplify algorithm description and presents hypotheses that establish the ground of our algorithm, followed by problem formulation in Section IV. Section V describes the details of our edit distance based instruction merging algorithm. Finally, Section VI presents experimental result and Section VII concludes the paper.

II RELATED WORKS

There has been handful of research on constructing flexible accelerators. [3] is a representative paper in automatic instruction set extension (ISE) field that discusses how a custom instruction can be generalized to support multiple applications. They proposed two schemes, one of which constructs bypass logic in a custom functional unit so that different computational pattern can be supported and the other of which adds redundancies of operations so that computational patterns with different operations but same interconnections can be supported. Although they basically try to support as many computational patterns as possible with a set of custom instructions, the guidance in which cases the schemes should be performed is unclear. In the worst case, they might bypass or add redundant operations that do not appear in future applications, which causes only overhead.

In loop accelerator (LA) domain, there have been multiple efforts to make a LA flexible. [9] discusses about construction of a LA to support multiple *given* applications. The LA is limited to support only the given applications. [4] discusses construction of a more flexible LA by supporting processor-like components such as generic functional units, global bus, rotating register files and etc. The addition of such generality may cause significant area and performance overhead.

[8] presented a set of algorithms to generate domain-specific coarse-grained arrays. Although they targeted a different platform from that ASIP, their algorithm is similar to ours in the sense that they tried to generate computing elements from the *given* applications for future applications instead of adding redundant operations and interconnections randomly. What is different from our approach is that they partition given computational patterns into fine grain simple paths and provide more interconnections than original computational patterns have. In that way, they can support other computational patterns that are not in the original ones. However, in an ASIP environment, partitioning computational patterns into finer grain ones may incur more overhead than using them as they are.

Our approach is significantly different from the previous works mentioned above in the sense that our technique is strictly guided by hypotheses that represent the characteristics of applications in the same domain instead of random guidance and targets future applications that would not be available at design time.

III PRELIMINARIES AND MOTIVATION

In this section, we make three hypotheses that ground our method to improve flexibility of custom instructions and then empirically verify them.

A Preliminaries

Definition 1 *Computational Pattern (CP)* of an application is a graph generated from control and data flow graph (CDFG) of the application that meets constraints of register file input and output and convexity. One can see an application as a set of computational patterns.

Definition 2 *Custom Instruction (CI)* is a set of computational patterns. In general, CI contains CPs that are different from one another.

Definition 3 *Coverage of a Custom Instruction (CCI)* Let CP_{all} denote a set of all possible CPs of a particular application enumerated by an ISE algorithm and CP_{CI} be a set of CPs associated with a custom instruction CI. Then CCI is defined as $\frac{|CP_{CI}|}{|CP_{all}|}$

Definition 4 *Coverage of a Set of Custom Instructions (CSCI)* Let SCI be a set of custom instructions whose elements are $\{ci_0, ci_1, ..., ci_{n-1}\}$. Then CSCI is defined as $\sum_{i=0}^{n-1} CCI(ci_i)$ assuming that $ci_i \bigcap ci_j = \emptyset$ where $i \neq j$.

Definition 5 *Current Application (CA)* is a set of applications that are considered at design time of CIs.

Definition 6 *Future Application (FA)* is a set of applications that are not considered at design time of CIs but in the same domain of CA. In other words, FA is a potential target of CIs generated for CA.

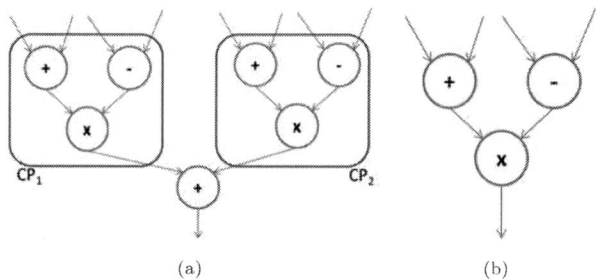

Fig. 1. (a) Computational Patterns in a CDFG (b) Custom Instruction for CP_1 and CP_2

Figure 1 shows an example of a computational pattern and a custom instruction. Figure 1(a) shows two computational patterns CP_1 and CP_2 in a CDFG and Figure 1(b) shows a custom instruction for CP_1 and CP_2. Although the custom instruction in the example includes two computational patterns that are exactly same, note that a custom instruction can contain CPs that are different.

B Hypotheses

It is significantly challenging to predict the characteristics of future applications that were not considered at design time of CIs. Thus, we need to construct a priori statement that holds for applications in the same domain and can guide us to develop precise objectives to make proper tradeoff between flexibility and other quality metrics such as performance, area, power and etc. To that end, we make three hypotheses presented below that we will use as the basis of our technique.

- Applications in the same domain exhibit significant similarity in terms of computational patterns (**H1**)

  - Applications in the same domain share the same computational patterns (**H1-1**)

– The difference of computational patterns of applications in the same domain is limited (**H1-2**)

• As CSCI increases for an application (A), it increases as well for other applications in the same domain as A (**H2**)

Our intuition is that applications in the same domain perform similar computations including types of operations and data dependency between operations, which leads to hypothesis H1. However, the similarity does not necessarily mean that computational patterns of the applications are exactly same. There are not only exactly same patterns (H1-1) but also different ones (H1-2). In case of different patterns, our hypothesis (H1-2) is that the difference is not significant. In other words, patterns are different by small amount of operations and data flows due to their similarity.

With regard to H2, it may seem to be trivial fact. However, for applications that exhibit little similarity to one another, the likelihood that increase of CSCI for one application is the case for other applications will be significantly low due to significant difference in their CPs.

In the next subsection, we empirically verify the hypotheses.

C   Validation of the Hypotheses

C.1   Validation Methodology

In this subsection, we describe how the hypotheses are verified. Before discussing what methods are used for the verification, let us present benchmark applications first. We simply chose three domains of applications in MIBENCH[5] such as security, audio and telecommunication. In addition, we chose medical imaging domain[2] with three popular computational kernels. Figure 2 shows applications in each domain with brief explanation on their functionality.

Domain	Application	Notes
Audio	mp3	Industrial mp3 decoder
	lame	Open-source mp3 encoder
	mad	Open-source mp3 decoder
Security	sha	Secure hash algorithm
	rijndael	Block cipher with 128-,192-,256-bit keys and blocks
	pgp	Pretty Good Privacy public key encryption algorithm
	blowfish	Symmetric block cipher with variable length key
Telecom	crc32	32-bit cyclic redundancy check
	gsm	GSM encoder/decoder
	adpcm	ADPCM encoder/decoder
Medical Imaging	denoise	Image denoise algorithm
	segmentation	Image segmentation algorithm
	registration	Image registration algorithm

Fig. 2. Benchmark applications

We select two applications in a domain one of which is assigned CA and the other of which is assigned FA. For example, in the case of security domain, when *sha* and *blow* are chosen, *sha* is CA and *blow* is FA. In the same way, we generate all possible combinations of application pairs for each domain and perform three measurement described in the next paragraph on each combination. Basically we generate two finite sets of CIs (CICA and CIFA) for CA and FA, respectively and then compare them together to understand the difference between them. For this particular experiment, we generate 10 CIs for CA and FA with input/output constraint of four inputs and two outputs by using our in-house ISE tool. Note that difference is quantified with the number of operations and data flows in data flow graphs associated with CIs.

In order to verify H1-1, we count the number of CIs in CICA that are exactly same as those in CIFA. By doing this, we can understand how many CPs are shared among applications in the same domain. We call the method as *Exact*. We also measure the difference between CICA and CIFA by means of the number of different operations and data flows to verify H1-2. We call this method as $Diff$. Lastly, in order to verify H2, we increase the size of CICA by two times and then compare the increased CICA to original CIFA in *Exact*, which is called as *Exact+*. Now we present the results of three methods discussed above in the next section.

C.2   Results

Table I shows experimental results of three methods. Due to the space limitation, we present only average number for each measurement . As shown in the first column of the table, about 16% of custom instructions in CICA is *exactly* same as those in CIFA on average. Based on the result, we can expect that CA and FA shares the same computational patterns to some extent (H1-1). When we increase the size of CICA by twice, then about 26% of custom instructions are same as shown in the second column of the table, which is significant increase (i.e. 62.5% increase). It verifies our hypothesis of H2 as well. Lastly, the difference between CICA and CIFA is only 1.6 operations or flows on average as shown in the third column of the table. This is important result because it means that we can support FA with CICA as well as CIFA by altering the operations or data flows of the elements of CICA by small amount.

TABLE I

Exact	Exact+	Diff
16.46%	25.90%	1.6

What the verification result means is that applications in the same domain actually shares the same CPs to some extent and difference is not significant, which leads us to believe that three hypotheses hold. In addition, flexibility of CIs can be improved by exploiting the hypotheses somehow with higher confidence than the techniques based on addition of redundancies in a random way. Therefore, as we expected in the beginning of this section, the hypotheses will play a role of a priori statement to develop a technique of flexibility improvement for CIs as discussed later sections.

IV   PROBLEM FORMULATION

In this section, we describe and formulate what problem we need to solve. Based on H1-1 and H2, we need to generate CIs targeting an input application (CA) that contain **maximum number of CPs of CA**. The reasons are as follows. 1) H1-1 implies that custom instructions targeting CA can also targets an application (FA) in the same domain as CA due to the similarity between CA and FA, and 2) H2 implies that the more CPs of CA are covered by CIs of CA (CICA) the more CPs of FA are covered by CICA as well. Putting both implications together, if we generate CIs for an application (CA) in such a way that the number of CPs of the application covered is maximized, then it is highly possible that the number of CPs of applications in the same domain as CA covered by the CIS is also maximized, which eventually maximizes the flexibility of the CIs.

978-1-4799-2817-0/14 $31.00 © 2014 IEEE

The best case solution would be to cover all possible CPs of CA with a set of CIs and then use them for other applications in the same domain. However, this is certainly not a practical solution due to area and latency overhead and intractable complexity of instruction selection that a compiler performs. Thus, **we need to limit the overhead while we maximize the number of CPs**. In this paper, we use two metrics as a proxy of area and timing overhead. One is the number of custom instructions finally generated. It has a strong correlation with total area of custom functional units that execute the custom instructions. The other is *Edit Distance* that represents the difference between two graphs. As briefly described in Section I, multiple CPs that may not be identical one another are merged into a single custom instruction. Depending on the difference among them, timing overhead could be large due to multiplexers. Thus, we limit the difference among the CPs by using *Edit Distance* to limit the number of multiplexers inserted.

In addition, while we maximize the flexibility of a set of custom instructions, we also need to maximize performance. Note that we also take traditional constraints such as register file input and output constraint and convexity constraint. Therefore, we describe the problem formally as follows.

Problem Given instruction count constraint (IC), edit distance (ED) constraint, register file input and output constraint (REGFIO), and a target application, generate a set of custom instructions (SCI) that maximizes both CSCI of SCI for the target application and performance improvement.

V EDIT DISTANCE BASED INSTRUCTION MERGING

In this section, we describe our approach to solving **Problem**. Before we present the details of our algorithm, we first discuss the limitation of traditional ISE in terms of flexibility of CIs. Figure 3 presents a flow of traditional ISE. It first enumerates all possible CPs in a given target application while they meet constraints of register input and output and convexity. Then, it merges isomorphic CPs into a single CI candidate. In this step, CCI of each CI candidate is determined. Since traditional merging procedure allows only isomorphic CPs to be merged together, the CCI is quite limited due to the small number of isomorphic CPs in an application. In addition, due to the strict constraint (i.e. isomorphic CPs), there is almost no controllability to make tradeoff between flexibility and any kind of quality metrics in the merging procedure. That eventually limits the flexibility of CIs generated by the traditional ISE. We introduce new merging procedure to allow the controllability for flexibility improvement. After merging, a set of CI candidates is generated and then, instruction selection is performed to choose CIs that maximizes performance improvement while they meet area constraint which is represented as instruction count (IC) in this paper. We apply Knapsack algorithm to choose best performing CIs by the amount of IC.

However, our approach increases the number of CPs associated with a CI by exploiting **H1-2** in Section B. That is, due to the small difference between CPs of CA and those of FA, we relax the constraint of *isomorphic* CPs in the traditional merging process. We allow small difference in merging CPs together. In addition, we do not add redundancies in a random way but add them by merging CPs in the same application. As described in **H1-1** and **H1-2**, CA and FA share significant similarity. That is why we add redundancies by merging similar CPs in the same application.

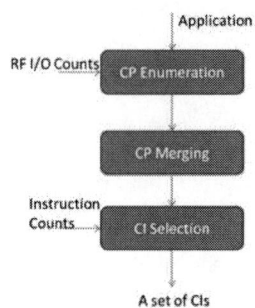

Fig. 3. A flow of traditional ISE

Then, one may ask how the *small difference* is quantified. In order to quantify it, we use *Edit Distance* concept. *Edit Distance* is the number of insertions and deletions of nodes and edges to convert a graph to another graph. Figure 4 shows an example of edit distance. Assuming that we convert Graph A to Graph B, edit distance is five because we have to delete one edge and add three edges and one vertex as shown as dotted ovals. In conventional algorithms of ISE, *isomorphic* CPs are considered as the same custom instruction. However, we relax the restriction by allowing small difference between two CPs. By doing that, we can consider more CPs that are different by a small amount as the same custom instruction, which improves flexibility of the set of CIs as well according to **H2**.

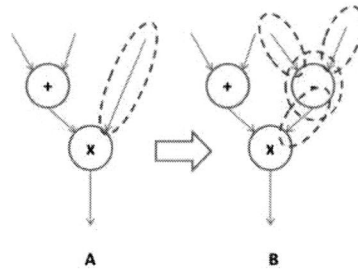

Fig. 4. Edit Distance is five when Graph A is converted to Graph B

Although we can increase the number of CPs to maximize flexibility of CIs, it is highly possible for the increase to increase area and latency overhead of resulting CIs. The reason is because we merge technically different CPs into the same CI, that causes additional functional units and multiplexers. Figure 5 presents an example of the overhead of additional functional units and multiplexers. We merge CP1 and CP2 together into merged CP as shown in the figure. As a result, we need to insert two 2-1 multiplexers. Therefore, we need a way to limit the overhead. In that sense, *Edit Distance* is also useful because it actually controls the amount of hardwares to be added by limiting difference among CPs.

Figure 6 shows our algorithm that maximizes the number of CPs associated with CIs by performing *Edit Distance* based CP merging. It accepts a set of CPs (**CIC**) that contains all possible CPs of a target application and *Edit Distance* (**ldist**) value. The output is a set of CI candidates each of which is a subset of CIC. Basically what the algorithm performs is to merge CPs in CIC whose edit distance is less than or equal to **ldist**. We traverse CPs in CIC one by one and greedily merge them together by computing edit distance in **line 1 -**

978-1-4799-2817-0/14 $31.00 © 2014 IEEE   222

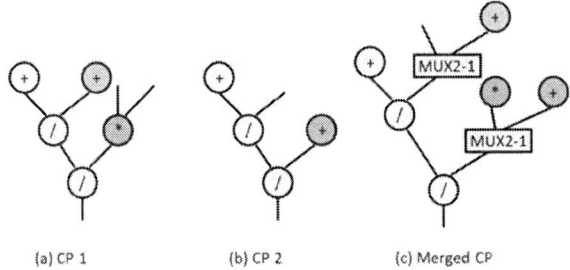

(a) CP 1   (b) CP 2   (c) Merged CP

Fig. 5. Additional hardware cost resulted from merging two CPs

```
Input  : a set of computational patterns (CIC)
         a constraint of edit distance (l_dist)
Output : a set of custom instruction candidates (MCIC)

 1: foreach cp1 in CIC
 2:    foreach cp2 in CIC
 3:       if (cp1 != cp2) then
 4:          if EditDistance(cp1,cp2) <= l_dist then
 5:             s = GetMergedSet(cp1);
 6:             if (s) then
 7:                if (EditDistance (s,cp2) <= l_dist) then
 8:                   insert (cp2, s);
 9:                endif
10:             else
11:                s = Create a new set for cp1;
12:                insert (s, MCIC);
13:                insert (cp1, s);
14:                insert (cp2, s);
15:             endif
16:          endif
17:       endif
18:    end foreach
19:    Remove CPs in s from CIC;
20: end foreach
21: UpdateCostFunction(MCIC);
22: return MCIC
```

Fig. 6. Edit Distance Based Merging Algorithm

line 18. Whenever one CI candidate is constructed, its CPs are removed from CIC to avoid overlap among CI candidates (**line 19**). This process is repeated until there is no CP in CIC. Once CI candidates have been generated (**line 1 - line 20**), we update cost of each CI by computing its latency (**line 21**). The updated latency is evaluated in choosing best performing CIs among the CI candidates generated by our merging procedure.

VI EXPERIMENTAL RESULTS

A Experimental Setup

In this section, we describe our in-house ISE tool that generates flexible CIs and what experiments we performed to evaluate our CP merging algorithm.

We have developed ISE tool based on LLVM[6] compilation framework. As shown in Figure 3, we first enumerate all possible CPs of a target application, based on the CP enumeration algorithm in [10]. Then, we perform our merging algorithm followed by dynamic programming based Knapsack problem solver to choose the best performing CIs with the constraint of instruction count.

In order to evaluate the benefit of our algorithm in terms of flexibility, we perform the same experiment described in Section C.1. Figure 7 shows flow of the experimentation. We consider input application of our ISE tool as CA and other applications in the same domain as FA. We generate CIs targetting CA (CICA) with our merging algorithm (*New Flow*)

and also generate CIs targetting FA (CIFA) with a conventional merging algorithm that merges only isomorphic CPs (*Conventional Flow*). Then, we compute how many CPs[2] in CICA are exactly same as CPs in CIFA. We show the number normalized with respect to the number of CPs in CIFA and call it as *coverage*.

We used the same benchmark applications as those in Section C.1 and categorized them into four domains as well. In addition, we set *Edit Distance* and instruction count constraints two and 10 for all experiments, respectively.

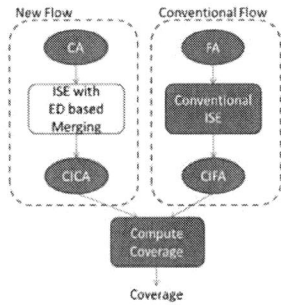

Fig. 7. Experiment flow to evaluate the amount of flexibility improvement of edit distance based merging algorithm

B Results

Figure 8 shows coverage result for audio domain. X-axis shows a pair of CA and FA. For example, in case of *lame−mad*, *mad* is CA and *lame* is FA. Y-axis represents the coverage. We computed coverage for three cases. One is *Exact* as explained in Section C.1. Another is our method called *Merging*. We generate CIs for CA by using our algorithm and compare them to CIs for FA generated by conventional algorithm to compute coverage. The last one is *Ideal* which computes coverage when all the CPs of CA are available in CICA. As shown in the Figure, our algorithm is always better than *Exact* because we included more CPs. We improved the coverage by 700% over *Exact* and the coverage of *Merging* is about 70.5% of *Ideal* on average for audio domain on average. We performed the same experimentation for the applications in domains of security, medical imaging and telecom as shown in Figures 9, 10 and 11, respectively. As shown in the Figures, our technique *Merging* is significantly better than *Exact* and approaching to *Ideal*. Similar to audio case, we have improved coverage on average by 106.7%, 36.8%, 66.7% for security, medical imaging and telecom domains, respectively. In addition, we have achieved 48.4%, 78.8% and 55.6% of *Ideal* coverage, respectively.

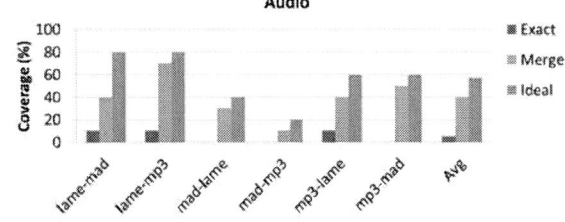

Fig. 8. Coverage result of applications in audio domain

---

[2]Note that single CI in CICA can have multiple CPs while a CI in CIFA has only one CP

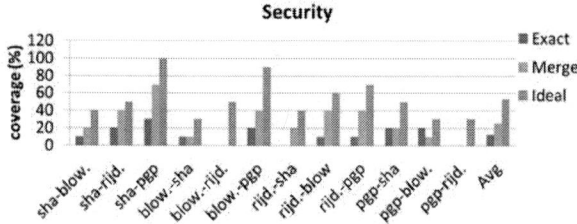

Fig. 9. Coverage result of applications in security domain

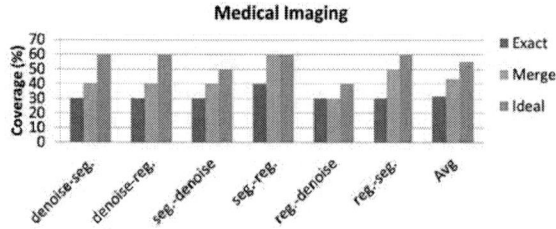

Fig. 10. Coverage result of applications in medical imaging domain

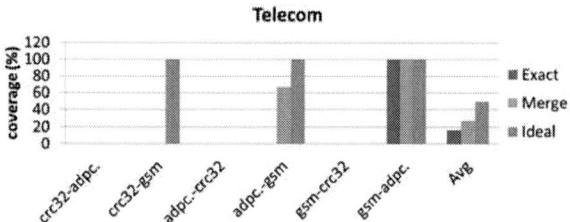

Fig. 11. Coverage result of applications in telecom domain

We improve coverage of CIs by adding redundancy, which certainly incurs latency overhead although we limit the overhead with *Edit Distance*. Thus, we have measured the latency increase due to the additional hardware such as multiplexers and functional units. Since we limit the overhead by *Edit Distance* of two, we can expect the increase is not significant. We compute the latency increase of a CI by computing the critical path of a CI with the consideration of additional 2-1 multiplexers and functional units. We have used delay numbers of functional units and multiplexers in [1]. In the case where delay numbers of components are not available in the table, we simply prorated with numbers in the table. For example, we divided the delay of 8:1 MUX by three to get that of 2:1 MUX. Figure 12 shows the results of latency increase. As shown in the Figure, the average latency increase of CIs generated by our merging algorithm is 23.3% over CIs generated by conventional merging algorithm. This result proves that *Edit Distance* indeed controls the overhead of merging similar CPs.

VII Conclusions

We have demonstrated the process of improving flexibility of custom instructions by first verifying three hypothesis on characteristics of applications in the same domain in terms of

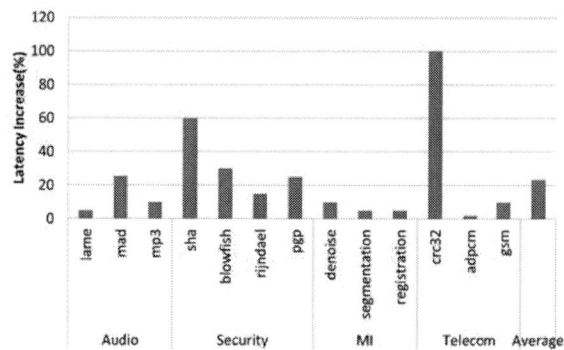

Fig. 12. Latency increase of our merging algorithm with respect to conventional merging algorithm

similarity of their computational patterns and then, proposing a new technique of instruction merging based on edit distance. We have shown that our technique improves coverage by custom instructions generated by our technique up to 7X as compared to conventional instruction merging technique. On the other hand, we have shown that our technique is able to control latency overhead by controlling edit distance. Our experimental result has demonstrated that latency of custom instructions increased 23.3% over the case where conventional instruction merging technique is used to generate custom instructions. Consequently, we can achieve significant flexibility improvement with our edit distance based instruction merging technique with limited latency overhead.

References

[1] http://www.design-reuse.com/articles/19171/ip-gate-count-estimation-micro-architecture-phase.html.

[2] A. Bui, K.-T. Cheng, J. Cong, L. Vese, Y.-C. Wang, B. Yuan, and Y. Zou. Platform characterization for domain-specific computing. In *Proceedings of Asian and South Pacific Design Automation Conference (ASPDAC)*, Jan 2012.

[3] N. T. Clark, H. Zhong, and S. Mahlke. Automated custom instruction generation for domain-specific processor acceleration. *IEEE Transactions on Computers*, 54(10):1258–1270, Oct 2005.

[4] K. Fan, M. Kudlur, G. Dasika, and S. Mahlke. Bridging the computation gap between programmable processors and hardwired accelerators. In *International Symposium on High Performance Computer Architecture (HPCA)*, pages 313–322. IEEE, Feb 2009.

[5] M. R. Guthaus, J. S. Ringenberg, D. Ernst, T. M. Austin, T. Mudge, and R. B. Brown. Mibench: A free, commercially representative embedded benchmark suite. In *IEEE 4th Annual Workshop on Workload Characterization*, pages 3–14. IEEE, Dec 2001.

[6] C. Lattner and V. Adve. LLVM: A Compilation Framework for Lifelong Program Analysis and Transformation. In *Proceedings of the International Symposium on Code Generation and Optimization (CGO)*, Mar 2004.

[7] B. T. Messmer. Efficient graph matching algorithms, 1995.

[8] M. Stojilovic, D. Novo, L. Saranovac, P. Brisk, and P. Ienne. Selective flexibility: Creating domain-specific reconfigurable arrays. *IEEE Transactions on Computer-Aided Design of Integrated Circuits and Systems*, 32(5):681–694, May 2013.

[9] S. Yehia, S. Girbal, H. Berry, and O. Temam. Reconciling specialization and flexibility through compound circuits. In *International Symposium on High Performance Computer Architecture (HPCA)*, pages 277–288. IEEE, Feb 2009.

[10] P. Yu and T. Mitra. Disjoint pattern enumeration for custom instructions identification. In *Proceedings of the International Conference on Field Programmable Logic and Applications (FPL)*, pages 273–278. IEEE, August 2007.

A Network-Flow-Based Optimal Sample Preparation Algorithm for Digital Microfluidic Biochips

Trung Anh Dinh[1]
trungda@ngc.is.ritsumei.ac.jp
[1]Ritsumeikan University, Japan

Shigeru Yamashita[1]
ger@cs.ritsumei.ac.jp

Tsung-Yi Ho[2]
tyho@csie.ncku.edu.tw
[2]National Cheng Kung University, Taiwan

Abstract—Sample preparation, which is a front-end process to produce droplets of the desired target concentrations from input reagents, plays a pivotal role in every assay, laboratory, and application in biomedical engineering and life science. The consumption of sample/buffer/waste is usually used to evaluate the effectiveness of a sample preparation process. In this paper, for the *first* time, we present an *optimal* sample preparation algorithm based on a minimum-cost maximum-flow model. By using the proposed model, we can obtain both the optimal cost of sample and buffer usage and the waste amount even for multiple-target concentrations. Experiments demonstrate that we can consistently achieve much better results not only in the consumption of sample and buffer but also the waste amount when compared with all the state-of-the-art of the previous approaches.

I. Introduction

Digital microfluidic biochips (DMFBs) have become the most promising technology in many biomedical fields. Compared with the conventional laboratory experiments, which are usually cumbersome and expensive, these miniaturized systems represent numerous advantages such as high portability, high throughput, low sample/reagent volume consumption, and so on. Recently, various on-chip biochemical applications such as immunoassay, DNA sequencing, and protein crystallization have been successfully demonstrated on DMFBs [1].

Before performing a biochemical application on a DMFB, *sample preparation* is a crucial preprocessing step, which dominates 90% of the cost and 95% of the analysis time in biochemical experiments [2]. In a sample preparation process, sample droplets are mixed with buffer droplets in an appropriate strategy in order to produce the droplets of the target concentration values. The main challenges of a sample preparation process is to minimize: (1) the cost of sample and buffer usage, which dominates most of the overall cost of the process; (2) the waste amount, which corresponds to the waste handling overhead. The above two concerns are even more critical in *multiple-target sample preparation* problem, in which multiple droplets of different concentrations need to be prepared at the same time. Given these concerns, we propose the *first optimal* algorithm for modern sample preparation process to minimize the total cost of buffer and sample us-

age. Our algorithm is based on a network-flow model that is generalized and capable of handling all the requirements simultaneously.

One of the first-generation methods for the sample preparation problem is based on a simple binary search [3]. After that, many heuristics have been proposed. Since the complexity of biochemical applications grows significantly in recent years, the need of multiple-target sample preparation becomes more and more urgent. However, the works in [4, 5, 6] cannot consider more than one different target concentration value at the same time, and thus they cannot solve multiple-target sample preparation problem effectively. Although the works in [2, 7, 8] can deal with multiple-target problem, all of them are based on heuristic strategies, and they sacrifice the optimality requirement of the problem. Moreover, it should be noted that all the previous works consider only one objective optimization, e.g., only the number of sample droplets ([2], [7]) or only the number of waste droplets ([5], [6]). In practical contexts, it is more desirable to consider all the objectives, i.e., the cost of sample and buffer usage and the number of waste droplets in one model; however, none of the previous works can do so.

To meet all the aforementioned challenges, in this paper, for the first time, we propose an optimal algorithm for the sample preparation problem. A *minimum-cost maximum-flow* model is introduced to formulate the problem. Since the proposed model is flexible to adjust the desired objective, we can obtain the *optimal* cost of sample and buffer usage as well as the *optimal* number of waste droplets even for multiple-target sample preparation.

The rest of this paper is organized as follows. We first give the formal definition of the general sample preparation problem in Section II. Then, Section III proposes an optimal solution for the problem, followed by Section IV that provides our experimental results. The paper is concluded in Section V.

II. Sample preparation on DMFBs

In laboratory, for one type of biochemical substance, the sample droplets and the buffer droplets, which have the concentration of 100% and 0%, respectively, are given. Mixing and splitting are the two fundamental operations, which are frequently used in order to alter the concentration value of a sample. General DMFBs can only perform serial dilution, in which samples are diluted using the same

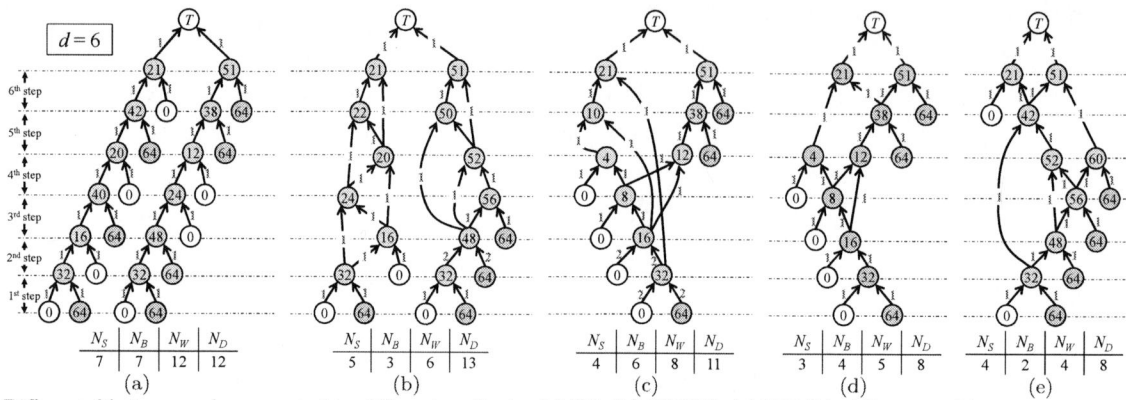

Fig. 1. Different dilution graphs generated by different methods: (a) BS, (b) DMRW, (c) REMIA, (d) $ours_S$, (e) $ours_W$.

mixing/splitting ratio, i.e., 1:1 [5]. Based on such speciality of DMFBs, a dilution method, a motivational example and a formal formulation of the sample preparation problem will be discussed in this section.

A. Dilution method

In a single dilution step, each of n droplets of concentration C_i ($0\% \leq C_i \leq 100\%$) is mixed with each of n droplets of concentration C_j ($0\% \leq C_j \leq 100\%$) to produce n two-unit-volume droplets of concentration $\frac{C_i+C_j}{2}$, and then each of the resultant mixture is split into two unit-volume droplets of concentration $\frac{C_i+C_j}{2}$. Therefore, the result of a dilution step is $2n$ droplets of concentration $\frac{C_i+C_j}{2}$. It can also be considered that n droplets of concentration C_i are diluted with n droplets of concentration C_j to produce $2n$ droplets of concentration $\frac{C_i+C_j}{2}$. A pair of mix and the corresponding split operations is often called *a dilution operation* for short. Note that besides the optimization objectives mentioned in Section I, in some cases, the number of dilution operations is also considered in a sample preparation process, since it illustrates the throughput, reliability of the process.

It can be noticed that a concentration value that is obtained by the above dilution method can always be expressed as the form $\frac{c_i}{2^d}$ ($0 < c_i < 2^d - 1$, $d > 0$). The number d, which is given as a constant factor in each sample preparation problem, represents the *precision level of concentration*. In this paper, for the sake of simplicity, we represent a concentration value only by its numerator (c_i) instead of the integral fraction ($C_i = \frac{c_i}{2^d}$). For example, when $d = 6$, in order to represent the concentration value $\frac{35}{64}$, we only use the number 35.

B. Motivational example

In order to illustrate a sample preparation process, we introduce a directed graph $DG = (\mathcal{V}_{DG}, \mathcal{E}_{DG})$, which is called a *dilution graph*. Actually, a dilution graph is a combination of several dilution operations, which represents a dilution strategy that can achieve the desired result. In a dilution graph, each vertex is labeled by a concentration value denoted only by its numerator as mentioned in the previous subsection.

For example, Fig. 1 describes and compares five different dilution graphs, which are generated by different methods (BS [4], DMRW [5], REMIA [7] and ours). The precision level d of the example is 6, and the objective is to produce two droplets of concentration 21 and 51. For the sake of convenience, this paper uses the notations N_S, N_B, N_W, and N_D to represent the number of sample droplets, buffer droplets, waste droplets, and dilution operations, respectively.

It can be seen in Fig. 1 (a) and (b) that BS and DMRW are not dedicated for multiple-target sample preparation, and thus they have to construct an individual subgraph for a concentration value, i.e., one subgraph for the concentration value 21 and one other for the concentration value 51. Therefore, they often consume more sample droplets than other methods. Compared to the graph $ours_S$ in Fig. 1 (d), which illustrates a sample preparation process with the optimal number of sample droplets ($N_S = 3$), the consumption of sample droplets of BS ($N_S = 7$) and DMRW ($N_S = 5$) is 57% and 40% higher, respectively.

On the other hand, the method proposed by REMIA can be applied to multiple-target sample preparation problem. It is based on the observation that the concentration values that have the form 2^k (1, 2, 4, 8, etc), can be obtained "easily" by diluting a sample droplet with buffer droplets repeatedly. Therefore, it tries to obtain the target concentrations by using such values greedily. For example, as shown in Fig. 1 (c), REMIA first generates droplets of concentrations 32, 16, 8, 4, and then uses such droplets to produce the desired result. Although this approach can reduce the number of sample droplets better than BS and DMRW, it totally ignores the roles of buffer and waste droplets. Therefore, REMIA necessitates a large number of buffer and waste droplets. For example, compared to $ours_W$ in Fig. 1 (e), which illustrates a process with the optimal number of waste droplets, REMIA consumes more than 66% buffer droplets and 50% waste droplets.

Compared to all the previous works, we adapt a minimum-cost maximum-flow model that can take a global and general view of the problem and treat sample and buffer equally. Moreover, by using the proposed approach, it is also easier to consider various concerns in one model. The proposed algorithm will be presented in the next section.

978-1-4799-2817-0/14 $31.00 © 2014 IEEE

C. Problem formulation

Given:

- Cost of one sample droplet: $cost_s$.
- Cost of one buffer droplet: $cost_b$.
- The required precision level of concentration values of the target samples: d.
- A set TC of N different target concentrations:
$$TC = \{c_1, c_2, ..., c_N\}, 1 \leq c_i \leq 2^d - 1.$$
- A set S of the required number of droplets for each target concentration:
$$S = \{s_1, s_2, ..., s_N\}$$
Each element $s_i \in S$ represents the required number of droplets of concentration c_i in TC. The total number of required droplets is represented by S_R.
$$S_R = \sum_{i=1}^{N} s_i$$

Output: Derive a sample preparation process, i.e., a dilution graph, which can generate all the target concentrations in TC with the corresponding numbers of droplets in S.

Objective: The resultant process should have the minimum cost of sample and buffer usage. Specifically, if u_s sample droplets and u_b buffer droplets are used, the following cost function:
$$\mathcal{F} = u_s \cdot cost_s + u_b \cdot cost_b$$
should be minimized.

An example with the following inputs is used to explain the proposed method.

- Cost of one sample droplet: $cost_s = 2$.
- Cost of one buffer droplet: $cost_b = 1$.
- The precision level: $d = 3$.
- The set of target concentrations: $TC = \{3, 5\}$.
- The set of required number of droplets for each target concentration: $S = \{4, 6\}$.
  Therefore, the total number of target droplets is $S_R = 4 + 6 = 10$.

This means that we need to generate four droplets of the concentration $\frac{3}{8}$, and six droplets of the concentration $\frac{5}{8}$.

Minimizing the number of sample droplets is just a special case of the cost function $\mathcal{F}$. In other words, when $cost_s$ and $cost_b$ are set to 1 and 0, respectively, the cost function $\mathcal{F}$ becomes u_s.

Since there is no other source of droplets rather than sample and buffer droplets, the number of waste droplets can be calculated by the following expression:
$$N_W = u_s + u_b - S_R$$
As S_R is a constant factor, the number of waste droplets (N_W) depends on the total number of sample and buffer droplets, i.e., $u_s + u_b$. Moreover, when both $cost_s$ and $cost_b$ are set to 1, the cost function expresses the total number of used sample and buffer droplets ($\mathcal{F} = u_s + u_b$). Therefore, minimizing $\mathcal{F}$ now leads to the minimization of the number of waste droplets as well.

III. PROPOSED METHOD

The proposed algorithm consists of three main steps: (1) formulate the input into a minimum-cost maximum-flow network model; (2) transform the original problem to an integer equal flows problem, which is solved by using ILP formulation; (3) derive the final dilution graph, which can obtain the required droplets with the optimal cost of sample and buffer usage.

For the sake of convenience, we introduce a function δ of a set of concentration values $SC = \{c_1, c_2, ...\}$ such that:
$$\delta(SC) = \{c_k | c_k = \frac{c_i + c_j}{2}, \forall c_i, c_j \in SC\}$$
For example, if $SC = \{0, 4, 8\}$, we have $\delta(SC) = \{0, 2, 4, 6, 8\}$ since $\frac{0+0}{2} = 0$, $\frac{0+4}{2} = 2$, and so on. The δ function implies that for any concentration value c_k in $\delta(SC)$, there exist two concentration values c_i and c_j in SC such that applying a dilution operation to two droplets of concentration c_i and c_j generates two droplets of concentration c_k.

A. Network-Flow Model

In order to achieve the optimal cost function $\mathcal{F}$, we construct a network-flow model $\mathcal{G} = (\mathcal{V}, \mathcal{A})$ by representing two construction rules and one constraint. The two rules describe the construction of the set of vertices $\mathcal{V}$ and the set of arcs $\mathcal{A}$. Since general DMFBs support only (1:1)-mix and (1:1)-split operations, the proposed model has an additional constraint, which differs from general network-flow problem. Figures 2 (a) and (b) are used to exemplify our network-flow model, and the intention of the flow values on the network's arcs, respectively. The two formulation rules and the constraint are detailed as follows.

A.1 Rule 1-Formulation of $\mathcal{V}$

The set of vertices in our network-flow model is constructed bottom-up from level 0. We use the notation $\mathcal{V}_l$ to describe the set of vertices at level l. As similar to the dilution graph (mentioned in Section II-B), each vertex is labeled by a concentration value. A vertex at level l, which is labeled by a concentration value c is denoted by $v_{l,c}$, e.g., in Fig. 2 (a), $v_{0,0}$ represents a concentration value of 0 at level 0. We denote a set of concentration values, which are used to label the vertices at level l, as SC_l. SC_l is constructed by the following intention: SC_l represents the set of concentration values that can be generated at level l by using the concentration values at level $l - 1$, i.e., $SC_l = \delta(SC_{l-1})$. For a precision level d, at the beginning of the sample preparation process, since we have only buffer and sample droplets, $SC_0 = \{0, 2^d\}$. After constructing the set of concentration values at level l, i.e., SC_l, we can obtain the set of vertices at level l by the following expression:
$$\mathcal{V}_l = \{v_{l,c_i} | c_i \in SC_l\}$$
For example, it can be seen in Fig. 2 (a) that at level 2, while the set of concentration values $SC_2 = \{0, 2, 4, 6, 8\}$, the set of vertices can be represented by $\mathcal{V}_2 = \{v_{2,0}, v_{2,2}, v_{2,4}, v_{2,6}, v_{2,8}\}$. The construction of the set of concentration values stops at level d. A key lemma, which will be presented at the end of this subsection, demonstrates that SC_d consists of all the concentration values from 0 to 2^d.

Besides the vertices in all levels, there are also three

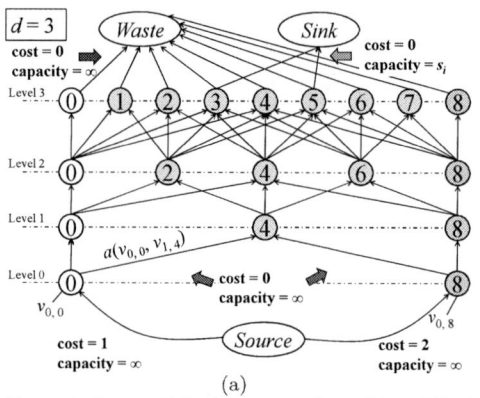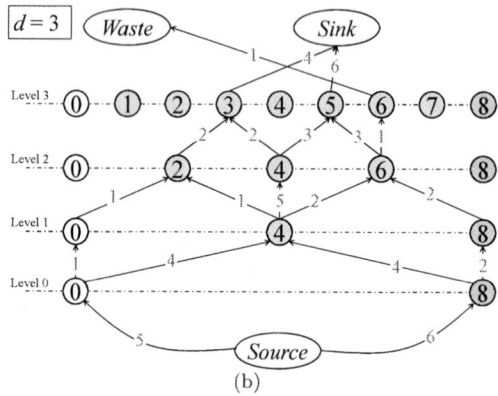

Fig. 2. (a) Network-flow model of the example problem (b) Minimum-cost maximum-flow of the example network-flow model.

special ones: *Source*, *Sink*, and *Waste* in our network-flow model. In short, the set of vertices $\mathcal{V}$ can be expressed by the following formulation:

$$\mathcal{V} = \bigcup_{l=0}^{l=d} \mathcal{V}_l \bigcup \{Source, Sink, Waste\}$$

Implementing the formulation rule of $\mathcal{V}$ for the above example, we can construct the set of vertices as shown in Fig. 2 (a). The input d does not only represent the precision level of concentration but also specify the number of levels in the proposed network-flow model, which is given by the following lemma. For the interest of space, the proof of this lemma is not discussed in detail.

Lemma 1. *In the proposed network-flow model, SC_d is the first set, which contains all the concentration values from 0 to 2^d.*

It can be concluded from Lemma 1 that the number of levels of the proposed network-flow model, i.e. d, is the minimal number of levels such that we can obtain any concentration value between 1 and $2^d - 1$.

A.2 Rule 2-Formulation of $\mathcal{A}$

An arc from v_x to v_y in the proposed network-flow model is denoted by $a(v_x, v_y)$. For example, the arc from $v_{0,0}$ to $v_{1,4}$ in Fig. 2 (a) is represented by $a(v_{0,0}, v_{1,4})$. The capacity and the flow value of an arc $a(v_x, v_y)$ are then denoted by $cap(v_x, v_y)$ and $f(v_x, v_y)$, respectively.

We classify the set of arcs in our proposed model into five different types, each of which is discussed in detail as follows.

- Type 1: There are two arcs from *Source* to $v_{0,0}$ and $v_{0,2^d}$. The arc $a(Source, v_{0,0})$ has infinite capacity and $cost_b$ cost per unit flow. The arc $a(Source, v_{0,2^d})$ has infinite capacity and $cost_s$ cost per unit flow.
- Type 2: There is an arc from $v_{l,c}$ to $v_{l+1,c}$ ($l < d, v_{l,c} \in \mathcal{V}_l, v_{l+1,c} \in \mathcal{V}_{l+1}$) with infinite capacity and zero cost per unit flow, e.g., $a(v_{1,4}, v_{2,4})$ in Fig. 2 (a).
- Type 3: If $\frac{c_i+c_j}{2} = c_k$, there are two arcs from v_{l,c_i} and v_{l,c_j} to v_{l+1,c_k} with infinite capacity and zero cost per unit flow ($l < d, v_{l,c_i}, v_{l,c_j} \in \mathcal{V}_l, v_{l+1,c_k} \in \mathcal{V}_{l+1}$). For example, in Fig. 2 (a), there are two arcs $a(v_{2,2}, v_{3,3})$ and $a(v_{2,4}, v_{3,3})$ since $\frac{2+4}{2} = 3$.
- Type 4: For each c_i in the set of target concentrations TC, there is an arc from v_{d,c_i} to $Sink$ with s_i

unit capacity and zero cost per unit flow. Recall that s_i is the number of required droplets of target concentration c_i. In our example, we need to generate four droplets of concentration 3, and thus there exists an arc $a(v_{3,3}, Sink)$ with four units capacity and zero cost per unit flow in Fig. 2 (a).

- Type 5: For each vertex at the dth level, there is an arc from v_{d,c_i} to *Waste* with infinite capacity and zero cost per unit flow, e.g., $a(v_{3,1}, Waste)$ in Fig. 2 (a). The arcs of this type are used to guarantee the validity of the flow values of the network-flow model.

Applying the above formulation rule of $\mathcal{A}$, the set of arcs of the proposed network-flow model is constructed as can be seen in Fig. 2 (a). The flow values of the proposed network-flow model have the same intention with the weight values of the dilution graph mentioned in Section II-B.

A.3 Constraint

Due to the speciality of current DMFBs, an additional constraint should be considered in the proposed network-flow model. Consider three vertices v_{l,c_i}, v_{l,c_j} and v_{l+1,c_k} such that $\frac{c_i+c_j}{2} = c_k$. According to the definition of the arcs of Type 3 as mentioned above, there exist two arcs $a(v_{l,c_i}, v_{l+1,c_k})$ and $a(v_{l,c_j}, v_{l+1,c_k})$. The flow values of the two arcs mean that each of $f(v_{l,c_i}, v_{l+1,c_k})$ droplets of concentration c_i will be diluted with each of $f(v_{l,c_j}, v_{l+1,c_k})$ droplets of concentration c_j to obtain droplets of concentration c_k. According to the definition of the dilution method in Section II-A, the two flow values must be equal, i.e., $f(v_{l,c_i}, v_{l+1,c_k}) = f(v_{l,c_j}, v_{l+1,c_k})$. For example, in the network-flow model in Fig. 2 (a), $f(v_{2,2}, v_{3,4}) = f(v_{2,6}, v_{3,4})$, $f(v_{2,0}, v_{3,4}) = f(v_{2,8}, v_{3,4})$, and so on.

A.4 Integer equal flows problem

The aforementioned two construction rules and constraint transform the sample preparation problem on DMFBs into the *integer equal flows* problem [10]. In an integer equal flows problem, besides the network-flow model, there exist some mutually disjoint sets of arcs: $R_1, R_2, ..., R_m$ such that all arcs in R_i ($i = 1, 2, ..., m$) must have equal flow values.

978-1-4799-2817-0/14 $31.00 © 2014 IEEE   228

It can be seen that a maximum-flow result of the proposed network-flow model represents a valid sample preparation process. Moreover, since all the arcs except $a(Source, v_{0,0})$ and $a(Source, v_{0,2^d})$ have zero cost per unit flow, the cost of the maximum-flow network is equivalent to the cost function $\mathcal{F}$. Therefore, by determining the minimum-cost maximum-flow result in the proposed network-flow model, we can obtain the sample preparation process with the optimal cost function $\mathcal{F}$.

For example, Fig. 2 (b) illustrates the flow values of the minimum-cost maximum-flow result of the example network-flow model in Fig. 2 (a). In Fig. 2 (b), at level 1, four droplets of concentration 0 are diluted with four droplets of concentration 8 to produce eight droplets of concentration 4. Moreover, as can be seen that $f(v_{3,3}, Sink) = 4$ and $f(v_{3,5}, Sink) = 6$, which means that at the end of the sample preparation, we can obtain the required four droplets of concentration 3 and six droplets of concentration 5. When it comes to the waste droplets, only one droplet of concentration 6 is considered as waste ($f(v_{3,6}, Waste) = 1$). Besides, $f(Source, v_{0,8})$ and $f(Source, v_{0,0})$ represents the number of sample droplets and buffer droplets, respectively. Since the cost of one sample droplet is $cost_s = 2$, and the cost of one buffer droplet is $cost_b = 1$, the optimal cost of sample and buffer usage in this case is $\mathcal{F} = 6 \times 2 + 5 \times 1 = 17$.

Based on the above analysis, the sample preparation problem is now transformed to the integer equal flows problem. Although the problem is NP-hard, if we specify the problem as ILP formulation, and employ an efficient ILP solver, solutions of the practical-size problems can be found in a reasonable computational time as will be seen in Section IV.

B. ILP Formulation

Using the flow values f of the proposed network-flow model as variables, the notations related to the sample preparation problem (as mentioned in Section II-C) and the notations related to the proposed network-flow model (as mentioned in Section III-A) as parameters for our ILP formulation, the sample preparation problem can be formulated as follows.

Minimize

$$\mathcal{F} = f(Source, v_{0,2^d}) \times cost_s + f(Source, v_{0,0}) \times cost_b$$

Subject to

1. *Capacity constraint:*
$$f(v_x, v_y) \le cap(v_x, v_y) \; \forall a(v_x, v_y) \in \mathcal{A}$$

2. *Target concentrations:*
$$f(v_{d,c_i}, Sink) = s_i \; (1 \le i \le N, c_i \in TC, s_i \in S)$$

3. *Network-flow conservation:*
   For each $v_x \in \mathcal{V}$:
$$\sum_{v_i : a(v_i, v_x) \in \mathcal{A}} f(v_i, v_x) = \sum_{v_o : a(v_x, v_o) \in \mathcal{A}} f(v_x, v_o)$$

4. *Integer equal flows constraint:*
$$f(v_{l,c_i}, v_{l+1,c_k}) = f(v_{l,c_j}, v_{l+1,c_k}) \text{ if } \frac{c_i + c_j}{2} = c_k$$
$$0 \le l < d$$

Although the main objective of the proposed method is to optimize the cost function $\mathcal{F}$, which includes the cost of both sample and buffer droplets, for a fair comparison with previous methods, the following two settings are used in our experiments:

TABLE I
COMPARISON FOR SINGLE-TARGET SAMPLE PREPARATION

	BS	DMRW	REMIA	$ours_S$	$ours_W$
N_S	5.00	3.52	2.41	**2.22**	2.49
N_B	4.01	3.50	6.09	9.68	2.50
N_W	8.01	6.02	7.50	10.90	**3.99**
N_D	8.01	12.52	10.13	15.80	9.85

- $ours_S$ ($cost_s = 1$, $cost_b = 0$): The cost function is now equal to the number of consumed sample droplets, i.e., $\mathcal{F} = u_s$, and thus the objective is to minimize the number of consumed sample droplets.
- $ours_W$ ($cost_s = 1$, $cost_b = 1$): The cost function is now equal to the total number of consumed sample and buffer droplets, i.e., $\mathcal{F} = u_s + u_b$. As discussed in Section II-C, this setting also leads to the minimization of the number of waste droplets.

IV. EXPERIMENTAL RESULTS

To demonstrate the effectiveness of the proposed method, two explicit experiments were performed. The first one is for single-target sample preparation, and the second one is for multiple-target sample preparation. The proposed algorithm was implemented by C++ in a Linux server (Intel(R) Core(TM) i7 CPU 920 2.67GHz 24GB Memory), and IBM CPLEX is used as our ILP solver [11].

A. Single-target sample preparation

The results of the three previous approaches (BS, DMRW, and REMIA) are compared with the ones achieved by the proposed method ($ours_S$ and $ours_W$). For a precision level $d = 10$, 1023 concentration values (from 1 to 1023) are considered as the target concentrations one by one. The average values of the numbers of sample droplets (N_S), buffer droplets (N_B), waste droplets (N_W), and dilution operations (N_D) are reported in Table I.

Note that it is guaranteed that $ours_S$ and $ours_W$ can obtain the optimal solutions with respect to the numbers of sample and waste droplets, respectively. In the following, we compare our optimal results (N_S and N_W) with the previous state-of-the-art methods, and also the other values (N_B and N_D) are compared to show the superiority of our method.

As shown in Table I, among the five results, $ours_S$ obtains the best number of sample droplets ($N_S = 2.22$), which is 7.8% better than the result of REMIA ($N_S = 2.41$), the best known method for sample usage minimization. As for the total number of sample and buffer droplets, $ours_W$ outperforms all the others. This leads to the fact that $ours_W$ also acquires a much better number of waste droplets ($N_W = 3.99$), which is 33% better than the one of DMRW (6.02), which is a waste minimization method. In terms of the number of dilution operations, the result achieved by $ours_W$ is only worse than BS's result (9.85 compared with 8.01). However, BS always produces the optimal results in terms of the dilution operation count for single-target sample preparation. Thus, we can observe that our method can obtain almost minimal solution even in terms of the number of operations.

TABLE II
Comparison for Multiple-Target Sample Preparation

	\multicolumn{6}{c}{$d = 9$}					
	\multicolumn{3}{c}{$N = 10$}			\multicolumn{3}{c}{$N = 100$}		
	REMIA	$ours_S$	$ours_W$	REMIA	$ours_S$	$ours_W$
N_S	19.59	**8.07**	8.95	203.98	**80.15**	80.43
N_B	31.19	12.67	**7.12**	277.60	76.44	**73.56**
N_W	40.78	10.74	**6.07**	381.58	56.59	**53.99**
N_D	65.90	43.21	**39.69**	654.44	276.67	**258.84**

	\multicolumn{6}{c}{$d = 10$}					
	\multicolumn{3}{c}{$N = 10$}			\multicolumn{3}{c}{$N = 100$}		
	REMIA	$ours_S$	$ours_W$	REMIA	$ours_S$	$ours_W$
N_S	17.33	**11.17**	13.96	182.73	**101.99**	103.41
N_B	43.35	19.79	**11.15**	320.25	157.19	**97.04**
N_W	50.68	20.96	**15.11**	402.98	159.18	**100.45**
N_D	89.77	73.12	**62.73**	720.64	438.36	**384.66**

B. Multiple-target sample preparation

To study the effectiveness of the proposed method, a comprehensive study using randomly generated test cases is carried out to compare the proposed method only with REMIA since it is reported in [7] as the best solution for multiple-target sample preparation in terms of the numbers of sample droplets, waste droplets and dilution operations. In the experiment, three frequently used precision levels of concentration are utilized ($d = 8, 9, 10$). As shown in [12], there are some applications which require up to one hundred different concentration values, and thus the number of target concentration values is set to 10, 20, 50 and 100 ($N = 10, 20, 50, 100$, respectively). For a fixed values of d and N, we generate 20 test cases, in each of which the set of target concentration values (TC) and the set of the required numbers of each target concentration value (S) are generated randomly. The average values of the results of the test cases are extracted for comparison. Due to the limitation of space, only a partial of the results ($d = 10$ and $N = 10, 100$) is reported in Table II.

Firstly, we compare the number of sample and buffer droplets during sample preparation. As shown in Table II, the proposed method can produce much better results in terms of the numbers of sample droplets (N_S) and buffer droplets (N_B). Specifically, $ours_S$ can achieve as high as 60% reduction on the number of sample droplets in comparison with REMIA. For example, when $d = 9$ and $N = 100$, the values of N_S achieved by REMIA and $ours_S$ are 203.98 and 80.15, respectively. As for the number of buffer droplets (N_B), we also observe more than 70% reduction compared to REMIA.

Secondly, when it comes to the number of waste droplets (N_W), it can be seen that $ours_W$ can achieve as high as 85% reduction on the number of waste droplets compared to REMIA. Less numbers of sample/buffer/waste droplets implies a low cost sample preparation process, and thus the proposed method is more promising than REMIA, the best-known method for multiple-sample preparation so far.

Finally, we compare the required number of the dilution operations (N_D). $ours_W$ produces the least number of dilution operations in all cases. Specifically, when $d = 9$ and $N = 100$, more than 60% reduction on the number of dilution operations when compared with REMIA is observed in Table II. This implies that a small number of the total number of sample and buffer droplets likely leads to a sample preparation process with less number of dilution operations. More importantly, this also guarantees a high-reliability chip design.

Compared with REMIA that uses simply greedy heuristic, although the runtime of our algorithm increases from second-scale to minute-scale for the largest test case ($352, 608$ variables and $175, 374$ constraints in the ILP formulation), the optimality and high reliability achieved by the proposed method are much essential for the sample preparation problem on DMFBs. Moreover, for a biochemical application, although the sample preparation algorithm is executed only once, its result can be used many times, i.e., one dilution graph can be used several times by varying different types of sample and buffer. For example, genealogical DNA test performs the same sample preparation process for different people. Therefore, the computational time of the proposed algorithm is obviously acceptable for practical situations.

V. Conclusion

In this paper, for the first time, we introduce a hybrid cost function, which represents the practical cost of sample and buffer usage of a general sample preparation process. A network-flow-based approach is proposed in order to obtain the *optimal* value of the cost function even for multiple-target cases. Note that it is the first method to solve the sample preparation problem optimally. Moreover, due to the flexibility of the proposed model, we can also minimize the number of the sample as well as the number of waste droplets. Experiments demonstrate that for multiple-target sample preparation, the proposed method can reduce 60% of the number of sample droplets, 70% of the number of buffer droplets, 85% of the number of waste droplets, and even 60% of the number of dilution operations compared to the state-of-the-art previous method, i.e., REMIA.

References

[1] T.-Y. Ho, J. Zeng and K. Chakrabarty, "Digital microfluidic biochips: A vision for functional diversity and more than Moore," in *IEEE/ACM ICCAD*, pp. 578-585, 2010.

[2] Y.-L Hsieh, T.-Y Ho and K. Chakrabarty, "A reagent-saving mixing algorithm for preparing multiple-target biochemical samples using digital microfluidic biochips," *IEEE TCAD*, vol. 31, no. 11, pp. 1656-1669, November 2012.

[3] E. J. Griffith, S. Akella and M. K. Goldberg, "Performance characterization of a reconfigurable planar-array digital microfluidic system," *IEEE TCAD*, vol. 24, no. 2, pp. 345-357, Feb. 2006.

[4] W. Thies et al., "Abstraction layers for scalable microfluidic biocomputing," *Natural Computing*, vol. 7, no. 2, pp. 255-275, June 2008.

[5] S. Roy, B. B. Bhattacharya and K. Chakrabarty, "Optimization of dilution and mixing of biochemical samples using digital microfluidic biochips," *IEEE TCAD*, vol. 29, no. 11, pp. 1696-1708, November 2010.

[6] S. Roy, B. B. Bhattacharya and K. Chakrabarty, "Waste-aware dilution and mixing of biochemical samples with digital microfludic biochips," in *IEEE/ACM DATE*, pp. 1059-1064, 2011.

[7] J.-D Huang, C.-H Liu and T.-W Chiang, "Reactant minimization during sample preparation on digital microfluidic biochips using skewed mixing trees," in *IEEE/ACM ICCAD*, pp. 377-383, 2012.

[8] D. Mitra et al., "On-chip sample preparation with multiple dilutions using digital microfluidics," in *ISVLSI*, pp. 314-319, 2012.

[9] S. Roy et al., "Layout-aware solution preparation for biochemical analysis on a digital microfluidic biochip," in *VLSID*, pp. 171-176, 2011.

[10] C. A. Meyers and A. S. Schulz, "Integer equal flows," in *Journal Operations Research Letters*, vol. 37, no. 4, pp. 245-249, July, 2009.

[11] "http://www.ibm.com/software/integration/optimization/cplex/,"

[12] T. Xu, K.Chakrabarty and V. K. Pamula, "Defect-tolerant design and optimization of a digital microfluidic biochip for protein crystallization," *IEEE TCAD*, Vol. 29, no. 4, pp. 552-565, April, 2010.

Exploring Speed and Energy Tradeoffs in Droplet Transport for Digital Microfluidic Biochips

Johnathan Fiske   Daniel Grissom   Philip Brisk
Department of Computer Science and Engineering
University of California, Riverside
Riverside, CA 92521, USA

Abstract - This paper transforms the problem of droplet routing for digital microfluidic biochips (DMFBs) from the discrete into the continuous domain, based on the observation that droplet transport velocity is a function of the actuation voltage applied to electrodes that control the devices. A new formulation of the DMFB droplet routing problem is introduced for the continuous domain, which attempts to minimize total energy consumption while meeting a timing constraint. Henceforth, DMFBs should be viewed as continuous, highly integrated cyber-physical systems that interact with and manipulate physical quantities, as opposed to inherently discrete and fully synchronized devices.

I. Introduction

Programmable *microfluidic* technology has the potential to revolutionize many subfields of (bio-)chemistry and bioengineering through miniaturization and automation [4, 5]. One such technology is the *Digital Microfluidic Biochip (DMFB)*, which manipulates discrete (hence the name "digital") droplets of liquid on a two-dimensional grid of electrodes [17], as shown in Fig. 1; Fig. 2 shows the basic operations supported: transport, merging and splitting, mixing, and storage. Over the past decade, there has been considerable interest in automating the process of compiling *assays* (biochemical protocols) into software programs that control the actuation of droplets on a DMFB [2, 10], shown in Fig. 3.

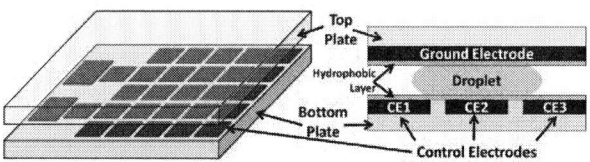

Fig. 1. A DMFB is a 2-dimensional electrode array, with I/O reservoirs on the periphery (a); a 1-dimensional cross-section; control electrodes (CEs) induce droplet motion according to the principles of electrowetting [17]; the hydrophobic layer provides insulation and prevents absorption (b).

Fig. 2. The instruction set of a DMFB: droplet transport, splitting, merging, mixing, and storage. Each of these operations is achieved by activating a specific set of electrodes in sequence in the vicinity of one or more droplets.

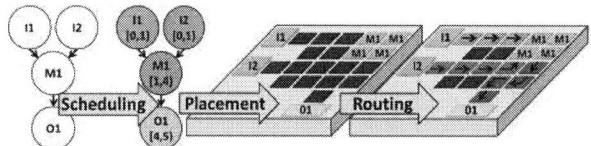

Fig. 3. The main steps of assay compilation; at present, assays are specified as directed acyclic graphs, without control flow. The focus of this paper is the routing stage.

A DMFB is a *cyber-physical system (CPS)*, meaning that it includes computational and physical components whose operation is tightly interleaved [12, 13]. One challenge in CPS design is that the computational domain is necessarily discrete and synchronized, i.e., actions occur on the granularity of *clock cycles*; in contrast, the physical domain is continuous. Prior work on DMFB compilation has imposed constraints on the system, which force it to operate in a discrete manner. One such example is the way that droplet transport, i.e., *routing* [1, 3, 11, 16, 18, 22, 25], is modeled.

A DMFB is assumed to operate synchronously: a typical actuation frequency is 100 Hz, [25], i.e., it takes 10 ms to transport one droplet to a neighboring electrode. By considering the state of the system only at 10 ms intervals, the entire system can be discretized, as shown in Fig. 4(a): a droplet can either stay at rest (e.g., Droplet 2 at 20ms), or move to a neighboring electrode (e.g., Droplet 2 at 10ms); if both electrodes are activated, then the droplet will stretch into an ovular shape that covers both electrodes.

In actuality, the velocity of a droplet traveling across the surface of a DMFB is a quadratic function of the voltage applied to each electrode [15, 17]. Thus, discretizing droplet transport through synchronization assumes that the same voltage value is always applied to each electrode. In practice, this does not need to be the case; moreover, there is no reason to believe that synchronization yields optimal or near-optimal performance (i.e., minimum droplet transport time) or energy consumption; but, considering continuous voltages and velocities transforms the problem of droplet routing from the discrete to the continuous domain, which changes the nature of the problem and the algorithmic solutions that are required, as shown in Fig. 4(b).

This paper introduces *the first* droplet transport algorithm for DMFBs that considers continuous voltages applied to electrodes, and continuous, rather than discrete, droplet velocities, thereby dropping the implicit assumption of synchronized droplet transport in prior work. A continuous time, voltage-aware formulation of the droplet routing problem is introduced that tries to minimize energy consumption during droplet transport given a timing constraint. This approach yields significant reductions in energy consumption without adversely affecting droplet routing times. Even greater energy savings could potentially be achieved by relaxing the timing constraint as well.

978-1-4799-2817-0/14 $31.00 © 2014 IEEE

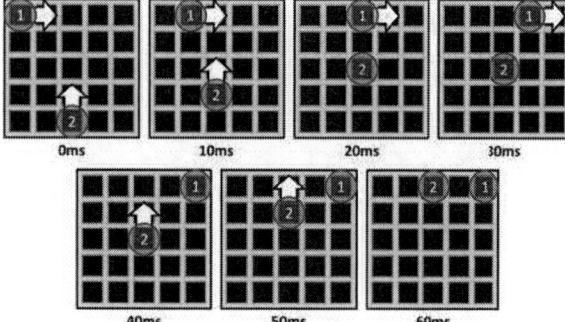

(a) Constant Discrete Voltage/Velocities

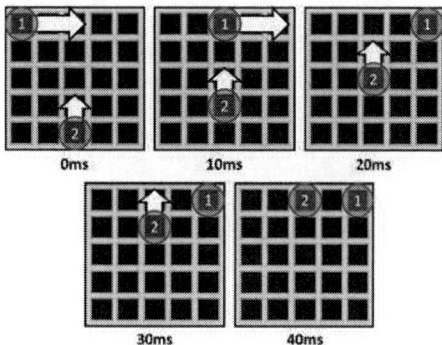

(b) Continuous Voltages/Velocities

Fig. 4. Droplets 1 and 2 cross a DMFB from one side to another (white arrows indicate the relative velocity) for (a) a DMFB driven by a single voltage, where all droplets move at the same velocity; and (b) a DMFB driven by multiple continuous voltages where droplets are allowed to move at varying velocities.

Minimizing energy consumption during execution of an assay on DMFB is an important problem, especially for portable point-of-care applications. One representative example would be to perform diagnostics as part of a health care effort targeting rural areas in the third-world, where societal infrastructure is lacking, and battery lifetime is limited.

II. Related Work

The routing algorithm presented in this paper is compatible with *direct addressing* [17] and *active-matrix* [9, 15] DMFBs, which provide independent control over all electrodes; compilation flows targeting these devices are mature [1, 3, 11, 16, 18, 20-22, 25]. It is not compatible with *pin-constrained* [2] and *cross-referencing* [6, 24] DMFBs, which provide more restricted forms of control.

The proposed algorithm would work best with DMFBs that have integrated *capacitive-touch sensors* into the substrate, which provide precise information about the position of droplets during transport [14, 19]. Otherwise, it would be too difficult to determine when a droplet completes its route at a given voltage/velocity.

Droplet routing is NP-complete [1]. The vast majority of routers that have been published, to date, are polynomial-time heuristics which cannot guarantee optimality; exceptions include optimal algorithms based on *A* search* [1] and *integer linear programming* [24], and a recent *iterative improvement* algorithm based on *particle swarm optimization* [16].

Several heuristics divide droplet routing into two parts [3, 11, 18]: (1) *path planning*, which determines the route that each droplet takes from its source to its destination, assuming that droplets are routed one-by-one; and (2) *route compaction*, which converts a the path planning result into a set of concurrent routes while adhering to proper spacing rules and trying to minimize the route of the longest droplet; thus. The algorithms presented in this paper take a similar approach: path planning followed by route compaction; in fact, any path planning algorithm can be used. We modified the compactor to consider voltage assignment to electrodes and the resulting droplet velocity in the continuous domain.

One recent paper varies the voltages applied to electrodes in a DMFB to improve reliability [23]. Another throttles droplet velocity by varying the frequency of electrode actuation [12]. To the best of our knowledge, no other papers have looked into the issues of varying voltage assignment and/or droplet velocity. Our objective is to reduce energy consumption, which is synergistic with, but not directly related to reliability.

III. Continuous-Time Droplet Routing: Preliminaries

A. Context and Spacing Rules

We assume that the assay has already been scheduled and placed on the DMFB; in principle, any appropriate algorithm could be used to accomplish these two steps. Droplet routing is decomposed into a series of *sub-problems* that must be solved [22].

As shown in Fig. 5, droplets must obey spacing rules to prevent inadvertent merging [22], which were originally introduced for discrete droplet transport. The spacing rules define an *interference region (IR)* around each droplet; if one droplet enters the IR of another, then the two will merge; to ensure correctness, the router must prevent this from occurring. The IR for a droplet resting at position $p = (x, y)$ is the 3x3 sub-array surrounding p.

The interference region for a droplet that is moving from position $p_1 = (x, y)$ to an adjacent position, p_2, which is $(x \pm 1, y)$ or $(x, y \pm 1)$, is the 3x4 sub-array surrounding p_1 and p_2. The former is called the *static* constraint, and the latter is called the *dynamic* constraint, as shown in Fig. 5. Consider a droplet at position p_1 at time t in the discrete domain. If the droplet remains at position p_1 at time $t+1$, then the static constraint must be satisfied; if it moves to an adjacent position at time $t+1$, then the dynamic constraint must be satisfied. Given the respective positions of each droplet in a larger set as a function of time, it is straightforward to determine whether *any* pair of droplets violates the static or dynamic constraint, as the number of constraint checks is finite.

In the continuous domain, at time t, a droplet is either at rest in position p, or in transport from position p_1 to adjacent position p_2: the static constraint must be satisfied in the former case, while the dynamic constraint must be satisfied in the latter case.

Droplets undergoing transport must also maintain a 1-position margin of separation with all other ongoing assay operations, such as mixing and storage; the size and location of these respective operations are fixed for the duration of each routing sub-problem.

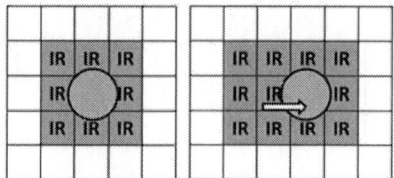

Fig. 5. The static interference region of a droplet at rest (left); the dynamic interference region of a droplet in motion (right).

B. Problem Statement

In a routing sub-problem, the input is a set of droplets to route, $D = \{d_1, d_2, ..., d_M\}$. The initial position of droplet d_i is called a *source*, denoted $src(d_i)$; d_i's final position is called a *sink*, denoted $sink(d_i)$. Each droplet must be routed from source to sink without violating static/dynamic interference constraints, while maintaining a 1-position separation with all ongoing assay operations. The objective is to minimize the completion time of all droplet routes.

C. Continuous-Time Routing Interference Constraints

The *path* taken by droplet d_i, is a sequence of positions taken by d_i along the route from $src(d_i)$ to $sink(d_i)$, and is denoted by the ordered sequence $P(d_i) = <p_{i,1}, p_{i,2}, ..., p_{i,N}>$, where $p_{i,1} = src(d_i)$, $p_{i,N} = sink(d_i)$, and $p_{i,j+1}$ is adjacent to $p_{i,j}$ for $1 \leq j \leq N-1$.

In the discrete model, d_i spends $z_{i,j} \geq 1$ time-steps at position $p_{i,j}$ along the path, as it may be necessary to wait while other droplets pass by. The number of time-steps spent during routing is

$$Z(d_i) = \sum_{j=1}^{N} z_{i,j}. \tag{1}$$

In the continuous model, d_i will spend a continuous amount of time $t_{i,j} \geq 0$ at position $p_{i,j}$, during which the static interference constraint must be satisfied, followed by time $u_{i,(j,j+1)} > 0$ while moving from $p_{i,j}$ to adjacent position $p_{i,j+1}$, during which the dynamic interference constraint must be satisfied. The total amount of time spent during routing is

$$T(d_i) = \sum_{j=1}^{N} \left(t_{i,j} + u_{i,(j,j+1)} \right), \tag{2}$$

and the total time spent routing d_i from $src(d_i)$ to position $p_{i,j}$, is

$$U(d_i, p_{i,j}) = \sum_{k=0}^{j-1} \left(t_{i,k} + u_{i,(k,k+1)} \right), \tag{3}$$

where $t_{i,0} = u_{i,(0,1)} = 0$.

The continuous time interval at which d_i waits at position $p_{i,j}$ is therefore given by the pair

$$I(d_i, p_{i,j}) = \left[U(d_i, p_{i,j}), U(d_i, p_{i,j}) + t_{i,j} \right], \tag{4}$$

and the continuous time interval at which d_i is spent in transportation from position $p_{i,j}$ to $p_{i,j+1}$ is given by the pair

$$J(d_i, p_{i,j}, p_{i,j+1}) = \left(U(d_i, p_{i,j}) + t_{i,j}, U(d_i, p_{i,j+1}) \right). \tag{5}$$

Thus, the timing information associated with each path is given by the ordered sequence of pairs

$$Q(d_i) = <I(d_i, p_{i,1}), J(d_i, p_{i,1}, p_{i,2}), ..., \\ I(d_i, p_{i,N-1}), J(d_i, p_{i,N-1}, p_{i,N})>. \tag{6}$$

Fig. 6 shows an example in which a droplet moves continuously across six positions; it waits at the fourth position for six time-units, and then increases its velocity as it travels to the sixth.

If droplet d_i moves across the chip with positive velocity, then it remains at position $p_{i,j}$ for an instant, which is assumed to consume no time. Hence, in the continuous model, $I(d_i, p_{i,j}) = 0$ except when

the droplet explicitly pauses at position $p_{i,j}$. A droplet may pause for two reasons: (1) to prevent the violation of an interference constraint; or (2) the droplet arrives at its respective sink and stops while waiting for other droplets to complete their routes.

Lemma 1. Any (static or dynamic) interference constraint between droplets d_i and d_j can be detected at continuous time $t \in \{t_1, t_2\}$ such that $(t_1, t_3) \in Q(d_i)$ and $(t_2, t_4) \in Q(d_j)$.

Proof: If d_i and d_j are initially placed at a location where a static constraint violation occurs (i.e., before droplets start to move), we are done, as $t = t_1 = t_2 = 0$. Now, assume that the constraint violation occurs at time $t > 0$. Either d_i or d_j must be in motion to cause the violation to occur; otherwise, it would have occurred at time $t' < t$. A constraint violation is an activation of an electrode at a position p^* inside the interference region of both d_i and d_j. The constraint violation occurs at the instant either droplet starts to move from an initial position p toward an adjacent position p^*. $\square$

Fig. 7 illustrates the proof of Lemma 1.

Lemma 1 shows how to determine if constraint violations occur in continuous time. If there are K droplets in the system, traverse $Q(d_i)$ for $i - 1$ to K; for each interval $(t_k, t_l) \in Q(d_i)$, identify the exact position of each other droplet d_j at time t_k. If d_i is moving toward position p^* at time t_k, then a constraint violating occurs if p^* is in the static or dynamic interference region of d_j, depending on whether d_j is at rest or in motion.

The total number of interference checks that must be made to verify that the routes of all droplets are interference-free is

$$C = (K - 1) \sum_{i=1}^{K} |Q(d_i)|. \tag{7}$$

Since droplets move at different velocities, the constraint check is *antisymmetric*. For example, suppose that droplets d_i and d_j have a constraint violation at time t, such that $t_1 < t < t_4$ for $(t_1, t_4) \in Q(d_i)$ and $t_2 < t < t_3$ for $(t_2, t_3) \in Q(d_j)$, such that $t_1 < t_2 < t_3 < t_4$. The constraint violation is observable at time t_2, but not at time t_1, in accordance with Lemma 1. In contrast, interference constraint violations in the discretized, synchronized case are symmetric.

D. Simplifying Assumptions

In DMFB routing, droplets move horizontally or vertically, but not diagonally [22]; we make the same assumption here.

Given a droplet path $P(d_i)$, as defined in the preceding section, a *segment* is defined to be a maximum-length contiguous subsequence $<p_{i,j}, p_{i,j+1}, ..., p_{i,k}>$ of $P(d_i)$ in which d_i travels in one direction (left, right, up, or down). Let $s_{i,j}$ denote the jth segment along path $P(d_i)$, and let $S(d_i) = <s_{i,1}, s_{i,2}, ..., s_{i,n}>$, $n \leq N$ (recall that N is the number of positions along the path) denote the decomposition of $P(d_i)$ into segments.

time											
Transport/Wait Times	$t_{i,1}$	$u_{i,(1,2)}$	$t_{i,2}$	$u_{i,(2,3)}$	$t_{i,3}$	$u_{i,(3,4)}$	$t_{i,4}$	$u_{i,(4,5)}$	$t_{i,5}$	$u_{i,(5,6)}$	$t_{i,6}$
	0	3	0	3	0	3	6	2	0	2	11
Arrival Time at Each Position	$U(d_i, p_{i,2})$		$U(d_i, p_{i,3})$		$U(d_i, p_{i,4})$		$U(d_i, p_{i,5})$		$U(d_i, p_{i,6})$		$U(d_i, p_{i,6}) + t_{i,6}$	
	3		6		9		17		19		30	
	$I(d_i, p_{i,1})$	$J(d_i, p_{i,2})$	$I(d_i, p_{i,2})$	$J(d_i, p_{i,3})$	$I(d_i, p_{i,3})$	$J(d_i, p_{i,4})$	$I(d_i, p_{i,4})$	$J(d_i, p_{i,5})$	$I(d_i, p_{i,5})$	$J(d_i, p_{i,6})$	$I(d_i, p_{i,6})$
Q(di)	<[0, 0],	(0, 3),	[3, 3],	(3, 6),	[6, 6],	(6, 9)	[9, 15],	(15, 17),	[17, 17],	(17, 19),	[19, 30]>

Fig. 6. Example illustrating continuous time-domain droplet routing parameters: droplet d_i starts at position $p_{i,1}$ at time zero. It travels to $p_{i,4}$ at a velocity of 1/3 of a position per time-unit. It then waits at $p_{i,4}$ for 6 time-units, and travels to $p_{i,6}$ at a velocity of 1/2 of a position per time-unit. It then waits at $p_{i,6}$ for 11 additional time-units. The total time taken, including transport and waiting, is 30 time-units.

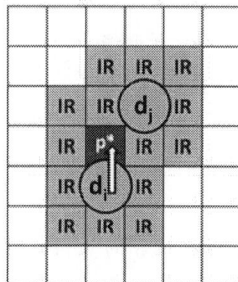 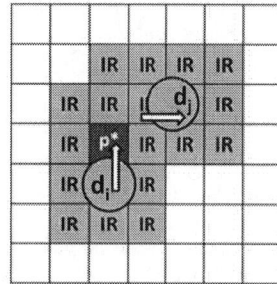

Fig. 7. Illustration of the proof of Lemma 1. Activating the electrode at position $p*$ transports droplet d_i into the interference region of droplet d_j; $p*$ is activated precisely at time instant t_k, which is the beginning of a time interval $(t_k, t_l) \in Q(d_i)$.

For example, suppose that droplet d_i travels from $src(d_i) = (1, 1)$ to $sink(d_i) = (4, 2)$ along path $P(d_i) = <(1, 1), (2, 1), (3, 1), (4, 1), (4,2)>$; $P(d_i)$ can be decomposed into a horizontal segment $s_{i,1} = <(1, 1), (2,1), (3, 1), (4,1)>$ and a vertical segment $s_{i,2} = <(4,1), (4,2)>$, so $S(d_i) = <s_{i,1}, s_{i,2}>$. It is important to note that the subsequence $<(1, 1), (2,1), (3, 1)>$ is not considered to be a segment here, because it is *not* maximal length.

We assume that each droplet travels at a constant velocity along each segment, and that the droplet may only stop at the initial and final segment positions. This means that voltage selection is performed on the granularity of segments, but interference constraints must still be resolved on the granularity of positions in continuous time, as discussed in the preceding section.

E. Velocity and Energy Models

Continuous time-domain droplet routing introduces, a new degree of freedom: to select the voltage applied to each electrode from a continuous range, as opposed to turning the electrode on (at a constant voltage) or off; the router must still adhere to continuous time-domain interference constraints (Section III.C).

This subsection summarizes the equations used to estimate the droplet velocity and power consumption that results from applying a voltage to an electrode in an active-matrix DMFB [9, 15]. To avoid confusion with notation established in previous sections, we use words, rather than letters, to represent these quantities. Droplet velocity (mm/s) as a function of voltage is derived from existing data [15, Fig. 3(a)] using a least-squares quadratic approximation:

$$Velocity = 0.005 \times Voltage^2 \tag{8}$$
$$+0.0358 \times Voltage - 0.9103$$

The energy consumption of each electrode depends on the voltage applied, the electrode's resistance, and the length of time that the voltage is applied:

$$Energy = Power \times Time = \frac{Voltage^2}{Resistance} \times Time \tag{9}$$

As noted previously, we assume that a droplet moves at a constant velocity along each segment; i.e., a constant voltage is applied to each electrode along the segment, and only one electrode (per droplet) is activated at any time.

F. Technology Parameters

The typical resistance of an electrode in an active-matrix DMFB is around $1G\Omega$, and the electrode pitch was reported to be $2.54mm$ [15], which we take to be the length of each square in the DMFB array. We assume that there are minimum and maximum voltage levels (13 and 70 Volts respectively) for electrode actuation.

Voltages below the minimum cannot actuate droplet motion, and voltages above the maximum will physically degrade the chip. As an example, consider the voltage applied to transport droplet d_i along segment $s_{i,j}$; let $|s_{i,j}|$ denote the number of positions that comprise the segment, so the total distance to travel along this segment will be $2.54mm \times (|s_{i,j}| - 1)$. The droplet velocity (and thus, the total time traveled, which we will denote $\Delta T_{i,j}$) depends on the constant voltage level $V_{i,j}$ chosen for this segment.

The energy consumed to transport d_i along segment $s_{i,j}$, denoted by $E_{i,j}$, depends on $V_{i,j}$ and $\Delta T_{i,j}$ as well:

$$E_{i,j} = \frac{V_{i,j}^2}{1G\Omega} \times \Delta T_{i,j} \tag{10}$$

As per Eq. (8), $\Delta T_{i,j}$ also depends on $V_{i,j}$ (via the classic mechanical equation: $distance = velocity \times time$). Thus, as long as droplet paths are determined a-priori, $E_{i,j}$ depends only on voltage.

We assume that 10 Volts are required to hold a droplet in-place; if no voltage is applied, the droplet will not be anchored and will drift aimlessly. Thus, a droplet at rest consumes energy regardless of whether a discrete or continuous routing model is used.

IV. Continuous-Time Droplet Routing Algorithms

Section III.A characterized the droplet routing problem; the objective was to finish all of the routes as quickly as possible, which is equivalent to minimizing the length (in time) of the longest route. This formulation of the droplet routing problem does not account for potential tradeoffs between droplet velocity and energy consumption of the device. Here, we assume that a time constraint is provided and the objective is to minimize energy consumption while meeting the timing constraint.

The algorithms that we introduce here employ the assumptions outlined in Section III.E. Our algorithm takes a two-part approach, in which droplet routing paths are computed a-priori, followed by a compaction step which enables concurrent droplet transport. Given the set of routing paths, compaction determines the sequence of timing intervals $Q(d_i)$ for each droplet d_i, as discussed in Section III.D. The compactor must prevent interference violations from occurring to ensure a legal routing solution. The voltage-aware formulation of the routing problem adds an additional constraint (timing) while defining a new optimization criterion (minimize energy) in addition to the basic criteria for an interference-free droplet route.

A. Compaction Strategy

Our implementation is based on an algorithm published by Roy et al. [18], which uses a Maze Router to compute droplet routes, and then applies a greedy heuristic to perform compaction. The heuristic sorts the droplets based on a priority function, and compact the routes one-by-one. A typical priority function, which we use in our implementation, is to prioritize the droplets by the length of their routes, such that droplets with the longest routes have the highest priority. A droplet d_i is *compacted* when the velocity along each segment in its route and the waiting time at each segment endpoint are known; i.e., once $Q(d_i)$ is computed.

In this algorithm, compacted routes are *never* revisited. When compacting droplet d_i, the compacted routes for all higher priority droplets impose constraints that must be satisfied to ensure a legal routing solution. If d_i's route crosses the compacted route of a higher priority droplet d_j, then the compactor must construct $Q(d_i)$ in a manner that prevents interference constraint violations involving any higher priority droplets from occurring; lower priority droplets are not given any consideration during compaction until they are routed themselves.

When compacting a droplet, the segments that compose the route are compacted one-by-one, in-order. To compact a segment along a route, the algorithm must choose a voltage (which, in turn, determines the droplet velocity) that leads to no interference constraint violations with any droplets whose routes are already compacted. The compaction process for the droplet completes when all segments along the route are successfully compacted.

B. Timing-Constrained Route Compaction Algorithm

This formulation of the routing problem includes a timing constraint, denoted T_C; all droplets must complete their routes in time no greater than T_C. The objective is to minimize the total energy consumption while doing so.

Droplets are processed for compaction, one-by-one, as described in the preceding subsection. The exact length $L(d_i)$ of the droplet path for droplet d_i is known, as it is the number of electrodes on the path multiplied by the electrode pitch (2.54mm in active-matrix technology [15]). The initial velocity for d_i is chosen to be $L(d_i)/T_C$, which is the minimum velocity required to meet the latency constraint, under the assumption that d_i does not pause along its path. The voltage level that produces this velocity can be determined using Eq. (8); if this voltage is lower than the minimum allowable voltage, then the minimum allowable voltage is chosen.

Starting at the first position in each segment the algorithm incrementally computes $Q(d_i)$ one electrode position at a time. As per the discussion in Section III.C, the compactor tests whether each advancement of $Q(d_i)$ at its chosen velocity violates an interference constraint each time that a new electrode is activated along the path. If a violation occurs, it is *not* possible to slow down d_i; otherwise, a timing violation would occur. Therefore, the only option is to increase d_i's velocity (by increasing the voltage) along the segment where the violation occurs. The increase should be the minimum necessary to suppress the constraint violation. Increasing the velocity along the segment will alter the timing intervals in $Q(d_i)$; therefore, the part of $Q(d_i)$ corresponding to the present segment being processed needs to be updated.

If a droplet is sped up along a segment, then it will arrive early at the destination if the initially computed velocity is used for the remainder of the route, and even earlier if its velocity increases along a future segment; in principle, the droplet could be slowed down along the remainder of the route in order to further reduce energy, as long as the timing constraint is not violated. Thus, whenever a droplet is sped up along a segment, the algorithm re-computes the minimum velocity required to complete the *remaining* route without violating the timing constraint. The remainder of the route is compacted using the voltage that corresponds to this newly computed velocity; if this voltage is lower than the minimum allowable voltage, then the minimum allowable voltage is used instead.

C. Limitations

In some cases, it may not be possible to find a compaction solution without exceeding the maximum allowable voltage. This is likely to occur if there exist a large number of droplets that need to be routed through a congested area of the DMFB. The only remedy for this situation is to re-schedule and re-place the assay, under the optimistic hope that the resulting routing sub-problems can be solved. The problem of unroutable routing sub-problems has been noted elsewhere [3, 18, 22, 25] and is not specific to the continuous time-domain formulation of the problem. We did not encounter any unroutable sub-problems in our experiments, although the theoretical possibility of their existence remains.

V. Simulation Results

A. Setup and Approach

We implemented the timing-constrained cyber-physical droplet routing algorithm in a publicly available open source framework for DMFB synthesis [8]. We modeled an active matrix DMFB [15], whose technology parameters are reported in Section III.F. We used a synthesis flow described in ref. [7] as our baseline; which includes Roy's algorithm for droplet routing [18]. We replaced the discrete compaction phase of Roy's algorithm with the continuous-time domain compactor described in Section IV.B, and compared the energy consumption with that of the baseline. We used a standard set of DMFB benchmarks that are publicly available and have been used in prior studies on DMFB synthesis and routing: polymerase chain reaction (PCR), multiplexed in-vitro diagnostics (In-vitro 1-5), and protein crystallization (ProteinSplit 1-4 and Protein).

Here, we focus on individual routing sub-problems in which multiple droplets are transported. For each routing sub-problem, we first route the droplets using the baseline routing algorithm using a constant voltage; the droplet transport velocity for a given voltage level is given by Eq. (8). We record the time t at which the last droplet finishes its route, along with the energy E_{Base} expended to transport all droplets, as modeled by Eq. (9). We then run the time-constrained droplet routing algorithm, described in Section IV, using t as a timing constraint; we record the energy expended to transport all droplets, denoted E_{TC}. We report the energy savings $E^* = (E_{Base} - E_{TC})/E_{Base}$ as a percentage.

Among the seven assays that we considered, PCR, Protein, and five variations of multiplexed in-vitro diagnostics, we identified 99 routing sub-problems of interest. We performed three comparisons per sub-problem, running the baseline routing algorithm with voltages of 30V, 50V, and 70V, yielding routing times denoted of t_{30V}, t_{50V}, and t_{70V}. We then ran the continuous time-domain routing algorithm using t_{30V}, t_{50V}, and t_{70V} as timing constraints, and report the energy savings E^*_{30V}, E^*_{50V}, and E^*_{70V}.

B. Results and Analysis

Fig. 8 reports the energy savings attained for all routing sub-problems at all three Voltage levels, and Fig. 9 concisely summarizes the results. The two key observations are: (1) for a given voltage level, the attainable energy savings varies significantly between routing sub-problems; and (2) greater energy savings are attainable at lower voltage levels.

The variation in energy savings across sub-problems is readily apparent in Fig. 9. The standard deviation in energy savings is equal to the average for the 30V experiments, and greater than the average for the 50V and 70V experiments. The reason for these high variations is not algorithmic inconsistency, but is actually an inherent property of the routing sub-problem instances. The experiment is set up so that the droplet that finishes last will take the same path and consume the same amount of energy in both the discrete and continuous-time models; whatever energy savings is accrued is due to other droplets slowing down by throttling the voltage. Droplets whose routes are significantly shorter than the longest route can travel at very low velocities without violating the timing constraint; thus, the greatest possible energy savings can be accrued for problem instances that have an inordinate number of droplets that fall into this category. In contrast, it is difficult to achieve significant energy savings for problem instances where all droplet routes have approximately equal length.

978-1-4799-2817-0/14 $31.00 © 2014 IEEE

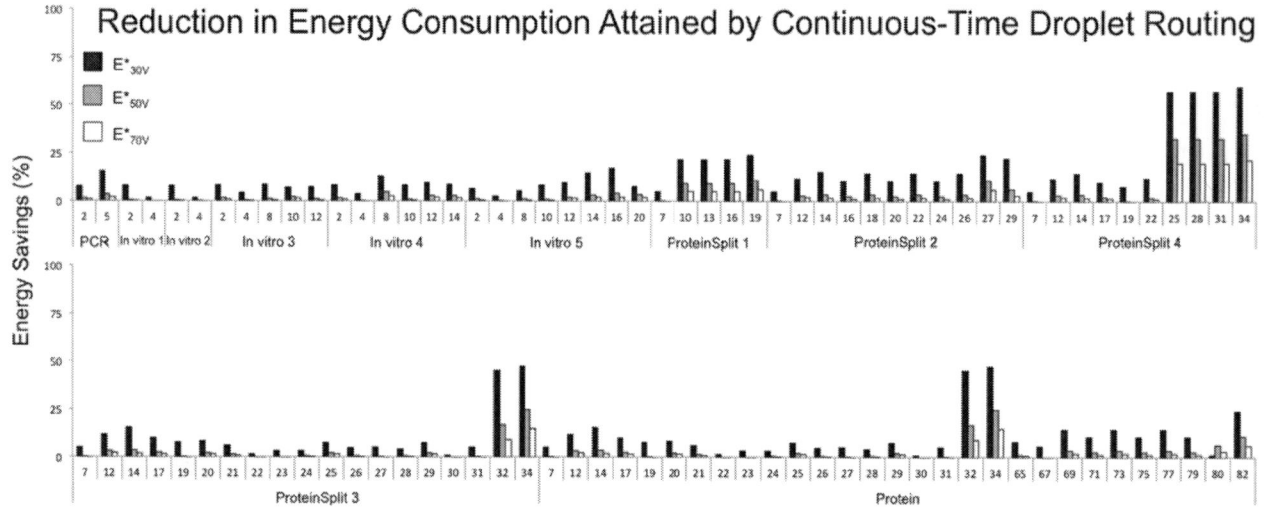

Fig. 8. Energy savings obtained by continuous-time droplet routing in comparison with discrete droplet routing at 30V, 50V, and 70V baselines. The experiments include 99 routing sub-problems; the ID number of each routing sub-problem is listed above each benchmark.

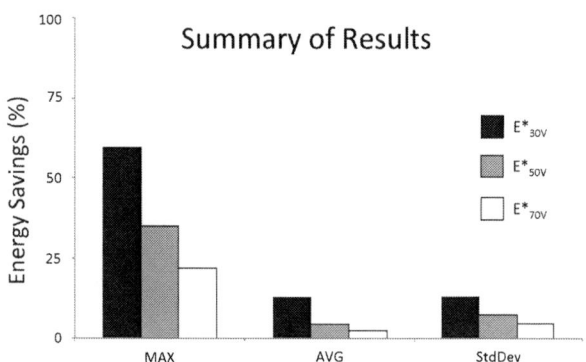

Fig. 9. Summary of the results reported in Fig. 8.

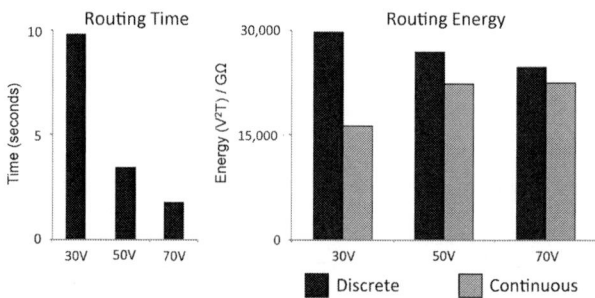

Fig. 10. Droplet routing times and energy consumption (as opposed to energy saving) for routing sub-problem 32 of ProteinSplit3.

Next, we address the issue of why the lower baseline voltage levels achieve greater energy savings than the higher voltage levels across the different routing sub-problems that were examined. As an example, Fig. 10 shows the droplet routing times and energy consumed for routing sub-problem 32 of ProteinSplit 3. The length of the longest droplet route at 30V is 2.9x longer than the length of the longest droplet route at 50V, and 5.6x longer than the length of the longest droplet route at 70V. As a consequence, there is a much greater window of time during which the compactor can slow down non-critical droplets to reduce energy at lower voltage levels.

Looking exclusively at the discrete routing model, increasing the baseline voltage yields a net energy savings because higher voltages yield higher droplet velocities, and shorter completion times. Thus, more energy is required to activate a sequence of electrodes at 30V for 9.8 seconds than to activate a sequence of electrodes at 50V for 3.4 seconds. Given the timing constraint provided by the discretized routing model at each voltage level, the continuous-time routing model reduces energy consumption; however, diminishing energy savings are observed at higher voltage levels due to shorter route completion times.

Altogether, these results demonstrate that varying the voltage applied to transport droplets at varying velocities in the continuous time-domain droplet routing model can lead to significant energy savings for favorable droplet routing sub-problems.

VI. Conclusion

This paper has developed *the first* algorithm by which synthesis tools can control droplet velocity and DMFB energy consumption during droplet transport by varying the voltage applied to electrodes. Both voltage and velocity are continuous quantities, so dealing with continuous, rather than synchronized and discrete, fluid flow creates new algorithmic challenges, especially to detect and prevent interference region constraint violations. Due to the discrete nature of the 2D electrode array, the interference regions themselves remain discrete, even though droplet transport is continuous in time. The proposed interference constraint detection algorithms leverages this observation to bound the number of distinct points in time at which constraints must be checked. This mechanism was integrated into a new continuous voltage-aware droplet routing algorithm, targeting activate-matrix actuation technology, which minimizes energy consumption while meeting timing constraints on droplet routes.

This initial foray into continuous droplet transport control is expected to launch future investigations on other aspects of DMFB synthesis that can be handled in the continuous domain. For example, it may be possible to leverage voltage assignment to explore trade-offs involving operation latency and energy consumption during scheduling.

978-1-4799-2817-0/14 $31.00 © 2014 IEEE

The tradeoffs explored in this paper can be extended to account for different problem formulations such as minimizing droplet transport time given an energy budget, and routing wash droplets to reduce cross-contamination. This work promotes understanding of DMFBs as cyber-physical systems that control continuous physical quantities with a hybrid discrete-continuous architecture.

Acknowledgment

This work was supported in part by NSF Grant CNS-1035603. D. Grissom was supported by an NSF Graduate Research Fellowship. J. Fiske was supported by a UC LEADS summer internship. Any opinions, findings, and conclusions or recommendations expressed in this material are those of the authors and do not necessarily reflect those of the NSF.

References

[1] Böhringer, K. F. 2006. Modelling and controlling parallel tasks in droplet-based microfluidic systems. *IEEE Trans. CAD* 25, 2 (Feb. 2006), 329-339. DOI= http://dx.doi.org/10.1109/TCAD.2005.855958

[2] Chakrabarty, K. 2010. Design automation and test solutions for digital microfluidic biochips. *IEEE Trans. Circuits and Systems I: Regular Papers*, 57, 1 (Jan. 2010) 4-17. DOI= http://dx.doi.org/10.1109/TCSI.2009.2038976

[3] Cho, M., and Pan, D. Z. 2008. A high-performance droplet routing algorithm for digital microfluidic biochips. *IEEE Trans. CAD* 27, 10 (Oct. 2008) 1714-1724. DOI= http://dx.doi.org/10.1109/TCAD.2008.2003282

[4] Fair, R. B. 2007. Digital microfluidics: is a true lab-on-a-chip possible? *Microfluidics and Nanofluidics* 3, 3 (Jun. 2007), 245-281. DOI= http://dx.doi.org/10.1007/s10404-007-0161-8

[5] Fair, R. B., et al. 2007. Chemical and biological applications of digital-microfluidic devices. *IEEE Design and Test of Computers* 24, 1 (Jan.-Feb. 2007), 10-24. DOI= http://dx.doi.org/10.1109/MDT.2007.8

[6] Fan, S-K., Hashi, C., and Kim C-J. 2003. Manipulation of multiple droplets on N×M grid by cross-reference EWOD driving scheme and pressure-contact packaging. In *Proceedings of the 16th IEEE International Conference on Micro Electro Mechanical Systems* (Kyoto, Japan, January 13 - 23, 2003). MEMS '03, 694-697. DOI= http://dx.doi.org/10.1109/MEMSYS.2003.1189844

[7] Grissom, D., and Brisk, P. Fast online synthesis of generally programmable digital microfluidic biochips. In *Proceedings of the ACM/IEEE International Conference on Hardware Software Codesign and System Synthesis* (Tampere, Finland, October 07 - 12, 2012). CODES-ISSS '12, 413-422. DOI= http://dx.doi.org/10.1145/2380445.2380510

[8] Grissom, D., O'Neal, K., Preciado, B., Patel, H., Doherty, R., Liao, N., and Brisk, P. A digital microfluidic biochip synthesis framework. In *Proceedings of the IEEE/IFIP International Conference on VLSI and System-on-a-Chip* (Santa Cruz, CA, USA, October 07 - 10, 2012). VLSI-SOC '12, 177-182, DOI= http://dx.doi.org/10.1109/VLSI-SoC.2012.6379026

[9] Hadwen, B., et al. 2012. Programmable large area digital microfluidic array with integrated droplet sensing for bioassays. *Lab-on-a-Chip* 12, 18 (May. 2012), 3305-3313. DOI= http://dx.doi.org/10.1039/c2lc40273d

[10] Ho, T-Y., Chakrabarty, K., and Pop, P. 2011. Digital microfluidic biochips: recent research and emerging challenges. In *Proceedings of the ACM/IEEE International Conference on Hardware Software Codesign and System Synthesis* (Taipei, Taiwan, October 09 - 14, 2011). CODES-ISSS '11, 335-343. DOI= http://dx.doi.org/10.1145/2039370.2039422

[11] Huang, T-W., and Ho, T-Y. 2009. A fast routability- and performance-driven droplet routing algorithm for digital microfluidic

biochips. In *Proceedings of the International Conference on Computer Design* (Lake Tahoe, CA, USA, October 04 – 07, 2009) ICCD '09, 445-450. DOI= http://dx.doi.org/10.1109/ICCD.2009.5413119

[12] Luo, Y., Chakrabarty, K., and Ho, T-Y. 2013. Design of cyberphysical digital microfluidic biochips under completion-time uncertainties in fluidic operations. In *Proceedings of the Design Conference* (Austin, TX, USA, June 02-06, 2013) DAC '13, article #44, DOI= http://dx.doi.org/10.1145/2463209.2488788

[13] Luo, Y., Chakrabarty, K., and Ho, T-Y. 2013. Error recovery in cyberphysical digital microfluidic biochips. *IEEE Trans. CAD* 32, 1 (Jan. 2013) 59-72. DOI= http://dx.doi.org/10.1109/TCAD.2012.2211104

[14] Murran, M. A., and Najjaran, H. 2012. Capacitance-based droplet position estimator for digital microfluidic devices. *Lab-on-a-Chip* 12, 11 (Mar. 2012) 2053-2059. DOI= http://dx.doi.org/10.1039/c2lc21241b

[15] Noh, J. H., Noh, J., Kreit, E., Heikenfeld, J., and Rack, P. D. 2012. Toward active-matrix lab-on-a-chip: programmable electrofluidic control enabled by arrayed oxide thin film transistors. *Lab-on-a-Chip* 12, 2 (Jan. 2012), 353-360. DOI= http://dx.doi.org/10.1039/c1lc20851a

[16] Pan, I., and Samanta, T. 2013. Efficient droplet router for digital microfluidic biochip using particle swarm optimizer. In *Proceedings of the SPIE 8760, International Conference on Communication and Electronics System Design 87601Z* (Jaipur, India, January 28, 2013) 87601Z-1 – 8760Z-10. DOI= http://dx.doi.org/10.1117/12.2012352

[17] Pollack, M. G., Shenderov, A. D., and Fair, R. B. 2002. Electrowetting-based actuation of droplets for integrated microfluidics. *Lab-on-a-Chip* 2, 2 (Mar. 2002), 96-101. DOI=http://dx.doi.org/10.1039/b110474h

[18] Roy, P., Rahaman, H., and Dasgupta, P. 2010. A novel droplet routing algorithm for digital microfluidic biochips. In *Proceedings of the 20th Great Lakes Symposium on VLSI* (Providence, RI, USA, May 16 - 18, 2010) GLSVLSI '10, 441-446. DOI= http://dx.doi.org/10.1145/1785481.1785583

[19] Shih, S. C. C., Fobel, R., Kumar, P., and Wheeler, A. R. 2011. A feedback control system for high-fidelity digital microfluidics. *Lab-on-a-Chip* 11, 3 (Feb. 2011) 535-540. DOI= http://dx.doi.org/10.1039/C0LC00223B

[20] Su, F., and Chakrabarty, K. 2008. High-level synthesis of digital microfluidic biochips. *ACM Journal on Emerging Technologies in Computing Systems* 3, 4 (Jan. 2008), article #16. DOI= http://dx.doi.org/10.1145/1324177.1324178

[21] Su, F., and Chakrabarty, K. 2006. Module placement for fault-tolerant microfluidics-based biochips. *ACM Trans. Design Automation of Electronic Systems* 11, 3 (Jul. 2006), 682-710. DOI= http://dx.doi.org/10.1145/1142980.1142987

[22] Su, F., Hwang, W., and Chakrabarty, K. 2006. Droplet routing in the synthesis of digital microfluidic biochips. In *Proceedings of Design Automation and Test in Europe* (Munich, Germany, March 06-10, 2006) DATE '06, 1-6. DOI= http://dx.doi.org/10.1109/DATE.2006.244177

[23] Yeh, S-H., Chang, J-W., Huang, T-W., and Ho, T-Y. 2012. Voltage-aware chip-level design for reliability-driven pin-constrained EWOD chips. In *Proceedings of the IEEE/ACM International Conference on Computer-Aided Design* (San Jose, CA, USA, November 05-08, 2012). ICCAD '12, 353-360. DOI= http://dx.doi.org/10.1145/2429384.2429461

[24] Yuh, P-H., Sapatnekar, S. S., Yang, C-L., and Chang, Y-W. 2009. A progressive-ILP-based routing algorithm for the synthesis of cross-referencing biochips. *IEEE Trans. CAD* 28, 9 (Sep. 2009), 1295-1306. DOI= http://dx.doi.org/10.1109/TCAD.2009.2023196

[25] Yuh, P-H., Yang, C-L., and Chang, Y-W. 2008. BioRoute: a network-flow-based routing algorithm for the synthesis of digital microfluidic biochips. *IEEE Trans CAD* 27, 11 (Nov. 2008), 1928-1943. DOI= http://dx.doi.org/10.1109/TCAD.2008.2006140

General Purpose Cross-Referencing Microfluidic Biochip with Reduced Pin-Count

Jackson H. C. Yeung

Department of Computer Science and Engineering
The Chinese University of Hong Kong
e-mail: jacksonyeung@gmail.com

Evangeline F. Y. Young

Department of Computer Science and Engineering
The Chinese University of Hong Kong
fyyoung@cse.cuhk.edu.hk

Abstract— The number of control pins used is a major factor affecting the manufacturing cost of Digital Microfluidic Biochip (DMFB). Pin-count on a DMFB can be reduced by sharing of control pins between electrodes. Most existing works on reducing pin-count are problem specific. Problem specific optimizations result in DMFB that can only perform certain specific bioassays. Cross-Referencing DMFB has a full array layout that is fully reconfigurable for any bioassay. Conventional Cross-Referencing DMFB uses $m+n$ number of pins. We have devised a non-problem specific pin assignment methodology that uses only $\sqrt{2}(\sqrt{m}+\sqrt{n})$ number of pins. The resulting DMFB are still fully reconfigurable. We have developed a droplet router specifically for cross-referencing DMFB with shared control pins. All real bioassay tested can be routed using a fixed and problem independent control pin mapping. Reduction on pin count ranges from 50% to 67%.

I. INTRODUCTION

Digital Microfluidic Biochip(DMFB) is receiving much attention in recent years. This technology allows traditional laboratory procedure to be conducted on a chip of the size of a few square centimeters. It is showing great promises in multiple disciplines such as molecular biology, medical, pharmaceutical and environmental monitoring.

A. Digital Microfluidic Biochip

Digital Microfluidic Biochip [1] enables the manipulation of liquid droplets on a 2-D array using electrowetting-on-dielectric (EWOD) technology. Movement of the liquid droplets is controlled by applying voltages to the electrodes. In the most conventional design, there is a bottom plate containing a patterned array of electrodes and a glass top plate containing a continuous electrode. To move a droplet, the target cell is activated by introducing a potential difference between the top plate and bottom plate of a cell. Basic operations such as mixing, detection and diluting can be carried out by manipulating the position of the droplets. In DMFB, locations can be re-used by different modules. This is unlike ASIC where the position of modules are fixed after fabrication. This property is in some way similar to FPGA and is referred as reconfigurability.

A.1 Direct Addressing and Pin Constrained DMFB

In direct addressing DMFB, each electrode is connected to a dedicated control pin. Such scheme maximizes flexibility at the expense of using a large number of control pins. Most productions of DMFB do not contain any active element. An external controller is used to control the electrodes. As a result the number of control pins is severely constrained by the cost and packaging issue, making direct address scheme is not viable in practice.

Various problem specific Pin-Constrained schemes are devised to reduce pin count on direct addressing DMFB. In a pin-constrained DMFB, each control pin addresses more than 1 electrode, reducing the number of control pins.

A.2 Cross-Referencing DMFB

Cross-referencing DMFB is first proposed in [2]. Rows and columns of electrodes are placed orthogonally on the top and bottom plates. A cell is activated by charging the row and column with opposite polarity. A major advantage of a cross-referencing design is the small pin-count. For a DMFB of size $n \times m$, a direct addressing layout requires $n \times m$ pins while a cross-referencing layout uses only $n+m$ pins. The low pin count of cross-referencing designs are due to its duo plate row/column electrode architecture rather than problem specific optimizations, as in the case of pin constrained design. As a result, a cross-referencing DMFB remains fully configurable.

B. Related Prior Work

B.1 Pin Constrained DMFB

Various pin-constrained methodologies have been proposed. They can significantly reduce the number of control pins by sharing of electrodes. An array partition scheme for DMFB is proposed in [5]. A chip is separated into partitions so that a control pin can address cells in multiple partitions. Another method that identify electrodes that can share a control pin from the actuation sequences is proposed in [9]. A method for co-optimization of routing and control pin mapping using ILP is proposed in [8]. All of the above methods are problem specific. It is not possible to reconfigure the chip for a new problem. An algorithm for producing a control pin layout for a general purpose chip layout is purposed in [3]. However, it does not consider electrode constraints between with diago-

978-1-4799-2817-0/14 $31.00 © 2014 IEEE

nally adjacent electrodes. So it is only useful in some specific layout.

B.2 Cross-Referencing DMFB

There are a number of works on solving the routing problem of cross-referencing DMFB. An ILP formulation for routing cross-referencing DMFB is proposed in [7]. Algorithms for placement and routing of cross-referencing DMFB is proposed in [4]. A SAT-based routing algorithm for cross-referencing DMFB is proposed in [6].

C. Our Contributions

By connecting a control pin to more than one electrodes, it is possible to reduce the number of control pins on a cross-referencing DMFB. Due to the unique row/ column electrode structure in cross-referencing DMFB methods of mapping control pins to electrodes for pin constrained DMFB cannot be applied here. We are not aware of any works on sharing control pin on cross-referencing DMFB. We have devised a method that maps the electrodes to a smaller number of control pins. Our method has the advantage of being problem independent and does not affect reconfigurability. The number of control pins is reduced from $(m+n)$ to $\sqrt{2}(\sqrt{m}+\sqrt{n})$.

The major contributions of this paper can be summarized as follows:
- We identified some unroutable conditions with control pin sharing in cross-referencing biochip. We found a lower bound on the number of control pins required to avoid those conditions.
- We devise a methodology for doing problem independent mapping of electrodes to control pins without affecting routability.
- We developed a router for cross-referencing DMFB with shared control pin. The router is able to route real world bioassay using only $\sqrt{2}(\sqrt{m}+\sqrt{n})$ number of pins.

Our methodology reduces pin count by 50 % to 67 % for the benchmark bioassays. Since the number of control pins required is asymptotically smaller than $(m+n)$, this method can realize huge savings on the number of pins for larger DMFB.

II. DROPLET ROUTING

Droplet routing in DMFB is fundamentally different from routing in VLSI. In droplet routing, the goal is to find a path for the droplet to travel from source to sink without violating any constraints. A path is not a permanent physical connection but rather the trajectory of the droplet through time. There are two important constraints which is explained below.

A. Fluidic and Electrode Constraint

To prevent unintended mixing, two droplets cannot occupy adjacent cells at any same moment, this is referred as fluidic constraint.

When a droplet is moving from cell Q to its adjacent cell P, cell P is activated by applying voltages of opposite polarity to its row and column electrodes. All other cells adjacent to P

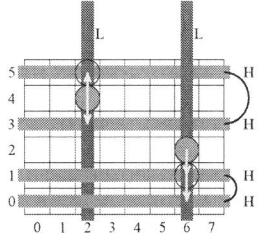 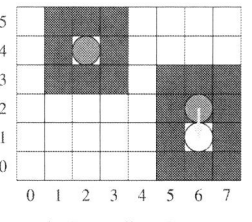

(a) Problematic pin sharing   (b) Bounding Box

Fig. 1.

and Q must not be activated. The adjacent cells of P and Q form a 4x3 bounding box as shown by the highlighted region on the right of Figure 1b. In this paper, this is referred as *dynamic electrode constraint*. For a stalling droplet, the 8 cells surrounding the droplet as shown in the left of Figure 1b must not be activated. In this paper, this is referred as *static electrode constraint*. When a cell is activated as a side effect of moving other droplets, such cell is said to be *extra activated*. The problem of assigning the correct voltage to electrodes so that the droplets are moved along the intended path is referred to as *electrode assignment*.

III. SHARING OF CONTROL PIN

Sharing of control pin on cross-referencing DMFB always put limitations on droplet movements. Most of these limitations can be avoided by the router by changing the droplet movement schedule or changing the droplet path. However, some conditions do make routing impossible. In this section, we identify the most common conditions that make routing impossible or significantly lengthens routing path.

Control pin is said to be shared when more then one electrodes are connected to the same control pin. For a sequence of electrodes $E = \{e_1, e_2, ..., e_n\}$, the control pin mapping can be represented by the sequence $Pin(E) = \{p_{\pi(1)}, p_{\pi(2)}, ..., p_{\pi(n)}\}$, where $p_{\pi(n)}$ is the control pin connected to electrode e_n. In this paper, the process of assigning control pins to electrodes is referred as *pin assignment*. $Pin(E)$ is referred as a control pin assignment sequence.

Charging a control pin also charges all electrodes connected to such control pin to the same polarity. If each row control pin and each column control pin is connected to k electrodes, charging a pair of control pin activates k^2 cells, creating $k^2 - 1$ extra activated cells. The extra activated cells can violate the constraint of other droplets. As a result, whenever control pins are shared we cannot guarantee that all droplets are able to move in all four possible directions without affecting other droplets. We will leave it to the router to find a path and movement schedule so that no extra activated cell causes problems.

A. Control pin spacing

When two electrodes sharing the same control pin have a distance less than or equal to two, electrode assignment to move a droplet may violate its own dynamic electrode constraint. This is illustrated in Figure 1a. The figure shows a DMFB where row electrode 3 and 5 are controlled by a single pin and row electrode 0 and 1 are again control by a single pin. The distance between row 3 and 5 is 2 and the distance between row 0 and 1 is 1. The droplet on the left is to be moved to cell (2,5). However, since row electrode 5 and 3 are connected, cell (2,3) is extra activated. This violates its dynamic electrode constraint. In fact, no droplet can ever be moved to any cell on row 5. The droplet on the right is to be moved to cell (6,1). Since row 1 and row 0 are connected, cell (6,0) is extra activated. This again violates the dynamic electrode constraint. As a result, no droplet can reach the cells on row 0 or row 1.

B. Bound Droplets

When the pin assignment sequence contains a repeating subsequence, it may happen that two droplets can enter a set of cell positions where electrode assignment to move one droplet to any of its four orthogonal neighbors produces the same movement in the other droplet. As a result, it is not possible to control the path of one droplet independently of the other one.

Definition 1. We define *bound droplets* as two droplets staying in cell positions where electrode assignment to move one droplet for a distance of two or more cells in any direction will also move the other droplet the same distance in the same or opposite direction.

It means that for droplet 1 in (x_1, y_1) and droplet 2 in (x_2, y_2). Moving droplet 1 to $(x_1 + 1, y_1)$ and then to $(x_1 + 2, y_1)$ will also move droplet 2 to $(x_2 + k, y_2)$ and then to $(x_2 + 2k, y_1)$, where $k = \pm 1$. Similar cases for vertical movement can be obtained by swapping x and y. The positions (x_1, y_1) and (x_2, y_2) are referred as bound positions.

The situation is illustrated in Figure 2a. In this figure, the control pin mapping is a repeating sequence $Pin(E) = \{0, 1, 2, 3, 4, 0, 1, 2, 3, 4\}$. Control pin number is labeled on the axises. The arrow in the figure shows the intended path of droplet 1. To prevent confusion, electrode number is not shown. We assume that the origin $(0,0)$ is on the lower left corner. By activating row control pin 3 and column control pin 0, droplet 1 is moved to position $(0, 8)$ and droplet 2 is moved to position $(5, 3)$. After this steps any attempt to move droplet 1 will move droplet 2 in the same direction and vice versa. This is because for all the subsequent positions, each of the 4 orthogonally adjacent cells of droplet 1 and the corresponding cell of droplet 2 are controlled by the same pair of control pins. Essentially, subsequent droplet paths mirror each other. This is shown by the arrowed line in the figure.

Given that droplet 1 is moving to $(0, 8)$, Figure 2b shows all possible positions at which a droplet will become bound with droplet 1. Any droplet residing in the highlighted cell when droplet 1 is moving to $(0, 8)$ will become bound with droplet 1.

It should be noted that this phenomenon is not limit to the case when the entire control pin assignment sequence contains a repeating subsequence. As long as there are some repeating sub-sequences, two droplet can become bound for a distance equals to the length of the subsequence. Although bound droplets do not always make routing impossible, it makes routing more difficult. It is very likely that the lengths of the routing paths have to be increased.

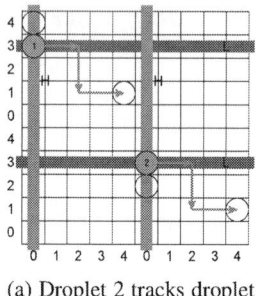 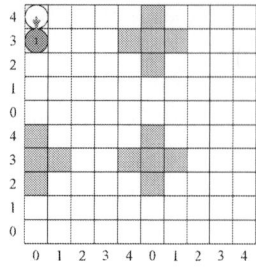

(a) Droplet 2 tracks droplet 1

(b) Cannot put droplet 2 in shaded area

Fig. 2. bound droplets

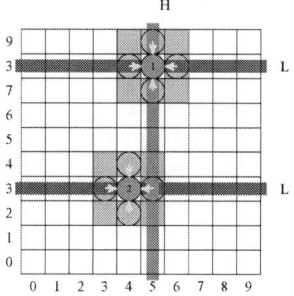

Fig. 3. Type 1 unreachable state

C. Unreachable States

Sharing of control pins can lead to a situation in which it is impossible to move two droplets to a certain combination of cell positions.

Definition 2. An *unreachable state* is a pair of cell positions (x_1, y_1) and (x_2, y_2), that for any droplet A not located at (x_1, y_1) and any droplet B not located at (x_2, y_2), no sequence of valid electrode assignment can produce a state in which A is in position (x_1, y_1) and B is in position (x_2, y_2).

We classifies this problem into two cases, which we refer as type 1 and type 2. An unreachable state is type 1 if the pair of cell positions is on adjacent rows and shares the same column control pin or Vice Versa. Otherwise it is of type 2. The problem is illustrated in the following subsections.

C.1 Type 1

In Figure 3, we want to move droplet 1 and droplet 2 to position 1 and position 2 respectively as indicated in the figure. We assume that each droplet can be in any of the four orthogonally adjacent cells of its destination, which covers all possible cases. The droplets can reach their destinations in 3 different order. The first case is for droplet 2 to reach its destination first, follow by droplet 1. The second case is the reverse of the first case. The third case is when both droplets reach their destinations at the same time-step.

For case 1, in order to move droplet 1 to its target position, row control pin 3 and column control pin 5 are charged with opposite voltages. The resulting electrode assignment violates the static electrode constraint for droplet 2. While droplet 1 will reach its destination, droplet 2 will move to position $(5, 3)$.

978-1-4799-2817-0/14 $31.00 © 2014 IEEE   240

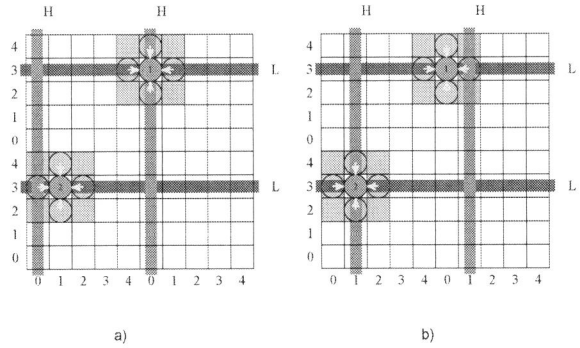

Fig. 4. Type 2 unreachable state (a) electrode assignment to move droplet 1 (b) electrode assignment to move droplet 2

For case 2, we have to charge row control pin 3 and column control pin 4 with opposite voltages. However, this will also move droplet 1 to $(4,8)$, away from its destination. For case 3, we have to charge column control pin 4 to high in addition to the electrode assignment shown in Figure 3. This electrode assignments will however activate two orthogonally adjacent cells for both droplet 1 and 2, which is illegal.

We refer the pair of droplet positions $(5,3)$ and $(4,8)$ as unreachable states. From the figure, we can see that two cell positions form an unreachable states if they share the same row control pin and have their column electrodes adjacent to each other, and vice versa. We refer this as type 1 unreachable state. Type 1 unreachable states exist whenever control pins are shared.

C.2 Type 2

The significance of type 2 unreachable state is that it is dependent on the control pin assignment sequence. In this case, neither the row electrodes or column electrodes of the target cells are adjacent to each other. This is illustrated in Figure 4. Similar to the type 1 unreachable state, moving droplet 1 to its target position violates the static electrode constraint for droplet 2, as illustrated in Figure 4a. Moving droplet 2 to its target position violates the static electrode constraint for droplet 1, as illustrated in Figure 4b. Moving the two droplets simultaneously will violating the dynamic electrode constraint for both moving droplets. Therefore, no two droplets can enter the combination of cell positions as indicated in the figure. This is referred as type 2 unreachable state.

IV. CONTROL PIN ASSIGNMENT

A. Control pin assignment constraint

In this section, we discuss the methodology for performing problem independent control pin assignment. We found that if a control pin assignment sequence $P = \{p_1, p_2, ..., p_n\}$ satisfies the following constraint, we can avoid the problems discussed in section A, section B and section C.2.

Constraint 1. *The control pin assignment sequence $P = \{p_1, p_2, ..., p_n\}$ should satisfies the following conditions:*

1. $\forall p_i \in P$ and $\forall p_j \in P$ where $p_i = p_j$ and $i \neq j$, $|j - i| \geq 3$

2. $\forall p_i \in P$ and $\forall p_j \in P$, if $p_i = p_j$ and $i \neq j$, $\{p_{i-1}, p_{i+1}\} \cap \{p_{j-1}, p_{j+1}\} = \emptyset$

The above constraint prescribes that any 2 numbers in the control pin sequence cannot be adjacent to each other in more than one place, which also implies that any 2 electrodes with distance apart less then 3 cannot share a control pin. For example, we have a sequence $\{1, 0, 1, 2, 3, 1, 2\}$. The subsequence $\{1, 0, 1\}$ violates condition 1. The subsequence $\{1, 2\}$ occurs twice in the sequence, violating condition 2. In the following subsections, we will describe how the above constraint can avoid the problems described in section III.

A.1 Control pin spacing

We will show that constraint 1 eliminates the problem discussed in section A, where extra activated cell produced by a moving droplet violates its own dynamic electrode constraint. When one pair of control pins are charged to move a droplet to its destination, case 1 of constraint 1 limits the minimum distance between any two activated cells to be 3. Since the boundary of the 4x3 bounding box of moving droplet is at most two cells away from its destination, constraint 1 ensures that any extra activated cell generated is outside the droplet's 4x3 bounding box. Therefore, the dynamic electrode constraint will not be violated.

A.2 Avoiding bound droplets and Unreachable state

If the control pin assignment sequence satisfies constraint 1, bound droplets and type 2 unreachable state cannot exist. Due to space limitation, the proof cannot be included in this paper. In this subsection, we give a brief explanation on how constraint 1 helps to avoid the above problems.

We will illustrate how constraint 1 avoid bound droplet with an example. Assume that droplet a at $(0,0)$ and droplet b at $(4,4)$ are bound, and moving droplet a 2 steps to $(2,0)$ will move droplet b to $(6,4)$. This implies that the column control pin assignment is $\{..., p_0, p_1, ..., p_0, p_1\}$, which violates condition 2 of constraint 1.

We will illustrate how constraint 1 avoids unreachable states using the example in Figure 4. In this example, column 0 and column 5 share the same control pin. Condition 2 of constraint 1 prevents column 1 from sharing a control pin with column 4 or column 6. Therefore, moving droplet 2 to its destination $(1,3)$ will not activate any adjacent cell of droplet 1. By moving droplet 1 to its destination first followed by moving droplet 2 to its destination, both droplets can reach their destinations.

B. Minimum Number of Pins

In this section, we will find the minimum number of control pins required to satisfy constraint 1.

Lemma 1. *The maximum length of a sequence P that can be formed by k distinct integer so that no pairs of integer can appear next to each other in more than 1 instance is represented by the following equation, where L is the maximum length of the sequence.*

When k is odd,

$$\frac{k(k-1)}{2}+1=L$$

When k is even,

$$\frac{(k-1)(k-2)}{2}+3=L$$

Proof. Consider an undirected graph $G(V,E)$ constructed by adding a vertex $v_i \in V$ for each of the k distinct integers i in the sequence P. For each pair of adjacent integers i and j in P, an edge $\{v_i, v_j\} \in E$ is added to $G(V,E)$. Note that constraint 1 is equivalent to constructing the sequence P as an Euler path in $G(V,E)$. An Euler path exists on an undirected graph if and only if at most 2 vertices have odd degree. When k is odd, the maximum length Euler path occurs when the graph is a complete graph. In this case the length of a maximal Euler path is equivalent to the number of edges, which is $\frac{k(k-1)}{2}$. Since each edge represents 2 adjacent integers in P, the maximum length of P is $\frac{k(k-1)}{2}+1$. When k is even, a graph containing a maximum length Euler path can be constructed by first forming a complete graph with $k-1$ vertices. Then two edges are added between the remaining vertex and any of those $k-1$ vertices. The number of edges in such a graph is $\frac{(k-1)(k-2)}{2}+2$ so the maximum length of P is $\frac{(k-1)(k-2)}{2}+3$.

$\square$

Theorem 2. *For a cross-referencing DMFB containing n electrodes on each axis, the number of control pins p required for each axis in order to satisfy constraint 1 is given by the following equation.*

If p is odd,

$$\frac{p(p-1)}{2}+1 \geq n$$

If p is even,

$$\frac{(p-1)(p-2)}{2}+3 \geq n$$

Proof. The sequence described in lemma 1 satisfies both conditions of constraint 1 for a control pin assignment sequence. This result thus follows from the proof of lemma 1. $\square$

When n is large the lower bound of p can be approximated by $p = \sqrt{2n}$.

C. Algorithm for Control Pin Assignment

The algorithm for generating the pin constraint sequence is shown in Figure 5. The algorithm generates a maximal length pin assignment sequence with p control pins by repeatedly generating subsequences of length p while making sure that constraint 1 is not violated.

D. Placement Constraint

As shown in previous section, type 1 unreachable state always exists when control pins are shared. The router will avoid using those combination of positions during routing. However, all droplets must reach their respective sink positions in each subproblem. We need to make sure that the sink positions for each subproblem do not have any type 1 unreachable states. This is done by the placement modification algorithm described below.

For each subproblem, we check all pairs of sink positions. If two sink positions share the same row control pin and have column electrode next to each other, or Vice Versa, it constitute

Fig. 5. Algorithm Generate Pin Assignment Sequence

```
1:  procedure GENPIN(p)
2:      for i = 1 → p do
3:          s₀[i] = i //initialize subsequence
4:          adj[i−1][i] = adj[i][i−1] = true
5:      end for
6:      current = s₀[p]
7:      repeat
8:          L = L + s₀; offset = 0
9:          for i = 1 → p do
10:             for j = offset → p + offset do
11:                 idx = j%p
12:                 if adj[current][s₀[idx]] is false and s₀[idx] ∉ s₁ then
13:                     s₁[i] = s₀[idx]
14:                     adj[current][s₀[idx]] = true
15:                     adj[s₀[idx]][current] = true
16:                     offset = idx; current = s₁[i]; break
17:                 end if
18:             end for
19:         end for
20:         s₀ = s₁; s₁ = ∅
21:     until s₀ = ∅ //stop if no new number found
22:     return L
23: end procedure
```

an type 1 unreachable state. In that case, we locate the module for the sink position and move the sink position to the nearest position along the module boundary. Then, we check all sink position again. This is repeated until no type 1 unreachable state is found.

V. ROUTER

Resolving electrode conflict in cross-referencing biochip is a difficult problem. Sharing of control pin significantly increases the chance of having electrode conflict. Traditional approach using Maze routing with rip-up and reroute does not have a global view of the problem and thus cannot in general resolve conflict between droplets in difficult situations.

The common approach to do routing in DMFB is to search for a viable route first. Then the router finds the correct voltage to assigned to each electrode at each time step in order to move the droplet along the chosen route. This approach is not efficient on cross-referencing DMFB because many combinations of droplet movements cannot be implemented using any valid electrode assignment. At each time step, our router searches on all the possible ways of assigning voltage to each control pin and then determines if the resulting droplet movement is useful. Our method eliminates the need to evaluate combination of droplet movement that cannot be carried out.

Our router considers up to two pairs of control pins activation in each time-step. Due to control pin sharing, charging two pairs of control pins will activate $\frac{2m \times n}{p^2}$ number of cells, where p is the number of control pins on each axis. For example, on a 21 by 21 cross-referencing DMFB using 7 control pins per axis, charging two pairs of control pins activate 18 cells, transporting a maximum of 18 droplets in each time-step.

In our router, a status $U_i = \{Q_i, C_i, t\}$ is used to represent the locations of all droplets and the control pins being activated at time t. $Q_i = \{(x_1,y_1),(x_2,y_2),...,(x_d,y_d)\}$ contains locations of all d droplets. $C_i = \{(c_{x1},c_{y1}),(c_{x2},c_{y2})\}$ contains up to two pairs of charged control pins. The weight of a status U_i is computed using the equation:

$$weight(U_i) = t + MD\_SUM(Q_i)$$

where U_i is a status and $MD\_SUM(Q)$ is the sum of Manhattan distances to sinks for all droplets. Although not discussed on this paper, the above equation can be easily modified to accomplish other optimization goals such as minimizing cross contamination.

First, a status representing the initial state $\{Q_0, \emptyset, 0\}$ is put into a priority queue. Q_0 contains the source positions. In each iteration, the router removes the status with the lowest weight from the priority queue $U_j = \{Q_j, C_j, t_j\}$. It then considers each possible pairs of control pin activations. If a pair of control pin activations $C_k = \{(c_{xk}, c_{yk})\}$ does not move any droplet or if it causes violation it is ignored. Otherwise new status $U_k = \{Q_k, C_k, t_j + 1\}$ is created and put into the priority queue, where Q_k contains the new droplet positions at time $t_1 + 1$. The status U_k is associated with a pointer that points to its parent U_j. If C_j contains only 1 pair of activated control pins, the router will also check if the droplet movements in U_j and U_k can be implemented in a single time-step. If combining the control pin activations in C_j and C_k do not violate any constraint, a new status $U_l = \{Q_l, C_l, t_j\}$, where $C_l = C_j \cup C_k$, is created and put into the priority queue.

This process is then repeated. The router terminates when it removes a status $\{Q_f, C_f, t_f\}$ where Q_f represents the sink positions for all the droplets and t_f is within the timing constraint. The router then performs a backtrack to find the routes for all droplets and the electrode assignments to implement such routes.

VI. EXPERIMENTAL RESULT

Our algorithm is tested on 4 real bioassays containing a total of 168 subproblems commonly use for testing cross-referencing DMFB. The benchmarks are obtained from [7]. The result is shown in Table I. All subproblems in benchmark In-vitro1, In-vitro2 and Protein1 are routed using the same control pin assignment sequence containing 7 pins. All subproblems in Protein2 are routed using the same control pin assignment sequence containing 6 pins. The number of pins required for cross-referencing DMFB is shown in column 4 and the number of pins required with control pin sharing is shown in column 5. The percentage difference in pin count is shown in column 6. The improvements are proportional to the size of the DMFB and ranges from 50 % to 67 %. The second column shows the number of subproblems in the benchmark and the third column shows the number of subproblems modified by our placement modification algorithm. Only 4 out of the 168 subproblem need placement modification. This testifies to the robustness of our control pin assignment scheme.

Table II compares our droplet routing time with Cross-router [4] without pin sharing. While the maximum routing time can increase significantly for some difficult subproblems, the average routing time increases by only 24% to 39%. We consider this a good result as our approach uses less than half of the original number of control pins without using any problem specific optimization.

TABLE I
Number of Control Pins

Bioassay	#P	#Mod	#Pin without sharing	#Pin with sharing	Diff. %
In-vitro1	11	0	32	14	-53%
In-vitro2	15	2	28	14	-50%
Protein1	64	2	42	14	-67%
Protein2	78	0	26	12	-54%

TABLE II
Routing Time

	Crossrouter		This Work		Diff.%
Bio-Assay	Max.	Avg.	Max.	Avg.	Avg.
In-vitro1	20	12.1	35	16.8	+39%
In-vitro2	19	10.7	38	13.7	+28%
Protein1	20	15.5	44	19.7	+27%
Protein2	20	9.9	45	12.3	+24%

VII. CONCLUSION

We studied the problem of control pin sharing in cross-referencing DMFB. We have developed a problem independent control pin mapping methodology for cross-referencing DMFB that uses only $\sqrt{2}(\sqrt{m} + \sqrt{n})$ number of control pins. Our approach dramatically reduces pin count of cross-referencing DMFB without sacrificing reconfigurability.

REFERENCES

[1] K. Chakrabarty and T. Xu. *Digital Microfluidic Biochips: Design and Optimization*. CRC Press, 2010.

[2] S.-K. Fan, C. Hashi, and C.-J. Kim. Manipulation of multiple droplets on n times;m grid by cross-reference ewod driving scheme and pressure-contact packaging. In *Micro Electro Mechanical Systems, 2003. MEMS-03 Kyoto. IEEE The Sixteenth Annual International Conference on*, pages 694 –697, jan. 2003.

[3] Y. Luo and K. Chakrabarty. Design of pin-constrained general-purpose digital microfluidic biochips. In *Design Automation Conference (DAC), 2012 49th ACM/EDAC/IEEE*, pages 18 –25, june 2012.

[4] Z. Xiao and E. Young. Placement and routing for cross-referencing digital microfluidic biochips. *Computer-Aided Design of Integrated Circuits and Systems, IEEE Transactions on*, 30(7):1000 –1010, july 2011.

[5] T. Xu, W. L. Hwang, F. Su, and K. Chakrabarty. Automated design of pin-constrained digital microfluidic biochips under droplet-interference constraints. *J. Emerg. Technol. Comput. Syst.*, 3(3), Nov. 2007.

[6] P.-H. Yuh, C. C.-Y. Lin, T.-W. Huang, T.-Y. Ho, C.-L. Yang, and Y.-W. Chang. A sat-based routing algorithm for cross-referencing biochips. In *Proceedings of the System Level Interconnect Prediction Workshop*, SLIP '11, pages 6:1–6:7, Piscataway, NJ, USA, 2011. IEEE Press.

[7] P.-H. Yuh, S. Sapatnekar, C.-L. Yang, and Y.-W. Chang. A progressive-ilp based routing algorithm for cross-referencing biochips. In *Proceedings of the 45th annual Design Automation Conference*, DAC '08, pages 284–289, New York, NY, USA, 2008. ACM.

[8] Y. Zhao and K. Chakrabarty. Simultaneous optimization of droplet routing and control-pin mapping to electrodes in digital microfluidic biochips. *Computer-Aided Design of Integrated Circuits and Systems, IEEE Transactions on*, 31(2):242 –254, feb. 2012.

[9] Y. Zhao, T. Xu, and K. Chakrabarty. Broadcast electrode-addressing and scheduling methods for pin-constrained digital microfluidic biochips. *Computer-Aided Design of Integrated Circuits and Systems, IEEE Transactions on*, 30(7):986 –999, july 2011.

978-1-4799-2817-0/14 $31.00 © 2014 IEEE

Wash Optimization for Cross-Contamination Removal in Flow-Based Microfluidic Biochips

Kai Hu[†], Tsung-Yi Ho[‡] and Krishnendu Chakrabarty[†]

[†]ECE Dept., Duke University, Durham, NC, USA
{kh131, krish}@duke.edu

[‡]National Cheng Kung University, Tainan, Taiwan
tyho@csie.ncku.edu.tw

Abstract—Recent advances in flow-based microfluidics have enabled the emergence of biochemistry-on-a-chip as a new paradigm in drug discovery and point-of-care disease diagnosis. However, these applications in biochemistry require high precision to avoid erroneous assay outcomes, and therefore are vulnerable to contamination between two fluidic flows with different biochemistries. Moreover, to wash contaminated sites, the buffer solution in flow-based biochips has to be guided along pre-etched channel networks. This constraint makes washing in flow-based microfluidics even harder. In this paper, we propose the first approach for automated wash optimization for contamination removal in flow-based microfluidic biochips. The proposed approach targets the generation of washing pathways to clean all contaminated microchannels with minimum execution time. A path dictionary is first established by pre-searching physically implementable paths in a given chip layout. When wash targets and occupied microchannels are defined, the proposed methods determine an optimized path set with the least washing time by calculating the priorities of wash targets. Two fabricated biochips are used to evaluate the proposed washing method. Compared to an ad hoc baseline method, the proposed approach leads to more efficient washing in all cases.

I. INTRODUCTION

Flow-based microfluidic biochips are an emerging technology that provides fluid-handling capability on a small scale [1]. These devices allow us to manipulate small volumes of fluid in a complex network of etched microchannels through thousands of integrated microvalves [2]. Recent developments in fabrication techniques, referred to as microfluidic Very Large Scale Integration [3], has compressed the typical sizes of valves to $6 \times 6 \, \mu m^2$. Increasing integration levels provide biochips with tremendous potential; hundreds of different assays, i.e., protocols for biochemistry, can be processed independently, simultaneously, and automatically [4]. As a result of these advances, flow-based biochips can be used for a variety of applications, such as drug discovery [5] and point-of-care disease diagnosis [6]. In 2011, Fluidigm, a biotech company that focuses on flow-based biochips, launched its initial public offering on NASDAQ, which is a significant milestone in the maturation of the microfluidic industry.

Despite the above developments, the problem of contamination removal in flow-based microfluidic biochips has not yet been addressed. This problem is important because bioassays always require high precision to avoid erroneous assay outcomes, and they are vulnerable to contamination between two fluidic flows with different biochemistries. Nevertheless, as feature sizes are scaled down, the increasing perimeter-to-area ratio of microchannels causes a considerable portion of fluid to be left in the microchannel walls. When two fluidic flows pass through the same microchannel one after the other,

the second flow can be contaminated by residue from the first flow, leading to erroneous assay outcomes. This problem is referred to as cross-contamination.

To avoid cross-contamination, wash operations are necessary to remove residue in microchannels. To accomplish this, a buffer fluid can be injected and flush microchannels in its flow path. Wash-time minimization is necessary because an increase in the time-to-result for a bioassay is detrimental to real-time detection and the analysis of biochemical samples that degenerate rapidly.

Recent advances in contamination avoidance for digital microfluidic biochips [7], [8] cannot be applied here because of the many differences in the underlying technologies. In digital microfluidic biochips, droplets are manipulated on a 2D electrode array. A wash droplet is dispensed and transported to contaminated spots for washing. Compared with digital devices, flow-based microfluidic biochips must consider the following additional constraints:

- The 2D grid layout of digital microfluidic biochips provides full freedom of wash droplet routing. Contaminated spots can be accessed from all the neighboring electrodes. However, for flow-based microfluidic biochips, wash fluid flows are restricted to the pre-etched microchannel networks. Contamination can be removed only along the direction of microchannels.
- In digital microfluidic biochips, multiple wash droplets can be dispensed at any time before washing. They can be stored at any positions on the chip and manipulated simultaneously for concurrent contamination removal at an appropriate time. In a flow-based device, fluid cannot be temporarily stored. Wash flows have to be generated on a path-by-path basis and should be executed immediately when a buffer fluid is injected.

The above constraints make wash optimization much harder for flow-based microfluidics than for digital microfluidics. To date, no systematic washing solution has been proposed for flow-based microfluidic biochips and washing has been neglected in chip design and assay execution.

In this paper, we propose the first approach for automated wash optimization for contamination removal in flow-based microfluidic biochips. The proposed approach targets the generation of washing pathways to clean all contaminated microchannels with minimum on-chip execution time. A path dictionary is first established by pre-searching physically implementable paths in a given chip layout. Once wash targets and occupied microchannels are defined by users, the proposed methods determine an optimized path set with the least washing time by calculating the priorities of wash targets. Two

978-1-4799-2817-0/14 $31.00 © 2014 IEEE

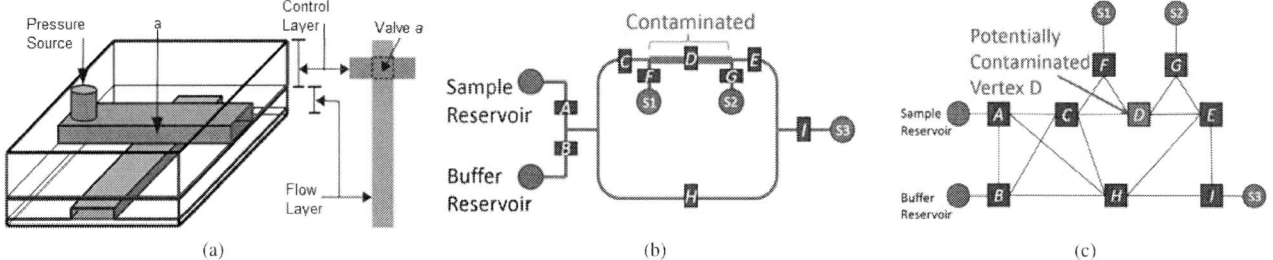

Fig. 1: (a) A schematic of a two-layer flow-based biochip. (b) Layout of a simple microfluidic biochip with a mixer (the circle) and a branch. The potentially contaminated microchannel is labeled in red. (c) A discretized graph corresponding to Fig. 1(b). The contaminated vertex is "D", marked in red.

fabricated biochips are used to evaluate the proposed washing method. The proposed approach leads to more efficient washing in all cases.

The rest of this paper is organized as follows. Section II formulates the contamination-washing problem for flow-based microfluidic biochips. The proposed algorithm for the search of washing paths is presented in Section III. Two fabricated flow-based devices are used to demonstrate the effectiveness of the proposed approach in Section IV. Conclusions are drawn in Section V.

II. PROBLEM DESCRIPTION AND FORMULATION

A basic microfluidic device is composed of two elastomer layers: a flow layer and a control layer (Fig. 1(a)). The flow layer is connected to a fluid reservoir through a pump that generates the fluid flow. The control layer is connected to an external air pressure source. A flexible membrane, working as a micro-valve, is formed at the overlapping area between channels of the two layers. When the pressure source is activated, high pressure deflects the membrane of the valve to pinch the underlying flow layer and block the fluid flow inside (closed valve). A simple layout of a flow-based biochip is shown in Fig. 1(b). The lines indicate flow channels. Rectangles indicate the positions of valves, which are connected to pumps via control channels (not shown in the figure). Components S1, S2 and S3 are outlet ports.

We can wash microchannels by generating buffer flows that cover all wash targets. In order to create a buffer flow, a pathway needs to be established from a buffer reservoir to a sink. A pump connected to the reservoir injects buffer into channel networks to generate a flow along the path. The desired path is generated by activating valves: valves along the path are opened and the rest are closed. For example, to clean the contaminated channel in Fig. 1(b), a washing path is established by closing any one of valves {A, F, H, E}, valves {A, F, H, G} and valves {A, C, G, I}.

For a given chip layout and a set of wash targets, we use a discretized graph to model a continuous fluid-flow topology and facilitate the analysis of biochip channel networks. Vertices in the graph represent not only valves, but also their downstream flow channels. All interconnection relationships between channels and valves are represented by edges. Fig. 1(c) illustrates an example for Fig. 1(b). Note that not all paths in the discretized graph are physically legal. For example, in Fig 1(c), path B⇒C⇒D⇒F⇒S1 is illegal because the opening of valves {B, C, D, F} cannot generate a

flow that passes through the contaminated channel; buffer will flow along B⇒C⇒F⇒S1 without passing D. There are four important criteria that must be considered to evaluate whether a washing path in graph is physically implementable:

1) Paths must start at a buffer reservoir and end at a sink.
2) Every vertex can be passed through only once. No loops are allowed. Each vertex in the graph is mapped to a microchannel (or node) in biochips. Therefore, flow can be injected into the same channel (or node) only once; otherwise there will be a conflict in fluid flow.
3) At most two valves can be open at each intersection. For example, if edge C↔F in Fig. 1(c) is enabled, edge C↔D and D↔F cannot be activated at the same time.
4) Washing and other operations should be processed in parallel in order to minimize waiting time. Accordingly, washing path must bypass occupied "busy" microchannels to avoid interruption of other concurrent tasks.

In wash optimization, our first goal is to generate a set of washing pathways that can clean all wash targets at least once. The second goal is to minimize the wash time, which is the sum of fluidic execution time for all selected washing paths. The washing time for a path consists of two parts: the time for path preparation and the time for buffer to flow through the path. The former is a constant for all paths and depends on how long it takes for valves to close (in parallel), while the latter is proportional to the length of the path. Therefore, we can model execution time for a path as $w = t + l/v$, where t represents the time for path preparation, l is the length of the path, and v is the velocity of fluidic flow. In a typical flow-based chip, the value of t is approximately 2 s [9] and the value of v is approximately 10 mm/s [10].

III. SEARCH FOR A SET OF WASHING PATHS

The problem of determining an efficient set of washing paths is difficult because of the following reasons:

- All paths must be physically implementable. They should satisfy all the constraints listed in Section II.
- The graph for a flow-based biochips consists of nodes with low degree because any intersection can have no more than four branches. Due to the lack of edges, it is difficult to find an implementable path that passes through a vertex corresponding to a specific contamination site.
- Wash optimization is considered after chip design. In order to retain generality, we assume that the graph generated for a biochip can have any arbitrary topology.

978-1-4799-2817-0/14 $31.00 © 2014 IEEE

Path ID	List of Vertices
1	B, C, F
2	B, C, D, G
3	B, C, D, E, I
4	B, H, I
5	B, H, E, G
6	B, H, E, D, F

a

Vertex ID	Path List	Vertex ID	Path List
B	1, 2, 3, 4, 5, 6	F	1, 6
C	1, 2, 3	G	2, 5
D	2, 3, 6	H	4, 5, 6
E	3, 5, 6	I	3, 4

b

Table I: Path dictionary corresponding to Fig. 1(b). (a) Initial dictionary format after path search. Data is organized path-by-path. (b) Reformatted path dictionary. Data is organized vertex-by-vertex.

Hence, techniques such as dynamic programming, which require specific problem substructure, cannot be exploited for wash optimization.

- Wash targets might be defined or updated dynamically by users when an experiment is set up or even during experiments. Therefore, for rapid response, an efficient search algorithm is necessary.

In this section, we describe an approach that is designed to identify a set of washing path with the least washing time. We first establish a path dictionary before the start of the experiment. Only physically legal pathways in the given chip layout are stored in the dictionary. This off-line data-preparation step needs to be performed only once but it can greatly accelerate washing-path search. After users define wash targets and occupied microchannels, we can simply look up in the dictionary the paths that include wash targets; they are known to be physically implementable. Finally, an efficient approach based on a weighted hitting-set problem formulation is used to determine path set with the least washing time.

A. Generation of Path Dictionary

The path dictionary is established before the execution of biochemical experiments, without any knowledge of wash targets and occupied channels. All paths in the dictionary must satisfy Criteria 1-3 of path implementability listed in Section II. We consider two methods for the generation of dictionary. All paths in the dictionary are candidate washing pathways.

The first approach for dictionary generation is based on depth-first search (DFS). The search starts at the buffer reservoir and transverse the entire graph until a sink is met (Constraint 1). A vertex list is created to register all "available" vertices and it is updated after each exploration: all neighbors of previously selected vertex are removed from the list to prevent violations of Constraint 2 and Constraint 3. A path is found and stored in the dictionary if a sink node is encountered. Backtracking occurs if none of the children nodes is "available". A path dictionary based on the graph in Fig. 1(c) is presented in Table I(a). Six paths are found and verified to be physically implementable.

The DFS method is able to find all physically implementable pathways. However, since DFS requires traversal of the complete graph, its worst-case complexity grows exponentially with the size of a biochip. Hence, it is only suitable for small chips. The second dictionary generation method, namely random path search, is more practical for large designs. In the random path-search method, a path is extended by randomly selecting an adjacent vertex in the "available" vertex list. Without traversing the complete graph, this search technique can potentially provide full coverage of vertices in the graph in low CPU time. Yet, this method suffers from the limitation that some valid candidate paths might be overlooked. An incomplete path dictionary may lead to longer washing paths and unsuccessful path search due to the lack of candidates. Hence, there is a trade-off between CPU time and the quality of the set of washing paths.

B. Storage of the Path Dictionary

After path search is completed, we obtain a path dictionary that can potentially be very large. Paths are listed sequentially in the dictionary; see Table I(a). However, only paths that contain a wash targets are candidates washing path. Other paths will not be considered for washing. When the wash targets and occupied channels are known during an actual experiment, we have to look up paths that include targets in the full dictionary. Obviously, this process is inefficient because usually only a few microchannels are contaminated and need to be washed in realistic scenarios.

We address this problem by reformatting the path dictionary. Instead of path-by-path organization, we store the dictionary entries in a vertex-by-vertex manner; see Table I(b). All paths that cover a vertex are listed explicitly. This data structure enables us to directly find washing path candidates (paths containing washing targets) without resorting to path filtering.

Moreover, if a number of wash targets, say p, are defined, a new dictionary with only paths containing wash targets is created and uploaded into memory. Other paths are deleted because they cannot be used for washing. For example, for the path dictionary in Table I, if wash targets $\mathcal{T} = \{C, E\}$ and occupied channel $\mathcal{O} = \{H\}$, the new dictionary is simply compressed as $\{C, E\} = \{\{1, 2, 3\}, \{3\}\}$, which considerably reduces both data volume and search complexity. Path 5 and Path 6 are deleted because it contains occupied channel H.

C. Identification of Washing-Path Set

In this subsection, we propose an efficient approach to determine an optimized path set with full-coverage of wash targets and minimum washing time. The approach needs to be computationally efficient because washing targets need to be updated during experiments. With the help of a path dictionary, the search is accelerated because all paths in the dictionary have been verified to be physically implementable.

1) Relevant Definitions: Recall that a flow-based microfluidic biochip can be modeled as a discrete graph. Vertices in the graph represent not only valves, but also their downstream flow microchannels. Let X denote the set of vertices in the graph. Similarly, each path in the dictionary, which is verified to be physically implementable, is indexed and let M denote the set of paths. For each vertex $j \in X$, we let the path-coverage set $S_j \subseteq M$ represent a set of paths that covers the jth vertex. Therefore, the generated path dictionary $\mathbb{S}$ is the collection of all path-coverage sets, that it, $\{S_j : j \in X\}$. Moreover, an additive weight function $w(p) = t + l(p)/v$ is defined to represent the washing time of path $p \in M$, where t

Symbol	Interpretation in Weighted Hitting-Set Problem	Interpretation in Wash-Optimization Problem
j	$j \in X$, the index of a set in collection $\mathbb{S}$	A vertex in the graph
X	An index set of collection $\mathbb{S}$	Set of all vertices in the graph
p	$p \in M$, the index of an element in the collection $\mathbb{S}$	A path in the graph
M	$M = \cup_{j \in X} S_j$, a set of all elements in the collection $\mathbb{S}$	Set of all paths in the graph
H	A hitting set	A wash path set
S_j	The jth set in the collection $\mathbb{S}$	Set of paths that covers the jth vertex
$\mathbb{S}$	$\{S_j : j \in X\}$, collection of S_j	Path dictionary
$\mathbb{S}'$	$\{S_j : j \in \mathcal{T}\}$, subcollection of collection $\mathbb{S}$	Compressed path dictionary
S	$S \in \mathbb{S}$, a set in the collection	A set in the path dictionary
$w(p)$	Weight of element $p \in M$	Wash time of path p
$\mathcal{T}$	$\mathcal{T} \subseteq X$, a set that specifies the sets required to be hit	Set of wash targets
$\mathcal{O}$	$\mathcal{O} \subseteq X$, a subset denoting all forbidden vertices	Set of occupied channels

Table II: Notation summary. All parameters of a variant of the weighted hitting-set problem can be mapped to that of the wash-optimization problem.

represents the time for path preparation, $l(p)$ is the length of the path p, and v is the velocity of fluidic flow.

2) Formulation as a weighted hitting-set problem: We next show that wash optimization can be described as a variant of the weighted hitting-set problem. A hitting set H of a collection $\mathbb{S}$ is a set that contains at least one element from each subset $S \in \mathbb{S}$, that is, $S \cap H \neq \emptyset, \forall S \in \mathbb{S}$. Each element in $\mathbb{S}$ is given a nonnegative weight, w, and the goal of the weighted hitting-set problem is to find the hitting set H with minimum total weight. This problem can be formalized as:

$$\underset{H}{\text{minimize:}} \sum_{p \in H} w(p) \qquad (1)$$

$$\text{subject to: } S \cap H \neq \emptyset, \forall S \in \mathbb{S} \qquad (2)$$

The standard hitting-set problem requires a non-empty intersection with every set $S \in \mathbb{S}$; see Equation (2). We relax this requirement by allowing the hitting-set to "hit" just a specified portion of collection $\mathbb{S}$, whose indices are recorded in a subset $\mathcal{T} \subseteq X$, instead of all sets S in $\mathbb{S}$. The new constraint can be formulated as:

$$S_j \cap H \neq \emptyset \text{ for every } j \in \mathcal{T} \qquad (3)$$

where S_j is the jth set in the collection $\mathbb{S}$. Moreover, an additional constraint can be placed on the standard hitting-set problem; some sets in $\mathbb{S}$ is forbidden to be "hit", i.e.,

$$S_i \cap H = \emptyset \text{ for every } i \in \mathcal{O}, \qquad (4)$$

where $\mathcal{O} \subseteq X$ is a subset of vertices denoting all the forbidden vertices corresponding to occupied microchannels.

Now we can establish the relationship between the wash-optimization problem and the weighted hitting-set problem. In wash optimization, our first goal is to find a set of washing path H that can cover all washing targets $\mathcal{T}$, i.e., satisfy Equation (3), and bypass all occupied microchannels $\mathcal{O}$, i.e., satisfy Equation (4). Washing paths in set H are executed one by one and therefore each target in $\mathcal{T}$ is washed at least once. Table II summarizes the parameters of the weighted hitting-set problems and their corresponding physical interpretation in the wash-optimization problem.

Therefore, the wash-optimization problem can be formulated as a variant of the weighted hitting-set problem:

$$\underset{H \subseteq M}{\text{minimize:}} \sum_{p \in H} w(p)$$

$$\text{subject to: } S_j \cap H \neq \emptyset \text{ for every } j \in \mathcal{T}$$

$$S_i \cap H = \emptyset \text{ for every } i \in \mathcal{O}.$$

where H is the optimal washing path set, $\mathbb{S}$ is the path dictionary, M is the set of all paths in the $\mathbb{S}$, S_j is a set of paths that covers the jth vertex, $\mathcal{T}$ is the set of wash targets, $\mathcal{O}$ is the set of occupied microchannels, and $w(p)$ is the wash time for the pth washing path.

3) Utility Function and Algorithm Design: A heuristic approach for solving this variant of hitting-set problem, which is known to be NP-complete, is a greedy approach that selects in each iteration the path that contains a maximum number of washing targets [11]. In this method, a path that contains a maximum of washing targets is selected in each step, and then the set of wash targets $\mathcal{T}$ is updated by removing vertices washed by the selected path. The washing path set is updated until all wash targets are covered. Due to its inherent simplicity, the above method is efficient. However, it is not always effective for solving the wash-optimization problem because it is based on an assumption that all vertices have equal probabilities to be visited. This assumption does not hold for flow-based microfluidic biochips. Graphs generated by flow-based biochips have low degree due to physical routing limits. Although there may be hundreds or even thousands of vertices in a graph, each vertex has no more than four neighbors. Low vertex degree leads to an unbalanced path-coverage set; some vertices are covered by more paths because they have larger degree in the graph, while others at branches are seldom visited. Therefore, vertices covered by fewer paths are more important and should be given higher priority in the search for a washing path.

For example, to wash targets {C, D, E, H, I} in Fig 1(b), the combination of Path 4 and Path 3 in Table I(b) is clearly the best. However, the approach discussed above for solving the hitting-set problems will choose Path 6 instead of Path 4 because less wash targets are covered by Path 4. Therefore, Path 6 and Path 3 will be finally selected as washing paths, which will clean Channel E and Channel D twice. On the other hand, based on the prioritization of wash targets, we conclude that Channel I is more important than Channel D and Channel E because it is covered by fewer paths. Channel D and Channel E are covered by more paths and therefore they can be targeted at a subsequent step. As a result, Path 4 is given higher priority and selected ahead of Path 6.

Due to the difference in effectiveness of path-coverage sets, different vertices need to be given different "wash priorities". We define a utility function, $f(j)$ to quantify the priority of a vertex j. A vertex with higher utility is more important and is given higher priority in the search for wash path. The utility function must have the following properties:

1) The wash priority of a vertex j is a function of $|S_j'|$, where S_j' is a path set that covers vertex j. Vertices covered by more paths need to be given lower priorities

Algorithm 1 The enhanced wash-optimization approach

1: Remove S_j in $\mathbb{S}$, if $j \notin \mathcal{T}$;
2: Remove all paths that include forbidden vertices from the $\mathbb{S}$;
3: Initialization H: $H = \emptyset$;
4: **while** $\mathbb{S} \neq \emptyset$ **do**
5:     Update $f(j)$: $f(j) = \log_a \frac{|S_j|-1}{|M|-1}$ for every vertex j, where $M = \cup_{j \in x} S_j$;
6:     Update $V(p)$: $V(p) = \sum f(j)$, $j \in \{j : p \in \mathcal{S}_j\}$;
7:     Find the path e with maximum value-to-cost ratio, $V(p)/w(p)$;
8:     $\mathbb{S} = \mathbb{S} - \mathcal{S}_j$ for every $j \in \{j : e \in S_j\}$;
9:     $H = H \cup e$;
10: **end while**
11: Return H

Fig. 2: Pseudocode for determining washing-path set H.

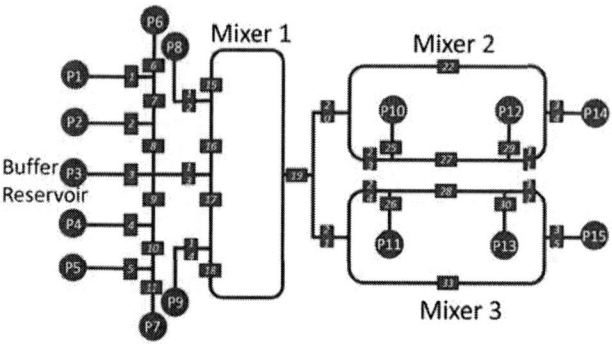

Fig. 3: Layout of a typical fabricated flow-based microfluidic biochip [12]. It contains 35 valves and 15 ports. A buffer reservoir is connected to P3.

because they can be targeted more easier at subsequent steps, i.e., The larger $|S'_j|$ is, the lower the value of vertex j must be.

2) If there exists a vertex that is covered by all the paths in $\mathbb{S}'$, i.e., $|S'_j| = |S'_1 \cup S'_2 \ldots \cup S'_m|$, this vertex need not be considered in wash optimization, i.e., $f(j) = 0$, such as Vertex B in Table I(b).

3) If there exists a vertex j that is covered by only one path in $\mathbb{S}'$, i.e., $|S'_j| = 1$, the path must be included in the H, i.e., $f(j) = \infty$.

4) The utility function should be nonlinear. It must be flatter when $|S'_j|$ is large, but decrease sharply when $|S'_j|$ is small to highlight the importance of vertices covered by only a few paths.

Guided by the above requirements, the utility function, $f(j)$ in this work is formulated as:

$$f(j) = \log_a \frac{|S'_j|-1}{|M'|-1}$$

where $M' = S'_1 \cup S'_2 \ldots \cup S'_m$, $0 < a < 1$. A log function is used for a nonlinear representation of the utility function and the parameter a is used to adjust its nonlinearity. With an increase in a, $f(j)$ becomes flatter and therefore vertices covered by fewer paths are less likely to be washed.

Based on the utilities of vertices, we can introduce the parameter $V(p)$ to evaluate the "wash value" of a path p. Here $V(p)$ is the sum of utilities of all the covered wash targets, i.e., $V(p) = \sum f(j)$, $j \in \{j : p \in S_j\}$. The path that covers more vertices with higher utilities, i.e., the path with higher "wash value" $V(p)$, should be given a higher priority during the selection of washing paths. In each iteration, to maximize efficiency, the path with highest value-to-cost ratio, $V(p)/w(p)$, is selected, where cost is the wash time $w(p)$. The pseudocode for the complete proposed wash-path set selection algorithm is shown in Fig. 2.

D. Complexity Analysis

Next we evaluate the worst-case computational complexity of Algorithm 1. Assume that the number of pre-defined wash targets is p, and the maximum number of paths covering each vertex is l. We analyze the algorithm step-by-step. Each iteration of the algorithm takes $O(pl)$ time for Step 5 and Step 6. In the worst case, the algorithm makes p iterations, which means that only one wash target is covered by each washing path. Step 7, Step 8 and Step 9 take $O(p)$, $O(pl)$ and $O(1)$,

respectively. Hence, the overall computational complexity is $O(pl) \cdot p + O(pl) = O(p^2 l)$.

IV. RESULTS: APPLICATIONS TO FABRICATED BIOCHIPS

In this section, we evaluate the wash-optimization approach using two fabricated flow-based biochips. The proposed approaches are compared with a baseline solution for algorithm evaluation. These methods are described below.

1) Baseline method (Method I): Buffer flows along the "longest" path to cover as many microchannels as possible. Without a dictionary, it is difficult to find a path that can cover a specific washing target. To reduce the number of wash paths, this baseline method simply chooses the paths that include the largest number of vertices, irrespective of whether they are wash targets.

2) Greedy method based on the proposed dictionary (Method II, proposed): The longest path in the compressed dictionary $\mathbb{S}'$ is selected in each iteration. With the help of a dictionary, washing-path search becomes more directed and the weighted hitting-set problem can be used. After removing all uncontaminated microchannels in the dictionary, we select the longest path (the path covering the maximum number of washing targets) at each iteration.

3) Enhanced wash-optimization approach (Method III, proposed): This method is based on the proposed approach illustrated in Fig. 2, where wash priorities of paths are calculated by the sum of utilities of covered washing targets. Compared to Method II, this method leads to less washing time (the time to execute all washing paths), but it requires more CPU time (the time to calculate washing path set).

A. Results for ChIP

The flow-based microfluidic biochip shown in Fig. 3 is targeted for chromatin immunoprecipitation (ChIP), an assay used to analyze DNA-protein interactions [12].

A path dictionary is first generated using DFS on the graph model corresponding to this chip. A total of 34 paths are found. The wash priorities for microchannels are evaluated in terms of the utility function, whose value is obtained by $f(j) = \log_a \frac{|S'_j|-1}{|M'|-1}$ with $a = 0.5$. The number of paths that cover a channel varies considerably for different microchannels. For example, Channel 3, which is near the reservoir, is covered by all paths. Therefore, its utility is 0. Channels at branches, such

978-1-4799-2817-0/14 $31.00 © 2014 IEEE

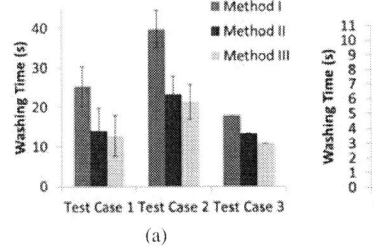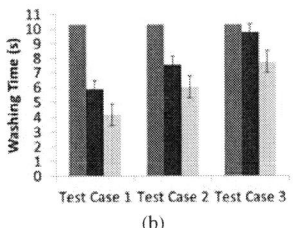

| | (a) | (b) |

Fig. 4: (a) Three test cases for the ChIP chip with different combinations of wash targets and occupied microchannels. Three approaches, a baseline method (Methods I), a greedy approach based on the proposed dictionary (Method II) and the enhanced wash-optimization approach method (Method III), are evaluated and compared. (b) Washing time for the programmable chip. Method III provides the shortest washing time in all cases.

	Method I (ms)	Method II (ms)	Method III (ms)
Test Case 1	15.95	2.78	7.61
Test Case 2	25.75	5.48	20.72
Test Case 3	10.23	1.87	3.41

Table III: The CPU time for the three methods and the different test cases

as Channel 6, are covered by only one path. Their utilities are infinity because as long as these channels are contaminated, the corresponding paths have to be selected. Three tests are performed to evaluate proposed methods:

1) Test Case 1: 10 wash targets and 4 occupied microchannels are randomly generated in the chip.
2) Test Case 2: 20 wash targets and 6 occupied microchannels are randomly generated in the chip.
3) Test Case 3: We simulate a real wash case in ChIP, where $\mathcal{T} = \{3, 13, 15, 16, 17, 18, 19, 20, 22, 23, 27, 31, 34\}$, and $\mathcal{O} = \{33\}$. The set of wash targets $\mathcal{T}$ is designed to clean a loading path for the second cell samples and the set of occupied channels $\mathcal{O}$ stores the DNA samples of the first cells. This chip is initially designed perform ChIP on the DNA fragments of one kind of cell. However, washing allows the chip to use DNAs from different cells without concerns about cross-contamination.

Case 1 and Case 2 are repeated 500 times, and results are reported in terms of mean and standard deviation of the wash times. The mean washing time and the standard deviation of all three methods are compared in Fig. 4(a). Methods III results in the most efficient washing pathways for all three test cases. Moreover, a path dictionary reduces the washing time by 50%.

The CPU time for the three methods are listed in Table III. Method II requires the lowest CPU time. Method III requires more computation than Method II, but it still takes only one-third of the CPU time of Method I. These simulations were carried out using a computer with a 4.2 GHz AMD-FX4170 processor and 8 GB memory.

B. A programmable microfluidic device with an 8-by-8 grid

Next we evaluate the wash-optimization approach using a programmable flow-based device, structured as an 8-by-8 grid [9]. Each of the nodes in this chip is surrounded by up to four valves, allowing variable interconnections and the implementation of arbitrary channel structures. All 112 on-chip valves are individually addressable.

A dictionary comprising of two million paths is first generated by random path search. Three tests are performed to evaluate the proposed method. In each test, we evaluate three methods using 500 independent simulations. In each simulation, a certain number of wash targets and occupied microchannels are randomly selected. The mean washing time and the standard deviation are reported. The following cases are considered:

1) 20 wash targets and 8 occupied microchannels
2) 40 wash targets and 10 occupied microchannels
3) 80 wash targets and 15 occupied microchannels

Fig. 4(b) indicates that the enhanced wash-optimization approach (Method III) outperforms the baseline (Method I) and Method II. The average CPU time is 48.5 s (Method I), 0.48 s (Method II) and 4.28 s (Method III).

V. CONCLUSIONS

We have described the first wash optimization approach for cross-contamination removal in flow-based microfluidic biochips. Given several wash targets and occupied microchannels, the proposed technique generates a set of washing pathways to clean all targets with minimum wash time. A path dictionary is first generated by searching *a priori* physically implementable paths in a given chip layout. Based on this path dictionary, the wash optimization problem can be formulated as a variant of hitting-set problem. Due to the low degree of the graph generated by flow-based biochips, we improve our algorithm by defining a utility function to evaluate the wash priorities of washing targets. Two fabricated biochips are used to evaluate the proposed methods. Compared with a baseline approach, the proposed approach leads to the most efficient washing for each of the test scenarios considered in this work.

REFERENCES

[1] T. Squires and S. Quake, "Microfluidics: Fluid physics at the nanoliter scale," *Reviews of Modern Physics*, vol. 77, pp. 977–1026, Oct. 2005.

[2] G. Whitesides et al., "Soft Lithography in Biology and Biochemistry," *Annual Review of Biomedical Engineering*, vol. 3, pp. 335–373, 2001.

[3] I. E. Araci and S. R. Quake, "Microfluidic very large scale integration (mVLSI) with integrated micromechanical valves," *Lab on a Chip*, vol. 12, pp. 2803–6, Aug. 2012.

[4] J. Melin and S. R. Quake, "Microfluidic Large-scale Integration: the Evolution of Design Rules for Biological Automation," *Annual Review of Biophysics and Biomolecular Structure*, vol. 36, pp. 213–31, 2007.

[5] S. Einav, et al., "Discovery of a hepatitis C target and its pharmacological inhibitors by microfluidic affinity analysis," *Nature biotechnology*, vol. 26, no. 9, pp. 1019–27, Sep. 2008.

[6] C. D. Chin, et al., "Microfluidics-based diagnostics of infectious diseases in the developing world," *Nature Medicine*, vol. 17, no. 8, 2011.

[7] Y. Zhao and K. Chakrabarty, "Cross-contamination avoidance for droplet routing in digital micro fluidic biochips," *IEEE/ACM DATE*, pp. 1290–1295, 2009.

[8] C. C.-Y. Lin and Y.-W. Chang, "Cross-contamination aware design methodology for pin-constrained digital microfluidic biochips," *IEEE/ACM DAC*, pp. 641–646, 2010.

[9] L. M. Fidalgo and S. J. Maerkl, "A software-programmable microfluidic device for automated biology," *Lab on a chip*, vol. 11, pp. 1612–9, 2011.

[10] Y. C. Lim, A. Z. Kouzani, and W. Duan, "Lab-on-a-chip: a component view," *Microsystem Technologies*, vol. 16, pp. 1995–2015, Sep. 2010.

[11] R. Niedermeier, et al., "An Efficient Fixed Parameter Algorithm for 3-Hitting Set," *Journal of Discrete Algorithm*, 1: 89-102, 2003.

[12] A. Wu et al., "Automated Microfluidic Chromatin Immunoprecipitation from 2,000 Cells," *Lab on a Chip*, vol. 9, pp. 1365–1730, 2009.

978-1-4799-2817-0/14 $31.00 © 2014 IEEE

ABCD-NL: Approximating Continuous Non-Linear Dynamical Systems using Purely Boolean Models for Analog/Mixed-Signal Verification

Aadithya V. Karthik[‡], Sayak Ray, Pierluigi Nuzzo, Alan Mishchenko, Robert Brayton, and Jaijeet Roychowdhury
The Department of Electrical Engineering and Computer Sciences, The University of California, Berkeley, CA, USA
[‡]Corresponding author. Email: `aadithya@berkeley.edu`

Abstract—**We present ABCD-NL, a technique that approximates non-linear analog circuits using purely Boolean models, to high accuracy. Given an analog/mixed-signal (AMS) system (*e.g.*, a SPICE netlist), ABCD-NL produces a Boolean circuit representation (*e.g.*, an And Inverter Graph, Finite State Machine, or Binary Decision Diagram) that captures the I/O behaviour of the given system, to near SPICE-level accuracy, without making any *apriori* simplifications. The Boolean models produced by ABCD-NL can be used for high-speed simulation and formal verification of AMS designs, by leveraging existing tools developed for Boolean/hybrid systems analysis (*e.g.*, ABC [1]). We apply ABCD-NL to a number of SPICE-level AMS circuits, including data converters, charge pumps, comparators, non-linear signaling/communications sub-systems, *etc.* Also, we formally verify the throughput of an AMS signaling system – modelled in SPICE using 22nm BSIM4 transistors, Booleanized with high accuracy using ABCD-NL, and property-checked using ABC.**

I. INTRODUCTION

AMS systems are becoming increasingly important in chip design. In recent times, AMS blocks have become key components that limit system-level performance [2]. Also, AMS components now account for a significant proportion of design bugs, designer time, and debugging cost. However, CAD tools for AMS verification have not kept pace with the rapid growth in complexity of these systems.

An important challenge for AMS verification is the *accurate modelling* of AMS components; because these components can introduce design flaws/loss of performance in a variety of subtle and non-obvious ways, it is important to model their behaviour at or near SPICE-level accuracy.

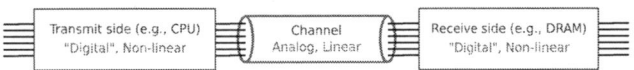

ISI, crosstalk, distortion, etc. need to be accurately modelled

Fig. 1. Schematic of a typical AMS signaling/communications sub-system that arises in signal integrity analysis.

For example, Fig. 1 depicts an AMS system that frequently arises in Signal Integrity (SI) applications. The system consists of digital components on the transmit side (*e.g.*, a CPU), whose outputs enter an analog channel. The channel introduces inter-symbol interference (ISI), crosstalk, *etc.* The other end of the channel (the receive side) has more digital components (*e.g.*, a DRAM/memory controller). A key figure of merit of this system is its throughput, *i.e.*, the maximum bitrate that can be reliably sustained. Guaranteeing the system's throughput is a non-trivial AMS verification problem, and to issue a meaningful guarantee, it is necessary to model this system at SPICE-level accuracy.

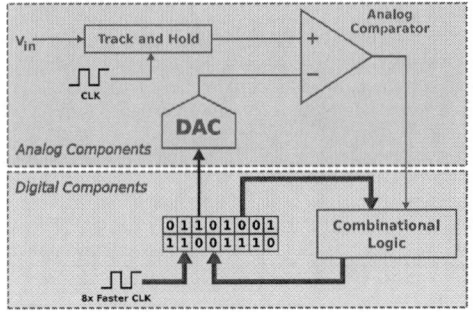

Fig. 2. Schematic of a successive approximation A/D converter (SAR-ADC).

To take another example, Fig. 2 depicts a successive approximation A/D converter (or SAR-ADC). This system contains a number of components, both analog and digital. The system's performance (*i.e.*, its speed, power consumption, *etc.*) is frequently limited by the analog components (*e.g.*, the speed of the DAC, the sensitivity/bandwidth of the comparator, *etc.*). Therefore, to verify the SAR-ADC, it is often necessary to model the analog components at SPICE-level accuracy (even though it may be sufficient to model digital components at the Boolean level).

Most existing approaches to AMS verification (*e.g.*, [3]–[8]), however, *do not* model AMS components at SPICE-level accuracy. The existing approaches are usually based on hybrid systems methods (*e.g.*, [9]–[14]); although these methods can reason about continuous analog quantities, they tend to be limited in terms of scalability. For example, they can fail even for relatively small AMS systems consisting of just 5-10 analog signals. Due to this limitation, existing verification flows usually adopt highly simplified "behavioural" macromodels for AMS components, which do not capture many performance-limiting analog effects (*e.g.*, non-linear operating regions, DC offsets, ISI, distortion, *etc.*). As a result, AMS designers often have to carry out extensive and time-consuming SPICE simulations; this is a tedious, expensive, and error-prone process.

In an attempt to address the above concerns, a technique called ABCD-L [15] was recently proposed. ABCD-L represents linear time invariant (LTI) AMS components using *purely Boolean* models, *i.e.*, without the need for any continuous variables[1]. ABCD-L has been shown to capture the dynamics of linear AMS components to almost SPICE-level accuracy [15]. Also, because ABCD-L models use only "cheap" Boolean variables, and not "expensive" continuous variables, the formal analysis and verification involving these models can be very efficient, using either the state-of-the-art Boolean techniques (*e.g.*, ABC [1]), or in conjunction with existing hybrid systems frameworks. The biggest drawback of ABCD-L, however, is that it applies only to linear systems. Therefore, ABCD-L cannot be used for verifying systems such as the communications sub-system of Fig. 1, or the SAR-ADC of Fig. 2, because these systems are strongly non-linear.

In this paper, we propose ABCD-NL[2], a new method that has all the advantages of ABCD-L, and in addition, works for a large class of non-linear systems. Given a non-linear AMS system (*e.g.*, a SPICE netlist), ABCD-NL produces a purely Boolean model (*e.g.*, as a Finite State Machine (FSM), an And Inverter Graph (AIG), or a Binary Decision Diagram (BDD)) that captures the dynamics of the given system to near SPICE-level accuracy. Furthermore, the Boolean model produced by ABCD-NL is well-suited for use with cutting-edge formal verification/model checking engines such as ABC [1], or with existing hybrid systems frameworks. Also, the Boolean model produced by ABCD-NL can be simulated very efficiently at the logic level (without needing to solve any differential equations), so ABCD-NL can be used as a much faster, almost-as-accurate, drop-in replacement for SPICE, in many applications.

ABCD-NL requires only one condition: a DC input to the given system should eventually result in a DC output. This condition is satisfied by almost all non-linear systems of interest to AMS designers, *e.g.*, D/A and A/D converters, amplifiers and comparators, linear and non-linear filters, equalizers, switches and multiplexers, charge pumps, I/O links, *etc.*

In the next section (§II), we describe the core techniques underlying

---

[1]Similar ideas have also been suggested in works like [16] and [17].

[2]**A**ccurate **B**ooleanization of **C**ontinuous **D**ynamics – **N**on-**L**inear

978-1-4799-2817-0/14 $31.00 © 2014 IEEE

ABCD-NL. We then apply ABCD-NL (in §III) to a number of non-linear systems that are of interest to AMS designers. For example, in §III-A, we Booleanize a charge pump/filter system using ABCD-NL. We also apply ABCD-NL to the analog components that make up the SAR-ADC of Fig. 2, such as the D/A converter (§III-C) and the comparator (§III-D). In all these cases, we show that ABCD-NL's Boolean model faithfully reproduces the circuit's behaviour.

We note that the primary focus of this paper is on *modelling AMS systems for verification*, as opposed to the verification itself. However, for completeness, we present (in §III-E) an example where we use the verification tool ABC [1], together with a Boolean model produced by ABCD-NL, to formally verify the throughput of a communications subsystem of the type shown in Fig. 1.

II. CORE TECHNIQUE: A NEW ALGORITHM FOR BOOLEANIZING NON-LINEAR ANALOG CIRCUITS

In this section, we describe the key ideas behind ABCD-NL. As outlined in §I, ABCD-NL takes a SPICE netlist as input, and it produces as output a purely Boolean model (*e.g.*, an FSM) of the given circuit.

The Boolean model represents the circuit's inputs and outputs using finite numbers of bits. The bits returned by the model are interpreted by the user as discretized analog waveforms. The goal is to preserve the behaviour of the original circuit as closely as possible in the Boolean model.

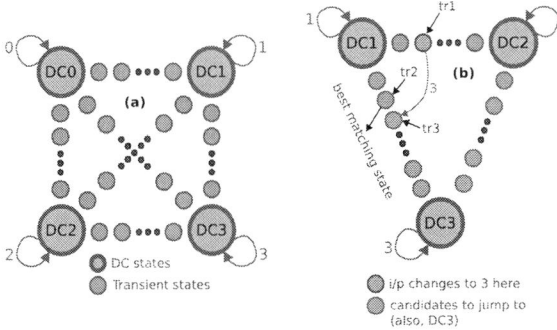

Fig. 3. (a) Structure of the FSM model derived by ABCD-NL from SPICE simulations, and (b) ABCD-NL's method of exploiting continuity to jump from one transient FSM state to another.

Algorithm 1: Converting an analog circuit into a purely Boolean ABCD-NL model

Inputs: SPICE netlist *cir*, Signal list *siglist*, Output signal *sigout*, FSM time step *tstepFSM*
Output: Purely Boolean FSM model *fsm* for the given circuit

1   *fsm* = new FSM()
2   *DCInputs* = enumerateDiscretizedCktInputs()
3   insertDCFSMStates (*fsm*, *DCInputs*)
4   **for** *initInput* in *DCInputs*:
5   *initFSMState* = encodeFSMState (*initInput*)
6   **for** *finalInput* ≠ *initInput* in *DCInputs*:
7   *finalFSMState* = encodeFSMState (*finalInput*)
8   *cirInputWaveforms* = generateStepFunctions (*initInput*, *finalInput*)
9   [*ts*, *vs*] = SPICESimulateTran (*cir*, *cirInputWaveforms*)
10   *tSettle* = estimateSettlingTime (*ts*, *vs*, *siglist*)
11   *numTranFSMStates* = ceil (*tSettle* ÷ *tstepFSM*) − 1
12   insertTranFSMStates (*fsm*, *initFSMState*, *finalFSMState*, *numTranFSMStates*)
13   annotateFSMArcs (*fsm*, *initFSMState*, *finalFSMState*, *ts*, *vs*, *sigout*)
14   annotateFSMStates (*fsm*, *initFSMState*, *finalFSMState*, *ts*, *vs*)

15   **foreach** transient FSM state *trst* in *fsm*:
16   **foreach** discrete input *inp* ≠ *trst.finalInput* in *DCInputs*:
      /* use state/output continuity to obtain new FSM arc */
17   *nextst* = estimateNextFSMState (*fsm*, *trst*, *inp*)
18   *outp* = estimateDiscreteOutputOnTransition (*fsm*, *trst*, *nextst*, *inp*)
19   insertFSMArc (*trst*, *nextst*, *inp*, *outp*)

20   **return** *fsm*

Fig. 3 (a) depicts the Boolean model produced by ABCD-NL. Each discretized input combination is associated with a *DC state* (that has a self loop) in an FSM. These DC states are akin to DC operating points of the circuit. For example, if the circuit has 2 inputs, and each input is

---

discretized using 3 bits (*i.e.*, 8 levels), the ABCD-NL FSM would have 64 DC states. These states capture the circuit's DC behaviour.

To capture the transient behaviour, ABCD-NL introduces additional *transient FSM states* between every pair of DC states described above. The exact number of transient states introduced depends on the dynamics of the circuit, *i.e.*, the time taken by the system to transition from one DC operating point to another, when a suitable step transition is applied at the input. Together, the DC states and the transient states capture the dynamics of the given system to high accuracy.

Algorithm 1 formally describes the FSM construction procedure used by ABCD-NL. Line 3 creates the DC states described above. For every pair of such DC states, ABCD-NL performs a transient SPICE simulation (Line 9), the results of which are used to create the transient FSM states (Lines 10–12). Note that, if needed, the user can explicitly specify a list of important signals (*siglist*) in the given circuit, which the algorithm takes into account while creating the transient states (by default, the algorithm considers all signals as important). Further, the algorithm uses the SPICE simulations above to label each FSM arc (Line 13) with an appropriate (discretized) output symbol.

Algorithm 2: Simulating ABCD-NL's Boolean model, and post-processing to the analog domain

Inputs: Boolean model *fsm*, Ckt. inputs *u(t)*, FSM step *tstepFSM*, simulation interval [t_0, t_f]
Output: Simulation trace [*ts*, *vs*] of the circuit's output *sigout* over the interval [t_0, t_f]

1   *ts* = [], *vs* = []
   /* start at the DC operating point for the input *u* at time t_0 */
2   t_{curr} = t_0
3   u_{curr} = discretizeInput (*u*(t_{curr}))
4   $state_{curr}$ = encodeFSMState (u_{curr})
5   **while** $t_{curr} \leq t_f$:
    /* simulate one time point by looking up the FSM's state transition table */
6   *transitionArc* = *fsm*.nextStateArc (t_{curr}, u_{curr})
    /* Look up the o/p on the transition arc, and post-process it from Boolean to analog */
7    $output_{curr}$ = *transitionArc.outputSymbol*
8    $analogOutput_{curr}$ = BooleanToAnalog ($output_{curr}$)
    /* record the output */
9   *ts*.append (t_{curr})
10   *vs*.append ($analogOutput_{curr}$)
    /* update the simulation variables for the next time point */
11   t_{curr} += *tstepFSM*
12   u_{curr} = discretizeInput (*u*(t_{curr}))
13   $state_{curr}$ = *transitionArc.finalState*

14   **return** [*ts*, *vs*]

By this time, all the states of the ABCD-NL FSM have been created, and all arcs have been specified for the DC states. However, not all arcs have been specified for the transient states. To fully specify the system, ABCD-NL uses an interpolation-based heuristic, as shown in Fig. 3 (b).

For example, suppose that the discretized version of the applied input switches from one Boolean-encoded value (say, 1), to another (say, 2). Corresponding to this, the FSM starts moving from state DC1 to DC2, as shown in Fig. 3 (b). However, before the FSM reaches DC2 (*i.e.*, when the FSM is in the state marked tr1), let us say the (discretized) input switches again, this time to 3. This forces the FSM to move towards state DC3, and we need a method to determine the next action of the FSM. For this, ABCD-NL takes advantage of the continuity of the underlying analog waveforms (Lines 17 to 19); it selects a transient FSM state that is, in some sense, "closest" to the current state tr1, along the paths DC1–DC3 and DC2–DC3 (this is possible because the ABCD-NL synthesis algorithm (Line 14) internally maintains an estimate of the *analog state* of the circuit at each *Boolean state* of the FSM). This completes the Boolean model generation, and the resulting FSM is returned, which, if necessary, can be transformed into an AIG or BDD using existing tools such as ABC [1]. (Our implementation of ABCD-NL ensures that its output can be directly read in by tools such as ABC.)

Algorithm 2 formalises how the ABCD-NL Boolean model can be simulated in the time-domain, at the logic level. Each time step in this simulation is simply a table lookup (Line 6), so there are no differential equations or Newton-Raphson iterations involved. Thus, ABCD-NL based simulation is much faster than SPICE. For example, even though we have implemented ABCD-NL in Python (a language not known for being fast), it was still 2x to 3x faster than HSPICE, for most examples presented in the next section. We believe that, if the code is re-written in C/C++, it would be quite straightforward to achieve ~30x speedup over HSPICE.

III. RESULTS

We now apply ABCD-NL to circuits that are of interest to AMS designers. These include, (1) a charge pump/filter system that is relevant to PLL design, (2) a signaling/communications sub-system that involves (non-linear) digital logic interfacing with an analog channel, (3) a D/A converter, and, (4) an analog comparator, the last two circuits being important analog components that make up a SAR-ADC. In each case, we show that the Boolean model produced by ABCD-NL is able to accurately reproduce the SPICE-level analog dynamics of the underlying circuit, including important performance-limiting non-ideal phenomena.

A. Charge pump driving an analog filter

Fig. 4 shows a charge pump driving an analog filter, a system that plays an important role in PLLs. The system works as follows: the transistors M1 and M2 form a current mirror that can pump a current I_0 into the load capacitor C_L, whereas transistors M3 and M4 form an opposing current mirror that can withdraw current I_0 from C_L. The circuit has two inputs, V_{up} and V_{down}. During normal operation, exactly one of these inputs is high; if V_{up} (V_{down}) is high, current is pumped in (out), driving the output voltage V_{out} higher (lower); this is called the charging (discharging) mode of the charge pump, and the system responds most quickly when in one of these modes (*e.g.*, using a 90nm process, response times are typically of the order of tens of nanoseconds).

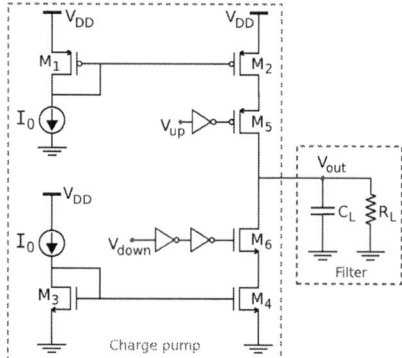

Fig. 4. Schematic of a charge pump driving an analog filter.

In addition to the normal mode of operation, it is also important to capture the behaviour of the charge pump under anamolous inputs; for example, if both inputs are high, the charge pump enters an imbalance-driven mode, which can either charge or discharge the load, depending on operating conditions. This type of charging/discharging is typically much slower than normal operation (*e.g.*, hundreds of nanoseconds response time), because only a small current flows through the load. Finally, if both inputs are off (cutoff mode), the output voltage, under ideal conditions, would remain constant; however, due to leakage currents and the resistive load R_L, the capacitor C_L slowly discharges to a DC voltage that is almost 0. The cutoff mode is the slowest mode of operation of the system, with response time in the microsecond range.

Since PLL performance critically depends on the non-linear charge pump/loop-filter dynamics, our goal is to use ABCD-NL to accurately model the behaviour of the above system under all possible operating conditions: charging, discharging, imbalance-driven, and cutoff. So we designed the system in 90nm CMOS, using BSIM4 device models. We then applied Algorithm 1 (of §II) to Booleanize this system (using 5 bits to encode the output waveform), and we simulated the resulting Boolean model using Algorithm 2, on a range of inputs that covered all four modes of operation. In each case, we compared the output predicted by ABCD-NL's Boolean model, against that predicted by HSPICE. Fig. 5 shows the results, where HSPICE waveforms are shown in blue, and ABCD-NL waveforms are in green. As the figure shows, in spite of the widely differing time scales involved in the four modes, the Boolean model produced by ABCD-NL closely matches the SPICE-level dynamics of the system in all its modes.

Fig. 6 further demonstrates the accuracy and robustness of the Boolean model produced by ABCD-NL. The figure shows a long pseudo-random bit sequence applied as input to the circuit, which switches the circuit in

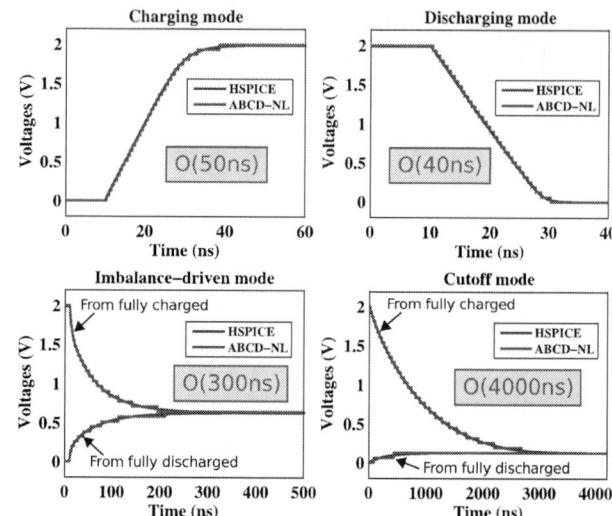

Fig. 5. ABCD-NL accurately captures the behaviour of the charge pump under all four modes of operation, in spite of the widely differing time scales involved.

and out of all 4 modes of operation over a long time frame. Throughout this time, it is seen that the Boolean model produced by ABCD-NL (the green waveform) closely tracks the SPICE-simulated output (the blue waveform) of the system. This indicates that ABCD-NL is indeed a powerful and accurate modelling technique, and one that can conceivably be used as a much faster, almost-as-accurate, drop-in replacement for SPICE over long transient runs.

B. Signaling system: Non-linear digital logic + analog channel

Fig. 7 shows a mixed-signal sub-system, of the type depicted in Fig. 1. As we mentioned in §I, such systems are often encountered in Signal Integrity (SI) applications. In this example, the transmit side takes 3 bits as input (A, B, and C_{in}), and adds them up using a full-adder, thereby producing 2 output bits: the sum S, and the output carry C_{out}. These 2 bits are then sent across an analog channel that consists of several RC stages, inter-linked with coupling capacitances. At the other end of the channel, the receiver cleans the arriving waveforms S_{ch} and C_{ch} using chains of inverters.

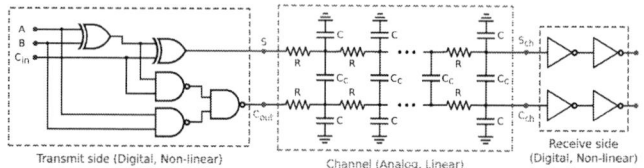

Fig. 7. A signaling/communications sub-system that arises in SI applications.

We have modelled each transistor in the above system using a 22nm BSIM4 analog SPICE model (obtained from [18]). Therefore, the system above exhibits several realistic non-linear analog effects, including leakage currents, loading effects, delays, channel-induced ISI/crosstalk, *etc.* Our goal is to use ABCD-NL to accurately reproduce the analog waveforms at the output of the channel (S_{ch} and C_{ch}), in the presence of these adverse analog effects. This is a crucial requirement for SI analysis.

Fig. 8 shows the results obtained by applying ABCD-NL to the above system. Part (a) shows three randomly generated 40-bit sequences (for A, B, and C_{in}, respectively), applied as inputs to the circuit. In part (b), these inputs are applied at a bitrate of 1 Gbps. At this bitrate, the system behaves in a fairly ideal manner, *i.e.*, distortion, crosstalk, *etc.* are minimal, as seen from the blue HSPICE waveforms S_{ch} and C_{ch} of Fig. 8 (b). Also, the green waveforms in the figure show the predictions made by ABCD-NL's purely Boolean model (using 4 bits to encode each output waveform); as the figure shows, ABCD-NL's purely Boolean model is able to accurately predict the system's dynamics at this bitrate.

Fig. 8 (c) shows the same bit pattern applied to the circuit, but at a higher bitrate (3.2 Gbps). At this bitrate, the system produces considerable

978-1-4799-2817-0/14 $31.00 © 2014 IEEE

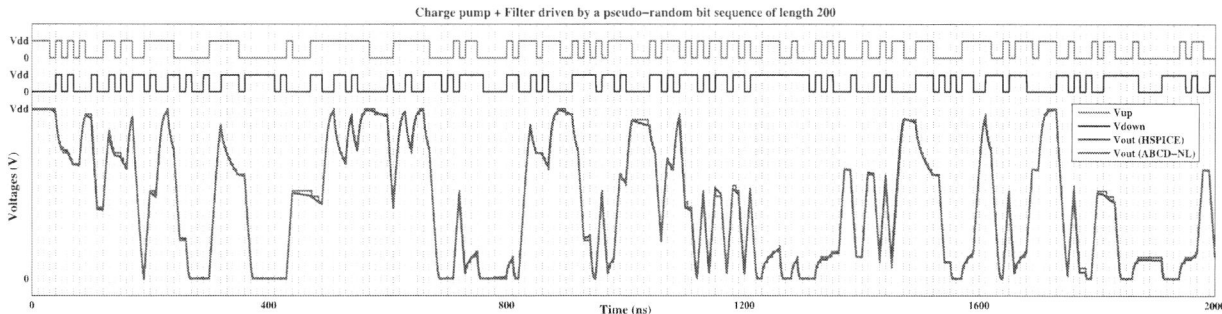

Fig. 6. ABCD-NL can predict the response of the charge pump/filter system, almost to SPICE-level accuracy, even over long time frames that involve rapid switching between all 4 modes of operation.

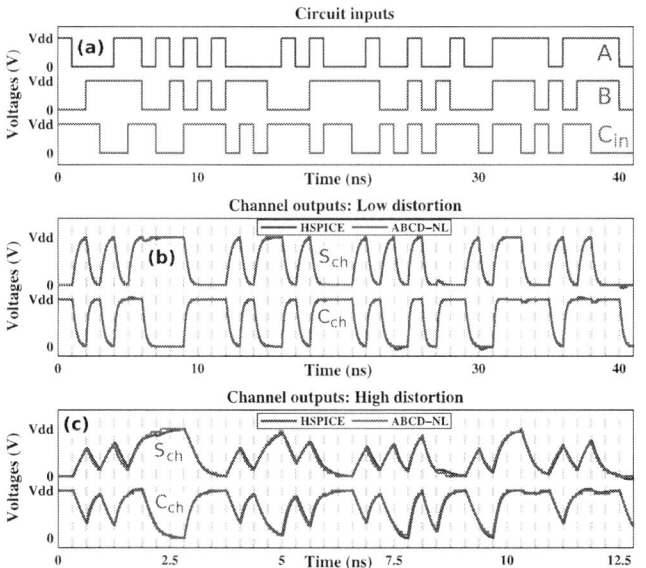

Fig. 8. ABCD-NL closely matches the SPICE-level analog behaviour of the signaling/communications sub-system of Fig. 7, at both high and low bitrates.

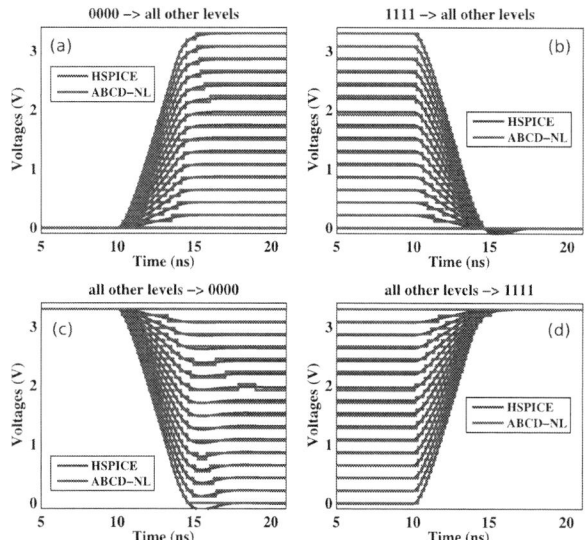

Fig. 10. ABCD-NL closely matches SPICE-level simulation of the D/A converter, for several input bit transitions.

distortion, with significant ISI and crosstalk. Here also, it is seen from the figure that ABCD-NL is able to accurately reproduce the analog dynamics exhibited by the system. This demonstrates that ABCD-NL is indeed a viable modelling technique for SI applications, even when the underlying system exhibits pronounced non-linear analog effects.

C. D/A converter (DAC)

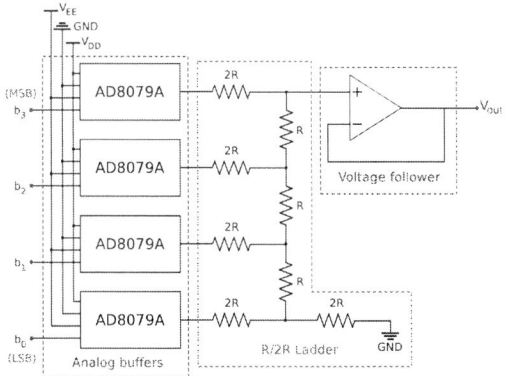

Fig. 9. Schematic of a D/A converter used within a SAR-ADC.

We now apply ABCD-NL to produce a purely Boolean model of a canonical mixed-signal system, a D/A converter used within a SAR-ADC. As Fig. 9 shows, we have a 4-bit D/A converter consisting of four Analog Devices AD8079A analog buffers (SPICE models available from [19]), and an R/2R ladder that feeds into a voltage follower.

A key figure of merit of a D/A converter is speed; so it is important to

accurately capture the delay of the system for all possible bit transitions at the input. Fig. 10 shows many of these transitions (due to space constraints, we are unable to show all the transitions), and indeed, it can be seen from the figure that ABCD-NL accurately captures the system's delay for all these inputs (using 6 bits to encode the D/A output).

Furthermore, because our D/A converter is embedded within a SAR-ADC, it is important to have our Boolean model reproduce the system's dynamics for input patterns that are typical to the SAR-ADC environment. Fig. 11 illustrates this environment for a 4-bit SAR-ADC. The red waveform shows a 150kHz sine wave, which is the ADC input. The ADC operates at about 8MHz, so each period of the input generates about 52 ADC samples. Over these samples, the input bits b_0 to b_3 of the D/A converter switch as shown in the top half of Fig. 11. The blue waveform at the bottom of the figure depicts the D/A output, as predicted by HSPICE. And as seen from the green waveform, ABCD-NL is able to reproduce this response very accurately. This shows that ABCD-NL is indeed an accurate and powerful way to Booleanize non-linear data converters for analysing mixed-signal systems.

D. Analog comparator

We now apply ABCD-NL to an analog comparator, another key component in a SAR-ADC. For this demonstration, we shall Booleanize an off-the-shelf comparator (LT1016 from Linear Technology, whose SPICE model is available online [20]), and deploy the resulting Boolean model in a SAR-ADC environment.

Booleanizing a SAR-ADC comparator is a particularly challenging problem because the circuit is highly non-linear, and very sensitive to its (large signal) inputs; for example, a differential input of 1mV elicits a very different response from the system compared to a 2mV (or 1V) differential, in terms of delay, the final steady state solution, etc. This necessitates very fine discretization of the input waveforms, which can

978-1-4799-2817-0/14 $31.00 © 2014 IEEE   253

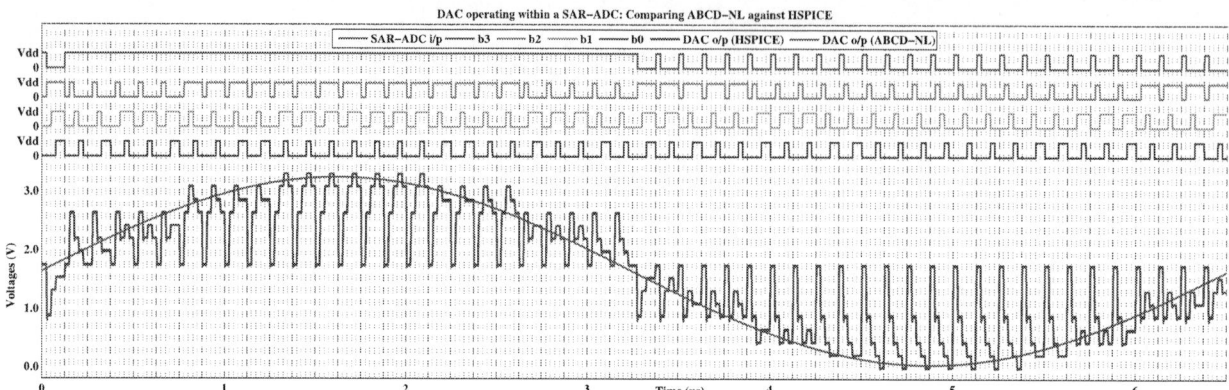

Fig. 11. ABCD-NL, using a purely Boolean model, is able to accurately capture the dynamics of a D/A converter embedded within a SAR-ADC, across a long time frame encompassing several ADC samples.

in turn result in a very large Boolean model unless domain knowledge is used (see below). Moreover, for each input sample of the SAR-ADC, the comparator usually begins at a large differential (of the order of 1V), and the closed-loop dynamics of the system uses feedback to progressively make this differential smaller, until it is of the order of 1mV. And the Boolean model must capture the dynamics of the comparator, to high accuracy, in spite of this large operating range.

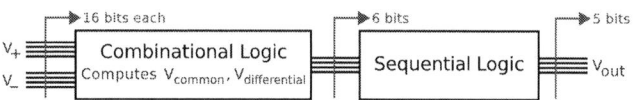

Fig. 12. ABCD-NL can use domain knowledge to achieve better efficiency. For a comparator, rather than directly operating on the input bits, it is significantly more efficient to first transform the input signals into common mode and differential mode components (using combinational logic), which can then be used to drive ABCD-NL's sequential Boolean model.

As indicated above, to reduce the size of the Boolean model, ABCD-NL supports the use of domain knowledge. Fig. 12 illustrates this point for the comparator. Because the input has to be discretized very finely (*e.g.*, using 16 bits or more per signal), it is inefficient to apply ABCD-NL directly to the discretized input. Instead, we use a combinational logic unit to compute the *common mode* and *differential* components of the input, which can be discretized relatively coarsely (*e.g.*, using just 6 bits for the differential mode, and 0-2 bits for the common mode). In particular, the differential component is discretized non-uniformly, placing more emphasis on small differentials (because they exert a powerful influence on system dynamics), and a smaller emphasis on large differentials (where the behaviour quickly saturates).

Fig. 13 illustrates the results. Part (b) of the figure shows that ABCD-NL, with the efficiency enhancements above, is able to accurately reproduce the SPICE-level analog behaviour of the comparator, for a wide range of input excitations. These excitations are generated as follows: initially, the input differential between the inputs V_+ and V_- is chosen to be either *large* (1V) or *small* (1mV). This choice has the effect of biasing the comparator either at a strongly polarized bias point, or a weak bias point. Now, a second choice is made: the input differential is suddenly reversed in polarity, using either a small driving strength (1mV differential), or a large driving strength (1V differential). This choice has a significant impact on response time; for example, a strong differential signal starting from a weakly polarized system state evokes a much faster response than a weak differential trying to flip the system from a strongly polarized state. The goal is to design the Boolean model to accurately capture all the different corner cases, and as Fig. 13 (b) shows, the Boolean model produced by ABCD-NL achieves this goal (using 5 bits to encode the output waveform).

Also, for verification purposes, it is important to model the comparator's departure from ideal behaviour. An important factor in this context is the DC sensing offset of the comparator. For example, Fig. 13 (c) shows a situation where the comparator behaves in a highly unexpected way: even though V_- is always higher than V_+ (*i.e.*, an ideal comparator's output would remain low throughout), the LT1016 actually switches from low to high. Such unexpected behaviour can potentially introduce bit-

errors in the context of a SAR-ADC, due to incorrect decisions made by the comparator within the feedback loop. Therefore, while analyzing a SAR-ADC that uses this comparator, it is important to use a comparator model that accurately accounts for such imperfections and shortcomings. And as Fig. 13 (c) shows, ABCD-NL's Boolean model does accurately reproduce the behaviour of the comparator. This is a powerful advantage offered by ABCD-NL – anything that SPICE can predict, the Boolean model can incorporate.

Finally, Fig. 13 (d) shows that ABCD-NL accurately reproduces the behaviour of the comparator when it is embedded in a typical SAR-ADC environment. The ADC, and the input to it, are the same as in Fig. 11. Over one time period of the input waveform (*i.e.*, 52 ADC samples), the top half of Fig. 13 (d) plots the comparator inputs V_+ and V_-. The bottom half of the figure shows that ABCD-NL is able to capture the system's response, almost to SPICE-level accuracy, over this entire time frame.

E. Formal verification with an ABCD-NL model

As we remarked before, the main focus of this paper is the *accurate modelling of AMS components for verification*, as opposed to the verification itself. However, for completeness, we now present an example where we Booleanized an AMS system using ABCD-NL, imported the resulting Boolean model into a verification engine (ABC, [1]), and carried out formal property checking of the model against an AMS-design relevant specification.

Fig. 14. Schematic of a system that was formally verified using a combination of ABCD-NL and ABC. The inverters were designed in a 22nm CMOS process, using BSIM4 models. The channel was modelled as a long RC chain.

Fig. 14 depicts the system that we formally verified. It follows the same pattern as the systems shown in Figs. 1 and 7. As we mentioned earlier, such systems play an important role in SI applications, where it is important to determine, and *formally verify*, the throughput of the system. We have used 22nm BSIM4 models for each transistor in the system, and an analog channel that consists of several RC units chained together.

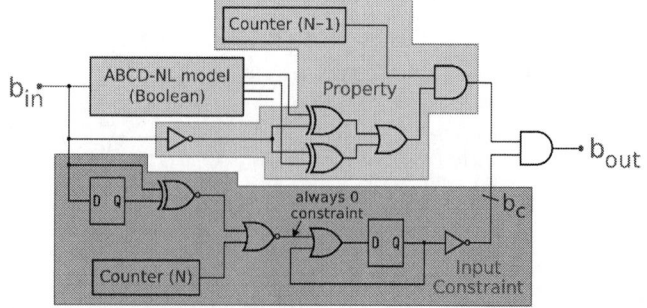

Fig. 15. Encoding the throughput property, along with constraints on the input, in a Boolean form so as to formally verify the ABCD-NL model using ABC [1].

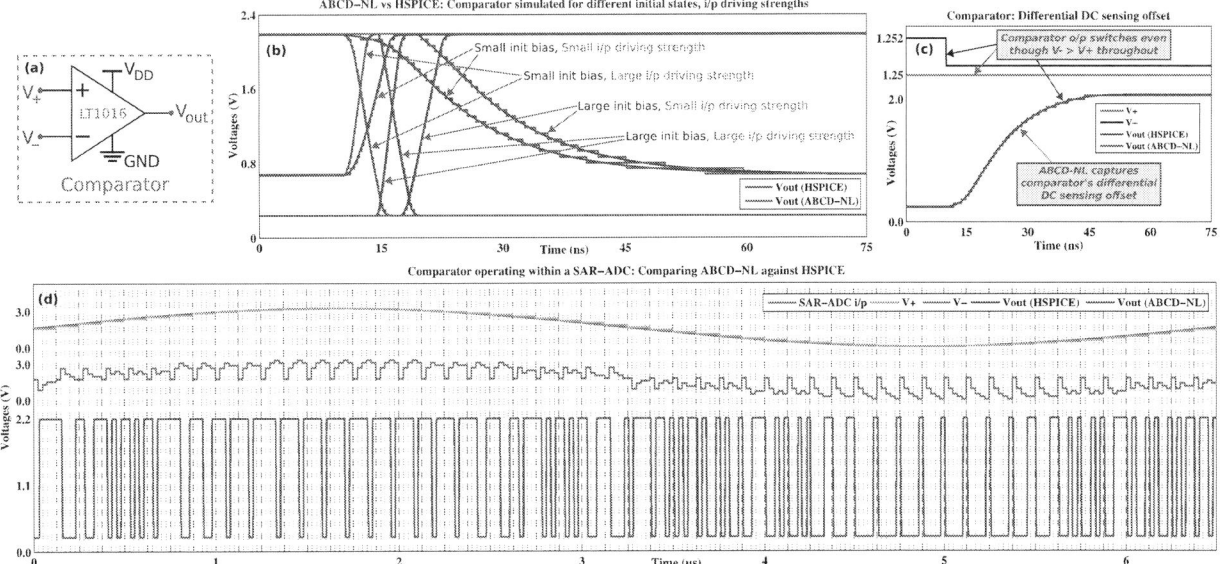

Fig. 13. ABCD-NL applied to a Linear Technology LT1016 comparator (part (a)). Part (b) shows that ABCD-NL is able to capture the dynamics of the comparator over a wide range of differential input excitations (from 1mV all the way to 1V). Part (c) shows that ABCD-NL can duplicate the SPICE-level dynamics of the comparator even when there is serious departure from ideal behaviour. Part (d) demonstrates that ABCD-NL is well-suited to model the comparator in the context of a SAR-ADC.

Fig. 15 shows the verification flow that we used. As the figure shows, the Boolean circuit that is verified consists of three parts, (1) the ABCD-NL Boolean model, (2) the Boolean logic that encodes the property to be checked, and (3) some Boolean logic to encode constraints on the inputs that can be applied to the AMS system. The circuit of Fig. 15 is constructed in such a way that the given AMS design fails to meet its throughput specification if and only if the bit b_{out} can somehow be asserted to 1 by choosing an appropriate sequence of bits applied at b_{in}.

The constraint on the input is that it can change only once per N clock periods of the ABCD-NL model. The time period of ABCD-NL's sequential Boolean model is 10ps (see Algorithm 1). The input constraint is modelled using a counter that outputs a 1 every N clock cycles (Fig. 15). If this constraint is violated, the bit marked b_c immediately becomes 0 and stays there forever, which makes it impossible to assert b_{out} to 1. This ensures that any counter-example returned by ABC would satisfy the input constraint.

The throughput property to be checked is that, given the above constraint on the input, the output should always reach an acceptable state before N clock cycles (i.e., before the input can change). This acceptable state is defined as being $\geq 0.8V$ for a 1, and being $\leq 0.2V$ for a 0.

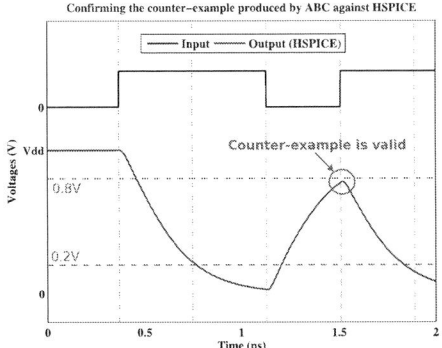

Fig. 16. Checking that the counter-example returned by ABCD-NL + ABC is valid in the analog domain.

Clearly, there is an N_0 such that the above throughput property will fail for all $N \leq N_0$. We can use ABC to quickly zero in on N_0, by incorporating ABC-based verification within a binary search loop. In this way, we were able to determine that $N_0 = 38$. This translates to a throughput of approximately 2.56Gbps. Furthermore, we were also able to confirm, using HSPICE, that for $N = 38$, the counter-example returned by ABC is a valid one in the analog domain (Fig. 16). Therefore, the throughput

bound obtained is tight and meaningful. Thus, ABCD-NL is a powerful and capable modelling technique that can be used for AMS verification.

IV. CONCLUSIONS

In conclusion, we have developed and demonstrated ABCD-NL, a new technique for producing purely Boolean models of non-linear analog circuits, suitable for AMS verification. We have applied ABCD-NL to several circuits that are of interest to AMS designers, including charge pumps, signaling/communications sub-systems, D/A converters and comparators used in SAR-ADCs, etc. In addition, we have demonstrated a formal verification example where we used ABCD-NL, in conjunction with ABC, to formally verify the throughput of an AMS system.

REFERENCES

[1] R. K. Brayton and A. Mishchenko. ABC: An academic industrial-strength verification tool. In CAV '10: Proceedings of the 22nd International Conference on Computer Aided Verification, pages 24–40, 2010.
[2] R. Parker. Analog design challenges in the new era of process scaling. At the 2012 International Workshop on Design Automation for AMS Circuits (co-located with ICCAD).
[3] W. Hartong, L. Hedrich, and E. Barke. Model checking algorithms for analog verification. In DAC '02: Proceedings of the 39th annual ACM Design Automation Conference, pages 542–547, 2002.
[4] S. Steinhorst and L. Hedrich. Model checking of analog systems using an analog specification language. In DATE '08: Proceedings of the ACM Conference on Design, Automation and Test in Europe, pages 324–329, 2008.
[5] G. Al-Sammane, M. H. Zaki, and S. Tahar. A symbolic methodology for the verification of AMS designs. In DATE '07: Proceedings of the ACM Conference on Design, Automation and Test in Europe, pages 249–254, 2007.
[6] S. Little. Efficient Modeling and Verification of Analog/Mixed-Signal Circuits Using Labeled Hybrid Petri Nets. PhD thesis, University of Utah, 2008.
[7] M. Althoff, A. Rajhans, B. H. Krogh, S. Yaldiz, X. Li, and L. Pileggi. Formal verification of Phase Locked Loops using reachability analysis and continuization. In ICCAD '10: Proceedings of the IEEE/ACM International Conference on Computer-Aided Design, pages 659–666, 2010.
[8] S. Steinhorst and L. Hedrich. Trajectory-directed discrete state space modeling for formal verification of nonlinear analog circuits. In ICCAD '12: Proceedings of the International Conference on Computer-Aided Design, pages 202–209, 2012.
[9] T. A. Henzinger, P. H. Ho, and H. Wong-Toi. HyTech: A model checker for hybrid systems. International Journal on Software Tools for Technology Transfer, 1(1):110–122, 1997.
[10] E. Asarin, T. Dang, and O. Maler. d/dt: A verification tool for hybrid systems. In CDC '01: Proceedings of the 40th IEEE Conference on Decision and Control, pages 2893–2898, 2001.
[11] A. Chutinan and B. H. Krogh. Computational techniques for hybrid system verification. IEEE Transactions on Automatic Control, 48(1):64–75, 2003.
[12] G. Frehse. PHAVer: Algorithmic verification of hybrid systems past HyTech. International Journal on Software Tools for Technology Transfer, 10(3):263–279, 2008.
[13] C. Tomlin, I. Mitchell, A. M. Bayen, and M. Oishi. Computational techniques for the verification of hybrid systems. Proceedings of the IEEE, 91(7):986–1001, 2003.
[14] M. Althoff, O. Stursberg, and M. Buss. Computing reachable sets of hybrid systems using a combination of zonotopes and polytopes. Nonlinear Analysis: Hybrid Systems, 4(2):233–249, 2010.
[15] A. V. Karthik and J. Roychowdhury. ABCD-L: approximating continuous linear systems using Boolean models. In DAC '13: Proceedings of the 50th Design Automation Conference, pages 63:1–63:9, 2013.
[16] C. Gu and J. Roychowdhury. FSM model abstraction for analog/mixed-signal circuits by learning from I/O trajectories. In ASP-DAC '11: Proceedings of the 16th Asia and South Pacific Design Automation Conference, pages 7–12, 2011.
[17] K. V. Aadithya and J. Roychowdhury. DAE2FSM: Automatic generation of accurate discrete-time logical abstractions for continuous-time circuit dynamics. In DAC '12: Proceedings of the 49th Design Automation Conference, pages 311–316, 2012.
[18] http://ptm.asu.edu/modelcard/HP/22nm_HP.pm.
[19] http://www.analog.com/Analog_Root/static/techSupport/designTools/spiceModels/license/spice_general.html?cir=AD8079A.cir.
[20] http://ltspice.linear.com/software/LTC.zip.

Toward Efficient Programming of Reconfigurable Radio Frequency (RF) Receivers

Jun Tao, Ying-Chih Wang, Minhee Jun, Xin Li, Rohit Negi, Tamal Mukherjee and Lawrence T. Pileggi

Electrical and Computer Engineering Department, Carnegie Mellon University, Pittsburgh, PA 15213

{juntao, yingchiw, mjun, xinli, negi, tamal, pileggi}@andrew.cmu.edu

Abstract – Reconfigurable radio frequency (RF) system is an emerging component to mitigate the growing engineering cost for wireless chip design. In this paper, we propose a new methodology for efficient programming of reconfigurable RF receiver. The proposed method is facilitated by two novel techniques: two-phase relaxation search and Pareto-based search space reduction. Our numerical experiments demonstrate that the proposed methodology is more robust (i.e., close to global optimum) and/or efficient (i.e., with low computational cost) than other traditional algorithms based on either local relaxation or simulated annealing.

I. Introduction

For most commercial wireless communication applications, rapid introduction of new wireless standards, increased autonomous applications and aggressive miniaturization result in a significant growth of engineering cost for developing various dedicated wireless chips. The advent of reprogrammable RF electronic systems provides a promising avenue for cost reduction by allowing circuit reuse via post-manufacturing reprogrammability [1].

Generally speaking, a reconfigurable RF system consists of a number of circuit blocks (e.g., low noise amplifier (LNA), mixer, filter, etc.) that can be adaptively reconfigured in terms of their block-level performance metrics (e.g., gain, bandwidth, nonlinearity, noise figure (NF), etc.) in order to meet a set of given performance specifications at system level (e.g., signal-to-noise ratio (SNR), bit error rate, etc.). Recently, several reconfigurable RF systems have been developed, including reconfigurable vector signal analyzer and software-defined receiver [2], multi-standard RF front-end [3] and ultra-low-power RF transmitter [4]. For a reconfigurable RF system, all circuit blocks must be carefully configured to ensure their desired functionality and performance. The process of finding the optimal configurations for all circuit blocks to meet the given system-level specifications is referred to as RF system *programming* in this paper. Our objective is to develop efficient algorithms to accomplish this programming task.

It is important to mention that most traditional algorithms developed for general-purpose analog optimization are not directly applicable to RF receiver programming. Traditionally, analog optimization often takes a fixed circuit topology and optimizes the device-level parameters, such as the widths and lengths of transistors, to meet a set of given performance specifications [5]. These traditional methods fall into two broad categories [6]: equation-based [7] and simulation-based [8]-[9]. Equation-based methods, such as geometric programming [7], rely on design equations that may require a lot of efforts to derive. Furthermore, these design equations may not be sufficiently accurate since various approximations are often made [5]. On the other hand, simulation-based methods, such as simulated annealing [8] and genetic programming [9], often take excessive computational time to extensively search the design space in hopes of reaching a solution close to global optimum [5]. For these reasons, there is a strong need to develop a new optimization technique that can solve our proposed problem of RF receiver programming both robustly (i.e., close to global optimum) and efficiently (i.e., with low computational cost).

Towards this goal, two novel techniques are proposed in this paper by exploiting the unique characteristics of RF receiver programming.

- *Two-phase relaxation search*: As will be discussed in Section II, RF receiver programming can be cast into an optimization problem that minimizes a cost function subject to a single constraint. Given this unique problem formulation, we propose to adopt a two-phase relaxation search algorithm. During the first phase, the constraint function is maximized without taking into account the cost function. Next, the cost function is minimized subject to the given constraint during the second phase. Such a two-phase approach was initially developed to explore the performance trade-offs between power and delay for gate sizing of digital circuits [10]. It has been demonstrated to be extremely robust (i.e., avoiding a large number of local optima) in the literature.

- *Pareto-driven search space reduction*: While most traditional algorithms were developed to optimize analog circuit blocks by tuning the device-level parameters, RF receiver programming should be performed at system level based on reconfigurable circuit blocks. In this case, we can take advantage of block-level Pareto optimal fronts (POFs) [11] to substantially reduce the search space. The POF of a circuit block captures the optimal trade-offs of the block-level performance metrics, such as the trade-off between nonlinearity and power of an LNA. For RF receiver programming, the block-level configurations on POFs are most likely to yield the optimal system-level performance. Therefore, it is desirable to search the optimal configurations among these candidates on POFs only for RF receiver programming.

The remainder of this paper is organized as follows. In Section II, we first describe the problem formulation of RF receiver programming and then develop the proposed programming algorithm in Section III. The efficiency of our proposed technique is demonstrated by a reconfigurable RF receiver example in Section IV. Finally, we draw conclusions in Section V.

II. Problem Formulation

The general goal of reconfigurable RF system design is to enable a common hardware architecture that reuses a set of circuit blocks for disparate wireless applications and standards through programming. As a result, the design cost can be reduced dramatically compared to the traditional non-configurable system. A reconfigurable RF system should be able to operate in the severely crowded and rapidly changing modern commercial environment, while maintaining the optimal performance. To arrive at an optimal reconfigurable RF system, it is crucial to minimize silicon area and power consumption for multiple wireless applications and standards by maximizing hardware sharing [3]. In addition, programming a reconfigurable RF system must be

978-1-4799-2817-0/14 $31.00 © 2014 IEEE

efficiently done by appropriately setting a number of programmable "knobs".

For a reconfigurable RF receiver, its programming can be performed through two possible avenues. First, the receiver can be programmed at system level by changing its architecture (e.g., transforming an RF receiver from superheterodyne to homodyne by using switch boxes). Second, the receiver can be programmed at block level by changing its block-level performances (e.g., varying the gain of an LNA by tuning its bias current) [2]. To successfully program a reconfigurable RF receiver, both system-level and block-level knobs must be appropriately set to achieve the desired functionality and performance. In this paper, we will first derive efficient algorithms for block-level programming, and then extend our proposed techniques to system-level programming for multiple receiver architectures.

As an example for system-level and block-level programming, a reconfigurable RF receiver is shown in Figure 1. In this system, the RF receiver can be programmed to two different architectures: superheterodyne and homodyne, by using system-level knobs, i.e., switch boxes. All circuit blocks, including LNAs, band-pass filter (BPF), low-pass filter (LPF), oscillators (LO1, LO2 and LO3), mixers, and variable gain amplifier (VGA), can be reconfigured by block-level programmable knobs, such as block-level switch boxes and/or tunable bias currents.

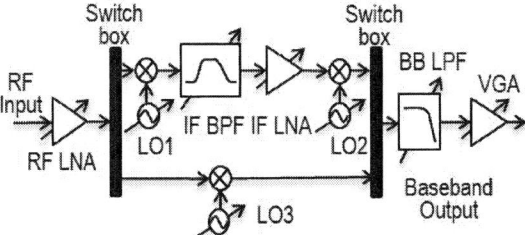

Figure 1. Simplified schematic is shown for a reconfigurable RF receiver which can be programmed to two different architectures: superheterodyne (the upper signal path) and homodyne (the lower signal path).

Mathematically, let $\mathbf{x}$ denote all block-level programmable knobs (e.g., the values of tunable bias currents used to control the performance metrics of circuit blocks) and $\mathbf{y}$ denote all system-level programmable knobs (e.g., the configurations of switch boxes used to set the receiver architecture). In general, the possible values of these programmable knobs are discrete and they are controlled by a finite number of digital bits (e.g., a finite set of possible bias current values controlled by a digital-to-analog converter). The problem of reconfigurable RF receiver programming can be formulated as an optimization with multiple constraints:

$$\min_{\mathbf{x},\mathbf{y}} \quad F(\mathbf{x},\mathbf{y})$$
$$\text{s.t.} \quad g_j(\mathbf{x},\mathbf{y}) \geq G_j \quad (j=1,2,\cdots,J) \quad , \quad (1)$$

where $F(\mathbf{x},\mathbf{y})$ (i.e., the cost function) and $g_j(\mathbf{x},\mathbf{y})$ (i.e., the constraint function) denote different system-level performances (e.g., power, area, SNR, etc.), and G_j is the given specification. In order to simplify the RF receiver programming problem, Eq. (1) can be rewritten as an optimization formulation that minimizes the cost function subject to a single constraint:

$$\min_{\mathbf{x},\mathbf{y}} \quad F(\mathbf{x},\mathbf{y})$$
$$\text{s.t.} \quad S(\mathbf{x},\mathbf{y}) = \min\left(g_1(\mathbf{x},\mathbf{y})-G_1,\cdots,g_J(\mathbf{x},\mathbf{y})-G_J\right) \geq 0 \quad , \quad (2)$$

where $S(\mathbf{x},\mathbf{y})$ is a new constraint function constructed by all the original constraints in (1). In the remainder of this paper, we will

take the formulation in (2) to derive the algorithm for efficient receiver programming.

The optimization problem in (2) is nonlinear, non-convex and discrete. There are two important technical challenges when developing efficient numerical algorithms to solve (2) for RF receiver programming.

- *Robustness*: The proposed optimization algorithm must be sufficiently robust to find the optimum of (2) that is at least close to global optimum. In other words, the optimization algorithm should not easily get stuck at local optima. Note that finding a robust solution for (2) is not trivial, since the problem formulation in (2) is non-convex and, hence, there may exist a large number of local optima in the search space.

- *Complexity*: The optimization algorithm for solving (2) must be computationally inexpensive so that programming an RF receiver can be done quickly. In particular, since evaluating the system-level performance metrics (i.e., $F(\mathbf{x},\mathbf{y})$ and $g_j(\mathbf{x},\mathbf{y})$ in (2)) requires expensive simulation (e.g., more than 10 minutes per simulation for our receiver example shown in Section IV), the proposed optimization algorithm must "smartly" explore the search space with few simulations. For this reason, even though many traditional analog optimization algorithms (e.g., simulated annealing [8]) can find a robust solution that is close to global optimum, they are not suitable for RF receiver programming due to their high computational complexity.

In what follows, we will propose a novel optimization framework that exploits the unique characteristics of RF receiver programming. It, in turn, facilitates us to find a robust solution of (2) with low computational cost.

III. Proposed Approach

As previously mentioned, there are two distinct types of variables in the optimization problem defined by (2): (i) the block-level knobs $\mathbf{x}$ to program the circuit implementation of each block, and (ii) the system-level knobs $\mathbf{y}$ to select the appropriate receiver architecture. In this paper, we propose to optimize $\mathbf{x}$ and $\mathbf{y}$ by applying a hierarchical approach. Namely, for each possible value of $\mathbf{y}$ that defines a particular receiver architecture, we optimize $\mathbf{x}$ to find out the optimal configurations of all circuit blocks. Next, the system-level performance metrics are compared among different architectures and the optimal architecture is selected to satisfy the given constraints with minimal cost function.

The aforementioned programming strategy is adopted because the optimal block-level implementations and their performance metrics are often substantially different for different receiver architectures. For instance, the performance requirements on noise and distortion are significantly different between superheterodyne and homodyne receivers [12]. Hence, it is not trivial to design an algorithm to optimize both $\mathbf{x}$ and $\mathbf{y}$ simultaneously.

In what follows, we first describe two novel ideas, two-phase relaxation search and Pareto-driven search space reduction, that facilitate us to efficiently find the optimal $\mathbf{x}$ (i.e., block-level configurations) for a given $\mathbf{y}$ (i.e., receiver architecture). Next, we further discuss the overall programming flow with optimal architecture selection.

A. Two-phase Relaxation Search

To find the optimal block-level implementations for a given receiver architecture, we re-write the optimization problem in (2) as:

$$\min_{\mathbf{x},\mathbf{y}} F\left(\mathbf{x},\mathbf{y}^{(q)}\right)$$
$$\text{s.t.}\quad S\left(\mathbf{x},\mathbf{y}^{(q)}\right)=\min\left(g_1\left(\mathbf{x},\mathbf{y}^{(q)}\right)-G_1,\cdots,g_J\left(\mathbf{x},\mathbf{y}^{(q)}\right)-G_J\right)\geq 0 \qquad (3)$$

where $\mathbf{x}$ is the optimization variable and $\mathbf{y}=\mathbf{y}^{(q)}$ corresponds to a particular receiver architecture. Without loss of generality, we assume that there are M programmable blocks and the configuration of the m-th block is defined by a row vector $\mathbf{x}_m$, where $m=1,2,\ldots,M$. Concatenating all these vectors $\{\mathbf{x}_m|m=1,2,\ldots,M\}$ forms the optimization variable $\mathbf{x}$, i.e., $\mathbf{x}=[\mathbf{x}_1 \quad \mathbf{x}_2 \quad \cdots \quad \mathbf{x}_M]$. We further denote the set of all possible values of $\mathbf{x}_m$ as $\left\{\mathbf{x}_m^{(i)}\big|i=1,2,\ldots,M_m\right\}$, where each $\mathbf{x}_m^{(i)}$ corresponds to one possible configuration and M_m is the total number of configurations for the m-th block.

When solving the optimization problem in (3), we must repeatedly evaluate the receiver performances for different values of $\mathbf{x}$. It requires a large number of expensive numerical simulations. To make the computational cost affordable, a local relaxation approach may be adopted to choose a single circuit block for optimization at one time.

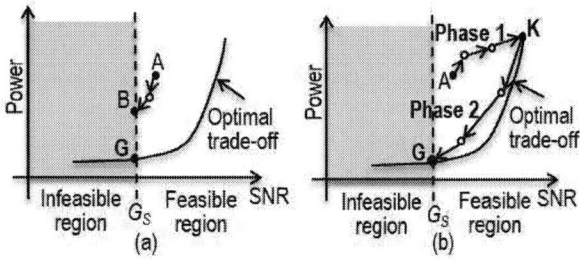

Figure 2. Two different optimization algorithms are compared with the objective to minimize power subject to a given SNR constraint. (a) The traditional algorithm of local relaxation starts from the initial point A and is stuck at a local optimum B. (b) The proposed algorithm of two-phase relaxation search starts from the initial point A and reaches the global optimum G.

Such a simple approach, however, often gets stuck at a local optimum, especially when the initial starting point is not appropriately set. To understand the reason, we consider a simple example shown in Figure 2(a) where the objective is to find the global optimum G by minimizing power subject to an SNR constraint. Suppose that the SNR specification is plotted as a vertical line G_S separating the feasible and infeasible regions in Figure 2. When local relaxation is applied, one circuit block is selected for optimization at each iteration step. The optimal configuration of the selected circuit block (say, the m-th block) should be found to minimize power while simultaneously satisfying the SNR constraint. All configurations of the m-th block that violate the SNR constraint are considered as "infeasible". Due to this reason, if the initial starting point is close to the constraint boundary G_S (e.g., the point A shown in Figure 2(a)), a large number of possible configurations of the m-th block may be considered to be infeasible because they violate the SNR constraint. In other words, starting from the initial point A in Figure 2(a) would prevent us from exploring many possible configurations of the m-th block that are infeasible in the current iteration step, but should be considered to be feasible if the configurations of other circuit blocks can be changed. The optimization, therefore, fails to converge to global optimum. It is one of the major limitations for the local relaxation algorithm.

To address this issue, we adopt a heuristic technique to solve

the optimization problem in (3) via two phases. During the first phase, local relaxation is applied to maximize the constraint function without considering the cost function:

$$\max_{\mathbf{x}} S\left(\mathbf{x},\mathbf{y}^{(q)}\right)=\min\left(g_1\left(\mathbf{x},\mathbf{y}^{(q)}\right)-G_1,\cdots,g_J\left(\mathbf{x},\mathbf{y}^{(q)}\right)-G_J\right). \quad (4)$$

Our goal is to find the configuration that is "farthest" away from the constraint boundaries $G_1,G_2,\ldots,G_J$, such as the point K in Figure 2(b), by maximizing the minimal distance between the constraint function $g_j\left(\mathbf{x},\mathbf{y}^{(q)}\right)$ and the given specification G_j. During the second phase, we further take the optimization result from the first phase as the initial starting point and apply local relaxation to solve (3), i.e., minimize the cost function subject to the given constraints. In this case, since the initial starting point is not close to the constraint boundaries, a large number of configurations should be "feasible" when optimizing a particular circuit block by local relaxation. In other words, the proposed two-phase approach is able to avoid a lot of local optima as compared to the simple local relaxation algorithm shown in Figure 2(a).

Algorithm 1: Constraint Function Maximization (Phase 1)
1. Start from the optimization formulation in (3) with a given $\mathbf{y}_q$ corresponding to a particular receiver architecture.
2. Let $\mathbf{x}^{(I)}$ denote the initial value of block-level knobs.
3. Simulate the receiver with $\mathbf{x}^{(I)}$ to obtain the value of the constraint function $S^{(I)}$.
4. Set $\mathbf{x}^{(OLD)}=\mathbf{x}^{(NEW)}=\mathbf{x}^{(I)}$ and $S^{(OPT)}=S^{(I)}$.
5. For $m=1,2,\ldots,M$
6. Set $\mathbf{x}_m$ to different values $\mathbf{x}_m=\mathbf{x}_m^{(i)}$, where $i=1,2,\ldots,M_m$, and simulate the receiver at each of these cases to obtain the values of the constraint function $\{S_m^{(i)}|i=1,2,\ldots,M_m\}$. The configurations of all other blocks are defined by $\left\{\mathbf{x}_i^{(NEW)}\big|i\neq m\right\}$ and they should be unchanged for these simulations.
7. Find the optimal value of $\mathbf{x}_m$ (say, $\mathbf{x}_m^{(OPT)}$) corresponding to the largest value of the constraint function (say, $S_m^{(OPT)}$) in the set $\left\{S_m^{(i)}\big|i=1,2,\ldots,M_m\right\}$.
8. If $S^{(OPT)}<S_m^{(OPT)}$, set $S^{(OPT)}=S_m^{(OPT)}$ and $\mathbf{x}^{(NEW)}=[\mathbf{x}_1^{(NEW)} \; \mathbf{x}_2^{(NEW)} \; \cdots \; \mathbf{x}_m^{(OPT)} \; \cdots \; \mathbf{x}_M^{(NEW)}]$.
9. End For
10. If $\mathbf{x}^{(NEW)}\neq\mathbf{x}^{(OLD)}$, set $\mathbf{x}^{(OLD)}=\mathbf{x}^{(NEW)}$ and go back to step 5. Otherwise, stop iteration and $\mathbf{x}^{(NEW)}$ is the optimal configuration with the largest value of the constraint function for the given receiver architecture.

Algorithm 2: Cost Function Minimization (Phase 2)
1. Start from the optimization formulation (2) and the optimal value $\mathbf{x}^{(NEW)}$ solved by Algorithm 1.
2. Calculate the cost function $F^{(OPT)}$ with the configuration $\mathbf{x}^{(NEW)}$.
3. Set $\mathbf{x}^{(OLD)}=\mathbf{x}^{(NEW)}$.
4. For $m=1,2,\ldots,M$
5. Set $\mathbf{x}_m$ to different values $\mathbf{x}_m=\mathbf{x}_m^{(i)}$, where $i=1,2,\ldots,M_m$, and simulate the receiver at each of these cases to obtain the values of the cost function $\{F_m^{(i)}|i=1,2,\ldots,M_m\}$ and the values of the constraint function $\{S_m^{(i)}|i=1,2,\ldots,M_m\}$. The configurations of all other blocks are defined by $\left\{\mathbf{x}_i^{(NEW)}\big|i\neq m\right\}$ and they should be unchanged for these simulations.

978-1-4799-2817-0/14 $31.00 © 2014 IEEE

6. Calculate the following evaluation function values $\{E_m^{(i)}|i=1,2,\dots,M_m\}$:

$$E_m^{(i)}=\begin{cases}F_m^{(i)} & if\quad S_m^{(i)}\geq 0\\ INF & if\quad S_m^{(i)}<0\end{cases}. \qquad (5)$$

7. Find the optimal value of $\mathbf{x}_m$ (say, $\mathbf{x}_m^{(OPT)}$) corresponding to the smallest evaluation function value (say, $E_m^{(OPT)}$) in the set $\{E_m^{(i)}|i=1,2,\dots,M_m\}$.

8. If $E_m^{(OPT)}<F^{(OPT)}$, set $F^{(OPT)}=E_m^{(OPT)}$ and $\mathbf{x}^{(NEW)}=[\mathbf{x}_1^{(NEW)}\ \mathbf{x}_2^{(NEW)}\ \cdots\ \mathbf{x}_m^{(OPT)}\ \cdots\ \mathbf{x}_M^{(NEW)}]$.

9. End For

10. If $\mathbf{x}^{(NEW)}\neq\mathbf{x}^{(OLD)}$, set $\mathbf{x}^{(OLD)}=\mathbf{x}^{(NEW)}$ and go back to step 4. Otherwise, stop iteration and $\mathbf{x}^{(NEW)}$ is the optimal configuration for the given receiver architecture.

Algorithm 1 and Algorithm 2 describe the detailed implementations of the aforementioned two phases. Several important clarifications should be made for these two algorithms. First, the aforementioned technique of two-phase relaxation search was initially developed for gate sizing of digital circuits where the total power should be minimized subject to a given delay constraint [10]. As demonstrated in the literature, it is able to avoid a large number of local optima for gate sizing. In this paper, we adopt the two-phase approach for RF receiver programming, since both gate sizing and RF receiver programming are discrete and they share a similar optimization formulation (i.e., minimizing a cost function subject to a given constraint).

Second, it is important to note that the computational cost of Algorithm 1 and Algorithm 2 is often dominated by the simulation time for performance evaluation, because accurately estimating system-level performances requires us to simulate the RF receiver over a long time period. As will be demonstrated by our numerical examples in Section IV, evaluating the system performances for a receiver with a given configuration takes more than 10 minutes, even if behavioral modeling is applied to speed up the simulation. Since the aforementioned performance evaluation must be repeated within the optimization loop of Algorithm 1 and Algorithm 2, it is extremely important to minimize the total number of required performance evaluations so that our proposed algorithm of two-phase relaxation search is computationally efficient for practical applications. For this reason, we will further develop a novel approach for search space reduction in order to minimize the computational cost of RF receiver programming.

B. Pareto-driven Search Space Reduction

Our key idea of search space reduction is to remove the configurations that cannot be the optimum of (3) or (4), before running Algorithm 1 or Algorithm 2. It, in turn, reduces the search space and, hence, the computational cost of receiver programming. Such a goal is facilitated by the concept of Pareto optimal front (POF) that was previously developed for analog system-level optimization in the literature [11].

Let $\mathbf{p}_m=[p_{m,1},p_{m,2},\dots,p_{m,D_m}]$ denote the performance metrics of interest for the m-th circuit block (e.g., gain, NF, IIP3 and power for an LNA). POF captures the "best" trade-off among these performance metrics. If a point $\mathbf{p}_m^{(i)}$ is on POF (i.e., referred to as a Pareto point), it cannot be dominated by any other point (say, $\mathbf{p}_m^{(j)}$) in the performance space:

$$\mathbf{p}_m^{(i)}\neq\mathbf{p}_m^{(j)}\text{ and }p_{m,k}^{(i)}\leq p_{m,k}^{(j)}\quad(k=1,2,\cdots,D_m). \qquad (6)$$

In (6), we assume that the value of a performance metric (say,

$p_{m,k}^{(j)}$) dominates another value (say, $p_{m,k}^{(i)}$), if $p_{m,k}^{(j)}$ is greater than $p_{m,k}^{(i)}$. In general, a linear transformation can be appropriately applied to any given performance metric to make this assumption valid. For instance, a small NF is often preferred for LNA design. In this case, we can define $-1\cdot$NF, instead of NF, as the performance of interest. Figure 3 shows a simple example of POF with three performance metrics: $p_{m,1}$, $p_{m,2}$ and $p_{m,3}$.

In our application of RF receiver programming, each possible configuration, $\mathbf{x}_m^{(i)}$ where $i=1,2,\dots,M_m$, is associated with a particular performance value $\mathbf{p}_m^{(i)}$ for the m-th block. To achieve the optimal performance for an RF receiver, we only need to consider the configurations on POF as "valid" programming options. Taking the RF LNA of the superheterodyne receiver shown in Figure 1 as an example, if the constraint function is set to SNR and we want to maximize it during the first phase, the RF LNA should be configured to have large gain, low NF and large IIP3. If a configuration option of the RF LNA is not on the POF with respect to gain, NF and IIP3, it is unlikely to reach the maximal SNR and can be simply ignored during the programming process.

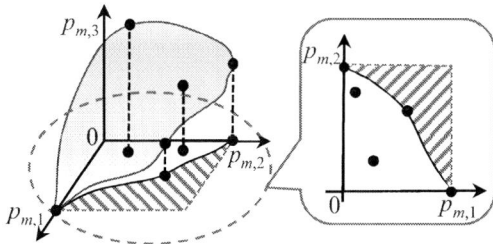

Figure 3. A 3-D POF is plotted for three performance metrics $p_{m,1}$, $p_{m,2}$ and $p_{m,3}$ and is projected to the 2-D plane defined by $p_{m,1}$ and $p_{m,2}$.

The aforementioned idea of Pareto-driven search space reduction is particularly useful when Algorithm 1 is applied to maximize the constraint function without considering the cost function during the first phase. Since the cost function $F(\mathbf{x},\mathbf{y}^{(q)})$ is ignored in Algorithm 1, the block-level performance metrics which only affect $F(\mathbf{x},\mathbf{y}^{(q)})$ can also be ignored. Therefore, the total number of block-level performance metrics and, hence, the dimensionality of the block-level performance space could be reduced for each circuit block. Considering the simple example shown in Figure 2(b) where the cost function is set to power and the constraint function is set to SNR, the block-level power consumption could be ignored when maximizing SNR during the first phase. Namely, the dimensionality of the block-level performance space is reduced by one. In this case, when we project the block-level performance points to a lower-dimensional space, a large number of configuration options may not be on POF and, therefore, can be simply ignored. To intuitively illustrate the reason, Figure 3 shows a simple 3-D POF example. If one of the performance metrics (i.e., $p_{m,3}$) is ignored and the 3-D performance points are projected onto the 2-D space defined by $p_{m,1}$ and $p_{m,2}$, many of these performance points are not on POF. Hence, the corresponding configuration options can be ignored.

Algorithm 3 summarizes the simplified flow of our proposed Pareto-driven search space reduction. It starts from a set of performance values $\Theta_m=\{\mathbf{p}_m^{(i)}|i=1,2,\dots,M_m\}$ corresponding to different configuration options of the m-th circuit block. Based on the definition of POF, a Pareto point $\mathbf{p}_m^{(i)}$ can be identified by choosing the performance point that has the greatest value for one of the performance metrics (say, $p_{m,1}$). Then all points in Θ_m that are dominated by $\mathbf{p}_m^{(i)}$ are removed. Next, the point $\mathbf{p}_m^{(j)}$ which has the second greatest value for the performance metric $p_{m,1}$ in Θ_m is found and it is taken as the second Pareto point because it is not dominated by $\mathbf{p}_m^{(i)}$ or any other point remaining in Θ_m. The aforementioned selection process is repeated until all Pareto points are found. Finally, the configuration options of the m-th circuit block corresponding to all selected Pareto points are identified and these configurations will be considered as valid options for RF receiver programming.

Algorithm 3: Pareto-driven Search Space Reduction

1. Start from a set $\Theta_m = \left\{ \mathbf{p}_m^{(i)} \middle| i = 1,2,\dots,M_m \right\}$ containing all performance values corresponding to M_m configuration options of the m-th circuit block.
2. Initialize the set $\Xi = \{\}$.
3. Find the Pareto point $\mathbf{p}_m$ from Θ_m that has the greatest value for the first performance metric $p_{m,1}$.
4. Remove all points in Θ_m that are dominated by $\mathbf{p}_m$.
5. Remove $\mathbf{p}_m$ from the set Θ_m, and add $\mathbf{p}_m$ to the set Ξ.
6. If the set Θ_m is not empty, go to Step 3. Otherwise, stop iteration and the set Ξ contains all Pareto points.
7. Determine the configuration options corresponding to the performance values belonging to the set Ξ.

C. Multi-Architecture Receiver Programming

The previous sub-sections describe several efficient algorithms to find the optimal value of the block-level programmable knobs $\mathbf{x}$ for a given receiver architecture defined by the system-level knobs $\mathbf{y}$. Suppose that the receiver system can be possibly implemented with Q different architectures: $\left\{ \mathbf{y}^{(q)} \middle| q = 1,2,\dots,Q \right\}$. For each architecture, the optimization problem in (3) is solved to determine the optimal block-level configurations. Next, all architectures are compared according to their system-level performance values, and the optimal architecture is selected to achieve the minimal cost while simultaneously satisfying the given constraints.

Algorithm 4: Reconfigurable RF Receiver Programming

1. Start from the optimization formulation (2) where $\mathbf{y} \in \left\{ \mathbf{y}^{(q)} \middle| q = 1,2,\dots,Q \right\}$.
2. For each $\mathbf{y}^{(q)}$ where $q = 1,2,\dots,Q$
3. Apply Algorithm 1~Algorithm 3 to find the optimal block-level configurations for the given receiver architecture defined by $\mathbf{y}^{(q)}$. Calculate the corresponding cost and constraint functions.
4. End For
5. Determine the optimal receiver architecture that can achieve the minimal cost subject to the given constraints.

Algorithm 4 summarizes the overall flow for RF receiver programming. By taking advantage of the efficient techniques described in this section, the proposed receiver programming only requires very few (e.g., less than 50) simulations to reach convergence, as will be demonstrated by our numerical examples in Section IV.

IV. Numerical Experiments

In this section, a reconfigurable RF receiver (including baseband signal processing) shown in Figure 1 is used as an example to demonstrate the efficiency of the proposed programming algorithm. This RF receiver is designed for the IEEE 802.11a WLAN standard where the channel bandwidth is 20MHz. In this example, the goal of RF receiver programming is to minimize power subject to a given SNR specification. At system level, the receiver can be programmed to two different architectures: superheterodyne (SH) and homodyne (HD). At block level, three circuit blocks are designed to be reconfigurable: RF LNA, IF LNA and IF BPF. Both the RF LNA and the IF LNA can be programmed to 11 configurations respectively, where the LNA performance metrics (i.e., gain, NF, IIP3 and power) are different for different configurations. The IF BPF can be programmed to 9 configurations with different bandwidth and unloaded quality factor. In total, there are 1100 different configurations for the aforementioned RF receiver. Each circuit block of the receiver is represented as a macromodel to facilitate efficient system-level simulation of the RF receiver in MATLAB SIMULINK.

A. Programming Results

Table 1. Programming results for the reconfigurable RF receiver with different SNR specifications

SNR Spec (dB)	Algorithm	SNR (dB)	Power (mW)	Selected Arch
6	Exhaustive	7.36	1.32	HD
	Relaxation	7.36	1.32	HD
	Annealing	7.36	1.32	HD
	Proposed	7.36	1.32	HD
8	Exhaustive	8.61	1.44	SH
	Relaxation	8.61	1.44	SH
	Annealing	8.61	1.44	SH
	Proposed	8.61	1.44	SH
9	Exhaustive	9.53	2.04	SH
	Relaxation	9.66	2.60	HD
	Annealing	9.53	2.04	SH
	Proposed	9.53	2.04	SH
10	Exhaustive	10.51	3.32	SH
	Relaxation	Infeasible	Infeasible	Infeasible
	Annealing	10.51	3.32	SH
	Proposed	10.51	3.32	SH

For testing and comparison purposes, four different programming algorithms are implemented: (i) exhaustive search (Exhaustive), (ii) local relaxation (Relaxation), (iii) simulated annealing (Annealing), and (iv) two-phase relaxation (Proposed). The exhaustive search method guarantees to find the globally optimal configuration by exploring all possible options. Hence, it is used to generate the "golden" results to evaluate the "robustness" of other programming methods.

Table 1 summarizes the programming results for different algorithms with different SNR specifications. Studying Table 1 reveals an important observation that the traditional algorithm of local relaxation fails to find the globally optimal configuration, when the SNR specification is 9 dB or 10 dB. On the other hand, the proposed algorithm of two-phase relaxation converges to the global optima in all test cases, thereby demonstrating its superior performance over the simple relaxation method. The simulated annealing algorithm is also able to find the global optima in this example; however, its computational cost is substantially more expensive than our proposed approach, as will be demonstrated in the next sub-section.

B. Computational Cost

During the programming process, system-level performances, such as SNR, must be repeatedly evaluated for different receiver configurations. For the RF receiver shown in Figure 1, a single SIMULINK simulation takes more than 10 minutes to evaluate its SNR for a given receiver configuration. Hence, the computational cost of receiver programming is completely dominated by the simulation time, and we can use the number of required simulations as a metric to compare the complexity of different programming algorithms, as shown in Table 2.

Table 2. The number of required simulations for different programming algorithms with different SNR specifications

SNR Spec (dB)	Exhaustive	Relaxation	Annealing	Proposed
6	1100	32	170	35
8	1100	32	179	34
9	1100	36	331	34
10	1100	36	259	40

Note that both exhaustive search and simulated annealing require a large number of simulations before reaching convergence. Our proposed algorithm of two-phase relaxation with Pareto-based search space reduction and the traditional algorithm of local relaxation can greatly reduce the computational cost by more than $5\times$. On the other hand, the proposed algorithm is more "robust" than local relaxation, as shown in Table 1. These observations demonstrate that our proposed method is superior over all other traditional approaches tested in this example.

In our experiments, we further observe that if the proposed heuristic method of Pareto-based search space reduction is not applied, more than 80 simulations are required to find the optimal configuration by using two-phase relaxation search. In other words, the proposed search space reduction successfully reduces the computational cost by more than $2\times$ in this example.

V. Conclusions

In this paper, we develop an efficient methodology for programming reconfigurable RF receivers to achieve minimal cost function value subject to a set of given constraints. It is mathematically formulated as a nonlinear, discrete optimization problem. Two novel techniques, two-phase relaxation search and Pareto-based search space reduction, are proposed to program the RF receiver both robustly (i.e., close to global optimum) and efficiently (i.e., with low computational cost). Our numerical experiments demonstrate that the proposed approach is superior over other traditional algorithms based on either local relaxation (in terms of robustness) or simulated annealing (in terms of complexity).

VI. Acknowledgements

This work is sponsored by the DARPA RF-FPGA (Radio Frequency-Field Programmable Gate Arrays) program under Grant HR0011-12-1-0005. The views expressed are those of the authors and do not reflect the official policy or position of the Department of Defense or the U.S. Government.

References

[1] H. Darabi, A. Mirzaei and M. Mikhemar, "Highly integrated and tunable RF front ends for reconfigurable multiband transceivers: A tutorial," *IEEE Trans. on. CAS-I,* vol. 58, no. 9, pp. 2038-2050, 2011.

[2] A. Goel, B. Analui and H. Hashemi, "A 130-nm CMOS 100-Hz–6-GHz reconfigurable vector signal analyzer and software-defined receiver," *IEEE Trans. on MTT,* vol. 60, no. 5, pp. 1375-1389, 2012.

[3] F. Agnelli, G. Albasini, I. Bietti, etc., "Wireless multi-standard terminals: System analysis and design of a reconfigurable RF front-end," *IEEE Circuits and Systems Magazine,* vol. 6, no. 1, pp. 38-59, 2006.

[4] F. Carrara and G. Palmisano, "High-efficiency reconfigurable RF transmitter for wireless sensor network applications," *IEEE RFIC,* pp. 179-182, 2010.

[5] H. Liu, A. Singhee, R. A. Rutenbar, etc., "Remembrance of circuits past: Macromodeling by data mining in large analog design spaces," *IEEE DAC,* pp. 437-472, 2002.

[6] G.G. Gielen and R.A. Rutenbar, "Computer-aided design of analog and mixed-signal integrated circuits," *Proceedings of the IEEE,* vol. 88, no. 12, pp. 1825-1854 , 2000.

[7] M. Hershenson, S. Boyd and T. Lee, "GPCAD: A tool for CMOS Op-Amp synthesis," *IEEE ICCAD,* pp. 296-303, 1998.

[8] G. Gielen, H. Walscharts and W. Sansen, "Analog circuit design optimization based on symblic simulation and simulated annealing," *IEEE JSSC,* vol. 25, no. 3, pp. 707-713, 1990.

[9] M. Krasnicki, R. Phelps, R.A. Rutenbar, etc., "MAELSTROM: Efficient simulation-based synthesis for analog cells," *IEEE DAC,* pp. 945-950, 1999.

[10] O. Coudert, "Gate sizing for constrained delay/power/area optimization," *IEEE Trans. on VLSI,* vol. 5, no. 4, pp. 465-472, 1997.

[11] T. Eeckelaert, T. McConaghy and G. Gielen, "Efficient multiobjective synthesis of analog circuits using hierarchical Pareto-optimal performance hypersurface," *IEEE DATE,* pp.1070-1075, 2005.

[12] B. Razavi, *RF Microelectronics,* Prentice Hall, 2011.

Efficient Matrix Exponential Method Based on Extended Krylov Subspace for Transient Simulation of Large-Scale Linear Circuits

Quan Chen, Wenhui Zhao and Ngai Wong

Department of Electrical and Electronic Engineering
The University of Hong Kong, Hong Kong
Emails: quanchen@eee.hku.hk, whzhao@eee.hku.hk, nwong@eee.hku.hk

Abstract— Matrix exponential (MEXP) method has been demonstrated to be a competitive candidate for transient simulation of very large-scale integrated circuits. Nevertheless, the performance of MEXP based on ordinary Krylov subspace is unsatisfactory for stiff circuits, wherein the underlying Arnoldi process tends to oversample the high magnitude part of the system spectrum while undersampling the low magnitude part that is important to the final accuracy. In this work we explore the use of extended Krylov subspace to generate more accurate and efficient approximation for MEXP. We also develop a formulation that allows unequal positive and negative dimensions in the generated Krylov subspace for better performance. Numerical results demonstrate the efficacy of the proposed method.

I. Introduction

Accurate yet fast transient simulation capability of large-scale integrated circuits has long been a challenge in electronic design automation. The work-horse approach so far is the direct method utilizing one-off sparse matrix factorization plus forward-backward substitution at each time step, adopted in most of the mainstream circuit simulators such as SPICE. Thanks to the recent developments in sparse matrix factorization technique, such as KLU [2], the direct methods have been demonstrated to be able to handle some millions-scale circuits.

However, there are still several shortcomings associated with the direct methods. First, the existing direct method is established on top of the traditional linear multi-step (LMS) schemes (trapezoidal, Gear's method, etc.) based on truncated polynomial expansion [7]. The allowable step size is usually small due to the limitation from the local truncation error (LTE). Second, the time and memory storage required in matrix factorization is highly sensitive to the sparsity pattern of the matrix; when parasitic coupling is included, i.e., the nonzero bandwidth increases and sparsity pattern becomes irregular, dealing with the excessive fill-ins generated during the factorization process remains a challenging task for direct solvers. Parallelization of matrix decomposition, though possible, is also more complicated. Thirdly, adaptive time-stepping is not favored in direct methods since in general the sys-

tem matrix has to be re-factorized once the step size is changed.

As an alternative to the direct methods, the matrix exponential method (MEXP) receives considerable attention in recent years [13–15]. The MEXP method differs from the traditional approaches in mainly two aspects: 1) it is based on the analytical solution of ordinary differential equation (ODE) using exponential operator (LMS methods can be viewed as polynomial approximation of the exponential operator); 2) the core computation is changed from matrix factorization to the approximation of the product of MEXP with a vector in the Krylov subspace.

$$\mathcal{K}_m = span\left\{v, Av, A^2v, ...A^{m-1}v\right\} \qquad (1)$$

where m is the subspace dimension and v the starting vector.

The first difference implies that the MEXP method is exact if the matrix exponential is calculated exactly, whereas the LMS methods remain approximative even when the matrix inversion is exact. Therefore, the step size in MEXP depends only on the quality of the Krylov subspace approximation other than the LTE. In other words, the "order of accuracy" of MEXP can be made very high by increasing the dimension of the Krylov subspace, in contrast to that in the LMS context such order is fixed once a particular scheme is chosen. Therefore, the step size in MEXP can be much larger than that allowed in LMS schemes even with a moderate subspace dimension [13]. The second aspect renders the computational cost less sensitive to the sparsity pattern of the matrices and allows convenient parallelization. The shift-invariant property of the Krylov subspace also greatly facilitates the adaptive time-stepping, enabling further computational advantages compared against tradition methods.

Despite the above merits, the MEXP method becomes less efficient when dealing with stiff circuits which have time constants differing by orders of magnitude. From spectral analysis perspective, large magnitude, well-separated, eigenvalues will appear in the spectrum of the numerical systems arising from stiff circuits. The underlying Arnoldi process, when used to generate bases of Krylov subspace, tends to capture these dominant eigenvalues first to minimize interpolation error, while leaving the spectrum in the vicinity of zero undersampled. For passive systems, all eigenvalues are negative, thus the large

978-1-4799-2817-0/14 $31.00 © 2014 IEEE

(magnitude) eigenvalues actually have negligible contribution to the computation of MEXP. The undersampling of small (magnitude) eigenvalues, on the other hand, introduces large error in the results, and accounts for the inefficiency of MEXP method for stiff systems.

The difficulty in finding small eigenvalues by ordinary Krylov subspace is a known issue in the context of eigenvalue solvers and iterative methods [4]. A natural remedy is to use the shift-invert strategy to transform the small eigenvalues into the dominant ones so that they can be more easily captured. This motivates to use, instead of the ordinary Krylov subspace (1), the extended Krylov subspace

$$\mathcal{K}_{l,m} = span\left\{A^{-l+1}v, ...A^{-1}v, v, Av, ...A^{m-1}v\right\} \quad (2)$$

to generate the MEXP approximation. Using the extended Krylov subspace for evaluating matrix functions was first proposed in [3], which proved that, when A is symmetric, the approximation quality of the exponential function in $K^{2m}(A, v)$ is the same as in $K^{\sqrt{2m}}(A, A^{-m}v)$. In [3], the negative dimension l has to be prespecified, so the subspace is augmented only in the positive direction. Simoncini [10] later developed a more flexible scheme to add two vectors at a time, one multiplied by A and one by A^{-1}, into the basis set to gradually expand the subspace in both the negative and positive directions, i.e.

$$\mathcal{K}_{m,m} = span\left\{v, A^{-1}v, Av, ..., A^{-m+1}v, A^{m-1}v\right\}. \quad (3)$$

However, since the computation of $A^{-1}v$ is usually more expensive than the computation of Av in circuit simulation, an equal number of positive and negative dimensions may not always be the best strategy in terms of total runtime when it comes to individual problems.

In this paper, we apply the extended Krylov subspace concept to improve the performance of MEXP method in transient simulation of stiff circuits. In particular, we develop an approach to generate the extended Krylov subspace with unequal positive and negative dimensions. In this way, one can enjoy the benefit of the extended Krylov subspace while avoiding unnecessary $A^{-1}v$ for faster simulation. Numerical experiments are conducted to demonstrate the effectiveness of the proposed approach.

II. MEXP BASED ON EXTENDED KRYLOV SUBSPACE

A. Formulation of MEXP Method

We first give a brief revisit to the MEXP method based on ordinary Krylov subspace method. For simplicity, we only consider linear circuits in this work. The numerical system to be solved in transient circuit analysis is a set of differential algebraic equations (DAE)

$$C\dot{x}(t) = Gx(t) + Bu(t). \quad (4)$$

where C, G and B denote the susceptance, conductance and input matrix, respectively, and $u(t)$ collects the voltage and current sources. The essence of MEXP lies in transforming (4) to an ODE

$$\dot{x}(t) = Ax(t) + b(t). \quad (5)$$

where $A = C^{-1}G \in \mathbb{R}^{N \times N}$ and $b(t) = C^{-1}Bu(t)$. Here we assume the C matrix is invertible. If C is singular, systematic regularization techniques [1] can be applied to produce a nonsingular system. With (5) and the common piece-wise linear (PWL) approximation of $u(t)$, the transient response of (4) can be analytically determined by

$$\begin{aligned} x(t+h) &= e^{Ah}x(t) \\ &+ (e^{Ah} - I)A^{-1}b(t) \\ &+ (e^{Ah} - (Ah+I))A^{-2}\frac{b(t+h) - b(t)}{h}, (6) \end{aligned}$$

where h is the step size. With some transformation (see [12], eq. 20), the three MEXP functions in (6) can be merged into one MEXP with a slightly bigger matrix, i.e.,

$$\mathbf{x}(t+h) = e^{Ah}x(t), \quad (7)$$

in which all vectors/matrices have been augmented by 2 dimensions.

The main computation in each time step is the application of MEXP to a vector $e^{Ah}v$ in (7). It is computed by the Krylov subspace method [5,9] using the Krylov subspace $\mathcal{K}_m(A, v)$ in (1) constructed by the Arnoldi process

$$AV_m = V_{m+1}\hat{T}_m \quad (8)$$

Here $V_m \in \mathbb{R}^{N \times m}$ is an orthonormal basis of $\mathcal{K}_m(A, v)$ and $\hat{T}_m \in \mathbb{R}^{(m+1) \times m}$ contains the orthonormalization coefficients. Then the MEXP-vec product is approximated in the computed subspace

$$e^{Ah}v \approx \beta V_m e^{T_m h}e_1, \quad (9)$$

where $\beta = \|v\|_2$, V_m, T_m is the $m \times m$ leading part of $\hat{T}_m$, and e_1 the 1-st column of identity matrix. A convenient posterior error estimate can be given for the approximation in (9) by [9]

$$err = \beta t_{m+1,m}\|e_m^{\mathbf{T}}e^{T_m h}e_1\|, \quad (10)$$

where $t_{m+1,m}$ is the bottom right element of $\hat{T}_m$.

The Arnoldi process itself is an iterative process. Since $Av = C^{-1}(Gv)$, in each iteration the main computation involves one sparse matrix-vector product and one sparse system solve (with C). In general, the cost per time step for MEXP is higher than that for LMS methods (except in the cases where C is much sparser than G). The benefit of using MEXP comes mainly from the possibility of using larger step sizes and thus fewer time steps for the same time interval than the low-order LMS methods. In addition, the ease in parallelization and adaptive time-stepping also add to the speedup over traditional methods, see, e.g., [13,14].

A moderate Krylov subspace dimension m for (9) to converge, e.g., < 100, is desired for the performance of MEXP. In million-scale problems, m is also better kept small to control the memory storage of the basis vectors. The necessary m generally depends on the eigenvalues distribution of A. It is known that the Arnoldi process for

978-1-4799-2817-0/14 $31.00 © 2014 IEEE

the ordinary Krylov subspace will prioritize the approximation of the eigenvalues that are large in magnitude and well-separated. In circuit analysis it implies the circuit contains many distinct fast modes, or is stiff (provided a slow mode also exist). If many such distinct outliers are present in the matrix spectrum, the Arnoldi process will spend most of its "dimension resources" building subspace close to the eigenspace associated with these eigenvalues. As a consequence, the part of spectrum in the neighborhood of zero is left inadequately sampled. Recall that the eigenvalues of A are with negative real parts, the exponential of these small (magnitude) eigenvalues are nonnegligible and thus the undersampling will induce substantial error in the approximation of $e^A v$. Then a larger subspace or a smaller step size (to compress the spectrum) is required to maintain the approximation accuracy, which significantly degrades the efficiency of MEXP.

B. Generalized Extended Krylov Subspace

The idea of using extended Krylov subspace is to generate the subspace containing the information of A^{-1} so that the MEXP approximation (9) can converge with fewer iterations. The technique was first proposed in [3], in which the number of negative dimensions l is fixed *a-priori* (only m is allowed to increase). Simoncini later developed an approach to increase the positive and negative dimensions of the subspace simultaneously by adding two basis vectors at a time [10], and showed that the extended Krylov subspace outperformed the standard Krylov subspace in a range of applications [6, 10, 11]. The technique is also known as the Krylov-plus-inverted-Krylov (KPIK) method [8]. In particular, it is shown that an Arnoldi-type relation analogous to (8) can be recovered from the orthonomalization coefficients without extra matrix-vector products [10]

$$AV_m = V_{m+2}\hat{T}_m, \tag{11}$$

where $\hat{T}_m \in \mathbb{R}^{(m+2)\times m}$ is a block Heisenbeig matrix.

Nevertheless, an equal number of positive and negative dimensions in KPIK may not always be the optimal choice. Since $A = C^{-1}G$, basis generation with Av requires a linear system solve with C, while that with $A^{-1}v$ requires one with G, which is usually denser than C in the MNA formulation. Thus expanding the subspace along negative dimension is more expensive, and the improvement in approximation quality as well as the increase in step size thereby allowed may not be able to pay off the extra cost resulted from the computation of $A^{-1}v$. Therefore, it is reasonable to allow an unequal number of dimensions in the extended Krylov subspace, i.e., we add one vector with A^{-1} after every k vectors with A.

$$\mathcal{K}_{m,km} = span\left\{ \begin{array}{l} v, A^1v, A^2v, ...A^kv, A^{-1}v, A^{k+1}v, ..., \\ A^{2k}v, A^{-2}v, ..., A^{km-1}v, A^{-m+1}v \end{array} \right\}. \tag{12}$$

Apparently KPIK is a special case of (12) for $k = 1$.

With the arrangement in (12), the recursive relation to obtain a new basis vector is given below

$$
\begin{array}{l}
\text{If } n = 1 \text{ or } \mod(n, k+1) = 2, ..., k \\
h_{n+1,n}v_{n+1} = Av_n - V_n h_{1:n,n}, \\
\text{If } \mod(n, k+1) = 0 \\
h_{n+2,n+1}v_{n+2} = Av_n - V_{n+1}h_{1:n+1,n+1}, \\
\text{If } n > k \text{ and } \mod(n, k+1) = 1 \\
h_{n,n-1}v_n = A^{-1}v_{n-k-1} - V_{n-1}h_{1:n-1,n-1}.
\end{array} \tag{13}
$$

where $h_{1:n+1,n}$ collects the $n+1$ orthonomalization coefficients against the previous basis vectors.

To obtain the relation in (11), we follow a similar strategy in [10] to recover $\hat{T}_m = V_{m+2}^T AV_m$ without really performing the matrix-vector products.

Proposition II.1 *Let* $\hat{T}_m = (t_{i,j})$, $i = 1, ..., 2m + 2$, $j = 1, ..., m$. *Then*

$$
\begin{array}{l}
\text{If } n = 1 \text{ or } \mod(n, k+1) = 2, ..., k-1 \\
t_{:,n} = h_{:,n} \\
\text{If } \mod(n, k+1) = 0 \\
t_{:,n} = h_{:,n+1} \\
\text{If } n > k \text{ and } \mod(n, k+1) = 1 \\
t_{:,n} = \frac{1}{h_{n,n-1}}\left(e_{n-k-1} - \left[\begin{array}{c} \hat{T}_{n-1}h_{1:n-1,n-1} \\ 0 \end{array} \right] \right)
\end{array} \tag{14}
$$

Proof Using the first equality of (13) and the orthogonality of V, for the first situations in (13), we have

$$
\begin{aligned}
t_{:,n} = V_{n+2}^T Av_n &= V_{n+2}^T v_{n+1}h_{n+1,n} + V_{n+2}^T V_n h_{1:n,n} \\
&= h_{1:n+1,n}.
\end{aligned}
$$

Similarly, using the second equality of (13), we obtain

$$
\begin{aligned}
t_{:,n} &= V_{n+2}^T Av_n \\
&= V_{n+2}^T v_{n+2}h_{n+2,n+1} + V_{n+2}^T V_{n+1}h_{1:n+1,n+1} \\
&= h_{1:n+2,n+1}.
\end{aligned}
$$

For the third situation, one can obtain from the last row of (13) that

$$h_{n,n-1}Av_n = v_{n-k-1} - AV_{n-1}h_{1:n-1,n-1}.$$

Then we can get

$$
\begin{aligned}
t_{:,n} &= V_{n+2}^T Av_n \\
&= \frac{1}{h_{n,n-1}}\left(V_{n+2}^T v_{n-k-1} - V_{n+2}^T AV_{n-1}h_{1:n-1,n-1} \right) \\
&= \frac{1}{h_{n,n-1}}\left(e_{n-k-1} - \left[\begin{array}{c} \hat{T}_{n-1}h_{1:n-1,n-1} \\ 0 \end{array} \right] \right),
\end{aligned}
$$

which completes the proof. ∎

Note that in (15) the current column in $\hat{T}$ cannot be filled until the next iteration is performed to generate $h_{:,n+1}$, so the evaluation of MEXP has also to be postponed to the next run.

A posterior error estimate can be given as

$$err = \beta\tau_{m+1,m}\|e_m^T e^{T_m h}e_1\|, \tag{15}$$

where τ is the 2×2 bottom right block of $\hat{T}_m$.

978-1-4799-2817-0/14 $31.00 © 2014 IEEE

III. NUMERICAL RESULTS

In this section, we compare the performance of the MEXP transient solver based on the standard Krylov subspace and the generalized extended Krylov subspace. The implementation was done in Matlab and the testing was performed with a 3.2GHz server with 32Gb memory.

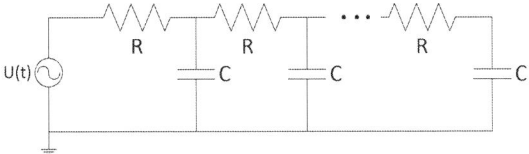

Fig. 1. RC ladder circuit

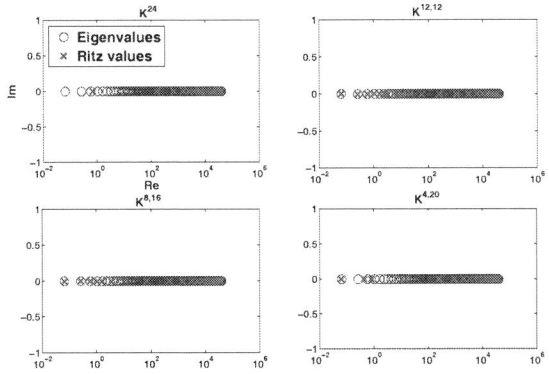

Fig. 2. Approximation of spectrum by standard Krylov subspace and extended Krylov subspace.

TABLE I
ERROR AND RUNTIME OF DIFFERENT KRYLOV SUBSPACES

Subspace	$\mathcal{K}_{24}$	$\mathcal{K}_{12,12}$	$\mathcal{K}_{8,16}$	$\mathcal{K}_{4,20}$
Error	4.2e-1	1.17e-6	3.6e-4	1.4e-2
Time (s)	0.09	0.37	0.17	0.11

We first use a simple RC ladder circuit (see the Fig. 1) to reveal the reason for the improvement led by the extended Krylov subspace. The order of the matrix is 1000, C is a diagonal matrix and G is simplified to the $[-1, 2, -1]$ pencil. The capacitor values are chosen uniformly distributed within $[1, 2] \times 10^{-15}F$ to produce a stiff circuit. We compute $e^{Ah}v$ by four Krylov subspaces with v being an all one vector and $h = 10^{-11}$. The dimension of the standard Krylov subspace is 24. The total dimensions of the three extended Krylov subspaces are also set to be 24, but with different negative-positive ratios $k = 1, 2, 5$, which correspond to the (negative, positive) dimensions of $(12, 12)$ (KPIK), $(8, 16)$ and $(4, 20)$. For notation convenience we denote the standard Krylov subspace as $k = 0$. Direct solution (backslash in Matlab) is applied in all sparse system solves. The eigenvalues of Ah and the Ritz values (the eigenvalues of $T_m h$) are plotted in Fig. 2. Note that the real parts are shown in log scale and the imaginary parts are in linear scale. The computation errors against the direct eigenvalue decomposition method and the runtime of the four cases are shown in Table I.

TABLE II
SPECIFICATIONS OF TEST CIRCUITS

Circuits	Category	Nodes	Matrix size	Stiffness
C1	Power grid	39K	54K	medium
C2	Power grid	164K	165K	high
C3	Trans. lines	5.6K	8.8K	high

Due to the very small capacitance, many large (magnitude) eigenvalues ($\sim 10^4$) are present in the spectrum of A. The method based on the standard Krylov subspace generates a good approximation to the right-end part of the spectrum, but a poor one for the left-end spectrum. This is expected since the basis is generated using positive power of A only, which amplifies the dominant eigenvalues. Failure in sampling the eigenvalues near the origin, however, induces substantial error as shown in Table I. On the other hand, the extended Krylov subspace all improve the approximation to the near-origin spectrum by including the information of A^{-1}. Table I suggests that using more negative dimensions improves the accuracy since the meaningful parts of spectrum locates at the left end. However, the computation is also more costly when more negative dimensions are included due to requiring system solves with the denser matrix G.

Next we evaluate the performance of MEXP based on different Krylov subspace with real circuit examples. Three linear circuits specified in Table II are tested. For a better comparison among subspaces, we run 100 time steps with a constant step size in each case, and allow the subspace dimension to vary dynamically to satisfy a prescribed tolerance of 10^{-6} (per step) using the error estimate (15). The input is a ramp signal with the final amplitude of 1V.

The results are reported in Table III. Apparently, a large subspace dimension is required for MEXP based on the standard Krylov subspace to converge, which is highly undesirable in terms of runtime and memory. The solvers based on extended Krylov subspace all requires a significantly smaller subspace dimension to achieve the same level of accuracy. This shows the advantage of using the extended Krylov subspace in the simulation of stiff systems.

The equal assignment of positive and negative dimensions ($k = 1$) results in the lowest total dimension, which, however, does not necessarily leads to the smallest runtime due to the extra cost in generating basis involving A^{-1}. For the relatively less stiff case (C1), the performance is the best for $k = 4$, i.e., one negative dimension is inserted after every four positive dimensions. For $k = 1$ it is even slower than the standard Krylov subspace method, since Av takes only $0.02s$ while $A^{-1}v$ requires $0.14s$. For

978-1-4799-2817-0/14 $31.00 © 2014 IEEE

TABLE III
PERFORMANCE OF MEXP BASED ON DIFFERENT KRYLOV SUBSPACE

Circuits	Step size (s)	Subspace dimensions				Total runtime (s)			
		k=0	1	3	4	k=0	1	3	4
C1	1e-12	(0,191)	(35,35)	(29,59)	(16,67)	591.8	641.5	595.1	**390.2**
C2	1e-11	(0,308)	(15,15)	(12,37)	(12,49)	5001.2	1459.1	**1002.7**	1256.3
C3	1e-12	(0,169)	(11,11)	(8,25)	(8,33))	6364.6	**1439.4**	1730.6	2014.7

the systems with stronger stiffness (C2 and C3), a larger portion of negative dimensions is desired. Therefore, the solver performs better with smaller values of k. The above experiments indicate that the best breakdown of positive and negative dimensions is generally problem dependent, and thus for maximum performance it is beneficial to allow a flexible basis generation along the two directions using the scheme developed in this work.

IV. CONCLUSION

We have investigated the use of extended Krylov subspace to enhance the accuracy of numerical approximation of MEXP-vector product, which in turn benefits the MEXP-based transient circuit simulation. The improvement is analyzed in terms of the quality of the approximation of the targeted part of matrix spectrum. In addition, we generalize the extended Krylov subspace to allow unequal positive/negative dimensions to maximize the overall performance in circuit simulation. Numerical results have confirmed the efficiency of the proposed method.

REFERENCES

[1] Q. Chen, S.-H. Weng, and C.-K. Cheng, "A practical regularization technique for modified nodal analysis in large-scale time-domain circuit simulation," *IEEE Transactions on Computer-Aided Design of Integrated Circuits and Systems*, vol. 31, no. 7, pp. 1031–1040, 2012.

[2] T. A. Davis and E. Palamadai Natarajan, "Algorithm 907: KLU, a direct sparse solver for circuit simulation problems," *ACM Transactions on Mathematical Software*, vol. 37, no. 3, p. 36, 2010.

[3] V. Druskin and L. Knizhnerman, "Extended Krylov subspaces: Approximation of the matrix square root and related functions," *SIAM J. Matrix Anal. Appl.*, vol. 19, no. 3, pp. 755–771, Jul 1998.

[4] M. Freitag and A. Spence, "Shift-invert Arnoldi's method with preconditioned iterative solves," *SIAM Journal on Matrix Analysis and Applications*, vol. 31, no. 3, pp. 942–969, 2010.

[5] M. Hochbruck and C. Lubich, "On Krylov subspace approximations to the matrix exponential operator," *SIAM J. Numer. Anal.*, vol. 34, no. 5, pp. 1911–1925, Oct 1997.

[6] L. Knizhnerman and V. Simoncini, "A new investigation of the extended Krylov subspace method for matrix function evaluations," *Numerical Linear Algebra with Applications*, vol. 17, no. 4, pp. 615–638, 2010.

[7] F. Najm, *Circuit Simulation*. Wiley, 2010.

[8] M. Popolizio and V. Simoncini, "Acceleration techniques for approximating the matrix exponential operator," *SIAM Journal on Matrix Analysis and Applications*, vol. 30, no. 2, pp. 657–683, 2008.

[9] Y. Saad, "Analysis of some Krylov subspace approximations to the matrix exponential operator," *SIAM Journal on Numerical Analysis*, vol. 29, no. 1, pp. 209–228, 1992.

[10] V. Simoncini, "A new iterative method for solving large-scale Lyapunov matrix equations," *SIAM J. Sci. Comput.*, vol. 29, no. 3, pp. 1268–1288, May 2007.

[11] ——, "Extended Krylov subspace for parameter dependent systems," *Appl. Numer. Math.*, vol. 60, no. 5, pp. 550–560, May 2010.

[12] S.-H. Weng, Q. Chen, and C.-K. Cheng, "Circuit simulation using matrix exponential method," in *2011 IEEE 9th International Conference on ASIC (ASICON)*, 2011, pp. 369–372.

[13] ——, "Time-domain analysis of large-scale circuits by matrix exponential method with adaptive control," *IEEE Transactions on Computer-Aided Design of Integrated Circuits and Systems*, vol. 31, no. 8, pp. 1180–1193, 2012.

[14] S.-H. Weng, Q. Chen, N. Wong, and C.-K. Cheng, "Circuit simulation via matrix exponential method for stiffness handling and parallel processing," in *2012 IEEE/ACM International Conference on Computer-Aided Design (ICCAD)*, 2012, pp. 407–414.

[15] S.-H. Weng, P. Du, and C.-K. Cheng, "A fast and stable explicit integration method by matrix exponential operator for large scale circuit simulation," in *2011 IEEE International Symposium on Circuits and Systems (ISCAS)*, 2011, pp. 1467–1470.

978-1-4799-2817-0/14 $31.00 © 2014 IEEE

SDG2KPN: System Dependency Graph to Function-level KPN generation of Legacy Code for MPSoCs

Jude Angelo Ambrose, Jorgen Peddersen,
Sri Parameswaran

School of Computer Science and Engineering
University of New South Wales
Sydney, Australia NSW 2052
e-mail: {ajangelo, jorgenp, sridevan}@cse.unsw.edu.au

Alvin Labios, Yusuke Yachide

Canon Information Systems Research Australia (CiSRA)
Australia

{Alvin.Labios,Yusuke.Yachide}@cisra.canon.com.au

ABSTRACT

The Multiprocessor System-on-Chip (MPSoC) paradigm as a viable implementation platform for parallel processing has expanded to encompass embedded devices. The ability to execute code in parallel gives MPSoCs the potential to achieve high performance with low power consumption. In order for sequential legacy code to take advantage of the MPSoC design paradigm, it must first be partitioned into data flow graphs (such as Kahn Process Networks — KPNs) to ensure the data elements can be correctly passed between the separate processing elements that operate on them. Existing techniques are inadequate for use in complex legacy code. This paper proposes SDG2KPN, a System Dependency Graph to KPN conversion methodology targeting the conversion of legacy code. By creating KPNs at the granularity of the function-/procedure-level, SDG2KPN is the first of its kind to support shared and global variables as well as many more program patterns/application types. We also provide a design flow which allows the creation of MPSoC systems utilizing the produced KPNs. We demonstrate the applicability of our approach by retargeting several sequential applications to the Tensilica MPSoC framework. Our system parallelized AES, an application of 950 lines, in 4.8 seconds, while H.264, of 57896 lines, took 164.9 seconds to parallelize.

1. INTRODUCTION

Parallel processing has been touted as a method to accommodate the demand for functional complexity by using increases in the availability of parallel resources. To exploit parallelism, the Multiprocessor System-on-Chip (MPSoC) paradigm has emerged as a viable implementation platform within embedded systems. In these MPSoC systems, multiple processors share the workload of an application by executing parts of the application (i.e., tasks) on separate processors to achieve better performance than executing the application on a single processor [5, 6, 7]. Applications designed for execution on an MPSoC are typically created with the tasks of each processor predetermined to communicate data between tasks executing on other processors. Thus, the data to be transmitted is known when the code is written [15].

Data communication between tasks is rarely considered when writing sequential code intended to execute on a single processor. Hence, code written in this manner (i.e., legacy code) requires analysis and/or a manual redesign before it can be split to execute on multiple processors, to determine which data must be transferred and when [4]. This analysis typically involves conversion to a data flow graph (DFG) [4, 15], where the application is partitioned into communicating tasks/functions. Tasks from these DFGs are mapped to multiple processors in the MPSoCs to improve performance, by exploiting parallelism [19], pipelining [6] and time multiplexing of tasks from different applications [15]. The Kahn Process Network (KPN) is considered a suitable and general model for a DFG [4, 19], consisting of concurrent processes/tasks/nodes that are able to communicate data. When modifying existing code to utilize parallelism, a designer is required to rewrite code to suit the system architecture and communication patterns, or they require an in-depth understanding of all parts of the original code and the interaction between its various components to construct the KPN of the application. This paper provides techniques to automate this process of creating a KPN from legacy code, making it easier and faster for a designer to modify existing code to make use of newer processing environments.

There are several key challenges that have been identified as being missing from previous methods for KPN generation from legacy code. 1), techniques [9, 13, 18] and tools (such as COMPAAN [9] and KPNGen [13]) exist to generate a KPN from a restricted piece of code. These techniques support static-affine nested loops and require manual modifications to the legacy code [4, 1]. This manual analysis is tedious and time consuming; 2), there was no method to efficiently detect variable dependencies across functions; 3), existing register and instruction level techniques to create dependence graphs [14] are restricted to specific hardware platforms, making them less portable; and 4), the state-of-the-art tools require significant manual modifications to the legacy code (i.e., completely rewriting most of the code and require heavy modification to the code by a talented designer in order for the code to work). Hence, reducing the amount of manual intervention will enable more legacy applications to be ported to MPSoCs.

For the first time, we propose the use of the System Dependency Graph (SDG) in order to detect dependencies between functions and/or tasks for KPN generation of legacy code. The SDG is a directed graph which represents inter-function and sub-function dependencies, at program point granularity, of legacy sequential code [20, 10]. SDG2KPN allows analysis of shared variables, unlike previous approaches. SDG2KPN performs automated analysis of the code and detects the usage of global variables between different functions. Finding variable usage between functions allows existing code to be implemented within a multi-processing environment in a fraction of the time of existing approaches, as the user only needs to identify potentially important functions and variables within the code rather than studying the interoperation of the functions and determining data flow manually, both of which would require the user to read and understand most or all of the legacy code. For simplicity, this paper focuses only on data dependencies between functions. The process can be extended to be applied at the granularity of basic blocks, although the extensions to achieve this are outside the scope of this paper.

Figure 1 depicts an example of our KPN generation methodology. The legacy code, shown in Figure 1(a), is received as an input.

978-1-4799-2817-0/14 $31.00 © 2014 IEEE

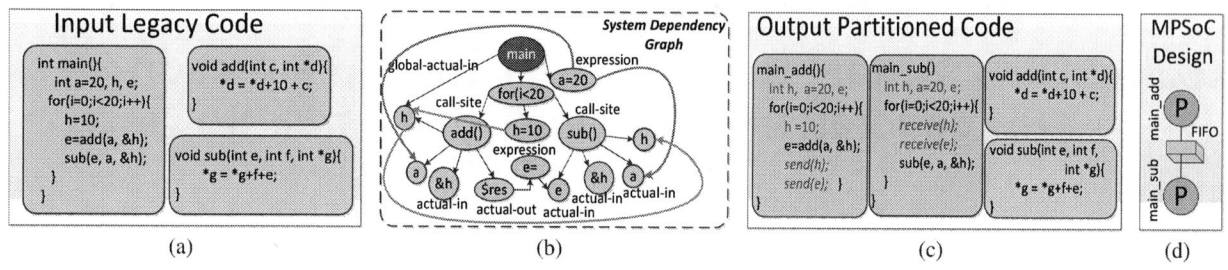

Figure 1: An Example KPN Generation Methodology

An intermediate representation of this code is produced in the form of a System Dependency Graph (SDG), as shown in Figure 1(b). As shown in the example SDG, the *main* function (referred to as the parent function) is decomposed into program points/vertices, representing all the statements of the code and their control and data dependencies. The goal of our SDG2KPN approach is to create a KPN for the parent function (i.e., *main*) by partitioning the *add()* function call and the *sub()* function call and identifying the variable dependencies between these function calls. The functions envisaged for partitioning are referred to as candidate functions. The *add()* and *sub()* candidate function calls (i.e., call-sites) are converted to KPN processes, *main_add* and *main_sub*, which transfer data within variables *h* and *e*, as shown in Figure 1(c). As shown in the example, the *sub()* call-site depends on the *add()* call-site for variables *e* and *h*. The variable *e* is updated by the add function and read by the sub function, whereas the variable *h* is passed by reference, modified by *add()* and passed to *sub()* for reading. Finding dependencies of shared global variables, such as *h*, across candidate functions is a challenge which has not been addressed before, with previous techniques only looking at explicitly defined data flow specified in the function's parameters and return value. We provide an automated rule based SDG analysis to detect such shared variable dependencies (including all the other variable dependency types). Such dependencies necessitate a communication link (such as FIFO, DMAs, shared memory, etc.) for transferring these variables from *add()* to *sub()*. Once a KPN is generated, we propose a design flow to generate partitioned code to execute in an MPSoC platform (as shown in Figure 1(d)). For example, the generated code shown in Figure 1(c), contains two main functions *main_add()* and *main_sub()* which will be executed in two different processors. Variables are transferred from *main_add()* using *send* commands, which write the contents of the variables to the FIFO, and are received at *main_sub* using *receive* commands, which copy the contents into appropriate variables. Figure 2 depicts an example streaming application (i.e., MPEG-2 encoding) where several candidate functions are generated into KPN processes (shown right). The initialization functions/code, i.e., those that appear outside of the candidate functions, are either mapped to the first processor or distributed to their respective processes of the KPN. The processes of the generated KPN are mapped to the processors in the MPSoC platform for parallel execution.

The rest of the paper is organized as follows; Section 2 and Section 3 detail the related work and methodology respectively. The results are presented in Section 4. Conclusions are provided in Section 5.

2. RELATED WORK

Several techniques and tools exist in literature for converting sequential code to partitioned concurrent code for MPSoCs.

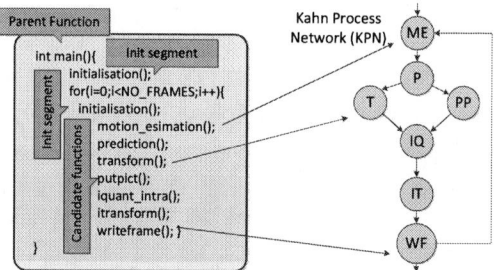

Figure 2: A Streaming Application Template

KPNGen [13] from the Daedalus framework allows creation of KPNs, where the user has to explicitly modify the function boundaries (parameters) to create the dependency links between the KPN processes. This puts much of the work in determining which variables are used onto the user. The passing of data via non-explicit techniques, such as global variables, is not handled by this approach.

COMPAAN [9] automatically creates a KPN process for each function of an Affine Nested Loop Program (ANLP) written in MATLAB syntax. Any legacy code written in other languages (such as in C) has to be manually converted to ANLP programs in MATLAB. An array data flow analysis is performed in COMPAAN [9], at the statement level, and therefore the task boundaries cannot reside across nonlinear or data-dependent conditions [2].

The authors in [4] suggest an overview for a sequential code to KPN transformation methodology, utilizing the call sequence graph and the control flow graph of sequential C code. Both these graphs were proposed for the analysis to find data dependencies between functions and assumed to have been created from a modified Abstract Syntax Tree (AST). No details were provided for the analysis methodology for task partitioning in [4] and it is explicitly stated that this is a suggested approach, which has not been implemented. The authors in [4] further suggest that they will modify the AST after task partitioning, which is quite a complicated and challenging endeavor.

Compared to these previous approaches, our methodology considers shared variables (such as globals and pointers) to detect data dependencies between functions, while requiring much less user input and does not require the legacy code to be rewritten or significantly modified. Our approach can be ported to any hardware platform and ISA, since it is performed at the code level. We do not require the code to be in a static affine nested loop form, as required by [9] and allow both pipelining and parallelization, in contrast to [14]. The closest approach to our methodology is [4], which proposes utilizing the AST, call sequence graph and control flow graph to create a KPN. The KPN is proposed to be generated

978-1-4799-2817-0/14 $31.00 © 2014 IEEE

from a modified AST. The AST is much finer-grained than SDG, hence including a significantly large amount of unnecessary information (i.e., much more nodes in the graph) which the analysis has to go through (if implemented). Hence our SDG based approach is less complicated and we demonstrate the feasibility and practicality of using the SDG in this paper.

2.1 Contributions

Some of our novel contributions are as follows:
- An SDG based function-level KPN generation methodology is proposed. This methodology is platform agnostic and supports shared variables.
- A rule-based traversal of the SDG is proposed to detect dependencies between functions. Such a traversal optimally detects variable dependencies.
- A complete design flow is demonstrated to generate application specific MPSoC systems.

3. SDG2KPN METHODOLOGY

Figure 3 depicts the SDG2KPN methodology, illustrating the integral components. The inputs are the legacy code and the application input data. An application specific MPSoC system is the final output. The Networker (detailed in Section 3.2) is the component which enables the novel contributions of this paper. The remaining portions of the methodology, such as the Abstractor, Mapper, Target Generator, Simulator, Annotator, Balance Checker, Graph Optimizer and Tuner, are developed based on existing state-of-the-art approaches.

The Abstractor (detailed in Section 3.1) receives the legacy code and creates a System Dependency Graph (SDG). An initial KPN is created by the Networker by analyzing the SDG. Each process/node in the initial KPN will correspond to a candidate function. The SDG is analyzed to determine the dependencies between candidate functions, by checking which variables are read or used by other candidate functions. Once dependencies are detected between two KPN processes, a corresponding communication link is established between them, to transfer the variables. The initial KPN is then passed to the Mapper (detailed in Section 3.3) which maps the KPN to processors. An MPSoC system for the analyzed legacy application is generated by the Target Generator (detailed in Section 3.4). MPSoC simulations are carried out to the generated MPSoC system with exemplary application input data (for example, image input files for JPEG).

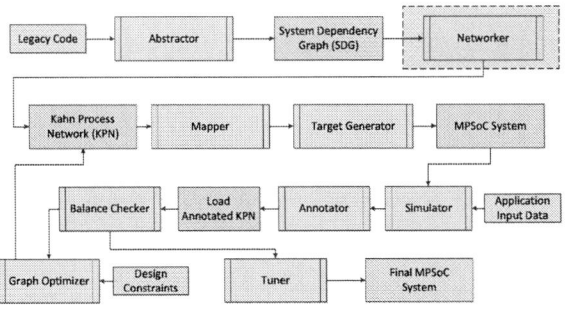

Figure 3: The SDG2KPN Flow

The Simulator (detailed in Section 3.5), performing cycle accurate simulation of the MPSoC system, reveals the total time consumed by each process which is then extracted by the Annotator (detailed in Section 3.6) to form the Load Annotated KPN. The Balance Checker (detailed in Section 3.7) evaluates the load and

considers any possible optimizations that can be made to the KPN to improve performance on the target architecture (such as load balancing for pipelined systems). Such optimizations are performed by the Graph Optimizer (detailed in Section 3.8), which then generates a modified KPN. This is an iterative step, which will terminate once there is no further performance improvement to be achieved. If the KPN is already verified by the Balance Checker to have no further identifiable improvements, the Tuner (detailed in Section 3.9) executes a fine grained adjustment to the hardware resources (e.g., processors) to further improve the performance of the MPSoC system, if possible. The final MPSoc system is output by the Tuner.

3.1 Abstractor

The Abstractor creates the SDG in three steps by: 1) creating an Abstract Syntax Tree (AST) [1] of the sequential legacy code; 2) creating the Program Dependency Graph (PDG) [2]; and then 3) creating the SDG by examining the dependencies across procedures/functions.

Our SDG2KPN methodology utilizes the commercial tool, Code-Surfer [3], as an Abstractor to produce the SDG. As shown in the example in Figure 1, the SDG contains vertices for each program point of the code. Each vertex has a type and is attached to a variable. For example, the a=20 vertex is of type *expression* and the variable attached is *a*. The *main* procedure has two candidate function call-sites *add()* and *sub()*. A parameter of a candidate function call-site is referred to as an *actual-in* and the return variable is referred to as an *actual-out*. For example, variable *a* is an *actual-in* of call-site *add()* whereas variable *$res* is an *actual-out* of the *add()* call-site. Any pointer variable or global variable will contain a global vertex, *global-actual-in* for variables being read and *global-actual-out* for variable being written using that call-site. For example, the variable *&h* being passed by reference creates a vertex *global-actual-in* for *h*. The arguments of the actual function are referred to as formal parameters (not shown in Figure 1). Inter- and intra- procedural edges are identifiable in the SDG (not shown). Readers are referred to [3] for more details about the SDG and its elements.

3.2 Networker

Figure 4 depicts the component flow of the Networker. The user specification is provided by the user, listing all the candidate function names and their line numbers from the legacy code. This allows the user to choose or override which functions would be converted to candidate functions. Candidate functions can also be automatically chosen if there are no user specifications. The Function Mapper component performs the extraction of these functions (which are either user specified or automatically chosen) from the SDG.

The Traverser defines the characteristics of the communication structures for the initial KPN. A Variable Dependency Graph (VDG) is generated to indicate the candidate functions and their variable dependencies to other candidate functions. A variable dependency exists when a variable set by one candidate function is used in another candidate function. Such a variable may be a global variable, passed to a candidate function as parameter of the function call or returned to the calling function at completion of the candidate function. Figure 5 provides an example for VDG, where the *sub()* call-site's variables *e*, *c* and *h* depend on *add()* call-site. These dependent variables form a communication link in the KPN, linking

---

[1] an AST is a graph representation of the syntactic structure of the code [4]

[2] PDG is a graphical representation of a logical order of execution of statements within each function

978-1-4799-2817-0/14 $31.00 © 2014 IEEE   269

the *add* and *sub* candidate functions. This data may be communicated by several techniques, such as FIFOs or shared memory. In this example, FIFOs have been chosen to be used for communication. Thus, the communication path is labelled *FIFO1* in Figure 5. The Traverser creates the VDG by analyzing the SDG for all the variable dependencies across candidate functions. The examination of the SDG involves the use of Traversal Rules, which describe methods to determine the variables shared between two candidate functions that have a dependency between them. For example, the dependency of returned variable *e* from the *add* candidate function in Figure 5 to the *sub* candidate function is detected using a traversal rule. Section 3.2.1 details the Traversal Rules and discusses further examples.

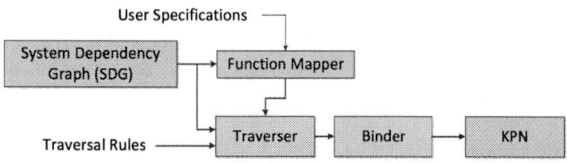

Figure 4: The Networker Flow

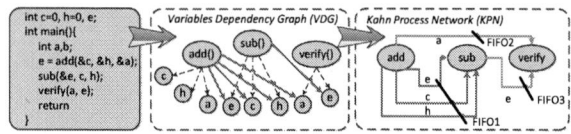

Figure 5: An Example for VDG to KPN conversion

The Binder receives the VDG and creates a corresponding KPN process for every candidate function in the VDG. Communication channels (FIFOs in the example) are created between candidate functions to establish communication for dependent variables, and are thus denoted in the VDG. Where maximum communication size is important, uch as when using FIFOs, these communication channels are given finite size and depth dimensions calculated to be at least as large as the total data transferred via that channel. For example, FIFO1, FIFO2 and FIFO3 of the KPN in Figure 5 are of size 12, 4 and 4 bytes respectively.

3.2.1 Traversal Rules for Dependencies

The SDG is analyzed using traversal rules to extract variable dependencies between candidate functions. Finding these dependencies allows the contributions of the paper to be realised.

As shown in the SDG in Figure 1, the *actual-in* of the *sub()* call-site has an intra-predecessor dependency to the *expression* "e=". This *expression* further depends on the *actual-out* $res. $res is the *actual-out* of another candidate function call-site *add()*. Such a dependency traversal identifies that variable *e* in the *sub* candidate function depends on the *add* candidate function.

To implement this example traversal to find the dependency (as a result, the VDG), the Traverser executes three nested loops. The outermost loop iterates through the candidate functions. The next loop iterates through each input variable of each candidate function. Detection of variable dependency using rule traversal is performed at the innermost loop. Table 1 summarizes all the major rules to create the VDG. Further rules can be added to consider complicated program patterns and dependencies.

Rule 1 is applied when the outermost loop iterates through each candidate function and finds multiple call-sites of the same candidate function. This step determines the first candidate function

and applies traversal Rule 2. The second step of the traversal process iterates through each *actual-in* and *global-actual-in* variable (also referred to as input variables) of a candidate function call-site. These input variables are analyzed using the traversal rules to determine which candidate functions and variables a particular input variable depends on. As shown in Table 1, non-global input variables are analyzed using Rules 3, 6 and 7, whereas the global input variables are analyzed using Rules 4, 5, 7 and 8.

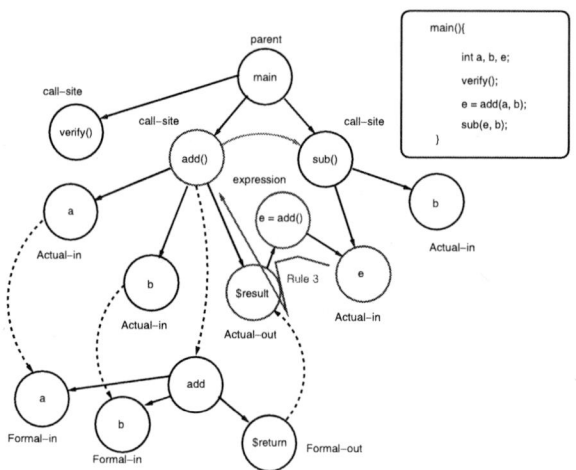

Figure 6: Example Rule 3 in Traverser

No.	Rule Name	Description
1	call-sites & KPN processes	- multiple call-sites per call - create unique KPN process per call
2	first candidate	- mapped to the first KPN
3	output of a call-site	- actual-in links to an intra-predecessor - the intra-predecessor is an actual-out
4	global-actual-in to global-actual-out	- global-actual-in links to global-actual-out - global-actual-out of a predecessor - traverse via intra-predecessor vertices
5	global-actual-in but not to global-actual-out	- none functions modify the global variable - dependency link from parent
6	actual variables modified using pointers	- variables passed as pointers - modified by the predecessors
7	C structures	- dependency in complicated types - entire type is considered as modified
8	Global variables only within function	- static variables also fall into this rule - none link betweens global actuals

Table 1: Traversal Rules Summary

Figure 6 depicts the SDG and its traversal path for Rule 3. Variable *e* is returned by the add function and passed as a parameter for the sub function, revealing a dependency which is captured by Rule 3. As mentioned above, the traversal first iterates for every candidate function call-sites (in this case, verify(), add() and sub()), and then the input variables of each candidate function's *call-site* is iterated (such as e, b in sub()). The third step is to traverse the SDG from an input variable to determine whether any rules are found, which in turn detects the dependency links. The SDG shown includes an *actual-in* vertex for *e* at the sub *call-site* which has an intra-predecessor (i.e., within the main procedure) link to an *expression* vertex (i.e., e=add()). An intra-predecessor link exists from the *expression* to an *actual-out*, which is labeled as *$result*. This *actual-out* belongs to the add candidate function call-site. Such a traversal path with the vertex types captures Rule 3, creating a link in the VDG between add function and sub function with variable *e*.

978-1-4799-2817-0/14 $31.00 © 2014 IEEE 270

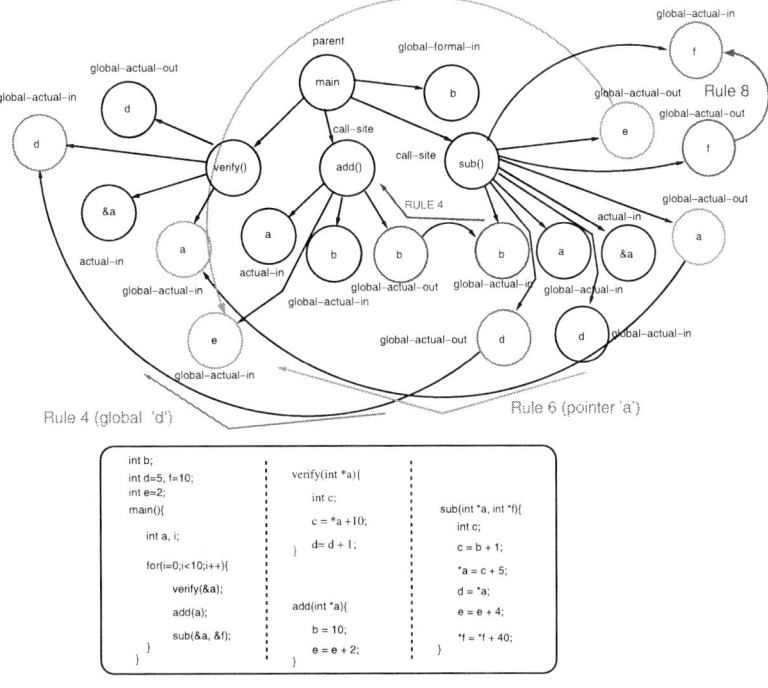

Figure 7: Examples for Rule Traversal

Figure 7 depicts a detailed example of Rule 4, 6 and 8 (i.e., traversing the SDG for dependencies on global variables and pointers). The sub function shown in Figure 7 modifies the global variable *d*, which creates a *global-actual-out* vertex in the SDG. As can be seen, the verify function is reading the modified variable *d* from the sub function, creating a *global-actual-in* for variable *d* in the SDG. When traversing from this *global-actual-in* vertex of *d* in an intra-predecessor fashion (i.e., within the procedure), an existence of a direct link to a *global-actual-out* will reveal a dependency link, captured by Rule 4. The global variables *e* and *b* are captured with the same rule to show a link between the sub and add functions (dependency links indicated in skyblue for variable *e* and blue for *b* in Figure 7).

Once the rules are analyzed and dependencies detected the Networker generates the initial KPN (using the Binder as shown in Figure 4) by combining the information gathered about the KPN processes and FIFOs from the SDG, VDG and the functions of the code. Figure 5 demonstrates an example of VDG to KPN conversion, where the dependency variables are combined to form FIFO links.

3.3 Mapper

The Mapper creates a resource map based on the current KPN. The resource map specifies the assignment of each node of the current KPN to one processor, amongst the plurality of processors in the target MPSoC system.

3.4 Target Generator

The Target Generator uses the resource map and the current KPN to create platform-specific target code. A set of individual build files, such as Makefiles and programs per KPN process are generated. The code for each processor is created by combining the code of the candidate function in the legacy sequential code that corresponds to the KPN process, with the appropriate FIFO read and write commands. A FIFO read command is used by an individual program to take the contents of a variable from the FIFO, while A FIFO write command is used to place the contents of a variable into a FIFO. Both these FIFO commands enable communication of variables between programs which are executed on individual processors.

3.5 Simulator

The Simulator configures a simulation platform, instantiating processors and memory blocks for FIFO communication. Individual executables are created for the target code to run on each processor. The executables dictate the operation of a processor during the simulation. Application specific input data, if required, is received by the simulator (stored in a predetermined block of memory).

3.6 Annotator

The Annotator takes the information regarding the load values (i.e., the results of the Simulator) of individual processors and creates a normalized representation as an estimate of the load performed by the corresponding KPN processes. It creates the load annotated KPN by appending the normalized load values to the description of each KPN process.

3.7 Balance Checker

The Balance Checker receives and evaluates the annotated KPN to determine whether or not the load distribution in the annotated KPN is sufficiently balanced. A KPN with a balanced load represents an efficient implementation of parallel pipelined code for an MPSoC system. This means that the partitioned application code enables maximum use of all respective processors, thereby minimizing the idle time.

3.8 Graph Optimizer

The Graph Optimizer creates a new optimized KPN from the current KPN. Graph optimization algorithms (such as the ones proposed in [12, 17]) are applied to either split or merge KPN pro-

978-1-4799-2817-0/14 $31.00 © 2014 IEEE   271

cesses after the evaluation of the load values. Design constraints, such as the number of processors in the MPSoC system, communication cost based on FIFO dimensions, performance and power budgets are manually entered by the user via an input file.

3.9 Tuner

When the Tuner receives the final KPN, a fine grained adjustment/optimization is made to the hardware resources of the MPSoC system. For example, if a KPN process has a relatively small load, the cache resources of the corresponding processor can be reduced in size to match the required load. Techniques presented in [6, 16] can be utilized for the Tuner. We utilize the automated tuning of code using custom instructions [6] from Tensilica.

4. RESULTS

We performed two separate experimentations for our SDG2KPN approach. 1), evaluation of MPSoC system generation using rule analysis, from legacy code to XTMP platform (for Tensilica setup) generation; and 2), evaluation of the generated MPSoC platform.

Table 2 depicts the SDG2KPN details for KPN generation and rule evaluation. Column one indicates the application benchmarks tested, whereas column two reveals the number of candidate functions (CFs) considered per application. Columns three through ten report the number of rules captured in each application. Column eleven presents the generation time (in seconds), time taken from reading the legacy code to the end of XTMP MPSoC platform generation using SDG2KPN. The lines of code (LOC) of each application benchmark (the legacy version) is reported in the twelfth column.

The main function in each application (freely available applications were used) was either used as is, if it already contained functions with substantial load, or converted into multiple functions by simply combining certain segments of the code. AES, MPEG-2 encoder, MJPEG and H.264 encoder were used as is, whereas the code segments related to each step in ADPCM encoder application was formed into a function.

A significant number of rules are captured for Rules 4 and 5 on the tested benchmarks. The H.264.enc application revealed 385 rules instances detected for Rule 4, showing that the application significantly exchanged data via global variables. H.264.enc further reveals that it captures 621 instances of Rule 5, which indicates that many global variables are not used to exchange data across candidate functions but defined as global. These variables were reported to the user by our SDG2KPN for manual modifications (moving the variable declaration and assignment locally to relevant candidate function). This will heavily optimize performance, eliminating unnecessary data being communicated. None of the tested applications have Rule 3, even though passing a returned variable is a widely used program pattern in other applications. Rule 1 has one to one connection with the number of candidate functions (i.e., each candidate function will have its unique *call-site*). H.264 has the highest count on Rule 7. This Rule 7 forces to send the entire structure via FIFOs instead of passing only a dependent element from the structure. Rule 2 constrains the first candidate function to the first processor hence no predecessor dependencies analyzed (only one occurrence per application).

As shown, MPEG-2.enc has the highest number of candidate functions considered (i.e., 10). Due to the huge amount of compilation time required, H.264 consumed approximately 164.9 seconds. The next longest one is MPEG-2 encoder, taking 21.4 seconds. The actual SDG2KPN time is much smaller than these values since

[3] No Balancing or Graph Optimization enforced, but one-to-one Mapping and Tuning applied

Apps.	CFs	No. of Rules								Gen. Time (sec)	LOC
		1	2	3	4	5	6	7	8		
AES	2	2	1	0	1	3	0	1	8	4.8	950
MPEG-2 (enc)	10	10	1	0	21	56	1	7	1	21.4	8k
MJPEG	6	6	1	0	8	55	0	3	2	8.0	2k
H.264 (enc)	9	9	1	0	385	621	0	47	1	164.9	58k
ADPCM (enc)	6	6	1	0	11	2	0	0	7	4.3	285

Table 2: KPN Generation and Rule Evaluation[3]

the generation time includes the compilation time and CodeSurfer SDG generation time as well (which are the major contributors to the overall generation time).

Our Tensilica MPSoC system used homogeneous processors, each containing 32kB instruction and data caches, 4-way set associative and 32-byte line size. 65nm technology was used with 1GHz frequency. Table 3 depicts the numbers from the XTMP simulations of the generated MPSoC platform. Column one shows the application benchmarks tested (MPEG-2 and H.264 were avoided due to their large XTMP simulation times). Column 2 indicates the number of processors used. Columns three, four and five report the latency, power and energy of the generated MPSoC system respectively. Columns six, seven and eight report the latency, power and energy of the application executing on a single Xtensa processor respectively (the same Xtensa processor from the MPSoC execution is utilized).

The MJPEG application is generated for both frame level (F) and macro block level (MB), by arranging the iterative loop boundary and the FIFO communication. Speedups of 3.4%, 16.8% and 5.15% are achieved for the XTMP executions of AES, MJPEG (Frames) and MJPEG (Macoblock) respectively, compared to the single processor execution. The ADPCM.enc has not gained any speedup but slowed down due to feedbacks from successive processes of the KPN to their predecessors. The code was not further partitioned to create more parallelism to improve speedup, since it was not the focus of this paper. The total power and energy have increased across all the bechmarks due to more processors in the MPSoC.

For fair comparison with SDG2KPN generation time, we manually created KPNs for several benchmarks, such as MPEG-2 and MJPEG. MPEG-2 manual generation took a few weeks due to the complexity in identifying global variables and their dependencies, whereas MJPEG manual generation took a few days. Our SDG2KPN took only a few minutes (worst case) to generate the code and consumed another few hours for manual modifications.

Table 4 depicts a feature comparison between our SDG2KPN tool and other state-of-the-art tools.

5. CONCLUSION

We propose an SDG to KPN conversion methodology using a rule based traversal of the SDG. All the variable constructs of the legacy code, including shared variables such as globals and pointers, are supported, which was hitherto not possible. We developed our technique in a complete design flow to demonstrate the practicality of the solution. The entire MPSoC platform was generated in a period measuring a few seconds to a few minutes for a simple AES application to a much complicated H.264 encoder application respectively. Our approach requires user modifications and inputs which are far less complicated than the state-of-the-art solutions.

Apps.	No. of Processors	Latency KPN (cycles)	Power KPN (mW)	Energy (KPN) KPN (uJ)	Latency single (cycles)	Power single (mW)	Energy single (uJ)
AES	2	130,631,568	180.54	23,573.59	135,289,982	123.34	16,687.53
MJPEG (F)	6	445,589,604	494.39	119,064.79	535,890,526	113.55	60,851.76
MJPEG (MB)	6	508,260,161	367.02	183,617.70	535,890,526	113.55	60,851.76
ADPCM.enc	6	223,964,173	579.19	129,718.42	113,307,334	131.25	14,872.71

Table 3: Application KPNs and MPSoC Executions

Features	SDG2KPN	COMPAAN[9]	KPNGen[13]	FP-MAP[8]	SPRINT[2]	DSWP[14]	Harmonic[11]	[4]
Shared variable support	Yes	No	No	No	No	No	No	No
Code rewrite/ modification	tool assisted minimal copying	convert to ANLP	manual	convert to nested loop	convert to single-entry single-exit tasks	convert to loop	convert to tasks	not implemented
Analysis granularity	SDG	code statement	code statement	code statement	CFG	loop threads	unknown	AST
Use KPN	Yes	Yes	Yes	No	Yes	No	Yes	Yes
Generates MPSoC platform	Yes	No	No	Yes (pipeline)	No (concurrent SystemC)	No (threads)	No (hardware units)	No (not implemented)

Table 4: Tools and Support

6. REFERENCES

[1] J. Ceng, J. Castrillon, W. Sheng, H. Scharwachter, R. Leupers, G. Ascheid, H. Meyr, T. Isshiki, and H. Kunieda. MAPS: An integrated framework for MPSoC application parallelization. In *DAC*, pages 754–759, 2008.

[2] J. Cockx, K. Denolf, B. Vanhoof, and R. Stahl. SPRINT: a tool to generate concurrent transaction-level models from sequential code. *EURASIP J. Appl. Signal Process.*, pages 213–213, 2007.

[3] CodeSurfer. http://www.grammatech.com/products/codesurfer/overview.html.

[4] V. K. Danish Ather, Raghuraj Singh. Transformation of sequential program to kpn - an overview. *International Journal of Computer Applications*, 2012.

[5] A. Hansson, K. Goossens, M. Bekooij, and J. Huisken. Compsoc: A template for composable and predictable multi-processor system on chips. *ACM Trans. Des. Autom. Electron. Syst.*, 2009.

[6] H. Javaid, A. Ignjatovic, and S. Parameswaran. Rapid design space exploration of application specific heterogeneous pipelined multiprocessor systems. *IEEE Trans. in CAD*, pages 1777–1789, 2010.

[7] H. Javaid, A. Janapsatya, M. S. Haque, and S. Parameswaran. Rapid runtime estimation methods for pipelined MPSoCs. In *DATE*, pages 363–368, 2010.

[8] I. Karkowski and H. Corporaal. FP-map-an approach to the functional pipelining of embedded programs. In *HiPC*, pages 415–420, 1997.

[9] B. Kienhuis, E. Rijpkema, and E. Deprettere. Compaan: deriving process networks from matlab for embedded signal processing architectures. In *CODES*, pages 13–17, 2000.

[10] D. Liang and M. Harrold. Slicing objects using system dependence graphs. In *Workshop on Source Code Analysis and Manipulation*, pages 358–367, 1998.

[11] W. Luk, J. Coutinho, T. Todman, Y. Lam, W. Osborne, K. Susanto, Q. Liu, and W. Wong. A high-level compilation toolchain for heterogeneous systems. In *SOC*, pages 9–18, 2009.

[12] S. Meijer, H. Nikolov, and T. Stefanov. Combining process splitting and merging transformations for Polyhedral Process Networks. In *ESTIMedia*, pages 97–106, 2010.

[13] H. Nikolov, M. Thompson, T. Stefanov, A. Pimentel, S. Polstra, R. Bose, C. Zissulescu, and E. Deprettere. Daedalus: toward composable multimedia MP-SoC design. In *DAC*, pages 574–579, 2008.

[14] G. Ottoni, R. Rangan, A. Stoler, M. J. Bridges, and D. I. August. From Sequential Programs to Concurrent Threads. *IEEE Comput. Archit. Lett.*, pages 2–9, 2006.

[15] S. Stuijk, M. Geilen, and T. Basten. SDF^3: SDF For Free. In *ACSD*, pages 276–278, 2006.

[16] F. Sun, S. Ravi, A. Raghunathan, and N. Jha. Custom-instruction synthesis for extensible-processor platforms. *IEEE Trans. on CAD*, pages 216–228, 2004.

[17] H. Toivonen, F. Zhou, A. Hartikainen, and A. Hinkka. Compression of weighted graphs. In *KDD*, pages 965–973, 2011.

[18] A. Turjan, B. Kienhuis, and E. Deprettere. Translating affine nested-loop programs to process networks. In *CASES*, pages 220–229, 2004.

[19] I. Viskic and D. Gajski. Modeling kahn process networks on mpsoc platforms. Technical Report CECS-08-08, Center for Embedded Computer Systems, University of California, Irvine, July 2008.

[20] N. Walkinshaw, M. Roper, and M. Wood. The Java system dependence graph. In *Workshop on Source Code Analysis and Manipulation*, pages 55–64, 2003.

Low Power Design of the Next-Generation High Efficiency Video Coding

Muhammad Shafique, Jörg Henkel

Chair for Embedded Systems (CES), Karlsruhe Institute of Technology (KIT), Germany
{muhammad.shafique, henkel} @ kit.edu

Abstract – This paper provides a comprehensive analysis of the computational complexity, power consumption, temperature, and memory access behavior for the next-generation High Efficiency Video Coding (HEVC) standard. We highlight the associated design challenges and present several low-power algorithmic and architectural techniques for developing power-efficient HEVC-based multimedia system. We explore the interplay between the algorithms and architectures to provide high power efficiency while leveraging the application-specific knowledge and video content characteristics.

***Index Terms*—low power design, power management, HEVC, video coding, architecture, algorithm, video memory, hardware accelerator, energy efficiency, analysis.**

I. Introduction and Related Work

Video coding based applications are ubiquitous, ranging from consumer, mobile devices, security, automotive, to the internet of things. Besides the increasing video resolutions (ultra-HD to Super-Vision), it is envisaged that video coding based services will cover 80%-90% of global consumer traffic by 2017 [1]. To address the massive video data rate challenges, the ITU-T/ISO/IEC Joint Collaborative Team on Video Coding (JCT-VC) have recently introduced the High Efficiency Video Coding (HEVC) as the next-generation video coding standard [2][3][1]. The HEVC provides ≈2x higher compression efficiency compared to state-of-the-art H.264/AVC video coding standard, while providing the same video quality [4][5]. It is achieved through several novel coding tools like: Coded Tree Unit (CTU) with Coded Tree Block (CTB) structure and recursive partitioning of the coded block, multiple transform and prediction unit sizes, numerous prediction modes for intra- and inter-prediction, complex interpolation filters, multiple in-loop filters, etc. [2][3][7]. However, this results in a vast mode decision space [6]-[10], thus incurring in 40%-70% higher computation effort and >2x more memory accesses compared to the H.264 encoder (see Fig. 1) that also corresponds to high power/energy consumption and temperature.

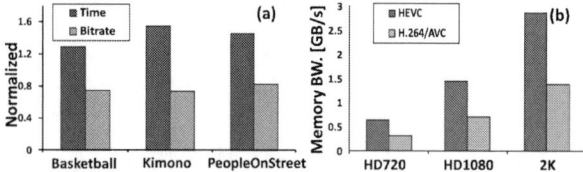

Fig. 1: (a) HEVC time and bitrate normalized to H.264 and (b) Memory bandwidth requirement for HEVC and H.264.

Advances in semiconductor technology complement such a high complexity through integrating 100s-1000s of cores on a single chip [13]-[15]. To exploit the plethora of cores for achieving high performance and power efficiency, HEVC provides inherent support for parallelism through so-called *video tiles* that can be encoded independently on different cores [3][16]. However, it comes with several challenges: (1) multiple cores concurrently processing independent video tiles leads to a significant increase in the energy consumption and memory pressure [17]; (2) integrating

billions of transistors in a tiny area results in escalated power densities and temperature/thermal issues that aggravate several reliability threats[2] and degrade chip lifetime[3] [14]; (3) workload balancing; etc.

Therefore, there is a need to develop low-power hardware and software solutions for advanced HEVC-based video coding systems with high performance-per-power efficiency while also accounting for temperature and reliability issues. The key to high performance-per-power efficiency is to optimize/adapt advanced HEVC algorithms towards embedded multimedia architectures and vice versa, i.e. exploring the potential of interplay of hardware and software.

The associated scientific challenges are:

- Developing hardware-software collaborative techniques for complexity management.
- Developing low-power hardware acceleration engines and HEVC-specific memory hierarchy.
- Developing efficient power and thermal management techniques.
- Developing efficient workload management and parallelization techniques.

The contributions of this paper: To address the above challenges, we have developed a low-power HEVC system with

- An adaptive complexity reduction scheme for fast intra-prediction;
- Low-power compute tiles with adaptive hardware accelerators and a low power video memory with application-driven DPM;
- Dynamic thermal management for HEVC
- Parallelization and workload balancing.

To facilitate an efficient system design, we have performed an extensive analysis of HEVC for performance/computational complexity, memory access behavior, and temperature.

Before proceeding further, we first present a short overview of the HEVC and its prominent novel coding tools.

II. Overview of High Efficiency Video Coding

The basic architecture of HEVC encoding is shown in Fig. 2. The input video is divided into block-shaped regions called *Coding Tree Units (CTUs)* and fed in raster-scan order to the HEVC encoder. The exact block partitioning sent to the decoder. HEVC is a hybrid encoding standard, i.e. both spatial (Intra) and temporal (Inter) redundancies are exploited for compressing the input video. The first picture is coded using only the *intra-prediction*, i.e. exploiting the spatial correlations and having no dependency on other pictures. For all the remaining pictures or between random access points, typically *inter-prediction* is employed using motion estimation and motion compensation. After searching for the best Intra and Inter predictions of the current CTU, the *residual* signal is computed as the difference of the current video data and the

---

[2] Like electromigration, HCI, TDDB, soft errors, etc. [14][18].
[3] E.g., a difference between 10ºC – 15ºC can result in a 2x difference in the MTTF (mean-time-to-failure) [19].

---

[1] Other parallel efforts are: Google's VP9 [11] and Daala [12].

978-1-4799-2817-0/14 $31.00 © 2014 IEEE

prediction. The residual is *transformed* by a linear spatial transform and *scaled, quantized, entropy coded* using the context adaptive binary arithmetic coder (CABAC), and transmitted to the decoder device along with the prediction information. HEVC employs uniform reconstruction quantization (URQ) with quantization scaling matrices for various transform block sizes. For generation of identical predictions of the subsequent pictures for future reference, *the encoder has a partial decoder in the reconstruction loop*. Therefore, the quantized coefficients are inverse quantized, inverse scaled, inverse transformed and added to the prediction to obtain the reconstructed signal. Afterwards, for reducing the blocking artifacts, two in-loop filters (*Deblocking and Sample Adaptive Offset (SAO) filters*) are employed and the reconstructed picture is stored in the *decoded picture buffer* and reused for future predictions.

In the following, we briefly describe important coding features of the HEVC standard (see more details in [2][3]).

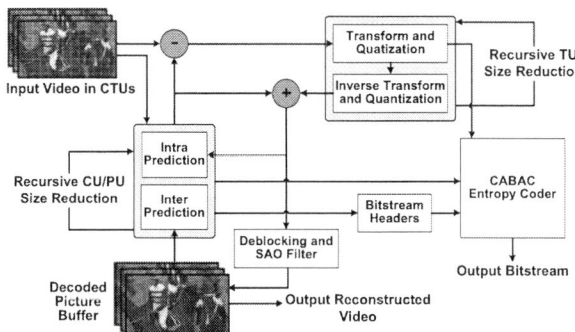

Fig. 2: Overview of the HEVC encoding systems.

A. Coded Tree Unit and Block Structure

Analogous to the 16x16 macroblock of H.264/AVC, the HEVC employs a large-sized block structure called *Coding Tree Unit* (CTU) which consists of lume and chroma *Coded Tree Blocks* (CTBs). The size of CTBs ranges typically from 64x64 to 16x16 [3], where large-sizes are better in terms of compression for higher resolutions. The CTBs are recursively partitioned using a *quadtree structure* (see Fig. 3 for a hierarchical partitioning) such that, the root of the quadtree is associated with the CTU. A CTB may contain one *Coding Unit* (CU) or may split into multiple CUs. At CU-level, HEVC decides whether to encode the current CU using Intra or Inter predictions. Therefore, each CU is partitioned into multiple *Prediction Units* (PUs) and *Transform Units* (TUs). The root of PU and TU is at the CU level and a further partitioning of PU and TU may be performed to achieve better compression. The PU sizes range from 64x64 to 4x4, while the TU sizes range from 32x32 to 4x4. To achieve high compression efficiency, HEVC employs a Rate-Distortion Optimized (RDO) mode decision that evaluates all possible combinations of CU, PU, and TU to determine the best quadtree partitioning and associated CU, PUT, and TU combinations. However, this incurs a significant computational load that we will illustrate in Section III.

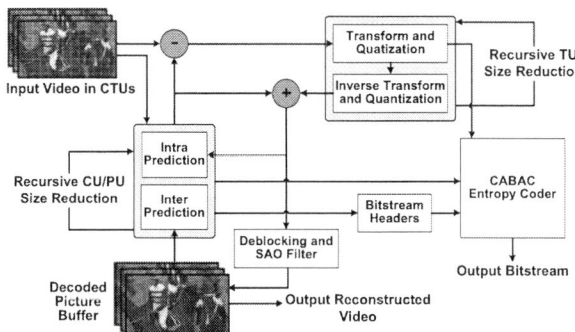

Fig. 3: An example CU splitting for CTUs of size 64×64.

B. Intra-Prediction

The HEVC employs 35 different modes (33 directional, 1 planar, and 1 DC) for intra-prediction as shown in Fig. 4. Each PU is evaluated for these prediction modes and the best one is selected after the RDO-optimized mode decision. A large set of predictors effectively eliminate the spatial redundancy. However, this comes at the cost of an increased search space and associate computational load.

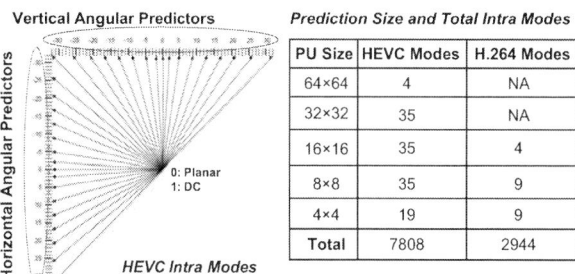

PU Size	HEVC Modes	H.264 Modes
64×64	4	NA
32×32	35	NA
16×16	35	4
8×8	35	9
4×4	19	9
Total	7808	2944

Fig. 4: Intra prediction modes in HEVC and H.264.

C. Inter-Prediction and Motion Compensation

The inter-prediction with motion estimation and compensation (considering hierarchical quadtree partitioning) is the most complex processing step in the HEVC. It employs 7-tap and 8-tap interpolation filters with quarter-sample precision for motion compensation and fractional-pixel motion estimation, multiple reference frames, uni- and bi-directional predictions. In the motion estimation step, a block is searched inside the so-called search window in the reference frame. The reference frame can be stored in an off-chip or an on-chip memory depending upon the throughput requirements and the architectural design constraints. This processing step incurs a huge amount of memory access for the block matching process and may spend >70% of the total motion estimation energy into them memory transfers [17][20]-[22]. Moreover, multiple video tiles may aggravate the memory pressure due to the data overlaps between different tiles and potentially more memory access [17].

D. Video Tiles and Slices

For improved error resilience, the picture can be partitioned into multiple *Slices* which are a sequence of CTUs that are processed in a raster scan order (see Fig. 5). The prediction is not performed across slice boundaries.

Unlike slices, the so-called *Video Tiles*[4] facilitate parallel encoding and decoding without any sophisticated thread synchronization. Each *video tile* is a rectangular-shaped group of CTUs. Fig. 5 presents an example for 6 tiles. The *video tiles* can be processed on different cores without any data dependencies with each other. Partitioning of a video frame into multiple tiles is a research challenge as different tiles may exhibit varying workload due to the texture and motion properties of their constituting CTUs. Therefore, mapping and processing of different tiles on different processing cores and hardware accelerators needs to account for workload, power density, temperature, and memory pressure balancing.

Parallelization in HEVC can also be achieved through *Wavefront Parallel Processing* (WPP) that divides a slice into rows

---

[4] To avoid confusion, we use the term "video tile" for the so-called tiles defined in HEVC and "compute tile" for traditional processing tiles in a many-core processor.

of CTUs such that the 2nd row of CTU can already start processing once the first two CTUs of the first row are encoded and so on.

Fig. 5: An example of Slices and Video Tiles in the HEVC.

III. Analysis of HEVC

In order to design an efficient HEVC system with optimized hardware/software components, we have performed an extensive analysis of computational complexity/processing load, memory accesses, and thermal behavior of the HEVC for different configurations.

A. Experimental Setup

Encoder Setup: The analysis is performed on the HEVC software [23]. For all experiments, the recommended test conditions (as provided by the JCT-VC) [24] are used, with a set of Quantization Parameter (QP)={ 22, 27, 32, 37}. Different test video sequences with diverse texture and motion properties are used for evaluation, ranging from 832x280 pixels resolution up to 2560x1600 pixels [25]-[30]. For Random Access experiments, the size of Group of Pictures (GOP) is 8, intra-period is 64, 48, and 32, frame rates are 60fps, 50fps, and 30fps. Most of the experiments were performed an Intel Core i7-2600 processor with 16GB RAM.

Temperature Analysis Setup (see Fig. 6) [32]: In the thermal experiments, an Intel Atom 45nm dual-core processor @ 1.8 GHz is employed. The chip temperature are measured using a DIAS pyroview 380L compact IR-thermal camera (temp. accuracy = ±1 °C, spatial resolution = 50 μm per pixel) [31] and the thermal images and readings are captured @50fps to a PC. To provide a clear thermal image, we deploy our in-house oil-free thermal measurement setup [32][33] that allows processor to work under normal/compliant operating conditions (see Fig. 6a). It uses a thermoelectric device that continuously cools down the chip from the bottom side (Fig. 6b) exploiting the Peltier effect, i.e. due to the thermal variations, one side of the device becomes very cold, while the other one very hot. The heat is dissipated using an additional water-based cooling system.

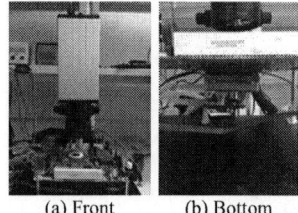 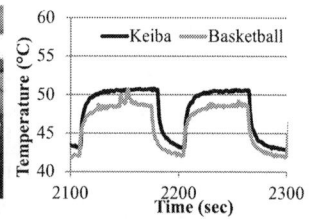

(a) Front   (b) Bottom

Fig. 6: Camera and board with measured processor

Fig. 7: Temperature cycles for two motion intensive sequences.

B. Computational Complexity Analysis

CTU Partitioning and Intra Prediction [6][9]: Considering the full RDO-optimized mode decision, for a CTU of size M×M pixels (divided into multiple CUs), the total number of predictions (β) can be given as $\sum_{i=0 \text{ to } \log2(M)-2}(2^{2i} \times N_i)$. N_i either denotes the total number of evaluated intra-prediction modes or the total number of motion search candidates of the ith CU size. All PUs (64×64→4×4) are generated and evaluated to find out the best CU structure for the given CTU. For the intra-prediction process, a 64x64 CTU will result in β=7808, ≈2.65× more predictions than that in the H.264.

Note that, recursive partitioning and motion search also requires additional interpolation effort, as discussed below. Fig. 8 illustrates an example partitioning after the full RDO-optimized mode decision. It shows that bigger partitions are selected for the homogeneous and low-variance regions, while smaller partitions are used for high variance regions. A more detailed distribution for different QP values is given in Fig. 9. It shows that for smaller QP value, more variations/texture is captured and the percentage of large-sized PUs is low. Increase the QP results in smoothening of the region and the number of large-sized PUs grows. This observation is leveraged to design a content-driven complexity reduction scheme at the application level.

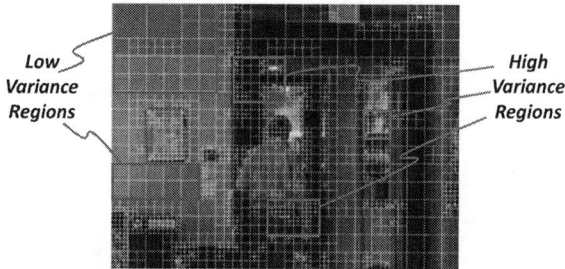

Fig. 8: PU borders on the 1st frame of Exit sequence with max CU of 64×64.

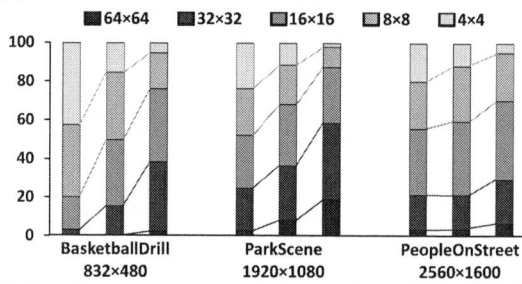

Fig. 9: Percentage image area occupied by a particular PU for QP of 22, 30 and 38 for different sequences.

Interpolation for Motion Compensation and Fractional-Pixel Motion Estimation [37]: Luma half- and quarter-pixels are generated using 8-tap and 7-tap filters, respectively while the chroma eighth-pixels are generated using a 4-tap filter. Recent analysis on HEVC complexity [34][35] demonstrates that these interpolation filters take 15%-38% of the total HEVC encoding and decoding time. Our analysis in Fig. 10 shows that the interpolation filter consumes ≈20%-30% of the total HEVC encoding execution time depending upon the video sequence and QP.

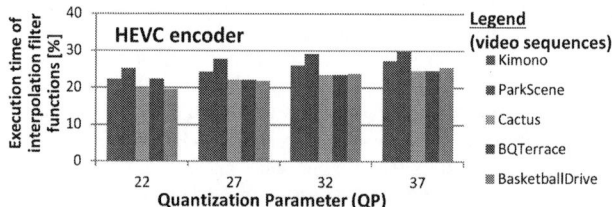

Fig. 10: Execution time (%) of interpolation filter in encoder and decoder.

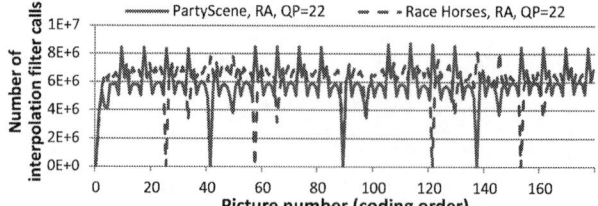

Fig. 11: Number of calls per picture to the interpolation filter basic function.

The frame-level analysis for two test video sequences ("Party Scene" and "Race Horses") in Fig. 11 shows variations in the

978-1-4799-2817-0/14 $31.00 © 2014 IEEE

interpolation filter calls for the first 180 frames at QP=22. This motivates the need for adaptive interpolation filters that adapt to the workload requirements depending upon the video sequence properties.

C. Memory Analysis

Figure illustrates the memory requirements for different test video sequences during the motion estimation process in HEVC for a single reference frame case. The box plots show the summary for three different search window sizes. Our analysis shows that the memory access requirements of the HEVC encoder are ≈3.86× more compared to that of the H.264/AVC encoder. This implies that the HEVC encoder puts a higher pressure on the memory sub-system compared to H.264. Fig. 12 also shows that the memory accesses variations for different test video sequences. This shows that leveraging the application-specific knowledge for video memory design, access management, and power management may provide a high potential for the corresponding savings. A more detailed analysis of HEVC memory access behavior can be found in [17][36]. Note that the quadtree structure with multiple CUs and multiple video tiles aggravate the memory problem.

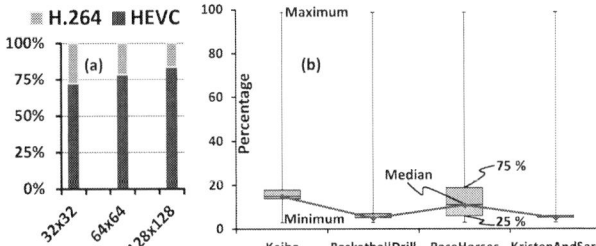

Fig. 12: Block matching memory access percentage statistics for TZ search in HEVC.

D. Temperature Analysis

Temperature Analysis for Different Video Sequences considering Core Idling: Fig. 7 shows the core temperature over time for the "Keiba" and "BasketballDrill" test video sequences considering the same arrival time for each frame of the both sequences. The "Keiba" sequence exhibits with high-motion content (dark line) while the "BasketballDrill" sequence exhibits low-motion content (gray line). In case, the encoding of a particular video frame is completed before the next frame's arrival time, the core is put into the idle mode, thus allowing for cooling down of th chip and temperature drop. Fig. 7 shows that for the "Keiba" sequence with high-motion incurs a high temperature for most of the time, while encoding of the "BasketballDrill" sequence experiences lower temperature and an earlier temperature drop due to a relatively longer idle time compared to the other case. The average temperature for the "BasketballDrill" sequence encoding is 2.5 °C lower compared to that of the "Keiba" sequence. It shows that workload of the high motion sequence need to be curtailed for lowering the temperature, while a workload balancing may contribute towards temperature balancing.

Temperature Analysis for Different Video Sequences under Frequency Scaling: When employing frequency scaling, the low-motion "BasketballDrill" sequence can be executed at a much lower frequency compared to the high-motion "Keiba" sequence while exploiting the slack between the next frame's arrival time and the encoding completion for the previous frame. The core is put to the idle mode if the encoding finishes before the next frame's arrival time. Fig. 13 shows that the frequency for the "BasketballDrill" sequence is reduced from 1.8 GHz to 1.35 GHz that results in the peak temperature decrease from 56.4 °C to 53.9 °C.

Temperature Analysis for Different Video Sequences using Multiple Concurrently Encoded Video Tiles: To meet the tight throughput constraints, the high-motion "Keiba" sequence is encoded using two video tiles processed on two cores concurrently. Fig. 14 shows that this leads to 5 °C higher temperature for the "Keiba" sequence compared to that for the "BasketballDrill" sequence.

A more detailed analysis of HEVC memory access behavior can be found in [33].

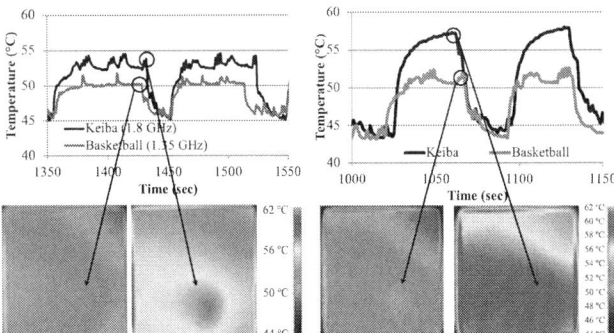

Fig. 13: Temperature analysis using FS for low motion sequence

Fig. 14: Temperature analysis of high motion sequence with 2 cores.

IV. Power-Efficient Design of HEVC

A. System Overview

Fig. 15 presents the overview of our low-power HEVC system with different hardware and software components.

At the software layer, the input video data is first pre-processed and a statistical analysis of the video data is performed. It is then characterized in different workload classes like high, medium, and low motion/texture frames, regions, and blocks [40][43]. Based on the statistical analysis and classification, workload budgeting is performed for different blocks, tiles, and frames considering the available compute capabilities from the underlying hardware platform and the user constraints (e.g., throughput in frame per second) [10]. Afterwards, different management algorithms are employed to curtail the computational load or to select and appropriate operational mode/configuration of the HEVC encoder application that fulfills the user and system constraints (e.g., available energy budget in a battery-power system) while incurring a minimal video quality loss [6][9]. Another important component at the software layer is the *video tile* formation scheme that applies sophisticated policies to partition the video frames into multiple *video tiles* such that their workload is balanced over different *compute tiles* [16]. This determines the computational demand of the HEVC video tiles from the underlying processing hardware.

At the hardware layer, we employ a low-power many-core processor that consists of multiple *compute tile*. A compute tile constitutes one or more general purpose cores to host the processing of the video tile, various hardware accelerators to expedite the processing of compute intensive functional blocks of HEVC like Sum of Absolute Differences, various intra-prediction modes, interpolation filters, modules for image processing (e.g., variance computation), etc. [6][17][22][36][37]. These accelerators are connected in a coprocessor fashion [37]. Fig. 16 presnets the hardware results for the interpolation accelerator synthesized for the Xilinx XC5VLX110T-2ff1136 FPGA device. Additionally, the compute tile has a specialized video memory with content-driven dynamic power management (DPM) [17][36]. The algorithm- and video content specific knowledge is passed from the application

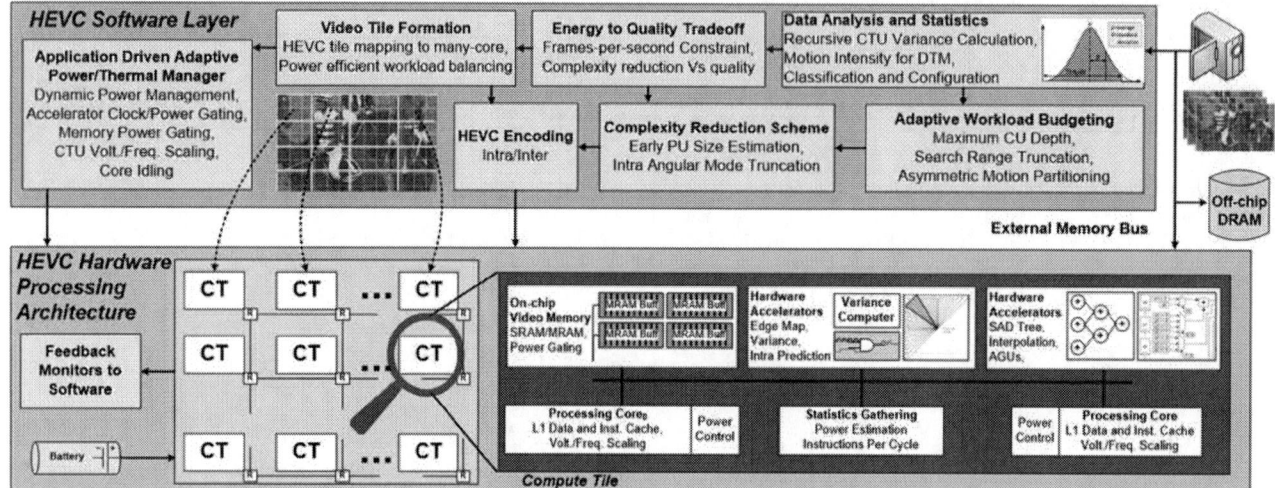

Fig. 15: Overview of our low-power HEVC system showing different components at the hardware and software layers.

software layer. Conceptually, our approach raises the abstraction level of the power management decisions to the application level, given the hardware-level power management infrastructure (like sleep transistors) [17][36]. In-depth details on various different case studies on application-driven DPM can be found in our previous works [20]-[22], [38]-[43].

The feedback on the achieved performance/throughput is provided to the software layer that adapts the computational requirements through adaptive workload budgeting, adaptive computational complexity reduction, energy-quality tradeoff triggers, and video tile formation with workload balancing.

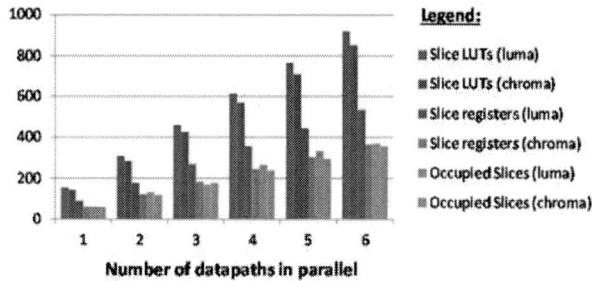

Fig. 16: Area results of interpolation filter datapath options for FPGA.

In the following, we briefly explain the key components of our low-power HEVC system.

B. Adaptive Complexity Reduction

The computational complexity of the HEVC encoder can be significantly reduced by restricting the total number of CU partitions for evaluation. In the following, we demonstrate the case for the HEVC intra encoding. Instead of performing the full RDO-optimized mode decision, our approach predicts a set of highly probable CU sizes depending upon the variance properties of different objects in the input video. Our analysis in Section III shows the relationship between CU size and the corresponding variance properties. Base on this observation, we develop a variance-aware CU/PU mode prediction as shown in Fig. 17 (see details in [6][9]). First, the variance at 4x4 sub-block level is computed. Afterwards, a recursive merge is performed to create bigger partitions depending upon the variance similarity between the current block and its neighboring blocks. The process is repeated unless no block for merging is available. This provides a so-called PU map which is used for as the PU size prediction for evaluation. Note that this completely eliminates the need for

multiple CU/PU size evaluations and only the sizes predicted using the PU map are evaluated. To lower the risk of miss-prediction, an additional PU map is generated which represents the one level above in the quadtree structure. An example of merging is shown in Fig. 17. Our experiments illustrate that our approach achieves an average speedup of 44% and energy reduction of 35% while incurring a negligible BD-PSNR [46] loss of -0.048dB. Our complexity reduction module facilitates integration of additional complexity reduction algorithms like early Intra angular direction prediction and early TU partitioning.

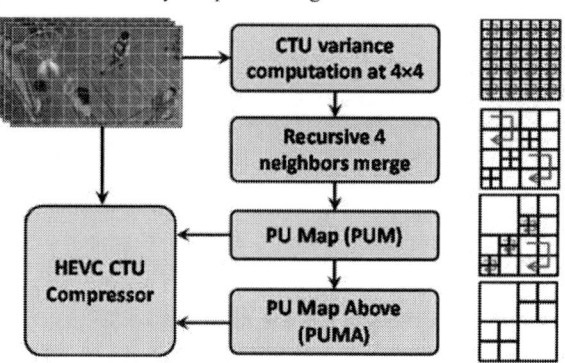

Fig. 17: An illustration of determining best CU sizes and locations using the variance of the video frame content.

C. Low-Power Video Memory Design

HEVC is a memory intensive application where the current and multiple reference frames are stored and used for inter-prediction in an intensive motion search and block matching operation. The increasing video resolutions aggravate this problem. For instance, Quad Full-HD resolution (3840×2160) at 30fps requires ≈12MB per frame. During the motion search process, large portions of the reference video frames are repeatedly brought on to an on-chip video memory. It results in high external memory access power and high leakage power of the on-chip video memory. These power-related issues can be alleviated through the following two means. (1) Reducing the off-chip memory access using intelligent data reuse, motion trajectory prediction, and leveraging this knowledge for designing and managing specialized on-chip video memories [17][20]-[22]. (2) Exploiting advanced memory technologies, i.e. Non-Volatile Memories (NVMs), like MRAM that overcome the limitations of SRAM by providing high density and very low leakage power (≈5.4× lower). However, these memories suffer from high dynamic write power (≈20.6× larger) and latency (≈2.6× larger).

978-1-4799-2817-0/14 $31.00 © 2014 IEEE

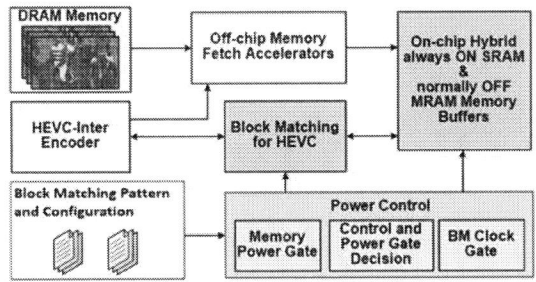

Fig. 18: Overview of AMBER system employed for HEVC-Inter encoding.

To address the above-mentioned memory power issues and to combine the benefits of both SRAM and NVMs, we have developed a novel hybrid video memory architecture along with an adaptive energy system called AMBER as shown in Fig. 18 (see details in [36]). It employs a high-speed but small SRAM-based memory to provide fast CTU reads and to hide the write latency of MRAMs. In parallel these CTUs are written into the high-density MRAM based on-chip reference frame buffers for a long-term use during the motion estimation of current or upcoming frames. The MRAM buffers are divided into sectors that can be selectively power-gated to save leakage and powered-on to fulfill the requirements of the current frame's motion estimation process. The power states of different sectors of the MRAM buffer are adaptively selected. Unlike SRAMs, the data is retained in the MRAM after power-gating due to its non-volatile nature. The power manager is a self-organizing map based learner that adapts to the changing memory search patterns without explicit supervision. Fig. 19 shows that our AMBER provides significant savings in the power consumption compared to a state-of-the-art search window based approach.

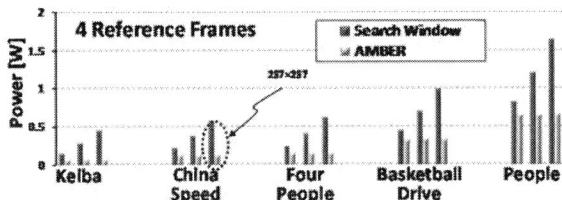

Fig. 19: AMBER power savings [36].

Note that AMBER supports a single-tile based HEVC implementation. Therefore, in order to support multiple tiles and concurrent memory access at high throughput, we have also designed a *distributed scratchpad based video memory design (dSVM)* for HEVC (see details in [17]). It employs a content-driven DPM and reduces the overall memory energy consumption by up to 51%-61% compared to parallelized state-of-the-art solutions. It leverages the application-specific knowledge to provide 54% on-chip leakage energy savings.

D. Workload Balancing for Multiple Video Tiles

The HEVC coding can be parallelized using multiple independent *video tiles* that are concurrently encoded on different *compute tiles*/cores. However, the workload of different video tiles varies depending upon their constituting CTUs (see Fig. 20). Therefore, *video tile* formation and determining a balanced distribution of the HEVC workload on multiple cores needs to take the underlying hardware features and application-specific properties into account. For instance, if the number of available cores and their maximum achievable frequency are ignored, the power consumption and achieved performance may not be sufficient or good enough. Moreover, it is important to consider that an increased number of *video tiles* may lead to a significant video quality loss. Therefore, the challenge is to determine the

tradeoff, i.e. finding an appropriate number of *video tiles* and their respective compositions.

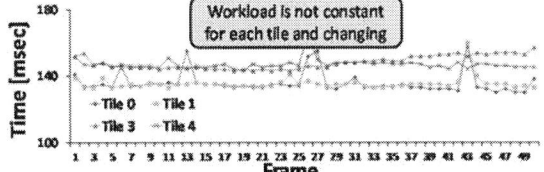

Fig. 20: Time consumption of each tile (4 tiles per frame, of 2×2 tile structure) of first 50 frames of RaceHorses sequence (832×480).

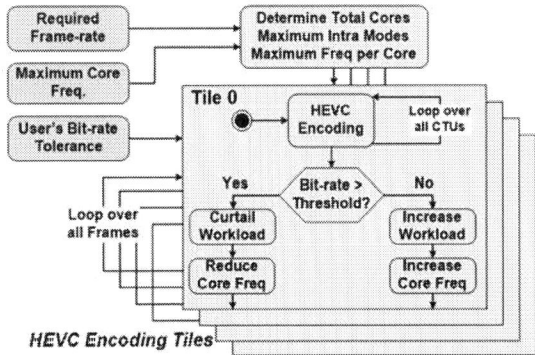

Fig. 21: HEVC adaptive workload balancing on many-core system.

We have developed a novel tile formation and workload balancing scheme for parallelized HEVC as shown in Fig. 21. Before starting the HEVC encoding process, the complete workload to be encoded is divided among different available cores. If the number of available cores is more than the required, the additional cores are power-gated. In case the available cores cannot meet the throughput requirements, the maximum workload for each core is curtailed using our complexity reduction scheme (see Section IV.B). Afterwards, the encoding is started. For each frame, we adapt the workload of each *video tile* (i.e. the set of CTUs that constitutes this *video tile*). Our analysis of video tile workload illustrates that *video tiles* may have a high correlation (in terms of workload, bit rate, etc.) with its temporal and collocated *video tiles*. This correlation is exploited to obtain a precise prediction of the estimated time to determine the *video tile* formation for the upcoming video frames. Additionally, we allow users to specify a level of tolerance to the encoded video quality in order to relax the encoder pressure and thereby allowing for a higher number of *video tiles*, i.e. moving towards massive multi-threading of HEVC. The power savings are proportional to the user-specified tolerance level. Further details can be found in [16]. Fig. 22 present different adaptations (i.e. frequency, bit rate, and intra modes) for two different tiles of the "Exit" test video sequence.

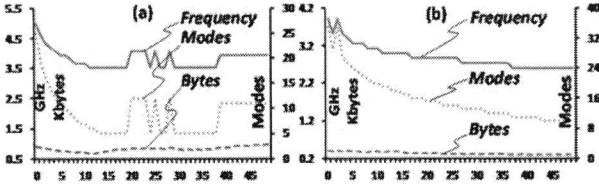

Fig. 22: Bit-rate, frequency and total intra modes adaptation of (a) tile 14 and (b) tile 17 for first 50 frames of "Exit" (640×480) video sequence [16].

E. Dynamic Thermal Management for HEVC

To efficiently manage the temperature while yet providing a high video quality, we have developed an application-driven dynamic thermal management (DTM) policy for the HEVC encoding [33]. The input is a design-time generated Pareto-Optimal frontier of different coding configurations providing temperature vs. video quality (in terms of PSNR – Peak Signal to Noise Ratio).

Our DTM policy determines an temperature-friendly coding configuration at frame level. At the start of each frame's encoding, it reads the temperature value from the underlying temperature sensors. In case, the temperature is approaching a pre-defined thermal threshold, it selects an appropriate coding configuration. It exploits the motion and texture information [40][43] to fine-tune the coding configuration at the frame and CU level. In case of thermal emergencies ($T_{Current} > T_{Th}$), our policy triggers the voltage/frequency scaling to rapidly cool down the core.

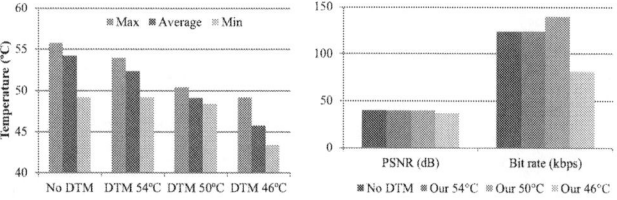

Fig. 23: Max, min and average temperature the "Keiba" test video sequence.

Fig. 24: PNSR and Bit Rate results for the "Keiba" test video sequence.

Fig. 23 shows the maximum, average and minimum temperature of the core for the complete encoding of the "Keiba" test video sequence encoding for without DTM and with our DTM under different thermal threshold values. It shows that our polidy significantly reduces the maximum, average and minimum temperature as the threshold decreases, which also reduces the thermal cycles intensity. Without consideration of DTM, the quality-wise best coding configuration is selected. However, to keep the temperature below the specified threshold another configuration is selected which may degrade the video quality. Still, our application-driven DPM policy incurs only a minimal video quality loss, i.e. an average PSNR loss is of 0.007 dB at $T_{Th}=54°C$, as shown in Fig. 24. The video quality degradation increases under tighter temperature constraints, i.e. at 46°C.

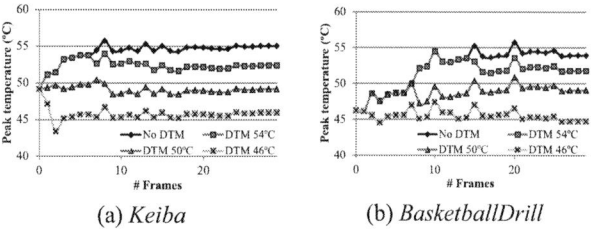

(a) *Keiba* (b) *BasketballDrill*

Fig. 25: Comparing temperature profile of our DTM at different T_{Th} with no DTM.

Fig. 25(a)-(b) show a detailed frame-level view in form of the temperature over time for encoding of two different test video sequences, i.e. "Keiba" and "BasketDrill" using different thermal thresholds.

Fig. 26 (a)-(d) show the thermal maps of the chip steady temperature when using no DTM policy and our DTM policy with different T_{Th} = 54°C, 50°C and 46°C for encoding the "Keiba" test video sequence. These thermal maps clearly show that our DTM policy achieves a much better temperature profile compared to when not considering an DTM.

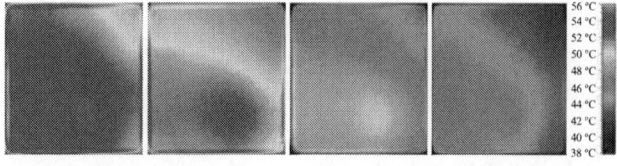

(a) no DTM (b) DTM 54°C (c) DTM 50°C (d) DTM 46°C

Fig. 26: Thermal maps of the die for different T_{Th} when encoding the "Keiba" test video sequence.

F. Our Multi-Threaded HEVC Software Architecture

In our lab, we have developed a multithreaded version of the HEVC Intra-only encoder, called CES265 ("CES" stands for the Chair for Embedded Systems, while "265" comes from the ITU's H.265 which is the standard specification name of the HEVC). The encoder is implemented in C++ and the sources can be compiled on both Linux and Windows based platforms.

The motivation behind the development of our customized multi-threaded HEVC encoder is:

- The HM reference encoders are quite slow and require a lot of memory. Hence, it is extremely difficult to export this software source for resource-limited embedded implementations and testing.
- There are not parallelization capabilities available in the reference implementation. It is very difficult to change the software to account for parallelization.
- Other implementations, like x265 are available. x265 is a C language based HEVC Intra only encoder. However, it doesn't offer parallelization capabilities.

The purpose of CES265 is to be efficient, and allow threading capabilities at GOP-, slice- and *video tile*-level. A single thread of CES265 is ≈13.2× faster than the HM-9.2 reference software.

V. Conclusion

In this work, we presented a comprehensive analysis of the next-generation High Efficiency Video Coding (HEVC) standard for computational load, memory accesses, and temperature. A low-power HEVC system is presented that leverages application-and video content specific knowledge and engages both hardware architecture and software layers. An adaptive complexity reduction scheme, a multimedia processor architecture with multiple compute tiles, a hybrid video memory with dynamic power management, video tile formation and workload balancing, and a dynamic thermal management scheme for HEVC are presented. The interplay of hardware and software provides a high potential of power savings and enables power-efficient multimedia systems for the next-generation HEVC standard.

Acknowledgement

This work was partly supported by the German Research Foundation (DFG) as part of the Transregional Collaborative Research Centre "Invasive Computing" (SFB/TR 89); http://invasic.de. The authors would like to thank Muhammad Usman Karim Khan, Daniel Palomino, Hussam Amrouch, and Claudio Diniz from CES, KIT for their efforts in conducting experiments and discussions.

References

[1] Cisco, "Cisco Visual Networking Index: Forecast and Methodology, 2012, 2017", May 2013.

[2] ITU-T, "SERIES H: AUDIOVISUAL AND MULTIMEDIA SYSTEMS - Infrastructure of Audiovisual Services - Coding of Moving Video – High Efficiency Video Coding." Apr. 2013.

[3] G. J. Sullivan, J. Ohm, W. Han, T. Wiegand, "Overview of the High Efficiency Video Coding," in IEEE Trans. Circuits Syst. Video Technol., vol. 22, no. 12, pp. 1649–1668, 2012.

[4] B. M. T. Pourazad, C. Doutre, M. Azimi, P. Nasiopoulos, "HEVC: The New Gold Standard for Video Compression: How Does HEVC Compare with H.264/AVC?," in IEEE Consumer Electronics Magazine,pp. 36–46, 2012.

[5] T. Nguyen, D. Marpe, "Performance analysis of HEVC-based intra coding for still image compression", Picture Coding Symposium (PCS), pp.233-236, 2012.

978-1-4799-2817-0/14 $31.00 © 2014 IEEE

[6] M. U. K. Khan, M. Shafique, M. Grellert, J. Henkel, "Hardware-Software Collaborative Complexity Reduction Scheme for the Emerging HEVC Intra Encoder", IEEE/ACM Design Automation and Test in Europe Conference (DATE), 2013.

[7] D. Marpe et al., "Video compression using nested quadtree structures, leaf merging, and improved techniques for motion representation and entropy coding," IEEE Trans. Circuits Syst. Video Technol., vol. 20, no. 12, pp. 1676–1687, 2010.

[8] B. M. T. Pourazad et al., "HEVC: The New Gold Standard for Video Compression: How Does HEVC Compare with H.264/AVC?," IEEE Consumer Electronics Magazine, pp. 36-46, 2012.

[9] M. U. K. Khan, M. Shafique, J. Henkel, "An Adaptive Complexity Reduction Scheme with Fast Prediction Unit Decision for HEVC Intra Encoding", 20th IEEE International Conference on Image Processing (ICIP), 2013.

[10] M. Grellert, M. Shafique, M. U. K. Khan, L. Agostini, J. C. B. Mattos, J. Henkel, "An Adaptive Workload Management Scheme for HEVC", 20th IEEE International Conference on Image Processing (ICIP), 2013.

[11] http://www.webmproject.org/vp9/. WebM VP9 project, accessed on 28.08.2013.

[12] http://people.xiph.org/~xiphmont/demo/daala/demo1.shtml. Xiph.Org Foundation's Daala Video Encoder, accessed on 28.08.2013.

[13] ITRS. http://www.itrs.net, 2013.

[14] J. Henkel, L. Bauer, N. Dutt, P. Gupta, S. Nassif, M. Shafique, M. Tahoori, N. Wehn, "Reliable on-chip systems in the nano-era: Lessons learnt and future trends", IEEE/ACM Design Automation Conference (DAC), 2013.

[15] A. Singh, M. Shafique, A. Kumar, J. Henkel, "Mapping on multi/many-core systems: Survey of current and emerging trends", IEEE/ACM Design Automation Conference (DAC), 2013.

[16] M. U. K. Khan, M. Shafique, J. Henkel, "Software Architecture of High Efficiency Video Coding for Many-core Systems with Power Efficient Workload Balancing", IEEE/ACM Design Automation and Test in Europe Conference (DATE), 2014.

[17] F. Sampaio, M. Shafique, B. Zatt, S. Bampi, J. Henkel, "dSVM: Energy-Efficient Distributed Scratchpad Video Memory Architecture for the Next-Generation High Efficiency Video Coding", IEEE/ACM Design Automation and Test in Europe Conference (DATE), 2014.

[18] S. Rehman, M. Shafique, F. Kriebel, J. Henkel, "Reliable software for unreliable hardware: embedded code generation aiming at reliability", CODES+ISSS, pp. 237–246, 2011.

[19] R. Viswanath, et al., "Thermal performance challenges from silicon to systems", Intel Technol. J., Q3, vol. 23, p. 16, 2000.

[20] M. Shafique, B. Zatt, F. L. Walter, S. Bampi, J. Henkel, "Adaptive Power Management of On-Chip Video Memory for Multiview Video Coding", IEEE/ACM Design Automation Conference (DAC), pp. 866-875, 2012.

[21] B. Zatt, M. Shafique, S. Bampi, J. Henkel, "A Low-Power Memory Architecture with Application-Aware Power Management for Motion & Disparity Estimation in Multiview Video Coding", IEEE International Conference on Computer-Aided Design (ICCAD), pp. 40-47, 2011.

[22] B. Zatt, M. Shafique, F. Sampaio, L. Agostini, S. Bampi, J. Henkel, "Run-Time Adaptive Energy-Aware Motion and Disparity Estimation in Multiview Video Coding IEEE/ACM Design Automation Conference (DAC), pp. 1026-1031, 2011.

[23] Joint Collaborative Team on Video Coding (JCT-VC), "HM 11.0 Reference Software" [Online]. Available: http://hevc.hhi.fraunhofer.de/.

[24] F. Bossen, "Common HM test conditions and software reference configurations", JCTVC-G1200, 2012.

[25] ftp://hvc:US88Hula@ftp.tnt.uni-hannover.de

[26] http://trace.eas.asu.edu/yuv/index.html

[27] http://media.xiph.org

[28] http://www.cdvl.org

[29] ftp://ftp.new.jp/KDDI/multiview

[30] ftp://ftop.merl.com/pub/avetro/mvc-testseq/orig-yuv

[31] "DIAS Infrared Camera," http://www.dias-infrared.com/pdf/pyroview380lcompact eng.pdf.

[32] Chair for Embedded Systems, KIT, Germany, http://ces.itec.kit.edu.

[33] D. Palomino, M. Shafique, H. Amrouch, A. Susin, J. Henkel, "hevcDTM: Application-Driven Dynamic Thermal Management for High Efficiency Video Coding", IEEE/ACM Design Automation and Test in Europe Conference (DATE), 2014.

[34] J. Vanne, M. Viitanen, T.D. Hamalainen, A. Hallapuro, "Comparative Rate-Distortion-Complexity Analysis of HEVC and AVC Video Codecs," IEEE Trans. Circuits Syst. Video Technol., vol.22, no.12, pp.1885,1898, Dec. 2012

[35] F. Bossen, B. Bross, K. Suhring, D. Flynn, "HEVC Complexity and Implementation Analysis", IEEE Trans. Circuits Syst. Video Technol., vol. 22, no. 12, 2012, pp. 1685-1696.

[36] M. U. K. Khan, M. Shafique, J. Henkel, "AMBER: Adaptive Energy Management for On-Chip Hybrid Video Memories", IEEE/ACM International Conference on Computer-Aided Design (ICCAD), 2013.

[37] C. M. Diniz, M. Shafique, S. Bampi, Jörg Henkel, "High-Throughput Interpolation Hardware Architecture with Coarse-Grained Reconfigurable Datapaths for HEVC", 20th IEEE International Conference on Image Processing (ICIP), 2013,.

[38] H. Javaid, M, Shafique, J. Henkel, S. Parameswaren, "System-Level Application-Aware Dynamic Power Management in Adaptive Pipelined MPSoCs for Multimedia", IEEE ICCAD, pp. 616-623, 2011.

[39] H. Javaid, M. Shafique, S. Parameswaran, J. Henkel, "Low-Power Adaptive Pipelined MPSoCs for Multimedia: An H.264 Video Encoder Case Study", IEEE/ACM Design Automation Conference (DAC), pp. 1032-1037, 2011.

[40] M. Shafique, B. Molkenthin, J. Henkel, "An HVS-based Adaptive Computational Complexity Reduction Scheme for H.264/AVC Video Encoder using Prognostic Early Mode Exclusion", IEEE/ACM Design Automation and Test in Europe Conference (DATE), pp. 1713-1718, 2010.

[41] M. Shafique, B. Vogel, J. Henkel, "Self-Adaptive Hybrid Dynamic Power Management Scheme for Many-Core Systems", IEEE/ACM Design Automation and Test in Europe Conference (DATE), 2013.

[42] M. Shafique, J. Henkel, "Agent-based distributed power management for kilo-core processors", International Conference on Computer-Aided Design (ICCAD), 2013.

[43] M. Shafique, L. Bauer, J. Henkel, "enBudget: A Run-Time Adaptive Predictive Energy-Budgeting Scheme for Energy-Aware Motion Estimation in H.264/MPEG-4 AVC Video Encoder", IEEE/ACM Design Automation and Test in Europe Conference (DATE), pp. 1725-1730, 2010.

[44] T.E. Carlson, W. Heirman, L. Eeckhout, "Sniper: Exploring the level of abstraction for scalable and accurate parallel multi-core simulation," in International Conference for High Performance Computing, Networking, Storage and Analysis, 2011.

[45] http://www.hpl.hp.com/research/mcpat/. McPAT simulator, HP labs, accessed on 29.08.2013.

[46] G. Bjontegaard, "Calculation of average PSNR differences between RD-curves", VCEG Contribution VCEG-M33, 2001.

Mapping Complex Algorithm into FPGA with High Level Synthesis

Reconfigurable chips with High Level Synthesis compared with CPU, GPGPU

Kazutoshi Wakabayashi   Takashi Takenaka   Hiroaki Inoue

Embedded System Solution Business Center   Green Platform Res. Labs

NEC Corp.

1753 Shimonumabe   Nakahara-ku, Kawasaki, 211-8666, Japan

Tel : +81-44-435-9416

Fax : +81-44-435-9491

e-mail : wakaba@bl.jp.nec.com   takenaka@aj.jp.nec.com   h-inoue@ce.jp.nec.com

Abstract - This presentation discusses on the comparison between "Reconfigurable Chip with High Level Synthesis" and "CPU, GPCPU with compiler such as CUDA" from the compiler perspective. Initially, we introduce several demands for acceleration with FPGA to achieve low latency calculation and control. As an application example, we show a High Frequency Trading. We accelerate it by FPGA NIC with C-based and SQL-based HLS, and show the necessity of high level language customizable reconfigurable chip. Then, we illustrate the difference of FPGA and processor (CPU, GPGPU) with the "FSM+Datapath" model and examine how the architecture difference affects delay and parallelism of operations. Next, we discuss parallelization of operations, threads with High Level Synthesis for FPGA and software compiler for processors. The main advantage of the former method is it is able to parallelize operations beyond control dependencies while the latter method has to obey control dependencies. Finally, some experimental results prove that "FPGA and HLS" generate better performance than a processor for control intensive algorithm.

I. Acceleration with FPGA and HLS

High Level Synthesis has been researched for more than twenty years. The authors have been developing a high level synthesis and high level verification tool suite called CyberWorkBench [1]. HLS become mature enough to design commercial chips for various area, such as computers, networking, digital appliance, printers, satellite. It has been used for LSI/FPGA design, and recently, it is used for a high level language compiler for FPGA to accelerate the high level language program, such as C, C++, Java, SQL which are originally running on the server. Currently, several FPGA NICs are used for accelerating stock trading as in Fig. 1, however FPGAs are designed in RTL language (Verilog HDL or VHDL) by a hardware designer. However, the micro level trading algorithm (strategy) should be changed every week, therefore the algorithm should be described in software programing language(e.g. C/C++). This easy customizability of "FPGA+HLS" is a reason why it is becoming more important for various areas which require low latency, such as Cyber-Physical-System, Complex Event Processing [2],[3].

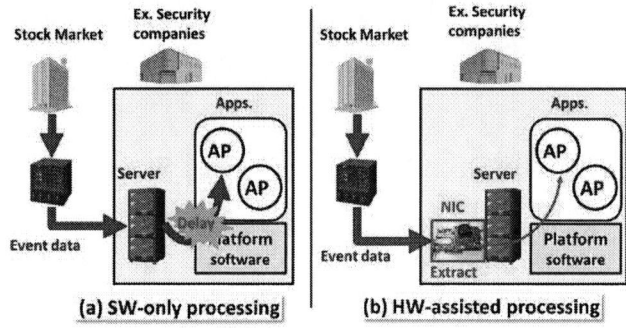

Fig. 1 Acceleration with FPGA-NIC for HFT

II. Architectural difference: FSM+Datapath model

This section discusses on the fundamental architectural difference of FPGA (Hardware) and Processor (Software) by using a "FSM+Datapath (FMSD)" model shown in Fig.2. A Program counter in a processor is modeled by a FSM which has only 2way branch (If-then branch) state transition. A controller for FPGA is a usual FSM which has an arbitrary number way of state transitions. Therefore, nested conditional branches or switch statement in a source code can be executed in one state (cycle) with FPGA, but multiple states (cycles) are necessary for a processor.

A Datapath of FPGA (Hardware) is flexible and its circuit's structure is tailored to a source code. GPGPU has many ALU's, and they are connected regularly, then SIMD operations (regular operations) can be executed parallel, but t not easy to use them parallel for control intensive (irregular operations). This flexibility of FPGA is attained by programmable wires, and the delay of it is relatively slower than fixed wires of a processor. Automatic Place and Route are tough for large circuit and takes a long computation period. A

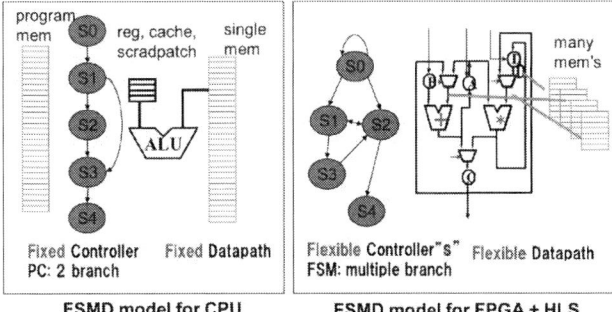

Fig. 2 FSMD models for a processor and FPGA

datapath of a processor is fixed, then the delay is much shorter than the delay of FPGA. GPGPU has several parallel ALU's for a SIMD operations, which works very well for a regular processing (straight line code) but does not work well for a control intensive code. Such flexibility difference of FSM and data path between FPGA and processor affects the degree of parallelization d of a compiler for FPGA and processor.

III. Parallelization for control dependencies and data dependencies

Traditionally, HW is used for regular processing and complex control is handled by a processor in a SoC. However, HLS allow a designer to use hardware even for control intensive application to take advantage of flexible FSM and datapath of FPAG.

A. Control Dependencies

As discussed in the previous section, control dependencies, such as conditional branches, jump, loops, are an obstacle to parallelize operations for a compiler for a processor. Even though CUDA for GPGPU is able to execute both IF-clause and Else-close together with many ALUs as a both branch speculation, it is not easy to parallelize operations in the different branches, because GPGPU have a limited number of branch unit. On the contrary, FPGA can generate arbitrary number of branching controls with random logics and parallelize multiple branches [4]. Fig.3 shows a simple example of scheduling for processor and FPGA. A source code in Fig.3 (a) contains two conditional branches. A usual software compiler for processor generates scheduling in Fig.3 (b) to keep the control dependencies. The latency is 4 or 3, 2 cycles to execute according to the conditionals. Fig.3(c) shows the scheduling of FPGA+HSL. The latency is always 2 cycles for the all conditionals. The four branch control signals control the execution of operations to keep the semantics of Fig.3 (a).

Also, FPGA can have multiple FSM, therefore several loops can be executed parallel and it can be exploited by HLS. This means that control dependencies is not an obstacle for parallelization with "FPGA+HLS", and only data dependency decides parallelism. This feature is a big advantage of FPGA+HLS compared to a processor.

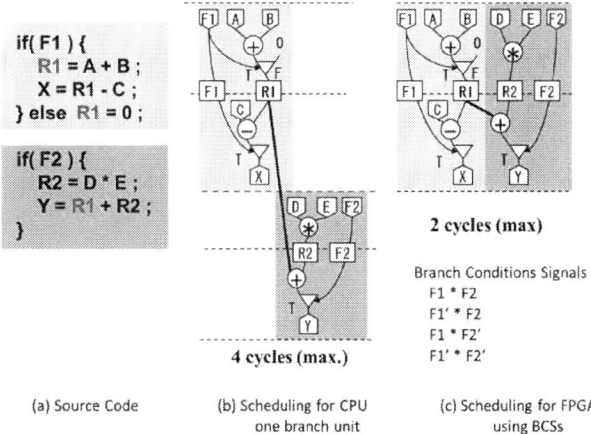

Fig. 3 Scheduling of multiple conditional branches

B. Data Dependencies

A compiler for a processor should hold the data dependency. For example, the compiler should not change the order of arithmetic operations to keep bit accuracy of operation (e.g. bit overflow). This is mainly because the datapath is fixed and the bit width of ALU is fixed (e.g 32, or 64). On the other hand, the datapath of FPGA is flexible and change the bit width of ALUs and HLS can adjust bit width of operation to avoid bit over flow. Fig.5 shows an example of Tree-Height-Reduction.

A loop in Fig.4 is unrolled and the Data Flow Graph (DFG) is scheduled as in Fig.4 (b) for a software compiler (e.g. C compiler). However, HLS is allowed to perform Tree Height Reduction because of the ability of automatic bit adjustment. For example, assume all input (a[0] through a[7]) be 8bit. The first stage additions require 8bit input and 9bit output adders and the second stage adders require 9bit input 10bit output, and the third stage addition requires 10bit input and 12 bit output to avoid bit overflow. HLS bind operations to appropriate bit width function unit, then it is allowed to transform DFG while keeping bit accuracy[1].

Also, FPGA can execute several sequential operations in a cycle (operation chaining), and has fast complex functional unit such as multiple input adders, combination of adder and multiplier with carry save adder logic. This sequential operation chaining makes the latency for FPGA smaller than that of processor, which does not execute operations sequentially in a cycle.

IV. Experimental Results

We compiled a graphical data processing program in C language in Fig.5 (a) into CPU, CPU+SIMD with C compiler and into Dynamic Reconfigurable Processor [5] (similar to FPGA) with HLS. Fig.5 (b) shows the comparison of processing period (latency x clock period) among them. Filter part is accelerated well by SIMD capability, however, decision

---

[1] This automatic bit adjustment is applicable for integer or fixed point arithmetic. It is not easy for floating point arithmetic to provide various bit width floating point units.

978-1-4799-2817-0/14 $31.00 © 2014 IEEE   283

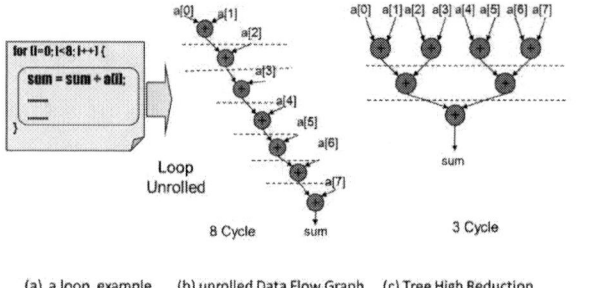

(a) a loop example   (b) unrolled Data Flow Graph   (c) Tree High Reduction
with "automatic bit adjusting"

Fig. 4 Tree height reduction with automatic bit width adjustment of HLS

or complex algorithm part cannot mapped to SIMD and they are still executed in CPU. Such complicated parts are compiled and mapped to DRP,FPGA with HSL. Then, the resulting processing period is 10 times shorter than CPU,CPU+SIMD.

V. Summary

This paper discuss on the performance period between processor and FPGA, when complex algorithm in C/C++/Java compiled by software C compiler or C-based HLS from the compiler perspective. Since FPGA has a flexible FSM controller and flexible datapath, HLS, or C compiler for FPGA has much more freedom to parallelize operations. Especially, it can parallelize operations beyond control dependencies, and somehow change even data dependencies, both of which are not easy for C compiler for a processor which has Program Counter and fixed datapath. One graphic example shows that FPGA+HLS allows shorter processing period. This method has been used as a hardware design tool, and also begun to be used for acceleration of servers in various area, such as financial trading, factory automations, database, data center, packet processing.

Acknowledgments

This work is partially supported by Dependable VLSI in CREST, JST. The authors thank all colleagues in CyberWorkBench research and development team.

Reference

[1] K. Wakabayashi and B. Carrion Schafer, "All-in-C" SoC Synthesis and Verification with CyberWorkBench., Springer, P. Coussy and A. Morawiec (Eds.), "High-Level Synthesis from Algorithm to Digital Circuit", XVI, ISBN: 978-1-4020-8587-1, Chapter 7, pp.113-127, 2008.

[2] Inoue, H., Takenaka, T., and Motomura, M. "C-Based Complex Event Processing on Reconfigurable Hardware," IEEE Transactions on Very Large Scale Integrated Systems (TVLSI), Vol.21, Issue.5, pp.971-974 (May 2013).

[3] Takenaka, T., Takagi, M., and Inoue, H. "A Scalable Complex Event Processing Framework For Combination of SQL-based Continuous Queries and C/C++ Functions," IEEE International Conference on Field Programmable Logic and Applications (FPL), pp.237-242 (August 2012).

[4] Kazutoshi Wakabayashi, "Unified Representation for Speculative Scheduling: General Condition Vector," IEICE Trans. Fundamentals of Electronics, Communications and Computer Sciences, vol. E89-A, No. 12, pp. 3408-3415 2006.

[5] T.Toi, N.Nakamura, Y.Kato, T.Awashima, K.Wakabayashi, L. Jing, "High-level synthesis challenges and solutions for a dynamically reconfigurable processor," Proc. of ICCAD, PP.702-708, 2006

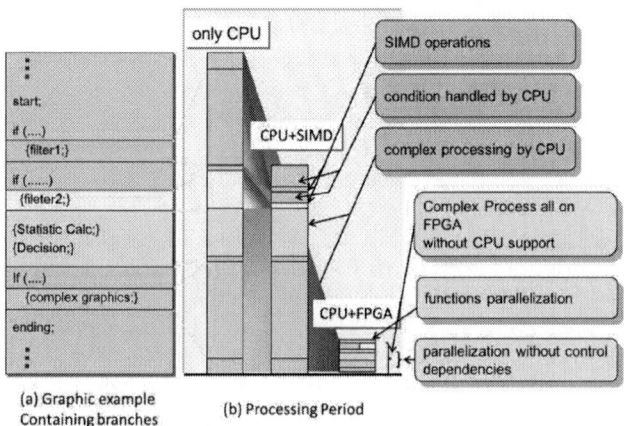

(a) Graphic example
Containing branches   (b) Processing Period

Fig. 5 Comparison of processing period

Leveraging Parallelism in the Presence of Control Flow on CGRAs

Jihyun Ryoo, Kyuseung Han, and Kiyoung Choi

Department of Electrical and Computer Engineering
Seoul National University, Republic of Korea
Tel: +82-2-880-1787   Fax: +82-2-882-4656
{jhshine, darprin, kchoi}@snu.ac.kr

Abstract— Coarse-Grained Reconfigurable Architectures (CGRAs) are suitable for accelerating data-intensive applications in embedded systems due to high performance and power efficiency. However, as application programs become complex having more control flows in them, it becomes harder to accelerate such programs on CGRAs. Previous researches on this issue have focused on correct execution of control flows rather than their acceleration. This paper reveals how control flows degrade the performance of programs and proposes a software approaches to accelerating control flows by exploiting parallelism residing in each conditionals as well as among conditionals. Experiments show that our proposed techniques improve performance by 2.51 times on average.

I. INTRODUCTION

A Coarse-Grained Reconfigurable Architecture (CGRA) is a viable solution for embedded systems because it can provide both high performance and power efficiency at the same time compared to full-blown instruction-set processors. A typical CGRA consists of tens of Processing Elements (PEs) and thus is quite suitable for highly parallelizable applications. Moreover, it has high power efficiency due to simplified control structures of PEs. By scheduling configuration (or instruction) fetching and data communication statically at compile time, it enables a single controller to orchestrate all PEs, thus minimizes power consumption not directly used for computation. Due to these benefits, there have been many researches proposing various kinds of CGRAs [1–10] and surveys on them [11, 12].

However, the simplified control structures make a CGRA vulnerable to control flows. Multiple PEs in a CGRA are controlled by a single, centralized control unit, and thus, the PEs should take the same branch (either if- or else-path) at each conditional branching point. This limitation prevents threads with *conditionals* (a *conditional* includes a conditional statement as well as the corresponding branches) from being parallelized on multiple PEs.

To overcome this limitation and exploit more parallelism, predicated execution technique (also called predication in short or guarded execution technique) [13] is adopted by parallel processors including CGRAs [14–20]. However, previous researches have just focused on the correct execution of control flows rather than their acceleration. In other words, control flows are just considered as an obstacle to exploiting other parallelism. In this paper, we reveal how control flows degrade the performance of programs and propose software approaches to accelerating control flows by exploiting parallelism among conditionals as well as within each conditional.

II. COARSE-GRAINED RECONFIGURABLE ARCHITECTURE

A. Architecture

A CGRA is generally composed of tens of PEs, which are tightly connected with each other in the form of a 1D or 2D array. A PE is equipped with functional units such as ALUs, shifters, and/or multipliers, and control signals reconfigure them at runtime to change the functionality of the PE itself as well as the interconnections among PEs. Although the concept of reconfiguration is similar to FPGA, its configuration has word-level granularity rather than bit-level one as in FPGA, thereby reducing the amount of configuration information and the time overhead of reconfiguration.

In addition to high performance and low power consumption, CGRAs also provide flexibility through the reconfigurability of PEs and their connections, which enables changing their functionality dynamically according to the change of applications. This feature makes CGRAs have the scalability in the number of PEs despite of the use of a single controller.

B. Mapping

To run an application on a CGRA, it is needed to generate configurations of the CGRA corresponding to the required functionality. The process of generating the configurations is called application mapping or just mapping in short. Since it is in general neither possible nor beneficial to execute the entire application on a CGRA, only data-intensive parts called kernels that have large parallelism are mapped to the CGRA. A typical kernel is a loop with multiple iterations. The iterations have no dependency between them and thus can be executed in parallel or have some dependency but can be pipelined with a short initiation interval.

The functionality of a kernel is generally expressed as a CDFG (Control Data Flow Graph) and the mapping is a process of generating configurations of the CGRA such that the PEs perform the operations in the CDFG while communications through the interconnections of PEs handle the data dependency among the operations. Specifically the process determines three components: when the operations are executed (scheduling),

978-1-4799-2817-0/14 $31.00 © 2014 IEEE   285

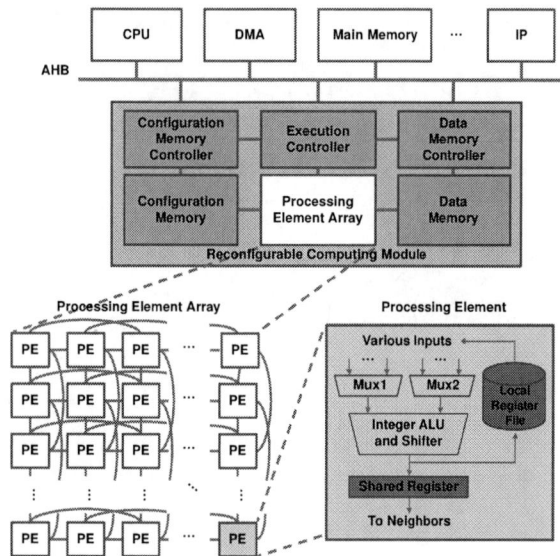

Fig. 1. The target architecture FloRA.

which PE performs the operations (binding), and which register stores the data (register allocation).

The CGRA mapping seems similar to compilation of conventional processors or High-Level Synthesis (HLS) in that they all need to determine the three components, but the internal framework is quite different. Since PEs on a CGRA are connected with each other with limited interconnect resources and use distributed register files which are not visible to other PEs, the mapping should consider routing for data communication. Because routing affects the three components, they should be considered at the same time when we map operations onto a CGRA.

C. Target CGRA

Our target architecture is a CGRA called FloRA [9, 21]. It has been developed mainly for multimedia applications such as video CODEC (MPEG4, H.264) and 3D graphics, which have instruction-level parallelism and/or data-level parallelism. Compared to other CGRAs, FloRA has merits of efficient resource sharing, low power configuration, and efficient floating-point operation supports. It has been implemented on a chip and its functionality and performance have already been verified [21, 22].

The overall architecture is shown in Fig. 1. It consists of four main components: the PE array, configuration memory, data memory, and a set of controllers. The PE array has 8×8 PEs, each of which is comprised of an integer ALU, a shifter, and a local register file and can be dynamically reconfigured every cycle if needed. The configuration memory contains configuration information used by the PE array and the data memory stores the input/output data used/generated by the PE array. It is accessible from the outside of the reconfigurable computing module through the bus, which enables host processors to provide data for the CGRA. The execution controller orchestrates all operations of these components.

III. CORRECT EXECUTION OF CONTROL FLOW

A. Limitation in Handling Control Flows

As mentioned in Section I, CGRAs have a limitation in handling control flows. This limitation prevents a program requiring different controls on different PEs from being parallelized. This problem is not actually limited to CGRAs but exists in other well-known parallel architectures such as VLIW or SIMD machines [14–19, 23].

B. Unique Solution: Predication

Predication [14–20] is the unique known solution for correct execution of control flows on CGRAs. It is a technique that converts control flows to data flows by modifying both architectures and compilers so that all instructions are fetched but executed selectively without branching. Since each PE has its own control of selecting instructions to be executed, the limitation of using a single controller can be overcome.

[14] classified predication techniques into partial (PARTIAL) and full predication. Recently, we proposed to further classify the full predication techniques into two groups: condition-based full predication (CONDFULL) and state-based full predication (STATEFULL) [20, 24]. This paper focuses on full predication, which does not need to execute unnecessary paths.

Fig. 2 shows an example of CONDFULL and STATEFULL. Fig. 2a is a simple example of a loop with an *if* structure written in the C language and Fig. 2b shows the assembly code of the loop body obtained by compiling the C code using CONDFULL. To support the predicated execution in Fig. 2b, each PE has a 1-bit *predicate* register (RP in Fig. 2b) for storing a boolean variable (true or false, a *predicate*), which is set by cmp instructions. Also, instructions are extended to hold an additional operand called *condition* (expressions in parentheses in Fig. 2b). The relation between the *predicate* register in a PE and the *condition* operand of each instruction determines whether the instruction is executed or not (*predicated*). The expressions (uc), (RP), and (!RP) used in Fig. 2b represent "unconditionally", "when RP is true", and "when RP is false", respectively. For example, if the value of R0 is zero, the cmp instruction in line 1 with ge (greater than or equal to) sets RP to zero (false). Then the instructions in line 2 and 3 are nullified and those in line 4 and 5 are executed.

STATEFULL inserts extra sleep instructions instead of making all instructions have the *condition* operand as shown in Fig. 2c. A sleep instruction makes a PE enter into a sleep mode for the cycles specified in the instruction. For example, if line 2 in Fig. 2c is activated, the PE will be in the sleep mode for three cycles so that line 3, 4, and 5 will be ignored. It is implemented by adding a state register and a sleep counter into each PE. STATEFULL can save power compared to CONDFULL since it does not lengthen instruction word and most parts of a PE can be turned off on untaken paths to avoid unnecessary bit switching. The readers are referred to [24] for more details of STATEFULL.

Several existing approaches adopted the predication techniques to CGRAs. [25] showed how to implement PARTIAL efficiently, and [26] maximized performance by using PARTIAL in a speculative way. [27] presented an automatic mapping

978-1-4799-2817-0/14 $31.00 © 2014 IEEE

```
1  for(i=0; i<8; i++){
2     if (c[i] >= 1){
3        x = x+1;
4        y = y+1;
5     }
6     else{
7        x = x-1;
8        y = y-1;
9     }
10 }
```

(a) C code

```
1  (uc)   cmp  ge RP R0 #1
2  (RP)   add     R1 R1 #1
3  (RP)   add     R2 R2 #1
4  (!RP)  sub     R1 R1 #1
5  (!RP)  sub     R2 R2 #1
```

(b) Assembly code using CONDFULL

```
1  cmp    ge RP R0 #1
2  sleep     (!RP) #3
3  add       R1 R1 #1
4  add       R2 R2 #1
5  sleep     (uc)  #2
6  sub       R1 R1 #1
7  sub       R2 R2 #1
```

(c) Assembly code using STATEFULL

Fig. 2. An example C code and its assembly code using predications.

framework integrating the method proposed in [26]. On the other hand, [23] proposed an automatic mapping to accelerate control-intensive kernels by using CONDFULL, and [20, 24] revealed that STATEFULL has merits in power consumption compared to CONDFULL.

IV. DIFFICULTIES IN HANDLING CONTROL FLOW WITH DISTRIBUTED REGISTER FILES

Many CGRAs including FloRA use distributed register files [1, 3, 5, 7–10]. It is natural in that centralized architecture generally has the scalability problem (register access latency increases as the size grows). Each PE has its own local register file, which is not visible to other PEs.[1] Not only (general) register files but also *predicate* registers are distributed.

However, the distributed architecture makes it difficult to map control flows. First, there can be heavy communications between PEs due to routing of *predicate* information, thereby degrading performance. For example, if a compare operation is executed on PE0 but the resulting *predicate* value is used for a control flow in PE1, then the value should be routed from PE0 to PE1. Moreover, since each PE generally has only a few *predicate* registers, the transfer of a *predicate* value to a PE may cause spilling of the *predicate* register (into a general register) if the existing value is alive. That happens frequently especially, when there is only one *predicate* register per PE.[2] Moreover, given such an architecture, if two parallelizable conditionals are

---

[1]For data communication, each PE has a few special registers or network logic accessible by other PE.

[2]Having only one *predicate* register helps simplify instruction encoding since cmp instruction does not need to specify the target register. In addition,

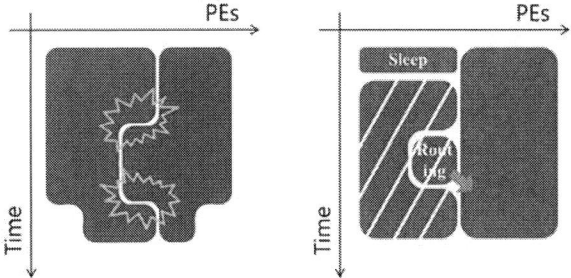

(a) Two conditionals are interleaved in their schedule on the same PE

(b) Routing is impossible due to sleep mode when STATEFULL is used

Fig. 3. Mapping problems with distributed register files.

mapped to the same PE and interleaved in their schedule like Fig. 3a, spilling will occur many times, resulting in performance degradation. In case of STATEFULL, the penalty is more serious because it requires extra instructions not only for spilling but also for updating state. Thus, it will be beneficial to allocate different conditionals to different PEs.

Secondly, if STATEFULL is adopted, a more critical issue arises from the fact that power-saving (sleep) mode of a PE renders almost all the resources of the PE including the local register file disabled. If a PE in a power-saving mode has a scalar variable stored in its register file, and the variable is needed by another PE, on-demand routing of the variable is impossible as shown in Fig. 3b. We must route the variable in advance before the PE goes into the power-saving mode. Or, the routing must be processed after exiting the power-saving mode, in which case, performance will be greatly degraded. Even worse, if another routing is required in the opposite direction (e.g., the PE in sleep mode is scheduled to wake up when the data from the other PE is available) at the same time, then the execution can go into deadlock and will be failed.

V. PROPOSED FRAMEWORK

To cope with the problems revealed in the previous section, we extend the framework proposed in [28]. In our framework, we map operations from different conditionals to different sets of PEs separated either in time or space in order to avoid wasteful instruction overhead for updating *predicate* registers (note that the two sets of operations in Fig. 3 have overlap in both schedule time and set of allocated PEs). Also we duplicate operations for calculating predicates and map them to different PEs to reduce the overhead of routing *predicate* values. The duplication of operations has also been adopted in [29]. However, it does not guarantee performance improvement since it is a CDFG-level approach. On the other hand, ours is a mapping-level approach that performs the duplication only when it improves performance.

Fig. 4 illustrates the overall flow of our framework for mapping loops with control flows on CGRAs with distributed register files. It starts from an Intermediate Representation (IR), which can be obtained easily by frontend tools. First, we convert the IR to a Control Data Flow Graph (CDFG), extracting

---

in the case of CONDFULL, the length of the *condition* operand can be reduced since it does not need to specify which *predicate* register is used.

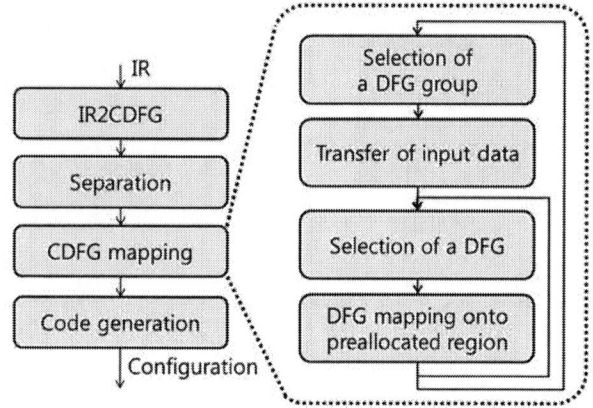

Fig. 4. The mapping framework for CGRAs with distributed register files.

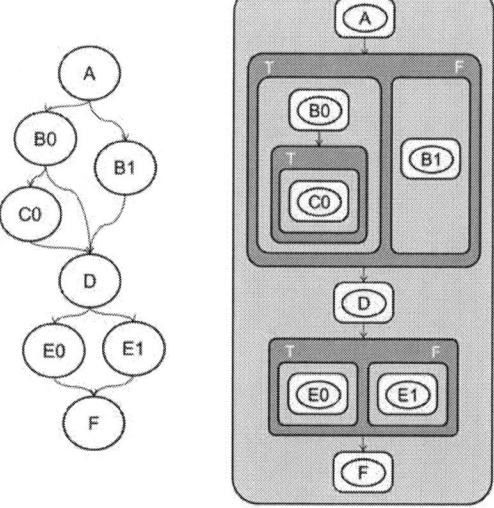

(a) A loop body represented as a CFG of DFGs

(b) Identifying conditionals (i.e., fork-join structures)

parallelism on the way. Then the next step takes the CDFG and allocates PEs to different parts of the CDFG so that each part (each DFG) can be mapped separately in a temporal or a spatial manner. Next, actual mapping of the CDFG is performed using the information generated from the previous step. Lastly, configuration codes are generated from the mapping results. We explain the steps in the following subsections except for the code generation step, which is straightforward.

A. From IR to CDFG

The IR for a loop body is generally given as a CFG, where each node is a DFG that represents a basic block. Fig. 5a illustrates the CFG of a loop body, which contains one *nested-if* construct followed by a simple *if-else*. However, such a flat structure cannot express how large each conditional is and how much parallelism exists between them. Thus, we first transform the CFG to our hierarchical CDFG representation so that the control structure and parallelism can be captured more explicitly. The hierarchical CDFG is defined as follows. Each node of the CDFG is a block of either of two types: *unipath* and *multipath*. A *unipath* block is simply a DFG, whereas a *multipath* block contains two CDFGs with a *condition* for each CDFG. Fig. 5b illustrates the CDFG corresponding to the IR in Fig. 5a. In the figure, ovals represents DFGs and round boxes colored white, dark gray, and light gray indicate *unipath* blocks, *multipath* blocks, and, and CDFGs, respectively. Directed edges indicate data dependency between two blocks. Note that the edges in Fig. 5b are obtained through data flow analysis and are different from those in Fig. 5a.

To transform an IR to a CDFG representation, we first identify conditionals to generate CDFGs at lower levels of hierarchy (see Fig. 5b). Since *multipath* blocks can contain other *multipath* blocks in such lower level CDFGs, *nested-if* structures can be naturally represented. We then update data dependency among blocks, which may reveal parallelism between blocks. In Fig. 5b, if DFGs D, $E0$, and $E1$ are not dependent on $B0$, $B1$, or $C0$, we can remove the dependency edge between them, making D an immediate successor of A as illustrated in Fig. 5c.

Lastly, we separate compare operations and their predecessors from original DFGs, use them to form new DFGs (called *fork DFGs*), and group them with *multipath* blocks that use the

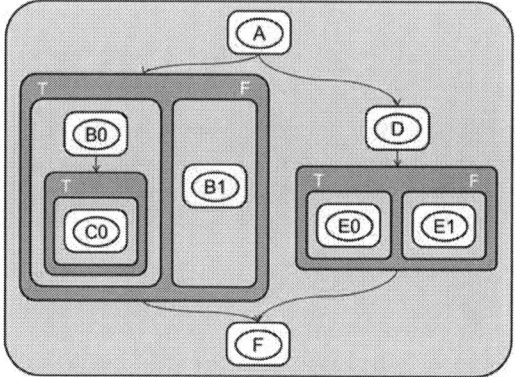

(c) Exploiting parallelism

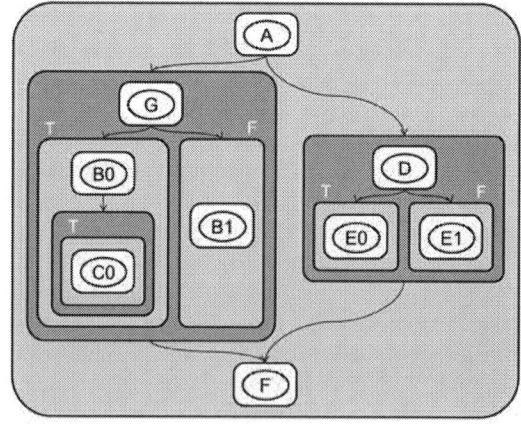

(d) Separating compare operations and their precedents

Fig. 5. Conversion process from IR to CDFG.

978-1-4799-2817-0/14 $31.00 © 2014 IEEE   288

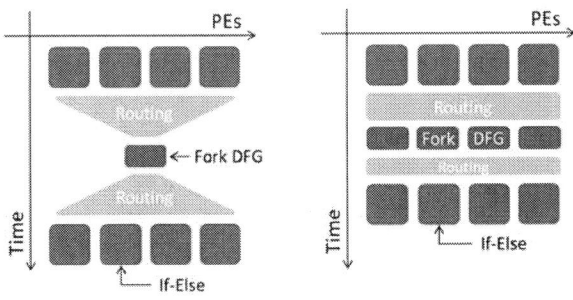

(a) Conventional mapping   (b) Proposed mapping which duplicates operations in fork DFGs

Fig. 6. Motivation for duplication of operations in fork DFGs.

results of the compare operations. This is to reduce the routing overhead of predicates. For instance, consider that DFG A in Fig. 5c has operations that create predicates used for the *multipath* block containing $B0$, $B1$, and $C0$. Thus we separate those operations out of A and use them to form a *fork DFG* G as shown in Fig. 5d. Also, assuming that the *predicate* for *multipath* block that holds $E0$ and $E1$ is generated by DFG D, we group it with the *multipath* block.

Fig. 6 illustrates how we can reduce routing overhead by duplicating such fork DFGs. Fig. 6a is the case of conventional mapping, where the *fork DFG* contains a compare operation and its predecessors. As shown in the figure, the compare operation can become a bottleneck since it needs to gather input data for the compare operation and route the resulting *predicate* to PEs executing *if* or *else* DFGs. Thus, centralized calculation of predicates can incur performance degradation. However, the routing overhead can be reduced by duplicating the *fork DFG* as shown in Fig. 6b.

B. Separation

To ensure correctness and maximize performance, we map operations from different conditionals separately either in time or space, which can be achieved by DFG grouping and PE-to-DFG allocation. We group DFGs into clusters such that the DFGs in a cluster can run in parallel (running in parallel means that the DFGs are not separated in time). Within a cluster, we allocate disjoint set of PEs to different DFGs to ensure spatial separation. The clusters are scheduled one after another (temporal separation) in the order of their generation considering data dependency (a DFG can be put into a cluster only when all predecessor DFGs are already in ohter clusters). When grouping DFGs, we need to consider two aspects. By putting many DFGs into one cluster, we can increase parallelism especially when each DFG has low instruction-level parallelism. However, if there are too many DFGs in a cluster, registers can be spilled, resulting in performance degradation. Thus our strategy is to allocate enough number of PEs to each DFG to avoid register spills and then group DFGs as long as PEs are available (refer to [28] for the details). The results of this step include mapping sequence of cluster, components (DFGs) of each group, allocated PEs per DFG, and allocated number of registers per DFG.

C. CDFG Mapping

The CDFG mapping step is divided into sub-steps as shown in Fig. 4. First, after selecting a cluster, for each DFG in the cluster, we route input data used by the DFG to the PEs allocated to that DFG if they are not already in the PEs, which prevents the problem depicted in Fig. 3b. After that, we move data that are in the PEs but are not used by the corresponding DFG to other PEs, if the number of registers available in the allocated PEs is less than the requirement. Then we map each DFG one by one onto the allocated PEs. In the case of STATEFULL, a state-update operation (or sleep instruction) is inserted at the entry of each conditionally executed DFG. For the DFG mapping itself, conventional algorithms can be easily integrated into our framework (e.g., [29–32]). In our experiments, we used a variant of list-scheduling-based mapping algorithm that performs scheduling, operation-to-PE binding, and register binding, all at once. It is similar to the approach in [30] except that we do not adopt modulo scheduling.

When mapping *multipath* blocks, we allow operations in *fork DFGs* to be mapped multiple times, which is different from the mapping of other parts. We duplicate the mapping only when it reduces the overhead of routing and thus improves the performance. In the example of Fig. 6b, the routing below the *fork DFG* can be eliminated by the duplication.

VI. EXPERIMENTS

A. Implementation

We use the Clang compiler [33] as a frontend tool in order to convert C code to IR. For the mapping of each DFG, we use integer linear programming (ILP) to model the problem of scheduling, binding, and register allocation in a flexible way. The ILP formulation makes it easy to model specific features of the target architecture and add/remove various constraints as needed. Also, we can optimize not only execution time but also other objectives such as the number of routing operations or that of register writes at the same time. Its modeling is similar as the one described in [34], but we additionally model lower level components like flag registers and multiplexers to reflect the characteristics of FloRA used by the proposed approach.

However, the increased complexity of the modeling prohibits us from mapping all operations together, and thus we mapped the operations one-by-one except those in *fork DFGs*. In case of *fork DFGs*, we need to map several operations together to see whether duplication gives a benefit or not. Since the mapping of *fork DFGs* could increase the complexity too much, we limit the size of each *fork DFG* to three operations when making a *fork DFG*, which corresponds to the process from Fig. 5c to Fig. 5d.

We implement the duplication of operations by leveraging the flexibility of ILP modeling. Only we have to do is the removal of the constraints on the number of mappings for the operations from the formulation. Moreover, we allow different number of duplications for different operations in a *fork DFG* so that the ILP solver could find optimal number of duplications for each operation to achieve the best performance. For the ILP solver, we use Gurobi optimizer 5.0.2 [35].

Our mapping framework is developed for STATEFULL since it has a merit in power consumption [20, 24]. However, the framework can be easily extended for other predication methods.

B. Applications for Experiments

We experiment with five applications:

- DCT 8x8 (*dct*): This is a widely used application that expresses data as the sum of cosine functions with different frequencies. We use the one in CUDA example.

- getANMS (*anms*): This is an application from the image stitch benchmark in the San Diego Vision Benchmark Suite (SD-VBS).

- Chromakey (*chromakey*): This is an application that replaces the concolorous background of an image to another image.

- SECDED decoding (*secded*): This is a widely used error correcting technique used in telecommunication. We use [8,4] Hamming code.

- Deblocking filter (*deblocking*): This is an application from H.264 video decoder. It operates at the boundary of transform blocks (macroblocks) and smoothens the edges.

In some of the kernels, we obtain parallelism among conditionals by unrolling loops that had control flows. However, some applications already have parallelizable conditionals inside.

C. Experimental Results

We compare the result of conventional serial mapping (SERIAL) of control flows with that of the parallel mapping (PARALLEL) and that obtained when multiple copies are allowed for *fork DFGs* (PARALLELMULTI). Fig. 7 shows relative performance normalized to SERIAL. For all five applications SERIAL has the lowest performance. It is quite clear since SERIAL treats control flows as if they have dependency between them. As a result, PARALLEL accelerates the performance by 2.37x on average.

By duplicating *fork DFGs*, PARALLELMULTI has 5.7% improvement on average compared to PARALLEL. In *dct* and *anms*, the conditionals are so small that each conditional is performed on one or two PEs, and thus there is no routing overhead that can be saved by duplicating *fork DFGs*. On the other hand, *chromakey*, *secded*, and *deblocking* are improved by about 9.7% on average since they have relatively large conditionals. Consequently, PARALLELMULTI have 2.51x speedup on average compared to SERIAL.

VII. CONCLUSION

In this paper, we present approaches to accelerating control flows on CGRAs. In particular, we present a new mapping framework that handles control flows efficiently on CGRAs having limited access to distributed register files. To solve the problem, we propose to duplicate operations and map DFGs

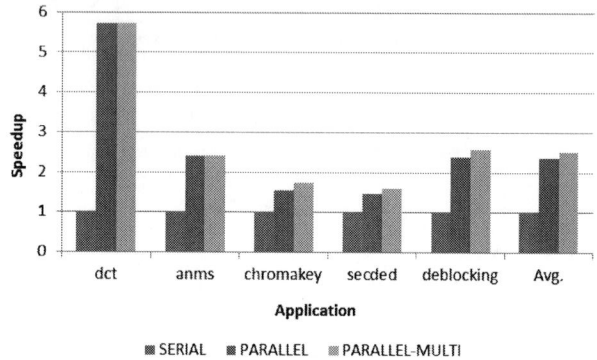

Fig. 7. Comparison of execution time. We calculate the speedup ratio by normalizing values using those of SERIAL.

with temporal or spatial separation. The framework uses a new CDFG structure and separates handling of control flows and data flows so that conventional mapping algorithms can be easily integrated into it. As a result, the framework achieves 2.51x performance improvement on average over conventional serial mapping.

ACKNOWLEDGMENTS

This work was supported by the National Research Foundation of Korea (NRF) grant funded by the Korea government (MEST) (No. 2012R1A2A2A0647297).

REFERENCES

[1] D. Chen and J. Rabaey, "A reconfigurable multiprocessor IC for rapid prototyping of algorithmic-specific high speed DSP data paths," *IEEE Journal of Solid-State Circuits*, vol. 27, pp. 1895–1904, Dec. 1992.

[2] C. Ebeling, D. C. Cronquist, and P. Franklin, "Configurable computing: the catalyst for high-performance architectures," in *Proceedings of the IEEE International Conference on Application-Specific Systems, Architectures and Processors*, 1997.

[3] T. Miyamori and K. Olukotun, "REMARC: reconfigurable multimedia array coprocessor," *IEICE Transactions on Information and Systems*, pp. 389–397, 1998.

[4] S. C. Goldstein, H. Schmit, M. Moe, M. Budiuy, S. Cadambi, R. R. Taylor, and R. Laufer, "PipeRench: a coprocessor for streaming multimedia acceleration," in *Proceedings of the International Symposium on Computer Architecture*, 1999.

[5] H. Singh, M. Lee, G. Lu, F. J. Kurdahi, N. Bagherzadeh, and M. C. Filho, "MorphoSys: an integrated reconfigurable system for data-parallel and computation-intensive applications," *IEEE Transactions on Computers*, vol. 49, pp. 465–481, May 2000.

[6] V. Baumgarte, G. Ehlers, F. May, A. Nückel, M. Vorbach, and M. Weinhardt, "PACT XPP–A self-reconfigurable data processing architecture," *Journal of Supercomputing*, vol. 26, pp. 167–184, Sept. 2003.

[7] B. Mei, S. Vernalde, D. Verkest, H. D. Man, and R. Lauwereins, "ADRES: an architecture with tightly coupled VLIW processor and coarse-grained reconfigurable matrix," in *Proceedings of the International Conference on Field Programmable Logic and Application*, 2003.

[8] F. Garzia, W. Hussain, and J. Nurmi, "CREMA: a coarse-grain reconfigurable array with mapping adaptiveness," in *Proceedings of the International Conference on Field Programmable Logic and Applications*, 2009.

[9] Y. Kim, M. Kiemb, C. Park, J. Jung, and K. Choi, "Resource sharing and pipelining in coarse-grained reconfigurable architecture for domain-specific optimization," in *Proceedings of the Design, Automation and Test in Europe Conference and Exhibition*, 2005.

[10] Y. Saito, T. Sano, M. Kato, V. Tunbunheng, Y. Yasuda, M. Kimura, and H. Amano, "MuCCRA-3: a low power dynamically reconfigurable processor array," in *Proceedings of the Asia and South Pacific Design Automation Conference*, 2010.

[11] R. Hartenstein, "A decade of reconfigurable computing: a visionary retrospective," in *Proceedings of the Design, Automation and Test in Europe Conference and Exhibition*, 2001.

[12] K. Choi, "Coarse-grained reconfigurable array: architecture and application mapping," *IPSJ Transactions on System LSI Design Methodology*, vol. 4, pp. 31–46, Feb. 2011.

[13] J. R. Allen, K. Kennedy, C. Porterfield, and J. Warren, "Conversion of control dependence to data dependence," in *Proceedings of the ACM SIGACT-SIGPLAN Symposium on Principles of Programming Languages*, 1983.

[14] S. A. Mahlke, R. E. Hank, J. E. McCormick, D. I. August, and W. W. Hwu, "A comparison of full and partial predicated execution support for ILP processors," in *Proceedings of the International Symposium on Computer Architecture*, 1995.

[15] M. L. Anido, A. Paar, and N. Bagherzadeh, "Improving the operation autonomy of SIMD processing elements by using guarded instructions and pseudo branches," in *Proceedings of the Euromicro Symposium on Digital System Design*, 2002.

[16] P. Dang, "An efficient implementation of in-loop deblocking filters for H.264 using VLIW architecture and predication," in *International Conference on Consumer Electronics Digest of Technical Papers*, 2005.

[17] J. Shin, M. Hall, and J. Chame, "Superword-level parallelism in the presence of control flow," in *Proceedings of the International Symposium on Code Generation and Optimization*, 2005.

[18] L. Huang, L. Shen, S. Ma, N. Xiao, and Z. Wang, "DM-SIMD: a new SIMD predication mechanism for exploiting superword level parallelism," in *Proceedings of the International Conference on ASIC*, 2009.

[19] K. Han, J. K. Paek, and K. Choi, "Acceleration of control flow on CGRA using advanced predicated execution," in *Proceedings of the International Conference on Field-Programmable Technology*, 2010.

[20] K. Han, S. Park, and K. Choi, "State-based full predication for low power coarse-grained reconfigurable architecture," in *Proceedings of the Design, Automation and Test in Europe Conference and Exhibition*, 2012.

[21] D. Lee, M. Jo, K. Han, and K. Choi, "FloRA: coarse-grained reconfigurable architecture with floating-point operation capability," in *Proceedings of the International Conference on Field-Programmable Technology*, 2009.

[22] M. Jo, D. Lee, K. Han, and K. Choi, "Design of a coarse-grained reconfigurable architecture with floating-point support and comparative study," *Integration, the VLSI Journal*, 2013, preprint.

[23] C. Arbelo, A. Kanstein, S. Lopez, J. Lopez, M. Berekovic, R. Sarmiento, and J. Y. Mignolet, "Mapping control-intensive video kernels onto a coarse-grain reconfigurable architecture: the H.264/AVC deblocking filter," in *Proceedings of the Design, Automation and Test in Europe Conference and Exhibition*, 2007.

[24] K. Han, J. Ahn, and K. Choi, "Power-efficient predication techniques for acceleration of control flow execution on cgra," *ACM Trans. Arch. and Code Opt.*, vol. 10, pp. 8:1–8:25, May 2013.

[25] J. Lee, Y. Kim, J. Jung, S. Kang, and K. Choi, "Reconfigurable ALU array architecture with conditional execution," in *Proceedings of the international SoC Design Conference*, 2004.

[26] K. Chang and K. Choi, "Mapping control intensive kernels onto coarse-grained reconfigurable array architecture," in *Proceedings of the International SoC Design Conference*, 2008.

[27] G. Lee, K. Chang, and K. Choi, "Automatic mapping of control-intensive kernels onto coarse-grained reconfigurable array architecture with speculative execution," in *Proceedings of the IEEE International Symposium on Parallel & Distributed Processing, Workshops and PhD Forum*, 2010.

[28] K. Han, K. Choi, and J. Lee, "Compiling control-intensive loops for CGRAs with state-based full predication," in *Proceedings of the Design, Automation and Test in Europe Conference and Exhibition*, 2013.

[29] M. Hamzeh, A. Shrivastava, and S. Vrudhula, "EPIMap: using epimorphism to map applications on CGRAs," in *Proceedings of the Design Automation Conference*, 2012.

[30] B. Mei, S. Vemaldet, D. Verkestt, H. D. Man, and R. Lauwereins, "DRESC: a retargetable compiler for coarse-grained reconfigurable architectures," in *Proceedings of the International Conference on Field-Programmable Technology*, 2002.

[31] T. Toi, N. Nakamura, Y. Kato, T. Awashima, and K. Wakabayashi, "High-level synthesis challenges and solutions for a dynamically reconfigurable processor," in *Proceedings of the International Conference on Computer-Aided Design*, 2008.

[32] J. W. Yoon, A. Shrivastava, S. Park, M. Ahn, and Y. Paek, "A graph drawing based spatial mapping algorithm for coarse-grained reconfigurable architectures," *IEEE Transactions on Very Large Scale Integration Systems*, vol. 17, pp. 1565–1578, Nov. 2009.

[33] http://clang.llvm.org/.

[34] G. Lee, K. Choi, and N. Dutt, "Mapping multi-domain applications onto coarse-grained reconfigurable architectures," *IEEE Transactions on Computer-Aided Design of Integrated Circuits and Systems*, vol. 30, pp. 637–650, May 2011.

[35] http://www.gurobi.com/.

978-1-4799-2817-0/14 $31.00 © 2014 IEEE

Physical-Aware Task Migration Algorithm for Dynamic Thermal Management of SMT Multi-core Processors

Bagher Salami †, Mohammadreza Baharani §, Hamid Noori †, Farhad Mehdipour ‡

† School of Engineering, Ferdowsi University of Mashhad, Mashhad, Iran

§ School of Electrical and Computer Engineering, University of Tehran, Tehran, Iran

‡ E-JUST Center, Graduate School of Information Science and Electrical Eng., Kyushu University, Fukuoka, Japan

Email: bagher.salami@stu-mail.um.ac.ir, m.baharani@ut.ac.ir, hnoori@um.ac.ir, farhad@ejust.kyushu-u.ac.jp

Abstract - This paper presents a task migration algorithm for dynamic thermal management of Simultaneous Multi-Threading (SMT) multi-core processors. The unique features of this algorithm include: 1) considering SMT capability of processors for dynamic thermal management via task scheduling, 2) using adaptive task migration threshold, and 3) considering cores physical features. This algorithm is evaluated on a commercial SMT quad-core processor. The experimental results indicate that our technique can significantly decrease the average and peak temperature compared to Linux standard scheduler, and two well-known thermal management techniques.

I. Introduction

As feature size is shrinking, the ability to have processors with larger number of cores is increasing. By the advent of Simultaneous Multi-Threading (SMT), the multi-core processors can exploit more thread-level parallelism by less hardware compared to non-SMT multi-core processors. SMT multi-cores are becoming the main trend in the new generations of processors. However, due to the increased density and complexity of these processors, the SMT multi-cores power consumption is increasing. The high power consumed in a small area die size results in increasing power density and generated temperature. Therefore, expensive processor packaging and cooling equipment are needed to remove hot spots. Moreover, increasing temperature potentially threatens system reliability, decreases both transistor age and transition speed and increases leakage current [1]. Therefore, thermal management at all levels of system design is crucial.

Dynamic Thermal Management (DTM) techniques are proposed to mitigate the aforementioned problems. DTM is a set of techniques that control processor temperature at run-time so that temperature does not go beyond a certain value known as critical temperature threshold. The DTM techniques are available at both hardware (HW) and software (SW) levels. Although HW approaches, such as stop-and-go and Dynamic Voltage and Frequency Scaling (DVFS), decrease temperature significantly, while degrades overall system performance due to longer execution time [1]. On the other hand, software-based DTM techniques such as task scheduling [2-3] and task migration [4-9] can reduce the temperature without significant performance degradation and any extra hardware.

Among DTM techniques, some of them are targeting temperature management of SMT processors [2, 3, 6, 7],

while others [4, 5, 8, 9] do not leverage SMT capability. DTM techniques for SMT processors can be divided into two categories: a) algorithms that are proposed and evaluated on simulators [2, 6], and b) algorithms that are proposed and evaluated on real platforms [3, 7].

The work of [2] introduces a simulation-based technique for parallel applications, called thread shuffling. This technique dynamically maps threads with similar criticality degrees into the same core and then applies DVFS to non-critical cores which execute fast threads. Their use of local DVFS restricts the proposed algorithm to only a few specific processors. [3] presents a DTM method for commercial single core SMT processors.

Lately, researchers proposed different instruments and algorithms for core and application thermal measurement and prediction to manage processors temperature efficiently. Two well-known methods, named as CMOS thermal sensors and performance-counter-based (software-based) sensors [1] are used to measure and predict processor thermal pattern. Alongside, application thermal profiling and performance counters are other two methods for application thermal categorization [1]. Since application thermal profiling is an offline method, it cannot reflect the real thermal pattern of the processor. Performance counters are usually used for online application temperature prediction, though they are inaccurate [7]. Moreover, reading different performance counters imposes significant overhead on application execution at run-time [1]. Therefore, recent proposed methods model overall core and application temperature with the aid of physical sensor and steady state temperature [8, 9]. Nevertheless, application temperature estimation of off-the-shelf SMT multi-core processors based on only physical temperature sensors is inaccurate, because generally, each core of an SMT multi-core processor has only one physical temperature sensor and it is hardly possible to know the real temperature of each thread.

Recent works predict future temperature of cores to reduce overheat temperature with negligible performance overhead. Their proactive task migration approaches predict the future temperature and manage workload to reduce and balance the temperature before reaching the temperature threshold. PDTM [8] is one of the first attempts that predicts core temperature. The prediction is based on both application thermal and core thermal models. PDTM migrates applications from the possible overheated core to the future coolest core in order to maintain the processor temperature below a threshold temperature. TAS [9]

978-1-4799-2817-0/14 $31.00 © 2014 IEEE

categorizes applications according to their thermal behavior and [4] considers the temperature of neighbor cores for improving the accuracy of temperature prediction.

Different cores of a processor do not have similar thermal behavior due to process variation [12], the temperature effect of neighbor components [4], and other physical issues [1]. The temperature difference between cores of a processor running the same application can be as much as $10\sim15°C$ [8]. In this paper, we name these phenomena as *physical features of cores*. It means that the cores of a multi-core processor show different thermal behavior for the same workload.

Motivated by these facts, we propose an algorithm which considers different thermal behavior of cores (*physical features of cores*) and uses both physical sensors and performance counters simultaneously to improve thermal management of SMT multi-core processors. We utilize physical sensors to estimate and predict the future temperature of the cores and performance counters to classify the applications thermal behavior at runtime. Another unique feature of proposed algorithm is that unlike all other previous algorithms, it has an adaptive migration threshold. To the best of authors' knowledge, no prior attempt has been made to implement a thermal-aware task scheduling on a commercial SMT quad-core product (Core i7-3770) under Linux environment considering SMT capability. The experimental results on Intel's Core i7-3770 running five to eight benchmarks indicate that our proposed method, PATM (Physical-Aware Task Migration), outperforms Standard Linux scheduler, PTDM, and TAS in reducing average and peak temperatures. The main contributions of this paper are summarized as follows:

- We propose a thermal-aware scheduling algorithm for multi-core SMT supported processors based on different thermal behavior of cores due to their *physical features*.
- Our experimental results on commercial processors indicate that our proposed approach, under full workloads, outperforms the Linux standard scheduler and two existing DTM techniques (PDTM, TAS).
- There is no additional hardware unit required for our prediction model and thermal-aware algorithm. It means that our approach is scalable for all the multicore systems and can be applied to off-the-shelf SMT multi-core products.

The remainder of this paper is organized as follows: The preliminaries of our algorithm are presented in Section II. Section III describes our proposed algorithm. Implementation and analysis results are shown in Section IV and final conclusions are drawn in Section V.

II. Preliminary

In this section, the preliminary of proposed algorithm is discussed.

A. Problem description

The system considered in this paper consists of an SMT multi-core processor with N cores, denoted as {$core_1$, $core_2$, ..., $core_N$} where each core can execute up to two threads simultaneously (two-context SMT cores). It is assumed that each of N cores has a physical thermal sensor. Since in this

paper we focus on SMT feature of processors, the number of tasks should be more than the number of the physical cores, thus it is assumed there are $N+1$ to $2\times N$ tasks for execution. The problem discussed in this paper is how to schedule these tasks among cores dynamically such that the average and peak temperature of the processor can be minimized under minimum performance loss and also temperature does not violate T_{max}. We propose a heuristic method to solve the above problem based on task migration and DVFS. It is assumed that the processor features global DVFS and performance counters. First, we introduce a new temperature prediction method, which predicts the future temperature of a core by considering both *cores physical features* and workload of the processor. Task migration is activated at *critical situations* i.e. when there is at least one core that reaches to T_{thr} in less than t_{res}, where T_{thr} is temperature threshold at which tasks are migrated to better cores in order to reduce the temperature, and t_{res} is the response time for the algorithm to decrease the core temperature.

B. Physical features of cores

As mentioned earlier, the temperature of each core of a processor is different from other cores. Table I summarizes our experimental results of running different applications of SPEC2006 benchmark suite, on four different cores of two Intel quad-core (Core i7-2600 and Core i7-3770) processors. This table shows the thermal behavior of cores (at a fixed fan speed) while one core executes an application and others are idle. The reported temperature is the maximum temperature among all four cores (e.g. $71°C$ is the peak temperature among all four cores of Core i7-2600, when core 3 executes the bzip2).

According to Table I, despite all four cores of Core i7-2600 have the same experimental setup, core 3 and core 1 are always the hottest and coolest cores, respectively. We tried the same experiment with Core i7-3770 and again observed this differential cores thermal behavior. As can be seen in Table I, core 2 and core 3 are the hottest and coolest cores respectively for Core i7-3770. This phenomenon, which we refer it as *physical features* of multi-core processors, motivated us for our proposed DTM algorithm. In the rest of paper, we fully explain how we take advantage of *physical features* to enhance the thermal management.

TABLE I
Temperature differential between cores.

Benchmark	Intel Core i7-2600			
	Executed on core 0	**Executed on core 1**	**Executed on core 2**	**Executed on core 3**
Gcc	59°C	58°C	61°C	64°C
Hmmer	66°C	62°C	63°C	66°C
bzip2	69°C	67°C	69°C	71°C
	Intel Core i7-3770			
Gcc	56°C	56°C	57°C	55°C
hmmer	60°C	60°C	62°C	58°C
bzip2	59°C	59°C	60°C	58°C

C. Temperature prediction

Our temperature predictor is modified version of [9]. Let assume T_{ss} as steady state temperature of an application (the steady state temperature of an application is defined as a

978-1-4799-2817-0/14 $31.00 © 2014 IEEE

temperature that the system would reach if the application is executed infinitely [8]). According to [9] the rate of temperature changes is proportional to difference between the current temperature and steady state temperature (Eq. 1):

$$\frac{dT}{dt} = c \times (T_{ss} - T), \tag{1}$$

where c is a core-specific constant. We add a new parameter w to Eq. 1 to extract Eq. 2:

$$\frac{dT}{dt} = c \times w \times (T_{ss} - T), \tag{2}$$

where w relates to core activity. w is added to reflect the thermal effects of other cores that are active (running applications) which has not been considered in [9]. The value of c, and w are determined offline using Eq. 2 by trying thermal curves corresponding to SPEC2006 benchmarks on the cores and running different number (one to eight applications) of them, simultaneously.

Solving Eq. 2, with $T(0) = T_{init}$ and $T(\infty) = T_{ss}$, we have:

$$T(t) = T_{ss} - (T_{ss} - T_{init}) \times e^{-c \times w \times t}. \tag{3}$$

Assigning $T(t) = T_{thr}$, we obtain:

$$t_r = \mu \times \ln\left(\frac{T_{ss} - T_{init}}{T_{thr} - T_{ss}}\right); \mu = \frac{1}{c \times w}, \tag{4}$$

where, t_r is the predicted time when the core reaches T_{thr}. According to our experiments the values of T_{ss} and c are different for each core. Therefore, the value of t_r should be calculated for each core. Based on the value of t_r the proposed algorithm decides when to start task migration.

III. Proposed Algorithm

This section discusses the proposed algorithm. In the following subsections, different parts of algorithm are fully explained. The flowchart of the proposed algorithm is depicted in Fig. 1 which briefly illustrates the intuituion behind the algorithm. The main parts of the algorithm are: *Threshold Management*, *Temperature Management*, and *Performance Management*. In *Threshold Management*, T_{thr} is tuned according to both migration frequency ($Migration_{\#}$) and migration limitation ($Migration_{limit}$). $Migration_{limit}$ is the maximum allowable task migration that can happen during specific iterations of the algorithm. Migrating more than $Migration_{limit}$ degrades performance and increases temperature due to the Ping-Pong effect [1]. In *Temperature Management* it is checked if cores are in *critical situations*. If so, the algorithm reschedules and migrates tasks, based on both application and core temperatures. After rescheduling, t_r for all cores are calculated, and if there is still any core in *critical situations*, it decreases the processor frequency (f_{cur}) to prevent violating T_{max}. In *Performance Management*, the goal is to minimize the performance degradation. In this phase, if the algorithm has not recently performed any migration and current processor frequency is lower than a predefined minimum frequency (f_{min}), it increases processor frequency to improve performance. In the following subsections, the aforementioned parts are thoroughly described.

A. Threshold Management

In all previous proposed task migration algorithms for DTM, T_{thr} is fixed [4-9]. Our experiments show that by having an adaptive T_{thr}, better results can be obtained (subsection IV.D). Therefore, we propose an adaptive T_{thr} unlike other algorithms. Finding a proper T_{thr} is crucial. In this subsection, it is explained how the algorithm adjusts T_{thr} based on changes of workload behavior. At first $T_{thr} = T_{max}$. During execution, if the total number of migrations in the last M iterations of the algorithm is higher than $Migration_{limit}$, T_{thr} increases (should not become greater than T_{max}) and if the total number of migrations in the last M iterations of the algorithm is zero, T_{thr} decreases. The higher migration frequency is, the more overall system performance degrades. Therefore, our proposed T_{thr} management tries to control migration frequency and prevent it from increasing. Note that rising T_{thr} results in decreasing migration frequency. However, increasing both T_{thr} and task migration deteriorate the overall system performance and temperature. Our proposed *Threshold Management* finds a trade-off between temperature threshold and task migration frequency regarding to workload and cores thermal behavior.

B. Temperature Management

A main challenge in task scheduling of SMT multi-core in

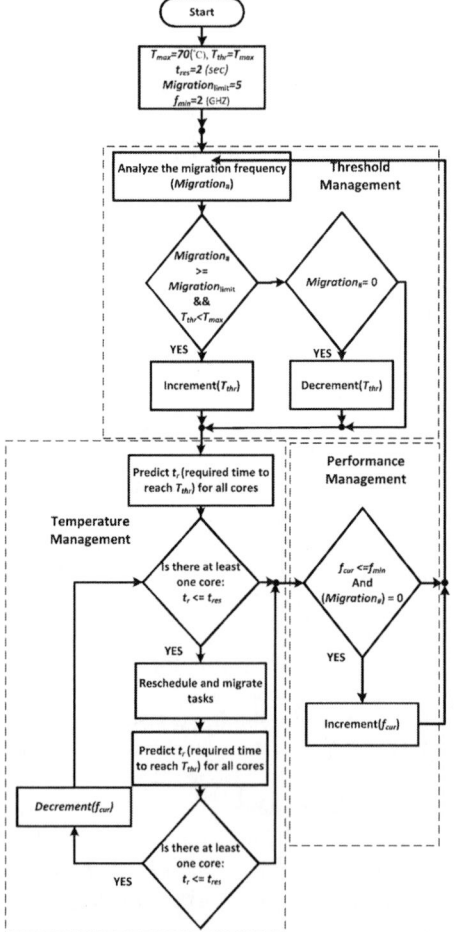

Figure 1- The flowchart of the proposed PATM algorithm.

order to improve performance, is co-scheduling of complementary threads on individual SMT cores [6] to make better use of shared pipeline resources. However, this way of scheduling causes higher heat generation due to more pipeline resources utilization [6]. To address this issue, we study five different strategies to find the most suitable pairs of tasks that should be co-scheduled to two-context SMT cores in order to minimize the average and peak temperature, while minimizing the performance degradation. Since, each core has one physical thermal sensor, we use performance counters to distinguish the cold thread from the hot thread on a two-context SMT core.

In the first strategy, cores and tasks are sorted according to their temperature. Physical sensors and performance counters are used to measure the temperature of the cores and tasks, respectively. After sorting cores and tasks, hottest and coolest tasks are paired and co-scheduled to the coldest core, then the second hottest and coolest tasks are paired and co-scheduled to the second coolest core and this process is continued. The second strategy is similar to the first one, except that cores are sorted according to their thermal behavior based on their *physical features* (e.g. for Core i7-2600, core 3 and core 1 are always the hottest and coolest cores, respectively). In our third strategy, after sorting cores according to their thermal behavior based on their *physical features*, the first two hottest tasks are co-scheduled to the coldest core. Fourth strategy is similar to the third strategy except that in this co-scheduling, the first two coldest tasks are assigned to the coldest core. In the second, third, and fourth strategies, sorting cores is based on their innate thermal behavior. Learning physical features of cores can be done offline and it is needed to be accomplished only once. Our fifth strategy reschedules tasks between only core that is in *critical situation* (the core that has $t_r<t_{res}$) and predicted coolest core (the core that $t_r>t_{res}$) instead of rescheduling all tasks among all cores as done for previous four strategies. In this strategy, the coolest core has the greatest t_r among all cores. The tasks in hot core will be moved to the coldest core and other cores are unchanged. The evaluations of these strategies are reported in Section IV. According to our results the second strategy is the best one. Fig. 2 illustrates selected task scheduling strategy.

After rescheduling, t_r is again predicted for all cores, and if there is still any core in critical situation, it means *Temperature Management* cannot perfectly manage core temperature at software level. At this moment, DVFS is used to decrease the processor frequency and hence, temperature.

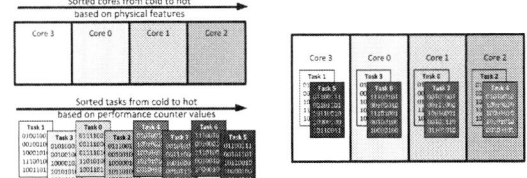

Figure 2 – Second task scheduling strategy.

C. Performance Management

As mentioned in the previous section, if *Temperature Management* cannot improve *critical situation*, it decreases the processor frequency. Although this action decreases temperature significantly, it ruins system performance. Our *Performance Management* function mitigates this problem with the aid of checking the workload of cores. If the number of migrations is zero in the last M iterations and current processor frequency is lower than f_{min}, algorithm increases the global frequency to enhance performance.

IV. Experimental Results

This section provides experimental results for different applications from SPEC CPU2006 benchmarks and an analysis on the obtained results.

A. Experimental Setup

The selected programs from SPEC2006 benchmark suite are summarized in Table II. The processor we use is an Intel Core i7-3770 while the SMT capability of processors is active. The size of the main memory of the system is 8 GB. The Linux kernel version is 3.2.0. The LM sensor [14] is used to read core temperatures. We use cpufreq to adjust the processor frequency and use perf subsystem in Linux for reading performance counters. In all of our experiments the fan speed has been fixed to a constant RPM. The value of t_{res}, and $migration_{limit}$ are set to two seconds and five, respectively. These values are selected empirically based on different experiments. f_{min} is set to 2 GHz because this is a frequency that if all cores are running applications, the maximum temperature will be less than T_{max}. The value of T_{thr} is adapted at run-time. At the start of the algorithm $T_{thr}=T_{max}$. The other constant is the number of iterations of the algorithm *(M)* for counting $migration_{\#}$ which is set to 10. The temperature threshold that we do not want to violate (T_{max}) is 70˚C.

B. Performance counter analysis

Our algorithm uses performance counters to *1)* distinguish the cold thread from the hot thread on a two-context SMT core and *2)* to sort applications according to their temperature. We need to find the performance counter which has the greatest correlation with temperature of programs. To do so, we first run different programs and profile all performance counters, then select one performance counter which has the greatest correlation with the temperature of programs using Pearson Product-Moment Correlation Coefficient (PPMCC) [13]. It is used as a criterion to measure the correlation between two variables X and Y. The r coefficient is calculated using Eq. 5:

$$r = \frac{\sum_{i=1}^{N}(X_i - \overline{X})(Y_i - \overline{Y})}{\sqrt{\sum_{i=1}^{N}(X_i - \overline{X})^2}\sqrt{\sum_{i=1}^{N}(Y_i - \overline{Y})^2}}, \quad (5)$$

TABLE II
SPEC CPU 2006 applications used in experimental results

Benchmarks	hmmer	libquantum	sjeng	perlbench	gobmk	gcc	mcf	bzip2
Avg. Temperature(˚C)	68.2	67	65.7	65	63.9	63.9	63	62.9

978-1-4799-2817-0/14 $31.00 © 2014 IEEE

where N is number of sampled data, $\overline{X}$, $\overline{Y}$ are the averages for X and Y variables, respectively. The relationship between X and Y is perfect when r is 1 or -1. Table III summarizes the average correlations between performance counters and applications temperature of ten programs: **astar, libquantum, gcc, bzip2, mcf, gobmk, sjeng, h264ref, perlbench, and hmmer.**

Acording to Table III *stalled-cycles-backend* event has the strongest correlation among other processor events, so our proposed algorithm uses this event as a metric to analyze the thermal behavior of applications. The negative value implies that if X variable increases, Y will decrease. Therefore, the larger the *stalled-cycles-backend* value for an application, the colder the application and vice versa.

TABLE III
Correlation between events and applications temperature.

Events	Correlation
stalled-cycles-backend	-0.37
cache-references	-0.35
stalled-cycles-frontend	-0.35
cache-misses	-0.33
Cycles	-0.29
task-clock	-0.24
context-switches	-0.03
branches	-0.03
page-faults	-0.01
branch-misses	0.02
CPU-migrations	0.04
Instructions	0.29
IPC	0.30

We set up an experiment to demonstrate the effect of choosing different events on results. Fig. 3 illustrates the average temperature of four cores while PATM once uses *stalled-cycles-backend* (highest correlation), and once uses *page-faults* (lowest correlation) as the events to measure application thermal behavior, respectively, and sort them from the hottest to the coldest. It is observed that using *stalled-cycle-backend* event causes peak temperature and average temperature improves about %6 (3°C) and %4. 7 (2.5°C) respectively compared to the case when the lowest correlation counter is used.

C. Task Scheduling analysis

The five strategies for task scheduling in *Temperature Management* phase (see section III.B) are tried and their results are compared against Linux scheduler. In each strategy we study four cases i.e. we execute from 5 to 8 different benchmarks simultaneously. The average results of

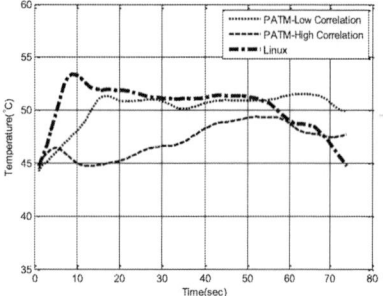

Figure 3 - PATM average temperature using high and low correlation counter for application ordering and Linux standard scheduler.

four cases are shown in Fig. 4. As can be seen, the second strategy has the best average and peak temperature improvement but there is about 0.38% performance overhead. The first strategy improves the average and peak temperature less than the second strategy but not only it does not degrade performance but also it improves it by about 0.76%. Since, the focus of this paper has been thermal management, we use the second strategy for the following experiments. The results show the effectiveness of considering *physical features* in the algorithm on the results.

D. Adaptive threshold analysis

In evaluating the effectiveness of having adaptive T_{thr} against fixed T_{thr}, we compare the results with the case where the *Threshold Management* section of the algorithm is disabled. Fig. 5 shows the results. Using an adaptive T_{thr}, 1.7% (0.9°C) and 6.3% (4°C) improvement is obtained on average and peak temperature compared to the fixed T_{thr}.

E. Temperature prediction analysis

Our temperature prediction model based on Eq. 2 predicts future temperature with less than 1°C mean absolute error on running different benchmarks. Fig. 6 illustrates the results of the prediction modell of TAS [9] vs. ours against the real core temperature. Using our predictor, mean absolute error (MAE) is 0.6 °C whereas MAE of TAS predictor is 0.8 °C on running gcc and hmmer programs simultaneously, which shows the effectiveness of new added w parameter to our predictor.

F. Thermal management results

Fig. 7 illustrates average and peak temperature and

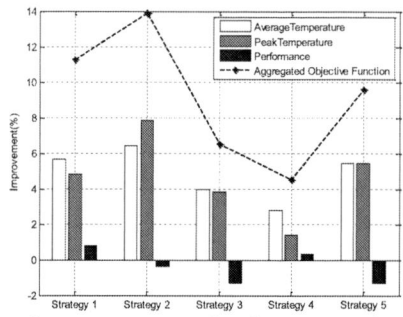

Figure 4 - performance, average, and peak temperature improvement of different strategies compared to Linux standard scheduler.

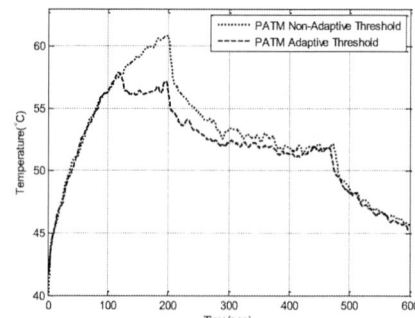

Figure 5 - Comparison of our proposed algorithms for two cases with adaptive and non-adaptive thresholds.

978-1-4799-2817-0/14 $31.00 © 2014 IEEE

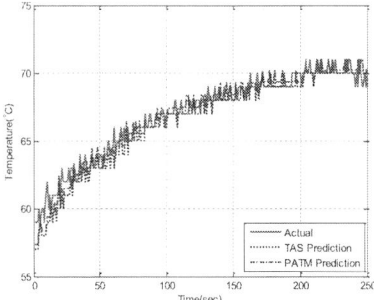

Figure 6 - The comparison of PATM and TAS predictors.

performance overhead for TAS, PDTM, Linux scheduler, and our proposed algorithm with various number of programs on an Intel core i7-3770.

After running a different set of programs on Intel Core i7-3770 processor, it is observable that our proposed technique (PATM) reduces the average temperature by about 7.7% (3.6C), and reduces peak temperature by almost 13.9% (7.8°C) with 1.7% performance (run-time) overhead compared to the standard Linux scheduler, on average. It should be noted that the average temperature is the mean of four cores temperature running programs simultaneously from beginning to the end. The experimental results also indicate that our proposed algorithm reduces the average temperature by about 1.1% (0.5°C) and 1.3% (0.6°C) compared to TAS and PDTM, respectively. Reduction in peak temperature through PATM is about 8.1% (4.5°C) and 5.8% (3.3°C) compared to both TAS and PTDM, correspondingly. The overall system performance (run-time) overhead is only 1.3% and 0.4% compared to TAS and PDTM, respectively. Table IV summarizes the comparison results for these four algorithms that are the mean values extracted at run-time for five to eight attempted applications. Hence, compared to Linux, PDTM, TAS, our proposed method indeed leads to more peak temperature reduction with negligible performance overhead.

V. Conclusion

In this paper, a dynamic thermal management algorithm with a future temperature predictor for SMT multicore processors is presented. The proposed algorithm manages processor temperature taking into account the workload and physical features of cores. As demonstrated, considering SMT, physical features of cores, applications thermal behavior, and adaptive migration threshold ability are extremely important in the DTM and they have significant influence on performance and temperature management. Experimental results obtained by SPEC CPU2006 running

on a desktop platform (Intel Core i7-3770) indicate that our algorithm outperforms Linux standard scheduler, TAS, and PDTM in terms of thermal management with negligible performance overhead.

TABLE IV
Comparison results of PATM against Linux, TAS, and PDTM.

DTM Algorithm	Average Temp.	Peak Temp.	Run Time(Second)
Linux	50.9(°C)	64(°C)	912.5(Sec)
PDTM	47.9(°C)	59(°C)	924.8(Sec)
TAS	47.8(°C)	60(°C)	916.8(Sec)
PATM	47.3(°C)	56(°C)	928.3(Sec)
Improvement PATM vs. PDTM	1.3%	5.8%	-0.4%
Improvement PATM vs. TAS	1.1%	8.1%	-1.3%
Improvement PATM vs. Linux	7.7%	13.9%	-1.7%

References

[1] J. Kong, S.W. Chung, and K. Skadron, "Recent thermal management techniques for microprocessors," *ACM Computer Survey*, vol. 44, no. 3, pp. 13:1-13:42, 2012.

[2] Q. Cai, J. Gonzalez, G. Magklis, P. Chaparro, and A. Gonzalez, "Thread shuffling: Combining DVFS and thread migration to reduce energy consumptions for multi-core systems," In *Low Power Electronics and Design (ISLPED)*, pp. 379-384, 2011.

[3] J. Choi, C.-Y. Cher, H. Franke, H. Hamann, A. Weger, and P. Bose, "Thermal-aware task scheduling at the system software level," In *Proc. of the 2007 international symposium on Low power electronics and design*, pp. 213-218, 2007.

[4] G. Liu, M. Fan, and G. Quan, "Neighbor-aware dynamic thermal management for multi-core platform," In *Design, Automation & Test in Europe Conference & Exhibition (DATE)*, pp. 187-192, 2012.

[5] Z. Liu, T. Xu, S.X.-D. Tan, and H. Wang "Dynamic thermal management for multi-core microprocessors considering transient thermal effects," In *18th Asia and South Pacific Design Automation Conference*, pp. 473-478, 2013.

[6] M. Gomaa, M.D. Powell, and T.N. Vijaykuma, "Heat-and-run: leveraging SMT and CMP to manage power density through the operating system," In *ACM SIGARCH Computer Architecture News*, pp. 260-270, 2004.

[7] A. Kumar, L. Shang, L. Peh, and N.K. Jha, "HybDTM: a coordinated hardw5are-software approach for dynamic thermal management," In *Proc. of the 43rd Design Automation Conference*, pp. 548-553, 2006.

[8] I. Yeo, C.C. Liu, and E.J. Kim, "Predictive dynamic thermal management for multicore systems," In *Proc. of the 45th annual Design Automation Conference*, pp. 734-739, 2008.

[9] I. Yeo, E.J. Kim, "Temperature-aware scheduler based on thermal behavior grouping in multicore systems," In *Proc. of the Conference on Design, Automation and Test in Europe*, pp. 946-951, 2009.

[10] Y. Li, K. Skadron, D. Brooks, and Z. Hu, "Performance, energy, and thermal considerations for SMT and CMP architectures," In *11th International Symposium on High-Performance Computer Architecture*, pp. 71-82, 2005.

[11] J. Donald, and M. Martonosi, "Temperature-aware design issues for SMT and CMP architectures," In *Proc. of the Workshop on Complexity-Effective Design. ACM Press*, 2004.

[12] E. Kursun, and C-Y. Cher, "Variation-aware thermal characterization and management of multi-core architectures," In *Proc. of International Conference on Computer Design(ICCD)*, pp. 280-285, 2008.

[13] J.L. Rodgers, and W.A. Nicewander. "Thirteen ways to look at the correlation coefficient," *The American Statistician* 42. no 1, pp. 59-66, 1988.

[14] Lm sensors linux hardware monitoring [Online]. Available: http://www.lm-sensors.org.

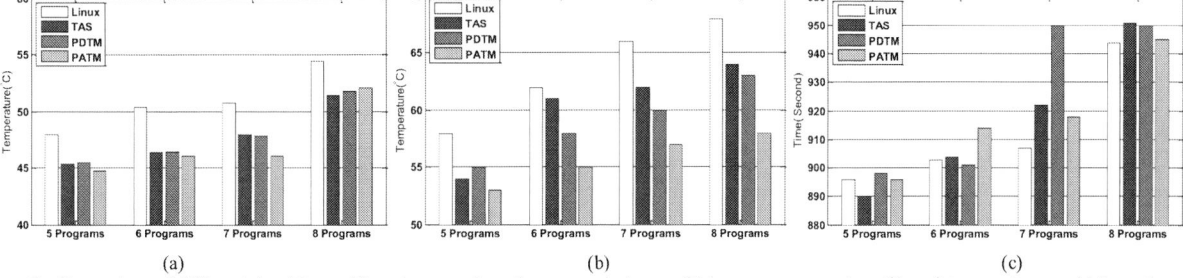

Figure 7 - Comparison of different algorithms with various number of programs in terms of (a) average temperature (b) peak temperature, and (c) run-time.

978-1-4799-2817-0/14 $31.00 © 2014 IEEE

Agile Frequency Scaling for Adaptive Power Allocation in Many-core Systems Powered by Renewable Energy Sources

Xiaohang Wang*, Zhiming Li*, Mei Yang[†], Yingtao Jiang[†], Masoud Daneshtalab[‡], Terrence Mak[§]*

*Guangzhou Institute of Advanced Technology, CAS, China
Email: {xh.wang, zm.li}@giat.ac.cn
[†]University of Nevada, Las Vegas, USA
Email: mei.yang@unlv.edu, yingtao@egr.unlv.edu
[‡]University of Turku, Finland
Email: masdan@utu.fi
[§]The Chinese University of Hong Kong, China
Email: stmak@cse.cuhk.edu.hk

Abstract— As low-power electronics and miniaturization conspire to populate the world with emerging devices, one appealing approach is to power these multi-core/many-core-based devices with energy harvested from various environments. Of the most important issues concerning these devices is how to effectively allocate power budget among the cores competing for power, which is formulated as one specific type of power-performance optimization problem in this paper. We attempt to solve this problem by proposing an Adaptive Power Allocation Technique (APAT) that uses a dynamic programming network. Our goal here is to maximize the overall system performance, taking into account a unique yet challenging fact that, available power budget might have to undergo a significant change when a renewable energy source is scavenging. APAT has a linear time complexity and low hardware overhead. Experiments have confirmed that APAT can reduce $20 \sim 30\%$ of execution time compared to other state-of-the-art power allocation algorithms. In addition, as APAT is quite insensitive to the changing rate of the power, lending itself well for power management in many-core systems powered by energy-harvesting sources.

I. INTRODUCTION

There is a general consensus that by the year 2020, the chip power density will increase as much as $10\times$ over the year 2012 [1]. A lot of these power hungry chips then will have to be powered by harvesting renewable energy. Renewable energy like solar energy is abundant to power electronic systems from low-power devices [2] to data centers [3]. One of the technical challenges to improve the energy efficiency of an energy harvesting based multi-/many-core chip is power budget allocation, which is to maximize the chip performance under a limited power or energy budget, *i.e.*, the power budgeting problem [4]. Frequency scaling can be used to solve the problem. However, this power budget problem is complicated by the following issues in an energy harvesting based many-core system.

1) In a many-core chip, the operating frequencies of the many resources are likely tunable. Consider an application mapped to a 16-core region, where the frequency of each core can take one of the four allowed values. The total frequency combination is as large as 4^{16}, which is

about 4×10^9. Heuristic-based approaches [5], [6] cannot find optimal solutions given such a large solution space.

2) The input power budget from the renewable energy might have high and rapid variations due to uncertainty in the environmental situations. Existing approaches [5], [6] need long run time, leading to unknown behavior or poor performance. Using energy buffers like supercapacitors or batteries can suppress the variation, which however, incurs significant energy loss and suffers from issues like battery aging, self-discharge, etc [7]. Thus, it would be beneficial to minimize the usage of energy buffers and use the renewable energy directly as much as possible.

In this paper, a novel frequency scaling scheme for power allocation is proposed, Adaptive Power Allocation Technique (APAT) using dynamic programming with renewable energy awareness to address the above aforementioned problems. The power budgeting problem is solved using dynamic programming network with linear time complexity. APAT has the following key features:

1) APAT can generate globally optimal power allocation solution under a given power budget from the harvested energy.
2) The run-time of APAT is much lower (*e.g.*, a few to dozens of cycles) than that of any other known power allocation algorithms [5], [6].
3) APAT can be used for many-core systems to control the power consumption of various on-chip resources at a finer grain.

Experimental results reveal that APAT can reduce the expectation time significantly compared with other three competing power allocation methods [5], [6] with much lower hardware and run-time overhead. Part of the work has been published in [8].

The paper is organized as follows. Section II reviews the related work. Section III introduces the performance-power model and problem definition, followed by the description of the dynamic programming based power allocation scheme in Section IV. Section V provides the experimental results and analysis of the proposed approach. Finally, Section VI concludes the paper.

This research program is supported by the Natural Science Foundation of China No. 61376024 and 61306024, Natural Science Foundation of Guangdong Province No. S2013040014366.

978-1-4799-2817-0/14 $31.00 © 2014 IEEE

II. RELATED WORK

One of the majaor challenges in renewable energy based systems is to allocate power to resources since the renewable energy source is not stable and might have high variation. Power management approaches for energy harvesting based multi-/many-core can be achieved by task scheduling [9], frequency/voltage scaling [10], etc. Similar approaches can be found to solve the power budgeting problem [4] in many-core systems. Frequency/voltage scaling can be performed for power allocation [4]–[6]. For example, [6] proposed a method where the number of cores that can be powered on is coarsely determined, after which fine-grained frequency adjustment is followed. Most of these approaches are heuristic-based [4]–[6], or linear programming-based [11]. Heuristic-based approaches cannot find the optimal solutions when the number of resources to be controlled is increased or when the applications' behavior becomes complicated. Linear programming-based approaches might lead to high run-time and power consumption overhead to find a good solution, which might waste the harvested energy.

III. MODELS AND PROBLEM FORMULATION

A. Performance model

Frequency scaling is used to balance between power consumption and performance. Suppose each application q occupies an N_q-tile region where the frequencies of the tiles are $f_1, \ldots, f_{Nq}$. Note that some of the regions can be overlapped, e.g., a cache bank might be shared by several applications. An application is assumed to be mapped to only one region. For a total of Q applications running simultaneously, we have $\sum_{q=1}^{Q} N_q = N$, where N is the total number of tiles. Performance of each application measured in execution cycles is modeled in terms of the frequencies of its region/tiles, as follows,

$$Cycle = g_{cycle}(f_1, \ldots, f_{Nq}) \qquad (1)$$

Regression models [12] are used to find the g_{cycle} in Eqn. 1 through curve fitting. We have proposed the following model in Eqn. 2 experimentally, which can result in lower regression error, and this model will be validated in the experimental part in Section V.B.

$$\ln Cycle = \sum_{i=1}^{N_q} a_i \cdot \sqrt{f_i} \qquad (2)$$

where a_i is the regression coefficient w.r.t. f_i. At the k-th time interval, the frequencies of an application's region are set randomly with a vector $\vec{f_k} = <f_1, \ldots, f_N>$ and measure the cycles $Cycle_k$. With K time intervals, the training data are collected. A linear regression model with the maximum likelihood estimator [12] will find the coefficients (a_i's) with the training data set. When some regions are overlapped, there are some resources shared by multiple applications, which implies resource contention. Thus, the performance models are trained for each application in parallel with possible contentions. These models remain accurate after the training.

B. Dynamic power model

Assuming all the cores are operating at the same voltage level, the dynamic power of an application with N_q tiles can be determined as follows:

$$P_q = \sum_{i=1}^{N_q} \alpha_i \cdot C_i \cdot f_i \cdot V^2 = \sum_{i=1}^{N_q} b_i \cdot f_i \qquad (3)$$

where α_i is the switching activity, C_i is the effective capacitance, V is the voltage, $b_i \equiv \alpha_i \cdot C_i \cdot V^2$.

If, besides the core frequencies, the core voltages can also be adjusted, the dynamic power can be calculated as

$$P_q = \sum_{i=1}^{N_q} \alpha_i \cdot C_i \cdot f_i{}^3 / K^2 = \sum_{i=1}^{N_q} d_i f_i{}^3 \qquad (4)$$

where K is a constant. Similarly, $d_i \equiv \alpha_i \cdot C_i / K^2$.

C. Problem Definition

With the above models, the power budgeting problem aims to minimize the overall execution time under the input power budget. With Q applications each occupying an N_q-tile region, we have

$$\sum_{q=1}^{Q} w_q \cdot P_q = P \qquad (5)$$

where each P_q is the power budget for application q at a given time t, and w_q are user defined priority weights for each application.

The power budgeting problem for each application q can be formulated as,

$$\min Cycle = g_{cycle}(f_1, \ldots, f_{N_q}) \qquad (6)$$

subject to

$$\sum_{i=1}^{N_q} b_i \cdot f_i \le P_q \qquad (7)$$

for each

$$f_i \in \{F_1, \ldots, F_M\} \qquad (8)$$

The power P_q in Eqn. 7 is determined by applying either Eqn. 3 or 4. Eqn. 8 specifies a discrete set of frequency values that each frequency variable f_i can take.

Dropping the ln notation in Eqn. 2 and converting the min operator to max in the objective equation, the above problem definition has the same form as a bounded knapsack problem, which is NP-hard. On the other hand, the power budget might have rapid variations, which requires the problem be solved with low run time and hardware overhead.

IV. THE APAT ALGORITHM

A. Overview of the APAT algorithm

Inspired by the dynamic programming approach to solve the knapsack problem, which is pseudo-polynomial time complexity, the problem in Section II.C can also be solved optimally by the following dynamic programming approach with polynomial time. Since the logarithmic function in the objective (Eqn. 2) is monotonic, minimizing Eqn. 2 is equivalent to minimizing $\sum_{i=1}^{N_q} a_i \cdot \sqrt{f_i}$. Let $C_{i,p}$ denote the minimum cycles of assigning $\{f_1, \ldots, f_i\}$ with $\sum_{j=1}^{i} b_j \cdot f_j \le p$ and $0 \le p \le P_q$. Thus, for each f_i,

- if $\sum_{j=1}^{i} b_j \cdot f_j > p$, $C_{i,p} = C_{i-1,p}$.
- Otherwise, $C_{i,p} = \min\{C_{i-1,p}, C_{i-1,p-b_i f_i} + a_i\sqrt{f_i}\}|_{f_i=F_j}$, where $F_j \in \{F_1, ..., F_M\}$ as in Eqn 8.

The above steps can be called recursively until the solution is found. The dynamic programming based algorithm can be summarized in Algorithm 1.

Algorithm 1: Dynamic Programming Based Frequency Assignment

Input: a_i, b_i: the coefficients in Eqns. 2 and 3
Output: $C_{i,p}$: the minimum cycles given the power budget $\leq p$.
Function: Find the minimum cycles.
begin
    Initialize all the $C_{i,p}$ to be 0;
    for *each f_i* **do**   /* $i = 1, ..., N_q$ */
        for *each F_j,* **do**   /* $\{F_1, ..., F_M\}$ */
            for *each $p \leq P_q$* **do**
                if $\sum_{j=1}^{i} b_j \cdot f_j > p$ **then**
                    $C_{i,p} = C_{i-1,p}$;
                else
                    $C_{i,p} = \min\{C_{i-1,p}, C_{i-1,p-b_i f_i} + a_i\sqrt{f_i}\}|_{f_i=F_j}$;
            end
        end
    end
end

The time complexity of the above algorithm can be further reduced to be linear [8]. To accelerate the computation, a multi-stage dynamic programming network (DPN) is designed to solve the problem for each application q with linear time complexity. It is first constructed by mapping the terms in the constraint (Eqn. 3) and objective (Eqn. 2) equations to the weights of the vertices or edges. Then the DPN is traversed to find a minimum weight path corresponding to the optimal solution.

In the DPN, each vertex represents a different power budget. An edge exists between two vertices in adjacent stages if the power consumption of assigning the frequency equals to the difference in the power budgets of the two vertices. Each edge is assigned a weight by the term in Eqn. 2, *i.e.*, minimum cycles by assigning frequency to a variable. If a stage (corresponding to a tile) is not in the application's region, the stage will be bypassed.

With the DPN, finding the minimum weight path is equivalent to an optimal frequency assignment. The traversal can be done in parallel. Each vertex at current stage selects the edge such that, the sum of the edge weight and the minimum cycles achieved by the later stage is minimum. This sum is transmitted back to vertices in the previous stage. In this manner, after reaching the source, the minimum weight path is found in linear time.

Fig. 1 shows the block diagram of APAT. The performance model in Section III.A for each application will be set up first with a regression model either online or offline. The DPN will be constructed based on the performance-power model and

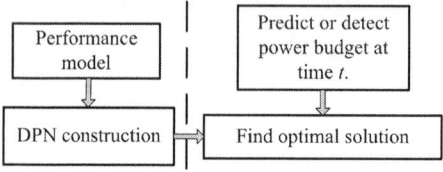

Fig. 1. The APAT system block diagram.

input power budget. Then the DPN is traversed to find the optimal solution.

B. The dynamic programming network

The dynamic programming approach in above sub-section can be run in a dynamic programming network [8]. Fig. 2 shows the transformation of the problem into a dynamic programming network. Once the DPN is constructed, it only needs minor update to remove some vertices and edges according to current power budget.

Definition 1. *Dynamic programming network. A dynamic programming network is denoted as a graph $DPN(V, E)$, with V and E represent the sets of vertices and edges, respectively.*

- Each vertex is assigned with two properties, $Cycle$ and P, where $Cycle\,(v_{i,j})$ denotes the minimum cycles given the power budget $P\,(v_{i,j})$ equal to j.
- An edge $e_{i,j,k}$ is added between vertices $(v_{i,j}, v_{i+1,k})$, if $j + b_i \cdot f_i|_{f_i=F_l} = k$, for $l = 1, ..., M$.
- The weight of an edge $e_{i,j,k} = (v_{i,j}, v_{i+1,k})$ is represented as $w_{i,j,k}$. $w_{i,j,k} = a_i\sqrt{f_i}|_{f_i=(j-k)/b_i}$ equals to the corresponding term in Eqn. 2.

The above edge connection means, there is a path between two adjacent vertices if the difference in the power budget corresponding to them equals to the power consumption by assigning frequency to the vertex between them.

Of the total of $P \times (N+1)$ vertices in a DPN, the network is organized as $N+1$ stages (as there are N frequency variables) with each group consists of P vertices (corresponding to the available power budget levels). Two dummy vertices, S and D, are added before stage 1 and after stage $N+1$, respectively, as in Fig. 2.

Note that if a stage i (corresponding to a tile) is not inside an application's region, it is bypassed, *i.e.*, connecting vertices directly to those in the previous stage with the edge weight set to be 0. This can be selected by a MUX as in Fig. 2.

C. Find the optimum solution

The DPN can be traversed to find an optimal solution as follows.

- The edge $e_{i,j,k}$ is assigned with weight

$$w_{i,j,k} = a_i\sqrt{f_i}|_{f_i=(j-k)/b_i} \qquad (9)$$

where a_i and b_i are defined in Eqns. 2 and 3.
- Find a minimum weight path from S to D, which corresponds to the optimal solution of the problem.

Note that f_m with $1 \leq m < i$ (of the stages proceeding stage i) are assigned a value at stage i, while f_n with $i < n \leq N$ (stages after stage i) are yet to be assigned.

Problem :

$$\min Cycle = e^{a_0 \cdot \sum_{i=1}^{N} a_i \sqrt{f_i}}$$

$$\sum_{i=1}^{N} b_i \cdot f_i \leq P$$

$$f_i \in \{F_1, ..., F_M\}$$

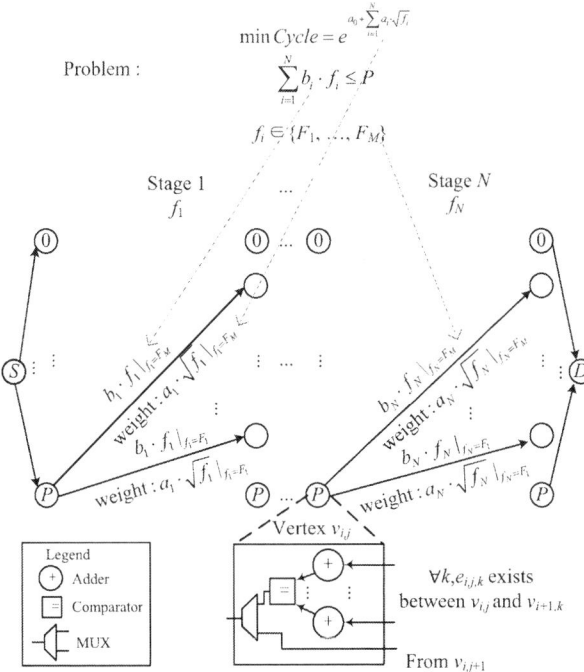

Fig. 2. Transformation of the optimization problem to a dynamic programming network. In total, there are $P \times (N+1)$ vertices.

The weight of each edge equals to the corresponding term in Eqn. 2 as shown in Fig. 2. When the power budget P_q changes, some vertices $v_{i,j}$ and the corresponding edges are removed if the power budget of $v_{i,j}$ (i.e., j) exceeds P_q. To find the minimum weight path under the power constraint in linear time, the following dynamic programming equations can be calculated within N iterations [13] following a backward moving procedure (from D back to S). At each stage, each vertex calculates the following value, a modified version of the Bellman equations [13].

$$Cycle_{MIN}(v_{i,j}) = \min_{\forall v_{i+1,k}, \exists \text{ an edge } e_{i,j,k} \text{ between}(v_{i,j},v_{i+1,k})} \{w_{i,j,k} + Cycle_{MIN}(v_{i+1,k})\} \quad (10)$$

where $Cycle_{MIN}(D) = 0$, and $w_{1,S,k} = w_{N,D,k} = 0$, with $k \in [1,P]$, i.e., the edges connecting S to the vertices in the first stage and the vertices in stage N to D have a weight of 0.

Thus, the calculation of Eqn. 10 will become to find the minimum weight path (optimal solution) under the power constraint in N steps as follows,

$$Cycle_{MIN}^*(PATH_{S,D}) = \min_{e_{i,j,k} \in PATH_{S,D}} \left\{ \sum_{i=1,j=1,k=1}^{i=N_q,j=P_q,k=P_q} w_{i,j,k} \right\}$$

$$= \min_{f_i \in \{F_1,...,F_M\}} \sum_{i=1}^{N_q} a_i \sqrt{f_i} \quad (11)$$

where $PATH_{S,D}$ is the set of all the paths from S to D. The optimal path from $v_{i,j}$ to $v_{i+1,\mu}$ at each vertex $v_{i,j}$ (corresponding to the assignment of each frequency variable f_i by $(j-\mu)/b_i$) along the minimum weight path can be

obtained as follows:

$$v_{i+1,\mu} = \arg\min_{\forall v_{i+1,k}, \exists \text{ an edge } e_{i,j,k} \text{ between}(v_{i,j},v_{i+1,k})} \{w_{i,j,k} + Cycle_{MIN}^*(v_{i+1,k})\} \quad (12)$$

where the optimal assignment of f_i is $F_l = (j - \mu)/b_i$. In this way, the optimal assignment of each frequency variable is found and the system can be tuned to run under this frequency configuration.

The pseudo-code in Algorithm 2 shows the traversal of the DPN to find the minimum weight path from S to D. Both DPN construction and traversal take N_q iterations for each application. Each iteration only involves add and compare operations which could be done in one cycle. Thus, the worst case run time is $2N_q$ cycles.

Algorithm 2: FindMinWeightPath

Input: $e_{i,j,k}$: the weight of each edge, for $i,j \in [1, N+1]$ and $k \in [1, P]$.
Output: $Cycle(v_{i,j})$:the minimum cycles of each vertex after assigning f_i.
Function: Find the minimum weight path & corresponding to the optimal solution.
begin
  Initialize all the $Cycle(v_{i,j})$ to be ∞, except $Cycle(D) = 0$;
  for *stages i from $N-1$ to 0* **do**
    for *each vertex $v_{i,j}$* **do**
      for *adjacent vertex $v_{i+1,k}$* **do**   /* an edge $e_{i,j,k}$ connecting $v_{i,j}$ and $v_{i+1,k}$ */
        if $Cycle(v_{i+1,j}) + w_{i,j,k} < Cycle(v_{i,j})$
        then
          $Cycle(v_{i,j}) = Cycle(v_{i+1,j}) + w_{i,j,k}$;
          $f_i = (j-k)/b_i$;
      end
    end
  end
end

V. EXPERIMENTAL RESULTS

A. Experimental setup

Experiments are performed using an in-house developed event-driven many-core simulator [14]. Table I lists the configuration of the many-core simulator and the mixes of the benchmarks selected from PARSEC and SPLASH-2. Each of the applications occupies a 4×4 mesh region in the many-core system. The weights of each application (Eqn. 5) is randomly selected whose sum equals to 1.

In all the experiments, we select a 8×8 2D mesh as the underline NoC topology. We compare the performance of the proposed APAT against three other best known schemes: (1) PGCapping [6], where both the number and the frequency of tiles can be adjusted, (2) PEPON [4], where the frequency of each processor and last-level cache bank can be adjusted, and (3) DPPC [11], a linear programming based approach. In what

978-1-4799-2817-0/14 $31.00 © 2014 IEEE

TABLE I.   Parameters Used in the Simulation

Number of processors	64 (MIPS ISA 32 compatible)
Fetch/Decode/Commit size	4 / 4 / 4
ROB size	64
L1 D cache (private)	16KB, 2-way, 32B line, 2 cycles, 2 ports, dual tags
L1 I cache (private)	32KB, 2-way, 64B line, 2 cycle
L2 cache (shared) MESI protocol	64KB slice/node, 64B line, 6 cycles, 2 ports
Main memory size	2GB
On-chip network parameters	
NoC flit size	72-bit
Data packet size	5 flits
Meta packet size	1 flit
NoC latency	router 2 cycles, link 1 cycle
NoC VC number	4
NoC buffer	5 × 12 flits
Workload mixes used as multiple applications running in a single NoC	
Mix-1	blackscholes, ferret, freqmine, swaptions
Mix-2	streamcluster, dedup, canneal, vips
Mix-3	barnes, raytrace, swaptions, vips
Mix-4	dedup, freqmine, fft, canneal

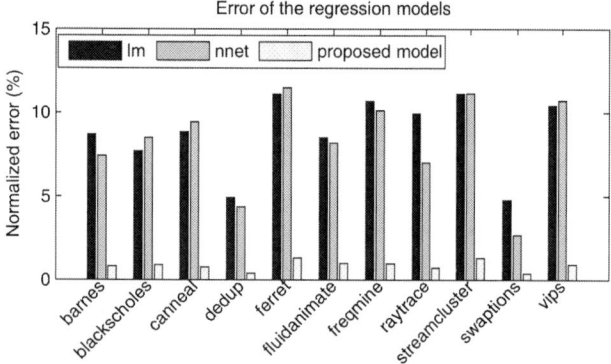

Fig. 3.   Comparison of the three models, linear regression (lm), neural network (NNet) and the proposed model as in Eqn. 2. Each of the application is runing on the 8x8 many-core system. All the errors (NRMSE) are normalized to the mean value of the cycles.

follows, we will present the verification of regression model of the performance vs. frequency first. Next the performance of the proposed APAT is compared against that of related approaches. In the end, the hardware cost and overhead of the DP-based optimization are analyzed.

B. Precision of the performance model

Suitable performance model plays an important role in the problem formulation. To justify the accuracy of the performance model in Eqn. 2, we compare it against two other approaches, the linear regression model (lm), and neural network model (nnet). In the lm model, the execution time is a linear combination of the frequencies of the tiles. In the nnet model, the relationship of the execution time and the frequencies of the tiles are found by an artificial neural network. We use the Matlab neural network toolbox to train a three stage neural network by varying the number of neurons in the hidden layer. The metric used here is the normalized root square mean squared error (NRMSE). Fig. 3 shows the errors of the three regression models, where the proposed regression model has shown significantly smaller error than the other two.

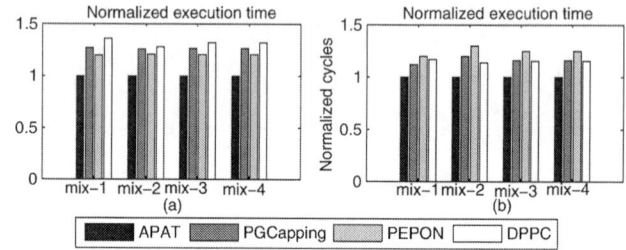

Fig. 4.   Execution time of the mixes under power budget of (a) 150W and (b) 70W.

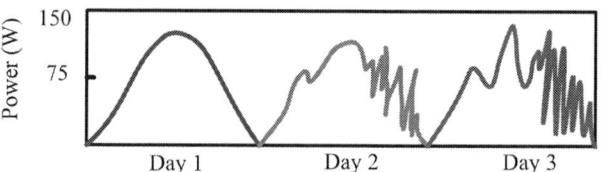

Fig. 5.   Solar energy data of three days in 2011 from ORNL website

C. Comparison of the power allocation methods

Fig. 4 shows the performance of the four methods when the power budget is high, 150W as in (a), and low, 70W as in (b). The execution time is measured as the total execution time of all the applications in one mix. From Fig. 4, one can see that APAT has the lowest execution time under both high and low power budgets. When the power budget is high, APAT records an average of 26 %, 20 % and 30 % less time than that of PGCapping, PEPON, and DPPC, respectively. When the power budget becomes low, on average, OPAD needs 19 %, 25 % and 16 % less execution time than that of PGCapping, PEPON, and DPPC, respectively.

To support solar energy powered system, a power budget estimator is used based on history to predict the power budget for the next period [9]. The prediction time interval is set to be 1 sec. The renewable energy can be smooth or highly variable, depending on the environment. The power data in Fig. 5 represent three days in different weather: a nice day, a day with nice morning and poor afternoon, and a terrible day. To see the real time power adaptiveness of the four methods, Fig. 6 compares the four methods with a variable power budget to mimic power budget from the renewable energy. The closer the actual power consumption is to the input power budget, the less the energy loss, and thus the better performance. In Fig. 6 the power budget is more smooth, representing a day of good weather. From Fig. 6 the power consumption of PGCapping, PEPON, and DPPC does not match the input power consumption and results in more energy loss. The energy losses (defined as the input power budget minus the power consumption and integrated over the time shown in Fig. 6) of PGCapping, PEPON, and DPPC over that of APAT are 11×, 4×, and 10×, respectively. In Fig. 7, lots of "jitters" or variations are made to mimic a day of poor weather. When the input budget variation is slow in Fig. 7 APAT still achieves better matching than the other three methods. The energy losses of PGCapping, PEPON, and DPPC over that of APAT are 11.3×, 3.5×, and 10×, respectively. Thus, APAT has the least energy loss.

Summary From the above experiments, APAT can have lower execution time given the same input power budget compared to three state-of-the-art methods, PGCapping, PEPON,

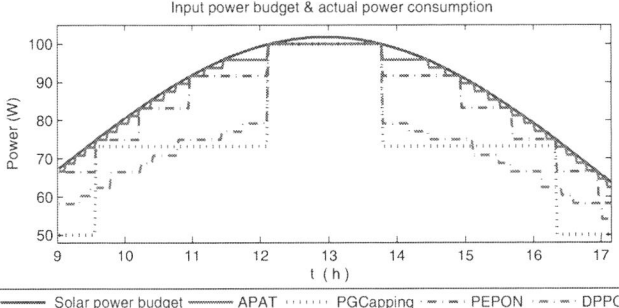

Fig. 6. Input power budget with low variation frequency mimicking a day of good weather and actual power consumption.

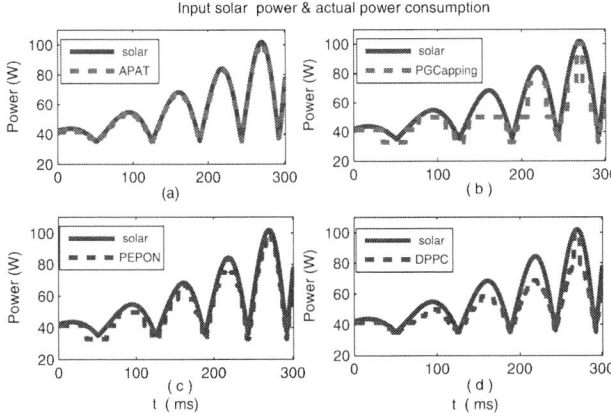

Fig. 7. Input power budget with variation frequency mimicking a day of bad weather and actual power consumption. The differences are shown between the input power budget and (a) APAT, (b) PGCapping, (c) PEPON and (d) DPPC.

and DPPC. On the other hand, as APAT has much lower overhead in power allocation, it is a beneficial method for many-core systems in which the power budget varies rapidly. Power consumption of APAT matches the input budget much better than the other three methods, resulting in less energy loss. Thus, APAT is suitable for online adaptive power allocation.

D. Hardware cost and runtime overhead

The hardware cost of the proposed APAT is mainly due to the nodes in the dynamic programming networks. Each node operation includes a 16-bit comparator and an adder. Each node has an area of 121 μm^2 and consumes 20 μW of power (assuming switching activity of 0.5) using Synopsys Design Compiler under 65nm TSMC library. As there are a total of $P \times (N + 1)$ vertices in DPN, for a system with 64 tiles and P in Eqn. 3 normalized to 10, the whole DPN area can be 78650 μm^2 and consumes 13 mW of power. For reference, a single 5×5 router with 2 virtual channels, flit size of 75 bits, and 4 flit-depth FIFO has an area of 145890 μm^2 and consumes 8 mW of power. That is, the total area and power consumption of the DPN is 53 % and 160 %, respectively, of a single router. The power consumption of the whole DPN is about 2% of the power consumed by the 64 routers.

The total run time of the APAT to allocate power online is $2N_q$ cycles, where N_q is the tile number (number of frequency variables) of application q. For an 16-tile application, only 32 cycles is needed. As a comparison, PGCapping, PEPON,

and DPPC each typically takes about $10 \sim 100$ M cycles, which is 6 to 7 orders of magnitude higher than that of APAT. In APAT, the regression model takes 10^5 cycles to compute. However, this is infrequently invoked; it only occurs at the initialization or at update interval for model error correction (typically, every 10 million cycles).

VI. CONCLUSIONS

In this paper, a power allocation algorithm was proposed to optimize performance with a power budget limit in energy harvesting based many-core systems. A dynamic programming approach is used which can run on a dynamic programming network to solve the problem with linear time complexity. The dynamic programming network is formed based on the performance-power model. The optimal solution can be found by traversing the network. This proposed algorithm can reduce execution time up to $20 \sim 30\%$ on average, compared to three existing approaches, PGCapping, PEPON, and DDPC with both smooth input power budget or power budget with rapid variations. The proposed algorithm also has very low run time, power and area overhead.

REFERENCES

[1] S. Borkar, "Thousand core chips: a technology perspective," in *Proc. Design Automation Conf.* pp. 746–749, ACM, 2007.

[2] M. Gorlatova, P. Kinget, I. Kymissis, D. Rubenstein, X. Wang, and G. Zussman, "Energy harvesting active networked tags (EnHANTs) for ubiquitous object networking," *IEEE Wireless Communications*, vol. 17, no. 6, pp. 18–25, 2010.

[3] C. Li, W. Zhang, C. Cho, and T. Li, "SolarCore: solar energy driven multi-core architecture power management," in *Proc. IEEE Int'l Symp. High Performance Computer Architecture*, pp. 205–216, IEEE, 2011.

[4] A. Sharifi, A. K. Mishra, S. Srikantaiah, M. Kandemir, and C. R. Das, "PEPON: performance-aware hierarchical power budgeting for NoC based multicores," in *Proc. Int'l Conf. Parallel Architectures and Compilation Techniques*, pp. 65–74, ACM, 2012.

[5] S. Imamura, H. Sasaki, N. Fukumoto, K. Inoue, and K. Murakami, "Optimizing power-performance trade-off for parallel applications through dynamic core and frequency scaling," *Proceedings of the RESoLVE*, 2012.

[6] K. Ma and X. Wang, "PGCapping: exploiting power gating for power capping and core lifetime balancing in CMPs," in *Proc. Int'l Conf. Parallel Architectures and Compilation Techniques*, pp. 13–22, 2012.

[7] C. Shen, X. Wang, W. Zhang, and F. Kang, "A high-performance three-dimensional micro supercapacitor based on self-supporting composite materials," *Journal of Power Sources*, vol. 196, no. 23, pp. 10465–10471, 2011.

[8] X. Wang, Z. Li, M. Yang, Y. Jiang, D. Masoud, and M. Terrence, "A low cost, high performance dynamic-programming-based adaptive power allocation scheme for many-core systems in the dark silicon era," in *Proc. ESTIMedia*, 2013.

[9] J. Lu, S. Liu, Q. Wu, and Q. Qiu, "Accurate modeling and prediction of energy availability in energy harvesting real-time embedded systems," in *Proc. Int'l Green Computing Conf.*, pp. 469–476, 2010.

[10] C. Li, A. Qouneh, and T. Li, "iSwitch: coordinating and optimizing renewable energy powered server clusters," in *Proc. Int'l Symp. Computer Architecture*, pp. 512–523, IEEE, 2012.

[11] K. Ma, X. Wang, and Y. Wang, "DPPC: dynamic power partitioning and control for improved chip multiprocessor performance," *IEEE Trans. Computers, in press*, 2013.

[12] C. M. Bishop, *Pattern recognition and machine learning*. Springer New York, 2006.

[13] T. Mak, P. Cheung, K. Lam, and W. Luk, "Adaptive routing in network-on-chips using a dynamic-programming network," *IEEE Trans. Industrial Electronics*, vol. 58, no. 8, pp. 3701–3716, 2011.

[14] J. Xue, A. Garg, B. Ciftcioglu, J. Hu, S. Wang, I. Savidis, M. Jain, R. Berman, P. Liu, M. Huang, H. Wu, E. G. Friedman, G. Wicks, and D. Moore, "An intra-chip free-space optical interconnect," in *Proc. Int'l Symp. Computer architecture*, pp. 94–105, ACM, 2010.

Variation-Aware Voltage Island Formation for Power Efficient Near-Threshold Manycore Architectures

Ioannis Stamelakos, Sotirios Xydis, Gianluca Palermo, Cristina Silvano

Politecnico di Milano - Dipartimento di Elettronica, Informazione e Bioingegneria
E-mail: {stamelakos, xydis, gpalermo, silvano}@elet.polimi.it

Abstract— **The power-wall problem and its dual utilization-wall problem are considered among the main barriers to feasible/efficient scaling in the manycore era. Several researchers have proposed the usage of aggressive voltage scaling techniques at the near-threshold voltage region, promising significant improvements in power efficiency at the expense of reduced performance values and higher sensitivity to process parametric variations. In this paper, we introduce a variability-aware framework for exploring the potential power-efficiency of the Near Threshold Computing (NTC) under performance constraints. We propose and analyze the usage of fine-grained voltage islands to cope with the increased effect of variability problem in the NTC region. For the considered workloads, we found that the power impact of fine-grained voltage islands formation can be up to 35% for a 128-core chip operating at NTC region, while the adoption of a variability aware technique can bring to a power reduction of up to 43% with respect to a variability unaware technique. Finally, we show that voltage regulator's complexity, in terms of voltage quantization levels, has a very low effect on the power efficiency at NTC, making in that way the usage of voltage islands a feasible solution for copying with variability.** [1]

I. INTRODUCTION

The continuous technology scaling following Moore's law [1], has recently raised the era of manycore architectures [2], [3], as the principal strategy for continuing performance growth. However, the end of Dennard's scaling [4] brings designers in front of the so called power/utilization wall. Projections show that the gap between the number of cores integrated on a chip and the number of cores that can be utilized will continue to widen on future technology nodes [5]. As a result, *dark silicon* - transistor count under-utilization due to power budget - has been recently emerged as major design challenge that jeopardizes the well established core count scaling path in current and future chip generations.

To address the dark silicon problem, researchers have proposed techniques at the micro-architectural level [6], [7], [8] down to physical and device level [9], [10]. Near-Threshold Voltage Computing (NTC) [11] has been proposed as a promising technique to mitigate the effects of dark silicon, allowing a large number of cores to operate under a given manycore power envelope. NTC takes advantage of the quadratic relation between the supply voltage (V_{dd}) and the consumed power, by lowering the operating V_{dd} to a region slightly larger than the transistors' threshold voltage (V_{th}).

[1]This work was partially supported by the EC under the grant HARPA FP7-612069

In comparison with the conventional Super-Threshold Voltage Computing (STC), computations at NTC regime are performed in a very energy efficient manner, unfortunately at the expense of reduced performance and high susceptibility to parametric process variations.

In this paper, we investigate the power efficiency potential of manycore architectures at the NTC regime, considering the process variations as well as a power delivery architecture supporting multiple V_{dd} domains, under strict performance constraints originated from multicore architectures at the STC regime. Unlike previous works on variation aware voltage allocation that target the STC regime [12], [13], [14], we propose the formation of voltage islands (VIs) for the minimization of the impact of within-die variations, which are more evident at NTC, in both performance and power. In particular, we developed a framework for manycore architectures operating at NTC to investigate the power efficiency under different workloads, while sustaining the performance when moving from the ST to the NT region. The framework has been parametrized in order to exploit different voltage island formation and to deal with process variability. Additionally, to generalize the analysis, we study four clustered manycore architectural organizations – differing on the number of cores per cluster.

Extensive experimental analysis showed that for the considered workloads, when moving to the NT regime for a 128-core architecture, average power gains close to 65% are delivered while sustaining the performance values obtained by a 16-core architecture at STC. The power impact of fine-grained voltage island formation can be up to 35% for a 128-core chip operating at NTC. Additionally, in comparison with variation unaware techniques, the proposed variation-aware NTC voltage island formation delivers power gain up to 8% considering a single VI per chip and up to 43% when considering the fine-grained multiple VI case, that is able to deal better with variability. Finally, analyzing the V_{dd} distributions at NTC, we demonstrated that the utilization of multiple VIs together with efficient integrated regulators can be considered a feasible option at NTC to efficiently deal with the process variability.

The remainder of the paper is organized as follows. Section II motivates this work by presenting the technical background regarding the NTC paradigm. Section III introduces the proposed framework for variation-aware voltage island formation

978-1-4799-2817-0/14 $31.00 © 2014 IEEE

at NTC under given performance constraints, while Section IV experimentally analyzes its effectiveness. Finally, Section V concludes the paper.

II. NTC: Motivation and Challenges

Near-threshold voltage operation relies on the aggressive tuning of the V_{dd} of the integrated circuit very close to the transistors' threshold voltage V_{th}, to a region where still $V_{dd} > V_{th}$. This decrement of the supply voltage increases the potential for energy efficient computation, e.g. by reducing V_{dd} from the nominal 1.1 V to 500 mV, energy gains of $10\times$ are reported [11]. While near-threshold region is not more power efficient than the theoretical limits provided in the sub-threshold/ultra-low voltage region, where $V_{dd} < V_{th}$, NTC has gained a lot of attention due to low energy operation at higher performance and easier adoption across multiple application platforms in comparison with the sub-threshold circuits. NT is the region that delivers interesting trade-offs regarding to energy efficiency and transistor delay, since super-threshold V_{dd} quickly reduces energy efficiency while sub-threshold V_{dd} leads to slower transistors. However, NTC comes together with two major drawbacks, i.e. reduced performance and increased sensitivity to process variations.

Performance reduction at NTC is exposed through the limited maximum achievable clock frequency. This is an implicit effect due to the reduction of the $V_{dd} - V_{th}$ difference, applied when moving to the NTC region. Performance degradation can be compensated by exploiting trade-off points of higher task parallelism at lower clock frequencies. Thus, an important open question for NTC to be examined is the following: *Is the inherent parallelism of the existing applications enough to retain the performance levels of super-threshold design with lower power consumption, thus making it worth going to near-threshold operation?* Pinckey et al. [15] studied the limits of voltage scaling together with task parallelization knobs to address the performance degradation at NTC by considering a clustered micro-architectural template with cores sharing the local cache memory [16]. They showed that under realistic application/architecture/technology features (i.e. parallelization efficiency, inter-core communication, V_{th} selection etc.) the theoretical energy optimum point ($\frac{dEnergy}{dV_{dd}} = 0$), moves from the sub-threshold to the near-threshold region. Considering a single supply voltage per die, the energy optimum point can be found within an interval of 200 mV higher V_{th}, thus implicitly defining the upper limits of the NTC region.

The second important challenge for manycore architectures operating at the NTC regime is their increased sensitivity to process variations. The transistor delay is heavily affected by the variation of V_{th} at NT voltages compared to the one in super-threshold voltages [17], [18]. In addition, failure rate of conventional SRAM cells is increased in low voltage operation [19]. As a consequence, the operating frequency of the cores varies considerably, reducing the yield. In addition, variation's effects on the total power of the chip have to be carefully considered, due to the exponential dependency of leakage current upon V_{th}. Karpuzcu et al. presented Various-NTV [20],

a micro-architectural model that adopts proper gate-delay and SRAM cell type models to capture the increased sensitivity of manycore chips to process variations at NTC. For mitigating variation effects on performance at NTC, the EnergySmart architecture and thread assignment methodology have been recently proposed in [21]. By assuming on-chip voltage regulators of low-efficiency, EnergySmart adopts single voltage - multiple frequency islands to cope with variability.

In this paper, we focus our study on the NTC design space defined though [11] and [21]. Specifically, we target power efficient NTC manycore architectures that sustain STC performance levels considering their increased sensitivity to process variation. While previous work [11] addresses the problem by considering single per chip voltage domain, in this paper we differentiate our approach by exploring multiple voltage domain NTC architectures through variation aware-voltage island formation techniques. Although in [21] some concerns are raised regarding the usage of on-chip regulators for multiple voltage islands in NTC, we motivated our research by the recent advancements in cost-effective fully-integrated voltage regulators (FIVR) [22]. FIVR regulators are expected to achieve very high voltage efficiency, enabling the integration of multiple power rails. We show that the workload characteristics at NTC enable fine-grained V_{dd} tuning, by delivering narrow V_{dd} distributions with very low power consumption (Section IV).

III. Framework for VI Formation at NTC

The overall exploration framework for variation-aware VI formation at NTC is shown in Figure 1. It accepts as main inputs the performance, power and area characterization curves of the targeted application at the STC regime. The STC characterization is performed by scaling the number of cores of the underlying manycore architecture template and then scaling accordingly the application's degree of parallelization. The Sniper multicore simulator [23] and the McPAT power modeling framework [24] have been used for the performance and power characterization, respectively. Designer/architect specific constraints are provided regarding the minimum allowed performance, L_{min} and the maximum core count constraint, C_{max} for the NTC manycore. Given the aforementioned inputs, the proposed exploration framework generates the VI configurations and the corresponding V_{dd} allocation decisions per VI, for a manycore architecture with C_{max} number of cores operating at the NTC regime and satisfying the performance constraint L_{min} under parametric process variation. In the remainder of this section, we describe the basic components of the proposed methodology.

A. NTC Frequency Calculation for Sustaining STC Performance

Core to core frequency and leakage variations can be mitigated by varying voltage/frequency levels of independent voltage islands. There are four different power management schemes supporting voltage/frequency islands, i.e. Single-Voltage/Single-Frequency (SVSF) for all cores,

978-1-4799-2817-0/14 $31.00 © 2014 IEEE   305

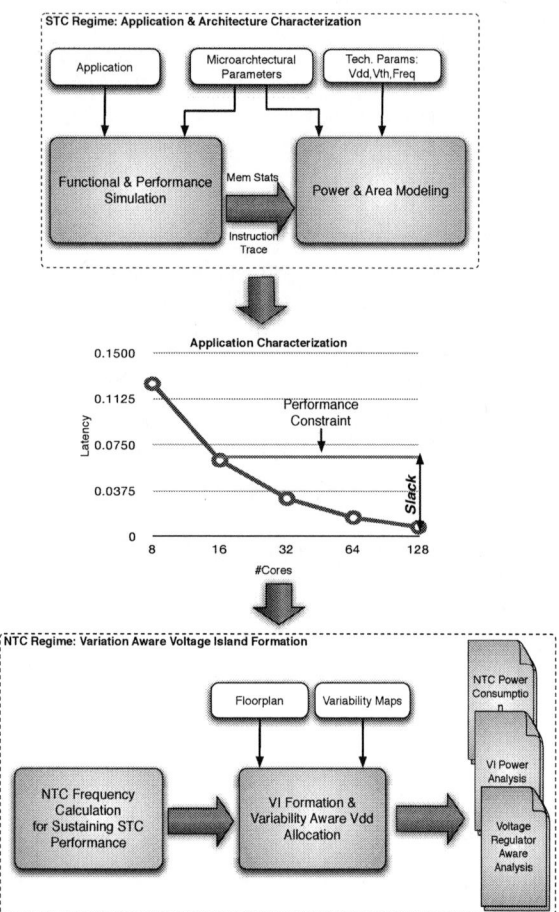

Fig. 1. Framework for variation-aware VI formation.

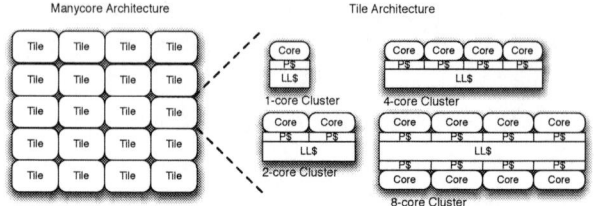

Fig. 2. Tile-based manycore architecture and the S1-S8 type of clusters.

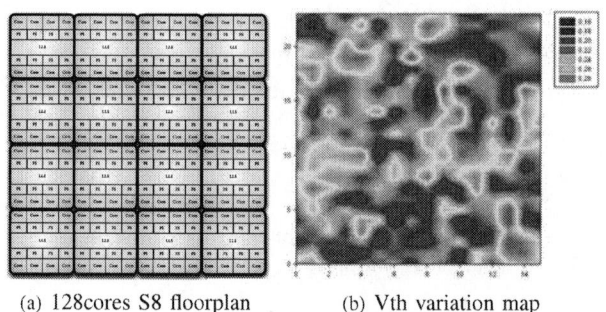

(a) 128cores S8 floorplan   (b) Vth variation map

Fig. 3. Manycore architecture: Floorplan and the corresponding V_{th} and timing variation maps.

lower frequency. Utilizing this positive slack, the f_{NTC} is calculated as follows:

$$f_{NTC} = \frac{L_{C_{max}}}{L_{min}} \times f_{STC} \qquad (1)$$

B. Tiled Manycore Architecture Description

While the framework considers the underlying manycore architecture as a free exploration parameter, we focus our study on tile-based architectures. Figure 2 shows an abstract view of the tile-based manycore architecture, as well as the different intra-tile organizations. We consider four intra-tile architectures by varying the number of cores per tile and the memory configuration of the last level cache (LLC) per tile. Each core owns a private instruction and data cache (P\$). The LLC (LL\$) is shared among the different cores composing a tile. The Intel Nehalem processor [25] configuration for the core and the P\$ has been adopted. While the P\$ size remains constant across the different intra-tile configurations, the size of the (LL\$) is scaled according to the number of cores in the tiles, keeping constant the total chip area. We use the following abbreviations for differentiating manycore architectures based on four tile types: (i) *S1:* each core owns a Last Level LL\$, (ii) *S2:* LL\$ is shared between 2 adjacent cores, (iii) *S4:* LL\$ is shared among 4 adjacent cores, (iv) *S8:* LL\$ is shared among 8 adjacent cores. While S4 and S8 resemble the cluster organizations proposed in [16], [21], we also explored more fine-grained clusters, i.e. S1 and S2. Tile's type defines the minimum VI granularity supported by each manycore configuration. Thus, for a *Si* manycore platform the finest granularity of a voltage domain is i core per VI.

Single-Voltage/Multiple-Frequencies (SVMF), Multiple-Voltages/Multiple-Frequencies (MVMF) and Multiple-Voltages/Single-Frequency (MVSF). While the SVSF scheme requires overdesigned power management decisions, the SVMF [21] and MVMF schemes mitigate variations by exposing the system heterogeneity to the programmer, i.e. application execution on different clock frequencies across the cores. In order to alleviate the aforementioned inefficiencies, we adopt the MVSF power management scheme that enables to mitigate process variations by tuning the V_{dd} allocation of the VIs while exposing to the programmer an iso-frequency view of the overall manycore platform.

Under this configuration, we first calculate the clock frequency of the platform at NTC regime, f_{NTC}, that satisfies the performance constraint, L_{min}. f_{NTC} will be then used by the VI formation and variability aware V_{dd} allocation module to configure the MVSF decisions. Let $L_{C_{max}}$ be the performance, in terms of latency, at the STC regime of a manycore architecture with C_{max} number of cores, running at f_{STC}. At STC region, $L_{min} - L_{C_{max}} > 0$ is the available slack in latency due to the higher parallelism degree of the architecture, that can be exploited to run the application in

C. Micro-architectural Process Variation Model at NTC

In order to capture the process variation at the NT regime, we integrate the Various-NTV [20] microarchitectural model within the proposed framework. While Various-NTV reuses the spherical distance function in [26] for modeling the intra-die spatial correlations, it heavily extends [26] by updating the STC micro-architectural delay and SRAM cell models to reflect in a more accurate manner the higher sensitivity of NTC on process variation. We used ArchFP [27] tool to automatically generate the floorplan of the targeted many-core architectures. Based on the provided manycore floorplan, Various-NTV generates the corresponding variation maps accounting for the within-die (WID) and die-to-die (D2D) process variations. Figure 3 shows the floorplan of the S8 manycore architecture with 128 cores (Figure 3(a)), together with a sample instance of its V_{th} variation map (Figure 3(b)). Assuming B as the set of component blocks found in the floorplan and D the set of dies, we now define $V_{th}^{(i,j)}, i \in B, j \in D$ that corresponds to the V_{th} of the architecture's component i in sample die j. Once extracted, $V_{th}^{(i,j)}$ is used for allocating to each component the lowest possible $V_{dd}^{(i,j)}$ for sustaining f_{NTC} frequency constraint given that:

$$f_{NTC} \propto \frac{(V_{dd}^{(i,j)} - V_{th}^{(i,j)})^{\beta}}{V_{dd}^{(i,j)}} \qquad (2)$$

where β is a technology-dependent constant (≈ 1.5).

D. Voltage Island Formation & Variability Aware V_{dd} Allocation

The final phase of the proposed framework first performs the generation of the VIs and then for each VI solution computes the per island V_{dd} assignment that satisfies the f_{NTC} derived by the latency performance constraint. The VI formation procedure explores all valid granularities of rectangular voltage islands in both vertical and horizontal directions, by grouping together adjacent cores. We use the notation $r \times c$, i.e. r rows - c cores per row, to indicate the type of the evaluated VI granularity. Depending on the type of the underlying tiled architecture (*S1, S2, S4, S8*), different constraints are included in the VI formation, since not all the core groupings are valid. For example, in the case of a *S8* tiled manycore, LL\$ is shared among 8 cores, thus the finest granularity that we can be evaluated is 2×4: 2 rows, 4 cores per row.

For the jth die, $j \in D$, each VI, $k \in VI$, operates in its own $V_{dd}^{(k,j)}$, tuned for the $VI_{k,j}$ group of processors and memories. In $VI_{k,j}$, the core with the highest $V_{th}^{(i,j)}, i \in B, j \in D$ determines the V_{dd} for the specific voltage island, to satisfy the VI_k's critical path timing. The trade-off by moving towards coarse grained VI granularities is that, we reduce area cost since less voltage regulation logic is allocated at the expense of degrading the power efficiency of the manycore in respect to the finest possible granularity. For $B_k, k \in VI$, the set of resources found in VI_k and from eq. 2, we calculate $V_{dd}^{(k,j)}$ according to the following relation:

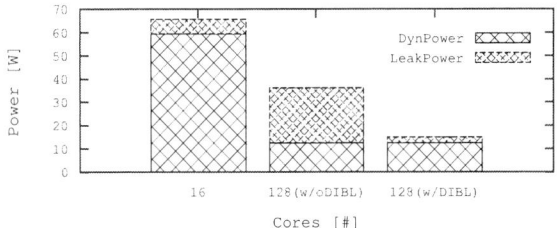

Fig. 4. Power breakdown for STC-16core and NTC-128core architectures with and without DIBL effect

$$V_{dd}^{(k,j)} = \max_{i \in B_k, j \in D} \left[V_{dd}^{(i,j)} \right] \qquad (3)$$

Given the V_{dd} allocation per VI, $V_{dd}^{(k,j)}, k \in VI, j \in D$, and the power characterization for the manycore with C_{max} number of cores at STC, we can calculate the power of each component in NTC. For $i \in B_k, j \in D, k \in VI$, the dynamic, DP and leakage, LP, power scaling factors are:

$$SF_{DP}^{(i,j,k)} = \left(\frac{V_{dd}^{(k,j)}}{V_{dd_{STC}}} \right)^2 \times \left(\frac{f_{NTC}}{f_{STC}} \right) \qquad (4)$$

$$SF_{LP}^{(i,j,k)} = \left(\frac{V_{dd}^{(k,j)}}{V_{dd_{STC}}} \right) \times \exp \left(\frac{V_{th_{STC}} - V_{th}^{(i,j)} + DIBL}{n \times V_{thermal}} \right) \qquad (5)$$

$$DIBL = \lambda (V_{dd}^{(k,j)} - V_{dd_{STC}}) \qquad (6)$$

where $DIBL$ is the coefficient modeling the Drain-Induced Barrier Lowering effect, $V_{thermal}$ is the thermal voltage, and n is the sub-threshold slope coefficient. The DIBL effect is a deep-submicron effect and is related to the reduction of the threshold voltage as a function of the drain voltage. DIBL is enhanced at higher drain voltage and tends to become more severe with process scaling to shorter gate lengths. Lowering supply voltage provides an exponential reduction in sub-threshold current resulting from the DIBL effect. Figure 4 shows the impact of DIBL effect on the reduction of leakage power in manycore architectures at NTC regime. As shown, by moving from STC multicore (16 cores) to NTC manycore (128 cores) architecture configurations, the DIBL effect accounts for a significant portion of the total power of the system.

IV. EXPERIMENTAL RESULTS

In this section, we experimentally evaluate the efficiency of the proposed framework. Without loss of generality, we consider that the performance L_{min} corresponds to a 16 core multicore in the STC regime, while the constraint C_{max} targets a 128 core many-core chip at NTC, at 22nm technology node. Maximum V_{dd} was set to 1.05V and the frequency to 3.2 GHz for the STC regime, according to parameter values derived from [28] for conservative technology scaling. From Various-NTV, we extract 100 different variation maps by using a 24x16 grid based on the core/cache granularity. The most significant parameters and their values are summarized in

TABLE I

EXPERIMENTAL SETUP

Parameters	Value
Process Technology	22nm
STC Frequency	3.2GHz
STC Supply Voltage	1.05V
Nominal $V_{th}/\sigma_{V_{th}}$	0.23V/0.025
Core Area	$6mm^2$
Private Cache Size/Area	$320KB/4.14mm^2$
Last Level Cache S_i Size – Area	$(2 \times i)$ MB / $(3.88 \times i)mm^2$

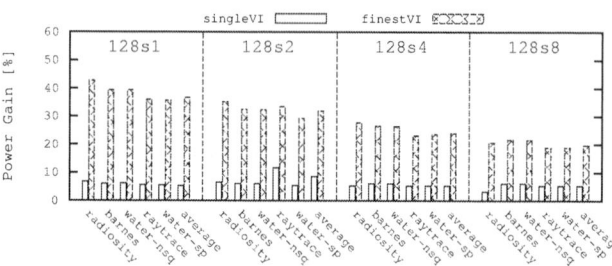

Fig. 6. Power gains of variability-aware NTC technique w.r.t. overdesign

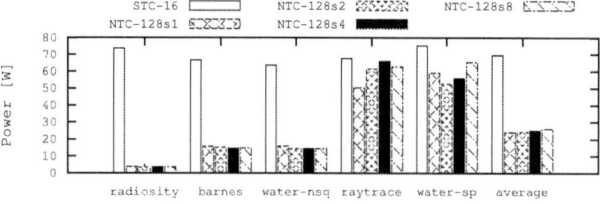

Fig. 5. Power reductions: 16-core STC chip versus 128-core NTC

Table I. For our experiments we used five applications from the SPLASH-2 benchmark suite [29] by using the "large dataset" provided within Sniper [23]. The applications exhibit different speedup scaling degrees, i.e. close to ideal (*radiosity*), medium (*barnes, water-nsq*) and limited (*raytrace, water-sp*). Additionally, we examined an *average* case workload, that aggregates in the execution sequence the five aforementioned applications. Specifically, this *average* case workload is like executing all the aforementioned applications, one after the other and then treating it as a single benchmark. In that way, we manage to see what happens in an average case since it includes benchmarks that scale well and others that don't scale well. We present results regarding the power efficiency delivered by adopting the proposed approach and we provide a sensitivity oriented analysis regarding parameters of the voltage regulation structure.

A. Analysis of Power Gains at NTC Regime

1) Moving from 16 STC Cores to 128 NTC Cores: Figure 5 shows the power consumption when moving from 16 cores at STC to 128 cores at NTC for each benchmark. Radiosity delivers the highest power gains since it scales almost ideally in terms of performance as the number of cores increases. For the radiosity benchmark we observed a 95% decrement in power while for the barnes and water_nsq that exhibit a medium scaling behavior, we observe a reduction of 77%. As shown, the scaling behavior of the applications with respect to increasing the number of cores, heavily affects the power efficiency at NTC, since the application takes advantage of the available performance slack when moving to a large number of cores. Thus, besides frequency we can aggressively scale down V_{dd} as well, and reduce power drastically, taking advantage of V_{dd}'s quadratic relation with dynamic power and its linear plus DIBL relation with leakage current. Raytrace and water_sp exhibit lower scaling degrees, limiting performance boost

when their load is split and distributed to 128 cores. In this case, V_{dd} assignment acquires higher values restricting power gains. The last column of Figure 5 depicts the power gains delivered in the average workload. Although benchmarks that don't scale well are included, a near threshold V_{dd} (∼0.4V) is acquired, delivering a 65% power reduction with respect to the 16 core STC manycore.

2) Variation Aware versus Overdesign NTC operation: We compared the power gains delivered by the proposed variation aware VI formation versus an overdesign approach to mitigate variation effects. From the V_{th} distribution, we calculate the V_{dd} of architectural components according to Eq. 2, with V_{th}'s overdesign value being equal to $\mu_{V_{th}} + 3\sigma_{V_{th}}$. Figure 6 reports the gains of the variability aware approach over the overdesign one. The histograms with the *singleVI* annotation represent power gains when having only one VI, and as a consequence one V_{dd} for the whole chip. Under a *singleVI* configuration, the variation aware approach achieves power gains around 5%, for all the available cluster architectures (S_i, $i \in \{1, 2, 4, 8\}$). On the contrary, the histograms with the *finestVI* annotation show the power gains achieved by considering the finest VI granularity possible for each architecture. Since S1 enables the finest 1×1 VI granularity to be exploited, it delivers the highest gains over the overdesign approach, that range between 34-42%. In the rest of the architectures, namely (S2, S4, S8), the gains vary between 29-34%, 25-28% and 18-23%, respectively.

3) Analysis of VI Granularity on NTC Power Efficiency: Figure 7 shows the impact of the different voltage island configurations in terms of power consumption at the NTC regime. The voltage island formation that has been analyzed includes all the possible combination in terms of power-of-two of the cores. We restrict the voltage island granularities to an aspect ratio between 1 and 1/4 (considering a voltage island configuration $c \times r$ the aspect ratio is c/r). In each case, we considered the tile size as the smallest possible voltage island. The constant trend over all the workloads and architectures, is that finer the granularity of the voltage island higher the power savings. In fact, selecting smaller voltage islands, we can cope with the variability in a more aggressive way by using a fine-grained tuning of the V_{dd} over the entire chip. The advantage passing from the single voltage island to the finest voltage island depends on the different architectural

978-1-4799-2817-0/14 $31.00 © 2014 IEEE

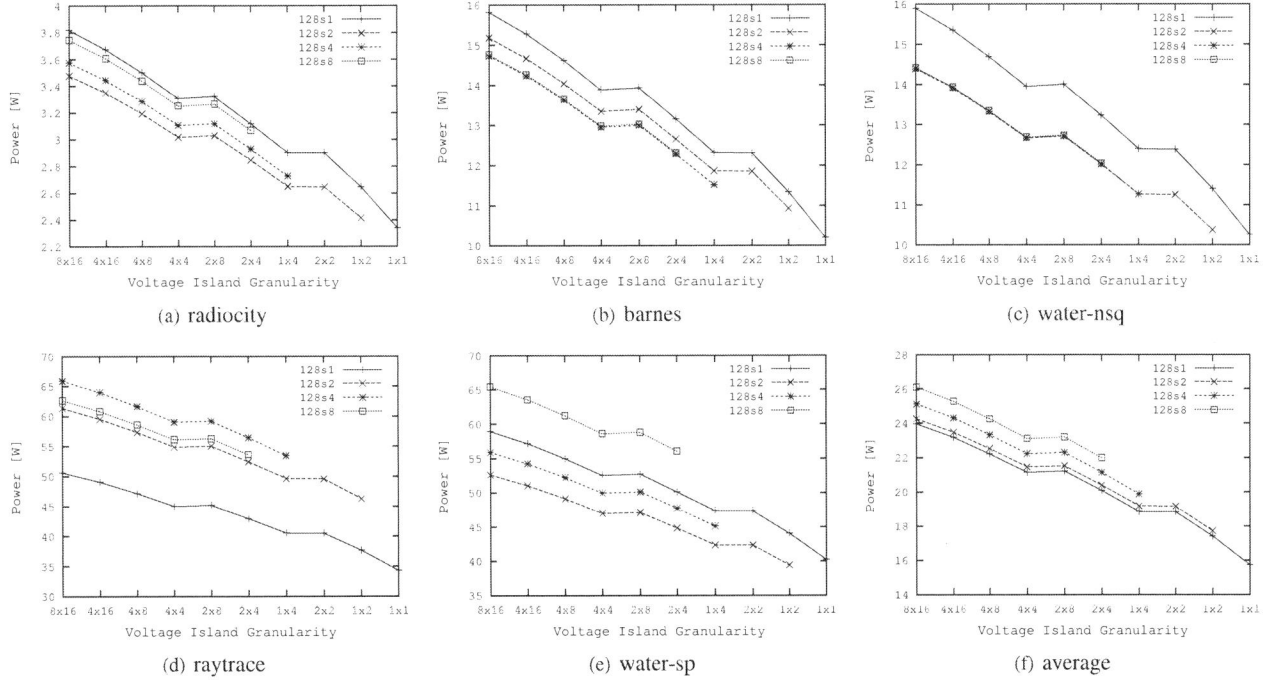

Fig. 7. Impact of voltage island granularity on power consumption

configuration: 30-35% for 128s1, 24-30% for 128s2, 19-24% for 128s4 and 14-18% for 128s8. In addition, the impact of the different voltage island configurations (when composed of the same number of cores), such as passing from 4×4 to 2×8 or passing from 1×4 to 2×2, is very limited. Despite the global trend, analyzing in more detail the power behavior over all the four clustered architectures, we noticed that the 128s1 architecture is not constantly the best configuration across the various VI granularities. In some cases the best configuration shifts over the 128s2 architecture (see water-sp in Figure 7). This phenomenon depends on the application scalability over the different clustered architectures.

B. Voltage Regulation Oriented Analysis

The analysis conducted so far considers the ability to ideally deliver all the requested voltage levels. Since this is not a realistic scenario according to current state-of-art power supply architectures, hereafter we analyze the impact of the on-chip voltage regulator resolution on power efficiency.

We analyzed three different voltage regulator resolutions, delivering voltage with a precision of (i) 12.5mV, (ii) 25mV and (iii) 50mV. Adopting the aforementioned schemes, we demonstrate the effect of allocating integrated regulators in the NTC region (from $[V_{th}] \longleftrightarrow [V_{th} + 200mV]$) that includes respectively 16, 8, and 4 voltage quantization levels. Figure 8 presents the average power overhead for each one of the voltage regulator precisions. Power overhead refers to the normalized difference between the power consumed in the ideal case (voltage regulator delivering arbitrary V_{dd} values) and the power with the specific value of voltage precision. The

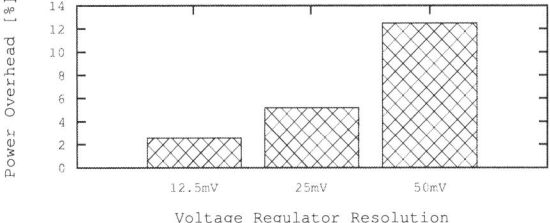

Fig. 8. Impact of voltage regulator resolution on power efficiency at NTC.

results are the average values for all the benchmarks and all the four architectures that we investigated. As expected, the higher is the resolution the smaller is the overhead since we are closer to the ideal case, passing from a 12% at 50mV to less than 3% at 12.5mV. For the applications that exhibit ideal or medium scaling with respect to increasing the number of cores, such as radiosity, barnes and water-nsq, the overhead of 12% can be efficiently compensated by the low-power consumption at NTC regime. On the opposite, for the case of applications with limited scaling, such as raytrace and water-sp, an integrated voltage regulation scheme that provides high resolution of the delivered V_{dd} is preferable.

Finally, Figure 9 shows the V_{dd} probability distribution considering the 12.5mV as the regulator's granularity for the *barnes* application running on the 128core architecture and operating at NTC, across all the examined *S1-S8* tile types. We can observe that the V_{dd} distribution is very concentrated across the mean value $\mu = 0.388V$ with $\sigma = 0.071V$. The

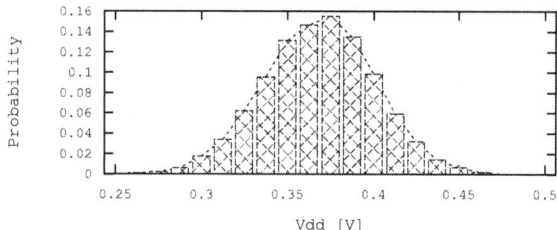

Fig. 9. Distribution of Vdd voltage at NT region.

V_{dd} distributions of the rest of the applications resemble the depicted one on Figure 9 with similar σ and slightly shifted μ values, according to the scaling behavior of the application. Narrow V_{dd} distributions together with low power consumption at NTC, instruct that the integrated voltage regulation circuitry can efficiently supply the requested power without the need of allocating multiple levels of voltage supplies, showing the efficiency of moving towards multiple VIs for supporting NTC operation on manycore chips.

V. CONCLUSION

In this paper, we presented a variability-aware framework for exploring the power-efficiency of Near-Threshold Computing. Motivated by recent advancement on power delivery systems, we proposed the utilization of voltage island formation combined with the operation at the near-threshold region as an effective technique for building power efficient manycore architectures that sustain performance values delivered by conventional super-threshold computing. Through extensive experimentation, we showed the optimization potentials of moving towards near-threshold voltage computation, exposing its high dependency on both workload characteristics and underlying architectural organization.

REFERENCES

[1] Gordon E Moore et al. Cramming more components onto integrated circuits, 1965.

[2] S. Dighe et al. Within-die variation-aware dynamic-voltage-frequency-scaling with optimal core allocation and thread hopping for the 80-core teraflops processor. *J. Solid-State Circuits*, 46(1):184–193, 2011.

[3] J. Howard et al. A 48-core ia-32 message-passing processor with dvfs in 45nm cmos. In *ISSCC*, pages 108–109. IEEE, 2010.

[4] R.H. Dennard, F.H. Gaensslen, V.L. Rideout, E. Bassous, and A.R. LeBlanc. Design of ion-implanted mosfet's with very small physical dimensions. *Solid-State Circuits, IEEE Journal of*, 9(5):256–268, 1974.

[5] Hadi Esmaeilzadeh, Emily Blem, Renee St. Amant, Karthikeyan Sankaralingam, and Doug Burger. Dark silicon and the end of multicore scaling. In *Proceedings of the 38th annual international symposium on Computer architecture*, ISCA '11, pages 365–376, 2011.

[6] N. Goulding-Hotta, J. Sampson, G. Venkatesh, S. Garcia, J. Auricchio, P. Huang, M. Arora, S. Nath, V. Bhatt, J. Babb, S. Swanson, and M.B. Taylor. The greendroid mobile application processor: An architecture for silicon's dark future. *Micro, IEEE*, 31(2):86–95, 2011.

[7] V. Govindaraju, Chen-Han Ho, and K. Sankaralingam. Dynamically specialized datapaths for energy efficient computing. In *High Performance Computer Architecture (HPCA), 2011 IEEE 17th International Symposium on*, pages 503–514, 2011.

[8] Yatish Turakhia, Bharathwaj Raghunathan, Siddharth Garg, and Diana Marculescu. Hades: architectural synthesis for heterogeneous dark silicon chip multi-processors. In *DAC*, page 173. ACM, 2013.

[9] Arun Raghavan, Yixin Luo, Anuj Chandawalla, Marios C. Papaefthymiou, Kevin P. Pipe, Thomas F. Wenisch, and Milo M. K. Martin. Computational sprinting. In *HPCA*, pages 249–260. IEEE, 2012.

[10] Francesco Paterna and Sherief Reda. Mitigating dark-silicon problems using superlattice-based thermoelectric coolers. In *Proceedings of the Conference on Design, Automation and Test in Europe*, DATE '13, pages 1391–1394, San Jose, CA, USA, 2013. EDA Consortium.

[11] Ronald G. Dreslinski, Michael Wieckowski, David Blaauw, Dennis Sylvester, and Trevor N. Mudge. Near-threshold computing: Reclaiming moore's law through energy efficient integrated circuits. *Proceedings of the IEEE*, 98(2):253–266, 2010.

[12] A. Das, S. Ozdemir, G. Memik, and A. Choudhary. Evaluating voltage islands in cmps under process variations. In *Computer Design, 2007. ICCD 2007. 25th International Conference on*, pages 129–136, 2007.

[13] Sohaib S. Majzoub, Resve A. Saleh, Steven J. E. Wilton, and Rabab K. Ward. Energy optimization for many-core platforms: communication and pvt aware voltage-island formation and voltage selection algorithm. *Trans. Comp.-Aided Des. Integ. Cir. Sys.*, 29(5):816–829, May 2010.

[14] Sebastian Herbert, Siddharth Garg, and Diana Marculescu. Exploiting process variability in voltage/frequency control. *IEEE Trans. VLSI Syst.*, 20(8):1392–1404, 2012.

[15] N. Pinckney, K. Sewell, R. G. Dreslinski, D. Fick, T. Mudge, D. Sylvester, and D. Blaauw. Assessing the performance limits of parallelized near-threshold computing. In *Proceedings of the 49th Design Automation Conference*, pages 1147–1152, 2012.

[16] Ronald G. Dreslinski, Bo Zhai, Trevor N. Mudge, David Blaauw, and Dennis Sylvester. An energy efficient parallel architecture using near threshold operation. In *PACT*, pages 175–188, 2007.

[17] M. Eisele, J. Berthold, D. Schmitt-Landsiedel, and R. Mahnkopf. The impact of intra-die device parameter variations on path delays and on the design for yield of low voltage digital circuits. *Very Large Scale Integration (VLSI) Systems, IEEE Transactions on*, 5(4):360–368, 1997.

[18] D. Markovic, C.C. Wang, L.P. Alarcon, Tsung-Te Liu, and J.M. Rabaey. Ultralow-power design in near-threshold region. *Proceedings of the IEEE*, 98(2):237–252, 2010.

[19] L. Chang, R.K. Montoye, Y. Nakamura, K.A. Batson, R.J. Eickemeyer, R.H. Dennard, W. Haensch, and D. Jamsek. An 8t-sram for variability tolerance and low-voltage operation in high-performance caches. *Solid-State Circuits, IEEE Journal of*, 43(4):956–963, 2008.

[20] Ulya R. Karpuzcu, Krishna B. Kolluru, Nam Sung Kim, and Josep Torrellas. Varius-ntv: A microarchitectural model to capture the increased sensitivity of manycores to process variations at near-threshold voltages. In *IEEE/IFIP International Conference on Dependable Systems and Networks, DSN*, pages 1–11, 2012.

[21] Ulya R. Karpuzcu, Abhishek A. Sinkar, Nam Sung Kim, and Josep Torrellas. Energysmart: Toward energy-efficient manycores for near-threshold computing. In *HPCA*, pages 542–553, 2013.

[22] Intels fourth generaton Core CPU Haswell. FIVR Fully Integrated Voltage Regulator, http://www.intel.com, 2013.

[23] Trevor E. Carlson, Wim Heirman, and Lieven Eeckhout. Sniper: Exploring the level of abstraction for scalable and accurate parallel multi-core simulations. In *International Conference for High Performance Computing, Networking, Storage and Analysis (SC)*, 2011.

[24] Sheng Li, Jung Ho Ahn, Richard D. Strong, Jay B. Brockman, Dean M. Tullsen, and Norman P. Jouppi. Mcpat: an integrated power, area, and timing modeling framework for multicore and manycore architectures. In *Proceedings of the 42nd Annual IEEE/ACM International Symposium on Microarchitecture*, MICRO 42, pages 469–480, 2009.

[25] D. Kanter. Inside nehalem: Intels future processor and system. http://www.realworldtech.com, 2008.

[26] S.R. Sarangi, B. Greskamp, R. Teodorescu, J. Nakano, A. Tiwari, and J. Torrellas. Varius: A model of process variation and resulting timing errors for microarchitects. *Semiconductor Manufacturing, IEEE Transactions on*, 21(1):3–13, 2008.

[27] Gregory G. Faust, Runjie Zhang, Kevin Skadron, Mircea R. Stan, and Brett H. Meyer. Archfp: Rapid prototyping of pre-rtl floorplans. In Srinivas Katkoori, Matthew R. Guthaus, Ayse Kivilcim Coskun, Andreas Burg, and Ricardo Reis, editors, *VLSI-SoC*, pages 183–188. IEEE, 2012.

[28] S. Borkar. The exascale challenge. In *VLSI Design Automation and Test (VLSI-DAT), 2010 International Symposium on*, pages 2–3, 2010.

[29] Steven Cameron Woo, Moriyoshi Ohara, Evan Torrie, Jaswinder Pal Singh, and Anoop Gupta. The splash-2 programs: characterization and methodological considerations. *SIGARCH Comput. Archit. News*, 23(2):24–36, May 1995.

An Evaluation of an Energy Efficient Many-Core SoC
with Parallelized Face Detection

Hiroyuki Usui, Jun Tanabe, Toru Sano, Hui Xu, Takashi Miyamori

Toshiba Corporation, Toshiba Semiconductor & Storage Products Company, Kawasaki, Japan
{hiroyuki1.usui, jun.tanabe, toru.sano, kei.jo, takashi.miyamori}@toshiba.co.jp

Abstract – New applications such as image recognition and augmented reality (AR) have become into practical on embedded systems. For these applications, we have developed a many-core SoC that includes two many-core clusters with 32 energy efficient processor cores connected by a low latency tree-based NoC. In this paper, we evaluate performance of many-core SoC by face detection as an example of real image recognition applications and discuss two parallelized implementations on the many-core clusters. By keeping balance of workloads on the cores, the performance scales up to 64 cores and the SoC consumes only 2.21W. The energy efficiency is several tens of times better than that of a high performance desk-top quad-core processor.

I. Introduction

New media processing applications such as image recognition, augment reality (AR), and computer vision, have become into practical on embedded systems, automotive, surveillance, digital-consumer, and mobile systems. These data-intensive applications require high performance with low power consumption. The power consumption is limited to roughly under 2-4W for cooling without fans, for example. One of effective approaches to meet these requirements is many-core architecture with energy efficient processor cores.

Accordingly, we have developed a many-core SoC designed for embedded applications [1]. Unlike many-core architecture composed of high performance cores [2-3] in High Performance Computing (HPC) area, our many-core SoC is composed of energy efficient processor cores. The many-core SoC consists of two 32-core clusters. The 32 cores in one of the clusters share an L2 cache and they are connected by a low latency tree-based Network-on-Chip (NoC). As realistic applications, we have evaluated full-HD H.264 video decoding and super resolution [1]. The power consumption of the many-core cluster is less than 1W. The many-core SoC with a similar concept has been proposed in [4]. However, they only evaluate the performance with kernels of image processing and recognition, corner detection and SIFT by simulation.

In this paper, we will show performance evaluation results of face detection, as an example of real image recognition applications, on actual many-core SoC chips. Face detection is widely utilized in embedded systems such as digital camera, automotive, surveillance, human-computer interface, and so on. Many kinds of algorithm to detect faces have been proposed to meet system requirements. For instance, surveillance systems need higher detection accuracy than others. Programmable many-core architecture has the advantage of being able to deal with various algorithms,

even ones proposed in the future, compared with hardware accelerators, which operate only specific algorithms. We implemented a face detection based on a method using Joint Haar-like features [5] that is high accuracy and meets requirements of surveillance systems.

In image recognition, an image includes many Regions-of-Interest (ROIs) and each ROI is generally processed independently. Therefore, it is efficient to exploit coarse grain thread level parallelism because overhead of scheduling thread can be minimized. This characteristic of image recognition is different from that of the video codec that has dependencies between macroblocks. However, there is an issue to parallelize face detection effectively. The workload of each parallelized task varies dynamically depending on whether faces exist or not in the ROIs. The total performance of face detection tends to be limited by the slowest task. To achieve high scalability, we have to allocate tasks to many cores balancing workloads among cores. Considering this point, we evaluate two parallelized methods; allocating scanned image lines cyclically and dividing an image equally. We also evaluate power consumption of the many-core SoC and show its energy efficiency.

II. Architecture of the Many-Core SoC

A. Overview of the Many-core SoC

Fig. 1 shows the block diagram of our many-core SoC. Unlike other proposed many-core LSIs [2-3] that consist of only the array of processor cores and minimum interfaces, the many-core SoC is highly integrated with two many-core clusters, two dynamic reconfigurable engines [6], image recognition and processing accelerators, integrated SRAMs, two ARM® Cortex™-A9 processors, and various interfaces. The many-core cluster consists of 32 cores called Media Processing Engine Block (MPB) and 2MB L2 cache. The cluster will be explained in more detail in the next section. The reconfigurable engines are designed for bitstream processing such as CAVLC and CABAC. The image recognition and processing accelerators are designed for commonly used fixed processing in image processing and image recognition, such as block matching, histogram, affine transformation, and filter processing [7]. Two channels of 32-bit DDR3 interfaces can provide peak memory bandwidth at 10.7GB/s. As shown in Fig. 2 and Table I, the chip is fabricated by 40nm CMOS process and the size of the chip and the many-core cluster are 15.0mm x 14.0mm and 7.4mm x 5.7mm respectively. All of the cores operate at 333MHz at

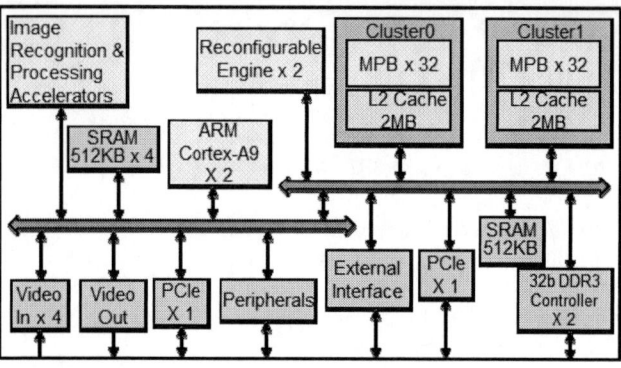

Fig. 1. Block diagram of the many-core SoC

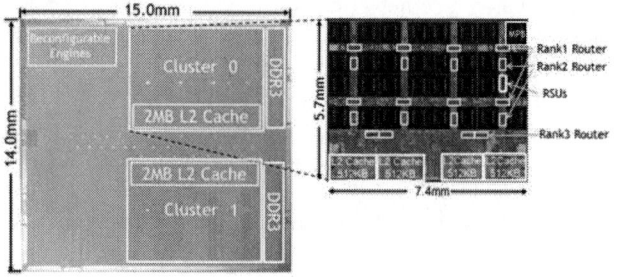

Fig. 2. Chip micrograph and features

1.1V. The total peak performance of the two clusters is 852GOPS.

B. Energy Efficient 32-Core Cluster

Fig. 3 shows the block diagram of the 32-core cluster. A processor core, Media Processing Engine (MPE), is a 3-way VLIW processor [8]. It consists of a 5-stage pipelined 32-bit RISC core and a 64-bit SIMD co-processor. Each core has private L1 instruction and data caches. The size of the L1 instruction cache is 32KB and the size of the L1 data cache is 16KB.

The cluster is based on a homogeneous architecture and a shared memory model. The 32 cores share the L2 cache. The programming model on the shared memory model is simple; however, many memory accesses concentrate at the shared memory when heavily memory-intensive applications are executed. To resolve this issue, we adopt a multi-banked L2 cache and a low latency tree-based NoC. To minimize the latency of the NoC router, source routing and a single bit on/off flow control are introduced. The router can operate at one cycle latency. The L2 cache is 2MB 32-way set-associative and interleaved to four 512KB banks, and its peak bandwidth is up to 42.6GB/s.

The NoC architecture of the cluster is also shown in Fig. 3. Unlike other published many cores based on mesh topology [2], we choose a tree-based topology. Tree-based topologies are suitable for shared memory models because all data transfers go through a shared memory, and there are no direct data transfers between cores. In case of other topologies like mesh, access latencies to a shared L2 cache would differ depending on positions of cores. In our tree-based topology, the L2 cache accesses from all of the cores have uniform latency. This is suitable for a homogeneous many-core architecture in which all cores are equivalent.

Each MPB has its own power switch to reduce leakage

TABLE I
Chip Specifications

Technology	40nm LP Process
Interconnect	8 metal (Cu)
Chip Size	15.0mm x 14.0mm
32-Core Cluster Size	7.4mm x 5.7mm
Transistors	87.5M
Cluster Frequency	333MHz, 1.1V

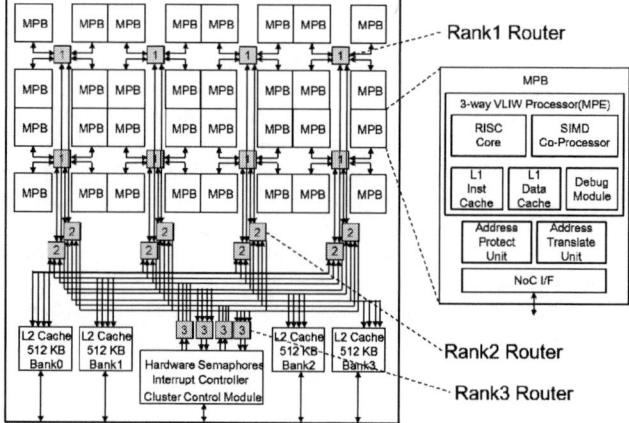

Fig. 3. Cluster architecture and NoC connection

power and the supply power can be gated independently of other cores. In addition, when all of the cores in the cluster are not in use, the supply power of the L2 cache and NoC can be gated by the dedicated power switch.

III. Face Detection using Joint Haar-Like Features

We used face detection using Joint Haar-like features [5] that is high accuracy and meets requirements of surveillance systems. This method can reduce the detection error by 37% compared with widely used Viola and Jones' method [9].

In general object detections, each ROI can be evaluated by a classifier, which detects a specific object independently. A classifier operates with a dictionary, which is trained for a specific object such as face detection.

The Joint Haar-like features are represented by combining binary values computed from multiple Haar-like features. Haar-like features are calculated with rectangular regions in an ROI. The procedure of face detection executed in the classifier is as follows:

First, the classifier calculates each of the Haar-like features in a ROI with a dictionary. A Haar-like feature is a value which is difference in average intensities between two groups of rectangular regions. Examples of Haar-like feature are shown in Fig. 4 and two groups of the rectangular regions are depicted as black and white rectangles. Fig. 5 is a pseudo code of calculating the leftmost Harr-like features in Fig. 4. To reduce calculation costs, an integral image is generally used. An integral image is a derived image whose pixel represents a sum of pixel values in upper left region of an original image. Using an integral image, the sum of each rectangular area (s0, s1) of a Haar-like feature can be calculated from the only four corners (e.g. v1, v2, v4 and v5 for s0) of the rectangular and the three arithmetic operations.

978-1-4799-2817-0/14 $31.00 © 2014 IEEE   312

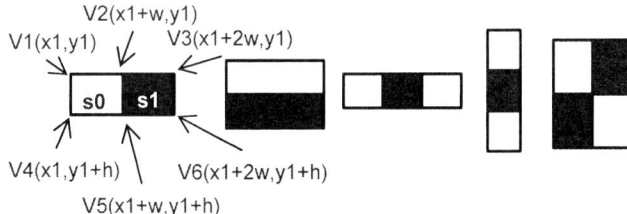

Fig. 4. Examples of Haar-like features

```
hs = h * stride;
// stride is width of image
p1 = img + y1 * stride + x1;
// img is pointer of the upper left
// pixel of the image
p2 = p1 + w;
p3 = p2 + w;
p4 = p1 + hs;
p5 = p2 + hs;
p6 = p3 + hs;
v1 = *p1;
v2 = *p2;
v3 = *p3;
v4 = *p4;
v5 = *p5;
v6 = *p6;
s0 = (v5 - v2) - (v4 - v1);
s1 = (v6 - v3) - (v5 - v2);
```

Fig. 5. A pseudo code of a Haar-like feature

Each value is compared with a predefined threshold and is converted into a binary output.

Then, the classifier combines binary outputs into one jointed N-bit value, where N is the number of Haar-like features. Using the jointed value as an index of tables, the classifier looks up two probabilities in two tables, probability of including faces and probability of not- including faces. If the probability of including faces is greater than that of not-including faces, the result of joint Haar-like features becomes 1, otherwise 0. These probability tables are included in the dictionary.

Next, the classifier accumulates the results of joint Haar-like features with weight. If the accumulated result is greater than another predefined threshold, the classifier judges that a face exists in the ROI of the image. During classification, if it is clear that no face exists in the ROI, the classifier terminates the remaining calculations.

In addition to lots of coarse grain parallelism based on ROIs, the algorithm has the following two important characteristics.

The first characteristic is that the algorithm does not have enough fine grain data parallelism. As shown in Fig. 5, accessed memory addresses (from v1 to v6) are not adjacent and memory accesses instructions are account for large percentage of the program. It means that it is inefficient to pack them into a SIMD register of CPUs or GPUs.

The second is imbalanced workloads. The workload of an ROI where a face exists is larger than that of an ROI without a face. It takes a longer time to evaluate an ROI with a face because the classifier needs to calculate more features.

These two characteristics cause random data accesses and

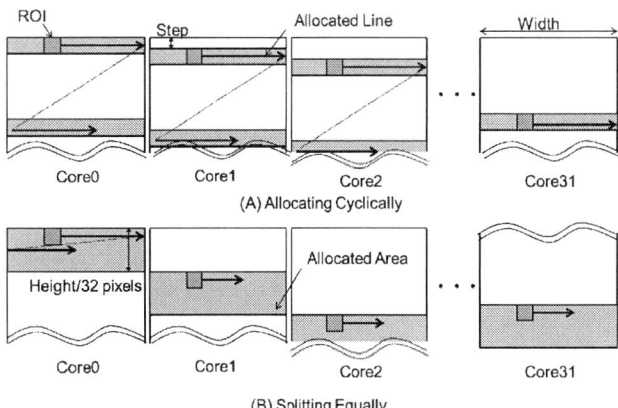

Fig. 6. How to allocate ROIs to the cores in a single cluster

branch divergence that are inefficient for current GPGPU architectures. For example, Sharma implements parallelized face detection based on Viola and Jones' method on a GPU using CUDA™ [10]. Although the workload of each thread is different from each other, all threads in Warp, a group of threads in a Streaming Processor, need to execute the same instructions. Therefore, the total performance is limited by the slowest thread and utilization of ALUs becomes low.

IV. Implementation of Face Detection on the Many-Core SoC

This section describes how to parallelize the face detection on the many-core SoC.

A. Parallelism of the Face Detection

As mentioned above, the face detection does not have enough fine grain data parallelism and we do not use the SIMD co-processor. We focus to exploit coarse grain thread level parallelism. When the resolution of an image is 4000 x 3200 pixels and the size of ROIs is 25 x 25 pixels and the step size is 2 pixels, there are about 3M ROIs in the image. There are lots of thread level parallelism to exploit by multiple processor cores. To minimize the overhead of scheduling threads, we decrease the number of threads and allocate one thread for each core. Because the workload of each ROI is different from each other, it is a consideration to allocate sets of ROIs to threads balancing the workload of each thread.

B. Implementation on the Single Cluster of the Many-Core

We parallelized the face detection and develop two implementations on the single cluster in the following two intra-cluster ways.

The first way, Allocating Cyclically, allocates lines to be scanned to each core cyclically. As shown in Fig. 6 (A), the scanned line 0 is allocated to the core 0, the scanned line 1 is to the core 1, and the line 31 is to the core 31. Then, the line 32 is allocated to the core 0 again and so on.

The second way, Splitting Equally, divides the image into N equal areas horizontally, where N is the number of operating cores, and the ROIs included in each split area are allocated to each core. As shown in Fig. 6 (B), when the

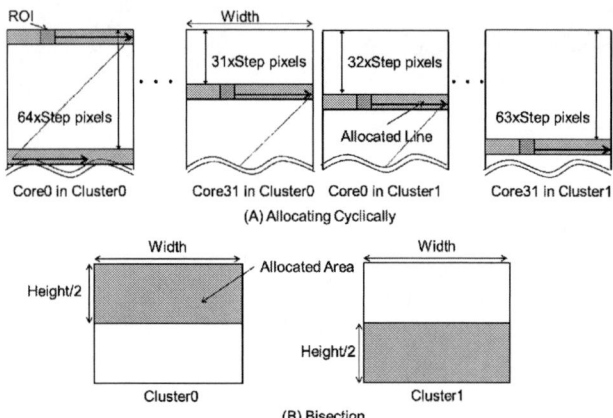

Fig. 7. How to allocate ROIs to the cores in two clusters

application is executed by 32 cores, a 4000 x 3200 image is divided into 32 areas whose size is 4000 x 100.

C. Implementation on the Dual Cluster of the Many-Core

When the application is executed on the two clusters, we have to take the bandwidth between the L2 cache and the DDR3 memory into consideration because it is narrower than that between the L1 and L2 caches. Thinking over it, we developed two implementations on the two clusters as the following inter-cluster ways. The first implementation, Allocating Cyclically, is the same as Allocating Cyclically in one cluster, and utilizes 64 cores across the clusters as shown in Fig. 7 (A). The second way, Bisection, divides the image into two blocks horizontally. As shown in Fig. 7 (B), the upper and lower blocks are assigned to the cluster 0 and the cluster 1 respectively. Then, the lines in each block are allocated to the 32 cores in the cluster by the Allocating Cyclically method.

D. Execution Platform and Parallel Execution Model

We parallelized the face detection program for the many-core cluster by using the thread-based parallel execution model which is compatible with the model of the multi-core architecture [11]. The thread model is much lighter than other thread models such as POSIX thread. The application is divided into threads and each thread is assigned to one core by a thread scheduler dynamically. It performs effectively on the many-core as well. In this program, all of the operations that one core executes is allocated to one thread. Because the thread scheduler manages the threads at only the beginning and end of the program, the overhead of synchronization is minimized.

V. Evaluation

A. Parallelism of the Face Detection

We evaluated the face detection with the pictures shown in Table II as target images. The sizes of the images are from 5.76M to 12.7M pixels, and the number of faces varies from 9 to 148.

The face detection programs are executed on the actual

TABLE II
Images for evaluation

No.	Resolution	# of detected face
0	4000 x 1440	30
1	3000 x 4082	37
2	4083 x 3062	78
3	4094 x 3107	148
4	3568 x 2568	9
5	3568 x 2568	10

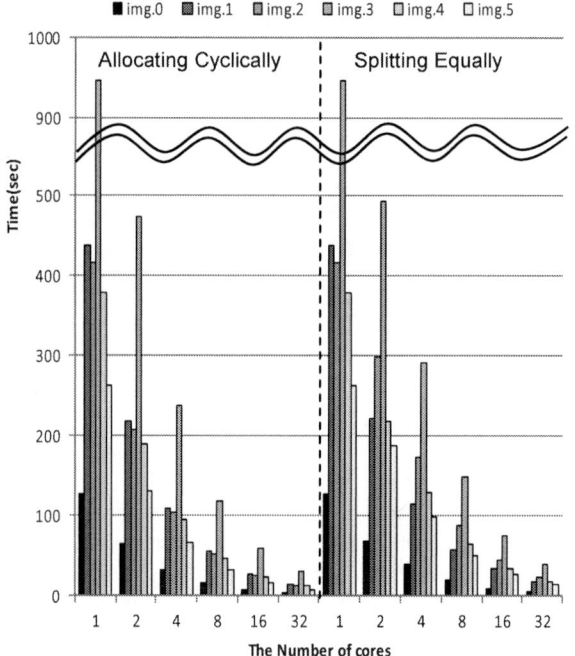

Fig. 8. Execution time of a single cluster

many-core SoC chips. We measured execution times and power consumption. Execution times to make integral images and image preprocessing are not included because these operations can be implemented on the filter accelerator and other accelerators in the SoC and over-wrapped with a main part of face detections executed by the many core clusters.

B. Evaluation Results on a Single Cluster

Fig. 8 shows the execution time in the single cluster of two intra-cluster ways: Allocating Cyclically and Splitting Cyclically. The execution times vary greatly depending on the images. The time of image 1 is under half of that of image 3 even though the resolutions of the images are almost the same. As indicated in Table II, the number of faces included in image 3 is larger than that in image 1. The ROI where a face exists has larger workload than that where no face exists. Therefore, it takes a long time to detect faces in image 3.

Fig. 9 shows the relative performance compared with the performance in one core. Allocating Cyclically scales up to 32 cores and accomplishes 15.5 times speed-up in 16 cores and 30 times speed-up in 32 cores. These performance improvements are close to ideal. On the other hand, in the

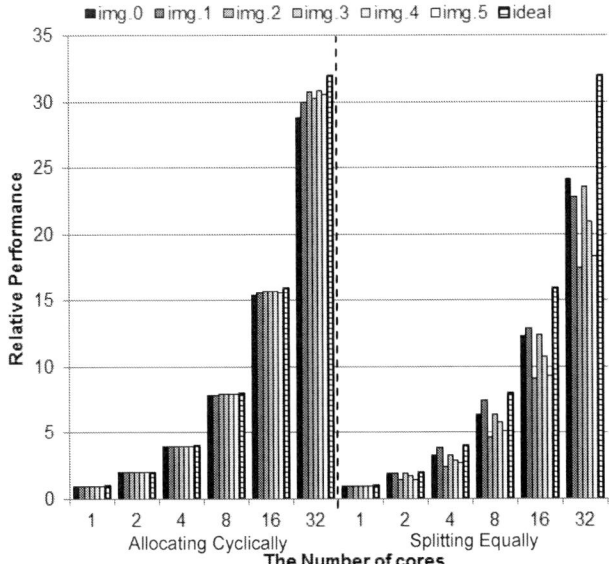

Fig. 9. Relative performance of a single cluster

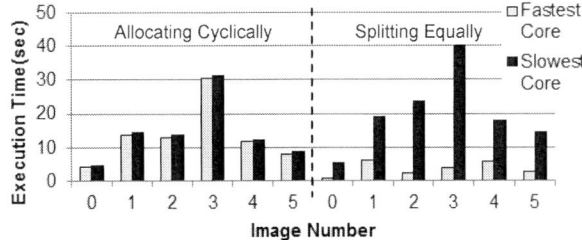

Fig. 10. Execution time of the fastest and the slowest cores when 32 cores are running in a single cluster

case of Splitting Equally, the speed-up is much lower than that of ideal case.

Fig. 10 shows the execution times of the fastest core and the slowest core when 32 cores are running. In Allocating Cyclically, the execution times of the fastest and slowest cores are almost equal, and the execution time of the slowest core is about 1.1 times slower than that of the fastest core. In Splitting Equally, on the other hand, the execution times of the fastest and slowest core are quite different, and the execution time of the slowest core is about 11 times slower than that of the fastest core. This causes large performance degradation. In image 2, the performance is much lower than ideal because the faces are gathered to a certain region of the image. In image 1, the relative performance is better than the case of image 2, because the faces are in scattered positions. However, even in image 1, the performance improvements are lower than those of Allocating Cyclically.

As the processor utilization, Allocating Cyclically achieves higher utilization ratio, from 90% to 95%, than those of Splitting Equally, from 54.3% to 76%. These results confirm that balance of workloads among threads is important and the parallelization by Allocation Cyclically is more efficient than that by Splitting Equally.

The bandwidth between the cores and the L2 cache is measured by the on-chip performance monitor. In both

parallelized cases, the utilized bandwidth is maximum 8.5GB/s at 32 cores and about 20% of maximum bandwidth (42.6GB/s). The NoC and the L2 cache are able to provide enough bandwidth to the 32 cores. The bandwidth between the L2 cache and the DDR3 memory in 32 cores is maximum 450MB/s, which is only 4.2% of the theoretical maximum bandwidth (10.7GB/s). This ensures that the L2 cache successfully reduces the transactions to the DDR3 memory. If the amount of transferred data between the L2 cache and the DDR3 memory are compared, the data size of Allocating Cyclically is from 70% to 85% of that of Splitting Equally in 32 cores. We confirm that Allocating Cyclically reduces L2 cache misses because most of the cores access a smaller area at the same time than that in Splitting Equally. This is another reason that Allocating Cyclically provides the better performance.

C. Evaluation Results on Two Clusters

The left graph of Fig. 11 shows the execution time in the 64 cores using two clusters. Allocating Cyclically achieves higher performance than Bisection in all of the images. The right graph of Fig. 11 is the relative performance of Allocating Cyclically normalized by the performance of one core. It scales up to 64 cores and the performance of 64 cores is from 56 to 60 times higher than that of one core.

Fig. 12 shows the execution time of the fastest core and the slowest core. In Allocating Cyclically, the execution time in the slowest core is at most 1.3 times slower than that in the fastest core (image 1), while in Bisection, the execution time in the slowest core is at most 2.8 times slower than that of the fastest core (image 2 and 5). The processor utilization of 64 cores is from 87% to 95% in Allocating Cyclically. On the other hand, in Bisection, the utilization is from 66% to 92%. This evaluation result ensures that Allocating

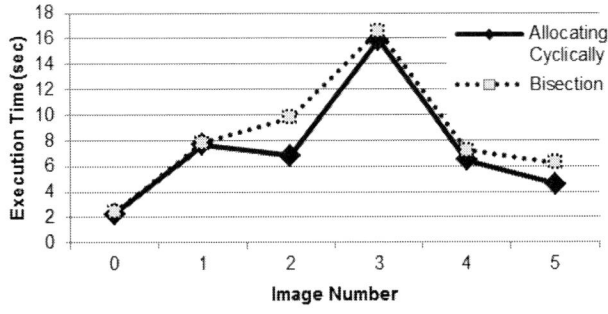

(A) Execution Time of Two Clusters

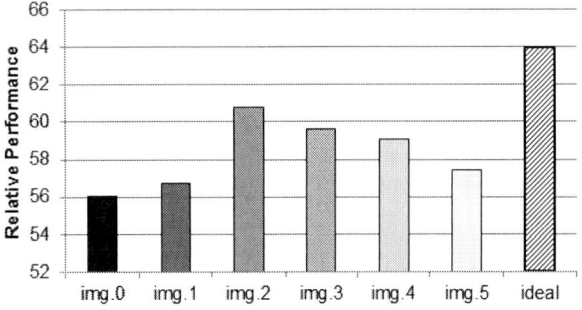

(B) Relative Performance (Allocating Cyclically)

Fig. 11. Performance of two clusters

978-1-4799-2817-0/14 $31.00 © 2014 IEEE   315

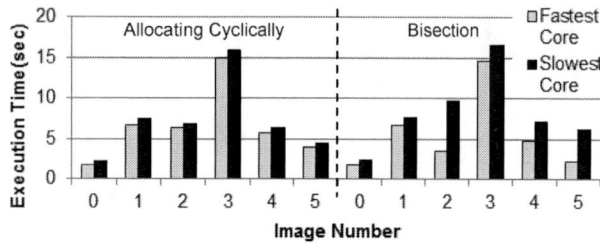

Fig. 12. Execution time of the fastest and the slowest cores when 64 cores are running in two clusters

Cyclically can balance the workloads among all of the cores across the clusters.

In Allocating Cyclically, the utilized bandwidth between the L2 cache and the DDR3 memory is maximum 750MB/s, which is about 7% of the theoretical maximum bandwidth (10.7GB/s). In this evaluation, the result ensures that the many-core SoC provides an adequate data bandwidth to the many cores even when the two clusters operate simultaneously.

D. Evaluation of Power Consumption

Fig. 13 shows the breakdown of the power consumption of the many-core SoC. When the number of the active cores is less than 64, the power supply of idle cores is gated and all power supply of the idle cluster is also gated. The power consumption of the clusters with the active 64 cores is 1.18W and the total power of the chip is 2.21W with a typical process sample at room temperature.

We compared the energy efficiency of many-core SoC with that of an Intel® Core™i7 3820 processor using the parallelized face detection with Allocating Cyclically method. The Core™i7 processor is a two hyper-threaded quad-core and its maximum operating frequency is 3.6GHz and its TDP is 130W. The performance of the many cores is from 30% to 60% of Core™i7 when it consumes about 2W. Although these energy values cannot be compared directly, the power efficiency of the many-core SoC is several tens of times better than that of desk-top CPUs.

VI. Conclusions

This paper presents the energy efficient many-core SoC and its evaluation using face detection as an example of realistic image recognition application. The many-core SoC is integrated with two 32-core clusters that consist of energy efficient processors. The algorithm has lots of coarse grain thread level parallelism that is suitable for many-core architecture. By choosing appropriate thread allocating policy, it achieves 30x speed-ups by 32 cores and 55x to 60x speed-ups by 64 cores. However, due to imbalance of workloads among threads, the performance decreases by 30% if the threads are not partitioned appropriately. The parallelized method that keeps workload balance of threads is important to achieve effective performance in many-core architecture. When the 64 cores are operating, the power consumption is only 2.21W. The energy efficiency is much better than that of the high performance quad-core processor. This result shows that our many-core SoC is remarkably

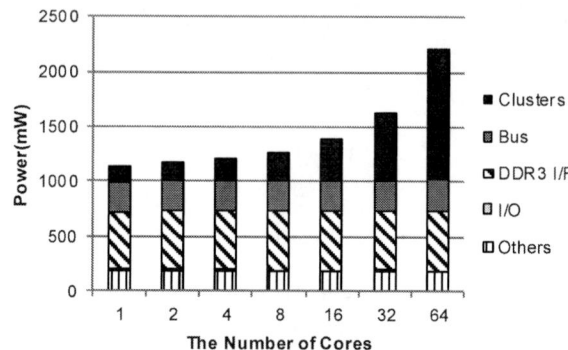

Fig. 13. Power consumption of the many-core SoC

energy efficient in image recognition applications.

As the next step of this study, we are going to implement other image processing and recognition applications on the many-core SoC and evaluate the advantage of the SoC in more detail. Furthermore, we plan to develop an improved many-core SoC for embedded applications to pursue higher performance and lower power consumption.

References

[1] H. Xu, J. Tanabe, H. Usui, S. Hosoda, T. Sano, et al. "A Low Power Many-Core SoC with Two 32-Core Clusters Connected by Tree based NoC for Multimedia Applications," VLSI Circuits, 2012, pp. 150-151.

[2] J. Howard, S. Dighe, Y. Hoskote, S. Vangal, D. Finan, et al., "A 48-Core IA-32 Message-Passing Processor with DVFS in 45nm CMOS," ISSCC, 2010, pp. 108-109.

[3] S. Bell, B. Edwards, J. Amann, R. Conlin, K. Joyce, et al., "TILE64-Processor: A 64-Core SoC with Mesh Interconnect," ISSCC, 2008, pp. 88-89.

[4] D. Melpignano, L. Benini, E. Flamand, B. Jego, T. Leply, et al., "Platform 2012, a Many-core Computing Accelerator for Embedded SoCs: Performance Evaluation of Visual Analytics Applications," DAC, 2012, pp.1137-1142.

[5] T. Mita, T. Kaneko, and O. Hori, "Joint Haar-like Features for Face Detection," ICCV, 2005, pp. 1619-1626.

[6] T. Yoshikawa, F. Hyuga, M. Tokunaga, Y. Yamada, S.Asano, "FlexGrip™: A small and high-performance programmable hardware for highly sequential application," Cool Chips XIV, 2011.

[7] Y. Tanabe, M. Sumioka, M. Nishiyama, I. Yamazaki, S. Fujii, et al., "A 464GOPS 620GOPS/W Heterogeneous Multi-Core SoC for Image-Recognition Applications," ISSCC, 2012, pp. 222-223.

[8] S. Nomura, F. Tachibana, T. Fujita, T. C. Kong, H. Usui, et al., "A 9.7mW AAC-Decoding, 620mW H.264 720p 60fps Decoding, 8-Core Media Processor with Embedded Forward-Body-Biasing and Power –Gating Circuit in 65nm CMOS Technology," ISSCC, 2008, pp. 262-263.

[9] P. Viola and M. Jones, "Rapid Object Detection using a Boosted Cascade of Simple Features", CVPR, 2001, pp. 511-518

[10] B. Sharma, R. Thota, N. Vydyanathan, and A. Kale, "Towards a Robust, Real-time Face Processing System using CUDA-enabled GPUs," HiPC, 2009, pp. 368-377

[11] T. Mori, Y. Ueda, N. Nonogaki, T. Terazawa, M. Sroka, et al., "A Power, Performance Scalable Eight-Cores Media Processor for Mobile Multimedia Applications," JSSCC, 2009, pp. 2957-2965.

Energy Aware Real-Time Scheduling Policy with Guaranteed Security Protection

Wei Jiang[1,3], Ke Jiang[2], Xia Zhang[3], and Yue Ma[4]

[1]*School of Computer Science and Engineering, University of Electronic Science and Technology of China, China*
[2]*Department of Computer and Information Science, Linköping University, Sweden*
[3]*School of Information and Software Engineering, University of Electronic Science and Technology of China, China*
[4]*Department of Computer Science and Engineering, University of Notre Dame, U.S.A.*
weijiang@uestc.edu.cn, ke.jiang@liu.se, zhangxia0414@163.com, yma1@nd.edu

Abstract—**In this work, we address the emerging scheduling problem existed in the design of secure and energy-efficient real-time embedded systems. The objective is to minimize the energy consumption subject to security and schedulability constraints. Due to the complexity of the problem, we propose a dynamic programming based approximation approach to find the near-optimal solutions with respect to predefined security constraint. The proposed technique has polynomial time complexity which is about half of traditional approximation approaches. The efficiency of our algorithm is validated by extensive experiments.**

I. INTRODUCTION

Real-time embedded systems are facing more and more severe security threats, e.g. due to the integration of new communication interfaces. One of the emerging needs is to protect sensitive data in critical embedded systems [1], [2]. Since snooping, spoofing and altering security-critical data can lead to significant losses or serious system failures, resulting in great loss of finance or human life. We refer to such systems as Security-Critical Real-Time Systems (SCRTS). Examples of SCRTS are flight control systems, satellite communication systems and radar tracking systems, which all have strict security requirements. To protect SCRTS against potential threats, a series of security services, i.e. integrity, confidentiality and authentication protection, need to be considered in the design process of SCRTS. With the most suitable security protections regarding the demands, SCRTS would be effectively protected via using the right amount of resources, e.g. CPU utilization.

Energy efficiency is another fundamental requirement in the context of SCRTS [3], [4], [5]. But security protections usually demand a significant amount of energy [2]. Quick energy depletion or early exhaustion of battery may cause failure to mission-critical tasks, resulting in unexpected losses, such as the energy incurred failure of Mars Pathfinder. Hence, how to design energy-efficient SCRTS becomes a great challenge.

Task scheduling policy plays an important role for achieving high performance in embedded systems. Unfortunately, traditional real-time scheduling approaches were mostly designed to guarantee timing requirements only [6]. Recently, security-aware real-time scheduling has become a hot research topic, e.g. [7], [8], [9], [10], [11]. However, all these works did not consider the energy factor in system-level designs, which may deliver solutions with unexpected energy consumptions.

In this paper, we identify the uniprocessor scheduling problem lying in many SCRTS designs considering energy, security and real-time dimensions. Specifically, we want to schedule a set of periodical real-time tasks with the objective of minimizing energy consumption, while satisfying security and timing constraints. The primary contributions of this paper are in two aspects. First, a typical security- and energy-aware

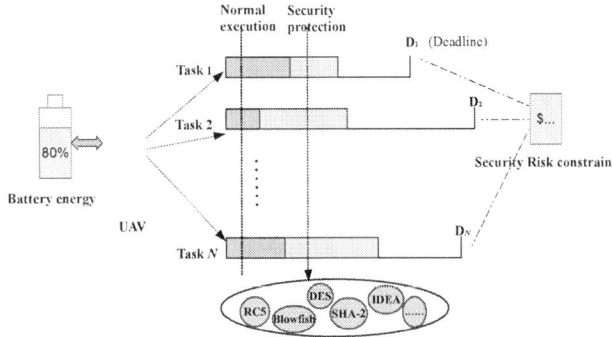

Fig. 1. A motivational application

SCRTS application is presented. Second, we study the existing policies and propose an efficient approximation approach that guarantees the security requirements. Our approach has polynomial time complexity, and requires bounded memory space. To the best of our knowledge, this is the first work that addresses the security- and energy-aware real-time task scheduling problem.

II. APPLICATION AND SYSTEM MODEL

A. Motivational application

In this paper we focus on the SCRTSs with limited energy budget, for example, an unmanned aerial vehicle (UAV) depicted in Fig. 1. The UAV is battery driven, and has limited resources, i.e. CPU and memory. It runs critical tasks that are periodically released, and exchanges information with other peers or service centers. Each task generates or receives some private data that needs to be transmitted over insecure environments. Different data has different requirements of security and deadline guarantee ratios. In order to make the communication secure (i.e. to protect the confidentiality or integrity of the messages), we need to perform cryptographic algorithms like RC5, DES and SHA-2, on the data before or after the normal executions of corresponding tasks. Thus, the energy consumption of each task consists of two parts that are from the normal execution and extra security protections, which will be discussed in Sec. II(B). Although there are many available cryptosystems, it is hard to obtain the best choices among different solutions having different execution and energy overhead, and the UAV only has limited energy and processing capability. Therefore, how to allocate resources to protect different data becomes an important design trade-off. In other words, we are aiming to schedule a set of periodic real-time tasks with the objective of minimizing energy consumption while satisfying security and schedulability constraints.

978-1-4799-2817-0/14 $31.00 © 2014 IEEE

B. Task model for security-critical real-time systems

In this paper, we consider a set of periodic security- and energy-aware tasks running on SCRTS. Each task T_i is modeled as a tuple $T_i = \{BE_i, L_i, S_i, S_i^{DM}, V_i, SR_i, P_i\}$. BE_i denotes the worst case execution time (WCET) of its non-security part. L_i is size of the data that is generated or received by T_i and needs to be protected using selected security service. S_i and S_i^{DM} are the chosen and designated security levels of T_i, respectively. If S_i^{DM} is achieved, this task is assumed to be absolutely secure. V_i is the security impact value of T_i representing the relevant importance of the messages processed by T_i. SR_i is the security risk of T_i, and means the potential loss of the security protection, which will be elaborated in Sec.II(D). P_i is the period and also the relative deadline of T_i.

C. Time and energy overhead of security critical tasks

It is known that security protections can be achieved by additional security services, which also compete resources with other executions. For example, doing AES encryption on one message may reduce the available CPU resource for protecting other messages. So it is indispensable to always allocate the right amount of resource to the security protections among tasks in order to reach the best global security protection.

It is still an open problem of quantifying the security strength of different cryptographic algorithms. So we quantify the security level of an algorithm based on its execution time similar to [8]. Then security levels of typical algorithms are enumerated as their relative strengths. Based on the measurement on a S3C2440 ARM board with 500MHz CPU and 64MB SDRAM [12], we obtain the time and energy overheads of six widely used confidentiality services for protecting 1 KB data as shown in Table I. For example, we assign security level 1 to the relatively weakest algorithm RC4 that has the shortest encryption time. In this paper, we only consider periodic tasks, so we assume that the key setup procedures for the security algorithms are prepared before the system starts, and thus, ignored, for the sake of simplicity. Note that this paper just gives a reasonable quantification of security level of cryptography algorithms. It can be changed by more reasonable data. If there exists newly developed algorithm with higher security strength while lower time overhead, we can use it to replace the ones with larger overhead and lower level.

If task T_i generates or receives L_i Bytes security sensitive data, then the execution time of T_i can be formulated as follows,

$$Exe_i = BE_i + \theta(S_i) * L_i \qquad (1)$$

$\theta(S_i)$ is the mapping function of security level S_i to unit execution time of the chosen algorithm. Taking $S_i = 3$ for example, $\theta(S_i)$ is thus the unit execution time overhead of BLOWFISH algorithm according to Table I. As can be observed from Table I, there is a close to linear relation between energy consumption and encryption time, i.e. approximately $320mJ/S$, which is recognized as the *power*. We also found that the *power* of task's basic execution is nearly the same situation in our measurement. Thus, the total energy consumption T_i including the data protections is the product of power (POW) and the whole execution time.

$$En_i = POW * Exe_i = POW * (BE_i + \theta(S_i) * L_i) \qquad (2)$$

TABLE I
TIME AND ENERGY OVERHEAD OF CONFIDENTIALITY ALGORITHMS

Ciphers	time(ms/KB)	Energy(mJ/KB)	Sec. Level
RC4	0.0063	2.0237	1
RC5	0.0125	4.0340	2
BLOWFISH	0.0170	5.4696	3
IDEA	0.0196	6.2822	4
SKIPJACK	0.0217	6.9658	5
3DES	0.0654	21.0914	6

D. Risk model for security-critical task

To quantify the security quality of tasks, it is necessary to introduce security risk model. In [8] and [10], linear profit model, which is the sum of weighted security levels, is used to evaluate the quality of each security-critical task. Obviously, this model is impractical to capture the real security quality. Since the potential risk is the product of security violation probability and consequence of security breach [13], we model the security risk (SR) as the expected security loss of T_i:

$$SR_i = V_i * Pro_i^{risk}, \qquad (3)$$

where V_i is the security impact value of T_i representing the relevant importance of its data. It could be finance loss or other metrics, e.g., 500\$ loss if T_i failed. Pro_i^{risk} is the failure probability of T_i with chosen security level S_i formulated as

$$Pro_i^{risk} = \begin{cases} 1 - e^{-\lambda_i(S_i^{DM} - S_i)}, & \text{if } S_i < S_i^{DM} \\ 0, & \text{if } S_i \geq S_i^{DM}. \end{cases} \qquad (4)$$

λ_i is the security risk coefficient of T_i, which can be adjusted by the designer based on different scenarios. As implied in Eq. 4, if the assigned security level is greater or equal to the demanded security, we assume no failure will occur when facing attacks. Inversely, the task has the probability to fail, and bigger security demand gap leads to higher probability of security violation. This paper just introduces a more reasonable and practicable metric, i.e. the security risk model, to quantify the quality of security-aware tasks, and uses it to assess the system performance.

III. PROBLEM FORMULATION

A. Original problem

In this paper, we consider a security-critical embedded system that has a set of N tasks. A system monitor gives the security Risk Bound (RB) that defines the current security requirement of the whole system. In another words, if the task executions lead to security violations, i.e. the system cannot satisfy the expected RB, then it is recognized as an unacceptable system. In certain situations, it is demanded that the security risk should not succeed $(1 + \beta)RB$, where β is the risk slack ratio and defined by designers according to the system requirement. Higher security-critical application has less value of β. So the problem is to assign the most suitable security services for tasks using the minimum energy consumption, and satisfy the strict real-time and security constraints.

Before going further, let us introduce the definition of *hyperperiod* (HP) that is the least common multiple of all tasks' periods. Thus, the purpose of this paper becomes to obtain the minimal long-term energy consumption for the task set within a hyperperiod. So our energy minimization scheduling problem can be formulated as

$$\text{Minimize } Energy = \sum_{i=1}^{N} (HP/P_i) * En_i \qquad (5)$$

978-1-4799-2817-0/14 \$31.00 © 2014 IEEE

Subject to

$$
\begin{cases}
\sum_{i=1}^{N}(HP/P_i)*SR_i \leq RB \\
\sum_{i=1}^{N}(BE_i+\theta(S_i)*L_i)/P_i \leq UB_x \\
S^{min} \leq S_i \leq S^{max},
\end{cases}
$$

where, UB_x denotes the utilization ratio bound of the x scheduling policy. A set of periodic tasks is schedulable on a processor using real-time scheduling policies, e.g. EDF ($UB_{EDF}=1$) and RM ($UB_{RM}=0.693$), if the processor utilization ratio is not more than the given bound. The first two constraints are the system security risk constraint and real-time constraint, respectively. The last constraint makes sure the correctness of the selected security protection method: the permitted security level must lie between S^{min} and S^{max}.

B. Reduced problem

The design problem that we are facing is a multi-constrained optimization problem, which can be transformed into a Multi-Dimensional Multiple Choice Knapsack Problem. It takes large computation overhead to get the optimal solution for large-scale designs. Thereby, we try to find an efficient assignment algorithm that can obtain good solution while satisfying the predefined constraints. In this section, we will reduce the fore-mentioned problem. Combining Eq. 2 and Eq. 5, we can rewrite our optimization objective as

$$
Energy = \sum_{i=1}^{N}(HP/P_i)*En_i
$$

$$
= HP*POW*\sum_{i=1}^{N}(BE_i+\theta(S_i)*L_i)/P_i \quad (6)
$$

where, for the fixed hardware platform and application, the hyperperperiod HP, power POW and task's non-security execution utilization $\sum_{i=1}^{N}BE_i/P_i$ are constant values. Consequently, we can rewrite the design problem as follows,

$$
\text{Minimize} \sum_{i=1}^{N}\theta(S_i)*L_i/P_i \quad (7)
$$

such that,

$$
\begin{cases}
\sum_{i=1}^{N}(HP/P_i)*SR_i \leq RB \\
\sum_{i=1}^{N}Exe_i/P_i \leq UB_x \\
S^{min} \leq S_i \leq S^{max}
\end{cases}
$$

Now with the reduced system problem, we only need to consider the CPU utilization ratio caused by security services and security risk for different level. In addition, this design optimization problem will be further transformed into a Markov decision-making procedure in the next section.

IV. APPROXIMATION BASED DYNAMIC PROGRAMMINGS

As the reduced system problem is still not easy to be solved, we must find an efficient approximation approach to solve it. In this section, we transform the problem to a multi-stage Markov decision-making procedure, and use approximate Dynamic Programming to address it.

A. Markov decision-making procedure

Considering a set of N periodic tasks, the utilization ratio minimization problem can be formed as an N-stage Markov decision-making procedure. The decision variable for the i-th stage is the chosen security service S_i that needs to be assigned for task T_i. Thus, the purpose can be transformed as to find a combination $S^*=(S_1,S_2,\cdots,S_N)$ with minimum system utilization ratio while satisfying the security risk constraint.

We denote a triple $(\xi_{ik},\gamma_{ik},S_{ik})$ to describe the k-th state in i-th stage. ξ_{ik} and γ_{ik} are two values which present the accumulated CPU utilization ratio and the accumulated security risk for the first i tasks, respectively. S_{ik} is the specific value of security decision variable S_i in this state.

The State Set Ω_i for i-th stage is defined as $\Omega_i = \{(\xi_{i1},\gamma_{i1},S_{i1}),(\xi_{i2},\gamma_{i2},S_{i2}),\cdots,(\xi_{ii^{max}},\gamma_{ii^{max}},S_{ii^{max}})\}$, where i^{max} is total number of states in the i-th stage. Given Ω_{i-1} in $(i-1)$-th stage, we can obtain the state subset Ω_{ik} by adding utilization ratio and security risk of task T_i under security level $S_{ik} \in [S^{min},S^{max}]$ to all states in Ω_{i-1} as follows,

$$
\Omega_{ik} = \Omega_{i-1} \biguplus (U_i(S_{ik}),SR_i(S_{ik}),S_{ik})
$$

$$
= \{(\xi_{i-1,1}+Exe_i(S_{ik})/P_i, \gamma_{i-1,1}+SR_i(S_{ik}),S_{ik}),
$$

$$
(\xi_{i-1,2}+Exe_i(S_{ik})/P_i, \gamma_{i-1,2}+SR_i(S_{ik}),S_{ik}),\cdots\}
$$

Thus, the state set in i-th stage is $\Omega_i = \bigcup_{k=1}^{|S_i|}\Omega_{ik}$. $|S_i|$ is the number of available options for task T_i. In the first stage, there are only $|S_1|$ states, and the number of states in i-th stage is the production of $|S_i|$ and the number of states in stage i-1. The maximal state space is:

$$
SS = \prod_{i=1}^{N}|S_i| \quad (8)
$$

As can be noticed from Eq. 8, the state space grows exponentially as the numbers of tasks and security choices grow. Thus it is infeasible to apply Dynamic Programming on large system designs. So, we must find methods to reduce the solution space. Inspired by the approximation algorithm of Knapsack problem [14], grouping the security risk into a series of discrete integers is a good approach to reduce the decision states on each stage. Setting Δ as the group factor, then the security risk of each task can be transformed to an integer decided by SR_i/Δ. Thus, in each stage, we just keep the state with minimal security risk when several states have the same risk value. Bigger Δ gives smaller scaled security risk values and smaller number of states in each decision-making stage. Since the number of states in each decision-making step can be maximized to a constant $M=\lceil RB/\Delta \rceil$, we can leverage a two-dimension table to show the states of all stages.

B. Four approximating approaches

1) Round to Ceiling approach: Round to Ceiling (RC) approach means that we round the divided value to the closest bigger integer. Most of the previous related works concerning energy used this policy to approximate the divided values, like [15] and [16]. For each T_i, the security risk value is divided by Δ. According to the RC policy, the result is scaled up to the closet bigger integer, i.e. $RC(SR_i)=\lceil \frac{SR_i}{\Delta} \rceil$.

2) Round to Floor approach: Round to Floor (RF) approach is similar to the fore-mentioned RC policy, but the obtained result is scaled down to its closest integer, i.e. $RF(SR_i)=\lfloor \frac{SR_i}{\Delta} \rfloor$.

3) Round Randomly: As can be observed, RC will bring positive deviation comparing to the real value, while RF leads to negative deviation. In this section, we introduce a Randoml Round (RR) policy, which round the divided value to the closest two integers with a certain probability i.e.

$$RR(SR_i) = \begin{cases} \lceil \frac{SR_i}{\Delta} \rceil \text{with prob. } \rho_1 = \frac{SR_i}{\Delta} - \lfloor \frac{SR_i}{\Delta} \rfloor \\ \lfloor \frac{SR_i}{\Delta} \rfloor \text{with prob. } \rho_2 = \lceil \frac{SR_i}{\Delta} \rceil - \frac{SR_i}{\Delta} \end{cases} \quad (9)$$

4) Round to Nearest approach: RR can erase the potential deviation for a set of tasks, but in extreme situations, the deviation is still big. Hence, we introduce a simple scaling policy, Round to Nearest (RN) integer, which scales the divided value to the closest integer with minimal deviation as follows.

$$RN(SR_i) = SR_i^{\Delta} = \begin{cases} \lceil \frac{SR_i}{\Delta} \rceil, \text{ if } \frac{SR_i}{\Delta} - \lfloor \frac{SR_i}{\Delta} \rfloor \geq 0.5 \\ \lfloor \frac{SR_i}{\Delta} \rfloor, \text{ if } \frac{SR_i}{\Delta} - \lfloor \frac{SR_i}{\Delta} \rfloor < 0.5 \end{cases} \quad (10)$$

C. $(1 + \beta)$ Approximating Analysis

As proposed in the last section, the approximation approach may return results deviating with the real security risks. Thus, in this section, we analyze the deviation of each approach in the whole optimization problem, and identify the most suitable policy to be used in our scheduling mechanism.

Let us assume that there exists N tasks, and RC policy is applied to discretize security risks, then the Overall Deviation (OD) between the real and approximated risk for the whole system is as follows.

$$OD^{RC} = \sum_{i=1}^{N}(SR_i - \Delta * RC(SR_i))$$
$$\geq \sum_{i=1}^{N}(SR_i - \Delta * (\frac{SR_i}{\Delta} + 1))$$
$$= -N\Delta \quad (11)$$

Similarly, we have

$$OD^{RF} \leq N\Delta \quad (12)$$

and

$$-N\Delta \leq OD^{RR} \leq N\Delta \quad (13)$$

RN policy has two extreme scenarios. The first one is that the fraction part of grouped security risk is always bigger than 0.5. Then, according to Eq. 10, we can obtain the risk deviation as

$$OD^{RN} = \sum_{i=1}^{N}(SR_i - \Delta * RN(SR_i))$$
$$\geq \sum_{i=1}^{N}(SR_i - \Delta * (\frac{SR_i}{\Delta} + 1/2))$$
$$= -N\Delta/2$$

For the other case when the fraction part of grouped security risk is always less than 0.5, the overall risk deviation is $OD^{RN} = N\Delta/2$. Hence, we can get that the risk deviation for all cases, since all scenarios are lying between the above two extreme cases, i.e.

$$-N\Delta/2 \leq OD^{RN} \leq N\Delta/2 \quad (14)$$

Bigger Δ reduces the states in the Markov decision procedure, but brings deviation comparing with the real security risk value. Based on the above analyses, bigger Δ will also increase the deviation for all four policies. If the approximation algorithm can find the near-optimal solution within polynomial time while satisfying the security risk deviation ratio β, then it is a good approach.

For RC policy, the real risk value is less than the scaled value, which means it won't exceed the security risk bound

---

Algorithm 1 RN-based approximation algorithm

1: //Step 1: Schedulability test
2: **if** $\sum_{i=1}^{N}(HP/P_i)SR_i(S^{max}) > RB$
    or $\sum_{i=1}^{N} Exe_i(S^{min})/P_i > UB_x$ **then**
3:    Return. /*Given task set is not schedulable*/
4: //Step 2: Initialization
5: Compute $\Delta = 2\beta RB/N$ and $M = \lceil RB/\Delta \rceil$
6: Initialize state matrix $\Omega_{N \times M}$ with each element $\Omega_{i,j} = (0,0,0)$
7: Initialize Ω_1 by calculate (ξ_1, γ_1, S_1) with each $S_1 \in [S^{min}, S^{max}]$
8: //Step 3: Update the state matrix in N-Stage decision procedure
9: **for** $i = 2$ to N **do**
10:   **while** $(\xi_{i-1}, \gamma_{i-1}, S_{i-1}) \neq (0,0,0)$ in Ω_{i-1} **do**
11:     **for** $S_i' = S^{min}$ to S^{max} **do**
12:      Calculate temporary state $(\xi_i', \gamma_i', S_i')$
13:      **if** $\xi_i' > UB_x$ or $\gamma_i' > RB$ **then**
14:       Ignore this state and break /*Schedulability or security violated*/
15:      **if** state $\Omega_{i,j}, (j = \gamma_i')$ is not existed **then**
16:       $\Omega_{i,j} = (\xi_i', \gamma_i', S_i')$ /*Store new state*/
17:      **else if** $\xi_i' < \xi_i$ in $\Omega_{i,j}$ **then**
18:       $\Omega_{i,j} = (\xi_i', \gamma_i', S_i')$ /*Keep state with smaller utilization*/
19: //Step 4: Find the minimal energy consumption solution
20: Find $\Omega_{N,j}^*$ with minimal utilization ratio ξ_N^*
21: Obtain the final security assignment decision $(S_1, S_2, \cdots, S_N)$ by backtracking
22: $Energy^* = \xi_N^* * HP * POW$ /*The minimal energy*/

(RB). To the satisfy the minus deviation ratio $-\beta$, it should satisfy following equation

$$-\beta * RB \leq -N * \Delta. \quad (15)$$

Then, the maximal Δ is $\Delta = \frac{\beta * RB}{N}$

For RF policy, the real risk value is more than the scaled value. Therefore, in order to satisfy the risk bound as $(1+\beta) * RB$, the maximal Δ is also $\frac{\beta * RB}{N}$. For RR policy, it has the deviation between RC and RF policies. Then, the maximal Δ is also the same as the above two policies to satisfy $(1+\beta) * RB$. Meanwhile, the case for RN policy is different. Given Δ, the deviation of RN policy is less than the other approaches. More specifically, the deviation is only one second of them according to Eq. 14. Hence, given risk slack ratio β, we can get the maximal Δ as

$$\Delta = \frac{2 * \beta * RB}{N} \quad (16)$$

From Eq. 16, we can notice that RN allows twice larger value of Δ than RF, RC and RR, while satisfying the given security risk deviation ratio. In another words, RN policy gives less time complexity, which is about half of other three policies. So RN is the best policy among the four studied methods, and we will present a RN based approximation algorithm for our system optimization problem in next section.

D. RN-based Approximating Algorithm

Based on the RN policy and the 2-dimensional states presentation, we proposed a Dynamic Programming based security-aware approximation solution. The main purpose of RN-based Approximation Algorithm (RNAA) is to assign the most suitable security level to each task using the minimum energy while satisfying the security requirements. The detailed optimization procedure is presented as pseudo-code in Algorithm 1.

RNAA is composed of four steps. In the first step, we test the schedulability of the given task set. If the minimal security risk is higher than the risk bound, i.e. all tasks have the maximal security levels, then the task set is not schedulable; if the minimal CPU utilization ratio is higher than the utilization bound even if only the minimal demanded security level are assigned on each task, then the task set is also not feasible. In Step 2, RNAA initializes the group factor Δ and the 2-dimensional state matrix.

Step 3, the core of RNAA, conducts the upgrading procedure of decision states for every decision-making stage. Based on each non-zero risk state in $(i-1)$-th stage, RNAA calculates every possible state for i-th stage. If a temporarily generated state obtains higher security risk than the given bound RB or more utilization ratio than UB, it is immediately ignored. Furthermore, if there is no such state with the same risk as the temporary state in N rows of the state matrix, or the temporary state has lower utilization ratio comparing with prior state in i-th stage with the same security risk, then RNAA replaces the old state with the new one that has lower utilization.

After all the decision states in each stage have been renewed, RNAA selects the state with the smallest utilization ratio in the N-th stage in Step 4. To obtain the final security protection decision, RNAA goes back from N to the first stage to get the vector of security assignments. In the end of Step 4, RNAA successfully obtains the minimal energy consumption.

Complexity Analysis: In step 1, it takes $O(n)$ to complete the schedulability test. In step 2, it takes $O(1)$. For step 3, there are $N-1$ decision stages (line 11). Based on each state at the $(i-1)$-th stage, it takes $O|S_i|$ to update all states at i-th stage (see lines 13-20). Due to the number of states can be at most M for $(i-1)$-th stage, it will take $O((n-1)*|S_i|*M)$ for the whole step 3. For step 4, it takes $O(n)$ to find the minimal energy solution. Therefore, the time complexity of RNAA can be inferred as $O(RNAA) = O(n) + O(1) + O((n-1)*|S_i|*M) + O(n) = O(n^2|S_i|/2\beta)$. According to the analysis in Sec.IV(D), it will take $O(n^2|S_i|/\beta)$ time overhead by using other policies (RF, RC, and RR). Thus, we conclude that RNAA is polynomial of tasks number n, security choices $|S_i|$ and $1/2\beta$ and has half time complexity of other traditional approximating policies.

V. EXPERIMENTAL RESULTS

In this section, we conducted synthetic experiments to verify the performance of the proposed algorithm. We implement a task scheduler that includes security assignment and scheduling in .Net environment. For evaluation purposes, we compare our RNAA algorithm with one group of approximation methods, i.e. RCAA and RSAA, and one group of heuristics, named GRDY and SEAS. RCAA is an approximation approach based on RC policy like in [16], while RRAA is the approximation approach based on RR policy. For GRDY scheme, the security levels are assigned in a greedy fashion. It provides the current highest security level to tasks step by step until all available energy slacks are depleted. SERS is a Security and Energy-aware Real-time Scheduling (similar to SASES [8]). SERS increase gradually the security level of tasks by comparing the risk-energy ratio among tasks, while satisfying the risk and utilization constraints, just like the benefit-cost ratio used in SASES.

The performance metrics in our experiments are energy consumption and risk deviation ratio. Risk deviation ratio is the deviation ratio of real system security risk to the risk bound. We performed two groups of simulations. For both groups, we generate three synthetic sets. Final results are obtained as the average of these three sets. For each task, the basic execution time is normally generated in the range of 5 ms to 10 ms, and the period is between 300 ms and 500 ms. The security demands are randomly assigned between level 6 and 8 for confidentiality protection. The impact value of each task is randomly generated between 5 and 10, and the size of sensitive data ranges from 100KB to 400KB uniformly. The security coefficient λ is set in $[1, 3]$ for all tasks. Other parameters are set correspondingly in each simulation group. For execution time and energy consumption of security services, we use the same values as Table I.

A. Impacts of Risk Bound over different approaches

In this group of simulation, we analyze the performances under different security bounds. Tasks are generated as discussed in the previous section, and are scheduled by EDF. Thus, the schedulability of the system can be tested by checking whether the utilization bound is not bigger than 1. For a set of tasks, the maximal security risk (MAR) and minimal security risk (MIR) can be calculated when each task is assumed to use the maximal security level or minimal level respectively. In this section, we refer to the risk bound as α. So the real risk bound can be obtained by $MIR + \alpha * (MAR - MIR)$. We set the risk slack ratio β to 0.05 and the risk bound varying from 0.4 to 0.9 with step size of 0.1. The overall energy cost is normalized to the energy consumption of all tasks (each with the maximal security level). Fig. 2 and 3 are the obtained results of the five different algorithms that were discussed at the beginning of this section under different security risk bounds.

There are several interesting phenomenons that can be observed in Fig. 2. For example, with the increase of security risk bound α, the energy costs of all approaches are gradually reducing. This is because that loose risk constraint gives larger search space for each algorithm to decrease their security protection levels, which can result in less energy consumption. GRDY has the largest energy cost while RNAA and RRAA have nearly lowest energy costs. The energy costs of RCAA and SERS lie between them. More specifically, RNAA saves 14.5%, 5.9% and 4.3% energy from GRDY, RCAA and SERS, respectively. RRAA has roughly the same average energy costs as RNAA.

From Fig. 3, we can see the real security risk deviation with different risk bound α. All RNAA, RRAA and RCAA can guarantee the risk slack ratio value 0.05. RCAA has the largest negative deviation ratio as -0.025 averagely. From this group of experiments, we can reach to the conclusion that our proposed RNAA uses the minimal energy cost within given security risk constraint among all the five solutions. Comparing with RRAA and RCAA, the time complexity of RNAA is about half of them according to the derivation in section IV. Therefore, RNAA is the most suitable approach for the scheduling optimization problem of this paper (refer to Section III).

B. Impacts of Risk Slack ratio

Different user or system designer may have different security risk deviation demands on approximating approaches. For example, users can tolerate larger risk deviation for less important applications but require smaller risk deviation for security-critical applications. Thus, the goal of this group of simulations is to evaluate the performance under different risk slack ratios β. We set the security risk bound as

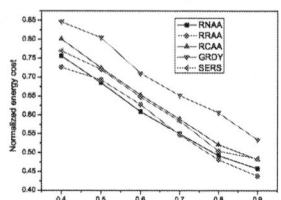

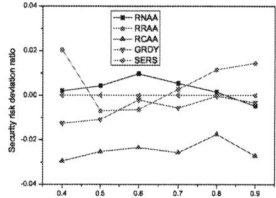

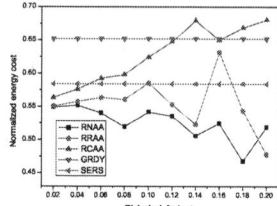

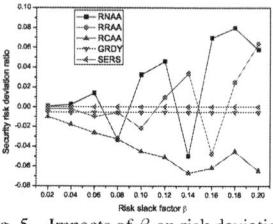

Fig. 2. Impacts of α on energy consumption

Fig. 3. Impacts of α on risk deviation ratio

Fig. 4. Impacts of β on energy consumption

Fig. 5. Impacts of β on risk deviation ratio

$MIR + 0.7 * (MAR - MIR)$. The risk slack ratio varies from 0.02 to 0.2 with step increment of 0.02. The task attributes are generated as discussed in the previous section. The simulation results of overall energy consumption and security risk deviation ratio are shown in Fig. 4 and Fig. 5, respectively.

For security risk in Fig. 4, we can see that RNAA also gives nearly the minimal energy cost on average, while GRDY approach results in the largest energy overhead. The energy costs of GRDY and SERS are constant in this experiment, because GRDY and SERS are not impacted by risk slack ratios. The energy cost of RCAA is increasing with bigger ratios, while the results of RNAA and RRAA demonstrate their random characteristics. RNAA can averagely reduce the energy cost than GRDY, SERS and RCAA by 19.3 percent, 10.0 percent and 16.3 percent, respectively.

Fig. 5 depicts the risk deviation ratios under different risk slack factors. The risk deviation ratios of GRDY and SERS are also constant as they are independent from risk slack factors. All RNAA, RRAA and RCAA can satisfy the given risk slack factor even if the slack factor is very small, e.g. 0.02 and 0.04. RCAA generally obtains negative deviation ratio. RNAA and RRAA have smaller (absolute) risk deviation ratio than RCAA. For example, RNAA can get smaller deviation ratio than RCAA, for example, when it becomes negative with the slack factor value 0.14. The reason is that RNAA and RRAA utilize random scaling policies which cancel out the positive and negative deviations. Based on these two figures, we can find that RNAA is the best algorithm among the five approaches which obtains the lowest energy cost with little security risk deviation.

VI. CONCLUSION

Energy and security are two important factors for designing mission-critical real-time embedded systems running on constrained resources. This paper addresses one common scheduling problem for security- and energy-critical real-time applications, in which minimizing energy consumption while satisfying the security and schedulability constraints is of central importance. This problem is a multi-dimensional knapsack problem which is proved to be NP-hard. Then, we reduce the problem based on the relationship between energy consumption and CPU utilization. To find the solution efficiently, we introduce and analyze four approximation policies, and then propose our dynamic programming based approximation algorithm to find the near-optimal solutions within predefined security risk constraints efficiently. The proposed algorithm has fully polynomial time complexity that is roughly half of the existing FPTAS approaches. Moreover, our algorithm has also low memory overhead, which is suitable to be used in resource limited embedded systems.

Finally, synthetic simulations demonstrate the advantages of our proposed scheduling framework.

VII. ACKNOWLEDGEMENT

This work was partly supported by the National Natural Science Foundation of China under Grant No. 61003032 and the Research Fund of National Key Laboratory of Computer Architecture under Grant No. CARCH201104.

REFERENCES

[1] A. Valenzano, L. Durante, and M. Cheminod, "Review of security issues in industrial networks," *IEEE Transactions on Industrial Informatics*, vol. 9, no. 1, pp. 277–293, 2012.

[2] R. Chandramouli, S. Bapatla, K. Subbalakshmi, and R. Uma, "Battery power-aware encryption," *ACM Transactions on Information and System Security (TISSEC)*, vol. 9, no. 2, pp. 162–180, 2006.

[3] V. Chaturvedi, H. Huang, S. Ren, and G. Quan, "On the fundamentals of leakage aware real-time dvs scheduling for peak temperature minimization," *Journal of Systems Architecture*, vol. 58, no. 10, pp. 387–397, 2012.

[4] Y. Wang, H. Liu, D. Liu, Z. Qin, Z. Shao, and E. H.-M. Sha, "Overhead-aware energy optimization for real-time streaming applications on multiprocessor system-on-chip," *ACM Trans. Des. Autom. Electron. Syst.*, vol. 16, no. 2, pp. 14:1–14:32, Apr. 2011.

[5] J. Gan, F. Gruian, P. Pop, and J. Madsen, "Energy/reliability trade-offs in fault-tolerant event-triggered distributed embedded systems," in *Proc. of ASP-DAC*, 2011, pp. 731–736.

[6] L. Sha, T. Abdelzaher, K.-E. Årzén, A. Cervin, T. Baker, A. Burns, G. Buttazzo, M. Caccamo, J. Lehoczky, and A. K. Mok, "Real time scheduling theory: A historical perspective," *Real-time systems*, vol. 28, no. 2, pp. 101–155, 2004.

[7] W. Jiang, K. Jiang, and Y. Ma, "Resource allocation of security-critical tasks with statistically guaranteed energy constraint," in *18th International Conference on Embedded and Real-Time Computing Systems and Applications*. IEEE, 2012, pp. 330–339.

[8] T. Xie and X. Qin, "Improving security for periodic tasks in embedded systems through scheduling," *ACM Transactions on Embedded Computing Systems*, vol. 6, no. 3, pp. 1–19, 2007.

[9] K. Jiang, P. Eles, and Z. Peng, "Optimization of message encryption for distributed embedded systems with real-time constraints," in *14th International Symposium on Design and Diagnostics of Electronic Circuits & Systems*. IEEE, 2011, pp. 243–248.

[10] M. Lin, L. Xu, L. T. Yang, X. Qin, N. Zheng, Z. Wu, and M. Qiu, "Static security optimization for real-time systems," *IEEE Transactions on Industrial Informatics*, vol. 5, no. 1, pp. 22–37, 2009.

[11] K. Jiang, P. Eles, and Z. Peng, "Optimization of secure embedded systems with dynamic task sets," in *Proc. of DATE*. IEEE, 2013, pp. 1765–1770.

[12] W. Jiang, Z. Guo, Y. Ma, and N. Sang, "Research on cryptographic algorithms for embedded real-time systems: A perspective of measurement-based analysis," in *Proc. of 9th IEEE International Conference on Embedded Software and Systems*. IEEE, 2012, pp. 1495–1501.

[13] B. Karabacak and I. Sogukpinar, "Isram: information security risk analysis method," *Computers & Security*, vol. 24, no. 2, pp. 147–159, 2005.

[14] H. Kellerer and U. Pferschy, "Improved dynamic programming in connection with an fptas for the knapsack problem," *Journal of Combinatorial Optimization*, vol. 8, no. 1, pp. 5–11, 2004.

[15] J. Gong, X. Zhong, and C.-Z. Xu, "Maximizing rewards in wireless networks with energy and timing constraints for periodic data streams," *IEEE Transactions on Mobile Computing*, vol. 9, no. 8, pp. 1187–1200, 2010.

[16] X. Zhong and C.-Z. Xu, "System-wide energy minimization for real-time tasks: Lower bound and approximation," *ACM Transactions on Embedded Computing Systems*, vol. 7, no. 3, p. 28, 2008.

978-1-4799-2817-0/14 $31.00 © 2014 IEEE

A Comprehensive and Accurate Latency Model for Network-on-Chip Performance Analysis

Zhiliang Qian[1], Da-Cheng Juan[2], Paul Bogdan[3], Chi-Ying Tsui[1], Diana Marculescu[2] and Radu Marculescu[2]

[1]Electronic and Computer Engineering Department, Hong Kong University of Science and Technology, Hong Kong
[2]Electrical and Computer Engineering Department, Carnegie Mellon University, Pittsburgh, U.S.A
[3]Ming Hsieh Department of Electrical Engineering, University of Southern California, Los Angeles, U.S.A

Abstract- In this work, we propose a new, accurate, and comprehensive analytical model for Network-on-Chip (NoC) performance analysis. Given the application communication graph, the NoC architecture, and the routing algorithm, the proposed framework analyzes the links dependency and then determines the ordering of queuing analysis for performance modeling. The channel waiting times in the links are estimated using a generalized G/G/1/K queuing model, which can tackle bursty traffic and dependent arrival times with general service time distributions. The proposed model is general and can be used to analyze various traffic scenarios for NoC platforms with arbitrary buffer and packet lengths. Experimental results on both synthetic and real applications demonstrate the accuracy and scalability of the newly proposed model.

I Introduction

Network-on-Chips (NoCs) have been proposed as an efficient and scalable solution to support complex on-chip communication for a variety of multi-core systems [1]. In order to cope with tight design constraints, design space exploration has been used extensively but turned out to be a major issue in many practical scenarios. More precisely, detailed simulation can suffer from long evaluation times and so only be used to estimate a small subset of alternatives [2,3]. Therefore, it is crucial to develop fast and accurate performance analysis tools that can guide the design space exploration during the overall optimization process.

Towards this end, several analytical models have been proposed for NoC performance estimation (Fig. 1) [2,3,4]. Based on the application and the target NoC architecture, the performance model typically works together with the design space exploration tool in the inner optimization loop. The traffic input to the performance model is first extracted from the current configuration (*e.g.*, task mapping, core floorplan and routing algorithm). Then, the analytical model estimates the performance metrics of interest (*e.g.*, latency, throughput) in order to guide the pruning of the design space.

This paper presents a new delay modeling methodology that can work for a variety of NoC configurations and traffic scenarios. More specifically, the proposed analytical model utilizes G/G/1/K queuing formalism and generalizes the NoC latency evaluation as follows: (1) The existing traffic arrival modeling using Poisson approximations [3,5] is extended to a generalized exponential (GE) packet inter-arrival distribution which can account for bursty and dependent traffic patterns.

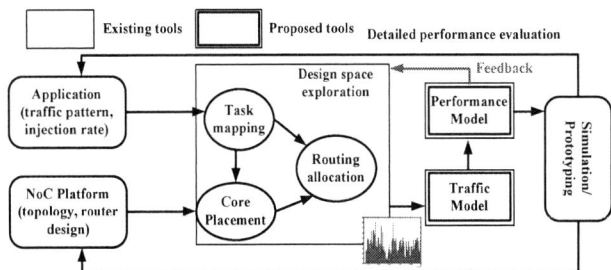

Figure 1 The proposed analytical performance model shown in the synthesis loop for design space exploration.

(2) The packet service process within each router is modeled by a general distribution to account for the service time correlation between routers and traffic flows. (3) The NoC architecture flexibility is enhanced to consider routers with finite buffer size and enable arbitrary buffer size and packet length combinations. 4) Finally, by examining the link dependencies, the proposed framework extracts the core communication graphs that can be used for further performance analysis. It is worth mentioning that this framework is completely generic and it can be applied to any NoC topology for which both the task mapping and routing algorithm are known.

The rest of this paper is organized as follows. In section II, we review related work and highlight the paper contributions. In section III, we present the proposed analytical model. Section IV shows the experimental results. Finally, we conclude by summarizing our main contributions in section V.

II. Related work and paper contributions

For the general class of network-based systems, much of prior work considers the modeling of wormhole (WH) routers under the assumption of Poisson arrival time and memoryless packet service time distributions. For example, in [6], an M/M/1 approximation for the link delay is used to analyze capacity and flow allocation. Although generally tractable, the accuracy of M/M/1 models can be significantly compromised as the assumption of exponential arrival and service time distributions may not hold in many real applications [7,9]. To address this shortcoming, several works have been proposed to improve the estimation accuracy by generalizing the arrival and service time distributions. More precisely, in [8], an M/G/1/K queue based analytical model is proposed to account for finite size input buffers in local area networks (LANs).

Table 1 Comparisons of NoC analytical models

	Previous NoC analytical models				This work
	[6]	[3,4]	[2]	[5]	
Application traffic modeled					
Queue	M/M/1	M/G/1/K	G/G/1	M/M/m/K	G/G/1/K
Arrival	Poisson	Poisson	General	Poisson	General
Service	Markov	General	General	Markov	General
NoC architecture modeled					
Buffer	Small	K packets	B flits	Small	B flits
PB[1]	m (≫ 1)	<1	arbitrary	m (≫ 1)	arbitrary
Arbiter	Round Robin	Round Robin	Fixed Priority	Round Robin	Round Robin

[1]PB ratio is defined as the ratio of average packet size (m flits) to the buffer depth (B flits)

Table 2 Parameters and notations in analytical modeling

Parameter	Explanation
H_s	Service time for header flit (including the link transfer)
m	Average packet size (flits)
B	Buffer size in each channel (flits)
$f_{s,d}$	Communication flow from source s to destination d
d_f	Length (number of hops) of flow f
l_{ab}	Link channel connecting router a and b
P_f $(P_{s,d})$	Set of links that form the routing path of flow $f_{s,d}$
$F_{l_{ab}}$	Aggregate set of flows sharing link l_{ab}
l_i^f	Link that resides in the ith hop of flow f
λ_f $(\lambda_{s,d})$	Mean packet arrival rate of flow f
C_f^2	SCV of the packet inter- arrival time of f
$\lambda_{l_{ab}}$	Mean packet arrival rate at link l_{ab}
$q_{l_i^f}$	Delay for a flit to reach the head of buffer in link l_i^f
$h_{l_i^f}$	Delay for a packet head to acquire the link l_i^f
$\eta_{l_i^f}$	Header flit transfer time over the link l_i^f
$s_{l_i^f}/s_{l_{ab}}$	Service time for a packet that travels link l_i^f / l_{ab}
$z_{l_i^f}$	Time that a header flit reaches the point where the accumulated buffer space can hold the whole packet
v_s	Source queuing time at tile s
L_f $(L_{s,d})$	Average latency of flow $f_{s,d}$ (cycles)

However, the model is based on the Laplace-Stieltjes transform and so is too complicated to be used in a typical NoC synthesis loop like in Fig.1. In [3], an M/G/1 based latency model for NoC analysis is proposed. It only assumes a Poisson process for the arrival rate of header flits (as opposed to the entire packet). In [2], a fixed-priority G/G/1 based NoC latency model has been proposed which further attempts at modeling the bursty arrival times with a 2-state Markov-modulated Poisson process (MMPP). However, this approach targets a specific priority-based router architecture, while many NoC routers can utilize a more fair arbitration such as the round robin (RR) scheme. In [5], an M/M/m/K queue-based analytical model is proposed to analyze the delay of NoCs with variable virtual channels per link. This approach assumes negligible flit buffers (*i.e.,* single flit buffer) such that a packet reaches its destination before its tail leaves the source host. In [4], an M/G/1/N queuing mode is proposed for both wormhole and virtual channel NoCs. However, this approach assumes that the granularity of the buffers is given in terms of packets instead of flits and therefore a single channel buffer can hold up to N packets during the analysis. This may not be the case for NoCs whose buffers are rather small (only several flits) to save area and power [2].

In this paper, we propose a new NoC latency model which generalizes the previous work by considering: (i) the arrival traffic burstiness, (ii) the general service time distribution, (iii) the finite buffer depth and arbitrary packet length. For clarity purposes, in Table I, we summarize and compare our proposed analytical frameworks against other models proposed to date. The newly proposed model offers a much broader coverage for a variety of temporal and spatial traffic patterns, as well as NoC architectures; this provides more flexibility when exploring the NoC design space.

III. NoC modeling for performance analysis

A. Basic assumptions and notations

We assume that the target applications have been scheduled and mapped onto the target NoC platform and the source and destination tile addresses for each specific flow f in the application are known. Also, borrowing from the idea of modeling bursty traffic in hyper-cube multicomputers [16], we assume that the packet inter-arrival times of flow f have been characterized using a general exponential (GE) distribution [10] (discussed later in Section III.D) with mean λ_f^{-1} and a square coefficient of variation (SCV) C_f^2. Therefore, the (λ_f^{-1}, C_f^2) characterization of the traffic model is an input to our analytical framework. Moreover, a deadlock-free and deterministic routing algorithm is used to guarantee that no cycles can be formed by link dependencies.

Without loss of generality, in this work we use X-Y routing, which can be generalized to other deadlock-free and deterministic routing schemes. We also adopt a wormhole router architecture, where there exists a single buffer at each input port. For simplicity, we assume that packets have a fixed size of m *(flits)* as in [9]. However, this assumption can be extended to arbitrary packet length distribution. Other assumptions include: the traffic source (*i.e.,* the source Processing elements) has an infinite queue size and the destination immediately consumes the arriving flits. To facilitate the discussion, the symbols in Table 2 are used consistently throughout this paper.

B. End-to-end delay calculation

In a wormhole routing network, the end-to-end flow latency $L_{s,d}$ (shown in Fig 2-a) of a specific flow $f_{s,d}$ consists of the queuing time at the source s (v_s), the packet transfer time ($\eta_{s,d}$) and the path acquisition time ($h_{s,d}$):

$$L_{s,d} = v_s + \eta_{s,d} + h_{s,d} \qquad (1)$$

In order to calculate $h_{s,d}$, we need to add up the path acquisition time of every link l_i^f in the routing path of f (Fig 2-b shows the links involved for flow $f_{0,8}$) and:

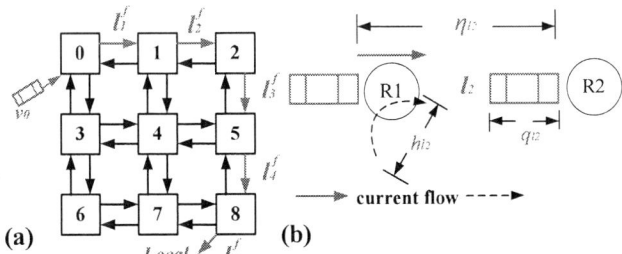
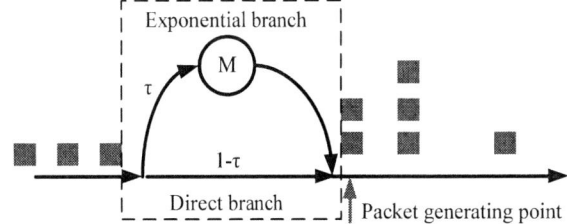

(a) **(b)**

Figure 2 Example showing the flow delay a) the links involved to calculate the end-to-end delay of flow $f_{0,8}$,b) the acquisition time h and transfer time η for link l_{12}

$$h_{s,d} = \sum_{i=1}^{d_f} h_{l_i^f} \qquad (2)$$

where $h_{l_i^f}$ is the time for a packet header to compete for a channel with other flows over link l_i^f.

If we denote $\eta_{l_i^f}$ as the time to transmit the header flit, then the packet transfer time $\eta_{s,d}$ can be rewritten as:

$$\eta_{s,d} = \sum_{i=1}^{d_f} \eta_{l_i^f} + (m-1) \qquad (3)$$

where the first term denotes the header flit transmission time over the network, and the second term approximates the packet serialization time of the body and tail flits. The notations of the queuing delay v_s, $h_{l_i^f}$ and $\eta_{l_i^f}$ in Eqn. 1-3 are also illustrated in Fig.2-a and -b.

In order to derive $L_{s,d}$, $\eta_{l_i^f}$ and $h_{l_i^f}$ for each link need to be obtained. Two issues hereby arise. First, it is important to determine the *order* of links that can be used for the queuing analysis. Second, a detailed queuing model is needed for obtaining η and h of each link. In the following, we first present the link dependency analysis to search for the correct order and then elaborate on the queuing formulation of a link to estimate η and h.

Algorithm 1 Link dependency analysis
_____
1: **Input**: F the application flow set
2: **Output**: G the ordered list of links for queuing analysis
3: **Container**: $LDG = (V, E)$ the link dependency graph
4: **for** any link l_{ab}
5:     $LDG.addnode(l_{ab})$;
6: **endfor**   // initialize LDG, $E = \emptyset$
7: **for** each flow $f \in F$
8:     $P_f = routing\_function(f)$ //set of links in the path
9:     $d_f = length(P_f)$
10:     **for** $i=1$: $d_f - 1$
11:         $l_{up} = P_f(i)$;   // the upstream link
12:         $l_{down} = P_f(i+1)$;   // the downstream link
13:         $LDG.addedge(l_{down}, l_{up})$
14:     **endfor**
15: **endfor**
16: **return** $G = topological\_sort(LDG)$;
_____

C. Link dependency analysis

To account for the impact of congestion on the succeeding hops, knowing the dependencies among all the links is important. For example, in Fig. 2-b, due to backpressure, the waiting times η_2^f and h_2^f of link l_{12} (i.e., l_2^f) will affect the time to serve a packet in the buffer head of link l_{01} (i.e., l_1^f).

Figure 3 Modeling of the GE packet generation process

Therefore, l_{01} is dependent on l_{12}. Iteratively, as η_2^f and h_2^f will be affected by the succeeding links of l_{12} in the same way, it is required to analyze all subsequent links of link l_{01} in the routing path of flow f *before* deriving the queuing service and waiting time of link l_{01}. In this work, we propose to build the link dependency graph (LDG) first and then utilize the topological sort algorithm as in [8] to order the links. The detail of link dependency analysis is presented in Algorithm 1. Please note that when the routing as well as the task mapping algorithm change, the LDG needs to be rebuilt. As shown in lines 7-15 of Algorithm 1, the LDG is built by checking every flow f in the application communication set F. An edge in the LDG is added if there exists a flow f whose routing path passes through the two links consecutively (line 13). After that, the links computation order G can be obtained (line 16).

D. GE-type traffic modeling

Many applications in NoCs show bursty patterns of traffic over a wide range of time scales [7]. Therefore, the Generalized Exponential (GE) distribution is utilized to model the arrival traffic at the source PEs and the links. The GE-type cumulative distribution function (cdf) of the inter-arrival time X is given by:

$$F(t) = P(X \leq t) = 1 - \tau e^{-\tau \lambda t}, t \geq 0 \qquad (4)$$

where $\tau = \frac{2}{1+C^2}$ and (λ^{-1}, C^2) are the mean and square coefficient of variation (SCV) of X.

We depict the GE packet generation process in Fig. 3. As shown in the figure, with probability $(1 - \tau)$, a packet experiences a zero service time to reach the departure point (*i.e.*, the point to generate the packet), while with probability τ, the packet needs to traverse a system with an exponentially distributed service time with mean $1/(\tau\lambda)$. The bulk of packets consist of a packet which comes from the exponential branch together with a number of successive packets arriving through the direct branch. The GE distribution is a versatile and simple distribution which helps to make the queuing formulation analytically tractable [10]. Moreover, it has been demonstrated that the GE distribution also provides an efficient approximation for both short and long-range dependent traffic in supercomputers [10,16] .

E. WH router modeling

In this section, we estimate the three key components in WH analytical models, *i.e.*, the path acquisition time h, the flit transfer time η, and the source queuing time v_s.

1) *Flit transfer time*: The flit transfer time η of link l_{ab} is

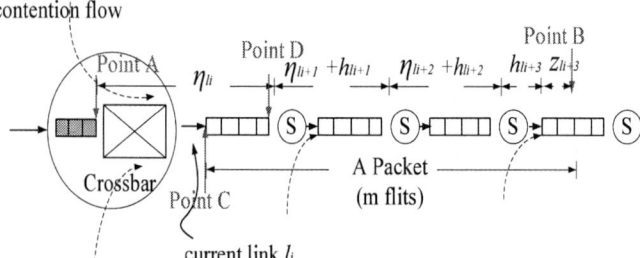

Figure 4 Illustration of service time of link l_i in WH routing

defined as the time taken for the header flit after being granted link access to leave the buffer head in the upstream node and reach the front of the buffer in current link l_{ab} (*e.g.*, from Point A to Point D in Fig. 4). It consists of two parts: the first part is the time to leave the upstream router (*i.e.*, from Point A to Point C in Fig. 4). For WH routing, this is a constant value of the router and link pipeline stages (H_s cycles). The second part accounts for the header flit to arrive at the buffer front of link l_i (*i.e.*, q_{l_i} in Table 2, from Point C to Point D). This time can be approximated as the waiting time of an M/M/1/K queue with system capacity being equal to the buffer size B (K = B). The mean flit rate arriving at the buffer is

$$\lambda_{flit}^{l_{ab}} = m \times \lambda_{packet}^{l_{ab}} = m \times \sum_{f \in F_{l_{ab}}} \lambda_f \qquad (5)$$

The mean time to serve a flit ($s_{flit}^{l_{ab}}$) in this queuing system is calculated as the weighted average of the service time from all flows passing through link l_{ab}, *i.e.*:

$$s_{flit}^{l_{ab}} = \frac{\sum_{\forall f \in F_{l_{ab}}} \left[\lambda_f \times \left(\frac{h_{l_{i+1}}^f}{m} + 1 \right) \right]}{\sum_{\forall f \in F_{l_{ab}}} \lambda_f} \qquad (6)$$

In Eqn. 6, for a packet with m flits, it takes $h_{l_{i+1}}^f$ cycles for a header in the buffer to win the next link of flow f (*i.e.*, $h_{l_{i+1}}^f$ + 1 cycles in service). After that, the body and the tail flits depart the buffer without any additional delay (*i.e.*, 1 cycle). Therefore, the mean flit service time over the whole packet is $h_{l_{i+1}}^f / m + 1$.

After $\lambda_{flit}^{l_{ab}}$ and $s_{flit}^{l_{ab}}$ are obtained, the M/M/1/B queuing formulation [15] can be applied to obtain q_{l_i} and thereby $\eta_{l_{ab}} = \eta_{l_f} = q_{l_i} + H_s$.

2) Path acquisition time: The path acquisition time h of flow f over its ith hop link l_{ab} (*i.e.*, l_i^f) is the delay for the header flit to be granted access to the buffers in l_{ab} after contention with other flows sharing the same link. It can be modeled as the waiting time of a G/G/1/K queuing system. More specifically, the system capacity K = k denotes the number of flows which come from the input ports in the same router and forward to the same output link (k = 3 as in Fig 4. for l_i^f). The arrival process is the merging of all traffic flows that route to l_{ab}. Therefore, the mean arrival rate is $\lambda_{l_{ab}} = \sum_{f \in F_{l_{ab}}} \lambda_f$, while the SCV can be approximated as:

$$C_{a_{l_{ab}}}^2 = \sum_{\forall f \in F_{l_{ab}}} (\lambda_f \times C_f^2) / \sum_{f \in F_{l_{ab}}} (\lambda_f) \qquad (7)$$

The service time of this queue is illustrated in Fig. 4. Without loss of generality, we consider the case when a packet with m flits is split over a limited number of downstream links along the path (assume the buffer depth is B flits). The service time accounts for the time that a packet occupies a shared link l_{ab}. For example, in Fig. 4, assume a header flit in Point A is granted access to the current link l_i. Then, the service process of this packet begins when the header flit leaves A and ends when the tail flit departs A so to release l_i for other flows. If the downstream links are not congested, this service time consists simply of m cycles as the entire packet can traverse downstream links smoothly. However, when a severe blockage exists along the path, the worst-case scenario needs the packet header to reach Point B (Fig.4) where the accumulated buffer spaces can hold the whole packet [8]. This time $x_{l_i}^f$ (*i.e.*, the delay from Point A to Point B) can be represented as in Eqn. 6 (assume z_{l_f} is small in Fig. 4):

$$x_{l_i}^f = \eta_{l_f} + \sum_{j=1}^{\Lambda_{l_i}^f} (\eta_{l_{i+j}^f} + h_{l_{i+j}^f}) + h_{l_{i+\Lambda_{l_i}^f}^f} \qquad (8)$$

where $\Lambda_{l_i}^f$ represents the effective number of hops that a packet may span. Specifically, if m/B is smaller than the remaining hops from the current link l_i^f to the destination of f (*i.e.*, $d_p - i$), then $\Lambda_{l_i}^f$ equals to $\lfloor m/B \rfloor$; otherwise, $\Lambda_{l_i}^f$ will be assigned the value $d_p - i$. The service time for flow f (*i.e.*, s_{l_f} in Table 2) is then bounded by m and $x_{l_i}^f$ under different congestion conditions and can be approximated according to [8,16]:

$$s_{l_f} = \begin{cases} \left[m(m + x_{l_i}^f) + 2x_{l_i}^f m \right] / (m + 2x_{l_i}^f) & \text{if } x_{l_i}^f < m \\ \left[m(m + x_{l_i}^f) + 2(x_{l_i}^f)^2 \right] / (m + 2x_{l_i}^f) & \text{otherwise} \end{cases} \qquad (9)$$

From Eqn. 8-9, once the downstream links transfer time η and contention time h are known, the overall link service time is then the weighted average of all flows passing through it, which yields the mean service time $\overline{s}_{l_{ab}}$ as:

$$\overline{s}_{l_{ab}} = \sum_{\forall f \in F_{l_{ab}}} (\lambda_f \times s_{l_f}) / \sum_{\forall f \in F_{l_{ab}}} \lambda_f \qquad (10)$$

Likewise, the SCV of the service time for link l_{ab} can be approximated by:

$$C_{s_{l_{ab}}}^2 = \frac{\overline{s_{l_{ab}}^2}}{(\overline{s}_{l_{ab}})^2} - 1 = \left(\frac{\sum_{\forall f \in F_{l_{ab}}} \lambda_f \times s_{l_f}^2}{\sum_{\forall f \in F_{l_{ab}}} \lambda_f} \right) / (\overline{s}_{l_{ab}})^2 - 1 \qquad (11)$$

Based on the mean arrival rate $\lambda_{l_{ab}}$, its SCV $C_{a_{l_{ab}}}^2$ in Eqn. 7 as well as the service time ($\overline{s}_{l_{ab}}, C_{s_{l_{ab}}}^2$) characterized by Eqn. 10-11, the GE/G/1/K queuing time $h_{l_{ab}}$ (*i.e.*, h_{l_f}, $\forall f \in F_{l_{ab}}$) can be approximated as:

$$h_{l_{ab}} = \frac{(C_{s_{l_{ab}}}^2 + C_{a_{l_{ab}}}^2)}{(1 + C_{s_{l_{ab}}}^2)} h'_{l_{ab}} \qquad (12)$$

where $h'_{l_{ab}}$ is the waiting time calculated according to the M/G/1/K queuing formula presented in [15].

3) Source queuing time: The source queue is modeled as a GE/G/1/∞ system [10]. Applying the GE/G/1 formula yields:

$$v_s = \frac{\overline{s}_{l_s}}{2} \left(1 + \frac{C_a^2 + \lambda_a \times \frac{(\overline{s}_{l_s} - m)^2}{\overline{s}_{l_s}}}{1 - \lambda_a \times \overline{s}_{l_s}} \right) - \overline{s}_{l_s} \qquad (13)$$

where the arrival process (λ_a, C_a^2) and the mean service time $\overline{s}_{l_s}$ are calculated in a similar way as the channel queues in Eqn 5-8.

F. The network performance analysis

The proposed analytical flow for NoC latency evaluation is

978-1-4799-2817-0/14 $31.00 © 2014 IEEE

summarized in Algorithm 2. We first apply the link dependency analysis presented in section III-B to determine the correct link ordering for performance analysis. Then, for each link l_{ab} in the ordered list G (line 3-15), we calculate the arrival traffic model $(\lambda_{l_{ab}}, C_{a_{l_{ab}}}^2)$ (line 4) first according to the routing algorithm and applications. As shown in Eqn. 6, the link transfer time $\eta_{l_{ab}}$ only depends on its downstream link contention time $h_{l_{i+1}^f}$, which should have been analyzed in the previous loop. On the other hand, the path acquisition time $h_{l_{ab}}$ depends not only on its downstream link contention time $h_{l_{i+1}^f}$ and transfer time $\eta_{l_i^f}$ but also on the current link $\eta_{l_i^f}$ (Eqn. 8). Therefore, $\eta_{l_{ab}}$ is calculated first in line 5. After $\eta_{l_{ab}}$ is obtained, then we calculate the mean and SCV of the link service time $(\overline{s}_{l_{ab}}, C_{s_{l_{ab}}}^2)$ (line 9). If this link connects two routers, we utilize GE/G/1/K queue to obtain $h_{l_{ab}}$ (line 11). Otherwise, we calculate the source queuing time v_a (line 13). Once all the (h, v, η) variables are known, Eqn. 1-3 can be applied to evaluate the delay for a specific flow $f_{s,d}$ from source s to destination d.

Algorithm 2 Analytical working flow

1: Input: F the application flow set; G ordered link list

2: Output: $L_{s,d}$ latency for each specific flow f

3: foreach $l_{ab} \in G$

4: $(\lambda_{l_{ab}}, C_{a_{l_{ab}}}^2)= traffic\_model\,(F_{l_{ab}})$ //Eqn. 5-7

5: $\eta_{l_{ab}}= calculate\_transfer\_time(\lambda_{l_{ab}}, s_{flit}^{l_{ab}}, m, B)$

6:   **foreach** $f \in F_{l_{ab}}$ and $l_i^f = l_{ab}$

7: $s_{l_i}^f = calculate\_link\_service\_time\,()$ //Eqn. 8-9

8:   **end**

9: $(\overline{s}_{l_{ab}}, C_{s_{l_{ab}}}^2) = service\_time\,()$ // Eqn. 10-11

10: **if** $a \neq b$ // the links between the routers

11: $h_{l_{ab}} = GE\_G\_1\_K\_queue\,(\lambda_{l_{ab}}, C_{a_{l_{ab}}}^2, \overline{s}_{l_{ab}}, C_{s_{l_{ab}}}^2, k)$

12: **else**   // the link is the source link

13: $v_a = GE\_G\_1\_queue\,(\lambda_{l_{ab}}, C_{a_{l_{ab}}}^2, \overline{s}_{l_{ab}}, C_{s_{l_{ab}}}^2)$

14: **endif**

15: endfor

16: foreach $f \in F$

17: $L_{s,d}=calculate\_flow\_latency()$ //Eqn. 1-3

18: end

IV. Experimental results

A. Basic assumptions and notations

We implement the proposed latency model in MATLAB and compare its accuracy with a cycle accurate NoC simulator called Booksim [13]. Without loss of generality, we adopt the mesh NoC as the target platform with two cycles to transmit a flit in a router and one cycle to cross the link (i.e., $H_s = 3$). Both synthetic traffic and real applications are utilized in the simulation. The real applications include *MMS* (multimedia system) [11] and *DVOPD* (Dual video object plane decoder)

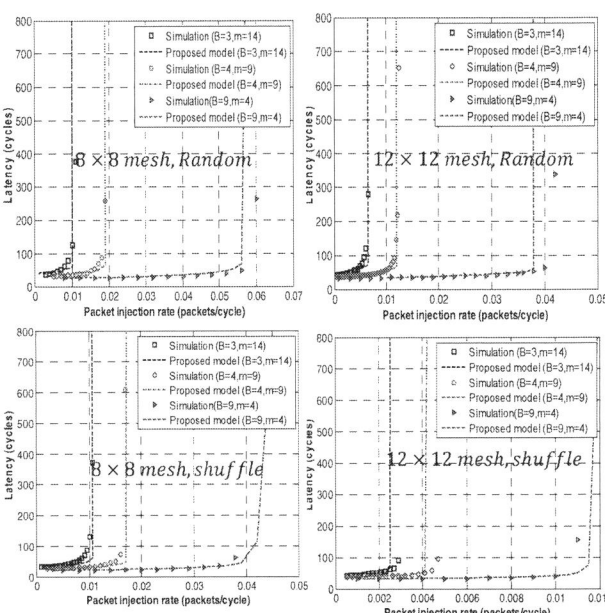

Figure 5 Comparison between proposed model and simulation

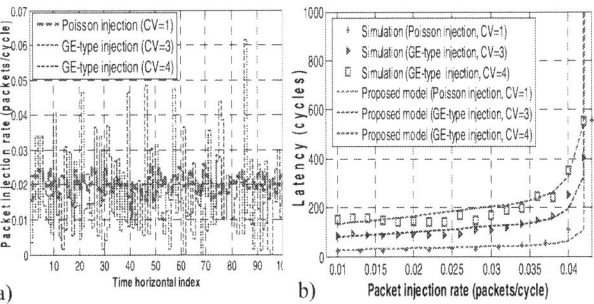

Figure 6 Comparison of GE-type and Poisson injection a) average injection rate , b) the latency comparison

[12] which are mapped onto a 5×5 mesh NoC and *MPEG4* (MPEG codec), *VOPD* (Video object plane decoder) [12] which are mapped onto a 4×4 mesh. Five traces extracted from SPECweb99 [14] are also adopted. They are 16-node multithreaded workloads for IBM and Oracle database server (i.e., *DVB2* and *Oracle*), the Apache HTTP server (i.e., *Apache*), scientific workloads of matrix factorization (i.e., *Sparse*) and the ocean dynamic simulation (i.e., *Ocean*).

B. Evaluation under synthetic traffic patterns

We first compare and evaluate the proposed analytical model using two synthetic traffic patterns (i.e., random and shuffle [13] traffic) with Poisson packet injection rates at each source node. Three packet length m and buffer depth B combinations are evaluated to cover different PB ratios in Table 1. For these (m, B) configurations, the models in [5,6] cannot be applied directly due to the finite buffer size (i.e., $B > 1$), while the models in [3,4] cannot be used as B is not a multiple of m. Moreover, because of the round-robin arbitration in the routers, the fixed-priority based model proposed in [2] is also not suitable for evaluating the target platform. The results for 8×8 and 12×12 mesh sizes are summarized in Fig. 5. As shown in Fig.5, the proposed analytical model achieves less than 12% error in predicting the network saturation point (i.e., the injection rate where the

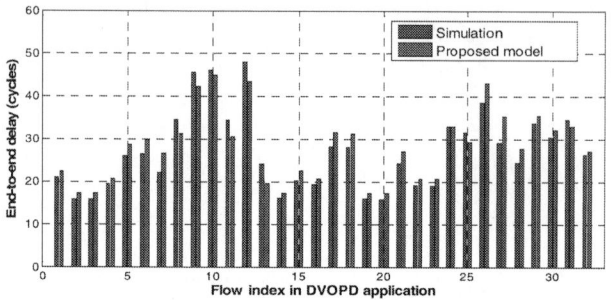

Figure 7 Flow latency comparisons for DVOPD application

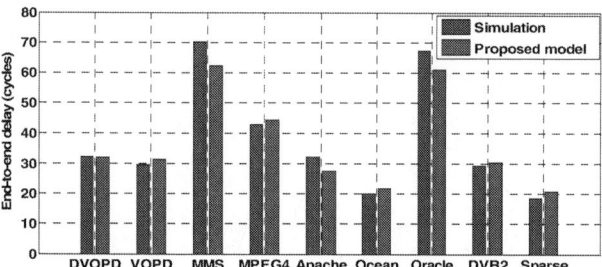

Figure 8 Accuracy comparisons for multiple NoC benchmarks

overall latency starts increasing dramatically) for all the synthetic traffic patterns.

Next, we consider the case where the traffic arrival times are bursty. Fig. 6-a depicts the average packet injection rate in a node against time for different values of coefficient of variation ($CV = C_a$). As shown in the figure, the Poisson arrival times cannot model the burstiness in the injection rates and greater CV values result in higher traffic injection intensities. In Fig. 6-b, we compare the proposed model against simulation where the burstiness is characterized by GE-traffic model for a 4×4 NoC with random traffic.

C. Evaluation under realistic benchmarks

We also evaluate the model accuracy using several real benchmarks. For *MMS*, *MPEG4*, *DVOPD* and *VOPD* benchmarks, we assume that packets are injected according to a Poisson process whose mean rate corresponds to the data communication rates on the edges of the task graphs. Moreover, the mapping of the task graphs is done manually utilizing the energy aware mapping algorithm in [11]. For *Apache*, *Ocean*, *Sparse*, *Oracle* and *DVB2* applications, the packets are injected according to the traces extracted. Based on the actual traces, we estimate the mean injection rate λ_f and its SCV C_f^2 in the analytical model. In Fig. 7, we show the per-flow latency comparison of the *DVOPD* application. As it can be seen, the proposed model achieves fairly good accuracy ($2.1\% - 8.4\%$ error) for each specific flow when compared against simulation. In Fig. 8, we show the latency predictions for a wide variety of applications. As shown in the figure, the proposed model achieves less than 13.5% error for all nine benchmarks evaluated.

D. Analytical and simulation time comparison

The average simulation time on a 12×12 mesh for random traffic is about thirty-five minutes while the analytical model only takes thirty seconds. Therefore, a 70X speedup over the simulation can be achieved. This can significantly benefit the NoC synthesis and optimization process since more potential solutions can be evaluated.

V. Conclusions

In this work, we have proposed a GE/G/1/K queue based analytical model for NoC performance analysis. Packet arrival burstiness has been taken into account by adopting a GE traffic model. In addition, different combinations of packet and buffer lengths have also been modeled. We have evaluated the accuracy and scalability of the model using both synthetic and real applications. As shown, the proposed model achieves less than 13% error for various traffic patterns, while providing about 70X speedup compared to simulation.

Acknowledgments

Z.-L. Qian and C.-Y. Tsui greatly acknowledge the support from Hong Kong Research Grant Council under GRF 619813. R.M. and D.M. greatly acknowledge partial NSF support under grant CNS-1128624. P.B. acknowledges support from USC. Finally, the authors would like to thank the anonymous reviewers for their valuable comments.

References

[1] Benini, L.; De Micheli, G., "Networks on chips: a new SoC paradigm," Computer , vol.35, no.1, pp.70,78, Jan 2002

[2] Kiasari, A.E.; Zhonghai L.,; Jantsch, A., "An Analytical Latency Model for Networks-on-Chip," Very Large Scale Integration (VLSI) Systems, IEEE Transactions on , vol.21, no.1, pp.113,123, Jan. 2013

[3] Ogras, U.Y.; Marculescu, R., "Analytical Router Modeling for Networks-on-Chip Performance Analysis," in proc. DATE '07 , vol., no., pp.1,6, 16-20 April 2007

[4] Mingche L.; Lei G.; Nong X.; Zhiying W., "An accurate and efficient performance analysis approach based on queuing model for network on chip," in proc. ICCAD'09., pp.563,570, 2-5 Nov. 2009

[5] Ben-Itzhak, Y.; Cidon, I.; Kolodny, A., "Delay analysis of wormhole based heterogeneous NoC," in proc. NoCs'11, pp.161,168, 1-4 May 2011

[6] Guz, Z.; Walter,I.; Bolotin, E.; Cidon, I.; Ginosar, R.; and Kolodny, A.; "Network delays and link capacities in application-specific wormhole NoCs" VLSI Design, 2007

[7] Bogdan, P.; Marculescu, R., "Non-Stationary Traffic Analysis and Its Implications on Multicore Platform Design," Computer-Aided Design of Integrated Circuits and Systems, IEEE Transactions on , vol.30, no.4, pp.508,519, April 2011

[8] Hu P. and Kleinrock L. "An Analytical Model for Wormhole Routing with Finite Size Input Buffers", 15th International Teletraffic Congress, 1997

[9] Nikitin, N.; Cortadella, J., "A performance analytical model for Network-on-Chip with constant service time routers," in proc. ICCAD'09. pp.571,578, 2-5 Nov. 2009

[10] Wu, Y.; Min,G.; Ould-Khaoua,M.; Yin, H.; and Wang. L; "Analytical modeling of networks in multicomputer systems under bursty and batch arrival traffic" The journal of Supercomputing, 51(2):115-130 , 2010

[11] Hu, J.; Marculescu, R., "Energy- and performance-aware mapping for regular NoC architectures," Computer-Aided Design of Integrated Circuits and Systems, IEEE Transactions on , 24(4): 551-562, 2005

[12] Bertozzi, D.; Jalabert, A.; Murali, S.; Tamhankar, R.; Stergiou, S.; Benini, L.; De Micheli, G., "NoC synthesis flow for customized domain specific multiprocessor systems-on-chip," Parallel and Distributed Systems, IEEE Transactions on , vol.16, no.2, pp.113,129, Feb. 2005

[13] Booksim simulator: https://nocs.stanford.edu

[14] SPEC benchmark: http://www.spec.org

[15] Tagagi H.; Queueing Analysis: "A Foundation of Performance evaluation: vol 2", North-Holland, New York, 1993

[16] Kouvatsos, D.D., SALAM, Assi and Mohamed ould-Khaoua, "Performance modeling of wormhole-routed hypercubes with bursty traffic and finite buffers" Int J Simul Pract Syst Sci Technol, vol. 6, pp. 69.81.2005

A Low-Latency Asynchronous Interconnection Network with Early Arbitration Resolution

Georgios Faldamis
Cavium Inc.
San Jose, CA 95131
gfaldamis@hotmail.com

Weiwei Jiang
Dept. of Computer Science
Columbia University
New York, NY 10027
wjiang@cs.columbia.edu

Gennette Gill
D.E. Shaw Research
New York, NY 10036
gennette@gmail.com

Steven M. Nowick
Dept. of Computer Science
Columbia University
New York, NY 10027
nowick@cs.columbia.edu

Abstract. *A new asynchronous arbitration node is introduced for use as a building block in an asynchronous interconnection network. The target network topology is a variant Mesh-of-Trees (MoT), combining a binary fan-out (i.e. routing) network and a binary fan-in (i.e. arbitration) network, which is becoming widely used for multi-core shared-memory interfaces. The two key features are: (i) each fan-in node can resolve its arbitration and pre-allocate the corresponding input channel, before the actual data arrives; and (ii) a lightweight shadow monitoring network fast forwards information as soon as data enters the network without synchronization to a fixed-rate clock, notifying each fan-in node on its path to enable the early arbitration. Simulations of the new arbitration node, using IBM 90nm technology and an ARM standard cell library, indicate latency reductions up to 54.4% over prior designs, while maintaining roughly comparable throughput. Network-level simulations were then performed on eight diverse synthetic benchmarks, comparing the new approach ("early arbitration") with two earlier alternative asynchronous MoT networks ("baseline" and "predictive"), using a mix of random and deterministic traffic. Considerable improvements in system latency were obtained on all benchmarks, ranging from 13.0% to 38.7%, with especially strong benefits for the two most adversarial benchmarks.*

1. Introduction

The introduction of networks-on-chip (NoC's) in recent years has been proposed to address several of the key challenges facing digital system designers [7], including design time, scalability, reliability and ease-of-integration. However, major challenges still remain in terms of system latency, throughput and power [23, 22].

The focus of the current work is on designing a flexible, high-performance and fully-asynchronous network, suitable for shared-memory chip multiprocessors (CMP's). The target network topology is a variant Mesh-of-Trees (MoT), combining a binary fan-out network (i.e. routing) and a binary fan-in network (i.e. arbitration) for each source-sink pair, as shown in Fig. 1(a). While this NoC topology is rarely used for embedded system-on-chip (SoC) platforms, it is receiving increasing attention as a foundation for shared-memory interface networks in high-performance parallel systems, to provide needed bandwidth for globally uniform memory access. The classic MoT, unlike the variant, which places the functional units at the leaves of the trees, also shows significant latency and throughput benefits over other topologies, such as 2D-mesh [17, 18], but is proved to have more contention and achieve lower saturation throughput than the variant. Several recent shared-memory parallel processors are using MoT, or close variants, for core-to-memory (or cache) interfaces [2, 24, 11]. Although MoT networks grow rapidly in size with the number of cores and memories, they are viable for medium-size parallel systems. In addition, extensions have been proposed to reduce area overhead through a hybrid MoT/butterfly topology, which maintains the throughput and latency benefits of MoT with the area advantages of butterfly [3].

There has been a surge of interest in recent years in asynchronous and globally-asynchronous locally-synchronous design (GALS) [28]. Several GALS NoC solutions have been proposed to enable structured system design. These approaches have been highly effective, especially for low- and moderate-performance distributed multicore systems [30, 1] which address a different point in the design space than the proposed work. Some have low throughput (e.g., 200-250 MHz)

This work was partially supported by NSF Grant CCF-1219013.

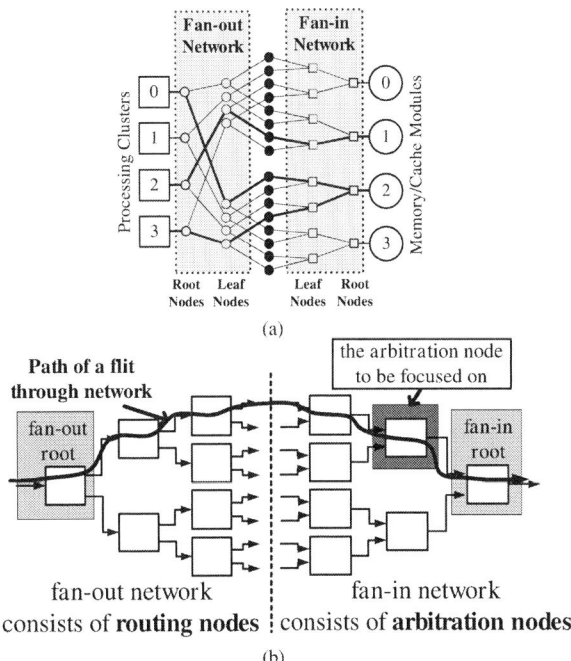

(a)

(b)

Figure 1: Mesh-of-trees network: (a) top-level topology; (b) block structure

[1], while those with moderate throughput (e.g., near 500 MHz [30, 8, 25, 4] often have large overhead in router node latency and area. Most are based on a standard 5-ported node architecture, and use *four-phase return-to-zero protocols,* involving two complete roundtrip channel communications per transaction rather than the single roundtrip communication targeted in our work (except for [14] which also uses *two-phase non-return-to-zero protocols*.) They also typically use *delay-insensitive data encoding*, resulting in a lower coding efficiency than the single-rail bundled encoding used in this paper [1, 25, 20, 4, 30, 27].

The proposed solution builds on our previous work of [12, 10, 13], which uses a transition-signaling single-rail bundled scheme for MoT networks. In [12, 13], two high-performance NOCs were introduced: one fully-asynchronous and the other GALS-style. Each used new lightweight asynchronous router node designs, targeting an MoT topology. These NOCs showed significant benefits under metrics of system latency, area and power. In particular the fully-asynchronous network, in 90nm technology, provided much lower system latency (by 1.7x) than a comparable synchronous network operating at 800MHz, with identical throughput over the latter's entire operating range. Our more recent approach [10] added limited *dynamic reconfiguration* capabilities to the network, using prediction based on local recent traffic history, resulting in further performance improvements for some benchmarks. The use of a transition-signaling single-rail bundled scheme has also been recently extended to implement a 5-port router, targeting a 2D-mesh topology [9].

However, both prior approaches ([12, 10, 13]) are shown to hit a latency wall, due to the overhead incurred by two steps: arbitration resolution and input channel allocation. At medium to high input rates, and with adversarial traffic patterns, these overheads create a

978-1-4799-2817-0/14 $31.00 © 2014 IEEE

ceiling or limit to the overall end-to-end system latency.

Contribution. In this paper, the **system-latency bottleneck** issue is addressed. The focus of the work is to provide support for high-performance memory (or cache) interfaces for shared-memory parallel processors [24, 11]. In this arena – even moreso than in the embedded system-on-chip (SoC) domain – system latency has been identified as the critical limitation. For example, the recent work on *NOC-OUT* [11] uses a tree-based NoC structure between processing cores and shared L2 cache in a so-called processing *pod*. In practice, the communication network between cores and cache typically has light traffic, and the key bottleneck is transport time.

Our contribution is a fundamentally different approach from previous high-performance asynchronous MoT networks of [12, 10, 13], aimed at overcoming the latency bottleneck, while still maintaining comparable throughput. In particular, a lightweight *monitoring network* is designed, which provides rapid advance notification to all destination nodes, whenever a new flit enters the network. The network information allows each router node on its path to complete arbitration and pre-allocate the corresponding input channel, *before* the actual flit arrival. As a result, in many traffic scenarios, the forward path through the network is directly optimized for end-to-end latency, and each arbitration node is configured to operate as a simple FIFO stage. A potential latency reduction of roughly 250 ps is obtained: 150 ps for mutex set and 100 ps for channel opening. The approach is supported by the design of a new arbitration node for the MoT network. Significantly improved latency results were obtained, across a diverse set of benchmarks.

Interestingly, the new arbitration node is much simpler than the predictive arbitration node proposed in [10]. While the latter contains 5 distinct mutual-exclusion elements, to resolve complex scenarios of mode change, the new node contains only 1 mutual-exclusion element. In addition, while [10] uses a monitoring network, it serves a secondary role: it identifies a timing window where nodes can safely revert from optimized ("biased") to conservative ("default") operating modes. In contrast, in our new network, the monitoring network is the key mechanism that supports pre-allocation of network paths, in advance of arriving packets.

The proposed monitoring concept can not only facilitate early arbitration and channel pre-allocation for performance improvements, but may be suitable as well to observe run-time workload characteristics, for both synchronous and asynchronous NoCs. This direction is becoming important for dynamic voltage and frequency adjustment in the field of power management [5].

Related Work. A number of acceleration techniques have been applied to enable better performance for NoCs. Express virtual channels [16] allows selective packets to use dedicated channels with accelerated routes through intermediate routers, resulting in significant performance improvements. However, the approach assumes baseline 5-ported routers with expensive 3-cycle operation, while our baseline uses low-radix routers with already heavily optimized performance (2.4 Gigaflits/sec throughput and 365 ps node latency in 90 nm technology [12, 13]) which is much more challenging to accelerate. The NoC in [19] uses prediction, a different approach from ours, and also is targeted to more complex 5-ported routers. The SMART NoC [6] uses low-swing clockless repeated links that allows packets to potentially traverse multiple hops in one cycle, but it requires advanced low-level circuit techniques with aggressive timing assumptions. The NoCs in [21, 29] combine a fast circuit-switched sub-network with a normal packet-switched sub-network, which allows the flits to alternate between the two sub-networks along the path to the destination. However, the link bandwidth and router resources are partitioned and statically allocated to one of the switching methods, leading to a low network utilization under low traffic rate. Also, circuit-switched techniques occur significant setup overhead, which is mitigated by some advanced low-latency circuit techniques. Closer to our work, the author of [15] uses "advanced bundles" (i.e. advanced information of flit arrival) for setting the router in advance, for high-performance 5-ported router designs. However, these bundles advance one cycle per hop, while our approach fast forwards monitoring signals not bound

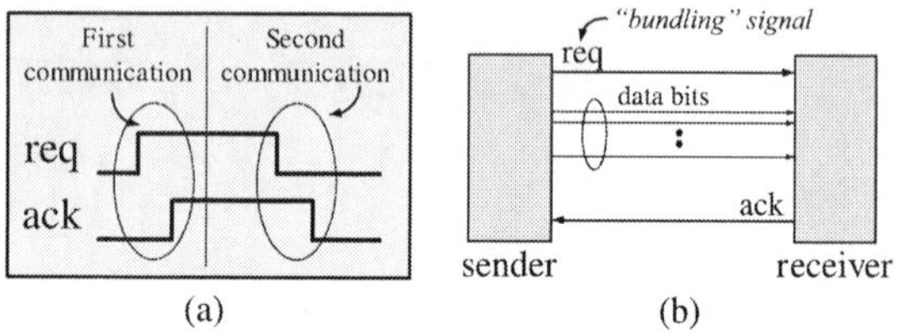

Figure 2: Asynchronous communication:
(a) Two-phase hand-shaking protocol; (b) Single-rail "bundled data"

to a clock, from network injection to destination nodes, with < 50 ps traversal time within each node.

A fast-forwarding technique was also proposed for asynchronous tree arbiters [31], in which an arbiter propagates the request to later stages in parallel with its own arbitration. However, this approach does not allow pipelined requests – which are critical for NoC design – which our design supports, and also does not include capabilities to interact with a complex NoC datapath.

2. Background

Asynchronous communication. Asynchronous protocols define channel communication between a sender and a receiver not synchronized to a master system clock, but rather locally and in an on-demand basis. As shown in Fig. 2(a) [26], in a *two-phase (NRZ) protocol*, only 2 events occur per transaction: a toggle on request followed by a toggle on acknowledge. In contrast, in a *four-phase (RZ) protocol*, both request and acknowledge are initially zero, and rising req/ack transitions are followed in turn by falling req/ack transitions. The paper uses *two-phase*, as it involves only a single roundtrip channel communication per transaction, whereas *four-phase* involves two roundtrip communications, and hence the former leads to better latency and throughput. *Single-rail bundled data* is used for data encoding in this paper. As shown in Fig. 2(b), the channel uses standard (synchronous style) single-rail data, with an accompanying worst-case "model delay" (i.e. the "bundling" *req* signal). In each communication, a hazard-free transition on *req* is asserted only *after* all data bits are valid and stable, indicating that data can be used.

Mesh-of-trees network. The microarchitecture of an MoT topology is shown in Fig. 1(b). It consists of a binary fan-out tree composed of *routing* nodes, which emanates from source root nodes and terminates in mid-network with leaf nodes; and a symmetric binary fan-in tree composed of *arbitration* nodes, which emanates from destination root nodes and also terminates in mid-network with leaf nodes. Leaf nodes of the two sub-networks are directly connected. Network routing is deterministic and packets are source routed. One useful feature for memory access in CMP's is that the number of hops from source to destination is always a small constant: log(n). The figure highlights a single flit traversal through the network, from a source fan-out root node to a destination fan-in root node. Each source-destination pair is associated with a unique path through the network. This feature significantly minimizes contention, since distinct fan-out sources are attached to different fan-out trees, and distinct fan-in destinations are attached to different fan-in trees. This separation guarantees that, unless the memory access traffic is extremely unbalanced, packets between different sources and destinations will not interfere (see [3, 2]).

Asynchronous primitives. Two asynchronous components were used to create the original asynchronous mesh-of-trees network by Horak *et al.* [12, 13]: *routing* and *arbitration* primitives. These components are based on an Mousetrap asynchronous pipelines [26]. The routing primitive receives one incoming stream and passes it to one of the two outgoing streams, while the arbitration primitive mediates between two incoming streams of flits – enforcing mutual exclusion – and merges the result into a single outgoing stream. The paper focuses on a new design for the arbitration node. The fan-out nodes in the new approach have almost the same basic function, but with minor modifications, and therefore are ignored due to space limitations.

Figure 3 shows the basic arbitration primitive by Horak *et al.* [12, 13]. When the arbitration primitive is empty, the control latches L1 and L2 are opaque while all the other latches, including the data register, are transparent. Following the bundled-data encoding, opera-

978-1-4799-2817-0/14 $31.00 © 2014 IEEE

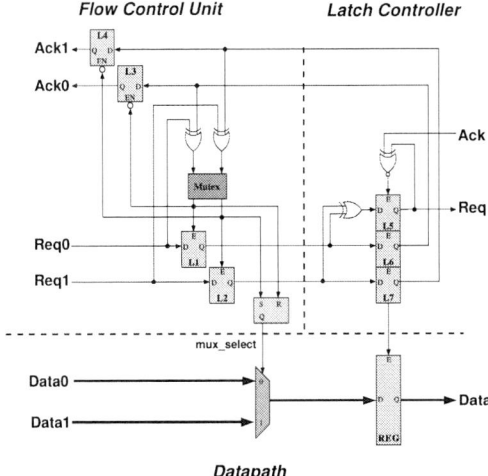

Flow Control Unit   *Latch Controller*

Datapath

Figure 3: Original arbitration node [12, 13]

tion begins when data appears at one of the input channels followed by a transition on corresponding req. To resolve contention between potentially concurrent incoming requests, the incoming request must first trigger a lock on the mutex. Next, two operations take place concurrently: 1) *mux_select* chooses the correct data input; and 2) either L1 or L2 is enabled, thereby forwarding the winning request. After the data is sent to the output channel, it is safely stored by making L5-L7 and the data register opaque. The mutex is then immediately reset and an acknowledge to the previous stage is generated. Finally, the right environment acknowledges the current stage and the entire operation is complete.

3. Overview of the Approach

The section introduces the basic strategy of the proposed new approach, by comparing it with two previous approaches: *baseline* [12, 13] and *predictive* [10].

The baseline design. The structure of a baseline asynchronous MoT network is shown in Fig. 1(b) [12, 13]. The flit is forwarded through the fan-out network followed by the fan-in network, with a path determined by the flit's source-routed address bits. After a flit arrives at an arbitration node, e.g. the shaded node highlighted in Fig. 1(b), the node first allocates and opens the channel. It then gets passed through and sent to the output. *Channel allocation only starts after the arrival of an actual flit.* If there is contention, with incoming flits on both input channels, arbitration grants to exactly one of the requesting flits based on their relative arrival time. Since the design is asynchronous, arbitration is not performed at discrete clock boundaries, but in continuous time, using an analog arbiter. Once the winning flit is processed, the other flit can win arbitration.

The predictive approach. The predictive approach [10] proposes enhanced arbitration nodes that can dynamically reconfigure themselves to accelerate the forward transmission of a stream of flits. In particular, if the node detects streaming activity on one input channel, and no incoming traffic on the other input channel, it *predicts* continued streaming activity (i.e. based on temporal locality) and puts the active input channel in an optimized state for improved performance.

Fig. 4(c) shows the network structure of the predictive approach, and Fig. 4(a) presents an arbitration node's operation. If all relevant nodes correctly predict and prepare for a flit in advance, an express path is created when the flit passes through the entire fan-in network. Prediction is performed locally, based solely on each arbitration node's recent observed traffic history. The node operates normally as the baseline design, i.e. default mode, but enters a *single-channel-biased mode* when two consecutive flits arrive on a single channel. Once in biased mode, the arbiter is completely ignored and the biased channel is held open. As a result, *arbitration and channel allocation steps are entirely eliminated.* Subsequent flits arriving on the biased channel will be directly sent to the output. The node reverts to default mode when *any flit* arrives on the inactive input channel. The challenge with this approach is that misprediction or thrashing

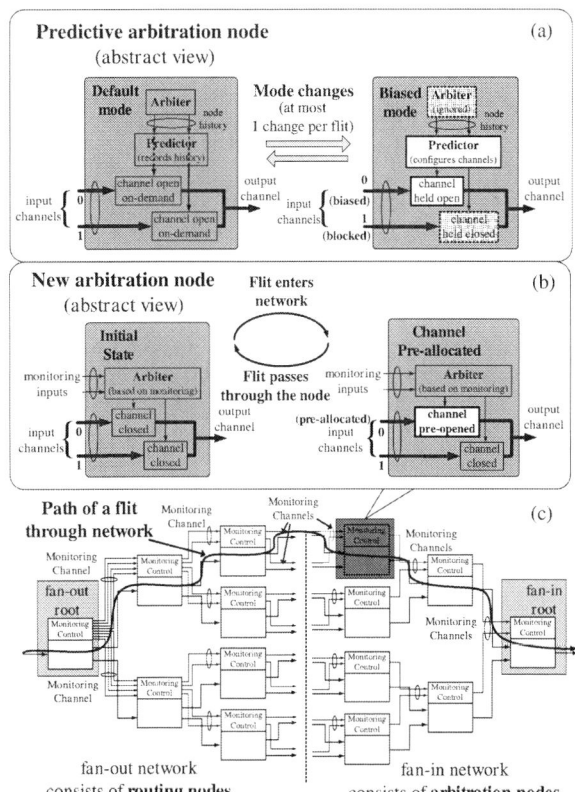

Figure 4: Block structure of MoT network: predictive and new approach
(a) Predictive node; (b) New node; (c) MoT network with monitoring

between modes may be induced, from adversarial traffic patterns.

A monitoring network is used to facilitate mode change operation for safety purposes only. It is only used to enable a reversion from optimized (i.e. biased) to unoptimized (i.e. default) mode.

The new approach. This approach proposes an alternative protocol for accelerating the forward transmission of flits through the fan-in network. New arbitration nodes are designed, which *anticipate* the arrival of incoming traffic by receiving early notification through a monitoring network. Once an arbitration node has been notified of a pending flit arrival, it rapidly performs early arbitration and pre-allocates the channel.

The overall network structure is the same as that of the predictive approach, but with fundamentally different operation. When a flit enters the network, injected by a source fan-out root node, advance notification is generated and fast forwarded to all the nodes along its path by the monitoring network. All arbitration nodes on the path *pre-allocate the corresponding input channel*, well before the actual flit arrives, as shown in Fig. 4(b). The input channel is then *kept open* until the flit passes through the node, as a combinational flow-through path with all latches transparent from a fan-in leaf to the fan-in root. When the flit arrives, it is directly sent through the pre-allocated channel to the output, *avoiding the overhead of a stop-and-go protocol.* The node then reverts back to the initial state and waits for the next operation. In the case of contention, arbitration is used to select one of the two input channels for pre-allocation; however, unlike the previous approaches, the arbitration is performed in advance, based on monitoring signals.

Several comparisons of the new vs. previous approaches are now highlighted. Similar to the baseline design, no history is recorded for the flits; this design decision allows a much simpler design. Similar to the predictive approach, in friendly cases, the arbitration and channel allocation are completely eliminated from the critical path, but using quite different strategies. However, the monitoring network has a major role in the new approach, as the node directly uses its advance notification to reconfigure in advance for higher performance. In contrast, in the predictive approach, monitoring serves a secondary

978-1-4799-2817-0/14 $31.00 © 2014 IEEE   331

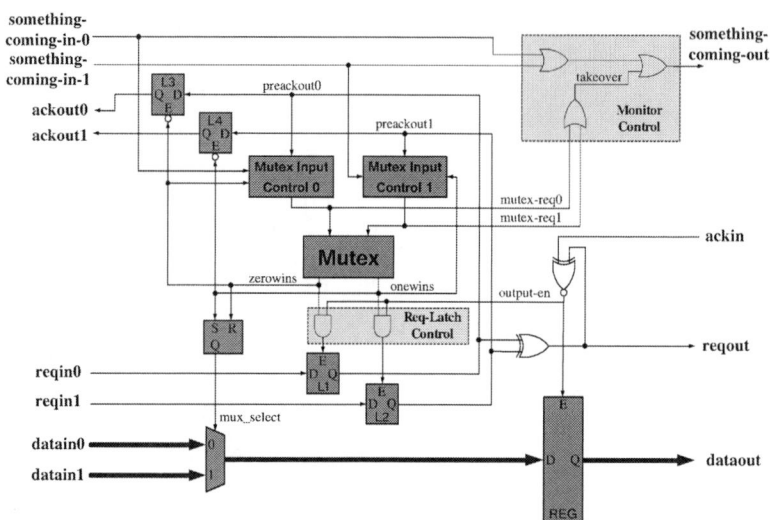

Figure 5: New arbitration primitive: single-flit design

role, to facilitate safe mode changes from optimized to default state.

4. The New Router Node Design

4.1 Arbitration Node Design and Operation

The structure and operation of the new arbitration node are first introduced, followed by details on several key components, for a basic design which handles single-flit packets. The section closes with timing analysis.

Structure. The structure of the new node is illustrated in Fig. 5. Compared to the baseline design [12, 13], additional monitoring channels are added to support early arbitration capability.

There exist three fundamental structural differences, compared to the baseline design: (i) two added *mutex input control* units, one for each input channel; (ii) a critical path from input to output with a single latch, while the previous designs use two latches; a new req-latch control is added to support this change; and (iii) a new *monitoring control* unit. These modifications facilitate early arbitration and allow the node to have extremely low latency, while still ensuring a simple design. Each contribution is presented in turn.

The *mutex-input controls* request and release the mutex during the operation. These controls initiate early arbitration, by asserting mutex request high whenever informed by early monitoring information. When a flit enters the network, *something-coming-in*, which is initially low, is asserted high, causing the control unit to request and win arbitration (i.e. mutex component). The control unit de-asserts its request, and releases the mutex, when the corresponding flit is sent to the output channel.

The newly-added *req-latch control* unit enables a dramatic optimization in the node's forward path, reducing it from two latches [12, 13, 10] to a single one. This latch now serves two purposes: (i) channel pre-allocation based on the winning of mutex, and (ii) flow control. In contrast, the earlier designs use separate registers to handle the two purposes. This requires a significant protocol modification on the *req-latch control* unit, which now opens the latch in synchronization with both left and right channels: it awaits early arbitration to complete (assertion of a mutex output) and an acknowledge of the previous flit by the right environment, for flow control (*output-en* signal asserted high).

The *monitoring control* unit is part of the entire new monitoring network. It enables a rapid propagation of the monitoring information from the node's input to output. In particular, *something-coming-out* is asserted high as soon as either of the two *something-coming-in* signals is asserted high with only a two-gate delay latency. Since *something-coming-in* signals are not persistent (i.e. they can go low early), a *takeover* is needed to maintain *something-coming-out* high until the actual flit arrives and is sent to the output channel.

Operation. Before showing detailed component designs, the operation of the entire node is illustrated by three simple scenarios.

(i) Single flit processing with no contention. This scenario illustrates how early arbitration works in a friendly case. The incoming flit is assumed to arrive on channel 0, without loss of generality, and with no traffic on the other input channel. Initially, latches L1 and L2 are opaque, while all others (including data register) are transparent. First, when the flit enters the network, early arbitration is performed and channel 0 is pre-allocated, well before the actual flit arrival: *something-coming-in-0* is asserted high, causing the *mutex-input-control-0* unit to request the mutex. After the mutex is won (*zerowins* asserted high), latch L1 is pre-allocated (i.e. becomes transparent), and the data mux selects the top input channel. Eventually, the actual flit arrives. Since input channel 0 is now pre-allocated, the flit passes directly to the output channel. A low forward latency is achieved: effectively one D-latch (L1) plus an XOR2. Finally, both L1 and the data register quickly close for data protection. In parallel, the mutex is released, allowing the left acknowledgment to be sent out through L3. In the end, the right environment acknowledges, and the entire operation is complete.

(ii) Two flits back-to-back on the same channel. This case illustrates when a data stream arrives on one input channel, e.g. channel 0. The first flit is processed similarly to the *single-flit* case above. In this scenario, though, the monitoring signal, *something-coming-in remains high*, indicating that there are more incoming flits. Note that the mutex is always briefly released between flits (i.e. *zerowins* de-asserted low), and *something-coming-in* is then re-sampled by the *mutex-input-control-0* unit for the second flit. The second flit on this input channel is then processed following the same procedure as the first one. Finally, *something-coming-in* is de-asserted low as no more flits are arriving. The mutex release between the two flits is forced to create a window for the other channel to have an opportunity to win the mutex. As a result, the arbitration is fair and starvation is avoided.

(iii) Contention between two channels. This scenario illustrates a simple case for the node to resolve contention. Two flits, one at each input channel, enter the network within a close time margin. Both *something-coming-in* inputs go high and the two *mutex-input controls* request the mutex almost simultaneously. Without loss of generality, assume channel 0 wins the mutex. The flit on channel 0 is processed normally by following the same procedure as in *single-flit* case, while the early arbitration on channel 1 cannot further proceed. As soon as channel 0 releases the mutex, channel 1 continues with its own early arbitration and the flit on channel 1 gets processed.

Mutex input control. A more detailed view of this control unit and its operation are now presented. The new arbitration node contains two *Mutex Input Control* units, one for each input channel, as shown in Fig. 6(a). They are the core components that initiate early arbitration whenever notified by monitoring signals.

In processing each flit, the *Mutex Input Control* unit requests and

978-1-4799-2817-0/14 $31.00 © 2014 IEEE

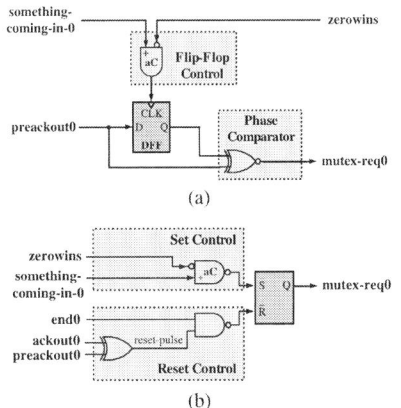

(a)

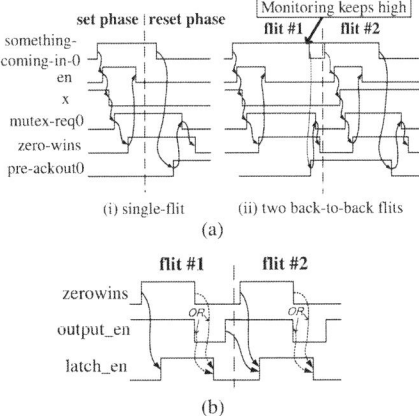

(b)

Figure 6: Mutex input control: (a) single-flit design (b) multi-flit design

Figure 7: Timing diagram: (a) Mutex input control; (b) req-latch control

releases the mutex exactly once, generating a clean pulse at its only output *mutex-req0*, as shown in Fig. 7(a)(i). An edge-triggered D-flipflop and a phase comparator are the core components for pulse generation. The phase comparator initially has different values at the two inputs and *mutex-req0* is zero. When a new flit enters the network and the local monitoring signal *something-coming-in-0* is asserted high, the *Flip-Flop Control* clocks the DFF. The two inputs of the phase comparator become equalized, and the control output (*mutex-req0*) is asserted high, resulting in a request to the mutex. (Once the mutex has been won, its *zerowins* output will then de-assert the *Flip-Flop Control* output.) Eventually, once the actual flit arrives at the node and is sent to the output channel, *pre-ackout* toggles. As a result, the bottom input to the phase comparator has a different value from the top input. Hence, the *mutex-req0* output is de-asserted low, thus releasing the mutex and also completing the output pulse.

A timing diagram for two back-to-back flits on the same channel is shown in Fig. 7(a)(ii). If flits are very close, *something-coming-in-0* stays high; alternatively, with a wider gap, it will briefly go low.

Req-latch control. This unit is implemented by two simple AND2 gates, one for each input channel. A timing diagram of the processing of two successive flits, widely spaced, is shown in Fig. 7(b).

The output of each of the AND2 gates, which opens the corresponding input channel, is asserted high based on a synchronization of its two inputs. For each AND2 gate, its left input (*zerowins/onewins*) serves as a *request for input channel allocation,* controlling the corresponding latch (L1 or L2). It is asserted high when the corresponding monitoring signal has arrived and wins early arbitration, and is de-asserted low after the actual flit arrives and is transmitted to the output channel. In particular, the corresponding *preackout0/1* signal toggles, causing the mutex to reset, and thereby de-asserting the corresponding mutex output (i.e., *zerowins/onewins*). The right input (*output-en*) has the role of flow control; it is a busy flag indicating whether the successor is free to accept a new data item. It is initially de-asserted high, indicating a free output channel. Once the flit arrives and is sent

out on this output channel, the *reqout* signal toggles and asserts the busy flag (*output-en*), as well as closing the data register. Once the right acknowledge *ackin* is toggled, the busy flag is de-asserted high.

For the case of downstream congestion, the right environment may take arbitrarily long to generate its acknowledgment. Even so, if the advance monitoring indicates a second flit heading to the same input channel, the node has sufficient parallelism to begin the pre-allocation step: it can win arbitration (i.e. assert *zerowins/onewins* high), and request to enable the input channel. At this point, as soon as the downstream congestion is cleared, the corresponding latch, L1 or L2, will be finally be allowed to open again.

Timing analysis. The three key timing constraints, which serve as design parameters for the physical layout level, are now briefly sketched.

Monitoring operation: re-sampling time. Once asserted high, the monitoring input (*something-coming-in*) must be de-asserted early enough to prevent a stale value from being re-sampled by the *mutex-input control* unit, otherwise the input channel will be re-allocated with no incoming flit. In practice, given the gate-level implementation, this timing condition has adequate margins. Soon after a flit arrives, the corresponding monitoring input signal (*something-coming-in*) is de-asserted low and immediately forwarded to the *mutex-input control* unit, thus nullifying the sampling signal. Concurrently, after the flit is transmitted to the output channel, an entire mutex release cycle is initiated, followed by a re-acquisition, before the *mutex-input control* unit can re-sample the monitoring input.

Hold time. Once data advances through the latch (L1/L2), it must be securely stored before new data arrives from the previous stage to prevent data overrun. In practice, L1/L2 closes nearly simultaneously with generating the left acknowledgment, because it involves small local paths. In contrast, the roundtrip time from generating the left acknowledgment to new data arrival involves one complete non-local channel traversal as well as paths through the left neighbor's control. The timing constraint, therefore, is easily satisfiable.

Pre-mature input channel opening. This timing constraint occurs whenever a flit is sent on the output channel. The right environment cannot acknowledge too quickly, before the mutex is released (*i.e. zerowins/onewins* goes low). Otherwise, L1/L2 will temporarily open for a short period; although no functional error can occur, possible glitches in control signals and other electrical issues may become a problem. In practice, as the right environment is normally another router of the same type, and the round-trip latency from generating req-out to receiving acknowledgment involves a complete non-local channel traversal, the timing constraint is also easily satisfiable.

4.2 Monitoring Network Design and Operation

Structure. The structure of the monitoring network is a replica of the entire MoT network. For modularity, it is combined into the existing MoT network. Each node – routing or arbitration – has an attached small *monitoring control* unit, implemented by several gates. The monitoring control units for fan-out primitives are similar to those in Gill *et. al* [10], with nearly identical functionality but using a slightly different protocol in monitoring communication between neighbors. They are not shown due to space limitations. The *monitoring control* unit for each arbitration node was shown in Fig. 5.[1]

System-level operation. Monitoring has distinct roles in the two halves of the network: in the fan-out part, monitoring signals only gets forwarded; in the fan-in part, monitoring signals get forwarded and are also used for channel pre-allocation at individual nodes. Whenever a flit enters the network through a root routing node, it initiates a monitoring signal transition. This signal is then rapidly forwarded to all the fan-out and fan-in nodes along the targeted path. A rapid wave of monitoring signals going high is detected as the relevant nodes get informed from the source to the destination only on this path. Then, monitoring signals are de-asserted low, node by node, as the actual flit proceeds through each router following the same path.

Node-level operation. A simple simulation is presented for an individual fan-in node for a single flit, as shown in Fig. 8(a). Assume

---

[1] The two top OR2 gates are combined using an OR3 gate in the actual implementation for improved performance.

978-1-4799-2817-0/14 $31.00 © 2014 IEEE   333

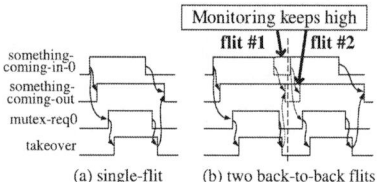

Figure 8: Timing diagram for fan-in monitoring control

Table 1: Area comparison for pre-layout arbitration nodes

Version	Area (μm²)	
	total	control only
baseline	869.1	190.5
predictive	1309.8	632.4
new	947.3	269.9

Table 2: Performance comparison: arbitration nodes

Version	Operating Mode	Latency (ps)	Max. Throughput (GFPS)	
			Single	All
baseline	N/A	416	1.35	2.28
predictive	Default	472	1.25	2.00
	Biased	264	1.77	N/A
new	N/A	215	1.42	2.43

the flit arrives on channel 0. When the flit enters the network, the node soon is notified: *something-coming-in-0* goes high and this information is quickly forwarded to *something-coming-out*. Concurrently, *takeover* is asserted, to maintain this output value. Eventually, the actual flit arrives, followed by *something-coming-in-0* going low. but *something-coming-out* stays high due to the *takeover* signal. Finally, after the flit is sent out on the output channel, the *takeover* is reset and allows *something-coming-out* to be de-asserted low.

In the case when there are two back-to-back flits approaching the node on channel 0, monitoring signals will stay high, until both incoming flits are processed. The timing diagram for this special case is illustrated by Fig. 8(b). Effectively, the monitoring signals are not normal handshaking signals, and no longer get reset when two consecutive flits are sufficiently close. It serves as a state bit, indicating the existence of incoming flit(s). Alternatively, runt pulses are possible for the monitoring signal, i.e. a brief de-assertion, when the flits are somewhat further apart. This aggressive protocol is much more efficient than a classical deterministic hand-shaking protocol where each asserted signal is required in turn to be de-asserted.

4.3 Multi-Flit Design

The previous subsections focused on a basic design to handle only single-flit packets. An enhanced version for the new arbitration node is now presented to support multi-flit capability, as in [12, 13]. In this design, once the header flit wins in early arbitration, the mutex is held until the entire packet is processed. In contrast, in [12, 13], the mutex cycles on a per-flit basis, being repeatedly won and released. The new approach has direct benefits in reduced switching activity, as well as improved throughput due to the complete elimination of this protocol overhead. The structure of the multi-flit fan-in node is largely identical to the single-flit one, shown in Fig. 5, except for a different *mutex-input-control* unit. The new control unit, shown in Fig. 6(b) replaces the one for *single-flit* design. It now takes extra tail flag (*end0/1*) information from each flit.

5. Experimental Evaluation

Detailed evaluations are now presented for technology-mapped new arbitration primitives as well as complete MoT networks. Results are compared to the *baseline* [12, 13] and *predictive* [10] designs. Both previous designs are re-implemented for fair comparison.

5.1 Asynchronous Arbitration Primitives

The new asynchronous arbitration primitive is evaluated in terms of area, latency and maximum throughput. The performance of monitoring signals is also evaluated. The same industrial cell library is used as in [10] for implementing the new and previous network primitives with a 32-bit wide datapath. Results are obtained using the Spectre simulator in Cadence Virtuoso environment at typical design corner with nominal temperature and supply voltage (1.0V, 27°C). As we use a different simulator from that of [10], reasonable results deviations are observed. All the results are obtained for the single-flit designs.

Area comparison. Table 1 presents pre-layout area results for the arbitration nodes, including total area and control only (which excludes the data path). The final layout node area is carefully estimated by summing up the cell areas and then dividing by a packing factor of 0.8. A 28% less area is observed for the new node when compared to the *predictive* one, due to a much simpler design. Even compared to the *baseline* arbitration primitive, the area overhead is less than 9%.

Latency and throughput. Table 2 provides the performance results for the new arbitration primitive. Latency is calculated assuming an empty primitive, from an input request transition to the time that the primitive toggles its output request. A complete 3-level fan-in tree

is implemented for evaluating maximum throughput under different steady-state traffic patterns in order to capture the realistic handshaking protocol overhead between neighboring nodes and avoid over-optimistic performance analysis. Then throughput is measured at the root primitive of the tree. Experiments are done under two different traffic patterns: *single* (i.e. traffic arrives at a single input channel) and *all* (i.e. traffic arrives at both inputs).

The new arbitration primitive has significant latency reduction comparing to the *baseline* (by 48%), and moderate to large latency improvement over the *predictive* arbitration primitive, depending on different operating modes that the *predictive* node operates in (54% improvement in *single* case and 19% in *all* case).

In terms of throughput, there is a mix of improvement and degradation for the new arbitration primitive, compared to the previous counterparts. In the *single* case, the new design has minor to moderate throughput improvements over the *baseline* and *predictive* (default mode), by 5.2% and 13.7% respectively, mostly due to a simpler design and an optimized forward path. However, about 20% worse throughput is observed, compared to the *predictive* (biased mode). This is because at high data input rate, there is not enough margin between two consecutive flits and the new design cannot fully complete channel pre-allocation before each flit arrival, while the *predictive* design operates in biased mode and the channel is always allocated. In the *all* case, the new design also has higher throughput, compared to both the *baseline* and *predictive* counterpart, in particular, 20% higher throughput for the latter comparison.

Monitoring signals. Two results are obtained to evaluate the latency of monitoring signals for the new arbitration node. First, the assertion latency is measured from the time that *something-coming-in* goes high to that of *something-coming-out*. The result is extremely fast, 41 ps, and proves the monitoring signal can be quickly forwarded. On the other hand, pre-allocation time is from *something-coming-in* assertion to the point when the channel is opened. The result, 463 ps, is fast enough for every fan-in node to complete channel pre-allocation before actual flit arrival when the network is lightly loaded.

Design Complexity. The *new* arbitration node is much simpler than the *predictive* one. It uses only a single analog mutex, while the *predictive* design requires five mutexes to solve complex mode change scenarios. The *predictive* node also contains complex *Policy* and *Safety* modules to store history information and maintain channel configuration. The *new* design requires none of these, and has simpler req- and ack-latch control units. However, the *new* design has two added small *mutex-input-controls* to allow channel pre-allocation.

5.2 Asynchronous Network

This section reports both network latency and throughput results for 8-by-8 mesh-of-trees networks. We conduct two sets of experiments: for single-flit and multi-flit designs. Single-flit experiments compare three networks: the *baseline*, *predictive* and *new*. For multi-flit experiments, only two networks are compared, *baseline* and *new*, since the *predictive* design does not support multi-flit packets.

Experimental setup. A similar network setup is used as in [10]. We use an asynchronous NoC simulator introduced in [10] developed by our group for testing, debugging and performance measurements, based on a PLI (Programming Language Interface) framework. The

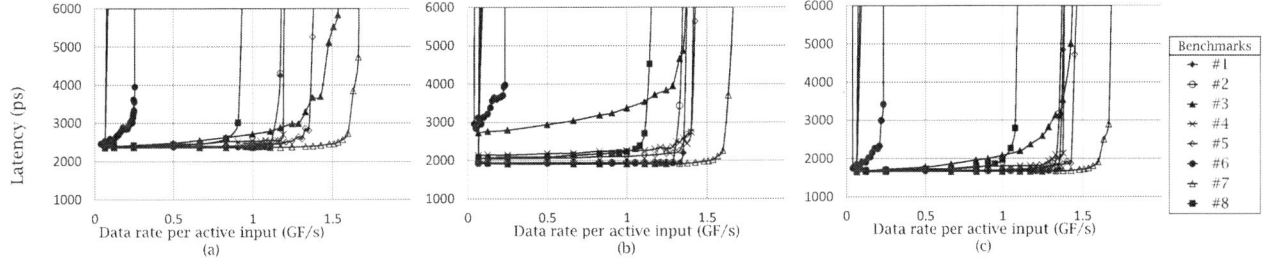

Figure 9: Network-level latency: (a) baseline network [12, 13]; (b) predictive network [10]; (c) new network

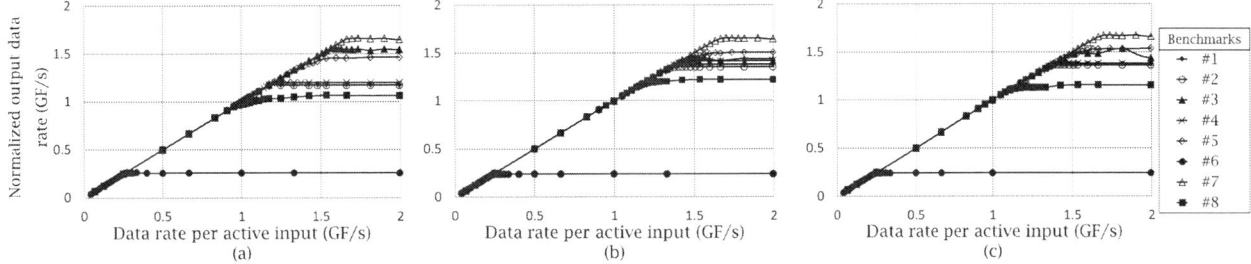

Figure 10: Network-level throughput: (a) baseline network [12, 13]; (b) predictive network [10]; (c) new network

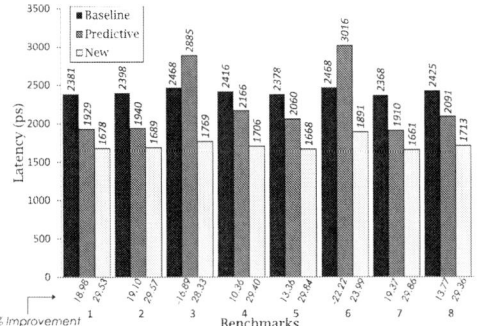

Figure 11: Latency comparison for 25% network load

Figure 12: Saturation throughput comparison

initial generic framework was developed by Prof. Simha Sethumadhavan (Columbia University) for synchronous applications only. Networks are modeled in structural Verilog using ARM 90nm standard cells while the test environment is written in C. Packet source queues are installed at network inputs ports for recording of latency [7]. For single-flit experiments, the input environment generates flits at random intervals that conforms to an exponential distribution. For multi-flit experiments, flits within a packet are put into the source queue almost simultaneously and the mean time between packet headers follows an exponential distribution. We follow the standard procedure to ensure a long enough warm-up and measurement time for each benchmark [7], where one simulation with given warm-up and measurement phases is compared to another with both of these periods doubled, to check whether results are comparable.

Gate modeling. Our previous work of [10] employs a fixed-latency gate model by taking the average of rising and falling transitions. In this paper, a more accurate model is extracted by Cadence simulations, which assigns distinct latencies to rising and falling transitions for each pair of I/O paths of each gate, rather than using a single fixed latency model.

Benchmarks. Experiments are conducted for the same eight synthetic benchmarks as in [10], chosen to represent a wide range of network conditions. Three benchmarks provide friendly scenarios with no contention, where each fan-in node receives data on only one input channel: **1)** *Bit permutation* [7], **2)** *Digit permutation* [7] and **7)** *Random single source broadcast*. Three benchmarks provide moderate contention, where some fan-in nodes have light or no contention, and others have moderate contention: **4)** *Simple alternation with overlap*, **5)** *Random restricted broadcast with partial overlap* and **8)** *Partial streaming*. The two remaining benchmarks are most adversarial, with

heavy contention at some fan-in nodes: **3)** *Uniform all-to-all random* [7] and **6)** *Hotspot8* [7]. The *predictive* network has particularly poor performance on these two, even worse than the *baseline* network.

Network-level simulation results: single-flit design. Overall, for the single-flit design experiments, the *new* network has moderate to high latency reductions across all benchmarks, over both *baseline* and *predictive* networks, ranging from 13 to 39%. For throughput, the *new* network out-performs the *baseline* network, in six of eight benchmarks, with improvements up to 17%, and only minor degradation in the other two benchmarks. Compared to the *predictive* network, the new network has almost identical throughput, within 6%.

Latency. Fig. 9 contains detailed latency results for the three networks. Results are plotted as the average end-to-end network latency for each flit vs. the mean offered data input rate [10]. When lightly loaded, network latency in the *baseline* network is nearly 2400 ps for every benchmark. The *predictive* network has a latency around 1900 ps for friendly benchmarks but up to 3000 ps for the adversarial ones. Like the *baseline,* the *new* network has a quite stable latency over all benchmarks; however, the latter is overall much lower, around 1700 ps. Fig. 11 highlights network latencies at identical input rates. An offered throughput of 25% saturation rate is chosen; this rate is high enough to differentiate benchmarks while still retaining an uncongested network. Compared to the *baseline* network, significant improvements are uniformly observed, ranging from 23% to 30%. Compared to the *predictive* network, the *new* network obtains improvements from 13% to 21% for friendly and moderately adversarial benchmarks, with even higher improvements – near 38% – for the most adversarial benchmarks.

The two key contributions – merging latches on the forward path and channel pre-allocation – are now separately evaluated, to observe

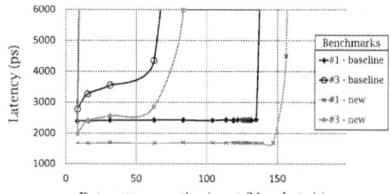

Figure 13: Latency for the networks with multi-flit capability

their impact. By merging latches, the network latency is improved by roughly 250 ps (13%), when comparing the *predictive* and the *new* network in the three friendly benchmarks (1, 2 and 7). These benchmarks have persistent source-sink pair-wise streaming traffic with no contention. Hence, the involved fan-in routers in the *predictive* network stay in optimal configuration (i.e. held in the biased mode), and therefore this differential identifies the improvement due solely to the latch removal. For other benchmarks, the latency gains are always higher (up to roughly 38% in benchmarks 3 and 6), showing the additional contribution of the channel pre-allocation scheme.

The *stability* of the latency is a key metric for memory access in shared-memory CMP's, capturing the degree of latency variation at fixed load under a variety of benchmarks. As highlighted in Fig. 11, while the *predictive* network has high latency variation, ranging from 1910 ps to 3016 ps, the *new* approach has *nearly uniform latency* over all benchmarks, which is another benefit of the proposed network.

Throughput. Fig. 10 explores network throughput, with results plotted as the output rate normalized to the number of active input sources vs. the mean offered data input rate [10]. Under fairly lightly-loaded traffic conditions, throughput tracks the input data rate. As the input rate increases, throughput results begin to level off to the *saturation throughput*. Fig. 12 provides a clean vision of the differences in saturation throughput values for the three networks. Compared to the *baseline,* the *new* network exhibits minor to moderate improvements – up to 17% – for six benchmarks, with up to 8% degradation on the two most adversarial ones. Compared to the *predictive* network, the *new* network exhibits nearly identical throughput – from 6% degradation to 6% improvement– across all benchmarks.

Monitoring Network Evaluation. Network-level simulations show that the monitoring network can rapidly forward monitoring information and allows enough time for each fan-in node on the path to complete channel pre-allocation before actual flit arrives. In a zero-load network, the tightest margin between monitoring signal arrival and actual flit arrival is 590 ps, which occurs at leaf fan-in nodes.

Network-level simulation results: multi-flit design. Finally, Fig. 13 shows network-level latency results, for *baseline* and *new* networks, using 8-flit packets. Only two representative benchmarks are selected. Benchmark 1 (shuffle) has no contention, and benchmark 3 (all-to-all random) is moderately adversarial. Significant improvements are obtained: 27% and 31%, respectively, for network latency; and 14% for both benchmarks, for saturation throughput. These strong results prove the effectiveness of the *new* approach in the multi-flit case.

6. Conclusions and Future Work

This paper introduces a novel approach to address the system-latency bottleneck for high-performance asynchronous interconnection networks. It targets an MoT topology for use in shared-memory chip multiprocessors. A lightweight shadow monitoring network is proposed to fast-forward information on arriving packets, in advance, allowing each node on its path to complete both arbitration and channel allocation before the data arrives. System latency improvements over 28% for 8 distinct benchmarks were obtained over our earlier network [12, 13], and up to 38% improvements were obtained over a more recent network [10], with no significant change in throughput.

Future work will be focused on narrowing the window between channel reservation and the actual flit arrival, which can potentially increase the network utilization. Instead of triggering the pre-allocation right after the flit enters the network, it can be triggered at the middle of the network, but still allowing enough time for completing pre-allocation in time. Also, mixed-timing interfaces will be created to connect the network with functional cores to form a GALS system for further experiments on real traffic benchmarks.

7. References

[1] J. Bainbridge and S. Furber. CHAIN: A delay-insensitive chip area interconnect. *IEEE Micro Magazine*, 22(5):16–23, 2002.

[2] A.O. Balkan, M.N. Horak, G. Qu, and U. Vishkin. Layout-accurate design and implementation of a high-throughput interconnection network for single-chip parallel processing. In *Hot Interconnects*, pages 21–28, 2007.

[3] A.O. Balkan, G. Qu, and U. Vishkin. An area-efficient high-throughput hybrid interconnection network for single-chip parallel processing. In *Proc. of ACM/IEEE DAC Conf.*, pages 435–440, 2008.

[4] T. Bjerregaard and J. Sparsoe. A router architecture for connection-oriented service guarantees in the MANGO clockless network-on-chip. In *Proc. Design, Automation and Test in Europe (DATE)*, pages 1226–1231, 2005.

[5] P. Bogdan, R. Marculescu, S. Jain, and R.T. Gavila. An optimal control approach to power management for multi-voltage and frequency islands multiprocessor platforms under highly variable workloads. In *Proc. of ACM NOCS Symp.*, pages 35–42, 2012.

[6] C.O. Chen, S. Park, T. Krishna, S. Subramanian, A.P. Chandrakasan, and L. Peh. SMART: A single-cycle reconfigurable NoC for SoC applications. In *Proc. Design, Automation and Test in Europe (DATE)*, pages 338–343, 2013.

[7] W. Dally and B. Towles. *Principles and Practices of Interconnection Networks.* Morgan Kaufmann Publishers, Inc., 2003.

[8] R. Dobkin, V. Vishnyakov, E. Friedman, and R. Ginosar. An asynchronous router for multiple service levels networks on chip. In *IEEE Int. Symp. on Asynchronous Circuits and Systems (ASYNC)*, 2005.

[9] A. Ghiribaldi, D. Bertozzi, and S.M. Nowick. A transition-signaling bundled data noc switch architecture for cost-effective GALS multicore systems. In *Proc. Design, Automation and Test in Europe (DATE)*, pages 332–337, 2013.

[10] G. Gill, S.S. Attarde, G. Lacourba, and S.M. Nowick. A low-latency adaptive asynchronous interconnection network using bi-modal router nodes. In *Proc. of ACM NOCS Symp.*, pages 193–200, 2011.

[11] B. Grot, D. Hardy, P. Lotfi-Kamran, B. Falsafi, C. Nicopoulos, and Y. Sazeides. Optimizing data-center TCO with scale-out processors. *IEEE Micro*, 32(5):52–63, 2012.

[12] M.N. Horak, S.M. Nowick, M. Carlberg, and U. Vishkin. A low-overhead asynchronous interconnection network for GALS chip multiprocessors. In *Proc. of ACM NOCS Symp.*, pages 43–50, 2010.

[13] M.N. Horak, S.M. Nowick, M. Carlberg, and U. Vishkin. A low-overhead asynchronous interconnection network for GALS chip multiprocessors. *IEEE Trans. on Computer-Aided Design of Integrated Circuits and Systems*, 30(4):494–507, 2011.

[14] M. Imai and T. Yoneda. Improving dependability and performance of fully asynchronous on-chip networks. In *IEEE Int. Symp. on Asynchronous Circuits and Systems (ASYNC)*, 2011.

[15] A. Kumar, P. Kundu, L.-S. Peh A.P. Singh, and N.K. Jha. A 4.6Tbits/s 3.6GHz single-cycle NoC router with a novel switch allocator in 65nm CMOS. In *IEEE Intl. Conf. on Computer Design*, pages 63–70, 2007.

[16] A. Kumar, L.-S. Peh, P. Kundu, and N.K. Jha. Express virtual channels: Towards the ideal interconnection fabric. In *Proc. of ISCA*, pages 150–161, 2007.

[17] S. Kundu and S. Chattopadhyay. Mesh-of-tree deterministic routing for network-on-chip architecture. In *Proc. of the ACM Great Lakes Symp. on VLSI (GLSVLSI)*, pages 343–346, 2008.

[18] S. Kundu, K. Manna, S. Gupta, K. Kumar, R. Parikh, and S. Chattopadhyay. A comparative performance evaluation of network-on-chip architectures under self-similar traffic. In *IEEE Intl. Conf. on Advances in Recent Technologies in Communication and Computing (ARTCom)*, pages 414–418, 2009.

[19] Z. Li, J. Wu, L. Shang, R.P. Dick, and Y. Sun. Latency criticality aware on-chip communication. In *Proc. Design, Automation and Test in Europe (DATE)*, pages 1052–1057, 2009.

[20] A. Lines. Asynchronous interconnect for synchronous SoC design. *IEEE Micro Magazine*, 24(1):32–41, 2004.

[21] M. Modarressi, H. Sarbazi-Azad, and M. Arjomand. A hybrid packet-circuit switched on-chip network based on SDM. In *Proc. Design, Automation and Test in Europe (DATE)*, pages 566–569, 2009.

[22] U.Y. Ogras, J. Hu, and R. Marculescu. Key research problems in NoC design: A holistic perspective. In *Proc. of CODES*, pages 69–74, 2005.

[23] J.D. Owens, W.J. Dally, R. Ho, D.N. Jayasimha, S.W. Keckler, and L.-S. Peh. Research challenges for on-chip interconnection networks. *IEEE Micro*, 27(5):96–108, 2007.

[24] A. Rahimi, I. Loi, M.R.Kakoee, and L. Benini. A fully-synthesizable single-cycle interconnection network for shared-L1 processor clusters. In *Proc. Design, Automation and Test in Europe (DATE)*, pages 1–6, 2011.

[25] A. Sheibanyrad, A. Greiner, and I. Miro-Panades. Multisynchronous and fully asynchronous NoCs for GALS architectures. *IEEE Design & Test*, 25(6):572–580, 2008.

[26] M. Singh and S. M. Nowick. MOUSETRAP: High-speed transition-signaling asynchronous pipelines. *IEEE Trans. on VLSI Systems*, 15(6):684–697, 2007.

[27] W. Song and D. Edwards. A low latency wormhole router for asynchronous on-chip networks. In *Asia and South Pacific Design Automation Conf. (ASP-DAC)*, pages 437–443, 2010.

[28] P. Teehan, M. Greenstreet, and G. Lemieux. A survey and taxonomy of GALS design styles. *IEEE Design & Test*, pages 418–428, 2007.

[29] N. Teimouri, M. Modarressi, and H. Sarbazi-Azad. Power and performance efficient partial circuits in packet-switched networks-on-chip. In *Euromicro Intl. Conf. on Parallel, Distributed and Network-Based Processing (PDP)*, pages 509–513, 2013.

[30] Y. Thonnart, P. Vivet, and F. Clermidy. A fully-asynchronous low-power framework for GALS NoC integration. In *Proc. of the Conference on Design, Automation and Test in Europe*, pages 33–38, 2010.

[31] A. Yakovlev, A. Petrov, and L. Lavagno. A low latency asynchronous arbitration circuit. *IEEE Trans. on VLSI Systems*, 2(3):372–377, 1994.

978-1-4799-2817-0/14 $31.00 © 2014 IEEE

A Vertically Integrated and Interoperable Multi-Vendor Synthesis Flow for Predictable NoC Design in Nanoscale Technologies

Alberto Ghiribaldi
University of Ferrara, Italy
alberto.ghiribaldi@unife.it

Hervé Tatenguem Fankem
University of Ferrara, Italy
herve.tatenguemfankem@unife.it

Federico Angiolini
iNoCs SàRL, Switzerland
angiolini@inocs.com

Mikkel Stensgaard
Teklatech A/S, Denmark
ms@teklatech.com

Tobias Bjerregaard
Teklatech A/S, Denmark
tb@teklatech.com

Davide Bertozzi
University of Ferrara, Italy
davide.bertozzi@unife.it

Abstract— We deliver a design flow for the synthesis and convergence of application-specific networks-on-chip. The flow comes with novel features that can better address nanoscale design challenges: front-end driven floorplanning, dynamic IR-drop minimization, fast and accurate system-level power grid modeling, predictable link design. Above all, such features are addressed by different prototype engines, even from different vendors, that can be smoothly integrated into the flow by means of a common specification format called Communication Exchange Format (CEF), that enables unprecedented tool interactions. This flow is validated by means of an extensive demonstration framework.

I. INTRODUCTION

As of today, there are few (actually, not too many) design methodologies and CAD tool flows for application-specific networks-on-chip (NOCs) [1, 2, 3]. Especially in industry, interconnect vendors complement their interconnect IP offering with tooling to automate the creation of the system interconnect based on the communication requirements of the SoC at hand. State-of-the-art design methodologies and toolflows typically cover a limited range of the whole design process, especially topology synthesis [9, 16, 13, 10, 15, 14], and rely on a library-based approach to NoC design, wherein predesigned soft macros are composed at instantiation time to build arbitrary topologies.

In all cases, when we look at current toolflows with respect to future system requirements, we can identify the following criticalities:

A. EDA flow interoperability. From the system designer's viewpoint, the integration of various tools is a common requirement throughout the whole NoC synthesis flow. Also, in a design flow that requires largely interdisciplinary skills, it is very common that such tools come from different vendors, each focused and specialized on a specific design step. If tool interoperability is not properly addressed in a multi-vendor flow, the design cycle may largely prolong.

B. Beyond composition: integration. NoC synthesis flows (possibly multi-vendor) today rely on the juxtaposition of different design steps rather than on their full integration, thus missing global visibility and optimization opportunities. This causes, among the other things, poor inter-play between front-end and back-end design, which is prone to a lengthy and possibly inefficient convergence process. In the future, integration of tools into the whole NoC synthesis flow will have to be mastered by a backbone design methodology with global visibility,

taking care of smart tool sequencing and of their interplay.

C. Evolving technology requirements. Today we are at a main transition point, where new features need to be there in design flows, all related to the intricacy of nanoscale designs. Issues such as process variations, power grid integrity, interconnect delay, etc. cannot be addressed as an afterthought any more, but from the ground up. At the same time, designs should be ranked with respect to these issues not late in the design flow, which may lead to lengthy iterations, but rather early, during a pruning step of the design space relying on technology awareness.

With respect to state-of-the-art, we develop a comprehensive and interoperable multi-vendor synthesis flow for application-specific nanoscale NoCs, and we present a structured validation framework of the novel features of this flow. In particular:

Claim A: *We deliver tool interoperability through the definition of a common and open specification format (named CEF), that provides a consistent representation of design data across all layers of the design process. Through the CEF format, design layers (and associated tools) exchange useful information, design intents, or directives. Therefore, CEF is the backbone of the proposed flow.*

Validation Means: Tool Interoperability is explicitly proved by running the proposed multi-vendor flow for the design of a multimedia system, and by proving convergence (section IV.B). The CEF format is the actual means of interoperability between mainstream and prototype tools, building up the flow.

Claim B: *We deliver a vertically integrated design flow, meaning that (i) the entire flow is addressed, and (ii) global scope of the flow allows us to anticipate floorplanning, a typical back-end design step, in the front-end, thus biasing topology synthesis toward the most efficient physical implementations. In addition, it is not just the length of wires that matters when synthesizing a topology: for the sake of low power, large amounts of traffic should be carried by short links overall; this is achieved by tightly coordinating floorplanning and topology synthesis, together with use-case specification, in the proposed flow.*

Validation means: Claim B is validated in section IV.C by proving the correlation between the communication cost metric used by the floorplanner tool to rank floorplans and the actual NoC dynamic power consumption of synthesized topologies, measured with post-layout analysis. This proves that topology synthesis has actually been driven to the most promising physical implementation.

978-1-4799-2817-0/14 $31.00 © 2014 IEEE

Claim C: *The proposed flow coherently integrates mainstream industrial tools with new prototype tools addressing the most daunting challenges of NoC design in nanoscale technologies.*

C1. In existing mainstream flows, analysis of the power delivery network is not available until after physical implementation. At this stage, the turn-around-time is often too long to justify changes to the floorplan. As a solution, the proposed flow includes a fast, early-phase power delivery network analysis engine and integrates this into the prototype front-end floorplanner tool. The user can thus create floorplans that are optimized for system-level communication (see Claim B) as well as for power integrity.

C2. NoC links are the major cause of lengthy design iterations between front-end and back-end designers. With the proposed flow it is possible to realistically project the actual length of NoC links, and even insert repeater stages upfront in order to meet predefined timing constraints, thus avoiding lengthy design iterations between front-end and back-end designers.

Validation means for claim C: The flow is put at work for the synthesis and physical convergence of an on-chip network for a media-rich mobile embedded system. The validation mean is the first-time-right design of this chip on a 40nm industrial technology library.

II. CEF FILE FORMAT

CEF (Communication Exchange Format) is an open design format which specifies SoC-level architecture and communication infrastructure. It is designed to ease tool interoperability and reduce the iterative exploitation of tools at different stages and abstraction levels of the design flow. CEF allows for the expression of both *design intent* (objectives and constraints) and *design implementation*, describing for example:

(i) system cores, including relevant interface parameters;

(ii) communication requirements across cores, with support for multiple use cases;

(iii) clock and power domains, including frequency and voltage scaling;

(iv) interconnect implementation, including key architectural parameters and routes;

(v) rough floorplan of the design, including wire length annotations.

The CEF format allows for iterative and incremental design steps. For example, communication requirements can be provided in a coarse way or modeled accurately. The floorplan of the design can be omitted, then added when available, and subsequently modified. New usage scenarios, updated NoC revisions, or even entirely new system cores can be added at any time, either reflecting engineering changes or the better understanding of the system as development proceeds.

CEF was conceived to ease the interoperability of tools involved in the development of on-chip interconnects. To this end, it has been designed to be expressive enough to capture the key design steps and abstraction layers involved in the design process. CEF does not enforce a strict order in which these tools must be used, instead promoting rich interaction between tools, back-annotation of parameters and successive refinements.

CEF has no aim to replace existing formats where established, useful standards exist, nor to model out-of-the-box ev-

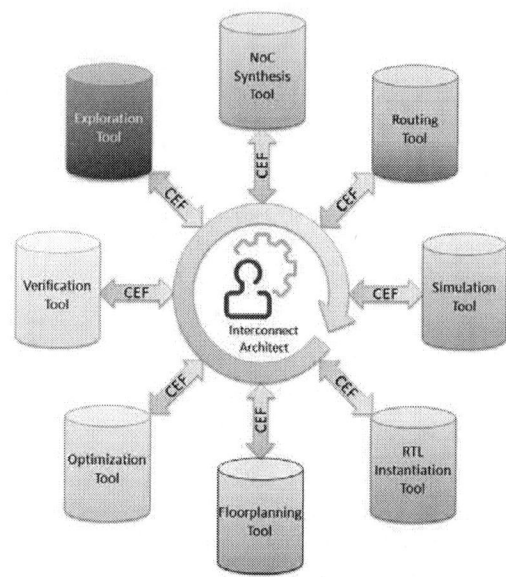

Fig. 1. CEF-enabled interoperability between NoC design tools.

ery nuance of the system. For example, at the architectural level, CEF does not replace existing NoC models written either in C++, SystemC, Verilog, or other languages, nor it models in detail the architectural parameters of a specific vendor's NoC library. Instead, CEF provides generic, broadly applicable, syntax to specify router connectivity, with only some select properties (e.g. virtual channels, flit widths, buffer depths) explicitly included. The set of built-in properties was carefully chosen so as to remain generic and vendor-independent, yet able to express the key parameters that impact in a major way the performance/area/power trade-offs.

Rather than as an intermediate representation format, the CEF format is devised as a data exchange format. The need to maximize the interoperability of different tools has motivated the choice of the XML to make the format easily consumable by machines while remaining readable by humans.

Figure 1 shows an example of the different interconnect design tools that can interoperate on a single target system specification thanks to CEF. The tools operate at different stages of a design process, but can now be seamlessly mixed and matched instead of being forced in a sequential flow, enriching the back-annotation and optimization opportunities. For instance:

- A NoC architectural simulator can now take into account wire propagation delays and variability effects identified by back-end tools;

- High-level exploration tools, optimizers and synthesizers can have a full view of constraints (e.g. communication requirements, block distances) and opportunities (e.g. architectural knobs such as buffering);

CEF was developed by the NaNoC consortium, including both academic and industrial partners with key expertise in system interconnect design. Since this paper is focused on the NoC synthesis flow as a whole, the interested reader is referred to [5] for CEF details.

III. THE FLOW AT A GLANCE

The design flow described in this paper revolves around physical design convergence. The main concepts are to (i) automate the floorplanning of such heterogeneous systems accord-

978-1-4799-2817-0/14 $31.00 © 2014 IEEE

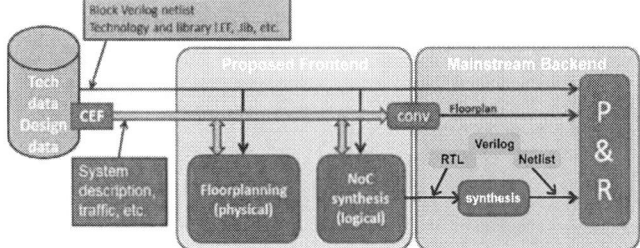

Fig. 2. Proposed flow for the implementation of application-specific NoCs targeting heterogeneous systems.

ing to their communication and power integrity requirements, (ii) perform NoC topology synthesis based on the resulting floorplan, (iii) a predictable physical synthesis process due to the early modeling and consideration of key technology-level convergence threats in the flow (namely interconnect delay and power grid integrity). A comprehensive overview of the flow is reported in Figure 2.

A. Front-End

The CEF file initially includes high-level specifications of the system, including communication bandwidth annotations between pairs of IP cores, specified on a use-case basis. A Liberty file characterizing the dynamic power usage of each IP core is also fed as input to the flow.

System floorplanning is performed *before* the synthesis of the NoC topology. Knowledge of the communication streams is used to direct the floorplanner tool to place closer together system blocks that communicate the most. Thus the total power of the NoC, once implemented, is lower, because the routing paths carrying high bandwidth streams are short. To enforce this behaviour, an *abstract communication cost metric* is defined by multiplying the bandwidth of communication streams by the distance between the associated communicating blocks in the floorplan. This metric is included in the objective function of the floorplanning tool.

The IR-drop of a design is greatly influenced by its coarse-grained placement, i.e., its floorplanning. Hence, in order to ensure power integrity closure, analysis of the power delivery network is warranted as early in the flow as possible [11]. Therefore, in the proposed flow the floorplanner not only reduces the communication cost function, but is also augmented to include an early-phase, dynamic IR-drop analysis directly in the optimization loop. This analysis leverages a prototype high-level power delivery network specification, and is fast enough (less than 50ms) for inclusion into the automated floorplanning process in the design front-end stage. As a result, the user can create floorplans that are optimized for system-level communication as well as for power integrity.

With respect to commercially available tool flows, floorplanning frameworks typically miss awareness of system-level design properties. Moreover, IR-drop analysis is not available until after physical implementation.

The above novel features of the proposed flow have been incorporated, for the sake of experimentation, into the existing Teklatech FloorDirector tool [4]. Plug-and-play of this tool into the flow was straightforward by making it support CEF as its input and output format.

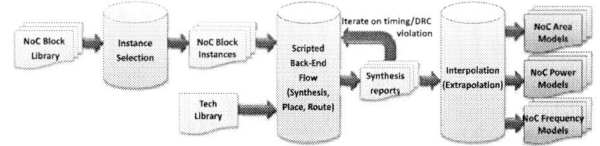

Fig. 3. NoC Topology Synthesizer: block characterization flow.

After the floorplanning stage the CEF file is refined to include the IP core placement. After this, it is handed over to the topology synthesis stage, where the Network-on-Chip is built according to the communication requirements of the system, described in the CEF too. The most relevant requirements for topology synthesis in the proposed flow are back-end modeling (parametric NoC component area and speed models, abstract component power models, interconnect delay models on the target technology), floorplan awareness for topology synthesis, and RTL automatic generation.

These features were to a large extent exposed by the commercial iNoCs NoC synthesis toolchain [3], which was then plugged into the whole flow by making it support the CEF format. The key novelty consisted of extending iNoCs framework to synthesize application-specific topologies on top of pre-assigned floorplans, delivered by a tool from a different vendor.

The final topology synthesis flow relies on NoC components characterization to compare different topology design points and block parameter settings. The characterization (Figure 3) relies on the complete synthesis, placement and routing of some instances of NoC RTL blocks, including Network Interfaces (NI), switches and links. This process yields power, area and maximum frequency values for those instances. Subsequently, by means of interpolation and extrapolation, predictions are made for all other possible instances of the library.

The baseline iNoCs' topology synthesizer was extended to accept a system floorplan from a CEF description as an input. The increasing impact of wiring, in terms of power and propagation delays, strongly suggests that it is essential to take into account the placement of the system cores in order to better estimate and optimize the interconnect [8, 12]. If such a flow is followed, it is possible to better estimate the power consumption of the interconnect, and to design the topology (including, where needed, automatically instantiating link repeaters to break critical timing paths) to meet timing constraints from the ground up. In particular, the topology synthesis tool:
- estimates wire length with Manhattan distance, or more complex routing when hard macros are in the way;
- takes decisions based on a pre-characterization of wire performance in the target technology library;
- meets timing constraints on links through smart switch floorplanning and link segmentation, thus making convergence of P&R more predictable.

In order to guarantee smooth integration of this step into the whole flow, the insertion of NoC components into the design does not lead to expanding the floorplan boundaries (Figure 4). In fact, power grid is designed before NoC generation, hence expanding the floorplan would impair the assumptions of the previous stage. To minimize the concern, while trying to avoid lengthy iterations, optimizations were performed in the iNoCs floorplanning engine to converge towards more compact floorplans.

978-1-4799-2817-0/14 $31.00 © 2014 IEEE

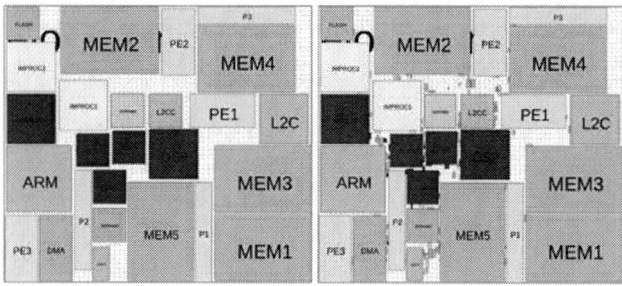

(a) Input floorplan for a SoC design (b) Output floorplan with NoC components instantiated

Fig. 4. The NoC synthesizer inserts the NoC blocks into input floorplan by minimally perturbing block positions.

RTL of a complete NoC topology, together with its succinct CEF description, are then handed over to the physical synthesis flow for P&R.

B. Back-End

The RTL design is translated into a gate-level netlist mapped to the target technology library (tool used: Synopsys Design Compiler), according to the enforced constraints (mainly frequency and operating conditions). Boundary timing constraints have been conservatively set to 25% of the fastest clock in the design in order to take into account the delay of inter-switch links, $100\times$ capacity of biggest inverter in the library has been used as output pin capacitance, and 0.4ns as transition time of signals at input pins.

The generated netlist is the starting point of the layout generation process, performed with Synopsys IC Compiler. We implemented a concurrent hierarchical Layout Generation methodology, exploiting a top-down approach. This flow is presented in Figure 5, and reflects mature industrial layout flows [6]. After the design has been imported, the floorplan definition (in particular die area and architectural blocks' position) is derived according to specifications extracted from CEF file. Then, power and ground grids are generated considering specifications from floorplanner tool. Next step is the partitioning of the Network-on-Chip into its fundamental components, thereby making it possible to perform place&route on each block concurrently and independently. This step is necessary to reduce layout generation complexity, machine runtime and to have tighter control on the layout generation process. Each block is then saved as a soft macro, so that only global routing between blocks is required in the top level. Once layout is completed, soft macros are "uncommitted", i.e. reverted back to their composing standard cells, and final global refinements (power and/or timing driven optimizations) are performed.

The outcome of this step is a completely placed and routed netlist, together with the associated parasitic effects. At this point, accurate power analysis can be performed, by annotating switching activity on the resulting design. This is done by injecting traffic patterns that reflect the predefined use cases using for instance the Modelsim tool, and performing timing and power analysis with proper tools (e.g., the Synopsys Prime-Time suite).

IV. EXPERIMENTAL RESULTS

The target experimental setting is a multimedia chips for mobile phones (or similar devices), which runs different use

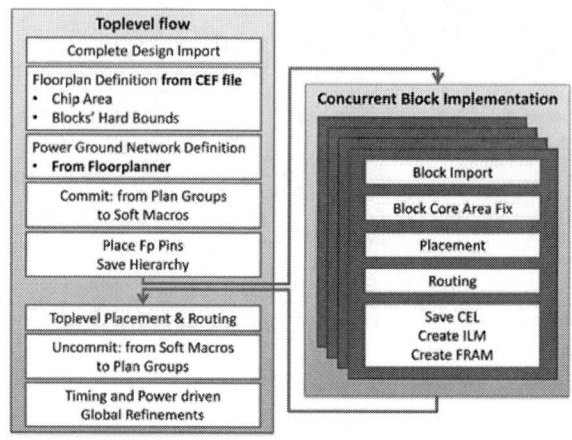

Fig. 5. Concurrent Hierarchical Layout Generation

Floorplan	CommCost	IR-Drop (mV)
Best/Worst	15.83	386
Best/Best	18.87	166
Worst/Worst	43.91	383
Worst/Best	44.52	155

TABLE I
FOUR SELECTED REFERENCE FLOORPLANS IN EXTREME CORNERS WITH REGARD TO COMMCOST AND IR-DROP.

cases. Its technical specifications reflect projected industrial trends of multimedia chip for the near future. This design under test features 25 IP Cores, which include CPU, Graphic Accelerator, various memory banks and peripherals. Further details in [7]. Extrapolating the usage scenarios of existing smart phones, communication requirements of CPU, hardware accelerators and memory for video playback have been scaled up to the high-end HD-TV resolution. The target technology library is an industrial 40nm 1.20V Low-Power CMOS Standard Cell library.

The goal of this section is to demonstrate that abstract metrics (the communication cost) and modeling frameworks (IR drops) used in the early steps of the proposed toolflow maintain a strong correlation with the physical properties of the design. In other words, we demonstrate that we are not pruning efficient points from the design space as an effect of abstract modeling of design properties or inaccurate/misleading objective functions in early-stage optimization tools. We thus selected 4 floorplan design points (Table I), representative of the design space, and fed them to the next steps of the flow to prove correlation: promising design points will actually outperform the others as the design is refined.

A. Floorplanning & Topology Synthesis, claims A and B

At this level, a correlation study was performed by synthesizing a large number of topology design points for the floorplans. The outcome can be seen in Figure 6, where dots of different colors represent topologies built on different reference floorplans. Recall that the communication cost computed by the floorplanning tool is expected to correlate with power consumption. The "red" and "maroon" dots represent solutions for the two floorplans with the lowest communication cost, and are very similar in actual power; the "blue" and "dark blue" dots

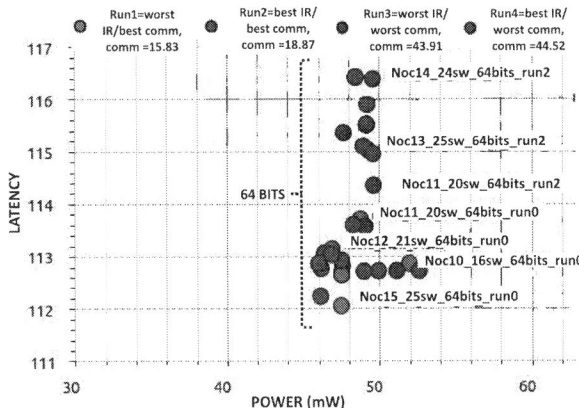

Fig. 6. Topologies generated by the Synthesizer engine for the various floorplans for different flit widths.

Corner Comm/IR	Switch Count	Flit Width (bits)	Reserved Area (mm^2)	NoC Power (mW)
Best /Worst	25	64	1.73	47.5
Best /Best	25	64	1.72	46.1
Worst /Worst	16	64	1.54	51.2
Worst /Best	16	64	1.54	52.7

TABLE II

MAIN METRICS OF THE FOUR SELECTED DESIGN POINTS, ONE PER INPUT FLOORPLAN, AS ESTIMATED BY THE NoC SYNTHESIZER.

represent solutions for the two floorplans with the highest communication cost, and are also clustered together. Indeed, there is a measurable difference (typically around 5 mW, or almost 10%) between "good" and "bad" floorplans. This separation is an average across use cases, and since this includes e.g. the Idle scenario in which the NoC is in fact inactive, it is clearly a conservative estimate; a separation of up to 15 mW or 25% was in fact noticed in traffic-intensive use cases. Moreover, the topologies built on a worse floorplan exhibit average flow latencies that are slightly higher, due to the need to insert, on average, a few more pipeline stages along links.

For the rest of the study, four specific design points among those produced by the Noc Synthesizer chosen, one per input floorplan. Without lack of generality, the flit width was fixed to 64 bits. Out of all the possible solutions at this width, we selected representative instances with low latency and typical power for the cloud of solutions based on the same floorplans. These four solutions are summarized in Table II. Please note that reserved area takes into account a target cell occupancy of 50%.

The design points at this stage consist of floorplans and of application-specific NoC topologies at the RTL level, custom-tailored for the floorplans and for the communication requirements of the application at hand. They have been derived by using tools from different vendors (Teklatech, iNoCs) coherently integrated into the same flow, and they have been showed to maintain correlation of power values across the hierarchy so far. With this respect, claim A is validated, and partly also claim B about correlation. The design points now undergo physical synthesis, involving the interplay with mainstream EDA tools and requiring to preserve correlation across the back-end design steps.

NoC Interconnect Timing Convergence			
Clock Domain	Target Frequency	Slack Pre-Opt	Slack Post-Opt
clk_Audio	100MHz	5,75ns	3,45ns
clk_CPU	500MHz	0,20ns	0,09ns
clk_DDR	250MHz	0,38ns	0,17ns
clk_DMA	200MHz	0,65ns	0,39ns
clk_DSP	300MHz	0,34ns	0,19ns
clk_Radio	150MHz	1,12ns	0,82ns
clk_USB	200MHz	0,53ns	0,23ns
clk_SPI	128MHz	6,33ns	2,27ns
clk_SRAM	500MHz	-0,19ns	0,14ns
clk_Video	300MHz	0,38ns	0,20ns

TABLE III

TIMING CONVERGENCE FOR ALL CLOCK DOMAINS WITH PROPOSED METHODOLOGY, BEFORE AND AFTER GLOBAL OPTIMIZATION

B. Physical Convergence, claims A and C2

The target design includes 10 clock domains. Since synchronization is taken care by Dual-Clock Fifos, clock signals have been defined as *asynchronous* to each other so the tool will not balance the sinks among the clock nets. In addition, a clock uncertainty of 10% has been enforced on every domain, to account for the possible clock skew.

During **Place&Route**, a hierarchical design flow is a must for an interconnect structure which is highly irregular, beyond being distributed. In order to have a tighter control of the results, Hard Bounds are created to define the areas where NoC building blocks will be placed.

Results: Total NoC area after layout turns out to be $946.152\mu m^2$ for Best Comm./Best IR drop topology and $898.744\mu m^2$ for Worst Comm./Worst IR drop topology, out of a chip area of $6855\mu m \times 6855\mu m$. Aggressive block boundary constraints applied to NoC switches led to timing convergence after the top level integration and global routing of the architectural blocks (3rd column of Table III). Indeed, every clock domain has a positive slack, and only one has a small violation due to the unmatched sizing of a few gates between intra-switch links and switch output cells. This confirms that the link delay prediction and design engine of the topology synthesis tool have done a good job. Given this excellent starting point, the tool does not struggle during the global optimization phase to meet the timing requirements. Therefore, it can afford relaxing the faster paths to reduce area and power consumption (4th column of Table III).

Overall, the described methodology actually delivers the fast convergence claimed by our flow (i.e., claim C2, or first-time right physical design), and completes the validation of claim A by proving effective integration of front-end and back-end multi-vendor tools.

C. Correlation of Dynamic Power, claim B

It is indeed possible to prove correlation between Sign-off and Early-phase total NoC power numbers in communication-intensive use cases, such as in "Video Playback", and "Video Capture" (see Figure 7). Deviations never exceed 8% for both "Best Comm" and "Worst Comm" floorplans.

978-1-4799-2817-0/14 $31.00 © 2014 IEEE

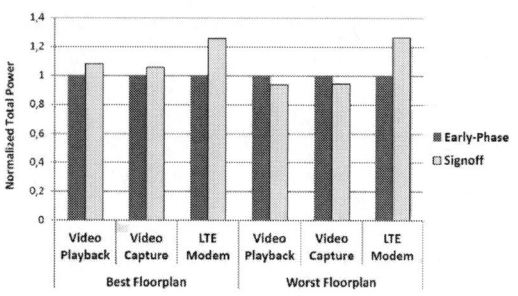

Fig. 7. Correlation of NoC power between early-phase and post-layout analysis.

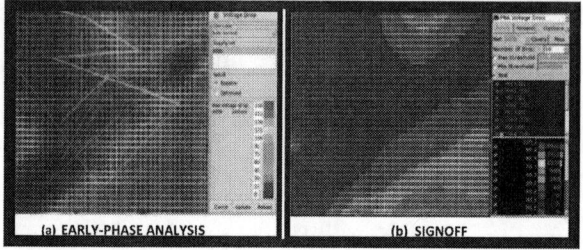

Fig. 8. Floorplan with best communication and best IR drop.

Only in those use cases where the network is heavily under-utilized ("LTE modem"), and approaches the idle use case, the power gap can be as large as 25%. This is due to the different choice of flip flops the synthesis tool made during the power model extraction methodology (Figure 3) and the actual physical synthesis of the design at hand. This difference is at such a low level of abstraction that it becomes extremely difficult to account for it upfront.

Given the large amount of specific timing and Design Rule optimizations each topology undergoes, all requiring different effort, and the necessarily more abstract power modeling performed during topology synthesis, these results ultimately validate claim B.

D. Correlation of IR Drop Maps, claim C1

This section needs to demonstrate that the IR drop maps derived by the front-end floorplanner (during the early phase analysis) and the same maps derived by the signoff tool indicate the same trends and critical hotspots. The In-Design Rail Analysis extension of Synopsys ICC was used as the reference sign-off tool.

It is worth observing that during early-phase analysis, only the current consumption of the IP cores is considered for IR drop analysis, in that the on-chip network has not been synthesized yet. To match (and later validate) this assumption, in the first place a special layout was inferred with the IP cores only.

In Figure 8 we report the IR drop maps displayed by Teklatech's early-phase floorplanning tool and the one displayed by Synopsys ICC for this design, with reference to the "Best Comm/Best IR drop" floorplan.

The correlation between the two maps is evident: they both agree on indicating the top-left quadrant of the design as the most critical in terms of IR drops. A similar correlation was observed for the "Worst Comm/Worst IR drop" floorplan.

When we compared the two floorplans (and their IR drop maps) with each other, we noticed that relative rankings between early-phase and signoff are in good agreement (degradations in the order of 2.3x and 1.5x respectively), thus confirming that the front-end floorplanner would have made the right choice when selecting the "Best Comm/Best IR drop" one.

Finally, we derived that the on-chip network has a maximum IR drop which is from 80 to 90% lower than that of IP cores, depending on the baseline floorplan. If we also look at the IR drop maps for the NoCs in isolation, we can see that they peak in almost the same physical spot of the IP core maps. This means that they do not even change the trend of the IP core maps. Therefore, it was hereby possible to validate the floor-plan tool's assumption to perform early-phase power integrity check by neglecting the on-chip network.

V. CONCLUSION

This paper reports a vertically integrated, interoperable and multi-vendor synthesis flow for NoCs. It ranges from application traffic specification to layout generation and physical convergence, and integrates prototype tools with mainstream industrial tools. Its main innovations include full vertical integration, global optimization of design steps for technology-aware design, flow extensions to deal with nanoscale designs. The validation strategy of the flow revolved around first-time-right design and the proof of correlation of early-phase analysis and design choices with signoff.

ACKNOWLEDGMENT

This work was supported by NaNoC and vIrtical European Projects (FP7-ICT-248972 and FP7-ICT-288574).

REFERENCES

[1] ARTERIS, "http://www.arteris.com",
[2] sonics, "http://sonicsinc.com",
[3] iNoCs, "http://www.inocs.com",
[4] Teklatech, "http://www.teklatech.com",
[5] Communication Exchange Format (CEF) format, "www.nanoc-project.eu",
[6] M.Igarashi, "Concurrent Hierarchical Design with IC Compiler. Real Life Application on Mobile Multi-Media Processor", DAC 2008.
[7] Tatenguem H. F., Ludovici D., Strano A., Bertozzi D., Reinig H. "Contrasting multi-synchronous MPSoC design styles for fine-grained clock domain partitioning: the full-HD video playback case study", NoCArc '11.
[8] Murali S., Meloni P., Angiolini F., Atienza D., Carta Salvatore, Benini L., De Micheli G., Raffo L. "Designing Application-Specific Networks on Chips with Floorplan Information" ICCAD '06
[9] Khan G.N., Tino, A. "Synthesis of NoC Interconnects for Multi-core Architectures" CISIS '12
[10] Binjie Song, Shan Zeng, Yuchun Ma, Ning Xu, Yu Wang "Tree-Based Partitioning Approach for Network-on-Chip Synthesis" CAD/Graphics '11
[11] Kapadia N., Pasricha S. "A Power Delivery Network Aware Framework for Synthesis of 3D Networks-on-Chip with Multiple Voltage Islands" VLSID '12
[12] Bo Huang, et.al., "Application-Specific Network-on-Chip synthesis with topology-aware floorplanning" SBCCI '12
[13] S.Murali, G.De Micheli, "SUNMAP: a tool for automatic topology selection and generation for NoCs", DAC 2004, pp.914-919, 2004.
[14] Vladimir Todorov, Daniel Mueller-Gritschneder, Helmut Reinig, Ulf Schlichtmann, "A spectral clustering approach to application-specific network-on-chip synthesis", DATE 2013: pp. 1783-1788
[15] S.Murali et al., "SUNFLOOR: Application-Specific Design of Networks-on-Chip", Poster at DATE 2006 University Booth.
[16] S.Murali et al., "NoC synthesis flow for customized domain specific multiprocessor systems-on-chip", IEEE Transactions on Parallel and Distributed Systems (Volume:16 , Issue: 2), pp.113-129, 2005.

Fuzzy Flow Regulation
for Network-on-Chip based Chip Multiprocessors Systems

Yuan Yao and Zhonghai Lu

Department of Electronic Systems, School for ICT

KTH Royal Institute of Technology, Stockholm, Sweden

{yuanyao, zhonghai}@kth.se

Abstract— Flow regulation is a traffic shaping technique, which can be used to improve communication performance with better utilization of network resources in chip multi-processors (CMPs). This paper presents fuzzy flow regulation. Being different from the static flow regulation policy, our system makes regulation decisions fully dynamically according to traffic dynamism and the state of interconnection network. The central idea is to use fuzzy logic to mimic the behavior of an expert that can recognize the network status and then intelligently control the admission of input flows. As the experiment results show, the maximum improvement in average delay reaches 53.0% against static regulation and 37.4% against no regulation. The maximum improvement in average throughput reaches 37.5% against static regulation and 23.8% against no regulation.

Keywords: Network-on-Chip, Chip Multiprocessor, Flow Regulation, Fuzzy Logic

I. INTRODUCTION

As large uniprocessors are no longer scaling in performance, chip multiprocessors (CMP) become the mainstream to build high-performance computers [2]. CMP chips integrate various components such as processing cores, L1 caches and L2 caches (some also contain L3 caches, for example, in the IBM Power7 multicore processor) together, and multiple CMP chips with external memory banks make up a CMP system. As discussed in [2, 3], CMP systems are rapidly becoming communication limited. Since buses (although long the mainstay of system interconnect) are unable to keep up with increasing performance requirements, network-on-chip (NoC) offers an attractive solution to this communication crisis and is becoming the pervasive interconnection network in CMPs.

As explained in [1, 5], in NoC based CMP systems, regulating traffic flows has been shown to be an effective means to improve communication performance and reduce buffer requirements. Fig. 1 illustrates the traffic flow regulation where

regulators are inserted between processing cores and the NoC to perform traffic shaping. To achieve an analyzable performance guarantee, leaky bucket flow control is used and the shaping function is expressed mathematically as (σ, ρ) shaped, with σ bounds a flow's burstiness and ρ long term sustainable rate. The (σ, ρ) flow regulation is based on *network calculus*, which has been studied thoroughly in [6, 4, 7]. It stipulates that during any time interval $(\tau, t]$, the amount of traffic denoted by $A_i(\tau, t)$, which is also called as regulated arrival function, from processor i entering the network is **upper-bounded** by :

$$A_i(\tau, t) \leq \sigma_i + \rho_i(t - \tau). \tag{1}$$

The constraint in inequality (1) is attractive since it is descriptive enough to model a wide variety of traffic patterns [8]. However, existing flow regulation policies are either static or partially dynamic. With the static regulation [7, 6], the parameters of the regulators are hard-coded during system configuration and remain constant. With the partially dynamic mechanism [1], the traffic dynamism is profiled to configure the regulation parameters online but does not take the dynamism of network congestion state into account.

In this paper, we propose a fuzzy flow regulation mechanism for NoC-based CMPs. Being different from the static and partially dynamic flow regulation policy, our system makes regulation decisions fully dynamically according to the traffic dynamism as well as to the state of interconnection network. As Fig. 1 shows, our mechanism uses fuzzy controllers (FCs) and network state recognizers (NSRs) to achieve network status awareness traffic. Based on a sampling window concept, FC, NSR, and the flow regulator form a feedback flow control loop for each processor. The goal of our design is to use fuzzy logic to mimic the behavior of an expert that validly controls the admission of input flows, *with the aim of making better use of on-chip resources and decreasing communication delays.* In our experiments, we show the fidelity and efficiency of our fuzzy flow control mechanism with both synthetic flows and traces from benchmark programs SPLASH2. We also demonstrate benefits that our fuzzy flow control brings in terms of average network delay and throughput.

The rest of the paper is organized as follows. Section II discusses related work. In Section III, we introduce (σ, ρ)–regulation and the basics of fuzzy logic. Section IV details the fuzzy flow control system from concept to design. In section V, we set up our experiment platform based on Multi-facets General Execution-driven Multiprocessor Simulator (GEMS, [13]) and report the results. Finally, we conclude in Section VI.

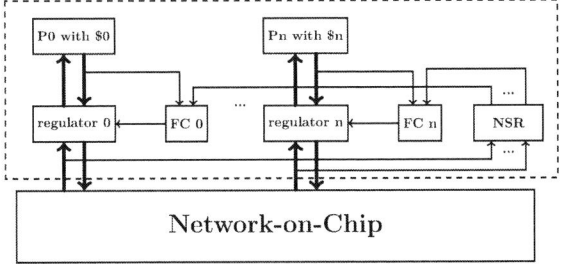

Fig. 1. Fuzzy flow regulation for CMP

978-1-4799-2817-0/14 $31.00 © 2014 IEEE

II. RELATED WORK

Since pioneered by Cruz [7] and Chang [4], network calculus has been successfully applied to design networks such as Internet with integrated/differentiated services [6], WSNs [9], etc. Based on network calculus, flow regulation is proposed to regulate the admission of traffic in order to minimize network overload, worst-case delay and buffer usage.

Most traditional networks (for example, WAN, LAN) used window-based flow regulation [11, 12], with advantages of low implementation cost and fast response time. However the effectiveness of this scheme in high speed networks is severely limited due to the increasing of performance requirements today [8].

To solve this problem, rate based flow regulation was introduced [6, 7, 4]. The most popular one of this kind is called leaky bucket flow regulation [7], where the admission of traffic is determined mathematically by a shaping function. Unlike window based flow control, no acknowledgement signals from the far-end or neighbor responser are used, thus this scheme can be implemented efficiently in fast speed networks. Works [6, 4] dealt with static regulation, requiring the characterization of traffic flows' (σ, ρ) values offline at design time. Though offline characterization methods are possible, they are inflexible and can not capture traffic dynamism. Work in [1] dealt with partially dynamic flow regulation. The regulator can capture and follow the characteristics of varying flows in an on-line fashion, but no network congestion status is considered.

The central idea for this work is to make *fully* dynamic regulation by considering the network congestion status, and then use such information to adaptively control the regulation strength. Unfortunately, there is no formal nor empirical models to precisely describe when a network is congested. Most of the time, network congestion is recognized through measurement metrics such as packet latency, link utilization, or throughput. With a terminal instrumentation measurement approach, Dally and Towles [3] inspected network status by examining average packet latency and average throughput against the offered traffics. In [2], Kunle et al. proposed a speculative parallel threads method to detect congestion status and to accelerate single-thread applications in CMP. Since these works aim to pinpoint the congestion status through measurement values, they are very difficult to be accurate in some situations. For example, suppose n cycles average packet delay is the indication for network congestion, can we say that the network is not saturated if the average packet delay reaches $n - 1$ cycles? The fundamental fuzzy property of the congestion recognition problem determines that the proposed methods in [2] and [3] are not always plausible.

This paper proposes a fuzzy flow regulation technique from concept, design to implementation. Our approach is based on a fuzzy flow regulation system which can recognize the network congestion status and make new flow regulation policies through fuzzy logics. Very different from the others, we aim to mimic the behavior of an expert that validly recognizes the network state and controls the admission of input flows (refer to Section IV). In section V, we implement the fuzzy flow regulation system within GEMS [13], which a full-system CMP simulation environment, and record the experiment results.

III. (σ, ρ)–REGULATION AND FUZZY LOGICS

A. (σ,ρ)–regulation Function

The (σ,ρ)–regulation function is a powerful characterization which helps to ensure QoS guarantees in networks [6, 4]. In reality, the leaky bucket model [7] can be employed to shape an incoming traffic flow and make it conform to a (σ,ρ)–regulation function.

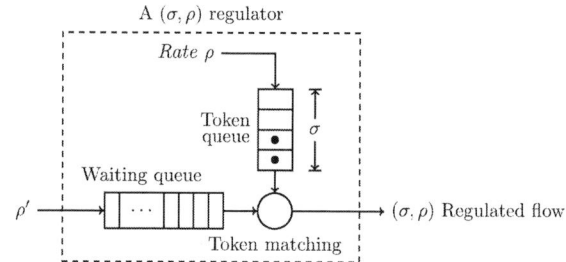

Fig. 2. Leaky bucket flow regulation

Fig. 2 shows a leaky bucket (σ, ρ)–regulator model. There is a token bucket into which tokens are fed at rate ρ. The bucket can hold up to σ tokens. Packet is allowed to depart from the regulator if and only if there are tokens available. For every departed packet, one token is consumed. If packets arrive and there are no tokens, then the data are buffered in the waiting queue until tokens are available. More about leaky bucket (σ, ρ)–regulation can be found in [7, 4, 3].

B. Fuzzy Logic Basics

First introduced by L. A. Zadeh [14, 15], fuzzy logic deals with many-valued logic or probabilistic logic problems, with the aim to reason about logics that are approximate rather than fixed and exact. To understand fuzzy logic clearly, we introduce some useful terminologies below.

- **Fuzzy set:** Fuzzy set is a pair $(\mathcal{U}, M)$, where exact values in $\mathcal{U}$ can be mapped to degree of truth values in $[0, 1]$ by membership functions in M.

- **Degree of truth:** Degree of truth represents that a proposition might be more or less true, rather than simply true or simply false. For instance, the proposition $1 + 1 = 2$ is simply true, while *The network is saturated* is neither simply true nor simply false. Degree of truth is in $[0, 1]$, with "1" indicates totally true and "0" totally false.

- **Membership Function (MF):** The MF is a generalization of the indicator function in classical sets. In fuzzy logic, it determines the degree of truth value of a proposition.

- **Fuzzy Rule:** A fuzzy rule is defined as a statement in the form: **IF** x is A **THEN** y is B, where x and y are linguistic variables; A and B are degree of truth values of x and y, respectively. The values of A and B are determined by proper membership functions (or consequence membership functions) invoked by x and y, respectively. In each rule, "y is B" is the consequence of "x is A". For example, **IF** *average packet delay* is $HIGH$, **THEN** *network* is $SATURATED$.

- **Consequence Membership Function (CMF):** The CMF indicates that if a rule's premise is true then the action indicated by its consequence should be quantified.

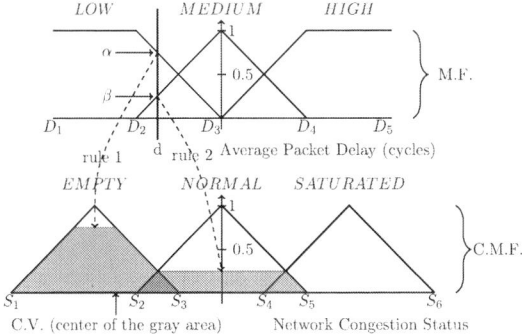

1. **IF** *avg. packet delay* is *LOW*, **THEN** *network status* is *EMPTY*
2. **IF** *avg. packet delay* is *MEDIUM*, **THEN** *network status* is *NORMAL* } Rule Base
3. **IF** *avg. packet delay* is *HIGH*, **THEN** *network status* is *SATURATED*

Fig. 3. Fuzzy network status recognization

Fig. 3 depicts the above concepts by showing the relationship between average packet delay and network congestion status. The fuzzy set is a pair $(\mathcal{U}, M)$ where $\mathcal{U}$ is a set of exact values (average packet delay ranges from D_1 to D_5 in this instance) and M the set of MFs with the mapping $m_i : \mathcal{U} \to [0, 1]$, $m_i \in M$. In the figure there are totally three MFs for measurement of low, medium and high delay, respectively. Each MF maps the measured delay value to a degree of truth value in the range [0, 1]. This progress is called *fuzzification*. Note that one exact value can invoke multiple MFs. For example average delay d is mapped to two degree of truth values (α and β) by MFs for low delay and medium delay, correspondingly. These degree of truth values further invoke rule 1 and rule 2 (illustrated by dashed lines), which show the consequences of the delay. For example, **IF** *average packet delay* is *LOW*, **THEN** *network* is *EMPTY*. This progress is called *inferencing using the rule base*.

To quantify the consequence, each rule is assigned a CMF, which can be treated as the "opposite" function of MF. CMF de-fuzzifies all the consequences of the invoked rules back to one Consequence Value (denoted by C.V. in the figure). This progress is called *defuzzification*. CMF can be the same or different shapes as the corresponding MF, but often with both ends closed [16]. There are dozens of approaches to implement defuzzification, each with various advantages or drawbacks [16]. One of the most popular is the "centroid" method, in which the "center of mass" of the results provide the consequence value. The "mass" (gray area in Fig. 3) of each CMF is determined by the degree of truth value from the corresponding MF. It is shown in Fig. 3 that given the average packet delay d, we can say the network is in $EMPTY$ state, since the final C.V. falls in the $EMPTY$ CMF. If the C.V. falls in the inter-crossed section $[S_2, S_3]$, the network status can be interpreted in either way. More about the "centroid" method will be introduced in Section IV. We also illustrate there what difference can be made if network status recognition is based on two metrics (average packet delay, average link utilization) instead of one.

IV. FUZZY FLOW REGULATION FOR CMP

A. Design Overview

To realize our fuzzy flow regulation mechanism requires: 1) module to recognize the network congestion status, and

2) module to make new flow regulation decisions and update the old ones. Fig. 1 shows the required modules as network state recognizer (NSR) and fuzzy controller (FC), respectively. Bold lines in the figure denote data paths and thin lines control paths. The NSR detects network congestion status from the performance statistics of memory transaction (link utilization and packet delay), and the FC updates new regulation policies based on both network congestion status and on unregulated flows of the local processing core. Note that the NSR is one per CMP chip, but the FC is attached to every core.

In this section, we present our fuzzy flow regulation from concept to design. We first introduce one problem of on-chip flow regulation, followed by design of the FC and NSR. As to design of leaky bucket flow regulator, please see [7, 1].

B. Concept: Sampling based Fuzzy Flow Regulation

One problem we face in CMP flow regulation is control latency, like any other control system introduced in [17] has. Control latency is the time between the generation of a control signal and the control signal begins to take effect. In our fuzzy flow control system, the flow regulation mechanism reacts to network status. This progress consumes time and this time is the control latency. Many factors can effect the control latency, such as the overhead of FC, response time of NSR, the system clock frequency, delays of the on-chip routers, etc. In order to tolerate the control latency problem, we design our fuzzy flow regulation mechanism based on sampling windows. The fuzzy controller does not make new flow regulation policy in every cycle. It samples different input signals during a period of time (called sampling window), and makes new policy after the sampling window is over. The new policy will make influence on the network status during the next sampling window.

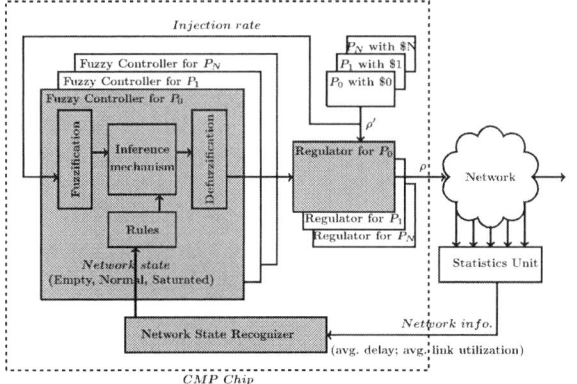

Fig. 4. Structure of fuzzy flow regulation system

As shown in Fig. 4, three components: fuzzy controller (FC), network state recognizer (NSR), and flow regulator work together to achieve fuzzy flow regulation. FC is very simple conceptually. At the end of each sampling window, three stages are invoked in each FC: an input stage (fuzzification), a processing stage (inferencing using rule base), and an output stage (defuzzification). In the input stage, inputs (injection rate, ρ', of the unregulated data flow from each processor) are mapped by the appropriate membership functions to degree of truth values. In the processing stage, appropriate rules are invoked and a result for each rule generated. In this step, the network status indicator K (has three values : *empty, normal, saturated*) generated by the NSR influences the mapping of each rule. The principle is that if the network is empty, the flow regulator casts a relaxing policy; if the network is saturated, the flow regulator tights up the incoming flow; if the network works normally, the flow regulation policy casts no effect. Finally, the output stage converts the combined results back into a new regulation policy. The new flow regulation policy updates the old one inside the regulator, which is leaky bucket based. The regulator forces the output flow to conform to the new injection rate ρ. We only restrict the per-window injection rate (ρ') of a data source, since in CMP the burst (σ) is determined by the data block size of memory transaction and that can not be changed.

The NSR is also based on fuzzy logic, due to the fact that it is hard to calculate NoC status by mathematical/empirical model. The approach we use is to monitor network state through network statistics information (average packet delay and average link utilization), as it is done in [3]. The network statistics information is collected by the statistics unit shown in Fig. 4, which is provided by GEMS [13]. Similar with the FC, NSR is also realized by three stages: fuzzification, inferencing using rule base, and defuzzification. The difference lies in that the NSR makes fuzzy decision on two inputs while the FC on one.

C. Design: Fuzzy Controller

The aim of the FC is to make flow regulation policy based on the status of network congestion. It takes the unregulated injection rate (ρ') as input, and generates the target injection rate (ρ) as output. Based on the window mechanism, the sampling of ρ' is straightforward. Within each sampling window, a flow is sampled at each time instant t_k for its traffic volume $f(t_k)$, where $t_k \in [L_0, L_0 + L_w]$ and $k \in [1, S]$, with L_0 indicating the start time of a window and L_w the length, and S is the total number of samples. At each t_k, ρ' is computed by $\rho'_k = f(t_k)/t_k$ to obtain the average rate of the flow. In this paper, we stipulate that all the sampling intervals are equal, which implies that all the t_k are distributed evenly within $[L_0, L_0 + L_w]$. Based on the sampled ρ'_k, the inferencing of each rule in the rule base is influenced by the network status indicator generated by NSR. Here we use K to symbolize the indicator, where $K \in [empty, normal, saturation]$.

Let m_i denote the ith membership function of the linguistic variable ρ' defined over the universe of discourse $\mathcal{P}'$. In our design, ρ' denotes "unregulated injection rate" and $\mathcal{P}'$ denotes all the possible values of ρ'. If we assume there are N membership functions, then the fuzzy set S of the fuzzy controller is a pair: $S = (\mathcal{P}', M)$, where $M = \{m_i : i = 1, 2, ..., N\}$. We assume that $\mathcal{P}' = [0, \infty]$ and $N = 7$. Similarly, let c_j denote the jth consequence membership function (CMF) of the linguistic variable ρ defined over the universe of discourse $\mathcal{P}$. In this paper, ρ denotes "regulated injection rate" and $\mathcal{P}$ all the possible values of ρ. Similarly, $j \in [1, 7]$ and $\mathcal{P} = [0, \infty]$.

As introduced in Section III, the mapping of the inputs to the outputs is characterized by a set of rule base, or formally, in modus ponens (If-Then) form. In FC, rules in the rule base has the form:

IF ρ' **is** $m_i(\rho')$, **THEN** ρ **is** $c_j(m_i(\rho'))$, $i, j \in [1, 2, ..., 7]$

By iterating i and j, we get a set of linguistic rules that specify on how to regulate the flows.

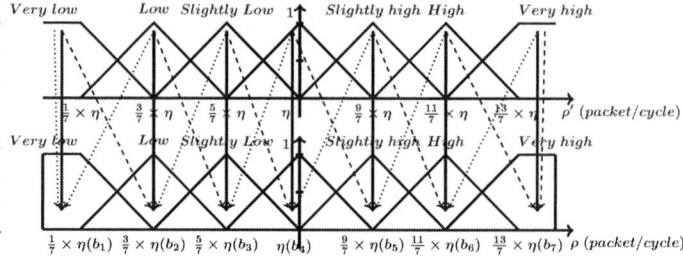

Fig. 5. Membership functions with different mapping ($\delta = 0, \pm 1$)

During $fuzzification$ and $inferencing$, S samples of ρ' are recored and denoted by ρ'_k, $k = 1, 2, ..., S$. Further, the degree of truth of each sample is denoted by $m_i(\rho'_k)$, where m_i is the ith membership function. The "mass" of each $m_i(\rho'_k)$ in the corresponding CMF is denoted by $c_j(m_i(\rho'_k))$, where c_j is the jth CMF. Different from static regulation and partly dynamic regulation [1], FC makes new regulation policy based on the network congestion status. The network status indicator K determines the extent to which each rule is relevant to the current situation. The mapping principles are as follows :

- When K is $normal$, mapping of $m_i(\rho'_k)$ to $c_j(m_i(\rho'_k))$ is done by $f : m_i \to c_j, i = j$, and is shown by the solid lines in Fig. 5. It shows that the network works fine with the current injection rate ρ', and the traffic just gets through without regulation.

- When K is $empty$, mapping of $m_i(\rho'_k)$ to $c_j(m_i(\rho'_k))$ is done by $f : m_i \to c_j, i = j + \delta, \delta \leq 0$. It shows that when empty, the network can accept more packets and increasing of ρ' is accepted. One example of $\delta = -1$ is shown by the dashed lines in Fig. 5.

- When K is $saturated$, mapping of $m_i(\rho'_k)$ to $c_j(m_i(\rho'_k))$ is done by $f : m_i \to c_j, i = j + \delta, \delta \geq 0$. It shows that when saturated, the network tights up the injection rate to recover from saturation. One example of $\delta = 1$ is shown by the dotted lines in Fig. 5.

Note that η in Fig. 5 denotes the average injection rate of each data source within a sampling window. In a CMP environment, data and control packets are injected if and only if cache miss happens. Thus η is treated as the average cache miss ratio (ACMR). Calculation of ACMR is beyond the topic of this paper. For more information please refer to [18].

During $defuzzification$, the centroid method is used to combine $c_j(m_i(\rho'_k)), i, j, \in [1, 2, ..., 7]$ into a single result by the following formula:

$$\rho = \frac{\sum_{j,i,k}^{N,N,S} b_j \, c_j(m_i(\rho'_k))}{\sum_{j,i,k}^{N,N,S} c_j(m_i(\rho'_k))}, \qquad (2)$$

where b_j is the center of c_j. The new ρ is passed to update the flow regulator, and the effects of the new regulation policy becomes effective during the next sampling window.

D. Design: Network State Recognizer

As mentioned in Section III, we also design the NSR with fuzzy logic approach, since the task of constructing and analyzing a mathematical model for complex systems such as network-on-chip is difficult, especially when resource contention issues are considered [1]. The structure of NSR is similar to that of FC (fuzzification, inferencing with rule base, and defuzzification), but differences exist in that NSR has two inputs (average link utilization and average packet delay) instead of one.

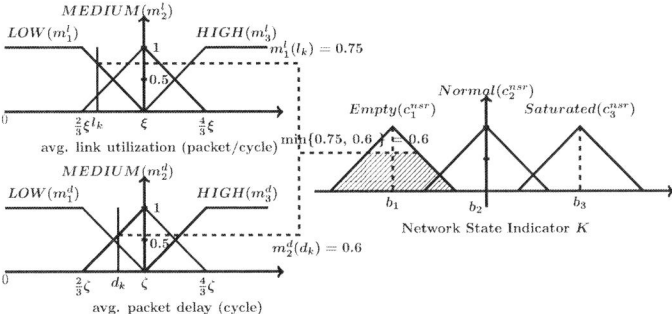

Fig. 6. Minimal intersection and defuzzification

Let l_k and d_k denote the kth samples of linguistic variables "average network link utilization" and "average network packet delay" in the corresponding universe of discourse $\mathcal{L}$ and $\mathcal{D}$, where $\mathcal{L}, \mathcal{D} \in [0, \infty]$, $k \in [1, S]$. Let m_i^l and m_j^d denote the ith and j^{th} membership functions invoked by l_k and d_k, respectively. During *fuzzification*, the membership functions map l_k and d_k to degree of truth values $m_i^l(l_k)$ and $m_j^d(d_k)$. Further, the fuzzy set for NSR is: $S = (\mathcal{D}, M^d) \times (\mathcal{L}, M^l) = (\mathcal{D} \times \mathcal{L}, M^d * M^l)$, where $M^d = \{m_j^d : j = 1, 2, ..., N^d\}$ and $M^l = \{m_i^l : i = 1, 2, ..., N^l\}$. The "$*$" operation symbolizes the intersection of the fuzzy sets, which achieves the mapping $M^d * M^l : \mathcal{D} \times \mathcal{L} \to [0, 1]$. In this paper, we manipulate that $N^d = N^l = 3$, and we use **and** as the intersection method ("$*$") of membership functions. There are many ways to implement the **and** intersection [16, 19]. In the NSR, we use the **minimal** method. It is based on the principle that if we are not very certain about the truth of one proposition, we can not be any more certain about the truth of that proposition "and" the other proposition. As shown in Fig. 6, two samples l_k and d_k generate four degree of truth values that are non-zero and the intersection of the ones generated by membership functions m_1^l and m_2^d is illustrated in dashed line. It is shown that by using minimal intersection, the degree of truth of the combined proposition is determined by the lower one.

The *inferencing* progress is determined by the rule base. We further manipulate that the network is in one of the three status: *empty, normal, saturated*. Thus, the number of CMF is also three, and each one is denoted by c_x^{nsr}, $x \in [1, 2, 3]$. Fig. 6 depicts both the MFs and the CMFs of NSR. Considering the mapping $M^d * M^l : \mathcal{D} \times \mathcal{L} \to [0, 1]$, according to the **and** intersection method, rules in NSR has the form:

IF l_k **is** $m_i^l(l_k)$ **and** d_k **is** $m_j^d(d_k)$,

THEN K **is** $c_x^{nsr}(min(m_i^l(l_k), m_j^d(d_k)))$. i, j, $x \in [1, 2, 3]$,

where K is the network status indicator as mentioned in Section IV.B. Since $N^d = N^l = 3$, there are totally nine rules for the combined propositions. For an unstable network [3], rule base can be designed pessimistically as shown in Table 1. To avoid performance drops, the unstable network is treated as saturated if any of the two inputs reaches high. For a stable network [3], more optimistic rules can be used.

TABLE I
RULE TABLE OF THE NETWORK STATE RECOGNIZER

Network status indicator K		Packet delay (cycle)		
		Low	Medium	High
Link utilization (packet/cycle)	Low	Empty	Normal	Saturated
	Medium	Normal	Normal	Saturated
	High	Saturated	Saturated	Saturated

During *defuzzification*, a consequence value K (network status indicator) is generated using the centroid method:

$$K = \frac{\sum_{j,i,x,k}^{N^d,N^l,3,S} b_x c_x^{nsr}(min(m_j^d(d_k), m_i^l(l_k)))}{\sum_{j,i,x,k}^{N^d,N^l,3,S} c_x^{nsr}(min(m_j^d(d_k), m_i^l(l_k)))}, \quad (3)$$

where b_x is the center of c_x^{nsr}. The network status indicator K then records in which section (empty, normal, saturated) itself falls and is passed to the fuzzy controller.

V. EXPERIMENT AND RESULTS

A. Experiment Setup

Our design is implemented using C++ in the cycle accurate full system simulator GEMS [13], which aims to characterize and evaluate the performance of multiprocessor systems.

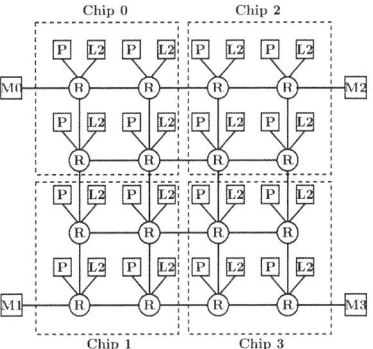

Fig. 7. Experiment CMP architecture

The architecture of the experimental CMP is constructed as shown in Fig 7. Due to space limitation, we do not show FC, NCR, and flow regulator. For their placement please refer to Fig. 1. Each CMP chip includes four processors (denoted by "P", with one L1 cache bank built in) and four L2 cache banks (denoted by "L2"). The L1 cache is private and the L2 cache shared. We simulate four CMP chips, with 16 L1 cache banks (64KB each bank) and 16 L2 cache banks (4MB each bank), totally. Outside the CMP chip exists four external memory banks (denoted by "M", with 1 GB each). The memory module mimics detailed DDR2/DDR3 SDRAMs. The memory coherence protocol is directory based MOESI protocol. The routing algorithm is "X-Y" routing and flow control type is wormhole. The experiments are done with both open-loop and closed-loop measurements [3]. The open-loop measurement focuses on the performance of NoC, while the closed-loop measurement focuses on the performance of the full system.

B. Open-loop Measurement: Results of Synthetic Flows

In the open-loop measurement, we use both uniform random traffic (to simulate load balanced situation) and permutation traffic (to simulate load unbalanced situation). Fig. 8

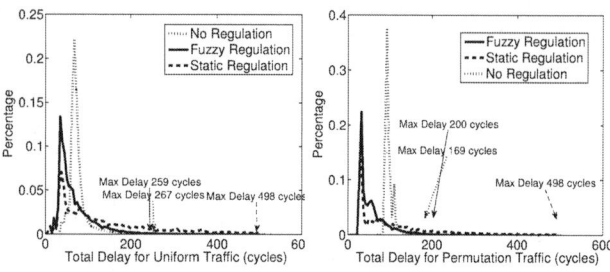

Fig. 8. Delay histogram for open-loop measurement

summarizes the global average packet delay histogram. The fuzzy flow regulation consistently improves the results of static regulation and no-regulation for both traffic patterns. In comparison with static regulation, the improvement in global average packet delay is 68 cycles (48.3% in percentage) for uniform random traffic and 73 cycles (53.0% in percentage) for permutation traffic. In comparison with no regulation, it is 10 cycles (12.6% in percentage) for uniform random traffic and 73 cycles (37.4% in percentage) for permutation traffic. Permutation traffic gets more improvement due to traffic unbalance which stresses the network more than uniform random traffic does. Fig. 8 further shows that the statically regulated flows have a large distribution in delay range from 100 cycles to 500 cycles, which is due to the queueing delay in the regulator's waiting queue. The unregulated flows have the narrowest delay span (from 50 cycles to 250 cycles approximately), but 95% of the packet delays fall in range from 70 to 100 cycles, which is caused by high network transaction delay (since the flow is unregulated). The delay of fuzzy regulated flow spans from 5 cycles to 300 cycles, and the distribution focuses on 50-150 cycles. This gain is due to fuzzy regulation system makes regulation policy based on the status of the network. Unnecessary queueing delay and large network transaction delay are both avoided.

C. Closed-loop Measurement: Results of Benchmarks

In the closed-loop measurement, we experimented with the Stanford ParalleL Applications for SHared Memory benchmark suite 2 (SPLASH-2) to confirm the performance benefits brought by the fuzzy flow regulation system. Of all the 14 benchmarks, we choose to report 4 of them as shown in Fig. 9, which depicts the throughput of the whole experiment CMP system. In benchmark Barnes, the improvement of fuzzy regulation against no regulation is 0.04 packet/cycle (22.2%), and against static regulation is 0.06 packet/cycle

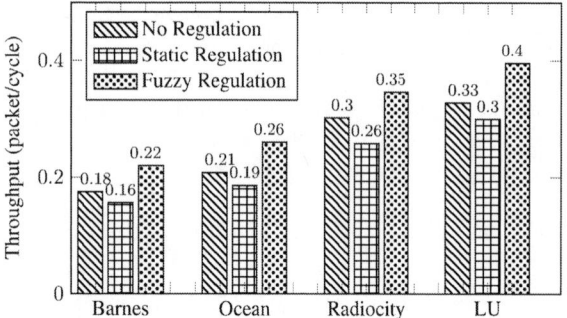

Fig. 9. Experiments with closed-loop measurement

(37.5%). In benchmark Ocean, the improvement of fuzzy regulation against no regulation is 0.05 packet/cycle (23.8%); and against static regulation is 0.07 packet/cycle (36.8%). In benchmark Radiocity, the improvement of fuzzy regulation against no regulation is 0.05 packet/cycle (16.6%), and against static regulation is 0.09 packet/cycle (34.6%). In benchmark LU, the improvement of fuzzy regulation against no regulation is 0.07 packet/cycle (21.2%), and against static regulation is 0.1 packet/cycle (33.3%).

VI. CONCLUSIONS

The central idea of this work is to make network status aware flow regulation through a fuzzy logic approach. On GEMS, our experiments with both synthetic traffic and SPLASH-2 benchmark traces show that the fuzzy regulation can flexibly adjust regulation strength on demand. As a result, it makes more effective use of the system interconnect, achieving significant improvement in average packet delay and throughput.

VII. ACKNOWLEDGMENT

The research is sponsored in part by Intel Corporation through a research gift.

REFERENCES

[1] Z. Lu, Y. Wang, "Dynamic Flow Regulation for IP Integration on Network-on-Chip," *Proceedings of the Sixth IEEE/ACM International Symposium on Network-on-Chip*, May, 2012.

[2] K. Olukotun, L. Hammond, J. Laudon, *Chip Multiprocessor Architecture: Techniques to Improve Throughput and Latency,* Morgan & Claypool Publishers, 2007.

[3] W. J. Dally, B. Towles, *Principles and Practices of Interconnection Networks,* Morgan Kaufman Publishers, 2004.

[4] C. Chang, *Performance Guarantees in Communication Networks,* Springer-Verlag, 2000.

[5] Z. Lu, et al., "Flow Regulation for On-Chip Communication," *Proceedings of the 2009 Design, Automation and Test in Europe Conference (DATE'09),* Nice, France, April 2009.

[6] J.-Y. L. Boudec, P. Thiran, *Network Calculus: A Theory of Deterministic Queueing Systems for the Internet,* No. 2050 in LNCS, 2004.

[7] R. L. Cruz, "A calculus for network delay, part I: Network elements in isolation; part II: Network analysis," *IEEE Transactions on Information Theory,* vol. 37, no. 1, January 1991.

[8] A. K. J. Parekh, "A Generalized Processor Sharing Approach to Flow Control," *PhD. Thesis of The Massachusetts Institute of Technology,* 1992.

[9] J. Schmitt, F. Zdarsky, and L. Thiele, "A comprehensive worst-case calculus for wireless sensor networks with in-network processing," *IEEE Real-Time Systems Symposium (RTSS07),* 2007.

[10] Y. Jiang, "A basic stochastic network calculus," *ACM SIGCOMM Comput. Commun. Rev.,* vol. 36, pp. 123134, August 2006.

[11] D. Bertsekas and R. Gallager, *Data Networks,* Prentice Hall, Englewood Cliffs, NJ, 1991.

[12] M. Gerla and L. Kleonock, "Flow control: A comparative survey," *IEEE Transactions on Communications,* COM-28(1980), pp. 553-574.

[13] M. M. K. Martin, et al., "Multifacet's General Execution-driven Multiprocessor Simulator (GEMS) Toolset," *Computer Architecture News (CAN),* September 2005.

[14] L.A. Zadeh, "Fuzzy Logic," *Stanford Encyclopedia of Philosophy,* Stanford University, 2006.

[15] L.A. Zadeh, "Fuzzy Sets," *Information and Control,* 1965.

[16] K. M. Passino, S. Yurkovich, *Fuzzy Control,* Addison-Wesley, 1997.

[17] G. C. Goodwin, S. F. Graebe, M. E. Salgado, *Control System Design,* Valparaiso, 2000.

[18] J. L. Hennessy, D. A. Patterson, *Computer Architecture: A Quantitative Approach, Fifth Edition,* Morgan Kaufmann, 2010.

[19] G. J. Klir, B. Yuan, *Fuzzy Sets and Fuzzy Logic: Theory and Applications,* Prentice-Hall, Englewood Cliffs, NJ, 1995.

Adjustable Contiguity of Run-Time Task Allocation in Networked Many-Core Systems

Mohammad Fattah, Pasi Liljeberg, Juha Plosila, Hannu Tenhunen
Department of Information Technology, University of Turku, Turku, Finland
Email: {*mofana, pakrli, juplos, hanten*}@*utu.fi*

Abstract—In this paper, we propose a run-time mapping algorithm, CASqA, for networked many-core systems. In this algorithm, the level of contiguousness of the allocated processors (α) can be adjusted in a fine-grained fashion. A strictly contiguous allocation ($\alpha = 0$) decreases the latency and power dissipation of the network and improves the applications execution time. However, it limits the achievable throughput and increases the turnaround time of the applications. As a result, recent works consider non-contiguous allocation ($\alpha = 1$) to improve the throughput traded off against applications execution time and network metrics. In contradiction, our experiments show that a higher throughput (by 3%) with improved network performance can be achieved when using intermediate α values. More precisely, up to 35% drop in the network costs can be gained by adjusting the level of contiguity compared to non-contiguous cases, while the achieved throughput is kept constant. Moreover, CASqA provides at least 32% energy saving in the network compared to other works.

Keywords—*Processor allocation; Application Mapping; Dynamic Many-Core Systems; Contiguous Task Mapping;*

I. INTRODUCTION

According to the International Technology Roadmap for Semiconductors [1], by 2020 Multi-Processor Systems-on-Chip (MPSoCs) will integrate hundreds of processing elements (PEs) connected by a Network on chip [2] (NoC) based communication infrastructure. NoCs provide a regular platform for connecting the system resources and make the communication architecture scalable and flexible compared to traditional bus or hierarchical bus architectures.

Such many-core systems will feature an extremely dynamic workload where an unpredictable sequence of different applications enter and leave the system at run-time. Applications are made up of a set of concurrent tasks which communicate to synchronize and exchange data among each other. This featured dynamic nature is handled through the mapping function of the system manager. This is to allocate available system resources to the tasks of applications at run-time.

Communication pattern of underlying NoC is directly influenced by the way communicating tasks are allocated to the PEs. The impact of mapping strategy on the performance of system is emphasized in many-core systems [3] as well as in recent supercomputers [4]. Placing communicating tasks in distant proximity will increase the power dissipation [5] and the congestion probability of the the network. In the presence of congestion, where different messages contend for the same network resources, the path length becomes an issue especially for large packet sizes [6]. Studies show up to eight fold increase in the message latency [4] and a doubled execution time of applications [7], in the absence of sophisticated allocation methods.

Accordingly, a contiguous allocation is preferred in which tasks of an application are settled on close proximity. Regarding the dynamic nature of many-core systems, however, a contiguous solution might not be always possible for an incoming application. Limiting to only contiguous solution, hence, keeps the applications waiting unnecessarily until the required contiguous area becomes available. This will degrade the achievable throughput (the total number of applications that complete their execution per time unit) and the turnaround time (total time between submission of an application and its completion) of each application. As a result, state-of-the-art dynamic mapping algorithms [8]–[11] allow a non-contiguous processor allocation with no limit on the level of dispersion.

In this paper we propose an allocation algorithm, CASqA, in which the level of desired contiguity (denoted by α) can be adjusted in fine-grain between strictly contiguous ($\alpha = 0$) to unlimited non-contiguous ($\alpha = 1$) solutions. When a contiguous solution is not possible, the application is mapped only if it can be mapped within the desired contiguity. This establishes a balance between the increased waiting time of applications (contiguous mapping) and the crippled performance (non-contiguous solutions). Interestingly, results show that the maximum throughput is not achieved in non-contiguous allocation ($\alpha = 1$), but when using intermediate α values ($0 < \alpha < 1$). Accordingly, the same throughput as non-contiguous mappings can be achieved with up to 30% improvement in the network costs. Note that in this paper we equivalently use the terms "the level of allowed dispersion" and "the level of desired contiguity" to express α.

In addition to the level of contiguity, the order in which tasks of an application are placed, within an area, affects the network costs and thus the application performance. Consequently, CASqA considers the pattern of the communication within an application tasks while allocating the processors. To this end, CASqA is equipped with a single evaluation metric representing both the power dissipation and congestion probability of the network incurred by the candidate mapping solutions.

The rest of the paper is organized as follows: In Section II the mapping problem as well as the contiguousness and power-congestion metrics are formally defined, while Section III discusses related work. Section IV describes our novel contiguity adjustable square allocation (CASqA) method. The simulation setups along with the effect of adjusting the level of desired contiguity (α) on the system performance is presented and discussed in Section V. Moreover, this Section illustrates a comparison with other state-of-the-art methods. Finally, Section VI concludes the paper and discusses some potential future works.

978-1-4799-2817-0/14 $31.00 © 2014 IEEE

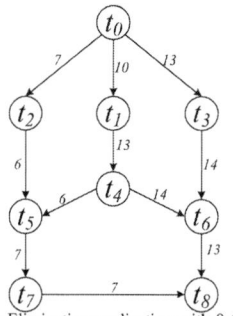

Fig. 1: A Gaussian Elimination application with 9 tasks and 11 edges.

II. DEFINITIONS

A. Problem Definition

Mapping algorithms try to optimally allocate system resources to the concurrent tasks of a requested application. Each application in the system is represented by a directed graph denoted as a task graph $Ap = TG(T, E)$. Each vertex $t_i \in T$ represents one task of the application, while the edge $e_{i,j} \in E$ stands for a communication between the source task t_i, and the destination task t_j. Task graph of a Gaussian Elimination application [12] which is extracted using TGG [13] is shown in Fig. 1. Note that $w_{i,j}$ of an edge $e_{i,j}$, i.e. the amount of data transferred from task t_i to t_j, is indicated on each edge.

An architecture graph $AG(N, L)$ describes the communication infrastructure of the processing elements. We consider a simple $W \times H$ 2D-mesh NoC with the XY wormhole routing (Fig. 2(a)). The AG contains a set of nodes $n_{w,h} \in N$, connected together through unidirectional links $l_k \in L$. Each node is the combination of a PE connected to the router.

Mapping of an application onto the system is defined as a one-to-one function from the set of application tasks to the set of nodes:

$$map : T \rightarrow N, s.t. map(t_i) = n_{w,h}; \forall t_i \in T, \exists n_{w,h} \in N \quad (1)$$

Fig. 2(a) illustrates a possible mapping of the application in Fig. 1, onto the described platform. For simplicity, we denote a node where a task, t_i, is mapped onto as nt_i and the routing path used by communication edge $e_{i,j}$ as $pck_{i,j}$; i.e. the traversed path by the packet sent from nt_i to nt_j. Note that due to XY routing assumption, the path can be determined once the application is mapped.

B. Evaluation Metrics

As mentioned before, traffic congestion and power dissipation are two main concerns of on-chip domain. Congestion increases the network latency dramatically [6] and cripples the application performance. The power dissipation and cooling issues are also known as the limiting walls [14], [15] of future silicon technologies.

The energy dissipation of the network related to an obtained mapping is a function of both the volume ($w_{i,j}$) and the traversed path ($pck_{i,j}$) of the communications [5]; while the congestion is related to the packets sharing the same path along their delivery. Contiguous mapping kills two birds with one stone. (i) It diminishes the network power consumption by decreasing the $pck_{i,j}$ lengths of the mapped application. (ii) It isolates the data communication of each application and hence the congestion among them (external congestion). As mentioned, however, limiting the mapping algorithm to only contiguous solutions will degrade the system throughput and applications turnaround time. Accordingly, recent works allow non-contiguous mapping.

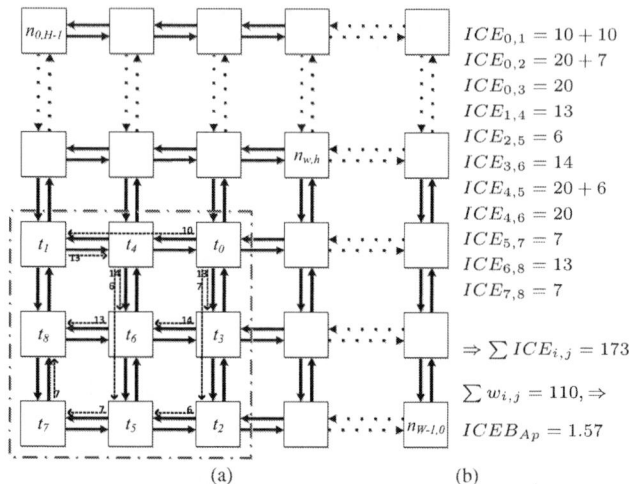

Fig. 2: (a) Mesh-Based platform with an application mapped onto it (the highlighted region.); and (b) $ICEB$ calculation of the mapped application

Several works (e.g. [9], [10], [16]–[18] etc.) considered the average pairwise Manhattan distance (MD, also known as L1 distance) of allocated nodes as a metric (called MRD in [19]) to assess the contiguity of a mapped applications. Bender et al. [20] showed that the most contiguous region, i.e. with the least MRD value, would be almost circular. Regarding the mesh topology of the network, however, a circular processor allocation will generate irregularity in remaining available nodes and more area fragmentation in long term.

As an alternative to circular; rectangular allocation forms regular regions, and isolates data communications of (and thus decreases the congestion probability between) different applications. Targeting both the contiguity and regularity of allocation method leads to square shape which is a rectangle with the smallest MRD. It is proved in [17] that a square allocation is a 7/4-approximation in the worst case in contrast to MRD minimization which is 2-approximation.

The preference of square allocation over circular is also observed in [19]. The $NormalizedMRD$ is provided as a metric to asses both the squareness and contiguousness of a mapped application, independent of the application size ($|T|$):

$$NMRD_{map(Ap)} = 1 + \frac{|MRD_{map(Ap)} - MRD_{SQ(|T|)}|}{MRD_{SQ(|T|)}}$$
$$(2)$$

where $MRD_{SQ(|T|)}$ is the MRD value for a square with $|T|$ nodes. The $NMRD$ value of 1 means a convex and squared area. $NMRD$ increases as the mapped area is getting more fragmented and/or less similar to a square shape.

As a result, CASqA collects required number of available PEs within a square shape as long as the allowed level of dispersion is respected. In addition to the level of allowed dispersion, the order in which tasks of an application are mapped onto a set of nodes will affect the energy dissipation and congestion probability of the network. This is referred to as internal congestion probability which stands for the network links shared among the data flows of the same application.

We define the *Internal Congestion and Energy (ICE)* of an edge, $e_{i,j} \in E$, in a mapped application as:

$$ICE_{i,j} = \sum_{\forall l_k \in pck_{i,j}} w_{l_k} \quad (3)$$

where, w_{l_k} is the total value of $w_{i,j}$ of edges which their $pck_{i,j}$ includes the link l_k. Let us follow the ICE calculation of $e_{0,2}$ for the mapped application in Fig. 2(a). The path between nt_0 and nt_2 ($pck_{0,2}$) is composed of two links:l_{nt_0,nt_3} and l_{nt_3,nt_2}. The first link is utilized by two edges: $e_{0,3}$ and $e_{0,2}$, so its w_{l_k} value is $w_{0,3} + w_{0,2} = 20$. The w_{l_k} for the second link is 7 as it is utilized only by $e_{0,2}$. The $ICE_{0,2} = 27$ is calculated by the summation of two w_{l_k} values.

According to the definition, the w_{l_k} value of a given link increases as being shared among several packets of an application. This reflects the increased probability of internal congestion. Moreover, the $ICE_{i,j}$ of a given data flow relates to the MD of corresponding nodes (nt_i and nt_j). This reflects the dissipated energy in the network. Consequently, *ICE per Bit* ($ICEB$) value for a mapped application is given by the sum of *ICE* values of all edges averaged by the total weights of the edges:

$$ICEB_{map(Ap)} = \frac{\sum ICE_{i,j}}{\sum w_{i,j}} \qquad (4)$$

The larger the $ICEB$ value, the higher energy dissipation and congestion probability. The $ICEB = 1$ means that all communications of the application are handled in one hop distance without sharing any common link. Fig. 2(b) shows the $ICEB$ evaluation of the example mapping of Fig. 2(a). Later in Section IV, CASqA uses $ICEB$ metric to select among available candidate nodes for allocating a task.

III. RELATED WORK

There are several works dealing with management of dynamic workload in massively-parallel systems. Early works in the supercomputer domain map applications only onto convex set of nodes. While recent works focus on non-contiguous processor allocation methods for supercomputers and many-core systems. In the following we study some of them in both domains. They start the allocation from an available *first node* and allow non-contiguous allocation.

Bender et al. [17] presented the MC1x1 method for processor allocation problem. Starting from an available node, MC1x1 explores the smallest square around that node; i.e. the square with radius 1; and collects all available nodes. The square radius is increased until the required number of available nodes is found. They start the same procedure from all available nodes, and finally select the *first node* which resulted in the smallest MRD. However, the communication pattern (task graph) of the application is not considered in their task allocation.

Carvalho et al. [21] presented Nearest Neighbor (NN) heuristic. NN maps each task of an application onto the nearest free neighbor of the task communicating with. This algorithm considers neither the energy dissipation nor the congestion. Later, they introduce Best Neighbor (BN) [8] which selects the best nearest free neighbor according to communication weights. They do not provide any method to converge the mapping area around the selected *first node* which significantly increases the congestion.

Chou et al. [10], in their incremental approach, broke down the mapping problem into two steps: region selection, and task allocation. First, in the region selection, they try to find a set of nodes to minimize MRD for both the application region and remaining nodes. Afterwards, the application tasks are mapped onto the chosen region to minimize the energy dissipation. Their method, however, results in increased dispersion, as they frequently might select an isolated *first node* in their region selection.

Algorithm 1 CASqA pseudo-code in expanding the current radius according to the level of allowed dispersion.

Inputs:
$TG(T, E)$: Task graph of the requested application, Ap.
t_f and n_f: The first task of the application and *first node* of the allocation process.
α: The allowed level of dispersion, $0 \leq \alpha \leq 1$.

Output:
$map : T \to N$.

Variables:
UNM, MET and MAP: Sets of unmet, met and mapped tasks.
R_{max}: The maximum radius the application is allowed to get dispersed around the n_f.
r: The current radius of the square around the n_f.

Body:
1: $nt_f \leftarrow n_f$;
2: $MAP \leftarrow t_f$;
3: $MET \leftarrow$ set of tasks connected to t_f in TG;
4: $UNM \leftarrow T - (MAP + MET)$;
5: **for** $r = 1 \to R_{max}$ **do**
6:     allocate the available nodes within the current radius (Algorithm 2);
7:     **if** $MAP = T$ **then**
8:         **return** *success*;
9:     **else if** $r = R_{max}$ **and** $|T| - |MAP| < \tau$ **then**
10:         $R_{max} \leftarrow R_{max} + 1$;
11:         $\tau \leftarrow \tau \times \alpha$;
12:     **end if**
13: **end for**
14: **return** *failure*;

Fattah et al. [9] aim at minimizing the internal congestion via CoNA approach. CoNA selects the node which has the largest number of free neighbors as the *first node* in the allocation. Then the task with largest degree is mapped onto that node. The task graph is traversed in breadth-first order, and neighbors making smaller squares are preferred. Later, the effect of *first node* selection method on the mapping performance is studied [19]. They show that a square shape is preferred for processor allocation. However, the utilized mapping algorithm, CoNA, does not collect the available nodes in a square shape.

There are several other works dealing with the dynamic workload management of many-core systems and/or supercomputers. They either limit the allocation to contiguous set of nodes or allow non-contiguous allocation with no limits. However, to the best of our knowledge, this is the first work allowing an adjustable level of allowed dispersion in the processor allocation. Results show that this improves the network costs while keeping the throughput as high as in non-contiguous solutions.

IV. CASqA MAPPING ALGORITHM

The pseudo-code of the proposed mapping algorithm with adjustable contiguity is listed in Algorithm 1[1]. The mapping is started by exploring the available nodes in the smallest square around the *first node*. CASqA expands the radius by one only after all the available nodes in the current radius are allocated.

Expanding the current radius continues up to the maximum radius, R_{max}, which is initialized according to (5). This is the minimum radius for the square which can fit $|T|$ nodes. The R_{max} value is then adapted at run-time (line 9–12) according to the level of allowed dispersion which is adjusted through the α. The α is a real value, where $\alpha = 0.0$ means a tight mapping while $\alpha = 1.0$ stands for no limit on R_{max}.

$$R_{max} = \left\lfloor \frac{\left\lceil \sqrt{|T|} \right\rceil}{2} \right\rfloor \qquad (5)$$

---

[1]The source code is available at http://users.utu.fi/mofana/CASqA.html

The maximum radius in which CASqA looks around for available nodes (R_{max}) is adapted through the τ threshold. It shows the maximum number of tasks that can be remained unmapped in order to allow the increase of R_{max}. In other words, once the current radius reaches the R_{max}, more dispersion can happen only if a limited number of tasks (max. τ) are remained for mapping (line 9). At each iteration of increasing the R_{max}, the τ threshold is tightened by multiplying it by α. Note that, the initial value for τ is $|T| \times \alpha$.

For instance, having $\alpha = 0.0$ keeps $\tau = 0$ from the beginning and the R_{max} is never increased from its initial value in (5). Hence, CASqA will map the application only if the required number of available nodes ($|T|$) can be collected within the smallest square. Otherwise, it will return a *failure* (line 14) and the mapping will be restarted from another *first node* or paused until the required area is available. On the other hand, when $\alpha = 1.0$ the τ is always kept to $|T|$ and thus the R_{max} can be increased with no limits until all the tasks are mapped onto the system.

As mentioned before, in addition to the provided adjustable contiguity, CASqA deals with the order in which tasks of the application are mapped onto the system nodes. The following subsection describes the CASqA approach in reducing the power dissipation and congestion probability of the network via $ICEB$ metric.

A. Task Mapping Details

CASqA assumes the task graph to be undirected, and meets and maps tasks through their predecessor tasks (called parents). For a given application, there are three sets of task: Tasks that are already mapped onto the system (MAP), tasks that are met through their mapped parent but are not mapped yet (MET), and the tasks that are not met yet and thus not mapped (UNM).

Through the initializing phase of the algorithm (lines 1–4), the *first node* is allocated to the first task of the application. The first task is added to MAP set and the tasks connected to it are added to MET set. The rest of the tasks are also moved to UNM set. Note that a basic random algorithm or a sophisticated heuristic, like SHiC [19], can be utilized for the *first node* selection. The task with the maximum degree is also selected as the first task.

The pseudo-code of CASqA for mapping application tasks onto the available nodes of the current radius, r, is listed in Algorithm 2. Within the current radius, MET tasks are preferred to get mapped onto the closest node to their parent. Accordingly, for each MD value, all MET tasks are examined for possible mappings. Given a MET task (t_c), list of available candidate nodes ($\tilde{N}$) in the current radius which are MD hop far from parent of t_c is extracted (line 3). If at least one candidate node exists to map t_c onto it (line 4), then the one which leads to minimum value of $ICEB$ is selected and the task is moved from the MET set to the MAP set (lines 5–6). Consequently, those of unmet tasks (UNM) that exchange data with the t_c will be moved to the MET set (lines 7–8). Now that new tasks are added to MET set, it might be possible to map them within one hop distance from their parent, t_c. Hence, the MD will be reset to one (line 9).

Let us follow the mapping process of the Gaussian Elimination application of Fig. 1 (with 9 tasks) onto the system configuration as shown in Fig. 3 (a)–(f). According to (5), the initial value for R_{max} will be 1. Since there are 8 nodes

Algorithm 2 *CASqA pseudocode in mapping application tasks within the available nodes of the current radius.*

```
 1: for MD = 1 → 4 × r do
 2:     foreach t_c ∈ MET do
 3:         Ñ ← available nodes in current radius with MD hop distance from t_c
                parent;
 4:         if Ñ ≠ ∅ then
 5:             nt_c ← the node of Ñ which results in a smaller ICEB value;
 6:             move t_c from MET to MAP;
 7:             if there are tasks in UNM connected to t_c then
 8:                 move tasks connected to t_c from UNM to MET;
 9:                 MD ← 1;
10:             end if;
11:         end if
12:     end for
13: end for
```

available within this radius around the selected *first node* CASqA will need to increase the search radius, R_{max}, to collect one available node beyond the first square. Hence, CASqA cannot map the application if one sets $\alpha \leq 1/9$; i.e. the initial value for $\tau \leq 1$. For the sake of simplicity, we assume $\alpha = 1.0$.

In the beginning, first task of the application (t_6) is mapped onto the *first node*, its connected tasks $\{t_3, t_4, t_8\}$ are added to MET set, and both r and MD are initialized to 1, Fig. 3 (a). As the next step, $t_3 \in MET$ can be mapped onto the smallest square ($r = 1$) around the *first node* with 1 hop distance ($MD = 1$) from its parent, t_6. All the candidate nodes (asterisk ones) result in the same $ICEB$. Hence, t_3 is mapped onto one of them alternatively and moved from MET set to MAP set, Fig. 3 (b). Consequently, t_0 is added to MET set as it is connected to t_3 in the task graph. Within next three iterations, MET tasks t_4, t_8, t_0 are mapped one by one with r and MD values equal to 1 and new tasks are MET through them, as shown in Fig. 3 (c). Note that r and MD values are still equal to 1.

Now, two nodes (asterisks in Fig. 3 (c)) can be selected for t_1, regarding the r and MD values. As t_1 also exchanges data with t_0, the bottom node results in less $ICEB$ value and thus is selected for t_1. Afterwards, t_5 is mapped onto the upper node of t_4, as it is the only node available regarding the current r and MD values. The system configuration after t_1 and t_5 being mapped is shown in Fig. 3 (d). Now that none of the MET tasks can be mapped onto 1 hop distance from their parents, the MD value is increased to 2. Consequently, t_2 is mapped within 2 hop distance from its parent (Fig. 3 (e)). Now that there is not any node available in the square with $r = R_{max} = 1$ and $\alpha = 1$, the r and R_{max} are increased to 2 and the MD is reset to 1. The only unmapped task is t_7 which should be mapped onto t_8 neighborhood. As it is not feasible to map t_7 within one hop distance from t_8, the MD is increased to 2 and t_7 is mapped in the last step, Fig. 3 (f).

Accordingly, the proposed CASqA algorithm lets adjust the level of expected contiguity of allocated nodes and tries to reduce the power dissipation and congestion probability of applications. In the next Section, we study the throughput and performance of the system, regarding different levels of contiguity.

V. RESULTS AND ANALYSIS

In this section, we assess our CASqA mapping algorithm in two main aspects. First we evaluate the impact of adjusting the level of desired contiguity on the system performance. Then we compare CASqA with other state-of-the-art methods. Several applications with 4 to 35 tasks are generated using TGG

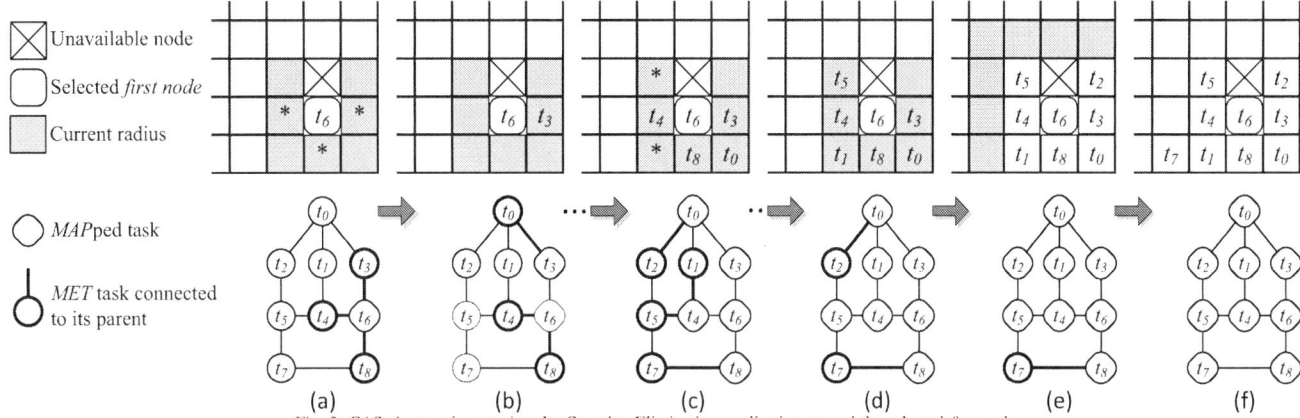

Fig. 3: CASqA steps in mapping the Gaussian Elimination application around the selected *first node*.

[13] where the communication volumes ($w_{i,j}$) are randomly distributed between 2 to 16 flits of data. Experiments are performed on our in-house cycle-accurate SystemC many-core platform which utilizes a pruned version of Noxim [22], as its communication architecture. A 16×16 network is instantiated, and each set of experiments are performed millions of cycles where thousands of applications enter and leave the system.

A random sequence of applications are entered the scheduler FIFO. This sequence is kept fixed in all experiments for the sake of fair comparison. Applications are scheduled based on First Come First Serve (FCFS) policy, and an application is scheduled if and only if there is enough available nodes in the system. An allocation request for the scheduled application is sent to the central manager (CM) of the platform residing in the node $n_{0,0}$. CM selects the *first node* using SHiC [19] method, and executes CASqA algorithm with the target α (denoted as CASqA$^\alpha$). Upon a mapping failure, new *first nodes* are selected until the current application being mapped. The successfully mapped application is allocated to the corresponding nodes, where each node emulates the behavior of its allocated task.

A. Contiguity Adjustment

In this subsection we study the effect of the level of allowed dispersion on the system performance. To this end, different experiments are done with α varying from 0 to 1 by 0.1 steps. The next application is pushed into the scheduler FIFO as soon as there is enough resources available for the application. This is to achieve the maximum possible throughput of the system. Accordingly, applications do not experience any waiting time in CASqA$^{1.0}$. Note that the turnaround time of an application is calculated from the moment it is pushed into the scheduler FIFO. While, the execution time is calculated from the moment the application tasks are allocated to the nodes.

A summary of the obtained results is demonstrated in Fig. 4. As can be seen, limiting the system to only contiguous allocations ($\alpha = 0$) minimizes the execution time of the applications. However, the achievable throughput is degraded by 17% as a result of the increased turnaround time.

Although releasing CASqA from contiguous allocations will improve the achievable throughput and decreases the turnaround time, this is not the case for $\alpha > 0.6$. Afterwards, the increased execution time will affect the turnaround time and the system throughput starts degrading. Note that the amount of the throughput degradation is less that 3%. In other

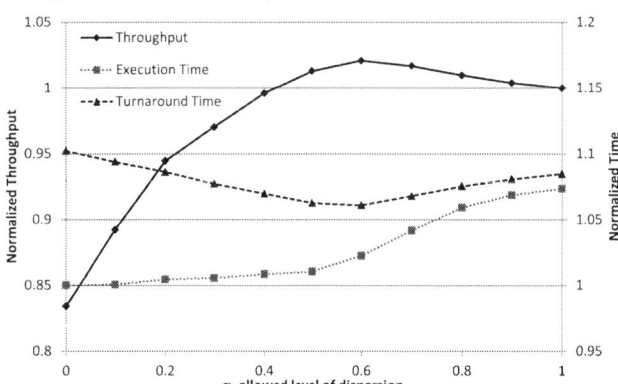

Fig. 4: Normalized achievable throughput across different α values. The normalized average execution and turnaround time of applications are also shown in the right axis.

words, the throughput gain of CASqA$^{0.6}$ over CASqA$^{1.0}$ is quite negligible.

The main conclusion is that the same throughput as CASqA$^{1.0}$ can be achieved when $\alpha \approx 0.4$. This equal throughput is achieved while there is almost 6% improvement in execution time, and 25 and 35 % gain in the latency and energy dissipation of the network, respectively, as shown in Fig. 5.

This is worth mentioning that in only 8% of the mapping failures of CASqA$^{0.4}$, trying new *first nodes* within the same configuration leads to mapping success. Put differently, in 92% of the mapping failures, the required contiguity is obtained only when some nodes get available (their application execution is completed). This proves that the obtained gains are not due to the change of the *first node*, but because of the adjusted

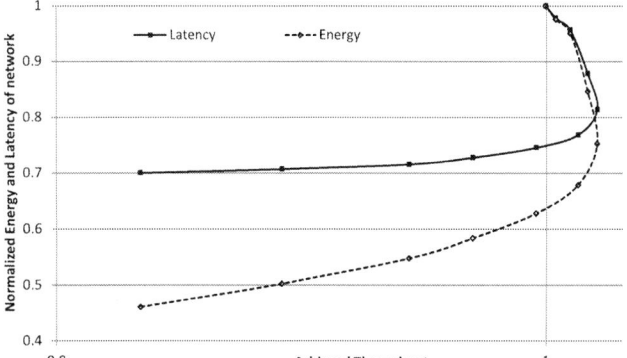

Fig. 5: The normalized dissipated energy and latency of the network versus the achieved throughput.

TABLE I: Results of Different Mapping Algorithms

Mapping	$NMRD$	$E_{norm.}$	L_{avg}	%$Congestion$
CASqA$^{0.5}$	1.13	1.00	41.00	30.52
CASqA$^{0.5*}$	1.13	1.25	48.23	37.70
CoNA	1.69	1.48	50.32	38.40
NN	1.97	1.56	55.53	41.98

contiguity. Moreover, according to Fig. 4, there is around 1% time overhead for mapping of an application and allocating its tasks to the selected nodes. This is the difference between the execution and turnaround time in CASqA$^{1.0}$.

B. ICEB Effect and Comparison

In this subsection, we make a comparison between CASqA and other state-of-the-art mapping algorithms and study the effect of the order of task placement within a area (Algorithm 2). Accordingly, several experiments are performed with exploiting different mapping algorithms: CASqA$^{0.5}$, NN [21] and CoNA [19]. Note that the experiments for CASqA case are executed twice, once (denoted as CASqA$^{0.5*}$) with arbitrary mapping of tasks onto the available nodes of the current radius. The rate in which the random sequence of applications are injected to the scheduler FIFO is set to the achieved throughput for $\alpha = 1.0$ in the previous subsection.

The extracted results are summarized in Table I. The normalized values for the dissipated energy of the network is shown in $E_{norm.}$ column, while the next column indicates the average latency, in clock cycles, of the network. The last column is the percentage of data packets experienced congestion along their path. As can be seen, thanks to the moderated dispersion, CASqA leads to at least 32% energy saving as well as 19% improvement in average latency of the network.

The comparison between two first rows, emphasizes the importance of tasks arrangement ($ICEB$ metric) within a area. However, the two middle rows show the superiority of contiguous mapping. Arbitrary task arrangement does not degrade the performance as non-contiguous allocation does.

Last but not least, CASqA$^{0.5}$ reduces the standard deviation and worst case latency of the network by 2 and 4 times, respectively, compared to CoNA. The reduction is 2.2 and 10.5 times for CASqA$^{0.0}$. This shows the potential of incorporating the run-time requirements into the selection of α values to deliver QoS at system-level. More details are left as future work.

VI. CONCLUSION AND FUTURE WORK

In this paper, we proposed a contiguity adjustable square allocation (CASqA) method. The proposed method attempts to form a square shaped area of available nodes around the selected *first node*. CASqA expands the exploration square according to the level of desired contiguity at run-time.

Experiments showed around 35% improvement versus non-contiguous approaches with the same throughput. Moreover, CASqA is equipped with $ICEB$ metric which resulted in improved network performance characteristics.

Regarding the variety of user demands and the application requirements, a different level of contiguousness can be suited for each run-time case. As a future work, we plan to incorporate the run-time criteria into α adjustment. Accordingly, a system-level adaptive approach, for e.g. QoS management, can be obtained. This will lead to increased performance while delivering the required functionality to each application.

REFERENCES

[1] S. I. Association et al., "International technology roadmap for semiconductors (ITRS), 2011 edition," 2011.

[2] W. Dally and B. Towles, "Route packets, not wires: on-chip interconnection networks," in *Design Automation Conference, 2001. Proceedings*, 2001. pp. 684–689.

[3] A. K. Singh et al., "Mapping on multi/many-core systems: survey of current and emerging trends," in *Proceedings of the 50th Annual Design Automation Conference*, 2013, pp. 1:1–1:10.

[4] A. Bhatele and L. Kale, "An evaluative study on the effect of contention on message latencies in large supercomputers," in *Parallel Distributed Processing, 2009. IPDPS 2009. IEEE International Symposium on*, 2009, pp. 1–8.

[5] T. Ye, L. Benini, and G. De Micheli, "Analysis of power consumption on switch fabrics in network routers," in *Design Automation Conference, 2002. Proceedings. 39th*, 2002, pp. 524–529.

[6] J. W. van den Brand et al., "Congestion-controlled best-effort communication for networks-on-chip," in *Proceedings of the conference on Design, automation and test in Europe*, 2007, pp. 948–953.

[7] V. Leung et al., "Processor allocation on Cplant: achieving general processor locality using one-dimensional allocation strategies," in *Cluster Computing, 2002. Proceedings. 2002 IEEE International Conference on*, 2002, pp. 296–304.

[8] E. de Souza Carvalho, N. Calazans, and F. Moraes, "Dynamic Task Mapping for MPSoCs," *Design Test of Computers, IEEE*, vol. 27, no. 5, pp. 26–35, 2010.

[9] M. Fattah et al., "CoNA: Dynamic Application Mapping for Congestion Reduction in Many-Core Systems," in *Computer Design (ICCD), 2012 IEEE 30th International Conference on*, 2012, pp. 364–370.

[10] C.-L. Chou, U. Ogras, and R. Marculescu, "Energy- and Performance-Aware Incremental Mapping for Networks on Chip With Multiple Voltage Levels," *Computer-Aided Design of Integrated Circuits and Systems, IEEE Transactions on*, vol. 27, no. 10, pp. 1866–1879, 2008.

[11] S. Kobbe et al., "DistRM: distributed resource management for on-chip many-core systems," in *Proceedings of the seventh IEEE/ACM/IFIP international conference on Hardware/software codesign and system synthesis*, 2011, pp. 119–128.

[12] A. Amoura, E. Bampis, and J.-C. Konig, "Scheduling algorithms for parallel Gaussian elimination with communication costs," *Parallel and Distributed Systems, IEEE Transactions on*, vol. 9, no. 7, pp. 679–686, 1998.

[13] "TGG: Task Graph Generator," *URL: http://sourceforge.net/projects/taskgraphgen/*, 2010.

[14] A. Agarwal and M. Levy, "The KILL Rule for Multicore," in *Design Automation Conference, 44th ACM/IEEE*, 2007, pp. 750–753.

[15] H. Esmaeilzadeh et al., "Dark silicon and the end of multicore scaling," in *38th Annual International Symposium on Computer Architecture (ISCA)*, 2011, pp. 365–376.

[16] J. Mache and V. Lo, "Dispersal Metrics for Non-Contiguous Processor Allocation," Tech. Rep., 1996.

[17] M. A. Bender et al., "Communication-Aware Processor Allocation for Supercomputers: Finding Point Sets of Small Average Distance," *Algorithmica*, vol. 50, no. 2, pp. 279–298, Jan. 2008.

[18] J. Mache, V. Lo, and K. Windisch, "Minimizing message-passing contention in fragmentation-free processor allocation," in *In Proceedings of the 10th International Conference on Parallel and Distributed Computing Systems*, 1997, pp. 120–124.

[19] M. Fattah et al., "Smart Hill Climbing for Agile Dynamic Mapping in Many-Core Systems," in *Proceedings of the 50th Annual Design Automation Conference*, 2013, pp. 39:1–39:6.

[20] C. M. Bender et al., "What is the optimal shape of a city?" *Journal of Physics A: Mathematical and General*, vol. 37, no. 1, p. 147, 2004.

[21] E. Carvalho, N. Calazans, and F. Moraes, "Heuristics for Dynamic Task Mapping in NoC-based Heterogeneous MPSoCs," in *Rapid System Prototyping, 2007. RSP 2007. 18th IEEE/IFIP International Workshop on*, 2007, pp. 34–40.

[22] F. Fazzino, M. Palesi, and D. Patti, "Noxim: Network-on-chip simulator," *URL: http://sourceforge.net/projects/noxim*, 2008.

978-1-4799-2817-0/14 $31.00 © 2014 IEEE

STD-TLB: A STT-RAM-based Dynamically-configurable Translation Lookaside Buffer for GPU Architectures*

Xiaoxiao Liu, Yong Li, Yaojun Zhang, Alex K. Jones, Yiran Chen

Department of Electrical and Computer Engineering, University of Pittsburgh, Pittsburgh, PA, 15261, USA

$\{xil116, yol26, yaz24, akjones, yic52\}$@pitt.edu

Abstract—Translation lookaside buffer (TLB) was recently introduced into modern graphics processing unit (GPU) architectures to support virtual memory addressing. Compared to CPUs, the performance of GPUs is more sensitive to the capacity of TLBs because of heavier memory accesses. However, large SRAM cell area greatly limits the implementable capacity of conventional SRAM-based TLBs. In this work, we propose using STT-RAM to construct TLBs in light of the unique memory access pattern in GPUs, i.e., infrequent data updates. STT-RAM TLB can replace its same-area SRAM counterpart with greater capacity, similar read performance and lower energy consumption. As an optimization of STT-RAM TLB, we further propose a STT-RAM-based dynamically-configurable TLB (STD-TLB) by leveraging differential sensing technique. STD-TLB can switch between high-capacity mode and high-performance mode on-the-fly based on real-time application needs. Our experiments show that compared to SRAM TLB, standard STT-RAM TLB improves the performance and energy delay product of GPU address translation by 32% and 75%, respectively, while STD-TLB achieves additional 15% and 13% improvements over standard STT-RAM TLB.

I. INTRODUCTION

Due to its outstanding parallel computing capability [1, 2], graphic processing unit (GPU) has been widely used not only in pure 3D graphic processing but also as the complimentary computing accelerators to CPUs. Virtual memory is implemented in GPUs to support a unified address space. The recent industry examples allow GPUs to be programmed more easily by releasing the software developer from the knowing the details on the memory management mechanism in parallel GPU computations, e.g., AMD Graphics Core Next (GCN) architecture [3] and NVIDIA Fermi architecture [4]. However, the advantages of using unified virtual address space in GPUs come with tradeoffs such as the need for virtual address translation and the deployment of translation lookaside buffer (TLB) on the critical path of memory accesses.

TLBs are used to store virtual-to-physical page addresses for address translation speedup in modern processor architectures. Because of the expensive miss penalty of TLBs [5–7], the performance of memory hierarchy is highly dependent on TLB hit ratio. Although TLBs were extensively studied in CPU designs [8–10], relatively few research efforts have made on the TLB design and optimization in GPUs, which require re-evaluations for the inherent difference between GPU and CPU. By handling a large volume of parallel processing threads, GPUs have much more frequent and heavier memory accesses than CPUs [11]. TLBs in a GPU, especially the last level TLB, must be sufficiently large to retain enough

physical page addresses and fast to achieve a high address translation performance. For example, Nvidia's *GT200* GPU, which was launched in 2008, has 4096 entries in L2 TLB for texture memory and 8096 L2 TLB entries for global memory, respectively [12]. As a comparison, in Intel's latest CPU Core i7 where each socket has 1 to 8 cores, each core has only 512 entries in the shared secondary DTLB [13]. TLBs are commonly implemented with SRAM cells to satisfy the capacity and performance needs of address translation. However, the capacity of TLB is often limited by the large cell area of SRAM and has become one of the major bottlenecks of GPU performance.

Spin-transfer torque random access memory (STT-RAM) is a promising new memory technology [14, 15] featuring high integration density, nanosecond read access time, and low leakage power consumption. However, the well-known STT-RAM drawbacks of long write latency and high write energy impose constraints on STT-RAM's application in last-level cache and main memory of CPUs [16–18] and GPUs [19]. In this work, we demonstrate that *TLB in GPUs is an ideal application of STT-RAM* due to relatively large memory footprints of GPU applications and infrequent updates of the stored contents. Hence, the large volume of address translation accesses could greatly benefit from the increased TLB capacity by using STT-RAM and achieve an improved hit rate. Additionally, the low write pressure typically observed by the TLB naturally mitigates the negative impact of STT-RAM in write performance and energy.

Furthermore, by leveraging *differential sensing* technique [20], we are able to improve the read performance of STT-RAM-based TLB whenever read access latency becomes critical. A novel TLB architecture – *STD-TLB*: STT-RAM-based dynamically-configurable translation lookaside buffer, is proposed to dynamically balance the access performance and capacity of the TLB upon run-time TLB access requirements. Compared to the existing research efforts on the application of STT-RAM in computing systems, our contributions are:

1) We propose using STT-RAM to implement GPU TLBs (namely, standard STT-RAM TLB) in which the data access patterns benefit from the unique STT-RAM access characteristics;

2) We leverage differential sensing technique to dynamically enhance the read speed of STT-RAM by trading off the available capacity;

3) We propose a novel TLB architecture – STD-TLB, which allows the TLB to dynamically switch between high-capacity and high-performance mode based on real-

---

*This work is supported in part by NSF awards CCF-1217947

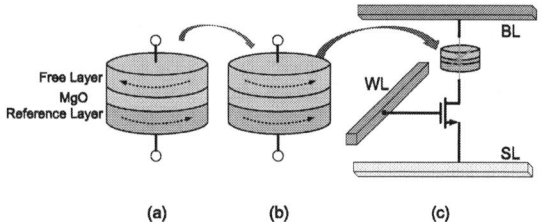

Fig. 1. MTJ Structure (a) Anti-parallel (high resistance state). (b) Parallel (low resistance state). (c) 1T1J STT-RAM cell structure.

time data access patterns of applications.

The rest of this paper is organized as follows: Section II introduces the backgrounds about STT-RAM designs and GPU TLBs as well as the motivation of this work; Section III describes the design details on two STT-RAM-based TLB designs, standard STT-RAM TLB and STD-TLB; Section IV presents the simulation results of different TLB designs in GPUs, including a detailed description of the evaluation methodology and experimental setup; Finally, Section V concludes our work.

II. BACKGROUND AND MOTIVATION

A. STT-RAM and Differential Sensing

1) STT-RAM basics: Data is stored in an STT-RAM cell as the resistance of a magnetic tunneling junction (MTJ) device. A MTJ consists of two ferromagnetic layers which sandwich a MgO layer. The magnetization direction of one ferromagnetic layer (fixed layer) is pinned while while that of the other ferromagnetic layer (free layer) can be flipped by passing a spin-polarized current. When the magnetization direction of the two layers are anti-parallel, the MTJ resistance is at high state (R_{high}); otherwise, the MTJ resistance is at low state (R_{low}), as shown in Fig. 1(a) and (b) [21], respectively. Fig. 1(c) depicts the popular "1T1J" STT-RAM cell structure where the MTJ switching current is supplied by a connected NMOS transistor. The STT-RAM cell area is mainly determined by the NMOS transistor size.

2) Differential sensing: A differential sensing STT-RAM architecture is recently proposed [20] to offer better read performance, same write latency and lower error rate than the 1T1J cell. Differential sensing technique can improve the readability of STT-RAM cells by trading off the memory capacity as follows: In addition to writing the data to one STT-RAM cell, the inversion of the data is written into an adjacent cell. Rather than comparing the cell states to a reference, the stored data is read out by comparing the resistance of these two complimentary cells. The sense margin is increased from $\frac{1}{2} \cdot (R_{high} - R_{low})$ to $(R_{high} - R_{low})$, considerably reducing the corresponding read latency with the improved read reliability. We refer to the cell structure of differential sensing as "2T2J", which is indeed composed of two 1T1J cells. Compared to an 1T1J cell, the dynamic read energy of a 2T2J cell remains fairly constant, while the write dynamic energy of a 2T2J cell essentially doubles as writing the data and its complement into neighboring cells.

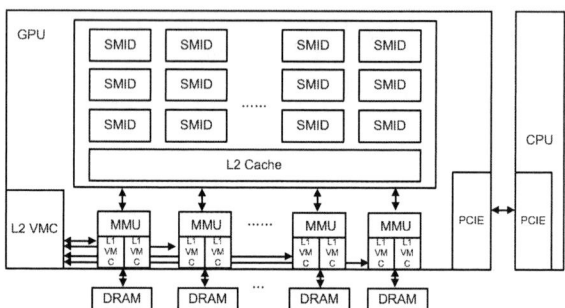

Fig. 2. Virtual memory with multi-level TLBs in GPUs.

B. Virtual Memory Systems and TLBs in GPUs

In 2009, Nvidia announced the first GPU to support unified address space – *Fermi* [4]. Later in 2010, AMD also announced *Graphic Core Next (GCN)* architecture with a virtual memory system to offer x86 addressing with unified address space for both CPU and GPU [3]. In these two architectures, GPU memories are mapped into a continuous 64-bit address space. General instructions are able to access local scratchpad memory, global memory and system memory addresses defined within the unified address space.

Fig. 2 illustrates the overview of the virtual memory system in GPU architectures. The translation from virtual address to physical address is conducted by the memory management unit (MMU) through a multi-level TLB hierarchy. Upon an L1 TLB miss, the corresponding page table entry (PTE) will be loaded from the L2 TLB, which is usually significantly larger than the L1 TLB and stores more cached PTEs.

C. Research Motivation

One important observation that motivates our work is that TLB entries are not modified as frequently as cache blocks. This difference is caused by the function of TLBs: A TLB primarily caches virtual to physical address mappings, which change only upon extremely infrequent OS events, e.g., page re-mapping, page swaps, context switches, and privilege changes. From an application's perspective, the TLB entry is updated only when an old mapping needs to be evicted and a new entry is filled into the TLB. The infrequent writing to the TLB entries makes TLBs particularly well suited to be built with STT-RAM, which usually has a long write latency but relatively short read latency. To verify this observation, we examine various GPU benchmarks and found that out of 13 benchmarks, only one benchmark (SAD) has similarly large write ratio (total number of writes divided by total number of reads) in both the data cache and TLBs. For all other benchmarks, the TLB write ratio is significantly lower than the cache write ratio, as illustrated in Fig. 3. For example,

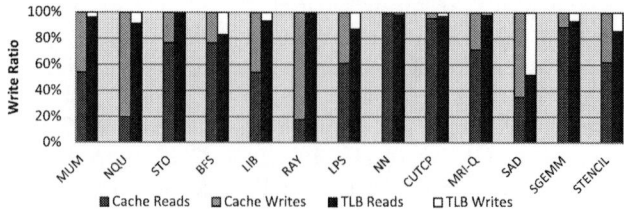

Fig. 3. Comparison of write ratio for data cache and TLB.

978-1-4799-2817-0/14 $31.00 © 2014 IEEE

in LIB, more than 40% of the data cache accesses are writes while less than 10% of the TLB accesses are writes. Based on this observation, a TLB implemented with STT-RAM is less harmed by the influence of the expensive writes but can still take advantage of the improved read speed from differential sensing and larger same-area replacement capacity to achieve better performance and lower energy consumption.

III. TLB DESIGN WITH STT-RAM

A. Standard STT-RAM TLB Design

Due to the structural simplicity, STT-RAM has much higher integration density than SRAM. Table I compares the design parameters of two SRAM-based and STT-RAM-based TLB configurations based on ITRS 2011 [14] and NVSIM [22] simulation results. In each configuration, the three designs with different memory technologies (i.e., SRAM, 1T1J and 2T2J) share similar area while the capacity of the standard STT-RAM TLB (1T1J) is $4\times$ as large as that of SRAM TLB. The capacity of an entirely differential sensing design (2T2J) is between that of the SRAM and standard STT-RAM since it trades half of its capacity for faster read access.

B. STT-RAM-based Dynamically-configurable TLB

A standard STT-RAM TLB configured for 1T1J access provides the best capacity and energy advantages among all TLB designs. The differential sensing design (2T2J) improves read performance, but increases miss rate due to the reduced capacity. As part of this work, we propose a STT-RAM-based dynamically-configurable TLB (STD-TLB) design that leverages the differential sensing technique to improve the read performance of selected memory blocks in a TLB while retaining the capacity advantage available in standard STT-RAM TLB design as much as possible.

Fig. 4 shows the memory structure of our STD-TLB design: In mode 0 (high-capacity mode), the stored data is read out by comparing the resistance of the memory cell to a reference cell; in mode 1 (high-performance mode), the resistances of the complimentary memory cells are compared to each other. Since the sense margin in mode 1 is $2\times$ that in mode 0, the read access time is reduced. However, two memory cells are now occupied to store a single entry in mode 1.

We note that the read accesses to TLB have very unbalanced patterns. We study different TLB entries and group them into "hot" and "cold" based on their frequency of accesses. As an

TABLE I
COMPARISON OF SRAM-BASED AND STT-RAM-BASED TLB

TLB Config.	TLB Config. 1			TLB Config. 2		
Technology	SRAM	STT-RAM		SRAM	STT-RAM	
	(1K)	1T1J(4K)	2T2J(2K)	(16K)	1T1J(64K)	2T2J(32K)
Memory Area(mm^2)	0.037	0.043	0.043	0.434	0.505	0.505
Sensing Time(ns)	0.49	1.13	0.71	0.49	1.13	0.71
Total Read time(ns)	1.78	1.81	1.39	7.24	6.78	6.36
Total Write time(ns)	2.48	11.37	11.37	6.89	14.02	14.02
Read energy (nJ)	0.12	0.06	0.06	0.13	0.09	0.09
Write energy (nJ)	0.13	0.39	0.79	0.15	0.49	0.98
Leakage power (mW)	62.49	21.86	21.86	522.67	157.48	157.48

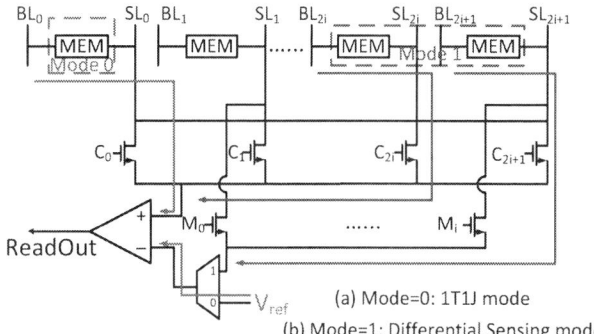

Fig. 4. Reconfigurable differential sensing circuit.

(a) Mode=0: 1T1J mode
(b) Mode=1: Differential Sensing mode

example shown in Fig. 5, the distributions of the read accesses across different TLB entries is highly skewed in benchmarks STO, NN, CUTCP and MRI-Q. Other benchmarks experience less but still noticeable uneven distributions. Generally there are far more accesses occurring on the hot TLB entries (more than 75% on average). Hence, it is desirable to switch the selected cache blocks of STD-TLB to high-performance mode only when the corresponding memory addresses are hot. As illustrated in Fig. 5, these active memory blocks experience frequent read accesses but only occupy a relatively small portion of the total memory blocks. Therefore, mode 1 can be effectively applied to a small number of TLB blocks during the operation, resulting in a marginal degradation on the total capacity and hit rate.

C. Organization of STD-TLB

Fig. 6 shows the organization of the proposed STD-TLB design. Each two adjacent cache blocks form a complimentary block pair when any of them switches to high-performance mode. Each cache block is augmented with a 1-bit mode flag (S) and a n-bit read counter (C) and can function independently in high-capacity mode (S = 0). When the paired cache blocks are in high-performance mode, both blocks' mode flags will be set to 1. The read counter is used to record the number of read accesses to the cache block.

D. TLB Working Mode Management

Through execution analysis we discovered that the page size of a data accessed by a GPU is directly linked to its access patterns. A large page size corresponding to an wide accessible data range from one PTE typically results in more accesses over time. Based on our observation on 13 GPU benchmarks, the PTEs of large pages always have more accesses than that of small pages. Therefore, our TLB working mode management is set heuristically as follows: when a new PTE is loaded

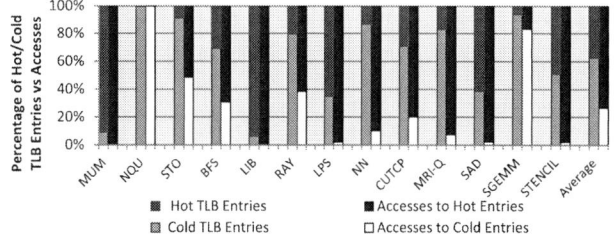

Fig. 5. Unbalanced TLB accesses in GPU.

TABLE II
SYSTEM CONFIGURATION

GPU Configuration		Memory Clock 1250MHz, Memory bandwidth 102.4GB/s, Fillrate Pixel 20.736GP/s		
	Technology	Level-1 TLB	Level-2 TLB	
TLB Configuration	SRAM baseline mode	4-way, 128 entries, 3-cycle latency	16-way, 2048 entries, 10-cycle latency	
	STT-RAM (1T1J mode)	4-way, 512 entries, 3-cycle latency	16-way, 8192 entries, 9-cycle latency	
	STT-RAM (2T2J mode)	4-way, 256 entries, 2-cycle latency	16-way, 4096 entries, 8-cycle latency	
Memory/Paging Parameters		4Kb small page and 128Kb large page, 300-cycle page table access latency		

into the TLB, we first check the wildcard bits of the PTE to retrieve the page size information. If the page size is bigger than 4KByte (i.e., wildcard > 0), the PTE will be written into the paired cache blocks in high-performance mode; otherwise, the PTE will be saved in the cache block in high-capacity mode.

Although this heuristic scheme works reasonably well, there are still some frequently accessed small pages that could not be covered by the scheme. To discover these pages, we utilize the augmented counter in each cache block to record how heavily the PTE is accessed. When a PTE is loaded into the TLB, the corresponding read counter is cleared to 0. The value of the counter increments upon each read access. The mode of the cache block is then dynamically reconfigured based on how intensively the page has been accessed: if the counter exceeds a threshold, the empty corresponding entry will be elevated to high-performance mode. In contrast, the entry set to high-performance mode (based on its page size) can be "demoted" to high-capacity mode during evictions: A high-performance TLB entry which is planned to be retired by the replacement policy (e.g., least-recently-used) will switch to high-capacity mode first rather than being evicted from the TLB immediately. Only the entries in high-capacity mode can be evicted from the TLB directly.

IV. EVALUATION

We created a STT-RAM device model based on an industrial prototype. We assume the MTJ switching time is 10ns [23] and the sense margins of the STT-RAM cell in high-capacity and high-performance modes are 40mV and 80mV, respectively. A modified CACTI [24] tool is used to derive the peripheral circuitry latencies of various TLB designs at 45nm technology [25]. The timing parameters of sense amplifiers are derived from SPICE simulations. All TLB design parameters are summarized in Table I.

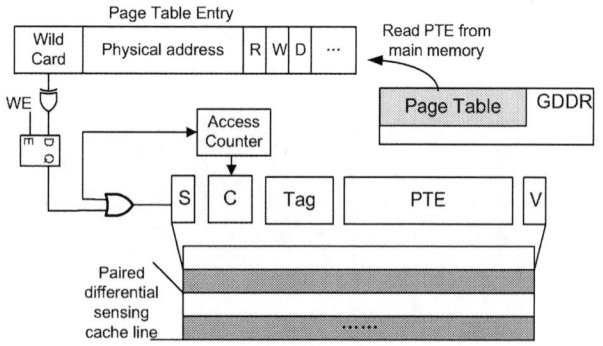

Fig. 6.   Organization of STD-TLB design.

A. System Configuration and Workloads

We verify our architectural design through functional and timing simulations on GPGPU-sim [26], a cycle accurate GPU performance simulator. We modified the baseline GPU architecture in GPGPU-sim based on Nvidia Quadro FX5800 [27] by adding virtual memory mapping support. In particular, we implemented a two-level TLB hierarchy including first-level private TLBs with small capacity and short translation latency, and a second-level TLB with much larger capacity and long translation latency, which is shared across all memory banks. We assume a dual-sized paging system with both small page (4K bytes) and large page (128K bytes). A large page is usually used for the applications with large memory footprint, e.g., enormous texture or color data.

We use SRAM TLB as the baseline implementation. Our proposed STT-RAM TLB implementations assume the same-area replacement associated with an increased capacity. The detailed parameters and system configuration are summarized in Table II. Because the cache blocks in STD-TLB can independently operate in high-capacity or high-performance mode, the maximum and minimum capacities of the STD-TLB are listed as 1T1J and 2T2J in the Table, respectively. In practice the effective capacity will be somewhere in between.

in our evaluations, we utilize a wide range of GPU workloads selected from the Parboil [28] and ISPASS2009 benchmark suite. Our intent is to primarily focus on the workloads which have high memory access intensity and thus could potentially benefit from STT-RAM-based TLBs. However, we also include some less access intensive workloads to test the potential impact of STT-RAM-based TLBs in different scenarios. The characteristics of the adopted workloads are listed in Table III.

B. Experimental Results

To demonstrate the influences of using STT-RAM to implement TLB in GPUs, we study the mode configuration accuracy, TLB miss rate, translation runtime performance, energy consumption, energy-delay product etc., and compare them with the SRAM baseline. We also evaluate the performance and energy improvement of STD-TLB over the standard STT-RAM TLB. Finally, we evaluate the overall system performance improvement and summarize the general rules of applying STD-TLB.

1) Mode Configuration Accuracy: We verify if the TLB entry of a small page is configured in an appropriate mode by using a counter to track its read accesses and comparing the value of the counter with a threshold. Ideally, the TLB entry corresponding to a small page shall be set to high-capacity mode and has a low counter value, i.e., smaller than a

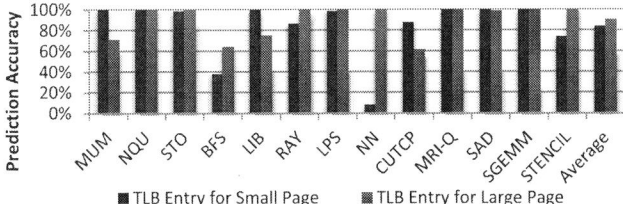

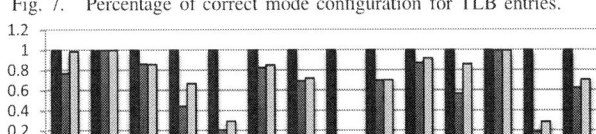

Fig. 7. Percentage of correct mode configuration for TLB entries.

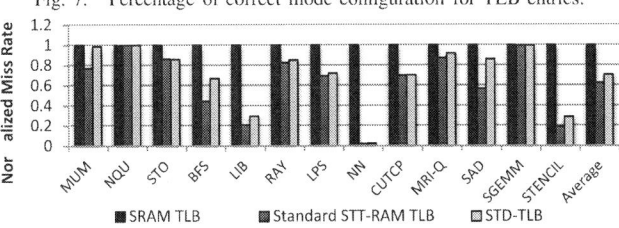

Fig. 8. TLB miss rate in form of off chip misses compared to all accesses.

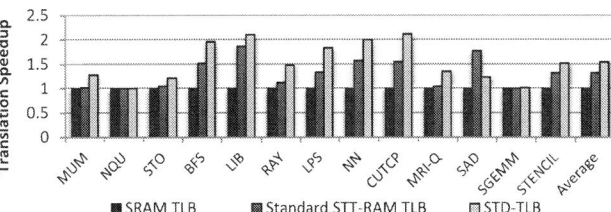

Fig. 9. TLB translation performance improvement.

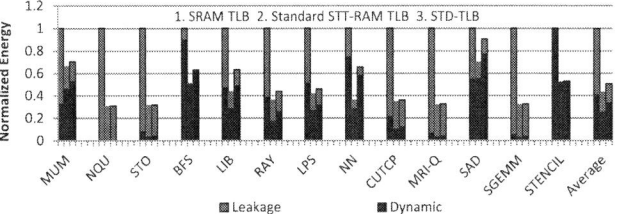

Fig. 10. Dynamic and leakage power consumptions for different TLBs.

threshold. Otherwise, the TLB performance may be adversely impacted because of its poor read speed. Heuristically, we set the threshold of defining a "hot" small page to 900. The configuration accuracy of the TLB entry of a large page is also measured by the probability of being demoted to high-capacity mode. Fig. 7 shows that our page-size-only mode selection mechanism (see Section III) captures averagely 83% of "cold" small pages and 90% of "hot" large pages. NN is an outlier and exhibits heavy reads for both small and large pages. On average, our page-size-only mode selection mechanism introduces a mode prediction accuracy from 40%~100% for both small and large pages. To improve the address translation efficiency, the promote/demote scheme is applied to STD-TLB atop the page-size-only mode selection mechanism, as presented in Section III-D.

2) TLB Miss Rate: Fig. 8 depicts the normalized overall miss rate of the TLB hierarchy for the tested benchmarks. Due to the larger capacity, standard STT-RAM TLB and STD-TLB exhibit significantly lower miss rates than SRAM TLB in 10 out of the 13 benchmarks. In NQU and SGEMM, all three designs perform closely because the kernels of these benchmarks exercise small working sets that can be effectively

TABLE III
CHARACTERISTICS OF GPU BENCHMARKS (INSTRUCTION NUMBER(IN))

	Abbrev.	Benchmark	IN
	MUM	MUMmerGPU, low cost ultra-fast sequence alignment	85M
	NQU	N-Queens Solver, finding an arrangement of queens	1M
ISPASS2009	STO	StoreGPU, accelerating numbers of hashing	125M
	BFS	Breadth First Search, a graph operation	17M
	LIB	LIBOR, calculating the evolution of 80 forward rates	998M
	RAY	Ray Tracing in 3D computer graphic	63M
	LPS	3D Laplace discretisation Solver	73M
	NN	a Neural Network Recognition of Handwritten Digits	84M
	CUTCP	Cutoff-limited Coulombic Potentiall	3G
Parboil	MRI-Q	Mri non-cartesian Q matrix calculation	1.8G
	SAD	Sum of Absolute Differences	300M
	SGEMM	SGEMM dense matrix operation	9M
	STENCIL	solving partial differential equations	2G

accommodated by a small TLB. For MUM, BFS, LIB and SAD, standard STT-RAM TLB drastically reduces the miss rate. STD-TLB also considerably reduces the miss rate of BFS, LIB, and SAD. But the reduction is less significant than standard STT-RAM TLB because of the smaller runtime effective capacity. Under these scenarios, the applications typically utilize more large pages with intensively accessed TLB entries. On average, standard STT-RAM TLB reduces the miss rate by 38% while STD-TLB reduces the miss rate by 30%, compared to the SRAM baseline.

3) Performance: The address translation runtime performance improvements [29] of different designs are presented in Fig. 9. Standard STT-RAM TLB achieves a significant speedup as a result of the prominent miss rate reduction. Although STD-TLB has a higher miss rate, STD-TLB further improves the runtime performance by performing a faster address translation for most accesses. We can see that other than two benchmarks – NQU and SGEMM, which have relatively small data sets, all other benchmarks experience remarkable runtime performance improvements under STD-TLB. On average, standard STT-RAM TLB improves the translation performance by 32% over SRAM baseline while STD-TLB further improves it by 55% and 15% compared to SRAM TLB and standard STT-RAM TLB, respectively.

4) Energy: The energy comparison of all designs is shown in Fig. 10. Both standard STT-RAM TLB and STD-TLB achieve over 70% leakage energy reduction compared to SRAM TLB. STD-TLB has slightly higher dynamic energy due to the higher write energy consumption in high-performance mode. For benchmarks that exhibit heavy write loads, e.g., NN and SAD, STD-TLB consumes considerably more dynamic energy than standard STT-RAM TLB. However, dynamic energy increases are negligible for other benchmarks. On average, standard STT-RAM TLB and STD-TLB achieve 57% and 49% energy savings over SRAM TLB, respectively.

5) Energy Delay Product: The overall benefit by both translation performance and energy is often denoted by energy delay product (EDP). As illustrated in Fig. 11, standard STT-RAM TLB achieves over 75% EDP reduction w.r.t. SRAM

978-1-4799-2817-0/14 $31.00 © 2014 IEEE

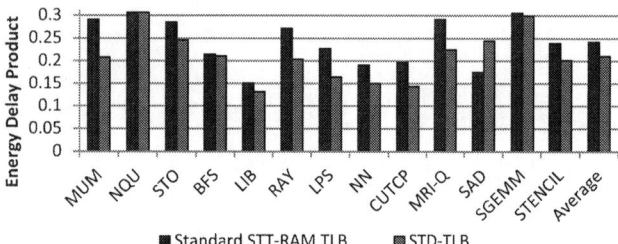

Fig. 11. Energy delay product improvements (normalized to SRAM TLB).

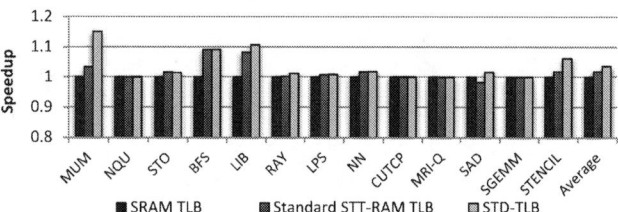

Fig. 12. Overall system performance speedup comparison of different TLBs.

TLB while STD-TLB further boosts it to nearly 80%. The significant improvement is mainly from: 1) low leakage power induced by STT-RAM technology; 2) low miss rate as a result of the increased TLB capacity; and 3) reduced translation latency. STD-TLB further achieves 13% EDP improvement over STT-RAM TLB regardless its nominally increased energy consumption. STD-TLB performs worse than standard STT-RAM TLB only in SAD because of the increased dynamic energy and translation latency induced by the heavy write loads and high miss rate.

6) System Performance: As shown in Fig. 12, standard STT-RAM TLB achieves very moderate performance improvement in many benchmarks. It is because modern GPU designs intent to hide memory latency, which is the main optimization goal of our proposed techniques. However, significant performance improvements are still achieved in some benchmarks like BFS (9.2%) and LIB (10.8%). After employing STD-TLB, significant GPU system performance improvement is achieved in MUM and STENCIL, say, 15% and 6%, respectively. On average, STD-TLB achieves 4% and 2% system performance speedup compared to SRAM TLB and standard STT-RAM TLB, respectively.

V. CONCLUSION

In this paper, we propose the use of STT-RAM in TLB for new virtually addressed GPUs. STT-RAM-based TLB introduces significant energy and performance advantages over SRAM implementation by realizing larger TLB capacity within the same area. We also present a novel STT-RAM-based dynamically-configurable TLB (STD-TLB) that leverages high read speed of STT-RAM differential sensing and dynamic configuration policy to retain the reduced miss rate from standard STT-RAM TLB while improving the translation performance of the heavily accessed pages. Compared to SRAM baseline, STD-TLB reduces TLB miss rate by 30%, boosts translation runtime performance by 55%, and improves the energy delay product by ~80%. As more systems employ on-board GPUs and more general purpose applications are mapped to these processors, alleviating the performance impact introduced by

address mapping will become very difficult in GPU system design. For these applications, we expect STD-TLB will provide a dramatic system benefit (e.g., 10% performance improvement over standard STT-RAM TLB in MUM) by dynamically switching between high-performance and high-capacity mode based on real-time application needs.

REFERENCES

[1] top500.org, *Supercomputers*, http://www.top500.org/list/2012/06/100.
[2] S. Che *et al.*, "Accelerating compute-intensive applications with gpus and fpgas," in *SASP*, 2008, pp. 101–107.
[3] AMD, *GCN Architecture*, http://www.amd.com/us/products/technologies-/gcn/Pages/gcn-architecture.aspx.
[4] Nvidia, *Fermi specifications*, http://www.nvidia.com/object/fermi-architecture.html.
[5] D. W. Clark and J. S. Emer, "Performance of the vax-11/780 translation buffers: Simulation and measurement," in *ACM Transactions on Computer Systems*, 1985, pp. 31–62.
[6] D. Nagle *et al.*, "Design tradeoffs for software managed tlbs," in *ISCA*, 1993, pp. 27–38.
[7] T. Barr *et al.*, "Translation caching: Skip, dont walk (the page table)," in *ISCA*, 2010, pp. 48–59.
[8] S. Srikantaiah *et al.*, "Sharp control: Controlled shared cache management in chip multiprocessors," in *MICRO-42*, 2009, pp. 517–528.
[9] G. Kandiraju and A. Sivasubramaniam, "Going the distance for tlb prefetching: An application-driven study," in *ISCA*, 2002, pp. 195–206.
[10] A. Saulsbury *et al.*, "Recency-based tlb preloading," in *ISCA*, 2000, pp. 117–127.
[11] A. Bhattacharjee *et al.*, "Shared last-level tlbs for chip multiprocessors," in *HPCA*, 2011, pp. 62–63.
[12] H. Wong *et al.*, "Demystifying gpu microarchitecture through microbenchmarking," in *ISPASS*, 2010, pp. 235–246.
[13] Intel, *Performance Analysis Guide for Intel Core i7 Processor and Intel Xeon 5500 processors*, http://software.intel.com/sites/products/collateral-/hpc/vtune/performance_analysis_guide.pdf.
[14] www.itrs.net, in *International Techonology Roadmap for Semiconductors*, 2011.
[15] W. Wen *et al.*, "PS3-RAM: A fast portable and scalable statistical STT-RAM reliability analysis method," 2012, pp. 1187–1192.
[16] Z. Sun *et al.*, "Multi retention level stt-ram cache designs with a dynamic refresh scheme," in *MICRO*, 2011, pp. 329–338.
[17] X. Wu *et al.*, "Hybrid cache architecture with disparate memory technologies," in *ISCA*, 2009, pp. 34–45.
[18] Y. Chen *et al.*, "On-chip caches built on multilevel spin-transfer torque RAM cells and its optimizations," *J. Emerg. Technol. Comput. Syst.*, vol. 9, no. 2, pp. 16:1–16:22, 2013.
[19] J. Zhao and Y. Xie, "Optimizing bandwidth and power of graphics memory with hybrid memory technologies and adaptive data migration," in *ICCAD*, 2012, pp. 81–87.
[20] Y. Zhang *et al.*, "Adams: Asymmetric differential STT-RAM Cell Structure For Reliable and High-performance Applications," in *ICCAD*, 2013, to appear.
[21] M. Hosomi *et al.*, "A novel nonvolatile memory with spin torque transfer magnetization switching: spin-ram," in *IEDM Technical Digest*, 2005, pp. 459–462.
[22] X. Dong *et al.*, *NVSim: A Circuit-Level Performance, Energy, and Area Model for Emerging Nonvolatile Memory*, 2012, vol. 31.
[23] C. Xu *et al.*, "Device-architecture co-optimization of stt-ram based memory for low power embedded systems," in *ICCAD*, 2011, pp. 463–470.
[24] P. Shivakumar and N. P. Jouppi, "Cacti 3.0: An integrated cache timing, power, and area model," hp, Tech. Rep., August 2001.
[25] W. Zhao and Y. Cao, *New Generation of Predictive Technology Model for Sub-45nm Early Design Exploration*, 2006, vol. 53.
[26] A. Bakhoda *et al.*, "Analyzing cuda workloads using a detailed gpu simulator," in *ISPASS*, 2009, pp. 163–174.
[27] Nvidia, *Quadro Specifications*, http://www.nvidia.com/object/product_quadro_fx_5800_us.html.
[28] J. A. Stratton *et al.*, "Parboil: A revised benchmark suite for scientic and commercial throughput computing," in *IMPACT Technical Report*, 2012.
[29] A. Bhattacharjee and M. Martonosi, "Inter-core cooperative tlb for chip multiprocessors," in *ASPLOS*, 2010, pp. 359–370.

Training Itself: Mixed-signal Training Acceleration for Memristor-based Neural Network

Boxun Li[1], Yuzhi Wang[1], Yu Wang[1], Yiran Chen[2], Huazhong Yang[1]

[1]Dept. of E.E., TNList, Tsinghua University, Beijing, China
[2]Dept. of E.C.E., University of Pittsburgh, Pittsburgh, USA
[1] Email: yu-wang@mail.tsinghua.edu.cn

Abstract—The artificial neural network (ANN) is among the most widely used methods in data processing applications. The memristor-based neural network further demonstrates a power efficient hardware realization of ANN. Training phase is the critical operation of memristor-based neural network. However, the traditional training method for memristor-based neural network is time consuming and energy inefficient. Users have to first work out the parameters of memristors through digital computing systems and then tune the memristor to the corresponding state. In this work, we introduce a mixed-signal training acceleration framework, which realizes the self-training of memristor-based neural network. We first modify the original stochastic gradient descent algorithm by approximating calculations and designing an alternative computing method. We then propose a mixed-signal acceleration architecture for the modified training algorithm by equipping the original memristor-based neural network architecture with the copy crossbar technique, weight update units, sign calculation units and other assistant units. The experiment on the MNIST database demonstrates that proposed mixed-signal acceleration is 3 orders of magnitude faster and 4 orders of magnitude more energy efficient than the CPU implementation counterpart at the cost of a slight decrease of the recognition accuracy ($< 5\%$).

Index Terms—Neural Network; Training; Memristor;

I. INTRODUCTION

Artificial neural networks have been widely employed in data processing applications, ranging from computer vision, speech recognition, pattern recognition to signal processing [1], [2]. On the top of that, more and more neuromorphic computing architectures have appeared and demonstrated a great potential for performance and energy efficiency gains [3].

In recent years, the innovation of memristor further explores the potential of neuromorphic computing systems. The memristor was first physical realized by HP Labs in 2008 and afterwards attracts significant research interest for its ultrahigh integration density, low power dissipation, and especially, the nonvolatile feature that the state could be tuned by the current passing through itself [4]. The similarity between the memristor and the biologic synapse makes the memristor a promising device to realize the neural network and mixed-signal neuromorphic systems [5]. For example, the memristor-based crossbar can efficiently perform the matrix-vector multiplication, which is the most common operation of ANN [6]. And the memristor-based neural network provides a promising solution to the low power on-chip approximate computing systems with power efficiency of more than 400 GFLOPS/W [7].

Training phase is the critical operation to obtain a neural network for a specific task. However, the process of training is usually time consuming and resource intensive [8]: it usually requires a large volume of memory and computing resources, while the time consumption can range from a few seconds to hundreds of hours, depending on the scale of the networks [9], [10]. Therefore, how to speed up the training process is one of the major concerns of the artificial neural network.

The training problem also exists on the memristor-based neural network: Although the memristor enables an efficient implementation of the operation phase of ANN, the training phase is still confined to the traditional CPU/GPU/FPGA systems because of the requirement of complicated calculation and caching a large amount of data. Users must first work out the parameters of the network through digital computing systems and then tune the memristor to the specific state [7]. However, the training process through digital systems is usually bulky and energy consuming [1], [10] and the tuning process of memristors is complicated, time consuming, hardware intensive and may even bring in unexpected errors [5], [11], [12]. Therefore, there urges an efficient acceleration for the memristor-based neural network.

Our object is to realize the self-training of memristor-based neural network through mixed-signal systems and achieve performance and energy gains. The analog unit of the mixed-signal training acceleration framework should be able to work out the training calculations efficiently and directly configure the memristor **without state tuning**. The digital unit only help control the state of training, instead of performing a large amount of complicate calculations. The following challenges must be overcome to realize this goal: 1) There're too many numerical calculations in the original training algorithm, which are hard for analog systems to realize. A modified algorithm is demanded to relieve the difficulty of analog numerical calculations. 2) Part of the results need to be back propagated when training, and therefore, there requires an efficient solution to the problem of caching analog signals. 3) The conversions between analog and digital in the mixed-signal system should be efficient and as few as possible in order to guarantee the performance and energy gains.

In this paper, we for the first time propose a mixed-signal training acceleration for memristor-based neural network and make it possible for memristor-based neural network to train itself. The contributions of this paper are:

1) We propose a mixed-signal training acceleration framework. The framework consists of a modified training algorithm and a mixed-signal training acceleration architecture, which can realize the self-training of memristor-based neural network. The experiment on the MNIST database demonstrates that our mixed-signal training acceleration framework is able to achieve a 3 orders of magnitude improvement of the training speed as well as a 4 orders of magnitude energy efficiency gains at the cost of a slight decrease of the recognition accuracy ($< 5\%$) compared with the CPU implementation counterpart.

2) We modify the original stochastic gradient descent algorithm by approximating the calculations of errors and decomposing the weight update calculations into sign calculations and numerical calculations. On top of that, we further replace the precious numerical calculations by an automatic adjustment of convergence rate. Such modifications reduce the difficulty of analog realization of the numerical calculations in the training process.

3) We propose *Copy Crossbar* technique and *Sign Calculation* unit to accomplish the analog numerical calculations and overcome the difficulty of caching analog signals. We configure the states of memristors through *Weight Update* units and vari-

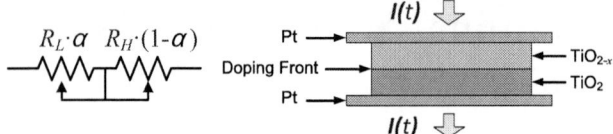

Fig. 1. Physical model of the HP memristor

able gain amplifiers (VGAs) directly without state tuning. In addition, we realize the automatic adjustment of convergence rate by controlling the factor gain of VGAs through digital unit.

The rest of this paper is organized as follows: Section 2 provides related background information. Section 3 introduces our modified training algorithm. The proposed mixed-signal training acceleration architecture is depicted in Section 4. Section 5 presents a case study of the MNIST database and Section 6 concludes this work.

II. PRELIMINARIES

A. Memristor

Fig. 1 shows the physical model of the HP memristor [13]. There is a two-layer thin film of TiO_2: One layer consists of intact TiO_2 and the other layer consists of TiO_{2-x}, which lacks of a small amount of oxygen. The two layers have different electrical conductivity and the overall resistance is the sum of the two layers. There is a boundary (doping front) between the two layers. When a current is applied to the device, the boundary will move and make the resistance of memristor change. The memristor also has other attractive features, such as the 'pinched hysteresis loop'. In this paper, we mainly take advantage of the variable resistance states of the memristor.

B. Implementation of Memristor-based Neural Network

An N-layer network has $N-1$ weight matrixes. The calculation between each pair of neighbour layers can be conceptually expressed as a combination of weighted input variations (matrix-vector multiplication) with a sigmoid active function ($f(x) = \frac{1}{1+e^{-x}}$). The operation of weighted summation in the neural network can be realized through memristor-based matrix-vector multiplication in analog [7] and the sigmoid activation function can be accomplished by the circuit similar to that in [14].

Fig. 2(a) illustrates a memristor crossbar array and Fig. 2(b) shows a complete implementation of the memristor-based matrix-vector multiplication. Each memristor-based matrix-vector multiplication needs two crossbars (positive crossbar and negative crossbar) to represent the positive and negative weights of a network, since M can only be positive [7]. The practical output of the implementation in Fig. 2(b) can be expressed as:

$$V_{oj} = \sum_k w_{kj} \cdot V_{ik} \tag{1}$$

where:

$$w_{kj} = R \cdot (g_{kj(pos)} - g_{kj(neg)}), g_{kj} = \frac{1}{M_{kj}} \tag{2}$$

where V_{ik} and V_{oj} represent the input and output voltage of each row and column, respectively. k and j are the index numbers of input and output voltages. R is the resistor at the end of the array. M_{kj} and g_{kj} are the resistance and admittance of each memristor.

The architecture shown in Fig. 2(b) realizes the weighted summation operation of ANN. By adding a series of the sigmoid circuits to the output ports of the memristor-based matrix-vector multiplication, a double-layer memristor-based neural network is realized. Finally, a multilayer network can be accomplished by combining several double-layer memristor-based neural networks together.

One thing needs to be aware of is that the polarities of each pair of memristors need to be set to the opposite direction in order to

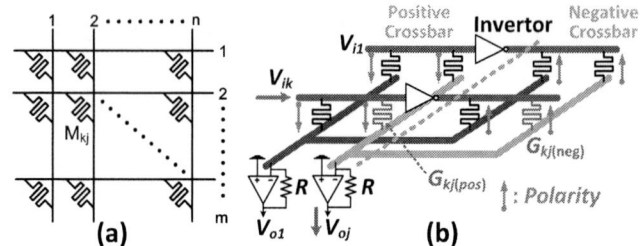

Fig. 2. (a) Memristor Crossbar; (b) Memristor Matrix-Vector Multiplication

make the deviations generated by the currents passing through the two memristors cancel each other, instead of accumulation [7]. This configuration will influence our circuit design in Section IV-C.

C. Stochastic Gradient Descent Algorithm

Stochastic gradient descent is one of the most common algorithms for neural network training [15]. The training process is realized by adjusting the weights in the network layer by layer. The update of each weight (w_{ji}) is:

$$w_{ji} \leftarrow w_{ji} + \eta \cdot \delta_j \cdot x_i \tag{3}$$

where η is the learning rate and δ_j is the error back propagated from the node j in the next neighbour layer. x_i is the input of the node i. For the node p in the output layer, δ_p is:

$$\delta_p = o_p \cdot (1 - o_p) \cdot (t_p - o_p) \tag{4}$$

where o_p and t_p are the actual and ideal output of the complete network, respectively. For the node h in the hidden layers:

$$\delta_h = o_h \cdot (1 - o_h) \cdot \sum_{k \in nextlayer} w_{kh}\delta_k \tag{5}$$

where o_h is the output of the node h in a hidden layer. w_{kh} is the weight between node h in this hidden layer and node k in the next neighbour layer. δ_k is the error of node k in the next neighbour layer.

The training algorithm is an iterative process and hard to be parallelized. For an N-layer network, where the amount of nodes in each layer is $N_1, N_2, ..., N_n$, the complexity of each iteration is $O(\sum_i^{n-1} N_i N_{i+1})$. In addition, it usually requires at least hundreds of thousands iterations to obtain an efficient network. Therefore, the time consumption of training process is usually quite large.

III. MODIFIED TRAINING ALGORITHM

There're two major modifications compared with the original algorithm: the approximation of the error calculations and accomplish the weight update calculating operations through sign calculations and automatically adjusting the convergence rate.

A. Error Approximation

It can be observed that there're too many operations of multiplication when calculating δ in Eq. (4) and Eq. (5) in the original algorithm. Such a large amount of multiplications is hard for analog circuits to realize.

We note that both $o_p \cdot (1 - o_p)$ and $o_h \cdot (1 - o_h)$ must be greater than zero because the output range of the sigmoid active function is $(0, 1)$. The polarities of δ only depend on $(t_p - o_p)$ in Eq. (4) and $\sum_k w_{kh}\delta_k$ in Eq. (5). Therefore, we decide to ignore $o_p \cdot (1 - o_p)$ and $o_h \cdot (1 - o_h)$ when calculating δ. In fact, we observe such an approximation of δ_p is efficient and has little impact on the training effect.

However, the neglect of $o_h \cdot (1 - o_h)$ may lead to a great error of training. The reason lies in the impact of o_h: The value of o_h is very likely to be close to 0 or 1, especially at the final state of the training process, and makes the value of δ_h close to zeros. Therefore, the complete neglect of $o_h \cdot (1 - o_h)$ will make the value of δ_h too

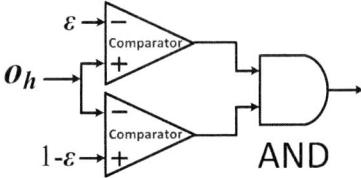

Fig. 3. Implementation of the Filter Operation

large and lead to a terrible training effect. To overcome this problem, we choose to filter o_h instead of approximating $o_h \cdot (1 - o_h)$ directly to 1: When $o_h > 1 - \epsilon$ or $o_h < \epsilon$, we will set $o_h \cdot (1 - o_h)$ to 0, or we set the value of it to 1. Here ϵ is a small value (e.g, $\epsilon = 0.1$) to filter o_h.

In conclusion, the approximation of δ_p in Eq. (4) is:

$$\delta_p = t_p - o_p \qquad (6)$$

And for δ_h in Eq. (5):

$$\delta_h = filter(o_h) \cdot \sum_{k \in nextlayer} w_{kh} \delta_k \qquad (7)$$

The calculations of δ_p can be realized through an analog subtractor. And the implementation of the filter operation can be accomplished through the unit shown in Fig. 3. The calculation of $\sum_k w_{kh} \delta_k$ will require the copy crossbar technique to be described in Section IV-B.

B. Calculation Decomposition

The error approximation technique just reduces the complexity of the calculations of δ, while the calculations of the complete weight updates require another technique called *Calculation Decomposition*. To be specific, we first decompose the weight update calculations into sign calculations and numerical calculations.

The sign calculations are used to extract the directions of the weight updates. The calculation of the updates of the weights in Eq. (3) can be expressed as:

$$w_{ji} \leftarrow w_{ji} + \eta \cdot sign(\delta_j) \cdot |\delta_j| \cdot x_i \qquad (8)$$

The advantage of the extraction of sign calculations is that the polarities of the weight updates can be achieved through zero-crossing detectors and analog comparators easily and quickly as shown in Fig. 4. And the computation between signs can be accomplished through digital logic efficiently. In addition, the digital data are able to get cached conveniently, which greatly reduces the difficulty of caching analog data.

As for the numerical calculations, in order to avoid the usage of the inefficient analog numerical computation, we accomplish the numerical calculation of $\eta \cdot |\delta_j| \cdot x_i$ in Eq. (8) by amplifying x_i signal with a gain factor equal to $\eta \cdot |\delta_j|$. However, the calculation of $\eta \cdot |\delta_j|$ is still hard to complete. Therefore, we choose to estimate the approximate value instead of performing the precise numerical calculations. Unfortunately, it is usually very difficult to make a good estimation of $\eta \cdot |\delta_j|$: A larger approximation will result in the error rebounding and a bad training effect, while a smaller approximation will make the speed of training too slow. In addition, the value of $\eta \cdot |\delta_j|$ also fluctuates frequently.

In order to make a better estimation of $\eta \cdot |\delta_j|$, we use the automatic convergence rate adjustment scheme: We set an initial convergence

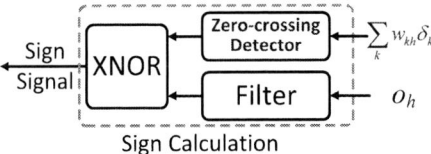

Fig. 4. Sign Calculation Unit

Algorithm 1: Modified Training Algorithm

Input: $Network$, $TrainingDataset$, η_{start}, η_{stop}, $DecayRate$, $MonitorPeriod$, $DetectNodes$
Output: $Network$

1 Initialize the weights in the $Network$ to small random values;
2 $\eta \leftarrow \eta_{start}$
3 $CurrentTrainState \leftarrow 0$
4 $LastTrainState \leftarrow Inf$
5 **while** $\eta > \eta_{stop}$ **do**
6   **for** $times = 0 \to MonitorPeriod$ **do**
7   Pick a sample from the $TrainingDataset$ and compute the $ActualOutput$ of the $Network$;
8       $\gamma \leftarrow$ Generate a random value between 0 and 1
9       $\delta_p \leftarrow t_p - o_p$
10      $\delta_h \leftarrow filter(o_h) \cdot \sum_{k \in nextlayer} w_{kh} \delta_k$
11      $w_{ji} \leftarrow w_{ji} + \gamma \cdot \eta \cdot sign(\delta_j) \cdot x_i$
12      $CurrentTrainState \leftarrow$ $CurrentTrainState + \sum_{p \in DetectNodes} abs(\delta_p)$
13   **end**
14   **if** $CurrentTrainState > LastTrainState$ **then**
15      $\eta \leftarrow \eta / DecayRate$
16   **end**
17    $LastTrainState \leftarrow CurrentTrainState$
18    $CurrentTrainState \leftarrow 0$
19 **end**

---

rate to represent the initial estimation of weight updates ($\eta \cdot |\delta_j|$). Then the training scheme will record the training states of several output nodes and periodically check the training results. Once the accumulation of the errors of the detected nodes rebound instead of decreasing after a period of monitoring, the convergence rate will decay automatically to guarantee the efficiency of the training process. In addition, if the convergence rate becomes lower than a certain small value, the scheme will assert that the training process is finished. We also multiply the convergence rate with a random value between 0 and 1 to imitate the fluctuation of the value. The complete modified updates of the weights are:

$$w_{ji} \leftarrow w_{ji} + \gamma \cdot ConvergenceRate_j \cdot sign(\delta_j) \cdot x_i \qquad (9)$$

where $ConvergenceRate$ is equal to $\eta \cdot |\delta_j|$ and γ is a random value between 0 and 1.

Finally, the automatic adjustment of convergence rate is performed by the digital assistant unit (e.g., FPGA) which requires only a small amount of registers, adders and other digital logic.

C. The Complete Modified Training Algorithm

Algorithm 1 demonstrates the complete process of the proposed modified training algorithm, where $Network$ is the neural network to get trained. $TrainingDataset$ contains the labeled data for training. The content of each labeled training data is $< \vec{x}, \vec{t} >$, where $\vec{x}$ is the input data of each training sample and $\vec{t}$ is the ideal output of the network. η_{start} and η_{stop} are the initial and final convergence rate, respectively. $DecayRate$ represents the decay rate of the convergence rate. $MonitorPeriod$ is the period of monitoring the states of the training to adjust the convergence rate. $DetectNodes$ represents the selected nodes in the output layer to estimate the training states.

Compared with the original training algorithm, only the calculations of the actual network outputs remains the same. Both the calculations of δ_p and δ_h are approximated. The calculation of weight update is replaced by the product of a random value, convergence rate and the result of sign calculation. In addition, there's an automatic adjustment of the convergence rate (η) to replace the precious numerical calculations of $|\delta|$.

978-1-4799-2817-0/14 $31.00 © 2014 IEEE

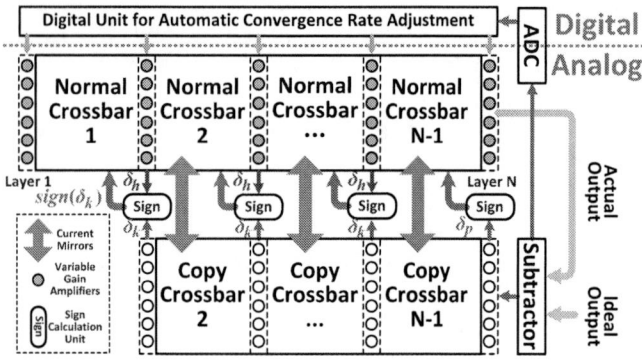

Fig. 5. Mixed-signal Training Acceleration Framework

IV. MIXED-SIGNAL TRAINING ACCELERATION ARCHITECTURE

Fig. 5 illustrates the overview of the proposed mixed-signal training acceleration framework for memristor-based neural network, which can realize the modified algorithm efficiently. For a training task of an N-layer neural network with $N-1$ weight matrixes, the mixed-signal training acceleration framework requires $N-1$ normal crossbar arrays and $N-2$ copy crossbar arrays. Each pair of normal crossbar and copy crossbar are connected by a series of mirror currents. A array of subtractors is used to work out the deviation (δ_p) between the actual and ideal output and δ_k will be calculated through the copy crossbar arrays. The details of the copy crossbar technique will be introduced later in Section IV-B. All the results of δ will be imported to the sign calculation units (mentioned in Section III-A and III-B). A array of VGAs is added to the input ports of each normal crossbar (mentioned in Section III-B). The outputs of the sign calculation units and VGAs will be imported to each normal crossbar to update the state of the memristor by the weight update unit equipped to each normal crossbar. The details of weight operation unit will be discussed later in Section IV-C. All of the above units are analog. At the same time, part of the results of δ_p will be converted to digital and then imported to the digital unit to control the gain factor of VGAs and perform the automatic adjustment of convergence rate as mentioned in Section III-B.

A. Operating Mechanism

Fig. 6 demonstrates the workflow of the mixed-signal training acceleration framework. There are two stages in each iteration in the training process: the forward computation state and the weight update state.

Before training, all the variables in the digital unit will be first initialized as the modified training algorithm. At the same time, all the crossbar arrays (both normal crossbar and copy crossbar) will be initialized to the same parameters. For example, all of the memristors in the network could be set to the same maximum state and then a series of pulses with the same random value will be set to each crossbar array.

Then the training process will begin. For an N-layer memristor-based neural network, there requires 1 phase for forward computation and $N-1$ phases for weight update in each training iteration:

- *Phase 1:* At the beginning of each iteration, the framework will first work at the forward computation state. The current mirrors are closed. The gain factor of all VGAs is set to 1. The framework will pick a sample *InputData* from the *TrainingDataset* and import it into the network. The normal crossbar arrays will work out the actual outputs of the network. At the same time, the ideal outputs of the network will be imported into the framework, too. A series of analog subtractors will compute the deviations (δ_p) of the actual and ideal outputs in parallel and import them directly to the copy crossbar to

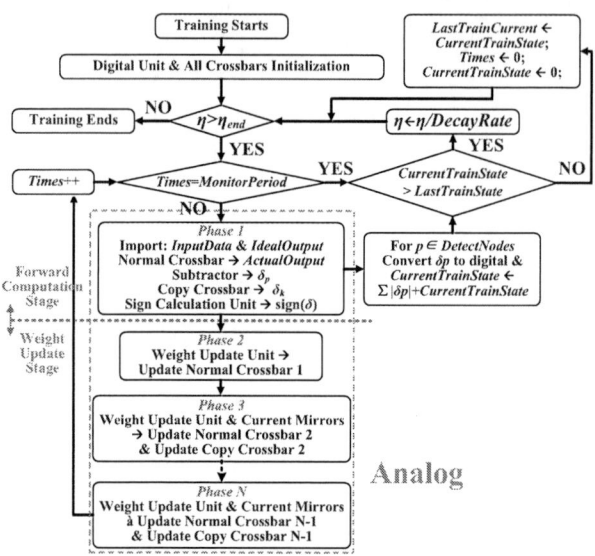

Fig. 6. Mixed-signal Training Flow

work out δ_k. The outputs of both the copy crossbar arrays and hidden layers will be imported into the sign calculation units to perform sign calculations. The sign calculation units will also convert the results of sign calculations to digital and cache them efficiently.

- *Phase 2:* After the results of sign calculations are cached, the framework will start to work at the weight update state. All the current mirrors will break over. As mentioned in Section II-C, the weights in the network get adjusted layer by layer, from the first normal crossbar array connected to the input layer (in *Phase 2*) to the last normal crossbar array connected to the output layer (in *Phase N*). In *Phase 2*, the *Normal Crossbar 1* between *Layer 1* and *Layer 2* will get updated. To be specific, the gain factor of the VGAs in *Layer 1* will be set to the corresponding convergence rate, and the VGAs in *Layer 2* will be closed to guarantee that only *Normal Crossbar 1* get updated. Then the weight update operation will be performed on the *Normal Crossbar 1*. The details of the weight update operation of each normal crossbar in each phase will be discussed later in Section IV-C.

- *Phase 3 ~ N:* After *Phase 2* is completed, *Normal Crossbar 1* will get be updated. Then the rest normal crossbars will get updated through the same operations: For *Phase X* ($3 \le X \le N$), the factor of all the VGAs from *Layer 1* to *Layer X-2* will be set back to 1 to generate the current inputs (x_i in Eq. (9)) of *Layer X-1*. The gain factor of the VGAs in *Layer X-1* will be set to the corresponding convergence rate, and the VGAs in *Layer X* will be closed to guarantee that only *Normal Crossbar X-1* get updated. Then the weight update operation will be performed on *Normal Crossbar X-1*. At the same time, the copy crossbar will get the same update with the help of current mirrors.

In addition, part of the results of deviations between the actual and ideal outputs of the network will be converted to digital and then imported to the digital system to monitor the training state of the network. The digital system will adjust the gain factor of VGAs according to the training state to realize automatic adjustment of the convergence rate.

B. Copy Crossbar Technique

Copy crossbar is a technique to duplicate the weight matrixes, which do not connect with the input layer. Therefore, an N-layer network will have $N-1$ weight matrixes will require $N-2$ copy crossbar arrays.

978-1-4799-2817-0/14 $31.00 © 2014 IEEE   364

The copy crossbar is aimed at directly calculating the following part of Eq. (7)

$$\sum_{k \in nextlayer} w_{kh} \delta_k \qquad (10)$$

instead of the scheme that first cache δ_k and then import it to the network in the reveres direction (from the output layer to the input layer).

In order to configure the copy crossbar, there will be a current mirror between each memristor in the normal crossbar and the corresponding memristor in the copy crossbar. Before training, all of the crossbar arrays (both normal crossbar and copy crossbar) will be initialized to the same parameters as mentioned in Section IV-A.

In each training iteration, the current mirrors will first be closed and let the copy crossbar complete the calculation of Eq. (10). Afterwards, the weight update operation will be executed to each normal crossbar. The current mirrors will break over and help realize the same update to the memristors in the corresponding copy crossbar by keep the current passing the pair of memristors at the same size.

In general, there will be a difference between each normal crossbar and the corresponding copy crossbar. However, we observe the difference rate is usually very small ($< 10\%$ in our simulations) and our latter experiment results will demonstrate that such a small difference rate only have a small effect on the training results.

C. Weight Update Unit

Fig. 7 demonstrates the weight update unit of our mixed-signal training framework. The unit is modified upon the memristor-based neural network described in Section II-B. The major modification is the addition of the training control unit between the output nodes of the crossbar arrays and the input nodes of sigmoid circuits. The Control Signal is used to switch between the forward computation state and the weight update state.

When Control Signal is ON, the network will work at the forward computation state. All the triodes at the end of the crossbar arrays will break over. And the current mirrors will close. All the crossbar arrays will carry out the multiplication between the weight matrix and the input voltages. And the complete network will generate the actual outputs.

When Control Signal is OFF, the network will work at the weight update state. In phase of the weight update stage, the conducting state of the triodes will depend on the Sign Signal. Sign Signal comes from the results of the sign calculations in Eq. (8). The hardware implementation of the sign calculation has been shown in Fig. 4.

When Sign Signal is ON, the memristor in the positive crossbar will be chosen. The memristor in the negative crossbar is insulated because of the invertor. A voltage pulse will be added to the input nodes of the crossbar array. According to the model in [13], when the current passing through the memristor is in the same direction of the memristor's polarity, the shift of the resistance of the memristor ($-\Delta_M$) caused by a pulse can be expressed as:

$$-\Delta_M = (R_H - R_L) \cdot \mu_v \cdot t_{pulse} \cdot V_{pulse} / h^2 \qquad (11)$$

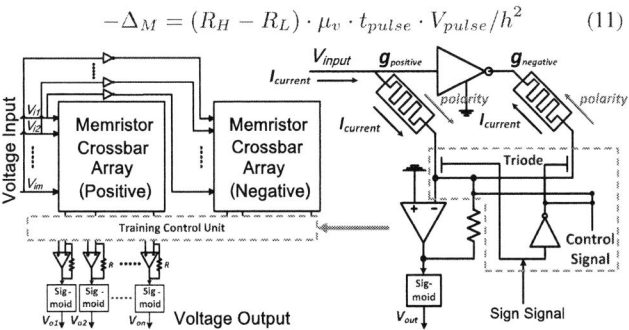

Fig. 7. Weight Update Unit

where R_L and R_H are the lowest and highest resistance of the memristor, respectively. μ_v is the equivalent mobility of dopants. t_{pulse} and V_{pulse} are the duration and value of each pulse. h is the thickness of the memristor.

As mentioned in Section III-B, the voltage pulse comes from the inputs of each layer amplified by VGAs, whose gain factor represents the convergence rate. Therefore, the shift of the resistance ($-\Delta_M$) is proportional to $\gamma \cdot ConvergenceRate_j \cdot x_i$. As the negative memristor is insulated, the total weight represented by this pair of memristors will shift as same as the positive memristor. Combining Eq. (2), the total update of the weight can be expressed as:

$$\Delta_w = R \cdot \left(\frac{1}{M_p - \Delta_{M_p}} - \frac{1}{M_p} \right) = R \cdot \frac{\Delta_{M_p}}{M_p(M_p - \Delta_{M_p})}$$
$$\approx R \cdot \frac{\Delta_{M_p}}{M_p^2} \propto \gamma \cdot ConvergenceRate_j \cdot x_i \qquad (12)$$

On the contrary, when Delta Signal is OFF, the positive memristor will be insulated and the state of the negative memristor will shift along with the input pulse because the polarities of memristors are the same as the current direction as mentioned in Section II-B. Therefore, the weight of the network will shift in the opposite direction of the pulse. In conclusion, the shift of each weight in the network will shift according to both Sign Signal and the input pulse, which realize the calculation of Eq. (9).

The weight update unit realizes an approximate configuration of the state of the memristor without memristor state tuning. In addition, we set each update of the weight to a very small value and a complete training task of the network can be realized through the accumulation of such small and approximate updates.

V. A CASE STUDY: MNIST

In order to test the effects of the proposed mixed-signal training framework, we use the MNIST database as a case study. MNIST is a widely used database with more than 60,000 handwritten digits for optical character recognition [16], [17].

A. Experimental Setup

We choose 20,000 examples of handwritten digits of '0'~'9' as the training set and 5,000 other examples for testing. The network used for pattern recognition is an 3-layer network with 784 input nodes (as the pixels of the input image is 28×28) with 300 hidden nodes. There're 10 output nodes in the network. The range of each input and output signal is $(0, 1)$. The amplitude of the output signal represents the similarity of the corresponding digit. We recognize the test image as the digit represented by the node with the biggest output.

The CPU implementation of the MNIST test is based on the Intel Math Kernel Library [18] and an Intel CPU core (i5-2320 @3.0GHz). The simulation of the mixed-signal training acceleration architecture is based on FPGA, MATLAB and SPICE, where the FPGA is used to implement and simulate the digital unit, the MATLAB is used to simulate the training process and the SPICE is used to simulate the power consumption.

The working frequency of the analog unit is 20MHz and each iteration of training costs 3 cycles. The lowest and highest resistance of the memristor (R_L and R_H) are set to 100Ω and $16k\Omega$, respectively. The equivalent mobility of dopants μ_v is set to $10^{-14} m^2 \cdot s^{-1} \cdot V^{-1}$ and the thickness (h) is set to $10nm$. The range of the input voltage V_{in} is $(0, 1)V$. As the analog unit works under 20MHz, t_{pulse} is equal to $50ns$. Taking the above value into Eq. (11), the shift of the resistance of the memristor (Δ_M) is equal to $0.0795V_{pulse}$. The ϵ in the Filter unit is set to 0.1. The $MonitorPeriod$ is set to 1,000. All the 10 output nodes ($DetectNodes$) are monitored through a 200MHz ADC. The resistance at the end the crossbar arrays (R) is

TABLE I

EXPERIMENT RESULTS OF THE MEMRISTOR IMPLEMENTATION OF TRAINING SCHEME

DecayRate	Noise Rate	Accuracy (%)	Iteration	Time (ms)	Energy (mJ)	Speed Up	Energy Saving	Accuracy Drop (%)
CPU		96.16	517,000	71067	6396030	-	-	-
1.2	0	94.84	140,000	21.0	146.89	3384	43544	1.37
	0.01	94.36	130,000	19.5	129.46	3644	49407	1.87
	0.05	94.40	134,000	20.1	132.97	3536	48102	1.83
	0.1	94.14	137,000	20.55	143.38	3458	44608	2.10
1.5	0	94.30	76,000	11.4	75.56	6234	84647	1.93
	0.01	93.96	77,000	11.55	81.43	6153	78543	2.29
	0.05	94.76	75,000	11.25	77.37	6317	82665	2.50
	0.1	93.34	66,000	9.9	67.31	7178	95021	2.93
1.8	0	93.02	50,000	7.5	49.53	9476	129128	3.27
	0.01	92.60	47,000	7.05	47.52	10080	134591	3.70
	0.05	92.70	46,000	6.9	47.04	10300	135984	3.60
	0.1	92.62	45,000	6.75	48.25	10528	132572	3.68

set to $1k\Omega$. The start and stop factor gains of the VGAs are set to 100 and 0.5.

B. Experiment Results

The simulation results of the power consumption of the analog unit is $\sim$1570.35mW. And the power of CPU and FPGA are $\sim$90W and $\sim$5W. The detailed time and energy consumption of the memristor implementation with DecayRate (in Algorithm 1) and noise rate is given in Table I. The noise in the experiment consists of both the noise of the analog signals and device deviations, such as the deviations between the copy crossbar and corresponding normal crossbar. The noise rate represents the maximum noise ratio added to the system. The accuracy represents the correct rate of the recognition of 5,000 test examples. Iteration stands for the times of iterations when the training is finished. The accuracy drop means the relative rate of the decrease of the recognition accuracy.

It can be seen that the effect of training depends both on the DecayRate and the noise rate. A lower DecayRate with a smaller noise rate will help achieve a better training result but the time and energy consumptions will become greater. Therefore, users must balance between the cost and the effect of the training.

Finally, the experiment results demonstrate that the mixed-signal acceleration framework is able to achieves a 3 orders of magnitude improvement of the training speed as well as a 4 orders of magnitude energy efficiency gains. It can be also observed that there's a slight decrease of the accuracy of recognition ($< 5\%$). However, such a cost is deserved compared with the huge performance and energy gains.

VI. SUMMARY AND CONCLUSIONS

In this work, we propose a mixed-signal training acceleration framework for memristor-based neural network. We first introduce a modified training algorithm, which enables the feasibility of the mixed-signal realization of the modified algorithm. We then propose a mixed-signal acceleration architecture for the modified algorithm, which can accomplish the training task of memristor-based neural network efficiently. Finally, we use the MNIST database as a case study to test the performance of our mixed-signal training acceleration framework for memristor-based neural network. The experiment results show that our training acceleration framework is able to realize a 3 orders of magnitude speed-up as well as a 4 orders of magnitude energy efficiency gains compared with the CPU implementation counterpart.

ACKNOWLEDGMENTS

This work was supported by National Natural Science Foundation of China (No. 61373026, No. 61261160501, No. 61028006), 973 project 2013CB329000, National Science and Technology Major Project (2013ZX03003013-003), youth talent development plan of Beijing (YETP0099) and NSF CAREER CNS-125324.

REFERENCES

[1] Jeffrey Dean et al. Large scale distributed deep networks. In P. Bartlett, F.C.N. Pereira, C.J.C. Burges, L. Bottou, and K.Q. Weinberger, editors, *Advances in Neural Information Processing Systems 25*, pages 1232–1240, 2012.

[2] Hadi Esmaeilzadeh, Adrian Sampson, Luis Ceze, and Doug Burger. Neural acceleration for general-purpose approximate programs. In *Proceedings of the 2012 45th Annual IEEE/ACM International Symposium on Microarchitecture*.

[3] Russell Beale and Tom Jackson. *Neural Computing-an introduction*. CRC Press, 2010.

[4] Dmitri B Strukov, Gregory S Snider, Duncan R Stewart, and R Stanley Williams. The missing memristor found. *Nature*, 453(7191):80–83, 2008.

[5] Beiye Liu, Miao Hu, Hai Li, Zhi-Hong Mao, Yiran Chen, Tingwen Huang, and Wei Zhang. Digital-assisted noise-eliminating training for memristor crossbar-based analog neuromorphic computing engine. In *Proceedings of the 50th Annual Design Automation Conference*, 2013.

[6] Miao Hu, Hai Li, Qing Wu, and Garrett S Rose. Hardware realization of bsb recall function using memristor crossbar arrays. In *Proceedings of the 49th Annual Design Automation Conference*, 2012.

[7] Li Boxun, Shan Yi, Hu Miao, Wang Yu, Chen Yiran, and Yang Huazhong. Memristor-based approximated computation. In *Proceedings of ISLPED 2013*.

[8] Rafael Menéndez de Llano and José Luis Bosque. Study of neural net training methods in parallel and distributed architectures. *Future Generation Computer Systems*, 26(2):267–275, 2010.

[9] Simon S Haykin, Simon S Haykin, Simon S Haykin, and Simon S Haykin. *Neural networks and learning machines*, volume 3. Prentice Hall New York, 2009.

[10] Jiquan Ngiam, Adam Coates, Ahbik Lahiri, Bobby Prochnow, Andrew Ng, and Quoc V Le. On optimization methods for deep learning. In *Proceedings of the 28th International Conference on Machine Learning (ICML-11)*, pages 265–272, 2011.

[11] Sieu D Ha and Shriram Ramanathan. Adaptive oxide electronics: A review. *Journal of Applied Physics*, 110(7):071101–071101, 2011.

[12] Wei Yi, Frederick Perner, et al. Feedback write scheme for memristive switching devices. *Applied Physics A*, 102:973–982, 2011.

[13] Robinson E Pino, Hai Li, Yiran Chen, Miao Hu, and Beiye Liu. Statistical memristor modeling and case study in neuromorphic computing. In *Design Automation Conference (DAC), 2012 49th ACM/EDAC/IEEE*, pages 585–590. IEEE, 2012.

[14] G. Khodabandehloo, M. Mirhassani, and M. Ahmadi. Analog implementation of a novel resistive-type sigmoidal neuron. *TVLSI*, 20(4):750–754, april 2012.

[15] Rainer Gemulla, Erik Nijkamp, Peter J Haas, and Yannis Sismanis. Large-scale matrix factorization with distributed stochastic gradient descent. In *Proceedings of the 17th ACM SIGKDD international conference on Knowledge discovery and data mining*, pages 69–77. ACM, 2011.

[16] Yann LeCun and Corinna Cortes. The mnist database of handwritten digits, 1998.

[17] Geoffrey E Hinton and Ruslan R Salakhutdinov. Reducing the dimensionality of data with neural networks. *Science*, 313(5786):504–507, 2006.

[18] Intel. Intel math kernel library. http://software.intel.com/en-us/intel-mkl/.

HDTV1080p HEVC Intra Encoder with Source Texture based CU/PU Mode Pre-decision

Jia Zhu, Zhenyu Liu, Dongsheng Wang*

IMETU and TNList, Tsinghua University
Beijing 100084, China.
E-mail: liuzhenyu73@tsinghua.edu.cn

Qingrui Han, Yang Song

Huawei Technologies Co., Ltd.
Bantian, Longgang District,
Shenzhen 518129, China

Abstract— HEVC doubles the coding efficiency with more than 4x coding complexity as compared to H.264/AVC. To alleviate the burden of Intra encoder, we estimate the RD-cost from the source image textures, and dynamically select two promising CU/PU mode candidates to execute exhaustive RDO processing. As integrated in our hardwired encoder, the averaged 61.7% computation complexity was saved with 4.53% rate augment. With TSMC 90nm technology, the real-time encoder for HDTV1080p at 44fps is implemented with 2269k-gate at 357MHz operating frequency.

I. INTRODUCTION

High Efficiency Video Coding (HEVC)[1] is the state-of-the-art video compression standard developed by Joint Collaborative Team on Video Coding (JCT-VC). HEVC aims to fulfill the growing requirements for higher quality and resolutions in video devices and applications, which are beyond the capabilities of H.264/AVC [2]. Thus, HEVC is devised to save around 35-40% bit-rate cost compared to H.264/AVC High Profile, while providing the equivalent object video quality[3].

To handle the large picture size, HEVC provides the flexible quad-tree structure based coding tree unit (CTU), which is composed of basic coding unit (CU), prediction unit (PU), and transform unit (TU). The $2N \times 2N$ ($N \in \{8, 16, 32\}$) CTU is the root of coding tree, which can be further split to four $N \times N$ CU, and this splitting procedure is feasible for each CU when its size is greater than 8×8. In the Intra prediction, PU size is equal to its $2N \times 2N$ CU when $N \geq 8$. On the other hand, 8×8 CU can use 8×8 or 4×4 PU sizes. The residue block of CU is further partitioned with a quad-tree structure, which is denoted as residue quad-tree (RQT), and the transform is processed on each leaf node, i.e., TU.

For each candidate parameter vector $\vec{p}$ (including CU/PU/TU structures), its coding performance is evaluated by the Lagrangian multiplier optimization technique,

$$J(\vec{p}) = D(\vec{p}) + \lambda \cdot R(\vec{p}), \qquad (1)$$

in which, $D(\vec{p})$ and $R(\vec{p})$ represent the distortion and rate costs, respectively, and λ is the Lagrangian multiplier controlling the rate-distortion trade-off. To derive the accurate values of $D(\vec{p})$

and $R(\vec{p})$, the encoder must carry out the time consuming transform, quantization, inverse quantization, inverse transform, and entropy coding procedures. The massive CU/PU/TU mode configurations lead to intensive computation of HEVC encoder. In HM reference software, the encoder exhaustively traverses all parameter vectors to find the best candidate

$$\vec{p}_o = \arg\min_{\vec{p}}\{D(\vec{p}) + \lambda \cdot R(\vec{p})\}. \qquad (2)$$

Experiments reveal that the coding time increase monotonically with the CU depth number.

To overcome the aforementioned obstacle, plethora algorithms have been proposed for fast intra coding, which fall into three primary categories: The first one substitutes the full RD-cost calculation with the low-complexity RD-cost estimation[4, 5]; The methods of the second kind rely on filtering out the most impossible candidate CU[6] or PU[7] modes; While, the third class of methods speed up the searching by early terminating the RDO execution during CU[8, 9], PU[10], or TU[11] search. However, the above fast algorithms are based on the software oriented recursive CU-depth search processing. As integrated in the real-time hardwired encoder systems, the CU level parallelism either degrades the performance or introduces considerable hardware overhead. In addition, the speedup efficiency of the early termination and the historical statistics based dynamic CU depth determination methods depends on the statistic features of the videos. Namely, the above fast algorithms could not ensure a stable speedup performance, which is essential for the real-time encoding system.

In this paper, we devise the VLSI friendly CU/PU mode decision algorithm for Intra prediction in HEVC hardwired encoder design. The proposed algorithm is primarily composed of the following three steps: First, the edge strength and direction of each $N \times N$ ($N \in 4, 8, 16, 32$) partitions in CTU are derived. Secondly, the linear model between predict error matrix of $N \times N$ partition and its edge strength matrix is constructed by using weighted least square linear regression. It should be noticed that the model is obtained by off-line learning. From the estimated prediction error, we can evaluate the RD-cost of $N \times N$ partition. Thirdly, our algorithm dynamically chooses one candidate for RDO from 16×16 and 32×32 CU modes. 8×8 CU mode RDO is always executed in our proposal. However, we dynamically determine its PU mode (8×8 or 4) according to their estimated RD-cost comparisons. We always discard 64×64 CU mode, which merely saves less

---

*This work is funded by Huawei Technologies, TNList cross-discipline foundation, the Nature Science Foundation of China (Grant No.60833004 and 60902101), and the National 863 High-Tech Programs of China (No.2012AA010905).

978-1-4799-2817-0/14 $31.00 © 2014 IEEE

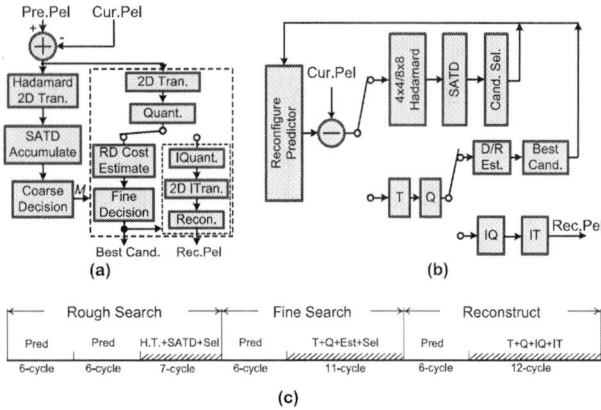

Fig. 1. HEVC prediction mode decisions for a $N \times N$ CU. (a)Flowchart for software. (b)Design for hardware. (c)Processing schedule of 4×4 CU. (Pred: Prediction signal generation; H.T.: Hadamard Transform; SATD: SATD computation; Sel: Select candidate modes; T: Transform; Q: Quantization; Est: RD-cost estimation; IQ: Inverse quantization; IT: Inverse transform)

than 0.22% rate in HD1080p and WQXGA (2560×1600) sequences. The hardwired Intra encoder is implemented integrated with the proposed fast algorithm and its performance is demonstrated.

The rest of this paper is organized as follows. In Section II, a brief introduction of our HEVC Intra encoding flow is described. In Section III, the hardware friendly mode decision algorithm is presented, including the theoretical analysis, parameters study method, and RD-cost estimation schemes. Our hardware design and experimental results are illustrated in Section IV, followed by the conclusions in Section V.

II. OVERVIEWS OF HEVC INTRA CODING

Carrying forward the advantages of H.264/AVC intra coding framework, HEVC standard also employs the block-based hybrid coding architecture, which is developed on the spatial prediction, followed by the transform coding and post processing steps. To handle the HDTV and UHDTV video sizes, HEVC brings in some advanced techniques contributing to the compression efficiency, including the quad-tree based coding unit structure, fine angular directional prediction, 16-bit length variable-block-size DCT/DST transforms, and prediction direction-based transform coefficient scanning.

The quad-tree coding block partition structure efficiently utilizes variable sizes of Coding, Prediction and Transform Units(CU/PU/TUs). As mentioned in Section I, RDO must be exhaustively executed to select the best coding parameters, i.e., CU quad-tree structure, PU and RQT partitions in each CU. In addition, to enhance the spatial prediction accuracy, HEVC employs 33 directional predictions as well as Planar and DC in PU mode search. The numerous Intra prediction modes impose a heavy burden to the encoder. The HM reference software employs the fast heuristic Intra prediction mode selection algorithm, which is composed of the low-complexity rough mode decision and the full RDO based fine mode search. Specifically, the rough decision select several candidates with the minimum SATD based RD-costs out of all 35 prediction modes. Fine

mode decision then choose the best Intra mode through the full RDO.

Even with the rough to fine heuristic RDO, the associated hardware and timing costs are both unfeasible for HDTV1080p real-time encoder implementations. The primary hindrance still comes from the full RDO. For the hardware consideration, even one variable block size DCT/DST accelerator consumes 320k-gate stand cells, let alone multiple engines required by parallelism RDO. On the other hand, the rate cost estimation in HM reference software is the CABAC alike algorithm, which is composed of the binarization and the bin-grain arithmetic coding. Even with the advanced ASIC accelerator, for one 32×32 block, the averaged 200-300 cycles are required in rate evaluation.

Our design simplifies the RD-cost estimation algorithm. Similar to literature [5], our design also obtains the distortion from quantization results. However, the original method merely saves the hardware of inverse transform and inverse quantization. Our contributions in the distortion evaluation algorithm include two aspects: First, we substitute the $N \times N$ DCT with the hardware saving $N \times N$ Hadamard counterparts as $N \geq 8$; Second, we restrict the bit-width of multipliers for SSE computation to 8-bit for saving chip area. By using binary classification and linear regression method, we develop the fast rate models, which are applied to estimate the rate according to the features of $N \times N$ quantization coefficient block. The latency of rate estimation is reduced to 3 cycles. In our proposal, the TU size is always equal to PU size, which introduces negligible coding quality loss. The flowchart of our PU mode decision algorithm and the corresponding system block are presented in Fig. 1(a) and (b), respectively. Figure. 1(c) provides the processing schedule of Luma 4×4 PU mode decision. Experiments illustrate that, on average, 1.27% rate increase is introduced by our simplified RDO.

In the VLSI implementation, parallel processing is imperative for the HDTV1080p@30fps real-time encoding. In detail, the processing latency of RDO for 8×8 CU with 4×4 PU mode is 248 clocks (220 for 4 4×4 Luma + 28 for 1 4×4 Chroma). Even for 4×4 PU mode RDO, 241MHz operating frequency is required. The serial processing of all CU modes needs 577MHz clock speed. On the other hand, with 90nm CMOS technology and traditional ASIC design flow, the typical clock speed is around 400MHz (our design is 357MHz), falling far behind the requirement.

The primitive parallel architecture, in which each PU mode search is equipped with the dedicated process engine, will consume 5282k-gates. To reduce the chip area in parallel Intra HEVC encoder, we discard the redundant CU/PU mode candidates during RDO by investigating the source image textures. Figure 2 shows the flowchart of our parallel RDO engine embedded with the proposed CU/PU mode filtering algorithm. The principle and detailed analysis of our CU/PU mode reduction will be explained in Section III. It should be noticed that, we devise the reconfigurable prediction accelerator, which is shared by all CU/PU RDO processing. The VLSI architecture and processing schedule will be described in Section IV.

978-1-4799-2817-0/14 $31.00 © 2014 IEEE

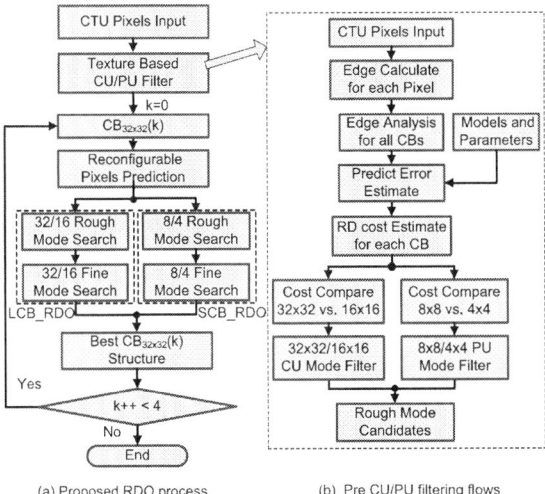

(a) Proposed RDO process   (b) Pre CU/PU filtering flows

Fig. 2. Propoed parallel RDO flowchart for a luma CTU sized 64×64, using edge based pre CU mode filtering.

III. CU/PU CANDIDATE REDUCTION USING RD-COST ESTIMATION FROM SOURCE IMAGE TEXTURE INVESTIGATION

A. Panoramic View of Proposed Method

As showed in Fig. 2, before the traditional coding configuration RDO, we first investigate the textures of source image and filter out the improper CU/PU candidates. This module is denoted as "Texture Based CU/PU Filter" in Fig. 2. As mentioned in Section II, 64×64 CU is always avoided, so the remaining CU candidates include 32×32, 16×16, and 8×8. For each 32×32 coding block ($CB_{32 \times 32}$), we dynamically discard 32×32 or 16×16 CU mode RDO from the texture investigation. 8×8 CU possesses two PU mode candidates, i.e., 8×8 and 4×4. The dedicated PU mode for each 8×8 CU is also defined during the texture preprocessing stage. In the following RDO processing, two modules (LCB_RDO and SCB_RDO) are employed to work in parallel. LCB_RDO is in charge of deriving the RD-cost of current 32×32 CB with the predefined 32×32 or 16×16 mode. Simultaneously, SCB_RDO obtains all RD-costs of 8×8 partitions with the specified PU mode. From the results of LCB_RDO and SCB_RDO, we then assemble the best coding tree structure.

The detailed flowchart of texture analysis is illustrated in Fig. 2(b). As the CTU source image is given, we can calculate the edge strength and edge direction on every pixel. The edge statistic features, such as the main edge direction, edge strength, and edge direction distributions, in any CU are obtained. According to the edge feature classification, we develop the linear relations between the pixel edge strength and its prediction error power, which will be unravel in Section B. Therefrom, we construct the RD-cost estimation method (as shown in Section C) to discard the improper CU/PU candidate modes. As our method is based on the source image investigation, and eliminates the constant computation burden, it neither degrades the pipeline performance nor incurs the encoding complexity oscillation, which are prohibited in the real-time encoding sys-

tem.

B. Linear Relation between Texture and Prediction Error

Prediction error is the primary factor that determines the coding cost. Generally, the more accurate the prediction is, the less the RD cost will be. The prediction error power at position (i, j) in a $N \times N$ CB is defined as

$$PE_k = (P_k - C_k)^2 \qquad (3)$$

where, i and j denotes the ordinate and abscissa of the pixel (left up corner is (0,0)), $k = i \cdot N + j$, P_k is the prediction pixel, and C_k is the source pixel. As we know, generating and searching the best prediction signal P_k require the intensive computation. Then, we strive for developing the model to fast estimate the power of PE_k.

The angular predictions in HEVC is on the basis of exploiting the local edge continuity. As expected, strong correlations exist between PE_k and its edge strength ES. Except for the texture features, PE_k is affected by the quantization noise QE. This is mainly because that the prediction signal P_k is deduced from the decoded pixels, which have been polluted by the quantization. From the above analysis, we suppose that the prediction error power (PE_k) stems from QE and ES, expressed as

$$PE_k \approx a \cdot Qs^2 + b_k \cdot ES_k. \qquad (4)$$

where a and b_k are linear regression parameters. The term $a \cdot Qs^2$ indicates that the quantization error power QE is linear with the square of quantization step (Qs^2).

$$Qs = 2^{Qp/6} \cdot Q[Qp\%6]$$
$$Q[i] \in \{0.625, 0.7031, 0.7969, 0.8906, 1, 1.125\} \qquad (5)$$

in which, Qp is the quantization parameter, $/$ and $\%$ are the quotient and remainder operators, respectively. ES_k at position (i, j) is proportional to the edge gradient, written as

$$ES_k = eh_k^2 + ev_k^2, \qquad (6)$$

where eh_k and ev_k denote the horizontal and vertical edge gradient, respectively.

For $N \times N$ CB, $N^2 + 1$ parameters, i.e., a and b_k ($k \in [0, N^2 - 1]$), are required. Through measurements, we can derive M sample groups, including the edge strength on each pixel $ES_k(\tau)$, the quantization error $QE(\tau)$, and the actual prediction error power $PE_k(\tau)$, with $\tau \in [0, M - 1]$. Let $\widetilde{PE}_k(\tau)$ represent the estimated prediction error power. Our target is that each prediction entry $\widetilde{PE}_k(\tau)$ approaches its actual value $PE_k(\tau)$ and the summary of $\widetilde{PE}_k(\tau)$ ($\sum_{k=1}^{N^2-1} \widetilde{PE}_k(\tau)$) also approaches the value ($\sum_{k=1}^{N^2-1} PE_k(\tau)$). Then, we have $N^2 + 1$ target functions. According to the weighted least squares estimation theory, the above question can be summarized as

$$\arg\min_{\{a, b_k\}} \left\{ \sum_{\tau=0}^{M-1} \left[\sum_{k=1}^{N^2-1} w_k \cdot (PE_k(\tau) - \widetilde{PE}_k(\tau))^2 \right. \right.$$
$$\left. \left. + w_{N^2} \cdot \left(\sum_{k=1}^{N^2-1} PE_k(\tau) - \sum_{k=1}^{N^2-1} \widetilde{PE}_k(\tau) \right)^2 \right] \right\} \qquad (7)$$

978-1-4799-2817-0/14 $31.00 © 2014 IEEE

Let assume that estimation error $\varepsilon_k = PE_k - \widetilde{PE_k}$ are uncorrelated with each other, and ε_k is a random variable with zero mean and constant variance(not changed with k). The optimum weighing vector $\vec{w}$ is

$$\begin{cases} w_k = 1 & 0 \le k \le N^2 - 1 \\ w_{N^2} = \dfrac{1}{N^2} \end{cases}. \qquad (8)$$

The solution $\vec{\theta} = (a, b_0, \cdots, b_{N^2-1})^T$ of (7) is formulated as

$$\vec{\theta} = (\boldsymbol{A}^T \boldsymbol{W} \boldsymbol{A})^{-1} \cdot \boldsymbol{A}^T \boldsymbol{W} \cdot \vec{PE}, \qquad (9)$$

in which, $\boldsymbol{A}$ is composed of M groups of measured Qs^2 and edge strengths, written as

$$\boldsymbol{A} = \begin{bmatrix} \boldsymbol{A}(0) \\ \boldsymbol{A}(1) \\ \vdots \\ \boldsymbol{A}(M-1) \end{bmatrix}. \qquad (10)$$

Each sub-matrix $\boldsymbol{A}(\tau)$ is defined as

$$\boldsymbol{A}(\tau) = \begin{bmatrix} Qs^2(\tau) & ES_0(\tau) & 0 & \cdots & 0 \\ Qs^2(\tau) & 0 & ES_1(\tau) & 0 & 0 \\ \vdots & \vdots & 0 & \ddots & \vdots \\ Qs^2(\tau) & 0 & 0 & \cdots & ES_{N^2-1}(\tau) \\ N^2 \cdot Qs^2(\tau) & ES_0(\tau) & ES_1(\tau) & \cdots & ES_{N^2-1}(\tau) \end{bmatrix}. \qquad (11)$$

The vector $\vec{PE}$ denominates the sampled prediction errors

$$\vec{PE} = \begin{bmatrix} \vec{PE}(0) \\ \vec{PE}(1) \\ \vdots \\ \vec{PE}(M-1) \end{bmatrix}, \qquad (12)$$

in which, each sub-vector is composed of all prediction error entries and their sum in one $N \times N$ CB .

$$\vec{PE}(\tau) = \begin{bmatrix} PE_0(\tau) \\ PE_1(\tau) \\ \vdots \\ PE_{N^2-1}(\tau) \\ \sum_{k=0}^{N^2-1} PE_k(\tau) \end{bmatrix}, \qquad (13)$$

$\boldsymbol{W}$ is the diagonal weighting matrix having the form of

$$\boldsymbol{W} = \mathrm{diag}(\boldsymbol{W}(0), \boldsymbol{W}(1), \cdots, \boldsymbol{W}(M-1)), \qquad (14)$$

and each sub-matrix $\boldsymbol{W}(\tau)$ is also a diagonal matrix $\boldsymbol{W}(\tau) = \mathrm{diag}(1, 1, \cdots, 1, 1/N^2)$.

To further improve the accuracy of our estimation algorithm, the sampled CBs are classified according to their texture distribution features, and each class possesses the dedicated model. As literature [12], for each CB, we can derive its edge direction histogram from its edge mapping. The edge direction histogram is composed of 33 cells corresponding to 33 prediction directions. From the distribution of histogram cells, the CB is classified with respect to edge direction homogeneity, prominent angle direction, and prominent angle strength.

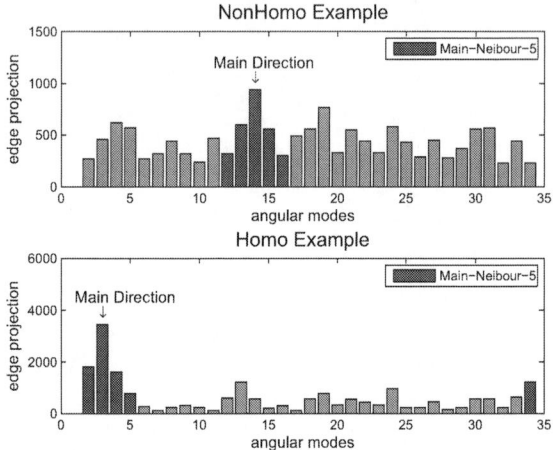

Fig. 3. Histogram of directional homogeneity examples

Fig. 4. Seven classes ($M0$ to $M6$) distinguished with the maximum edge strength.

Prominent angle direction In the histogram, the cell with the maximum amplitude points out the main angle direction, which is divided into 4 categories. The 33 prediction angles are indexed from 2 to 34 in HEVC. The first category (D0) ranges from 7 to 13, represents the horizontal alike directions; The second category (D1) includes modes from 23 to 29, represents the vertical alike directions; The third one (D2) includes -45 degree alike directions (modes 14 to 22); The other directions make up the fourth category (D3).

Directional homogeneity If the edges in a CB are concentrated in one direction, this CB is denoted as directional homogeneous; Otherwise, it is nonhomogeneous. In detail, let σ denote the sum of histogram cells of main direction and its 4 neighbors, Σ is the sum of all histogram cells, if $\sigma/\Sigma > 1.0 - 0.1 \log_2 N$, the $N \times N$ CB is directional homogeneous. Otherwise, it is labeled as nonhomogeneous. Figure 3 illustrates the homogeneous and nonhomogeneous examples.

Prominent angle strength Even though prediction error always increases monotonically with the edge amplitude, but the relationship is not linear. Therefore, we empirically develop 7 groups according to the prominent angle strength, which is depicted as Fig.4.

As employing the classification strategies, each $N \times N$ ($N \in \{4, 8, 16, 32\}$) CB has totally 56 linear model candidates. Parameters $\vec{\theta}$ are studied specially with respect to the CB size and the texture classifications. The parameters b_k of 8×8 CB with directional homogeneity and angle strength M0 are demonstrated by Fig.5. It can be observed parameter values monotonically increase with the ordinate and abscissa. This is because the degradation of prediction performance with the distance away from the boundary references. Another prominent feature is that the parameter values in D3 are much larger than those in other directional groups. With the classification based linear regression, the prediction performance is im-

978-1-4799-2817-0/14 $31.00 © 2014 IEEE

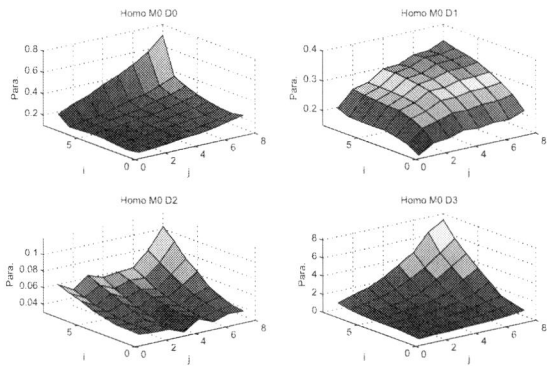

Fig. 5. Parameters b_k ($k = 0, 1, \cdots, 63$) of 8×8 CB with Homogeneous and M0 classifications.

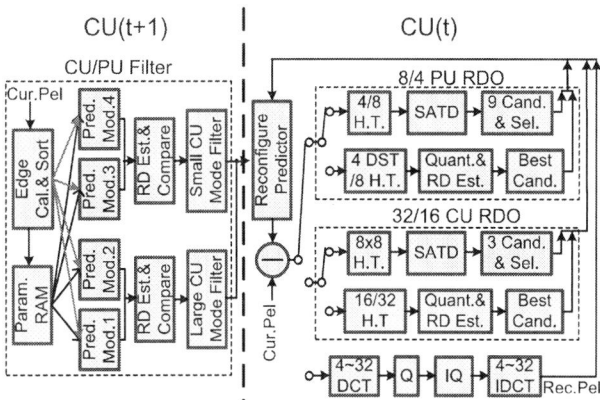

Fig. 6. Top Block Diagram of Proposed HEVC Intra Encoder.

is the additional side-information (prediction mode and code-block-flag) coding costs. In summary, $RD_{⊞N}$ is written as

$$RD_{⊞N} = \sum_{n=0}^{3} RD_{N/2}(n) + 3\frac{7}{64}(\gamma_{\mathrm{mode}} + \gamma_{\mathrm{cbf}}), \quad (18)$$

in which, $\gamma_{\mathrm{mode}} = 4$ represents the prediction mode bits, and $\gamma_{\mathrm{cbf}} = 1$ indicates the code-block-flag bit cost. It should be noticed that the ω_r value in $RD_{N/2}(n)$ is always smaller than that in RD_N.

If $RD_N \leq RD_{⊞N}$, the current CB is coded as a whole one; Otherwise, we chose sub-partition strategy in the following RDO search.

TABLE I
PARAMETER ω_r VALUES DEFINITION

$\widetilde{PE}_k/Qs^2$ \\ N	4	8	16	32
[0 , 1/8)	0	0	0	0
[1/8 , 1/4)	1/8	1/2	1/8	
[1/4 , 1/2)	1/4	1	1/4	1/2
[1/2 , 1)	1/2	4	1/2	2
[1 , 2)		16	1	8
[2 , 4)	1		2	32
[4 , 8)		32	4	64
[8 , ∞)			16	128

proved by the averaged 23%.

C. Prediction Error based RD Cost Estimation

The ultimate target is to derive the RD-cost estimation. Once the value of $\widetilde{PE}_k$ is obtained, we can evaluate the corresponding rate $\widetilde{R}_k$ and the distortion $\widetilde{D}_k$ overhead as follows.

$$\begin{cases} \widetilde{R}_k = \dfrac{7\omega_r \cdot \widetilde{PE}_k}{64} \\ \widetilde{D}_k = \omega_d \cdot \widetilde{PE}_k \end{cases} \quad (15)$$

ω_r and ω_d are the weighting factors for rate and distortion, respectively. Through theory analysis and experimental feedbacks, the values of ω_r are defined as showed in Table I, while ω_d is derived as

$$\omega_d = \begin{cases} 1 & , \quad \widetilde{PE}_k > Qs^2/16 \\ 0 & , \quad otherwies \end{cases} \quad (16)$$

For a $N \times N$ CB, if it is encoded as one partition, its RD-cost (RD_N) is estimated as

$$RD_N = \sum_{k=0}^{N^2-1} (\widetilde{R}_k + \widetilde{D}_k). \quad (17)$$

On the other hand, if the CB is partitioned into 4 sub-blocks, the corresponding RD-cost ($RD_{⊞N}$) is composed of two terms. The first one is $RD_{N/2}(n)$ for 4 sub-blocks, and second one

IV. EXPERIMENT RESULTS

The prototype HEVC Intra encoder integrated with our proposals is implemented with TSMC90nm technology. The top block diagram is shown as Fig. 6. The encoder is two CTU pipelined architecture. The first stage eliminates the redundant CU/PU modes by using our proposed methods. The directive information of the first stage is dispatched to the second stage for the RDO processing. The second stage is constituted of 4 prominent components: reconfigurable predictor, $8 \times 8/4 \times 4$ PU RDO engine, $32 \times 32/16 \times 16$ CU RDO engine, and the reconstructed datapath.

Each RDO engine is only assigned one mode, which is generated by the previous stage, to estimate its best RD-cost. It should be noticed that the predictor is shared by two RDO engines. This comes from two factors: First, the predictor is reconfigurable, which produces L modes $N \times 1$ row-wise prediction pixel in each cycle (N is the CB size and $L \times N = 128$); Second, as shown by Fig. 1, during the RD-cost estimation period (marked by the slash), $8 \times 8/4 \times 4$ RDO engine will not occupy the predictor. Therefore, the prediction engine can be used by two RDO engines alternatively. By using the variable block size unified DCT/IDCT architecture, one reconstruction datapath is shared by all CBs, when the best PU mode of a CB is decided.

All the modules for proposed RDO designs are described with Verilog HDL and synthesized with Design Compiler based on TSMC90nm 1P9M technology. The gate count and power dissipation of each primary component are shown in Table II. As compared with primitive parallel implementation (described in Section II), 57% hardware is saved. At the worst conditions (0.9v, 125°C), the maximum clock speed is 357MHz, which fulfills the real-time encoding of HD1080p@44fps. Accordingly, the power dissipation is 217.9mW.

TABLE II
HARDWARE CONSUMPTION OF PROPOSED HEVC INTRA ENCODER

Module	Pre-Mode Filter	Rcnf. Predictor	32/16 CU RDO	8/4 PU RDO	Rcns. Datapath	Total
Gates(K)	214.1	817.3	781.3	450.6	507.2	2269.0
Pwr(mW)	26.2	101.4	25.2	32.9	32.2	217.9

Modules in this talble could be referred to figure 6.

The coding performance analysis is illustrated by Table III. The original HM reference encoding (low-complexity configuration) is referred as the anchor. Twenty-two typical video sequences were tested with Intra coding configurations and $QP=\{22,27,32,37\}$. Bjontegaard Delta PSNR/Rate(BD-PSNR/Rate) is adopted to qualitatively measure the coding efficiency of our methods. The computational complexity reduction is measured in terms of the encoding time. As shown in Table III, the proposed methods averagely introduce 0.20dB BD-PSNR coding quality degradation, or equivalently 4.53% bit-rate increasing, while the complexity saving is up to 61.7% on average.

TABLE III
CODING PERFORMANCE COMPARED TO HM-10

Class	Sequence	BD-PSNR [dB]	BD-Rate [%]	Time Saved [%]
A	PeopleOnStreet	-0.21	4.61	61.4
	Traffic	-0.21	4.34	61.9
B	BasketballDrive	-0.17	6.73	61.7
	BQTerrace	-0.19	4.32	64.9
	Cactus	-0.14	4.28	72.6
	Kimono	-0.12	4.39	68.1
	ParkScene	-0.11	3.39	58.3
	Tennis	-0.18	5.92	62.0
C	BasketballDrill	-0.21	4.63	60.0
	BasketballDrillText	-0.21	4.75	60.6
	BQMall	-0.20	4.15	58.4
	RaceHorses	-0.19	3.38	58.6
D	BassketballPass	-0.24	4.80	58.8
	BlowingBubbles	-0.19	3.44	54.2
	BQSquare	-0.15	1.97	55.1
	Keiba	-0.21	4.02	60.3
E	SlideEditing	-0.41	2.94	61.1
	Vidyo1	-0.25	6.04	64.8
	Vidyo3	-0.23	5.35	63.4
	Vidyo4	-0.21	5.19	63.6
	Johnny	-0.21	5.15	63.5
	KristenAndSara	-0.24	5.86	63.6
Average		-0.20	4.53	61.7

V. CONCLUSIONS

This paper presents the fast HEVC Intra encoder architecture with the pre-CU/PU mode filtering algorithm friendly to parallel VLSI RDO processing. On average, 61.7% encoding complexity could be saved, while the incurred coding quality loss is 0.20dB BD-PSNR. Moreover, our pre-CU/PU filtering and parallel CU/PU searching design contributes to the realization of real-time encoding system, since it ensures a stable speedup performance without any computation complexity trembling. The encoder is implemented TSMC 90nm CMOS technology. With 357MHz clock speed, the proposed design supports the real-time encoding of 4:2:0 format HD1080p at the frame rate of 44fps.

REFERENCES

[1] Benjamin Bross, Woo-Jin Han, Jens-Rainer Ohm, Gary J. Sullivan, Ye-Kui Wang, and Thomas Wiegand, "High Efficiency Video Coding (HEVC) text specification draft 10," Geneva, CH, 2013, JCT-VC-L1003.

[2] T. Wiegand, G.J. Sullivan, G. Bjontegaard, and A. Luthra, "Overview of the H. 264/AVC video coding standard," *IEEE Trans. Circuits Syst. Video Technol.*, vol. 13, no. 7, pp. 560–576, 2003.

[3] J-R Ohm and Gary J Sullivan, "High Efficiency Video Coding: The Next Frontier in Video Compression [Standards in a Nutshell]," *IEEE Signal Processing Mag.*, vol. 30, no. 1, pp. 152–158, 2013.

[4] Mohammed Golam Sarwer and Lai-Man Po, "Fast bit rate estimation for mode decision of h. 264/avc," *IEEE Trans. Circuits Syst. Video Technol.*, vol. 17, no. 10, pp. 1402–1407, 2007.

[5] Yu-Kuang Tu, Jar-Ferr Yang, and Ming-Ting Sun, "Efficient rate-distortion estimation for H.264/AVC coders," *IEEE Trans. Circuits Syst. Video Technol.*, vol. 16, no. 5, pp. 600–611, 2006.

[6] Siwei Ma, Shiqi Wang, Shanshe Wang, Liang Zhao, Qin Yu, and Wen Gao, "Low complexity rate distortion optimization for hevc," in *Data Compression Conference (DCC)*. IEEE, 2013, pp. 73–82.

[7] Thaísa L da Silva, Luciano V Agostini, and Luis A da Silva Cruz, "Fast HEVC intra prediction mode decision based on edge direction information," in *Signal Processing Conference (EUSIPCO), 2012 Proceedings of the 20th European*. IEEE, 2012, pp. 1214–1218.

[8] Guilherme Correa, Pedro Assuncao, Luciano Agostini, and Luis A. da Silva Cruz, "Coding Tree Depth Estimation for Complexity Reduction of HEVC," in *Data Compression Conference (DCC)*. IEEE, 2013, pp. 43–52.

[9] Seunghyun Cho and Munchurl Kim, "Fast CU Splitting and Pruning for Suboptimal CU Partitioning in HEVC Intra Coding," *IEEE Trans. Circuits Syst. Video Technol.*, 2013(in press).

[10] Xiaolin Shen, Lu Yu, and Jie Chen, "Fast coding unit size selection for HEVC based on bayesian decision rule," in *Picture Coding Symposium (PCS), 2012*. IEEE, 2012, pp. 453–456.

[11] Su-Wei Teng, Hsueh-Ming Hang, and Yi-Fu Chen, "Fast mode decision algorithm for residual quadtree coding in HEVC," in *Visual Communications and Image Processing (VCIP), 2011 IEEE*. IEEE, 2011, pp. 1–4.

[12] Feng Pan, Xiao Lin, Susanto Rahardja, Keng Pang Lim, ZG Li, Dajun Wu, and Si Wu, "Fast mode decision algorithm for intraprediction in H. 264/AVC video coding," *IEEE Trans. Circuits Syst. Video Technol.*, vol. 15, no. 7, pp. 813–822, 2005.

Fast Large-scale Optimal Power Flow Analysis for Smart Grid through Network Reduction

Yi Liang, Deming Chen

Department of Electrical and Computer Engineering, University of Illinois at Urbana Champaign, USA
{yiliang2, dchen}@illinois.edu

Abstract— Optimal power flow (*OPF*) plays an important role in power system operation. The emerging smart grid aims to create an automated energy delivery system that enables two-way flows of electricity and information. As a result, it will be desirable if OPF can be solved in real time in order to allow the implementation of the time-sensitive applications, such as real-time pricing. In this paper, we present a novel algorithm to accelerate the computation of alternating current optimal power flow (*ACOPF*) through power system network reduction (*NR*). We formulate the *OPF* problem based on an equivalent reduced system and interpret its solution and the detailed optimal dispatch for the original power system is obtained afterwards using a distributed algorithm. Our results are compared with two widely used methods: full *ACOPF* and the linearized *OPF* with *DC* power flow and lossless network assumption, the so-called *DCOPF*. Experimental results show that for a large power system, our method achieves 7.01X speedup over *ACOPF* with only 1.72% error, and is 75.7% more accurate than the *DCOPF* solution. Our method is even 10% faster than *DCOPF*. Our experimental results demonstrate the unique strength of the proposed technique for fast, scalable, and accurate *OPF* computation. We also show that our method is effective for smaller benchmarks.

I. INTRODUCTION

The power system in United States is one of the largest and most complex cyber-physical systems in the world. In 2011, 40% of the total energy was consumed to generate electricity power by the United States [1]. To support its automation, power system needs to monitor, control and secure the grid in real time for efficient and reliable operation. Nowadays, the emerging smart grid aims to enable two-way flows of information and electricity to create an automated and advanced energy system with different decision makers involved. Timely and accurate analysis and control of such a large system are vitally important for its operating reliability and efficiency. Inaccurate or slow analysis of the power system may result in uneconomic operation of the grid, and potentially environmental pollution.

OPF has been widely used in power system planning and operation in the last 50 years, and seeks to optimize an objective function by adjusting a set of control variables subject to certain physical, operational and policy constraints. However, even today, full *ACOPF* has not been widely adopted in real-time operations for large-scale power systems because of the high computational requirement. In the smart grid paradigm, the problem size grows tremendously with the integration of renewable energy, energy storage, and demand response. In addition, more detailed model is needed to support various emerging applications, which further aggravates the computational burden.

With the advent of wholesale electricity market, *ACOPF* computation is now part of the core pricing mechanism for electricity pricing and trading. For example, an ultimate goal of Independent System Operator is the security-constrained, self-healing *ACOPF* with unit commitment over large-scale power system. Typically, this problem must be solved daily in 2 hours, hourly in 15 minutes, each five minutes in 1 minute by the Independent System Operator [2]. Currently, the problem is solved through various levels of approximations based on application and time sensitivity [2].

Although a highly nonlinear full *ACOPF* would provide the most accurate control settings in power system operation, due to the high computational demands of *ACOPF*, *DCOPF* is widely used. However, since *DCOPF* uses a linear approximation of the power flow equations and the lossless *DC* power flow assumption (the so-called *DC* power flow assumption), it is not accurate and the assumption of neglecting reactive power and power losses largely limits its application to real-world problems [3]. Currently, people use various approximation techniques and engineering judgments to explore reasonable solutions to the *ACOPF* problem. Today's inaccurate approximation may unnecessarily cost billions of dollars annually because of using inaccurate solutions [2]. It may also result in environmental pollution from unnecessary emissions and wasted energy. As a result, accelerating *ACOPF* computation and maintaining high accuracy are very important.

A wide variety of optimization techniques have been examined to solve the non-convex *ACOPF* problems, such as quadratic programming [4] and interior point method [5]. Alternative approaches include genetic algorithms (GA) [6] and particle swarm optimization [7]. However, these methods are not computationally efficient, and cannot be used in large-scale power system for real-time operation. Distributed algorithm for *ACOPF* problem was proposed in [8], where *OPF* problem for the original systems were decomposed into per-area instances. This approach assume the decoupling between different regions, which is not true for a densely interconnected power system. It can also result in very large border region which slows down the convergence and may even cause problem of non-convergence. In addition, the convergence is not guaranteed unless the objective function of the OPF problem is convex with respect to the border region variables, which is not always true in reality.

There are *NR* techniques to reduce the computational burden by finding an equivalent system. Some traditional reduction methods, such as Ward equivalent technique [9], are usually performed by computing the admittance and eliminating unnecessary elements that are not in the study area. The reduced model may lose sparsity and it may not yield the same power flow pattern as the original one. In addition, this technique is only used for power flow analysis. Alternatively, sensitivity matrix based methods, such as power transfer distribution factor (*PTDF*) based method, are used for *NR* [10] [11]. The method proposed in [10] preserves the same power flow pattern as that in the original system at the operation set point where the reduction is performed. The fact that the *NR* method depends on operation set point yields significant error when the system operates at a different set point. In [11], another *NR* method was proposed to derive an equivalent system that does not depend on set point. However, both [10] and [11] are proposed for power system long-term planning studies. The speed of generating equivalent system is not fast enough, which is not suitable for power system real-time operating purpose.

In this paper, a new method based on *NR* is proposed to solve *ACOPF* for large-scale power system. Our contributions are:

- We propose a novel method to partition the power network that can efficiently reduce the error brought by *NR* and a fast analytical approximation method to identify the parameters of the equivalent system without using *DC* power flow assumption.
- Instead of only considering the reduced equivalent system, we propose a distributed method to efficiently recover detailed solution for the original system. Congestions and transmission capacity of lines are considered in the algorithm to ensure the feasibility of *ACOPF* solution.
- This work provides an effective methodology for scalable computation of the *ACOPF* problems with high accuracy and speed.

The rest of this paper is organized as follows. In Section 2, we give the necessary background and the *ACOPF* formulation. Section 3 describes the framework and the algorithm of our *NR*-based *ACOPF* solution method. We present the numerical results in Section 4 and the conclusions in Section 5.

II. PRELIMINARIES

Over the past 50 years, steady-state *OPF* problem was well formulated and lots of variations of *ACOPF* formulations were studied. In this section, we begin with the background of power system analysis. We will introduce power flow analysis and *ACOPF* formulation.

A. Power Flow Analysis

The power flow equations constitute the steady-state model of the power system and are widely used to compute the system states once the injections and the withdrawals at each network node are specified.

We consider a power system with $N + 1$ buses and L lines. We denote by $\mathcal{N} \triangleq \{0, 1, 2, \cdots, N\}$ the set of buses, with the bus 0 being the slack bus, and by $\mathcal{L} \triangleq \{\ell_1, \ell_2, \cdots \ell_L\}$ the set of transmission lines that connect the buses in the set $\mathcal{N}$. We associate with each line $\ell \in \mathcal{L}$ the ordered pair $\ell = (i, j)$. The series admittance of line ℓ is denoted by $g_\ell - jb_\ell$. Each bus i is characterized by the voltage phasor $E_i = V_i e^{j\theta_i}$ and the net injected complex power $S_{net_i} = P_{net_i} - jQ_{net_i}$. Here V_i is the nodal voltage magnitude and θ_i is the nodal voltage phase angle. The net power injection at each node i is $P_{net_i} =$

$P_{g_i} - P_{L_i}$ and $Q_{net_i} = Q_{g_i} - Q_{L_i}$, where $P_{g_i} (Q_{g_i})$ is the real (reactive) power generated and $P_{L_i} (Q_{L_i})$ is the real (reactive) power consumed by the load at bus i. Equivalently, for each bus i, there are four real variables, P_{net_i}, Q_{net_i}, V_i, and θ_i. The power flow equations express the relationship that these variables must satisfy when the power system operates in the steady state. We denote by $\boldsymbol{Y}$ the $(N + 1) \times (N + 1)$ nodal admittance matrix, with Y_{ij} as the element in row $i + 1$ and column $j + 1$. We adopt the convention that $\boldsymbol{Y} = \boldsymbol{G} - j\boldsymbol{B}$, where $\boldsymbol{G}$ is the conductance matrix and $\boldsymbol{B}$ is the susceptance matrix. Then we have:

$$\boldsymbol{I} = \boldsymbol{Y}\boldsymbol{E}, \qquad (1)$$

where $\boldsymbol{I} = [\boldsymbol{I}_0, \boldsymbol{I}_1, \cdots, \boldsymbol{I}_N]^T$ is the vector of nodal current injection phasors; and $\boldsymbol{E} = [\boldsymbol{E}_0, \boldsymbol{E}_1, \cdots, \boldsymbol{E}_N]^T$ is the vector of nodal voltage phasors measured with respect to the ground node. In power system, we have 3 types of buses: (1) *slack bus* 0 with V_0 and θ_0 specified; (2) *P,V-bus* with P_{net_i} and V_i specified; and (3) *P,Q-bus* with P_{net_i} and Q_{net_i} specified. At each bus two of the four variables are known and the other two are unknown. At each bus i, the net complex power is given by

$$S_{net_i} = P_{net_i} - jQ_{net_i} = E_i^* I_i = E_i^* \sum_{k=0}^{N} Y_{ik} E_k. \qquad (2)$$

Therefore the power balance equations at each bus can be formulated as follows by separating the real and imaginary part.

$$P_{net_i} = \sum_{k=0}^{N} V_i V_k [G_{ik} \cos \theta_{ik} - B_{ik} \sin \theta_{ik}], \qquad (3)$$

$$Q_{net_i} = \sum_{k=0}^{N} V_i V_k [G_{ik} \sin \theta_{ik} + B_{ik} \cos \theta_{ik}], \qquad (4)$$

where $i \in \mathcal{N}$, and $\theta_{ik} = \theta_i - \theta_k$ is the voltage angle difference between bus i and k. The complex power flow in the transmission line $\ell = (i, j)$ can be formulated as

$$S_{ij} = E_i^* I_{ij}, \qquad (5)$$

The goal of power flow analysis is to solve the above nonlinear equations and obtain the voltage phasors and power flow in branches that represent the state of the system.

B. Optimal Power Flow

OPF is used to optimize the steady state performance of a power system in terms of an objective function under certain equality and inequality constraints. With specified reference bus angle, line admittance, shunt capacitances, P_{net_i} and Q_{net_i} at *P,Q-bus*, the *ACOPF* problem can be formulated as follows:

$$\min_{u} \quad f(\boldsymbol{x}, \boldsymbol{u})$$
$$s.t. \quad g(\boldsymbol{x}, \boldsymbol{u}) = \boldsymbol{0} \quad, \qquad (6)$$
$$h(\boldsymbol{x}, \boldsymbol{u}) \leq \boldsymbol{0}$$

where $\boldsymbol{u}$ is the vector of independent (or control) variables and $\boldsymbol{x}$ is the vector of dependent (or state) variables. Here,

$$\boldsymbol{u} = [P_m, V_m, t_\ell], \quad for \ \forall \ P,V\text{-}bus \ m, \qquad (7)$$
$$\boldsymbol{x} = [V_r, \theta_r, \theta_m], \quad for \ \forall \ P,V\text{-}bus \ m \ and \ \forall \ P,Q\text{-}bus \ r, \qquad (8)$$

where t_ℓ is vector of transformer tap settings. The equality constraints $g(\boldsymbol{x}, \boldsymbol{u}) = \boldsymbol{0}$ consist of nonlinear power balance equations in (3) and (4). The inequality constraints $h(\boldsymbol{x}, \boldsymbol{u}) \leq \boldsymbol{0}$ typically includes:

$$V_i^{min} \leq V_i \leq V_i^{max}, \qquad (9a)$$
$$P_{g_i}^{min} \leq P_{g_i} \leq P_{g_i}^{max}, \qquad (9b)$$
$$Q_{g_i}^{min} \leq Q_{g_i} \leq Q_{g_i}^{max}, \qquad (9c)$$
$$S_{\ell_k} \leq S_{\ell_k}^{max}, \qquad (9d)$$
$$t_{\ell_k}^{min} \leq t_{\ell_k} \leq t_{\ell_k}^{max}, \qquad (9e)$$

978-1-4799-2817-0/14 $31.00 © 2014 IEEE

for $\forall i \in \mathcal{N}$ and $\forall \ell_k \in \mathcal{L}$. Here, P_{g_i} and Q_{g_i} are the active power generation and reactive power generation of the generator at bus i. S_{ℓ_k} and t_{ℓ_k} are the power flow and the transformer tap setting on ℓ_k.

C. Applications

a) Minimization of Generation Cost

In the case of minimizing the generation cost, the objective function f is usually considered as the total active power generation cost:

$$f = \sum_{i \in \mathcal{N}_G} f^i(P_{g_i}), \qquad (10a)$$

where $\mathcal{N}_G = \{i \mid \text{bus } i \text{ is connected to a generator}\}$, and $f^i(P_{g_i})$ is the active power generation cost at bus i. $f^i(P_{g_i})$ is usually modeled by a quadratic function,

$$f^i(P_{g_i}) = a_i P_{g_i}^2 + b_i P_{g_i} + c_i, \qquad (10b)$$

where a_i, b_i, c_i are the cost coefficients. If this problem can be solved accurately in real time, optimal control operations will be updated timely to achieve the lowest generation cost and potentially a large amount of money can be saved.

b) Minimization of Line Loss

In this case, the objective function f is considered as the total loss on transmission lines [12].

$$f = \sum_{(i,j) \in \mathcal{L}} G_{ij}(V_i^2 + V_j^2 - 2V_i V_j \cos(\theta_i - \theta_j)). \qquad (11)$$

Solving this problem in real time will enable timely adjustment of control settings to reduce the line loss, which can improve the economic efficiency of power system operation. In 2011, around 7% of the total electricity was lost on the transmission line in U.S., which is worth about $3.23 billion. As a result, it is important to solve it quickly and accurately.

III. Solution Method

In this section, we present our algorithm for solving the *ACOPF* for large-scale power system. We will take the objective function of minimizing active power generation cost, which is shown in Equation (10), as example to illustrate our method.

A. Approach Overview

The idea of this approach is to accelerate the computation of the *ACOPF* solution by reducing the number of variables in the *ACOPF* thus reducing the size of the *ACOPF* problem. With *NR*, the size of $\boldsymbol{x}, \boldsymbol{u}$ and the admittance matrix $\boldsymbol{Y}$, are reduced down to the size of $\boldsymbol{x}^{eq}, \boldsymbol{u}^{eq}$, and $\boldsymbol{Y}^{eq}$ in the newly formulated *ACOPF* problem for the reduced equivalent system. We denote by $\mathcal{N}^{eq} \triangleq \{0, 1, 2, ..., N^{eq}\}$ the set of buses in the reduced system and by $\mathcal{L}^{eq} \triangleq \{\ell_1, \ell_2, \ell_3, ..., \ell_{L^{eq}}\}$ the set of transmission lines that connect the buses in set $\mathcal{N}^{eq}$. Similarly, power balance equations and line flow equations are formulated as:

$$\begin{aligned} S_{net_i}^{eq} = P_{net_i} - j Q_{net_i} &= E_i^{eq*} I_i^{eq} \\ &= E_i^{eq*} \sum_{j=0}^{N^{eq}} Y_{ij}^{eq} E_j^{eq}, \forall i \in \mathcal{N}^{eq}, \end{aligned} \qquad (12a)$$

$$S_{ij}^{eq} = E_i^{eq*} I_{ij}^{eq}, \quad \forall (i,j) = \ell \in \mathcal{L}^{eq}. \qquad (12b)$$

In order to keep the equivalence between the reduced *ACOPF* problem and the original *ACOPF* problem, power injection pattern and power flow pattern should be maintained. The goal of *NR* is to find an aggregation function, which maps the variables in the original system to the variables in the reduced system, that minimizes the mismatch between the original system and the reduced equivalent system. However, it

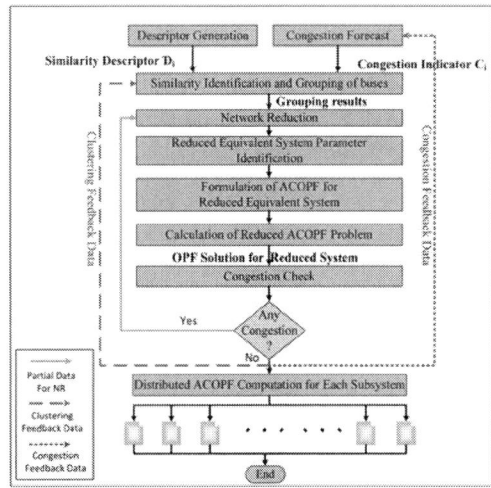

Fig. 1. Overall Algorithm Flow

is impossible to use existing methods to analytically or numerically find the exact aggregation function in real time for large-scale power system. In our approach, we approximate the aggregation function analytically by linearizing the *AC* power balance equations. Therefore, we propose our *NR*-based *ACOPF* computation algorithm.

The overall algorithm flow of the proposed method is shown in Figure 1. We first generate the similarity descriptors and the congestion indicators. By clustering, we group buses into subsystems. *NR* is performed to generate an equivalent reduced system. Then, We run *ACOPF* on the equivalent reduced system. After checking the feasibility of the solution, detailed control settings are recovered by solving *ACOPF* for each subsystem.

B. Generation of Similarity Descriptor

In order to identify the similarity between buses and group similar buses into one subsystem, a novel bus similarity descriptor containing voltage, load/generator model and surrounding network topology information is introduced here. Traditional descriptor in power system applications only considers the bus itself and ignores the interactions with its adjacent buses in its local network. Such isolation of buses cannot fully reflect bus features. We assign each bus $i \in \mathcal{N}$ the similarity descriptor $\mathcal{D}_i = \tau_i \cdot (V_i, \theta_i, M_{G_i}, M_{L_i}, \Gamma_i)$, where $\tau_i = (\tau_{V_i}, \tau_{\theta_i}, \tau_{M_{G_i}}, \tau_{M_{L_i}}, \tau_{\Gamma_i})$ is the weight vector; M_{G_i} and M_{L_i} represent the generator model and load model at bus i; $\Gamma_i = (\sum_{j,(i,j) \in \mathcal{L}} \frac{V_i}{V_j \cos \theta_{ij}}, \sum_{j,(i,j) \in \mathcal{L}} \frac{V_i}{V_j \sin \theta_{ij}})$ is the local topology descriptor. In our approach, we use polynomial "ZIP" load model [13] to describe the load model. The generators are modeled as synchronous generators and inductor generators [14] [15]. Similarity descriptor provides a measure of how "close" two buses are. It allows us to identify buses that can be merged together in the reduced network by using clustering algorithms. The generation of similarity descriptor is done off-line and will be updated when required.

C. Congestion Forecast

In power system, congestion occurs whenever the provision of transmission services required by the preferred generation/demand schedule exceeds the physical capability of the grid. In this paper, we only consider the restrictions imposed by the physical transmission capacity of the line. Congestion may increase the total generation cost because it may prevent cheap electricity generation from being dispatched. If we neglect congestion in the original system, it is possible that the calculated *ACOPF* solution after *NR* is not feasible. Therefore, it is very important to properly group buses in order to preserve the congestion profile. We propose a new method to ensure that the line flow constraints are not violated in the original system when applying the control settings derived from the *ACOPF* solution of the reduced network.

Congestion forecast is a heuristic method that takes power system field measurement data and load forecast result $S_{L_i}^f$ as input. In addition, it takes uncommitted transfer capability (*UTC*) of the successfully calculated *ACOPF* cases as feedback. *UTC* in line $\ell = (i, j)$ is defined by

$$u_{(i,j)} \triangleq S_{(i,j)}^{max} - |S_{(i,j)}|, \quad \text{where } \ell = (i,j) \in \mathcal{L}. \quad (13)$$

In order to predict congestion, we assign congestion indicator for bus $i \in \mathcal{N}$ based on the following heuristics:

(1) If the power generation capacity at bus i plus the total *UTC* of the transmission lines connected to bus i are larger than its demand, then bus i can either accommodate itself or import power from other generators.

(2) Motivated by *OPF*, power system will force cheap generators to generate as much power as they can and export it to reduce the overall cost until some factors, such as loss or congestion on the lines, limits the benefit of increasing generation output. We use the derivative of the cost function with respect to the power generation $\lambda_{P_{g_i}}^f = \frac{\partial f}{\partial P_{g_i}}$ at current operational state to evaluate it.

(3) Based on different loading and generating conditions, the system will update the control settings to the new optimal control settings by solving the *ACOPF*. Depending on the system condition, *UTC* changes correspondingly. We denote u_ℓ as the original *UTC* and $\tilde{u}_\ell$ as *UTC* after applying new optimal control settings. By comparing these two *UTC*s, we find the lines that became congested and accordingly predict which lines are going to get congested. Based on the above heuristics, we define

$$\phi_i^c = 1 - e^{\gamma_{\phi_i}(\sum_{(i,j) \in \mathcal{L}} u_{(i,j)} + S_{g_i}^{max} - S_{L_i}^f)}, \quad (14)$$

$$\alpha_i^c = \left(\frac{\lambda_{P_{g_i}}^f}{\max_i \lambda_{P_{g_i}}^f} \right)^{\gamma_{\alpha_i}}, \quad (15)$$

$$\beta_i^c = \min_{(i,j) \in \mathcal{L}} \beta_{ij}, \quad (16a)$$

$$\beta_{ij} = \begin{cases} 1, & u_{(i,j)} < \tilde{u}_{(i,j)} \\ (\frac{\tilde{u}_{(i,j)}}{u_{(i,j)}})^{\gamma_{\beta_i}}, & otherwise \end{cases}, \quad (16b)$$

where ϕ_i^c indicates the impact of supply demand balance on congestion; α_i^c reflects impact of power generation cost on congestion. β_i^c indicates the possibility of lines connected to bus i getting congested after applying the optimal control settings. $\boldsymbol{\gamma_i} = (\gamma_{\phi_i}, \gamma_{\alpha_i}, \gamma_{\beta_i})$ is the weight vector. Note that $\gamma_{\phi_i}, \gamma_{\alpha_i} > 0$ and $0 < \gamma_{\beta_i} < 1$. We define the congestion indicator C_i:

$$C_i = \phi_i^c * \alpha_i^c * \beta_i^c. \quad (17)$$

It is obvious that $\phi_i^c, \alpha_i^c, \beta_i^c \in [0, 1]$, thus $C_i \in [0, 1]$. The congestion indicator C_i for bus i is assigned to be 1 when bus i is connected to lines that are susceptible to congestion. We tend to isolate bus i if C_i is close to 0 and group i into a subsystem if C_i is close to 1.

D. Similarity Identification and Grouping of Buses

In the similarity identification process, each point is represented by Ψ_i, which is defined by

$$\Psi_i = \begin{cases} 0, & C_i < \delta \\ \mathcal{D}_i, & C_i \geq \delta \end{cases}, \quad (18)$$

where δ is a threshold for congestion indicator. System operator select δ to meet their accuracy and performance requirements. Hot start K-means algorithm is used to cluster the buses [15]. The most recent historical clustering result is used as the start point to improve the convergence speed of K-means algorithm. Due to the special physical features of slack bus and transformers, we isolate slack bus and make sure that lines with transformers are not grouped into subsystems unless its tap ratio is close to 1. After the clustering process, the system is then divided into S subsystems. The set $\mathcal{N}$ is divided into S subsets, where $\mathcal{N}_k \in \mathcal{N}$ and $\mathcal{N}_k \cap \mathcal{N}_m = \emptyset$ for $\forall k, m \leq S$. Subsystem k contains all the buses in $\mathcal{N}_k$. Let $c_k \in \mathcal{N}_k$ denote the centroid bus in subsystem k. Value at bus c_k is the average value of the cluster.

E. Network Reduction and Reduced System Generation

The *NR* process follows the following strategy:
a. Buses inside one subsystem are aggregated into one bus;
b. Lines between two subsystems are aggregated into one line;
c. Lines inside one subsystem are ignored.

The power network parameters are approximated to maintain the same power injection pattern and power flow pattern as the original system. We propose a fast method to approximate the aggregation function.

1. Power Demand and Generation in Equivalent System

Power demand $S_{L_k}^{eq}$ and power generation $S_{g_k}^{eq}$ at bus k in the equivalent system is calculated as follows.

$$S_{L_k}^{eq} = \sum_{i \in \mathcal{N}_k} S_{L_i} \quad \text{and} \quad S_{g_k}^{eq} = \sum_{i \in \mathcal{N}_k} S_{g_i}, \quad (19)$$

where S_{L_i} and S_{g_i} are the power demand and power generation at bus i in the original system.

2. Bus Voltage in Equivalent System

Since subsystem k is aggregated into bus k in the equivalent system, E_k^{eq} is approximated by the voltage of the centroid bus c_k in subsystem k.

$$E_k^{eq} = E_{c_k}. \quad (20)$$

3. Equivalent Line Admittance Approximation

Traditionally, many approaches identify the parameters of the equivalent system by calculating sensitivity matrix. However, that kind of approach is computationally expensive, especially for the large-scale power system, as the calculation of sensitivity matrix may take from minutes to hours to complete. In our approach, we approximate the parameter by the linearized power balance equations. We merge the power balance equations (3) and (4) of the buses in one subsystem and generate power balance equations for the single equivalent bus. Similarly, we merge the line flow equations from (5) where line flows in the equivalent system are

$$S_{mn} = \sum_{r \in \mathcal{N}_m, v \in \mathcal{N}_n} S_{rv}, \quad (21)$$

In order to maintain the same power injection pattern and the same power flow pattern, the line admittance matrix is approximated by

$$Y_{ij}^{eq} = \sum_{s \in \mathcal{N}_i, t \in \mathcal{N}_j} \left(\frac{V_s V_t \cos \theta_{st}}{V_{c_i} V_{c_j} \cos \theta_{c_i c_j}} G_{st} + j \frac{V_s V_t \sin \theta_{st}}{V_{c_i} V_{c_j} \sin \theta_{c_i c_j}} B_{st} \right). \quad (22)$$

F. Solve ACOPF for Reduced System

In order to perform *ACOPF* computation, new objective function and constraints after *NR* is generated based on (16)∼(19). Several generators are aggregated into a single bus in the equivalent system. The cost function of an equivalent generator is greedily changed to a piecewise function:

$$f^k(P_{g_k}^{eq}) = \min_{i \in \mathcal{N}_k} \sum_i f^i(P_{g_i})$$
$$s.t. \quad P_{g_k}^{eq} = \sum_{i \in \mathcal{N}_k} P_{g_i}, \quad \forall k \in \mathcal{N}^{eq}. \tag{23}$$

The equality constraints, which are the power balance equations are changed to

$$S_i^{eq} = P_i^{eq} - jQ_i^{eq} = E_i^{eq*} \sum_{k=0}^{N^{eq}} E_i^{eq} Y_{ik}^{eq}. \tag{24}$$

Inequality constraints are changed based on (16)∼(19). Line limits in the equivalent system are relaxed to the sum of the corresponding line limits in the original system. We find the optimal solution for the reduced equivalent system by performing *ACOPF* analysis. Further computation is needed to find the optimal solution to the original system.

G. Congestion Check

Based on the *ACOPF* solution for the reduced system, the feasibility of the solution is checked by performing power flow analysis while considering the constraints for each subsystem in parallel. As shown in Figure 1, if there are congested lines detected, we isolate the related buses, remove the congested lines out of the subsystem and go back the *NR* step.

H. Distributed ACOPF for Each Subsystem

The *ACOPF* solution for the reduced system gives the sum of the control variables inside each area. To decide the optimal dispatch inside the subsystem, it is still an *ACOPF* problem but with a smaller size and interchange power specified. Thus, *ACOPF* is computed to find the optimal settings for each subsystem. Finally, we obtain the detailed solution to the original *ACOPF* problem. With the nature of such a coarse-grained framework, we are able to distribute the computation of *ACOPF* for S subsystems to S processors to improve the speed.

We use primal-dual interior point method as the algorithm to solve the optimization problem. It is worth mentioning that our framework works for different solvers and is in parallel with the performance of optimization problem solver.

IV. NUMERICAL RESULTS

To test the proposed fast *ACOPF* computation algorithm for large-scale smart grid, we use two standard IEEE test power system and two modified large test system published in Matpower [16]. They are summarized in Table I. In our tests, we use the total active power generation cost as the objective function of the *ACOPF* problem. We run all the tests on a laptop, which has Intel Core2 Duo Processor of 2.26GHz and 2GB memory.

TABLE I
TEST BENCHMARK

Benchmark	Bus No.	Branch No.	Generator No.
IEEE 30-bus	30	41	6
IEEE 300-bus	300	411	69
Case 3120sp	3,120	3,693	505
Case 21k	21,084	25,001	2,692

TABLE II
30-BUS SYSTEM TEST RESULTS

	Standard 30-bus	With 5% DI	With 10% DI	With Congestion
Initial ($/h)	875.28	940.40	1008.05	875.28
ACOPF ($/h)	802.20	854.41	907.59	947.44
DCOPF ($/h)	806.97	859.70	913.44	967.67
PM ($/h)	802.35	854.54	907.69	947.63
DC error* ($/h)	4.77	5.26	5.75	20.23
DC error*	0.595%	0.612%	0.634%	2.135%
PM error* ($/h)	0.15	0.13	0.10	0.19
PM error*	0.018%	0.015%	0.011%	0.020%
Improvement**	96.86%	97.53%	98.26%	99.06%

PM: Proposed Method; DC error: DCOPF error; DI: Demand Increase.
*PM/DCOPF Error Compared to ACOPF
**PM Accuracy Improvement Compared to DCOPF

A. Numerical Results for IEEE 30-bus

IEEE 30-bus standard load-flow test system is used as a benchmark here. Figure 2 shows the network of IEEE 30 bus system and it is partitioned into 6 subsystems. Two modified IEEE 30-bus system with 5% more load demand and 10% more load demand and a modified IEEE 30-bus system with congestion are used to demonstrate the robustness of our proposed method including congestion forecast and congestion check. Real power costs for IEEE 30-bus test system were adapted from [10].

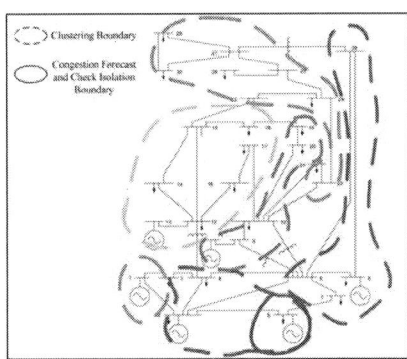

Fig. 2. Clustering Results - IEEE 30 Bus Test System

In this experiment, the congestion indicator threshold is set as 10% of the max value of all the congestion indicators C_is. For the non-congested test system, no additional bus is isolated. As shown in Figure 2, the dash lines are the boundaries of the subsystems. To illustrate the capability of our congestion forecast module, we set the transmission capacity in line 2-5 to be 32 MW while the active power flow in this line was 63.01 MW in standard case. Then bus 5 is isolated and the solid line shows the isolation.

Table II shows the experimental results for IEEE 30 bus test system. Initially, we set the power generation to be {260.9, 40.0, 0.0, 0.0, 0.0, 0.0} (MW), which is a feasible setting for the test system, the generation cost is 875.28 $/hr. By using the proposed method, the optimal setting is {178.91, 48.50, 21.18, 21.14, 11.93, 11.40} (MW), with the generation cost of 802.35 $/hr. Note that bus 1 is slack bus and we don't control the active power output. The total generation cost is reduced by 8.31%. The proposed method has 0.016% error on average compared to the most accurate full *ACOPF*. The proposed method reduces the error by 97.93% on average compared to *DCOPF*. The error of *DCOPF* is about 43 times larger than the proposed method in congestion free test systems. In the congestion case

978-1-4799-2817-0/14 $31.00 © 2014 IEEE

TABLE III
ACCURACY EVALUATION

	IEEE 30-bus	IEEE 300-bus	Case 3120sp	Case 21k
ACOPF ($/h)	802.20	719,725	2,142,704	2,732,880
DCOPF ($/h)	806.97	724,171	2,165,940	2,925,892
PM ($/h)	802.35	721,967	2,145,385	2,779,782
DC error ($/h)	4.77	4,446	23,236	193,012
DC error *	0.595%	0.618%	1.084%	7.06%
PM error ($/h)	0.15	2,242	2,681	46,902
PM error *	0.019%	0.311%	0.125%	1.72%
Improvement **	96.8%	49.6%	88.5%	75.70%

PM: Proposed Method; DC error: DCOPF error.
*PM/DCOPF Error Compared to ACOPF
**PM Accuracy Improvement Compared to DCOPF

TABLE IV
COMPUTATION TIME

	IEEE 30-bus	IEEE 300-bus	Case 3120sp	Case 21k
ACOPF (s)	0.6510	1.312	15.250	2552.8
DCOPF(s)	0.4720	0.5109	5.6130	400.7
PM (s)	0.4946	0.7966	7.2547	364.0
PM Speedup Compared to ACOPF	1.32X	1.63X	2.12X	7.01X
PM Speedup Compared to DCOPF	0.85X	0.64X	0.77X	1.10X

DCOPF has larger error which is about 2.1% and the error of DCOPF is about 106 times larger than the proposed method. Our proposed method handles congested systems much better than DCOPF.

B. Numerical Results for Larger Benchmarks

We also test our algorithm on larger benchmarks, including IEEE 300-bus test system, case 3120sp and case 21k from Matpower. In this approach, a 300-bus system is reduced to an 89 bus system with 112 lines; 3120-bus system is reduced to a 449 bus system with 565 lines; and 21k-bus system is reduced to a 4628 bus system with 5824 lines. Table III shows the accuracy of proposed method. The proposed method has 0.54% error on average (1.72% error for the 21k-bus system) compared to the most accurate full ACOPF. The proposed method reduces the error by 77.6% on average (75.7% for the 21k-bus system) compared to DCOPF. As the size of the power system increases, the error of obtaining optimal generation cost also increases. For 21k-bus system, 7.06% error was observed in DCOPF. The power system will unnecessarily lose 193,012 dollars per hour, which is 1.69 billion dollars per year. Our method can provide accurate solution to ACOPF problems that can reduce the error by 75.7% compared to DCOPF. Thus we can save 146,110 dollars per hour, which is 1.28 billion dollars per year.

Table IV shows the computation time of the proposed method. Compared to full ACOPF, the proposed method achieves 1.32X ~ 2.12X speedup for small benchmarks (30-bus, 300-bus, 3120-bus) and 7.01X speedup for large benchmark (21000-bus). The proposed method is slower than DCOPF for small benchmarks, but is faster than DCOPF for the large benchmark.

Our proposed method achieves better accuracy for all test systems compared to DCOPF. For large systems, the proposed method has advantage over DCOPF in terms of both accuracy and speed.

V. CONCLUSIONS

ACOPF is very important in power system operation. In some applications, it cannot be approximated by DCOPF because of the DC power flow assumption. In addition, the poor accuracy of DCOPF results in the great loss of social welfare. Therefore, faster ACOPF algorithm need to be developed for large-scale power system. In this work, we propose a fast ACOPF analysis framework through power system network reduction to speed up the computation of ACOPF problems. This distributed framework works with different ACOPF solvers, such as primal-dual interior point method. We demonstrate that our approach can achieve 1.32X to 7.01X speedup over full ACOPF while just introducing 0.54% error on average. With congestion forecast and check, as long as ACOPF can converge to the optimal solution, our proposed method can find an optimal solution, which demonstrate the robustness. Compared to the widely used DCOPF, we reduce the error by 77.6% on average. It can potentially save millions of dollars in smart grid operation. Also, experimental results show that the computation time of our algorithm grows almost linearly. The proposed method can be used to solve ACOPF for large-scale power system in many applications, such as operational reliability analysis and power market management.

REFERENCES

[1] U.S. Department of Energy, Annual Energy Review, 2011.
[2] Mary B. Cain, Richard P. ONeill, Anya Castillo, "History of Optimal Power Flow and Formulations," *FERC technical paper*, 2012
[3] B. Stott, J. Jardim, and O. Alsac, "Dc power flow revisited," *IEEE Trans. Power Syst.*, vol. 24, pp. 1290-1300, 2009.
[4] K. Aoki, A. Nishikori, and R. Yokoyama, "Constrained load flow using recursive quadratic programming," *IEEE Trans. Power Syst.*, vol. 2, no. 1, pp. 816, 1987.
[5] Y.-C. Wu, A. Debs, and R. Marsten, "A direct nonlinear predictor corrector primal-dual interior point algorithm for optimal power flows," *IEEE Trans. Power Syst.*, vol. 9, no. 2, pp. 876-883, 1994.
[6] L. Lai, J. Ma, R. Yokoyama, and M. Zhao, "Improved genetic algorithms for optimal power flow under both normal and contingent operation states," *Int. J. Elect. Power Energy Syst.*, vol. 19, no. 5, pp. 287-292, 1997.
[7] A. Abido, "Optimal power flow using particle swarm optimization," *Int. J. Elect. Power Energy Syst.*, vol. 24, no. 7, pp. 563-571, Oct. 2002.
[8] R. Baldick, B. H. Kim, C. Chase, and Y. Luo "A fast distributed implementation of optimal power flow," *IEEE Trans. Power Syst.*, vol. 14, pp. 858-863, Aug. 1999.
[9] J. B. Ward, "Equivalent circuits for power flow studies," *Electrical Engineering*, vol. 68, no. 9, pp. 794-794, 1949.
[10] X. Cheng and T. Overbye, "PTDF-based power system equivalents," *IEEE Trans. Power Syst.*, vol. 20, no. 4, pp. 1868-1876, 2005.
[11] H. Oh, "A new network reduction methodology for power system planning studies," *IEEE Trans. Power Syst.*, vol. 25, no. 2, pp. 677-684, 2010.
[12] U. Leeton, D. Uthitsunthorn, and U. Kwannetr, "Power loss minimization using optimal power flow based on particle swarm optimization," in *Proc. Int. Conf. Elect. Eng.*, 2010, pp. 440-444.
[13] I.T.F. on Load Representation for Dynamic Performance, "Bibliography on load models for power flow and dynamic simulation," *IEEE Trans. Power Syst.*, vol. 10, no. 1, pp.523-538, Feb. 1995.
[14] "IEEE Guide for Synchronous Generator Modeling Practices and Applications in Power System Stability Analyses," *IEEE Std 1110-2002 (Revision of IEEE Std 1110-1991)*, pp.1-72, 2003
[15] H. Li, Z. Chen, and L. Han, "Comparison and Evaluation of Induction Generator Models in Wind Turbine Systems for Transient Stability of Power System," in *IEEE, Int. Conf.Power Syst. Tech.*, Oct. 2006
[16] R. Zimmerman, C. Murillo-Sanchez, and R. Thomas, "MATPOWER's Extensible Optimal Power Flow Architecture," in *Power Energy Society General Meeting. 2009. PES '09. IEEE*, 2009, pp. 1-7.

978-1-4799-2817-0/14 $31.00 © 2014 IEEE

Storage-less and converter-less maximum power point tracking of photovoltaic cells for a nonvolatile microprocessor

Cong Wang*, Naehyuck Chang[†], Younghyun Kim[†], Sangyoung Park[†],
Yongpan Liu*, Hyung Gyu Lee[‡], Rong Luo* and Huazhong Yang*

Tsinghua National Laboratory for Information Science and Technology, Tsinghua University, China*
Department of EECS/CSE, Seoul National University, Korea[†]
School of Computer and Communication Engineering, Daegu University, Korea[‡]
ypliu@tsinghua.edu.cn*, naehyuck@elpl.snu.ac.kr[†], hgichon2@hotmail.com[‡]

Abstract—This paper pioneers the maximum power point tracking (MPPT) of photovoltaic (PV) cells that directly supply power to a microprocessor without an energy storage element (a battery or a large-size capacitor) nor power converters. The maximum power point tracking is conventionally performed by an MPPT charger that stores in the energy storage element, and a voltage regulator (typically a DC-DC converter) produces a proper voltage level for the microprocessor. The energy storage element is an energy buffer and makes it possible to perform MPPT of the PV cells and power management of the microprocessor independently. However, the energy storage element, MPPT charger and DC-DC converter cause seriously limited lifetime (when a typical battery is adopted), significant energy loss (typically over 20%), increased weight/volume and high cost, etc. The proposed method enables extremely fine-grain dynamic power management (DPM) in every a few hundred microseconds and performs the MPPT without using an MPPT charger and a DC-DC converter as well as an energy storage element. We achieve 84.5% of energy harvesting efficiency using the proposed setup with huge reduction in cost, weight and volume, and extended lifetime, which is not even numerically comparable with conventional MPPT methods.

I. INTRODUCTION

Energy harvesting gives a potential to make an electronics system sustainable and maintenance free. Any form of energy such as kinetic, thermal, etc., can be transformed to electrical energy, and photovoltaic (PV) cell is one of the most practical energy harvesting for electronics circuits in terms of power capacity, voltage and current magnitudes, cost, volume, weight, and so forth.

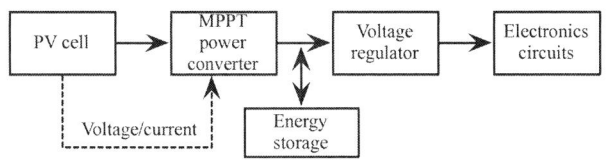

Fig. 1. Typical architecture of a solar energy harvesting system with MPPT.

Fig. 1 shows a typical architecture of a solar energy harvesting system with the maximum power point tracking (MPPT) [1] or maximum power transfer tracking (MPTT) [2]. A PV module requires MPPT because its internal impedance is not low enough to be a voltage source like conventional batteries. Furthermore, the MPP largely varies with the solar irradiance and temperature, which requires continuous tracking of the MPP. Common MPPT continuously maintains the MPP by the use of a perturb-and-observe method [3]. The MPP current can hardly match with the power supply current

This work was supported in part by the NSFC under grant 61271269, High-Tech Research and Development (863) Program under contract 2013AA01320, the Importation and Development of High-Caliber Talents Project of Beijing Municipal Institutions under contract YETP0102, the Center for Integrated Smart Sensors funded by the Ministry of Science, ICT & Future Planning as Global Frontier Project (CISS-201373718) and the NRF of Korea funded by the MEST (No. 2013035079). The ICT at Seoul National University provides research facilities for this study. Jaehyun Park of Seoul National University assisted with the evaluation board development.

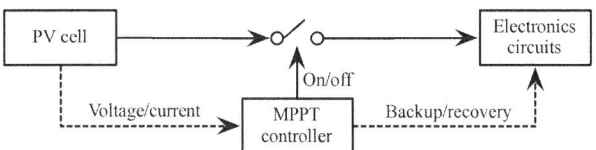

Fig. 2. Proposed converter-less and storage-less energy harvesting system architecture with MPPT.

of the electronics. Power supply current of a microprocessor continuously fluctuates. In other words, the PV energy harvesting system will generally fail MPPT if the electronics is directly connected to the PV cell without an energy buffer. The PV cell output voltage largely fluctuates as the microprocessor power supply current changes, which makes the PV cell output voltage change. In addition, the terminal voltage of typical energy storage devices generally does not match with the supply voltage of microprocessors. For example, Li-ion battery single cell voltage is commonly 3.7 V. The above reasons make a PV energy harvesting system to equip with two cascaded power converters (an MPPT charger and a DC-DC converter) and energy storage device, a battery or equivalent, in the middle. Such a setup has been widely used undoubtedly.

However, there are serious disadvantages from the power converters and the energy storage device though they were mandatory to make a PV energy harvesting system functional. First, the power converters and energy storage devices are typically the most expensive, heaviest, and largest components in a PV energy harvesting system. It is difficult and costly to integrate the expensive and bulky passive components, such as bulk capacitors, multilayer ceramic capacitors, inductors, etc. on chip. Second, the energy storage device, typically a rechargeable battery, is the primary component that significantly shortens the system lifetime or requires periodic maintenance of the system. These two factors seriously discourage the deployment of small, low-cost and self-sustainable ubiquitous smart sensors in various applications.

Elimination of the power converter and energy storage device gives a great freedom in size, cost, volume, lifetime, and you name it. However, there are obvious obstacles in such a radical approach. First, the MPP current and voltage should be exactly matched with the electronics supply voltage and current. Typical microprocessors cannot fulfill this requirement in general. Second, the electronics should be functional even after power interrupt when the solar irradiance is not strong enough to operate the electronics.

In this paper, we propose a converter-less and storage-less energy harvesting system that performs the MPPT. To the best of knowledge, this is the first paper that attempts storage-less and converter-less MPPT. We overcome the above mentioned obvious obstacles by the use of i) PV cell V-I characteristics and ii) a nonvolatile microprocessor. The PV cell MPP voltage is maintained within a narrow range regardless of the solar irradiance. In other words, the PV

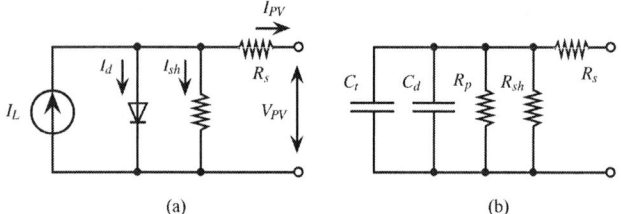

(a)   (b)

Fig. 3.   (a) A DC equivalent circuit of a PV module; (b) An AC equivalent circuit of a PV module.

cell produces an almost-constant voltage output as long as we keep track of the MPP current. The nonvolatile microprocessor provides extremely fast context saving and restoration before and after power interrupt, which makes a very fine-grain dynamic power management (DPM) feasible. Fast enough DPM makes the PV cell keep the MPP as if the electronics load current is a DC current as long as the average current is the same as the MPP current.

There are huge advantages from the proposed method as long as the target applications fit to the design philosophy. The advantages in terms of lifetime, cost, volume, weight, etc., are not even numerically comparable with conventional energy harvesting systems. The proposed method is not operational after sunset. However, the proposed PV energy harvesting systems are functional even with a very weak solar irradiance thanks to a very fine-grain DPM in a few hundred microseconds, which make a ultra low duty operation possible.

Experiment results show that the proposed storage-less and converter-less PV energy harvesting system with MPPT mitigates the high energy loss stemmed from power converters and state transitions, achieving an overall system efficiency of 84.5%. The proposed system executes $40\% - 2.5X$ more tasks than conventional baseline systems given the same PV cells and solar irradiance in a day.

II. COMPONENT MODELS

A. Photovoltaic Module

We use a well-known single diode equivalent circuit model to describe the DC characteristics of a PV module as shown in Fig. 3. The I-V characteristic of a PV module is represented by (1) and (2) [4].

$$I_{pv} = I_L - I_0 \left(e^{\frac{V_{pv}+I_{pv}R_s}{a}} - 1 \right) - \frac{V_{pv} + I_{pv}R_s}{R_{sh}}, \quad (1)$$

$$a \equiv \frac{N_s n_1 k T_c}{q}, \quad (2)$$

where N_s, n_1, k, T_c and q indicate the number of cells in series, the ideality factor of the diode, the Boltzmann's constant, the module temperature, and the electron charge, respectively.

We extract the five parameters (I_L, I_0, n_1, R_s, R_{sh}) with a curve-fitting method. We first measure a set of voltage-current points under a reference condition (solar irradiance of 200 W/m^2 and temperature of 27 $°C$), and then obtain the best-fit parameters that minimize the deviation between the curve provided by the model and the measured points. These parameter values are only valid for the reference condition. We use (3), (4) and (5) introduced in [5] to obtain PV characteristics at other temperature and solar irradiance conditions.

$$I_L = \frac{S}{S_{ref}} \left[I_{L,ref} + \alpha_{I_{sc}} (T_c - T_{c,ref}) \right], \quad (3)$$

$$\frac{I_0}{I_{0,ref}} = \frac{T_c}{T_{c,ref}} \exp\left[\frac{E_g}{k} \left(\frac{1}{T_c} - \frac{1}{T_{c,ref}} \right) \right], \quad (4)$$

$$\frac{R_{sh}}{R_{sh,ref}} = \frac{S_{ref}}{S}, \quad (5)$$

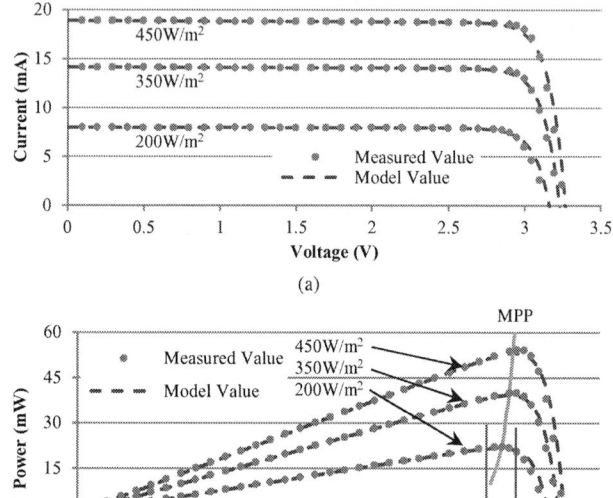

(a)

(b)

Fig. 4.   (a) I-V curve; (b) P-V curve of a 4.5×5.5 cm^2 experimental photovoltaic module under different solar insolation at the temperature of 300K. An MPPT window is shown on the curve for the irradiance of $200W/m^2$.

where S and E_g indicate the solar irradiance in W/m^2, and the material band gap energy, respectively. Symbols with subscript $[*]_{ref}$ are values under the reference condition. Following the investigation in [4], we assume the series resistance R_s and diode ideality factor n_1 are independent of temperature and solar irradiance in this paper.

TABLE I
MODEL PARAMETERS OF THE EXPERIMENTAL PV MODULE
(TEST CONDITION: $S = 200W/m^2$, $T_c = 300K$).

I_L	I_0	n_1	R_s	R_{sh}	N_s
8.01mA	$1.01^{-16}A$	1.885	0.17Ω	$22.36k\Omega$	2

Table I lists the five parameters of a PV module extracted from measurement results. As illustrated in Fig. 4(a), the model value fits well with the actual characteristic of the PV module.

The MPPT technique is widely used in PV systems. The MPP voltage is relatively stable regardless of the solar irradiance as shown in Fig. 4(b). We emphasize that the PV module output voltage variation is small enough even with different solar irradiance levels as long as we track the MPP. It is possible to remove the DC-DC converter between the PV module and the load device if the MPP voltage is also within the operating range of the load device. In other words, tracking the MPP implies output voltage regulation of the PV module.

B. DC-DC Converter

DC-DC converters convert and regulate the input voltage to a desired output voltage. Switching-mode DC-DC converters are generally used for voltage regulation as they are known to be more power efficient than linear voltage regulators [6, 7]. The efficiency of a switching DC-DC converter is defined as follows.

$$\eta_{conv} = \frac{P_{out}}{P_{in}} = 1 - \frac{P_{conv}}{P_{in}}, \quad (6)$$

where P_{in}, P_{out} and P_{conv} are input power, output power, and power loss in the converter, respectively.

The converter efficiency typically ranges from 10% to 80% due to the conduction power dissipation, gate drive power dissipation and controller power dissipation [8]. Moreover, conventional energy harvesting system is typically equipped with two stages DC-DC converters, as shown in Fig. 5(a). One is the input-stage power converter, which is typically an MPPT converter that charges the energy storage device with the PV module power. The other is a voltage regulator at the output stage. This two-stage converter system significantly degrades the power efficiency of the whole PV energy harvesting system. It would give a great potential to achieve tremendous power efficiency enhancement if we remove the power converters from a PV energy harvesting system.

C. Nonvolatile Processor

A nonvolatile processor is highly desired in a very frequent DPM because of its extremely low overheads in system states backup and recovery [9]. A recent work fabricated a nonvolatile microprocessor named THU1010N using a 130 *nm* ferroelectric technology from Rohm, where nonvolatile elements are incorporated into each conventional volatile flip-flop [10]. Data movement between flip-flops and their local nonvolatile elements are performed in parallel, thus expediting the backup and recovery processes.

TABLE II
SPECIFICATIONS OF THE NONVOLATILE MICROPROCESSOR USED IN EVALUATION PROTOTYPE.

Component	Part No.	Voltage Range	Power Consumption	Transition Overhead	
				Backup	Recovery
NVCPU	THU1010N	2.7 V – 3.6 V	5.78 *mW*	8 *μs*	3 *μs*

Table II summarizes the power consumption and transition overheads of the THU1010N nonvolatile microprocessor that we use in the proposed storage-less and converter-less PV energy harvesting system with MPPT. The backup and recovery time of the nonvolatile microprocessor are only 8 *μs* and 3 *μs*, respectively, being three to four orders of magnitude faster than typical microprocessors.

III. STORAGE-LESS AND CONVERTER-LESS MPPT

A. System Architecture

Fig. 5(a) illustrates the architecture of conventional solar energy harvesting system, which is in a typical two-tier setup. The MPPT charger maintains the PV operation at the MPP, extracting as much power as possible from the PV module, and stores it in the energy storage device. The harvested energy is retrieved from the storage device and delivered to the load device via power converters.

We mitigate the efficiency loss caused by the DC-DC converters, the degradation of lifetime, weight, volume stemmed from the energy storage unit, and the potential high cost incurred from both of them,

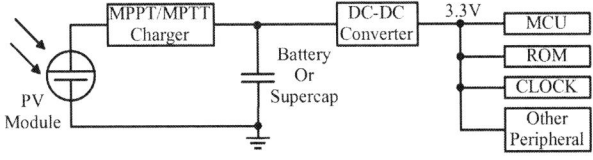

(a) Conventional energy harvesting system.

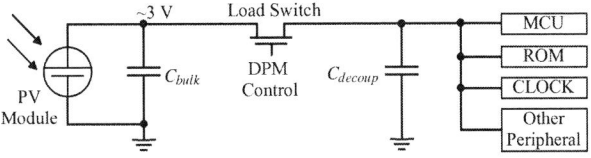

(b) The proposed storage-less and converter-less PV energy harvesting system.

Fig. 5. System diagram of the conventional and proposed architecture.

proposing a novel storage-less and converter-less architecture shown in Fig. 5(b). We connect the PV module to the electronics devices, the nonvolatile microprocessor, an OTP (one-time programming) ROM and peripherals, via a load switch. The MPP tracker (omitted in Fig. 5(b)) continuously turns on and off the load switch. The MPP controller controls the effective duty cycle of the nonvolatile microprocessor. This control process acts like the MPP tracking with the PV module current steering as the load switch turns on and off faster than the cut off frequency of the PV module. We keep the PV module output voltage close to the MPP voltage at all times. We carefully select the PV module that has the MPP voltage close to the legal operating voltage of the semiconductor devices. This makes the PV energy harvesting system operational without power converters nor an energy storage device while performing the MPPT.

There is a bulk capacitor connected in parallel with the PV cell. It extends the time constant of the PV module so that the PV module time constant may match with the feasible DPM period of the nonvolatile microprocessor. There is a decoupling capacitor on the load side, which maintains power integrity of the semiconductor devices.

Table III summarizes the power consumption and transition overheads of the components that we use in the evaluation prototype. The ROM is used as a code storage.

TABLE III
SPECIFICATIONS OF THE COMPONENTS USED IN EVALUATION PROTOTYPE.

Component	Part No.	Power Consumption	Transition Overhead	
			Backup	Recovery
ROM	AT27LV010A	16.50 *mW*	0	2 *μs*
Load Switch	TPS27081	0.31 *mW*	0	250 *ns*
Clock	74HC14	2.84 *mW*	0	2 *μs*
Other Peripheral	LCD and Sensors	14.56 *mW*	N/A	N/A

B. The MPPT with Fine-Grain Dynamic Power Management

We achieve the MPPT with fine-grain DPM of the nonvolatile microprocessor and associated peripherals. We adjust the average load current, as a result of the DPM, and attempt to match the load equivalent resistance with the internal resistance of the PV module. We turn on the nonvolatile microprocessor and its peripherals when the PV module output voltage is higher than the upper threshold, V_h (see Fig. 4). The PV module output voltage decreases as the load devices are connected to the PV module and the load current is higher than the MPP current. We assume the PV module size is over-designed if the load current is lower than the MPP current. The PV module output voltage becomes lower than the MPP voltage and then even lower than the lower threshold, V_l. The MPP controller switches off, and the PV module voltage recovers and goes back to V_h. This control policy maintains the PV module output voltage within $[V_l, V_h]$. We do not consider expensive dynamic voltage and frequency control in this paper.

A traditional microprocessor saves the data in volatile registers into a nonvolatile memory, for example a discrete FeRAM or Flash, before power-off and restores back to the their original places after power-on. Such context saving and restoration is done serially in conventional systems, and is very slow due to the limited IO bandwidth, resulting in large time and energy overheads. On the other hand, with a nonvolatile microprocessor, system states backup and recovery is no longer the bottleneck when we perform DPM.

IV. PROPERTIES OF STORAGE-LESS AND CONVERTER-LESS MPPT

We assume the load power consumption during the state transition ($P_{on,off}$ for backup and $P_{off,on}$ for recovery) is the same as the power consumption in ON state (P_{on}), i.e., $P_{on} = P_{on,off} = P_{off,on}$. The

MPPT window $[V_l, V_h]$ is narrow enough and the power dissipated in the load switch, P_{sw}, is a constant.

A. Load switch is OFF

The load devices are shut down, and the PV cell charges the bulk capacitor when the load switch is OFF. With the Kirchhoff's current law (KCL), we have

$$I_{pv} = C_{bulk} \frac{dV_{pv}}{dt}. \tag{7}$$

We derive the time to recover the bulk capacitor voltage from V_l to V_h as follows.

$$T_{l,h} = T_{off} = \int_{V_l}^{V_h} \frac{C_{bulk}}{I_{pv}} dV_{pv}, \tag{8}$$

where T_{off} represents how long the load switch and load device are kept in OFF state, and C_{bulk} is the size of the bulk capacitor.

The terminal voltage of the decoupling capacitors quickly drops down to 700 mV in our setup as soon as the load switch is OFF since there is no further power supply from the PV module. We assume the 700 mV is related to the threshold voltage of the transistors in the nonvolatile microprocessor. The terminal voltage continues to drop as time elapses, but it is relatively very slow compared with the DPM period we apply. We regard the terminal voltage of the decoupling capacitors, $V_{C_{decomp}}$ is 700 mV when the load switch is OFF, and we denote it as V_{th}.

B. Load switch is ON

The load switch would be turned ON once the bulk capacitor is charged to V_h.

1) Charge sharing: The charges in the bulk capacitor are shared with the decoupling capacitor C_{decoup} at the output stage immediately after the load switch is ON, resulting in an instant voltage drop of the bulk capacitor (PV cell) from V_h to V_{mid}. The value of V_{mid} is determined according to the law of charge conservation:

$$V_{mid} = \frac{C_{bulk} V_h + C_{decoup} V_{th}}{C_{bulk} + C_{decoup}}. \tag{9}$$

The resistance of the load switch in ON state is generally very low (tens to hundreds of $m\Omega$), and the charge sharing process is almost finished instantly. Therefore, we omit the time that bulk capacitor voltage drops from V_h to V_{mid}.

2) OFF-to-ON transition: The node goes into OFF-to-ON transition after the charge sharing process. The microprocessor restores its states and other components. We denote the OFF-to-ON transition time as $T_{off,on}$.

3) Task execution: Once the OFF-to-ON transition is over, the node goes into normal operation and start executing its tasks. We denote the time that allows the node to execute tasks as T_{task}.

4) ON-to-OFF transition: The voltage of the bulk capacitor drops during the task execution time. The MPP tracker keeps watching the voltage drop. When $V_{C_{bulk}}$ becomes close to V_l, it forces the node to backup its states and prepare to go to OFF mode. We denote the time reserved for system states backup before turning OFF the load switches as $T_{on,off}$.

We assume $P_{on} = P_{on,off} = P_{off,on}$, and during the load switch is ON, with the law of conservation of energy, we have

$$V_{pv} I_{pv} = P_{on} + P_{sw} + (C_{bulk} + C_{decoup}) V_{pv} \frac{dV_{pv}}{dt}. \tag{10}$$

We obtain the time it takes for PV cell voltage V_{pv} (also the bulk capacitor voltage $V_{C_{bulk}}$) to drop from V_h to V_l, denoted by $T_{h,l}$.

$$T_{h,l} = T_{off,on} + T_{task} + T_{on,off} = \int_{V_{mid}}^{V_l} \frac{(C_{bulk} + C_{decoup}) V_{pv}}{V_{pv} I_{pv} - P_{on} - P_{sw}} dV_{pv}. \tag{11}$$

The integral on the right side starts from V_{mid} and ends with V_l because we ignore the time of the instant charge sharing. The size of

the bulk capacitor and the operating voltage region $[V_l, V_h]$ should ensure that $T_{task} > 0$. Otherwise there is no time for actual task execution during time period $T_{h,l}$.

C. Fine-grain DPM Efficiency

We derive the energy efficiency of the proposed system η_{sys} in one DPM cycle with (8) and (11) as well as the transition overhead specifications of the components:

$$\eta_{sys} = \frac{E_{task}}{E_{mpp}} = \frac{P_{on} T_{task}}{P_{mpp}(T_{h,l} + T_{l,h})}, \quad (P_{mpp} < P_{on}), \tag{12}$$

where E_{task} is the effective energy used for task execution, and E_{mpp} and P_{mpp} are the maximum energy and maximum power that can be extracted from the PV cell, respectively. The load switches are always ON and the extra energy would be wasted if the harvested power exceeds the demand of the load device.

$$\eta_{sys} = \frac{P_{on}}{P_{mpp}}, \quad (P_{mpp} \geq P_{on}). \tag{13}$$

However, we do not assume this situation in this paper because it is over-designed.

The period of the DPM cycle is defined as

$$T_{dpm} = T_{h,l} + T_{l,h}, \tag{14}$$

where the duty ratio is

$$D_{dpm} = \frac{T_{h,l}}{T_{dpm}}. \tag{15}$$

D. Effects of the Bulk Capacitor

The sizing of bulk capacitor plays an important role because it determines the DPM time granularity, voltage drop during charge sharing, and the time and energy overhead of the DPM scheme. The DPM period can be estimated by (16), assuming that the PV module output power is close to its MPP in voltage range $[V_l, V_h]$.

$$T_{dpm} \approx \left(\frac{V_h - V_l}{I_{pv.mpp}} + \frac{V_{mid} - V_l}{I_{on} - I_{pv,mpp}} \right) C_{bulk}. \tag{16}$$

So the larger the bulk capacitor size, the longer the DPM period. Furthermore, if the bulk capacitor size is small, the terminal voltage decreases fast as the load current discharges the capacitor, and larger voltage drop occurs during charge sharing. If the bulk capacitor is large, the terminal voltage decreases slowly, and small voltage drop occurs during charge sharing. A larger capacitor makes it harder to track MPP of solar cell and perform finer-grain MPPT, but it lowers energy loss due to charge sharing and state transitions because the DPM is less frequent.

V. EXPERIMENTS

A. Experimental Setup

We compare the efficiency of the proposed system with two baseline systems. The architecture of the first baseline system, called the *volatile microprocessor baseline*, is shown in Fig. 5(b) assuming the microprocessor is a conventional volatile microprocessor with nonvolatile FRAM backup. The backup and recovery time overheads of this volatile microprocessor are 300 μs and 200 μs, respectively. The second baseline system, called the *conventional MPTT baseline*, is the system illustrated in Fig. 5(a).

The MPP tracking for the proposed system and the volatile microprocessor baseline system are based on the constant voltage principle, where the voltage window $[V_l, V_h]$ is set to be $[2.75\,V, 2.90\,V]$. The bulk capacitor size is 4.7 μF and decoupling capacitor is 20 nF. The conventional MPTT baseline has a 0.2 F supercapacitor as the energy storage device. We adopt the model parameters introduced in [2] for the converters in the conventional MPTT baseline.

978-1-4799-2817-0/14 $31.00 © 2014 IEEE

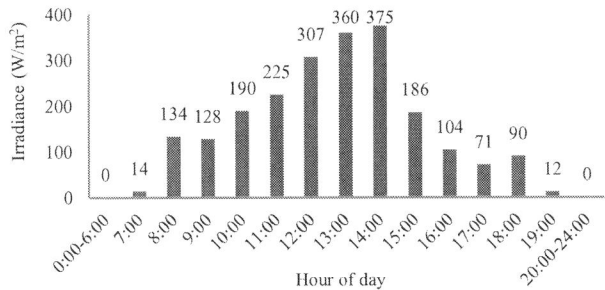

Fig. 6. Hourly solar irradiance during a day.

All the three systems are powered by the same PV module and have the same load power consumption. The PV module size is $4.5 \times 5.5\ cm^2$, with the nominal MPP voltage of 3 V. Model parameters for the PV module and component specifications for the load device are summarized in Table I and III, respectively.

B. System Efficiency

Fig. 6 depicts the hourly solar radiation data on a partly cloudy day that is chosen from the NSRDB (National Solar Radiation Data Base). The irradiance value is the average solar irradiance during the last one hour. We assume the solar irradiance changes every hour for simplicity in calculation, but actual deployment of the proposed method is not restricted by such assumption. The three systems are exposed to the solar irradiance characterized by Fig. 6, and we compare the overall energy efficiency of each system in a whole day.

TABLE IV
DYNAMIC POWER MANAGEMENT RESULTS.

Common DPM statistics				Proposed System			Volatile Microprocessor Baseline		
Time	V_{mpp} (V)	T_{dpm} (μs)	D_{dpm}	E_{mpp} (J)	Work	E_{task} (J)	η_{sys} (%)	Work	E_{task} (J)	η_{sys} (%)
7:00	2.50	N/A	N/A	4.83	No	0	0	No	0	0
8:00	2.73	218	31.6%	50.14	Yes	36.38	72.6	No	0	0
9:00	2.72	224	30.0%	47.81	Yes	34.30	71.7	No	0	0
10:00	2.76	191	46.9%	71.95	Yes	57.07	79.3	No	0	0
11:00	2.78	195	56.5%	85.69	Yes	71.25	83.1	No	0	0
12:00	2.80	301	79.7%	118.14	Yes	108.37	91.7	No	0	0
13:00	2.82	1100	95.1%	139.27	Yes	135.39	97.2	Yes	65.74	47.2
14:00	2.82	1360	99.6%	145.27	Yes	143.72	98.9	Yes	138.34	95.2
15:00	2.76	192	45.8%	70.38	Yes	55.51	78.9	No	0	0
16:00	2.70	260	23.6%	38.57	Yes	26.27	68.1	No	0	0
17:00	2.67	369	14.8%	25.98	Yes	15.90	61.2	No	0	0
18:00	2.69	294	19.9%	33.21	Yes	21.78	65.6	No	0	0
19:00	2.50	N/A	N/A	4.11	No	0	0	No	0	0
Overall				835.34		705.94	84.5		204.08	24.4

Table IV shows the simulation results for the proposed system and the volatile microprocessor baseline. The PV energy harvesting system is not continuously operational with the duty ratio less than 100% from 7:00 to 18:00 because the PV module we use is not overly large. The DPM period are mostly around $200-400\ \mu s$, except being over 1 ms during 12:00–14:00. The proposed system works in most of the time thanks to the low transition overhead of the nonvolatile microprocessor, achieving an overall system efficiency of 84.5%. On the other hand, the traditional volatile microprocessor suffers from its large transition overhead. The volatile microprocessor baseline only works during 12:00-14:00 because only during this period the load switch ON duration is longer than its transition

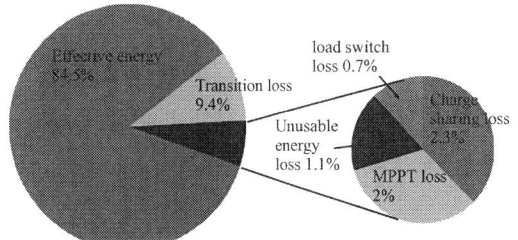

(a) Proposed system with nonvolatile microprocessor

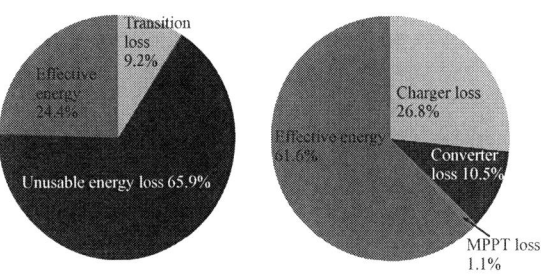

(b) Volatile microprocessor baseline   (c) Conventional MPTT baseline

Fig. 7. Energy loss breakdown in the three different PV energy harvesting systems.

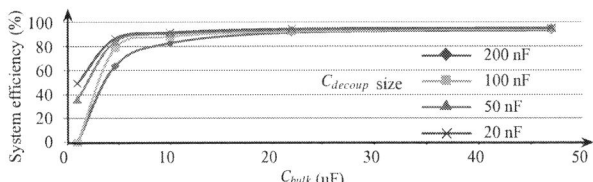

Fig. 8. Effects of the C_{bulk} and C_{decoup} on system efficiency.

overhead, providing time for task execution. The overall efficiency of the volatile microprocessor baseline is only 24.4%.

Fig. 7 illustrates the energy loss breakdown, where effective energy represents the energy utilized for task execution. The proposed method incurs unusable energy loss because the PV output voltage is not high enough to reach the legal operating range of the system when solar irradiance is very weak. The proposed method is not equipped with an energy storage element nor a DC-DC converter, and should waste the harvested energy in such a case.

The proposed system loses 9.4% of the PV energy due to transition overheads while loses a total of 6.1% of the harvested energy due to charge sharing, MPP tracking, etc. The volatile microprocessor baseline does not work during a large proportion of the time due to the large transition overhead, resulting in a 65.9% "unusable energy loss". The energy loss due to state transitions is also considerable at around noon, accounting for 9.2% of the energy loss. Energy loss due to charge sharing and MPP tracking is 0.5% for the volatile microprocessor baseline, which is omitted in the pie graph. The conventional MPTT system loses most of energy in the two-stage converters, resulting an overall 61.6% of the energy utilized for task execution.

C. Effects of C_{bulk} and C_{decoup}

Fig. 8 shows the system efficiency over a wide range of C_{bulk} and C_{decoup} selection. The energy loss stemmed from charge sharing decreases as C_{decoup} shrinks, and the overall efficiency goes up. Fig. 9 shows how C_{bulk} affects the DPM period and energy loss. The DPM period, T_{dpm}, increases as the bulk capacitor becomes larger as shown in Fig. 9(a). This makes the state transition and charge sharing less

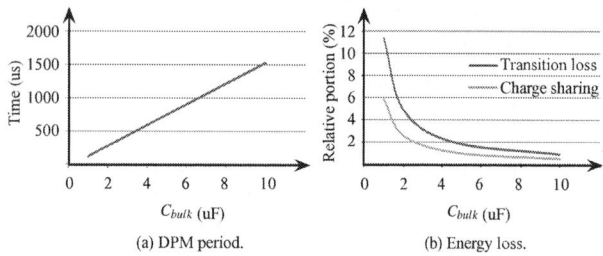

(a) DPM period.　　　　(b) Energy loss.

Fig. 9.　Effects of the C_{bulk} on (a) DPM period and (b) Energy loss.

frequent, and therefore lower the energy loss due to them. This is also why the system efficiency improves as C_{bulk} increases, as observed in Fig. 8.

The overall efficiency of the proposed system reaches up to 95.4% if we enlarge the bulk capacitor to 47 μF. Enlarging the bulk capacitor size also improves the efficiency of the volatile microprocessor baseline. Nevertheless, the proposed system would still outperform the volatile microprocessor baseline because of its low transition overheads.

D. Prototype Board

We implement an evaluation prototype following Fig. 5(b) to validate the functionality and analysis of the proposed system. Fig. 10 is a photograph of the proposed storage-less and converter-less PV energy harvesting system with MPPT. The board contains a PV module, load switches, a nonvolatile microprocessor, an instruction memory, a clock module, measuring interfaces, testing interfaces, voltage regulators and an MSP430 micro-controller for the DPM control. The MSP430 is only used for development purpose. It is powered separately from an external power supply. A low-power CMOS circuit will replace MSP430 in the final design. The voltage regulators are also used for development purpose. They are bypassed in the proposed system. The instruction memory for the nonvolatile microprocessor is a one time programmable (OTP) ROM with a start-up time of 2 μs.

The PV cell output voltage and DPM control signal waveform captured from the evaluation prototype board are shown in Fig. 11, with $C_{bulk} = 47\mu F$, $C_{decoup} = 20nF$ and the nonvolatile microprocessor THU1010N running a counting program. The PV cell output voltage, depicted by the upper curve, swings within the MPP window of [2.77 V, 2.97 V] under control of the DPM control signal.

VI. CONCLUSION

This paper introduces a breakthrough idea for maintenance-free long-lasting low-cost photovoltaic (PV) energy harvesting system. The proposed idea is a radical change of conventional design concept: use of the maximum power point tracking (MPPT) charger, an energy storage device, and a voltage regulator. These components are root causes of high cost, a short lifetime, large volume, and

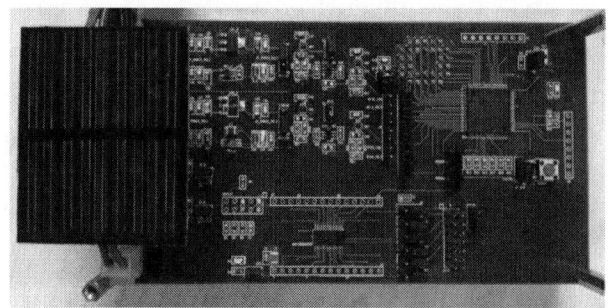

Fig. 10.　Photograph of the prototype board for evaluation.

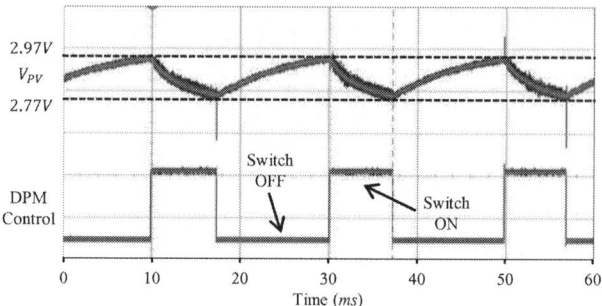

Fig. 11.　Waveform of the evaluation prototype board in operation.

heavy weight. We eliminate the energy storage device and power converters (the charger and regulator) and directly connect the PV cell to the microprocessor. We perform MPPT through a very fine-grain dynamic power management of the nonvolatile microprocessor in every a few hundred microseconds. The efficiency benefit is significant. We achieve an overall system efficiency of 84.5%, and the proposed method makes the PV energy harvesting system operational even with a very weak solar irradiance. The benefits in terms of cost, volume and weight are even higher, and lifetime extension is not even comparable.

REFERENCES

[1] T. Esram and P. Chapman, "Comparison of photovoltaic array maximum power point tracking techniques," *IEEE Trans. on Energy Conversion*, vol. 22, no. 2, pp. 439–449, 2007.

[2] Y. Kim, N. Chang, Y. Wang, and M. Pedram, "Maximum power transfer tracking for a photovoltaic-supercapacitor energy system," in *ISLPED*, 2010, pp. 307–312.

[3] N. Femia, G. Petrone, G. Spagnuolo, and M. Vitelli, "Optimization of perturb and observe maximum power point tracking method," *IEEE Trans. on Power Electronics*, vol. 20, no. 4, pp. 963–973, 2005.

[4] W. D. Soto, S. Klein, and W. Beckman, "Improvement and valida-tion of a model for photovoltaic array performance," *Solar Energy*, vol. 80, pp. 78–88, January 2006.

[5] R. A. Messenger and J. Ventre, *Photovoltaic Systems Engineering, Third Edition*.　CRC Press, 2010.

[6] D. Brunelli, C. Moser, L. Thiele, and L. Benini, "Design of a solar-harvesting circuit for batteryless embedded systems," *IEEE Trans. on Circuits and Systems I: Regular Papers*, vol. 56, pp. 2519–2528, 2009.

[7] F. Simjee and P. Chou, "Everlast: Long-life, supercapacitor-operated wireless sensor node," in *ISLPED*, 2006, pp. 197–202.

[8] Y. Choi, N. Chang, and T. Kim, "DC–DC converter-aware power management for low-power embedded systems," *IEEE Trans. on Computer-Aided Design of Integrated Circuits and Systems*, vol. 26, no. 8, pp. 1367–1381, 2007.

[9] Y. Liu, Y. Wang, H. Jia, S. Su, J. Wen, W. Zhang, L. Zhang, and H. Yang, "An energy harvesting nonvolatile sensor node and its application to distributed moving object detection," in *IPSN*, 2012, pp. 149–150.

[10] Y. Wang, Y. Liu, S. Li, D. Zhang, B. Zhao, M.-F. Chiang, Y. Yan, B. Sai, and H. Yang, "A 3us wake-up time nonvolatile processor based on ferroelectric flip-flops," in *Proceedings of the ESSCIRC*, 2012, pp. 149–152.

Soft Error Resiliency Characterization on IBM BlueGene/Q Processor

Chen-Yong Cher, K. Paul Muller, Ruud A. Haring, David L. Satterfield, Thomas E. Musta, Thomas M. Gooding, Kristan D. Davis, Marc B. Dombrowa, Gerard V. Kopcsay, Robert M. Senger, Yutaka Sugawara, Krishnan Sugavanam

IBM Research, and IBM Systems and Technology Group (STG)

1 Introduction

Soft Error Resiliency (SER) is a major concern for Petascale high performance computing (HPC) systems. In designing Blue Gene/Q (BG/Q) [8], many mechanisms were deployed to target SER including extensive use of Silicon-On-Insulator (SOI), radiation-hardened latches [7,13], detection and correction in on-chip arrays, and very low radiation packaging materials. On the other hand, it is well known that application behavior has major impacts on the masking (or "derating" factor) in system SER calculations. The principal goal of this project is to understand the interaction between BG/Q hardware and high-performance applications when it comes to SER by performing and evaluating a chip irradiation experiment.

Soft errors are transient fails that do not damage the hardware, and there is a short window of opportunity to detect them. Soft errors are caused by high energy particle incidence. These energetic particles have the ability to deposit charge in a transistor body. If the charge is sufficiently large, a non-conducting transistor can become conducting for a short period of time. This can have the effect of causing unwarranted bit flips in computation, control and data. For a terrestrial environment, we have to take cosmic neutrons, as well as alpha particles from packaging materials into account. Therefore, a detailed SER assessment is necessary to quantify a chip's reliability.

A detected soft error can be recoverable or unrecoverable. A recoverable error has typically no impact other than temporary performance degradation. If the soft error is unrecoverable, it leads to a check-stop which results in either an application hang, or unplanned job termination for which the usual recovery is restart from checkpoint. In order to reduce soft error rate at the scale of HPC systems, BG/Q compute chips are built using SRAM cells and latches in SOI technology that have a significantly reduced soft error rate compared to their bulk counterparts [6]. In addition, BG/Q uses a radiation-hardened stacked latch design in 45 nm SOI [5] and circuit detection and correction techniques [2] to enable high reliability at the HPC scale.

Fault injection through irradiation [3,4,9] (chip beaming) is an effective way to evaluate the overall soft error resiliency of microprocessors. By irradiating a microprocessor chip running an application with high-energy particles, an accurate assessment of fault masking and behavior under faults can be obtained without being biased by designer expectation or existing fault injection facilities. The chip beaming results can then be analyzed to project actual long term failure rate for larger-scale HPC systems. Chip beaming also enables integrated, cross-layer assessment of resiliency across the system stack. By collecting and analyzing the statistics of error behaviors under irradiation, failure events can be further classified and analyzed down to their sources, such as a specific on-chip hardware logic block or array located in caches, compute cores, interconnect or in the integrated network function. The beaming results and analysis help validating fault models for future designs and technologies.

In this project, we performed chip beaming on a BG/Q processor chip while running a variety of applications. Blue Gene/Q (BG/Q) is the third generation of IBM's massively parallel, energy efficient Blue Gene series of supercomputers. An overview of the BG/Q compute chip is given in [8].

We selected *AMG2006, UMT, LAMMPS, WUPWISE, QCD* and *LINPACK* as our applications because they are widely used in the HPC environments. Of the selected applications, *AMG2006, UMT* and *LAMMPS* were chosen from the full-application versions of the Sequoia Benchmark [10]. *AMG2006* (Algebraic MultiGrid) is a parallel algebraic multigrid solver for linear systems arising from problems on unstructured grids. *UMT* (Unstructured-Mesh deterministic radiation Transport) is a 3D, deterministic, multigroup, photon transport simulation program for unstructured meshes. *LAMMPS* (Large-scale Atomic and Molecular Massively Parallel Simulator) is a simulation program for classical molecular dynamics. *WUPWISE* (Wuppertal Wilson Fermion Solver) is a simulation program in the area of lattice gauge theory in quantum chromodynamics [1]. *QCD* (Quantum Chromo Dynamics) is an application for simulating the dynamics of quarks and gluons [11]. *LINPACK* is a benchmark that solves a dense system of linear equations and is widely used to measure performance of HPC systems [12]. In order to assess the resiliency of the BG/Q on-chip messaging unit (MU), we also include a separate program, *MU test*, for the BGQ on-chip messaging unit (MU) used in manufacturing tests that exercises both MU and processing cores of a single chip.

2 BGQ Chip 150 MeV Proton Beam Results

2.1 Background

Fault injection through chip beaming [3,4,9] is an effective, accelerated way to assess reliability and soft error resiliency of microprocessors and applications within a short period of time, typically a day. By placing a functional chip running applications under irradiation, the resiliency and behavior under errors of the system can be quantified, and the collected data can be extrapolated to model long term mean-time-between-failures (MTBF) for large-scale HPC systems.

With soft errors due to radioactive contaminants in packaging being controlled, the major sources of soft errors are energetic neutrons that result from high energy cosmic ray showers in the atmosphere. These neutrons can penetrate chip packaging and can create local induced charge events at sensitive circuits as a result of a nuclear reaction following neutron capture. The cosmic ray flux depends on the Earth's magnetic field (i.e. latitude, longitude, sunspot activity) and most critically on altitude (atmospheric depth). The resulting energetic neutron flux is customarily quoted for the reference location New York City (40.7 deg N, 74 deg W) at sea level, and is 12.9 neutrons per cm2 per hour. Separate experiments [3] have shown that the effects of the cosmic ray-induced neutron flux are modeled well by a 150 MeV proton beam. In the proton beam experiment, the average proton flux is approximately 1.0×10^{10} protons / cm^2 / hour, achieving an acceleration factor of 775 Million. To bridge the modeling gap of the neutron energy spectrum (from 10 MeV to 1 GeV) being replaced by a proton beam, we make use of a previous analysis [3] for 65nm SOI subjected to the same 150 MeV proton beam as we used for our experiment. The Qcrit, or minimum charge required to hold the state of a latch, is quite similar between the 45nm SOI of the BG/Q compute chip, and the 65nm SOI technology studied in the reference.

2.2 Fail case analysis and improvement

During the beaming experiment, we gathered 660GB of data, including application output files, chip diagnostic output files, system RAS logs, and scan dumps of every latched bit in the chip at the end of each failed run. We sought to identify the sources of the failures, often to a specific array or logic block on the chip, or to a specific piece of software code.

Applications	L1P and L1/L2 interface	L2 interconnect	Hung/ Crash	DCR	Unknown
MU Test	2	1	1	2	
AMG2006	5				
UMT		1			1
LAMMPS	1	1			
WUPWISE	4		1		
QCD			3		1
LINPACK	2				

Table 1: The remaining fail cases after assessment

Table 1 shows the distribution of the fail cases. The DCR (Device Control Register) errors were specific to the network test. However, recall that the other applications, being single node tests, did not use the MU. Thus we included the MU fails in projecting the system fail rate below.

2.3   System FIT rate projection

Assuming a 96 rack system with 1024 chip per rack, and equal workload mix of the measured applications, the results can be extrapolated to predict the MTBF for the larger systems. The expected MTBF for detectable-and-uncorrectable failures due to cosmic radiation and alpha particles from chip packaging materials is calculated to be 51 days for the sea-level at New York City. The worst-case expected detectable-and-uncorrectable MTBF is calculated to be 35 days assuming the system runs *QCD*, with 20% utilization of the messaging unit (MU).   Figure 1 shows the FIT rate per chip computed from analyzing the fail cases in Table 1.

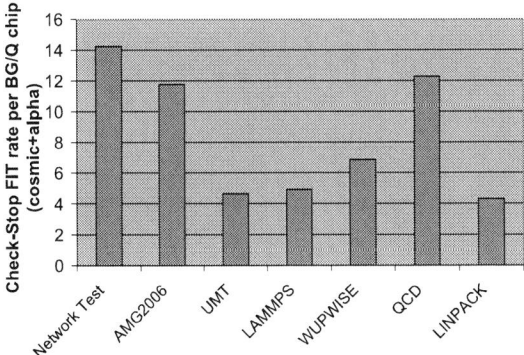

Figure 1: Check stop rate measured and extrapolated to FIT (Failure-in-Time, 1 FIT = 1 fail in 10^9 power-on hours) rate per chip for sea level, New York City

Summary

The principal goal of this project is to understand the soft error resiliency of BG/Q hardware when running high-performance applications. Based on the fails we encountered in the beaming experiment, we projected a mean-time-between-failure (MTBF) for a 20 PetaFLOP, 96 rack system running a comparable workload mix. The expected MTBF for detectable-and-uncorrectable failures is calculated to be 51 days for the sea-level. The worst-case expected detectable-and-uncorrectable MTBF is calculated to be 35 days assuming the system runs *QCD*, with 20% utilization of the messaging unit (MU). These are outstanding results for a machine of this magnitude. The beaming experiment and projected *MTBFs* significantly validate the necessity to include autonomous hardware detection and recovery techniques at the cost of design effort, silicon area and power.

ACKNOWLEDGEMENT

This work was supported in part by the Lawrence Livermore National Laboratory (LLNS) under the Blue Gene/Q Project [Subcontract #B554331]. The authors would also like to acknowledge Ethan Cascio and the Francis H. Burr Proton Therapy Center at Massachusetts General Hospital for their support and the use of their facilities.

REFERENCES

[1] Müller, Matthias et al, "SPEC HPG Benchmarks for Large Systems", High Performance Computing, Lecture Notes in Computer Science vol. 2858, Springer 2003.

[2] Michael Karl Gschwind and Robert Philhower, U.S. Patent 7512772 Soft error handling in microprocessors, Filing date: Jan 8, 2007, Issue date: Mar 31, 2009, Origional Assignee: IBM Corp.

[3] J. Kellington, R. McBeth, P. Sanda, and R. Kalla, "IBM POWER6 Processor Soft Error Tolerance Analysis Using Proton Irradiation," in Workshop on Silicon Effects of Logic - System Effects, 2007.

[4] P. Kudva et al, "Fault Injection Verification of IBM POWER6 Soft Error Resilience," in Workshop on Architectural Support for Gigascale Integration, 2007.

[5] Oldiges, P. et al., "Technologies to further reduce soft error susceptibility in SOI," Electron Devices Meeting (IEDM), 2009 IEEE International , vol., no., pp.1-4, 7-9 Dec. 2009.

[6] Loveless, T.D. et al, "Neutron- and Proton-Induced Single Event Upsets for D- and DICE-Flip/Flop Designs at a 40 nm Technology Node," IEEE Transactions on Nuclear Science, vol.58, no.3, pp.1008-1014, June 2011.

[7] Kenneth P. Rodbell, et al, "32 and 45 nm Radiation-Hardened-By-Design (RHBD) SOI Latches", IEEE Transactions on Nuclear Science, vol.58, no.6, pp.2702-2710, Dec. 2011; http://dx.doi.org/10.1109/TNS.2011.2171715

[8] R.A. Haring, et al, "The IBM Blue Gene/Q Compute Chip", IEEE Micro, vol. 32, no. 2, pp. 48-60, 2012; http://dx.doi.org/10.1109/MM.2011.108

[9] Michalak, S., et al, "Assessment of the Impact of Cosmic-Ray-Induced Neutrons on Hardware in the Roadrunner Supercomputer," IEEE Transactions on Device and Materials Reliability 2012 12:2, 445-454.

[10] ASC Sequoia Benchmark Codes: https://asc.llnl.gov/sequoia/benchmarks/

[11] US Lattice Quantum Chromodynamics http://www.usqcd.org/fnal

[12] The Linpack Benchmark http://www.top500.org/project/linpack.

[13] Ethan H. Cannon et al, U.S. Patent US 8354858 B2, Apparatus and method for hardening latches in SOI CMOS devices, Filing date: Jan 8, 2011, Issue date: Jan 15, 2013, Origional Assignee: IBM Corp.

Resiliency for Many-core System on a Chip

Tanay Karnik, James Tschanz, Nitin Borkar, Jason Howard, Sriram Vangal, Vivek De, Shekhar Borkar

Intel® Corporation

Hillsboro, OR 97124, U.S.A.

Tel : 503-712-4179

Fax : 503-264-6181

e-mail: tanay.karnik@intel.com

Abstract – **Resilient techniques are commonly employed for dynamic and static variation tolerance. In this paper, we present an adaptive clocking technique that achieves 31% throughput increase with 15% energy reduction, and an adaptive interconnect fabric technique that increases bandwidth by 63% with 14.6% energy reduction. We also discuss variations in many-core microprocessors and some techniques to enable a resilient many-core system on a chip.**

I Introduction

Resilient techniques for dynamic and static variation tolerance have been implemented in commercial and research microprocessors. This paper describes a circuit level technique, called adaptive clocking in Section II and an adaptive interconnect technique for modern systems-on-chip (SoC) in Section III. Section IV includes a discussion on system level techniques for many-core SoCs.

II. Circuit Level Techniques

Previous circuit techniques aim to reduce the effect of dynamic variations by explicit on-die sensing and adjusting the operating conditions [1], or by resilient timing-error detection and recovery by embedded error detectors [2-3]. This paper primarily aims at adaptive clocking techniques for dynamic variation tolerance. Previously, an adaptive frequency system was implemented by directly modulating the phase-locked-loop clock output with changes in the digital core V_{CC} to implicitly adapt F_{CLK} [4]. Figure 1 includes an experimental test chip that improves adaptive clocking beyond the modulation approach.

Test-Chip Block Diagram

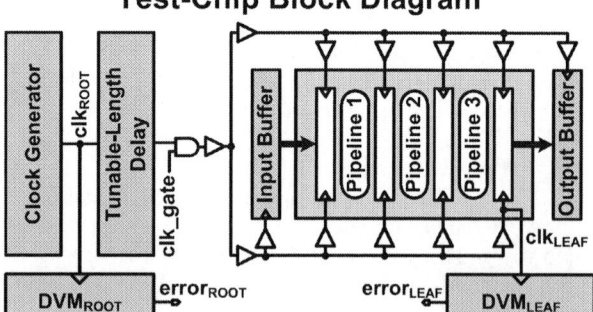

Figure 1: Adaptive Clocking Block Diagram

The design senses a supply droop to enable a sufficient response time to proactively gate the clock, thus mitigating the impact of V_{CC} droops on energy efficiency [5]. The adaptive clock distribution consists of a tunable-length delay, an on-die

dynamic variation monitor (DVM) at the clock distribution root node (DVM_{ROOT}), a clock-gating circuit and a DVM at a clock leaf node (DVM_{LEAF}). The DVM_{LEAF} captures both clock distribution and datapath delay variations. The tunable-length delay prevents path timing-margin degradation for multiple cycles after the VCC droop occurs to enable a sufficient response time for dynamic adaptation.

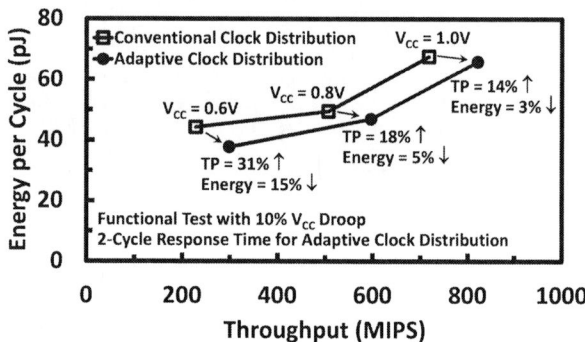

Figure 2: Energy Efficiency Improvement

In comparison to a conventional clock distribution, silicon measurements of the dynamically adaptive clock distribution from the 22 nm test chip demonstrate simultaneous throughput gains and energy reductions of 14% and 3% at 1.0 V, 18% and 5% at 0.8 V, and 31% and 15% at 0.6 V, respectively, for a 10% V_{CC} droop (Figure 2).

III. Interconnect Fabric Techniques

Packet-switched routers are communication systems of choice for modern SoC. Error-correction codes (ECC) have been previously used to mitigate transient failures in routers [6]. The associated energy overheads can be significant for detection and correction of multi-bit failures. We discuss a 6-port, 2-lane packet-switched input-buffered wormhole on-die interconnect router [7].

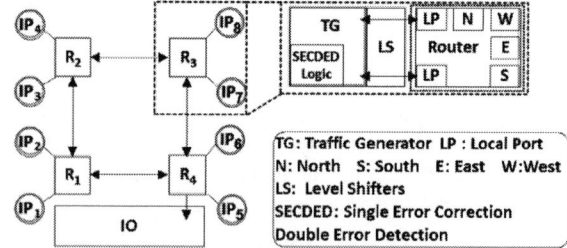

Figure 3: Resilient Router for on-die Interconnect Fabric

The resilient router incorporates an end-to-end forward error correction code and within router recovery from transient timing failures using error detection sequentials (EDS) and a flit replay scheme. The 2x2 2D mesh consists of a router, level-shifters and traffic generator (TG) at each node

978-1-4799-2817-0/14 $31.00 © 2014 IEEE

(Figure 3). The TG contains SECDED logic which appends or retrieves 9 ECC bits from a packet's tail FLIT, thus allowing end-to-end detection and correction of errors in the payload.

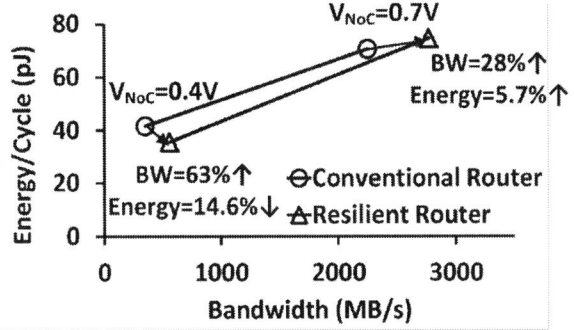

Figure 4: Measured Power, Performance, Energy across Wide Router Supply Voltage

Compared to a conventional router implementation, the resilient router offers 28% higher bandwidth for 5.7% energy overhead at 0.7V and 63% higher bandwidth with 14.6% energy improvement at 0.4V (Figure 4). The results demonstrate the scalability of the resiliency schemes down to near-threshold voltages (NTV).

IV. System Level Techniques

Multi- and many-core processors are available for various market segments. Recent large core count microprocessor chips were measured to have significant within-die variations as shown in Figure 5 [8, 9].

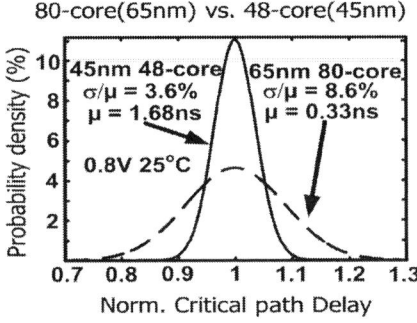

Figure 5: WID Variations across Processors/Processes

Dynamic voltage frequency scaling is employed across market segments for optimized energy efficient computing. Figure 6 shows F_{MAX} and Leakage variations as well as the dependence of variance on V_{CC}.

Chip #1 / Chip #2	Fmax		Chip #1 / Chip #2	Leakage @ 25°C	
Vcc(V)	μ (MHz)	σ/μ (%)	Vcc(V)	Min (A)	Max (A)
0.6	248 / 224	3.1 / 9.8	0.6	0.20 / 0.15	0.28 / 0.29
0.8	640 / 596	2.0 / 3.7	0.8	0.39 / 0.28	0.51 / 0.53
1.1	1179 / 1155	1.4 / 2.1	1.1	1.17 / 0.70	1.41 / 1.36

Figure 6: F_{MAX} and Leakage Variations across V_{CC}

Due to the significant variability across difference cores at different Voltage-Frequency pairs, many system level dynamic variation tolerant techniques have been proposed:

- **Core mapping:** If an application requires subset of the cores to be utilized then application-dependent core mapping can be done for minimum interconnect distances and maximum performance.
- **Core sparing:** After deployment a microprocessor can suffer from performance degradation due to aging and/or permanent faults. A core or set of cores can be maintained idle at the beginning of life. They can be brought on board as a replacement core[s] during mid-life of the processor.
- **Core hopping:** Dynamic variations, like thermal dependence can cause some core to run slower than usual. Threads can be migrated to cooler cores for faster results.
- **Core-cache adaptive interface:** For data intensive applications, threads can be run on cores physically local to the data cache banks.

All of these techniques have demonstrated significant energy efficiency improvements in commercial microprocessor systems [10].

V. Summary and Conclusions

This paper discussed circuit, interconnect and system level resilient design techniques for dynamic and static variation tolerance. The adaptive clocking technique was measured to increase throughput by 31% with 15% energy reduction, and the adaptive interconnect fabric technique was measured to increase bandwidth by 63% with 14.6% energy reduction.

References

[1] J. Tschanz, et al., "Adaptive Frequency and Biasing Techniques for Tolerance to Dynamic Temperature-Voltage Variations and Aging," in IEEE ISSCC Dig. Tech. Papers, pp. 292-293, Feb. 2007.

[2] S. Das, et al., "Razor II: In Situ Error Detection and Correction for PVT and SER Tolerance," IEEE J. Solid-State Circuits, pp. 32-48, Jan. 2009.

[3] K. A. Bowman, et al., "A 45nm Resilient Microprocessor Core for Dynamic Variation Tolerance," IEEE J. Solid-State Circuits, pp. 194-208, Jan. 2011.

[4] K. L. Wong, T. R. Arabi, M. Ma, and G. Taylor, "Enhancing Microprocessor Immunity to Power Supply Noise with Clock-Data Compensation," IEEE J. Solid-State Circuits, pp. 749-758, Apr. 2006.

[5] K. A. Bowman, C. Tokunaga, T. Karnik, V. K. De, and J. W. Tschanz, "A 22nm All-Digital Dynamically Adaptive Clock Distribution for Supply Voltage Droop Tolerance," IEEE J. Solid-State Circuits, pp. 907-916, April 2013.

[6] D. Rossi, C. Metra, A. K. Nieuwland, A. Katoch, "New ECC for Crosstalk Impact Minimization," IEEE Design & test of Comp., pp. 340-348, April 2005.

[7] S. Paul, et al., "A 3.6GB/s 1.3mW 400mV 0.051mm2 near threshold voltage resilient router in 22nm tri-gate CMOS", in IEEE Symp. VLSI Circuits, pp. C30-31, June 2013.

[8] S. Dighe, et al., "Within-Die Variation-Aware Dynamic Voltage Frequency Scaling Core Mapping and Thread Hopping for an 80-Core Processor," in IEEE ISSCC Dig. Tech. Papers, Feb. 2010.

[9] S. Dighe, et al., "A 45nm 48-Core IA Processor with Variation Aware Scheduling and Optical Core Mapping," in IEEE Symp. VLSI Circuits, pp. 250-251, June 2011.

[10] Q. Cai, et al., "Thread Shuffling," IEEE ISLPED, pp. 379-384, 2011.

978-1-4799-2817-0/14 $31.00 © 2014 IEEE

Rethinking Error Injection for Effective Resilience

Shahrzad Mirkhani

ECE Department
University of Texas at Austin
Austin, TX 78712
shahrzad@cerc.utexas.edu

Hyungmin Cho

EE Department
Stanford University
Stanford, CA 94305
hmcho@stanford.edu

Subhasish Mitra

EE and CS Departments
Stanford University
Stanford, CA 94305
subh@stanford.edu

Jacob A. Abraham

ECE Department
University of Texas at Austin
Austin, TX 78712
jaa@cerc.utexas.edu

Abstract— Soft errors, caused by radiation, have become a major challenge in today's computer systems and networking equipment, making it imperative that systems be designed to be resilient to errors. Error injection is a powerful approach to evaluate system resilience, and current practice is to inject errors in architectural registers of processors, program variables of applications, or storage elements in the hardware model. This paper, using answers to frequently asked questions, discusses the need for rethinking conventional approaches to error injection, showing data from recent research and our simulation results. Approaches to improving current error injections are also suggested.

I. INTRODUCTION

Utilizing more processing elements in current computing systems increases the likelihood of errors in computations due to transient hardware faults and environmental effects, such as cosmic rays. In particular, "soft errors" in internal states (i.e., flip-flops) need to be addressed in resilient systems. When evaluating the effects of these errors in a system, they are generally modeled by a single bit-flip in internal storage bits, or by a single bit change in architecture-level registers or memory elements. In this paper, we refer to the former as the "hardware-based model" and the latter as the "software-based model". Cho et al. [7] show that there are large inaccuracies in the outcomes of current software-based models compared to the hardware-based model.

Using answers to frequently asked questions, we discuss the need for rethinking the models used to evaluate the effects of soft errors. We describe the differences between software- and the hardware-based models using a detailed analysis of the propagation of single bit-flips at the flip-flops to architecture level. Statistics from this analysis can lead to a more accurate software-based model. The statistics shown in this paper are based on 40,000 error injections for an out-of-order Alpha based processor (IVM [24]) and 320,000 error injections on an in-order SPARC based processor (LEON3 [10]) mapped on the BEE3 FPGA emulation system [9].

II. FREQUENTLY ASKED QUESTIONS ABOUT SOFT ERRORS IN SYSTEMS

Q1: What is the big fuss about soft errors in computer systems? Can't we just re-execute the application to deal with these errors?

Soft errors can be critical in real-time systems (e.g., banking systems, avionics, etc.), where there may not be a chance to re-execute the application. In one reported case, errors in the interial reference unit in Qantas Airlines Flight 72 (Airbus A330) left many passengers and crew injured [3].

In addition, a more serious problem which could be caused by soft errors is silent data corruption (SDC). This is the case where incorrect data is produced, but we do not know that the data is incorrect. SDC has received attention lately due to its potential threat to large computing systems. Recently, researchers at Los Alamos National Laboratory have stated that "Silent Data Corruption has the potential to threaten the integrity of scientific calculations performed on high performance computing (HPC) platforms and other systems." [1]

Q2: What are viable ways of evaluating the resilience of a proposed design to soft errors?

Since the number of soft error candidates grows with hardware complexity and application run time, analyzing a system for all possible error candidates is not feasible. Several solutions have been proposed for evaluating the resilience of a design to soft errors. The two major approaches are architectural analysis ([17] [16]), and fault/error injection ([8] [24] [12] [11]).

Architectural analysis methodologies, in general, take advantage of certain characteristics of the architecture to estimate soft error vulnerabilities. Approaches such as [17] use architectural simulation statistics to estimate the vulnerability probability of each structure in the design. These methods mostly rely on occupancy-based analysis techniques and the performance models used in such techniques do not include all the hardware structures. In addition, these vulnerability analysis methods can be very difficult to apply to complex components.

On the other hand, in error injection methods ([12] [8]), errors can be injected in a chip using radiation [11], can be injected in an FPGA-based design using hardware-based error injectors [19], or can be injected in hardware models and then this model can be simulated. To overcome the slow speed of simulation-based methods, hierarchical simulation methodologies can accelerate the simulation process ([13] [24]). Quinn et al. [20] discuss the design, implementation, and validation of an error injector and also the current simulation and emulation based error injection systems in more detail.

Another solution to fault injection is to use software-based error models which flip the signal values at the higher levels of abstraction [5] [25] [18] [21] [26] [6]. These models flip a bit in an architectural component such as a program variable or architectural register and then execute the program with injected error in a high level simulator, such as an instruction set simulator. These software-based error models are prone to inaccuracies; this will be discussed in more detail in Q5.

Q3: What would be a good error model to use in error injection studies, and why? Is there any problem with this error model?

Single bit-flip on storage elements of a design is commonly used as the error model in many soft error injection studies. Since mak-

978-1-4799-2817-0/14 $31.00 © 2014 IEEE

ing system logic tolerant to soft errors is expensive (compared with memory arrays), it is important to be able to analyze the flip-flops carefully to design for optimized resilience vs. cost. Errors at the outputs of logic gates have not usually been considered in research. Most errors in logic will be masked before they reach a flip-flop, and the rest will change the states of the capture flip-flops. Recent results [23] indicate that combinational logic propagation in logic gates scales well with the technology, and logic gates are less susceptible to soft errors compared to storage elements.

The major problem with this approach is that the simulation speeds will be very low when analyzing large designs running realistic applications. This can be mitigated by developing hybrid/hierarchical techniques, approximating the effects of low-level bit flips at the architecture level, and by formulating accurate error models and statistics at the architecture level, based on simulations and analysis of the low-level behavior.

Q4: Why would we consider the single bit-flip at the hardware level the "gold standard" for modeling soft errors?

Research in [22] shows that the results for single bit-flips of flip-flops in the hardware implementation are very close to the results of errors due to actual radiation of a chip. Therefore, it can be considered as the gold standard for vulnerability analysis and evaluation of system resilience. As discussed in Q3, combinational logic is less susceptible to soft errors and considering storage elements in our error analysis is sufficient.

Q5: What are the benefits and problems with only considering single bit-flips in architectural (software-visible) registers and memory bits?

As mentioned in Q2, there are several models which consider a single bit flip in an architectural component. The errors injected this way can be evaluated at the software level, which can run much faster than simulation of the hardware model. Also, injecting software-based errors does not require information of the underlying hardware; errors can be injected easily by recompiling the application, for instance. Therefore, these errors are easier to inject and evaluate on realistic applications compared with bit-flips in the hardware.

Unfortunately, there are major drawbacks in using these software-based models. In some cases, these models have been used for studying the SDC occurrences in memory explicitly [15]; use of such models for evaluating system response to a faulty memory is sound. However, in cases where the software-level error injections are used to evaluate the effects of errors in logic elements, their effects on the computations are quite different from hardware-based injection and therefore, the results are inaccurate [7].

Q6: Why do we call the results of software-based error injection inaccurate?

One way to compare the software- and hardware-based error models is to compare the outcomes of a specific application using software- and hardware-based errors. Cho et al. [7] compared these outcomes for several standard models and showed that there are significant differences between the outcomes of a program under software-based and hardware-based errors. For example, they showed that using a software-based model [25], the error rates at the application level can range from 0.07x to 45x of the rates of hardware-level injection (Fig. 1). Such differences from the gold standard can lead to the design of either an expensive overdesigned or an unreliable underdesigned system.

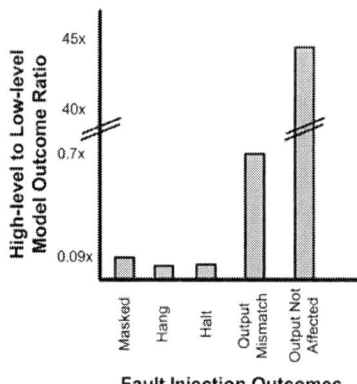

Fig. 1. Ratio of observed outcome rates at the high level versus the low level, for different outcomes

Q7: How do we perform hardware-based error injection?

The actual rate of bit-flips will depend on the type of radiation, the altitude at which the system would operate, and the cross-section of the flip-flops. However, we are primarily interested in the effect of a bit-flip on the results of a computation; injecting errors into the internal flip-flops, one at a time, will enable us to identify the resulting erroneous outcomes. If we need to determine the **rate** of flip-flop corruption, we would need more data on the environment (e.g., cosmic ray flux [4]) and the detailed layout of the flip-flops. For example in the mid 90s, IBM researchers reported error rates for different chips using radiation and field testing [27]. They estimated that a 16MB DRAM chip has a soft error rate of 52000 FIT (1 FIT= 1 failure in 10^9 hours). However, this rate can be different for other types of RAM (e.g., SRAMs) [2].

In order to manage the complexity, a sample of error candidates can be picked for error injection and simulation [14].

Q8: Is comparing hardware-based injection with software-based bit-flips really apples to apples?

As explained in Q5, if the architecture-level injections are used for modeling the effects of soft errors in a chip, these need to be compared with a gold standard (error injection in low-level flip-flops). The scientific process for evaluating a new technique is to compare it with existing techniques so that its benefits can be established.

Q9: Wouldn't the performance of hardware-based injection make this approach prohibitively expensive?

Since the analysis of hardware-based errors should be done (at least partially) at the register transfer level, or even the gate level of abstraction, it will take a very long time to complete the simulation of a reasonable sample of errors. On the other hand, we cannot use only software-based models for analyzing the vulnerability of a system, due to their inaccuracy. One way to compensate for the long execution time of hardware-based injection is to abstract the results of a small sample of hardware-based injections into more accurate higher-level models at the architecture level. This may be done by studying the effects of some hardware-level errors on architecture-level components, based on the structure of the circuit.

Although hardware-based injection is an expensive methodology, due to its relatively low speed, designers might need to pay this expense to meet design reliability goals. Therefore, we need to answer this question case by case, based on the risk of having an unreliable system.

978-1-4799-2817-0/14 $31.00 © 2014 IEEE

Q10: What would we do if we did not have access to the detailed hardware design, RTL or lower level descriptions?

In this case, the only possible way to analyze the system vulnerability is software-based injection. However, the patterns of bit-flips at the architecture level may be established from detailed experiments and past experience, so that the results are more accurate.

Q11: What are some of the statistics of the errors seen at the architecture level due to a single bit-flip at the hardware level?

The effect of each bit-flip in a storage element can be propagated to one or more architectural components in one or more cycles, while software-based error injections usually consider a single bit-flip in an architectural component in a single time frame. We have done a study of the propagation of low-level errors to the architecture level in two processors: the out-of-order, Alpha-based IVM [24], and the in-order SPARC-based LEON3 [10]. We have used SPECINT2000 benchmarks with MinneSpec input sets as our applications. Studying the patterns of hardware-based error propagation to architectural components is one way to improve the software-based model, so that the outcomes of the software-based model become closer to the hardware-based model outcomes.

In this study, we inject a hardware-based error in our system and estimate several parameters.

- Spatial effects such as,
  - error propagations to each register bit
  - the number of affected bits for each propagation
- Temporal effects such as
  - the number of propagations to an architectural register
  - the average time interval between each error propagation.

The results of these simulations (for propagation to architectural registers) are shown in Figures 2, 3, and 4. In these figures, the rate of each property that we are estimating (e.g., number of bit-flips in each register) is calculated by dividing the number of error injections with that property by the total number of injections. The rate is averaged for all registers for all applications. In this paper, we only demonstrate the error propagations to architectural registers. However, for a more precise analysis, the above parameters should be calculated for other components, like program variables, as well.

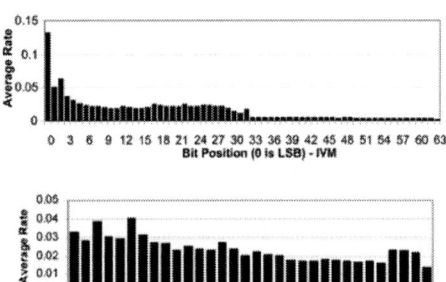

Fig. 2. The rate of hardware-based error propagation to each register bit

Figure 2 shows the average rate of injections that propagate to each bit of the registers (the rates are based on the average value for all the registers for all applications simulated). As can be seen in Fig. 2(a), the first 32 least significant bits observe more hardware-based error propagations than the 32 most significant bits in IVM (more than 13%

of the propagations are propagated to bit 0, around 18% of the propagations are propagated to bits 1 to 4, and less than 12% of the injections propagate to bits 33 to 63). In LEON3, the propagations to the bits are closer to a uniform distribution (Fig. 2(b)).

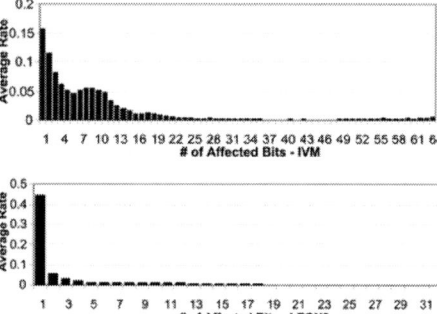

Fig. 3. The rate of the number of register bit flips for each propagation in each hardware-based error injection

Figure 3 shows the average number of register bits that are flipped for each propagation that is a result of a hardware-based error injection. As can be seen in Fig. 3(a), 16% of the hardware-based error injections result in one bit-flip in registers. Repectively, 11%, 8%, and 6% of the injections result in 2, 3, and 4 bit flips in registers. On the other hand, for the LEON3 processor (Fig. 3(b)), almost 45% of the hardware-based injections result in single bit flips in registers and a little over 10% of the injections result in 2, 3, and 4 bit flips in registers.

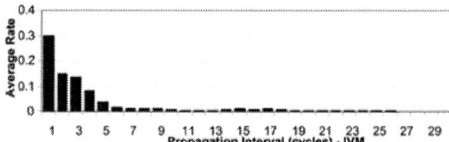

Fig. 4. The rate of cycle intervals between each propagation resulted from a hardware-based error injection in IVM

In our temporal analysis (done for IVM only), we have counted how many times each hardware-based error results in a propagation to registers during the application run-time. Our results show that (on average) around 65% of injections, which result in a propagation to the registers, propagate in one cycle and the remaining 35% of injections result in propagations in multiple cycles during the application run-time.

Our results show that most of the error injections that result in multi-cycle propagation end up propagating the errors to architectural registers over 2 to 8 cycles (for each injection).

For the injections that result in multi-cycle propagation, we have counted the intervals between each propagation. Figure 4 shows the results for average interval rates up to 30 cycles. As can be seen in this figure, almost 70% of the errors which result in multi-cycle propagation have 1 to 5 cycles between each propagation. Propagation intervals of more than 40 cycles happen rarely.

The statistics in our experimental results show that the behavior of hardware-based errors can be significantly different for different architectures, as we can see the differences in IVM and LEON3 processors. These results also show that software-based single bit-flip can model only a part of the hardware-based errors. Thus, in order to have a more accurate software-based error model, we need to think of a weighted (versus uniform) random generator which can consider the number of bits, bit positions, bit patterns, number of times when bits are flipped, and the time interval between each flip. The weights for these random generators would be based on the hardware structure as well as the applications running on the processor. Hardware-based error injections can provide these statistics before we perform the rest of the injections using a software-based error model.

III. CONCLUSIONS

In this paper, we discuss some frequently asked questions about modeling soft errors and their effects in applications, focusing on improving software-based error models using a small sample of hardware-based error simulations. Our analysis, on IVM and LEON3 processors, shows that the underlying architecture of a processor plays an important role in modeling software-based error injection. Study of low-level error propagation to architectural components shows that a single bit-flip software-based model can lead to non-realistic results. Using a sample of hardware-based error injections and studying their patterns of error propagation to architectural components can potentially lead us to improve the accuracy of software-based error models that we want to use for the rest of our soft error analysis.

IV. ACKNOWLEDGEMENT

The authors acknowledge the Texas Advanced Computing Center (TACC) at The University of Texas at Austin (http://www.tacc.utexas.edu) for providing HPC resources that have contributed to the simulation results for IVM processor reported within this paper. The Stanford researchers were supported in part by the National Science Foundation, the Defense Threat Reduction Agency, the Semiconductor Research Corporation, the Semiconductor Technology Advanced Research Network SONIC, and the Defense Advanced Research Projects Agency (Contract No. HR0011-13-C-0022). The views expressed are those of the authors and do not reflect the official policy or position of the Department of Defense or the U.S. Government.

REFERENCES

[1] HPC Wire. http://archive.hpcwire.com/hpcwire/2013-10-31/addressing_the_threat_of_silent_data_corruption.html.

[2] Soft Errors in Electronic Memory - A White Paper. http://www.tezzaron.com/about/papers/soft_errors_1_1_secure.pdf.

[3] The Soft Error Experts. http://thesofterrorexperts.blogspot.com/.

[4] JEDEC Standard, Measurement and reporting of alpha particle and terrestrial cosmic ray-induced soft errors in semiconductor devices. *Solid State Technology Association, JESD89A*, 2006.

[5] D. Chen, G. Jacques-Silva, Z. Kalbarczyk, R. K. Iyer, and B. Mealey. Error behavior comparison of multiple computing systems: A case study using linux on pentium, solaris on SPARC, and AIX on POWER. In *Dependable Computing, 2008. PRDC'08. 14th IEEE Pacific Rim International Symposium on*, pages 339–346. IEEE, 2008.

[6] G. Chen, M. Kandemir, N. Vijaykrishnan, and M. J. Irwin. Object duplication for improving reliability. In *Proceedings of the 2006 Asia and South Pacific Design Automation Conference*, pages 140–145. IEEE Press, 2006.

[7] H. Cho, S. Mirkhani, C.-Y. Cher, J. A. Abraham, and S. Mitra. Quantitative evaluation of soft error injection techniques for robust system design. In *Proceedings of the 50th Annual Design Automation Conference*, page 101. ACM, 2013.

[8] J. A. Clark and D. K. Pradhan. Fault injection: A method for validating computer-system dependability. *Computer*, 28(6):47–56, 1995.

[9] J. D. Davis, C. P. Thacker, C. Chang, C. Thacker, and J. Davis. BEE3: Revitalizing computer architecture research. *Microsoft Research*, 2009.

[10] A. Gaisler. Leon 3 processor. http://www.gaisler.com.

[11] U. Gunneflo, J. Karlsson, and J. Torin. Evaluation of error detection schemes using fault injection by heavy-ion radiation. In *Fault-Tolerant Computing, 1989. FTCS-19. Digest of Papers., Nineteenth International Symposium on*, pages 340–347. IEEE, 1989.

[12] M.-C. Hsueh, T. K. Tsai, and R. K. Iyer. Fault injection techniques and tools. *Computer*, 30(4):75–82, 1997.

[13] Z. Kalbarczyk, R. K. Iyer, G. L. Ries, J. U. Patel, M. S. Lee, and Y. Xiao. Hierarchical simulation approach to accurate fault modeling for system dependability evaluation. *Software Engineering, IEEE Transactions on*, 25(5):619–632, 1999.

[14] R. Leveugle, A. Calvez, P. Maistri, and P. Vanhauwaert. Statistical fault injection: quantified error and confidence. In *Design, Automation & Test in Europe Conference & Exhibition, 2009. DATE'09.*, pages 502–506. IEEE, 2009.

[15] D. Li, J. S. Vetter, and W. Yu. Classifying soft error vulnerabilities in extreme-scale scientific applications using a binary instrumentation tool. In *Proceedings of the International Conference on High Performance Computing, Networking, Storage and Analysis*, page 57. IEEE Computer Society Press, 2012.

[16] S. S. Mukherjee, J. Emer, and S. K. Reinhardt. The soft error problem: An architectural perspective. In *High-Performance Computer Architecture, 2005. HPCA-11. 11th International Symposium on*, pages 243–247. IEEE, 2005.

[17] S. S. Mukherjee, C. Weaver, J. Emer, S. K. Reinhardt, and T. Austin. A systematic methodology to compute the architectural vulnerability factors for a high-performance microprocessor. In *Microarchitecture, 2003. MICRO-36. Proceedings. 36th Annual IEEE/ACM International Symposium on*, pages 29–40. IEEE, 2003.

[18] K. Pattabiraman, G. P. Saggese, D. Chen, Z. Kalbarczyk, and R. Iyer. Automated derivation of application-specific error detectors using dynamic analysis. *Dependable and Secure Computing, IEEE Transactions on*, 8(5):640–655, 2011.

[19] A. Pellegrini, K. Constantinides, D. Zhang, S. Sudhakar, V. Bertacco, and T. Austin. Crashtest: A fast high-fidelity FPGA-based resiliency analysis framework. In *Computer Design, 2008. ICCD 2008. IEEE International Conference on*, pages 363–370. IEEE, 2008.

[20] H. M. Quinn, D. A. Black, W. H. Robinson, and S. P. Buchner. Fault simulation and emulation tools to augment radiation-hardness assurance testing. *IEEE Transactions on Nuclear Science*, 60(3):2119, 2013.

[21] P. Racunas, K. Constantinides, S. Manne, and S. S. Mukherjee. Perturbation-based fault screening. In *High Performance Computer Architecture, 2007. HPCA 2007. IEEE 13th International Symposium on*, pages 169–180. IEEE, 2007.

[22] P. N. Sanda, J. W. Kellington, P. Kudva, R. Kalla, R. B. McBeth, J. Ackaret, R. Lockwood, J. Schumann, and C. R. Jones. Soft-error resilience of the IBM POWER6 processor. *IBM Journal of Research and Development*, 52(3):275–284, 2008.

[23] N. Seifert, B. Gill, S. Jahinuzzaman, J. Basile, V. Ambrose, Q. Shi, R. Allmon, and A. Bramnik. Soft error susceptibilities of 22 nm trigate devices. *IEEE Transactions on Nuclear Science*, 59(6):2666–2673, 2012.

[24] N. J. Wang, J. Quek, T. M. Rafacz, and S. J. Patel. Characterizing the effects of transient faults on a high-performance processor pipeline. In *Dependable Systems and Networks (DSN), 2004 International Conference on*, pages 61–70. IEEE, 2004.

[25] K. S. Yim, Z. Kalbarczyk, and R. K. Iyer. Measurement-based analysis of fault and error sensitivities of dynamic memory. In *Dependable Systems and Networks (DSN), 2010 IEEE/IFIP International Conference on*, pages 431–436. IEEE, 2010.

[26] Y. Zhang, J. W. Lee, N. P. Johnson, and D. I. August. DAFT: decoupled acyclic fault tolerance. *International Journal of Parallel Programming*, 40(1):118–140, 2012.

[27] J. F. Ziegler, H. W. Curtis, H. P. Muhlfeld, C. J. Montrose, B. Chin, M. Nicewicz, C. Russell, W. Y. Wang, L. B. Freeman, P. Hosier, et al. IBM experiments in soft fails in computer electronics (1978–1994). *IBM journal of research and development*, 40(1):3–18, 1996.

Amphisbaena: Modeling Two Orthogonal Ways to Hunt on Heterogeneous Many-cores

Jun Ma[†‡], Guihai Yan[†], Yinhe Han[†] and Xiaowei Li[†]

[†]Institute of Computing Technology
Chinese Academy of Sciences, Beijing, China
[‡]School of Computer and Control Engineering
University of Chinese Academy of Sciences, Beijing, China
{majun, yan_guihai, yinhes, lxw}@ict.ac.cn

Abstract—Heterogeneous many-cores can deliver high performance or energy efficiency. There are two orthogonal ways to improve performance: 1) scale-out by exploiting thread-level parallelism, and 2) scale-up by enabling core heterogeneity. Predicting the performance of such architecture is increasingly challenging. We propose a comprehensive performance model Amphisbaena, or Φ, built from two orthogonal functions α and β. Function α describes the scale-out speedup and function β handles the scale-up speedup. The Φ model can clearly tell not only the overall speedup of a given multithreading and core mapping strategy, but also how to improve the multithreading and core mapping, hence should be a promising performance predictor for future heterogenous many-cores. The results show that Φ model's error rate is within 12%, which is lower than state-of-the-art methods. We demonstrate the application of Φ model by introducing a heuristic scheduling algorithm, which outperforms the baselines by 13% on average.

I. INTRODUCTION

Heterogeneous many-cores is a promising architecture to cater for increasing demand of computing. The successful stories have been witnessed in data centers and cloud computing [1][2][3]. Such architecture provides two ways to boost performance and energy efficiency: 1) highly thread-level parallelism supported by large core quantity, and 2) appropriate application-core mapping enabled by diversified core heterogeneity. Although such architecture promises attractive potential of performance/energy-efficiency, it is not easy to fully make the potential into reality. The main obstacle is the lack of high accurate and low complex perfo rmance model, which usually leads to sub-optimal runtime management.

To be an effective runtime management, one has to accurately predict and optimize the performance periodically. For heterogeneous many-cores, the Pareto Frontier of optimal performance depends on not only the number of active cores, but also the types of cores activated. However, given multi-threaded applications, there are still no quick answers to the both questions. A large body of prior work only focus on homogenous architectures [4][5]. For example, CPR [6] models the performance of multicore processor, but it is unable to provide hints for heterogeneous cores. FDT [7] and TR [8] investigate the upper bound of thread number by detecting the synchronization primitives and hardware contentions. However, without considering core heterogeneity, their methods cannot apply to the heterogeneous many-cores.

In terms of heterogenous architectures, many prior researches [9][10][11] only rely on memory related statistics to guide scheduling, and totally ignore the thread-level parallelism. The recently proposed PIE [12] model aiming to predict performance change between a pair of big and small cores. It claims that performance is determined by ILP and MLP; however, this model is still incomprehensive due to two reasons: 1) PIE cannot capture the performance impact from multithreading; 2) PIE is derived from only two type of cores and hence is unable to handle more complex heterogenous many-cores with more than two types of cores.

Based on the above analysis, we conclude that prior work can not model the heterogeneous many-cores well. In this paper, we propose a novel, comprehensive analytical model, called Amphisbaena[1], abbreviated as Φ in this paper. Φ model can simultaneously describe the performance speedup coming from two orthogonal ways: 1) large-scale multi-threading, or simply called "scale-out" speedup, and 2) core-heterogeneity, or "scale-up" speedup. The rationale behind the above two ways is that an application's performance is largely subject to either the number of threads spawned, or the heterogeneous cores' capabilities for executing those threads. Unlike prior performance models, Φ is deliberately designed from two orthogonal functions, α and β, to model the performance of two independent ways. By doing so, Φ model not only gives an overall performance speedup, more importantly, clearly indicates where the speedup comes from: scale-out or scale-up, or both of them.

In the runtime management, scale-out speedup indicates how many threads of an application should be spawned, and scale-up speedup indicates which type of cores should be activated. We will demonstrate that Φ can serve as the online performance predictor for different runtime scheduling. Experimental results show that the average error of Φ is within 12%, which is less than that of three state-of-the-art baselines. Moreover, we propose a dynamic scheduling to demonstrate the application of Φ model; the performance can outperform the baselines by 13% on average.

The rest of this paper is organized as follows: Section II illustrates our motivation. Section III presents the Φ model. Section IV proposes the runtime management based on Φ. Section V validates the Φ model and the scheduling algorithm. Section VI describes the related work. Section VII concludes this paper.

---

*The work was supported in part by National Basic Research Program of China (973) under grant No. 2011CB302503, in part by National Natural Science Foundation of China (NSFC) under grant No.(61076037, 60921002, 61100016).

[1]Amphisbaena is a mythological creature from Greek Mythology. It has a twin at the head end and one at the tail end, which metaphorically coincides with the two orthogonal ways in this paper.

978-1-4799-2817-0/14 $31.00 © 2014 IEEE

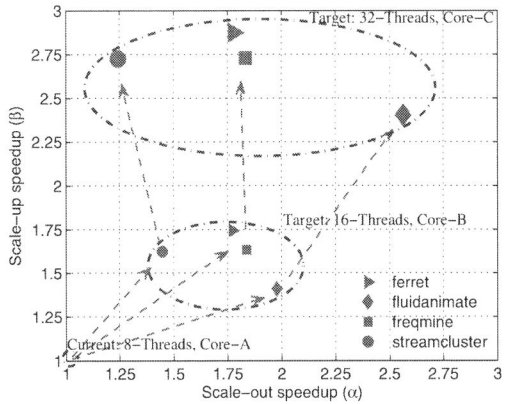

Fig. 1.   Φ space: scale-out speedup (α) and scale-up speedup (β)

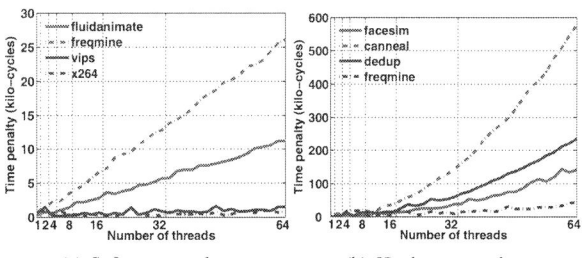

(a) Software penalty   (b) Hardware penalty

Fig. 2.   The time penalty of software synchronizations (a) and hardware contentions (b)

II. MOTIVATION

The motivation of predicting scale-out and scale-up speedup is better explained with an example. Scale-out speedup, modeled by α, is determined by the multithreading policy, i.e. how many threads of an application are spawned by the operating system. We defined α as the performance of target multithreading configuration to the current configuration on the same type of cores. Scale-up speedup, modeled by β, is determined by which type of cores is allocated to the application's threads. We defined β as the ratio of performance on target cores to current cores under the same multithreading configuration.

Figure 1 illustrates the α (X axis) and the β (Y axis) with four representative PARSEC benchmarks [13]. Supposing the current state is 8-thread configuration on core type A (issue width: 4; ROB size: 64), we then study α and β under target of 16 and 32 threads on the core type B (issue width: 6; ROB size: 96) and C (issue width: 8; ROB size: 128), respectively. The detailed core configurations are listed in Table I. The results in Figure 1 are called Φ space, which can profile how the performance responds to the scale-out and scale-up strategy respectively. Taking the `fluidanimate` for example, when the target state transits from current 8-thread with core A to target state of 16-thread with core B, the scale-out approach boosts performance by 2× while the scale-up speedup contributes to another 1.4×. Its overall performance speedup then can be estimated as 2.8× (i.e. 2×1.4).

Φ space can clearly indicate how to improve the overall performance of each application. First, spawning more threads not necessarily yields higher scale-out speedup, which is largely in line with the Amdahl's Law [14]. For example, multithreading from 16-thread to 32-thread brings no speedup for `ferret`, `freqmine`, and even hurts the performance of `streamcluster`, hence should be avoided. Second, choosing more powerful core usually brings scale-up speedup, but reveals different application-specific extents because of the different boundness to computing or memory intensity [12].

III. Φ MODEL

According to the definition of α and β, we can see that α and β are orthogonal to each other, which makes the overall performance modeling a trivial issue: the resultant performance speedup, denoted by $\Phi = Perf_{target}/Perf_{current}$, can be modeled by the product of α and β, i.e. $\Phi = \alpha \times \beta$. The orthogonality between α and β is confirmed by our experimental results shown in Section V.B.

A. Scale-out speedup: α

We use execution time as the performance metric in α modeling. According to the definition,

$$\alpha = \frac{T_{current}}{T_{target}}, \qquad (1)$$

where, T is the execution time on current and target threading configuration, respectively. Generally,

$$T = T_{serial} + T_{parellel} + T_{penalty}, \qquad (2)$$

where T_{serial} is the serial part, which determines the upper bound of scale-out speedup[15]. $T_{parallel}$ is the parallelizable part, which can be eaten away by spawning more threads. For online use, T_{serial} and $T_{parallel}$ can be accurately obtained by training method [7], or instrumentation technique [16]. For example, we can insert specific instructions at the entry and exit of serial and parallel sections to record the elapsed cycles. As the distinctive part of our α model, the multithreading overhead $T_{penalty}$ is indeed the bottleneck of scale-out speedup and determined by both 1) software, such as inter-thread contention due to synchronization with locks and barriers, and 2) hardware, such as contention for shared resources of LLC, memory controller, and memory bus. Unfortunately, accurately modeling $T_{penalty}$ is still an open question.

We tackle this problem by using the "univariate analysis". First, we assume a contention-free scenario by configuring hardware resources over-provisioning in our simulator, then study the performance impact only from software synchronization which results in thread waiting. Figure 2(a) shows the relationship between thread number and time penalty. We choose applications with the most amount of software synchronization (locks and barriers). No matter what type of synchronization happens, the penalty can be tracked and returned by bottleneck-identifying instructions (`BottleneckCall`, `BottleneckReturn` and `BottleneckWait`) which have been used in [16]. Basically, the penalty correlates linearly with the number of threads, even though with different slope due to different intensity of locks and barriers. The more synchronization operations, the steeper slope. For example, both contention-intensive `freqmine` and `fluidanimate` exhibit steeper slope. Hence, we reckon that $T_{penalty} \propto a_1 \times k_1 \times n$. n is the number of thread. k_1 represents software contention-intensity measured by synchronization waiting cycles per kilo-instructions (SPKI), which is accumulated by recording the penalties deriving from both locks and barriers. a_1 is a modulating constant.

Second, we tight hardware resource to match the reality and log the stall cycles due to hardware contentions. Because we can not hide the synchronization primitives of multithreaded applications, we just regard the incremental penalty of hardware resource-unlimited experiment as the hardware contention-induced penalty. Unlike software contention, we choose applications with the most intensive hardware contention (with largest working set). We found that the hardware

978-1-4799-2817-0/14 $31.00 © 2014 IEEE

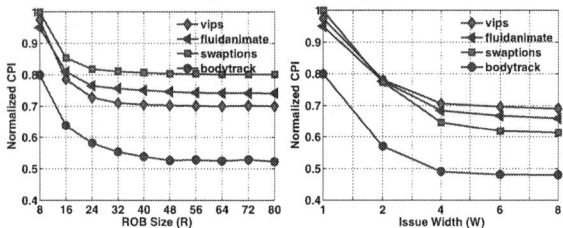

(a) Normalized CPI vs. ROB size (b) Normalized CPI vs. Issue width

Fig. 3.   The performance impacts from ROB size (R) and issue width (W)

contention-induced penalty increases much faster, roughly following a quadratic trend, as Figure 2(b) shows. Hence, we infer that $T_{penalty} \propto a_2 \times k_2 \times n^2$, where k_2 represents the contention-intensity measured by misses waiting cycles per kilo-instructions (MPKI). Since there are always misses happened before shared resource being accessed, MPKI just accounts for the hardware contention-intensity. MPKI is obtained by aggregating the penalty of to-LLC and to-MEM misses. Performance counters are needed to detect miss events at runtime. a_2 is also a modulating constant.

The overall penalty can be obtained by combining the above two components, i.e.

$$T_{penalty} = a_0 \times k_0 + a_1 \times k_1 \times n + a_2 \times k_2 \times n^2, \quad (3)$$

where k_0 is an application-specific bias derived from redundant computations [13], which is obtained on the fly and modulated by a_0.

B. Scale-up speedup: β

We use CPI as the performance metric in β modeling. Based on the definition,

$$\beta = \frac{CPI_{current}}{CPI_{target}}, \quad (4)$$

so given a thread, we need to accurately predict its CPI on various type of cores. Kenzo et al. correlated the CPI with two representative microarchitecture parameters: the frontend issue width (W) and backend ROB size (R) [12]. We also rely on the two parameters to differentiate heterogeneous cores, but build a more essential analytical function with them. We find that scale-up by core heterogeneity can be approximated with a power-exponent law, which is reasoned from the following observation. We first hold W and study the sensitivity of CPI to R. As Figure 3 (a) shows, the CPI improvement gets less significant with larger ROB capacity. Specifically, CPI initially reduces dramatically when ROB capacity is smaller than 32 entries, but quickly flats out when ROB capacity increases beyond 32 entries. This trend can be approximated well with a power law. We then hold R and study the sensitivity of CPI to W, as Figure 3(b) shows. The CPI also decreases along with the issue width growing. This trend fits an exponential function well (with negative exponent). Therefore, we build a new CPI prediction model for core heterogeneity, as Equation 5 shows.

$$CPI = r + s \times R^{-t \times W}. \quad (5)$$

Equation 5 is in accordance with prior knowledge: the marginal performance of scale-up speedup will decrease (lower slope) when running on more aggressive cores, because excessively large machine parallelism (larger issue width, ROB size, etc.) leads to mismatch with the available program parallelism (inherent instruction parallelism and concurrent memory access)[17]. Also, we find that the performance sensitivity to W and R shows very subtle difference. For instance, we find that when frontend capacity (represented by W) is too small, saying 1 or 2, then expanding backend will be ineffective. That's because the pipeline

becomes unbalance if the instructions issued in frontend underfeed the backend. So we put the W in the exponent function in Equation 5, which can show higher gradient than power function in most cases, for modeling the frontend impact to performance.

We find that s and t are also two application-specific constants, and are highly correlated with memory intensity and computing intensity respectively. For example, Figure 3(a) shows that, when comparing to fluidanimate, the CPI of vips is more sensitive to R. That's because vips is more memory intensive than fluidanimate. In Figure 3(b), swaptions, a financial analysis benchmark with more computing intensity, is more sensitive to W. Therefore, s and t can weight these kinds of sensitivities. Moreover, bodytrack, with approximate ratio of computing and memory intensity to fluidanimate, has a steady lower bias, which can be expressed by bias r.

These coefficients can be calculated as follows: given the CPI comes from two components: CPI_{base}, the base CPI without any stalls from memory access, and CPI_{mem}, the penalty of data waiting or pipeline stalls [18]. Let $\eta = \frac{CPI_{mem}}{CPI_{mem} + CPI_{base}}$, then s can be calculated by $s = b_1 \times \eta$, which highlights the memory intensity. It implies that the memory intensity linearly correlates with the ultimate CPI, no matter which type of cores is used. $t = b_2 \times (1 - \eta)$, which negatively weights the computing intensity on the exponent; this implies that the CPI of a computing intensive application can be reduced more by powerful frontend. The bias r reflects the intrinsic CPI of an application on the assumed oracle core with infinite R and W [17][19], which can be approximated by CPI_{base} in this paper. So r is calculated by $b_0 \times CPI_{base}$. Also, b_0, b_1 and b_2 are modulating constants.

The α and β model can be performed online, so does the Φ model. Based on any current phase of an application, we can use Φ model to predict its speedup of next phase on any other target configurations. The detailed model parameters are listed in Table II. In the next section, we will demonstrate the role of Φ model in runtime management.

IV.   RUNTIME MANAGEMENT

A generic runtime management can be divided into three steps: 1) Predict performance speedup coming from scale-out and scale-up at the beginning of each management interval. Then the predicted values serve as the input of the next step. 2) Invoke scheduling algorithm to figure out the optimal configuration in terms of, for example, maximizing performance. 3) The operating system enables the specified multithreading and application-core mapping indicated by the optimal configuration. Our Φ model takes charge of the first step.

First, we describe the implementation issue of Φ model. Those application-independent modulating constants such as a_0, a_1, a_2, b_0, b_1 and b_2, can be obtained by off-line regression. The other application-specific coefficients (k_1, k_2, s, t etc.) can be calculated online, by following the detailed calculation routines discussed in Section III.A and III.B respectively. Table II summaries their values and implementations. Note that SPKI and MPKI are calculated based on the bottleneck-identifying instructions [16] and the built-in performance counters [18], respectively. CPI_{base} and CPI_{mem} are derived from the CPI stack caculation [18]. The same method has been also used by PIE [12]. The bias of redundant computing derives from the cycles of instruction duplication, and the bias of intrinsic CPI is replaced by CPI_{base}.

To demonstrate the application of Φ model, we have to feed its output to a management algorithm, as the second step

978-1-4799-2817-0/14 $31.00 © 2014 IEEE

Algorithm 1 *How to scale-out(D_{out}) and scale-up(D_{up})*

1: Model α and β for each application;
2: Calculate expectations $E[\alpha]$ and $E[\beta]$;
3: Allocate thread number $n_i = \frac{E[\alpha_i]}{\sum_{i=1}^{N} E[\alpha_i]} \times N$;
4: Sort array $E[\beta_1..\beta_M]$ and set priority;
5: **repeat**
6:     $E_{max}[\beta]$ chooses the fastest cores;
7:     Pop $E_{max}[\beta]$ from array $E[\beta_1..\beta_M]$;
8: **until** array $E[\beta_1..\beta_M] = \emptyset$;

TABLE I. MICROARCHITECTURAL PARAMETERS

Parameter	OoO	In-order
Issue width	4(A), 6(B), 8(C)	1(D), 2(E), 4(F)
ROB size	64,96,128	—
I/D cache	64KB, 2-way	32KB, DM
I-L1/D-L1 access time	2ns/4ns	2ns/4ns
LLC access/miss time	12ns/54ns	12ns/54ns
Branch predictor	hybrid 2-level	static, 2k

TABLE II. COEFFICIENT DETAILS

Coefficients	Value	Implementation
a_0	1.837e-003	constant
a_1	0.05312	constant
a_2	-2.025e-005	constant
k_0	$Bias_{redundantComputing}$	online
k_1	$SPKI$	online
k_2	$MPKI$	online
b_0	0.2837	constant
b_1	1.1675	constant
b_2	1.8427	constant
r	$Bias_{intrinsicCPI}$	online
s	CPI_{mem}/CPI	online
t	CPI_{base}/CPI	online

described. Specifically, we introduce a heuristic algorithm, called Phi scheduling, as shown in Algorithm 1. The objective of Phi scheduling is to maximize performance. Noting for heterogenous many-cores, achieving such objective would be a NP-complete problem [20]. So, Phi scheduling actually is a greedy-based heuristic. Phi is divided into two sub-processes of decisions: D_{out} and D_{up}. First, D_{out} follows the policy that application with higher scale-out speedup should spawn more thread. So we make the number of threads proportional to its scale-out speedup:

$$n_i = \frac{E[\alpha_i]}{\sum_{i=1}^{N} E[\alpha_i]} \times N, \qquad (6)$$

where n_i is the thread number allocated for application i. N is the number of all available heterogeneous cores. $E[\alpha_i]$ is the scale-out expectation of application i. It accounts for the average level of scale-out speedup based on Equation 7.

$$E[\alpha_i] = \frac{1}{|U|} \sum_{j \in U} \alpha_{i,j}, \qquad (7)$$

where $\alpha_{i,j}$ is the scale-out speedup of application i at multithreading configuration j. U is the set of all possible number of threads for scale-out. The size of U ($|U|$) is 33 in this paper ($\{1\} \bigcup \{2 : 2 : 64\}$).

After D_{out}, D_{up} has to allocate available heterogeneous cores to the threads of each application. D_{up} follows the policy that application with largest $E[\beta]$ (calculated similar to $E[\alpha]$) will be allocated the fastest type of cores. So we rank the $E[\beta]$ of all applications in descending order, then follow the iterations: the application with maximal $E[\beta]$ always gets highest priority to choose the fastest cores to its threads, until all cores have been allocated.

The runtime interval is on order of sub-seconds. It depends on not only the penalty of runtime management, but also the effectiveness of management affected by interval granularity [21]. In this paper, the runtime interval is one second. The first two steps in charged by OS hypervisor can be finished within 10 ms. OS triggers OpenMP for changing thread number dynamically, and operates thread migration for application-core mapping. The former is on order of 10ms [22] and the latter costs about from 1ms to 20ms [12]. Therefore, the overhead of runtime management is no more than 50ms, which can be amortized well at runtime.

V. EVALUATION AND RESULTS ANALYSIS

A. Experimental Setup

We evaluate our work with gem5 simulator [23]. Similar to PIE [12], we choose heterogeneous cores which are obvious different in frontend issue width (W) and backend ROB capacity (R). We also implement three types of in-order cores to bring more types of heterogenous cores. The detailed configurations are listed in Table I. The cores are organized into clusters, and each cluster consists of 32 homogeneous cores. This layout method is similar to the baseline used in [3]. A distributed, cluster-level banked LLC is shared by all cores. Cache coherence is maintained by directory-based

MOEI protocol which supports heterogeneous transactions [24]. We use PARSEC [13] as the basic workloads, given it targets the general purpose processors and shows diverse preferences to heterogeneous cores.

B. Model Accuracy Validation

The accuracy of Φ model is determined by 1) the accuracy of α and β, and 2) the orthogonality between them. The following results are devoted to validate the accuracy and orthogonality, respectively.

Figure 4(a) shows the dispersion between the predicted scale-out speedup calculated by α model and the actual scale-out speedup by measurement. For each benchmark, there are 33 thread configurations (the threading is from 1, 2 to 64 in step of 2 threads, so the total space is $A_{33}^2 = 1056$). We randomly generate 600 validation points from the 12 benchmarks (50 phases per benchmark) to make sure the time-consuming simulation time acceptable. The result shows that the proposed α is fully competent to predict the scale-out speedup. The prediction error is under 5% on average, and the maximum error is no more than 8%. This merit mainly comes from the accurate model of penalty component ($T_{penalty}$ in Equation 3). As a comparison, we plot the predicted scale-out speedup according to the Amadahl's Law, which shows that the average error and maximum error reach up to 11.4% and 19.5%, respectively. Therefore, we believe the α model can serve as a good amendment to the classical Amadahl's Law.

Compared to the well-behavior α, predicting scale-up speedup β model from core heterogeneity is much harder. As Figure 4(b) shows (similar to α, 600 validation points are in the total space: $12 \times 50 \times A_6^2 = 18000$, where A_6^2 covers all cases of thread migrations between two types of cores), the worst case's prediction error reaches up to 25%. However, the average error is still kept below 8%, which makes β model still applicable in most cases. For comparison, we also study the prediction accuracy of recently proposed PIE model by applying it to the same samples, as shown in Figure 4(b). The result shows that PIE's average error and worst error are 12.2% and 33.7%, which agree with the prior work [12]. This comparison confirms that our β model is more accurate when predicting the performance impact from core heterogeneity.

The validation of α and β builds up high confidence in the

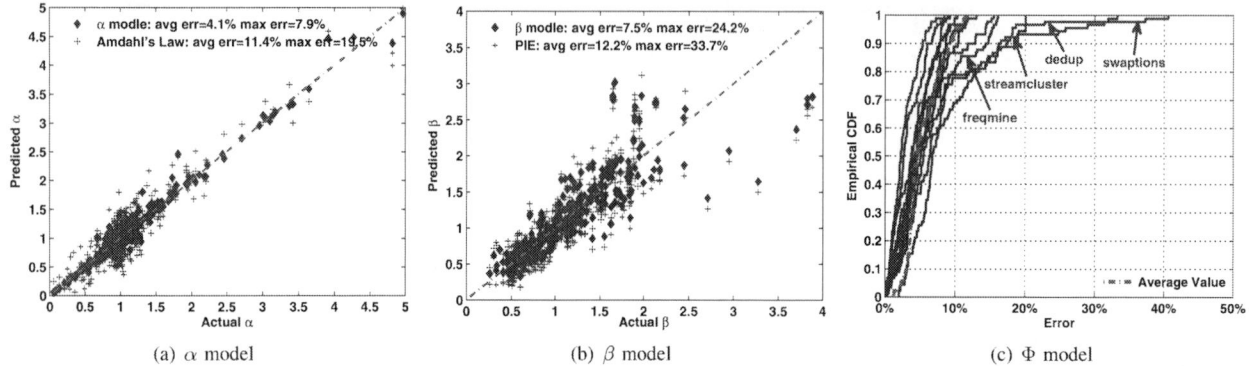

(a) α model   (b) β model   (c) Φ model

Fig. 4. Accuracy validation of α, β and Φ. We compare α to Amdahl's Law (a), β to PIE method (b), and shows the cumulative distribution function of Φ (c).

TABLE III.   WORKLOADS DESCRIPTION

Workload	Source
W1	fluidanimate, vips, ferret, canneal
W2	facesim, swaptions, blackscholes, bodytrack
W3	x264, fluidanimate, dedup, streamcluster
W4	swaptions, vips, ferret, blackscholes
W5	x264, dedup, freqmine, canneal
W6	facesim, vips, streamcluster, freqmine

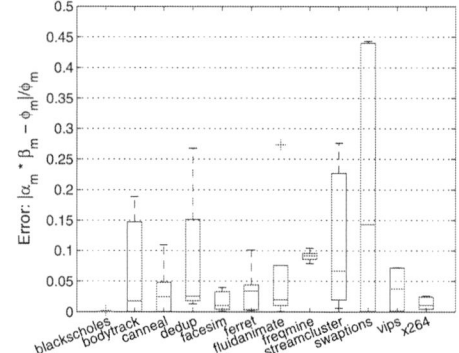

Fig. 5. Evaluate the orthogonality between α and β

proposed Φ model. Figure 4(c) illustrates the cumulative distribution of prediction error for Φ model. Up to 1080 samples are randomly picked out from a huge space (633600×18000). The results show that for most benchmarks, applying Φ model always results in less than 10% prediction error, or for more than 80% samples of `facesim`, `fluidanimate`, and 90% samples of `blackscholes`, `x264`. Although Φ has shown to be more comprehensive and accurate than many previous models, we admit that Φ is not perfect. In some corner cases, the prediction error is up to 30% and even higher, such as `dedup`, `swaptions`. That's largely because we make the tradeoff between model's complexity and accuracy. In particular, we deliberately assume that α and β are completely orthogonal to each other. We find that α and β are weakly correlated rather than totally orthogonal in most cases, because different number of threads will result in different average CPI of an application.

Figure 5 evaluates the orthogonality between α and β. In this experiment, we measure three *actual* values in each interval, i.e. α_m, β_m, and Φ_m. If α and β are orthogonal, then $|\alpha_m \times \beta_m - \Phi_m|$ should be zero. Hence, we use the normalized value $Cor = |\alpha_m \times \beta_m - \Phi_m|/\Phi_m$ to evaluate the orthogonality. First, we run one thread on baseline core A, and then T threads on core A. By measuring the running time, we can obtain the scale-out speedup α_m. Then, we run one thread on core B and measure the running time to obtain the scale-up speedup β_m. Lastly, measuring the speedup of T threads on core B can obtain the overall speedup Φ_m. Following this measurement, we build 2268 samples from the 12 benchmarks. The results are shown in Figure 5 with boxplot which can show central mark for median, interquartile range and possible outliers. These results can justify our orthogonality assumption. For most benchmarks, the Cor values are below 5%. The worst case happens on `swaptions`. This outlier reflects that scale-out speedup and scale-up speedup are not completely independent, because the thread CPI, or β can slightly change with different threads spawned. However, the results in Figure 5 confirm that our orthogonality assumption is a good approximation. As future work, it suggests a potential improvement to the current Φ model by involving the correlation between α and β.

C. Performance Comparison

We evaluate Phi scheduling by comparisons to three state-of-the-art algorithms. Bias [9] is indicated by memory related stalls, then schedules application that shows less stalls on big core or more stalls on small core. PIE [12] uses two analytical models to predict the performance impact between a pair of big and small cores, then moves the application with highest performance improvement to the big core. However, both Bias and PIE ignore the scale-out speedup. For fair comparison, we let their decision for thread number equal to Phi scheduling. Inversely, Static strategy allocates the same number of threads to each application, but lacks the indicator for heterogeneous core mapping. Therefore, we make its mapping policy the same to Phi. As a golden case, we also present an Oracle algorithm by exhaustively searching the huge solution space, though it is infeasible for online use.

In Figure 6, six workloads are selected and each workload is composed by four PARSEC benchmarks, as shown in Table III. The results are all normalized to Oracle. We can see that our Phi scheduling is the closest to Oracle. This result again justifies the efficacy of Φ model. Particularly, Phi averagely outperforms the other three baselines as high as 12.2% (Static), 13.3% (Bias) and 12.9% (PIE), even though Bias and PIE have the same multithreading configuration to Phi, or Static has the same heterogeneous core mapping to Phi.

VI. RELATED WORK

The performance model of heterogeneous architectures has been an active research topic for decades. Those methods are used to either facilitate early design stage to prune the huge design space [3], or guide runtime management [11]. Predicting performance becomes even more challenging with the advent of heterogeneous many-cores where performance is determined by not only scale-out, but also scale-up, or even

978-1-4799-2817-0/14 $31.00 © 2014 IEEE

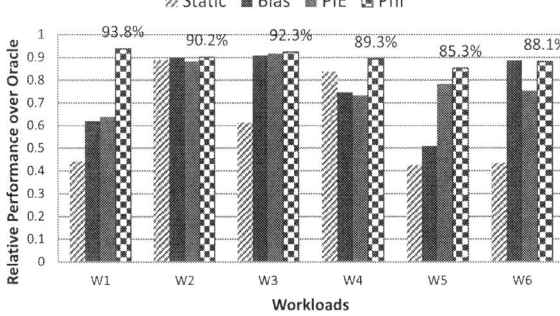

Fig. 6. Normalized performance comparing to three baselines

the complex interactions between them.

Predicting multithreading performance initially is accomplished with the classical Amdahl's Law [14]; however its beauty form fails to capture the inter-thread contentions such as data synchronization and hardware contention. Feedback-driven threading (FDT) [7] is designed to decide the upper bound of thread number. However, because FDT assumes a homogenous substrate, it becomes ineffective when applying to heterogeneous architectures.

Although heterogeneous architectures draw a lot of attentions [25][10][9], the work for performance prediction is far from mature. The recently proposed PIE [12] uses two models to predict performance change deriving from ILP and MLP. However, its cost of online implementation is very high. For example, PIE requires on-the-fly tracking a key parameter, called *dependency distance*, which involves a register table to track all the stream of dynamic instructions. By contrast, our model only needs limited runtime statistics which can be obtained from already built-in counters, and the results show that our model is more accurate than PIE. Furthermore, without considering performance impact from multithreading, PIE can not cope with the scale-out speedup.

Hill et al. [15] propose a performance prediction model for heterogeneous many-cores using the Amdahl's Law. That model considers both multithreading and core heterogeneity. However, it does not provide any analytical function or algorithm. Morad et al. [26] give a more comprehensive equation; however, it correlates the scale-up speedup to core's area, rather than micro-architectures, which results in large prediction error. Finally, both methods didn't clearly split scale-out speedup and scale-up speedup, so are hard to guide runtime management.

VII. CONCLUSION

In this paper, we propose a comprehensive analytical performance prediction model Φ for heterogeneous many-cores. Φ model can simultaneously capture the performance speedup from multithreading (scale-out) and core heterogeneity (scale-up), which are described with function α and β, respectively. Finally, we find that α and β are largely orthogonal, which greatly simplify the Φ model. The average error of Φ model is within 12%. We also present Phi scheduling to demonstrate the application of Φ model. The proposed Φ model highlights the tradeoff between scale-out speedup and scale-up speedup, hence we believe that it can serve as an ideal performance predictor for future runtime management.

REFERENCES

[1] Ripal Nathuji, Canturk Isci, Eugene Gorbatov, "Exploiting Platform Heterogeneity for Power Efficient Data Centers," *ICAC*, pp. 5–5, 2007.

[2] Christina Delimitrou, Christos Kozyrakis, "Paragon: QoS-aware Scheduling for Heterogeneous Datacenters," *ASPLOS*, pp. 77–88, 2013.

[3] Guevara, Marisabel, Lubin, Benjamin Lee, Benjamin C., "Navigating Heterogeneous Processors with Market Mechanisms," *HPCA*, pp. 95–106, 2013.

[4] Kai Ma, Xue Li, Ming Chen, Xiaorui Wang, "Scalable Power Control for Many-core Architectures Running Multi-threaded Applications," *ISCA*, pp. 449–460, 2011.

[5] Omid Azizi, Aqeel Mahesri, Benjamin C. Lee, Sanjay J. Patel, Mark Horowitz, "Energy-Performance Tradeoffs in Processor Architecture and Circuit design: A Marginal Cost Analysis," *ISCA*, pp. 26–36, 2010.

[6] Benjamin C. Lee, Jamison Collins, Hong Wang, David Brooks, "CPR: Composable Performance Regression for Scalable Multiprocessor Models," *MICRO*, pp. 270–281, 2008.

[7] M. Aater Suleman, Moinuddin K. Qureshi and Yale N. Patt, "Feedback-Driven Threading: Power-Efficient and High-Performance Execution of Multi-threaded Workloads on CMPs," *ASPLOS*, pp. 277–286, 2008.

[8] Kishore Kumar Pusukuri, Rajiv Gupta and Laxmi N. Bhuyan, "Thread Reinforcer: Dynamically Determining Number of Threads via OS Level Monitoring," *IISWC*, pp. 116–125, 2011.

[9] David Koufaty, Dheeraj Reddy and Scott Hahn, "Bias Scheduling in Heterogeneous Multi-core Architectures," *EuroSys*, pp. 125–138, 2010.

[10] Daniel Shelepov, Juan Carlos Saez Alcaide, Stacey Jeffery, Alexandra Fedorova et al., "HASS: a Scheduler for Heterogeneous Multicore Systems," *SIGOPS*, pp. 66–75, 2009.

[11] Rakesh Kumar, Dean M. Tullsen et al., "Single-ISA Heterogeneous Multi-core Architectures for Multithreaded Workload Performance," *ISCA*, pp. 64–75, 2004.

[12] Kenzo Van Craeynest, Aamer Jalee, Lieven Eeckhout, Paolo Narvaez and Joel Emer, "Scheduling Heterogeneous Multi-cores Through Performance Impact Estimation (PIE)," *ISCA*, pp. 213–224, 2012.

[13] Christian Bienia, Sanjeev Kumar, Jaswinder Pal Singh and Kai Li, "The PARSEC Benchmark Suite: Characterization and Architectural Implications," *PACT*, pp. 72–81, 2008.

[14] Dong Hyuk Woo, Hsien-Hsin S. Lee, "Extending Amdahl's Law for Energy-Efficient Computing in the Many-Core Era," *Computers*, pp. 24–31, 2008.

[15] Mark D. Hill, Michael R. Marty, "Amdahl's Law in the Multicore Era," *Computers*, vol. 41, no. 7, pp. 33–38, 2008.

[16] Jos A. Joao, M. Aater Suleman, Onur Mutlu, Yale N. Patt, "Bottleneck Identification and Scheduling in Multithreaded Applications ," *ASPLOS*, pp. 223–234, 2012.

[17] Norman P. Jouppi, "The Nonuniform Distribution of Instruction-level and Machine Parallelism and Its Effect on Performance," *Computers*, vol. 38, no. 12, pp. 1645–1658, 1989.

[18] Stijn Eyerman, Lieven Eeckhout, Tejas Karkhanis, James E. Smith, "A Performance Counter Architecture for Computing Accurate CPI Components," *ASPLOS*, pp. 175–184, 2006.

[19] Derek B. Noonburg, John P. Shen, "Theoretical Modeling of Superscalar Processor Performance," *MICRO*, pp. 52–62, 1994.

[20] Winter,J.A., Albonesi,D.H., Shoemaker,C.A., "Scalable Thread Scheduling and Global Power Management for Heterogeneous Many-Core Architectures," *PACT*, pp. 29–39, 2010.

[21] Canturk Isci, Gilberto Contreras, Margaret Martonosi, "Live, Runtime Phase Monitoring and Prediction on Real Systems with Application to Dynamic Power Management," *MICRO*, pp. 359–370, 2006.

[22] OPENMP, "OPENMP," *http://openmp.org/wp/*, 2013.

[23] gem5, "The gem5 Simulator System," *http://www.m5sim.org*, 2012.

[24] Suh, T., Blough, D. M., Lee, H. H. S., "Supporting Cache Coherence in Heterogeneous Multiprocessor Systems ," *DATE*, pp. 1150–1155, 2004.

[25] Youngjin Kwon, Changdae Kim, Seungryoul Maeng, and Jaehyuk Huh, "Virtualizing Performance Asymmetric Multi-core Systems," *ISCA*, pp. 213–224, 2011.

[26] Tomer Y. Morad, Uri C. Weiser, Avinoam Kolodny, Mateo Valero, Eduard A. E., "Performance, Power Efficiency and Scalability of Asymmetric Cluster Chip Multiprocessors," *Computer Architecture Letters*, vol. 5, no. 1, pp. 14–17, 2006.

Co-simulation Framework for Streamlining Microprocessor Development on Standard ASIC Design Flow

Tomoyuki Nakabayashi[†], Tomoyuki Sugiyama[†], Takahiro Sasaki[†], Eric Rotenberg[‡], and Toshio Kondo[†]

†Graduate School of Engineering, Mie University
Tsu, Mie, 514-8508, Japan
Tel: +81-59-231-9780, Fax: +81-59-231-9781
‡Department of Electrical and Computer Engineering, North Carolina State University
Raleigh, North Carolina, 27695-7911, USA
Email: {tomoyuki, sugiyama, sasaki, kondo}@arch.info.mie-u.ac.jp, ericro@ncsu.edu

Abstract— In this paper, we present a practical processor co-simulation framework for not only RTL simulation but also gate/transistor level simulation, and even chip evaluation with an LSI tester. Our framework includes an off-chip system call emulation mechanism, which handles system calls to evaluate and verify the processor design with general benchmark programs without pseudo-circuits in the processor design. Therefore, our framework can be consistently used from RTL design to chip fabrication. We also propose a checkpoint mechanism that resumes a program from a pre-created checkpoint. This mechanism is not affected by the non-deterministic problem on a multi-core processor. Moreover, we propose a cache warming mechanism when resuming from a checkpoint.

I. INTRODUCTION

As multi-core architecture has become commonly used to improve processor performance, designing a state-of-the-art multi-core chip in a short time has become essential for processor research. A development environment that contains useful mechanisms and can be used throughout the entire processor research provides efficient infrastructure to researchers. We classify the steps of fabricating a novel processor chip into five phases, 1) design space exploration using a simulator, 2) register transfer level (RTL), 3) gate level, 4) transistor level, and 5) fabricated chip. There are two challenges to streamline the processor development through the entire standard ASIC design flows.

1. *Emulating system calls in RTL through fabricated chip:*
   When researchers prototype a processor from RTL, to gate and transistor level, to ASIC, it is often desirable to focus on user level code because they are interested in the core part of the processor and not all of the system level support. They may be in this situation because they designed the RTL from scratch or because they are using open source toolsets (e.g., FabScalar) which provide a level of sophistication in the microarchitecture but do not currently feature system level support. As a matter of convenience, and a matter of research productivity, it is good to dispense with the issue of explicitly supporting system

calls in the processor design. While emulation is often used in simulators written in a high-level language, it is unwieldy to carry that over to RTL/gate/transistor simulations and not at all possible to emulate in the same way for a fabricated chip.

2. *Reducing turnaround time through sampled execution:*
   Except for the fabricated chip phase, all other simulation-based phases in particular gate/transistor level cannot simulate the entire workloads in a reasonable timeframe. Therefore, we need a checkpoint mechanism to resume a simulation from an arbitrary region of interest (ROI). Moreover, checkpoints are useful when hardware bugs are detected in the fabricated chip. A checkpoint allows for bypassing the bug (if it is infrequent) to get to another ROI. Therefore, it is also useful for validation in the fabricated chip phase.

In this paper, we propose a co-simulation framework to address these challenges. This paper makes the following contributions:

1. Our framework includes an off-chip system call emulation mechanism that handles system calls using general load and store instructions. This enables the processor design to involve the execution of a program using system calls without booting an OS on the target processor for evaluation and verification. The off-chip system call emulation mechanism enables a prototype processor that cannot handle system calls to execute a program using system calls. Therefore, researchers can improve research productivity.

2. Our framework also contains a checkpoint mechanism to reduce turnaround time for evaluation and verification. The checkpoint mechanism restores not only essential state (register file, program counter, inflight file/network operations, and memory state of the process) but also optional state (warming up the caches).

   - *Essential state restoration:* Our checkpoint mechanism resumes a program from an ROI even on a multi-core processor chip. With this mechanism, researchers can shorten turnaround time and obtain the result in the ROI.

978-1-4799-2817-0/14 $31.00 © 2014 IEEE

- *Optional state restoration:* The checkpoint mechanism also contains a start-up routine to warm up the cache with the cache replacement algorithm. The warm up mechanism reduces the time to achieve a peak performance and also improves simulation accuracy by diminishing the effect of cold started cache.

Our co-simulation framework can be consistently used in RTL, gate and transistor level simulations, and in fabricated chip evaluation because all the above mechanisms are implemented without pseudo-circuits such as direct programming interface-C (DPI-C) in the processor design. In addition, our framework does not depend on the microarchitecture of a processor. We introduced our framework into two processor design projects: a simple single pipeline processor and a complex out-of-order processor.

II. RELATED WORK

A. Processor simulators

Many processor simulators [1, 2] and system simulators [3, 4, 5] written in a high-level language are used for processor research. Researchers take advantage of such simulators in the early stage of research in accordance with their intended use. Since our main focus is from RTL to fabrication, in which researchers evaluate the precise hardware cost, energy efficiency, and circuit delay for their proposed approach, we describe three focused mechanisms: 1) system call emulation to simplify processor architecture, 2) checkpoint mechanisms to reduce simulation time, and 3) cache warming mechanism to achieve a highly accurate evaluation. These mechanisms, however, are used only in each simulator. Our goal is to use these mechanisms in all phases of standard ASIC design flows.

B. Synthesizable processors

Some open synthesizable processors can be used from RTL implementation to chip fabrication [5, 6, 7, 8]. Since FabScalar and OpenSPARC have a co-simulation environment, we describe these two processors in more detail.

FabScalar automatically generates synthesizable RTL designs of differently designed superscalar cores. FabScalar contains an instruction set simulator, called functional simulator, to verify RTL by concurrently running the same instructions in RTL design and the functional simulator, and cross-checking the architectural state instruction-by-instruction. The functional simulator is also used for emulating a system call, so RTL design can handle a system call as a one-cycle-instruction. Also FabScalar provides fast-skip and checkpoint mechanisms to avoid long simulation time and re-simulating up to a checkpoint. However, FabScalar currently has drawbacks in system call emulation (described in Section IV) and the checkpoint mechanism (described in Section V). Our aim was to improve the two mechanisms based on FabScalar.

OpenSPARC is the open-source version of UltraSPARC T1 and T2 processors. Currently, RTL design, simulation tools, and verification package are all available. OpenSPARC provides a complete RTL design to boot a full OS and useful tools

for simulation and verification including checkpoint and cache warming. However, there is the level of abstraction gap as the next step from a processor simulator, and this gap makes it difficult to advance the research phase beyond simulator-based exploration. By contrast, our off-chip system call emulation mechanism enables a designer to evaluate a prototype processor omitting touchy hardware for an OS with general benchmark programs. In addition, cache warming of OpenSPARC is implemented by programming language interface (PLI) in verilog, this limits the use to only in RTL simulation. Our cache warming mechanism is unique in that it is consistently used from RTL to fabricated chip.

III. CO-SIMULATION FRAMEWORK

A. Co-simulation overview

This subsection gives an overview of our co-simulation framework. The framework consists of a *functional simulator* written in a high-level language and *processor design*. Note that *processor design* refers to all designs of RTL, gate, transistor level, and fabricated chip. Fig. 1 shows how the functional simulator is used in our framework. The functional simulator assists in the verification and evaluation of processor design. The cross-checking architectural state guarantees instruction set level behavior of processor design, and fast-skip and checkpoint mechanisms reduce turnaround time. In addition, the functional simulator emulates system calls by calling the host OS according to a request from processor design.

B. Challenges

Three challenges in the co-simulation framework are described below.

System call emulation: System call emulation is significantly beneficial in that a designer can run a general program without booting an OS on the target processor. Our system call emulation mechanism described in Section IV can be used for every research phase. Our framework exploits general load and store instructions to communicate with the emulator; therefore, no special mechanism is necessary in the processor design.

Checkpoint mechanism: The checkpoint mechanism is used not only to reduce turnaround time but also to evaluate only an ROI. Checkpoint creation for a multiprocessor should consider the non-deterministic problem. In Section V we propose a checkpoint mechanism that solves this problem.

Cache warming: Our checkpoint mechanism contains cache warming mechanism as a part of optional checkpoints. Restoring a checkpoint is exposed to a large performance gap with a peak performance because the simulation is resumed with a cold started cache. In addition, in the gate and transistor level phases, it takes a long time to achieve the peak. Our cache warming mechanism described in Section VI warms up the cache in the shortest time and improves evaluation accuracy.

We introduced our framework into two processor design projects: an embedded processor [9] and FabScalar.

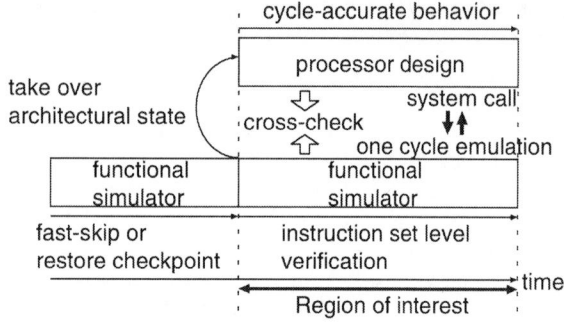

Fig. 1. Co-simulation framework.

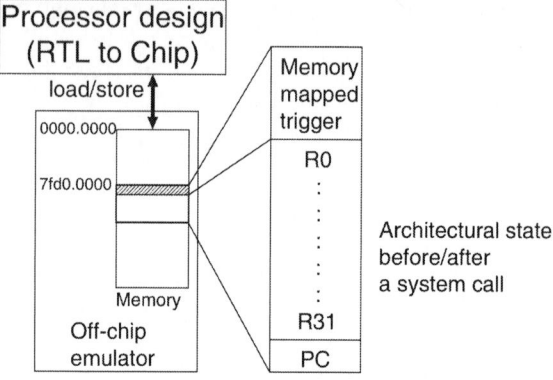

Fig. 2. Off-chip system call emulation mechanism.

IV. SYSTEM CALL EMULATION

In general benchmark programs such as SPEC, a processor must handle system calls (services from an OS kernel) to handle the file system, network, memory, process, thread, and security. For this reason, to evaluate a processor design with general benchmarks, the processor either boots an OS or uses an alternative stand-alone C library. There are two requirements. One is that researchers directly execute benchmarks on a full implemented processor to evaluate and verify in a short time. Since booting an OS takes a large amount of time, especially in gate and transistor level, it is difficult to evaluate or verify a processor design on a full system. Moreover, although using an LSI tester has an advantage of directly evaluating or testing a fabricated chip with input vectors, such LSI testers limit the execution cycle up to insufficient cycles for running a target application. The other requirement is that to evaluate a microarchitectural approach, researchers often require only primal instructions such as arithmetic, logical, branch, and memory access instructions, and tends to omit subsidiary hardware for an OS such as memory management unit and internal processor registers. Implementing such hardware requires researchers to have a deep understanding of such hardware.

Because of these two requirements, executing general programs without an OS is valuable. Newlib is a C library intended for use on embedded systems [10]. A processor can execute programs without an OS with the addition of a few low-level routines. However, another binary file using Newlib is needed. In addition, Newlib requires emulation of peripheral systems and does not support multiple processes and cores.

Emulating a system call as an instruction solves the above problems. When a system call occurs, the functional simulator detects the call and emulates it by calling the host OS. Later, the processor design continues execution after reflecting the result of the system call.

To emulate a system call, the processor design somehow notifies the emulator of the occurrence of the system call. Furthermore, the processor design must take over the system call result. In the RTL phase, this is not difficult because a test module can look into the submodule, which asserts a relative signal, and overwrite the architectural state. FabScalar currently uses this method; however, it is used only in the RTL phase and prevents regions including system calls from being evaluated on FPGA [11]. Therefore, our framework is necessary for emulat-

ing system calls beyond the RTL phase.

A. Implementation of off-chip system call emulator

The concept of our system call emulation mechanism is juggling a system call as consecutive stores and loads. We explain our emulation mechanism using Figs. 2 and 3. Fig. 2 shows memory mapping and how to trigger/reflect a system call emulation. We allocate a memory space (e.g., from address *7fd00000*) to interact with the off-chip emulator. First, when a system call occurs, the processor jumps to the system call trap routine like a real product. We use the routine shown in Fig. 3 instead of a true routine if an user wants to emulate system calls. Second, the processor design involves storing the architectural state, i.e., register file, to the prescribed space because the emulator requires the register file values to emulate the required system call. Next, the emulator emulates the system call when a store is executed into the probing address (*7fd00000*). Finally, the emulator writes the values updated by the system call into the same memory space to which the processor stored the register values, then the processor loads the modified values using load instructions. Note that if the processor includes a cache, the stores and loads should be non-cacheable memory access instructions. This emulation mechanism does not require any dedicated hardware in the processor design; therefore, the processor maintains a pure design. Since the processor interacts with the emulator using load and store instructions in our framework, the emulation mechanism can be consistently used from RTL to fabrication. This enables the processor design to execute a general program without booting an OS.

The off-chip system call emulation mechanism is of course used for evaluating and verifying a complete processor. Also, a prototype processor design which does not support system calls can be evaluated with general benchmarks. This aspect improves research productivity to evaluate a microarchitectural approach.

V. CHECKPOINT MECHANISM

A checkpoint mechanism saves the state of a simulation in an ROI and later continues the simulation from the ROI. A checkpoint mechanism goes through two phases: checkpoint

```
bfc003d0 <__trap_syscall>:
        /* Save architectural state */
        sw  $1,   0x104(k0)

        sw  $31,  0x17c(k0)
        /* Trigger for system call */
        sw  0x01, 0x0000(k0)
        /* Restore the result */
        lw  $1,   0x104(k0)

        lw  $31,  0x17c(k0)
```

Fig. 3. System call trigger routine

```
bfc00000 <__reset_handler>:
        lui k0,   0x7fd0
        lw  $1,   0x104(k0)

        lw  $31,  0x17c(k0)
        /* load program counter */
        lw  k1,   0x278(k0)
        jr  k1
```

Fig. 4. Reset routine

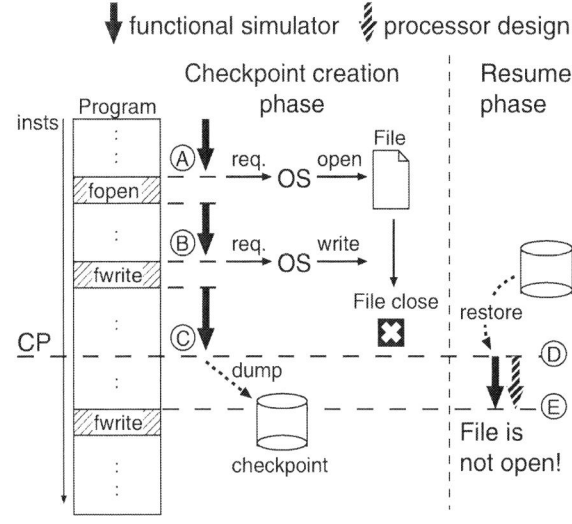

Fig. 5. Problem with checkpoint mechanism.

creation phase with only the functional simulator, and resume phase with the processor design. We can repeat the resume phase during verification and evaluation of a processor to reduce turnaround time.

In our co-simulation framework, we resume a program using a similar routine as the system call emulation for use in every design phase to restore the architectural state. We use the reset routine shown in Fig. 4. After the processor is reset, the program counter is initialized to *bfc00000*, which is the general start address of the reset routine. In the routine, the processor loads register file values and the program counter written into the prescribed memory space. The program counter indicates the starting point of a checkpoint.

However, if the co-simulator naively resumes a benchmark program from a checkpoint, a system call (file- and network-related) cannot be correctly executed. In the following explanation, we use a sequence of file system operations as an example to simplify the problem. The co-simulator leaves file input/output (I/O) to the OS running on a host computer. Fig. 5 shows the issue of resuming simulation from a checkpoint. When a file is opened, the off-chip system call emulation mechanism calls the OS to handle the file opening (Fig. 5.A). Once the file is opened, the co-simulator treats I/O operation to the file in the same way (Fig. 5.B). Once the simulation reaches at the start point of an ROI, the co-simulator creates the checkpoint, then the file is closed because the co-simulator quits (Fig. 5.C). For this reason, when the co-simulator resumes the simulation from the checkpoint (Fig. 5.D) and a file I/O occurs (Fig. 5.E), the co-simulator cannot handle the file I/O because the file is not open.

To solve this problem, FabScalar dumps the state not only at a checkpoint but also at file I/Os in an ROI as shown in Fig. 6. In the checkpoint creation phase, FabScalar executes a program beyond a checkpoint to dump the state after file I/Os in the ROI (Fig. 6.A). To resume a program, FabScalar restores the state at the checkpoint (Fig. 6.B). When a file I/O occurs during the resumed simulation, the dumped state after the file I/O is restored (Fig. 6.C); therefore, FabScalar reproduces the state after the file I/O. With this method, FabScalar provides a checkpoint mechanism. However, FabScalar can execute only file I/Os that were pre-executed in the checkpoint creation phase. In addition, this mechanism cannot be applied to a multiprocessor environment because we cannot create the preceding checkpoint. Because the execution order is non-deterministic in a multiprocessor, the order of system calls is also non-deterministic.

There are a few solutions to the problems. M5 uses the solution of dumping all necessary information into a checkpoint file, e.g., the offset of the file descriptor manipulated in the simulation. By contrast, we adopt another solution because M5's solution requires the dumping all inflight operations such as file system and network. In our solution, the simulator executes only related system calls to resume a program with inflight operations. Fig. 7 shows our solution. Our co-simulator dumps the difference in state between each file I/O up to a checkpoint in the checkpoint creation phase (Fig. 7.A). It skips the instructions between each file I/O using the dump file and executes only file I/Os (Fig. 7.B). After it reaches the checkpoint, it continues execution including file operations (Fig. 7.C). As a result, our co-simulator skips to a checkpoint at high speed without any restriction.

Although our solution has the advantage of handling all inflight operations in the same way, restoration speed depends on the number of system calls up to a checkpoint. To demonstrate that our solution is practical, we evaluated the restoring of speed using SPEC2000 INT benchmarks. We created a checkpoint in SimPoint [12] and resumed the checkpoint for each benchmark program. Table I lists the evaluation results. The upper half compares the time to forward each benchmark to the SimPoint. We evaluated all benchmarks on Intel Core i7-2600 CPU @ 3.40 GHz with 4 GB memory. We used a 4-width

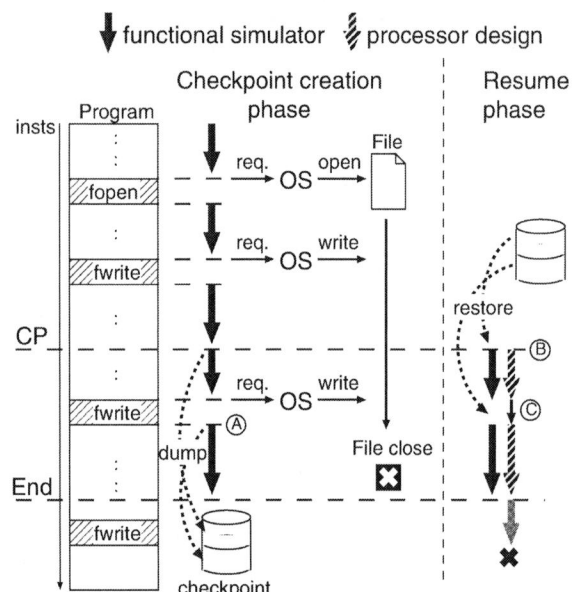

Fig. 6. Checkpoint mechanism of FabScalar.

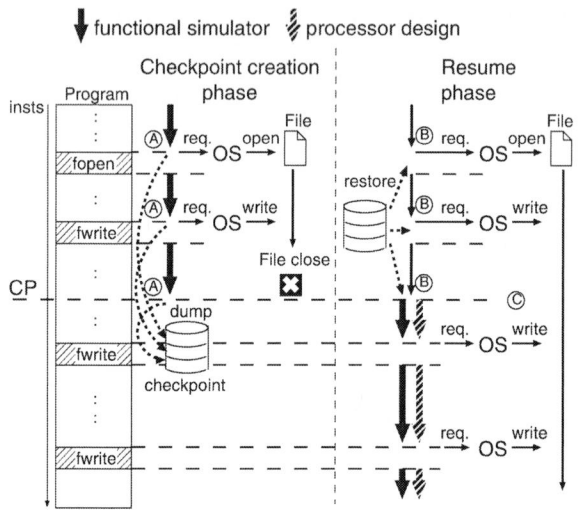

Fig. 7. Proposed checkpoint mechanism.

TABLE I
DEMONSTRATION OF CHECKPOINT MECHANISM.

	gzip	mcf	bzip	parser	twolf
Gate-level[a] (day)	13,757	6,389	11,297	13,260	12,319
RTL design[a] (day)	844	326	715	680	645
fast-skip (min.)	244	103	206	210	212
checkpoint (sec.)	0.50	0.68	0.77	3.84	0.53
skipped insts (100 million)	1,189	553	977	1,146	1,066
checkpoint file size (MB)	832	326	384	1780	119
system calls	65	116	101	1,027	133

[a]Estimated by million instructions per second (MIPS) value

fetch superscalar processor design as the RTL design and synthesized the RTL design for the gate-level estimation. We used Cadence NC-Verilog, version 09.20-s038, for simulation and Synopsys Design Compiler, version H-2013.03-SP2, for synthesis. We note that checkpoint restoration succeeded in both RTL and gate-level simulations. The lower half of the table summarizes the number of skipped instructions, the file size of the checkpoint, and the number of system calls up to the Sim-Point. The results show that restoring a checkpoint took a few seconds in the worst case and the file size of the checkpoint was not so large.

VI. CACHE WARMING MECHANISM

A cache system has a large impact on processor performance. When we resume a benchmark program from a checkpoint, a cold started cache incurs a performance gap with peak performance, as shown in Fig. 8. Therefore, the evaluation accuracy is degraded because of the performance gap. Further-

more, it takes a long simulation time to warm up a cache system to analyze the peak performance/energy. OpenSPARC has a cache warming mechanism using PLI in verilog HDL. This implementation limits the use to only in the RTL phase. By contrast, our cache warming mechanism can be used in all design phases. It is particularly effective in shortening the test vector for an LSI tester. In addition, our cache warming mechanism defines a certain time when the cache system is warmed up, this feature enables a designer to evaluate only a specified period after the processor achieves the peak in simulation, as shown in Fig. 9.

Fig. 9 shows our cache warming mechanism. Our co-simulator also has a cache simulator written in C language. When the co-simulator creates a checkpoint, the cache system dumps the cache warming routine (binary file, actually), as shown in Fig. 9. Lines that are accessed with the same index are dumped in order of the least recently used (LRU) value to restore the cache contents including the cache replacement algorithm. We depict that a lower LRU value has a higher priority for replacement, i.e., an entry whose LRU value is 0 will be replaced. The dumped routine is linked when the co-simulator starts restoration, and it is called in the reset routine before restoring the architectural state. This mechanism restores the cache contents by using a software level approach in the shortest time.

We briefly estimated the impact of our cache warming mechanism on an L1 data cache (total size: 16 KB and line size: 16 bytes). Even in the worst case (cache warming on blocking cache), our cache warming mechanism reduced 90% of the simulation time to stabilize performance compared with a simulation using cold started cache. As a result, we can cut 15 minutes in gate-level simulation; therefore, we will be enable to skip tens of hours for a peak performance evaluation in transistor-level simulation. Our cache warming mechanism reduces one more order of magnitude when we use a non-blocking cache for cache warming.

Currently, optional checkpoints are only for data cache warming. Expanding optional checkpoints such as instruction

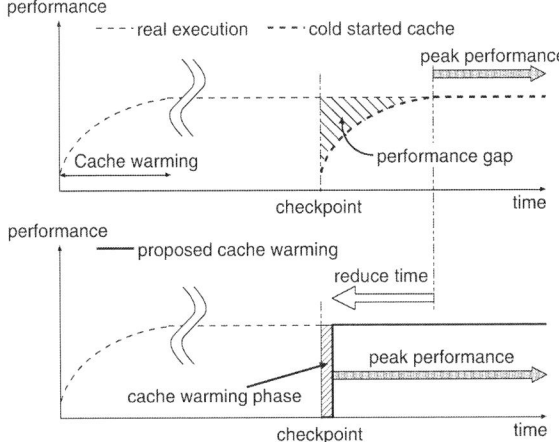

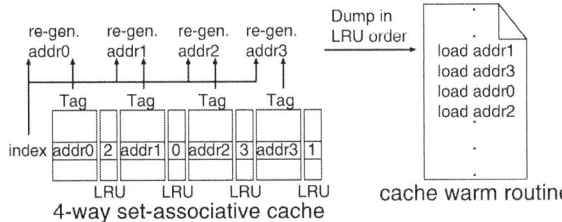

Fig. 8. Impact on performance using cache warming.

Fig. 9. Generating cache warming routine.

cache and branch predictors is left for future work.

VII. SUMMARY AND CONCLUSIONS

We proposed a practical processor co-simulation framework that provides system call emulation, checkpoint, and cache warming mechanisms through the RTL, gate level, transistor level, and fabricated chip phases. All the mechanisms were effective in two processor design projects.

For future work, we intend to demonstrate that our framework can be used for a fabricated chip we are currently designing. Also, the entire co-simulation environment will be available in the near future.

ACKNOWLEDGMENTS

This work was supported by JSPS KAKENHI Grant Number 24700047. This work is supported by the VDEC of the University of Tokyo in collaboration with Synopsys, Inc., Cadence Design Systems, Inc., and Rohm Corporation.

REFERENCES

[1] D. Burger and T. M. Austin, "The Simplescalar tool set, version 2.0. technical report" *CS-TR-1997-1342, University of Wisconsin-Madison*, 1997.

[2] R. Shioya, M. Goshima, and S. Sasaki, "The design and implementation of processor simulator Onikiri2", *the Annual Symposium on Advanced Computing Systems and Infrastructures, poster*, 2009.

[3] F. bellard. QEMU: open source processor emulator. http://bellard.org/qemu/.

[4] N. L. Binkert, R. G. Dreslinski, L. R. Hsu, K. T. Lim, A. G. Saidi, and S. K. Reinhardt, "The M5 simulator: modeling networked systems", IEEE Micro, 26:52-60, 2006.

[5] N. Fujieda, T. Miyoshi and K. Kise, "SimMips: A MIPS system simulator", *Workshop on Computer Architecture Education held in conjunction with MICRO-42*, pp. 32-39, December 2009.

[6] XUM Version 2.0: the eXtensible Utah Multicore Project, September, 2012. http://www.cs.utah.edu/formal_verification/XUM/.

[7] OpenSPARC: http://www.oracle.com/technetwork/systems/opensparc/index.html.

[8] N. K. Choudhary, S. V. Wadhavkar, T. A. Shah, H. Mayukh, J. Gandhi, B. H. Dwiel, S. Navada, H. H. Najaf-abadi, and E. Rotenberg, "Fab-Scalar: Composing synthesizable RTL designs of arbitrary cores within a canonical superscalar template", *Proceedings of the 38th IEEE/ACM International Symposium on Computer Architecture (ISCA-38)*, pp. 11-22, June 2011.

[9] T. Sugiyama, T. Sasaki, T. Nakabayashi, and T. Kondo, "Development of C++/RTL co-simulation environment for accelerating VLSI design of an embedded processor", *Proceedings of the 28th International Technical Conference on Circuits/Systems, Computers and Communications (ITC-CSCC 2013)*, pp. 281-284, July, 2013.

[10] Newlib http://sourceware.org/newlib/.

[11] B. H. Dwiel, N. K. Choudhary, and E. Rotenberg. "FPGA Modeling of Diverse Superscalar Processors", *Proceedings of the 2012 IEEE International Symposium on Performance Analysis of Systems and Software (ISPASS'12)*, pp. 188-199, April 2012.

[12] T. Sherwood, E. Perelman, G. Hamerly, and B. Calder, "Automatically Characterizing Large Scale Program Behavior", *10th International Conference on Architectural Support for Programming Languages and Operating Systems*, Oct. 2002, pp. 45-57.

Annotation and Analysis Combined Cache Modeling for Native Simulation

Rongjie Yan

State Key Laboratory of Computer Science
Institute of Software
Beijing, China,100190
yrj@ios.ac.cn

De Ma

Institute of Microelectronic CAD
Hangzhou Dianzi University
Hangzhou, China, 310013
made@hdu.edu.cn

Kai Huang Xiaoxu Zhang Siwen Xiu

Institute of VLSI Design
Zhejiang University
Hangzhou, China, 310013
{huangk,zhangxx,xiuswen}@vlsi.zju.edu.cn

Abstract— **To accelerate the speed of performance estimation and raise its accuracy for MPSoC, we propose a static analysis and dynamic annotation combined method to efficiently model cache mechanism in native simulation. We use a new cache model to statically analyze segmental profiling results to speed up simulation, and utilize a dynamic annotation technique to exactly trace the addresses of local variables. Experimental results show the efficiency of the proposed techniques for more accurate system performance estimation.**

I. INTRODUCTION

Rapid growth of silicon processing technology and embedded applications provide more opportunities and challenges for multi-processor system-on-chip (MPSoC) based design. The integration of more processor components brings high performance with concurrency capability and long market period with flexible programmability [10, 9]. The MPSoC design is naturally processor-centric, and thus software-centric. As a consequence, providing techniques to obtain fast performance estimation of software (SW) running on those MPSoC architectures becomes increasing important.

Nowadays, even equipped with powerful execution units and deep pipeline architecture, memory latency still impacts the accuracy of processor performance estimation greatly. Almost all advanced processors apply cache mechanism to overcome the gap between fast processor pipeline and slow memory access. Therefore, we have to take cache effects into account to exactly estimate the performance of MPSoCs. A great variety of estimation approaches have considered cache as a key element in system performance estimation. Almost all of the instruction set simulators (e.g., ConvergenSC [2], Realview[1], and MPARM [5]) have applied detailed cache models to give accurate SW execution cycles. The function of the cache models is to manage memory access performed by their processor simulators. Though the simulation using ISS could reflect the accurate execution cycles of a system, it is not suitable for early performance estimation for the low speed. Recently, native simulation based approaches have been proposed to achieve a good tradeoff among simulation speed, accuracy and retargeting ability [11, 8]. In native simulation, software runs on a host machine natively and its execution time on the target machine is calculated using annotation or analysis technique. Most native simulation techniques use tag-search based cache

models [15] or purely statistical models [11]. The method in [15] has adopted a classical cache model, with the binary code being divided into cache lines. During native simulation, whenever a line tag changes or a basic block terminates, the cache model is queried to check if a line is present. This model performs a sequential search over the whole set to check if it is hit or miss. Accessing and searching the cache model incur a much higher simulation cost. Castillo et. al. propose an improved fast Instruction Cache modeling technique [6] to avoid tag-searches by defining a two-dimension array structure for each basic block. During simulation, when an execution flow reaches the end of a basic block, we check the corresponding array element to know whether the basic block is in cache. This mechanism provides constant hit time detection. However, the cache updating granularity based on basic block level is quite small. It would become time-consuming for software codes with many branches. Even worse, few data cache modeling techniques have been considered in native hardware and software co-simulation. Compared with instruction cache model, it is difficult for data cache model to obtain the addresses of the data variables during pre-processing, for they are allocated dynamically. Pedram et. al. propose a method to obtain addresses for global data variables [13], which is only a small percentage of the total data variables. Improved techniques [7] trace addresses of all variables from native simulation for their data cache model. In this case, many annotations will be inserted into the source, and the cache model should be updated every time when data memory access happens, which cause great simulation time overhead. To reduce cache updating frequency, we have presented a segment-based technique to model instruction and data cache [12]. To model data cache accurately, it tries to establish the address offset tables for local variables in functions by analyzing the assembly code in the target processor and tracing the address of stack pointer in function callings according to the execution flow of native simulation. However, it is difficult to exactly trace the address of these local variables only with static analysis, and inaccurate address offset table may lead to wrong estimation results. Moreover, it does not handle the addressing problem of heap space.

To address the above issues, we propose a dynamic annotation and static analysis combined method to establish a fast and accurate cache model in native simulation. The main contribution is to analyze segmental profiling results statically to speed up simulation, and utilize a dynamic annotation tech-

978-1-4799-2817-0/14 $31.00 © 2014 IEEE

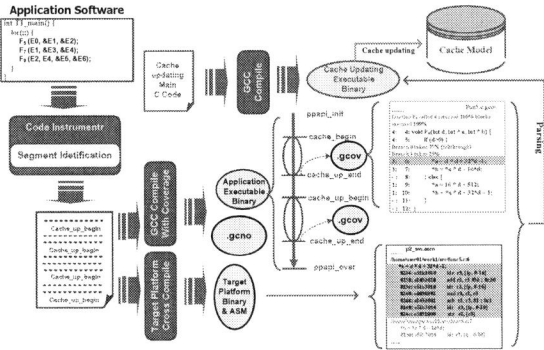

Fig. 1. The workflow of the cache modeling method

nique to trace the addresses of local variables exactly. Compared with existing works, we have mainly adopted three techniques in our cache model. First, we divide the code of an application into several segments according to the control flow of C language, and update the cache model in segments to avoid frequent cache model accessing. Second, for instructions and global or static variables, an address table of the target platform is established by statically analyzing the target-processor assembly code. Third, we insert annotation functions into GCC debugger script to trace the offset address of local variables. These annotation functions are used to establish the address table of dynamic variables by printing the variable hierarchy and their addresses in the host machine during native simulation.

II. THE PROPOSED CACHE MODELING METHOD

To achieve faster cache behavior simulation, we present a new cache modeling method basing on native simulation, as illustrated in Fig. 1. The basic idea is to use gcov [3] to obtain the profiling result of C statements during simulation. Then the status of a cache model is updated according to the analysis of the generated execution flow and execution times of each statement with its assembly code generated by the targeted platform assembler. There are three stages in our cache modeling process:

1. Code instrumentation stage to insert a cache update API at the boundary of each updating unit.
2. Profiling stage to collect profiling results of the code during simulation on a host machine.
3. Analysis stage to analyze profiling results with the assembly code in the target platform and update the cache status.

As Fig. 1 illustrates, to be executed in a host machine, we compile an application with GCC coverage parameters. Meanwhile, extra files suffixed by .gcno are generated by GCC with information to reconstruct the basic block graphs and assign the numbers of source lines to blocks. In Fig. 1, we also present a coverage file named fun5.c.gcov, which is generated by GCC profiling tool gcov during native simulation, with the total execution times of each statement and the percentage of executing different control branches. Based on the profiling results, there are still three problems to be solved in the cache modeling method:

1. What kind of cache model should be adopted to detect cache miss/ hit efficiently?

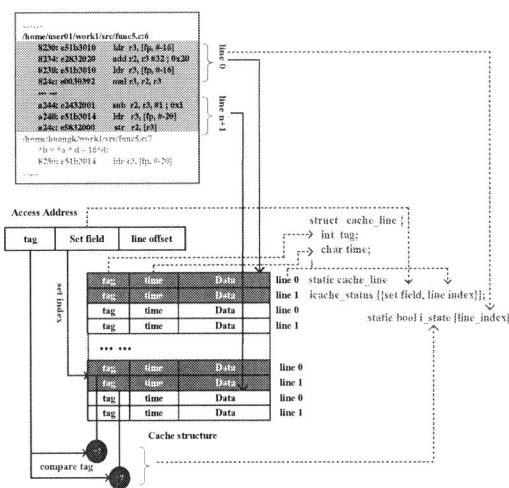

Fig. 2. An example of an instruction cache model

2. How to identify a basic cache updating unit to reduce cache updating frequency.
3. How to update the status of the cache model according to the profiling results.

In this section, we first propose our cache model. Then we review the process of segment identification. The third subsection presents the methods to trace instruction and data addresses, combined with dynamic annotation and static analysis techniques. Finally, we explain the updating mechanism of our cache model.

A. The proposed cache model

Cache memory is composed of multiple lines and each line consists of a certain number of bytes stored sequentially in memory, with a unique identity named tag and its replacement information. Cache lines are arranged in sets, depending on the association degree (e.g., for a 2-way association cache, there are two lines in one set), as shown in Fig. 2. The requested address is divided into three fields: *line offset, set field* and *tag*. The line offset is used to select data from the target line. Its width depends on the line size, e.g., for a cache with 32-byte line size, its width is 5 bits. The set field indicates at which line the set may be allocated. Within the target set, all the lines are equivalent. Thus, to determine if a required line is in a set, the most significant bits of an address are used to compare with all tags. With this mechanism, we can determine whether the access is a hit or a miss. When a memory access causes cache miss, the whole line will be fetched from the next level of memory. The updating scheme of the traditional tag-search based cache model depends on the number of cache lines. Using more lines leads to slower simulation speed, and it becomes worse with the increasing number of processors.

Our generic cache model contains two arrays, as illustrated in Fig. 2. One is declared as *bool* type with the size of the total number of instruction/data lines for the given application software, and each bit indicates whether the corresponding line is in cache. For example, if the code section of an embedded software is 1M and the cache line of the target processor is 32 bytes, the size of the first array is $2^{20}/2^5 = 32K$ bits. The

978-1-4799-2817-0/14 $31.00 © 2014 IEEE 407

other is an array indexed by the line number of the cache. The array consists of the tag of the instructions in the cache and the replacement information (time) for the corresponding line. The second element could be adjusted according to the replacement strategy of the cache. For example, it is ignored if random or round-robin replacement strategy is applied. However, we have to check the variable *time* when LRU (Least Recently Used) replacement strategy is in use. We do not introduce set index in our cache model, for it is indicated by the line index. Taking a four-way associative cache as an example, the line with index *line_num* belongs to set *line_num*/4.

Although the proposed model is memory consuming, the consumption can be ignored in modern workstations, which are equipped with several gigabytes of RAM. For processors with 32-bit data width, the maximum address range is 2^{32}. Considering a 16KB instruction or data cache of 32-byte line size [14], the maximum size of the first array will be $2^{32}/2^5 = 2^{27}$ bits while the second array is $5*(16/32) = 2.5$ KB. As the second array also exists in traditional tag-search based cache models, the memory overhead of our model is no more than 16MB. Therefore, even for the extreme condition that all the address space is covered, the amount of memory used for modeling this scenario is still quite far from the maximum capacities of modern workstations. The key advantages of the proposed cache model are as follows. For cache with random or round-robin replacement strategies, the time-consuming tag-search process could be avoided absolutely. And the cache hit is detected just by indexing the *bool* array, which always consumes constant time. However, when LRU is the selected strategy, we still have to search across the lines of the set to update the correct LRU information. If the status of a cache model is updated frequently, it will still be time-consuming for LRU cache replacement strategy. To reduce the frequency of cache updating, we adopt a segment-based cache updating strategy.

B. Segment-based cache updating strategy

We have proposed a segment-based cache updating method to reduce the updating frequency of cache without accuracy loss [12]. In this model, an application is divided into segments. Then the execution flow of statements contained in a segment could be analyzed statically according to the profiling result. The strategy allows cache effects of multiple statements to be applied in bulk.

The identification of a segment depends on the control flow of the corresponding C codes, which is either explicit or implicit. An explicit/implicit control block is composed of a set of basic blocks, whose size is constrained by the control structures in the C code. For example, the *for* loop in the code of Fig. 3 creates an explicit control block. And the two basic blocks involved in the *if-else* branch build up an implicit control block. Roughly speaking, the size of a segment is not bigger than the size of cache. A group of explicit control blocks can be in one segment if the sum of their sizes is smaller than the size of cache. For an implicit control block, it can be added to an existing segment if the size of the resulted segment is smaller than the size of cache. Otherwise, a new segment for this control block or a set of new segments for every basic block in this implicit control block will be created.

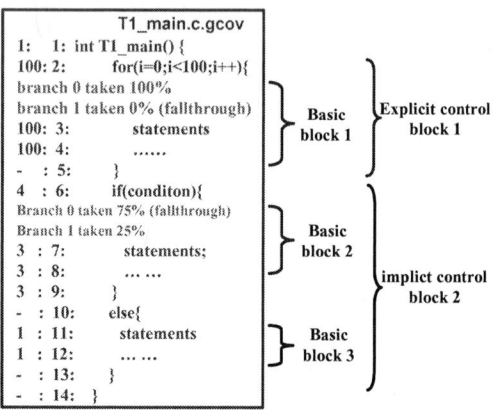

Fig. 3. An example of basic and control blocks

C. Tracing instruction and data addresses

In native-simulation, the source code is compiled with a host compiler. Different compiler optimization strategies and processor instruction sets may make the generated assembly code quite different from the real one running on the target platform. Therefore, to observe the exact instruction or data fetching behavior during native simulation, we compile the source code with two compilers: one is the host compiler to generate an application executable binary file for the native simulation and profiling; the other is the target cross-complier to generate a target assembly code. For instructions, the program counter is amenable to be analyzed statically based on the assembly code from the target platform. Using the target assembly code, we can easily find the relationship between a statement and its corresponding instructions, including the real addresses of the instructions in the target platform. In the example of Fig. 4, file p2_sw.asm shows the correspondence between C statements and their target assembly code. With this relationship, we can model the instruction memory access behavior on the target platform according to the profiling result.

Different from instruction cache modeling, data cache is actually more difficult to model. Tracing the address of a data variable is not easy, because it heavily depends on the dynamic behaviors with less regularity. The key idea to model data cache for native simulation is to derive the addresses of data variables to trace the data memory access behavior in the target platform. Source code annotation techniques are widely adopted to trace variables' addresses during native simulation, to indicate the type of data access and transfer the address to the cache model [7]. Whenever an annotation functions is called, the data cache model will be updated. However, this technique is not fit to segment-based cache updating model, for cache is only updated after the execution of a segment completes. Apart from that, the updating will be time-consuming for the following two reasons: 1) increased code size, where the code size after annotating will be several times bigger than the original; 2) frequent cache updating, where the cache model should be updated every time when data memory access happens.

To trace the addresses of data variables for our cache model, we further introduce a static analysis and dynamic annotation mixed method. According to the ELF standard [4], data allocation in the memory is easy to derive. Global variables are

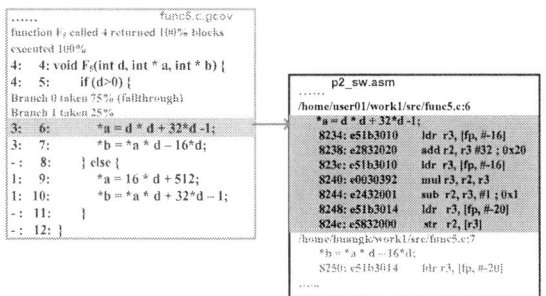

Fig. 4. Analyzing instruction addresses

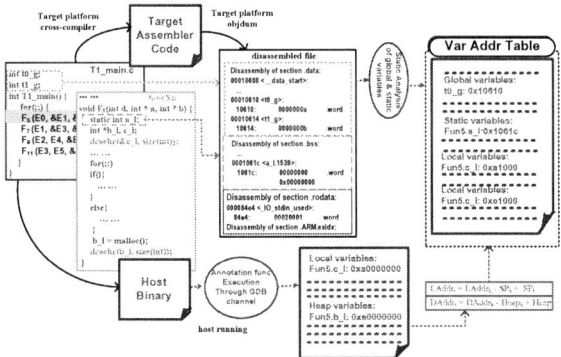

Fig. 5. Address tracing process of data variables

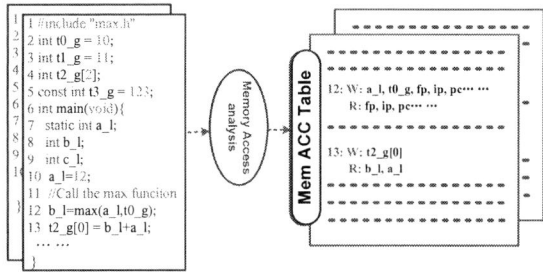

Fig. 6. An example of basic and control blocks

mainly allocated in *data*, *rodata* and *bss* sections. Local variables are allocated in the *stack*, and dynamic data are stored in the *heap*. Global or static variables are allocated statically during program code compilation. We can obtain their addresses by analyzing the disassembled file generated by target processor objdump tool, as shown in Fig. 5.

A local variable is declared in a function or a block and given in a local scope. The address of a local variable is allocated dynamically from the stack according to program running status. It is not possible to derive the absolute addresses of these variables directly from static analysis. Even worse, dynamic data is allocated from and released to the heap by *malloc* and *free* functions respectively during program running, and thus no regularity can be followed. Due to the complicity of heap, few performance estimation related works have considered it.

The idea of our dynamic annotation and static analysis combined address tracing method is as follows. For global or static variables, we establish a variable address table of the target platform by analyzing the cross-compiler ELF file statically. However, for dynamic variables allocated from stack or heap, we insert annotation functions into GCC debugger script to trace the offset address of these variables. In GCC debugger script, we set breakpoints where local variables are defined or memory is allocated for dynamic data. Then the debugger can print the addresses for these variables. The whole process is illustrated in Fig. 5. We use annotation functions to establish the address table of dynamic variables by printing the variable hierarchy and their addresses in the host machine during native simulation. For example, for a local variable *var* defined in *Fun5* with address 0x8a0000, the printing format is $Fun5.var : 0x8a0000$. Though the absolute addresses of the

host machine do not match those in the target platform, we can maintain the offset of the variables within each section. To keep the address offset in the host machine and its target platform the same, it is necessary to ensure that the types of variables in the host machine have the same size as those in the target. Otherwise, a slightly modification of the source code is necessary to covert the variable type. For example, if the type *long* in host machine is 64 bits while in the target processor it is 32 bits, the variable with type *long* should be converted to *int* type. With the printed result of annotation functions through native simulation, we can compute the addresses of the target platform by the following two formulas:

$$LAddr_t = LAddr_h - SP_h + SP_t \qquad (1)$$

$$DAddr_t = DAddr_h - Heap_h + Heap_t \qquad (2)$$

where $LAddr_t$, SP_t, $DAddr_t$ and $Heap_t$ represent the addresses of local variables, the stack pointer, dynamic data and heap base in the target platform respectively, while $LAddr_h$, SP_h, $DAddr_h$ and $Heap_h$ represent the corresponding variables in the host machine.

Besides the variable address table, we have also established a memory access table according to memory visits of each statement, as shown in Fig. 6. Before native simulation, each statement is analyzed statically to find all variable operations, each of which indicates a cacheable load or a store operation. A writing access is performed when a variable is on the left side of an equal expression (=, +=), or when there is an operator such as ++ or −−, or a function calling where stack push happens. A read access happens when a variable is on the right side of an equal expression, or comparing expression (>, <, ≥), or exiting a function where stack pop happens. Meanwhile, fp, ip, lr, and pc are system variables and should be reserved and recovered when calling and exiting functions respectively. The types of system variables and their corresponding operations are directly related to the type of a processor.

D. Cache updating

When a cache_up function is called in the profiling process, we need to update the cache model, and count the hit times of each instruction. The cache updating process consists of three stages: 1) Collect execution times of each statement by comparing current .gcov files with the one when the previous cache_up function was called. 2) Analyze the execution flow of statements in the segment according to the .gcov files. 3) Extract the target-platform assembly code of each statement for instruction cache, and the memory access information of each

TABLE I
DATA CACHE MISS COMPARISON BETWEEN DIFFERENT MODELS

cases	Data cache misses				
	ISS-based	Search-based	Annotation-based	Proposed method	Error(%)
Bubble 1000	127	127	126	119	6.2%
Bubble 10000	5199310	5207383	5207383	4980341	4.2%
Hanoi	38	44	44	40	5.2%
Factorial	375	500	500	410	9.3%

statement for data cache. And then update the instruction and data cache model respectively according to the execution flow.

To model instruction cache, both *i_state* and *icache_state* arrays are declared globally as static variables to remark the statuses of each instruction line in an application software and the instruction cache model respectively in cache update main code. In the initialization of a profiling task, all bits of the *i_state* array are initialized to be false while the time field of *icache_block* is initialized to zero. During simulation, when cache_up is invoked, the corresponding assembly code is analyzed to determine whether the cache is hit. This task is performed by checking the array *i_state* with the line address of the instruction. If a cache miss happens, both *i_state* and *icache_state* arrays are updated. If some cache line is empty in the set mapped by the instruction, we store the tag of the new instruction in the *icache_state* array directly, and change the bits of the corresponding *i_state*. Otherwise, one line in the cache will be replaced by the new with the replacement strategy. As most of the cache accesses result in a hit in normal operations such as loops and sequential statements, this solution minimizes the cache modeling overhead while keeps the estimation accuracy.

Data cache modeling in native simulation is similar to the instruction cache described above. Two arrays named *d_state* and *dcache_state* are declared globally as the static variables. *d_state*, indexed by the line addresses of variables, is used to indicate whether the variable is hit in the cache, while *dcache_state* is used to indicate the status of each cache block.

III. CASE STUDIES

To demonstrate the advantages of the proposed cache model, we take both small benchmarks (Bubble, Hanoi and Factorial) and a real application (H.264 decoder) as case studies. The simulation results have been compared with those from CK610 ISS, and results with ARM920T reported in [6, 7] and [14], to illustrate the efficiency of our techniques. The target processor CK610 includes 16KB instruction and data caches respectively with 2-way association and 16-Byte line size.

As small benchmarks are especially fit to analyze the corner cases, we first show the experimental results on data cache comparison between our techniques and others in Table I. The column for ISS (ARM920T Instruction Simulator) lists the accurate results, and the two subsequent columns present the results listed in [7] and [14], where the search-based method is traditional, and the annotation-based method is proposed in [7]

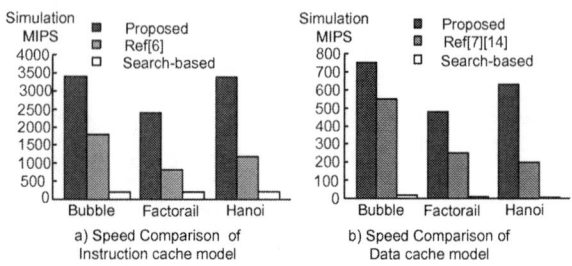

Fig. 7. Speed comparison between different techniques

and [14] (The two papers present same results). The code sizes of the benchmarks are so small that all the instructions are in the cache after the first execution.

In Table I, we can observe that for the bubble benchmarks, the annotation-based method has higher accuracy than ours. The reason is that our cache model is updated only after the execution of a segment, and the branch execution flow is analyzed statically, which will be different from the real one. For the annotation-based method, its cache model communicates with the annotation function whenever memory access happens during the native simulation. However, for the last two benchmarks, the accuracies of the annotation-based method reduce greatly with 25% and 33.3% error rates respectively, while the error rates of our model are only 5.2% and 9.3% respectively. The reason is that Hanoi and Factorial have a large percentage of function calls that cause a large amount of stack push/pop operations. We have considered both system variables (i.e., fp, ip, lr and pc) and formal parameters in our model, which are ignored in [7] and [14].

To compare simulation speed, we have executed the benchmarks with three techniques: search-based, annotation-based [6, 7, 14] modeling and our method. In Fig. 7, we show the speed comparison between the three techniques. The speed is described in simulation MIPS (Million Instructions Per Second, Millions of Instructions of the target platform could be simulated per second). In Fig. 7, we find that our cache model can achieve 2.5 times average speedup compared to the instruction cache model proposed in [6], and 2 times average speedup compared to the data cache model proposed in [7] and [14] respectively. All the models can avoid tag-search process for cache hit detection, which accelerates the speed greatly, compared to the search-based model. However, the instruction cache update in [6] is based on basic block, while the data cache proposed in [7] and [14] is updated whenever the memory access happens, both of which cause lots of simulation overhead for communication between cache models and the application. In our cache model, as the code sizes and address ranges for variables in all benchmarks are smaller than the sizes of instruction and data caches respectively, the whole routine is treated as a segment. Therefore, both instruction and data cache models are updated after the execution of the whole routine, and there is almost no overhead on simulation time.

To further analyze the efficiency of our cache model for different cache architectures, we choose a 30-frame Foreman QCIF H.264 decoder on a 4-processor hardware platform as the case study with different cache configurations, i.e., CK610 ISS, the model proposed in [12] and our model respectively. As shown in Fig. 8, the hit rate of instruction cache obtained from our model is almost the same as that from ISS, which means

978-1-4799-2817-0/14 $31.00 © 2014 IEEE   410

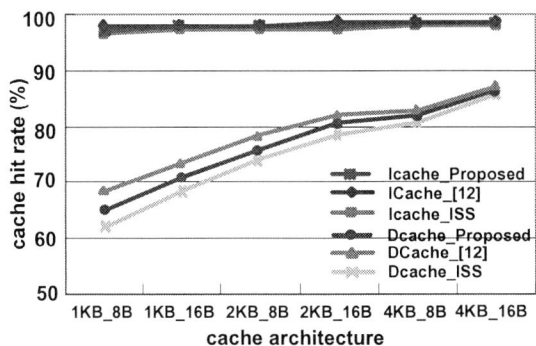

Fig. 8. Instruction and data cache hit rate comparison for different cache architectures

that our instruction cache model provides accurate estimation on instruction cache behavior. However, compared with ISS, the hit rate error of data cache estimated by our model is almost 4% in the case that CK610 processor is equipped with 1 KB cache and 8-Byte line size (1K_8B). This error is reduced to less than 1% if the cache size becomes 4 KB and its line size is 16 Bytes (4K_16B). The reduction comes from the increased size of cache. The average error rate of our model is reduced by half, compared with that of the model in [12]. Overall, the data cache hit rate curve of our model is similar to that of ISS, which indicates that our model is able to present the exact changing trends of cache performance through native simulation.

IV. CONCLUSION

We have presented a fast and accurate cache modeling method, which combines dynamic annotation and static analysis on the profiling results from native simulation. Meanwhile, we have considered more details in data cache model to obtain accurate cache hit information. In the proposed cache model, for global or static variables, we establish a variable address table of the target platform by analyzing the cross-compiler ELF file statically. For local variables and dynamic data, annotation functions are inserted to GCC debugger script to trace the offset address of local variables. Experimental results show the advantage of our model in simulation speed and estimation accuracy. For the future work, we will further investigate the techniques to improve the accuracy of cache model, and cache updating strategies to accelerate simulate speed with less loss of accuracy.

ACKNOWLEDGEMENTS

This work is supported in part by National Science Foundation of China under Grant No. 61100074, National Science and Technology Major Project of China under Grant No. 2012ZX01039-004 and Fundamental Research Funds for the Central Universities.

REFERENCES

[1] ARM, Inc. RealView MaxSim. http://www.arm.com/products/DevTools/MaxSim.html.

[2] Coware, Inc. ConvergenSC. http://www.coware.com/S.

[3] GNU, GCOV-a test coverage program. http://gcc.gnu.org/onlinedocs/gcc/Gcov.html.

[4] TIS committee, Executable and Linkable Format, version 1.2. http://www.parolamia.eu/ELF.pdf.

[5] L. Benini, D. Bertozzi, A. Bogliolo, F. Menichelli, and M. Olivieri. Mparm: Exploring the multi-processor soc design space with systemc. *J. VLSI Signal Process. Syst.*, 41(2):169–182, Sept. 2005.

[6] J. Castillo, H. Posadas, E. Villar, and M. Martinez. Fast instruction cache modeling for approximate timed HW/SW co-simulation. In *Proceedings of the 20th symposium on Great lakes symposium on VLSI*, pages 191–196, 2010.

[7] L. Diaz, H. Posadas, and E. Villar. Obtaining memory address traces from native co-simulation for data cache modeling in systemc. In *Conference on Design of Circuit and Integrated Systems*, 2010.

[8] P. Gerin, M. M. Hamayun, and F. Pétrot. Native mp-soc co-simulation environment for software performance estimation. In *Proceedings of the 7th IEEE/ACM international conference on Hardware/software codesign and system synthesis*, pages 403–412, 2009.

[9] A. Jerraya, H. Tenhunen, and W. Wolf. Guest editors' introduction: Multiprocessor systems-on-chips. *Computer*, 38(7):36–40, 2005.

[10] A. A. Jerraya and W. Wolf. Hardware/software interface codesign for embedded systems. *Computer*, 38(2):63–69, Feb. 2005.

[11] C. Kirchsteiger, H. Schweitzer, C. Trummer, C. Steger, R. Weiss, and M. Pistauer. A software performance simulation methodology for rapid system architecture exploration. In *Electronics, Circuits and Systems*, pages 494–497, 2008.

[12] D. Ma, R. Yan, K. Huang, M. Yu, H. Ge, X. Yan, and A. Jerraya. Performance estimation techniques with mpsoc transaction-accurate model. *Transaction on Computer-Aided Design of Integrated Circuits and Systems*, to appear.

[13] A. Pedram, D. Craven, and A. Gerstlauer. Modeling cache effects at the transaction level. In *Analysis, Architectures and Modelling of Embedded Systems*, volume 310, pages 89–101. 2009.

[14] H. Posadas, L. Daz, and E. Villar. Fast data-cache modeling for native co-simulation. In *16th Asia and South Pacific Design Automation Conference*, pages 425–430, 2011.

[15] J. Schnerr, O. Bringmann, A. Viehl, and W. Rosenstiel. High-performance timing simulation of embedded software. In *Proceedings of the 45th annual Design Automation Conference*, pages 290–295, 2008.

A Scorchingly Fast FPGA-Based Precise L1 LRU Cache Simulator

Josef Schneider   Jorgen Peddersen   Sri Parameswaran

School of Computer Science and Engineering, University of New South Wales, Sydney, Australia

{jschneider, jorgenp, sridevan}@cse.unsw.edu.au

Abstract—Judicious selection of cache configuration is critical in embedded systems as the cache design can impact power consumption and processor throughput. A large cache increases cache hits but requires more hardware and more power, and will be slower for each access. A smaller cache is more economical and faster per access, but may result in significantly more cache misses resulting in a slower system. For a given application or a class of applications on a given hardware system, the designer can aim to optimise cache configuration through cache simulation. We present here the first multiple cache simulator based on hardware. The FPGA implementation is characterised by a trace consumption rate of 100MHz making our cache simulation core up to 53x faster, for a set of benchmarks, than the fastest software based cache simulator. Our cache simulator can determine the hit rates of 308 cache configurations, of which it can determine the hit rates of 44 simultaneously.

I. INTRODUCTION

Ever since processors became faster than computer memory, much attention has been paid to maximising processor throughput despite the difference in speed between these components. One of the most obvious ways is to include a cache: a small, fast memory located close to the processor that temporarily stores copies of accessed data. It relies on two properties frequently found in software which are the likely reuse of recently used data and the spatial proximity of data accesses [1].

Ideally a designer wants to make the cache as big as possible in order to minimise the amount of cache misses. However, having a big cache increases die cost, makes the cache slower per access and increases its power consumption [2]. Such a large cache could cancel out the potential benefits of a higher hit rate as maximum hit rate is not synonymous with either minimum execution time or minimum energy consumption. Additionally, the memory access patterns fully depend on the software running on the system. An application or a set of applications can therefore have an *optimal cache configuration* where a larger cache is not necessary or where a suitable tradeoff between the conflicting timing, power and cost constraints can be found.

Much research has focused on simulating the cache behaviour under certain applications in order to find this optimal configuration. These simulators are software based and are either analytical [3][4][5][6] or precise [7][8][9][10][11]. Analytical cache simulators are characterised by fast simulation times but are approximate. Precise cache simulators keep track of every single memory access and determine precisely when a cache hit or miss occurs in a cache of given configuration. Despite many advancements these cache simulators can still be prohibitively slow, especially when simulating the latest high-power applications. For example, encoding 24 low resolution images into an MPEG2 video using a software encoder (i.e., without hardware acceleration) produces a trace file containing over 11 billion memory accesses. The fastest software cache simulators will take over an hour to process this file even though the output of the application is only *one second worth* of video. In an environment where design turnaround time is paramount, the speeding up of this simulation is vital.

Our contribution is as follows: We present the very first precise multiple cache simulator exclusively based on hardware and implemented on an FPGA.

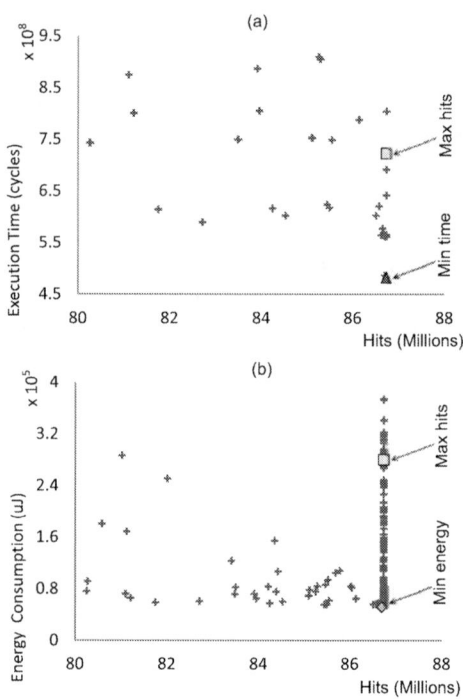

Fig. 1: Plots of the hit rates of different cache configurations vs. execution time (a) and energy consumption (b). It is easy to see that the configuration yielding the maximum hit count is far from being the configuration with minimum execution time or minimum energy consumption.

By exploiting the inherent parallelism found in FPGA fabric we can analyse many cache configurations at high speed. In this work we first describe the motivation for simulating cache hit and miss rates in Section II. The rest of this paper looks at previous implementations of cache simulators in Section III, and is followed by a reminder on how caches are configured in Section IV. A detailed description of the techniques used to minimise the hardware resources required for our cache simulator can be found in Section V. In Section VI we then compare the performance of our design with existing software cache simulators before looking into future work we intend to perform based on our implementation.

II. MOTIVATION

Many modern processor cores such as NIOS II, ARM and Xtensa provide the designer with the ability to configure the system cache or caches. Designers therefore need to know which cache configuration (set size, associativity and block size) would be optimal depending on the application they are running on the given processor. The task of choosing an optimal cache configuration is not trivial: we refer to work by Shwe et al. who clearly demonstrated [12] that a maximum cache hit rate has little to do with minimum execution time

978-1-4799-2817-0/14 $31.00 © 2014 IEEE   412

or energy consumption, as can be seen in Figure 1. To find the cache configuration yielding the minimum execution time and/or power consumption the cache hit and miss counts need to be determined for each unique cache configuration available. The obtained hit and miss values can directly be used to determine the timing and energy consumption of caches and even memory systems comprised of a cache and external memory device (e.g., a DRAM chip) by using equations such as those presented by Janapsatya et al. [9]. Armed with die area, timing and energy values, designers can then choose the best cache configuration for their purposes based on their design constraints.

In this paper we refer to a *multiple cache simulator* as a tool capable of establishing the cache hit and miss rates for many different cache configurations (different line sizes, set sizes and associativities) by analysing the memory access trace.

III. RELATED WORK

The ultimate goal when designing a cache simulator is to make it as exact and as fast as possible. In this section we cover the related work, principally in the field of simulation of caches with Least Recently Used (LRU) cache replacement policy. The category of analytical cache simulators mentioned earlier [3][4][5][6] uses probability to make guesses of the hit and miss rates of different cache configurations. They are limited by the fact that they are estimates. The assumption is made that memory accesses follow certain patterns. If the application is characterised by unruly address accesses, analytical cache simulators will yield large amounts of error.

Software-based precise cache simulators are inherently slower as they take into account every memory access. One of the most widely used precise cache simulators is DineroIV [8] by Jan Edler and Mark D. Hill. Though it is very flexible with instruction and data cache simulation of different configurations of both L1 and L2 caches, it is extremely slow for the purposes of multiple cache simulation. The same authors presented the forest simulation technique which relies on the *inclusion* and *set-refinement* properties [13].

Based on this same forest simulation technique, the cache simulator presented by Janapsatya et al. contains a collection of forests to compactly keep track of the states of many different cache configurations, searching the trees in a top-to-bottom manner [9]. Tojo et al. improved on this method, exploiting two supplementary cache properties they discovered for their CRCB algorithm [14].

Shortly after, Haque et al. tackled the problem from the other side with SuSeSim [10], determining that searching the trees in the opposite direction (bottom-to-top) showed a significant improvement in speed when combined with certain contrapositions of the inclusion cache properties. One of the latest precise cache simulators to have surfaced is HC-Sim by Chen et al. [11]. The simulation times cited for HC-Sim, however, are slow as their measured cache simulation times are up to one order of magnitude slower than SuSeSim. One cache simulator which has existed for many years is Cheetah [7] from Sugumar et al. which uses a binomial tree methodology.

As far as we are aware no hardware multiple cache simulators exist to this date. Cache hit and miss statistics could be extracted using expensive system monitoring hardware such as those provided by Lauterbach [15], yet this approach would only be able to analyse one single cache configuration at a time. The same shortcoming is seen in hardware accelerated simulators such as RAMP Gold [16], Protoflex [17], FAST [18] or HAsim [19]. These simulation systems cannot be compared with our work as they generally target timing analysis, multi-core systems and often rely on a front-end PC system. By comparison, our approach is a hardware-only, self-sufficient *multiple* cache simulator instance designed to determine the optimal cache configuration given a certain memory trace. It can easily be configured on any FPGA system to determine cache hit and miss rates for many different cache configurations by analysing a provided trace.

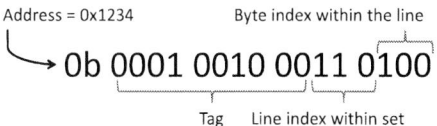

Fig. 2: Breakdown of an 16 bit address for a cache of line length 8 bytes and a set size of 8 lines

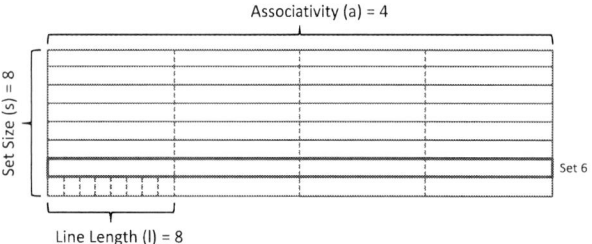

Fig. 3: Depiction of the data organisation within a cache with line length 8 bytes, set size 8 lines and associativity 4

IV. BACKGROUND

The three main configuration parameters that govern the manner in which data is stored in a cache are line length (l), set size (s) and associativity (a). The line length defines how many bytes are stored in one cache line and the set size how many sets are stored in the cache. When an address is accessed, the $log_2(l)$ lowest bits determine the desired byte, while the next lowest $log_2(s)$ bits point to the set index within the cache. The rest of the address bits are the Tag, which is held in the cache together with the associated data to determine whether a specific line is stored or not. Figure 2 shows how a 16 bit address access is broken down in a cache where $b = 8$ bytes and $s = 8$ sets, i.e., the lowest three bits are the byte index while the next lowest three bits are the set index.

Associativity defines the number of locations in which memory lines with the same set index can be stored. In a cache with associativity one[1] all addresses with set index $0b110$ could only reside in one location in the cache. On the off-chance that the software contains an access pattern where different lines with the same set index are frequently being accessed alternatively, we will get *cache thrashing*, leading to an excessive amount of cache misses. A set associative cache lowers the risk of such an occurrence as data can be stored in a different locations with other data of same set index. The data organisation of a 4-way associative cache can be seen in Figure 3. In the event that all four ways of a set are occupied and a new line needs to be stored, one of the older lines needs to be discarded to make room for the new line. The line to be discarded is determined by the cache replacement policy. In this paper we only investigate the LRU cache replacement policy which keeps track of the order in which lines were accessed and discards the one that was accessed the longest time ago.

V. MINIMISING HARDWARE RESOURCES

Different hardware implementations of the LRU replacement policy exist, but for our purposes it is convenient to represent a cache set with all levels of associativity as shift registers. This is similar to the linked list representation used in many software cache simulators [9][10]. In real cache implementations each entry would hold the address tag, the line data and a valid bit. As we are simulating the cache, only the address tag and the valid bit are needed. Figure 4 shows how different address accesses are processed by a cache set of a four way set associative cache.

---

[1]a.k.a. a direct mapped cache

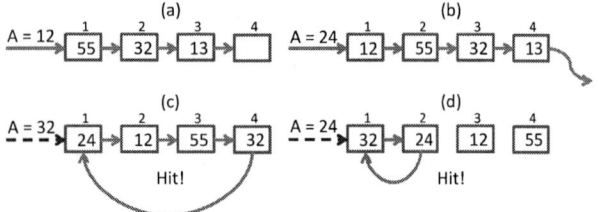

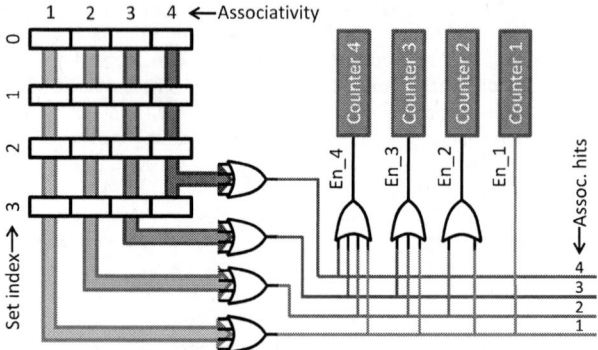

Fig. 4: A cache set with 4 levels of associativity implemented with shift registers and multiplexers. The solid arrows indicate in which direction data will be shifting at the next clock tick. (a) The address accessed, 12, is shifted in after which all four ways are full. (b) Another address token, 24, is shifted in. As all four ways are full the least recently used address token, 13, is discarded. (c) The accesses address matches the oldest address token, 32, moving it to the front of the queue. (d) Again a hit occurs as address 24 is already stored in the cache. The older address tokens, 12 and 55, do not move

Fig. 6: Cache simulator for caches of set size $s = 4$ and associativity $a \leq 4$

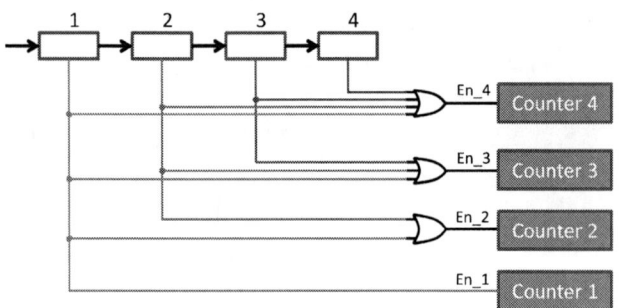

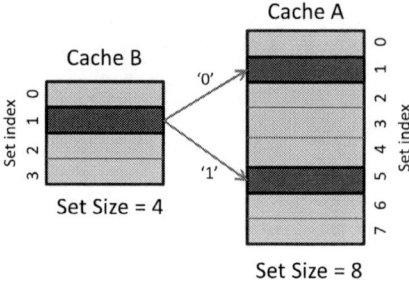

Fig. 5: One cache set of associativity 4 is all that is required to also simulate caches of same set size of associativity 3, 2 and 1. The signal coming out of each level of associativity (or line) indicates whether a hit has occurred at this level. The EN_x signals indicate when the counter for that level of associativity should increment

Fig. 7: Two caches of identical associativity and line size. Cached data stored at set index 1 in the smaller cache B would also always be stored at either of two locations at set indices 1 or 5 in cache A of double the set size. These two locations are denoted by '0', or *left* and '1', or *right*

A. Inclusion Property 1: Associativity

This is where we use the first LRU cache property to minimise the amount of hardware required: a cache of set size s, line length l and associativity a is a subset of a cache of set size s, line length l and associativity larger than a. This means that for multiple caches of identical set size, we only need to simulate the largest associativity in order to also simulate all lower levels of associativity with the same hardware.

Figure 5 depicts the design of a 4 way set associative cache (i.e., with four lines) containing only one set. If a hit occurs in the second line the En_2, En_3 and En_4 signals would enable the hit counters 2, 3 and 4 to increment. We would know that caches with associativity 2, 3 and 4 would have registered a hit, while a cache of associativity 1 would have suffered a cache miss. Caches usually have more than one set, in which case all the hit signals for a given level of associativity are 'or'-ed to drive the counters and to create a group of signals we call *assoc. hits*, as can be seen in Figure 6. This setup can be extended to simulate any group of caches of same set size s and associativity a and smaller, within the limits of the available FPGA resources.

To simulate multiple caches with different cache sizes it would be possible to implement multiple instances of this design. Yet such an approach would not be efficient in terms of resource usage: our current design utilises 71 LookUp Tables (LUTs) and 84 Registers for a cache set of associativity 4. These numbers hold true for a cache simulator capable of consuming a trace with an address width of 32 bits, but even with narrower address spans the resources required are very high. To overcome this issue, a second cache property is exploited.

B. Inclusion Property 2: Set Size

Caches with LRU replacement policy contain another important characteristic that can be used to reduce hardware utilisation of our

cache simulator. A cache of line length l, associativity a and set size s is always a subset of a cache of line length l, associativity a and set size larger than s. We can further refine this property as the data from a cache of given set size would be located in either of two sets of a cache of twice the set size. This can be seen in Figure 7, where data stored at set index 1 in a cache of set size 4 would also always be stored at either set index 1 or 5 in a cache of set size 8, given the same trace. *This means that caches of smaller set size can be simulated by keeping track of where their data is stored within a simulated cache of larger set size.*

For this purpose we designed hardware components we call *sub-sets*. Though their functionality is considerably more complex than the shift registers presented in Section V-A it does not need to manipulate, hold and compare as many data bits. As a result, one sub-set only requires 32 LUTs and 16 registers for an associativity of 4. To more clearly demonstrate the task of a sub-set, let us look at the example given in Figure 9 which shows the state the highlighted sets from Figure 7 could be in if the caches A and B had been subjected to the same trace. We can observe that:

1) all entries of the set in cache B are contained within the two given sets of cache A,
2) if the set from cache B contains x similar entries from one of the sets from cache A, then the matching entries will be the x entries of lowest associativity of that set from cache A, and
3) the order of the last four accesses was first 61, then 12 and 23, and finally 5.

Figure 10 shows the same cache set state as mentioned, only that the set of cache B is depicted by an extremely small shift register. Instead of containing a duplicate of all the address tokens held in Cache A, only two bits are shifted, a *source* bit and a *valid* bit. The valid bit is set when the associated source bit is not empty. The source bit determines if the address token it is keeping track of is located at set index 1 or at set index 5 of cache A, which is determined by a

978-1-4799-2817-0/14 $31.00 © 2014 IEEE

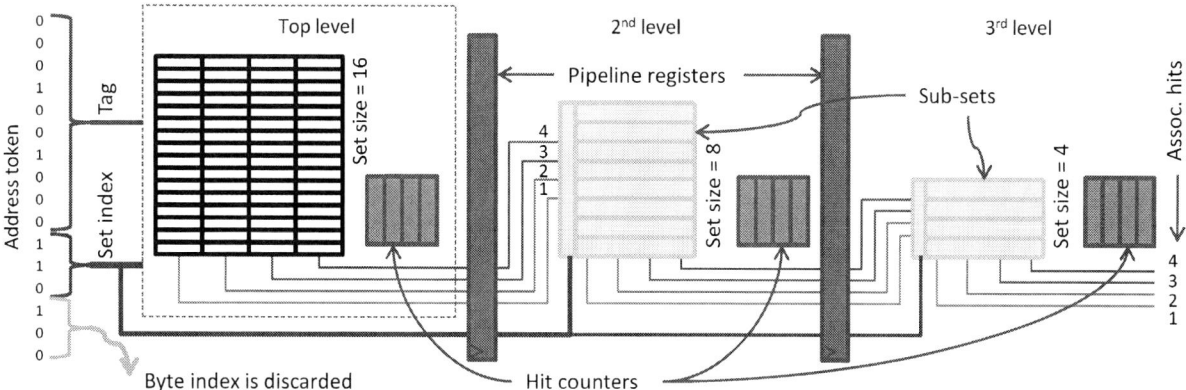

Fig. 8: Cache simulator for caches of set size $s = 16$, $s = 8$ and $s = 4$ and $a \leq 4$. Note that the top level of the cache simulator is identical to the simulator presented in Figure 6 except that it has a set size of 16, not 4. We re-use the address from Figure 2 to show how it is broken down. The assoc. hits and set index are passed down to the lower levels through registers, effectively pipelining the design for higher operating frequency.

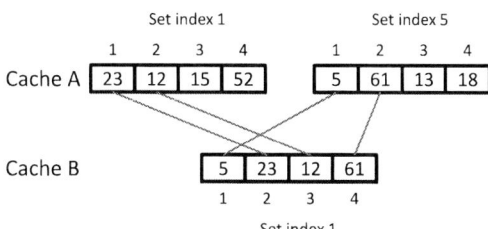

Fig. 9: Example of addresses that could be stored in the highlighted sets of Figure 7 if both caches A and B were administered the same trace. Each set is of associativity 4

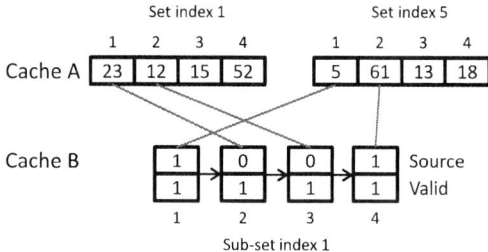

Fig. 10: Same example from Figure 9 with the set from cache B represented as a shift registers

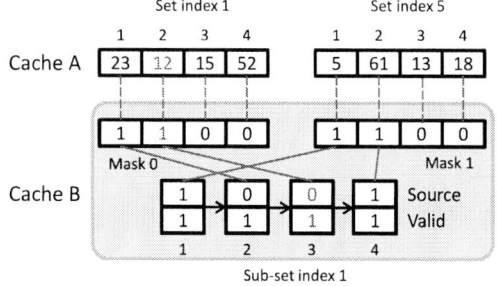

Fig. 11: One subset and its registers. If a hit occurs with address token 12 in cache A, set index 1, we would immediately be able to tell if a hit has also occurred in our subset thanks to Mask 0.

value of '0' or '1' respectively. In our example, the source bit order "1001" means that the first token (5) is stored at set index 5, the second (23) and third (12) tokens are stored at set index 1, and the last token (61) is stored at set index 5 of cache A.

Our sub-set is based on such a shift register but also includes two *mask registers* as seen in Figure 11. These registers indicate whether the address token at a certain level of associativity in cache A is held within cache B. In our current scenario, tokens 23 and 12, and 5 and 61 are tracked by the sub-set, which explains why the first two bits in Mask 0 and Mask 1 associated with those values are set. The purpose of these masks is to easily determine whether a hit has occurred in the sub-set, simplifying future operation.

At this point we can divide our cache simulator into *levels* where each level represents the hardware used to simulate caches of a given set size. The biggest set size (s^{top}) is simulated by the *top level*, like in the design depicted in Figure 6, and directly manipulates address tags. The assoc. hit vector of the top level is then passed down to the second level of the simulator which simulates caches of set size $s^{2nd} = s^{top}/2$, and is made up of s^{2nd} sub-sets. The three signal groups used to control the sub-sets are: an enable signal (determined from the set index), an address bit (the highest set index bit from the level above) and the assoc. hits vector from the level above. Each level of sub-sets also has its own assoc. hits output which is used to count the hits for this level and which can also be passed down to a lower level of the cache simulator. Figure 8 shows most of these signals, and more importantly, how they are connected within the cache simulator. Registers are introduced between the levels of the cache simulator to increase the maximum operating frequency (i.e., address token consumption rate), with each level increasing the latency by one clock cycle.

The entire cache simulator design is described in VHDL. The parameters of the simulating hardware determining the maximum and minimum set size and the largest associativity to be simulated are defined in one location. Lower levels are instantiated recursively, halving the set size at each level, until the smallest set size has been reached. The maximum address token width can also be modified to minimise the resources required by the top level if the range of the addresses is limited.

C. Line Length

For now our cache simulator can simulate different levels of associativity and different set sizes. Simulating different line lengths is trivial: when breaking down the address token, the size of the byte index is set to $log_2 l$ bits (where l is the line length in bytes), the set index is of constant size but is shifted so it is placed next to the byte index, and the remaining bits are used for the tag. This is demonstrated in Figure 12. However, an instance of the cache

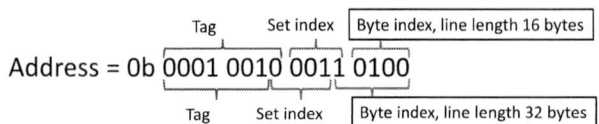

Fig. 12: Breakdown of the 16-bit address of value 0x1234 for the line lengths of 16 and 32 bytes and set size $s = 16$.

$addr_w$	a_{max}	set_count_{max}					
		32		256		1024	
		LUTs	Regs	LUTs	Regs	LUTs	Regs
17	4	4932	3702	25085	15326	79431	46077
	8	10529	7163	56859	29935		
	16	21434	14084	112098	59152		
22	4	5517	4342	26541	20446	87626	66557
	8	11644	8443	60321	40175		
	16	23672	16644	127876	79632		
27	4	5997	4982	31990	25566	105037	87037
	8	12655	9723	70216	50415		
	16	26203	19204	147526	100112		
32	4	6410	5622	35313	30686	126235	107517
	8	13500	11003	77732	60655		
	16	27829	21764				

TABLE I: Resource requirements (in LUTs and Registers) depending on the size of the largest cache simulated. Some parameter combinations could not be compiled as the resource usage of the resulting configuration would have been too large to fit on our FPGA (together with the SoC to control it)

simulator can only simulate one line length at any one time. This means that simulating a different line length requires us to re-run the simulator or to instantiate another cache simulator instance, in parallel, configured to simulate that line length.

VI. FPGA IMPLEMENTATION AND PERFORMANCE

A. Resource Usage

The amount of resources required in the form of LookUp Tables (LUTs) and Registers (Regs) by a hardware cache simulator instance is an important aspect to consider to make sure the design will fit on a given FPGA. This is entirely dependent on the size of the largest cache simulated. Three main cache simulator parameters defined at design time come into play: the address token width ($addr_w$), the maximum set count (set_count_{max}) and the maximum associativity simulated (a_{max}). Table I shows the required resources of a cache simulator with its CPU interface depending on these parameters. The numbers show that the size of our hardware instance grows linearly with the parameter values.

B. Performance

We tested our cache simulator on an Altera Stratix IV GX FPGA containing 230 thousand logic elements depicted in Figure 13. Given the amount of resources available, we were able to instantiate a simulator capable of simulating caches with four different levels of associativity ($a = 1, 2, 3$ and 4), eleven different set sizes ($s = 1$ to 1024) and seven different line lengths ($l = 4$ to 256). With these parameters 308 different cache configurations can be simulated, while 44 configurations can be simulated concurrently as the line length has to remain constant for each simulation run.

On our device, the resulting cache simulating hardware instance is capable of running at 100MHz with a latency of eleven clock cycles, i.e., it is capable of consuming an address trace at 100 million address tokens per second. As our hardware implementation is the first of its

Processing core count	32
Processor	Intel Xeon X7560 @ 2.27GHz
System memory	256 GB

TABLE II: Specifications of the setup used for software simulating purposes throughout this paper.

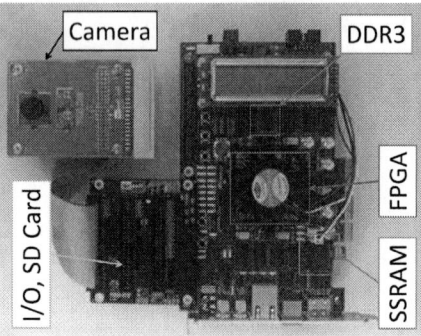

Fig. 13: FPGA hardware used to test our cache simulator. The devices outside the FPGA (Camera, I/O etc.) can be used in future work to implement a fully functional embedded system. A hardware cache simulator instance could then observe any memory bus to obtain the required memory trace from *within the system*.

kind, it is difficult to compare it with existing cache simulators. For static trace simulation such as that performed by the fastest precise software simulators, our current bottleneck lies in the method used to provide the cache simulator with such a trace as we are yet to implement a high-speed data interface. The correct functionality of the cache simulator was tested in Modelsim [20] and with a hardware implementation where static traces were provided through a JTAG UART connected to a PC.

Due to the extremely slow bandwidth of the JTAG UART interface we did not use it for our performance assessment. Instead, we compare the timing results of software-based cache simulators with the *throughput* of our cache simulating hardware. The traces simulated were obtained from both the SimpleScalar computer architecture simulator [21] and the Tensilica Xtensa processor simulator [22]. The traces of four Mediabench [23] applications and five SPEC CPU2000 [24] benchmarks were extracted using these two simulators. Every trace was fed to three different software-based cache simulators, namely DineroIV [8], Cheetah [7] and SuSeSim [10]. The simulations were performed on the high-end Intel system described in Table II. We exclusively simulated the 44 configurations our hardware is capable of simulating on Cheetah and SuSeSim. This was not possible with DineroIV as it does not allow for associativities that are not a power of two leading us to simulate only 33 cache configurations with this tool. The line length simulated was 8 bytes and the traces were purely made up of instruction memory accesses in order to simulate the instruction cache only.

Given the trace consumption rate that our FPGA implementation is capable of, the cache simulation times are given in Table III while the speedup with respect to software-based simulators can be seen in Figure 14. Compared with Cheetah, our hardware core displays a speedup of 32x to 53x, with an average speedup of 38x. When set against SuSeSim, our implementation is between 41x and 74x times faster, averaging 52x faster. DineroIV, despite only simulating 33 configurations, was still up to three orders of magnitude slower than our FPGA-based setup simulating 44 configurations.

VII. FUTURE WORK

This paper presents the first step in our hardware-based cache simulator design. Future work will include diversifying the design and making it more accessible for easy cache simulation. First, we will be considering different cache replacement policies such as FIFO and PLRU. A high-speed data interface will also be implemented to allow for the fast simulation of large static traces from a PC. Finally, the potential to integrate a multiple cache simulator instance within an embedded system will be considered. This would allow for a real-world embedded system to be configured together with a cache simulator onto an FPGA (using the I/O, SD Card and camera as seen in Figure 13) to run embedded applications. The hardware

Simulator	SimpleScalar				Tensilica Xtensa					
Application	JPEG encode	JPEG decode	MPEG2 decode	MPEG2 encode	bzip2 100	fma3d 100	gcc 100	gzip 100	mesa 100	MPEG2 encode
Trace token count	11.51M	2.84M	937.32M	2.36B	333.61M	312.60M	320.10M	331.70M	317.08M	11.15B
DineroIV (33)	92.12s	32.99s	6544s	16045s	2233s	2135s	2130s	2174s	2159s	83922s
SuSeSim (44)	6.37s	2.09s	593.92s	1476s	136.51s	151.71s	135.44s	150.31s	155.85s	3913s
Cheetah (44)	4.65s	1.51s	382.41s	1039s	108.34s	109.21s	105.92s	109.05s	113.87s	4794s
Our FPGA Sim (44)	0.12s	0.03s	9.37s	23.59s	3.34s	3.13s	3.20s	3.32s	3.17s	111.51s
Speedup w.r.t. Cheetah	40x	53x	41x	44x	32x	35x	33x	33x	36x	35x

TABLE III: Instruction cache simulation trace lengths and timings of 44 cache configurations (33 for DineroIV) based on ten application traces. The last row indicates the speedup that could be achieved by our cache simulation core as it can consume 100 million address tokens per second.

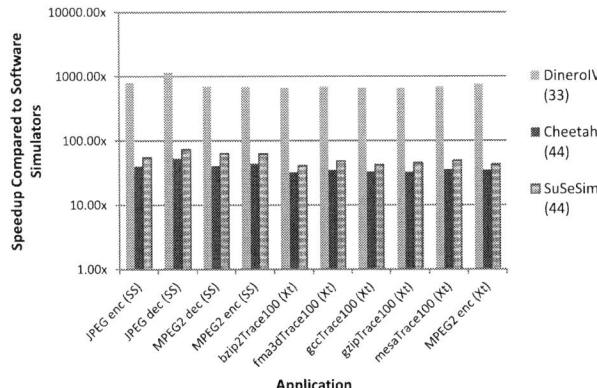

Fig. 14: Speedup of our FPGA-based cache simulator core when compared to existing software-based implementations. The application trace was extracted by either the SimpleScalar (SS) or the Tensilica Xtensa (Xt) simulator.

based cache simulator can then unobtrusively monitor any memory bus to extract the memory access trace required for cache simulation.

VIII. CONCLUSIONS

In this paper we have presented the first multiple cache simulator implemented in hardware on an FPGA. Its ability to process address tokens at 100MHz, concurrently simulating 44 different cache configurations, make it up to 53x faster than the fastest software-based cache simulators for a set of benchmarks, in our tests. Cache inclusion properties were exploited to optimise the amount of FPGA hardware resources required. As a result, our implementation of a cache simulator capable of simulating 308 different cache configurations requires over half the resources of a 230 thousand logic element Altera Stratix IV FPGA.

REFERENCES

[1] J. L. Hennessy and D. A. Patterson, *Computer Architecture: A Quantitative Approach*, 3rd ed. San Francisco, CA, USA: Morgan Kaufmann Publishers Inc., 2003.

[2] D. Brooks, V. Tiwari, and M. Martonosi, "Wattch: a framework for architectural-level power analysis and optimizations," in *Computer Architecture, 2000. Proceedings of the 27th International Symposium on*, june 2000, pp. 83 –94.

[3] Y. Liang and T. Mitra, "Static analysis for fast and accurate design space exploration of caches," in *Proceedings of the 6th IEEE/ACM/IFIP international conference on Hardware/Software codesign and system synthesis*, ser. CODES+ISSS '08. New York, NY, USA: ACM, 2008, pp. 103–108. [Online]. Available: http://doi.acm.org/10.1145/1450135.1450159

[4] A. Agarwal, J. Hennessy, and M. Horowitz, "An analytical cache model," *ACM Trans. Comput. Syst.*, vol. 7, no. 2, pp. 184–215, May 1989. [Online]. Available: http://doi.acm.org/10.1145/63404.63407

[5] J. J. Pieper, A. Mellan, J. M. Paul, D. E. Thomas, and F. Karim, "High level cache simulation for heterogeneous multiprocessors," in *Proceedings of the 41st annual Design Automation Conference*. ACM, 2004, pp. 287–292.

[6] A. Ghosh and T. Givargis, "Cache optimization for embedded processor cores: An analytical approach," *ACM Transactions on Design Automation of Electronic Systems (TODAES)*, vol. 9, no. 4, pp. 419–440, 2004.

[7] R. A. Sugumar and S. G. Abraham, "Set-associative cache simulation using generalized binomial trees," *ACM Trans. Comput. Syst.*, vol. 13, no. 1, pp. 32–56, Feb. 1995. [Online]. Available: http://doi.acm.org/10.1145/200912.200918

[8] J. Edler and M. D. Hill. Dinero iv trace-driven uniprocessor cache simulator. [Online]. Available: http://pages.cs.wisc.edu/~markhill/DineroIV/

[9] A. Janapsatya, A. Ignjatović, and S. Parameswaran, "Finding optimal l1 cache configuration for embedded systems," in *Proceedings of the 2006 Asia and South Pacific Design Automation Conference*, ser. ASP-DAC '06. Piscataway, NJ, USA: IEEE Press, 2006, pp. 796–801. [Online]. Available: http://dx.doi.org/10.1145/1118299.1118482

[10] M. S. Haque, A. Janapsatya, and S. Parameswaran, "Susesim: a fast simulation strategy to find optimal l1 cache configuration for embedded systems," in *Proceedings of the 7th IEEE/ACM international conference on Hardware/software codesign and system synthesis*, ser. CODES+ISSS '09. New York, NY, USA: ACM, 2009, pp. 295–304. [Online]. Available: http://doi.acm.org/10.1145/1629435.1629476

[11] Y.-T. Chen, J. Cong, and G. Reinman, "Hc-sim: A fast and exact l1 cache simulator with scratchpad memory co-simulation support," in *Hardware/Software Codesign and System Synthesis (CODES+ISSS), 2011 Proceedings of the 9th International Conference on*, oct. 2011, pp. 295 –304.

[12] S. M. M. Shwe, H. Javaid, and S. Parameswaran, "Rexcache: Rapid exploration of unied last-level cache," in *Proceedings of the 2013 Asia and South Pacific Design Automation Conference*, ser. ASP-DAC '13. IEEE Press, 2013.

[13] M. Hill and A. Smith, "Evaluating associativity in cpu caches," *Computers, IEEE Transactions on*, vol. 38, no. 12, pp. 1612 –1630, dec 1989.

[14] N. Tojo, N. Togawa, M. Yanagisawa, and T. Ohtsuki, "Exact and fast l1 cache simulation for embedded systems," in *Design Automation Conference, 2009. ASP-DAC 2009. Asia and South Pacific*, jan. 2009, pp. 817 –822.

[15] Lauterbach. (2013) Lauterbach development tools. [Online]. Available: http://www.lauterbach.com/

[16] Z. Tan, A. Waterman, R. Avizienis, Y. Lee, H. Cook, D. Patterson, and K. Asanović, "Ramp gold: an fpga-based architecture simulator for multiprocessors," in *Proceedings of the 47th Design Automation Conference*, ser. DAC '10. New York, NY, USA: ACM, 2010, pp. 463–468. [Online]. Available: http://doi.acm.org/10.1145/1837274.1837390

[17] E. S. Chung, M. K. Papamichael, E. Nurvitadhi, J. C. Hoe, K. Mai, and B. Falsafi, "Protoflex: Towards scalable, full-system multiprocessor simulations using fpgas," *ACM Trans. Reconfigurable Technol. Syst.*, vol. 2, no. 2, pp. 15:1–15:32, Jun. 2009. [Online]. Available: http://doi.acm.org/10.1145/1534916.1534925

[18] D. Chiou, D. Sunwoo, J. Kim, N. A. Patil, W. Reinhart, D. E. Johnson, J. Keefe, and H. Angepat, "Fpga-accelerated simulation technologies (fast): Fast, full-system, cycle-accurate simulators," in *Proceedings of the 40th Annual IEEE/ACM International Symposium on Microarchitecture*, ser. MICRO 40. Washington, DC, USA: IEEE Computer Society, 2007, pp. 249–261. [Online]. Available: http://dx.doi.org/10.1109/MICRO.2007.36

[19] M. Pellauer, M. Adler, M. Kinsy, A. Parashar, and J. Emer, "Hasim: Fpga-based high-detail multicore simulation using time-division multiplexing," in *High Performance Computer Architecture (HPCA), 2011 IEEE 17th International Symposium on*, 2011, pp. 406–417.

[20] M. Graphics. (2013) Modelsim. [Online]. Available: http://model.com/

[21] D. Burger and T. M. Austin, "The simplescalar tool set, version 2.0," *SIGARCH Comput. Archit. News*, vol. 25, no. 3, pp. 13–25, Jun. 1997. [Online]. Available: http://doi.acm.org/10.1145/268806.268810

[22] Tensilica. (2013) Tensilica. [Online]. Available: http://www.tensilica.com/

[23] C. Lee, M. Potkonjak, and W. H. Mangione-Smith, "Mediabench: a tool for evaluating and synthesizing multimedia and communicatons systems," in *Proceedings of the 30th annual ACM/IEEE international symposium on Microarchitecture*, ser. MICRO 30. Washington, DC, USA: IEEE Computer Society, 1997, pp. 330–335. [Online]. Available: http://dl.acm.org/citation.cfm?id=266800.266832

[24] J. L. Henning, "Spec cpu2000: Measuring cpu performance in the new millennium," *Computer*, vol. 33, no. 7, pp. 28–35, Jul. 2000. [Online]. Available: http://dx.doi.org/10.1109/2.869367

Redundant-Via-Aware ECO Routing*

Hsi-An Chien and Ting-Chi Wang

Department of Computer Science, National Tsing Hua University

Hsinchu, Taiwan

hsianchien@gmail.com, tcwang@cs.nthu.edu.tw

Abstract— Redundant via insertion (RVI) has become an inevitable means adopted in the routing or post-routing stage to enhance chip reliability and yield as feature size shrinks down to nanometer scale. The remaining routing resources, however, could become so limited after RVI, and make engineering change order (ECO) routing during a pre-mask stage or even a post-mask stage difficult to complete. In this paper, we study an ECO routing problem where redundant vias are present in the given layout but can be considered for replacement or removal to increase the routability and improve the routing quality. To find an ECO routing path, we construct only the necessary part of the routing graph on-the-fly, and develop an A* search based algorithm for achieving efficient path finding. We also take redundant via replacement, removal, and insertion into account when formulating the routing cost, and apply a state-of-the-art method to perform redundant via replacement and insertion. Experiments show that our algorithm not only successfully routes all test cases but also efficiently produces high-quality solutions.

I. INTRODUCTION

As technology goes into the nanometer era, the emerging design for manufacturing (DFM) issues result in difficulties in achieving design closure. During the last decade, techniques were proposed to deal with DFM issues such as antenna effect, single via failure, etc. [10]. With the advent of the manufacturing strategy that requires redundant via insertion (RVI) for tackling single via failure to enhance the chip reliability and yield, the RVI problem was recently and thoroughly studied in both the routing stage [2], [9], [12] and the post-routing stage [7] to minimize the number of dead vias[1]. Nowadays, the insertion rate of redundant vias tends to be high for a design [11]. Unfortunately, due to the large number of inserted redundant vias, a design could become too congested for engineering change order (ECO) routing to succeed.

For accelerating time-to-market while saving the design and fabrication costs, ECO has been widely used in complex ASIC chip design. In the chip design cycle, although ECO can be used any time, usually we do it after the design has almost been completed. For the purpose of fixing functional and/or timing problems effectively and efficiently after place-and-route or even tape-out, we can utilize pre-injected spare cells to change the original design and apply ECO routing to connect modified nets incrementally. This is called metal-only ECO, which only needs modifications on metal layers, and it could further reduce mask re-spin cost and shorten the manufacturing process cycle when the state of a design is post-mask [1], [4]. In general terms, ECO is performed to make a small functional or non-functional change of a design, and

ECO routing can be executed for broken nets, newly created nets, or fixing DRC violations. Unlike the conventional routing problem starting form scratch [2], [9], [10], [12], ECO routing is invoked in a more difficult situation which has to deal with a lot of blockages (e.g., existing interconnections) [3], [4], [8]. To solve an ECO routing problem, [3] introduced an appropriate routing graph based on an implicit representation, and [8] proposed a tile-based router combined with a graph reduction method to reduce the routing path search time without sacrificing solution quality. Recently, a redundant-wires-aware method for mask re-spin cost minimization was proposed [4], but the details of how to reuse redundant wires during ECO routing were not mentioned.

Clearly, a crowded design could make it harder to implement and accomplish ECO routing. Inserting redundant vias will further worsen congestion in layout. Therefore, ECO routing is definitely more difficult and challenging for a design with inserted redundant vias than the one without redundant vias. Nevertheless, if we are allowed to perturb a given layout a bit by properly replacing or removing existing redundant vias without changing the original routing connectivity, we are able to increase the routability and improve the routing quality for ECO routing. However, this feature is not provided in existing ECO routing approaches [3], [4], [8] which regard all existing layout objects as routing blockages. Let us consider an example, which is a routed design with inserted redundant vias and has two metal layers Metal1 and Metal2, and one via layer VIA1, as shown in Figure 1(a). Let rv_i $(1 \le i \le 5)$ denote the inserted redundant via of single via v_i. We assume that the net connecting two pins p_1 and p_2 is an ECO net. However, without disturbing the original design, we cannot obtain a feasible routing path for the ECO net. Now if we perturb the original layout a bit by allowing existing redundant vias to be considered for replacement or removal, a routing path without violating design rules can be found for the ECO net, as highlighted by the solid red line in Figure 1(b). As can be seen from Figure 1(b), the redundant via rv_1 is replaced

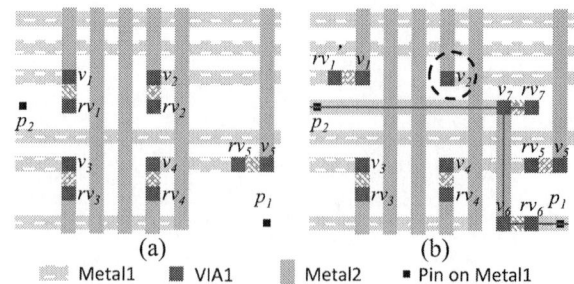

Fig. 1. Redundant via replacement/removal for routability enhancement. Dead via and ECO routing path are highlighted by dashed circle and solid red line, respectively.

---

*This work was supported in part by the National Science Council of Taiwan under Grant No. NSC-102-2220-E-007-012.

[1]A single via is said to be a dead via if and only if we cannot insert any redundant vias next to it without creating design rule violations.

with rv'_1 while the redundant via rv_2 is removed, making v_2 (highlighted by a dash circle) become a dead via. This example shows that a minor perturbation to existing redundant vias enables us to cope with the routing failure and thus to increase the routability.

Figure 2(a) shows another routed design with five single vias, v_1, v_2, ..., v_5 and their respective inserted redundant vias rv_1, rv_2, ..., rv_5. This example asks to route an ECO net from p_1 to p_2. By using a traditional ECO routing method, it is easy to get a feasible routing path that utilizes four vias, v_6, v_7, v_8, and v_9, as highlighted by the solid red line in Figure 2(b). But only redundant vias rv_6 and rv_8 can be inserted next to v_6 and v_8 while v_7 and v_9 become dead vias. With a careful observation on the original layout (i.e., Figure 2(a)), we find that one of the remaining vertical routing tracks within the minimum bounding box enclosing p_1 and p_2 is the one with redundant via rv_2 inserted on it, and therefore an ECO router needs to search a larger region and produces a scenic route as shown in Figure 2(b). Again, if we are allowed to replace or remove existing redundant vias, we may be able to reduce the routing wirelength. For example, if rv_2 is removed, a shorter routing path that has two vias, v_6 and v_7, and one inserted redundant via rv_6 for the ECO net can be found, as shown in Figure 2(c). In addition, it produces the same number of dead vias (v_2 and v_7) as that in Figure 2(b). To further reduce the amount of dead vias, we can replace rv_1 and rv_3 with rv'_1 and rv'_3, respectively, and construct another routing path with a small detour but only one dead via v_7, as shown in Figure 2(d). This example shows that by allowing existing redundant vias to be replaced or removed, we could enhance the routing quality.

Motivated by the above two examples shown in Figures 1 and 2, in this paper, we study an ECO routing problem where redundant vias are present in the given layout but can be considered for replacement or removal to increase the routability and reduce the routing cost. The cost we define considers several routing factors, including wirelength, vias, redundant vias/dead vias, and preferred routing directions. To allow replacement/removal for existing redundant vias, we employ a spatial search technique to construct only the necessary part of the routing graph on-the-fly. We develop an A* search based algorithm for achieving efficient path finding as well as apply a state-of-the-art method to perform redundant via replacement and insertion. We also present a technique to speed up the path finding algorithm, and extend the algorithm to handle more general problems. Experiments show that our algorithm not only successfully routes all test cases but also efficiently produces high-quality solutions.

The rest of this paper is organized as follows. In Section II, we give the problem formulation. The proposed approach and its extensions are presented in Sections III and IV, respectively. We report the experimental results in Section V, and conclude this paper in Section VI.

II. PROBLEM DEFINITION

In this section, we formulate our redundant-via-aware ECO routing problem, in which redundant vias are present in a

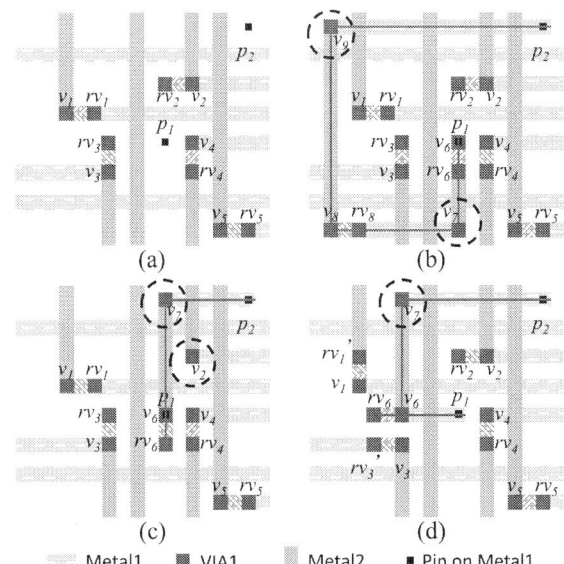

Metal1   ■ VIA1   Metal2   ■ Pin on Metal1

Fig. 2. Redundant via replacement/removal for routing quality improvement. Dead via and ECO routing path are highlighted by dashed circle and solid red line, respectively.

given layout but can be considered for replacement or removal to increase the routability and reduce the routing cost.[2] We assume that the given routed design consists of rectangles that correspond to layout objects, such as pins, wire segments, or vias. The geometric and connectivity information is also given for these layout objects. Such a routed design can be specified in, e.g., the LEF/DEF format. Note that since we do not have a 3D routing graph given as an input, our algorithm will construct only the necessary graph nodes on-the-fly according to wire width and spacing rules (details will be given in Section III-B). Same as previous RVI works [2], [7], [9], [12], for each single via in the original layout or on an ECO routing path, at most one out of four candidate positions can be inserted with a redundant via subject to design rules. Besides, if an existing redundant via is selected for replacement, it will be reinserted into one of the remaining locations if possible.

The ECO routing problem we consider is to find a path p that connects the source pin and target pin of a 2-pin ECO net with space reserved for subsequent redundant via insertion/replacement, and then to perform redundant via insertion for newly created vias on p and redundant via replacement for existing redundant vias that are removed to avoid design rule violations caused by p. The goal is to find a routing path with routing cost as small as possible (we will define the cost function in Section III-C). Note that this problem formulation is only for single ECO net with 2 pins. More general problems, including single ECO net with multiple pins and multiple ECO nets, will be discussed in Section IV.

III. THE PROPOSED APPROACH

In this section, we present our routing algorithm for single ECO 2-pin net. For an easy presentation, each routing layer

---

[2]Please note that if removing or replacing an existing redundant via will harm the timing of a design, it will not be a candidate for replacement or removal. To simplify our presentation, we assume all existing redundant vias can be considered for replacement or removal in the rest of this paper.

978-1-4799-2817-0/14 $31.00 © 2014 IEEE   419

is virtually superimposed with a grid to form the 3D routing graph and the source node s and target node t of the given ECO 2-pin net are physically mapped to two gridpoints. Each gridpoint is a square on a metal layer, whose width is the minimum wire width on that layer. To efficiently find the routing path to connect s and t, we exploit A* search [6].

A. A* Search Based Path Finding Algorithm

Instead of physically constructing the whole routing graph based on the huge amount of given layout objects at the beginning, we query the layout during A* search for getting necessary gridpoints whose query regions (the definition of the query region of a gridpoint will be defined in Section III-B) do not intersect with any layout objects or only intersect with existing redundant vias; we refer to these gridpoints as *feasible gridpoints*. Similar to [7], the layout objects of each layer are stored by an R-tree [5] for efficiently performing spatial query on the layout. Each feasible gridpoint will be assigned a routing cost, and the cost function will be defined in Section III-C. For simplicity, whenever we mention a gridpoint in the rest of this subsection, it means a feasible gridpoint.

The details of our routing algorithm are given in Algorithm 1. Each gridpoint we find will be added into the unvisited grid set at first. Next, a gridpoint with the smallest cost in the unvisited grid set will be moved into the visited grid set and be checked whether we reach target t (lines 4 and 5); if not, we perform spatial search to get its adjacent gridpoints (line 11). The details of how to search feasible neighboring gridpoints are given in next subsection. After that, the unvisited grid set will be updated according to the newly found gridpoints

---

Algorithm 1: Single ECO Net Routing Algorithm

Input : An ECO 2-pin net and the given layout
Output: The routing result

1 Set both unvisited and visited grid sets to be empty.
2 Add the source node s into unvisited grid set.
3 **while** unvisited grid set is not empty **do**
4   Remove an unvisited gridpoint n with minimal cost and add n into visited grid set.
5   **if** n == the target node t **then**
6   A routing path p from s to t can be obtained by following the parent gridpoints form t back to s.
7   Insert p into the layout and remove the redundant vias causing DRC violations with p.
8   Collect both the newly created vias on p and the old vias whose redundant vias have been removed, perform redundant via insertion/replacement for them, and update the layout accordingly.
9   **return** p
10   **end**
11   Search the feasible neighboring gridpoints of n.
12   **for** each feasible neighboring gridpoint m **do**
13   **if** m is visited **then**
14   Ignore it.
15   **end**
16   **if** m is already in unvisited grid set **then**
17   Update its cost and parent gridpoint when needed.
18   **else**
19   Insert m into unvisited grid set.
20   **end**
21   **end**
22 **end**
23 **return** routing failure

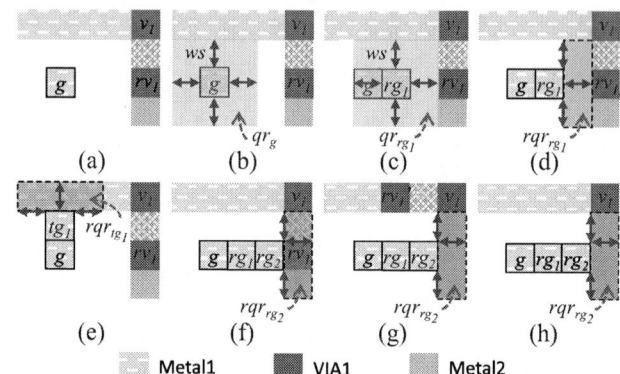

Fig. 3. Query regions for examining whether neighboring gridpoints on the same layer can be used for routing without causing DRC violations.

(line 16 to line 20). We continue the above procedure until a routing path is obtained or the unvisited grid set becomes empty. When the routing path is found, we can trace back from target t to source s to get the routing path (line 6). It is easy to see that the path finding algorithm iteratively explores the layout and gradually expands the search space. As a result, it constructs the routing graph on-the-fly, and only feasible gridpoints can become graph nodes. As for redundant via insertion/replacement, it is performed in lines 7 and 8, and the details will be given in Section III-E.

B. Feasible Gridpoint

In this subsection, we use examples to describe how to find feasible neighboring gridpoints. For each gridpoint, it has four neighboring gridpoints (top, bottom, left, and right) on the same metal layer and two neighboring gridpoints (upper and lower) on the adjacent metal layers. The R-tree structure is used to efficiently search feasible gridpoints in our method. The construction of the query regions for neighboring gridpoints on the same layer is illustrated in Figure 3. Given a feasible gridpoint g on Metal1 as shown in Figure 3(a), we want to find its same-layer neighboring gridpoints subject to the minimum wire spacing (denoted by ws) constraint. Basically, we can specify a query region for each gridpoint on Metal1 and claim that if the existing objects of Metal1 do not intersect with the query region, the gridpoint is feasible. Figure 3(b) shows the construction of the query region for g, which expands g by ws along each side to form the query region qr_g. The query region qr_{rg_1} of the right neighboring gridpoint rg_1 of g can be constructed in the same way as Figure 3(b), and can be used to determine whether rg_1 is a feasible gridpoint, as shown in Figure 3(c). But we can see that qr_g overlaps with qr_{rg_1}, so it is not necessary to query the overlapped region between qr_g and qr_{rg_1} again. Thus, the reduced query region rqr_{rg_1} shown in Figure 3(d) is used to lower the search effort when examining whether rg_1 is a feasible gridpoint. For the other same-layer neighboring gridpoints of g, we can also find their corresponding reduced query regions and check if they are feasible in a similar fashion. In Figure 3(e), the reduced query region rqr_{tg_1} of the top neighboring gridpoint tg_1 of g intersects with an existing wire segment on the same layer, and therefore tg_1 is not a feasible gridpoint. Since existing redundant vias can be replaced or removed, they will be ignored in any query region. Figure 3(f) illustrates that the redundant via rv_1 of via v_1 intersects with the reduced query region rqr_{rg_2} of the right neighboring gridpoint rg_2 of rg_1, but is allowed to be replaced with rv_1' (see Figure 3(g)) or removed out of rqr_{rg_2} (see Figure 3(h)), so rg_2 is treated as a feasible neighboring gridpoint for rg_1.

Connecting a gridpoint and its neighboring gridpoint on an adjacent (upper or lower) layer means that a via will be inserted between the two gridpoints. Figure 4 illustrates the examination of a neighboring gridpoint on the upper adjacent layer. To check the upper neighboring gridpoint ug on Metal2 of the gridpoint g as shown in Figure 4(a), we define a query region qr_{ug}^{Metal2} on the Metal2 layer (see Figure 4(b)). If no layout objects or only existing redundant vias intersect with the query region qr_{ug}^{Metal2}, ug is a feasible neighboring gridpoint on the adjacent upper layer. We can also deal with different wire widths and spacings on adjacent layers. Figure 5(a) shows that the adjacent layer Metal2 layer has a wire width w^{Metal2} different from the wire width w^{Metal1} of the Metal1 layer. When checking the upper neighboring gridpoint ug of g, due to the size mismatch of the two grids on Metal1 and Metal2, we pick the nearest upper neighbor ug and locate the via connecting g and ug at the intersection point v (see Figures 5(b) and 5(c)). The query regions qr_{ug}^{Metal1} and qr_{ug}^{Metal2} for Metal1 and Metal2 layers are shown in Figures 5(d) and 5(e), respectively.

C. Cost Function

Our path finding A* search algorithm uses the following cost function

$$cost(n) = g(n) + \epsilon h(n) \qquad (1)$$

where $g(n)$ is the actual routing cost from the source node s to the current node n and $h(n)$ is the estimated routing cost from n to the target node t. In our algorithm, $h(n)$ is measured by the Manhattan distance between n and t multiplied by a weighting coefficient ϵ. [6] indicates that A* search algorithm can find the minimum-cost solution when ϵ is 1. The weighting coefficient can be greater than 1 to trade off runtime and solution quality. The larger it is, the less runtime but worse solution quality we get. In our experiments, it is set to 1.5.

To consider various routing factors, when we visit a feasible gridpoint n, its actual routing cost $g(n)$ is the cumulative cost of the best path p from s to itself that has been found so far, and is calculated as follows:

$$g(n) = G(n) + \alpha V(n) + \beta I(n) + \gamma R(n) \qquad (2)$$

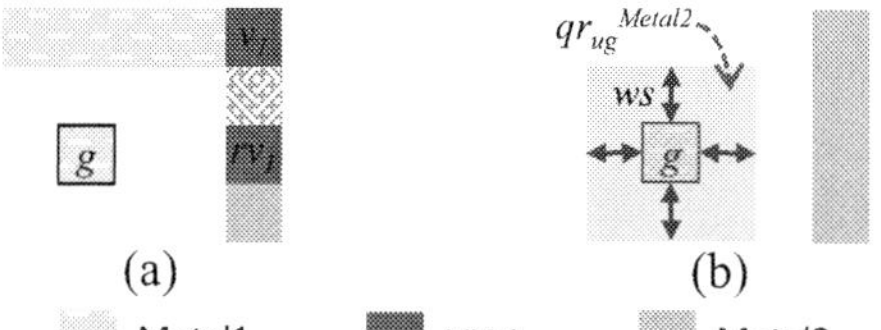

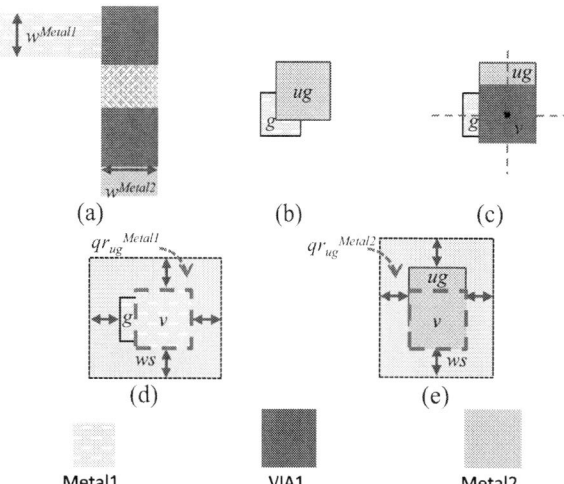

Fig. 5. Query regions for the neighboring gridpoint on an adjacent layer with a different metal width, where the wire spacing rules (denoted by ws) are set to be the same for all layers for simplicity.

where α, β, and γ are user-defined parameters. $G(n)$ is determined from the number of gridpoints on the routing path p. To consider both the wirelength and the preferred routing directions, we set $G(n)$ to be the summation of the number of gridpoints along preferred routing directions and twice the number of gridpoints along non-preferred routing directions. Doubling the cost for a gridpoint along a non-preferred direction could help prevent short jogs for better printability and reserve more longer routing tracks for other routes. To take the via cost into account, we add the amount of vias on the path p, denoted by $V(n)$, into $g(n)$. In addition, we also try to preserve space for these newly created vias to increase the chance of inserting redundant vias next to them. To do this, we add the cost $I(n)$, which is computed by the total number of "illegal redundant via insertions" among these new vias, into $g(n)$. Each single via has four possible positions to choose for inserting a redundant via, and if inserting a redundant via on a possible position causes design rule violations, it is counted as one illegal insertion. Each new via can have at most 4 illegal redundant via insertions. Clearly, the smaller $I(n)$ is, the higher chance these new vias can be inserted with redundant vias. In order to consider the perturbation on the original layout, we add $R(n)$ into $g(n)$ and compute it by the total number of "illegal redundant via replacements" among those redundant vias that need to be removed to avoid design rule violations caused by the routing path p. An illegal replacement is defined similarly to that of an illegal insertion, except that each removed redundant via has at most 3 illegal replacements because its associated single via has 3 remaining positions to choose for inserting a redundant via. Note that if all three possible replacements of an existing redundant via are illegal, then after removing the redundant via, its associated single via will become an additional dead via.

It is clear to see that the cost $g(n)$ considers several routing factors, including wirelength, vias, redundant vias/dead vias, and preferred routing directions. By properly adjusting the values of the three parameters, α, β, and γ, we can easily

Fig. 4. Query regions for examining whether the neighboring gridpoint on an adjacent layer can be used for routing without causing DRC violations.

handle the trade-off among these routing factors.

D. Speed-up for A* Search

The number of query operations and the amount of visited nodes will dominate the runtime of our path finding algorithm. To enhance the efficiency, our algorithm will search a certain amount of consecutive gridpoints along a preferred routing direction as a whole (instead of one gridpoint at a time). Let us illustrate this idea using the example shown in Figure 6. Suppose Metal1 is a horizontal layer and we are currently expanding search space from the gridpoint n. Along the preferred routing direction on Metal1, we examine the feasibilities of m left (right) consecutive gridpoints, lg_1 to lg_m (rg_1 to rg_m) of n by the corresponding reduced query region rqr_{left} (rqr_{right}), and return the maximum number of consecutive feasible gridpoints, lg_1 to lg_k (rg_1 to rg_m), where m is a user-defined parameter and is set to 20 in our current experiments. After finding a set of feasible neighboring gridpoints of n, all of them will be taken to update the unvisited grid set (line 12 to line 21 of Algorithm 1). A feasible gridpoint with less estimated routing cost will be visited earlier than others, and therefore the visited grid set of our path finding algorithm will be extended toward the target node rapidly.

E. Redundant Via Insertion/Replacement

After finding the path for an ECO net and removing those existing redundant vias that cause design rule violations (line 7 of Algorithm 1), our algorithm proceeds to perform redundant via insertion for the set NV of newly created vias on the path and redundant via replacement for the set OV of the old vias whose redundant vias have been removed. In order to minimize the total number of additional dead vias in the design, we perform redundant via insertion for NV and redundant via replacement for OV simultaneously (line 8 of Algorithm 1). This is done by employing an existing post-routing RVI approach [7], which guarantees to add the maximum number of redundant vias for the single vias in $NV \cup OV$.

IV. EXTENSIONS

In this section, we describe how to extend our algorithm to handle the cases of single multi-pin ECO net and multiple ECO nets. For single multi-pin ECO net, we first decompose the given net into a set of 2-pin subnets using a commonly used net decomposition method, such as minimum spanning tree or minimal Steiner tree. Since it is well known that determining a net order is a difficult problem, the set of 2-pin subnets is routed in an arbitrary order, one at a time by our routing algorithm. We also stop path finding for a 2-pin subnet when we reach any routed part of the multi-pin net.

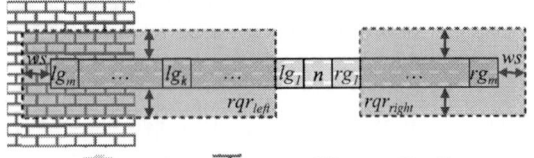

Metal1   Layout Objects on Metal1

Fig. 6. Reducing the number of query operations by searching consecutive gridpoints along a preferred routing direction as a whole.

As for multiple ECO nets, we can route them in an arbitrary order, and each of them is routed by our 2-pin or multi-pin routing algorithm.

V. EXPERIMENTAL RESULTS

We have implemented our ECO routing algorithm, including the redundant via insertion method proposed by [7], in C++ language. Our ECO router is capable of reading in a routed design in the LEF/DEF format and thus it can be easily integrated into an industry design flow. All experiments were conducted on a Linux workstation with an 2.2 GHz AMD CPU and 8G memory. To conduct the experiments, we used a real circuit whose redundant vias had been inserted by [7]. This circuit has four routing layers and 34376 vias with 94% insertion rate of redundant vias; moreover, the number of layout objects is 154551 and the virtual grid size is 6096×6096 per layer. The minimum wire widths and spacings of all metal layers are the same in this circuit. For this circuit, we randomly generated six test cases of single 2-pin ECO net and four test cases of multiple 2-pin ECO nets; as shall be seen later, these test cases include routable and unroutable cases when layout perturbation is prohibited.

To the best of our knowledge, there is no academic ECO routing method solving the same problem as ours. Thus, for making comparisons, we implemented two approaches, respectively named Router1 and Router2, to solve the problem. To see the impact of allowing redundant via replacement/removal on the routability and routing quality, Router1 was designed by disabling the option of redundant via replacement/removal of our algorithm. Router2 was designed to first remove all inserted redundant vias, then route ECO nets by using a variant of our algorithm which does not insert redundant vias for each routed ECO net, and finally insert redundant vias for the whole chip by [7].

We set the three parameters α, β, and γ of our algorithm to 2, 2, and 2, respectively. However, γ has no effect when the option of redundant via replacement/removal is turned off for Router1 and Router2. Tables I lists the results for the cases (eco1 to eco6) of single ECO net and the cases (eco7 to eco10) of multiple ECO nets. In the table, we report the number of ECO nets (in the "#Ns" column), the total wirelength of the routed ECO net(s) (in the "WL(μm)" column), the total number of newly created vias on the routed ECO net(s) (in the "#NVs" column), the total number of additional dead vias (in the "#DV" column), and the runtime measured in second (in the "T(s)" column). When an ECO net or more could not be routed (i.e., routing failure), we put an "NA" in each of its corresponding results.

Firstly, as compared with Router1, our router successfully constructed routing path(s) for each case. However, without removing existing redundant vias, Router1 could not achieve 100% routing completion rate for almost half of the test cases; note that one half and one sixth of the nets were not routable for the cases eco7 and eco9, respectively. So, it is clear to see that the routability of the given circuit was harmed by the existing redundant vias, such that we could

978-1-4799-2817-0/14 $31.00 © 2014 IEEE   422

TABLE I
THE ECO ROUTING RESULTS

Test Case	#Ns	Our router				Router1				Router2			
		WL(μm)	#NVs	#DVs	T(s)	WL(μm)	#NVs	#DVs	T(s)	WL(μm)	#NVs	#DVs	T(s)
eco1	1	300	0	0	0.005	1200	6	0	0.126	300	0	0	7.464
eco2	1	575	2	3	0.016	NA	NA	NA	NA	575	2	3	7.559
eco3	1	73000	6	0	0.437	73600	10	2	0.897	73000	6	0	7.858
eco4	1	42300	8	1	0.205	NA	NA	NA	NA	42300	8	1	9.294
eco5	1	90300	6	2	0.543	90925	12	3	0.75	90225	6	5	9.804
eco6	1	20275	6	0	0.086	20275	6	0	1.708	20275	6	0	8.083
eco7	8	323125	38	10	1.851	NA	NA	NA	NA	323075	38	13	8.684
eco8	8	281750	38	3	1.139	281925	44	3	0.994	281625	38	5	33.516
eco9	12	233000	54	13	1.342	NA	NA	NA	NA	232225	66	46	11.141
eco10	20	581025	84	0	7.623	581050	100	11	7.78	580700	80	29	13.016
Ratio		1.000	1.000	1.000	1.000	-	-	-	-	0.999	1.033	3.188	8.788

not easily accomplish the subsequent ECO routing without modifying its layout. In addition to routability, our router also produced better or same routing quality in terms of wirelength and the amount of newly created vias, and needed less runtime in almost all cases, due to the help from the replacement/removal scheme that effectively decreases search space. Even though redundant via removal is allowed in our router, the resulting number of additional dead vias produced by our router is still less then or equal to that of Router1, which means our cost function well guides our router to find a redundant-via-aware solution. Especially for case eco10, our router and Router1 produced approximately equal wirelength, but our router generated no additional dead vias while Router1 produced 11 dead vias.

Furthermore, when compared with Router2, our router has consistent improvements on the number of additional dead vias and runtime for almost all test cases. Without careful planning on reserving alternative space to replace those previously removed redundant vias, Router2 increased the number of additional dead vias by about 3 times averagely. Moreover, Router2 must perform a post-routing RVI for the whole chip, which also induces a runtime overhead. Apparently, Router2 also completed all ECO routing cases as expected, and its routing results regarding wirelength and number of vias were better than or as good as those of our router because it has a larger solution space to explore. Note that for case eco9, although our router got less number of vias and only a little longer wirelength, actually its routing cost is larger than that produced by Router2, because it generated more routing segments along non-preferred directions.

To further consider the mask re-spin cost in a post-mask ECO scenario, we used the same test cases given in Table I but imposed a layer restriction by which only Metal1 and Metal2 layers could be used for ECO routing, making these ECO routing test cases more difficult. With the layer restriction, half of the test cases were not completely routed without perturbing existing redundant vias (one more unroutable case than that without layer restriction). But again our router and Router2 successfully solved the ECO routing problem and produced a smaller wirelength for each case, and they also reduced the number of newly created vias in almost all cases. However, Router2 produced more additional dead vias and ran slower when compared with our router. Due to the limited space, we do not give the detailed experimental results in the paper.

No matter where the routing layer restriction is imposed or not, the experimental results demonstrate that in almost all cases our redundant-via-aware ECO routing algorithm not only effectively increased the routing completion rate using less routing resources as compared with Router1, but also produced less amount of additional dead vias with less runtime as compared with Router2.

VI. CONCLUSIONS

In this paper, we have presented an ECO routing approach that considers redundant via replacement, removal, and insertion. Our approach successfully increased the routing completion rate and improved the routing quality. We are currently testing our router on larger and more complex designs to further validate the robustness of our router.

REFERENCES

[1] H.-Y. Chang, I. H.-R. Jiang, and Y.-W. Chang. Simultaneous functional and timing eco. In *Proceedings of Design Automation Conference*, pages 140–145, 2011.

[2] H.-Y. Chen, M.-F. Chiang, Y.-W. Chang, L. Chen, and B. Han. Full-chip routing considering double-via insertion. *IEEE Transactions on Computer-Aided Design of Integrated Circuits and Systems*, 27(5):844–857, 2008.

[3] J. Cong, J. Fang, and K.-Y. Khoo. An implicit connection graph maze routing algorithm for eco routing. In *Proceedings of International Conference on Computer-Aided Design*, pages 163–167, 1999.

[4] S.-Y. Fang, T.-F. Chien, and Y.-W. Chang. Redundant-wires-aware eco timing and mask cost optimization. In *Proceedings of International Conference on Computer-Aided Design*, pages 381–386, 2010.

[5] A. Guttman. R-trees: a dynamic index structure for spatial searching. In *Proceedings of International Conference on Management of Data*, pages 47–57, 1984.

[6] P. Hart, N. Nilsson, and B. Raphael. A formal basis for the heuristic determination of minimum cost paths. *IEEE Transactions on Systems Science and Cybernetics*, 4(2):100–107, 1968.

[7] K.-Y. Lee, C.-K. Koh, T.-C. Wang, and K.-Y. Chao. Fast and optimal redundant via insertion. *IEEE Transactions on Computer-Aided Design of Integrated Circuits and Systems*, 27(12):2197–2208, 2008.

[8] Y.-L. Li, J.-Y. Li, and W.-B. Chen. An efficient tile-based eco router using routing graph reduction and enhanced global routing flow. *IEEE Transactions on Computer-Aided Design of Integrated Circuits and Systems*, 26(2):345–358, 2007.

[9] C.-T. Lin, Y.-H. Lin, G.-C. Su, and Y.-L. Li. Dead via minimization by simultaneous routing and redundant via insertion. In *Proceedings of Asia and South Pacific Design Automation Conference*, pages 657–662, 2010.

[10] D. Z. Pan, M. Cho, and K. Yuan. Manufacturability aware routing in nanometer vlsi. *Foundations and Trends in Electronic Design Automation*, 4(1):1–97, 2010.

[11] L. Scheffer. Recommended rules not recommended. Speech Presented at Electronic Design Processes Workshop, 2006.

[12] G. Xu, L.-D. Huang, D. Z. Pan, and M. D. F. Wong. Redundant-via enhanced maze routing for yield improvement. In *Proceedings of Asia and South Pacific Design Automation Conference*, pages 1148–1151, 2005.

A Fast and Provably Bounded Failure Analysis of Memory Circuits in High Dimensions

Wei Wu[†], Fang Gong[†], Gengsheng Chen[*], Lei He[†]

[†]Electrical Engineering Dept., UCLA
Los Angeles, CA 90095, US
{weiwu2011,gongfang}@ucla.edu
lhe@ee.ucla.edu

[*]School of Microelectronics, Fudan Univ.
Shanghai, 201203, China
gschen@fudan.edu.cn

Abstract— Memory circuits have become important components in today's IC designs which demands extremely high integration density and reliability under process variations. The most challenging task is how to accurately estimate the extremely small failure probability of memory circuits where the circuit failure is a "rare event". Classic importance sampling has been widely recognized to be inaccurate and unreliable in high dimensions. To address this issue, we propose a fast statistical analysis to estimate the probability of rare events in high dimensions and prove that the estimation is always bounded. This methodology has been successfully applied to the failure analysis of memory circuits with hundreds of variables, which was considered to be very intractable before. To the best of our knowledge, this is the first work that successfully solves high dimensional "rare event" problems without using expensive Monte Carlo and classic importance sampling methods. Experiments on a 54-dimensional SRAM cell circuit show that the proposed approach achieves 1150x speedup over Monte Carlo without compromising any accuracy. It also outperforms the classification based method (e.g., Statistical Blockade) by 204x and existing importance sampling method (e.g., Spherical Sampling) by 5x. On another 117-dimension circuit, the proposed approach yields 364x speedup over Monte Carlo while existing importance sampling methods completely fail to provide reasonable accuracy.

I. INTRODUCTION

Memory circuits (e.g., SRAM bit-cell, sense amplifier, delay chain, etc.) need to be replicated millions or even billions of times to achieve extremely high integration density in a smaller footprint, where the cutting-edge process technology is demanded. In this case, the stringent yield requirement of memory circuits can be translated into an extremely small failure probability of each component circuit, thereby making the circuit failure a "rare-event" [1].

In general, the probability estimation of "rare-event" is usually analytical intractable due to high complexity of memory circuits, therefore, sampling methods must be used. The most straightforward approach is the Monte Carlo (MC) method, which repeatedly draws samples and evaluates circuit performance with transistor-level SPICE simulation. However, MC is extremely time-consuming for rare-event estimation, because millions or even billions of samples are needed to capture one single failure.

To mitigate the complexity issue of the MC method, many statistical methodologies have been developed in past few years [2, 3, 4, 5, 6, 7, 8] which can be categorized into two groups:

(1) **Classification**: the approach in [2] makes use of a "classifier" to "block" those Monte Carlo samples that are unlikely to cause failures and simulates the remaining samples. However, this method has two limitations: first, a perfectly accurate classifier is usually unavailable. A safety margin is used in [2] to prevent the classifier error. Second, the imperfect classifier can easily incur large error beyond the safety margin for circuits with irregular failure region and strongly nonlinear behavior, which typically cannot be detected by the approach in [2].

(2) **Importance Sampling**: several approaches in [3, 4, 5, 6, 7] had been developed to construct a new "proposed" sampling distribution under which a "rare event" becomes "less rare" so that more failures can be easily captured. The critical issue is how to build an optimal proposed sampling distribution. Previous work investigated different approaches. For example, [3] mixes a uniform distribution, the original sampling distribution and a "shifted" distribution centering around the failure region. The approaches in [4, 5] simply shift the sampling distribution towards the point of failure region with a minimum L_2-norm. The work in [6] uses "particle filtering" to tilt more samples towards the failure region. The approach in [7] approximates the optimal sampling distribution with a parameterized sampling distribution by minimizing the Kullback-Leibler (KL) distance between them. These importance sampling based methods are plagued by the curse of high dimensionality [9, 10, 11]. In general, they can only be used in low-dimensional problems (e.g., those with a scope of 6-12 variables) but become very untrustworthy for high-dimensional problems.

Clearly, most of existing approaches can successfully be applied to low-dimensional problems with a few random variables but, in general, perform poorly in high dimensions. Therefore, an effective and low-complexity approach is still urgently needed for failure analysis of memory circuits in high dimensions.

In this paper, we proposed a novel statistical algorithm to efficiently estimate the failure probability of memory circuits in high dimension, where tens or hundreds of random variables are present. In details, the proposed methodology first constructs a new subset of the sampling space that dominates the failure region for memory circuits and can be efficiently estimated with a few samples. Then, the failure probability of memory circuits can be evaluated by the product rule of conditional probability within this sampling subset space. More importantly, the estimation from the proposed method is proved to be always bounded in high dimensions. Experiments on a 54-dimensional SRAM cell circuit show that the proposed approach achieves 1150X speedup over Monte Carlo without compromising any accuracy. It is also 204X faster than the classification based method (e.g., Statistical Blockade [2]) by and 5X faster than existing importance sampling method (e.g., Spherical Sampling [4, 5]). On another 117-dimension circuit, the classification based method fails to block "unlikely to fail" samples, and Spherical Sampling [4, 5] method completely fails to provide reasonable accuracy. Contrastingly, the proposed approach yields accurate result with 364X speedup over Monte Carlo.

The rest of this paper is organized as follows. In Section 2, we provide the necessary background on importance sampling and revisit the reasons for its failure in high dimensions. Section 3 describes the techniques underpinning the proposed algorithm in detail. The experiments are provided in Section 4 to validate the accuracy and efficiency of proposed method. This paper is concluded in Section 5.

II. BACKGROUND

A. Formulation of Probability Estimation

Let $f(X)$ be a probability density function (PDF) for a random variable X (e.g., any process or electronic variable parameters) which is the input of a measurement process as shown in (1); the output Y is an observation (e.g., voltage, amplitude, period, etc.) with input X:

$$\underbrace{X}_{\text{variable}} \Rightarrow \boxed{\text{Measurement, SPICE, etc.}} \Rightarrow \underbrace{Y}_{\text{observation}} \quad (1)$$

Usually, it is of great interest to estimate the probability of Y from a small subset $\mathcal{S}$ of the entire sampling space. For example, a small subset is the "failure region" for SRAM design and includes all failed samples where performance constraints cannot be satisfied. Therefore, the probability $p(Y \in \mathcal{S})$ can be estimated as:

$$p(Y \in \mathcal{S}) = \int I(X) \cdot f(X) dX. \quad (2)$$

$$I(X) = \begin{cases} 0 & \text{if } Y \notin \mathcal{S} \\ 1 & \text{if } Y \in \mathcal{S} \end{cases}$$

where Y is the observation/performance with the input variable X and the indicator function $I(\cdot)$ identifies whether $Y \in \mathcal{S}$ or not. Note that the integral in equation (3) is intractable because the analytical formula of $I(X)$ is unavailable. Therefore, sampling based method must be used. For example, the MC method enumerates as many samples of X as possible (e.g., $x_1, \cdots, x_n$) according to $f(X)$ and evaluates their indicator function values to estimate $p(Y \in \mathcal{S})$ as:

$$\tilde{p}(Y \in \mathcal{S}) = \frac{1}{n} \sum_{i=1}^{n} I(x_i) \xrightarrow[n \to +\infty]{a.s.} p(Y \in \mathcal{S}). \quad (3)$$

Here $\tilde{p}(X \in \mathcal{S})$ is an unbiased estimate from sampling method and can be very close to $p(X \in \mathcal{S})$ with a large number of samples.

B. Importance Sampling (IS)

When $Y \in \mathcal{S}$ is a *rare event*, the MC method becomes extremely inefficient because most $I(x_i)$ are zeros. Millions or billions of samples of X are needed to capture only one failed sample from the failure region $\mathcal{S}$.

To deal with this issue, the *importance sampling* (IS) has been introduced to sample from a "proposed" sampling distribution $g(X)$ that tilts towards $\mathcal{S}$ where a rare-event becomes more likely to happen:

$$p_{IS}(Y \in \mathcal{S}) = \int I(X) \cdot \frac{f(X)}{g(X)} \cdot g(X) dX$$
$$= \int I(X) \cdot w(X) \cdot g(X) dX. \quad (4)$$

Here, $w(X)$ is the "likelihood ratio" or the weight for each sample of X. $w(X)$ compensates for the discrepancy between $f(X)$ and $g(X)$ and unbiases the probability estimation under $g(X)$. Sampling based methods can be used to evaluate above integral as:

$$\tilde{p}_{IS}(Y \in \mathcal{S}) = \frac{1}{n} \sum_{j=1}^{n} w(\tilde{x}_j) \cdot I(\tilde{x}_j) \xrightarrow[n \to +\infty]{a.s.} p(Y \in \mathcal{S}). \quad (5)$$

$\tilde{x}_j$ $(j = 1, \cdots, n)$ follows the "proposed" sampling distribution $g(X)$ rather than the original distribution $f(X)$, because more *rare event* samples in the subset $\mathcal{S}$ can be easily chosen under the distribution $g(X)$.

Theoretically, $\tilde{p}_{IS}(Y \in \mathcal{S})$ is consistent with $p(Y \in \mathcal{S})$ in (3) if $\text{supp}(g(X)) \supset \text{supp}(I(X) \cdot f(X))$, where $\text{supp}(\cdot)$ denotes the support of a probabilistic distribution.

C. Failure Analysis of Importance Sampling

While importance sampling is, in principle, mathematically correct, the *degeneration* or *collapse* of the likelihood ratios leads to the failure of importance sampling in high dimensions as discussed in [10, 11].

Let us consider a classical case, as shown in Fig. 1, where $f(X)$ is the "original" sampling distribution and $g(X)$ is the "proposed" sampling distribution. The small circles with the same size within $g(X)$

are samples drawn from $g(X)$. In the bottom of Fig. 1, a few circles with different sizes represent the illustrative scales of the likelihood ratios corresponding to the samples on top of them. Clearly, if $g(X)$ has thinner tails than $f(X)$, the likelihood ratios $w(X) = f(X)/g(X)$ approach infinity in the tails of $g(X)$. Hence, the likelihood ratios vary dramatically and have extremely large variance that leads to unstable probability estimate.

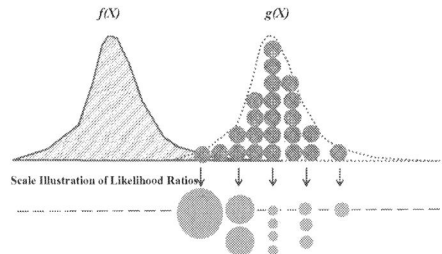

Fig. 1. The scale illustration of likelihood ratios in importance sampling.

Moreover, the reason for the collapse of likelihood ratio can be explained from another perspective: when importance sampling shifts $g(X)$ towards the rare-event region that is typically in the tails of $f(X)$, $f(X)$ and $g(X)$ become mutually singular and have "disjoint" support [10]. Therefore, IS fails to retain its accuracy.

This collapse issue of likelihood ratios becomes much worse in high dimensions because $w(X)$ is a product of probabilities for multiple parameters and consequently approaches infinity more quickly.

III. PROPOSED METHOD

A. Algorithm Overview

We consider a small subset $\mathcal{S}$ as the failure region in SRAM design under the given performance constraint (e.g., the performance of SRAM circuit Y should be greater than certain performance threshold t_c). Hence the subset $\mathcal{S} = \{Y | Y \geq t_c\}$ in Fig. 2 contains all failed samples that are "rare events".

The basic idea of the proposed algorithm is to construct a new subset $\mathcal{T}$ with a new threshold t (e.g., $t = 0.99$-quantile point). This new subset $\mathcal{T} = \{Y | Y \geq t\}$ includes "non-rare" events and dominates the "rare event" subset $\mathcal{S}$ (e.g., $\text{supp}(\mathcal{T}) \supset \text{supp}(\mathcal{S})$).

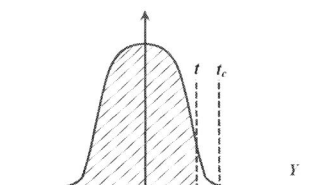

Fig. 2. Basic idea in proposed algorithm. (Noted that $\mathcal{T} = \{Y | Y \geq t\}$ contains $\mathcal{S} = \{Y | Y \geq t_c\}$).

In this way, the failure probability of SRAM design can be estimated by a product rule from the probability theory [12]:

$$P(Y \geq t_c) = P(Y \geq t) \cdot P(Y \geq t_c | Y \geq t). \quad (6)$$

The proposed algorithm has two stages and can be illustrated with Fig. 3:

1) Initial Sampling with MC: This step aims to evaluate the probability $P(Y \in \mathcal{T}) = P(Y \geq t)$ where t is the threshold, such as $t = 0.99$-quantile point shown in the left of Fig. 3. Since the samples in $\mathcal{T}$ are "non-rare" events, this evaluation needs only a few samples using standard MC method.

2) Conditional Probability Estimation: The most challenging task is to efficiently evaluate the conditional probability $P(Y \geq t_c | Y \geq t)$ where sampling method must be used. To expedite the convergence rate of estimation, a "proposed" sampling distribution $g(X)$ that is close to the failure region shall be constructed by *shifting* and *reshaping* the

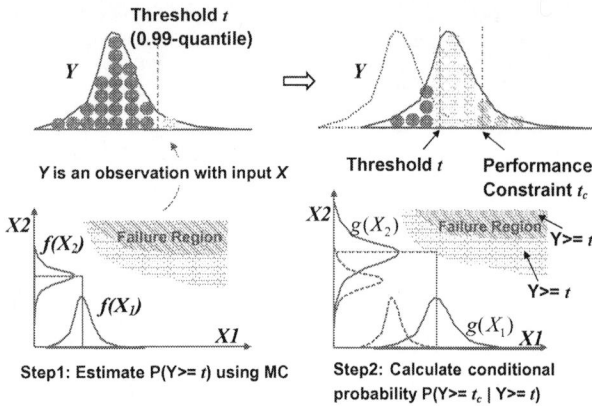

Fig. 3. Overall flow in proposed algorithm. (Noted that $\mathcal{T} = \{Y | Y \geq t\}$ contains $\mathcal{S} = \{Y | Y \geq t_c\}$).

Algorithm 1 Overall Algorithm

Input: random variables X with sampling distributions $f(X)$ and performance constraints $Y \geq t_c$.
Output: the estimation of failure probability $p_{IS}(Y \geq t_c)$.

1: /* 1: Initial Sampling with MC */
2: Use few MC samples to find the threshold value t of performance (e.g., t = 0.99-quantile point).
3: Run standard Monte Carlo method to calculate $P_{MC}(Y \geq t)$ with certain accuracy level.
4: /* 2: Conditional Probability Calculation */
5: Shift the original sampling distribution $f(X)$ towards the failure region.
6: Reshape the shifted $f(X)$ by changing its standard deviation to construct $g(X)$.
7: Generate samples from $g(X)$ and evaluate conditional probability $P(Y \geq t_c | Y \geq t)$.
8: /* 3: Failure Probability Estimation */
9: Solve for the failure probability $p_{IS}(Y \geq t)$ as

$$P_{IS}(Y \geq t_c) = P_{MC}(Y \geq t) \cdot P(Y \geq t_c | Y \geq t).$$

"original" sampling distribution (shown in the right of Fig. 3). More details will be discussed in the following section.

The overall algorithm flow is described in Algorithm(1). There are several issues that need to be resolved: 1) It is, at the moment, unclear how to *shift* and *reshape* the original sampling distribution $f(X)$ in order to build $g(X)$; 2) With the proposed sampling distribution $g(X)$, how to calculate the conditional probability; 3) It is of great interest to study whether the estimations of proposed algorithm is always bounded or not.

The following sections discuss how we solve these issues.

B. Shift and Reshape Sampling Distribution

B.1 Mean-Shift Vector Selection

Mean-shift is a typical way to move the sampling distribution towards the failure region where the failed samples are most likely to happen in previous works [3, 4, 5, 6, 7]. The key is to find the mean-shift vector for the original sampling distributions $f(X)$.

To this end, we propose to shift $f(X)$ towards a "non-rare" subset $\mathcal{T} = \{Y | Y \geq t\}$, because our target is to evaluate the conditional probability $P(Y \geq t_c | Y \geq t)$ around the subset $\mathcal{T}$. More importantly, as $\mathcal{T}$ is usually not far away from the mean of $f(X)$, the shifted distribution shares almost the *same* support with $f(X)$ so as to avoid the "disjoint support" issue.

In addition, we adopt the insights from [7] to find a close-to-optimal mean-shift vector in this work. Let us consider a 1-D problem as an

example. The algorithm in [7] starts with an initial parameterized distribution $\hat{f}(X, \hat{\mu})$ and tries to update the mean value iteratively to achieve a close-to-optimal sampling distribution $f^*(X, \mu^*)$ by an analytic formula:

$$\mu^* = \frac{\sum_{i=1}^{N} I(x_i) \cdot w(x_i) \cdot x_i}{\sum_{i=1}^{N} I(x_i) \cdot w(x_i)}. \quad (7)$$

Here x_i $(i = 1, \cdots, N)$ are samples drawn from $\hat{f}(X, \hat{\mu})$ and $w(x_i)$ are their likelihood ratios as $w(x_i) = f(x_i)/\hat{f}(x_i, \hat{\mu})$.

Intuitively, the updated mean value μ^* can be viewed as the coordinates of the *centroid point* in the failure region where the failed samples are most likely to happen. This interesting finding becomes more obvious if $\hat{f}(X, \hat{\mu})$ equals $f(X)$ and all likelihood ratios take on value 1. Hence, μ^* is:

$$\mu^* = \frac{\sum_{i=1}^{N} I(x_i) \cdot x_i}{\sum_{i=1}^{N} I(x_i)}. \quad (8)$$

Therefore, our mean-shift method tries to shift the sampling distribution towards the "centroid point" of the subset $\mathcal{T} = \{Y | Y \geq t\}$, which can be evaluated with available MC samples from the first step in Algorithm (1) and requires no extra sampling/simulation cost.

B.2 Standard Deviation Selection

Next, it is desired to *reshape* the shifted sampling distribution around the centroid of subset $\mathcal{T}$. In particular, the standard deviation for the proposed sampling distribution $g(X)$ must be properly chosen to reach the failure region $\mathcal{S} = \{Y | Y \geq t_c\}$, because the shifted and reshaped sampling distribution should *dominate* or completely cover the "rare-event" region $\mathcal{S}$.

As an illustration, let us consider a 2-D problem in Fig. 4. The problem now becomes how to choose the standard deviation of the proposed sampling distribution $g(X)$ to obtain the samples in the "rare-event" region $\mathcal{S} = \{Y | Y \geq t_c\}$.

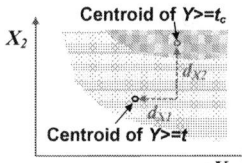

Fig. 4. The distance between centroid points of two subsets along each parameter axis.

The proposed algorithm first approximates the centroid point of $\mathcal{S} = \{Y | Y \geq t_c\}$ using uniformly-distributed samples and then calculates the distance between these two centroid points along each parameter axis (e.g., d_{X_1} and d_{X_2} shown in Fig.4). Then, we choose $max(d_{X_i}, \sigma_{(0, X_i)})$ as the standard deviation of $g(X_i)$ for the variable X_i, where $\sigma_{(0, X_i)}$ is the original standard deviation of $f(X_i)$. This choice can be intuitively explained as follows:

- $d_{X_i} > \sigma_{(0, X_i)}$: the failure region $\mathcal{S}$ is very far away from the subset $\mathcal{T}$, therefore, the larger value d_{X_i} is used to extend the range of $g(X_i)$ and obtain the rare-event samples in the failure region. In the meantime, $g(X_i)$ has almost the same supports with $f(X_i)$ because its mean position locates at the centroid point of $\mathcal{T}$ and is not far away from $f(X_i)$.

- $d_{X_i} < \sigma_{(0, X_i)}$: Suppose the smaller one, d_{X_i}, is chosen as the standard deviation of $g(X_i)$, the proposed sampling distribution $g(X)$ will have much smaller sampling space, thereby, making it fail to keep the same supports with $f(X_i)$ and suffer from "disjoint supports" issue. The proposed algorithm chooses $\sigma_{(0, X_i)}$ as the standard deviation of $g(X_i)$ in this case.

C. Conditional Probability Calcualtion

With the proposed sampling distribution $g(X)$, it is desired to efficiently estimate the conditional probability in Algorithm(1). We can start with the product rule in the probability theory [12]:

$$P(Y \geq t_c | Y \geq t) = \frac{P(Y \geq t_c, Y \geq t)}{P(Y \geq t)}. \quad (9)$$

In addition, when samples x_i $(i = 1, \cdots, N)$ are generated from $g(X)$, both $P(Y \geq t_c)$ and $P(Y \geq t)$ can be estimated mathematically with the indictor function and likelihood ratios. Thus, the equation (9) becomes:

$$
\begin{aligned}
P_{MIS}(Y \geq t_c | Y \geq t) &= \frac{P(Y \geq t_c)}{P(Y \geq t)} \\
&= \frac{\frac{1}{N} \sum_{i=1}^{N} w(x_i) \cdot I_{\{Y \geq t_c\}}(x_i)}{\frac{1}{N} \sum_{i=1}^{N} w(x_i) \cdot I_{\{Y \geq t\}}(x_i)}. \quad (10)
\end{aligned}
$$

where $I_{\{Y \geq t_c\}}(\cdot)$ and $I_{\{Y \geq t\}}(\cdot)$ are indicator functions for subsets $Y \geq t_c$ and $Y \geq t$, respectively. $w(x_i)$ are likelihood ratios for these samples. In this way, the conditional probability can be efficiently evaluated under proposed sampling distribution $g(X)$.

D. Boundedness Analysis

D.1 Importance Sampling
Let us first investigate the existing importance sampling and assume samples x_j $(j = 1, \cdots, M)$ are generated from the proposed sampling distribution $g(X)$.

We find the upper bound of probability estimate from the conventional importance sampling according to Boole's inequality (also known as the union bound from probability theory [12]) as:

$$
\begin{aligned}
P(Y \geq t_c) &= P_f(\sum_{j=1}^{M} I_{\{Y \geq t_c\}}(x_j)) \leq \sum_{j=1}^{M} P_f(x_j) \cdot I_{\{Y \geq t_c\}}(x_j) \\
&= \sum_{j=1}^{M} w(x_j) \cdot I_{\{Y \geq t_c\}}(x_j). \quad (11)
\end{aligned}
$$

In 11 P_f stands for the probability estimation under sampling distribution $f(X)$. As discussed in [10, 11], the likelihood ratios $w(x_j)$ can vary dramatically in high dimension and be any random quantities. Therefore, the union bound of the estimation $P(Y \geq t_c)$ in (11) approaches infinity and importance sampling becomes unreliable and untrustworthy.

D.2 Proposed Algorithm
The proposed algorithm constructs a subset $\mathcal{T} = \{Y | Y \geq t\}$ that *dominates* the failure region $\mathcal{S} = \{Y | Y \geq t_c\}$ (i.e., $\mathcal{T} \supset \mathcal{S}$). Therefore, the upper bound of conditional probability can be derived as:

$$
\begin{aligned}
P(Y \geq t_c | Y \geq t) &= \frac{P(Y \geq t_c)}{P(Y \geq t)} \\
&= \frac{\sum_{j=1}^{N} w(x_j) \cdot I_{\{Y \geq t_c\}}(x_j)}{\sum_{j=1}^{N} w(x_j) \cdot I_{\{Y \geq t\}}(x_j)} \leq 1. \quad (12)
\end{aligned}
$$

Note that no matter how likelihood ratios $w(x_j)$ vary, the same likelihood ratios for samples in the failure region $\mathcal{S} = \{Y | Y \geq t_c\}$ would appear in both numerator and denominator in (12) if and only if the calculations of both $P(Y \geq t_c)$ and $P(Y \geq t)$ utilize the *same* set of samples x_j $(j = 1, \cdots, M)$ drawn from $g(X)$. Clearly, the conditional probability estimation of proposed algorithm is always bounded by the upper bound 1. Thereby, the propose algorithm can reliably provide bounded estimation results.

IV. EXPERIMENTAL RESULTS

We investigate its performance of the proposed algorithm for failure analysis of memory circuits (e.g., SRAM bit-cell and sense amplifier) in this section. All experiments are performed using MATLAB and Hspice with BSIM4 transistor model. The proposed algorithm is named as HDIS (high-dimensional importance sampling) in this section. In addition, Monte Carlo (MC), statistical blockade (SB)[2], and spherical sampling (SS) [4, 5] have been implemented for comparison purpose.

TABLE I
PROCESS PARAMETERS OF MOSFETS.

Variable Name	σ/μ	unit
Flat-band Voltage (V_{fb})	0.1	V
Gate Oxide Thickness (t_{ox})	0.05	m
Mobility (μ_0)	0.1	m^2/Vs
Doping concentration at depletion (N_{dep})	0.1	cm^{-3}
Channel-length offset (ΔL)	0.05	m
Channel-width offset (ΔW)	0.05	m
Source/drain sheet resistance (R_{sh})	0.1	Ohm/mm^2
Source-gate overlap unit capacitance (C_{gso})	0.1	F/m
Drain-gate overlap unit capacitance (C_{gdo})	0.1	F/m

A. SRAM Circuit and Variation Modeling
A functional diagram of SRAM circuit with one bit-cell column is shown in Fig. 5, which consists of a decoder, bit-cells, a sense amplifier and a delay chain [13]. During the reading operation: the bit-cells store the data in forms of '0' or '1'; the decoder generates an address of a specific bit-cell and releases a read enable signal. Therefore, the chosen bit-cell starts to discharge the bit-lines (i.e., the lines that connect to all bit-cells) to produce a voltage difference between two bit-lines. the sense amplifier reads out the stored data by capturing and magnifying the voltage difference on bit-lines.

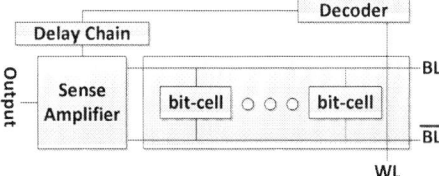

Fig. 5. Functional diagram of an SRAM circuit.

The process variations are introduced into each transistor of SRAM circuit, which are modeled by 9 process parameters shown in Table (I). The parameters are physically independent [14] and can be considered to be Gaussian random variables. Note that the threshold voltage V_{th} is not a process parameter and depends on V_{fb}, t_{ox}, ΔL and ΔW through related effects [14].

B. SRAM Cell with Reading Failure
A typical 6-transistor SRAM bit-cell is shown in Fig. 6: $Mn2$ and $Mn4$ control the accessing of the cell; the remaining four transistors form two inverters and use two stable states (either '0' or '1') to store the data in this memory cell. The *reading access failure* happens when the voltage difference between $\bar{BL}$ and BL is too small to be sensed by the sense amplifier at the end of reading operation [1].

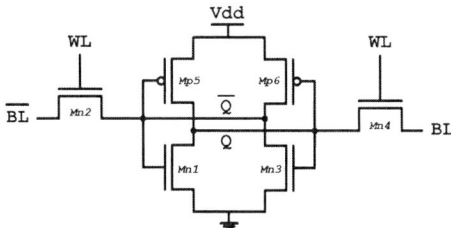

Fig. 6. The schematic of the 6T SRAM cell.

We perform different methods (MC, SS[5], SB[2], proposed) on this SRAM bit-cell example to predict the reading failure probability under process variations and the comparison results are shown in Table II.

B.1 Accuracy Comparison
At a first glance, we would be very surprised to find that SS[5] method based on conventional importance sampling framework can provide accurate failure rate predictions in this 54-dim problem!

TABLE II

COMPARISON FOR SRAM BIT-CELL ANALYSIS WITH 90% TARGET ACCURACY AND CONFIDENCE LEVEL.

	Monte Carlo (MC)	Spherical Sampling (SS)[5]	Statistical Blockade (SB)[2]	Proposed method (HDIS)
failure probability	2.413E-05 (0%)	2.8415E-05 (+17.7%)	2.7248e-05 (+12.9%)	2.4949E-05 (+3.39%)
#sim. runs	4.6e+6 (1150X)	2e+4 (5X)	8.16e+5 (204X)	4e+3 (1X)

However, this comparison cannot allow us to reach that conclusion, because this SRAM bit-cell example is a "pseudo" high-dimensional problem for two-fold reasons: (1) during the reading operation, not all transistors are active. In fact, both $Mp5$ and $Mn3$ are shut off, therefore, the process variations on these two transistors have no effect on discharge behavior of bit-lines at all; (2) without loss of generality, assuming $\bar{BL}$ = '0' and BL='1', the discharge current flows from $\bar{BL}$ to the ground through $Mn2$ and $Mn1$ so that to pull down the voltage of $\bar{BL}$. As such, the process variations in $Mn2$ and $Mn1$ have more significant effects on the discharge behavior of bit-lines and can potentially mask the variation effects in $Mp6$ and $Mn4$. In this way, there are only 18 "effective" variable parameters, which suggests that this example is a problem with modest dimension.

When compared with MC results, the proposed method provides the most accurate failure probability estimation with only 3.39% relative error, while the estimations from SS[5] and SB[2] have more than 10% relative error.

B.2 Efficiency Comparison

From Table II we also compare the efficiency of these methods: MC is very time-consuming and requires nearly 4.6 millions transistor-level SPICE simulations; SB[2] can provide 6X complexity reduction by screening out and simulating those "most-likely-to-fail"samples; SS[5] method is made more efficient (230X speedup over MC) by better choosing failed samples using importance sampling algorithm; the proposed algorithm achieves the best convergence rate (1150X faster than MC) by efficiently spreading more samples into the failure region using a sampling distribution with a large-standard-deviation in high dimensions.

C. Sense Amplifier for Target Gain

Next, we consider a sense amplifier example which includes 13 transistors as shown in Fig. 7.

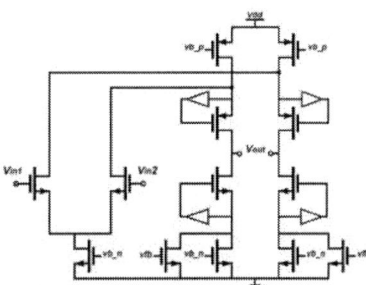

Fig. 7. The schematic of a sense amplifier circuit.

In a SRAM circuit, the sense amplifier is designed to magnify the voltage difference between $\bar{BL}$ and BL. If the gain is too small, the output of this amplifier might be too weak to be read by the decoder circuit. Therefore, a reading failure happens. With the variation modeling summarized in Table I, the sense amplifier example has 117 random variables in total. More importantly, all of these variable parameters are "effective" because the transistors are active and process variations on each transistor can significantly change the gain, which is a truly high-dimensional problem.

C.1 Accuracy Comparison

To validate the accuracy of the proposed algorithm, we apply different methods (MC, SS[5], SB[2] and proposed) on this 117-dim problem

to predict the timing failure probability. Here, MC serves as the "gold standard". SB[2], is not included in the further comparison, because the classifier used in SB[2] fail to block any Monte Carlo sample. Therefore, Considering the complexity of running the classifier, the SB [2] involves even higher computation complexity than MC method.

The evolution of the probability estimation in different methods are plotted in Fig. 8(a). Several observations can be made:

First, this figure shows the failure of conventional importance sampling (i.e., SS[5]). In fact, due to the degeneration or collapse of likelihood ratios, SS[4, 5] method converges to a random quantity which is obviously wrong and far away from the MC result. Moreover, SS[5] does not have a mechanism for improving accuracy even though more samples are added.

The proposed method builds an effective proposed sampling distribution to choose more failed samples easily and its estimation is theoretically bounded due to the proposed evaluation of conditional probability. Therefore, the proposed algorithm can reliably estimate the failure probability that matches with MC results.

C.2 Efficiency Comparison

Even though the Fig. 8(a) provides a rough comparison of efficiency, the detailed comparison can be shown in Fig. 8(b), where different methods try to achieve the "comparable" accuracy. Note that circuit simulation is the most time-consuming part and the runtime cost of the remaining computation becomes negligible. As such, the required number of circuit simulations for the same accuracy and confidence level serves as a measurement of the efficiency.

First, the Figure-Of-Merit (FOM) is used to quantify the accuracy of probability estimation as [4, 5]:

$$\rho = \frac{\sqrt{\sigma^2_{p(\text{fail})}}}{p(\text{fail})}. \tag{13}$$

where $p(\text{fail})$ is the failure probability and $\sigma_{p(\text{fail})}$ is the standard deviation of $p(\text{fail})$. In fact, the FOM can be viewed as a *relative error* so that lower FOM means higher accuracy of probability estimation.

We compare the evolutions of FOM for different methods in Fig. 8(b) and draw a dash line to indicate the 90% accuracy with 90% confidence ($\rho = 0.1$). And we can have following observations:

First, SS[5] has reached $\rho = 0.1$ but its estimation is completely wrong. Clearly, it cannot detect the failure at all. The same observation is applied to other existing importance sampling methods due to the boundedness analysis in Section 3.3.1.

Second, The proposed algorithm can provide the accurate estimation of failure probability with only a few thousands samples, which dramatically relieves the requirements of computing and storage efforts. As shown in this figure, the proposed method can achieve $708X$ speedup over Monte Carlo and be $17X$ faster than statistical blockade method [2].

C.3 Comparison for Different Failure Probabilities

We study various methods on the sense amplifier example with three different failure probabilities summarized in Table III. It is obvious that SS[5] method fails to achieve any reasonable accuracy in all these cases. This demonstrates the failure of conventional importance sampling method. On the contrary, the estimates from the proposed method match the MC result.

In addition, the table reveals that the proposed method provides the fastest convergence speed in all these cases and, more importantly, offers substantial complexity reduction as the failure probability becomes

978-1-4799-2817-0/14 $31.00 © 2014 IEEE

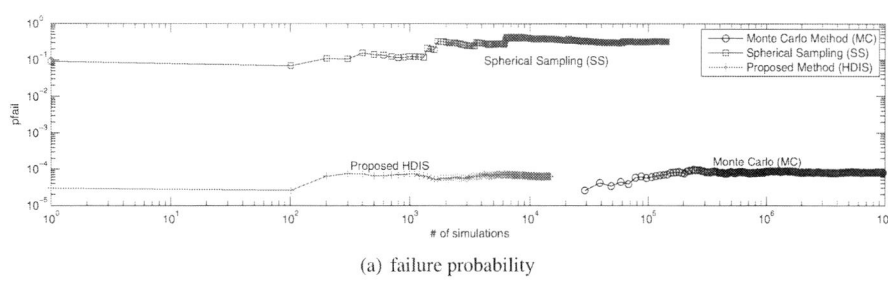

(a) failure probability

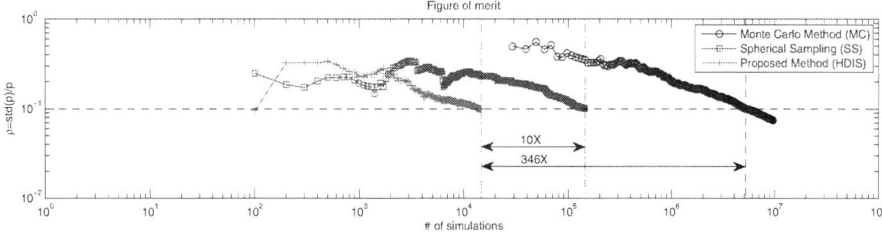

(b) figure of merit

Fig. 8. Evolution comparison of the failure probability estimation and figure of merit for different methods.

TABLE III

COMPARISON FOR SENSE AMPLIFIER ANALYSIS WITH 90% TARGET ACCURACY AND CONFIDENCE LEVEL

Target Failure Probability		Monte Carlo (MC)	Spherical Sampling (SS) [5]	Proposed Method (HDIS)
8e-3	prob:(failure)	8.136e-4	0.2603	7.861e-3 (3.4%)
	#sim. runs	4.800e+4 (24X)	16000 (8X)	2000
8e-4	prob:(failure)	8.044e-4	0.2541	8.787e-4 (9.2%)
	#sim. runs	4.750e+5 (36X)	8.330e4 (6.4X)	1.300e4
8e-5	prob:(failure)	8.089e-5	0.3103	8.186e-5 (1.2%)
	#sim. runs	5.156e+6 (346X)	1.430e+5 (10X)	1.500e+4

smaller. This property makes our proposed algorithm suitable for industrial problems where exist "rare events" with extremely small probability.

V. CONCLUSION

In this paper, we propose a fast statistical algorithm to estimate the extremely small probability of rare events in high dimensions which has proved to be bounded. The proposed algorithm has been successfully applied to failure probability prediction of memory circuits (e.g., SRAM bit-cell, sense amplifier) and demonstrates significant complexity reduction without compromising the accuracy. To the best of our knowledge, this is the first work that successfully handle the rare event estimation in high dimensions without using MC and classic importance sampling method. Experiments on a 54-dimensional SRAM cell circuit show that the proposed approach achieves 1150X speedup over Monte Carlo without compromising any accuracy. It is also 204X faster than the classification based method (e.g., Statistical Blockade [2]) by and 5X faster than existing importance sampling method (e.g., Spherical Sampling [4, 5]). On another 117-dimension circuit, the classification based method fails to improve the performance by blocking "unlikely to fail" samples, and Spherical Sampling [4, 5] method completely fails to provide reasonable accuracy. Contrastingly, the proposed approach yields accurate result with 364X speedup over Monte Carlo.

REFERENCES

[1] K. Agarwal and S. Nassif, "Statistical analysis of SRAM cell stability," ser. DAC '06, 2006, pp. 57–62.

[2] A. Singhee and R. A. Rutenbar, "Statistical blockade: very fast statistical simulation and modeling of rare circuit events and its application to memory design," *IEEE Tran. on CAD*, vol. 28, pp. 1176–1189, 2009.

[3] R. Kanj, R. Joshi, and S. Nassif, "Mixture importance sampling and its application to the analysis of SRAM designs in the presence of rare failure events," in *Proceedings of the 43rd annual Design Automation Conference*, ser. DAC'06, 2006, pp. 69–72.

[4] L. Dolecek, M. Qazi, D. Shah, and A. Chandrakasan, "Breaking the simulation barrier: SRAM evaluation through norm minimization," in *Proceedings of the 2008 IEEE/ACM International Conference on Computer-Aided Design*, ser. ICCAD '08, 2008, pp. 322–329.

[5] M. Qazi, M. Tikekar, L. Dolecek, D. Shah, and A. Chandrakasan, "Loop flattening and spherical sampling: Highly efficient model reduction techniques for SRAM yield analysis," in *Design, Automation Test in Europe Conference Exhibition (DATE), 2010*, 2010.

[6] K. Katayama, S. Hagiwara, H. Tsutsui, H. Ochi, and T. Sato, "Sequential importance sampling for low-probability and high-dimensional SRAM yield analysis," in *IEEE/ACM International Conference on Computer-Aided Design*, 2010.

[7] F. Gong, S. Basir-Kazeruni, L. Dolecek, and L. He, "A fast estimation of SRAM failure rate using probability collectives," in *Proc. ACM ISPD*, 2012, pp. 41–47.

[8] C. Dong and X. Li, "Efficient SRAM failure rate prediction via Gibbs sampling," in *Proceedings of the 43rd annual Design Automation Conference*, ser. DAC'11, 2011.

[9] S. Au and J. Beck, "Important sampling in high dimensions," *Structural Safety*, vol. 25, no. 2, pp. 139 – 163, 2003.

[10] T. B. B. Li and P. Bickel, "Curse-of-dimensionality revisited: Collapse of importance sampling in very high-dimensional systems," *Technical Report No.696, Department of Statistics, UC-Berkeley*, 2005.

[11] R. Y. Rubinstein and P. W. Glynn, "How to deal with the curse of dimensionality of likelihood ratios in monte carlo simulation," *Stochastic Models*, vol. 25, pp. 547 – 568, 2009.

[12] A. Papoulis and S. Pillai, "Probability, random variables and stochastic processes," *McGraw-Hill*, 2001.

[13] A. Pavlov and M. Sachdev, "CMOS SRAM circuit design and parametric test in nanoscaled technologies: Process-aware SRAM design and test," *Springer Publisher*, 2008.

[14] P. Drennan and C. McAndrew, "Understanding MOSFET mismatch for analog design," *IEEE J. of Solid State Circuits*, vol. 38, no. 3, pp. 450 – 456, 2003.

Predicting Circuit Aging Using Ring Oscillators

Deepashree Sengupta and Sachin S. Sapatnekar
Department of Electrical and Computer Engineering
University of Minnesota, Minneapolis, MN 55455, USA.

Abstract—This paper presents a method for inferring circuit delay shifts due to bias temperature instability using ring oscillator (ROSC) sensors. This procedure is based on presilicon analysis, postsilicon ROSC measurements, a new aging analysis model called the Upperbound on f_{Max} (UofM), and a look-up table that stores a precomputed *degradation ratio* that translates delay shifts in the ROSC to those in the circuits. This method not only yields delay estimates within 0.2% of the true values with very low runtime, but is also independent of temperature and supply voltage variations.

I. INTRODUCTION

Bias Temperature Instability (BTI) is a pressing reliability issue that degrades the threshold voltages (V_{th}) of nanometer-scale MOS devices during normal circuit operation under voltage and temperature stress. The degradation in PMOS [NMOS] is called Negative [Positive] Bias Temperature Instability, or NBTI [PBTI], and both are partially reversible on the removal of stress. Incorporating these recovery effects, the long-term degradation depends on the average duty cycle, but is independent of stressing signal frequency.

The overall effect of BTI is to reduce the maximum operating frequency, f_{Max}, of a circuit over its lifetime. To ensure that a chip meets its timing requirements over its lifetime, compensation techniques have been developed. In the presilicon design, appropriate delay guardbands may be added [1], [2], while the postsilicon phase may adapt the circuit during operation in the field [3], using sensors built in at the presilicon phase, by adjusting its clock frequency, supply voltage, or body bias. However, by definition, presilicon techniques are unaware of the runtime operating environment experienced by a chip and must consider worst-case scenarios by assuming pessimistic stress conditions for the circuits. Postsilicon techniques limit pessimism and deploy just enough adaptive compensation, based on monitors that periodically evaluate f_{Max} in the circuit under test (CUT). Two classes of monitors may be employed:

- *Surrogate circuit monitors*: These are test circuits used to estimate f_{Max} degradation in the CUT by trying to emulate the operating conditions/functionalities of the CUT. These range from simple ring oscillators (ROSCs) [4], [5] to more complex representative critical path (RCP) circuits [6], [7], [8].
- *CUT monitors*: In these methods, delay tests are directly performed on the CUT at predetermined intervals to measure its performance in terms of f_{Max} degradation [9], [10].

CUT monitors are accurate since they directly monitor the CUT, but may suffer from large hardware and test time overheads. Although such tests are required infrequently and their runtime can be reduced [11], the overheads of testing the entire chip may still be onerous. In this work we use surrogate circuit monitors (ROSCs to be specific) to characterize aging in the CUT.

Our contributions are summarized as follows. *First*, we propose a new *Upperbound on f_{Max} (UofM)* model to estimate a safe f_{Max} that the CUT can operate at. This model accounts for the possibility that critical paths may change over the lifetime of a chip due to nonuniform delay degradation on various circuit paths, by finding an envelope for the CUT delay. *Second*, we analyse the maximum pessimism in the UofM model and demonstrate it to be practically less than 0.2% in representative benchmark circuits. *Third*, we leverage the UofM model to present a novel aproach for inferring

the delay degradation of the CUT based on data from on-chip ROSCs. Our scheme involves an initial presilicon characterization that uses a compact on-chip look-up table to determine a calibration factor, which we call the *degradation ratio*, $\mathcal{D}$, that translates ROSC measurement data to CUT delay degradation while capturing process-voltage-temperature (PVT) fluctuations in the manufactured circuit. Our solution accounts for BTI recovery when the circuit is power-gated and is robust to changes in the on-chip temperature and DVFS-related supply voltage changes. Our approach also captures the effects of process variations on the sensors and the CUT.

The rest of the paper is organized as follows. We begin with an explanation of on-chip monitors in Section II. Next, Section III presents a brief background on BTI-induced delay degradation, followed by detailed overview of the UofM model in Section IV. Section V shows the maximum pessimism in delay estimation possibly incurred by the UofM model using a given library of gates. Section VI demonstrates the experimental setup and results, and we conclude in Section VII.

II. SURROGATE CIRCUITS FOR ON-CHIP MONITORING

ROSC-based surrogate circuits are widely used in industry to evaluate aging. In silicon odometers [4], they have been demonstrated to provide high resolution and remove common-mode disturbances. ROSCs have several advantages over RCPs. *First*, ROSCs are compact and uniform, and require the design and layout of only a single repeatable macrocell, as against RCPs, which must be designed and laid out individually. *Second*, the process of generating RCP circuits is computationally expensive (for circuit b15, RCP generation can take 30 minutes [7], as against this work, where the computational effort for ROSC characterization takes less than two seconds).

However, like RCPs, ROSCs are mere surrogates for the CUT. Therefore, measuring f_{Max} degradation in ROSC is not equivalent to measuring degradation in the CUT, for several reasons. *First*, since the gate types on a CUT path are not the same as those on the ROSC, the path delay sensisitvity to V_{th}-shifts under aging is different from the ROSC delay sensitivity. Although RCP circuits try to overcome this issue, they suffer from the limitations of irregularity in layout and high design effort, as pointed out above, and are less used in industrial designs than ROSCs. *Second*, the ROSC has a single path that ages along a constant profile through its lifetime; in contrast, the delay of a CUT is the maximum of all path delays. Since a set of near-critical paths may have different aging sensitivities, the critical path may change over the lifetime of the CUT, causing it to age at different rates at different times.

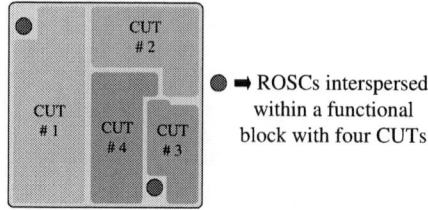

Fig. 1. Block diagram of a functional block with CUTs and ROSCs.

Within a larger circuit, ROSCs can be placed close to the CUT as illustrated in Fig. 1: since ROSCs are cheap and compact, many

978-1-4799-2817-0/14 $31.00 © 2014 IEEE

copies can be replicated within the chip. Small circuit blocks may share a ROSC, while a very large block could contain several ROSCs.

To correlate well to the CUT, a test structure should try to match its (a) temperature, (b) V_{dd}, (c) process parameter variations, and (d) signal stress probability. Spatial proximity of the ROSC and the CUT ensures similar temperatures and enables the design to connect the supply of the ROSC to that of the CUT (thus capturing the effects of V_{dd} change under DVFS and power gating). Further, due to proximity, the ROSCs face a similar set of systematic process variations and spatially correlated random variations, i.e., shifts within the CUT in any manufactured part are similar to those in the nearby ROSCs. The granularity at which these sensors are deployed reflects a trade-off between overhead and accuracy: in this work, we assume that ROSC placement is a user input. Thus, it is easy to match criteria (a) and (b) and some parts of (c) above.

Matching random parameter variations and signal stress probabilities is harder since ROSCs are surrogate circuits. Shifts in delay due to purely random process variations in the CUTs are, by definition, physically impossible to capture in any surrogate; however, these effects can be diluted by using ROSCs with many stages [4]. Within CUTs, which typically have ten or more stages of logic on their critical paths, there is also a natural level of dilution of random variations. The switching activity of the CUT is similarly impossible to capture in a surrogate ROSC circuit. Therefore it is common practice to assume pessimistic worst-case stress probabilities for the CUT that will guarantee correctness of a prediction. Note that this pessimism is not specific to our method and is an unavoidable guardband that must be built into any design methodology based on surrogates [2]. In addition, we assume that sufficient margin has been kept for the clock skew due to aging in the clock network, since CUT aging is interpreted in terms of timing violation.

In our work, we use identical ROSC-based sensors, thus simplifying layout and design effort to insert them in the chip, and derive analytical methods to predict delay degradation (and hence operable f_{Max}) in the CUT based on measurements from the ROSC. Our method captures PVT effects due to temperature, V_{dd}, and systematic/spatially correlated variations correctly, dilutes the effects of random variations, and uses routine pessimistic approaches to address the built-in inability of ROSCs in capturing CUT stress probabilities.

III. THE IMPACT OF BTI-INDUCED AGING ON DELAY

The precise mechanisms of BTI are a matter of debate within the research community. Two candidates have emerged as the most important: the reaction-diffusion (RD) model [12] and the charge trapping (CT) model [13]. The key difference is in the trap generation mechanism in the oxide layer: the former [latter] gives a power-law [logarithmic] dependence of the V_{th}-shift with time.

In general, the V_{th}-shift depends both on the stress time and the average duty cycle. If the precise switching probabilities in a circuit are known, they may be used to determine the average duty cycle. However, in many cases, it is impossible to guarantee that specific switching probabilities will be maintained, and instead a worst-case probability may be used. Common approaches include using a worst-case constant-stress assumption on each transistor for NBTI, or a stress probability (SP) of 0.95 on each transistor [2].

The magnitude of the V_{th}-shift in PMOS and NMOS device at time t, $\Delta V_{th,p}(t)$ and $\Delta V_{th,n}(t)$ respectively, is given by:

$$\Delta V_{th,p}(t) = K_1 \, \xi_1 \, f(t) = c_1 f(t)$$
$$\Delta V_{th,n}(t) = K_2 \, \xi_2 \, f(t) = c_2 f(t) \qquad (1)$$

where ξ_1 and ξ_2 are the SPs of the PMOS and NMOS device, respectively, and K_1 and K_2 are constants dependent on temperature and V_{dd} specified by the aging model ([12] or [13]). Since PMOS

[NMOS] devices are stressed when signal is low [high], ξ_1 [ξ_2] is the probability that the signal is at logic 0 [1]. The functions $f(t)$ are governed by the trap generation mechanisms for NBTI and PBTI. In principle, the functions $f(t)$ could be different for PMOS and NMOS devices, but these are experimentally observed to be the same, as documented in design manuals and the published literature [14]. Typically, $K_2 < K_1$.

It is particularly important to note that $f(t)$ is a *sublinear* and *monotonically increasing* function that captures BTI degradation: for the RD model, $f(t) \sim t^n$, where $n \sim 0.1 - 0.2$, and for the CT model, $f(t) \sim \log(t)$. Although the V_{th}-shifts through multiple stress-recovery cycles are not monotonic, $f(t)$ captures the envelope of the delay function, including recovery effects. The monotonicity property of the envelope is used later in Theorem 1.

For a given logic gate, let its delay $D(t)$ at time t be represented by the function $g(V_{th_x}(t))$ ($x : p$ or n depending on whether rise or fall delay is considered). For convenience, we henceforth drop subscripts p and n. Similarly, instead of c_1 and c_2, we use a general c and represent $\Delta V_{th} = c \, f(t)$. Under a small V_{th}-shift, the gate experiences a delay shift as:

$$\begin{aligned} D(t) &= g(V_{th}(t_0) + \Delta V_{th}) = g(V_{th}(t_0)) + \left.\frac{\partial g}{\partial V_{th}}\right|_{V_{th}(t_0)} \Delta V_{th} \\ &= D(t_0) + k \, f(t) \end{aligned} \qquad (2)$$

where $D(t_0)$ is the delay of the gate at time t_0[1] and k is a constant multiplicative factor. Here, $k = Sc$, where S is the sensitivity of delay with respect to the absolute value of V_{th}, calculated at the nominal $V_{th}(t_0)$. Thus, under fixed stress conditions of temperature, supply voltage, and duty cycle, the delay is a function of time and is easy to compute as long as the nominal delay and sensitivity to V_{th} variation for each gate have been characterized.

IV. DELAY ESTIMATION AND AGING PREDICTION

One of the primary difficulties in using a ROSC to predict the delay degradation of the CUT is that the CUT may have several near-critical paths that age at different rates and may become critical at various time points during its lifetime, while the frequency of the ROSC is determined by a single path. In this paper, we develop a method called the *Upperbound on f_{Max} (UofM)* procedure that provides a guaranteed upperbound on the delay (and thus the degraded maximum frequency of operation) of the CUT based on ROSC sensor data.

For an n-input gate, let D_i be the arrival time at input $i \in 1, \cdots, n$ of the gate, and $d_{i \to o}$ be the delay from input i to the output o of the gate; both parameters vary with time due to aging-related slowdowns. The arrival time at the gate output is given by:

$$D_o(t) = \max_{1 \le i \le n} (D_i(t) + d_{i \to o}(t)) \qquad (3)$$

where the form of $d_{i \to o}(t)$ is given by Equation (2). Therefore, the arrival time at the output of the gate is the envelope of a set of delay curves corresponding to each argument of the max operator above.

If we perform static timing analysis (STA) over the entire CUT, we can obtain the temporal delay of the CUT: this is an envelope of a set of path delays (exemplified in Fig. 2 for a CUT with four paths). The idea of the UofM method is simple: if the max operator could be pessimistically approximated by a smooth function with continuous derivatives, such as the red curve in the Fig. 2, then the unitary operation of finding a smooth approximation to $D_o(t)$ at any gate output could be repeated to find a smooth approximation to the delay of the CUT.

---

[1] Normally, one might consider $t_0 = 0$, but realistically, chips undergo a burn-in phase causing some level of accelerated aging. Thus, based on the $f(t)$ functions characterized on fresh devices, we begin with a general value of t_0 as 3 months (as assumed in several prior papers).

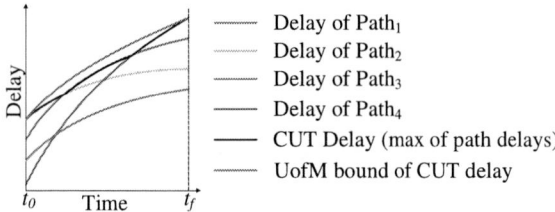

Fig. 2. Delay of CUT over its lifetime t_f.

A. An upper bound on the maximum delay

In this section, we lay the basis for the task of determining a smooth bound on $D_o(t)$ through the major theoretical result in our paper. Theorem 1 presents the case for upper-bounding the maximum of n aging curves. Pictorially, the theorem provides a precise expression for the red curve in Fig. 2, which is a continuous upper-bounding function for the maximum of n aging curves.

Theorem 1 In the interval $[t_0, t_f]$, an upperbound on the maximum of a set of monotonically increasing functions $x_1(t)$, $x_2(t)$, $\cdots$, $x_{n+1}(t)$ such that $x_i(t) = x_i(t_0) + k_i(f(t) - f(t_0))$, is given by

$$y_n(t) = x_M(t_0) + \left[\frac{x_M(t_f) - x_M(t_0)}{f(t_f) - f(t_0)} \right] (f(t) - f(t_0)) \quad (4)$$

where the function $x_M(t) = \max_{i \in \{1, \cdots, n+1\}}(x_i(t))$ represents the upper envelope of the functions x_1 through x_{n+1}.

Intuitively, the bound is simply the curve of the form in Equation (1) that matches the maximum curve at times t_0 and t_f. Note that the maximum at these two times could lie on different x_i curves, as illustrated in Fig. 2. The formal proof is deferred to Appendix I.

B. Applying the UofM bound to circuits using block-based analysis

In any circuit, the delay of a path is the sum of a set of gate delays whose temporal variations are each given by an equation of the form of Equation (2). Therefore, if $D_p(t)$ represents the delay of a path p in the circuit, the relationship between the path delays at any two times t_0 and t are given by a function of the form:

$$D_p(t) = D_p(t_0) + k(f(t) - f(t_0)) \quad (5)$$

From Equation (5), we can see that each path delay is similar to the form of the x_i functions in Theorem 1. Moreover, as discussed in Section III, the aging function $f(t)$ is a monotonically increasing function. Therefore, for a circuit with n paths, it is conceptually possible to obtain an upper bound, $y_n(t)$, on the delay of the circuit using the results of Theorem 1. As stated in Section III, while obtaining the UofM bound, we have considered SPs, $\xi_1 = 0.95$ and $\xi_2 = 0.95$, corresponding to NBTI and PBTI, respectively, at every gate of the CUT. Superficially, this may seem erroneous since a signal cannot be simultaneously high and low with the same probability. However, any input-output path through a gate goes through either the PMOS or the NMOS but *not* both. Therefore, either NBTI- or PBTI-based degradation is propagated, and our assignment of ξ_1 and ξ_2 correctly captures the worst-case delay for this worst-case path.

Theorem 1 allows the use of block-based STA by evaluating timing at the initial time t_0 and at the final time t_f, using the V_{th} aging model. In other words, only two timing analyses are needed to predict a safe upper bound on the delay over the entire lifetime.

The unitary operation in STA is to compute the maximum arrival time at the output of a gate, given the arrival times at the inputs. As an invariant, we assume that every arrival time is of the form in Equation (5): this invariant is preserved by writing the output arrival time in the same form. As is well known and apparent from Equation (3), STA involves two operations, "sum" and "max" and we now consider the preservation of the invariant under each operation.

- **Sum:** We add delay functions $D_1(t), \cdots, D_m(t)$, where $D_i(t) = D_i(t_0) + k_i(f(t) - f(t_0))$, to obtain

$$D_{sum}(t) = \sum_{i=1}^m D_i(t) = D_{sum}(t_0) + k_{sum}(f(t) - f(t_0))$$

where $D_{sum}(t_0) = \sum_{i=1}^m D_i(t_0)$ and $k_{sum} = \sum_{i=1}^m k_i$.

- **Max:** For the max over a set of delay functions $D_1(t), \cdots, D_m(t)$, each in the form of Equation (5), Theorem 1 immediately shows how the invariant is preserved.

Thus, block-based STA can compute the UofM function for the maximum arrival time at any node in the circuit, as well as the maximum delay of the circuit, in linear time in the number of gates.

C. Analyzing aging in the CUT based on ROSC data

Applying the analysis in Section IV-B to a given CUT, the temporal trend of the CUT delay can be represented using the UofM as:

$$D_{CUT}(t) = D_{CUT}(t_0) + k_{CUT}(f(t) - f(t_0)) \quad (6)$$

1) The delay of an aged ROSC: A ROSC is a chain of an odd number, $2l + 1$ of inverters connected in a closed loop. Assuming, for simplicity, that each inverter has a rise delay of d_r and a fall delay of d_f, the period of the ring oscillator is well known to be $(2l + 1)(d_r + d_f)$. We refer to the period of a ROSC as its delay, D_{ROSC}, and express it as:

$$D_{ROSC}(t) = D_{ROSC}(t_0) + k_{ROSC}(f(t) - f(t_0)) \quad (7)$$

The change in period of a ROSC can be measured easily on-the fly by prestablished methods such as the silicon odometer [4], which uses the notion of *beat frequencies* to measure delay variations in the ROSC to a very high degree of precision.

2) The Degradation Ratio, $\mathcal{D}$: We now examine how ROSC aging measurements can be used to predict aging in the CUT. Let the delay degradation in the CUT at time t be given by $\Delta D_{CUT}(t) = D_{CUT}(t) - D_{CUT}(t_0)$, and let the corresponding value for the ROSC be $\Delta D_{ROSC}(t) = D_{ROSC}(t) - D_{ROSC}(t_0)$. From (6) and (7), we define the CUT *degradation ratio*, $\mathcal{D}$, as:

$$\mathcal{D} = \frac{\Delta D_{CUT}(t)}{\Delta D_{ROSC}(t)} = \frac{k_{CUT}}{k_{ROSC}} \quad (8)$$

We make the following observations:

- From Equation (8), $\mathcal{D}$ for a CUT is a constant, independent of time t. Therefore, for any CUT, $\mathcal{D}$ can be precharacterized and stored in a look-up table to translate the ROSC delay degradation to the CUT delay degradation at various instants of time.
- The value of $\mathcal{D}$ may be different for different CUTs.
- If the CUT has one dominant critical path throughout its lifetime, the degradation ratio provides the true delay degradation of the CUT at any time (as does the UofM bound in this case).
- If the CUT has multiple critical paths that successively become dominant over its lifetime, $\mathcal{D}$ is based on the UofM bound, and Equation (8) provides a pessimistic bound on the CUT delay.
- PVT variations due to thermal and V_{dd} effects, systematic variations, and spatial correlations are all accounted for through the close spatial proximity of the ROSC and the CUT. Random variations in the ROSC can be reduced by using a larger number of stages [4].

D. Impact of temperature and V_{dd} on $\mathcal{D}$

As stated above, the effect of process variations is captured by the proximity of the ROSC and CUT, due to which their process parameters track each other. It is widely observed that a constant value of k in Equation (2) captures variations at all process corners. However, k depends on temperature, T and V_{dd} effects.

In this section, we investigate the impact of T and V_{dd} on $\mathcal{D}$ defined in Equation (8). To do so, it is necessary to analyse the $f(t)$

term in the aging model. The threshold voltage degradation with time for the RD model is given by [12] as:

$$\Delta V_{th}(t) = k_1 e^{\frac{E_{ox}}{E_0}} e^{\frac{-k_2}{T}} t^n \qquad (9)$$

For the CT model, this is given by [15] as:

$$\Delta V_{th}(t) = k_3 e^{\frac{-k_4 V_{dd}}{T}} e^{\frac{k_5}{T}} [A + \log(1 + Ct)] \qquad (10)$$

Here, $E_{ox} = \frac{V_{gs} - V_{th}}{t_{ox}}$, and $V_{gs} = \pm V_{dd}$ (during stressed/relaxed mode) and k_1, k_2, k_3, k_4, k_5, and E_0 are constants obtained from the aging model. Substituting $\Delta V_{th}(t)$ in Equation (2) and adding up gate delays to find the circuit delay degradation $\Delta D(t)$, we get:

$$\text{RD: } \Delta D(t) = k' e^{\left(\frac{-k_2}{T} + \frac{E_{ox}}{E_0}\right)} (t^n - t_0^n) \qquad (11)$$

$$\text{CT: } \Delta D(t) = k'' e^{\left(\frac{k_5 - k_4 V_{dd}}{T}\right)} \log\left(\frac{1 + Ct}{1 + Ct_0}\right) \qquad (12)$$

Here, k' and k'' represent the effect of adding the contributions of gate delays on a path during STA. The other terms are dependent on T, V_{dd} and time, which are identical for the CUT and ROSC by construction, and they cancel out when computing the ratio $\mathcal{D}$. Therefore, under both the RD and CT model, the value of $\mathcal{D}$ is:

$$\text{RD: } \mathcal{D} = \frac{k'_{CUT}}{k'_{ROSC}}; \qquad \text{CT: } \mathcal{D} = \frac{k''_{CUT}}{k''_{ROSC}} \qquad (13)$$

Since the right hand sides of both equations above are independent of T and V_{dd}, the degradation ratio $\mathcal{D}$ is independent of T and V_{dd}.

V. Bounding The Maximum Pessimism In UofM Model

The UofM bound is a pessimistic estimate of the delay of a CUT even when the actual switching activity of the CUT is known. In this section, we present Theorem 2, which bounds the maximum pessimism incurred by the proposed UofM model. The proof of the theorem is provided in Appendix II.

Theorem 2 If k_1 and k_2 are the aging sensitivities of the gates in the current library with minimum and maximum percentage degradation over their lifetime, respectively, the maximum error in delay estimation in a CUT incurred by the UofM model using this library is upper-bounded by E_{max} as:

$$E_{max} = \frac{(k_2 - k_1)(f(t_f) - f(t_0))}{4} \qquad (14)$$

where t_0, t_f, $f(t)$ and n have been defined earlier.

The maximum fractional error, E_{frac}, which is the ratio of E_{max} (Equation (14)) to $C_{top}(t')$ [or $C_{bot}(t')$] (delay of the two paths with maximum and minimum sensitivities when they cross over as shown in Fig. 5), is obtained by algebraic manipulation as:

$$E_{frac} = \frac{4(E_{max}^2)}{d_1 d_2 (\frac{k_2}{d_2} - \frac{k_1}{d_1})(f(t_f) - f(t_0))} \qquad (15)$$

where d_1 and d_2 are defined in the proof of Theorem 2. The values of k_1, k_2 and d_2 can be found from the current library and d_1 obtained by adjusting the number of gates with minimum percentage delay degradation such that $d_1 - d_2 = 2E_{max}$. Thus, E_{frac} depends on the gate library, the numerical value of which is shown in the Section VI.

VI. Experimental Setup And Results

The ideas in this paper are exercised on a set of representative ISCAS'89 and ITC'99 benchmarks. The circuits are implemented using a gate library that consists of the following functionalities: two- and three-input NAND and NOR gates, three- and four-input AOI gates, inverter, and buffer, each with drive strength X1, X2 and X4, from the NanGate 45nm Open Cell Library. Each gate is

characterized for nominal delay, output slew and delay sensitivities to V_{th} (for both rise and fall transitions) using the 45nm Predictive Technology Model (PTM). The benchmark circuits are synthesized using Synopsys Design Compiler.

The constant c_1 in Equation (1) is calibrated such that $V_{th,p}$ of the PMOS degrades by 25% in 10 years under a V_{dd} of 1.1V at an operating temperature of 85°C. The function $f(t)$ in Equation (2) follows the power-law model (with $n = \frac{1}{6}$), and the constant c_2 is chosen so that the $V_{th,n}$ degradation in NMOS due to PBTI is one-third of that due to NBTI [3]. To account for the initial transient and burn-in, we set $t_0 = 3$ months and constrain the circuit lifetime to be 10 years beyond this point. The choice of burn-in period does not however affect our proposed methodology. In addition, since the actual SPs are unknown, we use a pessimistic value of 0.95 for both ξ_1 and ξ_2 (in Equation (1)) at every gate input of the CUT to obtain the corresponding k_{CUT} values, similar to [2] (however, our results will look fundamentally similar even if we use a worst-case SP of 1.0 instead of 0.95). We obtain k_{ROSC} using a 33-stage inverter chain as in [4], by considering SP of each inverter in the chain as 0.5. It is to be noted that we do not solve any placement problem of the ROSC in this work and assume them to be in enough proximity to the CUT so that perfect correlation exists between the process parameters of the ROSC and the CUT.

Table I presents the results of our method on the representative benchmarks at $T = 85°C$ and $V_{dd} = 1.1$V. Each row corresponds to a single benchmark circuit associated with a single ROSC, except the circuit s_{mult}, which corresponds to a single ROSC that is shared by circuits s5378, s13207, and s15850. For each CUT, the second column in the table lists its gate count, $|G|$, and the third column shows the logical depth of the critical path d_{crit}: since the critical path may change over time, for convenience we consider the critical path at time $t = t_f$. We have observed that even if the critical path changes, the logical depth of the critical path does not vary appreciably over time. Generally speaking, larger values for d_{crit} correspond to larger values for the degradation ratio, $\mathcal{D}$, listed in the fourth column (note that $\mathcal{D}$ has no units).

Thus, we observe that extent of aging in a CUT is not necessarily dependent merely on the total number of gates, but also on the properties of the critical path(s), such as the logical depth. The sensitivity of a cell typically depends on its driving power and its load: cells with a larger driving power tend to have lower sensitivities, and those with larger loads have higher sensitivities. Critical paths are observed to contain large-sized cells, which have low sensitivity to V_{th}-shifts, but the cells that drive these large cells see large loads, particularly if they are smaller, making them sensitive to aging degradations. Thus, for critical paths with a larger number of stages, the impact of the larger cells is diluted by the smaller cells; this is less so for circuits with fewer stages. This explains why k_{CUT} (and hence $\mathcal{D}$, since k_{ROSC} is constant) generally increases with d_{crit}.

The next column represents the maximum percentage error of the UofM bound (over the entire lifetime): the estimation error due to the use of the upper bound is seen to be virtually negligible, even though each circuit has multiple near-critical paths. This is followed by a column that provides the percentage area overhead, ΔA, of the ROSC, i.e., the ratio of the ROSC area to the total area of the CUT+ROSC, as determined by Design Compiler. As expected, this area overhead is significant only for the smallest circuits, and is quite low when the CUT is larger. For the circuit, s_{mult}, in which the first three circuits in the table share a single ROSC, the overhead is small, and the estimation error is negligible. The final column lists the total CPU time, τ, that is required to compute the value of $\mathcal{D}$ for each CUT, evaluated on a 64-bit Ubuntu server (Intel® Core™2 Duo CPU E8400 3GHz). The modest runtimes (significantly faster than RCP procedures in, e.g., [7]) indicate the aptness of our method for handling very large CUTs. Thus, during design time, these $\mathcal{D}$ values can be computed cheaply and stored in look-up tables for each CUT, using which one can estimate its delay degradation at any point of time based on a cheap measurement of ROSC delay degradation.

TABLE I

RESULTS FOR ROSC-BASED ESTIMATE FOR $T = 85^oC$ AND $V_{dd} = 1.1V$

CUT	$|G|$ #	d_{crit} #	$\mathcal{D}$	Error (%)	ΔA (%)	τ (s)
s5378	690	11	1.51	0.17	4.63	0.50
s13207	590	18	1.88	0.00	5.74	0.46
s15850	336	18	1.95	0.00	10.41	0.36
s_{mult}	1616	18	1.95	0.00	2.06	0.78
s38417	4565	25	2.79	0.09	0.64	1.75
s38584	4585	26	2.61	0.00	0.67	1.69
b15	6311	63	6.35	0.00	0.49	1.99
b17	17882	62	6.28	0.00	0.17	5.18
b18	67776	130	12.78	0.00	0.04	16.95
b19	128494	120	11.89	0.00	0.02	31.64
b20	24080	128	12.25	0.00	0.13	6.09
b22	36149	128	12.23	0.00	0.09	8.38

To investigate the reason for the low errors, we further examined the characteristics of the critical paths. All of these circuits contain multiple near-critical paths, and although we do observe a crossover where the critical path changes as the circuit ages, the UofM bound is very close to the envelope of the maximum delay over time. We evaluate the bound in Theorem 2 to evaluate the theoretical maximum on the error. The maximum and minimum percentage delay degradation, $\left(\frac{D(t_f)-D(t_0)}{D(t_0)}\right)$, occur for the three-input NOR gate and buffer (having sensitivities k_2 and k_1), respectively, both with drive strength X4 and denoted as NOR3_X4 and BUF_X4. We synthesized a circuit with just two critical paths where each path was obtained by concatenating NOR3_X4 and BUF_X4 respectively. The nominal delays d_1 and d_2 of the paths were tuned by the number of stages in each path while keeping k_1 and k_2 unchanged, to obtain the maximum error scenario shown in Fig. 5. It can be proven that the maximum fractional error E_{frac} (as defined in Equation 15) for this circuit is the absolute maximum possible in this library (proof omitted due to space constraints). This error, found to be 3.59%, corresponds to the maximum possible pessimism of the UofM based delay estimate under our library.

In fact, achieving this bound requires a pathological case (since gates with small delays also tend to have small sensitivities) in which the critical path at t_0 has a low delay sensitivity, the critical path at t_f has a high delay sensitivity and is near-critical at t_0. The latter can be achieved if the critical path at t_0 has a small number of stages and the critical path at t_f has a large number of stages. This is unlikely to be seen in practice, and is not seen in any of our circuits. This is the reason why the UofM bound is even more accurate in practice than the already small bound on pessimism from Theorem 2.

Next, we evaluate the correctness of the notion that $\mathcal{D}$ is independent of the temperature, T, and V_{dd}, as claimed in Section IV-D. Simulations were run at various T (40°C, 85°C, 125°C) and V_{dd} (0.9V, 1.1V, 1.2V) values, and the dependence on the constants c_1 and c_2 in Equation (1) on V_{dd} and T was accounted for. Fig. 3(a) shows the k_{CUT} values of three CUTs, (s38585, b15, b22) for three values each of T on one axis and V_{dd} on another, normalized with respect to their baseline values at $T = 85°C$ and $V_{dd} = 1.1V$. For each (V_{dd}, T) point, the three bars correspond, from left to right, to s38584, b15, and b22, respectively. It can be seen that the bars at each such point are of equal height, indicating that each (V_{dd}, T) point experiences an equal multiplicative effect for each CUT.

Fig. 3(b) shows the $\mathcal{D}$ values, normalized to their corresponding baseline values in Table I, for the same three circuits and the T and V_{dd} values. When we examine the degradation ratio, $\mathcal{D}$, we find that all bars have a value that is very close to unity, i.e., $\mathcal{D}$ is independent

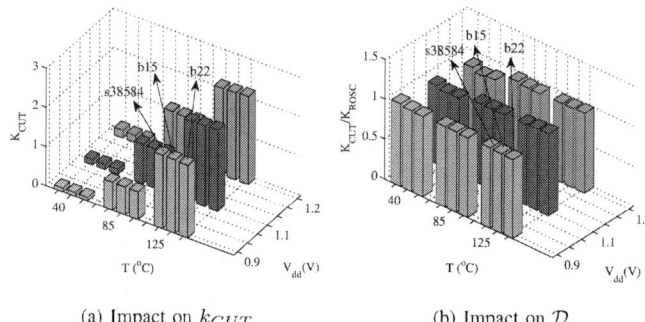

(a) Impact on k_{CUT} (b) Impact on $\mathcal{D}$

Fig. 3. Effect of change in temperature and V_{dd} on k_{CUT} and robustness of $\mathcal{D}$ to these changes

of V_{dd} or T. This may be understood by observing that k_{ROSC} also changes as T and V_{dd} are altered, and it tracks k_{CUT} very well at each point. The average absolute error of the normalized $\mathcal{D}$ from the ideal value of unity is only 3.79% for each of the three CUTs, across all T and V_{dd} values. Note that this error corresponds to a percentage of the already small delay shift (and not the delay), and is therefore negligible. This independence demonstrates that a single LUT serves the purpose of aiding delay degradation (and thus aging) estimation irrespective of variations in the operating temperature and V_{dd}, under our scheme where the CUT and the ROSC both operate in the same environment with regard to thermal changes and DVFS/power gating.

In this work, we are unable to show detailed comparisons with other approaches. To our knowledge, there is no other work that relates ROSC delays to CUT delays under aging. Existing work on RCPs for aging uses different methodologies, libraries and delay models. However, as pointed out earlier, the best proposed approach [7] requires significantly more computation than our method and entails complex layout issues. Moreover, our work is easier to implement in an industrial setting where ROSC-based methodologies have been in use for years.

VII. CONCLUSION

In this paper, we presented a technique to estimate delay degradation of a circuit using nearby ROSC-based aging sensors. We quantitatively determine how the data from the ROSC can be used to find the change in the circuit delay. Experimental results show that we can use the UofM metric to distill the relation between the CUT delay trend and the ROSC delay trend into a single degradation ratio, which can accurately predict the CUT delay degradation based on inexpensive measurements on the ROSC.

ACKNOWLEDGEMENT

This work was supported in part by the SRC (grant 2012-TJ-2234) and the NSF (award CCF-1162267).

REFERENCES

[1] S. V. Kumar, et al., "NBTI-aware synthesis of digital circuits," in Proc. DAC, pp. 370–375, 2007.
[2] M. Agarwal, et al., "Optimized circuit failure prediction for aging: Practicality and promise," in Proc. ITC, pp. 1–10, 2008.
[3] S. V. Kumar, et al., "Adaptive techniques for overcoming performance degradation due to aging in digital circuits," in Proc. ASP-DAC, pp. 284–289, 2009.
[4] T. H. Kim, et al., "Silicon odometer: An on-chip reliability monitor for measuring frequency degradation of digital circuits," IEEE J Solid-St. Circ., vol. 43, pp. 874–880, April 2008.
[5] T. B. Chan, et al., "DDRO: A novel performance monitoring methodology based on design-dependent ring oscillators," in Proc. ISQED, pp. 633–640, 2012.
[6] Q. Liu and S. S. Sapatnekar, "Synthesizing a representative critical path for post-silicon delay prediction," in Proc. ISPD, pp. 183–190, 2009.

[7] S. Wang, *et al.*, "Representative critical reliability paths for low-cost and accurate on-chip aging evaluation," in *Proc. ICCAD*, pp. 736–741, 2012.

[8] X. Wang, *et al.*, "Path-RO: a novel on-chip critical path delay measurement under process variations," in *Proc. ICCAD*, pp. 640–646, 2008.

[9] M. Agarwal, *et al.*, "Circuit failure prediction and its application to transistor aging," in *IEEE VLSI Test Symp.*, pp. 277–286, 2007.

[10] Y. Li, *et al.*, "CASP: concurrent autonomous chip self-test using stored test patterns," in *Proc. DATE*, pp. 885–890, 2008.

[11] Y. Li, *et al.*, "Concurrent autonomous self-test for uncore components in system-on-chips," in *IEEE VLSI Test Symp.*, pp. 232–237, 2010.

[12] S. Chakravarthi, *et al.*, "A comprehensive framework for predictive modeling of negative bias temperature instability," in *Proc. IRPS*, pp. 273–282, 2004.

[13] R. Da Silva and G. I. Wirth, "Logarithmic behavior of the degradation dynamics of metal oxide semiconductor devices," *J Stat. Mech.-Theory E.*, vol. P04025, pp. 1–12, April 2010.

[14] J. J. Kim, *et al.*, "PBTI/NBTI monitoring ring oscillator circuits with on-chip Vt characterization and high frequency AC stress capability," in *Proc. VLSIC*, pp. 224–225, 2011.

[15] J. B. Velamala, *et al.*, "Physics matters: statistical aging prediction under trapping/detrapping," in *Proc. DAC*, pp. 139–144, 2012.

APPENDIX I

In this section, we present a proof of Theorem 1. We begin by presenting a lemma for a simpler version of the theorem for just two paths, and then prove the theorem.

Lemma 1 Consider two monotonically increasing functions $x_1(t)$ and $x_2(t)$ in the interval $[t_0, t_f]$.

$$x_1(t) = x_1(t_0) + k_1(f(t) - f(t_0))$$
$$x_2(t) = x_2(t_0) + k_2(f(t) - f(t_0)) \quad (16)$$

An upper bound on maximum of $x_1(t)$ and $x_2(t)$ is given by:

$$y_1(t) = x_M(t_0) + \left[\frac{x_M(t_f) - x_M(t_0)}{f(t_f) - f(t_0)}\right](f(t) - f(t_0)) \quad (17)$$

where the function $x_M(t) = \max_{i \in \{1,2\}}(x_i(t))$ represents the upper envelope of the functions x_1 and x_2.

Proof: Without loss of generality, assume that $x_1(t_0) \leq x_2(t_0)$. Note that since both curves are montonically increasing, one of two possibilities must be satisfied, as illustrated in Fig. 4.
Case I: If $x_1(t_f) \leq x_2(t_f)$, then x_2 dominates x_1 over the interval.
Case II: If $x_1(t_f) \geq x_2(t_f)$, the curves cross over once in $[t_0, t_f]$.

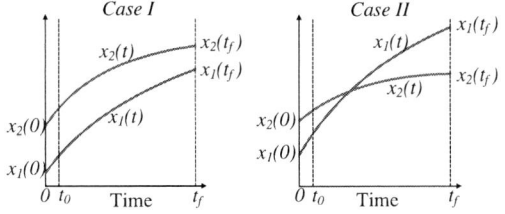

Fig. 4.   Possible trends for monotonically increasing $x_1(t)$ and $x_2(t)$.

In Case I, the result is trivially true, since the expression evaluates to the equation for $x_2(t)$. For Case II, $x_M(t_0) = x_2(t_0)$, $x_M(t_f) = x_1(t_f)$. Observing that $k_1 = \frac{x_1(t_f) - x_2(t_0)}{f(t_f) - f(t_0)}$, and performing some algebraic manipulation, we find that:

$$y_1(t) - x_1(t) = (x_2(t_0) - x_1(t_0))\left[\frac{f(t_f) - f(t)}{f(t_f) - f(t_0)}\right] > 0, \quad (18)$$

$$y_1(t) - x_2(t) = (x_1(t_f) - x_2(t_f))\left[\frac{f(t) - f(t_0)}{f(t_f) - f(t_0)}\right] > 0, \quad (19)$$

Each line above evaluates to be positive due to the monotonicity of f, i.e., $f(t_f) > f(t) > f(t_0) \ \forall \ t_0 < t < t_f$. Thus, $y_1(t)$ is an upper bound on $\max(x_1(t), x_2(t))$ over the interval $[t_0, t_f]$. $\square$

Proof of Theorem 1: Building upon Lemma 1, the proof is presented by mathematical induction over n. For the basis case, $n = 1$, Lemma 1 demonstrates that $y_1(t)$ forms an upper bound on $\max(x_1(t), x_2(t)) \ \forall t \in [t_0, t_f]$, and has the form:

$$y_1(t) = x_M(t_0) + \alpha_1(f(t) - f(t_0))$$

where α_1 is a constant of the form as in Equation (4).

For the inductive step, we assume that $y_{n-1}(t)$ is an upper bound on maximum of $x_1(t), \cdots, x_n(t)$, and attempt to show that $y_n(t)$ is an upper bound on maximum of $x_1(t), \cdots, x_{n+1}(t)$.

From the inductive hypothesis, $y_{n-1}(t)$ is an upper bound on the first n functions with the form:

$$y_{n-1}(t) = x_M(t_0) + \alpha_{n-1}(f(t) - f(t_0))$$

where α_{n-1} is a constant of the form in Equation (4). Therefore it is enough to prove that $y_n(t)$ is an upper bound on the maximum of $y_{n-1}(t)$ and $x_{n+1}(t)$ for $t \in [t_0, t_f]$. This result follows immediately from Lemma 1. In particular,

$$x_M(t_0) = \max(y_{n-1}(t_0), x_{n+1}(t_0)) = \max_{1 \leq i \leq n+1} x_i(t_0)$$

$$x_M(t_f) = \max(y_{n-1}(t_f), x_{n+1}(t_f)) = \max_{1 \leq i \leq n+1} x_i(t_f) \quad \square$$

APPENDIX II

Proof of Theorem 2: Consider a CUT with multiple critical paths over its lifetime. Let us represent the paths which are critical at $t = t_0$ and $t = t_f$ by P_1 and P_2 and denote them by the curves $C_{bot}(t)$ and $C_{top}(t)$, respectively, (Fig. 5 exemplifies this for a CUT with three critical paths) as:

$$C_{bot}(t) = d_1 + k_1(f(t) - f(t_0)) \ ; \ C_{top}(t) = d_2 + k_2(f(t) - f(t_0))$$

where d_1 and d_2 are the delays of P_1 and P_2 at $t = t_0$ respectively.

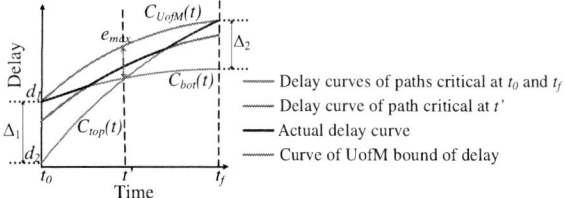

Fig. 5.   Error bound for the UofM model.

Evidently, $d_2 < d_1$. Using the UofM model, the estimated delay, denoted by $C_{UofM}(t)$, (e.g., the red curve in Fig. 5) is given by:

$$C_{UofM}(t) = d_1 + \left(k_2 + \frac{d_2 - d_1}{f(t_f) - f(t_0)}\right)(f(t) - f(t_0)) \quad (20)$$

The deviations of $C_{UofM}(t)$ from $C_{bot}(t)$ and $C_{top}(t)$ are given by $e_1(t)$ and $e_2(t)$, respectively, as:

$$e_1(t) = \frac{f(t) - f(t_0)}{f(t_f) - f(t_0)}\Delta_2 \ ; \ e_2(t) = \frac{f(t_f) - f(t)}{f(t_f) - f(t_0)}\Delta_1 \quad (21)$$

where Δ_1 and Δ_2 are the differences in the the two path delays at t_0 and t_f, respectively. The error (pessimism) of the UofM curve is bounded by minimum of $e_1(t)$ and $e_2(t)$ (as can be visualized from Fig. 5), which are monotonically increasing and decreasing, respectively. Therefore, the maximum error, e_{max}, occurs when both are equal, i.e., at $t = t'$ when the two curves cross over:

$$e_{max} = \Delta_1\left(1 - \frac{\Delta_1}{(k_2 - k_1)(f(t_f) - f(t_0))}\right) \quad (22)$$

Given the value of k_1 and k_2 as the sensitivities of the paths with minimum and maximum percentage degradation possible using the current gate library, the choice of Δ_1 can be optimized to Δ_1^{opt} that gives the maximum value of e_{max}, denoted by E_{max}. In other words, given fixed k_2 and d_2 for the path P_2, and k_1 for P_1, the number of gates in P_1 can be adjusted to obtain d_1 such that $d_1 - d_2 = \Delta_1^{opt}$. Note that changing number of cells in P_1 should not change k_1 (which can be ensured by concatenating same type of gates in the path). We obtain Δ_1^{opt} by differentiating e_{max} with respect to Δ_1, obtaining the maximum value of e_{max}, E_{max} as $(\Delta_1^{opt}/2)$, when $\Delta_1^{opt} = \frac{(k_2 - k_1)(f(t_f) - f(t_0))}{2}$. The result follows immediately.

At this optimum value, $\Delta_1 = \Delta_2$, so that the differences in the two path delays at time t_0 is identical (and the negative of) the difference at time t_f. $\square$

Statistical Analysis of Process Variation Based on Indirect Measurements for Electronic System Design

Ivan Ukhov
Linköping University
Sweden
ivan.ukhov@liu.se

Mattias Villani
Linköping University
Sweden
mattias.villani@liu.se

Petru Eles
Linköping University
Sweden
petru.eles@liu.se

Zebo Peng
Linköping University
Sweden
zebo.peng@liu.se

Abstract—**We present a framework for the analysis of process variation across semiconductor wafers. The framework is capable of quantifying the primary parameters affected by process variation, e.g., the effective channel length, which is in contrast with the former techniques wherein only secondary parameters were considered, e.g., the leakage current. Instead of taking direct measurements of the quantity of interest, we employ Bayesian inference to draw conclusions based on indirect observations, e.g., on temperature. The proposed approach has low costs since no deployment of expensive test structures might be needed or only a small subset of the test equipments already deployed for other purposes might need to be activated. The experimental results present an assessment of our framework for a wide range of configurations.**

I. INTRODUCTION AND PRIOR WORK

Process variation constitutes one of the major concerns of electronic system designs [1, 2]. A crucial implication of process variation is that it renders the key parameters of a technological process, e.g., the effective channel length and gate oxide thickness, as uncertain quantities. Therefore, the same workload applied to two "identical" dies can lead to two different power and, thus, temperature profiles since the dissipation of power and heat essentially depends on the aforementioned quantities. Consequently, process variation leads to performance degradation in the best case and to severe faults or burnt silicon in the worst scenario. Under these circumstances, uncertainty quantification has evolved into an indispensable asset of the fabrication workflows in order to provide guaranties on the efficiency and robustness of products.

An important target of uncertainty quantification is the characterization of the on-wafer distribution of a quantity of interest, deteriorated by process variation, based on measurements. The problem belongs to the class of inverse problems since the analyzed parameter can be seen as an input to the system and the measured data as the corresponding output. Such an inverse problem is addressed in this work: our goal is to characterize arbitrary process parameters with high accuracy and at low costs. The goal is accomplished by tracking supplementary quantities, which are more convenient and less expensive to be measured, and employing Bayesian statistics [3] to infer the needed parameters from the observed data.

Bayesian inference is utilized in [4] to identify the optimal set of locations on a wafer, in which the parameter under consideration should be measured in order to characterize it with the maximal accuracy. The expectation-maximization algorithm is considered in [5] in order to estimate missing test measurements. In [6], the authors consider an inverse problem focused on the inference of the power dissipation based on transient temperature maps using Markov random fields. Another temperature-based characterization of power is developed in [7] wherein a genetic algorithm is employed for the reconstruction of the power model. It should be noted that the procedures in [4, 5] operate on direct measurements, meaning that the output is the same quantity as the one being measured. In particular, [4, 5] rely heavily on the availability of adequate test structures on the dies and are practical only for the secondary quantities affected by process variation, such as delays and currents, but not for the primary ones, such as various geometrical properties. Hence, [4, 5]

often lead to excessive costs and have a limited range of application. The approaches [6, 7], on the other hand, concentrating on the power dissipation of a single die, are not concerned with process variation.

Our work makes the following main contribution. We propose a novel approach to the quantification of process variation based on indirect, incomplete, and noisy measurements. Moreover, we develop and implement a solid framework around the proposed idea and perform a thorough study of various aspects of our technique.

II. MOTIVATIONAL EXAMPLE

Let us consider an important application of the proposed technique: the characterization of the distribution, across a silicon wafer, of the effective channel length, denoted by u. The effective channel length has one of the strongest effects on the subthreshold leakage current and, consequently, on power and temperature [8]; at the same time, u is well known to be severely deteriorated by process variation [1, 2]. Assume the technological process imposes a lower bound u_* on u.[1] This bound separates defective dies ($u < u_*$) from those that are acceptable ($u \geq u_*$). In order to reduce costs, the manufacturer is interested in detecting the faulty dies and taking them out of the production process at early stages. Then the possible actions that they might take with respect to a single die on the wafer are: (a) keep the die if it conforms to the specification; (b) recycle the die otherwise. Let the distribution of u across the wafer be the one depicted on the left side of Fig. 1. The gradient from navy to dark red represents the transition of u from low to high values; hence, the navy regions have a high level of the power and heat dissipation.[2]

In order to quantify the uncertainty due to the variability of the effective channel length u, one can find the above-mentioned distribution by removing the top layer of (thus, destroying) the dies and measuring u directly. Alternatively, despite the fact that the knowledge of u is more preferable, one can step back and decide to characterize process variation using some other parameter that can be measured without the need of damaging the dies, e.g., the

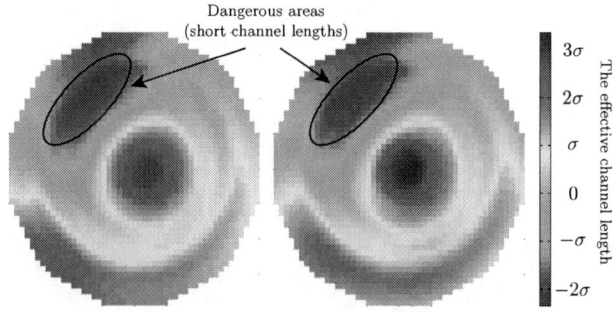

Figure 1. The true (on the left) and inferred (on the right) distributions of the effective channel length across the wafer. The color scheme shows the offset of u from the nominal value where σ stands for the standard deviation of u.

[1] For simplicity, a possible upper bound on the effective channel length is ignored in the motivational example.

[2] The experimental setup will be described in detail in Sec. VI.

978-1-4799-2817-0/14 $31.00 © 2014 IEEE

leakage current. It should be noted that, in this second case, the chosen surrogate is the final product, and u is left unknown. In either way, adequate test structures have to be present on the dies in order to take the corresponding measurements in sufficiently many points and at the desired level of granularity. Such a sophisticated test structure might not always be readily available, and its deployment might significantly increase production costs. Moreover, the first approach implies that the measured dies have to be recycled afterwards, and the second implies that the further design decisions will be based on a surrogate quantity instead of the primary source of uncertainty, which can compromise the reliability of the decisions. The latter concern is particularly urgent in the situations wherein the production process is not yet completely stable and, hence, the design decisions based on the primary subjects of process variation are desirable.

Our technique works differently. In order to characterize the effective channel length u, we monitor an auxiliary quantity q that depends on u and is more advantageous from the measurement perspective. The distribution of u across the whole wafer is then obtained via Bayesian inference [3] applied to the collected measurements of q. These measurements are taken only for a small number of locations on the wafer and can potentially be corrupted by the noise due to the imperfection of the measurement equipments.

Let us consider one particular helper q, which can be used to study the effective channel length u; specifically, let q be temperature (we elaborate further on this choice in Sec. VI). We can then apply a fixed workload (e.g., run the same application under the same conditions) to a few dies on the wafer and measure the corresponding temperature profiles. Since temperature does not require extra equipments to be deployed on the wafer and can be tracked using infrared cameras [7] or built-in facilities of the dies, our approach can reduce the costs associated with the analysis of process variation. The results of our framework applied to a set of noisy temperature profiles measured for only 7% of the dies on the wafer are shown on the right side of Fig. 1, and the locations of the selected dies are depicted in Fig. 2. It can be seen that the two maps in Fig. 1 closely match each other implying that our approach is able to reconstruct the distribution of the effective channel length with a high level of accuracy.

Another feature of the proposed framework is that probabilities of various events, e.g., $\mathbb{P}(u \geq u_*)$, can readily be estimated. This is important since, in reality, the true values are unknown for us (otherwise, we would not need to quantify them), and, therefore, we can rely on our decisions only up to a certain probability. We can then reformulate the decision rule defined earlier as follows: (a) keep the die if $\mathbb{P}(u \geq u_*)$ is larger than a certain threshold; (b) recycle the die otherwise. An illustration of this rule is given in Fig. 3 where the lower bound u_* is set to two standard deviations below the mean value of the effective channel length; the probability threshold of the action (a) is set to 0.9; the crosses mark both the true and inferred defective dies (they coincide); and the gradient from light gray to red corresponds to the inferred probability of a die to be defective. It can be seen that the inference accurately detects faulty regions.

In addition, we can introduce a trade-off action: (c) expose the die to a thorough inspection (e.g., via a test structure) if $\mathbb{P}(u \geq u_*)$ is smaller than the threshold of (a) and is larger than some other threshold, e.g., $0.1 < \mathbb{P}(u \geq u_*) < 0.9$. In this case, we can reduce costs by examining only those dies for which there is neither sufficiently strong evidence of their satisfactory nor unsatisfactory condition. Furthermore, one can take into consideration a so-called utility function, which, for each combination of an outcome of u and a taken action, returns the gain that the decision maker obtains. For example, such a function can favor a rare omission of malfunctioning dies to a frequent inspection of correct dies as the latter might involve much more costs. The optimal decision is given by the action that maximizes the expected utility with respect to both the observed

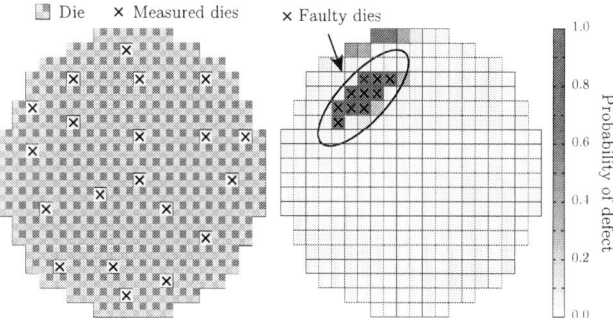

Figure 2. Measurements.   Figure 3. Probability of defect.

data and prior knowledge on u. Thus, all possible u weighted by their probabilities will be taken into account in the final decision, incorporating also the preferences of the user via the utility function.

Finally, we would like to emphasize that temperature is just one option. In certain situations, it might be preferable to perform the above inference based on measurements of some other auxiliary quantity q provided that it depends on the one that we wish to characterize, i.e., on u. For example, q can be the leakage current, which can be readily measured if adequate test structures have already been deployed on the wafer for other purposes.

III. PROBLEM FORMULATION

Consider a generic electronic system, which is fabricated on a silicon wafer hosting n_{d} dies. The system depends on a process parameter u, which we are interested in studying and shall refer to as the quantity of interest (QOI). Due to the presence of process variation, the value of u deviates from the nominal one, and this deviation can be different at different locations on the wafer. The QOI is assumed to be expensive/impractical for direct measurements.

The goal of this work is to develop a statistical framework targeted at the identification of the on-wafer distribution of u with the following properties: (a) low measurement costs; (b) high computational speed; (c) robustness to the measurement noise; (d) ability to accommodate prior knowledge on u; and (e) ability to assess the trustworthiness of the collected data and corresponding predictions.

In order to achieve the established goal, we propose the use of indirect measurements. Specifically, instead of u, we measure an auxiliary parameter q, which we shall refer to as the quantity of measurement (QOM). The observations of q are then processed via Bayesian inference in order to derive the distribution of the QOI, u. The QOM is chosen such that: (a) q is convenient and cheap to be tracked; (b) q depends on u, which is signified by $q = f(u)$; and (c) there is a way to compute q for a given u. The last means that f should be known; however, it does not have to be explicitly given: our framework treats f as a "black box." For example, f can be a piece of code or an output of an adequate simulator.

As the first step, the user of the proposed framework is supposed to harvest a set of observations of q at several locations on the wafer (recall Sec. II). Without loss of generality, we shall adhere to the following convention. One die corresponds to one potential measurement site, and $n'_{\mathrm{d}} \ll n_{\mathrm{d}}$ denotes the number of those sites that have been selected for measurements. Each site comprises n_{p} measurement points, and each point contains n_{t} data instances. For example, in Sec. II, each observation was an $n_{\mathrm{p}} \times n_{\mathrm{t}}$ matrix capturing temperature of n_{p} processing elements for n_{t} moments of time. Denote by $\mathcal{Q} = \{q_i^{\mathrm{msr}}\}_{i=1}^{n'_{\mathrm{d}}}$ the collected data set where $q_i^{\mathrm{msr}} \in \mathbb{R}^{n_{\mathrm{p}} \times n_{\mathrm{t}}}$ stands for one observation (one site) of the QOM. It is implied that the placement of each selected site is recorded along with $\mathcal{Q}$.

Note that, if f is the identity function, i.e., $q \equiv u$, the proposed technique will primarily focus on the reconstruction of any missing

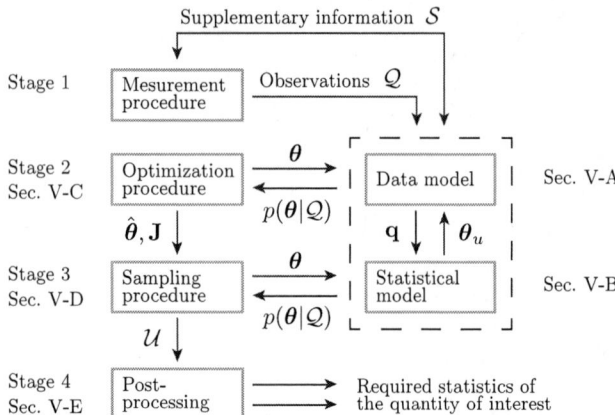

Figure 4. The proposed framework.

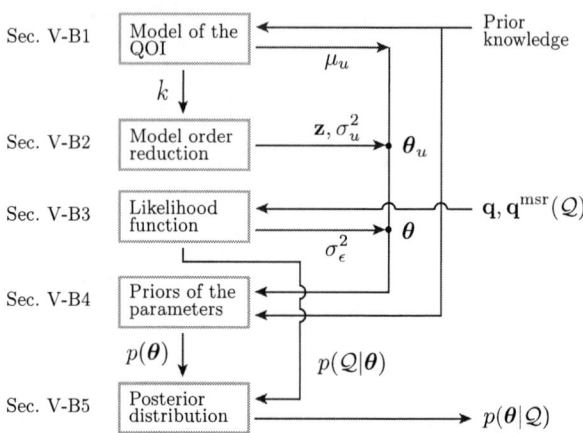

Figure 5. The statistical model.

observations (defined in Sec. V-B2) in $\mathcal{Q}$. From this standpoint, our approach is a generalization of those developed in [4, 5].

For convenience, we denote by $\mathcal{S}$ all the information relevant to the production and measurement processes including: (a) the layout of the wafer and (b) the floorplan of a die on the wafer.

IV. PRELIMINARIES

In order to give a clear presentation of the proposed technique, we first overview the basics of Bayesian inference [3]. Let θ be a set of unknown parameters (in our case, related to, e.g., the effective channel length), which we would like to characterize. Our arsenal to solve the problem includes: (a) a set of observations $\mathcal{Q}$ (in our case, e.g., temperature or current); (b) a data model connecting θ with $\mathcal{Q}$; and (c) prior beliefs on θ. A natural solution is Bayes' rule:

$$p(\theta|\mathcal{Q}) \propto p(\mathcal{Q}|\theta)\, p(\theta) \tag{1}$$

where $p(\cdot)$ denotes a probability density function. $p(\mathcal{Q}|\theta)$ is known as the likelihood function, which accommodates the data model and yields the probability of observing the data $\mathcal{Q}$ given the parameters θ. $p(\theta)$ is called the prior of θ, which represents our knowledge on θ prior to any observations. $p(\theta|\mathcal{Q})$ reads as the posterior of θ given $\mathcal{Q}$. Such a posterior is an exhaustive solution to our problem: having constructed $p(\theta|\mathcal{Q})$, all the needed characteristics of θ can be trivially estimated by drawing samples from this posterior.

Unfortunately, the posterior distribution often does not belong to any of the common families of probability distributions, which is primarily due to the data model involved in the likelihood function, and, therefore, the sampling procedure is not straightforward. To tackle the difficulty, one usually relies on such techniques as Markov Chain Monte Carlo (MCMC) sampling [3]. In this case, an ergodic Markov chain with the stationary distribution equal to the target posterior distribution is constructed and then utilized for the probability space exploration. A popular instantiation of MCMC sampling is the Metropolis-Hastings (MH) algorithm wherein such a Markov chain is attained via sampling from an auxiliary, computationally convenient distribution known as the proposal distribution. We shall further elaborate on this algorithm in Sec. V-B–Sec. V-D.

V. PROPOSED FRAMEWORK

In this section, we present our statistical framework for the characterization of process variation. The technique is divided into four major stages depicted in Fig. 4. Stage 1 is the data-harvesting stage wherein the user collects a set of observations of the QOM, q, forming the input set $\mathcal{Q}$. At Stage 2, we undertake an optimization procedure, which assists MCMC sampling at Stage 3 in the construction of an efficient proposal distribution. Stage 3 produces a collection of

samples of the QOI, u, such as the effective channel length, which is then processed at Stage 4 in order to estimate all the needed characteristics with respect to this QOI, e.g., the probability of the effective channel length to be smaller than a certain threshold as motivated in Sec. II. As it can be seen in Fig. 4, Stage 2 and Stage 3 actively communicate with the two models on the right, called the data and statistical models, which we discuss next.

A. Data Model

The data model is essentially a directed relation between the QOI, u, and the QOM, q, which we denote by the "black-box" transformation $q = f(u)$. f depends on the choice of q and is specified by the user according to the guidelines in Sec. III.

The data model is utilized to predict the values of the QOM at the same sites, at the same inner points, and with the same amount as the ones in $\mathcal{Q}$. The resulting data are then stacked into one vector with $n'_d n_p n_t$ elements (see Sec. III), which is denoted by $\mathbf{q}$. We also let $\mathbf{q}^{\mathrm{msr}} \in \mathbb{R}^{n'_d n_p n_t}$ be a stacked version of the data in $\mathcal{Q}$ such that the respective elements of $\mathbf{q}$ and $\mathbf{q}^{\mathrm{msr}}$ correspond to the same locations.

In order to acquire a better understanding of the data model, let us return to the setup considered in Sec. II. In this case, u stands for the effective channel length, and q stands for the temperature profile corresponding to a fixed workload. The data model $q = f(u)$ can be roughly divided into two transitions: (a) the effective channel length u to the leakage power p_{leak} and (b) the leakage power p_{leak} to the corresponding temperature profile q. The first transition is accomplished using one of the leakage models broadly available in the contemporary literature; see, e.g., [1, 2, 8]. In particular, a leakage model can be constructed via a fitting procedure applied to a data set of SPICE simulations of reference electrical circuits. The only requirement to such a model is that it should be parametrized by u. In addition, it can also be parametrized by temperature in order to account for the well-known interdependency between leakage and temperature. The second transition is undertaken by combining the leakage power p_{leak} with the dynamic power p_{dyn} that corresponds to the considered workload. The obtained total power along with the temperature-related information contained in $\mathcal{S}$ (mainly, the floorplan and thermal parameters of the die) are fed to a thermal simulator (see Sec. VI) in order to acquire the corresponding temperature q.

B. Statistical Model

Once the wafer has been fabricated, the values of u are fixed for all locations on the wafer; however, they remain unknown for us. In order to infer them, we employ the procedure, called the statistical model, developed in the current subsection and displayed in Fig. 5. The development consists of the five components described below.

1) Model of the QOI: The first step is to assign an adequate model to the unknown u. We model u as a Gaussian process [9] since: (a) it is flexible in capturing the correlation patterns induced by the manufacturing process; (b) it is computationally efficient; and (c) Gaussian distributions are often natural and accurate models for uncertainties due to process variation [2, 5, 8]. Thus, we have

$$u|\boldsymbol{\theta}_u \sim \mathcal{GP}\left(\mu, k\right) \qquad (2)$$

where $\mu(r)$ and $k(r,r')$ are the mean and covariance functions of u, respectively, and $r, r' \in \mathbb{R}^2$ denote coordinates on the wafer. Hereafter, the vertical bar, pronounced as "given," is used to mark the parameters that the probability distribution on the right-hand side depends on. In this case, such parameters are $\boldsymbol{\theta}_u$, which we shall identify later on. Prior to taking any measurements, u is assumed to be spatially unbiased; therefore, we let μ be a single location-independent parameter μ_u, i.e., $\mu(r) = \mu_u, \forall r \in \mathbb{R}^2$. The covariance function k is chosen to be the following composition:

$$k(r,r') = \sigma_u^2 \left(\eta\, k_{\mathrm{SE}}(r,r') + (1-\eta)k_{\mathrm{OU}}(r,r')\right) \qquad (3)$$

where

$$k_{\mathrm{SE}}(r,r') = \exp\left(-\frac{\|r - r'\|^2}{\ell_{\mathrm{SE}}^2}\right) \text{ and}$$

$$k_{\mathrm{OU}}(r,r') = \exp\left(-\frac{|\,\|r\| - \|r'\|\,|}{\ell_{\mathrm{OU}}}\right)$$

are the squared exponential and Ornstein-Uhlenbeck correlation functions [9], respectively; σ_u^2 represents the variance of u; $\eta \in [0, 1]$ is a weighting coefficient; ℓ_{SE} and $\ell_{\mathrm{OU}} > 0$ are so-called length-scale parameters; and $\|\cdot\|$ stands for the Euclidean distance. The choice of the covariance function k is guided by the observations of the correlation structures induced by the fabrication process [1, 10]: k_{SE} imposes similarities between the points on the wafer that are close to each other, and k_{OU} imposes similarities between the points that are at the same distance from the center of the wafer. ℓ_{SE} and ℓ_{OU} control the extend of these similarities, i.e., the range wherein the influence of one point on another is significant. Although all the above parameters of the model of u can be inferred from the data, for simplicity, we shall focus on μ_u and σ_u^2. The rest of the parameters, namely, η, ℓ_{SE}, and ℓ_{OU}, are assumed to be determined prior to our analysis based on the knowledge of the correlation patterns typical for the production process utilized (see [11] and references therein).

We have established a model for u given as a stochastic process. Now the model requires one additional treatment in order to make it computationally tractable, which we shall discuss next.

2) Model order reduction: The model of the QOI is an infinite-dimensional object as it characterizes a continuum of locations. For practical computations, however, it should be reduced to a finite-dimensional one. First, u is discretized with respect to the union of two sets of points: the first one is composed of the $n_{\mathrm{d}}'n_{\mathrm{p}}$ points where the observations in $\mathcal{Q}$ were made (n_{d}' selected sites with n_{p} inner locations each), and the other of the points where the user wishes to characterize u. For simplicity, we assume that the user is interested in all the sites, which is $n_{\mathrm{d}}n_{\mathrm{p}}$ points in total. Thus, we obtain an $n_{\mathrm{d}}n_{\mathrm{p}}$-dimensional representation of u denoted by $\mathbf{u} \in \mathbb{R}^{n_{\mathrm{d}}n_{\mathrm{p}}}$. Second, the dimensionality is reduced even further by applying the well-known principal component analysis to the covariance matrix of $\mathbf{u}$ computed via Eq. (3). More precisely, we factorize this matrix using the eigenvalue decomposition [12] and discard those eigenvalues (and their eigenvectors) whose contribution to the total sum of the eigenvalues is below a certain threshold. The result is

$$\mathbf{u} = \mu_u \mathbf{e} + \sigma_u \mathbf{L}\mathbf{z} \qquad (4)$$

where $\mathbf{e} = (e_i = 1) \in \mathbb{R}^{n_{\mathrm{d}}n_{\mathrm{p}}}$, $\mathbf{L} \in \mathbb{R}^{n_{\mathrm{d}}n_{\mathrm{p}} \times n_{\mathrm{v}}}$, and $\mathbf{z} = (z_i) \in \mathbb{R}^{n_{\mathrm{v}}}$ obey the standard Gaussian distribution. n_{v} is the final dimensionality

of the model of u; typically, $n_{\mathrm{v}} \ll n_{\mathrm{d}}n_{\mathrm{p}}$. Consequently, the QOI is now ready for practical computations. In what follows, the parameters of Eq. (2) are defined by $\boldsymbol{\theta}_u = \{\mathbf{z}, \mu_u, \sigma_u^2\}$ (see Fig. 5).

3) Likelihood function: In a Bayesian context, the observed information is taken into account via a likelihood function (see Sec. IV). In our case, the observed information is the measurements $\mathcal{Q}$ stacked into $\mathbf{q}^{\mathrm{msr}}$ as described in Sec. V-A. Since the measurement process is not perfect, we should also take into consideration the measurement noise. To this end, for a given u, the observed $\mathbf{q}^{\mathrm{msr}}$ is assumed to deviate from the data model prediction $\mathbf{q}$ as follows:

$$\mathbf{q}^{\mathrm{msr}} = \mathbf{q} + \boldsymbol{\epsilon} \qquad (5)$$

where $\boldsymbol{\epsilon}$ is an $n_{\mathrm{d}}'n_{\mathrm{p}}n_{\mathrm{t}}$-dimensional vector of noise, which is typically assumed to be a white Gaussian noise [9, 11]. Without loss of generality, the noise is assumed to be independent of u and to have the same magnitude for all measurements (characterized by the utilized instruments). Hence, the model of the noise is

$$\boldsymbol{\epsilon}|\sigma_\epsilon^2 \sim \mathcal{N}\left(\mathbf{0}, \sigma_\epsilon^2 \mathbf{I}\right) \qquad (6)$$

where σ_ϵ^2 is the variance of the noise, $\mathbf{0}$ is a vector of zeros, and $\mathbf{I}$ is the identity matrix. Let us denote the parameters of the inference by $\boldsymbol{\theta} = \boldsymbol{\theta}_u \cup \{\sigma_\epsilon^2\} = \{\mathbf{z}, \mu_u, \sigma_u^2, \sigma_\epsilon^2\}$ (observe this union in Fig. 5). Finally, combining Eq. (5) and Eq. (6), we obtain

$$\mathbf{q}^{\mathrm{msr}}|\boldsymbol{\theta} \sim \mathcal{N}\left(\mathbf{q}, \sigma_\epsilon^2 \mathbf{I}\right). \qquad (7)$$

The probability density function of this distribution is the likelihood function $p(\mathcal{Q}|\boldsymbol{\theta})$ of our statistical model, which is the first of the two components needed for the posterior given in Eq. (1).

4) Priors of the parameters: The second component of the posterior in Eq. (1) is the prior $p(\boldsymbol{\theta})$, which we now need to decide on. In this paper, we put the following priors on $\boldsymbol{\theta}$:

$$\mathbf{z} \sim \mathcal{N}\left(\mathbf{0}, \mathbf{I}\right), \qquad (8)$$

$$\mu_u \sim \mathcal{N}\left(\mu_0, \sigma_0^2\right), \qquad (9)$$

$$\sigma_u^2 \sim \text{Scale-inv-}\chi^2\left(\nu_u, \tau_u^2\right), \text{ and} \qquad (10)$$

$$\sigma_\epsilon^2 \sim \text{Scale-inv-}\chi^2\left(\nu_\epsilon, \tau_\epsilon^2\right). \qquad (11)$$

The prior for $\mathbf{z}$ is due to the properties of the decomposition in Eq. (4). The next three priors, i.e., a Gaussian and two scaled inverse chi-squared distributions, are a common choice for a Gaussian model with the mean and variance being unknown. The parameters μ_0, τ_u^2, and τ_ϵ^2 represent the presumable values of μ_u, σ_u^2, and σ_ϵ^2, respectively, and are set by the user based on the prior knowledge of the technological process and measurement instruments employed. The parameters σ_0, ν_u, and ν_ϵ reflect the precision of this prior information. When the prior knowledge is weak, non-informative priors can be utilized [3]. Taking the product of the densities in Eq. (8)–Eq. (11), we obtain the prior $p(\boldsymbol{\theta})$ completing Eq. (1).

5) Posterior: At this point, we have obtained the two pieces of the posterior shown in Eq. (1): the likelihood function, which is the density in Eq. (7), and the prior, which is the product of the four densities in Eq. (8)–Eq. (11). Thus, the posterior is

$$p(\boldsymbol{\theta}|\mathcal{Q}) \propto p(\mathbf{q}^{\mathrm{msr}}|\mathbf{z}, \mu_u, \sigma_u^2, \sigma_\epsilon^2)\, p(\mathbf{z})\, p(\mu_u)\, p(\sigma_u^2)\, p(\sigma_\epsilon^2). \qquad (12)$$

Provided that we have a way of drawing samples from Eq. (12), the QOI can be readily analyzed as we shall see in Sec. V-E. The problem, however, is that the direct sampling of the posterior is not possible due to the data model involved in the likelihood function via $\mathbf{q}$ (see Eq. (7) and Sec. V-A). In order to circumvent this problem, we utilize the Metropolis-Hastings (MH) algorithm [3] mentioned in Sec. IV. The algorithm operates on an auxiliary distribution called the proposal distribution, which is chosen to be convenient for sampling. Each sample, drawn from this proposal, is then used in Eq. (12) to evaluate the posterior probability of this sample and decide whether it should be accepted or rejected.[3] The acceptance/rejection strategy of the MH algorithm pushes the produced chain of samples towards regions of high posterior probability, which, after a sufficient number of steps, depending on the starting point of the chain and the efficiency of the moves, results in a good approximation of the target posterior distribution in Eq. (12). The preliminary computations needed for the proposal construction are discussed next, and the subsequent sampling procedure in Sec. V-D.

C. Optimization of the Proposal Distribution

In this section, we describe the objective of Stage 2 in Fig. 4. Although the requirements to the proposal distribution mentioned earlier are rather weak, it is often difficult to pick an efficient proposal, which would yield a good approximation with as few evaluations of the posterior in Eq. (12) and, thus, of the data model in Sec. V-A as possible. This choice is especially severe for high-dimensional problems, and our problem, involving around 30 parameters as we shall see in Sec. VI, is one them. Therefore, a careful construction of the proposal distribution is an essential component of our framework.[4] A common technique to construct a high-quality proposal is to perform an optimization of the posterior given by Eq. (12). More specifically, we seek for such a value $\hat{\boldsymbol{\theta}}$ of $\boldsymbol{\theta}$ that maximizes Eq. (12) and, hence, has the maximal posterior probability. We also compute the negative of the Hessian matrix at $\hat{\boldsymbol{\theta}}$, which is called the observed information matrix and denoted by $\mathbf{J}$ (see the output of Stage 2 in Fig. 4). Using $\hat{\boldsymbol{\theta}}$ and $\mathbf{J}$, we can now construct such a proposal, which will allow the MH algorithm (a) to start producing samples directly from the desired regions of high probability and (b) to explore those regions more rapidly.

D. Sampling via the Metropolis-Hastings Algorithm

Let us turn to Stage 3 in Fig. 4. We have at our disposal $\hat{\boldsymbol{\theta}}$ and $\mathbf{J}$ from Stage 2 in order to construct an adequate proposal and utilize it for sampling. A commonly used proposal is a multivariate Gaussian distribution wherein the mean is the current location of the chain of samples started at $\hat{\boldsymbol{\theta}}$, and the covariance matrix is the inverse of $\mathbf{J}$ [3]. In order to speed up the sampling process, we would like to make use of the potential of multicore parallelization. The above proposal, however, is purely sequential as the mean for the next sample draw is dependent on the previous sample. Therefore, we appeal to a variation of the MH algorithm known as the independence sampler [3]. In this case, a typical choice of the proposal is a multivariate t-distribution, independent of the current position of the chain:

$$\boldsymbol{\theta} \sim t_\nu\left(\hat{\boldsymbol{\theta}}, \alpha^2 \mathbf{J}^{-1}\right) \qquad (13)$$

where $\hat{\boldsymbol{\theta}}$ and $\mathbf{J}$ are as in Sec. V-C, ν is the number of degrees of freedom, and α is a tuning constant controlling the standard deviation of the proposal. Now the proposal samples and the time-consuming evaluation of their posterior in Eq. (12) can be computed for all samples in parallel. Then the precomputed samples can subsequently be accepted or rejected as in the usual MH algorithm.

Having completed the sampling procedure, we obtain a collection of samples of $\boldsymbol{\theta}$. The first portion of the drawn samples is typically discarded before the final computations as being unrepresentative; this portion is also known as the burn-in period. Each of the preserved samples of $\boldsymbol{\theta}$, comprising $\mathbf{z}$, μ_u, and σ_u^2, is then used in Eq. (4) to compute a sample of u, $\mathbf{u}_i \in \mathbb{R}^{n_d n_p}$. Denote such a data set with n_{mc} samples of the QOI by $\mathcal{U} = \{\mathbf{u}_i\}_{i=1}^{n_{\mathrm{mc}}}$.

E. Post-processing

At Stage 4 in Fig. 4, using the set of samples $\mathcal{U}$, the user computes the desired statistics of the QOI such as the most probable value of the effective channel length at some location of interest, the probability of a certain area on the wafer to be defective, etc. The computations boil down to the estimation of expected values with respect to the posterior distribution of $\boldsymbol{\theta}$, $p(\boldsymbol{\theta}|\mathcal{Q})$. This estimation is done in the standard sample-based fashion, that is, in order to compute some arbitrary quantity dependent on u, one needs to evaluate this quantity for each $\mathbf{u}_i$ in $\mathcal{U}$ and then take the average.

The strength of the Bayesian approach to inference really starts to shine when we are also interested in assessing the trustworthiness of the measured data and, therefore, the reliability of the estimates/decisions based on these data. Such an assessment can readily be undertaken using our framework since the delivered posterior distribution contains all the needed information about the QOI. This is especially helpful in decision making as exemplified in Sec. II.

VI. EXPERIMENTAL RESULTS

In this section, we assess our framework using the inference of the effective channel length u based on temperature q. This choice for illustration is dictated by the fact that such a high-level parameter as temperature constitutes a challenging task for the inference of such a low-level parameter as the effective channel length, which implies a strong assessment of the proposed technique. On the other hand, the effective channel length is an important target *per se* as it is strongly affected by process variation and considerably impacts the power/heat dissipation [1, 2, 8]; in particular, it also influences other process-related characteristics such as the threshold voltage. The performance of our approach is expected to only increase when the auxiliary parameter q resides "closer" to the target parameter u with respect to the transformation $q = f(u)$. For instance, such a "closer" quantity q can be the leakage current, which, however, might not always be the most preferable parameter to measure.

Now we shall describe the default configuration of our setup, which will be later adjusted according to the purpose of each particular experiment. We consider a 45-nanometer technological process. The diameter of the wafer is 20 dies, and the total number of dies n_d is 316. The number of measured dies n_d' is 20, and these dies are chosen by an algorithm, which pursues an even coverage of the wafer. The number of processing elements in each die is four, and they are the points of taking measurements, i.e., $n_p = 4$. The floorplans of the multiprocessor platforms are constructed in such a way that the processing elements form regular grids. The dynamic power profiles involved in the experiments are based on simulations of randomly generated task graphs via TGFF v3.5 [13]. The sampling interval of these profiles is 1 ms. The leakage model, parametrized by temperature and the effective channel length, is constructed by fitting to SPICE simulations of reference electrical circuits composed of BSIM4 v4.7 devices [14] configured according to the 45-nm PTM HP model [15]. The temperature calculations are undertaken using the approach described in [16], based on HotSpot v5.02 [17].[5] The input data set $\mathcal{Q}$ is obtained as follows: (a) draw a sample of u from a Gaussian distribution with the mean value equal to 17.5 nm, according to the considered technological process [15], and the covariance function given by Eq. (3) wherein the standard deviation is 2.25 nm; (b) perform one fine-grained temperature simulation per each of the n_d' selected dies under the corresponding dynamic power profile; (c) shrink the temperature profiles to keep only n_t, which is equal to 20 by default, evenly spaced moments of time; and (d) perturb the obtained data set using a white Gaussian noise with the standard deviation of 1 K (Kelvin).

---

[3] A reject means that the sequence of samples advances using the last accepted sample; therefore, the chain of samples is never interrupted.

[4] This has been also confirmed by our experiments. Without optimization, even for small examples, no adequate results were obtained in an affordable time. Therefore, all the experiments in Sec. VI include the optimization step.

[5] The floorplans of the platforms, task graphs of the applications, thermal configuration of HotSpot, etc. are available online at [18].

978-1-4799-2817-0/14 $31.00 © 2014 IEEE

Table I — MEASURED SITES n'_d							Table II — MEASURED POINTS PER SITE n_p					Table III — DATA AMOUNT PER POINT n_t						Table IV — NOISE DEVIATION σ_ϵ			
*	1	10	20	40	80	160	2	4	8	16	32	1	10	20	40	80	160	0 K	0.5 K	1 K	2 K
OT, m	0.41	2.49	3.34	4.59	7.33	10.29	2.67	3.34	5.20	7.37	13.85	1.12	3.02	3.34	3.62	3.64	4.20	5.08	3.73	3.34	3.19
ST, m	**2.40**	**3.99**	**4.60**	**5.79**	**8.49**	**12.96**	**3.71**	**4.60**	**6.03**	**8.92**	**14.77**	**2.40**	**4.38**	**4.60**	**4.67**	**4.80**	**4.97**	**4.76**	**4.70**	**4.60**	**4.71**
TT, m	2.81	6.47	7.94	10.38	15.81	23.25	6.38	7.94	11.23	16.29	28.62	3.52	7.40	7.94	8.29	8.44	9.16	9.84	8.43	7.94	7.90
PT, m	0.61	1.02	1.18	1.51	2.16	3.62	0.98	1.18	1.58	2.51	5.30	0.62	1.13	1.18	1.22	1.25	1.30	1.19	1.17	1.18	1.18
TT, m	1.02	3.50	4.52	6.10	9.49	13.91	3.65	4.52	6.78	9.88	19.15	1.74	4.16	4.52	4.84	4.89	5.50	6.27	4.91	4.52	4.37
E, %	30.49	4.40	3.42	1.09	0.85	0.67	4.71	3.42	3.68	2.73	1.94	7.48	2.72	3.42	1.83	2.34	1.32	0.02	2.71	3.42	4.05

* OT — optimization time, ST — sequential sampling time, PT — parallel sampling time, TT — total time (optimization plus sampling), and E — NRMSE.

Let us turn to the statistical model in Sec. V-B and summarize the intuition and our assignment for each parameter of this model. In the covariance function given by Eq. (3), the weight parameter η and the two length-scale parameters ℓ_SE and ℓ_OU should be set according to the correlation patterns typical for the production process at hand [1, 10]; we set η to 0.7 and ℓ_SE and ℓ_OU to half the radius of the wafer. The threshold parameter of the model order reduction procedure described in Sec. V-B2 and utilized in Eq. (4) should be set high enough to preserve a sufficiently large portion of the variance of the data and, thus, to keep the corresponding results accurate; we set it to 0.99 preserving 99% of this variance. The resulting dimensionality n_v of $\mathbf{z}$ in Eq. (4) was found to be 27–28. The parameters μ_0 and τ_u of the priors in Eq. (9) and Eq. (10), respectively, are specific to the considered technological process; we set μ_0 to 17.5 nm and τ_u to 2.25 nm. The parameters σ_0 and ν_u in Eq. (9) and Eq. (10), respectively, determine the precision of the information on μ_0 and τ_u and are set according to the beliefs of the user; we set σ_0 to 0.45 nm and ν_u to 10. The latter can be thought of as the number of imaginary observations that the choice of τ_u is based on. The parameter τ_ϵ in Eq. (11) represents the precision (deviation) of the equipments utilized to collect the data set $\mathcal{Q}$ and can be found in the technical specification of these equipments; we set τ_ϵ to 1 K. The parameter ν_ϵ in Eq. (11) has the same interpretation as ν_u in Eq. (10); we set it to 10 as well. In Eq. (13), ν and α are tuning parameters, which can be configured based on experiments; we set ν to eight and α to 0.5. The number of sample draws is another tuning parameter, which we set to 10^4; the first half of these samples is ascribed to the burn-in period leaving $5 \cdot 10^3$ effective samples n_mc. For the optimization in Sec. V-C, we use the Quasi-Newton algorithm [12]. For parallel computations, we utilize four processors. All the experiments are conducted on a GNU/Linux machine with Intel Core i7 2.66 GHz and 8 GB of RAM.

To ensure that the experimental setup is adequate, we first perform a detailed inspection of the results obtained for one particular example with the default configuration. The true and inferred distributions of the QOI are shown in Fig. 1 where the normalized root-mean-square error (NRMSE) is below 2.8%, and the absolute error is bounded by 1.4 nm, which suggests that the framework produces a close match to the true value of the QOI. We have also looked at the behavior of the constructed Markov chains and the quality of the proposal distribution; however, due to the shortage of space, these results are not presented here. All the observations suggest that the optimization and sampling procedures are properly configured.

Next we use the assessed configuration and alter only one parameter at a time: the number of measured sites/dies n'_d; the number of processing elements/measured points n_p on a site; the amount of data per measurement point n_t; and the noise deviation σ_ϵ.

A. Number of Measured Sites

Let us change the number of dies n'_d that have been measured. The considered scenarios are 1, 10, 20, 40, 80, and 160 measured dies, respectively. The results are reported in Tab. I. In this and the following tables, we report the optimization (Stage 2 in Fig. 4) and sampling (Stage 3 in Fig. 4) times separately (given in minutes). In addition, the sampling time is given for two cases: sequential and parallel computing, which is followed by the total time and error (NRMSE). The computational time of the post-processing phase (Stage 4 in Fig. 4) is not given as it is negligibly small. The sequential sampling time is the most representative indicator of the computational complexity scaling as the number of samples is always fixed, and there is no parallelization; thus, we shall watch this value in most of the discussions below (highlighted in bold).

We see in Tab. I that the more data the proposed framework needs to process, the longer the execution times, which is reasonable. The trend, however, is rather modest: with the doubling of n'_d, all the computational times increase less than two times. The error firmly decreases and drops below 4% with around 20 sites measured, which is only 6.3% of the total number of dies on the wafer.

B. Number of Measured Points Per Site

Here we consider five platforms with the number of processing elements/measurement points n_p on each die equal to 2, 4, 8, 16, and 32, respectively. The results are summarized in Tab. II. All the computational times grow with n_p. This behavior is expected as the granularity of the utilized thermal model (see Sec. V-A and [16]) is bound to the number of processing elements; therefore, the temperature simulations become more intensive. Nevertheless, even for large examples, the timing is readily acceptable, taking into account the complexity of the inference procedure behind and the yielded accuracy. An interesting observation can be made from the NRMSE: the error tends to decrease as n_p grows. The explanation is that, with each processing element, $\mathcal{Q}$ delivers more information to the inference to work with since the temperature profiles are collected for all the processing elements simultaneously.

C. Amount of Data Per Measured Point

In this subsection, we sweep the number of moments of time n_t captured by the measured temperature profiles. The scenarios are 1, 10, 20, 40, 80, and 160 time moments, respectively. The results are aggregated in Tab. III. As we see, the growth of the computational time is relatively small. One might have expected this growth due to n_t to be the same as the one due to n_p since, formally, the influence of n_p and n_t on the dimensionality of $\mathcal{Q}$ is identical (recall $\mathbf{q}^\mathrm{msr} \in \mathbb{R}^{n'_\mathrm{d} n_\mathrm{p} n_\mathrm{t}}$). However, the meaning of the two numbers, n_p and n_t, is completely different, and, therefore, the way they manifest themselves in the algorithm is also different. Therefore, the corresponding amounts of extra data are being treated differently leading to the discordant timing shown in Tab. II and Tab. III. The NRMSE in Tab. III has a decreasing trend; however, this trend is less steady than the ones discovered before. The finding can be explained as follows. The distribution of the time moments in $\mathcal{Q}$ changes since these moments are kept evenly spaced across the corresponding time spans of the input power profiles. Some moments of time can be more informative than the other. Hence, more or less representative

978-1-4799-2817-0/14 $31.00 © 2014 IEEE

samples can end up in $\mathcal{Q}$ helping or misleading the inference. We can also conclude that a larger number of spatial measurements is more advantageous than a larger number of temporal measurements.

D. Deviation of the Measurement Noise

Next we vary the standard deviation of the noise (in Kelvins), affecting the data $\mathcal{Q}$, within the set $\{0, 0.5, 1, 2\}$ coherent with the literature [7]. Note that the corresponding prior distribution in Eq. (11) is kept unchanged. The results are given in Tab. IV. The sampling time is approximately constant. However, we observe an increase of the optimization time with the decrease of the noise level, which can be ascribed to wider possibilities of perfection for the optimization procedure. A more important observation, revealed by this experiment, is that, in spite of the fact that the inference operates on indirect and drastically incomplete data, a thoroughly calibrated equipment can considerably improve the quality of predictions. However, even with a high level of noise of two degrees—meaning that measurements are dispersed over a wide band of 8 K with a large probability of more than 0.95—the NRMSE is still only 4%.

E. Sequential vs. Parallel Sampling

Let us summarize the results of the sequential and parallel sampling strategies. In the sequential MH algorithm, the optimization time is typically smaller than the time needed for drawing posterior samples. The situation changes when parallel computing is utilized. With four parallel processors, the sampling time decreases 3.81 times on average, which indicates good parallelization properties of the chosen sampling strategy. The overall speedup ranges from 1.49 to 2.75 with the average value of 1.77 times, which can be pushed even further employing more parallel processors.

VII. Conclusion

We proposed a framework for the analysis of process variation across semiconductor wafers based on cost-efficient, indirect measurements. The technique was exposed to an extensive study of various aspects concerning its implementation. The obtained results support the computational efficiency and accuracy of our approach.

We would like to note that, although the framework was demonstrated on the effective channel length and temperature, it can be readily utilized to analyze any other QOIs based on any other QOMs.

References

[1] A. Chandrakasan, F. Fox, W. Bowhill, and W. Bowhill. *Design of High-performance Microprocessor Circuits*. IEEE Press, 2001.

[2] A. Srivastava, D. Sylvester, and D. Blaauw. *Statistical Analysis and Optimization for VLSI: Timing and Power*. Springer, 2010.

[3] A. Gelman, J. Carlin, H. Stern, and D. Rubin. *Bayesian Data Analysis*. Chapman&Hall/CRC, 2004.

[4] W. Zhang, X. Li, and R. Rutenbar. "Bayesian Virtual Probe: Minimizing Variation Characterization Cost for Nanoscale IC Technologies via Bayesian Inference". In: *DAC*. 2010, pp. 262–267.

[5] S. Reda and S. R. Nassif. "Analyzing the impact of process variations on parametric measurements: novel models and applications". In: *DATE*. 2009, pp. 375–380.

[6] S. Paek, S.-H. Moon, W. Shin, J. Sim, and L.-S. Kim. "PowerField: A Transient Temperature-to-power Technique Based on MRF Theory". In: *DAC*. 2012, pp. 630–635.

[7] F. Mesa-Martinez, J. Nayfach-Battilana, and J. Renau. "Power Model Validation Through Thermal Measurements". In: *ISCA* (2007), pp. 302–311.

[8] D.-C. Juan, Y.-L. Chuang, D. Marculescu, and Y.-W. Chang. "Statistical Thermal Modeling and Optimization Considering Leakage Power Variations". In: *DATE*. 2012, pp. 605–610.

[9] C. Rasmussen and C. Williams. *Gaussian Processes for Machine Learning*. The MIT Press, 2006.

[10] L. Cheng, P. Gupta, C. Spanos, K. Qian, and L. He. "Physically Justifiable Die-level Modeling of Spatial Variation in View of Systematic Across Wafer Variability". In: *IEEE Transactions on CAD of ICs and Systems* 30.3 (2011), pp. 388–401.

[11] Y. Marzouk and H. Najm. "Dimensionality Reduction and Polynomial Chaos Acceleration of Bayesian Inference in Inverse Problems". In: *Journal of Computational Physics* 228.6 (2009), pp. 1862–1902.

[12] W. Press, S. Teukolsky, W. Vetterling, and B. Flannery. *Numerical Recipes: The Art of Scientific Computing*. Cambridge University Press, 2007.

[13] R. Dick, D. Rhodes, and W. Wolf. "TGFF: Task Graphs for Free". In: *CODES/CASHE*. 1998, pp. 97–101.

[14] *BSIM4*. Berkeley Short-channel IGFET Model Group at the University of California, Berkeley. URL: http://www-device.eecs.berkeley.edu/bsim/.

[15] *PTM*. Nanoscale Integration and Modeling Group at Arizona State University. URL: http://ptm.asu.edu/.

[16] I. Ukhov, M. Bao, P. Eles, and Z. Peng. "Steady-state Dynamic Temperature Analysis and Reliability Optimization for Embedded Multiprocessor Systems". In: *DAC*. 2012, pp. 197–204.

[17] K. Skadron, M. R. Stan, K. Sankaranarayanan, W. Huang, S. Velusamy, and D. Tarjan. "Temperature-aware Microarchitecture: Modeling and Implementation". In: *ACM Transactions on Architecture and Code Optimization* 1.1 (Mar. 2004), pp. 94–125.

[18] Embedded Systems Laboratory, Linköping University. URL: http://www.ida.liu.se/~ivauk83/research/SAPV.

Symbolic Computation of SNR for Variational Analysis of Sigma-Delta Modulator*

Jiandong Cheng and Guoyong Shi
School of Microelectronics, Shanghai Jiao Tong University,
Shanghai 200240, China
shiguoyong@ic.sjtu.edu.cn

Abstract—Signal-to-noise ratio (SNR) is an important design metric for switched-capacitor sigma-delta modulators (SC-SDMs). In an automatic synthesis environment, fast SNR computation is of paramount importance. So far the main SNR computation method has been behavioral simulation. Other less accurate methods are based on empirical formulas. These methods could not contribute too much to the enhancement of synthesis efficiency. In this work a highly efficient and purely symbolic SNR computation method is proposed. The difficulty in the computation of noise power (requiring integration of a rational function) is overcome by Taylor polynomial approximation. Together with a symbolic loop-transfer analysis tool, the SNR can be computed fully symbolically. This novel computation method is applied to variational SC-SDM analysis. The effectiveness and efficiency are compared to behavioral Monte Carlo simulation results.

Index Terms—Sigma-delta modulator (SDM), statistical analysis, switched-capacitor (SC), symbolic analysis, signal-to-noise ratio (SNR), Taylor approximation.

I. INTRODUCTION

$\Sigma\Delta$ modulator (SDM) is an important category of data converters widely applied in high performance analog signal processing [1]–[4]. Comparing to continuous-time implementation, switched-capacitor (SC) based discrete-time implementation is more popular due to its low fabrication cost. However, design and verification of SC-SDMs is not easy because of its mixed-mode (discrete and continuous-time) and mixed-signal nature. It is well-known that a fully transistor-level SPICE simulation of a practical SC-SDM design is extremely time-consuming. Hence, most of the automatic synthesis strategies proposed in the literature would not suggest fully SPICE-based simulation for the category of sigma-delta data converters.

A complete synthesis cycle would have to go through optimizations ranging from device sizes to modulator topology. Because of the importance of the related subjects, numerous publications have made attempts on studying many difficult problems from various angles [5]–[15].

Recently, the rising concern of process variation has caught the attention of researchers working on mixed-signal design automation [16]. Process variation causes severe variation in the SNR performance of SC-SDMs. From the design reliability perspective, optimizing the SNR performance during the synthesis cycle is highly important. So far only very few publications have addressed the SNR variation issues in the design of SDMs [12], [13], [17], [18].

Tang [17] proposed a symbolic SNR characterization method by calculating the SNR variance based on its first-order sensitivity with respect to capacitance variations. However, since the computation of noise power requires the integration of a noise transfer function (NTF), numerical integration is employed in the work [17].

Empirical approximations of SNR also are used in some publications. For example, extensive simulation is needed to fit parameters in [7] and rough pole-zero approximation is introduced in [19].

In this work a fully symbolic SNR computation strategy is developed with the following two innovations: 1) A symbolic tool is introduced for creating the loop transfer functions of modulators of arbitrary topology; and 2) a Taylor polynomial approximation method is proposed for the *symbolic* calculation of noise power. The new method eliminates the need of repeated numerical integration as required by the method proposed in [17]. Hence, this novel symbolic SNR computation method can be applied as an efficient computation engine for SDM synthesis considering SNR and/or, in particular, SNR variation.

In Section II we first make a preliminary review on the design issues involved with SC-SDM, then outline the steps for generating symbolic loop-transfer function. In Section III we present a Taylor approximate method for symbolically calculating SNR. Based on a symbolically generated SNR, we then present a procedure for calculating the sensitivity of SNR with respect to circuit parameters. SNR sensitivity has many applications. In this work we only present a first-order method for predicting SNR variation in Section IV due to space limit. The accuracy of the proposed Taylor approximation method for SNR computation and the reliability of using the first-order approximation for SNR variation estimation are validated by experiments reported in Section V. A conclusion is drawn in Section VI.

II. SDM DESIGN ISSUES AND SYMBOLIC SOLUTIONS

A. SDM Review

The idea of SDM is to achieve high quantization resolution by employing low-resolution internal ADCs via oversampling and filtering (noise-shaping). The resolution of internal ADC

---

*This research was supported in part by the National Natural Science Foundation of China (NSFC Grant No. 61176129) and by the SJTU-Synopsys Joint Research Project (2010-2013) sponsored by Synopsys, Inc.

978-1-4799-2817-0/14 $31.00 © 2014 IEEE

could be as low as one single bit, i.e. a two-level comparator [1]. SDM designers have to consider the choice of a proper oversampling ratio (OSR), optimal loop-filter design, setting the input dynamic range, together with the circuit defects (such as op-amp imperfection), etc. These interleaved design factors affect the overall system performance in a complicated way.

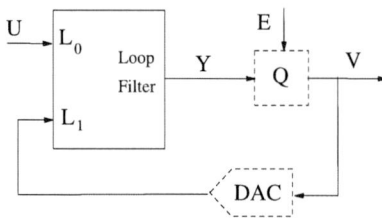

Fig. 1. Single-stage $\Sigma\Delta$ modulator.

Shown in Fig. 1 is the general structure of a single-stage SDM [4] with an analog input u, a digital output v, and a feedback connection that goes through a digital-to-analog converter (DAC). The "Q" block in the feedforward path stands for a quantizer. The quantization effect is modeled by an additive error signal e considered as the quantization noise. Simplified as a linear system, the output can be written in the following z-transformed form [4]

$$V(z) = \text{STF}(z)U(z) + \text{NTF}(z)E(z) \qquad (1)$$

where $\text{STF}(z)$ stands for the *signal transfer function* and $\text{NTF}(z)$ for *the noise transfer function*, defined by

$$\text{STF}(z) = \frac{L_0(z)}{1 + L_1(z)} \quad \text{and} \quad \text{NTF}(z) = \frac{1}{1 + L_1(z)}. \qquad (2)$$

The denominator $1 + L_1(z)$ is due to the negative feedback connection. The two transfer functions are interrelated, hence requiring careful design for both stability and noise shaping [4].

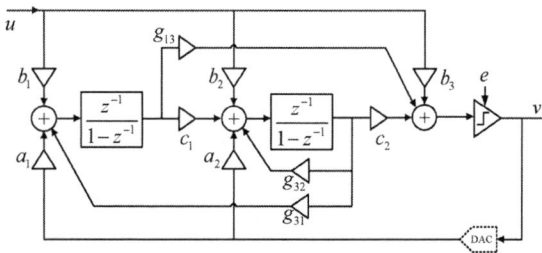

Fig. 2. A second-order SDM.

CMOS switched-capacitor circuits are commonly employed for SDM implementation considering speed, low-cost, and circuit reliability. The discrete-time realization requires the circuit analysis and simulation to be performed in the discrete-time domain. For methodology development, we assume that the SDM transfer functions STF and NTF take the following rational function form in general

$$H(z) = \frac{b_0 + b_1 z + b_2 z^2 + \ldots + b_m z^m}{a_0 + a_1 z + a_2 z^2 + \ldots + a_n z^n} \qquad (3)$$

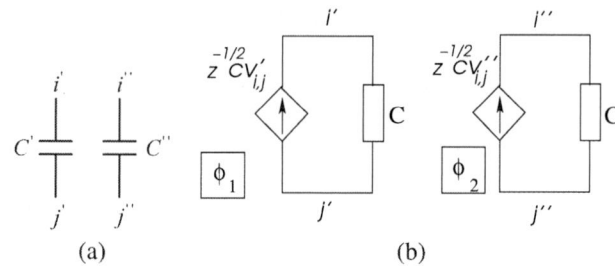

Fig. 3. Switched-capacitor equivalent circuit: (a) Capacitor symbol in two phases. (b) Equivalent circuit representation. It is assumed that $C' = C'' = C$, $V'_{ij} := V'_i - V'_j$, and $V''_{ij} := V''_i - V''_j$.

where z is the z-transform variable. The coefficients appear in the signal flow graph construction as that shown in Fig. 2. They can be realized by switched-capacitor ratios under the assumption that all op-amps in an SDM circuit are ideal.

B. Symbolic Analysis of SC Circuits

Discrete-time switched-capacitor circuits can be analyzed by a symbolic method, for which a variety of methods have been proposed, such as the one presented in [20] and the references therein. Typically, a signal-flow diagram (SFG) method or a nodal-analysis (NA) method is applied. Floberg and Mattisson [20] proposed a formulation to apply a continuous-time NA symbolic tool for analyzing discrete-time SC-circuits. We employ a similar approach in this work, hence explain a little more detail below.

A two-phase SC circuit can be divided into two subcircuits mutually dependent due to the charge transfer between two clock phases. By considering charge as "current" (mapped by time integration), the charge in a capacitor between two successive clock phases can be modeled by a *voltage controlled current source (VCCS)* as shown in Fig. 3. In the same vein, each capacitor in any clock phase is treated as a conductor (because charge has been treated as current). In Fig. 3 the quantities $z^{-1/2}C'V''_{ij}$ and $z^{-1/2}C''V'_{ij}$ (which are supposedly delayed charge quantities) are drawn as the *controlled currents*, where $z^{-1/2}$ indicates the half clock period latency (assuming one clock phase takes half of the clock cycle). This simple one-to-one conversion results in an equivalent continuous-time circuit, whose symbolic analysis can be performed by any existing continuous-time symbolic tool. After symbolic transfer functions are generated, converting them into discrete-time symbolic transfer functions is straightforward.

In this work we employ our continuous-time symbolic GPDD tool developed in our research laboratory [21]. This symbolic tool has the advantage of generating symbolic functions represented by a binary decision diagram (BDD) with easy correspondence between the circuit parameters and the manipulated symbols, rendering great ease in post-processing such as repeated evaluation, recomposition, or sensitivity generation [22]. Further details on the GPDD construction for discrete-time SC circuits can be found in [23].

III. Symbolic SNR Computation

A. SNR Computation and Complexity

SNR is one of the most important SDM design metrics. Let $\text{STF}(z; \xi)$ and $\text{NTF}(z; \xi)$ be parameter-dependent signal and noise transfer function, respectively, where ξ denotes the set of all circuit parameters considered as symbols in analysis. For a single-stage SDM, SNR can be calculated by the following formulas (assuming *sinusoidal* input):

$$\text{SNR}(\xi) := 10 \log \frac{P_S |\text{STF}(e^{j\omega_0}; \xi)|^2}{P_{N,in}(\xi)} \qquad (4)$$

and

$$P_{N,in}(\xi) := \frac{P_N}{\pi} \int_0^{\pi/R} |\text{NTF}(e^{j\omega}; \xi)|^2 d\omega, \qquad (5)$$

where $\omega_0 = 2\pi \frac{f_{in}}{f_s}$, $R = \frac{f_s}{2BW} = \text{OSR}$ (the oversampling ratio), BW is the bandwidth of the input signal, f_{in} and f_s are respectively the input sinusoidal wave frequency and the sampling frequency, P_S and P_N are constants defined by $P_S := A^2/2$ with A the amplitude of the sinusoidal input, and $P_N := \Delta^2/12$ with Δ the step size of the quantizer.

The SNR formula as given above requires integration of $\text{NTF}(z)$, $z = e^{j\omega}$, over the signal band. The SNR can be calculated directly by numerical integration, but it obviously requires the generation of the loop-transfer functions. Without a symbolic tool, repeated generation of such transfer functions is time-consuming in a synthesis loop. Many design tools do not have the facility of transfer function generation. In that case, **SNR is typically computed by time-domain behavioral simulation for the sake of speed**. The whole process requires simulating the time-domain waveforms, sampling and windowing the waveforms, then applying Fourier transform to obtain the SNR [4]. For the sake of accuracy, millions of samples might have to be processed, which is a big overhead. Therefore, the issue of SNR computation stands out as a serious bottleneck in the SDM synthesis flow.

B. Taylor Approximation

Given a rational function $\text{NTF}(z)$ (typically of equal degree in the numerator and the denominator). If the numerical values of the coefficients of $\text{NTF}(z)$ are known, a numerical integration can compute the noise power by (5), which is done in [17]. However, the computation efficiency becomes very low in variational analysis because the numerical integration has to be repeated for a huge number of times. Also, the numerical integration could not provide analytical relationship between SNR and the circuit parameters, whereas such correspondence is highly valuable for automatic SNR synthesis.

We therefore propose to solve this problem by using a *truncated Taylor polynomial* to approximate the rational function appearing in the integral. The coefficients of the Taylor polynomial can be determined by matching the polynomial coefficients.

Let $\text{NTF}(z; \xi)$ be a rational function involving all circuit parameters as the symbols. Expand it into a Taylor series centered at $z = 1$ (the dc point) up to the rth order,

$$\begin{aligned} \text{NTF}(z; \xi) &= \frac{b_0 + b_1 z + b_2 z^2 + \dots + b_n z^n}{a_0 + a_1 z + a_2 z^2 + \dots + a_n z^n} \\ &\approx m_0 + m_1(z - 1) + \dots + m_r(z - 1)^r \\ &= \widehat{\text{NTF}}(z; \xi), \end{aligned} \qquad (6)$$

where $\widehat{NTF}(z; \xi)$ denotes the truncated polynomial whose coefficients again are functions of all circuit parameters. Performing the Taylor expansion at $z = 1$ is for convenience because we have $z = e^{j\omega}$ and later an integration will be carried out from $\omega = 0$ to π/R.

Rewrite the numerator and denominator in polynomials of $(z - 1)^k$; i.e.,

$$\text{NTF}(z; \xi) = \frac{c_0 + c_1(z - 1) + \dots + c_n(z - 1)^n}{d_0 + d_1(z - 1) + \dots + d_n(z - 1)^n}. \qquad (7)$$

The coefficients m_i's of the Taylor polynomial are determined recursively by comparing the coefficients:

$$m_0 = \frac{c_0}{d_0}, \qquad m_i = \frac{c_i - \sum_{k=0}^{i-1} m_k d_{i-k}}{d_0}, \qquad (8)$$

for $i = 1, \cdots, r$, where $c_i = d_i = 0$ if $i > n$. These formulas can be implemented symbolically.

Knowing that $z = e^{j\omega}$, we change the integration variable ω to z and substitute $d\omega$ by $\frac{dz}{jz}$. Define indefinite integral

$$\begin{aligned} \mathcal{I}_N(z; \xi) &:= \int |\text{NTF}(e^{j\omega})|^2 d\omega \\ &= \frac{1}{j} \int \text{NTF}(z; \xi) \frac{NTF(z^{-1}; \xi)}{z} dz \\ &\approx -j \int \hat{g}_1(z - 1; \xi) \hat{g}_2(z - 1; \xi) dz \\ &= -j\hat{G}(z - 1; \xi), \end{aligned} \qquad (9)$$

where $\hat{g}_1(z - 1; \xi)$ is a truncated polynomial in $(z - 1)^k$ for approximating $\text{NTF}(z; \xi)$ and $\hat{g}_2(z - 1; \xi)$ is another truncated polynomial in $(z - 1)^k$ for approximating $\text{NTF}(z^{-1}; \xi)/z$. In the last step $\hat{G}(z - 1; \xi)$ denotes the polynomial in $(z - 1)^k$ after integration. Note that the two rational functions $\text{NTF}(z; \xi)$ and $\text{NTF}(z^{-1}; \xi)/z$ are separately approximated by two polynomials, $\hat{g}_1(z - 1; \xi)$ and $\hat{g}_2(z - 1; \xi)$, because the numerical stability is much better in implementation.

Both $\mathcal{I}_N(z; \xi)$ and $-j\hat{G}(z - 1; \xi)$ must be real functions. We then get

$$\begin{aligned} \mathcal{I}_N(\xi) &:= \int_0^{\pi/R} |\text{NTF}(e^{j\omega})|^2 d\omega = \mathcal{I}_N(q; \xi) - \mathcal{I}_N(1; \xi) \\ &\approx \text{Im}\left[\hat{G}(q - 1; \xi) - \hat{G}(0; \xi)\right], \end{aligned} \qquad (10)$$

where $q = e^{j\pi/R}$. Then the noise power is calculated as

$$P_{N,in}(\xi) \approx \frac{P_N}{\pi} \mathcal{I}_N(\xi), \qquad (11)$$

by the definition (5).

978-1-4799-2817-0/14 $31.00 © 2014 IEEE   445

IV. SNR SENSITIVITY AND VARIATION ANALYSIS

Define the semi-relative (semi-normalized) sensitivity of SNR with respect to a capacitance by

$$\mathcal{S}_C'^{\text{SNR}} = C \frac{\partial \text{SNR}}{\partial C}. \tag{12}$$

In view of (4), the expression for SNR also can be written in the following form

$$\text{SNR}(\xi) = 10 \log P_S + 10 \log | \text{STF}(e^{j\omega_0}; \xi)|^2$$
$$- 10 \log \frac{P_N}{\pi} - 10 \log \mathcal{I}_N(\xi). \tag{13}$$

Taking derivative leads to

$$\frac{\partial \text{SNR}(\xi)}{\partial C} = \frac{10}{\ln 10} \left(\frac{2 \operatorname{Re}\left[\frac{\partial \text{STF}(e^{j\omega_0};\xi)}{\partial C} \text{STF}(e^{-j\omega_0};\xi) \right]}{|STF(e^{j\omega_0};\xi)|^2} - \right.$$
$$\left. \frac{\int_0^{\pi/R} 2 \operatorname{Re}\left[\frac{\partial \text{NTF}(e^{j\omega};\xi)}{\partial C} \text{NTF}(e^{-j\omega};\xi) \right] d\omega}{\int_0^{\pi/R} | \text{NTF}(e^{j\omega};\xi)|^2 d\omega} \right) \tag{14}$$

Since both $\text{STF}(z;\xi)$ and $\text{NTF}(z;\xi)$ are symbolic functions, the derivatives of $\text{STF}(z;\xi)$ and $\text{NTF}(z;\xi)$ with respect to a capacitance can be generated by the GPDD tool with great ease [22]. The integration in the numerator of the second term in (14) can be calculated similarly to the approximate calculation of SNR because the integrand is again a rational polynomial.

A. Variational Analysis by Sensitivity

The SNR sensitivity can be used for approximate variational analysis by adopting the first-order approximation written as follows

$$\text{SNR}(\xi) \approx \text{SNR}(\xi_0) + \sum_i \frac{\partial \text{SNR}(\xi_0)}{\partial C_i} \cdot \Delta C_i, \tag{15}$$

where $\text{SNR}(\xi_0)$ represents the nominal SNR and $\text{SNR}(\xi)$ is the perturbed SNR. In SC-SDM circuits, we mainly consider the capacitors or the capacitor ratios as the design parameters. Hence, we may denote ξ to be the set of all capacitors C_i, i.e., $\xi := \{ C_i : i = 1, \cdots, N_C \}$, where N_C is the total number of capacitors. The perturbation of a capacitor is denoted by a small variation $\Delta C_i = C_i - C_{i,0}$ where $C_{i,0}$ is the nominal capacitance.

Typical relative variation of capacitors (in the sense of variance normalized by mean) in SC circuits is about $0.5 \sim 2\%$ [24]. We assume in this discussion that the random capacitor variations are independent variables subject to normal distributions $C_i \sim \mathcal{N}(\mu_{C_i}, \sigma_{C_i})$. (Note that the independence is just a technical assumption. We may always apply the technique of principle component analysis (PCA) to create a set of independent variables.)

By (15) the standard deviation of SNR can be predicted as

$$\sigma_{SNR}^2 \approx \sum_i \left\{ \frac{\partial \text{SNR}(\xi_0)}{\partial C_i} \right\}^2 \cdot \sigma_{C_i}^2$$
$$= \sum_i \left\{ C_{i,0} \frac{\partial \text{SNR}(\xi_0)}{\partial C_i} \right\}^2 \cdot \frac{\sigma_{C_i}^2}{C_{i,0}^2}$$
$$= \left(\sum_i \left\{ \mathcal{S}_{C_i}'^{\text{SNR}}(\xi_0) \right\}^2 \right) \left(\frac{\sigma_{C_i}}{\mu_{C_i}} \right)^2 \tag{16}$$

where we assume that the nominal capacitance is the mean of the capacitance. In the last equation we have written the scaled variance in the form of $\frac{\sigma_{C_i}}{\mu_{C_i}}$ which accounts for the capacitor fabrication reality. In real fabrication, capacitor variance is a relative quantity [24]. By normalizing variance to the mean, we are considering the variance of a unit capacitor. Moreover, it is reasonable to assume that the relative capacitor variance $\frac{\sigma_{C_i}}{\mu_{C_i}}$ is a constant independent of any specific capacitor.

V. EXPERIMENTAL RESULTS

We have implemented the approximate SNR computation method by Taylor approximation within our symbolic GPDD simulator. The experimental results are reported in two parts. Firstly, we demonstrate the numerical approximation accuracy by choosing different Taylor polynomial orders. Secondly, we report the capability of variational prediction of SNR by comparing to the numerical integration results and the behavioral simulation results. All behavioral simulations were conducted using the Simulink [†] toolbox by Schreier [25].

A. Test on Symbolic SNR Accuracy

The accuracy of the Taylor approximation method for SNR computation depends on the Taylor polynomial order (simply called *Taylor order*). The circuit used for test is a fifth-order SC-SDM from [4]. Shown in Fig. 4 is the effect of asymptotic approaching to the accurate SNR computed by numerical integration. We see that the approximation is sufficiently close to the numerical integration value as the Taylor order exceeds *six*. The numerical integration was implemented by using the method described in [26], [27]. We also plotted the SNR reference computed by behavioral simulation, which is slightly (about 1dB) higher than the numerically integrated value (due to the assumption made on the quantization error in linear analysis).

B. Tests on Variational SNR Accuracy

A statistical test also can validate the reliability of the symbolic SNR computation method. We chose adequately high Taylor orders for the calculation of SNR sensitivity. The symbolically calculated SNR histograms will be compared to those computed by direct numerical integration (that is considered exact) and by behavioral simulation. Different orders of SDMs, including the second up to the fifth-order modulators of different topologies and OSRs, were used in our experiment.

[†] Simulink is a product of the Mathworks, Inc.

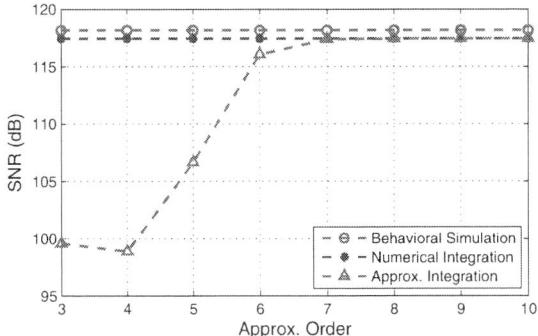

Fig. 4. SNR approximation accuracy for the 5th-order SDM versus Taylor expansion order.

Summarized in Tables I and II[‡] are the experimental results, where SDM2 is the second-order modulator shown in Fig. 2, SDM3 is a third-order modulator borrowed from [14, Fig. 9(a)], SDM4 is a fourth-order modulator generated by the delsig-toolbox [25] with OSR 64 and a CIFB structure, and SDM5 is a fifth-order modulator borrowed from [4, page 309]. We have mainly compared the means and variances of SNR computed by the three methods: 1) symbolic, 2) numerical integration, and 3) behavioral simulation. The input to all behavioral simulations is a sinusoidal signal of magnitude 0.2.

The modulator architectures such as CIFF (cascade-of-integrators in feedforward form) and CRFF (cascade-of-resonators in feedforward form) are defined in [4].

Table I shows the detailed results from our experiment by using the proposed symbolic method. μ_{SNR} is the computed mean, σ_{SNR} is the variance computed by Monte Carlo, μ_{SNR}^0 is the nominal mean for sensitivity analysis, and $\hat{\sigma}_{SNR}$ is the variance computed by sensitivity (whose computation requires only one round of sensitivity evaluation at the nominal capacitances which takes negligible time for the given circuit sizes.) The *GPDD size* (i.e., number of GPDD vertices) is an indication of the memory consumption for symbolic representation. The Taylor order was chosen by an empirical rule: *adding the modulator order by four*. The CPU time listed in the last column of Table I has accumulated all the time spent on symbolic construction of loop transfer functions, Taylor approximation, and sensitivity calculation.

The data given in Table II are collected for the other two SNR computation methods by numerical integration and behavioral simulation. The SNR results computed by the symbolic method in Table I should be compared to that by the numerical integration method. We see that the computed SNR values are extremely close to each other, which is an indication of good accuracy and reliability of the Taylor truncation method and the first-order approximation by symbolic sensitivity. The CPU time for Simulink behavioral simulation is the total time of 10,000 runs in the Monte Carlo simulation, each simulation samples 2^{14} (about 16K) time points for

[‡]All times were measured on an Intel Xeon server with four cores, 2.83GHz processor, and 16GB memory.

SNR computation. The SNR numbers computed behaviorally have around 1dB or so deviation, for which we have already explained the reason. The reader should observe that the behavioral simulation mostly took about 12 to 20 minutes, while the symbolic method took only *a small fraction of a second* for one round of variational SNR computation.

The histograms generated by Monte Carlo experiment of the three methods for the modulator SDM2 are shown in Fig. 5 for illustration. Similar histograms have been generated for other modulators. The dashed outlines in all histograms are the normal distribution curves fit by the computed means and variances.

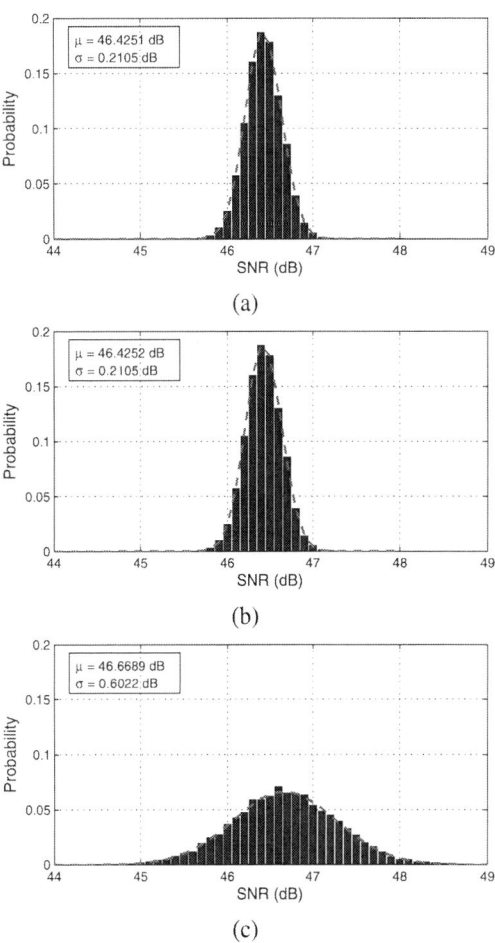

Fig. 5. Histograms of the second-order SDM (SDM2) computed by the three methods: (a) Symbolic SNR. (b) Numerical integration. (c) Behavioral simulation.

VI. CONCLUSION

A fully symbolic method has been developed for calculating the SNR of a switched-capacitor $\Sigma\Delta$ modulators as analytical (or symbolic) functions of the capacitor parameters. It turns out that Taylor approximation of a rational function followed by analytical integration is a highly reliable computation method for SNR computation. The proposed method in a sense has overcome the annoying bottleneck of time-consuming

TABLE I

VARIATIONAL SNR COMPUTED BY THE SYMBOLIC METHOD.

Modulator info				Symbolic						
Name	Topology	OSR	#C	μ_{SNR} (dB)	σ_{SNR} (dB)	μ^0_{SNR} (dB)	$\hat{\sigma}_{SNR}$ (dB)	GPDD size	Taylor order	CPU time
SDM2	CIFF	32	13	46.4251	0.2105	46.4252	0.2110	125	6	0.046 sec
SDM3	CIFF	32	14	58.7023	0.2433	58.7024	0.2425	158	7	0.061 sec
SDM4	CIFB	64	20	94.6307	0.2787	94.6518	0.2791	176	8	0.089 sec
SDM5	CRFF	80	20	118.0955	0.3033	118.1390	0.2952	479	9	0.256 sec

TABLE II

VARIATIONAL SNR COMPUTED BY NUMERICAL INTEGRATION AND BEHAVIORAL SIMULATION.

Modulator	Integration			Behavioral		
Name	μ_{SNR} (dB)	σ_{SNR} (dB)	CPU time	μ_{SNR} (dB)	σ_{SNR} (dB)	CPU time
SDM2	46.4252	0.2105	2.8 min	46.6689	0.6022	12.5 min
SDM3	58.6976	0.2435	3.9 min	59.2192	0.6417	14.2 min
SDM4	94.6209	0.2839	5.7 min	95.3173	0.8597	16.5 min
SDM5	118.0907	0.3041	16.0 min	119.0134	0.9719	19.0 min

SNR computation encountered in many applications requiring repeated SNR computations. As long as the computation accuracy is guaranteed, the remarkable speedup by using the proposed method is well expected. In this work the effectiveness of using the first-order approximation for variational SNR characterization is also demonstrated. A major future work is to apply the proposed method for automatic synthesis of SC-SDMs.

REFERENCES

[1] P. M. Aziz, H. V. Sorensen, and J. van der Spiegel, "An overview of sigma-delta converters," *IEEE Signal Processing Magazine*, vol. 13, no. 1, pp. 61–84, 1996.

[2] F. Medeiro, B. Pérez-Verdu, and A. Rodríguez-Vázquez, *Top-Down Design of High-Performance Sigma-Delta Modulators.* Boston, MA: Kluwer Academic Publishers, 1999.

[3] G. I. Bourdopoulos and A. Pnevmatikakis, *Delta-Sigma Modulators: Modeling, Design and Applications.* Imperial College Press, 2003.

[4] R. Schreier and G. C. Temes, *Understanding Delta-Sigma Data Converters.* Hoboken, NJ: John Wiley & Sons, 2005.

[5] F. Medeiro, B. P. Verdu, A. R. Vasquez, and J. L. Huertas, "A vertically integrated tool for automated design of $\Delta\Sigma$ modulators," *IEEE J. Solid-State Circuits*, vol. 30, pp. 762–777, July 1995.

[6] X. Haurie and G. W. Roberts, "A design, simulation and synthesis tool for delta-sigma-modulator-based signal sources," in *Proc. IEEE Int'l Symposium on Circuits and Systems (ISCAS)*, May 1996, pp. 715–718.

[7] A. Marques, V. Peluso, M. S. Steyaert, and W. M. Sansen, "Optimal parameters for $\Delta\Sigma$ modulator topologies," *IEEE Trans. on Circuits and Systems–II: Analog and Digital Signal Processing*, vol. 45, no. 9, pp. 1232–1241, September 1998.

[8] T. H. Kuo, K. D. Chen, and J. R. Chen, "Automatic coefficients design for high-order sigma-delta modulators," *IEEE Trans. on Circuits and Systems–II: Analog and Digital Signal Processing*, vol. 46, no. 1, pp. 6–15, January 1999.

[9] K. Francken, P. Vancorenland, and G. Gielen, "DAISY: A simulation-based high-level synthesis tool for delta-sigma modulators," in *Proc. IEEE/ACM Int. Conf. Computer-Aided Design (ICCAD)*, November 2000, pp. 188–192.

[10] K. Francken and G. Gielen, "A high-level simulation and synthesis environment for $\Delta\Sigma$ modulators," *IEEE Trans. Computer-Aided Design*, vol. 22, pp. 1049–1061, August 2003.

[11] O. Bajdechi, G. E. Gielen, and J. H. Huijsing, "Systematic design exploration of delta-sigma ADCs," *IEEE Trans. on Circuits and Systems–I: Regular Papers*, vol. 51, no. 1, pp. 86–95, January 2004.

[12] H. Tang and A. Doboli, "High-level synthesis of $\Delta\Sigma$ modulator topologies optimized for complexity, sensitivity, and power consumption," *IEEE Trans. on Computer-Aided Design of Integrated Circuits and Systems*, vol. 25, no. 3, pp. 597–607, March 2006.

[13] H. Wang and T. Kuo, "An automatic coefficient design methodology for high-order bandpass sigma-delta modulator with single-stage structure," *IEEE Trans. on Circuits and Systems–II: Express Briefs*, vol. 53, no. 7, pp. 580–584, July 2006.

[14] Y. H. Ho, H. K. A. Kwan, and K. L. Ho, "Designing globally optimal delta-sigma modulator topologies via signomial programming," *Int'l J. Circuit Theory and Applications*, vol. 37, no. 3, pp. 453–472, April 2009.

[15] S. Y. Lee, C.-Y. Chen, J. H. Hong, R. G. Chang, and M. P. H. Lin, "Automated synthesis of discrete-time sigma-delta modulators from system architecture to circuit netlist," *Microelectronics Journal*, vol. 42, no. 2, pp. 347–357, 2011.

[16] A. Graupner, W. Schwarz, and R. Schuffny, "Statistical analysis of analog structures through variance calculation," *IEEE Trans. on Circuits and Systems–I: Fundamental Theory and Applications*, vol. 49, no. 8, pp. 1071–1078, August 2002.

[17] H. Tang, "Symbolic statistical analysis of SNR variation for delta-sigma modulators," *IEEE Trans. on Circuits and Systems–II: Express Briefs*, vol. 54, no. 8, pp. 720–724, August 2007.

[18] H. Tang and M. Webb, "Optimal synthesis of delta-sigma modulator topologies considering SNR variation," in *Proc. 50th Midwest Symp. on Circuits and Systems*, 2007, pp. 730–733.

[19] V. Mladenov, P. Karampelas, G. Tsenov, and V. Vita, "Approximation formula for easy calculation of signal-to-noise ratio of sigma-delta modulators," *International Scholarly Research Network (ISRN) Signal Processing*, vol. 2011, pp. 1–7, August 2011, article ID 731989, DOI: 10.5402/2011/731989.

[20] H. Floberg and S. Mattisson, "Symbolic analysis of switched-capacitor networks using compacted nodal analysis in the s-domain," *IEEE Trans. on Computer-Aided Design of Integrated Circuits and Systems*, vol. 16, no. 10, pp. 1196–1199, October 1997.

[21] G. Shi, "Graph-pair decision diagram construction for topological symbolic circuit analysis," *IEEE Trans. on Computer-Aided Design of Integrated Circuits and Systems*, vol. 32, no. 2, pp. 275–288, February 2013.

[22] G. Shi and X. Meng, "Variational analog integrated circuit design by symbolic sensitivity analysis," in *Proc. Int'l Symposium on Circuits and Systems (ISCAS)*, Taiwan, China, May 2009, pp. 3002–3005.

[23] J. Cheng, G. Shi, A. Tai, and F. Lee, "Symbolic fault modeling for switched-capacitor circuits," in *IEEE Region Ten Conference (TENCON)*, Xi'an, China, 2013, accepted for publication.

[24] R. Aparicio and A. Hajimiri, "Capacity limits and matching properties of integrated capacitors," *IEEE J. Solid-State Circuits*, vol. 37, no. 3, pp. 384–393, March 2002.

[25] R. Schreier, "Delta sigma toolbox 6.0," 2004. [Online]. Available: http://www.mathworks.com/matlabcentral/fileexchange/

[26] M. Mori and M. Sugihara, "The double exponential transformation in numerical analysis," *Journal of Computational and Applied Mathematics*, vol. 127, no. 1-2, pp. 287–296, 2001.

[27] J. D. Cook, "Fast numerical integration," July 2009. [Online]. Available: http://www.codeproject.com/Articles/31550/Fast-Numerical-Integration/

Sparse Statistical Model Inference for Analog Circuits under Process Variations

Yan Zhang, Sriram Sankaranarayanan and Fabio Somenzi
ECEE, University of Colorado, Boulder, CO, USA
{yan.zhang, srirams, fabio}@colorado.edu

Abstract— In this paper, we address the problem of performance modeling for transistor-level circuits under process variations. A sparse regression technique is introduced to characterize the relationship between the process parameters and the output responses. This approach relies on repeated simulations to find polynomial approximations of response surfaces. It employs a heuristic to construct sparse polynomial expansions and a stepwise regression algorithm based on LASSO to find low degree polynomial approximations. The proposed technique is able to handle many tens of process parameters with a small number of simulations when compared to an earlier approach using ordinary least squares. We present our approach in the context of statistical model inference (SMI), a recently proposed statistical verification framework for transistor-level circuits. Our experimental evaluation compares percentage yields predicted by our approach with Monte-Carlo simulations and SMI using ordinary least squares on benchmarks with up to 30 process parameters. The sparse-SMI approach is shown to require significantly fewer simulations, achieving orders of magnitude improvement in the run times with small differences in the resulting yield estimates.

I. INTRODUCTION

Nano-CMOS technologies are widely used in modern VLSI design. While the move to nano-regime brings benefits to designers, many challenges emerge [1]. Among these challenges, one of the most critical issues is to design robust circuits under increasingly large *process variations*.

Statistical performance modeling is a common approach to understanding the effects of process variations. It has been studied extensively in the past to characterize the performance of circuits in terms of process parameters [2–6]. However, these techniques can only address problems in older technologies, wherein process variations were not a dominant issue and could be modeled by a few "inter-die" parameters using linear models. In nanoscale era, the process parameters have to be treated in a *per-transistor* fashion. For instance, the gate oxide thickness has to be considered separately for each transistor [1].

Furthermore, the variations are becoming larger and require nonlinear modeling techniques [7]. These challenges call for an approach that is able to efficiently construct nonlinear approximations in the face of many tens or hundreds of process parameters.

Due to the design challenges in nano-regime, performance modeling has received renewed attention. Li *et al.* [8] proposed a PROBE approach based on reduced rank regression. Their technique enables the use of quadratic models for nonlinear response surfaces modeling. A similar idea was proposed by Feng *et al.* [9] to handle the problem of interconnect modeling. One drawback of these two techniques is that they are limited to quadratic models, and thus cannot characterize higher-order response surfaces accurately. Singhee *et al.* [10] proposed a nonlinear regression approach based on latent variable regression and neural networks. For many practical circuits, they showed promising results. However, due to the complicated structure of neural networks, the resulting models can be prone to over-fitting. Furthermore, neural network models are hard to interpret. Li [11] used matching pursuit to find the "best" projection of the response surface onto an orthogonal polynomial basis. Although in principle, their technique can handle higher-order response surfaces, the experimental results are limited to quadratic models.

In this paper, we present a sparse regression technique to discover a polynomial approximation of an output response as a function of input process parameters. Such an approximation is used inside a statistical framework to construct a mathematical model that explains how the process variations affect the output response. Our approach collects data by sampling process parameters and simulating the circuit to find the output responses. Using the simulation data, we construct a low-degree and sparse polynomial in a stepwise fashion. The proposed technique consists of a heuristic that discovers relevant terms in the polynomial approximation, and a sparse regression algorithm based on LASSO to construct the actual approximation [12]. The heuristic efficiently discards terms that contribute the least to the output response. Then the regression algorithm computes the coefficients for the remaining terms under $\mathcal{L}_1$ regularization. This technique is used inside *statistical model inference (SMI)*,

---

This work was partially supported by the US National Science Foundation (NSF) under the award number 1016994 and CCF-1320069. All opinions expressed are those of the authors, and not necessarily of the NSF.

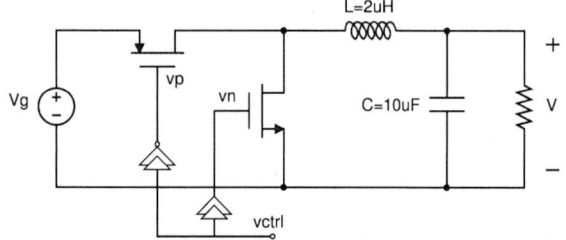

Fig. 1. A buck converter with L and C as process parameters.

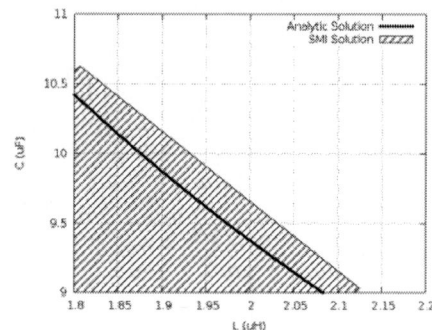

Fig. 2. Verification result of the buck converter. The region below the solid curve is shown to be unsafe. The shaded region is the statistically unsafe subset predicted by SMI.

a recently proposed statistical verification framework for transistor-level circuits [13]. SMI addresses the problem of estimating a safe region of the parameter space that guarantees a given design specification of a circuit, and examines specifications of the form $\phi(\mathbf{p}) \in [\ell, u]$ where ϕ is a performance metric, $\mathbf{p}$ a vector of process parameters, and ℓ and u are real-valued performance limit. We show that the proposed algorithm improves the scalability of SMI to handle circuits with many process parameters.

The rest of this paper is organized as follows. First, we provide an overview of SMI with a running example. Then we introduce the sparse regression algorithm in Section III. Finally, we demonstrate our technique by showing the verification of a ring oscillator and an operational amplifier under a large number of process variations.

II. OVERVIEW OF SMI

In this section, we review the statistical verification framework SMI with a running example [13]. Consider a circuit with process parameters $\mathbf{p}$ from a parameter space $\mathbb{P}$ with a joint distribution $\mathcal{F}$. Let $\phi = f(\mathbf{p})$ be a *output response* of interest. SMI handles design specifications of the form: $\phi \in [\ell, u]$, where ℓ and u are the tolerance limits of ϕ, and shows in the statistical sense whether a specification is satisfied under process variations.

Example II.1 (Buck Converter). *Figure 1 shows a simple buck converter. Assume that L and C are independent, uniform random variables:* [1] $L \in [1.8, 2.2]\mu H$ *and* $C \in [9, 11]\mu F$. *We verify that $\Delta v \leq 30mV$, where Δv is the ripple amplitude of the output voltage. For simplicity, the transistors are treated as ideal switches. Then we have*

$$\Delta v = f(L, C) = \frac{V_g - V}{16LC}DT_s^2, \qquad (1)$$

where V is the DC component of the output voltage, D is the duty cycle and T_s is the time period of the control voltage. Let $V = 3V$, $D = 0.25$ and $T_s = 2\mu s$. From (1), we derive the condition $LC \geq 18.75$ that guarantees the satisfaction of the specification.

In practice, ϕ is usually represented by a complicated function f with a unknown closed from. To verify a given

---

[1]Our approach can handle a larger variety of distributions that one may encounter in practice.

specification, SMI relies on repeated simulations to find a *statistical over-approximation* of f, which consists of a polynomial q and a tolerance interval I such that

$$\Pr\left(\phi \in q(\mathbf{p}) \oplus I\right) \geq \theta, \ \mathbf{p} \in \mathbb{P}, \qquad (2)$$

where $\oplus$ denotes the Minkowski sum and θ is a given probability. This means that the pair (q, I) over-approximates ϕ for at least θ portion of the parameter space $\mathbb{P}$. Such an approximation can be used to derive a *statistically* safe subset of $\mathbb{P}$ for the specification.

To obtain a statistical over-approximation of ϕ, SMI works in two phases: 1) a regression phase that finds a polynomial q to approximate ϕ, and 2) a bloating phase that derives a tolerance interval I. In the first phase, we construct a polynomial approximation q. A simple approach based on ordinary least squares (OLS) is used in Zhang *et al.* [13], which does not scale well to circuits with many parameters. Therefore, a novel algorithm is presented in Section III to improve SMI by using sparse regression techniques. The bloating phase remains unaltered and is briefly presented below.

Bloating. With an approximation q, we consider the problem of deriving a tolerance interval I for a given probability θ such that (2) holds. In SMI, it is formulated as a hypothesis testing problem that involves two hypotheses:

$$\mathcal{H}_0 : \Pr(\phi \in q \oplus I) \geq \theta \text{ and } \mathcal{H}_1 : \Pr(\phi \in q \oplus I) < \theta.$$

We wish to find an interval I so that we can be convinced of $\mathcal{H}_0$ as opposed to $\mathcal{H}_1$. In SMI, the interval I is derived using an algorithm based on Bayesian hypothesis testing. Starting with a zero interval $I = [0, 0]$, SMI repeatedly simulates the circuit with randomly sampled parameters and adjusts the interval I until for some $K > 0$ consecutive simulations, $\phi \in q \oplus I$. The threshold K is chosen so that when it is achieved, the observed data provide strong statistical evidence to accept $\mathcal{H}_0$, indicating that the interval I is sufficiently large. The rationale of SMI including the choice of K is discussed by Zhang *et al.* [13].

The two steps, regression and bloating, construct a model (q, I). Intuitively, such a model convinces a statistical model checker [14, 15] that (2) holds with a given

θ. With this model, identifying the safety of a point $\mathbf{p}$ reduces to checking whether $q(\mathbf{p}) \oplus I \subseteq [\ell, u]$. Furthermore, S, the safe subset of the parameter space, can be computed conservatively by nonlinear solvers, such as Z3 [16] and iSAT [17], or statistically by Monte-Carlo sampling.

Example II.2 (Verifying Buck Converter). *We use $N = 20$ simulations to construct a cubic polynomial $q_3(L, C)$. Using OLS, we obtain the following polynomial [2].*

$$\begin{pmatrix} 0.031 - 0.002L - 0.009C - 0.002L^2 - 0.005C^2 + 0.003LC+ \\ 0.004L^2C - 0.002LC^2 - 0.000L^3 + 0.010C^3 \end{pmatrix}. \quad (3)$$

As a comparison, using the proposed algorithm (as described below), we obtain a sparser approximation

$$0.028 - 0.001L - 0.001C. \quad (4)$$

Clearly, (4) provides a more compact representation than (3). Such a polynomial is bloated with $\theta = 0.95$, leading to a tolerance interval $I = [-3, 4]mV$. The model (q_3, I) is used to check the specification, $\Delta v \leq 30mV$. The result is shown in Figure 2. The solid curve plots of the equation $LC = 18.75$ so that the region below is known to be unsafe. The shaded region is the unsafe subset predicted by SMI. It can be shown that in this example, bloating (4) yields identical verification result as bloating (3).

III. A Sparse Regression Algorithm

In this section, we introduce a sparse regression algorithm. Let $\mathbf{p}$ be the process parameters and ϕ be a response of a circuit. Let $w(\mathbf{p})$ be the probability density function of $\mathbf{p}$. Our goal is to find a polynomial to approximate the response surface using a limited number of samples. The main challenge is that the number of parameters is often large, while the number of simulations is greatly limited by the available computational power. This means that the underlying linear system is underdetermined. Using conventional approaches, such as least squares fitting, can easily lead to over-fitting.

To solve this problem, we propose an approach shown in Algorithm 1, which uses LASSO [12], a $\mathcal{L}_1$ regularization technique, to solve the under-determined system. Let $\phi = f(\mathbf{p})$ where f is a square-integrable function with an unknown closed form. Given a set of n samples, $\mathbf{X}$ is an $n \times k$ matrix where k is the number of parameters, and $\mathbf{Y}$ consists of n observations of the property ϕ. The algorithm operates in an iterative fashion, wherein the approximation is constructed using a sequence of *latent variables*. The j-th iteration attempts to fit a polynomial of degree j to approximate the *residue* $\mathbf{R}_{j-1}$ from the previous iteration in terms of $\mathbf{X}$. Initially $\mathbf{R}_0 = \mathbf{Y}$. At each step, we discover a latent variable $z_j = q_j(\mathbf{p})$ that provides an approximation q_j of $\mathbf{R}_{j-1}$, and $\mathbf{R}_j = \mathbf{R}_{j-1} - q_j(\mathbf{p})$. The process is stopped when the degree limit D is reached or when $|R_j|$ is below a threshold. As a final step, the overall

---

Data: Matrices $\mathbf{X}, \mathbf{Y}$, Degree Limit D, Initial Degree d
Result: Approximation $q(\mathbf{p})$
latent_vars := empty set ;
$\mathbf{R} := \mathbf{Y}$;
while true **do**
  $S_d' :=$ choose_a_sparse_basis$(d, \mathbf{X}, \mathbf{R})$;
  $z, \mathbf{R} :=$ compute_latent_variables_by_lasso$(S_d', \mathbf{X}, \mathbf{R})$;
  latent_vars := latent_vars $\cup z$;
  $d := d + 1$;
  if $|\mathbf{R}| < \epsilon$ or $d == D$ **then**
    | $q(\mathbf{p}) :=$ do_least_squares_fitting(latent_vars, $\mathbf{X}, \mathbf{Y}$) ;
  end
end

Algorithm 1: Sparse Regression Algorithm.

model is constructed as a polynomial that approximates $\mathbf{Y}$ in terms of the latent variables $z_1, \ldots, z_k$. Our approach finds a hierarchical model that expresses the output response ϕ as a function of latent variables $z_1, \ldots, z_k$, each of which is a polynomial of increasing degree over $\mathbf{p}$.

A. Choosing a Sparse Basis

Polynomial chaos expansion (PCE) provides a mathematical basis for approximating functions of random variables by a polynomial obtained by a truncated generalized Fourier series

$$f(\mathbf{p}) \approx \sum_{i \in S_d} c_i Q_i(\mathbf{p}),$$

where S_d is an index set of terms, c_i a generalized Fourier coefficient, and $\{Q_i \mid i \in S_d\}$ is a set of orthogonal polynomials of degree d [18]. The polynomials Q_i are chosen based on the distribution $\mathcal{F}$ of process parameters using the standard Wiener-Askey scheme. We call each Q_i as a *term*. Intuitively, if c_i is sufficiently small, the corresponding term contributes little to f and can be dropped without losing much accuracy. The coefficients c_i can be computed from the following equation [18].

$$c_i = \frac{1}{\gamma_i} \langle f, Q_i \rangle = \frac{1}{\gamma_i} \int_{\mathbf{p} \in \mathbb{P}} f(\mathbf{p}) Q_i(\mathbf{p}) w(\mathbf{p}) d\mathbf{p}$$

where each γ_i is a normalization constant, $\langle, \rangle$ an inner product, and $w(\mathbf{p})$ a probability density function.

It can be difficult, if not impossible, to compute c_i exactly because (1) f may not have a closed form and (2) $\mathbf{p}$ may consist of a large number of parameters. As a possible solution, numerical quadrature methods suffer from their scalability. They also rely on the ability to sample at certain points, which is not always realizable in practice.

We use Monte-Carlo methods to estimate c_i,

$$\hat{c}_i = \frac{1}{\gamma_i n} \sum_{j=1}^n f(\mathbf{p}_j) Q_i(\mathbf{p}_j), \quad (5)$$

where n is the number of samples and each $\mathbf{p}_j$ is sampled according to $w(\mathbf{p})$. Note that such an estimation can at best capture the trend of c_i due to finite sample

---

[2]The parameter values for L and C are normalized to $[-1, 1]$.

978-1-4799-2817-0/14 $31.00 © 2014 IEEE

size. Using $\hat{c}_i$ usually leads to poor approximations of the response surface. However, the computation of (5) is efficient even with many parameters. Thus in our approach, we use (5) to construct a *filtering stage* such that terms with small $\hat{c}_i$ are dropped. In particular, we drop those terms whose corresponding $\hat{c}_i$ are in the lower k-quantile of the whole set, where k is controlled by the user. The index set of remaining terms is denoted by S_d'.

B. Computing Latent Variables by LASSO

We need to find an approximation consisting of Q_i, $i \in S_d'$. This is usually solved as a least squares problem

$$\min_{c_i, i \in S_d'} \|f(\mathbf{p}) - g(\mathbf{p})\|_2^2 \, , \; g(\mathbf{p}) = \sum_{i \in S_d'} c_i Q_i(\mathbf{p}) \, . \quad (6)$$

To avoid over-fitting, we use LASSO to solve the minimization problem in (6). LASSO handles the problem by adding an extra constraint on the coefficients c_i,

$$\min_{c_i, i \in S_d'} \|f(\mathbf{p}) - g(\mathbf{p})\|_2^2 + \alpha \sum_{i \in S_d'} |c_i| \, , \quad (7)$$

where α is a Lagrangian multiplier of the regularization term $\sum_{i \in S_d'} |c_i|$. Intuitively, the extra term forces the coefficients c_i to behave "regularly" so that they cannot range over many orders of magnitude. Furthermore, due to the nature of $\mathcal{L}_1$-norm minimization, proper choices of α result in sparse solutions. In general, a larger α leads to a sparser solution. When α approachs 0, LASSO reduces to ordinary least squares regression.

To apply LASSO in our algorithm, we construct a $n \times |S_d'|$ matrix $\mathbf{X}_{exp}$ that comprises observations mapped to each expansion term,

$$\mathbf{X}_{exp} = \begin{pmatrix} Q_{i_1}(\mathbf{X}_1) & \cdots & Q_{i_m}(\mathbf{X}_1) \\ Q_{i_1}(\mathbf{X}_2) & \cdots & Q_{i_m}(\mathbf{X}_2) \\ \vdots & \ddots & \vdots \\ Q_{i_1}(\mathbf{X}_n) & \cdots & Q_{i_m}(\mathbf{X}_n) \end{pmatrix} , \; i_k \in S_d', \quad (8)$$

where $\mathbf{X}_i$ is the i-th row of $\mathbf{X}$. Denote the coefficient vector $\mathbf{c} = (c_{i_1}, \ldots, c_{i_m})^T$. We solve the LASSO problem

$$\min_{\mathbf{c}} \|\mathbf{R}_{j-1} - \mathbf{X}_{exp}\mathbf{c}\|_2^2 + \alpha|\mathbf{c}| \, ,$$

where $\mathbf{R}_{j-1}$ is the residue from the previous iteration. The solution $\mathbf{c}$ yields a latent variable

$$z = c_{i_1}Q_{i_1} + \cdots + c_{i_m}Q_{i_m} \, , \; i_j \in S_d' \, ,$$

and the updated residue $\mathbf{R}_j = \mathbf{R}_{j-1} - \mathbf{X}_{exp}\mathbf{c}$.

C. Least Squares Fitting with Latent Variables

The iteration in Algorithm 1 terminates when either the residue $\mathbf{R}$ is smaller than some given ϵ, or the degree d reaches the limit D. The resulting model consists of the computed latent variables, each represented by a polynomial. In practice, since the number of latent variables is usually small, the fitting can be easily performed.

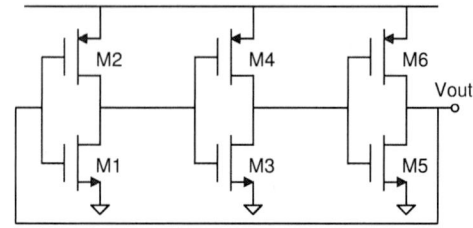

Fig. 3. A three-stage ring oscillator.

IV. EXPERIMENTAL EVALUATION

In this section, we experimentally evaluate the sparse-SMI approach proposed in this paper, comparing it with the original SMI proposed by Zhang *et al.* [13] and a yield estimation using Monte-Carlo simulation. Our comparison involves two analog circuits: a three-stage ring oscillator and a two-stage operational amplifier (opamp) over a range of output response functions (properties) for these circuits. The sparse-SMI technique presented in this paper is implemented in Python. We use LTSpice [19] as the circuit simulator. All experiments were performed on a quad-core 2.8GHz machine running Debian 6.0. Unless specified, all the times are measured in seconds.

The comparisons between SMI and sparse-SMI are done through a $\theta = 0.95$ and a threshold $K = 89$, indicating that the tolerance intervals covers at least 95% of the parameter space with a high statistical confidence. We compare the resulting models by using them to predict the yield, which is also computed using Monte-Carlo Simulations for comparison.

A. Ring Oscillator

Figure 3 shows the schematic of a three-stage ring oscillator. The circuit is designed to oscillate at 0.98GHz with a tolerance within ± 50MHz. The oscillation frequency is affected by various process parameters in the transistors $M_i, i = 1, \ldots, 6$. We select 24 process parameters to study, including the oxide thickness t_{ox}, threshold voltage under zero bias v_{t0}, channel width w and channel length l for each M_i. We assume that each variation follow a *truncated normal distribution* in the range $[\mu - 3\sigma, \mu + 3\sigma]$ where μ is the nominal value. The standard deviation of each variation is summarized in Table I.

TABLE I
STANDARD DEVIATION OF EACH PROCESS PARAMETER. μ IS THE NOMINAL VALUE OF THE CORRESPONDING PARAMETER.

	t_{oxi}	v_{t0i}	w_i	l_i
σ	0.05μ	0.05μ	0.1μ	0.1μ

Experiment results. The experiment results for the oscillator are presented in Table II. We choose a degree limit of 3 for SMI and sparse-SMI. The results for 10000 Monte-Carlo simulations are included. The

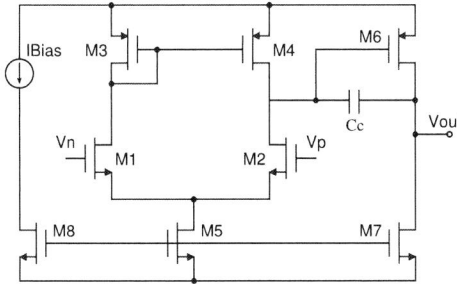

Fig. 4. A two-stage operational amplifier.

columns "Yield", "Var" and "SimTime" under "Monte-Carlo" show the sampling yield, variance of the sampling yield and time spent on simulations. Under the columns "SMI" and "Sparse-SMI", "#Sims" refers to the number of simulations performed in the regression and the bloating phase, respectively. Similarly, "SimTime" shows the simulation time and "Time" indicates the time spent in the regression and bloating phases of SMI, excluding the simulation time. "Safe" reports whether the specification is satisfied and if not, the predicted yield under the given variations. "Degree" is the final degree of the approximation in sparse-SMI.

Table II shows that under the given process variations, the specification $f_{osc} \in [0.93, 1.03]$MHz is not satisfied. First, let us compare SMI and sparse-SMI with the Monte-Carlo method. Since the SMI procedure procedure builds a statistically sound over-approximation for the underlying response surface, it tends to under-estimate the yield. It explains why the predicted yields from SMI and sparse-SMI are lower than the sampling yield from Monte-Carlo simulations. The results clearly show the advantage of using sparse-SMI as opposed to SMI. Whereas SMI requires 3200 simulations, sparse-SMI requires roughly 400 simulations in this case to result in almost identical yield estimates. The results reduce simulation time from roughly 1.5 hours for SMI to 12 minutes for sparse-SMI, a $7.5\times$ reduction. Comparing to Monte-Carlo simulations also show the benefits of a model building approach rather than direct yield estimation.

B. Two-stage Operational Amplifier

The schematic of a commonly used two-stage opamp is shown in Figure 4. The performance of an opamp is characterized by many properties, such as input offset voltage and slew rate. Usually, each property is measured using a specific type of simulation and circuit configuration. For example, input offset voltage is often measured by arranging the opamp in the unity-gain configuration and sweeping DC input voltage. In contrast, the measurement of slew rate requires transient simulation. Table III shows a list of specifications under verification and the types of simulation for the measurement. For a detailed description on how to simulate these properties, we refer

TABLE III
A LIST OF OPAMP SPECIFICATIONS UNDER VERIFICATION.

ID	Specification	Sim type
1	Input offset voltage $\leq$ 50mV	DC
2	DC voltage gain $\geq$ 60dB	AC
3	Unity-gain bandwidth $\geq$ 5MHz	AC
4	Phase margin $\geq 30°$	AC
5	CMRR $\geq$ 80dB	AC
6	PSRR (+) $\geq$ 80dB	AC
7	Slew rate (+) $\geq 10V/\mu s$	Transient

the interested reader to [20] (Chapter 6.6).

We select 4 process parameters, the oxide thickness t_{ox}, threshold voltage under zero bias v_{t0}, channel width w and channel length l, for each transistor $M_i, i = 1, \ldots, 8$. It leads to a total of 32 parameters. We assume the same variations for each parameter as in Table I.

Experiment results. The experimental results for the opamp are summarized in Table IV. The first column, "ID", indicates the specification under verification. Again, we include 10000 Monte-Carlo simulations for comparison. The degree limit for SMI and sparse-SMI is 3. For sparse-SMI, we use 200 simulations in the regression phase for all the properties.

First, we compare SMI and sparse-SMI with the Monte-Carlo method. As in the ring oscillator experiment, we observe that the predicted yield from the two approaches are similar. For all the properties, sparse-SMI requires fewer simulations and finishes much faster. Sparse-SMI reduces the running times from many hours to a few minutes, often resulting in $25\times$ or larger speed-ups.

The speed-up in terms of running time and number of simulations required is much more significant than in the ring oscillator verification. The regression time in SMI has become prohibitively large, while that in sparse-SMI is quite affordable. This demonstrates the scalability of sparse-SMI to handle problems that are too large for conventional approaches.

In the "Degree" column, we see that for some properties, sparse-SMI is able to construct accurate models with a degree lower than the limit. Correspondingly, for these properties, the regression times are significantly smaller. In fact, the major cost in regression lies in the computation of the generalized Fourier coefficients and LASSO. The former can be easily parallelized, leading to further performance improvements.

V. CONCLUSION

In this paper, we propose a sparse regression algorithm to address the problem of performance modeling for analog circuits under process variations. The algorithm uses a limited number of simulations to approximate response surfaces with many parameters. We have applied this technique to statistical model inference, a recently proposed statistical verification framework that aims to construct statistical over-approximations of the response sur-

TABLE II
VERIFICATION RESULTS FOR THE RING OSCILLATOR WITH 24 PROCESS PARAMETERS.

Monte-Carlo		SMI				Sparse-SMI				
Yield	SimTime	#Sims	SimTime	Time	Safe	#Sims	SimTime	Time	Degree	Safe
51%	4.8h	3000/213	1.5h	391s/17s	✗, 46%	200/197	693s	15s/4s	3	✗, 45%

TABLE IV
VERIFICATION RESULTS FOR THE OPAMP WITH 32 PROCESS PARAMETERS.

ID	Monte-Carlo		SMI				Sparse-SMI				
	Yield	SimTime	#Sims	SimTime	Times	Safe	#Sims	SimTime	Time	Degree	Safe
1	61%	1.8h	8500/201	1.6h	4.2h/45s	✗, 58%	200/193	264s	5s/3s	1	✗, 58%
2	65%	1.4h	8500/118	1.3h	4.9h/29s	✗, 61%	200/180	191s	96s/5s	3	✗, 60%
3	100%	1.9h	8500/241	1.7h	5.1h/56s	✓	200/161	271s	4s/3s	1	✓
4	98%	1.9h	8500/197	1.7h	4.4h/47s	✗, 94%	200/131	229s	101s/4s	3	✗, 93%
5	100%	2.2h	8500/194	1.7h	3.9h/51s	✓	200/281	381s	34s/7s	2	✓
6	62%	2.1h	8500/156	1.8h	4.9h/33s	✗, 55%	200/154	260s	91s/5s	3	✗, 54%
7	87%	2.8h	8500/123	2.5h	4.1h/28s	✗, 85%	200/201	413s	6s/5s	1	✗, 84%

face. We show that our sparse regression algorithm can significantly improve the scalability of SMI to handle circuits with many process parameters.

REFERENCES

[1] L. L. Lewyn, T. Ytterdal, C. Wulff, and K. Martin, "Analog circuit design in nanoscale CMOS technologies," *Proceedings of the IEEE*, vol. 97, no. 10, pp. 1687–1714, 2009.

[2] Z. Wang and S. W. Director, "An efficient yield optimization method using a two step linear approximation of circuit performance," in *IEEE EDAC*, 1994, pp. 567–571.

[3] G. Debyser and G. Gielen, "Efficient analog circuit synthesis with simultaneous yield and robustness optimization," in *ICCAD 98*, 1998, pp. 308–311.

[4] A. Dharchoudhury and S. Kang, "Worst-case analysis and optimization of VLSI circuit performances," *IEEE Trans. CAD*, vol. 14, no. 4, pp. 481–492, 1995.

[5] H. Yoon, P. Variyam, A. Chatterjee, and N. Nagi, "Hierarchical statistical inference model for specification based testing of analog circuits," in *VLSI Test Symposium*, 1998, pp. 145–150.

[6] H. Chang and S. S. Sapatnekar, "Statistical timing analysis considering spatial correlations using a single PERT-like traversal," in *ICCAD*, 2003, p. 621.

[7] S. R. Nassif, "Modeling and analysis of manufacturing variations," in *IEEE CICC*, 2001, pp. 223–228.

[8] X. Li, J. Le, L. T. Pileggi, and A. Strojwas, "Projection-based performance modeling for inter/intra-die variations," in *ICCAD*, 2005, pp. 721–727.

[9] Z. Feng and P. Li, "Performance-oriented statistical parameter reduction of parameterized systems via reduced rank regression," in *ICCAD*, 2006, pp. 868–875.

[10] A. Singhee and R. A. Rutenbar, "Beyond low-order statistical response surfaces: latent variable regression for efficient, highly nonlinear fitting," in *DAC*, 2007, pp. 256–261.

[11] X. Li, "Finding deterministic solution from underdetermined equation: large-scale performance variability modeling of analog/RF circuits," *IEEE Trans. CAD*, vol. 29, no. 11, pp. 1661–1668, 2010.

[12] R. Tibshirani, "Regression shrinkage and selection via the LASSO," *Journal of the Royal Statistical Society*, pp. 267–288, 1996.

[13] Y. Zhang, S. Sankaranarayanan, F. Somenzi, X. Chen, and E. Ábraham, "From statistical model checking to statistical model inference: Characterizing the effect of process variations in analog circuits," in *ICCAD*, 2013, preprint available on-line at http://bit.ly/12TH1WR.

[14] H. L. S. Younes and R. G. Simmons, "Probabilistic verification of discrete event systems using acceptance sampling," in *CAV*, 2002, pp. 223–235.

[15] S. K. Jha, E. M. Clarke, C. J. Langmead, A. Legay, A. Platzer, and P. Zuliani, "A Bayesian approach to model checking biological systems," in *CMSB*, 2009, pp. 218–234.

[16] L. De Moura and N. Bjørner, "Z3: an efficient smt solver," in *TACAS*, 2008, pp. 337–340.

[17] M. Fränzle, C. Herde, T. Teige, S. Ratschan, and T. Schubert, "Efficient solving of large non-linear arithmetic constraint systems with complex boolean structure," *JSAT*, vol. 1, no. 3-4, pp. 209–236, 2007.

[18] D. Xiu, *Numerical Methods for Stochastic Computation: A Spectral Method Approach.* Princeton university Press, 2010.

[19] "LTSpice." [Online]. Available: http://www.linear.com/designtools/software/

[20] P. E. Allen and D. R. Holberg, "CMOS analog circut design." Oxford University Press, Inc., 2002.

978-1-4799-2817-0/14 $31.00 © 2014 IEEE

Time-Domain Performance Bound Analysis for Analog and Interconnect Circuits Considering Process Variations

Tan Yu[*], Sheldon X.-D. Tan[*], Yici Cai[‡] and Puying Tang [§]

[*] Dept. Electrical Engineering, University of California, Riverside, CA 92521

[‡] Department of Computer Science and Technology, Tsinghua University, Beijing, China 10084

[§] School of Optoelectronic Information, University of Electronic Science and Technology of China, Chengdu, China 610054

Abstract—Time-Domain worst case or performance bound estimation for analog integrated circuits and interconnect circuits are crucial for both analog and digital circuit design and optimization in the presence of process variations. In this paper, we present a novel non-Monte-Carlo (MC) performance bound analysis technique in time domain. The new method consists of several steps. First the symbolic transient modified nodal analysis (MNA) formulation of the circuit matrices of (linearized) analog and interconnect circuits at a time step is formed. Then the closed-form expressions of the interested performance in terms of variational parameters of the circuit matrices of (linearized) analog and interconnect circuits are derived via a graph-based symbolic analysis method. Then time-domain performance response bound of current time step are obtained by a nonlinear constrained optimization process subject to the parameter variations and variational circuit state bounds computed from the previous time step. We study the bounds computed by the proposed against the different sigma bounds by the standard MC method, which shows that the proposed method is more efficient for computing high sigma bounds than the MC method. Experimental results show that the new method can deliver order of magnitudes speedup over the standard Monte Carlo simulation on some typical analog circuits and interconnect circuits with high accuracy.

1. Introduction

At the nano-scale, circuit parameters are no longer truly deterministic and most of the quantities of practical interests present themselves as probability distributions. Circuit designers must now contend with these variations and uncertainties to ensure the robustness of their circuit designs. Traditional corner-based verification can't meet the accuracy requirements. For statistical analysis of digital and analog integrated circuits under process variations, Monte-Carlo (MC) based statistical simulation are the most popular methods due to their advantage of generality and high accuracy [1], [2]. However, MC method is expensive and slow especially for rare events (high sigma estimations) as more samplings are required, which will lead to the bottleneck of analog circuit optimization. Many fast Monte Carlo methods have been proposed to improve the efficiency of classical Monte Carlo methods. Existing approaches include importance sampling [3], Latin hypercube sampling based method [4], [5], and quasi Monte Carlo based method [2], [6]. However, the importance sampling method is circuit specific, Latin hypercube sampling does not work for all the circuits, and quasi Monte Carlo method suffers the high-dimensional problems [5].

This research was supported in part by NSF grants under No. CCF-1116882, No. CCF-1017090, No. OISE-1130402, No. OISE-0929699 and UC MEXUS-CONACYT CN-11-575.

At the same time, some alternative non Monte Carlo methods have been proposed for statistical analysis. Among them, performance bound analysis methods emerged as attractive techniques for statistical analysis and yield estimation. Those techniques hold the promises that they are more scalable for high sigma and high dimensional statisitical analysis problems compared to existing Monte-Carlo method approaches. Bounding or worse case analysis of analog circuits under parameter variations has been studied in the past for fault-driven testing and tolerance analysis of analog circuits [7]–[9]. Recently, some frequency domain performance bound methods were proposed in [10]–[13] to compute the lower and upper bounds of transfer function's magnitude and phase. The work in [10] applied a control-based method [14] to obtain the performance bounds in frequency domain, and [11] applied an optimization based method to compute the bounds in time domain, in which the whole circuit equations are treated as constraints, which can be very expensive to enforce the constraints. This method has been improved by [12] where symbolic analysis approach was applied to derive exact transfer functions and affine interval method was used to compute variational transfer functions. However, the affine interval method can lead to over conservative results. This work has been extended to compute the time-domain performance bounds based on the frequency domain bounds [13]. Recently analog performance bound analysis in frequency domain based on the optimization method and symbolic analysis techniuqe was proposed in [15].

In this paper, we present a novel non-Monte-Carlo performance bound analysis technique in time domain. The new contributions of the paper lies in the following aspects:

- First, we develop a new general time-domain performance analysis method, which consists of several steps: First the time-domain symbolic modified nodal analysiss (MNA) formulation of (linearized) analog and interconnect circuits at a time step is formed. Then the closed-form expressions of the interested performance in terms of variational parameters of the circuit matrices of (linearized) analog and interconnect circuits are dervied via a graph-based symbolic analysis method. Then time-domain performance response bounds of current time step are obtained by finding the max/min values via a nonlinear constrained optimization process subject to the parameter variations and variational circuit state bounds computed from the previous time step.
- Second, we further study the bounds computed by the

proposed method against the different sigma bounds by the standard MC method, which shows that the proposed method is more efficient for computing high sigma bounds than the MC method, which will increase rapidly (almost exponentially) with increasing sigma. In contrast, the run time of the proposed method will remain the almost the same as it only deals with different parameter bounds with the same number of parameters.

Experimental results show that the new method can deliver one or two order of magnitudes speedup over standard Monte Carlo simulation on some typical analog circuits and interconnect circuits with very high accuracy.

2. New time-domain performance bound analysis technique

We first present the whole algorithm flow of the proposed new performance bound analysis algorithm in Alg. 1. Basically the proposed method consists of three major computing steps. The first step is to set up the symbolic circuit matrices in the time domain based on the companion models of the dynamic elements (Step 2). The second step is to compute the variational closed form expressions of interesting states from the variational circuit parameters, which will be done via DDD-based symbolic analysis method (Step 3). Third, we compute the time-domain response bounds via a constrained nonlinear optimization process in each time step (Step 6-7). We will present the computing steps in the following subsections.

Algorithm 1 New time-domain performance bound analysis

Input: Circuit netlist, bounds of selected parameters.
Output: Conservative performance bound of interests
1: Convert the circuit C and L elements into companion models
2: Generate symbolic expression of closed form expressions for interesting nodes
3: **for** each time step **do**   // *Perform transient analysis*
4:    Set bounds on process variational parameters.
5:    Set bounds on the voltage or current states from results of optimization of last time step.
6:    Run nonlinear constrained optimization (9) which uses closed form function as the objective. To find upper bound and lower bound.
7:    Save bound information for the optimization of next time step.
8: **end for**
9: Output the bound of voltage or current on every time step.

A. Symbolic transient analysis for analog circuits

In this subsection, we review a graph-based transient symbolic analysis for obtaining the exact symbolic closed form expressions of analog circuits. Graph-based symbolic technique is a viable tool for calculating the behavior or characteristic of analog circuits [16]. The introduction of determinant decision diagrams based symbolic analysis technique (DDD) allows exact symbolic analysis of much larger analog circuits than all the other existing approaches [17], [18]. Furthermore, with

hierarchical symbolic representations [19], exact symbolic analysis via DDD graphs essentially allows the analysis of arbitrary large analog circuits.

Existing symbolic analysis was mainly formed in the frequency domain to build the symbolic transfer functions [20]. Symbolic analysis in time domain is less investigated and will be explored in this paper. To better illustrate the proposed method, we would like to walk through one simple example. Fig. 1 shows a simple RC ladder circuit. To perform the transient analysis, we first convert capacitance into its companion models (using the Back-Euler method) as shown in Fig. 2.

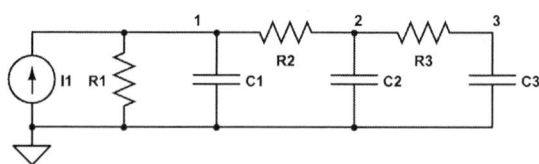

Fig. 1: RC ladder circuit

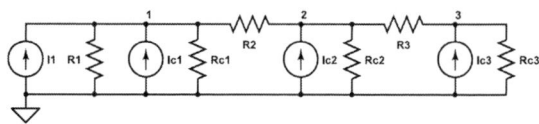

Fig. 2: RC ladder with companion models for capacitances

The corresponding modified nodal analysis (MNA) formulation of the circuit in time-domain at time step $n+1$ can be written as:

$$Y\vec{v}(n+1) = \vec{i}(n+1) \qquad (1)$$

where Y is the MNA matrix given by

$$\begin{bmatrix} \frac{1}{R_1} + \frac{1}{R_{c1}} & -\frac{1}{R_2} & 0 \\ -\frac{1}{R_2} & \frac{1}{R_2} + \frac{1}{R_{c2}} + \frac{1}{R_3} & -\frac{1}{R_3} \\ 0 & -\frac{1}{R_3} & \frac{1}{R_3} + \frac{1}{R_{c3}} \end{bmatrix} \qquad (2)$$

and

$$\vec{v}(n+1) = \begin{bmatrix} v_1(n+1) \\ v_2(n+1) \\ v_3(n+1) \end{bmatrix} \qquad (3)$$

and

$$\vec{i}(n+1) = \begin{bmatrix} i_1(n+1) + i_{c1}(n) \\ i_{c2}(n) \\ i_{c3}(n) \end{bmatrix} \qquad (4)$$

where $R_{c1} = \frac{C_1}{\Delta t}$, $R_{c1} = \frac{C_2}{\Delta t}$ $R_{c1} = \frac{C_3}{\Delta t}$, $i_{c1}(n) = \frac{v_1(n)*C_1}{\Delta t}$, $i_{c2}(n) = \frac{v_2(n)*C_2}{\Delta t}$, $i_{c3}(n) = \frac{v_3(n)*C_3}{\Delta t}$ and Δt is the time step size.

Then the unknown nodal voltage are solved using Cramer's rules.

$$v_i(n+1) = \frac{det(Y_i(n+1))}{det(Y)} \qquad (5)$$

where $Y_i(n+1)$ is the matrix formed by replacing the ith column of Y by vector $\vec{i}(n+1)$.

DDD is a very powerful tool to compute the symbolic determinant. Once the characteristics of circuits are presented by

DDDs, evaluation of DDDs, whose CPU time is proportional to the size of DDDs, will give exact numerical values.

We view each entry in the circuit matrix as one distinct symbol, and rewrite its system determinant in the left-hand side of Fig. 3. Then its DDD representation is shown in the right-hand side.

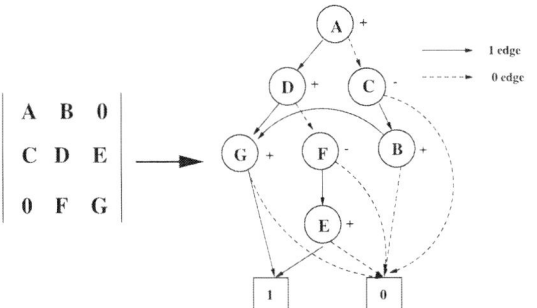

$$\begin{vmatrix} A & B & 0 \\ C & D & E \\ 0 & F & G \end{vmatrix} \longrightarrow$$

Fig. 3: A matrix determinant and its DDD representation.

Once a DDD has been constructed, the numerical values of the determinant it represents can be computed by performing the depth-first type search of the graph and performing one multiplication and addition at each node, whose time complexity is linear function of the size of the graphs (its number of nodes). The computing step is called *Evaluate(D)* where D is a DDD root. With proper node ordering and hierarchical approaches, DDD can be very efficient to compute transfer functions of large analog circuits [17], [19].

B. Variational symbolic closed-form expressions for transient states

To find the performance bounds of specific transient state variable, say $v_i(n+1)$ at time step $n+1$, DDD graphs are built for $det(Y_i(n+1))$ and $det(Y)$, we will obtain the following closed form symbolic expression for $v_i(n+1)$,

$$v_i(n+1) = f_i(p_1, \ldots, p_m, v_1(n), \ldots, v_k(n)) =$$
$$\frac{f_{n,i}(p_1, \ldots, p_m, v_1(n), \ldots, v_k(n))}{f_{d,i}(p_1, \ldots, p_m)} \quad (6)$$

where functions $f_{n,i}(p_1, \ldots, p_m, v_1(n), \ldots, v_k(n))$ and $f_{d,i}(p_1, \ldots, p_m)$ are represented by DDD graphs and $p_1, \ldots, p_m$ are m circuit variables and $v_1(n), \ldots, v_k(n)$ are the state variables computed from previous time step n. Notice that $v_i(n+1) = f_i(p_1, \ldots, p_m, v_1(n), \ldots, v_k(n))$ describes nonlinear functions in terms of $p_1, \ldots, p_m, v_1(n), \ldots, v_k(n)$. All the variables at current time step $n+1$ have variational bounds:

$$p_{il} \leq p_i \leq p_{iu} \quad (7)$$
$$v_{il}(n) \leq v_i(n) \leq v_{iu}(n) \quad (8)$$

Note that the variational bounds of state variable $v_i(n)$ are obtained form the previous time step n. In our presentation, we assume that the external voltage or current sources do not have variations to simplify our presentation. But this is not the limitation of the proposed method and we can trivially add this into our method.

To compute the numerical value of $v_i(n+1)$ for given specific values of $v_i(n+1) = f_i(p_1, \ldots, p_m, v_1(n), \ldots, v_k(n))$, this can be done by DDD *Evaluation* operation, which traverses the DDD in a depth-first style and performs one multiplication and one addition at each node.

Get back to the illustrative example, for voltage at node i at time step $n+1$, $v_i(n+1)$, we have

$$v_i(n+1) = f_i(C_1, C_2, C_3, R_1, R_2, R_3, v_1(n), v_2(n), v_3(n))$$

C. Variational bound analysis in time domain

To find the performance bounds subject to the parameter variations at time step $n+1$, we formulate the bound computing problem into a nonlinear constrained optimization problem. We use the lower bound of the voltage of node i on time step $n+1$ for an example. The symbolic expression of the voltage of node1, which has been obtained by DDD symbolic analysis, is used as the nonlinear objective function to be minimized:

$$\begin{aligned} \text{minimize} \quad & v_i(n+1)(\mathbf{x}) = f_i(\mathbf{x}) \\ \text{subject to} \quad & \mathbf{x}_{\text{lower}} \leq \mathbf{x} \leq \mathbf{x}_{\text{upper}}, \end{aligned} \quad (9)$$

where $\mathbf{x} = [\mathbf{p}, \mathbf{v}]$, in which, $\mathbf{p} = [p_1, \ldots, p_m]$ represents the circuit parameter variable vector, which is subjected to the optimization constraints $[\mathbf{p}_{\text{lower}}, \mathbf{p}_{\text{upper}}]$. In circuit design, foundries and cell library vendors supply these constraints. On the other hand, $\mathbf{v} = [v_1(n), \ldots, v_k(n)]$ represents the nodal voltage on the last time step, which are determined by the results of optimization of the last time step. Hence, after (9) is solved by an optimization engine, the lower bound of the v_1 on $(n+1)$th time step is returned and then serves as constrained condition for the optimization of voltage on $(n+2)$th time step.

The nonlinear optimization problem with simple upper and lower bounds given in (9) can be efficiently solved by several methods such as active-set, interior-point, and trust-region algorithms [21]–[23]. All those methods are iterative approaches starting with an initial feasible solution. In this work, we use the active-set method [23], as it turns to be the most robust nonlinear optimization method for our application. Active-set method is a two-phase iterative method that provides an estimate of the active set (which is the set of constraints that are satisfied with equality) at the solution. In the first phase, the objective is ignored while a feasible point is found for the constraints. In the second phase, the objective is minimized while feasibility is maintained. In this phase, starting from the feasible initial point $\mathbf{x}_0$, the method computes a sequence of feasible iterates $\{\mathbf{x}_k\}$ such that $\mathbf{x}_{k+1} = \mathbf{x}_k + \alpha_k \mathbf{d}_k$ and $f(\mathbf{x}_{k+1}) \leq f(\mathbf{x}_k)$ via methods like quadratic programming, where $\mathbf{d}_k$ is a nonzero search direction and α_k is a nonnegative step length.

Since the responses at two neighboring time step are usually close to each other, the starting point $\mathbf{x}$ for nth time step can be set using the solution on $(n-1)$th time step. This strategy tends to reduce the time required by the optimization to search its minimal or maximal point in the whole variable space, and thus speedup the calculation time of the bound analysis.

We remark that the active-set method is still a local optimization method, which finds the local optimal solutions. But find the true bound may come with more or much higher

978-1-4799-2817-0/14 $31.00 © 2014 IEEE   457

computing costs by performing many tries. In our approach, we still perform one optimization. Our experimental results show that the proposed method gives conservative bounds for given sigma values compared with Monte Carlo methods for the examples used.

3. Numerical results and discussions

In this section, we show experimental results of the proposed method on interconnect circuits and analog circuits. The DDD symbolic tool generates the exact transfer function expressions first [17], and all the follow-up optimization based bound calculation and yield estimation are done in MATLAB. The nonlinear constrained optimizations are solved by the *fmincon* function in MATLAB's Optimization Toolbox [24]. All running time are sampled from a Linux server with a 2.4 GHz Intel Xeon Quad-Core CPU, and 36 GB memory.

We compare proposed method with standard Monte Carlo analysis in terms of running time and accuracy using two examples. In all the examples, we assume that variational parameter has Gaussian distributions with the standard deviation σ. Their variational bound (3-sigma bound) will be $[-3\sigma + \mu, 3\sigma + \mu]$ where μ is the mean of the random process.

A. An interconnect RC tree circuit example

The first example is an interconnect RC tree example, which is driven by a voltage source as shown in Fig. 4.

The variational parameters are $R_i = 0.1\Omega$, $i = 1, 2, 3$, $C_j = 0.1pf$, $j = 1, 2, 3$. All parameters have 10% variations, which means that, for proposed method, the constrained condition is $(1 - 5\%) * p_{std} \le p \le (1 + 5\%) * p_{std}$, for Monte Carlo analysis, $\sigma = 1/6 * 10\% * p_{std}$ (3-sigma bound), in which, p represents the value of a certain varational parameter, and p_{std} is the standard value of the parameter.

Fig. 5 shows the transient step response 3-sigma bound of voltage of node 8 with that from the proposed method and simulation result from 5000 MC runs. This figure shows that the bounds from proposed method could safely cover the curves from the Monte Carlo simulation. Fig. 6 shows 3-sigma bounds from 2000 MC runs, 5000 MC runs and the proposed method at 0.5 ns.

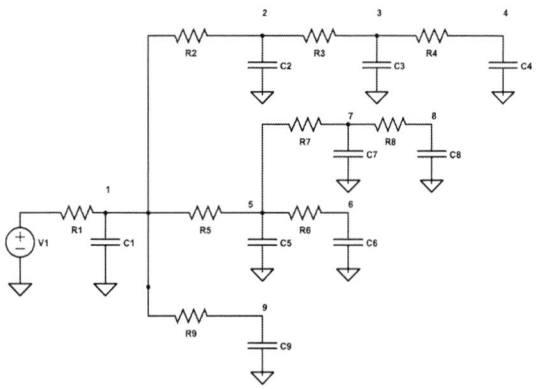

Fig. 4: A RC tree circuit

We have several observations: First, the bounds given by the proposed method matches with that given by the MC method

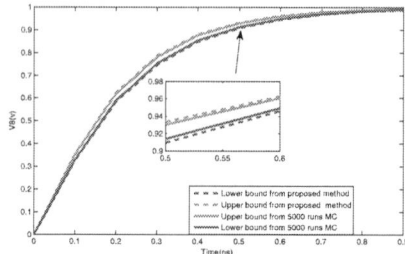

Fig. 5: The bounds of V8 obtained from 5000 MC runs and the proposed method on the RC tree circuit

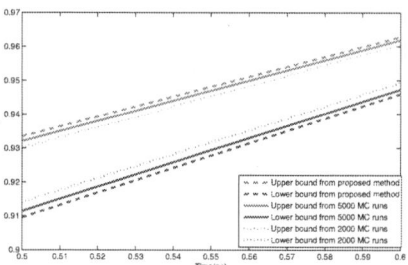

Fig. 6: Comparison of bounds of V8 from 2000 MC runs, 5000 MC runs and the proposed method on the RC tree circuit

very well. Since all of parameters take 3-sigma bounds, the bounds computed by the proposed method should be close to 3-sigma bounds as well. If the output bounds are Gaussian, then 3-sigma will cover 99.730% area under the probability density function (pdf) of Gaussian distribution, which means we need to take at least 370 MC runs to have event to reach the bound.

Table I compares the runtime, voltage values of the proposed method and that of the Monte Carlo method and also shows the error ratio of 2000 MC runs, 5000 MC runs. The table also shows that, our proposed method has 8.3x speedup over 5000 MC run simulation.

TABLE I: Comparison between the methods on lower bounds of V8 at t=0.5ns for the RC tree circuit

Method	Samplings#	CPU(sec)	Voltage(V)	% Error
Monte Carlo	2000	573.383	0.915	0.55
Monte Carlo	5000	1490.798	0.912	0.22
Proposed Method	1	180.432	0.910	N/A

To further study the bounds computed by the proposed method, we compared 3-sigma bounds given by 15K, 30K, 50K MC runs. Fig. 7a shows the 3-sigma upper bound of V8 from the 15K, 30K, 50K runs of MC simulation and that from our proposed method around 0.5ns. In this figure, we observe that 3-sigma bounds given by 30K and 50K now go outside the bound of the proposed method.

Fig. 7b shows the 3-sigma lower bound of V8 from 15K, 30K, 50K runs of MC simulation and the bound give by the proposed method. In this case, we observe that the the bound by the proposed method contain ALL the bounds by different

978-1-4799-2817-0/14 $31.00 © 2014 IEEE   458

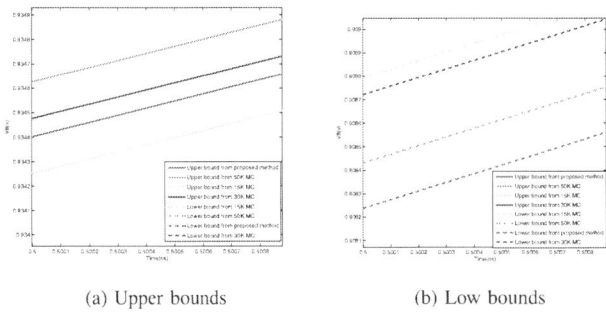

(a) Upper bounds   (b) Low bounds

Fig. 7: Comparison of 3-sigma upper and lower bounds of V8 from 15K MC runs, 30K MC runs, 50K MC runs and the proposed method on the RC tree circuit at 0.5ns

MC runs. This is a very interesting observation.

One possible explanation for the Fig. 7a and Fig. 7b is that the performance function may not be monotonic function for the some of parameter variables (called non-monotonic parameters here). In other words, the min/max values of performance function may not be reached at the edges of the bounds of those parameters. So the 3-sigma bound already includes the values to reach the min/max values of the functions. This can explain Fig. 7a, in which the maximum value is reached when most of variational parameter values found are not at edges of the bound. As a result, new approach will find more conservative bounds as those non-monotonic parameters reach their min/max values already. For Fig. 7b, on the other hand, the minimum value is reached when many variational parameters are at edges of the bound. However, it is almost impossible for all variational parameters are at edge of bounds at the same time considering the Gaussian distribution especially when the number of variational parameter is large. Therefore, the bound computed by new approach close to the true bound and more MC run can only get closer to the bound, but can't go beyond the true bound. As a result, we can see the new method tend to find the true bound more efficient than the MC method, especially for performance functions which achives min/max values when many parameters are at edge of bounds, as it requires quite a great amount of samplings to possibly get the maxmum or minimum.

To further study the behavior of the proposed method, we perform 4-sigma bound analysis in which the bounds of each parameters will be $[-4\sigma + \mu, 4\sigma + \mu]$. Fig. 8a shows the 4 sigma upper bounds from 100K MC runs, 200K MC runs and the proposed method at 0.1ns. Fig. 8b shows the 4-sigma lower bounds from 100K MC runs, 200K MC runs and that from proposed method at 0.1ns. From the two figures we can see, that in this case, the proposed method contain both upper bounds and lower bounds from the MC runs (even with 200K MC). It means that 4-sigma upper bound computed by our method is large than 4-sigma bounds of MC simulation. On the other hand, for the lower bounds, we observe the same results as 3-sigma bound results: the proposed method is always lowest bound among all the methods.

As a result, it can be seen that the proposed method is more efficient to find the high sigma bounds, as it takes almost the

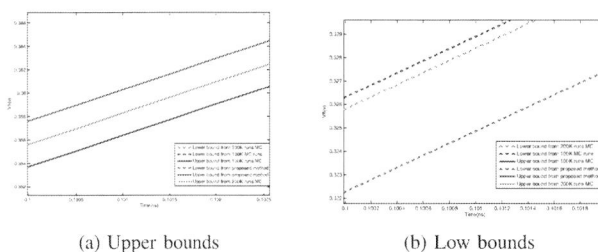

(a) Upper bounds   (b) Low bounds

Fig. 8: Comparison of 4-sigma upper and lower bounds of V8 from 100K MC runs, 200K MC runs and the proposed method on the RC tree circuit at 0.1ns

TABLE II: Comparison between the methods on lower bounds of Vout at t=1ms for the amplifier circuit

Method	Samplings#	CPU(sec)	Voltage(V)	% Error
Monte Carlo	2000	412.071	-0.942	3.7
Monte Carlo	5000	1112.597	-0.906	0.79
Proposed method	1	105.460	-0.899	N/A

same computational costs as computing low sigma bounds, than the standard MC methods, whose computational costs go up almost exponentially with high sigma bounds.

B. An opamp circuit example

The second example is an opamp circuit with 7 MOSFETs as shown in Fig. 9a. To perform the bound analysis, we use a linearized and simplified device models for the MOSFETs as shown in Fig. 9b. The variable parameters are $M_1.g_m = 1.5 * 10^{-5}$, $M_1.C_{gd} = 0.5fF$, $M_1.C_{gs} = 5fF$, $M_2.g_m = 1.5 * 10^{-5}\Omega^{-1}$, $M_2.C_{gd} = 0.5fF$, $M_2.C_{gs} = 5Ff$, $M_5.r_{ds} = 5 * 10^7\Omega$, $M_6.r_{ds} = 5 * 10^7$ Ω. Again all parameters have 10% variations.

Fig. 10 shows the transient response 3-sigma bound of Vout with sinusoidal wave input obtained from proposed method and simulation result from 5000 MC runs. Fig. 11 shows the 3-sigma bounds from 2000 MC runs, 5000 MC runs, and the proposed method at 1ms. In this case, we observe that the bounds from proposed method is still conservative such that it sill contain the bounds from all the MC runs. The possible reasons have been explained before.

Table II compares the runtime, voltage values of the proposed method and that of Monte Carlo method. It also shows the error ratio of 2000 MC runs and 5000 MC runs. It can been seen that the errors are quite small and get smaller as we take more MC runs, which is the consistent with the MC method.

The same table also shows that, our proposed method has 10.6x speedup over 5000 samplings MC simulation. We remark that, if high sigma (larger than 3 sigma) bounds, the standard MC runs will increase rapidly (almost exponentially), while the run time of the proposed method will remain the almost the same as it only deal with different parameter bounds with the same number of parameters. As a result, the proposed method indeed overcome the high sigma issues with the standard MC based method, which is the major advantage of the proposed method over MC based methods.

978-1-4799-2817-0/14 $31.00 © 2014 IEEE

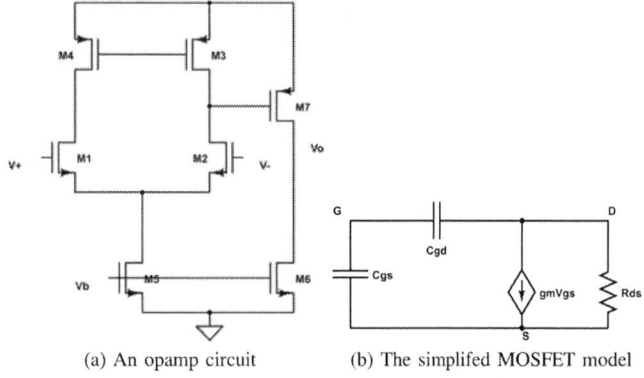

(a) An opamp circuit   (b) The simplifed MOSFET model

Fig. 9: The opamp circuit and its MOSFET model

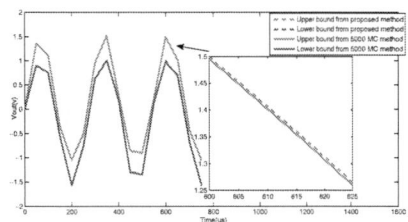

Fig. 10: The bounds from 5000 MC runs and the proposed method on the amplifier circuit

4. Conclusion

In this paper, we have presented a new non-Monte-Carlo performance bound analysis technique in time domain. The new method is based on the constrained non-linear optimization and advanced symbolic analysis techniques. We have shown that the proposed method is more efficient for computing high sigma bounds than the standard MC method. Experimental results have shown that the new method can deliver order of magnitudes speedup over standard Monte Carlo simulation on some typical analog circuits and interconnect circuits with very high accuracy.

References

[1] G. F. Fishman, *Monte Carlo, concepts, algorithms, and Applications.* Springer, 1996.

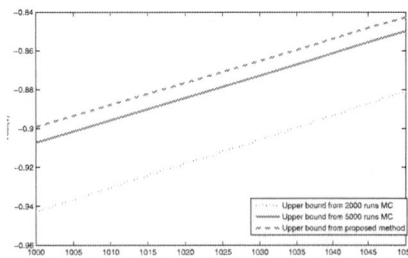

Fig. 11: Comparison of upper bounds of Vout from 2000 MC runs, 5000 MC runs and the proposed method on the the amplifier circuit

[2] A. Singhee and R. A. Rutenbar, "Why quasi-monte carlo is better than monte carlo or latin hypercube sampling for statistical circuit analysis," *IEEE Transactions on Computer-Aided Design of Integrated Circuits and Systems*, vol. 29, no. 11, pp. 1763–1776, 2010.

[3] T. S. Doorn, E. J. W. ter Maten, J. A. Croon, A. Di Bucchianico, and O. Wittich, "Important sampling monte carlo simulations for accurate estimation of sram yield," in *ESSCIRC 2008 - 34th European Solid-State Circuits Conference*, pp. 230–233, IEEE, 2008.

[4] J. F. Swidzinski, M. Keramat, and K. Chang, "A novel approach to efficient yield estimation for microwave integrated circuits," in *42nd Midwest Symposium on Circuits and Systems*, pp. 367–370, IEEE, 1999.

[5] A. B. Owen, "Latin supercube sampling for very high-dimensional simulations," *ACM Transactions on Modeling and Computer Simulation*, vol. 8, pp. 71–102, Jan. 1998.

[6] B. Liu, J. Messaoudi, and G. Gielen, "A fast analog circuit yield estimation method for medium and high dimensional problems," in *Proc. Design, Automation, and Test in Europe (DATE)*, pp. 751–756, 2012.

[7] L. Kolev, V. Mladenov, and S. Vladov, "Interval mathematics algorithms for tolerance analysis," *IEEE Trans. on Circuits and Systems*, vol. 35, pp. 967 –975, Aug. 1988.

[8] W. Tian, X.-T. Ling, and R.-W. Liu, "Novel methods for circuit worst-case tolerance analysis," *IEEE Trans. on Circuits and Systems I: Fundamental Theory and Applications*, vol. 43, pp. 272 –278, Apr. 1996.

[9] C.-J. R. Shi and M. W. Tian, "Simulation and sensitivity of linear analog circuits under parameter variations by robust interval analysis," *ACM Trans. Des. Autom. Electron. Syst.*, vol. 4, pp. 280–312, July 1999.

[10] L. Qian, D. Zhou, S. Wang, and X. Zeng, "Worst case analysis of linear analog circuit performance based on kharitonov's rectangle," in *Proc. IEEE Int. Conf. on Solid-State and Integrated Circuit Technology (ICSICT)*, Nov. 2010.

[11] S. Saibua, L. Qian, and D. Zhou, "Worst case analysis for evaluating VLSI circuit performance bounds using an optimization method," in *IEEE/IFIP 19th International Conference on VLSI and System-on-Chip*, pp. 102–105, 2011.

[12] Z. Hao, R. Shen, S. X.-D. Tan, and G. Shi, "Performance bound analysis of analog circuits considering process variations," in *Proc. Design Automation Conf. (DAC)*, pp. 310–315, July 2011.

[13] X.-X. Liu, S. X.-D. Tan, Z. Hao, and G. Shi, "Time-domain performance bound analysis of analog circuits considering process variations," in *Proc. Asia South Pacific Design Automation Conf. (ASPDAC)*, Jan. 2012.

[14] V. L. Kharitonov, "Asymptotic stability of an equilibrium position of a family of systems of linear differential equations," *Differential. Uravnen.*, vol. 14, pp. 2086 –2088, 1978.

[15] X. Liu, A. Palma-Rodriguez, S. Rodriguez-Chavez, S. X.-D. T. E. Tlelo-Cuautle, and Y. Cai, "Performance bound and yield analysis for analog circuits under process variations," in *Proc. Asia South Pacific Design Automation Conf. (ASPDAC)*, pp. 761–766, January 2013.

[16] G. Gielen, P. Wambacq, and W. Sansen, "Symbolic analysis methods and applications for analog circuits: A tutorial overview," *Proc. of IEEE*, vol. 82, pp. 287–304, Feb. 1994.

[17] C.-J. Shi and X.-D. Tan, "Canonical symbolic analysis of large analog circuits with determinant decision diagrams," *IEEE Trans. on Computer-Aided Design of Integrated Circuits and Systems*, vol. 19, pp. 1–18, Jan. 2000.

[18] C.-J. Shi and X.-D. Tan, "Compact representation and efficient generation of s-expanded symbolic network functions for computer-aided analog circuit design," *IEEE Trans. on Computer-Aided Design of Integrated Circuits and Systems*, vol. 20, pp. 813–827, April 2001.

[19] S. X.-D. Tan, W. Guo, and Z. Qi, "Hierarchical approach to exact symbolic analysis of large analog circuits," *IEEE Trans. on Computer-Aided Design of Integrated Circuits and Systems*, vol. 24, pp. 1241–1250, August 2005.

[20] Z. Qin, S. X.-D. Tan, and C. Cheng, *Symbolic Analysis and Reduction of VLSI Circuits*. Boston, MA: Kluwer Academic Publishers, 2005.

[21] R. H. Byrd, R. B. Schnabel, and G. A. Shultz, "A trust region algorithm for nonlinearly constrained optimization," *SIAM Journal on Numerical Analysis*, vol. 24, no. 5, pp. pp. 1152–1170, 1987.

[22] P. E. Gill, W. Murray, Michael, and M. A. Saunders, "Snopt: An sqp algorithm for large-scale constrained optimization," *SIAM Journal on Optimization*, vol. 12, pp. 979–1006, 1997.

[23] C. A. Floudas, *Nonlinear and Mixed-Integer Optimization: Fundamentals and Applications (Topics in Chemical Engineering)*. Oxford University Press, 1995.

[24] The Mathworks Inc., *MATLAB Optimization Toolbox*. http://www.mathworks.com/help/toolbox/optim/, 2012.

978-1-4799-2817-0/14 $31.00 © 2014 IEEE

A Robustness Optimization of SRAM Dynamic Stability by Sensitivity-based Reachability Analysis

Yang Song, Sai Manoj P D and Hao Yu

School of Electrical and Electronic Engineering,
Nanyang Technological University, Singapore 639798

haoyu@ntu.edu.sg

Abstract—**A robustness optimization of SRAM dynamic stability at nano-scale is developed in this paper by zonotope-based reachability analysis. A backward Euler method is developed to efficiently perform reachability analysis by zonotope to deal with multiple device parameters with tuning ranges. Moreover, a sensitivity calculation of zonotope is developed to optimize *safety distance* by simultaneously tuning multiple SRAM device parameters without multiple repeated computations. As such, sequential robustness optimizations can be performed such that the optimized SRAM designs can depart from unsafe region but converge into safe region. The proposed method is implemented inside a SPICE-like simulator. As shown by numerical experiments, the proposed method can achieve $600\times$ speedup on average compared to the traditional verification method by Monte-Carlo under the similar accuracy. In addition, compared to the traditional small-signal based sensitivity optimization, the proposed method can converge faster with high accuracy.**

I. INTRODUCTION

Stability verification and robustness optimization are emerging needs for integrated circuit (IC) designs at nano-scale. The stability challenge is pronounced for densely integrated SRAM circuits with minimum feature sizes. Note that static noise margin (SNM) [1] is traditionally deployed for SRAM stability characterization because of simple interpretation and measurement. As it may overestimate read-failure and underestimate write-failure, dynamic SRAM stability margin [2] is increasingly adopted by deploying critical word-line pulse-width that can produce a better estimation of failures. The verification of SRAM stability margin becomes harder at nano-scale. Firstly, due to the nonlinear dynamics, the SRAM characteristic behavior becomes less digital but more analog. Secondly, process variations such as threshold-voltage V_{th} [3], [4], [5], [6], [7] can further significantly suppress the SRAM stability margin, and result in higher failure rate during read/write operations.

Many recent works have been performed for SRAM stability characterization [8], [9], [10]. For example, Euler-Newton curve-tracing is utilized to find the boundary between safe and unsafe regions in the *parameter space* without brute-force exploration. But, this method is limited to considering two parameters, and the computational cost is prohibitive considering parameter variations from all transistors. The work in [10] formulates a dynamic stability margin to characterize the stability boundary, namely the *separatrix* [8] in the *state space*. The separatrix provides intuitive illustration of SRAM nonlinear behavior and is formed by combining two transient trajectories which start from the same equilibrium state on separatrix but move towards different directions. Yet, separatrix has to be recomputed every time when any parameter changes. What is more, it is unclear how to consider parameter tuning using separatrix for SRAM robustness optimization.

Note that reachability analysis has been widely deployed in stability verification of system dynamics by exploring potential trajectories of operating points in state space [11], [12]. It can accurately predict boundary of multiple trajectories with uncertain states by one-time simulation, in contrast to obtaining multiple trajectories in repeated simulations. The reachability analysis has been deployed for a number of hard analog circuit verifications [13], [14], [6]. A set of system trajectories in state space can be bounded by the zonotope-based over-approximation. A time-interval integrated reachability analysis with formed zonotope that can distinguish safe and unsafe regions is performed such that the failure in state space can be verified. However, it is unknown how to perform efficient reachability analysis that can be further applied for optimization with consideration of multiple device parameters.

In this paper, to consider multiple device parameters in one simulation, a zonotope-based reachability analysis is developed for robustness optimization of SRAM dynamic stability. Device parameters such as SRAM transistor widths are considered by zonotope matrices during reachability analysis. A corresponding sensitivity calculation is developed and deployed for the optimization of SRAM dynamic stability such that the SRAM designs can depart from unsafe regions. What is more, based on backward Euler method, zonotope-based verification and optimization procedures considering nonlinear device model of transistors are implemented in a SPICE-like simulator. Compared to the traditional Monte-Carlo based verification, our method achieves nearly up to $600\times$ speedup with similar accuracy. Moreover, as multiple-parameter large-signal sensitivity is generated for a safety distance, compared to the traditional single-parameter small-signal based sensitivity optimization, our method can converge faster with high accuracy.

II. PROBLEM FORMULATION OF SRAM FAILURE ANALYSIS AND OPTIMIZATION

Similar to [8], [9], [10], [6], the scope of this paper focuses on the transistor-level analytical approaches for SRAM dynamic stability optimization under V_{th} variations. When statistical distribution of V_{th} variation is known, one can efficiently generate yield statistics from the transistor-level verification results. In a 6T-SRAM there exists serious stability failure concern with V_{th} variation [6], which can lead to SRAM functional failures during read and write operations. Though transistor sizing may compensate the negative impact of V_{th} variations, it is unknown how to adjust transistor size for robustness optimization for the sake of SRAM dynamic stability.

In this section, we introduce the following definition to quantitatively describe the robustness of SRAM dynamic stability.

Definition 1:
Safety Distance is the Euclidean distance $\|p_{safe} - x\|_2$ in the state space between one operating point x and the safe state p_{safe}.

Note that different from separatrix based approaches [8], [10], the safety distance provides indication on the optimization direction of trajectory. As such, it can be conveniently leveraged within reachability analysis to consider parameter and also input variations at the same time by performing one-time transient simulation.

A. Failure Mechanisms

Physical mechanisms of SRAM failures i.e. read and write failures in terms of safety distance in the state space are described here. In addition, there exist two convergent regions in the state-variable space of SRAM [8]. Operating points on either region converge to the nearest equilibrium state.

1) Write Failure: A write-failure refers to the inability to write data properly into the SRAM cell. During write operation, both access transistors should be strong enough to change the voltage level at internal nodes. As shown in Fig.1, write operation can be described on the state-variable plane as the procedure of pulling the operating point from initial state (bottom right corner) to the target state (top left corner). Thus the safety distance refers to the distance between operating point and the target state. Given enough time, the operating point in any region will converge to the nearest stable equilibrium state. The write operation is aimed at pulling operating point to the target state and thus reducing the safety distance, as shown by point B in Fig.1.

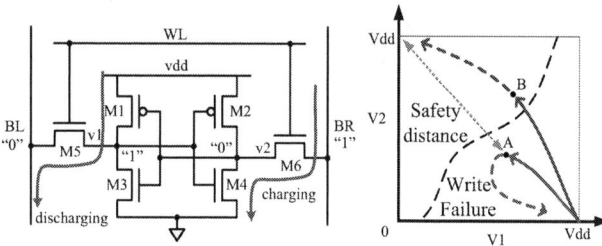

Fig. 1: Illustration of write failure.

V_{th} variations may cause write failure. An increase in V_{th} can reduce the strength of the transistor. For example, increase of V_{th} in M6 along with decrease of V_{th} in M4 can make it more difficult to pull up v_2. If operating point, which slowly moves towards target-state, cannot reach other convergent region before access transistors are closed, it will move back to the initial state implying a write failure. To resolve the failure, tuning width of M6 as increased while M4 as narrowed can help reduce safety distance and hence can mitigate the side effect from V_{th} variations.

2) Read Failure: A read-failure refers to the loss of the previously stored data in SRAM during read operation. Access transistors need careful sizing such that their pull-up strengths are not strong enough to pull digital "0" to "1" during read operation. On the state-variable plane, operating point of SRAM is inevitably perturbed and pulled towards the other convergent region. In this situation, the safety distance is from the operating point to its initial state. If read operation does not last too long, access transistors shut down before the operating point converges to the other region. The safety distance will converge to zero as the operating point returns to the initial state in the end, as shown by point A in Fig.2.

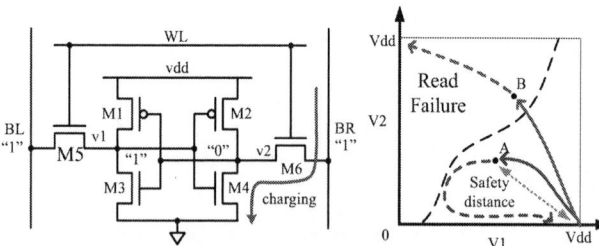

Fig. 2: Illustration of read failure.

V_{th} variations may also cause read failure. For example, variations caused by mismatch between M4 and M6 can result in unbalanced pulling strengths and v_2 can be pulled up more quickly. As a result, the operating point crosses to the other region resulting in read failure, as shown by point B in Fig.2. To resolve the failure, width of M6

needs to be scaled down to avoid excessive pulling strength, which may lead to write failure. In addition, V_{th} variations in M1-4 affect the locations of converging regions on the state-variable plane. As the opposite converging region migrates closer to the initial state, it becomes more likely for read-failure to happen.

Therefore, the problem to solve is to find an appropriate combination of sizing from all transistors to optimize the robustness of SRAM dynamic stability by circumventing potential hazards caused by V_{th} variations.

B. SRAM Dynamics

1) Nonlinearity: One primary challenge for SRAM dynamic stability optimization is from its nonlinear dynamic behavior. The time-evolution of safety distance depends on the nonlinear dynamics of SRAMs, which can be described by differential algebraic equation (DAE) as

$$\frac{d}{dt}q(x(t),t) + f(x(t),t) + u(t) = 0 \qquad (1)$$

where $x(t)$ is a state variable vector and $u(t)$ is input vector. Here $q(t)$, $u(t)$ contain charges and external sources; and $f(x,t)$ describes SRAM nonlinear dynamics. V_{th} variations of transistors can be introduced at input $u(t)$ as ad-hoc current sources [4].

After Newton iteration is performed at one selected operating point (or nominal point) x^*, $f(x(t),t)$ is linearized at this point as $\frac{\partial f}{\partial x}|_{x=x^*}$. Based on the mean-value theorem, the dynamic equation $f(x)$ at any neighbor operating point x can be expressed by a linear approximation with a 2nd-order residue i.e. the difference between nonlinear $f(t)$ and its linear approximation, called as *linearization error* denoted by L.

The SRAM dynamic equation (1) thereby can be depicted in a simplified form by

$$\frac{d}{dt}q(x,t) + f(x^*,t) + u^*(t) + G(x-x^*) + L = 0;$$

$with$

$$L = \frac{1}{2}(x-x^*)^T \frac{\partial^2 f}{\partial x^2}|_{x=\xi}(x-x^*); \quad G = \frac{\partial f}{\partial x}\Big|_{x=x^*} \qquad (2)$$

$$\xi \in \{x^* + \alpha(x-x^*)|0 \le \alpha \le 1\}.$$

Assume that $q(x,t)$ can be further decomposed into $q(x^*)$ and $C\Delta x$. Thus, we have (3)

$$\frac{d}{dt}q(x^*,t) + f(x^*,t) + u^*(t) = 0 \qquad (3a)$$

$$\frac{d}{dt}(C\Delta x) + G\Delta x + L = 0 \qquad (3b)$$

where $C = \frac{\partial q}{\partial x}|_{x=x^*}$.

(3a) is the nonlinear differential equation for the nominal point x^* and (3b) is the linear equation with the Euclidean distance from x^* to the neighbor point x.

On the basis of (3), reachability analysis can be deployed for SRAM dynamic stability verification and optimization in the state space. Reachability analysis can be performed on nonlinear trajectories with high accuracy by considering L.

2) Multiple Device Parameters: What is more, perturbations of multiple device parameters can be considered as well. Suppose that each transistor in SRAM has a width perturbation ΔW that affects transconductance g_m, namely Δg_m. One can have

$$\Delta g_m = \frac{\partial g_m}{\partial W}\Delta W. \qquad (4)$$

On basis of Δg_m, multiple device parameter perturbations can be included into conductance matrix by ΔG as follows

$$\Delta G = \begin{pmatrix} \ddots & & & \\ & \frac{\partial g_m}{\partial W} & -\frac{\partial g_m}{\partial W} & \\ & -\frac{\partial g_m}{\partial W} & \frac{\partial g_m}{\partial W} & \\ & & & \ddots \end{pmatrix} \Delta W. \qquad (5)$$

978-1-4799-2817-0/14 $31.00 © 2014 IEEE

Based on the above discussions to include multiple device parameters, one can deploy zonotope to form a set of region for multiple device parameters. With the further development of linear multi-step based integration for zonotope and its according sensitivity, one can develop reachability based robustness optimization, discussed in the later part of this paper.

C. Problem Formulation

Based on the aforementioned SRAM failure mechanisms and dynamics analysis, a robustness optimization for SRAM dynamic stability is proposed in terms of safety distance considering interval values of V_{th} variations from all transistors.

If the safety distance fails to converge to zero, a robust optimization of SRAM dynamic stability can be used to reduce safety distance, we call this as a verification oriented robustness optimization.

Problem of SRAM Robustness Optimization: To ensure the SRAM dynamic stability, one needs to minimize the safety distance measured at the final state of the system trajectory as follows

$$\min_{w} \quad F(w,t)$$
$$\text{subject to} \quad W_{min} < w_i < W_{max}, i = 1, 2, ..., m. \quad (6)$$

Here, w is the parameter or sizing vector for all m transistors with a defined range $[W_{min}, W_{max}]$. $F(w,t)$ is the objective function, weighted sum of safety distances for both read and write operations given by

$$F(w,t) = \begin{cases} D_w(w,t_w) + D_r(w,t_r), & \text{write and read failures} \\ D_w(w,t_w), & \text{write failure only} \\ D_r(w,t_r), & \text{read failure only} \end{cases} \quad (7)$$

where $D(w,t)$ is the safety distance and t is the pulse-width for read or write operation.

Due to the symmetrical structure, three transistor pairs are used to represent the 6T-SRAM. Thus, the robustness optimization task is performed in a three-dimensional parameter space, where a parameter-state point is denoted by $w \in \mathbf{R}^{3 \times 1}$. After the reachability analysis is performed, sensitivity of safety distance w.r.t. multiple device parameters can be obtained, which can guide the optimization routine to reduce or even eliminate failures caused by V_{th} variations with improved SRAM dynamic stability.

III. ZONOTOPE-BASED REACHABILITY ANALYSIS FOR SRAM DYNAMIC STABILITY VERIFICATION

Reachability analysis [11], [12], [13], [6] can efficiently determine a reachable region that a dynamic system evolves within a range of states. As such, one can perform one-time reachability analysis to form safe region with safe distance determined from the final state set as shown in Fig.3. With the linear multi-step implementation, the runtime cost or complexity of zonotope-based reachability analysis is similar with transient analysis in SPICE.

Here, we formulate of SRAM robustness optimization by *safety distance* in the framework of reachability analysis.

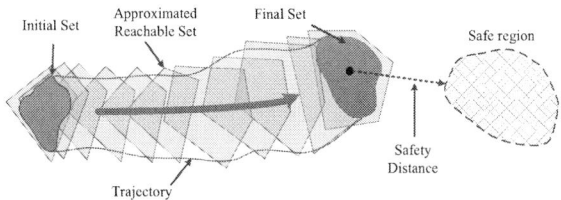

Fig. 3: System trajectory and safety distance with zonotopes.

A. Reachable Set and Zonotope

Interval-value analysis has been applied to model the uncertainties of state variables in [15], such as device parameters. For example, if Δx_1, Δx_2 model uncertainties in 2 different dimensions of state variable x with interval center c, then $x = c + [-1, 1]\Delta x_1 + [-1, 1]\Delta x_2$ is the neighboring point including these variations. However, there is no formal and efficient verification method developed to deal with multi-dimensional interval-value problem. Here, we model a multi-dimensional interval-value variable as a zonotope [16], [13], which is a convex polytope, to model multiple device parameters.

To start with, an important concept for reachability analysis is the reachable set.

Definition 2:

Reachable Set is the collection of all possible operating points or states in the state space that a system may visit, which can be approximated by an enclosing hypercube.

One simple and symmetrical type of hypercube, called *zonotope* [11] is defined as follows.

Definition 3:

Zonotope $\mathcal{X}$ is defined by

$$\mathcal{X} = \{x \in \mathbf{R}^{n \times 1} : x = c + \sum_{i=1}^{q} [-1, 1]g^{(i)}\} \quad (8)$$

where $c \in \mathbf{R}^{n \times 1}$ is the zonotope center; and $g^{(i)} \in \mathbf{R}^{n \times 1}$ is called as a zonotope generator.

One can observe from (8), a zonotope is essentially a multidimensional interval in affine form, with each generator $g^{(i)}$ in Fig.4 as a variation in a different direction, implying it can include device and parameter variations. Mathematically zonotope can also be expressed as *Minkowksi summation* [12] of two finite sets such that merged set preserves convexity. When reachability analysis for a nonlinear system is performed, the center of zonotope is utilized as the nominal point for the minimization of linearization error [16].

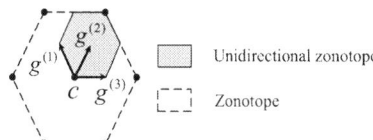

Fig. 4: Zonotope and unidirectional zonotope.

Note that in (8), range of scaling factors for generators is $[-1, 1]$ implying the difference vector from nominal point within a reachable set varies in two directions. For the calculation of sensitivity during robustness optimization, the scaling factor is defined within $[0, 1]$ to obtain a single-direction distance from the nominal point. We call this modified zonotope as *unidirectional zonotope* (Fig.4), determined by

$$\mathcal{X}^{uni} = \{x \in \mathbf{R}^{n \times 1} : x = c + \sum_{i=1}^{q} [0, 1]g^{(i)}\}. \quad (9)$$

Similar to (9), interval values for state matrices can also be modeled as zonotope [12]. Such an interval matrix can be described as *matrix zonotopes* $\mathcal{M}$.

$$\mathcal{M} = \{M \in \mathbf{R}^{n \times n} : M = M^{(0)} + \sum_{i=1}^{q} [0, 1]M^{(i)}\}. \quad (10)$$

Similar to zonotopes, the matrix $M^{(0)}$ is called *center matrix* and the matrix $M^{(i)}$ is called *generator matrix*, which contains the variation ranges of perturbed device parameters. Addition and multiplication rules for zonotopes and matrix zonotopes are defined in [12].

B. Reachability Analysis

To solve the dynamic equation of SRAM in (3a), a SPICE-like simulator is applied in this paper. Eq. (3b) with zonotope-based evolution can be solved by backward Euler method with discretized time-step h at kth-iteration by

$$\Delta x_k^{(i)} = A^{-1}(\frac{C}{h}\Delta x_{k-1}^{(i)} - L_k), k = 1, ..., K; i = 1, ..., q. \quad (11)$$

in which $A = \frac{C}{h} + G$ is Jacobian matrix, K and q represents number of time steps and zonotope generators respectively. These zonotope generators $\Delta x_k^{(i)}$ are the Euclidean distances in (2) from the nominal point (zonotope center) x^* to neighbor points x, with zonotopes formed as defined in (9).

The according iteration equation for reachability analysis is built after substituting $\Delta x_k^{(i)}$ by zonotope generator matrix $X_k = [\Delta x_k^{(1)}, ..., \Delta x_k^{(q)}]$, Jacobian matrix A by matrix zonotope $\mathcal{A}$, and capacitance matrix C by matrix zonotope $\mathcal{C}$. As such, one can have

$$X_k = \mathcal{A}^{-1}(\frac{\mathcal{C}}{h}X_{k-1} - L_k), k = 1, ..., K. \quad (12)$$

Here, $\mathcal{A}$ and $\mathcal{C}$ contain variations from multiple device parameters such as transistor width sizings in the case of SRAMs, whose solution is Minkowski summation [12] of individual matrices. In $\mathcal{A}$, interval conductance matrix ΔG can be computed using interval values of transistor width ΔW similar to (5). As such, the zonotope matrix in terms of interval-valued matrices given by

$$A \in [A^{(0)} - \sum_i |A^{(i)}|, A^{(0)} + \sum_i |A^{(i)}|]$$
$$A^{(i)} = \frac{\partial A^{(0)}}{\partial W}\Delta W^{(i)} = \Delta G^{(i)}. \quad (13)$$

Here, $A^{(0)}$ is the nominal state matrix without variations, and $A^{(i)}$ is the variation of state matrix caused by perturbation due to the i-th transistor width $\Delta W^{(i)}$. Similarly other interval conductances including g_{ds} and g_{mb} and capacitances can be derived.

Moreover, in (12) the reciprocal of matrix zonotope $\mathcal{A}$ with m width variations $\mathcal{A} = (A^{(0)}, ..., A^{(m)})$ can be evaluated by two steps. Firstly, $(A^{(0)})^{-1}$ is calculated by LU decomposition. And then, one can have $\mathcal{A}^{-1}$ expanded as

$$\mathcal{A}^{-1} = ((A^{(0)})^{-1}, ..., (A^{(0)})^{-1}A^{(m)}(A^{(0)})^{-1}). \quad (14)$$

This approach leads to an implementation of reachability analysis by SPICE-like simulator.

IV. ROBUSTNESS OPTIMIZATION OF SRAM DYNAMIC STABILITY

In this section, we first introduce safety distance under zonotope, and discuss the according sensitivity calculation of safety distance, which is applied for SRAM dynamic stability optimization by tuning multiple SRAM device parameters simultaneously.

A. Safety Distance

Assume that one safe state is located at p_{safe} in the state space. As for any zonotope in the form of (8), the safety distance for the reachable set can be expressed as

$$\mathcal{D} = \{d \in \mathbf{R}^{n \times 1} : d = p_{safe} - c - \sum_{i=1}^{q}[0, 1]g^{(i)}\}. \quad (15)$$

As shown in Fig.5, for one specific point inside a reachable set, safety distance D can be determined as

$$D = ||p_{safe} - c - \sum_{i=1}^{q}\varepsilon^{(i)}g^{(i)}||_2, 0 \le \varepsilon^{(i)} \le 1 \quad (16)$$

where $\varepsilon^{(i)}, i = 1, ...q$ is the coefficient of generators to determine the relative position of the point within zonotope. Note that the safety distance reduces to zero if zonotope settles in the safe region, which can be utilized to verify the dynamic stability of SRAM.

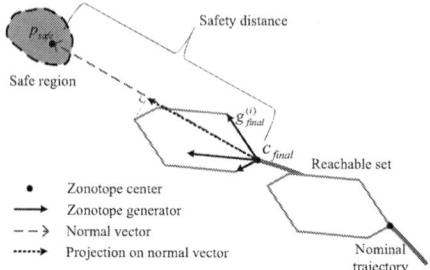

Fig. 5: Safety distance and its sensitivity in reachability analysis.

B. Sensitivity of Safety Distance

With the use of reachability analysis by zonotope, trajectory of SRAM is obtained at final reachable set x_{final} with center c_{final} as

$$x_{final} = c_{final} + \sum_{i=1}^{q}[0, 1]g_{final}^{(i)}.$$

Safety distance D for a reachable set can vary within a certain range as the perturbation of device parameters (13) can result in different operating points close-by.

Given the perturbation range of device parameters $[0, \Delta W]$ in form of interval entries of the matrix zonotope (10), the variation range of safety distance can be obtained from the final reachable set by

$$\Delta D = \sum_{i=1}^{q}\frac{(p_{safe} - c_{final})^T}{||p_{safe} - c_{final}||_2}g_{final}^{(i)}. \quad (17)$$

As shown in Fig.5, the perturbation of safety distance at the final reachable set is obtained by projecting zonotope generators $g_{final}^{(i)}$ to the normal vector $\frac{(p_{safe} - c_{final})^T}{||p_{safe} - c_{final}||_2}$, which is formed from zonotope center c_{final} to safe region p_{safe}.

As such, one can calculate the large-signal sensitivity $S(D, w)$ of the safety distance D w.r.t. device parameter w by

$$S(D, w) := \frac{\Delta D}{\Delta w}, \quad (18)$$

which becomes the ratio of their increment values ΔD and Δw for multiple device parameters simultaneously.

Note that different from the single-parameter small-signal sensitivity of state variable obtained by differentiating (11) with

$$s = \frac{\partial x_k}{\partial w} = -(\frac{C}{h} + G)^{-1}\frac{\partial G}{\partial w}(\frac{C}{h} + G)^{-1}(\frac{C}{h}x_{k-1} - u_k) \quad (19)$$

For the simplicity of presentation linearization error L_k and derivatives of capacitance matrices are omitted and state variable x_{k-1} is assumed as constant. Even though the single-parameter small-signal sensitivity is easy to obtain, compared to the large-signal sensitivity $S(D, w)$ in (18), s may fail to measure the accumulated variation from the previous states by multiple parameters. Moreover, small-signal sensitivity fails to provide accurate direction during optimization due to exclusion of nonlinearity. In contrast, the proposed multi-parameter large-signal sensitivity can be effectively utilized for SRAM dynamic stability optimization, which has faster convergence with higher accuracy as shown in experiment results.

C. Safety Distance Optimization by Sensitivity

The sensitivity of safety distance derived from reachability analysis can guide the optimization direction that departs from unsafe region and shorten safety distance by tuning multiple device parameters. So, it can be embedded in any gradient-based optimization algorithm to achieve a robust SRAM design under process variations. Moreover it can be applied for any circuits when safety distance is properly defined because no specific knowledge of circuit type is required. In case of robust stability optimization for SRAMs, transistor widths are

978-1-4799-2817-0/14 $31.00 © 2014 IEEE

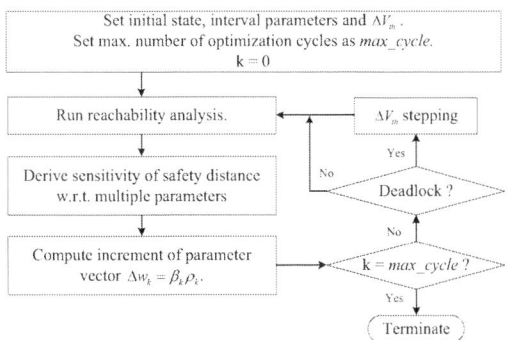

Fig. 6: Flow-chart of robust stability optimization of SRAMs.

sized to improve the dynamic stability. The complete flow of SRAM optimization is shown in Fig.6 with following detailed steps.

Firstly, in each searching step, increment Δw_k of parameter vector w_k is in the same direction with the sensitivity of safety distance.

$$\Delta w_k = \beta_k \rho_k \qquad (20)$$

where $\beta_k > 0$ is a scaling factor and ρ_k is the gradient of objective function, i.e. $S_k(F(w,t),w)$.

The parameter increment Δw_k for the next step is estimated by

$$F(w_k,t) + \Delta w_k^T \rho_k = 0. \qquad (21)$$

As such, one can obtain $\beta_k = -\frac{F(w_k,t)}{\rho_k^T \rho_k}$ after combining (20) with (21). Increment of parameter vector Δw (20) is obtained afterwards.

Though objective function $F(w,t)$ changes nonlinearly in the parameter space, its gradient $\frac{\partial F(w,t)}{\partial w}$ becomes small in magnitude around the safe region. As such, an empirical scaling factor γ can be utilized to resize the estimated increment of parameter vector as

$$\beta_k = -\gamma \frac{F(w_k,t)}{\rho_k^T \rho_k}, 0 < \gamma < 1 \qquad (22)$$

such that the convergence of optimization can be improved. What is more, to further improve the convergence, the initial value stepping can be used when the searching is stuck in the deadlock or out of the feasible range of device parameters ($W_{min} < w_{1,2,3} < W_{max}$).

V. EXPERIMENTAL RESULTS

The proposed optimization technique is implemented in a SPICE-like simulator by MATLAB. Manipulations of zonotopes are performed by MATLAB toolbox named Multi-Parametric Toolbox (MP-T) [17]. Experiment data is collected on a desktop with Intel Core i5 3.2GHz processor and 8GB memory.

SRAM circuit with $40nm$ CMOS is used as the technology node. Supply voltage of SRAM v_{dd} is set to $1V$ with initial states set as $v_1 = 1V$ and $v_2 = 0$. The transistor widths are varied in the range of $[100,600]nm$ with step of $1nm$. A budget of 30% threshold-voltage variation is considered for each transistor during the SRAM verification and optimization. For the robustness optimization, interval values of transistor widths are considered in zonotope matrix to derive sensitivity.

According to Section II, strong pulling strength of M1, M4 and other weak transistors lead to high probability of write failure. Thus the negative V_{th} variations are assumed for M1, M4 and positive variations for other transistors. Similarly for read operation, negative V_{th} variations in M2, M3, M6 and positive variations for other transistors are assumed. Variation magnitudes used for optimization are initially set as standard deviations.

A. Stability Optimization Results

1) Optimization of Read or Write Failure: Firstly, stability optimization for read operation only is performed with transistor widths $[W_1, W_3, W_5] = [200, 300, 300]nm$ and $9ns$ pulse width. The process of read stability optimization is shown in Fig.7(a), with trajectories plotted in light purple; and reachable sets (i.e. zonotopes) due to parameter changes in dark blue. Zonotopes are quite small due to small transistor width variation range $1nm$, resulting in limited variation range of trajectory. Multi-parameter large-signal sensitivity w.r.t. one zonotope set is calculated here, which is different from single-parameter small-signal sensitivity in (19).

At each nominal point 3 reachable sets are generated with different transistor widths and final sets are used to derive large-signal sensitivities (Fig.5). After 4 iterations, the optimized trajectory recovers from read failure. The optimized widths are $[148, 343, 217]nm$.

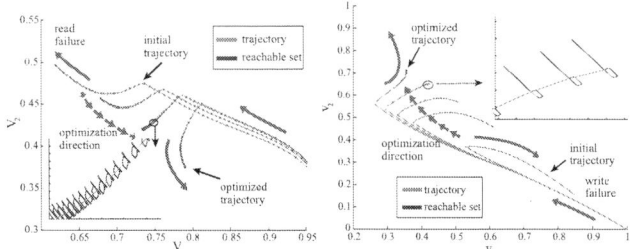

(a) optimization of read operation only (b) optimization of write operation only

Fig. 7: Optimization procedure for SRAM dynamic stability

Then, we perform stability optimization for write operation only with initial transistor widths $[400, 500, 350]nm$ and $0.050ns$ pulse width. The stability optimization by large-signal sensitivity calculated from reachability analysis can help the system trajectory to converge to safe region within 5 iterations (Fig.7(b)). The optimized transistor widths are $[381, 440, 497]nm$.

2) Optimization of Read and Write Failure: Initial transistor pair widths for read and write failure optimization are randomly chosen as $[200, 400, 400]nm$ with pulse width $9ns$ for read and $0.024ns$ for write operations.

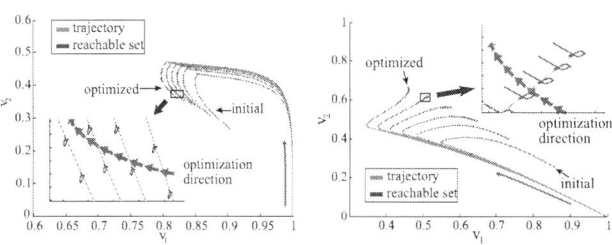

(a) optimization of read operation   (b) optimization of write operation

Fig. 8: Optimization procedure for read and write operation

The optimization directions of trajectory for read and write operations are shown in Fig.8(a) and Fig.8(b) respectively. The trajectory after performing initial optimization is represented as *initial*. From Fig.8(b), we can observe that at the beginning, write failure happens as the trajectory converges to initial state. With the use of proposed optimization technique, the trajectory of write operation moves towards target state and converges after 6 iterations. Meanwhile, if the read operation in Fig.8(a) is considered, initially read failure did not happen. As the write operation optimizes, trajectory for read operation moves upward. As such, the safety distance to the top-left corner (in this case) is decreased. In other words, the write operation is optimized at the expense of read operation to achieve a lower rate of failure for both cases.

The optimized transistor widths obtained by our approach is $[W_1, W_3, W_5] = [192, 330, 586]nm$. Statistical analysis by Monte-Carlo with 1000 samples is performed for SRAM before and after

978-1-4799-2817-0/14 $31.00 © 2014 IEEE

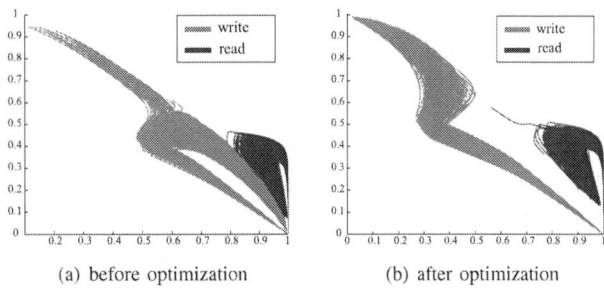

(a) before optimization   (b) after optimization

Fig. 9: Statistical yield calculation before and after optimization

optimization (Fig.9). Yield rate ($Y := 1 - \frac{N_{failure}}{N_{total}}$) considering both read and write functions is improved from 6.8% to 99.957%. Further improvement of yield rate can be achieved by introducing larger threshold-voltage variations during the optimization.

B. Comparisons

Time consumption of our proposed method and traditional Monte-Carlo method for read and write failure optimization shown in Fig.8 are compared here. Monte-Carlo based optimization is performed with 2000 samples. Runtime of optimization and transistor widths at each iteration is listed in Table I, and more than $600\times$ runtime speedup is achieved in our approach. *Iter* represents the iteration number; time consumed in *seconds* for optimization by proposed and traditional Monte-Carlo method are shown under *Sensitivity based RA* and *MC* columns respectively. Achieved speedup is listed under *speedup* column. For example, the proposed method takes about 9 seconds while Monte-Carlo based method needs nearly 2 hours for first iteration. The time consumption of reachability analysis is roughly the same with one transient simulation, since most computation is used on the simulation of the nominal trajectory. Variation of transistor widths at each iteration can also be observed. In our case, to derive large-signal sensitivity w.r.t. three transistor pairs, reachability analysis is performed 3 times.

TABLE I: Runtime comparison of SRAM stability optimization.

Iter.	Transistor Widths (nm)	Sensitivity based RA (s)	MC (s)	Speedup
1	[185, 371, 451]	9.37	5953.23	635.35×
2	[177, 359, 485]	9.69	5876.12	606.41×
3	[173, 349, 515]	9.53	5901.64	619.27×
4	[171, 340, 545]	9.34	5932.87	635.21×
5	[181, 329, 574]	9.58	5951.07	618.11×
6	[192, 330, 586]	9.51	5911.91	621.65×

Furthermore, we compare the our approach with another optimization routine by single-parameter small-signal sensitivity (19). For the same aforementioned test-case, the optimization result by small-signal sensitivity in shown in Fig.10. Unlike in Fig.8(b), the optimization routine by single-parameter small-signal sensitivity fails to find a feasible solution, and results in negative width after 3 iterations. Transistor pair widths, i.e. $[W_1, W_3, W_5]$ are shown in Fig.10. Note that W_5 fails to be tuned during optimization, because small-signal sensitivity w.r.t. W_5 is much smaller than the rest. Since the small-signal sensitivity only depends on the location of final state, the resulted gradient merely has local accuracy and changes irregularly as the trajectory moves and fails to converge. The proposed large-signal sensitivity by reachability analysis can achieve much higher accuracy for a faster converged SRAM optimization.

VI. CONCLUSIONS

To consider multiple device parameters during optimization, a zonotope-based reachability analysis is developed for robustness optimization of SRAM dynamic stability. Efficient backward Euler method is developed for zonotope-based reachability analysis for SRAM failure verification. Moreover, the proposed approach can generate sensitivity of zonotope by considering of multiple device

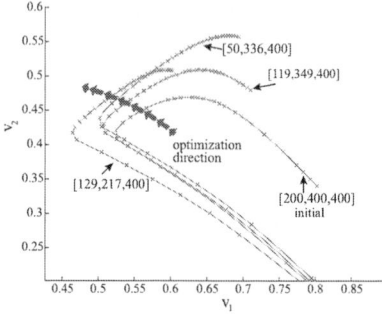

Fig. 10: Optimization of write operation by small-signal sensitivity

parameters with large-signal changes. The resulted sensitivity of safety distance during reachability analysis can be deployed during sequential optimizations to guide SRAM designs departing from unsafe region. Compared to the traditional Monte-Carlo based verification, our method is $600\times$ faster with similar accuracy. In addition, compared to the traditional single-parameter small-signal sensitivity based optimization, our method converges faster with high accuracy.

REFERENCES

[1] E. Grossar and et.al. Read stability and write-ability analysis of SRAM cells for nanometer technologies. *IEEE JSSC*, 41(11):2577–2588, Nov 2006.

[2] S. O. Toh and et.al. Dynamic SRAM stability characterization in 45nm CMOS. In *IEEE Symp. on VLSI Circuits*, 2010.

[3] F. Gong, H. Yu, and L. He. Fast non-monte-carlo transient noise analysis for high-precision analog/RF circuits by stochastic orthogonal polynomials. In *ACM/EDAC/IEEE DAC*, 2011.

[4] F. Gong and et.al. A fast non-monte-carlo yield analysis and optimization by stochastic orthogonal polynomials. *ACM TODAES*, 17(1):10:1–10:23, Jan 2012.

[5] H. Wang, H. Yu, and S. X-D. Tan. Fast timing analysis of clock networks considering environmental uncertainty. *Integration, the VLSI Journal*, 45(4):376 – 387, Sep 2012.

[6] Y. Song and et.al. SRAM dynamic stability verification by reachability analysis with consideration of threshold voltage variation. In *ACM ISPD*, 2013.

[7] S. B-Kazeruni and et.al. SPECO: Stochastic perturbation based clock tree optimization considering temperature uncertainty. *Elsevier Integration, the VLSI Journal*, 46(1):22 – 32, Jan 2013.

[8] G. M. Huang and et.al. Tracing SRAM separatrix for dynamic noise margin analysis under device mismatch. In *IEEE Int. BMAS Workshop*, 2007.

[9] C.J. Gu and J. Roychowdhury. An efficient, fully nonlinear, variability-aware non-monte-carlo yield estimation procedure with applications to SRAM cells and ring oscillators. In *IEEE/ACM ASP-DAC*, 2008.

[10] W. Dong and et.al. SRAM dynamic stability: theory, variability and analysis. In *ACM/IEEE ICCAD*, 2008.

[11] A. Girard. Reachability of uncertain linear systems using zonotopes. In *Int. conf. on Hybrid Systems: computation and control.* Springer, 2005.

[12] M. Althoff. Reachability analysis and its application to the safety assessment of autonomous cars. In *PhD Dissertation, TUM*, 2010.

[13] M. Althoff and et.al. Formal verification of phase-locked loops using reachability analysis and continuization. In *ACM/IEEE ICCAD*, 2011.

[14] Y. Song and et.al. Stable backward reachability correction for PLL verification with consideration of environmental noise induced jitter. In *IEEE/ACM ASP-DAC*, 2013.

[15] A. Singhee and et.al. Probabilistic interval-valued computation: Toward a practical surrogate for statistics inside CAD tools. *IEEE Trans. on CAD*, 27(12):2317–2330, Nov 2008.

[16] M. Althoff, O. Stursberg, and M. Buss. Reachability analysis of nonlinear systems with uncertain parameters using conservative linearization. In *IEEE Conf. on Decision and Control*, 2008.

[17] M. Kvasnica and et.al. Multi-parametric toolbox (mpt). MPT 2.6.3 is available at http://control.ee.ethz.ch/~mpt/.

978-1-4799-2817-0/14 $31.00 © 2014 IEEE

Accurate and Inexpensive Performance Monitoring for Variability-Aware Systems

Liangzhen Lai

Dept. of Electrical Engineering
University of California Los Angeles
Los Angeles, CA 90095
e-mail: liangzhen@ucla.edu

Puneet Gupta

Dept. of Electrical Engineering
University of California Los Angeles
Los Angeles, CA 90095
e-mail: puneet@ee.ucla.edu

Abstract—Designing reliable integrated systems has become a major challenge with shrinking geometries, increasing fault rates and devices which age substantially in their usage life. The proposed research is motivated by the observation that many of the in-field failures are delay failures and several variability signatures are also delay-related. The origins of temporal delay fluctuations include manufacturing variability, voltage/temperature changes, negative or positive bias temperature instability-related Vth degradation, etc. Since the actual delay changes depend on process variations as well as workload, on-chip monitoring may be the best way of predicting them. There is a need to monitor circuit performance during manufacturing as well as at runtime to predict achievable performance and warn against impending failures. Adaptive mechanisms in hardware and/or software can optimize the trade-off between errors, energy and performance based on the feedback from runtime circuit performance monitors.

This paper presents approaches for automated synthesis of design-dependent performance monitors. These monitors can be used to predict impending delay failures relatively inexpensively. For low-overhead monitoring, we propose multiple design-dependent ring oscillators (*DDROs*) as smart canary structures which can reliably predict achievable chip frequency but with margins for local variations. Early silicon results indicate that *DDROs* can reduce delay monitoring error by 35% compared to conventional ring oscillators. To further improve the prediction (albeit at a higher overhead), we propose in-situ slack monitors (*SlackProbe*) which can match local variations as well at overheads much smaller than monitoring all sequential elements. *SlackProbe* reduces the number of monitors required by over 15X with 5% additional delay margin in several commercial processor benchmarks. Finally, we show an example of software testbed that demonstrates a variability-aware system that utilizes the hardware monitors and operates with both hardware and software adaptation.

I. INTRODUCTION

With CMOS technology scaling, hardware variability continues to increase due to increasing amounts of manufacturing variability, ambient fluctuation, and circuit wear-out (e.g., NBTI, HCI etc.) degradation. These increased variations have led to increased margining in designing reliable integrated systems. If these variations can be captured and exposed to the software/system level during runtime, corresponding hardware and software adaptation can opportunistically reduce or even eliminate the design margin [1].

There are different approaches to capture hardware variability at runtime. On-line self-test and diagnostics allow a system to test itself concurrently during normal operation [2]–[4]. Software-based inference methods [5]–[7] use software-implemented test operations to capture variation and detect errors. Other than testing approaches, hardware monitors can also be used to capture variations. In this work, we focus on designing and utilizing hardware monitors to capture circuit performance variations. The proposed research is motivated

by the observation that many of the in-field failures are delay failures and several variability signatures are also delay-related.

There are mainly two classes of monitors that aim at monitoring circuit performance (i.e., measuring circuit critical path delay). They are replica monitors and in situ monitors.

Replica monitors, also known as canary circuits, [8]–[11] are stand-alone circuits which are designed to mimic the timing behavior of the original circuits. By measuring the replica circuit delay, one can estimate the delay of the original circuits. Typically, the replicas are simple circuits, e.g., simple paths or ring oscillators. Therefore, replicas are usually non-intrusive and with low overhead, but hard to cover the heterogeneity within the original circuits. Furthermore, replica monitors may fail to capture the variation that are local to real circuits such as random manufacturing variations and circuit aging.

In situ monitors measure the delay directly from the circuit paths [12]–[14].They can accurately capture the real path delay, but with significant overhead, especially when a large number of registers are timing critical.

In this work, we first introduce two circuit performance methodologies. Design-dependent ring oscillator (*DDRO*) [15] designs and leverages *multiple* replicas to accurately monitor circuit performance. *SlackProbe* [16] inserts in situ monitors at both path intermediate nets and path endpoints, which can greatly reduce the overhead of in situ monitoring. Then we present our implementation of an end-to-end variability-aware system [17] which utilizes the hardware monitors and operates with both hardware and software adaptation.

The rest of the paper is organized as follows: Section II presents *DDRO* replica monitoring methodology. Section III presents *SlackProbe* in situ monitoring methodology. The variability-aware system implementation and demonstration is described in Section IV, and Section V concludes the paper.

II. DESIGN-DEPENDENT RING OSCILLATOR (*DDRO*)

A. Motivation and Overview

In order to accurately estimate the circuit delay, replica monitors should be designed to follow the timing behavior of original circuits under variations. We know that the circuit's performance is determined by delay of the slowest path, i.e., critical path. Therefore, the replica should be designed to follow the timing behavior of the critical path.

Due to variations, there may be a number of potential critical paths and they may behave differently under variations, which makes a single replica monitor inadequate. The motivation of *DDRO* is shown in Fig. 1, in which each dot represents the delta delay of one critical path under variations of PMOS threshold voltage $Vthp$ (y-axis) and NMOS threshold voltage $Vthn$ (x-axis). The critical path delay sensitivities form natural

978-1-4799-2817-0/14 $31.00 © 2014 IEEE

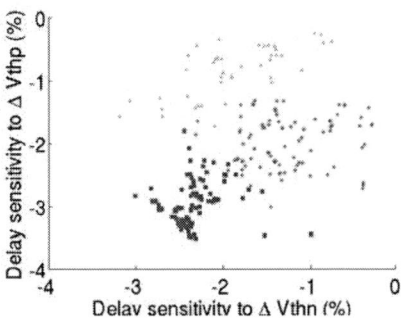

Fig. 1. Every dot represents delay sensitivities of a critical path in a design. In this example, we cluster the paths into 3 different clusters indicated by different colors. The results are based on SPICE simulation using commercial 45nm process technology. Critical paths are extracted from AES [18].

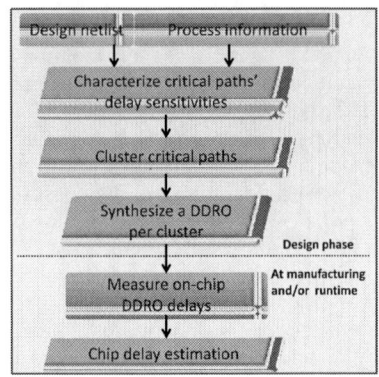

Fig. 2. Overview of *DDRO* design methodology.

clusters, which implies that *multiple* replica monitors can be designed to cover these clusters.

An overview of *DDRO* monitoring strategy is shown in Fig. 2. First, we extract critical paths of a design and characterize their delay sensitivities to variation sources. Second, we cluster the critical paths based on the sensitivities, and synthesize *DDROs* to match delay sensitivity of the clusters. By matching *DDRO* and cluster delay sensitivities, we ensure that the synthesized *DDROs* have good correlation with the critical paths. Since we use only standard cells (gates) to synthesize the *DDROs*, the design and placement of *DDROs* can be easily integrated with conventional implementation flows. Based on *DDRO* frequencies, we can estimate chip delay during manufacturing or runtime.

B. Path Sensitivity Extraction and Clustering

Delay sensitivity of path i (V_i^{path}) can be obtained using finite differences, i.e., taking the $\Delta delay$ by perturbating each variation source by 1σ. After characterizing all path delay sensitivities, we can cluster the critical paths. The objective function of the clustering is defined as

$$\text{minimize} \sum_{i=1}^{N} (w_i \times |\mathbf{V}_i^{path} - \mathbf{V}_k^{ro}|) \qquad (1)$$

where the summation is taken over paths i in cluster k, and w_i is the probability of a critical path delay exceeding the clock period of the design. The weight factor w_i is added so that we can impose a higher penalty for having mismatched delay sensitivities on a path with higher probability to fail (less timing slack). Detailed derivation of w_i can be found in [15].

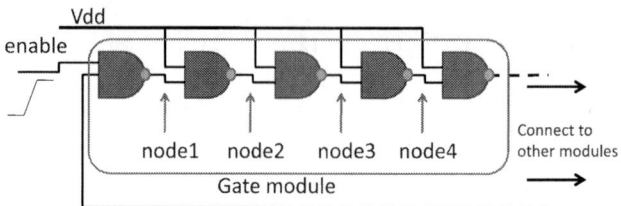

Fig. 3. Illustration of a gate module in a *DDRO*.

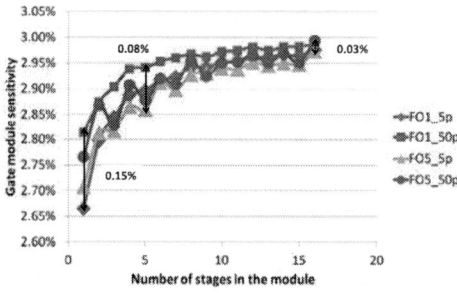

Fig. 4. Simulation results show that the sensitivities under different input slews {5ps, 50ps} and output loads {FO1, FO5} combinations converge as the number of stages in a gate module increases.

C. DDRO Synthesis

1) Gate Module: For the ease of synthesizing *DDRO*, we use gate module as basic building block. A gate module is constructed as several identical gates connected in series as illustrated in Fig. 3.

This repeated pattern can help decouple the load and slew interaction between gates. Simulation results in Fig. 4 show that the sensitivity difference due to different input slew and output load is reduced from 0.15% to 0.03%, as the number of stages in a gate module increases from 1 to 15. In this work, we use 5-stage gate modules as a trade-off between stability of sensitivity and total area of a gate module. For a gate with multiple input pins, gate delays through different input pins will have different delay sensitivities. Thus, each gate module type is defined with respect to a specific input pin. Extra input pins of a multi-input gate are assigned to high or low to make a gate module inverting or buffering (see Fig. 3).

Since the interconnect also affects path delay sensitivity, we use different wirelength in building our gate modules. Gate modules with different wirelength are considered as different instance types even if they have the same gate type.

2) ILP-based Synthesis: Given a delay sensitivity target ($\mathbf{V}^{ro}$) obtained from path clustering, we want to choose the number of each gate module in a *DDRO*, so that the delay sensitivities of the *DDRO* match the targeted delay sensitivities. Because of its unique structure, gate module delay variation is less sensitive to input slew and output load. Since each gate module type is instantiated a discrete number of times, we can formulate *DDRO* synthesis as an integer linear programming (ILP) problem, where the integer variable is the number of certain gate modules in the synthesized *DDRO*. Appropriate constraints are applied to limit the maximum RO length. Detailed description of gate module and ILP formulation can be found in [15].

D. Delay Estimation

1) Path-based Estimation: As shown in Fig. 2, at manufacturing and/or runtime, *DDRO* delay is measured and used

978-1-4799-2817-0/14 $31.00 © 2014 IEEE

to estimate chip delay.

Each critical path delay can be estimated by *DDROs*. Given M *DDROs*, we can represent delay sensitivity of each path ($\mathbf{V}_i^{path}$) as a linear combination of *DDRO* delay sensitivity($\mathbf{V}_k^{ro}$ ($k = 1, ..., M$)) as in Equation (2).

$$\mathbf{V}_i^{path} = \sum_{k=1}^{M} b_{ik} \cdot \mathbf{V}_k^{ro} + \mathbf{V}_i^{res\_path} \qquad (2)$$

where b_{ik} is a $[1 \times M]$ matrix containing constant coefficients and $\mathbf{V}_i^{res\_path}$ is a $[1 \times Q]$ matrix that represents the decomposition residue.

Using b_{ik}, we can estimate the path delay by:

$$d_i^{path} = d_i^{nom\_path}(1 + \sum_{k=1}^{M} \overbrace{(b_{ik}\mathbf{V}_k^{ro} \cdot \mathbf{g})}^{\text{measurable}} + \underbrace{u_i}_{\text{uncertainty}}) \qquad (3)$$

$$\text{where} \quad u_i \quad = l_i^{path} + \mathbf{V}_i^{res\_path} \cdot \mathbf{g}$$

where $\mathbf{g}$ is global variation vector, l_i^{path} is the local variation of the path. Equation (3) shows that d_i^{path} consists of a *measurable* term and an *uncertainty* term. While the value of the *measurable* term can be determined from the delays of *DDROs*, the value of the *uncertainty* term cannot be measured directly.

2) Cluster-based Estimation: The path-based estimation method requires one estimation for each critical path, which can cause significant computation and storage overhead. To reduce the overhead, we propose to utilize the clustering nature of critical paths and to have one estimation for each path cluster. We calculate the maximum delay of paths in each cluster using the method in [19], assuming that the means of path delays correspond to their nominal values. The outcome of this step gives us the expected maximum delay. But more importantly, it also extracts the sensitivity of the maximum delay to variation sources ($\mathbf{V}^{max}$). Similar to the path-based approach, we treat the cluster as a pseudo path with delay sensitivity $\mathbf{V}^{max}$. Simulation results show that the estimation error of this approximation approach compared to the reference method is very small.

E. Simulation and Silicon Result Highlights

To validate *DDRO* performance monitoring methodology, we synthesized, placed and routed processor benchmark circuits using a commercial 45nm SOI technology. Some simulation results of *DDRO* on Cortex-M0 [20] are highlighted in Fig. 5. For global variation only case, using more *DDROs* can dramatically decrease mean overestimation (i.e., required delay margin). But this benefit becomes less in presence of local variation. The results show that the estimation error of cluster-based approximation, compared to the path-based approach, is very small.

A testchip was taped out with *DDRO*-based performance monitoring using a 45nm IBM SOI technology with dual-V_{th} (RVT and HVT) libraries. The testchip has an ARM Cortex-M3 microprocessor [21] with five *DDROs*. For comparison, we also implemented several inverter-based ROs in the testchip. The silicon measurement results are shown in Fig. 7. The measurement results show that by using five *DDROs*, we can reduce the mean delay estimation error by 35% (from 2.3% to 1.5%) compared to generic inverter-based ROs.

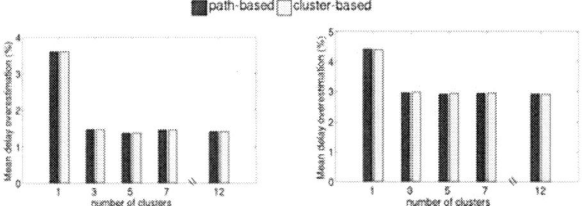

Fig. 5. Simulation results of *DDRO* on Cortex-M0 with global variation only (left) and both global and local variations (right).

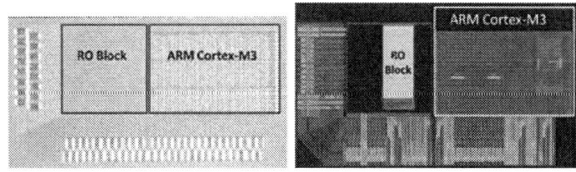

Fig. 6. Testchip die photo and layout illustration

III. *SlackProbe*: AN EFFICIENT AND FLEXIBLE IN SITU MONITORING METHODOLOGY

With *DDRO* methodology, replica monitors can predict global variation near perfectly. But the accuracy is still limited and constrained by local variation. In situ monitors can capture the actual path delay, including local variation, but usually incur significant overhead. We observe that most of existing work focuses on monitoring destination registers. In this section, we introduce a novel and flexible in situ monitoring methodology, *SlackProbe*, which inserts monitors at both path endpoints and path intermediate nets.

A. Monitor Working Principle and Overview

The monitor working principle is shown through an example in Fig. 8. If a monitor is inserted at an intermediate node A, a "probe", which consists of delay matching gates and a transition detector, is connected to A through a minimum size inverter. Signal transitions at node A are transferred through the delay chain to the transition detector and compared with incoming clock edge. If the transition is close to its required arrival time (RAT), i.e., within the margin window as in Fig. 8, a corresponding signal transition will arrive at node E after the clock edge. This triggers the transition detector and flags a signal indicating an impending delay failure.

The monitor inserted at node A is capable to monitor the delay of all critical paths passing through A. As shown in Fig. 8, in stead of monitoring all four destination registers,

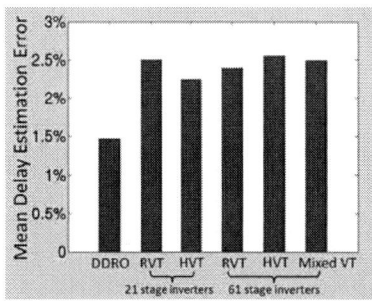

Fig. 7. Mean delay estimation error obtained from *DDROs* and inverter-based ROs. Estimation errors are calculated by taking the absolute difference between normalized estimation and normalized chip delay. RVT and HVT are the two Vt options for the inverters.

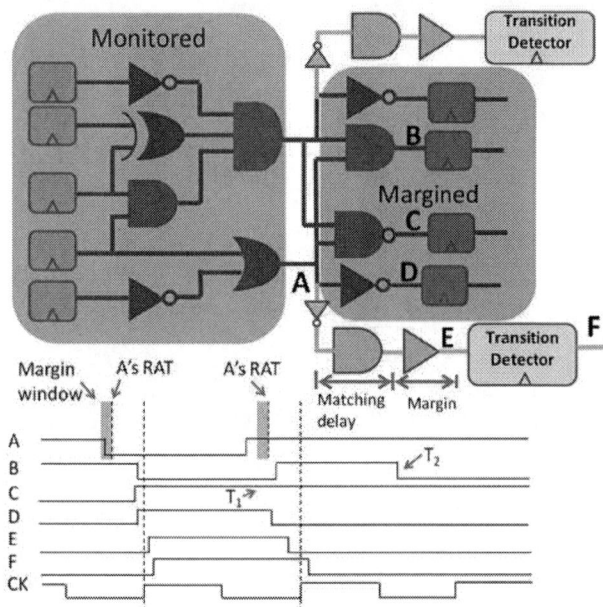

Fig. 8. *SlackProbe* working principle. As shown in the timing diagram, compared to inserting monitors at destination registers, the monitor inserted at A can monitor the path delay even when the transition does not propagate to the destination register (i.e., T_1 at C). But the monitor inserted at node A cannot capture transitions that do not pass through A (i.e., T_2 at B).

SlackProbe can use only two monitors while achieving the same path coverage.

Different transition detector designs as in [14], [22]. can be applied here. *SlackProbe* also allows monitors to be inserted at path endpoints where monitors as in [23]–[26] can be used as well. Since the additional margin makes the monitor detect an impending timing failure rather than an actual one, there is no datapath metastability issue as raised and discussed in [14]. The metastability issue of the monitor signal either results in a more pessimistic detection or is guardbanded by the monitor delay margin.

With the proposed monitoring strategy, the problem now becomes *when*, *where* and *how* to insert these monitors. In this work, we propose the monitor insertion flow as in Fig. 9. Different alternatives will also be discussed and compared against conventional approaches.

Monitor insertion starts with a placed and routed design, as the timing information is more accurate at this stage. Since we only care about timing-critical paths, a path selection process is applied to extract timing-critical paths and to construct the critical path graph. Then, monitor locations are picked from the graph using our proposed method. For each of the monitor locations, a delay cell path is synthesized. The final insertion flow is similar to Engineering Change Order (ECO) where the monitors are incrementally placed and routed. ECO metrics like those in [27] are applied when picking monitor locations to minimize the interference to the original design.

B. Monitor Delay Margin

If the monitor is placed at a path intermediate net, the path delay before the monitor can be captured. But some extra delay margin will be required for the remaining part of the path. As shown in Fig. 10, there are three types of relations between monitor and a path:

1) The path passes through the net, for example path A in Fig. 10. Since the delay up to G4 can be captured by

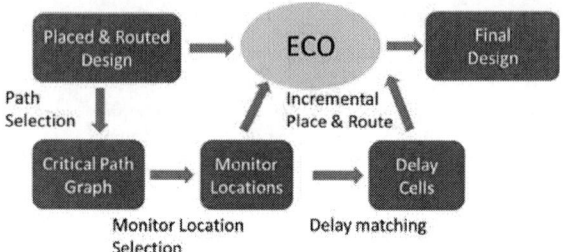

Fig. 9. Monitor insertion flow

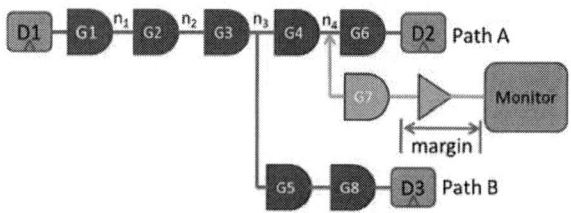

Fig. 10. Example of path-monitor pairs

the monitor, the delay path should account for the delay uncertainty of G6.

2) The path branches out at some net before the monitor, for example path B in Fig. 10. Depending on the application and gate type of G4, the monitor may be treated as being inserted between G3 and G5 with G4 as part of the delay matching. If the application is speed sensing, i.e., monitoring slow delay changes, path B can be considered as being monitored with delay uncertainty of G4, G5 and G8. If the application is event detection, only G4 with gate types that are transparent to signal transitions (e.g., inverter, buffer) are allowed.

3) None of the path instances fall in the fan-in cone of the monitor net. In this case, we consider that the path is not monitored.

Though different paths may require different delay margin, each monitor will have only one margin matching chain. The monitor margin should account for worst delay uncertainty after monitor insertion point and guarantee that the delay chain is always slower than margined part of monitored circuit paths.

In the example in Fig. 10, the best case delay of the delay chain (i.e., n_4 to the monitor) should match the worst case delay of the original path (i.e., $G6$). But this may be too pessimistic since the delay is likely to be correlated. Similar to on-chip variation modeling, in this work the delay chain is designed so that its delay at typical process corner matches the worst case delay of the original path. The equivalent delay margin in this case equals the delay of $G6$ at slow process corner (i.e., delay of the delay chain at typical process corner) minus the delay of $G6$ at typical process corner. This margin is considered as the delay uncertainty of $G6$.

The final delay margin for the entire circuit will be dominated by the monitor with the largest margin. Therefore, in this work, we define the delay margin cost as the maximum monitor delay margin constraint ϵ on each monitor. For a given ϵ, we can identify the feasible monitor candidate locations. Larger ϵ will give more flexibility in choosing monitor insertion location, which can potentially reduce the number of monitors required. But it will also reduce the monitoring benefit. The trade-off of delay margin ϵ will be discussed together with critical path selection in Section III-C.

978-1-4799-2817-0/14 $31.00 © 2014 IEEE

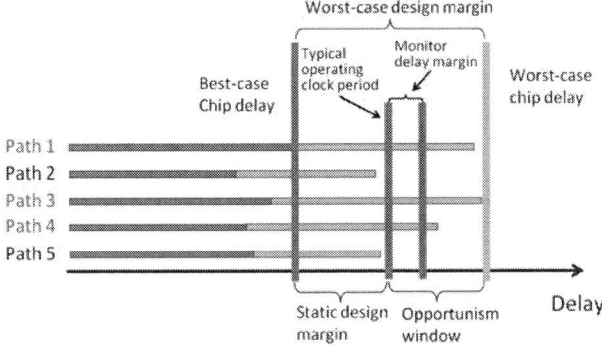

Fig. 11. Opportunism window is the margin saving comparing to worst-case design. Both monitoring benefit and overhead is affected by opportunism window size and monitor delay margin size.

C. Opportunism Window

As shown in Fig. 9, given a placed and routed design, the first step is to identify the part of the design that may be timing critical. Depending on the application, different criticality criteria may be applied for the selection process.

In this work, we propose a flexible path selection method by introducing user-defined *opportunism window*. As illustrated in Fig. 11, *opportunism window* and monitor margin will dictate the potential monitoring benefits. Typical worst-case design margins for the worst-case scenario, i.e., all chips will by default run at the worst-case chip delay regardless of what their actual delays are. In the presence of monitors, we may reduce the design margin and decrease the default operating clock period. The paths whose worst-case delay exceeds the default operating clock period should be selected and monitored. The amount of design margin reduction is called *opportunism window*, within which the circuit will operate opportunistically at its best effort.

This path selection method does not require any knowledge of correlation in the variations between the delay of different paths. Therefore, it can be used to select paths for applications like aging sensors, where exact delay degradations are context dependent with little pre-assumed correlation.

As illustrated in Fig. 11, there is a natural trade-off between monitoring benefits and monitoring overhead when deciding the *opportunism window* and monitor delay margin (ϵ). Larger opportunism window size and smaller monitor delay (ϵ) will increase the monitoring benefit, but also will increase the monitoring overhead.

After picking the critical paths, monitor location selection can be formulated as a Linear Programming (LP) problem, which resembles a max-flow problem. Detailed problem formulation and solution can be found in [16].

D. Experimental Result Highlights

To evaluate the effectiveness of our monitoring methodology, we apply *SlackProbe* on commercial processor benchmarks using a commercial sub-32nm process technology and libraries. For comparison purpose, three different monitoring methods are implemented:

- Baseline: A monitor is inserted at every critical path endpoint
- *SlackProbe* Event Detection: Monitors are inserted at both path intermediate nets and path endpoints. Monitors are inserted to capture every timing-critical signal switching events.

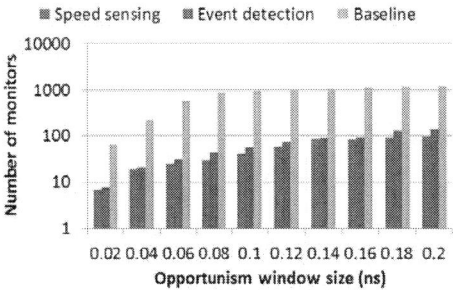

Fig. 12. Monitor count comparison between *SlackProbe* and baseline. The y-axis is plotted in log scale.

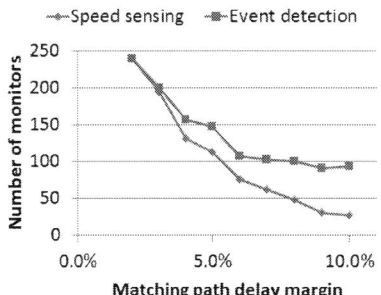

Fig. 13. Monitor count and cost vs. delay margin for processor A

- *SlackProbe* Speed Sensing: Monitors are inserted at both path intermediate nets and path endpoints. Monitor are inserted to capture the delay changes of all critical paths, given sufficient time of operations.

To evaluate our methodology and show the opportunism window trade-off, we implement all three methods on a processor benchmark. *SlackProbe* monitors are inserted with monitor delay margin ϵ as 5% of the clock period. The results are shown in Fig. 12. Compared to the baseline endpoint monitoring, with the 5% additional delay margin, *SlackProbe* can achieve up to 16X reduction in total number of monitors.

To show the trade-off between delay margin and monitor count, we also sweep the delay margin ϵ with opportunism window size of 0.2ns, i.e., equivalently 20% of the clock period. By allowing more delay margin and flexibility in selecting monitor candidate location, the number of monitor reduces for both *SlackProbe* methods (see Fig. 13).

IV. *RedCooper*: A TESTBED FOR VARIABILITY-AWARE SYSTEM

In this section, we will present our implementation of a complete end-to-end system [17] that demonstrates the use of hardware monitors with both hardware and software adaptation. We first describe the software adaptation concept of variability-aware software duty-cycling. Then we present our testbed implementation and variability-aware system demonstration.

A. Variability-Aware Software Duty-Cycling

For battery-powered embedded sensing system, the total lifetime energy is usually constrained. In order to meet the lifetime requirements, one particularly common power management techniques is duty cycling, where the system is at default in a sleep state and woken up periodically to attend to pending tasks and events. A system with higher duty cycle may, for example, sample sensors for longer intervals or at

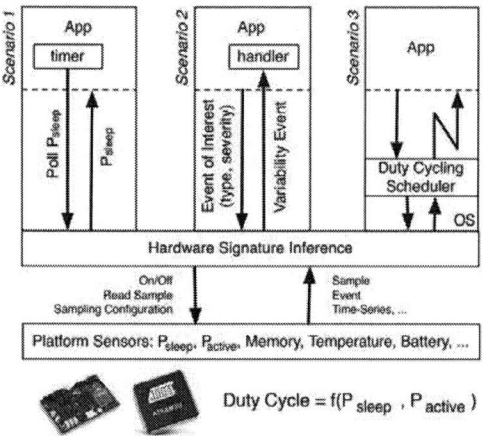

Fig. 14. Designing a software stack for variability-aware duty cycling [28]

higher rates, increasing data quality. A typical application-level goal is to maximize quality of data through higher duty cycles, while meeting a lifetime goal. Conventional approach determines duty cycles by either worst-case power specifications or datasheet power values, which may be heavily guardbanded. If the system's power consumption can be measured and exposed to the software, the application may be able to adapt its duty cycle rate and increase its QoS opportunistically according to the hardware power consumption status [28].

For example, for a fixed lifetime energy budget E and specified targeting lifetime constraints L, the software duty cycle rate DC can be calculated through the following equation [28]:

$$P_A \cdot DC + P_S \cdot (1 - DC) = \frac{E}{L}$$
$$DC = \frac{E - L \cdot P_S}{L \cdot (P_A - P_S)} \quad (4)$$

where P_A and P_S are the active and sleep power consumption respectively. By determining P_A and P_S on a per-instance basis, the duty cycle may be tailored to maximize active time for each individual sensor under a given deployment scenario (temperature profile, lifetime requirement, battery capacity).

There are several different ways that such an opportunistic stack may be organized as shown in Fig. 14. The scenarios differ in how the sense-and-adapt functionality is split between applications and the operating system. In this work, we use the last scenario, where variability is largely offloaded to the operating system.

B. Testbed Implementation and System Demonstration

The testbed is built upon *DDRO* testchip (Fig. 6). On the testchip, there are on-chip performance monitors (*DDRO*) and leakage sensors. For the purpose of demonstrating software duty-cycling, we also implement on-board power sensors to measure the power consumption of the testchip. the power of the Cortex-M3 processor and the (on-chip) SRAM memory are measured separately. A picture of the testbed board is shown in Fig. 17. Because *DDRO* and the processor are implemented as separate blocks when taping-out the testchip, we use on-board MCUs [29] to sample the monitor readings and feed into the on-chip SRAM.

The software running on the M3 core is shown in Fig. 16. The operating system (OS) is based on CoOS [30]. Three tasks are implemented within the OS:

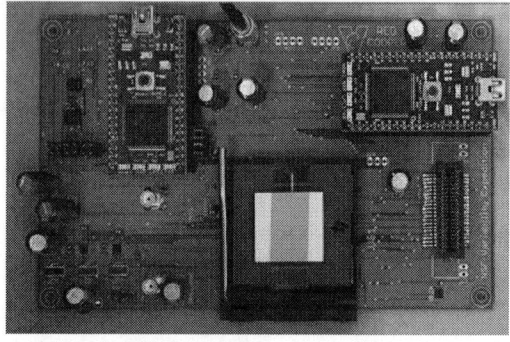

Fig. 15. *RedCooper* Testbed.

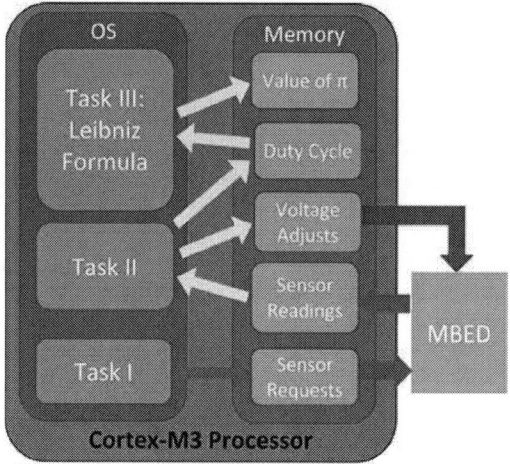

Fig. 16. Software Adaptation Illustration.

- Task I sends the sensing request by writing to a pre-specified memory location. Upon seeing the request, the on-board MCUs will start reading the sensor values and write the sensor readings directly to certain memory locations.
- Task II acts as the central adaptation center, which reads the sensor readings, including *DDRO* frequencies, current drawn by the M3 core, current drawn by the on-chip SRAM, and the on-chip leakage sensor. *DDRO* frequencies are used to calculate the performance slack and determine the adjusted voltage. The current and leakage sensor values are used to calculate the feasible duty cycle using Equation (4). The duty cycle rate is further translate to the number of iterations for Task III.
- Task III is the main application running in the OS, which calculates the value of π with its best effort under the constrained duty cycle.

In this demo, all three OS tasks are fired every 10 seconds. We set the processor to run at a fixed 600 MHz clock frequency. The active power includes both the core and SRAM power consumption. The sleep power includes the SRAM power and the projected leakage power of the core.

A snapshot of the entire hardware is shown in Fig. 17. We have two copies of *RedCooper* running side-by-side. Each testbed is equipped with an LCD reporting the system status. Some results are highlighted in Table I.[1] At room temperature, Instance B system has smaller sleep power than

---

[1]The complete demo video can be found at http://nanocad.ee.ucla.edu/Main/Codesign

978-1-4799-2817-0/14 $31.00 © 2014 IEEE

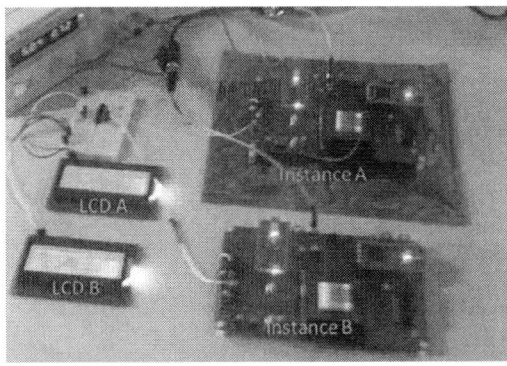

Fig. 17. A snapshot of the demo hardware.

TABLE I
SYSTEM DEMONSTRATION RESULT HIGHLIGHTS

	Instance A (room temperature)	Instance B (room temperature)	Instance A (after heat-up)
Supply Voltage	0.93 V	0.96 V	0.96 V
Active power	23.81 mW	24.29 mW	26.50 mW
Sleep power	4.74 mW	1.63 mW	7.58 mW
DC	27	49	18
Calculated π	3.1028	3.1206	3.2002
Calculation error	1.235%	0.668%	1.865%

Instance A, which implies the potential of achieving high duty cycle rate. Therefore, the adaptive software set the number of iterations (DC) at 49 for Instance B and the calculation error is smaller. After heating up Instance A from room temperature to about $45°C$, the system shows its capability for both runtime software and hardware adaptation (see last column in Table I). The voltage is boosted to compensate the performance loss. Software duty-cycle is reduced to compensate the increased power consumption. If the system is without hardware sensors and designed for the worst-case scenario (including process variations and temperature fluctuations), the number of iterations (DC), as in this demo, will be at most 18 (the case for Instance A after heat-up) with 1.865% calculation error. With the hardware sensors and adaptations, we are able to achieve 49 iterations and calculation error as small as 0.668%.

V. CONCLUSION

In this paper, we first present *DDRO*, an accurate replica performance monitoring methodology. Then we present *Slack-Probe*, an efficient and flexible in situ performance monitoring methodology. Last, we demonstrate an end-to-end variability-aware system that utilizes hardware monitors for both hardware and software adaptation.

ACKNOWLEDGMENTS

The work in Section II was jointly done with Tuck-Boon Chan and Andrew B. Kahng. The author would like to thank Vikas Chandra and Robert Aitken for collaboration for work presented in Section III. Finally, *RedCooper* testbed was jointly developed by the authors and Yuvraj Agarwal, Alex Bishop, Matt Fotjik, Paul Martin, Mani Srivastava, Dennis Sylvester, Lucas Wanner, and Bing Zhang.

This work is supported in part by NSF Variability Expedition grant CCF-1029030.

REFERENCES

[1] P. Gupta *et al.*, "Underdesigned and opportunistic computing in presence of hardware variability," *IEEE Transactions on Computer-Aided Design of Integrated Circuits and Systems*, 2012, Keynote Paper.

[2] Y. Li *et al.*, "Overcoming early-life failure and aging for robust systems," *IEEE Design and Test of Computers*, vol. 26, no. 6, pp. 28–39, 2009.

[3] H. Inoue *et al.*, "Vast: Virtualization-assisted concurrent autonomous self-test," in *Proc. IEEE International Test Conference.* IEEE, 2008, pp. 1–10.

[4] Y. Li *et al.*, "Casp: concurrent autonomous chip self-test using stored test patterns," in *IEEE/ACM Design, Automation and Test in Europe.* ACM, 2008, pp. 885–890.

[5] D. Lin *et al.*, "Quick detection of difficult bugs for effective post-silicon validation," in *Proc. ACM/IEEE Design Automation Conference.* IEEE, 2012, pp. 561–566.

[6] S. K. Sahoo *et al.*, "Using likely program invariants to detect hardware errors," in *Dependable Systems and Networks, IEEE International Conference on.* IEEE, 2008, pp. 70–79.

[7] M.-L. Li *et al.*, "Understanding the propagation of hard errors to software and implications for resilient system design," *ACM Sigplan Notices*, vol. 43, no. 3, pp. 265–276, 2008.

[8] A. Drake *et al.*, "A distributed critical-path timing monitor for a 65nm high performance microprocessor," in *Proc. IEEE International Solid State Circuits Conference*, feb. 2007.

[9] M. Bhushan *et al.*, "Ring oscillators for cmos process tuning and variability control," *IEEE Transactions on Semiconductor Manufacturing*, vol. 19, no. 1, pp. 10–18, 2006.

[10] Q. Liu *et al.*, "Capturing post-silicon variations using a representative critical path," *IEEE Transactions on Computer-Aided Design of Integrated Circuits and Systems*, vol. 29, no. 2, pp. 211–222, 2010.

[11] T.-B. Chan *et al.*, "Tunable sensors for process-aware voltage scaling," in *Proc. IEEE/ACM International Conference on Computer-Aided Design.* ACM, 2012, pp. 7–14.

[12] D. Fick *et al.*, "In situ delay-slack monitor for high-performance processors using an all-digital self-calibrating 5ps resolution time-to-digital converter," in *Proc. IEEE International Solid State Circuits Conference*, feb. 2010.

[13] S. Kim *et al.*, "Razor-lite: A side-channel error-detection register for timing-margin recovery in 45nm soi cmos," in *Proc. IEEE International Solid State Circuits Conference.* IEEE, 2013, pp. 264–265.

[14] K. Bowman *et al.*, "Energy-efficient and metastability-immune resilient circuits for dynamic variation tolerance," *IEEE Journal of Solid State Circuits*, jan. 2009.

[15] T.-B. Chan *et al.*, "Synthesis and analysis of design-dependent ring oscillator (ddro) performance monitors," *IEEE Transactions on Very Large Scale Integration Systems*, 2013, accpted for publication.

[16] L. Lai *et al.*, "Slackprobe: A low overhead in situ on-line timing slack monitoring methodology," in *IEEE/ACM Design, Automation and Test in Europe*, 2013, pp. 282–287.

[17] Y. Agarwal *et al.*, "Redcooper: Hardware sensor enabled variability software testbed for lifetime energy constrained application," Tech. Rep., http://nanocad.ee.ucla.edu/pub/Main/Codesign/red_cooper.pdf.

[18] [Online]. Available: http://opencores.org

[19] C. Visweswariah *et al.*, "First-order incremental block-based statistical timing analysis," *IEEE Transactions on Computer-Aided Design of Integrated Circuits and Systems*, vol. 25, no. 10, pp. 2170–2180, 2006.

[20] [Online]. Available: http://www.arm.com/products/processors/cortex-m/cortex-m0.php

[21] [Online]. Available: http://www.arm.com/products/processors/cortex-m/cortex-m3.php

[22] B. Rebaud *et al.*, "Digital timing slack monitors and their specific insertion flow for adaptive compensation of variabilities," in *International Conference on Integrated Circuit and System Design: Power and Timing Modeling, Optimization and Simulation*, ser. PATMOS'09, 2010.

[23] D. Ernst *et al.*, "Razor: a low-power pipeline based on circuit-level timing speculation," in *IEEE/ACM International Symposium on Microarchitecture*, dec. 2003.

[24] S. Das *et al.*, "RazorII: In situ error detection and correction for pvt and ser tolerance," *IEEE Journal of Solid State Circuits*, jan. 2009.

[25] H. Fuketa *et al.*, "Adaptive performance compensation with in-situ timing error prediction for subthreshold circuits," in *IEEE Custom Integrated Circuits Conference*, sept. 2009.

[26] M. Eireiner *et al.*, "In-situ delay characterization and local supply voltage adjustment for compensation of local parametric variations," *IEEE Journal of Solid State Circuits*, vol. 42, no. 7, pp. 1583–1592, 2007.

[27] J. Lee *et al.*, "Eco cost measurement and incremental gate sizing for late process changes," *ACM Transactions on Design Automation of Electronic Systems*, vol. 18, no. 1, p. 16, 2012.

[28] L. Wanner *et al.*, "Hardware variability-aware duty cycling for embedded sensors," *IEEE Transactions on Very Large Scale Integration Systems*, vol. 21, no. 6, pp. 1000–1012, 2013.

[29] [Online]. Available: http://mbed.org/

[30] [Online]. Available: http://www.coocox.org/CoOS.htm

978-1-4799-2817-0/14 $31.00 © 2014 IEEE

Quantifying Workload Dependent Reliability in Embedded Processors

Vikas Chandra
ARM R&D
San Jose, CA
vikas.chandra@arm.com

Abstract—With nearly three decades of continued CMOS scaling, the devices have now been pushed to their physical and reliability limits. Scaling to sub-20nm technology nodes changes the nature of reliability effects from abrupt functional problems to progressive degradation of the performance characteristics of devices and system components. The impact of unreliability results in time-dependent variability, directly translating into design uncertainty in manufactured chips. Further, application workloads can significantly affect the overall system reliability. In this work, we have analyzed aging effects on various design hierarchies of an embedded processor in 28nm running real-world applications. We have also quantified the dependencies of aging effects on switching-activity and power-state of workloads. Implementation results show that the processor timing degradation can vary from 2% to 11%, depending on the workload.

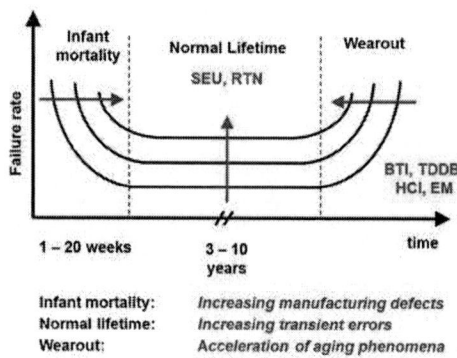

Fig. 1.   Failure rate during the design life-cycle

I. INTRODUCTION

When determining the reliability of a system, knowing the respective parameters of the underlying devices is mandatory. An accurate device-level evaluation requires a deep understanding of the fault mechanisms including aging, stress, process variation etc. The underlying devices of integrated systems are the gates, the transistors, and the physical connections. We will discuss the device level reliability mechanisms and focus on techniques to evaluate workload dependent reliability of a commercial embedded processor running real-world applications.

Degradation at the device level can be broadly divided into two groups: physical mechanisms, due to inherent defects and/or progressive degradation, and electrical mechanisms, which are transient in nature and do not permanently damage the device. Both must be responded to in a highly reliable system, but only the former needs a permanent workaround.

Figure 1 shows the failure rate versus time, typically referred to as the reliability "bathtub curve" due to its shape. The bathtub curve has three distinct regions:

- Infant mortality : Also known as early life failures. The high failure rate is due to random manufacturing defects like poor contacts/vias, gate-to-source shorts etc.
- Normal lifetime : The failures are due to transient errors (for example, single event upsets) and failure rate is fairly constant in this region
- Wearout/Aging: The sharp increase in the failure rate is due to aging mechanisms like NBTI and gate oxide breakdown [7], [10].

The rest of the paper is organized as follows. Section II explains the physical un-reliability mechanisms in devices.

Section III discusses the impact of application workloads on system reliability and the simulation flow to evaluate workload dependent reliability. Section IV analyzes the performance degradation results in an embedded processor. Section V summarizes the work and concludes.

II. PHYSICAL UN-RELIABILITY MECHANISMS

Reliability mechanisms which pose challenges to future scaling include Bias Temperature Instability (BTI), Time-dependent Dielectric Breakdown (TDDB), Hot Carrier Injection (HCI), Electromigration (EM), Soft errors and Random Telegraph Noise (RTN) [2].

Bias temperature instability occurs when a negatively-biased gate is stressed (PFET case for NBTI; for NFETs the effect happens with positive bias) at high temperature. Several models have been proposed to explain the precise mechanism [17]. Most of these involve the formation of traps in the gate oxide due to diffusion of hydrogen and subsequent increase in gate threshold voltage (Vt) leading to reduced performance over time. Unlike other aging effects, BTI can be partially offset by "healing" - when the device is oppositely biased, some of the traps collapse. Designers can thus mitigate the effects of BTI by balancing bias states. This can be quite challenging, depending on the portion of the design, but many efforts have been published [11], [15].

Time-dependent dielectric breakdown (TDDB), also known as soft oxide breakdown, is also trap related. FinFET devices can have improved TDDB due to vertical field reduction and increased barrier to tunneling [12], but 3D features must be carefully engineered and TDDB is still a critical reliability mechanism. With continued dimension scaling, in combination

978-1-4799-2817-0/14 $31.00 © 2014 IEEE   474

with pressure to employ lower dielectric constant films, increasing electric fields in the interconnect has extended TDDB concerns to the wiring [1].

Electromigration (EM) has moved from a problem of mild effect around high drive buffers to something that must be carefully checked throughout a design. The maximum DC current allowed through local metal has not been keeping pace with device current scaling. This trend, which should intensify into the FinFET era, necessitates more robust power delivery network (PDN) design, wider metal in critical nets, and even in some cases strapping outputs in M2. All of these remedies necessarily increase block area.

While the soft error rate (SER) per bit cell has stabilized in recent technologies, and with FinFETs may actually improve as compared to planar, the rate per area has been increasing [13]. SER has conventionally been most important in SRAM arrays and dealt with via error correction. With continued scaling, the flip-flop SER has become comparable to that of SRAMs [3], [9]. However, protecting (hardening) flip-flops against soft error is challenging due to the fact that they are spatially distributed. Further, soft error susceptibility of a design goes up with the onset of device aging due to TDDB [4].

Random telegraph noise (RTN), which causes time-varying threshold voltage [14], is a future reliability concern. While this effect has historically not been significant in digital design, the trend for RTN variation, $\sim (LW)^{-1}$, is steeper than that for random variability, $\sim (LW)^{-0.5}$.

III. Workload Dependent Reliability

Many of the mechanisms discussed in the previous section are activity dependent. BTI includes a "healing" component when a transistor is not active. TDDB happens only when an oxide is stressed. Neutron strikes will not cause soft errors when a memory is powered down. This activity dependence is very important when making reliability predictions for a given product. A product in constant operation is much more likely to suffer a reliability failure than one that is rarely used. Similarly, some portions of a design are much more likely to be stressed than others. In some cases, the stressed devices are obvious (clock networks, for example), while others are more subtle: In many designs (e.g. processors) memories tend to contain more 0s than 1s throughout normal operation, meaning that two of the six devices in each SRAM bit cell undergo much more TDDB and BTI related stress than the others, in the worst case leading to unbalanced bit cells and failing read/write operations. Consideration of application workloads can result in substantial changes to product reliability [8].

This work addresses the challenge of designing robust and energy-efficient systems in the presence of transistor aging. The aging mechanisms, Negative Bias Temperature Instability (NBTI) and Positive Bias Temperature Instability (PBTI), which are dominant in deeply scaled technology nodes, are considered. NBTI and PBTI gradually increase the magnitude of PMOS and NMOS threshold voltages over the lifetime, causing performance degradation. Because the sensitivity to aging is projected to increase at future process nodes and

lower supply voltages, delay degradations are expected to increase in the future. Hence, accurate estimation of aging effects is essential. This work can be extended to include other aging mechanisms. However, the HCI effect is typically much smaller than BTI, while the TDDB and EM affect the hard-breakdown time-to-failure.

This work presents analysis of workload-dependent aging effects in a large industrial design, with numerous complex combinational or sequential cell types, which runs typical applications. A simulation flow is presented to capture aging of all instances in the entire design with high accuracy, efficiency, and scalability, in a fully automated way. The flow creates a unique standard-cell library for each instance by aging each transistor according to its individual workload. Therefore, this flow can account for transistor-level aging variations due to switching-activity, power-state, process, voltage and temperature.

Figure 2 plots the power-state characteristics obtained from actual silicon measurements with full-system implementation running various real-world applications. The trace for each

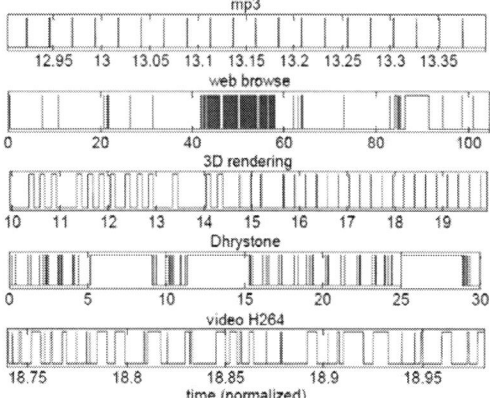

Fig. 2. Power-state profile for typical workloads in an embedded processor

application specifies the power-state (active or sleep) at each tick in time. Significant variations are clearly observed. Workloads such as "mp3" and "web-browsing" spend very little time in active-state, whereas workloads such as "Dhrystone" and "video H264" spend much more time in active-state. The average amount of time spent in the active-state is 0.54 for "video H264," 0.4 for "Dhrystone," 0.24 for "3D rendering," 0.07 for "web-browsing," and 0.03 for "mp3". Aging-induced threshold voltage degradation is increased during active-state and partially recovered in sleep-state. Long-term threshold voltage degradation depends not only on the average amount of time in the active-state, but also on the particular power-state waveform, due to its aperiodic nature with irregular active-sleep pattern. Using [5], [18], the degradation at the end of each tick can be obtained as a function of both degradation at the beginning of that tick and the power-state during that tick. Then, the long-term degradations for various workloads can be compared at a particular point in time.

A. Simulation Flow

A simulation flow is created to accurately and efficiently assess the impact of aging in a complex industrial processor core. We propose an "instance-based" approach (Figure 3), the main principle of which is to pair each instance in the design with a uniquely stressed standard-cell library, created by aging individual transistors according to the static and workload-dependent dynamic factors, such as switching activity, sleep mode, process, voltage, and temperature. By integrating transistor-level, circuit-level, and system-level simulations, the flow is designed to assess the impact of aging at the processor-level while taking into account aging variations between each transistor in the design.

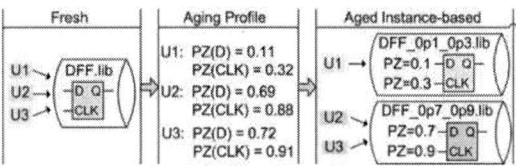

Fig. 3. Example of "instance based" approach. From instances (U1, U2, U3) that use standard-cell DFF (D Flip-Flop), bin and identify the unique sets of inputs logic probabilities (PZ), then create the aged DFF.lib

The simulation framework is illustrated in Figure 4 . RTL, synthesis, place and route (SPR), and STA are first performed with fresh standard-cell and reference libraries, to obtain the initial gate-level design, timing and power reports. This can be accomplished using existing industrial flows if available. To obtain the aged timing and power reports, the following

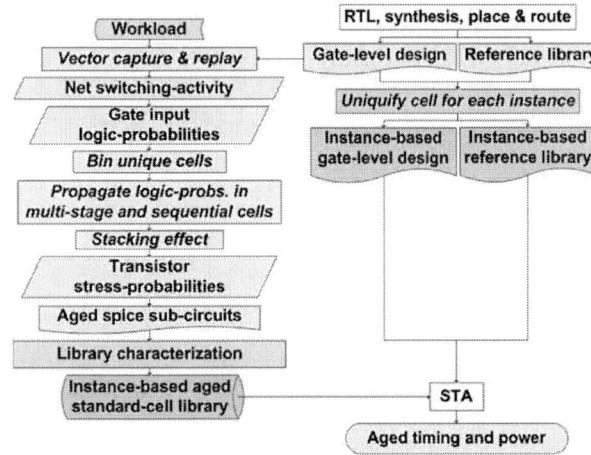

Fig. 4. Simulation flow with aged standard-cell library characterization

instance-based design and libraries are then created and used as new input files to the same STA tool:

1) *Instance-based gate-level design*, where each instance uses a unique standard-cell.
2) *Instance-based reference library*, which has physical information for each instance-based standard-cell, replicated from the original reference library.
3) *Instance-based aged standard-cell library*, which contains workload-dependent aged timing and power database for each instance-based standard-cell, e.g., in .lib format.

To create the "instance-based aged standard-cell library," the flow first runs system-level workload programs to obtain

the probability at logic zero (PZ) of every net in the design. The PZ at each input pin of every instance-based standard-cell can then be found by mapping the input pin to a corresponding net. The actual PZ, which determines the stress time, of every input is used, rather than a fixed value, e.g., the worst among them. The inputs PZ are then binned and discretized to identify the "unique instance-based standard-cells," such that all instance-based standard-cells with similar sets of inputs PZ are considered as one unique instance-based standard-cell. As illustrated in Figure 3, for example, DFF_U2 has PZ(D)=0.69 and PZ(CLK)=0.88; DFF_U3 has PZ(D)=0.72 and PZ(CLK)=0.91. Consequently, DFF_U2 and DFF_U3 can be identified as the same DFF_0p7_0p9. With a binning interval of 0.1, the number of unique standard-cells can be reduced by ~10-20X. For the case study presented here, binning reduces ~100K instances to ~6-11K unique instances, depending on the workload.

IV. RESULTS

The flow is demonstrated in a 28nm commercial processor, which has >100K instances with >10K sequential elements, and uses >700 cell-topologies with >11K timing-arcs. The design includes the parasitics. The nominal library characterization and aging stress conditions are 0.9V. 105C and 3 years. Two 28nm aging models, reaction-diffusion (RD) [5], [6], [18] and trapping-detrapping (TD) [16], which predict aging behavior under dynamic operation patterns, are used and scaled to match the desired process node and operating condition.

Figure 5 lists the fresh path ranks for the five worst aged paths, and the aged path ranks for the five worst fresh paths. The path rank is sorted by slack, so rank 1 means the worst, or

Path rank		% Timing degrad		Path rank		% Timing degrad
Fresh	Aged			Fresh	Aged	
179394	1	15.61		1	14084	7.64
145042	2	15.41		2	9781	7.94
134419	3	15.18		3	9329	8.02
1413427	4	17.57		4	12345	7.87
272323	5	15.67		5	6220	8.31

Fig. 5. Path rank shift in fresh versus aged

least, slack. Results show that path rank can shift significantly due to aging. Paths that are non-critical in the fresh design can become critical after aging due to much larger timing degradations, and paths that are critical in fresh design can become non-critical in aged design as they experience smaller timing degradations. The average degradation is 15.9% for the five worst aged paths, and 8% for the five worst fresh paths. Dhrystone workload and RD model are used for this analysis.

Figure 6 shows that timing degradations among paths are substantially non-uniform. For a sample of >660K critical paths, the largest degradation is 17.6%, the smallest is 4.9%, and the average is 9.5%.

Figure 7a shows that the timing degradation (left vertical axis) and aging-induced threshold-voltage shift (right vertical axis) of each instance along a path are correlated. Two paths, with nearly the largest and smallest timing degradations, are analyzed. Path A ranks 1413427 in fresh, ranks 4 in aged,

978-1-4799-2817-0/14 $31.00 © 2014 IEEE   476

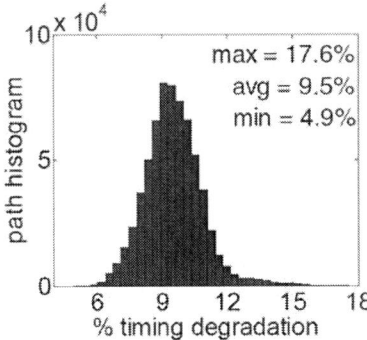

Fig. 6. Distribution of delay degradation for sampled 660K paths

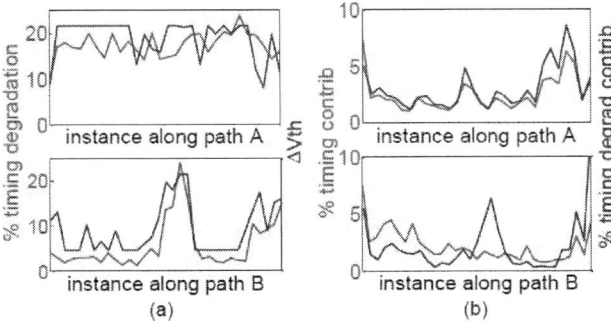

Fig. 7. Distribution of delay degradation for sampled 660K paths

and has 17.6% timing degradation. Path B ranks 1687 in fresh, ranks 1630836 in aged, and has 4.96% timing degradation. Comparison of the two paths shows that path timing degradation is closely affected by workload-dependent stress-probabilities along the path. Path A, which has higher stress-probabilities (hence higher aging-induced threshold-voltage shifts) along the path, has much larger timing degradation than path B, which has on average lower stress-probabilities. Figure 7b depicts the contribution of each instance to fresh path delay (left vertical axis), and the percentage of each instance delay increase to path delay degradation (right vertical axis). It shows that aging-induced path delay degradation can be dominated by some critical instances which may not be critical for fresh path delay.

Figure 8 analyzes the processor-level timing-degradations under various application workloads described earlier. It also

Workload	Switching-activity		Power-state		Switching-activity & power-state	
	RD	TD	RD	TD	RD	TD
mp3	10.0%	6.3%	3.6%	2.5%	2.3%	1.6%
web-browse	10.6%	7.3%	6.1%	4.2%	4.1%	2.8%
3Drendering	12.3%	8.4%	6.9%	4.7%	5.4%	3.7%
dhrystone	11.2%	7.7%	10.1%	6.9%	7.3%	5.0%
videoH264	12.5%	8.5%	11.4%	7.8%	9.1%	6.2%
all PZ = 0.5	6.8%	4.7%	15.6%	10.7%	6.8%	4.7%
worst-case	15.6%	10.7%	15.6%	10.7%	15.6%	10.7%

Fig. 8. Workload-dependent processor timing degradation

compares a synthetic workload which assumes all net have a logic-probability of 0.5, and a worst-case scenario, which assumes worst-case stress-probability and active-state at all times. The first column shows the switching activity depen-

dence, the second column shows the power-state dependence, and the third column shows the cumulative workload dependence. Strong workload-dependence is observed, ranging from 2.2% to 15.6%. The switching-activity dependence assumes active-state at all times, and similarly, power-state dependence assumes worst-case stress-probability during the active-state.

V. CONCLUSIONS

Reliability challenges in CMOS technologies are increasing due to dominance of spatial, temporal and dynamic variations in the nanoscale era. Further, application workload has a large impact on reliability and hence should be taken into account while evaluating the system reliability. Design for Reliability (DFR) flow should become an integral part of the complete technology offering to meet the power, performance and area requirements. DFR is a perfect example of design technology co-optimization, where close collaboration of innovation in devices, process, materials and circuit design can improve overall scaling as compared to past implementations (which generally simply guard-band around the unknown).

ACKNOWLEDGMENTS

The author would like to acknowledge Evelyn Mintarno and Subhasish Mitra (Stanford) and Rob Aitken and Dave Pietromonaco (ARM) for various insightful discussions.

REFERENCES

[1] R. Achanta et al, "Failure rates for interconnect dielectric breakdown: Trends determining technology reliability scaling limits," IEEE Transactions on Device and Materials Reliability, vol. 11 No.2, June 2011.

[2] R. Aitken, "Reliability Evaluation at the Device Level and its Impact on Design," IEEE Design Automation and Test in Europe (DATE), 2013.

[3] V. Chandra et al, "Impact of Technology and Voltage Scaling on the Soft Error Susceptibility of Nanoscale CMOS," IEEE Intl. Symposium on Defect and Fault Tolerance of VLSI Systems, 2008.

[4] V. Chandra et al, "Impact of Voltage Scaling on Nanoscale SRAM Reliability," IEEE Design Automation and Test in Europe (DATE), 2009.

[5] M. Chen et al, "A TDC-based Test Platform for Dynamic Circuit Aging Characterization," IEEE Intl. Reliability Physics Symposium (IRPS), 2011.

[6] G. Gielen et al, "Analog Circuit Reliability in Sub-32 Nanometer CMOS: Analysis and Mitigation," IEEE Design Automation and Test in Europe (DATE), 2011.

[7] B. Kaczer et al, "Impact of MOSFET gate oxide breakdown on digital circuit operation and reliability," IEEE Transactions on Electron Devices, Vol. 49, No. 3, pp. 500-506, Mar 2002.

[8] E. Mintarno et al, "Workload Dependent NBTI and PBTI Analysis for a sub-45nm Commercial Microprocessor," IEEE Intl. Reliability Physics Symposium (IRPS), 2013.

[9] A. Oates et al, "Reliability challenges for the continued scaling of IC technologies," IEEE Custom Integrated Circuits Conference (CICC), pp. 1-4, 2012.

[10] B. C. Paul et al, "Impact of NBTI on the temporal performance degradation of digital circuits," IEEE Electron Device Letters, Vol. 26, No. 8, pp. 560-562, Aug 2005.

[11] Z. Qi et al, "SRAM-based NBTI/PBTI sensor system design," Design Automation Conference (DAC), pp. 849-852, 2010.

[12] S. Ramey et al, "Intrinsic Transistor Reliability Improvements from 22nm Tri-Gate Technology," IEEE Intl. Reliability Physics Symposium (IRPS), 2013

[13] N. Seifert et al, "Soft error suceptibilities of 22 nm tri-gate devices," IEEE Transactions on Nuclear Science, Vol. 59, No. 6. Dec. 2012.

[14] K. Takeuchi et al, "Direct observation of RTN-induced SRAM failure by accelerated testing and its application to product reliability assessment," IEEE Symposium on VLSI Technology, 2010.

[15] R. Vattikonda et al, "Modeling and minimization of PMOS NBTI effect for robust nanometer design," Design Automation Conference (DAC), pp. 1047-1052, 2006.

[16] J. B. Velamala et al, "Statistical Aging under Dynamic Voltage Scaling: A Logarithmic Model Approach," IEEE Custom Integrated Circuits Conference (CICC), 2012.

[17] W. Wang et al, "The Impact of NBTI Effect on Combinational Circuit: Modeling, Simulation, and Analysis," IEEE Transactions on VLSI Systems, Vol. 18, pp. 173-183, 2010.

[18] R. Zheng et al, "Circuit Aging Prediction for Low Power Operation," IEEE Custom Integrated Circuits Conference (CICC), 2009.

QED Post-Silicon Validation and Debug: Frequently Asked Questions

David Lin
Dept. of Electrical Engineering
Stanford University
Stanford, CA, USA

Subhasish Mitra
Depts. of Electrical Engineering and Computer Science
Stanford University
Stanford, CA, USA

Abstract

During post-silicon validation and debug, one or more manufactured integrated circuits (ICs) are tested in actual system environments to detect and fix design flaws (bugs). According to several industrial reports, the costs of post-silicon validation and debug are rising faster than design costs. Hence, new techniques are essential to reverse this trend. QED, an acronym for Quick Error Detection, is such a technique that effectively overcomes several post-silicon validation and debug challenges. QED systematically creates a wide variety of validation tests to quickly detect bugs, not only inside processor cores, but also inside uncore components (i.e., components in an SoC that are neither processor cores nor co-processors) of multi-core SoCs. In this paper, we present a brief overview of QED through a series of frequently asked questions.

Keywords

Debug, Post-Silicon Validation, Quick Error Detection, Verification

I. FREQUENTLY ASKED QUESTIONS

Question 1. Why is post-silicon validation / debug important?

Post-silicon validation / debug is crucial because traditional pre-silicon verification alone is inadequate for today's complex integrated circuits (ICs). As a result, critical design bugs escape pre-silicon verification and are detected only during post-silicon validation [2], [3], [16], [23], [29]. Design bugs can be broadly classified into two categories:

(a) *Logic bugs* that are caused by design errors. In addition to bugs in the hardware implementation, this category includes incorrect interactions between the hardware implementation and the low-level system software (e.g., firmware).

(b) *Electrical bugs* that are caused by subtle interactions between a design and its "electrical" state. Electrical bugs often manifest themselves only under specific operating conditions, such as voltage, frequency, and temperature corners [36]. Examples of electrical bugs include signal integrity (e.g., cross-talk, power-supply noise), thermal effects, and process variations.

Traditional pre-silicon verification is too slow for "difficult" logic bugs, and it also does not adequately address electrical bugs that appear only after ICs are manufactured. These challenges get further magnified with the slowdown of the classical silicon CMOS (Dennard) scaling [8], as ICs incorporate tremendous design complexity to meet performance and energy-efficiency requirements. Examples include the use of multiple processor cores; co-processors and accelerators such as GPUs; uncore components such as on-chip cache controllers, memory controllers, and network controllers; adaptive power, thermal and reliability management; and, proliferation of heterogeneous integration. Here uncore components (also referred to as the nest or the northbridge) are defined as components in a system-on-chip (SoC) that are neither processor cores nor co-processors.

Existing post-silicon validation and debug practices are generally *ad-hoc*, and several sources [1], [44] indicate that their costs are rising faster than design costs. For example, it may take several weeks to debug a single bug during the post-silicon phase. Without systematic and scalable ways of taming such increasing levels of complexity, future systems can end up being highly vulnerable to bugs that may compromise correct operation and introduce security risks.

Question 2. Why is post-silicon validation and debug difficult?

During post-silicon validation and debug, manufactured ICs are tested in actual systems in order to detect and fix bugs. When a failure is detected, the bug that caused the failure is localized and root-caused (and then fixed). Several sources indicate that the effort to localize bugs from observed failures dominates the overall cost of post-silicon validation and debug [1], [6], [22], [23].

Post-silicon bug localization is difficult because existing approaches suffer from two major challenges:

(a) Reliance on failure reproduction, which involves returning the system to an error free state and re-executing the failure-causing stimuli (i.e. sequence of instructions, interrupts, voltage, temperature and frequency operating conditions). Failure reproduction is very difficult for complex SoCs with multiple clock domains and asynchronous I/Os.

(b) Reliance on system-level simulation to generate expected or golden responses. System-level simulation is several orders of magnitude slower than actual silicon (e.g., 1,000 cycles / second in simulation vs. 1 billion cycles / second for a 1GHz chip).

Question 3. What can be done to overcome post-silicon validation and debug challenges?

"Systematic" and "automated" techniques are essential to overcome the scalability challenges of existing post-silicon validation and debug approaches. Our analysis of bug databases for several commercial SoCs (enabled by our collaborations with several industrial partners) clearly indicates that these challenges are primarily caused by long error detection latencies [20], [26], [27], [38]. Error detection latency is defined as the time elapsed from when a test activates a bug and creates an error in the system to when the error manifests as an observable failure (e.g., system crash, timeout, deadlock, exceptions, or incorrect results). Ideally, error detection latencies should not exceed a few thousand clock cycles because most ICs contain trace buffers that can record on the order of 1,000 clock cycles of history [1]. However, during post-silicon validation and debug, error detection latencies for "difficult" bugs can exceed several millions or even billions of clock cycles, especially for bugs inside uncore components of multi-core SoCs [26]. Such long error detection latencies are highly challenging because it is extremely difficult to trace very far back into the history of system operation for debug.

We created the Quick Error Detection (QED) technique [20], [26], [27] to overcome the long error detection latency challenge. QED is organized as a variety of QED transformations that systematically transform existing tests (referred to as original tests) into new QED tests with bounded error detection latencies and improved coverage. A wide variety of tests, e.g., architecture-specific focused tests, random instruction tests, or end-user applications, can be transformed using QED. Examples of QED transformations include Error Detection using Duplicated Instructions for Validation (EDDI-V), EDDI-V with diversity, Control Flow Checking using Software Signatures for Validation (CFCSS-V), Control Flow Tracking using Software Signatures for Validation (CFTSS-V), and Proactive Load and Check (PLC). The details of these QED transformation techniques are presented in [20], [26], [27]. The QED transformations can be implemented entirely in software, in which case they are readily applicable during post-silicon validation of existing ICs. QED also can be augmented with hardware support to further improve error detection latency and coverage, and to reduce QED test execution time.

The basic principles of QED are rooted in Software Implemented Hardware Fault Tolerance (SIHFT) techniques [28], [30], [31], [32]. However, there are several major differences between QED for post-silicon validation and debug vs. SIHFT for fault-tolerant computing. For example, during post-silicon validation and debug, error detection latency is extremely important because debug time, rather than test execution time, dominates the overall cost [22]. Furthermore, QED tests must continue to detect bugs that are detected by the original tests [20], [26]. In contrast, performance impact (which is related to execution time) plays a very important role for fault-tolerant computing.

Question 4. What are the published QED results?

The following presents a summary of our published results.

(a) For the Intel® Core™ i7-based hardware platform (Fig. 1a), QED improves error detection latencies of electrical bugs by 6 orders of magnitude while simultaneously improving coverage 4-fold [20]. This is shown in Fig. 2, where the vertical axis represents the percentage of errors detected (normalized to QED, due to confidentiality reasons) and the horizontal axis represents the error detection latencies. The coverage improvement of QED is also demonstrated in Fig. 3; there is not a single operating point (frequency and voltage pair) for which the original test failed but the QED test passed. Furthermore, the operating point labeled with a star in Fig. 3 demonstrates coverage improvement using QED: for this operating point, the original test passed, whereas the QED test detected errors very quickly. QED creates valid tests that do not bring the system into any "illegal" states. Therefore, errors detected by QED are actual errors in the system. These results were also confirmed using the Intel® 48-core Single-Chip Cloud Computer [21] (Fig. 1b). For the processor core labeled with a star in Fig. 4, the original test passed but the QED test detected errors very quickly.

(b) We simulated an extensive list of "difficult" logic bugs (abstracted from actual bug databases) inside processor cores and uncore components using an OpenSPARC T2-like SoC [33]. The details of the logic bugs and our simulation setup are presented in [26]. The results are summarized in Fig. 5 (details in [26]), where the vertical axis represents the <u>absolute</u> cumulative percentage of bugs detected, and the horizontal axis shows the error detection latencies. QED improves error detection latencies by several orders of magnitude, from tens of millions or even billions of clock cycles to (mostly) a few hundred clock cycles, and simultaneously improves coverage up to 2-fold [26].

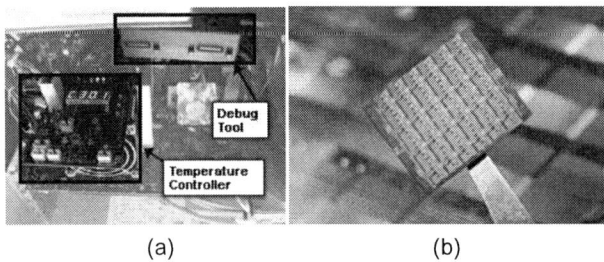

(a)   (b)

Figure 1. (a) Quad-core Intel® Core™ i7-based hardware platform with a DX58SO motherboard, a temperature controller, and a proprietary debug tool. (b) Intel® 48-core Single-Chip Cloud Computer [21].

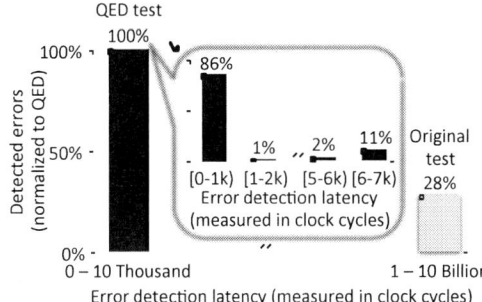

Figure 2. Distribution of error detection latencies for the Intel® Core™ i7-based hardware platform for the original Linpack test [14] with end result checks (Original test) and the Linpack test with QED (QED test). For confidentiality reasons, percentage of detected errors is normalized to QED. (The details are in [20]).

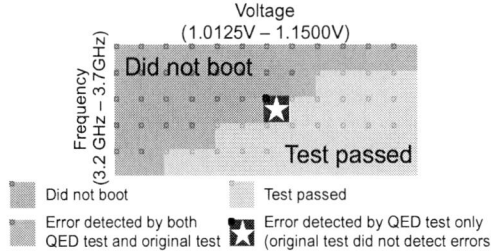

Figure 3. Shmoo plot for Intel® Core™ i7-based hardware platform running the Linpack test [14]. The operating point labeled with a star demonstrates coverage improvement using QED: the original test passed whereas the QED test detected errors very quickly.

978-1-4799-2817-0/14 $31.00 © 2014 IEEE

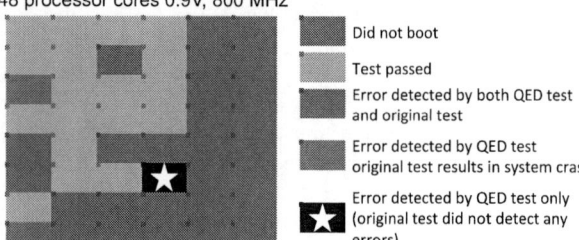

Figure 4. The Intel® 48-core Single-Chip Cloud Computer running the Linpack test [14]. Each box represents an individual processor core. The processor core labeled with a star demonstrates coverage improvement using QED: the original test passed whereas the QED test detected errors very quickly.

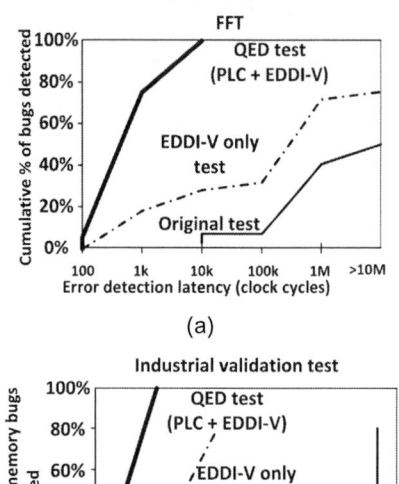

(a)

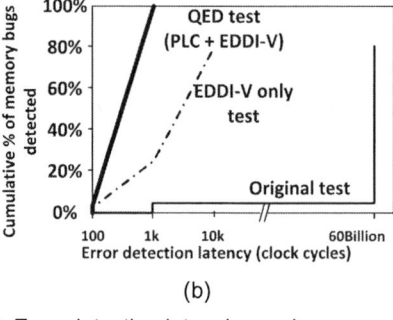

(b)

Figure 5. Error detection latencies and coverage of post-silicon validation tests on a simulated OpenSPARC T2-like SoC [33] using "difficult" logic bugs abstracted from actual bug databases. (a) FFT program from SPLASH-2 benchmark suite [42]. (b) Industrial validation test. The cumulative percentage of bugs detected represents absolute (and not normalized) percentage values.

Question 5. Is QED effective only for bugs inside processor cores?

QED targets bugs inside both processor cores and uncore components of SoCs. In [26], we simulated an extensive list of bugs, abstracted from actual "difficult" bugs that occurred in several commercial multi-core SoCs, to evaluate the effectiveness of QED. These are primarily logic bugs inside uncore components such as cache controllers, memory controllers, and on-chip interconnection networks (in addition to bugs inside processor cores). These bugs are considered to be difficult because the corresponding bug reports indicate very long debug times. Results in [26] (also summarized in Fig. 5) demonstrate that QED is highly effective for improving error detection latencies and coverage for bugs inside uncore components, as well as bugs inside processor cores.

Question 6. Is QED effective only for electrical bugs?

QED is effective for both electrical and logic bugs. [26] demonstrated that QED is highly effective for an extensive list of logic bugs abstracted from difficult logic bugs that occurred in several commercial multi-core SoCs (Fig. 5). [20] demonstrated that QED is highly effective for electrical bugs using the Intel® Core™ i7-based hardware platform (Fig 2. and Fig. 3).

Question 7. Is QED different from self-checking tests?

QED tests are different from self-checking tests that check the results of instructions by comparing the results against those derived using a golden model [5], or using instructions executed on processor cores [37], [43]. As demonstrated in [26], [27], such self-checking tests can incur extremely long error detection latencies, especially for bugs inside uncore components. In contrast, QED systematically inserts fine-grained and proactive checks to bound error detection latencies as well as to improve coverage for bugs inside both processor cores and uncore components.

Question 8. Is QED different from assertions?

QED is different from assertions used during post-silicon validation. The use of assertions during post-silicon validation suffers from several challenges:

1. Numerous assertions may be generated for a given design. These assertions have to be carefully crafted, and it is difficult to keep them up-to-date and to validate their correctness [7], [40]. While there exist techniques for automatically generating assertions (e.g., [15], [18], [39], [40]), it may be difficult to implement all such assertions in hardware.
2. Reconfigurable logic can somewhat ease the implementation of assertions in hardware [1], [9], [17]; however, one must be careful about selecting the relevant set of assertions for effective post-silicon bug localization.
3. An assertion may depend on signals that are located in two different regions of a chip; such an assertion must be decomposed into components that only use nearby signals [29].

In contrast, QED enables "structured" and "systematic" ways of performing extensive checks while avoiding the drawbacks of design-specific assertions.

Question 9. Doesn't the test execution time go up with QED?

QED does increase test execution time [26]. However, the costs of post-silicon validation and debug are dominated by the time required to localize bugs rather than the time required for running tests [22]. Since QED can detect bugs very quickly (e.g., a few hundred clock cycles after the bug is activated [20], [26]), bug localization can become significantly easier, resulting in a net improvement in productivity. Furthermore, the execution time impact of QED can be reduced using hardware support.

978-1-4799-2817-0/14 $31.00 © 2014 IEEE   480

Question 10. Since QED utilizes redundant execution, can QED miss bugs that affect both executions in the same way?

There are a wide variety of QED transformations, and time-redundant re-execution of instructions (EDDI-V) is only one of the techniques used by QED. For example, the EDDI-V with diversity transformation incorporates data diversity techniques based on the principles of ED^{4}I [31]. This ensures that the instructions inserted by QED execute differently vs. the original instructions, thereby minimizing the chances of a bug affecting the original instructions and the QED instructions identically. Moreover, for difficult bugs, time-redundant re-execution itself creates diversity (as shown by our results in [26]). Furthermore, QED techniques such as CFCSS-V and CFTSS-V do not rely on instruction re-execution.

Question 11. How do you localize bugs after quick detection using QED?

After quick detection using QED, many post-silicon bug localization techniques can be used: IFRA and BLoG [34], [35], simulation-based debug [10], [24], or debug techniques based on formal methods [11]. These techniques can directly benefit from the extremely short error detection latencies and improved coverage of QED. With the increasing complexity of uncore components in SoCs, new techniques for localizing bugs inside uncore components are required.

Question 12: Why does QED improve coverage?

QED improves the coverage of post-silicon validation tests by detecting errors that would otherwise be masked or undetected due to insufficient checking. For example, a bug may create an error in the value stored in a register, say register R1. Other instructions in the test that write to register R1 may mask that error by overwriting the erroneous value. Unless the value stored in register R1 is checked before it is overwritten, the error will be masked. QED inserts fine-grained and proactive checks to quickly detect such errors, as a result, coverage improves.

Question 13: Does QED introduce intrusiveness?

Since software-implemented QED transformations insert additional instructions, intrusiveness is possible; i.e., there may exist pathological examples for which the original test may detect a bug, but the corresponding QED test may not. To overcome this, QED transformations support special parameters, *Inst_min* and *Inst_max*, that enable systematic approaches to minimize possible intrusiveness due to QED transformations. *Inst_min* is defined as the minimum number of instructions in the original test that must execute consecutively before executing any instructions inserted by QED transformations. *Inst_max* is defined as the maximum number of instructions in the original test that must execute consecutively before executing any instructions inserted by QED transformations. For example, by increasing *Inst_min*, we can possibly reduce the intrusiveness introduced by QED. This is because a longer sequence of instructions from the original test will execute consecutively (uninterrupted by QED). However, excessively large *Inst_min* increases error detection latency. Strategies for adjusting these transformation parameters, e.g., through the creation of QED family tests with a range of *Inst_min* and *Inst_max* values, are discussed in [20], [27]. Moreover, hardware support for QED can minimize the intrusiveness introduced by QED transformations.

Question 14. Can QED be used for pre-silicon verification?

Pre-silicon verification uses emulation / acceleration platforms extensively [3], [4], [23], [25] to improve productivity. QED tests can be used to improve both error detection latencies and coverage in these environments as well.

Question 15. Are logic bug benchmarks available?

Thanks to our industrial collaborators, we received unprecedented access to bug reports for difficult bugs that occurred in several state-of-the-art commercial multi-core SoCs. From these bug reports, we compiled an extensive list of bug scenarios. We worked closely with the respective validation teams to abstract the bug reports into higher-level descriptions (and remove product-specific details). This allows the bugs to be simulated using various micro-architectural and RTL simulators. The list of bugs is presented in [26], and is also available at http://www.bugslist.org. Researchers can also submit their own bug scenarios on the website to benefit the post-silicon validation and debug research community.

Other researchers have also published logic bugs found in research chips, class projects, and errata pages [12], [13], [19], [41]. The list of bugs in [26] subsumes these bugs.

II. ACKNOWLEDGEMENT

This research was supported in part by NSF, SRC and SGF. The authors thank Eric Rentschler of AMD, Eswaran S. and Sharad Kumar of Freescale, Wisam Kadry, Ronny Morad, Amir Nahir and Avi Ziv of IBM, and Donald S. Gardner, Nagib Hakim, Jagannath Keshava, and Keshavan Tiruvallur of Intel for their valuable inputs.

III. References

[1] M. Abramovici, "A Reconfigurable Design-for-Debug Infrastructure for SoCs," *Proc. IEEE/ACM Design Automation Conf.*, pp. 7-12, 2006.

[2] A. Adir, A. Nahir, A. Ziv, C. Meissner, and J. Schumann, "Reaching Coverage Closure in Post-Silicon Validation," *Proc. Haifa Verification Conf.*, pp. 60-74, 2010.

[3] A. Adir, *et al.*, "A Unified Methodology for Pre-Silicon Verification and Post-silicon Validation," *Proc. IEEE/ACM Design, Automation and Test in Europe Conf.*, pp. 1-6, 2011.

[4] A. Adir, *et al.* "Leveraging Pre-Silicon Verification Resources for the Post-Silicon Validation of the IBM POWER7 Processor," *Proc. IEEE/ACM Design Automation Conf.* pp. 569-574. 2011.

[5] A. Aharon, *et al.*, "Test Program Generation for Functional Verification of PowerPC Processors in IBM," *Proc. IEEE/ACM Design Automation Conf.*, pp. 279-285, 1995.

[6] M. E. Amyeen, S. Venkataraman, and M. W. Mak, "Microprocessor System Failures Debug and Fault Isolation Methodology," *Proc. IEEE Intl. Test Conf.*, pp. 1-10, 2009.

[7] B. Bentley, and R. Gray, "Validating the Intel Pentium 4 Processor," *Intel Technology Journal*, Vol. 5 No. 1, pp. 1-8, 2001.

[8] M. Bohr, "The New Era of Scaling in an SoC World," *Proc. IEEE Solid-State Circuits Conf.*, pp. 23-28, 2009.

[9] M. Boule, J.-S. Chenard, and Z. Zilic, "Assertion Checkers in Verification, Silicon Debug and In-Field Diagnosis," *Proc. IEEE Intl. Symp. on Quality Electronic Design*, pp. 613-620, 2007.

[10] K.-H. Chang, I. L. Markov, and V. Bertacco, "Automated Post-Silicon Debugging and Repair," *Proc. IEEE/ACM Intl. Conf. on Computer-Aided Design*, pp. 91-98, 2007.

978-1-4799-2817-0/14 $31.00 © 2014 IEEE

[11] F. M. De Paula, A. J. Hu, and A. Nahir, "nuTAB-BackSpace: Rewriting to Normalize Non-Determinism in Post-Silicon Debug Traces," *Proc. Intl. Conf. on Computer Aided Verification*, pp. 513-531, 2012.

[12] A. DeOrio, A. Bauserman, and V. Bertacco, "Post-Silicon Verification for Cache Coherence," *Proc. IEEE Intl. Conf. Computer Design*, pp. 348-355, 2008.

[13] A. DeOrio, I. Wagner, and V. Bertacco, "DACOTA: Post-Silicon Validation of the Memory Subsystem in Multi-Core Designs," *Proc. IEEE Intl. Symp. High-Performance Computer Architecture*, pp. 405-416, 2009.

[14] J. Dongarra, P. Luszczek, and A. Petitet, "The LINPACK Benchmark: Past, Present and Future," *Concurrency and Computation: Practice and Experience*. Vol. 15, No. 9, pp. 803-820, 2003.

[15] M. D. Ernst, *et al.*, "The Daikon System for Dynamic Detection of Likely Invariants," *Science of Computer Programming*, Vol. 69, No. 1-3, pp. 35-45, Dec. 2007.

[16] T. J. Foster, D. L. Lastor, and P. Singh, "First Silicon Functional Validation and Debug of Multicore Microprocessors," *IEEE Trans. Very Large Scale Integration Systems*, Vol. 15, No. 5, 2007.

[17] M. Gao, H.-M. Chang, P. Lisherness, and K.-T. Cheng, "Time-Multiplexed Online Checking: A Feasibility Study," *Proc. IEEE Asian Test Symp.*, pp. 371-376, 2008.

[18] S. Hangal, S. Narayanan, N. Chandra, and S. Chakravorty, "IODINE: A Tool to Automatically Infer Dynamic Invariants," *Proc. IEEE/ACM Design Automation Conf.*, pp. 775-778, 2005.

[19] R. C. Ho, C. H. Yang, M. A. Horowitz, and D. L. Dill, "Architecture Validation for Processors," *Proc. ACM/IEEE Intl. Symp. Computer Architecture*, pp. 404-413, 1995.

[20] T. Hong, *et al.*, "QED: Quick Error Detection Tests for Effective Post-Silicon Validation," *Proc. IEEE Intl. Test Conf.*, pp. 1-10, 2010.

[21] J. Howard, *et al.*, "A 48-Core IA-32 Message-Passing Processor with DVFS in 45nm CMOS," *Proc. IEEE. Intl. Solid-State Circuits Conf.*, pp. 7-11, 2010.

[22] D. Josephson, "The Good, the Bad, and the Ugly of Silicon Debug," *Proc. IEEE/ACM Design Automation Conf.*, pp. 3-6, 2006.

[23] J. Keshava, N. Hakim, and C. Prudvi, "Post-silicon Validation Challenges: How EDA and Academia Can Help," *Proc. IEEE/ACM Design Automation Conf.*, pp. 3-7, 2010.

[24] A. Krstic, L.-C. Wang, K.-T. Cheng, and T. M. Mak, "Diagnosis-Based Post-Silicon Timing Validation Using Statistical Tools and Methodologies," *Proc. IEEE Intl. Test Conf.*, pp. 339-348, 2003.

[25] C. A. Krygowski, *et al.*, "Functional Verification of the IBM System z10 Processor Chipset," *IBM Journal of Research and Development*, Vol. 53, No. 1, pp. 1-11, 2009.

[26] D. Lin, T. Hong, F. Fallah, N. Hakim, and S. Mitra "Quick Detection of Difficult Bugs for Effective Post-Silicon Validation," *Proc. IEEE/ACM Design Automation Conf.*, pp. 561-566, 2012.

[27] D. Lin, *et al.*, "Overcoming Post-Silicon Validation Challenges Through Quick Error Detection (QED)," *Proc. IEEE/ACM Design, Automation and Test in Europe Conf.*, pp. 320-325, 2013.

[28] M. N. Lovellette, *et al.*, "Strategies for Fault-tolerant Space-based Computing: Lessons Learned from the ARGOS Testbed," *Proc. Aerospace Conf.*, pp. 5-2109-5-2119, 2002.

[29] S. Mitra, S. A. Seshia, and N. Nicolici, "Post-Silicon Validation Opportunities, Challenges and Recent Advances," *Proc. IEEE/ACM Design Automation Conf.*, pp. 12-17, 2010.

[30] N. Oh, P. P. Shirvani, and E. J. McCluskey, "Error Detection by Duplicated Instructions in Super-Scalar Processors," *IEEE Trans. on Reliability*, Vol. 51, No. 1, pp. 63-75, 2002.

[31] N. Oh, S. Mitra, and E. J. McCluskey, "ED^{4}I: Error Detection by Diverse Data and Duplicated Instructions," *IEEE Trans. on Computers*, Vol. 51, No. 2, pp. 180-199, 2002.

[32] N. Oh, P. P. Shirvani, and E. J. McCluskey, "Control Flow Checking by Software Signatures," *IEEE Trans. on Reliability*. Vol. 51, No. 1, pp. 111-122, 2002.

[33] "OpenSPARC: World's First Free 64-bit Microprocessor," http://www.opensparc.net.

[34] S.-B. Park, T. Hong, and S. Mitra, "Post-Silicon Bug Localization in Processors Using Instruction Footprint Recording and Analysis (IFRA)," *IEEE. Trans. Computer Aided Design Integrated Circuits Systems*, Vol. 28, No. 10, pp. 1545-1558, 2009.

[35] S.-B. Park, A. Bracy, P. Wang, and S. Mitra, "BLoG: Post-Silicon Bug Localization in Processors Using Bug Localization Graph", *Proc. IEEE/ACM Design Automation Conf.*, pp. 368-373, 2010.

[36] P. Patra, "On the Cusp of a Validation Wall," *IEEE Design & Test of Computers*, Vol. 24, No. 2, pp. 193-196, 2007.

[37] R. Raina, and R. Molyneaux, "Random Self-Test Method Applications on PowerPC^{TM} Microprocessor Cache," *Proc. ACM/IEEE Great Lakes Symp. VLSI*, pp. 222-229, 1998.

[38] K. Reick, "Post-Silicon Debug - DAC Workshop on Post-Silicon Debug: Technologies, Methodologies, and Best-Practices," *IEEE/ACM Design Automation Conf.*, 2012.

[39] F. Rogin, T. Klotz, G. Fey, R. Drechsler, and S. Rulke, "Automatic Generation of Complex Properties for Hardware Designs," *Proc. IEEE/ACM Design Automation Conf.*, pp. 545-548, 2008.

[40] S. Vasudevan, *et al.*, "GoldMine: Automatic Assertion Generation Using Data Mining and Static Analysis," *Proc. IEEE/ACM Design, Automation, Test in Europe Conf.*, pp. 626 – 629, 2010.

[41] M. N. Velev, "Collection of High-Level Microprocessor Bugs from Formal Verification of Pipelined and Superscalar Designs," *Proc. IEEE Intl. Test Conf.*, pp. 138-147, 2003.

[42] S. C. Woo, M. Ohara, E. Torrie, J. P. Singh, and A. Gupta, "The SPLASH-2 Programs: Characterization and Methodological Considerations," *Proc. ACM/IEEE Intl. Symp. Computer Architecture*, pp. 24-36, 1995.

[43] I. Wagner, and V. Bertacco, "Reversi: Post-Silicon Validation System for Modern Microprocessors," *Proc. IEEE Intl. Conf. Computer Design*, pp. 307-314, 2008.

[44] S. Yerramilli, "Addressing Post-Silicon Validation Challenges: Leverage Validation & Test Synergy," Keynote, *IEEE Intl. Test Conf.*, 2006.

978-1-4799-2817-0/14 $31.00 © 2014 IEEE

Efficient Synthesis of Quantum Circuits Implementing Clifford Group Operations

Philipp Niemann[1]

Robert Wille[1,2,3]

Rolf Drechsler[1,2]

[1]Institute of Computer Science
University of Bremen
28359 Bremen, Germany

[2]Cyber Physical Systems
DFKI GmbH
28359 Bremen, Germany

[3]Faculty of Computer Science
Technical University Dresden
01187 Dresden, Germany

e-mail: {pniemann,rwille,drechsle}@informatik.uni-bremen.de

Abstract— Quantum circuits established themselves as a promising emerging technology and, hence, attracted considerable attention in the domain of computer-aided design. As a result, many approaches for synthesis of corresponding netlists have been proposed in the last decade. However, as the design of quantum circuits faces serious obstacles caused by phenomena such as superposition, entanglement, and phase shifts, automatic synthesis still represents a significant challenge. In this paper, we propose an automatic synthesis approach for quantum circuits that implement Clifford Group operations. These circuits are essential for many quantum applications and cover core aspects of quantum functionality. The proposed approach exploits specific properties of the unitary transformation matrices that are associated to quantum operations. Furthermore, Quantum Multiple-Valued Decision Diagrams (QMDDs) are employed for an efficient representation of these matrices. Experimental results confirm that this enables a compact realization of the respective quantum functionality.

I. INTRODUCTION

Quantum computation provides a new way of computation based on so called qubits [1]. In contrast to conventional bits, qubits do not only allow to represent the (Boolean) basis states 0 and 1, but also superpositions of both. By this, qubits can represent multiple states at the same time which enables massive parallelism. By additionally exploiting further quantum mechanical phenomena such as phase shifts or entanglement, quantum computation enables asymptotical speed-ups for many problems (e.g. database search or integer factorization). Driven by these prospects as well as recent accomplishments in the physical design of corresponding devices (e.g. [2]), quantum circuits established themselves as a promising emerging technology and, hence, attracted considerable attention in the domain of computer-aided design.

However, in order to formalize the above mentioned phenomena, states of qubits are modelled as vectors in high-dimensional Hilbert spaces and are manipulated by quantum operations which can be described by unitary matrices – possibly including complex numbers. This already poses serious challenges to the representation, but even more to the development of proper and efficient methods for quantum circuit synthesis. By this, we mean the task of determining a cascade of building blocks (so called quantum gates) whose sequential application to the qubits realizes a given quantum operation.

As a special case, there are many existing synthesis approaches for Boolean reversible circuits (i.e. permutation matrices), see e.g. [3], [4], [5], [6], [7], [8], [9]. In the last decade, these approaches became very efficient and allow for synthesis of rather complex (Boolean) functionality. Since many important quantum algorithms such as Grover's search algorithm [10] or Shor's factorization algorithm [11] include

a substantial Boolean component, these are vital techniques for quantum circuit synthesis. Especially, since the resulting circuits can be mapped to elementary quantum gates using methods such as [12], [13], [14].

However, they do not allow for the realization of more general quantum functionality. For this matter, several approaches have been proposed, all of which rely on the fact that one-qubit gates together with the *controlled NOT* (CNOT) gate form a *universal* gate library that is sufficient to realize any given unitary matrix [12]. In fact, determining corresponding quantum circuits has a rich history (see e.g. [12], [15], [16], [17], [18], [19]). The drawback of these generic approaches is that they lead to a significant amount of gates (even for a small number of qubits) and that they rely on a set of arbitrary one-qubit gates. The latter poses a severe obstacle since, in physical realizations, these must be approximated by a restricted set of gates, in particular when fault tolerant methods are applied [1].

In this work, we provide an alternative synthesis approach that considers synthesis of quantum circuits implementing Clifford Group operations. By this, we avoid relying on a generic gate library and, instead, can realize quantum functionality with a precise and established set of gates including Hadamard, Phase, and CNOT gates which can be implemented in a fault tolerant way [20]. At the same time, this restricts the applicability of our approach (arbitrary unitary matrices are not supported). However, Clifford group circuits are essential for many quantum applications and cover core aspects of quantum functionality such as superposition, entanglement, and phase shifts [21]. Moreover, their functionality is sufficient for various quantum applications, particularly as stabilizer circuits for error-correcting codes [22], but also for the realization of the Greenberger-Horne-Zeilinger experiment [23], for quantum teleportation [24], or dense quantum coding [25].

This compromise on applicability enables a synthesis methodology allowing for the realization of considerably more compact quantum circuits. We explicitly exploit the effects of the clearly defined gate library in order to directly modify the given unitary matrices to be synthesized. In contrast to the previously proposed (generic) approaches and as confirmed by an experimental evaluation, this enables a reduction of the circuit sizes by several orders of magnitude. By additionally using a compact data-structure, namely *Quantum Multiple-valued Decision Diagrams* (QMDDs) [26], our approach enables an efficient processing of the respective unitary matrices.

The remainder of the paper is structured as follows. In Section II, the background of quantum computation, quantum circuits, as well as QMDDs is briefly reviewed. Section III introduces the main concepts of the proposed synthesis scheme, while the resulting synthesis algorithm is described in detail in Section IV. An experimental evaluation is presented in Section V and, finally, Section VI concludes the paper.

II. BACKGROUND

This section briefly reviews the basics on quantum computation and quantum circuits. Furthermore, we introduce the basic ideas of *Quantum Multiple-valued Decision Diagrams (QMDDs)*, a data-structure used to efficiently represent quantum functionality. For more detailed introduction, we refer to [1], [26].

A. Quantum Computation and Circuits

Quantum systems are composed of *qubits*. Analogously to conventional bits, a qubit can be in one of the computational *basis states* $|0\rangle$ and $|1\rangle$, but also in a so called *superposition* $\alpha|0\rangle + \beta|1\rangle$ for complex-valued α, β with $|\alpha|^2 + |\beta|^2 = 1$. An n-qubit quantum system can be in one of the 2^n basis states $(|0\ldots00\rangle, |0\ldots01\rangle, \ldots, |1\ldots11\rangle)$ or a superposition of these states. Accordingly, the state of a quantum system is represented by a *state vector* (of dimension 2^n).

By the postulates of quantum mechanics, the evolution of a quantum system due to a quantum operation can be described by a *unitary transformation matrix*, i.e. an invertible complex-valued matrix whose inverse is given by the adjoint matrix.

Example 1. *Commonly used quantum operations include the* Hadamard *operation H (setting a qubit into a superposition), the* phase shift *operations S (Phase gate) and $Z = S^2$, as well as the* NOT *operation X which flips the basis states $|0\rangle$ and $|1\rangle$. The corresponding unitary matrices are defined as*

$$H = \tfrac{1}{\sqrt{2}}\begin{pmatrix} 1 & 1 \\ 1 & -1 \end{pmatrix}, \; S = \begin{pmatrix} 1 & 0 \\ 0 & i \end{pmatrix}, \; Z = \begin{pmatrix} 1 & 0 \\ 0 & -1 \end{pmatrix}, \; X = \begin{pmatrix} 0 & 1 \\ 1 & 0 \end{pmatrix}.$$

Besides these operations that work on a single *target qubit, there are also controlled operations on multiple qubits. The state of the additional* control qubits *determines which operation is performed on the target qubit. An example is the* controlled NOT *(CNOT) operation on two qubits which is defined by*

$$\begin{pmatrix} 1 & 0 & 0 & 0 \\ 0 & 1 & 0 & 0 \\ 0 & 0 & 0 & 1 \\ 0 & 0 & 1 & 0 \end{pmatrix}$$

and applies the NOT *operation to the target if the control is in the $|1\rangle$-state.*

Realizations of quantum operations are represented by *quantum gates* g_i which eventually form a *quantum circuit* $G = g_1 \ldots g_d$ with $1 \leq i \leq d$. For the purpose of visualization each qubit is represented by a solid line and gates are arranged in a cascade from left to right indicating the application of the respective unitary matrices to the qubits over time.

Example 2. *Fig. 1 shows the corresponding gate representations of the quantum operations specified by the matrices from Example 1. Note that for the CNOT gate a black circle represents the control line connection, while a "$\oplus$" represents the target line connection.*

In this work, we consider the synthesis of quantum circuits implementing Clifford Group operations. It has been shown in [27] that each Clifford Group operation can be realized by

a cascade composed of Hadamard, Phase, and CNOT gates. Conversely, any cascade of these gates forms a Clifford Group operation. Therefore, these gates are also called *generators* of the Clifford Group.

B. Quantum Multiple-valued Decision Diagrams

QMDDs [26] have been introduced as a means for the efficient representation and manipulation of quantum gates and circuits. The fundamental idea is a recursive partitioning of the respective transformation matrix and the use of edge and vertex weights to represent various complex-valued matrix entries. More precisely, a transformation matrix of dimension $2^n \times 2^n$ is successively partitioned into four sub-matrices of dimension $2^{n-1} \times 2^{n-1}$. This partitioning is represented by a directed acyclic graph – the QMDD. The following example illustrates main aspects.

Example 3. *Fig. 2a shows a transformation matrix for which a QMDD as shown in Fig. 2b has been built. Starting with a single terminal vertex $\boxed{1}$ that represents the lowest partitioning level, i.e. single matrix entries, the next upper level of 2×2 matrices is represented by vertices labeled x_2. For each entry, there is an outgoing edge to the terminal vertex with an edge weight corresponding to the respective complex value. For simplicity, we omit edge weights equal to 1 and indicate edges with a weight of 0 by stubs. The vertices are normalized by dividing the weights of all outgoing edges by a normalization factor (here: such that the "leftmost" edge with a non-zero weight has weight 1). This factor is propagated to referencing edges, e.g. the factor $\frac{1}{2}$ is propagated upwards from the x_2-level to the x_0-level in Fig. 2b. By this, structurally equivalent sub-matrices are compressed to a shared vertex (highlighted in grey in Fig. 2a and 2b, respectively). This procedure is repeated for each level until a single vertex labeled by x_0 is created for the top level. This vertex is called the* root vertex*. Finally, a possible normalization factor of this vertex is assigned to the weight of the* root edge *which points to the root vertex, but has no source.*

To obtain the value of a particular matrix entry, one has to follow the corresponding path from the root to the terminal vertex and multiply all edge weights on this path. For example, the matrix entry $\frac{i}{2}$ from the top right sub-matrix of Fig. 2a (highlighted bold) can be determined as the product of the weights on the highlighted path of the QMDD in Fig. 2b.

It can be shown that normalization as described above enables canonical QMDD representations [26].

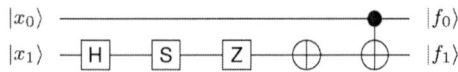

Fig. 1.   Quantum gates.

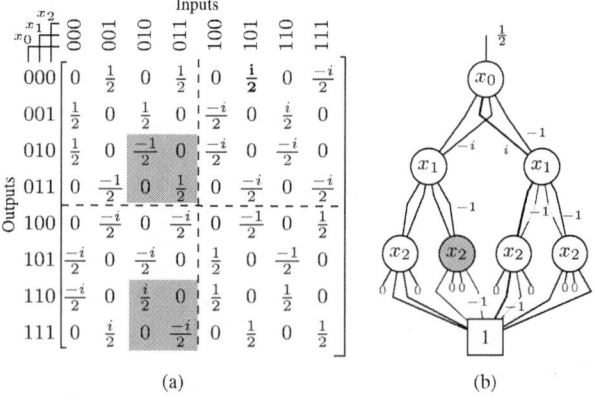

Fig. 2.   Matrix and QMDD representation of a 3-qubit quantum circuit.

III. MAIN CONCEPTS OF THE SYNTHESIS APPROACH

In this section, we introduce the general idea of the proposed synthesis approach. Furthermore, we describe the effect of the Clifford Group generators (Hadamard, Phase, CNOT) to a given transformation matrix. Based on that, the actual synthesis algorithm is afterwards described in Section IV.

A. General Idea

The task of synthesis is to determine a quantum circuit representing the desired quantum functionality F given in terms of a transformation matrix. We already know that all circuits considered in this work can be realized by a cascade composed of Hadamard, Phase, and CNOT gates. Therefore, applying transformation matrices representing these gates modifies F and for a distinct choice of these matrices we will eventually reach the identity matrix. Hence, the main goal of the proposed synthesis approach is to determine a sequence of quantum gates $g_1 \ldots g_d$ with this property. For this purpose, we identify the following three steps that are also illustrated in Fig. 3.

1) *Eliminate superposition*, i.e. apply quantum gates so that all multiple non-zero matrix entries in rows/columns are removed. This leads to a matrix that has a single non-zero entry per row/column (as illustrated in Fig. 3b), each of which has magnitude 1. Thus, the matrix is structurally equivalent to a permutation matrix. The corresponding circuit maps basis states to (possibly different) basis states and potentially applies phase shifts.
2) *Diagonalize*, i.e. establish the structure of a diagonal matrix as shown in Fig. 3c. By this, the structure of the transformation matrix is already equal to the structure of the identity matrix, i.e. each basis state is mapped onto itself. However, phase shifts still might be applied.
3) *Eliminate phase shifts* to eventually reach the identity matrix as shown in Fig. 3d.

All gates applied in order to perform these steps lead to a circuit realizing the inverse of the given transformaion matrix F. Since the Hadamard and CNOT gates are their own inverses and the inverse of the Phase gate is $S^3 = S \cdot Z$, the actually desired circuit can then be derived by simply inverting the order of all gates and replacing Phase gates by their inverses. Alternatively, we can convert F to its inverse in advance which is a simple operation using QMDDs.

B. Effect of the Clifford Group Generators

All synthesis steps sketched above can be accomplished by a clever sequential application of Hadamard, Phase, and CNOT gates, the generators of the Clifford Group. But before the respective algorithm is described in detail, we briefly investigate the effect of these gates to a transformation matrix F.

For simplicity, we illustrate all operations on a 4×4 transformation matrix F over two qubits x_0 and x_1. The generalization to larger matrices with more qubits is straightforward. In the following, we write U_{target} for the transformation matrix that represents the (uncontrolled) operation of a U-gate applied to a target qubit. Similarly, we use $U_{\text{target}}^{\text{control}}$ for a controlled U-gate, i.e. $X_{x_1}^{x_0}$ denotes the CNOT gate with control qubit x_0 and target x_1.

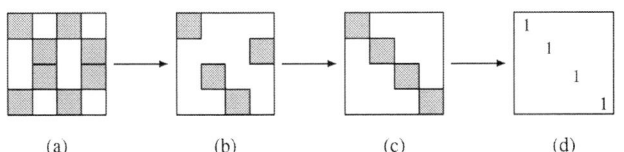

(a)   (b)   (c)   (d)

Fig. 3. General scheme of the synthesis procedure.

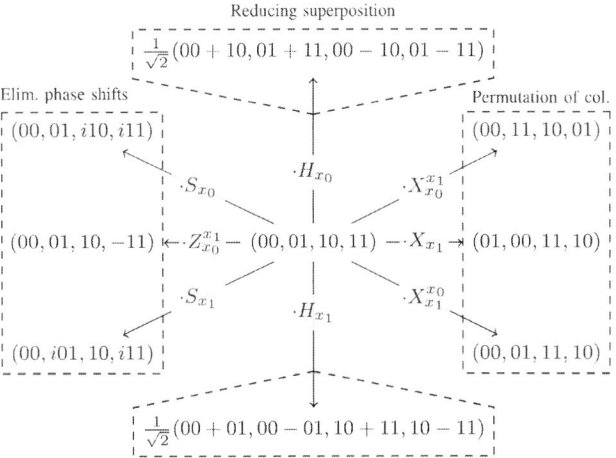

Fig. 4. Operation scheme of gate matrices.

Fig. 4 summarizes how the application of important Clifford operations affects a given transformation matrix F with columns $00, 01, 10$, and 11, i.e. $F = (00, 01, 10, 11)$. For our purposes, three main effects are important:

1) Permutation of columns: The CNOT gates $X_{x_1}^{x_0}$ and $X_{x_0}^{x_1}$ lead to matrices with columns 10 and 11 or 01 and 11 being interchanged, respectively. To additionally interchange column 00 with another one, an (uncontrolled) NOT gate (X) can be applied. This gate can be generated with Hadamard and Phase gates, more precisely $X = H \cdot S \cdot S \cdot H$. Hence, by composing gates $X_{x_1}^{x_0}$, $X_{x_0}^{x_1}$, and X_{x_1}, the columns of F can be re-arranged until any desired permutation is achieved. Note that for larger matrices we are no longer able to achieve any desired permutation of the columns, but we can still move a (single) column to any desired place.

2) Reducing superposition: Hadamard gates H_{x_0} and H_{x_1} link together pairs of columns as illustrated in Fig. 4, thereby allowing to create or reduce superposition. Hence, this operation is important for the first step of the synthesis approach sketched above. However, reduction of superposition is only possible for suitable pairs of columns (i.e. for 00 and 10 or 01 and 01) and only if the columns differ by no more than signs of the entries. If other pairs need to be linked together (e.g. 00 and 11), the corresponding columns have to be re-arranged before. For this purpose, permutation of columns as described above can be applied.

3) Eliminating phase shifts: Finally, Fig. 4 shows that Phase gates S_{x_0} and S_{x_1} apply a phase shift of i to a particular subset of columns of F. Beyond that, they do not alter the order of the columns. Applied again, they change this phase shift to -1 and $-i$ successively and, finally, remove the phase shift completely. Obviously, this is relevant for the third step of the synthesis approach sketched above. As Phase gates also only work on pairs of columns, we additionally might need so called controlled Z gates defined by

$$Z_{x_1}^{x_0} = Z_{x_0}^{x_1} = \begin{pmatrix} 1 & 0 & 0 & 0 \\ 0 & 1 & 0 & 0 \\ 0 & 0 & 1 & 0 \\ 0 & 0 & 0 & -1 \end{pmatrix}.$$

These apply a phase shift by -1 to a more restricted set of columns than Phase gates (here: a single column). This gate type can readily be constructed using Hadamard and CNOT gates, more precisely $Z_{x_1}^{x_0} = H_{x_0} \cdot X_{x_0}^{x_1} \cdot H_{x_0}$.

Exploiting these effects, the sketched synthesis approach can be realized as described next.

978-1-4799-2817-0/14 $31.00 © 2014 IEEE

IV. ALGORITHM

Based on the main concepts introduced in the previous section, we now describe the resulting synthesis approach in detail. The approach follows the three steps from Fig. 3. Furthermore, we employ QMDDs as reviewed in Section II-B as an efficient representation of transformation matrices on which all steps are conducted[1]. All steps are illustrated by a running example, namely the transformation matrix depicted in Fig. 2a and the corresponding QMDD depicted in Fig. 2b. We will exploit the specific property of Clifford group transformation matrices that all non-zero entries are multiples of a *basis weight*. More precisely, there is a complex number ω such that each non-zero entry is of the form $u \cdot \omega$ for $u \in \{\pm 1, \pm i\}$. This property follows from the theory of stabilizer circuits as discussed in [28].

A. Eliminating Superposition

As discussed in the previous section, superposition can be eliminated using Hadamard gates. More precisely, superposition involving a suitable pair of columns with no phase shifts can straightforwardly be tackled using a single Hadamard gate applied to the corresponding qubit.

Example 4. *In the running example from Fig. 2a, it can be seen that column 001 matches pairwise to column 011 (i.e. the entries differ only by sign). Hence, applying a Hadamard gate on qubit x_1 reduces the superposition of the matrix and leads to a matrix as shown in Fig. 5a.*

However, cases might be encountered where no Hadamard gates can be applied directly. Then, the columns have to be re-arranged using CNOT gates and possible phase shifts have to be removed using Phase gates. This is illustrated by the following example before the actually applied elimination scheme is described next.

Example 5. *The resulting matrix from Fig. 5a still includes superposition. In order to eliminate this, we need to derive a "valid" pair of columns for which a Hadamard gate can be applied. We choose columns 001 and 111 and first use a CNOT gate $X_{x_1}^{x_0}$ to move column 111 to 101. Then, we perform a Phase gate on qubit x_0 to remove the phase shift to column 001. Finally, the application of a Hadamard gate at qubit x_0 eventually eliminates all superposition in this matrix. The resulting matrix is shown in Fig. 6a.*

---

[1]Note that the proposed approach can as well be applied to alternative representations of unitary matrices.

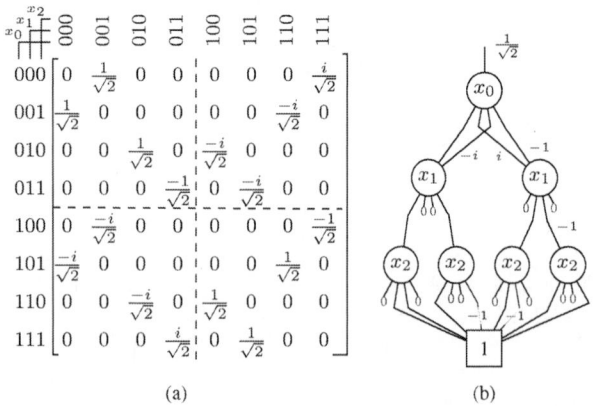

Fig. 5.   Running example after applying H_{x_1}.

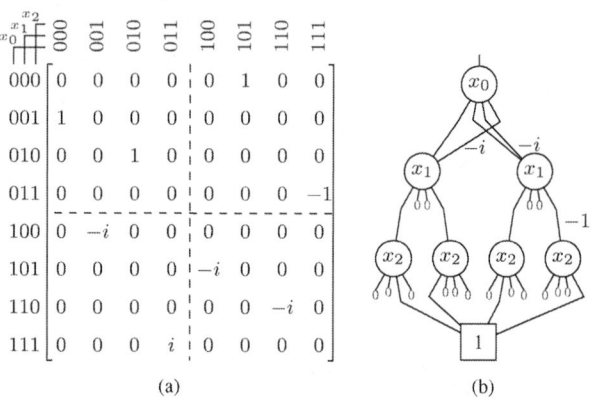

Fig. 6.   Running example after eliminating superposition.

Overall, superposition can gradually be eliminated by repeating the following steps.

1) Determine the first two non-zero entries in the first row and store their quotient q.
2) Rearrange the columns in such a way that their column index only differs in one place x_d. This can be done by choosing a place x_d where the indices initially differ and applying controlled NOT gates (controlled by x_d) on any other place where the indices differ.
3) If $q = \pm i$, perform a Phase gate on x_d.
4) Afterwards, perform a Hadamard gate on x_d.

Using QMDDs, these steps can be conducted efficiently. Since the basis weight ω of a matrix can be obtained through the weight of the root edge of its QMDD, it is easy to check whether there is superposition in the matrix or not. The resulting QMDD is shown in Fig. 6b.

B. Diagonalization

Once superposition has been removed from the matrix, there is a single non-zero entry per row and column (as in Fig. 6a). By unitarity, all of these entries and, hence, also the basis weight must have magnitude 1, i.e. the matrix is structurally equivalent to a permutation matrix and the corresponding circuit maps basis states to other basis states – possibly with phase shifts. Hence, in order to derive the structure of a diagonal matrix, the respective columns have to be permuted only. This can be done by applying suitable NOT and CNOT gates. In order to determine which gates to apply, we compute *line functions* f_{x_i} for each qubit x_i. These denote the (logic) formula that expresses for which inputs (columns) we get an output $|1\rangle$ on the respective qubit.

Example 6. *We read from the current matrix in Fig. 6a that*

$$
\begin{aligned}
f_{x_0} &= 001 \vee 011 \vee 100 \vee 110 &= x_0 \oplus x_2 \\
f_{x_1} &= 010 \vee 011 \vee 110 \vee 111 &= x_1 \\
f_{x_2} &= 000 \vee 011 \vee 100 \vee 111 &= \overline{x_1 \oplus x_2}
\end{aligned}
$$

In order to achieve a diagonal structure, we need $f_{x_i} = x_i$. To establish this for x_0, we need to XOR x_2 by applying $X_{x_0}^{x_2}$. For x_2, this can similarly be accomplished by applying $X_{x_2}^{x_1}$ and X_{x_2}. This leads to a matrix as shown in Fig. 7a.

It can be shown, that the line functions are always such XOR products of the qubits. However it may happen that x_i does not appear in f_{x_i} (but $x_j \neq x_i$ does). In this case, we firstly need to swap the qubits by applying $X_{x_j}^{x_i}$, $X_{x_i}^{x_j}$, and $X_{x_j}^{x_i}$.

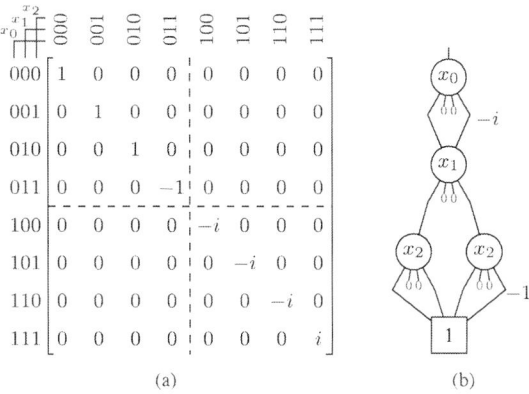

(a)

(b)

Fig. 7. Running example after diagonalization.

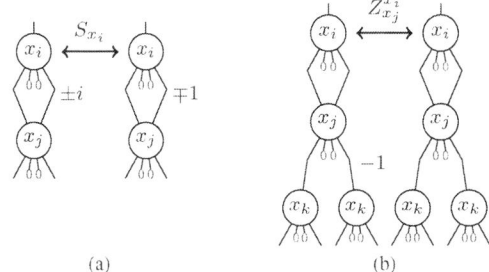

(a)

(b)

Fig. 8. Removing non-trivial edge weights from a QMDD.

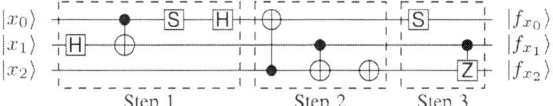

Fig. 9. Resulting quantum circuit.

Thus, the respective gates to be applied are derived by computing the line function for each qubit and, based on it, successively applying NOT/CNOT gates until the desired diagonal structure results. Using QMDDs, the line functions can easily be computed in a compressed form (as a Binary Decision Diagram) if the respective qubit is on the top level of the diagram.

More precisely, the following algorithm is applied (all variables are initially marked unvisited):

1) Pick an unvisited variable x and move it to the top of the QMDD by variable interchange.
2) Compute the output function f_x of x.
3) If f_x does not depend on x, i.e. if the top vertex of its compressed representation is not labeled by x (but by y), perform the CNOT gates X_y^x, X_x^y, and X_y^x.
4) For each variable $y \neq x$ on which f_x depends, perform a CNOT gate with target x and controlled by y, i.e. X_x^y.
5) Mark x as visited.

After all variables have been visited, each basis state is mapped onto itself and, thus, a diagonal matrix structure has been achieved. For the running example, this leads to the QMDD as shown in Fig. 7b.

C. Eliminating Phase Shifts

Finally, possible phase shifts are eliminated, i.e. all diagonal entries of the matrix are equalized. This is conducted by Phase gates (performing shifts by $\pm i$) and controlled Z gates (performing partial shifts by -1).

Example 7. *The phase shifts in the current matrix shown in Fig. 7a can be eliminated by applying a Phase gate at qubit x_0 (S_{x_0}). This transforms the values $\pm i$ to ± 1. To remove the phase shift -1 of $|011\rangle$, a controlled Z gate with target x_2 and controlled by x_1 ($Z_{x_2}^{x_1}$) is applied. This eventually leads to the identity matrix and, hence, terminates the synthesis.*

Note that we do not require the first entry of the diagonal (first row, first column) to have value $+1$. Any complex value of magnitude 1 is accepted and all other diagonal entries are transformed to the same value. This leads to a matrix that might not be the identity matrix itself, but that is equivalent up to *global phase* and, hence, physically indistinguishable [1].

Since phase shifts are indicated by edge weights $\pm i$ and -1 in the QMDD, they can easily be eliminated by applying the corresponding gate to qubits as illustrated in Fig. 8. In order to address edges on lower levels, re-ordering of the QMDD structure is applied. Eventually, all phase shifts can be eliminated by "moving" the respective qubit variable to the root level and applying the gates as shown in Fig. 8.

Overall, performing the respective steps as described above eventually leads to the quantum gate cascade depicted in Fig. 9. As explained in Section III-A, this realizes the inverse of the given unitary matrix of our running example from Fig. 2.

A detailed proof for the algorithm's convergence is available from the authors, but will not be given here due to page limitations.

V. EXPERIMENTAL RESULTS

The synthesis approach discussed above has been implemented in C on top of the original QMDD package presented in [26]. In this section, we evaluate the results obtained by the approach and compare them to synthesis schemes previously proposed in [18], [19]. To this end, arbitrary transformation matrices with up to 20 qubits (denoted *arbitrary*) as well as quantum functionality taken from [22] and realizing Shor's 9-qubit error correcting code (denoted by *9qubitN1* and *9qubitN2*), a 7-qubit encoding (denoted by *7qubitcode*), as well as an error syndrome measurement circuit for a 5-qubit code (denoted by *5qubitcode*) have been used.

The results are summarized in Table I. The first column provides the identifiers of the respective benchmarks followed by its number of qubits. In the remaining columns, the costs, i.e. the number of gates, of the resulting circuits are provided. It is a common understanding that (physical) implementations of CNOT gates are more error-prone and have greater delays than one-qubit gates. Therefore and in accordance with the evaluation of [18], [19], we distinguish between the number of one-qubit gates and the number of CNOTs in Table I. Furthermore, the best available results from [18], [19] are provided for comparison. Finally, the run-time of the proposed synthesis approach is provided in the last column. All experiments have been conducted on a 2.8 GHz Intel Core i7 machine with 8 GB of main memory running Linux.

First of all, we emphasize again that the approaches proposed in [18], [19] rely on a gate library composed of an arbitrary set of one-qubit gates together with a CNOT. This poses a severe obstacle since physical realizations and particularly fault tolerant methods impose limitations on the set of gates that may be used. In contrast, the approach presented here can realize the considered quantum functionality with a precise and established set of gates including only Hadamard, Phase, and CNOT gates[2].

---

[2]Note again that the controlled Z gates applied e.g. for eliminating phase shifts are eventually realized by a Hadamard–CNOT–Hadamard cascade.

978-1-4799-2817-0/14 $31.00 © 2014 IEEE

TABLE I
EXPERIMENTAL EVALUATION

Benchmark	#Qubits	Prev. Approach [18] #CNOTs	Prev. Approach [19]		Proposed Approach		
			#CNOTS	#one-qubit gates	#CNOTS	#one-qubit gates	Time (s)
Arbitrary transformation matrices							
arbitrary4	4	100	112	138	15	33	<0.01
arbitrary5	5	444	480	537	26	43	<0.01
arbitrary6	6	1 868	1 976	2 209	36	46	<0.01
arbitrary7	7	7 660	8 040	8 528	45	55	0.02
arbitrary8	8	31 020	32 456	33 455	61	72	<0.01
arbitrary9	9	124 844	130 408	134 415	68	61	0.03
arbitrary10	10	500 908	522 920	531 022	87	89	0.05
arbitrary11	11	2 006 700	2 094 376	2 110 669	102	98	0.10
arbitrary12	12	8 032 940	8 382 888	8 448 077	144	135	0.28
arbitrary15	15	$\approx 5.14 \cdot 10^8$	–	–	203	173	2.37
arbitrary20	20	$\approx 5.27 \cdot 10^{11}$	–	–	217	222	26.40
Quantum functionality taken from [22]							
5qubitcode	9	124 844	–	–	28	28	<0.01
7qubitcode	7	7 660	–	–	11	19	<0.01
9qubitN1	9	124 844	–	–	8	3	<0.01
9qubitN2	17	$\approx 8.23 \cdot 10^9$	–	–	34	4	<0.01

Benchmark: Name of benchmark – #Qubits: Number of qubits – #CNOTs: Number of CNOT gates – #one-qubit gates: Number of one-qubit gates

Besides that, Table I clearly shows that, using the proposed method, much more efficient quantum circuits can be realized for Clifford Group functionality compared to the more generic approaches presented before. In fact, reductions of several orders of magnitudes of CNOT gates can easily be obtained.

VI. CONCLUSIONS

In this work, we proposed a synthesis approach for quantum circuits implementing Clifford Group operations. Clifford group circuits are essential for many quantum applications such as stabilizer circuits, quantum teleportation, and more. In contrast to previous approaches for synthesis, we avoid relying on a generic gate library and, instead, exploit the specific effects of the Clifford Group generators to the unitary matrix to be synthesized. Experiments confirmed that, compared to the generic approaches presented before, quantum circuits with several orders of magnitude less gates can be realized using the proposed approach. Future work focuses on extending the proposed method to address more general quantum functionality.

REFERENCES

[1] M. Nielsen and I. Chuang, *Quantum Computation and Quantum Information.* Cambridge Univ. Press, 2000.
[2] R. V. Meter and M. Oskin, "Architectural implications of quantum computing technologies," *J. Emerg. Technol. Comput. Syst.*, vol. 2, no. 1, pp. 31–63, 2006.
[3] V. V. Shende, A. K. Prasad, I. L. Markov, and J. P. Hayes, "Synthesis of reversible logic circuits," *IEEE Trans. on CAD*, vol. 22, no. 6, pp. 710–722, 2003.
[4] G. Yang, X. Song, W. N. N. Hung, and M. A. Perkowski, "Fast synthesis of exact minimal reversible circuits using group theory," in *ASP Design Automation Conf.*, 2005, pp. 1002–1005.
[5] P. Gupta, A. Agrawal, and N. K. Jha, "An algorithm for synthesis of reversible logic circuits," *IEEE Trans. on CAD*, vol. 25, no. 11, pp. 2317–2330, 2006.
[6] M. Saeedi, M. S. Zamani, M. Sedighi, and Z. Sasanian, "Synthesis of reversible circuit using cycle-based approach," *J. Emerg. Technol. Comput. Syst.*, vol. 6, no. 4, 2010.
[7] R. Wille and R. Drechsler, "BDD-based synthesis of reversible logic for large functions," in *Design Automation Conf.*, 2009, pp. 270–275.
[8] R. Wille, S. Offermann, and R. Drechsler, "SyReC: A programming language for synthesis of reversible circuits," in *Forum on Specification and Design Languages*, 2010, pp. 184–189.
[9] M. Soeken, R. Wille, C. Hilken, N. Przigoda, and R. Drechsler, "Synthesis of reversible circuits with minimal lines for large functions," in *ASP Design Automation Conf.*, 2012, pp. 85–92.

[10] L. K. Grover, "A fast quantum mechanical algorithm for database search," in *Theory of computing*, 1996, pp. 212–219.
[11] P. W. Shor, "Algorithms for quantum computation: discrete logarithms and factoring," *Foundations of Computer Science*, pp. 124–134, 1994.
[12] A. Barenco, C. H. Bennett, R. Cleve, D. DiVinchenzo, N. Margolus, P. Shor, T. Sleator, J. Smolin, and H. Weinfurter, "Elementary gates for quantum computation," *The American Physical Society*, vol. 52, pp. 3457–3467, 1995.
[13] D. M. Miller, R. Wille, and Z. Sasanian, "Elementary quantum gate realizations for multiple-control Toffoli gates," in *Int'l Symp. on Multi-Valued Logic*, 2011, pp. 288–293.
[14] R. Wille, M. Soeken, C. Otterstedt, and R. Drechsler, "Improving the mapping of reversible circuits to quantum circuits using multiple target lines," in *ASP Design Automation Conf.*, 2013, pp. 85–92.
[15] J. J. Vartiainen, M. Mottonen, and M. M. Salomaa, "Efficient decomposition of quantum gates," *Phys. Rev. Lett.*, vol. 92, no. 177902, 2004.
[16] V. Bergholm, J. J. Vartiainen, M. Mottonen, and M. M. Salomaa, "Quantum circuits with uniformly controlled one-qubit gates," *Phys. Rev. A*, vol. 71, no. 052330, 2005.
[17] Y. Nakajima, Y. Kawano, and H. Sekigawa, "A new algorithm for producing quantum circuits using KAK decompositions," *Quantum Information & Computation*, vol. 6, no. 1, pp. 67–80, 2006.
[18] V. V. Shende, S. S. Bullock, and I. L. Markov, "Synthesis of quantum-logic circuits," *IEEE Trans. on CAD*, vol. 25, no. 6, pp. 1000–1010, 2006.
[19] M. Saeedi, M. Arabzadeh, M. S. Zamani, and M. Sedighi, "Block-based quantum-logic synthesis," *Quantum Information & Computation*, vol. 11, no. 3&4, pp. 262–277, 2011.
[20] P. O. Boykin, T. Mor, M. Pulver, V. Roychowdhury, and F. Vatan, "A new universal and fault-tolerant quantum basis," *Information Processing Letters*, vol. 75, no. 3, pp. 101–107, 2000.
[21] V. Kliuchnikov, D. Maslov, and M. Mosca, "Asymptotically optimal approximation of single qubit unitaries by Clifford and T circuits using a constant number of ancillary qubits," *CoRR*, vol. abs/1212.0822, 2012.
[22] N. D. Mermin, *Quantum Computer Science: An Introduction.* Cambridge University Press, 2007.
[23] D. M. Greenberger, M. A. Horne, and A. Zeilinger, *Bells Theorem, Quantum Theory, and Conceptions of the Universe.* Kluwer Academic Press, 1989.
[24] C. Bennett, G. Brassard, C. Crepeau, R. Jozsa, A. Peres, and W. Wootters, "Teleporting an unknown quantum state by dual classical and EPR channels," *Phys. Rev. Lett.*, vol. 70, no. 1895-1898, 1993.
[25] C. Bennett and S. J. Wiesner, "Communication via one- and two-particle operators on Einstein-Podolsky-Rosen states," *Phys. Rev. Lett.*, vol. 69, no. 2881, 1992.
[26] D. M. Miller and M. A. Thornton, "QMDD: A decision diagram structure for reversible and quantum circuits," in *Int'l Symp. on Multi-Valued Logic*, 2006, p. 6.
[27] D. Gottesman, "Stabilizer codes and quantum error correction," *arXiv preprint quant-ph/9705052*, 1997.
[28] ——, "The Heisenberg representation of quantum computers," *arXiv preprint quant-ph/9807006*, 1998.

Optimal SWAP Gate Insertion
for Nearest Neighbor Quantum Circuits

Robert Wille[1,2,3]
[1]Institute of Computer Science
University of Bremen
28359 Bremen, Germany

Aaron Lye[1]
[2]Cyber Physical Systems
DFKI GmbH
28359 Bremen, Germany

Rolf Drechsler[1,2]
[3]Faculty of Computer Science
Technical University Dresden
01187 Dresden, Germany

e-mail: {rwille,lye,drechsle}@informatik.uni-bremen.de

Abstract—Motivated by its promising applications e.g. for database search or factorization, significant progress has been made in the development of automated design methods for quantum circuits. But in order to keep up with recent physical developments in this domain, new technological constraints have to be considered. Limited interaction distance between gate qubits is one of the most common of these constraints. This led to the development of several strategies aiming at making a given quantum circuit nearest neighbor-compliant by inserting SWAP gates into the existing structure. Usually these strategies are of heuristic nature. In this work, we present an exact approach that enables nearest neighbor-compliance by inserting a *minimal* number of SWAP gates. Experiments demonstrate the applicability of the approach which enabled a comparison of results obtained by heuristic methods to the actual optimum.

I. INTRODUCTION

Quantum computation is an emerging technology that enables the computation of many relevant problems in less complexity than conventional computing paradigms [1]. Prominent examples how quantum circuits outperform conventional solutions include Shor's algorithm for factorization [2] and Groover's database search [3]. For these applications, the respective circuit netlists have manually been derived. But with the increasing interest in this technology, researchers also began to develop automatic solutions for the design of this kind of circuits. This led to a significant amount of contributions particularly for the automatic synthesis of corresponding circuit structures. Approaches addressing quantum circuits directly [4, 5, 6, 7, 8] or exploiting synthesis methods for reversible circuits [9, 10, 11] in combination with proper technology mapping schemes [12, 13] have been proposed for this purpose.

At the same time, the development of physical realizations for quantum computations continued. This led to new technological constraints which have to be considered by these synthesis methods. The emerge of so called *nearest neighbor* quantum circuits is one of the most common ones. The quantum gates of these circuits are restricted to work on *adjacent* circuit signals. While this can easily be achieved by inserting SWAP gates that move the respective gate signals together until they become adjacent, the precise fashion how this is accomplished has a significant effect on the overall costs of the resulting circuit. Accordingly, many approaches improving the SWAP gate insertion for nearest neighbor quantum circuits have been proposed [14, 15, 16, 17, 18]. However, almost all of them are of heuristic nature, i.e. do not guarantee an optimal solution. All this is discussed in more detail later in Section III.

In this work, we are addressing this problem. A SWAP gate insertion scheme for nearest neighbor quantum circuits is presented which keeps the actual number of SWAP gates to be inserted optimal. For this purpose, we are exploiting the deductive power of solvers for *Pseudo Boolean Optimization* (PBO). We are formulating the question which permutations of circuit signals shall be established in order to make all gates of a given quantum circuit adjacent as an satisfiability instance. Afterwards, we apply a cost function to be minimized which incorporates the respective costs of the respectively chosen permutation. By this, a PBO solver not only determines a possible SWAP insertion, but the one with the minimal number of SWAP gates.

Experiments illustrate that, considering the exponential complexity of the addressed problem, the proposed solution indeed is able to efficiently generate optimal results for many quantum circuits. Circuits composed of up to 6 circuit lines or up to 17 non-unary quantum gates can be handled. By this, we were able to prove that previously proposed (heuristic) approaches already determined the minimal number of SWAP gate insertions for some benchmarks. Nevertheless, also an example is shown where these heuristic results can still be further improved.

The remainder of this paper is structured as follows: Section II provides a basic introduction into quantum circuits and Pseudo Boolean Optimization before the SWAP gate insertion for nearest neighbor quantum circuits including existing approaches is reviewed in Section III. The proposed optimal approach is then presented in detail in Section IV and evaluated in Section V. Finally, the paper is concluded in Section VI.

II. BACKGROUND

In order to keep the paper self-contained, this section briefly reviews the basics on the considered circuit technology as well as the solving method utilized to tackle the addressed research problem.

II.A. Quantum Circuits

In contrast to conventional computation, *quantum computation* [1] works on qubits instead of bits. While bits allow binary values only, qubits may assume any superposition of them. More formally, a qubit is a two level quantum system, described by a two dimensional complex Hilbert space. The two orthogonal quantum states $|0\rangle \equiv \binom{1}{0}$ and $|1\rangle \equiv \binom{0}{1}$ are used to represent the Boolean values 0 and 1. Any state of a qubit may be written as $|x\rangle = \alpha |0\rangle + \beta |1\rangle$, where α and β are complex numbers with $|\alpha|^2 + |\beta|^2 = 1$.

Operations on n-qubits states are performed through multiplication of appropriate $2^n \times 2^n$ unitary matrices. Thus, each quantum computation is inherently reversible but manipulates qubits rather than pure logic values. At the end of the computation, a qubit can be measured. Then, depending on the current state of the qubit, either a 0 (with probability of $|\alpha|^2$) or a 1 (with probability of $|\beta|^2$) returns. After the measurement the state of the qubit is destroyed.

978-1-4799-2817-0/14 $31.00 © 2014 IEEE

TABLE I
QUANTUM GATES

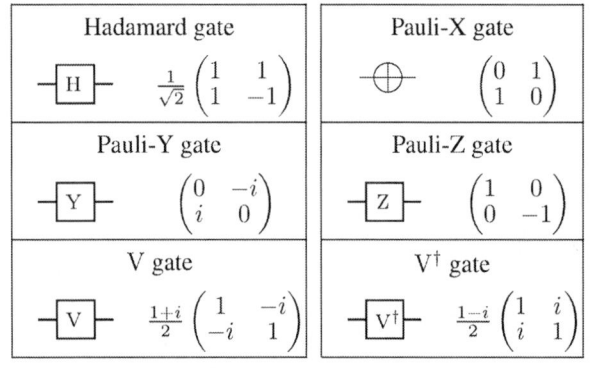

Hadamard gate	Pauli-X gate
$\frac{1}{\sqrt{2}}\begin{pmatrix} 1 & 1 \\ 1 & -1 \end{pmatrix}$	$\begin{pmatrix} 0 & 1 \\ 1 & 0 \end{pmatrix}$
Pauli-Y gate	Pauli-Z gate
$\begin{pmatrix} 0 & -i \\ i & 0 \end{pmatrix}$	$\begin{pmatrix} 1 & 0 \\ 0 & -1 \end{pmatrix}$
V gate	$V^\dagger$ gate
$\frac{1+i}{2}\begin{pmatrix} 1 & -i \\ -i & 1 \end{pmatrix}$	$\frac{1-i}{2}\begin{pmatrix} 1 & i \\ i & 1 \end{pmatrix}$

Fig. 1. Quantum circuit

Quantum computations are usually represented by *quantum circuits*. Here, the respective qubits are denoted by solid *circuit lines*. Operations are represented by *quantum gates*. Table I lists common quantum gates together with the corresponding unitary matrices describing their operation. In order to perform operations on more than one qubit, *controlled quantum gates* are applied. These gates are composed of a *target line* $|t\rangle$ and a *control line* $|c\rangle$ and realize the unitary operation represented by the matrix

$$ M = \begin{pmatrix} 1 & 0 & 0 & 0 \\ 0 & 1 & 0 & 0 \\ 0 & 0 & & \\ 0 & 0 & & U \end{pmatrix} $$

where U denotes the operation applied to the target line. That is, if $|c\rangle = |0\rangle$ all states remain unchanged and if $|c\rangle = |1\rangle$ the operation U is applied to the target line $|t\rangle$. In all other cases, the vector $(\alpha_{|c\rangle}\alpha_{|t\rangle}, \alpha_{|c\rangle}\beta_{|t\rangle}, \beta_{|c\rangle}\alpha_{|t\rangle}, \beta_{|c\rangle}\beta_{|t\rangle})$ is applied to M. In the remainder of this work, we use the following formal notation:

Definition 1 *A quantum circuit is denoted by the cascade* $G = g_1 g_2 \ldots g_{|G|}$, *where* $|G|$ *denotes the total number of gates. The number of qubits and, thus, the number of circuit lines is denoted by* n. *The costs of a quantum circuit are defined by the number* $|G|$ *of gates.*

Example 1 *Fig. 1 shows a quantum circuit composed of* $n = 2$ *circuit lines and* $|G| = 3$ *gates. This circuit gets* $|11\rangle$ *as input and transforms the qubits as indicated at the circuit signals.*

II.B. Pseudo Boolean Optimization

Solvers for *Boolean satsifiability* (SAT) and *pseudo-Boolean optimization* (PBO) are core technologies utilized in this work. Both problems are defined as follows:

Definition 2 *The* Boolean satisfiability problem *determines an assignment to the variables of a Boolean function* $\Phi : \{0,1\}^n \to \{0,1\}$ *such that* Φ *evaluates to 1 or proves that no such assignment exists. The function* Φ *is thereby given in*

Conjunctive Normal Form *(CNF). Each CNF is a conjunction of clauses where each clause is a disjunction of literals and each literal is a propositional variable or its negation.*

Definition 3 *The* pseudo-Boolean optimization problem *determines a satisfying solution for a pseudo-Boolean function* $\Psi : \{0,1\}^n \to \{0,1\}$ *which – at the same time – minimizes an objective function* $\mathcal{F}$. *The pseudo-Boolean function* Ψ *is thereby a conjunction of constraints defined by* $\sum_{i=1}^{n} c_i \dot{x}_i \geq c_n$, *where* $c_1 \ldots, c_n \in \mathbb{Z}$ *and* $\dot{x}_i$ *either is a positive or a negative literal. The objective function* $\mathcal{F}$ *is defined by* $\mathcal{F}(x_1, \ldots, x_n) = \sum_{i=1}^{n} m_i \dot{x}_i$ *with* $m_1, \ldots, m_n \in \mathbb{Z}$.

Example 2 *Let* $\Phi = (x_1 + x_2 + \overline{x}_3)(\overline{x}_1 + x_3)(\overline{x}_2 + x_3)$. *Then,* $x_1 = 1, x_2 = 1$, *and* $x_3 = 1$ *is a satisfying assignment solving the SAT problem.*

Accordingly, let $\Psi = (2x_1 + 3x_2 + \overline{x}_3 \geq 3)(2x_1 + x_2 \geq 2)$ *and* $\mathcal{F} = x_1 + x_2 + x_3$. *Then,* $x_1 = 1, x_2 = 0$, *and* $x_3 = 0$ *is a solution to the PBO problem, satisfying* Ψ *and, at the same time, minimizing* $\mathcal{F}$.

Both, SAT and PBO, are well investigated problems. In the past efficient solving algorithms (so called *SAT solvers* or *PBO solvers*, respectively) have been proposed (see e.g. [19, 20]). Instead of simply traversing the complete space of assignments, intelligent decision heuristics, powerful learning schemes, and efficient implication methods are thereby applied. In case of PBO, it is also common to translate the respective instance into a sequence of SAT instances in order to efficiently determine a solution. In the following, we apply these techniques as black boxes delivering a solution for the proposed problem formulation.

III. OPTIMIZING SWAP GATE INSERTION FOR NEAREST NEIGHBOR QUANTUM CIRCUITS

In the past, synthesis of quantum circuits has intensely been considered. While first quantum algorithms such as Grover's Search [3] or Shor's Algorithm [2] have manually been mapped into a circuit netlist, in the meantime also automatic synthesis approaches have been presented [4, 5, 6, 7, 8]. As every quantum circuit inherently is reversible, synthesis methods for reversible circuits [9, 10, 11] in combination with proper mapping methods [12, 13] are utilized for this purpose, too. However, these approaches usually assume a rather general quantum circuit model such as reviewed in Section II.A.

At the same time, new technological constraints emerged that should be considered by circuit designers and synthesis tools. Among the different technological constraints, the limited interaction distance between gate qubits is one of the most common ones. They are motivated by technologies such as ion traps [21], nitrogen-vacancy centers in diamonds [22], quantum dots emitting linear cluster states linked by linear optics [23], laser manipulated quantum dots in a cavity [24], and superconducting qubits [25] and assume that computations are only to be performed between adjacent (i.e. nearest neighbor) signals.

Accordingly, synthesis schemes addressing the nearest neighbor restriction have been introduced recently. While for specific quantum circuits such as quantum Fourier transformation [26], Shor's Algorithm [27], quantum addition [28], or error correction circuits [29] specialized realizations exist, general quantum circuits are usually made nearest neighbor-compliant by following a post-synthesis optimization scheme: First, the desired quantum functionality is realized using e.g. one of the synthesis approaches mentioned above. Afterwards, the resulting circuit is made nearest neighbor-compliant by inserting so called *SWAP gates*.

978-1-4799-2817-0/14 $31.00 © 2014 IEEE

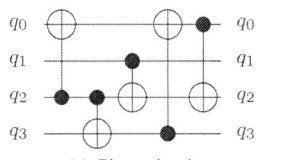 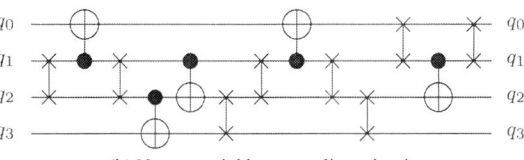 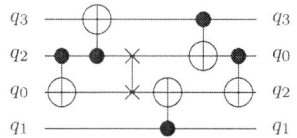

(a) Given circuit   (b) Nearest neighbor-compliant circuit   (c) Cheaper nearest neighbor-compliant circuit

Fig. 2. Establishing nearest neighbor-compliance

Definition 4 *A SWAP gate is a quantum gate $g(q_i, q_j)$ including two qubits q_i, q_j and maps $(q_0, \ldots, q_i, \ldots, q_j, \ldots, q_{n-1})$ to $(q_0, \ldots, q_j, \ldots, q_i, \ldots, q_{n-1})$. That is, a SWAP gate realizes the exchange of two quantum values.*

More precisely, adjacent SWAP gates are inserted before each gate g with non-adjacent circuit lines in order to "move" the control (target) line of g towards the target (control) line until they become adjacent. Afterwards, SWAP gates are inserted to restore the original ordering of circuit lines.

Example 3 *Consider the circuit depicted in Fig. 2(a). As can be seen, gates g_1, g_4, and g_5 are non-adjacent. Thus, in order to make this circuit nearest neighbor-compliant, SWAP gates before and behind all these gates are inserted as shown in Fig. 2(b).*

Obviously, inserting SWAP gates increases the cost of the resulting circuits[1]. Moreover, the fashion in which SWAP gates are inserted has a significant effect: The insertion of SWAP gates as shown in Fig. 2(c) (which additionally also exploits different orders of primary inputs/outputs) leads to a much cheaper nearest neighbor-compliant circuit for Example 3. Accordingly, different approaches for optimizing the insertion of SWAP gates have been proposed in the recent past (see e.g. [14, 15, 16, 17, 18]). For this purpose, strategies such as the re-ordering of circuit lines, window-based heuristics, or mapping the problem to a corresponding graph arrangement problem were applied and evaluated. However, all of them are of heuristic nature, i.e. do not guarantee an optimal solution. An exception might be [30] in which the determination of an optimal number of SWAP gate insertion has briefly been discussed by means of an exhaustive enumeration. But this approach has clearly been shown unfeasible[2] and was just used in order to motivate the application of heuristics. Furthermore, also the approach presented in [31] guarantees optimality. But this solution additionally allows for changing the order of the gates so that the results obtained there are not comparable to the previous work discussed above.

In this work, we are aiming to *exactly* determine the best, i.e. minimal, SWAP gate insertion in order to transform a given circuit into a nearest neighbor-compliant representation. For this purpose, the deductive power of solvers for Pseudo Boolean optimization is exploited.

IV. PROPOSED SOLUTION

This section presents an exact solution to the problem sketched above. For this purpose, first the general idea is discussed before details on the precise implementation are presented.

IV.A. General Idea

Given a quantum circuit G, we are looking for a minimal insertion of SWAP gates so that all gates g of G can adjacently be executed. As illustrated above, these SWAP gates are only applied in order to appropriately permute the order of the circuit lines so that all functional gates of the circuit can be executed adjacently. Hence, in order to determine the best possible insertion of SWAP gates, one has to consider

- all possible permutations of circuit lines that, in principle, can be established before each gate g of G and

- the costs (in terms of adjacent SWAP gates) that would be needed in order to create these particular permutations.

The precise cascade of adjacent SWAP gates and, by this, the costs for creating a particular permutation of circuit lines can thereby be calculated using *inversion vectors*. For any permutation π, an inversion vector $\vec{v} = (v_0 v_1 \ldots v_{n-1})$ is defined by $v_i = |\{e \in \pi \mid e > i\}|$ and $0 \leq i < n$, i.e. the ith element of $\vec{v}$ is the number of elements in π larger than i to the left of i. Inherently, this is also the number of (adjacent) SWAP operations to be performed in order to move $\pi(i)$ to i. Hence, the total amount of SWAP gates needed for creating a particular permutation of circuit lines can be calculated by summing up the respective entries in the inversion vector[3].

Example 4 *Assume a given circuit line order $(0, 1, 2, 3)$ shall be permuted to $(2, 3, 1, 0)$. The corresponding inversion vector is $\vec{v} = (3, 2, 0, 0)$. Hence, $3 + 2 + 0 + 0 = 5$ SWAP gates are required in order to create this permutation.*

Note thereby that this does not apply in order to permute the circuit lines before the first gate, i.e. before g_1. Here, in accordance with previous work (e.g. [17, 18]), we assume the circuit lines can arbitrarily be permuted with no additional costs just by re-arranging the primary inputs as necessary.

Taking all that into account, a naive approach that ensures minimality of SWAP gate insertion would work as follows:

1. Enumerately consider all possible permutations of circuit lines for all gates of the given circuit G.

2. For each set of permutations which lead to a circuit satisfying the nearest neighbor condition, calculate the costs according to the algorithm described above.

3. After all permutations have been considered, take the one with the smallest costs.

This requires to check all possible permutations for all gates of the circuit, i.e. $n!^{|G|}$ different combinations in total. Although it was shown in [30], that it is sufficient to only consider the permutations between the respective control and target lines of each gate, this remains an exponential complexity. Naive schemes as sketched here or discussed in [30] are

---

[1] In fact, different costs calculations are therefore applied in literature: SWAP gates are either treated as elementary gates with costs of 1 or realized through a cascade of three controlled Pauli-X gates (i.e. with costs of 3).

[2] The approach was aborted for almost all benchmarks due to time constraints.

[3] The same principle has been applied in [30] in order to show that a bubblesort-algorithm generates the minimum number of SWAP gates in order to construct an arbitrary permutation.

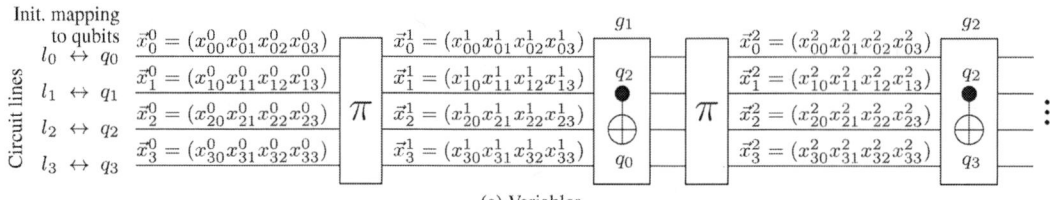

(a) Variables

Consistency-constraints:

$x^0_{00} + x^0_{01} + x^0_{02} + x^0_{03} = 1$
$\wedge x^0_{10} + x^0_{11} + x^0_{12} + x^0_{13} = 1$
$\wedge x^0_{20} + x^0_{21} + x^0_{22} + x^0_{23} = 1$
$\wedge x^0_{30} + x^0_{31} + x^0_{32} + x^0_{33} = 1$
$\wedge x^0_{00} + x^0_{10} + x^0_{20} + x^0_{30} = 1$
$\wedge x^0_{01} + x^0_{11} + x^0_{21} + x^0_{31} = 1$
$\wedge x^0_{02} + x^0_{12} + x^0_{22} + x^0_{32} = 1$
$\wedge x^0_{03} + x^0_{13} + x^0_{23} + x^0_{33} = 1$
$\dots$

Adjacency-constraints
(for g_1 with q_0 and q_2):

$(x^1_{00} \wedge x^1_{12})$
$\vee (x^1_{10} \wedge x^1_{22})$
$\vee (x^1_{20} \wedge x^1_{32})$
$\vee (x^1_{02} \wedge x^1_{10})$
$\vee (x^1_{12} \wedge x^1_{20})$
$\vee (x^1_{22} \wedge x^1_{30})$

Permutation-constraint (for $\pi = (2310)$ and $k = 1$)

$(\vec{x}^0_0 = \vec{x}^1_2 \;\wedge\; \vec{x}^0_1 = \vec{x}^1_3 \;\wedge\; \vec{x}^0_2 = \vec{x}^1_1 \;\wedge\; \vec{x}^0_3 = \vec{x}^1_0) \Leftrightarrow s^1_{2310}$

Objective function:

$\min((0 \cdot s^2_{0123} + 1 \cdot s^2_{0132} + 1 \cdot s^2_{0213} + 2 \cdot s^2_{0231} + 2 \cdot s^2_{0312} + 3 \cdot s^2_{0321} +$
$1 \cdot s^2_{1023} + 2 \cdot s^2_{1032} + 2 \cdot s^2_{1203} + 3 \cdot s^2_{1230} + 3 \cdot s^2_{1302} + 4 \cdot s^2_{1320} +$
$2 \cdot s^2_{2013} + 3 \cdot s^2_{2031} + 3 \cdot s^2_{2103} + 4 \cdot s^2_{2130} + 4 \cdot s^2_{2301} + 5 \cdot s^2_{2310} +$
$3 \cdot s^2_{3012} + 4 \cdot s^2_{3021} + 4 \cdot s^2_{3102} + 5 \cdot s^2_{3120} + 5 \cdot s^2_{3201} + 6 \cdot s^2_{3210})$
$+ \dots)$

(b) Constraints

Fig. 3. Resulting PBO encoding for circuit from Fig. 2(a)

infeasible due to their enumerative nature and the complexity of the problem. Hence, we propose an alternative approach which exploits the deductive power of the state-of-the-art PBO solvers described in Section II.B. Instead of naively enumerating all possible permutations, we are formulating the question which permutation shall be created as an problem of Boolean satisfiability. In addition to that, the costs of the respective permutations are incorporated in an objective function to be minimized. By this, a formulation results that can be passed to a PBO solver. From the solution of this PBO instance, the minimal SWAP insertions can be derived.

IV.B. Implementation

In order to encode the considered problem, we distinguish between the lines of a circuit G (denoted by $l_0, \dots, l_{n-1}$) and their corresponding qubits (denoted by $q_0, \dots, q_{n-1}$). Initially, each qubit corresponds to the circuit line with the same index, i.e. q_i corresponds to l_i for all $0 \le i < n$. Then, before each gate, we allow an arbitrary permutation (including the identity) which may lead to different mappings of qubits to circuit lines. Following this, Boolean variables are introduced to the PBO encoding representing which qubit currently corresponds to which line.

Definition 5 *Let $G = g_1 g_2 \dots g_{|G|}$ be a circuit over n qubits. Then, variables $\vec{x}^k_i = (x^k_{i0} x^k_{i1} \dots x^k_{in-1})$ with $0 \le k < |G|$ and $0 \le i < n$ are introduced representing which qubit corresponds to circuit line l_i initially (for $k = 0$) and before gate g_k (for $1 \le k < |G|$). More precisely, a variable x^k_{ij} states whether qubit q_j corresponds to the circuit line l_i ($x^k_{ij} = 1$) or not ($x^k_{ij} = 0$).*

Example 5 *Consider the circuit shown in Fig. 2(a) which works as running example throughout the remainder of this section. Fig. 3 sketches the resulting PBO encoding. The π-blocks denote the positions in which we allow an arbitrary permutation of circuit lines. This leads to a new qubit mapping which is represented by the corresponding $\vec{x}^k_i$-variables in Fig. 3. Here, e.g. the assignment $x^2_{21} = 1$ states that, before gate g_2, the qubit q_1 corresponds to circuit line l_2.*

Obviously, these mappings cannot arbitrarily be made. In fact, each circuit line must exactly correspond to one qubit and each qubit must exactly correspond to one circuit line. In order to ensure this, the following *consistency*-constraint is added to the PBO instance:

$$\bigwedge_{k=0}^{|G|-1} \left(\bigwedge_{i=0}^{n-1} (\sum_{j=0}^{n-1} x^k_{ij} = 1) \wedge \bigwedge_{i=0}^{n-1} (\sum_{j=0}^{n-1} x^k_{ji} = 1) \right)$$

The left part of this constraint states that, for each permutation position k ($0 \le k < |G|$) and for each circuit line l_i, the sum $x^k_{i0} + x^k_{i1} + \dots + x^k_{in-1}$ is fixed to 1, i.e. exactly one qubit corresponds to one circuit line. The right part of this constraint states that, for each permutation position k ($0 \le k < |G|$) and for each qubit q_i, the sum $x^k_{0i} + x^k_{1i} + \dots + x^k_{n-1i}$ is fixed to 1, i.e. exactly one circuit line corresponds to one qubit.

Example 6 *The bottom left of Fig. 3 sketches the consistency constraint for the example from Fig. 2(a).*

Next, we want to ensure that only permutations are applied which satisfy the nearest neighbor condition on all functional gates. As we know the control and target qubits of each gate, this can be enforced through the $\vec{x}^k_i$-variables and the following *adjacency*-constraint:

$$\bigwedge_{g_k(q_c,q_t) \in G} \left(\bigvee_{m=0}^{n-1} (x^k_{mc} \wedge x^k_{(m+1)t}) \vee \bigvee_{m=0}^{n-1} (x^k_{mt} \wedge x^k_{(m+1)c}) \right)$$

This constraint considers all gates $g_k(q_c, q_t)$ from a given circuit G with control qubit q_c and target qubit q_t. For each of these gates, a mapping of qubits to circuit lines is required so that either

- the control qubit q_c corresponds to a circuit line l_m and the target line q_t corresponds to a directly succeeding circuit line l_{m+1} (left part of the constraint) or

- the target qubit q_t corresponds to a circuit line l_m and the control line q_c corresponds to a directly succeeding circuit line l_{m+1} (right part of the constraint).

978-1-4799-2817-0/14 $31.00 © 2014 IEEE   492

That is, one of the possible adjacencies between the respective qubits of these gates has to be established.

Example 7 *Consider again the sketch of the encoding shown in Fig. 3. The gate blocks represent the qubits which must be adjacent (derived from Fig. 2(a)). Based on that, the bottom center of Fig. 3 exemplarily shows the resulting adjacency-constraint for gate g_1 with control qubit q_2 and target qubit q_0.*

Finally, the respectively chosen permutation of circuit lines at each position has to be extracted and the corresponding costs for creating it has to be linked to the objective function of the PBO instance. Again, the $\vec{x}_i^{\prime k}$-variables can be exploited for this purpose. Based on them, it can be derived what permutation is applied before gate g_k in order to change the previous circuit line order. This can be expressed by a *permutation*-constraint as follows:

$$\bigwedge_{k=1}^{|G|-1} \left(\bigwedge_{\pi \in \Pi} (\bigwedge_{i=0}^{n-1} \vec{x}_i^{\prime k-1} = \vec{x}_{\pi(i)}^{\prime k}) \Leftrightarrow s_\pi^k \right)$$

This constraint considers all possible permutations (denoted by Π) for each position k. If the assignments of $\vec{x}_i^{\prime k-1}$ and $\vec{x}_i^{\prime k}$ establish a particular permutation $\pi \in \Pi$, then a corresponding new free variable s_π^k is set to 1 (encoded through $\Leftrightarrow$). This states that this particular permutation π has been chosen before gate g_k and, hence, the corresponding costs for it have to be considered. This is eventually incorporated in the objective function

$$min\left(\sum_{k=1}^{|G|-1} \sum_{\pi \in \Pi} c_\pi s_\pi^k \right),$$

where c_π denotes the costs (in terms of adjacent SWAP gates) for creating a permutation π using the methods described in Section IV.A.

Example 8 *The permutation-constraint and the objective function for the running example from Fig. 2(a) are sketched at the bottom right of Fig. 3. In particular, the constraints for permutation $\pi = (2, 3, 1, 0)$ and $k = 2$ are shown. As already discussed in Example 4, creating this permutation requires 5 SWAP gates. Accordingly, costs of 5 are assumed in the objective function for this particular permutation. Furthermore, as it is assumed that circuit lines before gate g_1 can arbitrarily be permuted with no additional costs, all variables s_π^1 ($\pi \in \Pi$) are not part of the objective function.*

Combining all these constraints, a PBO instance results which is satisfiabile for all permutations of circuit lines that lead to a nearest neighbor-compliant circuit. The precise permutation to be created at position k can thereby be derived from the assignment to the s_π^k variables. If s_π^k has been assigned 1 by the PBO solver, a permutation π has to be created before gate g_k. By additionally optimizing the objective function, the PBO solver ensures a minimal number of SWAP gates.

Example 9 *Passing the PBO encoding presented above to a PBO solver, an optimal assignment with $s_{\pi_e}^0, s_{\pi_e}^1, s_{(0213)}^2, s_{\pi_e}^3, s_{\pi_e}^4$ set to 1 results (π_e represents the identity permutation). From that, the SWAP insertion as depicted in Fig. 2(c) can be derived. This represents an optimal solution to the SWAP insertion problem for the circuit given in Fig. 2(a).*

TABLE II
EXPERIMENTAL EVALUATION

Benchmark	n	$|G|$	$n!^{|G|}$	Swaps	Time
3_17_13	3	13	$1.3 \cdot 10^{10}$	2	0.1
4gt11_83	5	12	$8.9 \cdot 10^{24}$	6	630.1
4gt11_84	5	7	$3.6 \cdot 10^{14}$	1	8.9
4gt11-v1_85	5	7	$3.6 \cdot 10^{14}$	1	16.6
4gt13-v1_93	5	16	$1.8 \cdot 10^{33}$	6	9808.5
4mod5-v0_19	5	12	$8.9 \cdot 10^{24}$	6	489.2
4mod5-v0_20	5	8	$4.3 \cdot 10^{16}$	2	55.3
4mod5-v1_22	5	9	$5.2 \cdot 10^{18}$	3	45.5
4mod5-v1_24	5	12	$8.9 \cdot 10^{24}$	10	548.7
4mod5-v1_25	5	7	$3.6 \cdot 10^{14}$	1	11.9
alu-v0_27	5	13	$1.1 \cdot 10^{27}$	18	11705.3
alu-v1_29	5	13	$1.1 \cdot 10^{27}$	15	1685.5
alu-v4_37	5	14	$1.3 \cdot 10^{29}$	14	3669.9
decod24-v0_38	4	17	$2.9 \cdot 10^{23}$	4	23.6
decod24-v0_39	4	15	$5.0 \cdot 10^{20}$	5	19.2
decod24-v0_40	4	8	$1.1 \cdot 10^{11}$	3	0.4
decod24-v1_42	4	8	$1.1 \cdot 10^{11}$	2	0.4
decod24-v2_43	4	16	$1.2 \cdot 10^{22}$	3	7.6
decod24-v3_46	4	9	$2.6 \cdot 10^{12}$	3	0.4
graycode6_48	6	5	$1.9 \cdot 10^{14}$	0	4.5
mod5d1_63	5	11	$7.4 \cdot 10^{22}$	10	745.5
mod5d2_70	5	14	$1.3 \cdot 10^{29}$	18	3047.5
mod5mils_71	5	12	$8.9 \cdot 10^{24}$	7	735.7
QFT5	5	10	$6.2 \cdot 10^{20}$	6	2463.2
rd32-v0_66	4	12	$3.7 \cdot 10^{16}$	4	1.6
rd32-v0_67	4	8	$1.1 \cdot 10^{11}$	2	0.3
rd32-v1_68	4	12	$3.7 \cdot 10^{16}$	4	1.6
rd32-v1_69	4	8	$1.1 \cdot 10^{11}$	2	0.3

n: Number of lines $|G|$: Number of gates (does not include unary gates)
$n!^{|G|}$: Search space/complexity   Swaps: Min. number of SWAP gates
Time: Run-time in CPU seconds

V. EXPERIMENTAL EVALUATION

In this section, the results obtained with the proposed approach are presented and discussed. For this purpose, a PBO encoder has been implemented in C++ which takes a given quantum circuit and generates the PBO instance as described in the previous section. Afterwards, the PBO solver *clasp* [20] has been utilized for solving the resulting instance. To evaluate the performance of the proposed solution, quantum circuits from *RevLib* [32] as well as used in [18] have been applied. All evaluations have been conducted on an Intel E6700 Core2 CPU with 2.7 GHz and 4 GB of memory.

The results are summarized in Table II. The first three columns denote thereby the name of the considered benchmarks as well as their number n of circuit lines and their number $|G|$ of gates. Note that the latter does not include any unary gates. As unary gates are inherently nearest neighbor-compliant, they do not have to be considered for SWAP gate insertion and, thus, are ignored. Column $n!^{|G|}$ denotes the size of the search space and, by this, the precise complexity of the problem for the respective benchmark. Finally, the last two columns eventually provide the determined minimal number of SWAP gates to be inserted as well as the run-time (in CPU seconds) needed to obtain the result.

The results confirm the deductive power of the applied PBO solver that helped to determine the optimal SWAP gate insertion for many quantum circuits. In fact, circuits composed of up to 6 circuit lines or up to 17 non-unary quantum gates can be handled. In the most complex case (i.e. for *4gt13-v1_93*) $1.8 \cdot 10^{33}$ possible permutations have been considered – way too much to be tackled by an enumerative solution. Besides that, it can also be observed that the performance differs depending on the respective circuit. For example, *decod24-v2_43* and *mod5d1_63* have about the same complexity. How-

TABLE III
COMPARISON TO HEURISTIC APPROACHES

Benchmark	Minimal	[17]	[18]
3_17_13	2	12	4
4gt11_84	1	3	1
decod24-v3_46	3	4	3
QFT5	6	6	6
rd32-v0_67	2	2	2

ever, an optimal SWAP gate insertion can be determined for *decod24-v2_43* two orders of magnitude faster than for *mod5d1_63*. Such differences can be applied by the different fashions in which the PBO solver traverses the search space.

Using the proposed methodology, we were also able to prove that previously proposed (heuristic) approaches sometimes already determined the minimal number of SWAP gate insertions. Some numbers concerning this are shown in Table III (providing a comparison to some of the results from [17, 18][4]). As can be seen, particularly the approach recently presented in [18] was already able to determine minimal solutions for some benchmarks. However, minimality is not guaranteed in these approaches: As unveiled by our approach, even for the relatively less complex example *3_17_13*, a minimal solution with half the number of SWAP gate insertions exists.

VI. CONCLUSIONS

In this work, we proposed an approach for the optimal determination of SWAP gate insertions needed to make an arbitrary quantum circuit nearest neighbor-compliant. In order to handle the exponential complexity, the deductive power of PBO solvers has been exploited. That is, the given problem has been encoded as a PBO instance and, afterwards, solved by a proper solving method. Experiments confirmed the applicability of the proposed approach. By this, it was possible to compare results obtained by heuristic methods to the actual optimum. Future work focuses on determining the optimal number of SWAP gate insertions for alternative architectures, e.g. nearest neighbor quantum circuits based on 2D architectures [33] or relying on gate libraries particularly suited for nearest neighbor constraints [34].

ACKNOWLEDGMENTS

We would like to thank Mehdi Saeedi for making us some of his benchmarks available.

REFERENCES

[1] M. Nielsen and I. Chuang. *Quantum Computation and Quantum Information*. Cambridge Univ. Press, 2000.

[2] P. W. Shor. Algorithms for quantum computation: discrete logarithms and factoring. *Foundations of Computer Science*, pages 124–134, 1994.

[3] Lov K. Grover. A fast quantum mechanical algorithm for database search. In *Theory of computing*, pages 212–219, 1996.

[4] V. V. Shende, S. S. Bullock, and I. L. Markov. Synthesis of quantum-logic circuits. *IEEE Trans. on CAD*, 25(6):1000–1010, 2006.

[5] W.N.N. Hung, X. Song, G. Yang, J. Yang, and M. Perkowski. Optimal synthesis of multiple output Boolean functions using a set of quantum gates by symbolic reachability analysis. *IEEE Trans. on CAD*, 25(9):1652–1663, 2006.

[6] D. Große, R. Wille, G. W. Dueck, and R. Drechsler. Exact synthesis of elementary quantum gate circuits. *Multiple-Valued Logic and Soft Computing*, 15(4):270–275, 2009.

[7] Mehdi Saeedi, Mona Arabzadeh, Morteza Saheb Zamani, and Mehdi Sedighi. Block-based quantum-logic synthesis. *Quant. Info. Comput.*, 11(3&4):262–277, 2011.

[8] P. Niemann, R. Wille, and R. Drechsler. Efficient synthesis of quantum circuits implementing Clifford group operations. In *ASP Design Automation Conf.*, 2014.

[9] R. Wille, D. Große, G.W. Dueck, and R. Drechsler. Reversible logic synthesis with output permutation. In *VLSI Design*, pages 189–194, 2009.

[10] M. Saeedi, M. S. Zamani, M. Sedighi, and Z. Sasanian. Synthesis of reversible circuit using cycle-based approach. *J. Emerg. Technol. Comput. Syst.*, 6(4), 2010.

[11] Mathias Soeken, Robert Wille, Christoph Hilken, Nils Przigoda, and Rolf Drechsler. Synthesis of reversible circuits with minimal lines for large functions. In *ASP Design Automation Conf.*, pages 85–92, 2012.

[12] D. M. Miller, R. Wille, and Z. Sasanian. Elementary quantum gate realizations for multiple-control Toffoli gates. In *Int'l Symp. on Multi-Valued Logic*, pages 288–293, 2011.

[13] R. Wille, M. Soeken, C. Otterstedt, and R. Drechsler. Improving the mapping of reversible circuits to quantum circuits using multiple target lines. In *ASP Design Automation Conf.*, pages 85–92, 2013.

[14] M. Mottonen and J. J. Vartiainen. Decompositions of general quantum gates. *Ch. 7 in Trends in Quantum Computing Research, NOVA Publishers, New York*, 2006.

[15] A. Chakrabarti and S. Sur-Kolay. Nearest neighbour based synthesis of quantum boolean circuits. *Engineering Letters*, 15:356–361, December 2007.

[16] Mozammel H. A. Khan. Cost reduction in nearest neighbour based synthesis of quantum boolean circuits. *Engineering Letters*, 16:1–5, 2008.

[17] M. Saeedi, R. Wille, and R. Drechsler. Synthesis of quantum circuits for linear nearest neighbor architectures. *Quant. Info. Proc.*, 2010.

[18] A. Shafaei, M. Saeedi, and M. Pedram. Optimization of quantum circuits for interaction distance in linear nearest neighbor architectures. In *Design Automation Conf.*, 2013.

[19] N. Eén and N. Sörensson. An extensible SAT solver. In *SAT 2003*, volume 2919 of *LNCS*, pages 502–518, 2004.

[20] M. Gebser, B. Kaufmann, A. Neumann, and T. Schaub. Conflict-driven answer set solving. In *Int'l Joint Conference on Artificial Intelligence*, pages 386–392, 2007.

[21] N. H. Nickerson, Y. Li, and S. C. Benjamin. Topological quantum computing with a very noisy network and local error rates approaching one percent. *Nat Commun*, 4:1756, 2013.

[22] N. Y. Yao, Z.-X. Gong, C. R. Laumann, S. D. Bennett, L.-M. Duan, M. D. Lukin, L. Jiang, and A. V. Gorshkov. Quantum logic between remote quantum registers. *Phys. Rev. A*, 87:022306, 2013.

[23] David A. Herrera-Marti, Austin G. Fowler, David Jennings, and Terry Rudolph. Photonic implementation for the topological cluster-state quantum computer. *Phys. Rev. A*, 82:032332, 2010.

[24] N. Cody Jones, Rodney Van Meter, Austin G. Fowler, Peter L. McMahon, Jungsang Kim, Thaddeus D. Ladd, and Yoshihisa Yamamoto. Layered architecture for quantum computing. *Phys. Rev. X*, 2:031007, 2012.

[25] M. Ohliger and J. Eisert. Efficient measurement-based quantum computing with continuous-variable systems. *Phys. Rev. A*, 85:062318, 2012.

[26] Y. Takahashi, N. Kunihiro, and K. Ohta. The quantum fourier transform on a linear nearest neighbor architecture. *Quant. Info. Comput.*, 7(4):383–391, 2007.

[27] A. G. Fowler, S. J. Devitt, and L. C. L. Hollenberg. Implementation of Shor's algorithm on a linear nearest neighbour qubit array. *Quant. Info. and Comput.*, 4:237–245, 2004.

[28] R. Van Meter and M. Oskin. Architectural implications of quantum computing technologies. *J. Emerg. Technol. Comput. Syst.*, 2(1):31–63, 2006.

[29] A. G. Fowler, C. D. Hill, and L. Hollenberg. Quantum error correction on linear nearest neighbor qubit arrays. *Phys. Rev. A*, 2004.

[30] Yuichi Hirata, Masaki Nakanishi, Shigeru Yamashita, and Yasuhiko Nakashima. An efficient method to convert arbitrary quantum circuits to ones on a linear nearest neighbor architecture. In *International Conference on Quantum, Nano and Micro Technologies*, pages 26–33, Washington, DC, USA, 2009. IEEE Computer Society.

[31] A. Matsuo and S. Yamashita. Changing the gate order for optimal lnn conversion. In *Workshop on Reversible Computation*, pages 89–101, 2011.

[32] R. Wille, D. Große, L. Teuber, G. W. Dueck, and R. Drechsler. RevLib: an online resource for reversible functions and reversible circuits. In *Int'l Symp. on Multi-Valued Logic*, pages 220–225, 2008. RevLib is available at http://www.revlib.org.

[33] B.-S. Choi and R. Van Meter. A $\theta(\sqrt{n})$-depth quantum adder on the 2D NTC quantum computer architecture. *J. Emerg. Technol. Comput. Syst.*, 8(3):24, 2012.

[34] Z. Sasanian, R. Wille, and D. M. Miller. Realizing reversible circuits using a new class of quantum gates. In *Design Automation Conf.*, pages 36–41, 2012.

[4]As heuristics do not guarantee optimality, their run-time is usually negligible and, hence, not explicitly reported.

Qubit Placement to Minimize Communication Overhead in 2D Quantum Architectures

Alireza Shafaei, Mehdi Saeedi, and Massoud Pedram

Department of Electrical Engineering
University of Southern California
Los Angeles, CA 90089
{shafaeib,msaeedi,pedram}@usc.edu

Abstract— Regular, local-neighbor topologies of quantum architectures restrict interactions to adjacent qubits, which in turn increases the latency of quantum circuits mapped to these architectures. To alleviate this effect, optimization methods that consider qubit-to-qubit interactions in 2D grid architectures are presented in this paper. The proposed approaches benefit from Mixed Integer Programming (MIP) formulation for the qubit placement problem. Simulation results on various benchmarks show 27% on average reduction in communication overhead between qubits compared to best results of previous work.

I. Introduction

Quantum computing can offer significantly higher performance for a set of problems compared to what we have now, commonly-called classical computing. Quantum algorithms with superpolynomial speedup on a quantum computer include algorithms for number factoring, solving discrete-log and Pell's equation, and walk on a binary welded tree [1].

A well-known technique to implement a quantum algorithm on a quantum computer is to run a quantum physics experiment under the control of a classical computer. The experimental apparatus consists of physical qubits such as ions or photons where the quantum-mechanical properties of qubits are used to perform the required computation. A real-time classical computer directs the experiment by issuing instructions and reading out the quantum states. Final result may require post-process computation and answer checking.

A non-ideal quantum computer, however, is subject to noise and faces numerous limitations and constraints. Environmental disturbances and errors in the control systems are two common examples, which if ignored, can result in computational error. The error rate limits the computation length. In addition, current quantum technologies are subject to constraints on parallelism, *connectivity*, and bandwidth, which further limit the implementation of quantum algorithms.

Various proposals for quantum technologies with 1D, 2D and 3D interactions have been introduced. In general, 1D architectures with only two neighbors per qubit are highly restrictive, and 3D architectures with six neighbors per qubit are difficult to control. Hence, the most promising architecture for a quantum computing system is to arrange qubits in a 2D structure with four neighbors per qubit. Quantum technologies with 2D architectures include neutral atoms [2], superconductors [3], photonics [4], and quantum dots [5].

A physical realization of a quantum program can couple any distant two qubits with some communication overhead. However, this can result in a long sequence of operations, which in turn increases circuit latency and error rate. For instance, the nearest-neighbor communication overhead results in 175x reduction in error threshold for fault-tolerant error correction with a concatenated 7-qubit CSS code [6]. Improving error threshold is costly — it may require a more sophisticated control protocol to construct gates with higher fidelities or a more robust error correction code. Accordingly, optimization of quantum circuits are crucial in order to reduce the communication overhead.

During recent years, several techniques have been proposed to map arbitrary circuits to 1D quantum architectures [7]–[9], which as mentioned earlier have limited number of neighboring qubits. On the other hand, the few works on 2D architectures are hand-optimized techniques designed for special type of quantum circuits. Our focus, however, is to develop a design automation method to optimize qubit interactions considering the connectivity constraint in quantum technologies that use a 2D grid architecture.

Although conventional placement algorithms for VLSI physical design can be used for placement of qubits, the performance of such approaches is limited (Section IV). In this paper, we first characterize similarities and differences between the conventional placement algorithm and the one used in quantum technologies. Then, a Mixed Integer Programming (MIP) formulation is proposed as a standard grid placement algorithm to optimize qubit-to-qubit interaction. The proposed MIP formulation results in a valid placement for qubits. However, direct application of this formulation ignores specific properties of quantum technologies. Accordingly, the MIP formulation is improved by some heuristic techniques to properly capture the effects of quantum architectures.

The rest of this paper is organized as follows. We introduce basic concepts in Section II. Previous work is reviewed in Section III. Section IV discusses the proposed approach followed by experiments in Section V. Finally, paper is concluded in Section VI.

978-1-4799-2817-0/14 $31.00 © 2014 IEEE

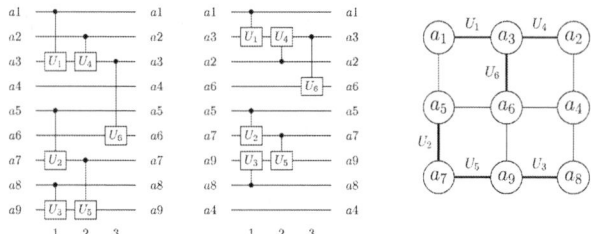

Fig. 1. A sample quantum circuit (left) and its implementation in 1D (middle) and 2D grid (right) architectures. The gate in time step 3 has a non-adjacent interaction in 1D architecture. However, all interactions in the 2D grid involve neighboring qubits.

II. BASIC CONCEPTS

In quantum computation, a quantum bit (qubit) is a unit of information which takes a linear superposition of the basis states $|0\rangle$ and $|1\rangle$. An n-qubit *quantum gate* performs a specific $2^n \times 2^n$ unitary operation on the selected n qubits. We do not use particular properties of any 1- or 2-qubit gates, except for 2-qubit SWAP gate. Therefore, we omit definitions. More information can be found in standard quantum computing textbooks and surveys [10], [11].

A quantum algorithm is described by a *quantum circuit* where a set of quantum gates is applied to transform the initial state of the quantum system into a final state. Each gate can involve an arbitrary number of qubits. The resulting circuit is then 'compiled' into another quantum circuit based on a library of primitive one- and two-qubit gates. This quantum circuit is an input to our problem.

Given a quantum circuit with one- and two-qubit gates, one should *map* the circuit into a quantum apparatus, which is a physical experiment that realizes the quantum circuit. The underlying quantum experiment is usually modeled as a *connectivity graph* with pre-defined connectivity patterns between graph nodes where nodes represent physical qubits. Therefore, a complete graph is an ideal quantum architecture with no limit on qubit interactions; and a path is a 1D architecture where only neighboring qubits on a line can interact (see Fig. 1 for an example). On the other hand, the quantum circuit is modeled with another graph, called *interaction graph*, where nodes denote qubits of the circuit. In this case, for each 2-qubit gate working on qubits i and j, an edge is added between nodes i and j in the graph.

Given an interaction graph and a connectivity graph, the mapping problem is a standard graph embedding problem with connectivity and interaction graphs as the *host* and *guest* graphs, respectively. The objective is then to minimize the total distance between adjacent nodes of the interaction graph. For a 2D grid connectivity graph, the mapping problem is NP-complete. Also, determining whether a given interaction graph can be embedded into a 2D grid is NP-complete [12].

If a solution for the grid-embedding problem is known, all circuit qubits have corresponding physical qubits. The next step is to apply quantum gates, which requires gates

to be adjacent. This means that all connected nodes in the interaction graph should be placed on adjacent grid nodes. For a qubit located at (i, j) in the grid, all qubits in locations $(i, j - 1)$ (left), $(i - 1, j)$ (up), $(i, j + 1)$ (right), $(i + 1, j)$ (down) are neighbors. Therefore for non-adjacent qubits (i, j) and (m, n) a connection should be made, which is achieved by applying a sequence of either MOVE or SWAP operations.

If MOVE operation is not supported by the quantum experiment, adjacent qubits should be exchanged step by step to transform a qubit from (i, j) to one of the four neighbors of (m, n). Since exchanging two neighboring qubits requires one SWAP gate, the total number of SWAP gates by this process is $|m - i| - 1$ if $n = j$ (same column), $|n - j| - 1$ if $m = i$ (same row), and $|m - i| + |n - j| - 1$ otherwise.

The added MOVE or SWAP operations are considered as communication overhead since such gates are not imposed by the original algorithm or circuit. Optimizations should be applied to reduce this overhead. In this paper, we focus on the quantum experiments that only support SWAP gates and not MOVE operations. Quantum technologies based on superconducting [3] and quantum dots [5] are examples of SWAP-based technologies.

III. PREVIOUS WORK

Certain circuits are amenable for specific interaction-cost optimizations. Examples include circuits for quantum Fourier transform [13], quantum adders [14], [15], modular exponentiation [16], [17], and error correction codes [6], [18]. A more general approach is developed in [19] where particular operations spanning n wires, e.g., rotation of n wires, were analyzed to be optimized for depth.

Optimization of arbitrary quantum circuits for 1D architectures is the topic of recent papers. Minimal number of SWAP gates required to transform one permutation of qubits in a line into another permutation was explored in [8]. Minimizing the number of SWAP gates by changing qubit locations dynamically was investigated in [7]. Minimum linear arrangement problem was employed in [9] to find (near-) optimal solutions, with respect to the number of SWAP gates, in different parts of an interaction graph. All these methods are based on 1D architectures. In [20], the authors considered qubit-to-qubit interaction optimization to map a circuit to a physical device where the underlying quantum device is a general graph (not a grid).

IV. THE PROPOSED METHOD

The conventional circuit placement problem in VLSI design starts with a (weighted) hypergraph where nodes represent standard cells and hyperedges denote connections among these cells. Each node of the hypergraph has a pre-defined size. Circuit placement determines center positions for nodes such that a specific objective function is optimized and some constraints are met (e.g., no overlap between cells). This is followed by a routing step that connects the placed cells via

978-1-4799-2817-0/14 $31.00 © 2014 IEEE

wires. Total wirelength, circuit delay, and power consumption are typical objectives in the VLSI physical design algorithms.

The qubit placement problem is similar to the conventional circuit placement problem, with some differences. In general, VLSI placement algorithms can be used for embedding a weighted undirected interaction graph. Also, similar to the minimized total wirelength in the conventional placement problem, we are interested in qubit placements with minimal total distance between connected nodes. However, in qubit placement, positions of instructions are not fixed, whereas in the conventional VLSI circuit placement gates (or instructions) are fixed. This time-variant nature of qubit placement imposes dynamic placement. Additionally, nodes (qubits) have no width and height in the qubit placement problem.

Dynamic placement of qubits can be used to reduce communication overhead. More precisely, after placing qubits in specific grid nodes in SWAP-based quantum technologies, one needs to exchange qubits step by step to 'route' two distant qubits towards each other in order to apply a gate. Location of other qubits on the path will change accordingly. To follow the placement solution, all moved qubits should return to their initial location by reversely applying the same sequence of SWAP gates. As used in e.g., [7] for 1D architectures, instead of returning qubits to their initial location, one may keep the current (updated) placement, and then apply the remaining gates based on the new locations of qubits.

Since VLSI designs include numerous gates, the most successful VLSI placement tools [21] apply several heuristics to avoid unbearable runtime. However, current quantum technologies are limited to a small number of qubits. Hence, we used an MIP-based grid-placement algorithm. Any other placement technique can also be used to solve the grid-embedding problem.

A. MIP-based Formulation

The MIP-based grid-embedding problem assigns each qubit to a unique location on the 2D grid such that frequently interacting qubits are placed as close together as possible. As a consequence, less number of SWAP gates are required in order to route qubits.

To mathematically formulate the problem, a binary variable x_{ij} is used which represents the assignment of $qubit_i$ (node i in the interaction graph) to $location_j$ in the grid. Moreover, w_{ik} denotes the weight between $qubit_i$ and $qubit_k$ in the interaction graph (i.e., the number of gates between them in the circuit), and $dist_{jl}$ represents the Manhattan distance between $location_j$ and $location_l$ in the grid. Hence, the cost of assigning $qubit_i$ to $location_j$ (i.e., x_{ij}) and $qubit_k$ to $location_l$ (i.e., x_{kl}) can be expressed as $c_{ijkl} = w_{ik} \times dist_{jl}$. The problem is then formulated as follows:

$$\min \quad \sum_{i=1}^{n}\sum_{j=1}^{n}\sum_{k=1}^{n}\sum_{l=1}^{n} c_{ijkl} x_{ij} x_{kl} \tag{1}$$

subject to

$$\sum_{j=1}^{n} x_{ij} = 1, \qquad i = 1, \ldots, n, \tag{2}$$

$$\sum_{i=1}^{n} x_{ij} = 1, \qquad j = 1, \ldots, n, \tag{3}$$

$$x_{ij} \in \{0, 1\}, \qquad i, j = 1, \ldots, n. \tag{4}$$

In this formulation, n is the number of grid nodes. More precisely, $n = hw$ for an $h \times w$ grid where h and w denote the number of rows and columns, respectively. Dummy nodes are also added to the interaction graph for the MIP formulation in cases where the number of qubits is less than n.

The objective function (1) is not linear; however, several equivalent formulations that linearize this objective function have been proposed. Among them, Kaufmann and Broeckx's linearization [22] has the smallest number of variables and constraints [23] which is described next. By defining $z_{ij} = x_{ij} \sum_{k=1}^{n}\sum_{l=1}^{n} c_{ijkl} x_{kl}$ for $i, j = 1, \ldots, n$, we can rewrite the objective function as $\sum_{i=1}^{n}\sum_{j=1}^{n} x_{ij} \sum_{k=1}^{n}\sum_{l=1}^{n} c_{ijkl} x_{kl} = \sum_{i=1}^{n}\sum_{j=1}^{n} z_{ij}$. Authors of [22] then proved that the following MIP formulation is equivalent to Eq. (1) - (4):

$$\min \quad \sum_{i=1}^{n}\sum_{j=1}^{n} z_{ij} \tag{5}$$

subject to

$$(2), (3), (4),$$

$$\alpha_{ij} x_{ij} + \sum_{k=1}^{n}\sum_{l=1}^{n} c_{ijkl} x_{kl} - z_{ij} \le \alpha_{ij}, \quad i, j = 1, \ldots, n, \tag{6}$$

$$z_{ij} \ge 0, \qquad\qquad i, j = 1, \ldots, n, \tag{7}$$

where $\alpha_{ij} = \sum_{k=1}^{n}\sum_{l=1}^{n} c_{ijkl}$ for $i, j = 1, \ldots, n$. This new formulation involves n^2 binary variables (x_{ij}'s), n^2 real variables (z_{ij}'s), and $n^2 + 2n$ constraints.

The above MIP formulation can find optimal placement solution with respect to the aforementioned objective and constraints. However, the resulting qubit placement may not be a valid solution for a SWAP-based quantum technology. In other words, the MIP formulation does not guarantee that all two-qubit gates become local; rather, it tends to place qubits that frequently interact with other as close as possible on the grid. Therefore, a mechanism is required to localize all two-qubit gates. For this purpose, after the MIP problem is solved, two-qubit gates are checked in order until a non-local gate is found. Afterwards, the corresponding control qubit is routed towards the target qubit based on xy routing algorithm (first along x-axis and then along y-axis) by inserting SWAP gates.

B. MIP-based Optimization Framework

The qubit placement discussed in Section A is obtained by applying the grid-embedding formulation on the whole interaction graph. Basically, the interaction graph has no view on scheduling of instructions. In other words, while $w_{i,k}$ reflects the number of interactions (or gates) between qubits i and k, qubits may interact in very different time steps. In this case, placing qubits i and k close to each other in the whole computation is not useful — one may place highly interacting qubits at different scheduling levels close to each other and move them to other locations when the corresponding qubits will not interact to leave space for other qubits.

In general, a small number of consecutive gates in a given circuit can be executed in parallel due to sharing control or target qubits. Accordingly, a given circuit is (almost) scheduled. Hence, working with gates at one scheduling level results in very few gates. As an alternative approach, we can apply m instances of the grid-embedding formulation on m subsets (subcircuits) of the interaction graph for a circuit with N gates. In this case, the interaction graph for subcircuit j is obtained by only considering consecutive gates between time steps $(j-1) * N/m + 1$ and $j * N/m$. Thus, each subcircuit can work on N/m gates simultaneously.

Using several instances of the grid-embedding problem, qubit placements for subcircuits j and $j + 1$ can be different. This requires a swapping network to align qubit arrangement of subcircuit j with qubit arrangement of subcircuit $j + 1$. For this purpose, we use the snake-like indexing (shown in Fig. 2(a)) with 2D bubble sort algorithm [24, Chapter 9]. While for 1D bubble sort, one can move the maximum element among unsorted items towards its proper location in one way, this is not the case in a 2D grid. If x and y are the row and column differences between an element and its proper location respectively, then the number of paths to move the element towards its proper location is $\frac{(x+y)!}{x!y!}$. Different paths for each element can affect other elements in the grid, which may result in very different number of SWAP gates. Moreover, considering the effect of moving minimum or maximum elements exacerbates the situation. Fig. 2(b) shows one example using two different strategies.

Fig. 3 illustrates the structure of the final circuit which is obtained from the following three steps. (1) MIP-based grid-embedding problem is solved for each subcircuit j in order to find the initial qubit placement of that subcircuit, P_j^i. (2) SWAP gates are inserted before non-local two-qubit gates of each subcircuit j which eventually leads to its final qubit placement, P_j^f. (3) In the end, swapping networks are added to change the final qubit placement of subcircuit j to initial qubit placement of subcircuit $j + 1$ ($P_j^f \rightarrow P_{j+1}^i$ for $1 \leq j \leq m - 1$).

V. EXPERIMENTAL RESULTS

Proposed methods were implemented in C++ and tested on a server machine with 4 Intel E7-8837 processors and 64GB memory. For MIP solver, we used Gurobi Optimizer Ver. 5.5.0 [25], which uses linear-programming relaxation techniques along with other heuristics in order to quickly solve large-scale MIP problems.

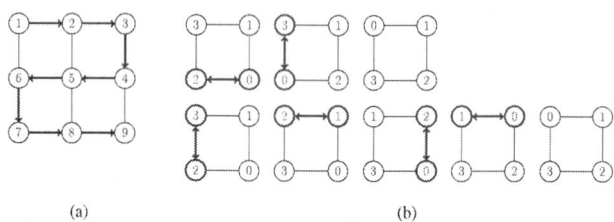

Fig. 2. (a) Snake-like indexing scheme for a 3×3 2D grid. It also shows how a path (traced by arrows) can be mapped on a 2D grid. (b) Sorting on a 2D grid based on different strategies can result in different number of SWAPs. In the first row, minimum element is always moved in XY direction (2 SWAPs). In the second row, maximum element is moved in XY direction (4 SWAPs).

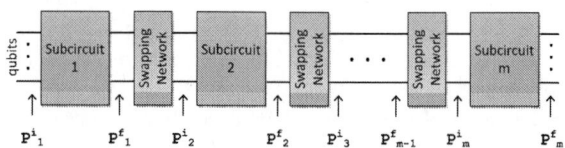

Fig. 3. Dynamic qubit placement via partitioning circuit into m subcircuits. P_j^i and P_j^f denote the initial and final qubit placements of subcircuit j, respectively. The swapping networks align qubit placements.

There are two methods in literature on optimization of communication overhead in 2D architectures for modular exponentiation [17] and adders [14], [15]. The method in [17] added $O(n^4)$ ancilla to reach $O(\log^2 n)$ depth for modular exponentiation. We do not add ancillae and our focus is on circuit size. Applying our techniques on log-size adders does not improve circuits in [15] (which improves [14]) in most cases. In particular, the optimizations in [15] are similar to our Method 2 (described next) while qubit placements for components were hand-optimized.

To evaluate the proposed methods we used reversible benchmarks in [7], [9] along with circuits for Shor's algorithm in [8]. The previous techniques [7]–[9], [16] are based on 1D interactions, which can also be mapped to 2D architectures[1]. However, using 2D interactions adds more flexibility and thus can lower the communication overhead. Accordingly, we do not intend to compare our results with 1D architectures. Instead, 1D results are reported to consider the effect of architectures on reducing overhead.

Runtime for the conventional placement algorithms in VLSI design is important, but as a secondary objective. In quantum computing, runtime for qubit placement is much less important, given that quantum technologies are in preliminary stages — there is no aggressive time to market. Accordingly, the main objective is still circuit quality. Additionally, due to the limitations of current quantum technologies to work with a large number of gates, investing runtime in favor of circuit quality is reasonable. Therefore, we used a time-limit of 30 minutes for each attempted benchmark. For small

---

[1]For mapping a 1D path onto a 2D grid, please see Fig. 2(a).

benchmarks, runtime varies from a few seconds to 5 minutes. For larger ones, we reported the best result after 30 minutes. For each circuit, we applied two methods:

- **Method 1:** This method uses a single grid-based MIP formulation on the whole interaction graph. When the *global* qubit placement solution is found, all 2-qubit gates are checked in order and SWAP gates are inserted before each non-adjacent gate. Placement of qubits will change accordingly. This new qubit placement is considered for the remaining gates.

- **Method 2:** Multiple instances of the grid-based placement problem are used in this method. For each instance, we used k consecutive gates. If a circuit includes $< k$ gates, the interaction graph is analyzed at once, same as Method 1. After finding a qubit placement for each instance, SWAP gates are applied before each non-adjacent gate. Swapping networks are also required between any two consecutive placements. We used a 2D bubble sort algorithm that moves (1) the maximum element in XY direction, (2) the maximum element in YX direction, (3) the minimum element in XY direction, (4) the minimum element in YX direction towards its proper location, and then selects the best network.

The results of applying the aforementioned methods are reported in Table I. In this table, for each benchmark we reported the number of qubits and the number of gates in the original circuit, as well as the number of two-qubit gates after decomposing the circuit based on [26] into one- and two-qubit gates. Columns 5-9 report the grid size ($h \times w$) that results in the smallest number of SWAPs along with its corresponding number of SWAP gates after applying the proposed methods. For Method 2, we also report the percentage of SWAPs in the swapping network as compared with total number of SWAPs (column 9). Column 10 shows the minimum number of SWAP gates achieved by our proposed methods (i.e., minimum of columns 6 and 8). Best prior result from different sources are also presented in Columns 11-12.

Comparing the best results for 2D architectures versus the best prior result for 1D architectures shows that the number of SWAP gates can be reduced extensively if one allows interactions in 2D architectures. As can be seen in Table I, our methods improve the best results of 1D architectures 27% on average and up to 67%. A sample circuit mapped to a 2D grid based on Method 1 is also illustrated in Fig. 4.

VI. CONCLUSION

We optimized qubit-to-qubit interactions in quantum technologies that allow 2D grid architectures. To achieve this, we formulated our problem by mixed integer programming as a grid-embedding problem. To consider scheduling of gates, the interaction graph is partitioned into several instances where the grid-embedding formulation is applied on each instance. To align qubit placement of one instance in the interaction graph with qubit placement of another instance, we applied a 2D bubble sort algorithm. Furthermore, for each

benchmark various grid sizes were examined to find the one with smallest number of SWAP gates. The proposed methods result in considerable reduction of communication overhead in 2D architectures. However, further heuristics can be applied to reduce the overhead more. For small circuits the prior methods for 1D architectures result in better circuits.

There are several lines for future researches. For very large circuits, conventional placement algorithms can be adopted to solve the graph-embedding problem. In addition, grid clustering and hierarchical qubit placement may be considered for large circuits, where our proposed method can be applied in each hierarchy level. Another topic for future research is to directly focus on depth of circuits. The method in [19] considered circuit depth for several basic operations in 1D architectures. The problem for 2D architectures is new and indeed interesting.

ACKNOWLEDGEMENTS

This research was supported by the Intelligence Advanced Research Projects Activity (IARPA) via Department of Interior National Business Center contract number D11PC20165. The U.S. Government is authorized to reproduce and distribute reprints for Governmental purposes notwithstanding any copyright annotation thereon. The views and conclusions contained herein are those of the authors and should not be interpreted as necessarily representing the official policies or endorsements, either expressed or implied, of IARPA, DoI/NBC, or the U.S. Government.

REFERENCES

[1] D. Bacon and W. van Dam. Recent progress in quantum algorithms. *Commun. ACM*, 53:84–93, Feb. 2010.

[2] M. Saffman, T. Walker, and K. Mølmer. Quantum information with rydberg atoms. *Rev. Mod. Phys.*, 82:2313–2363, Aug 2010.

[3] A. Blais et al. Quantum-information processing with circuit quantum electrodynamics. *Phys. Rev. A*, 75:032329, Mar 2007.

[4] E. Knill, R. Laflamme, G. Milburn. A scheme for efficient quantum computation with linear optics. *Nature*, 409:46–52, 2001.

[5] J. M. Taylor et al. Relaxation, dephasing, and quantum control of electron spins in double quantum dots. *Phys. Rev. B*, 76:035315, 2007.

[6] T. Szkopek et al. Threshold error penalty for fault-tolerant quantum computation with nearest neighbor communication. *IEEE Trans. Nanotechnol.*, 5(1):42 – 49, jan. 2006.

[7] M. Saeedi, R. Wille, R. Drechsler. Synthesis of quantum circuits for linear nearest neighbor architectures. *Quant. Inf. Proc.*, 10(3):355–377, 2011.

[8] Y. Hirata, M. Nakanishi, S. Yamashita, Y. Nakashima. An efficient conversion of quantum circuits to a linear nearest neighbor architecture. *Quant. Inf. Comput.*, 11(1–2):0142–0166, 2011.

[9] A. Shafaei, M. Saeedi, and M. Pedram. Optimization of quantum circuits for interaction distance in linear nearest neighbor architectures. *Design Autom. Conf.*, 2013.

[10] M. Nielsen and I. Chuang. *Quantum Computation and Quantum Information*. Cambridge Univ. Press, 2000.

[11] M. Saeedi and I. L. Markov. Synthesis and optimization of reversible circuits - a survey. *ACM Comput. Surv.*, 45(2):21:1–21:34, March 2013.

[12] S. N. Bhatt and S. S. Cosmadakis. The complexity of minimizing wire lengths in VLSI layouts. *Inform. Process. Lett.*, 25(4):263 – 267, 1987.

[13] Y. Takahashi, N. Kunihiro, and K. Ohta. The quantum Fourier transform on a linear nearest neighbor architecture. *Quant. Inf. Comput.*, 7:383–391, 2007.

978-1-4799-2817-0/14 $31.00 © 2014 IEEE

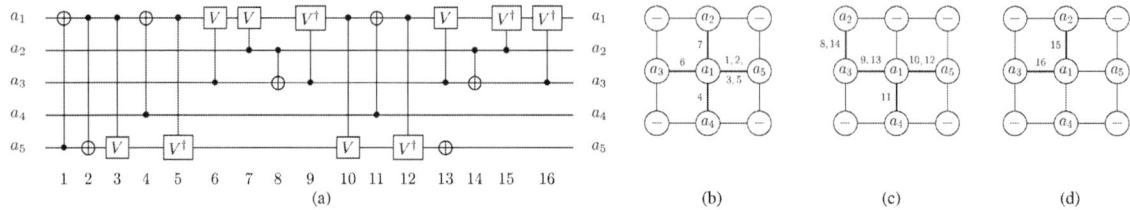

Fig. 4. The result of applying Method 1 on the 4gt13-v1_93 benchmark. Circuit in (a) is the original circuit with one- and two-qubit gates. The initial qubit placement is shown in (b). Since a_1 is adjacent with other qubits, gates in time steps 1-7 can be executed. In time step 8, non-adjacent qubits a_2 and a_3 should interact. A SWAP gate is applied to exchange the locations of qubit a_2 with its neighboring qubit. The result is shown in (c). Then, gates in time steps 8-14 can be executed. To execute the gate in time step 15, we return a_2 to its original location by applying another SWAP gate in (d). The remaining gates can be executed afterwards.

TABLE I

NUMBER OF SWAP GATES AFTER APPLYING PROPOSED METHODS AS WELL AS THE BEST PRIOR RESULT IN 1D ARCHITECTURES. FOR EACH BENCHMARK, THE BEST RESULT IS SHOWN IN BOLD FACE. ON AVERAGE, THE BEST RESULTS OF 1D ARCHITECTURES ARE IMPROVED BY 27%.

| Benchmarks | | | | Method 1 | | Method 2 | | | Our best result | Best results in 1D | | Imp. |
Name	n	# gates	# 2-qubit gates	grid size	#SWAPs	grid size	#SWAPs	SWAP Net		#SWAPs	Ref.	(%)
3_17_13	3	6	13	2x2	6	2x2	6	0%	6	4		-50
4_49_17	4	12	30	2x2	15	2x2	13	0%	13	12		-8
4gt10-v1_81	5	6	36	3x2	16	3x2	16	0%	16	20		20
4gt11_84	5	3	7	2x3	2	2x3	2	0%	2	1		-100
4gt12-v1_89	5	5	52	3x2	21	3x2	19	0%	19	35		46
4gt13-v1_93	5	4	16	3x3	2	3x3	2	0%	2	6		67
4gt4-v0_80	5	5	43	2x3	17	2x3	17	0%	17	34		50
4gt5_75	5	5	22	2x4	9	3x3	8	0%	8	12		33
4mod5-v1_23	5	8	24	2x3	11	2x3	11	0%	11	9		-22
4mod7-v0_95	5	6	40	3x3	13	3x3	13	0%	13	21		38
aj-e11_165	4	13	59	2x3	24	2x3	24	0%	24	36		33
alu-v4_36	5	7	31	2x3	10	2x3	10	0%	10	18		44
decod24-v3_46	4	9	9	3x2	3	2x2	3	0%	3	3		0
ham7_104	7	23	87	3x3	48	3x3	50	30%	48	68		29
lwb4_52	4	11	23	2x2	9	3x3	9	0%	9	10		10
hwb5_55	5	24	106	2x3	48	3x2	45	5%	45	63		29
hwb6_58	6	42	146	3x2	94	2x3	79	17%	79	118	[9]	33
hwb7_62	7	331	2659	3x3	1738	3x3	1688	2%	1688	2228		24
hwb8_118	8	633	16608	3x3	11133	3x3	11027	1%	11027	14361		23
hwb9_123	9	1959	20405	4x3	15063	4x3	15022	1%	15022	21166		29
mod5adder_128	6	15	81	3x2	44	3x2	41	15%	41	51		20
mod8-10_177	5	14	108	3x3	45	2x3	48	11%	45	72		38
rd32-v0_67	4	2	8	2x3	3	2x3	2	0%	2	2		0
rd53_135	7	16	78	5x2	39	3x3	42	15%	39	66		41
rd73_140	10	20	76	4x3	37	4x4	37	0%	37	56		34
sym9_148	10	210	4452	4x4	2363	4x4	2414	2%	2363	3415		31
sys6-v0_144	10	15	62	4x4	31	4x4	31	0%	31	59		47
urf1_149	9	11554	57770	3x3	38622	3x3	38555	1%	38555	44072		13
urf2_152	8	5030	25150	4x2	17175	2x4	16822	1%	16822	17670		5
urf5_158	9	10276	51380	3x3	34589	3x3	34406	1%	34406	39309		12
QFT5	5	10	10	3x2	5	3x2	7	0%	5	6		17
QFT6	6	15	15	5x2	6	2x3	6	0%	6	12		50
QFT7	7	21	21	2x4	18	5x2	18	0%	18	26		31
QFT8	8	28	28	5x2	25	4x2	18	50%	18	33		45
QFT9	9	36	36	3x4	40	3x3	34	0%	34	54		37
QFT10	10	45	45	4x3	54	5x3	53	0%	53	70		24
cnt3-5_180	16	20	125	3x6	69	4x4	82	30%	69	127		46
cycle10_2_110	12	19	1212	3x4	867	3x4	839	3%	839	2304		64
ham15_108	15	70	458	5x3	328	5x3	340	7%	328	715		54
plus127mod8192_162	13	910	65455	5x4	53598	5x4	53976	1%	53598	151794		65
plus63mod4096_163	12	429	29019	5x3	22118	3x5	22194	1%	22118	61556	[7]	64
plus63mod8192_164	13	492	37101	5x3	30358	5x3	29835	1%	29835	82492		64
rd84_142	15	28	112	5x3	54	5x3	68	28%	54	148		64
urf3_155	10	26468	132340	4x3	94017	4x3	94202	1%	94017	154672		39
urf6_160	15	10740	53700	5x3	43909	4x4	44394	1%	43909	88900		51
Shor3	10	2727	2076	3x5	1737	4x3	1710	1%	1710	1816		6
Shor4	12	6398	5002	3x6	4264	3x4	4339	4%	4264	5080	[8]	4
Shor5	14	12830	10265	5x4	8456	5x4	10890	2%	8456	10760		21
Shor6	16	23139	18885	4x6	20386	4x6	21418	1%	20386	20778		2

[14] B.-S. Choi and R. Van Meter. A $\sqrt{n}$-depth quantum adder on the 2D NTC quantum computer architecture. *J. Emerg. Technol. Comput. Syst.*, 8(3):24:1–24:22, August 2012.

[15] M. Saeedi, A. Shafaei, and M. Pedram. Constant-factor optimization of quantum adders on 2D quantum architectures. *Rev. Comp., arXiv:1304.0432*, 2013.

[16] A. G. Fowler, S. J. Devitt, and L. C. L. Hollenberg. Implementation of Shor's algorithm on a linear nearest neighbour qubit array. *Quant. Inf. Comput.*, 4:237–245, 2004.

[17] P. Pham and K. M. Svore. A 2D nearest-neighbor quantum architecture for factoring. *Rev. Comp., arXiv:1207.6655*, 2012.

[18] A. G. Fowler, C. D. Hill, and L. C. L. Hollenberg. Quantum error correction on linear nearest neighbor qubit arrays. *Phys. Rev. A*, 69:042314.1–042314.4, 2004.

[19] S. Kutin, D. Moulton, and L. Smithline. Computation at a distance. *Chicago J. of Theor. Comput. Sci.*, 2007.

[20] D. Maslov, S. M. Falconer, and M. Mosca. Quantum circuit placement. *IEEE Trans. CAD*, 27(4):752–763, Apr 2008.

[21] I. L. Markov, J. Hu, and M.-C. Kim. Progress and challenges in VLSI placement research. *ICCAD'12*, pages 275–282, 2012.

[22] L. Kaufmann and F. Broeckx. An Algorithm for the Quadratic Assignment Problem using Benders Decomposition. *European Journal of Operational Research*, 2(3):204–211, May 1978.

[23] R. E. Burkard, E. ela, P. M. Pardalos, and L. S. Pitsoulis. *The Quadratic Assignment Problem*. Handbook of Combinatorial Optimization, Kluwer Academic Publishers,1998, pp. 241-338..

[24] B. Parhami. *Introduction to Parallel Processing: Algorithms and Architectures*. Kluwer Academic Publishers, Norwell, MA, USA, 1999.

[25] Gurobi Optimization Inc. Gurobi optimizer reference manual, 2013.

[26] D. Maslov, G. W. Dueck, D. M. Miller, and C. Negrevergne. Quantum circuit simplification and level compaction. *IEEE Trans. CAD*, 27(3):436–444, March 2008.

A Novel Wirelength-Driven Packing Algorithm for FPGAs with Adaptive Logic Modules

Sheng-Kai Wu, Po-Yi Hsu, Wai-Kei Mak *

Dept. of CS, National Tsing Hua University, HsinChu, Taiwan

Email: u9562118@cs.nthu.edu.tw, boy.hsupaul@gmail.com, wkmak@cs.nthu.edu.tw

Abstract—Adaptive logic module (ALM) in modern field programmable gate array can serve as one 6-input lookup table (LUT) or two smaller lookup tables under certain constraints. In a typical design flow, a netlist of LUTs formed after technology mapping has to be merged into ALMs and then packed into coarse-grained logic blocks (CLBs) before placement and routing. How the LUTs are merged and the ALMs are packed has a significant impact on the quality of the placement. We propose a novel wirelength-driven algorithm to merge the LUTs and pack the ALMs to ensure that it will not adversely affect the final wirelength. Experimental results show that substituting AAPack [8] by our algorithm yields about 14.54% reduction in number of tracks required for routing and 17.97% wirelength improvement for ALM-based FPGA. Applying our algorithm to traditional FPGA, the minimum channel width and wirelength are reduced by 16.59% and 17.57%, respectively, compared to T-VPack.

I. INTRODUCTION

Most field programmable gate arrays (FPGAs) use lookup table (LUT) as the basic logic block. A k-input LUT can implement any Boolean logic function of up to k variables. Adaptive logic module (ALM) is a novel structure of logic block adopted by Altera's Stratix II [13] and later generations of Stratix family. One ALM can serve as a 6-input LUT or two smaller LUTs if the total distinct inputs is less or equal to 8 under certain constraints. This new logic module design has both performance and area advantages over conventional logic module. Hutton *et al.* [6] reported that the ALM architecture with ALM packing led to a 15% timing improvement after place and route and a 12% area reduction on average versus a standard BLE4 architecture for a set of 80 industrial designs.

As an ALM can implement two LUTs under certain constraints, how to merge two LUTs into one ALM so that it will not adversely affect the quality for later placement and routing is an important issue. [7] proposed an effective algorithm, ALMmap, to deal with the technology mapping problem for the ALM architecture considering area minimization with depth constraint. However, [7] uses a maximum cardinality matching algorithm to merge LUTs into ALMs to minimize the total ALM area without considering the impact on the quality of place and route. Merging LUTs into an ALM potentially has a big impact on placement and routing as it bolts two separate paths together. If we put two LUTs that should not be

merged together into one ALM, it will have an unfavorable impact on wirelength, routability, and performance.

The impact of fixed clustering on the wirelength and delay of placement solutions has been pointed out by [5]. And [5] proposed a simulated annealing-based simultaneous clustering and placement which allows LUTs to move between clusters. But a drawback of considering placement at the LUT-level is that it incurs significant overhead. In particular, modern FPGA architectures such as those with adaptive logic modules require expensive design rule checking before moving the LUTs. So, we propose a wirelength-driven packing algorithm suitable for ALM-based FPGA in this paper.

There are various works on FPGA clustering for simple traditional FPGA architecture. In [2], Betz et al. proposed a packing algorithm, VPack, which focuses on minimizing the number of clusters and the number of inputs for each cluster. Later, the same team introduced a new version of VPack, T-VPack [9], which is a timing-driven packing algorithm that clusters logic blocks on timing-critical paths to take advantage of fast internal connections. [4] proposed a clustering approach by performing a rough placement prior to clustering to obtain some physical information as guidance. [3] incorporates routability metrics into the clustering cost function. [11] presented a Rent's rule based algorithm. Recently, architecture-specific packing is considered in [1] and [8].

In this paper, we propose a novel packing algorithm for ALM-based FPGA targeting directly at wirelength minimization. Minimizing the total wirelength is of great importance. With smaller wirelength, the transition power of signal and the potential performance of the circuit will be improved. Also, in FPGA, the routing resource requirement can be reduced such that the routability of the design would increase. Although we focus on ALM-based architecture, the concept of our packing algorithm can be applied on any other FPGA architecture as well.

This paper is organized as follows. The preliminaries are presented in Section II. In section III, our proposed wirelength-driven packing algorithm is presented. Experimental results and conclusion are presented in Section IV and V, respectively.

II. PRELIMINARIES

In this section, we describe the concept of safe clustering that avoids wirelength degradation and define our problem.

---

*This work was supported in part by the National Science Council of Taiwan, under Grant NSC 102-2220-E-007-013.

978-1-4799-2817-0/14 $31.00 © 2014 IEEE

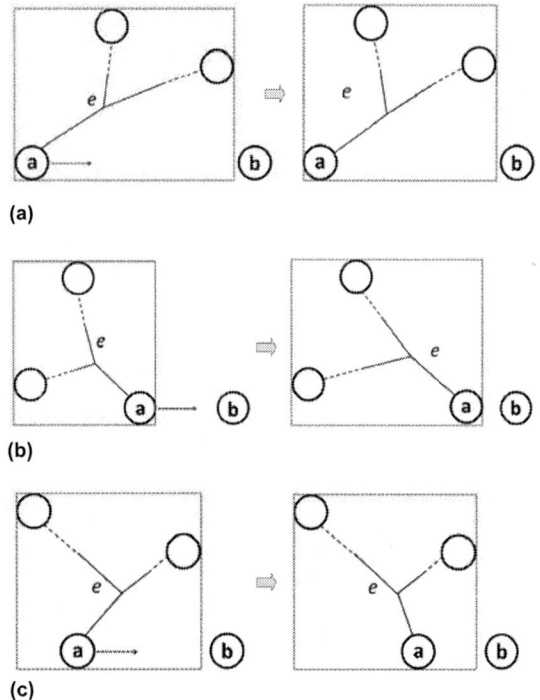

(a)

(b)

(c)

Fig. 1. Suppose a is on the left of b, and a is moved towards b. (a) If a is the only leftmost node of e, the wirelength of e will decrease. (b) If a is the rightmost node of e, the wirelength of e will increase. (c) Otherwise, the wirelength of e will not change.

A. The Idea of Safe Clustering

The concept of safe clustering for effectively reducing the problem size in ASIC placement was introduced in [12]. It guarantees that the clustering will not adversely affect the final total wirelength. Given a circuit modeled as a hypergraph $G(V, E)$, where V is the set of nodes and E is the set of hyperedges which correspond to the nets. Let P denote the set of all possible valid placements and E_v denote the set of hyperedges incident to v. Given a placement $p \in P$, a hyperedge $e \in E$, and a pair of nodes $\{a, b\}$ with a on the left of b. [12] defines $\Delta_a(p, e)$ as the gradient function of wirelength of e if a is moving toward b and $\Delta_b(p, e)$ as the gradient function of wirelength of e if b is moving toward a. Let $w_e (w_e \geq 0)$ be the weight of e, the gradient functions are given by:

$$\Delta_a(p, e) = \begin{cases} w_e, & \text{if } a \text{ is the rightmost vertex of } e \\ -w_e, & \text{if } a \text{ is the } only \text{ leftmost vertex of } e \\ 0, & \text{otherwise} \end{cases}$$

$$\Delta_b(p, e) = \begin{cases} w_e, & \text{if } b \text{ is the leftmost vertex of } e \\ -w_e, & \text{if } b \text{ is the } only \text{ rightmost vertex of } e \\ 0, & \text{otherwise} \end{cases}$$

When a is moved towards b, the wirelength of e will be reduced if $\Delta_a(p, e) < 0$; the wirelength of e will increase

if $\Delta_a(p, e) > 0$; the wirelength of e will not change if $\Delta_a(p, e) = 0$. An example is given in Fig.1. In Fig.1(a), a is the *only* leftmost node of e, so when a is moved towards b, the wirelength of e will decrease. In Fig.1(b), a is the rightmost vertex of e, so the wirelength of e will increase. Otherwise, the wirelength will not change as shown in Fig.1(c).

From the above two gradient functions, the total wirelength gradient function $F_{ab}(p)$ is defined as follows:

$$F_{ab}(p) = min(\sum_{e \in E_a} \Delta_a(p, e), \sum_{e \in E_b} \Delta_b(p, e)). \quad (1)$$

$F_{ab}(p)$ calculates the total wirelength gradients of placement p when a is moved towards b or when b is moved towards a, and take the smaller value. If $F_{ab}(p) > 0$, it means the total wirelength will increase if either a or b is moved towards the other.

To check the safeness of clustering a and b, the $F_{ab}(p_i)$ value needs to be computed under all possible placements $p_i \in P$. Suppose the number of all possible placements is l, then if $max(F_{ab}(p_1), F_{ab}(p_2), ..., F_{ab}(p_l)) \leq 0$, we can say it is safe to cluster a and b. However, it is not practical to generate all possible placements when considering clustering a and b. A selective placement enumeration approach is proposed in [2]. It only enumerates the placements that might generate worse wirelength if we merge a and b. Note that the increase or decrease of the total wirelength for merging a and b is only related to the nodes connected with at least one of a or b, we denote the set of these nodes as V_{ab}. For each node $v \in V_{ab}$, we have three possible locations for v, (1) v is on the left of a, (2) v is between a and b, and (3) v is on the right of b. [12] has proven that case(2) will never be worse than case(1) or case(3). Therefore, we can consider only two possible locations for v.

To further reduce the number of placements that need to be enumerated, [12] identified a subset of nodes in V_{ab} for which it is unnecessary to consider both possible positions for them. If there are α_{ab} nodes in this subset, then there are only $2^{|V_{ab}| - \alpha_{ab}}$ different placements that need to be enumerated to check the safeness of clustering a and b.

B. Problem formulation

The packing problem for ALM-based FPGA is stated as follows:

Given a mapped netlist of 6-input LUTs, merge the LUTs into ALMs under the ALM architecture constraint and cluster ALMs into CLBs so as to optimize the expected wirelength after place and route.

III. ALGORITHM

We introduce our wirelength-driven packing algorithm, ALMpack, in this section. Fig.2 shows the overall design flow. First, we perform technology mapping to generate a 6-LUT netlist. The LUT netlist can be generated by any technology mapper. Then we model merging LUTs into ALMs

978-1-4799-2817-0/14 $31.00 © 2014 IEEE

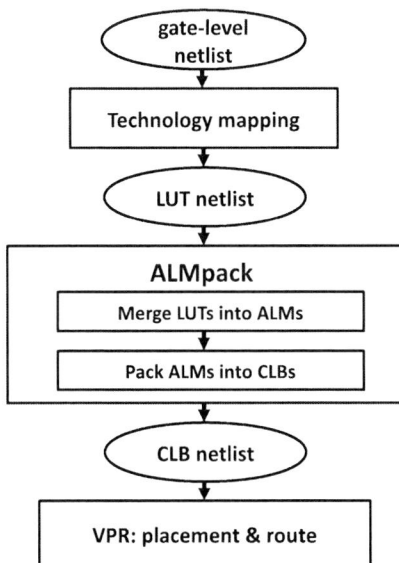

Fig. 2. The overall design flow.

as a minimum-weighted maximum matching problem on a weighted undirected graph and extend the same process for ALM packing when CLB size is 2^k. If CLB size is not 2^k, a heuristic seed-based wirelength-driven ALM packing will be adopted. We use VPR to conduct the remaining placement and routing.

A. Merge LUTs into ALMs

Since each ALM can serve as two LUTs to implement two functions for (1) each function has no more than five inputs each and the total number of inputs is not larger than 8, or (2) two 6-input functions that have the same logic operation sharing 4 inputs. We model the wirelength-driven ALM formation problem as a minimum weighted maximum matching problem. We construct a weighted undirected graph where each node correspondings to a LUT, and there is an edge between two nodes if and only if the corresponding LUTs can be merged into an ALM. We define the edge weight between two nodes a and b as follows:

$weight(a, b) =$

$$
\begin{cases}
0, & \text{if } 2^{|V_{ab}|-\alpha_{ab}} \leq enum\_bound \\
& \text{and } \forall i \; F_{ab}(p_i) \leq 0 \\
\frac{(\sum_{i=1 s.t. F_{ab}(p_i)>0}^{L_{ab}} F_{ab}(p_i))}{L_{ab}}, & \text{if } 2^{|V_{ab}|-\alpha_{ab}} \leq enum\_bound \\
& \text{and } \exists i \text{ s.t. } F_{ab}(p_i) > 0 \\
B - \beta \frac{|NET(a)\cap NET(b)|}{|NET(a)\cup NET(b)|}, & \text{if } 2^{|V_{ab}|-\alpha_{ab}} > enum\_bound
\end{cases}
\tag{2}
$$

where $L_{ab} = 2^{|V_{ab}|-\alpha_{ab}}$, $F_{ab}(p_i)$ is defined by equation(1), B is a big constant, $NET(v)$ is the set of nets incident to node

v, β is a user defined parameter, and a user defined enumeration bound, $enum\_bound$. In our experiments, B, β, and $enum\_bound$ are set to 1000, 10, and 2^{20}, respectively.

We perform enumeration only if the number of enumeration needed for checking the safeness of clustering a and b does not exceed $enum\_bound$. Otherwise, we will return $B - \beta \frac{|NET(a)\cap NET(b)|}{|NET(a)\cup NET(b)|}$ which accounts for less than 1% of the node pairs in our experiments. The big constant B is to distinguish these edges from other edges since their safeness is not sure. But if the node pair has more common nets, the edge weight between them will be smaller.

When enumeration is performed and for all enumerated placements the values of $F_{ab}(p_i)$ are all less than or equal to 0, we set 0 as the edge weight. We have preference for merging these pairs because they will not adversely affect the wirelength. On the other hand, if enumeration is performed and there exists a placement p_i such that $F_{ab}(p_i) > 0$, we compute the edge weight by $(\sum_{i=1 s.t. F_{ab}(p_i)>0}^{L_{ab}} F_{ab}(p_i))/L_{ab}$.

The reason why we only consider the situation where $F_{ab}(p_i) > 0$ is that if we use the formula of $(\sum_{i=1}^{L_{ab}} F_{ab}(p_i))/L_{ab}$ to compute the weight value for both safe pair and unsafe pair, some of them might have very close weight value. For example, for a pair of nodes (a, b), $F_{ab}(p_1) = -1, F_{ab}(p_2) = -2, F_{ab}(p_3) = -1$, and $F_{ab}(p_4) = 0$. For another pair of nodes (a, c), $F_{ac}(p_1) = 2, F_{ac}(p_2) = 1, F_{ac}(p_3) = 1$, and $F_{ac}(p_4) = -8$. As we can see, merging the pair (a, b) is safe while the pair of (a, c) is unsafe. Intuitively, it is better to merge a and b. However, $(\sum_{i=1}^{L_{ab}} F_{ab}(p_i))/L_{ab}$ has the same value as $(\sum_{i=1}^{L_{ac}} F_{ac}(p_i))/L_{ac}$ which means a has the same tendency to be merged with b or c. To avoid this, we ignore the negative $F_{ab}(p_i)$. After the modification, $(\sum_{i=1 s.t. F_{ab}(p_i)>0}^{L_{ab}} F_{ab}(p_i))/L_{ab}$ is 0 and $(\sum_{i=1 s.t. F_{ac}(p_i)>0}^{L_{ac}} F_{ac}(p_i))/L_{ac}$ becomes 1 so that a will have higher priority to be clustered with b. Moreover, we can ensure that the weight values of unsafe pairs are always larger than the weight values of safe pairs.

We also found out from experiments that better results[1] can be obtained if we modify equation(2) as follows:

$weight(a, b) =$

$$
\begin{cases}
0, & \text{if } 2^{|V_{ab}|-\alpha_{ab}} \leq enum\_bound \\
& \text{and } \forall i \; F_{ab}(p_i) \leq 0 \\
\frac{(\sum_{i=1 s.t. F_{ab}(p_i)>0}^{L_{ab}} F'_{ab}(p_i))}{L_{ab}}, & \text{if } 2^{|V_{ab}|-\alpha_{ab}} \leq enum\_bound \\
& \text{and } \exists i \text{ s.t. } F_{ab}(p_i) > 0 \\
B - \beta \frac{|NET(a)\cap NET(b)|}{|NET(a)\cup NET(b)|}, & \text{if } 2^{|V_{ab}|-\alpha_{ab}} > enum\_bound
\end{cases}
\tag{3}
$$

$$
F'_{ab}(p_i) = 1/2(\sum_{e \in E_a} \Delta_a(p_i, e) + \sum_{e \in E_b} \Delta_b(p_i, e)). \tag{4}
$$

After constructing the weighted undirected graph, we perform minimum-weighted maximum matching to find a

---

[1]It is likely due to the fact that non-optimal placement is always produced in practice even by the best placer.

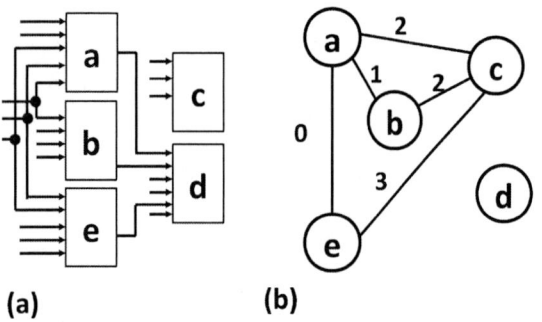

Fig. 3. (a) A netlist of LUTs. (b) A weighted undirected graph we construct from (a).

Algorithm 2 ComputeEdgeWeight

Input: A pair of nodes (a, b)
Output: The edge weight W between a and b
1: $L_{ab} = 2^{|V_{ab}| - \alpha_{ab}}$;
2: **if** $L_{ab} > enum\_bound$ **then**
3:     **return** $B - \beta \frac{|NET(a) \cap NET(b)|}{|NET(a) \cup NET(b)|}$;
4: **end if**
5: $W = 0$;
6: $safe = true$;
7: **for** $(i = 1; i \leq L_{ab}; i++)$ **do**
8:
$$F_{ab}(p_i) = min(\sum_{e \in E_a} \Delta_a(p_i, e), \sum_{e \in E_b} \Delta_b(p_i, e));$$
9:     **if** $F_{ab}(p_i) > 0$ **then**
10:         $safe = false$;
11:         $W += \frac{(\sum_{e \in E_a} \Delta_a(p_i, e) + \sum_{e \in E_b} \Delta_b(p_i, e))}{2}$;
12:     **end if**
13: **end for**
14: **if** $!safe$ **then**
15:     $W = W / L_{ab}$;
16: **end if**
17: **return** W;

Algorithm 3 Wirelength-driven ALM clustering

Input: An ALM netlist M
Output: A CLB netlist
1: N = cluster size;
2: **if** $N = 2^k$ **then**
3:     **for** $(i = 1; i \leq k; i++)$ **do**
4:         call wirelength-driven ALM packing similar
5:         to Algorithm 1;
6:     **end for**
7: **else**
8:     call Seed-based wirelength-driven ALM packing;
9: **end if**
10: **return** clustering result;

wirelength-driven LUT merging solution. The pseudo-code of our LUT merging algorithm is shown in Algorithm 1.

An example is demonstrated in Fig.3. Assume Fig.3(a) is a LUT netlist, we can construct a weighted undirected graph as shown in Fig.3(b). Since the total number of distinct inputs of LUT a and LUT b is 8, we construct an edge between them. However, the total number of distinct inputs of LUT b and LUT e is 9, there is no edge between them. The 6-input LUT d cannot be merged with any other LUT, so it is an isolated node. The matching result of Fig.3(b) is {a,e}, {b,c} and {d}.

B. CLB clustering

Algorithm 3 shows the pseudo-code of our CLB clustering algorithm. If the cluster size is 2^k, we will perform wirelength-driven ALM packing which is very similar to LUT merging algorithm described in Section III.A. The only difference between them is in the 4th line of Algorithm 1, 8 changes to the #inputs of CLB(I). The packing algorithm will repeat k times until reaching the target cluster size. If the cluster size is not 2^k, we use a heuristic seed-based approach to cluster ALMs into CLBs as shown in Algorithm 4.

Algorithm 1 Wirelength-driven LUT merging

Input: A LUT netlist M
Output: An ALM netlist
1: V = all nodes in M;
2: $E = \phi$;
3: **for** each pair of nodes (i, j) **do**
4:     **if** # distinct inputs of LUT_i and $LUT_j \leq 8$ **then**
5:         $w(e) = $ ComputeEdgeWeight(LUT_i, LUT_j);
6:         $E = E \cup e$;
7:     **end if**
8: **end for**
9: call minimum weighted maximum matching solver;
10: **return** matching result;

IV. EXPERIMENTAL RESULTS

We implemented our packing algorithm, ALMPack, using C++ on an Ubuntu workstation with 8 GB memory and 2.13 GHz CPU. We used the 20 largest MCNC benchmarks for the experiments. Our experiment settings are presented in TABLE I.

In the first experiment, we compare our packing algorithm with AAPack [8] and use VPR6.0 to perform placement and routing. The result is shown in TABLE II. Column 2 is the number of 6-LUTs that [7] generated for each MCNC benchmark, column 3 and column 4 show the total number of CLBs after applying the two different packing algorithms. The minimum channel width, total wirelength, and critical path delay after place and route are reported in columns 6 to 14. The last column is the total run time of ALMpack. The wirelength and delay values in column 9 and column 12 were obtained under the minimum channel width in column 6, while the wirelength and delay values in column 10 and column 13 were obtained

Algorithm 4 Seed-based wirelength-driven ALM packing

Input: An ALM netlist M
Output: A CLB netlist
1: N = cluster size;
2: **while** there exists any unclustered ALM in M **do**
3:     s = choose_seed(M);
4:     C = new_CLB(s);
5:     **while** C is able to accommodate ALMs **do**
6:         **for** each unclustered ALM a in M **do**
7:             Compute the $cost(C, a)$ by equation(3);
8:         **end for**
9:         Choose the feasible ALM with smallest cost;
10:        Add the chosen ALM to C;
11:        update_netlist(M);
12:    **end while**
13: **end while**
14: **return** clustering result;

TABLE I
ARCHITECTURE AND EXPERIMENTAL SETTINGS

	Exp. 1	Exp. 2
Logic element	ALM	4-LUT
Cluster size(N)	8	8
#inputs of CLB(I)	34	22
Placement Algorithm	Timing-driven	Timing-driven
Routing Algorithm	Timing-driven	Timing-driven
Segment length	4	4
Switch type	Buffered	Buffered
F_c(pad,input,output)	(1,0.5,0.25)	(1,0.5,0.25)
F_s	3	3

algorithm consistently outperforms AAPack for ALM-based FPGA and T-VPack for traditional FPGA by a large margin.

under the minimum channel width in column 7. It shows that the final wirelength is improved in all cases and is 17.97% shorter on average using ALMpack. In addition, using ALM-pack for packing reduced the minimum channel width in 18 of the 20 benchmarks and never increased the minimum channel width. This implies that packing produced by ALMpack can reduce the routing resource usage and increase the routability of design. We ran AAPack in default mode which optimizes both area and timing, but we still obtained 2.14% improvement for delay on average with comparable area. Moreover, our wirelength and delay improvement will increase to 24.00% and 10.19% on average if we route each design under the minimum channel width required by AAPack.

In the second experiment, we want to show the performance of our packing algorithm on traditional FPGAs. We use WireMap implemented in ABC [10] to generate the netlists of 4-LUTs and compare our algorithm with T-VPack [9]. As shown in TABLE III, we reduced the wirelength for all 20 benchmarks with 17.57% improvement on average. We achieved better minimum channel width in 19 of the 20 benchmarks with 16.59% improvement on average. Although ALM-pack does not directly target to minimize delay, it still obtained 3.62% delay improvement on average compared to the timing-driven packing algorithm, T-VPack, while reducing the channel width by 16.59%. We note that the wirelength and delay improvement will increase to 20.10% and 16.51% on average if we route each design under the minimum channel width required by T-VPack.

V. CONCLUSIONS

Minimizing the total wirelength is an important objective in VLSI design. For a FPGA-based design, smaller wirelength can improve both power and performance of the circuit and increase the routability of the design. In this paper we proposed a novel wirelength-driven algorithm to merge the LUTs and pack the ALMs to ensure that it will not adversely affect the final wirelength. The experimental results show that our packing

REFERENCES

[1] T. Ahmed, P.Kundarewich, J. Anderson, B. Taylor, and R. Aggarwal. "Architecture-Specific Packing for Virtex-5 FPGAs," in Proc. Intl. Symp. on FPGAs, pp. 5-13, 2008.

[2] V. Betz and J. Rose. "VPR: A new packing, placement and routing tool for FPGA research," in Proc. FPL, pp. 213-222, 1997.

[3] E. Bozorgzadeh, S. Memik, X. Yang, and M. Sarrafzadeh. "Routability-driven Packing: Metrics and Algorithms for Cluster-Based FPGAs," Journal of Circuits Systems and Computers, 13:77-100, 2004.

[4] D. Chen, K. Vorwerk, and A. Kennings. "Improving Timing-Driven FPGA Packing with Physical Information," in Proc. FPL, pp. 117-123, 2007.

[5] G. Chen and J. Cong, "Simultaneous Timing Driven Clustering and Placement for FPGAs," in Proc. FPL, pp. 158-167, 2004.

[6] M. Hutton, J. Schleicher, D. Lewis, B. Pedersen, R. Yuan, S. Kaptanoglu, G. Baeckler, B. Ratchev, K. Padalia, M. Bourgeault, A. Lee, H. Kim, and R. Saini, "Improving FPGA performance and area using an adaptive logic module," in Proc. FPL, pp. 135-144, 2004.

[7] Y.-Y. Liang, T.-Y. Kuo, S.-H. Wang, and W.-K. Mak, "ALMmap: Technology Mapping for FPGAs With Adaptive Logic Modules," IEEE Trans. Comput.-Aided Design, vol. 31, no. 7, pp. 1134-1139, Jul. 2012.

[8] J. Luu, J. Anderson, and J. Rose, "Architecture Description and Packing for Logic Blocks with Hierarchy, Modes and Complex Interconnect," in Proc. Intl. Symp. on FPGAs, pp. 227-236, 2011.

[9] A. Marquardt, V. Betz, and J. Rose, "Using cluster-based logic blocks and timing-driven packing to improve FPGA speed and density," in Proc. Intl. Symp. on FPGAs, Monterey, CA, pp. 37-46, 1999.

[10] A. Mishchenko et al. ABC: A System for Sequential Synthesis and Verification. http://www.eecs.berkeley.edu/ alanmi/abc, 2009

[11] A. Singh, G. Parthasarathy, and M. Marek-Sadowksa. "Efficient Circuit Clustering for Area and Power Reduction in FPGAs," ACM Trans. on Design Automation of Electronic Systems, 7(4):643-663, Nov 2002.

[12] J.Z Yan, C. Chu, and W.-K. Mak,"SafeChoice: A Novel Approach to Hypergraph Clustering for Wirelength-Driven Placement," IEEE Trans. Comput.-Aided Design, vol. 30, no. 7, pp. 1020-1033, Jul. 2011.

[13] Stratix II Device Family Overview. http://www.altera.com/literature/hb/stx2/stx2_sii51002.pdf

[14] Maximum weighted matching solver: http://lemon.cs.elte.hu/trac/lemon

978-1-4799-2817-0/14 $31.00 © 2014 IEEE

TABLE II
#CLBs, MINIMUM CHANNEL WIDTH, WIRELENGTH, AND CRITICAL PATH DELAY ON ALM-BASED ARCHITECTURE (EACH CLB CONSISTS OF 8 ALMs) COMPARED TO AAPACK[8].

Bench.	# LUTs	# CLBs			min. channel width			wirelength			delay(10^{-8}s)			ALMpack
		[8]	ALM pack	Impv. %	[8]	ALM pack	Impv. %	[8]	ALM pack	Impv. %	[8]	ALM pack	Impv. %	run time(s)
ex5p	745	54	57	-5.56	58	48	17.24	5372	4503	16.18	0.824	0.698	15.29	5.116
spla	2072	166	187	-12.65	106	86	18.87	27651	24123	12.76	1.4491	1.172	19.12	104.463
alu4	803	65	70	-7.69	62	58	6.45	7395	6956	5.94	0.9479	0.795	16.13	14.285
apex2	1058	88	85	3.41	74	72	2.70	11538	10554	8.53	1.073	1.002	6.62	16.977
apex4	787	65	63	3.08	72	72	0.00	8904	7198	19.16	1.1377	0.851	25.20	10.157
des	555	46	46	0.00	40	38	5.00	4999	4063	18.72	0.4674	0.461	1.37	2.728
ex1010	2703	224	221	1.34	114	94	17.54	35318	32048	9.26	1.1002	1.235	-12.25	85.801
misex3	818	65	66	-1.54	62	60	3.23	7349	7170	2.44	0.8506	0.905	-6.40	12.373
pdc	2417	195	206	-5.64	116	96	17.24	33328	31907	4.26	1.1725	1.341	-14.37	118.227
seq	960	77	76	1.30	74	74	0.00	9011	8804	2.30	0.6801	0.788	-15.87	19.073
bigkey	579	50	50	0.00	44	40	9.09	4560	2796	38.68	0.4346	0.293	32.58	6.308
clma	3911	330	328	0.61	88	74	15.91	41977	31467	25.04	1.399	1.247	10.86	306.903
dsip	689	43	43	0.00	38	34	10.53	3791	3055	19.41	0.3653	0.356	2.55	3.288
diffeq	660	55	54	1.82	44	30	31.82	4251	2628	38.18	0.5957	0.695	-16.67	6.84
elliptic	1795	140	138	1.43	74	50	32.43	14711	9946	32.39	0.8975	0.969	-7.97	62.224
frisc	1797	133	132	0.75	88	74	15.91	18962	14221	25.00	1.1101	1.175	-5.86	58.176
s298	780	62	62	0.00	60	58	3.33	6314	5859	7.21	1.158	1.075	7.17	10.373
s38417	2781	217	215	0.92	54	42	22.22	17742	15217	14.23	1.308	1.335	-2.06	103.922
s38584.1	2504	195	185	5.13	64	44	31.25	18278	14812	18.96	0.8522	0.824	3.31	97.722
tseng	660	51	49	3.92	40	28	30.00	3740	2217	40.72	0.5054	0.586	-15.95	7.448
Avg.				-0.47			14.54			17.97			2.14	

TABLE III
CLBs, MINIMUM CHANNEL WIDTH, WIRELENGTH, AND CRITICAL PATH DELAY ON TRADITIONAL ARCHITECTURE (EACH CLB CONSISTS OF 8 4-LUTs) COMPARED TO T-VPACK[9].

Bench.	# LUTs	# CLBs			min. channel width			wirelength			delay(10^{-8}s)			ALMpack
		[9]	ALM pack	Impv. %	[9]	ALM pack	Impv. %	[9]	ALM pack	Impv. %	[9]	ALM pack	Impv. %	run time(s)
ex5p	892	116	112	3.45	46	42	8.70	8214	7355	10.46	1.605	1.226	23.63	11.485
spla	3016	386	377	2.33	54	52	3.70	30214	28341	6.20	1.894	1.953	-3.14	203.053
alu4	1205	155	151	2.58	42	38	9.52	9842	9111	7.43	1.252	1.491	-19.10	27.938
apex2	1441	187	181	3.21	46	42	8.70	13004	11145	14.30	1.442	1.180	18.16	39.898
apex4	1061	138	135	2.17	50	44	12.0	9825	9006	8.34	1.438	1.410	1.283	10.83
des	1238	156	155	0.64	28	28	0.00	17292	16311	5.67	1.1487	1.722	-49.91	23.637
ex1010	3854	513	496	3.31	72	58	19.44	47525	37386	21.33	2.750	1.7329	36.99	203.12
misex3	1173	151	150	0.66	42	36	14.29	9401	8780	6.61	1.419	1.235	13.02	25.962
pdc	3435	442	430	2.71	62	58	6.45	38730	36819	4.93	2.349	2.694	-14.69	251.936
seq	1361	177	172	2.82	46	42	8.70	12430	11701	5.86	1.367	1.171	-2.80	36.554
bigkey	1146	144	144	0.00	18	12	33.33	14177	11482	19.01	0.5724	0.519	9.33	20.733
clma	5621	706	703	0.42	60	50	16.67	55226	40757	26.20	1.8288	1.796	1.79	958.74
dsip	1368	171	171	0.00	18	12	33.33	14155	9780	30.91	0.6116	0.610	0.26	17.045
diffeq	981	123	123	0.00	22	18	18.18	4671	3289	29.59	0.7795	0.871	-11.74	14.421
elliptic	2050	261	257	1.53	34	28	17.65	17088	11946	30.09	1.2081	1.255	-3.88	81.017
frisc	2282	288	286	0.69	50	40	20.00	19404	16656	14.16	1.3278	1.443	-8.68	102.042
s298	1053	134	133	0.75	38	36	5.26	8746	8347	4.56	1.5905	1.431	10.03	21.781
s38417	4978	623	623	0.00	32	20	37.50	23754	15717	33.83	0.9948	1.031	-3.64	419.22
s38584.1	4497	563	559	0.71	30	20	33.33	24731	13934	43.66	1.6917	0.815	51.82	393.43
tseng	779	98	98	0.00	16	12	25.00	4208	3017	28.30	0.7555	0.691	8.54	8.405
Avg.				1.40			16.59			17.57			3.62	

A Topology-based ECO Routing Methodology for Mask Cost Minimization

Po-Hsun Wu, Shang-Ya Bai, and Tsung-Yi Ho
Department of Computer Science and Information Engineering, National Cheng Kung University, Tainan, Taiwan

Abstract—**Although several Engineering Change Order (ECO) routers had been proposed to obtain a routing solution based on different design objectives, mask re-spin cost still cannot be effectively reduced because the ECO routing problem is handled in a sequential manner. This paper presents a three-stage ECO routing flow which can simultaneously route all ECO nets while considering routing layer minimization. Experimental results demonstrate that our proposed ECO routing flow can effectively reduce the number of changed masks with only negligible wirelength and via overhead.**

I. INTRODUCTION

With the rapid evolution of process technology over the last decade, more complicated design complexity becomes inevitable. As the design complexity becomes more complicated, more circuit failure and constraint violations may occur during later design stage due to increasing design rules. As a result, the circuit redesign iteration is required to correct circuit failure and meet all circuit constraints which is a very tedious and time-consuming procedure. Besides, if circuit redesign is made after the mask has been fabricated, then circuit redesign incurs mask re-spin which leads to very expansive Non-Recurring Engineering (NRE) cost. To reduce NRE cost, Engineering Change Order (ECO) technique is proposed to keep transistor-layer masks intact and only re-spin the less expensive metal-layer masks to perform incremental design changes.

Although many ECO optimization methodologies have been proposed in the literature [5], [8], [9], most of these works focus on how to choose desired spare cells to correct circuit functionality and to improve the circuit performance. After selecting spare cells, simple spare-cell rewiring (i.e., ECO routing) is performed to complete ECO procedure. In order to address the routing issue, some ECO routers have been proposed [5], [8], [9]. Li et al. [8] presented a tile-based ECO router to improve total wirelength and used vias. To address timing issue during ECO routing stage, Lin et al. [9] proposed a topology-aware buffer insertion and GPU-based massively parallel rerouting methodology to maintain circuit performance.

In addition to traditional metrics such as wirelength, via, and timing, it is important to reduce the number of changed masks after ECO procedure since mask re-spin cost is very expensive (each photomask may cost up to one million US dollars or even higher with technology advances [12]). To address this problem, Fang et al. [5] developed a redundant-wires-aware ECO approach to improve circuit timing and reduce the number of changed masks. By reusing all unused wires and dummy metals, rewiring nets has high possibility to cross over different obstacles in a layout such that better routability is achieved and less number of changed masks is also obtained. However, their routing approach is sequentially to route each ECO net which cannot effectively minimize the number of changed masks and the benefit of reducing the number of changed masks is mainly

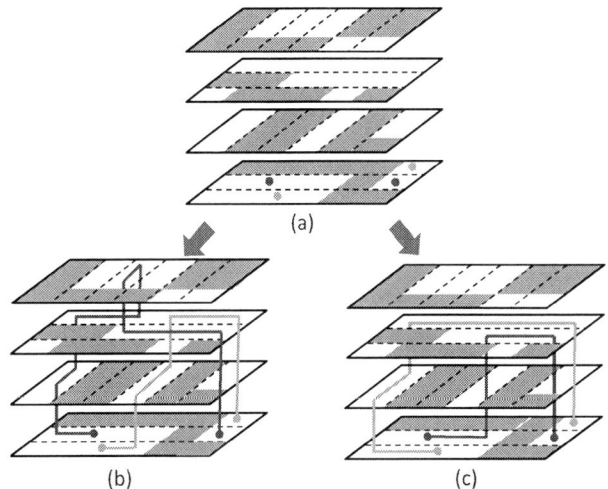

Fig. 1. Comparison of different routing style. (a) An ECO design with two ECO nets. (b) The routing result generated by sequentially routing these ECO nets. (c) The routing result generated by simultaneously routing these ECO nets.

due to the proper spare-cell selection in ECO stage.

Fig. 1 compares two ECO routing results which reroute all ECO nets in sequential and simultaneous manner, respectively. Given an ECO design with four layers and two ECO nets as shown in Fig. 1(a). By routing these ECO nets sequentially, a resulting routing solution using four layers (i.e., it needs to change four masks) is given in Fig. 1(b). However, a more desired routing solution with three changed masks can be obtained by simultaneously routing these ECO nets as shown in Fig. 1(c). In an ECO design, the routing space is mostly occupied by a great number of macro cells and pre-routed nets, simultaneous ECO routing is definitely required to effectively reduce the number of changed masks.

Since no ECO router can globally optimize different ECO nets to minimize the number of changed masks, we propose a new ECO routing flow that is capable to finish ECO routing while reducing total number of changed masks. Initially, a tile-based routing graph, which represents all available routing regions during ECO routing, is constructed. In ECO routing stage, several possible routing topologies for each ECO net are first generated while considering the occurrence of obstacles and pre-routed nets, then an Integer Linear Programming (ILP) model is facilitated to simultaneously determine the respective routing topology of each ECO net to reduce the number of changed masks. Finally, the multi-source multi-sink maze routing approach is used to complete the ECO routing. Since not all possible routing topologies of each ECO net are derived in routing topology generation step, the optimal routing result with the least number of changed masks may not be found. In order

---

978-1-4799-2817-0/14 $31.00 © 2014 IEEE

to further minimize the number of changed masks, a minimum-cost-maximum-flow (MCMF) model is developed to refine the routing result by adjusting the routing path of some ECO nets. In the end of our algorithm, a layout with all corrected ECO net is generated.

A. Our Contributions

In this paper, we proposed a new ECO routing flow considering the minimization of the number of changed masks. Our contributions can be summarized as follows:

- We have proposed a new ECO routing flow and algorithms for ECO routing problem. To our best knowledge, it is the *first* work to simultaneously optimize all ECO nets to reduce the number of changed masks. Based on the proposed ECO routing flow, an ECO routing result with less number of changed masks can be obtained.
- We have developed an ILP model to simultaneously determine the routing topology of each ECO net. Based on the proposed ILP model, the best routing topology of each ECO net can be obtained and routed with a minimum number of changed masks based on the set of derived routing topologies of each ECO net.
- In order to further minimize the number of changed masks, an ECO routing refinement approach is also developed based on the MCMF model. By facilitating the MCMF model, the routing path of some ECO nets can be properly adjusted between different masks to further reduce the number of changed masks. Experimental result demonstrates the proposed ECO routing refinement approach can further reduce the number of changed masks.

The rest of this paper is organized as follows. Section II describes how to generate the corresponding routing graph for a give layout. All detail ECO routing steps and the ECO routing refinement are explained in Section III and Section IV, respectively. Experimental results are given in Section V. Finally, Section VI concludes this paper.

II. TILE-BASED ROUTING GRAPH CONSTRUCTION

In order to efficiently determine the path of each routing net in the following routing steps, it first finds all available routing regions and represent these routing regions by a corresponding routing graph. Basically, there are two kinds of routing graphs: grid-based and gridless-based. In grid-based routing graph, a layout is divided into several routing regions, where each routing region is also called a grid and is represented by a node in the corresponding routing graph. Although a great number of grids can provide detailed routing paths and high feasible solution, it may lead to extreme high routing complexity when the number of nodes in the routing graph is dramatically increasing. In contrast, several adjacent grids are combined into a large block in a gridless-based routing graph and the graph complexity can been significantly decreased. Moreover, the gridless-based routing graph is more flexible to deal with different kinds of design rules [8]. In order to reduce the problem complexity of ECO routing, the gridless-based routing graph is used to represent all available routing regions.

Several kind of gridless-based routing graphs had been proposed in the literature. In [13], the routing graph is constructed by extending lines through the boundaries of each obstacle until they intersect with other obstacles or the boundaries of other routing regions. However, its drawback is inconvenient pre-construction and representation. Although [2], [3], [4] presented an implicit routing graph which can be efficiently constructed and can find the optimal path to represent all available routing regions, it will increase the problem complexity of ECO routing because the extended lines of each obstacle are allowed to cross other obstacles and thus more number of nodes and edges will be contained in the resulting routing graph.

In this paper, the widely-used tile-based routing graph [7] is adopted in our ECO routing flow and is implemented by using the corner-stitching data structure [11]. The tile-based routing graph can be constructed by extending lines through the boundaries of all obstacles until they intersect with other obstacles or the boundaries of other routing regions, where each tile is represented by a respective node and an edge between two nodes indicates two tiles are adjacent. To more accurately estimate the routing usage of each tile under the preferred direction consideration, all adjacent tiles with vertical (horizontal) routing direction are horizontally (vertically) merged, then the maximum horizontally/vertically stripped (MHS/MVS) tiles are applied for the layers with vertical/horizontal routing direction. If two adjacent tiles in the same layer, the capacity of the corresponding edge is estimated based on the abutted length of two tiles. As for two adjacent tiles in different layers, the capacity of the corresponding edge is estimated based on the intersection of projection area of two tiles. For example, the resulting layouts after applying MHS and MVS extraction are depicted in Fig. 2(a) and (b), and the corresponding routing graph of Fig. 2(b) is given in Fig. 2(c). Assume at most n routing paths are allowed to pass through the tile boundary between tiles A and B in Fig. 2(c), then a corresponding constraint is assigned to the edge connecting nodes A and B in the routing graph.

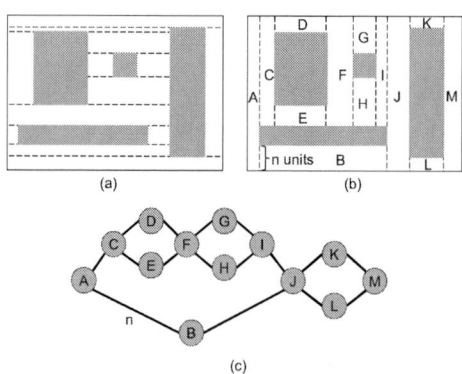

Fig. 2. (a) The resulting layout after applying MHS extraction. (b) The resulting layout after applying MVS extraction. (c) The corresponding routing graph of (b).

In order to further reduce the routing complexity, a Routing Graph Reduction (RGR) technique is used to eliminate all redundant tiles by iteratively reshaping and merging different neighboring tiles. Fig. 3 demonstrates the procedure of redun-

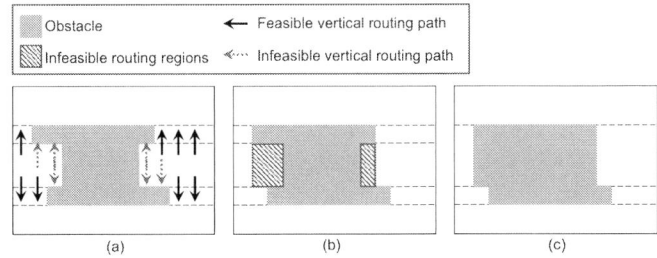

Fig. 3. The procedure of redundant tile elimination. (a) An initial tile layout. (b) The tile layout with infeasible routing regions highlighted. (c) The reduced tile layout.

dant tile elimination and vertical routing direction is assumed. Based on the vertical routing direction, some part of specific tiles are regarded as the infeasible routing regions because no vertical routing path is found to pass through the obstacles. As illustrated in Fig. 3(b), all infeasible routing regions blocked by the obstacles are highlighted. After reshaping these tiles and further merging them with adjacent tiles, two redundant tiles are eliminated as shown in Fig. 3(c) and the respective nodes and edges can be removed from the corresponding routing graph.

III. ECO ROUTING WITH MASK COST MINIMIZATION

The proposed ECO routing algorithm contains three major steps. Initially, all possible routing topologies of each ECO net are derived based on Obstacle-Avoiding Rectilinear Steiner Minimum Tree (OARSMT). Second, the proposed ILP model is used to optimally select the routing topology of each ECO net while minimizing the number of changed masks and total wirelength. Finally, the multi-source-multi-sink maze routing model is facilitated to connect all ECO nets.

A. Routing Topology Generation

Before performing ECO routing, the routing topology of each ECO routing should be efficiently estimated such that the routing congestion can be effectively eliminated to obtain better routability. For ECO routing, the routing path of each ECO net should avoid crossing all pre-placed macros and pre-routed nets, so the OARSMT routing model is adopted and the state-of-the-art OARSMT construction algorithm [10] is used to generate the routing topology of each ECO net. In order to further improve the routing quality, several OARSMTs are derived for each ECO net.

A naive approach of deriving multiple OARSMTs is to fast construct an OARSMT with shortest wirelength, then generate other OARSMTs by perturbing the routing path of the original derived OARSMT. However, the resulting OARSMTs may have the similar routing topologies and have worse routability in the occurrence of routing obstacles. To avoid the above drawbacks, we develop a congestion-aware OARSMT generation procedure which dynamically generates new obstacles during constructing different OARSMTs. Based on the proposed approach, the resulting OARSMTs have various kinds of routing topologies and have better routability in the occurrence of routing obstacles. In our proposed congestion-aware OARSMT generation approach,

we assign a corresponding parameter, t_i, with initial value 0 for each tile i. If the obtained OARSMT pass through tile i, then increase t_i by 1. Finally, the tile with the maximum parameter value is transformed into an pseudo-obstacle in later OARSMT generation procedure.

The detail steps of congestion-aware OARSMT generation procedure are described in the following. Initially, a respective OARSMT for each ECO net is generated. Based on the set of derived OARSMTs, the parameter value of every tile is updated accordingly and the tile with the maximum parameter value is transformed into an pseudo-obstacle in later OARSMT generation procedure. However, no new OARSMT is derived for some ECO nets if their previous OARSMT do not pass through the obstacles and pseudo-obstacle. In such case, some tiles are randomly selected and transformed into the random pseudo-obstacles to derive the new routing topologies of these ECO nets. The above steps are iteratively performed until no new routing topologies can be further generated. To generate various kind of routing topologies to more effectively alleviate the routing congestion and improve the routability, all pseudo-obstacles and random pseudo-obstacles are reserved after each iteration. It should be noted that the routing path and the number of changed masks of each routing topology are stored for further reference.

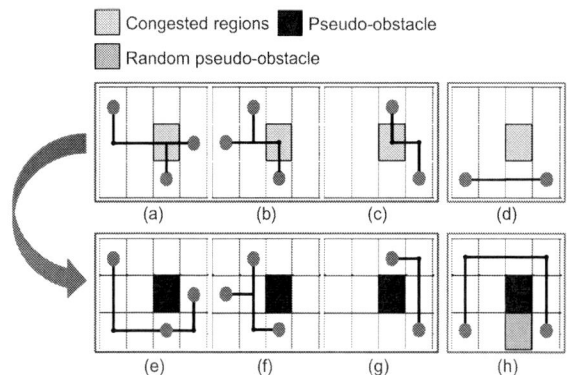

Fig. 4. The procedure of generating new routing topology for four given ECO nets. (a)-(d) The initial generated routing topologies of four ECO nets. (e)-(g) Three new derived routing topologies for the nets in (a)-(c) by avoiding crossing the pseudo-obstacle. (h) The new derived routing topology for the net in (d) by avoiding crossing both pseudo-obstacle and random pseudo-obstacle.

Fig. 4 shows the procedure of generating new routing topology for four given ECO nets as shown in Fig. 4(a)-(d) and the obstacles are ignored in these figures to give a more clear explanation. Based on the initial routing result, the most congested region can be estimated as highlighted in the figure and is transformed into the pseudo-obstacle. Three new routing topologies of (a)-(c) can be generated by avoiding passing through the pseudo-obstacle as illustrated in Fig. 4(e)-(g). However, the routing path in Fig. 4(d) does not pass through the pseudo-obstacle, then no new routing topology is thus derived. In order to derive new routing topology for the net in Fig. 4(d), a random pseudo-obstacle is formed beside the original pseudo-obstacle and a corresponding routing topology can be obtained as given in Fig. 4(h).

978-1-4799-2817-0/14 $31.00 © 2014 IEEE   509

B. Routing Topology Selection

After generating several possible routing topologies of each ECO net, the routing topology of each ECO net is determined based on the routing graph in Fig. 2(c) while minimizing the number of changed masks. Since the routing space have been occupied by a great number of pre-routed nets and pre-placed macros, only limited routing space can be facilitated to perform ECO routing. Besides, different routing orders and different selected routing topologies which significantly affect the routing quality should also be considered. In order to globally and optimally solve the routing topology selection problem, the ILP model is adopted as our global routing engine. Although several ILP model [1] is proposed to finish the global routing task, these works only focus on minimizing total wirelength. In order to solve the routing topology selection problem while minimizing the number of changed masks, the proposed version of objective function and several added constraints are used.

Given a set of n ECO nets $N_{ECO} = \{N_{ECO}^1, N_{ECO}^2, ..., N_{ECO}^n\}$ associated with m routing topologies, and a set of z tile boundaries $TB = \{TB_1, TB_2, ..., TB_z\}$ associated with a corresponding capacity constraint, the following objective function and ECO routing constraints can be derived to complete global routing while minimizing the number of changed masks:

Objective function:

$$Minimize\ L_{used} \tag{1}$$

ECO routing constraints:

$$\sum_{j=1}^{m} x_{ij} = 1, \forall N_{ECO}^i \in N_{ECO} \tag{2}$$

$$\sum_{i=1}^{n}\sum_{j=1}^{m}\sum_{k=1}^{z} (a_{ij}^k \times x_{ij}) \leq c_k, \forall N_{ECO}^i \in N_{ECO} \tag{3}$$

$$L_{used} \geq \max_{j=1 \to m} (l_{ij} \times x_{ij}), \forall N_{ECO}^i \in N_{ECO} \tag{4}$$

$$x_{ij} \in \{0, 1\} \tag{5}$$

where Equation (1) is the objective function for minimizing the number of changed masks, L_{used}, which is the maximum number of used layers of all routing nets. The remaining equations are designed for ECO routing constraints. Equation (2) is used to restrict only one routing topology of each ECO net is chosen, where x_{ij} is a Boolean variable which equals to 1 if the jth routing topology of ECO net N_{ECO}^i is chosen. Equation (3) is used to make sure all capacity constraints are satisfied, where a_{ij}^k is a Boolean constant which equals to 1 if the jth routing topology of ECO net N_{ECO}^i passes through the tile boundary TB_k and c_k is its corresponding capacity. Equation (4) is the function used to compute the number of changed masks for the current routing solution, where l_{ij} denotes the number of changed masks by choosing the jth routing topology of ECO net N_{ECO}^i. After solving the ILP model by using the above objective function and ECO routing constraints, a global routing result with mask cost minimization can be obtained.

C. Detailed Routing

After obtaining the global routing result, the detailed routing algorithm is performed to connect all I/O pins of spare cells. Various detail routing algorithms have been proposed to deal with detail the routing task. In order to effectively and efficiently generate a better routing result, the state-of-the-art multi-source-multi-sink maze routing algorithm is implemented.

IV. Network Flow-based ECO Routing Refinement

Despite the ILP model is used to generate the routing solution with the least number of changed masks based on a set of derived routing topologies of all ECO nets. However, not all possible routing topologies of each ECO routing are generated in routing topology generation step, the optimal routing solution with the least number of changed masks cannot be found even though the ILP model is adopted. Therefore, in this section, an ECO routing refinement approach is proposed to further reduce the number of changed masks.

Given a circuit design using L_{used} layers for ECO routing, the key idea of ECO routing refinement is to move each routing segment at the layer L_{used} (i.e., the source layer) to the routing tracks at the layer L_{used} - 2 (i.e., the target layer), where L_{used} must be greater than 2, then the number of changed masks can be further minimized if each routing segment at the source layer can be successfully placed at the target layer. For each segment at the source layer is regarded as a target segment to be moved to the target layer. Based on the routing topology (U-shaped, L-shaped, and I-shaped) of each target segment, a corresponding candidate region at the target layer is formed to find the routing tracks to place the target segment.

To minimize total routing wirelength and routing congestion in adjacent layer, the selected routing track at the target layer should locate in the candidate regionFor the U-shaped and L-shaped routing topologies , the candidate region should locate on the side, which has the connected segments, of the target segment such that shorter wirelength can be obtained after routing refinement. In contrast, for the Z-shaped routing topology, the candidate region can be on both sides of the target segment. Since the chosen routing track at the target layer may contain several pre-routed nets, these pre-routed nets have to be ripped up and rerouted to release the chosen routing track. Finally, the target segment can be moved to the chosen routing track.

A naive approach is to sequentially move each ECO routing segment to any one unoccupied routing track. However, the number of changed masks may not be effectively minimized by using the above method since all available routing tracks are unevenly distributed throughout the layout. To effectively reduce the number of changed masks, an MCMF model is developed to simultaneously distribute all target segments and all rerouted segments to all available routing tracks at the target layer.

The proposed MCMF model is shown in Fig. 5, where each node Seg_x represents a corresponding target segment (or rerouted segment) and each node T_y represents an available routing track at the target layer which can be easily calculated based on the space between different routing segments. There is an interconnection between two nodes, Seg_x and T_y, if T_y is located in the candidate region of Seg_x. When Seg_x is a rerouted segment, the respective candidate region can be

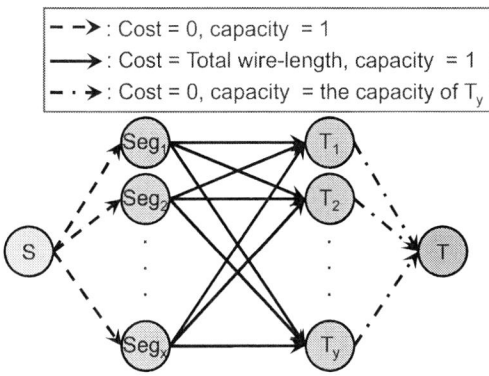

Fig. 5. The proposed MCMF model.

easily obtained. Since different routing wirelength are obtained by placing different segments Seg_x to different routing tracks T_y, the resulting wirelength of placing Seg_x to T_y is the cost of the respective edge between two nodes, Seg_x and T_y. Moreover, each segment is needed to be assigned to one of the routing tracks, so the capacity of the respective edge between S and Seg_x is 1. To avoid routing capacity violation, the available routing space of each routing track T_y is the capacity of the respective edge between T_y and T. After distributing all segments to different tracks at the target layer based on the MCMF model, the multi-source multi-sink maze routing algorithm can be applied to connect these segments. If some segments cannot be successfully routed, then the candidate region of some target segments are properly adjusted (i.e., add/delete an edge between two nodes, Seg_x and T_y) such that all segments can be successfully routed.

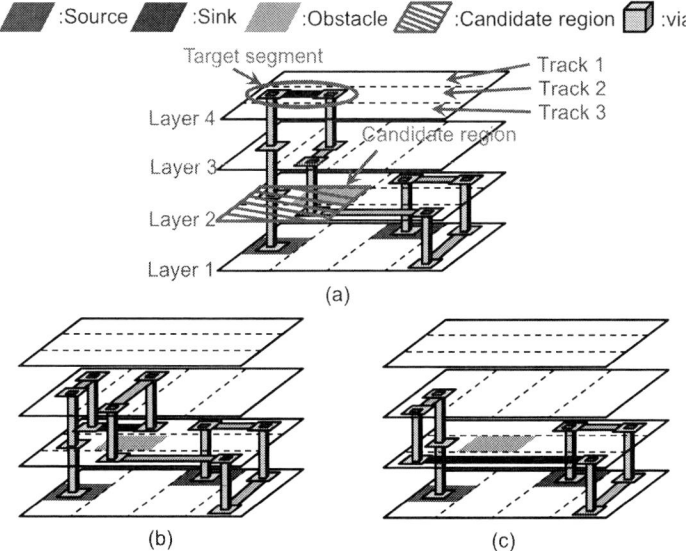

Fig. 6. (a) The original routing result. (b) The refinement result of (a) by moving the target segment to the track 1 at the target layer. (c) The refinement result of (a) by moving the target segment to the track 3 at the target layer.

Fig. 6 demonstrates an example of moving an ECO routing segment from the source layer to one of the routing tracks at the target layer. Initially, only one target segment is routed at the source layer (layer 4) and the target segment is tried to be moved to the target layer (layer 2) to further reduce the number of changed masks as shown in Fig. 6(a). Two possible refinement results are given in Fig. 6(b) and (c), respectively. As mentioned before, it tries to move the target segment to the target layer while maintaining similar wirelength, only the routing tracks in candidate region are considered. Therefore, the target segment is moved to the track 3 at the target layer as illustrated in Fig. 6(c).

V. EXPERIMENTAL RESULT

The proposed ECO router was implemented in the C++ programming language on a 2.0GHz Linux machine with 16GB memory. In our experiment, five industry routing testcases were used to test our ECO router. All testcases were originally routed by the state-of-the-art commercial tool, SoC Encounter [6], then 100 of pre-routed nets in each testcase was ripped up as the ECO nets and all remaining nets was regarded as the obstacles. TABLE I presents the routing comparison before and after applying our proposed ECO routing flow. As seen from the table, the number of used routing layers can be successfully reduced in three ($dsp1\_dfm$, $dsp2\_dfm$, $risc1\_dfm$) out of five testcases and all ECO nets are successfully routed with only negligible routing overhead.

To demonstrate the effectiveness of our developed routing refinement technique, we compared the routing result with and without refinement step as shown in TABLE II. Since the number of routing layers in three testcases ($dsp1\_dfm$, $dsp2\_dfm$, $risc1\_dfm$) can be further reduced as shown in TABLE I, we chose these testcases to perform our experiment. By facilitating the developed routing refinement technique, the number of used routing layers in two ($dsp1\_dfm$, $risc1\_dfm$) of these testcases can be further reduced with nearly the same routing wirelength and number of vias.

In industry practice, a common ECO routing methodology for minimizing the number of routing layers is try to route all ECO nets under a decreased number of routing layers. For example, given a circuit design with n routing layers, then it tries to route all ECO nets in $n - 1$ layers to further minimize the number of routing layers. In such methodology, the approach, which routes all ECO nets sequentially, cannot effectively reduce the number of routing layers. To show the effectiveness of our proposed algorithm in reducing the number of routing layers, we also compared our ECO router with a maze router under the same number of used routing layers as given in TABLE III. Since the number of routing layers in three testcases ($dsp1\_dfm$, $dsp2\_dfm$, $risc1\_dfm$) can be further reduced as shown in TABLE I, we perform our experiments in these testcases. Compared with our proposed ECO router, some nets in each testcase cannot be successfully routed by using the maze router as shown in TABLE III. Therefore, our proposed ECO router has better routability under the same routing resource such that less number of used routing layers can be obtained.

978-1-4799-2817-0/14 $31.00 © 2014 IEEE

TABLE I
ROUTING COMPARISON BEFORE AND AFTER APPLYING ECO ROUTING.

Testcases	# of Nets	# of ECO Nets	# of Total Layers	Before ECO Routing			After ECO Routing			
				# of Used Layers	Wirelength (μm)	# of Vias	# of Used Layers	Wirelength (μm)	# of Vias	Runtime (Sec)
dma_dfm	13156	100	6	4	3156000	268	4	3298400	283	537
dsp1_dfm	28347	100	6	5	1841600	370	4	2139600	406	924
dsp2_dfm	28331	100	6	6	1775200	376	5	1913200	391	1029
risc1_dfm	33934	100	6	6	1674800	409	5	1735600	415	1777
risc2_dfm	33934	100	6	4	1843600	390	4	2005200	421	2055
Norm.				1.00	1.00	1.00	0.89	1.06	1.05	-

TABLE II
COMPARISON OF ROUTING RESULTS WITHOUT AND WITH REFINEMENT.

Testcases	Without Refinement			With Refinement		
	# of Used Layers	Wirelength (μm)	# of Vias	# of Used Layers	Wirelength (μm)	# of Vias
dma_dfm	4	3298400	283	4	3298400	283
dsp1_dfm	5	2077800	389	4	2139600	406
dsp2_dfm	5	1913200	391	5	1913200	391
risc1_dfm	6	1693200	414	5	1735600	415
risc2_dfm	4	2005200	421	4	2005200	421
Norm.	1.00	1.00	1.00	0.87	1.01	1.01

TABLE III
COMPARISON OF ECO ROUTING UNDER LAYER LIMITATION
BETWEEN MAZE ROUTER AND OUR PROPOSED ECO ROUTER.

Testcases	# of Given Layers	Maze Router # of Failed Nets	Ours # of Failed Nets
dsp1_dfm	4	31	0
dsp2_dfm	5	27	0
risc1_dfm	5	44	0

VI. CONCLUSION

In this paper, we have proposed a new three-stage ECO routing flow and algorithms for ECO routing problem for mask cost minimization. First, we generate several possible routing topologies for each ECO net while considering the occurrence of obstacles and pre-routed nets. Then, the ILP model is facilitated to simultaneously determine the respective routing topology of each ECO net and the multi-source multi-sink maze routing approach is used to complete the ECO routing. Finally, the MCMF model is developed to further refine the routing result. Experimental result shows that the proposed ECO routing flow can effectively reduce the number of changed masks with only negligible routing overhead.

VII. ACKNOWLEDGMENT

This work was partially supported by the Taiwan National Science Council under grant No. NSC 102-2221-E-006-287-MY3 and 102-2220-E-006-007 and the Ministry of Education, Taiwan, R.O.C. under the NCKU Aim for the Top University Project Promoting Academic Excellence and Developing World Class Research Centers.

REFERENCES

[1] L. Behjat and A. Chiang. Fast integer linear programming based models for VLSI global routing. In *IEEE International Symposium on Circuits and Systems*, pages 6238–6243, 2005.

[2] J. Cong, J. Fang, and K.-Y. Khoo. An implicit connection graph maze routing algorithm for ECO routing. In *IEEE/ACM International Conference on Computer-Aided Design*, pages 163–167, 1999.

[3] J. Cong, J. Fang, and K.-Y. Khoo. DUNE: A multi-layer gridless routing system with wire planning. In *IEEE/ACM International Conference on Computer-Aided Design*, page 12, 2000.

[4] J. Cong, J. Fang, and K.-Y. Khoo. DUNE-a multilayer gridless routing system. *IEEE Transactions on Computer-Aided Design of Integrated Circuits and Systems*, 20(5):633–647, 2001.

[5] S.-Y. Fang, T.-F. Chien, and Y.-W. Chang. Redundant-wires-aware ECO timing and mask cost optimization. In *IEEE/ACM International Conference on Computer-Aided Design*, pages 381–386, 2010.

[6] Cadence, Inc. *SoC Encounter*. [Online]. Available: http://www.cadence.com/products/di/soc_encounter.

[7] Y.-L. Li, Chen H.-Y., and Lin C.-T. NEMO: A new implicit-connection-craph-based gridless router with multilayer planes and pseudo tile propagation. *IEEE Transactions on Computer-Aided Design of Integrated Circuits and Systems*, 26(4):705–718, 2007.

[8] Y.-L. Li, J.-Y. Li, and W.-B. Chen. An Efficient Tile-Based ECO Router Using Routing Graph Reduction and Enhanced Global Routing Flow. *IEEE Transactions on Computer-Aided Design of Integrated Circuits and Systems*, 26(2):345–358, 2007.

[9] Y.-H. Lin, Y.-J. Lo, J.-S. Tong, W.-H. Liu, and Y.-L. Li. Topology-aware buffer insertion and GPU-based massively parallel rerouting for ECO timing optimization. In *ACM/IEEE Asia and South Pacific Design Automation Conference*, pages 437–442, 2012.

[10] C.-H. Liu, I.-C. Chen, and D. T. Lee. An efficient algorithm for multi-layer obstacle-avoiding rectilinear Steiner tree construction. In *ACM/IEEE Design Automation Conference*, pages 613–622, 2012.

[11] J. K. Ousterhout. Corner Stitching: A Data-Structuring Technique for VLSI Layout Tools. *IEEE Transactions on Computer-Aided Design of Integrated Circuits and Systems*, 3(1):87–100, 1984.

[12] Reports of the International Technology Roadmap for Semiconductors,. 2012, http://www.itrs.net/.

[13] S. Q. Zheng, Joon Shink Lim, and S. S. Iyengar. Finding obstacle-avoiding shortest paths using implicit connection graphs. *IEEE Transactions on Computer-Aided Design of Integrated Circuits and Systems*, 15(1):103–110, 1996.

BOB-Router: A New Buffering-Aware Global Router with Over-the-Block Routing Resources Optimization

Yilin Zhang[1], Salim Chowdhury[2] and David Z. Pan[1]
[1] Department of ECE, University of Texas at Austin, Austin, TX, USA
[2] Oracle, Austin, TX, USA

Abstract—In this paper, we propose a new global router, BOB-Router, endowed with the ability to use over-the-block routing resources to the greatest extent in addition to traditional routing concepts of minimizing wirelength, via count and overflow. In previous global routing formulations, the routing resources over the IP blocks were either dealt as routing blockages leading to a significant waste, or simply treated in the same way as outside-the-block routing resources, which violates the slew constraints and thus fail buffering. Utilizing over-the-block routing resources could dramatically improve the routing solution, yet requires special attention, since the slew, affected by different RC on different metal layers, must be constrained by buffering and is easily violated. Moreover, even all nets are slew-legalized, the routing solution could still suffer from heavy congestion problem. For the first time, BOB-Router tries to solve the over-the-block global routing problem through minimizing overflows, wirelength and via count simultaneously without violating slew constraints. BOB-Router generates a slew-legalized initial solution followed by a Lagrangian-multiplier-based pricing phase and RC-constrained A* search to help explore new buffering-aware topologies on all metal layers. Our experimental results show that BOB-Router completely satisfies the slew constraints and significantly outperforms the obstacle-avoiding global routers in terms of wirelength, via count and overflows.

I. INTRODUCTION

As semiconductor technology keeps scaling into deeper sub-micron domain, interconnect delay becomes more critical in determining chip performance. Routing is one of the most important stages regarding performance of chip interconnection. The CEDA-sponsored ISPD Global Routing Contests [2] and [3] attract attention from dozens of academic and industrial participants. Inspired by the competitions, many high-performance global routers are published, including but not limited to, FastRoute 3.0 [22], FastRoute 4.0 [20], BoxRouter 2.0 [7], NTUgr [6], NTHU-Route [10], NTHU-Route 2.0 [5], GRIP [19], FGR [18], MaizeRouter [16], Archer [17] and NCTU-GR [9].

Those global routers can be roughly divided into two categories: sequential and concurrent algorithms. Sequential works [22], [20], [6], [10], [5], [16], [17], [9] route the nets based on heuristic rip-up and reroute (RNR) techniques, which tend to run 2D global routing followed by layer assignment. On the other hand, works such as [19] and [18] directly address the problem by running a full 3D global routing.

Meanwhile, extensively using IP-blocks to shorten turnaround time nowadays packs SOC designs with IP blocks or macros. To avoid routing over those blocks, obstacle-avoiding rectilinear Steiner minimum tree (OA-RSMT) problem has been actively studied over the years (e.g. [4],

[12]–[14]). However, completely avoiding those routing areas will result in significant underutilization of high-level metal layers which is the key to save power and close timing. To tackle that issue, new ideas of intelligently utilizing part of, instead of completely avoiding, the over-the-block routing resources with buffering awareness are proposed by [21] and [11] as BOB-RSMT [21] problem, as well as studied as scenic constraints in [15].

Since the guidance from two ISPD Global Routing Contests are similar, most published modern routers aim at the same problem: minimizing wirelength and via count in addition to alleviating congestion. However, the global routing problem has never been touched upon to not only consider wirelength, vias and overflows, but also properly use over-the-block routing resources. Studying this new problem is essential as to shorten the design cycle and improve the chip quality. If over-the-block routing resources are treated the same as that for outside-the-block, long nets over the block will fail buffering, leading to additional manual work; whereas if over-the-block routing resources are totally avoided, less remaining routing resources will significantly deteriorate the quality of the routing solution.

Our key contributions include:

1) We study the over-the-block global routing problem for the first time, providing global routing solution with overflows, wirelength, via count and buffering-awareness considered simultaneously.

2) We improve BOB-RSMT algorithm [21] by addressing its two limitations. Then we apply modified BOB-RSMT algorithm for our initial legal inside-tree generation.

3) For any block with overflow, in each iteration we evolve new topologies from inside trees confined within that block, with less cost associated with congestion, wirelength and via count,

4) We conduct Lagrangian-multipliers-based cost function to reflect the weighted impact from all generated topologies. It turns out that topologies with less cost will have more impact on determining the cost of covered edges.

5) An RC-constrained A* search is proposed to help incrementally evolve new topologies with minimum cost while meeting slew constraints.

The rest of paper is organized as follow. We first introduce basic concepts of inside trees, slew model and problem formulation in Section II. Our over-the-block routing algorithm will be presented in Section III, which includes three subsections.

978-1-4799-2817-0/14 $31.00 © 2014 IEEE

Section III-A discusses how to modify BOB-RSMT to generate initial legal inside topologies. Section III-B illustrates the process of incrementally evolving new topologies according to Lagrangian-multiplier-based cost function and RC-constrained A* search. Experimental results are shown and analyzed in Section IV, followed by conclusions in Section V.

II. PRELIMINARIES

A. Basic over-the-block concepts

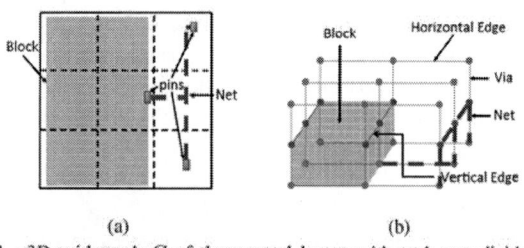

(a)   (b)

Fig. 1. 3D grid-graph G of three metal layers with each one divided into 3*3 global routing bins

In global routing, the chip is partitioned into rectangular global routing bins where a 3D grid-graph $G = (V, E)$ is used to model the multi-layer design. As depicted in Fig. 1(a) and Fig. 1(b), each global routing bin is a vertex $v \in V$. The boundary between two adjacent global routing bins on the same layer is modeled as an edge $e \in E$ with a capacity c_e reflecting the maximum routing resources between the cells.

After placement, the chip is packed with IP blocks or macros which occupy the low metal layers and forbid buffer insertion over the IP blocks. In our formulation, we set $B = \{b_1, b_2, \ldots, b_m\}$ as the set of blocks. Each block is modeled by a box in the 3D grid-graph G as the shadowed part in Fig. 1.

A set of multi-terminal nets $N = \{n_1, n_2, \ldots, n_k\}$ is required to be connected in the 3D graph G. The tree topology of each net n_i will enter and leave the blocks in the graph, which divides the whole tree topology into a set of outside trees TO_i and a set of inside trees TI_i.

For any inside tree t, the leaf nodes of t are on the boundaries of a block. Among all leaf nodes, one must be driving the signal and others are receiving. We name these leaf nodes that receive signals *escaping points* (EP), and the set of escaping points for t is $EP^t = \{EP_1^t, EP_2^t, \ldots, EP_{|EP^t|}^t\}$.

We use the same model as in [21] to check if any inside tree satisfies slew constraints. In our formulation, every inside tree is forced to be legal (i.e. satisfy slew constraints).

B. Problem Formulation

Matrices including wirelength, via cost and total overflow (TOF) are used to evaluate our routing solution. TOF is preferred to be zero since slightly overflowed global routing can still make detail routing considerably more difficult.

Our proposed buffering-aware global router will connect each net in N with the target of minimizing total wirelength in addition to reducing TOF. Over-the-block trees have to satisfy

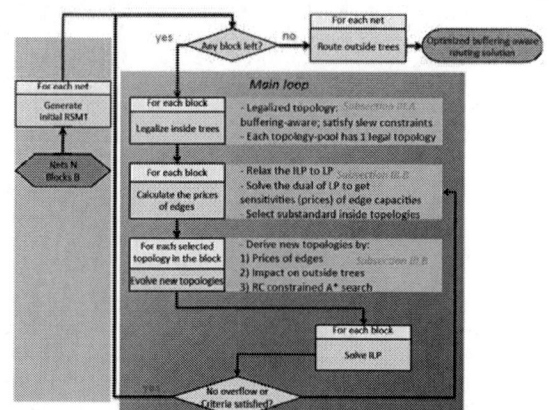

Fig. 2. Overall flow of BOB-Router

the slew constraints which ensure that every topology has feasible buffering solutions.

III. BOB-ROUTER ALGORITHMS

The overall flow of BOB-Router approach is depicted in Fig.2. The procedures in the "Main loop" frame consists of routing algorithm for inside trees while the rest part is composed of initial legal RSMT generation along with routing for outside trees.

In the BOB-Router problem formulation, any inside tree has to satisfy slew constraints to accomodate buffering. Due to this extra requirement, routing for inside trees becomes more challenging than that for outside trees. Our BOB-Router will route inside trees ahead of outside trees, as we algorithmically emphasize on inside-tree routing which prefers topologies with least downside or even betterment on the cost of outside-tree routing.

To avoid simultaneously coping with wirelength, via count, overflow and slew constraints in inside-tree-routing problem, we decouple the slew constraints by legalizing all topologies first and making sure every following step during the entire inside-tree routing will not violate the slew constraints. This decoupling process includes two steps. First, since the initial inside trees could violate slew constraints, we apply an EP-movement-based legalization procedure modified from [21] to legalize any illegal inside topology with minimum wirelength penalty. Second, during the "evolve new topologies" step (shown in Fig.2) in the inside-tree routing, we use an RC-constrained A* search to ensure that each operation during new topology evolution will not break the slew constraints.

A. Generate Legal Initial Topologies

We modify BOB-RSMT algorithm in [21] to generate legal initial inside trees. We will first briefly review BOB-RSMT approach, and then study two insufficiencies of BOB-RSMT. Lastly, we will present our modified BOB-RSMT (BOB-RSMT-m) to provide upgraded initial legal inside trees.

In BOB-RSMT, an initial RSMT is generated for each net by FLUTE [8]. Yet for any generated RSMT, it might violate

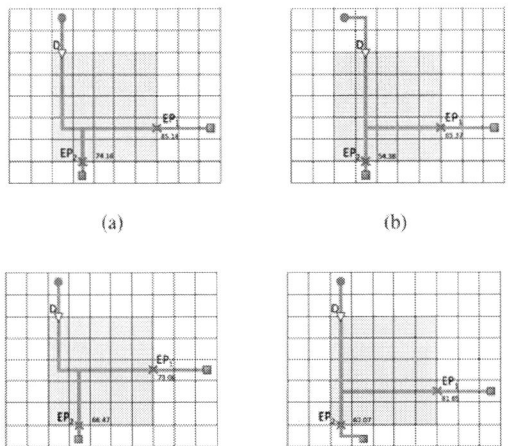

Fig. 3. Best move selection (a) shows an illegal inside tree. (b), (c) and (d) exhibit and evaluate the best single-unit move from the driver, EP_1 and EP_2 respectively.

slew constraints and become illegal. First of all, in order to legalize any illegal inside tree t in the illegal RSMT, all illegal escaping points (EP) of t will be sorted according to their slew violations. Next, in every iteration the first illegal escaping point EP_1^t with the worst slew violation based on sorting will be legalized. To legalize slew for EP_1^t, each escaping point from $\{EP_1^t, EP_2^t, \ldots EP_{|EP^t|}^t\}$ may slide to a different position by taking a combination of primitives with inside Steiner nodes updated as well. These primitives include parallel sliding which shifts the connected branch, perpendicular sliding which stretches the connected branch and EP merging with one of the neighboring EP. ILP is applied to select the final location for each EP from its *possible point set*. Finally, the location of EP_1^t will be fixed and next iteration is launched to legalize the next illegal escaping point EP_2^t. This process will be iterated until all escaping points satisfy slew constraints.

BOB-RSMT approach efficiently generates a topology satisfying slew constraints, however, it has two limitations. First, movement of the driver for an inside tree is not considered. Second, when two branches at the opposite end of the driver move simultaneously, slew improvement may be underestimated.

To address those two limitations, we keep the optimization primitives but replace ILP with a greedy approach. Instead of evaluating each possible point and applying ILP to selection, we assess every *single-unit move* from all EPs and the driver, and select the best move. For example Fig.3(a) presents an illegal inside tree, while Fig.3(b), Fig.3(c) and Fig.3(d) give the resulted topologies assuming the best moves from the driver, EP_1 and EP_2 are selected respectively. In order to directly show amount of slew improvement in Fig.3, we set one for unit R and unit C, along with zero for buffer-output resistance, buffer-input capacitance and output slew in our slew model. As we can see, the best slew improvement occurs in Fig.3(b), where the driver reduces worst slew by 19.88 slew units with single-unit move. However, in BOB-RSMT, this

solution cannot be found since the move of the driver is not considered.

The other benefit from proposed BOB-RSMT-m is that it accurately catches the slew difference when multiple branches at the opposite end of the driver moving simultaneously. In this rare case, BOB-RSMT approach might disregard slew improvement from antenna clearance and overestimate slew improvement from branch sliding. As is shown in Fig.4, moving EP_1 to the right by one unit (Fig.4(b)) can improve the worst slew for 9.89 slew units while moving EP_2 in the same way (Fig.4(c)) will improve the worst slew for 3.30 slew units. If the ILP in BOB-RSMT is applied to choose both slides, the total improvement will be summed up to 13.19 slew units in an incorrect way. The actual improvement in Fig.4(e) is 11.98 slew units which consists of the slew improvement from moving EP_1 to the right by one unit and removing of antenna segment circled in 4(d).

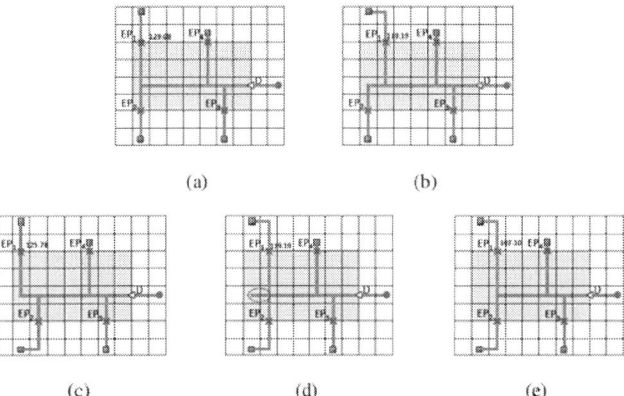

Fig. 4. Slew calculation method in BOB-RSMT and BOB-RSMT-m. (a) shows an illegal inside tree. (b) and (c) exhibit and evaluate the best single-unit move from EP_1 and EP_2 respectively. (d) and (e) illustrates if both EP_1 and EP_2 move.

B. Evolving Legal Congestion-Aware Min-Cost Topologies

The initial-inside-tree legalization guarantees one legalized topology for each inside tree. Placing these legal topologies simultaneously within each associated block could cause congestion problem. To resolve this issue, our approach uses the generated legalized topologies as seeds, giving birth to more legal congestion-aware topologies with less cost than current topologies. Finally, one topology will be chosen for each inside tree to achieve least overflow and cost.

We keep our topologies in Steiner tree structures instead of decomposing them into 2-pin nets in that (1) Steiner tree structures have more flexibility with unfixed Steiner points while 2-pin nets have to connect specified end points; (2) Steiner tree structure allows for tracking non-linear slew calculation over the entire tree, which is improbable for decomposed 2-pin nets. In normal global routing problem, it is non-trivial regarding how to come up with congestion-aware Steiner tree topologies with minimum cost. However, the Steiner tree topologies we demand for our inside trees have to satisfy additional slew constraints.

1) Formulations: First, we build an ILP formulation to describe the routing problem in each block. The ILP formulation contains no slew constraints, as every topology presented in the ILP formulation is legal.

$$\min. \sum_{i=1}^{n} \sum_{t \in \zeta_i} X_{it} W_{it} + M \sum_{i=1}^{n} S_i \qquad (1)$$

$$\text{s.t. } \forall i, S_i + \sum_{t \in \zeta_i} X_{it} = 1 \qquad (1a)$$

$$\sum_{i=1}^{n} \sum_{t \in \zeta_i} X_{ite} <= C_e \quad \forall e \prec b, b \text{ is block} \qquad (1b)$$

$$X_{it} \in \{0,1\} \qquad \forall i \in \{1,2,\ldots,n\} \qquad \forall t \in \zeta_i$$

$$S_i \in \{0,1\} \qquad \forall i \in \{1,2,\ldots,n\}$$

$TI = \{T_1, T_2, \ldots, T_n\}$ is the set of inside trees within block b, and ζ_i in the formulation is the collection of all topologies for T_i. For each Steiner tree topology $t \in \zeta_i$, parameter W_{it} represents the overall cost of the topology, including both wirelength and vias. Variable S_i denotes the routability of T_i; if S_i is positive, the inside tree T_i cannot be routed with available Steiner tree topologies. $\sum_{i=1}^{n} \sum_{t \in \zeta_i} X_{ite}$ contains the amount of routing resources demanded on every edge e, which is required to maintain under the edge capacity vector C_e.

In order to minimize overflow, MS_i is used in the objective function to penalize any unroutable inside tree T_i. Parameter M is a predefined large number which is greater than the wirelength of any possible Steiner tree in the chip. Solving ILP formulation (1) guarantees no inside-overflow and minimizes total cost with maximum number of routed inside trees.

The ILP formulation has the following two purposes in our approach: i) to select one topology for each inside tree to check if overflow-free solution could be achieved at the end of each iteration, ii) the dual problem of relaxed LP could provide cost of each edge.

2) Pricing the Edges: Before solving the ILP and fixing the topologies in current iteration, more Steiner tree topologies, instead of the initial one solely, are wanted to effectuate least TOF and cost routing solution. We use sensitivity analysis on the edge capacity constraints to price each edge, which provides a guidance for the evolution of new Steiner tree topologies from current topologies.

Different edges on different layers have various values in the routing since some of them are in congested area while some are not. We calculate *price* to describe the potential overflow on each edge. To obtain the prices for edges, we first relax the ILP formulation (1) into an LP formulation by relaxing binary variables $\{X_{ij}\}$.

The relaxation on binary variables $\{X_{ij}\}$ splits the constraint of choosing only one topology for each inside tree into a set of fractional numbers indicating several potentially preferred topologies. A topology avoiding congested area and costing less wirelength and vias will be preferred and assigned positive X_{ij} which depends on the quality of the topology.

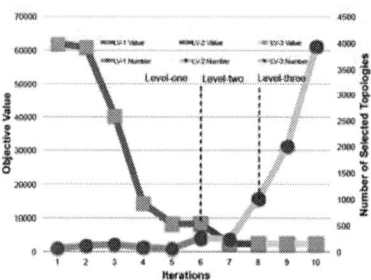

Fig. 5. Progression of objective value and number of selected "to-be-evolved" topologies over optimization rounds for one block on ADAPTEC1

The price of each edge comes from the dual of this LP formulation exhibited in (2). The variable λ_i is the Lagrange multiplier associated with relaxed topology-selection constraint (1a) for T_i and ρ_e is the Lagrange multiplier associated with relaxed capacity constraint (1b) for edge e. According to complementary slackness theorem, for optimal primal variables $X_i^*, i \in \{1,2,\ldots,n\}$ and optimal dual variable ρ_e^*, there exists $\rho_e^* * (\sum_{i=1}^{n} \sum_{t \in \zeta_i} X_{ite} - C_e) = 0$. When the ρ_e^* is positive, $\sum_{i=1}^{n} \sum_{t \in \zeta_i} X_{ite} - C_e = 0$ will be true, which means corresponding edge e has no "leftovers" in capacity. If the primal optimal solution exists, according to strong duality, the optimal dual variable ρ_e^* reflects how much improvement on the objective value we can make if the capacity of edge e increases by one. Therefore, we use the optimal dual variable ρ_e^* as the price for edge e. Compared with history-based cost in other routers, our price is more comprehensive because it considers all topologies we have and weights them according to their worth optimally.

$$\max. \sum_{i=1}^{n} (-\lambda_i) + \sum_{e \prec b} (-\rho_e) c_e \qquad (2)$$

$$\text{s.t. } \lambda_i + \sum_{e \prec t} \rho_e + W_{it} >= 0 \qquad (2a)$$

$$\lambda_i >= -M \qquad \forall i \in \{1,2,\ldots,n\}$$

$$\rho_e >= 0 \qquad \forall e \prec b$$

3) Three-level Topology Selection: If we evolve new topologies for each existing topology, the size of our topology pools will dramatically increase without corresponding speed of TOF mitigation. Therefore, we control the number of new evolved topologies in each iteration by only considering the most costly topologies.

We use a dynamic three-level topology-selection approach to determine certain topologies for evolution in each iteration. Only if current stage fails to further improve TOF, will next stage be launched. The following level selects topologies in a more broad way which enables further TOF reduction.

- Level-one: After we find all inside trees with positive S_i, all topologies associated with these unroutable inside trees will be evolved.
- Level-two: If evolution of topologies from level-one is unable to keep optimizing the LP formulation, we as-

semble an inside-tree-routing solution by selecting the topology with largest X_{ij} for each inside tree. Then the overflow of each edge could be counted. In addition to the topologies from level-one, any topology containing overflowed edge(s) will be added.

- Level-three: If the topology evolution in level-two fails to keep optimizing the LP formulation, we evolve topologies covering edges with positive price in addition.

We gradually loosen our requirement for topology evolution, pushing the optimization with control over the number of processed topologies. Fig.5 evaluates the three-level topology-selection during optimization iterations for one single block on ADAPTEC1. It shows that the first iterations in level-one increases the size of topologies slowly. As optimization halts during any iteration, next level will be launched to reduce TOF.

4) RC-constrained A* Search: After pricing and topology selection in every optimization iteration, we evolve new topologies with slew-aware rip-up and reroute. The pricy part will be ripped up and an RC-constrained A* search algorithm is applied to reroute disconnected parts without violating the slew constraints.

For any selected topology t, we find all wires with non-zero price, and sort them by their prices in descending order. After sorting, we sequentially rip-up and reroute each wire. For one wire w on t, signaling from U to V, we remove w from t first and notate the remaining part as $t\backslash w$. Then we calculate RC_p and C_p for each point $p \in t\backslash w$. RC_p and C_p are the maximum allowed RC and C connected to p without violation to the slew constraints. The maximum possible RC and C for all points on $t\backslash w$ will be:

$$RC^{max} = \max\{RC_p, p \in \{t\backslash w\}\} \qquad (3)$$

$$C^{max} = \max\{C_p, p \in \{t\backslash w\}\} \qquad (4)$$

Afterwards, an RC-constrained A* search is applied to reconnect V to the remaining part $t\backslash w$. We will only accept connections to point p with RC and C less than RC_p and C_p respectively. During RC-constrained A* search, any search path with RC exceeding RC^{max} or C exceeding C^{max} will be pruned away. The cost of each edge e in our A* search is the price of e plus one. The heuristic cost function we use is the 3-D Manhattan distance to the nearest point in $t\backslash w$, which clearly is a lower bound. This RC-constrained A* search guarantees least cost solution under slew constraints.

C. Outside-tree Routing

After topologies of inside trees are fixed, capacities of all edges within blocks are set to zero before blockage-avoiding outside-tree routing, which will be solved by existing academic routers.

IV. EXPERIMENTAL RESULTS

BOB-Router has been implemented in the C++ programming language. All experiments are conducted on an Intel Core 3.0GHz Linux machine with 16GB memory. We use 3D global routing benchmarks adaptec1 $\sim$ 4 and bigblue1 $\sim$ 4

from ISPD 2007 and 2008 Global Routing Contests for our experiments. Benchmarks from global routing contests are not annotated with blockage information explicitly. As far as we know, the block porosity information in the global routing benchmarks are derived from fixed macros in certain placement benchmarks. Owing to abutting blocks, it is arduous to directly retrieve geometric information of porosity areas from the routing benchmarks. Instead, we find the corresponding placement benchmarks, from which we are able to extract fixed macro geometric information. Besides, we remove nets containing pins inside blocks, which is beyond our formulation.

The wire resistance and capacitance for each metal layer are derived from ITRS [1], and we use 70ps as our maximal allowed slew.

We first evaluate the slew violation for each benchmark. Table II calibrates the slew numbers for all inside trees after RSMT topologies are generated by FLUTE and applied with a simple-layer-assignment heuristic. The heuristic will assign all inside trees on the lowest allowable pair of layers first. Then for all inside trees with slew violation, we will bring them to higher pair of metal layers according to extents of slew violations. From Table II, we can see that some benchmarks with no slew problem initially, such as bigblue2, may encounter slew problem because it is possible that most of inside trees have been promoted to the highest pair of metal layers.

TABLE II
SLEW DISTRIBUTION OF INSIDE TREES

Benchmarks	# nets	# inside trees	max slew	average slew
adaptec1	219794	57852	1713.8	36.9
adaptec2	260159	34769	494.4	28.5
adaptec3	466295	105137	23785.5	141.6
adaptec4	515304	86199	3986.7	65.8
bigblue1	282974	18763	380.1	22.1
bigblue2	576816	117259	69.9	4.0
bigblue3	1122340	79659	2025.1	22.1
bigblue4	2228930	234692	631.1	5.0

Since we eliminate all slew violation during initialization and keep slew under constraints, our final routing solution will not suffer from any slew problem. In Fig.6, we compare the slew distribution of inside trees from Table II with final routing solution for benchmark adpatec1. Initially, we observe the existence of inside trees with worst slew up to 1714ps. But after the benchmark is processed by BOB-Router, no inside tree has slew more than 70ps which is the maximum allowed slew rate in our slew constraints. The number of inside trees with slew between 60ps to 70ps is dramatically increased as most nets with slew violations originally are legalized to be just under 70ps.

If one router is not able to properly use the over-the-block routing resources, the safest way without breaking slew constraints thus involving manual work is to avoid the blocks completely by setting over-the-block routing capacity as zero (or large penalty). We compare our proposed BOB-Router with an obstacle-avoiding router (OA-Router) in terms of wirelength, via count and TOF. We modify NTHU-Router 2.0 [5] to be the OA-Router and the solver for outside-the-

TABLE I

COMPARISONS BETWEEN OUR PROPOSED BOB-ROUTER AND OA-ROUTER

Bench -marks	over-the-block				outside-the-block				overall				OA-Router			
	WL	Vias	TOF	cpu(s)	WL	Vias	TOF	cpu(s)	WL	Vias	TOF	cpu(s)	WL	Vias	TOF	cpu(s)
adaptec1	431886	138207	0	5690	2733837	1344218	199565	1421	3165723	1482425	199565	7111	3317320	1724765	450300	3463
adaptec2	261957	57838	265	4523	2615068	1258131	28847	1038	2877025	1315969	29112	5561	3371453	1836853	107498	4577
adaptec3	1235721	154123	1333	100210	8355049	2849048	639049	16527	9590770	3003171	640382	116737	10100613	3740726	1276779	18845
adaptec4	836840	105953	0	32718	8831370	2580484	329221	13202	9668210	2686437	329221	45920	11326871	3498262	438954	13455
bigblue1	98044	42090	0	55	3248498	1367350	22612	1637	3346542	1409440	22612	1692	3637249	1967568	70853	2232
bigblue2	258699	350385	0	520	3730497	2985365	3795	1131	3989196	3335750	3795	1651	4799773	3800398	5145	1346
bigblue3	522841	141885	0	2119	7800699	3847139	15148	2621	8323540	3989024	15148	4740	8961863	5267470	83416	8603
bigblue3	575639	731836	0	303	9358521	7489968	5266	2266	9934160	8221804	5266	2569	12363167	10444398	27939	5784
average	0.08	0.06	0.00	0.51	0.92	0.94	1.00	0.49	1.00	1.00	1.00	1.00	1.13	1.28	3.07	1.00

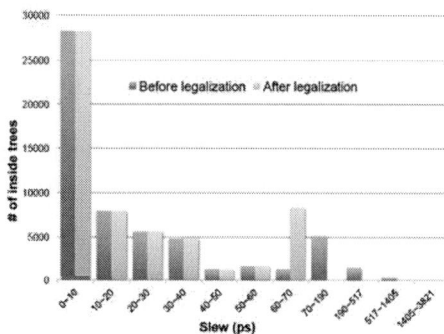

Fig. 6. Slew distribution of all inside trees in adpatec1 initially and finally. Each y coordinates number of inside trees with slew in the slot between current and previous x

block routing for its good performance. The results are shown in Table I. From the last row in the table, we can see that BOB-Router pushes about 8% of wirelength and 6% via count to the over-the-block part on average. The TOF of over-the-block routing is zero for most benchmarks. By using over-the-block routing resources, BOB-Router achieves about only 33% TOF, 88% wirelength and 78% via count of the OA-Router. We think more decrease of via count than wirelength is partially because BOB-Router performs full 3D routing for over-the-block part without layer assignment. Average runtime of BOB-Router is same with OA-Router. However, we notice that BOB-Router spends more time on bigger and tougher benchmarks, such as adaptec3, because more iterations and topologies are required.

V. CONCLUSION

In the past few years, traditional global routing has been extensively studied, which in turn makes it hard even to improve performance by 1%. We propose a new formulation of global routing problem from a different perspective. Solving this new BOB-Routing problem could keep shortening design cycle and improving routing quality. With our proposed approach, we can generate slew-violation-free solution with 66% less TOF, 12% less wirelength and 22% less via count compared with the obstacle-avoiding approach.

VI. ACKNOWLEDGMENT

This work is supported in part by NSF, SRC and Oracle.

REFERENCES

[1] 2012 Overall Roadmap Technology Characteristics (ORTC) Tables.
[2] ISPD 2007 Global Routing Contest and Benchmark Suite. http://archive.sigda.org/ispd2007/contest.html.
[3] ISPD 2008 Global Routing Contest and Benchmark Suite. http://archive.sigda.org/ispd2008/contests/ispd08rc.html.
[4] G. Ajwani, C. Chu, and W. Mak. FOARS: FLUTE Based Obstacle-Avoiding Rectilinear Steiner Tree Construction. In Proc. ISPD, pages 194–204, 2010.
[5] Y. Chang, Y. Lee, J. Gao, W. Wu, and T. Wang. NTHU-Route 2.0: A Robust Global Router for Modern Designs. In IEEE TCAD, volume 29(12), pages 1931–1944, 2010.
[6] H. Chen, C. Hsu, and Y. Chang. High-Performance Global Routing with Fast Overflow Reduction. In Proc. ASPDAC, pages 582–587, 2009.
[7] M. Cho, K. Lu, K. Yuan, and D. Z. Pan. BoxRouter 2.0: Architecture and implementation of a hybrid and robust global router. In Proc. ICCAD, pages 503–508, 2007.
[8] C. Chu and Y. Wong. FLUTE: Fast Loopup Table Based Rectilinear Steiner Minimal Tree Algorithm for VLSI Design. IEEE TCAD, 27(1):70–83, 2008.
[9] K. Dai, W. Liu, and Y. Li. Efficient Simulated Evolution Based Rerouting and Congestion-Relaxed Layer Assignment on 3-D Global Routing. In Proc. ASPDAC, pages 570–575, 2009.
[10] J. Gao, P. Wu, and T. Wang. A New Global Router for Modern Designs. In Proc. ASPDAC, pages 232–237, 2008.
[11] T. Huang and E. F.Y. Young. Construction of rectilinear Steiner minimum trees with slew constraints over obstacles. In Proc. ICCAD, pages 144–151, 2012.
[12] T. Huang and Evangeline F.Y. Young. An Exact Algorithm for the construction of Rectilinear Steiner Minimum Trees among Complex Obstacles. In Proc. DAC, pages 164–169, 2011.
[13] L. Li, Z. Qian, and Evangeline F.Y. Young. Generation of Optimal Obstacle-avoiding Rectilinear Steiner Minimum Tree. In Proc. ICCAD, pages 21–25, 2009.
[14] L. Li and Evangeline F.Y. Young. Obstacle-avoiding Rectilinear Steiner Tree Construction. In Proc. ICCAD, pages 523–528, 2008.
[15] W. Liu, Y. Wei, C. N. Sze, C. J. Alpert, Z. Li, Y. Li, and N. Viswanathan. Routing Congestion Estimation with Real Design Constraints. In Proc. DAC, 2013.
[16] M. D. Moffitt. MaizeRouter: engineering an effective global router. In Proc. ASPDAC, pages 226–231, 2008.
[17] M. Mustafa Ozdal and M. D. F. Wong. Archer: a history-driven global routing algorithm. In Proc. ICCAD, pages 488–495, 2007.
[18] J. A. Roy and I. L. Markov. High-Performance Routing at the Nanometer Scale. In IEEE TCAD, volume 27(6), pages 1066–1077, 2008.
[19] T. Wu, A. Davoodi, and J. T. Linderoth. GRIP: Scalable 3D Global Routing Using Integer Programming. In Proc. DAC, pages 320–325, 2009.
[20] Y. Xu, Y. Zhang, and C. Chu. FastRoute 4.0: Global Router with Efficient Via Minimization. In Proc. ASPDAC, pages 576–581, 2009.
[21] Y. Zhang, A. Chakraborty, S. Chowdhury, and D. Z. Pan. Reclaiming Over-the-IP-Block Routing Resources With Buffering-Aware Rectilinear Steiner Minimum Tree Construction. In Proc. ICCAD, pages 137–143, 2012.
[22] Y. Zhang, Y. Xu, and C. Chu. FastRoute3.0: A Fast and High Quality Global Router Based on Virtual Capacity. In Proc. ICCAD, pages 344–349, 2008.

978-1-4799-2817-0/14 $31.00 © 2014 IEEE

Routability-Driven Bump Assignment for Chip-Package Co-Design *

Meng-Ling Chen, Tu-Hsiung Tsai, Hung-Ming Chen
Institute of Electronics and SoC Center
National Chiao Tung U., Hsinchu, Taiwan
{souffle1219,thebear325}@gmail.com, hmchen@mail.nctu.edu.tw

Shi-Hao Chen
Global Unichip Corp.
Hsinchu Science Park, Taiwan
hockchen@globalunichip.com

Abstract—In current chip and package designs, it is a bottleneck to simultaneously optimize both pin assignment and pin routing for different design domains (chip, package, and board). Usually the whole process costs a huge manual effort and multiple iterations thus reducing profit margin. Therefore, we propose a fast heuristic chip-package co-design algorithm in order to automatically obtain a bump assignment which introduces high routability both in RDL routing and substrate routing (100% in our real case). Experimental results show that the proposed method (inspired by board escape routing algorithms) automatically finishes bump assignment, RDL routing and substrate routing in a short time, while the traditional co-design flow requires weeks even months.

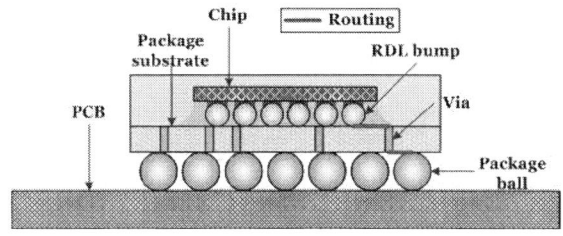

Fig. 1. The cross section of flip-chip. In this work, the problem structure is from I/O pads through bump pads to solder balls.

I. INTRODUCTION

Bump assignment plays an important role in chip and package designs. As illustrated in Fig. 1, flip-chip design consists of three domains: chip, package and board. In current industrial design flow, the designers first generate corresponding bump assignment based on experience according to the I/O sequence and the ball map given by customer, then perform RDL routing and substrate routing respectively. In addition, the assignment is composed of repeating pattern which satisfies power constraint but disregards routing information. It might consequently result in serious net congested problem on chip or package. Once the designers cannot find a legal routing solution, they have to reassign bumps then create chip and package layout all over again. Normally, this back and forth procedure has to be executed for many times, thus slowing down time to market and reducing the profits.

A. Previous Works

To improve the layout performance or to speed up design cycle, several previous works proposed cross-domain co-design methodology in various aspects such as placement [4, 5], routing [6, 7, 8], assignment [9, 10, 11] and design flow [12, 13]. For chip-package co-design problem, [4] proposed a multi-step algorithm based upon integer linear programming to find an I/O placement solution satisfying all design constraints. [5] addressed a block and I/O buffer placement method that optimizes wire length and signal skew. Some researches [6, 7] developed RDL routers for area-I/O to achieve better performance. [8] fast generated an estimation of wire planning in package and board for chip and board designs awareness. Since bump and finger are the interface between chip and package, [9, 10, 11] focused on the pin assignment to increase routability. Furthermore, the main reason that causes the bottleneck of co-design problem is the iteration of design process. Therefore, [12] provided a board-driven Λ-shaped co-design flow with true bi-directional information interactions and [13] offered a concurrent design flow to avoid much longer turn-around time.

However, the aforementioned previous works did not emphasize on substrate routing. Instead of physically connecting bumps and balls, some of them judge routability by fly-line or probabilistic prediction. Although the assignment proposed by

[9] considers routability on package, it is suitable only for wire-bonding package. As a result, they might still meet design difficulties.

B. Our Contributions

In this paper, we propose a fast heuristic in chip-package co-design in order to automatically obtain a bump assignment which introduces 100% routability both in RDL routing and substrate routing in our industrial case. Our approach also provides a practical RDL layout and a routing order that guides designers to easily finish net connection on package. Moreover, it can be used as a routing simulator as well. Since the results reflect the quality of initial I/O pin sequence and ball map, improper mapping can be fixed in the early stage for reducing design period and manual effort.

The rest of the paper is organized as follows. Section II introduces the framework of chip-package co-design problem. Section III discusses the proposed co-design flow then dilates on each stage. Section IV reports our experimental results on one real industrial case. Finally, we conclude this paper in Section V.

II. PROBLEM FORMULATION

Our chip-package co-design problem is to find a solution of bump assignment such that both RDL and substrate routing can meet the requirements without repeating the entire process over and over again. More specifically, we are given physical locations of peripheral I/O pads O, bump pads U, and solder balls B as illustrated in Fig. 2. Each I/O and ball are assigned to a specific net. The mapping between them is denoted by M_{OB}. Our goal is to appropriately assign bumps for the purpose of 100% routability in RDL routing and substrate routing under the corresponding I/O-bump mapping M_{OU} and bump-ball mapping M_{UB}. Note that the mapping derived from M_{OU} and M_{UB} should be exactly the same as M_{OB}. Furthermore, since we use [1] which guarantees 100% routability to complete RDL routing, our objective in chip domain is to minimize the number of routing tracks instead.

Input:
i) Chip domain:
  - Physical locations of each I/O pad O_i and each bump pad U_i on chip.
  - Peripheral I/O pin sequence is given by designer.
  - Routable region: top two layers in this research.
ii) Package domain:

---

*This work was partitially supported by the grant from National Science Council Taiwan No.102-2220-E-009-024.

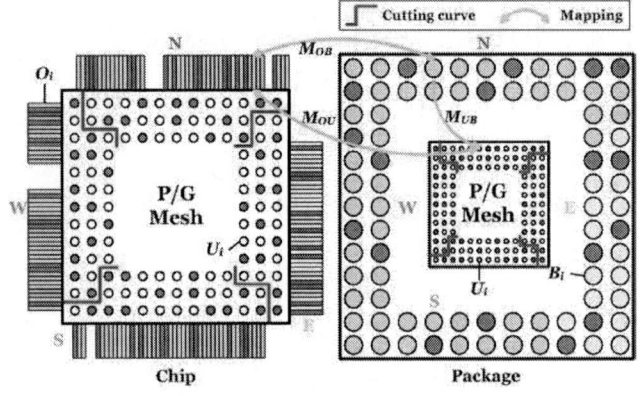

Fig. 2. Top view of chip and package. The whole model is partitioned into four portions: west, north, east and south. First, the I/O pads are divided into four groups by locality. Then, the solder balls are distinguished according to the groups of their corresponding I/O pads. Finally, the bump pads are sliced into four sectors by cutting curves derived from two diagonals of chip.

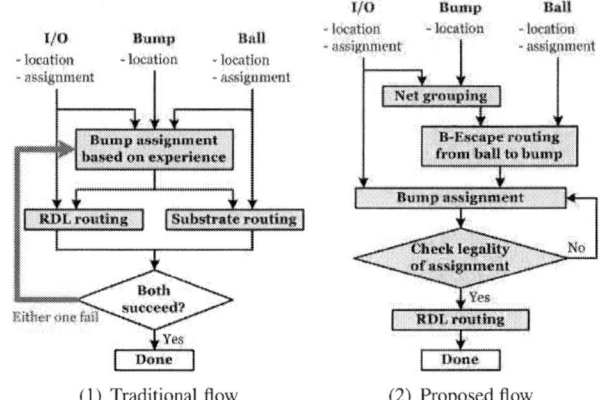

Fig. 3. The comparison between traditional flow and proposed flow. Both are consisted of three major domains marked with colors: bump assignment, RDL routing, and substrate routing. Traditional flow costs more human resource and time because of the iteration. Proposed flow automatically solves the problem in a short time.

- Physical locations of each bump pad U_i and each solder ball B_i on package.
- Solder ball map is given by designer.
- Routable region: single layer in this research.

iii) Design rules such as wire width, spacing constrain, size of O_i, U_i, B_i, and via.

Output:
i) A solution of bump assignment.
ii) Practical RDL routing result in pseudo single-layer [1].
iii) An illustration of planar substrate routing.

Objectives:
i) Chip domain: minimize the routing area borrowed from another existing metal layer (minimize number of tracks in channel routing) [1].
ii) Package domain: 100% routability.

In our implementation, the whole model is divided into four portions as shown in Fig. 2: west, north, east, and south by the following instructions. We first group I/O pads according to locality then distinguish their corresponding solder balls. The bump pads are partitioned by two diagonals of the chip and so does the routing area of RDL. Once solder balls and bump pads are sliced, four routing regions of package are generated. Here we transform coordinates into west coordinate system and consider west model only.

III. BUMP ASSIGNMENT AND PACKAGE PLANNING

In this section, we first compare between traditional co-design flow and the proposed co-design flow in Sec. III.A, then particularly describe each stage of our methodology in Sec. III.B-III.E. Since planar routing is still required (even though multiple routing layers are available for substrate routing) due to signal integrity and manufacturability as mentioned in [3], the substrate routing problem can be transformed into escape routing problem. Therefore, we reform B-Escape routing algorithm proposed by [2] with net grouping technique (Sec. III.B) to solve substrate routing (Sec. III.C). Based on the escape routing order of ball and the I/O pin sequence, a solution of bump assignment can be generated (Sec. III.D). Finally, we complete RDL routing by applying the work in [1] (Sec. III.E).

A. Design Flow

As shown in Fig. 3, the chip-package co-design problem consists of three major domains interacting with each other: bump assignment, RDL routing, and substrate routing. It is almost impossible to set compromising bump assignment between chip and package without any routing information. Therefore, designers first roughly assign bumps based on experience and specific patterns. Then they will iteratively revise

it depending on the routing results in the traditional design flow (as shown in Fig. 3(1)). In contrast, we propose a straightforward co-design flow to solve this problem by adjusting design order to be substrate routing, then bump assignment, and finally RDL routing, shown in Fig. 3(2).

Since the quality of an assignment is judged by routing results, we first detail our routing algorithm. For chip, we adopt [1] to solve congested RDL connection because of 100% routability and easy implementation. As shown in Fig. 4(1), it is intuitive to regard I/O pads as a pin track S_O. Additionally, a row-based projection is performed on bump matrix to build a virtual track S_{U_C}. Then we can apply classical channel routing algorithm for pin connection. Note that S_O does not have to be identical with S_{U_C} due to two routing layers.

For package, the routing area is decomposed into two parts: bump area and ball area. As shown in Fig. 4(2), S_{U_P} is obtained by the aforementioned projection and will be similar to S_{U_C}. According to [3], substrate routing should be planar although there are multiple available layers. Therefore the concept of escape routing, which is usually performed on PCB board, can be imported. Here we choose [2] to form S_B which must be exactly the same as S_{U_P} due to planar routing.

In conclusion, the ideal situation that concurrently achieves our two objectives mentioned in Section II occurs when $S_O \simeq S_{U_C} \simeq S_{U_P} = S_B$. For this purpose, we first divide nets into groups, which represents global assignment structure, depending on the given S_O and the above projection method. S_B will resemble S_O as much as possible by adding this information to the cost function of B-Escape algorithm [2]. In the end, we can determine the **detailed bump assignment** according to S_O and S_B, then generate RDL routing solution. The following subsections discuss the algorithm of each step.

B. Net Grouping

To simultaneously achieve high routability in both RDL routing and substrate routing, the escape routing order of ball should be similar to the given I/O pin sequence ($S_B \simeq S_O$) as discussed in last subsection. Since the original B-Escape routing algorithm [2] considers only routability, we introduce the I/O information by net grouping technique.

Fig. 4 demonstrates the way how a bump pin track S_{U_C} or S_{U_P} is constructed on chip or package. Since the bump pin track is generated by the row-based projection method, we group bumps row by row in top-down order with notation G_1, G_2, G_3 and so on. To reduce routing tracks in RDL, S_{U_C} should be quite similar to the given S_O. Consequently, each group is expected to be assigned with the corresponding nets. As shown in Fig. 4(1), group G_1 includes net N_1, N_2, and N_3, group G_2 includes N_4, and group G_3 includes N_5 and N_6.

978-1-4799-2817-0/14 $31.00 © 2014 IEEE

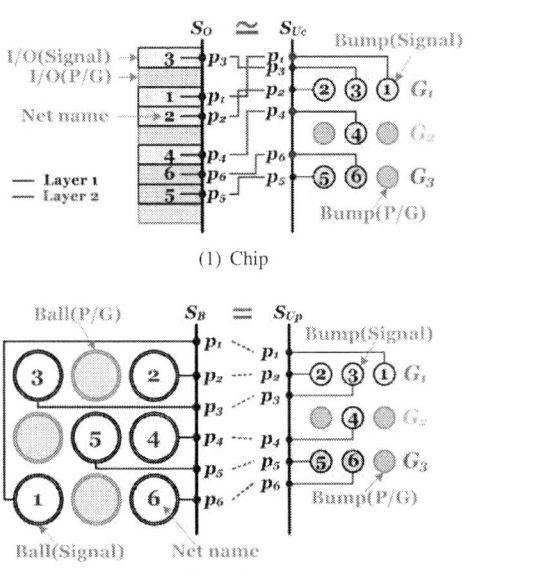

(1) Chip

(2) Package

Fig. 4. Four sequences, S_O, S_{U_C}, S_{U_P} and S_B, are built to specify chip-package co-design problem. S_O is the given I/O pin sequence. S_{U_C} and S_{U_P} are two virtual pin tracks constructed by performing a row-based projection for bump matrix on chip and package respectively. S_B symbolizes the escape pin ordering from ball matrix. In addition, $S_O \simeq S_{U_C}$ because we use channel routing algorithm to solve RDL routing and S_{U_P} must be exactly the same as S_B due to the constraint of single-layer routing on package. On the other hand, by grouping nets as G_1, G_2, G_3, we can offer flexible rules for bump assignment.

Algorithm 1 Reformation of B-Escape routing algorithm

1: **for** each of the six routing mode **do**
2:    **for** each unrouted net N_i **do**
3:       route net N_i from ball B_i to escape boundary S_B
4:       calculate the cost vector for net N_i
5:       clear the route generated for net N_i
6:    **end for**
7:    **if** upward mode **then**
8:       sort all net costs by group in non-decreasing order
9:    **else**
10:      sort all net costs by group in non-increasing order
11:    **end if**
12:    **for** each group **do**
13:      sort net costs by α and β in non-decreasing order
14:    **end for**
15:    choose the first net N_j
16:    **if** net N_j traps other nets **then**
17:      backtrack and reorder
18:    **else**
19:      route net N_j from ball B_j to escape boundary S_B
20:      remove net N_j from net cost order
21:    **end if**
22:    **until** all nets are routed or exceed the backtrack limit
23:    store the solution for this routing mode
24: **end for**
25: output the solution with the best routability

Net grouping symbolizes the idea of global bump assignment. If all the four pin tracks S_O, S_{U_C}, S_{U_P} and S_B are organized in the same group order, the number of net crosses is minimal. Furthermore, net grouping merely reveals the distribution, it does not specify the detailed assignment to each bump in row. In other words, S_{U_C} and S_{U_P} are uncertainly equal to each other in spite of the same group order. This can be seen in the following example: in Fig. 4(1), $S_{U_C} = (p_1 p_3 p_2)(p_4)(p_6 p_5) = G_1 G_2 G_3$; in Fig. 4(2), $S_{U_P} = (p_1 p_2 p_3)(p_4)(p_5 p_6) = G_1 G_2 G_3$. In summary, net grouping offers not only a standard but flexible rules for bump assignment.

C. Substrate Routing

As mentioned previously, we separate the routing area of bumps and balls on package. In this subsection, we focus on ball area since we use coherent routing method for bumps on chip and package which will be discussed in Sec. III.E. Considering signal integrity and manufacturability, substrate routing should be completed in single layer as described in [3]. For this reason, the connection from balls to S_B can be treated as *single-component escape routing problem* even though it acts on PCB board. The entire ball area split into four *components*. Each of them has an individual *escape boundary*, that is, S_B. In the next few paragraphs, we will describe how to solve this problem by modifying cost function in B-Escape routing algorithm [2] with net grouping.

Algorithm 1 shows our reformation of B-Escape routing algorithm [2]. The overall process comprises three steps (same as original algorithm): $Step1$ is to calculate routing cost of each net (Line 2-Line 6); $Step2$ is to sort net costs(Line 7-Line 14); $Step3$ is to route the first net or backtrack(Line 15-Line 21). However, the routing cost is calculated by new 3-element vector (α, β, γ). For this reason, the details of the three steps differ from [2] and will be individually specified in the next three paragraphs.

Here α and β follow the definition in [2] which respectively stands for the number of unroutable balls and the number of blocked balls caused by current routing net. These two elements dominate the routability (*Objective* ii in Sec. II). To optimize RDL routing (*Objective* i in Sec. II), S_O and S_{U_C} should be alike. According to the relationship among the four sequences described in Sec. III.A, $S_O \simeq S_{U_C} \simeq S_{U_P} = S_B$ is derived. Hence we define the third element γ as the group of net. By following the group order during escape routing, the escape order S_B will be similar to the given I/O pin sequence S_O.

Net cost ordering depends on *routing modes* clarified in [2]. In *upward* mode, a pin first goes straight up to the boundary then follows the boundary clockwise until it reaches S_B. All unrouted pins will be located lower than the current pin on S_B. Consequently, the net clustered in top group should be routed earlier. Thus, the net costs are arranged as $\{G_1, G_2, G_3, ...\}$. During *downward* mode, it is in reverse order $\{..., G_3, G_2, G_1\}$. Moreover, cost ordering inside each group is in non-decreasing order based on α and β.

Since the costs are organized by net group in the reformation, there are two situations: (a) the chosen net N_i and the next candidate net N_j are in the same group; (b) N_i and N_j are in different groups. Once all unrouted balls are trapped by each other ($\forall \alpha_i \neq 0$), it meets a *reordering point* and has to *backtrack* to route N_j first instead of N_i as described in [2]. Situation (a): as shown in Fig. 5(1), the exchange of routing order inside group does not influence the result of bump assignment. Therefore, it will not cause extra net crosses. Situation (b): the exchange of routing order between groups leads to the exchange of assignment between bump rows as shown in Fig. 5(2). As a result, net crosses are produced in RDL routing. Thus, situation (a) has higher priority when *backtracking*. According to aforementioned instructions, we can finally obtain an escape routing solution of balls whose S_B is the most similar to S_O with 100% routability.

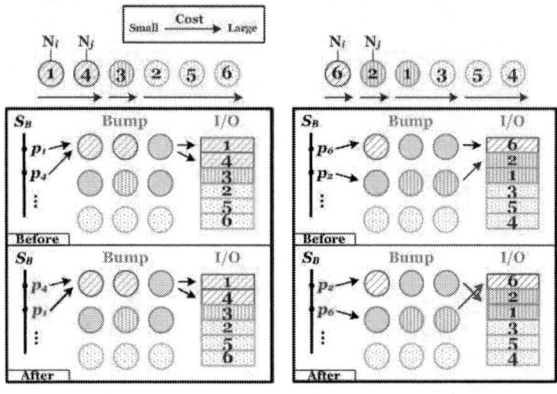

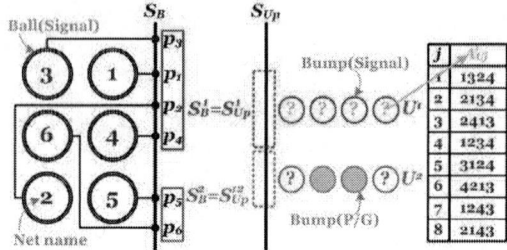

(1) Reordering inside group   (2) Reordering between groups

Fig. 5. Due to net grouping, there are two situations when reaching a reordering point of the reformation of B-Escape algorithm [2]: the chosen net N_i and the next candidate net N_j are in the same group (as shown in (1)) or not (as shown in (2)). The net cross caused by routing order exchange inside group could be solved by using different projection patterns of the bump row on chip and package. However, the net cross produced by routing order exchange between groups would never be solved since the projection method is row-based. Therefore, situation (1) has higher priority when backtracking.

D. Bump Assignment

Here we present three steps to determine an optimal solution of bump assignment by S_O and S_B. First, owing to single layer routing on package, S_{U_P} must be identical to S_B. For each bump row U^i, there is an assignment set A_U^i that can match the corresponding segment $S_{U_P}^i$ by performing the projection method, detailed in the next subsection. Then, we derive the candidates for $S_{U_C}^i$ from A_U^i in the same way. In addition, cost C_j^i is defined as the differences between candidate $S_{U_{C_j}}^i$ and S_O^i. Finally, we combine all A_U^i with minimum cost and verify it with design rules.

Fig. 6 demonstrates an example to clarify the process. As shown in Fig. 6(1), there are eight possible assignments $A_{U_1}^1$-$A_{U_8}^1$ for bump row U^1 to form $S_{U_P}^1 = p_3 p_1 p_2 p_4$. Fig. 6(2) illustrates the candidate $S_{U_{C_1}}^1$ projected from $A_{U_1}^1$. Comparing each $S_{U_{C_j}}^1$ and S_O^1, the cost C_j^1 is the difference between the length of S_O^1 and the length of the longest common sequence (LCS) of $S_{U_{C_j}}^1$ and S_O^1. Note that the LCS represents the similarity, the assignment which has lowest cost is denoted by $A_U^{1'}$. Repeat all steps above until all $A_U^{i'}$ are found.

In this work, we verify the sufficiency of routing resource for the solution composed of $A_U^{i'}$ instead of physically routing bumps to S_{U_P}. On chip, the number of tracks between two adjacent bump rows is larger than the number of bumps in a row. On package, it is completely the opposite. To improve routability, designers will relocate bump vias by connecting short wires from bumps in previous layer rather than dropping vias over bumps (vias are on top of the bumps). This method is quite similar to the *flexible via-staggering technique* proposed in [3]. For example, given $S_{U_P} = p_1 p_2 ... p_7$ and $A_U' = \{A_U^{1'}, A_U^{2'}, A_U^{3'}\}$ as shown in Fig. 7(1), if the vias are located right above the bumps, nets N_3 and N_7 will cause violation. It can be seen in Fig. 7(2) that A_U' is a legal solution by staggering and compacting vias based on DRC rules.

The formula (1) and (2) in Fig. 7(2)(2) are specified below:

$$F_a(n) = \begin{cases} \dfrac{max(d_v, d_{vw})}{2} + R & , n = 0 \\ \dfrac{max(d_w, d_{vw})}{2} + nW_w + (n-1)d_w + d_{vw} + R & , n \geq 1 \end{cases}$$

(1)

(1) S_{U_P} is the same as S_B due to single-layer substrate routing. For bump row U^1, there is an assignment set A_U^1 consisting of each candidate $A_{U_j}^1$ that produces corresponding projection $S_{U_P}^1$.

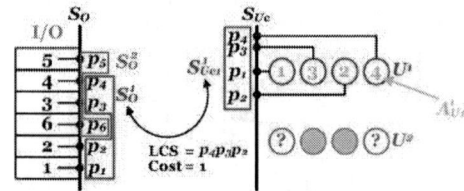

(2) First we draw the corresponding subsequence S_O^1 which is composed of net pins identical with that assigned to U^1. For each candidate $A_{U_j}^1$, perform the projection method with target S_O^1 to make the result $S_{U_{C_j}}^1$ as similar to S_O^1 as possible. Then we calculate the cost C_j^1 which is defined as the difference between the length of S_O^1 and the length of the longest common sequence (LCS) of $S_{U_{C_j}}^1$ and S_O^1.

Fig. 6. An example that demonstrates the process of bump assignment. (1) We derive all possible solutions of bump assignment based on ball escape routing result. (2)We choose the candidate which projeccts the most similar bump pin sequence to I/O pin sequence.

$$F_b(n) = \begin{cases} \sqrt{(2R + d_v)^2 - (\frac{D}{2})^2} & , n = 0 \\ \sqrt{(2R + 2d_{vw} + nW_w + (n-1)d_w)^2 - (\frac{D}{2})^2} & , n \geq 1 \end{cases}$$

(2)

Notations:
- n: the number of wires.
- D: the distance between two adjacent bump vias.
- R: the radius of a bump via.
- d_v: the minimum spacing between two adjacent vias.
- d_w: the minimum spacing between two adjacent wires.
- d_{vw}: the minimum spacing between a via and a wire.
- W_w: the minimum wire width in bump area of package.

Here we stagger and compact bump vias row by row from top to down. There are two cases: a via row is located between routing boundary and another via row; a via row is located between two via rows. Formula (1) is applied to the former and formula (2) is applied to the latter. It is evident to obtain the two formula according to Pythagorean theorem. After the lowest row is compacted, the legality of an assignment A_U' can be judged on whether the lowest row is located inside the routing region.

However, it is no guarantee of the existence of legal solution. Once A_U' fails, we will check and update $A_U^{i'}$. If all combinations of A_U' can not meet the requirement, it concludes that the routing constraint is extremely tight. Thus, the only way to solve this problem is to stagger vias manually or modify the given I/O pin sequence or ball map.

E. RDL Routing

After bump assignment, the methodology proposed in [1] is applied to solve RDL routing in pseudo single-layer. The problem is transformed into classical channel routing problem by considering I/O pads as a pin track S_O and projecting bump matrix to a virtual track S_{U_C}. As shown in Fig. 8(1), the projection in [1] is applied in only one direction (downward). The

978-1-4799-2817-0/14 $31.00 © 2014 IEEE   522

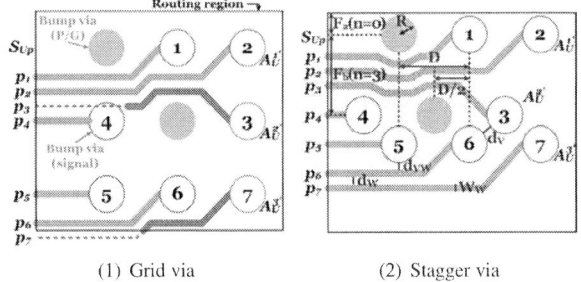

(1) Grid via   (2) Stagger via

Fig. 7. Comparison between grid via and stagger via. As shown in (1), an assignment A'_U consisting of $A_U^{1'}$, $A_U^{2'}$ and $A_U^{3'}$ is illegal since there is no routing space for nets N_3 and N_7 while dropping vias over bumps. However, (2) shows that A'_U is a legal solution by staggering and compacting vias based on DRC rules.

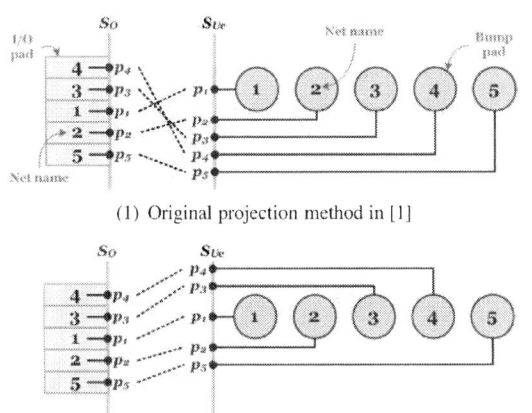

(1) Original projection method in [1]

(2) Proposed projection method

Fig. 8. Comparison between bump projection method in [1] and that in this paper. In (1), bumps can project in only one direction, it might cause more overlaps. Hence, a bi-direction projection method is developed to reduce net crossing as shown in (2).

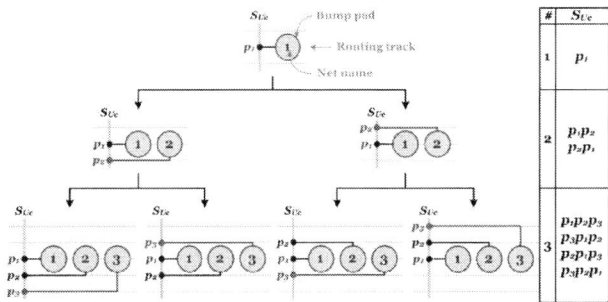

Fig. 9. List of bump pin track. Since the connection between bump pads and pins should be completed in single layer, the pin order can be derived as demonstrated. A new pin is located on either top or down of the original pin sequence. Thus, some particular patterns, for example, $S_{U_C} = p_2 p_3 p_1$, will never be generated.

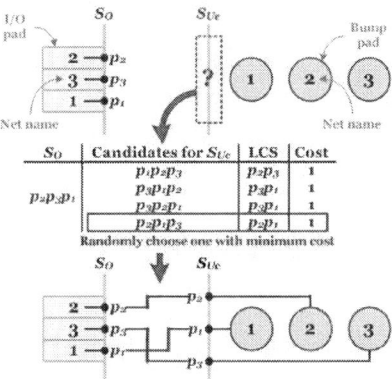

Fig. 10. An example for RDL routing. Given I/O pads and bump pads which are assigned, we first intuitively consider I/O pads as a pin track S_O. Then, we calculate the cost of each candidate for S_{U_C}. Finally, we choose the pattern with minimum cost and perform classical channel routing algorithm.

inflexibility of bump pin order will cause extra overlaps. Hence, we refine it by offering two projective directions (upward and downward) as shown in Fig. 8(2). Note that both techniques are row-based.

Fig. 9 lists all patterns of bump pin order generated by our projection method. Due to the non-detour routing in single-layer, some particular patterns, for example, $S_{U_C} = p_2 p_3 p_1$, will not be included. The cost of each pattern is equal to the difference between bump pin order and the corresponding I/O subsequence. It is similar to the definition of C_j^i described in Sec. III.D. With the minimum cost pattern, the usage of tracks will be reduced in RDL routing. The whole procedure is displayed in Fig. 10.

IV. EXPERIMENTAL RESULTS

The proposed algorithms are performed on a real and large-scale industrial case. This case contains chip domain and package domain. As shown in Fig. 11 and Fig. 12, the distributions of I/Os, bumps, and balls in south and east regions are uniform; in west region, balls are distributed uniformly but bumps and I/Os are distributed in corner; in north region, all I/Os, bumps, and balls are distributed in corner.[1] In this section, we first detail the execution flow, then compare the results of substrate routing, bump assignment and RDL routing with that of traditional co-design flow.

---

[1]The combination of various types of distribution in different design domain increases the routing difficulty. For example, in west region, EDI router can only complete less than half of nets in RDL routing.

A. Co-Design Flow

The whole methodology can be divided into three parts: (a) substrate routing in Sec. III.B-C, (b) bump assignment in Sec. III.D and (c) RDL routing in Sec. III.E. Parts (a) and (b) are implemented in C++; part(c) is implemented in tool command language Tcl. First, the data are fetched from the chip design in Encounter Digital Implementation (EDI) and the package design in Cadence Allegro. Then, a substrate routing order considering RDL routability with the corresponding simulation of single-layer substrate routing is generated by part(a). After bump assignment, part(c) connects I/O and bumps in pseudo single-layer and dumps scripts of commands in few minutes. Finally, the physical RDL layout will be obtained by sourcing these scripts in EDI.

B. Substrate Routability

Fig. 11 shows our results of bump assignment and substrate routing in a real industrial case with 507 signal nets (case1). Since we perform the reformation of B-Escape routing algorithm [2] on uniform grid, the routing paths produce 90 degree corner and cost more routing resource than 45 degree router. Therefore, it can be treated as the lower bound of routing performance. By following the escape order, designers can easily complete the physical substrate routing with 100% routability in Cadence Allegro. Although Allegro offers automatic router for package, designers still connect each bump and ball in manual because of the poor routability (40%-50%) and redundant detours.

TABLE I

THE COMPARISON OF CHIP-PACKAGE CO-DESIGN FLOW IN OUR CASE.

Method	Substrate routing			RDL routing		
	# routed nets	Rout.	Time	# routed nets	Rout.	Time
Ours	507/507 in 2nd layer	100%	909.52 (sec.)	507/507 in pseudo single-layer	100%	< 1 minute
Traditional	449/507 in 2nd layer 58/507 in 3rd layer	100%	a few days	507/507 in two layers	100%	> 2 weeks

(package size is 15.5mm*15.5mm)　　　　　　　　　　(chip size is 5.751mm*5.972mm)

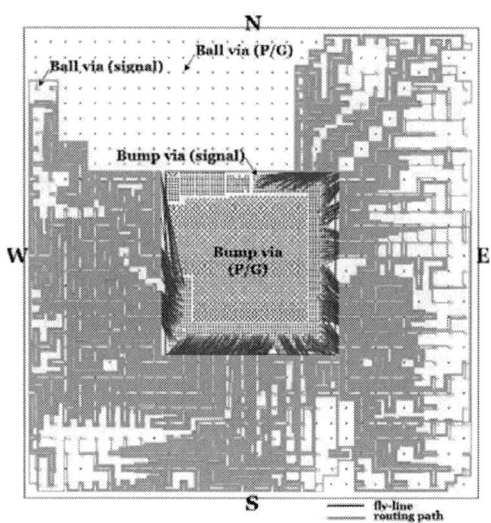

Fig. 11. A solution of bump assignment with single-layer substrate routing simulation that generates by the reformation of B-Escape algorithm [2] in up-down mode. Green paths are the escape routing result of ball; fly-lines colored in blue indicate the assignment between bumps and balls.

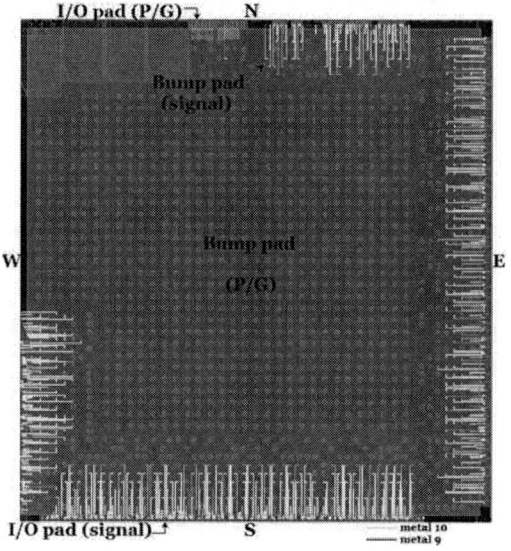

Fig. 12. Based on our bump assignment, we can achieve 100% routability in RDL routing by [1] in our case.

C. Bump Assignment

As shown in Table I, the substrate routing results reflect the quality of assignment method. By using the technique of via staggering mentioned in Sec. III.D during bump assignment, our co-design flow can achieve planar substrate routing. In contrast, designers assign bumps based on experience and iteratively revise the solution in traditional flow. Therefore, it requires more routing resource and more time to finish all connections.

D. RDL Routability

Fig. 12 shows the result of RDL routing by obtained [1] under our bump assignment. We first dump the assignment in script of command then source it in EDI to assign bumps. Because the assignment generated is based on I/O sequence, the routing area borrowed from another existing layer is quite equal to that of manual. Both flows produce 100% routability RDL routing, but our flow is much faster than the traditional co-design flow.

V. CONCLUSIONS

In this paper, we propose a chip-package co-design flow with routability-driven bump assignment based on [1] and [2]. By considering I/O pin sequence as a target order for ball escape routing, a compromise of bump assignment between chip and package can be generated. This technique offers information interactions to avoid the iterative revision in traditional design flow. The experimental results have shown that our approach achieves 100% routability in both RDL routing and substrate routing. In addition, since the whole process is automatic, it is much faster than the traditional design flow.

REFERENCES

[1] H. W. Hsu, M. L. Chen, H. M. Chen, H. C. Li, and S. H. Chen, "On effective flip-chip routing via pseudo single redistribution layer," in *Proc. of Design, Automation and Test in Europe Conference and Exhibition,* pp. 1597-1602, 2012.

[2] L. Luo, T. Yan, Q. Ma, M. D. F. Wong, and T. Shibuya, "B-escape: a simultaneous escape routing algorithm based on boundary routing," in *Proc. of International Symposium on Physical Design,* pp. 19-25, 2010.

[3] S. Liu, G. Chen, T. T. Jing, L. He, T. Zhang, R. Dutta, and X. L. Hong, "Substrate topological routing for high-density packages," in *IEEE Trans. on Computer-Aided Design of Integrated Circuits and Systems,* vol. 28, no. 2, pp. 207-216, Feb. 2009.

[4] J. Xiong, Y. C. Wong, E. Sarto, and L. He, "Constraint driven I/O planning and placement for chip-package co-design," in *Proc. of Asia and South Pacific Design Automation Conference,* pp. 207-212, 2006.

[5] M. F. Lai and H. M. Chen, "An implementation of performance-driven block and I/O placement for chip-package codesign," in *Proc. of International Symposium on Quality Electronic Design,* pp. 604-607, 2008.

[6] K. S. Lin, H. W. Hsu, R. J. Lee, and H. M. Chen, "Area-I/O RDL routing for chip-package codesign considering regional assignment," in *Proc. of IEEE Electrical Design of Advanced Packaging and Systems Symposium,* pp. 1-4, 2010.

[7] J. W. Fang and Y. W. Chang, "Area-I/O flip-chip routing for chip-package co-design considering signal skews," in *IEEE Trans. on Computer-Aided Design of Integrated Circuits and Systems,* vol. 29, no. 5, pp. 711-721, 2010.

[8] R. J. Lee, H. W. Hsu, and H. M. Chen, "Board- and chip-aware package wire planning," in *IEEE Trans. on Very Large Scale Integration Systems,* vol. 21, no. 8, pp. 1377-1387, Sept. 2012.

[9] C. H. Lu, H. M. Chen, C. N. J. Liu, and W. Y. Shih, "Package routability- and IR-drop-aware finger/pad assignment in chip-package co-design" in *Proc. of Design, Automation and Test in Europe Conference and Exhibition,* pp. 845-850, 2009.

[10] H. Han, W. Yin, W. Wang, and Z. Pang, "Auto-assign method for large scale flip-chip package design," in *Proc. of IEEE International Conference on ASIC (ASI-CON),* pp. 929-932, 2011.

[11] T. Meister, J. Lienig, and G. Thomke, "Interface optimization for improved routability in chip-package-board co-design," in *Proc. of International Workshop on System Level Interconnect Prediction,* pp. 1-8, 2011.

[12] H. C. Lee and Y. W. Chang, "A chip-package-board co-design methodology," in *Proc. of ACM/IEEE Design Automation Conference,* pp. 1082-1087, 2012.

[13] R. J. Lee and H. M. Chen, "A study of row-based area-array I/O design planning in concurrent chip-package design flow," in *Proc. of ACM Trans. on Design Automation of Electronic Systems,* vol. 18, no. 2, pp. 1-19, 2013.

978-1-4799-2817-0/14 $31.00 © 2014 IEEE

VFGR: A Very Fast Parallel Global Router with Accurate Congestion Modeling *

Zhongdong Qi[1], Yici Cai[1], Qiang Zhou[1], Zhuoyuan Li[2], Mike Chen[2]

[1] Tsinghua National Laboratory for Information Science and Technology,
Tsinghua University, Beijing, China
[2] Nimbus Automation Technologies, Shanghai, China
e-mail : zhongdongqi@gmail.com

Abstract - With the rapid growth of design size and complexity, global routing has always been a hard problem. Several new factors contribute to global routing congestion and can only be measured and optimized in 3-D global routing rather than 2-D routing. We propose an enhanced congestion model in global routing to capture local congestion and more accurately reflect modern design rule requirements. To achieve better global and detailed routing solution quality, we propose a 3-D global router VFGR with parallel computing using this congestion model. Experimental results show that VFGR can achieve comparable or better global routing solution quality with two start-of-the-art global routers in shorter runtime. It is also demonstrated that adopting proposed congestion model in global routing, higher solution quality and much shorter runtime can be achieved in detailed routing stage.

I. Introduction

Global routing is one of the most important stages in back-end VLSI design. It can serve as a congestion estimator to guide routability optimization in placement, and can also provide start point for detailed router to construct final layout. The chip's routability, timing, power and time-to-market are all heavily affected by global routing.

In global routing (GR), the routing region is tessellated into rectangular global routing cells (gcells), and a 2-D or 3-D routing graph is constructed to measure the routing congestion. In the graph, typically edge capacity is measured by the number of routing tracks between two gcells, while edge demand is the number of wire segments across the gcell boundary. Global router plans paths of nets on this graph. Guided by GR results, detailed routing (DR) searches and realizes real wires and vias inside all gcells.

As technology advances, several new factors are introduced, such as varying metal widths over the routing layers, via resource consumption and increase of design rules volume. They all enlarge the gap between GR congestion model and DR resource consumption situation, and contribute to complexity of global routing problem.

From 90nm technology node, varying width and thickness metal layers are available in VLSI design. At the 65nm technology node, typically three different metal widths, namely 1X, 2X and 4X are used across 9 metal layers (illustrated in Fig. 1). Even more metal widths and thickness are used across more metal layers in successive technology nodes. This makes routing truly a 3-D problem [1]. As shown in Fig. 1, varying metal widths and thicknesses introduce fat vias into interconnections, which consume much more

resources than normal vias. In addition, more stacked vias are generated in global routing to form multi-layer connections as the number of routing layers increases. Resources consumed by fat vias and stacked vias can only be measured and optimized in 3-D global routing. Varying width metal layer stack and via resource consumption make 3-D global routing crucial.

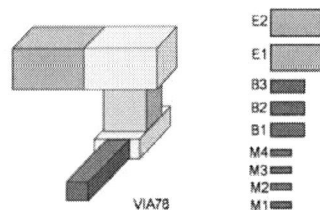

Fig. 1. A fat via and a metal layer stack with varying widths over different layers.

Due to sub-resolution lithography, chemical-mechanical polishing (CMP) and yield enhancement requirements, the number of design rules becomes larger in successive technology nodes. The design rules greatly affect amount of routing resources consumed by local connections and vias. Vias and local connections affected by design rules contribute a major mismatch between global routing congestion model and detailed routing resource consumption.

During global routing, an inaccurate congestion model misleads path search and enlarges the gap between global routing and detailed routing. The conventional congestion model cannot take above factors into account and underestimates resource consumption of global routes. Using the model, a global routing path could be unroutable in detailed routing stage, causing costly path rerouting in dense detailed routing grid. This makes a practical global congestion model essential.

In addition, 3-D global routing has a significantly larger search space than 2-D global routing, which would lead to much longer runtime. Today's multi-core architecture makes multi-threaded computing available, which can be used to accelerate this computational intensive problem.

In this paper, we propose an accurate and practical congestion model to capture some of resource consuming factors in modern technology. Using 3-D global routing graph based on the model, we develop a multi-threaded 3-D global routing algorithm named VFGR, which can produce high quality solutions using short runtime.

Our contributions are as follows:

1) An accurate congestion model capturing local congestion due to vias and local nets resource consumption along with

---

* This work is supported by National Natural Science Foundation of China (NSFC) No.61274031.

978-1-4799-2817-0/14 $31.00 © 2014 IEEE

related design rules. We demonstrate that using proposed congestion model in global routing can lead to great reduction of detailed routing runtime and the number of design rule violations.

2) A constructive hierarchical global routing framework and detailed techniques which are performed on 3-D routing graph. Experimental results demonstrate that good solution quality can be achieved.

3) Efficient parallelization of proposed global routing framework. A 6X speed up can be achieved using 8 threads on an 8-core CPU.

The rest of paper is organized as follows. Section II gives a brief statement of related work. Section III presents the details of proposed congestion model. The framework of VFGR and routing techniques are described in section IV, followed by experimental results. We conclude the paper in section VI.

II. Related Work

A. Congestion Modeling

In recent years, some pioneer works in global routing or routability estimation were done to remedy the inconsistency between GR congestion model and DR resource consumption. Hsu et al. [4] proposed concepts of via capacity and via overflow to take stacked vias into account. As the technology node becomes smaller, this model is not practical and out dated. After that, in work of Taghavi et al. [21], pin geometries and density was used to measure detailed routing difficulty in routability-driven placement. Wei et al. [23] used pin density factor to measure local congestion. Local resource usage is considered in by increasing edge demand. Most recently, Shojaei et al. [20] used vertex demand to measure local nets resource consumption.

The above works improved the accuracy of routability estimation during placement or congestion modeling in global routing. However, fat vias were not modeled, and design rules were not modeled explicitly. The vias and local nets were not measured in a uniform model.

B. Global Routers & Parallel Global Routing

Significant progress on global routing algorithms has been made in recent years. A number of high quality global routers were developed, which can be roughly categorized into two classes, sequential ones and concurrent ones, according to how the nets are routed.

In sequential class, FGR [19], MaizeRouter [16], Archer [17], NTHU-Route 2.0 [3], NTUgr [4], FastRoute [18], MGR [26], BFG-R [11], NCTU-GR 2.0 [14] and several other high performance global routers use ripping-up and rerouting technique to get overflow minimized solutions. Among these routers, FGR, BFG-R and MGR can perform 3-D routing directly, while others comprise of 2-D global routing and layer assignment. In concurrent class, routers typically use mathematical programming methods to route a group of nets at the same time. BoxRouter [5] and GRIP [24] use integer programming to find routes for nets in a routing region. There has been very few works on parallel global routing to date. PGR [15] introduces a net-level collision concept to multi-threaded global routing. PGRIP [25] uses region

partition to generate sub-problems which solved using integer programming in a distributed computing cluster. Although PGRIP is parallel, the usage of integer programming makes the runtime still relatively long. The work in [10] adds computing with GPU into parallel global routing. In this paper, we try to explore region-level and net-level parallelism in shorter runtime using multi-threaded computing.

III. Accurate Congestion Modeling

The explosion of design rules makes modeling all the rules impossible. Many rules are seldom triggered in practical design. Two major design rules affecting local congestion are minimum area rule and end-of-line spacing rule (depicted in Fig. 2). Minimum Area Rule (*MinArea*) specifies the minimum metal area required for polygons on the layer. All polygons must have an area that is greater than or equal to a minimum area value. End-of-Line Spacing Rule (*EOL-Spacing*) indicates that a polygon edge that is shorter than certain length requires spacing greater than or equal to a spacing value beyond the end-of-line within some distance.

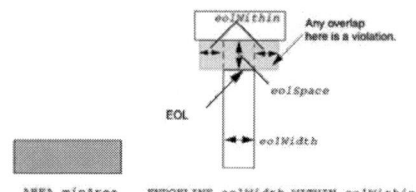

Fig. 2. Illustration of MinArea and EOL Spacing rules [13]

To practically model detailed routing resource consumption, let's first look at an example of a gcell with detailed routing results shown in Fig. 3. This gcell has 8 wire segments and two fat vias (one stacked and one non-stacked) in it. For inter-gcell connections, it has 7 and 6 wire segments on left and right gcell boundary respectively, resulting edge congestion as 7/12 and 6/12 on corresponding left and right global routing graph edges. Measured by these values, the gcell is not congested. However, considering intra-gcell connections and design rules, the region is too congested to let an extra wire segment cross the gcell. The fat via enclosure's width is wider than the minimal width on the layer. Considering the spacing rule, adjacent two tracks are not available for segments to pass through. As for a stacked via enclosure, its area is smaller than the minimal area value in MinArea rule, so the enclosure is enlarged with an extra wire to obey the rule.

From this case, we can see that (1): the fat vias and stacked vias can cause serious routability problems in detailed routing if they are not treated carefully at global routing stage; (local nets contribute to congestion in a similar way) (2): only inter-gcell congestion (edge congestion in global routing grid graph) is not capable to measure real detailed routing congestion. So we use pass-through capacity and demand to measure intra-gcell congestion, which is embedded in vertices of 3-D global routing graph. A simple example is depicted in Fig. 4.

In this graph, inter-gcell congestion is measured by edge capacity and demand, and intra-gcell congestion is measured by pass-through capacity and demand.

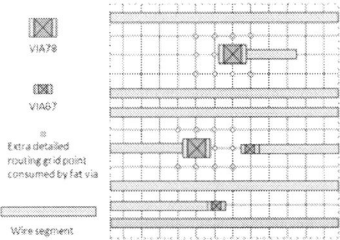

Fig. 3. A gcell with 2 fat via enclosures and 8 wire segments.

To avoid design rules checking surprise in detailed routing stage, we need to assign enough amounts of resources to each feature of net routes in global routing.

A **fat via** enclosure on thin metal layer is as wide as minimum width of the upper thick metal layer which it connects. For example, in a typical $65nm$ technology, due to minimum spacing rule on thin metal layer, the via enclosure makes neighboring two wiring tracks unavailable for placing wire segments to pass through. So a fat via contributes 3 to pass-through demand of this gcell on thin metal layer.

A **stacked via** have enclosures on connected metal layers. An enclosure may trigger MinArea and EOL-Spacing rule. To avoid violations, the enclosure is enlarged to *minArea* with spacing *eolSpace* on both ends. As a results, for a stacked via enclosure, the contributed pass-through demand is *minArea / enclosure_width + 2 * eolSpace*.

A **local net** (or subnet of a global net) has its pins all in one gcell. In a detailed routed design, a local net is typically routed using Metal2 and Metal3. Each local net is decomposed to segments using rectilinear minimum spanning tree construction. Since a local net is relatively small, a wire segment can easily violate MinArea and EOL-Spacing rules. The way of computing pass-through demand contributed by a local net segment is similar to that of a stacked via enclosure.

Inter-gcell wire segments also contribute to pass-through demand. A wire passing through the gcell increases pass-through demand by 1. The wire segments connecting to the gcell but not passing through it (denoted as side wire segments) contribute $\max\{N_l, N_r\}$ to pass-through demand, where N_l and N_r are number of left/lower and right/upper side wire segments connecting to the gcell, respectively. For example, $N_l = 2$ and $N_r = 1$ for the gcell in Fig. 3.

This congestion model is compatible with widely-used path search algorithms in global routing, such as pattern routing and maze routing. Conventionally, these path search algorithms are carried out on 2-D/3-D global routing grid graph edges. The cost function is generally related to edge congestion, wirelength, via count etc. Since vertices are connected by edges in the routing graph, it is natural to perform these algorithms on the routing graph to utilize both edge and vertex information. The only revision for the path search algorithms is to take vertex (pass-through) congestion into account.

IV. Multi-threaded 3-D Global Routing

A. Proposed Global Routing Framework

In global routing problem, nets compete for limited routing resources to achieve their own routes. In our observation on designs routed by academic and commercial routers,

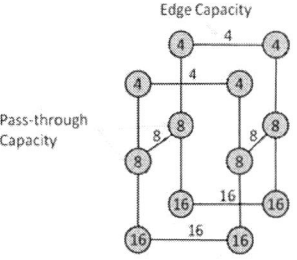

Fig. 4. Proposed congestion model in 3-D global routing graph.

characteristics of the resources consumed by a net are highly correlated with the size of the net. Generally, smaller nets consume resources in smaller region on lower routing layers, and larger nets take resources in a larger scope on higher routing layer. Smaller net generally has less routing flexibility, while larger net has more. This principle was used in several previous works [2] [7].

From this observation of resource distribution characteristics, we propose a hierarchical 3-D global routing approach. In this approach, multiple levels of hierarchy are constructed on the routing region, in which each hierarchy level has different size of global routing cell. Nets are fitted into different hierarchy level according to their bounding box size. Global routing is carried out from bottom hierarchical level to top level. In each hierarchy level, parallelism in region-level or in net-level is used, which is presented in subsection IV-C.

The design flow of proposed global routing is depicted in Fig. 5. Detailed techniques employed in the flow as well as parallelization of different level routing are presented in the following sub-sections.

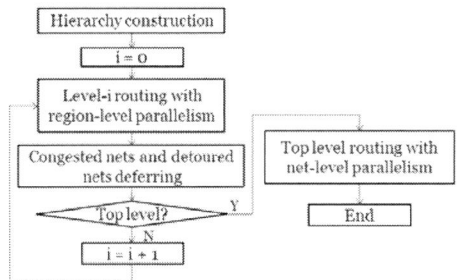

Fig. 5. Flow of proposed global routing.

For hierarchical global routing, hierarchical 3-D grid graphs are constructed beyond original global routing grid graph. We denote the original global routing grid graph as level 0 graph. Each level-$(i+1)$ graph is constructed by merging level-i graph neighboring N x N gcells (N is 4 in our implementation). The graph coarsening procedure stops when current level coarse graph has row or column number less than predetermined threshold.

All the nets are decomposed into 2-pin subnets (as described in subsection IV-B1), and are fitted to each hierarchy level according to bounding box size. A subnet is mapped to level-i if its bounding box is not larger than level-i graph gcell but is larger than level-$(i-1)$ graph gcell. Nets in each hierarchy level include nets fully embedded in a gcell (denoted as gcell nets) and nets across gcell boundary (denoted as boundary nets).

The hierarchical global routing proceeds from bottom level

to top level. In each level routing, gcell nets are routed in region of one gcell. Carried out in each region of 2 x 2 gcells in current level graph, boundary nets are then routed. Because the boundary nets cross different boundaries with even and odd indices, four passes of 2 x 2 gcell routing is performed with different start row and start column index.

In each regional routing problem, 3-D path search is directly performed on original global rouging grid graph (details are in subsection IV-B2). We do not utilize coarsened graph to route nets due to difficulty of modeling congestion on the coarsened graph.

After routing in each region, congested nets and detoured nets are deferred to higher level routing, which is presented in subsection IV-B3.

B. Detailed Techniques

(1). Flexible Net Topology and Resource Sharing

In global routing, net decomposition method, which impacts net tree topology, has strong impact on global routing solution quality and runtime. Rectilinear Steiner minimal trees (RSMT) and rectilinear minimum spanning tree (RMST) are two choices for decomposition. RSMT has shorter wirelength but less flexibility for rerouting, while RMST has better flexibility for rerouting but relatively longer wirelength. In our router, both RSMT and RMST are utilized to deliver advantages of both kinds of net topology. We use minimum spanning tree (MST) to decompose each multiple pin net to 2-pin subnets, and use RMST generated by FLUTE [6] to guide the path search of each subnet. Guided by RMST, a path search has zero wirelength cost for an edge connected to a Steiner point. This helps wirelength reduction during subnet routing. An example is shown in Fig. 6.

A necessary partner technique with net decomposition using MST is resource sharing between subnets of each net [19]. In global routing grid graph, each edge records which nets are utilizing it. A path search has zero wirelength cost using an edge which is already utilized by another subnet of the net.

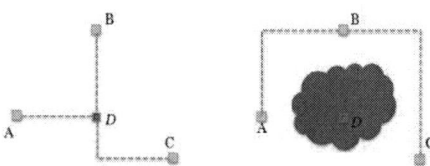

Fig. 6. A net with three pins *A*, *B*, and *C*. Point *D* is Steiner point. The net is decomposed into two subnets *A-B* and *B-C*. If there is no congestion around Steiner point *D*, paths passing *D* is created for both subnets, with resource sharing between them. If congestion exists in region around *D*, overflow-free paths can also be created.

(2). Regional Routing with Bounded Box Path Search

In each hierarchical level, routing of nets in each sub-region is proceeded using ripping-up and rerouting (RRR). Specifically, negotiated-congestion routing [27] is used. Each subnet is rerouted by a 3-D A* search engine. The iterative RRR finishes when overflow is reduced to zero or iteration count reaches a threshold.

An important aspect of hierarchical routing is to ensure appropriate resource consumption by nets in different levels. In regional routing of a certain level, routing path of a subnet should be constrained into a bounded box. If a subnet path

occupies much bigger region than bounding box of its pins, it consumes resources belonged to remaining unrouted nets. In order to limit resource usage, A* search is performed in a bounded box to facilitate bounding box control of routed path. During several rounds of rerouting, bounding box of A* search is gradually enlarged to find a less congested feasible solution.

(3). Deferring Congested and Detoured Nets to Higher Level

In regional routing, a net path could be congested after iterative RRR, which implies current region restricts the path search of the net. A net path could also be much longer than its optimal path wirelength (an example is shown in Fig. 7), which consumes the resources of other unrouted nets. These nets should be routed in a larger region. When routing in a region finishes, congested nets as well as nets with detour ratio larger than a threshold are deferred to higher level routing and rerouted in higher level. This enables a more natural resource utilization.

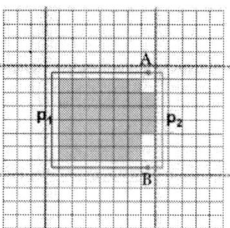

Fig. 7. A subnet *A-B* with long detoured path p_1. Blue dashed lines are boundaries of gcells on this level. Black solid lines are boundaries of gcells of original routing graph (level-0 gcells). Orange colored gcells are 100% full. It's proper to have path p_2 for the subnet, which can be achieved by deferring this subnet to higher level routing.

C. Parallelism in Different Level Routing

We now present method of parallelization in different level routing, using multi-threaded computing, in shared memory multiple-instruction multiple-data (MIMD) computing platform.

In proposed algorithm framework, the most time consuming part is regional routing in each level. Because all regions are independent in data, i.e. subnets to be routed and portion of global routing grid graph used in regional routing, *region-level* parallelism can be performed to speed up entire global routing.

However, higher levels which contain relatively small amount of gcells could have poor load balance during routing. If the amount of regions to perform routing is less than the amount of available CPU cores, some cores are inevitably idle while other cores are working. Similar situation occurs when number of nets to be routed varies greatly over different regions and amount of regions is close to the amount of CPU cores.

To handle this issue, we utilize *net-level* parallelization for higher level routing. During hierarchy construction phase of algorithm, graph coarsening is set to stop when row or column number of next level coarse graph is less than 2 * *K*, where *K* is the amount of cores. The reason of choosing 2 * *K* is that boundary nets are routed in a region with size of 2*2 gcells. This level is set to be top level of hierarchical routing, and contains all remaining unmapped nets. These nets are routed in parallel in a path search bounding-box overlap free manner.

(1). Region-level parallelism in lower level routing

For a lower hierarchical level, routing of each region is constructed as a computing task. The routing procedure presented in subsection IV-B2 needs no revision. A routing space containing a partial global routing grid graph covered by the region, and all subnets to be routed in the region are prepared as task data. All the tasks are pushed into task queue and are consumed by available CPU cores. When a CPU core finishes a task, it obtains a new task from task queue.

(2). Net-level parallelism in top level routing

In top level, routing of each subnet is a computing task. A pointer to common routing space containing the entire global routing grid graph, and pins of the subnet are constructed as task data. All the tasks are constructed are pushed into task queue.

Routing procedure in each subnet task is to gradually enlarge path search bounding box (BBox) and search for a path using bounded box A* search engine. Each subnet in routing task has a record of a *path search BBox* (not subnet BBox) for current path search. To avoid multiple threads access or modify a global routing grid graph edge or vertex congestion information in the same time, all the path search BBoxes of tasks proceeded in a certain time are guaranteed to have no overlap. This is achieved by judging a task's path search BBox. If a task has a path search BBox not overlapped with those of current processing tasks, it's compatible with current tasks and will be consumed by a thread, else it will be pushed back to the task queue.

V. Experimental Results

We implemented VFGR using C++. The C++ code is compiled and run on a server with Intel 8-core 2.40GHz CPU and 24GB memory.

The benchmarks we used are the DAC 2012 placement and routing benchmark suite which contains several challenges in modern technology [22]. These designs contains massive hard macros, varying metal layers width and spacing, pins on higher metal layers. The placement solutions we used are produced by the winner placer NTUplace4 in DAC 2012 routability-driven placement contest (achieved from [8]).

Two flows using different settings of the benchmark suite are performed to demonstrate effectiveness of our global router and proposed congestion model respectively (subsection V-A and V-B). The first one uses the original gcell size and routing capacity setting in DAC 2012 benchmark suite to only perform global routing. The second one uses modified smaller gcell size and routing capacity to run full-flow global routing and detailed routing. The details are as follows. To facilitate detailed routing, we change the gcell size from 32x40 to 9x9 (9 is the placement row site height, in unit of 1 pitch) in second setting. We map one unit in placement image to one pitch (i.e. 180nm in 65nm technology node) in Metal1. We also add the 65nm design rules to the designs and construct the pin figures and obstacles in standard cells. The benchmark files are transformed to LEF and DEF files and also OpenAccess database. This set of benchmarks can be used for both global routing and detailed routing.

A. Performance of VFGR

To demonstrate the performance of our global router, we compare the runtime and solution quality of BFG-R and NCTU-GR 2.0 with that of VFGR under traditional congestion model (by turning off vertex capacity and design rules related congestion factors) on original DAC 2012 benchmarks.

BFG-R runs in multi-layer mode and performs 3-D routing. NCTU-GR 2.0 comprises of 2-D global routing and layer assignment, and run in regular mode. Since the released NCTU-GR 2.0 version does not support multi-threaded computing, it is performed using a single CPU core. VFGR does parallel routing using 8 threads. Three routers run on the same machine with an 8-core CPU.

The results are listed in Table 1. "OF" represents number of total edge overflow. "WL" is routed wirelength, which is in unit of 10^{10} nm. "via" is total via count in unit of 10^6. "E-CPU" denotes elapsed runtime of global routing, in unit of minutes, while "CPU" is the total CPU time of 8 cores.

As a 3-D global router, VFGR is 1.24x and 7.29x faster than another 3-D router BFG-R in term of CPU time and elapsed time, respectively. Since NCTU-GR 2.0 uses 2-D routing followed by layer assignment, which has smaller search space than 3-D router, NCTU-GR 2.0 spends shorter CPU time than VFGR. Using multi-threaded computing, VFGR is faster than NCTU-GR 2.0 measured by elapsed runtime. In term of overflow, wirelength and via count, VFGR achieves comparable or better solution quality with NCTU-GR 2.0 and BFG-R.

The speed up of parallel computing in VFGR is about 6X on average, which is relatively high for a multi-threaded program run on 8-core CPU. This implies data dependency in global routing is controlled to a limited scope using proposed hierarchical global routing framework.

B. Effectiveness of Proposed Congestion Model

We use two full-flow routing configurations to show the influence of different congestion models. In these two configurations, global routing solutions are achieved by VFGR with proposed congestion model and BFG-R, respectively. Then a commercial detailed router supporting 130nm to 45nm design rules is used to perform detailed routing. The detailed routing results are checked by a commercial design rule checker. BFG-R uses traditional congestion model only considering wire congestion. The benchmark suite is the one with smaller gcell size to adapt with the detailed router. We compare the results of global routing and detailed routing phases which are listed in Table 2. "OF" represents number of total edge overflow (both edge and vertex overflow for VFGR). "WL" is routed wirelength, which is in unit of 10^{10} nm. "via" is total via count in unit of 10^6. "CPU" denotes runtime in unit of minutes. "DRC" is total number of design rule violations, in unit of 10^3.

Using proposed congestion model, VFGR produced slightly better wirelength and vias in global routing stage than BFG-R, with much shorter elapsed runtime. Guided by VFGR solutions, detailed router produced 59% fewer DRC violations and 6% and 9% smaller wirelength and via count in detailed routing solutions with 51% shorter runtime compared to the one using BFG-R global routing results. VFGR reported many edge and pass-through overflows which indicated detailed routing congestion (i.e. DRC violations).

978-1-4799-2817-0/14 $31.00 © 2014 IEEE

These results validate the effectiveness of proposed global congestion model, which is better correlated to detailed routing. This indicates that proposed congestion model using both edge and vertex capacity captures detailed routing congestion much more accurately than conventional model does. The global routing paths guided by proposed model can guide detailed router finding feasible interconnections with less design rule violations and avoids massive rerouting in dense detailed routing grid.

VI. Conclusion

Facing the challenges brought by successive technology nodes for global routing, we propose enhancements to current congestion models, to capture local congestion and more accurately reflect modern design rule requirements. We propose a hierarchical 3-D global router with parallel computing named VFGR. Experimental results show the effectiveness of proposed parallel global routing algorithm. It is also demonstrated that proposed congestion model is better correlated to detailed routing resource consumption than conventional model.

References

[1] Charles J. Alpert, et al. "What makes a design difficult to route." in Proc. of ISPD, 2010, pp. 7 - 12.
[2] Yao-Wen Chang and Shih-Ping Lin. "MR: a new framework for multilevel full-chip routing." IEEE TCAD 23.5 (2004): 793 - 800.
[3] Yen-Jung Chang, et al. "NTHU-Route 2.0: a robust global router for modern designs." IEEE TCAD 29.12 (2010): 1931 - 1944.
[4] Chin-Hsiung Hsu, Huang-Yu Chen, and Yao-Wen Chang. "Multi-layer global routing considering via and wire capacities." in Proc. of ICCAD, 2008, pp. 350 - 355.
[5] Minsik Cho and David Z. Pan. "BoxRouter: a new global router based on box expansion and progressive ILP." IEEE TCAD 26.12 (2007): 2130 - 2143.
[6] Chris Chu and Yiu-Chung Wong. "FLUTE: Fast lookup table based rectilinear steiner minimal tree algorithm for VLSI design." IEEE TCAD 27.1 (2008): 70 - 83.
[7] Jason Cong, Jie Fang, and Yan Zhang. "Multilevel approach to full-chip gridless routing." in Proc. of ICCAD, 2001, pp. 396 - 403.

[8] http://archive.sigda.org/dac2012/contest/dac2012_contest.html
[9] Ke-Ren Dai, Wen-Hao Liu, and Yih-Lang Li. "NCTU-GR: efficient simulated evolution-based rerouting and congestion-relaxed layer assignment on 3-D global routing." IEEE TVLSI 20.3 (2012): 459 - 472.
[10] Yiding Han, et al. "Exploring high throughput computing paradigm for global routing." in Proc. of ICCAD, 2011, pp. 298 - 305.
[11] Jin Hu, Jarrod A. Roy, and Igor L. Markov. "Completing high-quality global routes." in Proc. of ISPD, 2010, pp. 35 - 41.
[12] Ryan Kastner, et al. "Pattern routing: use and theory for increasing predictability and avoiding coupling." IEEE TCAD 21.7 (2002): 777 - 790
[13] LEF/DEF reference 5.7, Cadence, Nov. 2009
[14] Wen-Hao Liu, et al. "NCTU-GR 2.0: Multithreaded Collision-Aware Global Routing With Bounded-Length Maze Routing." IEEE TCAD 32.5 (2013): 709 - 722.
[15] Wen-Hao Liu, et al. "Multi-threaded collision-aware global routing with bounded-length maze routing." in Proc. of DAC, 2010, pp. 200 - 205.
[16] Michael D. Moffitt. "MaizeRouter: Engineering an Effective Global Router." IEEE TCAD 27.11 (2008): 2017 - 2026.
[17] Muhammet M. Ozdal and Martin D.F. Wong. "Archer: a history-based global routing algorithm." IEEE TCAD 28.4 (2009): 528 - 540.
[18] Min Pan, et al. "FastRoute: an efficient and high-quality global router." VLSI Design 2012 (2012): 14.
[19] Jarrod A. Roy and Igor L. Markov. "High-performance routing at the nanometer scale." IEEE TCAD 27.6 (2008): 1066 - 1077.
[20] Hamid Shojaei, Azadeh Davoodi, and Jeffrey Linderoth. "Planning for local net congestion in global routing." in Proc. of ISPD, 2013, pp. 85-92.
[21] Taraneh Taghavi, et al. "New placement prediction and mitigation techniques for local routing congestion." in Proc. of ICCAD, 2010, pp. 621 - 624.
[22] Natarajan Viswanathan, et al. "The DAC 2012 routability-driven placement contest and benchmark suite." in Proc. of DAC, 2012, pp. 774 - 782.
[23] Yaoguang Wei, et al. "GLARE: global and local wiring aware routability evaluation." in Proc. of DAC, 2012, pp. 768 - 773.
[24] Tai-Hsuan Wu, Azadeh Davoodi, and Jeffrey T. Linderoth. "GRIP: Global routing via integer programming." IEEE TCAD 30.1 (2011): 72 - 84.
[25] Tai-Hsuan Wu, Azadeh Davoodi, and Jeffrey T. Linderoth. "A parallel integer programming approach to global routing." in Proc. of DAC, 2010, pp. 194 - 199.
[26] Yue Xu and Chris Chu. "MGR: Multi-level global router." in Proc. of ICCAD, 2010, pp. 250 - 255.
[27] Larry McMurchie and Carl Ebeling, "PathFinder: a negotiation-based performance-driven router for FPGAs," in Proc. of ACM Int. Symp. on FPGAs, pp. 111–117, 1995.

TABLE I. Global routing results comparison under conventional congestion model.

Testcase	VFGR					NCTU-GR 2.0				BFG-R			
	OF	WL	via	E-CPU	CPU	OF	WL	via	CPU	OF	WL	via	CPU
superblue2	0	9.32	5.25	6.93	45.07	0	9.21	5.29	16.40	0	9.61	5.35	97.48
superblue3	0	5.5	4.82	5.48	33.98	0	5.42	4.81	11.05	0	5.99	5.17	49.83
superblue6	0	5.36	4.93	3.23	19.40	0	5.3	4.94	6.55	0	5.54	5.21	19.60
superblue7	0	6.48	7.29	3.87	22.42	0	6.46	7.31	5.90	0	7.33	7.92	25.32
superblue9	0	3.95	3.96	2.17	12.35	0	3.91	4.01	4.23	0	4.40	4.23	13.65
superblue11	0	5.22	4.2	2.15	12.03	0	5.21	4.37	2.82	0	5.49	4.37	14.68
superblue12	0	5.55	6.71	2.20	12.53	0	5.49	6.68	4.08	0	6.05	6.79	11.67
superblue14	0	3.54	3.32	3.50	17.85	0	3.51	3.30	3.22	0	3.69	3.61	18.48
superblue19	0	2.34	2.22	1.02	5.50	0	2.35	2.29	1.13	0	2.37	2.40	6.12
Average	0	1.000	1.000	0.17	1.00	0	0.992	1.009	0.29	0	1.066	1.058	1.24

TABLE II. Full-flow routing results using different global routers.

Testcase	BFG-R								VFGR with Proposed Congestion Model							
Stage	GR stage				DR stage				GR stage				DR stage			
	OF	WL	via	CPU	DRC	WL	via	CPU	OF	WL	via	E-CPU	DRC	WL	via	CPU
superblue2	0	10.01	7.58	306.3	2355	11.4	12.54	484.8	2056	9.92	7.51	39.5	713	10.7	11.5	271.5
superblue3	0	5.77	7.28	137.9	1312	6.69	11.45	516.2	2408	5.75	7.13	16.1	602	6.24	10.9	242.6
superblue6	0	5.54	7.42	122.7	1026	6.32	9.39	351.6	1516	5.49	7.27	17.9	513	5.93	9.12	179.3
superblue7	0	7.48	11.0	82.6	685	8.48	15.61	392.8	994	7.25	10.8	9.1	305	7.92	14.3	157.1
superblue9	0	4.39	6.31	51.7	316	5.13	10.58	308.9	330	4.34	6.19	6.5	116	4.77	9.53	139.0
superblue11	0	5.58	6.61	100.2	2561	6.41	10.21	329.6	4708	5.51	6.60	11.8	1256	6.01	9.91	168.1
superblue12	0	6.15	10.7	69.9	812	7.11	14.70	251.2	1302	6.10	10.5	7.6	319	6.62	14.0	145.7
superblue14	0	3.71	4.72	74.0	5910	4.29	8.40	340.2	10716	3.69	4.61	9.2	2294	4.01	7.5	170.1
superblue19	0	2.43	3.55	31.5	6627	2.73	7.76	279.6	9407	2.42	2.51	5.9	2198	2.64	5.69	136.7
Average	0	1.00	1.00	1.00	1.00	1.00	1.00	1.00	3715	0.99	0.98	0.13	0.41	0.94	0.91	0.49

Efficient Simulation-Based Optimization of Power Grid with On-Chip Voltage Regulator

Ting Yu and Martin.D.F. Wong

Department of ECE, University of Illinois at Urbana-Champaign

tingyu1@illinois.edu, mdfwong@illinois.edu

Abstract— IR-drop values of power grid can be reduced through inserting on-chip low-dropout voltage regulators (LDO). In this paper, we explore the optimization of LDOs to meet the IR-drop constraint, where the maximum IR-drop value is less than 10% of power supply. With Cholesky direct solver and SPICE, we propose a method to simulate power grid with LDOs. Based on the simulation method, we develop an efficient flow to optimize the number and locations of the LDOs. Effectiveness of the proposed method is verified by the experimental results. To the best of our knowledge, this is the first work optimizing the number and locations of LDOs to meet the IR-drop constraint.

I. INTRODUCTION

Voltage regulators are essential components in power delivery system. Traditionally, they are off-chip and very large. In recent years, researchers have developed low-dropout on-chip voltage regulators (LDO) [1]. It significantly improves load regulation, reduces crosstalk, eliminates load-transient spikes and saves board space [2]. It is very promising to integrate a large amount of on-chip LDOs into power distribution network to improve its performance.

There are a few works developed for integrating on-chip LDOs into power delivery system. Suming et al. analyzed the stability of integrating LDOs into power delivery system [3]. Zhiyu et al. demonstrated the significant benefits of on-chip LDOs in suppressing both high-frequency switching noises and mid-frequency voltage drops caused by resonance [4]. They also discussed the number of LDOs based on the assumption of even distribution of the LDOs. However, the assumption did not count in the effect of uneven current distributions and may result in more LDOs than the case of uneven distribution of the LDOs. Besides, restricted by the locations of whitespace, it is not always possible to evenly distribute the LDOs.

Based on previous work, we can see that adding a set of on-chip LDOs reduces IR-drops of power grid. To generate a reliable circuit, maximum IR-drop value should be less than 10% of the supply voltage. In the rest of this paper, this criterion is referred as the IR-drop constraint.

The number of LDOs that are added to meet the IR-drop constraint depends on multiple factors, including power grid structure, current distribution and devices' switching activities. Because the LDOs occupy chip area and routing resources, it is preferable to add fewer LDOs. Besides, the locations of the LDOs depend on the area and distribution of whitespace. Based on these considerations, we need to optimize the number and locations of the LDOs. In this paper, we define the optimization problem as: meeting the IR-drop constraint by adding a reasonable amount of LDOs at the whitespace locations, where the IR-drop values are effectively reduced. We will explore the optimization problem through simulation.

The optimization problem is very challenging, especially considering devices' switching activities. We need to develop an efficient flow to find out how many LDOs are required to meet the IR-drop constraint of all the time-steps. These LDOs have to be set at the locations that can effectively reduce the IR-drop values. Besides, we need to check whether the IR-drop constraint is met by adding a set of LDOs. To obtain accurate results, the verification should be done through simulation. However, lacking an effective numerical model of LDO, efficient power grid solvers such as Cholesky solver, random walks or Multigrid method are not able to simulate power grid with LDOs.

To address these issues, we first propose an efficient method to simulate power grid with LDOs. It is utilized to check whether IR-drop constraint is met with the LDOs or not. Based on the simulation method, we propose an efficient flow to find a reasonable solution of the optimization problem. We divide power grid into global and local power grids. Only local power grid is utilized to optimize the LDOs to meet the IR-drop constraint. The LDOs are settled at the locations where the IR-drop values are reduced very effectively.

Our major contributions include:

1. We propose an efficient method to simulate power grid with on-chip LDOs. Cholesky solver is utilized to simulate power grid, while SPICE is utilized to capture the transient responses of LDOs.

2. We propose an efficient flow to optimize the LDOs to meet the IR-drop constraint. Only local power grid is utilized to perform the LDO optimization. We

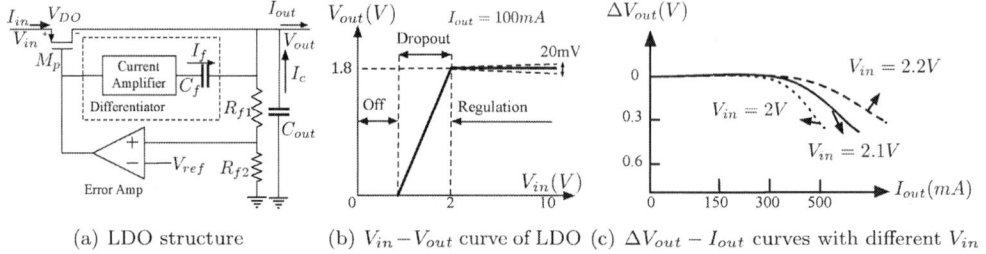

(a) LDO structure   (b) $V_{in} - V_{out}$ curve of LDO   (c) $\Delta V_{out} - I_{out}$ curves with different V_{in}

Fig. 1. Structure and electrical characteristic curves of a LDO

develop a method to set each LDO at the location where the IR-drop values are reduced very effectively.

3. We compare the optimization flow with two other methods, one of which utilizes even distribution of LDOs, while the other one adds the LDOs with a different strategy. The comparisons verify the effectiveness of the proposed flow.

In the following, Section II and Section III show the structure of LDO and the model of power grid with LDOs, respectively. Section IV illustrates the simulation method, while Section V presents the optimization flow. Experimental results and final conclusions are given in Section VI and Section VII, respectively.

II. ON-CHIP LOW-DROPOUT VOLTAGE REGULATOR

Figure 1(a) shows the model of an external capacitor free low-dropout voltage regulator (LDO) [1]. It is composed with a pass element (M_p), sampling resistors (R_{f1}) and (R_{f2}), reference voltage (V_{ref}), an error amplifier and a differentiator. We adopt the LDO topology from [1] and modify the design to generate an output voltage of $1.8V$. With $100mA$ output current, the $V_{in} - V_{out}$ curve of the LDO is shown in Figure 1(b).

As shown in Figure 1(b), LDO works in three different regions: off region, dropout region and regulation region. In dropout region, the output voltage of LDO is a linear function of input voltage. As the input voltage increases, LDO starts to work in regulation region, where the output voltage is constant and is almost not affected by the variation of input voltage.

Due to the resistance of M_p, there is a voltage drop between the input and output voltage of LDO when it starts to work in regulation region. For example, Figure 1(b) shows that when the LDO starts to work in regulation region, $V_{in} = 2V$ and $V_{out} = 1.8V$. The voltage drop between V_{in} and V_{out} is $200mV$.

LDO's output voltage depends on its input voltage and output current. Utilizing $V_{out} = 1.8V$ as a reference, Figure 1(c) shows the changes of LDO's output voltages with different output currents and input voltages. As output current increases, output voltage starts to decrease. The larger the output current, the less the output voltage. Increasing input voltage helps keeping output voltage from decreasing.

With the above observations, we should increase LDO's

input voltage to make it less affected by big output currents. However, input voltage cannot be too high, because higher input voltage reduces LDO's power efficiency [4]. Combining these considerations, we set $V_{in} = 2.2V$.

III. POWER GRID MODEL WITH LDOs

We adopt the power grid structure with LDOs from [4], which is shown in Figure 2(a). Power grid is represented as a RLC network. The dotted line segments represent resistors of the global or local power grid. LDOs connect the global and local power grids. The benefit of utilizing LDO to connect the global and local grids is to reduce both high-frequency on-chip power grid noises and mid-frequency noises caused by package variation [4]. In this paper, we focus on decreasing the on-chip noises of VDD network with flip chip technology. Only a single power domain is considered.

As stated in Section II, input voltage and output voltage of the LDO are $2.2V$ and $1.8V$, respectively. Because global grid supplies power to the LDOs, VDD values of the global grid should be at least $2.2V$. In this paper, we set $VDD = 2.2V$. Ideally, local power grid obtains $1.8V$ voltages from the LDOs.

With this power grid structure, voltage values of the local grid are directly affected by the distribution of the LDOs. In order to satisfy the IR-drop constraint, we need to optimize the number and locations of the LDOs.

IV. SIMULATING POWER GRID WITH LDOs

As mentioned before, it is very challenging to simulate power grid with LDOs. Based on Cholesky direct solver and SPICE, we propose a method to perform the simulation.

The simulation models of power grid and LDO are shown in Figure 2(b). In the power grid simulation model, LDOs are treated as voltage sources. The values of the voltage sources are the LDOs' input and output voltages. LDOs are simulated by SPICE with the model in Figure 2(b). In Figure 2(b), V_{ildo} and V_{oldo} are the input and output voltages of LDOs. V_A and V_B are the voltage values of nodes in global and local power grids.

The simulation flow is illustrated in Figure 2(c). Starting with $V_{ildo} = 2.2V$ and $V_{oldo} = 1.8V$, we simulate the global and local power grids to obtain V_As and V_Bs. In the simulation, companion models of capacitors and in-

978-1-4799-2817-0/14 $31.00 © 2014 IEEE   532

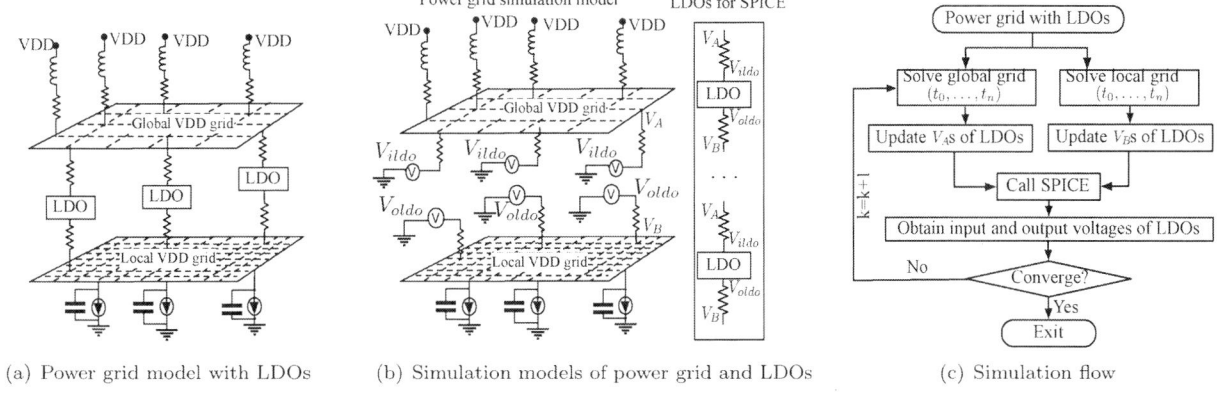

(a) Power grid model with LDOs   (b) Simulation models of power grid and LDOs   (c) Simulation flow

Fig. 2. Power grid structure and simulation flow with LDO integration

ductors are adopted from [5]. "CHOLMOD" is utilized as the Cholesky solver. After this, SPICE is utilized to simulate the LDOs and update V_{ildo}s and V_{oldo}s. This process repeats until the voltage values of power grid nodes are converged.

V. OPTIMIZING POWER GRID WITH LDOS

Based on the simulation method, we propose an effective flow to optimize the LDOs to meet the IR-drop constraint.

A. Optimization flow

The optimization flow is shown in Figure 3. With the geometrical information of device blocks, we extract candidate locations of the LDOs. For power grid, initially, we assume there is a single randomly settled LDO connecting the global and local power grids. Based on the DC analysis of local power grid, we develop a method to set the LDO at a location, where the IR-drop values can be effectively reduced. The optimization improves DC operation point of the circuit, which helps reduce IR-drop values of the transient time steps.

After optimizing the single LDO, we perform transient analysis of local power grid. A set of LDOs is added to eliminate the IR-drop violations. We develop an efficient method to find the locations of the new LDOs. Adding LDOs at those locations reduces IR-drop values very effectively. Each newly added LDO is assumed to provide an ideal output voltage to local power grid. Every LDO is connected to global power grid through the closest node of global power grid to the LDO. After adding the LDOs, we verify power grid with LDOs with our simulation method. Accurate output voltages of LDOs are obtained at this step. New iteration starts if there is still an IR-drop violation.

In the flow, we only simulate local power grid to optimize the LDOs. The insertion of the new LDOs depends on the voltage values of power grid nodes, which are obtained through transient analysis. Because both the voltages of global and local power grid nodes are af-

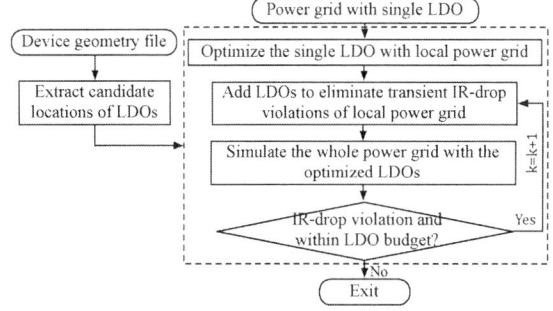

Fig. 3. Power grid optimization flow with LDOs

fected by the LDOs, we have to go through iterations of transient analysis of the whole power grid and SPICE simulations to obtain accurate voltage values of power grid nodes and LDOs. In our flow, we avoid the iterations of transient analysis and SPICE simulation by approximating the voltage values of the LDOs. LDOs that are new to the optimization loop have ideal output voltages. Other LDOs utilize their output voltages from previous iteration of the optimization loop. New LDOs are inserted by only performing transient analysis of local power grid, which makes the optimization much faster. Besides, by assuming that each new LDO provides an ideal output voltage to the local power grid, we are conservative about the optimization results. Because real output voltages of the LDOs are always lower than the ideal ones, the risk of adding excessive LDOs is avoided.

B. Candidate locations of LDOs

Assuming a LDO as a rectangle, we divide the chip into grids by the unit of LDO's width and height. The grids are shown as the dotted small rectangles in Figure 4. Grids that are not occupied by device blocks are the candidate locations of LDOs.

We represent a grid with its bottom left corner node. For example, in Figure 4, node A and B represent the rectangles which are occupied by the LDOs. The star symbols around the LDOs represent the nodes of local power grid. The lines connecting the LDOs and the nodes are the metal wires. LDOs should be close to the nodes

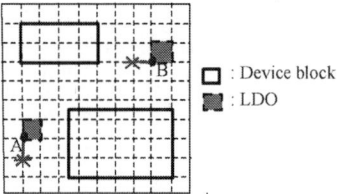

Fig. 4.   Candidate locations of LDOs

to reduce the wire lengths.

C. Optimizing the location of the single LDO

The process of optimizing the single LDO is shown in Figure 5. We develop an effective method to calculate the new location of the LDO. DC analysis of local power grid is performed after moving the LDO to the new location. After the DC analysis, we record the maximum IR-drop value of local power grid and the new location of the LDO. After repeating this process for M iterations, we recover the LDO to the location that results in smallest maximum IR-drop value. In our experiment, $M = 5$.

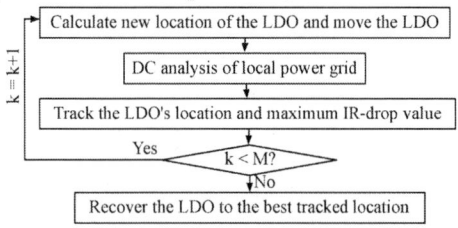

Fig. 5.   Optimization flow of the single LDO

The new location of the LDO is calculated as follows. With DC IR-drop values of local power grid nodes, similar to [6], we assign each node a weight, which is proportional to its IR-drop value. We then calculate the weighted geometrical center of local power grid nodes. The LDO should be moved to the closest candidate location to the geometrical center. In order to shorten the wire length from the LDO to local power grid, the LDO should connect to the local grid node that is closest to the geometrical center.

The above method effectively counts in the severity of different IR-drop values, which are weighted by their values and the number of nodes with the IR-drop values. Setting the LDO at the new location not only reduces the maximum IR-drop value, but also results in a more even IR-drop distribution. By moving the LDO for a few times, the single LDO can be quickly settled at a location where IR-drop values of local power grid are effectively reduced.

D. Adding LDOs to eliminate IR-drop violations

In previous step, we optimize the location of the single LDO. Depends on devices' switching activities and current extractions, more LDOs may be required to meet the IR-drop constraint.

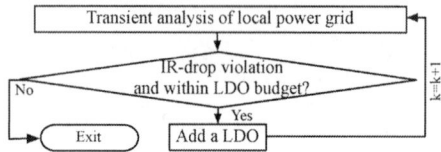

Fig. 6.   Flow of adding LDOs to eliminate IR-drop violations

Our flow of adding LDOs is shown in Figure 6. First, we perform transient analysis of local power grid. If there is an IR-drop violation, one LDO is added. This process repeats until the IR-drop constraint is met or there is no space to insert another LDO.

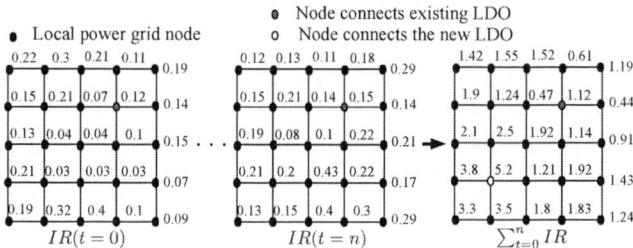

Fig. 7.   Accumulating IR-drop values for local power grid nodes

We develop a very effective method to add the new LDO. As shown in Figure 7, while performing transient analysis, we record the IR-drop values of the nodes of local power grid. By the end of the analysis, we obtain the accumulated IR-drop value of each node. The new LDO is settled at the closest candidate location to the node with maximum accumulated IR-drop value, such as the white node of Figure 7.

Setting the LDO based on the accumulated IR-drop values counts in the severity of both devices' switching activities and current extractions along all the time steps. Each added LDO is able to effectively reduce the IR-drop values of local power grid.

The pseudo code of the whole optimization process is listed in Algorithm 1. V_{oldo_ideal} is the ideal output voltage value of LDO. $V_{oldo\_ideal} = 1.8V$.

VI. EXPERIMENTAL RESULTS

The program is developed with C++ and SPICE. It is executed on a single CPU with a frequency of 2.13GHz, 512K cache and 32G RAM. We generate several test cases to test the optimization flow. Each test case includes a power grid file and a device geometry file. The device geometry file specifies the shape and locations of device blocks and initial location of the single LDO.

The test cases have similar parameters as that of IBM power grid benchmarks [7]. There are 1000 time steps in the transient analysis. We generate pulse current sources to represent devices' switching. The period of the currents varies from $1ns$ to $10ns$. We also assign small decoupling capacitors with the devices, whose values vary from $0.01\mu F$ to $0.1\mu F$. These capacitors are to compen-

978-1-4799-2817-0/14 $31.00 © 2014 IEEE   534

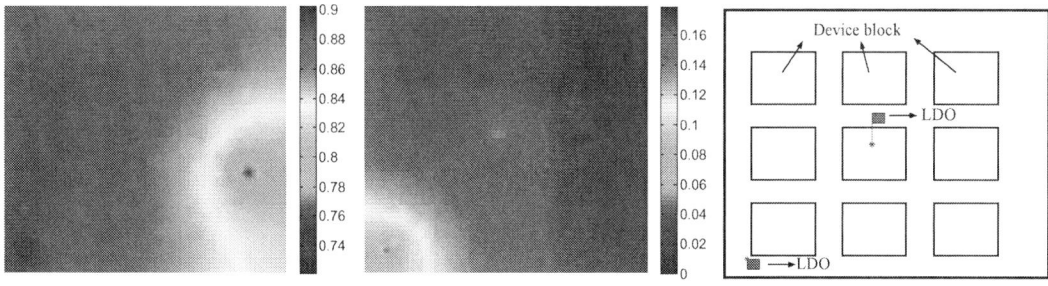

(a) Worst-case IR-drop values before optimization   (b) Worst-case IR-drop values after optimization   (c) Locations of optimized LDOs

Fig. 8. Worst-case IR-drop values and locations of LDOs with the proposed optimization method

---

Algorithm 1: Optimize power grid with LDOs

1 **begin**
2   Optimize the initial LDO with DC local power grid analysis;
3   **while** *(1)* **do**
4   **while** *(1)* **do**
5   Perform transient analysis of local power grid;
6              $\Delta V_{max} \leftarrow$ worst-case maximum IR-drop value;
7   **if** *($\Delta V_{max} \leq 10\% \times V_{oldo\_ideal}$ or no space for new LDO)* **then**
8   break;
9   **end**
10   Add a LDO;
11   **end**
12   Solve the whole power grid and the LDOs with the simulation method;
13   **if** *($\Delta V_{max} \leq 10\% \times V_{oldo\_ideal}$ or no space for new LDO)* **then**
14   break;
15   **end**
16   **end**
17 **end**

---

sate for the fast switching of devices.

In our experiment, the convergence of the proposed simulation method happens when the worst-case voltage differences of all the nodes between two continuous iterations are less than $3e^{-4}V$. Because LDOs provide local power grid power with an ideal output voltage of $1.8V$, the IR-drop constraint is:

$$\Delta V_{max} \leq 0.1 \times 1.8V = 0.18V \tag{1}$$

A. Effect of the proposed optimization method

We test the effect of the optimization flow on a small size power grid. There are around 17K nodes in local power grid. Global power grid has similar number of nodes. The initial LDO is located at the right hand side of the chip. The maximum DC IR-drop value is $0.206V$. With the initial LDO, worst-case IR-drop values are shown in Figure 8(a). The maximum IR-drop value is $0.913V$, which seriously violates the IR-drop constraint.

In the optimization, the LDO is moved to the center of the chip, which decreases the DC maximum IR-drop value to $0.158V$. The IR-drop constraint is met by adding only one more LDO. The locations of two LDOs are shown in Figure 8(c). With the two LDOs, worst-case IR-drop values are shown in Figure 8(b). The maximum IR-drop value is $0.179V$.

We randomly pick up a node of local power grid and record its voltage values before and after the optimization, which are shown as the red and blue curves in Figure 9, respectively. The node's voltage values are greatly increased by performing the optimization.

By comparing the IR-drop values and nodes' voltage values before and after the optimization, we can see that our method is very efficient in meeting the IR-drop constraint.

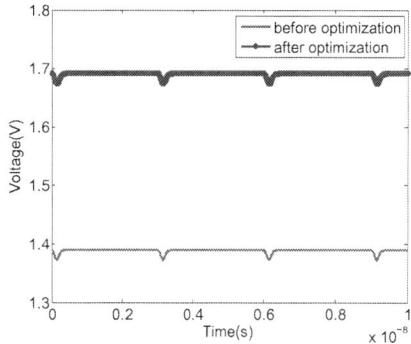

Fig. 9. Voltage values of a node before and after optimization

B. Comparison of the proposed method and others

Utilizing the above test case, we compare our proposed method with two other methods. The first one assumes even distribution of LDOs. The second method adds LDOs with a different strategy.

We first compare the proposed method with an even distribution of LDOs. Proven by our proposed method, two LDOs are sufficient to meet the IR-drop constraint. For fairness, we generate two LDOs and evenly distribute them in the whitespace. The maximum IR-drop value is $0.692V$, which is much higher than our method. Worst-case IR-drop values are shown in Figure 10(a). Comparing the IR-drop values to our optimization results, which are shown in Figure 8(b), we can see that our method achieves much less IR-drop values by allowing uneven distribution of LDOs.

In the second comparison, we develop another method to add the LDOs. Instead of adding a LDO after the complete transient analysis of local power grid, the LDO is added when there is an IR-drop violation at any time step. The new LDO is settled at the closest candidate location

978-1-4799-2817-0/14 $31.00 © 2014 IEEE   535

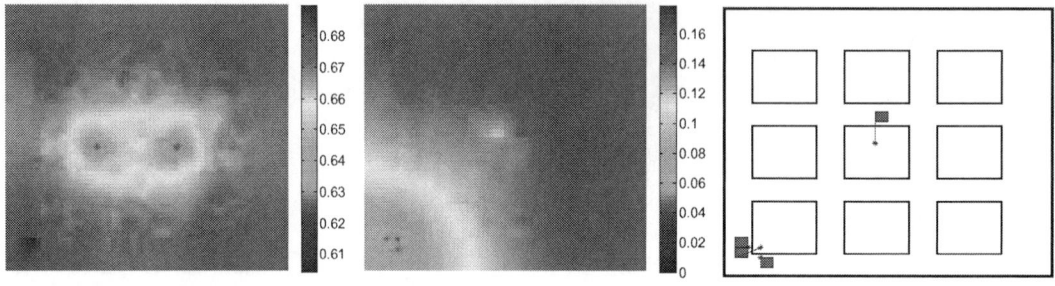

(a) Worst-case IR-drop values with even distribution of LDOs

(b) Worst-case IR-drop values of the second comparison method

(c) Optimized LDOs of the second comparison method

Fig. 10. Worst-case IR-drop values and locations of LDOs of the comparison methods

to the local power grid node, which has the maximum IR-drop value of the time step. With the method, three LDOs are added, which are shown in Figure 10(c). Worst-case IR-drop values are shown in Figure 10(b). The optimized maximum IR-drop value is $0.163V$.

Comparing Figure 10(b) and Figure 8(b), we see that the second method requires more LDOs than ours to meet the IR-drop constraint. The reason is as follows. In the second method, a LDO is added with the appearance of IR-drop violation at any time step. Adding the LDO improves the DC operation point of the transient analysis, which results in higher real voltages than the ones obtained by continuing simulating the later time steps. The method does not count in this factor and continues adding LDOs based on worse IR-drop values than real ones, which results in excessive LDOs to meet the IR-drop constraint.

C. Simulation results of more test cases

We test our optimization flow on several bigger benchmarks. A summary of the simulation results is listed in Table I. The names of the test cases indicate the number of nodes in the local power grid. Global grid has similar number of nodes. Meanings of other items are listed under Table I.

TABLE I
SIMULATION RESULTS OF THE PROPOSED OPTIMIZATION FLOW

Case	N_{ldo}	N_B	ΔV_{imax}	ΔV_{omax}	$T(h:m:s)$	K
C17K	2	301	0.913V	0.179V	00:49:20	98
C525K	3	580	0.935V	0.155V	08:54:50	53
C1M	13	139	1.534V	0.167V	34:38:24	92
C2M	21	621	1.192V	0.173V	62:30:12	103

1. N_{ldo}: number of LDOs; 2. N_B: number of LDO candidates; 3. ΔV_{imax} and ΔV_{omax}: maximum IR-drop value before and after optimization; 4. $T(h:m:s)$: runtime (hour:minute:second); 5. K: total number of transient analysis of global or local power grid

With the initial LDO, maximum IR-drop values of the test cases vary from $0.913V$ to $1.534V$, which greatly violate the IR-drop constraint. The violations are effectively removed by adding a set of LDOs with the optimization method. Optimized IR-drop values all meet the IR-drop constraint.

Table I also shows the runtime increases a lot with bigger test cases. Over 90% of runtime is cost on the local and global power grid transient analysis, which are utilized to insert the LDOs and to obtain the voltage values of power grid nodes and LDOs. Considering there are 1000 time steps, performing the multiple times of transient analysis results in long runtime, especially for bigger test cases. Runtime can be reduced by accelerating the transient analysis.

VII. CONCLUSIONS

In this paper, we propose an efficient method to meet IR-drop constraint of power grid by optimizing the number and locations of on-chip low-dropout voltage regulators (LDO). The optimization is based on a simulation method, which is composed by Cholesky solver and SPICE. Experimental results verify the effectiveness of the proposed method. To the best of our knowledge, this is the first work optimizing the number and locations of LDOs to meet the IR-drop constraint.

REFERENCES

[1] R. Milliken, J. Silva Martinez, and E. Sanchez Sinencio, "Full on-chip cmos low-dropout voltage regulator," *IEEE Transactions on Circuits and Systems I: Regular Papers*, vol. 54, no. 9, pp. 1879 –1890, September 2007.

[2] K. N. Leung and P. Mok, "A capacitor-free cmos low-dropout regulator with damping-factor-control frequency compensation," *IEEE Journal of Solid-State Circuits*, vol. 38, no. 10, pp. 1691–1702, October 2003.

[3] B. Y. Suming Lai and P. Li., "Stability assurance and design optimization of large power delivery networks with multiple on-chip voltage regulators," in *International Conference On Computer-Aided Design (ICCAD).*, November 2012, pp. 247–253.

[4] Z. Zeng, X. Ye, Z. Feng, and P. Li, "Tradeoff analysis and optimization of power delivery networks with on-chip voltage regulation," in *47th ACM/IEEE Design Automation Conference (DAC)*, June 2010, pp. 831–836.

[5] T. Yu and M. Wong, "PGT_SOLVER: An efficient solver for power grid transient analysis," in *IEEE/ACM International Conference on Computer-Aided Design (ICCAD)*, 2012, pp. 647–652.

[6] T. YU and M. Wong, "A novel and efficient method for power pad placement optimization," in *International Symposium on Quality Electronic Design (ISQED)*, 2013, pp. 158–163.

[7] "IBM Power Grid Benchmarks," http://dropzone.tamu.edu / pli/PGBench.

978-1-4799-2817-0/14 $31.00 © 2014 IEEE

Walking Pads: Fast Power-Supply Pad-Placement Optimization

Ke Wang[*], Brett H. Meyer[†], Runjie Zhang[*], Kevin Skadron[*], and Mircea Stan[‡]

[*]Dept. of Computer Science
University of Virginia
Charlottesville, VA, 22904, USA
{kewang, runjie}@virginia.edu,
skadron@cs.virginia.edu

[†]Dept. of Elec. and Comp. Engineering
McGill University
Montréal, Québec, H3A 0E9, Canada
brett.meyer@mcgill.ca

[‡]Dept. of Elec. and Comp. Engineering
University of Virginia
Charlottesville, VA, 22904, USA
mircea@virginia.edu

Abstract— **We propose a novel C4 pad placement optimization framework for 2D power delivery grids: Walking Pads (WP). WP optimizes pad locations by moving pads according to the "virtual forces" exerted on them by other pads and current sources in the system. WP algorithms achieve the same IR drop as state-of-the-art techniques, but are up to 634X faster. We further propose an analytical model relating pad count and IR drop for determining the optimal pad count for a given IR drop budget.**

I. INTRODUCTION

In modern system-on-chip design, supply-voltage-noise induced reliability issues are becoming increasingly challenging due to increasing current density [1]. Among the various sources of voltage noise, IR drop refers to the resistive drop across metal wires in the power delivery network (PDN). Typical design rules tolerate an IR drop ratio no more than 5% of supply voltage; violations can lead to timing errors.

In a flip-chip design, because the underlying silicon chip has a non-uniform power dissipation, the number and locations of controlled-collapse-chip-connection (C4) pads connecting to the on-chip PDN have a large impact on IR drop. Thus optimizing both the number and location of power supply C4 pads becomes critical to guarantee the desired IR drop target. Moreover, given the fact that both power supply and signal I/O share the same physical interface—C4 pads—determining the minimum number of power pads required for a given chip design through such optimization can help a designer to determine the available I/O bandwidth, or even perform tradeoffs between I/O bandwidth and the IR drop target.

Previous works have addressed pad placement optimization for the purpose of minimizing IR drop [2, 3, 4]. However, their approaches have scalability limitations, and as a result are not suitable for the large pad placement design space of modern systems. Some other works provide analytical methods to estimate max IR drop when pad number and pad locations are given [5, 6]. To the best of our knowledge, no existing work investigates the minimum number of C4 pads required to satisfy a target IR drop in a 2D PDN grid.

In this paper, we propose a fast method to obtain the minimum pad number for a target IR drop and corresponding optimized pad locations. First, we introduce a new method of power pad placement optimization, *Walking Pads* (WP). The key idea behind WP is to convert a global optimization problem, the placement of n pads given m candidate locations, into

a local balance problem, the placement of individual pads (current sources) with respect to various nearby current demands. Treating pads as "mobile positive charges" and the on-chip PDN grid as a 2D electrostatic voltage field, WP optimizes pad location by letting pads "walk" in the direction of the total virtual force exerted upon them to achieve local force balance.

WP achieves significant speedup over existing methods in the literature because it has two significant advantages:

1. WP leverages the underlying voltage gradients to quickly identify promising pad locations.
2. WP allows all pads to step toward their balanced positions simultaneously, reducing algorithm complexity significantly as a function of target pad count.

Second, we derive an analytical formula to describe the relationship between IR drop and pad number based on optimized pad locations. While not a closed-form model, our analytical formula only requires that three coefficients be fit to a curve, and can identify the optimal pad count to within two pads for systems with 128-1024 pads. When combined with WP, our analytical formula can quickly and accurately predict the minimum required pad count.

This paper makes two principal contributions:

1. We propose WP and demonstrate that it achieves at least 100X speedup with respect to the classical simulated annealing (SA) methods in the literature, while sacrificing no more than 0.1% VDD in steady-state IR drop.
2. We propose an analytical formula that describes the relationship between the number of pads and the expected maximum IR drop assuming optimized pad locations.

Together, the analytical model and WP algorithm are positioned to significantly accelerate the optimization of power pad count and placement, and therefore create new opportunities for joint optimization.

II. RELATED WORK

Sato et al. proposed the Successive Pad Assignment (SPA) method of power pad location optimization for pad ring allocation [3]. Zhao et al. provided a solution of mixed integer linear program (MILP) for pad ring allocation [2]. The computational complexities of both SPA and MILP grow quickly as problem size increases. As a result, they are not tractable for large scale 2D C4 arrays. Zhong and Wong proposed a fast power pad placement optimization algorithm within the framework of simulated annealing (SA) [4]. This method localizes the ef-

978-1-4799-2817-0/14 $31.00 © 2014 IEEE

fect of pad movement using a node-based iterative method and therefore improves the performance of each SA iteration. However, the localization is based on the hypothesis that the voltages of pad-PDN connection points cannot affect each other. This is not true when the package circuit and pad resistance are considered. Furthermore, their approach sacrifices accuracy when accelerating calculations [4], and cannot work with other efficient numerical methods like preconditioned Krylov subspace methods [7].

Shakeri proposed a theoretical method of accurate IR drop estimation for uniform power consumption floorplans with uniformly distributed pads [5]. Rius extended this work to a closed-form expression for non-uniform power consumption floorplans with arbitrary pad counts and locations [6]. However, Rius' work is based on the assumption that power pads are uniformly distributed on a rectangular 2D array. As shown in Section VII, IR drop is systematically overestimated in this case relative to the expected IR drop of optimally placed pads.

Walking Pads and the analytical model we have developed, unlike any prior work, enable designers to efficiently determine the relationship between pad count and IR drop, and therefore optimal pad allocation. Such an approach is critical for pre-RTL design, as the number of pads required for power delivery affects the number of pads available for I/O, and therefore has implications for system architecture and microarchitecture.

III. PROBLEM FORMULATION

A. Power Delivery Network Model

The typical regularity of the on-chip PDN's physical structure makes compact PDN modeling feasible. A well accepted methodology models the multi-layer metal stack as a 2D resistor mesh [8]. C4 pads are modeled as individual resistors attached to on-chip grid nodes, and the relative locations of those connection points in the grid represent the actual locations of the C4 pads on the silicon die. Ideal current sources are used to model the load (*i.e.* switching transistors). Off-chip components like the package or printed circuit board (PCB) are lumped into single resistors. To the best of our knowledge, lumped package models are adopted in most current related work. We adopt this methodology and build the model skeleton as in Fig. 1 [9]. We assume the PCB represents an ideal power supply and simultaneously model lumped package resistors, pad resistors and on-chip 2D resistor mesh; the steady state equations we solve therefore capture not only the on-chip 2D resistor mesh, but the package and pad resistors as well, with the latter elements changing as pads move from one candidate location to another.

To solve for voltage and current values in the model circuit, we employ sparse LU decomposition with pivoting, using SuperLU [10]. A direct solver with pivoting is generally considered a numerically stable and accurate method, and protects optimization quality from numerical errors. When implemented using advanced reordering techniques [11], sparse LU reduces memory usage significantly and achieves adequate performance for use in our experiments. It is worth noting that the proposed Walking Pad algorithm framework is a high level optimization framework, is thus not restricted to a particular nu-

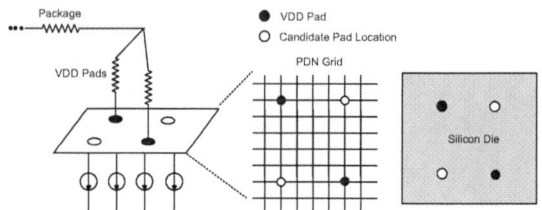

Fig. 1. Model of 2D PDN.

merical method, and can therefore take advantage of ongoing advances in numerical methods [7, 12].

B. Power Pad Location Optimization

Given the system floorplan, the number of power pads to place, and system power trace, the objective of power pad location optimization is to identify grid locations at which to place pads in order to minimize the maximum observed IR drop. The size of C4 bumps restricts the locations where they may be placed. We assume that power pads can be allocated on a coarse pad grid that depends on the ratio of pad pitch and metal pitch. Each possible allocation of power pad to grid locations is called a *configuration*. The total number of configurations is the binomial coefficient of the number of pad locations and number of pads, and is larger than 10^{200} in the case studies considered in this paper (and larger than 10^{1400} for a scaled system in Section VI.B). In this context, effective and computationally efficient search techniques are needed to rapidly identify pad allocations that achieve near-optimal IR drop.

IV. WALKING PADS

The key idea behind WP is to convert a global optimization problem, the placement of n pads given m candidate locations, into a local balance problem, the placement of individual pads (current sources) with respect to various nearby current demands. To find the proper virtual force for local balance, we first observe that there is a similarity between the 2D PDN on-chip voltage field and a 2D electrostatic voltage field. The steady state equation of a voltage field can be regarded as the finite-difference version of the 2D Poisson equation [5, 13]:

$$\frac{\partial^2 V}{\partial x^2} + \frac{\partial^2 V}{\partial y^2} = I_{xy} R, \qquad (1)$$

where V is the on-chip voltage field, I_{xy} is the workload current density at point (x, y) and R is the resistance per unit length in the x and y directions. Gauss's law of electrostatic systems can be similarly described [14]:

$$\frac{\partial^2 \tilde{V}}{\partial x^2} + \frac{\partial^2 \tilde{V}}{\partial y^2} = \frac{\rho_{xy}}{\varepsilon_{xy}}, \qquad (2)$$

where $\tilde{V}$ is the electrostatic field, and ρ_{xy} and ε_{xy} are the charge density and permittivity at point (x, y). Note that in this paper we only consider the case where R is the same in the x and y directions; WP algorithms are also suitable if on-chip resistance is anisotropic.

When viewing pad placement as a 2D electrostatic voltage field problem, the current in the PDN is analogous to the electric flux lines in an electrostatic system, which are proportional to the voltage gradient. In this way, power pads can be regarded as "positive point charges" that source currents, and the underlying architectural blocks in the processor system can be regarded as "negative surface charges" that sink currents. Like charges repel each other, while unlike charges attract each other. We therefore define the voltage gradient at a pad location as the virtual force to direct pad movement.

In this context, Walking Pads allows pads to move in reaction to the forces exerted on them by current sources and other pads in the PDN; the pads "walk," toward the locations where these forces balance. No matter where the pads are placed, the total current through all pads is invariant. However, when pads reach their balanced positions, the gradient of the voltage field (directly proportional to the current) in each direction is equalized and reduced. Therefore, IR drop (the integral of voltage gradients) is minimized.

WP also minimizes max on-chip current density and PDN metal power dissipation at the same time. On-chip max current always occurs in those wires directly connected to a pad; max on-chip current density is therefore also minimized by WP because WP minimizes the current through these wires. PDN metal power dissipation is an analogue to the total energy of the electrostatic system. Therefore, the PDN metal power dissipation is also reduced when pads move under virtual forces, and is minimized when all forces on surface charges are balanced.

A. Walking Pads Algorithm Framework

An iteration of a Walking Pads algorithm uses three steps to incrementally move all pads toward their balanced positions:
1. Solve steady state equations.
2. Calculate virtual forces and decide the direction and distance of movement for each pad based on total forces.
3. Move pads.

Grid voltage and current values are determined in step 1. In step 2, current values are used to guide pad movement. Step 3 moves all pads simultaneously. WP achieves a significant performance improvement over SA by employing a deterministic approach to the selection of pad movement direction and distance in step 2 and allowing all pads to move simultaneously in step 3. As more optimization is achieved with each iteration, fewer iterations are needed.

B. Efficient Total Force Calculation

Once steady state current and voltages have been calculated for each node in the PDN, WP must determine in which direction to move each pad by computing virtual forces.

A intuitive way to determine the total virtual force on each pad is to apply the law of superposition and sum the contributions of virtual force from all other pads and current sources together. Some previous work uses this approach [15]. However, such methods are inherently inefficient due to their complexity. Using Gauss's Law, the force on a pad in one direction can be calculated from the voltage gradient in that direction. In the case of 2D PDN, one pad connects to four lines in the

east, north, west and south directions. The resultant force is the vector summation of these four currents.

C. Walking Pads Algorithm Variants

We propose three variants of Walking Pads. The first, Walking Pads - Neighbor (WP-N), only allows the pads to move to neighboring locations based on a comparison of the strength of vertical and horizontal forces: the stronger force determines the direction the pad moves, either up/down or left/right. Because all pads move at the same time and traverse a constant distance—one pad candidate location in the direction of motion—this algorithm results in the oscillation of pad locations around balanced positions. In practice, WP-N regards oscillation as convergence: when oscillation is detected, the algorithm terminates. As a result, WP-N does not perform well, but remains useful for quick, but low-quality, optimization.

The second variant, Walking Pads - Freezing (WP-F), is shown in Algorithm 1. WP-F allows pads to move in an arbitrary direction defined by the normalized virtual force $\vec{F}/\left\|\vec{F}\right\|$. Large move distances are also adopted in early iterations. To force pads to stop at approximately balanced positions, we introduce a freezing process which gradually decreases the move distance of each pad. The distance a pad moves D_i decreases with the constant freezing rate γ. WP-F terminates when pads no longer move. The large-step stage of WP-F helps pads to jump out of local minima, while the small-step stage helps pads gradually freeze in their balanced positions.

Set: initial move distance D_0, freezing rate γ
repeat
    Solve steady state;
    foreach *pad* **do**
        $\vec{F} = (I_{north} - I_{south}, I_{east} - I_{west})$
        $\vec{Disp} = \vec{F}/\left\|\vec{F}\right\| * D_i$
    end
    $D_{i+1} = D_i * \gamma$
until *check_converge() == True*;

Algorithm 1: Walking Pads - Freezing (WP-F) algorithm.

Walking Pads - Refined (WP-R), is shown in Algorithm 2. The first two versions of WP take advantage of the simultaneous movements of all pads. Simultaneous movements reduce the quality of the solution to some extent, however, because the forces on one pad may change when other pads move. To address this, WP-R performs a greedy search: it moves pads one by one and only accepts movements that decreases the max IR drop. For a 2D grid, we assume that moving pads near the location of max IR drop has greater effect than moving distant ones. To improve efficiency, WP-R sorts the pads by their distances to the max IR drop location and lets nearby pads move first. When the location or the value of maximum IR drop changes, WP-R re-sorts the pads and continues. The algorithm terminates when no pad movement improves IR drop. Because of its algorithm complexity, WP-R is used to supplement WP-F or WP-N to further improve the results when high optimization quality is required.

Set: $D_0 = PadPitch$, initial $maxIRDrop$
repeat
  Sort pads by distance to max IR place $\rightarrow$ PadList;
  foreach *pad in PadList* **do**
    $\vec{F} = (I_{north} - I_{south}, I_{east} - I_{west})$
    $\vec{Disp} = \vec{F} / \left\|\vec{F}\right\| * D_0$
    Solve steady state and get $new\_maxIRDrop$;
    if $new\_maxIRDrop < maxIRRrop$ **then**
      accept the movement;
      $maxIRRrop = new\_maxIRDrop$; break;
    else
      reject the movement;
    end
  end
until *check_converge() == True*;

Algorithm 2: Walking Pads - Refine (WP-R) algorithm.

D. Algorithm Complexity Analysis

The worst-case complexity of WP algorithms occurs when a pad must move from an initial position in one corner of the chip (e.g., the left-top corner) to a balanced position in the opposite corner (e.g., the right-bottom). In this case, WP-N requires $\#grid_{row} + \#grid_{column} - 2$ iterations to converge. For the practical cases of randomly initialized pad positions, the average number of iterations required is on the order of $B_0(\#grid_{row} + \#grid_{column} - 2)/\#pad$. B_0 is larger than 1 for the case that a pad does not move directly from its initial to the balanced position (i.e., it takes a *detour*).

For WP-F, the convergence speed is controlled by freezing rate γ. The approximate traveling distance of one pad before being frozen is $(D_0 - 0.5pad\_pitch)/(1 - \gamma)$, where D_0 is the initial move distance. Again, to beat the worst case, D_0 and γ are chosen to make the travel distance of each pad larger than the diagonal length of the grid. In our experiments, starting from roughly uniform pad locations results in much faster convergence than this theoretical upper bound. Detours are also possible in WP-F. In practice, we add a safety coefficient C_0 in the range of $2.0 \sim 4.0$ to balance the effect of detours and the speedup due to uniform initial positions and get:

$$\frac{D_0 - 0.5pad\_pitch}{1 - \gamma} = C_0 * \sqrt{\#grid_{row}^2 + \#grid_{column}^2}. \tag{3}$$

We choose an initial move distance $D_0 = 3 * pad\_pitch$ and freezing rate $\gamma = 0.99$ for our case studies; this results in 180 WP-F iterations. The total number of iterations required is independent of the number of pads to be placed.

V. EXPERIMENTAL SETUP

To evaluate our WP algorithms, we compare their convergence speed and solution quality with the simulated annealing (SA) algorithm proposed by Zhong and Wong [4]. For SA, we evaluate two cooling rates, 0.98 (practical cooling speed, SA-P) and 0.999 (very slow cooling speed, SA-S) for efficiency and quality comparison respectively; we have observed that the cooling rate of 0.85 proposed by Zhong and Wong is too fast to produce high-quality results. In our SA implementation, we maximize the square of the worst node voltage and implement

the movement window shrinking strategy proposed in the literature [4]. The SA algorithm is considered converged when the movement window is too small for pads to move.

We begin by comparing SA with WP-N and WP-F, and compare SA with WP-F+WP-R then. To compare WP-R and SA, we need to terminate WP-R iteration to get results of similar quality as those from SA, and then compare the speedup. To compare with SA-P and SA-S respectively, WP-F+WP-R-T1, terminates after #pad/2 iterations of WP-R, and WP-F+WP-R-T2 terminates after #pad*8 iterations of WP-R. These cutoffs were determined heuristically to yield similar quality.

We select a 24-core, Intel Penryn-like multiprocessor at 16nm technology as the platform to evaluate the above optimization algorithms. To estimate the power consumption for each functional block, we use McPAT, an architecture-level power model [16]. To model the worst-case power dissipation in the system, we assume that each architectural unit dissipates 85% of its max power [17]. We assume a supply voltage of 0.7V; architectural floorplans were generated using an architecture-level tool, ArchFP [18]. We assume that the top metal pitch is $30\mu m$ top layer metal pitch, and that wires in this layer are $6\mu m$ wide and $4\mu m$ thick; this results in a PDN model consisting of a 236 by 296 resistor grid, where each resistor has a resistance of $41m\Omega$. We assume that the C4 pad pitch is 285 μm, resulting in a grid with 2880 pad candidate locations for our 24-core system. According to ITRS projections, C4 pad density will be held constant in the foreseeable future [19]; we adopt the ITRS projection for pad density in our experiments. All our experiments are conducted on an Intel Xeon E5-1650 3.20 GHz CPU with 32 GB memory.

VI. RESULTS

A. WP Speedup and Result Quality

We first compare two basic WP algorithms, WP-N and WP-F, with SA-P; the results of this comparison are illustrated in Fig. 2. Fig. 2 plots algorithm convergence and solution quality for WP-F (dotted line), WP-N (dashed line), and SA-P (solid line) with respect to IR drop, max current density and power consumed in PDN metal; iteration count is plotted on the x axis. We use iteration counts alone to compare the efficiency of each approach because solving for steady state voltage and current values—required by, and equivalent in, each approach—requires over 99.9% of the total time to complete a single iteration in each case. SA, WP-N, WP-F and WP-R have about the same runtime *per iteration* and memory usage (approximately 0.3s and 220MB for the case of 512 pads on 24-core floorplan).

In Fig. 2, VDD pads are initially allocated uniformly to every fourth pad candidate location in the vertical and horizontal directions, representing 180 pads among 2880 pad candidate locations. We summarize the IR drop (IR), max on-chip current density (J), metal power dissipation (P) and required iteration (Iter) for each pad allocation method in Table I.

We observe that uniform pad allocation does not produce good results: SA reduces IR drop by 45% with respect to that from uniform pad location. Furthermore, we observe that all three algorithms jointly optimize all three metrics, if at different rates, and with differing effectiveness. WP-N converges the

978-1-4799-2817-0/14 $31.00 © 2014 IEEE

TABLE I
COMPARISON OF DIFFERENT ALLOCATION METHODS

Method	IR (% VDD)	J $(10^{10}A/m^2)$	P (W)	Iter
Uniform	12.5	2.246	10.11	–
WP-N	10.2	1.903	8.752	36
WP-F	7.5	1.543	8.365	180
SA-P	6.9	1.530	8.571	28,261

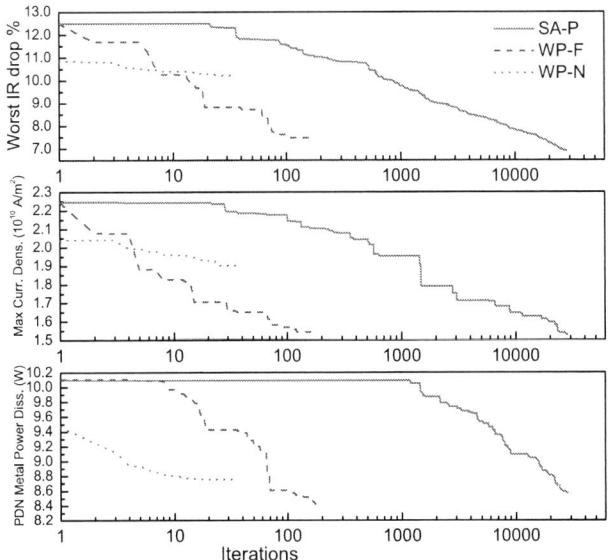

Fig. 2. WP-F (dotted), WP-N (dashed) and SA (solid) all jointly optimize IR drop, max current density and power dissipated in on-chip PDN metal, but at different rates and with different effectiveness. In practice, the techniques above do not monotonically improve each figure of merit; for clarity, we plot the results for the best explored configuration so far at a given iteration count.

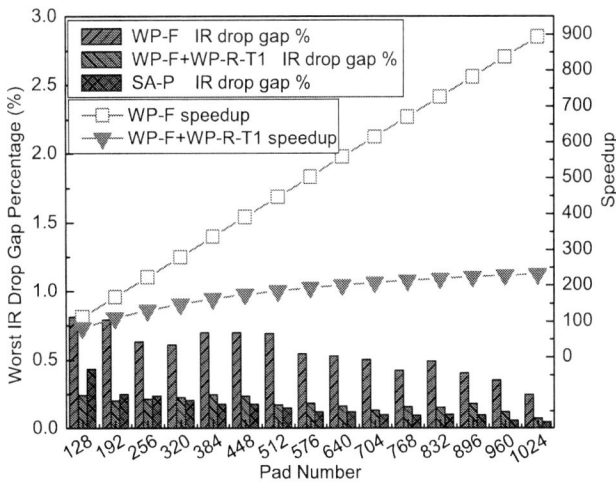

Fig. 3. Comparision of Walking Pads and simulated annealing: differences in worst IR drop and speedup. WP-R-T1 terminates after $\#pad/2$ iterations.

TABLE II
COMPARISON OF DIFFERENT WALKING PADS ALGORITHMS

WP Str.	Speedup (X)		Max Gap in %VDD	
	vs SA-P	vs SA-S	vs SA-P	vs SA-S
WP-F	112-893	–	0.54	0.81
F+R-T1	82-232	–	0.09	0.25
F+R-T2	–	337-388	–	0.12
F+R	–	20-220	–	0.10

fastest, finishing in 20% of the time required for WP-F; however, WP-N converges too quickly to get high-quality results, resulting in an IR drop 48% higher than that produced by SA. WP-F only sacrifices 0.6% VDD in IR drop, but obtains a 157X speedup when compared with SA.

We next evaluate the effect of combining WP-F and WP-R to achieve better optimization quality. Fig. 3 plots the IR drop gap and convergence efficiency of WP-F, WP-F+WP-R-T1 (terminates after #pad/2), and SA-P for varying pad counts, relative to the results from SA-S. The pad allocations selected by SA-S are considered the global optimal and are used to evaluate the result quality of other methods. SA-S, which cools at a rate of 0.999 instead of 0.98, needs $3176 \times \#pad$ iterations to converge while SA-P needs $157 \times \#pad$ to converge.

In Table II we summarize the quality and speedup on a 24-core floorplan with 128 to 1024 pads. Four different WP strategies (WP Str.) - WP-F, WP-F+WP-R-T1 (F+R-T1, WP-R-T1 terminates at $\#pad/2$), WP-F+WP-R-T2 (F+R-T2, WP-R-T2 terminates at $\#pad * 8$), and WP-F+WP-R (F+R, no early termination), are investigated. WP-F achieves up to 893X speedup with respect to SA-P, but sacrifices too much quality (0.54 %VDD). When refined with WP-R, WP-F+WP-R-T1

achieves up to 232X speedup with respect to SA-P, but produces results matching those from SA-P with a gap less than 0.1% VDD. We therefore think WP-F+WP-R-T1 can replace SA-P to obtain optimized pad locations with practical quality. In the case of 832 pads, WP-F+WP-R-T1 requires less than four minutes to achieve results of comparable quality to SA-P after 15 hours. For the same reason, we think WP-F+WP-R-T2 can replace SA-S to obtain intensively optimized pad locations with a speedup in the range of 337-388X. We have not compared WP-F and WP-R-T1 with SA-S and WP-R-T2 and WP-R with SA-P.

B. Synthetic and Scaled System Benchmarks

To demonstrate that WP performs well under a variety of scenarios, we developed a series of benchmarks including (a) six synthetic floorplans (Fig. 4) and (b) three variants of the 24-core system with 16, 32, and 48 cores. Our results are summarized in Table III. For each benchmark (Bench.), we report the number of pads allocated (# pads), the number of candidate locations (# loc), and the corresponding speedup (Speedup) of WP-F (F), WP-F+WP-R-T1 (R-T1) and the IR drop gap (% Gap) of WP-F (F), WP-F+WP-R-T1 (R-T1) and WP-F+WP-R (R), each relative to SA-P. The IR drop gap between SA-P and WP is calculated as $(IR_{WP} - IR_{SA-P})/VDD$. A negative gap means WP outperforms SA-P.

For the synthetic benchmarks, we observe that WP-F and

978-1-4799-2817-0/14 $31.00 © 2014 IEEE

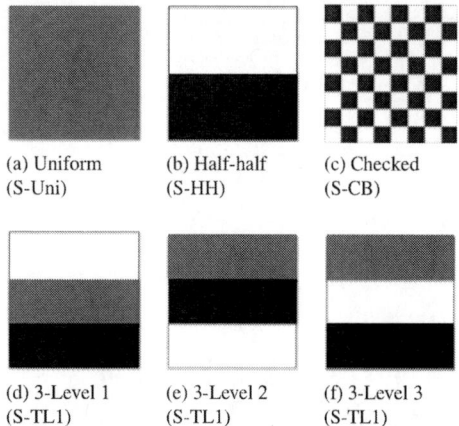

(a) Uniform (S-Uni) (b) Half-half (S-HH) (c) Checked (S-CB)

(d) 3-Level 1 (S-TL1) (e) 3-Level 2 (S-TL1) (f) 3-Level 3 (S-TL1)

Fig. 4. The floorplan of each synthetic model is 20 x 20 mm^2. 512 pads are allocated to deliver a total of 150 W. In (b), the power density ratio of black to white is 4:1. In (d), (e) and (f) the power density ratio of black, gray and white is 3:2:1.

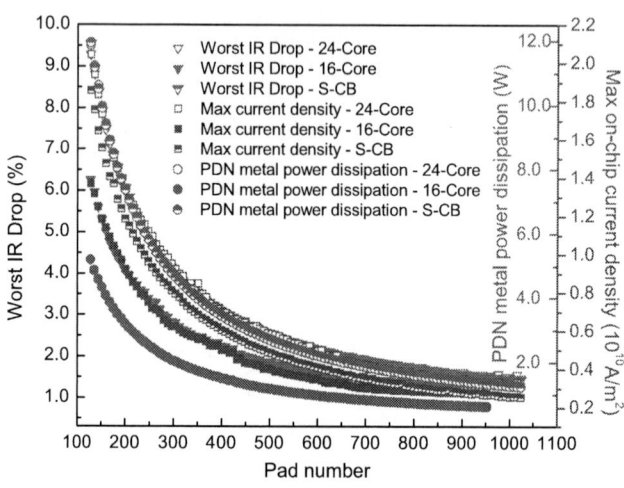

Fig. 5. Pad number effect on IR drop, max current density and PDN metal power dissipation based on optimized pad locations. Optimization uses WP-F+WP-R (no early termination) and starts from randomly allocated pads. Points are plotted at a interval of 8 pads in this figure.

TABLE III
WP RESULTS FOR SYNTHETIC AND MULTI-CORE MODELS

Bench.	# pads	# loc	Speedup		% Gap		
			F	R-T1	F	R-T1	R
S-Uni	512	4900	498	206	0.18	-0.03	-0.11
S-HH	512	4900	498	206	0.23	-0.03	-0.11
S-CB	512	4900	498	206	0.20	-0.06	-0.12
S-TL1	512	4900	498	206	0.15	-0.03	-0.10
S-TL2	512	4900	498	206	0.24	0.01	-0.12
S-TL3	512	4900	498	206	0.19	-0.03	-0.11
16-Core	512	1914	375	155	0.42	0.16	-0.07
24-Core	768	2880	670	277	0.33	0.071	-0.04
32-Core	1024	3844	961	397	0.41	0.070	-0.07
48-Core	1536	5776	1536	634	0.39	0.055	-0.09

WP-R-T1 achieve a speedup of 498 and 206X relative to SA-P. WP-F and WP-R-T1 further achieve IR drops within 0.25% and 0.01% of SA-P. For the Penryn-like variants, the speedup advantage of WP-F and WP-R-T1 increases as the chip grows, up to 634X, and the IR drop gap for WP-R-T1 shrinks marginally; the IR drop gap for WP-F is relatively constant across chip sizes.

VII. ANALYTICAL MODEL

While the above results show that WP efficiently places a given number of pads, naively determining the appropriate pad count to meet a given IR drop budget requires many WP executions, one for each pad count. We therefore developed an analytical model capable of predicting the appropriate pad count, significantly reducing the number of required WP executions.

Fig. 5 illustrates the relationship between pad count, IR drop, max current density and PDN metal power when pad locations are optimized with WP-R. As the pad count increases, each of the three metrics decreases in a similar way.

To model the relationship between pad count and IR drop, we begin with several simplifying assumptions:

1. The load current density ρ is uniform.
2. All pad currents are equal.
3. Each pad serves a circular area around it with radius r_0.

From Gauss's Law we have:

$$\frac{\partial V}{\partial r} = \frac{\pi r_0^2 \rho - \pi r^2 \rho}{2\pi r} * R. \quad (4)$$

Integrating V from (r_ε) to r_0, the IR drop at r_0 is:

$$V|_{r_0} = \frac{\rho R r_0^2}{2} ln\frac{r_0}{r_\varepsilon} - \frac{\rho R}{4}(r_0^2 - r_\varepsilon^2) + \frac{I_0 R_p}{N_p} + V_{packagedrop}. \quad (5)$$

where r_ε is the effective radius of pad, and R is the resistance per unit length of on-chip resistor grid. Substituting $r_0 = \sqrt{\frac{I_0}{\pi \rho N_p}}$ and substituting for the constant coefficients with a, b, and c, we have:

$$V_{drop} = a \frac{1}{N_p} log(\frac{1}{N_p}) + b\frac{1}{N_p} + c. \quad (6)$$

To validate Eq. (6), we performed curve fitting against the IR drop data in Fig. 5, and find that $R^2 = 0.998$ and 0.9998 for the 16-core and 24-core models respectively. Furthermore, when used to derive max on-chip current density and PDN metal power, fitting Eq. (6) results in $R^2 = 0.998$ and 0.9999 respectively for the 16-core model, and $R^2 = 0.9995$ and 0.99997 respectively for the 24-core model. Eq. (6) clearly is effective at predicting each metric as a function of pad count.

To explore the predictive power of our analytical model, we select four different IR drop budgets for the 24-core system, use Eq. (6) to estimate the appropriate number of pads, and compare this with the minimum pad count satisfying the budget. The parameters of Eq. (6) are fitted using three randomly selected pad counts: 200, 520, and 840. The results of this experiment are summarized in Table IV. We observe that the predicted pad count (Pred.) is within two of the optimal pad count (Optimal) in each case. It is worth noting that even if all

978-1-4799-2817-0/14 $31.00 © 2014 IEEE

TABLE IV
PREDICTED AND OPTIMAL PAD COUNT FOR 24-CORE MODEL

IR Drop Budget	Pred.	Optimal	Actual IR Drop
5%, 35mV	240	238	34.63mV
4%, 28mV	304	306	27.97mV
3%, 21mV	416	418	20.77mV
2%, 14mV	673	672	13.99mV

pad counts in $\{128, 136, ..., 1024\}$ are used for curve fitting, the predicted number of pads does not change.

While validating our analytical model, we noticed that there is a significant difference between the worst-case IR drop experienced under uniform pad distribution and that experienced when pad locations are optimized. For example, the worst IR drops with uniform pads allocations on a rectangular 2D array are 12.0%, 7.0% and 3.3% for the cases of 180, 320 and 720 VDD pads in our 24-core model. The corresponding worst IR drops with WP-optimized pad allocations are 6.6%, 3.8% and 1.9% respectively. This suggests that previous analytical models based on uniform pad allocations (e.g., [6]) systematically overestimate worst-case IR drop.

VIII. CONCLUSIONS AND FUTURE WORK

In this paper we describe a fast method for determining the minimum number of pads required to satisfy an IR drop constraint and their corresponding optimized locations. We introduce a novel pad placement optimization framework for 2D grids: Walking Pads (WP). Three algorithms are proposed in the WP framework to meet the conflicting requirements of results quality and optimization time. The experimental results show that combining the Walking Pads - Freezing (WP-F) and Walking Pads - Refined (WP-R) algorithms achieves up to 634X speedup when compared with simulated annealing (SA), without sacrificing more than 0.1% VDD in IR drop. Our scalability test also shows that speedup and result quality of WP increase as the chip grows. We also propose an analytical model to describe the relationship between the number of allocated, optimized pads and resulting IR drop. This model matches WP results well and leads to fast minimum-pad-number determination when working with WP algorithms.

In this paper we take the first step of demonstrating the viability of the WP paradigm. There are several directions for future research using the WP framework: (1) The joint optimization of VDD and GND pad placement should be considered to make further IR drop optimization across both the VDD and GND layers; (2) Spatial constraints in the 2D pad candidate location grid should be considered in WP for the placement of signal pads; (3) WP could be used to support IR-drop-aware floorplanning, by moving 'negative charges' (functional units or standard cells) instead of 'positive changes' (power pads); (4) WP algorithms could be simply extended for through-silicon via (TSV) placement in 3D IC; (5) WP algorithms can be easily extended to temperature-aware placement by replacing the voltage field with a temperature field.

ACKNOWLEDGMENTS

This work was supported in part by NSF grants CNS-0916908, CCF-0903471 and CCF-1116673, and C-FAR, one of six centers of STARnet, a Semiconductor Research Corporation program sponsored by MARCO and DARPA.

REFERENCES

[1] Mikhail Popovich, Andrey V. Mezhiba, and Eby G. Friedman. *Power distribution networks with on-chip decoupling capacitors*. Springer, New York; London, 2008.

[2] Min Zhao, Yuhong Fu, Vladimir Zolotov, Savithri Sundareswaran, and Rajendran Panda. Optimal placement of power supply pads and pins. *Proc. DAC '04*, pp. 165–170, New York, NY, USA, 2004. ACM.

[3] T. Sato, Hidetoshi Onodera, and M. Hashimoto. Successive pad assignment algorithm to optimize number and location of power supply pad using incremental matrix inversion. *Proc. ASP-DAC '05*, vol. 2, pp. 723–728, 2005.

[4] Yu Zhong and Martin D. F. Wong. Fast placement optimization of power supply pads. *Proc. ASP-DAC '07*, pp. 763–767, Washington, DC, USA, 2007.

[5] K. Shakeri and J.D. Meindl. Compact physical IR-drop models for chip/package co-design of gigascale integration (GSI). *IEEE Transactions on Electron Devices*, vol. 52(6), pp. 1087–1096, 2005.

[6] J. Rius. IR-Drop in on-chip power distribution networks of ICs with nonuniform power consumption. *IEEE Transactions on Very Large Scale Integration (VLSI) Systems*, vol. 21(3), pp. 512–522, 2013.

[7] Jianlei Yang, Zuowei Li, Yici Cai, and Qiang Zhou. PowerRush: a linear simulator for power grid. In *ICCAD '11*, pp. 482–487, 2011.

[8] Meeta S. Gupta, Jarod L. Oatley, Russ Joseph, Gu-Yeon Wei, and David M. Brooks. Understanding voltage variations in chip multiprocessors using a distributed power-delivery network. In *Proc. DATE '07*, pp. 624–629, San Jose, CA, USA, 2007.

[9] Runjie Zhang, Brett H. Meyer, Wei Huang, Kevin Skadron, and Mircea R. Stan. Some limits of power delivery in the multicore era. *WEED*, Oregon, USA, 2012.

[10] Xiaoye S. Li. An overview of SuperLU: algorithms, implementation, and user interface. *ACM Trans. Math. Softw.*, vol. 31(3), pp. 302–325, 2005.

[11] Joseph W. H. Liu. Modification of the minimum-degree algorithm by multiple elimination. *ACM Trans. Math. Softw.*, vol. 11(2), pp. 141–153, 1985.

[12] Zhuo Li, Raju Balasubramanian, Frank Liu, and Sani Nassif. 2011 TAU power grid simulation contest: benchmark suite and results. In *Proc. ICCAD '11*, pp. 478–481, Piscataway, NJ, USA, 2011.

[13] Andrew B. Kahng, Bao Liu, and Qinke Wang. Stochastic power/ground supply voltage prediction and optimization via analytical placement. *IEEE Trans. Very Large Scale Integr. Syst.*, vol. 15(8), pp. 904–912, 2007.

[14] David J. Griffiths. *Introduction to Electrodynamics*. Addison-Wesley, 4 edition, October 2012.

[15] Yi-Lin Chuang, Po-Wei Lee, and Yao-Wen Chang. Voltage-drop aware analytical placement by global power spreading for mixed-size circuit designs. *IEEE Transactions on Computer-Aided Design of Integrated Circuits and Systems*, 30(11):1649–1662, 2011.

[16] Sheng Li, Jung-Ho Ahn, R.D. Strong, J.B. Brockman, D.M. Tullsen, and N.P. Jouppi. McPAT: an integrated power, area, and timing modeling framework for multicore and manycore architectures. *MICRO-42*, pp. 469–480, 2009.

[17] A.M. Joshi, L. Eeckhout, L.K. John, and C. Isen. Automated microprocessor stressmark generation. *HPCA 2008*, pp. 229–239, 2008.

[18] Gregory G. Faust, Runjie Zhang, Kevin Skadron, Mircea R. Stan, and Brett H. Meyer. ArchFP: rapid prototyping of pre-RTL floorplans. *VLSI-SoC*, pp. 183–188. IEEE, 2012.

[19] ITRS, 2011. http://www.itrs.net.

978-1-4799-2817-0/14 $31.00 © 2014 IEEE

Power Supply Noise-Aware Workload Assignments for Homogeneous 3D MPSoCs with Thermal Consideration

Yuanqing Cheng, Aida Todri-Sanial, Alberto Bosio,
Luigi Dilillo, Patrick Girard, Arnaud Virazel
LIRMM – University of Montpellier 2 / CNRS, Montpellier, France
{yuanqing.cheng, aida.todri, bosio, dillio, girard, virazel}@lirmm.fr

Abstract— In order to improve performance and reduce cost, multi-processor system on chip (MPSoC) is increasingly becoming attractive. At the same time, 3D integration emerges as a promising technology for high density integration. 3D homogeneous MPSoCs combine the benefits of both. However, high current demand and large on-chip switching activity variations introduce severe power supply noises (PSN) for 3D MPSoCs, which can increase critical path delay, and degrade chip performance and reliability. Meanwhile, thermal gradient should also be considered for 3D MPSoCs to avoid hot spots. In the paper, we investigate the PSN effects of different workloads and propose an effective PSN estimation method. Then, a heuristic workload assignment algorithm is proposed to suppress PSN under the given thermal constraint. The experimental results show that PSNs can be reduced significantly compared with thermal-balanced workload assignment scheme, and the system performance can be improved as well.

I. Introduction

Due to aggressive technology scaling, billions of transistors can be fabricated on a single die. However, interconnects on chip can not scale very well with transistors, and it is difficult to increase frequency further when processor operating frequency approaches several giga Hertz. Moreover, power consumption also becomes a bottleneck of processor design. As a result, single processor can not afford to increasing performance and functionality requirements. Multiprocessor System-on-Chip (MPSoC), which integrates IP module, FPGA, hardware accelerator and processors together, provides a cost-effective way to tackle this problem [1]. At the same time, three dimensional integration technology emerges to reduce interconnect delay and power consumption [2]. Combining both technologies can provide more functionalities and higher performance within given power budget. In addition, 3D MPSoC can integrate modules fabricated by disparate processes effectively. Therefore, 3D MPSoCs draw much attention from both academia and industrial fields.

3D MPSoCs, however, also bring several new challenges. Among others, power supply noise (PSN) is a big concern threatening signal integrity, performance and reliability, especially for shared power delivery networks [3]. Compared with 2D counterparts, 3D MPSoC power supply current becomes much larger, and causes more severe IR drop. Moreover, due to power/ground TSV parasitics, power supply noise manifests large variations among different tiers, and can propagate from one tier to other ones through TSVs [4]. These distinguish characteristics require further investigations to suppress PSN magnitudes for 3D MPSoCs. On the other hand, thermal dissipation has become a prominent issue for 3D MPSoCs. Stacking structure prevents effective heat removal of bottom tiers from the heat sink. High temperature can degrade system performance, cause thermal runaway, and increase package cost [5].

There are many existing research works involving suppressing PSN and thermal optimization for both 2D and 3D ICs. Todri et al. investigated PSN interactions for 2D CMPs and proposed some useful workload assignment guidelines [3]. Gupta et al. explored the voltage variation impact on chip multiprocessors using a distributed power delivery network model [6]. Xu et al. claimed that PSN can cause severe variations of skitter of clock networks and degrade system performance dramatically [7]. Lung et al. explored the online task scheduling technique to maximize system throughput with the thermal constraint for 3D MPSoCs [8]. Coskun et al. proposed both static and dynamic task scheduling schemes to reduce thermal gradient and thermal cycles. Zhou et al. takes advantage of strong thermal correlations among vertical tiers, and proposed a thermal-aware task scheduling algorithm for 3D CMPs, which can reduce both peak temperature and occurrences of thermal emergency [9].

However, these works are either on purely circuit level or high system level, and can not fill the gap between different design levels and capture the close relations among power supply noise, thermal gradient and workload characteristics. In this paper, we integrate circuit-level PSN estimation, system level thermal evaluation and workload assignment together effectively. Our contributions are listed as follows:

- To the best of our knowledge, it is the first work to

978-1-4799-2817-0/14 $31.00 © 2014 IEEE

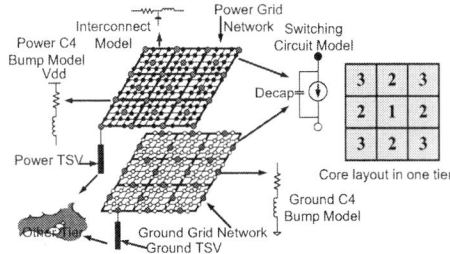

Fig. 1. Power delivery network model of 3D MPSoCs and core layout within one tier

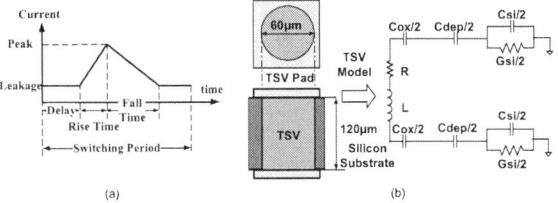

Fig. 2. (a) Workload current waveform, (b) TSV model [10]

consider both PSN effect and thermal issue for 3D MPSoCs from the workload assignment perspective.

- Through extensive circuit-level simulations on the 3D power delivery network, an effective and efficient P-SN estimation method is proposed to guide workload assignment.

- An effective heuristic algorithm is proposed to optimize system performance by suppressing PSN under the given thermal constraint.

The rest of the paper is organized as follows. Section II briefly introduces both power delivery network and thermal modeling techniques used in our paper. Section III motivates our research work through an illustrative example. The problem investigated is formulated in Section IV. Section V explores the PSN estimation method, and proposes a heuristic workload assignment algorithm to attack the problem. Section VI presents the experimental results to validate the effectiveness of our algorithm, and Section VII concludes the paper.

II. PRELIMINARIES

A. Power delivery network model and PSN estimation

Since we explore PSN interactions at core level, it is sufficient to model global power supply interconnects. In the paper, we assume the power grid topology is mesh as in [6]. Each interconnect segment is modeled as a RLC T model based on PTM interconnect model [11]. In a single tier, both power and ground power grid are considered to capture voltage drop and ground bounce simultaneously.

TABLE I
ELECTRICAL PARAMETERS OF PDN MODEL

INTERCONNECT		TSV	
segment length	$100\mu m$	diameter	$10\mu m$
global metal width	$15\mu m$	aspect ratio	$1:8$
global metal space	$15\mu m$	resistance	$20m\Omega$
segment resistance	$122m\Omega$	inductance	$25pH$
segment capacitance	$152.7fF$		
segment inductance	$60pH$		
C4 BUMP		CORE	
inductance	$9.52m\Omega$	size	$1.6mm \times 1.6mm$
resistance	$12.65pH$	mesh grid	$16 \times 16/core$

PDNs residing on different tiers are connected by TSVs as illustrated in Fig. 1. Both decoupling and intrinsic capacitances are modeled as a lump decap in parallel with a current source representing the workload. Workload switching activities are represented by triangular waveforms as in [12], which can be characterized by several shape parameters, i.e., peak current, leakage, delay time, rise/fall time and period. The workload current waveform is shown in Fig. 2 along with the TSV model. The physical parameters of the PDN model are listed in Table I, and the workload current waveform can be derived by architecture level power simulation tools as done in [6].

As stated in [13], critical path delay and system performance are affected by average PSN instead of peak PSN. Thus, we calculate average voltage drop as follows [14]:

$$V_{pnoise} = \int_{t_s}^{t_e} max\{V_{dd} - V_p(t), 0\}dt/t_{switching} \quad (1)$$

$$V_{gnoise} = \int_{t_s}^{t_e} max\{V_{ss}(t), 0\}dt/t_{switching} \quad (2)$$

$$V_{noise} = V_{pnoise} + V_{gnoise} \quad (3)$$

where t_s/t_e denotes the start/end of switching window. V_{dd} is the normal power supply voltage. $t_{switching}$ denotes the length of switching window. $V_p(t)$ and $V_{ss}(t)$ capture the power supply variation and ground bounce respectively.

B. Workload performance modeling

We use the formula proposed by [13] to model the relationship between average PSN and critical path delay, i.e.,

$$\frac{D}{D_0} = 1 - k_1\frac{\Delta V}{V_{dd} - V_t} + k_2(\frac{\Delta V}{V_{dd} - V_t})^2 \quad (4)$$

where V_{dd} is nominal supply voltage. V_t is the transistor threshold voltage. k_1 and k_2 are process dependent constants. D_0 is the ideal critical path delay. Using Eq. (4), we can derive the clock frequency taking PSN into account. Then, MIPS (million instructions per second) is used to evaluate the system performance:

$$MIPS_i = \frac{IPC_i \times f_i}{10^6} \quad (5)$$

978-1-4799-2817-0/14 $31.00 © 2014 IEEE

$$Performance = \sum_{i=1}^{p} MIPS_i \qquad (6)$$

In the above formula, f_i is the running frequency of workload i considering PSN effect. IPC_i is instructions per cycle of workload i, and p is the number of workloads.

C. Thermal evaluation

Thermal analysis generally takes advantage of duality between thermal and electrical properties [8]. Chip temperature can be derived by solving differential thermal equations. However, it is too time-consuming to be applied for iterative workload assignment optimization. In this paper, we adopt the *stack power gradient* concept proposed by [9] to estimate thermal gradient across the chip. Assume that the MPSoC has t tiers, and each tier has $m \times n$ cores. To compute stack power gradient, we need to obtain stack power firstly as follows:

$$P_{stack}^{i,j} = \sum_{k=1}^{t} P_{k,i,j} \qquad (7)$$

$$1 \leq i \leq m, 1 \leq j \leq n \qquad (8)$$

Then, the stack power gradient of the chip can be calculated as:

$$P_{Gradient} = max\{| P_{stack}^{i,j} - P_{stack}^{i',j'} |\} \qquad (9)$$

$$1 \leq i, i' \leq m, 1 \leq j, j' \leq n, (i,j) \neq (i',j') \qquad (10)$$

In the paper, we will use this metric as the thermal constraint of workload assignment problem.

III. MOTIVATION

As heat sink is often attached to the top tier via thermal interface material, workloads should be intuitively allocated to the upper tier from the heat removal perspective. On the other hand, package is attached to the bottom tier while power supplies for upper tiers are provided by the bottom tier through TSVs. As a result, upper tiers may suffer larger PSNs, and may not be optimal positions for workload assignment from the PSN perspective. We use an illustrative example to show that with the same stack power gradient, PSN can be suppressed significantly by adjusting workload positions.

In the example, we assume the 3D MPSoC has 2 layers. There are 2×2 cores on each layer. The two layers are integrated using face-back bonding technique. The PDN model used for PSN estimation is presented in Section II. There are two types of workloads waiting for assignment, and their current profiles are: workload type I: {0.15A, 1.5A, 75ps, 75ps, 75ps, 300ps, 0.2W}, workload type II: {0.12A, 1.2A, 325ps, 325ps, 325ps, 1.3ns, 0.1W}. The meanings of the first six parameters are described in Section II. The last parameter denotes the workload's average power consumption. Assume that the workload set

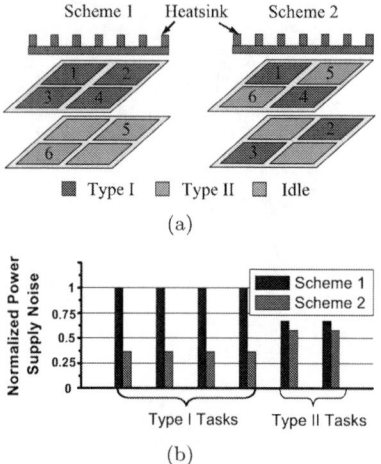

Fig. 3. (a) Two different workload assignments with the same stack power gradient (b) PSN distributions of all workloads corresponding to two workload assignment schemes

consists of 4 type I and 2 type II workloads. In the paper, we ignore power consumptions and time overheads of data communications among running workloads because data transfer can be done very efficiently by TSVs.

First of all, we perform workload assignment using scheme 1 shown in the left part of Fig.3. If we exchange task 2 and 5, and task 3 and 6, the stack power gradient remains the same but the power supply noise reduces as shown in Fig. 3(b). Since workloads with high activities are assigned near the package, the PSNs are reduced. Then, we use Hotspot [15] to evaluate the MPSoC peak temperature. The simulations results show that peak temperature of scheme 1 is 319.42 while that of scheme 2 is 319.62, which are almost the same. This example indicates PSN can be reduced significantly without affecting thermal dissipation through deliberate task assignment strategy. It also validates the effectiveness of taking stack power gradient as the thermal evaluation metric.

IV. PROBLEM FORMULATION

Given a 3D MPSoC architecture $\mathbf{Ar}(t,m,n)$, which has t tiers. Each tier has $m \times n$ cores. There are p workloads waiting for assignment. Every workload is characterized by a set of parameters, i.e., $\{L, P, D, R, F, T, W\}$. The objective is to assign p workloads to $\mathbf{Ar}$ such that their performance can be maximized under the given thermal constraint, i.e.,

$$Maximize : Performance = \sum_{i=1}^{p} MIPS_i \qquad (11)$$

s.t.

$$P_{Gradient}(\mathbf{Ar}) \leq C \qquad (12)$$

where formula (12) specifies stack power gradient can not exceed the given threshold value C to avoid local hot spots. Eq. (11), (4) and (5) indicate that to calculate performance, we must obtain the PSN magnitude firstly. Although PSN analysis can be performed by HSPICE simulation or other power grid analysis methods as in [16], the solving procedure is very time-consuming due to the large scale of the power grids. In the next section, we propose an effective and efficient method to estimate PSN impacts through investigating the PSN characteristics of different type of workloads.

V. PROPOSED METHOD

In the paper, we assume a $3 \times 3 \times 3$ 3D homogeneous MPSoC platform, and there are three type workloads for assignment, i.e., high, middle and low activity workloads. Their current profiles are listed as follows: High Activity: {0.1A, 1A, 50ps, 50ps, 50ps, 200ps, 2.09W} , Middle Acitivity: {0.07A, 0.7A, 250ps, 250ps, 250ps, 1ns, 1.375W}, Low Acitivity: {0.03A, 0.3A, 1ns, 1ns, 1ns, 4ns, 0.55W}. The idle power consumption is assumed to be 0.1W. In the rest of the section, we introduce our PSN estimation method at first. Then, we propose a heuristic algorithm to optimize PSN under the thermal constraint.

A. PSN estimation method

Our PSN estimation consists of two stages. First, we perform off-line pre-characterization for each type of workload using HSPICE simulation. In the first stage, to account for the PSN impact of the idle core, we obtain the PSN distribution when no workload is assigned, and denote it as PSN_{idle}. Then, we assign a specific type of workload individually to a specific position each time. As shown in Fig. 1, due to symmetry, there are three distinct positions within one tier (i.e., position 1, 2, and 3). The PSN increase due to the workload can be calculated as,

$$\Delta PSN_c^{k,i,j} = PSN_c^{k,i,j} - PSN_{idle}^{k,i,j} \qquad (13)$$

$$1 \le k \le t, 1 \le i \le m, 1 \le j \le n \qquad (14)$$

$$c \in \{High, Middle, Low\} \qquad (15)$$

where the triple k, i, j specifies the core position within 3D MPSoC, c denotes the type of workload, t is the number of tiers, m is the number of rows of cores within one tier, and n is the number of columns of cores within one tier. Using Eq. (13), we can obtain the PSN magnitude increase caused by a single workload. The pre-characterization results are stored in the MPSoC memory as a table for reference during online task assignment.

During online task assignment stage, we estimate the PSN distributions induced by multiple running workloads by the following formula:

$$PSN^{(k,i,j)} = PSN_{idle} + \sum_{t=1}^{p} \Delta PSN_t^{k,i,j} \qquad (16)$$

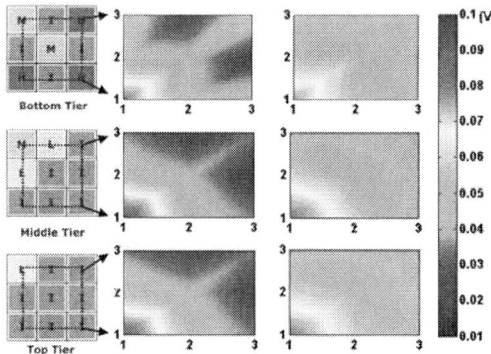

Fig. 4. Validation of the proposed method compared to HSPICE simulation results

The equation implies that the PSN impacts caused by multiple workloads are the sum of the PSN impacts when no workload is assigned and the PSN impact increase caused by each single workload. Note that our method is different from the common superposition rule in electronic circuits. We just use the formula to estimate the relative PSN impacts caused by multiple workloads, and do not attempt to obtain the accurate absolute value of PSN magnitudes.

To validate the correctness of our method, we plot the PSN distributions predicted by our method and HSPICE simulations respectively in Fig. 4. As shown in the figure, the left part shows the workload assignment scenario, and the right part shows the PSN distributions corresponding to the dashed line rectangle area. The PSN distributions on different tiers are derived by our estimation method and HSPICE respectively. It shows that although our method has some errors on predicting the PSN magnitudes, it can predict the hot spot area correctly from the power supply noise perspective, and provide enough guidelines for our online task assignment algorithm. We simulate different workload assignment scenarios and get the similar results. The prominent advantage of our method is that it only involves very simple calculations based on the pre-characterization results, and PSN impact evaluation can be performed very efficiently for online workload assignment. The detailed algorithm is described next.

B. Proposed Workload Assignment Algorithms

As workload assignment problem is well known as a NP-hard problem [17], we propose a simulated annealing (SA) based heuristic algorithm to attack it. The inputs of the algorithm are the number of cores(n), initial assignment (S^*), PSN impact of each kind of workload derived from subsection A (PSN_Table), iteration count set for simulated annealing procedure($IterCnt$), the given stack power gradient constraint ($\mathcal{P}$) and instruction per cycle (IPC) value of each workload (IPC). The pseudo-code of

Algorithm 1 The proposed heuristic algorithm

1: **procedure** PSN_AWARE_ASSIGN($n, \mathcal{S}^*, PSN\_Table,$ $\mathcal{P}, IPC$)
2: $\quad \mathcal{S} \leftarrow \mathcal{S}^*$ // set initial assignment
3: $\quad maxMIPS \leftarrow CalcMIPS(\mathcal{S}, IPC, PSN\_Table, \mathcal{P})$
4: $\quad Q \leftarrow IterCnt, K \leftarrow n(n-1)/2, L_0 \leftarrow QK$
5: $\quad oscillation \leftarrow false,$ set initial temperature $t \leftarrow t_0$
6: $\quad$ BestAssignment=Simulated_Annealing($S, \quad IPC,$ $PSN_Table, \mathcal{P}, Q, K, L_0, oscillation, t$) // $Q, K, L_0,$ $oscillation,$ t are parameters of simulated annealing procedure
7: $\quad S_{opt} \leftarrow TABOO\_Search(S, IPC, PSN\_Table, \mathcal{P})$

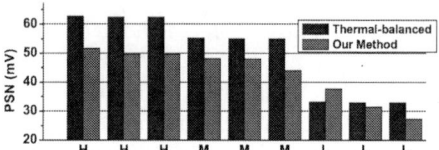

Fig. 5. PSN comparisons of all workloads of case I

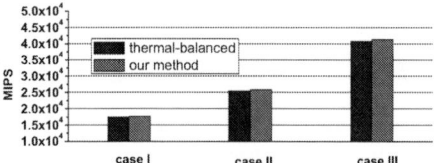

Fig. 6. system performance comparisons

the algorithm is shown in Algorithm. 1. Line 2-6 is the initialization of the algorithm, including computing performance corresponding to the initial assignment (based on Eq. (5) and Eq. (6)) and initialization of simulated annealing parameters. Then the simulated annealing procedure is invoked, which is based on the algorithm proposed by [18]. It exchanges the workload positions in the neighborhood and evaluate the performance corresponding to the exchange to find the optimal solution. During the procedure, the power gradient constraint is checked consistently. If the new assignment violate the constraint, it is dropped. At last, the obtained solution is further improved by TABOO search technique [18].

The time complexity of the algorithm is mainly determined by simulated annealing procedure, which is $O(IterCnt * InnerCnt * n^2)$. IterCnt, InnerCnt are outer and inner iterative counts of SA algorithm respectively. n is the number of cores of the 3D MPSoC.

VI. Experimental Results

The simulation platform used in the experiments is a $3 \times 3 \times 3$ homogeneous 3D MPSoCs. The workload PSN distributions are obtained by the PDN model described in Section II. Our workload assignment algorithm is implemented by C++ and runs on a machine with Intel 2.60G dual core and 4GB memory. Our simulation results are compared with thermal-balanced algorithm proposed by [9]. This algorithm can obtain the workload assignment with the minimum stack power gradient. To make the paper self-contained, we briefly describe its main idea. First, all workloads are sorted by their power consumptions in descending order. Each time, a workload is selected and assigned to the stack with the minimum power. Within each stack, the workload with high power consumption is assigned to the position near the heat sink. This procedure is repeated until all workloads are assigned. In the rest of paper, we call it thermal-balanced algorithm. In the comparisons, we firstly apply thermal-balanced algorithm to obtain the minimum stack power gradient and take it as the stack power gradient constraint. Then we apply our proposed method to optimize the PSN and sys-

tem performance.

For simplicity, we assume the IPC of each type workload is 1. The performance degradation due to PSN is derived by Eq. 4. If the PSN reduces the high activity frequency by α ($\alpha > 1$), then it will decrease the frequencies of low and middle workloads by $(\alpha + 4)/5$ and $(\alpha + 19)/20$ respectively.

A. Case I: 3 high, 3 middle and 3 low activity workloads

In case I, we assume the workloads set consisting of 3 high, 3 middle and 3 low activity workloads. The minimum stack power gradient is 1.54W, which is set as the stack power gradient constraint. Fig. 5 plots the PSN comparisons of all workloads when use different workload assignment schemes. As the figure shown, PSNs introduced by thermal-balanced algorithm are larger for almost all workloads compared with results of our method, especially for high activity workloads, which have the largest effects on system performance. On average, our method can reduce PSN by 12.42% compared to the thermal-balanced algorithm. For high activity workloads, our method can reduce 19.63% PSN on average. The left two bars of Fig. 6 shows the system performance (MIPS) comparisons considering PSN. Our method can increase system performance by 262 MIPS.

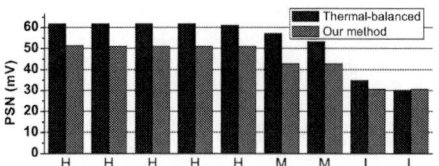

Fig. 7. PSN comparisons of all workloads of case II

978-1-4799-2817-0/14 $31.00 © 2014 IEEE   548

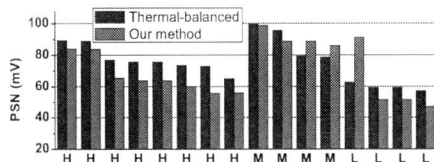

Fig. 8. PSN comparisons of all workloads of case III

REFERENCES

[1] Wayne Wolf. The future of multiprocessor systems-on-chips. In *DAC'04*, pages 681–685, 2004.

[2] Banerjee, K. et al. 3-D ICs: a novel chip design for improving deep-submicrometer interconnect performance and systems-on-chip integration. *Proc. of the IEEE*. 89(5):602–633, 2001.

[3] Todri, Aida et al. Power supply noise aware workload assignment for multi-core systems. In *ICCAD'08*, pages 330–337, 2008.

[4] Huang, Gang et al. Power Delivery for 3D Chip Stacks: Physical Modeling and Design Implication. In *EPEP'07*, pages 205–208, 2007.

[5] Coskun, A.K. et al. Temperature aware task scheduling in mpsocs. In *DATE '07*, pages 1–6, 2007.

[6] Gupta, Meeta S et al. Understanding voltage variations in chip multiprocessors using a distributed power-delivery network. In *DATE'07*, pages 624–629, 2007.

[7] Xu, Hu et al. Timing Uncertainty in 3-D Clock Trees Due to Process Variations and Power Supply Noise. *IEEE Trans. on VLSI*, 2013.

[8] Lung, Chiao-Ling et al. Thermal-aware on-line task allocation for 3D multi-core processor throughput optimization. In *DATE'11*, pages 1–6, 2011.

[9] Zhou, Xiuyi et al. Thermal-Aware Task Scheduling for 3D Multicore Processors. *TPDS*, 21(1):60–71, 2010.

[10] Cadix, L. et al. Integration and frequency dependent electrical modeling of Through Silicon Vias (TSV) for high density 3DICs. In *IITC'10*, pages 1–3, 2010.

[11] PTM Interconnect Model, http://ptm.asu.edu/.

[12] Chen, Howard H et al. Power Supply Noise Analysis Methodology for Deep-Submicron VLSI Chip Design. In *DAC'01*, pages 638–643, 1997.

[13] Saint-laurent, Martin et al. Impact of Power-Supply Noise on Timing in High-Frequency Microprocessors. *IEEE Trans. on Advanced Packaging*, 27(1):135–144, 2004.

[14] Conn, Andrew R. et al. Noise considerations in circuit optimization. In *ICCAD '98*, pages 220–227, 1998.

[15] Wei Huang, S. Ghosh, S. Velusamy, K. Sankaranarayanan, K. Skadron, and M. R. Stan. Hotspot: a compact thermal modeling methodology for early-stage vlsi design. *IEEE Transactions on Very Large Scale Integration (VLSI) Systems*, 14(5):501–513, 2006.

[16] Zhao, Shiyou et al. Decoupling capacitance allocation for power supply noise suppression. In *ISPD'01*, pages 66–71, 2001.

[17] Garey, M. R. et al. *Computers and Intractability: A Guide to the Theory of NP-completeness*. WH Freeman & Co. New York, NY, USA, 1979.

[18] A. Misevicius. A modified simulated annealing algorithm for the quadratic assignment problem. *Informatica*, 14(4):497–514, 2003.

B. Case II: 5 high, 2 middle and 2 low activity workloads

In this scenario, we assume there are 5 high, 2 middle and 2 low activity workloads to be assigned to 3D MP-SoCs. The minimum stack power gradient is still 1.54W. The PSN comparisons of all workloads are plotted in Fig. 7. Our method can reduce PSN by 15.6% compared to thermal-balanced method on average. For high activity workload, our method can reduce PSN by 17.14%. The system performance comparisons are shown with the middle two bars in Fig. 6. As shown in the figure, our method can increase 366 MIPS compared with thermal-balanced method.

C. Case III: 8 high, 4 middle and 4 low activity workloads

In this case, we assume there are 8 high, 4 middle and 4 low activity workloads to be assigned on 3D MPSoCs. The minimum stack power gradient is 1.275W. The PSN comparisons of all workloads are plotted in Fig. 8. As the figure shown, our method can reduce PSN by 6.14% on average. For high activitly workloads, our method can reduce PSN by 14.23% on average. The system performance comparisons are shown as the right two bars in Fig. 6. Although our method increases PSN of some workloads, the system performance increases 582 MIPS compared with thermal-balanced method because our method can reduce PSNs of high activity workloads effectively, which dominate the system performance.

VII. CONCLUSIONS

3D MPSoCs combine the benefits of 3D integration and many-core technologies, and are becoming more attractive. However, due to dense stacking structure and high current delivery demand, the PSN and thermal issues become big concerns. In the paper, we investigate the PSN effects induced by different workloads. Then, we propose an effective PSN-aware heuristic to suppress PSN under given thermal constraint. For $3 \times 3 \times 3$ 3D homogenous MPSoCs, the experimental results show that the proposed algorithm can reduce PSN significantly compared with thermal-balanced algorithm. Due to reduced PSNs, system performance can also be improved as well.

978-1-4799-2817-0/14 $31.00 © 2014 IEEE

SwimmingLane: a Composite Approach to Mitigate Voltage Droop Effects in 3D Power Delivery Network

Xing Hu[†‡], Yi Xu[§¶], Yu Hu[†] and Yuan Xie[¶‖]

[†]Institute of Computing Technology, Chinese Academy of Sciences,Beijing, China
[‡] University of Chinese Academy of Sciences, Beijing, China
[§]Space Science Institute, Macau University of Science and Technology
[¶]AMD Research China Lab
[‖]Pennsylvania State University, USA

Abstract—One of the design challenges for the emerging 3D ICs is the power integrity. With multiple dies stacked vertically, the voltage droop may result in severe power integrity issues. In this paper, we first analyze the impact of application behaviors on voltage droop in a 3D power supply network (PDN) and observe that voltage droop is extremely imbalanced either across different layers or among the cores in the same layer. Based on the observation, we propose *Swimming Lane*, a hardware/software co-design method with two key schemes: (1) Mitigating the interference among different dies via a layer-independent scheme, and (2) balancing the intra-layer voltage droop and reducing the worst-case margin via OS scheduling. Compared to conventional designs, our method can reduce power consumption by 18%, worst-case voltage droops by 13%, and the number of voltage violations by 40%.

I. INTRODUCTION

The integration of multiple cores on a single die is expected to accentuate the already daunting memory-bandwidth problem. Supplying enough data to a many-core chip will become a major challenge for performance scalability. Three-dimensional integrated circuits (3D ICs) are attractive options for overcoming the barriers in interconnect scaling, thereby offering an opportunity to continue performance improvements using CMOS technology. Consequently, 3D integration is envisioned as a solution for future multi-core design to mitigate the interconnect crisis and the memory wall problem [1]. One of the design challenges for the emerging 3D ICs is the power integrity. With multiple dies stacked vertically, the voltage droop may result in severe power integrity issues, due to imperfect parasitic impedance of the power delivery network (PDN) and current fluctuation of circuits [10]. Compared to 2D chips, 3D chips have higher device density, with multiple dies connected vertically via through-silicon-vias (TSVs), which makes power delivery more challenging due to larger load current and longer power delivery paths [6].

Prior works focused on the impact of physical design [7][8] and the floorplan [9] on voltage droop in 3D PDN and observed that increasing decoupling capacitance or TSV density alleviates voltage droops [10]. Nevertheless, it incurs

---

This work is supported in part by National Natural Science Foundation of China (NSFC Program) under Grant No.(61076018, 61274030), National Basic Research Program of China (973 Program) under Grant No.2011CB302503. Yuan Xie is in part supported by NSF 0905365 and 1017277, and SRC grants.

large area overhead to place many TSVs and on-chip decoupling capacitors to overcome the power integrity issue. In addition, decoupling capacitors should be placed next to the active circuits [5] to reduce the voltage noises effectively. Hence, a static solution may not be efficient because the status of the circuits changes dynamically.

In this paper, we study the voltage droop issues in the context of 3D multi-core design where multi-thread applications are distributed to different cores at different layers of a 3D stacked chip. We first analyze the causes of voltage droop from two different perspectives, application behavior and thread distribution, and observe significant *temporal and spatial variations* in voltage droop for 3D multicore architecture: (1) *temporal variations:* The amplitude of voltage droop varies significantly during the program execution. For example, the worst-case voltage droop is almost four times larger than the average case during the execution of the same application; (2) *spatial variations:* The voltage droop within a layer and across layers also vary significantly. Such observations imply that allocating a conservative margin for the worst case is not efficient due to significant spatial and temporal variations in voltage droop. In addition, there are three key observations motivating our thread-mapping strategy: 1. Leveraging thread diversity can eliminate large voltage droop by avoiding destructive horizontal interferences among cores. 2. The top layer (i.e. Die 4 in Fig. 1) always has larger voltage droop than the remaining layers; in addition, the vertical interaction exerts more influence on voltage droop than horizontal resonance; 3. The placement of voltage-violent threads is critical to the amplitude of worst-case voltage droop.

In viewing this phenomenon, we come up with a composite approach: *Swimming Lane*, which allocates an exclusive independent frequency domain, or "lane", for every die of chip. To improve the performance per layer by balancing the intra-layer voltage droop, we propose an OS-scheduling strategy to decide the threads, or "swimmers", for each lane. To summarize, we make the following contributions:

1) We analyze the impact of application behavior and find that thread distribution cause significant spatial and temporal variations in voltage droop in 3D PDN.

2) Based on the characteristics analysis, we propose an OS-scheduling strategy to mitigate the worst-case voltage droops and smooth the differences of voltage margins within one layer.

3) In viewing that voltage droops are asymmetrically

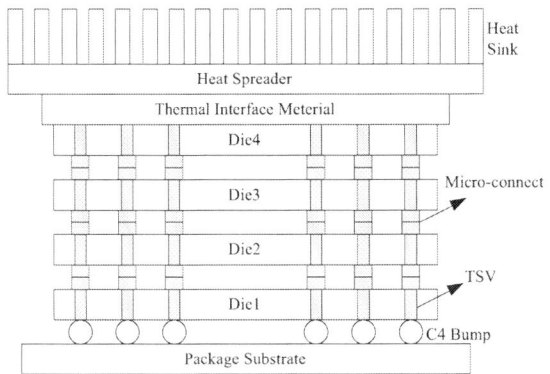

Fig. 1. Overview of 3D integration technology

distributed in different layers, we propose a layer-independent design to alleviate the penalty for achieving failsafe operation under timing errors.

In the rest of the paper, Section II introduces the background of power delivery systems in 3D stacked chips and the related work. Section III describes the simulation set-up and analyzes the impact of thread behavior and thread distribution on voltage droop in 3D chips, followed by the elaboration of the proposed Swimming Lane design. Section V compares the Swimming Lane with conventional designs and shows the results. Section VI concludes the paper.

II. BACKGROUND AND RELATED WORK

In this section, we first give a brief introduction on power delivery network in 3D stacked chips, and then describe related work.

A. PDN in 3D stacked chips

The conceptual structure of 3D chips in flip-chip packaging technology is illustrated in Fig. 1 [10]. In the figure, four dies are stacked on top of the package substrate by micro-connects. TSV is used for both electrical and power signals. Power is delivered from an off-chip voltage regulator, through controlled-collapse-chip-connection (C4) bumps to the bottom die (i.e. Die1 in Fig. 1), then to the upper dies via PG TSV. Hence the power delivery path of each layer is naturally different which causes different voltage droop level at different layers.

The power delivery network consists of two parts: off-chip paths and on-chip networks. The off-chip paths are referred to the power delivery path from voltage source, package to the chip. The on-chip network refers to the R(L)C network inside the chip, consisting of parasitic resistance and inductance in the delivery path, and decoupling capacitance for eliminating the transient voltage noise. The running cores are acting as current sources which fluctuate along with the application execution. Such fluctuation would bring voltage droop in the R(L)C circuit.

B. Related Work

Application behavior exerts significant influence on voltage droop. In 2D chips, clock-gating techniques – a wildly used power-saving scheme in modern CPU products – stalls

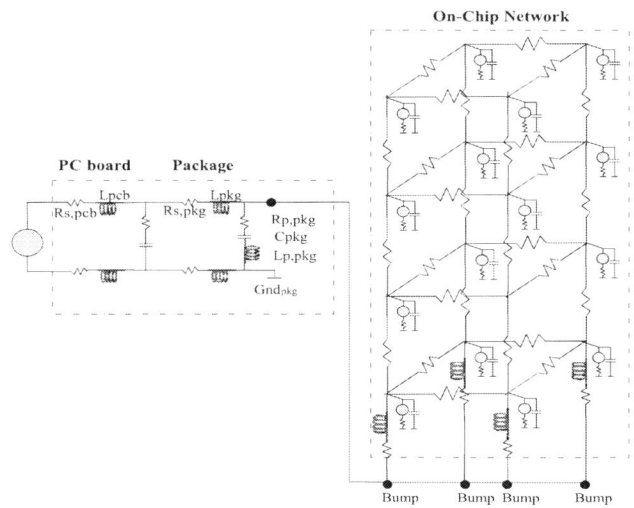

Fig. 2. Four-layer processor baseline and PDN model

or activates the pipeline upon micro-architectural events such as branch mis-prediction, long latency operation, cache misses or TLB misses. Previous work showed that such changes of pipeline status incur current fluctuation within a single core and cause voltage droop [23]. In multi-core processors, interferences between adjacent cores are the major causes of large voltage droop [24], especially when the threads invoked by the same application have similar current traces [16]. Cores sharing similar current traces occasionally cause destructive interferences. In 3D chips, interferences among cores are extended from the horizontal plane to the vertical dimension, which is not considered in the 3D circuitry design or 2D solutions. Hence, we analyze the impact of thread distribution and application behaviors in 3D PDN in the following section.

III. VOLTAGE DROOP CHARACTERISTICS IN 3D CHIPS

With 3D stacking, voltage of cores is affected by vertical and horizontal interaction and the length of delivery path. In this section, we first introduce the simulation setup for the analysis of voltage droop characteristics, then analyze the influences of application behavior and thread distributions on voltage droops based on the experimental results, which motivates the design of Swimming Lane.

We use SPLASH-2 benchmark suite [12] which covers most computation and communication patterns as workloads. The full-system simulator GEMS and power model Wattch [13] are used to estimate current traces. A 16-core processor is simulated as the baseline, which has four layers with each layer hosting four cores, shown in Fig. 2. The floorplan of this processor is adopted from [22]. The core configuration is listed in Table I. After obtaining current traces, the HSPICE is used to simulate the voltage traces of each core. Fig. 2 depicts the PDN utilized in this paper. The off-chip model is consistent with Xeon package information [18][19] and the on-chip network of every layer is based on the multi-core processors model in [19]. The resistance of TSV connecting each layer is slightly larger than the bump resistance, as mentioned in [4]: $R_{via} = 0.5\ mohm$ [5]. The inductance

TABLE I
CORE CONFIGURATION

Parameters	Configuration
Clock Frequency	2.0 GHz
Fetch/Decode Width	4 instructions/cycle
Branch-Predictor Type	64 KB bimodal gshare/chooser, 1K entries
Reorder Buffer Size	128
Unified Load/Store Queue Size	64
Physical Register File	32-entry INT, 32-entry FP
INT ALU, INT Mul/Div, FP ALU, FP Mul/Div	4/2/4/2
L1 Data Cache	16KB, 2-way, 32B line-size, 1-cycle latency
L1 Instruction Cache	16KB, 2-way, 32B line-size, 1-cycle latency
L2 Unified Cache	1MB, 4-way, 64B line-size, 16-cycle latency
I-TLB/D-TLB	64-entry, fully-associative

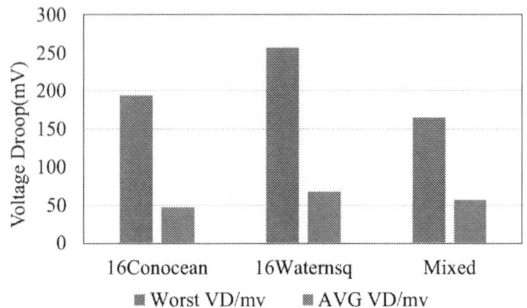

Fig. 3. Voltage droop of diverse applications

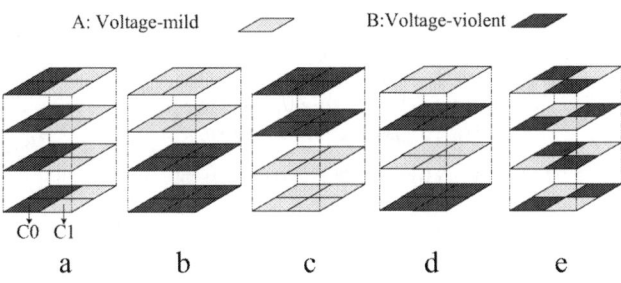

Fig. 4. Thread-mapping strategies

of a via is very small and can be neglected [3]. Based on the experimental setup, we ran the simulations and have following observations:

(1). Worst-case voltage droop is much more serious but rarely occurs. The worst-case voltage droop is much more serious than the average cases. Taking *conocean* and *watersp* as examples, Fig. 3 shows the maximum and average voltage droops by running 16 threads of each application for 1 million cycles. The amplitude of the worst-case droop is almost three times greater than the average cases. However, based on our experimental data, the probability of such a large voltage droop ($\leq$ -130 mv) is only 0.6%. Therefore, it is inefficient to allocate margin conservatively for the worst case.

(2). Thread diversity can mitigate destructive horizontal interference. When eight *conocean* and *watersnq* threads are running simultaneously in the chip, as shown in the "Mixed" case of Fig. 3, the occurrences of the worst-case voltage droop turn out to be much less than the previous two scenarios. This is mainly because of the interference among the neighboring threads in the same plane. Threads with similar switching activity introduce voltage resonance among the active cores, which is referred as destructive interference. When threads with diverse behaviors are arranged together, the voltage-mild thread helps stabilize the voltage-violent one. Such interaction can reduce the worst voltage droop and is referred to as constructive interference. Threads invoked by the same application often have similar power traces and are highly likely to produce destructive interference. Thus, the first two scenarios in Fig. 3 generate higher voltage droop than the mixed scenario. Leveraging thread diversity can eliminate large voltage droop by avoiding destructive interference among cores.

(3). Vertical interference can be damaging. The third dimensional interaction introduced in 3D chip is vertical voltage interaction. Here, we use a simplified example to investigate how vertical interaction and horizontal interference affect voltage droop. Two types of synthetic current traces are employed to generate voltage traces of each core. Thread A represents voltage-mild thread, while thread B is voltage-

violent with twice the sinusoidal amplitude of A. Different thread-mapping strategies are depicted in Fig. 4. In scenario (a), threads invoked by the same application are allocated in the same vertical stack, while (b) illustrates a scenario in which the most violent threads are arranged in the bottom layers.

The corresponding voltage droops of the two scenarios are shown in Fig. 5. C0 and C1 stand for the two cores in each layer. The worst-case voltage droop in scenario (a) is 14% larger than that in (b). It implies that the vertical interference is more serious than the horizontal resonance effect and would amplify the voltage droop. Meanwhile, scenario (a) has larger intra-layer gap among cores than (b). The worst-case voltage droop of C0 is 13% larger than C1, which implies that vertical interference could potentially enlarge the intra-layer imbalance.

(4). Violent threads arranged in top layer always hurt the vertical neighbors. Scenario (b), (c), (d) and (e) are compared to illustrate the influence of violent thread location on voltage droop. The strategy in (c) in Fig. 4 is to place the most violent threads in the topmost layer. Strategy in (d) is to interleave voltage-violent and voltage-mild threads along the vertical direction. The strategy in (e) is to interleave different types of threads both in the horizontal and vertical directions.

As we can observe in Fig. 5, scenario (c) results in the worst voltage droop among the four scheduling methods. Scenario (b) outperforms other cases in terms of the worst-case voltage droop. It implies that arranging violent threads farther from the voltage source will cause larger voltage droop. Even though scenario (e) tries to alleviate both horizontal and vertical core resonance, the location of voltage-violent threads exerts more impact on voltage droop, which is considered as the first-order factor for thread-mapping.

978-1-4799-2817-0/14 $31.00 © 2014 IEEE

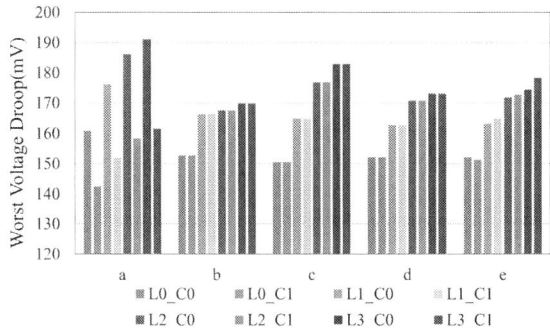

Fig. 5. Worst voltage droop in different scenarios

In summary, we can derive several implications from these results:

1) There is a large gap between common-case and the worst-case voltage droop. Hence it would be inefficient to allocate conservative voltage margin.
2) Thread diversity can mitigate destructive horizontal interference. The voltage-mild threads can offer current relief to voltage-violent threads.
3) Vertical interference is more serious than the horizontal resonance. The vertical location of voltage-violent threads exerts an important impact on voltage droop, which is the dominant factor for thread scheduling.

These implications indicate that voltage margin is distributed unevenly during the execution. Both application behavior and thread distribution affect voltage droops in 3D chips. Based on these observations, we first propose a violent-first thread mapping strategy to alleviate voltage droop effect, and then propose a hardware design to reduce the margin gaps between and inside layers as well.

IV. VIOLENT-FIRST THREAD MAPPING STRATEGY

As shown in Fig. 5, diverse thread distributions have different impacts on voltage droop. We aim to minimize the worst-case voltage droop and the number of voltage violations. To achieve these goals, we need to quantify the voltage droop characteristics of each thread and develop an algorithm to guide thread scheduling based on the predicted threads' voltage features.

Voltage feature characterization. Previous work used the metric of intrinsic droop intensity (IDI) to indicate whether a thread is voltage-violent or voltage-mild [16]. Voltage-violent threads often have higher IDI, while voltage-mild threads have lower IDI. They correlate relationship between IDI and micro-architectural events, such as cache miss, TLB miss and branch mis-prediction. We adopt this practice in this paper, using an on-line predictor to estimate IDI of every thread by capturing performance counter information.

Scheduling strategies. There are three guidelines for thread scheduling: (1). Arrange the violent threads in the lower layers, otherwise it will induce serious droop in the vertical stack. (2). Cluster the threads with close IDI into one layer. There are two advantages of doing this. First, it can mitigate vertical interference. As mentioned in Section III.D, vertical resonance can arouse larger voltage droop than horizontal

direction, so it is reasonable to put similar threads in the same die instead of the same vertical stack. Second, it smooths the intra-layer margin gap among cores. Since the whole layer shares one frequency-monitoring and actuation system, timing margin needs to tolerate the worst thread. Therefore it is important to minimize inter-layer margin gap, otherwise lots of margin would be wasted. (3). Place threads of different applications in the neighborhood in every die. This strategy reduces horizontal voltage resonance induced by similar pipeline activities.

The detailed scheduling algorithm that determines the location of each thread is shown as follows.

1) Estimate IDI of each thread according to micro-architectural events of them. The detailed estimation process is described in [16].
2) Sort threads in descending order in terms of IDI and enqueue them. If multiple threads have the same IDI, a round-robin algorithm is employed to choose threads from different applications, so that threads from the same application would not be arranged close to each other.
3) Select the thread at the head of the queue and place it in the idle core of the available bottom layer.
4) Check if the queue is empty. If not, go back to Step3.

V. CONSTRUCTION OF THE SWIMMING LANE

With the violent-thread first scheduling strategy, we further develop a composite design to avoid wasting margin that includes both hardware design and thread scheduling. It equips the 3D chip with a power delivery system but multiple frequency domains referred to as *swimming lanes*; every layer can work at an individual frequency and be controlled separately.

The hardware design of *Swimming Lane* is illustrated in Fig. 6. Power is delivered from the bottom layer and goes through the middle layers to the upper layers. Swimming Lane equips every die with one monitor and actuation system to isolate the effect of voltage droop within one die. As analysis shown in Section III, the upper layers have larger voltage droop than the other layers. The gap between the worst cases of the bottom layer and the top layer can reach 30mv. As a result, the voltage violations in the upper layer occur more frequent than in the lower layers. To reduce failsafe penalties of timing errors, the layer-independent design allows each layer to work at its appropriate frequency and make adjustments or recovery locally without affecting others. For example, when a core reduces the local frequency to avoid a potential timing error, other layers will not be interrupted and can continue full-speed execution.

To solve voltage violation issues, margin-monitor and actuation mechanism are necessary. The output of critical-path monitor (CPM) [21] indicates the timing margin of the monitored critical paths. Timing delay is affected by multiple factors: process, thermal, aging and voltage. The thermal and aging influences are sluggish; thus, the transient timing violations are related to voltage variation. Therefore, transient variation of CPM can be interpreted as voltage variation. We set the voltage margin for common cases instead of the worst case to save power. Once CPM senses voltage violation, the digital-phase locked loop (DPLL) can

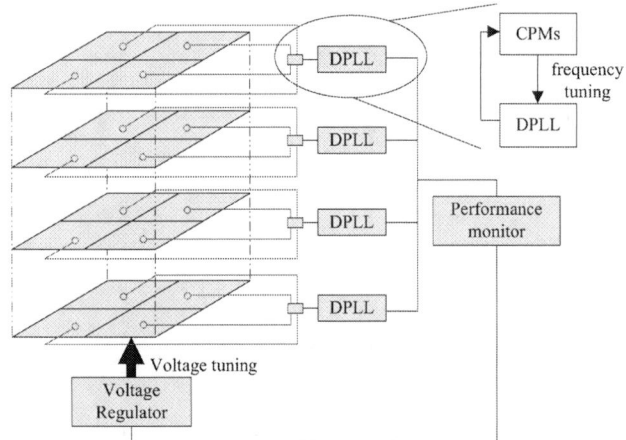

Fig. 6. Hardware design of Swimming Lane

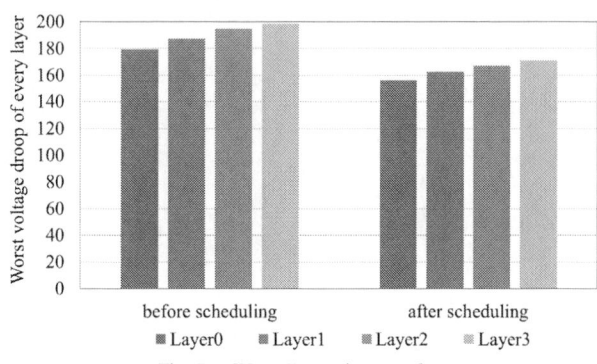

Fig. 7. Worst Droop in every layer

tune the frequency quickly to avoid a timing error. Studies have shown that fast DPLL is able to turn down 7% of peak frequency in several cycles [14]. The CPM circuits and DPLL changes add little penalty in processors, as shown in [14], only 0.046 mm^2 of silicon.

A performance monitor is adopted to supervise the working frequency of the system periodically. When the frequency exceeds the upper threshold, it indicates voltage margin is over-provisioned for this period of time. Then the voltage regulator starts to decrease voltage level. When frequency is lower than low bound, it indicates the supplied voltage is insufficient and needs to be raised up. The voltage tuning resolution is 6.25 mV, and it takes several micro-seconds to complete the entire tuning process [14].

The working process of *Swimming Lane* is as follows. The *Swimming Lane* predicts if threads are voltage-violent or voltage-mild according to their performance counter information, using the solution in previous work [16]. Then the threads are mapped to cores according to the violent-first thread mapping strategy. During threads execution, CPM senses the voltage variation and DPLL is triggered to reduce working frequency if voltage emergency is about to occur. The voltage regulator tunes voltage if the frequency of system is lower than the preset values in a period of time. In this way, *Swimming Lane* can smooth voltage emergencies and alleviate performance penalty induced by voltage emergencies as well.

VI. Experimental Results

We use C_Margin to represent the margin for common-case voltage droop which assures system with no voltage emergencies most of times. Compared to the solution using worst-case margin (W_Margin), the supply voltage can be decreased for ($\mid W\_Margin \mid - \mid C\_Margin \mid$) mV. In our experimental work, W_Margin is set at -260mV and the C_Margin varies from 95 to 130mV with different applications. The nominal voltage is 1.4V; thus, the voltage is about 1.27V with C_Margin. The power consumption of the common-margin scheme is about 82% of the conservative solution.

When we use the common-margin scheme, voltage violations may occur in some cases, followed by rapid frequency tuning. In this session, we validate the effectiveness of *Swimming Lane* in eliminating voltage violations, alleviating the worst voltage droop and reducing differentiation of voltage margins across chip.

A. Voltage violation reduction in whole chip

Nine scientific applications in SPLASH-benchmark suite are used in our simulations: barnes, cholesky, conocean, fft, lu, radix, radiosity, water-nsq and water-sp. Every application, invoking four threads, has been executed for 1B instructions which are cut into program slices of 1M cycles. Every sample selects 16 program slices from four applications to build the start point of workload. We build 100 samples from Splash2 benchmarks to validate the effectiveness of Swimming Lane. To simplify the discussion, we assumed that cores do not support simultaneous multi-threading, so each core can run only one thread at a time. Compared to the case that schedules threads in the vertical stack, the Violent-first scheduling strategy can reduce voltage violations by 40%.

B. Worst voltage droop in each layer

The following experimental results mainly focus on the voltage-dominated workloads. The *Swimming Lane* can mitigate voltage violations in every layer and reduce the worst voltage droop by 13%, as shown in Fig. 7. This is because it makes voltage feature prediction for threads and proactively schedules threads to minimize large voltage droops.

In addition to reducing voltage violations, this scheduling method can reduce C_Margin by 9%. Besides, it reduces the gap of voltage droops inside one layer. The gap of the voltage droop in layer0 was about 20mv; after scheduling, the gap of layer0 was only 6mv.

C. Frequency boost

When we use the same supply voltage as the conservative solution, the performance of *Swimming Lane* can be boosted by utilizing additional voltage margin. Fig. 8 shows the boosted frequency taken by several techniques. Assuming a 1.5 voltage-to-frequency scaling factor [17][15], $A\%$ of voltage increase would transfer to $1.5 * A\%$ of frequency

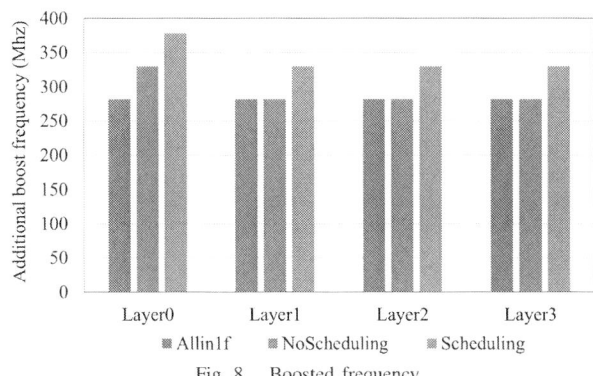

Fig. 8. Boosted frequency

increase. By allocating C_Margin instead of W_Margin, the frequency can be boosted up for 380MHz. When using layer-independent design, the bottom layer can be boosted up for 48MHz. *Swimming Lane* can boost frequency up for 96MHz for the bottom layer and 48MHz for the other layers.

VII. CONCLUSIONS

3D integration technology is a promising solution to improve performance and decrease power. However, it also faces many challenges, such as power integrity, which becomes more serious due to the long delivery path and heavier load in 3D chips. This paper analyzes the impact of application behaviors on voltage droop in 3D power delivery network and proposes a hardware and software co-design based on the observations. Results reveal that compared to conservative margin solution, our scheme can save 18% of power. In addition, our scheme can mitigate the worst-case voltage droop and the number of voltage violations by 40%.

REFERENCES

[1] Gabriel H. Loh and Yuan Xie, "3D Stacked Microprocessor: Are We There Yet?," *IEEE Micro* 30(3): pp. 60-64, 2010.

[2] Michael B. Healy and Sung Kyu Lim, "Power-Supply-Network Design in 3D Integrated Systems," *IEEE Design, Automation Test in Europe Conference (DATE)*, pp. 223-228, 2011.

[3] J. S. Pak, J. Cho, J. Kim, J. Lee, H. Lee, K. Park, and J. Kim. "Slow Wave and Dielectric Quasi-TEM Modes of Metal-Insulator-Semiconductor Structure Through Silicon Via in Signal Propagation and Power Delivery in 3D Chip Package," *IEEE Proceedings of Electronic Components and Technology Conference (ECTC)*, pp.667-672, 2010.

[4] Michael B. Healy, and Sung Kyu Lim, "Distributed TSV Topolog for 3-D Power-Supply Networks", *IEEE Transacations on Very Large Scale Intergration Systems (TVLSI)*, vol. 20, pp. 2066-2079, 2012.

[5] Zheng Xu, Christopher Putnamy, Xiaoxiong Gux, Michael Scheuermannx, Kenneth Rose, Buckwell Webbx, John Knickerbockerx, and Jian-Qiang Lu,"Decoupling Capacitor Modeling and Characterization for Power Supply Noise in 3D Systems," *Advanced Semiconductor Manufacturing Conference (ASMC)*, pp 414-419, 2012.

[6] Pulkit Jain, Tae-Hyoung Kim, John Keane, and Chris H. Kim, "A Multi-Story Power Delivery Technique for 3D Integrated Circuits," *International Symposium on Low Power Electronics and Design(ISLPED)*, pp 57-62 , 2008.

[7] Taigon Song and Sung Kyu Lim, "A Fine-Grained Co-Simulation Methodology for IR-drop Noise in Silicon Interposer and TSV-based 3D IC," *IEEE Conference on Electrical Performance of Electronic Packaging and Systems (EPEPS)*, pp. 239-242 , 2011.

[8] Nauman H. Khan, Syed M. Alam, and Soha Hassoun, "Power Delivery Design for 3-D ICs Using Different Through-Silicon Via (TSV) Technologies," *IEEE Transactions on Very Large Scale Integration Systems (TVLSI)*, vol. 19, NO. 4, pp. 647-658, 2011.

[9] Zuowei Li, Yuchun Ma, Qiang Zhou, Yici Cai, Yu Wang, Tingting Huang and Yuan Xie,"Thermal-aware Power Network Design for IR Drop Reduction in 3D ICs," *Asia and South Pacific Design Automation Conference (ASPDAC)*,pp.47-52, 2012.

[10] Gang Huang, Muhannad Bakir, Azad Naeemi, Howard Chen and James D. Meindl,"Power Delivery for 3D Chip Stacks: Physical Modeling and Design Implication," *IEEE Conference on Electrical Performance of Electronic Packaging (EPEP)*,pp.205-208, 2007.

[11] Milo M. K. Martin, Daniel J. Sorin, Bradford M. Beckmann, Michael R. Marty, Min Xu, Alaa R. Alameldeen, Kevin E. Moore, Mark D. Hill, and David A. Wood, "Multifacet's general execution-driven multiprocessor simulator (GEMS) toolset," *SIGARCH Comput. Archit. News*, vol. 33, pp. 92-99, 2005.

[12] S.C. Woo, M. Ohara, E. Torrie, J.P. Singh, and A. Gupta, "The SPLASH-2 programs: characterization and methodological considerations," *International Symposium on Computer Architecture (ISCA)*, pp. 24-36 1995.

[13] D. Brooks, V. Tiwari and M. Martonosi, "Wattch: a Framework for Architectural-level Power Analysis and Optimizations," *International Symposium on Computer Architecture (ISCA)*, pp.83-94, 2000;

[14] Charles R. Lefurgy, Alan J. Drake, Michael S. Floyd, Malcolm S. Allen-Ware, Bishop Brock, Jose A. Tierno, and John B. Carter, "Active Management of Timing Guardband to Save Energy in POWER7," *International Symposium on Microarchitecture (MICRO)*, pp. 1-11, 2011.

[15] Vijay Janapa Reddi and David Brooks, "Resilient Architectures via Collaborative Design: Maximizing Commodity Processor Performance in the Presence of Variations," *IEEE Transactions on Computer-Aided Design of Integrated Circuits and Systems (TCAD)*, vol. 30, pp 1429-1445, 2011.

[16] Xing Hu, Guihai Yan, Yu Hu and Xiaowei Li, "Orchestrator: a low-cost solution to reduce voltage emergencies for multi-threaded applications," *IEEE Design, Automation Test in Europe Conference (DATE)*, pp. 208-213, 2013.

[17] K.A. Bowman, J.W. Tschanz, Nam Sung Kim, J.C. Lee, C.B. Wilkerson, S.L. Lu, T. Karnik, and V.K De, "Energy efficient and metastability immune timing error detection and instruction replay-based recovery circuits for dynamic variation tolerance," *ISSCC*, pp. 401-403, 2008.

[18] Arun Chandrasekhar, David Ayers, Farzaneh Yahyaei-Moayyed, and Chung-Chi Huang, "Chip-package-board co-design of a 45nm 8-core enterprise Xeon processor," *Electronic Components and Technology Conference (ECTC)*, pp. 536-542, 2010.

[19] Guihai Yan, Xiaoyao Liang, Yinhe Han, and Xiaowei Li, "Leveraging the core-level complementary effects of PVT variations to reduce timing emergencies in multi-core processors," *International Symposium on Computer Architecture (ISCA)*, pp. 485-496, 2010.

[20] Meeta S. Gupta, Jarod L. Oatley, Russ Joseph, Gu-Yeon Wei, and David M. Brooks, "Understanding Voltage Variations in Chip Multiprocessors using a Distributed Power-Delivery Network," *IEEE Design, Automation Test in Europe Conference (DATE)*, pp. 1-6, 2007.

[21] A. Drake, R. Senger, H. Deogun, G. Carpenter, S. Ghiasi, T. Nguyen, N. James, M. Floyd, and V. Pokala, "A Distributed Critical-Path Timing Monitor for a 65nm High-Performance Microprocessor," *ISSCC*, pp. 398-399, 2007.

[22] Xiuyi Zhou, Jun Yang, Yi Xu, Youtao Zhang, and Jianhua Zhao, "Thermal-Aware Task Scheduling for 3D Multicore Processors," *IEEE Transactions on Parallel and Distributed Systems (TPDS)*, vol. 21, pp. 60-71, 2010.

[23] Meeta S. Gupta, Krishna K. Rangan, Michael D. Smith, Gu-Yeon Wei, and David Brooks, "Towards a software approach to mitigate voltage emergencies," *International Symposium on Low Power Electronics and Design (ISLPED)*, pp. 123-128, 2007.

[24] V.J. Reddi, S. Kanev, Wonyoung Kim, S. Campanoni, M.D. Smith, Gu-Yeon Wei and D. Brooks, "Voltage Smoothing: Characterizing and Mitigating Voltage Noise in Production Processors via Software-Guided Thread," *International symposium on Microarchitecture (MICRO)*, pp. 77–88, 2010.

Spiking Brain Models: Computation, Memory and Communication Constraints for Custom Hardware Implementation

Anders Lansner
Dept. of Computational Biology,
Stockholm university and CSC, KTH,
Sweden
Email: ala@csc.kth.se

Ahmed Hemani, Nasim Farahini
Dept. of Electronic Systems,
School of ICT, KTH, Sweden
Email: [hemani,farahini]@kth.se

Abstract – We estimate the computational capacity required to simulate in real time the neural information processing in the human brain. We show that the computational demands of a detailed implementation are beyond reach of current technology, but that some biologically plausible reductions of problem complexity can give performance gains between two and six orders of magnitude, which put implementations within reach of tomorrow's technology.

I. Introduction

The human brain is an extremely complex and intricate structure which outperforms all human made artifacts when it comes to real-time adaptation and sensory processing of complex and uncertain information as well as advanced motor control. This performance is achieved using slow and unreliable computational elements, neurons and synapses, in large numbers. Though our understanding has made some progress, the precise underlying mechanisms are still mostly unknown. Some basic information about the human brain reveals an almost alien technology (Table 1).

Table 1. Human cortex – some estimated numbers	
Dimensions	$0.3 \text{ cm} \times 2400 \text{ cm}^2$
N:o neurons	$2 \ 10^{10}$
Fan-out (synapses/neuron)	10^4
N:o synapses	$2 \ 10^{14}$
Power dissipation	30 W
Wiring length	10^{10} m
Wiring length/cm^2	$4 \ 10^6$ m
Conduction delays	1-20 ms
N:o spikes generated/s	$3 \ 10^9$
N:o communication events/s	$3 \ 10^{13}$

The main functions of the brain can be characterized as perception, memory, action selection ("decision making"), and production of motor output, i.e. behavior. The brain is capable of continuously operating on-line learning which mainly occurs through synaptic plasticity. In humans, the brain is dominated by cortex and many functions are highly dependent on cortex, for instance language which we use extensively to represent and reason about the world.

In addition to cortex, other parts of the mammalian brain and central nervous system are critically important. Subcortical divisions of the brain as well as the brain stem and spinal cord, host critical "lower-level" largely automated functions like breathing, sneezing, coughing, swallowing, and locomotion. Higher functions like stimulus-response learning and decision making, involve the basal ganglia located centrally in the brain. The cerebellum which has as many

neurons as the rest of the brain, seems to be mostly involved in fine-tuning and coordination of motor output. Also the motivational and emotional aspects of behavior and their influence on, for instance, attention and learning involve to a large extent subcortical structures like amygdala and neuro-modulatory systems and nuclei.

The amount of experimental data about the structure and function of the brain is increasing at a rapid pace and new sophisticated technologies that enable novel measurements are developing fast. Yet, the brain is in many ways an inaccessible structure, in particular if we want to study it under natural conditions *in vivo* and non-invasively as is mostly necessary in human subjects. Thus, despite the already enormous and expanding amount of data, a mechanistic understanding of the brain's normal function and what goes wrong in neurological and psychiatric conditions is still largely lacking.

In recent years, mathematical modeling and computer simulation has gained interest as a tool in brain science. It has the unique ability to bring together in a coherent fashion experimental data about the brain obtained with different measurement technologies and relating to different scales of observation, i.e. from molecules over neurons, synapses and microcircuits, to macroscopic structures like cortical areas and the whole brain. Computational modeling is thus an enabling technology that helps to extract the maximal knowledge from data generated e.g. in animal experiments, and to assist the planning of new experiments. Once we understand the fundamental mechanisms behind the amazing capabilities of the brain, we will also be able to design brain-like machines that mimic aspects of these functions.

Large-scale neural network models of the brain are highly complex and to handle them, high-performance computers as well as accelerators like GPGPUs and dedicated neuromorphic hardware are increasingly used. Supercomputers are becoming increasingly used tools in computational neuroscience and brain simulation [1]. They can execute very large network models, though typically much slower than real time and with a high power consumption. Currently ongoing European projects aim to reduce size and power dissipation by the design of dedicated neural simulation hardware [2,3] and major companies like IBM have similar in-house activities [4]. Designs using more exotic components and elements like memristors [5] and carbon nano-wires [6] are also under way.

Since electronic components and systems typically operate on a much faster time scale than the biological system, and conduction delays are much longer in biology, running in biological real time allows for much time-multiplexing in communication as well as in computation, which can be exploited in digital solutions.

978-1-4799-2817-0/14 $31.00 © 2014 IEEE

The aim of this paper is to estimate the computation, memory, and communication demands for full-scale brain simulation. We first consider a network model that explicitly represents the same number of neurons and synapses as in the brain. We thereafter discuss possible reductions of complexity which are implementation friendly on digital machines, but largely maintain computational capabilities and we finally estimate the capacity and power dissipation gains of such algorithmic reductions.

II. Computation and Communication in the Brain

A. Neuron and Synapse Level Detailed Network Models

In order to estimate the amount of memory, computation and communication needed to simulate the information processing in the cortex we need to start from a hypothetical network model formulated mathematically by billions of coupled differential equations and state variables. Simulation would involve solving these equations by numerical integration on a large cluster supercomputer with a typical time step of 0.1 milliseconds, as well as communicating all the spikes generated as messages between the processors.

A complicating fact is that there is not yet a real consensus within the computational neuroscience and brain simulation community with regard to how detailed the component neurons and synapses need to be.

B. Neurons and Neuron Models

The most numerous neuron type in the cortex is the pyramidal cell, comprising about 80% of all neurons there. In addition, there are some other types of excitatory neurons and a number of different kinds of inhibitory interneurons. Neuron models of different levels of complexity are used when modeling the brain, from simple graded output units over integrate-and-fire point neuron models to complex multi-compartmental ones based on the Hodgkin-Huxley formalism [7].

In fact, for estimation of capacity requirements, the type of neuron model and its computational complexity is of minor importance, since the synapses in the brain are three to four orders of magnitude as numerous. Thus, the communication as well as memory and computation requirements are dominated by these tiny structures.

C. Communication in the Brain – Spiking and Neuromodulation

Communication in the brain is dominated by spikes, i.e. binary messages sent via physical fibers, the axons. Other ways of communication are via direct electric contact, so called gap junctions, or via neuromodulators and hormones. From a computational view, the spike communication is dominant and will be the focus here. Each generated spike typically fan-out to on the order of several thousand post-synaptic target neurons. The average spiking frequency in the brain is low; it has been estimated to 0.16 spikes/s/neuron based on metabolic constraints [8]. In the active brain it is likely that less than 1% of cells fire around $10 - 50$ s^{-1}, and the rest about 0.05 s^{-1}.

Conduction speed along axons vary between 0.1 and 10 m/s, the faster ones achieved by myelinated fibers. In addition, there may be a significant delay in transmission along dendrites of the receiving neuron. The total delay between spike initiation zones of synaptically connected neurons in the brain are on the order of one to a few tens of milliseconds, which is comparable to the time scale of the synaptic and membrane time constants and thus cannot be ignored.

D. Synaptic Transmission and Plasticity in the Brain

Synaptic input via incoming spikes to a neuron is acting either via fast transmittors or via slower neuromodulatory ones. The former type of input generates excitatory or inhibitory postsynaptic potentials lasting on the order of five to hundred milliseconds. If this input excites the cell beyond its firing threshold an output spike is generated. Neuro-modulators typically regulate neuron and synaptic properties including synaptic plasticity on a slower time scale. We will not further consider neuromodulation here.

Synapses models have typically a weight value which is a conductance. Some also feature voltage and Ca-gated conductances as well as different forms of synaptic plasticity, like e.g. fast synaptic depression.

A very important function of the brain is learning and adaptation and the most prominent underlying mechanism is changes in the strength of the synaptic connections between neurons, so called synaptic plasticity, i.e. long-term potentiation and depression. There are a number of forms of plasticity and they occur on several different time scales. A biological synapse is a very complicated molecular machine comprising about a thousand different proteins intricately interacting to support synaptic transmission and plasticity. In models of neural information processing this complexity is typically dramatically reduced, to a handful of multiply-and-add (muladd) operations per time step.

Memories are stored by strengthened synapses and also reinforcement type stimulus-response learning is based on reward modulated synaptic plasticity, mediated by e.g. the neuromodulator dopamine.

Structural plasticity refers to the reorganization of synaptic connections, i.e. their formation and removal. This is important in particular during development and as a response to and for repair after injury. Although important for the development of proper function, considering structural plasticity is probably not critical when it comes to performance requirements, since it runs on a very slow "developmental" time scale. Moreover, this kind of plasticity is rarely present in contemporary network models and will not be treated here, except in the discussion.

978-1-4799-2817-0/14 $31.00 © 2014 IEEE

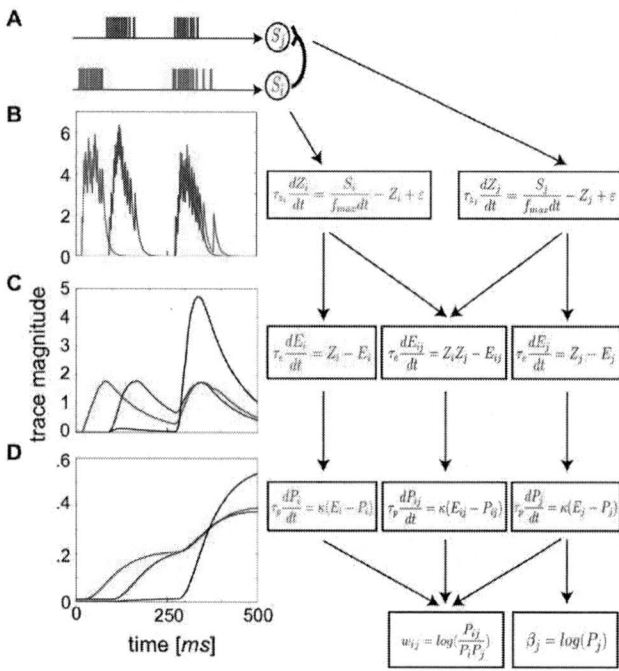

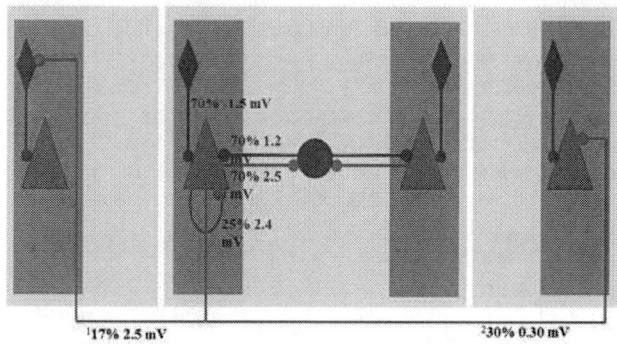

Fig. 2. Microcircuit of the KTH layer 2/3 cortex model. Minicolumns shown in dark grey, comprise 30 pyramidal cells (red, connected 25%) and 2 dendritic targeting, vertically projecting inhibitory interneurons (blue diamond). Hypercolumns (soft WTA modules) are shown in lighter grey with a pool of inhibitory basket cells (blue disc). Large network models have 100 minicolumns and 200 basket cells per hypercolumn. The model has only rudimentary layers 4 and 5.

Fig. 1: Schematic flow of the on-line spike based BCPNN learning rule. (A) The S_i pre- (A-D, red) and S_j postsynaptic (A-D, blue) neuron spike trains are presented as input patterns. Each subsequent row (B-D) corresponds to a single stage in the Bayesian-Hebbian learning rule used for the incremental Bayesian weight update. (B) Primary traces low pass filter input spike trains with time constants τ_{zi} and τ_{zj}. (C) Eligibility traces compute a low pass filtered representation of the primary traces at time scale τ_e. (D) Secondary traces are fed into final estimated probabilities bounded in [0,1], which are used to compute the weight (w_{ij}) and bias (β_j) terms. These tertiary traces have the longest and slowest time course, established by τ_p.

There is a lack of detailed information about the learning rules realized in the brain. Thus, as with the neuron models, there is little consensus on the form and complexity of the learning rules. Different types of spike-timing- dependent plasticity (STDP) rules have been proposed [9]. Here, however, we use a slightly more complex Bayesian- Hebbian online learning rule, the Bayesian Confidence Propagation Neural Network (BCPNN) [10], which has recently been transformed into a spiking version (Fig. 1). Its differential equations can be integrated in every time step using an Euler forward numerical method, but it also allows for lazy evaluation (see below, section V). The BCPNN builds on the idea of a naive Bayesian classifier. The activity of network units represents stochastic events, and unit bias and weights are calculated by applying Bayes' rule as measures of correlation between unit activities in the past [11]–[13]. These statistics are collected during training, and in inference mode the output activity of a unit represents the *a posteriori* probability of the corresponding event given other activated units, i.e. events that have occurred recently. Despite the extreme independence assumptions made in the derivation, this learning rule works very well in practice.

III. A First Estimate of Memory, Computation and Communication demands

We have previously developed and simulated quite detailed and large-scale associative memory models of mammalian cortex and found that many phenomena related to memory retrieval seen experimentally also can be reproduced and explained by such models [14]–[16]. These networks have a modular structure in terms of minicolumns and hypercolumns and feature three or more neuron types (Fig. 2). We will next estimate the computational capacity and power requirements necessary for real-time simulation of a such a cortex network model of the same size as the human brain (Table 1) on a large (today non-existing) supercomputer.

It is reasonable to assume that a synapse has a handful of state variables and that a synaptic transmission operation takes about two muladd instructions per time step and the synaptic plasticity some more such instructions. More precisely, from the equations in Fig. 1, it can be seen that there are 10 plasticity state variables and update equations and that each update is around two muladd operations. The human brain has at least 10^{14} synapses and if we assume four byte (B) per state variable, we end up with a memory demand of about 8 petabyte (PB). The time step in these neural simulations is typically a tenth of a millisecond, resulting in a required computational capacity of around 40 exaflop/s (XF/s) (Table 2) if all updates are performed locally by each synapse. The amount of spiking communication per second would be around 32 TB with each message about 4B using Address Event Representation (AER) [17]. From Table 2 it is quite clear that synaptic transmission is less demanding than plasticity. This is mainly because only one connection state variable (Z_i) require updating in this case.

978-1-4799-2817-0/14 $31.00 © 2014 IEEE

Table 2. Capacity requirements for real time full brain model

	Transmission	Plasticity
Memory	2 PB	8 PB
Computation	4 XF/s	40 XF/s
Memory bandwidth	8 XB/s	80 XB/s
Spike communication	32 TB/s	32 TB/s
Power dissipation	2 GW	20 GW

Table 2. Capacity requirement for real time simulation of a full size brain model. Numbers are given separately for transmission and plasticity. Power dissipation is estimated from the Blue Gene value of 2 Gflop/W.

It is quite clear that to simulate this kind of brain model in biological real time at this level of quite moderate detail surpasses most if not all computational technologies of today. A cat brain sized model has been run on a Blue Gene supercomputer at 1/100 of real time [4]. But given that roadmaps exist for Exaflop in 2018 and Zetaflops in 2030[1], real time execution of this kind of full-scale cortex model will eventually be achieved.

So, is there some way to achieve functionally equivalent full-scale cortex-like computation in a reduced brain model at a significantly lower computational cost? The rest of this paper will explore these possibilities and discuss possible assumptions and reductions based on a reduced spiking cortex model that has already been studied in supercomputer simulation, showing interesting computational capabilities.

IV. A Reduced Full-scale Cortex model

As stated above, the memory and computation requirements for a single neuron and synapse level model of the brain are such that simulation on today's supercomputers as well as dedicated hardware is impossible. However, if we aim for a lower complexity but functionally equivalent model reduced based on some neurobiologically plausible hypotheses of modularity and patchy connectivity the situation is somewhat more bright.

We will describe one case of a functional abstraction of the brain's structure and discuss what the computational units may be, what modular structure there might be in terms of minicolumns and hypercolumns, and how brain areas can be seen as composed of arrays of hypercolumns, and finally the cortex as composed of some hundred such areas. We will further consider how the characteristic patchy long-range connectivity within cortex can reduce the demands for communication significantly.

A. What is the Computational Unit in the Cortex – the Single Neuron or the Minicolumn?

Functional minicolumns like e.g. orientation columns in the primary visual cortex are composed of a local population of some hundred neurons, possibly somewhat intermingled with

---

[1] http://www.itproportal.com/2011/06/20/knights-corners-beginning-intel-aims-1000-petaflop-supercomputer-2018/

---

nearby columns [18]. Very recent information indicate that these columns originate from the same stem cell during brain development and that they are interconnected with a probability of about 25%, so they presumably have quite correlated output [19]. Thus, they likely represent the same external entity (feature) and collaborate to establish links between it and other information in distant cortical locations. Our assumption here is that such a minicolumn unit (MCU) and not the single neuron is the smallest functional unit in the human brain.

B. Hypercolumns as Aggregates of Minicolumns

Hypercolumns are aggregates of the order of hundred functional columns or minicolumns [20]. They operate as normalizing winner-take-all modules [21] due to the competitive feedback inhibition between the minicolumns in the hypercolumn, mediated by local inhibitory interneurons [18,19]. A hypercolumn unit (HCU) can thus be regarded as representing a discrete coded attribute with up to hundred discrete values, each represented by one MCU. Here we regard this columnar structure as generic for the entire cortex, and the human cortex then consists of about two million hypercolumns. The spatial extent of a real cortical hypercolumn is about 0.5x0.5 mm and it contains some tens of thousands of neurons and 10^3-10^4 as many incoming connections. The incoming synaptic connections to the hypercolumn are to about 50% originating locally and 50% from other neurons outside the hypercolumn via long-range connections. In our reduced model, the function of the internal connections have been hidden within the MCU units. Each hypercolumn unit (HCU) is represented by its constituent spiking MCUs and the normalization of activity has been hardwired within the HCU (Fig. 3), so only external connect-ions are represented explicitly. The MCU spikes are generated by a Poisson process as shown by the state update equations in Fig. 3. These equations together with the on-line plasticity update rules in Fig. 1 define our model. It could be noted that the spiking MCUs feature spike frequency adaptation, which is a prominent feature of many real cortical neuron types. This effect is often mediated by Ca-gated K-channels and acts on a time scale of hundreds of milliseconds to gradually slow down the spiking rate of an active neuron.

In our reduced modular cortex model each MCU is assumed to deliver spikes to 10000 different target MCUs, about ten times more than any single neuron and maybe close to the number of global connections from a real minicolumn. We assume that about 1/10 of its hundred neurons are large layer V pyramidal cells that provide long-range connections to distant parts of cortex. As a consequence, the average and maximum spiking rates of an MCU are assumed to be about 10 times that of a single neuron *in vivo*, i.e. 1 s^{-1} and 100 s^{-1} respectively.

This modular abstraction and reduction of model complexity contributes to a significantly more hardware implementation friendly network model and also sets the stage for the use of patchy connectivity.

C. Patchy Long-range Connectivity

A universal feature of cortical neurons is their tendency to form terminal clusters or "patches" of synapses, targeting far away sites in the horizontal direction [24], [25]. The diameter of such a cluster corresponds approximately to the size of a hypercolumn. In our reduced model, all long-range connections are patchy and each patch originates from an MCU in the source HCU and contacts all MCUs in the target HCU, though each one with an individual plastic weight, modifiable

$$s_j = \beta_j + \sum_i w_{ij} S_j + I_j + \gamma_a \alpha_j$$

$$\frac{d\alpha_j}{dt} = -\frac{(S_j - \alpha_j)}{\tau_a}$$

$$\frac{d\hat{s}_j}{dt} = \frac{(s_j - \hat{s}_j)}{\tau_m}$$

$$o_j = \begin{cases} \dfrac{e^{\gamma \hat{s}_j}}{\sum_k e^{\gamma \hat{s}_k}} & if \ \sum_k e^{\gamma \hat{s}_k} > 1 \end{cases}$$

Fig. 3. Equations for the updating of minicolumn unit state. Variables are as follows: s is support, β is a bias, S is spike output, w_{ij} is a weight from i to j, I is input, α is adaptation, $\hat{s}$ is dynamic support, γ_x are gains and τ_x are time constants.

according to the equations in Fig. 1. In real cortex, it has been found that one large pyramidal cell generates on the order of ten such patches. The MCUs in our model are therefore the origin of one hundred patches. The BCPNN learning rule can yield excitatory (positive) as well as inhibitory (negative) connections. In cortex, the former ones are assumed to target pyramidal cells in the target minicolumn directly, whereas the latter influence indirectly via inhibitory interneurons. In a human cortex sized network model, each of the two million hypercolumns is the source of on the order of 10000 patches, and also the target of the same number of incoming patches. The same arrangement in our reduced model generates one million incoming MCU-to-MCU connections. The corresponding biological number would be around 100 million synapses.

We will now give a rough estimate of the memory, computation, memory bandwidth, and spike communication requirements for this reduced functional model.

V. Capacity Requirements for a Reduced Cortex Model

Here, we will present two very similar cases, based on assumptions of modularity and patchy connectivity. The first one employs a straightforward periodic updating method, while the other is based on lazy evaluation. Since the HCU is a natural building block of our network, the estimates will often be given per such module. An HCU is also a natural unit for parallel implementation, on supercomputers using MPI as well as on custom VLSI.

A critical variable for the simulation of a brain model is the simulation time step. It varies in neural simulations depending on accuracy demands from about 10^{-5} to 10^{-3} seconds. Our simulations of the reduced model typically use the latter longer time step. Given that neuronal passive membrane time

constants as well as the time course of postsynaptic potentials are on the order of 5-10 milliseconds, it is unlikely that a higher time resolution is required.

B. Memory Requirements

The connections are involved in storing memory as well as in "synaptic" transmission and plasticity. This implies that the memory requirements and also computation is dominated by the connections in the network. Of the synaptic state variables (Fig. 1), three (E_{ij}, P_{ij}, W_{ij}) need to be represented per connection (10^6 per HCU), three (Z_i, E_i, P_i) per incoming terminal cluster (10^4 per HCU), and four (Z_j, E_j, P_j, β_j) per MCU in the HCU (100 per HCU). Thus, the HCU connection state matrix typically has 10^4 rows and 100 columns. The total amount of memory required per HCU is about 12 MB.

C. Computation and Memory Bandwidth

For our reduced model, computation is necessary for synaptic transmission as well as for the plasticity and, as previously, the transmission is very cheap and uncomplicated compared to the plasticity.

Transmission. Each time a spike is received access of the weight of 100 connections on the corresponding row is needed. This happens on average about 10 times per time step in a HCU. The weight summation will take about one muladd per connection. The exponential decay of the 10^4 postsynaptic potentials in each time step will require additional computation. The required memory bandwidth is moderate since a spike only requires access to weight values from one row of the connection matrix.

Plasticity. The differential equations in Fig. 1 can be solved using a simple difference approximation. The updating of connection variables involves an exponential decay of all variables in the connection matrix in every time step, which becomes expensive. This also means that the entire memory need to be accessed in every time step.

Table 3. Capacity requirements for reduced brain model (Periodic update)

	Full	Per HCU	Gain
Memory	24 TB	12 MB	333
Computation	48 PF/s	24 GF/s	833
Memory bandwidth	48 PB/s	24 GB/s	1667
Spike communication	88 GB/s	44 kB/s	364

The capacity demands for transmission together with plasticity update is given in Table 3.

However, it is possible to do better, since this particular learning rule can be subjected to lazy evaluation. The plasticity equations for one connection can be solved analytically over the interval where there is no incoming or outgoing spike. For each incoming spike, it is necessary to update a row in the connection matrix and a generated outgoing spike triggers update of the corresponding column in the connection state matrix. Updating requires both to read from and store back into memory. Though the numerical expressions for deferred updating are quite complicated this will still save quite a lot of computation since spikes are

978-1-4799-2817-0/14 $31.00 © 2014 IEEE

relatively infrequent. But the lazy evaluation requires slightly more total memory, since it is necessary to store a time stamp to keep track of the time elapsed since last update. The most dramatic saving is, however, that the required memory bandwidth reduces by almost three orders of magnitude (Table 4)!

D. Spiking Communication

In spiking neural network models, spiking communication is typically represented as an event using AER (Address Event Representation) [17] and each spike is delivered after a delay depending on the conduction speed and distance between source and target location. In our model, the spike generation is not balanced at the single minicolumn level, but more so, on the hypercolumn level where typically one unit is dominating spike production and the others are silent, due to the winner-take-all internal dynamics.

The patchy connectivity in our network means that only one message is required to contact all minicolumns in a target hypercolumn, giving a moderate 88 GB/s of messaging assuming 37b packets.

Table 4. Capacity requirements for reduced brain model (Lazy update)

	Full	Per HCU	Gain
Memory	28 TB	14 MB	1
Computation	732 TF/s	366 MF/s	66
Memory bandwidth	112 TB/s	56 MB/s	429
Spike communication	88 GB/s	44 kB/s	1

VI. Summary and Concluding Remarks

Real-time simulation of full-scale detailed brain models are out of reach today and several years ahead and this is also the case for quite simplified network models matching the number of neurons and synaptic connections in the brain. However, design and execution of reduced models with a function in principle comparable to the human brain will soon be within reach. Their design is, however, still resting on many controversial assumptions. We have here demonstrated a specific case based on the BCPNN model.

The reduced capacity demands demonstrated here comes from four key optimizations: a) The smallest computational element is an MCU unit which is an aggregation of 100 neurons; b) Variables and operations that depend only on pre- or postsynaptic activity (single indexed variables in Fig. 1) are shared for entire rows and columns respectively in the connection matrix; c) a perfect patchy connectivity is assumed, which matches modularity and decreases communication demands; d) Periodic updating of plasticity equations in every time step is replaced by a lazy evaluation method. Together, these algorithmic innovations significantly reduce the computational, power, and memory bandwidth requirements compared to a naive neuron and synapse level implementation (Table 5). Future work may include, for instance, exploitation of lower precision state variables [26] and less frequent updating of processes with long time constants.

This reduced model still puts high demand on the hardware

implementation but in a companion paper [27] we demonstrate that a highly customized and optimized architecture, parallelized memory access and 3D integration technology handles successfully the fundamental performance and power bottlenecks. We estimate that in the technology of 2018 this kind of reduced full-scale cortex model can be contained within a form factor of a pizza box and a power envelop of 6 kW [27].

Table 5. Capacity comparison

	Full brain model	Reduced	Gain
Memory	8 PB	28 TB	286
Computation	40 XF/s	732 TF/s	54645
Memory bandwidth	80 XB/s	112 TB/s	714286
Spike communication	32 TB/s	88 GB/s	364
Power dissipation	20 GW	6 kW	$3 \cdot 10^6$

Table 5. Comparison of capacity and power requirements for full brain model and the reduced model. For the power dissipation, please note that the "Full model" value is calculated based on 2 Gflop/W whereas the "Reduced" is based on the power dissipation of the VLSI design described in the companion paper [27].

The basic BCPNN algorithm investigated here is limited to "L2/3 capability" representing in an abstract manner the holistic pattern processing supported by lateral and feed-back projections within and between cortical areas, mostly involving cortical layers 2/3 and 5. To reach a more complete brain model, this basic scheme needs to be extended with "L4 capability" supporting the operations of mainly layer 4 of cortex, involved in transforming representations from one stage to the other in the processing hierarchy. The resulting connectivity links two areas by taking the output from layers 2/3,5 in the source area, sparsifying and decorrelating it by local feature extraction in layer 4 of the target area, then forwarding this transformed activation to layer 2/3,5 in the target area. This operation can be applied recursively at the next higher level of the hierarchy. We are currently working on adding such mechanisms to our reduced cortex-like network model. The goal is to do this without increasing dramatically the complexity and capacity requirements of the computational model as presented here.

References

[1] A. Lansner and M. Diesmann, "Virtues, pitfalls, and methodology of neuronal network modeling and simulation on supercomputers," in in Computational Systems Neurobiology, N. Le Novère, Ed. Springer, 2012, pp. 283–315.

[2] J. Schemmel, D. Brüderle, A. Grübl, M. Hock, K. Meier, and S. Millner, "A wafer-scale neuromorphic hardware system for large-scale neural modeling," in ISCAS 2010 - IEEE International Symposium on Circuits and Systems: Nano-Bio Circuit Fabrics and Systems, 2010, pp. 1947–1950.

[3] F. Galluppi, S. Davies, A. Rast, T. Sharp, L. A. Plana, and S. Furber, "A hierachical configuration system for a massively parallel neural hardware platform," in CF '12 Proceedings of the 9th conference on Computing Frontiers, Cagliari, 2012, pp. 183–192.

[4] R. Ananthanarayanan, S. K. Esser, H. D. Simon, and D. S. Modha, "The Cat is Out of The Bag: Cortical Simulations with 10^9 neurons and 10^13 synapses," in SC09, 2009.

[5] S. H. Jo, T. Chang, I. Ebong, B. B. Bhadviya, P. Mazumder, and W. Lu, "Nanoscale Memristor Device as Synapse in Neuromorphic Systems," Nano Lett, vol. 10, pp. 1297–1301, 2010.

[6] M. S. Zaveri and D. Hammerstrom, "Performance/price estimates for cortex-scale hardware: a design space exploration.," Neural Netw., vol. 24, no. 3, pp. 291–304, Apr. 2011.

[7] C. Koch and I. Segev, "Methods in Neuronal Modeling: From Ions to Networks." MIT Press, Massachusetts, 1998.

[8] P. Lennie, "The Cost of Cortical Computation," Curr Biol, vol. 13, pp. 493–497, 2003.

[9] A. Morrison, M. Diesmann, and W. Gerstner, "Phenomenological models of synaptic plasticity based on spike-timing," vol. 98, pp. 459–478, 2008.

[10] N. Wahlgren and A. Lansner, "Biological evaluation of a Hebbian-Bayesian learning rule," Neurocomputing, vol. 38–40, pp. 433–438, 2001.

[11] A. Lansner and A. Holst, "A higher order Bayesian neural network with spiking units," Int. J. Neural Syst., vol. 7, no. 2, pp. 115–128, 1996.

[12] A. Sandberg, A. Lansner, K. M. M. Petersson, Ö. Ekeberg, and O. Ekeberg, "A Bayesian attractor network with incremental learning," Netw. Comput. Neural Syst., vol. 13, no. 2, pp. 79–194, May 2002.

[13] C. Johansson and A. Lansner, "Towards Cortex Sized Artificial Neural Systems," Neural Networks, vol. 20, pp. 48–61, 2007.

[14] M. Djurfeldt, M. Lundqvist, C. Johansson, M. Rehn, Ö. Ekeberg, and A. Lansner, "Brain-scale simulation of the neocortex on the IBM Blue Gene/L supercomputer," IBM J Res Dev., vol. 52, pp. 31–41, 2008.

[15] M. Lundqvist, A. Compte, and A. Lansner, "Bistable, Irregular Firing and Population Oscillations in a Modular Attractor Memory Network," PLoS Comput Biol, vol. 6, no. 6, pp. 1–12, 2010.

[16] M. Lundqvist, P. Herman, and A. Lansner, "Variability of spike firing during theta-coupled replay of memories in a simulated attractor network," Brain Res, vol. 1434, pp. 152–161, 2012.

[17] K. A. Boahen, "Point-to-point connectivity between neuromorphic chips using address events," IEEE Trans. Circuits Syst. II Analog Digit. Signal Process., vol. 47, no. 5, pp. 416–434, 2000.

[18] V. B. Mountcastle, "The columnar organization of the neocortex," Brain, vol. 120, pp. 701–722, 1997.

[19] Y. Li, H. Lu, P. Cheng, S. Ge, H. Xu, S.-H. Shi, and Y. Dan, "Clonally related visual cortical neurons show similar stimulus feature selectivity.," Nature, vol. 486, no. 7401, pp. 118–21, Jul. 2012.

[20] D. Hubel and T. N. Wiesel, "The functional architecture of the macaque visual cortex. The Ferrier lecture.," Proc. R. Soc. B, vol. 198, pp. 1–59, 1977.

[21] M. Oster, R. Douglas, and S.-C. Liu, "Computation with spikes in a winner-take-all network.," Neural Comput., vol. 21, no. 9, pp. 2437–65, Sep. 2009.

[22] M. Carandini, D. J. Heeger, and J. A. Movshon, "Linearity and Normalization in Simple Cells of the Macaque Primary Visual Cortex," J. Neurosci., vol. 17, pp. 8621–8644, 1997.

[23] C. Meli and A. Lansner, "A modular attractor associative memory with patchy connectivity and weight pruning," Netw. Comput. Neural Syst., 2013.

[24] J. C. Houzel, C. Milleret, and G. Innocenti, "Morphology of Callosal Axons Interconnecting Areas 17 and 18 of the Cat," Eur. J. Neurosci., vol. 6, pp. 898–917, 1994.

[25] T. Binzegger, R. J. Douglas, and K. A. C. Martin, "Stereotypical Bouton Clustering of Individual Neurons in Cat Primary Visual Cortex," J Neurosci, vol. 27, pp. 12242–12254, 2007.

[26] C. Johansson and A. Lansner, "Implementing plastic weights in neural networks using low precision arithmetic," Neurocomputing, vol. 72, pp. 968–972, 2009.

[27] N. Farahini, A. Hemani, A. Lansner, F. Clermidy, and C. Svensson, "A Scalable Custom Simulation Machine for the Bayesian Confidence Propagation Neural Network model of the Brain," in Proc ASP-DAC 2014, 2014.

Advanced Technologies for Brain-Inspired Computing

Fabien Clermidy[*], Rodolphe Heliot[*], Alexandre Valentian[*], Christian Gamrat[**],
Olivier Bichler[**], Marc Duranton[**], Bilel Blehadj[#] and Olivier Temam[#]
[*]CEA-LETI, Grenoble, FRANCE, fabien.clermidy@cea.fr
[**]CEA-LIST, Paris, FRANCE, christian.gamrat@cea.fr
[#]INRIA, Paris, FRANCE, olivier.temam@inria.fr

Abstract – **This paper aims at presenting how new technologies can overcome classical implementation issues of Neural Networks. Resistive memories such as Phase Change Memories and Conductive-Bridge RAM can be used for obtaining low-area synapses thanks to programmable resistance also called Memristors. Similarly, the high capacitance of Through Silicon Vias can be used to greatly improve analog neurons and reduce their area. The very same devices can also be used for improving connectivity of Neural Networks as demonstrated by an application. Finally, some perspectives are given on the usage of 3D monolithic integration for better exploiting the third dimension and thus obtaining systems closer to the brain.**

I. Introduction

Brain-inspired computing has long been a nice theoretical research topic facing implementation issues. Even with powerful computing capabilities on some of the hardest problems, very few industrial applications have appeared.

As largely known, implementation limitations are linked to two main elements:
- Difficulty to "emulate" the behavior of synapses and neurons with transistors. Classical CMOS technology is not aimed at providing devices compatible with neurons requirements.
- Difficulty to connect neurons. This is linked to 2D-layered implementation used for simplifying CMOS integration.

Some solutions have been proposed to cope with these issues. Analog neuron is an elegant implementation solution to reduce the number of transistors needed for a neuron. However, while reducing the number of transistors, large capacitances are needed for obtaining a good RC delay.

On the connection purpose, time multiplexing is currently used, but it limits the overall performance while increasing the power consumption due to high frequency links. Moreover, the increasing capacitance of wires in advanced technologies limits this solution.

A global conclusion of all these works is the relative inexpediency of standard CMOS with Neural Networks (NN) implementation.

In recent years, new technologies have appeared that can change the game. Some of these technologies and their application or potential of application to NN are presented in this paper.

In section II, resistive memories are used to improve implementation of synapse thanks to their nice properties.

Both Phase Change Memories (PCM) and Conductive-Bridge Random Access Memories (CBRAM) are studied for this purpose.

In section III, analog neurons are revisited through the usage of high capacitances of Through Silicon Vias (TSV) used for 3D stacking. TSV also opens the way towards 3D integration and interconnection length reduction which is discussed in section IV.

Finally, emerging 3D monolithic technology is discussed in section V. This technology allows high vertical interconnects density, thus becoming closer to the brain and opening the way to a high range of possibilities to NN implementation.

II. Exploiting Resistive Memories

Memristive memories can be defined as two-terminal devices whose conductance can be modulated by flowing current though it. In this section, we show how memristor can be used to produce brain-like circuits and in particular arrays of artificial synapses. The functional analogy between memristive technologies and the behavior of a synapse has been anticipated by Chua [1] and later popularized by Strukov [2].

Since then several authors have shown that it was a fruitful idea. Most notably, Snider [3] showed the concept of using memristive memories to implement the Spike Timing Dependent Plasticity (STDP) learning rule, by virtue of their own physics. This idea has since been experimentally demonstrated to be actually working by several groups: University of Michigan using Si-Ag devices [4], University of Aachen with nano-ionics devices [5] and Alibart et al. with Organic transistors (NOMFET) [6].

However, there is a bit more than the basic functionality of a single device to yield an actual working memristive based brain-like circuit. As a matter of fact, the circuit architecture allowing the exploitation of such devices is of utmost importance. Depending on the polarity and electrical characteristics of investigated devices, three types of circuits have been identified which are now described in the following paragraphs.

A. Circuits for Bipolar Memristors

Most of the works on memristive devices that have been published over the last couple of years focus on bipolar resistive switching devices [2] [3] [4] [5]. Indeed, these devices exhibit characteristics that are the closest to the original Memristor predicted by Chua. Their resistance can be gradually increased or decreased with opposite polarity voltage pulses and the resistance change is cumulative with the previous state of the device, which make them particularly suitable to implement synaptic-like functionality.

Several (simplified) models have been proposed to develop nano-architectures capable of learning with memristive devices. In [7], a new behavioral model especially tailored at modeling the conductance evolution of some popular memristive devices in pulse regime for synaptic-like applications is introduced. This model demonstrates high tolerance of NN architectures to Memristor variability.

A biologically-inspired spiking NN-based computing paradigm which exploits the specific physic of those devices is presented in [8]. In this approach, CMOS input and output neurons are connected by bipolar memristive devices used as synapses. It is natural to lay out the nanodevices in the widely studied crossbar as illustrated on Figure 1, where CMOS silicon neurons and their associated synaptic driving circuitry are the dots, the squares being the nanodevices. Learning is competitive thanks to lateral inhibition and fully unsupervised using a simplified form of STDP.

Using this topology, comparable performance to traditional supervised networks has been measured [8] on the textbook case of character recognition, despite extreme variations of various memristive devices' parameters. With the same approach, unsupervised learning of temporally correlated patterns from a spiking silicon retina has also been demonstrated. When tested with real-life data, the system is able to extract complex and overlapping temporally correlated features such as car trajectories on a freeway [9].

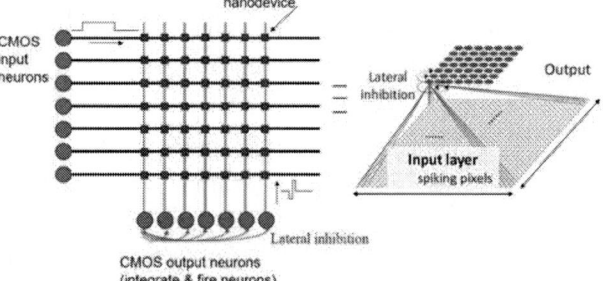

Figure 1: Basic circuit topology. Wires originate from CMOS input layer (horizontal black wires) and from the CMOS output layer (vertical gray wires). Memristive nanodevices are at the intersection of the horizontal and vertical wires [8].

B. Circuits for Unipolar Memristors (PCM)

Among the resistive technologies, Phase-Change Memory (PCM) has good maturity, scaling capability, high endurance, and good reliability [10]. PCM resistance can be modified by applying a temperature gradient modifying the material organization between an amorphous and a crystalline phase. The amorphous region inside the phase change layer can be crystallized by applying Set pulses, thus increasing device conductance. It was shown that the magnitude of the relative increase in conductance can be controlled by the pulse amplitude and by the equivalent pulse width [11]. Amorphization, on the other hand is a more power-hungry process and is not progressive with identical pulses. The current required for amorphization is typically 5–10 times higher than for crystallization, even for state-of-the art devices.

To overcome these issues, a novel low-power architecture "2-PCM Synapse" was introduced in [12]. The idea is to emulate synaptic functions in large scale neural networks thanks to two PCM devices constituting one synapse as shown in Figure 2. The two devices have an opposite contribution to the neuron's integration. When the synapse needs to be potentiated, the Long Term Potentiation (LTP) PCM device undergoes a partial crystallization, increasing the equivalent weight of the synapse. Similarly, when the synapse must be depressed, the Long Term Depression (LTD) PCM device is crystallized. As the LTD device has a negative contribution to the neuron's integration, the equivalent weight of the synapse is reduced. Furthermore, because gradual crystallization is achieved with successive identical voltage pulses, the pulse generation is greatly simplified.

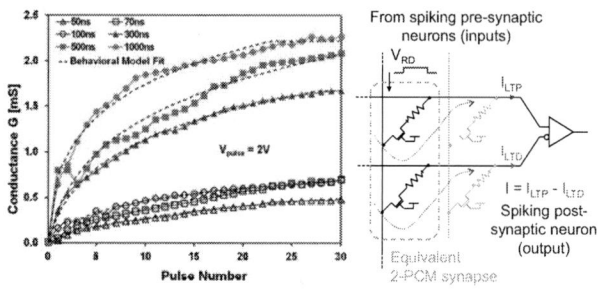

Figure 2: (Left) Experimental LTP characteristics of Ge2Sb2Te5 (GST) PCM devices. For each curve, first, a reset pulse (7 V, 100 ns) is applied followed by 30 consecutive identical potentiating pulses (2 V). Dotted lines correspond to the behavioral model fit used in our simulations. (Right) 2-PCM synapse principle.

C. Using Bipolar Memories (CBRAM)

1T-1R CBRAM multi-level programming was also proposed to emulate biological synaptic-plasticity. Like PCM, it is difficult to emulate a gradual synaptic depression effect using CBRAM, due to the abrupt nature of the set-to-reset transition in those devices. Moreover, LTP behavior is possible, but requires to gradually increasing the select transistor gate voltage. This implies keeping a history of the previous state of the synaptic device, thus leading to additional overhead in the programming circuitry.

In [13], it is shown that when weak SET programming conditions are used immediately after a RESET, a probabilistic switching of the device appears, as illustrated in Figure 3. Then, the switching probability can be tuned by using the right combination of programming conditions. This opens the way to exploit the intrinsic stochasticity of CBRAM to implement synapses. Similarly to the system level, a functional equivalence [14] exists between multi-level deterministic synapses and binary probabilistic synapses. Using this property, a low-power stochastic neuromorphic system for auditory (cochlea) and visual (retina) cognitive processing applications was proposed in [13].

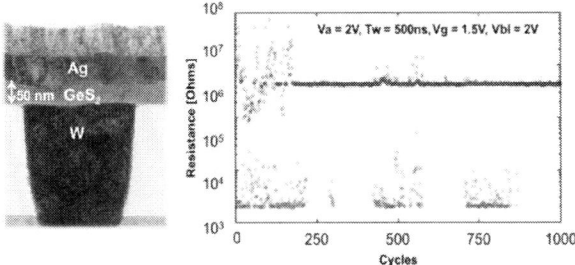

Figure 3: (Left) TEM of the CBRAM resistive element. (Right) Stochastic switching of 1T-1R device during 1000 cycles using 100 ns pulse shows switching probability of around 0.5.

D. Conclusion

This section has demonstrated how synaptic weight adaptation rules: LTP, LTD can be implemented using various memristive devices. Thanks to this property, such devices can lead to intrinsic implementation of STDP based learning rule when coupled with a spike coding scheme [15]. It is safe to say that memristive technology is a promising way for implementing synaptic arrays for brain-inspired computing while neurons circuits will be implemented using advanced CMOS technology as described in the following section. This hybrid CMOS-memristive approach is particularly appropriate for embedded neuromorphic devices.

III. Analog neurons and 3D-TSV

Beyond pure technological advantages, a mix between advanced design and new technologies is an appealing way to succeed in integrating more neurons and greatly increase the overall performance of NN. In this section, we discuss the advantages of analog neurons design style and their combination with Through-Silicon-Vias (TSV) used in 3D stacking technologies.

Analog neurons are considered powerful computational operators thanks to 1) compact design, 2) low power consumption, 3) ability to interface sensors directly with the processing part, and 4) computational efficiency. Even if some analog neuron designs have already been proposed, most are geared towards fast emulation of biological neural networks [16] [17]. Consequently, these hardware neurons contain a feedback loop for learning purposes [19] which is irrelevant when learning can be done off-line such as many current applications. Still, some of these designs have been applied to computing applications [18], but they come with the aforementioned overhead of their bio-inspired motivation. Recently, we implemented and fabricated a

Leaky Integrate and Fire (LIF) analog neuron in 65nm technology node (Figure 4) without feedback loop. This neuron is capable of harnessing input signals ranging from low (1kHz) to medium (1MHz) frequency, compatible with most of the signal processing applications [20]. Note that the aforementioned frequencies (1kHz to 1MHz) correspond to the rate at which the neuron and I/O circuits can process spikes per second, not the overall update frequency of the input data. This frequency can potentially scale up to much higher values by leveraging the massive parallelism of the architecture: an "x" MHz input signal can be de-multiplexed and processed in parallel using x input circuits.

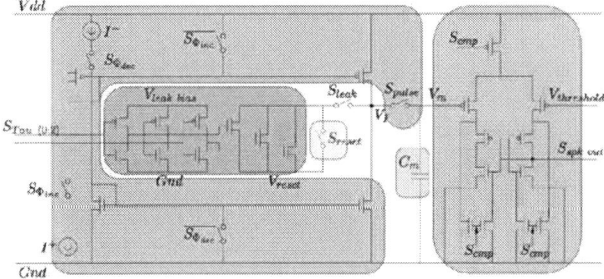

Figure 4: Circuit diagram of the analog neuron (injection in purple, leak in blue, reset in yellow, capacitance in orange, comparator in red).

For instance, the motion estimation application processing a black and white SVGA (800x600) image at 100Hz corresponds to a maximum spike frequency of 48MHz (800*600*100). This application requires $4*N_p + 3$ neurons for N_p pixels. Thus, (4*800*600+3)*100 spikes must be processed per second. Since the maximum processing rate of a neuron is 1Mspikes/s in our design, we need $((4*800*600+3)*100)/10^6 = 193$ neurons to process images at a speed compatible with the input rate.

However, an analog neuron has both assets and drawbacks. The main drawback is the need for a large capacitance. The capacitance has a central role as it accumulates spikes, i.e., it performs the addition of inputs. In spite of this drawback, the analog neuron can remain a very low-cost device: our analog neuron has an area of $120\mu m^2$ at 65nm. This area accounts for the capacitance size and the 34 transistors. Since the capacitance is implemented using only two metal layers, most of the transistor logic can fit underneath. As a result, most of the area actually corresponds to the capacitance.

The analog neuron of Figure 4 operates as follows. When a spike arrives at the input of a neuron, before the synapse, it triggers the $S\varphi_{inc}$ switch (see bottom left), which is bringing the current to the capacitance via V_I. The different transistors on the path from $S\varphi_{inc}$ to V_I form a current mirror, which aims at stabilizing the current before injecting it. Now, if the synapse has value V, this spike will be converted into a train of V pulses. These pulses are emulated by switching on/off S_{pulse} V times, inducing V current injections in the capacitance. We have physically measured the neuron energy at the very low value of 1.4pJ per spike (we used the maximum value of V = 127 for the measurement). The

978-1-4799-2817-0/14 $31.00 © 2014 IEEE

group of transistors on the center left part of the figure implements the variable leakage of the capacitance (which occurs when S_{leak} is closed) necessary to tolerate signal frequencies ranging from 1kHz to 1MHz. After the pulses have been injected, the neuron potential (corresponding to the capacitance charge) is compared against the threshold by closing S_{cmp}, and an output spike is generated on S_{spkout} if the potential is higher than the threshold.

As explained before, analog spiking neurons require large capacitances to store the internal membrane voltage of the neuron. Typical capacitances values are in the order of 0.5−1 pF, which corresponds to 50−200 μm² capacitors in 32−65 nm technologies, and up to 50% of the total neuron area.

3D stacking is a technique that can provide massive parallelism between layers by stacking them and directly connecting them via a large number of Through-Silicon Vias (TSVs). TSVs are used to create vertical interconnections in 3-D stacked chips. As shown in Figure 5, they are composed of a metal wire isolated from the substrate. They are not perfect wires however, and act as MOS capacitors.

In traditional digital circuits, these TSVs are a significant limitation of 3D stacking: they consume on-chip area and suffer from their parasitic capacitance behavior [21]. However, we can actually take advantage of these capacitances to implement the capacitors of spiking neurons, turning a weakness into a useful feature.

An RLC π-shaped electrical model of TSVs was used, as described in [21] (Figure 5, middle). The TSV model features a resistance R_{tsv}, and an inductance L_{tsv}. It is isolated from the substrate by three capacitors C_{ox}, C_{dep} and C_{si}, while parasitic silicon substrate losses are represented by a conductance G_{si}. All of these values are process dependent and can change with TSV density, height, diameter, operating frequency and oxide thickness.

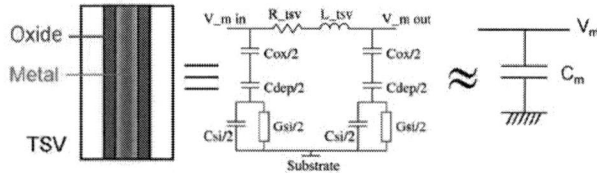

Figure 5: Through Silicon Vias structure and electrical model

We designed an analog LIF neuron using the above TSV model and compared it through Spice simulations against the previously designed analog neuron. Figure 6 shows the behavior of the neuron internal potential V_m for standard (top) and TSV (bottom) neurons (for different densities). When V_m reaches a threshold voltage V_{th}, V_m is reset to a resting potential. With a medium density TSV (400/mm2), it can be seen that the TSV-neuron exhibits the exact same behavior as the standard neuron. TSVs with other densities exhibit similar behavior, albeit with different time constants [22].

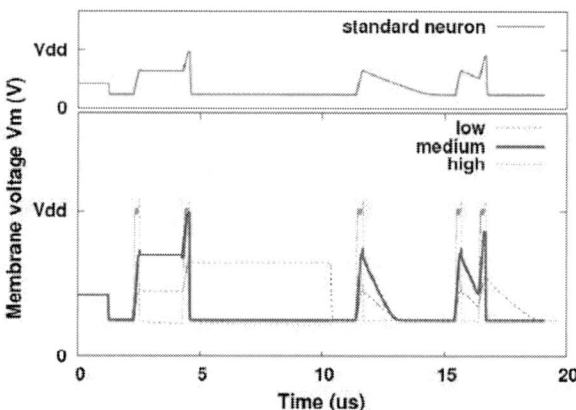

Figure 6: Membrane potential of standard- and TSV-based (several TSV densities) neurons.

3D stacking neuromorphic architectures with TSVs offers massive tremendous opportunities. Figure 7 summarizes some of the options that are offered. These different setups come with different gains in terms of area or with additional connectivity. Some of these applications are discussed in next section.

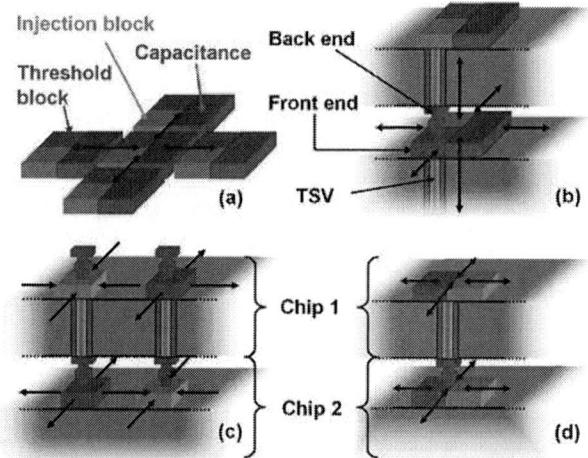

Figure 7. Illustrations of (a) a standard 2D neuromorphic architecture (b) a 3D-IC with standard 2D neurons, and (c) (d) a 3D-IC with TSV-based neurons

IV. 3D staking integration

Neuromorphic architectures are fundamentally 3D structures with massive parallelism, but the 2D planar circuits on which they are implemented considerably limit the bandwidth between layers (or significantly increase the area and energy cost required to achieve sufficient bandwidth). Indeed, neural networks typically exhibit a high level of connectivity. Especially when biologically relevant neural networks are considered, a given neuron can receive information from up to 10.000 neurons. Deep computation and cognitive functions may be reached with thousands of neurons and millions of synapses [21][24][25][26].

In this section, we assume that the neural network is multilayer and densely-connected, and neurons are point-to-point connected (hardwired connections). Dense hardware integration is, therefore, required for advanced processing tasks. Analog neurons are compact circuits but connecting thousands of them leads to routing and throughput problems. The 3D architectures are expected to reduce the routing congestion problem related to neuron interconnect, increase the scalability of the network, reduce critical paths length, and finally save circuit area and power.

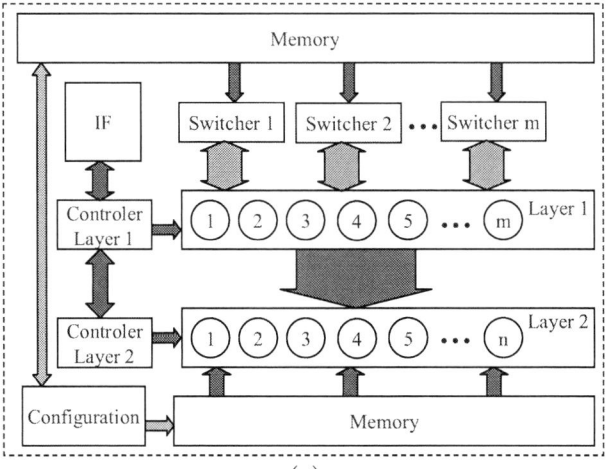

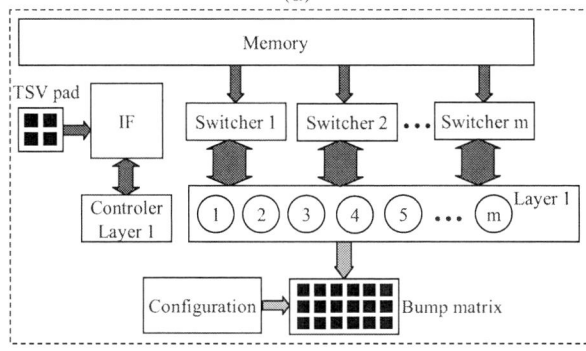

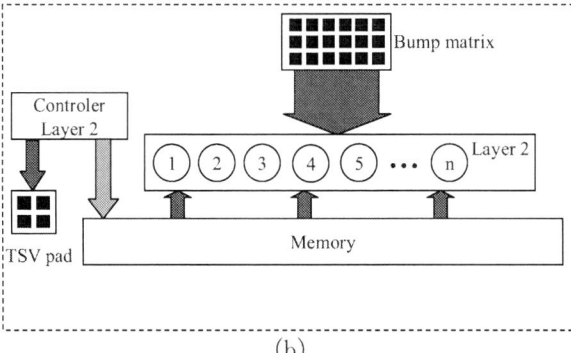

Figure 8: 2D (a) and 3D (b) designs of a neural processor. The two layers of the neural network are mapped onto separate silicon layers. The mean wire-length is reduced, as well as the routing complexity and the inter-neuron throughput.

The case study presented in this paper is a neural processor with a densely connected 2-layers neural network and aiming to recognize objects appearing in a 1000frame/s video stream. The final objective is to execute complex tasks,

such as real-time surveillance (motion detection + image filtering + shape recognition) with low-power and compact circuitry. The 3D design offers more flexibility to meet timing, area and power constraints. Figure 8 shows the block diagrams of both 2D and 3D partitioning of the neural processor design. The intensity of inter-block communication is represented by the thickness of edges (thick edges for intense communication). The routing congestion risk is represented using colors (red edges for higher congestion risk to green edges for the opposite). In 2D design, the communication between the two neuron layers is the most intensive in terms of data exchange and throughput requirements. This complexity is mitigated after mapping the network layers onto separate silicon layers for 3D design. Throughput and congestion problems are, interestingly, alleviated by the increasing number of metal layers and short connections created by 3D routing.

The design methodology we used consists on designing two separate circuits for face-to-face staking fabrication process. We use two types of TSV to link both circuits: TSVs that are created by bonding the last metal layers of both circuits (bumps), and TSVs that cross all the metal layers to output pad signals of the hidden circuit (IO TSV). Bumps have a size of $5x5\mu m^2$ and may be placed everywhere in the circuit body, while IO TSVs have a size of a $60x90\mu m^2$ and can only be placed in the corona area of both circuits.

A 2-tier 3D layout has been built in a 130nm technology. Layouts show a decrease in the mean wire-length. Short connections reduce the footprint area in the 3D design, which is largely related to the buffer count reduction because of shorter wire-length and hence better timing. Routing demand is quite different between 2D and 3D designs. As for the 2D case, a large number of inter-neuron and over-neuron connections are required, and this increases both total and average wire-length. Thus, more high metal layers are necessary to complete inter-neuron routing. In the other hand, many wires in the 3D design are connected to nearby bumps, and this reduces wiring demand significantly.

Figure 9 shows the mapping results of the same neural network in 2D (left) and 3D (right). 3D mapping allows to tremendously decrease the total connections length, and hence the associated power consumption.

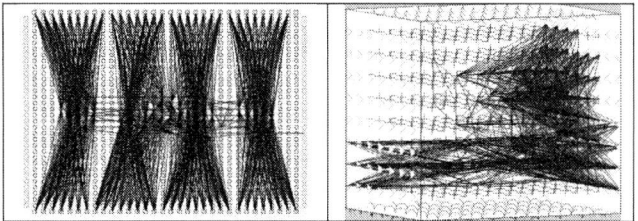

Figure 9: mapping results of the same neural network (184 neurons, 1181 connections) in 2D (left) and 3D (right). Total connections length is reduced by 3x.

978-1-4799-2817-0/14 $31.00 © 2014 IEEE

V. 3D monolithic integration

3D-TSV opens the way towards more integration thanks to new functions (3D-capacitances) and reduced wire lengths as demonstrated in previous sections. However, recent ITRS [27] roadmap shows that TSV alignment will be limited between ~ 0.5μm and 1μm to guarantee correct operation (Table I). This is due to the TSV process which is performed outside the foundry, thus not benefiting from advanced lithography.

Due to the restriction in the alignment, the pitch between two TSVs cannot be smaller than ~ 4 - 8μm. In addition, height of TSV (~ 20 - 50 μm) implies performance impact when signals crossing TSV are in the chip critical paths. As we have seen before, high capacitance can be used for neurons design. However, the limited number of vertical TSV per mm^2 prevent from using this technology for ultra-dense wires.

Recently, a new 3D technology called 3D Monolithic Integration (3DMI) has appeared. With 3DMI, transistor layers are fabricated one after another on the same die using high-end lithography equipment. It results in an improved alignment performance of ~10nm [28].

Therefore, vertical connections can be placed with a very small footprint of less than 100nm diameter in 65nm technology [29].

Table 1. 3DMI vs 3D-TSV comparison

	Alignment (μm)	Diameter (μm)	Pitch (μm)	Minimum height (μm)
TSV	0.5 - 1	2 - 4	4 - 8	20 – 50
3DMI	0.01	0.1	0.2*	0.1
3DMI vs TSV gain	50x to 100x	20x to 40x	20x to 40x	200x to 500x

* Pitch is assumed to be at least two times larger than the diameter.

Moreover, the distance between the top and bottom layers can be reduced to 100nm [28]. Consequently the performance while passing through the vertical via is greatly improved. As a result, efficient, very fine grain partitioning is achievable with 3DMI. Figure 10 shows some partitioning possibilities based on 3DMI integration. Contrary to TSV, the 3D connections are on the same size than classical 2D vias. No guard interval is required and classical design rules can be used. This characteristic allows ultra-fine grain partitioning: as shown in Figure 10.a, a classical library cell can be split in two layers. Figure 10.b shows a so-called "cell-on-cell" approach requiring new 3D place & route tools with a fine grain partitioning [30]. In all cases, 3DMI becomes an ideal choice for highly integrated 3D circuits.

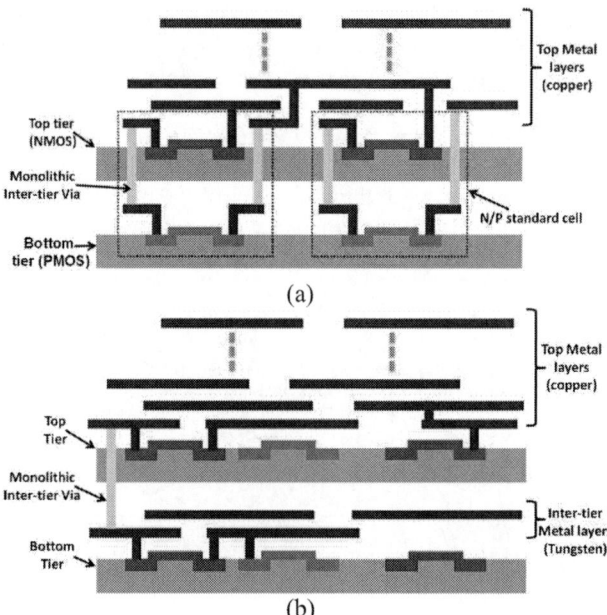

Figure 10 Monolithic 3D integration approaches, (a) transistor-level (N/P), (b) gate-level (cell-on-cell)

3DMI implementation of NN has not been demonstrated yet. However, compared to 3D TSV, a high degree of improvement is expected, especially when considering multi-layers of neurons which is not an issue with 3DMI. If we consider that NN are working at quite low frequency compared to classical devices, thermal issues should not be a problem.

VI. Conclusion

In this paper, we have presented the usage of emerging technologies for greatly improve the implementation of Neural Networks: Synapses effects can be obtained by Memristors thanks to PCM or CBRAM. Neurons cost can be reduced by mixing analog design and high TSV capacitances. Finally, neurons connections issues can be partially solved through 3D integration, monolithic integration being the most advanced case.

All these technologies appear at the same time, and when mixed with new "smart" and low-power applications requirements, we believe we have here a unique opportunity of pushing NN forward.

References

[1] L. Chua, "Memristor-The missing circuit element," *Circuit Theory, IEEE Transactions on*, vol. 18, no. 5, pp. 507-519, 1971.

[2] D. B. Strukov, G. S. Snider, D. R. Stewart, and R. S. Williams, "The missing memristor found," *Nature*, vol. 453, no. 7191, pp. 80-83, 2008.

[3] G. S. Snider, "Spike-timing-dependent learning in memristive nanodevices," *Prof. of IEEE International Symposium on Nanoscale Architectures 2008 (NANOARCH)*, 2008, pp. 85-92.

[4] S. H. Jo, T. Chang, I. Ebong, B. B. Bhadviya, P. Mazumder, and W. Lu, "Nanoscale Memristor Device as Synapse in Neuromorphic Systems," *Nano Letters*, vol. 10, no. 4, pp. 1297-1301, 2010.

[5] R. Waser and M. Aono, "Nanoionics-based resistive switching memories," *Nature Materials*, vol. 6, no. 11, pp. 833–840, 2007.

[6] F. Alibart, S. Pleutin, D. Guérin, C. Novembre, S. Lenfant, K. Lmimouni, C. Gamrat, and D. Vuillaume, "An organic nanoparticle transistor behaving as a biological spiking synapse," *Advanced Functional Materials*, vol. 20, no. 2, pp. 330–337, 2010.

[7] D. Querlioz, P. Dollfus, O. Bichler, and C. Gamrat, "Learning with memristive devices: How should we model their behavior?," *Nanoscale Architectures (NANOARCH), 2011 IEEE/ACM International Symposium on*, pp. 150-156, 2011.

[8] D. Querlioz, O. Bichler, P. Dollfus, C. Gamrat, "Immunity to Device Variations in a Spiking Neural Network with Memristive Nanodevices," *Nanotechnology, IEEE Transactions on*, vol. 12, no. 3, pp. 288-295, 2013.

[9] O. Bichler, D. Querlioz, S. J. Thorpe, J.-P. Bourgoin, and C. Gamrat, "Extraction of temporally correlated features from dynamic vision sensors with spike-timing-dependent plasticity," *Neural Networks*, vol. 32, pp. 339–348, 2012.

[10] A. Fantini et al, "N-doped GeTe as performance booster for embedded phase-change memories," *Proc. IEDM*, 2010, pp. 29.1.1–29.1.4.

[11] D. Kuzum, R. G. D. Jeyasingh, B. Lee, and H.-S. P. Wong, "Nanoelectronic programmable synapses based on phase change materials for braininspired computing," *Nano Lett.*, vol. 12, no. 5, pp. 2179–2186, 2012.

[12] O. Bichler, M. Suri, D. Querlioz, D. Vuillaume, B. DeSalvo, and C. Gamrat, "Visual pattern extraction using energy-efficient '2-PCM synapse' neuromorphic architecture," *Electron Devices, IEEE Transactions on*, vol. 59, no. 8, pp. 2206–2214, 2012.

[13] M. Suri, D. Querlioz, O. Bichler, G. Palma, E. Vianello, D. Vuillaume, C. Gamrat, B. DeSalvo, "Bio-Inspired Stochastic Computing Using Binary CBRAM Synapses," *Electron Devices, IEEE Transactions on*, vol. 60, no. 7, pp. 2402–2409, 2013.

[14] D. H. Goldberg, G. Cauwenberghs, and A. G. Andreou, "Probabilistic synaptic weighting in a reconfigurable network of VLSI integrate-and-fire neurons," *Neural Networks*, vol. 14, pp. 781–793, 2001.

[15] Alibart, F., Pleutin, S., Bichler, O., Gamrat, C., Serrano-Gotarredona, T., Linares-Barranco, B. and Vuillaume, D. A "Memristive Nanoparticle/Organic Hybrid Synapstor for Neuroinspired Computing". *Advanced Functional Materials*, vol. 22-3, pp. 609-616, 2012.

[16] J. V. Arthur and K. Boahen, "Silicon-Neuron Design: A Dynamical Systems Approach," *Circuits and Systems I: Regular Papers, IEEE Transactions on*, vol. 58, no. 99, p. 1, 2011.

[17] G. Venkatesh, J. Sampson, N. Goulding-hotta, S. K. Venkata, M. B. Taylor, and S. Swanson, "QsCORES : Trading Dark Silicon for Scalable Energy Efficiency with Quasi-Specific

Cores Categories and Subject Descriptors," in International Symposium on Microarchitecture, 2011.

[18] R. J. Vogelstein, U. Mallik, J. T. Vogelstein, and G. Cauwenberghs, "Dynamically reconfigurable silicon array of spiking neurons with conductance-based synapses," IEEE Transactions on Neural Networks, vol. 18, no. 1, pp. 253–265, 2007.

[19] A. Hashmi, A. Nere, J. J. Thomas, and M. Lipasti, "A case for neuromorphic ISAs," in International Conference on Architectural Support for Programming Languages and Operating Systems. New York, NY: ACM, 2011.

[20] A. Joubert, B. Belhadj, and R. Heliot, "A robust and compact 65 nm LIF analog neuron for computational purposes," in IEEE NEWCAS conference, vol. 10, pp. 9–12, 2011.

[21] C. Fuchs, J. Charbonnier, and S. Cheramy. "Process and RF modelling of TSV last approach for 3D RF interposer", IITC/MAM 2011, pages 9–11, 2011.

[22] A. Joubert, M. Duranton, B. Belhadj, O. Temam and R. Heliot, "Capacitance of TSVs in 3-D stacked chips a problem? Not for neuromorphic systems", IEEE/ACM Design and Automation Conference, DAC'12, June 2012.

[23] W. Gerstner and W. M. Kistler, Spiking Neuron Models. Cambridge University Press, 2002.

[24] P. Merolla, J. Arthur, F. Akopyan, N. Imam, R. Manohar, and D. Modha, "A digital neurosynaptic core using embedded crossbar memory with 45pJ per spike in 45nm," in IEEE Custom Integrated Circuits Conference. IEEE, Sep. 2011, pp. 1–4.

[25] U. Rutishauser and R. J. Douglas, "State-dependent computation using coupled recurrent networks," Neural computation, vol. 21, no. 2, pp. 478–509, 2009.

[26] J. Schemmel, J. Fieres, and K. Meier, "Wafer-scale integration of analog neural networks," in International Joint Conference on Neural Networks. Ieee, Jun. 2008, pp. 431–438.

[27] Semiconductor Industry Association, "The International Technology Roadmap for Semiconductors (ITRS)", 2011 Edition.

[28] P. Batude et al., "GeOI and SOI 3D Monolithic Cell integrations for High Density Applications" VLSI Technology (VLSIT), symposium on, IEEE, 2009.

[29] Soon-Moon Jung; et al, "Highly cost effective and high performance 65nm S3 (stacked single-crystal Si) SRAM technology with 25F2, 0.16um2 cell and doubly stacked SSTFT cell transistors for ultra high density and high speed applications," VLSI Technology, 2005. Digest of Technical Papers. 2005 Symposium on , vol., no., pp.220,221, 14-16 June 2005.

[30] B. Shashikanth et al., "CELONCEL: Effective Design Technique for 3-D Monolithic Integration targeting High Performance Integrated Circuits", proceedings of the 16th Asia and South Pacific Design Automation Conference, IEEE Press, 2011.

GPGPU Accelerated Simulation and Parameter Tuning for Neuromorphic Applications

Kristofor D. Carlson[1], Michael Beyeler[2], Nikil Dutt[2], Jeffrey L. Krichmar[1,2]

Department of Cognitive Sciences[1]
School of Social Sciences
University of California, Irvine
Irvine, CA 92697
{kdcarlso} {jkrichma}@uci.edu

Department of Computer Science[2]
Bren School of Information and Computer Sciences
University of California, Irvine
Irvine, CA 92697
{mbeyeler} {dutt}@uci.edu

Abstract - Neuromorphic engineering takes inspiration from biology to design brain-like systems that are extremely low-power, fault-tolerant, and capable of adaptation to complex environments. The design of these artificial nervous systems involves both the development of neuromorphic hardware devices and the development neuromorphic simulation tools. In this paper, we describe a simulation environment that can be used to design, construct, and run spiking neural networks (SNNs) quickly and efficiently using graphics processing units (GPUs). We then explain how the design of the simulation environment utilizes the parallel processing power of GPUs to simulate large-scale SNNs and describe recent modeling experiments performed using the simulator. Finally, we present an automated parameter tuning framework that utilizes the simulation environment and evolutionary algorithms to tune SNNs. We believe the simulation environment and associated parameter tuning framework presented here can accelerate the development of neuromorphic software and hardware applications by making the design, construction, and tuning of SNNs an easier task.

I Introduction

Neuromorphic systems are gaining importance as traditional scaling in CMOS technology begins to reach its physical limits. These systems aim to mimic the biological structure of the nervous system; potentially both for solving engineering applications as well as understanding neural computation, which is one of the grand challenges of the 21st century [1], [2]. Biological information processing systems operate at performance levels set by fundamental physical limits, and do so under severe constraints of size, weight, and energy resources. As a result of these constraints, biological nervous systems are extremely energy-efficient (8–9 orders of magnitude better than digital computation [3]). In addition, brain networks employ learning at all levels of computation, are capable of adapting to complex environments, and possess remarkable fault tolerance by maintaining excellent performance even after the loss of many neurons. Thus investigating the computational mechanisms and engineering strategies that give rise to these system properties may not only further our understanding of the brain, but may also lead to novel algorithmic and architectural approaches that can overcome the limits of Moore's law.

A key aspect to the computing power of brain circuits is their massively parallel architecture. Brain systems are organized into highly interconnected modules that operate in concert with one another to carry out intrinsically parallel algorithms, rather than parallelizations of inherently serial procedures. For example, a certain computation in the brain might be carried out by millions of low-precision processing elements (neurons) in less than 100 serial steps [4]. On the other hand, parallelization of serial code typically leads to very limited speedup due to Amdahl's law [5]. The resulting energy efficiency of brain architectures is remarkable: For example, although there are approximately 20 billion neurons and 240 trillion synapses in the human cortex alone, the power consumption of the human brain is estimated to be no more than 13–15 watts [6].

Another key aspect to the computing power of brain circuits is the use of an event-driven communication protocol. Generally speaking, neurons employ relatively infrequent (~10–100 Hz) brief electrical pulses (called action potentials or spikes) as their main communication means. These pulses travel along a wire (axon) to the connection site (synapse) of another neuron, where they cause a small change in the electric potential of the receiving neuron. Neurons integrate these small changes and spike when their voltage reaches a threshold value [7]. Neuromorphic systems have successfully modeled this type of communication using address-event representation (AER), which is a communication protocol that represents each spike by its location (that is, the neuron that fired; explicitly encoded as an address) and the time at which it occurred (implicitly encoded) ([8], [9]). Using AER, it is possible to emulate massive connectivity in an efficient way.

A powerful framework in the development of neuromorphic applications that captures both of these key aspects is the use of spiking neural network (SNN) models in combination with highly parallel, off-the-shelf graphics processing units (GPUs). SNN models provide detailed neuronal dynamics [10] while utilizing the digital AER protocol for efficient communication, which makes them amenable to hardware application development. Furthermore, recent developments in high-performance GPUs enable the simulation of large-scale SNNs in real-time on affordable, programmable platforms. In order for the field of neuromorphic engineering to produce results and applications of practical value, such large-scale networks will be necessary. However, the tuning and stabilization of these large-scale dynamical systems is challenging, due to the large number of state variables and open parameters. Incorrect values in the parameter landscape lead to unstable, chaotic or undesired network operations (e.g.,

978-1-4799-2817-0/14 $31.00 © 2014 IEEE

epileptic oscillations). Thus there is a need for both hardware and software tools that can aid the modeler in the otherwise extremely tedious and possibly error-prone process of tuning complex, large-scale dynamical systems.

In this paper we introduce a software environment for the efficient simulation of SNN models on general-purpose graphics processing units (GPGPUs) as well as an automated parameter tuning framework that uses evolutionary algorithms (EAs) to tune SNN models in parallel. The remainder of the paper is organized as follows. In Section II, we briefly discuss recent hardware and software efforts aimed at building neuromorphic applications. Section III then introduces a GPU-accelerated SNN simulator called CARLsim, and the accompanying parameter tuning interface (PTI) for use in the neuromorphic engineering community.

We believe that the simulation environment and parameter tuning framework presented here will allow neuromorphic engineers to more easily construct larger, more complex SNNs, leading to the development of more powerful neuromorphic applications that may offer practical solutions to currently unsolved real-world problems.

II. Neuromorphic Hardware Devices and Software Tools

A. Neuromorphic Hardware Devices

Neuromorphic engineers have made significant progress developing neuromorphic devices to emulate both sensory systems and cognitive architectures. We briefly review recent advances in the development of neuromorphic cognitive architectures from research teams in the USA and Europe. We refer the reader to the following review on neuromorphic sensory systems [11] and focus our discussion on neuromorphic devices that emulate cognitive architectures.

The construction of neuromorphic chips and devices is an active area of research, and has spawned major research initiatives. The Neurogrid board at Stanford University is a neuromorphic device that emulates ion channels with analog circuit components but handles synaptic addressing with digital circuit components. It is capable of simulating 1 million neurons and 6 billion synaptic connections in real-time using only 5 watts and is an impressive example of the speed and power that can be achieved with neuromorphic devices [12]. Both IBM and HRL Laboratories, LLC (HRL) participated in the DARPA-funded systems of neuromorphic adaptive plastic scalable electronics (SyNAPSE) project, where the goal was to build highly scalable neuromorphic devices. The Cognitive Computing Group at IBM recently unveiled its True North architecture, which features a hierarchical design of neurosynaptic cores, each with 256 neurons and approximately 256k synapses [13]. These neurosynaptic cores are built from silicon neurons that can perform many realistic biophysical behaviors but lack synaptic plasticity (and thus lack learning capabilities). HRL also released a general purpose neural chip which has 576 neurons and 70k time multiplexed virtual synapses. The neural chip implements simple spiking neural models and plasticity in the form of spike-timing-dependent plasticity (STDP) [14], a learning paradigm which modulates the weight of synapses according to their degree of causality.

European researchers have also made advances in the design of neuromorphic chips due to funding from a number of initiatives that include two projects called fast analog computing with emergent transient states (FACETS) and brain-inspired multiscale computation in neuromorphic hybrid systems (BrainScaleS). The FACETS/BrainScaleS projects have produced two neuromorphic hardware devices to date. The first neuromorphic device is a single chip system called Spikey, which simulates 384 spiking neurons and 256 synapses per neuron [15]. The second neuromorphic device is more ambitious and is referred to as a wafer-scale neuromorphic hardware system. This system is constructed from 352 separate analog network cores (ANCs), in which each contains 512 spiking neurons and 16k synapses per neuron. The 352 ANCs can fit onto a single 20 cm wafer with a total of 180k neurons and $4 \cdot 10^7$ synapses [16].

Neuromorphic architectures can be either digital, analog, or a hybrid. The SpiNNaker (a contraction of spiking neural network architecture) project uses a digital design that has resulted from a unique collaboration between UK universities and industry partners. The ultimate goal of the SpiNNaker project is to build a computing engine that consists of 1,036,800 ARM9 processor cores capable of simulating 1 billion neurons in real-time. The SpiNNaker computing engine consists of an array of nodes, each containing 18 ARM9 cores, which communicate via packets using a custom interconnect fabric. Each ARM9 core can model 100 neurons and approximately 1M synapses or 10,000 inputs per neuron [17]. The projected power dissipation for the full million-core machine is 90 kW. SpiNNaker can implement spiking and non-spiking neurons models and synaptic plasticity with a novel form of STDP [18]. On the other hand, research teams from the University and ETH Zurich have recently developed a hybrid analog/digital VLSI implementation of an SNN with programmable synaptic weights. The chip has 32 silicon neuron circuits, 128 virtual synaptic weights per neuron, and STDP learning. Because the synaptic weights can be changed on-line and saved offline, the chip can be used to explore spike-based learning rules [19].

B. Neuromorphic Software Tools: SNN Simulators

Many research groups have developed SNN simulators to study brain function [20]–[22] or develop neuromorphic applications [23], [24]; we briefly review SNN simulators produced from these research efforts. Both NEURON [25] and GENESIS [26] began as simulation environments originally designed for detailed neuronal modeling at the ionic channel level, but both have the capability to run network models. These simulation environments have a large user-base, extensively-tested code, and parallelized versions that can be executed using MPI on super-computing clusters. However, the computation cost for solving all the equations governing channel activity and propagation of signals makes these models difficult to use in large network applications. Other simulation environments like CSIM/PCSIM, NCS, XPPAUT, SPLIT, Brian, NEST, and Mvaspike were built specifically to run SNNs and have been optimized with SNNs in mind; see [27] for more information. Of particular note are Brian [20] and NEST [21], which have Python interfaces, multiple spiking neuron model implementations, and distributed parallel implementations. Brian is flexible and easily extensible, partially because it is written in Python. However, there is a slight performance penalty, as it is about 25% slower than similar C implementations. NEST has an optional Python front-end but avoids the potential decrease in performance by using a kernel written in C++. Although NEST has an MPI implementation, it does not currently have a parallel GPU implementation. Because NEST is a larger, more mature software project, the implementation of new features like accelerated GPU implementations may take more time.

There are SNN environments implemented for specific hardware and simulation environments. For example, IBM's C2 SNN simulator is massively parallelized and designed to run on powerful Blue Gene/P supercomputing clusters [28]. The C2 simulator ran a large-scale SNN simulation that consisted of 1.6 billion neurons and 8.87 trillion synapses on a Blue Gene/P with 147,456 central processing units (CPUs) and 144 TB of memory, which is one of the largest SNN simulations to date [29]. NENGO is an SNN simulator that uses a control theory oriented approach called the neural engineering framework (NEF) to specify the synaptic weights required to achieve a desired computation and has been used to build a large-scale brain model with impressive functionality [22]. There are also SNN simulators designed to mimic neuromorphic hardware computing architectures, such as HRLsim [23] developed at HRL and Compass [24] developed at IBM. However, these classes of simulators are currently not available for public use.

C. GPU-Enabled SNN Simulators

Many SNN modelers are turning to GPGPUs for developing application software. Modern GPUs are a low-cost alternative to traditional supercomputing clusters for applications in scientific computing [30] and theoretical neuroscience [31]. A number of groups have developed parallel implementations of SNN simulators that run on GPUs [23], [32]–[38] . For a more comprehensive review see [39].

Whereas some of these software tools are still under heavy development, there are a number of fully functional SNN simulators that feature a whole range of detailed neuronal and synaptic dynamics (such as spike-rate adaptation, plasticity, homeostasis, and specific ion channels), routines that enable the construction of arbitrary connection topologies, and an optimized GPU implementation. Among them are HRLsim [23], NeMo [33], and CARLsim [36], [37]. Although HRLsim is the only simulator to offer parallelization across a GPU cluster using MPI and CUDA, it is currently unavailable for public use. NeMo is a C++ library that simulates networks of Izhikevich neurons on multiple CUDA-enabled GPUs, with a frontend in C/C++, Matlab, and Python. NeMo also features synaptic plasticity in the form of STDP and axonal delays. CARLsim has specifications that are similar to NeMo, in that it simulates networks of Izhikevich neurons on a single CUDA-enabled GPU. Multi-GPU support is planned in the future. Additionally, CARLsim offers optimal computational efficiency by utilizing a reduced AER protocol, different kinds of synaptic plasticity, a method to stabilize synaptic dynamics, the simulation of specific ion channels, and the option to either run the network on a CPU or a GPU. Moreover, CARLsim has been used in a variety of computational studies

to simulate detailed large-scale models of cortical processing. The next section will discuss these features in more detail.

III. CARLsim: An SNN Simulator

The Cognitive Anteater Robotics Laboratory Simulator (CARLsim) was designed to make large-scale SNN modeling readily available and is intended for use by the computational neuroscience and neuromorphic engineering communities [36], [37]. CARLsim is written in C/C++ and has both a single-threaded CPU implementation and parallelized GPU implementation. To maximize accessibility, CARLsim runs on both generic x86 CPU architectures and the widely used NVIDIA CUDA GPU architecture under both the Windows and Linux operating systems. We provide a user-friendly programming interface similar to that of PyNN [40] and allow the user to specify a number of preprogrammed connection topologies along with a mechanism to allow for user-defined connection topologies. CARLsim is publicly available at http://www.socsci.uci.edu/~jkrichma/CARLsim/.

A. CARLsim Features

CARLsim uses the Izhikevich spiking neuron model [41], which is well-suited for large-scale neuromorphic applications, because it is computationally efficient yet allows for complex neuronal dynamics that closely mimic biological neurons. CARLsim includes expressions to model specific ion channels (such as AMPA, GABA, and NMDA), which play an important role in neuronal excitability and plasticity.

CARLsim employs a reduced AER protocol for efficient encoding of neuronal communication, which helps reduce both memory usage and memory bandwidth limitations. Recall that AER stores a spike event by representing it as an address-time pair. If many neurons have the same time step, however, this approach leads to high memory overhead due to duplicate storing of time for each address. We overcome this limitation by removing the duplicate time entry for each address, and instead store the cumulative count of fired neurons during each time step [37]. Additionally, SNN state variables are compactly stored in memory and ordered in a manner that minimizes the state update process.

CARLsim includes descriptions of synaptic plasticity at different time scales. Short-term plasticity (STP) occurs over timescales of milliseconds to minutes, whereas long-term plasticity (LTP) occurs over time steps of minutes or longer [42]. STP changes the synaptic weight over time in a way that reflects the history of presynaptic activity. It can be used to model phenomena such as synaptic facilitation or synaptic depression (fatigue). LTP can be induced through STDP, which has been shown to play a crucial role in unsupervised learning [43], forming sparse representations of temporal sequences [44], and computing with neural synchrony [45]. CARLsim offers routines to implement both STP and LTP.

Learning rules in SNNs can often lead to unstable, runaway synaptic dynamics and completely disrupt learning and neural function [46]. To cope with this challenge, CARLsim implements a biologically plausible weight update rule that promotes stable learning and homeostasis that mimics experimental observations [47]. For more information on the details of the STDP and homeostasis update rules please see reference [48].

B. CARLsim Parallel GPU Implementation

The GPU-accelerated NVIDIA CUDA implementation of CARLsim is an important aspect of the simulation framework and allows for a significant speed up over the single-threaded CPU implementation, as we will see later. The basic structure of the CUDA GPU architecture is shown in Figure 1. Each GPU consists of multiple streaming multiprocessors (SMs) and a global memory accessible by all SMs. Each SM is built from multiple floating-point scalar processors, a cache/shared memory, and one or more special functions units (SFU), which execute transcendental functions such as sine, cosine, and square root operations. CUDA groups parallel threads into 'warps', where the number of threads per warp varies depending on the specific CUDA architecture. Each SM also has at least one warp scheduler that is built to maximize the number of threads running concurrently. The warp scheduler monitors threads within a warp and automatically switches to another warp when a thread makes a time-consuming memory access.

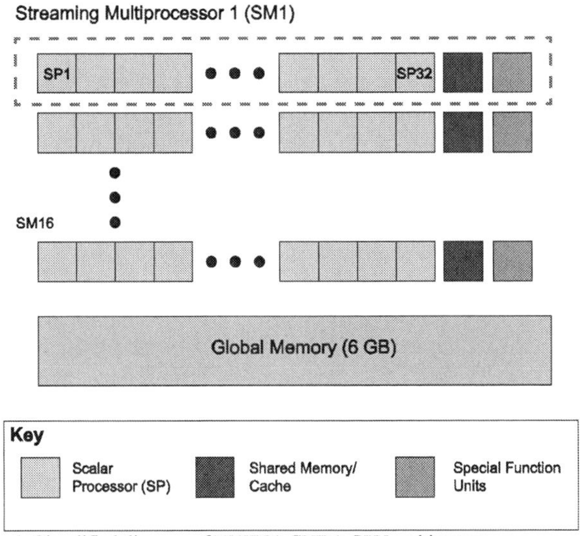

Streaming Multiprocessor 1 (SM1)

SP1 ••• SP32

SM16 •

Global Memory (6 GB)

Key
- Scalar Processor (SP)
- Shared Memory/ Cache
- Special Function Units

Fig. 1. Simplified diagram of NVIDIA CUDA GPU architecture.

CARLsim simulations are primarily run on the M2090 Tesla GPU that utilizes the CUDA Fermi architecture. The Tesla M2090 has 6 GB of global memory, 512 cores (each operating at 1.30 GHz) grouped into 16 SMs, and a single precision compute power of 1331.2 GFLOPS. Each SM is composed of 32 SPs, 16 load/store units, 4 special function units, and two warp schedulers [49].

The GPU implementation of CARLsim utilizes a number of approaches to maximize the degree of parallelization, minimize memory usage, reduce the effects of memory bandwidth limitations, and avoid thread/warp divergence. There are number of ways to assign SNN calculations to the GPU threads for parallelization. N-parallelism describes organizing the computations assigned to GPU threads by neuron while S-parallelism organizes the computations assigned to GPU threads by synapse. CARLsim uses both of these approaches by using N-parallelism for neuron state

variable updates and S-parallelism for synapse variable updates. Thread/warp divergence occurs when a thread executes a different operation than other threads in a warp causing the other threads to wait for its completion. CARLsim prevents thread/warp divergence by buffering data until all threads are ready to execute the same operation. More information about the CARLsim GPU implementation can be found in reference [36].

C. CARLsim Applications

CARLsim has been used to construct large-scale simulations of cognitive processes on the order of 10k–100k neurons and millions of synapses, with examples that include models of visual processing [37], neuromodulation [50], and neural plasticity [48]. Although their efficient implementation is challenging due to the associated computational cost, investigating large-scale models of cortical networks in more biological detail is widely regarded as crucial in order to understand brain function [51].

One of our most recent works [52] concerned the ability of a large-scale spiking neural network model to rapidly categorize highly correlated patterns of neural activity such as handwritten digits from the MNIST database [53]. Although many studies have focused on the design and optimization of neural networks to solve visual recognition tasks, most of them either lack neurobiologically plausible learning rules or decision-making processes. In contrast, our model demonstrated how a low-level memory encoding mechanism based on synaptic plasticity could be integrated with a higher-level decision-making paradigm to perform the visual classification task in real-time.

The model consisted of Izhikevich neurons and conductance-based synapses for realistic approximation of neuronal dynamics, an STDP-like synaptic learning rule for memory encoding (previously described in [54]), and an accumulator model for memory retrieval and categorization [55]. Grayscale input images were fed through a feed-forward network consisting of visual cortical areas V1 and V2 (selective to one of four spatial orientations, in 45° increments), which then projected to a layer of downstream classifier neurons through plastic synapses that implement the STDP-like learning rule mentioned above. Decision neurons were equally divided into ten pools, each of which would develop selectivity to one class of input stimuli from the MNIST dataset (i.e., one of the ten digits) as a result of training. Population responses of these classifier neurons were then integrated over time to make a perceptual decision about the presented stimulus.

The model constitutes an important proof of concept; that is, (i) to show how considerably hard problems such as visual pattern recognition and perceptual decision-making can be solved by general-purpose neurobiologically inspired cortical models solely relying on local learning rules that operate on the abstraction level of a synapse, and (ii) to do it in real-time. The network achieved 92% correct classifications on MNIST in 100 rounds of random sub-sampling, which provides a conservative performance metric, yet is comparable to other SNN approaches ([54], [56]). Additionally, the model correctly predicted both qualitative and quantitative properties of reaction time distributions reported in psychophysical experiments. The full network, which comprised 71,026 neurons and approximately 133 million synapses, ran in real-time on a single NVIDIA Tesla M2090, which demonstrates the efficiency of the CARLsim implementation. Moreover, because of the scalability of the approach and its neurobiological fidelity, the model can be extended to an efficient neuromorphic implementation that supports more generalized object recognition and decision-making architectures found in the brain.

III. CARLsim Parameter Tuning Interface

The size and complexity of the SNN models used by the neuromorphic engineering community has been steadily increasing in an effort to capture more biological realism [57] and underlying functionality [22]. This complexity has taken the form of more detailed neuron models, the inclusion of plasticity rules, and connection topologies that include recurrent connections. The integration of these features into SNN models comes at a cost: it produces SNNs with less stable neuronal and synaptic dynamics. Both the size and instability of this new generation of SNNs have made the task of constructing and tuning them a difficult one. To meet this challenge, we have developed an automated parameter tuning framework that uses evolutionary algorithms (EAs) and CARLsim to construct and tune SNNs in parallel with GPUs.

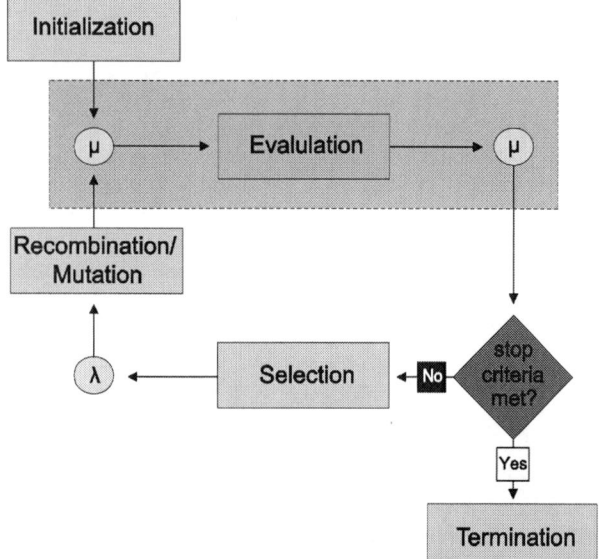

Fig. 2. Diagram illustrating the overall approach of the parameter tuning framework using evolutionary algorithms. The dotted gray box indicates processes that are done in parallel using the GPU implementation of CARLsim. The rest of the processes are done sequentially using Evolving Objects.

The process of tuning an SNN with the automated parameter tuning framework is detailed in Figure 2: (1) a population of SNNs, each with a different set of parameters, is created. (2) Each SNN is evaluated by a fitness function and assigned a fitness score based on how well the behavior of the SNN matches a target behavior. (3) The highest scoring SNNs are selected to produce the next generation of offspring via recombination and mutation while the remaining individuals are discarded. (4) This evolutionary process continues until the desired fitness is reached or another termination condition

has been met. Using this approach, an SNN can be tuned to produce suitable firing patterns, stable learning dynamics, and behaviors that closely match experimental data.

The automated parameter tuning framework has three components: the CARLsim SNN simulator, an open source EA library called Evolving Objects [58], and a program to pass information between CARLsim and Evolving Objects we call the parameter tuning interface (PTI). Evolving Objects handles all EA computations shown with light blue boxes in Fig. 2. CARLsim executes the most computationally expensive portion of the tuning algorithm, evaluating the fitness of each SNN, in parallel indicated by the light brown box. During each new EA generation, Evolving Objects assigns parameters from each offspring individual in the population to CARLsim SNNs using the PTI. CARLsim then runs these SNNs, each with a different set of parameters, in parallel and assigns them a fitness which is passed backed to Evolving Objects via the PTI. Evolving Objects then selects those individuals with the best fitness and produces the new parent generation via recombination and mutation until a termination condition is reached.

A. Preliminary Tuning Framework Results

As a proof of concept, an SNN with 4,104 Izhikevich neurons, two forms of synaptic plasticity, and feedback connections was successfully tuned to reproduce neuronal responses found in the visual cortex. Each generation consisted of 10 SNNs and the simulation framework took 127.2 hours of wall-clock time to complete 287 generations. The parallelized GPU CARLsim implementation was performed on an NVIDIA Tesla M2090 GPU card with 6 GB of memory and 512 cores. The single-threaded CPU implementation was performed on a system with an Intel Core i7 2.67 GHz quad-core processor with 6 GB of memory.

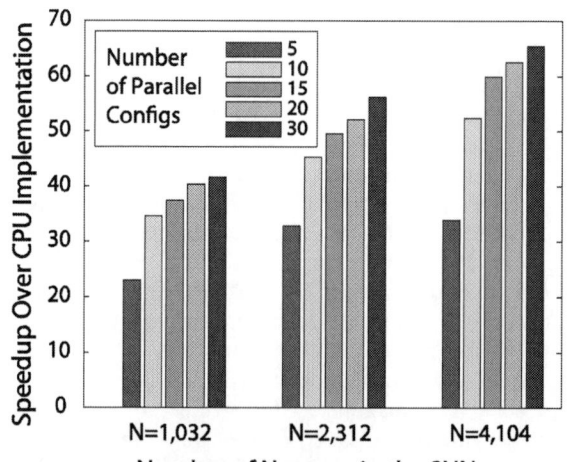

Fig. 3. Comparison of speedups of the parallel GPU CARLsim implementations over the single-threaded CPU CARLsim implementation. The different colored bars represent the number of configurations run in parallel for each SNN network size. The maximum speedup occurs at the largest network size (4,104 neurons) that ran 30 SNN configurations in parallel.

978-1-4799-2817-0/14 $31.00 © 2014 IEEE

To further characterize the relationship between the speedup of the GPU implementation over the CPU implementation, both the SNN network size and number of SNN configurations were varied. The results are shown in Figure 3. Three SNNs with 4104, 2312, and 1032 neurons were run for 10 simulated minutes each. The number of configurations each SNN ran in parallel was varied from 5 to 30. There were significant speedups over the single threaded CPU implementation for all three network sizes when 5 or more SNNs were executed in parallel. The largest speedup was approximately (65x) and occurred when 30 configurations of 4,104-neuron SNN were simulated in parallel, as shown in Figure 3. The automated parameter tuning framework presented here efficiently searches SNN parameter spaces using EAs and performs fitness evaluations in parallel using GPUs. The tuning approach presented here uses these two techniques to build a powerful tool for researchers to design, construct, and test complex SNNs for neuromorphic applications.

IV. Summary and Conclusions

Neuromorphic applications have the potential to provide insight into brain function and create low-power, fault tolerant neuromorphic devices for use in sensory systems and cognitive computing architectures. However, due to the massive number of computing elements and the unstable neuronal and synaptic dynamics inherent in these models, the design and construction of neuromorphic applications is a difficult task. We presented the CARLsim SNN simulation environment, which features Izhikevich spiking neurons, three plasticity mechanisms, and GPU acceleration for use in the computational neuroscience and neuromorphic engineering communities. We also presented an automated parameter tuning framework which integrates CARLsim and an EA library to efficiently tune SNNs in parallel using GPU acceleration. We believe these software tools will accelerate the design and construction of large-scale neural models and neuromorphic applications, potentially offering practical solutions to currently unsolved real-world problems.

Acknowledgments

This work was supported by the Defense Advanced Research Projects Agency (DARPA) subcontract 801888-BS and by NSF Award IIS/RI-1302125. We thank Micah Richert for his work developing CARLsim 2.0 and Jayram Nageswaran for his work developing CARLsim 1.0 and the parameter tuning framework.

References

[1] J. M. Nageswaran, M. Richert, N. Dutt, and J. L. Krichmar, "Towards reverse engineering the brain: Modeling abstractions and simulation frameworks," in *VLSI System on Chip Conference (VLSI-SoC), 2010 18th IEEE/IFIP*, 2010, pp. 1–6.

[2] "NAE Grand challenges for engineering -www.engineeringchallenges.org."

[3] J. Hasler and B. Marr, "Finding a roadmap to achieve large neuromorphic hardware systems," *Front. Neuromorphic Eng.*, vol. 7, p. 118, 2013.

[4] G. Lynch, G. S. Lynch, and R. Granger, *Big Brain: the origins and future of human intelligence*. Macmillan, 2008.

[5] G. M. Amdahl, "Validity of the single processor approach to achieving large scale computing capabilities," in *Proceedings of the April 18-20, 1967, spring joint computer conference*, 1967, pp. 483–485.

[6] C. Koch, *Biophysics of computation: information processing in single neurons*. Oxford university press, 2004.

[7] E. R. Kandel, J. H. Schwartz, T. M. Jessell, and others, *Principles of neural science*, vol. 4. McGraw-Hill New York, 2000.

[8] J. Lazzaro, J. Wawrzynek, M. Mahowald, M. Sivilotti, and D. Gillespie, "Silicon auditory processors as computer peripherals," *Neural Netw. IEEE Trans. On*, vol. 4, no. 3, pp. 523–528, 1993.

[9] M. Mahowald, "An Analog VLSI System for Stereoscopic Vision," 1994.

[10] W. Gerstner and W. M. Kistler, *Spiking neuron models: Single neurons, populations, plasticity*. Cambridge university press, 2002.

[11] S.-C. Liu and T. Delbruck, "Neuromorphic sensory systems," *Curr. Opin. Neurobiol.*, vol. 20, no. 3, pp. 288–295, Jun. 2010.

[12] R. Silver, K. Boahen, S. Grillner, N. Kopell, and K. L. Olsen, "Neurotech for neuroscience: unifying concepts, organizing principles, and emerging tools," *J. Neurosci.*, vol. 27, no. 44, pp. 11807–11819, 2007.

[13] S. K. Esser, A. Andreopoulus, R. Appuswamy, P. Datta, D. Barch, A. Amir, J. Arthur, A. Cassidy, M. Flickner, P. Merolla, S. Chandra, N. Basilico, S. Carpin, T. Zimmerman, F. Zee, R. Alvarez-Icaza, J. A. Kusnitz, T. M. Wong, W. P. Risk, E. McQuinn, T. K. Nayak, R. Singh, and D. S. Modha, "Cognitive Computing Systems: Algorithms and Applications for Networks of Neurosynaptic Cores," in *The 2013 International Joint Conference on Neural Networks (IJCNN)*, 2013.

[14] J. M. Cruz-Albrecht, T. Derosier, and N. Srinivasa, "A scalable neural chip with synaptic electronics using CMOS integrated memristors," *Nanotechnology*, vol. 24, no. 38, p. 384011, 2013.

[15] T. Pfeil, A. Grübl, S. Jeltsch, E. Müller, P. Müller, M. Schmuker, D. Brüderle, J. Schemmel, and K. Meier,

"Six networks on a universal neuromorphic computing substrate," *Front. Neuromorphic Eng.*, vol. 7, p. 11, 2013.

[16] J. Schemmel, D. Brüderle, A. Grübl, M. Hock, K. Meier, and S. Millner, "A wafer-scale neuromorphic hardware system for large-scale neural modeling," in *Proceedings of 2010 IEEE International Symposium on Circuits and Systems (ISCAS)*, 2010, pp. 1947–1950.

[17] S. B. Furber, D. R. Lester, L. A. Plana, J. D. Garside, E. Painkras, S. Temple, and A. D. Brown, "Overview of the SpiNNaker System Architecture," *IEEE Trans. Comput.*, vol. 99, no. PrePrints, 2012.

[18] S. Davies, F. Galluppi, A. D. Rast, and S. B. Furber, "A forecast-based STDP rule suitable for neuromorphic implementation," *Neural Netw.*, vol. 32, pp. 3–14, Aug. 2012.

[19] S. Moradi and G. Indiveri, "An Event-Based Neural Network Architecture With an Asynchronous Programmable Synaptic Memory," *IEEE Trans. Biomed. Circuits Syst.*, vol. Early Access Online, 2013.

[20] D. Goodman and R. Brette, "Brian: a simulator for spiking neural networks in Python," *Front. Neuroinformatics*, vol. 2, p. 5, 2008.

[21] M.-O. Gewaltig and M. Diesmann, "NEST (NEural Simulation Tool)," *Scholarpedia*, vol. 2, no. 4, p. 1430, 2007.

[22] C. Eliasmith, T. C. Stewart, X. Choo, T. Bekolay, T. DeWolf, Y. Tang, and D. Rasmussen, "A Large-Scale Model of the Functioning Brain," *Science*, vol. 338, no. 6111, pp. 1202–1205, Nov. 2012.

[23] C. M. Thibeault, "Computational Neuroscience: Theory, Development and Applications in Modeling The Basal Ganglia," Ph.D., University of Nevada, Reno, United States -- Nevada, 2012.

[24] R. Preissl, T. M. Wong, P. Datta, M. Flickner, R. Singh, S. K. Esser, W. P. Risk, H. D. Simon, and D. S. Modha, "Compass: a scalable simulator for an architecture for cognitive computing," in *Proceedings of the International Conference on High Performance Computing, Networking, Storage and Analysis*, Los Alamitos, CA, USA, 2012, pp. 54:1–54:11.

[25] M. L. Hines and N. T. Carnevale, "The NEURON simulation environment," *Neural Comput.*, vol. 9, no. 6, pp. 1179–1209, 1997.

[26] J. M. Bower, D. Beeman, and A. M. Wylde, *The book of GENESIS: exploring realistic neural models with the GEneral NEural SImulation System.* Telos Santa Clara, Calif, 1998.

[27] R. Brette, M. Rudolph, T. Carnevale, M. Hines, D.

Beeman, J. M. Bower, M. Diesmann, A. Morrison, P. H. Goodman, F. C. Harris Jr, M. Zirpe, T. Natschläger, D. Pecevski, B. Ermentrout, M. Djurfeldt, A. Lansner, O. Rochel, T. Vieville, E. Muller, A. P. Davison, S. El Boustani, and A. Destexhe, "Simulation of networks of spiking neurons: a review of tools and strategies," *J. Comput. Neurosci.*, vol. 23, no. 3, pp. 349–398, Dec. 2007.

[28] R. Ananthanarayanan and D. S. Modha, "Anatomy of a cortical simulator," in *Proceedings of the 2007 ACM/IEEE conference on Supercomputing*, New York, NY, USA, 2007, pp. 3:1–3:12.

[29] R. Ananthanarayanan, S. K. Esser, H. D. Simon, and D. S. Modha, "The cat is out of the bag: cortical simulations with 109 neurons, 1013 synapses," in *Proceedings of the Conference on High Performance Computing Networking, Storage and Analysis*, New York, NY, USA, 2009, pp. 63:1–63:12.

[30] J. D. Owens, D. Luebke, N. Govindaraju, M. Harris, J. Krüger, A. E. Lefohn, and T. J. Purcell, "A Survey of General-Purpose Computation on Graphics Hardware," in *Computer graphics forum*, 2007, vol. 26, pp. 80–113.

[31] J. Baladron, D. Fasoli, and O. Faugeras, "Three Applications of GPU Computing in Neuroscience," *Comput. Sci. Eng.*, vol. 14, no. 3, pp. 40–47, Jun. 2012.

[32] D. Yudanov, M. Shaaban, R. Melton, and L. Reznik, "GPU-based simulation of spiking neural networks with real-time performance & high accuracy," in *Neural Networks (IJCNN), The 2010 International Joint Conference on*, 2010, pp. 1–8.

[33] A. K. Fidjeland, E. B. Roesch, M. P. Shanahan, and W. Luk, "NeMo: A Platform for Neural Modelling of Spiking Neurons Using GPUs," in *Application-specific Systems, Architectures and Processors, 2009. ASAP 2009. 20th IEEE International Conference on*, 2009, pp. 137–144.

[34] V. K. Pallipuram, M. C. Smith, N. Raut, and X. Ren, "Exploring Multi-level Parallelism for Large-Scale Spiking Neural Networks," 2012.

[35] T. Nowotny, "Flexible neuronal network simulation framework using code generation for NVidia(R) CUDATM," *BMC Neurosci.*, vol. 12, no. Suppl 1, p. P239, 2011.

[36] J. M. Nageswaran, N. Dutt, J. L. Krichmar, A. Nicolau, and A. V. Veidenbaum, "A configurable simulation environment for the efficient simulation of large-scale spiking neural networks on graphics processors," *Neural Netw. Off. J. Int. Neural Netw. Soc.*, vol. 22, no. 5–6, pp. 791–800, Aug. 2009.

[37] M. Richert, J. M. Nageswaran, N. Dutt, and J. L. Krichmar, "An efficient simulation environment for modeling large-scale cortical processing," *Front. Neuroinformatics*, vol. 5, no. 19, 2011.

[38] "NeoCortical Simulator - http://www.cse.unr.edu/brain/ncs." .

[39] R. Brette and D. F. M. Goodman, "Simulating spiking neural networks on GPU," *Netw.-Comput. Neural Syst.*, vol. 23, no. 4, pp. 167–182, 2012.

[40] A. P. Davison, D. Bruderle, J. Eppler, J. Kremkow, E. Muller, D. Pecevski, L. Perrinet, and P. Yger, "PyNN: A Common Interface for Neuronal Network Simulators," *Front. Neuroinformatics*, vol. 2, Jan. 2009.

[41] E. M. Izhikevich, "Simple model of spiking neurons," *IEEE Trans. Neural Netw.*, vol. 14, no. 6, pp. 1569 – 1572, Nov. 2003.

[42] P. Dayan and L. F. Abbott, *Theoretical neuroscience*, vol. 31. MIT press Cambridge, MA, 2001.

[43] T. Masquelier and S. J. Thorpe, "Unsupervised learning of visual features through spike timing dependent plasticity," *PLoS Comput. Biol.*, vol. 3, no. 2, p. e31, 2007.

[44] S. Byrnes, A. N. Burkitt, D. B. Grayden, and H. Meffin, "Learning a sparse code for temporal sequences using STDP and sequence compression," *Neural Comput.*, vol. 23, no. 10, pp. 2567–2598, Oct. 2011.

[45] R. Brette, "Computing with neural synchrony," *PLoS Comput. Biol.*, vol. 8, no. 6, p. e1002561, Jun. 2012.

[46] L. F. Abbott and S. B. Nelson, "Synaptic plasticity: taming the beast," *Nat. Neurosci.*, vol. 3, no. 11, pp. 1178–1183, Nov. 2000.

[47] G. Turrigiano, "Homeostatic synaptic plasticity: local and global mechanisms for stabilizing neuronal function," *Cold Spring Harb. Perspect. Biol.*, vol. 4, no. 1, Jan. 2012.

[48] K. D. Carlson, M. Richert, N. Dutt, and J. L. Krichmar, "Biologically Plausible Models of Homeostasis and STDP: Stability and Learning in Spiking Neural Networks," in *The 2013 International Joint Conference on Neural Networks (IJCNN)*, Dallas, Texas, 2013.

[49] "NVIDIA's Next Generation CUDA Compute Architecture: Fermi," NVIDIA Corp., white paper, 2009.

[50] M. C. Avery, D. A. Nitz, and J. L. Krichmar, "Simulation of cholinergic and noradrenergic modulation of behavior in uncertain environments," *Front. Comput. Neurosci.*, vol. 6, p. 5, 2012.

[51] C. J. Honey, R. Kötter, M. Breakspear, and O. Sporns, "Network structure of cerebral cortex shapes functional connectivity on multiple time scales," *Proc. Natl. Acad. Sci.*, vol. 104, no. 24, pp. 10240–10245, Jun. 2007.

[52] M. Beyeler, N. D. Dutt, and J. L. Krichmar, "Categorization and decision-making in a neurobiologically plausible spiking network using a STDP-like learning rule," *Neural Netw. Off. J. Int. Neural Netw. Soc.*, vol. 48C, pp. 109–124, Aug. 2013.

[53] Y. LeCun, L. Bottou, Y. Bengio, and P. Haffner, "Gradient-based learning applied to document recognition," *Proc. IEEE*, vol. 86, no. 11, pp. 2278–2324, 1998.

[54] J. M. Brader, W. Senn, and S. Fusi, "Learning Real-World Stimuli in a Neural Network with Spike-Driven Synaptic Dynamics," *Neural Comput.*, vol. 19, no. 11, pp. 2881–2912, Sep. 2007.

[55] P. L. Smith and R. Ratcliff, "Psychology and neurobiology of simple decisions," *Trends Neurosci.*, vol. 27, no. 3, pp. 161–168, Mar. 2004.

[56] D. Querlioz, O. Bichler, and C. Gamrat, "Simulation of a memristor-based spiking neural network immune to device variations," in *The 2011 International Joint Conference on Neural Networks (IJCNN)*, 2011, pp. 1775–1781.

[57] E. M. Izhikevich and G. M. Edelman, "Large-scale model of mammalian thalamocortical systems," *Proc. Natl. Acad. Sci. U. S. A.*, vol. 105, no. 9, pp. 3593–3598, Mar. 2008.

[58] M. Keijzer, J. J. Merelo, G. Romero, and M. Schoenauer, "Evolving objects: a general purpose evolutionary computation library," in *Artficial Evolution*, vol. 2310, P. Collet, C. Fonlupt, J. K. Hao, E. Lutton, and M. Schoenauer, Eds. Berlin: Springer-Verlag Berlin, 2002, pp. 231–242.

A Scalable Custom Simulation Machine for the Bayesian Confidence Propagation Neural Network model of the Brain

Nasim Farahini[*], Ahmed Hemani[*], Anders Lansner[†], Fabian Clermidy[‡], Christer Svensson[**]

*Dept. of Electronic Systems, School of ICT, KTH, Sweden; [farahini|hemani]@kth.se
†Dept. of Computational Biology, School of CSC, KTH, Sweden; ala@csc.kth.se
‡Digital design and Architecture Lab, Cea-Leti, Grenoble, France; Fabian.clermidy@cea.fr
**Dept. of Electrical Engineering, Linköping University, christer@isy.liu.se

Abstract - A multi-chip custom digital super-computer called eBrain for simulating Bayesian Confidence Propagation Neural Network (BCPNN) model of the human brain has been proposed. It uses Hybrid Memory Cube (HMC), the 3D stacked DRAM memories for storing synaptic weights that are integrated with a custom designed logic chip that implements the BCPNN model. In 22nm node, eBrain executes BCPNN in real time with 740 TFlops/s while accessing 30 TBs synaptic weights with a bandwidth of 112 TBs/s while consuming less than 6 kWs power for the typical case. This efficiency is three orders better than general purpose supercomputers in the same technology node.

I. Introduction

The dramatic developments in brain science and neuroscience over the past few decades, together with the formidable developments in VLSI technology, have brought us to the edge of building functioning brain-like devices and systems. Developing and deploying such devices in real-world information processing has been a dream and vision for quite some time, though success has so far been limited. In recent years, computational neuroscience has developed rapidly, and researchers are currently designing and studying large-scale complex brain models, thus helping to integrate the vast amounts of information about the brain from different sources and levels of description into working models of the brain. Supercomputers have become an enabling tool for simulating complex brain models with a scale approaching that of small and medium-sized mammals. As a reference human brain has 20 billion neurons and more than 100 trillion synapses. While the use of supercomputers to simulate complex brain models has the benefit of generality, the capital and the running costs of supercomputers can be prohibitive and beyond the reach of many research groups besides being harmful to the environment. Moreover, they lack the capacity to simulate the complete human brain in real-time. Lastly, these machines are not scalable in cost, form factor and energy consumption to brain-like computers that could be mass manufactured and deployed in industry and society as part of intelligent embedded systems. The objective of the research presented in this paper is to demonstrate with a concrete custom multi-chip digital design how to simulate a model of the human brain in real time that overcomes the limitations identified above. The proposed custom design that we call *eBrain* for Electronic Brain implements a specific scalable spiking neural network model called Bayesian Confidence Propagation Neural Network (BCPNN) [1]. This algorithm with its aggregation of neuronal behavior, spiking communication and embarrassingly parallel nature is hardware friendly. We further improve the efficiency by introducing a lazy evaluation model

[6] that reduces the computation and bandwidth requirement by more than two orders of magnitude. By using custom design for computation, spike communication infrastructure and I/O to external storage architecture, eBrain's total power consumption is less than 6 kWs in 22nm node. These numbers show three orders of magnitude improvement in silicon and computational efficiencies compared to general purpose supercomputers like BlueGene and Cray. These improvements will have two disruptive impacts: It will make available an affordable real time brain simulation engine to many more neuroscientists than what is feasible with supercomputers and it will enable building industrial machines, with human like intelligence for perception, motor control and decision making as a complement to the prevalent von-Neumann computing paradigm.

II. BCPNN Computation Model

BCPNN builds on the idea of a naive Bayesian classifier. The activity of network units represents stochastic events, and unit bias and weights are calculated by applying Bayes' rule as measures of correlation between unit activities in the past. These statistics are collected during training, and in inference mode the output activity of a unit represents *a posteriori* probability of the corresponding event given other activated units, i.e. events that have occurred recently. Despite the extreme independence assumptions made in the derivation, this learning rule works very well in practice [14].

This algorithm has a modular structure in terms of hypercolumns units (HCUs) and minicolumns units (MCUs) inspired by a generalization of observed cortical micro-circuitry [14]. Relating to biological neuronal networks, an MCU represents a local population of some hundreds of neurons (e.g. a cortical minicolumn). HCUs are aggregates of 100 MCUs and have 10 000 incoming connections from other MCUs. Each incoming connection targets all MCUs in the HCU. 2 million HCUs are needed to reach a size of the human cortex. The state variables of the HCU are held in a matrix of 10 000×100 MCU state vector elements each one representing represents the status of a connection between a pair of MCUs. HCUs are triggered by circa 10 000 spikes per second and produce 100 spikes per second on an average; each of these 100 spikes is fanned out to 100 HCUs. When an input spike arrives, it identifies one of the 10 000 MCU rows corresponding to the input connections. For the lazy evaluation method of BCPNN [6] the computational, storage and the bandwidth requirements are 370 MFlops/s/HCU and 15MBs/HCU and 56 MBs/s/HCU respectively. The storage

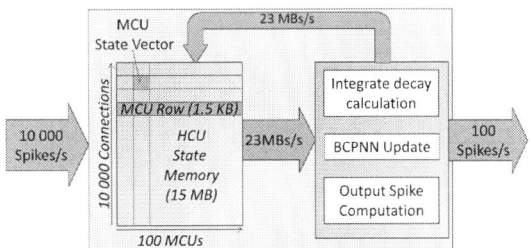

Fig. 1. BCPNN computational model and the HCU requirements using lazy evaluation method

includes time stamps that are necessary in the lazy evaluation model to calculate and apply the integrated decay corresponding to the elapsed time since last activation of MCU records. The aggregate bandwidth required for spike propagation in BCPNN is relatively modest. Each HCU gets 10K spikes/s and as there are 2 million HCUs, the aggregate spike propagation bandwidth is 100 GBs/s. The requirements above have been used to dimension the eBrain. Though the power numbers reported are for the typical case when 50% of the HCUs are active.

III.   The eBrain Architecture

Even after adopting the lazy evaluation method, the requirements of BCPNN in the foreseeable future will remain beyond the integration capacity of a single chip traditional 2D VLSI. For this reason, eBrain has been designed as a regular fabric of multiple identical chips called Brain Computational Units or BCUs that are interconnected by an inter-BCU Spike Propagation Network (SPN) as shown in Fig. 2.a. BCUs are integrated units with 3D stacked DRAM memories. The integration can be 2.5D or multi-chip package.

BCPNN has two levels of parallelism. One is the embarrassing Thread Level Parallelism (*TLP*); each HCU is an independent thread. The second is HCU computation that has rich Arithmetic Level Parallelism (*ALP*) potential. While other spiking brain models also have TLP, BCPNN because of its aggregation of MCUs in an HCU offers additional ALP.

BCUs are organized as a regular fabric of design units called H-Tiles as shown in Fig. 2.b. An H-Tile is a collection of computation, storage, interconnect and control resources to service a cluster of HCUs. If there are L BCUs, M H-Tiles per BCU and N HCUs per H-Tile, $L \cdot M \cdot N = 2$ million HCUs that are needed to model the human cortex according to the BCPNN. In terms of the eBrain architectural dimensions, the $L \cdot M$ H-Tiles operate in embarrassingly TLP manner and the N HCUs in each H-Tile are serviced sequentially, though servicing of each HCU exploits the rich ALP in the HCU computation. Smaller L implies fewer BCUs and thus smaller system form factor and minimal off-chip communication. The 30 TBs required to store all HCU state memories and the 112 TBs/s bandwidth required to access them are the dominant factors that have influenced the architectural and dimensioning decisions of the eBrain design. L is constrained by a) the amount of the HCU state memory that the technology allows to integrate in a single chip and b) the available bandwidth between BCU and the HCU state storage.

If S_{MAX} is the maximum storage that can be integrated in a single BCU and if B_{MIN} is the minimum bandwidth between BCPNN computation and the HCU state storage then $L = \lceil 30 \ TBs/S_{MAX} \rceil$ $B_{MIN} \geq M \cdot N \cdot 56$ MBs/s where 56MBs/s is

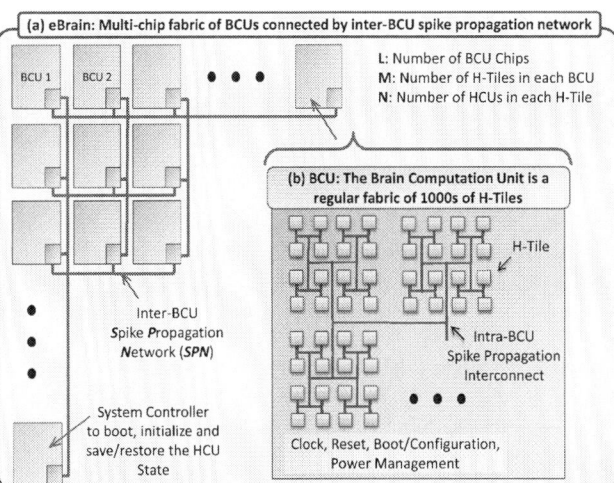

Fig. 2. a) Conceptual eBrain organizations, b) Brain Computation Unit Structure

the bandwidth required by each HCU. This implies that each BCU implements $M \cdot N = (2 \ million \ / \ L)$ HCUs.

eBrain has two levels of spike propagation interconnect schemes. One is an inter-BCU Spike Propagation Network (SPN) that transports spikes amongst the BCUs and the second is an on-chip intra-BCU spike distribution interconnect. Since every spike that is generated needs to discriminate amongst 2 million HCUs and once it arrives in an HCU it needs to further discriminate amongst the 10K MCU rows, we have dimensioned the address space to 35 bits. A spike is a boolean biological event that has a logic one value when a spike is generated and a logic zero value otherwise. Since a spike is in essence an address, in spirit this is an Address Event Representation scheme. However, unlike [3] the spike communication in eBrain involves no request/acknowledge handshake. We rely on the robustness of the communication infrastructure to deliver the spikes with negligible loss. This is substantiated in section V.B when we concretely dimension the communication infrastructure.

Reducing L reduces the off-chip spike propagation bandwidth requirement and increases the on-chip spike distribution requirement which is more easily manageable. Though the inter-BCU SPN is not the focus of this paper, we have made some preliminary design decisions regarding it as they influence the design and dimensions of the BCU which is the focus of this paper. The inter-BCU SPN will be a mesh topology network with static minimal path routing with both *phit (physical unit of transmission)* and *flit (logical unit of data transmission, e.g., packet/datagram)* size of a single spike (35b). Though the routing is static, it is possible to reconfigure it at runtime to make the connectivity between communicating MCUs adaptive as would be required in the biological brain. We plan to use Serdes based physical links that delivers 1-2 GBs/s for the inter-BCU SPN. In section V, where we dimension the BCU, we show that this bandwidth would be sufficient.

One of the units in the eBrain fabric serves as a system controller (Fig. 2) that initializes the fabric, can stop the execution, save the state in an external storage and restore it as part of the initialization and startup. The inter-BCU SPN that is normally used to distribute spikes doubles as the HCU state distribution network for both saving and restoring tasks.

IV. The Brain Computation Unit (BCU)

The organization of BCU in terms of design units is shown in Fig 3. We next elaborate the role and design of these units.

A. H-Tile

The role of an H-Tile in eBrain is to implement N HCUs, i.e., provide storage, computation and spike communication needs for them. H-Tiles are organized as two parts; one part holds the state memories of the N HCUs in the stacked DRAM/SRAM dies and the other part that implements the computation and spiking communication for the N HCUs in the logic die. This sub-section focuses on the logic part of the H-Tile, hereafter referred to as simply H-Tile for brevity. The architectural schematic of the H-Tile is shown in Fig. 4. The memory interface block in the H-Tile is responsible for accessing the HCU state memory in the 3D stacked memory and fetching the required state memory records for the BCPNN computation. The H-Tile controller triggers the data to be fetched based on the incoming spike that is being serviced. Two scratchpad memories have been architected to serve as ping-pong buffers to pipeline the fetching of data from the 3D stacked HCU state memory with the BCPNN computation. The size of each scratchpad memory is 1.5 KB corresponding to a row in the HCU state memory identified by the incoming spike (Fig. 1.). A VLIW style computation unit customized for BCPNN computation with multiple single precision floating point units implements the lazy evaluation variant of the BCPNN algorithm under the control of a simple sequencer and a small (~1 KB) local program memory. Being customized to the BCPNN computation that is straight line code, the computation unit has negligible overhead of typical general purpose VLIWs in the form branch prediction, managing deep pipelines, detecting and avoiding hazards etc.

The incoming spikes are queued and serviced in the order of their arrival. Each of the 100 MCUs in an HCU can potentially generate an output spike that is fanned out to 100 HCUs, i.e., the 100 output spikes become 100×100=10K spikes per second to be dealt with in terms of architectural requirements. The addresses of the 100 target HCUs to which these output spikes are distributed are stored as part of the MCU record and defines the MCU to HCU connectivity in the eBrain. These addresses can be adaptive and changed as a result of the computation. Outgoing spike packets are queued in a delay buffer to mimic the conduction delay. The dimensioning these queues for incoming and outgoing spikes is elaborated in section V.B

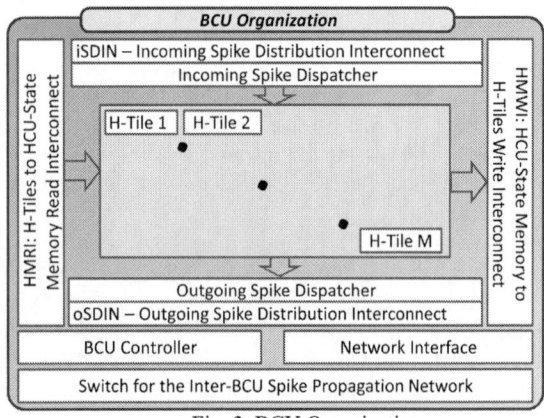

Fig. 3. BCU Organization

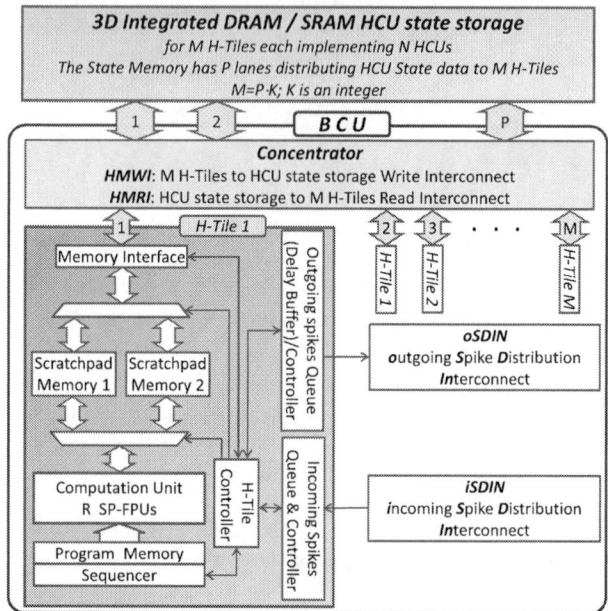

Fig. 4. H-Tile organization

The entire HCU state memory composed of MCU records is updated every second. Three HCU operations are time-dependent with an accuracy of 1 ms. To fulfill this, each H-Tile has an independent 10b time-counter that is incremented every ms. The three time-dependent HCU operations are as follows: 1) When an MCU record is updated and written back it is time stamped. This enables measurement of the time elapsed between two updates to calculate the integrated decay that should be applied to the synaptic weights. 2) The 10K output spikes generated every second are time stamped so that their individual delays (typically 2 – 20 ms) can be measured and ejected when the delay has elapsed. 3) The spikes of each MCU are generated by a Poisson process driven by the MCU's activation variable, which is time dependent on a millisecond time scale. The Intra-BCU Spike Distribution Interconnects

Two spike distribution interconnects exist in each BCU, one for the incoming spikes, the *iSDIN* (Fig. 5.a) and the other for the outgoing spikes, the *oSDIN* (Fig. 5.b). The *iSDIN* is a pipelined hierarchical spike distribution interconnect realized using registered de-multiplexors. Pipelining is a necessity because of the long wires and also enables multiple spike transmission to proceed in a pipelined parallel manner. The advantage of this design is its simplicity resulting in a low implementation cost and an equal log latency spike distribution to all the H-Tiles. Being a hierarchical interconnect from the *Incoming Spike Dispatcher* where the spikes enter the BCU and terminate in the H-Tiles it does not provide any *direct* intra-BCU H-Tile to H-Tile spike propagation. When an output spike from one H-Tile is destined for another H-Tile in the same BCU, the *oSDIN* propagates the output spike to the Switch which then sends it to the *Incoming Spike Dispatcher* to be forwarded to the destination H-Tile. Another aspect of the *iSDIN* is that only one spike can enter it every clock tick. Note that, the eBrain clock tick is of the order of 10 ns, so even if the spikes are delivered sequentially, the delay is inconsequential as the

978-1-4799-2817-0/14 $31.00 © 2014 IEEE   580

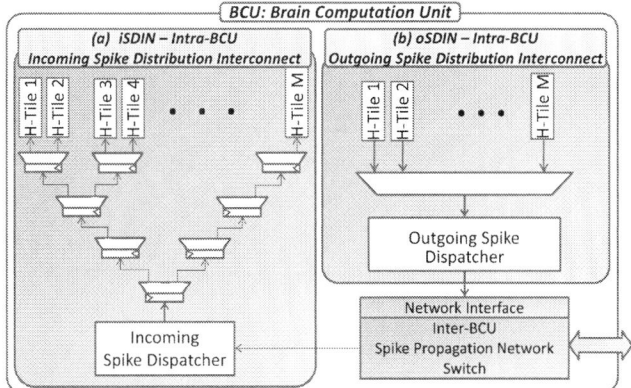

Fig. 5. The two intra-BCU spike distribution interconnects

biological time constants are of the order of 1ms. Finally, *iSDIN* does not have queues; the incoming spikes are queued in the individual H-Tiles. While Fig 5.a depicts the *iSDIN* conceptually as a pyramid like tree, the physical layout of the network is like an H-Tree akin to clock distribution network in VLSI and also shown in Fig 2.b.

The 100×100=10K outgoing spikes are queued locally in each H-Tile in a delay buffer and ejected from this queue when their respective delays have elapsed. The ejection happens as a result of round robin polling by the outgoing spike dispatcher of the delay buffers in each H-Tile (Fig 5.b). The ejected spike is then sent out via the Network Interface/inter-BCU SPN Switch to an HCU in another BCU or in the same BCU.

B. HCU State Distribution Interconnects

Each BCU has two parallel interconnects, *HMWI* and *HMRI* (Fig. 4), for distributing the HCU state data stored in the external 3D stacked DRAM memories to the *M* H-Tiles. We assume that the 3D HCU storage would have *P* parallel lanes; $P \leq M$ and in the extreme case *P=1*. These lanes are connected to the *M* H-Tiles via a concentrator that services their requests in a round robin fashion. This is feasible because the aggregate bandwidth of *P* lanes is greater than bandwidth required by the *M* H-Tiles. Like *oSDIN*, these two interconnects (*HMWI* and *HMRI*) are two sets of parallel pipelined paths between the H-Tiles and the concentrator.

C. BCU Controller, Network Interface and the Switch

The BCU controller is responsible for booting up and initializing and configuring the BCU. It works under the supervision of the eBrain's system controller with whom it communicates via the inter-BCU SPN. This network in the normal operational mode is used to propagate spikes but in the boot up mode it can be used to communicate commands and initialization/configuration data. When the eBrain is powered down, the BCU controller is again involved in saving the HCU state memory in an external non-volatile storage, again using the inter-BCU SPN.

The Network Interface and the Switch for the inter-BCU SPN enable the other design units in the BCU to communicate with the external world including other BCUs and the eBrain system controller.

D. BCU Power Management

BCU has a power management architecture that has been customized for the BCPNN model and it relies on power

gating as the principle low power technique besides using the standard clock gating and operand isolation options when synthesizing the logic. Dynamic Voltage Frequency Scaling has not been used because all H-Tiles have the same performance need. BCU has been divided into two power domains: always-on and H-domains; part of each H-Tile is an independent H-domain. The always-on domain includes a) all the logic outside H-Tiles, b) the SRAM macro for program memory,the incoming queue controller in each H-Tile and the time counter. The H-domains (one per H-Tile) covers all the remaining logic and SRAM macros in the H-Tiles. The incoming spike queue controller in each H-Tile also acts as a simple power management controller; when the queue is empty it first waits for the ongoing operations to complete and then waits for a certain number of cycles for the next incoming spike and if no input spike comes, it powers down the local H-domain and then powers it up when it has a few input spikes in the queue.

Even though the BCU die is expected to be quite large (~100 mm^2), the clocking problem is manageable because the operating frequency is expected to be low in 100s of MHz and the communicating design units in BCU are nearest neighbors, inducing manageable skew between them. For instance, H-Tiles do not communicate with each other, only with the registered de-multiplexors. Likewise, in the *oSDIN* the path from H-Tiles to the incoming queue controller is pipelined.

V. BCU Design Space Exploration & Dimensioning

In this section we elaborate the design space exploration (DSE) and the dimensioning of the BCU. The HCU state memory in 3D stacked DRAM/SRAM dies largely decides the number of HCUs that can be serviced by each BCU and thereby decides the number of BCUs required for eBrain, i.e. the *L* parameter. BCUs are organized in terms of *M* H-Tiles with each H-Tile clustering *N* HCUs that are serviced sequentially; note that *M* and *N* are not independent because their product is a constant for a given *L*.

The degree of TLP is decided by *M* (and thus also *N*) and these parameters strongly influence the ALP and thereby the voltage frequency operating point, power density, area and utilization. Let *R* represent the degree of ALP in every thread, i.e., the number of Floating Point Units (FPUs) per H-Tile. Given that each HCU requires 370 Mflops/s, the required operational frequency can be expressed in terms of ratio of *N* and *R*:

$$f_{reqd} = \frac{N}{R} 370 \ MHz \qquad (1)$$

The ratio *(N/R)* expresses a *composite* thread and arithmetic level parallelism per HCU and decides if the f_{reqd} would be less than, equal to or more than 370 MHz. Lowering *N*, thereby increasing *M* implies increased TLP and increasing *R* implies increased ALP. Increasing parallelism of either or both types does not change the number of operations performed but it lowers the unit energy cost of operations by enabling a lower voltage frequency operating point according to (Eq. 2) [15] where α is the velocity saturation index, V_t is the threshold voltage and f_{max} is the maximum frequency at which a design can operate when operating at max V_{dd_max}.

$$f_{reqd} = K \frac{\left(V_{DD\_f_{reqd}} - V_t\right)^\alpha}{V_{DD\_max}} f_{max} \qquad (2)$$

HCU Operations and the hardware resources in BCU on

which they are performed can be divided into three categories: *arithmetic, SRAM* and *infrastructural*. Arithmetic operations including fetching and storing operands from the scratchpad memories and instructions from program memory are performed by the computational unit (Fig. 4) consisting of R FPUs. Storage needs like scratchpad memory, program memory and queues are mapped to the SRAM macros. Finally all other operations queuing and de-queuing spikes, spike and HCU state data distribution etc. are mapped to various controllers, interconnects, dispatchers, concentrators etc. as described in section IV and are collectively referred to as infrastructural resources.

While increasing the TLP has the benefit of improving the computational efficiency (Gflops/s/watt) by lowering the voltage frequency operating point but it comes at the expense of lowering the utilization of the infrastructural resources thus increasing the area required for the infrastructural resources. For instance, the two scratchpad memories, the program memory and the sequencer are needed in every H-Tile independent of the number of HCUs clustered in it as they are serviced sequentially. This contributes to increased area and static power consumption. Additionally, there will be larger spike distribution and HCU state memory distribution interconnects. On the positive side, the same amount of activity distributed over larger area contributes to lower power density and the sparse activity in the larger number of H-Tiles provides more opportunity to power gate them when they are inactive and better matches the dark silicon era constraints.

To systematically explore the BCU design space for a given technology, we express the power and power density in terms of technology, architectural and dimensioning parameters in the Eqs (3-7) and sweep them over plausible ranges to find an optimal solution. Eq. 3 specifies the power consumption in each H-Tile. Note that the TLP and ALP parameters, M/N and R are indirectly expressed in f_{reqd} and V_{DD_freqd} besides being also explicitly present in the Eqs (3-7). As the area and power consumption in the infrastructural resources inside an H-Tile are negligible, the $P_{infrastructure}$ and the $A_{infrastructure}$ parameters in Eqs 5-6 are the infrastructure resources outside H-Tile at the BCU level.

$$P_{H-Tile} = P_{comp} + P_{SRAM} \qquad (3)$$

$$P_{comp} = \frac{1}{2} f_{reqd} \cdot V_{DD\_freqd}{}^2 \cdot R \cdot C_{FPU} \qquad (4)$$

$$P_{BCU} = M \cdot P_{H-Tile} + P_{infrastructure} \qquad (5)$$

$$A_{BCU} = M[R \cdot A_{FPU} + A_{SRAM}] + A_{infrastructure} \qquad (6)$$

$$Power\ Density_{BCU} = \frac{P_{BCU}}{A_{BCU}} \qquad (7)$$

A. Design case for 8 GB Hybrid Memory Cube and 22nm node

In this section, we present concrete BCU design dimensions when using the Hybrid Memory Cube [16] for the HCU state memory and 22nm technology node for implementing the BCU logic chip or BCU for brevity. HMC is a Micron led standard that exploits 3D stacking of DRAM dies to achieve high density, bandwidth and low power consumption. Micron has recently announced engineering samples of 2 GB and announced plans for 4GB next year. In line with realistic schedule for manufacturing the BCU logic chips, we have assumed availability of 8 GB HMCs by 2015.

Number of BCUs (L) and the Bandwidth: In the baseline eBrain design, we use 8×8 GB HMCs surrounding a single

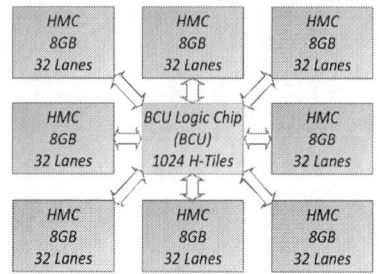

Fig. 6. BCU Module: 8×8 GB HMCs and 1 BCU logic chip

BCU logic chip in a 3×3 matrix organization as shown in Fig 6. BCPNN's requirement is 15MB/HCU for state memory. This results in a requirement that each BCU will implement 8×8 GB/15MB ≥ 4096 HCUs and with 2 million HCUs this translates to 490 BCU Modules; 2 million being the number of HCUs in a human cortex in the BCPNN model.

In BCPNN's lazy evaluation method, the bandwidth requirement is 56 MB/s/HCU. This translates into 4096×56MB/s = 230 GB/s aggregate bandwidth requirement for the 4096 HCUs. Since each HMC has 32 lanes and each lane can provide 1.25 GB/s to 1.875GB/s, the total aggregate bandwidth of the 8 HMCs in the BCU module is 320 GBs/s. This being raw bandwidth, we need a minimum of 230/320=72% utilization. This can be achieved because BCPNN accesses are large and can utilize the maximum 128 byte packet size that the HMC protocol allows. In conclusion, a BCU module (Fig 6) has sufficient storage and bandwidth to support 4096 HCUs and this translates into L=490.

For the 50nm node micron reports HMC efficiency of 13.7 pJ/bit [17]. We scaled down this value by 30% for 22nm based on data in Shekhar Borkar's IPDPS 2013 Keynote [10] to get 9.6 pJ/bit. Since the aggregate HCU state bandwidth requirement is 112 TBs/s, the power consumed by the HMCs for the entire eBrain fabric would be 10 kWs for the theoretical worst case when all the HCUs are active and 5 kWs for the typical case when 50% of the HCUs are active.

TLP and ALP (M and R): Having settled the dimensions of the BCU module for the HCU state memory and the bandwidth for 4096 HCUs, we next decide the dimensions of the BCU logic chip (BCU for brevity) that implements the computation for the 4096 HCUs in terms of degree of TLP and ALP, the M (# of H-Tiles) and the R (# of FPUs per H-Tile) parameters. We use equations 1-7 to explore the design space by sweeping M from 256 to 4096 and R from 4 to 32 to gauge the impact on Power, Area and Power Density. M and R decides the f_{reqd} (Eq. 1) that in turn influences the V_{dd_reqd} (Eq. 2). Fig 7 shows results of the DSE in terms of total power consumption in the 490 BCUs and power density in each BCU.

The area per BCU is also shown for a few select interesting design points. These numbers were calculated based on synthesis results for a 40 nm technology node and data from an embedded SRAM compiler in the same node. These values were then scaled down to 22nm node using the conservative scaling scenario presented in [18]. We prioritized minimizing power and considered solutions that had power density less than 3 watts/cm^2 to enable forced air cooling of eBrain.

Notice that there are quite many designs in the range of 1 to 1.3 kWs but they dramatically differ in their area cost from 375 mm^2 to 10 mm^2, with implication on power density. This justifies the DSE to decide on the optimal degree of TLP and

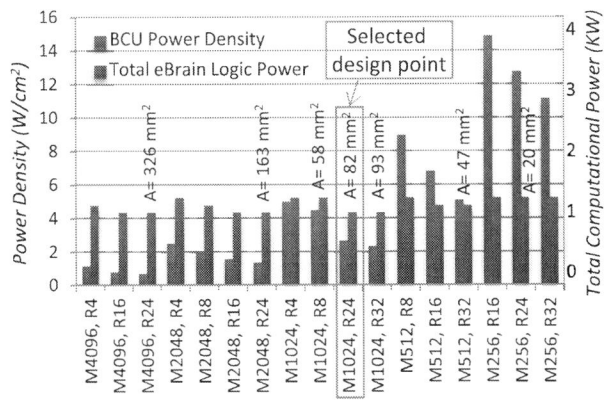

Fig. 7. Design Space Exploration of BCU

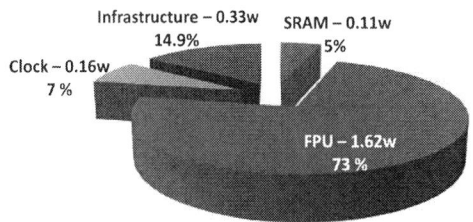

Fig 8. Breakdown of BCU Logic Chip's Power Consumption

ALP. For the selected solution: M=1024, R=24, f_{reqd}=62 MHz the breakdown in power among the various BCU components is shown in Fig. 8. The f_{reqd} = 62 MHz is the minimum frequency but operating at this frequency keeps the logic on for a greater percentage of the 1s BCPNN computation cycle increasing the leakage current. To mitigate this, we operate at 100 MHz to give power management more opportunity to power gate the H-Domains. The typical power consumption for logic computation is 2.2 W per BCU or ~1 kW for the entire eBrain. FPU dominates because each HCU in one second requires 370 million FPU operations but only 56 million accesses to the on chip scratchpad SRAM macros (introduced in Fig. 4). The FPU component includes fetching data and instructions from the SRAM macros. The low percentage for clock is explained by the fact the design is flip-flop scarce owing to low frequency and little or no pipelining. Though not shown separately, the infrastructure component is dominated by the HMRI and HMWI's heavy data movement between the chip periphery and the H-Tiles.

B. Dimensioning Incoming and Outgoing Spike Queues

Incoming and outgoing spike queues also influence the dimensions of the BCU logic chip; deeper queues allow, computing at a more relaxed pace and vice versa. However, given the sparse spiking activity of BCPNN, the overall impact of spike queues is not significant. The dimension of these queues is significant because it decides the robustness of the spike communication infrastructure so that negligible numbers of spikes are dropped. This is especially important because in eBrain, we have adopted a send and forget policy for spike propagation.

The incoming spikes arrive in the H-Tiles with Poisson distributed inter-spike arrival time. Each H-Tile with 4 HCUs, based on the dimensions decided in Section V.A, gets 40K spikes/s (10K spikes/s/HCU) that are buffered in a queue managed by its controller (Fig. 4). The worst case scenario has an infinitesimal small probability of occurring and results in an unreasonably large size, 9KB, of the queue. To have a robust dimension with a reasonable queue size we simulated the queue build up with spikes arriving with random Poisson distributed inter-spike arrival times and emptying at a steady rate of 17 μs per spike as per the ALP and TLP dimensions and the operating frequency decided in Section V.A. This simulation was performed five million times and we plotted the distribution of the required queue size, which as expected is Gaussian. By picking up a queue size that is at 6σ, we arrive

at a dimension of 55 spikes with a probability of dropping a spike once a month in each H-Tile.

Each H-Tile with 4 HCUs in it generates 400 outgoing spikes/s that enters the queue with Poisson distributed inter-arrival time. Each of these output spikes is fanned out to 100 HCUs. Effectively, the queue needs to deal with 40K spikes/s, each with an individual delay. The difference compared to incoming spikes is that these spikes arrive sparsely but in chunks of 100s and have an individual delay associated with them for which they stay in the queue. When the delay becomes zero they are moved to the top of the queue ready for ejection by the polling outgoing spike dispatcher (Fig. 5). Based on our preliminary calculations, we are able to inject one spike, i.e., a *phit* every 10 ns into the inter-BCU SPN. As there are 1024 H-Tiles to poll by the dispatcher, this means that each H-Tile can eject a spike every 1024×10 ns = ~10 μs. With 40K outgoing spikes in each H-Tile to dispatch every second, the total time required is ~400 ms, well within the 1s BCPNN computation cycle in which the 40K×1024 = ~40 mega output spikes in the BCU that are generated and must be dealt with.

To dimension the outgoing queue in each H-Tile, we again simulated 5 million times the queue build up with 400 spikes arriving with Poisson distributed inter-arrival time and queue being emptied at the rate of 10 μs. The 6σ value of the queue size on the resulting Gaussian distribution turned out to be 20 spikes; i.e. 20×100 spikes as each spike is fanned out to 100 addresses. In conclusion, in each H-Tile we need an incoming spike queue of size 55 and outgoing spike queue of size 20 to have a robust spike distribution infrastructure with very reasonable cost.

VI. Related Work

Supercomputers like Blue Gene have been widely used to model human brain in varying degrees of complexity starting from Blue Brain Project at EPFL [5]. It has also been used by Prof. Lansner's group to model a spiking cortex model with BCPNN type modular structure and simulated on IBM Blue Gene/L supercomputer reaching a model size of 22 million neurons and 11 billion synapses [2]. In this experiment, every hypercolumn unit was implemented on one BG/L node, which is composed of two PowerPC-440s. The C2S2 SyNAPSE project at IBM Almaden research lab used IBM's Dawn Blue Gene/P supercomputer with 147,456 CPUs and 144TBs of main memory to simulate 1.6 billion cortical neurons with simpler models than Blue Brain and 9 trillion synapses with a speed 1/83rd of the real time. Compass [7] is a massively parallel, multi-threaded functional simulator developed by the same group to simulate the functionality of 65 billion neurons and 16 trillion synapses running at 1/388th of the real time on

978-1-4799-2817-0/14 $31.00 © 2014 IEEE

a 16 rack IBM Blue Gene/Q (262144 CPUs and 256 TBs of memory) with an average spiking rate of 8.1 Hz. GPGPUs have also been used to simulate neuron level (Izhikevich model) large-scale spiking neural network [8,9].

Besides the super-computers and high-performance computing platforms discussed above, there have been attempts to build dedicated machines for simulating brain. The FACETS [4] and its follow up, the BrainScaleS project [11] aims at designing a spiking neural network with 200k integrate-and-fire neurons and 50 million modifiable synapses in a mixed signal CMOS design on wafer scale integration with the possibility to stack multiple wafers to make design scalable. The SpiNNaker system is a digital design built using large number of general purpose ARM cores communicating via a custom designed spike communication fabric. Each chip has 18 cores and the entire system has the size of a refrigerator with 1 million cores running 1 billion spiking I-F neuron.. A human cortex sized system would be 20 times bigger than this. The IBMs cognitive chip prototype [13] emulates 256 silicon neurons with 262,144 programmable weighted synapses.

VII.   Discussion and Analysis of the eBrain design

eBrain delivers the required 740 TFLOPs/s, 30 TBs of external storage accessed with a bandwidth of 112 TBs/s for the real time execution of the BCPNN model of human cortex while consuming 6 kWs of power for the typical case. This is three orders of magnitude better than other general purpose super-computers like BlueGene and Cray. In an experiment performed in Prof. Lansner's group, BCPNN has been mapped to Cray with one HCU per core that results in 7.44 W/HCU (scaled to 22 nm technology using [18]) as opposed to 6mW/HCU for eBrain, i.e. eBrain is 1240 times more efficient. This three orders of magnitude improvement in efficiency comes from customization of eBrain in three architectural aspects: 1) I/O architecture to enable high-bandwidth energy efficient access to the external HCU state storage access, 2) computational architecture whose TLP and ALP has been customized and optimized for BCPNN computation and 3) custom infrastructure for managing and distributing the spikes. It is the *combined* effect of these three customizations that gives the three orders of magnitude improvement and not just the customization of computation and improvements in performance (GFlops/s) and computational and silicon efficiencies (GFlops/s/watt and GFlops/s/mm^2) that are very often the comparison metrics.

The lack of these three customizations in the general purpose supercomputers degrades their efficiencies (computational and silicon) compared to eBrain forces them to use the same complex computing unit and sequencer designed for heavy number crunching for also performing the infrastructural chores like spike distribution, queuing, delay management, data movement between the external HCU state storage and the computational units etc. This sequentializes computation and infrastructural operations that results in loss of performance and underutilizes the complex computing unit and sequencer that results in loss of silicon and computational efficiencies. In contrast, the BCU in eBrain has customized resources as described in section IV for managing the aforementioned infrastructural operations that not only operate in parallel to the computational resources but also take

significantly less area and energy that results in the three orders of magnitude improvement in the efficiencies.

SpiNNaker, unlike BlueGene and CRAY, is a dedicated brain simulation machine that has a customized spike propagation infrastructure. The eBrain differs from SpiNNaker in two key aspects. The first is that eBrain has dedicated resources for both spike and HCU state distribution and management, whereas SpiNNaker has only dedicated resource for spike distribution; for HCU state distribution it uses standard ARM infrastructural IPs. The second is that the BCU in eBrain distributed dedicated hardwired controllers to implement the infrastructural operations. Additionally, the hardwired controllers are the masters (the BCU and the H-Tile Controllers) and the sequencer implementing the BCPNN computation is the slave. In SpiNNaker, the ARM9 cores remain the master controllers forcing time multiplexing of computation and infrastructure management. In conclusion, we expect SpiNNaker to have better efficiencies compared to BlueGene and Cray but we expect eBrain to have significantly better efficiencies compared to SpiNNaker; we plan to map BCPNN to SpiNNaker in near future to be able to concretely compare the efficiencies of the two systems.

There are some general purpose multi/many core high-performance (**MCHP**) parallel computing machines that report impressive performance and computational and silicon efficiency numbers as shown in Table 1. These numbers are for the case when the computation is performed with data in on-chip SRAM. In eBrain, the bulk of power consumption, about 80%, goes into the external storage for the HCU state memory and access to it. For fair comparison we only compare the computational and silicon efficiencies of the BCU Logic Chip to these machines. The results are shown in Table 1; we have scaled the values reported in [19] to 22nm using the more conservative scaling scenario presented in [18]. Results in Table 1 might suggest that if BCPNN were to be mapped to these MCHP machines, the efficiency would be only 5-20 times worse off. We argue that the degradation in efficiencies would be significantly more because like BlueGene and Cray, these machines have the same sequencer(s) and computational units that would time multiplex between infrastructural operations and computational operations degrading both the performance and the silicon and computational efficiencies.

Table 1. Silicon and Computational Efficiencies of general purpose multi-cores high-performance *(MCHP)* compared to BCU

	ARM A9Cortex	Adapteva	Vivante GC2000	eBrain (BCU Logic Chip)
# of Cores	4	16	4	1024
Total SRAM (KB)	256	512	128	10,000
Die Area (mm^2)	2.3	1.03	3.2	81.74
Frequency (Hz)	1.27 G	636 M	1.59 G	100 M
Peak Flop (GFlops)	20.35	20.35	25.44	2088.9
GFlops/s/watt	25.53	97.83	50.83	585.9
GFlops/mm^2	8.90	19.93	18.23	25.56

We further point out that, eBrain has custom hardwired resources for infrastructural operations but the BCPNN computation is still fully programmable and also uses the single precision floating point units. Keeping this in mind the 5-20 times improvement in computational efficiency is still significant and as we argued is amplified by the custom hardwired distributed control and resources for infrastructural operations. A common argument against custom design is the large engineering cost associated with it. We counter this with the observation that two factors significantly lower the engineering cost of the eBrain design. One is the regularity, which requires designing and verifying a tile that is then replicated. The second is the simplicity that comes with the custom design that avoids the complexity associated with the generality; it is harder to design and verify SOCs built with general purpose IPs compared to designing and verifying custom hardware that implements a single functionality.

VIII. Conclusions and Future Work

We have presented a scalable multi-chip digital design called eBrain that implements the BCPNN model in real time with 740 TFlops/s, providing 30 TBs for HCU state storage in HMC and accessing it with a bandwidth of 112 TBs/s. eBrain consumes only 6 kWs of power for the typical performance case in 22nm technology node which is three orders of magnitude lower than executing BCPNN in a general purpose supercomputer like Cray XE in the same node. This paper has focused mainly on designing the BCU logic chip.

The next steps for us in the near future are to complete the design by making decisions on the inter-BCU SPN and deciding on the integration form for the BCU module. Our inclination is for 2.5D and we plan to refine the eBrain design for this option.

We are developing a follow up design to eBrain where the computation unit is based on integer arithmetic that will significantly improve the computational efficiency of eBrain. More critically, the three customizations in eBrain address power components that constitute only 20% of the total power. HMC, while more power efficient compared to its competitors is not customized for BCPNN. We plan to replace the DRAM based HCU state storage with 3D stacked SRAMs. SRAM is predicted to overtake the conventional DRAM in density and provides 2-3 orders better energy efficiency [20]. Critically it also allows significant customization in how the HCU state is organized in terms of its MCU records and with incorporation of smart streaming address generation units [21-22]. These customizations will significantly lower the most dominant component of eBrain's power consumption.

Long term, we plan to explore use of emerging memory technologies like memristors to further lower the power consumption and possibly also incorporate some of the computations in the memristors themselves [23]. We further plan to investigate implementing a scaled down version of BCPNN using less integers for embedded system applications in robotics and vision.

References

[1] A. Lansner , A. Holst, "A higher order Bayesian neural network with spiking units," *in Int. J. of Neural Systems*, vol. 7, 1996.

[2] M. Lundqvist, C. Johansson, M. Rehn, O. Ekeberg, A. Lansner, "Brain-Scale Simulation of the Neocortex on the IBM Blue Gene/L Supercomputer," *in IBM Journal of Research and Development*, vol. 52, 2008.

[3] K.A Boahen, "Point-to-point connectivity between neuromorphic chips using address events," *IEEE Trans. Circuits Syst. II Analog Digit. Signal Process.* vol. 47, no. 5, 2000.

[4] D. Brüderle, et al. "A comprehensive workflow for general-purpose neural modeling with highly configurable neuromorphic hardware systems," *Biological cybernetics*, 2011.

[5] H. Markram, "The Blue Brain Project", in *Nature Reviews*, Vol. 7, February 2006, pp 153-160.

[6] A. Lansner, A. Hemani, N. Farahini, "Spiking Brain Models: Computation, Memory and Communication Constraints for Custom Hardware Implementation," in *IEEE ASP-DAC*, 2014.

[7] R. Preissl, T. Wong, P. Datta, M. Flickner, R. Singh, S. Esser, W. Risk, H. Simon , and D.Modha "Compass: A scalable simulator for an architecture for Cognitive Computing," in *Proc. of the International Conference for High Performance Computing, Networking, Storage, and Analysis*, 2012.

[8] M. Richert, J. Nageswaran, N. Dutt, and J. Krichmar, "An efficient simulation environment for modeling large-scale cortical processing," *Frontiers in Neuroinformatics*, vol. 5, 2011.

[9] A. Fidjeland, M. Shanahan, "Accelerated simulation of spiking neural networks using GPUs," on *Intr. Joint Conf. on Neural Networks (IJCNN)*, 2010.

[10] Shekhar Borkar. "ExaScale Computing – Fact or a Fiction?, Slide 8," Keynote at IPDPS, May 21, 2013.

[11] J. Schemmel, et al. "Live demonstration: A scaled-down version of the BrainScaleS wafer-scale neuromorphic system," in *Proceedings of IEEE International Symposium on Circuits and Systems (ISCAS)*, 2012.

[12] E. Painkras, L. Plana, J. Garside, S. Temple, F. Galluppi, C. Patterson, S. Furber, "SpiNNaker: A 1-W 18-Core System-on-Chip for Massively-Parallel Neural Network Simulation," in *IEEE Journal of Solid-State Circuits*, 2013.

[13] D. Modha, R. Ananthanarayanan, S. Esser, A. Ndirango, A. Sherbondy, and R. Singh, "Cognitive Computing: Unite neuroscience, supercomputing, and nanotechnology to discover, demonstrate, and deliver the brain's core algorithms," *Communications of the ACM*, vol. 54, 2011.

[14] C. Johansson, A. Lansner, "Towards Cortex Sized Artificial Neural Systems," in *Neural Networks (Elsevier)*, vol. 20, 2007.

[15] M. Flynn, P. Hung, K. Rudd, "Deep submicron microprocessor design issues," *IEEE Micro*, vol.19, no.4, 1999.

[16] *Hybrid Memory Cube Specification 1.0*, Hybrid Memory Cube Consortium, 2013.

[17] J. Thomas Pawlowski, "Hybrid Memory Cube (HMC)," in *Hot Chips 23*, Aug. 4, 2011.

[18] H. Esmaeilzadeh, E. Blem, R. Amant, K. Sankaralingam, D. Burger, "Dark silicon and the end of multicore scaling," in *Proc. International Symposium on Computer Architecture*, 2011.

[19] Linley Gwennap, "Adapteva: More Flops, Less Watts. Epiphany Offers Floating Point Accelerator for Mobile Processors," *Microporcessor Report*, The Linely Group, 2011.

[20] B. Hoefflinger, "Chips 2020: A Guide to the Future of Nanoelectronics," *Springer Publishers*, 2012.

[21] M.A. Shami, A. Hemani, "Address generation scheme for a coarse grain reconfigurable architecture," *Application-Specific Systems, Architectures and Processors (ASAP)*, 2011.

[22] N. Farahini, A. Hemani, K. Paul, "Distributed Runtime Computation of Constraints for Multiple Inner Loops," in *Euromicro Conference on Digital System Design*, 2013.

[23] S. H. Jo, T. Chang, I. Ebong, B. B. Bhadviya, P. Mazumder, and W. Lu, "Nanoscale Memristor Device as Synapse in Neuromorphic Systems," *Nano Lett*, vol. 10, 2010.

978-1-4799-2817-0/14 $31.00 © 2014 IEEE

NoΔ: Leveraging Delta Compression for End-to-End Memory Access in NoC Based Multicores

Jia Zhan[*], Matt Poremba[*], Yi Xu[†], Yuan Xie[*†]

[*]The Pennsylvania State University, {juz145,poremba,yuanxie}@cse.psu.edu
[†]AMD Research China Lab, {yi1.xu,yuan.xie}@amd.com

Abstract—As the number of on-chip processing elements increases, the interconnection backbone bears bursty traffic from memory and cache accesses. In this paper, we propose a compression technique called *NoΔ*, which leverages *delta compression* to compress network traffic. Specifically, it conducts data encoding prior to packet injection and decoding before ejection in the *network interface*. The key idea of *NoΔ* is *to store a data packet in the Network-on-Chip as a common base value plus an array of relative differences (Δ).* It can improve the overall network performance and achieve energy savings because of the decreased network load. Moreover, this scheme does not require modifications of the cache storage design and can be seamlessly integrated with any optimization techniques for the on-chip interconnect. Our experiments reveal that the proposed *NoΔ* incurs negligible hardware overhead and outperforms state-of-the-art zero-content compression and frequent-value compression.

I. Introduction

The last few years we have witnessed a growing trend of accommodating many cores and large caches in chip multiprocessors. Conventional shared-memory multi-core design adopts a single bus with limited bandwidth as the communication backbone. Consequently, the bus bears enormous strain and even becomes the performance bottleneck due to frequent packet transmission. To provide an efficient and scalable interconnect for memory and cache accesses, researchers proposed Network-on-Chip (NoC) [1] to replace conventional bus-based architectures. However, NoCs also have limitations such as long end-to-end communication latency and high network power dissipation. Therefore, many previous studies [2]–[4] explored optimization on router microarchitecture, network topologies, routing algorithms, etc. However, these techniques inevitably increase design complexity and even incur significant performance or power overhead.

Instead of seeking new NoC architectures, an alternative yet complementary approach is *data compression*, also known as *source coding*, which compresses data messages into more compact bits before they are stored or transmitted. Data compression has been utilized in hardware design to help enhance system performance as well as diminish power dissipation. For example, *cache compression* expands the cache capacity by packing in more data blocks than given by the space. Dusser *et al.* [5] proposed Zero-Content Augmented (ZCA) cache design that is able to represent zero cache lines. Zhang *et al.* [6] discovered the frequent value locality in many application programs, i.e., a few values appear repeatedly and engage in a large portion of memory accesses. They proposed a compression cache [7] that employs a value-centric mechanism with regard to such a phenomenon. Additionally, Alameldeen and Wood [8] exploited some frequent patterns in programs such as all bytes in a 4-byte word are the same. Recently, Pekhimenko *et al.* [9] introduced a new compression algorithm called Base-Delta-Immediate (BΔI) compression, which stores cache lines as a set of differences (Δ) to the base values. However, all of these cache compression techniques require fundamental changes of the cache architecture to support variable sizes of cache lines. As a result, cache compression may not be applicable to modern SoC implementation that may comprise several unmodifiable *hard* IP blocks from various vendors.

Alternatively, simple compression/decompression modules can be integrated in the *network interface* (NI) and triggered to compress network traffic. In this case, there is no need to modify the cache structure, allowing for a plug-and-play capability. Additionally, the whole process is conducted in the NI, and thus it is orthogonal and complementary to any further optimization techniques for the switch microarchitecture or network topology. Recently, data compression is explored in the research domain of Network-on-Chip. For instance, Das *et al.* [10] proposed to compress network messages based on zero bits in a word. Subsequently, Jin *et al.* [11] and Zhou *et al.* [12] adopted frequent value compression and design table-based schemes to manage the compression/decompression process.

Delta compression is a method to transmit data in the form of differences rather than the complete data set. It is widely adopted in video compression codecs to considerably reduce frame size, or in the networking domain to reduce Internet traffic by allowing HTTP servers to send updated Web pages based on differences between versions. For memory/cache data, they also display a low dynamic range due to several reasons: programs tend to change register numbers or memory addresses and the changed values are the same across many instructions; similar data values and types are grouped together in the memory; arrays are commonly used to represent large piece of data, etc. This indicates that delta compression can be a promising candidate to compress memory/cache traffic flowing through the on-chip interconnect.

Therefore, in this work, we investigate a new compression scheme called *NoΔ*, which leverages the *delta compression* technique in NoC-based multi-core by transmitting values of bytes as differences (Δs) between sequential values, rather than the complete values themselves. We demonstrate how delta compression can be applied to the NoC domain with specific architecture support and evaluate the corresponding performance gain and power savings. Overall, this paper makes the following major contributions:

• To the best of our knowledge, this is the first work that exploits the usage of *delta compression* in NoC-based architectures. We run full system simulation with SPEC CPU2006 benchmark suite and analyze data patterns to get the compression ratios of different applications.

• A lightweight compression/decompression module is designed which incurs minimum area and power overhead. We integrate these modules in the Network Interfaces and provide detailed synthesis results to validate the architecture design.

• We compare the proposed *NoΔ* technique over previous zero-content compression and frequent-value compression. Experimental results show that our scheme outperforms the other two in terms of both performance and power.

This work is in part supported by NSF CCF-0903432, CNS-0905365, and SRC grant.

978-1-4799-2817-0/14 $31.00 © 2014 IEEE

II. Data Compression in On-Chip Networks

A. NoC-based Multi-core Architecture

The revolution of multicore/Chip-Multiprocessor (CMP) opens up opportunities to circumvent the power wall, but it also poses challenges on the design of the cache hierarchy. Consequently, the last level cache (L2, or where applicable, L3 caches) becomes oversized to accommodate the needs of all cores, and gets fragmented into numerous banks due to the increased capacity. However, it becomes clearly inefficient to demand that every cache access incurs the delay penalty of accessing the furthest bank. To this end, some researchers proposed innovations of network fabric that enable caches to support non-uniform cache bank accesses, which are referred to as *Non-Uniform Cache Access (NUCA)* architectures [13].

Fig.1 depicts the high-level view of our NoC-based multi-core system in this paper. This 4×4 tile-based system forms a 2D mesh network interconnected through routers. In this architecture, each node consists of a CPU core, a private L1 instruction and data cache, and a separate bank of a shared last level cache (L2). In this work we adopt a SNUCA [14] architecture for the L2 where the mapping of data blocks is unique and cannot be dynamically moved. These processing elements are coupled with the router through the Network Interface (NI). The main memory is attached to this multi-core chip via memory controllers (MCs) on the corners. Each MC controls a single memory channel with possibly many DIMMs.

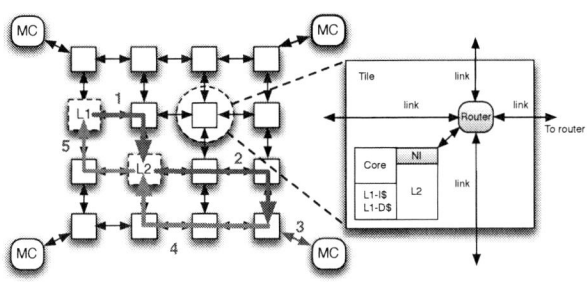

Fig. 1: A 4×4 NoC-based multicore with SNUCA architecture.

B. Memory Access and Packet Switching

Fig.1 also demonstrates the detailed flow of a memory access that produces network traffic. Specifically, once a core issues a memory request, its private L1 cache is first checked to see whether the data is locally available. If the data is not available in the L1, the request is then injected into the on-chip network and forwarded to an L2 bank (path 1). Subsequently, if an L2 miss is encountered, the request message is further delivered to the appropriate memory controller (path 2), and the off-chip memory access (path 3) is triggered consequently. Correspondingly, the response message from memory goes back on chip and is sent to the L2 cache bank (path 4) for updates. It is finally transferred to the L1 (path 5) that can be accessed by the core. Note that in this example, a private L1 cache hit would not generate any network traffic, and an L2 cache hit would not further lead to message flow in paths 2 to 4 which are directly used for off-chip memory request/response.

The memory request/response messages together with the cache coherence messages infuse into the on-chip network through the Network Interface (NI). A NI is responsible for the packetization/depacketization of these messages, and the flit fragmentation/assembly for flow control. When running some memory-intensive applications

on the cores, there will be many packets competing for the shared resource in the network. As a result, network latency plays a vital role in memory accesses to deliver the required Quality-of-Service. In addition, network power dissipation will increase significantly due to frequent packet accesses. In this work, we analyze the end-to-end memory access patterns in NoC-based multi-core and explore data compression to reduce the packet size, and hence relieve network load for performance improvement as well as power savings.

C. $No\Delta$ Mechanism

In general, data packets are generated by *load* and *store* instruction misses in the processor cache. These misses are converted into *read request* and *write request* packets and forwarded into the network. Each read request contains the memory address to be read where data will be fetched in a *read reply* message. The write request contains not only the address information but also the actual data to be stored. Fig.2 illustrates two packet formats, where the header flit includes the address and other packet identification information, and the body flits contain the actual data.

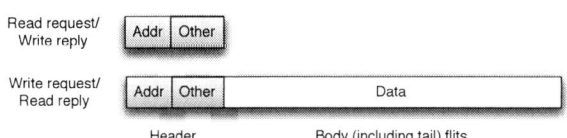

Fig. 2: The two packet formats required for end-to-end memory accesses through on-chip interconnect

The goal of data compression is to explore the internal redundancy of data values and squeeze packets into more compact units. However, data compression in on-chip networks may result in extra queuing delays of packets in the network interfaces as well as hardware overhead from those compression/decompression modules. In this work, we apply a lightweight compression mechanism called $No\Delta$ to the on-chip interconnect domain. It succeeds to achieve a significant compression ratio with minimal extra overhead.

$No\Delta$ is adapted from *delta compression* and leverages a key observation: the data values stored in the packets that are traversing the network have a low dynamic range, i.e., when the whole packet is divided into multiple fixed-length units, the relative value differences between different units are small. Therefore, the original data can be reorganized as *a common base value of fixed size plus an array of relative differences (Δs)*. Then the packet size information will be updated in the header flit. Assume that the data portion of a packet has D bytes, the size of base value is B bytes, and the other values can be represented by a set of Δs: $\{\Delta_1, \Delta_2, ..., \Delta_n\}$, where $n = D/B$. Then we can claim that the packet is compressible only if:

$$\max\{size(\Delta_i)\} < B, \forall i \in \{1, 2, ..., n\} \quad (1)$$

Then the size of delta (Δ) chosen for compression is:

$$size(\Delta) = \max\{size(\Delta_i)\}, \forall i \in \{1, 2, ..., n\} \quad (2)$$

For different data patterns, Fig.3 plots the percentage of packets that have a specific data pattern over the total number of network packets. SPEC CPU2006 benchmark suite [15] is used for our observations. We compare the ratio of zero-content packets (Denoted as *ZeroCmpr*), frequent value packets (Denoted as *FreqCmpr*), and

---

The detailed evaluation methodology is described in Section IV.

978-1-4799-2817-0/14 $31.00 © 2014 IEEE   587

packets that have $No\Delta$ compression capabilities. As the figure demonstrates, data packets that have "Δ" patterns constitute a big portion of the total network packets. On average, the ratios of compressible packets for *ZeroCmpr*, *FreqCmpr*, and $No\Delta$ are 13.5%, 14.6%, and 21.9%, respectively. Therefore, leveraging these "Δ" patterns in the NoC compression can potentially achieve good performance. Note that these results only show the rough ratios of compressible packets that motivate our work; further investigation of how much traffic can be reduced is required.

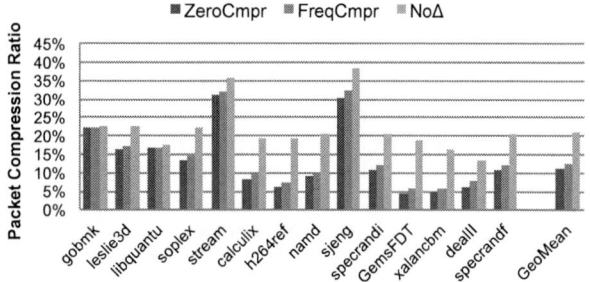

Fig. 3: Ratios of compressible packets for different patterns

As an example, Fig.4 demonstrates how $No\Delta$ works for a data packet that consists of a header flit and four data flits. Here we assume that a flit is only four-byte long and the first data flit is treated as the base. As the figure indicates, the data flits can be represented using a four-byte base value, 0xC0D45800, and an array of three two-byte deltas. As a result, the deltas can be stored comfortably into one flit and hence the whole packet is compressed into three-flit long. We refer to this compression scheme as *flit-based compression* because the base size is exactly equal to one flit.

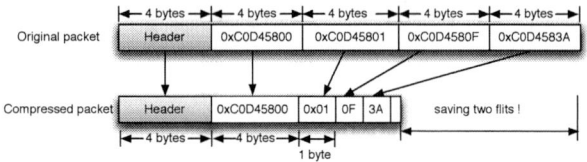

Fig. 4: Packet compression demonstration

Additionally, the above flit-based compression algorithm can be extended in two ways to further improve the compression ratio:

Variable-length base: Instead of fixing the size of the base to one flit, a data packet can be represented by different sets of fixed-size values. For example, a 64-byte data can be represented as 4 16-byte, 8 8-byte, or 16 4-byte values. Then the base value can be of size 16B, 8B, or 4B accordingly. In this case, multiple choices are provided to accommodate various data localities. In contrast, the flit-based compression, as indicated in Fig.4, has two major limitations:

• The base is always fixed as one-flit long, and thus it would be difficult to find another flit whose value has a relative difference that falls within a small range of Δ.

• Since a flit is the minimal flow control unit, a packet can be at most encoded into three flits: the header flit, the base flit, and another flit that stores all Δs. Alternatively, variable-length base compression can aggressively squeeze a data packet into two-flit long, or even single-flit long if there is extra space in the header flit that can be utilized.

We use a four-byte long flit in the example to save space. In real experiments, our baseline flit size is 16 bytes.

Multiple bases: Our observation shows that a single packet may mix different data patterns. Therefore applying delta compression with multiple bases can increase the compression ratio. To improve compression efficiency without sacrificing design simplicity, we use two bases: one is always 0; the other is the *first* arbitrary value from the set of fixed-size data values, just like the example in Fig.4.

Table I shows an encoding table that summarizes the bases and deltas we analyze in this work. Note that all potential compressed sizes are known statically. The smallest compressed packet size will be chosen in case multiple candidates are applicable for the same packet. The encoding bits are used to identify which base-delta pair has been applied for a compressed packet. For a packet that contains all zeros, only one extra bit "0" will be added to the header.

TABLE I: Encoding table for the $No\Delta$. All sizes are in bytes.

Name	Base	Δ	Enc.	Name	Base	Δ	Enc.
Zero	NA	NA	0	B8Δ4	8	4	0101
B16Δ8	16	8	0001	B8Δ2	8	2	0110
B16Δ4	16	4	0010	B8Δ1	8	1	0111
B16Δ2	16	2	0011	B4Δ2	4	2	1000
B16Δ1	16	1	0100	B4Δ1	4	1	1001

III. Architecture Support

To support compression and decompression in a network-on-chip, modules for both functions need to be placed in all of the NIs. When data is injected into the network, the compression module tests if compression is possible. Likewise, the decompression module checks if the ejected data is compressed. Therefore, the data is only compressed when it is injected into the network, and does not need to be modified between hops.

A. Compression Module

A diagram of the compression modules is shown in Fig.5. Here we use the same packet from Fig.4. The size of the base is 4 bytes, and each delta is a 1-byte difference between the base and the subsequent 4-byte segments. The figure contains 3 subtractors to calculate the differences between each of the four 4-byte packet segments and the base. In general, N-1 subtractors are needed for a packet with N segments. The subtractors must have a bit width of the segment size, since we need to check *all* of the most significant bits to check for sign extensions. In the example, the subtractor would have 32-bit inputs and outputs, and the 24 most significant bits are checked for sign extension. Checking for sign extension is done by simply using AND and OR gates on all of the most significant bits. The output of the AND is high if all the bits are ones and the output of the OR is zero if all the bits are zero. When all the bits are zeros or ones, the output of the subtractor is considered to be sign extended. If *any* of the subtractor outputs are sign extended, the packet can be compressed. The outputs of the sign extension are placed in the header flit to be used as enable signals for the subtractors in the decompression unit.

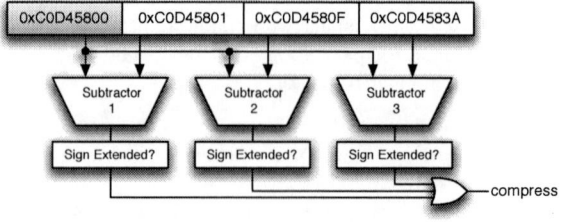

Fig. 5: A single $No\Delta$ compression module

978-1-4799-2817-0/14 $31.00 © 2014 IEEE

B. Decompression Module

Decompression modules use subtractors to regenerate the original uncompressed packet via two's complement addition. An example of a decompression module is given in Fig.6. The base value is passed through directly to the output and each of the subtractors. Each of the following 1-byte segments following the 4-byte base are passed to the subtractors. An enable signal on the subtractor is tied to the sign extension data placed in the header flit by the compression module. When a subtractor is disabled, a value of zero is used in place of the base and the output is the original uncompressed data. The sign extension bits from the header are also used to select the secondary input to the subtractor via a multiplexor. This selects either Δ bits if the segment is compressed, or an entire segment if it is uncompressed. The output of the subtractors is the reconstructed packet data. In this module, the subtractors only need to be the width of the delta. In this example, the subtractor inputs and output are 8-bit signals. The remaining upper bits, 24 bits in the example, can be hard wired to the most significant bits of the base when compression is enabled.

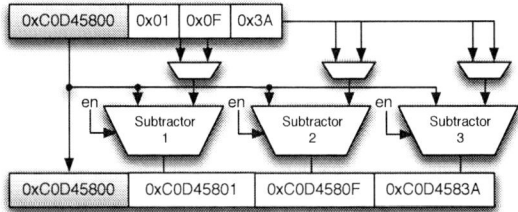

Fig. 6: A single NoΔ decompression module

C. NoΔ Architecture

In order to accommodate compression of different base or Δ sizes, multiple compression modules are needed. Our design includes compression modules for base and delta sizes listed in Table I, organized in parallel as shown in Fig.7. These modules execute simultaneously and output a `compress` signal specifying whether the packet can be compressed with the given base size. It is possible that multiple modules will indicate that the packet is compressible. In this case, we arbitrarily choose a compression with the smallest compressed number of body flits. Furthermore, we can break ties where the flit length is the same by choosing the encoding with the least decompression power, based on results in section IV-D. The compressed sizes and number of *body* flits are shown in Table II. The priority in this table indicates the order in which compression techniques are chosen. Lower priority values have larger compression ratios and are chosen first, if possible.

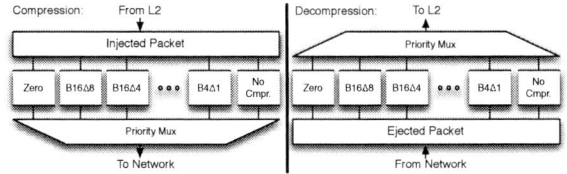

Fig. 7: NoΔ architecture with multiple compression and decompression modules

IV. Experimental results

A. Platform Setup

We use GEM5 [16] full system simulator to setup the basic system platform. Ruby [17] and Garnet [18] are enabled to model a detailed

TABLE II: Compressed packet sizes and priorities for the *NoΔ* encodings

Name	Packet Size (Flits)	Priority	Name	Packet Size (Flits)	Priority
Zero	0 bits (0)	1	B8Δ4	288 bits (3)	9
B16Δ8	320 bits (3)	8	B8Δ2	176 bits (2)	6
B16Δ4	224 bits (2)	5	B8Δ1	120 bits (1)	2
B16Δ2	176 bits (2)	4	B4Δ2	272 bits (3)	10
B16Δ1	152 bits (2)	3	B4Δ1	152 bits (2)	7

memory system and interconnection network. We use DSENT [19] for detailed network power analysis.

We model a 16-core multicore architecture at 45nm technology, as shown in Fig.1, Section II. The detailed system configurations are listed in Table III. Each core has a split private L1 cache (instruction and data cache each 8KB in size). All the cores share a distributed L2 cache that comprises 16 banks. Each bank of the L2 cache has a size of 1MB and is allocated to a tile. From a high-level point of view, these 16 tiles form a 2D-Mesh topology that employs a dimension-order routing algorithm and wormhole flow control. Each network packet contains five flits and each flit is set to be 16-byte long. As for the router microarchitecture, we adopt a classic router design with five-stage pipelines. Each input port of the router contains two virtual channels (VCs). Each VC has a buffer depth of 4 flits.

TABLE III: System configurations

core count	16
clock frequency	2GHz
L1 I & D cache	2-way, 8KB, 1cycle
L2 cache	8-way, 16×1MB, 5 cycles per bank
memory	4GB DRAM
# memory controllers	4 (on the corner)
cache-coherency	MESI protocol
network topology	4×4 mesh
routing	dimension-order
router pipeline	classic five-stage
buffer	2 VCs per port, 4-flit depth per VC
Packet length	5 flits, each flit 16 bytes

B. Performance and Energy Evaluation

We use a diverse set of workloads from SPEC CPU2006 [15] benchmark suite for performance and power analysis. Given each workload, the benchmark program is duplicated to 16 instances and mapped to the corresponding 16 cores. Note that most of the workloads we selected are memory-intensive which relatively generate more network traffic.

Meanwhile, we also implement two other state-of-the-art compression techniques adopted in the NoC compression domain: zero-content compression (*ZeroCmpr*) [10] and frequent value compression (*FreqCmpr*) [11]. Specifically, for a zero-content packet, it can be represented by the header flit plus an extra encoding bit "0"; For frequent-value compression, we explore repeated data patterns at different granularity: 16 bytes, 8 bytes, 4 bytes, and 2 bytes. We compare different compression schemes with the baseline that does not utilize any compression techniques.

Intuitively, NoC compression decreases the number of flits injected to the network. This can be observed in Fig.8a, which shows the normalized total flit count in the network for different applications. It also reflects the compression ratio: the less number of flits traversing the network, the higher the compression ratio. Among the three compression techniques, *NoΔ* has the highest average compression

978-1-4799-2817-0/14 $31.00 © 2014 IEEE   589

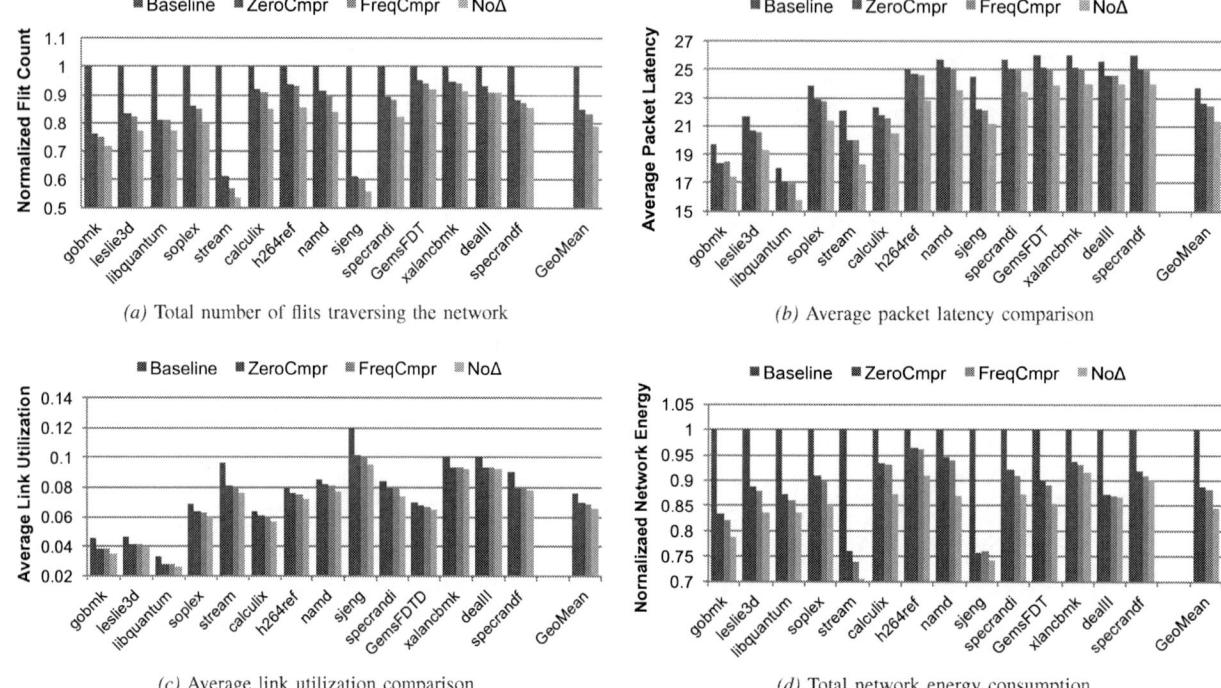

(a) Total number of flits traversing the network

(b) Average packet latency comparison

(c) Average link utilization comparison

(d) Total network energy consumption

Fig. 8: Comparisons of different compression techniques on flit count, packet latency, link utilization and network energy.

ratio of 21.1% for all applications. However, the compression ratio varies for different applications. For instance, *sjeng* and *stream* indicate significant space for compression (more than 40%), while some applications like *GemsFDTD*, *xalancbmk*, and *dealII* display weak compressibility (less than 10%).

Note that the number of packets remains the same after compression despite the decrease of flit count. However, since the basic flow control unit is a flit, fewer flits cause less network congestion. As a result, packets will be delivered smoothly without heavily suffering from head-of-the-line blocking. In order to analyze the impact of different compression techniques on the network performance, we collect average packet latencies after running different applications on different compression architectures. As implied in Fig.8b, *No*Δ outperforms all the other compression methods. It can lower the packet latency by up to 17.5% (*stream*) with an average reduction of 10.1%. This phenomenon can be further explained through Fig. 8c: the average link utilization, which reflects the network load, drops significantly when compression is utilized. *No*Δ again reduces the load to the highest degree, i.e, up to 20.8% for *sjeng* and *stream* and on average 13.1% for all applications. The decrease of network load reduces network congestion. Therefore, individual packets tend to progressively propagate to their destinations with less contention with each other for shared resources.

Furthermore, another primary goal of using data compression in NoC is to conserve energy. Since packet compression will reduce the number of network operations, such as buffer write and read, allocation and arbitration, crossbar traversal, link traversal, etc., the dynamic energy consumption of the communication fabric will be reduced accordingly. On the other hand, the shrink of average packet latency reduces the total system execution time and thus can potentially save the static energy as well.

Fig.8d demonstrates the normalized total network energy con-

sumption using different encoding techniques. Compared to the baseline, *No*Δ for *stream* benchmark achieves the highest energy savings — 29.5%. On average, *ZeroCmpr*, *FreqCmpr*, and *No*Δ cut down the total network energy by 11.1%, 11.9%, and 15.3%, respectively. *No*Δ proves to be the most competitive candidate for NoC compression.

C. Sensitivity Study

Fig.4 demonstrates *No*Δ with ten different compression/decompression modules. However, depending on the application, real design may only utilize a few of them to reduce power/area overhead. To evaluate how the selection of encodings affect *No*Δ, we test SPEC 2006 with only 16-byte bases, 8-byte bases, or 4-byte bases, respectively. Fig.9 shows the compression ratio of flits. The baseline design (Base_all) is to use all the ten different encodings.

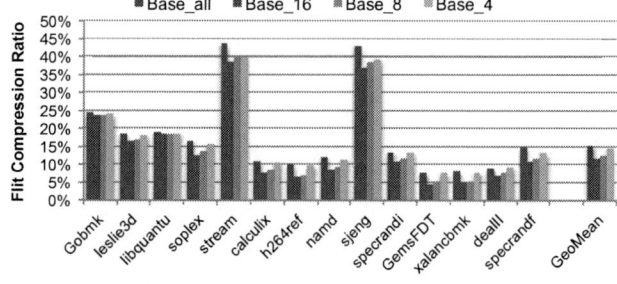

Fig. 9: Flit compression ratios for a subset of encoders/decoders. Base_n means only 16-byte base is utilized.

There are three important observations from Fig.9: (1) Different applications may prefer different encodings. (2) For each application, the best encoding design may yield a very close compression ratio as the baseline. (3) On average, Base_4 generates the highest compression ratio, which only incurs 1.0% compression loss compared to the baseline design. Therefore, we may choose Base_4 encodings only while still obtaining satisfiable compression ratios.

D. Overhead Estimation

We verified our design by implementing the modules in behavioral Verilog, creating a testbench, and simulating using Mentor Graphics Modelsim. We further investigate the overheads in terms of power, area, and critical path by synthesizing our Verilog using the Synopsys Design Compiler. We choose a 45nm technology node using the worst-case low-power library. The results of individual compression and decompression module pairs are shown in Fig.10. The average dynamic power for one pair of compression and decompression units is just over 1mW. Leakage power is negligibly low, adding only an additional 9.6 μW on average. In contrast, the total power for a single router based on the DSENT power model ranges from 55mW—144mW depending on benchmark. Therefore, a single compression and decompression unit pair consumes 1.81%—0.69% of the power of a router.

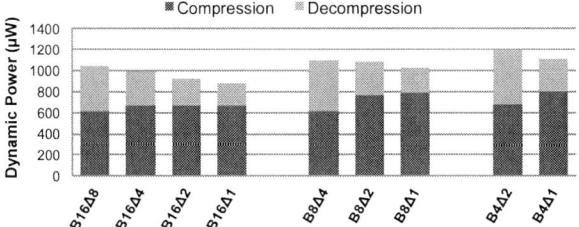

Fig. 10: Dynamic power breakdown for the compression/decompression modules

From the power results we can also observe a trend in the dynamic power for compression and decompression modules. Compression and decompression module power increases as the base decreases. This is due to additional adders needed to calculate the Δ values. The power also increases as the delta shrinks due to larger fan-in AND and OR gates needed for the sign extended check logic. Conversely, the dynamic power for the decompression module decreases as the delta shrinks, since the subtraction units in the decompression module have input sizes of Δ bits. Based on this observation of the decompression power, we can break ties when multiple compressions are possible. Specifically, we prefer smaller Δ values and larger bases. This decision is reflected by the priorities in Table II in Section III-C.

NoΔ has modest hardware overhead and implementation complexity because the compression and decompression algorithms involve only simple vector addition, subtraction, and comparison operations. The average area of a compression and decompression module pair is 0.006 mm^2. Based on the area of the network interface and router, this area is relatively small. Furthermore, we analyze the impact of these modules on the critical path. We successfully synthesized our design with a clock rate of 1 GHz, meaning the design can compress a packet in a single network clock cycle in our simulated system. *This additional latency is accounted for during simulation by adding extra queuing latency to packets.*

Finally, we compare the power/area overhead of *NoΔ* with *ZeroCmpr* and *FreqCmpr* under 45nm technology, as shown in Table IV. For frequent-value compression, we consider one design from Reference [12] with adaptive compression (*FreqCmpr_a*) and another two designs from Reference [11]: One with private tables (*FreqCmpr_p*) and the other with a shared table (*FreqCmpr_s*). For *NoΔ*, we consider Base_4 only as explained in Fig.9. We can see *NoΔ* incurs modest area overhead (only larger than *FreqCmpr_s*) and the least power overhead.

TABLE IV: Power/Area overhead for different compression schemes

	Area(mm^2)			Power(mW)		
	Encoder	Decoder	Total	Encoder	Decoder	Total
ZeroCmpr [10]	NA	NA	0.183	NA	NA	273
FreqCmpr_p [11]	0.00526	0.00523	0.17	NA	NA	1626.1
FreqCmpr_s [11]	NA	NA	0.023	NA	NA	220
FreqCmpr_a [12]	0.00443	0.00432	0.14	4.26	3.95	131.36
NoΔ	0.00474	0.00173	0.10	1.47	0.81	36.48

V. Conclusion

In this paper, we propose *NoΔ* that conducts data encoding/decoding in the network interface prior to NoC packet injection/reception. The encoder/decoder module incurs minimal area/power overhead. We evaluate *NoΔ* using a full system simulator with real benchmarks. Experimental results show an average compression ratio of 21.1%, which results in 10.1% performance improvement and 15.3% energy savings. In addition, a comprehensive comparison verifies that our method outperforms other compression techniques like zero-content compression and frequent-value compression.

References

[1] W. J. Dally and B. Towles, "Route packets, not wires: On-chip interconnection networks," in *DAC*, 2001, pp. 684–689.

[2] J. Kim *et al.*, "A novel dimensionally-decomposed router for on-chip communication in 3D architectures," *ACM SIGARCH Computer Architecture News*, vol. 35, no. 2, pp. 138–149, 2007.

[3] B. Grot, J. Hestness, S. W. Keckler, and O. Mutlu, "Express cube topologies for on-chip interconnects," in *HPCA*, 2009, pp. 163–174.

[4] S. Ma, N. E. Jerger, and Z. Wang, "Whole packet forwarding: Efficient design of fully adaptive routing algorithms for networks-on-chip," in *HPCA*, 2012, pp. 1–12.

[5] J. Dusser, T. Piquet, and A. Seznec, "Zero-content augmented caches," in *ICS*, 2009, pp. 46–55.

[6] Y. Zhang, J. Yang, and R. Gupta, "Frequent value locality and value-centric data cache design," in *ACM SIGOPS Operating Systems Review*, vol. 34, no. 5, 2000, pp. 150–159.

[7] J. Yang, Y. Zhang, and R. Gupta, "Frequent value compression in data caches," in *MICRO*, 2000, pp. 258–265.

[8] A. R. Alameldeen and D. A. Wood, "Adaptive cache compression for high-performance processors," in *ISCA*, 2004, pp. 212–223.

[9] G. Pekhimenko *et al.*, "Base-delta-immediate compression: practical data compression for on-chip caches," in *PACT*, 2012, pp. 377–388.

[10] R. Das *et al.*, "Performance and power optimization through data compression in network-on-chip architectures," in *HPCA*, 2008, pp. 215–225.

[11] Y. Jin, K. H. Yum, and E. J. Kim, "Adaptive data compression for high-performance low-power on-chip networks," in *MICRO*, 2008, pp. 354–363.

[12] P. Zhou *et al.*, "Frequent value compression in packet-based NoC architectures," in *ASPDAC*, 2009, pp. 13–18.

[13] B. M. Beckmann and D. A. Wood, "Managing wire delay in large chip-multiprocessor caches," in *MICRO*, 2004, pp. 319–330.

[14] J. Huh *et al.*, "A NUCA substrate for flexible CMP cache sharing," in *ICS*, 2005, pp. 31–40.

[15] J. L. Henning, "SPEC CPU2006 benchmark descriptions," *ACM SIGARCH Computer Architecture News*, vol. 34, no. 4, pp. 1–17, 2006.

[16] N. Binkert *et al.*, "The gem5 simulator," *ACM SIGARCH Computer Architecture News*, vol. 39, no. 2, pp. 1–7, 2011.

[17] M. M. Martin *et al.*, "Multifacet's general execution-driven multiprocessor simulator (GEMS) toolset," *ACM SIGARCH Computer Architecture News*, vol. 33, no. 4, pp. 92–99, 2005.

[18] N. Agarwal, T. Krishna, L.-S. Peh, and N. K. Jha, "GARNET: A detailed on-chip network model inside a full-system simulator," in *ISPASS*, 2009, pp. 33–42.

[19] C. Sun *et al.*, "DSENT-a tool connecting emerging photonics with electronics for opto-electronic networks-on-chip modeling," in *NOCS*, 2012, pp. 201–210.

DPA: A Data Pattern Aware Error Prevention Technique for NAND Flash Lifetime Extension

Jie Guo[1], Zhijie Chen[1], Danghui Wang[2], Zili Shao[3], Yiran Chen[1]

[1]Department of Electrical and Computer Engineering, University of Pittsburgh, Pittsburgh, PA USA 15261

[2]Northwestern Polytechnical University, Xi'an, China 710072

[3]The Hong Kong Polytechnic University, Kowloon, Hong Kong

{$jig26, zhc31, yic52$}@pitt.edu, $wangdh$@nwpu.edu.cn, $cszlshao$@comp.polyu.edu.hk

Abstract—**The recent research reveals that the bit error rate of a NAND flash cell is highly dependent on the stored data patterns. In this work, we propose Data Pattern Aware (DPA) error protection technique to extend the lifespan of NAND flash based storage systems (NFSS). DPA manipulates the ratio of 1's and 0's in the stored data to minimize occurrence of the data patterns which are susceptible to bit error noise. Consequently, the NAND flash cell bit error rate is reduced, leading to system endurance extension. Our simulation result shows that, with marginal hardware and power overhead, DPA scheme can increase the NFSS lifetime by up to $4\times$, offering a complementing solution to other lifetime enhancement techniques like wear-leveling.**

I. INTRODUCTION

NAND flash has emerged as a promising storage device in the embedded system due to good scalability, low power consumption and excellent random access performance. A NAND flash cell consists of a floating gate transistor. Logic bits are represented by transistor threshold voltage (V_{th}) levels, which can be programmed by injecting the electrons into the floating gate. In MLC NAND flash, a cell stores two logic bits which are denoted by four V_{th} levels. With technology scaling down, the noise margin on each V_{th} level shrinks. Consequently, MLC cells become more susceptible to external disturbs and bit error rate quickly grows.

Program disturb, read disturb and retention time limit have been identified as the major contributors to MLC NAND flash bit error. As P/E cycle count increases, the effect of these three noises is aggravated, eventually resulting in uncorrectable bit error. Hence, BCH Error Correction Code (ECC) is widely employed in NAND flash based storage systems (NFSS) to mitigate this reliability issue. However, the incurred hardware cost quickly becomes prohibitively high as the required correcting capability increases. Some low-cost schemes, e.g., refreshing the MLC cells for retention time extension, have been also proposed [1][2]. However, these schemes can only prevent power-on errors but not be able to handle the error occurring during power-down period. Other signal processing techniques are also adopted in NAND flash designs to improve the system reliability [3][4][5][6].

The recent works reveal that bit error rate is dependent on the programmed V_{th} level [2][7]. If the vulnerable V_{th}

levels can be avoided, bit error rate will decrease. Based on this observation, we propose Data-Pattern-Aware (DPA) error prevention technique to protect the data integrity in NFSS. The major contributions of this work are:

- We propose the Pattern Probability Unbalance (**DPA-PPU**) technique to reduce the probability of the data patterns sensitive to noise. DPA-PPU checks the data inter-correlation and peforms de-correlation or scrambling accordingly to skew the ratio of 1's and 0's. By minimizing the number of cells programmed to the vulnerable V_{th} levels, the bit error rate is substantially reduced.

- We propose the Data-Redundancy Management (**DPA-DRM**) scheme to mitigate DPA-PPU induced performance degradation without impairing system reliability. DPA-DRM adjusts DPA-PPU efficiency based on P/E cycle count to avoid unnecessary data redundancy. It also employs several schemes to minimize redundant bits access. Due to random data pattern, we apply a stronger protection to redundant bits than normal data bits.

II. PRELIMINARY

A. MLC NAND Flash

An MLC NAND flash chip is composed of multiple blocks as shown in Fig. 1(a). Each block is an array of NAND flash cells, or floating gate transistors. Each wordline (WL) represents one NAND flash page, which includes both data area and Out-of-Band (OOB) zone. In each cell, 2-bit data 11, 10, 01, 00, can be represented by four V_{th} levels of the floating gate transistor, i.e., L0 (the lowest level), L1, L2, L3 (the highest level), respectively. An erase operation needs to be conducted before a cell is re-programmed.

However, V_{th} is susceptible to extrinsic noises, i.e., program disturb, read disturb and retention time limit [8]. The noise effect is aggravated when Program/Erase (P/E) cycle counts increase.

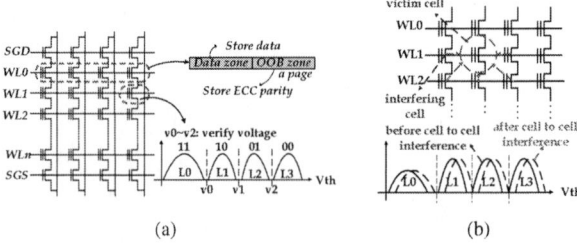

Fig. 1. (a) Device structure and V_{th} distribution of MLC NAND flash. (b) Cell-to-cell interference.

---

[1]This research was sponsored by National Science Foundation under contract no. CNS1116171, National Natural Science Foundation of China under contract no. 61272122 and Basic Research Project of Northwestern Polytechnical University of China under contract no. JC201256.

978-1-4799-2817-0/14 $31.00 © 2014 IEEE

B. V_{th} Distortion Noise

Programming one cell can induce V_{th} shift of its neighboring cells via parasitic capacitance-coupling (see Fig. 1(b)). This is called cell-to-cell interference, one major factor of *program disturb*. The V_{th} shift of the victim cell ΔV_{c2c} can be modeled by [2]:

$$\Delta V_{c2c} = \sum_{k=0} \Delta V_p^{(k)} \times \gamma^{(k)}. \tag{1}$$

Here, $\Delta V_p^{(k)}$ and $\gamma^{(k)}$ denote V_{th} shift of the interfering cell after programming and the coupling ratio, respectively. In [9], Moon *et al.* show that the effect of cell-to-cell interference is pattern dependent: the largest V_{th} shift of the victim cell occurs when the interfering cell is programmed to L1 or L3.

Read operation can also lead to V_{th} shift due to Fowler-Nordheim tunneling and stress induced leakage current (SILC). This is called *read disturb*. As pointed out in [7], the lowest V_{th} level L0 demonstrates the highest susceptibility to read disturb. *Retention time limit* comes from electron detrapping and SILC, which cause charge loss and therefore V_{th} reduction [2]. The retention time limit induced V_{th} shift distribution can be modeled as Gaussian distribution $N(\mu_d, \sigma_d^2)$. Here we have [2]:

$$\begin{cases} \mu_d = K_s(x - x_0)K_d N_{PE}^{0.3}ln(1 + t/t_0), & (2) \\ \sigma_d^2 = K_s(x - x_0)K_m N_{PE}^{0.4}ln(1 + t/t_0). & (3) \end{cases}$$

K_s, K_d, K_m and t_0 are fitting constants. N_{PE} is P/E cycle count; x_0 and x are V_{th} at L0 and the initial V_{th} after programming, respectively; t is the data storage time. Eq. (2) and (3) show that L3 has the highest susceptibility to retention time errors among all V_{th} levels.

III. MOTIVATIONS

BCH ECC code is prevalently employed in NFSS to prevent bit error. The lifetime of a NAND flash block is defined as the P/E cycle count at which the ECC failure rate reaches a threshold T_{uber}. Assume n-bit BCH ECC (m,l,n) is performed on a l-bit data block with a total codeword length of m, the block yield $Y_{ber}(n)$ can be calculated by

$$Y_{ber}(n) = \sum_{k=0}^{n} C_m^k p_c^k (1 - p_c)^{(m-k)}. \tag{4}$$

Here p_c is the error rate of a single NAND flash cell. Assume p_i and p_{li} ($i = 0, 1, 2, 3$) denote the probability of programming a flash cell to the threshold voltage Li and the corresponding error rate, respectively. p_c can be rewritten as

$$p_c = \sum_{i=0}^{3} p_{li} \times p_i. \tag{5}$$

TABLE I
PARAMETERS IN CELL RELIABILITY SIMULATIONS

K_r	11.9 [10], [7]	γ_x	0.08 [5]
K_d	4×10^{-5} [12]	γ_y	0.06 [5]
γ	1.1	γ_{xy}	0.048 [5]
t_s	$8\mu s$	K_m	3×10^{-6} [12]
α	5.3×10^{-2} [10], [7]	β	8.8×10^5

p_i is related to the ratio of 1's and 0's in the stored data. If we use $p(0)$ and $p(1)$ to denote the probability of storing 1's and 0's, respectively, then $p_0 \sim p_3$ can be calculated by $p(1)p(1)$, $p(0)p(1)$, $p(1)p(0)$ and $p(0)p(0)$.

p_{li} induced by program disturb and retention time limit at different V_{th}'s can be derived from Eq. (1),(2),(3). By assuming the V_{th} fluctuation caused by read disturb follows Gaussian distribution, we derive the mean μ_{rd} and the standard deviation σ_{rd} of the V_{th} fluctuation as [7][10]:

$$\begin{cases} \mu_{rd} = \frac{1}{\gamma}In[1 + \gamma\beta t_s e^{\gamma(V_{CG} - V_{T,0})} N_{PE}^{0.3} K_r] & (6) \\ \sigma_{rd} = \sqrt{\alpha(1 - e^{\gamma\mu_{rd}})}. & (7) \end{cases}$$

Here γ, β, K_r and α are the fitted coefficients. V_{CG} and $V_{T,0}$ represent the voltages applied to the wordline and the floating gate transistor, respectively. t_s is the read pulse width.

For the illustration purpose, we simulated the ECC failure rates of a 512B data block with 8-bit BCH ECC under different P/E cycle counts by assuming $p(0) = p(1) = 0.5$ [11]. T_{uber} is set to 10^{-13} [8]. We also assume MLC Flash cell adopts even/odd bitline structure [5] and the coupling ratios between the cells at three directions are defined as γ_x, γ_y and γ_{xy}. The ECC failure rate resulting from read disturb and retention time limitation are obtained by assuming a read count of 5K and 1 year data storage time, respectively. The simulation parameters and setup are depicted in TABLE I. As shown in Fig. 2, error rate resulting from these three noises are below T_{uber} only below 7.5K P/E cycle count.

IV. DATA PATTERN AWARE ERROR PREVENTION

A. Overview of DPA

The efficacy of our DPA technique in reducing NAND flash cell error rate is based on the following two observations:

- Program errors are prone to occur when the interfering cell is programmed to L1 (bits 10) or L3 (bits 00).
- Retention time errors are prone to occur when the cell is programmed to L3 (bits 00).

DPA intuitively aims to increase the probability of programming NAND flash cells to L0 by maximizing the ratio of 1's in the stored data. DPA system architecture is depicted in Fig. 3, including two modules DPA-PPU and DPA-DRM. A data buffer with simple LRU replacement policy is deployed in front of an array of flash chips. Data from the host is first stored in the data buffer; when the data is evicted from the buffer, its data pattern is converted to a favorable one

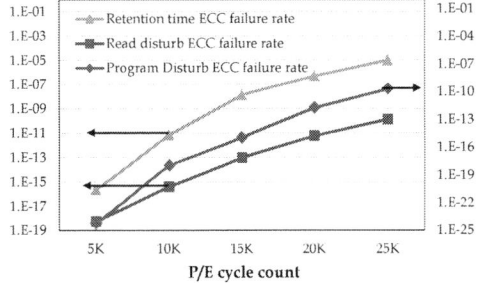

Fig. 2. The ECC failure rates under different P/E cycle counts.

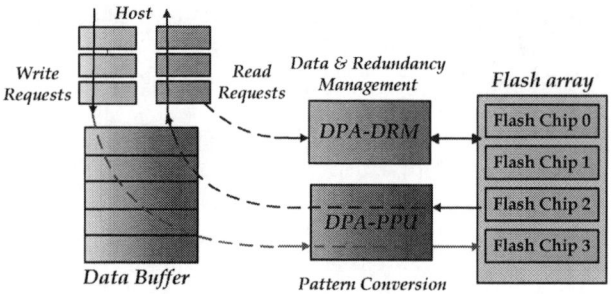

Fig. 3. The system architecture of DPA.

by DPA-PPU before being flushed to the flash chips. DPA-DRM performs data management to reduce the performance overhead induced by DPA-PPU.

B. DPA-PPU: Pattern Probability Unbalance

DPA-PPU performs data pattern conversion to increase the ratio of 1's in the stored data by performing de-correlation or scrambling, followed by bit inversion operation. If the data demonstrates a strong correlation, the de-correlation scheme in [11] can be adopted to decrease the number of 1's in the data by applying XOR operations between neighboring bytes. If the data is weakly correlated, we propose a scheme based on scrambling coding to skew the ratio of 1's and 0's. After the de-correlation or scrambling, all data bits are flipped to obtain the codeword of which the majority are 1's.

The architecture of DPA-PPU is presented in Fig. 4. A data page is divided into multiple data chunks. The data chunks are processed by de-correlation and scrambling circuits in parallel. Scrambling circuit performs modulo 2 division, i.e., cyclic XOR, on the data by using a polynomial as divider. If the pattern of the polynomial overlaps with the data much, the number of 1's can be effectively reduced. DPA-PPU implements n different scrambling circuits which are differentiated by the polynomial tags from 0 to $n-1$. The data is scrambled by different polynomials and the result that has the fewest 1's will be selected as the output of scrambling circuit. Then the number of 1's in the output data from de-correlation and scrambling circuits are checked by a comparator. The one with the fewer 1's is selected and inverted before being flushed into the flash chips. If the de-correlated data is selected then the correlation bit is set to 1. Otherwise, the correlation bit is set to 0. The corresponding polynomial tag (if applicable) and the correlation bit must be also stored in the flash chip as they are required when re-correlation or descrambling is performed for data recovery at read operations.

The scrambling circuit efficiency is determined by the polynomial pattern, the data chunk size, the number of polynomials and the polynomial order. For example, the number of 1's can be effectively reduced by maximizing the overlapping bits between the data and the polynomial. In normal applications, the occurrence probabilities of 1's and 0's in the stored data are approximately equal [11]. Hence, the polynomials with a 1-to-0 ratio of 1:1 are generally preferred. The increases of data chunk size diversifies the data pattern and thus potentially decreases the overlapping area between the data chunk and the polynomial. Hence, the scrambling efficiency degrades

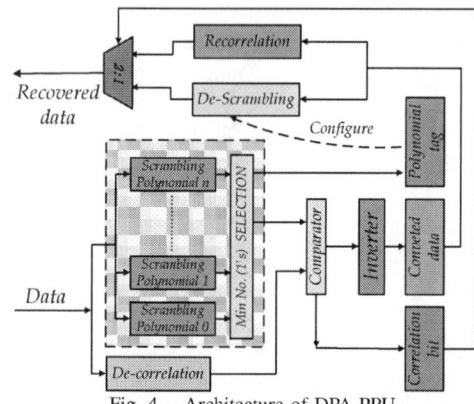

Fig. 4. Architecture of DPA-PPU.

when the data chunk size rises. Similarly, increasing the number of polynomials may enhance the scrambling efficiency by maximizing pattern overlapping probability. Polynomial order itself does not directly affect the scrambling efficiency. However, the polynomial with a higher order can offer more available polynomials which may improve the pattern overlapping probability.

The scrambling-descrambling circuits can be implemented with simple linear feedback shift registers (LFSRs) with very marginal hardware cost. At 65nm technology node, the power of the scrambling-descrambling circuit is around 9mW, which is negligible compared to the large power consumption of the read and write operations of NFSS (which is at $\geq$ 5W for a 256GB system). Since only one correlation bit is required by one data page, the incurred hardware cost is negligible. Similarly, the hardware overhead of polynomial tags is also marginal: assume the data chunk size is 8B and total 64 16-order polynomials are included in the scrambling-descrambling circuits, the extra space required to store the polynomial tags in a 256GB NFSS is only 16GB (6%) because one third of the tags can be stored in OOB zone.

C. DPA-DRM: Data-Redundancy Management

DPA-PPU needs redundant bits to store polynomial tags. Compared to the scrambled data, the redundant bits have a more random pattern. Hence, the redundant bits are more vulnerable to bit errors than the scrambled data. Also, programming and reading the extra pages that stores the redundant bits incur performance degradation. In DPA scheme, we deal with the scrambled data and the redundant bits in different ways: for a data page consisting of multiple data chunks with different polynomial tags, only a part of the polynomial tags is stored in the OOB zone of the same data page while the rest are stored in a separate page (we refer to it as redundant page). Redundant pages and data pages are stored in different blocks. A mapping table is utilized to track the corresponding redundant page of the data page, as shown in Fig. 5.

In DPA scheme, the data page and the redundant bits stored in the same page are protected by applying the n-bit BCH ECC to every 526-byte data block. For a redundant page, we divide it into multiple data blocks, which are smaller than 512 bytes. The reliability of the redundant page can be enhanced by

978-1-4799-2817-0/14 $31.00 © 2014 IEEE

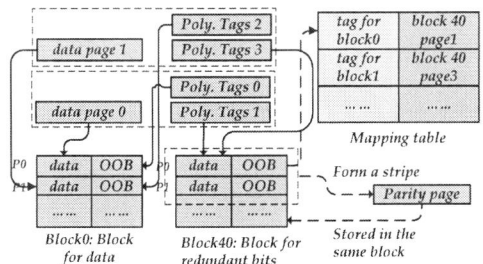

Fig. 5. Redundant pages and data pages.

applying stronger BCH ECC to the divided data blocks. Such a scheme leads to more ECC parity bits and we can store the extra parity bits in the rest data zone of the redundant pages. RAID-5 is also applied to the redundant pages to minimize ECC failure rate: unlike conventional RAID-5 [13] which forms the stripe based on logic page number, we form the stripe based on physical page number. The parity page is stored in the same block to eliminate the extra mapping information. In the example shown in Fig. 5, the data page 0 have two polynomial tags: tag 0 and tag 1. Since the OOB zone can hold only one tag and ECC parity, only tag 0 is stored in the OOB zone and tag 1 is stored in the redundant page. The data page and the redundant page are included in different flash blocks, i.e., block 0 and block 40, respectively. Two redundant pages, P0 and P1 in block 40, form a stripe of RAID-5. A parity page is calculated and also stored in block 40. Physical RAID-5 eliminates the necessities of parity mapping table for conventional RAID-5 scheme.

The performance degradation induced by DPA-PPU can be minimized by adjusting the data chuck size based on the P/E cycle count: at early post-cycling stage, the NAND flash cell error rate is relatively low; a large data chuck size is adopted to reduce the number of scrambling operations by sacrificing the scrambling efficiency. The induced access overhead of the redundant bits is low. As P/E cycle count increases, the error rate grows; a small data chunk size can be adopted to improve the scrambling efficiency. However, the volume of the generated redundant bits also increases, leading to the rising of the access overhead.

To further reduce the performance degradation, we adopt a delayed write scheme: when a redundant page is ready, it is retained in the data buffer first and then flushed into flash chips only during system idle time. In the read operations, the redundant page is read out together with the data page to configure the de-scrambling circuit for data recovery. Therefore, redundant pages and data pages are stored in different chips so that they can be accessed in parallel. The mapping table,

TABLE II
WORKLOAD CHARACTERISTICS.

Disk trace	Write ratio	Seq.wr.	Application
WIN 7	42%	15.2%	p2p, office and web serfing
RHEL	93%	2.3%	Server access
TPC-C	99%	0.9%	OLTP application
financial [14]	98%	1.9%	OLTP application
web search [14]	0.02%	0	Access to search engines

TABLE III
THE PARAMETERS OF MLC NAND FLASH

Capacity	Block Size	Block Number	Page Size
	512KB	4096	4KB+218Bytes
Timing	Program Latency	Read Latency	Erase Latency
	900 μs	50 μs	3.5ms

the size of which is usually small (e.g., 15.2MB for a 256G NFSS), can reside in data buffer for fast access.

We note that DPA-PPU places more cells at L0 and exposes the cells to a high risk of read disturb. However, increasing the number of cells at L0 also mitigates the cell-to-cell interference and enlarges the noise margin to read disturb. The impact of DPA on read disturb will be analyzed in Section V.

V. EVALUATION

We evaluate the efficacy of DPA on NAND flash lifetime enhancement through running experiments on Flashsim simulator [15]. The simulator is modified by adding multichip access capability and incorporating our DPA technique. The NFSS capacity is set to 256GB. An ideal wear-leveling technique is adopted to guarantee that all flash blocks wear out uniformly. The efficiency and performance overhead of DPA are evaluated under 5 representative applications listed in Table II. The specification of the MLC NAND flash chips used in our experiments is shown in Table III.

A. Evaluation of DPA-PPU Efficiency

We first study the efficiency of the de-correlation and scrambling circuits on reducing the ratio of 1's in the data (or reducing the ratio of 0's in the data after the bit inversion in Fig. 4) for the seven data types listed in TABLE IV, Fig. 6 shows the ratio of 1's in the data before and after processed by de-correlation circuit. Among all data types, only *system metadata* exhibits good correlation where the ratio of 1's decreases from 0.45 to 0.27. Other data types, however, display a more random pattern and the reduction of the ratio of 1's is very small after the de-correlation.

In the evaluations of different scrambling schemes, the default values of data chunk size, polynomial number and polynomial order are set to 8B, 64 and 16, respectively. We only change the value of one parameter each time to evaluate its impact on scrambling efficiency. As shown in Fig. 7(a), the average ratio of 1's decreases from 0.42 to 0.34 in the scrambling scheme as the polynomial order rises from 4 to 16. Such improvement is due to increase of the available polynomials. Fig. 7(b) shows that the average ratio of 1's

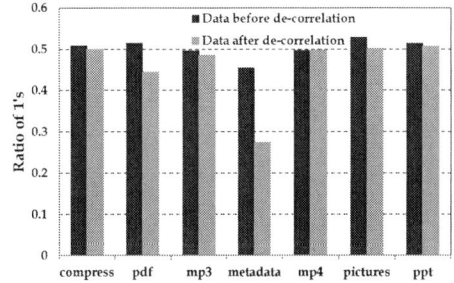

Fig. 6. The ratio of 1's before and after de-correlation.

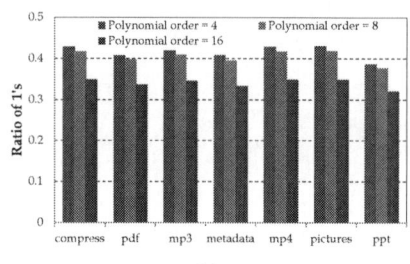 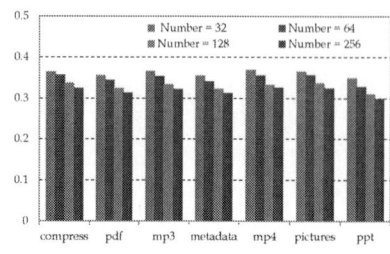 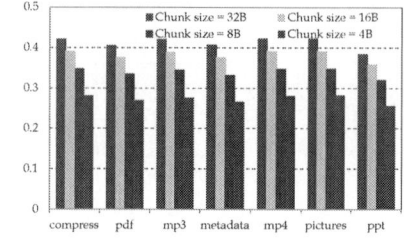

(a)   (b)   (c)

Fig. 7. (a) The efficiency of scrambling under 4,8 and 16-order polynomials. (b) The efficiency of scrambling when the polynomial number changes from 32 to 256. (c) The efficiency of scrambling when the data chunk size changes from 4B to 32B.

TABLE IV
FILE DATA CHARACTERISTICS

File type	File Number	File size
mp3	2	7.5MB, 6.3MB
mp4	1	101MB
compressed file (tar.gz)	2	5.3MB,4.2MB
pictures (.jpg)	2	1.42MB, 1.37MB
pdf	2	8.6MB,4.2MB
office (.ppt)	2	797KB, 1.2MB
system matadata	5	total 2MB

decreases by 11.5% as the polynomial number changes from 32 to 256. Fig. 7(c) shows that the ratio of 1's increases from 0.27 to 0.42 on average by increasing the chunk size from 4B to 32B due to the reduced overlap pattern between the data chunk and the polynomial.

Finally, we compare the ratios of 0's in the outputs of the DPA-PPU's with a data chunk size of 4B and 8B, respectively. Note that the data from the DPA-PPU is the inversion of one output from the de-correlation or scrambling circuits. All other parameters are set to the default values. As shown in Fig. 8, the efficiency of DPA-PPU is generally better when the data chunk size is small. The only exception is *system metadata* under 8B data chunk size where the data is mainly processed by de-correlation circuit.

B. DPA Error Failure Rate

Fig. 9 shows the simulated V_{th} programming probability of NAND flash cells under WIN 7 workload before and after applying DPA-PPU. The results of different data chunk sizes, i.e., 4B and 8B, are also simulated. The probability of L0 increases from 24.5% to 48% when the DPA-PPU with a data chuck size of 8B is applied. When reducing the data chunk sizes down to 4B, the probability of L0 further raises to 59%.

We also simulated the ECC failure rates of NAND flash generating from program disturb, read disturb and retention time limit, respectively, by using the model in Section III.

DPA applies 8-bit BCH ECC (4312,4208,8) to every 526-byte data block. Fig. 10(a) shows that the ECC failure rate induced by program disturb keeps below 2×10^{-18} when the P/E cycle count increases from 5K to 40K as the cell-to-cell interference is significantly minimized by DPA-PPU. Similarly, Fig. 10(b) shows that the ECC failure rate induced by retention time limit is also substantially suppressed by DPA-PPU compared to the system without DPA-PPU. Fig. 10(c) depicts the ECC failure rate induced by read disturb at the read count of 5K. ECC failure rate reduction is still achieved by DPA-PPU even though more cells are placed on L0. It is because the minimization of cell-to-cell interference improves the cell noise margin and more reads can be tolerated. We note that the ECC failure incurred by retention time limit dominates all errors and primarily determines the NAND flash lifetime. When a 8B data chunk size is applied, ECC failure rate reaches the reliability threshold $T_{uber} = 10^{-13}$ when P/E cycle count = 23K; when a 4B data chunk size is applied, the ECC failure rate reaches the T_{uber} when P/E cycle count = 30K. Compared to the system without DPA-PPU where the maximum P/E cycle count = 7.5K (see Fig. 2), the NAND flash lifetime is improved by $3\times$ and $4\times$, respectively. In all scenarios, reducing the data chuck size always improves the NAND flash reliability.

Fig. 11 demonstrates the tradeoff between the maximum tolerable read count and P/E cycle count when the T_{uber} is fixed, say, 10^{-13}. The maximum read count of the system without DPA-PPU quickly drops down to zero when the P/E cycle count raises above 18K. By applying DPA-PPU scheme, however, the working range of flash cells is dramatically expanded.

C. Overheads of DPA

We also compared the performance of DPA scheme under the data chuck size of 4B and 8B. The data chunk size is switched from 8B to 4B when the P/E cycle count reaches

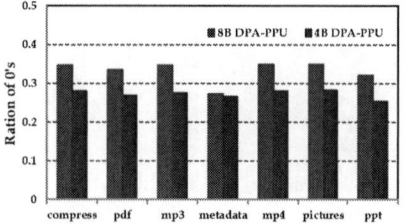

Fig. 8. Ratio of 0's under DPA-PPU.

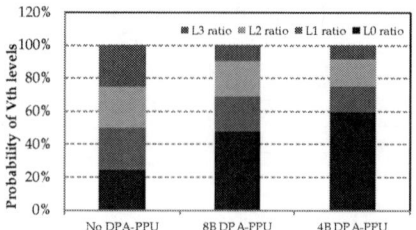

Fig. 9. V_{th} distribution after DPA-PPU.

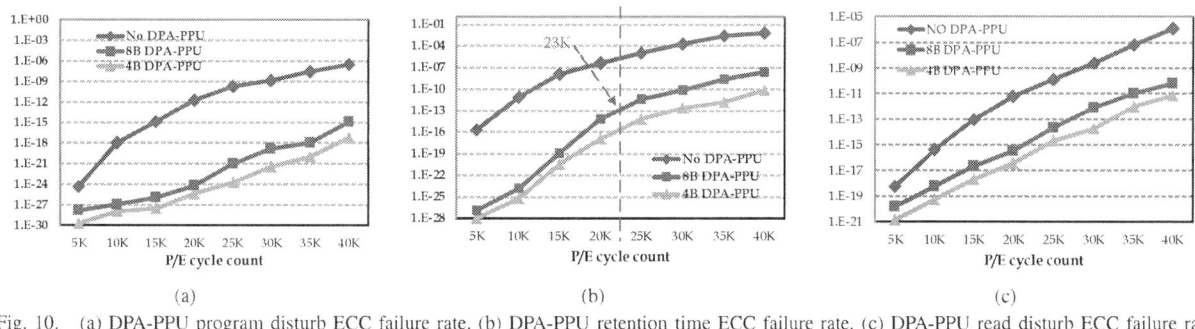

(a)   (b)   (c)

Fig. 10.   (a) DPA-PPU program disturb ECC failure rate. (b) DPA-PPU retention time ECC failure rate. (c) DPA-PPU read disturb ECC failure rate.

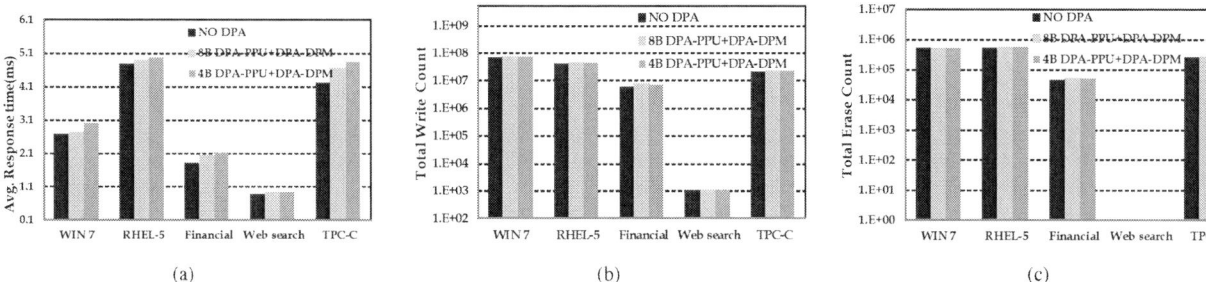

(a)   (b)   (c)

Fig. 12.   (a) The average response time of DPA-DRM. (b) Write counts of DPA-DRM. (c) Erase counts of DPA-DRM.

23K (the reliability limit for 8B data chunk size). When the data chunk size is 8B, one third of the redundant bits can be stored in OOB zone. When the data chunk size changes to 4B, the redundant bits increase and only 1/7 of them can be stored in OOB zone, inducing more page accesses overhead. 8-bit ECC is applied to the data blocks of 170B in the redundant pages and 4 redundant pages form one RAID-5 stripe.

Fig. 12(a) shows that after 23K P/E cycle, compared with no DPA system, the response time of DPA+DRM with 8B and 4B data chunk size degrades by $\sim 9\%$ and $\sim 13\%$ on average, respectively. The maximum performance degradation (12.4% and 15.5%) occurs at *TPC-C*. Fig. 12(b) and Fig. 12(c) show the write counts and erase counts of two DPA schemes, respectively. Averagely, under 4B data chuck, DPA+DRM increases the write count and erase count by 8% and 4%, respectively. Interestingly, DPA+DRM shows 4.1% performance degradation in *websearch* which primarily includes only read accesses.

VI. CONCLUSION

In this work, we proposed a data pattern aware error prevention technique named DPA to extend the lifespan of NAND flash storage systems. We observe that the V_{th} level L0 is resilient to retention time error and cell-to-cell interference. Based on this observation, we propose Pattern Probability Unbalance (DPA-PPU) scheme to skew the ratio of 1's and

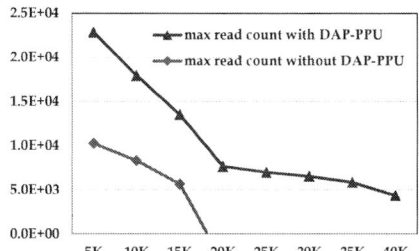

Fig. 11.   Tradeoff between read count and P/E cycle count.

0's in the stored data so as to place more cells on L0. We also employ Data Redundancy Management (DPA-DRM) scheme to mitigate the performance overhead induced by DPA-PPU. Experimental results show that DPA can prolong NAND flash lifetime by $3\times$ or $4\times$ with $\sim 13\%$ performance overhead, respectively. Only very marginal hardware and power overhead are introduced.

REFERENCES

[1] Y. Cai and et al, "Flash correct-and-refresh: Retention-aware error management for increased flash memory lifetime," in *ICCD*, 2012, pp. 94 – 101.

[2] Y. Pan and et al, "Quasi-nonvolatile ssd: Trading flash memory non-volatility to improve storage system performance for enterprise applications," in *HPCA*, 2012, pp. 1–10.

[3] S. Li and et al, "Improving multi-level nand flash memory storage reliability using concatenated bch-tcm coding," *TVLSI*, vol. 18, no. 10, pp. 1412–1420, 2010.

[4] Y. Maeda and et al, "Error control coding for multilevel cell flash memories using nonbinary low-density parity-check codes," in *DFT*, 2009, pp. 367–375.

[5] G. Dong and et al., "Using data postcompensation and predistortion to tolerate cell-to-cell interference in mlc nand flash memory," *TCAS*, vol. 57, no. 10, pp. 2718–2728, 2010.

[6] B. Shin and et al., "Error control coding and signal processing for flash memories," in *ISCAS*, 2012, pp. 409–412.

[7] L. Cola and et al, "Read disturb on flash memories: Study on temperature annealing effect," *Microelectronics Reliability*, 2012.

[8] N. Mielke and et al, "Bit error rate in nand flash memories," in *IRPS*, 2008, pp. 9–19.

[9] J. Moon and et al, "Noise and interference characterization for mlc flash memories," in *ICNC*, 2012, pp. 588–592.

[10] M. Compagnoni and et al, "Analytical model for the electron-injection statistics during programming of nanoscale nand flash memories," *T-ED*, vol. 55, no. 11, pp. 3192–3199, 2008.

[11] M. R. Stan and et al, "Low-power encodings for global communication in cmos vlsi," *TVLSI*, vol. 5, no. 4, pp. 444–455, 1997.

[12] H. Sun and et al, "Quantifying reliability of solid-state storage from multiple aspects," 2011.

[13] Y. Lee and et al, "Fra: a flash-aware redundancy array of flash storage devices," in *CODES+ISSS*, 2009, pp. 163–172.

[14] "Oltp application i/o and search engine i/o," http://traces.cs.umass.edu/index.php/storage/storage.

[15] "A simulator for various ftl scheme," http://csl.cse.psu.edu/?q=node/322.

Scattered Refresh: An Alternative Refresh Mechanism to Reduce Refresh Cycle Time

T. Venkata Kalyan, Kasha Ravi and Madhu Mutyam

Programming languages, Architecture and Compiler Education (PACE) Laboratory,
Department of Computer Science and Engineering,
Indian Institute of Technology - Madras,
Chennai - 600 036, India.
Email - {kalyantv, kasha, madhu}@cse.iitm.ac.in

Abstract— **With realization of high density DRAM devices, the amount of time spent in refreshing a DRAM bank is increasing. This reduces the availability of the bank to the requests from the processing cores, leading to degradation in performance. In this work we target to reduce the** *refresh cycle time* **of the DRAM device by** *scattering* **the rows in a refresh operation to different subarrays and leveraging the available parallelism in their access. Considering** $8Gb$ **devices, we show that** *Scattered Refresh* **achieves up to** 10.2% **of overall system performance improvement.** *Scattered Refresh*, **being orthogonal to the existing refresh handling techniques, can be employed along with any of them, boosting their effectiveness further.**

I. INTRODUCTION

Dynamic Random Access Memory (DRAM) technology forms the core of main memory systems, finding wide range of applicability from smartphones to high-end servers. The one-transistor-one-capacitor circuit, able to store a single bit of data, is orders of magnitude faster than the secondary memory, and being highly dense provides high capacities. The demand for large main memory is propelled by the integration of multiple processing cores on a single chip. To meet the increased demand, commodity DRAMs with the help of advances in fabrication technology, are able to realize high-density DRAM devices operating at high speeds. For example, the recently announced JEDEC DDR4 standard [1] hints that device densities and frequencies can be as high as $32Gb$ and $3200MT/s$, respectively.

The charge in the capacitor of the DRAM cell is representative of the data it stores. The cell loses the charge because of leakage and so, the charge level needs to be restored periodically if the cell is not used. JEDEC standard specifies that a DRAM cell can retain data for $64ms$ under normal temperature ($< 85^0$C) and the retention time reduces to $32ms$ for high temperatures ($> 85^0$C). This data retention time is normally denoted as refresh period, and the process of charge restoration in DRAM cells is termed as a refresh operation. Refresh operations for every refresh period help the DRAM based main memory systems to maintain data integrity of the cells and are clearly unavoidable. The down side of a refresh operation is that during an on-going refresh, the DRAM device is unavailable to the memory controller for normal operation (for issuing read and write requests from the processing cores). This increases the average memory latency considerably, affecting the overall system performance.

During the initial stages of the DRAM development, when the DRAM devices were small (with few Mb capacity), entire DRAM device was refreshed with a single refresh command (REF) issued by the memory controller. This scheme was termed as burst refresh. With increase in device size distributed refresh scheme was implemented, wherein the memory controller issues multiple refresh commands, each command refreshing only a small part of the device. Within a refresh period, these multiple refresh commands are spaced evenly, and a refresh command is issued after every refresh interval (t_{REFI}). During a refresh operation either the memory controller (*RAS-only* type) or the refresh controller (*CAS-before-RAS* type) can supply the address of the row to be refreshed [4]. CAS before RAS (CBR) is preferred to RAS-only refresh because of its ease of implementation. In this work, we consider CBR refresh with distributed refresh scheme.

JEDEC standard specifies that within a single refresh period 8K REF commands have to be issued with t_{REFI} equals to $7.8\mu s$ ($3.9\mu s$ for high temperatures). For every REF command issued, a subset of rows (called refresh bundle [11]) of the bank will be refreshed. The size of the refresh bundle is the size of each group when all the rows in a bank are divided into 8K groups. The time spent in refreshing a refresh bundle is termed as refresh cycle time, given as t_{RFC} in the JEDEC specification.

In high-density DRAM devices the refresh bundle size is increasing resulting in dramatic increase in t_{RFC}. Table I provides the value of t_{RFC} for different DRAM device densities. Addressing refresh cycle time, DDR4 JEDEC standard [1] introduced fine granularity 2x and 4x refresh modes, wherein, the refresh interval is reduced to half and one-fourth of the normal (1x) mode, respectively. This implies that more number of refresh commands are sent (16K and 32K, respectively), each consuming t_{RFC_2x} and t_{RFC_4x} time for a single refresh, and refreshing half and one-fourth of the normal refresh bundle, respectively.

If we consider $3.2GHz$ core frequency, for an $8Gb$ device, in 1x mode at high temperatures, almost 9% of the time is spent in refreshing (complete experimental setup is described in Section IV), during which the device is unavailable. For a variety of workloads, it is observed that for a system that needs periodic refresh of main memory, the overall performance degrades by 18.8% when compared to that of a system with main memory that does not need refresh (ideal but impractical). This is mainly because of increase in the average memory latency for read requests due to refresh. For an ideal memory system, the average read request latency is 183 processor cycles. The latency increases to 218 cycles when refresh operations are considered.

Motivated by the observation that a DRAM bank is made up of small independent subarrays sharing some global circuitry, we propose Scattered Refresh, where we apply a new mapping policy for the refresh controller of the DRAM device. With this new mapping, different rows in a refresh bundle are *scattered* so as to target different subarrays, opening an opportunity to overlap part of refresh operation of each row with another and leading to reduction in the refresh cycle time. Implementing the proposed technique on a detailed memory simulator we observe 10.2% improvement in the overall system performance.

978-1-4799-2817-0/14 $31.00 © 2014 IEEE

TABLE I

REFRESH CYCLE TIMES (IN ns) FOR DIFFERENT DDR4 DRAM DEVICES.
VALUES FOR LARGE CHIPS ARE EXTRAPOLATED.

Device density	t_{RFC}	t_{RFC_2x}	t_{RFC_4x}
4 Gb	260	160	110
8 Gb	350	260	160
16 Gb	480	350	260
32 Gb	640	480	350

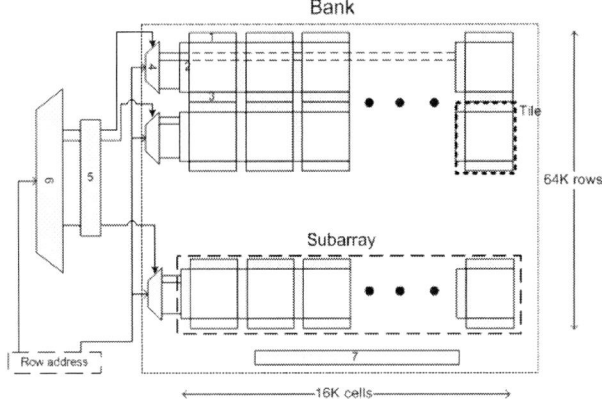

Fig. 1. Physical organization of a DRAM bank. 1, 2 and 3 represent local precharge, local wordline and local sense amplifier circuitry, respectively. 4 and 6 represent local and global row-decoders, respectively. 5 represents global row address latch and finally, 7 represents global row-buffer.

Because of its impact on overall system performance, optimizations targeting refresh operations have been proposed [6, 5, 10, 11, 12]. Among these, Elastic Refresh [6] tries to schedule the refresh operations by using the feature provided by JEDEC standard that eight refresh commands can be postponed or preponed. RAIDR [5] and Smart Refresh [10] propose to reduce the number of refresh commands issued by exploiting variable retention time of DRAM cells and avoiding refreshes to recently accessed rows, respectively. Refresh Pausing [11] identifies specific points to pause an on-going refresh operation and issues any pending read request. Techniques like [12] try to reduce the energy consumption of refresh operations by refreshing rows with critical data only and that of non-critical data infrequently. One aspect of Scattered Refresh is that it is orthogonal to the existing techniques targeting to reduce the impact of refresh on the overall system performance and/or power. While those techniques try to reduce the effect of refresh on performance by refresh scheduling [6] or pausing [11] or eliminating unnecessary refreshes [5, 10, 12], Scattered Refresh targets to reduce the refresh cycle time. Scattered Refresh can be applied in combination of any of these techniques. We show that Scattered Refresh boosts the performance of the best refresh handling technique, Refresh Pausing [11], further by almost 2.3%.

II. BACKGROUND

In Subsection A we first provide the logical organization of a DRAM based main memory system, and then explain the important DRAM commands and their basic functionality by considering the logical organization of a bank. Physical organization of a DRAM bank and the working of each of the DRAM commands will be explained in Subsection B.

A. Logical Organization of Main Memory and Bank

DRAM based main memory systems are arranged as channels, ranks, banks, rows and columns. Column is the smallest granularity of access for the memory. A row comprises of multiple columns. Bank is a collection of these rows. At any given time, only one row of a bank can be accessed. Two or more banks can be accessed in parallel. All the banks in a rank share a single address and data bus. Channel is a set of ranks that share a common address and data bus with the processor. Each channel is controlled by a memory controller and provides the highest level of parallelism. At the abstract level, a bank can be looked at as the smallest structure that can be accessed in parallel.

A memory controller can issue four types of commands to a DRAM bank. Activate (ACT), Read/Write (R/W CAS), Precharge (PRE) and Refresh (REF). ACT, CAS and PRE commands are issued by the memory controller to service a memory request. When an ACT command is issued to the bank, a row of cells is selected and read into a temporary storage called row-buffer. Row-buffer is generally made up of sense amplifiers which also perform latching. Upon receiving a read (write) CAS command, few columns are read-out (written) from

(to) the row-buffer. A PRE command is sent by the memory controller to decouple the sense amplifiers from the activated row. When a REF command is issued by the memory controller, for each row in the refresh bundle the refresh controller activates a row of the bank and then precharges it.

B. Physical Organization of a DRAM Bank

DRAM bank is not implemented as a single monolithic array because of large capacitance (resulting in large delay) associated with the long bitlines and wordlines. In practise, it is implemented as a two dimentional arrangement of small arrays called tiles [4, 7, 8, 13]. Typically a tile is of size 256Kb, wherein, the DRAM cells are arranged in a 512 x 512 array (Figure 1). Apart from the cell array, a tile consists of wordline drivers, precharge circuits and sense amplifiers [7, 8, 13]. Between the tiles, the wordline drivers form a vertical strip, and the sense amplifiers and the precharge circuits form the horizontal strips.

We refer to a row of tiles in horizontal direction as a subarray. A subarray has a single wordline (referred to as global wordline) driver, i.e., the wordline drivers within the tiles of a subarray are driven by these global wordline drivers. The global wordline driver strengthens the output of a row-decoder which is unique for a subarray. The sense amplifiers of all the tiles in a subarray are together called as local row-buffer.

Figure 1 shows a block level representation of various circuits in the physical organization of a bank. In order not to clutter the figure, global bitlines, local bitlines and global precharge circuits are not shown. The subarray row-decoders (4) are driven by a single global row address latch (5) that is part of row pre-decode logic along with global row decoder (6), and stores the partially decoded row address. The local row buffers of all the subarrays are connected to a single global row-buffer (7) through a set of global bitlines and column select logic. The global row-buffer strengthens the changes in global bitlines when they are selectively connected to the local row-buffers. The global row buffer is inturn connected to the I/O circuitry of the device. Global and local (1) precharge circuits precharge the global and the local bitlines, respectively. For further details on tiles, subarrays and their arrangement along with the peripheral circuitry within the device the reader can refer [4, 7].

For each of the ACT, CAS and PRE commands from the memory controller, a subset of above circuits is used. When an ACT is issued, the row address is taken up by the global row-decoder, and partially decoded row address is latched by the global row address latch. The global row address latch selects a single subarray row-decoder that decodes the rest of the row address bits and drives a single global wordline of the subarray. The global wordline drives the local wordline drivers of the tiles, and the selected row is connected to the local bitlines (already precharged to $V_{dd}/2$). The perturbations on the local bitlines are captured by the local row-buffers of the subarray. Now, when a CAS command is issued, using the column address bits, a subset of local row buffers is connected to the global row buffer through global bitlines (precharged to $V_{dd}/2$). The global row-buffer then drives the selected column bits to the I/O drivers and the output latches of the device. Working of the CAS command is not affected with our proposal and henceforth, we do not consider it further.

For a PRE command, the memory controller does not specify a row address and hence, all the bitlines of *all* the subarrays are disconnected from the wordlines and precharged to $V_{dd}/2$. For a PRE command the address bus is driven to an invalid value.

Two important observations are in order here. Firstly, for an ACT, apart from the subarray access, only the global address decoder and the global address latch are used. They alone limit the amount of parallelism possible in accessing two subarrays. Secondly, a PRE operation is common to all the subarrays. We intend to exploit these observations during a REF operation and modify its working, as explained in Section III.

III. SCATTERED REFRESH

A. Conventional Refresh Operation

Once the memory controller issues a REF command to the DRAM device, the internal refresh controller takes over the device operation. The refresh controller consumes the row address supplied by the internal address counter (IAC) and issues an activate (ACT_{rc}) and a precharge (PRE_{rc}) abiding by their respective timing constraints (suffix rc is used to distinguish the commands that are issued by the refresh controller). After issuing an ACT_{rc}, the refresh controller issues a PRE_{rc} after t_{RAS} time period, and the next ACT_{rc} is sent after t_{RP} time. Once a PRE_{rc} has been issued by the refresh controller for the current row, the IAC value is incremented by one and the value denotes the next row address. For each row, the operations of ACT_{rc} and PRE_{rc} as described in the earlier section utilize the global circuitry and the subarrays for sometime.

Conventionally, the IAC increments the row address by one each time. Due to this, during a single refresh operation, all the row addresses map to a single subarray. For example, consider an 8Gb DRAM device with 8 banks. Each bank consists of 64K rows of size 16K bits each and the refresh bundle size is 8. If the subarray size is 512 rows, the bank contains 128 subarrays. Now, during a refresh period, let us assume that the memory controller has issued 2048 REF commands so far. When the 2049th REF command is issued, the rows to be refreshed are 16385 to 16392, all belong to the 32nd subarray. So, during the current REF operation, the refresh controller issues eight pairs of ACT_{rc}-PRE_{rc}'s (as shown in Figure 2), which are mainly intended for the 32nd subarray.

The problem with the conventional IAC address generation is that all the rows during a refresh operation map to a single subarray, making it mandatory to issue a PRE_{rc} for every ACT_{rc} issued to the subarray. This is required because within a subarray, only one row can be active at any given moment and to activate another row the wordlines have to be detached from the previously opened row, and the bitlines have to be precharged. For this reason the refresh cycle time t_{RFC} of the device generally equals to the refresh bundle size times the row cycle time (t_{RC}) (with some additional recovery time

t_{REC}) [4, 7]. With increase in the refresh bundle size, t_{RFC} increases (from one generation of DDR to another, there has been no major improvement in t_{RC} time), making the DRAM device unavailable for large periods of time for the memory controller to service the read requests. If within a refresh operation multiple subarrays are engaged and the available parallelism is extracted, the effective time spent in refreshing a row reduces, decreasing the t_{RFC}. We next explore the impact of mapping a refresh bundle to multiple subarrays.

B. Scattered Refresh - Concept

The requirement for the refresh controller to wait for t_{RAS} and t_{RP} for every ACT_{rc} and PRE_{rc}, respectively, is mainly because all the rows in a refresh bundle belong to a single subarray. But if the refresh bundle is seen as a collection of rows each from a different subarray, the refresh controller can issue commands to the subarrays so that their operations can overlap in time and hence, reduce the refresh cycle time. If rows in a refresh bundle are mapped to different subarrays, immediate issue of PRE_{rc} is no longer a necessity, but an option to the refresh controller. Every single PRE_{rc} avoided reduces the t_{RFC} by t_{RP} cycles. But in Scattered Refresh, for every ACT_{rc} a PRE_{rc} is issued at the appropriate time so as to avoid the peak power consumption of multiple activated rows in multiple subarrays.

Our idea of *scattering* the rows in a single refresh operation to different subarrays is motivated by the observation that two different subarrays can work almost independently with few constraints from the global circuitry they share. In this subsection the effect of refresh controller using multiple subarrays is studied. Changes to the refresh controller, limitations posed by the global circuitry and the modifications required in the periphery of the bank to overcome these limitations are dealt in Subsection III-C.

During each refresh, the rows that are to be refreshed can be scattered to different subarrays if the IAC increments by any value greater than the subarray size. Without loss of generality we assign 8K as the increment value because in distributed refresh 8K REF commands are sent by the memory controller independent of the DRAM device density. The choice of 8K also simplifies the process to determine the next row to refresh without need of any extra hardware to track either the rows refreshed or to be refreshed. Now, let us consider the previous example of the 8Gb device with 8 banks and 128 subarrays per bank (each subarray contains 512 rows). Two row addresses generated consecutively by the IAC go to two subarrays that are 16 subarrays apart. During the 2049th REF command sent by the memory controller the refresh bundle is collection of rows - 2049, 10241, 18433, 26625, 34817, 43009, 51201 and 59393, and their corresponding subarrays are 5, 21, 37, 53, 69, 85, 101 and 117, respectively. The starting address of the IAC for each REF command is the number of REF commands sent by the memory controller in the present refresh period. Figure 3 gives a pictorial representation of the commands issued by the refresh controller adopting Scattered Refresh.

Because the addresses generated using the modified IAC increment value target different subarrays, the refresh controller need not wait for t_{RAS} and t_{RP} time to send a PRE_{rc} and ACT_{rc}, respectively. Consider rows 2049 and 10241 during the present refresh operation, they belong to subarrays 5 (SA5) and 21 (SA21), respectively. The refresh controller begins the refresh operation by issuing an ACT_{rc} to SA5 and after t_{RAS} time it issues another ACT_{rc} to SA21. Unlike the traditional refresh controller, for the refresh controller employing Scattered Refresh PRE_{rc} to SA5 is not in the critical path, saving t_{RP} amount of time. In Scattered Refresh, the refresh controller issues PRE_{rc} to SA5 only after t_{RCD} time (reason to be explained in Subsection III-C) after ACT_{rc} has been issued to SA21. In this way, the refresh controller can overlap the activation and precharge operations directed to two different subarrays SA5 and SA21 successfully.

978-1-4799-2817-0/14 $31.00 © 2014 IEEE

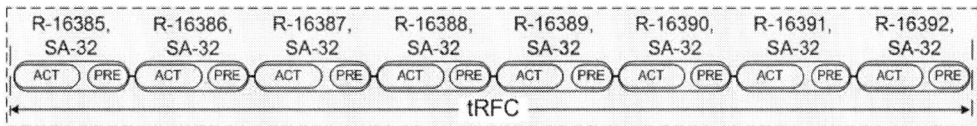

Fig. 2. Refresh operation in traditional DRAM devices with IAC generating contiguous row addresses. Row 16385 and subarray 32 is represented as R-16385, SA-32.

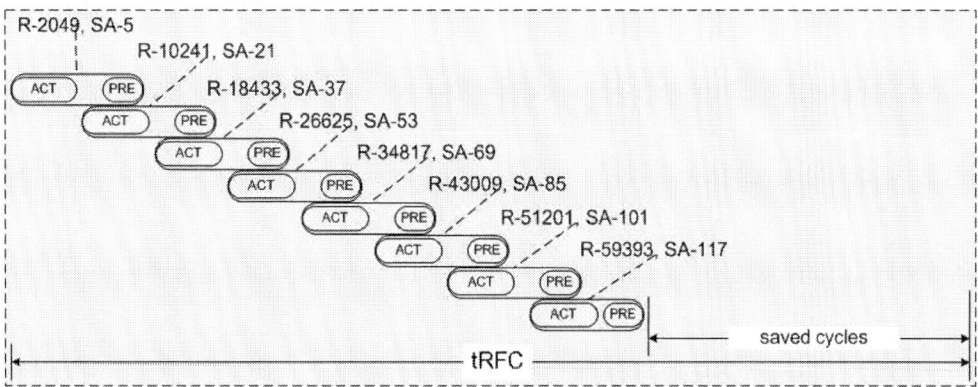

Fig. 3. Refresh operation in DRAM devices with Scattered Refresh, IAC scattering rows to different subarrays.

With this possible overlap of operations to different subarrays, the effective time spent in refreshing a bundle reduces by a significant amount. For a refresh bundle size of eight, time taken by seven out of eight precharge operations is overlapped by the activate operations, reducing the refresh cycle time t_{RFC} by 7 * t_{RP} amount. For a 11-11-11 DRAM device the savings come to 77 cycles, freeing-up significant time for the memory controller's usage and hence, improving the overall performance.

C. Implementation

Few modifications are needed to IAC of the traditional refresh controller to implement Scattered Refresh.

For every REF command, the starting value of IAC is the number of REF commands sent so far in the present refresh period. Hence, a 13-bit counter is needed. Changes are required to the physical organization of the DRAM bank to be able to handle commands to more than one subarray. To be precise, one row in each subarray can be activated and precharged if a latch is chosen per subarray instead of a single global address latch. We adopt the design proposed by Kim, et al., [9] and introduce the per-subarray row-address latches for each subarray. As pointed out by the authors, the area and power overheads of these additional latches are negligible when compared to those of the DRAM device. Now, during an ACT_{rc} the local row decoder is driven by the local row address latch instead of global row address latch. To precharge a specific subarray, with these changes the refresh controller needs to send its subarray ID along with each PRE_{rc}.

The refresh controller in the proposed Scattered Refresh issues a PRE_{rc} to the subarray after issuing an ACT_{rc} to another subarray with some delay to avoid simultaneous usage of global row decoder by the two commands. We choose the delay to be t_{RCD} so as not to overlap the peak power consumption of ACT_{rc} and PRE_{rc}, leading to increased peak power consumption of device. Usually t_{RAS} is much higher than t_{RP} of the device, and it takes t_{RCD} time for the row address to be consumed by the global row decoder, for the global wordlines to drive the local wordlines within the subarray and

for the sense amplifiers to sense the changes in the bitlines. During this period, peak operating current is drawn by the device. After t_{RCD} time, the sense amplifiers in local row-buffer stabilize the bitlines and recharge the DRAM cells, the current drawn during this period is minimal and hence, PRE_{rc} can be issued without overshooting the peak power profile of the device if issued after t_{RCD} time.

IV. EXPERIMENTAL METHODOLODY

In order to evaluate the proposed technique, we use USIMM [3] from the Memory Scheduling Championship (MSC) [2]. The USIMM simulator models a DRAM based memory system in detail with the required timing constraints and, memory request queues and scheduling policies. We extensively modify the refresh model in the tool for our study. In every cycle, a refresh operation can contend with read and write requests. In our baseline (Forced Refresh), refresh operations are scheduled only when eight of them have been pending. That is, our baseline configuration gives higher priority to read requests and schedules a refresh when no further delay is possible. In all the refresh techniques we study, at most eight refresh operations can be delayed (as per the JEDEC standards). We implement fine grained granularity (2x and 4x) refresh policies, wherein, at most 16 and 32 refresh commands can be delayed. Refresh intervals t_{REFI_2x} and t_{REFI_4x} are equal to half and one-fourth of t_{REFI}, respectively. t_{RFC} for 2x and 4x modes is chosen accordingly [1]. To employ Scattered Refresh over 2x and 4x modes we choose the IAC increment value to be 16K and 32K, respectively.

Table IV gives the base system configuration, with four processing cores operating at 3.2GHz and memory operating at 800MHz frequency. By default, the read requests are prioritized over the writes. Write requests are scheduled only when the occupancy of the requests in write queue reaches high watermark or when the read queue is empty and no refresh is pending. We opt for close-page policy (with FR-FCFS) as our scheduling policy, since it is better suited for multi-programmed setup. We consider 8Gb 11-11-11 DDR4 DRAM devices

978-1-4799-2817-0/14 $31.00 © 2014 IEEE

TABLE II
SYSTEM CONFIGURATION

Number of cores	4
Processor frequency	3.2GHz
Pipeline depth	10
ROB size	160
Fetch / Retire width	4 / 4
Last level cache	512KB per core (Private)
Cache block size	64B
Memory clock frequency	800MHz
Num. of channels	4
Num. of ranks	2 per channel
Num. of banks	8 per rank
Num. of rows	64K per bank
Num. of columns	128 cache blocks per row
Write queue size	64
Write queue high and	40 and
low watermarks	20

TABLE III
DRAM TIMING PARAMETERS

Parameter	DRAM cycles
$tRCD$	11
tRP	11
$tCAS$	11
$tRAS$	28
tRC	39
$tRRD$	6
$tFAW$	32
tWR	12
$tWTR$	6
$tRTP$	6
$tCCD$	4
$tRFC$	280
$tREFI$	3120

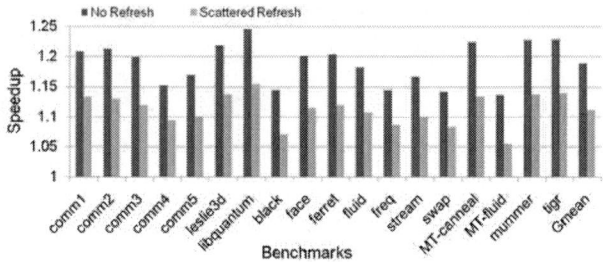

Fig. 4. Performance improvement achieved by Scattered Refresh.

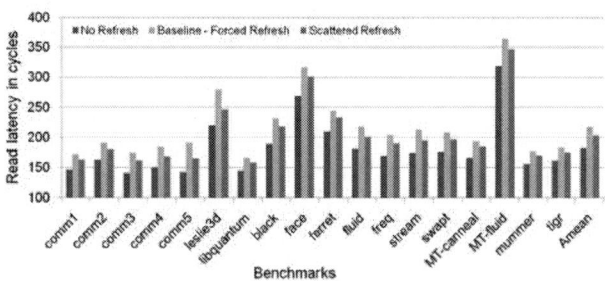

Fig. 5. Average read latency for all the benchmarks (lower the better).

operating at 800MHz with t_{RFC} and t_{REFI} as $350ns$ and $3.9\mu s$, respectively. Important timing parameters are listed in Table III.

To evaluate our technique for a wide range of workloads, we use the workload set from MSC [2]. The MSC workload suite contains 18 benchmarks in total, comprising of five commercial applications (*comm1* to *comm5*), nine benchmarks from the PARSEC suite (*black*, *face*, *ferret*, *fluid*, *freq*, *stream*, *swapt*, *MT-canneal* and *MT-fluid*), two benchmarks from SPEC2006 suite (*leslie3d*, *libquantum*)) and lastly, two from Biobench suite (*mummer*, *tigr*). *MT-canneal* and *MT-fluid* are multithreaded versions of *canneal* and *fluid*, respectively, from the PARSEC suite. For our four core setup, each simulation consists of running four instances of the benchmark, one on each core simultaneously. In each simulation, the run continues till the end of the last benchmark.

V. RESULTS AND ANALYSIS

Figure 4 shows the performance improvement obtained by Scattered Refresh over the baseline (i.e., Forced Refresh). No Refresh case gives the potential improvement possible (18.8%). Scattered Refresh obtains 10.2% improvement, equals to almost 59.2% of the potential performance gain. Benchmarks that have higher scope for improve-

ment (> 20%) for No Refresh (eg. *comm1*, *comm2*, *leslie3d*, *libquantum*, *face*, *ferret*, *MT-canneal*, *mummer* and *tigr*) are clearly the ones benefited most (> 10% improvement) with Scattered Refresh. For these benchmarks the impact of refresh on performance is high and hence, reduction in the *refresh cycle time* reflects on overall performance.

The speedup obtained can be understood better by the reduction in the average read latency of the system. Figure 5 shows the average read latency (in terms of processor cycles) of ideal No Refresh, baseline and Scattered Refresh techniques. For the baseline configuration, the read latency is 218 cycles, 35 cycles more than No Refresh case. This clearly shows that refresh has a significant impact on the overall performance. Scattered Refresh reduces the read latency by 10 cycles, to 208 cycles on an average, and hence, the improvement in performance of the system.

To understand the variation in the savings obtained by Scattered Refresh among different benchmarks, we observe the frequency of read requests at the memory controller. Figure 6 presents the average number of read requests received at the memory controller per 1K cycles. A direct implication of higher number of average read requests is that time saved by the Scattered Refresh during evey refresh operation can be utilized effectively for servicing them. This is clearly observed from comparing Figures 4 and 6. For example, benchmarks like *libq*, *mummer* and *tigr* have more than 54 read requests (on an average) for every 1K cycles, and hence, experience more than 12% of improvement in performance when Scattered Refresh is applied. On the other hand, benchmarks like *comm5* and *MT-fluid* show relatively less improvement because the frequency of read requests for these benchmarks is less than 17 (on an average).

MT-fluid is the only benchmark that has performance improvement less than 5.5%. We observe that because of high average read latency (approx. 365 cycles), reduction in the refresh cycle time alone is not effective even though it reduces the latency to (approx.) 347 cycles. Even in No Refresh case the average read latency is 318 cycles, suggesting that techniques handling memory contention would be more effective for this benchmark.

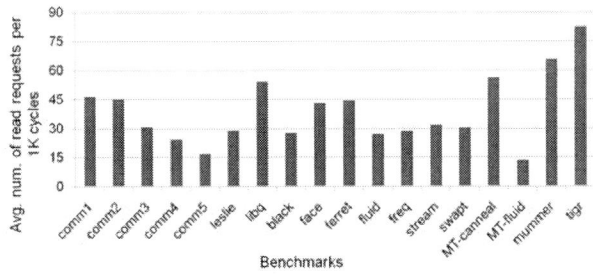

Fig. 6. Benchmark wise average number of read requests per 1K cycles.

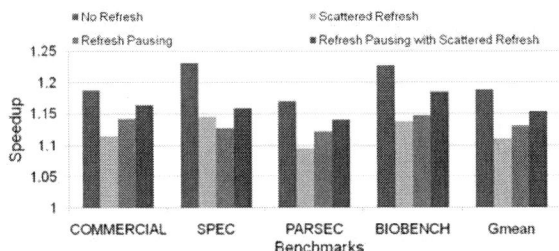

Fig. 7. Performance improvement for Scattered Refresh, Refresh Pausing and when both are applied together.

When Scattered Refresh is applied over 2x and 4x refresh modes, performance improved by 5.5% and 1.5%, respectively. Decreased effectiveness of Scattered Refresh is because of increased fraction of t_{REC} in t_{RFC} for 2x and 4x modes.

A. Comparison with Refresh Pausing Technique

In [11], the authors show that Refresh Pausing outperforms the existing refresh scheduling techniques. Hence we chose to study our technique in comparison with Refresh Pausing. Figure 7 shows the speedup observed from Refresh Pausing and Scattered Refresh. Refresh Pausing clearly performs better than Scattered Refresh (except for *libquantum*), by achieving 13.1% improvement over the baseline. This is because when Refresh Pausing is employed the maximum time a request waits during a refresh is equal to the row cycle time. Comparatively, the requests wait for longer time during Scattered Refresh. The average read latency incurred supports this observation. For Refresh Pausing the average latency for reads is observed to be 194 cycles, almost 10 cycles less than that of Scattered Refresh.

In the case of *libquantum*, Refresh Pausing achieves only 8.3%, whereas Scattered Refresh achieves 15.3% improvement in performance. The effectiveness of Refresh Pausing is limited for *libquantum* because of frequent pausing of refreshes. This leads to refreshes eventually getting done in Forced mode where no more pausing is possible.

Given that Scattered Refresh introduces changes only to the refresh controller, it is orthogonal to other techniques addressing refresh. Scattered Refresh can be employed with any of the available techniques without any disruption. We study the combined behaviour of Refresh Pausing and Scattered Refresh, and observe that speedup has improved by 2.3% over Refresh Pausing. For each benchmark category, the rightmost bar in Figure 7 shows the result of combining the two techniques. Together, Refresh Pausing and Scattered Refresh are able to reduce the average read latency 4 cycles further, when compared to Refresh Pausing alone.

VI. CONCLUSION

Given that for high-density DRAM devices time spent in refreshing is increasing significantly, in this work we targeted to reduce refresh cycle time. We proposed *Scattered Refresh*, a refresh mechanism wherein the refresh controller *scatters* rows in a refresh bundle to different subarrays and exploits the parallelism inherently available in accessing them. With few modifications to the refresh controller and to the peripheral circuitry of the DRAM bank, Scattered Refresh recovers up to 59.2% of performance loss incurred by traditional refresh. It is also shown that Scattered Refresh being orthogonal, when employed with other existing techniques improves their performance further. Given that there is no change in the interface between the memory and the refresh controllers, Scattered Refresh can be implemented in future systems without any disruptive changes.

ACKNOWLEDGMENTS

The authors would like to thank Abhilash Bhandari, Sudharsan J and Tripti S Warrier, and anonymous reviewers for their valuable feedback. This work is supported in part by grant from British Council under UKIERI project (Contract No. IND/CONT/E/11-12/099).

REFERENCES

[1] JESD79-4, JEDEC Committee JC-42.3 Std. DDR4, Sept. 2012.

[2] Memory scheduling championship (msc), 2012. Available: http://www.cs.utah.edu/ rajeev/jwac12/

[3] N. Chatterjee, et al., "USIMM: the utah simulated memory module", University of Utah, Tech. Report, 2012, uUCS-12-002.

[4] Itoh K, "VLSI Memory Chip Design", Springer, 2001.

[5] Jamie Liu, et al., "RAIDR: retention-aware intelligent DRAM refresh", *International Symposium on Computer Architecture* (ISCA), 2012, pp. 1 - 12.

[6] Jeffrey Stuecheli, Dimitris Kaseridis, Hillery C. Hunter and Lizy K. John, "Elastic refresh: techniques to mitigate refresh penalties in high density memory", *International Symposium on Microarchitecture* (MICRO), 2010, pp. 375 - 384.

[7] Keeth B, Jacob Baker R, Brian Johnson and Feng Lin, "DRAM circuit design - Fundamental and high-speed topics", Wiley-Interscience and IEEE press, 2007.

[8] Keeth B, Bunker Layne G., Beffa Raymond J. and Ross Frank F. "256Meg dynamic random access memory". United States Patent Application No. 7477556B2, 2009.

[9] Kim Y, Seshadri V, Lee D, Liu Jamie and Mutlu O, "A case for exploiting subarray-level parallelism (salp) in DRAM", *International Symposium on Computer Architecture* (ISCA), 2012, pp. 368 - 379.

[10] Mrinmoy Ghosh and Hsien-Hsin S. Lee, "Smart refresh: an enhanced memory controller design for reducing energy in conventional and 3D die-stacked DRAMs", *International Symposium on Microarchitecture* (MICRO), 2007, pp. 134 - 145.

[11] Prashant Nair, Chia-Chen Chou and Moinuddin K. Qureshi, "A case for refresh pausing in DRAM memory systems". *International Symposium on High Performance Computer Architecture* (HPCA), 2013, pp. 627 - 638.

[12] Song L, Karthik P, Thomas M and Benjamin Z, "Flikker: saving DRAM refresh-power through critical data partitioning", International Conference on Architectural Support for Programming Lanaguages and Operating Systems (ASPLOS), 2011, pp. 213 - 224.

[13] Thomas Vogelsang, "Understanding the energy consumption of dynamic random access memories", *International Symposium on Microarchitecture* (MICRO), 2010, pp. 363 - 374.

978-1-4799-2817-0/14 $31.00 © 2014 IEEE

A Read-Write Aware DRAM Scheduling for Power Reduction in Multi-Core Systems

Chih-Yen Lai*, Gung-Yu Pan*, Hsien-Kai Kuo*, Jing-Yang Jou*†

*Department of Electronics Engineering and Institute of Electronics, National Chiao Tung University

{yoshilai, astro, hsienkai.kuo}@eda.ee.nctu.edu.tw

†Department of Electrical Engineering, National Central University

jyjou@faculty.nctu.edu.tw

Abstract— The demand of high performance and low power has increased the importance of power efficiency in multi-core systems. In modern multi-core architectures, DRAM has dominated the power consumption and therefore reordering based DRAM scheduling has been intensively studied to reduce the power. However, the benefit of reordering is not fully explored by the previous studies. To further reduce the power, this paper proposes the read-write reordering and the read-write aware throttling. When compared to the existing work, the proposed techniques reduce 10% more DRAM power with less performance degradation.

I. INTRODUCTION

In the latest multi-core systems, the increasing performance comes at the cost of the higher power consumption [1]. Within a multi-core architecture, it has been shown that main memory contributes up to 40% of the system power consumption [2][3][4]. As the result, reducing the power consumption of main memory has become the main issue for chip designers.

Among existing memory circuits, *dynamic random access memory* (DRAM) is by far the mainstream of main memory used in modern multi-core systems. Hence, this paper focuses on reducing DRAM power consumption. To generalize the discussion, this paper chooses to work on the widely used JEDEC standard DRAM architecture [5].

A. DRAM Basics

The JEDEC standard DRAM is composed of multiple dual-in-line memory modules (*DIMMs*) and communicates with the processors through the *memory controller*. The memory controller receives memory commands from the processors and issues them to DIMMs. Sets of DRAM chips lie in a DIMM, and each set of the DRAM chips is called a *rank*. Each rank can be further partitioned into *banks*, which spread across all DRAM chips within a rank.

For each memory operation, a rank is activated to carry out the read or write access. The memory command identifies the bank within the rank it is destined to and that bank reads out an entire row of data. This row of data is stored in the *row buffer*, which is a private buffer owned by each bank. The memory command then identifies the column that it needs from the row in the row buffer and returns them to the processors.

Since a rank is the minimum number of DRAM chips that need to be activated for a memory access, a rank is

the smallest set of DRAM chips that can be turned off. The JEDEC standard DRAM supports rank level power mode control and allows users to turn on or turn off each rank [5][6]. The power mode transition commands are also sent to the DIMMs by the memory controller. An activated rank consumes active power, burst power for read and write requests, refresh power and background power. If a rank is turned off, it only consumes refresh power and background power that is lower than the active power.

Although turning off idle ranks reduces DRAM power consumption, switching power mode of a rank takes a transition delay of time. The delays of switching power mode of ranks bring extra latencies to the system performance. Therefore, one of the main goals for DRAM power management is to design a policy that determines when to turn off idle ranks. An immature policy leads to loss of power saving or serious degradation to the system performance.

B. Related Works and Motivation

Some approaches to DRAM power management includes re-designing the physical structure of DRAM, migrating data in DRAM and implementing power management policy on the modern DRAMs. Re-designing the physical structure of DRAM increases the potential of turning off idle DRAM chips. Some proposed to separate ranks into mini-ranks by adding mini-rank buffers inside each rank [7]. Others proposed to change the arrangement of arrays in each bank [2]. These solutions may have good performance, but they require modification to the modern DRAM architecture. Automatic data migration tends to make ranks empty and turns them into the low power mode for a period of time [8]. However, it requires a large hardware overhead to implement. This paper targets on DRAMs that are currently available on the market.

An intuitive power management policy is *time-out power-down* policy. The time-out power-down policy turns off a rank once it is idle for a fixed period of cycles, regardless of the upcoming commands. This results in inflexible power mode transitions and may harm the system performance dramatically.

Besides time-out power-down policy, *queue-aware power-down* policy is proposed [9]. If a rank is idle, this policy checks commands in the memory controller to see if any of the pending commands is destined for the idle rank. Unless there is no pending commands heading to the idle rank, it will not be turned off. Queue-aware power-down

978-1-4799-2817-0/14 $31.00 © 2014 IEEE

policy does not affect the system performance as much as the time-out power-down policy does, but its effect on reducing DRAM power consumption is limited. The *power-aware memory scheduler* and *memory throttling* mechanism are also proposed and combined with the queue-aware power-down policy in the previous work [9]. The power-aware memory scheduler utilizes the fact that the memory controller is able to reorder the received memory commands. It clusters memory commands destined for the same rank together. While the memory concentrates on accessing one rank, other ranks can be switched to the low power mode to save power. Since only one rank is activated at a time, other ranks stay in the low power mode for longer periods. Memory throttling mechanism further increases the power saving by blocking memory commands in the memory controller. No command will be issued by the memory controller before it has blocked commands for a fixed period of cycles. This fixed period of cycles is called *throttle delay*. With the queue-aware power-down policy, power-aware scheduler and the memory throttling mechanism, the previous work achieves a power saving closed to time-out power-down policy with a moderate system performance degradation.

However, it does not take into consideration that read requests are more critical for the system performance than write requests [10]. Utilizing this fact, one can further reduce DRAM power consumption with a slight system performance overhead.

This paper chooses to add new techniques based on the power-aware scheduler and the memory throttling mechanism. However, since all these related works are orthogonal to each other, one can integrate our work with other related works.

C. Our Contributions

To reduce DRAM power consumption, this paper proposes the read-write aware DRAM scheduling. This work includes two techniques, the *read-write aware throttling* mechanism and the *rank level read-write reordering*. The read-write aware throttling mechanism effectively cuts down DRAM power consumption. The rank level read-write reordering is employed to significantly reduces the system performance degradation caused by DRAM power management while maintaining the power saving. Our work achieves a superior 75.34% power saving from the DRAM with no power management. When compared to the existing work, the proposed techniques reduce 10% more DRAM power with less performance degradation. Moreover, by evaluating and comparing with an oracle policy, it has shown that our work can reduce more power at the expense of a slight system performance degradation.

The remainder of this paper is organized as follows. The next section describes the problem formulation. Section III places the proposed algorithm. The experiment results are presented in Section IV. Finally, this paper concludes in Section V.

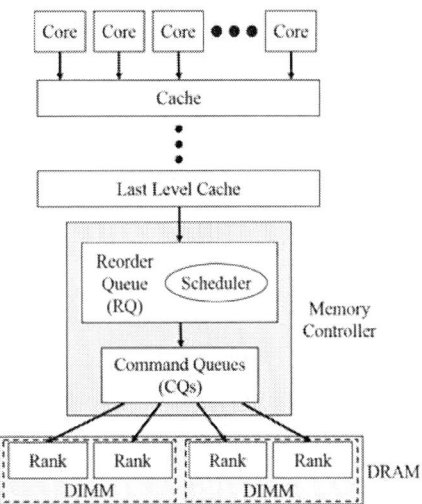

Fig. 1. System hierarchy diagram.

II. PROBLEM DESCRIPTION

As shown in Fig. 1, given a processor with one or more cores, these cores are connected to a multi-level cache and are equipped with write buffers. A DRAM main memory is connected to the last level cache. A memory controller lies between the last level cache and the DRAM structure. The memory controller receives memory commands from the processors and issues these commands to DIMMs inside the DRAM. These memory commands contain information including its target address, access type and the time it enters the memory controller.

Within the memory controller, there are two queues. Memory access commands from the last level cache are first stored in the *reorder queue (RQ)*. These commands are mapped to certain DRAM ranks and banks according to their target addresses in the RQ. A scheduler inside the RQ reorders the commands in the RQ while keeping the data hazard-free. After address mapping and reordering, the memory commands on top of the RQ are sent to the *command queues (CQs)* one at a cycle. Each CQ is assigned to a certain rank. A CQ handles all kinds of commands destined for its corresponding rank. These commands includes access commands from the RQ and other commands such as precharge, refresh, power-up, and power-down commands generated by the memory controller. The CQs issue these commands to the DRAM ranks in first-in-first-out (FIFO) order.

The memory controller supports throttling mechanism. It blocks memory commands in the RQ for throttle delay of cycles. When the throttle delay is reached, the RQ starts sending the commands to the CQs one at a cycle until the RQ is empty.

The DRAM supports rank level power-mode control. That is, ranks can be turned on or turned off by the memory controller. The memory controller puts power-up and power-down commands into CQs to switch the power mode of DRAM ranks. Switching the power mode of a rank is at the

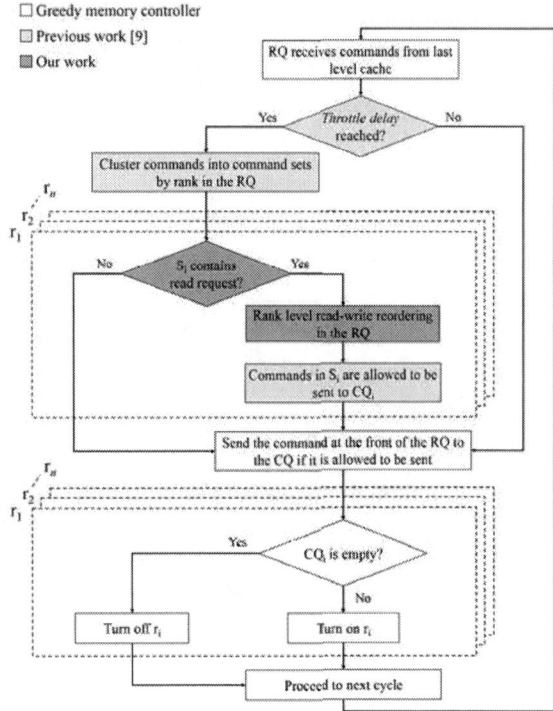

☐ Greedy memory controller
☐ Previous work [9]
▨ Our work

Fig. 2. Flow chart of the proposed power management policy.

cost of transition delays.

The goal of this paper is to find a delicate DRAM scheduling scheme. The scheme should reduce the power consumption of DRAM with minor system performance degradation.

III. THE PROPOSED TECHNIQUES

A. Overview

The proposed DRAM power reduction techniques address on lowering DRAM power consumption with slight system performance degradation. Fig. 2 depicts the overall flow of the proposed power management policy. The white rectangles and decision box are the procedures of the greedy memory controller, which employs the greedy power-down policy that turns off a rank whenever it is empty. The light-gray rectangles and decision boxes are power management policies added in the previous work [9]. The dark-gray rectangle and box are the proposed techniques. The memory is assumed to have n ranks. Notation r_i is used to represent the ith rank in the memory, and CQ_i denotes the command queue for the ith rank.

Since the throttling mechanism is employed, all the main memory commands from the last level cache are blocked in the RQ for throttle delay of cycles. DRAM ranks stay in the low power mode until the throttle delay is reached. Once the throttle delay is reached, the scheduler in the RQ clusters the memory commands into *command sets* by the ranks they are destined for. Command set composed of commands destined for the ith rank is represented by S_i in Fig. 2. The policy goes down to rank level after command sets are formed and handles each command set separately. If the command set to

a certain rank is empty, that memory controller turns off that rank.

On the other hand, the read-write aware throttling mechanism, which is depicted by the dark-gray decision box, checks for the existence of read commands in the nonempty command sets. Only the command sets containing read requests are sent to the CQs and only their target ranks are turned on. The other ranks stay in the low power mode for another throttle delay to reduce the DRAM power consumption.

The dark-gray rectangle shows that the commands in each command set are reordered by the rank level read-write reordering before they are sent to the CQs. The read requests in each command set get higher priorities than write requests. The commands with higher priorities enter the CQs earlier. Since the CQs send commands to the DIMMs in a FIFO order, the read requests reach their target ranks as soon as possible. This makes read requests, which are critical to system performance, to be served by the DIMMs earlier and thus the system performance degradation caused by the throttling mechanism is reduced.

These two techniques are described in the following subsections.

B. Read-write Aware Throttling

The read-write aware throttling mechanism determines which rank should be turned on whenever the throttle delay is reached. It checks on each command set, which is composed of memory commands destined for the same rank in the RQ, for the existence of read requests. If there is no read request in the command set, its target rank is turned off and the command set remains in the RQ for another throttle delay. The rank is otherwise turned on and the commands in the command sets are sent to the CQ.

The read-write aware throttling utilizes the fact that read requests affect system performance more than write requests [10]. It is performed on rank level and checks the existence of critical read requests in every command set whenever the throttle delay is reached. If a read request appears in a command set, the memory controller sets the target rank of this command set to be *urgent*. Ranks with no pending read requests are set to be *trivial*. The commands destined for urgent ranks are sent to the CQs, while other commands remain in the RQ for another throttle delay. This allows the memory controller to only turns on urgent ranks and keeps trivial ranks in the low power mode, contributing to a large DRAM power saving.

Notice that when a rank is set to be urgent, all the commands destined for it, including write requests that have been blocked in the RQ, are sent to its corresponding CQ. If there is no read request to any rank for a very long period, which rarely happens to DRAM, and the RQ is overloaded with blocked write requests, it simply flushes all the requests to the CQs when the throttle delay is reached.

The hardware overhead for read-write aware throttling is tiny. Only a one bit flag identifying the urgency of each rank is required.

C. Rank Level Read-write Reordering

The rank level read-write reordering is a reorder policy for the command sets containing read requests in the RQ. It gives read requests higher priority over write requests. The commands in a command set are sent to the CQ in descending order of priority. Since the CQ issues commands to the DIMM in FIFO order, read requests are issued to the DIMM prior to write requests. This forces the DIMM inside the DRAM to process read requests, which are critical to system performance, as soon as possible. The rank level read-write reordering effectively relieves the system performance degradation caused by the DRAM power management policy.

The system performance degradation is greatly reduced if the read requests are sent to the CQ prior to all the write requests. However, reordering memory commands blindly can lead to a data hazard issue since the memory commands are no longer handled by the DIMM in the same order as the processors sent out. If a read request enters the RQ after a write request and they are both destined for a same address, a read-after-write (RAW) data hazard occurs if the DIMM returns data for this read request prior to the write request. To avoid RAW hazard, the rank level read-write reordering performs a check before reordering. For each read request in the a command set, the rank level read-write reordering checks all the write requests that entered the RQ earlier than this read request in the same command set. If one or more write requests target to the same address as the read request does, they are combined in their original order to form a *command group*. All the command groups are then reordered according to the read requests inside them. The earlier the read request enters the RQ, the higher priority will be set for the command set.

After reordering, the reordered command sets are sent to the CQs. The target ranks of these command sets are turned on to process these command. It is switched back to the low power mode once all the commands are finished and is kept in the low power mode until the throttle delay is once again reached.

An example of how the rank level read-write reordering works is illustrated in Fig. 3, which depicts one of the command sets in the RQ. Each box in Fig. 3 is a memory command with its access type represented by W (Write) or R (Read) followed by an index and the address it is destined for. The command at the top of the command set has the highest priority while the command at the bottom has the lowest priority. When the throttle delay is reached, the commands are sent to the CQ in descending order of priority one at a cycle. The rank was originally in the low power state because all the commands from last throttle period are finished. Suppose that the throttle delay is reached at cycle k and there are 5 commands in this command set. The rank level read-write reordering checks through the command set and finds that W2 and W4 target to the same address as R5. These three commands are then combined to form a command group and reordered to the top of the command set. The commands are sent to the CQ and their target rank is turned on. The

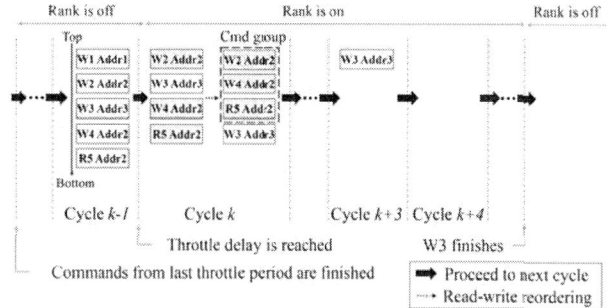

Fig. 3. Example for rank level read-write reordering.

rank is turned off again when all the command finish. The example shows that the rank level read-write reordering forces the DRAM to process read requests as early as possible and thus reduces the system performance degradation.

IV. EXPERIMENTAL RESULTS

A. Simulation Environment

The performance of our work is evaluated with Multi2Sim [11], a widely used cycle-accurate system simulator. Multi2Sim provides detailed simulation of single core or multicore processors and gives us the statistics of the system performance. Our evaluation integrates DRAMSim2 [12] into Multi2Sim to obtain more accurate statistics of DRAM, including the power consumption. DRAMSim2 is a cycle accurate, JEDEC DDRx memory system simulator, which models the memory controller, memory channels, DRAM ranks, and banks. DRAMSim2 also models the power mode transitions of DRAM, including the transition delay and the transition power. The evaluation acquires the system throughput statistics from Multi2Sim and observes the power consumption and access delay of main memory simultaneously with DRAMSim2.

The baseline system in the simulations has four cores and two levels caches. In order to evaluate our work by comparing to the previous work [9], our simulations use the same cache sizes as in the previous work. The detail parameters of the baseline system are presented in Table I. The main memory used in the evaluation is a DDR2 SDRAM, which is one of the JEDEC standard memory available on market [6].

The workload for our simulations is SPEC CPU2006 benchmark suite [13]. In order to compare the power management policies, only sixteen memory-intensive benchmarks from CPU2006 are selected, while all the policies can obtain acceptable results on benchmarks which are not memory intensive. Therefore, each of our simulations randomly chooses four benchmarks from floating point benchmarks and assigns them to the cores. Notice that the proposed techniques works well on integer benchmarks too. However, since integer benchmarks are not memory intensive, one can simply uses any naive power management policy and gets an acceptable result. The benchmark combinations are listed in Table II. Each benchmark combination is run for one billion CPU cycles in our evaluation.

978-1-4799-2817-0/14 $31.00 © 2014 IEEE

TABLE I
BASELINE SYSTEM PARAMETERS

Parameter	Value
Number of cores	4
Processor frequency	2.132 GHz
Size of L1D cache	32 KB
Set associativity of L1D cache	4-way
Size of L1I cache	64 KB
Set associativity of L1I cache	2-way
Size of L2 cache	2 MB
Set associativity of L2 cache	8-way
DRAM frequency	**533 MHz**
Number of DRAM ports	**2**
DRAM device width	**8**
Number of DRAM ranks	**4**
Number of DRAM banks	**4**
Number of DRAM rows	**8192**
Number of DRAM columns	**4096**

TABLE II
BENCHMARK COMBINATIONS

Comb.	Benchmarks			
fp1	410.bwaves	416.gamess	433.milc	434.zeusmp
fp2	435.gromacs	436.cactusADM	437.leslie3d	444.namd
fp3	447.dealII	450.soplex	453.povray	454.calculix
fp4	459.GemsFDTD	465.tonto	470.lbm	481.wrf

In the evaluation, the power consumption is measured in Watts. The system performance is measured in million instructions per second (MIPS), which represents the throughput of the system. All the results are normalized to the DRAM with no power management policy.

B. Analysis on Different Techniques

In the evaluation, our work is compared to the previous work and an oracle policy. In the oracle policy, the order of the memory accesses is transparent so that the DRAM ranks can be ideally turned on and off when needed. Furthermore, there is no transition delay and transition power in the oracle policy. The power reductions of different policies are compared with the throttle delay set to 100, at which both our work and the previous work performs well. The evaluation results are shown in Fig. 4. The term *RwAT* is used for the read-write aware throttling and the term *RwR* is used for the read-write reordering.

The evaluation results show that when the read-write aware throttling is employed alone, it reduces the DRAM power consumption 10%–15% more than the previous work but causes around 1% more system performance degradation. On the other hand, when the read-write reordering is used alone, it improves the system performance by around 1% but the DRAM power consumption remains the same. By combining these two techniques, our work saves 10%–15% more power than the previous work with the same, or even slighter, system performance degradation. More importantly, our work reduces DRAM power consumption to below the oracle solution. The reason is that the oracle solution stands for the maximum power reduction when there is no system performance degradation. Our work trades a reasonable system performance for more power reduction. The only exception is the benchmark combination fp3, in which the number of read requests to DRAM is much greater than the number of write requests.

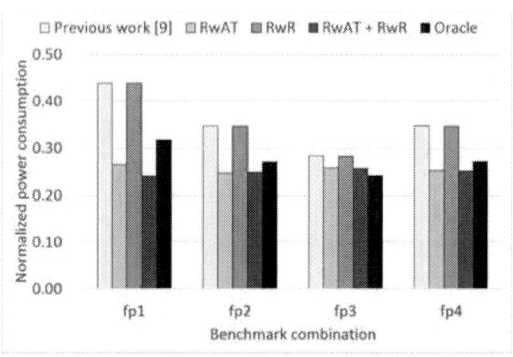

(a) Normalized DRAM power consumption

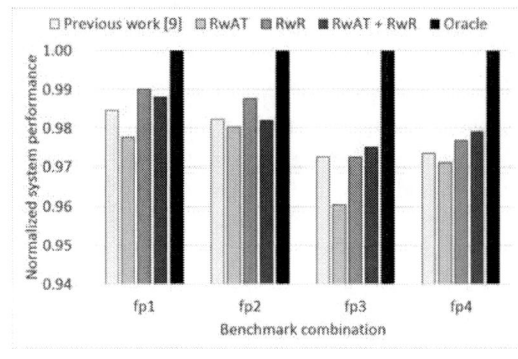

(b) Normalized system performance

Fig. 4. Power and performance of different policies.

TABLE III
EFFECT OF DIFFERENT THROTTLE DELAYS

Throttle delay	Previous work		Our work		Improvement	
	PRP	Overhead	PRP	Overhead	PRP	Overhead
100	64.67%	1.03%	75.46%	0.61%	10.79%	-0.42%
200	64.54%	0.87%	75.26%	0.72%	10.72%	-0.15%
400	64.58%	1.98%	75.10%	1.73%	10.52%	-0.25%
800	65.32%	5.03%	75.14%	4.88%	9.82%	-0.15%
1600	65.49%	6.46%	74.96%	6.39%	9.47%	-0.07%
3200	68.24%	7.63%	75.17%	7.46%	6.93%	-0.17%
6400	71.02%	7.92%	75.42%	7.88%	4.4%	-0.04%

This limits the effect of the read-write aware throttling in our work. Nevertheless, our work still manages to save 5% more DRAM power consumption than the previous work.

C. Stability with Different Throttle Delays

The *throttle delay* used in the throttling mechanism is critical to both the DRAM power consumption and the system performance. Long throttle delay harms the system performance but brings more potential to power reduction. To see how the throttle delay affects the performance of our work, an evaluation on different throttle delays is carried out. The evaluation uses all benchmark combinations fp1, fp2, fp3 and fp4. The average simulation results are shown in Table III. The term *PRP* is used as an abbreviation of power reduction percentage. The improvements are the differences between our work and the previous work.

The evaluation results shows that our work is stable with different throttle delays. It steadily reduces around 75% of DRAM power consumption, which is around 10% better

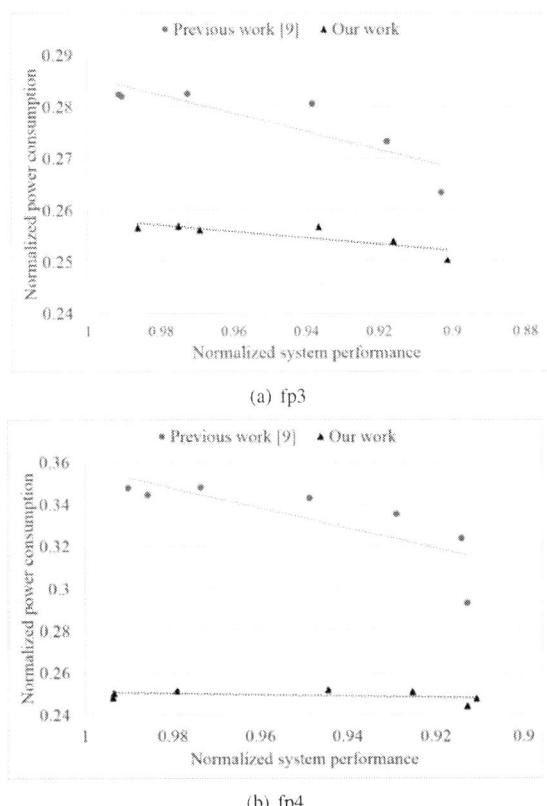

(a) fp3

(b) fp4

Fig. 5. Power to performance trade-off.

than the previous work. Moreover, the system performance degradation of our work is better than the previous work. This shows that the read-write aware reordering mechanism effectively relieves the impact on the system performance. The evaluation results also shows that when the throttle delay is too large, the difference between our work and the previous work is small. It is because that when the throttle delay is too large, all the DRAM ranks are in the low power mode for most of the time. Therefore the power consumption is low and the performance degradation is dramatic.

D. Power and Performance Trade-off

With the results in Table III, this subsection further illustrates the trade-off characteristic between power and performance. By varying the throttle delay, the evaluation shows how our work reacts to different system performance degradations. The evaluation results of benchmark combinations fp3 and fp4 are shown in Fig. 5 as examples. The number of read requests is larger than write requests in fp3, while write requests dominates over read requests in fp4. The results show that when the application is more read intensive, our work dramatically reduces more power when a slight more system performance degradation is allowed. Nevertheless, under the same system performance degradation, our work reduces around 10% more of DRAM power than the previous work.

V. CONCLUSIONS AND FUTURE WORKS

This paper proposes a DRAM scheduling policy to magnificently reduce the power consumption of DRAM. The read-write aware throttling mechanism allows the DRAM ranks to stay in the low power mode for a longer period of cycles. It improves the power saving by 10%–15% on average. Read-write reordering forces DRAM to handle read requests, which are critical to the system performance, as soon as it can. It reduces the system performance degradation caused by the power management without sacrificing much power saving. From the experiments, our work reduces the DRAM power consumption by around 75%, which is better than the previous work and the oracle solution. Meanwhile, it causes only 1%–3% system performance degradation, which is smaller than the existing power management policy.

As for the future, the proposed techniques can be improved by designing a controller that dynamically adjusts the throttle delay in run-time. The system performance degradation can be further reduced with the flexible throttling mechanism. It is also worth exploring the potential of combining the techniques with other related works such as the automatic data migration, which creates empty ranks that can be shut off until the throttle delay is reached.

REFERENCES

[1] G. Zhang, H. Wang, X. Chen, S. Huang, and P. Li, "Heterogeneous multi-channel: Fine-grained DRAM control for both system performance and power efficiency," in *Proceedings of the 49th Design Automation Conference*, 2012.

[2] A. N. Udipi, N. Muralimanohar, N. Chatterjee, R. Balasubramonian, A. Davis, and N. P. Jouppi, "Rethinking DRAM design and organization for energy-constrained multi-cores," in *Proceedings of the 37th International Symposium on Computer Architecture*, 2010.

[3] V. Delaluz, M. Kandemir, N. Vijaykrishnan, A. Sivasubramaniam, and M. J. Irwin, "DRAM energy management using software and hardware directed power mode control," in *Proceedings of the 7th International Symposium on High-Performance Computer*, 2001.

[4] K. Chandrasekar, B. Akesson, and K. G. W. Goossens, "Run-time power-down strategies for real-time SDRAM memory controllers," in *Proceedings of the 49th Design Automation Conference*, 2012.

[5] *Jedec Standard: DDR2 SDRAM Specification*, JEDEC Solid State Technology Association, 2009.

[6] *512Mb: x4, x8, x16 DDR2 SDRAM*, Micron Technology, Inc., 2012.

[7] H. Zheng, Z. Zhang, E. Gorbatov, and Z. Zhu, "Mini-rank: Adaptive DRAM architecture for improving memory power efficiency," in *Proceedings of the 40th International Symposium on Microarchitecture*, 2008.

[8] V. D. L. Luz, M. Kandemir, and I. Kolcu, "Automatic data migration for reducing energy consumption in multi-bank memory systems," in *Proceedings of the 39th Design Automation Conference*, 2002.

[9] I. Hur and C. Lin, "A comprehensive approach to DRAM power management," in *Proceedings of the 14th International Symposium on High-Performance Computer*, 2008.

[10] C. J. Lee, V. Narasiman, E. Ebrahimi, O. Mutlu, and Y. N. Patt, "DRAM-aware last-level cache writeback: Reducing write-caused interference in memory system," Tech. Rep., Apr. 2010.

[11] R. Ubal, B. Jang, P. Mistry, D. Schaa, and D. Kaeli, "Multi2Sim: A Simulation Framework for CPU-GPU Computing," in *Proceedings of the 21st International Conference on Parallel Architectures and Compilation Techniques*, 2012.

[12] P. Rosenfeld, E. Cooper-Balis, and B. Jacob, "DRAMSim2: A Cycle Accurate Memory System Simulator," *Computer Architecture Letters*, vol. 10, no. 1, pp. 16 –19, Jan.–Jun. 2011.

[13] J. L. Henning, "SPEC CPU2006 benchmark descriptions," *SIGARCH Computer Architecture News*, vol. 34, no. 4, pp. 1–17, Sep. 2006.

978-1-4799-2817-0/14 $31.00 © 2014 IEEE

A Coherent Hybrid SRAM and STT-RAM L1 Cache Architecture for Shared Memory Multicores

Jianxing Wang, Yenni Tim
Weng-Fai Wong, Zhong-Liang Ong

School of Computing
National University of Singapore
Singapore, 117417
wangjx, timyn, wongwf, zoo@nus.edu.sg

Zhenyu Sun, Hai (Helen) Li

Swanson School of Engineering
University of Pittsburgh
Pittsburgh, PA 15261
zhs25, hal66@pitt.edu

Abstract— STT-RAM is an emerging NVRAM technology that promises high density, low energy and a comparable access speed to conventional SRAM. This paper proposes a hybrid L1 cache architecture that incorporates both SRAM and STT-RAM. The key novelty of the proposal is the exploition of the MESI cache coherence protocol to perform dynamic block reallocation between different cache partitions. Compared to the pure SRAM-based design, our hybrid scheme achieves 38% of energy saving with a mere 0.8% IPC degradation while extending the lifespan of STT-RAM partition at the same time.

I. INTRODUCTION

The concern about energy consumption in the current technology nodes has driven researchers to examine alternative solutions. To reduce the increasing leakage energy in conventional SRAM-based caches, several non-volatile random-access memory (NVRAM) technologies [1] are being investigated as candidates for future on-chip processor caches, and off-chip primary memory. Among these NVRAM solutions, *spin-torque transfer RAM* (STT-RAM) [2] has emerged as one of the promising candidates. Besides its non-volatility, STT-RAM offers an access (read) speed comparable to SRAM, a high integration density comparable to DRAM, and is compatible with the CMOS process.

However, there are three main challenges in deploying STT-RAM as on-chip caches. The first two are its long write latency and high dynamic write energy. Many architectural solutions to mitigate these issues [3, 4, 5, 6, 7, 8, 9] have been investigated, mostly with the STT-RAM used in last level caches. Sun et. al. [10] has recently proposed architectures for pure STT-RAM L1 caches. Still, there is a third challenge that few of the earlier work had considered: reliability. In this paper, we would like to address all three challenges in using STT-RAM in the highest level of the cache hierarchy.

The key insights of the paper are that under a shared memory multicore environment, most of the private L1 cache blocks have predictable behaviors during an application's execution, and the transition between different cache coherency states provides a good approximation of a block's future behavior. Based on these observations, we propose a novel hybrid private L1 cache design scheme that operates with a small SRAM partition, and a larger STT-RAM partition. Our approach not only

brings energy benefits to processors, but also significantly reduces the performance impact as well as the risk of STT-RAM cell failure. In particular, this paper makes the following contributions:

- We present a hybrid L1 cache architecture design that attempts to mitigate the performance and energy penalty induced by the slow and energy-consuming write operation of STT-RAM.

- We propose three different cache block migration schemes that are based on the MESI protocol. Simulation shows that while compared to a conventional pure SRAM-based cache design, our approach achieved significant energy savings with minor IPC degradation.

- We investigate the write endurance issue of STT-RAM and evaluate the improvement achieved by our hybrid cache architecture. We found that the write endurance of STT-RAM cache in our hybrid design is significantly improved compared to a pure STT-RAM-based design.

II. STT-RAM BASICS

The data storage unit inside a STT-RAM cell is the *magnetic tunnel junction* (MTJ). Fig. 1 shows a typical one transistor, one MTJ (1T-1MTJ) cell structure. The MTJ has two ferromagnetic layers that are separated by an oxide tunnel barrier. The ferromagnetic layer at the bottom (the reference layer) comes with a fixed magnetization direction while that for the top layer (the free layer) can be changed. When the magnetization direction of the free layer is parallel (anti-parallel) to the reference

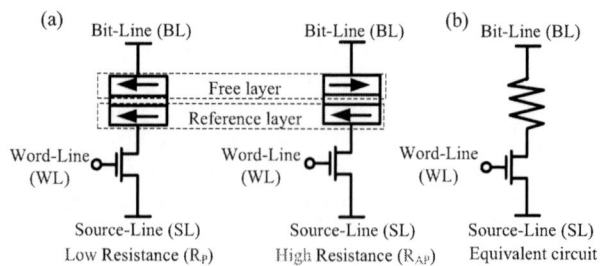

Fig. 1. STT-RAM cell structure

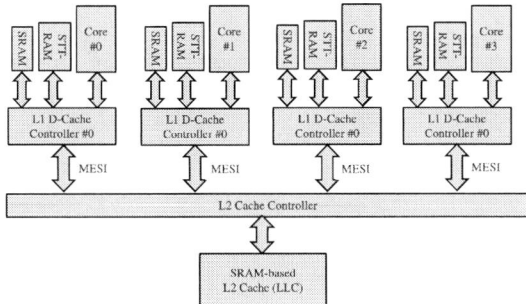

Fig. 2. The hybrid cache hierarchy.

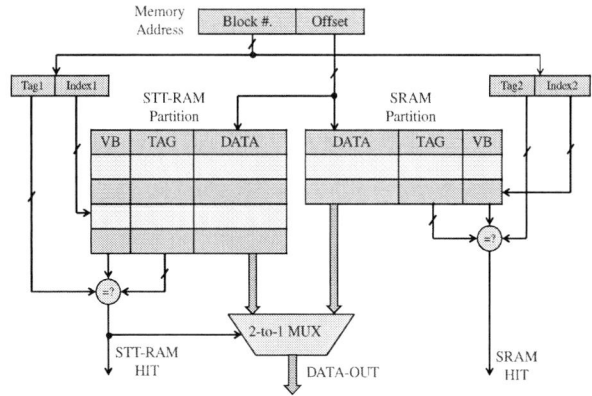

Fig. 3. Read from the hybrid cache.

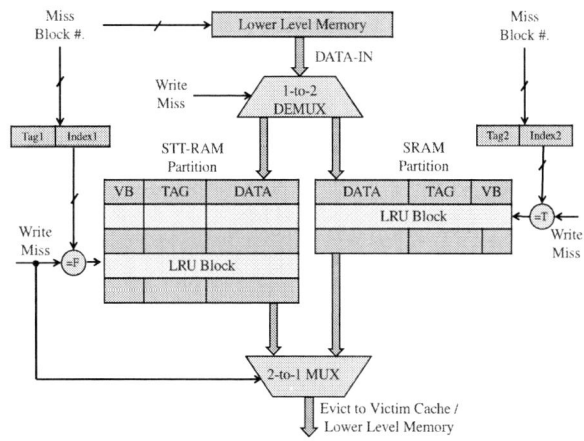

Fig. 4. Miss handling for the hybrid cache.

layer, the overall resistance of the device is in low (high) state, which can be read out by injecting a sensing current. Conceptually, logical '1' is represented by low conductivity while logical '0' corresponds to the high conductive state. To write the device, a strong polarized current is applied from source-line to bit-line, which switches the magnetization direction of the free layer and turns the cell into parallel or anti-parallel state.

The *data retention time* (the expected time until a bit flip appears) of the STT-RAM cell can vary from 10+ years to microsecond level, depending on the MTJ designs [10].

III. HYBRID L1 CACHE ARCHITECTURE

A. The Proposed Hybrid Cache Hierarchy

Fig. 2 illustrates our proposed hybrid cache hierarchy for a quad-core machine. Each of the processor core has a private L1 cache comprising of a SRAM and a STT-RAM partition while sharing a unified L2 cache. The L1 cache controller is responsible for handling cache access operations, and transferring cache blocks between the two partitions when necessary.

B. A Naïve Solution

Due the high write latency and write energy of STT-RAM, it is desirable to move writes to the SRAM partition, and only read from the STT-RAM partition. However, the difficulty lies in predicting a cache block's access pattern. A naïve solution would allocate all read-miss blocks to STT-RAM, and write-miss blocks to SRAM when a cache miss happens. This method assumes for every cache block, the first operation on it would be the dominant one for this block until it is evicted. However, such an assumption does not guarantee to hold all the time. For example, during an application's warming up phase, cache behavior is generally unstable, and would easily invalidate this assumption.

Fig. 3 illustrates a read operation under the naïve implementation. Both cache partitions are accessed simultaneously, and at most one partition would produce a cache hit. On a cache hit, the corresponding data would be sent via a 2-to-1 MUX to the processor. The MUX is selected by the hit signal. In any case, the additional delay caused by the MUX is negligible when compared to a pure SRAM design as the two accesses are done in parallel.

Under this hybrid model, cache miss is defined by none of the partitions produces a hit. When it happens, a prediction is made and the data block is loaded into the partition that potentially can maximize the performance and energy benefits. Fig. 4 illustrates this process. A DEMUX is added at the top to direct the input data from lower level memory to corresponding cache partition based on the prediction. Since STT-RAM has a longer write latency than SRAM, read-misses require more time to handle than write-misses. In addition, the *hit ratio* (HR) of each partition can be calculated separately by

$$HR_{SRAM} = \frac{\#Hits_{SRAM}}{\#Hits_{SRAM} + \#Misses_{write}}$$

$$HR_{STT-RAM} = \frac{\#Hits_{STT-RAM}}{\#Hits_{STT-RAM} + \#Misses_{read}}$$

C. Analysis of Cache Block Behavior under MESI

Although the naïve prediction method does reduce the energy consumption by a certain amount when compared to the pure SRAM baseline, it incurs a relatively large impact on performance (Section IV.B). The critical point is to mitigate the performance penalty induced by those ill-placed blocks which reside in a less beneficial cache partition.

978-1-4799-2817-0/14 $31.00 © 2014 IEEE   611

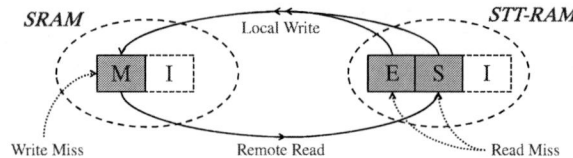

Fig. 5. Immediate transfer (IT) diagram.

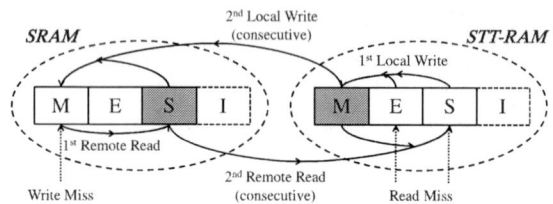

Fig. 6. Delayed transfer (DT) diagram.

Under the MESI protocol, if at one point in time a cache block is in the MODIFIED (M) state, it must have been written to before, and more importantly, it probably will be written to *again* exclusively by the current owner in the near future. In other words, such blocks are likely to be overwritten many times, and might benefit more by staying in the SRAM partition rather than in STT-RAM partition. On the other hand, EXCLUSIVE (E) and SHARED (S) states indicate that the block is a clean copy for now, and is likely to remain clean as a read-only exclusive/shared copy. Hence, it is better to keep such blocks in the STT-RAM partition of the cache. In addition, during a program's runtime, it can happen that a cache block may have to be moved from one partition to the other in order to take the advantages of both caches. Thus, some mechanisms are required to take care of the block transfer decision.

D. Transfer Mechanism 1 (Immediate Transfer)

Our first proposed block transfer mechanism is called the *immediate transfer policy* (IT). Fig. 5 shows a block's transition state and movement under this policy. While the naïve solution is based entirely on temporal locality, IT takes things a step further. A remote read operation hit on SRAM partition would cause the corresponding block to be evicted, and transferred to the STT-RAM partition in the new S state, while a local write hit on a STT-RAM partition makes the block dirty, and necessitate its transfer to the SRAM partition of the cache under its new M state. Note that here we only consider those operations that can change the coherent state, so there is no extra information required. Besides, the way of handling cache misses is the same as in the naïve solution.

Note that the order of handling the actual memory request and transferring the affected cache blocks can potentially cause differences in performance and energy saving, but we found that the variations is generally small, and can be safely ignored. Throughout this paper, we assume a block transfer completes before serving the memory request.

E. Transfer Mechanism 2 (Delayed Transfer)

Although IT does result in a better overall performance than the naïve approach, it fails to consider certain cases. In a typical scenario of a single producer with multiple consumers, some cache blocks may keep changing their states between M and S. IT may therefore be too aggressive under this circumstance. The *delayed transfer policy* (DT) differs from IT by delaying the transfer triggered by the first operation. Fig. 6 illustrates the transfer decision making process. Block migration only happens when two consecutive operations satisfying the transfer condition are encountered. In particular, to move a cache block

from SRAM partition to STT-RAM partition, we need to have two remote reads with no intervening writes, while in order to move a block from STT-RAM partition to SRAM partition, two consecutive local writes must have be encountered.

The flowchart of DT is shown in Fig. 7. In the implementation, an extra *transfer determine* bit (TD bit) is added to each cache blocks. All newly-allocated blocks will have the TD bit reset to '0', but the runtime set and reset policy is different between SRAM partition and STT-RAM partition. In the SRAM partition, a remote read will set the TD bit to '1' while any write will have it reset to '0'. On the other hand, in the STT-RAM partition, only local writes set the TD bit to '1' and any other reads will reset it to '0'. Hence, when a potential transfer-trigger operation is encountered in a cache partition, TD bit can be used to determine whether a transfer is actually needed. Unlike IT, this transfer policy does not impose any MESI state restriction of a cache block in the caches, hence we may encounter blocks with E or S state in the SRAM partition, as well as M blocks in STT-RAM partition. The main objective here is to eliminate the thrashing effect where a block keep moving between the two partitions without receiving enough references.

IV. EVALUATIONS

A. Experiment Setup

To evaluate our proposed design, we use the cycle-accurate simulator MARSSx86 [11] to model a conventional multicore x86-64 processor with a 2-level cache hierarchy as shown in Fig.2. Ten diverse multi-threaded workloads from PARSEC 2.0 [12] benchmark suite were chosen for the experiment. The complete list of simulator parameters is given in Table I. Both

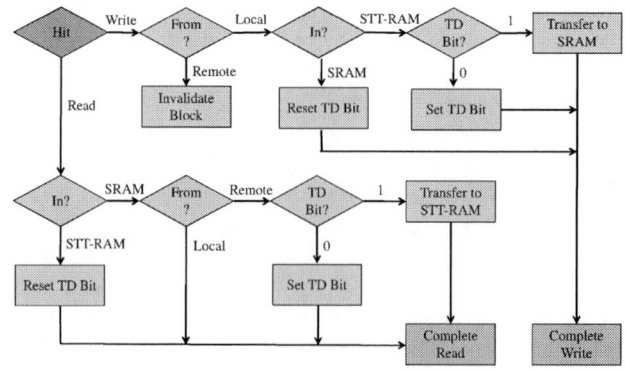

Fig. 7. Flowchart of DT under cache hit.

TABLE I
SIMULATION PLATFORM.

Processor Core	3GHz frequency, 4 cores
Pipeline Width	4
Functional Units	2 ALUs, 2 FPUs, 2 LUs, 1 SU
ROB / LSQ Entries	128 / 80
In-flight Branches	24
ITLB/DTLB Entries	32 / 32
Cache Block Size	64-byte
L1 I-Cache (SRAM)	64KB, 8-way, 3-cycle latency
L1 D-Cache (SRAM)	4-64KB, 8-way, 3-cycle latency
L1 D-Cache (STT-RAM)	64/128KB, 8-way, 3-cycle read latency, 9-cycle write latency
Coherent Protocol	Illinois MESI
Shared L2 Cache (SRAM)	2MB, 16-way, 15-cycle latency

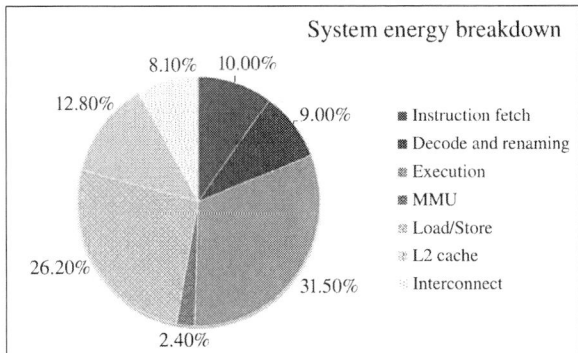

Fig. 8. Energy breakdown for each pipeline stage.

of the SRAM and STT-RAM cache energy and latency numbers were generated using NVSim [13] with a 32nm technology node assumed. To improve the write performance of STT-RAM, its non-volatility is relaxed with a retention time of $26.5\mu s$ [10]. This assumes that a conventional DRAM-style refresh scheme is used to maintain data integrity. The baseline reference is a pure 64KB SRAM cache and we took care not to exceed the silicon area of this when exploring possible hybrid cache size configurations.

Before evaluating the energy consumption of our hybrid cache architecture, we shall first show the energy distribution for each component within the processor. Fig. 8 presents a typical breakdown of energy for each pipeline stage during runtime using the McPAT tool [14] with our processor configuration (Table I) as input. The statistics were generated according to the event log of MARSSx86's cycle-accurate simulation. In particular, the load/store unit, and the L1 data cache in particular (which is part of the load/store unit) consumes more than 1/4 and 1/5 of the total energy, respectively. This result justify our effort in reducing the energy consumption of L1 cache. To simplify the discussion, we will only focus on the energy consumption of L1 data cache for the rest of the paper.

B. Performance

Fig. 9 shows the normalized *instructions-per-cycle* (IPC) for the different hybrid cache configurations running under the IT policy. All the hybrid cache configurations maintain a comparable performance to the pure SRAM baseline. In the worst case, only an average 1.5% performance degradation was recorded for the smallest size configuration. This is even better than the best case in the naïve solution (Fig. 11).

However, a particularly bad degradation of nearly 8% loss in IPC was recorded in `swaptions`. Unlike the others, `swaptions` requires more exclusive blocks for writing during its execution. This is why there is a continuous performance improvement as the size of the SRAM partition increases. Without a sufficiently large SRAM partition, doubling the size of STT-RAM partition will not compensate for the performance loss.

Fig. 11 compares the performance for the different transfer scenarios. Across all the cache size combinations, both IT and DT perform better than the naïve solution. On the other hand, the IPC differences between IT and DT are marginal. The largest one recorded is in the smallest cache size configuration (4KB SRAM + 64KB STT-RAM), where DT outperforms IT by around 1%. Notice that a serious IPC drop of 5.7% is recorded when the baseline is directly replaced with pure STT-RAM. This proves the importance of our hybrid design.

C. Energy

The total energy consumption is modeled by the sum of three components: *leakage energy*, *refresh energy* and *dynamic energy*. Besides the usual runtime read/write energy, dynamic energy in our context also includes the energy for probing both cache partitions, and the *transfer energy* when one of the block transfer policies is applied.

Fig. 10 shows the normalized energy consumption for the hybrid cache with IT policy. In general, our design saves more energy in those read-intensive benchmarks such as `vips` and `swaptions`. Even for the most write-intensive benchmark, `raytrace`, around 30% of energy saving is achieved in the best case. When taking IPC into consideration, a medium cache size combination (8KB SRAM + 64KB STT-RAM) provides the best trade-off among the four configurations tested.

Within the three transfer scenarios (Fig. 11), the largest energy saving occurs in the smallest cache configuration, though

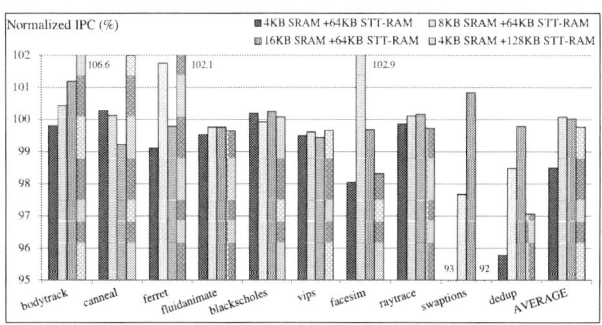

Fig. 9. Normalized IPC for hybrid cache with IT policy.

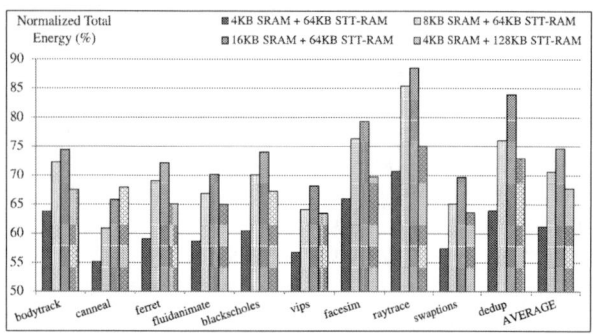

Fig. 10. Normalized total energy for hybrid cache with IT policy.

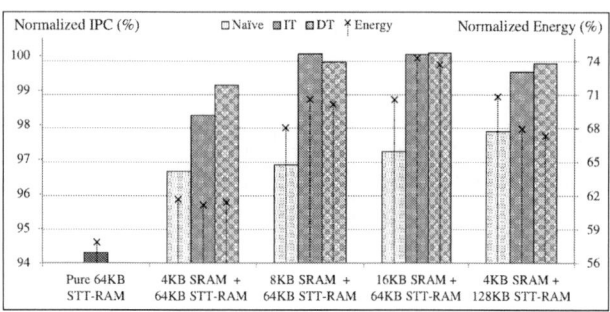

Fig. 11. Normalized IPC and total energy averaged from all workloads.

the lowest IPC rates were recorded for all of them. This is because the leakage and refresh energy consumed in a larger cache is more significant than the energy spent in extra dynamic operations that are caused by either a higher miss ratio (smaller cache) or ill-placed blocks (naïve solution). Although the pure STT-RAM solution consumes the least energy, the difference between it and the best hybrid scheme is merely 3%, while the IPC drop is significant. Overall, we consider DT(4KB SRAM + 64KB STT-RAM) to be the best hybrid cache configuration since it provides the best energy-IPC trade-off (saves 38% energy under a mere 0.8% IPC degradation).

D. Impact of Retention Time

The retention time of STT-RAM cells is related to the thermal barrier Δ of an MTJ, which can be expressed as $t = C \times e^{k\Delta}$ where t is the retention time, and C and k are fitting constants [9]. Any variations in the planar area and the thickness of the MTJ affect the thermal barrier, and thus impact the retention time. However, given a particular set of read/write latency and the cache size, the lowest possible retention time under a DRAM-style refresh scheme is bounded by $\#cache\,blocks \times (read\,latency + write\,latency) \times cycle\,time$.

Fig.12 shows the IPC and energy consumption under various retention time values for a 4KB SRAM + 64KB STT-RAM hybrid cache with IT policy. The data are normalized to the perfect case when the refresh is completely eliminated (inf). In general, a lower retention time causes more refresh conflicts and thus increases the latency for memory requests. It also consumes more energy due to a more frequent refresh schedule, but the impact become marginal while the refresh period

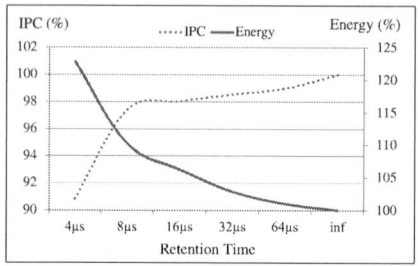

Fig. 12. IPC and energy consumption for various retention time values of a 4KB SRAM + 64KB STT-RAM hybrid cache with IT policy.

is larger than $32\mu s$. We demonstrate these results is because STT-RAM are not yet in commercial deployment and the manufacturing parameters are not fully known. An extreme short retention period can seriously downgrade performance and lead to a higher energy consumption.

E. STT-RAM Endurance Study

The *write endurance* issue for STT-RAM has so far been only considered in the context of last level caches [15]. However, the issue is even more pressing when STT-RAM technology is employed in L1 caches as they have a much higher write activity than last level caches. Although a prediction of 10^{15} programming cycles [16] is often cited as the write endurance for STT-RAM, real experiments thus far have showed that the achieved endurance is only about 10^{13} write cycles [2]. This would be a severe constraint on any pure STT-RAM L1 solution, and is one of the key motivations for our hybrid design.

Fig. 13 demonstrates the average number of writes to STT-RAM partition happened within each CPU cycle. Among all the benchmarks, `facesim` has the highest average writes per cycle. Consider the case when the writes are perfectly distributed to all cache blocks in `facesim` for a pure STT-RAM design, each block would need to stand 2.4×10^5 writes per second. Under a conservative estimate, each block has a lifespan of about 1.3 years.

However, further analysis shows that writes are not at all evenly distributed. This is true within a single cache as well as across different private caches. We observed that some blocks are seldom used while several other blocks suffer from

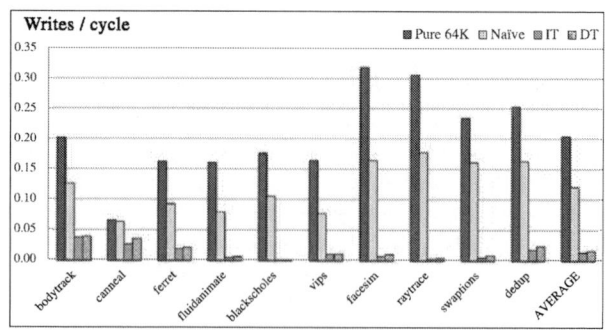

Fig. 13. Average number of STT-RAM writes occurs in each CPU cycle. All hybrid caches use 4K SRAM + 64K STT-RAM with IT policy.

TABLE II
STT-RAM CACHE LIFESPAN ESTIMATION BASED ON THE WRITE
FREQUENCY OF FACESIM.

	Average partition	Worst partition	Worst block
Pure STT-RAM	1.3 years	0.3 years	< 22 mins
Hybrid Naïve	3.5 years	1.0 year	0.9 hr
Hybrid IT	41.2 years	6.9 years	51.6 hrs
Hybrid DT	32.9 years	7.0 years	54.3 hrs

All hybrid caches use 4K SRAM + 64K STT-RAM.

a huge amount of writes. For example, about 15% writes target at a particular block in facesim, while the corresponding cache partition receives nearly 50% writes among the four private caches. In another words, assuming a 3 GHz clock and a write endurance of 10^{13} write cycles, that STT-RAM block may fail within half an hour if no proper measure is taken. Due to the low-latency requirement and high access frequency of L1 caches, existing wear-levelling methods for last level caches [15] are not feasible.

Table II lists the average and worst case lifespan of the STT-RAM cache for facesim under the write endurance assumption of 10^{13} programming cycles. The "average partition" column assumes that the writes are perfectly distributed to all of the private caches while the "worst partition" only assumes the writes are evenly distributed within each private cache partition. In the "worst block" case, the actual number of writes per cache block is computed. Compared to a pure 64KB STT-RAM solution, the lifespan of the most write-intensive cache block in a hybrid configuration with the DT policy is increased by up to $150\times$, while the worst cache partition lifespan is increased by 2333%. Although the DT policy performed slightly better than IT policy in the worst case endurance, it suffered from a lower average lifespan because the blocks in STT-RAM recieve more writes by the delayed migration. Note that the naïve solution for hybrid cache is still insufficient for actual deployment.

In practice, caches would come with redundancy or error correction code to improve reliability. Also, it is very unlikely to see such a sustained high write frequency on a particular block especially for personal workload. Thus the worst case block lifespan should be much longer. Nonetheless, the risk exists, especially for the current state-of-the-art in STT-RAM technology. Our experiments showed clearly that the proposed hybrid architecture can significantly reduce this risk.

V. CONCLUSION

This paper proposed a hybrid L1 cache architecture that uses both conventional SRAM as well as the new STT-RAM technology. By exploiting the MESI cache coherence protocol, our design mitigates the impact of STT-RAM's high write latency on IPC performance, while reducing the overall energy consumption compared to a pure SRAM-based design. In addition, our proposal enhances the reliability of STT-RAM, making it more suitable for actual deployment. As the CMOS technology continues to scale, increasing energy consumption and design complexity demands for more power-efficient designs. We believe that our hybrid cache is an attractive architecture for the next-generation non-volatile computing.

ACKNOWLEDGMENTS

This research was supported in part by the Singapore Ministry of Education Research Grant MOE2010-T2-1-075.

REFERENCES

[1] R. Bez and A. Pirovano, "Non-volatile memory technologies: emerging concepts and new materials," *Materials Science in Semiconductor Processing*, vol. 7, no. 4, pp. 349–355, 2004.

[2] Z. Diao, Z. Li, S. Wang, Y. Ding, A. Panchula, E. Chen, L. Wang, and Y. Huai, "Spin-transfer torque switching in magnetic tunnel junctions and spin-transfer torque random access memory," *Journal of Physics: Condensed Matter*, vol. 19, no. 16, 2007.

[3] W. Xu, H. Sun, X. Wang, Y. Chen, and T. Zhang, "Design of Last-Level On-Chip cache using Spin-Torque transfer RAM (STT RAM)," *IEEE Transactions on Very Large Scale Integration Systems*, vol. 19, no. 3, pp. 483–493, 2011.

[4] C. Smullen, V. Mohan, A. Nigam, S. Gurumurthi, and M. Stan, "Relaxing non-volatility for fast and energy-efficient STT-RAM caches," in *IEEE 17th International Symposium on High Performance Computer Architecture*, pp. 50–61, 2011.

[5] A. Jadidi, M. Arjomand, and H. Sarbazi-Azad, "High-endurance and performance-efficient design of hybrid cache architectures through adaptive line replacement," in *International Symposium on Low Power Electronics and Design*, pp. 79–84, 2011.

[6] M. Rasquinha, D. Choudhary, S. Chatterjee, S. Mukhopadhyay, and S. Yalamanchili, "An energy efficient cache design using spin torque transfer (STT) RAM," in *ACM/IEEE International Symposium on Low-Power Electronics and Design*, pp. 389–394, 2010.

[7] G. Sun, X. Dong, Y. Xie, J. Li, and Y. Chen, "A novel architecture of the 3D stacked MRAM l2 cache for CMPs," in *IEEE 15th International Symposium on High Performance Computer Architecture*, pp. 239–249, 2009.

[8] J. Li, C. Xue, and Y. Xu, "STT-RAM based energy-efficiency hybrid cache for CMPs," in *IEEE/IFIP 19th International Conference on VLSI and System-on-Chip*, pp. 31–36, 2011.

[9] A. Jog, A. K. Mishra, C. Xu, Y. Xie, V. Narayanan, R. Iyer, and C. R. Das, "Cache revive: Architecting volatile stt-ram caches for enhanced performance in cmps," in *Design Automation Conference (DAC), 2012 49th ACM/EDAC/IEEE*, pp. 243–252, 2012.

[10] Z. Sun, X. Bi, H. H. Li, W. Wong, Z. Ong, X. Zhu, and W. Wu, "Multi retention level STT-RAM cache designs with a dynamic refresh scheme," in *Proceedings of the 44th Annual IEEE/ACM International Symposium on Microarchitecture*, pp. 329–338, 2011.

[11] A. Patel, F. Afram, S. Chen, and K. Ghose, "MARSS: a full system simulator for multicore x86 CPUs," in *Proceedings of the 48th Design Automation Conference*, DAC '11, pp. 1050–1055, 2011.

[12] C. Bienia, S. Kumar, J. P. Singh, and K. Li, "The PARSEC benchmark suite: characterization and architectural implications," in *International Conference on Parallel Architectures and Compilation Techniques*, pp. 72–81, 2008.

[13] X. Dong, C. Xu, Y. Xie, and N. Jouppi, "NVSim: a Circuit-Level performance, energy, and area model for emerging nonvolatile memory," *IEEE Transactions on Computer-Aided Design of Integrated Circuits and Systems*, vol. 31, no. 7, pp. 994–1007, 2012.

[14] S. Li, J. H. Ahn, R. D. Strong, J. B. Brockman, D. M. Tullsen, and N. P. Jouppi, "Mcpat: an integrated power, area, and timing modeling framework for multicore and manycore architectures," in *42nd Annual IEEE/ACM International Symposium on Microarchitecture.*, pp. 469–480, 2009.

[15] Y. Chen, W. Wong, L. Hai, and C. Koh, "Processor caches built using multi-level spin-transfer torque ram cells," in *Proceedings of the 2011 ACM/IEEE international symposium on Low power electronics and design*, ISLPED '11, pp. 73–78, 2011.

[16] F. Tabrizi, "The future of scalable stt-ram as a universal embedded memory," *Embedded.com*, 2007.

Allocation of FPGA DSP-Macros in Multi-Process High-Level Synthesis Systems

Benjamin Carrion Schafer

Department of Electronic and Information Engineering

The Hong Kong Polytechnic University

(b.carrionschafer@polyu.edu.hk)

Abstract— High-Level Synthesis (HLS) is a single process synthesis method that has shown to produce very good results compared to hand coded RTL, especially for DSP-related applications. At the same time FPGAs are reaching capacities that allow entire systems to be implemented on them. Most of these systems are also DSP-related and make intensive use of the FPGAs' embedded hardmacros (e.g. DSP-blocks). This works presents a method to efficiently allocate DSP-macros in multi-process systems created using HLS in order to minimize the overall area. The proposed method calculates the area sensitivity of each process when its multiply-accumulate (MAC) operations are either mapped onto the FPGA's hardmacro or its configurable resources and allocates the available hardmacros across all processes. Experimental results show that our method creates very good results compared to the optimal solution at a negligible running time.

I. Introduction

High-Level Synthesis (HLS) is finally being used extensively in commercial designs. The increase in productivity combined with the improvement in the quality of the results of commercial HLS tools has lead many design teams to make the transition. Most of the applications are Digital Signal Processing (DSP) related, which contain large amounts of Multiply-Accumulate(MAC)operations. For these types of applications, also called data-intensive applications, a well known optimization technique called resource sharing [1] can be used in order to reduce the area. In resource sharing a single functional unit (FU) is re-used among different computational operations in the behavioral description. The total design area can be significantly reduced by using resource sharing in HLS compared to a manually optimized RTL implementation [2]. Although RTL designers can also manually create architectures which shared resources, they generally do not like to perform manual resource sharing because it makes it much harder to design an especially to verify (a Finite State Machine (FSM) is needed in order to generate the control signals for the muxes and the timing can easily become very complicated). In contrast, HLS tools can easily maximize the amount of resource sharing and create au-

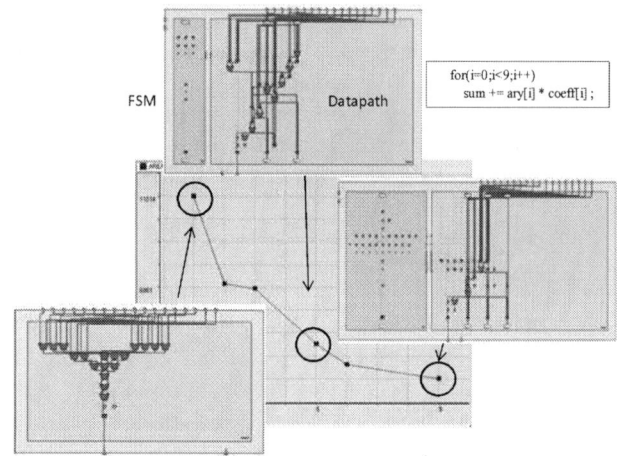

Fig. 1. FIR filter Functional Unit (FU) Design Space Exploration (DSE) results targeting ASIC 45nm(x-axis=Area, y-axis=Latency)

tomatic testbenches to verify their correct behavior.

A typical approach to achieve different degrees of resource sharing in HLS is by constraining the maximum number of FUs that the high-level synthesizer can instantiate. The high-level synthesizer will maximize the amount of resource sharing based on the number and type of available FUs specified in the resource constraint file. In traditional ASIC designs this leads to the generation of designs with very different area vs. performance trade-offs. Fig. 1 shows the trade-off curve obtained by a commercial HLS tool [9] for a 9-tap FIR filter when the number of MACs FUs are reduced from 9 to 1 targeting an ASIC 45nm technology (area vs. latency). Three different architectures have been highlighted. The left-most is the highest performance architecture with no resource sharing at all. It can be observed that the architecture has no FSM and that it is only composed of the datapath containing the 9 MAC. The schematic view of this design also reveals that the filtering will be executed in a single clock cycle and hence the circuit latency=1. The other two examples present different degrees of resource sharing. It is easy to observe from the schematic diagram that the higher the amount resource sharing the larger the FSM logic becomes. This implies that the number of states increases and so does the design latency, decreasing

978-1-4799-2817-0/14 $31.00 © 2014 IEEE

at the same time the overall design area as shown in the trade-off curve.

At the same time, FPGA device capacities keep increasing and currently entire systems can now be implemented on single devices. It is therefore important to understand the differences to efficiently synthesize designs using HLS when targeting ASICs vs.FPGAs. When re-synthesizing the FIR filter used previously targeting a Xilinx Virtex 6 FPGA a very different result is observed as shown in Fig. 2. By default the trade-off curve labled DSP Macros is obtained. It can be observed that in contrary to the ASIC case, the area grows when the number of FUs is reduced. The main reason as observed in [6] is that the MAC operations of the FIR filter are now being mapped onto the FPGA's embedded DSP-macros, which effectively make them *free*, while the cost of the multiplexers increases when the amount of resource sharing increases. Another important observation is that if the same designs are re-synthesized mapping the MAC operations onto the configurable logic (LUTs), then a similar response to the one obtained in the ASIC case is observed (Fig. 2 - LUTs). Finally Fig. 2 also shows the overlapped synthesis result when all MAC operations are either mapped on the FPGA's DSP-hardmacros or on the LUTs. It can be observe that in all cases the implementation using DSP-macros is superior (dominates) the implementation when the MACs are implemented on the LUTs.

A quick conclusion of these observations would be to always use the FPGA's hardmacros to implement MAC operations and to reduce the amount of resource sharing as much as possible. The main problem with this conclusion is that it obviates the fact that HLS is a *single* process synthesis. As FPGAs capacities increase, complete multi-process systems can now be implemented on them. A method to allocated the DSP-macros effectively across multi-processes systems is therefore needed. It could also be counter-argued that state of the art FPGAs have large amount of DSP-macros and that hence the designer does not need to care about their allocation. These is true for high-end FPGAs, e.g. Xilinx Virtex 7 contains between 1,260 to 3,600 DSP-macros or Altera's Stratix 5 containing 512 to 3,926 DSP-macros. Unfortunately none of these devices can be used to target consumer products as their unit price is prohibitively expensive and therefore these devices are normally only used for ASIC prototyping. Low-end FPGAs, e.g. Xilinx Artix 7 FPGAs and Altera's Cyclone 5 FPGAs, contain between 60-740 (25x18 multiplier) and 50-684 (18x18 multiplier) DSP-macros respectively depending on the selected device with an average of 218 DSP-macros for the Artix 7 FPGA family and 243 for the Cyclone 5 family. One last argument defending the need to allocate the FPGAs' DSP-macros in a systematic way is that for the 9-tap FIR filter case without resource sharing shown in the previous example 9 DSP48 were consumed, while only 16 LUT slices were needed. It could easily happen that for larger DSP sys-

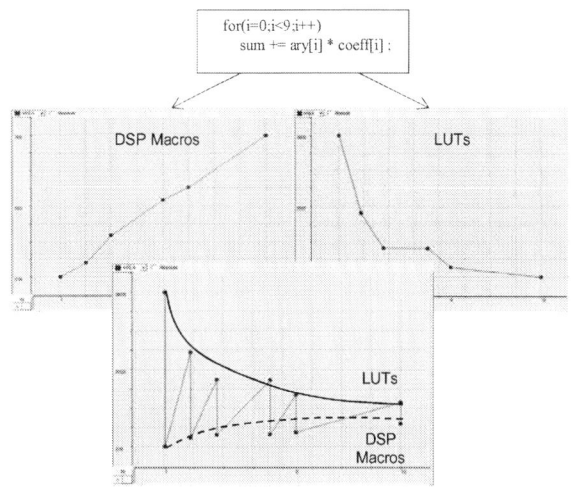

Fig. 2. FIR filter resource sharing exploration results (DSPs vs.LUTs) targeting FPGA .

tems (e.g. decimation filters) the DSP-macros could be quickly consumed, while much of the configurable logic would be kept unused.

In this work, we therefore study the effect of instantiating MAC operations onto DSP-macros or onto the FPGAs configurable logic in systems composed of different DSP related applications and propose a method to efficiently allocate these FPGA's DSP-macros. In particular the novelty of this work is twofold: Firstly, we introduce an automatic HLS Design Space Explorer (DSE) for FPGAs which explores not only the number of Functional Units(FUs) increasing/decreasing the amount of resource sharing for behavioral descriptions given in ANSI-C/SystemC but also mapping their MAC operations to either DSP-macros or the configurable logic. Secondly, we propose a method called *Allocation of DSP-macros for Multiple Processes* (ADSP_MULTP) to efficiently map a maximum number of given DSP-macros to multi-process FPGA designs reducing the total design area under given latency constraints.

II. BACKGROUND AND RELATED WORK

Resource sharing is a typical technique used in High-Level Synthesis(HLS) to reduce area in which a single Functional Unit (FU) is re-used among different computational operations in the behavioral description. Nevertheless when targeting FPGAs resource sharing often leads to larger designs, as shown in the previous section, due to the availability of embedded DSP-Macros and due to the high cost of multiplexers (MUXes) in FPGAs.

In [6] it was shown that in some cases, resource sharing of some non-multiplication operations in FPGAs can lead to smaller designs for larger LUT sized FPGAs e.g. Stratix IV vs. traditional 4-input LUT FPGAs due to the ability to combine portions of the sharing muxes to be combined into the same LUTs that implement the operators themselves. Resource sharing that target particulary FPGAs has been studied previously in [3], [4] and [5]. In

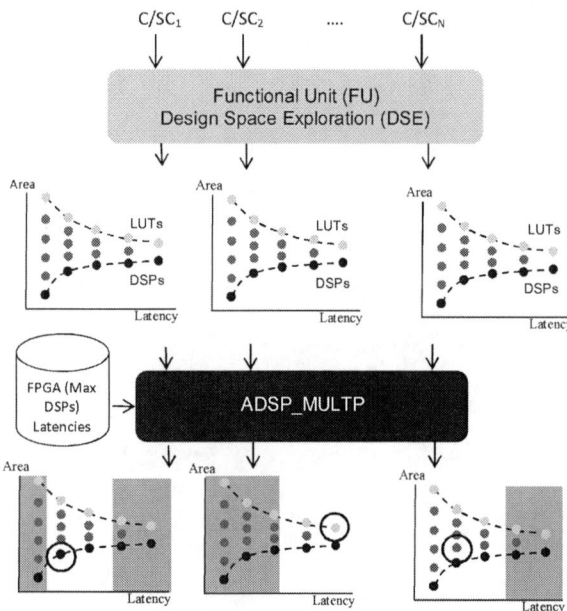

Fig. 3. Overall design flow overview.

all cases the effect of resource sharing is only studied at the single process level, but not on multi-process systems.

On the other hand, the design space exploration (DSE) of FUs for HLS has been mainly targeted to the number and type of FUs (e.g. non-pipelined vs. pipelined). Smita et al. [7] introduced the concept of component selection for DSE. More recent work target specifically FPGAs [8]. In this work the authors perform a combined DSE, modifying the number of FUs and performing different component selections. The effect of FU number exploration and implementation type (on the FPGA's DSP-macros) or configurable logic has not been studied so far. To the best of our knowledge no previous work has attempted to study this effect over multiple processes given the result of a DSE with different degrees of resource sharing and implementations of the MAC operations on either the configurable logic or the FPGA's DSP-macros.

III. PROPOSED METHOD

The DSP-macro allocation method proposed in this work is called *Allocation of DSPs for Multiple Processes* (ADSP_MULTP). Fig. 3 shows an overview of the complete flow. The method starts by performing a DSE for each individual process, specified by its behavioral description in ANSI-C/SystemC. The explorer, as we will see in the next subsection in detail, explores the number of FUs and their implementation in either the FPGA's DSP-macros or LUTs. Once the designs space for each process is explored, our method continuous by selecting the implementation that minimizes the total system area given a fixed global DSP-macro budget and a latency constraints for each of the processes. The next sub-sections describe in detail how the DSE and the DSP-macro allocator work.

A. Design Space Explorer

The DSE starts by parsing each individual process's behavioral description given in ANSI-C or SystemC and performing HLS and logic synthesis on it in order to report accurate results. No FU constraint is set at this first synthesis. This leads to the HLS tool to instantiate the maximum number of FUs needed to parallelize the behavioral description as much as possible and hence resulting in the fastest of all the possible implementations as no resource sharing is performed at this initial design. By default this design will also use the FPGA's DSP-macros and hence lead to a small design(left-most bottom design in Fig. 4 labeled as *Max FUs*).

The DSE then continues by retrieving the number and type of FUs used for this first design and by reducing it automatically by a fixed percentage. It was found empirically that 20% provides a good compromise between running time and design space covered. This will increase the amount of resource sharing and hence increase the latency of the generated circuit. As shown at the introduction, the area for most of the DSP applications observed increases with the reduction of MAC operators due to the increasing size of the muxes needed to support the increased sharing of these resources. Nevertheless, it was also observed that for some other applications the trade-off curve followed a parabolic curve, where designs with fewer FUs actually lead to smaller designs than the initial design with no resource sharing at all. This explorations step finishes when all the FUs are set to one, which maximizes the amount of resources sharing and could potentially lead to the smallest possible design, but also the slowest of all the designs. Fig. 4(a) shows the result of this very first exploration step.

The DSE then continues by revisiting all the different designs and mapping the multipliers increasingly from the FPGA's hardmacros to its LUT's. As shown in Fig. 4(b) the area for a given design grows as more multiplications are migrated onto the configurable logic. This continues until the the entire design space has been explored as shown in Fig. 4(c).The problem here is that the HLS tool does only allow all multipliers to be mapped onto DSP-macros or LUTs, but not partially assigning some to DSPs-macros and some to LUTs, e.g. how to specify which MAC operation should be mapped to DSP-macros or LUTs at a for-loop calculating a sum or products as shown in the introductional example? To circumvent this limitations the DSE directly annotates at the RTL code generated by the HLS the implementation type using FPGA vendors' synthesis attributes. This gives the DSE full control on where the multiplier should be instantiated. E.g. for Xilinx:

```
attribute use_dsp48 : string;
attribute use_dsp48 of mul16s9ot : signal is "no";
attribute use_dsp48 of mul16s8ot : signal is "yes";
```

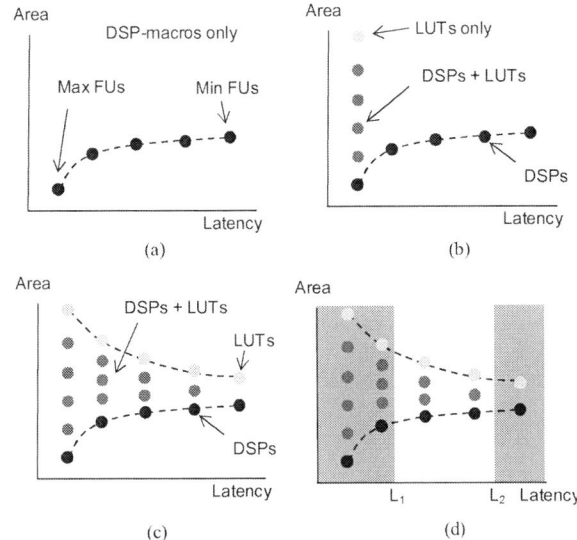

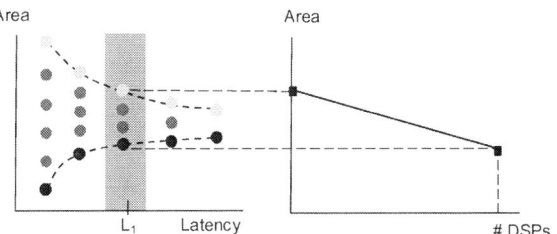

Fig. 5. Example of sensitivity for a given design at a given latency.

Fig. 4. Functional Unit (FU) Design Space Exploration (a) Resource sharing exploration (b) Multiplication instantiation exploration (DSP-macros vs. LUTs) (c) Final exploration result (d)Latency constraint result

Although most of the designs returned by the DSE are not Pareto-optimal, they cannot be discarded because very often the design latency is a global constraint. This constraint can vary during different project stages e.g. when the process is integrated into the system or when it is re-used in later projects. Therefore the full exploration results are stored and the most efficient implementation is selected when the latency or latency interval constraint is specified. Only those designs within the specified latency interval are considered by our method as shown in Fig. 4(d).

B. DSP-macros Allocation Method

Once the DSE finishes, our proposed method continues by allocating selected multiplications of the different processes onto the FPGAs DSP-macros. Fig. 6 shows a summary of our proposed method ADSP_MULTP, which consists of 4 main steps. Our method starts by searching the smallest design for a given latency L_n or latency range $[L\_n, L\_m]$. If no latency constraint is given then the entire design space is considered. The method then continues by calculating the sensitivity S of each process to implementing MAC operations on either DSP-macros or on the FPGA's configurable logic. The sensitivity is basically the degree of change of the area vs.the number of DSPs and is defined as: $S = {}^{\Delta Area}/_{\Delta \#DSPs}$.

Fig. 5 shows an example of sensitivity calculation for a design implemented completely on the FPGAs DSP-macros, a mixture of DSP-macros and LUTs all the way to a full LUT implementation for a specific latency. The sensitivity S is calculated for each process P_i given in the Process List $PList = P_1, P_2, ..., P_n$ for the smallest Design Family DF (see definition below) found in the first step. ADSP_MULTP then continues by sorting all the

processes from highest to lowest sensitivity. The last step involves allocating the DSP-macros to the process with highest sensitivity consecutively (their area is reduced the most when mapped to DSP-macros), until the DSP-macro budget is exhausted or the process does not need any more DSP-macros, in which case the next process with next highest sensitivity is processed. Fig. 6 summarizes the procedure of our DSP allocation method. It consists of 4 main steps:

Step1: Smallest Design Latency Search: For the set of designs D returned by the DSE, $DSE = \{D_1, D_2, ..., D_n\}$ for each process $P = \{P_1(DSE_1), P_2(DSE_2), .., P_n(DSE_x)\}$ with each design characterized by its Area (A) and Latency(L) the Design Family (DF) is a group of D with different degrees of DSP-macros usage $DF_i = \{(A_l, L_m, DSP_n), (A_p, L_m, DSP_{n-1}), .., (A_q, L_m, 0)\}$, find

---

ADSP-MULTP: <u>DSP Allocation</u>(PList{DSE$_i$, L$_i$}, DSP)
  PList: List of all the synthesized processes
  DSE$_i$: Design Space Exploration result for each process
  L$_i$: Latency constraint for each process
  DSP: Total DSP-macros budget

/* **Step 1** Find design family with smallest area in L_i */
foreach(P_n in $PList$) do
  foreach(D in DSE of P_n) do /* go through all designs*/
    • Find smallest design in the given
      latency range $[L_n, L_m]$;
    • Assign Design Family (DF) to Process List (PList)
  endforloop;
endforloop;

/* **Step 2** Calculate Sensitivity of each Design Family (DF)*/
foreach(P_n in $PList$) do
  foreach(D_i of DF) do /* For each design impl.*/
    • Find smallest Design (annotate Area min and DSPs);
    • Find largest Design(annotate Area max and DSPs);
    • Δ Area= Area_max - Area_min;
    • Δ DSP= DSPs_max - DSPs_min;
    • S=Δ Area/Δ DSP;
  endforloop;
endforloop;

/* **Step 3** Sort processes by sensitivity*/
sort_processes(PList);

/* **Step 4** Assign DSP-macros to processes*/
foreach(P_n in $PList$) do
  • PList(D) <= DSP-macro[n]; /* assign DSP-macro */
  • DSP -= DSP-macro[n]; /* decrease total DSP-macro budget
    if(DSP <= 0) /* if DSP-macro budget allocated
      break;
endforloop;
return Design with DSP allocation for each Process
  PList(P$_1$(D$_n$), P$_2$(D$_m$), ..., P$_x$(D$_q$));

Fig. 6. A summary of ADSP_MULTP method.

978-1-4799-2817-0/14 $31.00 © 2014 IEEE

the design D_x with the smallest latency L_y for all the designs within the specified latency interval $D = [L_1..L_2]$. We define the following terms:

Definition: *Design family DF of the explored design space DSE of a single process P_i is a subset D_i of $\forall D \in DSE$ that satisfies: for each $D_i = \{D_1, D_2, ..., D_x\}$ the latency L_i for each design is the same and each design $D_x = \{(A_p, L_i, DSP_n)\}$ is defined by its unique area A_p and number of DSP-macros used DSP_n.*

Step2: Sensitivity Calculation S: For the Design Family DF found in step1 for each process compute the sensitivity S using the sensitivity equation indicated previously, where S is the difference between the largest design $D_{largest} = (A_n, DSP_m)$ (using the smallest amount of DSP-macros DSP_m and the smallest design $D_{smallest} = A_p, DSP_q$ using the largest amount of DSP-macros DSP_q.

Step3: Sensitivity Based Process Sorting: Once the S has been computed for each DF the entire process list $PList$, $PList = \{P_1(S_1, DF_1), P_2(S_2, DF_2), ..., P_n(S_n, DF_n)\}$ the method sorts the processes with decreasing sensitivities, so that $S_{P_1} > S_{P_2} > S_{P_n}$. A quick sort is used in this step as it is easy to implement and although in the worst case it needs $O(n^2)$ comparisons in practice is only needs $O(nlog(n))$.

Step4: DSP-macro allocations: The last step allocates the DSP-macro budget among each process $P \in PList$, starting with the P with highest S. For each $DF \in P$ allocate enough DSP-macros so that the D_i with the smallest area A can be selected.

The main weakness of ADSP_MULTP is that it assumes that the effect of mapping a multiplication onto a DSP-macro grows linearly within the same design family (DF). This is not always the case. As shown in [5] the size of the muxes can grow in a none-linear way and hence the sensitivity S is not constant within a DF. To better understand the impact of the non-linearity in the sensitivity a variation of our proposed method was implemented. We call this method ADSP_MULTP fast_brute. This method follows the exact same steps as the original method except for the very last step (step 4), where instead of assigning DSP-macros to DFs with highest sensitivities in a linear way, it performs a brute force search trying all possible DSP-macro assignments within the selected DFs only.

IV. EXPERIMENTAL RESULTS

Different DSP intensive application were selected and grouped together into complex systems in order to test our proposed method. Table I shows the benchmarks selected and how the complex system benchmarks were created. FIR is a 9-tap FIR filter, CVIDIQ is an image filter, interp is a 3-stages FIR interpolation filter, decim is a 5-stages decimation filter and forces is the forces update

TABLE I
COMPLEX SYSTEM BENCHMARKS.

Bench	S1	S2	S3	S4	S5	S6	S7	S8
FIR	1	1	1	1	1	2	2	2
CVIDIQ	1	1		1	1	2	2	2
FFT	1	1	1		1	2	2	2
Interp	1	1	1	1	1	2	2	2
Decim		1	1	1			2	2
Forces			1	1				2
Max mults.	68	103	89	116	118	136	206	236

unit of the Discrete Element Method (DEM)-method to simulated the behavior or particle assemblies [11]. The FIR, FFT and Forces were given in ANSI-C and the rest in synthesizable SystemC. The number in the table indicates how many instantiations of that particular benchmark are instantiated into each system. E.g. benchmark S1 consists of 4 individual processes that need to be mapped onto the same FPGA, sharing the FPGA's DSP-macros. In particular test case FIR, CVIDIQ, FFT and the Interpolation filter. On the other hand, the most complex benchmark (S8) uses 2 instantiations of each of the test cases. Complex benchmarks of different sizes were chosen with smaller ones having known optimal solutions used for clarity. The last row indicates the total number of multiplications of these complex benchmarks when they are synthesized targeting maximum performance (smallest latency). It should be noted that the size of the multiplication differs between the different testcases, e.g. FIR uses 8-bit signed multipliers, while the interpolation filter requires 17-bit signed multipliers and the decimation filter 19-bit signed multipliers.

The experiments were run on an Intel Pentium 4 running at 3.20GHz machine with 1 GBytes of RAM running Linux SUSE version 3.0.13-0.27. The HLS tool used is CyberWorkBench v.5.2 [9] from NEC. The number of LUTs and registers reported are extracted from Xilinxs ISE 14.2 [10] synthesis report for the RTL generated by the HLS tool. The targeted FPGA is a Xilinx Virtex 6 VCX130T and the target HLS frequency is 75MHz. The running time given comprises only the DSP allocation part of our proposed method as all methods share the same initial DSE exploration results.

Table II(a) shows the results of the two variations of our propose method compared to a brute force approach without any latency constraint. Table II(b) on the other hand shows the comparison between all three methods when a latency constraint is given on each of the individual processes that form the complex system benchmark. A random latency range that covers less than $1/4$ of the total latency range was chosen for each of the processes. In Table II the first column specifies the complex benchmark name indicated in Table I. The second column show the DSP-macro budget. In both cases the DSP-macro budget is set to 75% of the total number of multiplications that each complex benchmark would need in order to maximize its parallelism and hence achieve the fastest design, as shown in Table I. This would ensure that other processes still have enough DSP-macros if needed. This

978-1-4799-2817-0/14 $31.00 © 2014 IEEE

TABLE II

EXPERIMENTAL RESULTS1 (WITHOUT LATENCY CONSTRAINT) (A)

	#DSPs	Brute Force(1)		ADSPMULTP fast_brute(2)		ADSPMULTP(3)		Δ LUT Slices		
		Run[s]	LUTs	Run[s]	LUTs	Run[s]	LUTs	Δ LUTs 1-2	Δ LUTs 1-3	Δ LUTs 2-3
S1	51	3	8,610	<1	8,610	<1	10,190	0.00	15.51	15.51
S2	77	564	12,137	<1	14,124	<1	15,365	14.07	21.01	8.08
S3	67	302	9,901	<1	10,192	<1	10,192	2.86	2.86	0.00
S4	87	10,810	10,810	<1	12,598	<1	12,598	14.19	14.19	0.00
S5	89	27,324	13,047	2	14,283	<1	14,651	8.65	10.95	2.51
S6	102	NA	NA	13	19,644	<1	21,010	NA	NA	6.50
S7	155	NA	NA	186	21,883	<1	23,347	NA	NA	6.27
S8	1 177	NA	NA	139,307	28,566	<1	31,127	NA	NA	8.23
Avg.								7.95	12.90	5.89

EXPERIMENTAL RESULTS2 (WITH LATENCY CONSTRAINT)(B)

	#DSPs	Brute Force(1)		ADSPMULTP fast_brute(2)		ADSPMULTP(3)		Δ LUT Slices		
		Run[s]	LUTs	Run[s]	LUTs	Run[s]	LUTs	Δ LUTs 1-2	Δ LUTs 1-3	Δ LUTs 2-3
S1	51	1	8,776	<1	8,776	<1	8,776	0.00	0.00	0.00
S2	77	20	12,598	<1	12,598	<1	12,731	0.00	1.04	1.04
S3	67	11	10,268	<1	10,268	<1	10,268	0.00	0.00	0.00
S4	87	28	11,278	<1	11,278	<1	12,071	0.00	6.57	6.57
S5	89	824	13,515	<1	13,515	<1	13,515	0.00	0.00	0.00
S6	102	NA	NA	1	17,552	<1	18,013	NA	NA	2.56
S7	155	NA	NA	3	21,374	<1	22,781	NA	NA	6.18
S8	177	NA	NA	12	27,030	<1	27800	NA	NA	2.77
Avg.								0.00	1.52	2.39

means that the DSP-macro budget for the first benchmark (S1)is only 36 (=48 × 0.75). Column 3 to 8 indicate the running time and the total area in LUT slices after logic synthesis of the 3 different methods. Brute force, ADSP_MULTP fast_brute and ADSP_MULTP. The last 3 columns compare the area achieved by all three methods, where the brute force is taking as a reference for the cases in which it could achieve a result. For S6, S7, and S8 the brute force method could not find a solution after 5 days.

From the result it can be seen that our methods achieve good results on average 7.95% to 12.90% worse than the optimal solution for either the fast brute force allocation method (2) or the direct DSP-macro allocation method (3) respectively when no latency constraint is specified and therefore the global minimum solution is searched. The on average 5.41 % difference between ADSP_MULTP fast_brute and ADSP_MULTP indicates that the usage of process sensitivity to allocated DSP-macros works well, but that the non-linearity effect leads to a small result degradation. In respect to running time, the ADSP_MULTP method is extremely fast finishing in all cases in less than 1 second and even the ADSP_MULTP fast_brute is very reasonable. For the latency constraint case, the results shown in Table II(b) indicate that our proposed method in all cases found the optimal solution for (2), and that it is only on average 2.39% worse for (3). It is therefore save to conclude that our method works well and achieves very good results extremely quickly.

V. SUMMARY AND CONCLUSIONS

In this work we have presented a method to allocate FPGA's DSP-macros efficiently across multiple processes synthesized using HLS. HLS is becoming common in most VLSI design flows as it has shown to increase the design productivity and achieves very good results compared to

hand-coded RTL, especially for DSP-related application. At the same time low-cost FPGAs are reaching capacities that allow entire systems to be implemented on them. Most of these systems are DSP related and make use of the FPGAs' embedded hardmacros (e.g. DSP blocks). Our proposed method (*Allocation of DSP-macros for Multiple Processes* - ADSP_MULTP) achieves very good results compared to the brute force optimal solutions extremely quick. We have introduced the concept of sensitivity S to allocate DSP-macros across the different processes and are currently investigating the usage or non-linear sensitivity methods to more accurately assign these hardmacros across the processes. Directions for future work include enhancing our proposed method to deal with non-linear sensitivities and studying the effect of our method on FPGAs with different underlying architectures.

REFERENCES

[1] S. Raje and R. Bergamaschi, "Generalized Resource Sharing," *IEEE/ACM ICCAD Dig. Tech.*,pp. 326–332,1997.

[2] P. Coussy and A. Moraweic, "High-Level Synthesis from Algorithm Digital Circuit," *Springer, ISBN 978-1-4020-8587-1*, ch.7,pp. 113–127, 2008.

[3] J. Cong and W. Jiang "Pattern-based behavior synthesis for FPGA resource reduction," *ACM FPGA*, pp. 107-116, 2008.

[4] D. Chen et al., "Optimality study of resource binding with multi-Vdds.," *IEEE/ACM DAC*,pp. 580-585, 2006.

[5] E. Casseau and B. Le Gal., "High-level synthesis for the design of FPGA-based signal processing systems," *Symp. on Systems, Architectures, Modeling, and Simulation*,pp. 25–32, 2009.

[6] S. Hadjis et al., "Impact of FPGA Architecture on Resource Sharing in High-Level Synthesis," *FPGA*,pp. 111-114, 2012.

[7] S. Bakshi and D. Gajski, "Component Selection for High-Performanc Pipelines," *TVLSI*,Vol.4–2pp. 181-194, 1996.

[8] W. Sun et al. "FPGA Pipeline Synthesis Design Exploration Using Module Selection and Resource Sharing." *IEEE TCAD*,Vol.26, No.2 pp. 254-265, 2007.

[9] NEC CyberWorkBench 5.2, *www.cyberworkbench.com*

[10] Xilinx ISE 14.2, *www.xilinx.com*

[11] P.A. Cundall and O.D.L. Strack, "A discrete numerical model for granular assemblies," *Geotechnique*,no.29, 4765, 1979.

Array Scalarization in High Level Synthesis

Preeti Ranjan Panda, Namita Sharma

Indian Institute of Technology Delhi
New Delhi, India

Arun Kumar Pilania, Gummidipudi Krishnaiah,
Sreenivas Subramoney, Ashok Jagannathan

Intel Technology India Pvt. Ltd.
Bangalore, India

Abstract— Parallelism across loop iterations present in behavioral specifications can typically be exposed and optimized using well known techniques such as Loop Unrolling. However, since behavioral arrays are usually mapped to memories (SRAM) during synthesis, performance bottlenecks arise due to memory port constraints. We study *array scalarization*, the transformation of an array into a group of scalar variables. We propose a technique for selectively scalarizing arrays for improving the performance of synthesized designs by taking into consideration the latency benefits as well as the area overhead caused by using discrete registers for storing array elements instead of denser SRAM. Our experiments on several benchmark examples indicate promising speedups of more than 10x for several designs due to scalarization.

I. Introduction

Arrays in behavioral specifications are typically mapped to on-chip memory structures such as SRAM during high-level synthesis (HLS or Behavioral Synthesis). Since array accesses occurring within loops account for a significant fraction of the execution latency, several classical compiler optimizations involving arrays and loops have been adapted into HLS. Loop transformations such as unrolling and pipelining fall in this category. Even though conceptually similar to the compiler scenario these transformations can have different consequences due to differences in the underlying architecture – for a compiler, the target processor architecture is fixed, whereas in HLS, the architecture is the output of the synthesis process.

We propose a technique for automating the decision of when to perform an important high-level synthesis transformation – *Array Scalarization*. In this transformation, array variables are converted into a group of scalar variables, with a discrete scalar variable replacing each array element. Such a transformation can potentially improve performance by removing bottlenecks arising out of memory port constraints. This is illustrated in Figure 1. A simple loop that updates an array element in each iteration is shown in Figure 1(a). The schedule of the loop body is shown in Figure 1(d), assuming that the addition and memory accesses (load/store operations being indicated by LD and ST) require one clock cycle each; and address computation is simplified to form part of the memory access operation. The sequential dependences in the loop body inhibit any performance improvement even if additional resources might be present. The four iterations require 12

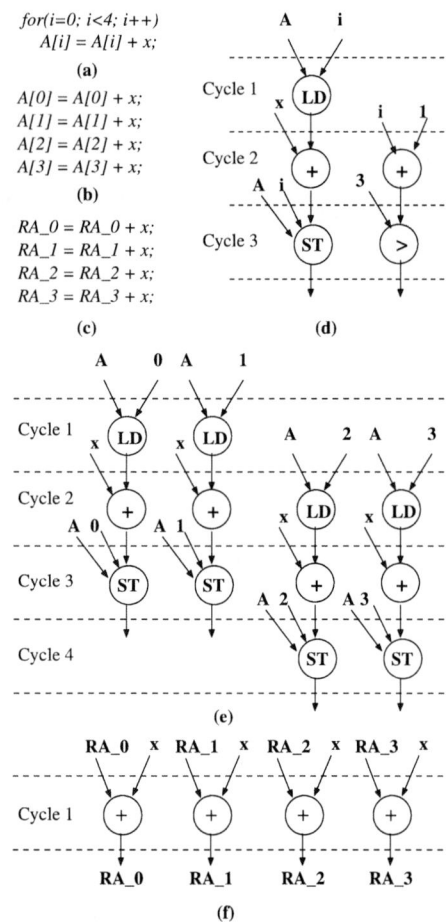

Fig. 1. Illustration of Array Scalarization

cycles. The well known loop unrolling transformation results in the code of Figure 1(b). The loop structure is eliminated; the loop index increment and comparison with the bound are unnecessary; and the array indices are now constants. Assuming we have enough adders and that array A is stored in a dual-port memory with two independent Read/Write ports, the resulting schedule is shown in Figure 1(e). The latency is reduced to 4 cycles. The improved performance is essentially due to the unrolling transformation exposing parallelism across loop iterations, which is now recognized and exploited by the scheduler. Nevertheless, the memory port constraint does limit the performance improvement extracted by the transformed code, since only two simultaneous memory operations are allowed. Figure 1(c) shows the code resulting

978-1-4799-2817-0/14 $31.00 © 2014 IEEE   622

from array scalarization applied to the unrolled loop. Each array element now becomes a scalar variable. The entire array will now be stored in discrete registers, and there is no SRAM structure inferred. Since any number of discrete registers can be simultaneously accessed, the memory port constraint vanishes, and the resulting schedule is shown in Figure 1(f). The total latency is 1 cycle, a 92% reduction over the original loop, and a 75% reduction over the unrolled loop.

The array scalarization transformation is conceptually simple and attractive, but cannot be arbitrarily applied during synthesis because it entails area overheads and is subject to other constraints such as static mapping after analysis of array accesses across multiple loops. We outline a technique that automates the array scalarization decision by analyzing the input behavior and selectively applying the transformation in a way that is aware of these overheads.

II. RELATED WORK

High-level synthesis algorithms need to be aware of a variety of different memory architecture parameters such as ports, size (word count and word width), and banks (independent physical memories). Memory port constraints are handled by essentially treating each port as equivalent to a classical resource that needs to be scheduled and bound, with each array access translating to an address computation and a memory read/write operation. Several related problems arise in this context, such as memory allocation, bank assignment, memory partitioning, and port assignment [18], [21], [22], [4], [15], [16], [9], [19], [8], [24].

Transformations such as scalar expansion of arrays [3] in the vectorizing compiler domain proceed in the reverse direction – scalar variables in loops sometimes lead to false dependencies across loop iterations, and are converted into arrays that are conveniently mapped to vector registers. While vectorizing compilers share our high-level objectives of eliminating sequentialization of loop iterations, they differ significantly from synthesis tools with respect to important details such as the processing and memory architecture not being fixed in synthesis, area overheads, and absence of cache in synthesis, all of which lead to a very different analysis (the example in Figure 1(c) would not be considered good code in a vector processor).

Loop transformations from the mainstream compiler domain have been investigated in the context of embedded systems and HLS, for their impact on area, performance and power/energy [11]. Partitioning of the memory data space has been of recent research interest, in order to accommodate banked memory available on FPGAs [5], [12], [23]. Compiler optimizations such as register pipelining [14], scalar replacement of arrays [7], and in-place optimization of arrays [8] use the idea of replacing array accesses by scalar variables to eliminate redundant memory accesses, but the HLS environment provides a very different context – the presence of a large number of scalar variables actually increases parallelism, unlike in a processor. Our scalarization is done in conjunction with the loop unrolling transformation. The effects of the loop unrolling transformation has been studied in different contexts

such as: I-Cache effects and code compaction [20], and balancing memory and floating point operations [7]. Unrolling has been incorporated in several HLS systems such as [10] and investigated for its effect on FSM delay [13].

Commercial high-level synthesis tools such as SystemC Compiler did include the facility of scalarizing an array through a compiler directive specified by the designer. However, as we show later, the decision of whether to scalarize or not is non-trivial for large complex specifications, and the designer needs to be aware of several consequences, which are better captured by our automated analysis technique.

III. ISSUES IN ARRAY SCALARIZATION

The application of the array scalarization optimization requires a careful analysis of its implications. While the potential performance improvement is intuitive, the transformation is not universally applicable primarily due to area overheads/constraints and possibly limited benefits due to dependencies. The issues are discussed below.

A. Area Overheads

The primary deterrent to array scalarization is that the scalars would have to be stored in discrete registers instead of the more compact SRAM cells, leading to an area overhead, not just from the larger area occupied per bit, but also due to additional routing space required in comparison to the regular SRAM structure. When operating under area constraints, arrays eligible for scalarization need to be prioritized in terms of the resulting performance benefits and the area overhead incurred by scalarization. Consider the three loops below:

```
for(i=0;i < 128; i++)
  a[i] = a[i] * some_param;
for(j=0; j < 128; j++)
  b[j] = b[j] << some_shift;
for(k = 0; k < 128; k++)
  b[k] = b[k] + some_val;
```

Arrays a and b have equal size, but scalarizing b results in more latency reduction than scalarizing a, for an equivalent area overhead. If the additional area constraint imposed on the synthesis tool were to accommodate only one scalarized array, then scalarizing b would be preferable.

B. Sparse Array Accesses

Scalarization may be detrimental when array accesses are sparse or occur outside loops, a frequently occurring situation. Consider the code below:

```
int MetricMap[128];
...
current_metric = MetricMap[3];
```

Scalarizing arrays such as *MetricMap* would result in insignificant performance benefits (because it is accessed only once) while incurring a large area penalty. Similar arguments apply against scalarization of arrays accessed inside loops but with access count relatively small compared to their sizes.

978-1-4799-2817-0/14 $31.00 © 2014 IEEE   623

C. Loop Carried Dependencies and Resource Constraints

Sequentialization of a schedule can be forced by constraints inherent in the loop specifications and external resource constraints. For example, in the code below:

```
for(int i=0;i<50;i++)
  A[i] = A[i] + A[i-1] * 7;
```

the loop carried dependency forces a sequentialization in the execution of the loop iterations, leading to a limited number of memory accesses per clock cycle, even if scalarization were to expose parallel access to all array elements. Similarly, in the example of Figure 1, scalarization may not be useful if the number of adder resources were constrained to be very small. A good scalarization algorithm must, therefore, be aware of the potential benefits of scalarization, and perform the transformation only when evidence points to its usefulness.

D. Complex or Data-dependent Array Indexing

Scalarization may not be attractive if array index expressions are data dependent or dynamically determined. Consider the code below:

```
int a[150]; b[100];
...
for(int i=0;i<150;i++){
  index = some_calc(i) % 100;
  shift_val = b[index];
  a[i] <<= shift_val;
}
```

Array a can be scalarized, but the inputs to the registers containing the a elements have to be read from array b, which incurs a large cost upon scalarization. All the 100 b values could possibly be an input to each $a[i]$, so if we scalarize b, then for parallel updates to all $a[i]$, we have to create paths from every element of b to the register corresponding to $a[i]$. This can be done by introducing a 100-to-1 MUX at the input of all the 150 $a[i]$ registers, whose select input is connected to *index*, but there are inherent scalability problems with this approach – the area overhead is large and the access delays are significantly larger. These costs need to be taken into account in the scalarization decision.

E. Obligatory Scalarization

Scalarization of an array accessed in a loop may imply the scalarization of other arrays accessed in the same loop, if the true performance benefits are to be exploited. Consider the code below:

```
for(int i=0;i<50;i++)
  A[i] = B[i] + 3;
```

Suppose array A is scalarized. In order to exploit the parallelism across loop iterations, array B also needs to be scalarized, otherwise accesses to B get sequentialized, resulting in limited advantage from A's scalarization in this loop because the memory bottleneck continues to remain (there could still be a benefit in other loops where A is accessed). A scalarization algorithm needs to be aware of this implication.

The above issues – all of which are derived from our experiments in applying the transformation to real applications – make the decision of whether or not to apply array scalarization a challenging one. We propose and evaluate a systematic method for analyzing a behavioral specification and selectively applying the transformation on specific arrays.

IV. PROBLEM FORMULATION AND APPROACH

A. Problem Definition

The array scalarization problem may be stated as follows. Given a behavioral description with loops $L_1, ..., L_n$ accessing arrays $A_1, ..., A_m$, with each loop L_j accessing a subset S_j of the arrays; array size for A_i being $Size(i)$; and an allowed area overhead constraint OV; compute the Scalarization Vector $SV[1...m]$ (where $SV[i]$ = TRUE means A_i is scalarized and = FALSE means A_i is not scalarized), such that the overall latency (cycle count) reduction is maximized while the total area overhead $< OV$.

We present a heuristic approach based on estimates of latency reduction and area overhead, which is conceptually similar to the *value density* heuristic used in the Knapsack Problem.

We perform the array scalarization transformation in conjunction with loop unrolling, because scalarization, in general, does not save on latency if the sequential loop structure is maintained. The handling of multi-dimensional arrays accessed in loop nests is similar to that of single-dimensional arrays – multiple loop levels need to be unrolled. If array indices are data-dependent or otherwise unresolvable statically, then the MUX-based architecture indicated in Section III-D is assumed, and the associated costs are incorporated into the area and delay estimates. We also need to either compute array access counts statically, or rely on external profiling to give us branch probabilities of conditionals so that access counts can be estimated. Finally, in this work we assume full scalarization; the possibility of partitioning an array and storing it in multiple memory banks is a topic of future research.

B. Array Access Graph

We model array accesses within loops using a hypergraph structure where each node represents an array, and each hyperedge connecting multiple nodes represents a loop accessing the arrays represented by the nodes. The edge captures the observation made earlier that, if one node in the edge is scalarized, then all nodes of the edge must be scalarized to achieve the full latency benefit of scalarization. The graph construction is illustrated in Figure 2. If we scalarize B, C, and D, but not A, then only loop $L2$ will benefit significantly from scalarization, but not $L1$ and $L3$, since the memory port bottleneck introduces an inherent sequentialization while accessing A. However, if the area overhead constraint permits the scalarization of only 2 arrays, then $\{A, B\}$ (or $\{A, D\}$) would be a better scalarization choice; this improves the performance of only $L1$ (or $L3$). If there is space for scalarizing only one array, then there is no good choice.

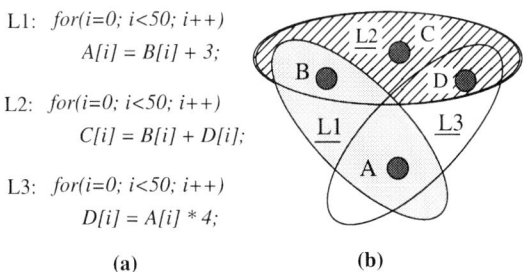

L1: `for(i=0; i<50; i++)`
    `A[i] = B[i] + 3;`

L2: `for(i=0; i<50; i++)`
    `C[i] = B[i] + D[i];`

L3: `for(i=0; i<50; i++)`
    `D[i] = A[i] * 4;`

(a)   **(b)**

Fig. 2. (a) Example code (b) Hypergraph representing array accesses

C. Cost Estimation

We consider the cost of scalarizing an array to be the additional area that results from two sources: (1) storing the array elements in discrete registers instead of SRAM cells; and (2) MUX costs due to scalarization. We have:

$$Cost(A_i) = Storage\_Area(i) + MUX\_Area(i) \quad (1)$$

where $Storage\_Area(i)$ and $MUX\_Area(i)$ are the additional storage- and MUX-related area due to scalarization of A_i. The storage overhead is given by:

$$Storage\_Area(i) = Size(i) \times ElementSize(i) \times (RArea - SArea) \quad (2)$$

where $Size(i)$ is the number of array elements in array A_i, $ElementSize(i)$ is the number of bits in each element, $SArea$ is the area of a single SRAM cell, and $RArea$ is the area of a flip flop. $SArea$ and $RArea$ are technology dependent. The MUX_Area cost arises due to two factors: (1) wider muxes at the FU inputs due to additional paths from the scalarized registers (instead of just the memory data bus) and (2) data-dependent or complex accesses to the array in any loop, if any. In the first case, the muxes are of average width $Size(i)/R$ where R is the number of resources of the corresponding operation type with array A_i as operand, and in the second case, the width is $Size(i)$.

$Cost(A_i)$ represents the approximate additional datapath area. Further area overheads are incurred due to routing during the physical design phase, but we treat the routing phase as contributing a fixed overhead (such as a 30% overhead over cell area often assumed in early stages by CAD tools) to the entire design. Area overhead for a hyper-edge in the access graph is just the sum of the scalarization costs of the individual nodes connected to the edge, representing the total cost of scalarizing all arrays accessed in a loop.

$$Cost(L_i) = \sum Cost(A_j), A_j \in S_i \quad (3)$$

D. Benefit Estimation and Priority Function

In our approach, we estimate the benefit due to scalarization as the potential reduction in a loop's latency after unrolling and scalarization. Such reduction depends on several factors: loop carried dependencies, resource constraints, operation count in the loop body, and the loop iteration count. We use an estimate similar to the ones used in [2], [13] for this computation. We first compute the maximum Initiation Interval (*II*) for the loop, which gives the highest rate at which new iterations can be initiated. This is a function of the loop carried dependencies and resource constraints, and is given by:

$$II = max\left(max\left\lceil \frac{ops(i) \times lat(i)}{FU(i)} \right\rceil \forall i, max\left\lceil \frac{dep(t)}{it(t)} \right\rceil \forall t\right) \quad (4)$$

where $ops(i)$ is the number of operations of type i; $lat(i)$ is the latency of the function unit implementing this operation; $FU(i)$ is the number of resources of type i; $dep(t)$ is the maximum distance between dependent nodes in the loop body's schedule; $it(t)$ is the iteration distance for these nodes. The first term gives the bound imposed by resource constraints, and the second term gives the bound imposed by dependencies. Since memory resources are no longer used for the scalarized arrays, these resources are omitted from the above computation.

The latency of the unrolled and scalarized loop L_i is now computed as:

$$Lat(i) = Body(i) + II \times (iter(i) - 1) \quad (5)$$

where $Body(i)$ is the schedule length of a single iteration of the loop body and $iter(i)$ is the iteration count of the loop. The latency reduction is now:

$$Red(L_i) = Body(i) \times iter(i) - Lat(i) \quad (6)$$

where the first term above gives the total latency of the original loop. The case where there are multiple basic blocks in a loop body involves a simple variation. The priority function for performing the unrolling and scalarization transformation for arrays in a loop L_i can now be defined as follows:

$$Priority(L_i) = Red(L_i)/Cost(L_i) \quad (7)$$

The priority of a loop is, thus, the ratio of the latency benefits to the additional area incurred when its arrays are scalarized. Note that if some arrays in loop L_i have already been scalarized, then the corresponding area overhead has already been considered, and the cost computation for the loop excludes the contribution of that array.

E. The Scalarization Algorithm

The scalarization technique is outlined in Algorithm 1. We start by creating a hyper-graph modeling the array accesses in loops. For each edge E_i in the graph representing loop L_i, we first compute the $Cost(L_i)$, latency reduction $Red(L_i)$, and initial priority. A candidate list CL is created, sorted according to the loop priority. We then consider the loops in order of priority, applying scalarization to arrays in the loop if the overheads are less than OV. When we decide to scalarize arrays in a loop, we remove E_i from CL; update the $Cost$ values of overlapping edges; recompute their priorities; and re-insert them into CL (because the priority has changed). If a loop cannot be subject to scalarization, then the edge E_i is removed from consideration. The algorithm terminates when CL is empty or the area overhead constraint is consumed.

V. EXPERIMENTS AND RESULTS

We validated our scalarization algorithm on several designs from the DSP and multimedia domain, including *JFD-CTint* (forward Discrete Cosine Transform), *Wavelet* (Debaucles 4-Coefficient Wavelet filter used in image compression), *ADPCM* (implementation of adaptive differential pulse-code modulation), *DHRC* (differential heat release computation), *Laplace* (for edge enhancement in an image) and *GSM_param* (from GSM specification for computing LTP gain and LTP lag for the long term analysis filter). We also evaluated our approach on DCT based channel estimation kernel in WLAN

978-1-4799-2817-0/14 $31.00 © 2014 IEEE

Algorithm 1 Array Scalarization

Input: Loops $L_1...L_n$, Arrays $A_1...A_m$, Allowed area overhead OV
Output: Scalarization Vector $SV[1...m]$
1: Create hyper-graph G. Initialize $SV =$ all FALSE
2: **for all** hyper-edge E_i corresponding to loop L_i **do**
3:     Compute area overhead $Cost(L_i)$ due to unrolling and scalarization
4:     Compute latency reduction $Red(L_i)$ and priority $Priority(L_i)$
5:     Insert E_i into candidate list CL sorted by priority
6: **end for**
7: **while** $OV > 0$ and $CL \neq \phi$ **do**
8:     remove head E_i from CL
9:     **if** $OV > Cost(L_i)$ **then**
10:         $SV[j] =$ TRUE $\forall$ nodes A_j on which E_i is incident
11:         /* loop is unrolled and arrays in it scalarized */
12:         **for all** j such that $E_j \cap E_i \neq \phi, i \neq j$ **do**
13:             $Cost(L_j) = Cost(L_j) - \sum(Cost(A_p)), p \in E_i \cap E_j$
14:             $Priority(L_j) = \frac{Red(L_j)}{Cost(L_j)}$
15:             $CL = CL - E_j$; Re-insert E_j into CL
16:             Reset $Cost(A_p) = 0 \ \forall p \in E_i \cap E_j$
17:         **end for**
18:         $OV = OV - Cost(L_i)$
19:     **else**
20:         /* area constraint too tight to scalarize all arrays in loop */
21:     **end if**
22: **end while**
23: return SV

Design	Orig (cycles)	unroll (cycles)	unroll, scal (cycles)	Speedup	Area Ovhd
JFDCTint	339	157	41	8.3x	19.59%
Wavelet	267	107	26	10.3x	9.87%
DHRC	1795	652	83	21.6x	7.66%
Scal_array	63	30	7	9x	5.89%
ADPCM	44	40	38	1.15x	17.29%
ChannelEst	832	408	39	21.33x	40.73%
Laplace	1601	501	53	31.05x	47.34%
GSM_param	7325	5287	556	13.2x	20.53%

TABLE I
SUMMARY OF RESULTS

11n for the 20MHz channel bandwidth. We used an in-house high level synthesis tool written in C++. From a CDFG structure, the synthesis tool performs typical HLS transformations and optimizations before the standard scheduling, binding, and register allocation steps, and ultimately generates RTL-VHDL output, which is then taken through a commercial synthesis tool flow using a 90nm standard cell library.

Table I shows the number of cycles needed for complete execution of each design under three scenarios – in original form (no unroll and scalarization); full unroll and no scalarization; and full unroll combined with the most suitable scalarization of the selected array subset. When arrays are stored in memory, we use a 2-port SRAM capable of supporting one memory read and one memory write operation in parallel. In cases of designs that operate continuously on data samples, such as *ADPCM*, the cycle count represents the latency of one iteration of the outermost loop. For this experiment, we have assumed the availability of 10 resources of each type (adder, multiplier, comparator, shifter, logical AND/OR) and as much additional memory area as needed for exploiting the full parallelism discovered by scalarization.

As seen from Table I, the speedup ranges from 1.15x to 31.05x, revealing a huge potential for performance improvement by the scalarization transformation. Designs with larger number of memory accesses and larger loop iteration count (such as *GSM_param*) result in more impressive improvements. Designs with a higher degree of sequential dependence and nested conditional structures (such as *ADPCM*) yield only marginal benefits. Table I also notes the memory area overhead (as a fraction of the total area) from the scalarization for the different examples. The data is derived from the *CACTI* estimation tool [17] and from synthesis results.

In Figure 3 and 4 we investigate the performance improvement obtained for the same examples, over the original code, for a range of different resource constraints. The resources are varied from 2 to 50 instances of each resource type. In Figure 3, we plot the performance difference when we unroll the loops but do not apply array scalarization. We notice that the performance saturates quickly, and does not vary too much with increasing resources. This can be attributed to the memory bottleneck, which causes sequentialization of array accesses in spite of the availability of computation resources. In Figure 4, we plot the performance difference when we apply both unrolling and array scalarization. We observe that removing the memory bottleneck by scalarization allows the exploitation of the available parallelism by the increased computation resources. The large speedups in some designs are due to large numbers of loop iterations being executed in parallel, as well as due to the total elimination of memory accesses, address computations, and loop variable increments and comparisons, leading to reduced latency for the loop body. In *Scal_array*, no further improvement is observed for higher resource counts because the loop has a small iteration count ($= 12$).

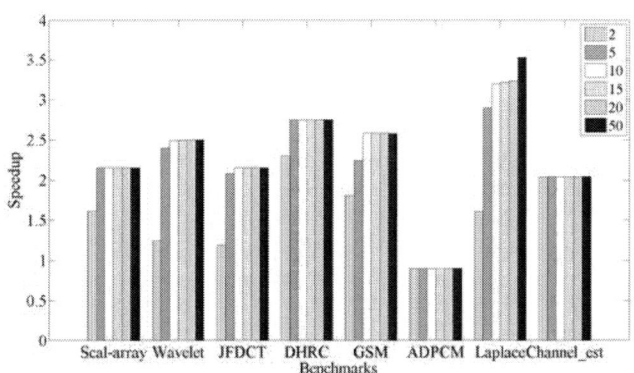

Fig. 3. Latency reduction on unrolling without scalarization

The *GSM_param* design offers some further insight into the scalarization problem involving multiple loops and arrays. We study the effect of unrolling and scalarizing the arrays accessed in two loops in the design, *Loop-1* and *Loop-2*. In Figure 5(a) we show the performance results where the transformation is applied on *Loop-1* (with an area overhead of 11.93%), and *Loop-2* (with an area overhead of 20.53%) individually, and both together (with an area overhead still being 20.53%). *Loop-1* is an attractive candidate for unroll/scalarization. Despite incurring a significant area overhead, *Loop-2* offers neg-

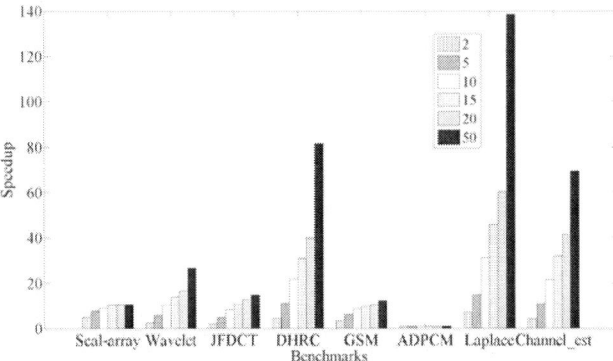

Fig. 4.   Latency reduction on unrolling and scalarization

ligible improvement over *Loop-1*. Our prioritization function recognizes this difference and helps choose the right order for scalarization when there are area overhead constraints.

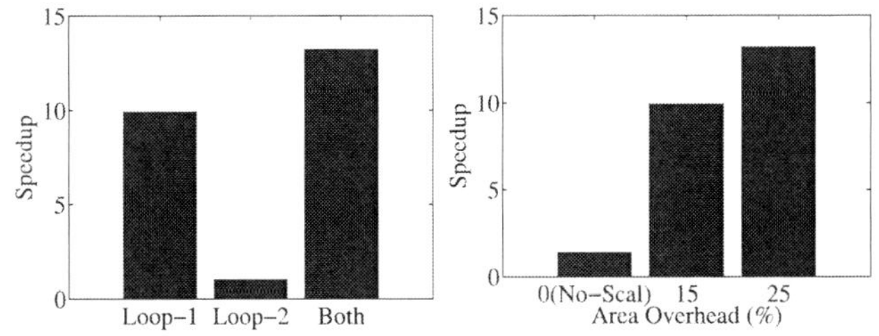

Fig. 5.   (a)Unroll and Scalarization on different loops (b)Performance variation with different area overhead constraints for *GSM_param*

We finally study the impact of imposing different area overhead constraints for the *GSM_param* design keeping the computation resources constant. In Figure 5(b) we present a comparison of the performance improvements under different area overhead constraints. In one case with a 15% overhead, we obtain a speedup of 9.92x. When the allowed overhead is increased to 25%, the speedup is 13.2x.

Designs such as *Laplace* and *ADPCM* distinguish our approach from a simple algorithm (sometimes used in FPGA synthesis tools) that scalarizes all arrays that are smaller than a pre-determined threshold value. In Laplace, such an algorithm (using, say, a threshold value of 10 32-bit elements) would not scalarize the arrays as their size exceeds 100. However, their scalarization results in 31.05x speedup with an area overhead of 47.34%, which leads to an attractive design point (for example, with a far superior *(area × delay)* metric value). In case of *ADPCM*, the simple algorithm's decision to scalarize the arrays (array sizes being small), is not a good option and results in a lower *(area × delay)* product. This demonstrates the need for a more intelligent analysis for scalarization based on estimates of changes in both performance and area.

VI. CONCLUSIONS AND FUTURE WORK

We investigated array scalarization, a behavioral synthesis optimization that selectively transforms arrays accessed in loops to scalar variables to improve performance under an area overhead constraint. In complex designs, it is not trivial to make scalarization decisions manually, and our scalarization algorithm automates this process by first prioritizing

the loops in the behavioral code, taking into account the expected performance benefits and the resulting area overhead. Our experimental studies reveal the potential for significant performance improvements resulting from the removal of memory port related bottlenecks. In the future, we plan to have this transformation interact more intelligently with other loop transformations, and with arrays mapped to multi-bank and multi-port memories. Metrics such as power can be incorporated into the algorithm.

ACKNOWLEDGMENT

This work was partially supported by a research grant from Intel. We are grateful for their support. We also thank Reeno Joseph for his help with the experiments.

REFERENCES

[1] A. Agarwal, D. A. Kranz, and V. Natarajan "Automatic partitioning of parallel loops and data arrays for distributed shared memory multiprocessors", ICPP, 1993.
[2] V. H. Allan, R. B. Jones, R. M. Lee, and S. J. Allan, "Software pipelining," ACM Computing Surveys, 27(3), 1995.
[3] R. Allen and K. Kennedy, "Optimizing Compilers for Modern Architectures", Morgan Kaufman, 2001.
[4] P. Athanasopoulos, P. Brisk, Y. Leblebici, P. Ienne, "Memory organization and data layout for instruction set extensions with architecturally visible storage", ICCAD, 2009.
[5] Y. Ben-Asher, N. Rotem, "Automatic Memory Partitioning: Increasing Memory Parallelism via Data Structure Partitioning", CODES+ISSS 2010.
[6] M. F. P. O'Boyle and P. M. W. Knijnenburg, "Integrating loop and data transformations for global optimization", PACT, 1998.
[7] S. Carr and K. Kennedy, "Improving the ratio of memory operations to floating-point operations in loops," ACM TOPLAS, 16(6), 1994.
[8] F. Catthoor, S. Wuytack, E. De Greef, F. Balasa, L. Nachtergaele, A. Vandecappelle, "Custom memory management methodology", 1998.
[9] C. Chavet and P. Coussy, "A memory mapping approach for parallel interleaver design with multiples read and write accesses", ISCAS 2010.
[10] S. Gupta, R. K. Gupta, N. D. Dutt, and A. Nicolau, "Coordinated Parallelizing Compiler Optimizations and High-Level Synthesis", ACM TODAES, 9(4), 2004.
[11] M. Kandemir, M., N. Vijaykrishnan, M. J. Irwin, and Y. Wu, "Influence of compiler optimizations on system power", IEEE TVLSI, 9(6), 2002.
[12] P. Li, Y. Wang, P. Zhang, G. Luo, T. Wang, and J. Cong, "Memory Partitioning and Scheduling Co-optimization in Behavioral Synthesis", ICCAD 2012.
[13] S. Kurra, N. K. Singh, and P. R. Panda, "The impact of loop unrolling on controller delay in high level synthesis", DATE, 2007.
[14] H. Falk and P. Marwedel, "Source Code Optimization Techniques for Data Flow Dominated Embedded Software", Springer, 2004.
[15] S. Meftali, F. Gharsalli, F. Rousseau, A. A. Jerraya "An optimal memory allocation for application-specific multiprocessor SoC", ISSS, 2001.
[16] R. Nalluri, R. Garg, P. R. Panda, "Customization of Register File Banking Architecture for Low Power," VLSI Design, 2007.
[17] N. Muralimanohar, R. Balasubramonian, N. P. Jouppi, "CACTI 6.0: A Tool to Model Large Caches", HP Labs., Tech. Rep. HPL-2009-85, 2009.
[18] P. R. Panda, N. D. Dutt, "Low-power mapping of behavioral arrays to multiple memories," ISLPED, 1996.
[19] P. R. Panda, F. Catthoor, N. D. Dutt, K. Danckaert, E. Brockmeyer, C. Kulkarni, A. Vandecappelle, P. G. Kjeldsberg "Data and Memory Optimization Techniques for Embedded Systems", ACM TODAES 2001.
[20] V. Sarkar, "Optimized unrolling of nested loops," ICS, 2000.
[21] J. Seo, T. Kim, and P. R. Panda, "An Integrated Algorithm for Memory Allocation and Assignment in High-level Synthesis", DAC, 2002.
[22] J. Seo, T. Kim, P. R. Panda, "Memory allocation and mapping in high-level synthesis - an integrated approach", TVLSI, 11(5), 2003.
[23] Y. Wang, P. Zhang, X. Cheng, and J. Cong, "An Integrated and Automated Memory Optimization Flow for FPGA Behavioral Synthesis," ASP-DAC 2012.
[24] Y. Wang, P. Li, P. Zhang, C. Zhang, and J. Cong, "Memory partitioning for multidimensional arrays in high-level synthesis", DAC, 2013.

978-1-4799-2817-0/14 $31.00 © 2014 IEEE

Data Compression via Logic Synthesis

Luca Amarú[1], Pierre-Emmanuel Gaillardon[1], Andreas Burg[2] and Giovanni De Micheli[1]

[1]Integrated Systems Laboratory (LSI), EPFL, Switzerland
[2]Telecommunication Circuits Laboratory (TCL), EPFL, Switzerland

Abstract—Nowadays, most software and hardware applications are committed to reduce the footprint and resource usage of data. In this general context, lossless data compression is a beneficial technique that encodes information using fewer (or at most equal number of) bits as compared to the original representation. A traditional compression flow consists of two phases: data decorrelation and entropy encoding. Data decorrelation, also called entropy reduction, aims at reducing the autocorrelation of the input data stream to be compressed in order to enhance the efficiency of entropy encoding. Entropy encoding reduces the size of the previously decorrelated data by using techniques such as Huffman coding, arithmetic coding, and others. When the data decorrelation is optimal, entropy encoding produces the strongest lossless compression possible. While efficient solutions for entropy encoding exist, data decorrelation is still a challenging problem limiting ultimate lossless compression opportunities. In this paper, we use logic synthesis to remove redundancy in binary data aiming to unlock the full potential of lossless compression. Embedded in a complete lossless compression flow, our logic synthesis based methodology is capable to identify the underlying function correlating a data set. Experimental results on data sets deriving from different causal processes show that the proposed approach achieves the highest compression ratio compared to state-of-art compression tools such as ZIP, bzip2 and 7zip.

I. INTRODUCTION

The maturity of *Electronic Design Automation* (EDA) tools enables today's billions transistors chip to be designed and tested efficiently. Such great success drives the application of EDA algorithms to non-traditional EDA fields, e.g., cure of cancer [1], smart water [2], cure of genetic diseases [3], smart grid [4], etc. In this work, we use EDA synthesis techniques to reduce the footprint and resource usage of binary data.

It is estimated that the total amount of data stored in world's digital devices could be compressed by a factor of 4.5x without any loss of information [5]. The efficiency of data compression techniques is key to securing a profitable usage of physical resources. For this reason, data compression is a widely studied and active research field. Original compression methods, such as Huffman and arithmetic coding, are based on the Shannon entropy [6]. The Shannon entropy quantifies the information contained in data. When the considered data is a sequence of *independent and identical distributed* (i.i.d.) *Random Variables* (RVs), Shannon entropy provides the upper bound on the best possible lossless compression. In this condition, Huffman and arithmetic coding methods are optimal since they are proven to asymptotically achieve the Shannon entropy. Unfortunately, most of the real-world information cannot be modeled as sequence of i.i.d. RVs. Indeed, digital data is often strongly correlated. In order to overcome this limitation, succeeding techniques have been developed embedding a decorrelation pre-processing step prior (or integrated) to entropy encoding based compression [6].

Notably, dictionary techniques incorporate the structure of correlated data, by building a list of frequently occurring patterns, to increase the compression ratio. This compression approach has been used in microprocessor architectures, e.g., Thumb from Arm, where the instruction stream is compressed in I-memory and decompressed in the processor [7]. The *Lempel-Ziv* (LZ) method [8], and its evolutions, represents a very successful dictionary based compression technique employed in most practical compression standards. However, LZ-based methods imply some assumptions on the pattern recurrence locality. Moreover with dynamic dictionaries the efficiency of the compression converges slowly with the size of the data. For this reason, specialized decorrelation transforms have been successively studied with the aim at further improving the compression ratio [6]. Among existing decorrelation transforms, fixed-basis (heuristic) transforms, such as discrete cosine transform, are used in practice while variable-basis (optimal) transforms, such as KLT [6], are used to bound the best theoretical transform performance. Indeed, the variability of the transform basis is not a desirable property in data compression. Fixed-basis transforms usually target a specific class of data, for example the discrete cosine transform is advantageous for imaging (JPEG standard [10]).

A major aim for today's compression tools is to overcome the limits of traditional decorrelations techniques in order to unlock the full potential of lossless data compression.

In this paper, we use logic synthesis to compact the size of causal data sets enabling novel lossless compression opportunities. Logic synthesis is the process by which an abstract Boolean function description is transformed into a corresponding minimized logic circuit [11]. In the data compression context, the capability of modern logic synthesis tools to identify and remove redundancy, while preserving the initial behavior of the circuit, paves the way for a general lossless compression approach with the aim to find the underlying logic function generating the data we want to compress. Experimental results on data sets derived from causal processes show that our compression method based on logic synthesis reduces the data size by 1~2 orders of magnitudes compared to state-of-art compression tools such as ZIP, bzip2 and 7zip.

The remainder of this paper is organized as follows. Section II first provides background on lossless data compression and logic synthesis, then it discusses the contributions of this work compared to prior art on compression techniques performed via Boolean minimization. In Section III, the proposed logic synthesis based compression methodology is presented. Experimental results for our compression approach are presented and compared with state-of-art compression standards in Section IV. We conclude the paper in Section V.

978-1-4799-2817-0/14 $31.00 © 2014 IEEE

II. Background and Motivation

This section first presents background on data compression and logic synthesis. Later, it discusses the contributions of this work compared to prior art on compression techniques performed via Boolean minimization. Please note that, in this work, we focus only on lossless data compression.

A. Data Compression

A *lossless* compression scheme refers actually to two algorithms: (i) a compression algorithm that takes in input a data sequence X and reduces it to X_c (that requires fewer, or at most equal, bits than X), and (ii) a reconstruction algorithm that recovers *exactly* X from X_c, i.e., with no loss of information. In other words, in a lossless compression scheme the compression algorithm must be *reversible* [6]. Given the dual nature of *lossless* compression schemes, we concentrate hereafter only on compression algorithms provided that they satisfy the reversibility requirement.

Nowadays, practical compression approaches are based on two subsequent phases: first, the input data is decorrelated and then entropy encoding techniques are applied as final compression step.

1) Data Decorrelation (Entropy reduction): Data decorrelation, also called entropy reduction, is a data preprocessing phase aiming at reducing the autocorrelation of the input data.

Linear transformations are efficient means to accomplish this task [6]. Among all the decorrelation transforms, the *best* coding gain, in term of compression ratio, is provided by the *Karhunen-Loeve Transform* (KLT). The KLT is a transform having signal-dependent basis and random variables as coefficients. In the discrete (binary) domain, the KLT is described by a matrix having as columns the eigenvectors of the autocorrelation matrix of the input (binary) sequence considered [6]. The signal-dependency of the KLT makes its use inefficient in practical applications where the basis (or the data autocorrelation itself) must be provided to the decoding algorithm which is unaware of the properties of the compressed data. Such overhead removes the advantage of the KLT theoretical optimality. For this reason, the KLT is typically used as theoretical bound for the best coding gain achievable by decorrelation transforms. On the other hand, practical decorrelation techniques of interest make assumptions on the original data properties resulting in fixed basis transforms. A well-known example is the *Discrete Cosine Transform* (DCT), widely used in image compression methods, e.g., JPEG [10], which employs cosine functions as fixed transform basis. Other specialized transforms have been developed in literature, e.g., BCJ/BCJ-2 [20] for binary executables, Burrows-Wheeler [9] especially efficient in the field of bio-informatics etc.

Another notable approach to decorrelate data is based on dictionary techniques. The core idea of dictionary techniques is to build a dictionary (static or dynamic) of recurring patterns in the input data. Such patterns are directly encoded by their indexes in the dictionary. The *Lempel-Ziv* (LZ) method [8] is one of the most recognized dictionary based compression technique used in many practical compression standards. The general efficiency of existing dictionary techniques is limited by (i) assumptions on pattern recurrence locality and (ii) issues related to dictionary flexibility.

We refer the interested reader to [6] for an extended and complete discussion of data decorrelation.

2) Entropy Encoding: Entropy encoding refers to a class of (lossless) coding techniques able to compress an input data down to its entropy. When the entropy information is defined according to the exact probabilistic model, entropy encoding achieves the optimum compression for any input data. However, the correlation and the (often complex) underlying function that produced the data set is typically not known. Hence, the choice of the right probabilistic model is a difficult problem usually simplified by decorrelating transforms. Once the data is (fully) decorrelated, a simple stochastic model is reliably employable to define the information entropy. Then, entropy encoding methods can be applied successfully. Original entropy encoding techniques are Huffman coding [12] and arithmetic coding [6] that form the basis of current compression software and standards. For a review of such methods, we refer again to [6].

B. Logic Synthesis

Logic synthesis is the process by which (virtually) all digital integrated circuits are designed [11]. Logic synthesis aims to transform a general description of a Boolean function into its minimal logic circuit implementation. The development of synthesis algorithms have been driven by the exponential growth of digital electronics, requiring contemporary tools to be scalable, efficient and capable to produce near-optimal results. Logic synthesis shares optimization criteria with many other problems, consequently its application to non-traditional (non-EDA) fields is attracting the interest of researchers in several science communities.

For the sake of brevity, we do not review basic concepts and notation for logic synthesis, please see [11] for a review.

C. Contribution to Prior Art

Previous works in [13]–[15] explored the possibility to compress data using Boolean minimization. Their method consist in treating a binary sequence of 2^N elements as a complete truth table for a N-input single output Boolean function. Two-level minimization is applied in [13]–[15] to lossless compress the information stored in such truth table (representing the initial sequence). We differentiate from these methods by (i) the use of a more sophisticated initial SOP representation in place of exhaustive truth tables, (ii) extension from single output to multioutput function representation to enhance the manipulation flexibility, (iii) introduction of encoding/decoding *arrow of time* in binary strings to facilitate the logic synthesis task and (iv) identification and selective manipulation of uncorrelated data with traditional entropy encoding methods.

III. Logic Model for Data Compression

Binary data is generated, stored and transmitted by digital electronic systems. Given its intrinsic nature, binary data often derived from a set of logic operations performed in the logic core of an electronic system and repeated over different combinations of logic operands. Consequently, a sensible gain in data compression efficiency is achievable if the kernel set of logic operations (logic function) generating the data is known and transmitted/stored in place of the binary data itself. Kolmogorov complexity [16] is a theoretical generalization of this idea: it considers the *shortest* program for a universal computer that outputs the sequence. Unfortunately, Kolmogorov complexity is incomputable [16] limiting its applicability to theory. Nevertheless, relaxing the *shortest* property requirement, the search for such a sub-optimal program (logic function) is still of interest for efficient data compression.

Motivated by this intuition, we study in this section a methodology able to identify a valid logic function that can generate back the initial binary data. Then, such logic function is minimized by logic synthesis techniques effectively eliminating redundancy. We integrate this approach in a lossless compression flow and then we show its reversibility for exact data decompression.

A. Description of a Logic Function Generating a Specific Binary String

The input of general data compression tools is a string of bits, say B. Data decorrelation techniques begin by partitioning B in M sub-blocks $\{S_0, S_1, ..., S_{M-1}\}$ of length $L = \lceil |B|/M \rceil$[1] [6]. We follow this approach to describe a logic function G that outputs B. In this context, the requirement for G is to produce an output S_i when stimulated by the *Binary Representation* (BR) of the partition index i (BR(i) has $N = \lceil log_2(M) \rceil$ bits). More formally, G is an L-output, N-input Boolean function implementing the relation $G(\mathrm{BR}(i)) = S_i$. The original string B can be generated back by stimulating G with consecutive values of BR(i) and concatenating the corresponding output. Note that, the choice for the partitioning number M is dictated by practical considerations on the binary data to be compressed. The same argument also holds for general decorrelation transforms [6].

Algorithm 1 creates a *Sum Of Products* (SOP) representation for G given the partition $\{S_0, S_1, ..., S_{M-1}\}$. Note that G is initially represented in SOP form but can be minimized later in various ways as long as the final logic circuit for G is small. Two promising candidate techniques to minimize G are multi-level synthesis and binary decision diagrams. On the one hand, multi-level synthesis is efficient to manipulate a large amount of primes that are present in the SOP of G for large data. On the other hand, binary decision diagrams based methods can map G in a program with nested if then else to regenerate data. Regardless of what is the best approach, the use of logic synthesis is orthogonal to our compression method enabling a high degree of flexibility to choose the most appropriate minimization technique. The process in Algorithm 1 to generate the SOP for G consists of two nested for loops, the first considers all the L outputs of G while the second one considers all the M partitioned sub-blocks. The rationale is the following: when a sub-block S_i assumes the logic 1 value at the k-th bit, the k-th output of G is updated accordingly by adding the cube BR(i) to its SOP. At the end of this procedure, a valid description for the Boolean function G is obtained.

[1]For the sake of simplicity, fixed-length partitions are considered.

Algorithm 1 G function description.

INPUT: binary strings $\{S_0, S_1, ..., S_{M-1}\}$ (L-bits per each)
OUTPUT: SOP representation for G function
FUNCTION: Construct G$(\{S_0, S_1, ..., S_{M-1}\})$

    for all $k = 0 : L - 1$ **do**
        for all $i = 0 : M - 1$ **do**
            if $(S_i(k) == 1)$ **then**
                add cube BR(i) to SOP for the k-th output of G
            end if
        end for
    end for

G function example: $B =$ 000001010011000001110111, partition coefficients $M = 8$ ($N = 3$), $L = 3$, partition set $\{S_0, S_1, .., S_7\} = \{000, 001, 010, 011, 000, 001, 110, 111\}$. Consider the first bit highlighted in bold (G_0). The SOP of G_0 is $G_0 = $ BR(6)+BR(7). With BR$(i) = \{I_0, I_1, I_2\}$, the first (0) bit function becomes $G_0 = I_0 I_1 \overline{I_2} + I_0 I_1 I_2$.

By using Algorithm 1, it is possible to describe a function G for every partitioned binary string B. It can be easily verified that the worst case size for the SOP of G is $O(L \cdot M) = O(|B|)$ cubes of N bits each. Thus, the SOP description complexity is linear in the size of the initial string B. However, we want a logic circuit representation for G much smaller than $O(|B|)$ in order to have advantageous compression for B. Logic synthesis techniques are capable to shrink down the size of G preserving its functionality. In the previous example, $G_0 = I_0 I_1 \overline{I_2} + I_0 I_1 I_2$ is minimized in $G_0 = I_0 I_1$ by logic synthesis. On the one hand, optimal synthesis techniques give the best result in terms of synthesized circuit size but require a long runtime. On the other hand, synthesis heuristics produce near-optimal results with efficient runtime. Since data compression targets the size reduction of large files, heuristic techniques are key to have affordable runtime. Then, the capability of synthesis heuristic to produce satisfactory results heavily depends on the initial logic representation. Moreover, there exist logic functions too complex to be recognized and minimized by traditional synthesis heuristics, e.g., arithmetic operations such as exponential, logarithm etc. When one of these complex functions describes the underlying correlation in the data set that we want to compress, the direct synthesis of G via heuristics may reveal to be unfruitful.

We propose here to improve the efficiency the (heuristic) synthesis process by providing additional information about the function G. While G can be fully described by the binary indexes BR(i) for the partitions of B, G also admits some more flexibility exploiting the sequential nature of the decoding process, i.e., G can be stimulated by BR(i) *if and only if* it has been previously stimulated by BR$(i - 1)$. Therefore, introducing $S_{i-1} = G(\mathrm{BR}(i - 1))$ as additional input, on top of BR(i), some further simplification for G are enabled. Please note that by introducing S_{i-1} we are not moving to a sequential synthesis approach, instead the synthesis process is made unaware of the provenience of the input S_{i-1} hence remaining in a combinational context. To exploit the S_{i-1} information it is necessary to consider two different cases: (i) $S_{i-1} = G(\mathrm{BR}(i-1))$ is unique in $\{S_0, S_1, ..., S_{M-1}\}$ and (ii)

$S_{i-1} = G(\mathrm{BR}(i-1))$ repeats in $\{S_0, S_1, ..., S_{M-1}\}$. In the first case, the L bits for $S_{i-1} = G(\mathrm{BR}(i-1))$ are sufficient information to uniquely determine $G(\mathrm{BR}(i-1)) = S_i$, consequently the information for S_{i-1} is ORed to the original description of G. In the second case, the information of S_{i-1} has to be ANDed with $\mathrm{BR}(i)$ to uniquely determine the output, this condition is not added to G since it does not contain additional (non redundant) information. The updated procedure is presented in Algorithm 2.

Algorithm 2 Synthesis-facilitated description of G.

INPUT: binary strings $\{S_0, S_1, ..., S_{M-1}\}$ (L-bits per each)
OUTPUT: SOP representation for G function
FUNCTION: Construct G($\{S_0, S_1, ..., S_{M-1}\}$)

>**for all** $k = 0 : L - 1$ **do**
>>**for all** $i = 0 : M - 1$ **do**
>>>**if** ($S_i(k) == 1$) **then**
>>>>add cube $\mathrm{BR}(i)$ to SOP for the k-th output of G
>>>>**if** (S_{i-1} is unique in $\{S_0, S_1, ..., S_{M-1}\}$) **then**
>>>>>add cube S_{i-1} to SOP for the k-th output of G
>>>>**end if**
>>>**end if**
>>**end for**
>**end for**

The key improvement with respect to Algorithm 1 is the following: if the $i\text{-}th$ output for S_i has a unique predecessor S_{i-1} this can be used as alternative (logical or with $\mathrm{BR}(i)$) information to describe G. In other words, we can uniquely determine $G_i = G(\mathrm{BR}(i)) = S_i$ by giving the index i and its binary representation $\mathrm{BR}(i)$ but also by specifying the predecessor of S_i, S_{i-1}, provided that it does not repeat in the initial partition. Heuristic synthesis techniques have new minimization opportunities for G, thanks to the additional disjunctive information on $G(\mathrm{BR}(i))$ deriving from S_{i-1} (exploiting the causal nature of G). It is clear that Algorithm 2 can be extended to deal with S_{i-2}, S_{i-3} etc., the choice of the numbers of previous outputs to be employed is then a tradeoff between the effectiveness of the additional information and the increase in representation size. Practical discussion on this topic is given in Section IV.

B. Compression Flow

Our proposed compression flow employs logic synthesis to identify and remove redundancy in binary data. If logic synthesis fails at discovering the underlying function for a certain portion of data, entropy encoding is used to complete the compression process. Procedure details are given in Algorithm 3. First, the input binary string B is partitioned in substrings $\{S_0, S_1, ..., S_{M-1}\}$ as in usual compression flows [6]. A logic function G is then constructed for $\{S_0, S_1, ..., S_{M-1}\}$ using the methodology presented in Algorithm 2. G is synthesized via traditional heuristics considering each output separately. It is worth to stress that even if G has an initial description in SOP (2-level) form, synthesis heuristics can produce general logic networks potentially more compact than 2-level circuits. When the synthesis for a certain output G_i is not effective (it generates time-out or results in a large circuit) then the

---

Algorithm 3 Lossless compression of binary string B.

INPUT: Binary string B
OUTPUT: Compressed string C
FUNCTION: Compress B

>$R = \emptyset$ (set of integers – indexes)
>$K = \emptyset$ (set of logic functions)
>$W = \emptyset$ (set of bits – string)
>partition B in L-bit long $\{S_0, S_1, ..., S_{M-1}\}$ substrings
>construct logic function G for $\{S_0, S_1, ..., S_{M-1}\}$ (Alg. 2)
>**for all** output i of G **do**
>>synthesize G_i
>>**if** ((synthesis time-out) **or** (size $G_i >$ threshold)) **then**
>>>$R \leftarrow i$
>>**else**
>>>store G_i's synthesized logic circuit in K
>>**end if**
>**end for**
>logic sharing extraction in K
>**for all** $i \in R$ **do**
>>**for all** $j = 0 : M - 1$ **do**
>>>$W \leftarrow S_j(i)$
>>**end for**
>**end for**
>$C =$ binary representation of K + entropy encoding of W

---

index of the corresponding input is stored in R for successive treatment. Otherwise, when the synthesis is effective, the logic circuit for G_i is stored in a common optimized logic circuit K. After all the outputs of G have been considered, the logic circuits stored in K may contain redundancy and therefore a sharing extraction algorithm is applied to K to reduce its size. Considering then the indexes in R, they represent difficult functions for which synthesis heuristics have not been able to produce satisfactory (small) logic circuits. We assume that the words (collection of bits) corresponding to such indexes in R have an *uncorrelated nature* and hence are suited to be compressed with entropy encoding techniques rather than to be synthesized in a circuit. Entropy encoding is applied to W (concatenated bits corresponding to the indexes in R). Finally, the compressed string is obtained by concatenating the binary representation for the logic circuit K and the entropy encoded sequence W. The integer set R and the integer numbers $\{M, L\}$ must be provided to the decompression stage in order to fully reconstruct the original string B.

C. Decompression Flow

The compression method in Algorithm 2 is lossless and therefore exactly reversible. The corresponding decompression method is depicted by Algorithm 4. The input of the decompression method are: the previously compressed string C (logic circuit K + binary string W), the set of integer–indexes R, and the integer values $\{M, L\}$.

The first decompression step is achieved by simulating the logic circuit K with consecutive values of $\mathrm{BR}(i)$ in input. Note that the logic circuit K is designed to be stimulated only by incremental values of i, otherwise the functionality of G is lost. If G is described using Algorithm 1 the decoding model

Algorithm 4 Lossless decompression of binary string C.

INPUT: Compressed string C (logic circuit K + binary string W), integer set R, integer values $\{M, L\}$
OUTPUT: Original string B
FUNCTION: Decompress C

    $X = \emptyset$ (set of bits – string)
    $Y = \emptyset$ (set of bits – string)
    for all $i = 0 : M - 1$ **do**
        $X \leftarrow K(\mathrm{BR}(i))$
    end for
    $Y \leftarrow$ entropy-decode(W)
    $B =$ interleave X and Y according to $\{R, L\}$

---

is just a combinational logic circuit producing as output the (partial) strings $\{S_0, S_1, ..., S_{M-1}\}$ in sequence. Otherwise, when Algorithm 2 is employed to facilitate the synthesis of G, the decoding model needs to be updated. The Mealy FSM model in Fig.1 is a valid extension to decompress logic circuits described by Algorithm 2. The previous output S_{i-1} becomes

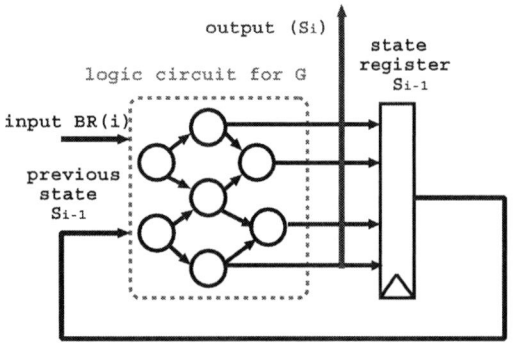

Fig. 1. Decompression FSM (Mealy) model.

the state of the Mealy machine and together with the current input $\mathrm{BR}(i)$ determines the output value. Denote by X the concatenation of binary outputs produced by the consecutive simulation of G (either FSM or combinational model).

The second decompression step is perfomed via entropy decoding of the binary string W, produced by Algorithm 3. Denote by Y the entropy decoded string.

The final step to recover the original string B consists of interleaving strings X and Y. Indeed, the initial partition $\{S_0, S_1, ..., S_{M-1}\}$ can be built back from X and Y by (i) partitioning in M sub-strings X and Y and (ii) merging pair-wise each X_i, Y_i sub-strings in unique L-bit long sub-string $\{X_i, Y_i\} \rightarrow S_i$. Such merged substring S_i has Y_i elements placed in the indexes pointed by the R set and X_i elements placed consecutively in the rest of the positions. At the end of this procedure the initial partition $\{S_0, S_1, ..., S_{M-1}\}$ is obtained, B follows by direct concatenation of S_i strings one after the other.

X-Y interleaving example: $X = 000111010$, $Y = 101$, $M = 3$ ($N = 2$), $L = 4$, $R = \{2\}$. Given M (number of partitions) X and Y can be seen as $X = \{000, 111, 010\}$, $Y = \{1, 0, 1\}$. R says that the 2nd index of each partition

originally belongs to Y while, by duality, the rest of indexes belong to X. Consequently, $B = \{0100, 1011, 0110\}$ which corresponds to $B = 010010110110$.

Note that the overall decompression technique requires M simulations of the logic circuit for G, plus M interleaving operations. Thus, it can be shown that the decompression runtime is $O(M \cdot |G|)$, where $|G|$ represents the size of the logic circuit G to be simulated.

IV. EXPERIMENTAL RESULTS

In this section, we evaluate the advantage of the proposed lossless compression technique. Decompression runtime is not reported for brevity. Comparisons with state-of-art lossless compression tools are also given.

A. Methodology

The top module of the lossless compression method proposed is described by Algorithm 3. Then, its core constituents are (i) algorithm(s) for G description, (ii) logic synthesis heuristics to minimize G and (iii) entropy encoding for the string portion not represented by G. The top module is implemented in PERL language and manages the interaction between the three major sub-modules. Algorithm 2 for G description is implemented in C language. Only the previous value S_{i-1} is considered in the implementation of Algorithm 2 since, for the considered data sets, the use of additional previous values do not carry synthesis improvements but representation size overhead. Logic synthesis is accomplished using ABC [17] academic synthesis tool. Entropy encoding for the portion of the input string not represented by G is achieved via the standard-de-facto ZIP tool [18].

In order to validate the proposed compression method, we consider data set benchmarks deriving from causal processes: 1) a perfect line measurement, 2) a line measurement affected by white noise, 3) a parabolic measurement and 4) apparently random data but automatically generated by a XOR-intensive logic circuit. Benchmarks 1-3 are data-set observable in physics experiments (Electronics, Mechanics, Astronomy etc.) while benchmark 4 represents the output of a computer program, or integrated-circuit, running over several operation cycles. In all 4 cases, we consider only data sets larger than 1 MB. The focus on such particular data sets, in place of the general benchmark suite, is motivated by the nature of the proposed compression, which is intended, and designed, for high-correlated data sets.

For these benchmarks (still binary strings B), the number of partitions considered (M) is $\lceil |B|/20 \rceil$. Therefore, G in our experiments is a 20-output, $\lceil log_2(\lceil |B|/20 \rceil) \rceil$-input Boolean function. The synthesis timeout threshold in Algorithm 3 is set as 1.0s while the size threshold is 10% of the initial logic representation size.

The counterpart compression tools considered are: (i) ZIP [18] (based on LZ [8] technique), (ii) DCT transform + ZIP, (iii) bzip2 [19] (based on Burrows-Wheeler transform [9]) and (iv) 7zip [20] (evolutions of LZ [8] technique).

B. Results

Table I shows data compression results. Among all the compression techniques considered, our approach shows the

TABLE I
DATA COMPRESSION RESULTS

Benchmark	Original Data Size	ZIP	DCT+ZIP	bzip2	7zip	Our Approach	ZIP Runtime
Linear Data	2.2 MB	208 KB	868 KB	316 KB	60 KB	8 KB	0.3 s
	25 MB	2.1 MB	8.3 MB	3.1 MB	888 KB	8 KB	2.1 s
	287 MB	21 MB	81 MB	31 MB	3.4 MB	302 KB	32 s
Linear Data + Noise	2.2 MB	264 KB	872 KB	258 KB	212 KB	80 KB	0.4 s
	25 MB	2.7 MB	8.4 MB	2.6 MB	2.4 MB	700 KB	3.0 s
	287 MB	27 MB	84 MB	30 MB	23 MB	7.1 MB	43 s
Quadratic Data	3.3 MB	484 KB	816 KB	532 KB	272 KB	8 KB	1.0 s
	39 MB	5.3 MB	7.6 MB	6.1 MB	3.3 MB	16 KB	6.1 s
	449 MB	59 MB	71 MB	67 MB	40 MB	566 KB	64 s
Random (XOR-intensive network)	1.6 MB	116 KB	304 KB	124 KB	44 KB	8 KB	0.1 s
	20 MB	1.2 MB	3.2 MB	1.5 MB	796 KB	8 KB	1.2 s
	230 MB	12 MB	31 MB	15 MB	3.8 MB	234 KB	10 s
Average runtime (normalized to ZIP)	–	1	–	1.5 x	8 x	12 x	–

highest compression gain ranging from 1 to 3 order of magnitues. The best compression ratio among counterpart tools is achieved by 7zip, being still 1~2 order of magnitudes smaller than our approach. The considerable compression gain improvement of our method comes from the capability of logic synthesis to recognize (by eliminating redundancy) the core function underlying in the data we want to compress. For example, the first benchmark representing linear data is reduced to the function $G(\mathrm{BR}(i))=\mathrm{BR}(i)$, which corresponds to a logic circuit where input and output are just connected by a wire. Then, the third benchmark (square function) is reduced to the function $G(\mathrm{BR}(i))=\mathrm{BR}(i) + G(\mathrm{BR}(i-1))$ (where $+$ is the binary addition operation), which corresponds to a two-operands binary adder circuit. It is then obvious why such large compression improvement is possible: we store only the basic function (or program) generating the data while other traditional compression tools cannot rely on this opportunity. About the runtime, ZIP is the fastest tool while our approach is on average 12x slower. This is due to the current implementation of the compression software calling external synthesis (ABC) and entropy encoding tools (ZIP). We expect that the runtime can be improved in a fully integrated software.

It is interesting to notice that in the data set 2 (linear data + noise) our proposed approach is able to identify the random portion of the data and then isolate it. Indeed, for the noisy bits of the partitions, the synthesis process always produced timeout. Then, the identified complex bits are treated as un-correlated and compressed with entropy-encoding technique. As a result, the size of the compressed data tends to the size of the random noise superposed on the linear data.

The largest data set compressed is 449 MB evidencing the scalability of our method. This is thanks to the $O(|B|)$ size of the initial G description and to the efficiency of logic synthesis heuristics employed (and-inverter-graphs based techniques in ABC [17]).

V. CONCLUSIONS

Resource usage is a key factor of today's software and hardware applications. To reduce resource usage, in terms of data storage or transmission capacity, lossless data compression

techniques are widely employed. In this paper, we use logic synthesis to compact the size of causal data sets enabling novel lossless compression opportunities. In the data compression context, the capability of modern logic synthesis tools to identify and remove redundancy, while preserving the initial behavior of the circuit, paves the way for a general lossless compression approach with the aim to find the underlying logic function generating the input data. Experimental results on data sets derived from causal processes show that our compression method based on logic synthesis reduces by 1~2 orders of magnitudes the data size compared to state-of-art compression tools such as ZIP, bzip2 and 7zip.

REFERENCES

[1] S. Nassif, *From Circuits to Cancer*, Keynote, ASPDAC 2013.
[2] F. Liu, B.R. Hodges, *Dynamic River Network Simulation at Large Scale*, Proc. DAC 2012.
[3] P. Lin, S.P. Khatri, *Application of Logic Synthesis to the Understanding and Cure of Genetic Diseases*, Proc. DAC 2012.
[4] Byron Washom, *Smart Grid for the 21st Century*, ICCAD 2009.
[5] M. Hilbert, P. Lopez, *The World's Technological Capacity to Store, Communicate, and Compute Information*, Science 332 (6025): 60-65, March 2013.
[6] K. Sayood, *Introduction to Data Compression*, 3rd ed., Elsevier, 2006.
[7] C. Lefurgy *et al.*, *Improving code density using compression techniques*, 30th ACM/IEEE Proc. on Microarchitectures, 1997.
[8] J. Ziv, A. Lempel, *A Universal Algorithm for Sequential Data Compression*, IEEE Trans. on Information Theory 23 (3): 337-343, May 1977.
[9] M. Burrows, D. Wheeler, *A block sorting lossless data compression algorithm* Technical Report 124, Digital Equipment Corporation, 1994.
[10] Joint Photographic Experts Group standard, http://www.jpeg.org.
[11] G. De Micheli, *Synthesis and Optimization of Digital Circuits*, McGraw-Hill, New York, 1994.
[12] D.A. Huffman, *A Method for the Construction of Minimum-Redundancy Codes*, PhD thesis, MIT, 1952.
[13] J. Augustinea, *et al.*, *Switching theoretic approach to image compression*, Signal Processing, 44, 243-246, 1995.
[14] D. Pramanik, *et al.*, *Lossless compression of images using minterm coding*, Proc. ICICS, 1997.
[15] J. Yang, *et al.*, *Lossless compression using 2-level and multilevel Boolean minimization*, Proc. SIPS, 2006.
[16] A.N. Kolmogorov, *On Tables of Random Numbers*, Theoretical Computer Science 207 (2): 387395, 1963.
[17] ABC Synthesis Tool [Online] http://www.eecs.berkeley.edu/alanmi/abc/
[18] ZIP Compression Tool [Online] http://www.pkware.com/
[19] bzip2 Compression Tool [Online] http://www.bzip.org/
[20] 7ZIP Compression Tool [Online] http://www.7-zip.org/

Synthesis of Power- and Area-Efficient Binary Machines for Incompletely Specified Sequences

Nan Li

Royal Institute of Technology
Stockholm, Sweden
nan3@kth.se

Elena Dubrova

Royal Institute of Technology
Stockholm, Sweden
dubrova@kth.se

Abstract— Binary Machines (BMs) are a generalization of Linear Feedback Shift Registers (LFSRs) in which a current state is a nonlinear function of the previous state. It is known how to construct a BM generating a given completely specified binary sequence. In this paper, we present an algorithm which can efficiently handle the case of incompletely specified sequences. Our experimental results show that it significantly outperforms the approaches based on all-0 or random fill in both area and power dissipation. On average, it reduces dynamic power dissipation twice compared to all-0 fill approach and 6 times compared to random fill approach. The presented algorithm can potentially be useful for many applications, including Logic Built-In Self Test (LBIST).

I. INTRODUCTION

The importance of low-power optimization techniques for our society and business is hard to overestimate. At the end of 2011, one billion people in the world were connected to mobile broadband. The wireless Internet will continue growing. Industry foresees that over 50 billion people and devices will be connected into an Internet of Things by 2020. Newer generations of products and applications (e.g. 4G telephony) will significantly increase the required computational effort, data rates and supported transmission bandwidth.

Power consumption is often the main limiting factor for the amount of functionality which can be placed on a chip. The continuous growth of power consumption is caused by the exponential increase of the integration density of CMOS circuits combined with a reduction of the clock periods. Two components of active power dissipation are:

1) Dynamic power, which accounts for the energy needed to repeatedly re-charge internal circuit nets during the circuit operation. Dynamic power is a function of the switching activity of the circuit nets, the individual gate parameters and interconnection loads, the operation frequency, and supply voltage of the circuit.
2) Static power, which is mainly caused by leakage currents of the individual gates.

In the past years, multiple design techniques have been developed that can significantly reduce the dynamic and static power consumption. These techniques simultaneously adjust supply voltages, transistor threshold voltages and transistor sizing for achieving a power-optimal design point for given performance requirements. To achieve this, coarse-grain methods work on larger blocks of circuitry whereas fine-grain methods perform adjustments for individual gates or transistors. These techniques are shown to be highly effective in reducing power consumption during functional mode of operation. However, reducing power consumption during testing remains an open problem. During testing, the sequential elements of the entire circuit are typically configured into the scan chains. The test vectors are first shifted into the scan chains (shift) and then applied to simulate the circuit behavior (capture). The loading and unloading of scan chains translate into increased level of switching during shift. In addition, during capture, the circuit operates in the functional mode for one or more clock periods. The shift and capture together may significantly increase power dissipation during testing. Increased power may negatively affect yield, cause local hot spots, or even damage the circuit.

Major techniques for reducing power dissipation during test are lowering of scan shift frequency, hierarchical design partitioning and using power-friendly test patterns which meet a specified switching threshold on a shift cycle [1]. Such patterns can be generated by a power-aware Automatic Test Pattern Generation (ATPG)-tool and then stored in a memory of an external tester. However, it is much harder to control test patterns in Logic Built-In Self Test (LBIST).

LBIST incorporates test generation and response-capture logic on-chip [2]. It typically uses a Pseudo-Random Pattern Generator (PRPG) to generate pseudo-random test patterns that are applied to the scan chains. PRPG is often complemented with an on-chip Deterministic Pattern Generator (DPG) which stores pre-computed deterministic top-off patterns. Such a combination makes possible achieving a higher fault coverage in a shorter time. The switching threshold of pseudo-random patterns can be reduced by modifying the PRNG to generate sequences with a biased distribution of 0s and 1s [3]. However, no solution for controlling the level of switching of top-off patterns is known. Using a power-aware ATPG-tool for specifying don't care bits in top-off patterns is not a good idea, because a DPG constructed for the resulting fully specified patterns is typically quite large.

In this paper, we present an approach for constructing DPGs for top-off patterns with unspecified don't care bits which minimizes both, the level of switching of top-off patterns and the area of the DPG. Our experimental results show that it significantly outperforms the approaches based on all-0(1) or random fill. On average, it reduces dynamic power twice compared to all-0 fill and 6 times compared to random fill. It is particularly efficient for test patterns with many don't cares. This is important because test patterns for large designs typically contain 95%-99% don't-cares [4].

We generate deterministic test patterns using a type of state machines known as Binary Machine (BM) [5]. An n-stage BM consists of n binary stages, n updating functions, and a clock. At each clock cycle, the current values of all stages are synchronously updated to the next values computed by the updating functions.

Several heuristic algorithms for constructing a minimal BM generating a given binary sequence have been presented [6], [7], [8], [9], [10]. However, these algorithms are applicable to completely specified sequences only. In this paper, we present an algorithm which can efficiently handle the case of incompletely specified sequences.

The rest of the paper is organized as follows. Section II gives an introduction to BMs. Section III describes the previous work. In Section IV, the algorithm for constructing BMs for incompletely specified sequences is presented. Section VI presents the experimental results. Section VII concludes the paper and discusses open problems.

II. PRELIMINARIES

Throughout the paper, we use "·", "+" and "⊕" to denote Boolean AND, OR and XOR, respectively. We use $\bar{x}$ to denote the complement of x, defined by $\bar{x} = 1 \oplus x$.

A *Binary Machine (BM)* consists of n binary storage elements, called *stages*. Each stage $i \in \{0, 1, \ldots, n-1\}$ has an associated *state variable* x_i which represents the current value of the stage i and a *updating function* $f_i : \{0,1\}^n \to \{0,1\}$ which determines how the value of i is updated (see Figure 1).

If $f_i = x_{i+1}$ for $i = 0, 1, \cdots, n-2$, then a BM is called an *Non-Linear Feedback Shift Register* (NLFSR). If updating function f_{n-1} of an NLFSR is linear, then it is called a *Linear Feedback Shift Register* (LFSR). LFSRs are the most popular type of shift registers today. They have been extensively studied and the theory behind them is well-understood [5].

A *state* of a BM is a vector of values of its state variables. At every clock cycle, the next state of a BM is determined from its current state by simultaneously updating the value of each stage i to the value of f_i, for $i \in \{0, 1, \ldots, n-1\}$. The *period* of a BM is the length of the longest cyclic output sequence it produces [5].

The *degree of parallelization* of an n-stage BM, k, is the number of output bits generated at each clock cycle, $1 \leq k \leq n$.

BMs are typically smaller and faster than NLFSRs generating the same sequence [11]. For example, consider the 4-stage NLFSR with the updating functions

$$
\begin{aligned}
f_3(x_0,x_1,x_2,x_3) &= x_0 \oplus x_2 \oplus x_3 \oplus x_1x_2 \oplus x_1x_3 \oplus x_2x_3 \\
f_2(x_3) &= x_3 \\
f_1(x_2) &= x_2 \\
f_0(x_1) &= x_1
\end{aligned}
$$

If this NLFSR is initialized to the state $(x_0x_1x_2x_3) = (1000)$, it generates the output sequence 100011110110010 with the

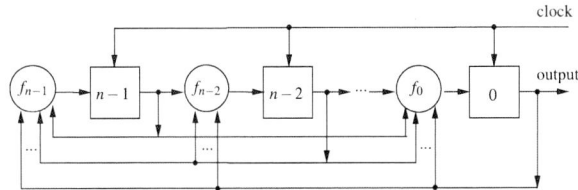

Fig. 1. General structure of an n-stage binary machine.

period 15. The same sequence can be generated by the 4-stage BM with the updating functions

$$
\begin{aligned}
f_3(x_0,x_2,x_3) &= x_0 \oplus x_2x_3 \\
f_2(x_3) &= x_3 \\
f_1(x_1,x_2) &= x_2 \oplus x_1 \\
f_0(x_1) &= x_1
\end{aligned}
$$

using the same initial state.

We can see that the BM uses 3 binary operations, while the NLFSR uses 8 binary operations. Furthermore, the depth of updating functions of the BM is smaller than the depth of the updating function of the NLFSR. Thus, the BM has a smaller propagation delay than the NLFSR.

III. RELATED WORK

LFSRs are one of the most popular devices for generating pseudo-random binary sequences today. A minimal LFSR generating a given binary sequence can be constructed using the *Berlekamp-Massey algorithm*. This algorithm was originally invented by Berlekamp for decoding Bose-Chaudhuri-Hocquenghem (BCH) codes [12]. Massey [13] linked the Berlekamp's algorithm to LFSR synthesis and simplified it.

The Berlekamp-Massey algorithm constructs classical LFSRs, which generate one bit of the output sequence per clock cycle. A number of techniques have been developed for constructing LFSRs with the degree of parallelization p, generating p bits of the output sequence per clock cycle. Two main approaches are: (1) synthesis of subsequences representing p decimation of some phase shift of the original LFSR sequence [14] and (2) computation of the set of states reachable from any state in p steps. The latter is usually done by computing pth power of the connection matrix of the LFSR [15]. Parallel LFSRs are used in applications where high data rate is important, such as Cyclic Redundancy Check (CRC), widely used in data transmission and storage for detecting burst errors [16].

NLFSRs and BMs have been much less studied compared to LFSRs. The problem which received most attention is synthesis of full-length NLFSRs whose period is of length 2^n (see [17] for an excellent overview). Algorithms for constructing an NLFSR generating a given binary sequence were presented in [6], [7]. For BMs (also called (n,k)-*NLFSRs* [18]), algorithms were presented in [9] and [10].

The algorithm [9] constructs BMs with the guaranteed minimum number of stages for a given binary sequence. This algorithm exploits the unique property of BMs that *any* binary

978-1-4799-2817-0/14 $31.00 © 2014 IEEE

n-tuple can be the next state of a given current state. The algorithm assigns every 0 of a sequence a unique even integer and every 1 of a sequence a unique odd integer. Integers are assigned in an increasing order starting from 0. For example, if an 8-bit sequence 00101101 is given, the sequence of integers 0,2,1,4,3,5,6,7 can be used. This sequence of integers is interpreted as a sequence of states of a BM. The largest integer in the sequence of states determines the number of stages. In the example above, $\lceil \log_2 7 \rceil = 3$, thus the resulting BM has 3 stages. The feedback functions f_0, f_1, f_2 implementing the resulting current-to-next state mapping are derived using the traditional logic synthesis techniques [19]. In [10], the algorithm [9] was extended to BMs generating p bits of the output sequence per clock cycle. The main idea is to encode a binary sequence into an 2^p-ary sequence which can be generated in a simpler way. As an example, suppose that we use the 4-ary encoding $(00) = 0$, $(01) = 1$, $(10) = 2$, $(11) = 3$ to encode the binary sequence 00101101 from the example above into the quaternary sequence 0231. Then, we can construct a parallel BM generating 00101101 2-bits per clock cycle with a sequence of states 0, 2, 3, 1. Note that $\lceil \log_2 3 \rceil = 2$, so the resulting parallel BM has one stage less than the BM constructed above.

IV. ALGORITHM FOR SYNTHESIS OF BMs FOR INCOMPLETELY SPECIFIED SEQUENCES

An *incompletely specified* binary sequence contains don't-care bits which can either be a logic 1 or a logic 0 in the specified sequence.

In general, it is possible to deal with incompletely specified sequences in two ways:

1) To develop a BM synthesis algorithm capable of handling don't-cares, or
2) To specify the don't-care bits first and then to use an existing BM synthesis algorithm for completely specified sequences.

In this paper, we take the second approach. We specify don't-care bits using *Inherit* algorithm presented in this section and then apply a modified version of the algorithm [10] to construct a BM for the resulting completely specified sequence.

A. Algorithm for Specifying Don't-Care Bits

The easiest way to specify don't-care bits is to fill them with either all zeros or all ones. Such a strategy is good for reducing the number of transitions $0 \rightarrow 1$ and $1 \rightarrow 0$ and thus decreasing the dynamic power dissipation. However, it results in a biased 0-to-1 ratio which increases the number of stages in the resulting BM and, potentially, its total area.

A random fill can be used to create a balanced sequence. However, by doing so we increase the number of transitions $0 \rightarrow 1$ and $1 \rightarrow 0$, which negatively affects the dynamic power dissipation. Furthermore, we increase the overall entropy of the sequence. It is known that the entropy of data puts a theoretical limit on the size of the minimal representation for the data [20]. A random sequence has the highest entropy.

Algorithm 1 Inherit(S, p): Specify don't-care bits in a binary sequence $A = (a_0, a_1, \ldots, a_{|A|-1})$ for the degree of parallelization p.

1: **for** $i \leftarrow 0$ to $p - 1$ **do**
2:     **if** a_i is don't-care **then**
3:         $a_i \leftarrow \text{rand}()$
4:     **end if**
5: **end for**
6: **for** $i \leftarrow p$ to $|A| - 1$ **do**
7:     **if** a_i is don't-care **then**
8:         $a_i \leftarrow a_{i-p}$
9:     **end if**
10: **end for**

Therefore, a random fill typically results in large BMs, especially when the percentage of don't-care bits is high.

Our goal was to find an algorithm which reduces the number of transitions $0 \rightarrow 1$ and $1 \rightarrow 0$ while keeping the ratio 0-to-1 unbiased. After trying many different strategies including

- specify all don't-cares to 0's (or all to 1's),
- specify to random values,
- specify odd positions to 1's, even positions to 0's (or vice versa),
- specify to corresponding bits in a pseudo-random sequence generated by e.g. an LFSR,

we found that a simple approach based on "inheriting" values of previously specified bits of the sequence works best. On one hand, by creating a dependence between the previous and next bits in a sequence, we reduce randomness of a sequence. On the other hand, since both zeros and ones are inherited, the ratio 0-to-1 remains relatively unbiased. In addition, by "inheriting" previous values, we minimize the number of transitions $0 \rightarrow 1$ and $1 \rightarrow 0$, reducing the dynamic power.

The pseudocode of *Inherit* algorithm is shown as Algorithm 1. The input of *Inherit* is an incompletely specified binary sequence $A = (a_0, a_1, \ldots, a_{|A|-1})$ to be generated and the degree of parallelization p. First, the sequence A is partitioned into p-tuples. Each don't-care bit is assigned an identical value to the bit with the same position in the previous p-tuple. The don't-care bits in the first p-tuple are chosen at random.

B. Algorithm for Synthesis of BMs

In order to generate a given binary sequence A with the degree of parallelization p, the algorithm [10] first partitions A into p-tuples and then appends $m = \lceil log_2 N_{max} \rceil$ extra bits to each p-tuple, where N_{max} is the largest number of identical p-tuples in the sequence A. The extra bits are required in order to make current-to-next state mapping of the resulting BM unique.

These extra bits are assigned values according to the binary expansions of integers from 0 to $2^m - 1$ in the increasing order. For example, if $N_{max} = 4$ and a p-tuple b occurs in the sequence four times, then b is expanded into $(p+2)$-tuples $(b, 0, 0), (b, 1, 0), (b, 0, 1), (b, 1, 1)$. The extra bits for p-tuples which occur less than N_{max} are assigned in the same way. For

978-1-4799-2817-0/14 $31.00 © 2014 IEEE

example, if a p-tuple c occurs only twice in the sequence, then it is expanded into $(c,0,0)$ and $(c,1,0)$.

In [21], the algorithm [10] is modified to exploit a possibility of using don't-care values for the p-tuples which occur less than N_{max} number of times in the sequence. So, in the example above, the p-tuple c is expanded into $(p+2)$-tuples $(c,0,x),(c,1,x)$, where "x" stands for a don't-care value. The extra bits are assigned values in correspondence with a Gray code [22]. In the example above, the p-tuple b is expanded into $(p+2)$-tuples $(b,0,0),(b,1,0),(b,1,1),(b,0,1)$.

The modified BM synthesis algorithm takes as its input a completely specified binary sequence $A=(a_0,a_1,\ldots,a_{|A|-1})$ and the desired degree of parallelization p. It constructs a BM for this sequence as follows:

1) Partition the input sequence into p-tuples.
2) Compute the largest number of identical p-tuples in the sequence, N_{max}.
3) Append to each p-tuple $m=\lceil log_2 N_{max}\rceil$ extra bits as follows. If the p-tuple occurs in the sequence N times, assign the first $\lceil log_2 N\rceil$ extra bits of the N identical p-tuples values corresponding to N consecutive $\lceil log_2 N\rceil$-bit codewords of a Gray code. Assign the remaining $m-\lceil log_2 N\rceil$ extra bits don't-care values.
4) The resulting sequence $S=(s_0,s_1,\ldots,s_{\lceil|A|/p\rceil-1})$ of $(p+m)$-tuples s_i represents the sequence of states of the BM.
5) Define the current-to-next state mapping as $s_i\mapsto s_{i+1}$, for $i\in\{0,1,\ldots,s_{\lceil|A|/p\rceil-2}\}$. All remaining states are mapped to don't-care values.
6) Synthesize the functions $f_0,f_1,\ldots,f_{p+m-1}$ implementing the resulting current-to-next state mapping using the traditional logic synthesis techniques [19].

If the input sequence length $|A|$ is not a multiple of p, the sequence is padded with don't-care values.

Since, by construction, the first p bits of each $(p+m)$-tuple s_i in $S=(s_0,s_1,\ldots,s_{\lceil|A|/p\rceil-1})$ correspond to the ith p-tuple of the input sequence, the resulting BM generates the input sequence A with the degree of parallelization p.

C. An Example

As an example, let us construct a BM which generates the following 40-bit sequence with the degree of parallelization 4:

$$A = \text{x0x0010x10xx01x1101x1xx00x01xx1x1xx00x01}.$$

First, we partition A into ten 4-tuples:

$$\text{x0x0},\text{010x},\text{10xx},\text{01x1},\text{101x},\text{1xx0},\text{0x01},\text{xx1x},\text{1xx0},\text{0x01}.$$

Then, we assign don't-cares in the first 4-tuple at random. Suppose that the resulting assignment is 1010. Then, for the second and other 4-tuples, we specify the don't-cares according to the value in the previous 4-tuple above. The resulting assignment is:

$$1010,0100,1000,0101,1011,1010,0001,0011,1010,0001.$$

Since the most frequent 4-tuple 1010 occurs 3 times, $N_{max}=3$. Thus, we need to add $m=2$ bits to each 4-tuple. After

assigning values to the extra bits according to the modified synthesis algorithm, we get the following sequence of BM states:

$$S = (101000,0100xx,1000xx,0101xx,1011xx,101010,$$
$$00010x,0011xx,101011,00011x).$$

The functions implementing the resulting current-to-next state mapping are defined by the following table:

$x_0x_1x_2x_3x_4x_5$	$f_0f_1f_2f_3f_4f_5$
1 0 1 0 0 0	0 1 0 0 x x
0 1 0 0 x x	1 0 0 0 x x
1 0 0 0 x x	0 1 0 1 x x
0 1 0 1 x x	1 0 1 1 x x
1 0 1 1 x x	1 0 1 0 1 0
1 0 1 0 1 0	0 0 0 1 0 x
0 0 0 1 0 x	0 0 1 1 x x
0 0 1 1 x x	1 0 1 0 1 1
1 0 1 0 1 1	0 0 0 1 1 x
0 0 0 1 1 x	x x x x x x

The remaining 54 input assignments are mapped to don't-cares. We can implement the above functions as:

$$f_5 = x_0 \oplus x_3$$
$$f_4 = x_5$$
$$f_3 = x_4 \oplus x_3 x_5$$
$$f_2 = x_3$$
$$f_1 = x_2 \oplus x_4$$
$$f_0 = x_1 \oplus x_2 x_3.$$

As we can see, the modified synthesis algorithm introduces don't-cares into both, input and output, parts of the defining table of updating functions. This helps minimizing the size of updating functions.

If instead of using *Inherit* algorithm we used all-0 fill, we would get the following updating functions:

$$f_5 = \bar{x}_0 x_2$$
$$f_4 = x_1 x_5 + x_2$$
$$f_3 = x_0 \bar{x}_2$$
$$f_2 = x_3$$
$$f_1 = \bar{x}_2 \cdot (\bar{x}_0 \bar{x}_1 \bar{x}_3 + x_0 \bar{x}_4)$$
$$f_0 = \bar{x}_0 x_2 + x_1.$$

As we can see, the resulting BM uses 11 binary operations, while the BM obtained with *Inherit* algorithm, uses only 6 binary operations.

V. SWITCHING ACTIVITY ANALYSIS

In this section, we perform a probabilistic analysis on the number of transitions $0\to 1$ and $1\to 0$ in output sequences generated by BMs for different methods for specifying don't cares. If these sequences are used as test patterns for LBIST, then fewer transitions imply smaller dynamic power dissipation in circuit under test during shift, as well as smaller dynamic power dissipation in the BM itself during test generation.

Let A be an incompletely specified binary sequence of length n in which every specified element is selected independently and uniformly at random from $\{0,1\}$. Suppose that, for each bit of A, the probability that it is specified is s, $0\le s\le 1$.

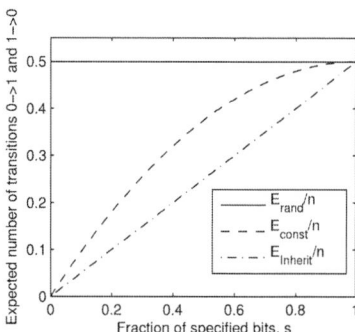

Fig. 2. Expected number of transitions $0 \to 1$ and $1 \to 0$ for different specifying methods.

Then, the bits of A have values $\{0, 1, x\}$ with the following probabilities:

$$\begin{cases} 0 & : & s/2 \\ 1 & : & s/2 \\ x & : & 1-s \end{cases} \qquad (1)$$

For a BM with degree of parallelization p, we compare three methods for specifying don't-cares:

1) *Random fill* specifies all don't-cares to either 0 or 1 with equal probability.
2) *Constant fill* specifies all don't-cares to the same value (0 or 1).
3) *Inherit fill* specifies all don't-cares to be equal to their pth prior positions in the sequence, as described in Section IV-A.

For a sequence $A = (a_0, a_1, \ldots, a_{|A|-1})$ generated with the degree of parallelization p, the number of transitions $0 \to 1$ and $1 \to 0$ is determined by the number of pairs (a_i, a_{i+p-1}) in A such that either $a_i = 0$ and $a_{i+p-1} = 1$, or $a_i = 1$ and $a_{i+p-1} = 0$. We call the 1st element of the pair, a_i, the *prior bit* of the 2nd element of the pair, a_{i+p-1}.

After a random fill, every bit is different from its prior bit with 50% probability. Therefore, the expected number of transitions in A after a random fill, E_{rand}, is:

$$E_{rand} = \frac{n}{2}.$$

Let $P(c)$ be the probability that a bit of A has the value $c \in \{0, 1\}$ after all don't-care bits of A are specified to c. From (1), it follows that:

$$P(c) = 1 - s/2.$$

Then, the probability that a bit of A is specified to the value $\bar{c} = c \oplus 1$ is:

$$P(\bar{c}) = s/2.$$

So, the expected number of transitions in A after a constant fill with c, E_{const}, is:

$$\begin{aligned} E_{const} &= nP(c)P(\bar{c}) + nP(\bar{c}) \cdot P(c) \\ &= 2nP(c)(1 - P(c)) \\ &= s(1 - \frac{s}{2})n. \end{aligned}$$

TABLE I
AREA AFTER MAPPING FOR RANDOM SEQUENCES OF LENGTH 2^{16}.

% of don't cares	Area after mapping			Improvement		Runtime, sec		
	random A_1	all-0 A_2	proposed A_3	$\frac{A_1}{A_3}$	$\frac{A_2}{A_3}$	random t_1	all-0 t_2	proposed t_3
0	45765.0	45765.0	45765.0	1.00	1.00	4.44	4.42	4.43
25	45764.7	46908.0	44919.1	1.02	1.04	4.43	4.71	4.36
50	45907.0	46015.5	41652.9	1.10	1.10	4.42	4.82	4.08
75	45836.7	40573.2	34890.8	1.31	1.16	4.42	4.34	3.56
90	45813.8	32160.5	25988.2	1.76	1.24	4.42	3.69	2.86
95	45924.4	24993.2	19399.5	2.37	1.29	4.67	3.38	2.40
99	45899.8	10550.3	8418.4	5.45	1.25	4.53	2.12	1.45
avg	45844.5	35280.8	31576.3	2.00	1.16	4.48	3.93	3.31

TABLE II
AREA AFTER MAPPING FOR RANDOM SEQUENCES WITH 90%
DON'T-CARES.

$\|S\|$	Area after mapping			Improvement		Runtime, sec		
	random A_1	all-0 A_2	proposed A_3	$\frac{A_1}{A_3}$	$\frac{A_2}{A_3}$	random t_1	all-0 t_2	proposed t_3
2^{10}	926.4	562.0	438.0	2.12	1.28	0.17	0.15	0.13
2^{11}	1863.6	1193.7	963.0	1.94	1.24	0.23	0.20	0.17
2^{12}	3607.4	2384.8	1979.1	1.82	1.20	0.37	0.30	0.26
2^{13}	6932.5	4620.5	3783.9	1.83	1.22	0.64	0.52	0.43
2^{14}	12924.5	8890.5	7423.9	1.74	1.20	1.18	0.97	0.79
2^{15}	24380.9	16842.3	13961.0	1.75	1.21	2.26	1.88	1.49
2^{16}	45937.9	32034.9	26090.0	1.76	1.23	4.43	3.70	2.88
2^{17}	86030.0	60648.6	47913.1	1.80	1.27	9.18	7.53	5.63
2^{18}	159315.6	114281.7	86154.0	1.85	1.33	19.74	16.40	11.49
2^{19}	290470.5	215869.3	152749.8	1.90	1.41	41.50	35.34	23.26
avg	63238.9	45733.7	34145.6	1.85	1.26	7.97	6.70	4.65

For the Inherit fill method, a transition occurs only if a bit is specified and it is different from its prior bit. This happens with probability $s/2$. Thus, the expected number of transitions in A after an Inherit fill, $E_{Inherit}$, is:

$$E_{Inherit} = \frac{sn}{2}.$$

We can conclude that, for $0 < s < 1$, $E_{Inherit} < E_{const} < E_{rand}$. Figure 2 illustrates the difference in the expected number of transitions for the three methods.

VI. EXPERIMENTAL RESULTS

In this section, we compare *Inherit* algorithm to other possible strategies for specifying don't-care bits, such as

1) Specifying don't-cares at random
2) Specifying all don't-cares to 0 and or all to 1

All experiments were carried out on a PC with Intel i5 3.20 GHz quad-core CPU and 4GB RAM. BMs for the specified sequences were constructed using our modified version of the algorithm [10]. Circuits for updating functions were synthesized using *resyn2* script [23] of UC Berkeley's tool ABC [24], and were mapped into *mcnc.genlib* library.

Table I shows results for random sequences of length 2^{16} with a different percentage of don't-cares for the degree of parallelization 16. For each percentage of don't-cares, an average result for 10 random sequences is given. For all-0 and all-1 fills, the results were nearly identical. Therefore, in the table we show the case of all-0 fill only.

As we can see, *Inherit* algorithm significantly outperforms all-0 and random fill approaches. It is particularly efficient for sequences with many don't-cares. For example, for the case

978-1-4799-2817-0/14 $31.00 © 2014 IEEE

TABLE III
DYNAMIC POWER DISSIPATION FOR RANDOM SEQUENCES WITH 90% DON'T-CARES.

$|S|$	Dynamic power dissipation, uW/MHz			Improvement	
	random	all-0	proposed	$\frac{P_1}{P_3}$	$\frac{P_2}{P_3}$
	P_1	P_2	P_3		
2^{10}	1.15	0.59	0.38	3.02	1.54
2^{11}	1.80	0.81	0.50	3.58	1.60
2^{12}	3.33	1.21	0.63	5.26	1.91
2^{13}	5.87	2.11	1.04	5.64	2.03
2^{14}	9.99	3.25	1.63	6.12	1.99
2^{15}	18.37	5.78	2.38	7.71	2.42
2^{16}	33.90	11.71	4.34	7.82	2.70
2^{17}	63.60	17.85	7.10	8.96	2.52
2^{18}	121.19	27.33	12.88	9.41	2.12
2^{19}	250.57	48.48	26.60	9.42	1.82
avg	50.98	11.91	5.75	6.69	2.07

of 99% don't-cares, it constructs BMs which are 5.45 times smaller than the ones obtained using random fill method. This result can be important for testing because test patterns for large designs typically contain 95%-99% don't-care bits [4].

To investigate how the efficiency of different specifying methods changes with the growth of sequence length, we applied them to random sequences of lengths form 2^{10} to 2^{19}. Table II shows the results for the case of 90% of don't-cares and the degree of parallelization 16. For each sequence length, an average result for 100 random sequences is given.

We can see that *Inherit* algorithm outperforms all-0 filling strategy by at least 20%, and random fill by at least 74%.

In the last columns of Tables I and II we show the runtime required to synthesize BMs using different approaches. It includes all steps of the flow, from sequence reading to technology mapping. The proposed approach is the fastest in all cases.

In the third experiment, BMs generating random sequences of length ranging from 2^{10} up to 2^{19} with 90% don't-cares, are synthesized and mapped to a 90nm commercial library. The dynamic power dissipation of the BMs synthesized with the proposed approach is measured under 1.0V core voltage using a commercial tool, and compared to all-0 and random fill methods. The results are shown in Table III. The proposed method constructs BMs which, on average, dissipate twice less dynamic power compared to all-0 fill method and 6 times less dynamic power compared to random fill method.

VII. CONCLUSION

In this paper, we present an efficient method for synthesis of BMs for incompletely specified binary sequences. Our experimental results show that the presented strategy for specifying don't-cares significantly outperforms the approaches based on all-0, all-1, or random fill, in both chip area and dynamic power dissipation.

Future work includes evaluating the presented method on test patterns of real designs.

ACKNOWLEDGEMENT

This work was supported in part by the project No 2011-03336 from Swedish Governmental Agency for Innovation

Systems (VINNOVA) and in part by the research grant No SM12-0005 from the Swedish Foundation for Strategic Research.

REFERENCES

[1] S. Goel and K. Chakrabarty, "Power-aware test data compression and bist," in *Power-Aware Testing and Test Strategies for Low Power Devices* (P. Girard, N. Nicolici, and X. Wen, eds.), pp. 147–173, Springer US, 2010.

[2] E. McCluskey, "Built-in self-test techniques," *IEEE Design and Test of Computers*, vol. 2, pp. 21–28, 1985.

[3] C. Chin and E. J. McCluskey, "Weighted pattern generation for built-in self test," Tech. Rep. TR - 84-7, Stanford Center for Reliable Computing, Aug. 1984.

[4] Z. Wang and K. Chakrabarty, "Test data compression for IP embedded cores using selective encoding of scan slices," in *Proceedings of International Test Conference (ITC'2005)*, pp. 10 pp. –590, Nov. 2005.

[5] S. Golomb, *Shift Register Sequences*. Aegean Park Press, 1982.

[6] C. J. A. Jansen, "The maximum order complexity of sequence ensembles," *Lecture Notes in Computer Science*, vol. 547, pp. 153–159, 1991. Adv. Cryptology-Eupocrypt'1991, Berlin, Germany.

[7] D. Linardatos and N. Kalouptsidis, "Synthesis of minimal cost nonlinear feedback shift registers," *Signal Process.*, vol. 82, no. 2, pp. 157–176, 2002.

[8] K. Limniotis, N. Kolokotronis, and N. Kalouptsidis, "On the nonlinear complexity and Lempel-Ziv complexity of finite length sequences," *IEEE Transactions on Information Theory*, vol. 53, no. 11, pp. 4293–4302, 2007.

[9] E. Dubrova, "Synthesis of binary machines," *IEEE Transactions on Information Theory*, vol. 57, pp. 6890 – 6893, 2011.

[10] E. Dubrova, "Synthesis of parallel binary machines," in *Proceedings of International Conference of Computer-Aided Design (ICCAD'2011)*, (San Jose, CA, USA), pp. 200–206, Nov. 2011.

[11] E. Dubrova, "A transformation from the Fibonacci to the Galois NLFSRs," *IEEE Transactions on Information Theory*, pp. 5263–5271, November 2009.

[12] E. R. Berlekamp, "Nonbinary BCH decoding," in *International Symposium on Information Theory*, (San Remo, Italy), 1967.

[13] J. Massey, "Shift-register synthesis and BCH decoding," *IEEE Transactions on Information Theory*, vol. 15, pp. 122–127, 1969.

[14] A. Lempel and W. L. Eastman, "High speed generation of maximal length sequences," *IEEE Trans. Comput.*, vol. 20, pp. 227–229, February 1971.

[15] S. Mukhopadhyay and P. Sarkar, "Application of LFSRs for parallel sequence generation in cryptologic algorithms," in *Computational Science and Its Applications - ICCSA 2006*, vol. 3982 of *Lecture Notes in Computer Science*, pp. 436–445, Springer Berlin / Heidelberg, 2006.

[16] J. McCluskey, "High speed calculation of cyclic redundancy codes," in *Proc. of the 1999 ACM/SIGDA seventh international symposium on Field programmable gate arrays*, FPGA '99, (New York, NY, USA), pp. 250–256, ACM, 1999.

[17] H. Fredricksen, "A survey of full length nonlinear shift register cycle algorithms," *SIAM Review*, vol. 24, no. 2, pp. 195–221, 1982.

[18] E. Dubrova, M. Teslenko, and H. Tenhunen, "On analysis and synthesis of (n,k)-non-linear feedback shift registers," in *Design and Test in Europe Conference (DATE'08)*, pp. 133–137, March 2008.

[19] R. K. Brayton, C. McMullen, G. Hatchel, and A. Sangiovanni-Vincentelli, *Logic Minimization Algorithms For VLSI Synthesis*. Kluwer Academic Publishers, 1984.

[20] C. E. Shannon, "A mathematical theory of communications," *The Bell System Technical Journal*, vol. 27, pp. 379–423,623–656, July,October 1948.

[21] N. Li, S. S. Mansouri, and E. Dubrova, "Secure key storage using state machines," in *Multiple-Valued Logic (ISMVL), 2013 IEEE 43rd International Symposium on*, pp. 290–295, 2013.

[22] F. Gray, "Pulse code communication," 1953. U.S. patent No. 2,632,058, 1953.

[23] A. Mishchenko, S. Chatterjee, and R. Brayton, "DAG-aware AIG rewriting: a fresh look at combinational logic synthesis," in *Design Automation Conference, 2006 43rd ACM/IEEE*, pp. 532 –535, 2006.

[24] Berkeley Logic Synthesis and Verification Group, "ABC: A system for sequential synthesis and verification, release 70930."

978-1-4799-2817-0/14 $31.00 © 2014 IEEE

Multi-Mode Trace Signal Selection for Post-Silicon Debug

Min Li and Azadeh Davoodi

Department of Electrical and Computer Engineering

University of Wisconsin at Madison, USA

Email: {mli46, adavoodi}@wisc.edu

Abstract— Trace buffers are used during post-silicon debug to increase the visibility to the internal signals of a chip via online tracing of a few state elements within a capture window. Due to the small bandwidth of the trace buffer, only a few state elements can be selected or tracing in order to restore the states of the remaining state elements as many as possible. In this work, we show that the quality of restoration corresponding to a set of trace signals selected for a single operating mode, may significantly degrade over the remaining operating modes of a design; an operating mode refers to specific values taken by control signals such as signals for mode selection and scan enable. This is the first work to study the *multi-mode* trace signal selection problem in order to maximize the restoration over all the operating modes of a design. We propose algorithmic strategies for this problem as well as a procedure to reduce the number of modes by merging the ones with "similar" restoration maps; merging improves the runtime scalability of our multi-mode trace selection algorithm with increase in the number of modes, without much loss in the solution quality.

I. INTRODUCTION

Due to the high complexity in modern designs, comprehensive verification prior to fabrication is not possible. Later on, at the post-silicon stage, lack of visibility to the signals inside the chip has made validation and debug a cumbersome task.

An on-chip trace buffer allows tracing a few state elements of a design within an capture window corresponding to the buffer depth [1]. Tracing is ignited by a trigger condition at runtime. These trace signals can be arbitrarily selected from the set of state elements of the design but the selection is done once during the design stage. At the post-silicon stage, the collected traces are analyzed off-line within the capture window to debug logic errors [6], [17]. Specifically the traces are first used to restore as many other state elements for each cycle of the capture window before analyzing a bug.

Due to the complexity of modern IC designs, selection of trace signals can no longer be done manually. Recently, many automated trace signal selection algorithms have been proposed in order to maximize the restoration ratio of the untraced signals. These can be categorized into faster (but lower quality) metric-based algorithms [8], [12], [15] versus slower (but higher quality) simulation-based algorithms [5]. The works [3], [10] propose quick procedures with better balance in the solution quality and execution runtime. These previous works have been based on gradual selection/elimination of signals (i.e., forward greedy strategy in [3], [8], [10], [12], [15] and backward pruning strategy in [5]). Other procedures for simultaneous selection based on Integer Linear Programming (ILP) [16] and circuit Satisfiability [17] have also been proposed. However, such procedures may not be scalable with increase in the design size. ILP-based selection strategy is subject to higher error in approximating the restoration ratio, (as opposed to computing using simulation [5]).

A design may operate in various modes based on the values taken by its control signals. Examples of control signals include signals for mode selection, scan enable, power gating and clock gating, encryption, etc. Some of these control signals may also be introduced

This research was supported by National Science Foundation under Grant 1053496.

by the CAD tools for example based on automated strategies for controlling sleep/wakeup in power-gated designs (e.g., [9]).

Control signals typically have a much lower switching activity compared to most of the other primary inputs in a design. However, selection of the trace signals based on unrealistic changing of the control signals (e.g., random assignment in simulation) may yield to a significantly poor restoration for any of the individual operating modes. The simplest example are set and reset signals which should be appropriately set to active or inactive when evaluating the restoration within the trace signal selection process. Otherwise, by assuming such signals can take any values using random assignment, the selected traces may have a lower restoration quality than reality.

To acknowledge the impact of control signals during the trace signal selection process, previous works have considered generating *separate* solutions for each operating mode [3], [5], [8], [10], [12], [15]; this means the control signals are kept constant during each trace signal selection process. However as we show in this work, by selecting trace signals which are optimized for a single operating mode, the restoration ratio of the remaining modes can be significantly degraded for that same solution. This issue may significantly bring down the value of trace buffers, if bug analysis at the post-silicon stage needs to be conducted using signals traces which were optimized for an operating mode different than where the bug is observed. Moreover, while it is possible to find different sets of trace signals for each operating mode, and feed them all to an interconnect network [11], [14] which selects only one set for tracing at each time, however, the overhead of such a network (including the routing to feed in various signals to the network) may significantly increase with increase in the number of operating modes.

In this work, we propose the multi-mode trace signal selection (MMTS) problem which aims to maximize an objective corresponding to the restoration ratio over all the operating modes. Our contributions to solve the MMTS problem are summarized below.

1. A procedure to reduce the number of modes by merging the modes with "similar" restoration maps;
2. A procedure based on perturbing an initial single-mode-optimized solution, selected from a suitable "start" mode, to improve the restorability over all the modes.

Our algorithm has a non-greedy nature and at each iteration tries to improve the previous solution within a gradually-increasing perturbation radius. It is driven by a metric which quickly reflects the attainable restorability by a group of candidate trace signals, over all the operating modes. In our simulation results, we use large-sized designs from IWLS05 and ISPD12 gate sizing contest benchmarks with up to 150K gates. We measure the quality of our algorithm with respect to a defined point of reference providing an upper bound, and four other alternative strategies.

We start by reviewing some preliminaries in Section II and define the MMTS problem in Section III. Our mode merging strategy is given in Section IV. Our procedure for multi-mode trace signal selection is given in Section V. Simulation results are discussed in Section VI followed by conclusions in Section VII.

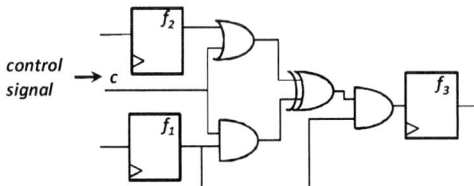

Fig. 1. Example circuit for explaining the notations

II. PRELIMINARIES

A. Single-Mode Trace Signal Selection (SMTS)

Consider a trace buffer of size $B \times N$, where B designates the buffer bandwidth corresponding to the number of signals that can be traced, and N designates the buffer depth corresponding to the size of a *capture window* of N clock cycles for tracing.

Moreover, for a sequential circuit, we have already identified the control signals denoted by the set $\mathcal{S}_C$. When the circuit operates in a single mode, we assume the control signals take constant and known (0 or 1) values (one out of at most $2^{|\mathcal{S}_C|}$ combinations) which do not change within the capture window of the trace buffer.

For a trace buffer of size $B \times N$ and control signal set $\mathcal{S}_C$ specified for a single mode m, the single-mode trace signal selection (SMTS) problem aims to select B state elements in order to maximize restoration of the untraced signals expressed by the equation below.

$$SRR^m = \frac{B \times N + \sum_{n=1}^{N} k_n}{B \times N} \quad (1)$$

Here SRR^m expresses the *State Restoration Ratio* (SRR) of mode m. k_n indicates the number of untraced signals in cycle n which can be restored using the trace signals and the control signals in mode m. SRR^m measures the average restoration attainable per trace signal per clock cycle within the capture window.

B. Overview of X-Simulation

In Equation 1, computing the value of k_n at cycle n is done using X-Simulation. During X-Simulation for a single clock cycle, the values of the trace signals and control signals are known and fixed. The rest of the signals may be 0, 1, or unknown (x). The circuit is iteratively simulated from the selected trace signals to the primary outputs (i.e., a forward traversal) and then backwards towards the primary inputs (i.e., a backward justification). As the iterations proceed, the values of some of the unknown state elements will be restored which will be used to restore the values of the remaining ones in the future iterations as many as possible. For example in Figure 1, assume the control signal takes value $c = 0$. If flipflop f_3 is getting traced and takes value 1, we can restore f_1 and f_2 both to value 1 using backward justification. The X-Simulation iterations continue for each clock cycle until no more signals can be restored. X-Simulation is a time-consuming process and must be implemented as efficient as possible for effective utilization in the trace selection procedures [5], [8], [10].

III. MULTI-MODE TRACE SELECTION PROBLEM

A. Problem Definition

Given a trace buffer of size $B \times N$, and a set of control signals defining M operating modes, the Multi-Mode Trace Selection (MMTS) problem selects B state elements in order to maximize the *summation* of State Restoration Ratios over all the modes (denoted by $MSRR$), and given by the equation below.

$$MSRR = \sum_{m=1}^{M} SRR^m \quad (2)$$

In the above equation, SRR^m is the state restoration ratio of mode m, given by Equation 1. To evaluate SRR^m for a single mode m, X-Simulation is used to compute restoration of the remaining signals using the (same) selected trace signals but with (different) control signal values corresponding to mode m.

We note the single mode state restoration ratio SRR^m has widely been used as the metric for measuring the quality of the SMTS problem [3], [5], [8], [10], [12], [15]. So it is intuitive to define the multi-mode objective by building upon it.

B. A Point of Reference for A High Quality $MSRR$

Given a solution for MMTS, we discuss a point of reference in order to get a sense for the quality of the corresponding $MSRR$. Given recent advancements in the algorithms to solve the SMTS problem (e.g., [3], [5], [8], [10], [12], [15]), a natural point of reference is to (parallel-) solve SMTS for each of the M modes independently and then add the corresponding SRR^m values. (Note this procedure generates M distinct solutions for M modes.) In practice, this will likely be an upper bound for $MSRR$ because it is obtained from solutions each of which is individually optimized for one mode. We elaborate using a simple example of S38584 with only two modes. We use the SMTS procedure in [10]. The results are shown in the table below. The two rows are when solving SMTS for modes 0 and 1. For each trace selection solution, SRR^0 and SRR^1 are computed for modes 0 and 1 as reported in columns 2 and 3. The reference $MSRR$ here is $17 + 8.2 = 25.2$.

TABLE I
USING SMTS SOLUTION FOR MMTS

	SRR^0	SRR^1	$MSRR$	$MSRR$ (%REF)
SMTS^0	17.0	4.3	21.3	85%
SMTS^1	14.3	8.2	22.5	89%

Consider the example in Table I for S38584 again. We report the ratio of the $MSRR$ of each single-mode solution with respect to the reference case in column 5. In this example, we observe that the solution of SMTS^1 has a higher $MSRR$ of 89% than of SMTS^0. (Notice that as expected, SRR^0 is lower in SMTS^1 than SMTS^0.) However, we show using our proposed strategy which emphasizes multi-mode selection throughout the procedure that it is possible to achieve a solution with $MSRR$ that is 99% of the reference case. This is partially due to the observation that there is a small (and sometimes no) sharing in the selected trace signals of the single-mode solutions (i.e., SMTS^0 and SMTS^1 in this example).

IV. IDENTIFYING AND MERGING "SIMILAR" MODES

We observe that some combinations of the values of control inputs have similar impact in restoring the other signals, i.e., they result in similar "restoration maps". A restoration map shows the signals which can be restored with corresponding values of 0 or 1, if restored *only* using the control signals. With this observation, we propose to merge the operating modes with similar restoration maps into a single one. Mode merging reduces the number of modes considered in MMTS and as a result, directly translates into speed improvement of our MMTS procedure without much loss in its solution quality.

We explain our strategies using an example. Figure 2 shows the restoration maps of ISCAS89 benchmark S35932 for four different modes corresponding to the values taken by the two control signals given in [10] which exist in this benchmark. Each restoration map shows the signals which can be restored, *only* when the two control signals are applied and the remaining primary inputs do not take any values (except an inactive reset). In each plot of Figure 2, the green dots indicate the components (i.e., gates or state elements)

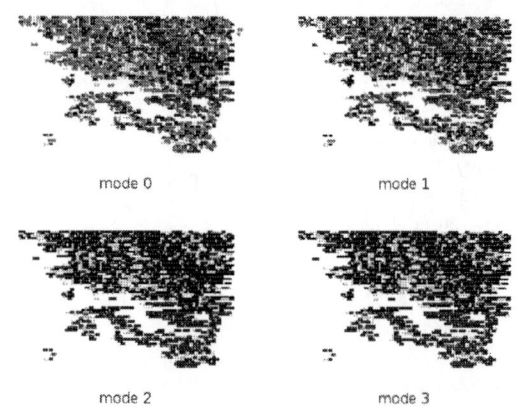

mode 0   mode 1

mode 2   mode 3

Fig. 2. Restoration maps of S35932 under different working modes; modes with similar restoration maps can be merged into a single mode.

which are restored to value 0 while the black dots indicate the components which are restored to value 1. The red dots show the components which were not restored. (Note, Figure 2 only shows about 20% of this benchmark; the restoration maps of the remaining 80% were similar in the four modes and contained mostly unrestored components.) The above restoration maps indicate that modes 0 and 1 may be suitable to be merged into one mode. (Similarly modes 2 and 3 are suitable to be merged.) So the MMTS problem can be solved for two modes after merging instead of four.

Using the observations from the above example, we now propose a merging strategy for combining *two* modes into a single mode if they have a "similar" restoration map. Consider two modes m_i and m_j. The number of restored components in the two modes are denoted by R_i and R_j. Assume the number of common components restored by both modes is denoted by C_{ij}. (This is without considering the values that they are restored to.) We define similarity between the restoration maps of modes m_i and m_j using the equation below.

$$S_{ij} = \frac{C_{ij}}{\max(R_i, R_j)} \qquad (3)$$

Note in the above equation, the similarity S_{ij} is a ratio between 0 and 1. We consider two modes can be merged if $S_{ij} \geq \alpha$, where α is a constant threshold set to 0.7 in our simulations.

When merging two modes, we replace them with a single representative mode. It is the mode which has a restoration map with the highest number of restored components. Next, we extend the strategy for combining two modes in order to combine multiple modes. We start by visiting each mode exactly once and try to combine it with *one* of the existing modes. Here an existing mode can be an unvisited mode, or a mode representing merging of previously-visited modes. Specifically, when visiting mode m_i, if we find it can be merged with more than one of the existing modes, we only merge it with the mode m_j with the highest similarity with m_i (i.e., S_{ij} is highest compared to the other modes which can be merged with m_i). The process terminates when all (unmerged) modes are visited once.

V. ALGORITHM FOR THE MMTS PROBLEM

We first define a set of metrics which drives our algorithm to identify the top candidates for multi-mode restoration, and then discuss the selection procedure using these metrics.

A. Metrics to Find the Top Candidates for Multi-Mode Restoration

Our metrics are defined by extending the modeless metrics in [10].

1) Reachability List in Mode m: For a state element f and single operating mode m, we define two reachability lists L_{fv}^m when f takes value $v = 0$ and $v = 1$ and the control signals take the values corresponding to mode m. Here L_{fv}^m is the subset of state elements which can be restored by f without the help of any other state elements but accounting for values of the control signals. For example in Figure 1, assume the control signal is set to value $c = 0$. For f_3 we have $L_{f_3^0}^{c=0} = \emptyset$ and $L_{f_3^1}^{c=0} = \{f_1, f_2\}$.

We also define a single (merged) reachability list for each operating mode $m = 1, \ldots, M$ and each state element $f \in F$ as given below.

$$L_f^m = L_{f0}^m \cup L_{f1}^m \qquad (4)$$

The merged reachability list of a state element in one operating mode may be different than another mode. For example for the circuit in Figure 1, we have $L_{f_3}^{c=0} = \{f_1, f_2\}$ and $L_{f_3}^{c=1} = \{f_1\}$.

Computation of the reachability lists of all the state elements (for a single operating mode) is done only once as a pre-processing step prior to trace selection. This computation is fast and can be ignored compared to the runtime of the trace selection procedure [10].

2) Restoration Demand in Mode m: Restoration demand (denoted by $d_{i,fv}^m$) is defined between state elements i and f for mode m. Here i is an untraced state element and f is a candidate considered for tracing. Furthermore, a subset of the required trace signals may have already been selected which (partially) contribute to restoration of i. The restoration demand metric approximates how much of the remaining restoration need of i can be provided by f, and is computed by the equation below [10]:

$$d_{i,fv}^m = \min(1 - r_i^m, a_{fv}^m), \ \forall i \in L_{fv}^m \qquad (5)$$

In the above equation r_i^m is the restorability rate of state element i which is a number between 0 and 1 reflecting the rate that i can be restored using the currently-selected trace signals in mode m. (The work [10] shows that these restorability rates can be computed for all the state elements quickly, using *one round* of X-Simulation in an *observation window* (of 64 cycles) which is much smaller than the trace buffer's capture window of typically a few thousand cycles.) Also a_{fv}^m is the rate that state element f takes value v obtained using simulation with sufficient amount of randomly-selected primary input patterns. Note even if state element i is in L_{fv}^m, the maximum restoration which it can get from f (when f takes v) is at most a_{fv}^m.

3) Multi-Mode Impact Weight: Using the demands, impact weights in mode m are computed for each untraced state element using the equation below.

$$W_f^m = \sum_{v=0,1} \sum_{\forall i \in L_{i,fv}^m} d_{i,fv}^m \qquad (6)$$

The impact weight W_f^m of state element f reflects how much a state element can contribute to restoring the remaining untraced elements (in its reachability list) in mode m. The state elements with higher impact weights are more suitable candidates for tracing [10].

Finally, we define a multi-mode impact weight (MW_f) as below.

$$MW_f = \sum_{m=1}^{M} W_f^m \quad \forall f \in F \setminus T \qquad (7)$$

It represents the impact weight of state element f over all the modes.

B. An Iterative Multi-Mode Algorithm (IteM)

1) Overview: Our algorithm is denoted by IteM (for Iterative Multi-Mode). IteM starts with a trace signal solution that is optimized for a single-mode, identified as a suitable "start" mode (m_{start}) which is more likely to yield to the highest $MSSR$. IteM then perturbs this solution iteratively by replacing some of the existing

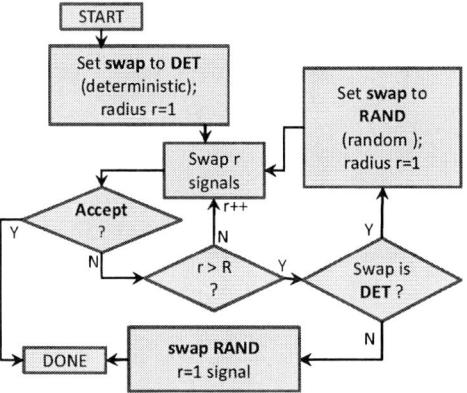

Fig. 3. Procedure to swap up to R trace signals per iteration, with gradual increase in perturbation radius r

TABLE II
BENCHMARK INFORMATION

	#FF	#Gates	M	M_{merged}	
S38584	1166	10552	2	2	ISCAS89
S35932	1728	11032	4	2	ISCAS89
b17	1317	33888	4	4	IWLS05
b18	3020	119762	2	2	IWLS05
dsp	3605	54730	8	2	IWLS05
DMA	2192	36556	8	4	ISPD12
des_perf	8802	149066	2	2	ISPD12

trace signals with the other unselected state elements, in order to improve restoration over all the modes. It has a non-greedy nature based on a gradually-increasing perturbation radius within each iteration. Specifically, at each iteration, up to R number of trace signals in the current solution may be swapped. The process terminates upon observing no improvements in $MSSR$ in 20 consecutive iterations.

One motivation behind IteM is that one operating mode m may have significantly higher potential for restoration compared to many of the other modes. Moreover, working with the initial solution generated by an SMTS procedure for m_{start} allows a faster pace to maximize multi-mode restorability in an iterative procedure.

The specific strengths of IteM are (1) quickly identifying a suitable starting mode and finding its corresponding single-mode solution; (2) considering non-greedy and randomized moves with gradual degree of perturbation when swapping at each iteration; (3) explicit consideration of multi-mode restoration at each swap using the multi-mode impact weight introduced in the previous subsection.

2) Identifying the Start Mode and Generating the Initial Solution:
To find m_{start}, we first compute the reachability list of each state element for each mode (L_f^m given by Equation 4). Recall if a state element is selected as a trace signal, then all the state elements in its reachability list can be restored by it, without the help of any other trace signal. We then compute m_{start} using the following equation.

$$m_{start} = \operatorname{argmax}_m \left(\sum_{\forall f \in F} |L_f^m| \right) \qquad (8)$$

where the sizes of the reachability lists of the state elements in a single mode are added, and the mode with the maximum aggregate size is selected. Note, this process requires pre-computing $M \times |F|$ reachability lists (for each state element and each mode), however in practice it is quite fast, e.g., negligible runtime to compute all the reachability lists for a single mode compared to solving SMTS for that mode, as we show in our experiments. In case of mode merging, a representative mode is used instead. So the reachability lists of the state elements are only computed for the representative mode.

After m_{start} is identified, we apply the SMTS algorithm in [10] for initial solution generation due to its fast runtime and relatively good solution quality. This initial solution is used as the basis for the following perturbation steps in order to gradually improve $MSRR$.

3) Iterative Perturbation of the Current Solution: Figure 3 shows the process of swapping up to R trace signals at a single iteration of IteM. We start by the smallest *radius* of perturbation of $r = 1$, corresponding to swapping a single trace signal with one of the remaining state elements. A probabilistic acceptance criteria is used to evaluate the swap. If the acceptance criteria is satisfied, the swap will be made and the iteration terminates. Otherwise, the radius of perturbation is gradually increased up to $r = R$, as shown in Figure

3. This gradual increase in the radius of perturbation increases the likelihood to satisfy the acceptance criteria. If a suitable swap can not be found at the maximum radius, the radius is set back to 1 and a new search starts based on switching to swapping using a random strategy. The process repeats itself using random swapping up to the maximum radius. In the end, if the acceptance criteria is still not satisfied, the random swap strategy is used to swap one signal (without further evaluating the criteria).

Below we explain further details of the algorithms.

Each swap is made of two steps: 1) eliminating the least promising trace signal from the current solution, and 2) adding the most promising of the remaining state elements as a trace signal. In the case of random swap, the elimination part is based on random elimination of state elements (and not the least promising one) from the existing trace signals but the addition step remains the same.

Elimination of the least promising trace signal is done as follows. For a given trace signal solution with B traces, we consider eliminating each trace signal and compute the corresponding $MSSR$ for the remaining $B - 1$ trace signals. (Recall $MSSR$ is given by Equation 2 and computed using X-Simulation.) The elimination candidates are ranked with respect to the corresponding $MSSR$ and the top candidate for elimination is the one with the highest $MSSR$ (associated with the remaining signals).

Computing the elimination ranking of B candidates requires B number of $MSSR$ computations involving X-Simulation. To speedup the process, once a (ranked) elimination queue is created and the swaps in one iteration are made, the same queue will continue to be used for a total of 8 iterations. The eliminations in the remaining 7 iterations is done by just selecting the trace from the top of the list. We observed the above procedure to be a reasonable approximation while providing significant speedup in our simulations.

Addition of the most promising trace signal is done by ranking the unselected trace signals using their multi-mode impact weights ($MW_f \forall F \in T \setminus T$) given by Equation 7. Next a small number of top candidates (i.e., 3% of state elements) with highest MW_f are selected. The best candidate is then found by computing $MSSR$ among the top candidates.

Applying $r > 1$ number of swaps is done as follows. First r eliminations are done by selecting the r trace signals from the top of the elimination queue. Next r new trace signals with the highest multi-mode impact weights (MW_f) are sequentially selected. In this case, every time, a new trace signal is added, the multi-mode impact weights MW_fs are also updated.

To accept one (or more) swaps, first, we consider the $MSSR$ for the set of trace signals if the swaps are made. (Note, this $MSSR$ is already computed during the addition step when the best candidate is identified.) We denote this by $MSSR_{cur}$. If the $MSSR_{cur}$ is better (higher) than $MSSR_{prv}$ (i.e., the $MSSR$ of the previous solution), then the swap is accepted. Otherwise, it is accepted by following Boltzmann's criteria [7] ($e^{\frac{MSRR_{cur} - MSRR_{prv}}{T}} > rand$ with $rand$ a randomly generated number between 0 and 1). The parameter T is updated at each iteration to be 0.95 of its value in the previous iteration with an initial value of 10K.

VI. SIMULATION RESULTS

We verify the IteM algorithm using a subset of ISCAS89 [4], IWLS05 [2], and ISPD12 gate sizing contest [13] benchmarks. These are the benchmarks we identified to include control signals. Moreover, the IWLS05 and ISPD12 benchmarks are much larger in size than the benchmark used in previous published works; Table II shows the number of state elements and combinational gates are listed in columns 2 and 3, respectively. The largest one is des_perf with about 150K gates with 8K sequential components. Therefore, due to the larger number of sequential components we use a trace buffer bandwidth of $B = 64$ in our experiments. (Previous works used $B = 32$ for much smaller-sized benchmarks.) We also used a capture window of 2048 cycles which corresponds to the buffer bandwidth. We note both this buffer bandwidth and capture window are considered as reasonable values for on-chip trace buffer [17]. Column 4 lists the number of modes defined by the control signals. (In defining the modes, we excluded obvious control signals such as reset by assuming reset is always inactive.)

When implementing IteM, we applied the mode merging procedure explained in Section IV with a similarity threshold of 0.7. The number of modes after merging is reported in column 5. Mode merging reduced the number of modes in 3 of the benchmarks which are shown by highlighted rows. For example, in dsp mode merging allowed reducing the number of modes from 8 to 2. We verified that the identified signals are control signals in different ways. First, using the similarity metric given by Equation 3, we observed difference in the restoration maps in the operating modes defined by the identified control signals, as can be seen from Table II columns 4 and 5. Moreover, by solving the single-mode trace selection problem (SMTS) for each mode, we observed either significant variation in the corresponding single-mode restoration ratio SRR^m, and a significant difference in the generated solutions (i.e., the selected trace signals).

We compared the following techniques in our simulations:
- IteM: Our proposed multi-mode selection procedure
- HYBR: Single-mode selection procedure of [10]
- RATS: Single-mode selection procedure of [3]
- SimF: Single-mode simulation-based procedure of [5]
- HYBRM: Extension of HYBR for multi-mode signal selection

IteM was implemented as explained in Section V-B with a maximum radius of $R = 3$ and includes the mode merging step given in Section IV. HYBR, RATS and SimF are based on solving the SMTS problem for the same (promising) start mode identified for IteM. SimF is a variation of the simulation-based approach of [5] which selects the trace signals based on a forward greedy selection strategy. (We were not able to generate a solution within a 24-hour runtime limit for larger benchmarks, using the original selection procedure of [5] based on backward elimination of not-promising signals.)

HYBRM is an alternative multi-mode selection procedure. It follows forward greedy trace selection and runs for B iterations, selecting one trace signal at each iteration. At each iteration, first, the top 3% of state elements with the highest value of the multi-mode impact weight given by Equation 7 are selected. Among these candidates, X-Simulation is used to find the best one which provides the highest $MSRR$ value. HYBRM is a direct extension of HYBR incorporating multi-mode impact weights and $MSRR$.

In addition, to generate a reference as an upper bound for the attainable $MSRR$, we also used the procedure given in Section III-B by generating M distinct single-mode optimized solutions and adding their (individual) SRR^m $\forall m = 1, \ldots, M$. To obtain the single-mode solution in each mode m for each benchmark, we ran (in parallel), each of the single-mode techniques (i.e., HYBR, RATS, SimF) and selected the best solution to compute the highest SRR^m for mode m. We denote this reference case by REF.

TABLE III
COMPARISON OF $MSRR$ NORMALIZED TO THE REFERENCE CASE (REF)

	REF	RATS	HYBR	SimF	HYBRM	IteM
S38584	25.20	0.86	0.85	0.95	0.95	1.00
S35932	66.40	0.64	0.74	0.65	0.91	0.91
b17	7.90	N/A	0.62	0.58	0.76	0.94
b18	5.90	N/A	0.50	0.92	0.61	0.80
dsp	42.80	N/A	0.41	0.88	0.37	0.92
DMA	50.67	0.76	0.88	0.89	0.84	0.92
des_perf	77.60	N/A	0.97	0.98	0.98	0.99
Average	**1.0**		**0.71**	**0.83**	**0.77**	**0.93**

TABLE IV
COMPARISON OF RUNTIME (MIN)

	RATS	HYBR	SimF	HYBRM	IteM
S38584	0.1	2	19	4	13
S35932	0.1	2	14	5	15
b17	24+	1	19	4	24
b18	24+	4	2151	119	90
dsp	24+	2	92	28	251
DMA	5	7	99	38	125
des_perf	24+	16	469	24	94

A. Comparison of Solution Quality

We compare the solution quality of the described techniques using the multi-mode restoration ratio $MSRR$ given by Equation 2. The $MSRR$ of a trace selection solution is computed using X-Simulation over M modes for the capture window of 2048 cycles. The results are reported in Table III. $MSRR$ of REF is given in column 2. For the remaining columns, an $MSRR$ ratio, normalized to REF is given.

As can be seen, IteM consistently performs better than all the other techniques except in b18 where it performs worse than SimF but it still has an $MSRR$ of 0.8 of REF. Also SimF takes 20X longer to generate a solution in this benchmark compared to IteM. The $MSRR$ values of IteM are on average 0.93 of the $MSRR$ of REF.

HYBRM which explicitly considers multi-mode selection, often performs better than HYBR, but it is sometimes worse than SimF, except in S35932 and b17 in which it performs significantly better than SimF. However, we note HYBRM also has a runtime much faster than SimF, as we report in the next experiment.

We only reported results of RATS in 3 benchmarks as shown in the table. These results for RATS are similar to or worse than HYBR. We note, even though RATS could quickly generate a solution in these 3 benchmarks, it was not able to generate a solution for the other ones, which were all from ISPD12 and IWLS05 benchmarks, after imposing a 24-hour runtime limit. (We note while implementing RATS, the procedure for setting a custom parameter used in a step to extend the graph edge weights was not provided in [3]. We therefore tuned this parameter for each benchmark to achieve the best possible implementation of RATS for that benchmark, for the purpose of comparison in our experiments.)

B. Comparison of Runtime

Here we compare the runtime of different techniques. All techniques were implemented in C++ and ran on a quad-core CPU using multi-threading. Specifically, our setup allowed up to 8 parallel threads to run simultaneously. This parallel implementation was utilized when running X-Simulation using bit-wise parallelism procedure given in [8]. This same implementation of X-Simulation was utilized in the source codes of all our trace selection strategies.

The runtime results are reported in Table IV given in minutes. As can be seen, RATS has the fastest runtime (fraction of a minute) for the two ISCAS89 benchmarks and only 5 minutes for DMA. However, it was not able to generate a solution for the remaining benchmarks after a 24-hour runtime limit. We found out that in these benchmarks,

978-1-4799-2817-0/14 $31.00 © 2014 IEEE

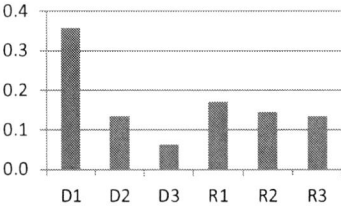

Fig. 4. Ratio of number of swaps (Y-Axis) and swap types (X-Axis)

the time spent by RATS on finding "the exact paths" of all pairs of state elements to compute the edge weights create the runtime bottlenecks. Among the rest, HYBR runs much faster due to its hybrid nature which incorporates fast metrics with a small number of X-Simulations. In contrast SimF which is purely based on X-Simulation usually has the highest runtimes in most of the benchmarks. HYBRM has a runtime between the HYBR and SimF in all benchmarks. It takes more than HYBR due to the extensions to evaluate the multi-mode impact weight and compute $MSRR$.

The runtime of IteM is reported in column 6 which includes the runtime to generate the initial solution. The amount of time to generate the initial solution in IteM has already been reported in column 3 which is the runtime of HYBR. (Recall, we used HYBR to generate the single-mode optimized initial solution in IteM.) We note, our implementation of HYBR includes an improvement to incrementally update consecutive restoration maps during the forward-greedy selection process which significantly improves the runtimes of HYBR.

When looking into the individual runtime of IteM for each benchmark, it can be seen that dsp has a much longer runtime than the others. This is because the initial solution generated by HYBR is not as good as the other benchmarks, as can be verified from the solution quality reported in Table III for this benchmark. Therefore IteM takes more iterations to improve upon the initial solution results.

Overall, we consider the runtime of IteM reasonable given the large size of most of the benchmarks and emphasize that IteM provides significantly better solution quality compared to the faster techniques.

C. Further Analyzing the IteM Algorithm

Recall that each iteration of IteM considers swapping with increasing perturbation of radius $R = 1$ up to a maximum radius of $R = 3$. The swaps are first considered using (deterministic) elimination of the least promising trace, and if the acceptance criteria is not satisfied, the process repeats but this time random elimination is used in each swap. The iteration stops as soon as the acceptance criteria is satisfied and the corresponding swap (with the corresponding radius) is enforced.

To show these strategies were all useful in IteM, in Figure 4 we reported the percentage of the iterations which stopped at a given radius (further divided into deterministic and random cases) in benchmark S35932. The X-axis shows the corresponding 6 cases. For example, D2 and R2 show the percentages of iterations in which two trace signals were swapped using deterministic and random eliminations, respectively. As can be seen, the maximum number of swaps was using D1 (one state element/deterministic) which happened in about 36% of the iterations. The minimum was D3 in 6% of the iterations, and interestingly the remaining cases (including random swaps) each had higher than 10% of the iterations.

VII. Conclusions

This paper is the first work on the multi-mode trace signal selection problem. We provided an objective to optimize the multi-mode restoration, introduced a strategy to merge the modes with similar restoration maps, and proposed a procedure to solve the problem based on iterative perturbation of an initial solution for a single but suitable start mode. Our experiments show that our procedure performs better than various single-mode procedures. It also performs better than a procedure based on direct extension of an existing single-mode procedure to optimize restoration over multiple modes. We also showed high solution quality compared to a reference case which can be considered to provide an upper bound to the best attainable solution. Our procedure runs in a reasonable time while experimenting with benchmark sizes of up to 150K gates.

References

[1] M. Abramovici, P. Bradley, K. N. Dwarakanath, P. Levin, G. Memmi, and D. Miller. A reconfigurable design-for-debug infrastructure for SoCs. In *DAC*, pages 7–12, 2006.

[2] C. Albrecht. IWLS 2005 benchmarks. In *IWLS*, 2005.

[3] K. Basu and P. Mishra. RATS: restoration-aware trace signal selection for post-silicon validation. *IEEE TVLSI*, 21(4):605–613, 2013.

[4] F. Brglez, D. Bryan, and K. Kozminski. Combinational profiles of sequential benchmark circuits. In *ISCAS*, 1989.

[5] D. Chatterjee, C. McCarter, and V. Bertacco. Simulation-based signal selection for state restoration in silicon debug. In *ICCAD*, pages 595–601, 2011.

[6] M. Gort, F. M. de Paula, J. J. W. Kuan, T. M. Aamodt, A. J. Hu, S. J. E. Wilton, and J. Yang. Formal-analysis-based trace computation for post-silicon debug. *IEEE Trans. VLSI*, 20(11):1997–2010, 2012.

[7] S. Kirkpatrick, C. D. Gelatt, and M. P. Vecchi. Optimization by simulated annealing. *Science*, 220:671–680, 1983.

[8] H. F. Ko and N. Nicolici. Algorithms for state restoration and trace-signal selection for data acquisition in silicon debug. *IEEE Trans. on CAD*, 28(2):285–297, 2009.

[9] M.-C. Lee, Y. Shi, and S.-C. Chang. Efficient wakeup scheduling considering both resource usage and timing budget for power gating designs. *IEEE Trans. on CAD*, 31(7):1041–1049, 2012.

[10] M. Li and A. Davoodi. A hybrid approach for fast and accurate trace signal selection for post-silicon debug. In *DATE*, pages 485–490, 2013.

[11] X. Liu and Q. Xu. On efficient silicon debug with flexible trace interconnection fabric. In *ITC*, pages 1–9, 2012.

[12] X. Liu and Q. Xu. On signal selection for visibility enhancement in trace-based post-silicon validation. *IEEE Trans. on CAD*, 31(8):1263–1274, 2012.

[13] M. M. Ozdal, C. Amin, A. Ayupov, S. Burns, G. Wilke, and C. Zhuo. The ISPD-2012 discrete cell sizing contest and benchmark suite. In *ISPD*, pages 161–164, 2012.

[14] S. Prabhakar, R. Sethuram, and M. S. Hsiao. Trace buffer-based silicon debug with lossless compression. In *VLSI Design*, pages 358–363, 2011.

[15] H. Shojaei and A. Davoodi. Trace signal selection to enhance timing and logic visibility in post-silicon validation. In *ICCAD*, pages 168–172, 2010.

[16] J.-S. Yang and N. A. Touba. Efficient trace signal selection for silicon debug by error transmission analysis. *IEEE Trans. on CAD*, 31(3):442–446, 2012.

[17] Y.-S. Yang, A. G. Veneris, and N. Nicolici. Automating data analysis and acquisition setup in a silicon debug environment. *IEEE Trans. VLSI*, 20(6):1118–1131, 2012.

Implicit Intermittent Fault Detection in Distributed Systems

Peter Waszecki[1], Matthias Kauer[1], Martin Lukasiewycz[1], Samarjit Chakraborty[2]

[1] TUM CREATE, Singapore, Email: peter.waszecki@tum-create.edu.sg
[2] TU Munich, Germany, Email: samarjit@tum.de

Abstract—**This paper presents a novel approach to detect resources in distributed systems with an increased occurrence of intermittent faults that exceed the amount of unavoidable transient faults caused by environmental phenomena. Intermittent faults occur due to stressed resources and often are a precursor of permanent faults. The proposed early fault detection and diagnosis allows the use of precautionary measures before the permanent failure of a component in a distributed system occurs. In this paper, we present four methods that can implicitly detect intermittent faults by taking the distributed applications and their dependencies into account. Thus, explicit tests are not required which would lead to additional costs and resource load. On the other hand, the implicit approach may considerably reduce the number of plausibility tests compared to the conservative solution with one test per resource. We analyzed and evaluated implementations of the proposed fault detection principle. The experimental results give evidence of the feasibility of our approach and show a comparison of the implemented methods in terms of runtime and detection rate.**

I. Introduction

Today, the reliability of embedded systems and the associated safety aspects are of high relevance in many domains with strict real-time requirements such as in avionics and automotive. Also for non-safety critical applications, for example, in consumer electronics, efficient fault detection and fault tolerance mechanisms are important to fulfill the customers' quality expectations. On the other hand, ever-growing Very Large Scale Integration (VLSI) processes with shrinking geometries and decreasing power supply voltages result in devices which are increasingly susceptible to transient faults and, hence, might have a negative impact on the system reliability [1]. In distributed systems, a failure of a single component can influence the behavior of a multitude of applications. It is therefore desirable to detect potential failures of components before they actually happen and apply precautionary measures which can vary from graceful degradation to a replacement of the affected component. For such an early detection, an increased number of non-permanent faults is a suitable indicator to determine stressed components.

Contributions of the paper. In this paper, we propose an approach to implicitly detect components in a distributed system that show an increased number of non-permanent faults. Assuming that deficient hardware causes the presence and accumulation of so-called *intermittent* faults which lead to software errors, then a potential imminent failure of a specific resource can be projected by analyzing the results of a set of plausibility tests running within regular tasks or as discrete applications. A major objective of the proposed approach is to perform such a detection implicitly to keep the additional costs and resource utilization low. In contrast to an explicit detection

This work was financially supported in part by the Singapore National Research Foundation under its Campus for Research Excellence And Technological Enterprise (CREATE) programme.

which would require additional test tasks for each component, the proposed fault detection relies on existing plausibility tests which are part of the distributed applications. Also, due to the architectural constraints of the resources it is often only possible to check the consistency of test results at specific points, e.g., on particular Electronic Control Units (ECUs). For this purpose, we take the distribution of applications, the runtimes of tasks, and their data-dependencies into account to implicitly determine the component with an increased number of intermittent faults.

We propose four different implementations for an implicit intermittent fault detection in distributed systems. The first two methods are based on the analysis of linear dependencies within the system model, while the other two methods use Integer Linear Programming (ILP) formulations for an optimization-based approach. In the experimental results, we show the feasibility of the proposed fault detection and compare the different methods in terms of their runtime and detectability. For the sake of simplicity, in this paper we consider a system with at most one stressed resource. However, ILP approaches innately support the detection of multiple stressed resources and the methods based on the analysis of linear dependencies might be adapted appropriately. The extension of our approach to a concurrent detection of multiple faulty resources is part of future work.

In this paper, we consider the automotive domain, working on the level of Electrical and Electronic (E/E) architectures, i.e., we assume ECUs, sensors, and actuators as the basic components. However, the proposed methodology is also applicable to other levels of granularity, e.g., in case of a Multiprocessor System-on-Chip (MPSoC), basic components might be computing units, buses, switches, or memories. While for E/E architectures the number of basic components stays in a double-digit or lower three-digit range, other domains might easily reach ranges of 10^4 or even 10^5, which makes the scalability of our approach an important objective. Since the detection might also be performed offline by analyzing the detected failed tests within a certain time interval, the approaches do not have to satisfy strict runtime requirements, but have to be within a reasonable range with a good scalability.

Organization of the paper. The remainder of the paper is organized as follows: Section II discusses the related work. Section III describes the system model, including a formal definition of the system and problem. The detection approaches are proposed in Section IV. Experimental results are presented in Section V before the paper is concluded in Section VI.

II. Related Work

In the considered distributed systems a resource can execute one or multiple periodic tasks (or generally speaking manipulate data), while these tasks might have data-dependencies across different resources by using messages or shared memory for communication. The mapping of tasks to resources is a crucial point and is discussed, e.g., in [2]. To deal with errors in the system, the most common fault tolerance strategies are, on the one hand, re-mapping of tasks as discussed in [3] and [4] for the use of Network-on-Chip (NoC)-based MPSoCs. On the other hand, when intact resources do not have enough free capacity to take over tasks of the affected resources, a graceful degradation mechanism is necessary as, for example, described in [5]. Our approach is set at the fault detection stage and, thus, before the fault tolerance mechanisms come into action. For that reason, it builds on an existing schedule and a given task distribution.

A number of approaches are dealing with the reliability and fault diagnosis in distributed systems. In [6], a hardware based monitoring is proposed that enables the detection of transient faults in a Real-Time Operating System (RTOS). Three on-line self-testing policies for multi-processors are investigated by Héron et al. [7] in terms of performance and detection probability. Approaches to the problem on a system design level are presented in [8] and [9] where the system is modeled by a Data Flow Graph (DFG) and the authors maximize the reliability by exchanging the resources until the defined constraints are met. In contrast, [10] considers not only reliability but also all other design objectives simultaneously.

However, none of these approaches consider the problem that an accumulation of non-permanent faults might finally lead to a permanent failure of a resource or, in the worst case, the entire system. To the best of our knowledge, this is the first work analyzing and comparing test failure rates to implicitly detect resources tending to fail through the steady increase of intermittent faults.

III. SYSTEM MODEL

In our system model, we assume that a resource can be affected by two types of faults: permanent faults, which durably damage the system and non-permanent faults, which can further be divided into transient and intermittent faults, and only temporarily affect the system [11]. Although permanent faults are going along with enduring physical changes of the affected hardware, they are often preceded by an increased number of intermittent faults which themselves are caused by malfunctioning hardware and occur with a high arbitrary frequency, see [1]. In contrast to intermittent faults, transient faults are caused by temporary environmental phenomena, like cosmic rays, electromagnetic interference, electrostatic discharge or radiation from lightning. The main purpose of this work is to identify intermittent faults early before a permanent fault occurs in order to apply appropriate precautions. Besides this, an important aspect is the *implicit* nature of the fault detection to avoid additional testing overhead. The occurrence of intermittent faults can be validated with the help of plausibility tests which are evaluating the outcome of particular periodic tasks. According to the hypothesis of the Resilience Articulation Point (RAP), all faults originating from a physical phenomenon, if not masked, will manifest as a permanent or non-permanent single- or multi-bit flip [12]. We assume that the errors caused by these bit flips are propagated between data-dependent tasks and finally result in plausibility test failures, with the probabilistic distribution of the intermittent faults remaining unaltered. Thus, a failed plausibility test indicates an error in an own or a preceding task originally caused by a fault on a resource. To model the

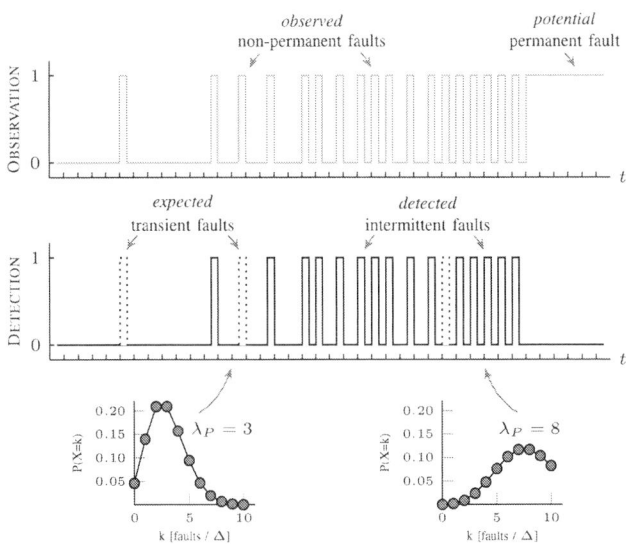

Fig. 1. Illustrative example for the occurrence of faults on a resource. In the beginning the operation is affected by transient faults only. Then, gradually, more and more intermittent faults occur which finally result in a permanent fault. The probabilistic distribution of the intermittent faults depicts an early ($\lambda_P = 3$) and a late ($\lambda_P = 8$) stage of the fault period.

probability of the fault occurrence, the Poisson distribution is used, however, our approach is flexible enough to adopt other probabilistic error models, if required. Figure 1 shows how non-permanent faults can be split into a transient and an intermittent part, with the latter becoming more and more predominant over time as illustrated by a higher mean value λ_P of the Poisson distribution. It is important to consider transient faults as possible noise in our analysis approach to avoid false positive fault detections as far as possible.

Generally, due to architectural and cost reasons not every resource can be equipped with specific tests to detect intermittent faults. Therefore, our implicit detection makes use of existing plausibility tests which are part of the user applications. In our approach, we analyze a number of plausibility tests at the end of particular task sequences and compare the expected and observed results of several tests. This allows us to draw the right conclusion about the fault-causing resource. A motivating example is shown in Figure 2, where eight application tasks τ_{x_i/y_i} and two test tasks $t_{x/y}$ are mapped to three ECUs as indicated by the gray background areas. Depending on the particular assignment and utilization of the tasks, a higher rate of intermittent faults on a resource can be detected by analyzing the failure ratio of the test tasks. Given an equal utilization among all application tasks and a consistent error propagation towards the test tasks, a faulty ECU_1 will cause a failure ratio between t_x and t_y of $\frac{2}{1}$, since it is running two tasks from the t_x-task chain and only one from the t_y-task chain. Note that correspondingly a failure ratio of $\frac{1}{2}$ indicates a faulty ECU_2 and a ratio of $\frac{1}{1}$ a faulty ECU_3.

Formal Problem Description. In the following, a mathematical description shall formally define the problem, which is mainly based on the sets and functions listed below.

T	set of available tests
R	set of resources to be considered
$\mathcal{T}_r$	set of tasks on a resource $r \in R$
Δ	time interval for test failure observation
O_t	observed failures of test $t \in T$ in Δ

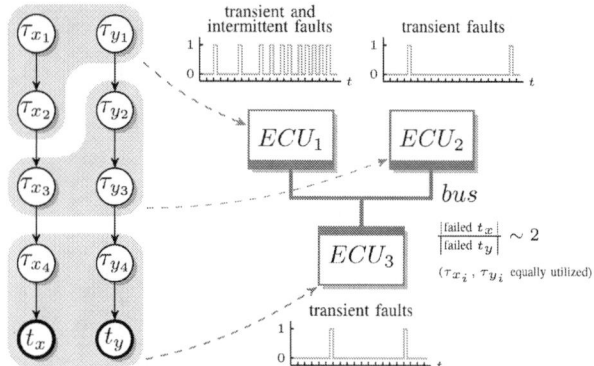

Fig. 2. Simplified example for the detection of a faulty resource. The application tasks τ_{x_1}, τ_{x_2}, τ_{y_1} are mapped to ECU_1 and τ_{x_3}, τ_{y_2}, τ_{y_3} are mapped to ECU_2 whereas the remaining τ_{x_4} and τ_{y_4} as well as the plausibility tests t_x and t_y are assigned to ECU_3. The increased intermittent fault rate on ECU_1 results in a test failure ratio of 2 between t_x and t_y.

E_t expected failures of test $t \in T$ in Δ

$\lambda : T \times R \to \mathbb{R}$ frequency of a test $t \in T$ failing due to a fault of $r \in R$

The mere quantitative analysis of plausibility test failures is not sufficient to detect a faulty resource. This is particularly the case, when considering the non-trivial example of having a system consisting of less tests than resources (see the example in Figure 2). Consequently, for each test $t \in T$ and each resource $r \in R$ we define a $\lambda(t,r)$ which represents the number of expected test failures per time interval caused by the respective resource.

The actual fault rate (i.e., number of faults per time interval) for a specific resource is a random variable following the Poisson distribution with a known average value. Figure 1 illustrates the increasing fault occurrence over time for a hypothetical resource with an increasing expectation value λ_P towards the permanent fault. In Equation (1) the fault rate is represented as an independent random variable X_r of the Poisson distribution with an expected value $E[X_r]$. The expected value is strongly dependent on the resource's susceptibility to intermittent faults and must be investigated experimentally or determined from the manufacturer's hardware description. In the scope of this work $E[X_r]$ shall be assumed to be known.

$$X_r \sim \text{Poi}(E[X_r]) \qquad (1)$$

With e_τ describing the average execution time of a periodic task $\tau \in \mathcal{T}_r$ and h_τ describing its period, $\lambda(t,r)$ is defined as in Equation (2). Here, $pred(t)$ represents the set of tasks that are predecessors of the test t.

$$\lambda(t,r) = \sum_{\tau \in \mathcal{T}_r \cap pred(t)} \frac{e_\tau}{h_\tau} \cdot E[X_r] \qquad (2)$$

The equation indicates whether the faulty resource runs any task in which an error would lead to a failure of test t. That can be the test task itself, or any of its predecessors.

The allocation of each test to the resources which can cause its failure leads to a test expectation matrix Λ which comprises all $\lambda(t,r)$ values as its elements (see Equation (3)).

$$\Lambda = \big(\lambda(t,r)\big)_{t,r} \qquad (3)$$

This matrix describes the expected fault distribution under normal operation for transient faults only. In case of intermittent

faults, the observations will deviate from this matrix and enable an implicit fault detection as described in Section IV.

Detectability Analysis. Before the actual detection process, it is important to initially ensure that the given distribution of test tasks leads to an unambiguous recognition of the affected resource. Given the presumption that only one resource can fail in the considered time interval a mutual comparison of the tests related to a specific resource enables a conclusion about the system's detectability. In detail, the so-called cosine similarity in Equation (4a) between two column vectors v_{r_i} and v_{r_j} in the Λ matrix will result in a number proportionally approximating 1 the smaller the angle θ between these two vectors is. It is lower than 1 otherwise, thus, indicating a better distinctness between two potentially faulty resources. Equation (4b) is illustrating this correlation in which ϵ_{det} is the maximum acceptable deviation of the cosine value from 1.

$$cos(\theta) = \left| \frac{\mathbf{v}_{r_i} \cdot \mathbf{v}_{r_j}}{\|\mathbf{v}_{r_i}\| \cdot \|\mathbf{v}_{r_j}\|} - 1 \right| \qquad (4a)$$

$$cos(\theta) \begin{cases} \leq \epsilon_{det} \Rightarrow r_i \approx r_j, & \theta = \angle(\mathbf{v}_{r_i}, \mathbf{v}_{r_j}) \\ > \epsilon_{det} \Rightarrow r_i \neq r_j, & \theta = \angle(\mathbf{v}_{r_i}, \mathbf{v}_{r_j}) \end{cases} \qquad (4b)$$

A valid definition of detectability can be formulated as follows.

Definition 1 (Fully Detecting Implementation): We call the allocation of plausibility tests to resources fully detecting, if each sub-matrix $\Lambda_s = \big(\mathbf{v}_{r_i}, \mathbf{v}_{r_j}\big)$ of the expectation matrix Λ consists of linearly independent columns, i.e. has full rank (Equation (5)).

$$rank(\Lambda_s^{|T| \times 2}) = min(|T|, 2) \qquad (5)$$

A trivial way to a fully detecting test setup would be to attach a plausibility test to every single resource. This leads to a diagonal matrix Λ_d which can always be solved and for this reason can be regarded as the reference solution. Therefore, one of the goal of this work is to beat the reference solution by reducing the number of required tests.

IV. DETECTION APPROACHES

Based on the presumption that the system to analyze is fully detectable, a potentially faulty resource shall be located. As a matter of principle, the solution of the problem can be abstracted to an analysis of the ratio between observed and expected test failures. Given the expectation matrix Λ, four possible realizations of the general fault detection principle shall be described and evaluated with respect to their correctness and performance. In the following, the sets O_t and E_t are representing observed and expected failures of a test $t \in T$, respectively, which occur within a time interval Δ. We assume that at least one resource $r \in R$ is faulty if the number of observed test failures originating from all possible resources is significantly higher than the number of the expected test failures within the time interval Δ, as shown in Equation (6).

$$O_t \gg \Delta \cdot \sum_{r \in R} \lambda(t,r) \qquad (6)$$

A. Method I: COSINE SIMILARITY

The first detection method uses a similar approach as the detectability analysis in Section III. We regard the expectation matrix as a set of column vectors, where each vector $\mathbf{v}_{r_i}^{\ *}$ is assigned to one resource and contains the corresponding utilization dependent expected test fault rates (Equation (7a)).

978-1-4799-2817-0/14 $31.00 © 2014 IEEE

In addition, an observation vector $\mathbf{v}_{O_T}$ which consists of the observed failures of each plausibility test is used (Equation (7b)).

$$\mathbf{v}_{r_i} = [\lambda(t_1, r_i), \lambda(t_2, r_i), \cdots, \lambda(t_m, r_i)]^T \quad (7a)$$

$$\mathbf{v}_{O_T} = [O_{t_1}, O_{t_2}, \cdots, O_{t_m}]^T \quad (7b)$$

A cosine similarity analysis between the observation vector and the expectation vectors can detect the resource responsible for the failed tests, when the corresponding cosine value is close to 1. This is illustrated in Equation (8), where θ is the angle enclosed by the examined vectors and $\epsilon_{\cos}$ is the maximum acceptable deviation from 1.

$$cos(\theta) \begin{cases} \leq \epsilon_{\cos} \Rightarrow O_T \text{ from } r_i, & \theta = \angle(\mathbf{v}_{O_T}, \mathbf{v}_{r_i}) \\ > \epsilon_{\cos} \Rightarrow O_T \text{ not from } r_i, & \theta = \angle(\mathbf{v}_{O_T}, \mathbf{v}_{r_i}) \end{cases} \quad (8)$$

B. Method II: SVD-Test

Alternatively, the second method calculates the Singular Value Decomposition (SVD) of the sub-matrix $\Lambda_s = (\mathbf{v}_{O_T}, \mathbf{v}_{r_i})$ to determine the linear dependence, as shown in Equation (9). The observation O_T is thought of being caused by resource r_i, if the rank of the diagonal matrix Σ containing the singular values σ_1 and σ_2 is less than 2 (Equation (10)).

$$\Lambda_s^{|T| \times 2} = U\Sigma V, \text{ with } \Sigma = \begin{bmatrix} \sigma_1 & 0 \\ 0 & \sigma_2 \end{bmatrix} \text{ and } |T| \geq 2 \quad (9)$$

$$rank(\Sigma) \begin{cases} < 2 \Rightarrow O_T \text{ from } r_i, & \text{if } \exists \sigma \leq \epsilon_{\text{svd}} \\ = 2 \Rightarrow O_T \text{ not from } r_i, & \text{if } \forall \sigma > \epsilon_{\text{svd}} \end{cases} \quad (10)$$

Regarding the performance and correctness, both methods lead to comparable results when using appropriate threshold values and, thus, are highly dependable on the correct choice of the tolerance interval ϵ. Numerous test runs proved $\epsilon_{\cos} \sim 10^{-3}$ and $\epsilon_{\text{svd}} \sim 10^{-5}$ as good ranges for the analyzed system sizes.

C. Method III: Confidence Interval

A fundamentally different detection approach can be applied by making use of the non-deterministic nature of fault occurrences. To mirror the stochastic uncertainty of the expected intermittent faults caused by a resource, we can create a confidence interval around E_t inside which an observation is thought to conform to its corresponding λ-values. Assuming, that the faults are following a Poisson distribution and that during normal operation all observed failures O_t lie within three standard deviations from the expected value E_t, we can define the interval according to the 3σ-rule as seen in Equation (11). Here, σ represents the standard deviation and μ the mean of the Poisson-distributed variable x.

$$P_r(\mu - 3\sigma \leq x \leq \mu + 3\sigma) \approx 0.9973 \quad (11)$$

Given the Cumulative Distribution Function (CDF) of x as $F_\lambda(x \leq k)$, the values λ_{lo} and λ_{hi} can be determined by approximating the corresponding CDFs to the limits resulting from the 3σ-rule, as shown in Equations (12a) and (12b).

---

*Note, that the expectation vectors $\mathbf{v}_{r_i}$ are not scaled with the observation interval Δ, and hence do not represent the actual expected values E_t. This has no influence on the analytical result of the cosine similarity analysis, as it is magnitude independent.

Accordingly, this leads to a confidence interval $[\lambda_{lo}, \lambda_{hi}]$ which is located around E_t.

$$\underset{\lambda_{lo} \in \mathbb{R}_0^+}{\text{minimize}} \ \lambda_{lo}, \quad \text{subject to } F_{\lambda_{lo}}(x \leq O_t) \leq \frac{P_r + 1}{2} \quad (12a)$$

$$\underset{\lambda_{hi} \in \mathbb{R}_0^+}{\text{maximize}} \ \lambda_{hi}, \quad \text{subject to } F_{\lambda_{hi}}(x \leq O_t) \geq \frac{1 - P_r}{2} \quad (12b)$$

As a next step, introducing a stress factor allows us to formulate an optimization problem with the confidence interval as one constraint (Equation (13b)). We define the expected observations E_t as sum of weighted λ-values over all resources, where the stress factor is x_r (Equation (13c)). Additionally, in Equations (13d) and (13e) we use y_r as a switch variable to decide whether x_r is significantly large ($y_r = 1$) or still in an acceptable range ($y_r = 0$), so that E_t stays inside the confidence interval. The threshold for this decision is set by x_{var}. The objective function (Equation (13a)) tries to find cases where only a minimal number of stressed resources is responsible for a failed test.

$$\underset{y_r \in \{0,1\}}{\text{minimize}} \ \sum_{r \in R} y_r \quad (13a)$$

subject to:

$$\forall t \in T: \quad \lambda_{lo}(O_t) \leq \frac{E_t}{\Delta} \leq \lambda_{hi}(O_t) \quad (13b)$$

$$\forall t \in T: \quad E_t = \sum_{r \in R} x_r \cdot \lambda(t, r) \cdot \Delta \quad (13c)$$

$$\forall r \in R: \quad x_r \geq x_{\text{var}} \cdot y_r \quad (13d)$$

$$\forall r \in R: \quad x_r \leq x_{\text{var}} + 10^{10} \cdot y_r \quad (13e)$$

For the optimization algorithm $\lambda(t, r)$ and O_t are considered to be constants from the set of positive rational numbers $\mathbb{R}_0^+$. Similarly, the variables E_t and x_r are defined in $\mathbb{R}_0^+$, whereas y_r contains only two values, $\{0, 1\}$.

To rule out additional resources being responsible for the failed tests, the optimization algorithm must be started a second time with the first solution excluded from the test. Only if the latter test reveals no solution, one can be sure about the correctness of the first run.

D. Method IV: χ^2-Test

The fourth method to detect faulty resources is based on a statistical hypothesis test. To determine how well the expectation fits an observation the *Pearson's χ^2-Test* shall be used as a statistical model to describe the goodness of fit (see Equation (14)).

$$\chi^2 = \sum_{t \in T} \frac{(O_t - E_t)^2}{E_t} \quad (14)$$

The null hypothesis H_0 for the test is defined in Equation (15) and it indicates that all *observed* test failures O_t result from the *expected* test failures E_t.

$$H_0: E_t \rightarrow O_t \quad (15)$$

H_0 shall be accepted if the p-value of the χ^2-Test is higher than a predefined significance level and rejected otherwise. Equation (16) denotes this inequality where $F(\chi^2, |T| - 1)$ is the CDF of the χ^2-distribution with $|T| - 1$ degrees of freedom.

$$1 - F(\chi^2, |T| - 1) \geq 0.05 \quad (16)$$

To find solutions for the χ^2-test we can formulate an optimization problem analogue to that, used for the confidence interval approach. Again, we try to find cases where the sum of all switch variables y_r is minimal. However, the χ^2-Test puts E_t in a quotient term and hence, leads to a non-linear problem. In order to still be able to use linear solvers, the variable R_t serves as a binary representation of the reciprocal value E_t^{-1}. This is defined in the Equations (17b) and (17c) for the optimization problem formulated below.

$$\underset{y_r \in \{0,1\}}{minimize} \quad \sum_{r \in R} y_r \tag{17a}$$

subject to:

$$\forall t \in T: \quad \chi^2 = \sum_{t \in T} O_t^2 \cdot R_t - 2 \cdot O_t + E_t \tag{17b}$$

$$\forall t \in T: \quad E_t \cdot R_t = 1 \tag{17c}$$

$$\forall t \in T: \quad E_t = \sum_{r \in R} x_r \cdot \lambda(t,r) \cdot \Delta \tag{17d}$$

$$\forall r \in R: \quad x_r \geq x_{\mathrm{var}} \cdot y_r \tag{17e}$$

$$\forall r \in R: \quad x_r \leq x_{\mathrm{var}} + 10^{10} \cdot y_r \tag{17f}$$

Similarly to the previous fault detection method, the test has to be repeated to rule out alternative solutions.

V. EXPERIMENTAL RESULTS

In this section the experimental results are presented. All experiments were carried out on an Intel Core i5 with 2.6 GHz and 4GB RAM. For the methods that required an ILP solver, GUROBI [13] was used. In the following, the implementation details and the test cases will be presented. The results will be discussed afterwards.

Implementation and Test Case Modeling. Expected test failure rates for all resources are represented as λ-values of a Poisson distribution and implemented as a Λ-matrix, as defined in Section III. To specify a stressed resource, its λ-value is weighted with a factor s in the range of 10^3, which is equivalent to an increase of its initial fault rate. The implementation of the application, which consists of interdependent tasks, and architecture, which consists of interconnected resources, is based on a previously presented graph-based model [14]. Figure 3 shows an exemplary system specification with two functions and five resources as well as the corresponding expectation matrix. In this example, resource $r3$ is considered faulty, which affects the highlighted λ-values in the expectation matrix.

240 tests cases of realistic sizes and topologies have been generated. These test cases comprise various system specifications with 10-100 resources and 3, 5 and 10 tasks executed on each resource, respectively. The ratio of tests to resources varies between $\frac{1}{4}$ and 1. For each test case, the observation time was as well varied. Here, we assume that only half of the test cases have actually a stressed resource to detect also false positive results.

Results. All test cases have been applied to the fault detection methods presented in Section IV with each being subject to three different values of the parameters x_{var} in case of the two ILP approaches and ϵ_{cos} and ϵ_{svd} in case of the two linear approaches. Table I shows the parameter values used for the tests.

First, we investigate the absolute detectability in Figure 4. We make the distinction between *correct*, *false positive*,

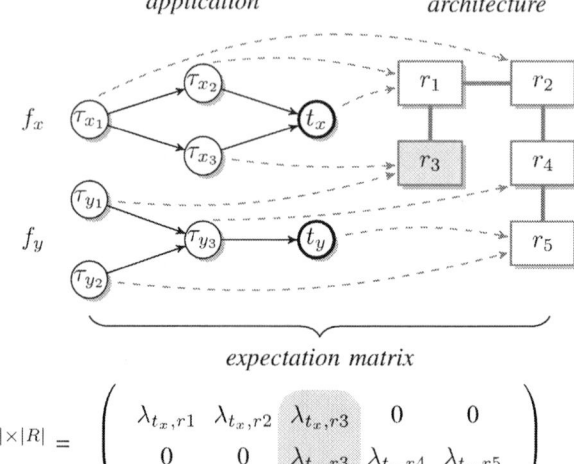

Fig. 3. The figure illustrates an exemplary systems specification consisting of the application functions (f_x, f_y), the resources (r_1 - r_5) and mappings between tasks and resources ($--\rightarrow$). The expectation matrix derived from the specification highlights the affected λ-values of the faulty resource $r3$.

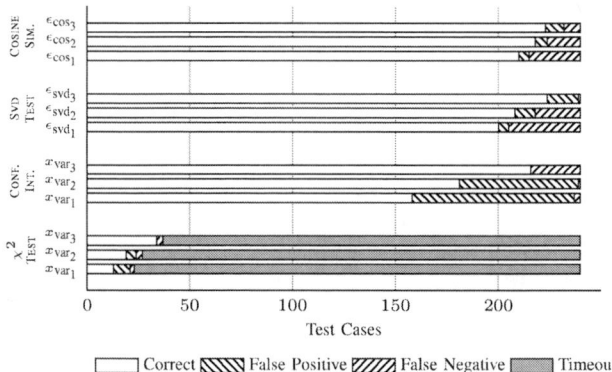

Fig. 4. The graph compares the absolute numbers of the test case results. Each bar represents the result for one fault detection method with one specific parameter value. False positive and false negative results refer to test cases which could not be solved correctly. The timeout for finding a solution for one single test case was set to 60 seconds.

false negative, and *timeout*. *Correct* results include proper detection of stressed resources as well as the right detection of systems without a stressed resource. In the case of a *false positive* result, a different resource than the stressed one has been detected or one or several other resources, including the stressed one, have been found. A test result is *false negative* when a stressed resource has not been detected. A *timeout* aborts the run after 60 seconds marking the corresponding test case.

Figure 5 visualizes the test runtimes. Each cross stands for the computation time of one single test case where the lowest

TABLE I
LIST OF PARAMETER VALUES USED FOR THE TESTS

	Value 1	*Value 2*	*Value 3*
ϵ_{cos}	$3.0 \cdot 10^{-3}$	$6.0 \cdot 10^{-3}$	$1.2 \cdot 10^{-2}$
ϵ_{svd}	$4.3 \cdot 10^{-5}$	$8.50 \cdot 10^{-5}$	$1.70 \cdot 10^{-4}$
x_{var}	1.1	1.5	1000

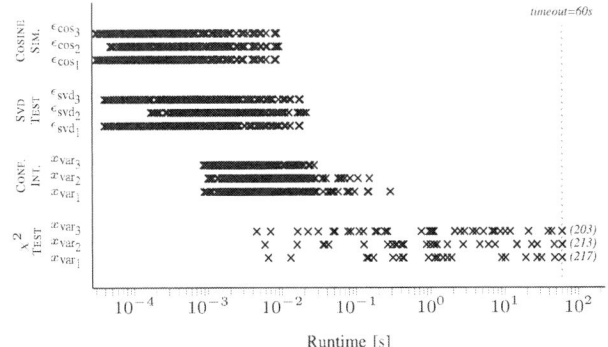

Fig. 5. The figure shows the computation times of each test case with a timeout at 60 seconds. The numbers of runs that are aborted due to a timeout are annotated in brackets.

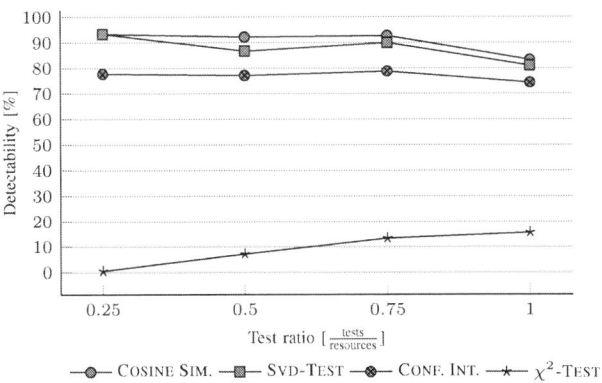

Fig. 6. The graph illustrates the percentage of detectability for each method depending on the relative number of tests.

values are in the range of microseconds and the timeout is set to 60 seconds.

Finally, the detectability of the system with respect to the relative number of tests is shown in Figure 6. Here, the y-axis represents the number of correct test results as percentage values and the x-axis marks four points with different test to resource ratios between 0.25 and 1. For this plot, the results of all three parameter values have been combined.

Discussion. The presented test case results clearly give evidence of the feasibility of our intermittent fault detection approach. The *Cosine Similarity* and the *SVD-Test* show a very good overall detection rate when using an appropriate tolerance interval ϵ. On the other hand, the *Confidence Interval* method can achieve similar detectability results with the right choice of the decision limit x_{var}. The results from the χ^2-*Test* look promising, when not taking the timeout into account. It is expectable that it would deliver detection rates comparable to the other methods or even better with a higher timeout for finding the solution.

The computation time of each test is in a millisecond range for the two linear approaches and still in a sub-second range for the *Confidence Interval* method. The runtime for the χ^2-*Test* might be reduced significantly by using a non-linear solver, as this would eliminate the additional computationally expensive constraint for the expectation variable E_t. Additionally, it should be noted that the runtime does not have to support any detection within real-time since already the observation

time of a system can be rather long and compared to that the presented methods are very fast.

When splitting the results with reference to different tests to resources ratios all methods except for the χ^2-*Test* show a good detectability in a high two-digit percentage range with only small deviations between the different ratios. The decline of the linear methods curves at the $1:1$ ratio derives from a lightly increased number of false positive detections, when many tests are applied, and might be reduced by adapting the parameter values accordingly. As for the χ^2-*Test*, a higher number of tests seems to improve the detectability. However, it must be considered that this correlation might be distorted due to the limited number of test results caused by the timeout, which is also responsible for the low detection rate. It therefore appears, that the choice of the right parameter values has a greater impact on the fault detection than the number of plausibility tests, proving that our approach works well also for a low number of tests with respect to the system size.

VI. CONCLUSION

In this paper, an approach was proposed that enables an early and implicit detection of resources in a distributed system affected by intermittent faults . The experimental results give evidence of the efficiency and effectiveness of the proposed methods. While we presented only results for at most one faulty resource, the methods either support or can be extended to detect also multiple faults. The investigation of multiple faulty resources is a part of the future work.

REFERENCES

[1] Cristian Constantinescu. Trends and challenges in VLSI circuit reliability. *Micro, IEEE*, 23(4):14–19, 2003.

[2] Martin Lukasiewycz and Samarjit Chakraborty. Concurrent architecture and schedule optimization of time-triggered automotive systems. In *Proc. of CODES*, pages 383–392, 2012.

[3] Alexandre Amory, César Marcon, Fernando Moraes, and Marcelo Lubaszewski. Task mapping on NoC-based MPSoCs with faulty tiles: Evaluating the energy consumption and the application execution time. In *Proc. of RSP*, pages 164–170, 2011.

[4] Anup Das and Akash Kumar. Fault-aware task re-mapping for throughput constrained multimedia applications on NoC-based MPSoCs. In *Proc. of RSP*, pages 149–155, 2012.

[5] Michael Glaß, Martin Lukasiewycz, Christian Haubelt, and Jürgen Teich. Incorporating graceful degradation into embedded system design. In *Proc. of DATE*, pages 320–323, 2009.

[6] Dhiego Silva, Letícia Bolzani, and Fabian Vargas. An intellectual property core to detect task schedulling-related faults in RTOS-based embedded systems. In *Proc. of IOLTS*, pages 19–24, 2011.

[7] Olivier Héron, Julien Guilhemsang, Nicolas Ventroux, and Alain Giulieri. Analysis of on-line self-testing policies for real-time embedded multiprocessors in DSM technologies. In *Proc. of IOLTS*, pages 49–55, 2010.

[8] Yuan Xie, Lin Li, Mahmut Kandemir, Narayanan Vijaykrishnan, and Mary Jane Irwin. Reliability-aware co-synthesis for embedded systems. In *Proc. of ASAP*, pages 41–50, 2004.

[9] Suleyman Tosun, Nazanin Mansouri, Ercument Arvas, Mahmut Kandemir, and Yuan Xie. Reliability-centric high-level synthesis. In *Proc. of DATE*, pages 1258–1263, 2005.

[10] Michael Glaß, Martin Lukasiewycz, Thilo Streichert, Christian Haubelt, and Jürgen Teich. Reliability-aware system synthesis. In *Proc. of DATE*, pages 409–414, 2007.

[11] Veerle Desmet, Yiannakis Sazeides, and Costas Vrioni. Performance implications of hard-faults in non-architectural structures. In *Proc. of RAAW*, 2007.

[12] Andreas Herkersdorf, Michael Engel, Michael Glaß, Jörg Henkel, Veit B Kleeberger, et al. Cross-layer dependability modeling and abstraction in system on chip. In *Proc. of SELSE*, 2013.

[13] Gurobi Optimizer. Gurobi 5.0. http://www.gurobi.com/.

[14] Martin Lukasiewycz, Michael Glaß, Christian Haubelt, and Jürgen Teich. Efficient symbolic multi-objective design space exploration. In *Proc. of ASP-DAC*, pages 691–696, 2008.

A Segmentation-Based BISR Scheme

Georgios Zervakis[1], Nikolaos Eftaxiopoulos[2], Kostas Tsoumanis[3], Nicholas Axelos[4], Kiamal Pekmestzi[5]

Department of Computer Science
School of Electrical & Computer Engineering
National Technical University of Athens
Athens, Greece
Email: zervakis@microlab.ntua.gr [1], eftaxiopoulos@microlab.ntua.gr [2], kostastsoumanis@microlab.ntua.gr [3],
njaxel@microlab.ntua.gr [4], pekmes@microlab.ntua.gr [5]

Abstract – With memory estate increasing in System-On-Chips and highly integrated products, memory defects and wearout effects are the determining factor in the chip's yield loss and reliability. In this paper, a multiple cache-based Built-In Self-Repair scheme is proposed that is able to repair from the word level down to the bit level. Moreover, it is proved that the level of segmentation does not affect the repair efficiency. An exploration is then conducted to find the optimal scheme in terms of area overhead.

I. Introduction

In modern SOC (System-On-Chip) and highly integrated products, where memory estate is dominant over logic, the determining factor pertaining to yield loss and reliability problems throughout the lifetime of the product, is SRAM memory [1]. Manufacturing defects and process variation effects are particularly exacerbated in SRAM memories, due to the small geometries utilized to increase memory density and the continuous area and voltage scaling. Fault-tolerance techniques are therefore necessary to improve chip yield loss and negate memory failures during its lifespan.

Regarding manufacturing defects, the conventional approach is the use of spare rows and columns in the cell array. External test equipment captures the stimulus response and processes it to create a fault map. The detected faulty elements are then deactivated and replaced by spare ones [2]. However, this process in modern dense memories is costly and time-consuming [3].

BISR (Built-In Self-Repair) techniques move the repair process on-chip, with the repairing circuitry intervening in the decoding operation to reroute the access of faulty elements to spare ones. Localization of the faulty cells is provided by a BIST (Built-In Self-Test) circuitry. Examples of such techniques can be found in [4]-[12]. In [4], a column-based approach is followed, using hardwired remapping. A 2D redundancy structure which utilizes spare columns/rows and a heuristic algorithm for redundancy analysis is proposed in [5]. A scheme that uses direct fault mapping via fused-based logic and operates at word level is proposed in [6]. In [7], remapping is performed using a dedicated CAM, while in [8] the row/column approach is improved by segmenting the spare resources into banks. A soft-fused technique in the row

decoding circuit for remapping the faulty row to the next available is proposed in [9]. In [10], a FLASH programmable BIST/BISR scheme utilizes spare columns for repairs. In [11] a BISR solution for multi-port RAMs is proposed, using additional rows and IOs. A complete platform for test and repair of multiple memories using spare rows/columns is given in [12].

The granularity at which the repair takes place affects the remapping/routing circuitry overhead and the utilization degree of the spare elements [13], [14]. Row/column-based techniques require simpler circuitry but a complete row/column may be spent on a single faulty cell or word. On the other hand, block/bit-based techniques allocate more efficiently the spare resources but require complex remapping circuitry. Considering this tradeoff, in [15] the authors proposed a cache-based architecture with multiple cache-banks that repair faulty elements at word level. The faulty words of the MUR (Memory Under Repair) are directly mapped to redundant cache-banks using a part of the faulty address as an index pointer to a spare word and the remaining part as a tag. According to a statistical analysis presented, ~100% reparability can be achieved using this technique. In [16], a first attempt by the authors to reduce the area overheads of the BISR circuitry by decreasing the remapping granularity at bit level is presented, with promising results.

In this paper, based on the BISR architectures of [15] and [16], we propose a parameterized (i.e. various segmentations of the MUR word) BISR scheme and conduct an exploration in order to reach an optimal segmentation where area overheads are minimized. Based on mathematical analysis, we prove that by repairing the MUR at any segmentation level does not affect the ~100% reparability achieved by the technique of [15].

The rest of the paper is organized as follows: In Section II we describe in detail the proposed segmentation-based BISR scheme as well as its interaction with the BIST circuitry. In Section III we present a detailed mathematical analysis for the MUR reparability based on various segmentations of the MUR word. In Section IV we calculate the area overhead with respect to the selected segmentation level and explore for the optimal segmentation. Section V concludes our work.

978-1-4799-2817-0/14 $31.00 © 2014 IEEE

II. Architecture

A. Proposed Segmentation-Based BISR Scheme

In the proposed scheme (Fig. 1) an $N \times M$ MUR (i.e. a MUR with N M-bit words) is considered. Each M-bit MUR word is segmented in p parts of d bits each ($d=M/p$). In order to repair λ faulty parts of the MUR, the primary cache-bank provides $n=2\lambda$ words and each next bank half the words of its previous one [15]. Thus, there are $b=\log_2(n)+1$ banks in total. In a cache-bank of k words, N/k words of the MUR are directly mapped in the same position using a $\log_2 k$-bit address referred as index (DI). This address is formed by the least significant $\log_2 k$ bits of the $\log_2 N$-bit address of the MUR. The remaining $\log_2(N/k)$ bits are referred as tag (TI) and are used to identify in which of the N/k words, that are mapped in the same cache-bank word, belongs the repaired part. In order to identify which part of the MUR word is repaired by the cache-bank word, an extra field of $e=\log_2 p$ bits referred as part is used as an offset. Moreover, a valid bit is used in each cache-bank word to indicate if it is in use ($V=1$) or not ($V=0$). Therefore, each cache-bank word provides a d-bit spare for the repaired part along with a valid bit, a tag field and a part field. As a result a cache-bank with k words requires $k \times (1+e+\log_2(N/k)+d)$ bits.

Except for the cache-banks, there is a remapping and routing circuitry which identifies if there are parts of the requested word that are stored in any of the cache-banks. In that case, the respective parts are driven to the data bus from the cache-banks instead of the MUR, through the routing circuit.

When a memory access request is performed, the respective word in each of the cache-banks (which operate in parallel) is activated according to the index (least significant $\log_2 k$ bits of the $\log_2 N$-bit address) using the respective decoder. In each cache-bank the tag field of the activated word is compared to the $\log_2(N/k)$ most significant bits of the address (TI) through a comparator. The result is driven along with the

valid bit to a NAND gate which generates a local enable signal (En_i). This signal indicates if there is a repaired part of the requested word in the specific cache-bank. Moreover, it operates as an enable signal for a decoder which takes the part field as input. The decoder of the i-th cache-bank has p outputs ($L_{i,1} - L_{i,p}$) and only one of them can be set to "1" when the respective part is repaired by the specific cache-bank. The $L_{i,j}$ signals are called local hit signals and indicate if the j-th part of the requested MUR word is stored in the i-th cache bank.

The local hit signals $L_{1,j} - L_{b,j}$ are driven to an NOR gate to generate the global hit signal hit_j which indicates if the j-th part of the requested word is repaired by any of the cache-banks ($hit_j=1$) or not ($hit_j=0$). Thus, there are p NOR gates to generate the global hit signals.

Each local hit signal $L_{i,j}$ controls d transmission gates and if the j-th part of the requested word is stored in the i-th cache-bank, it drives its data field ($cdata_i$) to the final group of multiplexers. There are p final multiplexers controlled by the global hit signals hit_j which have two d-bit inputs. The first one is the respective part of the MUR word and the second one is the repaired part (if existing in any cache-bank). Depending on the global hit signals, each part of the requested word is connected to the data bus either from the MUR ($hit_j=0$) or from the cache-banks ($hit_j=1$).

B. Interaction with BIST

The BISR circuitry described in II.A considers the existence of a BIST circuitry which has previously identified all the faulty parts of the MUR and has distributed them in the spare cache-banks. When a faulty part is detected, the BIST circuitry stores the necessary information (tag, part) in the first available cache-bank. Depending on the index, the BIST circuitry selects the appropriate line in the first cache-bank. If the valid bit is not set, the tag and part fields are stored, the flag bit is set and the respective part is considered repairable.

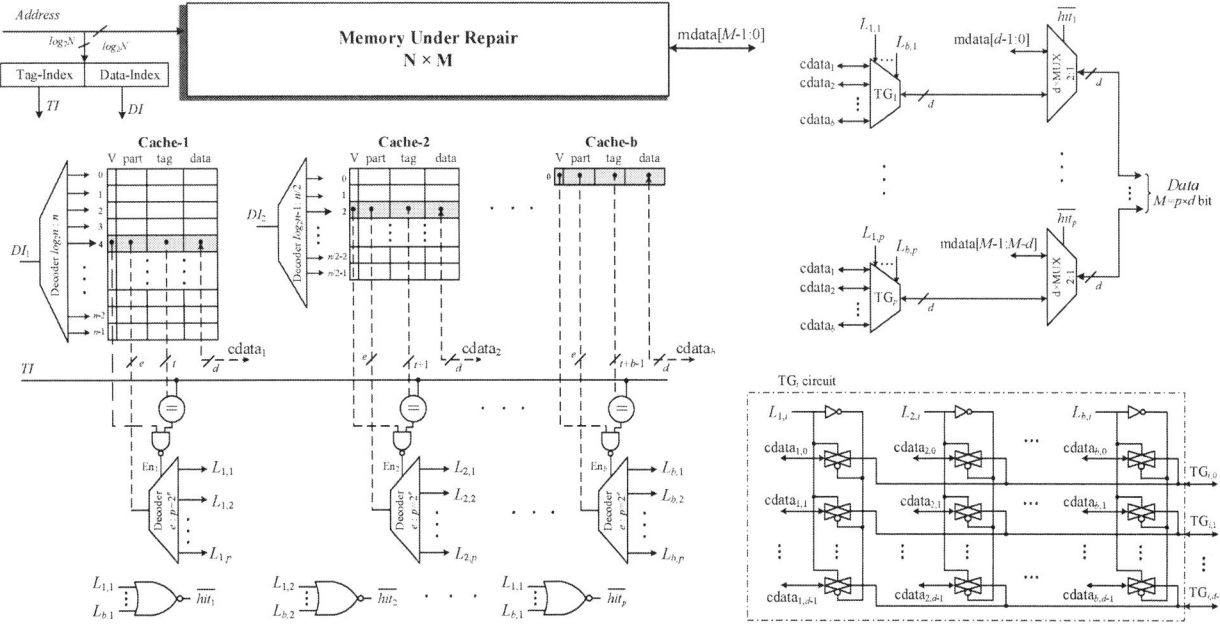

Fig. 1. Proposed BISR scheme

978-1-4799-2817-0/14 $31.00 © 2014 IEEE

If the valid bit is already set, the specific line repairs another faulty part and cannot be used. The BIST circuitry repeats the aforementioned procedure until all cache-banks are examined. If there is no available cache line to repair the faulty part, the MUR is considered as non-repairable.

III. Mathematical analysis

A MUR with N words of M bits and random fault distribution is considered, as well as a BISR circuitry with word level repair [15], with n spare words in the primary cache-bank (bank-1) and half spare words in each secondary one ($n/2$, $n/4...1$ spare words respectively). As presented in [15] a spare word of the primary bank is mapped onto $m=N/n$ words of the MUR. This group of m MUR words will be onwards referred as a block. A word of bank-2 is mapped onto $2m=N/(n/2)$ MUR words, which correspond to two MUR blocks or two spare words of bank-1. Similarly, a spare word of bank-3 is mapped onto $4m$ MUR words etc. Assuming that the MUR has f_w faulty words, the probability of a MUR word being fault-free is:

$$p_w = 1 - f_w / N \tag{1}$$

A spare word of bank-1 can repair a block if it contains at most one faulty word. As a result, the block must have at least m-1 faulty-free words and the probability of being repairable by a spare word of the primary bank is:

$$p_1(x \leq 1) = p_w^m + \binom{m}{1} p_w^{m-1}(1-p_w) \tag{2}$$

Therefore, a MUR is repairable by the primary bank only if all the n blocks mapped to its spare words are repairable. This probability is given by the following equation.

$$P_1 = p_1^n(x \leq 1) \tag{3}$$

A spare word of bank-2 maps two spare words of bank-1 or two MUR blocks. The block pair is repairable by a spare word of bank-2 if only each block contains at most one faulty word or one of them contains exactly two and the other one at most one. The probability of that case is:

$$p_2(x \leq 1) = p_1^2(x \leq 1) + \binom{2}{1} p_1(x \leq 1)p_1(x = 2) \tag{4}$$

where the probability of a block having exactly 2 faulty words is:

$$p_1(x = 2) = \binom{m}{2} p_w^{m-2}(1-p_w)^2 \tag{5}$$

As a result the probability of the MUR being fully repairable by the first and the second bank is:

$$P_2 = p_2^{n/2}(x \leq 1) \tag{6}$$

Similarly, the probability of a spare word of bank-q providing full repair to its respective blocks is:

$$p_q(x \leq 1) = p_{q-1}^2(x \leq 1) + \binom{2}{1} p_{q-1}(x \leq 1)p_{q-1}(x = 2) \tag{7}$$

The probability of r faulty words mapped onto a spare word of bank-q is given by:

$$p_q(x = r) = \sum_{i=2}^{\lfloor r/2 \rfloor + 1} c_i p_{q-1}(x = i) p_{q-1}(x = r + 2 - i)$$
$$+ \binom{2}{1} p_{q-1}(x \leq 1) p_{q-1}(x = r + 1) \tag{8}$$

where $c_i=1$ if r is even and $i=1+r/2$ else $c_i=2$

and $$p_1(x = r) = \binom{m}{r} p_w^{m-r}(1-p_w)^r \tag{9}$$

As the BISR circuitry has $b=\log_2 n+1$ banks, the probability of the MUR being fully repaired while having f_w faulty words is:

$$P_b = p_b(x \leq 1) \tag{10}$$

and is calculated using (1), (2), (7), (8), (9).

In the proposed scheme a BISR circuitry which repairs parts of the MUR word is considered. Each word has p parts of M/p bits and these parts share the same index. Therefore, a spare word of bank-1 is mapped onto $m'=pm=pN/n$ MUR parts. Assuming that the MUR has f_s faulty parts, the probability of a part being fault-free is:

$$p_s = 1 - f_s / (pN) \tag{11}$$

Thus, replacing m with m' and p_w with p_s in (2) and (9), the probability of the MUR being fully repaired by the proposed segmented BISR scheme can be calculated.

In this case f_s refers on how many faulty parts exist and p_s is the probability of a part being fault-free. In order to compare the reparability provided by the segmented and non-segmented BISR circuitries, a relation between the probability of a non-segmented word being fault free and the number of faulty parts f_s is required. Assuming a MUR word with p parts WP_1, WP_2 ... WP_p, the probability p_w of being fault-free is equal to:

$$p_w = p_s(\bigcap_{i=1}^{p} WP_i) = p_s^p \tag{12}$$

Therefore, the probability of the MUR, being fully repaired by the non-segmented BISR scheme, while having f_s faulty M/p-bit parts can be calculated using (12), (2), (7), (8), (9), (10). Hence, the reparability of the segmented and non-segmented BISR schemes can be compared for any f_s value. Moreover, using the same equations, BISR circuitries of different repair level can be evaluated. For example, if one BISR circuitry segments the MUR word into p_1 parts and another into p_2 parts with $p_1 < p_2$, then if pN is replaced by $p_2 N$ in (11) and the exponent p is replaced by p_2/p_1 in (12), the probability that the MUR is fault free in both cases if it has f_s faulty word parts of M/p_2 bits can be calculated.

The probability of a 1M×64 MUR (without any loss of generality) being fully repaired by BISR schemes with different repair levels when it has f_s faulty bits is depicted in Fig. 2. In this figure there are graphs for all the possible segmentation schemes of the 64-bit MUR word (1, 2, 4 ... 32-bit word parts), as well as for the non-segmented scheme. All the aforementioned schemes have a primary bank with 64 lines

(n=64). As illustrated in Fig. 2 the probability of the MUR being fully repaired by the BISR circuitry does not depend on the level of segmentation and is almost equal to the probability of the non-segmented scheme (the graphs are overlapped). In Fig. 3 the ratio of the repair probability provided by the segmented schemes to the non-segmented one is illustrated in order to observe the negligible differences between the probability graphs of Fig. 2. The ratio is equal to 1 up to 32 faulty bits where the probability of repair is equal to 1 for all schemes and decreases insignificantly only for higher number of faulty bits where ~100% reparability is not provided even from the non-segmented scheme. Therefore, the reparability is independent of the segmentation level and the only requirement to achieve ~100% reparability is that the primary cache-bank has double lines of the number of faulty parts being repaired [15].

IV. Area Overhead and Exploration of Optimal Segmentation

In Section II a BISR scheme with parametric repair level was proposed (Fig. 1). Using the proposed architecture, each MUR word can be virtually segmented in p parts and be repaired in that level. In Section III a mathematical analysis was conducted and proved that for a fixed-size MUR, in order to achieve ~100% reparability, the number of lines of the primary cache-bank (and consequently of all the secondary ones) is independent of the segmentation level. Therefore, an exploration for the optimal segmentation has to be conducted. Since the reparability remains unaffected, one of the basic criteria is the area occupied by the BISR circuitry. Segmenting in bit-level seems an obvious solution but the additional remapping and routing circuitry becomes significantly larger. Thus, in this section there is a detailed analysis of the area

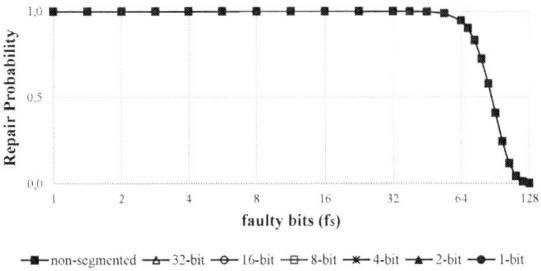

Fig. 2. Repair probability for the non-segmented and segmented schemes

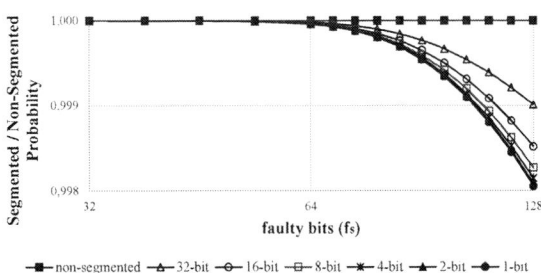

Fig. 3. Ratio of the repair probability provided by the segmented schemes to the non-segmented one

occupied by the proposed BISR scheme for all segmentations in terms of transistor count assuming an $N \times M$ MUR and b cache-banks with n lines in the primary one. Each M-bit word is segmented in p parts of $d=M/p$ bits. In Table I the transistor count of the basic units used is presented. The cell of the cache banks is considered to be the common 6T cell.

A. Transistor Count

1. Caches and Decoders

The first cache-bank has n spare lines and each secondary one has half the lines of the previous one until the last cache-bank which has only one cache-line. Therefore, there are $2n$-1 total cache lines and each one has a valid bit, a part field of $e=\log_2 p$ bits, a tag field and a data field of $d=M/p$ bits. The tag field of the primary bank has $t=\log_2 N - \log_2 n$ bits and is increased by one bit for each secondary bank. The transistor count of the caches is given in (13).

$$T_{Caches} = \sum_{i=0}^{b-1} (1 + e + (t+i) + d) \cdot \frac{n}{2^i} \cdot 6 \quad (13)$$

For each cache bank with k lines there is a $\log_2 k$-to-k decoder. The decoder used is the Lyon-Schediwy decoder [17]. The transistor count of all the decoders is:

$$T_{Decoders} = \sum_{i=0}^{b-1} Dec[\log_2(\frac{n}{2^i})] \quad (14)$$

2. Local-Hit Circuitry

Each cache-bank requires a comparator for the cache tag and the address tag, a NAND gate which is driven by the output of the comparator and the valid bit to generate the enable signal En_i and a Lyon-Schediwy decoder [17] with enable signal, driven by the part field, which generates the local hit signals ($L_{i,j}$). A $\log_2 k$-to-k Lyon-Schediwy decoder requires $k+1$ additional transistors for the enable circuitry. For k-bit tags, the comparator requires k 2-input XNOR gates and a k-input AND gate. The transistor count of the local-hit circuitry is given in (15).

$$T_{LHit} = \sum_{i=0}^{b-1} \left[nand_2 + (t+i) \cdot xnor_2 + and_{(t+i)} + DecEn(e) \right] \quad (15)$$

TABLE I
Transistor Count of Basic Units [17]

Unit	No of Inputs	Transistor count
xnor	2	8
and	*tag*	$2 \times tag + 2$
nand	2	4
not	1	2
nor	*banks*	$2 \times banks$
mux	2:1	6
transmission gate (tg)	1	2
decoder without enable	n	$n \times 2^n + 2(n+2^n) - 2$
decoder with enable	n	$n \times 2^n + 2(n+2^n) - 2 - 1 + 2^n$

3. Global-Hit Circuitry

A b-input NOR gate driven by the local hit signals is required for each part to produce the $\overline{hit}_j$ signal. The transistor count for the global-hit circuitry is given in (16).

$$T_{Ghit} = p \cdot nor_b \qquad (16)$$

4. Transmission Gate Circuitry

There are p TG_i (transmission gate) circuits and each one requires d transmission gates for each cache-bank. Moreover, for each TG_i circuit there are b NOT gates to generate the $\overline{L}_{i,j}$ signals. Therefore, the transistor count of the TG circuitry is the one given in (17).

$$T_{TGC} = p \cdot TG_i = p \cdot (d \cdot b \cdot tg + b \cdot not) \qquad (17)$$

5. Final Multiplexing Circuitry

A 2-to-1 multiplexer is required for each bit of the requested word. Therefore, M 2-to-1 multiplexers are required and the transistor count is:

$$T_{FMC} = p \cdot d \cdot mux_{2:1} \qquad (18)$$

B. Results and Comparison

The parameters which affect the occupied area of the BISR circuitry are the configuration of the MUR (i.e. the N and M values) as well as the lines of the primary bank (n). Initially an exhaustive exploration about the effect of the MUR lines (N) in the area overhead was conducted and according to the results the impact is insignificant. This is expected as only the length of the tag field is modified. Therefore, in the following exploration, without any loss of generality, the number of MUR lines is set to 2^{20}.

The most usual memory word lengths (M) are 8, 16, 32 and 64 bits. For each one of these values an exploration of the total transistor count is conducted for a range of lines of the primary cache-bank (n) from 4 to 2048 and all the possible segmentations. It has to be mentioned that for $p=1$ the proposed parametric segmented scheme is degenerated in the non-segmented scheme [15]. The graphs of the transistor count of the proposed BISR scheme for 8, 16, 32 and 64 bits word length are presented in Fig. 4, Fig. 5, Fig. 6 and Fig. 7 respectively. In each graph, except of all the segmentations, the non-segmented scheme is illustrated as well. Moreover, the results for all the segmentation schemes are normalized to the values of the non-segmented.

According to Fig. 4, for 8-bit word length the area overhead is minimized when segmenting in 2-bit groups except of the case where $n=4$. As illustrated in Fig. 5, for 16-bit word length the minimum area overhead for n up to 16 is achieved when segmenting in 4-bit groups and for n from 32 to 2048 when segmenting in 2-bit groups. For 32-bit word length (Fig. 6), the 4-bit segmentation has the minimum area overhead for n up to 32 and for the remaining values the optimal segmentation is the 2-bit one. When the word length is set to 64-bit (Fig. 7), the 8-bit segmentation is the optimal for n from 4 to 8, the 4-bit segmentation for n from 16 to 64 and the 2-bit segmen-

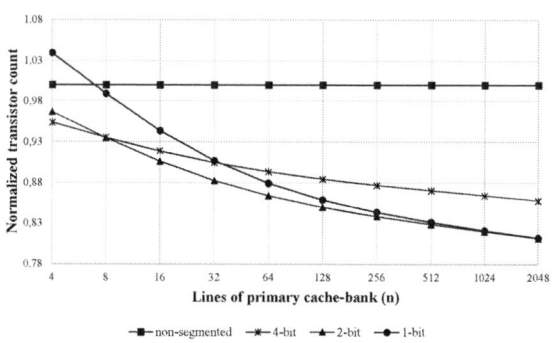

Fig. 4. Transistor count of the proposed BISR scheme for 8-bit MUR word length and all the segmentations

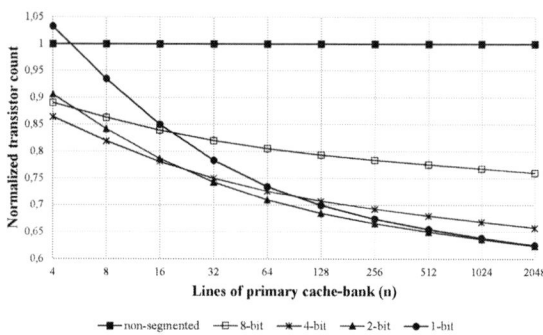

Fig. 5. Transistor count of the proposed BISR scheme for 16-bit MUR word length and all the segmentations

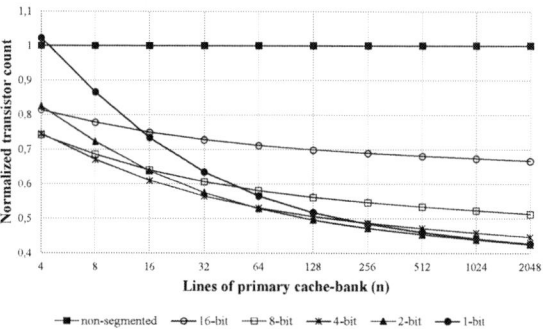

Fig. 6. Transistor count of the proposed BISR scheme for 32-bit MUR word length and all the segmentations

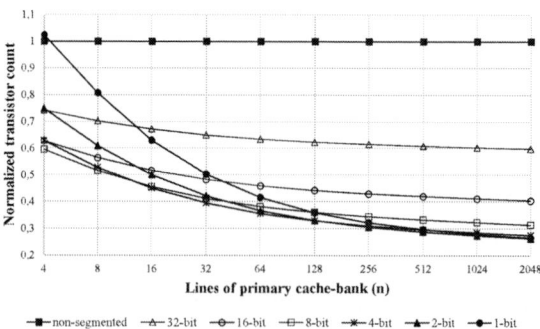

Fig. 7. Transistor count of the proposed BISR scheme for 64-bit MUR word length and all the segmentations

tation for n from 128 to 2048.

From an overall observation of the results, the non-segmented scheme is the one with the highest area overhead and it must be avoided. The 1-bit segmentation, although it might seem as the obvious solution, never proves to be the best option. It always begins with a high area overhead and ends up being close to the optimal segmentation. The segmentation in 2-bit groups seems to be the optimal choice when the MUR has a high fault rate and as this rate becomes smaller and the word length increases, segmentation in bigger groups (e.g. 4-bit or 8-bit) should be preferred.

V. Conclusion

In this paper a new BISR scheme with parametric repair level is proposed. The proposed scheme can operate for all segmentations of the MUR word and according to a mathematical analysis the probability of the MUR being fully repaired does not depend on the level of segmentation. Therefore, an exploration about the optimal repair level was conducted for 8, 16, 32 and 64 bit word length. This exploration was based on minimizing the occupied area in terms of transistor count, which is one of the most significant overheads. According to the experimental results, the non-segmented scheme proved to have the higher area overhead, while the 2-bit one proved to have the minimum overhead for MURs with high fault rate.

Acknowledgements

This research has been co-financed by the European Union (European Social Fund – ESF) and Greek national funds through the Operational Program "Education and Lifelong Learning" of the National Strategic Reference Framework (NSRF) - Research Funding Program: Thales -UOA- HOLISTIC. Investing in knowledge society through the European Social Fund.

References

[1] E. J. Marinissen, B. Prince, D. Keltel-Schulz, and Y. Zorian, "Challenges in embedded memory design and test," in *Proc. Design Automation and Test in Europe*, 2005, pp. 722–727.

[2] S. Hamdioui, G. Gaydadjiev, and A. van de Goor, "The state-of-art and future trends in testing embedded memories," in

Rec. Int. Workshop Memory Technology Design and Testing, 2004, pp. 54–59.

[3] Y. Zorian and S. Shoukourian, "Embedded-memory test and repair: infrastructure ip for soc yield," in *IEEE Design & Test of Computers*, vol. 20, no. 3, pp. 58–66, May-June 2003.

[4] I. Kim, Y. Zorian, G. Komoriya, H. Pham, F. Higgins, and J. Lewandowski, "Built in self repair for embedded high density sram," in *Proc. Int. Test Conf.*, 1998, pp. 1112–1119.

[5] H.-C. Kim, D.-S. Yi, J.-Y. Park, and C.-H. Cho, "A bisr (built-in selfrepair) circuit for embedded memory with multiple redundancies," in *Proc.Int. Conf. VLSI and CAD*, 1999, pp. 602–605.

[6] V. Schober, S. Paul, and O. Picot, "Memory built-in self-repair using redundant words," in *Proc. Int. Test Conf.*, 2001, pp. 995–1001.

[7] A. Benso, S. Chiusano, G. Di Natale, and P. Prinetto, "An on-line bist ram architecture with self-repair capabilities," in *IEEE Trans. Reliability*, vol. 51, no. 1, pp. 123–128, 2002.

[8] J. F. Li, J. Yeh, R. F. Huang, and C. W. Wu, "A built-in self-repair design for rams with 2-d redundancy," in *IEEE Trans. VLSI Systems*, vol. 13, no. 6, pp. 742–745, 2005.

[9] J. Lee, Y. J. Lee, and Y. B. Kim, "Sram word-oriented redundancy methodology using built in self-repair," in *IEEE Proc. Int. SOC Conf.*, 2004, pp. 219–222.

[10] R. Zappa, C. Selva, D. Rimondi, C. Torelli, M. Crestan, G. Mastrodomenico, and L. Albani, "Micro programmable built-in self-repair for srams," in *Rec. Int. Workshop on Memory Technology Design and Testing*, 2004, pp. 72–77.

[11] T.-W. Tseng, C.-H. Wu, Y.-J. Huang, J.-F. Li, A. Pao, K. Chiu, and E. Chen, "A built-in self-repair scheme for multiport rams," in *IEEE Proc. VLSI Test Symp.*, 2007, pp. 355–360.

[12] C.-D. Huang, J.-F. Li, and T.-W. Tseng, "Protar: An infrastructure ip for repairing rams in system-on-chips," in *IEEE Trans. VLSI Systems*, vol. 15, no. 10, pp. 1135–1143, 2007.

[13] J. Segal, S. Bakarian, J. Colburn, M. Kumar, C. Hong, and A. Shubat, "Determining redundancy requirements for memory arrays with critical area analysis," in *Rec. IEEE Int. Workshop Memory Technology, Design and Testing*, 1999, pp. 48–53.

[14] S. Shoukourian, V. Vardanian, and Y. Zorian, "An approach for evaluation of redundancy analysis algorithms," in *IEEE Int. Workshop Memory Technology, Design and Testing*, 2001, pp. 51–55.

[15] N. Axelos, K. Pekmestzi and D. Gizopoulos, "Efficient Memory Repair Using Cache-Based Redundancy," in *IEEE Trans. VLSI Systems*, vol. 20, no. 12, pp. 2278–2288, 2012.

[16] N. Axelos and K. Pekmestzi, "A bit level area aware cache-based architecture for memory repairs," in *2010 IEEE Int. On-Line Testing Symp.*, July 2010, pp. 154–158.

[17] Ivan E. Sutherland, Bob F. Sproull and David L. Harris, "Wide Structures," in Logical Effort: Designing Fast CMOS Circuits, Morgan Kaufmann, 1999, ch. 11, sec. 3, pp. 179-180

Fault-tolerant TSV by Using Scan-chain Test TSV

Fu-Wei Chen, Hui-Ling Ting and TingTing Hwang

Department of Computer Science

National Tsing Hua University, Hsinchu, Taiwan 30013, R.O.C.

fwchen@cs.nthu.edu.tw, yummy99100@gmail.com, tingting@cs.nthu.edu.tw

Abstract— In order to increase the yield of 3-D IC, fault-tolerance technique to recover failed TSV is essential. In this paper, an architecture of TSV recovery by using scan-chain test TSV is proposed. With the architecture, only a small amount of redundant TSVs is required to be inserted. Extra TSV area that occurs by our method is much less than that of other methods. Moreover, a 3-D IC scan-chain optimization algorithm is proposed taking into consideration the locations of functional TSVs as well as test TSVs, so that the number of total TSVs including test TSV and extra redundant TSV of a 3-D IC design is effectively reduced.

I. Introduction

Three-dimension Integrated Circuit (3-D IC) provides smaller interconnect delay and higher device density by using Through-Silicon-Vias (TSVs) linking signals among multiple stacked dies [1].

TSV provides communication links for dies in vertical direction and is a critical design issue in 3-D integration. Just like other components, the fabrication and bonding of TSVs may fail [2], [3]. A failed TSV can severely increase the cost and decrease the yield. To improve the yield, some recovery mechanism for faulty TSV is needed.

In order to increase the yield of 3-D IC, numerous techniques to recover TSV were proposed. Hiesh et al. first proposed a redundant TSV architecture to recover faulty TSVs by using redundant TSVs [4]. Then, Wu et al. presented a built-in self repair (BISR) circuit for recovering faulty TSV in 3-D Random Access Memory by using global redundant TSVs [6]. Ye and Chakrabarty proposed spare TSVs to replace failed TSVs in considering physical locations [5]. Although these TSV recovery techniques can successfully recover the failed TSV, they need extra chip area to place these spare or redundant TSVs. Apart from these recovery techniques using extra TSVs, repairing failed TSV without using extra TSVs was proposed by sharing circuits and signal wires of pre-bond clock tree [7]. However, this method can only be applied to TSV in 3-D IC clock tree.

In recent years, research studies of 3-D IC testing have been also raised. Testing in 3-D ICs includes pre-bond testing and post-bond testing where pre-bond testing tests individual die before stacking and post-bond testing tests the whole circuit after stacking.

Pre-bond testing of 3-D ICs reduces yield-loss by preventing defective dies and/or wafers from stacking for 3-D IC. Marinissen et. al., proposed an optimization of 3-D DFT containing TSVs for pre-bond testing [9]. Panth et. al., studied the scan chain and power delivery network synthesis for pre-bond testing of 3-D ICs [9]. Kumar et. al., proposed a hyper-graph-based partitioning for pre-bond 3-D IC testing to reduce design-for-test (DfT) cost [10]. Chen et. al., used sense amplifier and write buffer to implement on-chip TSV monitoring before bonding [15].

Post-bond testing presents some unique technical issues that are different from testing in 2-D design. For 3-D IC scan-chaining, Wu et. al., proposed several scan-chain ordering optimization methods [12].

This work is supported in part by the National Science Council of Taiwan under Grant NSC-100-2221-E-007-043-MY3.

Liao et. al., developed a fast scan-chain ordering algorithm for 3-D IC designs under TSV constraints [13]. Then, based on the test method used in SoC design, Lewis and Lee proposed the first paper addressed to testing of 3-D IC using scan-chain structure [11].

For a scan path, if two scan flip-flops are located at different tiers and chained next to each other, TSV is constructed for connection. These TSVs regarded as *test TSVs* are earmarked for shifting data in scan-in and scan-out phases during test mode. These *test TSVs* shifting test data are used only during post-pond testing but not used during normal mode. Hence, these *test TSVs* can be regarded as redundant TSVs in normal mode.

In this paper, we propose a new architecture to recover failed TSV by using *test TSV*. Table I summarizes different TSV recovery techniques. In this table, we can see that the first three schemes need extra TSVs area for all their redundant TSVs. The fourth method requires no extra area for TSV but is good for clock signal only. Our method only requires small amount area for redundant TSVs and can be applied to all types of signal.

The rest of this paper is organized as follows. In Section II, we introduce our architecture for TSV recovery and problem definition. Section III presents our algorithm for TSV recovery using *test TSV*. Section IV shows our experimental results. Finally, the conclusion of this paper is given in Section V.

II. Preliminary and Problem Definition

In this section, first, the architecture for TSV recovery is described in Section II-A. Next, the fault model is explained in Section II-B. Then, in Section II-C, 3-D IC scan-chain structure is presented. Finally, TSV recovery using TSVs of 3-D IC scan-chain and our problem formulation is introduced in Section II-D.

A. Architecture for TSV Recovery

The architecture of TSV recovery using redundant TSV is adopted [4] and shown in Figure 1(a). For each TSV, two configuration MUXs are added at two ends to shift the signal to neighboring TSV when one TSV fails. The TSVs are connected as a chain where the redundant TSV is placed at the end of the chain. When no TSV fails, all signals are transferred by original TSVs. When a TSV fails, the signal of the failed TSV needs to be shifted. This in term causes all signals between the failed TSV and the redundant TSV to be shifted. For example, let TSV_1 fail as shown in Figure 1(b). The selection inputs of configuration MUXs for TSV_2, TSV_3, and TSV_4 are set to 1. The signal paths after shifting are shown in Figure 1(b). This recovery architecture is very effective.

We will adopt this architecture to recover failed TSV in this paper. In the following, the term TSV-chain will be used to refer to the structure of this redundant TSV architecture.

B. Fault Model

The recovery rate for failed TSVs is analyzed based on probabilistic models. According to the result presented in Hiesh's work [4], assuming that single-fault model is used and the failure rate of a single TSV ranges between 10^{-4} to 10^{-5}, the recovery rate is 90%

TABLE I
COMPARISON OF THE CHARATERISITCS OF DIFFERENT TSV RECOVERY TECHNIQUES

Recovery Scheme	Main Recovery Device	Extra TSV Area	Type of TSV Signal
Hsieh's [4]	redundant TSV	all	all
Ye's [5]	spare TSV	all	all
Wu's [6]	redundant TSV	all	memory
Lung's [7]	clock wires for pre-bond test	no	clock
Ours	*test* and redundant TSV	small amount	all

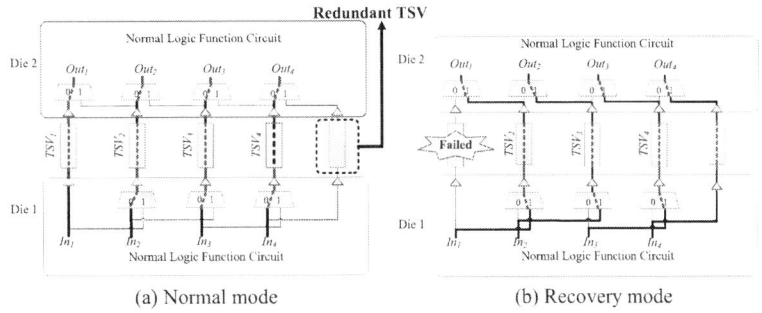

(a) Normal mode   (b) Recovery mode

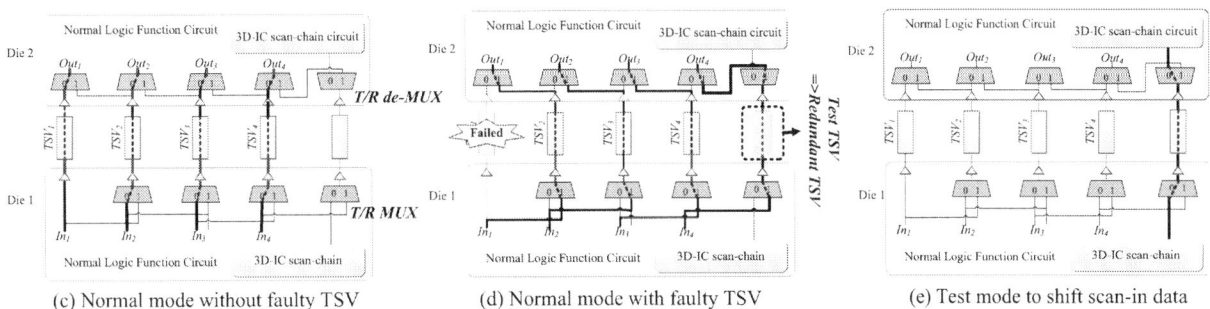

(c) Normal mode without faulty TSV   (d) Normal mode with faulty TSV   (e) Test mode to shift scan-in data

Fig. 1. Architecture of TSV recovery (a) normal mode, (b) recovery mode, (c) normal mode without faulty TSV, (d) normal mode with faulty TSV, and (e) test mode to shift scan-in data

when the number of TSVs in a chain is no greater than 50, and 95% when the number of TSVs in a chain is no greater than 25.

C. 3-D IC Scan Chain Structure

Scan chain improves testability and provides fault diagnosis of integrated circuits. In general, after circuit has been logic synthesized, the sequential components, flip-flops, can be replaced with scan flip-flops. Then, all scan flip-flops are connected to form a chain.

Figure 2 shows the netlist of a scan chain. Each scan flip-flop has two input sources, backward scan flip-flop/test data input (TDI), and combinational circuit, and two output targets, forward scan flip-flop/test data output (TDO), and combinational circuit. During normal mode, the input source and output target are outputs and inputs of combinational circuit, respectively. And during test mode, the input source and output target are backward scan flip-flop/TDI and forward scan flip-flop/TDO, respectively.

In 3-D IC, after placement, scan flip-flops are distributed in different tiers. If there are two scan flip-flops in different tiers being connected in scan-chain, *test TSV* is required to provide connection between these scan flip-flops.

D. TSV Redundancy using Test TSV and MUX Configuration

Based on TSV redundancy and 3-D IC scan-chain structure, our proposed architecture will use *test TSV* as redundant TSV. A *test TSV* acting as a redundant TSV in a TSV-chain is shown in Figure 1(c)-(d). In order to play two roles, the circuit of *test TSV* (**redundant TSV**) is modified. A MUX, denoted as *T/R MUX*, is connected to source end, and a de-MUX, denoted as *T/R de-MUX*, is connected to the destination end. During normal mode, the selection inputs of

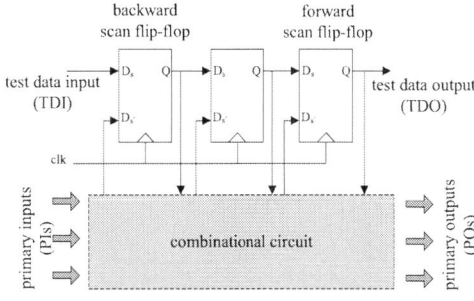

Fig. 2. 3-D IC scan chain

T/R MUX and *T/R de-MUX* are configured to 1 and the TSV is a redundant TSV, while during scan-in and scan-out of test mode, the selection inputs of *T/R MUX* and *T/R de-MUX* are configured to 0 and the TSV is a *test TSV*.

In normal mode, since *test TSVs* do not transport test data, *test TSVs* can be regarded as redundant TSVs. These redundant TSVs can be used to recover failed TSV. If there is no faulty TSV in the TSV-chain, the data is transmitted as shown in bold line of Figure 1(c). If there is a faulty TSV in the TSV-chain, the last input (i.e., In_4 in Figure 1(d)) will be shifted passing through redundant TSV as shown in bold line of Figure 1(d).

In test mode, when the TSV is used as a *test TSV* to shift scan flip-flop data, *T/R MUX* and *T/R de-MUX* are re-configured to 0 as shown in Figure 1(e). Figure 3 shows our proposed flow of scan test in 3-D IC. In testing, test access is only through external pins of the bottom die. Before shifting test data in 3-D IC scan-chain, the selection inputs of *T/R MUXs* and *T/R de-MUX* is configured to 0.

978-1-4799-2817-0/14 $31.00 © 2014 IEEE

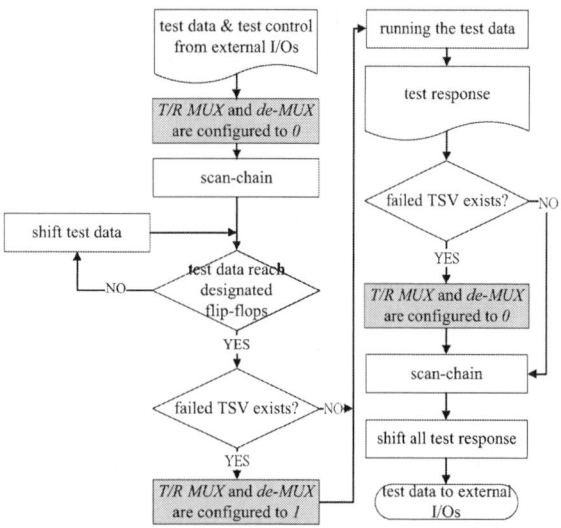

Fig. 3. Flow of testing 3-D designs

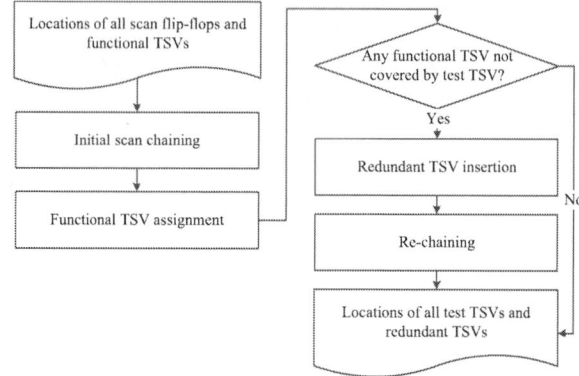

Fig. 4. The flow of functional TSVs repairing algorithm

In order to send test data to all designated flip-flops in each die for testing, the test data is transmitted through 3-D IC scan-chain till all test data arrive their designated scan flip-flops. Then, before running the test data, *T/R MUX* and *T/R de-MUX* are re-configured to 1 acting as a redundant TSV in normal mode if there is one faulty TSV on the TSV-chain. Next, to shift out the test response, *T/R MUX* and *T/R de-MUX* are configured to 0 and the test response can be shifted out. Note that testing of TSV is usually performed before testing the whole design [5], [6], [14], [15]. Whether there is any failed TSV is known before the testing. If no failed TSV is found, the second and the third configuring steps for *T/R MUX* and *de-MUX* are bypassed.

Given a set of placed module blocks, logic gates, and a scan flip-flops. Our objective is to chain scan flip-flops and construct TSV-chain so that minimum redundant TSV is required and the total wire length of 3-D IC scan-chain and TSV-chain is minimized.

III. Algorithm

A. Overview of Our Proposed Method

Figure 4 is the overview of our proposed algorithm. The flow starts with *initial_scan_chaining*. In this step, an initial scan chain will be determined. The chaining will take into consideration *wire length*, *congestion* and *repairing cost*. Once the sequence of chaining is determined, the number and locations of *test TSVs* are also set.

Then, *functional_TSV_assignment* step is to actually assign each functional TSV to *test TSV* so that each group of functional TSVs has one *test TSV* as their repairing TSV. This assignment is modeled as a network-flow optimization problem. After this step, if any functional TSV is remained uncovered due to conditions such as distance constraint, and covering constraint, *redundant_TSV_insertion*, will insert real redundant TSVs to cover these unassigned functional TSVs. Finally, *re_chaining* step reroutes scan chain to minimize the wire length of the scan chain. We will show the details in Section III-B to Section III-E.

B. Initial Scan Chaining

Chaining of scan cells requires routing resource. On the other hand, in our proposed architecture, *test TSVs* will also be used as redundant TSVs in normal mode to repair faulty TSV. Connecting *test TSV* and functional TSV also requires routing resource. (Note that if two scan flip-flops are chained next to each other and they are in different tiers, *test TSVs* are inserted.)

Therefore, the decision of the scan chain sequence should consider the above-mentioned factors. The chaining of scan flip-flops is

modeled as a complete graph, $G = (V, E, C)$, where $r \in V$ denotes a flip flop, $e_{ij} \in E$ the chaining of nodes r_i and r_j, $c_{ij} \in C$ the cost of chaining nodes r_i and r_j. The minimum cost scan chain solution is to find a minimum distance *Traveling Salesman Problem* (TSP). To consider wire length, congestion and repairing cost, we define the cost function, c_{ij}, for each pair of scan flip-flops, r_i and r_j, $i \neq j$,

$$c_{ij} = \alpha \cdot WL_{ij} + \beta \cdot cong_{ij} + \gamma \cdot repair_{ij} \qquad (1)$$

The details of each term is explained in the following Subsections from III-B1 to III-B3.

1) Wire Length: As mentioned above, the wire length of a scan chain should keep short. We define WL_{ij} as following. If two scan flip-flops, r_i and r_j, are at the same tier, we have

$$WL_{ij} = HPWL(r_i, r_j) \qquad \text{if } l_{r_i} = l_{r_j},$$

where $HPWL$ is Half-Perimeter Wire Length, and l_{r_i} is the tier of scan flip-flop r_i. On the other hand, if r_i and r_j are in different tiers and assume l_{r_i} is lower than l_{r_j}, denoted by $l_{r_i} < l_{r_j}$, we have

$$WL_{ij} = HPWL(r_i, r_j) + (l_{r_j} - l_{r_i}) \times h_{tsv} \qquad \text{if } l_{r_i} \neq l_{r_j},$$

where h_{tsv} represents the height of a TSV.

2) Congestion Cost: This cost only works when r_i and r_j are in different tiers. To insert *test TSVs* for scan chain between different tiers in 3-D IC, we should take into consideration the placement congestion in the bounding box of r_i and r_j. We restrict the *test TSV* not to be inserted out of the bounding box of r_i and r_j so that the wire length of the scan chain does not increase. Our congestion term is defined as

$$cong_{ij} = \frac{1}{WS_{ij}},$$

where WS_{ij} is the white space in the bounding box of r_i and r_j.

3) Repairing Cost: The repairing capabilities of all candidate *test TSVs* will be evaluated. The repairing capability of a *test TSV* is defined as the number of functional TSVs it can repair under a distance constraint.

The distance constraint, d, is used to prevent long distance between a functional TSV and a *test TSV*. A long distance means long repairing wire length. Since the exact location of candidate *test TSV* is not known yet, the center of bounding box of r_i and r_j is taken as its location for computation. We denote the repairing cost by $repair_{ij}$ for r_i and r_j.

We want to compute the repairing cost of a *test TSV* which will be created if two scan flip-flops are chained and they are in different tiers. Let this *test TSV* be denoted as $ttsv_{ij}$. If $ttsv_{ij}$ is allocated, all functional TSVs, $ftsv_{ijk}$, within the distance d can be repaired by $ttsv_{ij}$. However, a $ftsv_{ijk}$ may also be repaired by other *test TSV*. If there are n *test TSVs* which can repair $ftsv_{ijk}$, then the probability of $ftsv_{ijk}$ to be repaired by *test TSV*, $ttsv_{ij}$, is $\frac{1}{n}$. Let

978-1-4799-2817-0/14 $31.00 © 2014 IEEE

the probability of $ftsv_{ijk}$ repaired by $ttsv_{ij}$ be denoted as $prob_{ijk}$. Then, the repairing cost of $ttsv_{ij}$

$$repair_{ij} = \frac{WL_{repair}}{\frac{tn}{\sum_{k=1}^{tn} prob_{ijk}}},$$

where tn is the number of functional TSVs that can be repaired by $ttsv_{ij}$ (i.e. within the distance d), WL_{repair} is the $HPWL$ of the bounding box of these tn functional TSVs, and $prob_{ijk}$ is the probability of $ftsv_{ijk}$ repaired by $ttsv_{ij}$.

C. Functional TSV Assignment

After the scan chain sequence is determined by our minimum cost TSP, we know the actual allocated *test TSV* (i.e. if two scan flip-flops are chained and they are at different tiers, a *test TSV* is allocated). With a *test TSV*, we are to perform actual covering of functional TSVs. To consider yield, one *test TSV* can not cover more than N functional TSVs as we described in Section II-B. This is the covering constraint of a *test TSV*. On the other hand, a functional TSV may be covered by more than one *test TSVs*. Moreover, wire length needs to be considered when this assignment problem is performed. We model the problem as a network-flow optimization problem. At first, we denote the flow network by $G(V, E)$, where node set $V = s + TSV_F + TSV_T + t$ and edge set $E = IE + FE + OE$. Node s and node t represent the source and sink node of the flow network, respectively. Nodes in sets TSV_F and TSV_T represent sets of functional TSVs and *test TSVs*. IE, FE, OE represent sets of incoming edges, assignment edges, outgoing edges. Incoming edges are constructed from source to TSV_F, and outgoing edges are constructed from TSV_T to sink node.

An assignment edge $e_{ij} \in FE$ is constructed if a node $v_{f_i} \in TSV_F$ representing a functional TSV tsv_f can be repaired by a node $v_{t_j} \in TSV_T$ representing a *test TSV* tsv_t. The capacity of incoming edges and assignment edges are all set to 1 while that of an outgoing edge is set to N. The capacity of outgoing edge ensures that the *test TSV* can repair N functional TSVs at most (covering constraint).

To consider the wire length to connect a functional TSV and a *test TSV*, cost is modeled on the edge. All incoming edges and outgoing edges will set to 0. For assignment edges, the cost is defined as follows. For a functional TSV, tsv_f, assigned to a *test TSV*, tsv_t, we define the cost as

$$cost(tsv_f, tsv_t).$$

We first define *essential repaired functional tsv*. A *test TSV* can repair a functional TSV as long as this functional TSV is within d distance to the *test TSV*. That is, a functional TSV may be within the distance d to no *test TSVs*, one *test TSV*, or many *test TSVs*.

If a functional TSV is within the distance to only one *test TSV*, we call it *essential repaired functional tsv*, and *essential repaired functional tsv* will be assigned to its only *test TSV*.

Now, we are to define $cost(tsv_f, tsv_t)$. If tsv_f is an *essential repaired functional tsv*, the cost on the edge of tsv_f and tsv_t is set to 0 because this assignment edge must be selected. For a *test TSV*, tsv_t, if there is any *essential repaired functional tsv$_j$* at the distance d to tsv_t, then

$$cost(tsv_f, tsv_t) = min\{dist(tsv_f, essential\ repaired\ functional\ tsv_j),$$
$$dist(tsv_f, tsv_t)\},$$
(2)

where $dist(tsv_f, essential\ repaired\ functional\ tsv_j)$ is the manhattan distance from tsv_f to the closest *essential repaired functional tsv$_j$* that is within d distance to tsv_t, and $dist(tsv_f, tsv_t)$ is the manhattan distance from tsv_f to tsv_t. Otherwise, if there is no *essential repaired*

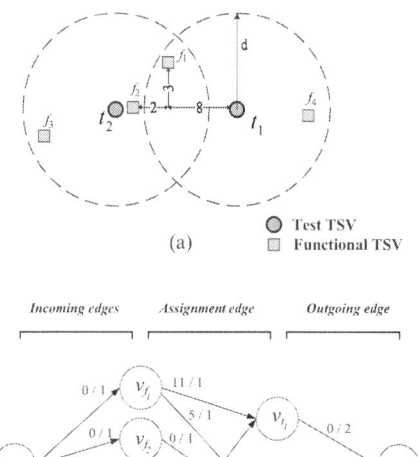

Fig. 5. The flow network example

functional tsv, then

$$cost(tsv_f, tsv_t) = dist(tsv_f, tsv_t).$$

The reason why the cost function is defined is because our repairing architecture is a chain-style interconnection as shown in Figure 1. After modeling the problem as a flow network graph, we find the minimum cost maximum flow of the network as our solution. Figure 5 shows an example of the flow network modeling. Figure 5(a) shows that there are two *test TSVs*, t_1 and t_2, and four functional TSVs, f_1, f_2, f_3 and f_4. Functional TSVs, f_1 and f_4, are within the distance d to *test TSV* t_1, and functional TSVs, f_1, f_2, and f_3 are within the distance d to *test TSV* t_2. f_2, f_3 and f_4 all are covered by only one *test TSV*, and hence *essential repaired functional tsv*. The flow network is modeled in Figure 5(b).

In Figure 5(b), there are four nodes, v_{f_1}, v_{f_2}, v_{f_3}, v_{f_4}, and two nodes, v_{t_1}, v_{t_2} in TSV_F and TSV_T, respectively. Functional TSV, f_1, is covered by *test TSVs*, t_1 and t_2, and thus there are one assignment edge, e_{11}, from v_{f_1} to v_{t_1} and one assignment edge, e_{12}, from v_{f_1} to v_{t_2}. For f_1, the closest *essential repaired functional tsv* from it is f_4 in the covering region of t_1, and the manhattan distance from f_1 to t_1 is shorter than that from f_1 to f_4. Hence, as defined in Equation (2), the cost of e_{11} is set to 11 computed by manhattan distance between f_1 and t_1. On the other hand, the closest *essential repaired functional tsv* from f_1 is f_2 in the covering region of t_2, and the manhattan distance from f_1 to f_2 is shorter than that from f_1 to t_2. The cost of e_{12} is set to 5 which is the manhattan distance between f_1 and f_2. For f_2, it is an *essential repaired functional tsv* and covered only by t_2. Thus there is an assignment edge, e_{22}, from v_{f_2} to v_{t_2} and the cost of e_{22} is set to 0 because it has to be selected in the flow model. For f_3 and f_4, they are also *essential repaired functional tsvs* and hence are defined similarly as f_2.

To consider yield of TSVs, the covering constraint for one *test TSV* must be set. Let the covering constraint be 2. Then the capacity of each outgoing edge will be set to 2. We solve the covering problem by minimum cost maximum flow such that the total flow from s to t is maximal, and the total cost of the flow is minimal. The solution to the network flow problem will ensure that one functional TSV will be repaired by one and only one *test TSV*. In this example, a minimum cost maximum flow from v_{f_1}, v_{f_4} to v_{t_1}, and from v_{f_2}, v_{f_3} to v_{t_2} is found which corresponds to assigning functional TSVs,

$\{f_1, f_4\}$, and $\{f_2, f_3\}$ to test TSVs, t_1 and t_2, respectively.

D. Redundant TSV Insertion

As pointed out in the previous section, a functional TSV may not be at the distance of d to any *test TSV*. In this case, we will insert actual redundant TSV for repairing.

First, we identify all functional TSVs that are not covered by any *test TSV*. Then, we cluster all these remaining functional TSVs into groups by their distance to each other.

The clustering algorithm is performed by K-means [16] tier by tier for 3-D IC. Before calling the K-means algorithm, we need to decide the value of K. We will use the distance constraint to compute K. To ensure the distance of functional TSVs in one cluster consistent with the distance using *test TSV* for repairing, we reuse the distance d (i.e. the longest distance between a *test TSV* and a functional TSV if the functional TSV is repaired by the *test TSV*) to calculate the number of clusters K. Let the width and height of a die be W and H. Then, the number of clusters, K, is computed as

$$\lfloor \frac{W}{2 \cdot d} \rfloor \times \lfloor \frac{H}{2 \cdot d} \rfloor.$$

The calculation is based on the observation that if a redundant TSV, tsv_r, is assigned at the center of a circle, all functional TSVs in the circle at distance of d can be repaired by it. For example, if the width and height of a die are $5d$ and $4d$, $K = \lfloor 5d/2d \rfloor \times \lfloor 4d/2d \rfloor$. Therefore, we will perform 4-means to cluster all the remaining functional TSVs, and the coordinate of clustering seeds will be set to be $(d, d), (3d, d), (d, 3d), (3d, 3d)$, respectively.

After K-means algorithm is performed, we check if there is any cluster whose functional TSVs exceeds the covering constraint, N. Assume there is one such cluster containing n functional TSVs, $n > N$, we set $K = \lceil \frac{n}{N} \rceil$ and select seeds randomly from the functional TSVs in the cluster. Then, redo the K-means for functional TSVs in this cluster only. K-means algorithm is recursively called till the number of functional TSVs in a cluster is under N.

Finally, a redundant TSV is inserted to empty space close to the center of the cluster.

E. Re-chaining

Now, All the functional TSVs can be repaired by a *test TSV* or an actual allocated redundant TSV. In order to reduce wire length of the scan chain, we re-chaining the scan chain.

Note that a *test TSV* is used as a redundant TSV in normal mode. Similarly, a redundant TSV can be used as a *test TSV* in test mode.

In the *re_chaining* step, *test TSVs* and redundant TSVs are all regarded as potential link for two scan flip-flops to connect. Minimum cost TSP is modeled again. However, only the wire length is modeled as the cost on the edge. After chaining order is determined, those TSVs connected scan flip-flops are viewed as *test TSVs* and all others are viewed as redundant TSVs.

IV. EXPERIMENTAL RESULTS

Our experiment is developed on 3.0 GHz Linux environment with 64 GB memory. We implemented the proposed algorithm using C++/STL programming language with LEDA library [19] and performed experiments on the OpenCore benchmark circuits [17], which are circuits described in behavioral level. These benchmark circuits are synthesized to gate level format by using Design Compiler [21] with 45-nm technology library. Next, 3-D IC placement using the tool developed by UCLA laboratory [20] is performed to place the circuits to four tiers. In this experiment, TSV diameter and height are set to 5μm and 10μm, respectively. Finally, the locations of all cells and functional TSVs are produced and fed to our algorithm as

TABLE II
BENCHMARK STATISTICS

circuit name	$\#TSV_f$	$\#DFF$
openMSP430	747	684
tv80_core	1123	347
fft	441	992
lcd_ctrl	1201	892
pci_controller	1173	424
ac97	575	2229
usb2.0	882	1746
sub_x86_cpu	1099	517

input. In our proposed algorithm, our weight factors, α, β, and γ of Equation (1) are set to 1, 1, and 0.01, respectively. In our experiment, the scan chain modeled as TSP is solved by standard 2-opt and 3-opt heuristics adopted by most scan chain ordering [18] methods. Our covering constraint is set to 50 to achieve 90% recovering rate.

At first, we compare the traditional architecture, denoted as ***single-use TSV***, with our proposed architecture, denoted as ***dual-use TSV***. In the traditional architecture, the *test TSVs* and redundant TSVs are used independently. The *test TSVs* are used to connect the scan flip-flops, and the redundant TSVs are inserted to recover faulty functional TSVs. On the contrary, in our architecture, both *test TSVs* and redundant TSVs can be used to recover faulty functional TSVs.

Then, we compare two different algorithms for our ***dual-use TSV*** architecture. The two algorithms have different considerations. The first one, denoted as *repairing oriented*. It considers repairing, congestion and wire length cost when connecting the scan flip-flops as proposed in Section III. The second one, denoted as *wirelength oriented*, optimizes the wire length without considering repairing cost. The algorithm proceeds as follows. It will first optimally connect the scan flip-flops taking wire into consideration only (i.e. TSP with only wire length modeled on the edge). Then, the second step is to utilize *test TSVs* which are inserted during scan chaining to repair functional TSVs. Clustering functional TSVs uses the same algorithm as proposed in *repairing oriented* method. Then, each formed cluster of functional TSVs is assigned a *test TSV* nearby and will be repaired by the *test TSV* if there is any faulty functional TSV. Finally, if there are functional TSVs not covered, actual redundant TSVs are inserted.

Table II shows profiling of benchmark circuits, where $\#TSV_f$ and $\#DFF$ represent the number of functional TSVs and scan flip-flops, respectively.

The comparison results of the two experiments are shown in Table III and Table IV, where WL_{sc}, $\#TSV_r$, $\#TSV_t$, and $\#TSV_{r/t}$, represent the wire length (2000 unit = 1μm) of 3-D IC scan-chain, the number of single-use redundant TSVs, the number of single-use *test TSVs*, and the number of dual-use TSVs, respectively. Note that single-use TSV means that the TSV is use eigher as a *test TSV* for 3-D IC scan-chain or a redundant TSV for repairing while dual-use TSV means that the TSV is not only a *test TSV* but also a redundant TSV. The column labeled $\#TSV_{total}$ is the total number of TSVs (i.e. $\#TSV_{total} = \#TSV_r + \#TSV_t + \#TSV_{r/t}$).

In Table III, we compare our ***dual-use TSV*** architecture with the ***single-use TSV*** architecture. The results show that the average improvement of the total number of TSVs is 40%, and the routing resource overhead of 3-D IC scan-chain is 3%. In ***dual-use TSV*** architecture, column labeled $\#TSV_t$ shows that the number of single-use *test TSVs* is close to 0. It means that almost all *test TSVs* are also used as redundant TSVs for repairing. Hence, it does not need so many redundant TSVs like ***single-use TSV*** architecture to repair functional TSVs. Because the traditional architecture considers wire length cost of 3-D IC scan-chain only, the ***single-use TSV*** architecture produces shorter wire length of 3-D IC scan-chain than that of ***dual-use TSV*** architecture.

To understand the TSV usage and routing overhead of different algorithms, we conduct an experiment to compare the *wirelength*

978-1-4799-2817-0/14 $31.00 © 2014 IEEE

TABLE III
COMPARISONS OF ARCHITECTURE

circuit name	single-use TSV					dual-use TSV				
	WL_{sc}	$\#TSV_r$	$\#TSV_t$	$\#TSV_{r/t}$	$\#TSV total$	WL_{sc}	$\#TSV_r$	$\#TSV_t$	$\#TSV_{r/t}$	$\#TSV total$
openMSP430	8680261	19	8	0	27	9034431	10	0	8	18
tv80_core	5008458	31	13	0	44	5151862	20	0	9	29
fft	10486698	12	10	0	22	10852080	5	0	10	15
lcd_ctrl	10512128	34	19	0	53	10580498	21	0	9	30
pci_controller	11795747	13	10	0	23	12735565	1	1	9	11
ac97	20376028	18	12	0	30	20415944	9	0	8	17
usb2.0	18473980	23	9	0	32	18554764	13	0	7	20
sub_x86_cpu	4610518	22	16	0	38	4825646	10	0	10	20
ratio	1.00				1.00	1.03				0.60

TABLE IV
COMPARISONS OF ALGORITHM

circuit name	repairing oriented					wirelength oriented				
	WL_{sc}	$\#TSV_r$	$\#TSV_t$	$\#TSV_{r/t}$	$\#TSV_{total}$	WL_{sc}	$\#TSV_r$	$\#TSV_t$	$\#TSV_{r/t}$	$\#TSV_{total}$
openMSP430	9034431	10	0	8	18	8577055	12	3	7	22
tv80_core	5151862	20	0	9	29	4936040	21	5	10	36
fft	10852080	5	0	10	15	10432260	6	6	6	18
lcd_ctrl	10580498	21	0	9	30	10383456	24	3	10	37
pci_controller	12735565	1	1	9	11	11761163	7	4	6	17
ac97	20415944	9	0	8	17	20349586	6	6	12	24
usb2.0	18554764	13	0	7	20	18343500	14	2	9	25
sub_x86_cpu	4825646	10	0	10	20	4637212	12	6	10	28
ratio	1.03				0.60	0.99				0.75

oriented with *repairing oriented* methods based on **dual-use TSV** architecture. The results produced by **single-use TSV** architecture is our baseline result (i.e. reference result). Table IV shows the results produced by two methods as compared to the baseline result.

First, compared with the baseline result, the ratio of the total number of TSVs by *wirelength oriented* method is 25% better, and the wire length of the scan chain is almost the same. There is 1% decrease in wire length which is caused by different insertion sequences of *test TSVs* and redundant TSVs. (Note that **single-use TSV** architecture inserts redundant TSVs first.)

Second, compared with *repairing oriented* method, in terms of ratio of the total number of TSVs, *wirelength oriented* method is 15% worse, but in terms of the ratio of the wire length, it is 4% better. The *wirelength oriented* method does not take repairing factor into consideration during scan chaining step. It results in shorter wire length of the scan chain, but more single-use *test TSVs*. Thus, it needs more redundant TSVs to repair the functional TSVs.

To sum up, we obtain a significant reduction of the total number of TSVs in our **dual-use TSV** architecture with little routing resource overhead. In our proposed method, *repairing oriented* method can generate *test TSVs* that are dual-use during scan chaining. Hence, fewer extra redundant TSVs are needed. Meanwhile, the *wirelength oriented* method provides another choice to generate dual-use TSVs that results in shorter wire length of 3-D IC scan-chain.

V. CONCLUSIONS

In this paper, we have proposed an architecture of TSV recovery by using scan-chain *test TSV*. Extra area that incurs by our proposed architecture is much less than that of other methods. The number of total TSVs based on *test TSV* as compared with that of single-use redundant TSVs is 40% better in average with 3% wiring overhead.

REFERENCES

[1] C. S. Tan, Ronald J. Gutmann, and L. Rafael Reif, "Wafer Level 3-D ICs Process Technology," *Springer* , 2008

[2] P. R. Morrow , M. J. Kobrinsky , S. Ramanathan , C.-M. Park , M. Harmes , V. Ramachandrarao , H.-M. Park , G. Kloster , S. List and S. Kim, "Wafer-level 3D interconnects via Cu bonding," *Proceeding of Advanced Metallization Conference (AMC)*, pp.125-130, 2004

[3] P. Garrou, C. Bower, and P. Ramm, "Handbook of 3D Integration: Technology and Application of 3D Integrated Circuits," Wiley-VCH Verlag GmbH & Co. KGaA, Weinheim, vol.1-2, 2008

[4] Ang-Chih Hsieh, TingTing Hwang, "TSV Redundancy: Architecture and Design Issues in 3-D IC," *Very Large Scale Integration (VLSI) Systems, IEEE Transactions on*, vol.20, no.4, pp.711-722, April 2012

[5] Fangming Ye and Krishnendu Chakrabarty, "TSV Open Defects in 3D Integrated Circuits: Characterization, Test, and Optimal Spare Allocation," *Design Automation Conference*, pp.1024-1030, 2012

[6] Cheng-Wen Wu, Shyue-Kun Lu and Jin-Fu Li, "On Test and Repair of 3D Random Access Memory," *Design Automation Conference, 2012 17th Asia and South Pacific (ASP-DAC)*, pp.744-749, 2012

[7] Chiao-Ling Lung, Yu-Shih Su, Shih-Hsiu Huang, Yiyu Shi and Shih-Chieh Chang, "Fault-tolerant 3D clock network," *Design Automation Conference (DAC), 2011 48th*, pp.645-651, 2011

[8] E. J. Marinissen, C.-C. Chi, J. Verbree, and M. Konijnenburg, "3D DFT architecture for pre-bond and post-bond testing," International 3D System Integration Conference, pp.1-8, Nov. 2010

[9] S. Panth, and S. K. Lim, "Scan Chain and Power Delivery Network synthesis for pre-bond test of 3D ICs," VLSI Test Symposium (VTS), pp.26-31, 2011

[10] A. Kumar, S. M. Reddy, I. Pomeranz, and B. Becker, "Hyper-graph based partitioning to reduce DFT cost for pre-bond 3D-IC testing," Design, Automation and Test in Europe Conference and Exhibition (DATE), pp.1-6, March 2011

[11] Dean L. Lewis and Hsien-Hsin S. Lee, "A Scan-Island Based Design Enabling Prebond Testability in Die Stacked Microprocessors," *In Proceedings IEEE International Test Conference (ITC)*, pp.1-8 October 2007.

[12] X. Wu, P. Falkenstern, K. Chakrabarty, and Y. Xie, "Scan-Chain Design and Optimization for Three-Dimensional Integrated Circuits," ACM Journal on Emerging Technologies in Computing systems (JETC), vol. 5, Issue 2, Article 9, July 2009

[13] C.-H. Liao, W.-T. Chen, Y.-Z. Lin, and H.-P. Wen, "Fast Scan-Chain Ordering for 3-D-IC Designs Under Through-Silicon-Via (TSV) Constraints," IEEE Transaction on Very Large Scale Integration System (TVLSI), vol. 21, no. 6, pp.1170-1174, 2012

[14] D. L. Lewis and H. S. Lee, "A scanisland based design enabling prebond testability in die-stacked microprocessors," Test Conference, IEEE International (ITC) , pp. 1-8, 2007

[15] Po-Yuan Chen, Cheng-Wen Wu and Ding-Ming Kwai, "On-chip testing of blind and open-sleeve TSVs for 3D IC before bonding," VLSI Test Symposium (VTS), pp.263-268, 2010

[16] J. B. MacQueen, "Some Methods for classification and Analysis of Multivariate Observations," Proceedings of 5-th Berkeley Symposium on Mathematical Statistics and Probability, Berkeley, University of California Press, vol. 1, pp. 281-297, 1967

[17] "OpenCore", [Online]. Available: http://opencores.org/

[18] K. D. Boese, A. B. Kahng and R. Tsay, "Scan Chain Optimization: Heuristic and Optimal Solutions", MARCO Gigascale Research Center and Cadence Design Systems, Oct. 1994. [Online]. Available: http://vlsicad.ucsd.edu/GSRC/Bookshelf/Slots/ScanOpt/

[19] "LEDA Library", [Online]. Available: http://www.algorithmic-solutions.com/leda/

[20] "UCLA 3D Physical Design Flow", [Online]. Available: http://cadlab.cs.ucla.edu/three_d/3dic.html

[21] "Design Compiler", Synopsys.

Suppressing Test Inflation in Shared-Memory Parallel Automatic Test Pattern Generation

Jerry C. Y. Ku, Ryan H.-M. Huang, Louis Y. -Z. Lin, and Charles H.-P. Wen
Dept. of Elec. Comp. Engr., National Chiao Tung University, Hsinchu, Taiwan 300
E-mail: jerrycyku.cm99g@nctu.edu.tw, hmhuang.eed00g@g2.nctu.edu.tw,
louislin.eed99g@g2.nctu.edu.tw, opwen@g2.nctu.edu.tw

Abstract — Multi-core machines enable the possibility of parallel computing in Automatic Test Pattern Generation (ATPG). With sufficient computing power, previously proposed parallel ATPG has reached near linear speedup. However, test inflation in parallel ATPG yet arises as a critical problem and limits its practicality. Therefore, we developed a parallel ATPG system that incorporates (1) *concurrent interruption (CI)*, (2) *ripple compaction (RC)* and (3) *fan-in-cone based fault ordering (FIC)* to deal with such problem. Concurrent interruption aborts test generation on simultaneously detected faults by fault simulation. Ripple compaction combines tests for different faults while fan-in-cone based fault ordering strategically arranges the fault list to reduce the number of test generations and thus speeds up the ATPG process. According to our experiments, the proposed parallel ATPG system effectively reduces 11% pattern count and achieves ~0% test inflation while maintaining an average of 6.5X speedup with no attenuation in fault coverage on experimental circuits.

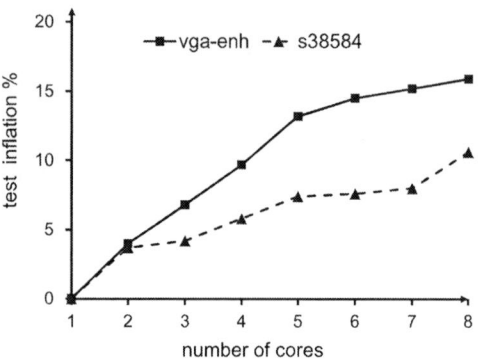

Fig. 1. Test-inflation problem in parallel ATPG

I. INTRODUCTION

As VLSI designs continue to grow in size and complexity, the demand for faster ATPG arises for their verification testing. The conventional ATPG algorithm running on single processor has reached its bottleneck and become unsustainable to generate quality tests effectively for modern designs. The rapid development of multi-core processors enables the possibility of **parallel computing** and emerges to be a solution for scaled designs. By its communication protocol, a **parallel computing** architecture can be classified into: *shared-memory system* and *message-passing system*. Both parallel computing systems provide additional computing power and thus are used in [1][2][10] to speed up ATPG recently.

Common strategies used in parallel ATPG include *fault parallelism, heuristic parallelism, search-space parallelism, algorithmic parallelism,* and *circuit parallelism*. Fault parallelism [4] divides all faults among available processors and each processor generates tests for its corresponding faults. For heuristic parallelism [5], each processor employs a distinctive ATPG algorithm to generate a test for the same fault. Search-space parallelism [5] allows the processors to work collectively in finding a test for a single fault by dividing search space into discrete pieces and let each processor work on these search spaces simultaneously. ATPG algorithms are divided into fine-

grained tasks and processors work in parallel on each task in algorithmic parallelism [6]. For circuit parallelism [7], the circuit is divided into disjoint sub-circuits and each processor will perform ATPG operations on its respective part.

Although various parallelism approaches were proposed, the core of parallel ATPG does not vary with these approaches and involves only two major operations: *test generation* and *fault simulation*. Krishnaswamy et al. in [8] parallelized fault simulation but overlooked test generation. Yeh et al. in [2] demonstrated parallelism in test generation as well as fault simulation, achieving sub-linear speedup, but suffers from severe test inflation. Wolf et al. in [10] parallelized ATPG completely but did not implement test compaction to suppress test inflation. Aguado et al. in [3] implemented a serial test compaction to reduce excessive pattern counts. Such technique did not correspond to parallel structures. A new parallel ATPG with parallelized test compaction was proposed in [1], recently. It applied depth-first-search (DFS) compaction as well as dynamic fault partitioning and achieved near linear speedup.

However, one fatal limitation for a parallel ATPG is *test inflation* [1] [2], which is illustrated by Figure 1. Test inflation is an inherent problem of parallel ATPG and to overcome test inflation, a conventional technique, *fault broadcasting*[13][14], is often used. Fault broadcasting passes the information of the generated test patterns or newly detected faults to all processors, thus, reducing the probability of generating a test pattern for identical faults. However, even with fault broadcasting, such problem remains serious. For example, a parallel ATPG executing on eight threads still suffers from 15.9% and 10.8%

978-1-4799-2817-0/14 $31.00 © 2014 IEEE

inflation on pattern count for two sample circuits (*s38584* from ISCAS'89 and *vga-enh* from IWLS'05) as Figure 1 illustrates. According to [11], an increase (15.9%) in test patterns can cause a 100% increase in test cost per unit at the worst-case scenario, not to mention the additional test time and the financial loss due to delayed time to market. Hence, test inflation has become a critical problem that limits the practicality of parallel ATPG.

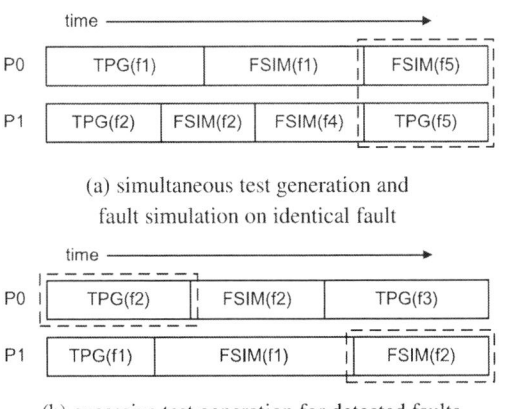

(a) simultaneous test generation and
fault simulation on identical fault

(b) excessive test generation for detected faults

Fig. 2. Two reasons for the test-inflation problem

Figure 2(a) shows one cause of test inflation, which is stemmed from redundant test generation of identical faults by different processors and simultaneous test pattern generation and fault simulation on identical fault. Figure 2(b) shows another cause of test inflation where test inflation is caused by excessive test generations for faults that could have been detected by fault simulation from other processes. TPG, FSIM, P, and f in Figure 2 denote test-pattern generation, fault simulation, processor, and a target fault, respectively.

Upon observing the causes of test inflation, in this work, we propose **concurrent interruption** to prevent simultaneous test generation and fault simulation on identical faults. In addition, we improve DFS compaction to ripple compaction, which mechanism of ripple compaction is also implemented in parallel along with a strategic **fault ordering** specially customized for improving parallel ATPG process and suppressing test inflation. The fault ordering and test compaction complement each other and needs carefully design so that they can work together. Three aforementioned approaches are implemented onto a parallel ATPG algorithm on a shared-memory system in a fault-parallelism fashion. As a result, the enhanced parallel ATPG reduces total pattern count by 11% and successfully suppresses test inflation to 0% while achieving a 6.5X speedup in average for benchmark circuits as in [1].

The rest of this paper is organized as follows: Section II describes the architecture of our baseline parallel ATPG and three techniques, *fault ordering*, *concurrent interruption*, and *ripple compaction*, used to suppress test inflation are elaborated. Experimental results and data analysis are presented in section III. Finally, we draw the conclusions and outline future works in section IV.

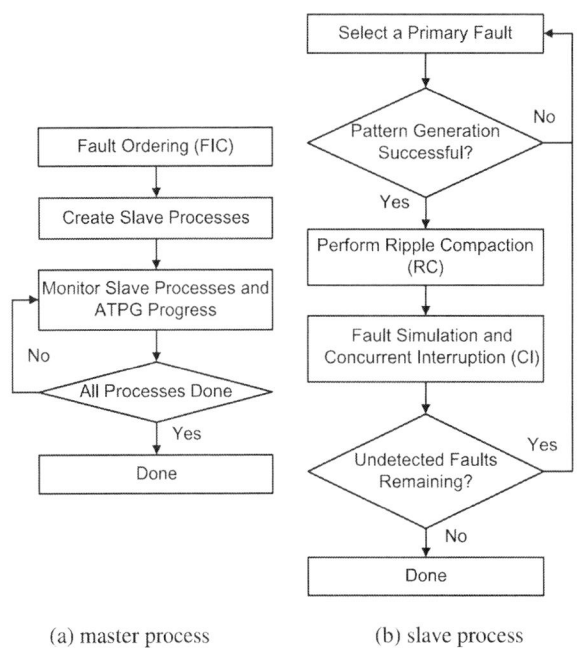

(a) master process   (b) slave process

Fig. 3. The flow of master and slave process

II. CORE TECHNIQUES FOR SUPPRESSING TEST INFLATION IN PARALLEL ATPG

Our baseline parallel ATPG algorithm uses a master-slave architecture where the only master process creates multiple slave processes on a shared-memory machine. Dynamic fault parallelism is employed to ensure that workloads distributed among slave processes are balanced. The master process computes a fault order into a list, prior to the creation of slave processes and monitors the progresses of slave processes as illustrated in Figure 3(a).

A slave process is proposed as Figure 3(b), in which the top-priority fault is picked from the fault list, set as the primary target, and then sent to test generation. Once a fault is processed by a slave process, the slave process updates the fault status in the global memory so that none of other slave processes can access such fault, simultaneously. Ripple compaction is performed later to combine tests of different faults. Upon the completion of ripple compaction, the slave process begins fault simulation and checks for the necessity of concurrent interruption. Such slave process repeats the aforementioned procedure until no more faults can be detected, and returns to join the master process.

In the proposed parallel ATPG structure, three core techniques: (1) **concurrent interruption**, (2) **fault ordering**, and (3) **ripple compaction**, are incorporated. Concurrent interruption is only applicable in the parallel ATPG. Fault ordering and ripple compaction yield better performance on parallel ATPG than single-core ATPG in terms of total pattern count and runtime speedup. The rest of the section will explain three techniques, respective. At first, we re-examine the causes of test inflation in parallel ATPG in detail and then introduce how each

978-1-4799-2817-0/14 $31.00 © 2014 IEEE   665

technique is applied to suppress test inflation and reduce test patterns, collectively.

A. Concurrent Interruption

As mentioned before, many previous works like [1][2] have used *fault broadcasting* to deal with the test-inflation problem. However, it is not sufficient when fault dropping and test generation happen simultaneously on an identical fault as shown in Figure 2(a). Fault broadcasting only passes information; it does not stop on-going test generation for newly detected faults. This scenario happens frequently in parallel ATPG and thus test patterns increase. As a result, ***concurrent interruption*** is proposed to overcome this problem. When a fault is detected, the slave process informs other slave processes about its latest discovery and interrupts other slave process if it is generating a test for the same fault, concurrently.

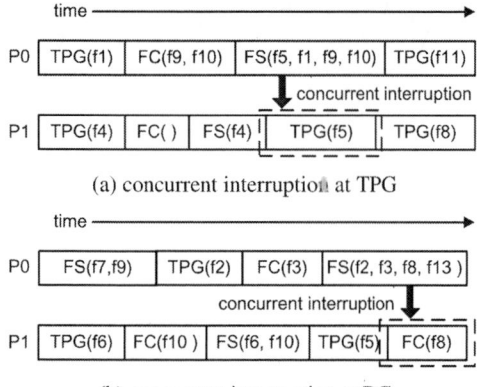

(a) concurrent interruption at TPG

(b) concurrent interruption at RC

Fig. 4. Two types of concurrent interruption

In Figure 4, a simple example uses two processors (P0 and P1) to illustrate two types for concurrent interruption. Here TPG, FC and FS represents the stages of test pattern generation, fault compaction and fault simulation in ATPG, respectively. Figure 4(a) represents the first type where target fault *f5* is processed by FS on P0 and TPG on P1 at the same time. Hence, TPG on P1 is redundant and can be interrupted. Similarly, Figure 4(b) shows the second type that target fault *f8* is simultaneously processed by FS on P0 and FC on P1 where the latter operation can be interrupted. The interrupted slave process will stop immediately, terminate its current operation, and perform test generation for a new fault. By instantly aborting test generation for newly detected faults, type (a) in Figure 2 can be solved, meanwhile saving computation time of ATPG and an excessive test pattern.

B. Fan-in-cone (FIC) based Fault Ordering

The technique of fault ordering refers to how faults are arranged in a fault list that will later be executed sequentially during the test generation. An ATPG engine like Podem or Atalanta sorts faults into a list where faults with a lower topological order will be executed first. This fault list fails to lower the number of test generation. Previous parallel ATPG only

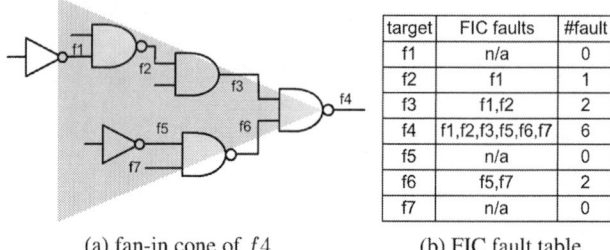

(a) fan-in cone of $f4$ (b) FIC fault table

Fig. 5. An example for fan-in-cone (FIC) based fault ordering

emphasizes on distributing faults evenly to processors to balance their workloads but completely neglects the importance of the fault order for pattern-count reduction. Krauss et al. in [12] proposed that a fault list should be sorted according to the detection probability where the hardest-to-detect faults are treated first. Since many faults are covered by tests generated for hard-to-detect faults, this method ensures that the number of tests generated decreases. Therefore, strategically arranging faults into a proper order can shorten ATPG runtime and reduce pattern count, simultaneously.

Test inflation caused by excessive test generation for faults that *could have been* detected by fault simulation by other processes occurs frequently in parallel ATPG. This situation happens when test generation of a *could-have-been* detected fault ends before it is detected by another slave process at the fault-simulation stage. Both concurrent interruption and fault broadcasting cannot solve this problem. Hard-to-detect faults have a lower probability to be detected during fault simulation according to [12]. Therefore, a fault list with mostly hard-to-detect faults remaining is *less likely* to encounter test inflation. By maximizing the detection on easy-to-detect faults at the beginning of ATPG will leave a fault list that is populated by hard-to-detect faults, thus alleviating the test-inflation problem. In addition, the number of test generation will decrease as well, resulting in a shorter ATPG runtime.

We observed that a fault with considerable quantities of faults in its fan-in cone is effective for fault dropping. Namely, the test for this fault is able to detect a vast number of faults. Therefore, the proposed fault list is arranged by sorting faults according to the number of faults that have not yet been detected (denoted as *not-yet-detected*) as weights in their respective fan-in cones. Figure 5(a) shows *not-yet-detected* faults (including $f1, f2, f3, f5, f6$ and $f7$) in the fan-in cone of $f4$, making $f4$'s weight 6. FIC fault lists of each fault of this example are listed in 5(b) where three columns denote target faults, FIC faults and FIC weights, respectively. According to such FIC table, the final FIC fault ordering is $f4 \rightarrow f3 \rightarrow f6 \rightarrow f2 \rightarrow f5 \rightarrow f1 \rightarrow f7$.

Figure 6 further compares the orders in [9] and [12] with our FIC fault order by showing the fault-dropping numbers for parallel ATPG on an 8-core machine for a benchmark circuit-vga-enh. Such result suggests that, our FIC fault order drops faults in parallel ATPG more quickly over the other two orders, [9] and [12], and leaves mostly hard-to-detect faults, thus preventing test inflation effectively.

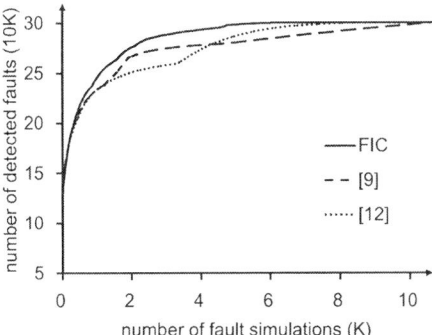

Fig. 6. Comparing the effect of fault dropping by different fault orderings on 8 core parallel ATPG for vga-enh

C. Ripple Compaction

ATPG uses test compaction to reduce pattern counts by combining tests for different faults into one test. In test compaction, faults can be classified into two types of faults: *primary* and *secondary* faults. Secondary faults fall into two statuses: incompatible and compatible. The work [1] uses a DFS compaction to combine tests of different faults and to search for compatible secondary faults for the primary fault. We observed that *structurally-correlated* faults have a higher tendency of being compatible to each other. Hence, a compaction strategy named ***ripple compaction*** which searches structurally-correlated compatible faults in the fan-in and fan-out cone is proposed instead.

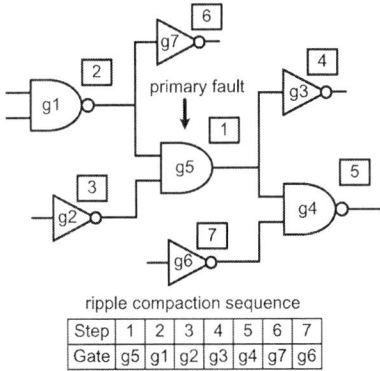

Fig. 7. An example for ripple compaction

Compaction sequence refers to the order of secondary faults being compacted with the primary fault. The compaction sequence affects the compaction success rate since the test of a compatible secondary fault becomes the constraint in test generation of the successive faults. Therefore, two different compaction sequences of the neighboring faults by breadth-first search (BFS) and depth-first search (DFS) were assessed in ripple compaction. Results show that the BFS search yields better runtime and compaction success rates.

Figure 7 shows a sample circuit for the ripple compaction. Assume that the primary fault is targeted at gate $g5$. As mentioned before, the BFS search is used for finding compatible secondary faults. At first, faults from the input cone (including those at gate $g1$ and $g2$) are explored followed by faults from the output cone (including those at gate $g3$ and $g4$). After that, input/output-related faults (in order of faults just being added: $g1 \rightarrow g2 \rightarrow g3 \rightarrow g4$) are also searched and starts from I/O cones of respective gates. As a result, the fault at gate $g7$ and the fault at gate $g6$ are added. Therefore, the final compaction sequence is $g5 \rightarrow g1 \rightarrow g2 \rightarrow g3 \rightarrow g4 \rightarrow g7 \rightarrow g6$.

Hence, our ripple compaction uses the BFS search for finding compatible secondary faults in the proximity of the primary fault. According to our experimental results, the **ripple compaction** can achieve a higher success rate (56.1%) of test compaction than that (47.6%) of the DFS compaction. During **ripple compaction**, test generation is performed for the secondary fault if the test of the primary fault still contains unspecified bits (i.e. X's). The known bits of the primary test become the constraints of successive test generation. The test of the secondary fault can only be generated when it satisfies the constraints of the primary fault. A compatible secondary fault refers to faults that have the same known bits as primary fault while the unknown bits can be 0,1, or remain as unknown as Figure 8 (a). In Figure 8 (b), a conflict value (0 versus 1) in the gray boxes causes the primary and secondary faults become incompatible and cannot be compacted.

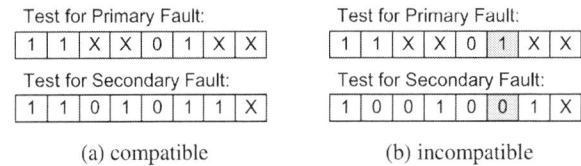

Fig. 8. Compatible and incompatible secondary faults

Setting a constraint on the secondary fault limits the search space for a test solution and speeds up the ATPG process. If a test for the secondary fault is found, this test becomes the new constraint. The search will iterate until all undetected faults in the proximity of the primary fault are exhausted. Once all faults have been explored, the slave process enters the fault-simulation stage. This test is guaranteed to detect the primary fault, all compatible secondary faults, as well as some un-targeted faults.

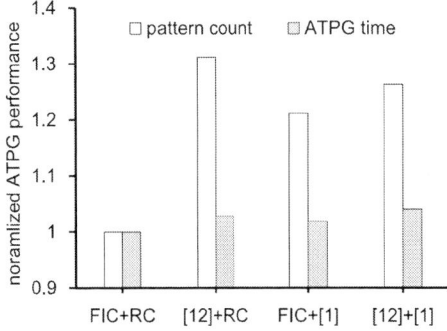

Fig. 9. Comparing fault orders and compaction methods on vga-enh

A comparison on pattern count and ATPG runtime of different combinations of fault-ordering strategies (including [12]

and ours) and test-compaction approaches (including [1] and ours) for a benchmark circuit-vga-enh is illustrated in Figure 9. *FIC* and *RC* denote fan-in-cone based fault ordering and ripple compaction proposed in this paper, respectively. Experimental results show that combining our **FIC fault ordering** and **ripple compaction (RC)** is the best strategy which yields the least amount of ATPG runtime and lowest number of test patterns.

TABLE I
CIRCUIT INFOMRATION

Circuit	# PI	# PO	# Gates	# SA Faults
s35932	35	320	16,065	39,094
s38584	38	304	19,253	36,303
s38417	28	106	22,179	31,180
ethernet	10,640	10.657	72,609	190,924
des3	9,041	8,872	86,309	218,524
vga-enh	17,160	17,180	115,011	300,406
des	9,035	8,942	135,555	270,762
des-ex	11,657	5,154	406,665	754,692

III. EXPERIMENTAL RESULTS

Three ISCAS'89 and five IWLS'05 circuits with gate counts from 16K to 410K are used as benchmark cases in our experiment. Moreover, time-frame expansion by three times is performed on the biggest IWLS'05 circuit (*des*) to create a bigger circuit denoted as *des-ex*. The underlying ATPG kernel is ATALANTA which is a Fan-based algorithm [9]. The proposed parallel ATPG was implemented in C and Pthreads. Table I lists the number of gates and the number of stuck-at faults on benchmark circuits. Experiments were conducted on a Multi-core SMP Server that comprises four Intel Xeon X5650 2.66GHz CPUs and 96 GB memory. DFS compaction proposed in [1] was also implemented on ATALANTA to make a comparison between the proposed ripple compaction to DFS compaction.

Figure 10(a) compares the runtime speedup of parallel ATPG between the proposed approach with our implementation of work [1] over eight benchmark circuits. As a result, our approach has an average speedup of 6.46X while the implementation of work [1] has an average of 5.73X. Figure 10(b) shows the pattern-count reduction of parallel ATPGs. The proposed approach manifests no test inflation with an average of 11% pattern reduction. In contrast, the work [1] suffers from the test inflation on two particular benchmark circuits and achieved no pattern reduction in average.

Table II further lists the percentage of pattern reduction by the three proposed techniques individually on 1, 2, 4, and 8 cores. Here *CI*, *FIC* and *RC* denote concurrent interruption, fan-in-cone based fault ordering and ripple compaction, respectively. The average pattern reduction by concurrent interruption is 0.77%, 1.24%, and 3.22% in 2, 4, and 8 cores of parallel ATPG, respectively. This result suggests that test inflation due to simultaneous test generation and fault simulation on identical fault deteriorates as the number of cores grows and concurrent interruption is capable of suppressing this effect from occurring.

Particularly, our FIC fault ordering causes an additional 4.80% and 0.17% of test patterns on *s38417* and *ethernet* during single core ATPG operation. However, this effect is exceptional and only limited to the single-core APTG operation. In general, pattern reduction by fault ordering is 5.85%, 6.25%, 7.21%, and 7.53% in 1, 2, 4, and 8 cores, respectively. FIC fault ordering becomes more efficient in pattern-count reduction as number of cores grows. This trend suggests that our fault ordering is capable of preventing test inflation in multi-core ATPG operation by lowering the probability of excessive test generations on faults that could have been detected.

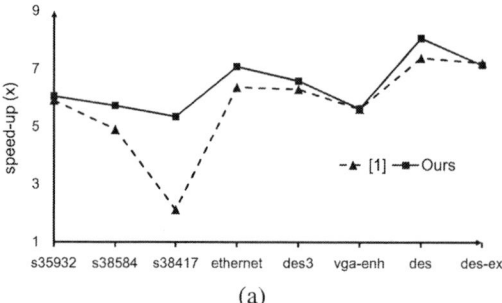

(a)

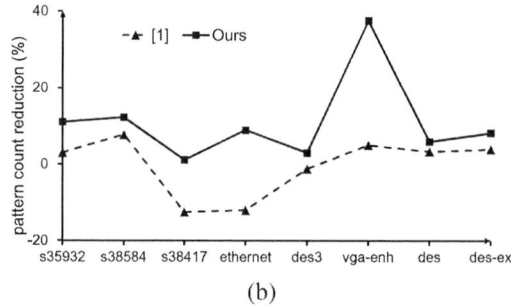

(b)

Fig. 10. (a) Speedup on 8-core parallel ATPG (b) Pattern reduction on 8-core parallel ATPG

Ripple compaction's effect on pattern reduction does not vary with the number of cores. Namely, it does not make direct contribution to suppressing test inflation. However, test compaction is still an effective approach in pattern-count reduction (4.38%, 4.57%, 4.14% , and 4.38% on 1, 2, 4, and 8 cores) for maintaining the practicality of parallel ATPG.

Test-inflation suppression is shown in Figure 11(a). Results indicate that test inflation in parallel ATPG has been greatly suppressed. The most severe test-inflation case is only 0.70% on *ethernet* circuit. In contrast, the reimplementation of work [2] showed a test inflation of 50%. The original average test inflation was 6.81% and is decreased to 0% after applying the proposed techniques. Moreover, the best effect of test-inflation suppression on *vga-enh* with respect to different cores used for parallel ATPG is illustrated in Figure 11(b). Test inflation has been succesfully suppressed by 4.0%, 9.7% and 15.3% on 2, 4, and 8 cores respectively. This trend proves that our techniques can scale with number of cores and suppresses test inflation effectively. Note that our parallel ATPG is also efficient in memory usage which only used an additional 1.26%, 3.75%, and 13.5% memory in 2, 4, and 8 cores, which is similar to the result reported by work [1].

TABLE II
PATTERN REDUCTION OF EACH BENCHMARK CIRCUITS

	1 core			2 core				4 core				8 core			
	pattern reduction (%)			pattern reduction (%)				pattern reduction (%)				pattern reduction (%)			
name	FIC	RC	total	CI	FIC	RC	Total	CI	FIC	RC	total	CI	FIC	RC	total
s38584	8.23	2.67	10.90	1.33	7.18	4.20	12.71	3.22	7.62	3.28	14.12	4.14	9.62	3.92	17.68
s35932	6.25	1.25	7.50	0.60	3.80	8.22	12.82	1.30	8.23	1.90	11.43	8.14	2.33	4.65	15.12
s38417	-4.80	6.21	1.41	0.71	0.47	1.26	2.44	0.70	1.24	0.80	2.74	7.10	0.54	2.33	9.97
ethernet	-0.17	9.50	9.33	0.33	0.97	9.31	10.61	0.91	2.76	9.23	12.90	0.60	4.76	9.08	14.44
des3	1.47	1.76	3.23	0.91	0.31	0.31	1.53	1.20	0.30	0.00	1.50	1.76	0.88	1.76	4.40
vga-enh	28.00	9.72	37.72	0.14	30.90	9.46	40.50	0.35	34.30	7.90	42.55	1.38	36.30	7.89	45.57
des	3.87	1.03	4.90	1.30	3.10	0.52	4.92	0.76	1.53	5.87	8.16	1.02	2.05	1.98	5.05
des-ex	3.91	2.88	6.79	0.85	3.25	3.28	7.38	1.44	1.64	4.11	7.19	1.61	3.82	3.41	8.84
Average			10.20				11.61				12.57				15.13

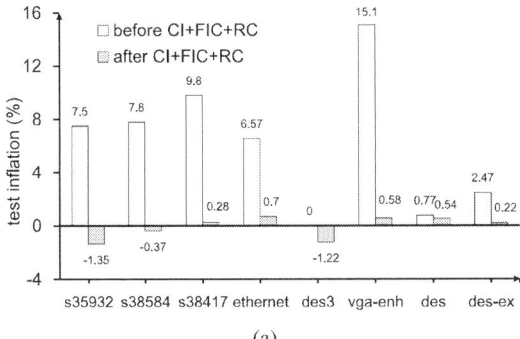

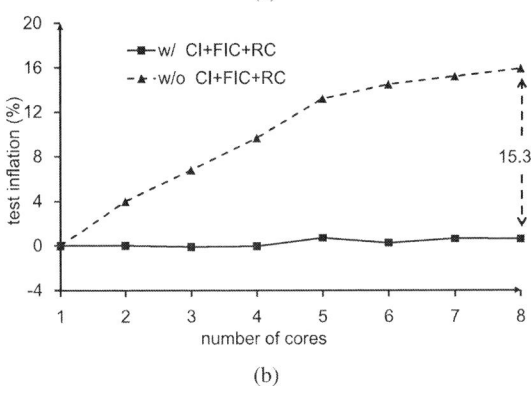

Fig. 11. (a) Test inflation suppression on 8-core parallel ATPG (b) Break down of test inflation suppression on 8-core parallel ATPG on vga-enh

IV. CONCLUSION

Runtime and test inflation are two most critical issues in parallel ATPG for modern VLSI testing. Previous works mainly emphasize on load-balancing and runtime while neglecting the deterioration in test inflation. Test inflation incurs severe financial loss due to excessive test cost and delayed time to market. These drawbacks severely limit the practicality of parallel ATPG. Therefore, three techniques are proposed to suppress test inflation in parallel ATPG on shared memory that employs dynamic fault parallelism. **Concurrent interruption (CI)** is implemented to intervene test generation of other slave process when necessary. **Fan-in-cone based fault ordering (FIC)** is proposed, especially for parallel structures, to prevent excessive test generations for detected faults. **Ripple compaction**

(RC) shortens ATPG runtime by limiting search space of compatible faults. Experimental results show that the proposed parallel ATPG effectively reduces test-patterns by 11% in average and successfully suppresses test inflations from 6.81% to 0% while achieving a 6.5X speedup on benchmark circuits

REFERENCES

[1] X. Cai, P. Wohl, J. A. Waicukauski and P. Notiyath, "Highly efficient parallel ATPG based on shared memory," *Proc. Intl Test Conf.*, 2010, pp.1-7.

[2] K.-W. Yeh, M.-F. Wu and J.-L. Huang, "A low communication overhead and load balanced parallel ATPG with improved static fault partition method," *Proc. Intl Conf. on Algorithms and Architectures for Parallel Processing*, 2009, pp. 362-371.

[3] M. 1. Aguado, E. de la Torre, M. A. Miranda and C. Lopez-Barrio, "Distributed implementation of an ATPG system using dynamic fault allocation," *Proc. of Intl Test Conf.*, 1993, pp. 409-418.

[4] S. Patil and P. Banerjee, "Fault partitioning issues in an integrated parallel test generation fault simulation environment," *Proc. Intl Test Conf.*, 1989, pp. 718-726.

[5] R.H. Klenke, R.D. Williams and J. H. Aylor, "Parallel-processing techniques for automatic test pattern generation," *IEEE Computer*, vol. 25, Issue 1, 1992, pp. 71 - 84.

[6] A. Motohara, K. Nishimura, H. Fujiwara and I. Shirakawa, "A parallel scheme for test-pattern generation," *Proc. Intl Conf. on Computer-Aided Design*, 1986, pp. 156-159.

[7] R. H. Klenke, R. D. Williams and J. H. Aylor, "Parallelization methods for circuit partitioning based parallel automatic test pattern generation," *Proc. VLSI Test Symposium*, 1993, pp. 71-78.

[8] D. Krishnaswamy, E. M. Rudnick, J. H. Patel and P. Banerjee, "SPIT-FIRE: scalable parallel algorithms for test set partitioned fault simulation," *Proc. VLSI Test Symposium*, 1997, pp. 274-281.

[9] H. Fujiwara and T. Shimono, "On the acceleration of test generation algorithms," *IEEE Trans. on Computers*, vol. C-32, 1983, pp. 1137-1144.

[10] M. Wolf, L. M. Kaufman, R. H. Klenke, J. H. Aylor and R. Waxman, "An analysis of fault partitioned parallel test generation," *IEEE Trans. on Computer-Aided Design of Integrated Circuits and Systems*, vol. 15, no. 5, 1996, pp. 517-534.

[11] I. D. Dear, C. Dislis, A. P. Ambler, and J. Dick, "Economic effects in design and test," *Design & Test of Computers*, vol.8, no.4, 1991, pp. 64-77.

[12] P. A. Krauss and M. Henftling, "Efficient fault ordering for automatic test pattern generation for sequential circuits," *Proc. of Third Asian Test Symposium*, 1994, pp. 113-118.

[13] Butler, R., Keller, B., Paliwal, S., "Schoonover, R., Swenton, J.:Design and Implementation of a parallel Automatic Test Pattern Generation Algorithm with Low Test Vector Count," *Proc. Intl Test Conf.*, 2000, pp.530-537.

[14] Klenke, R.H., Kaufman, L., Aylor, J.H., Waxman, R., Narayan, "P. Workstation based parallel test generation," *Proc. Intl Test Conf.*, 1993, pp.419-428.

A Volume Diagnosis Method for Identifying Systematic Faults
in Lower-yield Wafer Occurring during Mass Production

Tsutomu Ishida[1], Izumi Nitta[1], Koji Banno[2], Yuzi Kanazawa[1]

[1]Fujitsu Laboratories Limited
4-1-1 Kamikodanaka, Nakaharak-ku, Kawasaki, Japan

[2]Fujitsu Semiconductor Limited
50 Fuchigami, Akiruno-shi, Tokyo, Japan

{ishi.tsutomu.11, nitta.izumi, banno.kouji, ykanazawa}@jp.fujitsu.com

Abstract - This work focuses on volume diagnosis for identifying systematic faults in lower-yield wafers, whose yields are lower than baseline level due to systematic faults during mass production. We develop a model-based volume diagnosis method. To diagnose accurately using the fail data with one lower-yield wafer, we apply modeling techniques for handling pseudo-faults and random faults in the fail data. Experimental results show our method's efficiency; we succeeded in identifying the failure layer for 20/22 data sets with actual lower-yield wafers.

I. Introduction

Continuous CMOS process scaling has made it more and more difficult to achieve and maintain high yield in integrated circuit (IC) manufacturing. IC manufacturers analyze failing chips screened out by scan-based logic test, find the defect sites and identify the root causes so as to accomplish high yield for making their products more profitable. Defects can be classified into random or systematic faults. The systematic faults are especially critical for yield enhancement. Hence, IC manufacturers eliminate the systematic faults by modifying the layout and, in some cases, the logic circuit to achieve yield improvement.

In this work, we focus on identifying systematic faults in lower-yield wafers, whose yields are lower than baseline level due to systematic faults during mass production. The lower-yield wafers are caused by setting changes of manufacturing equipment, electronic supply problems, defects caused by design problems, process defects, and so on. To improve the lower yield as soon as possible, IC manufacturers have to identify systematic faults using the fail data based on limited number of lower-yield wafers.

In recent years, many researches on volume diagnosis were presented to show their effectiveness for identifying systematic faults [1-8]. Volume diagnosis requires *failure reports*, each of which contains fault candidates traced by logic diagnosis [11-13] for a failing chip. Previous works on volume diagnosis generally use a lot of failure reports for yield improvement in various situations (including lower-yield wafers) by statistical analysis such as iterative learning [3,4], chi-square test [1] and clustering [7].

Model-based statistical diagnosis, which was proposed in [14] for speed-path analysis to identify systematic timing effects with design-silicon timing mismatch, can also be applied to volume diagnosis. The overview of the model-based volume diagnosis (MVD) method is shown in Fig.

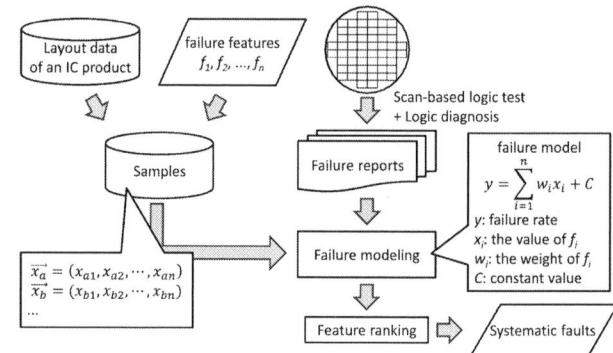

Fig. 1. Overview of model-based volume diagnosis (MVD) method.

1. In this work, our model calculates a failure rate based on layout features which likely cause systematic faults, and we will refer to the model and the layout features as *failure model* and *failure features*, respectively. For example, when a net a is represented with failure features as $\vec{x_a}$, a failure model $(\vec{w}, C)$ calculates the failure rate of a as $\vec{w} \cdot \vec{x_a} + C$. MVD method constructs such a failure model statistically and then identifies failure features whose sensitivity is larger in the model as systematic faults. We employ and develop MVD method so that we can identify systematic faults using the failure reports obtained from one lower-yield wafer.

In order to diagnose accurately using the limited fail data, we have to properly handle two causes of erroneous diagnosis; one is pseudo-faults in failure reports and the other is random faults due to manufacturing residuals. Pseudo-faults are often reported as fault candidates because logic diagnosis generally cannot identify fault sites definitely due to various limitations such as logic equivalence and lack of layout information. Fig. 2 shows an example of pseudo-fault with its failure report. The volume diagnosis methods in [1,7] can be applied to analysis of lower-yield wafers. However, the authors in [7] used exclusively the failure reports in each of which there is only one fault candidate to avoid the uncertainty due to the pseudo-faults. Although the authors in [1] proposed a method for handling the pseudo-faults, the method needs hundreds of failure reports to be efficient. In this work, to diminish the noise of the pseudo-faults and the random faults, we apply three modeling techniques: *likelihood selection* [16], regression learning using epsilon-support vector regression (ε-SVR) [17] and a grouping method. Likelihood selection, a combinatorial optimization technique, selects the most likely fault from among fault candidates in each failure report. At this point, we assume a

978-1-4799-2817-0/14 $31.00 © 2014 IEEE
670

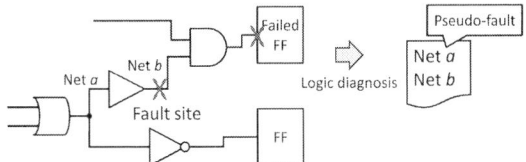

Fig. 2. An example of pseudo-fault with a failure report.

situation where defects are attributed to a single dominant defect mechanism. Since lower-yield wafers are most commonly caused by a single dominant factor, likelihood selection suits the situation discussed in this work. In experiments, we used fail data sets with practical IC products. Each data set consists of the failure reports obtained from an actual lower-yield wafer, with which certain layout patterns on a particular layer are systematically fractured in failing ICs. Experimental results show that MVD method could identify the failure layer correctly for 20/22 data sets.

The main contributions of this work are as follows:

- We develop MVD method that can identify systematic faults using the failure reports obtained from one lower-yield wafer. MVD method can also be utilized for mature yield enhancement where a dominant defect mechanism causes some dies to fail.
- In order to properly handle the pseudo-faults and the random faults in fail data, we employ three key techniques: likelihood selection, regression learning using ε-SVR and a grouping method. Especially, we first apply combinatorial optimization (used in likelihood selection) to volume diagnosis and confirm its effectiveness.
- We consider that likelihood selection can be general diagnosis for problems where there are multiple fault candidates and the true one is unclear. This work is one example for confirming the correctness of this view.

In this paper, we deal with defects related to nets (more precisely defects in metal layers) because defects with lower-yield wafers during mass production are actually almost always caused not in transistor layers but in metal layers. In addition, just focusing on nets simplifies discussions. MVD method can be extended theoretically to handle defects in transistor layers.

The rest of the paper is organized as follows. Section II gives the overview of MVD method. In Section III, we explain the ideas of the three key techniques for accurate diagnosis. Section IV describes the overall MVD method. Section V presents experimental setup and results. Finally, conclusions are summarized in Section VI.

II. Overview of MVD Method

The goal of this work is to find some layout features which have relatively strong correlation with fault nets. A small amount of fail data makes it difficult to obtain the exact correlation, but it is not necessary. We believe that the model-based statistical diagnosis is the best strategy for our situation where the number of input failure reports is small.

Fig. 1 shows the overview of MVD method. The inputs are layout data of an analyzed IC product, user-defined fail-

ure features and the failure reports obtained from a lower-yield wafer. The outputs are one or more failure features identified as systematic faults. MVD method is mainly formed from sample set generation phase and failure modeling phase. A sample set for failure modeling is generated only once for each IC product. Therefore, once a sample set is built, an analysis with a new set of failure reports begins with failure modeling.

In sample set generation phase, *net grouping* groups a certain amount of nets and generates net groups as samples for failure modeling. The reason for the grouping is that an exact failure rate of each net cannot be obtained with the limited fail data due to variation of defects. Each net group is encoded with failure features to a group feature vector.

In failure modeling phase, MVD method constructs linear failure model for net group. In this work, we assume that defects are caused by certain failure features independently, and therefore, we apply linear model. Likelihood selection selects the most likely fault net for each failure report, and a failure rate is computed for each net group. When all m net groups gain their own feature vector and failure rate, a failure model is statistically obtained by solving the following regression problem.

$$y_k = \sum_{i=1}^{n} w_i x_{ki} + C \quad (1 \le k \le m). \quad (1)$$

y_k is the normalized failure rate for the net group k; w_i is the weight of the failure feature f_i; x_{ki} is the normalized value of f_i for the net group k; C is the constant value. $w_1, w_2, \cdots, w_n$ and C are obtained by ε-SVR. A feature ranking method computes each failure feature's sensitivity in the obtained model and ranks all failure features. Finally, the higher-ranked failure features are identified as systematic faults.

III. Ideas of Key Techniques

This section explains the ideas of the three key techniques: likelihood selection, regression learning using ε-SVR and net grouping. As described in Section I, we have to properly handle the pseudo-faults and the random faults in fail data. Likelihood selection is applied for avoiding that the pseudo-faults are used for failure modeling. The noise of the random faults is handled by regression learning using ε-SVR and net grouping. The following subsections describe the idea of each key technique.

A. Likelihood Selection

Likelihood selection is a combinatorial optimization process developed in [16] for speed-path analysis. In speed-path analysis, there is an ambiguity that a failing path, a path causing a FF to fail, is unclear. Fig. 3 illustrates an example of the ambiguity.

Fig. 4 shows the overview of likelihood selection for speed-path analysis. The key component of likelihood selection is an index measuring model accuracy, and we call it *model-accuracy index*. In Fig. 4, a failing path suspect is selected from failing path candidates for each failed FF.

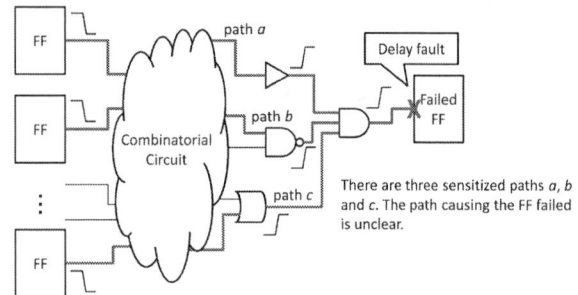

Fig. 3. An example of the ambiguity with a failed FF.

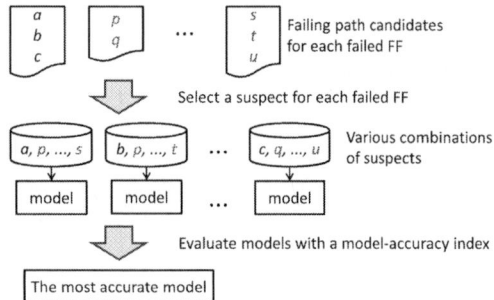

Fig. 4. Overview of likelihood selection.

Next, the combination of the selected suspects is used to construct a model which estimates a path delay difference between pre- and post-silicon. A large number of the combinations construct the corresponding number of models, and these models are evaluated with the model-accuracy index. Finally, the most accurate model is determined and used for identifying systematic timing effects with design-silicon timing mismatch. The most accurate model gives the most likely failing path for each failed FF. Therefore, we call this process "likelihood selection". Likelihood selection is formulated into a combinatorial optimization problem with the model-accuracy index.

Likelihood selection can be also applied to volume diagnosis. It selects the true fault net for each failure report under the assumption that there is one true fault net in each failure report. There may exist more than one true fault nets in a failure report. In such a case, selecting one true fault net is enough for accurate diagnosis if all the true fault nets are caused by a dominant defect mechanism. Likelihood selection for volume diagnosis is almost same procedure in Fig. 4. A fault suspect net is selected for each failure report. Next, the combination of the suspects is used to formulate the equations (1), and a failure model is obtained by solving the regression problem. Many failure models are created and evaluated by a model-accuracy index. Finally, the most accurate failure model is determined and used for identifying systematic faults.

An appropriate model-accuracy index is required for accurate diagnosis with likelihood selection. The followings are valuable conclusions obtained in [16] with model-accuracy index:

- Coefficient of determination (R^2) can be used for model-accuracy index. When the R^2 of a model is low, the

diagnosis result is not reliable.
- The locally optimized model by Hill Climbing (HC) with R^2 can be used to identify systematic timing effects with design-silicon timing mismatch. At this point, an initial condition for HC should be pretty close. In speed-path analysis, the slowest path with static timing analysis (STA) is extracted for each failed FF and they are used as the initial condition for HC.

In this work, we use R^2 and HC. In order for accurate diagnosis, we need to get a proper initial condition. We can obtain it by looking at *match* and *conflict* indices returned by logic diagnosis. Suppose logic simulation gives n failed FFs under the assumption that a given fault candidate net is broken; m and c is the number of truly-failed and non-failed FFs in the n FFs, respectively. m is the match value and c is the conflict value. We select the net that has the minimum conflict value from among the nets having the maximum match value for each failure report as the initial condition.

B. Regression Learning using ε-SVR

ε-SVR is an appropriate regression learning method to solve the regression problem (1) because the epsilon parameter in ε-SVR can handle the noise of the random faults. The epsilon parameter indicates a margin for model-prediction error. Hence, a proper epsilon value can decrease the noise of the random faults. The discussion of the proper epsilon value is described in the next subsection because it is closely linked to net grouping.

C. Net Grouping

Net grouping groups a certain amount of nets which have similar layout feature into a net group as a sample for failure modeling. The method sorts all nets by the value of an item which is a layout feature likely correlated to systematic faults, and it divides the sorted nets into n_{div} equal groups so that similar nets in terms of the item are grouped into a same net group. This net grouping is based on a grouping method in [1].

The division number n_{div} should be determined properly according to the number of input failure reports for accurate diagnosis. We consider that variation of defects includes the random faults and also the pseudo-faults not eliminated by likelihood selection. The n_{div} determines a standard deviation for the average number of fault nets in a net group. Since the average standard deviation shows how much variation exists from the average number of fault nets in a net group, we seek to use the average standard deviation to avoid the effect of the variation of defects for a net group in failure modeling. Therefore, we calculate the average standard deviation value σ_{ave} under binomial distribution and utilize it to determine a proper epsilon value in ε-SVR. A larger epsilon value degrades the sensitivity of regression learning but a smaller epsilon value cannot eliminate the effect of the variation of defects. In this work, we chose the n_{div} that leads the σ_{ave} to be at least one and less than two or three nets for each data set because it gives good diagnosis results in the experiments.

978-1-4799-2817-0/14 $31.00 © 2014 IEEE

IV. Overall MVD Method

This section explains the implementation of the overall MVD method. Fig. 5 illustrates the flow of MVD method. The following subsections describe the components in the flow.

A. User-defined Failure Features

At the first step in MVD method, users have to define layout features which are candidates for systematic faults as failure features. We see no need to define proper detailed failure features. For example, finding the layer related to systematic faults is of assistance for physical failure analysis (PFA). In this work, we extract layout patterns which likely cause open and/or short defects and then define them for each layer as failure features. For instance, when sparse-wiring pattern is extracted as a cause of interconnect open defects, we define the following failure features: Layer1-sparse-wiring, Layer2-sparse-wiring and so on. The layout patterns are extracted based on accumulated fail data and knowledge of expert engineers. Automation of failure feature extraction remains one of the problems for MVD method. The work [7], which automatically discovers important layout features, could help to solve our problem.

B. Net Encoding

Each net is encoded with failure features to a net feature vector. In this paper, we use a window-based net encoding method. It divides the IC layout of each layer into windows and calculates the value of each failure feature for each window. Fig. 6 shows an example of a meshed window. Each failure feature value of a net is obtained by summing the feature values of the windows through which the net passes. Suppose we define failure features f_1, f_2, f_3. Net a passes through the window p on layer 2 and the window q on layer 3. The net feature vector $\overrightarrow{n_a}$ is calculated as follows:

$$\overrightarrow{n_a} = (x'_{p1}l_{pa} + x'_{q1}l_{qa}, x'_{p2}l_{pa} + x'_{q2}l_{qa}, \\ x'_{p3}l_{pa} + x'_{q3}l_{qa}). \quad (2)$$

x'_{ki} denotes the feature value of f_i with the window k; l_{ka} denotes the wire length of a in the window k. This window-based net encoding method is one of the methods for encoding a net with failure features. Another net encoding method can be also applied to MVD method.

C. Net Grouping

The pseudocode described in Fig. 7 generates net groups. After net grouping, each group feature vector is obtained by summing all the net feature vectors involved in the group.

D. Failure Modeling

The flow of failure modeling is illustrated in Fig. 5. First, a fault suspect net is selected for each failure report. We select the net that has the minimum conflict value from among the nets having the maximum match value as the initial fault suspect net. Second, the fault suspect nets are used to com-

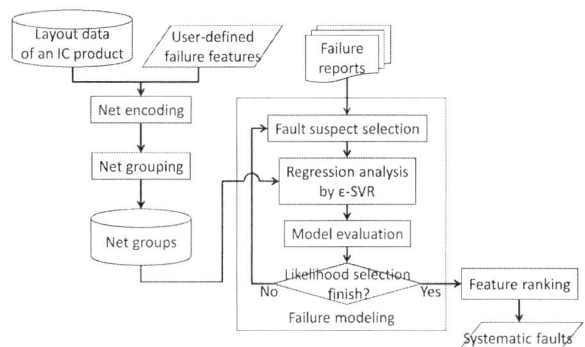

Fig. 5. Flow of MVD method.

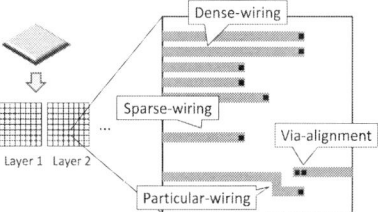

Fig. 6. An example of a meshed window.

pute each net group's failure rate. Specifically, we obtain the failure rate of a net group NG_a by dividing the number of all fault suspect nets in NG_a by the number of all nets in NG_a. When all net groups gain their own failure rate and group feature vector, the net groups are used to construct the regression equations (1). At this point, we cluster failure features which are correlated to each other into a new feature to avoid multicollinearity in regression analysis. We also normalize failure rates and failure feature values for accurate diagnosis. Third, a failure model is built by ε-SVR and evaluated with R^2. A failure model which is built with the initial fault suspect nets is the initial failure model. The failure model is iteratively updated beginning with the initial failure model in likelihood selection described in Fig. 8, and finally, the optimized failure model is obtained.

E. Feature Ranking

In order to identify systematic faults, a feature ranking method ranks all failure features using their sensitivities in the optimized failure model. There are many feature ranking methods [14-16]. In this work, we calculate an *importance*

Net grouping

1: **Input:** all nets *NETS*, layout features *ITEMS*;
2: **Output:** *GROUPS*;
3: *GROUPS* = ();
4: Determine n_{div};
5: **FOR** *item* **IN** *ITEMS*
6: Sort *NETS* in descending order with the value of *item*;
7: Divide the sorted *NETS* into n_{div} equal net groups: $GP_1^{item}, GP_2^{item}, \cdots, GP_{n_{div}}^{item}$, such that the top $|NETS|/n_{div}$ nets go into GP_1^{item}, then next $|NETS|/n_{div}$ nets go into GP_2^{item} and so on;
8: Push $GP_1^{item}, GP_2^{item}, \cdots, GP_{n_{div}}^{item}$ into *GROUPS*;
9: **END FOR**

Fig. 7. Pseudocode of net grouping method.

```
          Likelihood selection for failure modeling
 1:  Input: failure reports REPORTS, net groups GROUPS;
 2:  Output: optimized_failure_model;
 3:  Select an initial fault suspect net for each failure report;
 4:  initial_failure_model = the failure model built with the initial
     fault suspect nets using GROUPS;
 5:  optimized_failure_model = initial_failure_model;
 6:  best_R2 = the R² of initial_failure_model;
 7:  DO
 8:    pre_best_R2 = best_R2;
 9:    FOR report IN REPORTS
10:      FOR suspect_net IN all fault candidate nets with report
11:        current_model = a failure model built under the as-
           sumption that suspect_net is the fault suspect net of re-
           port using GROUPS;
12:        current_R2 = the R² of current_model;
13:        IF (current_R2 > best_R2) THEN
14:          optimized_failure_model = current_model;
15:          best_R2 = current_R2;
16:          Replace the fault suspect net of report to suspect_net;
17:        END IF
18:      END FOR
19:    END FOR
20:  WHILE (best_R2 > pre_best_R2)
```

Fig. 8. Algorithm of likelihood selection.

Table 1. Information of IC product A and B.

IC product	A	B
Process technology	28nm CMOS	65nm CMOS
#Gates	60 million	10 million
#Layers	L1 - L9	L1 - L6

value for each failure feature. Suppose there is a failure model $(w_1, w_2, \cdots, w_n, C)$ and top m net groups with failure rate. The importance value of a failure feature f_i is obtained as follows:

$$imp(f_i) = w_i x_{1i} + w_i x_{2i} + \cdots + w_i x_{mi}. \qquad (3)$$

$imp(f_i)$ denotes the importance value of f_i; x_{ki} denotes the value of f_i with the k-th net group. The failure features with the higher importance values are extracted as systematic faults.

V. Experiments

In this section, we describe experimental setup and results. We evaluated MVD method using actual fail data taken from lower-yield wafers with practical IC products.

A. Setup

We used actual fail data with practical IC product A and B. The information of A and B is summarized in Table 1. "#Gates" shows the approximate number of the logic gates; "#Layers" shows the analyzed metal layers. We have 22 data sets; one is a data set with A and the others are with B. Each data set consists of the failure reports obtained from a lower-yield wafer. Table 2 shows 5 data sets of them. Columns 1 and 2 give the data set name (Product-Lot-Wafer) and the number of the failure reports, respectively. In each lower-yield wafer, certain layout patterns on a particular layer are fractured systematically in failing ICs. For the data set A-lot1-wf1, the defect sites were confirmed by PFA; for the other data sets, the failure layer was identified with difficulty using IC test results and equipment history by expert engineers. Column 3 shows the failure layer.

Experimental specifications of MVD method are as follows:

- Failure features

For each layer, we defined the following four failure features: dense-wiring (dw), sparse-wiring (sw), particular-wiring (pw) and via-alignment (via). Fig. 6 shows an example with these features. The particular-wiring feature consists of critical wiring patterns obtained from accumulated defective patterns. We extracted the defined feature values by a design rule checking (DRC) tool. It took about one day for A and less than a half day for B using 4 CPUs on a Linux server with 3.33GHz CPUs and 192GB RAM. We did not require post-optical proximity correction (post-OPC) layout data to extract the feature values.

- Net grouping

We determined the division number $n_{div} = 20$ and $n_{div} = 40$ for the data consisting of less than 80 and more than or equal to 80 failure reports, respectively.

- ε-SVR

We calculated the σ_{ave} and set $2\sigma_{ave}$ as the epsilon value in ε-SVR. We used libsvm [18] for ε-SVR.

- Feature ranking

We extracted the top 3 failure features whose importance values are larger than a certain threshold as systematic faults.

We implemented MVD method in C++ and all the analyses were performed using 1 CPU on a Linux server with 3.33GHz CPUs and 96GB RAM.

B. Results

MVD method could identify the failure layer correctly for 20/22 data sets. Some experimental results are summarized in Table 2. Columns 4, 5 and 6 show the results with the optimized failure model by likelihood selection (OptModel); columns 7 and 8 show the results with the initial failure model for likelihood selection (IniModel). Columns 4 and 7 show the failure features identified as systematic faults. "L6via,L7via", for example, indicates that they are clustered into one feature to avoid multicollinearity in regression analysis.

OptModel finds the failure features with the failure layer for 20/22 data sets, while IniModel finds for 17/22 data sets. The reason for the reasonably good results with IniModel comes from that the initial fault suspect nets selected by match and conflict indices are pretty close. For the data sets with good R^2 in IniModel, the failure layer is identified, as the B-lot1-wf2, 3 results show. This shows that good R^2 with IniModel indicates that the initial fault suspect nets are set close to the true fault nets. On the other hand, when the R^2 with IniModel is poor, the failure layer is not identified. In such cases, likelihood selection is further effective for identifying the systematic faults, as the results with A-lot1-wf1 and B-lot3-wf4 show. Fig. 9 shows the normalized importance values of the top 5 ranked failure features with

978-1-4799-2817-0/14 $31.00 © 2014 IEEE

Table 2. Comparison of feature rankings with the optimized failure model (OptModel) and the initial failure model (IniModel).

Data set	#Reports	Failure layer	OptModel (obtained by likelihood selection)			IniModel (without likelihood selection)	
			Ranking (1st/2nd/3rd)	R^2	time [sec]	Ranking (1st/2nd/3rd)	R^2
A-lot1-wf1	35	L7	**L7dw**/L6via,**L7via**	0.751	704	L5via,L6sw,L6pw,L6dw/L5sw,L5dw	0.217
B-lot1-wf2	60	L4	**L4pw/L4dw**	0.916	162	L3sw/L3via/**L4sw**	0.748
B-lot1-wf3	46	L4	L1dw/L3sw/L1sw	0.889	142	**L4sw**	0.642
B-lot2-wf2	73	L2	L4dw/L4pw	0.892	210	L3via/L3sw/L1via,**L2via**	0.531
B-lot3-wf4	44	L4	**L4via**/L5sw,L5pw,L5dw/**L4pw**	0.740	206	-	0.141

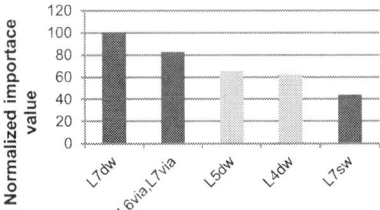

Fig. 9. Top 5 ranked failure features obtained from OptModel with A-lot1-wf1. L7 failure features are ranked higher.

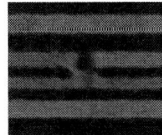

Fig. 10. SEM image of a defect on L7 with A-lot1-wf1.

OptModel for A-lot1-wf1. The L7dw is top-ranked. We confirmed by PFA for two failing ICs in A-lot1-wf1 that the dense-wiring patterns on L7 layer are systematically damaged. Fig. 10 shows a scanning electron microscope (SEM) image of the PFA result. However, there are two data sets (B-lot1-wf3, B-lot2-wf2) whose failure layers are not identified by OptModel. For the two data sets, although IniModel finds the failure features with the failure layer, OptModel does not. We think that overfitting may be caused in likelihood selection because R^2, as we have known, is not always the best model-accuracy index for likelihood selection. Hence, to develop other model-accuracy indices is part of our future work. Or since real defects are not confirmed by PFA for the data sets with B, there might be real defects on the layer estimated by OptModel.

VI. Conclusions

In this paper, we developed a model-based volume diagnosis (MVD) method. When lower-yield wafers are produced, MVD method statistically identifies systematic faults using the fail data taken from one lower-yield wafer. Likelihood selection is applied to reduce the noise of pseudo-faults in the input fail data. In experiments, MVD method could identify the failure layer correctly for 20/22 data sets with practical IC products. However, MVD method could not identify the failure layer for the remaining two data sets. We think that overfitting may occur in likelihood selection for the two data sets. To develop a technique to avoid overfitting in likelihood selection is part of our future work.

Acknowledgements

We thank Mitsuhiro Nishigaki and Akihisa Takechi of Fu-

jitsu Semiconductor Limited for the valuable support.

References

[1] M. Sharma, et al., "Efficiently Performing Yield Enhancements by Identifying Dominant Physical Root Cause from Test Fail Data", International Test Conference, p. 14.3, 2008

[2] C. Hora, R. Segers, S. Eichenberger, and M. Lousberg, "An Effective Diagnosis Method to Support Yield Improvement", International Test Conference, p. 9.4, 2002

[3] H. Tang, S. Manish, J. Rajski, M. Keim, and B. Benware, "Analyzing Volume Diagnosis Results with Statistical Learning for Yield Improvement", European Test Conference, pp. 145-150, 2007

[4] B. Benware, C. Schuermyer, M. Sharma, and T. Herrmann, "Determining a Failure Root Cause Distribution From a Population of Layout-Aware Scan Diagnosis Results", IEEE Design & Test of Computers, vol. 29, no.1, pp. 8-18, Feb. 2012

[5] W. Yang, T.-P. Tai, T. Chandilya, C. Hao, and D. Carder, "Enabling Baseline Yield Improvement with Diagnosis Driven Yield Analysis", ECS Trans. 2012, Volume 44, Issue 1, pp. 1029-1035, 2012

[6] R. Turakhia, M. Ward, S. K. Goel, and B. Benware, "Bridging DFM and Volume Diagnosis for Yield Leraning – A Case Study", VLSI Test Symposium, pp. 167-172, 2009

[7] W. C. Tam, O. Poku, and R. D. Blanton, "Systematic Defect Identification through Layout Snippet Clustering", International Test Conference, p. 13.2, 2010

[8] S. Wang, and W. Wei, "Machine Learning-Based Volume Diagnosis", Design, Automation & Test in Europe Conference, pp. 902-905, 2009

[9] W. C. Tam, O. Poku, and R. D. Blanton, "Precise Failure Localization using Automated Layout Analysis of Diagnosis Candidates", Design Automation Conference, pp. 367-372, 2008

[10] M. Keim, N. Tamarapalli, H. Tang, M. Sharma, and J. Rajski, "A Rapid Yield Learning Flow Based on Production Integrated Layout-Aware Diagnosis", International Test Conference, p. 7.1, 2006

[11] I. Yamazaki, H. Yamanaka, T. Ikeda, M. Takakura, and Y. Sato, "An Approach to Improve the Resolution of Defect-Based Diagnosis", Asian Test Symposium, pp. 123-128, 2001

[12] R. Desineni, O. Poku, and R. D. Blanton, "A Logic Diagnosis Methodology for Improved Localization and Extraction of Accurate Defect Behavior", International Test Conference, p. 12.3, 2006

[13] S. Venkataraman and S. B. Drmmonds, "POIROT: A Logic Fault Diagnosis Tool and Its Applications", International Test Conference, p. 9.4, 2000

[14] P. Bastani, N. Callegari, Li-C. Wang, and M. S. Abadir, "Statistical Diagnosis of Unmodeled Systematic Timing Effects", Design Automation Conference, pp. 355-360, 2008

[15] P. Bastani, N. Callegari, Li-C. Wang, and M. S. Abadir, "Diagnosis of design-silicon timing mismatch with feature encoding and importance ranking – the methodology explained", International Test Conference, p. 14.2, 2008

[16] T. Ishida, I. Nitta, K. Homma, Y. Kanazawa, and H. Komatsu, "Speed-path Analysis for Multi-path Failed Latches with Random Variation", International Symposium on Quality Electronic Design, pp. 545-552, 2012

[17] H. Drucker, C. J. C. Burges, L. Kaufman, A. Smola, and V. Vapnik, "Support Vector Regression Machines", Advances in Neural Information Processing Systems 9, pp. 155-161, 1997

[18] Chih-Chung Chang and Chih-Jen Lin, "LIBSVM: a library of support vector machines", ACM Trans. Intel. Syst. Technol. 2, 3, Article 27 (May 2011), 27 pages, 2011

An Overview of Spin-based Integrated Circuits

Wang Kang[1,2], Weisheng Zhao[1*], Zhaohao Wang[1], Jacques-Olivier Klein[1], Yue Zhang[1], Djaafar Chabi[1], Youguang Zhang[2], Dafiné Ravelosona[1], and Claude Chappert[1]

1, Institute d'Electronique Fondamentale (IEF),
Univ. Paris-Sud, CNRS,
Orsay, 91405, France
Tel : (+33)169156292
Fax : (+33)169154000
e-mail : weisheng.zhao@u-psud.fr

2, Electronics and information engineering,
Univ. Beihang
Beijing, 100191, China
Tel : +86 01082314978
Fax : +81 01082339478
e-mail : kanebuaa@gmail.com

Abstract - Conventional CMOS integrated circuits suffer from serve power and scalability challenges as technology node scales into ultra-deep-micron technology nodes. Alternative approaches beyond charge-only based circuits. In particular, spin-based devices or integrated circuits show promising merits to overcome these issues by adding the spin freedom of electrons to the electronic circuits. Spintronics has now become a hot topic in both academics and industrials. This paper overviews the status and prospects of spin-based integrated circuits under intense investigation and address particularly their merits and challenges for practical applications.

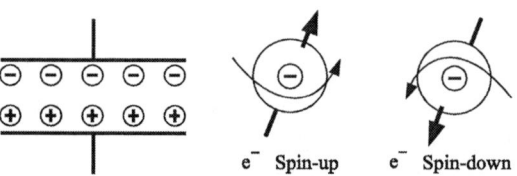

Fig. 1. Comparison between the principles of conventional electronics and Spintronics [7]. (a) Conventional electronics uses only the electrical charges; (b) Spintronics exploits the spin degrees of freedom of electrons.

I Introduction

Thanks to the good characteristics, such as high speed and small size, metal-oxide-semiconductor field-effect transistors (MOSFETs) and complementary MOS (CMOS) devices [1-3] (hereafter, both are referred to as MOS devices) are the fundamental technologies for mainstream integrated circuits. However the rapid and continual technology scaling progress of the MOS devices below 40 nm drives more and more issues and challenges, e.g., intrinsic leakage currents, dynamic power and process variability etc, indicating the end of MOS scaling in the near future [2, 3]. Many new advanced solutions, e.g., silicon on insulator (SOI) [4], have been proposed recently in order to alleviate this dilemma. They achieve one order of enhancement in some aspects, but they cannot overcome all the problems induced in the deep-micron technology nodes. In addition, the production cost including facility investment becomes very huge. Therefore alternative scaling-independent technologies, i.e., spin-based devices and circuits [5-7], for improving the integrated circuit performance have attracted considerable attention to sustain the Moore's Law beyond the MOS scaling limit [2, 3]. Many research groups, including academia and industries are undergoing for this purpose.

Spintronics (or spin-electronics) is a rapidly emerging R&D area and it shows great promise in the future integrated circuits (both memory and logic computing) [5-7]. The basic concept of Spintronics is to control the spin property (besides charge) of electrons, as shown in Fig. 1, in ferromagnetic thin films based solid state nano-devices, such as spin-valve and magnetic tunnel junction (MTJ). It provides new concepts for future electronics, namely spin-based integrated circuits. The founding step of Spintronics triggered the discovery of the giant magnetoresistance (GMR) in 1988 (awarded Nobel Prize Physics 2007 for A. Fert and P. A. Grunberg) [8, 9], and soon

triggered the definition of the spin-valve [10]. The spin-valve sensor was firstly commercialized by IBM in 1997 to replace the anisotropic magnetoresistance (AMR) sensor for hard disk drive (HDD) read heads and drove immediately the storage areal density growth rate of 100% per year. Another big step forward for Spintronics came from replacing the metallic spacer layer of the spin-valve with a thin non-magnetic insulating oxide layer (e.g., Al_2O_3 or MgO), thus creating the magnetic tunnel junction (MTJ) [11-12]. In such configuration, much higher resistance difference, denoted as tunnel magnetoresistance (TMR) ratio, can be obtained at room temperature. The discovery of MTJ triggered intensive research and indicated a rapid development era on spin-based integrated circuits, such as magnetic random access memory (MRAM) and spin-transistors [5-7]. This paper overviews the status and prospects of the spin-based integrated circuits (including memory and logic circuits), and focuses mainly on the period after the discovery of the MTJ.

The rest of this paper is organized as follows: section II introduces the fundamentals of Spintronics. The spin-based memory and logic designs are described in section III and section IV respectively. Finally conclusions and perspectives are discussed in section V.

II. Fundamental of Spintronics

Spintronics is an emerging technology exploiting both the intrinsic spin degree of freedom and the magnetic moment of the electron, in addition to its fundamental electrical charge, in solid state devices. The origins of Spintronics can be traced back to the 1970s [13] and the research of Spintronics widely

978-1-4799-2817-0/14 $31.00 © 2014 IEEE

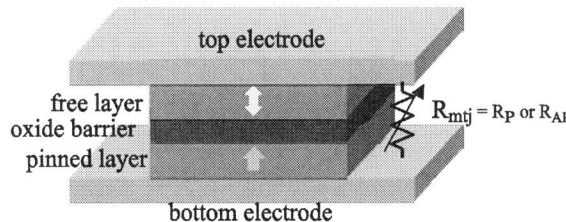

Fig. 2. Schematic structure of PMA magnetic tunnel junction (MTJ) nanopillar: It is composed of an oxide barrier layer sandwiched between two ferromagnetic (FM) layers. According to the different configuration (Parallel or Anti-parallel) of the two FM layers, the MTJ shows low or high (R_P or R_{AP}) resistance property and the resistance difference value is denoted by TMR=(R_{AP}-R_P)/R_P [5].

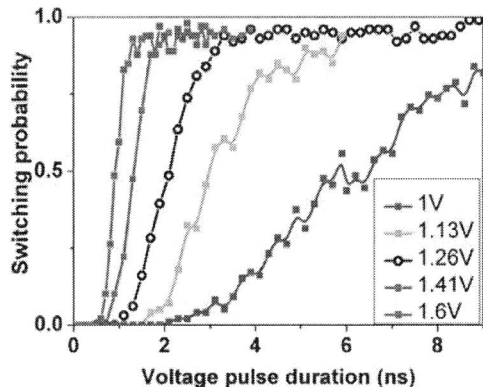

Fig. 3. Experimental measurement of STT stochastic switching behaviours [18].

emerged from the discovery of spin-dependent electron transport phenomena in solid-state devices in the 1980s [8, 9]. The discovery of spin valve or GMR in 1988 opens a new era in Spintronics, and later in 1995, the discovery of MTJ or TMR boosts its intensive research and applications [11, 12]. The MTJ nanopillar, as shown in Fig. 2, is one of the most important devices of current spin-based integrated circuits. Both perpendicular magnetic anisotropy (PMA) and in-plane shape anisotropy MTJs can be designed with different thin film composition.

The MTJ states, i.e., R_P or R_{AP}, can be switched by applying either a magnetic field or a spin polarized current. However conventional field-induced magnetic switching (FIMS) approaches [14] require writing current in the order of few mA to generate the required magnetic field and induce consequently high power consumption and hardware area. Much of academic and industrial research efforts have been focused on developing efficient strategies for switching the magnetization of the MTJ. One promising method relies on the spin transfer torque (STT), which was firstly proposed independently by Berger [15] and Slonczewski [16] in 1996, shows good performance. Only a small bidirectional current is required for the MTJ switching, which simplifies greatly the integration with the peripheral MOS circuits and therefore allows much higher density. Unfortunately according to the experimental measurements and theoretical models, the STT switching mechanism is normally stochastic (see Fig. 3 and Eqs. (2)-(5)) [17-19], thus resulting in reliability issues, which should be addressed in the practical spin-based applications.

Depending on the relative magnitude between the write current I_{write} and the critical current I_{C0} (see Eq. (1)), the STT switching behaviour of the MTJ can be categorized into two regions: precessional switching region ($I_{write} > I_{C0}$) and thermal activation region ($I_{write} < I_{C0}$).

$$I_{C0} = \alpha \frac{\gamma e}{\mu_B g}(\mu_0 M_S)H_K V = 2\alpha \frac{\gamma e}{\mu_B g}E \quad (1)$$

where α is Gilbert damping coefficient, γ the Gyro-magnetic constant, e the magnitude of the electron charge, μ_B the Bohr magneton constant, g the spin polarization efficiency factor, $\mu_0 M_s$ is the saturation field in the free layer, H_K the anisotropy field, V the volume of the free layer and $E = (\mu_0 M_S)H_K V/2$ is the barrier energy.

When the current flowing through the MTJ exceeds the critical current I_{C0}, the MTJ experiences a fast precessional switching regime, and the switching duration and probability can be expressed as,

$$\Pr(t_{pulse}) = 1 - \exp(-\frac{t_{pulse}}{\tau_1}) \quad (2)$$

$$\frac{1}{\tau_1} = \left[\frac{2}{C+ln(\pi^2\Delta)}\right]\frac{\mu_B P}{em(1+P^2)}(I_{write} - I_{C0}) \quad (3)$$

where t_{pulse} is the driver current pulse duration, τ_1 is the mean duration for precessional switching regime, $C = 0.577$ is the Euler's constant, $\Delta = E/k_B T$ is the thermal stability factor, k_B the Boltzmann constant, T the temperature, m the free layer magnetic moment, P the tunneling spin polarization of the ferromagnetic layers. Otherwise, if the current is lower than I_{C0}, the switching process can still occur with a long pulse thanks to the thermal activation. The switching probability and pulse duration for the thermal activation regime can be expressed as,

$$\frac{d\Pr(t_{pulse})}{(1-\Pr(t_{pulse}))\,dt} = \frac{1}{\tau_2} \quad (4)$$

$$\tau_2 = \tau_0 exp(\frac{E}{k_B T}(1 - \frac{I_{write}}{I_{C0}})) \quad (5)$$

where τ_0 is the attempt period, τ_2 the mean pulse duration for the thermal activation regime. In the practical spin-based integrated circuits, we generally require the MTJ operates at precessional switching regime and choose short pulse current with amplitude larger than I_{C0} to get high speed operations.

The MTJ resistance states can be sensed with either a bias voltage V_{bias} or a directional current I_{read} due to the TMR effect. Generally a relative larger V_{bias} or I_{read} is required to improve the sense margin (SM). However it is worth noting that the sensed current or I_{read} should be sufficiently less than I_{C0} to avoid any read disturbance (RD) (see Eq. (6)) for the STT-based MTJ [19-21]. In addition, larger V_{bias} may lead to TMR loss (see Eq. (7)) [21].

$$\Pr_{dis}(t_{read}) = 1 - \exp(-\frac{t_{read}}{\tau_0}\exp(-\Delta(1 - \frac{I_{read}}{I_{C0}}))) \quad (6)$$

$$TMR_{real} = \frac{TMR(0)}{1+V_{bias}^2/V_h^2} \quad (7)$$

where I_{read} and t_{read} are the amplitude and accumulated duration of the sensing current, TMR_{real} is the TMR ratio after adding bias voltage, $TMR(0)$ is the TMR ratio with 0V

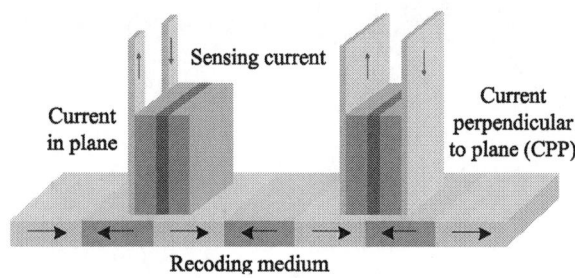

Fig. 4. The spin-valve read head for hard-disk recording. (Left) The schematic configuration of the 'current in plane' geometry; (Right) The 'current perpendicular to plane (CPP)' geometry [5].

voltage, V_h is the bias voltage as $\mathrm{TMR_{real}} = 0.5\,\mathrm{TMR}(0)$.

The STT-based MTJ has been widely used and become the major candidate device, at least in the short term, in the spin-based integrated circuits. Many STT-MTJ based memory and logic circuits or prototypes have been presented in the last years, such as STT-MRAM and Magnetic Flip-Flop (MFF) etc. They show many great merits, such as high speed, low power, non-volatility and instant on/off capability, especially the good compatibility and scalability performance to continue the Moore's scaling law.

III. Spin-based memory circuits

There are different methods to categorize these spin-based memory circuits. In this section, we will overview and classify them from the device structure point of view.

A. Spin-based HDD read head

Spin-valve sensor for HDDs [5], as shown in Fig. 4, utilized in 1997 by IBM, is the first commercialized application of Spintronics, which provides a sensitive and scalable read technique for HDD and brought huge economic interests. Both geometries, i.e., current in plane and current perpendicular to plane (CPP) were proposed, but the CPP geometry is much better for integration and achieving minimum dimension. In addition, the CPP configuration is also more favorable for reducing the recoding track width. The spin-valve sensor increased the HDD areal recording density by three orders of magnitude (from ~0.1 to ~100 Gbit in^{-2}) between 1997 and 2003. However, this growth rate started to slow down after 2003, when other problems joined the limiting spin-valve head. The MTJ read head, with the same structure as spin-valve shown in Fig. 4, except replacing the metallic spacer layer M with a thin non-magnetic insulating layer (e.g., Al_2O_3 or MgO), was then commercialized in 2004 by Seagate. The MTJ read head promises to achieve over 200 Gbit in^{-2} recording density and 1 Gbit/second data rate.

B. Magnetoresistive Random Access Memory (MRAM)

MRAM is mainly based on the hybrid structure, i.e., the magnetic memory elements (generally referred to the MTJ) integrated with the MOS devices. There exist two basic array architectures, i.e., one-transistor with one MTJ (1T-1MTJ) and cross-point architectures [5, 22-23]. In the 1T-1MTJ array

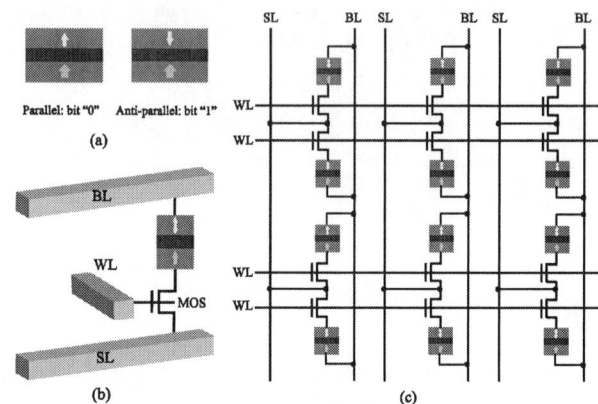

Fig. 5. Schematic of MRAM structure [5]. (a) Data representation with different MTJ configurations; (b) Schematic of the 1T-1MTJ cell structure; (c) Schematic of the 1T-1MTJ array architecture.

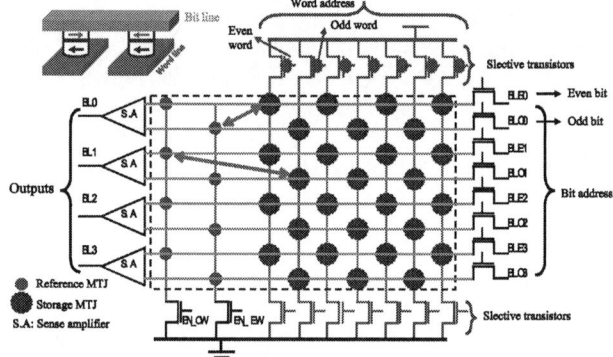

Fig. 6. Schematic of the cross-point architecture for MRAM [22].

architecture, as shown in Fig. 5, each MTJ is connected in series with an MOS transistor, where the gate of the transistor is connected to the word-line (WL), drain to the bit-line (BL) crossing the MTJ and source to the source-line (SL). It is convenient for selecting the cell for the write/read operations. However the density of the 1T-1MTJ array architecture is less than the cross-point architecture due to the transistor required for each memory cell. In the cross-point array architecture, the MTJs lie at the intersection of the WLs and BLs, as shown in Fig. 6, where the pinned layer and free layer of the MTJ are connected directly to the BLs and WLs respectively. This arrangement allows for a considerable packing density, since no contact is made to the silicon die within the cell. However it involves several significant design challenges, such as low data access speed and sneak currents, leading to poor write/read performance.

The read operation of the MRAM utilizes a bias voltage on the BL to measure the effective current, or applies a current flowing through the cell to measure the voltage. There are three basic decision-making methods to determine the data value stored in the memory cell: the reference-cell method, complementary-cells comparison method and self-referenced method [23-26]. In the reference-cell method, each sensed value of the memory cell is compared to a reference cell, with resistance value $R_{ref} = (R_P + R_{AP})/2$. This method should be designed with consideration of the process variations, especially as technology scales down to the ultra-deep-micron

978-1-4799-2817-0/14 $31.00 © 2014 IEEE

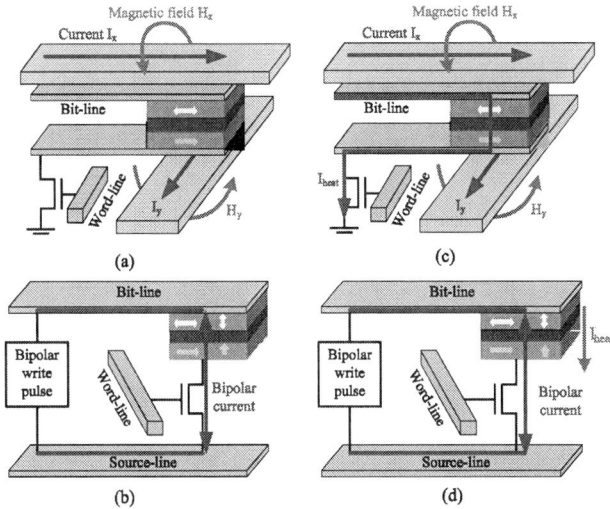

Fig. 7. 1T-1MTJ cell structures of various MRAM families. (a) Magnetic field-only FIMS-MRAM; (b) STT-MRAM; (c) TAS FIMS-MRAM; (d) TAS STT-MRAM.

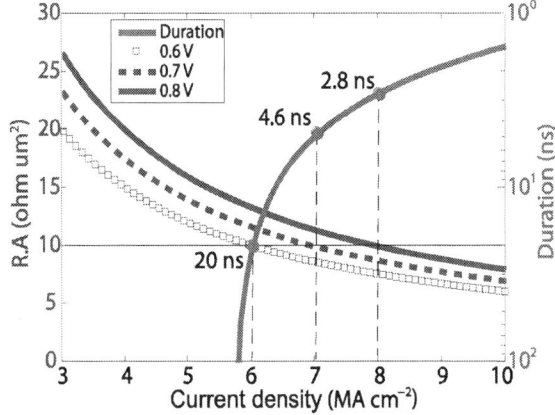

Fig. 8. To obtain high I_{write} for fast speed, either high V_{bias} or low R.A is required, leading to poor endurance of the MTJ [21].

technology nodes. The second method uses two MTJs, which are always written to opposite states, to store one bit of data. This method can improve greatly the sense margin (SM) compared with the first one, but with lower storage density. The self-referenced method accesses the memory cell twice, and makes the decision based on a known value written into the cell at the second time or based on the physical characteristics of the MTJ R-I curves. This method requires no hardware for the reference cells, and it is insensitive to process variations. However the repeated sense cycles result in considerable read latency and power consumption. In addition, it also leads to endurance reliability issues.

The write operation of the MRAM is different based on the witching mechanism of the MTJ, and we can classify the MRAM family into several categories as following (see also Fig. 7): (a) Filed induced magnetic switching (FIMS) MRAM; (b) spin transfer torque (STT) MRAM; (c) Thermally-assisted switching (TAS) MRAM.

FIMS-MRAM

FIMS-MRAM [14] is the first generation of MRAM that based on the field-only writing mechanism, as shown in Fig. 7 (a), where the magnetic fields are generated by two orthogonal current lines. In such approach, the write selectivity is based on the combination of two perpendicular pulses of magnetic fields, i.e., H_x and H_y, which may result in narrow write margin and half-selectivity issues. In addition, high power consumption (magnetic fields requiring write currents of about 7-10 mA) and poor scalability (selectivity problems caused by magnetic field dispersion) limit its commercialization. Later in 2003, the selectivity problem is solved by introducing a new write method, named toggle-switching [27], which is based on the use of a synthetic ferrimagnetic (SF) free layer and is commercialized in 2006 by Everspin. However this method needs to read before write and it also consumes high write power. Most importantly, the general downscale scalability problem (cell size about 30 F^2) with field writing mechanism cannot be well overcame.

STT-MRAM

STT-MRAM [16-17, 28], as shown in Fig. 7 (b), uses a bi-directional low current (~1-2mA/cm^2) to switch the MTJ, and this simplifies greatly the integration with the peripheral MOS circuits, thus achieving very good scalability (down to 10 nm diameter) and high integrated density (with minimum cell size of 6 F^2). The key challenge for STT-MRAM is to achieve low write current, high speed, good endurance as well as long retention simultaneously, because (a) both I_{C0} and the thermal stability factor Δ is proportional to the barrier energy E (see Eq. (1) and (3)), there exists a conflict between low current and long retention; (b) to get high speed (i.e., short write duration t_{pulse}), a high write current I_{write} is required (see Eq. (3) and (5)), therefore a high bias voltage V_{bias} or low resistance.area product (R.A) is needed to achieve this purpose, as shown in Fig. 8. However, both the two solutions lead to the oxide barrier breakdown and thus poor endurance of the MTJ.

TAS-MRAM

To solve the dilemma between write performance and retention, and meanwhile to continue the MRAM downsize scalability, a new concept, named thermally assisted switching (TAS) was proposed [29, 30]. As shown in Fig. 7 (c) and (d), this approach uses firstly a current pulse flowing through the cell to heat temporarily the free layer of the MTJ above its magnetic ordering temperature by Joule dissipation, greatly reducing the required magnetic fields or STT current for the MTJ switching. Then a magnetic field or current is applied for the magnetic switching of the MTJ. Finally the cell is rapidly cooled down to the room temperature and the magnetization of the MTJ subsequently remains frozen in the new direction. In such technique, the switching field or current can be fixed whatever the size of the MTJ, allowing a largest integration density and resolving the selectivity problem (only the addressed MTJ is heated). Moreover, the MTJ was written at elevated temperature and stored/read at room temperature, ensuring a high thermal stability and long data retention.

MRAM is an emerging non-volatile random access memory technology under development since the 1990s and is still under intensive research. Many prototypes or chips have been proposed or commercialized in markets currently. We believe that MRAM will eventually become dominant for all types of

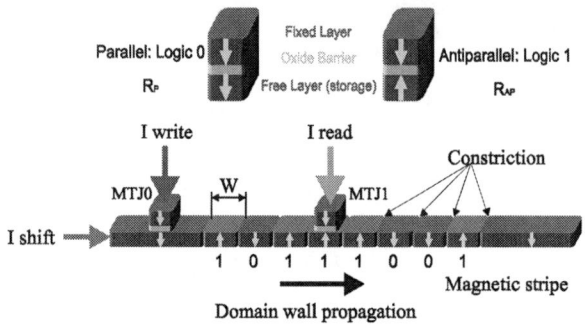

Fig. 9. Schematic of domain-wall (DW) racetrack memory (RM) based on current induced DW motion.

TABLE I
Comparison of Various MRAM Technologies

Technology	FIMS MRAM	STT MRAM	TAS-FIMS MRAM	TAS-STT MRAM
Scalability	Poor (>60 nm)	Very good (<10 nm)	Good (>40 nm)	The best (<8 nm)
Min cell size	Large (~30 F^2)	Very small (~6 F^2)	Small (~10 F^2)	The best (~4 F^2)
Endurance	10^{16}	10^{16}	10^{12}	10^{12}
Writability	Poor	Very good	Good	The best
Power	Very high	Low	High	The best
Latency	Very long (>20 ns)	Short (<10 ns)	Long (>10 ns)	The best (<8 ns)

memory due to its overwhelming merits, becoming a universal memory one day. In summary, we compare the performance of various MRAM families, as shown in Table I.

C. Domain-wall (DW) racetrack memory

The observation of the current-induced domain wall (DW) motion in magnetic nanowires triggered the 3D storage devices, i.e., racetrack memory (RM) [31-32], as shown in Fig. 9, which promises to offer ultra-high storage density by storing the data in a U-shaped nanowire normal to the plane of the substrate. Combining with the MTJ nanopillars as the read and write heads, good CMOS integrability and fast data access can be achieved. The first RM prototype fabricated on 90 nm node has been presented recently; however, it is based on the in-plane magnetic anisotropy in NiFe nanowires with intrinsic low energy barrier, leading to insufficient data retention. Recent progress demonstrates that the perpendicular magnetic anisotropy (PMA) materials (e.g., CoFeB) can further improve the storage density, access speed and power consumption of the RM. Nevertheless many challenges should be addressed before this device can be used in industry. One of the key challenges to build RM is to avoid any pinning defects in the magnetic strips, because even one single pinning defect may prevent the DW motion of the whole track.

D. Advanced spin-based memories

With the continual academic and industrial progresses on Spintronics, some new advanced spin-based memory concepts, such as voltage-controlled (VC) MRAM [33], multi-level-cell (MLC) MRAM [34] and multi-terminal structures [35, 36] etc, are proposed recently. The VC-MRAM uses the electrical filed

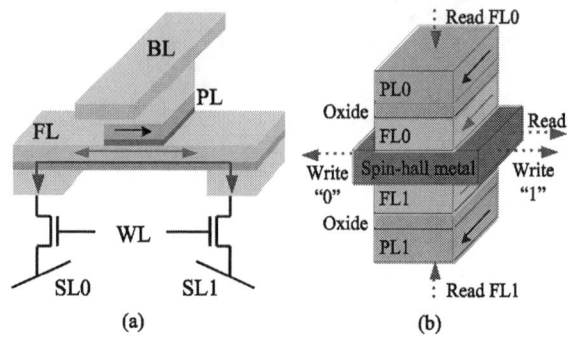

Fig. 10. Schematic of the multi-terminal cell structures based on the spin-orbit torque effect. (a) 3-terminal cell structure; (b) 4-terminal differential spin-hall cell structure.

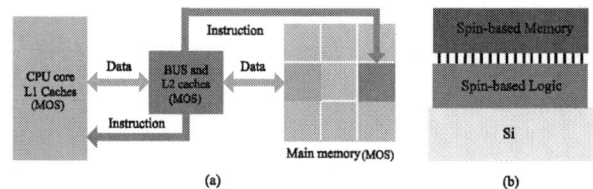

Fig. 11. Schematic of logic computing architectures. (a) Classic Von-Neumann architecture; (b) The 3D integration of spin-based logic architecture.

through a voltage to assist or accomplish the switching of the MTJ, resulting in lower power consumption. MLC-MRAM formed by modifying the free layer structure of the MTJ or by stacking multiple MTJs in parallel or series, can store multiple bits in one memory cell, thus doubling or tripling the storage density. The multi-terminal cell structures based on spin-orbit torque effect: spin-hall or Rashba, as shown in Fig. 10, can indeed provide some aspects of advantages, e.g., improved endurance by reducing the electrical stress on the oxide barrier, but they also induce some drawbacks, e.g., larger cell size due to the two access MOS transistors. In summary, all these advanced concepts are still far away to be considered as promising approaches for practical applications due to their maturities and technological feasibilities.

IV. Spin-based logic circuits

Classic computing model of microprocessor is based on the Von-Neumann architecture [37], as shown in Fig. 11 (a), which consumes both high static power (due to the volatile cache memories based on MOS devices) and dynamic power (due to the data traffic between the CPU core and the main memories), limiting its further downsize scalability. For instance, the power to access the memory for fetching the instructions and reading/writing the data (e.g. ~1 pJ/bit/mm) is much higher than that for performing logic operations (e.g. ~ 1 fJ at 22 nm node). Therefore the spin-based logic devices and circuits, such as hybrid MTJ/CMOS logic structures, all-spin logic devices and spin-based transistors [38-41] etc, are under intensive investigation. They are expected to bring the non-volatility into the MOS circuits and then allow them to achieve instant on/off operations (i.e., powered off when unused and retrieved instantly on active state), reducing

978-1-4799-2817-0/14 $31.00 © 2014 IEEE   680

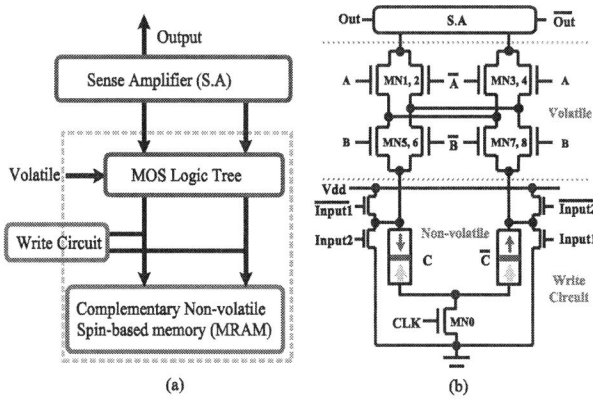

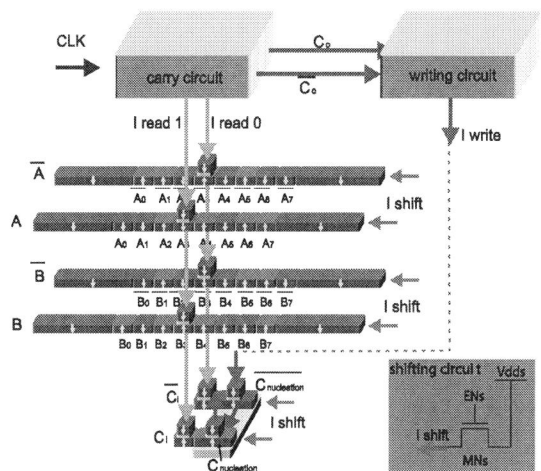

Fig. 12. Schematic of the hybrid MTJ/CMOS logic circuits. (a) Logic-in-memory structure; (b) An example of "XOR" gate.

Fig. 13. Schematic of a magnetic full adder (MFA) based on the current induced domain wall motion in magnetic nanowires.

greatly the static power. In addition, thanks to the vertical structure of the spin-based storage devices (e.g., MTJ), they can be fabricated above the MOS circuits at the back-end process. This 3D integration structure [7], as shown in Fig. 11 (b), shortens greatly the data traffic distance between the memory and logic chips, thus accelerating greatly the logic computing speed and saving significantly the dynamic power. In the following, we will overview several spin-based logic devices and structures that draw much attention currently.

A. Hybrid MTJ/CMOS logic circuits

The hybrid MTJ/CMOS logic circuits are mainly based on the logic-in-memory structure [38, 39], as shown in Fig. 12. It is mainly composed of four parts: a sense amplifier (S.A) to evaluate the logic result, a non-volatile memory block (e.g., MRAM), a write circuit, and a volatile MOS logic block. This type of spin-based logic circuits is the most popular one currently and draws considerable attention due to its instant on/off capability, zero standby power, good compatibility with conventional computing architectures and easy integration with the existing MOS technology process. Many hybrid MTJ/CMOS logic circuits or prototypes have been presented in the past few years, such as magnetic flip-flop (MFF), magnetic full adder (MFA) and magnetic look-up table (MLUT) etc. However they also suffer from some challenges which should be addressed, e.g., the switching latency (several ns) of the MTJ is much larger than those of the conventional MOS logic circuits, which limits the computing frequency to the order of GHz. Another severe issue is the poor sense reliability caused mainly by the device mismatch (both MOS and MTJs devices) of the S.As and the intrinsic stochastic switching effects of the MTJs. Different from the memory circuits where complex error correction circuits (ECCs) can be employed [42], it is difficult to embed them in the logic circuits while keeping fast speed, low area and high power efficiency. Therefore the current efforts that concentrate on this topic are fast-access MTJ development, high-performance S.A design, low-cost and reliable integration process etc.

B. Domain wall based logic circuits

Besides memory circuits, domain wall (DW) motion in magnetic nanowires provides also the ability of logic designs.

One of the demonstrations is to use geometry dependence of the DW motion to perform logic computation [43], which uses no MOS transistors and exhibits ultra-low power consumption. However, these circuits utilize magnetic field to drive the DW movement and have some critical shortcomings, such as low speed (<100 KHz), magnetic field dissipation, scalability and reliability issues etc, for practical applications. The current induced magnetic DW motion becomes an effective solution to overcome these issues [31-32]. An alternative design is based on the DW motion racetrack memory (RM), as shown in Fig. 13 is an example of a MFA circuit [44]. It employs also the general logic-in-memory structure to perform the logic computing operations. Unlike the hybrid MTJ/CMOS logic structure, where parts of the input data are volatile provided by the MOS circuits (see Fig. 12), in the DW motion based logic circuits, all the input and output data are stored in the non-volatile RMs, overcoming completely the standby power issue and achieving a real non-volatile logic circuit. However, all the data inputs are stored in non-volatile state, which is difficult to be realized with a reasonable area and latency overheads. In addition, it faces the same challenges (e.g., pinning defects) as introduced in the RM circuits.

C. All-spin logics

All spin logic devices (ASLD) [40], as shown in Fig. 14 (a), which employs nano-magnets as digital spin capacitors to store data information and spin currents (through spin transfer torque) to communicate, realizing logic gates based on the spin majority evaluation. As shown in Fig. 14 (b) is an example to demonstrate the possible layouts for constructing cascadable ASLD logic gates. The magnetization directions of the nano-magnets can be switched between the stable states if enough torque is exerted on them. Information stored in the input magnet is used to generate a spin current that can be routed along a spin-coherent channel to the output magnet, determining its state based on the spin transfer torque effect. The key feature of ASLD is its compactness and completeness, because no MOS transistor is needed for the logic operations and all the logic functions can be constructed with a minimal set of Boolean logic gates. With such design, a full spin

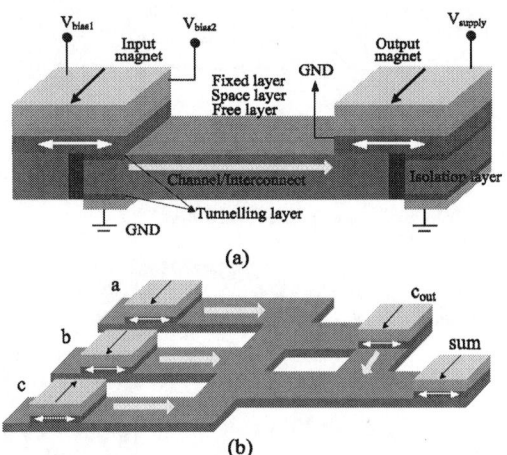

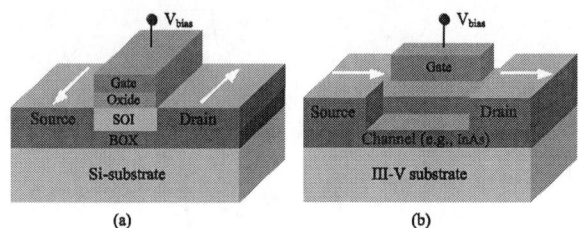

Fig. 15. Schematic device structures of the spin transistors. (a) spin-MOSFET device; (b) spin-FET device [41].

Fig. 14. (a) Schematic of all-spin logic device (ASLD) structures. (b) An example of full adder layout based on ASLDs.

E. Other spin-based logic devices and circuits

There are many other spin-based logic devices and circuits, such as spin-valve logic gates and graphene-based transistors etc [48-52]. Among them, graphene (awarded Nobel Prize in Physics in 2010 [51]) has been widely studied recently due to its ultra-high charge carrier mobility and long spin diffusion lengths, allowing it to be used as a promising channel material in the spin transistor. In 2008, the smallest graphene-based transistor was reported [50] to be successfully fabricated with only one atom thick and 10 atoms wide, and later in 2011, IBM announced that they had succeeded in creating the first graphene-based integrated circuit [52]. Graphene has proved generally the potentiality and capability of replacing silicon as a mainstream semiconductor. However many issues and challenges should be addressed before it can be widely used in industries. One of the key challenge is that single sheet of graphene is very hard to produce, and even harder to make on top of an appropriate substrate.

computing system can be expected with extremely low switching power. However this is still a theoretical prospect currently and many issues, such as reliability and clock control, are remaining unresolved. The most critical challenge for this ASLD is the "magic" material with strong spin orbit interaction and there are few experimental prototypes can produce successfully the theoretical results so far.

D. Spin-transistors

The concept of spin-transistor has been predicted early in the 1990s [45], but it was experimentally developed recently thanks to the rapid progress of ferromagnetic study on spin orbit interaction, and then a wide variety of spin transistors based on various operating principles have been proposed so far [46-47]. Spin-transistors, including "spin-MOSFET" and "spin-FET" devices, benefit from the similar structure as MOS transistors, as shown in Fig. 15. In the spin-FET and its related devices, the source and drain have the same spin alignment, and the on/off switching operation can be achieved by spin precession of the spin-polarized carriers injected in the channel through spin-orbit interaction. Here the spin-orbit interaction is controlled by the gate voltage. It is worth noting that particular channel materials with strong spin-orbit coupling, such as InGaAs, InAs and other III-V compounds, are required to sufficiently induce the spin-orbit interaction. However, the channel region of the spin-MOSFET requires a material with low spin-orbit coupling, since spin-MOSFET requires no spin precession of spin-polarized electrons in the channel. In the spin-MOSFET devices, the alignment of the drain magnetization is fixed, while that of the source can be changed, so the gate allows current to flow from the source to the drain without modulation. In contrast to the spin-FETs, the cutoff state of the spin-MOSFET is simply achieved by a gate bias condition in the same manner as an ordinary MOS transistor. In both type of devices, the spin is injected from the ferromagnetic source, and then transported through the channel to the drain and electrons with spin aligned with the drain are passed and generate current. Spin transistors provide a potential building element for novel integrated circuits and open a promising path for achieving real all-spin based integrated circuits.

V. Summary and Conclusions

In this invited paper, we presented a general overview of the spin-based integrated circuits and addressed particularly its using as memories and logics by adding the spin freedom of electrons. The rapid progress in both physics and electronics makes them promising to overcome the power and scalability bottlenecks of conventional MOS-based integrated circuits, sustaining the Moore's Law beyond the MOS scaling limit. Recently many prototypes or small-scale products have been successfully presented or commercialized, however many challenges should be relieved before they can be widely used in practical applications. For this purpose, many efforts have been done or are undergoing in both academics and industries. We believe that spin-based integrated circuits will become mainstream solution to build up the next-generation storage and computing systems.

Acknowledgments

This work was supported by ANR-DIPMEM, ANR-MARS, NVCPU projects and the CSC exchange program.

References

[1] S. Kang, and Y. Leblebici, *CMOS digital integrated circuits*, Third Edition, McGrawHill publisher, 2002.

[2] International technology roadmap for semiconductor (ITRS), 2010. http://www.itrs.net.

978-1-4799-2817-0/14 $31.00 © 2014 IEEE

[3] S. E. Thompson, S. Parthasarathy, "Moore's law: the future of Si microelectronics," *Mater. Today*, Vol. 9 No. 6, pp. 20-25, 2006.

[4] A. Marshall, and S. Natarajan, *SOI design: analog, memory and digital techniques*, Springer, 2002.

[5] C. Chappert, A. Fert and F. Dau, "The emergence of spin electronics in data storage", *Nature Mater.*, Vol. 6, pp. 813-823, 2007.

[6] I. Žutić, J. Fabian, and S. D. Sarma, "Spintronics: fundamentals and applications," Rev. Mod. Phys., Vol. 76, PP. 323-410, 2004.

[7] W. S. Zhao, et al, "Spin-electronics based logic fabrics," in *Proc. IEEE VLSI-SOC*, 2013, in press.

[8] M. N. Baibich, et al, "Giant magnetoresistance of (001)Fe/(001)Cr magnetic superlattices," *Phys. Rev. Lett.* Vol. 61, pp. 2472–2475, 1988.

[9] G. Binasch, et al, "Enhanced magnetoresistance in layered magnetic structures with antiferromagnetic interlayer exchange," *Phys. Rev. B*, Vol. 39, PP. 4828-4830, 1989.

[10] B. Dieny, et al. Giant magnetoresistance in soft ferromagnetic multilayers. *Phys. Rev. B*, Vol. 43, pp. 1297-1300, 1991

[11] J. S. Moodera, L. R. Kinder, T. M. Wong, and R. Meservey, "Large magnetoresistance at room temperature in ferromagnetic thin film tunnel junctions," *Phys. Rev. Lett*, Vol. 74, PP. 3273-3276, 1995.

[12] S. S. P. Parkin, et al, "Giant tunnelling magnetoresistance at room temperature with MgO (100) tunnel barriers," *Nature Mater*, Vol. 3, PP. 862–867, 2004.

[13] M. Julliere, "Tunneling between ferromagnetic films," *Phys. Lett. A*, Vol. 54, pp. 225-201, 1975.

[14] B. N. Engel et al., "A 4-Mb toggle MRAM based on a novel bit and switching method," *IEEE Trans. Magn.* Vol. 41, pp. 132-136, 2005.

[15] L. Berger, "Emission of spin waves by a magnetic multilayer traversed by a current," *Phys. Rev. B.*, Vol. 54, PP. 9533, 1996.

[16] J. C. Slonczewski, "Current-driven excitation of magnetic multi-layers", *J. Magn. Magn. Mater.*, Vol. 159, pp. L1–L7. 1996

[17] D. C. Worledge, et al., "Spin torque switching of perpendicular Ta|CoFeB|MgO based magnetic tunnel junctions," *Appl. Phys. Lett.*, Vol. 98, no. 2, pp. 022501, 2011.

[18] Y. Lakys, et al., "Self-enabled "error-eree" switching circuit for spin transfer torque MRAM and logic," *IEEE Trans. Magn.*, Vol. 48, no. 9, pp.2403-2406, Sept. 2012.

[19] L. Faber et al., "Dynamic compact model of spin-transfer torque based magnetic tunnel junction (MTJ)," in *Proc. IEEE DTIS*, pp. 130–135, 2009.

[20] W. Kang, et al. "High reliability sensing circuit for deep submicron spin transfer torque magnetic random access memory," *Electron. Lett.*, Vol. 49, PP. 1283-1285, 2013.

[21] W. S. Zhao, et al, "Failure and reliability analysis of STT-MRAM," *Microelectron. Reliab.*, Vol. 52, pp. 1848-1852, 2012.

[22] W. Zhao, et al, "Cross-point architecture for spin transfer torque magnetic random access memory," *IEEE Trans. Nanotech.*, Vol. 11, pp. 907-917, 2012.

[23] T. M. Maffitt, et al, "Design considerations for MRAM," *IBM J. Res. Devel.*, Vol. 50, pp. 25-39, 2006.

[24] W. Zhao, C. Chappert, V. Javerliac and J. P. Noziere, "High stability and low power sensing amplifier for MTJ/CMOS hybrid logic circuits," *IEEE Trans. Magn.*, Vol. 45, no. 10, pp. 3784-3787, 2009.

[25] Z. Sun, H. Li, Y. Chen, and X. Wang, "Voltage driven nondestructive self-reference sensing scheme of spin-transfer torque memory," *IEEE Trans. VLSI. Syst.*, Vol. 20, pp. 2020-2030, 2012.

[26] G. Jeong, et al, "A 0.24-m 2.0-V 1T1MTJ 16-kb nonvolatile magneto-resistance RAM with self-reference sensing scheme," *IEEE J. Solid-State Circuits*, Vol. 38, no. 11, pp. 1906–1910, 2003.

[27] L. Savtchenko, B. N. Engel, N. D. Rizzo, M. DeHerrera, and J. Janesky, "Method of writing to scalable magnetoresistance random access memory element," *U. S Patent 6545906*, 2003.

[28] Y. Huai, and P. P.Nguyen, "Magnetic element utilizing spin transfer and an MRAM device using the magnetic element," *U.S. Patent 6920063*, 2005.

[29] I. L. Prejbeanu, et al., "Thermally assisted MRAM," *Jour. Phys. Conden. Matter*, Vol. 19, PP. 165218, 2007.

[30] Z. Li, and S. Zhang, "Thermally assisted magnetization reversal in the presence of a spin-transfer torque," *Phys. Rev. B*, Vol. 69, pp. 134416, 2004.

[31] S. S. P. Parkin, M. Hayashi, and L. Thomas, "Magnetic domain-wall racetrack memory," *Science*, Vol. 320, no. 5873, pp. 190-194, 2008.

[32] Y. Zhang, et al., "Perpendicular-magnetic-anisotropy CoFeB racetrack memory," *J. Appl. Phys.*, Vol. 111, pp. 093925, 2012.

[33] N. Lei, et al., "Strain-controlled magnetic domain wall propagation in hybrid piezoelectric/ferromagnetic structures," *Nature Commun.*, Vol. 4, pp. 1378, 2013.

[34] M. Aoki, et al., "Novel highly scalable multi-level cell for STT-MRAM with stacked perpendicular MTJs," in *Symp. VLSIT*, pp. 134-135, 2013.

[35] I. M. Miron, et al., "Perpendicular switching of a single ferromagnetic layer induced by in-plane current injection," *Nature*, Vol. 476, pp. 189-193, 2011.

[36] Y. Kim, S. H. Choday, and K. Roy, "DSTT-MRAM: differential spin hall MRAM for on-chip memories," *arXiv:1305.4085v1*, 2013.

[37] G. J. Myers. *Advances in computer architecture*. John Wiley and Sons, 1982.

[38] H. Dery, and P. Dalal, "Spin-based logic in semiconductors for reconfigurable large-scale circuits," *Nature*, Vol. 447, PP. 573-576, 2007.

[39] W. Zhao, et al., "High performance SoC design using magnetic logic and memory," in *VLSI-SoC: Advanced Res. Syst. Chip*, Vol. 379, pp. 10-33, 2012.

[40] B. Behin-Aein, D. Datta, S. Salahuddin, and S. Datta, "Proposal for an all-spin logic device with built-in memory," *Nature nanotech*, Vol. 5, pp. 266-270, 2010.

[41] S. Sugahara, J. Nitta, "Spin-transistor electronics: an overview and outlook," in *Proc. IEEE*, Vol. 98, pp. 2124-2154, 2010.

[42] W. Kang, et al, "A low-cost built-in error correction circuit design for STT-MRAM reliability improvement," *Microelectron. Reliab*, Vol. 53, pp. 1224-1229, 2013.

[43] D. A. Allwood, et al., "Magnetic domain-wall logic," *Science*, Vol. 309, pp. 1688-1692, 2005.

[44] H. P. Trinh, et al., "Domain wall motion based magnetic adder," *Electron. lett.*, Vol. 48, pp. 1049-1051, 2012.

[45] S. Datta, and B. Das, "Electronic analog of the electrooptic modulator," *Appl. Phys. Lett.*, Vol. 56, pp. 665–667, 1990.

[46] J. Schliemann, J. C. Egues, and D. Loss, "Nonballistic spin field effect transistor," *Phys. Rev. Lett.*, Vol. 90, pp. 146801, 2003.

[47] S. Sugahara and M. Tanaka, "spin metal-oxide-semiconductor field effect transistor using half-metallic-ferromagnet contacts for the source and drain," *Appl. Phys. Lett.*, Vol. 84, pp. 2307–2309, 2004.

[48] T. Schneider et al., "Realization of spin-wave logic gates", *Appl. Phys. Lett*, Vol.92, 022505, 2008.

[49] B. Dlubak et al., "Highly efficient spin transport in epitaxial graphene on SiC," *Nature Phys.*, Vol.8, pp.557-561, 2012.

[50] L. A. Ponomarenko, et al., "Chaotic dirac billiard in graphene quantum dots," *Science*, Vol. 320, pp. 356- 358, 2008.

[51] K. S. Novoselov, et al., "Electric field effect in atomically thin carbon films," *Science,* Vol. 306. pp. 666-669, 2004.

[52] Y. M. Lin, et al., "Wafer-scale graphene integrated circuit," *Science*, Vol. 332, pp. 1294-1297, 2011.

Advances in Spintronics Devices for Microelectronics – From Spin-transfer Torque to Spin-orbit Torque

S. Fukami[1*], H. Sato[1], M. Yamanouchi[1,2], S. Ikeda[1,2,3], F. Matsukura[1,2,4], and H. Ohno[1,2,3,4]

[1] Center for Spintronics Integrated Systems, Tohoku University, Sendai, 980-8577, Japan
[2] Laboratory for Nanoelectronics and Spintronics, RIEC, Tohoku University, Sendai 980-8577, Japan
[3] Center for Innovative Integrated Electronic Systems, Tohoku University, Sendai 980-0845, Japan
[4] WPI Advanced Institute for Materials Research, Tohoku University, Sendai 980-8577, Japan
* e-mail: s-fukami@csis.tohoku.ac.jp

Abstract - Recent advances in spintronics devices make it possible to open a new era of microelectronics. In this paper, we review the spintronics devices utilizing spin-transfer torques (STTs) and spin-orbit torques (SOTs) developed in recent years. The progresses of two-terminal STT device with CoFeB-MgO based magnetic tunnel junction (MTJ), three-terminal magnetic domain wall (DW) motion device with Co/Ni multilayer, and three-terminal SOT device with Cu-based channel are described. Integrated circuits with the developed spintronics devices are also reviewed.

I. Introduction

Increase of power consumption and interconnection delay are major issues in the future microelectronics such as very large-scale integrated circuits (VLSIs) technology [1,2]. These issues are attributed to the architecture adopted in the current microelectronics; volatile memories, *e.g.* static random access memories (SRAMs) and dynamic random access memories (DRAMs), are employed for cache and main memories, and are located separately from logic circuits with global wiring. Replacing the volatile memory cells by nonvolatile ones and applying a power-gating technique [3] allow one to significantly reduce standby power which occupies a large portion of the total power consumption. Furthermore, by adopting logic-in-memory architecture where the memory devices are distributed over the logic circuits [4], one can drastically reduce the power dissipation for data transmission and interconnection delay. Among several nonvolatile devices, spintronics devices are the most promising because of its high-speed capability, virtually unlimited endurance, and low operation voltage as the same as logic circuits [5].

The spintronics devices are equipped with magnetic tunnel junction (MTJ) that comprises a tunnel barrier sandwiched by two ferromagnetic layers: one is a recording layer with reversible magnetization direction and the other is a reference layer with fixed magnetization direction. Information is stored as the magnetization direction of the recording layer. For reading, a tunnel magnetoresistance (TMR) effect is used, where the resistance of the MTJ changes according to the stored information. For writing, various methods are utilized. One established method is magnetic field induced switching, which is utilized for the first product of magnetic random access memory (MRAM) that is one of the representative applications of the spintronics devices [6]. However, since the field-induced switching is not scalable, alternative writing

schemes with scalability have been intensively explored. That is why the manipulation of magnetization using spin-transfer torque (STT) and spin-orbit torque (SOT) has attracted a great deal of attention.

In this paper, we review the recent advances in spintronics devices utilizing STT and/or SOT. First we give an overview of the spintronics devices and describe the characteristics of each type of devices. We then review our recent studies of individual devices and finally refer to the integrated circuits with the developed spintronics devices.

II. Overview of Spintronics Devices

Table I summarizes the device structure, cell circuit, operation, and characteristics of three kinds of spintronics devices described in this paper: two-terminal STT device, three-terminal domain wall (DW) motion device, and three-terminal SOT device, all of which utilize scalable writing techniques. The cell circuit structure is categorized into two-terminal and three-terminal architectures. In the two-terminal STT device, STT induced magnetization switching [7,8] is used for the writing method. Its primitive cell has one cell transistor and one MTJ (1T1MTJ), achieving a relatively small cell size of ideally $6F^2$, where F is the feature size of the MTJ layer. The write current passes through the tunnel barrier as is also the case with the read current. Accordingly, the read current should be small enough that the write event, *i.e.* magnetization switching, does not take place, and the write current should be small enough that it does not give rise to a barrier breakdown (BD). On the other hand, in the three-terminal DW-motion device and SOT device, current-induced DW motion [9] and magnetization switching induced by SOT [10-12], respectively, are utilized. Their primitive cell has two transistors and one MTJ (2T1MTJ), and ideal cell size is 12 or $18F^2$, larger than the two-terminal device. The write and read current paths are different from each other; the former is applied to the metallic wire through the two cell transistors and the latter is applied to the MTJ. Thus the upper limit of the read and write currents are determined from the BD of tunnel barrier and metallic wire, respectively. In general, it is possible to obtain a larger margin for read and write operation in the three-terminal devices than in the two-terminal device. Namely, the two-terminal cell architecture is suitable for large capacity application, typically Gb scale, whereas the three-terminal cell architecture is

TABLE I
Summary of the spintronics devices described in this paper.

Device	Two-terminal STT	Three-terminal DW-motion	Three-terminal SOT
Circuit diagram	SL, MTJ, BL, WL	WBL, MTJ, RBL, $\overline{\text{WBL}}$, WL	
Write	STT switching	DW motion	SOT
Read	TMR		
Ideal cell size (F^2)	6	12 or 18	12 or 18
Operation margin	Read/Write Write/Barrier BD	Read/Barrier BD Write/Wire BD	Read/Barrier BD Write/Wire BD

suitable for high-speed operation, typically sub GHz class. In the followings, we focus on the individual devices. We describe the background of each technology and review our recent studies.

III. Two-terminal STT device

A. Background

The STT induced magnetization switching in MTJ is utilized for the writing method of the two-terminal STT device. The magnetization switching results from a transfer of spin angular momentum from conduction electrons to the magnetization of the recording layer [7,8]. While early studies employed MTJ with in-plane magnetic easy axis [13-18], now the technology has shifted to MTJ with perpendicular easy axis (pMTJ) [19-22] because of an advantage to achieving efficient current utilization to switch the magnetization [19].

The TMR effect is utilized for the reading method of the STT device. The magnitude of TMR effect is represented by the TMR ratio defined as $(R_{AP} - R_P) / R_P$, where R_{AP} (R_P) denotes the resistance of MTJ at antiparallel (parallel) magnetic configuration. The TMR effect originates from a spin-dependent tunneling [23]. MTJ with MgO tunnel barrier was found to be able to achieve a high TMR ratio of more than 100% due to a coherent tunneling of Δ_1 band electrons which has fully-polarized spin [24-27]. By using CoFeB-MgO based MTJ, a large TMR ratio over 600% was observed at room temperature [28].

In order to achieve the non-volatility, one has to ensure the stability of magnetization in the recording layer from thermal agitation. The thermal stability is represented by a thermal stability factor (Δ) defined as $\Delta = E / k_B T$, where E denotes the energy barrier between two stable states, k_B the Boltzmann constant, and T the absolute temperature. The required value of Δ depends on the capacity, for example Δ of more than 67

is necessary for 32-Mb capacity [29].

The two-terminal STT device needs to satisfy the following requirements: (1) a small write current I_W to achieve small cell area, i.e., large capacity, (approximately $I_W < F$ µA, where F is in nm), (2) a high TMR ratio to achieve fast sensing (> 100%), and (3) a high Δ ($\Delta > 60-70$). In addition to them, durability against thermal treatment over 350 °C is desirable for application into the VLSIs.

B. CoFeB-MgO MTJ with perpendicular easy axis

CoFeB-MgO not only shows a high TMR ratio but it has been found to exhibit perpendicular magnetic anisotropy at the interface, large enough that, when thickness of the CoFeB layer is reduced to about 1 nm, its easy axis becomes out-of-plane [30]. The interface anisotropy has its origin in the hybridization of Fe 3d and O 2p orbitals [31].

The first demonstration of p-MTJ using CoFeB-MgO was reported in [21]. The stack structure was, from the substrate side, Ta(5)/ Ru(10)/ Ta(5)/ $Co_{20}Fe_{60}B_{20}$(0.9-1.3)/ MgO(0.85-1.05)/ $Co_{20}Fe_{60}B_{20}$(1.5-1.7)/ Ta(5)/ Ru(5) (numbers in parenthesis are nominal thickness in nm). A reasonably small intrinsic switching current (I_{C0}) of 49 µA was obtained at the MTJ size of 40 nm in diameter. The TMR ratio was 124%, which did not degrade after an annealing at 350 °C. The Δ was determined to be 43, which was high enough to ensure the retention of stored information of a single bit but not enough for a large number of bits for 10 years. The highlight of this work was that the CoFeB-MgO pMTJ has a high potential in meeting major requirements for the practical application at the same time, and thus triggered numerous studies [32-39].

C. Double-MgO MTJ

As described above, Δ of more than 60 is required to realize non-volatile VLSIs, which was not realized by the first CoFeB-MgO pMTJ [21]. For CoFeB-MgO MTJs, it was

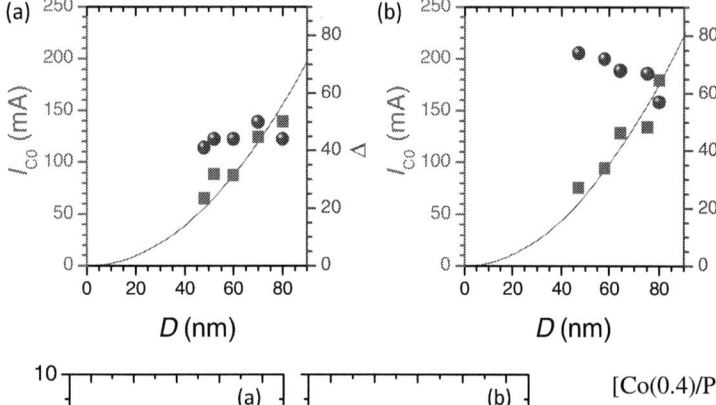

Fig. 1. Intrinsic critical current (I_{C0}) and thermal stability factor (Δ) as a function of MTJ diameter (D) for single- (a) and double-(b) MgO MTJ. Red solid lines represent fitting curve with quadratic function. Double-MgO pMTJ shows larger Δ without increase in I_{C0}.

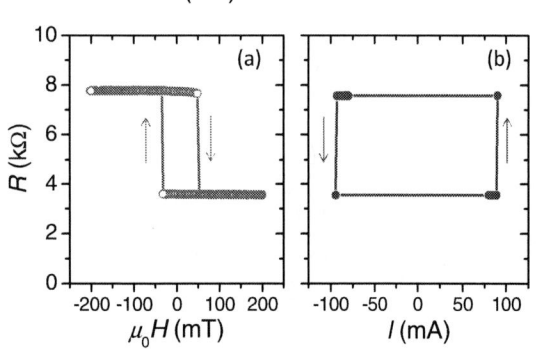

Fig. 2. *R-H* (a) and *R-I* (b) properties of two-terminal STT device with CoFeB-MgO pMTJ where double MgO recording layer and SyF reference layer are employed.

found that the Δ shows almost constant value down to a critical size, indicating that the magnetization switching of CoFeB layer is initiated by a reversal of nucleation embryo which determines the Δ of the system [33,34]. In this case, the Δ is proportional to the thickness of recording layer (t_{rec}) [35]. We found that double-MgO MTJ structure, where the recording layer consisting of CoFeB/Ta/CoFeB was sandwiched by MgO layers allowed us to increase the t_{rec} with keeping the perpendicular easy axis, resulting in the Δ twice as large as that of the previous structure, i.e., single-MgO MTJ [37]. The Ta insertion layer acted as an absorber of B, without which the magnetic easy axis became in-plane. Figure 1 shows the I_{C0} and Δ as a function of MTJ diameter (D) for single MgO interface (a) and double MgO interface (b) structures. The stack structure of them are CoFeB(1.15) and CoFeB(1.6)/ Ta(0.4)/ CoFeB(1.0), respectively. The double-MgO pMTJ exhibits higher Δ without an increase in I_{C0}. The higher Δ is attributed to the increase in total t_{rec}, whereas no increase in I_{C0} is speculated to be due to a decrease of damping constant upon increasing the t_{rec}.

D. SyF reference layer

In the first demonstration of CoFeB-MgO pMTJ [21], pseudo spin valve structure was used. This leads to an asymmetric thermal stability between parallel and antiparallel magnetic configurations, which is unfavorable for application. To overcome this issue, we developed a synthetic ferrimagnetic (SyF) reference layer structure [39]. Figure 2 shows the MTJ resistance versus applied magnetic field (*R-H*) and current (*R-I*) for the CoFeB-MgO pMTJ with SyF reference layer. The stack structures of recording layer and reference layer are CoFeB(1.6)/ Ta(0.4)/ CoFeB(1.0) and

[Co(0.4)/Pt(0.4)]$_6$/ Co(0.4)/ Ru(0.4)/ [Co(0.4)/Pt(0.4)]$_2$/ Co(0.4)/ Ta(0.3)/ CoFeB(1.0), respectively. A well-centered *R-H* curve is obtained. The shift field of *R-H* curve is 10 mT whereas that of pseudo spin valve structure was 25 mT. Also, the intrinsic critical current density was 3.5 MA/cm^2 for both samples. The SyF reference layer and double-MgO recording layer are particularly effective when one reduced the MTJ size. With these technologies, we have achieved $\Delta = 59$ at $D = 29$ nm. [39]

IV. Three-terminal DW-motion device

A. Background

The key ingredient of the three-terminal DW-motion device [40,41] is a current-induced DW motion, which is also associated with the STT [9]. In 2004, a motion of single DW by an electric current was demonstrated [42-44]. While materials with in-plane easy axis were widely used in early stage of researches [42,43,45], those with perpendicular easy axis were found to have a potential to achieve both small critical current density and large thermal stability from a theoretical study [46]. Later, a Co/Ni multilayer with perpendicular easy axis was found to be a promising material system for the practical use [47,48]. One of the advantages of the three-terminal DW motion device is that one can design the thermal stability independently of the critical current [49-51], resulting in a high scalability. In addition, the DW-motion device with the Co/Ni multilayer has high tolerance to a wide range of temperature [52,53] and external field [54]. High reliability in write endurance was also confirmed in the Co/Ni based DW-motion device recently [55]. We also note that, besides the three-terminal DW motion device, a racetrack memory utilizing the current-induced DW motion, which can potentially replace the hard-disk drive (HDD) with drastically improved access time, has also gathered considerable attention [56].

B. Depinning probability of DW by nanosecond pulses

Since the target of the three-terminal DW motion device is applications for high-speed and high-reliability memory circuits, e.g. replacement of logic-embedded SRAMs and non-volatile latch in logic circuits, it is crucial to investigate the probability of DW motion, especially a depinning of DW from a potential well, driven by nano or subnanosecond-long current pulses. To this end, the DW depinning probability from an artificially-prepared pinning site in the Co/Ni wire

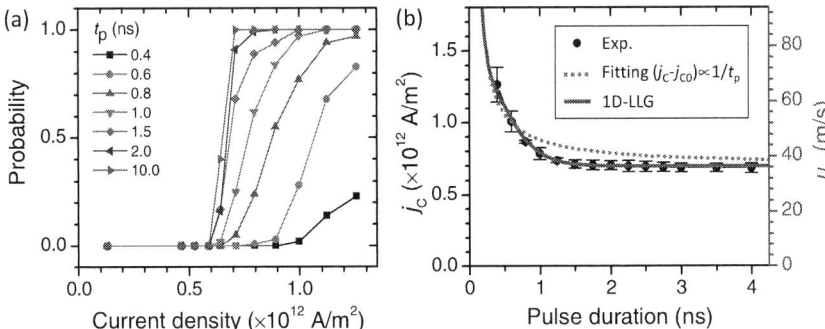

Fig. 3. DW depinning probability versus current density (a) and critical current density (j_C) versus pulse duration (b). Critical current density used in 1D-LLG calculation (u_C) is defined as $u_C = g\mu_B Pj / 2eM_S$, where g the g factor, μ_B the Bohr magnetron, P the spin polarization, j the critical current density, e the elementary charge, M_S the saturation magnetization.

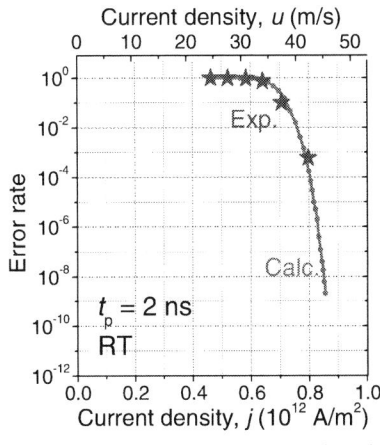

Fig. 4. Experimental and calculation results of depinning error rate of DW as a function of current density.

was examined with various amplitudes and durations [57].

The stack structure used in the work was Ta(3)/ Pt(2)/ [Co(0.3)/Ni(0.6)]$_4$/ Co(0.3)/ Pt(1.5)/ Ta(3). The film was patterned into a 160-nm-wide wire with a pair of Hall probe, which was used for the detection of DW depinning through the anomalous Hall effect and at the same time acted as a pinning site for the DW. Measurement was repeated 100 times for each pulse condition.

Figure 3 shows the DW depinning probability versus current density for various pulse durations (t_p) (a) and critical current density (j_C) versus the t_p (b). The j_C does not increase with decreasing the t_p down to about 2 ns and it is not fitted by an inverse relation, i.e., $j_C - j_{C0} \propto 1/t_p$, which is the case for the STT switching. This presents a promising potential in terms of high-speed operation. The depinning error rate was also measured from 10^4-times iteration, as shown in Fig. 4. It was found that the error rate steeply decreased above a certain threshold with increasing the applied current density. This feature is also promising to achieve an extremely small write error rate.

These experimental results were reproduced by a theoretical calculation, in which one-dimensional Landau-Lifshitz-Gilbert (1D-LLG) equation including the effect of thermal fluctuation was used. In Figs. 3(b) and 4, the numerical calculation results are also shown, which describe well the relation between j_C and t_p (blue solid line in Fig. 3(b)) and that between the error rate and current density (red solid line in Fig. 4). The required error rate for practical application is less than about 10^{-12} for the three-terminal cell architecture with a specific redundancy [58]. From the numerical calculation shown in Fig. 4, it was found that this requirement could be satisfied at the current density of about 1.3-times larger than the critical current density.

V. Three-terminal SOT device

A. Background

It is now widely known that, when an electric current is applied to a heavy metal/ ferromagnet heterostructure, the SOTs are exerted on the magnetization of ferromagnetic layer [10-12, 59-62]. The SOT has its origin in a spin accumulation due to the spin-orbit interaction. The possible mechanism of the SOT is still under debate but the Rashba effect [63] and the spin Hall effect (SHE) [64] have been proposed. The magnetization reversal by the SOT has been observed in ferromagnetic materials with perpendicular [10-12] and in-plane [11] easy axis.

B. SOT device with Cu:Ir channel

In order to apply the three-terminal SOT device to VLSIs, it is desirable for the metallic layer adjacent to the ferromagnetic recording layer to have process compatibility with standard VLSI interconnection technology. From this viewpoint, Ir doped Cu (Cu:Ir), which also shows the SHE [65], is a material of interest because Cu is widely used as interconnection in VLSIs. We fabricated a three-terminal SOT device with Cu:Ir channel and evaluated its switching properties [66].

The stack structure of the fabricated device was Cu$_{90}$Ir$_{10}$(10)/ CoFeB(1.5)/ MgO(1.7)/ CoFeB(3)/ Ru(0.8)/ CoFe(2.5)/ Ta(5)/ Ru(5), where the ferromagnetic stack above the MgO was the reference layer with SyF coupling. The easy axis of the magnetic layers was in-plane. The stack was processed into 100×350 nm^2 rectangular MTJ on a 1-μm-wide Cu:Ir channel and annealed at 300 °C for 1 h under an in-plane magnetic field of 0.4 T along the long axis of the MTJ, which is orthogonal to the channel direction. The MTJ resistance R was measured by a lock-in technique.

Figure 5 (a) and (b) show dependence of R on H (R-H) and current pulse I (R-I) with $t_p = 10$ ms for the fabricated device, respectively. The positive (negative) current switches the resistance from low (high) to high (low). Assuming that the switching is driven by the SHE, this direction of switching indicates that the spin Hall angle of the Cu:Ir is positive, which is consistent with the previous study [65].

To analyze the contribution of SOT (anti-damping torque)

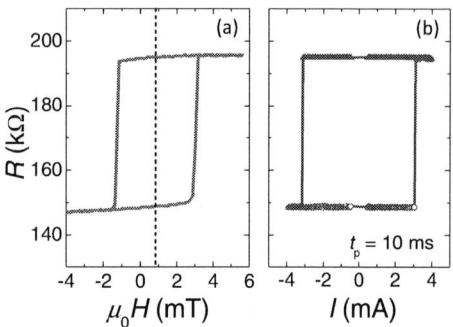

Fig. 5. *R-H* (a) and *R-I* (b) curves for three-terminal SOT device with Cu-Ir channel.

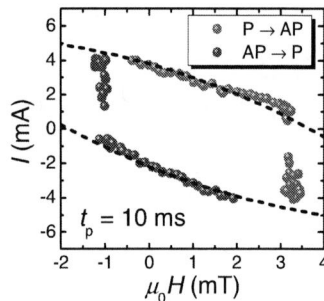

Fig. 6. Switching current (*I*) versus magnetic field (*H*) phase diagram for the three-terminal SOT device. Broken lines are fitting curve based on a theoretical equation.

and field torque, we measured switching current versus applied in-plane magnetic field phase diagram as shown in Fig. 6. The obtained phase diagram was fitted by a theoretical equation [67] with taking into account the magnetic fields generated from channel current, from which we obtained $I_{C0} = 7.4$ mA ($j_{C0} = 8.5 \times 10^{11}$ A/m^2) and $\Delta = 42$. The derived I_{C0} is about half of the expected current to switch the magnetization solely by the magnetic fields, indicating that the SOT played a decisive role for the magnetization reversal. Assuming that the SHE was only the source of the SOT, we determined the lower bound of the spin Hall angle to be 0.03, which was close to the reported value [65]. Although a reduction of the critical current is necessary for applications in terms of both write endurance and cell area, these results show a potential to implement the three-terminal SOT device with Cu-based channel into the VLSIs.

VI. Spintronics VLSIs

In this section, we review integrated circuits with the developed spintronics devices. The MRAM is the most typical implementation of the spintronics devices for VLSI with non-volatile working memory. Current VLSI uses SRAM cache, DRAM main memory, and storage (HDD or solid-state drive (SSD)), and the speed gap between them limits the performance of today's computing systems. Since the spintronics devices have capability for realizing high-speed operation and large-capacity in principle, it is possible to construct a new memory hierarchy where the MRAM with STT or SOT is employed and thus the speed gap issue can be

solved [68]. As described in section II, the primitive memory cell with two-terminal STT device has 1T1MTJ structure. This is suitable for large-capacity memory but not suitable for high-speed operation like embedded SRAMs. To overcome it, four transistors and two MTJs (4T2MTJ) cell was proposed [69], with which a fast wake-up operation from a standby mode within 1 ns was demonstrated. Also, by using a cell structure with six transistors and two MTJs (6T2MTJ), data writing within 2.1 ns was achieved in 1-Mb MRAM [70].

Nonvolatile logic-in-memories with the spintronics devices were also demonstrated. For example, content-addressable memories (CAMs) have been designed and fabricated [71,72]. Thanks to the nonvolatile feature of the logic-in-memory architecture with non-volatile spintronics devices, power dissipation required for search operation was reduced to 1/100 [71]. Nonvolatile look-up table (LUT) circuit for instant power-on/off field programmable gate array (FPGA) is another example. In [73], by using a nonvolatile LUT with six inputs, 62%-reduction of the number of transistor compared to the previous proposal was demonstrated.

The spintronics device-oriented design flows have also been developed [74,75]. It is vitally important to develop a CAD tool that can semi-automatically design the spintronics-based VLSIs with the logic-in-memory architecture. A practical SPICE incorporating MTJ model parameters was developed [75]. By using the MTJ SPICE model and the semi-automated design flow, MTJ/MOS-hybrid video coding hardware has been successfully designed, and reduction of power dissipation was demonstrated [74].

VII. Summary

We reviewed the recent advances in three types of spintronics devices utilizing STT or SOT devoted to future microelectronics. Taking advantage of nonvolatile feature as well as high-speed capability and virtually unlimited endurance of the spintronics devices, one can obtain nonvolatile VLSIs where drastic reduction of power consumption and high performance are achieved together. As the spintronics devices, we described two-terminal STT device, three-terminal DW motion device, and three-terminal SOT device. The CoFeB-MgO pMTJ enabled the two-terminal STT device to meet major requirement for the applications. In particular, double-MgO MTJ and SyF reference layer showed promising features to reduce the MTJ size with maintaining the required properties. It was revealed that the three-terminal DW motion device with Co/Ni wire had a high potential to apply for the high-speed and high-reliability memory device and logic circuits. Also, a three-terminal SOT device with Cu-based channel was fabricated and its basic operation was demonstrated. We hope that these advances in spintronics devices will establish a new paradigm of microelectronics, where a human- and environment-friendly computer and communication are realized.

Acknowledgments

This work was supported by the Japan Society for the Promotion of Science (JSPS) through its "Funding Program for World-Leading Innovative R&D on Science and

Technology (FIRST Program)," and Research and Development for Next-Generation Information Technology of the Ministry of Education, Culture, Sports, Science and Technology. The authors would like to thank N. Kasai, L. Chen, J. Kim, M. Hayashi, I. Morita, T. Hirata, H. Iwanuma, Y. Kawato, K. Goto, M. Murahata, C. Igarashi for fluitiful discussion and technical support.

References

[1] N. S. Kim, T. Austin, D. Blaauw, T. Mudge, K. Flautner, J. S. Hu, M. J. Irwin, M. Kandemir, and V. Narayanan, "Leakage current: Moore's law meets static power," IEEE Computer, vol. 36, no. 12, pp. 68-75, Dec. 2003.

[2] M. T. Bohr, "Interconnect scaling – The real limiter to high performance ULSI," in IEDM Tech. Dig., Dec. 10-13, 1995, pp. 241-244.

[3] K. Roy, S. Mukhopadhyay, and H. Mahmoodi, "Leakage current mechanism and leakage reduction techniques in deep-submicrometer CMOS Circuits," Proc. IEEE, vol. 91, no. 2, pp. 305-327, Feb. 2003.

[4] W. H. Kautz, "Cellular logic-in-memory arrays," IEEE Trans. Comuputers, vol. C-18, no. 8, pp. 719-727, Aug. 1969.

[5] H. Ohno, T. Endo, T. Hanyu, N. Kasai, and S. Ikeda, "Magnetic tunnel junction for nonvolatile CMOS logic," in IEDM Tech. Dig., Dec. 6-8, 2010, pp. 9.4.1-9.4.4.

[6] http://www.everspin.com/

[7] J. C. Sloncewski, "Current-driven excitation of magnetic multilayers," J. Magn. Magn. Mater., vol. 159, no. 1-2, pp. L1-L7, Jun. 1996.

[8] L. Berger, "Emission of spin waves by a magnetic multilayer traversed by a current," Phys. Rev. B, vol. 54, no. 13, pp. 9353-9358, Oct. 1996.

[9] L. Berger, "Exchange interaction between ferromagnetic domain wall and electric current in very thin metallic films," J. Appl. Phys., vol. 55, no. 6, pp. 1954-1956, Mar. 1984.

[10] I. M. Miron, K. Garello, G. Gaudin, P.-J. Zermatten, M. V. Costache, S. Auffret, S. Bandiera, B. Rodmacq, A. Schuhl, and P. Gambardella, "Perpendicular switching of a single ferromagnetic layer induced by in-plane current injection," Nature, vol. 475, pp. 189-194, Aug. 2011.

[11] L. Liu, C.-F. Pai, Y. Li, H. W. Tseng, D. C. Ralph, and R. A. Buhrman, "Spin-torque switching with the giant spin hall effect of Tantalum," Science, vol. 336, pp. 555-558, May 2012.

[12] L. Liu, O. J. Lee, T. J. Gudmundsen, D. C. Ralph, and R. A. Buhrman, "Current-induced switching of perpendicular magnetized magnetic layers using spin torque from the spin Hall effect," Phys. Rev. Lett., vol. 109, no. 9, 096602, Aug. 2012.

[13] M. Tsoi, A. G. M. Jansen, J. Bass, W.-C. Chiang, M. Seck, V. Tsoi, and P. Wyder, "Excitation of a magnetic multilayer by an electric current," Phys. Rev. Lett., vol. 80, no. 19, pp. 4281-4284, May 1998.

[14] E. B. Myers, D. C. Ralph, J. A. Katine, R. N. Louie, and R. A. Buhrman, "Current-induced switching of domains in magnetic multilayer devices," Science, vol. 285, pp. 867-870, Aug. 1999.

[15] J. A. Katine, F. J. Albert, R. A. Buhrman, E. B. Myers, and D. C. Ralph, "Current-driven magnetization reversal and spin-wave excitations in Co/Cu/Co pillars," Phys. Rev. Lett., vol. 84, no. 14, pp. 3149-3152, Apr. 2000.

[16] M. Hosomi, H. Yamagishi, T. Yamamoto, K. Bessho, Y. Higo, K. Yamane, H. Yamada, M. Shoji, H. Hachino, C. Fukumoto, H. Nagao, and H. Kano, "A novel nonvolatile memory with spin torque transfer magnetization switching : Spin-RAM," in IEDM Tech. Dig., Dec. 5-7 2005, pp. 459-462.

[17] T. Kawahara, R. Takemura, K. Miura, J. Hayakawa, S. Ikeda, Y. M. Lee, R. Sasaki, Y. Goto, K. Ito, T. Meguro, F. Matsukura, H. Takahashi, H. Matsuoka, and H. Ohno, "2 Mb SPRAM (SPin-transfer torque RAM) with bit-by-bit bi-directional current write and parallelizing-direction current read," J. Solid-State Circuit, vol. 43, no. 1, pp. 109-120, Jan. 2008.

[18] J. M. Slaughter, N. D. Rizzo, J. Janesky, R. Whig, F. B. Mancoff, D. Houssameddine, J. J. Sun, S. Aggarwal, K. Nagel, S. Deshpande, S. M. Alam, T. Andre and P. LoPresti, "High Density ST-MRAM Technology," in IEDM Tech. Dig., Dec. 10-13 2012, pp. 29.3.1-29.3.4.

[19] S. Mangin, D. Ravelosona, J. A. Katine, M. J. Carey, B. D. Terris, and E. E. Fullerton, "Current-induced magnetization reversal in nanopillars with perpendicular anisotropy," Nature Mater., vol. 5, pp. 210-215, Mar. 2006.

[20] T. Kishi, H. Yoda, T. Kai, T. Nagase, E. Kitagawa, M. Yoshikawa, K. Nishiyama, T. Daibou, M. Nagamine, M. Amano, S. Takahashi, M. Nakayama, N. Shimomura, H. Aikawa, S. Ikegawa, S. Yuasa, K. Yakushiji, H. Kubota, A. Fukushima, M. Oogane, T. Miyazaki and K. Ando, "Lower-current and fast switching of a perpendicular TMR for high speed and high density spin-transfer-torque MRAM," in IEDM Tech. Dig., Dec. 15-17 2008, pp. 309-312.

[21] S. Ikeda, K. Miura, H. Yamamoto, K. Mizunuma, H. D. Gan, M. Endoh, S. Kanai, J. Hayakawa, F. Matsukura, and H. Ohno, "A perpendicular-anisotropy CoFeB-MgO magnetic tunnel junction," Nature Mater. vol. 9, pp. 721-724, Sept. 2010.

[22] D. C. Worledge, G. Hu, P. L. Trouilloud, D. W. Abraham, S. Brown, M. C. Gaidis, J. Nowak, E. J. O'Sullivan, R. P. Robertazzi, J. Z. Sun, and W. J. Gallagher, "Switching distributions and write reliability of perpendicular spin torque MRAM," in IEDM Tech. Dig., Dec. 6-8 2010, pp. 12.5.1-12.5.4.

[23] M. Julliere, "Tunneling between ferromagnetic films," Phys. Lett., vol. 54A, no. 3, pp. 225-226, Sept. 1975.

[24] W. H. Butler, X.-G. Zhang, T. C. Schulthess and J. M. MacLaren, "Spin-dependent tunneling conductance of Fe|MgO|Fe sandwiches," Phys. Rev. B, vol. 63, no. 5, 054416, Jan. 2001.

[25] J. Mathon and A. Umerski, "Theory of tunneling magnetoresistance of an epitaxial Fe/MgO/Fe(001) junction," Phys. Rev. B, vol. 63, no. 22, 220403(R), May 2001.

[26] S. S. P. Parkin, C. Kaiser, A. Panchula, P. M. Rice, B. Hughes, M. Samant and S. H. Yang, Nature Mater. vol. 3, pp. 862-864, Dec. 2004

[27] S. Yuasa, T. Nagahama, A. Fukushima, Y. Suzuki and K. Ando, "Giant room-temperature magnetoresistance in single-crystal Fe/MgO/Fe magnetic tunnel junctions," Nature Mater. vol. 3, pp. 868-871, Dec. 2004.

[28] S. Ikeda, J. Hayakawa, Y. Ashizawa, Y. M. Lee, K. Miura, H. Hasegawa, M. Tsunoda, F. Matsukura, and H. Ohno, "Tunnel magnetoresistance of 604% at 300 K by suppression of Ta diffusion in CoFeB/MgO/CoFeB pseudo-spin-valves annealed at high temperature," Appl. Phys. Lett. vol. 93, no. 8, 082508, Aug. 2008.

[29] R. Takemura, T. Kawahara, K. Miura, H. Yamamoto, J. Hayakawa, N. Matsuzaki, K. Ono, M. Yamanouchi, K. Ito, H. Takahashi, S. Ikeda, H. Hasegawa, H. Matsuoka and H. Ohno, "A 32-Mb SPRAM with 2T1R memory cell, localized bi-directional write driver and '1'/'0' dual-array equalized reference scheme," IEEE J. Solid-State Circuits, vol. 45, pp. 869-879 Apr. 2010.

[30] M. Endo, S. Kanai, S. Ikeda, F. Matsukura, and H. Ohno, "Electric-field effects on thickness dependent magnetic anisotropy of sputtered MgO/Co40Fe40B20/Ta structures," Appl. Phys. Lett., vol. 96, no. 21, 212503, May 2010.

[31] R. Shimabukuro, K. Nakamura, T. Akiyama, and T. Ito, "Electric field effects on magnetocrystalline anisotropy in ferromagnetic Fe monolayers," Physica E, vol. 42, no. 4, pp.

1014-1017, Feb. 2010.

[32] D. C. Worledge, G. Hu, D. W. Abraham, J. Z. Sun, P. L. Trouilloud, J. Nowak, S. Brown, M. C. Gaidis, E. J. O'Sullivan, and R. P. Robertazzi, "Spin torque switching of perpendicular TaCoFeBMgO-based magnetic tunnel junctions," Appl. Phys. Lett., vol. 98, no. 2, 022501, Jan. 2011.

[33] H. Sato, M. Yamanouchi, K. Miura, S. Ikeda, H. D. Gan, K. Mizunuma, R. Koizumi, F. Matsukura and H. Ohno, "Junction size effect on switching current and thermal stability in CoFeB/MgO perpendicular magnetic tunnel junctions," Appl. Phys. Lett. vol. 99, no. 4, 042501, Jul. 2011.

[34] J. Z. Sun, R. P. Robertazzi, J. Nowak, P. L. Trouilloud, G. Hu, D. W. Abraham, M. C. Gaidis, S. L. Brown, E. J. O'Sullivan, W. J. Gallagher, and D. C. Worledge, "Effect of subvolume excitation and spin-torque efficiency on magnetic switching," Phys. Rev. B, vol. 84, no. 6, 064413 Aug. 2011.

[35] H. Sato, M. Yamanouchi, K. Miura, S. Ikeda, R. Koizumi, F. Matsukura and H. Ohno, "CoFeB Thickness Dependence of Thermal Stability Factor in CoFeB/MgO Perpendicular Magnetic Tunnel Junctions," IEEE Magn. Lett., vol. 3, 3000204, Apr. 2012.

[36] J. H. Park, Y. Kim, W. Lim, J. Kim, S. Park, J. Kim, W. Kim, K. Kim, J. Jeong, K. S. Kim, H. Kim, Y. J. Lee, S. Oh, J. E. Lee, S. O. Park, S. Watts, D. Apalkov, V. Nikitin, M. Krounbi, S. Jeong, S. Choi, H. Kang and C. Chung, "Enhancement of data retention and write current scaling for sub-20nm STT-MRAM by utilizing dual interfaces for perpendicular magnetic anisotropy," in IEEE Symp. on VLSI Tech., Dig. Tech. Pap. Jun. 12-14, 2012, pp. 57-58.

[37] H. Sato, M. Yamanouchi, S. Ikeda, S. Fukami, F. Matsukura and H. Ohno, "Perpendicular-anisotropy CoFeB-MgO magnetic tunnel junctions with a MgO/CoFeB/Ta/CoFeB/MgO recording structure," Appl. Phys. Lett., vol. 101, no. 2, 022414, Jul. 2012.

[38] G. Jan, Y.-J. Wang, T. Moriyama, Y.-J. Lee, M. Lin, T. Zhong, R.-Y. Tong, T. Torng, and P.-K. Wang, "High Spin Torque Efficiency of Magnetic Tunnel Junctions with MgO/CoFeB/MgO Free Layer," Appl. Phys. Express, vol. 5, 093008, Sept. 2012.

[39] H. Sato, M. Yamanouchi, S. Ikeda, S. Fukami, F. Matsukura, and H. Ohno, "MgO/CoFeB/Ta/CoFeB/MgO Recording Structure in Magnetic Tunnel Junctions With Perpendicular Easy Axis," IEEE Trans. Magn., vol. 49, no. 7, pp. 4437-4440, Jul. 2013.

[40] S. Fukami, T. Suzuki, K. Nagahara, N. Ohshima, Y. Ozaki, S. Saito, R. Nebashi, N. Sakimura, H. Honjo, K. Mori, C. Igarashi, S. Miura, N. Ishiwata, and T. Sugibayashi, "Low-Current Perpendicular Domain Wall Motion Cell for Scalable High-Speed MRAM," in Symp. VLSI Tech., Dig. Tech. Pap., Jun. 16-18, 2009, pp. 230-231.

[41] S. Fukami, M. Yamanouchi, T. Koyama, K. Ueda, Y. Yoshimura, K.-J. Kim, D. Chiba, H. Honjo, N. Sakimura, R. Nebashi, Y. Kato, Y. Tsuji, A. Morioka, K. Kinoshita, S. Miura, T. Suzuki, H. Tanigawa, S. Ikeda, T. Sugibayashi, N. Kasai, T. Ono, and H. Ohno, "High-speed and reliable domain wall motion device: Material design for embedded memory and logic application," in Symp. VLSI Tech., Dig. Tech. Pap., Jun. 12-14, 2012, pp. 61-62.

[42] A. Yamaguchi, T. Ono, S. Nasu, K. Miyake, K. Mibu, and T. Shinjo, "Real-space observation of current-driven domain wall motion in submicron magnetic wires," Phys. Rev. Lett., vol. 92, no. 7, 077205, Feb. 2004.

[43] N. Vernier, D. A. Allwood, D. Atkinson, M. D. Cooke, and R. P. Cowburn, "Domain wall propagation in magnetic nanowires by spin-polarized current injection," Europhys. Lett., vol. 65, no. 4, pp. 526–532, Feb. 2004.

[44] M. Yamanouchi, D. Chiba, F. Matsukura, and H. Ohno, "Current-induced domain-wall switching in a ferromagnetic

semiconductor structure," Nature, vol. 428, no. 1, pp. 539–542, Apr. 2004.

[45] L. Thomas, M. Hayashi, X. Jiang, R. Moriya, C. Rettner, and S. S. P. Parkin, "Oscillatory dependence of current-driven magnetic domain wall motion on current pulse length," Nature, vol. 443, pp. 197-200, Sept. 2006.

[46] S. Fukami, T. Suzuki, N. Ohshima, K. Nagahara, and N. Ishiwata, "Micromagnetic analysis of current driven domain wall motion in nanostrips with perpendicular magnetic anisotropy," J. Appl. Phys., vol. 103, no. 7, 07E718, Jan. 2008.

[47] T. Koyama, G. Yamada, H. Tanigawa, S. Kasai, N. Ohshima, S. Fukami, N. Ishiwata, Y. Nakatani, and T. Ono, "Control of domain wall position by electrical current in structured Co/Ni wire with perpendicular magnetic anisotropy," Appl. Phys. Express, vol. 1, 101303, Oct. 2008.

[48] S. Fukami, Y. Nakatani, T. Suzuki, K. Nagahara, N. Ohshima, and N. Ishiwata, "Relation between critical current of domain wall motion and wire dimension in perpendicularly magnetized Co/Ni nanowires," vol. 95, no. 23, 232504, Dec. 2009.

[49] T. Suzuki, S. Fukami, N. Ohshima, K. Nagahara, and N. Ishiwata, "Analysis of current-driven domain wall motion from pinning sites in nanostrips with perpendicular magnetic anisotropy," J. Appl. Phys., vol. 103, no. 11, 113913, Jun. 2008.

[50] S. Fukami, T. Suzuki, K. Nagahara, N. Ohshima, and N. Ishiwata, "Large thermal stability independent of critical current of domain wall motion in Co/Ni nanowires with step pinning sites," J. Appl. Phys., vol. 108, no. 11, 113914, Dec. 2010.

[51] K.-J. Kim, R. Hiramatsu, T. Koyama, K. Ueda, Y. Yoshimura, D. Chiba, K. Kobayashi, Y. Nakatani, S. Fukami, M. Yamanouchi, H. Ohno, H. Kohno, G. Tatara, and T. Ono, "Two-barrier stability that allows low-power operation in current-induced domain-wall motion," Nature Commun., vol. 4, 2011, Jun. 2013.

[52] H. Tanigawa, K. Suemitsu, S. Fukami, N. Ohshima, T. Suzuki, E. Kariyada, and N. Ishiwata, "Effect of device temperature on domain wall motion in a perpendicularly magnetized Co/Ni wire," Appl. Phys. Express, vol. 4, 013007, Jan. 2011.

[53] K. Ueda, T. Koyama, D. Chiba, K. Shimamura, H. Tanigawa, S. Fukami, T. Suzuki, N. Ohshima, N. Ishiwata, Y. Nakatani, and T. Ono, "Current-Induced Magnetic Domain Wall Motion in Co/Ni Nanowire at Low Temperature," Appl. Phys. Express, vol. 4, 063003, May 2011.

[54] Y. Yoshimura, T. Koyama, D. Chiba, Y. Nakatani, S. Fukami, M. Yamanouchi, H. Ohno, and T. Ono, "Current-Induced Domain Wall Motion in Perpendicularly Magnetized Co/Ni Nanowire under In-Plane Magnetic Fields," Appl. Phys. Express, vol. 5, 063001, May 2012.

[55] S. Fukami, M. Yamanouchi, H. Honjo, K. Kinoshita, K. Tokutome, S. Miura, S. Ikeda, N. Kasai, and H. Ohno, "Electrical endurance of Co/Ni wire for magnetic domain wall motion device," Appl. Phys. Lett., vol. 102, no. 22, 222410, Jun. 2013.

[56] S. S. P. Parkin, M. Hayashi, and L. Thomas, "Magnetic domain-wall racetrack memory," Science, vol. 320, pp. 190-194, Apr. 2008.

[57] S. Fukami, M. Yamanouchi, S. Ikeda, and H. Ohno, "Depinning probability of a magnetic domain wall in nanowires by spin-polarized currents," Nature Commun., vol. 4, 2293, Aug. 2013.

[58] Y. Tsuji, R. Nebashi, N. Sakimura, A. Morioka, H. Honjo, K. Tokutome, S. Miura, T. Suzuki, S. Fukami, K. Kinoshita, T. Hanyu, T. Endoh, N. Kasai, H. Ohno, and T. Sugibayashi, "Spintronics Primitive Gate with High Error Correction Efficiency $6(P_{error})^2$ for Logic-in Memory Architecture," in Symp. VLSI Tech., Dig. Tech. Pap., Jun. 12-14, 2012, pp. 63-64.

[59] K. Obata, and G. Tatara, "Current-induced domain wall motion

in Rashba spin-orbit system," Phys. Rev. B, vol. 77, 214429, Apr. 2008.

[60] A. Monchon and S. Zhang, "Theory of nonequilibrium intrinsic spin torque in a single nanomagnet," Phys. Rev. B, vol. 78, 212405, Dec. 2008.

[61] A. Chemyshow, M. Overby, X. Liu, J. K. Furdyna, Y. Lyanda-Geller, and L. P. Rokhinson, "Evidence for reversible control of magnetization in a ferromagnetic material by means of spin–orbit magnetic field," Nature Phys., vol. 5, pp. 656-659, Sep. 2009.

[62] I. M. Miron, G. Gaudin, S. Auffret, B. Rodmacq, A. Schuhl, S. Pizzini, J. Vogel, and P. Gambardella, "Current-driven spin torque induced by the Rashba effect in a ferromagnetic metal layer," Nature Mater., vol. 9, pp. 230-234, Mar. 2010.

[63] E. I. Rashba, Sov. Phys. Solid State, vol. 2, p. 1109, 1960.

[64] M. I. Dyakonov, V. I. Perel, "Current-induced spin orientation of electrons in semiconductors," Phys. Lett. vol. 35A, no. 6, pp. 459-460, Mar. 1971.

[65] Y. Niimi, M. Morota, D. H. Wei, C. Deranlot, M. Basletic, A. Hamzic, A. Fert, and Y. Otani, "Extrinsic Spin Hall Effect Induced by Iridium Impurities in Copper," Phys. Rev. Lett. vol. 106, no. 12, 126601, Mar. 2011.

[66] M. Yamanouchi, L. Chen, J. Kim, M. Hayashi, H. Sato, S. Fukami, S. Ikeda, F. Matsukura, and H. Ohno, "Three terminal magnetic tunnel junction utilizing the spin Hall effect of iridium-doped copper," Appl. Phys. Lett., vol. 102, no. 21, 212408, May 2013.

[67] Z. Li, and S. Zhang, "Thermally assisted magnetization reversal in the presence of a spin-transfer torque," Phys. Rev. B, vol. 69, no. 13, 134416, Apr. 2004.

[68] T. Endoh, T. Ohsawa, H. Koike, T. Hanyu, and H. Ohno, "Restructuring of memory hierarchy in computing system with spintronics-based technologies," in Symp. VLSI Tech., Dig. Tech. Pap., Jun. 12-14, 2012, pp. 89-90.

[69] T. Ohsawa, H. Koike, S. Miura, H. Honjo, K. Tokutome, S. Ikeda, T. Hanyu, H. Ohno, and T. Endoh, "1Mb 4T-2MTJ nonvolatile STT-MRAM for embedded memories using 32b fine-grained power-gating technique with 1.0ns/200ps wake-up/power-off times," in Symp. VLSI Cir., Dig. Tech. Pap., Jun. 12-14, 2012, pp. 46-47.

[70] T. Ohsawa, S. Miura, K. Kinoshita, H. Honjo, S. Ikeda, T. Hanyu, H. Ohno, and T. Endoh, "A 1.5nsec/2.1nsec random read/write cycle 1Mb STT-RAM using 6T2MTJ cell with background write for nonvolatile e-memories," in Symp. VLSI Cir., Dig. Tech. Pap., Jun. 11-13, 2013, pp. C110-C111.

[71] S. Matsunaga, N. Sakimura, R. Nebashi, Y. Tsuji, A. Morioka, T. Sugibayashi, S. Miura, H. Honjo, K. Kinoshita, H. Sato, S. Fukami, M. Natsui, A. Mochizuki, S. Ikeda, T. Endoh, H. Ohno, and T. Hanyu, "Fabrication of a 99%-Energy-Less Nonvolatile Multi-Functional CAM Chip Using Hierarchical Power Gating for a Massively-Parallel Full-Text-Search Engine," in Symp. VLSI Cir., Dig. Tech. Pap., Jun. 11-13, 2013, pp. C106-C107.

[72] R. Nebashi, N. Sakimura, Y. Tsuji, S. Fukami, H. Honjo, S. Saito, S. Miura, N. Ishiwata, K. Kinoshita, T. Hanyu, T. Endoh, N. Kasai, H. Ohno, and T. Sugibayashi, "A Content Addressable Memory Using Magnetic Domain Wall Motion Cells," in Symp. VLSI Cir., Dig. Tech. Pap., Jun. 15-17, 2011, pp. 300-301.

[73] D. Suzuki, M. Natsui, T. Endoh, H. Ohno, and T. Hanyu, "Six-input lookup table circuit with 62% fewer transistors using nonvolatile logic-in-memory architecture with series/parallel-connected magnetic tunnel junctions," J. Appl. Phys., vol. 111, no. 7, 07E318, Feb. 2012.

[74] M. Natsui, D. Suzuki, N. Sakimura, R. Nebashi, Y. Tsuji, A. Morioka, T. Sugibayashi, S. Miura, H. Honjo, K. Kinoshita, S. Ikeda, T. Endoh, H. Ohno, and T. Hanyu, "Nonvolatile Logic-in-Memory Array Processor in 90nm MTJ/MOS Achieving 75% Leakage Reduction Using Cycle-Based Power Gating," in IEEE Int. Solid-State Cir. Conf. (ISSCC), Dig. Tech. Pap., Feb. 17-21, 2013, pp. 194-195.

[75] N. Sakimura, R. Nebashi, Y. Tsuji, H. Honjo, T. Sugibayashi, H. Koike, T. Ohsawa, S. Fukami, T. Hanyu, H. Ohno, and T. Endoh, "High-Speed Simulator including Accurate MTJ Models for Spintronics Integrated Circuit Design," in IEEE Int. Symp. on Cir. and Sys (ISCAS), May 20-23 2012, pp. 1971-1974.

Hybrid CMOS/Magnetic Process Design Kit and SOT-based Non-volatile Standard Cell Architectures

Gregory Di Pendina

Spintec Laboratory
CEA-INAC/CNRS/UJF/G-INP
Grenoble, France
Tel: +33 4 38 78 47 46
Fax: +33 4 38 78 21 27
e-mail: gregory.dipendina@cea.fr

Kotb Jabeur

Spintec Laboratory
CEA-INAC/CNRS/UJF/G-INP
Grenoble, France
Tel: +33 4 38 78 23 83
Fax: +33 4 38 78 21 27
kotb.jabeur@cea.fr

Guillaume Prenat

Spintec Laboratory
CEA-INAC/CNRS/UJF/G-INP
Grenoble, France
Tel: +33 4 38 78 63 15
Fax: +33 4 38 78 21 27
guillaume.prenat@cea.fr

Abstract—This paper gives an overview of hybrid CMOS/magnetic logic circuit design. We describe the magnetic devices, the expected advantages of using them beside CMOS to help to circumvent the incoming limits of VLSI circuits and the tools required to design such circuits, including Process Design Kit (PDK) and Standard Cells (SC). As a case of study, we particularly focus on a new and promising device technology based on Spin Orbit Torque (SOT) effect.

I. INTRODUCTION

For 40 years, microelectronics has been following the Moore's law, stating that the complexity and performance of integrated circuits would double every 18 months. But today, this trend is facing insurmountable physical limits. Due to decreasing devices size, leakage current is becoming a significant contributor to power dissipation of CMOS. The increased number of transistors per chip and reduction in die size lead to heat dissipation and reliability issues. Moreover, the dynamic power keeps on growing up since both clock frequency and capacitance still increase, while the power supply is not scaled down accordingly. Several solutions are studied to try to push forward these limits at technology, circuit or architecture levels. Among them, the use of non-volatile devices is seen as a promising solution to reduce power consumption, improve reliability and offer new functionalities. Several technologies are intensively investigated like Phase Change Random Access Memory (PCRAM), Ferroelectric RAM (FeRAM), RedoxRAM and Magnetic RAM (MRAM). In its 2010 report, ITRS identified RedoxRAM and MRAM as the two most promising technologies for embedded memories at technology nodes below 16nm. In this paper, we give an overview of the use of the magnetic technology beside CMOS in logic circuits, with a focus on a new device based on Spin Orbit Torque (SOT) effect. In section II, we describe the magnetic technologies and devices. In section III, we see how the use of this devices at different levels of the memory hierarchy can dramatically improve the characteristics of logic systems. In section IV, we present the tools developed to allow designing hybrid CMOS/Magnetic circuits, at different abstraction levels.

II. MAGNETIC TECHNOLOGIES AND DEVICES

A. Spintronics

Spintronics is a discipline at the frontier between electronics and magnetism, which consists in using the spin of the electron in addition to its charge to conceive innovative devices like Spin Valves (SV) or Magnetic Tunnel Junctions (MTJ). These devices have been widely used in read heads of hard drives. But they can be used for other applications, for example as memory elements for logic circuits, RF oscillators or receivers and current or magnetic field sensors. MTJs combine intrinsic non-volatility, hardness to radiations, high writing and reading speed compared to others NV technologies, high density and quasi-unlimited endurance, which make them particularly suitable for many logic applications. They can be fabricated in post-process above CMOS devices since both processes are fully compatible.

B. Magnetic Tunnel Jonction

The MTJ is the basic element of MRAMs [1]. It is a nanostructure basically composed of two FerroMagnetic (FM) layers sandwiching a thin insulating tunnel barrier. The magnetization of one of the layers, called Reference Layer (RL), is pinned and acts as a reference. The magnetization of the second layer, called Storage Layer (SL), can be switched between two directions, Parallel (P) or Anti-Parallel (AP) to the RL, with an hysteretic behavior (Fig.1). According to the relative orientations of the two layers, the electrical resistance changes: it is bigger at AP state than at P state. This phenomenon is called Tunnel Magneto-Resistance (TMR) [2]. The relative resistance variation is given by the TMR ratio $(R_{AP} - R_P)/R_P$ whose value is typically 200% and up to 600% in laboratory environment [3]. The digital information (logic '0' or '1') is then represented by the resistance of the device and is intrinsically non-volatile. Since the information is no more coded by an electric charge but by a magnetic state, the device itself is totally immune to radiations. Reading information consists in reading this resistance (for example by polarizing the MTJ at a given voltage and measuring the current). Writing information consists in switching magnetic state of the MTJ, in a P or AP configuration. Several writing schemes, corresponding to different MRAM generations, exist:

978-1-4799-2817-0/14 $31.00 © 2014 IEEE

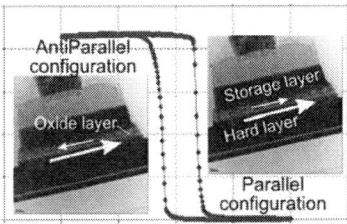

Fig. 1. MTJ device description

B.1 Field Induced Magnetic Switching

In the first generation of MRAM, called FIMS for Field Induced Magnetic Switching, the magnetization of the SL was switched using an external magnetic field, generated by an electrical current passing through lines close to the MTJ (Fig.2). For use in a memory arrays, two currents lines are required, to select the addressed line and row of the memory. Only the MTJ at the intersection of the two active lines sees a field large enough to switch its magnetization. In this approach, the stability of the magnetization in standby mode is ensured by the shape anisotropy: in a magnetic structure, the magnetic field tends to spontaneously align in the longuest dimension. If the jonction has an elliptical shape for instance, the magnetization will stand along the longuest axis. The shape factor is the ratio between the lengths of the axis. The larger the ratio, the better stability and so the larger energy required to write MTJ. In this technology, the writing time is typically about few nanoseconds and writing using only magnetic fields does not create a stress on the tunnel barrier: the endurance is considered unlimited.

The main limitation of this technology is its selectivity: indeed, for advanced technology nodes, the stability of the MTJ state decreases because of the reduced volume of the magnetic materials. Thus, due to the increasing density and process variations, the field generated to write a MTJ can accidentally write the neighbour cells (issue of half selected bits). Moreover, in this technology, the value of the writing magnetic field is important and does not scale properly with the size of the device. The current required to generate this field is typically around 16mA per current line and per bit. This leads to electromigration problems in advanced technology nodes. For all these reasons, this writing scheme is not scalable below the 90nm technology node.

B.2 Toggle switching

A solution to solve the selectivity issues was proposed in [4]. It consists in replacing the SL by an Synthetic Anti-Ferromagnetic (SAF) layer, composed of two FM layers antiparallely coupled. Instead of combining simultaneously two static magnetic fields to write, a sequence of currents is applied in both current lines to generate a rotating field. This writing scheme is used in the only commercial MRAM product existing today. It solves the selectivity issues, but remains power consumming when writing and hardly scalable for advanced technology nodes due to the large writing fields. Moreover,

each write destroys the previous state. However, this technology has been proven to be mature for industrialisation and used for instance in the flight controllers of airbus aircrafts.

B.3 Thermally Assisted Switching

Another solution was proposed to solve the limitations of magnetic field writing [5]. It is called TAS for Thermally Assisted Switching and relies on the strong dependancy of the magnetic parameters upon the temperature. In this technology, the stability of the SL is ensured by exchange interaction with an Anti-FerroMagnetic (AFM) layer. In this case, the shape of the junction can be circular (no shape factor). The idea is to apply a current through the MTJ to heat by joule effect. When temperature exceeds a so-called "blocking temperature", the exchange energy disapears, releasing the magnetization which can then be written at lower field (Fig.3). This allows decorrelating the stability in standy mode and the writing energy. In this approach, only one current line is required since half of the selection in made by the heating current. For the same reason, the magnetic field can be shared between several bits (the typical current is then around 5mA per word). The per bit writing energy is then dominated by the heating current which is relatively small and proportional to the surface of the MTJ, making the technology more scalable. In comparison with the FIMS technology, the selectivity issues are removed, the power consumption is reduced and the density improved. It is scalable down to 45nm node. The main limitation is the writing time which is limited by the heating and cooling phases durations, around 30 nanoseconds.

B.4 Spin Transfer Torque switching

The STT (Spin Transfer Torque) effect was first predicted in [6]. If TMR is the effect of a magnetization on the electrical conduction, the STT effect can be seen as the opposite: when a spin-polarized current flows through a FM layer, there is an interaction between the spin of the conducting electrons and the magnetization. In practical implementations [7], when a current flows through a MTJ, the electrons get polarized by the RL's magnetization and apply in return a torque to the SL's magnetization. Above a critical current, switching occurs. The value written in the MTJ depends on the direction of the current (Fig.4). This approach does not require external magnetic fields, but the generation of a bidirectional current. It has intrinsically no selectivity issues and is scalable, since the current

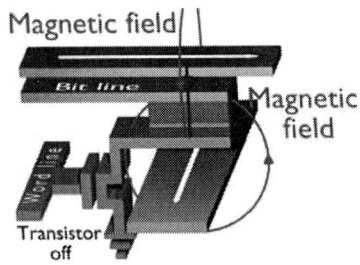

Fig. 2. FIMS writing scheme

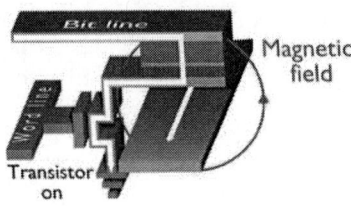

Fig. 3. TAS writing scheme

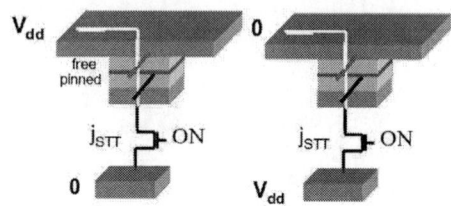

Fig. 4. STT writing scheme

decreases proportionnaly to the surface of the device. The order of magnitude of the writting current density is $10^6 A/cm^2$. Two configurations exist for STT MTJ (Fig.5): [8]:

- In the standard configuration, refered as planar STT, the magnetization is parallel to the plane of the magnetic layers. The stability is ensured by the shape factor. According to the application, it is possible to choose between a long retention (10 years for standalone memories for instance) with a relatively large writing current and a degraded stability (for cache applications for example) but with a smaller writing energy. However, due to this trade-off, stability problems appear for advance technology nodes and this technology is not suitable below 65 nm.

- In the second configuration, refered as perpendicular STT, interface interaction is used to maintain the magnetizations of the SL and RL perpendicular to the layers. In this case, the stability is ensured by the interface anisotropy. No shape factor is required. This technology allows maintaining the thermal stability down to 20 nm with a reasonnable writing current density, comparable to that of planar STT.

STT technology is currently considered as the most promising for logic applications. The writing time is a few nanoseconds, the density is very good because a memory cell only consists of one MTJ and one selection transistor. MRAM based on this technology are often refered as STT-RAM and a lot of major compagnies are interested in working on it. Everspin is about to commercialize a 64Mb product based on this technology. However, it still suffers from some limitations: first, since a current is directly injected in the MTJ to write, the same path is used for reading and writing. It is then very important to make sure that the reading current is far lower than the writing current to avoid accidental writting when reading. Second point is related to the endurance: when writing, a current is applied through the MTJ and so a voltage on the tunnel barrier. If this voltage is too close to the breaking voltage of the barrier, the endurance is affected. This limits the acceptable MTJ's value, leading to more tricky reading, because the nominal resistance of the MTJ is close to that of the transistors in their "on" state. Typically, a ratio of 3 is chosen between the breakdown voltage and the writing voltage as well as between the writing and reading voltages. Another issue is the thermal noise affecting the STT writing, resulting in some stochasticity on the switching occurrence.

B.5 Spin Orbit Torque switching

More recently, a new effect was discoverd, leading to new devices with three terminals. This effect is called Spin Orbit Torque (SOT), a generic term encompassing in reallity two effects, the Rashba effect and the Spin Hall Effect (SHE) [9]. It allows, by interface interaction, switching the magnetization of a magnetic layer by passing a current in the plane of the conductive contact line below the SL, instead of through the MTJ itself (Fig.6). This allows separating the reading and writing paths with great benefits: firstly, since no large writing current is passing through the tunnel barrier (which generates aging), the endurance can be considered infinite. Secondly, there is no risk of accidental writing during reading anymore and thirdly, it is possible to choose a higher resistance of the junction to ease the reading if required. Moreover, the switching duration using this scheme is related to the value of the current and can be very fast, typically some hundreds of picoseconds. The current density is higher than for STT (typically $5.10^7 A/cm^2$), but since it flows through a very thin line (10nm), the resulting writing current can be very small. The main limitation of this technology compared to STT is the density because of the additionnal terminal and corresponding contacts. Since the writing current decreases proportionnally to the lateral dimension of the device and not with its surface, the scalability is weaker than for STT writing, but remains much better than for magnetic field writing. The thermal stability of the device is ensured down to 20nm. Another specificity of this writing scheme is that a small static magnetic field is required to avoid stochastic switching: without it, for a given direction of the writing current, the magnetization can switch in either P or AP configuration. Adding an external magnetic field make the switching deterministic, with a direction of the current writing the P state and the opposite direction the AP state. This field can be generated by a magnetic layer with a static magnetization, acting like a per-

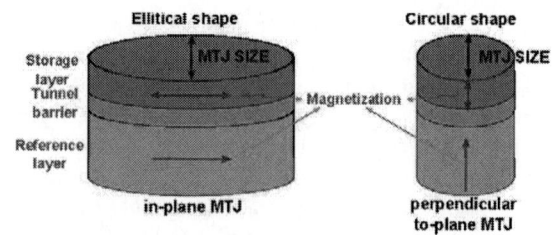

Fig. 5. Comparison between planar and perpendicular STT MTJ structures

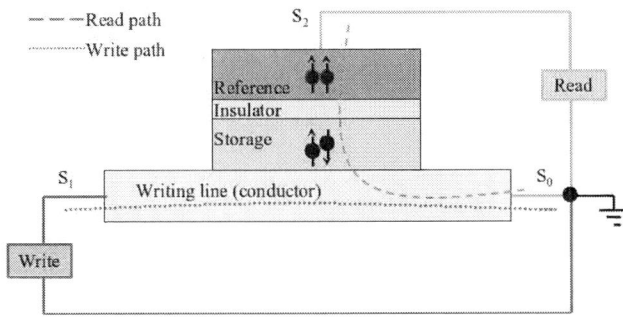

Fig. 6. SOT writing scheme

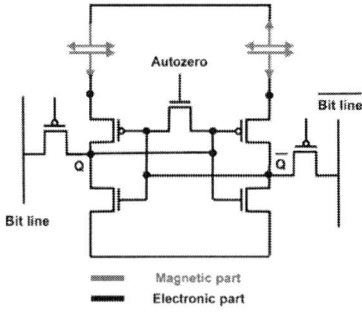

Fig. 7. Black and Das non-volatile SRAM latch

manent magnet, with no additionnal power consumption. This particularity can be turned into an advantage. Indeed, if the direction of the magnetic field is switched, the operation of the device is inverted: the direction of the current previously used to write a logic '0' will write a logic '1' and vice-versa. It can be useful for reconfigurable computing. In this case, the magnetization of the layer generating the magnetic field could be chosen using a current line, like in FIMS writing. The corresponding power consumption would then depend on the reconfiguration frequency.

Although this technology is at very early stage development, these properties make it particularly interesting for high speed applications that does not require very high densities. Typically, it can be suitable for cache 1 replacement, which is not possible with classical STT. Since this technology solves some of the major issues of STT due to separate reading and writing paths, a lot of academic organisms are intensively working on it and we have good confidence that it can have a predominant place in the futur of spintronics, for a lot of applications.

III. NON-VOLATILE HYBRID CMOS/MAGNETIC LOGIC CIRCUITS

A. Standalone or embedded memories

As introduced in section II, MRAM combines a unique set of performance compared to existing memories: it is intrinsically non-volatile, immune to radiations. Its writing speed and energy are much better than Flash and could become comparable to SRAM for advanced writing schemes like SOT. Its density can be close to DRAM for perpendicular MTJ in advanced technology nodes, since writing can be performed using a minimum size transistor. Thus, MRAM could replace some part of the memory hierachy in systems to reduce the power consumption, improve the performance or allow new functionalities. For instance, perpendicular STT-RAM could be a candidate for DRAM replacement thanks to its density or SRAM replacement in high levels of cache, while SOT-RAM could target SRAM replacement in fast caches. Better yet, even if MRAM is not as performant as existing memories for each feature taken individually, it is possible to totally rething the architectures of systems to take advantage of its combined qualities cite6081627: for instance, thanks to its density 4-5 times better than SRAM, STT-RAM could be used as cache memory, its

much bigger size compensating its relative slowness for some applications. The non-volatility allows easing the widely used power gating technique, which consists in cutting-off the power supply of unused parts of the memory to save static power consumption: indeed, their content can be backed-up at any moment locally in the magnetic part, without using distant non-volatile or very low-leakage memories and restored in a few hundreds of picoseconds.

B. Non-volatile registers or Flip-Flops

Another interesting application of MTJs consists in making SRAM cells non-volatile. This idea was first proposed by Black and Das (B&D) in [10] (Fig.7). It is based on a classical 6 transistors SRAM cell, with two MTJs operating in differential mode, meaning that when one MTJ is in P state, the other one is in AP state and vice-versa. An additional reading transistor, called AutoZero (AZ) is added between the inputs. Two differential data are coded in this latch: one in the CMOS part (Q and $\overline{Q}$) and one in the MTJs. When the AZ transistor is open, the two data are independant and the latch can be used in standard mode at CMOS speed. When applying an AZ signal, the corresponding transistor is closed and the latch is shorten. Thus, transistors in the two branches of the latch enter the conduction mode and a current is flowing through the branches. Due to the difference of resistance of the MTJs, the latch reaches a "metastable" state where the two outputs have slightly different voltages. When the AZ signal is removed, the latch switches in a stable state corresponding to the value stored in the MTJs. During this AZ phase, the latch operates as a sense amplifier to read the content of the MTJs. Such a NV latch can be associated with a standard one to give a Non-Volatile Flip-Flop (NVFF). The introduction of non-volatility in FFs containing the active data of the calculations allows simplifying the power gating technique inside the logic itself. They can also be used to implement a checkpoint for rollback operation in processors, to restore the system state with a snapshot taken in the past.

C. Logic-in-memory

Although the intrinsic non-volatility of MTJs naturally encourage to use them as memory elements, it is also possible to intrinsically mix logic and memory to shorten and multiply the communication between them, leading to a new paradigm

978-1-4799-2817-0/14 $31.00 © 2014 IEEE

in logic architectures. This concept, called "logic-in-memory" has been proposed in [11]. Examples of circuits based on Current Mode Logic (CML) have been proposed and implemented in silicon demonstrators giving encouraging results [12].

IV. DESIGN TOOLS

In order to design hybrid CMOS/magnetic circuits, it is necessary to introduce the magnetic devices and technlogy in standard design tools and flows of microelectronics. We present here the ongoing work aiming at developping a full design methodology, from device to system level, for the hybrid technology.

A. Compact models of the magnetic devices for full custom design

To design elementary circuits using MTJs, like NVFFs or memory arrays, it is necessary to develop a compact model of the device's behavior for electrical simulations. Several compact models has been proposed for MTJs, in particular for STT writing. Some of them are mainly based on physical equations [13, 14, 15, 16, 17, 18], others more behavioral to fit characterization results [19]. The first approach describes the behavior of the device more realistically and is generally more generic and flexible. The second one is often more accurate for a given technology and faster in terms of simulation time. They can be implemented using a generic language like C and compiled for a given simulator [16, 17, 18], or using a generic language like verilogA [15, 19] or VHDL-AMS [14]. Interestingly, a library of STT-MTJ models has been made available for the community [20].

More recently, a compact model of the SOT MTJ has been proposed by SPINTEC Lab [21]. It is written in VerilogA language, based on physical equations modeling the behavior of the device. It takes into account in particular the dynamics of the magnetization, based on the Landau-Lifshitz-Gilbert (LLG) equation [22], subject to SOT and the dependance of the resistance upon the magnetic configuration. In this model, all the physical quantities, like the coordinates of the magnetization, m_x, m_y and m_z are represented by voltages. Fig.8 shows a block diagram of the model, with two modules, one representing the dynamics of the magnetization and the other representing the TMR. Fig.9 shows a simulation result using the SPECTRE electrical simulator of Cadence. It illustrates the operation of the device, by showing the dynamics of the m_z coordinate of the magnetization for different values of the writing current densities J_{app}. In this particular case, the MTJ has a magnetization perpendicular to the layers, so the value of m_z is 1 for P state and -1 for AP state. Intermediate values correspond to transient steps during the switching process. The duration of the writing pulse is 3ns. We see that during the application of the current pulse, the magnetization alignes in the plane of the layers ($m_z=0$). Once the magnetization is stabilized, the current is removed and the magnetization switches in its final state according to the direction of the current. For $J_{app} = 10^{12} A/m^2$, the SOT is not strong enough to initiate the switching and the jonction remains in its P state. For $J_{app} = 2.10^{12} A/m^2$, the 3ns pulse is not long enough for the

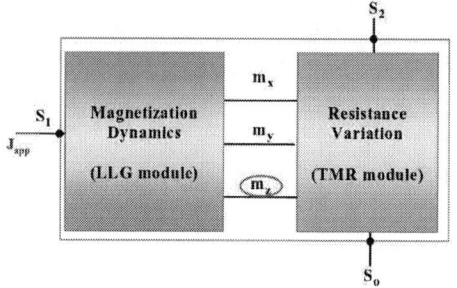

Fig. 8. Block diagram of the SOT-MTJ compact model

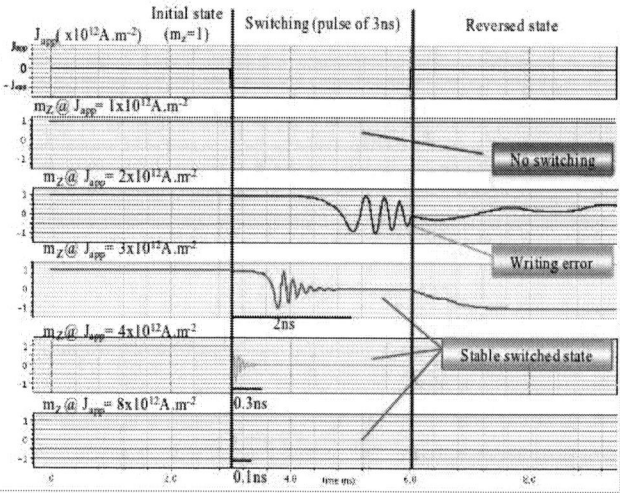

Fig. 9. Simulation of the dynamic behavior of the SOT-MTJ perpendicular magnetization mz according to the current

magnetization to stabilize and the MTJ switches back to its initial P state. For higher values of J_{app}, switching occurs, with a duration decreasing with the increase of the current density, allowing a trade-off between speed and power consumption. We see that for high current values ($4.10^{12} A/m^2$ for instance), the switching time can be less than 1ns. In Fig.10, we see the influence of an external magnetic field on the switching. Without it, the switching is stochastic. When the field is applied, switching becomes deterministic, with a switching time duration decreasing when the magnitude of the field increases.

B. Non-volatile standard cells and integration in the digital design flow

The design of a complex digital circuit requires automatically interconnecting standard cells, performing more or less elementary logic functions. As seen in section III, NV devices are advantageously introduced in registers or Flip-Flops. Based on the B&D approach, a lot of NVFF have been proposed in the litterature, optimized for area, speed, or reliability against process variations [23, 24, 25]. For example, in [26], the authors present a ultra-compact NV latch based on the loadless 4 transistors SRAM cell.

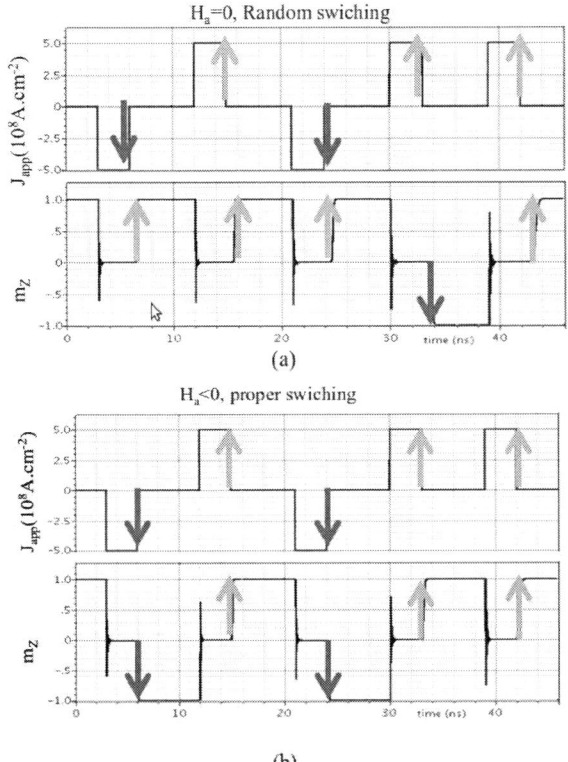

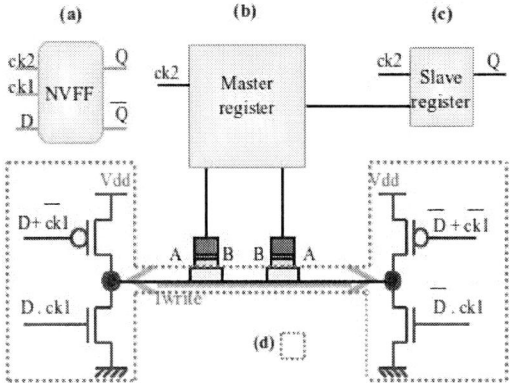

Fig. 11. SOT-MRAM based NVMFF (a) symbol, (b) master register architecture (c) slave register, (d) writing circuit

(left column figures)

H$_a$=0, Random swiching

H$_a$<0, proper swiching

(b)

Fig. 10. Study of the dynamic behavior of the perpendicular magnetization mz according to the value of the external applied field H_a

means that with such a device, the power supply can be cut-off at any moment, without further action and restarted in the same state. This opens the door to the concept of "normally-off" or "instant-on" circuits [27].

Once designed and characterized, these circuits has to be integrated in the design flow with behavioral description (in verilog or VHDL) and timing libraries for logic synthesis and layout for place and route. Such a work has been performed for the TAS technology in post-process above a STMicroelectronics 130nm CMOS technology [28] and can be easily adapted to other technologies as soon as specific standard cells has been designed. It is currently developed for the SOT technology in the framework of a European Research project called spOt [29].

C. System level investigations

To evaluate the gain that can be expected of using MRAM in real applications, it is necessary to carry out system level

In order to illustrate the potential of SOT, we describe here a NVFF architecture based on this technolgy (Fig.11), developped at Spintec. It is composed of a master latch made non-volatile using a pair of SOT MTJs (Fig.11(b)) connected to a standard slave latch (Fig.11(c)). The operation is managed by a non-overlaping two-phases clock ($ck1$, $ck2$). Four transistors are used to generate a bidirectional current flowing in the both SLs of the SOT devices in series. The directions of the current is determined by the input data D. The generation of the current is enabled during the first phase clock $ck1$. During the second phase clock $ck2$, the master latch reads the data of the SOT MTJs and transfer it to the output Q. This operation is illustrated in Fig.12, showing a simulation of the FF using the compact model presented in section 8. In this work, the technolgy node is 40nm, for both CMOS and SOT MTJ. In these conditions, the writing current is 60uA, comparable to perpendicular STT-RAM. But as seen in inserts, the switching time is 250 ps, four times faster than STT, dividing the energy by four. Moreover, the current is only flowing through a current line whose resistance is much smaller than the MTJ's resistance, reducing all the more the power consumption.

The operation of this FF fondamentally differs from those presented so far: indeed, the magnetic part is written at every clock cycle contrary to the previous cases where the data was backed up in the magnetic part only when required, on demand. This is made possible because of the high writing speed and low-power consumption of the SOT technology. It

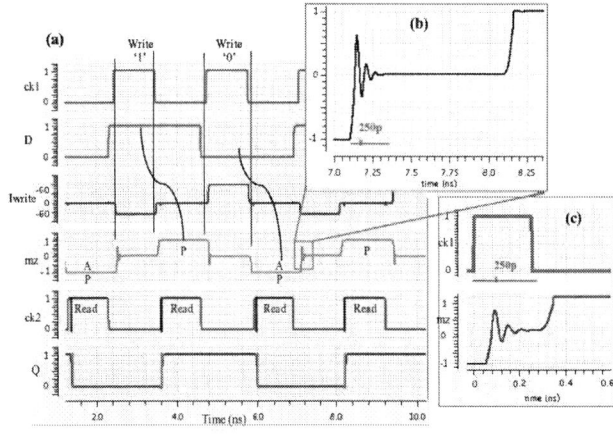

Fig. 12. SOT-MRAM based NVMFF simulation results a) write and read operations (b) perpendicular magnetization mz behavior of the SOT-MRAM during a 1ns current pulse to switch from AP (mz=-1) to P (mz=1) (c) SOT-MRAM switching with only a 250ps pulse (exactly after oscillations stabilization)

978-1-4799-2817-0/14 $31.00 © 2014 IEEE

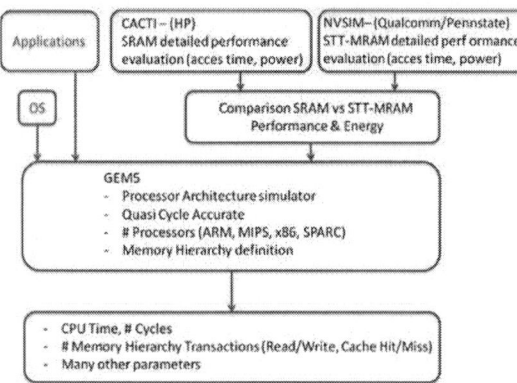

Fig. 13. System level design flow to evaluate the performance of processors embedding MRAM

investigations. Indeed, the possible benefit strongly depends on the application and the system architecture. Some ongoing works aim at evaluating the performance of NV processors [30] (Fig.13). GEM5 is a quasi-cycle accurate processor architecture simulator, developed by ARM, which takes as inputs the architecutre of the processor and the memory hierarchy definition. It provides a lot of performance information like the CPU time, the number of cycles, of memory transactions (read/write, cache hit/miss) for a given running application. To operate, GEM5 needs a model of the cache memories. These models can be obtained by simulation/characterization of an existing memory, or using memory simulators which provide detailed performance of a memory, without having to design it. CACTI, from HP, is a widely-used SRAM simulator which gives very accurate results. NVSIM is an extension of CACTI, developped by Qualcomm in collaboration with Pennstate University, for resistive memories. It can be used to predict the performance of MRAM (ongoing work, for example in the framework of the spOt project [29]). However, although it gives preliminary interesting results, it remains very generic for non-volatile memories and should be adapted to take into account the specificities of MRAM and to get really accurate. The outputs of the simulator processor can in return provide useful feedback to help defining and designing the memory.

V. CONCLUSION

The emergence of new MRAM generations has led to a keen interest all around the world for this technology as a solution to push forward the limits of microelectronics. A lot of tools are developed to allow exploring new circuits and system architectures using MRAM devices. As seen in this article, each MRAM technology has its own strenghs and limits and is thus suitable for a given set of applications. A wide field of investigation is open which covers technology developments to improve the existing devices and discover new ones, and design of innovative circuits and systems. Two ways must be followed in parallel: identifying relevant application targets for each technology and adapt the architectures of the systems to gain maximum benefits from it. The performance of the most

recent technologies and in particular SOT, in terms of speed and power consumption which could compete with SRAM, paves the way towards a true "universal" memory.

ACKNOWLEDGMENTS

This work has been partially funded by the European Commission under the spOt project (grant agreement n318144) in the framework of the Seventh Framework Program, and from the French National Research Agency (ANR) under the DIP-MEM project (contract ANR-12-NANO-0010) in the framework of the P2N program.

REFERENCES

[1] S. Tehrani, B. Engel, E. Chen, M. DeHerrera, M. Durlam, and J. Slaughter, "Recent developments in magnetic tunnel junction mram," in *Magnetics Conference, 2000. INTERMAG 2000 Digest of Technical Papers. 2000 IEEE International*, 2000, pp. 282–282.

[2] M. Julliere, "Tunneling between ferromagnetic films," *Physics Letters A*, vol. 54, pp. 225–226, 1975.

[3] S. Ikeda, J. Hayakawa, Y. Ashizawa, Y. Lee, K. Miura, H. Hasegawa, M. Tsunoda, F. Matsukura, and H. Ohno, "Tunnel magnetoresistance of 604% at 300 k by suppression of ta diffusion in cofeb/mgo/cofeb pseudo-spin-valves annealed at high temperature," *Applied Physics Letters*, vol. 93, no. 8, pp. 082 508–082 508–3, 2008.

[4] L. Savtchenko, B. Engel, N. Rizzo, M. Deherrera, and J. Janesky, "Method of writing to scalable magnetoresistance random access memory element," Patent US6 545 906, 2003.

[5] I. L. Prejbeanu, W. Kula, K. Ounadjela, R. Sousa, O. Redon, B. Dieny, and J.-P. Nozieres, "Thermally assisted switching in exchange-biased storage layer magnetic tunnel junctions," *Magnetics, IEEE Transactions on*, vol. 40, no. 4, pp. 2625–2627, 2004.

[6] J. Slonczewski, "Currents and torques in metallic magnetic multilayers," *Journal of Magnetism and Magnetic Materials*, vol. 159, 1996.

[7] G. Fuchs, N. Emley, I. Krivorotov, P. Braganca, E. M. Ryan, S. I. Kiselev, J. C. Sankey, D. C. Ralph, R. Buhrman, and J. Katine, "Spin-transfer effects in nanoscale magnetic tunnel junctions," *Applied Physics Letters*, vol. 85, no. 7, pp. 1205–1207, 2004.

[8] K. C. Chun, H. Zhao, J. Harms, T.-H. Kim, J.-P. Wang, and C. Kim, "A scaling roadmap and performance evaluation of in-plane and perpendicular mtj based stt-mrams for high-density cache memory," *Solid-State Circuits, IEEE Journal of*, vol. 48, no. 2, pp. 598–610, 2013.

[9] I. M. Miron, K. Garello, G. Gaudin, P.-J. Zermatten, M. V. Costache, S. Auffret, S. Bandiera, B. Rodmacq, A. Schuhl, and P. Gambardella, "Perpendicular

switching of a single ferromagnetic layer induced by in-plane current injection," *Nature*, vol. 476, no. 7359, pp. 189–193, Aug. 2011. [Online]. Available: http://dx.doi.org/10.1038/nature10309

[10] W. C. Black and B. Das, "Programmable logic using giant-magnetoresistance and spin-dependent tunneling devices (invited)," *Journal of Applied Physics*, vol. 87, no. 9, pp. 6674–6679, 2000.

[11] A. Mochizuki, H. Kimura, M. Ibuki, and T. Hanyu, "Tmr-based logic-in-memory circuit for low-power vlsi*," *IEICE Trans. Fundam. Electron. Commun. Comput. Sci.*, vol. E88-A, no. 6, pp. 1408–1415, Jun. 2005. [Online]. Available: http://dx.doi.org/10.1093/ietfec/e88-a.6.1408

[12] S. Matsunaga, J. Hayakawa, S. Ikeda, K. Miura, H. Hasegawa, T. Endoh, H. Ohno, and T. Hanyu, "Fabrication of a nonvolatile full adder based on logic-in-memory architecture using magnetic tunnel junctions," *Applied Physics Express*, vol. 1, no. 9, p. 091301, 2008. [Online]. Available: http://apex.jsap.jp/link?APEX/1/091301/

[13] S. Lee, S. Lee, H. Shin, and D. Kim, "Advanced hspice macromodel for magnetic tunnel junction," *Japanese Journal of Applied Physics*, vol. 44, no. 4B, pp. 2696–2700, 2005. [Online]. Available: http://jjap.jsap.jp/link?JJAP/44/2696/

[14] M. Madec, J. Kammerer, F. Pregaldiny, L. Hebrard, and C. Lallement, "Compact modeling of magnetic tunnel junction," in *Circuits and Systems and TAISA Conference, 2008. NEWCAS-TAISA 2008. 2008 Joint 6th International IEEE Northeast Workshop on*, 2008, pp. 229–232.

[15] A. Jander, L. Engelbrecht, and P. Dhagat, "Dynamic verilog-a model of a magnetoresistive spin valve," in *Behavioral Modeling and Simulation Workshop, 2008. BMAS 2008. IEEE International*, 2008, pp. 50–54.

[16] W. Guo, G. Prenat, V. Javerliac, M. E. Baraji, N. de Mestier, C. Baraduc, and B. Diny, "Spice modelling of magnetic tunnel junctions written by spin-transfer torque," *Journal of Physics D: Applied Physics*, vol. 43, no. 21, p. 215001, 2010. [Online]. Available: http://stacks.iop.org/0022-3727/43/i=21/a=215001

[17] G. Prenat, B. Dieny, W. Guo, M. El Baraji, V. Javerliac, and J.-P. Nozieres, "Beyond mram, cmos/mtj integration for logic components," *Magnetics, IEEE Transactions on*, vol. 45, no. 10, pp. 3400–3405, 2009.

[18] M. El Baraji, V. Javerliac, W. Guo, G. Prenat, and B. Dieny, "Dynamic compact model of thermally assisted switching magnetic tunnel junctions," *Journal of Applied Physics*, vol. 106, no. 12, pp. 123 906–123 906–6, 2009.

[19] Y. Zhang, W. Zhao, G. Prenat, T. Devolder, J.-O. Klein, C. Chappert, B. Dieny, and D. Ravelosona, "Electrical modeling of stochastic spin transfer torque writing in magnetic tunnel junctions for memory and logic applications," *Magnetics, IEEE Transactions on*, vol. 49, no. 7, pp. 4375–4378, 2013.

[20] W. Zhao. Spinlib — spintronics nanodevice spice-compatible compact model libarary. [Online]. Available: http://www.ief.u-psud.fr/ zhao/spinlib.html

[21] K. Jabeur, G. Prenat, G. Di Pendina, L. Buda-Prejbeanu, I. Prejbeanu, and B. Dieny, "Compact model of a three-terminal mram device based on spin orbit torque switching," in *Semiconductor Conference Dresden-Grenoble (ISCDG), 2013 International*, 2013, pp. 1–4.

[22] L. D. Landau and E. Lifshitz, "On the theory of the dispersion of magnetic permeability in ferromagnetic bodies," *Phys. Z. Sowjetunion*, vol. 8, no. 153, pp. 101–114, 1935.

[23] W. Zhao, E. Belhaire, V. Javerliac, C. Chappert, and B. Dieny, "A non-volatile flip-flop in magnetic fpga chip," in *Design and Test of Integrated Systems in Nanoscale Technology, 2006. DTIS 2006. International Conference on*, 2006, pp. 323–326.

[24] N. Sakimura, T. Sugibayashi, R. Nebashi, and N. Kasai, "Nonvolatile magnetic flip-flop for standby-power-free socs," in *Custom Integrated Circuits Conference, 2008. CICC 2008. IEEE*, 2008, pp. 355–358.

[25] Y. Guillemenet, L. Torres, G. Sassatelli, and N. Bruchon, "On the use of magnetic rams in field-programmable gate arrays," *Int. J. Reconfig. Comput.*, vol. 2008, pp. 1:1–1:9, Jan. 2008. [Online]. Available: http://dx.doi.org/10.1155/2008/723950

[26] G. Di Pendina, K. Torki, G. Prenat, Y. Guillemenet, and L. Torres, "Ultra compact non-volatile flip-flop for low power digital circuits based on hybrid cmos/magnetic technology," in *Integrated Circuit and System Design. Power and Timing Modeling, Optimization, and Simulation.* Springer, 2011, pp. 83–91.

[27] T. Kawahara, "Scalable spin-transfer torque ram technology for normally-off computing," *Design & Test of Computers, IEEE*, vol. 28, no. 1, pp. 52–63, 2011.

[28] G. Di Pendina, G. Prenat, B. Dieny, and K. Torki, "A hybrid magnetic/complementary metal oxide semiconductor process design kit for the design of low-power nonvolatile logic circuits," *Journal of Applied Physics*, vol. 111, no. 7, pp. 07E350–07E350–3, 2012.

[29] spot - spin orbit torque memory for cache and multicore processor applications. [Online]. Available: http://www.spot-research.eu/

[30] L. Torres, R. Brum, L. Cargnini, and G. Sassatelli, "Trends on the application of emerging nonvolatile memory to processors and programmable devices," in *Circuits and Systems (ISCAS), 2013 IEEE International Symposium on*, 2013, pp. 101–104.

Architectural Aspects in Design and Analysis of SOT-based Memories

Rajendra Bishnoi, Mojtaba Ebrahimi, Fabian Oboril and Mehdi B. Tahoori

Karlsruhe Institute of Technology (KIT)

· Chair of Dependable Nano Computing (CDNC)

Karlsruhe, Germany

e-mails: {rajendra.bishnoi, mojtaba.ebrahimi, fabian.oboril, mehdi.tahoori}@kit.edu

Abstract—Magnetic Random Access Memory (MRAM) is a very promising emerging memory technology because of its various advantages such as non-volatility, high density and scalability. In particular, Spin Orbit Torque (SOT) MRAM is gaining interest as it comes along with all the benefits of its predecessor Spin Transfer Torque (STT) MRAM, but is supposed to eliminate some of its shortcomings. Especially the split of read and write paths in SOT-MRAM promises faster access times and lower energy consumption compared to STT-MRAM. In this work, we provide a very detailed analysis of SOT-MRAM at both circuit- and architecture-level. We present a detailed evaluation of performance and energy related parameters and compare the novel SOT-MRAM with several other memory technologies. Our architecture-level analysis shows that with a *hybrid*-combination of SRAM for the L1-cache and SOT-MRAM for the L2-cache the energy consumption can be reduced by 63 % in average while the performance can be increased by 1 %. In addition, the memory area is 43 % lower compared to an SRAM-only configuration.

I. INTRODUCTION

As the continuous downscaling of CMOS technology becomes more and more challenging, there has been a great deal of efforts to find feasible alternatives. For random access memory (RAM), *nano-magnetic storage devices (MRAM)* are very promising candidates to replace the traditional CMOS-based memory solutions. Especially the non-volatility of MRAM is a major advantage, which minimizes static power consumption and paves the way towards normally-off/instant-on computing. In particular, MRAM based on *Magnetic Tunnel Junction[1] (MTJ)* [26,27] storage devices is one of the most interesting candidates as identified by the ITRS [12]. Among these memory technologies, *Spin Transfer Torque MRAM (STT-MRAM)* [1] gains a lot of attention as it is non-volatile, scalable, and has a low read access time [4, 10, 27]. In addition, due to the high resistance of the MTJ storage elements, STT-MRAM is compatible with the CMOS process. Furthermore, the magnetization of the storage layer, and hence the stored data, can be switched without requiring an external magnetic field. Instead, a spin polarized current flowing through the MTJ device is employed.

Despite all these advantages, STT-MRAM also faces various challenges. First, although the write current is much lower than in many other MRAM technologies [10], it is still very high, leading to a high energy consumption (10x more energy

per write operation than SRAM) [6, 24]. In addition, the high current through the MTJ imposes a severe stress for the memory cell. As a result, it leads to a time dependent degradation of the MTJ performance parameters such as tunneling magneto resistance, write current, and write latency. Moreover, also the lifetime is reduced, as the MTJ oxide is threatened by time dependent dielectric breakdown [20, 28]. Second, beside the high write current, also the write path itself is a challenge. In STT-MRAM, the read and write operations share the same access path (through the junction) which can impair the reliability, i.e. a read operation can by mistake lead to a bit flip (magnetization of the storage layer is switched). Third, the long write latencies usually prohibit the use of STT-MRAM in first level caches [4].

To mitigate these issues, *Spin Orbit Torque MRAM (SOT-MRAM)* has been recently proposed [7, 13, 18]. SOT-MRAM uses a three terminal MTJ-based concept to isolate the read and the write path compared to the two terminal concept of STT-MRAM. As a result, in SOT-MRAM the read and the write path are perpendicular to each other which significantly improves the read stability [13]. Moreover, the write current is much lower and also the write access is supposed to be much faster, as the write path can now be optimized independently.

In this paper, we provide a detailed circuit- and architecture-level analysis of the SOT-MRAM in both memory array design and its implications for a hybrid memory hierarchy in an advanced computing system. As we will show, the read and write latencies of SOT-MRAM are comparable to those of SRAM. In addition, SOT-MRAM offers a much higher density, lower energy consumption, is radiation immune and non-volatile. All of these aspects make SOT-MRAM a viable candidate for on-chip memory, not only for the last-level cache, but also for lower levels of cache to replace SRAM. To illustrate these benefits, we perform both circuit-level and architecture-level evaluations in which we compare SOT-MRAM with SRAM and STT-MRAM as L1- and L2-cache memory. This analysis shows that a *hybrid*-combination of SRAM for the L1-cache and SOT-MRAM for the L2-cache can reduce the energy consumption 63 % in average, while it even increases the performance slightly by 1 %. In addition, the area occupied by the memory units is 43 % lower compared to an SRAM-only solution. Even more energy savings are possible, if SOT-MRAM is used in both cache levels. However, this incurs a small performance penalty (up to 2 %).

The rest of this paper is organized as follows. In Section II, the basics of SOT-MRAM are introduced. Section III explains the details of the memory architecture using SOT-MRAM and the resulting memory characteristics such as access latencies,

---

[1]memory component consisting of the two magnetic layers and a barrier oxide in between storing a logic value in form of a resistance state (see Fig 1)

978-1-4799-2817-0/14 $31.00 © 2014 IEEE

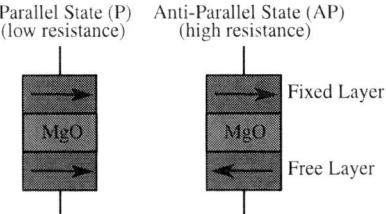

Fig. 1. MTJ resistance according to the magnetization of the free layer

energy consumption and density. Furthermore, the extracted data is compared with various other memory technologies. In addition, this information is used in Section IV to analyze the advantages and disadvantages of SOT-MRAM as a possible replacement of SRAM inside a classical memory hierarchy. Finally, Section V concludes the paper.

II. BACKGROUND

A. Magnetic Tunnel Junction Device

The storage devices in Spin Orbit Torque memories are Magnetic Tunnel Junction (MTJ) cells in which data is stored as a resistance state value. An MTJ device, as shown in Figure 1, consists of two independent ferromagnetic layers (e.g. CoFeB) separated by a very thin ($\approx$ 1 nm) barrier oxide layer such as magnesium oxide (MgO) [13]. One of the two ferromagnetic layers has a fixed magnetization, i.e. the orientation of its magnetic field is fixed. Hence, this layer is known as *fixed* or *reference layer*. In contrast, in the second magnetic layer the magnetization can be freely rotated based on the direction of the current (i.e. spin of the electric particles) flowing through the MTJ device. Therefore, this layer is referred to as *free layer*.

When the direction of the magnetic field of the free layer is *parallel (P)* to the fixed layer, i.e. the magnetic field orientations in both layers are the same, the MTJ cell has a low resistance value. On contrary, when the magnetization of the free layer is opposite or *anti-parallel (AP)* to the fixed layer, the MTJ cell has a high resistance value. This high and low resistance values are used to represent logic '1' and '0' values.

B. SOT-MRAM Structure

The MTJ cell is the core part of a bit-cell in SOT-based memories as well as in STT-MRAM as shown in Figure 2. How-

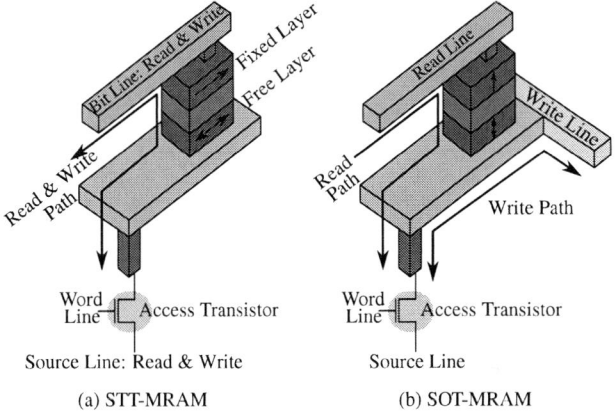

(a) STT-MRAM   (b) SOT-MRAM

Fig. 2. Comparison of a standard bit-cell for STT-MRAM and SOT-MRAM

ever, to eliminate the shortcomings of STT-MRAM, the SOT-MRAM bit-cell has an additional terminal to separate the (uni-directional) read and the (bidirectional) write path which are perpendicular to each other. The terminals comprise a *read line*, a *write line*, a *source line* and a *word line*. The word line is used to access the required bit-cell during memory accesses via the NMOS-based access transistor. If such an access is a read operation, the source line is connected to the ground and the read line is used to measure the MTJ resistance by sensing the current flowing through the MTJ cell. During the write operation the current flows between the source line and the write line. In fact, The current direction is determined by the potentials of the source line and the write line (i.e. the write path is bidirectional). The current direction in turn affects the magnetization of the free layer and hence the value stored in the bit cell. If the current flows from the source line to the write line, the MTJ resistance will be low. To achieve a high MTJ resistance, i.e. anti-parallel state, the current needs to flow from write to source line (high potential for the write line). However, the underlying physical relation between the current and the magnetic field orientation is still under discussion. On the one hand, the *Rashba effect* is said to be responsible for the current-induced magnetization switch [7, 19]. On the other hand, many people explain this phenomenon with the *Spin Hall Effect* [18]. Due to this reason, some also refer to SOT-MRAM as "Giant Spin Hall Effect" MRAM. Nevertheless, in both cases the spin-orbit-torque is responsible for the status of the free layer magnetization, which is the origin of the name SOT-MRAM.

Since the read and write paths are independent of each other in SOT-MRAM, they can be also optimized separately. This is used to reduce the write current and write latency in SOT-MRAM compared to STT-MRAM. As we will show later, this is the reason why SOT-MRAM can achieve access times similar to SRAM, while STT-MRAM suffers from high write latencies. In addition, also the asymmetry between read and write operations can be significantly reduced, such that in SOT-MRAM read and write operations have the same access times, while in STT-MRAM a write access requires considerably more time.

It can be inferred from Figure 2 that a bit-cell consists of two different technologies, namely CMOS for the transistor and a nano-magnetic technology for the MTJ device. Therefore, the MTJ cells require additional layers in the layout and more processing steps during the fabrication process.

III. CIRCUIT-LEVEL EVALUATION OF SOT-MRAM

A. Details of the SOT-MRAM Architecture

The architecture of an SOT-MRAM memory array is shown in Figure 3. As it can be seen, similar to the SRAM memory architecture, it has a decoder which is responsible for the activation of the word line indicated by the memory address. The major difference with SRAM is in the write and read circuitry. As mentioned in Section II, the SOT bit-cell is a four terminal device which has different paths for write and read operations. In case the write enable signal is inactive, a read operation is performed by connecting the read line of the desired bit-cell to a current sense amplifier. The current sensed on the read line is

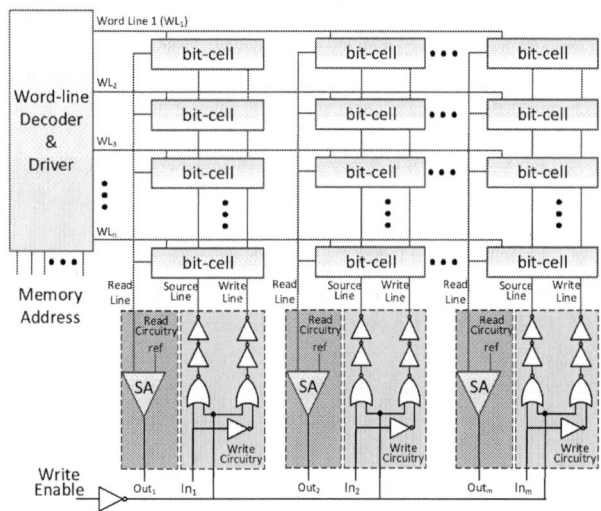

Fig. 3. Read and write operation using SOT-based bit-cell

	SOT-MRAM	STT-MRAM
Read Latency [ps]	221	226
Write Latency [ps]	266	10.500 (reset) / 3,700 (set)
Write Current [uA]	100	525 (reset) / 616 (set)
Read Energy [pJ]	1.8	1.8
Write Energy [pJ]	0.1	3.9 (reset) / 3.4 (set)

TABLE I

COMPARISON OF SOT-MRAM AND STT-MRAM FOR A SINGLE BIT-CELL

compared with a reference value to distinguish the value stored in the bit-cell.

For the write operation, the write enable signal has to be activated. In fact, the write operation in SOT-MRAM is bidirectional, i.e. the data stored in the bit-cell depends on the direction of the current which in-turn is determined by the input data value. As a result, the write circuitry can be designed in such a way that the high resistance state of the MTJ cell represents either a logic '1' or a logic '0'. For the write circuitry shown in Figure 3, it is assumed that the anti-parallel state (high resistance) represents a logical value of '1'. When the write enable signal is active and the input data has a logical value of '1', the current flows from the write line to the source line in the MTJ cell resulting in high resistance.

B. Comparison with other Memory Technologies

To investigate the SOT based memory architecture and compare it with other memory technologies, we use a multi-level approach. First, we analyze the behavior for a single bit-cell only. Afterwards, this information is used to extract the data for an entire memory array.

B.1 Circuit-Level Memory Evaluation Platform

For the bit-cell analysis of SOT-MRAM, we use the framework proposed in [13] in combination with the TSMC 65 nm general purpose library for the CMOS elements. For STT-MRAM we apply the model from [8], which employs in-plane magnetization, whereas the model for SOT-MRAM uses a perpendicular magnetic anisotropy. For both technologies, the magnetic switching dynamics for the free layers are described by the Landau-Lifshitz-Gilbert model [16].

The results of this analysis are summarized in Table I and underline the benefits of SOT-MRAM over STT-MRAM. In SOT-MRAM, the write access latency for a single bit-cell is similar to that of a read operation. However, SOT-MRAM has the same latency for the two possible write operations, i.e. write '1' (set) and write '0' (reset), while there is a huge difference for STT-MRAM. Hence, the significant asymmetry of STT-MRAM is no longer an issue for SOT-MRAM. This is mainly due to the

fact that the write path in SOT-MRAM can be optimized separately as explained in Section II. Moreover, also the per-access energy and the write current of SOT-MRAM is much lower. Therefore, the access transistor of an SOT-MRAM bit-cell can be designed much smaller. This in turn leads to a lower leakage power for SOT-MRAM.

Based on the results obtained from a single bit-cell, we extracted the area, read and write latency, per access energy and leakage power for a complete memory array using NVSim [5]. NVSim contains circuit-level performance, energy, and area models for various non-volatile memory technologies such as SRAM, PC-RAM, R-RAM, NAND-Flash and in particular STT-MRAM. However, the standard models used in thus tool for STT-MRAM do not consider its asynchronous write behavior (set vs. reset). Therefore, we modified NVSim to support this effect. Beside these necessary modifications for STT-MRAM, we adapt this model also for SOT-MRAM, which is possible as both technologies are very related. Moreover, we assume that the additional terminal of SOT-MRAM does not affect the bit-cell footprint. Therefore, the actual area numbers for SOT-MRAM could be higher than those reported here. For all memory technologies expect MRAM (i.e. SOT-MRAM and STT-MRAM) the default parameters of NVSim, which are based on the ITRS data, are applied for this study. For STT- and SOT-MRAM we use the previously extracted bit-cell information to feed the modified NVSim models as these are more accurate than the data provided by NVSim for STT-MRAM. All NVSim evaluations use the latency-optimized parameter set.

B.2 Comparison of SOT-MRAM with other Memory Technologies

To compare various memory technologies, we use a 512 KByte memory as a case study for which the results are summarized in Table II. For NAND-Flash, we consider the size of one page as 256 Byte and the write access energy number is reported per page. Furthermore, we report only the worst-case write latency and energy for all memories. As the results show, SOT-MRAM is competitive with SRAM in terms of performance and is even superior when it comes to energy consumption and cell density. In addition, unlike SRAM, SOT-MRAM does not have scalability limitations [1] and is also radiation immune. Although PC-RAM and R-RAM are comparable to SOT-RAM in terms of area and read latency, these memory technologies suffer significantly from their high write latency and write energy [11]. NAND-Flash has the smallest area and leakage, however it has problems with a high write energy, scalability and endurance.

Please note that for every memory technology different ways of implementation are possible, e.g. low-power, high-performance or high-density optimized versions. As a consequence, also the absolute numbers presented in Table II would change for other implementations. However, the major trends

978-1-4799-2817-0/14 $31.00 © 2014 IEEE

	6T-SRAM [25]	NAND-FLASH [17,23]	STT-MRAM [15,21]	SOT-MRAM [7,13,18]	PC-RAM [22,29]	R-RAM [14]
Data Storage	Latch	Floating Gate Device	Magnetization	Magnetization	Resistance	Resistance
Non-Volatility	no	yes	yes	yes	yes	yes
Area [mm^2]	2.78	0.17	1.63	1.51	0.31	0.66
Read Latency [ns]	2.17	565.365	1.2	1.13	0.55	1.15
Write Latency [ns]	2.07	2×10^5	11.22	1.36	150.4	20.66
Read Access Energy [pJ]	587	3921	260	247	363.4	193
Write Access Energy [pJ]	355	6902	2337	334	63670	592
Leakage Power [mW]	932	77	387	254	153	115
Process	CMOS	Floating Gate Device	CMOS + STT-MTJ	CMOS + SOT-MTJ	CMOS + GST^2	CMOS + MIM^3
Features (based on ITRS [12])	(−) Scalability (++) Endurance (-) Radiation vulnerable	(-) Scalability (−) Endurance (-) Radiation vulnerable	(+) Scalability (+) Endurance (+) Radiation immune (-) Bit Failure Rate	(+) Scalability (+) Endurance (+) Radiation immune (-) Bit Failure Rate	($\pm$) Scalability (-) Endurance (+) Radiation immune	(+) Scalability (-) Endurance (+) Radiation immune (-) Bit Failure Rate (-) Retention

TABLE II

COMPARISON OF VARIOUS MEMORY TECHNOLOGIES FOR A 512 KBYTE MEMORY BASED ON THE FLOW FROM SECTION III.B.1

will remain the same. Therefore, the main purpose of this analysis, as summarized in Table II, is a comparative analysis of the trends for several memory technologies and their usabilities for the on-chip memory hierarchy, rather than the actual numbers.

B.3 SOT-MRAM Scaling for Various Memory Sizes

Beside the analysis for a single memory size of 512 KByte, we also evaluated the most important memory parameters for SRAM (6T), STT-MRAM and in particular SOT-MRAM for various other memory sizes in the range between 16 KByte and 4 MByte using the same methodology as in the previous subsection. The results are summarized in Figure 4 as well as Figure 5 and are discussed in the following paragraphs. Please note that the actual numbers can differ based on the particular memory architecture, but the overall trends discussed here will remain the same.

Area: The first interesting observation of our analysis is the scaling behavior of the area occupied by the memory (Figure 4(a)). As it can be seen, for large memory capacities all three memory technologies show the same trend, i.e. with duplicated memory capacity also the area increases by a factor of almost 2. However, for sizes smaller than 512 KByte, the area

[2]GST: An alloy for Phase change material Ge$_2$Sb$_2$Te$_5$
[3]MIM : Metal-Insulator-Metal component

of STT-MRAM and SOT-MRAM increases slower than the capacity. In contrast, SRAM still scales with the same trend. As a result, SRAM offers better area usage for small memory capacities, while SOT-MRAM is superior for larger sizes (here starting from 256 KByte).

To explain this phenomenon it is necessary to decompose the memory area into the total bit-cell area and the area of the periphery (i.e. write circuitry, decoder, sense amplifier). In this regard, the bit-cells for SOT-MRAM and STT-MRAM are much smaller than those for SRAM. In contrast the periphery for MRAM is larger, due to the higher write current. Both aspects together lead to the fact that, in case of MRAM, the size of the periphery dominates or is similar to the total bit-cell area for memory capacities below 64 KByte. Furthermore, the size of the periphery does not scale linearly with the memory capacity, while the total bit-cell area does. Hence, for small memory capacities, the scaling of SOT-MRAM and STT-MRAM is limited by the size of the memory periphery, while for SRAM the total bit-cell area is the limiting factor and thus it scales better.

Please note that the actual area numbers for SOT-MRAM could be higher than those reported here, since we assume that the additional bit-cell terminal does not increase the bit-cell footprint compared to STT-MRAM. In fact, the overhead due to the additional terminal depends on various aspects, e.g. design rules and size of the access transistor. Therefore, as SOT-

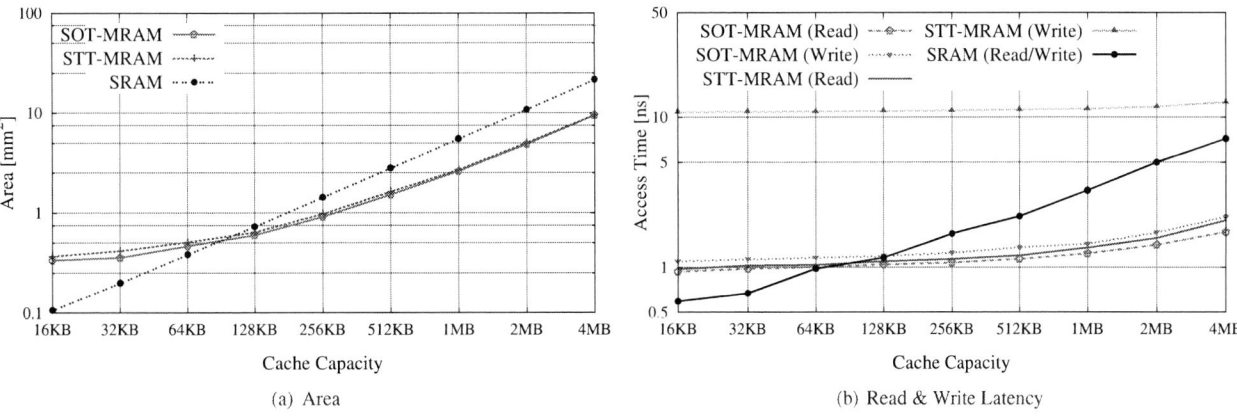

Fig. 4. Area and latency scaling behavior for SRAM, STT-MRAM and SOT-MRAM for various memory sizes

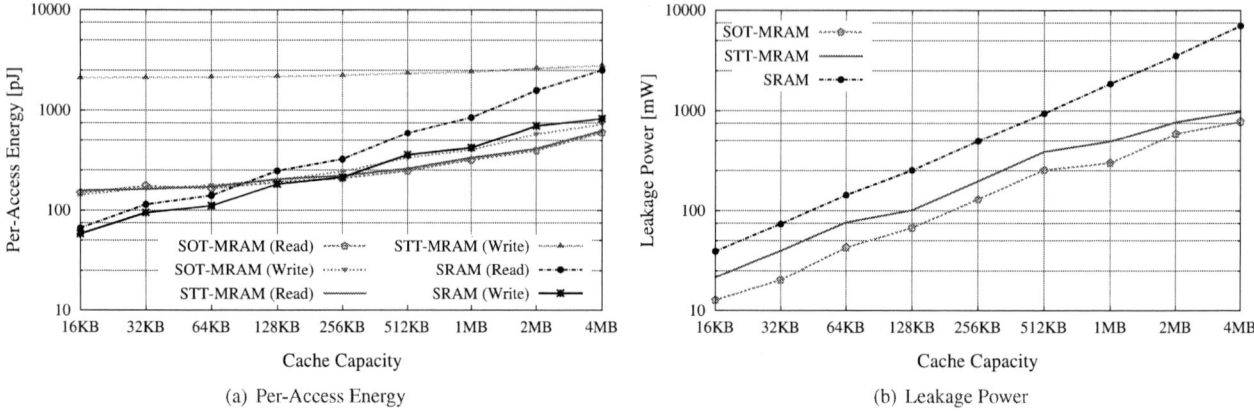

Fig. 5. Access energy and leakage scaling behavior for SRAM, STT-MRAM and SOT-MRAM for various memory sizes

MRAM is not yet in production, it is not possible to quantify the additional area required due to the fourth terminal.

Access Latencies: Another interesting phenomenon can be observed for the scaling behavior of the access latencies (see Figure 4(b)). Since the load capacitance of an SRAM-based bit-cell is much higher than that of an MTJ-based bit-cell, as the latter is much smaller, the access latencies of SRAM are stronger correlated to the number of bit-cells than those of SOT-MRAM or STT-MRAM. For MRAM memories, in the evaluated size range, the major contributor is the latency of the periphery circuitry and the routing delay. Thus, the access latencies of SOT-MRAM and STT-MRAM do not increase as much as those of SRAM with increasing memory size. As a result, although SRAM is the fastest memory technology for very small memory sizes, it is slower than SOT-MRAM for both read and write operations for larger memory sizes. While STT-MRAM is comparable to SOT-MRAM in terms of read latency, it suffers from its very long write latency. This underlines how effective the separation of read and write paths and hence their independent optimization in SOT-MRAM is. As a result, the asynchronous access behavior (almost) disappears for SOT-MRAM.

Per-Access Energy: The per-access energy shows a similar behavior as the access latencies as shown in Figure 5(a). Thereby, the reasons are the same as explained in the previous paragraph. As a result, SRAM is again the best choice for very small memories, but for larger memories (here: starting with 256 KByte) SOT-MRAM starts to become the better solution. In contrast, STT-MRAM has a very high write-access energy, due to the high write current required [3].

Leakage Power: In terms of leakage power SOT-MRAM is superior compared to STT-MRAM and SRAM. The reason for the high leakage power of SRAM is its CMOS nature. For STT-MRAM it is the larger access transistor compared to SOT-

MRAM which is the reason why its leakage power is worse than that of SOT-MRAM.

Summary: In summary, based on our observations, SOT-MRAM is a very good replacement for SRAM in cache memories. However, its suitability for an L1-cache compared to SRAM strongly depends on the size of this cache and the clock frequency. For slower clock frequencies or larger cache sizes SOT-MRAM could be a viable choice even for L1 cache. However, the real cache performance depends not only on these parameters but also the application and its characteristics, e.g. read to write ratio or hit rate. Therefore, in the following section, we present a detailed study of SOT-MRAM as a candidate in various levels of the cache hierarchy in a real system.

IV. EVALUATION OF SOT-MRAM AS CACHE MEMORY

Based on the comparison of various memory technologies presented in Section III.B, SOT-MRAM is a promising candidate to (partially) replace SRAM as the memory technology for caches in microprocessors. Therefore, we analyze the advantages and disadvantages of SOT-MRAM as L1- and L2-cache memory technology in terms of performance, energy consumption as well as area. For this reason, various *"hybrid"* cache configurations are evaluated in which different memory technologies (SRAM, STT-MRAM, and SOT-MRAM) are used for different levels of cache hierarchy.

A. Hybrid-Memory Evaluation Platform

Our evaluation uses gem5 [2], a full-system, cycle-accurate performance simulator that supports various memory configurations and allows to configure all relevant cache parameters such as capacity, associativity, latency, block size and policy. However, to model the asymmetric behavior of STT- and SOT-

Processor	Single-core @ 3 GHz, out-of-order, 4-issue
L1-Cache	32 KByte, 2-way set associative, 64 B line size, 1 bank, MESI cache (SRAM: 0.7 ns, SOT: 1.0 ns/1.1 ns, STT: 1.0 ns/10.9 ns)
L2-Cache	512 KByte, 16-way set associative, 64 B line size, 1 bank, MESI cache (SRAM: 2.1 ns, SOT: 1.1 ns/1.4 ns, STT: 1.1 ns/11.2 ns)
Execution Units	2x ALU, 2x CALU, 2x FPU
MiBench applications	BasicMath, BitCount, QSort, Dijkstra, Patricia, StringSearch, SHA, CRC, FFT

TABLE III
CONFIGURATION DETAILS FOR THE EXPERIMENTS

978-1-4799-2817-0/14 $31.00 © 2014 IEEE

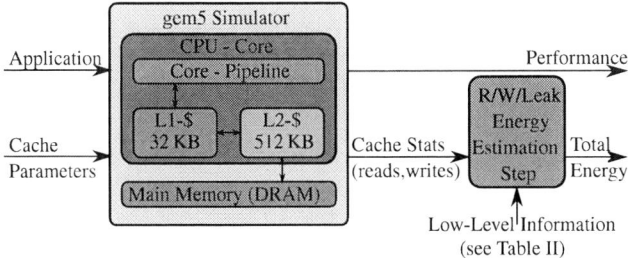

Fig. 6. Analysis flow to obtain performance & energy consumption for different cache configurations

MRAM we had to extend gem5 to support different read and write latencies for each cache.

The baseline configuration for our study is summarized in Table III. It is based on a single-core processor with a clock frequency of 3 GHz and an out-of-order pipeline based on the Alpha 21264 processor. Furthermore, the processor has an L1-cache with a capacity of 32 KByte and its L2-cache is 512 KByte large. For each memory technology we extracted the read and write access latencies for the L1- and L2-cache according to the methodology presented in Section III.B. Please note that due to the chosen clock frequency of 3 GHz the latencies correspond to 3 cycles and 7 cycles for SRAM, 3 (4) and 4 (5) cycles for SOT-MRAM for read (write) accesses, and 3 (33) and 4 (34) cycles for STT-MRAM for read (write accesses), for L1 and L2 caches respectively. This indicates already that the final performance does not only depend on the memory technology for each cache level, but also the clock frequency for each cache.

To evaluate the benefits and shortcomings of SOT-MRAM as cache memory, we use nine workloads out of the MiBench benchmark suite [9] as detailed in Table III. All workloads are simulated completely, including the initialization phase, to be as close as possible to the real world. Afterwards, the performance and cache statistics obtained from gem5 are used to estimate the dynamic (read & write) and static energy (leakage) for each memory configuration. Therefore, for each memory technology the per access energy and leakage power are taken into consideration. By considering also the number of read (write) accesses and the runtime, the total energy consumption for every application can be estimated as shown in Figure 6.

B. Main Results

The main results of our analysis are summarized in Figure 7. For this figure and all further discussions a configuration such as SRAM+SOT means that SRAM is used for the L1-cache, while SOT-MRAM is used for the L2-cache.

To explain these, we first focus on the area, afterwards on the performance and then discuss the energy consumption.

Area: As expected the usage of SOT-MRAM or STT-MRAM significantly reduces the cache area, which is due to the fact that both technologies have much smaller bit-cells than SRAM. In this regard, the major savings can be achieved, if SOT-MRAM or STT-MRAM are used for the L2-cache, since it occupies much more area due to the higher capacity. If SOT-MRAM instead of SRAM is employed for the L2-cache, area can be reduced by more than 40%. As the L1-cache in our study is quite small, the phenomenon discussed in Section B.3 occurs, i.e. for this cache size SRAM is smaller than SOT-MRAM. Hence, if the L1-cache uses SOT-MRAM as memory technology, the size increases by 5%.

Performance: The results also show that SOT-MRAM can replace SRAM in terms of performance, while STT-MRAM suffers from its long write latency as expected based on the analysis presented in Section III.B.3. However, the benefits strongly depend on the cache-level. For the L1-cache, SRAM offers in average 1% more performance than SOT-MRAM, while for the L2-cache SOT-MRAM is slightly faster, which results in a slight performance increase (i.e. runtime reduction) of 1% in average. Since the write access latency of the L2-cache is not so important, even STT-MRAM can be used for this cache-level [4]. However, for the L1-cache it is not feasible (i.e. runtime increase of 30% in average).

Energy: To analyze the energy consumption, let us first focus on the L2-cache. As Figure 7 shows, SRAM is not competitive

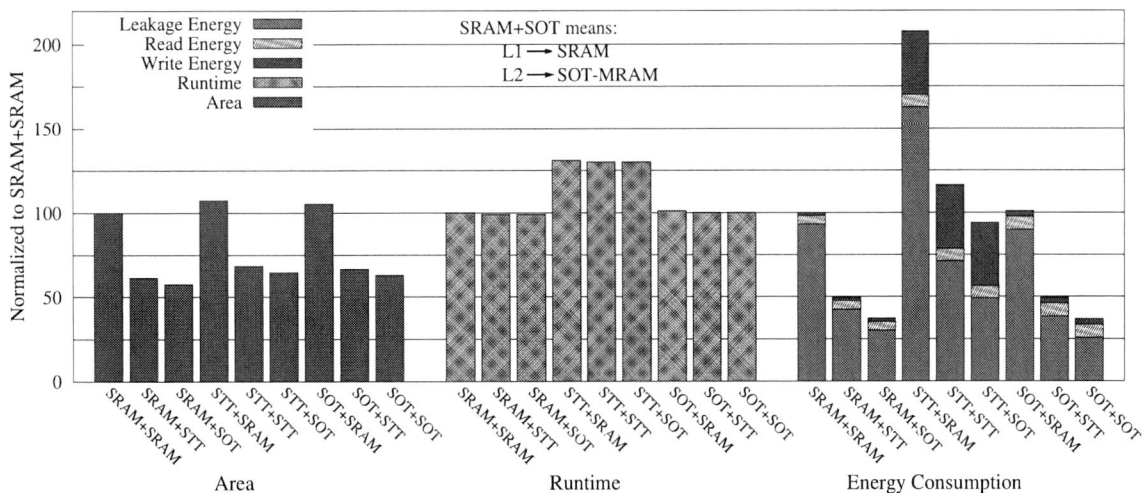

Fig. 7. Comparison of various cache configurations in terms of occupied area, average application runtime and average energy consumption (normalized to the standard configuration, i.e. SRAM for L1- and L2-cache)

978-1-4799-2817-0/14 $31.00 © 2014 IEEE   705

	Runtime [ms]				Energy [mJ]			
	SRAM+SRAM	SRAM+STT	SRAM+SOT	SOT+SOT	SRAM+SRAM	SRAM+STT	SRAM+SOT	SOT+SOT
BasicMath	61.4	**59.8**	**59.8**	60.5	66.4	31.6	23.6	**22.8**
BitCount	130.1	**130.1**	**130.1**	130.1	133.8	63.0	45.6	**40.4**
CRC	998.8	**998.8**	**998.8**	1025.5	1075	531.5	398.1	**395.7**
Dijkstra	62.7	**62.4**	**62.4**	62.6	75.5	41.2	**32.9**	36.8
FFT	176.1	175.4	**175.3**	176.1	191.9	95.5	72	**71.6**
Patricia	49.1	46.7	**46.7**	47.6	54.6	25.8	19.5	**19.4**
QSort	35.2	34.9	**34.9**	34.9	36.7	17.6	12.7	**11.6**
SHA	23.3	**23.3**	**23.3**	23.3	26.1	13.4	**10.3**	10.7
StringSearch	1.5	**1.5**	**1.5**	1.5	1.7	0.9	0.7	**0.7**
Average	170.9 (100 %)	170.0 (99 %)	**170.0 (99 %)**	173.3 (101 %)	184.6 (100 %)	91.2 (49 %)	68.4 (37 %)	**67.7 (36 %)**

TABLE IV

PER BENCHMARK ANALYSIS OF DIFFERENT "HYBRID" CACHE CONFIGURATION (BOLD NUMBERS REPRESENT THE BEST VALUE)

with STT-MRAM or SOT-MRAM for this cache-level. This is due to the high leakage power of SRAM and the memory capacity of 512 KByte. If instead MRAM is used, the energy consumption can be reduced by more than 50 %, in average. Furthermore, as explained in Section B.3 the leakage power of SOT-MRAM is also smaller than that of STT-MRAM, due to smaller access transistors. As a result, SOT-MRAM offers the least power hungry solution for the L2-cache.

In contrast, for the small L1-cache (just 32 KByte), SRAM requires less energy than STT-MRAM. This is due to two facts. First, the per-access energy of STT-MRAM is much higher than that of SRAM, especially due to the high write current required by STT-MRAM. Second, leakage power is less important for this small memory. Both aspects together lead to a 40 % increase energy consumption compared to SRAM in average (if SOT-MRAM is used for the L2-cache). Since SOT-MRAM has a much lower per-access energy consumption than STT-MRAM, it eliminates a major shortcoming and thus allows to even reduce the energy consumption of the L1-cache compared to SRAM.

Summary: In summary, SOT-MRAM is a viable candidate to replace SRAM as memory technology for some levels of the cache hierarchy. It does not only offer a higher density and lower energy consumption but has also a similar performance. However, for smaller cache sizes such benefits reduces accordingly. As a consequence, the per-access energy gains importance and in turn SOT-MRAM looses advantages. Based on our observations, SOT-MRAM is a viable SRAM replacement for the L2-cache and in some cases even for the L1-cache, if it is large enough (in our setup, at least 64 KBytes). In other words, when the L1-cache size is small enough, SRAM is still a better choice. Moreover, for register files, due to their small sizes, SOT-MRAM is not a suitable choice.

C. In-Depth Evaluation

In Table IV the results per benchmark for the hybrid cache-configurations SRAM+SRAM, SRAM+STT, SRAM+SOT and SOT+SOT are shown. As it can be seen, SOT+SOT is in average the solution with the lowest energy consumption and hence is the best choice for low power systems. However, for some applications the combination of SRAM for the L1-cache and SOT-MRAM for the L2-cache offers a slightly better energy efficiency (e.g. Dijkstra or SHA). This is due to the fact that the combination of SRAM and SOT-MRAM is often faster and

hence has the advantage of a lower runtime. In addition, the per-access energy for SRAM as L1-cache is lower than that of SOT-MRAM as memory technology for the L1-cache.

Furthermore, it can be seen that often SRAM+STT and SRAM+SOT deliver the same performance. This is the case for applications that have a low write access rate to the L2-cache (e.g. StringSearch or BasicMath). If the ratio of write access to the L2-cache is higher (e.g. FFT) SRAM+SOT is a better solution in terms of performance as STT-MRAM has much higher write access times. In terms of energy consumption SRAM+SOT is always much better than SRAM+STT.

The combination of SRAM+SRAM is neither the fastest nor the most energy saving solution for any benchmark. Hence, this configuration is, at least for our setup, not a viable choice. Instead a hybrid solution or SOT-MRAM-only is favorable. However, considering all aspects, i.e. performance, energy consumption and area, the hybrid solutions offers the best trade-off for our processor configuration.

V. CONCLUSIONS

For shrinking technologies, non-volatile memories are promising storage technologies due to their low static power. In this paper, we evaluated a novel nano-magnetic memory technology called Spin Orbit Torque (SOT-MRAM). It is related to Spin Transfer Torque MRAM (STT-MRAM), but has independent read and write paths. As a result SOT-MRAM can achieve access latencies similar to SRAM which makes SOT-MRAM a viable candidate for on-chip memory, not only for the last-level cache, but also for lower levels of cache to replace SRAM. Depending on the cache size, SOT-MRAM can even replace SRAM as memory technology for the L1-cache. In fact, our detailed architecture-level analysis shows that an SOT-only solution is the best choice for low power systems. We also found out that for very small memory blocks, such as register files or small L1-caches, SRAM is still superior to SOT-MRAM in terms of area and performance. Therefore, the best combination of performance, energy efficiency and area cost is offered by a "hybrid" solution composed of SRAM for the small L1-cache (32 KByte) and SOT-MRAM for the larger L2-cache (512 KByte). Compared to an SRAM-only configuration this allows to reduce the energy consumption by 63 %, the area by 43 % and in addition the performance will increase by 1 %.

978-1-4799-2817-0/14 $31.00 © 2014 IEEE

VI. ACKNOWLEDGEMENT

This work was partly supported by the European Commission under the Seventh Framework Program as part of the spOt project (http://www.spot-research.eu/). Furthermore, the authors would like to thank Gregory di Pendina, Kotb Jabeur and Guillaume Prenat from SPINTEC for their valuable feedback and discussion during this work.

REFERENCES

[1] D. Apalkov, A. Khvalkovskiy, S. Watts, V. Nikitin, X. Tang, D. Lottis, K. Moon, X. Luo, E. Chen, A. Ong, A. Driskill-Smith, and M. Krounbi, "Spin-transfer torque magnetic random access memory (STT-MRAM)," *ACM Journal on Emerging Technologies in Computing Systems*, pp. 13:1–13:35, May 2013.

[2] N. L. Binkert, R. G. Dreslinski, L. R. Hsu, K. T. Lim, A. G. Saidi, and S. K. Reinhardt, "The M5 Simulator: modeling networked systems," *IEEE Micro*, pp. 52–60, Jul. 2006.

[3] R. Bishnoi, M. Ebrahimi, F. Oboril, and M. Tahoori, "Asynchronous asymmetrical write termination (AAWT) for a low power STT-MRAM," in *Design, Automation and Test in Europe*, Mar. 2014.

[4] M.-T. Chang, P. Rosenfeld, S.-L. Lu, and B. Jacob, "Technology comparison for large last-level caches (L^3Cs): low-leakage SRAM, low write-energy STT-RAM, and refresh-optimized eDRAM," in *High Performance Computer Architecture*, Feb. 2013, pp. 143–154.

[5] X. Dong, C. Xu, Y. Xie, and N. P. Jouppi, "NVSIM: A circuit-level performance, energy, and area model for emerging nonvolatile memory," *Computer-Aided Design of Integrated Circuits and Systems, IEEE Transactions on*, pp. 994–1007, 2012.

[6] X. Dong, X. Wu, G. Sun, Y. Xie, H. Li, and Y. Chen, "Circuit and microarchitecture evaluation of 3D stacking magnetic RAM (MRAM) as a universal memory replacement," in *Design Automation Conference*, Jun. 2008, pp. 554–559.

[7] D. A. P. Gambardella and I. M. Miron, "Current-induced spin-orbit torques," *Philosophical Transactions of the Royal Society A: Mathematical, Physical and Engineering Sciences*, pp. 3175–3197, 2011.

[8] W. Guo, G. Prenat, V. Javerliac, M. E. Baraji, N. de Mestier, C. Baraduc, and B. Diny, "SPICE modelling of magnetic tunnel junctions written by spin-transfer torque," *Journal of Physics D: Applied Physics*, p. 215001, May 2010.

[9] M. R. Guthaus, J. S. Ringenberg, D. Ernst, T. M. Austin, T. Mudge, and R. B. Brown, "MiBench: a free, commercially representative embedded benchmark suite," in *Workshop on Workload Characterization*, 2001, pp. 3–14.

[10] M. Hosomi, H. Yamagishi, T. Yamamoto, K. Bessho, Y. Higo, K. Yamane, H. Yamada, M. Shoji, H. Hachino, C. Fukumoto *et al.*, "A novel nonvolatile memory with spin torque transfer magnetization switching: Spin-RAM," in *International Electron Devices Meeting*, 2005, pp. 459–462.

[11] J. Hutchby and M. Garner, "Assessment of the potential & maturity of selected emerging research memory technologies," in *Workshop and ERD/ERM working group meeting*, Apr. 2010.

[12] International Technology Roadmap for Semiconductors, http://www.itrs.net, 2012.

[13] K. Jabeur, L. D. Buda-Prejbeanu, G. Prenat, , and G. D. Pendina, "Study of two writing schemes for a magnetic tunnel junction based on spin orbit torque," *International Journal of Electronics Science and Engineering*, pp. 501–507, 2013.

[14] D. S. Jeong, R. Thomas, R. S. Katiyar, J. F. Scott, H. Kohlstedt, A. Petraru, and C. S. Hwang, "Emerging memories: resistive switching mechanisms and current status," *Reports on Progress in Physics*, p. 076502, 2012.

[15] A. Khvalkovskiy, D. Apalkov, S. Watts, R. Chepulskii, R. Beach, A. Ong, X. Tang, A. Driskill-Smith, W. Butler, P. Visscher *et al.*, "Basic principles of STT-MRAM cell operation in memory arrays," *Journal of Physics D: Applied Physics*, pp. 74001–74020, 2013.

[16] L. D. Landau and E. Lifshitz, "On the theory of the dispersion of magnetic permeability in ferromagnetic bodies," *Phys. Zeitsch. der Sow.*, pp. 153–169, 1935.

[17] Y. Li and K. N. Quader, "NAND Flash memory: challenges and opportunities," *Computer*, pp. 23–29, 2013.

[18] L. Liu, C.-F. Pai, Y. Li, H. W. Tseng, D. C. Ralph, and R. A. Buhrman, "Spin-torque switching with the giant spin hall effect of tantalum," *Science*, pp. 555–558, 2012.

[19] I. M. Miron, K. Garello, G. Gaudin, P. Zermatten, M. Costache, S. Auffret, S. Bandiera, B. Rodmacq, A. Schuhl, and P. Gambardella, "Perpendicular switching of a single ferromagnetic layer induced by in-plane current injection," *Nature*, pp. 189–193, 2011.

[20] G. Panagopoulos, C. Augustine, and K. Roy, "Modeling of dielectric breakdown-induced time-dependent STT-MRAM performance degradation," in *Device Research Conference*, 2011, pp. 125–126.

[21] D. Ralph and M. D. Stiles, "Spin transfer torques," *Journal of Magnetism and Magnetic Materials*, pp. 1190–1216, 2008.

[22] S. Raoux, G. W. Burr, M. J. Breitwisch, C. T. Rettner, Y.-C. Chen, R. M. Shelby, M. Salinga, D. Krebs, S.-H. Chen, H.-L. Lung *et al.*, "Phase-change random access memory: a scalable technology," *IBM Journal of Research and Development*, pp. 465–479, 2008.

[23] N. Shibata, H. Maejima, K. Isobe, K. Iwasa, M. Nakagawa, M. Fujiu, T. Shimizu, M. Honma, S. Hoshi, T. Kawaai *et al.*, "A 70 nm 16 Gb 16-level-cell NAND flash memory," *Solid-State Circuits, IEEE Journal of*, pp. 929–937, 2008.

[24] G. Sun, X. Dong, Y. Xie, J. Li, and Y. Chen, "A novel architecture of the 3D stacked MRAM L2 cache for CMPs," in *High Performance Computer Architecture*, 2009, pp. 239–249.

[25] N. Weste and D. Harris, *CMOS VLSI Design: a circuits and systems perspective*, 4th ed. USA: Addison-Wesley Publishing Company, 2010.

[26] S. A. Wolf, D. D. Awschalom, R. A. Buhrman, J. M. Daughton, S. von Molnr, M. L. Roukes, A. Y. Chtchelkanova, and D. M. Treger, "Spintronics: a spin-based electronics vision for the future," *Science*, pp. 1488–1495, 2001.

[27] S. A. Wolf, J. Lu, M. R. Stan, E. Chen, and D. M. Treger, "The promise of nanomagnetics and spintronics for future logic and universal memory," *Proceedings of the IEEE*, pp. 2155–2168, 2010.

[28] C. Yoshida, M. Kurasawa, Y. M. Lee, K. Tsunoda, M. Aoki, and Y. Sugiyama, "A study of dielectric breakdown mechanism in CoFeB/MgO/CoFeB magnetic tunnel junction," in *International Reliability Physics Symposium*, 2009, pp. 139–142.

[29] P. Zhou, B. Zhao, J. Yang, and Y. Zhang, "A durable and energy efficient main memory using phase change memory technology," in *ACM SIGARCH Computer Architecture News*, vol. 37, no. 3, 2009, pp. 14–23.

Timing Anomalies in Multi-core Architectures due to the Interference on the Shared Resources

Hardik Shah, Kai Huang and Alois Knoll

Department of Informatics VI, Technical University Munich,

85748 Garching, Germany.

{shah,huangk,knoll}@in.tum.de

Abstract— Timing anomalies in single-core processors have been theoretically explained and well understood phenomenon. This paper presents new timing anomalies which occur in multi-core architectures due to the interference on the shared resources. We derive formulation to capture these anomalies and provide practical evidences using real applications from the Mälardalen WCET benchmark suit executing on NIOS II multi-core architecture on an Altera FPGA.

I. INTRODUCTION

Timing anomaly is a *counter intuitive* timing behavior. The term was coined by Lundqvist & Stenström [1]. They observed that in a dynamically scheduled processor, a cache hit at certain execution point could lead to longer execution time than a cache miss at the same point. Thus, the timing anomaly for processor architectures is defined as, *A processor architecture is said to be timing anomalous when a locally favorable event (e.g. cache hit) could result in a globally unfavorable event (e.g. longer execution time) and vice versa.* The formal definition of timing anomaly is provided by Reineke et al [2].

Simple processors could also exhibit timing anomalous behavior [3]. The occurrence of timing anomalies can be analyzed using static worst case execution time (WCET) analysis techniques [4]. However, these events are rarely (or never) observed in real life. This argument is often used against the static WCET analysis techniques. Hence, it is important to present real-life evidences of the presence of timing anomalies.

In multi-core architectures, the shared resource interference caused by accesses from the co-existing applications increases shared resource access latencies for the application-under-test. This leads to longer execution time of the application. It is intuitive that if the co-existing applications are very aggressive in accessing the shared resource then the application-under-test experiences higher latencies to the shared resource and the execution time increases accordingly. The other intuition is, the higher the number of aggressive co-existing applications, the higher the latencies to the shared resource. In this paper, we show that certain applications behave counter intuitively to these intuitions.

The major contributions of the paper are as follows. i) We prove that in the presence of aggressive accesses from the co-existing applications, some applications-under-test

benefit and experience less than the average case latencies to the shared resource. ii) We prove that in the presence of the aggressive accesses, for some applications-under-test, shared resource latencies under less number of interfering applications could be more than the shared resource latencies under more number of interfering applications. iii) We also provide practical evidence of these timing anomalies using real applications from the Mälardalen WCET benchmark suit executing on the NIOS II based multi-core architecture on Altera Cyclone III FPGA. The paper focuses on the round robin arbiter which is one of the most popular *starvation free* arbiters and a default component of many off-the-shelf interconnect architectures [5, 6]. However, other *starvation free* arbiters are discussed in Sec. VI.

The paper is organized as follows. Sec. II provides existing work related to this paper. Sec. III provides necessary background information. Sec. IV provides formalism from which the timing anomalous behavior is inferred. Sec. V provides practical evidences of timing anomalous behavior. Sec. VI discusses the timing anomalies in other *starvation free* arbiters and Sec. VII concludes the paper.

II. RELATED WORK

This paper has two dimensions, i) Interference analysis in multi-core architectures and its effect on the WCET. ii) Timing anomalies. We address the related work on both the topics in this section.

Lv et al [7] propose to build Timed Automata (TA) models of concurrently executing applications using abstract interpretation. These TA models are then combined with the shared resource arbiter's TA model. The combination is analyzed using a model checker to find the WCET of each application considering the maximum interference among them. Pellizzoni et al [8] use cache activity trace of the application-under-test and upper bound on I/O traffic. These inputs are analyzed using real-time calculus to calculate WCET of the application-under-test considering maximum interference from the I/O traffic. Both of these approaches take the knowledge of the co-existing applications into the account.

On the contrary, our previous work [9, 10, 11] and Paolieri et al [12] analyze applications in isolation for shared SDRAM interference. Here, the worst possible behavior from co-existing applications is assumed, including the faulty behavior. All the above mentioned works assume timing-anomaly-free architecture.

978-1-4799-2817-0/14 $31.00 © 2014 IEEE

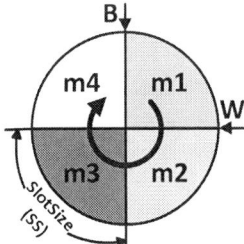

Fig. 1.: Graphical View of the Round Robin Arbiter

Li et al [13] model timing anomalous processor for the WCET analysis. They analyze the interaction between Basic Blocks (BB) and its effect on the instruction cache state. Kirner et al [14] identify a *new* timing anomaly that arises due to the parallel decomposition. The parallel decomposition is used for hardware state space reduction. These approaches target single-core architectures.

Recently, there has been some work on the WCET analysis in multi-cores in the presence of timing anomalies. Chattopadhyay et al [15] provides the first unified analysis that takes various micro-architecture components into account. Interaction among these components is analyzed to estimate the WCET of applications on a timing anomalous multi-core architecture. Kelter et al [16] investigate when a BB will start executing with respect to the statically assigned slot to access the shared bus. The shared bus access latency is estimated by analyzing whether the access from the BB lays in the current slot or in next slots. Both, [15] and [16] use TDMA as a shared bus arbiter.

All the above mentioned work related to timing anomalies explain it hypothetically without real evidences. In this paper, we derive formulation which is used to infer timing anomalies due to the shared resource interference in multi-core architectures. Moreover, we provide real-life evidences of their presence.

III. BACKGROUND

In this section, we provide some background information in order to facilitate discussion in the later sections.

A. Work Conserving Round Robin Arbitration

The Fig. 1 depicts the Round Robin (RR) arbitration graphically. Under the RR scheme, the shared resource contenders are assigned fixed number of slots in a virtual ring depending on their bandwidth requirements. Although, our analysis is valid for any slot allotment, for simplicity, we consider one slot per contender without loss of generality. The figure shows four contenders (master1, master2, master3 and master4) in the ring. Here, we assume that these contenders are processor cores executing independent applications and the shared resource is shared main memory. These cores access the shared main memory when a cache miss occurs. *Throughout the paper, we use master and core terms interchangeably. Similarly, we use memory and shared memory interchangeably.*

The arbiter continuously searches for a master which wants to access the memory in a clock-wise direction. We call this master an active master. As soon as an active

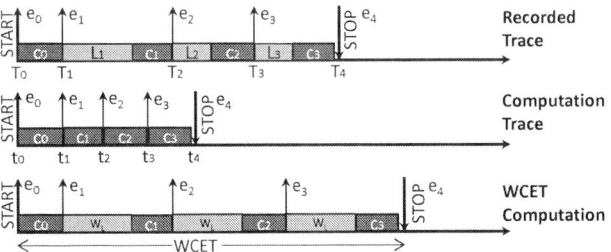

W_L = Worst case latency, L_x = Measured latency, c_x = Computation time, e_x = x^{th} event, T_x = Time in recorded trace, t_x = Time in computation trace

Fig. 2.: WCET Computation using the Computation Trace

master is encountered, it is granted the memory for a pre-defined maximum number of clock cycles (SlotSize - SS). The SS is big enough to accommodate the burst issued for one cache-line fill. After the granted master finishes its burst access, the search process resumes from the next slot in the ring. Thus, the memory is always occupied as long as there is at least one active master (hence the name, "*work conserving*" or "*greedy TDMA*").

Now, let us assume that the application-under-test is executing on m1 and the SS is same for all masters. For this architecture, an access request from m1 experiences the worst case completion latency ($W_L = 4 \times SS$) if it is issued when the arbiter pointer is at **W** in Fig. 1 *AND* all other masters utilize their slots. Similarly, an access request experiences the best case completion latency ($B_L = 1 \times SS$) if it is issued when the arbiter pointer is at **B** in the figure. If the exact location of the arbiter pointer *OR* activity of other masters are unknown, the completion latency could be any number in the range $[B_L, W_L]$.

Intuitively, the average case latency, $A_L = (B_L + W_L)/2$. *In this paper, by latency of an access, we mean completion latency of an access (scheduling latency + time required to complete the access). Moreover, we also assume that the application-under-test executes on m1.*

B. Computation Trace

The computation trace is an execution trace of an application path where cache misses are denoted by timeless events. The motivation behind the timeless events is the following. In the shared memory architecture, the shared main memory is accessed when a cache miss occurs. The contention on the shared memory delays service to this memory access. Typically, the collision of cache misses on the shared memory is extremely difficult to predict. Moreover, the delay in service also delays the subsequent cache misses (memory accesses) of the application-under-test by the same amount. This causes difficulties in estimating the worst case interference and its impact on the WCET. To avoid these difficulties, at first, we remove all the latencies related to the memory accesses. This means, there is no interference at all and the shared memory takes zero cycles to respond. Later, theoretically calculated worst possible latency is added for each cache miss.

Computation trace, depicted in the Fig. 2, can be easily obtained through simulation. At first, occurrence time ($T_0, T_1, ...$) of each cache miss event ($e_0, e_1, ...$), and its

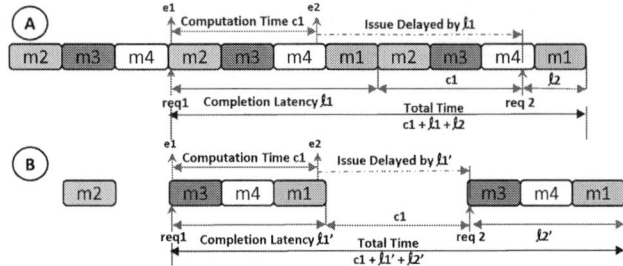

Fig. 3.: In **case A**, over all execution time is lower than in **case B** although the accesses generated by the co-existing application is more aggressive in **case A** than in **case B**

experienced latency $(L_0, L_1, ...)$ are recorded in a trace by executing the application on cycle accurate simulation models of processor and memory. Later, these latencies are removed and each event is shifted towards left in time. The resulting trace is the computation trace. Now, to compute the WCET considering the worst case interference, each cache miss event in the computation trace is annotated by the worst possible latency (W_L) and all the subsequent accesses are shifted to the right. Now, the computed WCET contains the effect of the worst possible interference the application may experience.

It must be noted that the latencies in the recorded trace depend on shared memory interference at the time of measurement. However, the computation trace remains unchanged[1] when the same path is executed multiple times, provided that each time we start the application from the same cache state[2] and use the same data as an input.

Note that the WCET computed using this method is actually WCET of the *path* being executed on the application-under-test. Moreover, the method considers only shared resource interference as an execution time modifying source. Practically, applications have multiple paths through execution and caches, pipelines, branch predictors etc. contribute heavily to the execution time deviation. The core contribution of the paper is to identify the timing anomalies originating from the shared resource interference and to provide practical evidences of its presence. Therefore, we do not present analysis of caches, pipelines, branch predictors, execution path etc in this paper and fully concentrate on interference analysis.

C. Latencies under round Robin Arbitration

In this subsection, we analyze different latency scenarios for the computation trace. As explained before, the computation trace remains constant for multiple runs of the application, however, the interference scenarios can be different resulting in different execution times. Fig. 3 depicts two different interference scenarios. In scenario A, the co-existing applications generate aggressive accesses (un-

interrupted accesses/interference) to the shared memory while in scenario B, the co-existing applications generate sparse accesses. The computation trace in the figure contains two cache miss events **e1** and **e2**. The computation time between these events is **c1**.

In scenario A, the request 1 (corresponding to event **e1**) is issued just after the arbiter has scheduled **m2**. Since cores **m2**, **m3** and **m4** are generating uninterrupted accesses, the request 1 from **m1** can be scheduled only after $3 \times SS$. After request 1 is scheduled, it needs another SS to complete. Hence, request 1 has latency of $l1 = 4 \times SS$, which is the worst case latency (W_L). After requests 1 is served, the **m1** does computation for **c1** amount of clock cycles. During this time, the **m1** executes from caches and on-chip registers and does not send any request to the shared memory. However, the co-existing cores send uninterrupted accesses to the shared memory and keep on rotating the arbiter pointer. Thus, when request 2 (corresponding to **e2**) is issued, the arbiter pointer is close to the next scheduling opportunity of **m1**. Hence, the latency of request 2, $l2 << W_L$.

In scenario B, although the co-existing applications generate sparse accesses, the total execution time is longer than that of scenario A. Similarly, another scenario can be presented where both the accesses experience the worst case latencies which results into the real WCET[3].

IV. LATENCY ANALYSIS UNDER α INTERFERENCE

This section focuses on the uninterrupted interference and derives equations for experienced latencies under it. The first subsection defines the α interference and the second subsection provides analysis of experienced latencies.

A. α Interference

Definition: *The α interference is defined as the uninterrupted interference produced by α number of co-existing masters.* Under α interference, the rotation of the arbiter pointer becomes deterministic since accesses from the co-existing masters are deterministic (uninterrupted or no access at all). Here, except **m1**, on which the application is being executed, other masters either continuously utilize their slots or they do not utilize their slots at all.

The α interference occurs in real life in the following scenarios, i) Immediately after reset or after a new task is scheduled on co-existing applications, there are high number of cache misses. At this point, the co-existing applications send many accesses to the shared resource in a relatively short period of time. ii) When some of the co-existing applications generate aggressive traffic e.g. DMA and the remaining are idle for a short period of time. iii) When some of the co-existing masters have a stuck-at-fault on the request line and other masters are idle for a short time. It is clear that the α interference could occur randomly for a short period of time in real life. However, applications with relatively short life times (typical

---

[1]In this paper, we assume absence of jitter in occurrence of cache misses due to operating system, floating point unit, pipeline etc.

[2]In order to fully concentrate on the *new* timing anomalies originating from interference, we assume classical timing-anomaly-free architecture in this paper. Certainly, these *new* timing anomalies are valid in the presence of the classical ones as well.

---

[3]In this paper, we focus only on the effect of the interference on WCET and consider the effects of other components such as caches, pipe-lines, branch predictors etc as constant.

978-1-4799-2817-0/14 $31.00 © 2014 IEEE

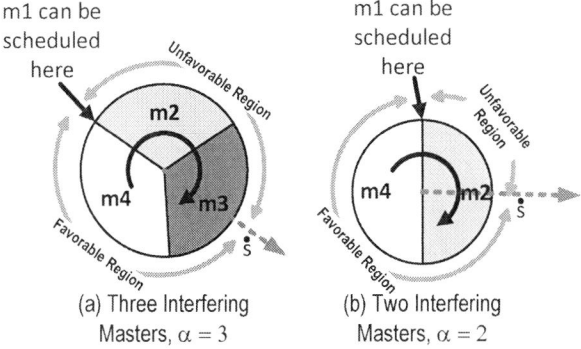

Fig. 4.: Deterministic Rotation of the Arbiter Pointer (`SlotSize` - SS not to the scale)

hard real-time control applications) could experience it during their entire execution. In the remaining sections, we will show that, counter intuitively, these uninterrupted accesses from co-existing masters could also be beneficial to some applications.

B. Analysis

Fig. 4 depicts two scenarios, $\alpha = 3$ and $\alpha = 2$. In $\alpha = 3$ scenario (same as Fig. 3 **scenario A**), all other masters m2, m3 and m4 do uninterrupted accesses to the shared resource. In $\alpha = 2$ scenario, only masters m4 and m2 do uninterrupted accesses; m3 is idle. Hence, there are only two slots in Fig. 4(b). Note that there are total four masters in the system and the theoretical values of B_L, W_L and A_L are derived considering all masters in the system, *irrespective* of number of active masters.

We denote the latency of ith access under α interference as deterministic latency (DL_α^i) since it is derived assuming the deterministic rotation of the arbiter pointer. Its value can be obtained using the following equation.

$$DL_\alpha^i = (\alpha + 1) \times SS - \{c_{(i-1)} \bmod (\alpha \times SS)\} \quad (1)$$

Here, $c_{(i-1)}$ is the computation time between i^{th} and $(i-1)^{th}$ cache miss events in the computation trace (Fig. 2). Let $\Theta_\alpha^{(i-1)} = \{c_{(i-1)} \bmod (\alpha \times SS)\}$,

$$DL_\alpha^i = (\alpha + 1) \times SS - \Theta_\alpha^{(i-1)} \quad (2)$$

Consider average of all DL_α^i values is $\overline{DL_\alpha}$ and the average of all Θ_α^i values is $\overline{\Theta_\alpha}$ (note that $\overline{DL_\alpha}$ and $\overline{\Theta_\alpha}$ represent average values over entire execution path). Hence, equation (2) can be re-written as,

$$\overline{DL_\alpha} = (\alpha + 1) \times SS - \overline{\Theta_\alpha} \quad (3)$$

$\overline{DL_\alpha}$ is the average experienced latency when the application is executed in the presence of α interference.

Recall that $A_L = (B_L + W_L)/2$, $B_L = 1 \times SS$ and $W_L = N \times SS$, N is total number of master in the system. From this information, the average-case latency of the ith access, A_L^i, is given by the following equation,

$$A_L^i = \frac{N+1}{2} \times SS \quad (4)$$

Since value of A_L in (4) depends only on constant numbers, the average of all A_L^i values is,

$$\overline{A_L} = \frac{N+1}{2} \times SS \quad (5)$$

Equations (1) to (5) are used to infer counter intuitive timing behavior. First, let us prove couple of lemmas.

Lemma 1 $\exists c_{(i-1)}, \alpha : A_L^i > DL_\alpha^i$. *In other words, for a combination of α, and $c_{(i-1)}$, it is possible that the observed latency of ith access (DL_α^i) is less than the theoretical average-case latency of the ith access (A_L^i).*

Proof: Since $\Theta_\alpha^{(i-1)} = \{c_{(i-1)} \bmod (\alpha \times SS)\}$, $\Theta_\alpha^{(i-1)} \in [0, (\alpha \times SS - 1)]$. Thus, if $c_{(i-1)} = n(\alpha \times SS) - 1$, $\Theta_\alpha^{(i-1)} = \alpha \times SS - 1$, $n \in \mathbb{N}^+$.

Putting $\Theta_\alpha^{(i-1)} = \alpha \times SS - 1$ in (2), $DL_\alpha^i = SS + 1$. Here, $DL_\alpha^i < A_L^i$, $\forall N > 1, SS > 2$. Hence, the lemma holds. □

The lemma 1 leads to a counter intuitive observation. It proves that an access from an application could experience less than the average-case latency under α interference. Moreover, the latency does not depend on the absolute value of $c_{(i-1)}$, rather on the $\Theta_\alpha^{(i-1)}$. This phenomenon is explained graphically in Fig. 4. Here, favorable and unfavorable regions are depicted. If the application has all c_i such that $\overline{\Theta_\alpha}$ lies in the favorable region, the average experienced latency ($\overline{DL_\alpha}$) is less than the average-case latency ($\overline{A_L}$). The sweet point ($\dot{s}$) between the boundaries of the regions is derived by the following equation.

$$\dot{s} = W_L - A_L \quad (6)$$

The proof of lemma 1 can also be used to derive condition for experiencing the worst case latency for all accesses is, $\forall i, c_i = n(\alpha \times SS), n \in \mathbb{N}^0$ AND $\alpha = (N - 1)$. The application fulfilling this condition always requests exactly when the next master to it in the ring is scheduled (at point **W** in Fig. 1) and all other masters in the system utilize their allocated slots.

In real-life, it is difficult to find an application which fulfills the above mentioned conditions for the worst case latencies. However, applications that experience less than the average-case latencies under α interference are not rare (see Sec. V).

Lemma 2 $\exists c_{(i-1)} : DL_{\check{\alpha}}^i > DL_{\hat{\alpha}}^i$, $\check{\alpha} < \hat{\alpha}$. *In other words, under α interference, for a particular value of $c_{(i-1)}$, less number of interfering masters could result in longer latency than more number of interfering masters.*

Proof: Let, $c_{(i-1)} = (\check{\alpha} \times SS)$. From equation (1),

$$DL_{\check{\alpha}}^i = (\check{\alpha} + 1) \times SS = \check{\alpha} \times SS + SS \quad (7)$$

Again using equation (1) and since $\check{\alpha} < \hat{\alpha}$, $c_{(i-1)} \bmod (\hat{\alpha} \times SS) = \check{\alpha} \times SS$. Hence, the value of $DL_{\hat{\alpha}}^i$ can be given by the following equation,

$$DL_{\hat{\alpha}}^i = (\hat{\alpha} + 1) \times SS - (\check{\alpha} \times SS) \quad (8)$$

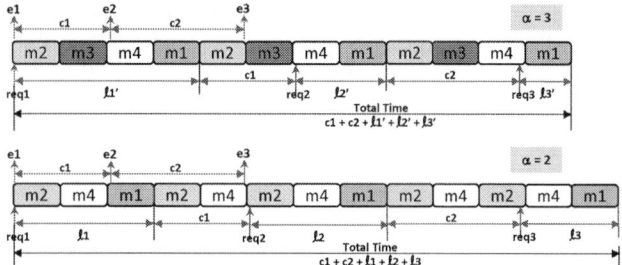

Fig. 5.: Experienced Latencies under $\alpha = 3$ and $\alpha = 2$ Interference

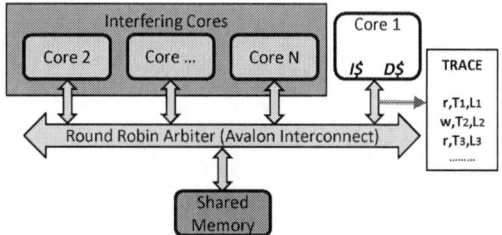

Fig. 6.: Test Set-up

$$DL_\alpha^i = (\hat{\alpha} - \check{\alpha}) \times SS + SS \qquad (9)$$

From equation (7) and (9), $DL_{\hat{\alpha}}^i > DL_{\check{\alpha}}^i, \forall \hat{\alpha} : \check{\alpha} < \hat{\alpha} \leq 2\check{\alpha}$. Hence, the lemma holds. □

The Fig. 5 provides supporting example for the lemma 2. Here, for the given application, $\alpha = 2$ interference produces higher latencies than $\alpha = 3$ interference.

V. Test Cases

The goal of this section is to provide real life evidences of timing anomalies inferred by the lemmas of previous section. We did intensive testing by executing applications from the Mälardalen WCET benchmark suit[4] [17] on Altera Nios II multi-core architecture as depicted in Fig. 6. We experimented with different cache sizes and different number of cores. Note that with the variation in cache size, the computation trace also varies (c_i and total number of cache miss). Hence, for each new cache configuration, new set of traces for each application was created using cycle accurate simulation models.

We did experiments on Altera Cyclone III FPGA Development board. Altera provides cycle accurate simulation models of processor and memory. These models were used to capture the recorded trace of each application under different cache configurations. For recording trace, connection point between core 1 and the shared memory was probed as depicted in the figure. Here, core 1 executes one of the applications and all other cores execute a dummy application (similar to [18, 19, 20]) that uninterruptedly accesses the shared memory. We started with total 4 cores in the system and step by step increased the total number of cores to 8. Note that, unlike variation in cache size,

---

[4]We chose only single path applications from the suit to avoid path analysis since the path analysis is not our contribution.

Benchmark	Oet	Acet	Wcet	$\overline{DL_3}$
cover	12207	11586	14274	24.95
crc	104331	101196	108396	26.57
duff	6777	5936	7364	28.63
edn	**360574**	**361342**	**391834**	**19.91**
expint	16573	16469	16781	23
fac	1129	1107	1227	24
fdct	**22079**	**24173**	**32453**	**18**
fibcall	1110	1098	1182	24
jane	832	813	921	25
jfdcint	28209	28107	33891	21.97
minver	158910	142289	189893	25.95
prime	196676	186533	228197	25
quart	224508	204366	271530	24.99
ud	38248	34815	46443	24.93

TABLE I

: Execution times in clock cycles under $\alpha = 3$ Interference

cover	α	$\alpha = 3$	$\alpha = 4$	$\alpha = 5$	$\alpha = 6$	$\alpha = 7$
	Oet	10717	**11686**	**11046**	11942	13662
	$\overline{DL}$	23	**30**	**27**	32	42

TABLE II

: Execution times in Clock Cycles under varying α interference

variation in number of cores does not change the computation trace since computation trace of an application is independent of experienced latency.

A. Test 1

In this experiment, we used instruction and data caches of 512 Bytes each. Total of $N = 4$ cores were used, hence, $\alpha = N-1 = 3$. As depicted in Fig. 2, we inserted the worst case (W_L) and the average-case (A_L) latencies for each cache miss to obtain the WCET and the ACET (Average Case Execution Time) of the applications, respectively.

Table I depicts the results. The first column depicts Observed Execution Time (OET) of the applications. It is clear from the table that most of the applications experienced more than the average-case latencies. However, the edn and the fdct applications experienced less than the average-case latencies under this hardware configuration. These applications did majority of accesses in the favorable region of Fig. 4 (note the lower values of $\overline{DL_3}$ for these applications). Thus, the results provide evidence for the lemma 1 of the previous section.

B. Test 2

The evidence of the lemma 2 was captured when we increased the cache size to 1 KB. We started from total 4 cores ($\alpha = 3$) and step-by-step increased the total number of cores to 8 ($\alpha = 7$). After adding each core, the OET was measured. For the cover application, execution times are listed in the Table II. This application exhibits higher execution time under $\alpha = 4$ interference than under $\alpha = 5$ interference. Due to its shared memory access pattern (cache miss pattern), it experiences longer latencies in the presence of less number of aggressive masters than presence of more number of aggressive masters, which is counter intuitive.

Note that the cache configuration is different in this experiment and the previous experiment. Hence, the cover application has different OET in the tables for $\alpha = 3$. Except the cover application, other applications followed intuition and experienced more latencies as α was increased.

VI. DISCUSSION

These anomalies are not limited to the round robin arbiter. In the budget based arbiters, such as Credit Controlled Static Priority (CCSP), Priority-based Budget Scheduler (PBS) and Dynamic Priority Queue (DPQ) this phenomenon can also be observed. Under these arbiters, all masters are assigned a unique budget to access the shared resource per unit time. If a master consumes its budget, it is termed ineligible and cannot access the shared resource until the unit time expires [9, 11]. Thus, due to the aggressive accesses, if co-existing applications become ineligible, shared resource accesses from the application-under-test experience low latencies. Again, it depends on access pattern of the application-under-test. If the application-under-test itself is aggressive then it quickly becomes ineligible and following accesses experience high latencies. Thus, similar to classical timing anomalies, these anomalies are application dependent. Only under the TDMA and the Priority Division [21] arbiters, these timing anomalies are absent.

The timing anomalies presented in this paper depend on the shared resource access pattern (cache miss pattern) of the application. Modification in cache line size, associativity, cache size etc modifies the shared resource access pattern. The modification can either remove these timing anomalies or introduce them. This makes the measured execution time in the presence of uninterrupted interference a highly unreliable WCET candidate.

VII. CONCLUSION

This paper has identified two new timing anomalies which occur due to the interference on shared resources in multi-core architectures. The anomalies are as follows: i) Some applications could benefit from aggressive co-existing applications and experience less than the average-case latencies while accessing the shared resource. ii) Some applications experience more latencies in the presence of less number of aggressive co-existing applications than in the presence of more number of aggressive co-existing applications while accessing the shared resource. The anomalies are inferred from formulation and the real-life evidences of their presence are provided using applications from the Mälardalen WCET benchmark suit. The experiments are conducted on the Altera NIOS II multi-core with shared memory architecture implemented on Altera Cyclone III FPGA development board.

Collisions of cache misses of concurrently executing applications on a shared main memory is extremely difficult to predict. Moreover, such prediction is limited to the particular set of application execution paths and precise phase of their starting time. Thus, this prediction is not useful for WCET analysis, practically. To estimate WCET in the presence of unpredictable interference, it is intuitive to let the co-existing applications generate uninterrupted accesses to the shared memory and assume that this is the highest possible interference. However, this paper concludes that the measured execution time of application-under-test in the presence of uninter-

rupted shared resource accesses from co-existing applications could be highly optimistic and well below the WCET considering the theoretical worst case interference.

ACKNOWLEDGMENT

This work was funded by German BMBF projects ECU (13N11936) and Car2X (13N11933).

REFERENCES

[1] T. Lundqvist and P. Stenstrom. Timing anomalies in dynamically scheduled microprocessors. In *Proc. RTSS*, 1999.

[2] J. Reineke, *et al.* A definition and classification of timing anomalies, wcet 2006.

[3] Gernot Gebhard. Timing Anomalies Reloaded. In *WCET 2010*, Dagstuhl, Germany.

[4] Reinhard Wilhelm *et al.* The worst-case execution-time problem—overview of methods and survey of tools. *ACM Trans. Embed. Comput. Syst.*, 7(3), 2008.

[5] Advance microcontroller bus architecture (amba).

[6] Avalon interface specifications.

[7] Mingsong Lv *et al.* Combining abstract interpretation with model checking for timing analysis of multicore software. In *Proc. RTSS*, 2010.

[8] R. Pellizzoni *et al.* Impact of peripheral-processor interference on wcet analysis of real-time embedded systems. *IEEE Transactions on Computers*, 2010.

[9] H. Shah *et al.* Bounding WCET of Applications Using SDRAM with Priority Based Budget Scheduling in MPSoCs. In *Proc. DATE*, 2012.

[10] Dynamic priority queue: An SDRAM arbiter with bounded access latencies for tight WCET calculation. Technical report, 2012.

[11] Hardik Shah, Alois Knoll, and Benny Akesson. Bounding sdram interference: detailed analysis vs. latency-rate analysis. In *Date '13*, Grenoble, France.

[12] Marco Paolieri *et al.* An Analyzable Memory Controller for Hard Real-Time CMPs. *Embedded Systems Letters, IEEE*, 2009.

[13] Li. Xianfeng *et al.* Modeling out-of-order processors for software timing analysis. In *RTSS, 2004*.

[14] R. Kirner *et al.* Precise worst-case execution time analysis for processors with timing anomalies. In *ECRTS '09*.

[15] S. Chattopadhyay *et al.* A unified wcet analysis framework for multi-core platforms. In *RTAS 2012*.

[16] T. Kelter *et al.* Bus-aware multicore wcet analysis through tdma offset bounds. In *ECRTS*, 2011.

[17] J. Gustafsson *et al.* The Mälardalen WCET benchmarks – past, present and future.

[18] M. Fernández *et al.* Assessing the suitability of the ngmp multi-core processor in the space domain. In *EMSOFT '12*.

[19] P. Radojković *et al.* On the evaluation of the impact of shared resources in multithreaded cots processors in time-critical environments. *TACO*, 2012.

[20] J. Nowotsch *et al.* Leveraging multi-core computing architectures in avionics. In *EDCC '12*.

[21] H. Shah, A. Raabe, and A. Knoll. Priority division: A high-speed shared-memory bus arbitration with bounded latency. In *Proc. DATE*, 2011.

A Unified Online Directed Acyclic Graph Flow Manager for Multicore Schedulers

Karim Kanoun[†], David Atienza[†], Nicholas Mastronarde[‡], and Mihaela van der Schaar[⋆]

[†] Ecole Polytechnique Fédérale de Lausanne (EPFL), Switzerland. [‡] State University of New York at Buffalo (UB), U.S.A. [⋆] University of California, Los Angeles (UCLA), U.S.A
email: {karim.kanoun, david.atienza}@epfl.ch, nmastron@buffalo.edu, mihaela@ee.ucla.edu.

Abstract— Numerous Directed-Acyclic Graph (DAG) schedulers have been developed to improve the energy efficiency of various multi-core systems. However, the DAG monitoring modules proposed by these schedulers make a priori assumptions about the workload and relationship between the task dependencies. Thus, schedulers are limited to work on a limited subset of DAG models. To address this problem, we propose a unified online DAG monitoring solution independent from the connected scheduler and able to handle all possible DAG models. Our novel low-complexity solution processes online the DAG of the application and provides relevant information about each task that can be used by any scheduler connected to it. Using H.264/AVC video decoding as an illustrative application and multiple configurations of complex synthetic DAGs, we demonstrate that our solution connected to an external simple energy-efficient scheduler is able to achieve significant improvements in energy-efficiency and deadline miss rates compared to existing approaches.

I. INTRODUCTION

Emerging real-time video processing applications such as video data mining, video search, and streaming multimedia (e.g., H.264 video streaming [17]) have stringent delay constraints, complex Directed Acyclic Graph (DAG) dependencies among tasks, time-varying and stochastic workloads, and are highly demanding in terms of parallel data computation. Multimedia applications are in general modeled with DAGs where each node denotes a task, each edge from node j to node k indicates that task k depends on task j and each group of tasks has a common deadline d_i. As illustrated in Fig. 1, DAG models for applications with dependent tasks can be roughly classified into 4 types depending on the relationship between the task dependencies and task deadlines.

We define a DAG monitoring solution as the module used to process and analyze these DAGs before scheduling the tasks. This module is different from the scheduler and it is responsible for finding parallelization opportunities, tracking the execution of the DAG and providing relevant information to a connected external scheduler. Numerous offline [15][16] and online [6][7][8] DAG monitoring solutions have been proposed to assist schedulers for multimedia applications. However, these solutions are usually closely related to their connected schedulers, and their output cannot be directly exploited by other schedulers. Thus, the problem of finding a generic online DAG

This work was supported in part by a Joint Research Grant for ESL-EPFL by CSEM, and the Spanish Government Research Grant TIN2008-00508.

monitoring solution to assist online energy-efficient schedulers, making the DAG processing problem independent from the connected scheduler, is becoming increasingly important.

Moreover, we also contend that none of these existing online DAG monitoring solutions have considered the general DAG model in which a task's children can have different deadlines (e.g., model 4 in Fig. 1). We present an example of this problem in the remainder of this paragraph. Fig. 2 illustrates an example of two different DAGs modeling the same H.264 video decoder application. The DAG of Fig. 2a preserves the original dependencies between I, P and B frames, where I-frames are compressed independently of the other frames, P-frames are predicted from previous frames, and B-frames are predicted from previous and future frames [1]. Each frame is composed of three type of tasks, namely, initialization (e.g., I_1), slice decoding (e.g., S_2, S_3 and S_4) and the deblocking filter (e.g., F_5). In the example shown in Fig. 2 with 4 frames (I-B-P-B), there are 2 deadlines corresponding to the display deadlines of the two B frames. These deadlines are imposed by the frame rate and the underlying dependency structure. In fact, if frame k depends on frame $k + l$ (with $l > 0$), then both frames will have their deadlines set to the minimum one, i.e., $k/30$ seconds. Finally, in this DAG model, a task's children may have different deadlines (e.g., $F_5 \rightarrow (I_6, I_{16})$). Existing online DAG monitoring approaches (cf. Section II) are not able to handle the additional dependencies between the tasks with different deadlines. Instead, they are forced to convert the original DAG to the fork-join DAG model as presented in Fig. 2b where critical edges (i.e., edges linking 2 tasks with different deadlines) are removed and replaced by a single join edge that links the last task with deadline d_i to the first task with deadline d_{i+1}. Although the fork-join model (which is defined as a sequence of sequential and parallel segments [2]) preserves the dependency coherency between tasks, it restricts the scheduler to operate on tasks belonging to one deadline at a time (i.e., the earliest deadline). Hence, several parallelization opportunities are missed (e.g., frame B with a deadline d_i and frame P with a deadline d_{i+1} can be decoded in parallel, once frame I is decoded).

To summarize, each existing scheduler implements its own DAG monitoring solution with several restrictions on the DAG model. Moreover, none of the existing solutions are able to handle the general DAG illustrated with model 4 in Fig. 1, which allows a task's children to have different deadlines.

In this paper, we propose a novel unified DAG monitoring solution, which we call DAG Flow Manager (DFM). The key contributions of this work are as follows:

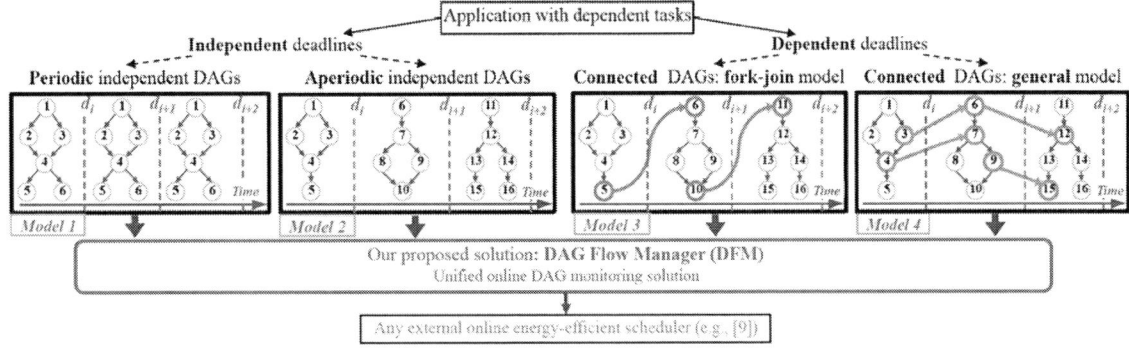

Fig. 1. Classification of DAG models of applications with dependent tasks. Our proposed solution handles all possible DAG models

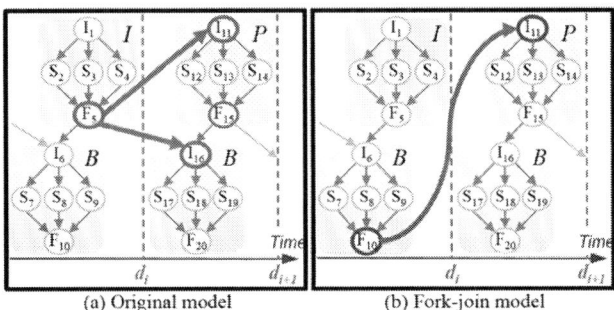

(a) Original model   (b) Fork-join model

Fig. 2. H.264 decoder DAG model [1]

• A low-complexity online DAG monitoring solution that is fully independent of the scheduler that it is connected to it.

• Our DFM does not impose any restrictions on the DAG (e.g., restrictions on deadline dependencies as in the fork-join model). Our DFM covers online all DAG models (Fig. 1).

• Our DFM provides detailed information about the execution status of tasks and deadlines within a look-ahead window, allowing simple connected schedulers to have optimal control of the core assignment and DVFS selection of each task.

Our results for the H.264 video decoder and different configurations of synthetic DAGs demonstrate that our proposed DFM allows a connected online scheduler based on [9] to reach over 50% reduction in energy consumption and over 80% reduction in deadline miss rates compared to existing DAG monitoring solutions [7][8] connected to the same scheduler.

The remainder of this paper is organized as follows. In Section II, we describe the limitations of current offline and online DAG monitoring approaches. In Section A, we introduce the system and application model. In Sections B, C, D and E, we describe our online DFM algorithm in detail. In Section IV, we present our experimental results. Finally, we summarize the main conclusions in Section V.

II. RELATED WORK

In Table I, we summarize different application models considered by existing DAG analyzers and we compare them to the DAG modeled by our solution.

Existing static approaches [15][16] model the application as a DAG with periodic dependent tasks as shown in Fig. 1 (model 1). They propose a coarse-grained task-level software pipelining algorithm to transform periodic dependent tasks into a set of independent tasks based on a retiming technique. However, these approaches are unsuitable for multimedia applications with dynamic DAGs. In fact, the assumption of a periodic DAG

limits the applicability of static approaches because highly optimized modern and emerging video coders do not always have periodic task-graphs (e.g., they may use adaptive group of pictures structures). Hence, applying techniques such as pipelining is not possible, especially for applications that do not adopt a fixed task-graph structure but instead adapt their task-graphs on the fly (e.g., stream mining applications [18]). Moreover, for H.264 video decoding, such pipelining techniques require buffering delays that are proportional to the Group of Pictures (GOP) size, which may be large. Finally, static approaches rely on worst-case execution time estimates to generate the schedulers input. These approaches are efficient if the workload and the starting time of each task is fixed and known. However, they are unsuitable for multimedia applications with dynamic workloads because modeling a non-deterministic workload with its worst-case execution time leads a connected scheduler to create significant slack time and utilize resource inefficiently.

Few online DAG monitoring solutions [6][7][8] have been recently proposed for scheduling problems. In [6][7][8], they consider periodic tasks where each task is represented as an independent DAG with a single deadline (i.e., a task is modeled as a group of jobs or threads having the same deadline similar to models 1-2 of Fig. 1). Therefore, in this approach, monitoring n deadlines simultaneously requires n independent DAGs. Then, each DAG (i.e., each deadline) is decomposed into segments in order to identify future parallelization opportunities. Therefore, if we consider the general case of DAG of applications with dependent deadlines illustrated with model 4 of Fig. 1 and Fig. 2a where a job's children may have different deadlines, the solutions presented in [7][8] are then forced to convert these DAGs into the fork-join model (e.g., model 3 of Fig. 1 and Fig. 2b). Although the fork-join model preserves the dependencies between tasks, it restricts schedulers to scheduling tasks belonging to only one deadline at a time. Hence, several parallelization opportunities are missed (e.g., Fig. 2: frames B and P with deadline d_i and d_{i+1} respectively). The solu-

TABLE I

COMPARISON OF EXISTING DAG MONITORING APPROACHES TO OUR DFM: APPLICATION MODEL AND ANALYSIS TYPE

DAG-monitoring solution	type of deadlines	type of DAG	analysis
[15, 16]	independent	periodic	offline
[6, 7, 8]	independent	periodic	online
[13]	dependent	H.264	online
Our solution	dependent	general	online

978-1-4799-2817-0/14 $31.00 © 2014 IEEE

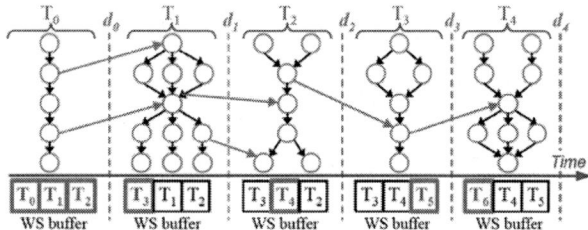

Fig. 3. Evolution of the WS buffer during the execution

tions in [10][9] also suffer from the same limitation. Only one existing solution accounts for dependencies among tasks with different deadlines [13], however, it is restricted to the H.264 DAG model.

III. ONLINE DAG FLOW MANAGER (DFM)

A. System Model

We model our target computationally intensive application as a DAG $G =< \mathcal{N}, \mathcal{E} >$ of dependent tasks t_j with non-deterministic workload w_j and coarse-grained deadlines. $\mathcal{N}$ is the node set containing all the tasks. $\mathcal{E}$ is the edge set, which models the dependencies between the tasks. Each node in the DAG denotes a task t_j. e_j^k denotes that an edge is pointing from t_j to t_k indicating that task k depends on task j. Each task t_j is characterized with its index j and a deadline d_i (i is the deadline index). Each deadline can be assigned to a subset of tasks. Our model covers all general DAG models (e.g., Fig. 1) including the general case where a task's children may have different deadlines than the task itself and its other children (e.g., model 4 of Fig. 1). Finally, we assume that our target multicore platform has M cores with DVFS capability to trade off energy consumption and delay. Each processor can operate at a different frequency $f_i \in F$, where F denotes the set of available operating frequencies and $f_i < f_{i+1}$.

B. Overview of the proposed DFM

We define T_i as the subset of tasks having the same deadline d_i. We also define the Working Set WS as a look-ahead window buffer of n T_is. Each time all the tasks inside a T_i finish their execution, we request the next T_i input from the application. For each new added T_i to the WS buffer, our approach requires from the application its adjacency matrix, its deadline value and the list of edges connecting this T_i with T_{i+l} (with $l \neq 0$). In our *DAG Flow Manager (DFM)*, we propose then to process the full DAG of the application using this WS buffer where only a limited number of deadlines are processed at a time. Analyzing the full DAG of an application by subset of n deadlines (i.e., T_is) is the key for having a low complexity online DAG monitoring solution. Thus, it will be possible to efficiently adapt to applications that have a highly variable workload and adapt their task-graphs on the fly. Fig. 3 illustrates an example of our DFM processing a general DAG model using a WS buffer of 3 deadline slots. In this example, for the sake of clarity, we assume that each T_i finishes its execution at d_i. First, at the beginning of the execution, the WS buffer is filled with T_0, T_1 and T_2. Next, when a deadline finishes its execution, a slot becomes available and it gets automatically filled with the next available deadline. In Section C, we describe the initialization phase that we apply on each new

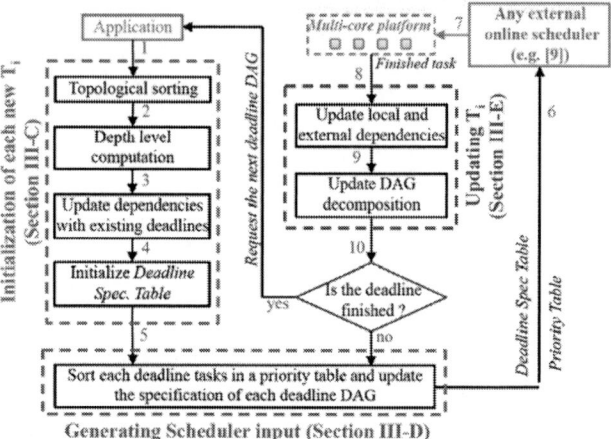

Fig. 4. A complete overview of our DFM

added T_i to the WS buffer and how we handle online the additional dependencies between these T_is. In Section D, we then describe different data structures that our DFM provides to assist an external scheduler and how they can be used efficiently. Finally, we explain how we update online the generated data structures and synchronize it with recent scheduling decisions.

C. Initialization of each new added deadline to the WS

The full initialization process is illustrated in Fig. 4 (arrows 1, 2, 3 and 4). First, we sort the tasks of each new added T_i in the WS buffer, in topological order. An optimal topological sorting algorithm with a complexity of $\mathcal{O}(|\mathcal{N}| + |\mathcal{E}|)$ has been proposed in [12]. Moreover, in [14], it was demonstrated that a DAG with n nodes has a worst case of $n * (n-1)/2$ edges. The topological sorting step is applied only once for each new T_i. This reduces the complexity of computing the depth δ_i^j of each task t_j with deadline d_i in the WS buffer. While sorting the tasks, the list of direct parent tasks $\mathcal{L}_j^P$ (i.e., list of tasks t_k linked to ingoing edges e_k^j) and the list of direct children tasks $\mathcal{L}_j^C$ (i.e., list of tasks t_k linked to outgoing edges e_j^k) for each task t_j are also generated. $\mathcal{L}_j^P$ is used to compute the critical path workload in Section D, while $\mathcal{L}_j^C$ is used for the depth level computations of each available task in the WS buffer. In fact, for each outgoing edge e_j^k (i.e., edges connecting t_j to $\mathcal{L}_j^C$ tasks) of each remaining task t_j in T_i, our solution updates the depth value δ_i^k of the node t_k with $\delta_i^k \leftarrow max(\delta_i^j + 1, \delta_i^k)$. The complexity of the graph traversal algorithm that we apply to compute δ_i^k is then $\mathcal{O}(|\mathcal{E}|)$.

We denote by $l_{i,k}$ the group of tasks t_j having the same depth level $\delta_i^j = k$ in T_i. Note that, for all tasks in T_i, tasks at depth level $k + 1$ (i.e., $l_{i,k+1}$) can only be scheduled after tasks at depth level k (i.e., $l_{i,k}$) are finished. Fig. 5 illustrates in detail the difference between t_j, T_i, $l_{i,k}$ and δ_i^j after applying our algorithm on a WS buffer containing part of the DAG of the H.264 video decoder.

While traversing the DAG for the first time for each new T_i, the number of unfinished direct parent tasks t_k for each task t_j, that we call the dependency status r_j, is also computed. r_j is used to assist the scheduler in detecting available parallelization opportunities from the remaining unscheduled tasks in the WS (i.e., entry nodes $r_j = 0$). In fact, the dependency status r_j is the key idea to track existing dependencies between tasks

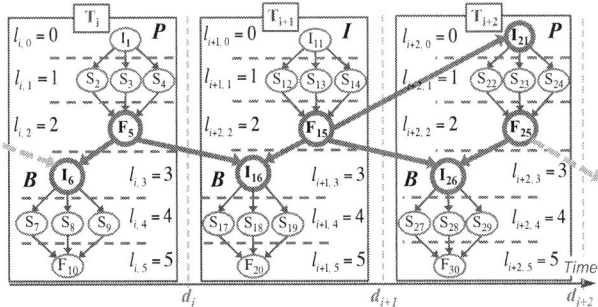

Fig. 5. The decomposition applied by our DFM on H.264 DAG [1]

with different deadlines (i.e., between T_is). Once the initial dependency status r_j is computed for each task within the newly added T_i, we check for its dependencies with other deadlines. To this end, for each task t_j in T_i with an incoming edge from tasks with earlier deadlines (i.e., T_{i+l} with $l < 0$), we increment its r_j value. Then, for each task t_j in T_i with an outgoing edge to other deadlines tasks (i.e., T_{i+l} with $l > 0$) we update its list of direct children tasks $\mathcal{L}_j^C$. This will be used later for the update phase in Section E for clearing the dependencies between the T_is. Finally, it may happen that a task finishes its execution and the other deadline that is connected to does not yet exist in the WS buffer. In this case, we need to make sure that when this new deadline is added to the buffer, its dependency with this finished task is cleared correctly. Therefore, we store this dependency in a new list, that we call the *list of non-cleared dependencies*. This list is generated during the update process (cf. Section E). We use the list of non-cleared dependencies in the last step of this initialization phase in order to check if a dependency has to be cleared by decrementing the r_j value of the concerned task. Each time a dependency is cleared, we remove its corresponding entry from the list.

D. Generation of the scheduler input

To assist online schedulers with immediate parallelization opportunities and the priorities of the available tasks, we define the first output structure, called *Priority Table*, that we provide to any connected scheduler. Each entry in the Priority Table corresponds to a task t_j and it is characterized by: estimated workload w_j, earliest release time s_j, expected ending time z_j, fixed deadline d_i^j where i refers to T_i and $d_i^j = d_i \ \forall j$, critical path workload to its deadline $w_j^{d_i}$ and the dependency status r_j (i.e., the total number of parent tasks that it still depends on). w_j, s_j, z_j and $w_j^{d_i}$ are in clock cycles and d_i^j is in seconds. The tasks in the Priority Table are sorted by our DFM according first to their deadline d_i, then refined to their depth level $l_{i,k}$ in T_i and finally to their estimated workload in case of a tie. We use the Quicksort algorithm with an average complexity of $\mathcal{O}(|\mathcal{N}|.log(|\mathcal{N}|))$. Several workload estimation methods [3][4] with a negligible overhead have been proposed for multimedia applications. Full overhead measurements on the H.264 decoder application are provided in Section IV.

Regarding the computation of this output, z_j and s_j are calculated while the WS buffer is traversed by the depth level computation algorithm (cf. Section C) with $z_j \leftarrow w_j + s_j$ and $s_k \leftarrow max(z_j, s_k)$ for $\forall t_k \in \mathcal{L}_j^C$. If the task is currently running at frequency f_j then we use $z_j \leftarrow w_j * \frac{f_{max}}{f_j} + s_j$ to take into account the applied frequency. Finally, for the calculation

of the critical path workload value $w_j^{d_i}$ starting from each task t_j to its local sink node in T_i, the method used to calculate $w_j^{d_i}$ is the same as the depth level computation algorithm except that we traverse the topological graph in the inverse order and we replace $\mathcal{L}_j^C$ with $\mathcal{L}_j^P$. Our solution indicates to the scheduler which tasks are entry nodes in the remaining tasks set using the dependency status r_j of each task t_j. Nodes with $r_j = 0$ in the Priority Table are potential starting tasks for parallelization. Even though the Priority Table gives a straightforward solution to select the task to schedule, it does not give the scheduler general information related to the execution status of each remaining T_i in the buffer, such as their currently running tasks progress. Having such information, a scheduler will be able to efficiently tune the deadlines values for a more energy efficient execution [9]. Moreover, such information could be exploited by any connected scheduler to set the priority of each T_i, which can be efficiently used in parallel with the priority table. To this end, we then propose a second output to online schedulers in order to track the execution status of future deadlines tasks that we call the *DeadlineSpec Table*. This latter output is used to track the overall progress of each group of tasks T_i. Each entry in the DeadlineSpec Table stores relevant computed information related to each T_i namely: total workload, executed workload, scheduled workload and finally a depth table. Each entry k in the depth table corresponds to a $l_{i,k}$ in T_i and it stores relevant computed information namely: total workload, maximum number of allowed cores and minimum amount of parallelizable workload. The full DeadlineSpec Table is initially generated using previously computed information related to each task. The complexity of creating this table is then linear to the number of nodes in the considered WS.

E. Updating the DAG decomposition and the scheduler input

As shown in Fig. 4 (arrows 8, 9 and 10), the working set decomposition and the generated output are updated each time a task finishes its execution. First, we remove the finished task t_j from the list of remaining tasks and re-estimate the workload of similar tasks. Then, we decrement r_k of each task t_k in $\mathcal{L}_j^C$ (i.e., its direct children tasks t_k). However, it may happen that one direct child has a different deadline that is not available yet in the WS buffer. In this case, we save this dependency in a list that contains all the non-cleared dependencies. This list is used during the initialization phase to clear future dependencies when a new T_i linked to this edge is added to the WS buffer as described in Section C. This first step allows instantly detecting new entry nodes (i.e., when $r_j = 0$) among all the available tasks of all the T_is in the Priority Table output. Then, we set the earliest starting time of each task t_j depending on its execution status. If the task t_j is an entry node and did not start yet, then the current execution time is assigned as its earliest starting time. However, if the task t_j is currently executing then some values from the DeadlineSpec Table, namely, the scheduled workload and executed workload, are updated from values of the currently running task t_j. Finally, we update the information related to each remaining task t_j in T_i by applying the same graph traversal algorithm of Section C on the remaining unscheduled nodes of T_i. Therefore, the total number of times the graph traversal algorithm is applied for the full execution of a T_i will be equal to twice the number of tasks in the

978-1-4799-2817-0/14 $31.00 © 2014 IEEE

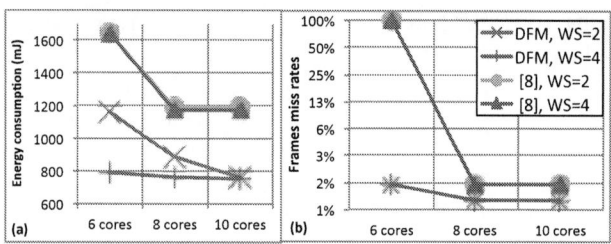

Fig. 6. H.264 decoder: comparison between our DFM and [8], connected to the same scheduler. (a)Energy consumption. (b)Deadline miss rates.

T_i and during each update it will be executed with one node less (i.e., without the finished tasks). Indeed, each time a task finishes its execution the algorithm will be executed first during the depth computation phase and then for the critical path workload. This guarantees real-time information regarding future available parallelization each time before a scheduler has to make a decision.

IV. EXPERIMENTAL RESULTS

We demonstrate the advantages of our DFM, which we have implemented in C, to assist an energy-efficient online scheduler on two experimental benchmarks namely, the H.264 video decoder [17], and multiple configurations of synthetic DAGs generated with GGEN tool [11]. We explain our experimental setup in Section A. Then, we present our results for the two sets of experiments in Sections B and C.

A. Experimental setup

As described in Sections I and II, existing online DAG monitoring solutions [6][7][8] do not consider DAG models where a task's children can have different deadlines. The only way for these approaches to handle such a DAG online is to process it deadline by deadline (i.e., fork-join parallelism model) where critical dependencies between the deadlines are replaced with a single join edge as shown in Fig. 2. However and since our DFM applies similar task decompositions technique to existing solutions for DAGs without critical dependencies between the deadlines, we can perfectly simulate existing approaches by first converting general DAG model to the fork-join model then by applying our DFM on the converted DAGs.

We connect the simulated existing solution [8] ([8] uses the same decomposition technique as in [7]) and our DFM to an external online scheduler based on [9], which applies the least possible restrictions on its application model compared to other schedulers. In [9], an online scheduling approach called MLTF was proposed for multimedia application, where the earliest deadline is scheduled with limited consideration of future tasks' deadlines and workloads. Indeed, based on the derived estimated duration of all pending tasks, a new virtual deadline is set in order to have more balanced workload distribution over the time. Then, a Largest Task First (LTF) schedule is applied. In [9], the algorithm applied for frequency selection does not take into account the dependencies between the tasks. Therefore, we made the scheduler select the frequency of each task based on the remaining critical path workload and the available amount of clock cycles. Due to dependencies between tasks, gaps (i.e., when a core is idle and waiting for another task to finish) may occur during the schedule. Therefore, we add another simple module on top of the MLTF scheduler presented in [9]

to fill the gap. This module compares the amount of available gap (which can be easily estimated from the schedule generated by [9]) to the workload estimation of the available task t_j with $t_j \in T_{e+l}$ (i.e., having deadline d_{e+l}) to compute the minimum frequency $f_{d_e}^j$ that allows t_j to fit the available gap occurring before d_e. Then, the module applies again the MLTF schedule but this time on T_{e+l} between d_{e+l} and d_{e+l-1} to estimate the minimum frequency $f_{d_{e+l}}^j$ used when scheduling task t_j during its allocated time (i.e., between d_{e+l} and d_{e+l-1}). Finally, if $f_{d_e}^j \leq f_{d_{e+l}}^j$, then the gap is filled. All these computations are then based on MLTF [9].

B. H.264 decoder - Energy, deadline miss rates and overhead

For our multimedia benchmark, we have used the Joint Model reference software version 17.2 (JM 17.2) of an H.264 encoder [17]. The DAG model and the deadlines configuration that we consider for our benchmark are similar to the one shown in Fig. 2 with an IBPB GOP structure, 8 slices per frame and 30 frames/second for CIF (352x288) resolution video sequences. We use accurate statistics generated from an H.264 decoder that we have parallelized and executed on a sophisticated multiprocessor virtual platform simulator. In fact, in this work, we use the multiprocessor ARM (MPARM) virtual platform simulator [5], which is a complete SystemC simulation environment for MPSoC architectural design and exploration. MPARM provides cycle-accurate and bus signal-accurate simulation for different processors. In our experiments, we have generated with MPARM the workloads and the dynamic power consumption statistics of each task (i.e., frame initialization, slice decoding and deblocking filter) using ARM9 power consumption figures with DVFS support (300MHZ at 1.07V, 400MHZ at 1.24V and 500MHZ at 1.6V).

In Fig. 6, we show the energy consumption and frame miss rates when decoding the Foreman sequence. We have used the same scheduler connected first to our DFM and second to existing DAG monitoring solution [8] as described in the previous section. The results of Fig. 6a show that [8] does not allow the connected scheduler to use more than 8 cores as the maximum number of parallelizable tasks within a single T_i is 8 in the considered DAG (i.e., 8 slices per frame). However, our solution allows the connected scheduler to take advantage of the additional number of cores as demonstrated by the relative energy decreasing with the number of cores. The results show also that our DFM can efficiently exploit the increased size of the WS buffer. In fact, our DFM allows the connected scheduler to reduce the energy consumption by up to 52% compared to [8] thanks to the information provided by our DFM regarding each of the available deadlines in the buffer to the connected scheduler. For the frame miss rates, the decomposition technique and the information provided by existing DAG monitoring solution do not allow the connected scheduler to efficiently schedule the tasks before their deadlines. In fact, as shown in Fig. 6b, all the frames are missed if the number of cores is less than 8 cores due to heavy workloads. However, by exploiting our DFM output related to each of the available deadlines in the buffer and the detected parallelization among the deadlines tasks, the connected scheduler is then able to achieve less than 1.5% miss rates starting from only 6 cores as the gaps are filled with available tasks with future deadlines.

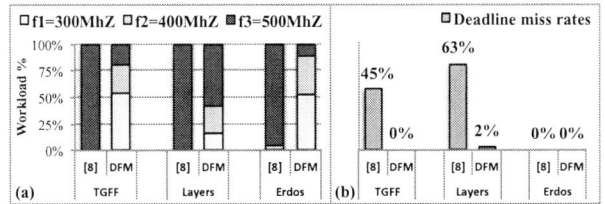

Fig. 7. Synthetic DAGs (deadlines values are only 1% greater than the critical path workload) : comparison between our DFM and [8]. connected to the same scheduler. (a)Frequency usage. (b)Deadline miss rates.

Finally, we present the overhead in terms of processor clock cycles with respect to the workload of the Foreman video sequence. We measured an overhead of 0.82%, 0.93% and 1.04% of our proposed DFM when using 2, 3 and 4 deadlines per WS respectively (with 20 tasks per deadline). The overhead of the connected scheduler varies between 0.7% and 1.24% depending on the number of cores and the WS buffer size.

C. Generalizing the results to the general DAG model

We use the GGEN [11] tool to model an application with 100 connected generated DAGs. We connect these DAGs by randomly adding m edges in a way that some tasks in DAG g depend on some other tasks in DAG $g-1$. For the workload, we assume that an application of n tasks has k types of workloads. We assign then each task t_j with a random type number a_j between 1 and k, and we compute the workload with $w_j = w(a_j) + x$ where $w(a_j)$ is the minimum workload value of all the tasks with type a_j, and x is a random number between 0 and $(w(a_j) * \alpha)$ with $\alpha \in [0, 0.5]$. x represents the workload variation of each task with respect to its type. Finally, to assign a deadline to each DAG, we compute the critical path workload w_i^{cp} of each DAG i and the final deadlines values (in seconds) are assigned with $d_i = d_{i-1} + w_i^{cp} * \frac{1+\beta}{f_{max}}$ with $\beta \in [0, 0.5]$.

We generate the synthetics DAGs with Erdos, TGFF and Layers DAG generation methods as presented in [11]. We set the number of cores to 6 and previously described parameters to $n = 25$, $k = 5$, $\beta = 0.01$ (i.e., deadlines values are only 1% greater than the critical path workload), $\alpha = 0.4$ and $m \in [10, 15]$. For TGFF method, we set the maximum number of ingoing and outgoing edges per node to 4, for Erdos and Layer we set the probability of an edge to appear in each DAG to 0.5 and for the Layer method we set the number of layers to 4. We choose these parameters in order to simulate a congested system. Fig. 7a shows the distribution of the frequency usage of the total workload assigned by the scheduler exploiting our DFM output compared to the same scheduler exploiting [8] output. A higher fraction of workload processed at lower frequencies is desirable because it indicates lower dynamic energy consumption. We also compare the deadlines miss rates in Fig. 7b. Our DFM significantly reduces the usage of the maximum frequency by up to 84% and the deadline miss rates by up to 61% (with 0% to 2% miss rates overall). Our solution provides the information related to the execution status of each deadline which allows a connected scheduler to take foresighted decisions that target the lowest possible frequency, reducing then the dynamic energy consumption. Moreover, connecting the deadlines together with multiple edges restricts existing solutions [8] to provide the scheduler with only parallelization within a single deadline. Therefore, our DFM allows any con-

nected scheduler to exploit more parallelization opportunities resulting into a more balanced workload distribution and more optimal usage of the available resources.

V. CONCLUSION

In this paper we have proposed a novel unified DAG monitoring solution, that we called DAG Flow Manager (DFM). The key contributions of this work were as follows: (i) A low-complexity online DAG monitoring solution that is fully independent of the scheduler that it is connected to it; (ii) Our DFM does not impose any restrictions on the DAG and covers online all DAG models (Fig. 1); (iii) Our DFM provides detailed information about the execution status of tasks and deadlines within a look-ahead window, allowing simple connected schedulers to have optimal control of the core assignment and DVFS selection of each task. Our results for the H.264 decoder have demonstrated that our proposed DFM solution allowed a connected online scheduler to reach up to 52% reduction in energy consumption and over 80% reduction in deadline miss rates compared to the schedule generated by the same scheduler but relying on existing DAG monitoring solutions. The overhead in terms of processor clock cycles for the proposed DFM, for the aforementioned results, is less than 1% with respect to the total workload of the foreman video sequence. Finally we have generalized these results on synthetic DAGs with different DAG configurations.

REFERENCES

[1] T. Wiegand, et al., "Overview of the h.264/avc video coding standard." in *IEEE TCSVT*, vol. 13, no. 7, pp. 560–576, July 2003.

[2] "OpenMP," http://openmp.org.

[3] Y. Andreopoulos, et al., "Adaptive linear prediction for resource estimation of video decoding," in *IEEE TCSVT*, vol. 17, no. 6, pp. 751–764, June 2007.

[4] S.-Y. Bang, et al., "Run-time adaptive workload estimation for dynamic voltage scaling," in *IEEE TCAD*, vol. 28, no. 9, pp.1334 –1347, Sept. 2009.

[5] L. Benini, et al., "Mparm: Exploring the multi-processor soc design space with systemc," *J. VLSI Signal Process. Syst.*, vol. 41, no. 2, pp. 169–182, Sept. 2005.

[6] M. Qamhieh, et al., "A Parallelizing Algorithm for Real-Time Tasks of Directed Acyclic Graphs Model," in Proc. *RTAS*, Apr. 2012.

[7] A. Saifullah, et al., "Real-time scheduling of parallel tasks under general dag model," online: http://www.cse.wustl. edu/~ saifullaha/BIB_Files/dag_parallel. pdf, 2012. tech. report.

[8] J. Li, et al., "Outstanding Paper Award: Analysis of Global EDF for Parallel Tasks," in Proc. *ECRTS*, 2013..

[9] Y.-H. Wei, et al., "Energy-efficient real-time scheduling of multimedia tasks on multi-core processors," in Proc. *ACM SAC*, 2010.

[10] J. Cong, et al., "Energy efficient multiprocessor task scheduling under input-dependent variation," In Proc *DATE*, 2009.

[11] D. Cordeiro, et al., "Random graph generation for scheduling simulations," in Proc. *SIMUTools*. 2010.

[12] T. H. Cormen, et al., "Introduction To Algorithms," MIT Press, 2001.

[13] N. Mastronarde, et al., "Markov decision process based energy-efficient on-line scheduling for slice- parallel video decoders on multicore systems," in *IEEE TMM*, vol. 15, no. 2, pp. 268–278, Feb, 2013.

[14] O. Sinnen, "Task scheduling for parallel systems," in Wiley, 2007.

[15] Y. Wang, et al., "Overhead-aware energy optimization for real-time streaming applications on multiprocessor system-on-chip, in " *ACM TODAES*, vol. 16, no. 2, pp. 14:1–14:32, Apr. 2011.

[16] Y. Wang, et al., "Optimal task scheduling by removing inter-core communication overhead for streaming applications on mpsoc," in Proc. *RTAS*, 2010.

[17] *H.264/14496-10 AVC Reference Software Manual (revised for JM 17.1)*.

[18] R. Ducasse, et al., "Adaptive topologic optimization for large-scale stream mining," in *IEEE JSTSP*, vol. 4, no. 3, pp. 620–636, June 2010.

978-1-4799-2817-0/14 $31.00 © 2014 IEEE

Variation-aware Statistical Energy Optimization on Voltage-Frequency Island based MPSoCs under Performance Yield Constraints

Song Jin*   Yinhe Han[#]   Songwei Pei[†]

*Department of Electronic and Communication Engineering, School of Electrical and Electronic Engineering,
North China Electric Power University, P. R. China

[#]State Key Laboratory of Computer Architecture, Institute of Computing Technology,
Chinese Academy of Sciences, P. R. China

[†]Department of Computer Science and Technology, Beijing University of Chemical Technology, P. R. China

jinsong@ncepu.edu.cn   yinhes@ict.ac.cn   peisw@mail.buct.edu.cn

Abstract— Energy efficiency is a primary design concern for embedded multiprocessor system-on-chips (MPSoCs). Recently, Voltage-Frequency Island (VFI) -based design paradigm was introduced for fine-grained power management, which can seamlessly combine with the task scheduling algorithm to optimize system energy. However, the ever-increasing variabilities cause large uncertainty on delay and power. Such statistical nature in performance parameters easily makes deterministic energy optimization hard to achieve desirable performance yield, defined as the probability of the design meeting timing constraints of the system. In this paper, we propose a variation-aware statistical energy optimization framework, which takes account of performance yield constraints in energy-aware task scheduling, voltage assignment and VFI partitioning process. Energy optimization sensitivity, defined as energy variations of the task under voltage scaling, combines with the statistical slack of the task to guide the overall optimization flow. Experimental results demonstrate the effectiveness of the proposed scheme.

I. Introduction

Technology scaling dramatically increases integration capacity of the transistors, enabling design of embedded systems transform into Multiprocessor System-on-chip (MPSoCs) [1]. Typically, an MPSoC can provide an entire system function for high-performance applications by integrating various processing elements (PEs) into a single chip, including microprocessor, DSP and memory, etc. For the embedded MPSoCs, energy efficiency has become a primary design concern since the power supply commonly relies on the battery. It is desirable to execute the applications with minimum energy meanwhile still meeting the real-time constraints of the system.

Recently, concept of Voltage-Frequency Island (VFI) was introduced into the MPSoC design for fine-grained power management [2, 3]. VFI-based design partitions the chip into multiple voltage islands and each island operates at its own voltage and frequency. Combining with task allocation and scheduling, such design paradigm manifests great potential in system energy reduction. As a result, various energy optimization schemes targeting VFI-based MPSoCs were proposed

[4, 10, 11, 12]. Generally speaking, the existing schemes first perform an energy-aware deterministic task scheduling algorithm. After the task schedule, the static voltage assignment process assigns appropriate supply voltages to individual PEs for total energy minimization. At last, VFI construction is performed by merging adjacent PEs and adjusting the supply voltages. The total optimization flow aims at executing tasks with minimum energy while still meeting the deadline constraints.

However, process variation (PV) exacerbated with the relentless scaling of transistor's feature size [7], posing significant challenges on the deterministic energy optimization solutions. With PV effects, performance parameters like operating frequency and power of the PE always deviate from the designated values, and should be treated as random variables. Correspondingly, execution time and power of the scheduled task also manifest statistical characteristics. Facing such scenario, it is essential for any energy optimization scheme to meet the deadline constraint in maximum possibility, therefore minimizing the yield loss. Unfortunately, under serious variations, the deterministic energy optimizations relied on nominal value hard to meet the predefined deadline constraints. While designing for worst case scenarios is costly and may not be a viable choice. Given above discussions, it is obvious that the statistical nature in performance parameters should be taken into account in the overall VFI-based energy optimization processes.

In this paper, we propose a variation-aware statistical energy optimization framework for VFI-based MPSoCs. To meet deadline constraints in different process corners, performance yield [9] is used to quantify timing characteristic throughout the optimization process, defined as the probability of the design meeting timing constraints of the system. Unlike prior work, we perform energy-aware task scheduling, voltage assignment and VFI partition in a statistical manner by considering probability feature in the execution time and power of the task. We combine energy optimization sensitivity, defined as energy variations of the task under voltage scaling, with the statistical slack of the task to guide the overall optimization flow. Experimental results demonstrate that our scheme achieves on average 54% improvement in performance yield over the deterministic solution, with comparable energy optimization results.

978-1-4799-2817-0/14 $31.00 © 2014 IEEE

Comparing to the worst case -based optimization, on average more than 38% energy reduction is obtained from our scheme, under the same performance yield constraints.

The rest of the paper is as follows. Section 2 introduces preliminaries. Section 3 details the proposed statistical energy optimization framework. Section 4 presents the experimental results. We conclude in Section 5.

II. PRELIMINARIES

The target hardware platform is defined as below. Without loss generality, the target MPSoC platform is tile-based structure with various heterogeneous PEs. Inter-tile communication is supported by Networks-on-Chip (NoC). Mixed-voltage/mixed-frequency FIFOs [3] located at border between VFIs are used for data synchronization.

A. Deterministic Energy Optimization on VFI-based MPSoCs

The large amount of literature has been conducted on energy optimization for VFI-based hardware platform. Orgas et al. [4] proposed an energy-aware VFI partition solution which performs the deterministic task scheduling algorithm (EAS) [10] and VFI partition in a unified way. Zhang et al. [11] formulated task scheduling and voltage selection as an integer linear programming problem (ILP) to optimize energy for MPSoCs. Jang et al. [12] proposed an energy optimization framework by combing VFI partition with NoC mapping. However, under serious variations, above deterministic schemes hard to achieve desirable performance yield. Marculescu et al. [13] analyzed deterministic bounds for application completion times running on single and multiple VFI-based platforms. Their work, however, only focused on timing yield analysis rather than system energy.

The deterministic energy and latency models are commonly used to guide the existing energy optimization schemes. Given the target task graph and VFI-based MPSoC platform with NoC infrastructure, the system energy E_{sys} can be expressed as:

$$E_{sys} = E_{comp} + E_{comm} \tag{1}$$

where E_{comp} and E_{comm} denote the total computation and communication energies, respectively. The total computation energy comes from the execution of task graph and can be expressed as $E_{comp} = \sum_{i=1}^{n} (NC_j \cdot C_i \cdot V_i^2)$. Here, NC_j denotes the execution cycles of task j. C_i stands for the totally switching capacitance per cycle. V_i denotes the voltage that the PE i operated on. n is the total number of the PEs in MPSoC.

To calculate the total communication energy, as in [20] and [21], we first define bit energy consumed on transmitting one bit from PE i to PE j as:

$$E_{bit} = \sum_{i}^{n_V} \left(E_{bit}^R(i) + E_{bit}^{Link}(i) + E_{bit}^{FIFO}(i) \right) \frac{V_i^2}{V_{DD}^2} \tag{2}$$

where E_{bit}^R, E_{bit}^{link} and E_{bit}^{FIFO} are bit energies consumed on the router, the link and the mixed-voltage/mixed-frequency FIFOs, respectively. n_V is the number of VFIs. V_i is the supply voltage that $VFI(i)$ operated on.

Based on E_{bit}, the total communication energy consumed by a task graph with m communication transactions can be formulated as:

$$E_{comm} = \sum_{k=1}^{m} E_{bit} \times Q_k \tag{3}$$

where Q_k denotes the data volume in communication transaction k between PE i and PE j.

As for timing quantities in the task graph, we define latest finish time of a task FT as

$$FT(i) = T_{exe}(i) + T_{comm}(i,j) \tag{4}$$

where $T_{exe}(i)$ denotes execution time of the task on PE i. For the task with execution cycles NC, $T_{exe}(i) = NC/f_i$. Here, f_i is the operating frequency of PE i. $T_{comm}(i,j)$ denotes the communication latency between two tasks executed on PEs i and j, respectively.

For a communication transaction k between two PEs, the communication latency T_{comm} can be formulated as:

$$T_{comm} = \sum_{i=1}^{n_V} \frac{NC_R}{f_i} + \sum_{j=1}^{n_V-1} \frac{NC_{FIFO}}{f_j} + \frac{Q_k}{W} \tag{5}$$

where NC_R is the number of cycles that a flit traverses a single router and a outgoing link. NC_{FIFO} is the number of cycles that a flit traverses the mixed-voltage/mixed-frequency FIFO. W is the channel width. Obviously, the first two terms in Eq.5 denote header flit latency, while the last term stands for the serialization latency.

B. Statistical Task Scheduling

With PV effects, operating frequency of the PE should be treated as a random variable. Correspondingly, above mentioned timing quantities, such as FT, T_{exe} and T_{comm}, also become the statistical variables. As a result, statistical static timing analysis (SSTA) [14] is commonly conducted for task graph analysis. Two atomic functions *sum* and *max* are used to calculate the statistical timing quantities. For two random timing quantities X and Y following Guassian distribution with means μ_X and μ_Y, sum of X and Y is still a Guassian random variable with a mean of $\mu_X + \mu_Y$ and variance of $\sqrt{\sigma_X^2 + \sigma_Y^2 - 2\rho\sigma_X\sigma_Y}$. ρ is the correlation coefficient. As for *max* function, the tight probability [15] is used for comparison of two random variables, defined as the probability of one random variable being larger than the other. Clark equation [15] can then be exploited to calculate the maximum of two random variables.

As for statistical task scheduling, Wang et al. [9] proposed *performance yield* to quantify the scheduling effectiveness, defined as the probability that a task schedule meets the predefined deadline constraints. Huang et al. [16] proposed a variation aware quasi-static task scheduling algorithm and adopted Monte Carlo simulation to calculate the performance yield. Singhal et al. [17] proposed a non-probabilistic approach to reduce computation complexity of performance yield. Chon et al. [18] presented a task allocation and scheduling method that takes the impact of resource sharing into consideration. Above literature, however, only focused on improvement in performance yield whereas ignoring the energy optimization. In [19],

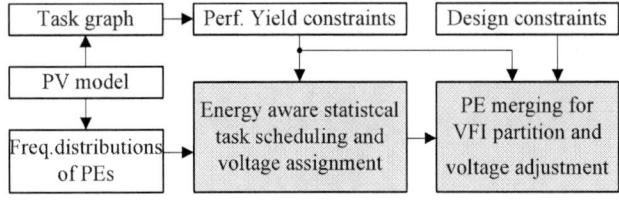

Fig. 1. Overview of our scheme.

Ghorbani proposed the concept of energy yield and formulated yield-aware task scheduling as an ILP problem. The work in [19] does not fit to VFI-based platform. Mover, rationality of energy yield remains as an open problem since it is desirable for an embedded system to consume energy as low as possible.

III. STATISTICAL ENERGY OPTIMIZATION

A. Problem Formulation

The problem studied in this paper can be formulated as: Given

- A directed acyclic task graph including a set of tasks and directed arcs. Each task contains related information, such as execution cycles, power and deadline constraint. While the arcs denote control and data dependencies between tasks as well as the communication quantities.

- An MPSoC platform consists of various heterogeneous PEs. Frequency distributions of the PEs under PV effects with respect to different voltage levels.

- A set of design constraints, such as the maximum allowable VFI number, performance yield constraints.

To determine an energy-aware statistical task schedule and VFI partition with appropriate voltage assignment such that the total system energy is minimized meanwhile the designated performance yield constraints are satisfied.

B. Overview of Our Scheme

Figure 1 sketches the proposed statistical energy optimization framework. Our scheme performs task scheduling, voltage assignment and VFI partition in a statistical manner. To summarize, we first perform statistical task scheduling. The task with largest energy optimization sensitivity is assigned with the highest priority and scheduled onto the PE which provides the maximum statistical slack. After task schedule, appropriate supply voltages are assigned to each PE for total energy minimization meanwhile still meeting the performance yield constraints. At this time, each PE is recognized as a single VFI. Next, we merge adjacent PEs to form the VFIs given design constraints, such as maximum allowable VFI number. The merging process carefully adjusts supply voltages of the formed VFIs, aiming at achieving the performance yield constraints with minimum increase in system energy.

C. Energy-aware Statistical Task Scheduling and Voltage Assignment

Statistical task scheduling on MPSoCs for energy minimization has been proven to be an NP-hard problem [10]. In this paper, we combine energy optimization sensitivity and statistical slack of the task to efficiently guild the task scheduling algorithm. Generally speaking, we always try to schedule the task with the most potential in energy optimization onto the processor which provides largest slack.

A. Energy optimization sensitivity: for a task, the energy optimization sensitivity is defined as the energy variations under voltage scaling. Assume energy consumed by the task at nominal voltage is E_{norm}, energy optimization sensitivity of the task S_E w.r.t n voltage levels is defined as:

$$S_E = E_{norm} \sum_{i=1}^{n} \frac{V_i^2}{V_{DD}^2} \qquad (6)$$

where V_i is the available voltage levels for voltage scaling.

B. Statistical slack: we determine statistical slack of the task as follows. For each task t_i, μ_i, mean of execution time of t_i over different PEs is calculated. Then we calculate the total slack for each path in the task graph based on $\mu's$ and the deadline constraints. At this time, the total slack is a deterministic value. Then, the total slack of a path will be proportional allocated to the tasks on this path. The task with larger S_E will be allocated with more slack. Finally, with $\mu's$ and the slacks, an initial deadline for each task can be determined correspondingly. As a result, for each task we can calculate a set of statistical slacks corresponding to all of the task-PE pairs by using the initial deadline in statistical timing analysis. The tight probability [15] can then be used to compare these statistical slacks.

C. Statistical task scheduling and voltage assignment: Figure 2 presents the pseudo-code of the statistical task scheduling and voltage assignment processes. The required inputs include full task list (FTL) constructed from the task graph and full PE list (FPL) constructed from MPSoC platform. The statistical task scheduling starts with the construction of ready task list (RTL) and available PE list (APL) (line 4). The ready task denotes a task for which the preceding tasks have already been scheduled. While the available PE denotes a PE is idle at the present scheduling time. After RTL constructed, the task in the RTL with highest S_E is selected. Then a set of statistical finish times and statistical slacks are calculated assuming that this task is scheduled onto all of the available PEs (lines 6–8). Next, the tight probability is exploited for comparison of the set of statistical slacks. Finally, the task will be assigned onto the PE which provides the largest statistical slack (lines $9-10$). After this time of scheduling, the scheduled task and the occupied PE are removed from RTL and APL (line 11). The next round of task schedule will start by pick out a new task in the RTL with highest S_E. The whole statistical task scheduling terminates until all of the tasks in the task graph have been scheduled.

In the following voltage assignment process, initially each PE is recognized as a single VFI. The initial voltages are assigned to all the PEs for total energy minimization based on Eq.1-Eq.3. Then SSTA is performed to calculate the achieved performance yield. If the designated performance yield constraints are not satisfied, the assigned voltages on the PEs will

978-1-4799-2817-0/14 $31.00 © 2014 IEEE

Procedure of Statistical Task Scheduling and Voltage Assignment
1.	//Energy-aware Task Scheduling Process
2.	// Input: Full Task List FTL, Full PE List FPL;
3.	// Output: Task schedule: t_i in FTL $\rightarrow P_j$ in FPL;
4.	Construct Ready Task List RTL and Available PE List APL;
5.	while(RTL is not empty && APL is not empty){
6.	Select the task t_i with largest S_E in RTL;
7.	For all the PEs in APL
8.	Calculate $FT(t_i, P_j)$ and $SSlack(FT(t_i), P_j)$;
9.	Apply tight probability to calculate $max(SSlack(FT(t_i), P_j))$;
10.	Schedule t_i onto P_j given $max(SSlack(FT(t_i), P_j))$;
11.	Remove t_i from RTL and P_j from APL;}
12.	Jump to step 4;
13.	**Voltage Assignment Process**
14.	// Input: Task Schedule; Available voltage levels; FTL;FPL;
15.	// Output: Assigned voltages for PEs in FPL;
16.	Assign initial voltages to PEs in FPL to minimize total energy;
17.	Perform SSTA to calculate Perf. yield;
18.	while(Perf. Yield constraints are not satisfied){
19.	Select P_j in FPL with minimum $\sum S_E$(scheduled tasks);
20.	Relax assigned voltage of P_j and flag P_j as adjusted PE;
21.	Perform SSTA to calculate Perf. Yield;}

Fig. 2. Procedure of statistical task scheduling and voltage assignment.

be adjusted. Each time we choose the PE on which the scheduled tasks have the minimum sum of energy optimization sensitivities. Then we relax (increase) the operating voltage of this PE. Such adjustment can increase statistical slacks of the scheduled tasks and improve the overall performance yield, meanwhile posing least impact on system energy. The whole adjustment process terminates until the performance yield constraints are satisfied.

D. VFI Partition

Given the design constraints such as the maximum allowable VFI number, we may need to perform VFI partition by merging adjacent PEs and adjusting the supply voltages for the constructed VFIs. In fact, the preceding voltage assignment process has assigned the lowest voltages to the PEs to minimize system energy while meeting performance yield constraints. Therefore, voltage adjustment for a VFI at this time is inclined to increase voltage for the PE which originally has lower voltage before merging. As a result, energy may increase correspondingly. To minimize the increase in energy, we once again use the energy optimization sensitivity of the task to guide the merging process. When two adjacent PEs are considered to be merged together, we always increase the voltage of the PE on which the scheduled tasks having the minimum sum of energy optimization sensitivities. Each time a merging process finishes, SSTA is performed to re-check the performance yield constraints. We evaluate all the VFI partitions from 1-VFI to maximum allowable VFI number. The VFI partition with minimum energy while still meeting the performance yield constraints is selected as the final partitioning decision.

IV. EXPERIMENT

A. Experimental Setup

A. Simulation platform: experiments are performed on a hypothetical heterogeneous MPSoC which is tile-based NoC structure with 4×4 topology. PEs in the platform include various microprocessors, DSPs and lower-power processing elements. Frequency ranges from $350MHz$ to $750MHz$. All of the PEs are assumed to support voltage scaling with five discrete levels $[0.7V, 0.8V, 0.9V, 1.0V, 1.1V]$. Besides PEs, each tile contains a router for task communication. The deterministic x-y routing algorithm is exploited to avoid livelock and deadlock. Mixed-voltage/mixed-frequency FIFOs are used at border between VFIs for data synchronization. The bit energy used in Eq.2 is retrieved from [20, 21].

B. Frequency distribution under PV: we simulate different kinds of PEs based on critical path model [22] with different number of logic levels, ranging from 12 to 20 levels. HSPICE Monte Carlo simulation is performed with the PTM $65nm$ [23] transistor model to obtain frequency distributions of the PEs under PV effects, corresponding to all the discrete voltage levels. The total variation of 10% on process parameters consists of die-to-die variation of 6% and within-die variation of 8%. The within-die variation is further evenly divided into system and random components.

C. Task graph: Two sets of task graphs are used as target applications. The first set is selected from industrial benchmark E3S [24]. Since the number of tasks in E3S benchmark suite commonly less than the number of PEs in the experimental platform, we group multiple task graphs in each benchmark together and treat them as a new task graph. The second set of 6 task graphs is randomly generated using TGFF [25], which include a number of tasks ranging from 80 to 100. We implemented statistical task scheduling algorithm proposed in [9] to obtain the performance yields for all the task graphs. The obtained performance yields are then used as the constraint in the following experiments. Note that with large variations, performance yield will not necessarily be 100% for some tasks with tight deadline constraints. Table I lists the related statistic for the task graphs and the obtained performance yield constraints.

B. Experimental Results

The proposed statistical energy optimization scheme is applied to the target MPSoC platform and the task graphs. For comparison purpose, we also implement the deterministic scheme proposed in [4] based on nominal and worst case scenarios, respectively. In nominal cases (called D-NC), mean of frequency distribution of the PE under nominal supply voltage ($1V$) is adopted. While in worst case (called D-WC), $\mu+3\sigma$ of frequency distribution of the PE is adopted in the experiment. Considering the design constraints, the maximum allowable VFI number is set to 8. For clarity, energy optimization results are normalized to the ones obtained from D-NC performed on 1-VFI.

Table II lists the statistic of optimization results for three kinds of schemes. Columns $2-4$ denote the optimal number of VFIs after partition. Columns $5-7$ denote the normalized energy after optimization. Columns $8-10$ denote the achieved

TABLE II
STATISTIC OF EXPERIMENTAL RESULTS

Bench	Optimal VFI Number			Energy Normalized			Performance Yield			CPU time (mins)
	D-NC	D-WC	ours	D-NC	D-WC	ours	D-NC	D-WC	ours	
Consumer	3	3	5	0.56	0.89	0.52	52%	100%	100%	3
Auto-indust	4	5	6	0.59	0.91	0.60	60%	100%	100%	7
Networks	3	3	4	0.75	0.96	0.72	56%	100%	100%	3
Telecomm	4	4	4	0.51	0.87	0.52	60%	100%	100%	8
TG1	5	6	6	0.43	0.83	0.42	57%	100%	100%	15
TG2	4	5	6	0.46	0.90	0.46	61%	100%	100%	17
TG3	4	4	5	0.40	0.87	0.41	56%	100%	100%	17
TG4	4	4	6	0.53	0.91	0.61	51%	97%	98%	16
TG5	4	3	6	0.58	0.86	0.66	53%	97%	97%	18
TG6	4	6	5	0.52	0.93	0.58	53%	96%	96%	18

Energy normalized: normalized energy to the one in 1-VFI; D-NC&D-WC: deterministic optimizations on nominal value and worst case

TABLE I
STATISTIC OF TASK GRAPHS

	Task	Communication.	Perf. yield
Consumer	11	low	100%
Auto-indust	24	medium	100%
Networks	13	high	100%
Telecomm	30	high	100%
TG1	88	medium	100%
TG2	96	medium	100%
TG3	101	high	100%
TG4	91	medium	98%
TG5	96	medium	97%
TG6	100	high	96%

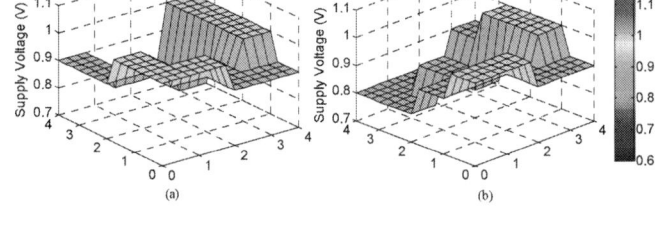

Fig. 3. Impact of slack on voltage assignment. (a) original version of TG6 with tight deadline, and (b) modified version of TG6 with loose deadline.

performance yield. The last column presents the CPU times consumed by our scheme.

The data in Table II can be divided into two parts for analysis. The first part consists of the results in Rows $3-9$ while the second part covers the results in the last three rows. Benchmarks in the first part generally have relative loose deadline constraints, i.e., large slacks. The performance yields of these benchmarks always achieve 100% (the last column in Table I). For these benchmarks, our scheme outperforms much better in improving performance yield over D-NC. On average about 54% improvement in performance yield is observed (Column 8 vs. Column 10) with almost the same energy optimization results (Column 5 vs. Column 7). Although D-WC also achieves desirable performance yield, the resulted energies after optimization are far more larger than the ones in our scheme (Column 6 vs. Column 7).

However, the benchmarks in the second part (Rows $10-12$) have more tight deadline constraints (i.e. small slacks) compared to the ones in the first part. As shown in the last column in Table I, performance yields of these benchmarks cannot

achieve 100%. Such tight deadline constraints in these benchmarks obviously degrade the energy optimization, as shown in Column 5 and Column 7 in the last three rows in Table II. To achieve the designated performance yield constraints, our scheme has to relax the energy optimization effort, resulting in less energy reduction. Nevertheless, our scheme still performs much better than D-WC. On average our scheme achieves about 32% energy reduction over D-WC.

We now analyze the impact of the slack on statistical energy optimization in detail. We choose benchmark TG6 and relax its deadline constraint. After deadline relaxing, TG6 also achieves 100% performance yield. Once again our scheme is applied to TG6 for energy optimization. At this time, the normalized energy achieved by our scheme is about 0.54, very close to the result obtained by D-NC. Figure 3 illustrates the voltage assignments for the twice applications of our scheme on TG6. As shown in Figure 3(a), for the original version of TG6 with the tight deadline (small slack), more PEs have to operate at higher voltage ($1V$) for performance yield consideration. On the contrary, for the modified version of TG6 with the loose deadline (large slack), more PEs can operate at lower voltage, as shown in Figure 3(b). Such observation indicates that for

978-1-4799-2817-0/14 $31.00 © 2014 IEEE

the application with tight deadline constraint, other energy optimization scheme like DVFS may need to be conducted for more aggressive energy optimization at runtime.

V. CONCLUSION

In this paper, we proposed a statistical energy optimization framework for VFI-based MPSoC. Unlike existing deterministic solutions, our scheme takes account of performance yield constraints in the overall statistical task scheduling, voltage assignment and VFI partitioning process. Energy optimization sensitivity, combining with the statistical slack of the task, is used to guide the overall optimization flow. Experimental results demonstrate the effectiveness of the proposed scheme.

ACKNOWLEDGMENTS

The work was supported in part by National Basic Research Program of China (973) under grant No.2011CB302503, in part by National Natural Science Foundation of China (NSFC) under grant No.(61204027, 61076037, 60921002), in part by Natural Science Foundation of Hebei province of China under grant No. F2013502274, in part by the Fundamental Research Funds for the Central Universities under Grant No.12MS123, in part by China Postdoctoral Science Foundation funded project under grant No.2011M500321 and No.2012T50092.

REFERENCES

[1] A. Jerraya, W. Wolf. Multiprocessor Systems-on-Chips. Morgan Kaufmann Publishers, 2005.

[2] J. Dorsey, S. Searles, M. Ciraula et al. An Integrated Quad-core Opteron Processor. IEEE International Solid-State Circuits Conference (ISSCC), pp.102-103, 2007.

[3] J. Howard, S. Dighe, S. R. Vangal et al. A 48-Core IA-32 Processor in 45 nm CMOS Using On-Die Message-Passing and DVFS for Performance and Power Scaling. IEEE Journal of Solid-State Circuits (JSSC), Vol.46, No.1, pp.173-183, 2011.

[4] U. Y. Ogras, R. Marculescu, P. Choudhary et al. Voltage-Frequency Island Partitioning for GALS-based Networks-on-Chip. ACM/IEEE Design Automation Conference (DAC), pp.110-115, 2007.

[5] D. Marculescu, S. Garg. Process-Driven Variability Analysis of Single and Multiple Voltage-Frequency Island Latency-Constrained Systems. IEEE Transactions on Computer-Aided Design of Integrated Circuits and Systems, Vol.27, No.5, pp.893-905, 2008.

[6] C. Seiculescu, S. Murali, L. Benini et al. Comparative Analysis of NoCs for Two-Dimensional versus Three-Dimensional SoCs Supporting Multiple Voltage and Frequency Islands. IEEE Transactions on Circuits and Systems, Vol.57, No.5, pp.364-368, 2010.

[7] O. S. Unsal, J. W. Tschanz, K. Bowman et al. Impact of Parameter Variations on Circuits and Microarchitecture. IEEE MICRO, Vol.26, No.6, pp.30-39, 2006.

[8] S. Herbert, D. Marculescu. Characterizing Chip-Multiprocessor Variability-Tolerance. ACM/IEEE Design Automation Conference (DAC), pp.313-318, 2008.

[9] F. Wang, C. Nicopoulos, X. Wu et al. Variation-aware Task Allocation and Scheduling for MPSoC. ACM/IEEE International Conference on Computer-Aided Design (ICCAD), pp.598-603, 2007.

[10] J. Hu, R. Marculescu. Communication and Task Scheduling of Application-specific Networks-on-chip. Computers and Digital Techniques, Vol.152, No.5, pp.643-651, 2005.

[11] Y. Zhang, S. Hu, Z. Danny. Task Scheduling and Voltage Selection for Energy Minimization. ACM/IEEE Design Automation Conference (DAC), pp.183-188, 2002.

[12] W. Jang, D. Duo, D. Z. Pan. A Voltage-Frequency Island Aware Energy Optimization Framework for Networks-on-chip. ACM/IEEE International Conference on Computer-Aided Design (ICCAD), pp.264-269, 2008.

[13] D. Marculescu, S. Garg. Process-Driven Variability Analysis of Single and Multiple VoltageCFrequency Island Latency-Constrained Systems. IEEE Transactions on Computer-Aided Design of Integrated Circuits and Systems (TCAD), Vol.27, No.5, pp.893-905, 2008.

[14] C. Hongliang, S. S. Sapatnekar. Statistical Timing Analysis Considering Spatial Correlations Using A Single PERT-like Traversal. ACM/IEEE International Conference on Computer-Aided Design (ICCAD), pp.621-625, 2003.

[15] C. Clark. The Greatest of A Finite Set of Random Variables. Operations Research, PP.145-162, 1961.

[16] L. Huang, Q. Xu. Performance Yield-Driven Task Allocation and Scheduling for MPSoCs under Process Variation. ACM/IEEE Design Automation Conference (DAC), pp.326-331, 2007.

[17] L. Singhal, E. Bozorgzadeh. Process Variation Aware System-Level Task Allocation Using Stochastic Ordering of Delay Distributions. ACM/IEEE International Conference on Computer-Aided Design (ICCAD), pp.570-574, 2008.

[18] H. Chon, T. Kim. Timing Variation-Aware Task Scheduling and Binding for MPSoC. IEEE Asia and South Pacific Design Automation Conference (ASP-DAC), pp.137-142, 2009.

[19] M. Ghorbani. A Variation and Energy aware ILP Formulation for Task Scheduling in MPSoC. IEEE International Symposium on Quality Electronic Design (ISQED), pp.772-777, 2012.

[20] U. Y. Ogras, R. Marculescu, D. Marculescu. Design and Management of Voltage-Frequency Island Partitioned Networks-on-Chip. IEEE Transactions on Very Large Scale Integration (VLSI) Systems, Vol.17, No.3, pp.330-341, 2009.

[21] Y. Cheng, L. Zhang, Y. Han et al. Thermal-constrained Task Allocation for Interconnect Energy Reduction in 3-D Homogeneous MPSoCs. IEEE Transactions on Very Large Scale Integration (VLSI) Systems, Vol.21, No.2, pp.239-249, 2013.

[22] K. A. Bowman, A. R. Alameldeen, S. T. Srinivasan et al. Impact of Die-to-Die and Within-Die Parameter Variations on the Clock Frequency and Throughput of Multi-Core Processors. IEEE Transactions on Very Large Scale Integration (VLSI) Systems, Vol.17, No.12, pp.1679-1690, 2009.

[23] W. Zhao and Y. Cao. New generation of predictive technology model for sub-45nm early design explorations. Available at http://www.eas.asu.edu/_ptm. TED, 53(11):2816-2823, Nov. 2006.

[24] R. Dick. Embedded System Synthesis Benchmarks Suites (E3S). Available: http://ziyang.eecs.umich.edu/ dickrp/e3s/

[25] Available: http://ziyang.eecs.umich.edu/ dickrp/tgff/

QoS-Aware Dynamic Resource Allocation for Spatial-Multitasking GPUs

Paula Aguilera   Katherine Morrow   Nam Sung Kim
University of Wisconsin - Madison
paguilera@wisc.edu, kati@engr.wisc.edu, nskim3@wisc.edu

Abstract - General-purpose computing on GPUs (GPGPU computing) is becoming widely adopted; however, some GPGPU applications fail to fully utilize GPU resources. In these cases, spatial multitasking better exploits the parallelism offered by GPUs by partitioning the GPU resources among simultaneously-running applications. When one or more such applications have quality-of-service (QoS) requirements, enough resources must be allocated for those applications to satisfy their requirements. Remaining resources can be either disabled to reduce power consumption or used to accelerate other applications. However, we observe that the amount of resources for a QoS application to satisfy its performance requirement is dependent in part upon the co-executing applications. In this paper, we propose a runtime technique to dynamically partition GPU resources between concurrently running applications—at least one of which has a QoS requirement. We demonstrate that the proposed technique can satisfy a 100% QoS requirement while also achieving either a 7W power consumption reduction or a 17.57% performance improvement for co-executing best-effort applications.

I. Introduction

Consumer expectations are putting ever-growing demands on today's electronic devices—particularly those used for multimedia applications. It has become extremely important to provide both high performance and energy efficiency for mobile devices running such applications. As new languages such as OpenCL [1] and CUDA [2] have enabled general-purpose computing on the GPU (GPGPU), GPUs have emerged as alternative computational platforms that can offer both high performance and energy efficiency.

Yet, many GPGPU applications fail to fully utilize the GPU—their performance increases sub-linearly with increasing allocation of GPU resources [3]. We can use the GPU more effectively by simultaneously executing two or more applications, each on a subset of the GPU resources. We refer to this form of multitasking—dividing space (resources) rather than time (as with cooperative or preemptive multitasking)—as spatial multitasking. As commercial GPUs begin to support concurrent execution of multiple kernels [4, 5], partitioning GPU resources amongst multiple kernels or applications to maximize performance and/or energy efficiency has become a key challenge.

Meanwhile, many multimedia and other applications have certain quality-of-service (QoS) requirements. Performing below the application's QoS requirement is considered undesirable, whereas performance above it may not offer any benefit. These applications may concurrently run on the GPU with other applications that do not have QoS requirements, but do benefit from increased performance. We refer to this latter class of applications as best-effort applications.

QoS requirements complicate the task of GPU resource allocation for GPUs using spatial multitasking, but they also provide an opportunity. If QoS requirements can be satisfied without using all GPU resources, the remaining resources can either be disabled to reduce power consumption, or allocated to any co-executing best-effort applications to increase their performance.

This paper examines dynamic GPU resource allocation when at least one application has a QoS requirement. To the best of our knowledge, this is the first work to study QoS on a spatial-multitasking GPU. The key contributions are:

- We show that GPU resources that remain after QoS requirements are satisfied can be disabled to reduce power consumption or allocated to a co-executing best-effort application to increase its performance (Section VI).

- We demonstrate that the resource requirement to satisfy an application's QoS depends both on its characteristics and those of any co-executing applications. A runtime dynamic resource allocation algorithm is thus needed to maximize the benefit of a spatial-multitasking GPU running at least one QoS application (Section VII).

- We propose a runtime dynamic resource allocation algorithm and evaluate its efficacy. (Section VIII).

Using our proposed technique, we show that we can reduce power consumption by an average of 7W by disabling resources unnecessary for meeting QoS requirements of two co-executing QoS applications on the GPU. Alternately, when a QoS application is running with a best-effort application, performance of the best-effort application is increased by an average of 17.57% while satisfying the performance requirements of the QoS applications.

II. Related Work

How to allocate limited resources to multiple applications to satisfy their QoS requirements has been studied in other circumstances. Usually such research takes into account low-level system details such as arrival rate, period, service time, bandwidth and buffer size (e.g., [6], [7]). For our work, we use a single QoS dimension, which is the average IPC achieved by the QoS application. Several other QoS requirement types could also be represented in this form by profiling the application to determine the average IPC required to meet the alternate requirement.

Some work in embedded real-time systems has focused on reducing energy consumption by adapting the QoS of the different tasks (e.g., [8]). We also examine reducing energy consumption, but without the need to adapt the QoS. Other work proposed scheduling algorithms that modify the supply voltage and frequency on DVFS-capable CPUs to maximize

978-1-4799-2817-0/14 $31.00 © 2014 IEEE

QoS and minimize energy consumption (e.g., [9]). We pursue the same objective for a different style of device by attempting to use the minimum number of GPU resources to meet QoS requirements.

Hong and Kim developed an analytical GPU performance and power model to estimate the number of GPU cores that provide the highest performance per Watt for one application [10]. In contrast, we aim to partition GPU resources between multiple QoS and best-effort applications running on the same GPU to minimize power consumption or maximize performance of a co-scheduled best-effort application.

QoS has also been studied on the GPU, TimeGraph is a real time GPU scheduler for multi-tasking environments that focuses on prioritization and isolation of competing GPU applications [11]. Our approach satisfies QoS by partitioning the GPU resources, instead of the execution time, among applications running in a spatial multitasking GPU.

III. Overview and Motivation

A GPU is comprised of various resources, such as multiple compute cores (e.g., streaming multiprocessors (SMs) in NVIDIA's GPUs), on-chip memory/interconnect, and off-chip DRAM. GPUs are thus powerful computing platforms for not only graphics processing but also general parallel computations. GPGPU applications combine GPU and CPU execution, with the most compute-intensive portions accelerated by the GPU. However, scheduling the use of the GPU resources among multitasking GPGPU applications is a recent area of research. Furthermore, as these types of applications become more common in mobile applications, meeting QoS requirements will become an important aspect of scheduling in this form of computing.

Past research showed that some GPGPU applications fail to take full advantage of all GPU resources [3]—a trend likely to continue with increasing GPU compute capabilities from advances in GPU design. This motivates the use of spatial multitasking, which divides GPU resources amongst simultaneously-running applications, each of which might otherwise under-utilize the GPU. This approach improves overall system performance over cooperative multitasking, which instead gives each application a slice of time to execute on the full GPU [3]. QoS applications executing on the GPU must be allocated enough resources to meet their QoS requirements. Remaining resources can then be disabled to reduce power consumption, or allocated to a co-executing best-effort application to improve its performance.

However, as we will show, resource allocation to meet QoS requirements is not necessarily straightforward. The exact number of SMs required to meet QoS requirements with spatial multitasking is not static; it depends in part on any other applications simultaneously executing on the GPU. If we must statically choose the number of SMs to allocate to an application, it must be enough for the application to meet its QoS requirements in the worst case—when executing with applications that most degrade its performance.

A dynamic approach that reacts to the actual executing workload can allocate SMs more precisely, increasing the benefit (i.e., lower power consumption and/or higher

performance) of spatial multitasking. In this paper, we propose a runtime dynamic resource allocation algorithm, and evaluate its benefit for a spatial-multitasking GPU running at least one QoS application.

Dynamic SM allocation could be done by the GPU hardware, which already assigns computations to resources (e.g., thread blocks to SMs) without OS intervention. This is because many threads with short execution time run on the GPU, and frequent OS intervention would incur excessive overhead. For a spatial-multitasking GPU, its hardware is also responsible for assigning thread blocks to the SMs and thus it tracks which SMs are allocated to which applications. For QoS support, the OS now must inform the GPU of an application's performance target so the GPU allocates enough SMs to satisfy the application's QoS requirement.

IV. Evaluation Platform

We use a modified version of GPGPU-Sim 2.1.1 [12] that implements spatial multitasking [3]. We further modified this simulator to consider QoS and to dynamically assign SMs to applications. For our experiments, we model the NVIDIA Quadro FX 5800 GPU (GT200 architecture) with 30 SMs and 8 memory controllers [13]. Although this architecture models an NVIDIA GPU, these results can be generalized to other architectures. TABLE I lists the parameters used with GPGPU-Sim to model this architecture.

We use CUDA benchmarks [3] for our work. AES Decoding (AES-D), JPEG Decoding (JPEG-D) and SHA1 are our QoS applications; they have a variety of characteristics and represent applications that may have a QoS requirement: Encryption algorithms may need to keep pace with communication or media playback, and JPEG-D might be required to decode frames at a certain rate. Image Denoising (ID), Ray Tracing (RAY), Dirac Video Codec (DVC), Sum of Absolute Differences (SAD), and Fractals represent our *best-effort* applications. TABLE II summarizes their most relevant characteristics.

TABLE I: GPGPU-Sim parameters.

# SMs	30	CUDA Cores	240
SM Frequency	1300 MHz	Warp Size	32
# Memory Controllers	8	Interconnect Topology	Crossbar
Memory Frequency	800 MHz	Interconnect Frequency	650 MHz

TABLE II: Application characteristics

App	Category	Kernels	Thread Blocks	Threads / Block	Blocks / SM
AES-D	Encryption	1	4097	256	5
JPEG-D	Image Processing	1	13824	64	8
SHA1	Encryption	1	65	64	3
DVC	Video	14	130 (avg.)	231 (avg.)	4 (avg.)
Fractals	Graphics	2	512, 336	64, 256	8,4
ID	Image Processing	1	110592	64	8
RAY	Graphics	1	2048	128	8
SAD	Video	3	4800, 300, 300	61, 128, 32	8, 8, 8

978-1-4799-2817-0/14 $31.00 © 2014 IEEE

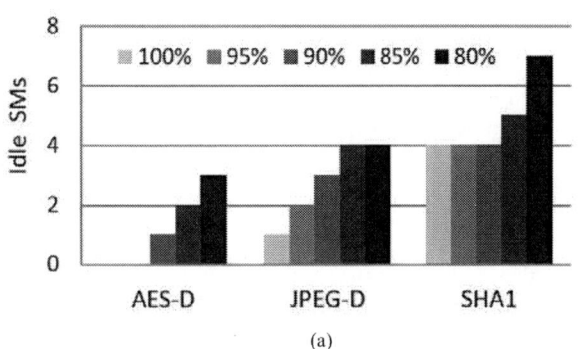
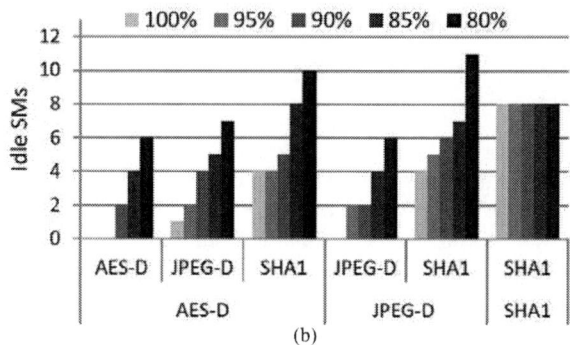

(a)   (b)

Fig. 1. (a) SMs not required by each QoS application running in isolation on the GPU. (b) SMs not required for two QoS applications sharing the GPU.

V. Methodology

We compare the power and performance effects of using spatial vs. cooperative multitasking while meeting QoS requirements. We measure power differences by noting the different number of SMs that must be enabled for each method. We measure performance differences in terms of the number of instructions completed during the same time span [14]. For our experiments, we assume that the QoS applications meet their QoS using cooperative multitasking. We then evaluate if spatial multitasking can also meet these requirements, and if so, how many SMs can be left idle to save power or assigned to best-effort applications to improve performance.

To establish a QoS requirement for each application, we measure the work (number of instructions) it completes when allocated 100% of the GPU resources (30 SMs) for 50% of the simulated execution time (i.e., cooperative multitasking). This same instruction count must then be met by a partial allocation of SMs for 100% of the simulated time. Different QoS requirements do not change the fundamental approach, merely the result of applying it. To test the effects of "tighter" vs. "looser" requirements, we also examine 95%, 90%, 85% and 80% of the calculated QoS.

To evaluate the benefits of spatial multitasking over cooperative multitasking for QoS GPGPU applications, we examine three execution scenarios. First, one QoS application is executing in isolation, and we determine the minimum number of SMs needed to satisfy its QoS. Second, we examine two concurrently-executing QoS applications, and determine how many SMs are not needed to meet the applications' QoS requirements—which depends in part on the co-executing applications. Last, we examine combinations of one best-effort and one QoS application to determine how allocating the "extra" SMs (those not needed by the QoS application) to the best-effort application can improve performance.

VI. Profile-Based Resource Allocation

In this section, we present application profile information showing how many SMs each QoS application needs to meet its requirement. First, we run each QoS application in isolation using up to 50% of the GPU resources (15 SMs) for 100% of simulated time and measure the number of SMs (out of the 15) that can be disabled to reduce power consumption while meeting QoS requirements (Fig. 1 (a)). We can disable 4 SMs for SHA1's 100% of execution time but none for AES-D at 100% of the calculated QoS. As the QoS requirement is relaxed, more SMs can be disabled.

Second, we concurrently run two QoS applications using spatial multitasking on the GPU, and evaluate the number of SMs that can be disabled while still meeting their QoS (Fig. 1 (b)). Spatial multitasking allows the GPU to disable 11 SMs when concurrently executing SHA1 and JPEG-D; with cooperative multitasking there would be no idle SMs—each application uses 100% of the resources for 50% of the time.

Finally, we examine all possible SM allocations when one QoS and one best-effort application co-execute on the GPU. We determine, for each pairing, the minimum number of SMs for the QoS application to satisfy its QoS requirement, then measure the performance improvement of the best-effort application when it is allocated the extra SMs. Fig. 2 shows the throughput increase factor of best-effort

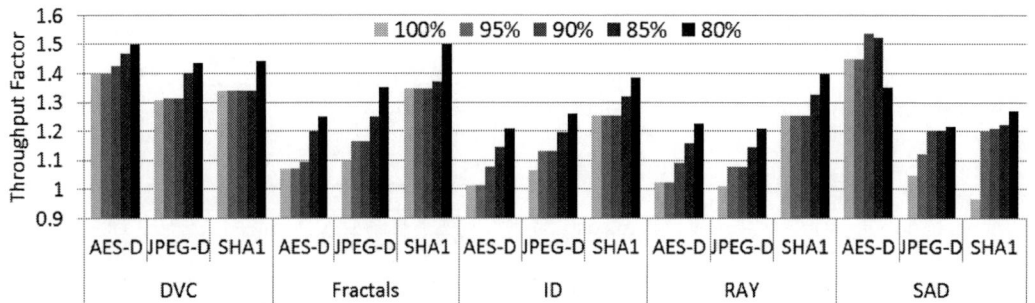

Fig. 2. The throughput increase of best-effort applications when spatially-multitasked with one QoS application for 100% of the simulated time as compared to running in cooperative multitasking using all the resources (30 SMs) for 50% of the simulated time.

978-1-4799-2817-0/14 $31.00 © 2014 IEEE

applications when they are allocated the "extra" SMs with spatial multitasking. This is computed by dividing the amount of work that the best-effort application completes when spatially-multitasked with the QoS application, divided by the amount of work that the best-effort application completes when cooperatively-multitasked using all the resources (30 SMs) for half of the time. When a best-effort application such as Fractals, ID and RAY concurrently runs with SHA1, their performance increases at least 20% for the 100% QoS requirement. Cooperative multitasking, however, cannot allocate unused SMs to best-effort applications to increase performance.

When we only run QoS applications, any SMs unneeded for QoS is left idle; idle SMs contribute to power savings because they consume less dynamic power than active SMs. Current GPUs do not employ core power gating, but future ones could as a result of the increasing number of cores and the increasing importance of power consumption especially in embedded devices. Using GPUWattch [15], we determine the power consumption per SM for every application and then compute the power savings obtained by leaving SMs idle as well as by power gating the otherwise-idle SMs when running two QoS applications simultaneously (using spatial multitasking) on the GPU (TABLE III).

VII. Impact of Co-Scheduled Applications on Performance of QoS Applications

Section VI showed that the number of SMs required by QoS applications to meet their performance requirements with spatial multitasking depends on any co-executing applications. For each QoS application, we determined which best-effort application most degraded the QoS application's performance. TABLE IV reports this performance degradation for each QoS application relative to its performance in isolation with the same number of SMs. We show results for both 10 SMs (1/3 of the GPU) and 15 SMs (1/2 of the GPU). AES-D does not have notable performance loss, but JPEG-D and SHA1 are more sensitive to co-executing applications. Even with 50% (15) of the SMs, JPEG-D and SHA1 exhibit up to 16% and 18% performance loss (respectively) in the worst case.

It is thus difficult to statically determine QoS applications' required SM allocation. If the application(s) that might co-execute with it are not known a priori, we must conservatively allocate SMs to ensure the QoS is met for the worst case. This could considerably reduce or completely negate the performance benefit of spatial multitasking over cooperative multitasking. We further examine this problem in the next section, and propose a method to address it.

VIII. Dynamic Resource Allocation

As previously discussed, a QoS application's performance depends on any co-executing applications due to contention in shared resources such as memory and interconnect. Thus, we propose a runtime algorithm to estimate QoS application performance and dynamically allocate the minimum required

TABLE III: Power savings when spatially-multitasking two QoS applications, if unneeded SMs are left idle vs. are power-gated

QoS	Idle SMs(W)	Power GatedSMs (W)
100%	2.6	6.9
95%	5.8	11.2
90%	8.8	15.8
85%	13	22.2
80%	18.1	28.6

TABLE IV: Maximum performance loss of each QoS application when sharing the GPU with another application vs. alone with the same # of SMs

Application	Max. performance loss (15 SMs)	Max. performance loss (10 SMs)
AES-D	0.3%	0.3%
JPEG-D	16%	23.2%
SHA1	18.2%	24%

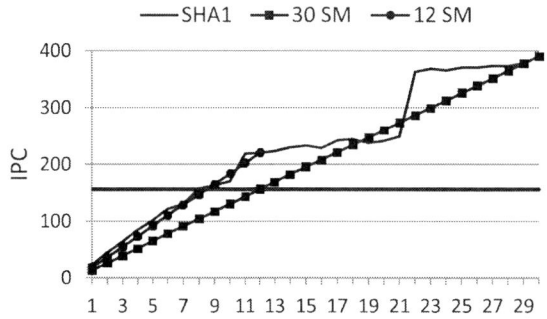

Fig. 3. Linear approximation of SHA1, running in isolation on the GPU, from 30 SM for a target QoS of 160 IPC. X-axes are the number of SMs.

SMs to them to maximize the benefit (power reduction and/or performance increase) of spatial multitasking.

For performance estimation, we assume that application performance increases linearly with the number of SMs. We assign an initial number of resources to the QoS application and measure its performance at that point; our performance estimation is the line that contains that point and a slope equal to the performance divided by the number of SMs at that point. We can plot this estimation on a graph of performance vs. SM allocation by drawing a line between the current measurement point (# SMs, performance) and a point at 1 SM and a performance level equal to the measured performance divided by the SM count for that performance. We can use this line to estimate the potential performance impact of changing the number of allocated SMs. Although the actual slope of a measured (not estimated) performance graph depends on the characteristics of both QoS and co-executing applications, most applications show sub-linear speedups with the number of allocated SMs. Hence, when used as a guide primarily for estimating a *reduction* in SMs, this performance estimation is generally pessimistic. Fig. 3 demonstrates this by showing the performance of SHA1 running in isolation for allocations of 1 to 30 SMs.

Since this estimation method is not 100% accurate, however, we use an iterative approach as follows:
1. We initially assign 50% of the SMs to a QoS application and measure its performance after it has run on those

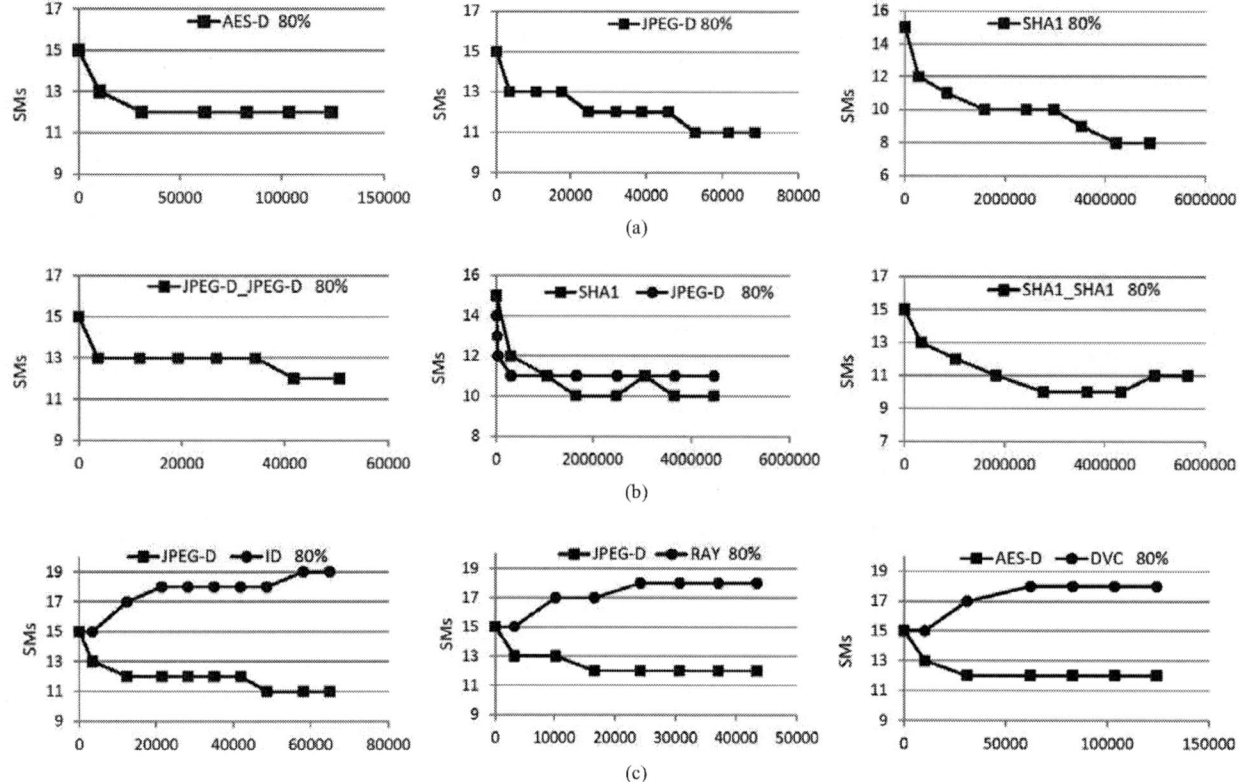

Fig. 4. (a) Convergence of each QoS application, in isolation, to the minimum # of SMs to satisfy its QoS. (b) Convergence of two QoS applications spatially-multitasked on the GPU to the minimum # of SMs to satisfy their QoS. (c) Convergence of one QoS application to the minimum # of SMs to satisfy its QoS when spatially-multitasked with a best-effort application. The x-axes represent the # of GPU cycles (passage of time).

SMs for at least the duration of one thread block. We use this sample point to create a linear performance model.

2. If the current performance exceeds the QoS requirement, we determine whether or not the QoS would still be satisfied with fewer SMs based on the model. If so, we decrease the allocated SMs to the minimum number needed to meet the QoS, as predicted by the linear model. On the other hand, if the QoS is not satisfied, we increase the SM allocation according to the model.

3. We re-evaluate the performance after running the application with the adjusted number of SMs, and repeat step two if appropriate.

To reallocate an SM from one application to another, we first need to make sure that the SM is no longer executing work. There are three ways to do this: (i) migrate the remaining work to a different SM; (ii) terminate the work and later restart it on another SM; or (iii) wait for the SM to finish executing. As each SM on a GPU has many threads executing simultaneously, each keeping their intermediate results in large register files and local memory, migrating work across SMs requires moving a large amount of state and thus incur a great deal of overhead. Terminating the work and later restarting it on another SM can be very costly because some threads perform a large amount of work that might then be lost. For this paper, we choose the third option: wait for the SM to finish executing the remaining work it had been assigned from the first application; then, allocate the SM to the new application.

Fig. 3 demonstrates how our algorithm works. It shows how to determine the minimum number of SMs that SHA1 needs to satisfy a QoS of 160 IPC, shown as a horizontal line. We start at 30 SMs, and after executing the application for the duration of a thread block, our algorithm estimates that 12 SMs are needed to satisfy the QoS. We wait for the extra SMs to finish execution of any remaining work that was assigned to them, then we reduce the number of SHA1's SMs to 12. After executing our application now on 12 SMs for the duration of a thread block, the algorithm runs again and estimates that only 9 SMs are needed. With 9 SMs the QoS is satisfied, and if we run the algorithm again it determines that we cannot further reduce the SM allocation. This process is done for any QoS application running on the system, depending on the other co-executing applications the results will change.

For our experiments, our main concern is to satisfy the QoS requirements instead of converging fast to the number of SMs. Thus, we conservatively choose the center point of the line for our next SM allocation, intentionally attempting to over-estimate the required SM allocation. This requires more iterations, but reduces the probability of backtracking to a higher allocation (if the QoS is not met). Although we always want to satisfy the QoS requirements, the proposed iterative approach may still violate QoS requirements for short time periods due to inaccurate estimations. However, such violations can be tolerable for soft QoS requirements, as long as the performance returns to an acceptable level.

978-1-4799-2817-0/14 $31.00 © 2014 IEEE   730

To show the efficacy of our iterative algorithm, we first study the case where only one QoS application runs on the GPU. Fig. 4 (a) shows how the number of SMs converges to the minimum over time, the extra SMs are left idle. The x-axes represent the cycle count. AES-D converges to 12 SMs, JPEG-D to 11 SMs and SHA1 to 8 SMs for 80% QoS.

Fig. 4 (b) shows how the number of SMs needed by two QoS applications running simultaneously using spatial multitasking converges to the minimum required to satisfy 80% QoS, with the unused SMs left idle. Two copies of JPEG-D running on the GPU need 12 SMs each, as opposed to the 11 SMs that would have been required for one copy of JPEG-D in isolation. Two copies of SHA1 need 11 SMs each, whereas a single copy needs only 8 SMs. When SHA1 and JPEG-D co-execute on the GPU, JPEG-D needs 12 SMs if SHA1 has more than 12 SMs, but when SHA1 is reduced to 12 SMs or less, JPEG-D can satisfy its QoS with 11 SMs.

Fig. 4 (c) shows combinations of one QoS and one best-effort applications. As the number of SMs assigned to the QoS application decreases, the number of SMs assigned to the best-effort application increases (the SMs not needed by the QoS application are reallocated to the best-effort one). When JPEG-D runs with ID it only needs 11 SMs to satisfy its QoS; when running with RAY it needs 12 SMs.

We observe that all the QoS applications converge to the minimum number of SMs needed to satisfy their QoS. Each application takes a different number of cycles to converge, depending on the duration of their thread blocks. This is because we execute the application for a complete thread block to obtain performance information before running our algorithm. If we decide to reassign an SM, we also wait for that SM to complete its in-progress work before re-allocating it. When running just QoS applications on the GPU, the SMs that are not needed by the QoS applications could also be power gated during the search process of the algorithm. Since we start from a conservatively high number of SMs and reduce it from there, we rarely suffer from overhead related to waking-up the power gated SMs during the (re-)allocation process.

IX. Conclusions

Spatial multitasking on the GPU can provide power and performance benefits while satisfying QoS application requirements. With spatial multitasking, a QoS application can be assigned the minimum resource count to meet its requirements, and remaining SMs can be allocated to best-effort applications improving overall system performance—or left idle to reduce power consumption. By using spatial multitasking with QoS applications on the GPU we obtain 7W power consumption reduction or 17.57% performance improvement for co-executing best-effort applications.

However, allocating the minimum number of SMs to a QoS application is complicated by the fact that QoS application performance depends at least in part on what, if any, other applications are co-executing (spatially multitasked) on the GPU. Contention in shared resources such as memory and interconnect cause performance degradation sufficient to affect the number of SMs needed to meet a QoS requirement. Thus, a static allocation of resources is not a good approach.

In this paper, we demonstrated that a linear approximation algorithm allows us to dynamically allocate the minimum number of resources to the QoS application so that it satisfies its performance requirements no matter what other applications run simultaneously on the GPU. Dynamic GPU resource allocation for spatial multitasking, including allocation based on QoS requirements, will continue to increase in importance as GPGPU computing becomes more pervasive in compute-intensive embedded systems.

Acknowledgements

This work was supported in part by an SRC grant (Task ID: 2080.001), NSF grants (CNS-0952425 and CNS-1217102), and generous gift from AMD. N. Kim has a financial interest in AMD.

References

[1] K. Group, "OpenCL," [Online]. Available: http://www.khronos.org/opencl/.

[2] NVIDIA, "CUDA C Programming Guide v4.2," 2012.

[3] J. Adriaens, K. Compton, N. S. Kim and M. Schulte, "The case for GPGPU spatial multitasking," in *HPCA*, 2012.

[4] NVIDIA, "NVIDIA's Next Generation CUDA Compute Architecture: Fermi," 2009.

[5] NVIDIA, "NVIDIA's Next Generation CUDA Compute Architecture: Kepler GK110," 2012.

[6] K. Jeffay, D. Stone and F. Smith, "Kernel support for live digital audio and video," in *Computer Communications*, 1992.

[7] D. Clark, S. Shenker and L. Zhang, "Supporting Real-Time Applications in an Integrated Services Packet Network: Architecture and Mechanism," in *SIGCOMM*, 1992.

[8] F. Harada, T. Ushio and Y. Nakamoto, "Adaptive Resource Allocation Control for Fair QoS Management," in *IEEE Transactions on Computers*, 2007.

[9] C. Poellabauer, L. Singleton and K. Schwan, "Feedback-based dynamic voltage and frequency scaling for memory-bound real-time applications," in *RTAS*, 2005.

[10] S. Hong and H. Kim, "An integrated GPU power and performance model," in *ISCA*, 2010.

[11] S. Kato, K. Lakshmanan, R. Rajkumar and Y. Ishikawa, "TimeGraph: GPU scheduling for real-time multi-tasking environments," in *USENIXATC*, 2011.

[12] A. Bakhoda, G. Yuan, W. Fung, H. Wong and T. Aamodt, "Analyzing CUDA workloads using a detailed GPU simulator," in *ISPASS*, 2009.

[13] NVIDIA, "NVIDIA Quadro FX 5800," [Online]. Available: http://www.nvidia.com/object/product_quadro_fx_5800_us.html.

[14] K. Rupnow, J. Adriaens, W. Fu and K. Compton, "Performance Metrics for Hybrid Computation in Multi-Tasking Systems," in *FPL*, 2009.

[15] J. Leng, T. Hetherington, A. ElTantawy, S. Gilani, N. S. Kim, T. M. Aamodt and V. J. Reddi, "GPUWattch: enabling energy optimizations in GPGPUs," in *ISCA*, 2013.

Automated Debugging of Missing Assumptions

Brian Keng[1], Evean Qin[3], Andreas Veneris[1,2], Bao Le[1]

Abstract—**Formal verification has increased efficiency by detecting corner case design bugs but it has also introduced new challenges when failures are detected. Once a counter-example is returned by a formal tool, the user typically does not know if the failure is caused by a design bug, an incorrectly written assertion, or a missing assumption. Previous work in debug automation has focused on the former two cases. This paper introduces a novel methodology to automatically debug missing assumptions. It begins by generating multiple formal counter-examples for the error. Next, a function is extracted from these counter-examples that encodes the input combinations that cause the assertion to fail. This function is later used to generate a list of fixed cycle assumptions that prevent failures similar to the generated counter-examples. These filtered assumptions can then be used as hints for the actual missing assumption. Further, if a missing assumption is not the cause of the failure, the method offers the additional benefit that the counter-examples it generates can be utilized to debug the RTL and/or the assertion. An extensive set of experimental results on OpenCores designs and assertions show that the number of generated assumptions can be reduced by an average of 38% using ten counter-examples, while an average of 28 assumptions is returned to the user.**

I. INTRODUCTION

Functional debugging today has become a bottleneck taking up 60% of the total verification time [1]. To cope with this burden, many debugging techniques [2]–[4] have been introduced to automatically localize design errors and improve debugging efficiency. At the same time, techniques such as formal property checking and assertion-based verification [5] have grown in popularity, leading to new challenges that extend beyond traditional design error debugging.

Formal property checkers [6] aim to increase verification efficiency by exhaustively verifying an assertion encoding the design intent. In the ideal case, if an assertion is violated, the formal tool returns a single counter-example allowing detection and debugging of corner case design bugs. However, as documented in industry reports [7], debugging formal counter-examples can be challenging, as the engineer does not have confidence whether the observed failure is due to a design bug, an incorrectly written assertion, or a missing assumption. Despite the need for designer intervention to determine the actual root-cause of the failure, previous work [4], [8] in debug automation has shown to be effective in aiding the designer in the former case. Today, diagnosing missing assumptions still poses a significant bottleneck in formal verification flow as they can cause up to 50% of the formal failures [7].

Assumptions are necessary in formal verification as they model the design's intended environment and ensure that Register Transfer Level (RTL) bugs can be detected. Debugging missing assumptions can be a challenging task because – unlike assertions – they are rarely explicitly documented. Instead, they are expressed implicitly by either the design specification or the functionality of adjacent design blocks. For the engineer, this can lead to a tedious "guess-and-check" iterative debugging process, introducing multiple time-consuming calls to the formal tool. To alleviate this pain and

make formal technology effective to its full potential, more debug automation is needed to help analyze the behavior of the counter-example and identify candidate missing assumptions.

Identifying missing assumptions has been researched before in the context of compositional verification, software model checking and reactive system synthesis [9]–[11]. More recently, a technique to tackle this problem in a hardware formal verification context has shown promising results [12]. Given a counter-example due to a missing assumption, this technique generates a list of fixed cycle assumptions that prevent the assertion failure under conditions similar to the original counter-example. In essence, the *goal* for these added assumptions is to be used as hints by the engineer to identify the missing assumption or to provide confidence that the problem is most likely into the RTL and/or the assertion and not into the assumption(s) itself. Results from that paper [12] on a handful of instances are encouraging despite their dependence on a single counter-example, a fact which may severely limit the quality of these generated assumptions.

In this work, we present an automated assumption debugging methodology. The proposed work is based upon the framework developed in [12] but overcomes its limitations as it provides higher quality assumptions that can be used during debugging. It also complements other debugging techniques as it provides additional information through new counter-examples. In detail, the contributions of the work are twofold. At first, a novel algorithm is presented to generate multiple distinct counter-examples from a single assertion failure by iteratively extracting constraints from previous counter-examples and re-running the formal tool. As an added benefit, when a missing assumption is not the root-cause of the observed failure, the generated counter-examples can be used by the engineer to improve the resolution of existing automated tools when debugging incorrect assertions [8] and/or RTL design errors [3]. Next, a method of using multiple distinct counter-examples to improve the quality of results of the assumption debugging methodology is also presented.

An extensive set of experiments is performed over a wide variety of `OpenCores` [13] designs with SystemVerilog assertions written from their specification documents. Multiple counter-examples are shown to reduce the number of generated assumptions by 38% on average while an average of 28 assumptions are returned to the user. This confirms the benefit of the proposed approach.

The remaining paper is as follows. Section II presents background material. Section III describes an overview of the assumption debugging methodology, while Section IV present the details of the proposed work. Section V presents the experiments and Section VI contains the conclusion.

II. PRELIMINARIES

A. Minimal Correction Sets and Unsatisfiable Cores

For a given unsatisfiable (UNSAT) Boolean formula ϕ in conjunctive normal form (CNF), an *UNSAT core* is a subset of clauses of ϕ that are unsatisfiable. A *Minimal Unsatisfiable Subset* (MUS) is an UNSAT core where every proper subset is satisfiable (SAT). A *Minimal Correction Set* (MCS) is a

---

[1] University of Toronto, ECE Department, Toronto, ON M5S 3G4 ({briank, veneris, lebao}@eecg.toronto.edu)

[2] University of Toronto, CS Department, Toronto, ON M5S 3G4

[3] Vennsa Technologies, Inc., Toronto, ON M5V 3B1 (evean@vennsa.com)

978-1-4799-2817-0/14 $31.00 © 2014 IEEE

minimal set of clauses of ϕ such that removing them will result in ϕ being SAT. There exists a duality relationship between MUSs and MCSs such that, if one has all MUSs, then all MCSs can be computed and vice versa [14].

MCSs of ϕ can be computed by introducing a fresh *relaxation variable* to each clause. If the variable is active, then the clause is effectively removed from the problem. By additionally introducing cardinality constraints on these relaxation variables, one can find all minimal sets of relaxation variables which will result in ϕ being SAT. Each one of these solutions represents an MCS corresponding to the associated relaxation variables. This idea has been used extensively in modern Max-SAT solvers [15], [16] to compute MCSs as well as design debugging [3], [4] applications.

B. Minimally Unsatisfiable Input Sets and Debugging Missing Assumptions

Let I represent the initial state, X the counter-example input vector, T the unrolled circuit transition relation, and P the property to be checked. The unrolled counter-example in CNF, denoted by ϕ, is given by:

$$\phi = I \cdot X \cdot T \cdot P \tag{1}$$

Equation 1 is UNSAT by construction because the counter-example exposes an assertion failure.

Let C^k denote the kth *Minimal Correction Input Set* (MCIS), defined as a minimal set of input unit clauses of X that when removed, will result in ϕ being SAT *i.e.*, C^k is the minimal set of input clauses to remove to correct the failure. This is analogous to the idea of a MCSs except with respect to only input unit clauses.

Let U^k denote the kth *Minimal Unsatisfiable Input Subset* (MUIS), defined as a minimal unsatisfiable set of input unit clauses such that $I \cdot T \cdot P \cdot U^k$ is still UNSAT *i.e.*, U^k is the minimal set of input clauses needed to expose the failure. Similarly, U^k is analogous to MUSs except with respect to input unit clauses.

Each U^k represents a minimal set of input combinations from the counter-example that can directly excite the assertion failure. The disjunction of all U^k represents all combinations of inputs from the counter-example that can lead to the assertion failure, given by:

$$F = U^0 + ... + U^k \tag{2}$$

As shown in [12], this function can be computed by first calculating all C^k using relaxation variables, and then building F using a duality relationship analogous to the one between MUSs and MCSs. Thus, F can be re-written in terms of C^k, where c_i^k represents the ith input unit clause of C^k:

$$F = \overline{\overline{c_0^0} \cdot ... \cdot \overline{c_{|C^0|}^0} + ... + \overline{c_0^k} \cdot ... \cdot \overline{c_{|C^k|}^k}} \tag{3}$$

Given an assertion failure due to missing assumptions and its associated counter-example, the technique in [12] uses the function from Equation 3 to aid in debugging missing assumptions. This is accomplished in several steps. First, F from Equation 3 is extracted from the given counter-example. Next, a list of candidate fixed-cycle assumptions on primary inputs are generated from a dictionary model. Each candidate input assumption, A, can be filtered out by creating a SAT instance, ψ, as such:

$$\psi = F \cdot A \tag{4}$$

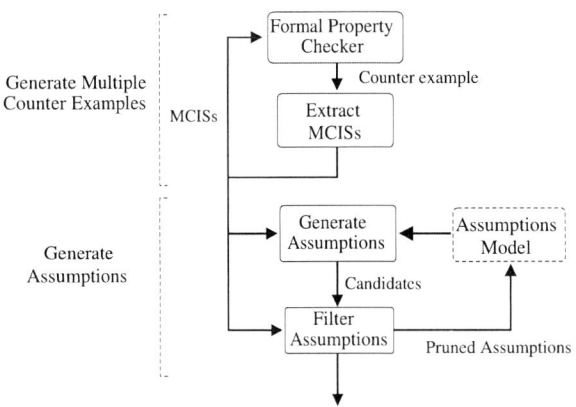

Fig. 1. Assumption Debugging Methodology

If ψ is SAT, then the assumption cannot prevent the assertion failure in the given counter-example (implicitly encoded in F) and should be pruned. Otherwise, A prevents a failure similar to the given counter-example and can be returned to the user. The end result is a list of assumptions that can prevent failures similar to those by the given counter-example, allowing the engineer to either potentially use the assumption directly, or build a strong intuition for the actual missing assumption.

III. ASSUMPTION DEBUGGING FLOW

This section presents an overview of our methodology for debugging missing assumptions in a formal property checking environment. Although the overall flow is presented in the context of debugging missing input assumptions, as noted earlier, the work here can still be valuable in debugging other types of formal failures such as RTL design errors or incorrectly written assertions. This is because a key by-product of our flow (*i.e.*, generating multiple formal counter-examples) can be utilized by "traditional" automated techniques [3], [4], [8] to aid debugging for all types of failures.

The overall methodology is shown in Figure 1 and consists of two major phases. Given an assertion failure and its associated counter-example, the *first phase* attempts to iteratively generate multiple formal counter-examples. For each counter-example, MCISs are extracted and used to generate constraints which are then passed back into the formal tool. These constraints ensure that another distinct counter-example is found. This process is repeated until the formal tool cannot return any more counter-examples, or a desired number of vectors has been reached.

The *second phase* iteratively generates input assumptions that can prevent failures similar to those seen in the existing counter-examples. This is accomplished by using a model of simple assumption structures and filtering them based on the MCIS constraints extracted from the counter-examples. The resulting filtered assumptions are returned to the user and can either be used as suggestions for the missing assumptions, or as hints to which signals and expressions might be needed.

This flow improves debugging efficiency in two ways. First, multiple counter-examples can greatly improve debugging of formal failures regardless of their type because they provide a more general representation of the assertion failure, benefiting both manual and automated debugging [3], [4], [8]. Second, the assumptions returned by the methodology improve upon previous work by generating easy-to-understand properties based upon common assumption structures. The next section describes the phases of our methodology in detail.

IV. GENERATING MULTIPLE COUNTER-EXAMPLES

In a typical formal flow when an assertion fails, the formal tool generates a single counter-example to be used later in debugging. Multiple counter-examples are beneficial for debugging because they allow a broader view of the root-cause of failure and may improve the resolution of automated debugging techniques [3], [8]. Despite the benefits of multiple counter-examples, existing formal property checkers do not support this feature. In the past, there has been some work to generate diverse SAT solutions [17], although there is guarantee as to how different the counter-example will be.

The difficulty in this process is not simply generating a second counter-example, but rather generating a *useful* second counter-example that causes the assertion to fail in a different manner. The following sub-sections describe a method to generate multiple formal counter-examples that are quantitatively different from each other. It also outlines how to apply them to filter candidate input assumptions and improve quality of the final result.

A. Minimal Correction Input Sets as Blocking Constraints

In the context of debugging, a failure can be viewed as a counter-example exciting an error, propagating its effect through design components, and causing an assertion to fail. This corresponds to the unrolled (in time) CNF of the counter-example from Equation 1. The initial states and input vector propagate through the clauses that model the design, and cause a conflict with the modeled property. The corresponding clauses can be abstractly viewed as a set of MUSs. As such, a natural way to quantify two counter-examples as being different is when the observed failures occur with no identical MUSs. This leads to the following definition of distinct counter-examples:

Definition 1 *Given two counter-examples R and S, and their respective unrolled CNF instances from Equation 1, ϕ_R and ϕ_S, let M_R and M_S represent the set of all MUSs from ϕ_R and ϕ_S, respectively. Counter-examples R and S are said to be distinct iff $M_r \cap M_s = \emptyset$.*

Using this definition, we can generate multiple distinct counter-examples by preventing previously seen MUSs from occurring again. To prevent a MUS, we need to ensure at least one of its clauses is not present. Since the circuit behavior should not change, only the clauses corresponding to the primary input vector should be blocked to prevent previously found MUSs. This corresponds directly to generating a blocking constraint on the inputs to prevent previously found MUISs. Using the duality between MUISs and MCISs, this constraint can be computed from a single MCIS.

In more detail, for the unrolled counter-example ϕ from Equation 1 and MCIS $C^k = \{c_0, ..., c_{|C^k|}\}$, removing C^k will break all MUISs (and thus all MUSs) in ϕ since their removal will make the instance SAT. Since C^k is minimal, this is equivalent to reversing the polarity of the corresponding input unit clauses in ϕ, and it can be expressed as the following blocking constraint B^k for the kth MCIS:

$$B^k = \overline{c_0^k} \cdot \overline{c_1^k} \cdot .... \cdot \overline{c_{|C^k|}^k} \qquad (5)$$

This blocking constraint can be used in conjunction with the design and assertion to generate another distinct counter-example using an additional call to the formal tool. The following lemma describes this idea:

Lemma 1 *For counter-example R, let ϕ_R be the unrolled counter-example in CNF. If B^k is the kth blocking constraint* of ϕ_R, *then any counter-example that satisfies B_k is distinct from R.*

Proof: From Equation 5, B^k is the conjunction of all the negations of the kth MCIS, C^k. By definition, removing the literals of C^k from the ϕ_R will result in the instance being SAT, effectively breaking all the MUSs from ϕ_R. Since C^k is minimal and no proper subset has the property of being a correction set, any SAT assignment will contain the negation of all the literals of C^k, precisely the expression B^k. It follows then that any counter-example that contains the assignment from B^k will necessarily not contain any MUSs from ϕ_R, and therefore it is distinct. ∎

As such, to generate a new distinct counter-example we can use Lemma 1 and pass a blocking constraint in the form of Equation 5 to the formal tool. If the formal tool returns a counter-example, it implicitly guarantees that B^k is satisfied, resulting in a distinct counter-example.

B. A Practical Algorithm

Algorithm 1 shows the pseudo-code for generating multiple counter-examples. The algorithm begins by generating the first counter-example from the formal property checker and extracting all MCISs from it (lines 2-5). The loop from line 6-14 generates multiple counter-examples. For a given MCIS, it will attempt to find a new counter-example. If successful (line 9), this MCIS is saved in *blocking* to ensure that future counter-examples remain distinct. This process greedily selects new MCISs to add to *blocking* only when it can find a new counter-example. Once the new counter-example is saved, a new set of MCISs are extracted (lines 11-7), and the process repeats using this new set of MCISs. The loop stops when either none of MCISs combined with the existing constraints in *blocking* can generate another counter-example, or when the maximum number of user-specified counter-examples has been reached. The following theorem confirms the benefits of these counter-examples:

Theorem 1 *All counter-examples returned by Algorithm 1 are mutually distinct.*

Proof: Each counter-example generated from the run of the formal tool on line 8 will run under the set of blocking constraints, $blocking \cup C$. By Lemma 1, any counter-examples that derived any of the constraints in $blocking \cup C$ will be distinct from the newly generated one. Since blocking constraints are added only when a new counter-example is found, $blocking \cup C$ maintains a set of MCISs from each of the previously seen counter-examples. Therefore, each newly generated counter-example will respect these blocking constraints and it will be mutually distinct. ∎

One important aspect of Algorithm 1 is that it iteratively adds a blocking constraint in the form of Equation 5. This could have alternatively been implemented using the disjunction of all blocking clauses from a single counter-example *i.e.*, the negation of Equation 3. However, our experience with an industrial formal property checker shows that this latter approach significantly slows down the tool causing time-outs or bounded proofs, an observation that can be explained as follows. Our blocking constraints are just unit clauses, which are easily modeled within many different model checking algorithms. Whereas, the disjunction of multiple MCISs can be significantly more complicated to model (or at least require specialized optimizations). This allows for a more generic method without any need to use a specialized property checker.

Algorithm 1 Generating Multiple Counter-Examples

1: **procedure** MULTIPLECOUNTEREXAMPLES(max)
2:    $blocking = \emptyset$
3:    c-ex = RUNFORMAL($blocking$)
4:    CEX = {c-ex}
5:    MCIS = EXTRACTALLMCIS(c-ex)
6:    **while** MCIS $\neq \emptyset$ and —CEX— $< max$ **do**
7:      $C =$EXTRACTBLOCKING(MCIS)
8:      c-ex = RUNFORMAL($blocking \cup C$)
9:      **if** c-ex $\neq \emptyset$ **then**
10:        $blocking = blocking \cup C$
11:        CEX = CEX $\cup$ c-ex
12:        MCIS = EXTRACTALLMCIS(c-ex)
13:      **end if**
14:    **end while**
15:    **return** CEX
16: **end procedure**

C. Applications for Debugging Missing Input Assumptions

As mentioned in Section II-B, a single counter-example can be used to filter a candidate assumption A using Equation 4. The filtering function F used to rule out candidate assumptions is derived directly from a set of MCISs. If Algorithm 1 is used to derive multiple counter-examples, all MCISs from each counter-example are indirectly generated as a by-product. These can be used to generate a set of filtering functions $F_1, ..., F_N$ for N counter-examples, respectively, which can naturally be combined to extend the filtering function and generate the following instance:

$$\psi = (F_1 + ... + F_N) \cdot A \qquad (6)$$

The disjunction of all the F_i in Equation 6 correspond to all the MUISs for each of the counter-examples. This implicitly encodes all the input behaviors that led to the observed assertion failures in the given counter-examples. Similar to Equation 4, if the instance is SAT, then the assumption is not strong enough to prevent at least one of the observed failures. Otherwise, the assumption is generalized enough to prevent all the observed failures and should be returned to the user.

It should be noted that although we present this filtering function in the context of pruning generated assumptions, it is equally valid to say that this function can be used to test manually generated assumptions by the engineer. Thus, the filtering function can provide quick feedback to determine if a given assumption can prevent the failure(s) present in the current counter-example(s).

D. Assumption Model

Table I shows a summary of the model used to generate candidate missing input assumptions that is similar to [12]. Each row corresponds to one of four categories of properties presented in SystemVerilog. These categories correspond to simple unit Booleans, combined Boolean operators, one-hot operators, and stability expressions. An assumption is generated by taking the property and using the same clock and reset as the target failing assertion. Each assumption is then checked against the filtering function to determine if it should be returned to the user. In the table, `input` refers to a single bit primary input pin, while `bus` refers to a semantic grouping of primary input pins.

TABLE I
ASSUMPTION MODEL

Category	Model
Unit Booleans (unit)	`input, !input`
Combined Booleans	`<unit> & <unit>,` `<unit> & <unit> & <unit>,` `<unit>	<unit>,` `<unit>	<unit>	<unit>`
One-hot	`$onehot(bus), $onehot0(bus),` `$onehot({<unit>, <unit>}),` `$onehot({<unit>, <unit>, <unit>})`
Stability	`$stable(bus), bus == 0` `input |=> !input, !input |=> input`

TABLE II
DESIGN INFORMATION

Design Name	# Gates (k)	# Flops	# Inputs
cpu	50.9	1270	51
ddr2	55.5	2475	431
hpdmc	9.8	431	210
mips	51.1	2250	82
mrisc	9.9	1372	69
pci	60.3	3886	162
spi	1.7	133	16
usbf	33.2	1954	128
wb	4.0	98	143

V. EXPERIMENTAL RESULTS

This section presents experimental results for the proposed methodology. All experiments are performed on a single core of an Intel Core i5 3.1 GHz quad-core workstation with 16 GB of RAM. A commercial property checker [18] is used with default settings to perform all formal checks, while the extraction of MCISs as well as generation and filtering of candidate assumptions are all implemented in C++, using Minisat [19] as the SAT engine. Nine designs are selected for evaluation from OpenCores [13] with assertions written based upon their specification documents.

For each of the gathered designs and assertions, the formal property checker is run and any failure is considered to be an instance of a missing assumption. The instances listed in the following tables correspond to a single failing assertion and are labeled by appending a number to the design name.

Table II presents information for each of the designs used in the experimental results. The columns of the table list the design name, number of gates (including state elements), number of state elements, and number of primary input pins.

A. Generating Multiple Counter-Examples

This subsection presents experimental results for the proposed approach to generate multiple counter-examples from Section IV. Experiments in this subsection are conducted for each instance by running Algorithm 1 to generate as many counter-examples as possible within 1800 seconds to a maximum of 15. Next, using either 1, 5, 10, or 15 counter-examples, candidate assumptions are generated and filtered to examine if multiple counter-examples are useful in reducing the number of generated assumptions. To simplify the experiments, combined and one-hot type properties from Table I are omitted when generating assumptions. Additionally, each pin of an input bus is also used in unit Boolean properties so that we have a sufficient set of properties across all input pins. Note that a cone of influence [6] optimization is run on the failing assertion of each instance, resulting in a potentially different number of total generated assumptions

TABLE III
MULTIPLE COUNTER-EXAMPLE EXPERIMENTS

Instance Name	# CE	MCIS Time (s)	Form Time (s)	Tot Can	Filt Using n CE			
					1	5	10	15
cpu_1	15	653	356	154	2	2	2	2
cpu_2	10	778	616	154	3	3	3	-
ddr2_1	3	625	86	226	68	-	-	-
ddr2_2	9	383	1395	257	15	2	-	-
hpdmc_1	15	112	148	97	17	16	11	11
hpdmc_2	15	123	200	97	25	16	13	11
mips_1	4	278	93	163	36	-	-	-
mips_2	12	813	959	163	13	13	13	-
mrisc_1	8	88	1126	92	11	5	-	-
mrisc_2	7	190	1339	92	8	8	-	-
pci_1	8	611	761	267	9	9	-	-
pci_2	8	648	723	267	11	11	-	-
spi_1	15	8	177	22	6	6	6	6
spi_2	15	47	214	22	19	10	7	7
usbf_1	12	901	858	131	61	28	26	-
usbf_2	10	186	1152	131	22	0	0	-
wb_1	15	6	846	13	9	7	6	6
wb_2	15	8	344	41	10	2	2	2

between instances of the same design. Table III shows the results of these experiments.

The first five columns list the instance name, number of counter-examples generated, run-time to extract MCISs from the counter-examples, run-time of the formal tool to generate that many counter-examples, and the total number of candidate assumptions for that instance. The last four columns show how many of the candidate assumptions remain after filtering using the technique from Section IV-C with 1, 5, 10, and 15 counter-examples, respectively.

Overall the last four columns show that using more counter-examples can effectively reduce the number of filtered assumptions. On average, for instances that are able to generate either 5, 10, or 15 counter-examples, the number of filtered assumptions are reduced by 30.4%, 37.9% and 38.3%, respectively, compared to a single counter-example. This confirms that the additional counter-examples generated do generalize the assertion failure, which shows that our definition of distinct is indeed a valid one.

The ability of the proposed technique to filter candidate assumptions works well in most of the instances (such as `ddr2_2` and `usbf_1`), but not all (such as `cpu_1` and `mips_2`). This can be explained as follows. The former case is the ideal behavior where the second counter-example does indeed find a different way to excite the design and cause the assertion to fail. While the latter case finds a counter-example similar to the original one but shifted in time. One needs to note that in this situation, it is often the case that there may only be one way to cause the assertion fail.

When analyzing the run-time, there are two main contributors. The first is the extraction of MCISs, which depends on the size of the design, number of input pins, and the length of the counter-example. For many cases, such as `mrisc_1` and `wb_1`, it is relatively fast. In the case of `ddr2_1`, however, the excessive number of inputs (431) cause the run-time of extracting the MCISs to be large. The other contributor to overall run-time is multiple iterations of the loop in Algorithm 1, which may require many calls to the formal tool before a counter-example is found. However within the 1800 second timeout, 7 out of the 18 instances were able to generate 15 counter-examples, and 16 out of the 18 were able to generate at least 5. This shows that this technique is effective in generating multiple counter-examples within a short amount of time.

B. Assumption Debugging Methodology

This section presents experimental results for the overall assumption debugging methodology from Section III. For each instance, counter-examples are generated within a time limit of 1800 seconds up to a maximum of 10, which was chosen qualitatively to be a good balance between filtering and run-time. Additionally, the full assumption model from Section IV-D is used to generate and filter assumptions. Note that this may result in a different number of candidate assumptions compared to the previous subsection. Similarly, a cone of influence [6] optimization is run on the failing assertion for each instance. Table IV shows the quantitative results of these experiments.

Table IV is divided into two parallel sections. The columns in each section list the instance name, number of counter-examples generated, time to extract MCISs from the counter-examples, time of the formal tool to generate that many counter-examples, time to generate and filter candidate assumptions, total number of candidate assumptions, and the number of assumptions after filtering.

From columns 7 and 14, the absolute number of filtered assumptions returned to the user is relatively small with an average of 28. It is important that this number is not too large, or else the list of assumptions may become overwhelming for a user to analyze. Although most of the instances fall close to this average, there is one outlier `ddr2_2` with 333 returned assumptions. This is due to the large number of input pins which generates a significant number of candidate assumptions (4094). However as described in Section V-C, the different categories of properties allow one to narrow down the analysis. In this case, only analyzing the unit Booleans assumptions proved most useful.

When analyzing run-time of generating and filtering candidates in columns 5 and 12, in most instances the time is relatively small and both tasks can be completed within 60 seconds. However, `ddr_1` and `ddr_2` are again outliers, where the former hit a time limit of 1800 seconds. Here, the excessive number of input pins cause an exponential number of generated assumptions in the more complex properties. In these cases, it may be more prudent to only generate simpler properties or limit the number of pins used to generate assumptions. Both these solutions are easily implementable within the proposed flow.

C. Qualitative Analysis of Generated Assumptions

The following is a detailed discussion on the qualitative aspects of how the proposed flow can be used to aid debugging of missing assumptions. In the proposed flow, the engineer will analyze the assumptions and decide which assumption is appropriate. We selected two instances from Table IV to illustrate the benefits.

a) mips_1: The assertion for this instance checks to see that a finite state machine transition occurs with the correct conditions:

```
P: (CurrState == `IDLE) && (irq && ~iack)
   |=> (CurrState == `IRQ)
```

The approach generated 22 assumptions, a sample of which is listed here:

```
A1: !pause
A2: pause
A3: $onehot(zz_ins_i[31:0])
A4: $onehot0(zz_ins_i[31:0])
```

In this case, both `pause` and its negation are suggested by the technique. This is because holding pause either high or low for

TABLE IV
ASSUMPTION DEBUGGING METHODOLOGY EXPERIMENTS

Instance Name	# CE	MCIS Time (s)	Form Time (s)	Gen Time (s)	Tot Can	Filt Can	Instance Name	# CE	MCIS Time (s)	Form Time (s)	Gen Time (s)	Tot Can	Filt Can
cpu_1	10	255	100	5	31	3	mrisc_4	9	116	898	5	39	14
cpu_2	10	778	616	7	28	5	pci_1	8	611	761	7	25	10
ddr2_1	3	625	86	TO	857	21	pci_2	8	648	723	7	22	11
ddr2_2	9	383	1395	1504	4094	333	pci_3	8	564	518	8	25	10
hpdmc_1	10	70	60	4	90	33	pci_4	2	466	60	27	261	82
hpdmc_2	10	77	65	8	65	18	spi_1	10	4	48	1	20	9
hpdmc_3	10	6	77	1	8	3	spi_2	10	28	60	15	74	29
mips_1	4	278	93	9	59	22	usbf_1	10	737	334	14	148	58
mips_2	10	455	276	8	39	10	usbf_2	10	186	1152	132	1135	44
mips_3	5	134	458	7	39	6	usbf_3	10	18	244	2	16	7
mips_4	10	589	631	10	59	7	wb_1	10	3	123	1	16	5
mrisc_1	8	88	1126	5	39	10	wb_2	10	4	111	1	81	2
mrisc_2	7	190	1339	6	34	9	wb_3	10	5	79	1	19	2
mrisc_3	5	79	169	4	20	9	wb_4	10	4	98	1	81	2

the entire trace will prevent the assertion from failing. This is a common occurrence in many of the generated assumptions, however, usually only one of the two stuck-at assumptions will be relevant.

By tracing the relationship between `pause` and the state machine, it is clear that when `pause` is asserted, the state transition will be stopped. In this case, A1 is precisely the needed constraint. Interestingly, the specification does not explicitly mention that the state transition will be stopped by the `pause` signal, a common omission that causes counter-examples due to missing assumptions.

Finally, the last two assumptions are on the primary input bus corresponding to the CPU's instruction. They are obviously not very meaningful and correspond to a vacuous fix (*i.e.*, a case where the antecedent is always false).

b) usbf_1: The assertion for this instance checks the property where a buffer overflow occurs when a packet has been received that does not fit into the buffer. The packet will be discarded and a `NACK` will be sent to the host.

```
P: buffer_overflow ##0 send_token[->1]
   |-> (token_pid_sel == NACK)
```

The approach generated 58 assumptions, several of which are listed here:

```
A1: !wb_stb_i
A2: !wb_cyc_i
A3: !wb_addr_i[17] & wb_cyc_i & wb_stb_i
A4: $stable(DataIn_pad_i[7:0])
```

The first two assumptions that pull down `wb_stb_i` and `wb_cyc_i` are vacuous fixes. But the assumptions with both of these signals high (A3) are more interesting. In this assumption, the 17th bit in `wb_addr_i` controls the source of the data to be sent. This assumption tells the user that during the assertion, the data should be selected from the register file instead of memory, avoiding the cause for the failure.

The next assumption `$stable(DataIn_pad_i[7:0])` provides a useful hint. This signal controls which data will be selected from the endpoint. It tells the user that during the assertion, the same endpoint should be selected, providing another way to avoid the failure.

VI. CONCLUSION

In this work, a novel debug automation methodology for missing input assumptions is presented. It begins by generating multiple formal counter-examples for the failure along with a function that encodes the input combinations that caused the assertion to fail. This function is later used to generate a list of fixed cycle assumptions that prevent the failures

seen in the counter-examples, which can then be used as hints for the actual missing assumption. An extensive set of experimental results on OpenCores designs and assertions show the efficacy and usability of the approach in an industrial formal verification environment.

REFERENCES

[1] H. Foster, "Applied assertion-based verification: An industry perspective," *Foundations and Trends in Electronic Design Automation*, vol. 3, no. 1, pp. 1–95, 2009.

[2] S. Huang and K. Cheng, *Formal Equivalence Checking and Design Debugging*. Kluwer Academic Publisher, 1998.

[3] A. Smith, A. Veneris, M. F. Ali, and A. Viglas, "Fault diagnosis and logic debugging using Boolean satisfiability," *IEEE Trans. on CAD*, vol. 24, no. 10, pp. 1606–1621, 2005.

[4] G. Fey, S. Staber, R. Bloem, and R. Drechsler, "Automatic fault localization for property checking," *IEEE Trans. on CAD*, vol. 27, no. 6, pp. 1138–1149, 2008.

[5] H. Foster, A. Krolnik, and D. Lacey, *Assertion-Based Design*. Kluwer Academic Publishers, 2003.

[6] E. Clarke, O. Grumberg, and D. Peled, *Model Checking*. MIT Press, 1999.

[7] A. Matsuda. (2011, May.) Overcoming the challenges of formal verification and debug. [Online]. Available: http://www.eetimes.com/design/eda-design/4216119/ Overcoming-the-challenges-of-formal-verification-and-debug

[8] B. Keng, S. Safarpour, and A. Veneris, "Automated debugging of SystemVerilog assertions," in *Design, Automation and Test in Europe*, 2011, pp. 323–328.

[9] J. M. Cobleigh, D. Giannakopoulou, and C. S. Pasareanu, "Learning assumptions for compositional verification," in *Tools and Algorithms for the Construction and Analysis of Systems*. Springer-Verlag, 2003, pp. 331–346.

[10] S. Joshi, S. K. Lahiri, and A. Lal, "Underspecified harnesses and interleaved bugs," in *Principles of Programming Languages*, 2012, pp. 19–30.

[11] W. Li, L. Dworkin, and S. A. Seshia, "Mining assumptions for synthesis," in *Int'l Conf. on Formal Methods and Models for Codesign*, 2011.

[12] B. Keng and A. Veneris, "Automated debugging of missing input constraints in a formal verification environment," in *Formal Methods in CAD*, 2012.

[13] OpenCores.org, 2007. [Online]. Available: http://www.opencores.org

[14] M. H. Liffiton and K. A. Sakallah, "On Finding All Minimally Unsatisfiable Subformulas," in *Int'l Conf. on Theory and Applications of Satisfiability Testing*, 2005, pp. 173–186.

[15] J. Marques-Silva and J. Planes, "Algorithms for maximum satisfiability using unsatisfiable cores," in *Design, Automation and Test in Europe*, 2008, pp. 408–413.

[16] M. H. Liffiton and K. A. Sakallah, "Generalizing Core-Guided Max-SAT," in *Int'l Conf. on Theory and Applications of Satisfiability Testing*, 2009, pp. 481–494.

[17] A. Nadel, "Generating Diverse Solutions in SAT," in *Int'l Conf. on Theory and Applications of Satisfiability Testing*, 2011, pp. 287–301.

[18] Cadence Design Systems, "Incisive Formal Verifier," 2012. [Online]. Available: http://www.cadence.com/products/ld/formal_verifier/pages/ default.aspx

[19] N. Eén and N. Sörensson, "An extensible SAT-solver," in *Int'l Conf. on Theory and Applications of Satisfiability Testing*, 2003, pp. 502–518.

978-1-4799-2817-0/14 $31.00 © 2014 IEEE

Property Directed Reachability for QF_BV with Mixed Type Atomic Reasoning Units

Tobias Welp[1] Andreas Kuehlmann[1,2]

[1] University of California at Berkeley, CA, USA
[2] Coverity, Inc., San Francisco, CA, USA

Abstract

The generalization of Property Directed Reachability (PDR) for the theory QF_BV presented in [1] outperforms the original formulation if the required inductive invariant can be represented efficiently as a set of polytopes. However, many QF_BV model checking instances do not belong in this class and can be solved quickly with the original PDR algorithm. In this paper, we present a hybrid approach which uses both polytopes and Boolean cubes as atomic reasoning units combining the advantages of either homogeneous approach. We discuss theoretic properties of the presented algorithm and report experimental results demonstrating its effectiveness.

1 Introduction

Property Directed Reachability (PDR, also known as IC^3) was initially proposed in [2] and demonstrated remarkable good performance (3rd place) in the Hardware Model Checking Competition (HWMCC) 2011. The authors of [3] have shown that a more efficient implementation of the algorithm would have won the HWMCC 2011.

In addition to the excellent runtime performance, PDR has additional favorable properties. For instance, it excels in finding counterexamples and in proving that none exists alike, it has modest memory requirements, it is parallelizable, and it is incremental.

These convincing algorithmic properties of PDR have instigated interest in generalizations of the binary formulation to richer logics. For instance, the authors of [4] present generalizations for pushdown systems and to the theory of quantifier-free formulae over linear arithmetic (QF_LA), the authors of [5] an extension for the verification of timed systems, and the authors of [6] propose the use of PDR for verifying safety properties in well-structured transition systems.

In [1], a generalization of PDR to the theory of quantifier-free formulae over bitvectors (QF_BV) was proposed. In contrast to the binary formulation of the algorithm that uses Boolean cubes as atomic reasoning units (ARUs), the presented approach uses polytopes as ARUs. The main incentive of polytopes is that they can capture linear relations between bitvectors that are interpreted as integers efficiently. As a consequence, the PDR algorithm with polytopes outperforms the original algorithm on instances which require inductive invariants composed of linear relations for their solution.

Although QF_BV model checking instances with relevant piecewise linear relations as inductive invariants are frequent in application domains such as program verification, many instances also require inference of logic invariants that cannot be efficiently represented as piecewise linear invariants. For instances of this type, the Boolean PDR algorithm [2] often outperforms its generalization with polytopes. Whereas one could use the original version for these instances, none of the homogeneous approaches is successful in cases where invariants are required which combine logic parts with piecewise linear parts.

In this paper, we present a novel hybrid approach that uses both Boolean cubes and polytopes as ARUs. As such, the presented approach overcomes the limitations of exclusive use of polytopes or Boolean cubes and experiments with a set of benchmarks derived from program verification suggest that the proposed algorithm with mixed type ARUs outperforms either homogeneous approach.

The remainder of this paper is structured as follows: To keep the paper self-contained, we provide a cursory description of a generic PDR algorithm in the following section. Next, in Section 3, we present details of the proposed hybrid approach and discuss some of its characteristics followed by a description of our implementation and experimental results in Section 4. Finally, we draw conclusions and describe future work in Section 5.

2 Property Directed Reachability for QF_BV

In this section, we present aspects of the QF_BV PDR algorithm relevant to this paper. For conciseness, many details, some of them fundamental for the performance of the algorithm, are omitted. We refer the reader to [3] for a detailed discussion of the algorithmic skeleton and to [1] for a detailed discussion of the specifics associated to polytopes.

Given a state space spanned by the domain of n bitvector variables $\mathbf{x} = x_1, x_2, \ldots, x_n$, the input of the PDR algorithm is a model checking instance $M = (I, T, B)$ where the QF_BV formula I defines a set of initial states in $\mathbf{x}$, B a set of bad states in $\mathbf{x}$ and T a transition relation over $\mathbf{x}$ and $\mathbf{x}'$, where $\mathbf{x}'$ is a copy of $\mathbf{x}$ which corresponds to the same variables one time step later. T is true iff $\mathbf{x} \rightarrow \mathbf{x}'$ represents a valid transition of the system. The model checking problem is to decide whether a state in B can be reached from a state in I by using only valid transitions in T.

The conceptual strategy of PDR to solve a given model checking problem is to iteratively find small truth statements and combine all these lemmata to obtain the desired proof for the given problem. The motivation for this strategy is that solving a large number of small problems may be more tractable than attempting to solve one big problem at once.

More concretely, PDR constructs a trace t consisting of frames $f_0, f_1, \cdots$. Each frame contains a set of ARUs $\{a_i\}$. For the original Boolean algorithm, each ARU a is a Boolean cube and for the formulation with polytopes, each ARU a is a set of linear inequalities $\sum_i c_i x_i \leq b$. If there is an ARU a in frame f_i, the semantic meaning of this is that all states contained in a cannot be reached within i steps from the initial states. In this case, we say that the states in

978-1-4799-2817-0/14 $31.00 © 2014 IEEE

a are *covered* in frame *i* and we will call all ARUs in frame f_i the *cover*. The inverse of the cover in f_i represents an overapproximation of the states that are reachable in *i* steps. In frame f_0, a state is covered if it is not in the initial set *I*.

Algorithm 1 PDR(*I, T, B*)

1: **while true do**
2:    ARU *a* = FINDBADARU()
3:    int *l* = LENGTHSTRACE()
4:    **if** *a* **then**
5:       **if** !RECCOVERARU(*a, l*) **then return** "property fails"
6:    **else**
7:       PUSHNEWFRAME()
8:       **if** PROPAGATEARUS() **then return** "property holds"

Algorithm 2 RECCOVERARU(ARU *a*, int *l*)

1: **if** *l* = 0 **then return false**
2: **while** *a* reachable from *ã* in one transition **do**
3:    **if** !RECCOVERARU(*ã, l* − 1) **then return false**
4: EXPAND(*a, l*) ; PROPAGATE(*a, l*)
5: add *a* to F_l
6: **return true**

Algorithm 1 shows the overall PDR algorithm. In each iteration, the algorithm searches for an ARU in the last frame of the trace that is in *B* and not yet covered using FINDBADARU(). If such an ARU *a* exists, the algorithm tries to recursively cover the proof obligation *a* (RECCOVERARU(), see Algorithm 2). To this end, the routine checks if *a* is reachable from the previous frame. Assume for now that this is the case and denote with *ã* an ARU in the previous frame that is not covered and from which *a* can be reached in one step. Then RECCOVERARU() calls itself recursively on *ã*. If such a sequence of recursive calls reaches back to frame f_0, the corresponding call stack effectively proves that *a* can be reached from *I*, i.e. that the property fails. Otherwise, if an ARU *a* is proved to be unreachable from the previous frame, it can be added to the cover of the current frame. For efficiency, however, the algorithm attempts to expand and to propagate the ARU to later frames before doing so. Continuing the discussion of Algorithm 1, if FINDBADARU() returns without successfully finding an uncovered ARU in *B*, we know that *B* is covered in the last frame f_l. As we maintain the invariant that the cover in frame f_i is an underapproximation of the space not reachable within *i* steps, we can conclude that *B* is not reachable within *l* steps and we push a new frame to the end of the trace. Afterwards, we attempt to propagate ARUs from frame *l* to frame *l* + 1. If this is successful for all ARUs, we have shown that from an overapproximation of the reachable states in frame f_l we cannot reach any state outside this overapproximation in frame f_{l+1}. In other words, we have found an inductive invariant. Moreover, the set of bad states *B* is disjoint of this inductive invariant. This proves that the property holds.

3 Hybrid Approach with Mixed Type ARUs

We start this section by discussing a motivating example for the approach with mixed type ARUs which illustrates the shortcomings of using either Boolean cubes or polytopes exclusively. Next, we will define the hybrid approach and discuss some of its properties. Finally, in Subsection 3.3, we discuss an extension of simulation based expansion of proof obligations suitable for the hybrid approach.

3.1 Example: Hybrid Invariant

Consider the following model checking problem

$$I \quad := \quad (2 \times x_1 \equiv x_2) \wedge (x_2 + x_1 \leq 3)$$
$$T \quad := \quad (x_1' \equiv x_1 + 1) \wedge (x_2' \equiv x_2 - 2) \wedge (x_1' > x_1) \wedge (x_2' < x_2)$$
$$B \quad := \quad (x_2 + x_1 \geq 4) \vee (x_2 \bmod 2 \equiv 1)$$

The snapshot of the PDR trace in Figure 1 illustrates the issue the PDR algorithm with polytopes encounters when attempting to solve this model checking instance.

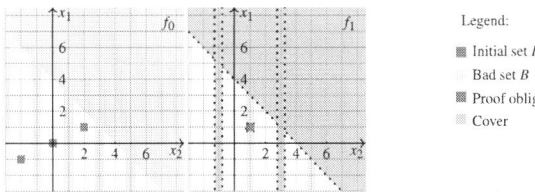

Figure 1: Attempt to Solve Example Hybrid Invariant

The inductive invariant required to solve this instance is $(x_2 + x_1 \leq 3) \wedge (x_2 \text{ even})$. The corresponding cover in the PDR trace includes two parts, '$x_2 + x_1 \geq 4$' and 'x_2 odd'. The first part can be covered using a polytope. However, the other aspect can not be efficiently represented using polytopes. Instead, the algorithm would add a polytope for each odd number representable by x_2. Assuming that x_2 is a 32-bit bitvector, this would require 2^{31} polytopes, causing the overall algorithm to be very inefficient.

In contrast, the Boolean PDR algorithm would efficiently find this second part of the inductive invariant as it can be represented by a single Boolean cube but fail in finding the first part efficiently.

3.2 Probabilistic Specialization

The shortcomings of using either exclusively Boolean cubes or polytopes motivate a hybrid approach. To this end, we define the ARU to be either a Boolean cube or a polytope. Each ARU is created either in FINDBADARU() in line 2 of Algorithm 1 as a point in bad that is not yet covered or in line 2 of Algorithm 2 as point from which another proof obligation is reachable. In both cases, the point is found by a theory solver and can be interpreted as a polytope or a Boolean cube. A specialization to a specific kind of ARU becomes necessary before the proof obligation is expanded because the expansion sequence is different for polytopes (relaxing individual inequalities) and Boolean cubes (chopping of literals). We propose to specialize a point probabilistically. Therefore, we define probability *c* and specialize to a Boolean cube with probability *c* and to a polytope with probability 1 − *c*.

This probabilistic choice guarantees that a favorable specialization can be expected to be chosen in a constant number of attempts. To see why, consider the example in Figure 1 for another time. With probability *c*, the marked proof obligation is specialized to a Boolean cube. In this case, the desired part 'x_2 even' of the inductive invariant would be found immediately. Otherwise, the proof obligation would be covered by the polytope $x_2 = 1$. In this case, after the next call to FINDBADARU(), a new proof obligation with '**x** odd' is found followed by another probabilistic decision for specialization.

The probability that a choice for a Boolean cube is made in the ith iteration is $c(1-c)^{i-1}$. Hence, the expected number of trials until the favorable specialization is found can be calculated to be

$$E\left\{\text{trials until Boolean cube specialization}\right\} = c\sum_{i=1}^{\infty} i(1-c)^{i-1} = \frac{1}{c}$$

Analogously, one calculates

$$E\left\{\text{trials until polytope specialization}\right\} = (1-c)\sum_{i=1}^{\infty} ic^{i-1} = \frac{1}{1-c}$$

Consequently, as long as $c = (0,1)$, one can expect a favorable decision within a constant number of times. The optimal choice for c depends on the concrete model checking instance. We will investigate the impact of c in the experimental section of this paper.

3.3 Expansion of Proof Obligations

The excellent performance of the binary version of PDR as documented in [3] can be partly attributed to its ability to efficiently and effectively expand proof obligations using ternary simulation. In [1], this idea was generalized to interval simulation for the QF_BV PDR model checker with polytopes. In this section, we describe a suitable generalization for the presented QF_BV PDR model checker with mixed type ARUs.

Proof obligations are expanded after their specialization. At this time, a proof obligation a consists of a single point in the state space. The aim of the expansion is to add additional points to a so that the resulting set can be processed as a union in the sequel, stimulating more abstract reasoning.

Note that expansion of proof obligations, albeit practically critical for performance, is not required for the correctness of the algorithm. The goal is to find a close underapproximation of the maximum possible expansion that can be calculated efficiently.

The expansion is achieved using an expansion sequence suitable for the type of the proof obligation. At each step of the sequence, an expansion $\hat{a}$ is proposed. The proposal will be accepted if we can reach a point in B from any point in $\hat{a}$.

The backbone for this test is simulation of expressions. In the following, we will denote with $\Phi_e(\hat{a})$ an overapproximation of the values expression e can take under the constraint $\hat{a}$ on the variables.

Expansion of proof obligations is used in two contexts. First, we use it for expanding ARUs in B that are not yet covered (line 2 in Algorithm 1). To this end, we test whether a proposed expansion $\hat{a}$ is valid by checking if

$$\Phi_{b \wedge u}(\hat{a}) \equiv \text{true} \tag{1}$$

where b is an expression describing the bad set B of the model checking instance and u is the expression describing the points that are not yet covered. If the equation holds, the expansion is valid.

Second, in the context of expanding an ARU $\tilde{a}$ from which another proof obligation a is reachable in one step (line 2 in Algorithm 2), we check whether the possible valuations of the variables after one step starting from $\hat{\tilde{a}}$ are included in the possible valuations of the corresponding variables under a, formally

$$\forall x_i . \Phi_{\text{next}_{x_i}}(\hat{\tilde{a}}) \subseteq \Phi_{x_i}(a). \tag{2}$$

where we denote with next_{x_i} the expression which captures the computation of the next value of x_i. Note that this requires that the transition relation is in the format of transition functions solvable for each variable. In many applications, this is naturally the case.

Under the assumption that the simulation values $\Phi_e(\hat{a})$ represent overapproximations, the checks (1) and (2) are conservative in the sense that an expansion to $\hat{a}$ or $\hat{\tilde{a}}$, respectively, is accepted only if a bad state is reachable from all points in the proposed expansion.

3.3.1 Ternary Simulation

The original PDR algorithm uses ternary simulation to calculate simulation values for expressions. Representing binary variables as intervals $[1,1]$, $[0,1]$, and $[0,0]$ which stand for true, true or false (X), and false, respectively, one can use the rules in Figure 2 to obtain simulation values for any expression composed of AND- and INV-operations. As one can decompose any bitvector expression

$$\frac{}{\Phi_c = [c,c]} \text{[const]} \qquad \frac{}{\Phi_X = [0,1]} \text{[x]} \qquad \frac{l \le v \le u}{\Phi_v = [l,u]} \text{[var]}$$

$$\frac{\Phi_{e_1} = [l_1,u_1] \quad \Phi_{e_2} = [l_2,u_2]}{\Phi_{e_1 \wedge e_2} = [\min\{l_1,l_2\},\max\{u_1,u_2\}]} \text{[and]} \qquad \frac{\Phi_e = [l,u]}{\Phi_{\neg e} = [1-u,1-l]} \text{[inv]}$$

Figure 2: Simulation Rules for Ternary Simulation

into equivalent AND-Inverter expressions using their circuit representations, this allows to calculate the simulation values required in checks (1) and (2).

It remains to discuss how the range of variables is restricted given an ARU $\hat{a}$. If the ARU is a Boolean cube, this is straightforward. Otherwise, if the ARU is a polytope, the situation is slightly more complicated: It is instructive to make two observations. First, note that at the beginning of the expansion move, $\hat{a}$ represents a point. Second, any proposed expansion for a polytope is derived by relaxing individual inequalities of the polytope. Combining these two observations, we can conclude that at any point of the expansion sequence, the polytope encodes a static integer interval for each variable. For a bitvector variable x of m bits, let $[l,u]$ be the interval x can take while staying in $\hat{a}$. With i being the index of the bit we are interested in, we can use SELECT(l,u,i) in Algorithm 3 to obtain the corresponding range of values the bit can take. Note that in Algorithm 3, we denote with $l[i:m-1]$ the vector of the $m-i$ most significant bits of l if l was stored in a bitvector of length m. The definition of $u[i:m-1]$ is analog.

Algorithm 3 SELECT(l,u,i)

1: **if** $l[i:m-1] \equiv u[i:m-1]$ **then return** $[l[i],l[i]]$
2: **else return** $[0,1]$

We can conclude that ternary simulation can be used for the expansion of proof obligations for the QF_BV generalization of the PDR algorithm. However, expansion of proof obligations using ternary simulation cannot be expected to perform effectively if used for a model checking instance where the transition function captures high-level constraints where bitvectors are interpreted as integers. In particular, using ternary simulation, it is not possible to represent any intervals of integers that contain both positive and negative values in the two's complement representation causing gross overapproximation of simulation values in practice.

3.3.2 Interval Simulation

These limitations of ternary simulation suggests a simulation method that captures arithmetic operations more naturally. One way to achieve this is interval simulation as proposed in [1]. Consider the representative choice of simulation rules in Figure 3 where we denote the minimally and maximally representable numbers for a given expression with $-\infty$ and ∞, respectively.

In the case of interval simulation, restricting the range of variables given an ARU $\hat{a}$ is immediate if $\hat{a}$ is a polytope. In the case $\hat{a}$

$$\frac{\Phi_{e_1} = [l_1, u_1] \quad \Phi_{e_2} = [l_2, u_2] \quad l_1 + l_2 \geq -\infty \quad u_1 + u_2 \leq \infty}{\Phi_{e_1 + e_2} = [l_1 + l_2, u_1 + u_2]} \text{ [plus-r]}$$

$$\frac{\Phi_{e_1} = [l_1, u_1] \quad \Phi_{e_2} = [l_2, u_2] \quad l_1 + l_2 < -\infty \vee u_1 + u_2 > \infty}{\Phi_{e_1 + e_2} = [-\infty, \infty]} \text{ [plus-o]}$$

$$\frac{\Phi_{e_1} = [l_1, u_1] \quad \Phi_{e_2} = [l_2, u_2] \quad u_1 < l_2}{\Phi_{e_1 < e_2} = [1, 1]} \text{ [lt-1]}$$

$$\frac{\Phi_{e_1} = [l_1, u_1] \quad \Phi_{e_2} = [l_2, u_2] \quad l_1 > u_2}{\Phi_{e_1 < e_2} = [0, 0]} \text{ [lt-0]}$$

$$\frac{\Phi_{e_1} = [l_1, u_1] \quad \Phi_{e_2} = [l_2, u_2] \quad u_1 \geq l_2 \quad l_1 \leq u_2}{\Phi_{e_1 < e_2} = [0, 1]} \text{ [lt-x]}$$

$$\frac{\Phi_{e_1} = [l_1, u_1] \quad l_1 = 1}{\Phi_{\text{if } e_1 \text{ then } e_2 \text{ else } e_3} = \Phi_{e_2}} \text{ [ite-t]} \qquad \frac{\Phi_{e_1} = [l_1, u_1] \quad u_1 = 0}{\Phi_{\text{if } e_1 \text{ then } e_2 \text{ else } e_3} = \Phi_{e_3}} \text{ [ite-e]}$$

$$\frac{\Phi_{e_1} = [l_1, u_1] \quad \Phi_{e_2} = [l_2, u_2] \quad \Phi_{e_3} = [l_3, u_3] \quad l_1 \neq u_1}{\Phi_{\text{if } e_1 \text{ then } e_2 \text{ else } e_3} = [\min\{l_2, l_3\}, \max\{u_2, u_3\}]} \text{ [ite-x]}$$

Figure 3: Choice of Simulation Rules for Interval Simulation

is a Boolean cube constraining the m bits of x such as $x[i] \subseteq [l_i, u_i]$, we can use procedure $\text{CONJOIN}(l_0, u_0, \ldots, l_{m-1}, u_{m-1})$ in Algorithm 4 to calculate the interval for bitvector x.

Algorithm 4 $\text{CONJOIN}(l_0, u_0, \ldots, l_{m-1}, u_{m-1})$

1: l, u = bitvector with length m
2: **for** $i = 0$ to $m - 1$ **do**
3:    **if** $l_i \equiv u_i$ **then** $u[i] = l[i] = l_i$
4:    **else**
5:      $l[i] = \begin{cases} 1 & \text{if } i \equiv m-1 \\ 0 & \text{otherwise} \end{cases} \; ; \quad u[i] = 1 - l[i]$
6: **return** $[\text{INTEGERVALUE}(l), \text{INTEGERVALUE}(u)]$

Though these rules capture the meaning of arithmetic operations well, analogous bitvector rules for logic operations such as AND are difficult to formulate and representing their results as integer intervals can lose a lot of precision. Consider, e.g. we wanted to calculate $\Phi_{e_1 \wedge e_2}$ given $\Phi_{e_1} = [-1, 0]$ and $\Phi_{e_2} = [-8, -7]$ where all expressions are 4-bit integers. Using the two's complement representation of integers, this corresponds to the following calculation

$$\wedge \begin{array}{r} [-1, \; 0] \\ [-8, -7] \\ \hline [-8, \; 1] \end{array} \binom{\text{interval}}{\text{simulation}} \quad \leftrightarrow \quad \wedge \begin{array}{r} - - - - \\ 1\,0\,0\,- \\ \hline -\,0\,0\,- \end{array} \binom{\text{ternary}}{\text{simulation}}$$

Even the optimal (tightest overapproximation) result representable as an integer interval $[-8, 1]$ contains more points than the exact result on the right which does e.g. not include -2.

3.3.3 Hybrid Simulation

The strengths and weaknesses of each homogeneous simulation strategy suggests a hybrid approach. To this end, we define two additional simulation rules [select] and [conjoin] as in Figure 4 that use Algorithms 3 and 4, respectively. The rule [select] is for extracting individual bits from a bitvector and the rule [conjoin] serves for combining bits to a bitvector. Combining these rules with the rules from ternary simulation in Figure 2 and the rules of interval simulation as in Figure 3 allows for a simulation where ternary simulation is used for the logic operations and interval simulation for the arithmetic operations. At the interfaces between the

$$\frac{\Phi_e = [l, u]}{\Phi_{e[i]} = \text{SELECT}(l, u, i)} \text{ [select]}$$

$$\frac{\Phi_{e[0]} = [l_0, u_0] \quad \ldots \quad \Phi_{e[m-1]} = [l_{m-1}, u_{m-1}]}{\Phi_e = \text{CONJOIN}(l_0, u_0, \ldots, l_{m-1}, u_{m-1})} \text{ [conjoin]}$$

Figure 4: Conjoin and Select Simulation Rules

two kinds of operations act the transformation rules [select] and [conjoin].

As such, the choice between interval simulation and ternary simulation is driven by the specific expressions and an operation is treated by the kind of simulation which captures its conceptional idea the best. The rules [conjoin] and [select] which are susceptible to losing precision are only applied when necessary at the interface between operations of different kinds.

3.3.4 Example: Simulation

We conclude the discussion of the simulation based expansion of proof obligations with an example. Assume we attempt to simulate $\Phi_e(\hat{a})$ with

$$\begin{array}{llll} e & := & (e_1 < 2) \wedge (0 \leq e_1) & e_1 := e_2 \wedge e_3 \\ e_2 & := & x_1 - x_2 + 2 & e_3 :- y_1 \vee y_2 \end{array}$$

and $\hat{a} := (1 \leq x_1 \leq 5) \wedge (0 \leq x_2 \leq 3) \wedge (y_1 \in \text{-00-}) \wedge (y_2 \in \text{100-})$

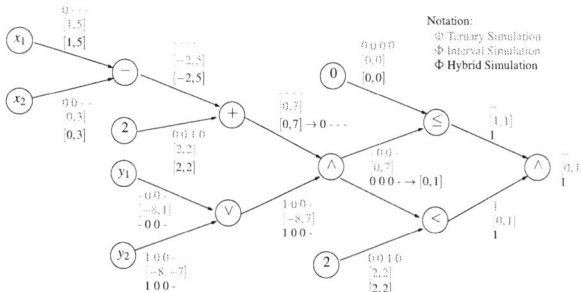

Figure 5: Example of a Hybrid Simulation

Figure 5 shows the expression DAG of e and the simulation results for all (sub)expressions using ternary, interval, and hybrid simulation. On the one hand, ternary simulation works well for the simulation of subexpression e_3 sustaining the information that the second and third bits of the expression are low whereas interval simulation looses this information entirely. On the other hand, interval simulation performs well for the simulation of subexpression e_2 pertaining the information that the value must be positive while ternary simulation looses all information for this expression. However, only hybrid simulation is able to maintain both pieces of information and to combine them to $\Phi_{e_1}(\hat{a}) = [0, 1]$. Hence, hybrid simulation yields the correct result that $\Phi_e(\hat{a}) = \text{true}$ while both ternary and interval simulation fail due to their individual shortcomings. If e corresponded to $B \wedge u$ in check (1), the expansion to $\hat{a}$ would only succeed with hybrid simulation.

4 Implementation and Experimentation

We implemented the proposed hybrid approach. The skeleton of our implementation is similar to the one suggested in [3]. As backend solver, we use the SMT-solver Z3 [9] instead of a SAT-solver. This choice allowed us to implement all of the suggested optimization techniques described in [3], in particular to consider the unsatisfiable core for fast expansion and reusing learned clauses by

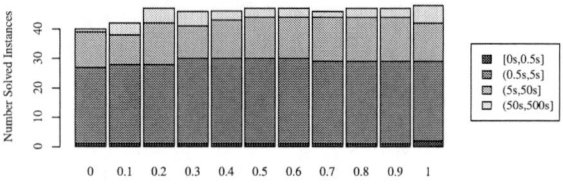

(a) Instances extracted from the SV-COMP [7] bitvector benchmark set.

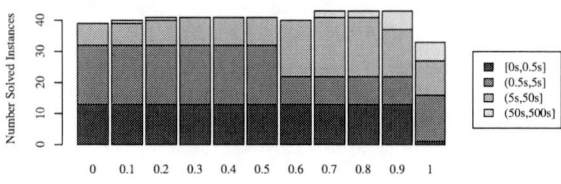

(b) Instances extracted from the INVGEN [8] benchmark set.

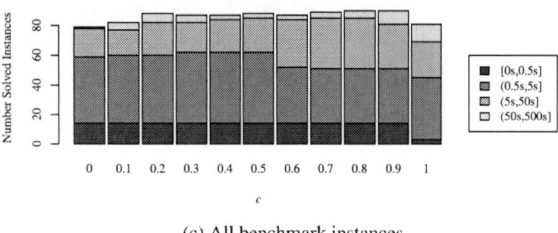

(c) All benchmark instances.

Figure 6: Impact of c on Number of Solved Benchmark Instances

incremental solving with retractable assertions (hot solving). To generalize proof obligations, we use hybrid simulation as discussed in Section 3.3.

For experimentation, we compiled a set of roughly 100 benchmark problems derived from the regression test suite of INVGEN [8] and the bitvector set of the SV-COMP software competition benchmarks [7]. In all model checking problems, the property holds, i.e. no bad states are reachable from the initial states.

4.1 Impact of Parameter c on Quality of Results

As discussed in Section 3, the choice of a specific kind of ARU is important for the success of the QF_BV PDR algorithm. In the proposed hybrid algorithm with polytopes and Boolean cubes, parameter c controls the probability that a point in the state space is specialized to a Boolean cube. The point is specialized to a polytope with probability $1 - c$.

Figure 6 documents the impact of parameter c on the efficacy of the overall algorithm. The extreme cases $c = 1$ and $c = 0$ are the configurations of the algorithm with exclusive use of Boolean cubes (corresponding to the approach presented in [3]) and polytopes (corresponding to the approach presented in [1]), respectively. Figure 6a shows the results for a subset of the entire set of instances derived from the bitvector set of the SV-COMP [7] benchmarks. On this subset, the algorithm with Boolean cubes only performs the best. Figure 6b displays the same plot with instances derived from the INVGEN [8] benchmark set. On this subset, the configuration with Boolean cubes only performs the worst. The varied result can be explained by the characteristics of the benchmark sets. On the one hand, the benchmarks derived from the SV-COMP benchmarks focus on bit-level characteristics. For instance, the inductive

invariants required to solve these problems often specify a specific bit having a specific value or two bits having the same value. On the other hand, in the INVGEN benchmarks, bitvectors are usually interpreted as integers and the relevant inductive invariants often represent linear relations between different variables using this interpretation. In this setting, the polytope interpretation is particularly suitable. In Figure 6c, we illustrate the impact of c for the entire set of model checking instances. Here, the algorithm performs similarly for all configurations but the two pure versions (only Boolean cubes or only polytopes) for which the algorithm perform worse. This validates the theoretic discussion in 3 in so far as that the specific value for c does not matter much in practice unless c is chosen at the extreme values.

4.2 Impact of Simulation Type on Quality of Results

In our second experiment, we investigated the impact of the simulation type on the performance of the algorithm. To this end, we attempt to solve the set of model checking instances with ternary simulation, interval simulation, and hybrid simulation.

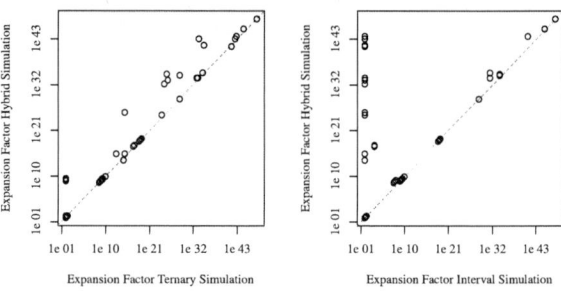

Figure 7: Impact of Simulation Type on Expansion Factor

Figure 7 shows the *expansion factor*, the quotient of the volumes of ARUs after and before the expansion move, with ternary simulation versus hybrid simulation and with interval simulation versus hybrid simulation, respectively, accumulated for several individual expansion attempts. As can be seen, hybrid simulation outperforms the pure simulation types in many cases, often by several orders of magnitude.

Figure 8 illustrates how this increase in simulation efficacy impacts the overall performance of the model checker when run with $c = 0.5$. As in Subsection 4.1, we show the result both for the complete set of benchmarks and for the two subsets separately. All plots suggest that hybrid simulation performs best. Interestingly, for the model checking instances derived from the SV-COMP benchmarks, ternary simulation outperforms interval simulation. For the benchmarks originating from the INVGEN benchmarks, one would draw the converse conclusion. Similar as discussed in Subsection 4.1, the different observations can be explained by the focus on logic operations in the first subset of benchmarks rather than on arithmetic operations which dominate in the second subset.

4.3 Performance Comparison vs ABC PDR

In the last experiment, we compare the performance of our generalized PDR algorithm with the original Boolean version contained in the logic synthesis and verification tool ABC [10]. To this end, we translated the benchmark model checking instances into BTOR [11], a format for specifying word-level model checking problems, and used the tool SYNTHEBTOR, which is part of the distribution of the BOOLECTOR SMT-solver [12], to bit-blast

978-1-4799-2817-0/14 $31.00 © 2014 IEEE

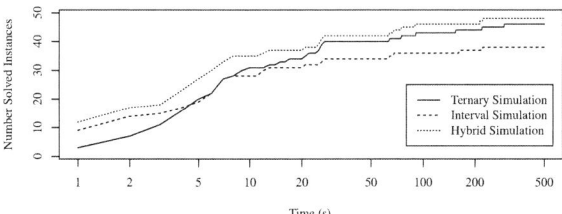

(a) Instances extracted from the SV-COMP [7] bitvector benchmark set.

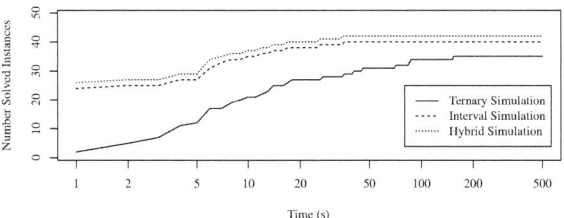

(b) Instances extracted from the INVGEN [8] benchmark set.

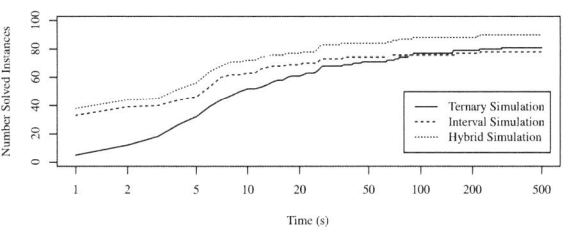

(c) All benchmark instances.

Figure 8: Impact of Simulation Type on Number of Solved Benchmark Instances

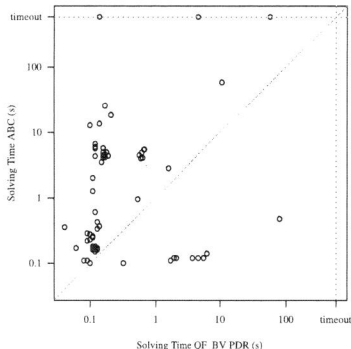

Figure 9: Comparison of QOR of QF_BV PDR vs ABC PDR.

core of the hybrid approach rest (1) a probabilistic specialization of proof obligations after being found by the theory solver and (2) hybrid simulation for the expansion of proof obligations. We implemented the novel algorithm and reported experimental results that show that the PDR algorithm with hybrid ARUs is dominating both homogeneous alternatives on a set of benchmarks extracted from program verification benchmarks.

As future work, it would be interesting to investigate the potential of adding other types of ARUs to the algorithm.

the problems into AIGER [13], a standard format for specifying hardware-model checking problems supported by ABC.

Figure 9 shows a comparison of running times to solve the given benchmark model checking instances with the presented QF_BV generalization of PDR versus the Boolean PDR model checker in ABC. For roughly 95% of the benchmarks, the presented QF_BV generalization outperforms the Boolean PDR algorithm when run on the bit-blasted versions of the benchmarks, often by more than an order of magnitude. This demonstrates that the QF_BV algorithm is able to use the additional information encoded in the word-level formulation of the model checking instances to speed up the solving. In the other roughly 5% of the benchmarks, the inductive invariant required to solve the model checking instance can simply be represented as a set of Boolean cubes. In this case, the generalized algorithm has no conceptional advantage over the Boolean algorithm and attempting to solve the model checking instance using polytopes is not purposeful, hence one cannot expect the generalized algorithm to outperform the original one.

5 Conclusions and Future Work

PDR originated as an efficient and effective algorithm for solving hardware model checking problems and its excellent algorithmic properties motivate research to find generalizations for richer logics. In this paper, we presented an extension to the generalization of PDR to the theory QF_BV with polytopes presented in [1]. Instead of only using polytopes, we propose a hybrid approach with Boolean cubes and polytopes as ARUs that overcomes some of the limitations associated with either homogeneous approach. At the

References

[1] T. Welp and A. Kuehlmann, "QF_BV Model Checking with Property Directed Reachability," in *Proc. of the 16th Int'l Conf. on Design, Automation and Test in Europe*, DATE '13, (Washington, DC, USA), IEEE Computer Society, 2013.

[2] A. R. Bradley, "SAT-Based Model Checking without Unrolling," in *Proc. of the 12th Int'l Conf. on Verification, Model Checking, and Abstract Interpretation*, VMCAI'11, (Berlin, Heidelberg), pp. 70–87, Springer-Verlag, 2011.

[3] N. Eén, A. Mishchenko, and R. Brayton, "Efficient Implementation of Property Directed Reachability," in *Proc. of the Int'l Conf. on Formal Methods in Computer-Aided Design*, FMCAD '11, (Austin, TX), pp. 125–134, 2011.

[4] K. Hoder and N. Bjørner, "Generalized Property Directed Reachability," in *Proc. of the 15th Int'l Conf. Theory & Appl. Satisfiability Testing (SAT)*, (Berlin, Heidelberg), pp. 157–171, Springer-Verlag, 2012.

[5] R. Kindermann, T. Junttila, and I. Niemelä, "SMT-based Induction Methods for Timed Systems," in *Proc. of the 10th Int'l Conf. on Formal Modeling and Analysis of Timed Systems*, FORMATS'12, (Berlin, Heidelberg), pp. 171–187, Springer-Verlag, 2012.

[6] J. Kloos, R. Majumdar, F. Niksic, and R. Piskac, "Incremental, Inductive Coverability," in *Proc. of Computer Aided Verification*, vol. 8044 of *Lecture Notes in Computer Science*, pp. 158–173, Springer-Verlag, 2013.

[7] D. Beyer, "Competition on Software Verification," in *Proc. of the 18th Int'l Conf. on Tools and Algorithms for the Construction and Analysis of Systems*, TACAS'12, (Berlin, Heidelberg), pp. 504–524, Springer-Verlag, 2012.

[8] A. Gupta and A. Rybalchenko, "InvGen: An Efficient Invariant Generator," in *Proc. of Computer Aided Verification*, vol. 5643 of *Lecture Notes in Computer Science*, pp. 634–640, 2009.

[9] L. De Moura and N. Bjørner, "Z3: An Efficient SMT Solver," in *Proc. of the Theory and Practice of Software, 14th Int'l Conf. on Tools and Algorithms for the Construction and Analysis of Systems*, TACAS'08/ETAPS'08, (Berlin, Heidelberg), pp. 337–340, Springer-Verlag, 2008.

[10] R. Brayton and A. Mishchenko, "ABC: An Academic Industrial-Strength Verification Tool," in *Proc. of Computer Aided Verification*, vol. 6174 of *Lecture Notes in Computer Science*, pp. 24–40, Springer-Verlag, 2010.

[11] R. Brummayer, A. Biere, and F. Lonsing, "BTOR: Bit-Precise Modelling of Word-Level Problems for Model Checking," in *Proc. of the Joint Workshop of the 6th Int'l Workshop on SMT & 1st Int'l Workshop on Bit-Precise Reasoning*, SMT '08/BPR '08, (New York, NY, USA), pp. 33–38, ACM, 2008.

[12] R. Brummayer and A. Biere, "Boolector: An Efficient SMT Solver for Bit-Vectors and Arrays," in *Proc. of the 15th Int'l Conf. on Tools & Algorithms for the Construction & Analysis of Systems*, TACAS '09, (Berlin, Heidelberg), pp. 174–177, Springer-Verlag, 2009.

[13] A. Biere, "The AIGER And-Inverter Graph format." http://fmv.jku.at/aiger, 2006.

978-1-4799-2817-0/14 $31.00 © 2014 IEEE

Adaptive Interpolation-Based Model Checking

Chien-Yu Lai
d02943034@ntu.edu.tw

Cheng-Yin Wu
gro070916@yahoo.com.tw

Chung-Yang (Ric) Huang
cyhuang@ntu.edu.tw

Graduate Institution of Electronic Engineering,
National Taiwan University

Abstract - Interpolation-based model checking (IMC) is an important technique in modern formal verification tools. In essence, it relies on an abstraction and refinement process to derive an adequate image approximation for the reachability analysis. However, previous IMC algorithms only offer fixed degrees of abstraction and thus may fail in the proofs if the abstraction is too coarse- or fine-grained. In this paper, we propose an adaptive interpolation-based model checking algorithm in which the degree of abstraction can be adjusted on demand. That is, during the proof process, we closely monitor the effectiveness of the interpolation-based over-approximated image computation and thus adjust the degree of abstraction for the best performance. The experimental results confirm that our flexible interpolation indeed leads to an adequate degree of abstraction as our IMC algorithm outperforms previous ones in various aspects.

I. Introduction

Interpolation-based model checking (IMC) has been an important unbounded model checking technique to verify safety properties since McMillan proposed it in 2003[1]. As shown in Fig. 1, IMC can disprove a safety property (i.e. concludes that the property is unsafe) if it finds a counter-example in the bounded model checking (BMC)[2] phase. On the other hand, it can formally validate the property (i.e. proves the property safe) if it converges on a fixed point during the over-approximated state reachability analysis in the interpolation (ITP) phase.

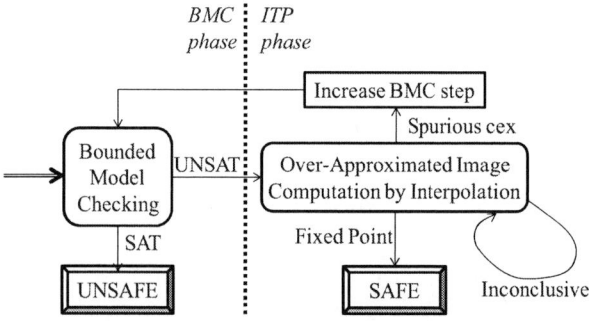

Fig. 1. The classical IMC algorithm flow

The iteration between these two phases can be viewed as an "abstraction and refinement" process. That is, when the BMC phase fails to identify the counterexample at certain depth k (i.e. SAT engine returns unsatisfiability), the ITP phase will then utilize the unsatisfiable proof model as the abstraction guidance to compute the over-approximated image of the current set of reachable states. This over-approximated state reachability analysis will iterate until either the fixed point is found, or a (spurious) counterexample is encountered. In the latter case, since the counterexample is from the over-approximated image and

thus may be spurious, the IMC will have to go back to the BMC phase to refine the proof model with increased depth.

Conceptually, the iteration between the BMC and ITP phases is trying to refine the proof model to an adequate abstraction so that the proof engine can either identify a true counterexample or reach a fixed point in the over-approximated state reachability analysis (say, with a depth $k = d$). Consequently, the best case for the IMC algorithm is that an adequate abstraction is quickly found and thus the proof can be efficiently concluded. On the contrary, the worst case is either that in the first $(d-1)$ iterations, the abstraction is too *fine-grained* so that it takes a long time to yield the spurious counterexamples, or that in the depth (d) where the fixed point can be met, the abstraction is too *coarse-grained* and thus the proof process fails to converge. In short, it will be ideal that the proof engine has the "intelligence" to adjust the granularity of the abstraction, making it promptly jump out of the ITP phase when the spurious counterexample is bound to occur, and efficiently conclude the proof with the adequate image computation in the reachability analysis.

However, in the previous IMC algorithms, there is no such flexibility in adjusting the abstraction granularity. In the original IMC [1], the over-approximated next-state image is generated from the Craig interpolation [3][4] of the unsatisfiable BMC model. Since the interpolant is derived from the resolution graph of the unsatisfiability proof, it is pre-determined by the decision procedure and thus the IMC algorithm has no effective way to control it. On the other hand, the NewITP algorithm proposed in [5] utilizes both SAT and UNSAT generalizations to derive cubes as the over-approximated reachability. There is no guidance to refine the UNSAT-generalized cubes except the minterms of spurious counter-examples in NewITP. It cannot affect the granularity of abstraction but only prune a very small number of states from the interpolant.

In this paper, we propose an adaptive IMC in which an adjustable over-approximated imagine computation engine is included. With the mechanism to evaluate the effectiveness of the abstraction and thus the flexibility to generate proper interpolants for the over-approximated reachability, our IMC algorithm can intelligently settle down to an adequate abstraction for the efficient conclusion of the proof. The experimental results indeed show that our proposed IMC algorithm outperforms previous ones in various aspects.

The remaining of this paper is organized as follows: In Section II, we provide some background knowledge for this paper and in Section III, we overview our adaptive IMC algorithm. The details of the proposed adaptive over-approximated image computation techniques are presented in Sections IV and V. Finally, Section VI demonstrates the experimental results and Section VII concludes the paper.

II. Preliminaries

Without loss of generality, the circuits we are verifying with model checkers can be viewed as a finite-state concurrent system [6]. That is, the set of states that are reachable from the initial state(s) is finite. For an assertion property to be formally verified, we need to make sure that it is evaluated true for all the reachable states under all the input combinations.

The set of reachable states can be computed by the *reachability analysis* process. First, the *transition relation* (TR) function that characterizes the relationship between the current and the next sets of states can be derived from the circuit functionality. Then from the set of initial states, the set of next reachable states (also called the *image* of the current states) can be computed from the TR. This *image computation* process is iterated and the set of the reachable states is accumulated until no more new states can be found in the process. In such case, we call it a *fixed point* of the reachability analysis.

For the assertion checking, we check whether the current set of reachable states can falsify the assertion property. If yes, a counterexample can be derived. Otherwise, if no counterexample is found before the fixed point is reached, we can conclude that the assertion is formally verified.

In general, the set of reachable states, although finite, is exponential to the number of state variables (i.e. the registers of the circuit) and thus the *exact* reachability analysis is mostly intractable for real designs. For practicality reason, people usually apply certain *abstraction* techniques with much lower computational cost to derive the *over-approximated* set of reachable states. To prove an assertion property, if no counterexample is found in the over-approximated reachability analysis, we can guarantee that no counterexample can be found in the exact reachability analysis and thus the property is asserted. On the other hand, if a counterexample is encountered for certain over-approximated set of states, we cannot judge whether it is a valid counterexample and thus usually treat it as *spurious*. In such case, a *refinement* process is required to adjust the approximation of the reachability analysis.

The abstraction techniques applied in the over-approximated reachability analysis generally involve the abstractions on the circuit model, on the TR function, or on the image computation. For the previous IMC algorithms, the utilized abstraction techniques only include the abstraction on the image computation, while in our proposed IMC, we apply all the techniques to realize the adaptive abstractions and thus achieve the best performance.

III. Overview of Our Adaptive Interpolation-Based Model Checker

Our IMC has the capability to adjust the granularities of abstraction in order to improve the proving capability of the model checking algorithms. In essence, our IMC flow is similar to the conventional ones as shown in Fig. 1. However, the key differentiator is that we make the over-approximated image computation *adaptive* so that it can adjust to an *adequate abstraction* that is sufficient to conclude the proof. In other words, we make it flexible to choose either a coarser-grained abstraction that can spot the spurious counterexample earlier, or to a finer-grained abstraction that can find a fixed point more quickly.

More specifically for the over-approximated image computation step, Fig. 2 illustrates the differences between the conventional and our IMC algorithms. As described in Section I, the conventional IMC algorithm generates the interpolation from the refutation graph of the unsatisfiability proof and thus the over-approximated image is pre-determined by the decision procedure. Consequently, there is no flexibility in the reachability analysis and thus the room for improvement on the proving capability is limited.

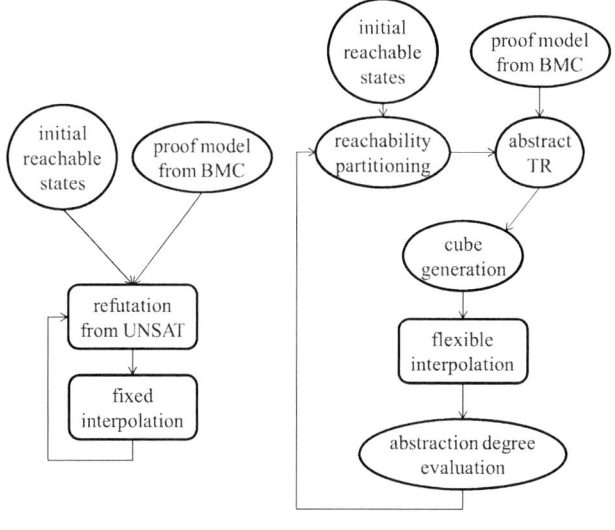

Fig. 2. Comparison between conventional and proposed IMC on over-approximated image computation

On the other hand, in our IMC, to achieve the flexibility in image computation, we devise several techniques to adjust the abstraction model and evaluate the effectiveness of the abstraction. We first define the notation "abstract transition relationship (ATR)" as a subset of the original TR constraints (see Section IV-A). As the ATR represents a looser constraint for the transition system, the interpolant derived from the ATR and the BMC proof model will be an over-approximated image of the current states.

In addition, we apply "cube enumeration" to generate the interpolant as the over-approximated image (see Section IV-C). This enables us to easily partition the set of reachable states as subsets of cubes. As each of the subset of cubes can render an over-approximated image of the corresponding subset of states, the disjunction of these sub-images will also be an over-approximated image of the current set of states. As we will describe in Section IV-B, this disjunction forms a larger image (i.e. coarser-grained abstraction) than the one

978-1-4799-2817-0/14 $31.00 © 2014 IEEE   745

computed from the un-partitioned reachability. Therefore, by choosing different cardinalities in partitioning, we can fulfill the flexibility in controlling the size of the over-approximated image.

In our implementation, we further design an "abstraction degree evaluator" to evaluate the effectiveness of the over-approximated image computation for the current abstraction (see Section V). It is mainly based on the number of cubes in the representation of the over-approximated image. Our algorithm will use this evaluation result to adjust the granularity in the reachable state partitioning.

In addition, at the beginning of the ITP phase, we determine the granularity of the initial abstraction based on the depth of the previous spurious counterexample. This is mainly to avoid too coarse-grained abstraction. From our observation, if the spurious counterexample is identified with a small numbers of steps in the previous ITP phase, it is highly possible that the interpolation generated in the current round with the same degree of abstraction will also be too coarse even if the proof model is refined by the incremented BMC step. Therefore, we will adjust our over-approximated computation to a finer-grained initial abstraction.

IV. Image Computation by Flexible Interpolation

Instead of computing over-approximated image directly, our flexible interpolation computes abstract transition relation first followed by image computation. The flexibility is realized by controlling the size of USNAT core and utilizing it as abstract transition relation, In this way, the degree of transition relation abstraction gets flexible and so does the corresponding over-approximated image.

A. Abstract Transition Relation From UNSAT Core

Given initial states I_0, a transition relation T and a reachability, R, consider BMC proof instance $I_0 \wedge T^k \wedge R$. Let the off-set be $I_0 \wedge T^{k-1}$ and the on-set be $T \wedge R$. The backward interpolation process is able to continue the over-approximated reachability analysis as long as the on-set does not intersect with the off-set. Therefore, other than using the interpolant of the on- and off-sets to over-approximate the set of reachable states, we can also perform the over-approximated reachability by abstracting the transition relation as long as this abstract transition relation (ATR) yields an over-approximated image (i.e. $ATR \wedge R$) that does not intersect with the off-set $I_0 \wedge T^{k-1}$.

Clearly, the UNSAT core of the BMC proof instance offers a good example of the ATR. Figure 3 illustrates such concept. Let the red solid line be the exact set of the reachable states $T \wedge R$. Since T is encoded as a set of clauses [7], where every clause is a transition constraint, the UNSAT core can abstract T into ATR as a subset of calsues/transition constraints. In other words, with weaker constraints, such ATR can lead to a larger set of reachable states while still assures the inconsistency with the off-set $I_0 \wedge T^{k-1}$.

Let the green dashed line be the over-approximated set of reachable states (i.e. $ATR \wedge R$). As ATR implies certain absence of transition constraints, we can symbolize the

effect of the absent transitions/clauses as *dashed arrows*, that is, as certain forces to enlarge the state space of the reachable states.

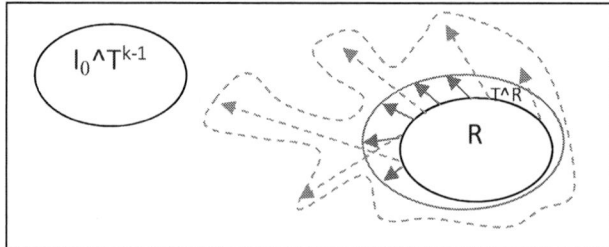

Fig. 3. The Boolean space of the UNSAT instance $I_0 \wedge T^{k-1} \wedge T \wedge R$

B. Flexible Interpolation by Reachability Partitioning

Fig. 4 illustrates a similar case to Fig. 3, except that we are considering a smaller subset of R, denoted as R_s, where s refers to *"small"*. Obviously, smaller reachability implies more room for abstraction. Therefore, the ATR for R_s can contain fewer transition constraints from the UNSAT core.

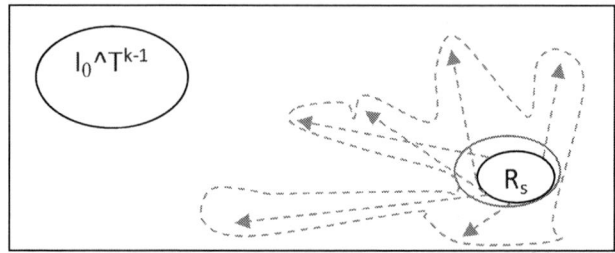

Fig. 4. The Boolean space of the UNSAT instance $I_0 \wedge T^{k-1} \wedge T \wedge R_s$

Based on the observation from the correlation between the reachability and the corresponding UNSAT core as the abstract transition relation, we propose an *Flexible Interpolation by Reachability Partitioning (FIRP)* algorithm to manipulate the degree of abstraction. As shown in Alg. 1, the input of *FIRP* is the reachability R, and the number of slices n. The output is the over-approximated pre-image of R. The key operation is partitioning R into n slices, $R_1 \sim R_n$ (line 1). The final interpolant is the disjunction of the sub-interpolants (line 5) computed from the corresponding abstract transition relation of partitioned sub-reachability R_i (line 3-4). The degree of abstraction is controlled by the input number n. If n is larger, then the final interpolant gets coarser.

FIRP(Reachability R, int n)

// n ↑ , degree of abstraction ↑
1. Partition R into R_1, R_2, ... , R_n
2. For each slices R_i
3. ATR_i = abstract transition relation from UNSAT core of BMC (R_i)
4. Itp_i = image(R_i, ATR_i)
5. return $\cup Itp_i$

Alg. 1 Flexible Interpolation by Reachability Partitioning

The following example explains how *FIRP* works. In Fig. 5, R is partitioned into R_1 and R_2. R_1 or R_2 needs fewer constraints from UNSAT proof than R does to be inconsistent with the off-set $I_0 \wedge T^{k-1}$ because the reachability becomes smaller. The disjunction of the over-approximated pre-images of R_1 and R_2 (Fig. 6) is therefore coarser than the over-approximated pre-image of R in Fig. 3.

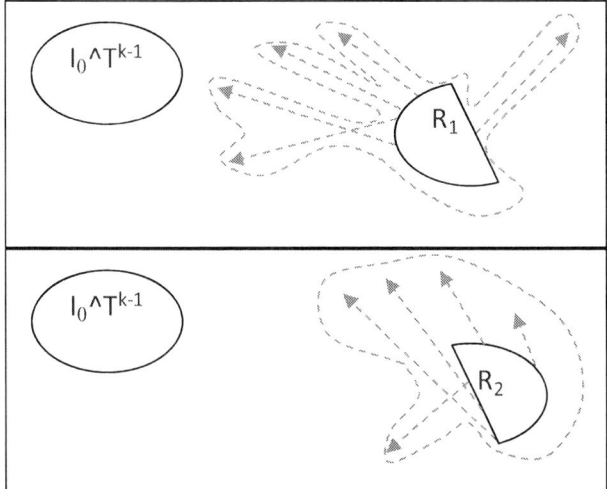

Fig. 5. The over-approximated pre-images of R1 and R2

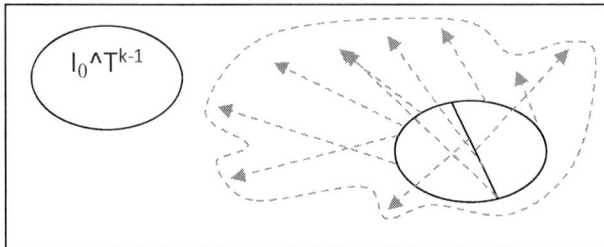

Fig. 6. The disjunction of the over-approximated pre-images of R1 and R2 from Fig. 5

C. Efficient Image Computation from Abstract Transition Relation

McMillan's IMC and NewITP are hard to adopt FIRP to be flexible for several reasons. In McMillan's IMC, the interpolant is represented with a circuit, of which the Boolean space is hard to partition into slices. In NewITP, although the interpolant is a set of cubes and is easy to partition, the interpolation process always solves on concrete transition relation for generalization. To compute the image in FIRP, we propose an efficient algorithm to compute backward image(i.e. pre-image) represented with cubes.

Fig. 7 shows our image computation algorithm, of which the efficiency is from two features: simple iterative algorithm flow and high-speed generalization with pure circuit simulation. The inputs are the reachability R and the corresponding abstract transition relation T' extracted from UNSAT core. The iterative operation in Fig. 7 can be decomposed into two steps: SAT query for $T'^{\wedge}R$ as state probing and *ATR Circuit Simulation (ATRCktSim)* as state generalization. In each iteration, a state s_i in $T'^{\wedge}R$ is returned by the SAT solver and then generalized by *ATRCktSim*, of which details are revealed soon. The generalized cubes are collected iteratively until all states that can reach R are collected and the collection of generalized cubes is the over-approximated pre-image of R.

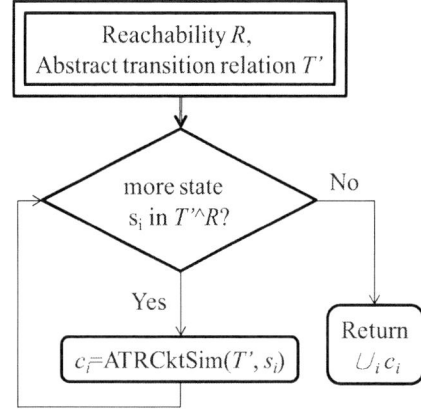

Fig. 7. Image computation by ATR Circuit

To manifest the technique of *ATRCktSim* in Fig. 7, we present the definitions of *endcoding status* and *ATR circuit*.

Definition 1. An encoding status of a gate g in the abstract transition relation T' is the presence of clauses encoding g in T'.

For example, an AND gate c = a&b is encoded in concrete transition as (c+a'+b')(c'+a)(c'+b)[7], while in an abstract transition relation, it is possible that only (c+a'+b') exists because (c'+a) and (c'+b) are not needed for UNSAT proof. In this case, the encoding status of the gate c is (c+a'+b').

Definition 2. An ATR circuit from an abstract transition relation, T', is a circuit where the encoding status of each gate in T' is recorded on the corresponding gate.

Similar with ternary simulation[8], the proposed *ATR circuit simulation* explores don't-care terms by setting the state variable as X(i.e. unknown signal). In original ternary simulation, X is blocked on a gate by a side input with controlling value and propagated through the gate if its side inputs are all non-controlling. For example, consider an AND gate c = a & b with b = X. If a=1, then X propagates to output c, makes c=X and fails in generalizing b as don't-care. In *ATR circuit simulation*, however, X may be blocked by incomplete encoding, too. Take encoding status of c as (c+a'+b')(c'+a) for example. Because the constraint "c→b" is absent, b can be X when c=1. That is, even when ab = 1X, the signal c can be 1 and b is treated as don't-care because the abstract transition relation is still satisfied. Every clause absent in the abstract transition relation allows additional X to be blocked (Take AND gate for example. See Table 1).

By taking the encoding status in abstract transition relation into consideration, ATR Circuit Simulation is a SAT&UNSAT one-way generalization method because the simulation runs on the abstract transition relation. Blocking X by side inputs with controlling value explores don't-care in the concrete transition relation as SAT generalization and blocking Xs by absence of clauses constraints in proof explores over-approximation as UNSAT generalization. By doing simulation once, we can generalize a state to an over-approximated cube. Compared with NewITP, our interpolation engine saves much time when BMC step grows

978-1-4799-2817-0/14 $31.00 © 2014 IEEE   747

because our generalization only needs simulation and the time cost for generalization is linear in the BMC step.

Encoding status of AND	X-blocking rule
(c+a'+b')(c'+a)(c'+b)	abc = 111, 0X0, X00
(c+a'+b')(c'+a)	abc = 1X1, 0X0, X00
(c+a'+b')(c'+b)	abc = X11, 0X0, X00
(c'+a)(c'+b)	abc = 111, XX0
(c'+b)	abc = X11, XX0
(c'+a)	abc = 1X1, XX0
(c+a'+b')	abc = XX1, 0X0, X00
None	abc = XX1, XX0

Table 1. X-blocking rules of an AND gate

V. Abstraction Degree Evaluation

Influencing the effectiveness of the adaptivity, abstraction degree evaluation plays an important role in our IMC (see Fig. 2). If the evaluation itself is ineffective, the adaptive mechanisms may fail in guiding the flexible interpolation for adequate abstraction. To have an effective evaluation, we propose an efficient method by taking advantage of representing reachability in cubes.

In essence, the proposed method evaluates the degree of abstraction by simply watching the number of cubes in an interpolant. Consider reachability R in Fig. 8. To contain all minterms in R, its over-approximation can be composed of finer-grained (right side) or less coarse (left side) cubes. Given an interpolant represented with cubes, if the number of cubes is large (small), we evaluate the degree as fine(coarse). In this way, the degree of abstraction can be evaluated effectively and efficiently.

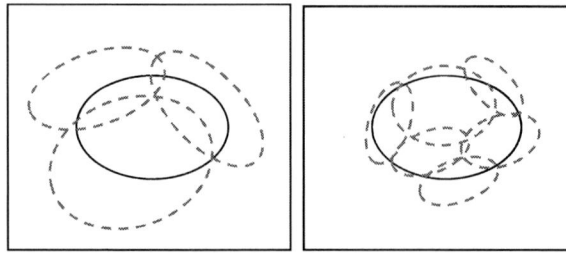

Fig. 8. An reachability (central circle) and its over-approximations in different degree of abstraction

VI. Experimental Results

Our experiment runs on Intel(R) Xeon(R) CPU E5405, 2.00GHz. The Number of available CPU core is 1. The memory limit is 7GB and time limit is 15 minutes, just as Hardware Model Checking'12 (HWMCC'12) [9]. The testcase suit is 'hwmcc11nointel.7z' downloaded from HWMCC official website, which contains 405 instances. Our experiment compares our adaptive IMC to conventional McMillan's IMC and NewITP. We let McMillan's ITP compute forward over-approximated image from initial states for fair comparison since McMillan's ITP works better in this way instead of computing backward over-approximated image from the property.

To demonstrate the adaptivity, the implementation of our adaptive algorithm in the experiment sets the degree of abstraction as finest for the first call of ITP phase. That is, the number of partitioned slices on reachability is 1, just like the traditional McMillan's method. If our algorithm is not adaptive, it may be worse than McMillan's ITP since our interpolation technique has to generate cubes in one interpolation process. With the flexibility of interpolation, we expect the adaptivity to adjust the degree of abstraction and lead IMC to prove more instances.

A. Comparison of the Number of Solved Instances

Fig. 9 shows our experimental results, where the x-axis is run-time and the y-axis is the number of solved instances. Our algorithm, adaptive IMC (AIMC), solves 255 instances, while McMillan's ITP solves only 217 and NewITP solves only 244 instances.

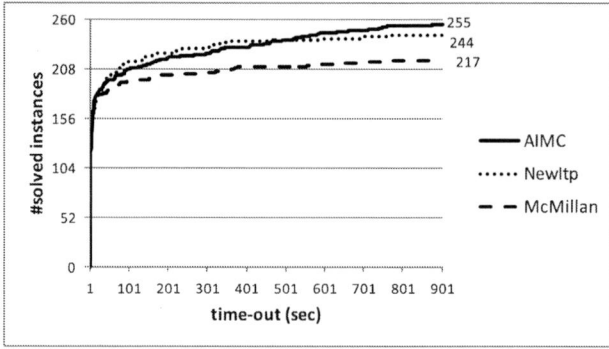

Fig. 9. The accumulated solved instance with respect to runtime

McMillan's ITP can solve easy instances fast. Nevertheless, most of the instances unsolved in early BMC steps are still unsolved after the several refinements of BMC step. In consequence, McMillan's ITP hardly solves new instances after 300seconds.

Although NewITP solves much more instances than McMillan's ITP, the refinement of NewITP is still inefficient. We can see that the easier instances can be solved faster by NewITP than by AIMC because in this experiment, our degree of abstraction is set as McMillan's IMC. However, NewITP hardly solves new instances after 300 seconds, too. The reason is the guidance from counterexamples only refines the local Boolean space of generalized cubes. Once the granularity is not adequate for the instance, NewITP hardly proves it even with enough run-time for several refinements of BMC steps and guidance from counterexamples. This result indicates these two refinements are ineffective and inadequate.

In addition to BMC step and guidance from spurious counterexamples, our adaptive IMC refines the degree of abstraction directly. Therefore, the number of instances solved by our adaptive IMC grows steadily with run-time since the adequate abstraction is found by the adaptive mechanisms. Eventually, our adaptive IMC computes adequate abstraction and solves the most instances.

B. Statistics Analysis

Table 2. shows the number of solved and unsolved instances by IMC algorithms. In addition to 116 instances which are too hard for interpolation-based method and 179 trivial instances solved by all IMCs in 405 instances, we analyze the result for the remaining 110 instances. The row "Solved only" shows the number of instances solved only by the respective IMCs. Because the adequate abstraction is found by our adaptive mechanisms, our method solves 20 instances which are not solved by the other two previous works. NewITP solves only 14 and McMillan's IMC solves only 7 instances which cannot solve by the other two IMCs. The row "Unsolved only" shows the number of instances unsolved only by the respective IMCs. Our adaptive IMC cannot solve only 13 instances which are solved by one of previous works. The statistics in these two rows show Our proposed adaptive IMC explores more granularities than previous works to solve 20 instances hard for previous IMCs and keep good quality when solving those solved by NewITP or McMillan's IMC.

405 instances in total			
	AIMC	NewITP	McMillan
All Solved	179		
Solved only	20	14	7
Unsolved only	13	18	38
All Unsolved	116		
100 instances unsolved by PDR			
	AIMC	NewITP	McMillan
Solved	15	7	12

Table 2. Detailed number of experimental statistics

C. Comparison of Test Instances Hard for PDR[8][10]

PDR is another kind of model checking, and it is known as the most powerful single engine which can solve 305 of 405 instances. We extract the remaining 100 instances and solve them by IMCs. In the last row of Table 2, the statistics show the numbers of instances which are unsolved by PDR while solved by receptive IMCs. While NewITP solves more instances in total than McMillan's ITP, it solves only 7 instances which PDR cannot solve. Because MaMillan's IMC is fundamentally different from PDR, MaMillan's IMC can solve 12 instances in this category, even more than NewITP. NewITP uses generalization techniques in PDR without the incremental refinement on reachability. When finding counterexamples, NewITP does not refine the reachability but turns to BMC instead. The method itself and the result indicate NewITP is actually similar to PDR while weakened refinement from counterexamples makes NewITP unable to compete with PDR. In consequence, PDR almost dominates the advantage of NewITP.

On the other hand, our adaptive IMC solves 15 instances that PDR cannot solve. Our flexible interpolation engine considers the abstract transition relation for reachability and our IMC is adaptive, while PDR does not have these two features. That makes our method solve 15 more instances. Compared to NewITP, our method solves more instances while keeps the feature of conventional IMC algorithm flow. As a result, we solve the most instances hard for PDR.

VII. Conclusion

Based on the observation that abstraction can be achieved by reachability partitioning in IMC, we have proposed an adaptive IMC algorithm. With effective and efficient evaluation of degree of abstraction, our adaptive IMC refines the granularity of the abstraction directly in an efficient and effective way. The experimental results shows that our adaptive IMC outperforms the best existing IMC algorithm and explores the possibility to solve hard instances for interpolation-based model checking.

This work is partially supported by the National Science Council of Taiwan ROC under Grant No. NSC-102-2221-E-002 -233-.

Reference

[1] K. McMillan. Interpolation and sat-based model checking. In *Computer Aided Verification*, pages 1--13. Springer, 2003.

[2] Biere, A. Cimatti, E. Clarke, and Y. Zhu. Symbolic model checking without bdds. *Tools and Algorithms for the Construction and Analysis of Systems*, pages 193--207, 1999.

[3] W. Craig. Linear reasoning. a new form of the herbrand-gentzen theorem. *Journal of Symbolic Logic*, pages 250--268, 1957.

[4] P. Pudlák. Lower bounds for resolution and cutting plane proofs and monotone computations. *Journal of Symbolic Logic*, pages 981--998, 1997.

[5] C. Y. Wu, C. A. Wu, C. Y. Lai, and C. Y. Huang, and W.-L. Hung. A counterexample-guided interpolant generation algorithm for SAT-based model checking, In *Design Automation Conference (DAC), 2013 50th.*

[6] Edmund M. Clarke, Jr. , Orna Grumberg , Doron A. Peled, Model checking, MIT Press, Cambridge, MA, 2000

[7] G. S. Tseitin. On the complexity of derivation in propositional calculus. In J. Siekmann and G. Wrightson, editors, *Automation of Reasoning 2: Classical Papers on Computational Logic 1967-1970*, pages 466--483. Springer, Berlin, Heidelberg, 1983.

[8] N. Een, A. Mishchenko, and R. Brayton. Efficient implementation of property directed reachability. In *Formal Methods in Computer-Aided Design (FMCAD), 2011*, pages 125--134. IEEE, 2011.

[9] Hardware Model Checking Competition. *http://fmv.jku.at/hwmcc12/benchmarks.html.*

[10] A. Bradley. Sat-based model checking without unrolling. In *Verification, Model Checking, and Abstract Interpretation*, pages 70--87. Springer, 2011.

[11] N. Eén and N. Sörensson. An extensible sat-solver. In *Theory and Applications of Satisfiability Testing*, pages 333--336. Springer, 2004.

[12] M. Sheeran, S. Singh, and G. Stålmarck. Checking safety properties using induction and a sat-solver. In *Formal Methods in Computer-Aided Design*, pages 127--144. Springer, 2000.

Efficient Parallel GPU Algorithms for BDD Manipulation[*]

Miroslav N. Velev

Aries Design Automation, LLC
Chicago, IL, U.S.A.

e-mail: miroslav.velev@aries-da.com

Ping Gao

Aries Design Automation, LLC
Chicago, IL, U.S.A.

Abstract—We present parallel algorithms for Binary Decision Diagram (BDD) manipulation optimized for efficient execution on Graphics Processing Units (GPUs). Compared to a sequential CPU-based BDD package with the same capabilities, our GPU implementation achieves at least 5 orders of magnitude speedup. To the best of our knowledge, this is the first work on using GPUs to accelerate a BDD package.

Keywords—Binary Decision Diagrams (BDDs); Boolean Satisfiability; Formal Verification; Graphics Processing Unit (GPU); Parallel Execution.

I. Introduction

Since their invention, Binary Decision Diagrams (BDDs) [4, 5] have been used in a wide range of computer science applications, and have had a significant impact on the field of Electronic Design Automation (EDA). Previous efforts to parallelize BDD packages have been based on CPU execution, e.g., [2, 8, 11, 16]. However, Graphics Processing Units (GPUs) have had at least an order of magnitude performance and memory bandwidth advantage relative to CPUs so far [6], and that will continue by Moore's law [13, 15].

GPU-based processing of a variant of BDDs, Zero-Suppressed BDDs (ZBDDs), was presented by Minato et al. [12], who reported a speedup of 6 times relative to sequential execution on a CPU. However, they do not give many implementation details. Other authors explored the acceleration of Boolean Satisfiability (SAT) solvers on GPUs, and reported speedups of 1 order of magnitude for random instances [7, 10], but up to 3.4× for real-world formulas [10].

The contribution of this paper are efficient algorithms for BDD manipulation on GPUs, including efficient implementation of recursion for the ITE operator, as well as exploiting cuckoo hash tables for parallel hashing on the GPU. Compared to a sequential CPU-based BDD package with the same capabilities, our GPU implementation achieves at least 5 orders of magnitude speedup, which is increasing with the complexity of the Boolean formulas. To the best of our knowledge, we present the first work on using GPUs to accelerate a BDD package.

II. Background

A. Binary Decision Diagrams

Binary Decision Diagram (BDD) is a directed acyclic graph (DAG). The graph has two sink nodes labeled 0 and 1, representing the Boolean functions 0 and 1. Each non-sink node is labeled with a Boolean variable v and has two out-edges labeled 1 (or then), and 0 (or else), pointing to the 1-child and 0-child BDD nodes, respectively. Each non-sink node represents the Boolean function corresponding to its 1 edge if $v = 1$, or the Boolean function corresponding to its 0 edge if $v = 0$.

An Ordered Binary Decision Diagram (OBDD) is a BDD with the constraint that the input variables are ordered and every source to sink path in the OBDD visits the input variables in ascending order. A reduced ordered binary decision diagram (ROBDD) is an OBDD where each node represents a distinct logic function. Bryant [4] was the first to prove that the ROBDD is well-defined. Bryant also showed the ROBDD is a canonical form for a logic function, i.e., two functions are equivalent if, and only if, the ROBDDs for each function are isomorphic. It is well known that the size of the ROBDD for a given function depends on the variable order chosen for the function. This paper is not concerned with the variable ordering problem.

As in previous BDD packages, we exploit the observation that all operations on two argument Boolean functions can be represented as an instance of the ITE operator with a suitable arrangement of the two Boolean functions or their negations, and the sink nodes 0 and 1. For example, AND(F, G), where F and G are Boolean functions, can be computed as ITE(F, G, 0); OR(F, G) as ITE(F, 1, G); NOT(F) as ITE(F, 0, 1); and XOR(F, G) as ITE(F, NOT(G), G).

Fig. 1 presents the pseudocode for an implementation of the ITE operator for sequential CPU execution [3]. The ITE function takes three arguments, which are pointers to BDDs representing the if-, then-, and else-Boolean functions, and returns a pointer to BDD for the result, ITE(I, T, E).

---

[*] Research partially funded by NASA under various contracts.

978-1-4799-2817-0/14 $31.00 © 2014 IEEE

```
bdd_ptr ITE(bdd_ptr I, bdd_ptr T, bdd_ptr E) {
if (is_terminal_case(I, T, E))
  return (result_terminal_case(I, T, E));
else {
  p = findHash(computed_hash, I, T, E);
  if (p) return (p);
  else {
    RootVar = getSmallestRootVar(I, T, E);
    PosFactor = ITE(RESTRICT(I, RootVar, 1),
                    RESTRICT(T, RootVar, 1),
                    RESTRICT(E, RootVar, 1));
    NegFactor = ITE(RESTRICT(I, RootVar, 0),
                    RESTRICT(T, RootVar, 0),
                    RESTRICT(E, RootVar, 0));
    if (PosFactor == NegFactor)
      return (PosFactor);
    p = findOrCreateBDD(RootVar,
                    NegFactor, PosFactor);
    insertHash(computed_hash, I, T, E, p);
    return (p);
}}}
```

Fig. 1. The ITE operator for sequential CPU execution.

The algorithm starts by checking if a terminal case applies, where the result can be easily computed given the values for the I, T, and E arguments; if so, the result is returned. Else, the algorithm checks the computed hash table, `computed_hash` (that maps previously computed ITE operators and their arguments to their results) if the results has already been computed, and if so returns it. Else, the smallest root variable for the three arguments is determined, `RootVar`, and the positive and negative cofactors of the original ITE operator are computed. (This is based on Shannon's decomposition theorem.) If they are equal, one of them is returned as the result. Otherwise, function `findOrCreateBDD()` is used to check a unique hash table if the new BDD node has been created—if so, the pointer to it is returned; if not, the BDD node is created, and the information for it is inserted in the unique hash table. The result is also inserted in the computed hash table to avoid future recomputations, and is returned.

B. GPU Architecture

Programming NVIDIA's Graphics Processing Units (GPUs) for general-purpose applications is enabled through the Compute Unified Device Architecture (CUDA) [14]. It allows a programmer to use a GPU's thousands of cores that can collectively run millions of threads. Each core, called a Scalar Processor (SP), belongs to a set of SPs that collectively form a *multiprocessor* on the same device. The SPs in a multiprocessor share resources, such as registers and memory. The on-chip shared memory allows the parallel tasks running on the SPs to share data without the need of sending it over the system memory bus.

CUDA provides an API as a set of library functions that extend the standard ANSI C. The programming model is based on an abstraction of the GPU parallel architecture, using a minimal set of programming constructs, such as hierarchy of threads, hierarchy of memories, and synchronization primitives. A compiler generates executable code for the GPU. A CUDA program comprises of a host program which is executed on the CPU or host, and a set of CUDA kernels that are launched from the host program on the GPU device. A *kernel* is a function that is executed in parallel as a set of threads, each operating on different data. The threads are organized into groups called *blocks*. The threads within a block synchronize among themselves through barrier synchronization primitives in CUDA, and communicate through a shared memory space that is available to the block. A kernel comprises of a *grid* of one or more blocks—see Fig. 2. Each thread in a block is uniquely identified by its thread id (threadIdx) within its block, and each block is uniquely identified by its block id (blockIdx). Each CUDA thread has access to various memories at different levels in the hierarchy. The threads have a private local memory space and register space. The threads in a block use a shared memory space. The GPU DRAM is accessible by all threads in a kernel. The threads of a block are executed on the same multiprocessor in groups of 32 threads called *warps*.

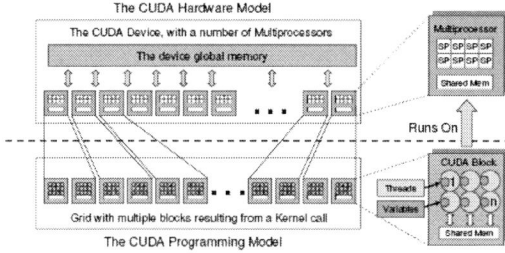

Fig. 2. CUDA hardware model (top), programming model (bottom), and block-to-multiprocessor mapping.

C. Cuckoo Hash Tables for Efficient Parallel Hashing on the GPU

As seen from the sequential implementation of the ITE operator in Fig. 1, critical building blocks are the two hash tables, the computed and unique hash tables. An efficient GPU-based implementation of the ITE operator requires efficient GPU implementations of the two hash tables with support for many parallel accesses, which are necessary in a breadth-first recursive processing of an ITE operator. For this purpose, we use a modified version of GPU-based cuckoo hashing. The original GPU implementation of cuckoo hashing—see Chapter 4 of [9] and [1]—is summarized next, and our modified version is presented in Sect. III.C.

In a GPU implementation of cuckoo hashing, a hash table is represented as an array of 64-bit unsigned integers in the memory of a GPU. Half of each element stores a 32-bit key, and the other half a 32-bit value. The

978-1-4799-2817-0/14 $31.00 © 2014 IEEE   751

efficiency is based on the restriction that a (`key`, `value`) pair can be placed in one out of a limited set of locations in the array, where each such location is computed by application of a different hash function to the `key` of the pair. The two main operations of a GPU implementation of cuckoo hashing are: 1) parallel search of multiple keys that may contain duplicates; and 2) parallel insertion of multiple (`key`, `value`) pairs that do not contain duplicate keys. The array is initialized with pairs (`EMPTY_KEY`, 0), where `EMPTY_KEY` is a 32-bit integer that is reserved and is used only to indicate that an array element does not contain a valid (`key`, `value`) pair. We call this array a cuckoo array.

The parallel search of multiple keys is implemented with a GPU kernel that is run in parallel for all keys. Each invocation of the kernel processes one key by computing all locations in the array that the key may map to by means of the set of hash functions, and searching those locations one by one to find the first location that contains a key equal to the one searched for by this kernel invocation. If such a location is found, then its (`key`, `value`) pair is returned, or else the pair (`EMPTY_KEY`, 0) is returned.

The parallel insertion of multiple (`key`, `value`) pairs that do not contain duplicate keys is done with another GPU kernel that is run in parallel for all such pairs. Each invocation of the kernel processes one (`key`, `value`) pair by computing the first location in the array where the pair can be placed, and then performing an atomic exchange operation, `atomicExch()`, to swap the given (`key`, `value`) pair with the one that was previously at that location, (`old_key`, `old_value`) pair. If `old_key` is different from `EMPTY_KEY`, then the location had been updated and contained useful information. The old pair (`old_key`, `old_value`) is then atomically exchanged with the pair at the next location for `old_key`. This is repeated up to a fixed number of times, `INSERTION_ATTEMPTS` until an empty location is found, which means that all replaced elements were rearranged successfully. If no empty location is found within the given limit of such atomic exchanges, then the algorithm signals failure, which can be corrected in several ways or combinations thereof: 1) increasing the number of hash functions that map each key to a location in the array; 2) modifying these hash functions; 3) increasing the value of `INSERTION_ATTEMPTS`; 4) using a larger array; or 5) rebuilding the array by reinserting the initial (`key`, `value`) pairs in a different permutation allowed by the hash functions.

III. Efficient GPU-Based Algorithms for BDD Manipulation

A. BDD Node Representation in the GPU Memory

To achieve scalability for large BDDs in our GPU-based parallel BDD package, we made several design decisions. First, we decided to represent the data in the GPU in sets of arrays, as opposed to data structures that are allocated dynamically. This would save time from many allocations. Furthermore, this would allow us to split the arrays into multiple disjoint sections that can be either placed in the memories of different GPU cards on the same host computer, or situated in their entirety in the memory of the host computer with only the active sections of the arrays moved to/from GPU cards when needed to perform computations on active sections of large BDDs (this functionality will be implemented later).

Second, because of the representation of the data structures with sets of arrays, we decided to identify each BDD node with its index of the elements in the arrays that store the information for that BDD node, as opposed to with a pointer to a BDD data structures as done in a typical BDD package for sequential execution on a CPU (see Sect. II.A). Using indices will simplify the implementation of partitioned arrays, where sections of the same large array may be situated in the memories of different GPU cards for efficient GPU-based parallel processing. Note that if multiple GPU cards are used, a value of a pointer to a data structure in the memory of one GPU card would also exist as a location in the memory of another GPU card, and that would complicate the distributed processing of large BDDs by using several GPU cards (this will be implemented later).

We used the following arrays of elements of type `unsigned int` to store the BDD nodes on the GPU: `g_bdd_variable`, `g_bdd_uid_low`, and `g_bdd_uid_high`, where elements with the same index contain the decision variable, 0-child, and 1-child of a BDD node, respectively. We call such an index a "uid" (for unique identifier) of a BDD node.

B. Representing the Recursion for Computing ITEs

To efficiently implement recursion for the ITE operator on the GPU, we modeled a recursion stack as a collection of six arrays, as well as separate GPU kernels `ITE_down` and `ITE_up` that process the ITE operators at one level of the recursion in the direction, respectively, down towards the leaves of the recursion (where each ITE operator can be computed as a terminal case, based on its arguments), and up towards the result given the results for all ITE operators at lower levels of recursion.

The six arrays included five arrays of type `unsigned int`, where the elements with the same index in these arrays represent, respectively, the uid for the I, T, and E arguments of an ITE operation, as well as the 0-child and the 1-child. The elements in the last two arrays contain either an index to an element of the recursion stack that represents the corresponding child ITE operation at a lower recursion level (where the arguments I, T, and E are the negative and positive cofactors, respectively, of the arguments of the current ITE operator), or the uid of the result BDD, as returned by the computed hash table (see Sect. III.C). The sixth array, `valid`, is of type `unsigned char`, and is used for bookkeeping, such as to indicate a valid ITE operator that has to be computed, and what information is

978-1-4799-2817-0/14 $31.00 © 2014 IEEE   752

contained in the 0-child and 1-child. Additionally, we used two arrays to record, respectively, the number of ITE operators in each level of recursion, and the location of the first of these ITE operators on the recursion stack, with the rest situated in consecutive locations. In our prototype implementation, all arrays had fixed sizes that were set at compile time, and were given values that were sufficiently large to allow all experiments to complete.

The GPU kernel `ITE_down` is applied in parallel to all ITE operations at the same level of recursion. This results in breadth-first traversal of the recursion space, which exposes significantly more GPU parallelism, compared to depth-first traversal. The initial recursion level has only the ITE operator that is to be computed. The kernel checks if the argument BDDs for I, T, and E represent a terminal case, and if so computes the result and stores it in the 0-child element for that ITE operation. Else, computed are the positive and negative cofactors of the I, T, and E arguments. The computed hash table is checked for availability of results for the ITE operations with arguments the positive and negative cofactors, respectively, and if found those results are placed in the 1-child and 0-child elements, respectively, with corresponding bookkeeping done on element `valid`. Else, the cofactors of the I, T, and E arguments that were not found in the computed hash table are placed as arguments for new ITE operations in the next level of recursion, with corresponding bookkeeping done for the current ITE operation. In the final level of recursion, all ITE operations at that level are computed as terminal cases of their corresponding arguments.

The results are composed by applying the GPU kernel `ITE_up` in parallel to all ITE operations at the same level of recursion, when the results for all lower levels of recursion have been computed. This kernel identifies duplicate BDD candidates from each recursion level, and ensures that only one of them is inserted in the unique hash table.

C. Modified Cuckoo Hashing

We implemented an extension of the cuckoo hashing scheme from Chapter 4 of [9] (see Sect. II.C)—where the original algorithm works on pairs consisting of a 32-bit key and a 32-bit value—in order to adapt it to our application, where the keys have 3 fields, corresponding to the unique identifiers for the I, T, and E arguments for an ITE operator in the case of the computed hash table, and the decision-variable index, 0-child, and 1-child in the case of the unique hash table. Namely, we used the cuckoo arrays to store signature-value pairs (instead of key-value pairs), where the signature and the value are of 32 bits each as in the original cuckoo hashing approach, except that the signature is computed from the complete key consisting of its three 32-bit integer parts, and the value is an index to the elements in the

four arrays that contain the complete key and the actual value, respectively.

Thus, the search is done by checking the signature-value pairs stored in the limited number of locations in the cuckoo array that correspond to that 32-bit signature (we experimented with 4 and 5 hash functions that map each signature to 4 and 5 locations, respectively). If the 32-bit signature at one of these locations equals the 32-bit signature of the complete key for the search, then we check for equality between the three parts of the complete key for the search, and the three parts of the complete key of the hashed element that is located in the arrays with the complete information and is pointed to by the 32-bit value extracted from the found element in the cuckoo array. The signature of a 3-part key is computed in the following way for the computed cuckoo hash table, given the 32-bit unique identifiers for the BDDs of the I, T, and E arguments of an ITE operator:

```
temp_key = (I*6000011 + T)*6000023 + E;
key = (unsigned int) temp_key;
```

where the two constants, 6000011 and 6000023, are primes that are greater than the maximum number of BDDs of 5000000 that we had sized the arrays in our code for.

In the above, `temp_key`, is a 64-bit unsigned integer, and `key` is a 32-bit unsigned integer that represents the signature of the complete key. Note that aliasing in the cuckoo array in our implementation—different complete keys evaluating to the same signature, and/or different signatures mapping to the same location—will not result in returning wrong values, because if the signatures match, then we check if the complete key stored for that element also matches the complete key for the search, and only then return the corresponding value. However, aliasing results in longer search times.

IV. Results

We compared our GPU-based BDD package with a base CPU BDD package (see Sect. II.A) that is executed sequentially on a single-core of the CPU, and uses function `malloc()` to dynamically allocate memory for new data structures, including BDD nodes, and then uses pointers to identify each BDD node, as done in a typical BDD package. Both packages have comparable capabilities in that they implement all Boolean operators in terms of the ITE operator, and are otherwise each a minimal implementation without optimizations such as negated pointers [3], garbage collection, etc.

The comparison was done on gate-level implementations of a ripple-carry adder that we generated for width of the two operands and the result of between 8 and 8,192 bits. The experiments were conducted using a single NVIDIA Tesla K20 card with 5 GB of memory on a workstation with a 3.47-GHz Intel Xeon W3690 CPU and 24 GB of 1,333-MHz ECC memory, running Red Hat Enterprise Linux v6.3. For the GPU-based experiments, we used NVIDIA CUDA v5. The results are shown in Table 1.

Table 1. Experimental results from building the BDDs that represent the functionality of ripple-carry adders with between 8 and 8,196 bits for each of their two operands and result. As can be seen from the results for the 21-bit adder, the GPU-based BDD package results in 5 orders of magnitude speedup. Furthermore, the speedup is increasing with the size of the benchmarks.

Number of bits in each operand and the result	BDD Statistics				Time [s]		Speedup
	BDD Variables	BDD nodes in most significant bit of carry	BDD nodes in most significant bit of sum	Total BDD nodes	Base CPU BDD package	GPU-based BDD package	
8	16	25	45	117	0.003	0.573	0.005×
16	32	49	93	245	33.9	0.614	55×
17	34	52	99	261	225	0.610	369×
18	36	55	105	277	1,133	0.638	1,776×
19	38	58	111	293	4,959	0.652	7,606×
20	40	61	117	309	20,560	0.650	31,631×
21	42	64	123	325	83,774	0.657	**127,510×**
32	64	97	189	501	————	0.712	————
64	128	193	381	1,013	————	1.1	————
128	256	385	765	2,037	————	1.4	————
256	512	769	1,533	4,085	————	2.2	————
512	1,024	1,537	3,069	8,181	————	3.6	————
1,024	2,048	3,073	6,141	16,373	————	6.4	————
2,048	4,096	6,145	12,285	32,757	————	12.4	————
4,096	8,192	12,289	24,573	65,525	————	24.7	————
8,192	16,384	24,577	49,149	131,061	————	50.8	————

As can be seen from the table, the formulas required: between 16 and 16,384 BDD variables; between 117 and 131,061 total BDD nodes. The most significant bit of the carry was represented with between 25 and 24,577 BDD nodes, and the most significant bit of the sum with between 45 and 49,149 BDD nodes. The longer execution time of the GPU-based BDD package on the adder with 8-bit operands and result, compared to the base BDD package, is due to the overhead for creating large arrays on the GPU before the actual execution begins.

Using our GPU-based BDD package, we achieved **at least 5 orders of magnitude (127,510×) speedup** for the adder with 21-bit operands and result, compared with the base BDD package, executed on a single-core of the same CPU. Furthermore, the speedup is increasing with the size of the Boolean formula.

The experiments were conducted with 4 hash functions for mapping of a 32-bit signature (of a key consisting of three 32-bit integer parts) to a location in the cuckoo arrays for the computed and unique hash tables. Increasing the number of hash functions to 5 did not result in addition-al speedup. We found that setting parameter INSERTION_ATTEMPTS to 120, as used in the insertion kernel for the cuckoo arrays of both the computed and unique hash tables, allowed all experiments to complete.

V. Conclusion

We presented parallel algorithms for BDD manipulation optimized for efficient execution on Graphics Processing Units (GPUs). Compared to a sequential CPU-based BDD package with the same capabilities, our GPU implementation achieves at least 5 orders of magnitude speedup. Furthermore, the speedup is increasing with the size of the Boolean formula. To the best of our knowledge, this is the first work on using GPUs to accelerate a BDD package. Our future work will focus on fine-tuning the presented implementation.

978-1-4799-2817-0/14 $31.00 © 2014 IEEE

References

[1] D.A.F. Alcantara, *Efficient Hash Tables on the GPU*, Ph.D. thesis, Department of Computer Science, University of California at Davis, 2011.

[2] F. Bianchi, F. Corno, M. Rebaudengo, M. Sonza Reorda, and R. Ansaloni, "Boolean Function Manipulation on a Parallel System Using BDDs," *High-Performance Computing and Networking*, LNCS 1225, Springer, 1997, pp. 916–928.

[3] K.S. Brace, R.L. Rudell, and R.E. Bryant, "Efficient Implementation of a BDD Package," *Design Automation Conference (DAC'90)*, June 1990, pp. 40–45.

[4] R.E. Bryant, "Graph-Based Algorithms for Boolean Function Manipulation," IEEE Transactions on Computers, Vol. C-35, No. 8 (1986), pp. 677–691.

[5] R.E. Bryant, "Symbolic Boolean Manipulation with Ordered Binary-Decision Diagrams," ACM Computing Surveys, Vol. 24, No. 3 (1992), pp. 293–318.

[6] J.F. Croix, and S.P. Khatri, "Introduction to GPU Programming for EDA," *International Conference on Computer-Aided Design (ICCAD'09)*, November 2009, pp. 276–280.

[7] K. Gulati, and S.P. Khatri, "Boolean Satisfiability on a Graphics Processor," *Great Lakes Symposium on VLSI (GLS-VLSI'10)*, May 2010, pp. 123–126.

[8] Y. He, "Multicore-Enabling a Binary Decision Diagram Algorithm," 2009. Available from: http://software.intel.com/en-us/articles/Multicore-enabling-a-Binary-Decision-Diagram-algorithm

[9] W.-M.W. Hwu, *GPU Computing Gems Jade Edition*, Morgan Kaufmann, 2011.

[10] P. Manolios, and Y. Zhang, "Implementing Survey Propagation on Graphics Processing Units," *International Conference on Theory and Applications of Satisfiability Testing (SAT '06)*, August 2006, pp. 311–324.

[11] K. Milvang-Jensen, and A.J. Hu, "BDDNOW: A Parallel BDD Package," *Formal Methods in Computer-Aided Design (FMCAD '98)*, LNCS 1522, Springer, November 1998, pp. 501–507.

[12] S.-I. Minato, M. Onsjo, and O. Watanabe, "Faster Evaluation of ZBDD Compressed Multi-Linear Functions with GPU Parallelism," Research Report C-274, Dept. of Math. & Comp. Science, Tokyo Institute of Technology, January 2011.

[13] G.E. Moore, "Cramming More Components onto Integrated Circuits," Electronics, Vol. 38, No. 8, April 1965.

[14] *NVIDIA CUDA Architecture: Introduction & Overview.* http://developer.download.nvidia.com/compute/cuda/docs/CUDA_Architecture_Overview.pdf

[15] SOA News Desk, "Moore's Law: 'We See No End in Sight,' Says Intel's Pat Gelsinger," May 2008. http://www.sys-con.com/node/557154

[16] T. Stornetta, and F. Brewer, "Implementation of an Efficient Parallel BDD Package," *Design Automation Conference (DAC '96)*, June 1996, pp. 641–644.

978-1-4799-2817-0/14 $31.00 © 2014 IEEE

Efficient Techniques for the Capacitance Extraction of Chip-Scale VLSI Interconnects Using Floating Random Walk Algorithm

Chao Zhang   Wenjian Yu

Tsinghua National Laboratory for Information Science and Technology
Department of Computer Science & Technology
Tsinghua University, Beijing 100084, China
Tel : (8610)-62773440
e-mail : eric.3zc@gmail.com, yu-wj@tsinghua.edu.cn

Abstract – To enable the capacitance extraction of chip-scale large VLSI layout using the floating random walk (FRW) algorithm, two techniques are proposed. The first one is a virtual Gaussian surface sampling technique. It makes efficient random sampling on the Gaussian surface for complex nets with vias, and optimizes the sampling scheme to reduce the time of random walk. The other one is a parallelized, improved construction approach for Octree based space management structure. It can be over 5000X faster than the existing approach and provides same convenience to the FRW procedure. Numerical experiments on large cases with up to half million conductors validate the proposed techniques, and demonstrate a fast FRW solver for chip-scale extraction task.

I. Introduction

With the increase of the density of integrated circuit (IC), capacitances among the interconnects in IC have the growing impact on circuit performance. So, accurately and quickly extracting the interconnect capacitances is more and more important. The three-dimensional (3-D) field solver which directly solves the electrostatic field possesses the highest degree of accuracy. Despite the boundary element method (BEM) based field solvers [1-3] have been proposed, they are only able to handle small structures due to the bottleneck of runtime and memory. Differing from these deterministic methods, the floating random walk (FRW) algorithm was also proposed for capacitance extraction [4-9]. It is based on the Monte Carlo (MC) method, and on the field solver level of accuracy. Compared with the deterministic solvers, the FRW algorithm has the advantages of using less memory, better parallelism and tunable accuracy. More importantly, it is suitable for the chip-scale large interconnect structures, since its cost of computing time barely depends on the number of conductors involved.

The FRW algorithm has been developed into commercial software (such as *QuickCap*), despite with approximating approaches to adapt to complex VLSI structures and obtain high efficiency. An efficient FRW algorithm (called *RWCap*) was proposed to accurately extract the multi-dielectric VLSI structures, by pre-characterizing the transition domain including multiple dielectric layers and quickening the convergence of MC procedure with a comprehensive variance reduction scheme [6]. Efficient techniques were also proposed

---

This work was supported in part by NSFC under Grant 61076034, the Beijing Natural Science Foundation under Grant 4132047, the Opening Foundation of ASIC and System State Key Laboratory (Fudan University, No. 12KF009), and the Tsinghua University Initiative Scientific Research Program.
*W. Yu is the corresponding author.

to parallelize the FRW based capacitance extraction on GPUs [9]. However, only medium-scale interconnect structures was tested in [6], and the technical details of the FRW algorithm for chip-scale large structures have not yet been reported in literature. In actual scenario, the distributed capacitances for a net (a number of connected interconnect wires) are needed. The environment of the nets, which constitutes chip-scale large structures, should also be handled to preserve high accuracy. This brings challenges to existing FRW algorithms, for the difficulties of handling the Gaussian surface of a whole net and millions of conductor blocks while executing the FRW procedure.

In this paper, two techniques are proposed to make the FRW algorithm effective for the chip-scale capacitance extraction task. An efficient virtual Gaussian surface sampling (VGSS) technique for a whole net with vertical vias is firstly proposed. It is versatile, and incorporates with the importance sampling on the Gaussian surface. The approach to optimize the placement of Gaussian surface for the FRW efficiency is also discussed. Then, an improved Octree construction algorithm is proposed to largely reduce the time for handling a large quantity of conductor blocks.

Experiments are carried out on several large scale VLSI layouts, and distributed full-net capacitances are extracted with 1% 1-σ error of total capacitance. Experimental results validate the effectiveness of the VGSS technique, and reveal that the optimized placement and sampling scheme of Gaussian surface could bring **1.4X** speedup to the FRW procedure. Compared with the existing approach, the improved Octree construction algorithm is over **5000X** faster. For a layout with 3037 nets including vias, the enhanced FRW solver is able to accomplish the extraction of all nets in about half an hour, on an 8-core CPU machine.

II. Background

The electric potential of a point r can be expressed as an integral of the potential on the surface S surrounding it:

$$\phi(r) = \oint_S P(r, r^{(1)})\phi(r^{(1)})dr^{(1)}, \qquad (1)$$

where $\phi(r)$ is the potential of point r and $P(r, r^{(1)})$ is called the surface Green's function. The domain enclosed by S is often called the transition domain. $P(r, r^{(1)})$ is non-negative for any point $r^{(1)}$ on S, and can be regarded as the probability density function (PDF) for selecting a random point on S. With the principle of Monte Carlo (MC) simulation, $\phi(r)$ can be estimated as the statistical mean of $\phi(r^{(1)})$.

To compute the capacitances related with conductor i (called master conductor), a Gaussian surface G is firstly constructed to enclose it (see Fig. 1). According to the

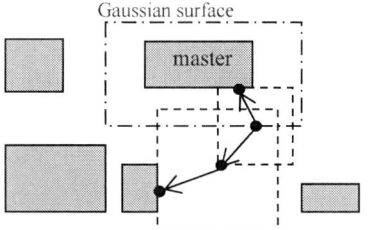

Fig. 1. Two examples of the random walk in the FRW algorithm for capacitance extraction (a 2-D cross-section view).

Gaussian theorem, charge Q of conductor i becomes

$$Q = \oint_G F(\boldsymbol{r}) \int_{S^{(1)}} \omega(\boldsymbol{r}, \boldsymbol{r}^{(1)}) q(\boldsymbol{r}, \boldsymbol{r}^{(1)}) \phi(\boldsymbol{r}^{(1)}) d\boldsymbol{r}^{(1)} d\boldsymbol{r} \quad . \quad (2)$$

where $F(\boldsymbol{r})$ is the dielectric permittivity at point $\boldsymbol{r}$, $q(\boldsymbol{r}, \boldsymbol{r}^{(1)})$ is the PDF for sampling on $S^{(1)}$ which may be different from $P(\boldsymbol{r}, \boldsymbol{r}^{(1)})$, and $\omega(\boldsymbol{r}, \boldsymbol{r}^{(1)})$ is the weight value. Thus, Q can be estimated as the statistical mean of sampled values on G, which is further the mean of sampled potentials on $S^{(1)}$ multiplying the weight value. When $\phi(\boldsymbol{r}^{(1)})$ is unknown, (1) can be applied recursively to (2), which means the sampling procedure repeats until the potential of a sample point is known. This can be interpreted as a floating random walk (FRW) procedure: for the ith hop of a walk, a transition domain centered at $\boldsymbol{r}^{(i-1)}$ is constructed and then a point $\boldsymbol{r}^{(i)}$ is randomly selected on its boundary according to the discrete probabilities obtained with $P(\boldsymbol{r}^{(i-1)}, \boldsymbol{r}^{(i)})$. The walk terminates after k hops if $\phi(\boldsymbol{r}^{(k)})$ is known, e.g. it is on the surface of a conductor with known potential (see Fig. 1). With the techniques in [6], the weight value varies slightly, which reduces the random walks for the convergence of MC.

Although the surface Green's function for a spherical transition domain has simple analytical expression, we only consider the cubic transition domain that is well suited to the Manhattan-shaped interconnects in VLSI circuit. The surface Green's function only depends on the relative position of $\boldsymbol{r}^{(1)}$. So, we can pre-calculate and tabulate the sampling probability and weigh value for a unit-size cube. In multi-dielectric environment, the cubic transition domain filled by multiple dielectrics can also be pre-characterized [6].

The above deduction reveals that the statistical mean of the weight values for the walks terminating at conductor j approximates the capacitance C_{ij} between conductors i and j (if $j \neq i$), or the self-capacitance C_{ii} of master conductor i.

III. Techniques for Full-Net Capacitance Extraction

Calculating the delay of a full signal path, for which the interconnect delay caused by parasitic resistance and capacitance (RC) is a dominant factor, is a critical task in digital circuit design. To achieve accuracy, the interconnect wires (including vertical vias) in a net should be segmented and converted to a distributed RC network with parasitic extraction techniques. Due to the simplicity of resistance calculation, the capacitance extraction is the major challenge. The existing FRW algorithm should be enhanced to consider the existence of vias and produce the distributed capacitances, accurately and efficiently. This is also required for the accurate crosstalk analysis and circuit simulation.

Below, we will propose a virtual Gaussian surface sampling technique for the FRW-based full-net capacitance extraction. Considering the influence on the runtime of FRW algorithm, we will also discuss how to optimize the placement of Gaussian surface.

A. The basic idea

To obtain the capacitances for a net, which is made of a number of metal blocks, a simple approach is to construct one Gaussian surface for the whole net as if we only extract the capacitances related with the whole net. The key to get the distributed capacitances is to split the Gaussian surface into pieces and assign each block a piece. Thus, counting the walks starting from the Gaussian piece for a block, we can get the capacitances related to the block. In this way, the statistical error of a block capacitance could be larger than that of the full-net total capacitance, since the former is obtained with a smaller number of walks. However, due to the statistical cancellation, the error of derived net delay would be comparable to the error associated with the full-net total capacitance [7]. An example has shown that the statistical error of Elmore delay is ±2.3%, while the error of the full-net total capacitance is ±2% [7]. Therefore, setting a moderate accuracy criterion for the full-net total capacitance can ensure the accuracy we need for the sequent analysis.

It is a problem to generate the Gaussian surface for a net with complicated geometries of blocks and vias. A straightforward idea is to generate the Gaussian surface for each block and then calculate the envelope of them with geometric operations. However, this could cost a lot of computing time and bring complication to the representation and sampling of the Gaussian surface, which is not necessary. The FRW algorithm actually needs not to know what the Gaussian surface exactly is. All it needs is to make random sampling on it. So, we propose a virtual Gaussian surface sampling (VGSS) technique, which avoids calculating the envelope of the block's Gaussian surface (BGS). By suitably discarding invalid samples, the VGSS technique ensures the correctness and high efficiency.

B. The virtual Gaussian surface sampling technique

For each block in a net, we can construct its BGS as if its adjacent blocks in the net do not exist. If getting the BGSs for all blocks in a net, they may intersect with each other. This makes some parts of BGSs illegal for sampling. The idea of VGSS is selecting points on the BGSs and discarding illegal or invalid samples, so that the selected points can be treated as they are on the net's Gaussian surface which is not really generated. Take Fig. 2 as an example. Points like P_1 and P_2 are legal points, while P_3 within the BGSs of block 1 and the via is an illegal point. P_4 is also on legal area, which is however a place where two BGSs coincide. If we want to make uniform sampling on the Gaussian surface by sampling the BGSs, P_4 should be accept with probability of 1/2. With these considerations, the uniform sampling on the Gaussian surface is easily obtained.

It is not required that the Gaussian surface is sampled uniformly. Actually, some different sampling PDF may be

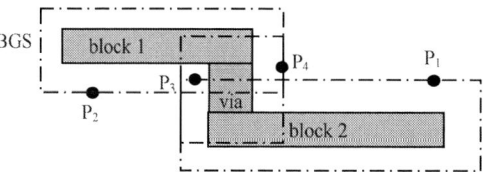

Fig. 2. Side view of two blocks and a via' BGSs.

used to take advantage of the importance sampling technique. With a non-negative function $p(r)$ that satisfies

$$\oint_G p(r)\mathrm{d}r = H \neq 0 , \qquad (3)$$

Eq. (2) can be transformed to

$$Q = H\oint_G \frac{p(r)}{H}\oint_{S^{(1)}} \frac{F(r)}{p(r)}\omega(r,r^{(1)})q(r,r^{(1)})\phi(r^{(1)})\mathrm{d}r^{(1)}\mathrm{d}r \quad .(4)$$

This means that we can use $p(r)/H$ as the PDF to sample the Gaussian surface. To achieve this, we can firstly make a uniform sampling with the aforementioned approach, and then accept the sample with probability of $p(r)/U$, where U is a constant value no less than the maximum of $p(r)$.

The efficiency of the FRW algorithm, i.e. the number of walks needed to reach a specified accuracy, is affected by the choice of $p(r)$. The basic option is $p(r)=1$, which means uniformly sampling. The second choice is $p(r) = F(r)$, which could remove the variance on the weight value brought by (4). In [8], an IS technique based on electric field is proposed, which samples the Gaussian surface with a PDF proportional to $1/D(r)$. $D(r)$ denotes the distance of point r to the master conductor along Gaussian surface's normal direction. Following this strategy, the third choice of $p(r)$ could be $1/D(r)$. Finally, we can also try the choice $p(r) = F(r)/D(r)$. For the four choices of the importance sampling on Gaussian surface, experiments are presented in Section V.A to see which one produces the best performance.

In (4), the weight value for each random walk should be multiplied by $HF(r)/p(r)$. So, the left problem is calculating H. If we apply the MC method to (3), H can be estimated with $S_G \cdot \overline{p}(r)$, where S_G is the area of Gaussian surface and $\overline{p}(r)$ is the statistical mean of $p(r)$. Suppose N_G points r on the Gaussian surface are generated during the VGSS sampling procedure. With them we get $\overline{p}(r)$. Suppose we totally generate N_T points on the BGSs, in which N_G points forms the uniform sampling of the Gaussian surface. Because the sampling on the BGSs is also a uniform one, N_G/N_T approximates the area ratio of the Gaussian surface to the sum of areas of BGSs. Therefore, $S_G \approx S_{all} \cdot N_G/N_T$, where S_{all} is the sum of areas of BGSs. Note that the samples on the Gaussian surface, equal to the number of walks, is usually at least several thousands. So, we can calculate H, and thus the weight value for (4) very accurately.

The following Algorithm describes how to get a valid sample on the Gaussian surface with the VGSS technique. $n_c(r)$ denotes the number of coinciding BGSs on point r.

Algorithm 1 SampleOnVGSS (BGS list)
---
1. Randomly select BGS_i from the BGS list according to their areas;
2. Randomly select a point r on BGS_i uniformly;
3. If r is in the domain enclosed by other BGS **then goto** 1;
4. **Elseif** r is also on some other BGS whose outer normal direction is opposite to that of BGS_i **then goto** 1; **Endif**
5. Get two uniform random numbers in (0, 1): x_1, x_2;
6. If $x_1 > 1/n_c(r)$ **then goto** 1 **Endif**
7. If $x_2 > p(r)/U$ **then goto** 1 **Endif**
8. **Return** r.

The VGSS is a rejection sampling procedure. When we sample each BGS separately and uniformly, it actually give samples on G_0 with PDF $f_1(r) = n_c(r)/N_c$, where $G_0 = \{r \mid r$ is on a BGS's surface$\}$ and $N_c = \oint_{G_0} n_c(r)\mathrm{d}r$. In the VGSS, each of the samples on BGSs is only accepted with a probability

$P_{ac}(r)$. When $r \in G$, $P_{ac}(r) = 1/n_c(r) \cdot p(r)/U$. Otherwise, $P_{ac}(r) = 0$. According to the principle of rejection sampling [11], the accepted samples obey the PDF proportional to $f_0(r) = f_1(r) \cdot P_{ac}(r) = p(r)/(UN_c)$, for $r \in G$. It is otherwise 0. This is the function what we want for sampling the Gaussian surface as in (4), since U and N_c are both constants.

Because points may be discarded, the selection will repeat several times until get a valid point. The time cost of selecting one point is about 1% of the cost of a random walk. So, as long as the repetition time of sampling is not very large, it will not harm the performance. The experiments reveal that the average repetition time is always less than 5.

In line 3 and 4 of Algorithm 1, we need to judge if a point is within or on some BGS. This can be done by comparing the point with all the BGSs. But it could be slow since a net may contain more than 50 blocks. To do this quickly, we pre-calculate each BGS's neighbor BGS list. Two BGSs are neighbors if they intersect. Only the BGSs in the neighbor list of the source BGS, which the point is selected from, need to be tested. This neighbor list is normally very short, so the efficiency of the judging can be high enough.

C. Optimize the placement of Gaussian surface

Before discussing how to determine the optimal placement of a BGS, we first give some definitions about distances. Symbols A and B denote geometric objects which may be cuboids, points or faces. The last two are the special cases of a cuboid. D denotes one of directions X, Y and Z. S denotes one of signs P (positive) and N (negative).

Definition 1: PD-distance from A to B is B's minimum D-coordinate minus A's maximum D-coordinate which is also ND-distance from B to A. $d_{PD}(A, B)$ and $d_{ND}(A, B)$ denote the PD- and ND-distance from A to B respectively.

Definition 2: The distance between A and B, denoted by $d(A, B)$ is the maximum of $d_{PX}(A, B)$, $d_{NX}(A, B)$, $d_{PY}(A, B)$, $d_{NY}(A, B)$, $d_{PZ}(A, B)$, $d_{NZ}(A, B)$.

The placement of a BGS is determined by six distances between it and the enclosed conductor block along six directions (PX, PY, PZ, NX, NY and NZ). The placement of Gaussian surface has an impact on the performance of FRW algorithm. In [8], each face of the Gaussian surface is set to be exactly at the middle of the block and its nearest block. Therefore, points on the Gaussian surface are equidistant from the master and from an adjacent block, and the first cubic transition domain of random walk can touch two blocks. This gives the random walk more probability to stop after the first hop. However, this strategy is not suitable for the master conductor with rare blocks around, where the positions of Gaussian surface in different directions may have very large disparity. And, the larger distance from the Gaussian surface to the conductor also worsens the convergence behavior of the FRW procedure. The authors of [6] firstly find the six equidistant positions from the master, and then use the minimum of the six distances to construct the Gaussian surface. However, this may cause a very small BGS, which is also not good to the efficiency of FRW.

Below, we present a general strategy to determine the Gaussian surface, which compromises those used in [6] and [8]. The effect of multiple dielectric layers is also considered. The position of Gaussian surface is adjusted to guarantee that there is probability of terminating the random walk just after one hop in a two-dielectric transition cube. Algorithm 2 describes how to determine the placement of a BGS. A *scale_factor* not less than 1 is used, whose best value will be discussed in Section V.A. If *scale_factor*=1, the approach is that in [6]; if *scale_factor* is infinite, it is that in [8].

Algorithm 2 GenerateBGS (master block A, other blocks)
---
1. **For** signed direction K in {PX, PY, PZ, NX, NY, NZ} **do**
2. dm := inifinty;
3.   **For** each other block B **do**
4.   **If** $d_k(A, B) = d(A, B)$ **then**
5. dm := min(dm, $d(A, B)$)
6.   **Endif**
7.   **Endfor**
8. $D[K]$:= dm / 2;
9. **Endfor**
10. **For** K in {PZ, NZ} **do**
11.   Start from A's K-face and find the second dielectric interface I on A's K-direction.
12.   **If** $d(A, I) < D[K]$ **then** $D[K]$:= $d(A, I)$ / 2 **Endif**
13. **Endfor**
14. d := min($D[PX]$, $D[PY]$, $D[PZ]$, $D[NX]$, $D[NY]$, $D[NZ]$);
15. d := d * *scale_factor*
16. **For** K in { PX, PY, PZ, NX, NY, NZ} **do**
17. $D[K]$:= min(d, $D[K]$);
18. **Endfor**
19. Use six distances $D[PX]$, $D[PY]$, $D[PZ]$, $D[NX]$, $D[NY]$ and $D[NZ]$ to construct the BGS.

IV. Parallel Space Management Techniques

For each hop of a walk, a conductor-free cubic transition domain is needed. And, we hope it could be as large as possible. Thus, we need to find the distance from current point to its nearest block as the cubic transition domain's radius. A space management structure which indexes all the blocks will helpfully accelerate this. With such approach, only the distances to a few of blocks, which are in the so-called "candidate list" are calculated for each hop.

Octree is a widely used space management structure. The basic idea of Octree structure is dividing the problem domain into small subdomains and organizing them as a tree [10]. Each non-leaf node has 8 children and each leaf node has a candidate list which contains every possible nearest block from any point in the node (subdomain). To find the nearest block from a specified point, we can firstly find the leaf node that contains the point and then compare every block in that node's candidate list to find the nearest one. The candidate list is usually not long, so the Octree has very high performance in finding the nearest block.

Construction of Octree can be done by inserting all blocks into an empty Octree's root one by one. There are two thresholds in Octree. One is called n_t. When a node's candidate list contains more than n_t blocks, the node should be divided into 8 subnodes. To avoid the tree growing too high, we set a minimal size limit l_t for node. The primary operation for constructing the Octree is checking if a block should be added into a node's candidate list. This is judged with the domination relationship defined as follows [6].

Definition 3: T is a node, and B_1, B_2 are two blocks. If for any point $P \in T$, and $P \notin B_1 \cup B_2$, $d(P, B_1) \leq d(P, B_2)$, we say B_1 *dominates* B_2 regarding T.

A block will not be inserted into a node's candidate list unless it's not dominated by any other blocks in that candidate

list. Thus, the domination relationship should be judged between the block and every block in the node's candidate list. This could cost much runtime while handling a structure with a lot of blocks. For an example structure with 37 thousand blocks, the original Octree construction approach can take half an hour. Note that extracting the capacitances of a net in this structure only costs several seconds. In this section, we will propose the techniques to accelerate the Octree construction.

A. The pruning skills for judging domination relationships

We firstly give two definitions.

Definition 4: The *size* of a cuboid node is defined as the maximum of the node's length, width and height.

Definition 5: The *distance limit* $L(T)$ of an Octree node T is the minimum distance between it and a block in its candidate list, plus the node's size.

Actually, the distance limit $L(T)$ is an upper bound of the distance to nearest block from any point in node T. For a new block B, the candidate list of a leaf node T and block B_0 in the list, if $d(B, T) \geq L(T) = d(B_0, T)+Size(T)$, then B is dominated by B_0. Because for any point P in the conductor-free space within T, we have $d(P, B_0) \leq d(T, B_0)+Size(T) = L(T) \leq d(T, B) \leq d(P, B)$. When a new block is added to a node's candidate list, the value of distance limit dynamically changes. Algorithm 3 describes the candidate checking procedure. For most blocks, it returns at beginning instead of after testing all blocks in the candidate list.

Algorithm 3 CandidateCheck(block B, node T)
---
1. d := $d(B, T)$; l is the size of T;
2. **If** $d \geq L(T)$ **then return** false;
3. **For** each b in the candidate list of T **do**
4.   **If** b dominate B **then return** false;
5.   **Elseif** B dominate b **then**
6.   Remove b from the candidate list of T;
7.   **Endif**
8. **Endfor**
9. Add B to the candidate list of T;
10. **If** $(d + l) < L(T)$ **then** $L(T)$:= $d + l$; **Endif**
11. **Return** true.

The distance limit of a node is initialized with infinity since the first tried block should be added to the candidate list. However, we can set it to be a finite value d_{nb}, to prune more block testing. In this way, the candidate list no longer contains all of possible nearest block but only the blocks in the node's *neighbor region* (see Fig. 3). When the candidate list of a node is incomplete, more operation is needed after finding the minimum distance from current point to the candidates. Because maybe a block out of the neighbor region is the nearest block to current point, the minimum distance should be compared with the distance between the point and the boundary of neighbor region, and the smaller one (such as d_b rather than d_3, in Fig. 3) should be used to guarantee a conductor-free transition domain.

Because the transition domain obtained by inquiring the

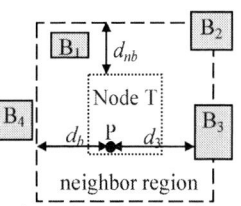

Fig. 3. The illustration of a point in node, node's neighbor region and calculating the nearest distance.

incomplete candidate list is not the largest, it would increase the number of hops in a walk. But with an appropriate value of d_{nb}, this shortcoming is negligible. Our experiments tell that the distance limit and setting the neighbor region can accelerate the construction by thousands times.

B. Parallel construction of the Octree

The Octree structure has great potential in parallelization. Finding the nearest block is embarrassingly parallel, since all the operations on Octree are read-only. The construction of Octree can also be parallelized with a little modification. Here, we come up with two different ways to parallelize it with multi-thread model.

In the parallel construction of Octree, we need to avoid different threads accessing same tree node simultaneously. One way for parallelization is adding a lock to each node and all the accesses need to acquire the lock. Because the insertion of candidate to Octree is performed recursively, the lock of parent node should be released before trying its children nodes. Otherwise, the entire tree would be locked. The cost of this method has two parts: the cost of lock operation itself and the cost of waiting same lock. The first part depends on the implementation of lock. Since only basic functions are needed for the lock, an atomic variable is enough and able to keep the cost of lock at the lowest. The second part can be reduced by two ways: traversing the children nodes in random order or temporarily skipping the occupied node. They both reduce the probability that two threads accessing same node. However, they are not very helpful, because with the distance limit the operation on a node is usually simple and fast, and they won't help to avoid frequently waiting for the root.

Another way of parallelization is letting different threads inserting all blocks into different parts of the Octree. Thus, the lock will be needed any more. We split the root node into 8 subnodes. Each of them represents a subtree and can be further split to get more subtrees if necessary. It's better not to split the tree too many times since the splitting times decide the minimal height of the Octree which affects the efficiency to locate a leaf node.

Each subtree will be assigned to a thread and all the blocks will be tried for insertion. With above pruning skills, most blocks will only be tested with the subtree's root and affirmed not in any node's candidate list of this subtree. This guarantees each thread only has few work to do. Because different region contains different number of blocks, the workload of different subtree may be very different. To balance all the threads, the number of subtrees should be much larger than the number of threads.

Our experiments tell that the second method with several subtrees works much better than the first one with lock. 4.5X speedup can be achieved on a machine with 8-core CPU.

V. Numerical results

We have implemented the FRW algorithm [6], and the proposed techniques in C++. The obtained enhanced FRW algorithm is used to extract the full-net distributed capacitances for several large-scale VLSI layouts.

The first test case is an artificially generated complex structure, which includes a net formed by 11 blocks and 6 vias in 3 metal layers, along with other 12 nets. The second case is an actual design called "FreeCPU" based on the *180-nm technology* with the minimum wire width of 200nm. It includes 37,062 conductor blocks in 5 metal layers, which forms 3037 nets with vias. There are 12 dielectric layers. The lateral and vertical dimensions of the case are about 700μm and 9.4μm, respectively. The third case is an artificially created layout based on the *45-nm technology* with the minimum wire width of 70nm. It includes 101,595 conductor blocks in 3 metal layers, with random widths, spacing, lengths and positions. And, it contains 9 dielectric layers. The lateral size of Case 3 is about 1000μm. The fourth case is an even larger case with 484,441 conductor blocks, based on the 180-nm technology. Its parameters are similar to the "FreeCPU" case.

We firstly verify the correctness of the VGSS technique, and reveal the best sampling PDF and placement of the Gaussian surface. Then, the efficiency of the pruning skills and parallelization techniques for constructing the Octree is demonstrated. Experiments are carried out on a Linux server with Intel Xeon E5-2650 8-core CPU of 2.0 GHz, while the accuracy criterion of FRW algorithm is 1% 1-σ error.

A. Validate the VGSS technique and optimized parameters

For Case 1, we have manually generated the real Gaussian surface, which is represented by several rectangular planes. Random walks are executed with the VGSS technique and on the real Gaussian surface, respectively. The results show the both programs run same number of walks, and the resulting capacitances have less than 1% discrepancy, consistent to the set accuracy criterion.

The *scale_factor* in Algorithm 2 decides the placement of Gaussian surface and affects both the number of walks and the number of hops per walk. We set it to different values to find out the best one. We also compare the different Gaussian surface sampling PDFs. 33 randomly selected nets in each of Case 2 and 3 are extracted. The total time for performing FRW procedure is plotted in Fig. 4.

From Fig. 4, we see that setting *scale_factor*=1.25 and $p(r)=F(r)$ brings the best efficiency. A larger *scale_factor* increases the number of walks for convergence. And, the caused larger transition cube causes smaller probability to terminate after the first hop. Meanwhile, a very small factor also causes more hops in a walk. Comparing with the situation with *scale_factor*=1 (used by [6]), the optimized placement of Gaussian surface could bring **1.4X** speedup to the FRW procedure, as shown in Fig. 4(a). Compared with the strategy in [8], the speedup could be even larger. As for the sampling scheme on the Gaussian surface, the advantage of $p(r)=F(r)$ over other PDFs are obvious. This again validates

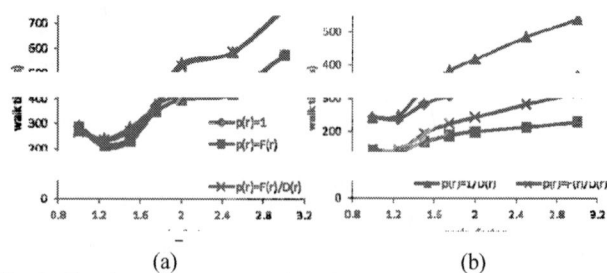

Fig. 4. The time of FRW procedure for extracting 33 nets in Case 2 (a) and Case 3 (b), with varied Gaussian surface placement and $p(r)$.

the importance of the variance of the weight values to the convergence rate. For Case 3, with $p(r)=F(r)$ the average number of walks is 245182, whereas it is 422909 (72% larger) if we make uniform sampling ($p(r)=1$). Finally, we have counted the repetition time of sampling for each PDF choices. The results reveal that the repetition time is always less than 5, which means the overhead of the VGSS technique is negligible, and hardly affects the total runtime.

B. Validate the improved Octree construction algorithm

We firstly use Case 2 to find out suitable values for thresholds n_t and l_t considering both the Octree construction time and the performance of FRW procedure. Because cases with different process technology may need different value of l_t, we use l_t/w_{min} as the variable, where w_{min} represents the minimum wire width. The Octree construction time and the time for performing a million FRW walks under different settings of n_t and l_t are shown in Fig. 5. From it, we see the construction time decreases when l_t/w_{min} or n_t increases. It is because the Octree includes fewer nodes with larger l_t/w_{min} or n_t. The time for performing walks changes little when n_t increases. This is because larger n_t increases the time for traversing the candidate list, while reducing the height of Octree. As for the l_t/w_{min}, it affects only when it is very large (≥ 128 in this experiment) causing a very long candidate list. Therefore, to balance the efficiency of Octree construction and the FRW procedure, we set l_t/w_{min} to 32, and n_t to 25.

In Table I, the Octree construction times for the three large cases are listed. It reveals that, with the proposed pruning techniques, our approach can be over **5000X** faster than the approach in [6]. The improved approach costs only 12 seconds for a large case with half a million blocks.

We use the largest Case 4 to test the parallel Octree construction technique. The approach using lock achieves 2.2X speedup of parallelization, while running 8 threads on the machine with 8-core CPU. On the contrary, the approach with several subtrees performs better. It achieves 4.4X and **4.5X** speedups for 8 and 12 threads, respectively. The experiments demonstrate that the improved parallel Octree

construction costs very short time, and greatly facilitates the capacitance extraction of even larger layout structures.

C. Results of full-chip capacitance extraction

Full-chip capacitance extraction producing all distributed capacitances for all nets is performed. With the proposed techniques, optimized parameter settings and 8-thread multi-core parallelization, the enhanced FRW solver completes the full-chip extraction for Case 2 in 35.2 minutes. This means an average extraction speed of about **0.7 second per net**, under the accurate 1% error criterion. Similar high speed of extraction is also observed on other test cases.

VI. Conclusions

The proposed techniques provide efficient solutions to the problems faced in extracting full-net distributed capacitances and handling chip-scale large interconnect structures using the FRW algorithm. With them, the major advantage of the FRW solver over the deterministic solvers can therefore become reality, which makes executing full-chip capacitance extraction directly with field solver possible. The techniques are crucial to satisfy the increasing accuracy demand in the capacitance extraction of VLSI designs.

References

[1] K. Nabors and J. White, "FastCap: A multipole accelerated 3-D capacitance extraction program," *IEEE Trans. Computer-Aided Design*, Vol. 10, pp. 1447–1459, Nov. 1991.

[2] W. Yu and Z. Wang, "Enhanced QMM-BEM solver for three-dimensional multiple-dielectric capacitance extraction within the finite domain," *IEEE Trans. Microwave Theory Tech.*, Vol. 52, No. 2, pp.560-566, 2004.

[3] W. Yu, X. Wang, Z. Ye, and Z. Wang, "Efficient extraction of frequencydependent substrate parasitics using direct boundary element method," *IEEE Trans. Computer-Aided Design*, Vol. 27, No. 8, pp. 1508–1513, Aug. 2008.

[4] Y. Le Coz and R. B. Iverson, "A stochastic algorithm for high speed capacitance extraction in integrated circuits," *Solid-State Electron.*, Vol. 35, no. 7, pp. 1005–1012, Jul. 1992.

[5] T. A. El-Moselhy, I. M. Elfadel, and L. Daniel, "A hierarchical floating random walk algorithm for fabric-aware 3D capacitance extraction," in *Proc. ICCAD*, 2009, pp. 752-758.

[6] W. Yu, H. Zhuang, C. Zhang, G. Hu and Z. Liu, "RWCap: A floating random walk solver for 3-D capacitance extraction of VLSI interconnects," *IEEE Trans. Computer-Aided Design*, Vol. 32, no. 3, pp. 353–366, Mar. 2013.

[7] M. Kamon and R. Iverson, "High-accuracy parasitic extraction," in *EDA for IC Implementation, Circuit Design, and Process Technology*, CRC Press, Boca Raton, FL, 2006.

[8] S. H. Batterywala and M. P. Desai, "Variance reduction in Monte Carlo capacitance extraction," in *Proc. 18th Int. Conf. VLSI Design*, Jan. 2005, pp. 85–90.

[9] W. Yu, K. Zhai, H. Zhuang, and J. Chen, "Accelerated floating random walk algorithm for the electrostatic computation with 3-D rectilinear-shaped conductors," *Simulation Modelling Practice and Theory*, Vol. 34, No. 5, pp. 20-36, 2013.

[10] H. Samet, *Applications of Spatial Data Structure*, Addison-Wesley, Reading, MA. 1990.

[11] W. H. Press, S. A. Teukolsky, W. T. Vetterling, and B. P. Flannery, *Numerical Recipes in C*, 2nd ed. Cambridge, U.K.: Cambridge Univ. Press, 1992.

TABLE I
The Construction Time of the Octree (in unit of second)

Case	#block	Approach of [5]	Proposed approach	Speedup
2	37062	1757.9	0.53	3316
3	101595	16595.6	3.12	5319
4	484441	> 2 days	12.27	--

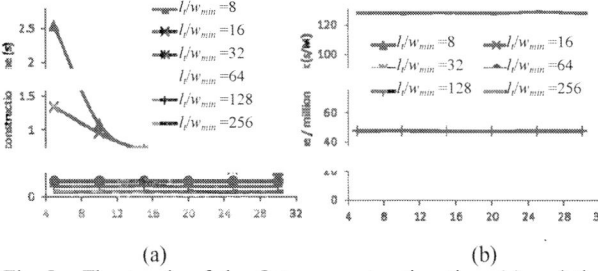

(a)   (b)

Fig. 5. The trends of the Octree construction time (a) and the time for performing a walk (b), with different n_t and l_t/w_{min}.

3DLAT: TSV-based 3D ICs crosstalk minimization utilizing Less Adjacent Transition code

Qiaosha Zou [*], Dimin Niu [*], Yan Cao [*], Yuan Xie [*][†]

[*] The Pennsylvania State University, USA

[†] AMD Research China Lab

Email: {qszou, dun118, yancao, yuanxie}@psu.edu

Abstract—**3D integration is one of the promising solutions to overcome the interconnect bottleneck with vertical interconnect through-silicon vias (TSVs). This paper investigates the crosstalk in 3D IC designs, especially the capacitive crosstalk in TSV interconnects. We propose a novel ω-LAT coding scheme to reduce the capacitive crosstalk and minimize the power consumption overhead in the TSV array. Combining with the Transition Signaling, the LAT coding scheme restricts the number of transitions in every transmission cycle to minimize the crosstalk and power consumption. Compared to other 3D crosstalk minimization coding schemes, the proposed coding can provide the same delay reduction with more affordable overhead. The performance and power analysis show that when ω is 4, the proposed LAT coding scheme can achieve 38% interconnect crosstalk delay reduction compared to the data transmission without coding. By reducing the value of ω, further reduction can be achieved.** [1]

I. INTRODUCTION

With technology scaling, the interconnect delay becomes one of the major bottlenecks due to the increasing delay gap between interconnect and transistors. The interconnect delay problem can be alleviated through migrating designs to the third dimension [2], [3]. The vertical interconnect can dramatically reduce the routing length and therefore result in smaller transmission delay and power consumption. Nevertheless, the capacitive crosstalk, which is one of the major contributors to delay and power consumption of on-chip interconnect [1], is not eliminated in 3D designs.

As the key interconnect component in 3D ICs, through-silicon vias (TSVs) substitute the long global wires in 2D designs and enable the vertical stacking of dies for smaller footprint and the capability of heterogeneous designs. In spite of these benefits, the signal integrity issues in TSVs become the major challenges in 3D designs [4], [5]. Studies show that the coupling problem is not negligible in TSVs because of the relatively large diameter, which results in TSV-to-TSV coupling, TSV-to-device coupling, and TSV landing pad to wire/device coupling. In these coupling effects, the TSV-to-TSV coupling attracts lots of research interests because of the relatively large coupling noise and the lack of sufficient mitigation methods. Elaborated analysis and modeling methods are provided for the TSV-to-TSV coupling [5]–[8]. They claim that TSV-to-TSV crosstalk problem is an important signal reliability issue and should be taken into consideration during design phase.

Several crosstalk minimization techniques have been proposed in 2D designs, such as active shielding [9], data coding [10]–[12], and wire spacing [13]. Other techniques to migrate the crosstalk delay with negligible area overhead are proposed, such as alleviating the crosstalk effect through transmission cycle tuning [14] by observing that not all the transmissions result in large crosstalk delay. Nevertheless, these techniques can't be directly applied in 3D designs. The additional dimension results in significant difference in crosstalk problem analysis between 2D and 3D designs. In most 2D cases, the

coupled wires are usually considered placed in the same planar and each victim has only at most two aggressors, however, each victim TSV is surrounded by at most eight aggressors in 3D and the crosstalk noises come from the coupling capacitance network. Consequently, the crosstalk analysis and elimination become complex in 3D designs.

In this paper, we propose a coding mechanism called ω-Less Adjacent Transition (LAT) to reduce the crosstalk delay and transmission power in a TSV array. This mechanism encodes the input data to a codeword which contains limited number of 1s (indicated by ω) in every 3×3 TSV array. In the analysis, we show that by applying our coding mechanism, the crosstalk delay can be reduced by 38% and the power consumption overhead is minimized, when we constrain at most four 1s in each 3×3 array (ω=4). Further delay reduction can be obtained with a smaller value of ω with the cost of area overhead. The contribution of this paper are as follows:

- A novel crosstalk and power consumption overhead minimization mechanisms, ω-LAT, is proposed. This coding scheme takes the horizontal, vertical, and diagonal TSVs of a victim into consideration.
- Both the normal coding scheme and overhead optimization schemes are explained. Their efficiency and advantage over other crosstalk minimization schemes are shown through evaluations.
- Compared to previous 3D crosstalk avoidance coding mechanism, LAT can maintain affordable overhead when the crosstalk delay is aggressively reduced.

The rest of paper is organized as follows. Section II introduces the preliminaries on 3D capacitive crosstalk and the related work. The framework of ω-LAT mechanism and the corresponding optimization scheme with overhead analysis are described in Section III. The comparison and the analysis of power consumption and transmission delay are shown in Section IV. At last, the conclusion is drawn in Section V.

II. PRELIMINARIES

In this section, we first introduce the capacitance crosstalk model that is used in our mechanism. According to the crosstalk model, we classify the crosstalk into 10 classes for the convenience of analysis in the following sections. After that, the previous studies about crosstalk reduction and elimination in both 2D and 3D designs are briefly introduced.

A. Crosstalk in TSV Array

The coupling capacitance network among adjacent TSVs is shown in Figure 1. In this model, the middle TSV i is the victim and proximate eight TSVs are aggressors. Due to the different distance, we use C_d and C_c to represent the coupling capacitance of diagonal TSVs and vertical/horizontal TSVs, respectively. C_L is the capacitance between TSV and bulk silicon, also known as the self-capacitance. With the notation above, the approximate signal delay considering the

---

[1]Zou, Niu, and Xie's work is supported in part by NSF 0903432, 1017277 and SRC grants.

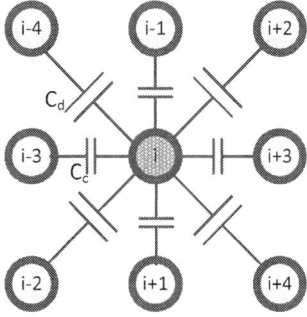

Fig. 1. The capacitance crosstalk model in a 3×3 TSV array.

capacitance crosstalk can be expressed in the following equation [1], [15]:

$$\tau_i(\alpha) = -RC_L(1 + \lambda_1(\delta_{i,i-3} + \delta_{i,i-1} + \delta_{i,i+1} + \delta_{i,i+3}) + \\ \lambda_2(\delta_{i,i-4} + \delta_{i,i-2} + \delta_{i,i+2} + \delta_{i,i+4})) \quad (1)$$

where

$$\begin{cases} \lambda_1 = \frac{C_c}{C_L} \\ \lambda_2 = \frac{C_d}{C_L} \\ \Delta V_i = V_i(t^+) - V_i(t^-) \\ \delta_{i,k} = abs(\frac{\Delta V_i - \Delta V_k}{V_{dd}}) \end{cases}$$

In the above equations, $V_i(t^+)$ and $V_i(t^-)$ denote the voltages before and after the transition. The value of $\delta_{i,j}$ can only be 0, 1, or 2 to represent the relative signal transition direction between victim and aggressor. For example, if two signals in TSV i and k are switching in the opposite directions (i.e. $0 \rightarrow V_{DD}$ and $V_{DD} \rightarrow 0$), $\delta_{i,j}$ equals to 2 because $\Delta V_i = V_{DD}$ and $\Delta V_k = -V_{DD}$. If two signals are switching in the same direction, $\delta_{i,j}$ takes the value of 0. Note that Equation 1 is applied when a transition happens on victim TSV i, otherwise, there is no signal delay since signal i stays unchanged. Assuming the ratio between C_d and C_c is 1/4 [16], the effective crosstalk capacitance $C_{eff,i}$ on victim TSV i is defined as follows:

$$C_{eff,i} = C_L(1 + \lambda_1(\delta_{i,i-3} + \delta_{i,i-1} + \delta_{i,i+1} + \delta_{i,i+3}) + \\ \frac{\lambda_1}{4}(\delta_{i,i-4} + \delta_{i,i-2} + \delta_{i,i+2} + \delta_{i,i+4})) \quad (2)$$

based on the effective coupling capacitance, we classify the crosstalk into 10 categories . In addition to **0C** to **8C** classes that have been defined in [16], we take the diagonal TSVs into consideration and extend the crosstalk classes to **9C** and **10C**. In brief, the classification can be done in the following steps:

- Initialize crosstalk class to **0C**;
- Determine the transition direction of TSV i;
- Determine the accumulated $\delta_{i,k}$ for each vertical/horizontal TSV according to the definition and increase the crosstalk class by $\sum \delta_{i,k}\mathbf{C}$;
- Determine the accumulated $\delta_{i,d}$ for each diagonal TSV, if the value exceeds 0.5, increase the crosstalk class by **1C**.

B. Related Work

Plenty of research have been done to minimize or eliminate the crosstalk delay in 2D design to close the increasing delay gap between interconnect and transistor. Generally, there are mainly two types of methods to handle the crosstalk problem: static methods and data coding (CAC) [9]–[13]. Static methods include data shielding and wire spacing in design time. The benefit of static methods is that

data input doesn't need to be changed. However, large overhead is introduced. Coding schemes try to minimize crosstalk by coding the data input to avoid patterns that cause large crosstalk delay during data transmission. These methods have been proposed and their effectiveness is proved in the 2D regime.

Even though the crosstalk handling methods are mature in 2D designs, they can't be directly applied on 3D designs. Unlike in the 2D bus, where usually only two adjacent wires are considered as aggressors, the additional dimension in 3D interconnect increases the complexity in crosstalk minimization. Shielding and coding methods have been initially explored, focusing this unique feature of 3D designs.

The crosstalk problem and influence factors have been examined in 3D designs [7], [8], [17]. These studies prove that by increasing the pitch or inserting power/ground TSVs between signal TSVs, the crosstalk noise can be significantly reduced. Nevertheless, this kind of shielding method introduces large area overhead and increases the routing complexity during design time. ShieldUS [15] suggests to use data signals as shields to minimize the crosstalk during data transmission. This approach remaps the bits that are relatively stable as shields to separate the more active TSVs. The area overhead is negligible in this method. However, the worst case delay cannot be guaranteed and dynamic remapping circuitry introduces overhead. 3D Crosstalk avoidance code (CAC) is proposed and analyzed with different levels of crosstalk minimizations [16]. The benefit of such a coding scheme is that it can provide the desired level of crosstalk minimization, therefore, the transmission delay can be guaranteed. The disadvantage is that it requires significant extra interconnect resources and the crosstalk from diagonal TSVs is not considered.

In this work, we propose a coding mechanism for 3D ICs, called Less Adjacent Transmission. In addition to the consideration of diagonal TSV, the proposed approach can guarantee the crosstalk delay in different levels while reduce the power consumption overhead in data transmission.

III. 3D LAT CODING MECHANISM

In this section, the Less Adjacent Transition (LAT) coding mechanism is introduced. Since the LAT code is derived from 2D No Adjacent Transition (NAT) code [10], we briefly review the NAT coding scheme which is used on 2D design and can reduce the bus crosstalk delay. Through analytical analysis, we show that the NAT code is impractical to apply in 3D ICs. Then we propose the LAT code for 3D designs which can minimize the crosstalk delay in TSV arrays while maintain the coding overhead within reasonable range. Considering the non-negligible overhead, we further propose the overhead optimization schemes. Finally, the heuristic strategy of encoder and decoder (CODEC) designs are given.

A. Preliminaries in 2D NAT Code

In the coding scheme, the original input is defined as *dataword* and the data after coding is called *codeword*. The coding scheme is used to design a sequence of codeword that satisfy several constraints and map each dataword into the corresponding codeword. Due to the constraints, usually the codeword width is longer than that of the dataword, resulting in high coding overhead and potentially larger power consumption. In order to minimize both crosstalk and power, the NAT coding scheme is proposed [10]. In addition to the Transition Signaling and the Limited Weighted Code, NAT coding scheme constrains that no adjacent transition is allowed.

Transition Signaling defines that the transition takes place only when the input bit is a 1. For example, if the codeword is 00100, then the middle wire in the bus changes to its compliment value while

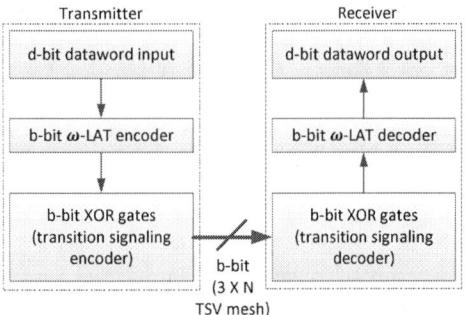

Fig. 2. Framework overview of signal transmission with LAT coding scheme.

others remain unchanged. The encoder and decoder can be simply constructed with XOR gates. The number of 1's in the dataword is called the data weight. Limited Weighted Code is a coding method that limits the maximum weight in the codeword. Therefore, the number of transitions is bounded by the maximum weight. Combining these two transmission techniques, the NAT coding mechanism restricts that no adjacent 1s are allowed in the codeword, therefore, crosstalk is eliminated because no adjacent transitions are allowed. Furthermore, power consumption with NAT is reduced because it is proportional to the maximum weight.

This coding scheme can be easily applied to a 2D bus since in crosstalk analysis, each wire usually only considers two adjacent wires. However, the adjacency concept in 3D designs has changed. Therefore, if we still apply the rule that no adjacent 1s are allowed in the codeword, it would mean that every 1 requires eight surrounding 0s. In consequence, the codeword overhead is extremely large. A Less Adjacent Transition (LAT) coding scheme is proposed to handle this unique 3D crosstalk challenge.

B. 3D ω-LAT Code

The framework overview of signal transmission with LAT code is shown in Figure 2. The data input first goes through the LAT encoder to generate the corresponding codeword. Before transmission, the transition signaling encoder which is constructed with XOR gates is applied. At the receiver side, the decoding steps (transition signaling decoder and LAT decoder) are performed to generate the final data output.

Before introducing the LAT coding scheme, we need to clarify two concepts that are used in the analysis.

Definition For each single TSV i, all the surrounding eight TSVs, including diagonal TSVs and horizontal/vertical TSVs are called i's adjacent TSVs. The victim TSV and its eight adjacent TSVs construct a 3×3 TSV array, which is called TSV subarray throughout this paper.

As illustrated in previous section, it is impossible to have no adjacent transitions in a 3D TSV array. Therefore, we relax this requirement and propose a ω-LAT code. ω represents the maximum weight for every TSV subarray, that is, every TSV and its eight adjacencies. By restricting the maximum weight, we can minimize the worst case crosstalk delay within each subarray as shown in the following.

Lemma 1: For each ω-LAT code when $\omega \leq 5$, the crosstalk class of the transmission will not exceed $(\omega - 1)*\mathbf{2C}$.

Proof: From the definition, we can see that crosstalk class is determined from $\delta_{i,j}$. In a TSV subarray, if the transition direction of the middle TSV i is opposite to the transition of its direct neighbor j, then $\delta_{i,j}$ equals to 2. Therefore, for a ω-LAT code, the worst case of crosstalk delay happens when the signal for middle TSV is 1 and signals for $\omega - 1$ of its horizontal/vertical neighbors are 1s,

meanwhile, the transition directions of these TSVs are in opposite. In this case, the crosstalk is maximum and the class is $(\omega - 1)*\mathbf{2C}$. Since each subarray follows this rule, this conclusion can be easily applied to a N×N TSV array which contains multiple subarrays. ∎

Next, we will introduce how to calculate the code cardinality of a ω-LAT code that can satisfy the one-to-one mapping between codeword and dataword. Due to the calculation complexity, we only show the analytical analysis of a 3×N TSV array. The terms that are used in the following analysis are shown below:

d = dataword bitwidth

N = the number of TSV columns in the codeword, therefore, the codeword bitwidth is $3 * N$

α_c = the weight of column c, since we only consider 3 rows, α_c cannot exceed 3

$T(\beta, c)$ = the number of codewords (code cardinality) when the TSV array has c columns and the weight of each subarray is exactly β

$T_\omega(c)$ = the number of codewords with c TSV columns when the maximum weight of each subarray is ω

Based on the definitions and known conditions, the code cardinality with maximum weight ω can be calculated as follows:

$$T_\omega(N) = \sum_{\forall c, \alpha_c + \alpha_{c+1} + \alpha_{c+2} \leq \omega} \prod_{c=0}^{N} \binom{3}{\alpha_c} \tag{3}$$

In this equation, we assume that each column has weight α_c, and the weight in consecutive three columns cannot exceed ω. When N is small, the value of $T_\omega(N)$ can be generated with enumeration. However, when size N increases, it is infeasible to solve this equation within polynomial time. Alternatively, we consider the feasible lower bound of code cardinality which can be described with the following equation:

$$T_\omega(N) \geq \sum_{\forall c, \alpha_c + \alpha_{c+1} + \alpha_{c+2} = 0}^{\omega} \prod_{c=0}^{N} \binom{3}{\alpha_c} = \sum_{\beta=0}^{\omega} T(\beta, N) \tag{4}$$

where β is the total weight for any consecutive three columns as defined. The above equation can calculate the code cardinality's lower bound because it sacrifices the coding flexibility.

In Equation 4, we constrain that the total weight in any consecutive three columns (consecutive subarray) are the same to provide the lower bound. The calculation can be done in inductive procedure. Since each subarray contains the same weight, the equation $\alpha_{N-3} + \alpha_{N-2} + \alpha_{N-1} = \alpha_{N-2} + \alpha_{N-1} + \alpha_N = \beta$ is held. Therefore, the weight in column N should be the same as in column $N - 3$, meaning the weights in every three columns are the same. The following equation is constructed assuming N is an integer multiple of 3:

$$T(\beta, N) = \sum_{\alpha_1 + \alpha_2 + \alpha_3 = \beta} \left[\binom{3}{\alpha_1}\binom{3}{\alpha_2}\binom{3}{\alpha_3} \right]^{\frac{N}{3}}$$

$$T(\beta, N+1) = \sum_{\alpha_1 + \alpha_2 + \alpha_3 = \beta} \left[\binom{3}{\alpha_1}\binom{3}{\alpha_2}\binom{3}{\alpha_3} \right]^{\frac{N}{3}} \binom{3}{\alpha_1}$$

$$T(\beta, N+2) = \sum_{\alpha_1 + \alpha_2 + \alpha_3 = \beta} \left[\binom{3}{\alpha_1}\binom{3}{\alpha_2}\binom{3}{\alpha_3} \right]^{\frac{N}{3}} \binom{3}{\alpha_1}\binom{3}{\alpha_2} \tag{5}$$

Note that the previous equations can be used to derive the codeword count when $N \geq 3$. For N smaller than 3, enumeration is applied

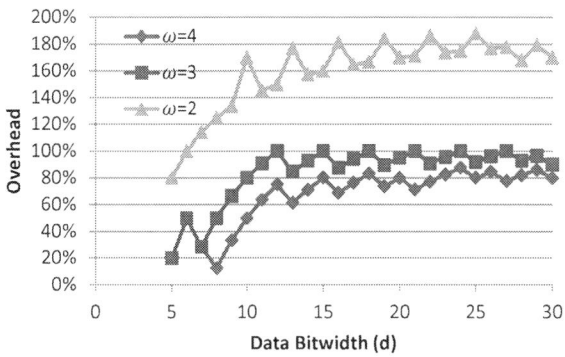

Fig. 3. The overhead caused by redundant TSVs with different ω. The horizontal axis is the bitwidth of dataword and the vertical axis is the overhead percentage.

to get the code cardinality. Given the value of β, all the possible combinations of α_1, α_2, and α_3 subjected to $\alpha_1 + \alpha_2 + \alpha_3 = \beta$ are enumerated and the corresponding code cardinality is calculated.

In order to satisfy the mapping between dataword and codeword, we should find the minimal N such that $T_\omega(N) \geq 2^d$. Therefore, the overhead can be calculated from $\frac{3*N-d}{d}$. The coding overhead with respect to the input data bitwidth is shown in Figure 3.

When the ω is reduced, the overhead increases to achieve lower crosstalk delay. The upper bounds of ω =2, 3, and 4 are 190%, 100%, and 90%, respectively. The overhead of ω=2, which means the maximum crosstalk in transmission is 2C, is significantly larger than the other two cases. When the data bitwidth is smaller than 15, the overhead grows proportionally with the increased data input. After that, it varies within a small range and approaches the asymptotic upper bound. Note that the calculated cardinality is the lower bound of ω-LAT code, therefore, the overhead is the upper limit. Even though the overhead in LAT is larger than 3D CAC [16] with 4C and 6C crosstalk reduction, it is significantly smaller when the crosstalk is aggressively reduced to 2C. Moreover, the estimated overhead in 3D CAC is a lower bound while our overhead is the upper bound.

C. LAT Code Optimization

In order to reduce the code overhead, the first simple technique we can apply is bus inverting [18]. A global weight detector is used to determine if the dataword's weight exceeds $d/2$. If the weight is larger than $d/2$, we change the input data to its complement. The bus inverting bit is set at the transmitter side which will indicate the reverse operation at the receiver side. Therefore, the codeword cardinality only needs to be larger than $2^{(d-1)}$. The weight detector can be implemented with hamming distance circuits by comparing the hamming distance between data input and all 0's.

The main purpose of LAT coding scheme is to restrict the weight in the TSV subarray for crosstalk minimization. Therefore, if the dataword doesn't violate this restriction, we don't need to perform the coding. By excluding the qualified dataword, we can reduce the number of inputs that needs to be encoded, and further reduce the codeword length and overhead. In this case, a local weight detector is needed to determine if every TSV subarray has its weight smaller than ω. If the dataword doesn't need the encoding, the encoding bit is reset and the encoder is disabled. The data input is directly sent to the receiver through TSV arrays. In total, $d/3 - 2$ number of 9-bit weight detectors are required.

The block diagram of the optimized coding mechanism at the transmitter side is shown in Figure 4. In the optimization scheme, two extra bits are needed: a bus inverting bit and an encoding bit. The receiver side performs decoding and bus inverting based on these

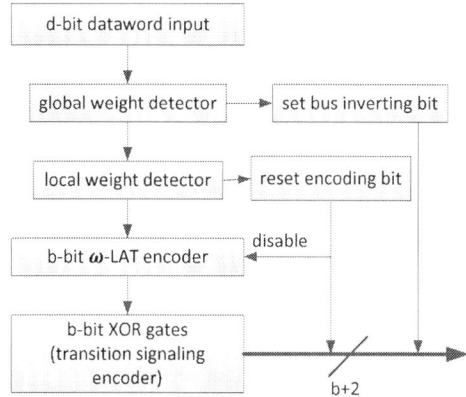

Fig. 4. The overview of signal transmission procedure after overhead optimization.

two signal bits accordingly. According to the previous analysis, the codeword length is decided by finding the minimal N to satisfy the following condition:

$$T_\omega(N) \geq 2^{(d-1)} - T_\omega(d/3) \tag{6}$$

Equation 4 is used to calculate the lower bound of $T_\omega(N)$. Table I show the coding overhead comparison of LAT code with and without optimization when $\omega = 4$. From the results, for small input size, large overhead saving can be obtained. For example, if we need to encode a 5-bit data input, after optimization, only 3 TSVs would be needed. Considering the two added signal bits, there is no overhead. Nevertheless, as the data input size increases, the overhead saving is reduced. It is because when the input size increases, the percentage of qualified patterns (ratio of $T_\omega(d/3)$ to 2^d) is reduced. Meanwhile, in the optimization scheme, the size of both global and local weight detectors grows linearly with increasing dataword length. The corresponding power consumption and detection delay increase with longer input, therefore, the power and delay saving from crosstalk minimization is sacrifices. The design tradeoff should be carefully considered to obtain the optimal power and performance.

D. Heuristic LAT CODEC Design

In LAT design, each subarray should contain weight no more than ω. The codeword cardinality changes with different ω values, therefore, it is hard to design a universal encoder and decoder for every possible ω and data bitwidth. The CODEC can be designed only after the codeword length is determined through data bitwidth and ω during design time. In general, two levels of comparators are needed in the encoder. The first level is used to decide the weight in each TSV subarray and the second level is used to select the combination of α_1 to α_3. In the following analysis, we assume ω=4, data bitwidth equals to 16, and a data input value of 1024 as an example to explain the CODEC design.

Based on the configuration, the required TSV column to encode 16-bit data is 9 and the total codeword bitwidth is 27. The codeword cardinalities with weights equal to 0, 1, 2, 3, 4 are 1, 81, 2268, 24060, 61398, respectively. Therefore, we need five comparators on the first level, where each has a value 1, 82, 2350, 26410, and 87808, respectively. These comparators operate in parallel to reduce timing overhead. Data 1024 is within the range of 82 to 2350, therefore, the weight in each subarray should be 2.

The possible number of combinations of α_1 to α_3 is fixed when the subarray weight is known. For maximum weight of 4, 12 combinations are generated, therefore in the second level, 12 comparators are needed. For each combination, we can calculate the codeword

TABLE I
CODING OVERHEAD COMPARISON RESULTS OF WITH AND WITHOUT OPTIMIZATION.

Data Bitwidth	Optimized		original		reduced ratio ($\frac{T_\omega(d/3)}{2^d}$)	overhead reduction (%)
	column	overhead (%)	column	overhead (%)		
5	1	-40	2	20	25	60
10	3	-10	5	50	25	60
15	8	60	9	80	4.04	20
20	11	65	12	80	0.38	15
25	15	80	15	80	0.02	0

cardinality and determine the α_1 to α_3 combination. For example, the input 1024 should be encoded with $\alpha_1 = 1$, $\alpha_2 = 0$ and $\alpha_3 = 1$.

Next, we need to decide the row position of 1s. The total weight is 6 from the value of α. For column containing 1s, we define a coefficient k, and k is 0, 1, or 2 when the 1-valued bit is on row 0, 1, or 2. Therefore, the data value can be represented in $k_0 * 3^0 + k_1 * 3^1 + ... + k_5 * 3^5$. In the last step, we calculate the value of these coefficients and generate the codeword. For example, the coefficients of k_0 to k_5 for data 1024 are 0, 2, 2, 2, 1, and 0, respectively.

The 16-bit comparator is implemented and synthesized with Nangate 45nm library using Synopsys Design Compiler. The total area consumed by the two-level comparators is about $4264\mu m^2$ while the signal delay of each level is about $0.86ns$. Pipeline design can be used to reduce the timing overhead from CODEC.

At the decoder side, we first use one weight detector to examine the weight of one TSV subarray. After that, three weight detectors are applied on each column in the subarray to determine α_1 to α_3 combination. And based on the row position of the 1-valued bit, we can get the coefficient and the data input.

IV. EVALUATION

In this section, we perform the evaluation of our proposed ω-LAT coding scheme and compare our scheme with two 3D crosstalk elimination mechanisms: ShieldUS [15] and 3D CAC code [16]. First, we introduce the power model to analyze the interconnect power consumption of these three mechanisms. After that, the crosstalk reduction analysis and effectiveness on real benchmark traces are shown.

A. Interconnect Power Analysis

Because of the capacitive crosstalk in interconnects, the dynamic power consumption of each TSV P_k mainly comes from two parts: the switching power P_k^s caused by wire transition and the coupling power P_k^c caused by inter-wire transition [19]. The total power consumption can be obtained by $P = \sum_{k=1}^{3N} P_k^s + \sum_{k=1}^{3N-1} P_k^c$. In order to avoid redundant summation of same item, we consider three adjacencies, the south, east and southeast of each TSV in the crosstalk power. Similar to the analysis in [10], the equation of P_k^s is shown as follows:

$$P_k^s = \frac{1}{2}C_L V_{DD}^2 * Pr(trans) \qquad (7)$$

where $Pr(trans)$ represents the probability of transition in TSV k. Due to the increased number of neighbors, the value of P_k^c can be calculated from the following equation:

$$P_k^c = C_c V_{DD}^2 * Pr(V_k(t^+) \neq V_{k+1}(t^+)) * E_t(k, k+1)$$
$$+ C_c V_{DD}^2 * Pr(V_k(t^+) \neq V_{k+3}(t^+)) * E_t(k, k+3)$$
$$+ C_d V_{DD}^2 * Pr(V_k(t^+) \neq V_{k+4}(t^+)) * E_t(k, k+4) \qquad (8)$$

where $k+1$, $k+3$, and $k+4$ denote the south, east and southeast adjacencies of TSV k. $Pr(V_k(t^+) \neq V_{k+1}(t^+))$ represents the probability of different voltage between TSV k and its

TABLE II
THE POWER CONSUMPTION RELATED PARAMETERS COMPARISON BETWEEN UNCODED, SHIELDUS, 3D 6C CAC, AND 3-LAT SCHEMES.

code	$Pr(trans)$	$Pr(V_k(t^+) \neq V_{k+1}(t^+))$	$E_t c(k, k+1)$
uncoded	0.5	0.5	1
ShieldUS	0.5	0.5	≤ 1
6C CAC	0.5	0.367	1
4-LAT	0.4079	0.5	0.8159

neighbor after transition. $E_t c(k, k+1)$ is the expected number of transitions in $(V_k(t^-), V_{k+1}(t^-)) \rightarrow (V_k(t^+), V_{k+1}(t^+))$ when $V_k(t^+) \neq V_{k+1}(t^+)$.

Compared to uncoded data transmission, the coding scheme changes the probability of transition and the expected value of transition count between two coupled TSVs. Table II shows the parameter values when the input bitwidth is 15 in uncoded cases, ShieldUS, 3D CAC, and the proposed ω-LAT schemes. Due to space limitation, the difference between boundary and middle TSVs are not listed. For fair comparison, we choose ω as 4 and the 3D 6C CAC, which means that the maximum crosstalk is 6C. Since 3D CAC doesn't consider the diagonal TSVs, we hereby assume that the parameters in diagonal TSVs are the same as in direct neighbors.

For uncoded and ShieldUS, the input data doesn't have any constraints, therefore, $Pr(trans)$ and $Pr(V_k(t^+) \neq V_{k+1}(t^+))$ are all $\frac{1}{2}$. The transition count between two TSVs is hard to predict in ShieldUS since it is still highly dependent on the input data. Nevertheless, the expected transition should be smaller than uncoded cases due to the shielding effect.

For 6C CAC, we implement and calculate the valid patterns count to code a 16 bit data. The $Pr(V_k(t^+) \neq V_{k+1}(t^+))$ is smaller than 0.5 in 6C CAC because this coding scheme only contains the valid pattern. Valid pattern is determined by the complemental value count in the four neighbors. For example, the definition of a valid 2C pattern is that for the central TSV, only one out of four direct neighbors can has its complement value. Therefore, the probability that two nearby bits have different values reduces from $\frac{1}{2}$ to $\frac{1}{4}$.

In 4-LAT cases, since transition signaling is applied, therefore, transition happens only when the encoded data is 1. By limiting the weight in each subarray, we can reduce the probability of transition. $Pr(trans)$ can be calculated as $\sum_{\beta=0}^{\omega} \frac{\beta}{9} * \frac{T(\beta,N)}{T_\omega(N)}$. When the value of ω reduces, the transition probability also reduces. If other conditions are the same and ω changes from 4 to 3, the probability of transition reduces from 0.4079 to 0.3251.

Note that bit 1 represents transition in LAT, not the voltage value of the corresponding wire, while in Equation 8, V_k is the voltage value. We assume in the initial state, $Pr(V_k(t^+) \neq V_{k+1}(t^+))$ is $\frac{1}{2}$, then after the transition, the probability of inequality can be calculated considering two cases: the two initial voltages are the same, and one goes through a transition; the two initial voltages are different, both go through transitions or remain unchanged. In each case, $Pr(trans)$ is used for transition probability. Similarly, the parameter $E_t c(k, k+1)$ can be expressed as the expected number of 1s in TSV k and $k+1$.

When the parameter λ_1 is set as 5.54 [15], for single TSV, the power consumption is $8.56C_L V_{DD}^2$ in uncoded cases and

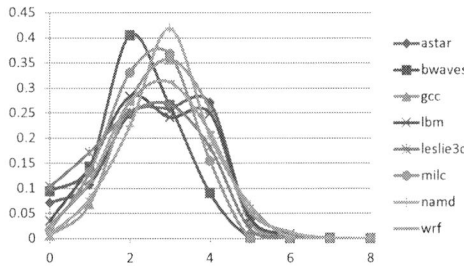

Fig. 5. The benchmark crosstalk characteristic in 3D cases.

$6.98 C_L V_{DD}^2$ in 4-LAT cases, respectively. Due to the TSV overhead after coding, the total power consumption of 4-LAT is slightly larger than uncoded cases, however, the average power consumption for each TSV is 18.46% smaller than uncoded cases.

B. Crosstalk Delay Analysis

In the crosstalk minimization evaluation, we use the extracted data trace from benchmark SPEC2006 and simulate the transmission time of ShieldUS, 4C CAC, and 3-LAT. The architecture simulator GEM5 is used for memory data trace extraction. Each benchmark is executed with four cores and the Ruby memory model is used to connect cores to memory. The 3D crosstalk classification of the data trace is analyzed with our implemented crosstalk analyzer and the results are shown in Figure 5. Most of the data transmissions fall into the range of 2C to 4C. The data bitwidth is 64 and we divide every 16 bits into one group. Therefore, for every group, we use 4×4 TSVs in both uncoded and ShieldUS cases, while the TSV arrangements are 3×9 and 3×10 in 4C CAC and 3-LAT, respectively.

When λ_1 is set as 5.54, the transmission delays in 4C CAC and 3-LAT code are the same as $23.16 RC_L$. For fair comparison, in ShieldUS, we set the delay time of 4C as one clock cycle and multiple cycle transmission method is used. The sampling interval for ShieldUS is 100 transmissions. For uncoded cases, the transmission cycle is determined by the longest crosstalk delay (**10C** class), which is $56.4 RC_L$. We also simulate the transmission delay in the ideal case, which means the clock cycle is flexible and determined only by the crosstalk delay in each transmission.

Figure 6 shows the average benchmark data transmission delay of ShieldUS, 4C CAC/3-LAT, and ideal cases normalized to the uncoded case. Based on the benchmark characteristics, the most transmissions are in the range of 2C to 4C, therefore, the performance of ShieldUS is close to the 4C crosstalk minimization schemes (4C CAC and 3-LAT). The 4C crosstalk minimization schemes can guarantee the transmission time, however, ShieldUS can only reduce the delay when plenty of data bits are unchanged with no transmission time guaranteed. Moreover, from the experiment, we find that ShieldUS sometimes increase the transmission crosstalk.

V. CONCLUSION

Due to the complexity in 3D capacitive crosstalk minimization, we propose a novel ω-LAT coding scheme to minimize both the crosstalk and interconnect power consumption overhead in vertical interconnects. Combining transition signaling and adjacent weight limitation, the LAT coding can reduce the maximum capacitive crosstalk to 6C, 4C, or 2C when ω is 4, 3, or 2. An optimized mechanism is also introduced to reduce the codeword overhead.

Compared with uncoded cases, our proposed coding scheme can achieve 38% and 58.9% interconnect delay improvement with ω equal to 4 and 3, respectively. The LAT code can reduce crosstalk to 2C with affordable overhead compared to 3D crosstalk avoidance code.

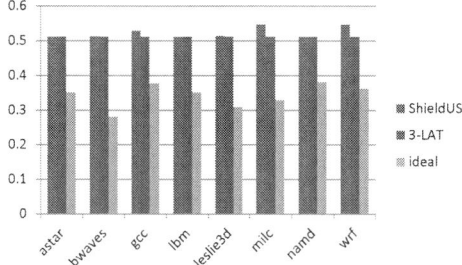

Fig. 6. The transmission time results of ideal case, ShieldUS, and 6C crosstalk minimization schemes normalized to uncoded case.

REFERENCES

[1] C. Duan, B. J. LaMeres, S. P. Khatri, *On and Off-Chip Crosstalk Avoidance in VLSI Design*, C. Duan, Ed. Springer, 2010.

[2] V. Pavlidis and E. Friedman, *Three-Dimensional Integrated Circuit Design*, V. Pavlidis and E. Friedman, Eds. Morgan Kaufmann, 2009.

[3] C. S. Tan, *3D Integration for VLSI Systems*, C. S. Tan, K.-N. Chen, and S. J. Koester, Eds. Pan Stanford Publishing, 2011.

[4] K. Tu, "Reliability challenges in 3D IC packaging technology," *Microelectronics Reliability*, vol. 51, pp. 517–523, 2011.

[5] T. Song, C. Liu, D. H. Kim, S. K. Lim, J. Cho, J. Kim, J. S. Pak, S. Ahn, J. Kim, and K. Yoon, "Analysis of TSV-to-TSV coupling with high-impedance termination in 3D ICs," in *International Symposium on Quality Electronic Design*, 2011.

[6] D. H. Kim, S. Mukhopadhyay, and S. K. Lim, "Fast and accurate analytical modeling of through-silicon-via capacitive coupling," *IEEE Transactions on Components, Packaging and Manufacturing Technology*, vol. 1, pp. 168–180, 2011.

[7] C. Liu, T. Song, J. Cho, J. Kim, J. Kim, and S. K. Lim, "Full-chip TSV-to-TSV coupling analysis and optimization in 3D IC," in *Design Automation Conference*, 2011.

[8] T. Song, C. Liu, Y. Peng, and S. K. Lim, "Full-chip multiple TSV-to-TSV coupling extraction and optimization in 3D ICs," in *Design Automation Conference*, 2013.

[9] H. Kaul, D. Sylvester, and D. Blaauw, "Active shields: a new approach to shielding global wires," in *Great Lakes Symposium on VLSI*, 2002.

[10] P. Subrahmanya, R. Manimegalai, V. Kamakoti, and M. Mutyam, "A bus encoding technique for power and cross-talk minimization," in *International Conference on VLSI Design*, 2004.

[11] B. Victor and K. Keutzer, "Bus encoding to prevent crosstalk delay," in *International Conference on Computer Aided Design*, 2001.

[12] C. Duan, A. Tirumala, and S. Khatri, "Analysis and avoidance of crosstalk in on-chip buses," in *Hot Interconnects*, 2001.

[13] R. Arunachalam, E. Acar, and S. Nassif, "Optimal shielding/spacing metrics for low power design," in *Computer Society Annual Symposium on VLSI*, 2003.

[14] L. Li, N. Vijaykrishnan, M. Kandemir, and M. Irwin, "A crosstalk aware interconnect with variable cycle transmission," in *Design, Automation and Test in Europe*, 2004.

[15] Y. Y. Chang, Y. C. Huang, V. Narayanan, and C. T. King, "ShieldUS: A novel design of dynamic shielding for eliminating 3D TSV crosstalk coupling noise," in *Asia and South Pacific Design Automation Conference*, 2013.

[16] R. Kumar and S. P. Khatri, "Crosstalk avoidance codes for 3D VLSI," in *Design, Automation and Test in Europe*, 2013.

[17] Z. Xu, A. Beece, D. Zhang, Q. Chen, K. N. Chen, K. Rose, and J. Q. Lu, "Crosstalk evaluation, suppression and modeling in 3D through-strata-via (TSV) network," in *International 3D Systems Integration Conference*, 2010.

[18] M. Stan and W. Burleson, "Bus-invert coding for low-power I/O," *IEEE Transactions on Very Large Scale Integration Systems*, vol. 3, pp. 49–58, 1995.

[19] P. Sotiriadis and A. Chandrakasan, "Low power bus coding techniques considering inter-wire capacitances," in *Custom Integrated Circuits Conference*, 2000.

978-1-4799-2817-0/14 $31.00 © 2014 IEEE

Tackling Close-to-band Passivity Violations in Passive Macro-Modeling

Moning Zhang and Zuochang Ye, *Senior Member, IEEE*

Abstract—Passivity enforcement is important for macro-modeling for passive systems from measured or simulated S-parameter data. State-space systems generated from vector fitting usually present strong passivity violation outside the frequency bandwidth especially in the close-to-band (CTB) region. Removing such close-to-band violation is very difficult with existing passivity enforcement techniques without severely sacrificing the accuracy of the model. In this paper we propose a frequency data extension method which aims to reduce or even eliminate such close-to-band violations without sacrificing model accuracy. The generated model can be used in a later stage for further passivity enforcement. Experiments show that with applying the proposed method, the accuracy of the generated model can be significantly improved.

Index Terms—S-parameter, state-space model, passivity enforcement

I. Introduction

Passivity-enforced macro-modeling for linear dynamic system has been studied for many years. Vector fitting [1] and rational fitting [2] have been widely used in macro-modeling of passive circuit elements in RF circuit design as well as package modeling for signal/power integrity analysis based on tabulated s-parameter data obtained either by measurement or numerical simulation. In such macro-modeling methods, an important issue is the passivity, as it affects the numerical stability in circuit simulation when the macro-model is connected to other networks. Thus to ensure passivity in macro-modeling is a must.

Existing methods for passive macro-modeling can generally be divided into two stages. The first stage employs robust macro-modeling methods, such as the vector fitting [1], to generate models without considering passivity issues. The second stage is to enforce the passivity of the model. Existing passivity enforcement methods can be categorized as convex optimization methods and find-and-fix methods. Convex programming methods [3] generates passive models with optimal accuracy. However, such methods are very expensive. Popular find-and-fix passivity enforcement methods include Eigenvalue Perturbation Method [4]–[6], Frequency Residual Perturbation [7], Local compensation [8]. The common idea in these methods is to locate the passivity violations and fix them either

Moning Zhang is with the Institute of Microelectronics, Tsinghua University, Beijing, 100084, China (Email: zhangmn11@mails.tsinghua.edu.cn.)

Zuochang Ye is with the Institute of Microelectronics, Tsinghua University, Beijing, 100084, China (Email: zuochang@tsinghua.edu.cn.)

This work is supported by National Natural Science Foundation of China (No. 61106031), National 973 Program under Grant 2011CBA00604, 2010CB327403 and Tsinghua University Initiative Scientific Research Program.

by perturbing the residual or by introducing additional poles and residues.

In practice we found that almost all existing methods based on find-and-fix strategy encounter a common problem for realistic industrial cases relating to large CTB passivity violations. A CTB violation is a violation that is locate outside yet very close to the original modeling band. Such violations are very difficult to remove by existing methods without significantly degrading the accuracy.

To address the issues with CTB violations, we propose a robust frequency data extension method. In the proposed method, a set of artificial data at augmented frequency points are created. These artificial data allows vector fitting to generate systems with much less close-to-band violations. The generated system can then be fed into existing passivity enforcement methods to produce passive models. Comparing with passivity enforcement without the data extension, the proposed method generates passive systems with much better in-band accuracy.

The rest of this paper is organized as follows. In Section II we introduce the mathematical background of the problem, and introduce the traditional framework for passivity constrained macro-modeling. Section III we revise the traditional framework adding our passivity extension method. Section IV we give a typical example to illustrate the effect of our proposed extension method and compare it with another possible extension method.

II. Background

A. Macro-modeling Passive System

A general linear system can be described using state-space representation as

$$\dot{x} = Ax + Bu \qquad (1)$$

$$y = Cx + Du \qquad (2)$$

The corresponding frequency domain transfer function can be written as

$$H(s) = C(sI - A)^{-1}B + D \qquad (3)$$

Suppose we are given a set of tabulated data $H_{origin}(s_k), 1 \leq k \leq N_o$ for a passive system obtained via measurement or simulation, and N_o refers to the number of the frequency samples and s_k represents each complex sampling frequency. If this system is passive and described using hybrid representation, then these original data should satisfy the following inequalities

$$H(s_k) + H(s_k)^H \geq 0, 1 \leq k \leq N_o \qquad (4)$$

978-1-4799-2817-0/14 $31.00 © 2014 IEEE

Macro-modeling this passive system is to find a state-space model to fit the data via minimizing the following error metric

$$Err = \sum_{k=1}^{N_o} |H_{origin}(s_k) - H(s_k)|^2, \qquad (5)$$

The state-space model should satisfy the passive condition

$$H(s) + H(s)^H \geq 0, s = j\omega \text{ from } -\infty \text{ to } +\infty \qquad (6)$$

B. Existing Methods for Passive Modeling

Most of existing modeling algorithms adopt the two-stage framework as follows. The first stage employs robust rational approximation methods, such as VF (Vector-Fitting) to generate an initial model as the start point, at this stage we temporarily ignore the passive constraints and only ensure the accuracy. The second stage performs iterative enforcement to fix the passivity. Except for algorithms that performs global optimizations which are computationally expensive, such as the SDP method [3], the passivity enforcement procedure can generally be divided as finding and fixing the violations as described in Fig 1. The common problem of such methods is that they are all suitable only for fixing small violations. Applying such methods for problems with large violations will often lead to convergence issue and/or significant accuracy degradation.

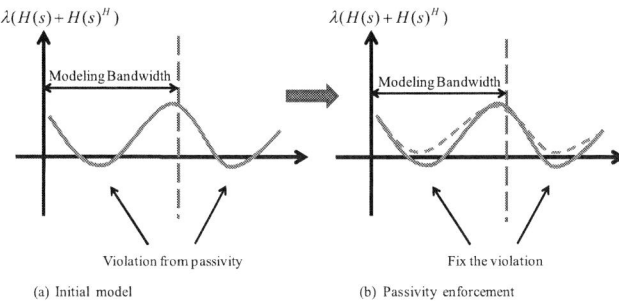

Fig. 1.   Illustration of traditional two-stage framework.

III. CTB VIOLATION AND PASSIVE DATA EXTENSION

In this section, we will first introduce a common issue in traditional passive macro-modeling methods caused by large CTB passivity violation, and then we will propose a method to address this issue by passive data extension.

A. The Issue of Large CTB Violation

Consider a system generated by fitting techniques like vector fitting from tabulated data measured for a passive system. The modeling frequency band is

$$f_{min} \leq f \leq f_{max} \qquad (7)$$

A passivity violation is defined as any frequency region where the passivity condition is violated. Passivity violations can be divided as in-band violations and out-of-band violations, depending on whether the violation is inside or outside the

modeling band. CTB violations are those out-of-band violations which are close to the modeling band.

Comparing to in-band violations and other out-of-band violations, CTB violations are much more difficult to remove. This is because in-band passivity violations are usually small provided that the original data is passive and the vector fitting is done properly, and out-of-band violations which are far from the modeling band can be removed by local compensation with low-order high-pass filters [8], as illustrated in Fig 2(a). In order to use the same compensation strategy to fix a CTB violation, the high-pass filter must be a sharp-edge high-order filter. Unfortunately, it is not known so far how to implement a high order filter in local compensation method.

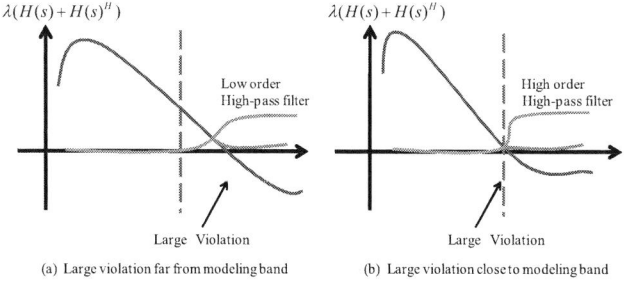

(a) Large violation far from modeling band   (b) Large violation close to modeling band

Fig. 2.   Two different cases with large out-band violation.

B. Frequency Data Extension

The basic idea to fix the CTB violations is to artificially create data points to extend the original data to a higher frequency, and use the augmented data to perform vector fitting, and hopefully the CTB violations in resulting system is reduced or eliminated. The artificial data should be properly designed such that 1) the system generated by vector fitting has less CTB violations, and 2) the accuracy of vector fitting, measured by comparing with the original data should be acceptable. After the data extension, local compensation with low-order high-pass filter can be applied to fix the remaining out-of-band passivity violations illustrated in Fig 3.

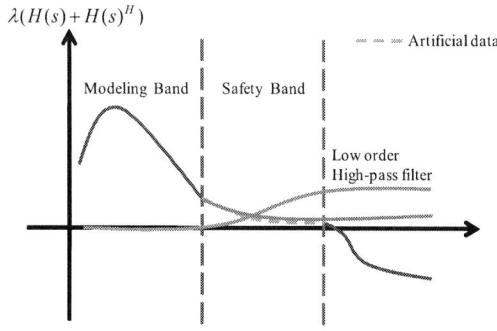

Fig. 3.   Fix original framework by adding safety bandwidth.

The question now becomes how to design the artificial data to fulfill the above two conditions. Firstly, we divide the whole positive frequency axis into three parts,

$$0 \leq f \leq f_{max}, \text{ modeling band}$$
$$f_{max} < f \leq f_{max} + BW_{safety}, \text{ safety band}$$
$$f_{max} + BW_{safety} < f < +\infty, \text{ far band}$$

The most concerning part is the second part, i.e. the safety band. The task now becomes to design an artificial transfer function matrix $\widetilde{H}(s_l)$ at pre-given frequencies s_l in the safety band. Now we can formulate the frequency extension problem as an optimization problem shown below.

$$
\begin{aligned}
\text{Variable}: \ & A, B, C, D \\
& \widetilde{H}(s_l), \ s_l = j2\pi f_l, f_l \text{ travel through} \\
& \text{safety band} \\
\text{Object}: \ & \text{minimize} \sum_l |H(s_l) - \widetilde{H}(s_l)|^2, s_l \text{ is same} \\
& \text{as above} \\
\text{Constraint}: \ & \widetilde{H}(s_l) + \widetilde{H}(s_l)^H \geq 0 \\
& Err < \delta_{max}
\end{aligned}
\tag{8}
$$

Where $A, B, C, D, H(s)$ and Err are the same as that defined in (3), (4) and (5) and δ_{max} is the predefined modeling accurracy tolerance.

To be clear, we also formulate the original global passivity enforcement problem in an optimization problem form.

$$
\left\{
\begin{aligned}
\text{Variable}: \ & A, B, C, D \\
\text{Object}: \ & \text{minimize } Err \\
\text{Constraint}: \ & \text{eq (6)}
\end{aligned}
\right.
\tag{9}
$$

In general, the global passivity condition, namely eq (6) is a very strong constraint, in contrast the two constraints in problem (8) hold in two separate finite bandwidth, and thus it is relatively easier to solve.

However, the optimization problem (8) is not convex thus it is difficult to find a global optimal solution. So in the proposed algorithm we employ a greedy scheme shown in Algorithm 1 to reach an approximate optimal solution.

Algorithm 1 Framework of Frequency Data Extension

---

initialize A, B, C, D by fitting the modeling band;
r=0;
while $Err < \delta_{max}$ and $r < MaxRun$ **do**
    r=r+1;
    Step 1: Detect passivity violations of Model $\{A, B, C, D\}$ in safety band;
    Step 2:Find $\widetilde{H}(s_l)$ which minimize sum $\sum_l |H(s_l) - \widetilde{H}(s_l)|^2$ by fixing passivity violations in safety band;
    Step 3:Find $\{A_1, B_1, C_1, D_1\}$ which minimize sum $\sum_l |H_1(s_l) - \widetilde{H}(s_l)|^2$ by fitting the extended dataset $\{H(s_k), \widetilde{H}(s_l)\}$;
    $\{A, B, C, D\} := \{A_1, B_1, C_1, D_1\}$;
end while

It's not difficult to verify the above pseduo code ensures convergency. It also indicates that this framework would

require two kinds of algorithms, namely a fitting method and a passivity fixing method. As there already exist many robust fitting methods like VF, our task is then to find an efficient passivity fixing algorithm used in Step 2. Theoretically it can be proved that the best fixing (achieve minimum $\sum_l |H(s_l) - \widetilde{H}(s_l)|^2$) is by directly fixing the negative eigenvalues to zero, however this would introduce large discontinuity which will lead to large accuracy degradation in Step 3. So approximate algorithm must be used. We'll discuss this issue in detail in the next section.

C. Passivity Fixing Method

The Hamiltonian matrix [4] or Extended Hamiltonian Pencil [9] can be used to detect passivity violations. After locating the passivity violation region, we need to fix those violations in safety-band. It is necessary to emphasize that with all possible fixing strategy, one must preserve continuity otherwise it may cause the followed fitting to produce large error. This can be illustrated as Fig 4, in which although the original data part can be fitted accurately by methods like VF, after extending an inappropriate part the whole data set can no longer been accurately approximated by rational function.

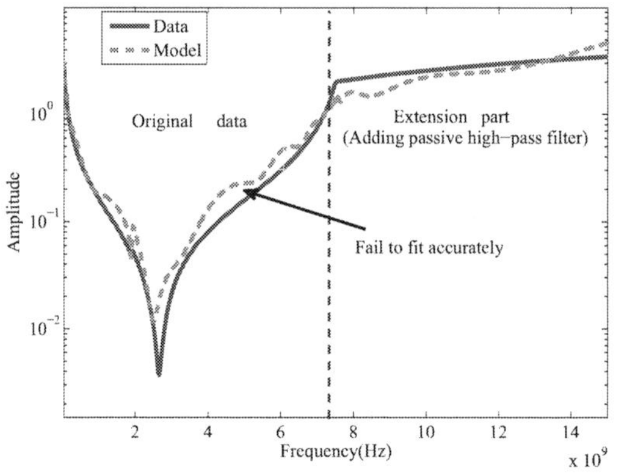

Fig. 4. In-band accuracy greatly affected by extension data.

For each single passivity violation, first-order derivative can be preserved by using a second-order rational function

$$G_{fix}(s) = K \frac{s^2 + as + b}{s^2 + cs + d} \tag{10}$$

The real part of this rational function can be written as

$$\text{Re}(G_{fix}(j\omega)) = K \frac{\omega^4 + (ac - b - d)\omega^2 + ac}{(d - \omega^2)^2 + c^2\omega^2} \tag{11}$$

Thus in order to ensure passivity, when K is set to be positive, it is sufficient to set

$$\Delta = (ac - b - d)^2 - 4ac \leq 0 \tag{12}$$

First derivative condition of two sides as illuatrated in Fig 5. makes another four equations,

$$
\left\{
\begin{aligned}
G_{fix}(P_1) &= f(P_1) \\
G_{fix}(P_2) &= f(P_2) \\
G'_{fix}(P_1) &= f'(P_1) \\
G'_{fix}(P_2) &= f'(P_2)
\end{aligned}
\right.
\tag{13}
$$

By properly setting Δ, we can numerically calculate all coefficients. Although higher-order rational function may also be used, our experiment showed that this second-order fixing can work well enough in most cases.

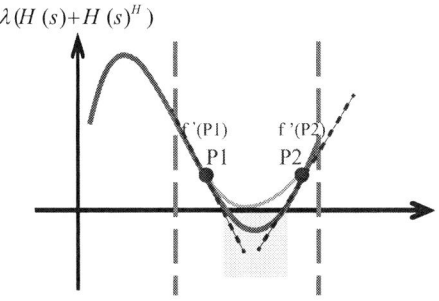

Fig. 5. Fixing with first derivative condition.

D. Asymptotic Passivity

[8] has shown that transfer matrix realized in below form can have high-pass $\mathrm{Re}(\lambda)$ when the coefficients are appropriately given.

$$G_{c0}(s) = \frac{K_1 s^2 + K_2 s}{s^2 + as + b} \qquad (14)$$

$$G_c(s) = \begin{bmatrix} G_{c0}(s) & 0 & \cdots & 0 \\ 0 & G_{c0}(s) & \cdots & 0 \\ \vdots & \vdots & \ddots & \vdots \\ 0 & 0 & \cdots & G_{c0}(s) \end{bmatrix}$$
$$= \mathrm{diag}(G_{c0}, G_{c0}, \cdots, G_{c0}) \qquad (15)$$

By choosing the bandwidth of this high-pass matrix at the end of our extended region, we can achieve asymptotic passivity with acceptable in-band error by simply cascading it.

E. Summarize

Finally, we can summarize the whole realization of the proposed Frequency Data Extension method in Algorithm 2.

It's necessary to emphasize that the proposed scheme is not a passivity enforcement method, i.e. the resulting model does not necessarily be passive. The goal is to make the following enforcement methods easier by removing large CTB violations.

IV. RESULTS

In this section, we verify the proposed algorithm using an example with 14 ports. The modeling bandwith is from 0 to $7.5GHz$.

A. Large Out-band Violation

We first use vector fitting to generate the initial model. The eigenvalue plot of two initial model with different orders are presented in Fig 6.

Both models have large CTB passivity violations at about $8GHz$, which is very close to the modeling bandwidth

---

Algorithm 2 Realization of Frequency Data Extension

Input: $H_{origin}(s_k), 1 \le k \le N_0$, extend length BW_{safety}
Output: A, B, C, D
  $\{A, B, C, D, Err\}$=VF(H_{origin});
  r=0;
  $s_{ext} = s_{N_0} + j2\pi BW_{safety}$;
  $\tilde{H}(s_{N_0} : s_{ext}) = H(s_{N_0} : s_{ext})$;
  while $Err < \delta_{max}$ and $r < MaxRun$ **do**
    r=r+1;
    array$[\omega_1, \omega_2]$=DetectViolation(A, B, C, D);
    for each $[\omega_1, \omega_2] \subseteq [s_{N_0}, s_{ext}]$ **do**
      $P_1 = \omega_1 - \epsilon, P_2 = \omega_2 + \epsilon$;
      $[a, b, c, d, K]$=fsolve(eq.(12)(13));
      for $P_1 \le \omega_l \le P_2$ **do**
        $U\Lambda U^H$=EVD($H(s_l) + H^H(s_l)$);
        $\Lambda(violation) = G_{fix}(s_l)$;
        $\tilde{H}(s_l) = \frac{1}{2}(H(s_l) - H^H(S_l)) + \frac{1}{2}U\Lambda U^H$;
      end for
    end for
    $\{A,B,C,D,Err\}$=VF($[H_{origin}, \tilde{H}]$);
  end while
  $\{A, B, C, D\}$=HPC(A, B, C, D, s_{ext});
  * VF: Vector Fitting; EVD: EigenValue-Decomposition;
  HPC: High-Pass Compensation

$7.5GHz$. Various methods have been tried using these initial model as the initial model. As shown in Table I, they either fail to converge or bring very large error within the modeling band. It can be seen from the result that when increasing the order of VF, the in-band accuracy is improved. However, the CTB violations do not reduce at all.

TABLE I
DIFFERENT PASSIVITY ENFORCEMENT METHOD.

Method	iterations	fitting error
EPM	> 40	
SDP	> 40	
FRP	> 40	
LC	26	large error shown in Fig 11(a)
LC + Freq. Extension	10	small error shown in Fig 11(b)

B. Extension Method

1) Extension by Adding Arbitrary Passive Data: In order to show that our proposed iteration process is necessary, we first test a simple extension method by directly adding a sequence of passive data. We use response of the high-pass local compensating system from [8] as the extension data, and perform window-mean smoothing at the junction region between the original frequency samples and the augmented frequencies in order to ensure enough continuity. The original data, the augmented data and the whole data with smoothing are shown in Fig 7. Fig 8 shows the result for the system generated with the augmented data by vector fitting. Unacceptably large in-band errors are observed. This indicates that only guaranteeing the zeroth-order continuity is not enough to guarantee the accuracy.

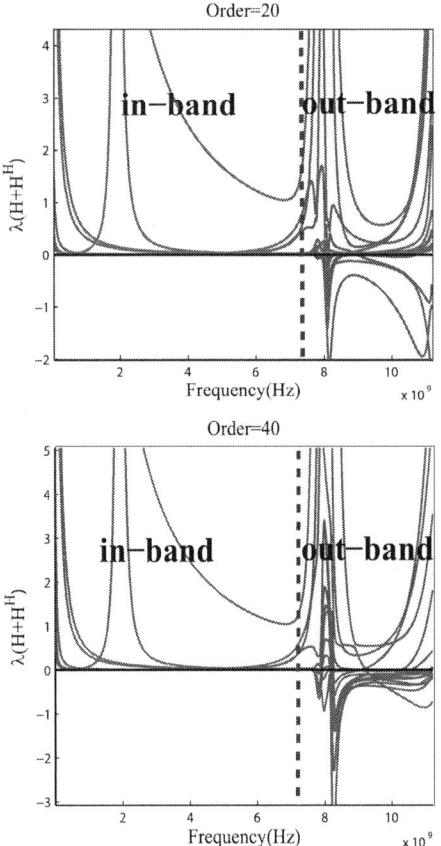

Fig. 6. Eigenvalue plot of two initial model by VF using different order.

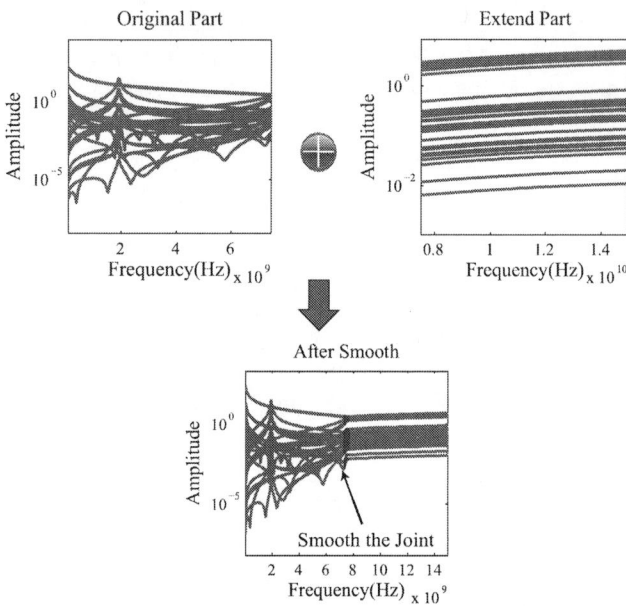

Fig. 7. Entries of the transfer matrix for the original data and new initial model.

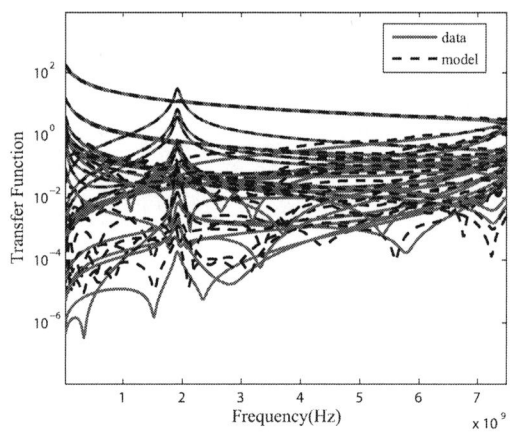

Fig. 8. Entries of the transfer matrix for the original data and initial model generated by comparison extension method.

2) Proposed Extension Method: Fig 9 shows the result after employing the proposed extension method and then use VF to fit the whole data set. The in-band accuracy is satisfactory.

The eigenvalue plot of the new initial model is shown in Fig 10. Compared with Fig 6, the passivity violations around $8GHz$ to $10GHz$ have been ceased to only about $\frac{1}{20}$ compared with the old initial model. Thus this initial model can be considered much easier to be handled by a following passivity enforcing method.

C. Using Local Compensation

Now we can employ passivity enforcement method to fix the passivity iteratively. Here we choose local compensation [8] as the enforcement method. As shown in Fig 11(a), if we do not employ the proposed extension method, local compensation took 26 iterations to produce a passive model. However, in-band modeling error is too large due to the compensation for out-band violation near $8GHz$. In contrast, with the proposed data extension, local compensation took 10 runs to produce a passive model. And the comparison of $\lambda(H + H^H)$ is shown in Fig 11(b). The accuracy is greately improved.

V. CONCLUSIONS

In this paper we address a common issue in traditional passivity macro-modeling methods caused by large close-to-band (CTB) passivity violations. Such violations are very difficult to remove with most of existing methods. We proposed a frequency data extension method to significantly alleviate this issue. The resulting model can be further fed into any of existing passivity enforcement algorithm to generate passive models. The generated passive model is much more accurate compared with passivity enforcement without using the proposed methods when there are large CTB violations.

REFERENCES

[1] B. Gustavsen and A. Semlyen, "Rational approximation of frequency domain responses by vector fitting," *Power Delivery, IEEE Transactions on*, vol. 14, no. 3, pp. 1052–1061, Jul 1999.

[2] C. Coelho, J. Phillips, and L. Silveira, "Robust rational function approximation algorithm for model generation," *Design Automation Conference, 1999. Proceedings. 36th*, pp. 207–212, 1999.

[3] ——, "A convex programming approach for generating guaranteed passive approximations to tabulated frequency-data," *Computer-Aided Design*

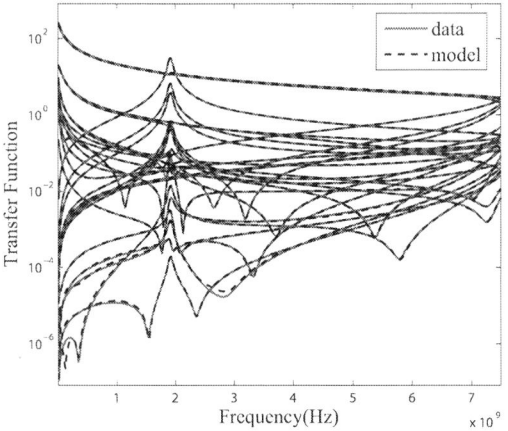

Fig. 9. Entries of the transfer matrix for the original data and initial model generated by proposed extension method.

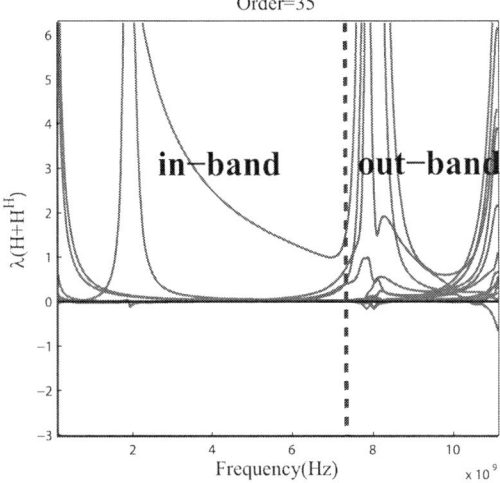

Fig. 10. Eigenvalue plot of the original data and new initial model.

of Integrated Circuits and Systems, IEEE Transactions on, vol. 23, no. 2, pp. 293–301, Feb. 2004.
[4] S. Grivet-Talocia, "Passivity enforcement via perturbation of hamiltonian matrices," Circuits and Systems I: Regular Papers, IEEE Transactions on, vol. 51, no. 9, pp. 1755–1769, Sept. 2004.
[5] S. Grivet-Talocia and A. Ubolli, "On the generation of large passive macromodels for complex interconnect structures," Advanced Packaging, IEEE Transactions on, vol. 29, no. 1, pp. 39–54, Feb. 2006.

[6] S. Grivet-Talocia, "An adaptive sampling technique for passivity characterization and enforcement of large interconnect macromodels," Advanced Packaging, IEEE Transactions on, vol. 30, no. 2, pp. 226–237, May 2007.
[7] B. Gustavsen, "Fast passivity enforcement for pole-residue models by perturbation of residue matrix eigenvalues," Power Delivery, IEEE Transactions on, vol. 23, no. 4, pp. 2278 –2285, oct. 2008.
[8] T. Wang and Z. Ye, "Robust passive macro-model generation with local compensation," Microwave Theory and Techniques, IEEE Transactions on, vol. 60, no. 8, 2012.
[9] Z. Ye, L. M. Silveira, and J. R. Phillips, "Fast and reliable passivity assessment and enforcement with extended hamiltonian pencil," in ICCAD '09: Proceedings of the 2009 International Conference on Computer-Aided Design, 2009, pp. 774–778.

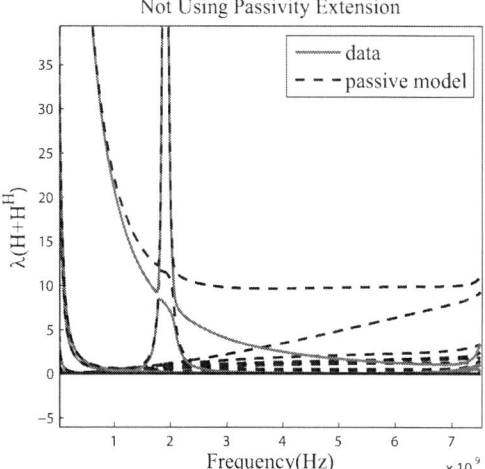

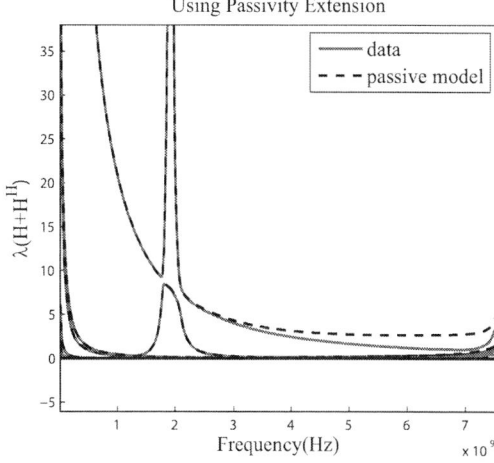

Fig. 11. Eigenvalue plot of the original data and macro-model generate via LC (a) Not using extension method (b) Using extension method

HIE-Block Latency Insertion Method for Fast Transient Simulation of Nonuniform Multiconductor Transmission Lines

Takahiro Takasaki

Dept. of Systems Eng.,
Graduate School of Eng.,
Shizuoka University
3-5-1, Johoku, Naka-ku,
Hamamatsu-shi, 432-8561 Japan
Tel: +81-53-478-1237
Fax: +81-53-478-1269
takasaki@tzasai7.sys.eng.shizuoka.ac.jp

Tadatoshi Sekine

Dept. of Mechanical Eng.,
Shizuoka University

3-5-1, Johoku, Naka-ku,
Hamamatsu-shi, 432-8561 Japan
Tel: +81-53-478-1237
Fax: +81-53-478-1269
sekine@tzasai7.sys.eng.shizuoka.ac.jp

Hideki Asai

Nanovision Research Division,
Research Institute of Electronics,
Sizuoka University
3-5-1, Johoku, Naka-ku,
Hamamatsu-shi, 432-8561 Japan
Tel: +81-53-478-1237
Fax: +81-53-478-1269
hideasai@sys.eng.shizuoka.ac.jp

Abstract— **This paper describes a hybrid implicit-explicit block latency insertion method (HIE-block-LIM) for fast simulation of nonuniform multiconductor transmission lines. In the HIE-block-LIM, an implicit difference method is used with respect to the current variables in one direction, and an explicit method is adopted to update the other variables. The HIE-block-LIM can alleviate a time step size limitation of the existing block-LIM by taking both advantages of the explicit and implicit difference methods.**

I. INTRODUCTION

With the rapid progress of integrated circuit (IC) technology, high-density and high-frequency electronic circuits have been designed. Therefore, signal delay, reflection, and cross-talk occur notably and cause unexpected errors on tightly coupled multiconductor transmission lines (MTLs) [1, 2]. In order to estimate those effects in a design stage, circuit-based modeling and simulation techniques are adequate in terms of accuracy and efficiency. Although general-purpose SPICE-like simulators are applicable to such an estimation, their algorithms based on a direct matrix solver are not efficient in transient simulation of a large network.

Recently, the latency insertion method (LIM) has been proposed as one of the fast simulation techniques [3]. LIM is based on an explicit leapfrog scheme and can be simulated more than ten times faster than SPICE-like simulators. However, the time step size used in the method is limited by the minimum value of reactive elements such as capacitance and inductance in the circuit because the leapfrog scheme is an explicit difference method which has a strict numerical stability condition [4–6]. On the other hand, the block-LIM has been proposed to extend LIM to simulations of the MTLs. The time step size of the block-LIM is still limited as with LIM. Meanwhile, in the field of computational electromagnetics, the finite-difference time-domain (FDTD) method is well-known and adopts the leapfrog scheme. Instead of the values of reactive elements, the maximum time step size used in the FDTD method is restricted by the minimum cell size due to the Courant-Friedrichs-Lewy (CFL) condition [7]. It is important to note that there is an anal-

ogy between the time step size limitations of the LIM-based algorithm and FDTD method: small values of the reactive elements are extracted indeed from the fine meshes, and the small reactances and fine meshes induce the small time step sizes. Therefore, if an object to be analyzed includes extremely fine components and they result in the small reactances, the computational cost of the explicit leapfrog scheme increases significantly.

In order to alleviate the CFL condition and generate a weakly conditionally stable electromagnetic field solver, a hybrid implicit-explicit (HIE)-FDTD has been proposed [8–11]. In the HIE formulation, if there exists an object which needs the fine mesh only in one direction, e.g., vertical direction, as the case of PCBs, an implicit difference method is used with respect to the differential equations associated with that direction; the other equations along with the horizontal directions are solved using the explicit method. Because the implicit method is unconditionally stable, the time step size, which is used in the explicit method, depends only on the large cell sizes in the horizontal directions.

In this paper, we propose the HIE-block-LIM by combining the block-LIM and HIE formulation for the fast transient simulation of the equivalent circuit of nonuniform MTLs. The proposed method applies the implicit scheme to a local area which is composed of inductance elements with small values. By using the proposed method, the time step size can be chosen depending only on the relatively-large inductance component, and the larger time step size than that of the conventional block-LIM can be used.

The remainder part of the paper is organized as follows. In Section II, we explain the nouniform MTLs. We describe the basic LIM and block-LIM in Section III. Section IV describes the formulation of the HIE-block-LIM. Section V shows some numerical results, and the conclusion is given in Section VI.

II. NONUNIFORM MTLs

The cross-section view of the nonuniform MTLs focused in this paper is shown in Fig. 1. The MTL in Fig. 1 has the three metal layers: the top and bottom layers have the signal

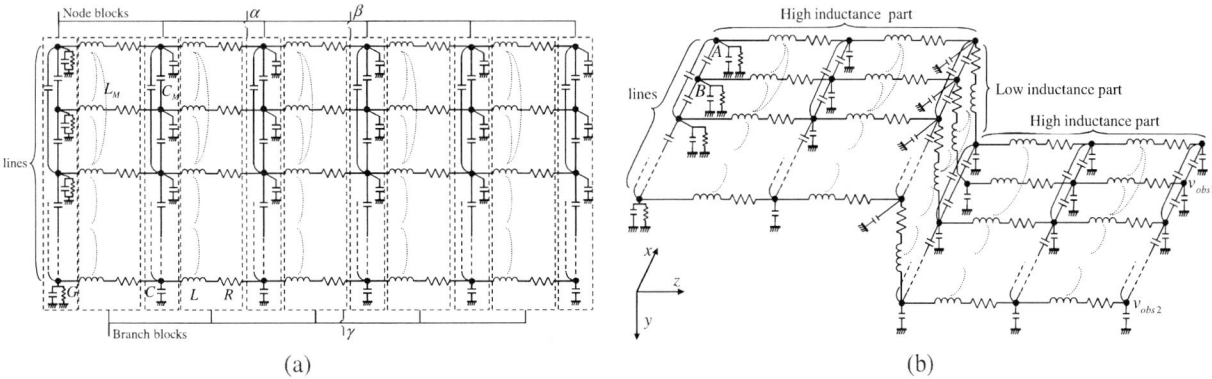

Fig. 2. The equivalent circuit of a nonuniform MTL which is composed of 128 transmission lines.

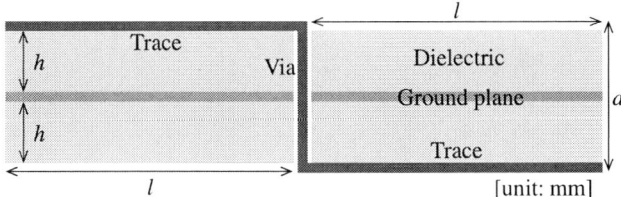

Fig. 1. The structure and dimension of the nonuniform MTL.

traces, the middle layer is the ground plane, and each space between the metal layers is filled with the dielectric. The ground plane has the holes through which the top and bottom traces are connected to each other by using the vertical vias to provide a signal paths from top to bottom. Although we assume that there are actually many signal traces in parallel, only one of them is depicted in Fig. 1 for simplicity. The lengths of the traces in the top and bottom layers are l mm, the vertical via is d mm in length, the thickness of each dielectric layer is h mm, and the relative permittivity of the dielectric is $\varepsilon_r = 4.2$. The equivalent circuit model of the entire MTL is illustrated in Fig. 2(a). In Fig. 2(a), R, L, C, G, L_M, and C_M are the resistance, inductance, capacitance, conductance, mutual inductance, and mutual capacitance. For the explanation of the proposed method, the equivalent circuit in Fig. 2(a) is redescribed in a three-dimensional equivalent circuit shown in Fig. 2(b). In addition, the high and low inductance parts mean the sub-circuits with large and small inductance components, respectively, and correspond to the vias and traces. From the viewpoint of the three-dimensional coordinate system, the low inductance part is allocated in the y-direction, and the high inductance parts are allocated in the z-direction.

III. BACIC LIM AND BLOCK-LIM FORMULATION

A. Basic LIM

In the basic LIM, it is assumed that the circuit to be analyzed is composed of a number of branch and node topologies shown in Fig. 3(a) and (b). The branch topology, where the

branch current i_{ab} flows, is represented as serial combination of the inductance L_{ab}, resistance R_{ab}, and independent voltage source E_{ab}. On the other hand, the node topology includes the voltage variable v_a at the node a and is represented as parallel combination of grounded elements, namely, capacitance C_a, conductance G_a, and independent current source H_a, where N_b^a is the number of branch topologies connected to the node a. In order to derive the updating formulas of the basic LIM, we first apply Kirchhoff's current law (KCL) to the node and Kirchhoff's voltage law (KVL) to the branch in Fig. 3. Then applying the leapfrog scheme to the KCL and KVL equations leads to the updating formulas

$$v_a^{n+\frac{1}{2}} = \frac{C_a}{C_a + \Delta t G_a} v_a^{n-\frac{1}{2}} + \frac{\Delta t}{C_a + \Delta t G_a} \left(-\sum_{k=1}^{N_b^a} i_{ak}^n + H_a \right), \quad (1)$$

$$i_{ab}^{n+1} = \frac{L_{ab} - \Delta t R_{ab}}{L_{ab}} i_{ab}^n + \frac{\Delta t}{L_{ab}} \left(v_a^{n+\frac{1}{2}} - v_b^{n+\frac{1}{2}} + E_{ab}^{n+\frac{1}{2}} \right), \quad (2)$$

where Δt is the time step size, and n denotes the index of the time step. In addition, the maximum time step size $\Delta t_{\max}$ used in LIM is limited by

$$\Delta t_{\max} < \sqrt{2} \min_{a=1}^{N_n} \left(\sqrt{\frac{C_a}{N_b^a} \min_{m=1}^{N_b^a} (L_{am})} \right) \quad (3)$$

where N_n is the number of the node topologies in the network [4]. From (3), it is confirmed that $\Delta t_{\max}$ depends on the small reactive elements. Thus, if the small reactive elements exist in the circuit, the efficiency of the basic LIM is reduced significantly.

B. Block-LIM

The block-LIM is suitable for the fast transient analysis of a tightly coupled circuit constructed by connecting a number

978-1-4799-2817-0/14 $31.00 © 2014 IEEE

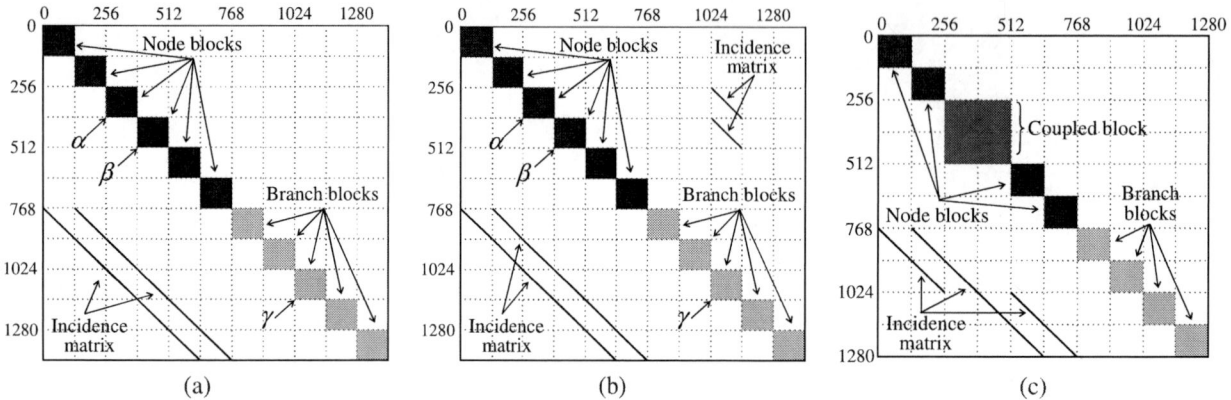

Fig. 5. Structural representations of the coefficient matrices associated with the circuit shown in Fig. 2. (a) For the block-LIM. (b) For the HIE-block-LIM. (c) For the HIE-block-LIM eliminating the block sub-matrix γ.

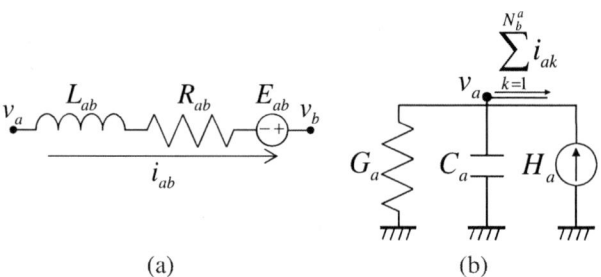

Fig. 3. Circuit topologies required for LIM. (a) Branch topology. (b) Node topology.

of node and branch blocks shown in Fig. 4. The branch block includes n_b branches, each of which is composed of the series-connected resistance R_m, self inductance L_m, and voltage source E_m, and the mutual inductances $L_{p,q}$ exist among the self inductances of the branches, where $m, p, q = 1, 2, ..., n_b$, and $p < q$. On the other hand, in the node block with n_n nodes, the conductance G_m, capacitance C_m, and current source H_m are connected in parallel between a node m and the ground, and the mutual capacitances $C_{p,q}$ exist among the nodes, where $m, p, q = 1, 2, ..., n_n$, and $p < q$. We define the node voltages v_m and the branch currents i_m as the unknown variables in the circuit. Note that the network of the nonuniform MTLs shown in Fig. 2 is composed of the node and branch blocks which are connected alternately.

In order to derive the updating formulas of the block-LIM, we first apply KCL to the node block and KVL to the branch block. After that, we get

$$\mathbf{G}_a \mathbf{v}_a + \mathbf{i}_a + \mathbf{C}_a \frac{d}{dt} \mathbf{v}_a = \mathbf{h}_a \qquad (4)$$

$$-\mathbf{v}_{ab} + \mathbf{R}_{ab}\mathbf{i}_{ab} + \mathbf{L}_{ab}\frac{d}{dt}\mathbf{i}_{ab} = \mathbf{e}_{ab}. \qquad (5)$$

The matrices and vectors in (4) and (5) are defined in [12]. Finally, applying the leapfog scheme to (4) and (5) leads to the

updating formulas

$$\left(\frac{1}{\Delta t}\mathbf{C}_a + \mathbf{G}_a\right)\mathbf{v}_a^{n+\frac{1}{2}} = \frac{1}{\Delta t}\mathbf{C}_a\mathbf{v}_a^{n-\frac{1}{2}} - \mathbf{i}_a^n + \mathbf{h}_a^n \qquad (6)$$

$$\frac{1}{\Delta t}\mathbf{L}_{ab}\mathbf{i}_{ab}^{n+1} = \left(\frac{1}{\Delta t}\mathbf{L}_{ab} - \mathbf{R}_{ab}\right)\mathbf{i}_{ab}^n + \mathbf{v}_{ab}^{n+\frac{1}{2}} + \mathbf{e}_{ab}^{n+\frac{1}{2}}. \qquad (7)$$

The block-LIM is based on the leapfrog time-marching procedure as with the basic LIM, and the amount of the updating calculations can be substantially reduced by using the LU factorization individually within each block. The maximum time step size used in the block-LIM can be estimated by (3) [12]. Therefore, for the simulation of the nonuniform MTLs in Fig. 2, the maximum time step size of the block-LIM is restricted due directly to the small inductance in the low inductance part.

IV. HIE-BLOCK-LIM

If we apply the block-LIM formulation to the MTLs shown in Fig. 2, the coefficient matrix of the entire circuit equation constructed by assembling the updating formulas of all blocks has the block diagonal structure illustrated in Fig. 5(a). In this case, we assume that the number of the signal traces is 128, $l = 12$, and the traces and vias are divided into five branch blocks and six node blocks. In Fig. 5(a), the block sub-matrix γ is the branch block related to the current variables in the y-direction, the block sub-matrices α and β denote the node blocks connected to the y-directional branch block. And the dimension of each block sub-matrix equals the number of the transmission lines, i.e., 128.

In order to derive the updating formulas of the proposed method, we first deal with the KCL equations associated with the sub-matrices α and β. By splitting the current vector in each KCL equation into the vectors $\mathbf{i}_\gamma$ and $\tilde{\mathbf{i}}_\alpha$, which contain the currents in the y and z -directions, respectively, the KCL equations are rewritten as

$$\mathbf{G}_\alpha \mathbf{v}_\alpha + \tilde{\mathbf{i}}_\alpha + \mathbf{i}_\gamma + \mathbf{C}_\alpha \frac{d}{dt}\mathbf{v}_\alpha = \mathbf{h}_\alpha \qquad (8)$$

978-1-4799-2817-0/14 $31.00 © 2014 IEEE

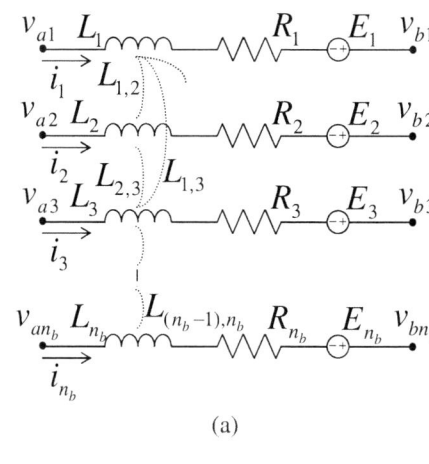

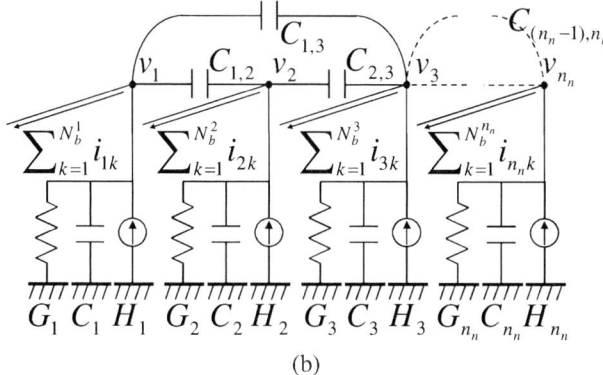

Fig. 4. Circuit topologies required for the block-LIM. (a) Branch block. (b) Node block.

$$\mathbf{G}_\beta \mathbf{v}_\beta + \tilde{\mathbf{i}}_\beta + \mathbf{i}_\gamma + \mathbf{C}_\beta \frac{d}{dt}\mathbf{v}_\beta = \mathbf{h}_\beta \qquad (9)$$

where the vector with tilde denotes the current vector containing the z-directional currents. Next, applying the backward difference method with respect to $\mathbf{i}_\gamma$ and the leapfrog scheme with respect to the other variables leads to

$$\left(\frac{1}{\Delta t}\mathbf{C}_\alpha + \mathbf{G}_\alpha\right)\mathbf{v}_\alpha^{n+\frac{1}{2}} = \frac{1}{\Delta t}\mathbf{C}_\alpha \mathbf{v}_\alpha^{n-\frac{1}{2}} - \tilde{\mathbf{i}}_\alpha^n - \mathbf{i}_\gamma^{n+\frac{1}{2}} + \mathbf{h}_\alpha^n \quad (10)$$

$$\left(\frac{1}{\Delta t}\mathbf{C}_\beta + \mathbf{G}_\beta\right)\mathbf{v}_\beta^{n+\frac{1}{2}} = \frac{1}{\Delta t}\mathbf{C}_\beta \mathbf{v}_\beta^{n-\frac{1}{2}} - \tilde{\mathbf{i}}_\beta^n - \mathbf{i}_\gamma^{n+\frac{1}{2}} + \mathbf{h}_\beta^n. \quad (11)$$

On the other hand, applying the backward difference method to (2) leads to

$$\mathbf{i}_\gamma^{n+\frac{1}{2}} = \left(\frac{1}{\Delta t}\mathbf{L}_\gamma + \mathbf{R}_\gamma\right)^{-1}$$
$$\left(\mathbf{v}_\alpha^{n+\frac{1}{2}} - \mathbf{v}_\beta^{n+\frac{1}{2}} + \frac{1}{\Delta t}\mathbf{L}_\gamma \mathbf{i}_\gamma^{n-\frac{1}{2}} + \mathbf{e}_\gamma^{n+\frac{1}{2}}\right). \quad (12)$$

Then, the coefficient matrix of the entire circuit equation in the block-LIM is changed to that of the HIE-block-LIM shown in 5(b). In (10) and (11), there exist the node voltages and branch currents that are arranged in the same time point. Specifically, in Fig. 5(b), the block sub-matrices α and β are coupled by the

block sub-matrix γ. In order to eliminate the current variables at the $(n+1/2)$-th step, (12) is substituted into (10) and (11), and we get the difference equations in which only the node voltages at the $(n+1/2)$-th step are unknowns:

$$\mathbf{T}_\alpha \mathbf{v}_\alpha^{n+\frac{1}{2}} - \hat{\mathbf{K}}\mathbf{v}_\beta^{n+\frac{1}{2}} = \frac{1}{\Delta t}\mathbf{C}_\alpha \mathbf{v}_\alpha^{n-\frac{1}{2}} - \tilde{\mathbf{i}}_\alpha^n - \hat{\mathbf{K}}\frac{1}{\Delta t}\mathbf{L}_\gamma \mathbf{i}_\gamma^{n-\frac{1}{2}} + \mathbf{s}_\alpha \quad (13)$$

$$\mathbf{T}_\beta \mathbf{v}_\beta^{n+\frac{1}{2}} - \hat{\mathbf{K}}\mathbf{v}_\alpha^{n+\frac{1}{2}} = \frac{1}{\Delta t}\mathbf{C}_\beta \mathbf{v}_\beta^{n-\frac{1}{2}} - \tilde{\mathbf{i}}_\beta^n - \hat{\mathbf{K}}\frac{1}{\Delta t}\mathbf{L}_\gamma \mathbf{i}_\gamma^{n-\frac{1}{2}} + \mathbf{s}_\beta \quad (14)$$

where

$$\hat{\mathbf{K}} \equiv \left(\frac{1}{\Delta t}\mathbf{L}_\gamma + \mathbf{R}_\gamma\right)^{-1}, \quad \mathbf{T}_a \equiv \frac{1}{\Delta t}\mathbf{C}_a + \mathbf{G}_a + \hat{\mathbf{K}}$$

$$\mathbf{s}_a \equiv -\hat{\mathbf{K}}\mathbf{e}_\gamma^{n+\frac{1}{2}} + \mathbf{h}_a^n, \quad a = \alpha, \beta.$$

In this case, because $\mathbf{L}_\gamma$ and $\mathbf{R}_\gamma$ are relatively small block sub-matrices, the calculation cost to derive $\hat{\mathbf{K}}$ is not large. After the above formulation, the coefficient matrix of the entire circuit equation has the structure illustrated in Fig. 5(c). In Fig. 5(c), the node blocks connected to the branch block of the low inductance part are united with each other and become the single and locally dense block. This structure is defined as the coupled block circuit. The updating formula of the voltage variables in the coupled block circuit is derived by combining (13) and (14) in a matrix-vector form as

$$\begin{bmatrix} \mathbf{T}_\alpha & -\hat{\mathbf{K}} \\ -\hat{\mathbf{K}} & \mathbf{T}_\beta \end{bmatrix} \begin{bmatrix} \mathbf{v}_\alpha^{n+\frac{1}{2}} \\ \mathbf{v}_\beta^{n+\frac{1}{2}} \end{bmatrix} = \frac{1}{\Delta t} \begin{bmatrix} \mathbf{C}_\alpha & \mathbf{0} \\ \mathbf{0} & \mathbf{C}_\beta \end{bmatrix} \begin{bmatrix} \mathbf{v}_\alpha^{n-\frac{1}{2}} \\ \mathbf{v}_\beta^{n-\frac{1}{2}} \end{bmatrix}$$
$$- \begin{bmatrix} \tilde{\mathbf{i}}_\alpha^n \\ \tilde{\mathbf{i}}_\beta^n \end{bmatrix} - \begin{bmatrix} \hat{\mathbf{K}}\frac{1}{\Delta t}\mathbf{L}_\gamma \mathbf{i}_\gamma^{n-\frac{1}{2}} \\ \hat{\mathbf{K}}\frac{1}{\Delta t}\mathbf{L}_\gamma \mathbf{i}_\gamma^{n-\frac{1}{2}} \end{bmatrix} + \begin{bmatrix} \mathbf{s}_\alpha \\ \mathbf{s}_\beta \end{bmatrix}. \quad (15)$$

In order to solve (15), we use the direct matrix solver. After that, the current variables in the branch block γ are updated using (12). Finally, the other unknown variables are updated using (6) and (7) as with the block-LIM.

The voltage variables included in the coupled block are connected to the current variables defined in the y-direction. Therefore, in the proposed updating process, the y-directional unknown currents are solved implicitly, and the other unknown variables are solved explicitly. As a result, the time step size used in the HIE-block-LIM depends only on the reactive elements in the high inductance parts because the unknown variables in the low inductance part are solved stably by using the implicit method. Therefore, the HIE-block-LIM can adopt a larger time step size and perform the faster analysis than the block-LIM.

V. NUMERICAL RESULTS

In order to verify the accuracy and efficiency of the HIE-block-LIM, we perform transient analyses of the circuit shown in Fig. 2 by using the HIE-block-LIM and block-LIM. Then, we set the MTL's parameters as follows: $l = 48$, $d = 0.221$ and $h = 0.1016$, and we assume that the number of the signal traces is 256. The traces and vias are divided into 17 branch blocks

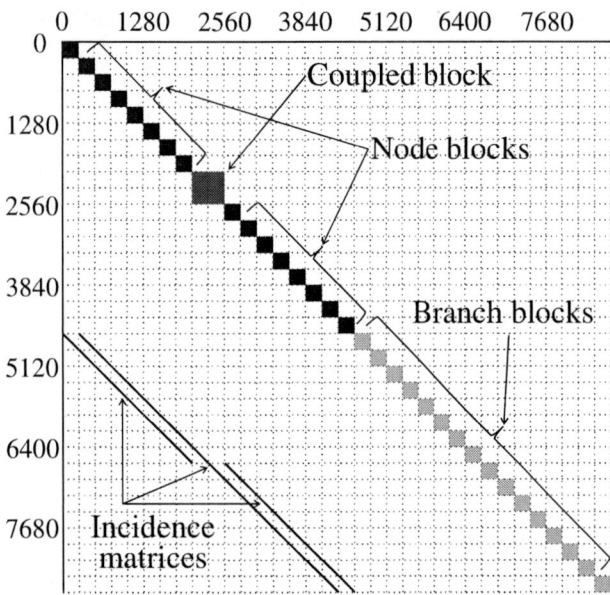

Fig. 6. Structural representations of the coefficient matrix structure of the unknown voltages and currents generated by (6), (7) and (15).

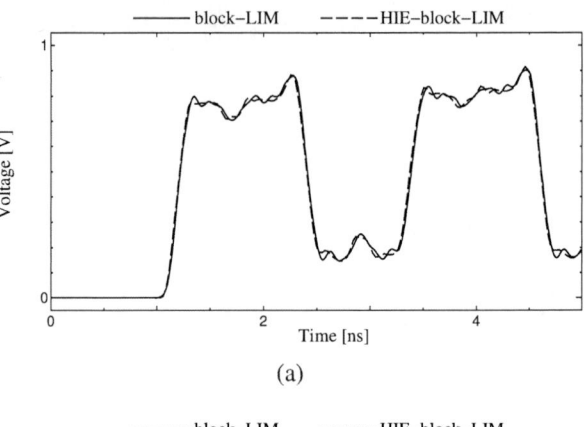

(a)

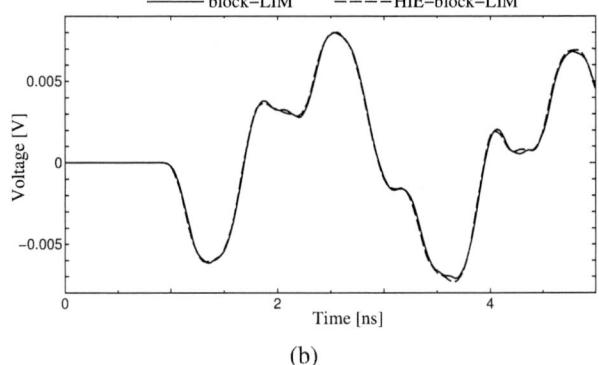

(b)

Fig. 7. Voltage waveforms at the observation points. (a) v_{obs1} and (b) v_{obs2}.

TABLE I
CPU TIME

Method	CPU Time [s]	Speed-up
block-LIM	18.21	1.00
HIE-block-LIM	4.66	3.91
HSPICE	6242.04	2.91e-3

and 18 node blocks. Also, we set the element values as follows: $C = 0.01$ pF and $G = 0.02$ S. Furthermore, we use $L = 0.936$ nH and $R = 33.1$ mΩ for the high inductance parts, and $L = 21.4$ pH and $R = 1.12$ mΩ for the low inductance part. These element values can be obtained by using techniques in [13, 14]. However, the coupling coefficient k of the mutual inductance between the braches is set to 0.015, and the mutual capacitance C_m between the nodes is set to 0.543 fF for simplicity. The maximum time step sizes used in the HIE-block-LIM and the block-LIM can be approximated by (3) [12], and we use $\Delta t = 26.6$ ps with $L = 0.936$ nH in the HIE-block-LIM and $\Delta t = 4.02$ ps with $L = 21.4$ pH in the block-LIM. Although the time step size used in the HIE-block-LIM is larger compared with that in the block-LIM, it is small enough to obtain the accurate waveform. The input current sources H_{in1} and H_{in2} are connected to A and B in Fig. 2. These sources are the same trapezoidal pulse of which the initial value is 0 A, the pulse value is 0.02 A, the rise and fall times are 0.2 ns, the pulse width is 1.0 ns, and the period is 2.2 ns with the exception of the delay times, which are 0.5 ns and 0.6 ns, respectively. Then, we observe the node voltages v_{obs1} and v_{obs2} in Fig. 2(b). The simulation interval is from 0 ns to 5.0 ns.

Fig. 6 shows the coefficient matrix structure of the unknown voltages and currents generated by (6), (7) and (15). The waveform results of v_{obs1} and v_{obs2} are illustrated in Fig. 7. In Fig. 7, it is confirmed that the proposed method can provide nu-

merically stable solution even if the larger time step size than that of the block-LIM is used. The results of the HIE-block-LIM can be evaluated by the feature selective validation (FSV), and the confidence histograms in terms of the global difference measure (GDM) are shown in Fig. 8 [15, 16]. In Fig. 8, it is confirmed that the waveform obtained from the block-LIM and HIE-block-LIM agree well with each other. The CPU times and speed-up ratios of the HIE-block-LIM to the block-LIM are shown in Table I. In Table I, the CPU times of the block-LIM and the HIE-block-LIM are 18.21 s and 4.66 s, and the HIE-block-LIM is about 3.91 times faster than the block-LIM without losing accuracy. Note that the proposed method is substantially fast because the block-LIM is about 343 times faster than HSPICE, and the waveform of the block-LIM completely agrees with that of HSPICE.

Furthermore, it is expected that the proposed method can be used to simulate the power distribution network (PDN). Actually, it has been proven that locally implicit (LI) LIM [17–19] and block-LILIM [20], which are similar to the proposed method, are much faster than the basic LIM and useful for the analysis of PDN.

VI. CONCLUSION

In this paper, we have proposed the efficient circuit simulation technique, the HIE-block-LIM, based on the block-LIM and HIE formulation. In the proposed method, the limit of the time step size could be alleviated by applying the implicit dif-

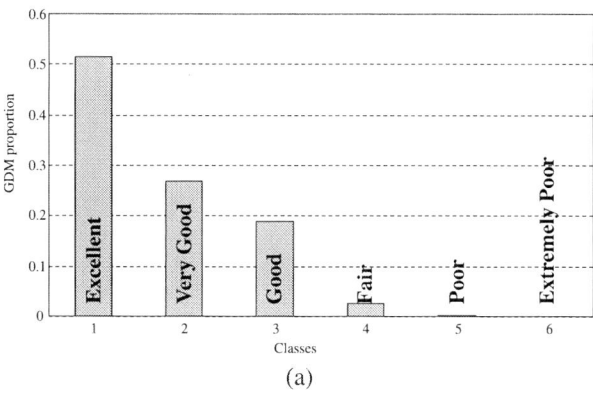

(a)

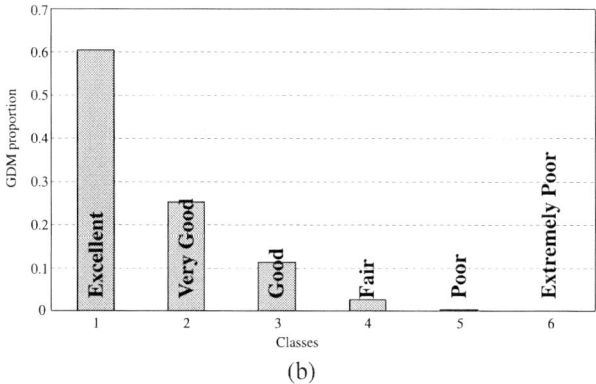

(b)

Fig. 8. The GDM confidence histograms generated by FSV in the cases of (a) v_{obs1} and (b) v_{obs2}.

ference method with respect to the current variables in the y-direction and the explicit difference method with respect to the other variables. Therefore, the proposed method was suitable for the simulation of the nonuniform MTLs. The numerical results showed that the proposed method was about 3.91 times faster than the block-LIM with appropriate accuracy.

ACKNOWLEDGMENTS

This work is partially supported by The Semiconductor Technology Academic Research Center (STARC).

REFERENCES

[1] H. Asai, "Present status and future trend of power/signal integrity problems in chip/package/board codesign," *Journal of signal processing*, vol. 2, no. 6, pp. 433–438, Nov. 2007.

[2] ——, "Advanced PI/SI/EMI simulation technology for 3D co-design," in *Tutorials in ASP-DAC 2011*, Jan 2011, pp. 179–215.

[3] J. E. Schutt-Ainé, "Latency insertion method (LIM) for the fast transient simulation of large networks," *IEEE Trans. Circuits Syst. I*, vol. 48, pp. 81–89, Jan. 2001.

[4] S. N. Lalgudi, M. Swaminathan, and Y. Kretchmer, "On-chip power-grid simulation using latency insertion method," *IEEE Trans. Circuits Syst. I*, vol. 55, no. 3, pp. 914–931, Apr. 2008.

[5] J. E. Schutt-Ainé, "Stability analysis for semi-implicit LIM algorithm," in *Proc. IEEE APMC 2009*, Dec. 2009, pp. 1270–1272.

[6] Z. Deng and J. E. Schutt-Ainé, "Stability analysis of latency insertion method," in *Proc. IEEE EPEP 2004*, Oct. 2004, pp. 167–170.

[7] A. Taflove and S. C. Hagness, *Computational Electrodynamics: The Finite-Difference Time-Domain Method*, 3rd ed. Boston: Artech House Inc., Jun. 2005.

[8] J. Chen and J. Wang, "Numerical simulation using HIE-FDTD method to estimate various antennas with fine scale structures," *IEEE Trans. Antennas Propag.*, vol. 55, no. 12, pp. 3603–3612, Dec. 2007.

[9] ——, "A 3d hybrid implicit-explicit FDTD scheme with weakly conditional stability," *Microw. Opt. Tech. Lett.*, vol. 48, no. 11, pp. 2291–2294, Nov. 2006.

[10] ——, "A three-dimensional semi–implicit FDTD scheme for calculation of shielding effectiveness of enclosure with thin slots," *IEEE Trans. Electromagn. Compat.*, vol. 49, no. 2, pp. 354–360, May 2007.

[11] M. Unno and H. Asai, "HIE-FDTD method for hybrid system with lumped elements and conductive media," *IEEE Microw. Wireless Compon. Lett.*, vol. 21, no. 9, pp. 453–455, Sep. 2011.

[12] T. Sekine and H. Asai, "Block latency insertion method (block-LIM) for fast transient simulation of tightly coupled transmission lines," *IEEE Trans. Electromagn. Compat.*, vol. 53, pp. 193–201, Feb. 2011.

[13] J. D. Kraus, *Electromagnetics with Applications*, 5th ed. New York: McGraw-Hill, Jun. 2005.

[14] C. R. Paul, *Inductance: Loop and Partial*, 1st ed. Hoboken, NJ: Hoboken, NJ: Wiley-IEEE Press, Dec. 2009.

[15] A. P. Duffy, A. J. M. Martin, G. A. A. Orlandi, T. M. Benson, and M. S. Woolfson, "Feature selective validation (FSV) for validation of computational electromagnetics (CEM). part i – the FSV method," *IEEE Trans. Electromagn. Compat.*, vol. 48, no. 3, pp. 449–459, Aug. 2006.

[16] A. Orlandi, A. P. Duffy, G. Antonini, B. Archambeault, G. Antonini, D. E. Coleby, and S. Connor, "Feature selective validation (FSV) for validation of computational electromagnetics (CEM). part ii – assessment of FSV performance," *IEEE Trans. Electromagn. Compat.*, vol. 48, no. 3, pp. 460–467, Aug. 2006.

[17] H. Kurobe, T. Sekine, and H. Asai, "Locally implicit LIM for the simulation of PDN modeled by triangular meshes," *IEEE Microw. Wireless Compon. Lett.*, vol. 22, pp. 291–293, Jun. 2012.

[18] S. Okada, H. Kurobe, T. Sekine, and H. Asai, "Fast transient analysis of power distribution network modeled by unstructured meshes by using locally implicit latency insertion method," in *Proc. IEEE EDAPS 2012*, Dec. 2012, pp. 22–25.

[19] T. Takasaki, T. Sekine, and H. Asai, "Efficient PDN simulaion by locally implicit latency insertion method based on rectangular meshes," in *Proc. IEEE EDAPS 2013*, Dec. 2013.

[20] S. Okada, T. Sekine, and H. Asai, "Locally implicit block-LIM for the simulation of multilayered PDN modeled by triangular meshes," in *Proc. AP-RASC 2013*, Sep. 2013, pp. 1–6.

978-1-4799-2817-0/14 $31.00 © 2014 IEEE

The Role of Photons in Cryptanalysis

Juliane Krämer
Technische Universität Berlin
Security in Telecommunications
10587 Berlin, Germany
juliane@sec.t-labs.tu-berlin.de

Michael Kasper
Fraunhofer Institute for
Secure Information Technology (SIT)
64295 Darmstadt, Germany
michael.kasper@sit.fraunhofer.de

Jean-Pierre Seifert
Technische Universität Berlin
Security in Telecommunications
10587 Berlin, Germany
jpseifert@sec.t-labs.tu-berlin.de

Abstract— Photons can be exploited to reveal secrets of security ICs like smartcards, secure microcontrollers, and cryptographic coprocessors. One such secret is the secret key of cryptographic algorithms. This work gives an overview about current research on revealing these secret keys by exploiting the photonic side channel. Different analysis methods are presented. It is shown that the analysis of photonic emissions also helps to gain knowledge about the attacked device and thus poses a threat to modern security ICs. The presented results illustrate the differences between the photonic and other side channels, which do not provide fine-grained spatial information. It is shown that the photonic side channel has to be addressed by software engineers and during chip design.

I. INTRODUCTION

The role of photons in harming or increasing security has not been considered by manufacturers of smart cards until recently. Current research results on photonic emission analysis, however, reveal that photonic emissions of an integrated circuit can help an attacker to steal secret information [5, 10, 13, 18, 19]. In a photonic side channel attack, the attacker analyses photonic emissions while a device encrypts or decrypts data to reveal the secret key of the cryptographic algorithm.

Side channel attacks are a significant research area since the seminal papers of Kocher in 1996 and 1999, which introduced the timing [8] and the power side channel [9]. Since then, other side channels, e.g., electromagnetic (EM) radiation [14], and various analysis methods, such as template attacks [2] and mutual information analysis [1], have been developed. Most side channel attacks focus on system-wide information leakage, whereas the photonic side channel, which was first introduced in 2008 [5], also allows selective analysis of specific parts of the hardware. Since attacks targeting single transistors are possible, the selectivity of photonic emission analysis greatly exceeds the selectivity of EM analysis. Targeting specific elements of an integrated circuit (IC) results in significantly better signal-to-noise ratios and potentially, signals can be captured that consist entirely of side channel leakage. However, due to the huge cost and complexity of the necessary equipment used in [5], the photonic side channel was not regarded as a realistic threat at that time. Since then, new research has introduced new applications and has demonstrated that photonic side channel attacks can be realized with low-cost equipment [10, 18, 19].

This work gives a thorough overview on the research on the photonic side channel. Known results are recapitulated and

new results are presented. All in all, these research results demonstrate that photonic side channel attacks pose a serious threat to modern security ICs and thus have to be addressed as well by software engineers as during chip design.

Organization.

The remainder of this paper is organized as follows: In Section II, we will go through the related work and explain the AES algorithm, which is the target of all presented attacks. In Section III we introduce Photonic Emission Analysis and present two variants, Simple Photonic Emission Analysis and Differential Photonic Emission Analysis, in Sections IV and V. We conclude in Section VI.

II. BACKGROUND

A. Related Work

The first use of photonic emissions in CMOS for a side channel attack was presented in 2008 [5]. The authors utilized Picosecond Imaging Circuit Analysis (PICA), one of the most complex detector technologies in use today, to spatially recover information about exclusive or operations ($\oplus$). More recently, an integrated PICA system and laser stimulation techniques were used to attack a cryptographic algorithm on an FPGA [3]. The authors partiallly recovered the secret key using temporally resolved measurements. However, the use of equipment worth more than two million Euros did not make such analysis particularly relevant. Most recently, a novel low-cost optoelectronic setup for time- and spatially resolved analysis of photonic emissions was presented [18]. The authors also introduced a corresponding methodology, named Simple Photonic Emission Analysis. They successfully performed such analysis of a proof-of-concept AES implementation and were able to recover the full AES-128 key by monitoring memory accesses. This work was extended to AES-192 and AES-256 [19]. The authors also introduced Differential Photonic Emission Analysis and presented an attack against AES-128 [10]. They successfully revealed the secret key.

In the field of electromagnetic side channel analysis, location-dependent leakage was successfully exploited in an attack on an elliptic curve scalar multiplication implementation on an FPGA using a near-field EM probe [6]. The authors demonstrated that location-dependent leakage can be used in a template attack and countermeasures against system-wide leakage can thus be circumvented.

978-1-4799-2817-0/14 $31.00 © 2014 IEEE

```
1   subBytes:
2     cbi   PORTB, Pin5  ; Set trigger
3     ldi   r24, 0x00    ; i = 0
4   do_subBytes:
5     ld    r30, X       ; Load &state[i]
6     ldi   r31, 0
7     subi  r30, 0xC1
      ; Add SBox low address byte (0x3F)
8     sbci  r31, 0xFD
      ; Add SBox high address byte (0x02)
9     ld    r25, Z       ; Load &SBox + &state[i]
10    st    X+, r25      ; Store new state[i]
11    subi  r24, 0xFF    ; i++
12    cpi   r24, 16      ; i < 16?
13    brne  do_subBytes
14    sbi   PORTB, Pin5  ; Clear trigger
```

Fig. 1. `SubBytes` operation of the attacked AES implementation.

B. The AES Algorithm

The presented work focuses on the Advanced Encryption Standard (AES), since its memory access patterns are particularly susceptible to cryptanalysis. AES is a 128-bit block cipher ratified as a standard by National Institute of Standards and Technology of the USA (NIST). AES operates on a 4×4 matrix of bytes, named the state. Depending on the length of the key, which is 128, 192, or 256 bits, the cipher is termed AES-128, AES-192, or AES-256. The algorithm is specified as a number of rounds that transform the 128-bit input plaintext into the ciphertext. Each round consists of 4 different operations (`SubBytes`, `ShiftRows`, `MixColumns`, `AddRoundKey`), except for the final round, which skips the `MixColumns` operation. Additionally, there is an initial `AddRoundKey` operation before the first round. Regarding AES-128, the secret 128-bit key is used for this initial `AddRoundKey` operation. Each 128-bit round key is derived deterministically from this original secret key.

Photonic emission attacks do not exploit mathematical weaknesses of the AES. Thus, we present only a high level description of the operations in order of appearance. The `AddRoundKey` step combines each byte of the state with a byte of the (round) key using the exclusive or operation ($\oplus$). In the non-linear `SubBytes` step, each byte of the state is substituted by an affine transformation of its multiplicative inverse over the Galois field $GF(2^8)$. Since this deterministic operation is very costly, an 8-bit lookup table is often used. This is called S-Box. Photonic emissions related to the S-Box will be exploited in Sections IV and V. In the `ShiftRows` step, each row of the matrix is shifted to the left by 0, 1, 2 or 3 bytes and in the `MixColumns` step, the four bytes of each column of the state are combined using an invertible linear transformation, resulting in another four bytes.

III. PHOTONIC EMISSION ANALYSIS

In this section, we describe the general methodology for a Photonic Emission Analysis (PEA). A PEA reveals the se-

cret key of a cryptographic algorithm based on an analysis of photonic emissions, measured during the repeated execution of this algorithm. The experimental setup necessary for such attack is also presented in this section.

Photonic emissions pose a side channel since CMOS transistors emit near-infrared light, so-called Hot-Carrier Luminescence, when current flows through the conductive channel. For a standard CMOS-inverter, this creates data-dependent photonic emissions in the following way: If the input is changed from 0 to 1, the n-type transistor will carry a current, emitting photons. For the inverse case the p-type transistor will carry the current. Consequently, optical emissions of CMOS logic show a data-dependent behaviour. Photonic emissions are comparable to power consumption and electromagnetic field emissions. However, in contrast to those, photonic emission is a statistical process. Measurements result in discrete count numbers and the absolute number of detectable photons is very low. A single measurement is only a sequence of binary digits and thus does not carry enough information for an analysis. Therefore, a large number of measurements of the same data has to be collected. For efficiency reasons, we average over many switching operations before storing the data. We call such averaged measurement trace. The number of measurements per trace depends on the optoelectronic setup. For the analyses presented in this paper, depending on the hardware architecture, we averaged 250,000 to 30,000,000 measurements for a single trace, whereas our latest research indicates that as little as 10,000 measurements will be sufficient.

In side channel analysis, the cryptographic key is guessed in small portions, e.g., single bytes, referred to as subkeys. The sets of admissible subkeys and plaintexts, respectively, are denoted with $\mathcal{K}$ and $\mathcal{D}$. In the case of AES, both $\mathcal{K}$ and $\mathcal{D}$ equate to the set $\mathcal{K} = \mathcal{D} = \{0, \ldots, 2^{8-1}\}$. A plaintext byte is given by $d \in \mathcal{D}$ and a subkey candidate by $k \in \mathcal{K}$. The correct subkey is denoted by k°. In general, each byte $b \in \{0, 1, \ldots, 2^8 - 1\}$ is of the binary form $b = b_7|b_6|b_5|b_4|b_3|b_2|b_1|b_0$ with $b_i \in \{0, 1\} \forall i \in \{0, \ldots, 7\}$, i.e., we count the bits starting with the least significant bit. We denote an observation trace of photonic emissions by $\vec{\omega}$. In the case of the attacks presented in this paper, the trace $\vec{\omega}_d$ is recorded while the device encrypts under the fixed secret key the plaintext in which all bytes equate to $d \in \mathcal{D}$, i.e., we collected 256 traces. Each trace consists of T sample points, each of which corresponds to one point in time, i.e., T is the length of the traces and $\vec{\omega}_d = (\omega_{d,1}, \ldots, \omega_{d,T})$. The traces $\vec{\omega}_d$ and their components $\omega_{d,t}$, with $t \in \{1, \ldots, T\}$, refer to averaged real photonic emissions and thus, correspond to a number of count events.

A. Experimental Setup

The experimental setup for the presented attacks is identical to the one employed in [10, 18, 19].

Device under Test To maximize detection efficiency, modern security ICs are best observed from the backside as interconnect layers obstruct observation from the frontside. Detection efficiency, specifically for silicon detector technologies, can be further boosted by thinning the IC substrate with standard backside polishing machines. The device under test

(DUT) used in the presented attacks consists of an inversely soldered ATmega328P supplied with 5 V operating voltage, a 16 MHz quartz oscillator as well as decoupling capacitors and an I/O header for external communications and programming. The backside of the ATmega328P was mechanically polished with an automated backside sample preparation machine. The remaining substrate was approximately 25 μm thick.

The proof-of-concept implementation running on the ATmega328P consists of a software AES implementation, as seen in Fig. 1. To increase the frequency of the execution, only the initial `AddRoundKey` operation and the subsequent `SubBytes` operation were computed on the chip after which the input was reset and the measurement restarted. The AES S-Box was implemented in the microcontroller's data memory, i.e., the SRAM. The ATmega328P microcontroller is based on the AVR architecture. This is an 8-bit architecture with a 16-bit or 32-bit fetch and 16-bit data memory addresses.

Photonic Emission Detector System Detection of hot-carrier luminescence from the backside, i.e., single near-infrared photons, must overcome the issue that charge coupled devices (CCD) exhibit high spatial resolution, but only allow slow frame rates. Single pixel detectors like Photo Multiplier Tubes and Avalanche Photo Diodes (APD), however, offer picosecond timing resolution, but only for one small detection area. The experimental setup, therefore, consists of two detectors optically and electrically connected to the DUT via a custom-built near-infrared microscope. The first detector is a Si-CCD which captures NIR photons below the silicon bandgap energy. The second detector is a single InGaAs/InP Avalanche Photo Diode commonly found in telecom applications (Telcordia GR-468-CORE). The diode is coupled to the microscope via an optical fiber. The fiber's aperture can be freely positioned in the image plane. Areas of interest, identified in an emission image taken by the Si-CCD, can thus be selected for temporal analysis with high spatial selectivity. In gated operation, the APD is rendered sensitive only for a short window in time, called the detection gate, in every signal cycle. To reconstruct the complete signal temporally, the detection gate has to be synchronized and shifted relative to the signal with every signal cycle iteration, similar to a sampling oscilloscope. Provided the gate delay can be controlled with high resolution, the time resolution and inversely the measurement time depend only on the minimal gate width. To implement this detection scheme we used an FPGA-based controller phase-locked to the DUT clock. As the DUT executed the target program code, the phase-locked FPGA digitally delayed and triggered the APD detection gate. Detection events were sent back to the FPGA and counted in the corresponding time bins. The experimental system was constructed with off-the-shelf components and employs readily available technical solutions. In 2013, its price is approximately sixty thousand Euros, which is comparable to that of a mid-range oscilloscope.

IV. SIMPLE PHOTONIC EMISSION ANALYSIS

In this section, we explain the general methodology for a Simple Photonic Emission Analysis and present a practical result against AES which allows total key recovery. The attacks

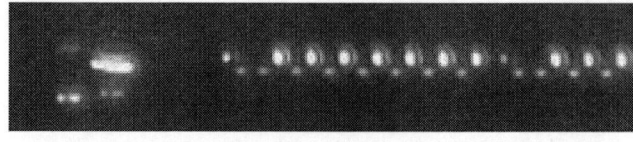

(a) Access to 0x308

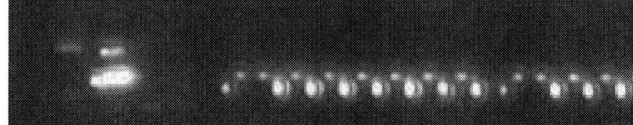

(b) Access to 0x300

Fig. 2. 120 s emission images of memory accesses to two adjacent memory rows of the ATmega328P obtained with the Si-CCD detector.

presented in this section were first described in [18, 19]. In addition to the ATmega328P, we also conducted these attacks on an ATxmega128A1. Though this newer IC includes a hardware AES implementation, it executed the same software AES routine as the ATmega328P.

Simple Photonic Emission Analysis (SPEA) reveals the secret key of a cryptographic device based on traces of photonic emissions that have been recorded while the device encrypts or decrypts different data using the same secret key. The photonic emissions are recorded at specific components of the chips logic, which were previously identified in an emission image. Due to the spatial separation of logic circuits, the traces may even consist entirely of side channel leakage. The photonic emissions whithin the traces show a strong data dependency. Statistical analyses are not required. The traces require little to no additional analysis to recover the secret key.

A. Monitoring SRAM Access

By studying emission images of our AES implementation, we identified the S-Box within memory. It is contained within the SRAM, which is structured in rows and columns. Each of the four 512-byte SRAM banks of the ATmega328P is made up of 64 rows and 64 columns. Thus, each row stores a total of 64 bits, i.e., 8 bytes. Any SRAM cell read begins with assertion of the word line for the entire row [15]. This row select signal is driven by an inverter which is part of the row decode logic [21]. This address and access logic is implemented similarly across all platforms and types of memory. Since the S-Box was implemented as an array of bytes within data memory, its start address depends on the state of the SRAM before the S-Box is loaded. It is therefore not necessarily aligned to the beginning of a memory row. Emission images thus reveal that the S-Box spans 33 rows of the SRAM and not 32 as expected.

The emissions of the row drivers are clearly visible to the left of the individual memory lines. Thus, a row access results in a clearly defined photon detection peak. Expecting input-dependent accesses to the memory rows, the APD detector measures the photonic emissions from the row driver inverter corresponding to a fixed memory row with S-Box elements. With spatial resolution, accesses between even two adjacent rows of SRAM can be clearly differentiated, cf. Fig. 2, which is reused from [18]. The same kind of analysis can also

978-1-4799-2817-0/14 $31.00 © 2014 IEEE

Offset o	Remaining Candidates per Key Byte		Unresolved Bits of the 128-Bit Secret Key	
	$r = 1$ or $r = 33$	$r \in \{2, ..., 32\}$	$r = 1$ or $r = 33$	$r \in \{2, ..., 32\}$
0	8	8	48	48
$\in \{1, 3, 5, 7\}$	**1**	**1**	**0**	**0**
$\in \{2, 6\}$	2	2	16	16
4	4	8	32	48

TABLE I

NUMBER OF CANDIDATES PER KEY BYTE AND UNRESOLVED BITS OF THE WHOLE 128-BIT KEY, DEPENDING ON THE OFFSET AND ROW, WHEN EACH SRAM ROW STORES 8 BYTES. FOR AN OFFSET OF 0, THERE ARE ONLY 32 ROWS CONTAINING S-BOX ELEMENTS.

be applied to the column addresses. The capture of an emission trace $\vec{\omega}_d$ for a row driver on the ATmega328P, being the average of 250,000 measurements, required an acquisition time of 90 seconds. The time to capture the photonic emissions for all 256 input messages amounted to just over six hours.

The ATmega328P and the ATxmega128A1 differ in the size and layout of their SRAM. The ATxmega128A1 has eight 1024-byte SRAM banks and a total of 8 Kilobytes of SRAM. Each bank is made up of 64 rows, but 128 columns, corresponding to 16 bytes per row. For an acceptable trace of a row driver, a higher number of measurements and detection events, respectively, was necessary. This is due to the lower supply voltage and smaller feature size of this architecture as compared to the ATmega328P. For the ATxmega128A1, a single trace took just over two hours and consists of 30,000,000 measurements.

B. SPEA Key Recovery

Since AES operates on bytes, this SPEA works sequentially for all 16 key bytes. As described in Secion III, for an arbitrary but fixed key byte k° we use all corresponding plaintext bytes $d \in \{0, \ldots, 255\}$ as input values. For a fixed secret key byte, the AddRoundKey operation is bijective. Thus, due to these plaintexts, each of the 256 elements in the S-Box will be accessed exactly once. For an S-Box starting at the beginning of a memory row, i.e., with offset $o = 0$, this means that w memory accesses can be observed for each row, with w denoting the memory line width. For an S-Box offset of $o \geq 1$ bytes, the first row of the S-Box in memory will exhibit $w - o$ accesses, and the last row will show o accesses. For a given input value each observed access can be assigned to a number of key candidates: For this access, there are exactly as many key candidates as there are S-Box elements in the observed row; those that could have called an address within that memory line. However, since k° is used during all accesses for all input values, the correct byte k° will be in the intersection of up to w sets of possible key candidates. Table I shows the number of remaining candidates for the whole 128-bit AES key when the attack is conducted against the ATmega328P with a memory line width of $w = 8$ bytes. The respective numbers for the ATxmega128A1 and for an SRAM storing 16 bytes per row, respectively, can be found in Table 2 in [19]. Both tables show that for an uneven offset, this SPEA always results in full key recovery. They also show that the number of remaining candidates per key byte equates to the greatest common divisor of the memory line width w and the offset o, independent of the observed row. The only exception is the case $o = \frac{w}{2}$, when the number of remaining candidates per key byte equates to w for all rows except for the first and last one.

In this SPEA attack, both DUTs had a maximum offset $o = w - 1$, leaving just a single element of the S-Box in the first row of SRAM containing S-Box elements. Thus, all 16 key bytes could be directly inferred from the corresponding single accesses and known plaintext bytes. As soon as an attacker observes as many accesses to a certain row as there are S-Box elements stored, there is no need for further measurements.

Attacking AES-192 or AES-256 The attack can be directly transferred to the first 16 key bytes of AES-192 and AES-256. For the remaining bytes, the S-Box accesses during the SubBytes operation of the second round have to be observed as well. Here, the attack using only those plaintexts with all bytes equal reveals less bits per key byte than expected. This has two different sources.

The first effect only occurs when the S-Box exhibits an even offset and $x > 1$ candidates remain for each of the first 16 key bytes. It stems from the operation MixColumns, during which four state bytes are combined. Thus, after the MixColumns operation, each state byte depends on four former state bytes. In case the first 16 key bytes have not been fully revealed, however, an attacker does not know these four state bytes exactly, but has x candidates for each of them. Consequently, the attacker has up to x^4 candidates for each state byte after the MixColumns operation. The success of the attack against the first SubBytes operation, however, is based on the exact knowledge of the state bytes before the initial AddRoundKey operation. This is ensured since the attacker chooses the plaintexts, which are exactly these state bytes. Since the attacker does not know the value of the state bytes directly before the second AddRoundKey operation, the observed S-Box accesses have to be related to each of these candidates. Thus, if y candidates remain per key byte used during the analysis of the leakage in the second round, there are up to $x^4 \times y$ candidates altogether for each of the key bytes used during the second round.

The second effect generally leads to $y > 1$ and $y > x$: During the revelation of the first 16 key bytes, due to the chosen input messages and the bijective AddRoundKey operation, the S-Box gets accessed at 256 pairwise different addresses for each of the 16 bytes, i.e., all of its elements are accessed exactly once. Regarding the second round, however, the state values leading to the S-Box accesses are the result of the initial `AddRoundKey` operation and the complete first round of AES. Considering only an arbitrary state byte, all operations except `MixColumns` are injective, whereas the `MixColumns` operation maps always four bytes to one byte by mixing these four bytes. Therefore, using just the input messages with all bytes equal, an attacker does not have 256 pairwise different values per state byte after the first round, but considerably less. In fact, we can expect approximately 162 pairwise different accesses [19] and thus, we observe less accesses to a fixed row. Consequently, less key candidates get ruled out. Thus, more key candidates are left for the key bytes used during the second `AddRoundKey` operation, as compared to the first 16 key bytes. However, an attacker can solve this problem by using additional plaintexts.

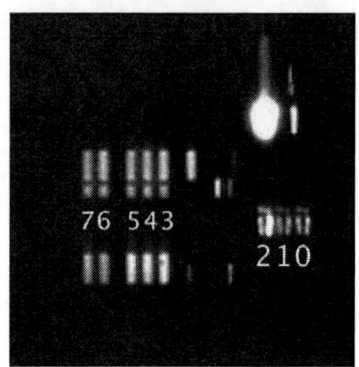

Fig. 3. Emission image of the driving inverters for the second SRAM bank on the ATmega328P. The bit order of the driving inverters is also shown.

C. Countermeasures

The concrete attack presented in this section could be prevented by calculating the byte substitution, rather than using a lookup table during the `SubBytes` operation. However, this would increase the necessary time for encryption and decryption considerably. In addition to this caveat, this would allow for a Differential Photonic Emission Analysis, cf. Section V, since the states before and after the `SubBytes` operation would be stored in the same register directly after each other. However, since the photonic side channel requires averaging many measurements to get a single trace, properly implemented randomized masking schemes can prevent the mere capture of emission traces, since they make the emissions dependent on some unknown, random source.

V. DIFFERENTIAL PHOTONIC EMISSION ANALYSIS

In this section, we explain the general methodology for a Differential Photonic Emission Analysis and present exemplary results of three different distinguishers, which all reveal the whole secret key of a PoC-AES-128 implementation on the ATmega328P. One of the attacks presented in this section was already described in [10].

Differential Photonic Emission Analysis (DPEA) reveals the secret key of a cryptographic device based on a large number of traces of photonic emissions that have been recorded while the device encrypts or decrypts different data using the same secret key. The photonic emissions are recorded at specific components of the chips logic, which were previously identified in an emission image. The spatial separation of logic circuits leads to a low noise level. The data dependency of the intensity of the photonic emissions at certain points in time, which do not have to be known in advance, is exploited by a statistical distinguisher. Such a distinguisher typically is based on methods for statistical analyses.

A. Monitoring Driving Inverters for an SRAM bank

Since a DPEA requires an intermediate result which depends on the input data and on the secret key, we chose to analyze the first `SubBytes` operation, where the absolute S-Box address is read first and next, the S-Box output is loaded and stored as new state. Thus, we attacked the AVR architecture's datapath to recover photonic side channel leakage from the AES calculation. We refer to the notation and pseudocode of the AES subroutine, which was introduced in Fig. 1. The 16 state bytes and the AES S-Box were located in the SRAM of the microcontroller. On the ATmega328P SRAM is mapped to the data memory and is accessed via load (`ld`) and store (`st`) instructions in conjunction with the registers X, Y and Z for data indirect memory addressing. The 8-bit registers `r26` and `r27`, `r28` and `r29`, and `r30` and `r31` form the low and high bytes of 16-bit registers X, Y, and Z, respectively. In the `SubBytes` function register X points to the address of the 16 state bytes. To perform the `SubBytes` operation, a state byte is read (`ld r30, X`). The value of this state byte is the result of the initial `AddRoundKey` operation. Next, this value is used to index the S-Box by adding an offset to this value, i.e., the base address of the S-Box, which was `0x23F`. The S-Box output is loaded and stored and the X pointer is incremented to point to the next state byte.

The ATmega328P has four 512-byte memory banks. Each bank is individually connected to the rest of the datapath. This connection consists of very large driving inverters. The bit order of these was determined by analyzing emission images with the techniques introduced in [13], see Fig. 3, which is reused from [10]. Considering the IC's layout, the transistors form two groups, the five most significant bits (MSB) and the three least significant bits (LSB). Because of the distance between the two groups and the additional enable and clock signals that lie between them, it is impractical to measure the emissions of both groups in a single trace. For this reason we chose to measure the 5 MSB and the 3 LSB separately. The corresponding observations are denoted by $\vec{\omega}_d^\top$ and $\vec{\omega}_d^\perp$ for the MSB and LSB measurements. The union set is given by $\vec{\omega}_d^\cup$. We averaged 1,000,000 measurements for every input value.

B. DPEA Key Recovery

We present three ways to perform a DPEA. First, we present the first published DPEA that led to the recovery of the complete secret key of an AES-128 implementation [10]. For this DPEA, we used the Difference of Means method based on results gained during a previous correlation analysis. Next, we discuss the Pearson correlation, as another classical side channel distinguisher. Finally, we present the first application of the Stochastic Approach as statistical distinguisher for a DPEA.

In order to compare the efficiency of the employed distinguishers, we consider the global success rate (GSR) as quality metric. A DPEA is declared as successful, if all correct key bytes k° are ranked as first key hypotheses. Fig. 4 depicts the GSR of the corresponding distinguishers. The x-axis shows the number of necessary traces according to the GSR, which is mapped on the y-axis and was computed on the basis of 32 experiments, which were performed on random selection of traces from the input data sets. Since all presented methods need considerably less than 256 input data to reveal the whole 128-bit key, all presented analyses can be directly transferred to AES-192 and AES-256 and longer keys, respectively.

Since AES operates on bytes, we attacked and revealed each of the key bytes separately in all presented analyses. Therefore, unless otherwise stated, the descriptions in this section always refer to a fixed but arbitrary byte.

Difference of Means We started to analyze the measured data using the Difference of Means method (DoM), which was already used in Kocher's first work on DPA [9] and has since occasionally been used, e.g., [11, 12]. The DoM method belongs to the partition distinguishers [20]. It requires and exploits reliable information of just a single bit to reveal the whole key if a nonlinear function can be attacked. The general approach is to partition for each key hypothesis $k \in \mathcal{K}$ the traces according to the value of a certain bit (i.e., 1-bit partition) after a nonlinear function has been calculated. For each $k \in \mathcal{K}$, the attacker thus gets two sets of traces, and calculates a mean for both. Afterwards, he computes the difference of these two mean traces - hence the name, Difference of Means. The assumption is that in case a key candidate is wrong, the partition of the two sets is more or less random, so that both mean traces are approximately equal and thus, the difference trace gets drowned out by the noise. However, in case the traces were partitioned according to k° and the emissions of the weighted bit influence the measurements, there is a significant difference in the two mean traces at some point in time and thus, their difference trace will exhibit a peak at this point.

We applied this method to `SubBytes`$(d \oplus k)$. During a preceding correlation analysis, we learned that especially bit 2 is a perfect discriminator. This can also be seen in Fig. 3, where the emissions of bit 2 visibly exceed those of bit 1 or

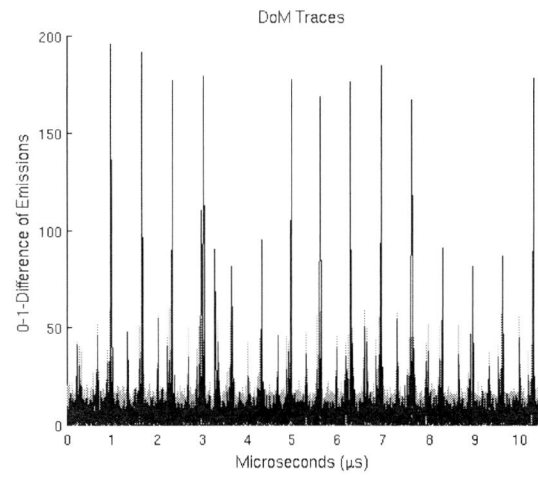

Fig. 5. All 256 DoM traces. The equidistant black peaks correspond to the key bytes. The traces for the remaining key candidates are plotted in gray.

0. Thus, we partitioned for each $k \in \mathcal{K} = \{0, 1, \ldots, 255\}$ the 256 traces $\vec{\omega}_d^\perp$ according to the value `SubBytes`$(d \oplus k)_2$. As can be seen in Fig. 5, we were able to extract the subkeys with a good distinguishability. The explicit equidistant peaks indicate the correct subkey. They show that emissions of a single transistor can be sufficient for a successful DPEA. We also analyzed the whole trace $\vec{\omega}_d^\perp$ and the combined dataset $\vec{\omega}_d^\cup$ and also got perfect results. These results can also be obtained with less than 256 emission traces, see Fig. 4. Using the same sets of traces, but only a random selection of these, we achieved GSR = 1 for only 44 and 56 random inputs for the traces $\vec{\omega}_d^\perp$ and $\vec{\omega}_d^\cup$, respectively.

Pearson Correlation We also analyzed the measured data using Pearson correlation as means of statistical analysis, which proved to be effective in former side channel attacks [11]. By calculating for each possible key the correlation between the leakage function $m(d, k) =$ HammingWeight(`SubBytes`$(d \oplus k)$) for all input data and the actual leakage, the attacker determines the unknown subkey k° by selecting the value k with maximum correlation coefficient ρ_k. Given the leakages $\vec{\omega}_t(\cdot)$, denoting the emissions of all considered input plaintexts at time t, and the corresponding hypotheses vectors of the leakage model $m(d, k)$, the calculation of the coefficient ρ_k is done as follows.

$$\rho_k = \frac{\text{Cov}(\vec{\omega}_t(\cdot), \vec{m}(\cdot, k))}{\sqrt{\text{Var}(\vec{\omega}_t(\cdot)) \cdot \text{Var}(\vec{m}(\cdot, k))}}$$

This distinguisher was applied using the traces $\vec{\omega}_d^\perp$ as well as the combined data set $\vec{\omega}_d^\cup$. We revealed the whole 128-bit secret key. Again, this could also be achieved with only 44 and 56, respectively, different input plaintexts, cf. Fig. 4.

Stochastic Approach We also analyzed the traces with the Stochastic Approach. The Stochastic Approach, which was introduced in [17], approximates the real leakage function of the exploited side channel as a linear combination of basis functions in a suitable vector space. For a mathematical and more

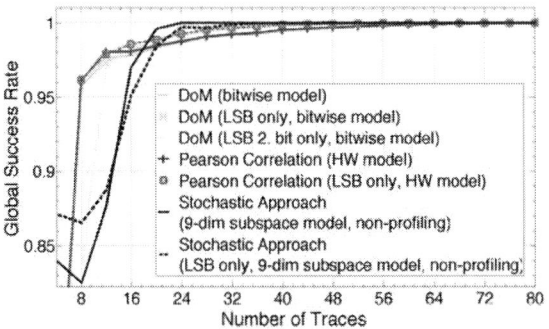

Fig. 4. Global Success Rate for extraction of all subkeys using Difference of Means, Pearson correlation or the Stochastic Approach as distinguisher.

detailed treatment, we refer to [7, 16]. The Stochastic Approach can be applied in two scenarios: The Stoachastic Approach with Profiling requires additional traces from the DUT, measured while the device encrypts data with another secret key, different from k°. Additionally, a variant of the Stochastic Approach can handle non-profiling based scenarios. Schindler et al. [16] first mentioned that the Stochastic Approach also allows key extraction without explicit profiling. Doget et al. substantiate this assumption and give an explicit description, referred to as Linear Regression [4]. We applied the latter approach to the analysis of photonic emissions.

For $k \in \mathcal{K}$ we define the leakage function $h_{t;k} : \mathcal{D} \times k \to \mathbb{R}$, describing the real data- and key-dependent part of the photonic emissions. The Stochastic Approach does not aim at the unknown function $h_{t;k}(\cdot, k)$, but at its best approximator $h_{t;k}^*(\cdot, k)$ in the vector subspace $\mathcal{F}_{u,t;k}$, which is spanned by some basis functions $g_{0,t;k}, \ldots, g_{u-1,t;k} : \mathcal{D} \times k \to \mathbb{R}$, which are selected by the attacker with regard to a sound leakage model. Again, we used the model $\texttt{SubBytes}(d \oplus k)$ and considered the bitwise model, i.e., $u = 9$. In particular, $g_{0,t;k}(d, k) = 1$ and $g_{j,t;k}(d, k) := \texttt{SubBytes}(d \oplus k)_{8-j}$ for $d \in \mathcal{D}$ and $j = 1, \ldots, 8$, where we interprete all bits as integers. The information leakage is estimated with $h_{t;k}^*(\cdot, k)$ for each key candidate $k \in \{0,1\}^8$. In particular, we estimate

$$h_{t;k}^*(\cdot, k) = \sum_{j=0}^{u-1} \beta_{j,t;k}^* \cdot g_{j,t;k}(\cdot, k)$$

with suitable coefficients $\beta_{j,t;k}^* \in \mathbb{R}$. To estimate the $\beta_{j,t;k}^*$, the attacker builds the real-valued $N \times u$-matrix

$$M := \begin{pmatrix} 1 & g_{1,t;k}(d_1, k) & \ldots & g_{u-1,t;k}(d_1, k) \\ \vdots & \vdots & \ddots & \vdots \\ 1 & g_{1,t;k}(d_N, k) & \ldots & g_{u-1,t;k}(d_N, k) \end{pmatrix}.$$

The square matrix $M^T M$ generally is regular, and the equation $M^T M \vec{\beta} = M^T \vec{\omega}_t$ has the unique solution $\vec{\beta}^* := (\beta_{0,t;k}^*, \ldots, \beta_{u-1,t;k}^*)$. Least square estimation provides an estimate $h_{t;k}^*(\cdot, k)$ on basis of the photon emission traces $\vec{\omega}_t(\cdot)$, denoting the emissions of all considered input plaintexts at time t. All estimated functions $h_{t;k}^*(\cdot, k)$ with $k \in \mathcal{K} \backslash k^\circ$ are not related to the deterministic part of the photonic emissions. For the correct subkey k°, however, it is to be expected that the subspace $\mathcal{F}_{u,t;k^\circ}$ is closer to $h_{t;k^\circ}$ than $\mathcal{F}_{u,t;k}$ is for all $k \in \mathcal{K} \backslash k^\circ$. Consequently, we decide for that $k \in \mathcal{K}$ that minimizes $\|\vec{\omega}_t(\cdot) - h_{t,k}^*(\cdot, k)\|$ among all candidates and all time instants t within the interval in which the respective key byte is processed. The Stochastic Approach shows the best performance of all presented methods, with only 24 required random traces $\vec{\omega}_d^\cup$ for a GSR = 1, cf. Fig. 4, although it is less efficient than the other presented distinguishers for less than 16 traces.

C. Constructiveness

For a side channel attack which aims to further improve the design and to implement architecture specific countermeasures, the characterization of the leakage is very interesting. Regarding the protection of cryptographic implementations and to guarantee a desired level of side channel resistance, leakage characterization is of the upmost importance. While

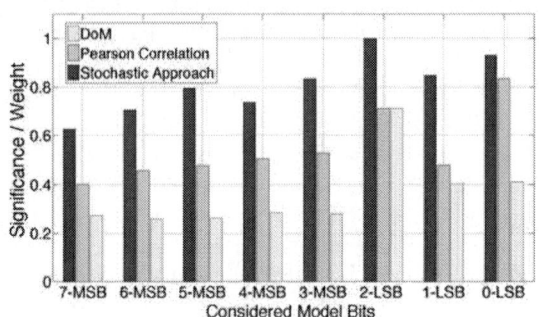

Fig. 6. Constructiveness Information from each of the 8 Bits.

our work on a comprehensive constructiveness analysis is still in process, the analyses presented in this paper already provide relevant constructiveness information.

Fig. 6 shows the scaled photon leakage characteristic for DoM, Pearson correlation and the Stochastic Approach on the ATmega328P. The x-axis represents bit j of the register bank, i.e., $j \in \{7, 6, 5, 4, 3, 2, 1, 0\}$. The analyses presented in this paper led to slightly different leakage characteristics for the applied distinguishers, cf. Fig. 6. However, for all distinguishers, the LSB bits $\{2, 0\}$ show higher absolute values than all remaining bits. In particular, the significance of bit 2 is significantly larger for the DoM distinguisher and also shows a significantly higher contribution for the Stochastic Approach. This is not due to any mathematical reason, but directly related to the chip itself. The Stochastic Approach additionally offers an inherent leakage characteristic through the regression coefficients, the so-called β-characteristic, which is the collection of the absolute values $|\beta_{j,t;k^\circ}|$ for $j > 0$. This characteristic corresponds to the basis vector $g_{j,t;k^\circ}(d, k)$.

This exposes that DPEA helps to gain constructive knowledge about potential vulnerabilities and design flaws of the DUT, since such information supports the chip designer and cryptographic engineers in finding weaknesses of the implementation and attack vectors.

D. Countermeasures

To remove potential attack vectors for photonic side channel attacks, known countermeasures against system-wide leakage might also be successfull. However, the specific characteristics of PEA have to be considered. Hardware countermeasures include the distribution of bits that compose the same byte over different components of the SRAM, preventing a joint signal from them with a good SNR. However, as soon as an attacker finds out the relevant transistors, DPEA attacks can be applied to measurements of even a single transistor. By properly implementing randomized known countermeasures, DPEA can be prevented, since DPEA requires averaging thousands of measurements with the same data for a single trace. However, the lifecycle of an IC generation has to be considered, since advances in optical technologies will lead to better measurement tools, which will help attackers to circumvent countermeasures.

978-1-4799-2817-0/14 $31.00 © 2014 IEEE

VI. CONCLUSION

This work summarizes the research about the exploitation of photonic emissions to reveal the secret key of a cryptographic device up to now. Simple Photonic Emission Analysis and Differential Photonic Emission Analysis are presented. Since photonic side channel attacks are able to exploit photonic emissions even of selected components of the attacked device, they pose a huge threat to modern security ICs and thus have to be addressed by chip designers, chip manufacturers, and software engineers. So they do not only have to consider global side channels, such as power analysis, but also all local attack vectors. From the view of an attacker, the need for averaging thousands of measurements for a single trace is an obvious disadvantage. However, this helps chip designers and software engineers to prevent such analyses by properly implementing randomized countermeasures. To design an implementation resistant to photonic side channel attacks, all potential attack vectors have to be considered and known countermeasures have to be adjusted. When designing implementations with such countermeasures, the continuous development in optical technologies also has to be considered: For the first results on the photonic side channel, one million measurements were needed for a single trace, whereas latest research shows that as little as ten thousand measurements are sufficient.

REFERENCES

[1] BATINA, L. ; GIERLICHS, B. ; PROUFF, E. ; RIVAIN, M. ; STANDAERT, F.-X. ; VEYRAT-CHARVILLON, N. : Mutual Information Analysis: a Comprehensive Study. In: *J. Cryptology* 24 (2011), Nr. 2, S. 269–291

[2] CHARI, S. ; RAO, J. R. ; ROHATGI, P. : Template Attacks. In: *Cryptographic Hardware and Embedded Systems CHES 2002*, 2002, S. 13–28

[3] DI-BATTISTA, J. ; COURREGE, J.-C. ; ROUZEYRE, B. ; TORRES, L. ; PERDU, P. : When Failure Analysis Meets Side-Channel Attacks. In: *Cryptographic Hardware and Embedded Systems CHES 2010*

[4] DOGET, J. ; PROUFF, E. ; RIVAIN, M. ; STANDAERT, F.-X. : Univariate side channel attacks and leakage modeling. In: *J. Cryptographic Engineering* 1 (2011), Nr. 2, S. 123–144

[5] FERRIGNO, J. ; HLAVÁČ, M. : When AES blinks: introducing optical side channel. In: *Information Security, IET* 2 (2008), Nr. 3, S. 94 –98

[6] HEYSZL, J. ; MANGARD, S. ; HEINZ, B. ; STUMPF, F. ; SIGL, G. : Localized Electromagnetic Analysis of Cryptographic Implementations. In: *Topics in Cryptology CT-RSA 2012*. Springer Berlin / Heidelberg, 2012, S. 231–244

[7] KASPER, M. ; SCHINDLER, W. ; STÖTTINGER, M. : A stochastic method for security evaluation of cryptographic FPGA implementations. In: *FPT*, IEEE, 2010, S. 146–153

[8] KOCHER, P. C.: Timing Attacks on Implementations of Diffie-Hellman, RSA, DSS, and Other Systems. In: *CRYPTO*, 1996, S. 104–113

[9] KOCHER, P. C. ; JAFFE, J. ; JUN, B. : Differential Power Analysis. In: *CRYPTO*, 1999, S. 388–397

[10] KRÄMER, J. ; NEDOSPASOV, D. ; SCHLÖSSER, A. ; SEIFERT, J.-P. : Differential Photonic Emission Analysis. In: *COSADE*, 2013, S. 1–16

[11] MANGARD, S. ; OSWALD, E. ; POPP, T. : *Power Analysis Attacks – Revealing the Secrets of Smart Cards*. Springer, 2007

[12] MESSERGES, T. S. ; DABBISH, E. A. ; SLOAN, R. H.: Examining Smart-Card Security under the Threat of Power Analysis Attacks. In: *IEEE Trans. Computers* 51 (2002), Nr. 5, S. 541–552

[13] NEDOSPASOV, D. ; SCHLÖSSER, A. ; SEIFERT, J. ; ORLIC, S. : Functional Integrated Circuit Analysis. In: *Hardware-Oriented Security and Trust (HOST)*, 2012

[14] QUISQUATER, J.-J. ; SAMYDE, D. : ElectroMagnetic Analysis (EMA): Measures and Counter-Measures for Smart Cards. In: *E-smart*, 2001, S. 200–210

[15] RABAEY, J. M. ; CHANDRAKASAN, A. ; SODINI, C. G. (Hrsg.): *Digital Integrated Circuits*. Second Edition. Pearson Education, 2003 (A Design Perspective)

[16] SCHINDLER, W. : Advanced stochastic methods in side channel analysis on block ciphers in the presence of masking. In: *J. Mathematical Cryptology* 2 (2008), Nr. 3, S. 291–310

[17] SCHINDLER, W. ; LEMKE, K. ; PAAR, C. : A Stochastic Model for Differential Side Channel Cryptanalysis. In: *CHES*, Springer Berlin Heidelberg (LNCS), S. 30–46

[18] SCHLÖSSER, A. ; NEDOSPASOV, D. ; KRÄMER, J. ; ORLIC, S. ; SEIFERT, J.-P. : Simple Photonic Emission Analysis of AES - Photonic side channel analysis for the rest of us. In: *Cryptographic Hardware and Embedded Systems CHES 2012*, 2012

[19] SCHLÖSSER, A. ; NEDOSPASOV, D. ; KRÄMER, J. ; ORLIC, S. ; SEIFERT, J.-P. : Simple photonic emission analysis of AES. In: *J. Cryptographic Engineering* 3 (2013), Nr. 1, S. 3–15

[20] STANDAERT, F.-X. ; GIERLICHS, B. ; VERBAUWHEDE, I. : Partition vs. Comparison Side-Channel Distinguishers: An Empirical Evaluation of Statistical Tests for Univariate Side-Channel Attacks against Two Unprotected CMOS Devices. In: *ICISC*, 2008, S. 253–267

[21] WESTE, N. H. E. ; HARRIS, D. : *CMOS VLSI Design: A Circuits and Systems Perspective*. Fourth Edition. Addison Wesley, 2010

SPADs for Quantum Random Number Generators and beyond

Samuel Burri[1] Damien Stucki[2] Yuki Maruyama[3]
Claudio Bruschini[1] Edoardo Charbon[1,3] Francesco Regazzoni[4]

[1]School of Computer &
Communication Sciences
1015 Lausanne
Switzerland
samuel.burri@epfl.ch

[2]IdQuantique

1227 Carouge
Switzerland
damien.stucki@idquantique.com

[3]Delft University
of Technology
2628 Delft
Netherlands
e.charbon@tudelft.nl

[4]ALaRI
USI
6900 Lugano
Switzerland
regazzoni@alari.ch

Abstract—Single-Photon Avalanche Diodes (SPADs) are solid-state photo-detectors capable of detecting single photons by exploiting the avalanche effect that occurs in the breakdown of a p-n junction biased above breakdown voltage. By this effect, a SPAD translates an incoming photon to a macroscopic current pulse.
These devices are currently used for building medical devices characterized by a very high time resolution. An appealing application of SPAD is to use them as a basic block for building the entropy source of true random number generators.
In this paper we focus on such application, and we explore the design challenges behind the realization of a quantum random number generator based on a massively parallel array of SPADs. The matrix under investigation comprises 512x128 independent cells that convert photons onto a raw bit-stream, which, as ensured by the properties of quantum physics, is characterized by a very high level of randomness. The sequences are read out in a 128-bit parallel bus, concatenated, and pipelined onto a de-biasing filter.
Subsequently, we fabricated the proposed chip using a standard CMOS process. Our results, achieved on the manufactured device and coupling two matrices, show that our architecture can reach up to 5 Gbit/s while consuming 25pJ/bit, thus demonstrating scalability and performance for any random number generators based on SPADs.

I. INTRODUCTION

Random numbers are required in many applications, ranging from password or cryptographic key generation to gaming (e.g. winning number drawing or card deck shuffling). Although for certain applications pseudorandom numbers are sufficient and even desirable, true random numbers are increasingly used either for security or regulatory reasons. The emergence of quantum key distribution as a technique to enable secure key exchange according to information theory and the pervasive diffusion of privacy sensitive applications, such as web services for e-commerce and e-health, push for the development of low cost true random number generators in the multi-Mb/s range for clients and in the multi-Gb/s range for servers.

High speed True Random Number Generators (TRNGs) have been proposed based on mechanisms, such as thermally induced jitter from ring oscillators, block RAM write collisions, flip-flop metastability, etc. on FPGAs [12] and ASICs [8]. TRNGs may also exploit optical effects. In [11] and [4], the use of superluminescent LEDs and lasers was proposed as a source of physical entropy achieving rates of up

to 300Gb/s, however, both TRNGs were implemented in non-standard processes.

An effective way to create an optical TRNG is to use the quantum nature of photons. Reference [13] for instance measured the quantum phase noise of a laser operating at a low intensity levels for rates up to 6.25 Gbits/s. However, the system is fabricated in a custom process and the operating conditions to achieve stable, high-quality random numbers are hard to achieve and/or to maintain. To date, commercial quantum random generators can only reach speeds of 150Mb/s and are often built in expensive custom processes. Alternatively, CMOS quantum random number generators have been proposed by a number of authors, usually in the multi-Mb/s. However, a complete and exhaustive study of the scalability of this approach was still missing thus the use of massively parallel quantum random number generators is so far mainly unexplored.

In this paper, we explore the use of a large matrix of SPADs to realize a fully scalable quantum CMOS TRNG. The generator uses the quantum process of single-photon detection, implemented using a standard CMOS technology, whereby a large number of detectors, organized in a regular array, are used in parallel to increase the overall throughput [1]. This approach is sound because each detector tends to respond independently from the others, assuming near-zero crosstalk. However, in real SPAD array crosstalk is non-zero; in additon, another non ideality, namely afterpulsing, emerges. Afterpulsing, as we will see later, relates to spurious pulses generated after a photon detection, resulting in false photon counts correlated to real ones.

Our design exploits SPADs as detectors and a LED as photon source. If properly designed, SPADs exhibit the needed low optical and electrical crosstalk, while, the bit-stream of each SPAD can be considered a random process, assuming zero afterpulsing. Afterpulsing, in fact, introduces a correlation between subsequent pulses, thus degrading the quality of the randomness in a similar way as crosstalk.

In order to evaluate the quality of our design, we manufactured the proposed quantum CMOS TRNG using a standard CMOS technology. We studied the effects of detector- and source-related properties, while varying the number of activated pixels, as well as supply voltage and temperature. Finally, the throughput and the quality of the TRNG was validated using the NIST and diehard test suites. Overall, our TRNG achieves a throughput of 5Gb/s with an energy requirement of 25pJ/bit, to the best of our knowledge the lowest to date for a TRNG. With the proposed quantum CMOS TRNG

978-1-4799-2817-0/14 $31.00 © 2014 IEEE

we demonstrate the scalability of our approach from a single pair of single-photon detectors to 65,536 pairs in a dual chip configuration, paving the way to the use of massively parallel arrays of detector as high performance random number generators.

The reminder of the paper is organized as follows. Section II introduces the concepts of SPADs. Section III describes the architecture of our TRNG. Section IV reports the performance of the proposed chip.

II. SINGLE PHOTON AVALANCHE DIODE

A SPAD is a pn junction biased above breakdown, so as to operate in Geiger mode. In this mode of operation, the SPAD is capable of detecting single photons with a probability known as photon detection probability (PDP). The PDP is a function of the excess bias voltage, i.e. the voltage above breakdown at which the diode is biased, and wavelength. SPADs have been known for a long time and their characteristics are the topic of several publications [2, 6].

Only recently, however, SPADs have been integrated together with their driving circuits in a standard CMOS fabrication process [9]. At the beginning, there were only single diodes (pixels) and small arrays on a chip, while nowadays it is possible to integrate thousands of pixels into much more complex systems [10].

In order to fabricate SPADs using a standard CMOS processes, it is necessary to create planar diodes between layers commonly available for transistor layout. This approach allows to place a large number of electronic components needed to drive and read out the SPADs next to the pixels with limited parasitics. It must be noticed however that standard fabrication procedures are not ideal manufacturing of SPAD. Thus, CMOS SPADs are generally noisier then their counterpart implemented in dedicated technologies.

The p+ anode to deep n-well cathode junction forms the active part of the device where the avalanches is triggered. A p-well guard ring is fabricated around the p+ anode to prevent premature edge breakdown. Putting the whole structure in the p-substrate allows independent voltages and provides additional isolation [7].

As an example Figure 1 reports the cross-section of the SPAD used for the random number generator which will be described in the following sections. The p+ anode to deep n-well cathode junction forms the active part of the device where the avalanches occur. A p-well guard ring is fabricated around the p+ anode to prevent premature edge breakdown. Putting the whole structure in the p-substrate allows independent voltages and provides additional isolation [7].

A SPAD has several sources of non-idealities: dead time, dark counts, afterpulsing, and crosstalk. Dead time relates to the time required to return to the initial state after an avalanche has occurred; it is the minimum time between detection pulses. Dark counts relate to a SPAD's activity in the dark. Dark counts are spurious thermal events due to two quantum mechanisms, i.e. trap-assisted and band-to-band tunneling. It is generally a Poissonian process and it is characterized through dark count rate (DCR), a mean or median event rate, which a function of excess bias, temperature, and the active area of the SPAD.

Afterpulsing is a phenomenon by which spurious events occur after a primary photon absorption. These pulses are due to

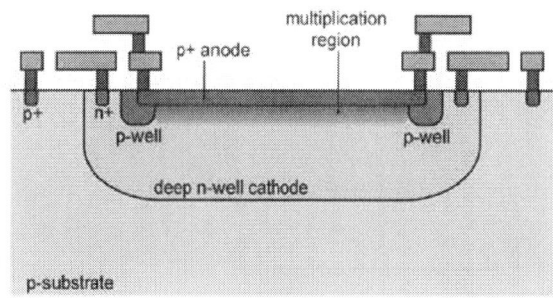

Fig. 1.: Cross-section of a SPAD standard CMOS fabrication process. The p+ anode to deep n-well cathode junction forms the active part of the device where the avalanches occur. A p-well guard ring is fabricated around the p+ anode to prevent premature breakdown. Putting the whole structure in the p-substrate allows independent voltages and provides additional isolation[7].

secondary avalanches triggered by a primary avalanche that, in turn, is caused by a photon or a dark count. Secondary avalanches are due to trapped carriers that are released at a random time after the primary avalanche. Afterpulsing is characterized by afterpulsing probability, a parameter that relates the probability of secondary and higher-order avalanches to excess bias and dead time. It is generally derived by inter-arrival-time characterization in the dark and in controlled illumination. Figure 2 plots the typical afterpulsing probability in a SPAD as

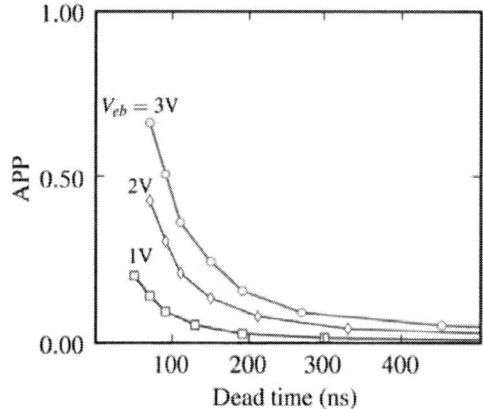

Fig. 2.: Afterpulsing probability as a function of the dead time between readout operations, equivalent to the read cycle time [3].

a function of dead time and excess bias at room temperature.

Crosstalk is a SPAD's spurious activity due to the activity of adjacent SPADs by way of optical and electrical coupling. Electrical coupling occurs due to substrate and supply line spurious pulses that propagate from a SPAD to others via metal lines. Optical crosstalk is due to photons emitted during an avalanche by impact ionization that are captured by victim SPADs that are triggered consequently. Crosstalk is a function of the radiation intensity to the overall sensor and it rapidly decays with SPAD to SPAD separation. We assume an exponential decay based on observation. This behavior is due to photon absorption and transport in silicon, the dominant mechanism supporting crosstalk.

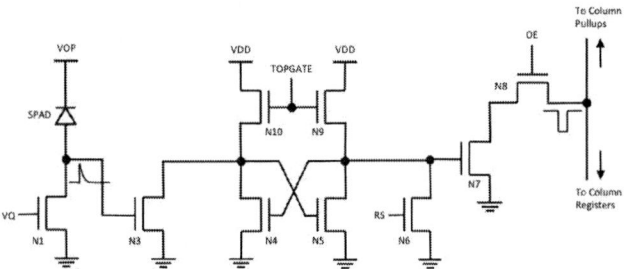

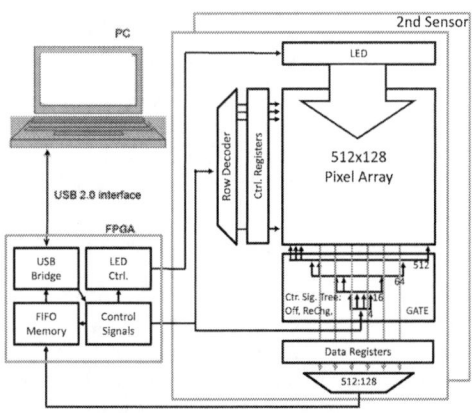

Fig. 4.: Block diagram of the proposed true random number generator. The pulsed light is produced by a LED with peak emission at 830nm, which is placed on the center of the array at a distance of 2cm to allow homogeneous illumination of the whole matrix. The internal memory bank (512 memory elements) is connected with external memory elements, which are read out in parallel and concatenated to produce the bit stream. The bit-stream is input to a filter to remove the bias of the sequence and final stream is output outside.

Fig. 3.: Schematic of the 9T pixel in the proposed TRNG. The SPAD is quenched via N1; the cathode drives N3 which in turn sets the latch formed by N4-N5-N9-N10, upon photon detection. The NMOS latch is controlled by TOPGATE that can also be used to save power, and reset by RS via N6. The output of the latch controls the pulldown N7 that is used to change the column line via select transistor N8 controlled by signal OE. The latter is pulled up at the top of the column and read out at the bottom.

A complete sensor contains pixels formed by a circuit made of a SPAD and several transistors. The sensor circuit used for realizing the TRNG which will be described in the following sections is depicted in Figure 3 and it is composed of nine transistors. The SPAD is quenched via N1; the cathode drives N3 which in turn sets the latch formed by N4-N5-N9-N10, upon photon detection. The NMOS latch is controlled by TOPGATE that can also be used to save power, and reset by RS via N6. The output of the latch controls the pulldown N7 that is used to change the column line via select transistor N8 controlled by signal OE. The latter is pulled up at the top of the column and read out at the bottom.

Each pixel contains local shutter transistors for fast response and a memory where photon detections during the active time are registered. Through selection and reset transistors a full line of the sensor will be read out and reset in one operation.

III. TRNG ARCHITECTURE

The block diagram of the system, shown in Figure 4, it comprises three main units: the photon source, the detector array, and an algorithmic post-processing unit. The photon source is a pulsed LED with peak emission at 830nm, and it is placed on the center of the array at a distance of 2cm to allow homogeneous illumination of the whole matrix.

The detector array consists of a dual 512x128 pixel array. Each pixel comprises a SPAD, implemented as described in Section II, and a one-bit memory element. The chip supports two possible acquisition modes: global-shutter mode, in which all the sensors have the same shutter signals, and rolling shutter

mode, in which the integration time of each line is equal to the time which passes between one read-out of the line and the next read-out of the same line, resulting in integration periods which roll across time. The row decoder signal enables the desired row in the array at the time of the read-out after which the memories in the row are reset.

Every column is read out independently via a fast memory and a serializer. The entire content of one array (65,536 bits) is completely read out in 6.4μs (frame duration) via a 128-bit bus; note that a bit-stream of 10.2Gb/s is achieved by the circuit, irrespectively of the sequence and the number of rows read out in the frame duration. As mentioned before, our system comprises a dual detector array operating independently, thus doubling the throughput.

To acquire one frame of random data the memories in the sensor are reset and the SPADs are activated by applying the excess voltage which brings them in the Geiger regime. The LED is then activated for a duration which will give each SPAD a 50% chance of being triggered by a photon. After deactivation of the SPAD frontend circuit, the resulting random bits are read out and the memories reset again for the next acquisition.

The bit-streams are pipelined onto the algorithmic post-processing unit, which implements a von Neumann filter to de-bias the sequence thereby suppressing any residual optical and electrical crosstalk. More in details, the von Neumann filter considers pairs of bits at a time and performs one of the following three actions: if two successive bits are equal, they are discarded; if the sequence is 1;0, the output will be 1 and the sequence if 0;1, the output will be 0. The used filter is expected to reduce the throughput from a raw bit-stream of 20.4Gb/s to 5Gb/s of a de-biased bit stream.

The overall system is controlled by a dedicated FPGA which takes care of uploading the streams of bits to the computer us-

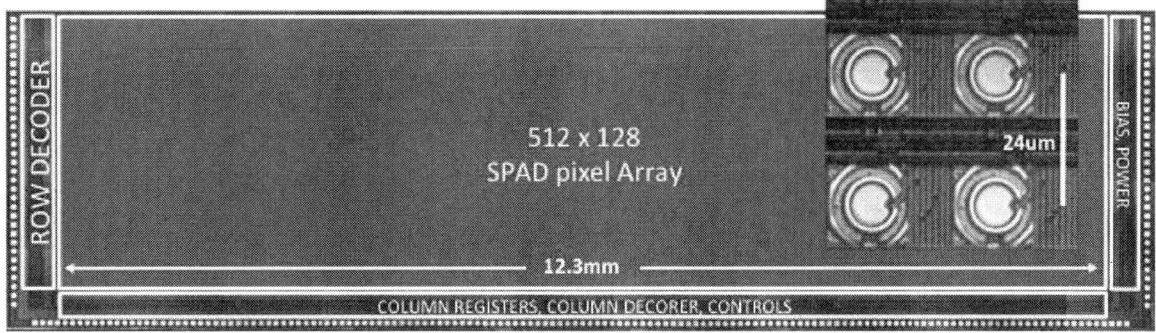

Fig. 5.: Chip micrograph of the true random number generator. The inset shows the detail of the pixels, which compose the complete array. The chip was fabricated in standard 0.35m CMOS technology. It measures 12.3mm x 3.3mm.

Fig. 6.: Photograph of the chip mounted and wire-bonded on a PCB.

ing th USB 2.0 interface.

The chip was implemented using a standard 0.35μm CMOS technology and it measures 12.3mm x 3.3mm. The micrograph of one of the two pixel arrays is captured on Figure 5. The figure also reports the detail of the pixel used to build the complete true random number generator. The photograph of both the pixel arrays mounted and wire-bonded on a PCB is reported in Figure 6.

IV. RESULTS

In this section we report the experiments we carried out to demonstrate the scalability of quantum random number generator based on CMOS and we analyze its performance.

The architecture of the chip was designed to investigate the trade-offs between architectural parameters, source parameters, and throughput. We define Random Bit Efficiency (RBE) as the ratio between de-biased bit-streams and the raw bit-streams. RBE is 100% when the raw bit-streams are perfectly biased and it is zero when no random content can be found in the raw bit-stream. RBE was computed on the output bit-stream of the chip for a range of temperatures from -25C to 70C biasing the SPADs in the pixels at an excess bias voltage from 2.0V to 5.5V. The LED was biased at an average power of 100μW and pulsed at 156kHz with a duty cycle of 0-15%, i.e. a pulse length from 0 to 900ns.

The RBE is plotted in Figure 7 as a function of excess bias voltage and LED pulse length. The plots, obtained at the indicated temperature range, demonstrate that an optimum is found in a large region of operation, showing the insensitivity of the quality of the random sequences from temperature, voltage, and LED parameters.

Next, a variable number of pixels was activated on the chip, so as to analyze the impact of the number of independent pixels operating at the same time to the quality of the random bit-stream. The plot in Figure 8 shows the throughput of the TNRG plotted as a function of activated pixels before and after de-biasing; with fewer pixels, the minimum readout cycle of 50ns is used. At this speed, afterpulsing degrades the quality of the TRNG sequences and thus the usable throughput after de-biasing, is relatively low. Increasing the number of activated pixels has the effect of increasing the readout cycle, thereby reducing afterpulsing and thus increasing RBE and thus the overall throughput. The relation between afterpulsing and readout cycle time is complex however it becomes negligible at readout cycles in the order of μs [3].

De-biasing is effective in improving the quality of our TRNG that is already below 50% at any time for an excess bias of less than 2.5V; this explains why the measured curve approaches the theoretical one even at low pixel counts. The maximum theoretical throughput is never achieved but it is approached at less than 1% (1-sigma) with 4096 pixels, far fewer than the maximum of 65,536 pixels. Above this count a raw throughput of 20.0Gb/s is always guaranteed, irrespective of temperature (in the -25C to 70C range) and supply voltage (10%). After de-biasing, the raw throughput is reduced to 5Gb/s and it is guaranteed to pass all NIST and diehard tests.

A performance comparison between the proposed TRNG and the literature is illustrated in Table I. To the best of our knowledge, the proposed TRNG is the fastest implemented in a standard CMOS process, while higher throughput is only achieved by Wei et al [11], using a non-quantum process in a custom, non-CMOS technology. Our architecture is expected to outperform the performance obtained by Wei et al [11] if fabricated in standard 65nm CMOS, a process that is widely available since 2009 and for which SPADs are already available. Finally, Table II lists the tests passed by the TRNG and the conditions in which these tests were conducted.

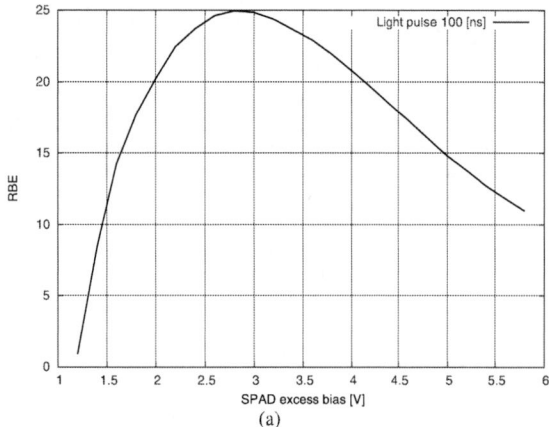 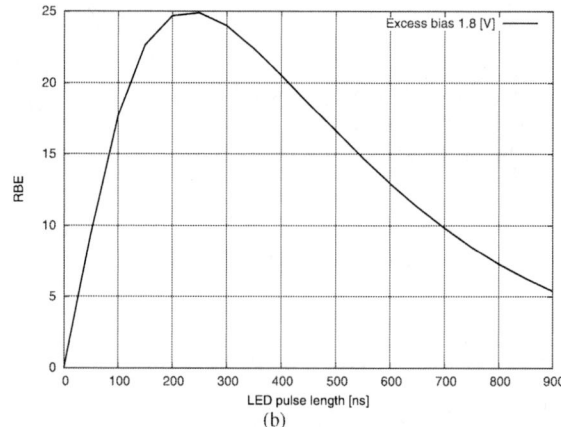

Fig. 7.: Measured random bit efficiency (RBE) vs. (a) excess bias voltage and (b) LED pulse length. The plot shows a maximum efficiency of 25%. These measurements were repeated in the temperature range with no statistical deviation.

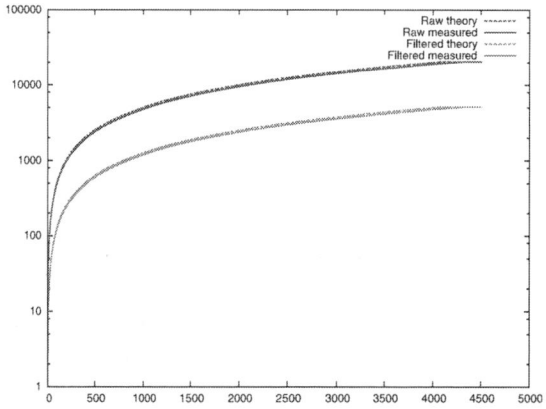

Fig. 8.: Measured and theoretical throughput vs. number of pixels activated in an experiment with 2.8V of excess bias. The throughput is shown before and after de-biasing, whereas a 4x reduction is caused by von Neumann based de-biasing. The plot shows that the theoretical linear relation between throughput and the number of active pixels is achieved with the increase of the number of active pixels. De-biasing is effective in eliminating the effects of afterpulsing. However, due to 5% of the pixels that are non-functional, a certain level of redundancy must always be added and thus the measured and theoretical curves (before and after de-biasing) do not perfectly overlap.

978-1-4799-2817-0/14 $31.00 © 2014 IEEE

TABLE I

: Comparison of the proposed TRNG performance and the state-of-the-art. *) The area refers to the active core. **) Data not available.

Measure	Min	Typ	Max							Unit	
Reference	This work			[13]	[5]	[12]	[8]	[11]	[4]		
Raw Throughput	10	15	20	6.4	0.02	0.3	0.04	280	300	Gb/s	
Temp. Range	-25	27	70	**	**	**	**	**	**	C	
Vdd	3.0	3.3	3.6	N/A	N/A	N/A	N/A	N/A	N/A	V	
Excess Bias	1.2	1.8	4.0	N/A	N/A	N/A	N/A	N/A	N/A	V	
LED Pulse Length	50	100	500	N/A	N/A	N/A	N/A	N/A	N/A	ns	
LED Duty Cycle	0.8	2	10	N/A	N/A	N/A	N/A	N/A	N/A	%	
Power		500		**	1.9	**	29	**	**	mW	
Area		7.7		**	0.012	**	0.752	**	**	mm	
Energy/bit		25		**	950	**	725	**	**	pJ/bit	
Technology	0.35μm CMOS			Custom (InGaAs)	SiN MOSFET	CMOS (FPGA)	0.35μm CMOS	Custom	Custom		

TABLE II

: Results of the NIST tests applied to a sequence generated with a LED pulse length of 100ns and an excess bias voltage of 2.8V. The tests were run on the data from the de-biasing filter.

Test	Accept Threshold	Von Neumann	Pass / No Pass
Frequency	0.951464	0.9833	Y
BlockFrequency	0.951464	0.9833	Y
CumulativeSum	0.951464	0.9833	Y
Runs	0.951464	1.0000	Y
LongestRun	0.951464	1.0000	Y
Rank	0.951464	1.0000	Y
FFT	0.951464	0.9833	Y
NonOverlapping Template	0.951464	0.9667	Y
Universal	0.951464	1.0000	Y
ApproximateEntropy	0.951464	1.0000	Y
RandomExcursion	0.951464	0.9744	Y
RandomExcursion Variant	0.942202	0.9744	Y
Serial	0.951464	1.0000	Y
LinearComplexity	0.951464	1.0000	Y

VI. Acknowledgements

The authors are grateful to Xilinx for providing the FPGAs used in the readout system. This work has been partially supported by the Swiss NCCR-QP, NCCR-QSIT, NCCR-MICS, Swiss Experiment, and EMRP (project IND06-MIQC). The EMRP is jointly funded by the EMRP participating countries within EURAMET and the European Union

References

[1] S. Burri, D. Stucki, Y. Maruyama, C. Bruschini, E. Charbon, and F. Regazzoni. Jailbreak Imagers: Transforming a Single-Photon Image Sensor into a True Random Number Generator. In *International Image Sensors Workshop*, Snowbird Resort, Utah, USA, June 2013.

[2] S. Cova, A. Longoni, and A. A. Towards picosecond resolution with single-photon avalanche diodes". *Rev. Sci. Instrum.*, 52:408–412, 1981.

[3] M. Fishburn. *Fundamentals of CMOS Single-Photon Avalanche Diodes*. PhD Thesis. TU Delft, Sep 2012.

[4] I. Kanter, Y. Aviad, I. Reidler, E. Cohen, and M. Rosenbluh. An Optical Ultrafast Random Bit Generator. *Nature Photonics*, 4:58–61, 2010.

[5] M. Matsumoto, S. Yasuda, R. Ohta, K. Ikegami, T. Tanamoto, and S. Fujita. $1200\mu m^2$ Physical Random-Number Generators Based on SiN MOSFET for Secure Smart-Card Application. In *IEEE International Solid-State Circuits Conference, 2008. ISSCC 2008. Proceedings of the*, pages 414–415, Feb 2008.

[6] R. J. McIntyre. Recent developments in silicon avalanche photodiodes. *Measurment*, 3:6, 1985.

[7] C. L. Niclass. *Single-Photon Image Sensors in CMOS: Picosencond Resolution for Three-Dimensional Imaging*, volume 4161 of *PhD Thesis*. Ecole Polytechnique Fdrale de Lausanne (EPFL), 2008.

V. Conclusion

In this paper we explored the use of SPADs as security building block, focusing in particular on the scalability of one of the most appealing applications for quantum CMOS: true random number generation. In particular, we discussed the two coupled matrices each consisting of 512x128 independent cells that convert photons onto a raw bit-stream. This TRNG is characterized by a very high level of randomness.

The performance is measured on a device manufactured in standard CMOS process. Measurements show that our architecture can reach up to 5 Gbit/s while consuming 25pJ/bit. Our results prove the scalability and performance for any random number generators based on SPADs, while achieving the lowest power consumption to date.

[8] F. Pareschi, G. Setti, and R. Rovatti. Implementation and Testing of High-Speed CMOS True Random Number Generators Based in Chaotic Systems. In *IEEE Trans. Circ. & Sys.*, volume 57–I(12), pages 3124–3137, Oct 2010.

[9] A. Rochas, M. Gosch, A. Serov, P. A. Besse, R. S. Popovic, T. Lasser, and R. Rigler. First fully integrated 2-D array of single-photon detectors in standard CMOS technology. *IEEE Photonics Technology Letters*, 15:963–965, Jul 2003.

[10] C. Veerappan, J. Richardson, R. Walker, D.-U. Li, M. W. Fishburn, Y. Maruyama, D. Stoppa, F. Borghetti, M. Gersbach, R. K. Henderson, and E. Charbon. A 160x128 single-photon image sensor with on-pixel 55ps 10b time-to-digital converter. In *IEEE International Solid-State Circuits Conference, 2011. ISSCC 2011. Proceedings of the*, pages 312–314, Feb 2011.

[11] W. Wei, G. Xie, A. Dang, and H. Guo. High-Speed and Bias-Free Optical Random Number Generator. *Photonics Technology Letters*, 24(6):437–439, June 2012.

[12] K. Wold and S. Petrovic. Optimizing Speed of a True Random Number Generator in FPGA by Spectral Analysis. *ICCIT*, pages 1105–1110, Nov 2009.

[13] F. Xu, B. Qi, X. Ma, H. Xu, H. Zheng, and H.-K. Lo. Ultrafast Quantum Random Number Generation based on Quantum Phase Fluctuations. *Optics Express*, 20(11):12366–12377, Nov 2012.

Quantum key distribution with integrated optics

Mirko Lobino

Centre for Quantum Dynamics
Griffith Univeristy
Brisbane, Australia 4111
e-mail: m.lobino@griffith.edu.au

Pei Zhang

Department of Applied Physics
Xian Jiaotong University
Xi'an, China 710049
e-mail: zhangpei@mail.ustc.edu.cn

Enrique Martín-López

Centre for Quantum Photonics
University of Bristol
Bristol, United Kingdom BS8 1UB
e-mail: em9657@bristol.ac.uk

Richard W. Nock

Department of Electrical and Electronic Engineering
University of Bristol
Bristol, United Kingdom BS8 1UB
e-mail: richard.nock@bristol.ac.uk

Damien Bonneau

Centre for Quantum Photonics
University of Bristol
Bristol, United Kingdom BS8 1UB
e-mail: Damien.Bonneau@bristol.ac.uk

Hong Wei Li

Nokia Research Centre
Cambridge, United Kingdom CB3 0FA
e-mail: hongwei.3.li@nokia.com

Antti O. Niskanen

Nokia Research Centre
Cambridge, United Kingdom CB3 0FA
e-mail: antti.niskanen@nokia.com

Jeremy L. O'Brien

Centre for Quantum Photonics
University of Bristol
Bristol, United Kingdom BS8 1UB
e-mail: jeremy.l.obrien@me.com

Anthony Laing

Centre for Quantum Photonics
University of Bristol
Bristol, United Kingdom BS8 1UB
e-mail: Anthony.Laing@bristol.ac.uk

Kanin Aungskunsiri

Centre for Quantum Photonics
University of Bristol
Bristol, United Kingdom BS8 1UB
e-mail: phxka@bristol.ac.uk

Joachim Wabnig

Nokia Research Centre
Cambridge, United Kingdom CB3 0FA
e-mail: joachim.wabnig@nokia.com

Jack Munns

Bristol Centre for Functional Nanomaterials
University of Bristol
Bristol, United Kingdom BS8 1UB
e-mail: Jack.Munns.07@bristol.ac.uk

Pisu Jiang

Centre for Quantum Photonics
University of Bristol
Bristol, United Kingdom BS8 1UB
e-mail: p.jiang@bristol.ac.uk

John G. Rarity

Department of Electrical and Electronic Engineering
University of Bristol
Bristol, United Kingdom BS8 1UB
e-mail: john.rarity@bristol.ac.uk

Mark G. Thompson

Centre for Quantum Photonics
University of Bristol
Bristol, United Kingdom BS8 1UB
e-mail: Mark.Thompson@bristol.ac.uk

Abstract— We report on a quantum key distribution (QKD) experiment where a client with an on-chip polarisation rotator can access a server through a telecom-fibre link. Large resources such as photon source and detectors are situated at server-side. We employ a reference frame independent QKD protocol for polarisation qubits and show that it overcomes detrimental effects of drifting fibre birefringence in a polarisation maintaining fibre.

I. INTRODUCTION

Quantum key distribution (QKD) exploits the quantum mechanical principle that requires microscopic systems to be changed upon observation [1, 2, 3, 4, 5, 6, 7, 8, 9, 10, 11, 12, 13] for enhancing the security in communication. In particular the quantum properties of single photons are commonly used for the generation the random binary key used in one-time pad type protocols. Here we demonstrate a compact implementation of QKD based in integrated optical devices and telecom optical fibers that is particularly suited for equipping mobile communication devices.

The transmission of quantum information, suitable for QKD, normally requires that a state sent by *Alice* is faithfully received by *Bob*, unless an eavesdropper, *Eve*, observes the state and reveals her presence as a disturbance in the communication. Yet even in the absence of Eve, an unstable fibre communication link or instability in the sending and receiving apparatus, is equivalent to an unknown or varying reference frame, and has the effect of unhelpfully transforming the states that Bob receives [18].

Attempts to overcome potential reference frame misalignment used qubits encoding over larger systems [19, 20, 21] requiring the creation, manipulation, and detection of many-photon entangled states, which is technically challenging and very loss-sensitive. Other protocols that use transver spatial modes of light [22, 23, 24, 25] may facilitate communication between misaligned parties but suffer from problems such as mode dispersion when transmitting through fibre. Protocols that exploit the arrival time of photons as a logical basis have been implemented [26] and in particular the *plug and play* system which sends each light pulse back and forth along the same fibre, with the aid of a Faraday mirror, to cancel the effects of birefringence [28, 29, 30, 31, 32]. Interferometric stability results form both halves of the time-split pulse retracing each others path. A drawback to this double-pass arrangement is the potential for an increased error rate due to Rayleigh backscattering, which requires the addition of a storage line to hold a train of pulses which must complete a round trip before the next train is sent. A further stability constraint, related to the length of transmission, is that fluctuations in the fibre and interferometer should be slow on the time-scale of the double pass.

II. THE REFERENCE FRAME INDEPENDENT PROTOCOL

The reference frame independent QKD protocol (rfiQKD) [33, 34, 36, 35], deployed here, operates between unknown and changing reference frames, is independent of any particular choice of information encoding, requires no entanglement, and is implementable with two-level systems encoded onto single photons that may be approximated with weak coherent laser

pulses, making it intrinsically practical. The protocol will generate a secret key as long as the rate of change between reference frames is slow on the rate of particle repetition.

Here we show the implementation of a rfiQKD protocol using photon-polarisation qubits over an unstablised fibre link where the photons are manipulated using commercially available lithium niobate integrated polarisation controllers [37]. Larger resources, such as the photon source and superconducting detectors [39, 38], are situated on Bob's side, which can be regarded as the *server side*, while Alice, as the *client*, requires only the capability to perform single qubit operations. The scenario is one in which the client tethers a hand held device, with an integrated photonic chip, to a telecom-fibre to receive dim laser pulses from the QKD server, which the client attenuates to the single photon level, before encoding each pulse with a qubit of information for return transmission to the server, along a different fibre. We demonstrate a stable, constant, and continuously positive secret key rate over the unstablised fibre link using the rfiQKD protocol, while the rate for the BB84 QKD protocol falls. We go on to show that our rfiQKD system automatically and passively recovers from the deliberate introduction of large amounts of noise in the form of rapid fluctuations.

Although formulated as an entanglement-driven protocol, rfiQKD can be implemented with weak coherent states that sufficiently approximate single photons, where Alice randomly prepares and sends one of six polarisation states $\{D/A, R/L, H/V\}$ corresponding to the eigenvectors of the Pauli matrices, which we label as $\{X, Y, Z\}$ [43]. In our experiment, the requirement of one known and stable basis, used to encode the key, is fulfilled with the horizontal (H) and vertical (V) polarisation states, that are preserved throughout transmission in polarisation maintaining fibre (PMF). The other four states, superpositions of H and V used to guarantee security, are unhelpfully transformed by phase fluctuations in PMF due to environmental influences on the birefringence of the fibre; while this effect is troublesome for other protocols, rfiQKD operates in the presence of phase drifts that are slow on the repetition rate of sent photons. For fluctuations sufficiently rapid to force protocol failure, rfiQKD will automatically recover in calmer periods without the need for re-alignment, as we demonstrate.

The operation of the protocol, with its phase invariant security measure, works as follows: Expressing Alice and Bob's measurement bases as $Z_A = Z_B$, $X_B = \cos(\beta)X_A + \sin(\beta)Y_A$ and $Y_B = \cos(\beta)Y_A - \sin(\beta)X_A$, where β slowly changes with time in an unknown way, Alice randomly prepares quantum states which she sends to Bob, who measures in his randomly chosen basis; later they publicly reveal their choice of bases. The raw key is obtained when they both measure in the Z direction, providing a quantum bit-error rate

$$Q = \frac{1 - \langle Z_A Z_B \rangle}{2}. \tag{1}$$

The other two slowly rotating bases are used to estimate the knowledge of a potential Eve. The quantity

$$C = \langle X_A X_B \rangle^2 + \langle X_A Y_B \rangle^2 + \langle Y_A X_B \rangle^2 + \langle Y_A Y_B \rangle^2, \tag{2}$$

is independent of the relative angle β. If there is no Eve and the communication channel is ideal with a fixed (although unknown) phase, then the correlation function $\langle Z_A Z_B \rangle$ is equal to 1 whereas $\langle X_A X_B \rangle$, $\langle X_A Y_B \rangle$, $\langle Y_A X_B \rangle$ and $\langle Y_A Y_B \rangle$ each take a constant value between -1 and 1 determined by β. Also, $\langle Z_A X_B \rangle$, $\langle Z_A Y_B \rangle$, $\langle X_A Z_B \rangle$ and $\langle Y_A Z_B \rangle$ should be zero as X, Y, and Z are mutually unbiased. With Z bases aligned, $Q = 0$ and $C = 2$ will be achieved, but in a realistic implementation, Q will be greater than zero and C will be less than 2.

The correlation functions involved in (1) and (2) are calculated from the rates of photon detections by assigning positive or negative signs to correlated or anti-correlated detections respectively. For example, if Bob labels his pair of detectors as $b = \{0, 1\}$, while Alice labels the pair of states she sends (within a particular basis) as $a = \{0, 1\}$ then the $\langle AB \rangle$ expectation value is calculated from the number of detector clicks n_{ab} as $(n_{00} + n_{11} - n_{01} - n_{10})/(n_{00} + n_{11} + n_{01} + n_{10})$, which is essentially the normalised difference of correlated and anti correlated detections.

The security proof of the rfiQKD protocol [33] shows that when $Q \lesssim 15.9\%$, Eve's information is given by

$$E(Q, C) = \frac{h}{2}\left(1 + (1 - Q)u_{max} + Q\,v(u_{max})\right) \quad (3)$$

where

$$u_{max} = \min\left[\frac{\sqrt{C/2}}{1 - Q}, 1\right],$$

$$v = \frac{1}{Q}\sqrt{C/2 - (1 - Q)^2 u_{max}^2},$$

and the binary entropy

$$h(x) = -x\log_2(x) - (1 - x)\log_2(1 - x).$$

The secret key rate is given by

$$r = 1 - h(Q) - E(Q, C). \quad (4)$$

III. EXPERIMENTAL RESULTS

The experimental setup is shown in Fig. 1. At the server side, light from a 1550 nm continuous-wave laser source is sent through a pulse generator (PG) to produce pulses of 100 ns width with a repetition rate of 1 MHz [44], which are filtered by a horizontal polariser and transmitted to the client through PMF. At the client side, a fibre beam splitter and photodetector expose hypothetical attacks (for example, [40]). A ≈ 75 dB attenuator reduces the light intensity to the single photon level of ≈ 0.1 photons per pulse so that the probability of more than one photon per pulse ≈ 0.005. The client randomly prepares among the six polarisation states $\{D, A, R, L, H, V\}$ using a LiNbO$_3$ [41, 42] polarisation controller (PC) commanded from a field programmable gate array (FPGA) and associated driver circuits. Photonic qubits are returned to the server, along another unstabilised 5 m length of PMF, where a similar PC and FPGA, together with a fibre polarisation beam splitter and superconducting single photon detectors, perform projective measurements chosen randomly among the three relevant bases.

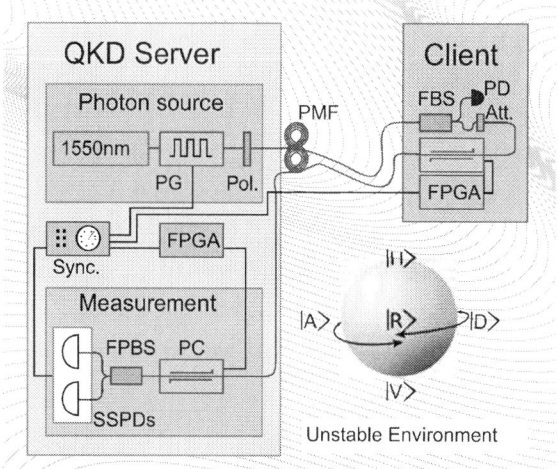

Fig. 1. Experimental set-up for the rfiQKD protocol. The server side generates the light pulses with a telecom wavelength (1550 nm) laser and a 1 MHz pulse generator (PG) and fixed polariser, to send light pulses to the client through a polarisation maintaining fibre (PMF). At the client side, an integrated polarisation controller (PC) encodes qubits into the polarisation of the attenuated (Att.) light. A fibre beam splitter (FBS) and photodetector (PD) continuously monitor power for malicious attacks. Qubits received back at the server side are measured with a similar PC, fibre polarising beamsplitter (FPBS), and superconducting single photon detectors (SSPDs), all controlled by an electronic board synchronisation (Sync.), function programmable gate array (FPGA), and processor. The Bloch sphere illustrates the effects of an unstable environment on polarisation.

All system elements are synchronised to the SYNC-FPGA platform, which allows for precise timing of all stages within the transmission period. Each repetition begins with an optical pulse from the server side PG; when the pulse arrives at the client PC, it is set to prepare a particular polarisation state; then, timed appropriately for the return of the pulse, the server PC is set to measure in a particular basis; finally, the state of the detectors is recorded by the FPGA.

In addition to demonstrating the feasibility of QKD between telecom-fibre linked integrated photonic devices in the described client-server scheme, we aim to show two features of the rfiQKD protocol that are particularly relevant: robustness to phase drift, inevitable in long range fibre; and automatic passive recovery from rapid noise.

We also present analysis for the uncalibrated variant urfiQKD protocol which assumes not only unaligned reference frames, but also removes the assumption for alignment within a reference frame, allowing for non-orthogonality within a basis and mutual bias between bases, and differing detector efficiencies that arise in a real world implementation [35]. This is achieved by using an explicit device model and minimising the key rate over possible model parameters.

We demonstrated drift robustness of the rfiQKD and urfiQKD generated key rates in comparison with that of BB84, beginning key exchange with well aligned client-server reference frames, so that states prepared by the client PC had a high fidelity with projectors determined by the server PC. The PMF quantum channel was unfixed but undisturbed for the duration of the key exchange, subject only to ambient environmental influences. Figure. 2 shows the secret key rate fraction, r, as a function of time for the BB84, rfiQKD, and urfiQKD protocols. The duration of key exchange is 240 s, so each of the 24 points corresponds to data integrated over 10 seconds, which is long

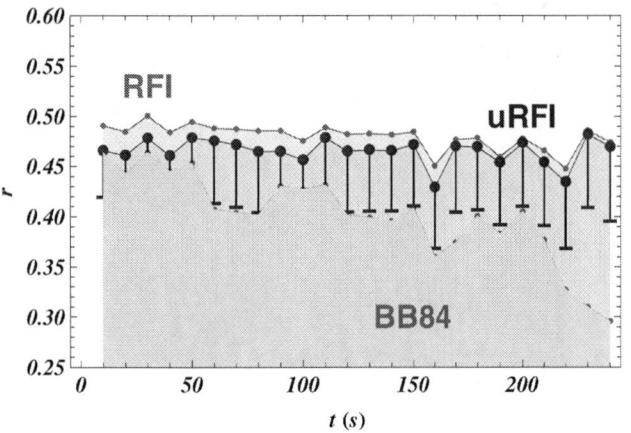

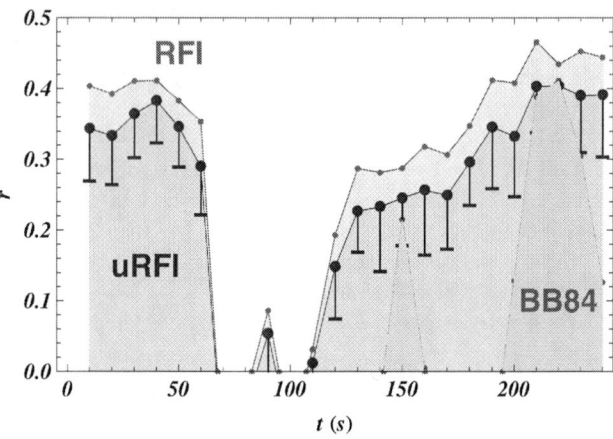

Fig. 2. Experimental data for secret key rate fraction r showing robustness to drift. Data is initially collected in the situation of well aligned client-server reference frames, but the unfixed PMF quantum channel is subject to ambient environmental influences, which effect a slowly varying reference frame. While the key rate for the rfiQKD protocols is constant, for BB84 it suffers a fall as the alignment drifts. Lower bounds on the secret key rate from the urfiQKD analysis are shown.

Fig. 3. Experimental data for secret key rate fraction r showing automatic recovery from rapid noise. During the initial 60s, the unaligned and slowly varying reference frames result in BB84 failure while the rfiQKD protocols operate. Between $t = 60$s and $t = 120$s we deliberately introduced rapid PMF deformations force an rfiQKD failure. However, an automatic and passive revival of r for the rfiQKD protocols is observed during the subsequent calm period.

enough to collect sufficient data to produce small error bars, corresponding to a precision of 3 standard deviations, but short enough to avoid significant depletion of the value of r, from integration over largely different values β. For clarity, these error bars are not displayed, instead we show the lower bound on r from the urfiQKD analysis. It can be seen that while the secret key rate for BB84 falls as a function of time, the rfiQKD protocols do not.

We demonstrated the *automatic, passive recovery* capability of our system after periods of rapid and substantial noise that force a protocol failure. Figure. 3 shows r as a function of time for 24 points, each corresponding to 10 seconds of data, as before. During the initial 60 s of key exchange, the PMF quantum channel is unfixed and undisturbed but unaligned so that the BB84 protocol immediately fails. However the changes resulting from the ambient environment are slow enough for the rfiQKD protocols to operate successfully. Between $t = 60$ s and $t = 120$ s we deliberately introduced a large amount of noise by continually and significantly deforming the PMF to simulate a rapidly changing reference frame, forcing r to fall below zero. At $t = 120$ s, the noise ceases and as the PMF relaxes from the mechanical strains, a positive key rate automatically returns and achieves initial values for the rfiQKD protocols. In contrast, BB84 achieves a positive r only in brief transitionary periods of near-alignment. Again, the lower bound on r for the urfiQKD analysis is displayed.

IV. CONCLUSIONS

In conclusion, we demonstrated a client-server QKD protocol where all large resources reside at server side, and the client requires only an integrated photonic device that could be further integrated into a hand held communication device. The key is exchanged though a PMF telecom-fibre tether using the rfiQKD protocol, that is shown to be passively robust to typical environmental drift effects and can automatically recover

from large noise levels to re-establish QKD in calmer periods with no requirement for alignment. The results significantly broaden the operating potential for QKD outside of the laboratory and pave the way for quantum enhanced security for the general public with handheld mobile devices.

Future directions are to adapt the system for time multiplexing with the rfiQKD protocol which avoids the requirement for temperature-stabilised on-chip asymmetric Mach Zehnder interferometers. Such a system would be well suited to take advantage of existing (non PMF) telecom-fibre networks, without the need for double passing and the inheritance of associated problems. The challenge is to then increase the distance over which secret key may be exchanged in telecom-fibre.

V. ACKNOWLEDGEMENTS

This work was supported by EPSRC, ERC, QUANTIP, PHORBITEC and NSQI. P. Z. acknowledges support from the Fundamental Research Funds for the Central Universities, the Special Prophase Project on the National Basic Research Program of China (Grant No.2011CB311807), and the National Natural Science Foundation of China (Grant No. 11004158). J.M. acknowledges EPSRC grant code EP/G036780/1. J.L.OB. acknowledges a Royal Society Wolfson Merit Award and a Royal Academy of Engineering Chair in Emerging Technologies.

REFERENCES

[1] C. H. Bennett and G. Brassard, *Proceedings of IEEE International Conference on Computers, Systems, and Signal Processing* (IEEE, New York, 1984), pp. 175.

[2] A. K. Ekert, Phys. Rev. Lett. **67**, 661 (1991).

[3] D. Bruß, Phys. Rev. Lett. **81**, 3018 (1998).

978-1-4799-2817-0/14 $31.00 © 2014 IEEE

[4] H. Bechmann-Pasquinucci, and N. Gisin, Phys. Rev. A **59**, 4238 (1999).

[5] H.-K. Lo and H. F. Chau, Science **283**, 2050 (1999).

[6] P. W. Shor and J. Preskill, Phys. Rev. Lett. **85**, 441 (2000).

[7] N. Gisin, G. Ribordy, W. Tittel, and H. Zbinden, Rev. Mod. Phys., **74**, 145 (2002).

[8] C. Kurtsiefer, P. Zarda, M. Halder, H. Weinfurter, P. M. Gorman, P. R. Tapster, and J. G. Rarity, Nature (London) **419**, 450 (2002).

[9] T. Honjo, K. Inoue, and H. Takahashi, Opt. Lett. **29**, 2797 (2004).

[10] H. Takesue and K. Inoue, Phys. Rev. A **72**, 041804(R) (2005).

[11] E.-L. Miao, Z.-F. Han, T. Zhang, and G.-C. Guo, Phys. Lett. A **361**, 29 (2007).

[12] C. Bonato, A. Tomaello, V. D. Deppo, G. Naletto, and P. Villoresi, New J. Phys. **11**, 045017 (2009).

[13] M. Fujiwara, M. Toyoshima, M. Sasaki, K. Yoshino, Y. Nambu, and A. Tomita, Appl. Phys. Lett. **95**, 261103 (2009).

[14] V. Scarani, H. Bechmann-Pasquinucci, N. J. Cerf, M. Dušek, N. Lütkenhaus, and M. Peev, Rev. Mod. Phys. **81**, 1301 (2009).

[15] M. Dušek, N. Lütkenhaus, and M. Hendrych, Prog. Opt. 49, **381** (2006).

[16] H.-K. Lo and Y. Zhao, Quantum cryptography in Encyclopedia of Complexity and System Science (Springer, New York, 2009), Vol. 8, p. 7265.

[17] N. Gisin and R. Thew, Nature Photon., **1**, 165 (2007).

[18] Z. Tang, Z. Liao, F. Xu, B. Qi, L. Qian, and H.-K. Lo, arXiv:1306.6134, (2013).

[19] S. D. Bartlett, T. Rudolph, and R. W. Spekkens, Phys. Rev. Lett. **91**, 027901 (2003).

[20] J.-C. Boileau, D. Gottesman, R. Laflamme, D. Poulin, and R. W. Spekkens, Phys. Rev. Lett. **92**, 017901 (2004).

[21] T.-Y. Chen, J. Zhang, J.-C. Boileau, X.-M. Jin, B. Yang, Q. Zhang, T. Yang, R. Laflamme, and J.-W. Pan, Phys. Rev. Lett. **96**, 150504 (2006).

[22] F. M. Spedalieri, Opt. Commun. **260**, 340 (2006).

[23] L. Aolita and S. P. Walborn, Phys. Rev. Lett. **98**, 100501 (2007).

[24] C. E. R. Souza, C. V. S. Borges, A. Z. Khoury, J. A. O. Huguenin, L. Aolita, and S. P. Walborn, Phys. Rev. A **77**, 032345 (2008).

[25] V. D'Ambrosio, E. Nagali, S. P. Walborn, L. Aolita, S. Slussarenko, L. Marrucci, L. Lorenzo, and F. Sciarrino, Nat. Commun. **3**, 961 (2012).

[26] I. Marcikic, H. de Riedmatten, W. Tittel, H. Zbinden, M. Legré, and N. Gisin, Phys. Rev. Lett. **93** 180502 (2004).

[27] A. Tanaka, M. Fujiwara, S. Woo Nam, Y. Nambu, S. Takahashi, W. Maeda, K. Yoshino, S. Miki, B. Baek, Z. Wang, A. Tajima, M. Sasaki, and A. Tomita, Opt. Express. **16**, 11354 (2008).

[28] A. Muller, T. Herzog, B. Huttner, W. Tittel, H. Zbinden, and N. Gisin, Appl. Phys. Lett. **70**, 793 (1997).

[29] H. Zbinden, J.D. Gautier, N. Gisin, B. Huttner, A. Muller and W. Tittel, Electron. Lett. **33**, 586 (1997).

[30] G. Ribordy, J. D. Gautier, N. Gisin, O. Guinnard, and H. Zbinden, Electron. Lett. **34**, 2116 (1998).

[31] D. Stucki, N. Gisin, O. Guinnard, G. Ribordy, and H. Zbinden, New J. Phys. **4**, 41.1 (2002).

[32] P. Eraerds, N. Walenta, M. Legré, N. Gisin, and H. Zbinden, New J. Phys. **12**, 063027 (2010).

[33] A. Laing, V. Scarani, J. G. Rarity, and J. L. O'Brien, Phys. Rev. A **82**, 012304 (2010).

[34] L. Sheridan, T. P. Le, and V. Scarani, New J. Phys. **12**, 123019 (2010).

[35] J. Wabnig, D. Bitauld, H. W. Li, A. Laing, J. L. O'Brien, and A. O. Niskanen, New J. Phys. **15**, 073001 (2013).

[36] T. P. Le, L. Sheridan, and V. Scarani, Int. J. Quantum Inf., **10**, 1250035 (2012).

[37] D. Bonneau, M. Lobino, P. Jiang, C. M. Natarajan, M. G. Tanner, R. H. Hadfield, S.N. Dorenbos, V. Zwiller, M. G. Thompson, and J. L. O'Brien, Phys. Rev. Lett. **108**, 053601 (2012).

[38] S. N. Dorenbos, E. M. Reiger, U. Perinetti, V. Zwiller, T. Zijlstra, and T.M. Klapwijk, Appl. Phys. Lett. **93**, 131101 (2008).

[39] M. G. Tanner, C. M. Natarajan, V. K. Pottapenjara, J. A. O'Connor, R. J. Warburton, R. H. Hadfield, B. Baek, S. Nam, S. N. Dorenbos, E. Bermudez Urena, T. Zijlstra, T. M. Klapwijk, and V. Zwiller, Appl. Phys. Lett. **96**, 221109 (2010).

[40] L. Lydersen, C. Wiechers, C. Wittmann, D. Elser, J. Skaar, and V. Makarov, Nature Photon. **4**, 686 (2010).

[41] S. Thaniyavarn, Appl. Phys. Lett. **47**, 674 (1985).

[42] S. Thaniyavarn, Opt. Lett. **11**, 39 (1986).

[43] As is typical, $\{D, A, R, L, H, V\} \equiv \{\text{diagonal, anti-diagonal, right circular, left circular, horizontal, vertical}\}$.

[44] The initial 1MHz pulse rate is attenuated at client-side to an average intensity of ≈ 0.1 photon/pulse; this and the efficiency of the SSPDs, at 10%, give a final average repetition rate of 10kHz.

978-1-4799-2817-0/14 $31.00 © 2014 IEEE   799

Constraint-based Platform Variants Specification for Early System Verification

Andreas Burger[1], Alexander Viehl[1], Andreas Braun[1], Finn Haedicke[2,3], Daniel Große[2]
Oliver Bringmann[1,4], Wolfgang Rosenstiel[1,4]

[1]FZI Research Center for Information Technology, Haid-und-Neu-Str 10-14, 76131 Karlsruhe, Germany
[2]solvertec GmbH, Anne-Conway-Str. 1, 28359 Bremen, Germany
[3]Institute of Computer Science, University of Bremen, 28359 Bremen, Germany
[4]University of Tuebingen, Sand 13, 72076 Tuebingen, Germany

[1][aburger,viehl,abraun]@fzi.de   [2][grosse,haedicke]@solvertec.de
[4][bringman,rosenstiel]@informatik.uni-tuebingen.de

To overcome the verification gap arising from significantly increased external IP integration and reuse during electronic platform design and composition, we present a model-based approach to specify platform variants. The variants specification is processed automatically by formalizing and solving the integrated constraint sets to derive valid platforms. These constraint sets enable a precise specification of the required platform variants for verification, exploration and test. Experimental results demonstrate the applicability, versatility and scalability of our novel model-based approach.

I. INTRODUCTION

The increase in complexity of distributed embedded systems and systems-on-chip (SoC) platforms over the last decade in combination with the decrease in acceptable time-to-market are issuing tremendous challenges for the platform development. Besides the growing complexity, other problems are the increase in integration of external IPs and the rising usage of reused logic in the platform design composition [1]. Due to these facts, the integration and parameterization of platform components and their configurations grow significantly which causes huge platform variant and configuration spaces.
In [2] for example, an automatic gear shifting application from Daimler Trucks is described which consists of 6.4 million valid variants. Each of them can be integrated in trucks and should be implemented, verified and tested. Another important example can be given by automotive networks like *FlexRay* [3] or *Media Oriented System Transport* [4] (MOST). A MOST network can integrate from 2 up to 64 communication devices as well as a set of different protocol parameters for each of them. This leads to a space of more than 10^{21} valid MOST network variants.
Due to these facts and the given examples, the verification of IP blocks in different platform variants and the verification of the interaction between multiple IP block instances is gaining importance. Further, the relevance of the verification of platform characteristics (e.g. network topology, component instances) and different component variants (e.g. software versions, component parameters) is growing. Therefore the focus in early verification and test is moving away from fixed virtual prototype platforms with variable test scenarios to virtual prototype platforms which are variable in the number and kind of components considering complex interdependencies, constraints and requirements.
Hence it is inevitable to devise methods which allow focusing on the model-based description and generation of feasible platform variants. This could reduce the manual effort in verification, exploration and test significantly and enables verification of IPs within different platform variants.
Therefore we demonstrate our novel approach on virtual prototyping but it can also be applied for real tests (e.g. *Hardware-in-the-loop* (HIL), *Software-in-the-loop* (SIL)). The structure of the platform variant specification is defined by UML-based [5] templates, so-called *Constrained-Structural Templates (CST)*. The hierarchically structured templates enable a high degree of structural flexibility. The usage of constraints allows to specify requirements and to realize complex interdependencies between different IPs and platform components. Equally they enable to control the huge platform variant and configuration space efficiently. This allows generating required variants for verification, exploration or test. The constraints are specified by an extended subset of the *Object Constraint Language (OCL)* [6]. They are encoded as a SAT instance on bit-vector-level to solve them by metaSMT [7]. The platform variant framework is implemented parallel and multi-threaded to overcome the overhead of generation and solving. In summary, the major contributions are:

- Precise specification of valid and feasible platform variant space
- Complete generation of required variants for verification, exploration and test
- High structural flexibility in the platform variants specification
- Fast reconfiguration capabilities of the specified platform variants space

By experimental evaluations we demonstrate our approach for verifying a MOST network using virtual prototyping. We also adopted our approach for an exploration of a FlexRay-based traffic sign recognition application. The examples demonstrate the necessity to specify the required variants for verification, exploration and test as well as the wide application field of our novel methodology.

978-1-4799-2817-0/14 $31.00 © 2014 IEEE

II. Related Work

For variant and alternative modeling several approaches have been proposed (e.g. [8–12]). In [8] the *Common Variability Language* (CVL), a generic language for modeling variability in different domain models is presented. In [9] feature models are introduced for modeling software product line variants.
Feature models are also used in [10] to describe variants. They are combined with an XML-based language (XVCL) to add much more flexibility to the feature models. The approach presented in [11] allows to model variants of industrial automation systems within a product line. Another approach for variant handling is described in AUTOSAR 4.0 [12, 13]. In [14] the authors give a good overview on the current state of the art in platform-based product design and development which contain product variant management approaches.
The main weaknesses of these approaches are that they are mostly used for software or product design and have no capability to specify platforms and especially their variants. Furthermore the generation of variants is not supported and they even provide very limited methods to specify complex interdependencies between different components and variants.
In the area of stimuli generation for test and simulation of platforms via SMT/SAT solving, several approaches have been proposed. For instance, in [15] the SystemC Verification (SCV) [16] library has been extended for stimulus generation based on SMT. The authors in [17] present a sampling algorithm combining concepts from the Metropolis-Hastings algorithm, Gibbs sampling and WalkSAT to generate solutions with an approximately uniform distribution. In both approaches the generated solutions for the constraints are stimuli parameter that can be classified on a lower system level in comparison to our approach.
In the verification and validation of large design components constraint-based random simulation remains an integral part. In [18] entropy metrics to define coverage targets of internal signals is combined with a framework which uses small randomized XOR constraints to generate stimuli for inadequately stimulated circuit regions. The authors in [19] present a methodology to analyze contradictory constraints that occur in constraint-based random simulation. Actually all of the approaches in constraint-based random simulation are used to generate test and stimuli for fixed platforms. In contradiction to our novel methodology which lifted up verification, exploration and test to variable virtual prototype platforms.
The authors in [20] describe how OCL constraints and UML models can be encoded in a SAT instance. In comparison, we support the encoding of more OCL features (e.g. if-expressions, allInstances() operation, extended OCL operators). In summary, to the best of our knowledge, no model- and constraint-based generation of platform variants has been considered so far.

III. Platform Variants Specification and Generation

This section introduces our platform variants specification approach based on constraints and structural templates. Before we describe our approach in detail we give an overview. The elaborated framework, presented in Fig. 1, is divided in three parts: the template-based platform variants specification, the automated encoding of template-attached constraints including the SMT/SAT-solving process and the virtual prototype simulation. The platform variants specification contains the platform templates based on UML, see Subsection A, as well as the attached constraint sets. The UML platform templates are structured to three libraries. Within these libraries the templates are structured hierarchically by composition structure diagrams. Hence a hierarchical structure for every component of the libraries can be built. So every component can be abstracted as complex or detailed as required.

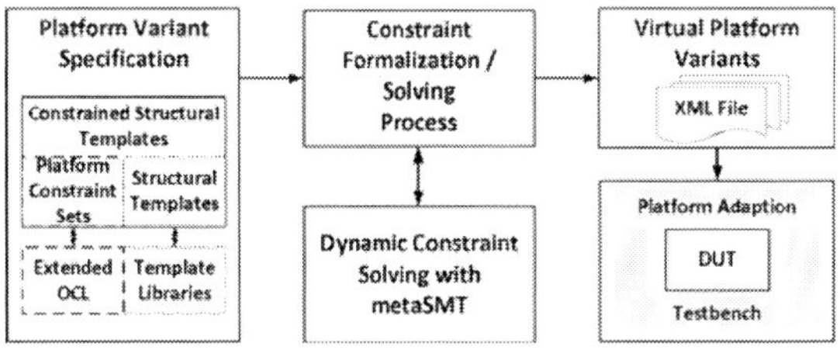

Fig. 1. Overview of the platform variants generation

The user-defined constraint sets are attached at the different hierarchical template layers to describe the variability, the dependencies and the configuration capabilities of different platform variants. Currently we attach these constraints by using UML standard constraints [21]. The constraint sets in combination with the structural templates allow a precise specification of complex dependencies among platform templates. For example more restrictive constraint sets enable an effective reduction of the huge variant space to the relevant platforms for verification.
We encode the constraint sets to a SAT instance using metaSMT. The constraint solutions are derived for generating valid and feasible platform variants. These platform variants are described as XML. This enables to link the established tools for platform modeling with our platform variant approach to specify, generate and simulate platform variants. We pass the XML to our platform execution tool which registers instantiates and links the modules of the platform variant automatically [22].

A. UML Platform Templates

Within this section we present the UML-based structural platform templates and a small tailored UML platform profile. This profile and the platform templates form the basis for the structural specification of different variants of virtual prototype platforms (e.g. SystemC platforms). The UML platform profile consists of four relevant stereotypes:

1. *<platform_module>* is a virtual prototype module described by an instance of a class definition.
2. *<platform_template>* is a component which abstracts a part of a system for simplicity, variability or structural reasons.
3. *<platform_or_template>* is a component which contains more than one not related platform templates or modules to realize possible alternatives.
4. *<platform_constraint>* marks a template/module attached constraint.

Beside this UML profile, the UML standard profile for Primitive Types [5] is used for data types (String, Float, etc.).

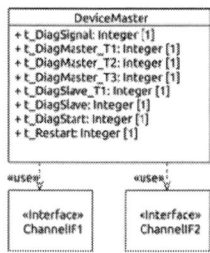
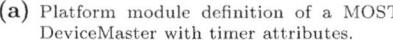

(a) Platform module definition of a MOST DeviceMaster with timer attributes.

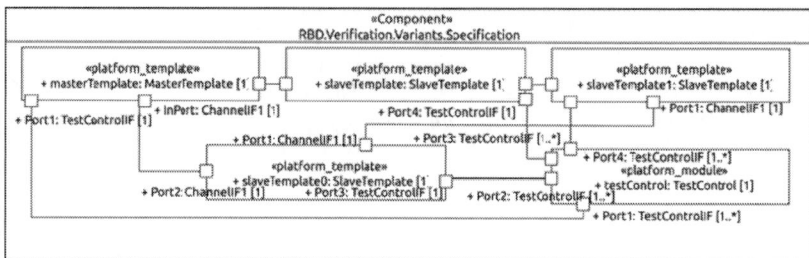

(b) Platform variant specification describing all relevant variants for the MOST Ring Break Diagnosis (RBD) process.

Fig. 2. Platform module DeviceMaster specifing a MOST master node and its timer parameters as well as the used and required interfaces. Platform variant specification defines different MOST ring variants for the verification of the Ring Break Diagnosis (RBD).

As mentioned before, the templates and modules of the platform variants specification are structured to three libraries. This well-established library approach provides clarity, supports reusability as well as reintegration of specified platform templates/modules in new platform variants specifications. The first library contains the platform modules representing virtual prototype modules (e.g. SystemC modules). Every virtual prototype module is described by a platform module definition, see Fig. 2a. Within this definition the module parameters are specified. Every parameter must have a data type and has to be marked with *public (+)*, *private (-)* or *protected (#)*. In the platform variants specification only parameters which are specified as *public (+)*, are allowed to be configured. Optionally a composition structure diagram is associated with the platform module to define a hierarchical structure.

The second element type of the first library is the platform template (*<platform_template>*). A platform template abstracts a part of a system for simplicity, variability or structural reasons. It consists of several platform modules and templates which are described in a composition structure diagram, see Fig. 2b. The last type of platform elements is the or-template (*<platform_or_template>*). They are integrated in the platform variants specification representing alternatives for specific platform components. The or-templates consist of two or more not related platform templates or modules. A valid platform variant contains always one element of them.

In the second library, interfaces are defined to guarantee a correct linking between templates and modules. Fig. 2a demonstrates the annotation of interfaces to a module definition. The linking requirements will be explained in detail in Section C. The last library contains enumerations which enable the specification of user-defined data types.

B. Platform OCL Constraints

This section introduces the constraint sets which are used to define variability, configuration capabilities and dependencies among platform elements. The constraint sets are attached to the corresponding platform modules/templates and are specified in OCL. Commonly OCL is used to define constraints at the M1 layer of the *Meta Object Facility (MOF)* Standard [23]. Hence concrete instances, respectively distinctive data (M0), of the M1 models can be verified by these constraints. Table I presents the deviation of our Platform Meta Layers to the standard MOF layers. The M3 layer is represented by the UML 2.4.1 Meta model. The elements and concepts of the UML-Meta Model are derived to build the plat-

TABLE I
MOF LAYERS

	Standard MOF Layers	Platform Meta Layers
M3	Meta Meta Model	UML-Meta Model
M2	UML-Meta Model	Platform Templates/Profile
M1	User-defined UML-/Object-models	Platform Variants Specification
M0	Distinctive Data	Platform Variant Space

form templates and the platform profile. These two platform concepts build the M2 layer in our approach. The M1 layer is formed by the platform variants specification itself with the attached OCL constraints. The OCL constraints combined with the platform variants specification span the M0 layer which consists of the valid platform variant space. We deploy a subset of OCL to specify the platform constraints. We build this particular subset by adding new operators to define platform constraints more accurately. In the following the subset will be called *Platform-OCL* (P-OCL). P-OCL supports several OCL standard types and their operators like: Collection-Related Types (*Sequence*, *Bag*), Collection-Related Operators/Operations (*.,->, includes, size, ...*), Primitive Types (*Integer*, *Real*, *Boolean*) or Object Model Types (*Class*, *Attribute*).

At first we present an OCL extension which supports probability distributions for the OCL Collection-Type *Sequence*. We implemented the *gaussian* and the *invGaussian* probability operator because these distributions are often used in validation and test. Definition 1 introduces the gaussian probability distribution for *Sequences*.

Definition 1 (P-OCL Gaussian Sequence-Operator (**gaussian()**)). Let $S = Sequence\{min..max\}$, $O = S.gaussian()-> includes(Class|Attribute)$ the OCL Expression and R is the result set of O then follows: $\forall X_i \in R$ and X_i are continuous random variables with $|R| = 0 \leq i < (max - min)$ because duplicates are removed from R. These random variables correspond a gaussian probability distribution with the density function $f(x) = \frac{1}{\sigma\sqrt{2\pi}}e^{-\frac{1}{2}(\frac{x-\mu}{\sigma})^2}$ with $\mu = \frac{max-min}{2}$ and $\sigma = \frac{\mu+min}{3}$. Thereby μ defines the mean and σ represents the standard deviation.

Additionally to the gaussian probability distribution operator an inverse gaussian probability distribution operator is added.

Definition 2 (P-OCL Inverse-Gaussian Sequence-Operator (**invGaussian()**)). Let $S = Sequence\{min..max\}$, $O = S.invGaussian()-> includes(Class|Attribute)$ the OCL Expression and R is the result set of O then follows: $\forall X_i \in R$ and X_i are continuous random variables with $|R| = 0 \leq i < (max - min)$ because duplicates are removed from

R. These random variables correspond an inverse-gaussian probability distribution with the density function $f(x) =$
$$\begin{cases} (\frac{\lambda}{2\pi x^3})^{\frac{1}{2}} e^{-\frac{\lambda(x-\mu)^2}{2\mu^2 x}} & x > 0 \\ 0 & x \leq 0 \end{cases} \text{ with } \mu = \frac{max-min}{2} \text{ and } \lambda > 0.$$
Thereby μ defines the mean and λ represents the shape parameter.

Both operators enable to specify critical boundary conditions of attribute value configurations or critical numbers of platform templates/modules. Likewise they can be used to reduce the feasible number of valid variants.

The next extension, the *active* operator, supports the constraining of or-templates.

Definition 3 (P-OCL active-Operator (**active()**)). Let $R_i = Class|Component$ with $i = 0, ..., n$ and $n \in \mathbb{N}$, $OT = or - template$, $R_i \in OT$ and $O = R_1.active()$ a P-OCL expression. Then follows for the resulting variants V: R_1 is integrated in all variants V.

The *active* operator defines which template or module of an or-template should be integrated in the generated platform variants. This enables to define constraint sets which integrate different platform templates/modules in valid solutions.

The following rules show the formalization of different types of P-OCL constraints in SMT/SAT expressions. Rule 1 formalizes a P-OCL *Sequence* expression. This expression optionally allows to define a step size, e.g. to reduce huge parameter ranges. Therefore the step size functionality, which is not contained in OCL Collection-Types standardly, is provided semantically by an iterator definition. This iterator definition is specified by the OCL operation *select*. In the formalization process integer values are represented as metaSMT bit-vectors. Subsequently bit-vectors are represented in rules as vectors (e.g. $\vec{b}$). In the following a natural number is converted to a bit-vector and vice versa by *conv* and $conv^{-1}$.

Rule 1. Let $rmin$, $rmax$, $stepsize \in \mathbb{N}$, then follows $\vec{a} \equiv conv^{-1}(rmin)$, $\vec{b} \equiv conv^{-1}(rmax)$ and $\vec{s} \equiv conv^{-1}(stepsize)$. The current bitwidth is defined by $bw \in \mathbb{N}$. The *OCL Attribute/Class Reference* is mapped to y. Then a formalization of a P-OCL Sequence definition is defined as follows: $Sequence\{rmin..rmax\}->select(e : Integer \mid e \ / \ stepsize = 0)->includes(Attribute/Class)$,
is mapped to: $y \geq \vec{a} \ \& \ y \leq \vec{b} \ \& \ (y - \vec{a}) \ \% \ \vec{s} == \vec{0}$

The first step of the formalization is to represent the *Attribute/Class-Reference* by a variable, here y. This reference-to-variable mapping has to be unique to allocate the different *Sequences* distinctly. In metaSMT, bit-vectors are represented by function $bv\_uint()[bw]$ with bit-width bw. Rule 2 demonstrates the formalization of a list definition.

Rule 2. Let be *Class/Attribute-Reference* x and $n \in \mathbb{N}$, then: $Bag\{b_0, ..., b_n\}->includes(Attribute/Class)$ is mapped to $x < \vec{m}$, with $\vec{m} \equiv conv^{-1}(n+1)$.

A list definition in P-OCL can contain values of different *Primitive Types*. These lists are specified by the OCL standard type *Bags*. In order to represent a list definition in metaSMT, the list index values are formalized as bit-vectors to assign each list entry a unique index. Regarding to Rule 2 it is obvious that the variable, respectively the Class/Attribute reference, has to be smaller than the highest index

number plus one. This rule is also adopted for floating point value sequences. Due to the standard definition of *Sequences* in OCL it is necessary to define floating point *Sequences* with explicit values. These results from the fact that a floating point *Sequence* interval specification is be determined infinite. One last example for the SMT/SAT formalization is given in Rule 3. It shows the formalization of an if-expression. The if-expression, subsequently described by if-condition (c), then-branch (t) and else-branch (e), is defined semantically equivalent to the Boolean formula: *(c implies t) and (not c implies e)*.

Rule 3. Let c, t and e be P-OCL constraints and the function *form* is defined by $form(\lambda) \in$ Boolean expression, with $\lambda \in$ P-OCL. Then the formalization of a P-OCL if-expression is defined by: if c then t else e is mapped to $logic\_ite(\ form(c)$, $form(t)$, $form(e))$

C. Template Linking

This structural Section demonstrates the versatility in defining different system topologies, e.g. ring-, bus- or mesh-topologies, with *CST*, linking requirements and port cardinalities. The Template Linking takes place, once a solution is determined by the SMT/SAT solving process. Then the number of newly generated platform module/template instances has to be derived from the solution. Afterwards they are connected due to specific linking requirements. The linking requirements are based on templates, port cardinalities, interfaces and connectors.

1. Every newly generated platform module/template instance can be linked to their target platform modules/templates as long as target port instances are left. The number of possible port instances is defined by cardinality tags annotated to the port definitions, see Fig. 2b or 3.

2. If no target port instance is left the newly generated platform module/template instance will be connected to the previously generated instance of the same type, see the ring topology example in Fig. 3.

3. Platform modules/templates which are specified variably by constraints are only allowed to connect to non-variable platform modules/templates.

4. If two platform modules/templates are connected to each other and both are specified variably by constraints it is necessary to abstract them by a *CST*.

Fig. 3 demonstrates the linking process for two examples regarding to the above defined linking requirements. In both cases template *T1* has to be integrated three times in the final platform variant. This can be enforced by the constraint $Sequence\{1..3\}->includes(self.allInstances()->size())$ whereby *self* refers to *T1*. *S* and *S1* are the target platform modules/templates to which *T1* is connected to. In the ring topology example has the target port instance of connector *C2* the cardinality of *[1]*. Therefore every newly generated instance of *T1* has to be connected to the next newly generated one. The bus topology is generated because the cardinality of the target port at *S* of *C1* is unlimited *[1..*]*. Hence every newly generated instance of *T1* can be connected to *S*. The given examples in this section are synthetic and should only demonstrate the definition of different system topologies by using *CSTs*, linking requirements and port cardinalities. Due

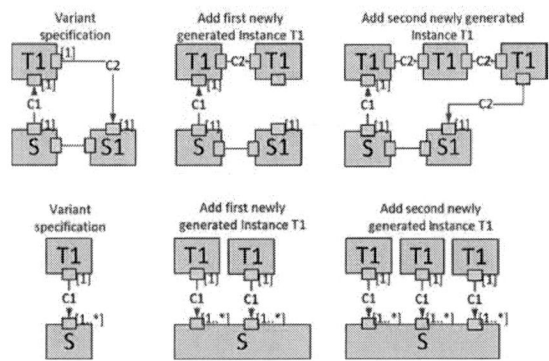

Fig. 3. Linking process for two platform variants specifications regarding to defined linking requirements.

to the limited space in this paper the example is visualized in an abstract way.

IV. Experimental Results

This section discusses our experimental results from platform variants specification for the verification of an industrial MOST application and for the exploration of a FlexRay traffic sign recognition example. All experiments have been carried out on an Intel Core 2 Quad 2.50 GHz with 4 GB of RAM using Ubuntu 12.04 64bit.

A. Ring Break Diagnosis Application (RBD)

We demonstrate the flexibility, changeability and necessity of our methodology for the verification of virtual prototypes variants by means of a MOST Ring Break Diagnosis [24] verification example. Ring Break Diagnosis is classified within ISO-OSI Data Link Layer of the MOST network interface section (NetInterface). It serves the purpose of localizing a fatal error in the network. The RBD process can be started by various triggers, which must be chosen and implemented by the System Integrator [24]. In our MOST model we implemented all different MOST layers as well as the entire functionality of RBD regarding to the state chart description of the algorithm given by the MOST specification.

In order to verify the RBD algorithm reasonably it is necessary to specify a feasible subspace of the $54*10^{21}$ valid MOST platform variants. In [22] six different verification scenarios turned out to be suggestive: *error free, ring break, excessive attenuation, multi master, all slave* and *combination*. Every scenario is specified by a different number of templates and constraints. Table II summarizes the generated variants, used templates and specified constraints for each scenario. The number of constraints and templates remained fairly

TABLE II
GENERATED VARIANTS FOR RBD

Scenario	Variants	Templates	Constraints
error free	25133	8	38
ring break	24478	8	37
excessive attenuation	24564	8	37
multi master	25231	8	37
all slave	25117	7	29
combination	24756	8	38

constant because often one constraint or template limits the boundary to another scenario. This fast reconfiguration capability is demonstrated by changing the *multi master* to the *all slave* scenario. Therefore it is just necessary to remove

the MasterTemplate from the variant specification and to re-link the remaining SlaveTemplates.

In Fig. 2b the top level description of the ring break variant specification scenario is displayed. For the ring break scenario we use 8 templates and 37 constraints. Four of the 8 templates are displayed in the top level description. Each of these four templates contain one or-template and one MOST Device node. Each of the four top level templates contain a minimum of 8 constraints to specify the different timer of MOST masters or slaves as well as to trigger which channel should be used. Listing 1 shows some of these constraints. The if-constraint in line 11 to 18 take care that only one ring

```
1  ---  These Sequence Constraints are attached
2  ---  at MasterTemplate and slaveTemplate1.
3  Sequence{1..63}->includes(self);
4  Sequence{1..63}->includes(self);
5  slaveTemplate.allInstances()->size() +
6     slaveTemplate0.allInstances()->size() +
7     MasterTemplate.allInstances()->size() <= 63
8  ---  Constraints defined within MasterTemplate
9  Sequence{2000..2500}.invGaussian()->
10    includes(self.t_Config);
11 if self.deviceChannelRB.active()
12   then SlaveTemplate.allInstances->
13     forAll(e | !e.deviceChannel.active())
14   else if slaveTemplate0.deviceChannel.active()
15   then self.allInstances
16     ->forAll(m | !m.deviceChannel.active())
17   endif
18 endif
```

Listing 1 Small subset of ring break scenario constraints

break is injected in a final platform variant. This is necessary because the RBD algorithm is designed to detect and identify the position of one ring break. For every scenario we generate approximately 25000 variant instances.

Altogether we generated for the six scenarios over 150000 variants. Thereby the solving and generation process has been done in about 2 hours. In order to ensure a suggestive verification regarding to RBD, we configured the maximum simulation time to 3500 msec. The parallelized implementation of our platform variants specification framework enables to process simulations parallel. Therefore the entire solving, generation and simulation process can be done in a few hours in case of sufficient resources. Owing to our limited resources the simulation process needs up to 5 days.

B. Traffic Sign Recognition Example

In our second example we demonstrate the versatility of our novel constraint-based platform variants specification approach. We applied our approach for an exploration use case within a heterogeneous system consisting of virtual prototype modules and modules containing hardware target code. We explored different camera modules for an automotive traffic sign recognition (TSR) platform and different hardware parallelization options. The TSR platform is implemented by a FlexRay bus system.

The top level of the variants specification of the system is defined by four structural platform templates which encapsulate the functionality of the TSR. These four templates include: the camera, the circle detection application, the traffic sign classification and the display. Each of these four templates abstracts up to 25 platform modules which provide the functionality of the camera, the circle detection or the classification. The camera exploration constraints in Listing 2 specify different camera types by display resolution,

978-1-4799-2817-0/14 $31.00 © 2014 IEEE

grayscale- or colored-camera and scale factor. The second constraint set in Listing 3 specifies the different hardware parallelization options for the circle detection. The circle detection is implemented in target code for a Tilera board to explore these options on the board. The options enable to

```
1  Bag{320,480,576}->includes(self.roiHeight);
2  Bag{480,640,720}->includes(self.roiWidth);
3  Bag{1,3}->includes(self.unBytesPerPixel);
4  Bag{0.3, 0.6, 0.1}->includes(self.scale);
```

Listing 2 Camera exploration constraints

trigger a number of threads on the board which provides up to 53 cores. Thereby the circle detection module is started

```
1  Sequence{1..53}.includes(self.noOfCores);
2  Sequence{5..10}.includes(self.minRadius);
3  Sequence{75..80}.includes(self.maxRadius);
```

Listing 3 Exploration constraints for parallization

on a certain number of cores with a calculated sub range corresponding to a given minRadius and maxRadius. Hence different circle radii can be detecting parallel.

Due to the complete variants generation process of our approach, 71150 valid platform variants are generated. They are evaluated with regard to the frame rate and the number of recognized traffic signs. The evaluation video stream contains 6 traffic signs.

The results showed that only one camera module recognized all 6 traffic signs correctly and has no classification errors. It is defined by the parameter set: *(320, 640, 1, 0.5)* (*roiHeight, roiWidth, unBytesPerPixel, scale*). The best results for the parallelization are: *minRadius = 5, maxRadius = 80* and *noOfCores = 20*. Within this example we highlighted the versatile and the flexible usage of our variants specification approach for heterogenous systems.

V. CONCLUSION

In this paper we proposed a novel constraint-based specification approach to enable variability for virtual prototype platforms. Applying this methodology for verification, exploration and test of virtual prototype platforms enables to specify and generate precisely, plausibly and comprehensibly valid platform variants. The usability, applicability and flexibility of our approach are demonstrated for the verification of an industrial MOST application and the exploration of a traffic sign recognition application on a FlexRay bus. Likewise the MOST variant space consists of more than $54 * 10^{21}$ valid variants which points out the scalability regarding to huge variant spaces. This huge variant space can be reduced efficiently to feasible and required platform variants by applying our methodology. Next steps are a model-to-model mapping to import IP-XACT descriptions and to integrate a model-based design flow for reliability assessment of safety-relevant automotive systems [25].

VI. ACKNOWLEDGEMENTS

This work has been funded by the German Federal Ministry of Education and Research (BMBF) within project 16M3195E.

REFERENCES

[1] W. R. Group. (2010) The 2010 Wilson Research Group Functional Verification Study. Wilson Research Group. [Online]. Available: http://goo.gl/rIyYu

[2] P. Montag and S. Altmeyer, "Precise WCET calculation in highly variant real-time systems," in *Design, Automation Test in Europe Conference Exhibition (DATE), 2011*, march 2011, pp. 1 –6.

[3] Altran GmbH. FlexRay Consortium. Altran GmbH. [Online]. Available: http://www.flexray.com/

[4] M. Cooperation, *MOST Specification*, Most Cooperation Specification, Rev. 3.0 E2, 07 2010. [Online]. Available: http://goo.gl/nRqj2

[5] OMG. (2011, August) Unified Modeling Language. Object Management Group. [Online]. Available: http://www.uml.org/#UML2.0

[6] ——. (2012, January) Object Constraint Language (OCL). ISO Release. [Online]. Available: http://www.omg.org/spec/OCL/ISO/19507/PDF

[7] F. Haedicke, S. Frehse, G. Fey, D. Große, and R. Drechsler, "metaSMT: Focus On Your Application Not On Solver Integration," in *Program Proceedings of the 1st International Workshop on Desigen and Implementation of Formal Tools and Systems (Austin, TX/USA)*, M. K. Ganai and A. Biere, Eds., 2011.

[8] OMG. (2009, March) Common Variability Language. Object Management Group. [Online]. Available: http://goo.gl/yPW7j

[9] K. Kang, S. Cohen, J. Hess, W. Novak, and S. Peterson, "Feature Oriented Domain Analysis (FODA) Feasibility Study," nov 1990.

[10] S. Jarzabek and H. Zhang, "XML-based method and tool for handling variant requirements in domain models," in *Requirements Engineering, 2001. Proceedings. Fifth IEEE Int. Symposium on*, 2001, pp. 166 –173.

[11] C. Maga and N. Jazdi, "An approach for modeling variants of industrial automation systems," in *Automation Quality and Testing Robotics (AQTR), 2010 IEEE Int. Conference on*, vol. 1, may 2010, pp. 1 –6.

[12] AUTOSAR. (2012) About. AUTOSAR. [Online]. Available: http://goo.gl/KO4cq

[13] ——. (2011, 10) Generic Structure Template. AUTOSAR. [Online]. Available: http://goo.gl/ccByb

[14] X. F. Zha and R. D. Sriram, "Platform-based product design and development: A knowledge-intensive support approach," *Know.-Based Syst.*, vol. 19, no. 7, pp. 524–543, Nov. 2006. [Online]. Available: http://dx.doi.org/10.1016/j.knosys.2006.04.004

[15] R. Wille, D. Große, F. Haedicke, and R. Drechsler, "SMT-based stimuli generation in the SystemC Verification library," in *Specification Design Languages, 2009. FDL 2009. Forum on*, sept. 2009, pp. 1 –6.

[16] IEEE, "IEEE Standard for Standard SystemC Language Reference Manual," *IEEE Std 1666-2011 (Revision of IEEE Std 1666-2005)*, vol. 1, pp. 1 –638, 9 2012.

[17] N. Kitchen and A. Kuehlmann, "Stimulus generation for constrained random simulation," in *Computer-Aided Design, 2007. ICCAD 2007. IEEE/ACM Int. Conference on*, nov. 2007, pp. 258 –265.

[18] S. Plaza, I. Markov, and V. Bertacco, "Random Stimulus Generation using Entropy and XOR Constraints," in *Design, Automation and Test in Europe, 2008. DATE '08*, march 2008, pp. 664 –669.

[19] D. Grosse, R. Wille, R. Siegmund, and R. Drechsler, "Contradiction analysis for constraint-based random simulation," in *Specification, Verification and Design Languages, 2008. FDL 2008. Forum on*, sept. 2008, pp. 130 –135.

[20] M. Soeken, R. Wille, M. Kuhlmann, M. Gogolla, and R. Drechsler, "Verifying UML/OCL models using Boolean satisfiability," in *Design, Automation Test in Europe Conference Exhibition (DATE), 2010*, March, pp. 1341–1344.

[21] OMG, *Constraints Package*. OMG, August 2011, vol. 2.4.1, ch. 9.6, pp. 40–43. [Online]. Available: http://www.omg.org/spec/UML/2.4.1/Infrastructure

[22] A. Braun, O. Bringmann, D. Lettnin, and W. Rosenstiel, "Simulation-based verification of the most netinterface specification revision 3.0," in *Design, Automation Test in Europe Conference Exhibition (DATE), 2010*, march 2010, pp. 538 –543.

[23] OMG. OMG Meta Object Facility (MOF) Core Specification. OMG. [Online]. Available: http://www.omg.org/spec/MOF/2.4.1/PDF/

[24] MOST, *MOST Media Oriented Systems Transport Multimedia and Control Networking Technology*, MOST Cooperation Std., Rev. 3.0, 05 2008.

[25] S. Reiter, M. Pressler, A. Viehl, O. Bringmann, and W. Rosenstiel, "Reliability assessment of safety-relevant automotive systems in a model-based design flow," in *Design Automation Conference (ASP-DAC), 2013 18th Asia and South Pacific*, 2013, pp. 417–422.

A Transaction-Oriented UVM-Based Library for Verification of Analog Behavior

Alexander W. Rath*, Volkan Esen* and Wolfgang Ecker*
*Infineon Technologies AG
Email: *Firstname.Lastname*@infineon.com

Abstract—The Universal Verification Methodology (UVM) has become a de facto standard in today's functional verification of digital designs. However, it is rarely used for the verification of Designs Under Test containing Real Number Models. This paper presents a new technique using UVM that can be used in order to compare models of analog circuitry on different levels of abstraction. It makes use of statistic metrics. The presented technique enables us to ensure that Real Number Models used in chip projects match the transistor level circuitry during the whole life cycle of the project.

I. Introduction

In today's IC designs more and more parts of the analog implementation are shifted to the digital domain, since digital circuits scale better with new technologies. This trend leads to mixed signals designs. Their analog and digital parts interface with each other as well as with the outside world.

The functional verification of the analog parts is different compared to the verification of the digital parts:

- Digital parts are functionally verified on register transfer level (RTL). Very sophisticated transaction-based methodologies like OVM [1] or UVM [2] are used in order to accomplish this task. The key concepts of these methodologies are the generation of constrained-random stimulus [3], automated checking mechanisms [4] and the collection of functional coverage [5].
- The analog parts of mixed signal designs are usually verified on SPICE level by using network simulators [6]. This approach covers mainly the verification of electrical parameters, e. g. input resistance and amplification. However, it is also used to verify the functional behavior of the block.

In the verification of the whole chip (chip level verification), where the system-level behavior as well as the interconnectivity of the blocks are to be checked, the detailedness of the SPICE models is often not required. Also, they slow down the simulation speed drastically.

In consequence, it is a common practice not to use the SPICE models in chip level simulations. Instead, so called real number models (RNM) are used [7], [8]. They purely reflect the functional behavior of the analog parts and are developed by using a hardware description language, e. g. VHDL, Verilog or SystemVerilog. The advantage of this approach is that a regular event driven simulator can be used in order to perform the chip level verification.

As the specification of the SPICE modules changes often during the project development cycle, due to new findings, it becomes a challenge to keep the aforementioned RNMs consistent to the their SPICE counterparts.

A big portion of this challenge is to verify the consistency of RNMs and their SPICE counterparts, because the evaluation is usually done by visual inspection of simulator wave forms over a limited set of test stimuli. On one hand, this approach is error prone and on the other hand very time consuming.

In this paper, we present a verification approach which enables automatic reevaluation of the consistency between the SPICE models and the RNMs, while allowing a much broader coverage of tested functionality.

This approach is based on an extension of the UVM, which is state of the art [9] in automated digital verification.

In the following chapters we present an outline of the approach and how it is mapped to the UVM. Furthermore, we describe the application of this approach to a typical example and provide an overview of experimental results. In connection to that, an analysis with regard to related work in this area is given, followed by a conclusion and an outline of steps that are to be addressed next.

II. Approach

In order to ensure consistency between the RNM and its SPICE counterpart, it is mandatory to stimulate both models with the same stimuli and automatically compare the results of both models for equality.

In principal, UVM offers these mandatory features. However, it is originally intended for digital stimulus generation and the comparison of digital signal sequences. Our purpose, however, requires the generation of a wide range of analog stimuli and the comparison of analog signal sequences.

While digital behavior represents a sequence of binary values at specific points in time, analog behavior represents a continuous progression of arbitrary values, i.e. a function $t \mapsto f(t)$. Hence, it is necessary to provide a framework which on hand can create stimulus in such a way and on the other hand, provides methods to compare such kind of behavior.

In addition to this an RNM is typically more abstract and hence, creates a simplified behavior compared to the SPICE counterpart. Therefore, the framework also needs to provide mechanisms, which allow tolerances in the comparison. The acceptable tolerances are to be defined upfront in order to enable a clear pass/fail criterion.

978-1-4799-2817-0/14 $31.00 © 2014 IEEE

In the following sections, we describe how we enhance the UVM framework in order to support these requirements. Our enhancement to UVM, we call A-UVM (analog UVM).

A. Transactions

The core elements in UVM-based testbenches are the transactions. In the following, we shortly explain that claim and the consequences, it implies for testbenches. Furthermore, we extend the concept of transactions towards the analog world.

1) Transactions in UVM: Not only in a UVM context, a transaction is a set of parameters, which describes a certain protocol in an abstract manner above the signal level. For example, a transaction for a simple serial protocol may contain a parameter "address" and another parameter "data". However, it will not contain information about how exactly the transaction will look like on pin level. Thus, the same transaction could potentially be used to abstractly describe a simple parallel protocol. The motivation of modeling protocols using transactions is that a verification engineer is usually not interested in the signal level behavior of the design. For example, if it is to be checked that a certain register in the DUT holds the correct value, only the address and data information of the respective communication is of interest. There is no reason for carrying the information about what exactly happened on pin level. However, in order to communicate with a DUT, the transaction has to be translated from transaction level to pin level and vice versa. In a UVM testbench so-called drivers and monitors are used to accomplish this task. Together, a driver and a monitor form an agent, which is also called a verification component (VC). Hence, the transactions that are used to communicate with the DUT determine the structure of the testbench.

In UVM, the code to describe the example transaction mentioned would look like this:

```
class my_item extends uvm_sequence_item;
  int address;
  int data;
  'uvm_object_utils_begin(my_item)
    'uvm_field_int(address, UVM_ALL_ON)
    'uvm_field_int(data,    UVM_ALL_ON)
  'uvm_object_utils_end
endclass
```

2) Analog Transactions: Analog signals are different compared to digital signals, as their co-domain is practically unlimited. That allows single analog signals to adopt different shapes, whereas a single digital signal is always of rectangular shape. Despite this, it is possible to classify the shape of an analog signal. For example, an analog signal can be of a linear, harmonic or cubic spline shape or of any other shape as well. Obviously, in order to precisely describe an analog signal, it is not sufficient to simply name its shape. Additional parameters are required. For example, in order to describe a linearly shaped signal, its slope as well as one value at a certain point in time are to be specified.

In A-UVM, we identify the term "shape" with the term "protocol" known from purely digital interfaces. The term

"transaction" stands for a data structure which contains the parameters needed to specify an analog signal.

For example, a signal with a sinusoidal shape would be described by the two real-valued parameters "amplitude" and "frequency".

However, in our approach we found it necessary to add some meta data to the transactions. Thus, we separated the transaction parameters from the sequence item, by defining a new class `a_uvm_data_structure`. This class serves as a base class for a new class containing the analog transaction parameters.

```
class sinus extends a_uvm_data_structure;
  real ampl;
  real freq;
  'uvm_object_utils_begin(sinus)
    'uvm_field_real(ampl, UVM_ALL_ON)
    'uvm_field_real(freq, UVM_ALL_ON)
  'uvm_object_utils_end
endclass
```

The transaction itself references the data structure—the three meta data fields, we explain in the next sections.

```
class a_uvm_sequence_item extends
uvm_sequence_item;
  // transactions parameters
  rand a_uvm_data_structure data_str;
  // meta data
  string algorithm_name;
  protected a_uvm_tlm_time
    duration;
  protected a_uvm_tlm_time
    sample_rate;

  'uvm_object_utils_begin
    (a_uvm_sequence_item)
    ...
  'uvm_object_utils_end
endclass
```

B. Generating constraint random analog stimulus

In UVM-based testbenches, drivers are used to transform transactions to signal level activity. In this section we show how A-UVM accomplishes this task regarding the analog transactions defined in the previous section.

In purely digital environments, a protocol is engraved into the digital driver using an FSM. According to the previous paragraph, a signal shape must be engraved into the analog driver. However, this cannot be done using an FSM. Instead a numerical algorithm must be used.

A-UVM uses a predefined interface for the communication between a generic driver and the algorithms. This interface allows new – potentially project specific – algorithms to be plugged in. It also allows to exchange the algorithm to be used during runtime. This plug-in mechanism is realized using the so-called "strategy pattern" from [10]. Regarding algorithms,

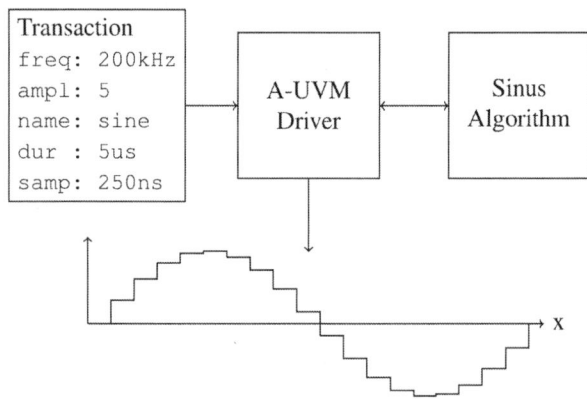

Figure 1. Process of driving an analog transaction onto a signal in A-UVM.

A-UVM is not restricted to SystemVerilog. Algorithms written in C, Matlab or other languages can be plugged in as well.

The interface that is to be provided by every algorithm consists of the following methods. We explain them in the following paragraphs.

pure virtual function void pre_process(a_uvm_data_structure data_str);

pure virtual function real get_real(**real** x);

virtual function void post_process();

When the driver receives a potentially randomized transaction, it reads the meta field algorithm_name. Then the driver selects this algorithms from its data base and passes the field data_str to the algorithm by calling its method pre_process. This algorithm can use this in order to open a connection to external tools, e.g. to Matlab, if required.

After that the driver starts to call get_real repeatedly. The interval in which the driver calls this method is determined by the meta field sample_rate in the transaction. The argument x of the method represents the time elapsed since the start of the transaction. This information is used by the algorithm in order to compute the actual signal value. It is returned by the method get_real. The whole process lasts until the time specified by the meta filed duration in the transaction is elapsed. After that, the driver calls post_process. Upon this call, the algorithm can perform finalization tasks, e.g. closing the connection to an external tool. Now, the A-UVM driver is ready for the next transaction, which can be processed by the same or another algorithm.

Figure 1 visualizes the whole process at the example of a sine wave. However, A-UVM is not restricted to sine waves. It features algorithms for FOURIER synthesis, cubic spline interpolation, ramps, jumps and others.

C. Monitoring analog behavior

In this section we show, how we extend the concepts presented in the previous sections towards monitoring of analog signals.

In principle, the definitions from the previous sections can be used directly for monitoring as well: An analog monitor must be able to extract the parameters of a signal of a given shape. In order to do so, the monitor uses a – potentially project dependent – algorithm reflecting the signal shape. Because of the potential project dependency of the algorithm, we separated the monitor from the monitoring algorithm it deploys. This allows us to keep all the management required for monitoring in our library. Also, it allows the verification engineer to exchange the monitoring algorithm during run time. This is useful, if the shapes that are to be monitored vary over time because of the DUT being in different states. In the following subsections we present the plug-in approach in more detail as well as some complications that need be taken into account which we discuss in the following sections. Since the complications have an impact on the plug-in mechanism, we go for them first.

1) Triggering: When stimulating a DUT's interface, the driver is not responsible of determining the point in time, when exactly a transaction is supposed to start. Instead, it just starts to drive the DUT's interface once it receives the transaction from the sequencer. The sequencer in turn is controlled by higher testbench facilities, e.g. by a test sequence.

When monitoring a DUT's interface, this circumstance is substantially different. There is no testbench facility that can tell the monitor when to start exactly. Instead, the monitor has to determine the start of a transaction by recognizing a characteristic activity in the DUT's output.

For analog signals, typical start activities could be e.g.

- discontinuities,
- discontinuities in the first derivative,
- changes in frequency or
- crossing a certain value.

Of course, other characteristics are possible as well.

In order to enable the monitor to start the construction of a transaction at one of the aforementioned activities, we added the concept of triggers to A-UVM. A trigger is an object that raises an event upon one of the aforementioned activities. In A-UVM, we defined triggers covering these activities. However, additional user-defined triggers can be plugged in via a callback mechanism. Hence, the approach provides the flexibility required to be useful across different projects.

In the previous paragraph, we focused on a start activity *within* the signal to be monitored. However, there can be cases, where the start is actually not an activity within the signal to be monitored. The start activity could be an activity in another signal – analog or digital. Hence, we designed A-UVM in such a way that the trigger can be sensitive not only to the signal to be monitored (see fig. 2). Furthermore, we designed it in a way, such that triggers can be combined. This way, triggers can form logic interconnections.

2) Single threaded vs. Multi threaded: The monitor in A-UVM knows two operating modes:

- Single threaded, i.e. serial operation mode and
- Multi threaded, i.e. parallel operation mode

978-1-4799-2817-0/14 $31.00 © 2014 IEEE

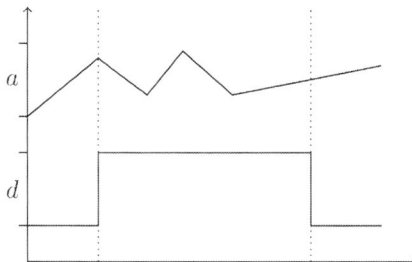

Figure 2. Arbitrary analog signal a and arbitrary digital signal d. If the monitor algorithm is supposed to determine the greatest value of a within the interval where d is high, it needs the help of a start and a stop trigger, in order to determine start and end time.

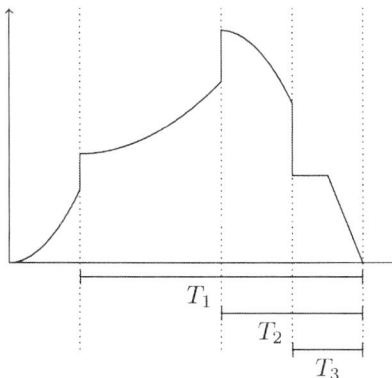

Figure 3. Arbitrary analog signal with discontinuities. If the purpose of our monitor is to measure the time between the discontinuities and the zero, it must operate in parallel mode in order to capture all three times.

In the example of fig. 2, a rising edge and a falling edge of d can never occur at the same point in time. Therefore, the transactions in this example are not interleaved. Hence, the monitor must operate in sequential mode, i.e. only one instance of the algorithm is active at the same time.

However, in the example fig. 3, the situation is different. Let us assume that the characteristic to be monitored is the time between a discontinuity and a zero. The actual time T_i between a discontinuity and a zero is measured by the monitoring algorithm. The discontinuities are detected by a start trigger and the zero is detected by a stop trigger (or – alternatively – by the algorithm itself). In this example, the monitor has to spawn three algorithms, because there are three discontinuities before the zero. Therefore, the monitor must operate in parallel mode. If, instead, the monitor was operating in sequential mode, the transactions belonging to the intervals T_2 and T_3 in fig. 3 would not be generated, since the monitor would be blocked by determining T_1.

3) Plug-in approach for monitoring: The A-UVM monitor communicates with one or more attached monitoring algorithms via a predefined API. A verification engineer who designs a new monitoring algorithm has to create a class whose methods comply with this API. Thus, the API serves as an abstraction layer. In this subsection we present the plug-in API in more detail.

All monitoring algorithms have to extend the virtual class `a_uvm_monitor_algorithm` which is part of our library. This class defines three virtual methods which are to be overridden by the concrete monitor algorithms:

pure virtual task
    start(a_uvm_sequence_item trans);

This method is called by the generic monitor, once a transaction is started. This happens once a start trigger attached to the monitor has encountered a respective event. Since the method is pure virtual, the user has to provide a project dependent implementation, which monitors the transaction. The monitoring algorithm stores the results it extracted in a data structure that we call `a_uvm_data_structure`. The concrete layout of this data structure depends on the monitoring algorithm. Thus, we provide it as a virtual class as well. After having filled the data structure, the monitoring algorithm attaches it to the field `data_str` in the argument `trans`.

4) The generic monitor: In this subsection we present the generic monitor in more detail. Its task is to perform all the required management for the communication between triggers and the actual monitoring algorithm. The class is a library element. Hence, it is not required to override it.

However, it can be configured dynamically through a small set of methods, which we will present within this subsection as well. Since it is not required to override these methods, they are non-virtual.

function void register_monitor_algorithm
    (a_uvm_monitor_algorithm algorithm);

Using this method, algorithms according to the previous subsection can be registered. Being able to register more than one algorithm is useful, since it is sometimes desired to be able to monitor different signal shapes within one simulation.

function void set_monitor_algorithm
    (**string** algorithm_name);

This method selects one of the registered algorithms. This is the algorithm used by the generic monitor from now on.

function void set_monitor_operation_mode
    (a_uvm_operation_mode_type
        operation_mode);

Through this function one can select whether the monitor should run in sequential or parallel mode. Additionally, one can completely shut down the monitor through this method.

function void set_start_trigger
    (a_uvm_monitor_trigger trigger);

Using this method, the triggers described above can be registered. Based on the configurations set through the aforementioned methods, the A-UVM monitor provides all the required management. This includes

– Reacting on triggers
– Starting of algorithms

– Pushing monitored transaction through a TLM port to higher testbench facilities

D. Checking analog transactions

In UVM-based testbenches, the transactions produced by monitors and reference models are to be compared in order to determine, whether the DUT behaves correctly.

Comparing transactions in UVM is done by comparing their fields bit-wise. However, analog transactions have real-valued fields. Due the fact that real value operations are not exact [11], a bit-wise comparison of the fields of an analog transaction will almost never lead to accordance.

Thus, A-UVM provides a comparison mechanism based on cosine similarity [12]: [1]

$$r(X,Y) = \frac{\sum_{i=0}^{n-1} X_i Y_i}{\sqrt{\sum_{i=0}^{n-1} (X_i)^2 \sum_{i=0}^{n-1} (Y_i)^2}}, \qquad (1)$$

where X_i and Y_i are the parameters of the transactions to compare and n their size, i.e. the number of parameters, such a transaction has.

This coefficient satisfies $|r| \leq 1$ and has the following interpretation:

– A value of 1 means that there is a scalar $a > 0$ such that

$$Y = aX. \qquad (2)$$

That just means that X and Y are linearly dependent.
– A value of -1 means the same, but with $a < 0$.
– Any other value between -1 and 1 means that there is no such linear relationship between X and Y.

Consider the following examples with $X = \begin{pmatrix} 1 \\ 2 \\ 3 \end{pmatrix}$:

$$Y = \begin{pmatrix} 2 \\ 4 \\ 6 \end{pmatrix} \Rightarrow r(X,Y) = 1, \qquad (3a)$$

$$Y = -\begin{pmatrix} 1 \\ 2 \\ 3 \end{pmatrix} \Rightarrow r(X,Y) = -1, \qquad (3b)$$

$$Y = \begin{pmatrix} 3 \\ 0 \\ -1 \end{pmatrix} \Rightarrow r(X,Y) = 0, \qquad (3c)$$

$$Y = \begin{pmatrix} 1.2 \\ 1.8 \\ 3.3 \end{pmatrix} \Rightarrow r(X,Y) \approx 0.996. \qquad (3d)$$

For the examples 3a and 3b equation 2 is easily solved and, thus, $|r| = 1$. For the examples 3c and 3d equation 2 cannot be solved. However, in example 3d the value of r is very close to 1, indicating that the vectors X and Y are not far away of being linearly dependent. Hence, it can be assumed that the models producing the transactions X and Y are at least *somehow* similar.

[1]The name of this metric is motivated by the fact that $\arccos(r(X,Y))$ is the angle between the vectors X and Y in an n-dimensional coordinate system.

III. RELATED WORK

UVM [2] is the emerging de facto standard for creating reusable testbenches and verification environments. Released by Accellera, this standard defines a class library, which allows verification engineers to build verification components (VCs) and environments in a standardized way. Further, the UVM class library provides a callback mechanism, which enables VCs and system models to communicate via TLM (Transaction Level Modeling) [13].

For the analog domain, no such abstract communication technique is available. However, several different approaches to extend modern hardware, verification and system description languages with the ability to describe analog behavior have been developed; the newest being SystemC-AMS, presented in [14]. SystemC-AMS allows modeling engineers to describe analog behavior in frequency and time domain.

The drawback is that no verification library, that is such sophisticated as UVM, exists for SystemC or SystemC-AMS.

Another new approach is UVM-MS presented in [15] and [16]. However, this approach focuses mainly on the direct stimulation of the pins of the DUT using UVM and an additional Verilog-AMS layer.

All these newer approaches for the analog domain enable the verification of AMS models. The difference between AMS models and the aforementioned RNMs is that AMS models aim more at a higher level of electrical accuracy that is often not required for chip level verification, since in a chip level verification the over all functionalities rather than the electrical parameters are of interest. In consequence, none of the approaches mentioned in this section fits to our verification problem described in section I.

IV. APPLICATION AND DISCUSSION

In this section we show an application of the approach presented in the previous paragraphs.

The application is a voltage regulator circuit for which an RNM written in SystemVerilog has been developed, in order to speed up chip level simulation. Since jump stimuli cause the most significant output for regulators, we have built a testbench (see figure 4) that stimulates both regulators with jumps of random height and evaluates the responses of both models for consistency. The step responses upon a unit jump of both models are shown in figure 5.

The testbench's driver produces jumps. The height of one jump is the only parameter of the incoming transaction. The

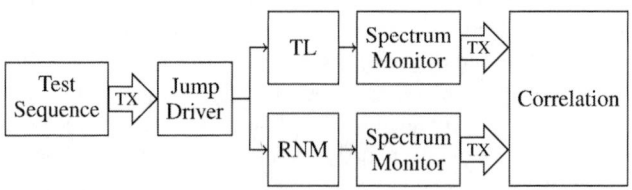

Figure 4.   The testbench used in order to compare two different abstraction levels of a voltage regulator

978-1-4799-2817-0/14 $31.00 © 2014 IEEE

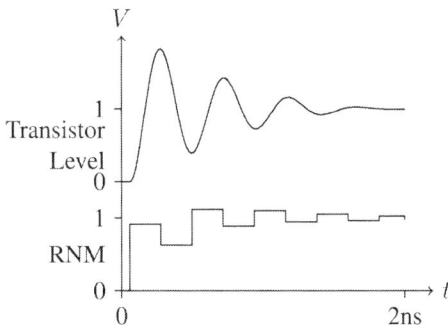

Figure 5. Step responses of the two voltage regulator models. The frequency of the transient oscillation is the same for both models. That indicates, that the frequency spectra of both models can be correlated, in order to evaluate the consistency. If the oscillation frequency of the transistor level circuit was changing due to a change in its design, the correlation would become much smaller, indicating that the RNM is to be updated.

height is randomized by the test sequence. The monitors calculate the frequency spectra of the model's outputs within an interval of 2ns. One transaction produced by a monitor carries the frequency spectrum of one step response, i.e. a list of complex values. After that, the comparison component of the testbench calculates the correlation of a pair of transactions.

Our test sequence produces 1000 random jumps in a row. The first version of the RNM simulated together with its SPICE counterpart leaded to a correlation coefficient greater than 0.89 for every produced frequency spectrum pair. We ran this test sequence in a nighty regression. After a design change in the transistor level regulator, the oscillation frequency was reduced by a factor of 2. Since the RNM was not updated, the correlation dropped down to a value smaller than 0.24 for every transaction pair. This was a clear indication that an update of the RNM is required. After updating the RNM, the correlation went up to 0.9 again.

The effort for constructing the testbench was about one week. In contrast, the effort for building a testbench relying on directed tests and manual wave form checking is slightly smaller. However, such a directed testbench can not be run in nightly regressions and the effort that is needed in order to check the consistency between the two models manually over and over again during the project exceeds the effort spent for our approach by far. Furthermore, the A-UVM-based approach presented in this paper features randomization. Thus, it covers corner cases that are easily forgotten in a directed testbench.

V. CONCLUSION AND OUTLOOK

In this paper we introduced a strategy for verifying analog behavior and highlighted its key features. The technique tackles the necessity of being able to compare models with analog behavior on different levels of abstractions. Our future work

will focus on the extensions of the presented methodology, regarding usability and flexibility. The goal is to provide a UVM based building box that covers the need of verification engineers to simulate and verify designs containing real number models. This building box shall include methods and techniques for driving, monitoring and checking of analog signals, as well as for coverage collection and reference modeling.

REFERENCES

[1] "OVM User Guide – Version 2.1.2," June 2011. [Online]. Available: www.verificationacademy.com

[2] "Universal Verification Methodology (UVM) 1.1 User's Guide," May 2011. [Online]. Available: www.uvmworld.org

[3] N. Kitchen and A. Kuehlmann, "Stimulus generation for constrained random simulation," in *Proceedings of the 2007 IEEE/ACM international conference on Computer-aided design*, ser. ICCAD '07. Piscataway, NJ, USA: IEEE Press, 2007, pp. 258–265. [Online]. Available: http://dl.acm.org/citation.cfm?id=1326073.1326127

[4] G. Allan, "Architectural considerations of scoreboard design," in *Proceedings of the 2012 DVCON international conference*, ser. DVCON '12, 2012. [Online]. Available: http://events.dvcon.org/2012/proceedings/papers/01P_7.pdf

[5] H. Carter and S. Hemmady, *Metric Driven Design Verification: An Engineer's and Executive's Guide to First Pass Success*. Springer, 2010. [Online]. Available: http://books.google.de/books?id=sAWtcQAACAAJ

[6] P. Li, L. Silveira, and P. Feldmann, *Simulation and Verification of Electronic and Biological Systems*. Springer, 2011. [Online]. Available: http://books.google.de/books?id=N48SiipG8LkC

[7] A. Elzeftawi, "CDNLive! – Real Number Model Development and Application in Mixed-Signal SoC Verification," April 2012. [Online]. Available: http://www.cadence.com/Community/blogs/ms/archive/2012/04/09/cdnlive-real-number-model-development-and-application-in-mixed-signal-soc-verification.aspx

[8] W. Hartong and S. Cranston, "Real Valued Modeling for Mixed Signal Simulation," January 2009. [Online]. Available: http://www.cadence.com/rl/Resources/application_notes/real_number_appNote.pdf

[9] T. Poikela, J. Plosila, T. Westerlund, J. Buytaert, M. Campbell, X. Llopart, R. Plackett, K. Wyllie, M. van Beuzekom, V. Gromov, R. Kluit, F. Zappon, V. Zivkovic, C. Brezina, K. Desch, X. Fang, and A. Kruth, "Architectural modeling of pixel readout chips velopix and timepix3," *Journal of Instrumentation*, vol. 7, no. 01, p. C01093, 2012. [Online]. Available: http://stacks.iop.org/1748-0221/7/i=01/a=C01093

[10] E. Gamma, *Design Patterns: Elements of Reusable Object-Oriented Software*, ser. Addison-Wesley Professional Computing Series. Addison-Wesley, 1995. [Online]. Available: http://books.google.de/books?id=6oHuKQe3TjQC

[11] *The Art Of Computer Programming, Volume 2: Seminumerical Algorithms, 3/E*. Pearson Education, 1998.

[12] A. Rath, V. Esen, and W. Ecker, "Comparison of Analog Transactions using Statitics," in *Proceedings of International Syposium on System-on-Chip*, 2013.

[13] M. Glasser and J. Bergeron, "Tlm-2.0 in systemverilog," in *Proceedings of the 2011 DVCON international conference*, ser. DVCON '11, 2011. [Online]. Available: http://events.dvcon.org/2011/proceedings/papers/04_2.pdf

[14] "OSCI SystemC-AMS extensions," March 2010. [Online]. Available: www.systemc-ams.org

[15] N. Khan, Y. Kashai, and H. Fang, "Metric Driven Verification of Mixed-Signal Designs," March 2011.

[16] N. Khan and Y. Kashai, "From Spec to Verification Closure: A Case Study of Applying UVM-MS for First Pass Success to a Complex Mixed-Signal SoC Design," February 2012.

Automata-Theoretic Modeling of Fixed-Priority Non-Preemptive Scheduling for Formal Timing Verification

Matthias Kauer[1], Sebastian Steinhorst[1], Reinhard Schneider[2],
Martin Lukasiewycz[1], Samarjit Chakraborty[2]

[1] TUM CREATE, Singapore, Email: matthias.kauer@tum-create.edu.sg
[2] TU Munich, Germany, Email: samarjit@tum.de

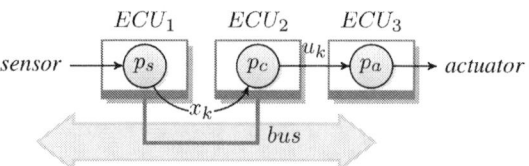

Fig. 1: A distributed control application (p_s, p_c and p_a are sensor, controller and actuator task, respectively).

Abstract—The design process of safety-critical systems requires formal analysis methods to ensure their correct functionality without over-sized safety margins and extensive testing. For architectures with state-based events or scheduling, such as load-dependent frequency scaling, model checking has emerged as a promising tool. It formally verifies timing behavior of real-time systems with minimal over-approximation of the worst case delays. In this context, Event Count Automata (ECAs) have become a valuable modeling approach because they are specifically designed to handle typical arrival patterns and integrate well with analytic techniques.

In this work, we propose an extension of the ECA framework's semantics and use it in a Fixed-Priority Non-preemptive Scheduling (FPNS) model that correctly abstracts the intra-slot behavior in the slotted-time model of the ECA. This is challenging because straightforward implementations cannot capture the full behavior of event-triggered scheduling with such a time model that the ECA shares with most model checking based methods. In a case study, we obtain bounds via model checking a basic model and then our proposed model. We compare these bounds with a SystemC simulation. This shows that the bounds from the basic model are too optimistic – and exceeded in practice – because it does not capture the full behavior, while the bounds from the proposed extended model are both safe and reasonably tight.

I. INTRODUCTION

Modern embedded systems are becoming increasingly distributed and heterogeneous at the same time. Typically, they are running two kinds of applications. Best effort applications (such as streaming) need to meet a certain average quality criterion and often require high throughput. Safety-critical tasks, on the other hand, interact with the real world via sensors and actuators and therefore operate under hard real-time requirements. A typical example is the distributed control application in Fig. 1. Here, the sensor reading of the state x_k is transmitted over a bus before the controller computes an appropriate input and communicates it to the actuator which applies it to the system. On the one hand, the sensor-to-actuator delay needs to fulfill a strict threshold, but, on the other hand, high cost sensitivity (especially in the automotive industry) can lead to progressively shrinking safety margins.

Thus, during the design phase, the behavior of the system has to be analyzed in order to guarantee its timing requirements. Since the individual components and shared resources follow diverse scheduling and arbitration policies, this analysis becomes challenging. At the same time it is impossible to give guarantees using simulation, as the worst-case pattern is often obscure and might only rarely occur in practice. In a complex system, it is therefore never certain that the behavior has been sufficiently simulated. As a remedy, the system can be analyzed using formal verification approaches such as model checking. The results of such a formal analysis can yield conservative, guaranteed bounds

This work was financially supported in part by the Singapore National Research Foundation under its Campus for Research Excellence And Technological Enterprise (CREATE) programme.

for the timing behavior with minimal over-approximation. There exist many formalisms and associated model checking tools to represent communication architectures. The applied modeling formalisms, however, determine whether the system behavior can be fully captured in the model.

Usually, there is a trade-off between time-granularity and scalability of the modeling. Timed Automata (TAs), for instance, intend to track the system's behavior with high precision and often result in models that cannot be solved in practice. Event Count Automata (ECAs), on the other hand, have been developed for streaming applications and their inherent uncertainty. As a result, they naturally model data arrival patterns that need to be scheduled and integrate well with analytic approaches. With their slotted-time model – common among most model checking based approaches – they are well-suited for time-triggered systems.

Since communication architectures are often mixed, it is of interest to use ECAs as effectively as possible for event-triggered communication policies such as FPNS. The application of ECAs to these systems has not been investigated yet, however. Towards this, we construct an ECA that is aware of all the possible scenarios that FPNS can lead to under different, but indistinguishable arrival patterns within a time slot.

Our contributions. In this work, we focus on the analysis of FPNS scheduling and, in particular, on one of its most common implementations, the Controller Area Network (CAN) bus, which is the most widely used bus in the automotive industry and other domains.

We (1) analyze why a basic FPNS model does not work if it relies on a slotted timing assumption, using the ECA framework as an example. We then (2) extend the semantics of the ECA framework to explicitly keep track of newly arrived messages and allow non-deterministic update functions. Finally, we (3) propose a FPNS model that utilizes our extended framework to accurately capture all intra-slot scenarios. We implement both the basic and the newly proposed model in the model checking tool Symbolic Analysis Laboratory (SAL) [1]. We then compare the bounds obtained by formally verifying both implementations to an analytic solution and a SystemC simulation. These experiments show that only the proposed model leads to conservative bounds without requiring increased time granularity and therefore helps to avoid state space explosion.

Overview of the paper. The remainder of this paper is organized as follows. Related approaches to delay analysis are discussed in Section II. We then give the problem description in Section III and the basics of the ECA framework in Section IV. This serves as

978-1-4799-2817-0/14 $31.00 © 2014 IEEE

background information for the introduction of a straightforward FPNS model in Section V. After we analyzed its shortcomings, we then extend the ECA's semantics and propose advanced FPNS models in Section VI. Finally, we demonstrate their value with a case study in Section VII. Section VIII concludes.

II. RELATED WORK

Commonly studied schedulability analysis techniques such as those based on demand bound or response time analysis, e.g., [2] for CAN, become pessimistic for the non-preemptive case. Furthermore, they are not compositional in nature. This has led to more general compositional models and techniques such as the Real-Time Calculus (RTC) framework [3], [4] – a purely analytic technique built upon the interaction of upper and lower arrival curves with corresponding service curves that contain information of the system from an interval-based, sliding window perspective. Often, RTC delivers precise results, but it cannot accurately capture *state* information. Efforts in this direction have been the analysis of correlated streams in [5] and lightweight state mechanics in [6]. A FPNS model has been proposed for RTC in [7], but it cannot directly interact with state-based subsystems.

The need for state-based and non-preemptive scheduling has motivated automata-based methods, e.g., [8], that can be used to obtain timing guarantees by delegating to powerful, established model checking tools originating from the digital design area. These tools find an arrival pattern that violates a certain bound if it cannot be guaranteed. Their results are therefore tight by design within the precision of the model. However, they all suffer from state space explosion once the architecture under consideration becomes too large.

Approaches to FPNS analysis using TAs from [9] have been presented in [10] and [11]. Because these approaches employ a continuous time model, they track the evolution of the system in a much more fine-grained fashion than necessary in most cases. In addition, they are difficult to interface with analytical methods, such that scalability often becomes an issue. For instance, the technique from [11] is not applicable when arrival times are given in intervals.

Consequently, previous efforts have supplemented the RTC framework with a model checking based extension, the ECA [12]. It strikes a balance between the high precision tracking of the TA and the scalability of RTC. Large parts of the hardware architecture under consideration can be modeled in plain RTC and then composed with an ECA-modeled part [13]. This considerably improves scalability. Modeling state-based scheduling becomes straightforward as with all automata-based frameworks and they have shown to be a good model for capturing the timing properties of streaming applications. At the same time, they can test and verify richer interfaces [14], as they might arise from feedback control systems [15].

However, ECA models of FPNS have not been examined yet. In order to enable FPNS modeling with ECA, we capture behavior arising from the slotted-time model by taking all possible scenarios into account and delegating the decision to the model checker. Hence, our proposed model does neither require an overly fine time scale nor complicated message tracking mechanics.

III. PROBLEM DESCRIPTION

We consider a distributed control application as shown in Fig. 1 that is communicating over a shared FPNS bus. It is operating with a period T, that acts as a deadline for the *sensor-to-actuator delay*. This kind of feedback control loops typically fulfill safety-critical tasks in physical systems such as cars or airplanes. Since *the physical world does not wait* for the control system to finish processing, the consequences of a missed deadline can be severe.

We focus our analysis on the FPNS bus and try to ensure that the message x_k needs to wait a maximum of T_m there, with T_m being sufficiently small to guarantee that p_s and p_a can be executed in time. This analysis integrates well with larger ECA networks and carries over to more flexible requirements, such as m, k-firm deadlines as discussed in [14].

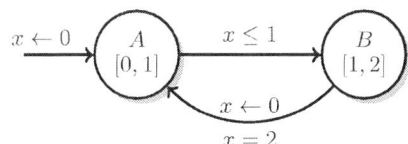

Fig. 2: Example ECA showing the framework's basic functionality.

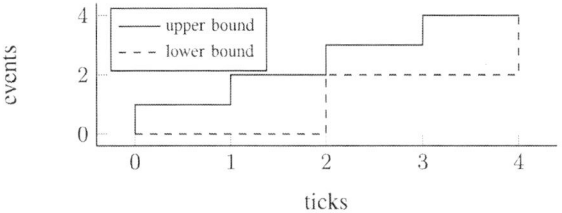

Fig. 3: Upper and lower bounds for the events of the example ECA in Fig. 2.

IV. EVENT COUNT AUTOMATA

Before we describe the issues surrounding an event-triggered system such as FPNS modeled in discrete time and the proposed FPNS implementation to capture them, we briefly introduce the basic ECA behavior. We illustrate the formal definition using a simple example in Section IV-A, describe how they combine to form larger networks in Section IV-B, and show how we can derive a message's maximum delay from an ECA network in Section IV-C. A complete description of ECAs can be found in [12].

A. Individual ECA

Individual Event Count Automata (ECAs) are given as tuples

$$\mathcal{A} = (S, s_{in}, X, V_{in}, Inv, \rho, \rightarrow). \qquad (1)$$

In this formulation:

- S is a set of states and s_{in} is the initial state.
- X is a set of count variables.
- V_{in} is the initial valuation of the count variables.
- $Inv : S \rightarrow \Phi(X)$ is the Invariant Constraint Function that assigns constraints to states with

$$\Phi(X) = x \leq K | x < K | x \geq K | x > K | \varphi_1 \wedge \varphi_2$$

  where K is a lower/upper bound enforced on a count variable x and $\varphi_l \wedge \varphi_2$ allows combinations of such constraints.
- $\rho : S \rightarrow \mathbb{N} \times \mathbb{N}$ is the rate function.

$$\rho(s) = [l, u]$$

  Every state is assigned an upper bound u and a lower bound l for the arrival or service rate in that state.
- $\rightarrow \subset S \times \Phi(X) \times 2^X \times S$ is the transition relation. Here $\Phi(X)$ indicates that transitions can have guards of the same form as the invariant constraints of the states. 2^X indicates the set of count variables that will be reset on a transition, typically written as $x \leftarrow 0$.

Consider the simple example depicted in Fig. 2. It has a single count variable $X = \{x\}$. It further consists of two states $S = \{A, B\}$ with associated rate intervals $\rho(A) = [0, 1]$ and $\rho(B) = [1, 2]$. In other words, while in B, one or two events can occur.

The states are connected by transitions with guards such as $x = 1$ and $x \leq 1$ and reset actions $x \leftarrow 0$. The ECA starts in the initial configuration $(s_{in}, V_{in}) = (A, [0])$ where V corresponds to the valuation of the count variable x. Since all transitions are considered urgent by default, we then move to either $(B, [0])$ or $(B, [1])$. In B, the subsequent guard enforces an amount of events such that $x = 2$ before the automaton resets. This models a jitter in the system.

The transitions in the system deliver the event counts $c \in [C] \stackrel{\text{def}}{=} \{1, \ldots, C\}$ (with C being a finite upper bound that is required to keep the automaton tractable). These automaton-wide event counts increment all the count variables. They represent data arrival or processing during a single step and form the alphabet for the language of the ECA : $[C]$. It contains sequences (or "strings") of event counts $\sigma = (c_1 c_2 c_3 \cdots)$ that represent arrival or processing patterns.

Fig. 3 shows the upper and lower bound on the events that can arrive in this automaton. Note that they are in the time domain, but in this case they could easily be converted to the interval-based interpretation of RTC. In general, ECAs allow more complicated patterns than RTC's service and arrival curves can accurately capture.

B. ECA Networks

Individual ECAs can be connected to form an ECA network which is a structure

$$\mathcal{N} = \left(\{\mathcal{A}_p\}_{p \in \mathcal{P}}, \{U_p\}_{p \in \mathcal{P}}, \mathcal{B}, b, C, IN, OUT \right) \quad (2)$$

In this definition:

- $\mathcal{P}$ is a finite set of nodes that the various ECAs $\mathcal{A}_p$ are associated with.
- U_p is the update function that defines how an ECA consumes and deposits items from its buffers.
- $\mathcal{B}$ is a finite set of buffers of capacity $B_{\max}$.
- C is the maximum number of items that any ECA in the network can handle in a single step. This needs to be finite for the state space underlying the ECA that we need to explore to remain finite.
- $IN : \mathcal{P} \to 2^{\mathcal{B}}$ and $OUT : \mathcal{P} \to 2^{\mathcal{B}}$ link the buffers to the ECA. It is assumed that every $\mathcal{A}_p$ has at least one buffer and that its input and output buffers are disjoint. Furthermore, buffers cannot be shared and merging of streams has to be handled in the automatons.

Consider the following, more intuitive description: Arrival ECAs deposit messages into initial buffers and service ECAs process messages from their input to their output buffers. Each ECA $\mathcal{A}_p$ is characterized by an associated update function U_p that describes from which input to which output it processes data. With N_p, the total amount of buffers attached to $\mathcal{A}_p$, separating into $N_p = N_p^{in} + N_p^{out} \stackrel{\text{def}}{=} |\text{IN}(p)| + |\text{OUT}(p)|$, $U_p : \mathbb{N}^{N_p} \to \mathbb{N}^{N_p}$ maps input and output buffer levels to *changes* in input and output buffer levels.

Data is preserved within individual ECAs. It must therefore hold for all vectors of associated buffer levels $\mathbf{b} = \left[(b_i)_{i \in \text{IN}(p)} \quad (b'_i)_{i \in \text{OUT}(p)} \right] \in \mathbb{N}^m$, event counts $c \in [C]$ and states $s \in S$:

$$\sum_{i \in \text{IN}(p)} U_p^{(i)}(s, c, \mathbf{b}) = \sum_{j \in \text{OUT}(p)} U_p^{(j)}(s, c, \mathbf{b})$$

Data can however be overwritten by the buffer mechanics in (3) where b_i is the fill level of the i-th buffer and the $(\cdot)^+$ operator is a temporal next operator[1].

$$b_i^+ = \max \left[0, \min \left(b_i + \sum_{p : i \in \text{OUT}(p)} U_p^{(i)} - \sum_{p : i \in \text{IN}(p)} U_p^{(i)}, B_{\max} \right) \right] \quad (3)$$

---

[1]More precisely, $b_i = b_i^k$ refers to the content of the buffer at the end of slot k, $b_i^+ = b_i^{k+1}$ refers to the content one time step later. This is a convenient notation because there is no interaction between more distant time steps and it translates well into the implementation language for SAL[1] – a well-known Binary Decision Diagram (BDD)-based model-checking tool that has often been selected for ECA verification.

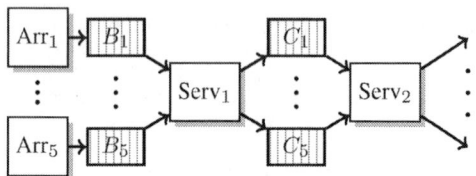

Fig. 4: Example ECA network demonstrating the interconnection using buffers.

Consider the example in Fig. 4. It begins on the left with a set of arrival ECAs. These automata use the trivial update function that deposits all the events into the only attached buffer:

$$U(s, c, \mathbf{b}) = c$$

From the first buffers, the data is further processed by Serv_1 and Serv_2. These service automata could be implementing a Time Division Multiple Access (TDMA) scheduling policy, i.e., they could be ECAs with a state for every TDMA slot and associated update function

$$U^{(i)}(s_i, c, \mathbf{b}) = U^{(i')}(s_i, c, \mathbf{b}) = c$$

for the input buffer b_i and the output buffer b'_i that correspond to the current state s_i and

$$U^{(j)}(s_i, c, \mathbf{b}) = U^{(j')}(s_i, c, \mathbf{b}) = 0$$

otherwise for $j \neq i$.

C. Delay calculation

Once the ECA network is set up, the maximum delay for a certain message stream can be calculated in the following way. Initialize an integer array $D_j = 0$. D_j tracks how many messages have been in the system for j units of time or more. Update D_j to D_j^+ – cf. b_i^+ in equation (3) – as follows[2]:

$$D_j^+ = \begin{cases} \text{inc}(b_1) + D_1 - \text{dec}(b_{\text{last}}) & , \text{if } j = 1 \\ \max\left(0, D_{j-1} - \text{dec}(b_{\text{last}}) \right) & , \text{otherwise} \end{cases}$$

The maximum delay is then $\max\{j : D_j > 0$ at any time$\}$ and can be obtained by initializing $d = 0$ and updating d via:

$$d^+ = \begin{cases} d & , \text{if } D_{d+1} = 0 \\ d + 1 & , \text{otherwise} \end{cases}$$

Note that this model does not delete buffer overwrites from the delay count. Any overwrite will therefore cause an unbounded worst case delay for the concerned message stream which is the correct interpretation in most cases.

V. STRAIGHTFORWARD FPNS MODEL

We will now describe a basic FPNS model implemented in the ECA framework. Its mechanics are very close to those of a simulation implementation in, e.g., SystemC. To avoid state space explosion, the time granularity of an ECA model must be chosen much coarser than that of a simulation. Therefore, this intuitive model can not capture all the intricacies of a real-world CAN bus. In this section, we will explore these effects and see how they can lead to overly optimistic bounds in detail. The presented modeling might be used in a typical verification scenario. Relying on its resulting bounds, however, can and will lead to deadline misses and the aforementioned, potentially severe consequences in practice.

---

[2]$\text{inc}(b_1)$ refers to the increment of the first buffer, i.e., the new arrivals; $\text{dec}(b_{\text{last}})$ to the decrement of the last buffer, i.e., the messages leaving the system. Both are to be regarded with respect to one specific message stream.

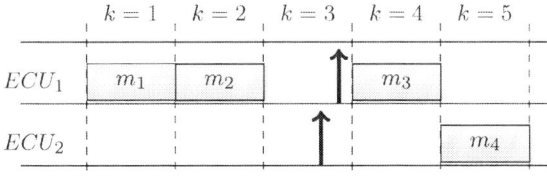

Fig. 5: **Perfectly aligned timing as assumed by the straightforward FPNS model ignores blocking by lower-priority message. Table I shows the corresponding ECA trace. The arrows denote the actual message arrival in the send-buffer.**

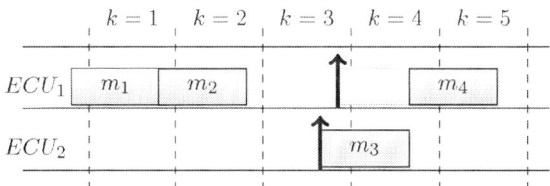

Fig. 6: **Timing diagram illustrating how a higher priority-message can be blocked by a lower-priority one in FPNS scheduling. Fig. 5 and Table I show the corresponding sequence in the basic ECA implementation that fails to capture this effect.**

A. Implementation Description

The most basic implementation of FPNS, that is closest to how a simulation tool would implement an abstract CAN bus, would employ a constant service automaton (one state A with constant rate $\rho(A) = [c, c]$) and combine it with a priority-based update function where $\pi(b)$ is the priority of a certain message buffer. We obtain $U^{(i)}(s_i, c, \mathbf{b}) = U^{(i')}(s_i, c, \mathbf{b}) = g_i$ with

$$g_i = \max\left(0, \min\left(b_i, c - \sum_{\substack{j \in \text{IN}(p) \\ \pi(b_j) > \pi(b_i)}} b_j\right)\right) \quad (4)$$

This ensures that the Processing Element (PE) handles all the data in the higher priority buffers before moving on to lower priority items. Note that for the highest priority buffer it holds $U = \min(c, b_i)$, meaning it always gets the full service unless it doesn't have enough data in the buffer.

B. Timing Issues of the Straightforward FPNS Implementation

The straightforward FPNS model described in Section V-A assumes that the messages on the FPNS bus are aligned to the ticks of the ECA as illustrated in Fig. 5. The arbitration in the model takes place at the beginning of every tick and it is impossible to resolve timing issues within a slot. More precisely, if two messages arrive in the same slot, it is impossible to say which one was first. The alignment of the ECA's ticks to the real life timing of, e.g., the CAN bus is an overly simplistic assumption that cannot hold in practice, however. When the bus is idle, it does not wait for the next time slot to begin before transmitting once a message is in the buffer ready to be sent. Such a behavior might make verification easier, but it is certainly undesirable when optimizing for throughput.

It can therefore occur in the real CAN bus that higher priority messages are blocked by lower priority messages. This is not captured by the straightforward model and leads to over-optimistic bounds when analyzing it in a model checker. To see this issue in more detail, let us assume for simplicity that there are only two message streams and that they are indexed in order of descending priority (1 being the highest priority). Further let the underlying constant service automaton process one message per time slot, i.e., $c = 1$.

As shown in Fig. 6, after two messages, m_1 and m_2 from the buffer for ECU_1 are sent, all the buffers are empty and the bus idles at the beginning of $k = 3$. Then, a message arrives for ECU_2 (indicated by the black arrow) and the bus starts processing it immediately since ECU_2 is the highest priority accessing it at that moment. When ECU_1 wants to transmit data shortly after (black arrow), the bus is already utilized and ECU_1 is blocked (transparent) until ECU_2 finishes its uninterruptible message m_3. Only then is m_4 transmitted. This blocking is a well-known phenomenon and forms the basis of, e.g., the FPNS analysis for RTC [7]. To the best of our knowledge, it has till now not been addressed in the context of ECA or other discrete-time approaches, however.

Let us compare this to how the straightforward FPNS model would process this arrival pattern. Fig. 5 contains the same

arrivals, but the messages are processed according to the straightforward model. The evolution of the two buffers b_1 and b_2 under this model is further detailed in Table I. It starts out just like the real pattern and handles the two items in the buffer of ECU_1. At the end of $k = 2$, the buffers are empty, so nothing is processed in $k = 3$ but there are two arrivals.

The model starts to differ from the observed pattern in step $k = 4$. Here, it would just process the higher priority message because it performs arbitration at the beginning of the slot only. This has been marked in bold in Table I. The bus, however, would have started processing the message of ECU_2 already before the beginning of slot $k = 3$ as described above. This would correspond to $[b_1 \ b_2] = [1 \ 0]$ at the end of $k = 4$. Calculating the maximum delay for ECU_1 from this model, e.g., by using the method from Section IV-C, and using it as a bound would therefore be overly optimistic. As we will see in the case study in Section VII, such a bound will be violated by some of the messages.

VI. PROPOSED FPNS MODEL

Blocking by lower-priority messages as described in Section V-B does occur and needs to be taken care of at the ECA modeling level. We will start by showing how a model could look like for an FPNS element that processes one message per tick (or less, e.g., one message every 3 steps) and then extend it to a multi message model.

A. Single Message Model

In this section, we will introduce a model that takes the possible blocking by lower priority messages into account when a transmission starts before the slot it ends in. Towards this, we need to consider the various paths the intra-slot behavior could take due to the underlying slotted-time mechanics. These paths then need to be included into the update function that governs the arbitration between the individual message streams. The example from Fig. 6 and Table I showed that we cannot just take the highest priority from the buffer and process it. Instead, we have to consider that any newly arrived message with even higher priority could take its place. Fig. 7 illustrates this in greater detail. With one message in the buffer for ECU_3 at the end of $k = 1$ and one message arriving for both ECU_1 and ECU_2 during $k = 2$, the discrete-time semantics allow for the scenarios discussed in the following. Each scenario consists of two arrivals (arrows) and one message (labeled a, b, c) and is further described in the following paragraphs:

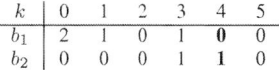

k	0	1	2	3	4	5
b_1	2	1	0	1	**0**	0
b_2	0	0	0	1	**1**	0

TABLE I: **Buffer fill levels at the end of individual steps k showing the behavior of the basic FPNS model for the arrival pattern in Fig. 5 and Fig. 6.**

978-1-4799-2817-0/14 $31.00 © 2014 IEEE   815

(a) The message from ECU_1 arrives before the transmission of m_1 ends and it therefore gets to send next. It is irrelevant when the message of ECU_2 arrives in this case.

(b) The message from ECU_2 arrives before m_1 is entirely transmitted and ECU_1 is ready to send only afterwards. ECU_2 therefore sends next.

(c) Both ECU_1 and ECU_2 get ready to send after the transmission of m_1 finishes. ECU_3 therefore transmits next.

Since the slotted time does not hold enough information to resolve which message arrives first within the time slot, all the possibilities need to be accounted for.

To calculate these possibilities, we need to track which messages were already present at the beginning of the slot and which have just arrived during the last step. We therefore extend the basic buffer semantics from Eq. (3) where only the total buffer content was tracked. In addition to b_i, the total elements in the buffer at the end of the slot, we now track n_i, the elements that newly arrived during this slot. This results in the following equations:

$$n_i^+ = \sum_{p:i \in \text{OUT}(p)} U_p^{(i)}$$

$$b_i^+ = \max\left(0, b_i + \sum_{p:i \in \text{OUT}(p)} U_p^{(i)} - \sum_{p:i \in \text{IN}(p)} U_p^{(i)}\right) \quad (5)$$

The elements that remained in the buffer after all the messages from the last step were sent are then given by

$$r_i^+ = b_i^+ - n_i^+. \quad (6)$$

With this additional and more fine-grained information we can now select the message i_0 that the bus starts sending at the end of the last step and finishes in the current step. As shown in Fig. 7, this can be either the message with the highest priority that was already in the buffer and whose transmission therefore would have started immediately after the bus was idle (scenario (c)) or any of the newly arrived messages that have higher priority (scenarios (a) and (b)). This sums up to the following options:

$$i_0 \in \Big\{ i : \underbrace{b_i > 0}_{\text{has message to send}} \quad \text{AND} \quad \underbrace{r_j = 0 \quad \forall j > i}_{\text{no higher priority is waiting}} \Big\} \quad (7)$$

With regard to the update function, after setting

$$g_i \stackrel{\text{def}}{=} \begin{cases} 1 & , \text{if } i = i_0 \\ 0 & , \text{otherwise} \end{cases} \quad (8)$$

we then let $U^{(i)}(s, c, \mathbf{b}) = U^{(i')}(s, c, \mathbf{b}) = g_i$, where i and i' refer to the in- and output buffer of stream i, respectively.

The single message model extends well into a more fine-grained time discretization, where one message is transmitted every couple of time steps. Here, one would just change the underlying service automaton and leave the arbitration as described when there is service to distribute. Additional delay could be modeled by introducing new buffer stages that the message has to travel through.

B. Multi Message Model

As discussed in Section VI-A, extension of the single message model towards more fine-grained timing is straightforward. Establishing a similar model for even less fine-grained timing where multiple messages are to be sent during a single time slot still poses a challenge, however. One possibility for modeling the behavior in this case is to iterate the single message model from Section VI-A.

We start by assuming that i_0 was chosen in the previous time step using the rule from Eq. (7). The next message to be transmitted faces a situation that is similar to the single message case shown in Fig. 7. The difference in the selection of i_1, the ECU to send after i_0, lies in the available messages. Firstly, the messages that were in the buffer at the end of slot $k = 1$ are available if they were not sent already, i.e., minus i_0. Additionally,

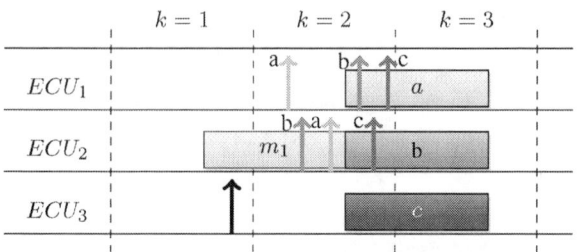

Fig. 7: Three scenarios (a,b,c) with identical arrival signature from the ECA's point of view lead to different outcomes when the CAN's arbitration is not aligned to the ECA's ticks in the single message case.

all the messages with higher priority that might have arrived by then could be ready to send. Since, again, it is impossible to distinguish arrivals within a single time step, this refers to all the messages arriving in the respective slot. Therefore i_1 has to be selected as

$$i_1^+ \in \Big\{ i : \underbrace{b_i + n_i^+ - \delta_{ii_0} > 0}_{\text{potentially has message}} \text{ AND } \underbrace{b_j - \delta_{ji_0} = 0 \quad \forall j > i}_{\text{no higher priority}} \Big\}$$

where δ_{ij} refers to the Kronecker delta:

$$\delta ij = \begin{cases} 0 & , \text{if } i \neq j \\ 1 & , \text{if } i = j \end{cases} \quad (9)$$

After i_1 has been determined, further i_l can then be calculated iteratively:

$$i_l^+ \in \Big\{ i : \quad b_i + n_i^+ - \sum_{m=0}^{l-1} \delta_{ii_m} > 0$$

$$\text{AND } b_j - \sum_{m=0}^{l-1} \delta_{ii_m} = 0 \quad \forall j > i \Big\} \quad (10)$$

VII. CASE STUDY

To demonstrate the advantages of the model developed in Section VI-A and extended in VI-B, we performed a case study using Symbolic Analysis Laboratory (SAL) [1] and SystemC. We converted the abstract ECA networks from both the straightforward model and the proposed extended models to their transition system representation. We then implemented them in the modeling language of SAL. Finally, we employed its BDD-based model-checker to verify the respective system behavior. We calculated bounds on the delay for every stream via the method outlined in Section IV-C. These will be referred to as *StraightForw*, *SingleMsg* and *MultiMsg*, respectively. Additionally, we constructed a SystemC model for the CAN bus and obtained the average delay *Avg* and the maximum delay *Max* as observed in a simulation for $T = 10^7 \text{ms} \approx 2.7\text{h}$. Finally, we performed a manual analysis of the worst case scenario where all messages start at the same time and are initially blocked by the longest lower priority message. This is denoted by *Manual*. It is both tedious and error-prone and unlike ECAs and SystemC, this analysis cannot be applied to state-based or even just larger systems.

We assumed 4 message streams numbered #1 to #4 from highest to lowest priority with periods

$$P = \begin{bmatrix} 1\text{ms} & 2\text{ms} & 4\text{ms} & 5\text{ms} \end{bmatrix}$$

and a common jitter of $j = 0.5\text{ms}$ where jitter refers to an arrival in the interval $[k \times P_i, k \times P_i + j]$.

The message streams share a CAN bus with transmission time of 0.5ms per message. No state dependence was modeled in order to have readily available reference solutions for comparison.

Runtime Evaluation. The model checking runs for the following case study took 16s for the single message model and 85s for the multi message model on an Intel i7 at 3.4GHz with 16GB RAM. Note that while the multi message model took longer to verify in this case, it might be the preferred or only choice in case another PE dictates a certain slot time.

Single message model. To test the single message model, we set the slot time to 0.5ms in accordance with the transmission time. The resulting bounds created by the formal verification of the ECA models via the model checking tool SAL are listed in Table II as *Straightforw* and *SingleMsg* respectively.

Subsequently, we took the same architecture and modeled it in SystemC. We simulated the model and recorded the delays of the individual messages. Fig. 8 shows the delay distribution of Stream #3. It illustrates that some messages do in fact incur a larger delay than 2.5ms, the bound calculated using the straightforward model. In contrast, no message exceeded 4.5ms, the bound guaranteed by SAL when verifying the newly developed single message model.

Because of their similarity, we do not show histograms for the other message streams. Instead, we have summed up their average and maximum delay in Table II. All the bounds from the single message model shown there prove to be tight within the resolution of the ECA. To clarify, consider stream #3 in Table II as an example. There, we observed a maximum delay of 3.63ms (equivalent to 4ms or 8 slots, rounded up for the ECA) and our manual analysis gave a worst-case delay of 4.0ms that can occur in practice. Verifying the ECA model with SAL leads to an upper bound of 8 slots. This corresponds to the same 4.0ms and would therefore be tight as well. However, we additionally account for the fact that this worst-case message could have arrived early in slot $k = k_0$ and was transmitted late in slot $k = k_0 + 8$, increasing the upper bound *SingleMsg* by an additional slot to 4.5ms.

Multi message model. We used the same case study with the multi message model. Toward this, we changed the slot time to 1ms, such that two messages can be transmitted per slot. By transferring this model to SAL and running its BDD-based model checker again, we obtained the bounds *StraightForw* and *MultiMsg* from Table III. These bounds appear to be very conservative with respect to the simulation results and the analysis from Table II. This is due to the higher implied jitter in the multi message case. Since the slot time is 1ms, the model now assumes that all the messages suffer from a jitter of 1ms as well. Repeating the simulation and the analysis with this jitter in mind, we obtain the columns *Avg*, *Max* and *Manual* in Table III. Again, we observe that the straightforward model is overly optimistic. In contrast, the proposed model successfully bounds the transmission delay and it is tight within the selected resolution of the ECA framework when compared to the analytical worst case. We could, however, not observe the worst case for all message streams in the SystemC simulation.

VIII. CONCLUDING REMARKS

This paper points out and thoroughly analyzes the challenges inherent in modeling FPNS on top of a slotted-time model. A basic

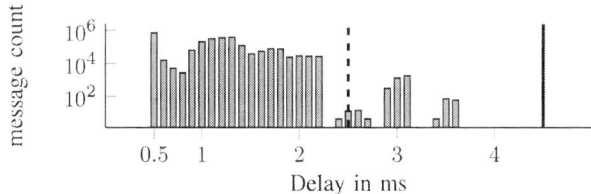

Fig. 8: Histogram of delays experienced by message stream #3 as measured in the SystemC simulation. The bound from the straightforward model (dashed, Section V) is exceeded by a significant part of the messages while the bound from the single message model (solid, Section VI-A) holds.

implementation does not account for blocking by lower priority messages and can thus lead to bounds that do not hold in practice. Instead of simply refining the time scale to capture these effects – a futile approach due to state space explosion – we propose a FPNS model that can be created with reasonable modifications to the semantics of the well-established ECA framework. We compare the bounds obtained by formally verifying a straight-forward model as well as our proposed model with results from a SystemC simulation. This case study demonstrates the importance of the improved modeling, as the bounds from the straightforward approach are in fact violated by a significant amount of messages.

With the proposed extensions to the ECA semantics and the developed FPNS model, an accurate and efficient verification of timing bounds becomes possible.

REFERENCES

[1] L. de Moura, S. Owre, H. Rue, J. Rushby, N. Shankar, M. Sorea, and A. Tiwari, "SAL 2," in *Computer Aided Verification*, ser. Lecture Notes in Computer Science. Springer Berlin / Heidelberg, 2004, vol. 3114, pp. 251–254.

[2] R. I. Davis, A. Burns, R. J. Bril, and J. J. Lukkien, "Controller Area Network (CAN) Schedulability Analysis: Refuted, Revisited and Revised," *Real-Time Systems*, vol. 35, no. 3, p. 239272, 2007.

[3] L. Thiele, S. Chakraborty, and M. Naedele, "Real-time Calculus for Scheduling Hard Real-Time Systems," in *ISCAS*, 2000.

[4] E. Wandeler and L. Thiele, "Real-Time Calculus (RTC) Toolbox," http://www.mpa.ethz.ch/Rtctoolbox, 2006.

[5] K. Huang, L. Thiele, T. Stefanov, and E. Deprettere, "Performance Analysis of Multimedia Applications Using Correlated Streams," in *DATE*, 2007.

[6] A. Bouillard, L. Phan, and S. Chakraborty, "Lightweight Modeling of Complex State Dependencies in Stream Processing Systems," in *RTAS*, 2009.

[7] D. B. Chokshi and P. Bhaduri, "Modeling Fixed Priority Non-Preemptive Scheduling with Real-Time Calculus," in *RTCSA*, 2008.

[8] K. G. Larsen, P. Pettersson, and W. Yi, "Compositional and Symbolic Model-Checking of Real-Time Systems," in *RTSS*, 1995.

[9] R. Alur and D. L. Dill, "A Theory of Timed Automata," *Theoretical computer science*, vol. 126, no. 2, p. 183235, 1994.

[10] J. Krakora and Z. Hanzalek, "Timed Automata Approach to CAN Verification," in *INCOM*, 2005.

[11] E. Fersman, L. Mokrushin, P. Pettersson, and W. Yi, "Schedulability Analysis of Fixed-Priority Systems using Timed Automata," *Theoretical Computer Science*, vol. 354, no. 2, p. 301317, 2006.

[12] S. Chakraborty, L. Phan, and P. Thiagarajan, "Event Count Automata: a State-Based Model for Stream Processing Systems," in *RTSS*, 2005.

[13] L. Phan, S. Chakraborty, P. Thiagarajan, and L. Thiele, "Composing Functional and State-Based Performance Models for Analyzing Heterogeneous Real-Time Systems," in *RTSS*, 2007.

[14] M. Kauer, S. Steinhorst, D. Goswami, R. Schneider, M. Lukasiewycz, and S. Chakraborty, "Formal Verification of Distributed Controllers using Time-Stamped Event Count Automata," in *ASP-DAC*, 2013.

[15] G. Weiss and R. Alur, "Automata Based Interfaces for Control and Scheduling," *HSCC*, 2007.

Stream #	Delay from SystemC		Analytic Delay Bounds		
	Avg	Max	Manual	Straightforw	SingleMsg
1	0.66ms	0.98ms	1.0ms	1.0ms	1.5ms
2	0.83ms	1.83ms	2.0ms	1.5ms	2.5ms
3	1.10ms	3.63ms	4.0ms	2.5ms	4.5ms
4	1.17ms	3.65ms	4.0ms	4.5ms	4.5ms

TABLE II: Single message model case study results.

Stream #	Delay from SystemC		Analytic Delay Bounds		
	Avg	Max	Manual	Straightforw	MultiMsg
1	0.65ms	1.4ms	1.5ms	2ms	2ms
2	0.78ms	2.3ms	2.5ms	2ms	4ms
3	1.02ms	4.5ms	5.5ms	3ms	7ms
4	1.14ms	5.45ms	7.5ms	5ms	9ms

TABLE III: Multi message model case study results.

PROCEED: A Pareto Optimization-based Circuit-level Evaluator for Emerging Devices

Shaodi Wang, Andrew Pan, Chi On Chui, and Puneet Gupta

Department of Electrical Engineering
University of California, Los Angeles
Los Angeles, CA 90095
E-mail: shaodiwang@g.ucla.edu

Abstract—**Evaluation of novel devices in a circuit context is crucial to identifying and maximizing their value. We propose a new framework, PROCEED, and metrics for accurate device-circuit co-evaluation through proper optimization of digital circuit benchmarks. PROCEED assesses technology suitability over a wide operating region (MHz to GHz) by leveraging available circuit knobs (V_t assignment, power management, sizing, etc.) and improves accuracy by 3X to 115X compared to existing methods while offering orders of magnitude improvements in runtime over full physical design implementation flows. To illustrate PROCEED's capabilities, we deploy it to assess novel tunneling transistors (TFETs) compared to conventional CMOS.**

Index Terms—**Tunneling transistor (TFET), silicon-on-insulator (SOI), circuit-level device evaluation, Pareto optimization, simulation-based optimization.**

I. INTRODUCTION

As traditional silicon devices approach their fundamental limits, it is important to explore additions or alternatives to CMOS. To do so, it is essential to systematically compare emerging devices in the context of the circuits they would be used to build. Many technology benchmarking methods have been proposed to meet this need [1]-[9]; unfortunately, as summarized in Table I, all those methods are inadequate due to neglect of various essential circuit features, any one of which can dramatically alter the benchmarking conclusions. Because of the variety and complexity of modern circuits, devices and circuit designs must be carefully chosen to complement each other before assessing viability; this requires a level of flexibility in the benchmarking process that has not existed until now.

Device/circuit assessments must consider several factors to draw realistic conclusions. For instance, *effective evaluations*

should examine the power-delay (PD) tradeoff over several orders of magnitude since modern circuits' performances span a wide range from KHz to GHz frequencies. For a particular circuit to be properly used, *crucial tuning knobs such as logic gate sizing or supply voltage (V_{dd}) or threshold voltage (V_t) selection must be optimized*. In addition, since circuit performance depends critically on the chosen device operating point, *benchmarks should consider the full device I-V characteristics rather than only simple device metrics* like saturation current I_{on} or off-state leakage I_{off}. A given device may not be suitable for all circuit architectures because of *variations in logic depth histogram (LDH) patterns and logical or physical structure. Such adaptivity and circuit topology must be considered in any assessment*. Meanwhile, as technologies scale down, *device variability from ambient process fluctuations becomes ever more important* and impacts circuit viability. Such complexities might seem to require a complete circuit design flow, but that is impractically time-consuming. Thus, alternative evaluation method must be used *which accounts for the above factors with reasonable computational run time*.

To meet these needs, we propose a new device evaluation framework, PROCEED (PaReto Optimization-based Circuit-level Evaluator for Emerging Devices), for fully circuit-aware benchmarking. It incorporates typical circuit design flow flexibilities and tunes physically adjustable device and circuit parameters to generate realistic conclusions about the combined device-circuit performance. PROCEED remedies the flaws enumerated above in several ways:

(1) We use Pareto curves to analyze PD tradeoff over a realistically wide range of power and performance.

(2) The range and number of V_t as well as range of logic gate sizes are inputs to PROCEED, and the evaluation circuit benchmarks can use one or several V_{dd} supply voltages, in accord with realistic designs.

Table 1 COMPARISON OF VARIABLES CONSIDERED IN BENCHMARK METHODOLOGIES IN THE LITERATURE

Methodologies:		Ref. [1]	Ref. [2,4]	Ref. [3-4]	Ref. [5]	Ref. [6]	Ref. [7]	Ref. [8]	Ref. [9]	PROCEED
Metrics		CV/I, CV^2, I_{on}/I_{off}	*PD* Pareto Curves, I_{on}/I_{off}	*PD* Pareto Curves	Clock Frequency	SS, I_{on}	Energy, Clock Frequency	CV/I, CV^2, Model Power, Delay	CPI	*PD Pareto Curves*
Benchmark Circuit		Latch, Inverter Chain	μP	μP	μP	Device	Inverter Chain	Small Logic Elements	μP	Arbitrary Circuit (μP here)
Power management		✗	✗	✗	✗	✗	✗	✗	✗	✓
Optimization Knobs	V_{DD}, V_t	✓	✓	✓	✗	✗	✓	✗	V_{DD} only	✓
	Size	✗	✗	✗	✓	✗	✗	✗	✗	✓
	Multiple V_{DD}, V_t	✗	✗	✗	✓	✗	✗	✗	✗	✓
Circuit Conditions	Interconnect	✓	✓	✓	✓	✗	✓	✓	✓	✓
	LDH*	✗	✗	✗	✗	✗	✗	✗	✗	✓
	Activity	✓	✓	✗	✓	✗	✓	✓	✗	✓
Device Model	Current	I_{on}, I_{off}	I_{on}, I_{half}, I_{off}	I_{on}, I_{half}, I_{off}	I_{on}, I_{off}	TCAD	Model	Model	R_{on}, R_{off}	Compact Model
	Capacitance	Fixed	Fixed	Full *C-V*	Fixed	N/A	Fixed	Full *C-V*	-	Full *C-V*

*LDH: Logic depth histogram (Using slack histogram to estimate)

978-1-4799-2817-0/14 $31.00 © 2014 IEEE   818

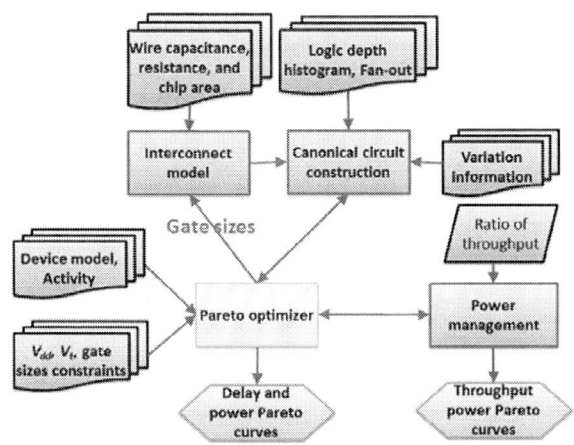

Fig. 1. Overview of PROCEED framework.

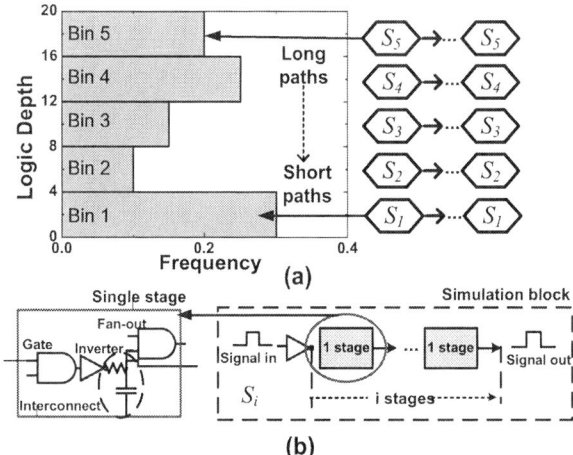

Fig. 3 Circuit schematic for simulation and optimization.

(3) To properly account for device operation at each bias, we utilize compact or lookup table-based full device models.

(4) To assess circuit topology, the full chip characteristics are considered including LDH, interconnect loads, activity factor (i.e. average gate toggle rate) and average fan-out.

(5) We analyze the circuit impact of device variability due to factors like random dopant fluctuation (RDF) and parasitic voltage drops by calculating delay for logic gates evaluated at different variation corners.

(6) For computational efficiency, we adopt scalable Pareto optimization techniques.

(7) Power gating and dynamic voltage and frequency scaling (DVFS) are modeled to assess power management and scaling.

In this paper, we describe the PROCEED framework and, as a case study, deploy it to compare a traditional technology, silicon-on-insulator (SOI), with the novel tunneling FET (TFET). The TFET is a new device concept currently drawing intense interest because of its potential for highly energy efficient operation due to its steep subthreshold switching [10]. However, comprehensive assessments of its system-level performance are still lacking; therefore we perform a microprocessor-level study of the SOI benchmark technologies and elucidate their respective strengths and disadvantages. We outline the methodology behind PROCEED in section II, and explain details of the Pareto optimization procedure in section III. We present results of our PROCEED study on TFET and SOI devices in section IV and summarize our conclusions in section V.

II. OVERVIEW OF PROCEED FRAMEWORK

As shown in Fig. 1, typical inputs to PROCEED include interconnect information (including average wire resistance and capacitance (RC) and chip size), benchmark design (i.e. design LDH and average fan-out), variability (through supply voltage

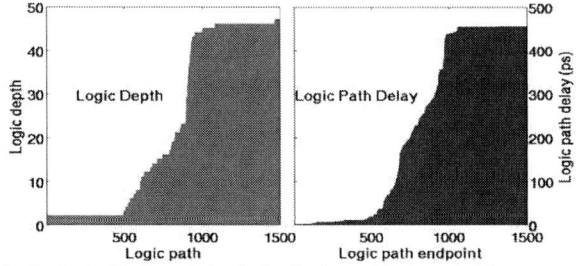

Fig. 2 Typical logic path depth distribution and logic path delay extracted from a synthesized CortexM0.

drops, threshold voltage shifts, etc.), a full device model, operating activity, and optional constraints on V_{dd}, V_t, chip area, and ratio of average to peak throughput. With input and feedback from the Pareto optimizer (through tuning parameters like V_{dd}, V_t, and gate sizes), the needed simulation blocks with interconnect loads are created in the canonical circuit construction process. Optimized results are generated in the form of the *PD* Pareto curve. Finally, power management analysis including DVFS and power gating is performed based on this Pareto curve. As presently implemented, PROCEED is capable of evaluating an arbitrary device candidate as long as it does not cause a dramatic change in circuit topology. For instance, multistate logic devices fall outside PROCEED's present scope of use because of the unconventional circuit architectures within which they must operate.

A. Canonical Circuit Construction

Full, exact optimization is an impossible job for large digital circuits. Since the goal of our approach is to predict the best performance and power tradeoffs for emerging devices, detailed circuit design is thus not our target and contributes little to evaluation. We utilize therefore only essential design information to maximize performance and determine the optimal V_{dd}, V_t, and gate sizes at a given power. A typical circuit design contains both long and short logic paths and the path delay is usually proportional to the logic depth, as shown in Fig. 2. Hence we derive the LDH by extracting endpoint slacks from benchmark designs and estimating logic paths. In Fig. 3, we show an example of the simulation blocks used to construct a specific circuit. For simplicity, we first divide logic paths into *n* bins based on logic depth; in Fig. 3(a), for instance, *n* = 5. More bins improve accuracy at the expense of computation time. Each bin is modeled by the corresponding simulation blocks S_i (S_1-S_5 in Fig. 3(a)), which are in turn made of *i* gate stages. We use the gate design for S_i to construct logic paths belonging to a given bin *i*. The LDH is divided such that the longest path in each bin has the same delay if all these blocks have the same delay. Fig. 3 shows an example of this, with five evenly spaced bins for logic paths from one to twenty stages such that the first bin contains one to four stage paths, the second holds paths with five to eight stages, and so forth. The delay weight W_D is the number of copies of S_i needed to construct the longest path in bin *i* (W_D is 4 in Fig. 3). The logic gate and interconnect used for a single stage in the simulation blocks is shown in Fig. 3(b). The gate can be NAND, NOR, or a more complicated gate like XNOR, depending on the average number of transistors per gate in a given benchmark. The gate choice can also differ from bin to bin, though in this paper's examples we will

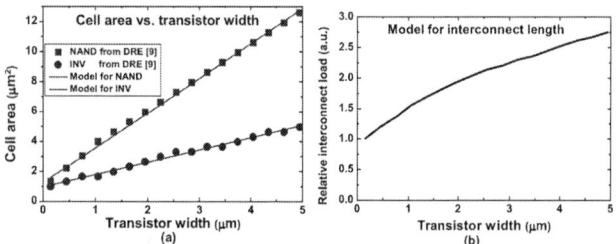

Fig. 4. (a) Cell area and (b) interconnect load as a function of transistor width. In (b), transistor width is the same in Inverter and NAND gate.

use NAND gates for all bins. An inverter or buffer is inserted after the gate to drive the fan-out (which is a replica of the chosen gate sized to average fan-out) as well as wires represented by interconnect RC elements. We have verified the reliability of the PROCEED results through comparison with commercial synthesis tools, as discussed in Section IV.A.

B. Process Variation and Voltage Drop

As devices scale to ever smaller technology nodes, device variations due to process and ambient variations are becoming more important and should not be neglected in PD evaluation. In circuit design, slow corner devices are commonly used to estimate the upper bound on delay and create a "safe" design with sufficient delay margin. We define the slow corner as a device with reduced effective V_{dd} and increased V_t due to variability and parasitic effects; these voltage shifts are inputs to PROCEED. Separate models for other variability effects may be incorporated as needed. During circuit optimization, delay is calculated using the slow corner device while power is simulated with the normal device to model the worst-case scenario.

C. Interconnect Load

We model interconnect loads using a series RC circuit. To construct load as a function of gate width, we use UCLADRE[1] [11] to find a relation between cell area and gate width, and then fit linear models to each cell used in PROCEED. The model accuracy is demonstrated in Fig. 4(a). We assume R and C are linear with interconnect length and chip area is linear to cell area, so the load will be proportional to the square root of the average cell area [12], and can be dynamically changed based on average gate width. Shown in Fig. 4(b) is an example of interconnect load as a function of transistor width, using a combined NAND and INV cell to estimate the cell area. The average RC and extracted gate width are then fed into PROCEED.

D. Pareto-Based Optimization

Following logic canonical circuit construction, all logic paths are replaced by simulation blocks (S_i) which will be optimized. However, these blocks cannot be optimized separately because they usually share a common V_{dd} and V_t, complicating the procedure. To perform the optimization we use a modified form of a general simulation-based Pareto technique [13] which is discussed in more detail in Section III. The simulation target is regarded as a black box with two optimization objectives: design power P and critical delay D (or minimum working clock period).

E. Power Management Modeling

Current technologies usually allow circuits to operate in at least three modes: normal, power saving, and sleep mode. Previous evaluation works only considered the normal mode when devices

[1]Freely available for download at
http://nanocad.ee.ucla.edu/Main/DownloadForm

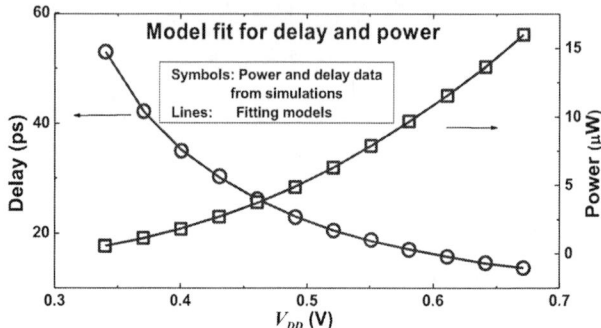

Fig. 5. Model fitting for simulation block's delay and power as a function of V_{dd}.

continuously work at peak performance. PROCEED allows devices to also operate at a second, lower supply V_{dd2} (DVFS) as well as in the off state (power gating). This allows us to evaluate device PD scalability as a function of V_{dd}, an important feature which, to the best of our knowledge, has been ignored in all previous evaluations.

The ratio of average to peak throughput is another input for PROCEED. To study power management, we choose all designs from the generated Pareto points which achieve the lowest power and peak throughput. From this, the optimizer selects the best choice for the second power rail and divides the time spent operating at high V_{dd1} (the original supply) and the new lower V_{dd2}. This is done as follows. Starting from the optimized design (with maximized peak throughput), we carry out circuit simulations by sweeping voltages lower than the original V_{dd1}. The original design may even have multiple supply voltages, in which case different blocks can use different V_{dd2} values. Delay and power models for every simulation block S_i as functions of V_{dd} are constructed using polynomial functions, as in Fig. 5:

$$D_{Si}(V) = \sum_{j=-2}^{5} a_{i,j} V^i, \quad P_{Si}(V) = \sum_{j=-1}^{5} b_{i,j} V^i \quad (1)$$

We have tested and found this model to be sufficiently accurate; for instance, in our experiments the relative error of the polynomial fittings is less than 2%. We then optimize for the weighted power sum $f_1 P_1 + f_2 P_2$, subject to

$$D_2 >= W_D D_{Si}(V_{i2}), \; P_2 >= \sum_{i=1}^{n} W_{Pi} P_{Si}(V_{i2}) \quad i=1,2,...,n$$
$$f_1 \cdot 1/D_1 + f_2 \cdot 1/D_2 >= T_{Ave}, \; 0 \le f_1 + f_2 \le 1 \quad (2)$$

Here $D_{1,2}$ and $P_{1,2}$ are the delay and power using $V_{dd1,2}$, W_D and W_P are the delay and power weight mapping from simulation blocks to the design, and f_1 and f_2 are the fractions of time spent operating with V_{dd1} and V_{dd2} with any remaining time assumed to be spent in the off state. Typically this step is not a feasible convex optimization problem; however, by using the fitted model of Eq. (1), an enumeration approach can solve this problem very efficiently with acceptable accuracy.

F. Activity Factor

Activity varies widely with application: in embedded sensing, for instance, factors below 1% are observed in car-park management [14], while those for systems like VigilNet exceed 50% [15]. Activity factor can therefore dramatically change evaluation results and is included as an input to PROCEED. In circuit simulations, the dynamic and leakage power are separately extracted and the total power is their weighted sum. From this the circuit can be optimized for a known activity factor.

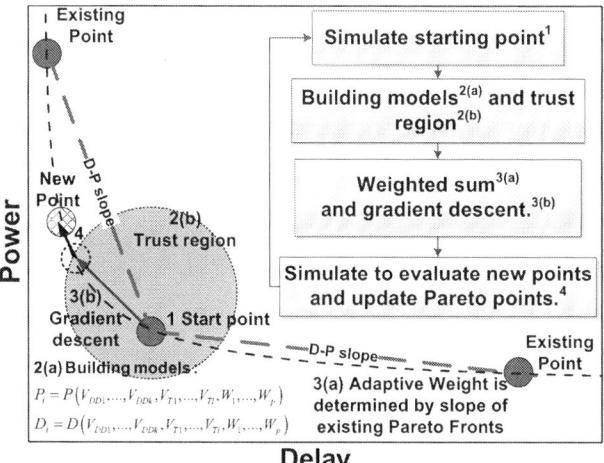

Fig. 6 Optimizer overview. Adaptive weight is chosen by slope of existing fronts. Based on starting point, meta-modeling is built and gradient descent is used to find potential points. Simulate potential points to get new Pareto points.

G. Multiple V_{dd} and V_t

In modern circuit designs, multiple V_{dd} and V_t values are used. In our scheme, transistors in each simulation block S_i must be assigned the same voltages, so to optimize a design with integer m different V_{dd} or V_t biases, the number of simulation blocks must be greater than m. In addition, our optimization is an iterative process whereby Pareto points are updated and improved based on previous iterations. Therefore, if the same V_{dd} or V_t is shared by multiple simulation blocks, this assignment cannot be changed during the optimization. A full optimization for multiple V_{dd} and V_t is implemented by considering designs with all sets of reasonable voltage assignments in parallel. For example, if we have five simulation blocks S_1-S_5 and two available threshold voltages, then for i from 1 to 4, blocks S_1 to S_i use the high V_t and S_{i+1} to S_5 use the low V_t. This comprises the set of useful voltage assignments, since simulation blocks with longer logic paths require higher performance (lower V_t).

III. PARETO OPTIMIZATION

Fig. 6 presents an overview of our Pareto optimization process. PROCEED treats circuit simulations as a black box and uses models to optimize tuning parameters based on the simulation results. Gradient descent is utilized to find minimal objectives in the trust region. Final simulations are performed on designs outputted by the model-based optimization. The vector of tuning parameters X for optimization is represented as:

$$X = (x_1, x_2, ..., x_m) = (y_1, y_2, ..., y_n)$$
$$y_i = (V_{dd,i}, V_{t,i}, W_{i1}, W_{i2}, ..., W_{i,2i}), i = 1, ...n \quad (3)$$

where $V_{dd,i}$ and $V_{t,i}$ are the supply and threshold voltages for simulation block S_i, W_{ij} are sizes for gates and inverters in S_i, x_j are the variables of X, and y_i are vectors of the tuning parameter variables for S_i. The optimization entails the following steps:

(1) *Picking a starting point*: each iteration of the optimization process uses a starting set of variables X_0 around which to explore. For the first iteration, any reasonable X_0 may be inputted. The choice of the initial point may affect runtime but not final accuracy, since "bad points" will gradually be eliminated by the optimization process and converge to the true answer. Subsequently X_0 is determined from already existing Pareto points by computing the Euclidean distance between all neighboring points in delay/power coordinates, as shown in Fig. 6. The point with the

largest total distance from its two neighbors is chosen as X_0 since it lies in the sparse region, which is usually suboptimal.

(2) *Building a local model around X_0*: To accelerate the optimization process, second-order delay and power models are constructed based on simulation results. The delay and power models D_{Si} and P_{Si} for each block S_i are calculated separately and then combined to reduce the number of simulations, as determined by the size of the Hessian matrix (proportional to the number of variables squared). D_{Si} and P_{Si} are represented by the gradient vector G_{Di} and Hessian matrix H_D as

$$D_{Si}(y_{i,0} + \Delta y_i) = D_{Si,0} + G_{Di}^T \Delta y_i + \frac{1}{2}\Delta y_i^T H_{Di} \Delta y_i$$
$$P_{Si}(y_{i,0} + \Delta y_i) = P_{Si,0} + G_{Pi}^T \Delta y_i + \frac{1}{2}\Delta y_i^T H_{Pi} \Delta y_i \quad (4)$$

This second-order model is a local estimation near the starting point. To guarantee validity, an adaptive trust region is applied as shown in Fig. 6, limiting the model range inside the region

$$X_0 - \lambda(r) < X < X_0 + \lambda(r) \quad (5)$$

where r is the radius of this "trust region" and λ is the range of the tuning parameters X and is a linear function of r.

(3) *Model-based optimization*: In this step, four metrics are used in optimization: D, P, $W_{dl} \times D + W_{pl} \times P$, and $W_{dr} \times D + W_{pr} \times P$. Minimization of D and P yields the fastest and lowest power designs in the local region, while the weighted sums of delay and power are used to populate the phase space by finding two Pareto points between the starting point and its neighbors. Since the problem may not be convex, gradient descent with the logarithmic barrier method [16] is used to find these optimal points. The model's region of validity lies in the intersection of the trust region and the inputted bounds for the tuning parameters. The objective function is performed as follows:

$$\text{Minimize } W_D D(X) + W_P P(X) - t\left(\sum_{j=1}^{m}\log(-x_j + x_{j,u}) - \sum_{j=1}^{m}\log(x_j - x_{j,l})\right)$$
$$D(X) = D(X_0) + G_D(X_0)^T(X - X_0) + (X - X_0)^T H_D(X_0)(X - X_0) \quad (6)$$
$$P(X) = P(X_0) + G_P(X_0)^T(X - X_0) + (X - X_0)^T H_P(X_0)(X - X_0)$$

where $x_{j,l}$ and $x_{j,u}$ are the upper and lower bounds for variable x_j, and D and P are delay and power for the entire design, respectively. The weights for delay and power are defined as follows:

$$W_{dl(r)} = (P_{l(r)} - P_0) / \sqrt{(P_{l(r)} - P_0)^2 + (D_{l(r)} - D_0)^2}$$
$$W_{pl(r)} = (D_0 - D_{l(r)}) / \sqrt{(P_{l(r)} - P_0)^2 + (D_{l(r)} - D_0)^2} \quad (7)$$

where (D_0, P_0) is the starting point and (D_l, P_l) and (D_r, P_r) are the left and right neighbor points, respectively. The solid points in Fig. 6 are examples of such points. The direction vectors (W_{dl}, W_{pl}) and (W_{dr}, W_{pr}) of the weighted sum of objectives are calculated so as to be perpendicular to the connecting lines between the starting point and its neighbors, as illustrated by the dashed line in Fig. 6. D and P are given by

$$D(X) = W_D \cdot \max\left((D_{S1}(y_1), D_{S2}(y_2), ..., D_{Sn}(y_n))\right), P = \sum_{i=1}^{n} W_i \cdot P_{Si} \quad (8)$$

where W_D is the delay weight discussed in Section II.A and W_i is the number of S_i used in the canonical circuit construction. Because the maximizing function does not have a continuous derivative, we use higher order norms to estimate the maximum, so the elements of gradient vector and Hessian matrix for delay are derived as follows:

$$D(X) \approx \|D\|_K, D = (D_{S1}(y_1), D_{S2}(y_2), ..., D_{Sn}(y_n))$$
$$G_{D,j}(X) = \frac{\partial D(X)}{\partial x_j} \approx \frac{\partial \|D\|_K}{\partial x_j}, H_{D,jk}(X) = \frac{\partial^2 D(X)}{\partial x_j \partial x_k} \approx \frac{\partial^2 \|D\|_K}{\partial x_j \partial x_k} \quad (9)$$

978-1-4799-2817-0/14 $31.00 © 2014 IEEE

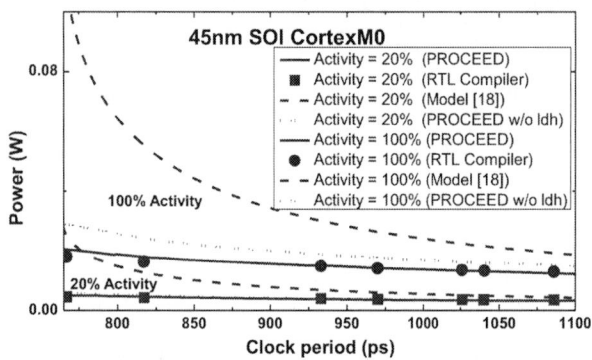

Fig. 7. Comparison between commercial synthesis tool, Model [4], and PROCEED. V_{dd} and V_t are constants and only size is a variable.

where K is the order of the norm. Higher K results in more accurate results (we use $K = 100$ in our simulations). Similarly, the elements of the gradient vector and Hessian matrix for power are given as

$$G_{P,j} = \frac{\partial P(X)}{\partial x_j} = \sum_{i=1}^{n} W_i \cdot \frac{\partial P_{Si}(y_i)}{\partial x_j}, \quad H_{P,jk} = \frac{\partial^2 P(X)}{\partial x_j \partial x_k} = \sum_{i=1}^{n} W_i \cdot \frac{\partial^2 P_{Si}(y_i)}{\partial x_j \partial x_k} \quad (10)$$

(4) *Addition of new Pareto points*: To correct for model errors, circuit simulations are performed to evaluate D and P for all remaining potential Pareto points found by the optimization. In Fig. 6, this process is illustrated by the shift of the hatched point to the dotted circle. Finally, points not on the Pareto frontier (such that at least one other point with both lower delay and power exists) are filtered out.

(5) *Iteration termination*: For each iteration, when choosing the starting point for each step, the radius of trust region around this point is decreased by a factor of p ($p > 1$). Two termination conditions are applied: 1) existence of a sufficient Pareto point density in the region of interest, defined by the largest gap between any two neighboring points being smaller than a given criterion. This condition is usually used for devices with large operating regions (i.e. suitable for both high speed and low power applications). 2) Reduction of the radius of trust below a given criteria. This usually occurs due to limitations on the device operating region or device model discontinuities.

The PROCEED runtime is of order $O(r \times m^2) + O(r)$, where r is the resolution constraint (number of points in a unit Pareto curve), m is the total number of tuning parameters, $O(r \times m^2)$ is the complexity of the simulations for gradient and Hessian matrix calculation, and $O(r)$ is the complexity of simulating potential Pareto points. In our experiments, runtimes are mainly dominated by the resolution constraint; however, for large m, the $O(r \times m^2)$ term will dominate. The average PROCEED runtime to generate a full Pareto curve over three orders of magnitude in performance is about 4 hours on a single CPU. We use MATLAB in the optimization process and HSPICE for circuit simulations.

IV. EXPERIMENT RESULTS

To illustrate PROCEED's capabilities, we compare it with existing evaluation methods and use it to assess SOI and silicon TFET devices at the 45 nm node. Because of their use of interband tunneling, TFETs are capable of very low leakage and extremely steep subthreshold swing, making them well-suited for low voltage operation [8]. Currently, however, nonidealities in experimental devices and low on-current limit their performance. We examine the viability of currently achievable TFETs using a de-

vice compact model [17]-[18] calibrated against TCAD simulations and experimental SOI devices [19]. While this does not represent the best possible TFET, which may require a different channel material or device structure, it have the advantages of being experimentally validated and structurally comparable to conventional SOI devices and represents a realistic lower bound. 45 nm SOI MOSFETs are modeled using commercial characteristics and compact model. Unless otherwise specified, all circuit results are generated with one V_{dd} and two V_t. To easily compare devices, we will refer to the Pareto crossover, defined as the delay above which the optimized novel device (here, the TFET) consumes less power than the established technology (SOI); lower Pareto crossover means the novel device is more promising for a given case.

A. Framework Evaluation

To validate our PROCEED framework, we use the widely employed evaluation model of Ref. [4] (hereafter Model [4]), and a commercial synthesis tool to evaluate the PD Pareto curve for a CortexM0 microprocessor with a commercial 45 nm SOI library and model. The information needed for PROCEED and Model [4] (LDH, average fan-out and interconnect load) is extracted from a synthesized, placed, and routed netlist at a clock period of 933 ps. Only one constant V_{dd} and one constant V_t are used, as Model [4] does not support multiple voltages and the commercial library has only constant supply and threshold voltages. As shown in Fig. 7, the PROCEED predictions are in much better agreement with the comprehensive optimized results from the RTL compiler compared to Model [4], which is frequently used for device evaluation [2]-[3]. The operating range for comparison is chosen by the synthesis results with the commercial library with one V_{dd} and V_t. We note that using the compiler for evaluation purposes is completely impracticable, since generating a Pareto curve from kHz to GHz speeds necessitates libraries with V_{dd} and V_t varying from 0.5V to 1.2V and 0.1V to 0.5V respectively. However, the generation and optimization of these libraries would consume months of runtime, whereas we completed the same study in hours using PROCEED. Meanwhile, the computationally simple Model [4] takes seconds to complete such Pareto curves but grossly overestimates power for two reasons: the neglect of LDH in assuming all gates have the same (large) size used for the critical path, and the use of analytical *PD* models rather than circuit simulations using full device characteristics. The dotted line is the Pareto curve generated by PROCEED while neglecting LDH, illustrating the accuracy improvement contributed by the two foregoing points. We further note that Model [4] cannot account for adaptivity, variability, or multiple V_{dd} and V_t effects. By benchmarking to the RTL results in Fig. 7, we observe *PROCEED improves accuracy by 3X to 115X compared to the current standard Model [4]*.

B. Impact of Multiple V_{dd}, V_t, and Gate Sizing

More tuning parameters create a larger phase space for design optimization, as illustrated in Fig. 8 for a 45 nm SOI CortexM0 topology. As more LDH bin divisions are introduced, power is increasingly optimized because of a greater range of gate sizes with which to construct the design. Similarly, the introduction of additional supplies and threshold voltages substantially improves performance. The result does not account for the overhead consumed by the voltage shifter used in multiple V_{dd} design. Overall, however, we observe that *the evaluated optimal power at a given delay may change by over 50% as gate size tuning and multiple V_{dd} and V_t are introduced, demonstrating the necessity of including these effects in any quantitative comparison.*

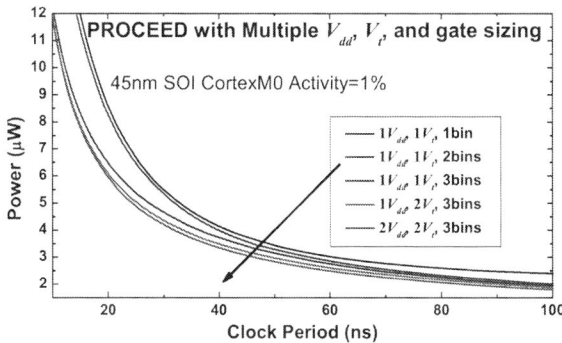

Fig. 8. 45nm SOI CortexM0 power-clock period as tuning parameters are increased.

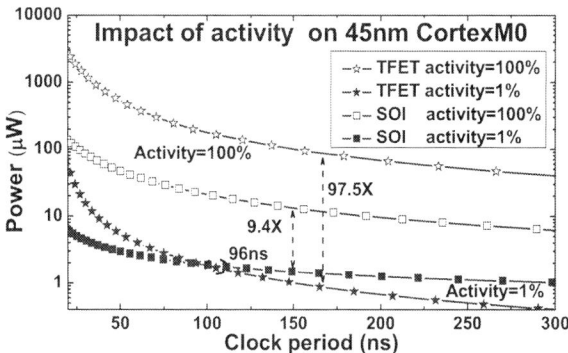

Fig. 10. Activity impact on 45nm Si SOI and Si TFET power-clock period.

C. Impact of Benchmarks on Evaluation – SOI vs. TFET

To show the impact of benchmark selection, we compare the performance of two microprocessors, CortexM0 and MIPS, using SOI and TFET devices and two supply rails and two threshold voltages. We choose these benchmarks because, as shown in Fig. 9(a), they have a similar number of critical path stages (56 in CortexM0 vs. 62 in MIPS) and total gates (8990 vs. 9248), but the CortexM0 has a more evenly distributed LDH. The power consumption in MIPS is dominated by short paths, which means it will be more accommodating of slow devices compared to the CortexM0. Accordingly, in Fig. 9(b), both SOI and TFET achieve better power efficiency in MIPS designs because the second V_{dd} and V_t can be optimized to save power along the short paths. The crossover points where the Pareto curves for different devices intersect define their advantageous operating regions; a device changes from being less power efficient on one side of the crossover to being more efficient on the other side. If multiple crossovers are found, then the Pareto curve can be divided into several regions (high performance, low power, etc.) such that in each one, there is only a single crossover point. This allows us to demarcate the (possibly multiple) favorable operating ranges for each device. The Pareto crossover occurs at 73 ns and 106 ns for MIPS and CortexM0, respectively, showing that TFETs are more acceptable for applications like MIPS which tolerate slower devices. However, TFET drive currents must be increased if they are to be usable at higher clock rates. Previous evaluations, like those in Table 1, which ignore LDH, are not able to distinguish between benchmarks in this way. *These results show how the choice of circuit topology strongly impacts the suitability of emerging devices.*

D. Impact of Activity Factor – SOI vs. TFET

We next examine how activity factor affects SOI- and TFET-based CortexM0 processors in Fig. 10. As activity reduces

from 100% to 1%, TFET circuit power scales in lockstep by 97.6X due to low device leakage. However, the corresponding SOI designs only see power reduction of 9.4X because of its higher off-current. We see that *TFETs change from being completely impracticable at 100% activity to being superior to SOI beyond the 96 ns delay point at 1% activity; thus activity factor, and hence system use contexts, can drastically alter the device evaluation and must be considered.*

E. Power Management Modeling

The results of the previous subsections make clear that there is no panacea device and that device-circuit evaluation must be done with specific applications and operating windows in mind. DVFS and power gating are crucial ingredients for such usage-mindful evaluation. In Fig. 11, we show PROCEED- generated Pareto curves at different ratios of average to peak throughputs for SOI and TFET CortexM0 using DVFS and power gating. Power is reduced by operating at the lower supply rail or turned off by power gating; the achievable power reduction differs with device and operating region. The peak throughput crossover point for TFETs shifts from 10.9M to 21.5M operations per second as the ratio of average to peak throughput reduces from 100% to 10%; *the relative performance of TFETs effectively doubles as throughput requirements become less aggressive, emphasizing the importance of incorporating power management into device benchmarking.*

F. Variation-Aware Evaluation

To illustrate how variability might impact conclusions drawn using nominal devices, we show in Fig. 12 how the SOI and TFET Pareto curves are changed when slow corner devices are used. We define the slow corner as a device with 10% effective voltage

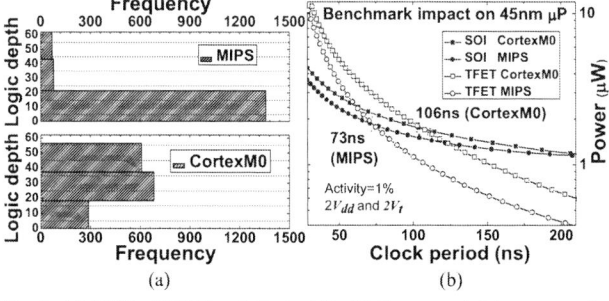

Fig. 9. (a) LDH of MIPS and CortexM0. (b) Power and delay curves for MIPS and CortexM0 designed with TFET and SOI respectively. Activity is 1% and two V_{dd} and two V_t are applied.

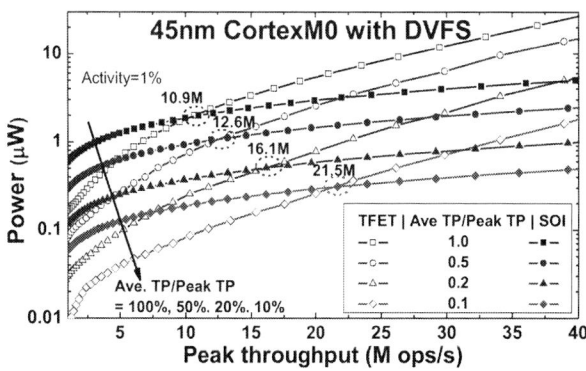

Fig. 11. 45nm SOI and TFET CortexM0 microprocessors with power management. The ratio of average to peak throughputs are 10%, 20%, 50% and 100%. Curves with ratios of 100% are designs outputted from Pareto optimizer.

978-1-4799-2817-0/14 $31.00 © 2014 IEEE

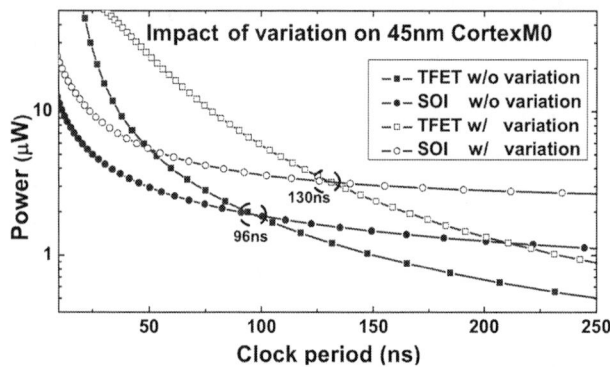

Fig. 12. Variation-aware evaluations of 45nm technologies. Assumed voltage drop is 90% and V_t shift is 50mV.

reduction and 50 mV V_t shift; total power is simulated using the nominal device, while delay is evaluated with the slow corner. We observe that the TFET is more sensitive to variability effects than SOI, as the Pareto crossover shifts from 96 ns to 130 ns. This is due to the TFET's steep subthreshold swing around the crossover, leading a high sensitivity of drive current to voltage [20]-[21] *This suggests that TFETs need to show substantial **nominal device** advantages in order to buffer this sensitivity and demonstrates that even a simple consideration of variability is important in device evaluation and selection.*

V. CONCLUSION

The proposed circuit-device co-evaluation framework [2] accounts for circuit topology, adaptivity, variability and use context using efficient Pareto optimization heuristic. Previous device evaluation frameworks ignore one or more crucial factors like multiple supply and threshold voltages, power management, logic depth, variability, etc., which can easily lead to misleading results. For instance, we find that including power management in our evaluation can effectively double the usable operating range for TFETs, and that choice of activity factor can dictate whether TFETs are acceptable at all in a given application. These observations are made possible by PROCEED's scope and computational efficiency in studying several orders of magnitude in possible device/circuit performance, and demonstrate the power and flexibility of our new methodology.

VI. ACKNOWLEDGEMENT

We would like to acknowledge the generous support of IMPACT UC Discovery Grant (http://www.impact.berkeley.edu/) in accomplishing this work.

REFERENCES

[1] L. Wei, S. Oh, and H.-S. P. Wong, "Performance benchmarks for Si, III–V, TFET, and carbon nanotube FET-re-thinking the technology assessment methodology for complementary logic applications," *Proc. IEDM*, pp.16.2.1-4, 2010.

[2] L. Wei and D. A. Antoniadis, "CMOS device design and optimization from a perspective of circuit-level energy-delay optimization," *Proc. IEDM*, pp.15.3.1-4, 2011.

[3] P. M. Solomon, D. J. Frank, and S. O. Koswatta, "Compact model and performance estimation for tunneling nanowire FET," *Proc. DRC*, pp.197-198, 2011.

[4] D. J. Frank, W. Haensch, G. Shahidi, and O. H. Dokumaci, "Optimizing CMOS technology for maximum performance," *IBM J. Research and Dev.*, vol. 50, no. 4/5, pp.419-431, Jul.-Sep. 2006.

[5] D. Sylvester and K. Keutzer, "System-Level Performance Modeling with BACPAC – Berkeley Advanced Chip Performance Calculator," *Proc. SLIP*, pp. 109-114, 1999.

[6] M. Luisier, M. Lundstrom, D. A. Antoniadis, and J. Bokor, "Ultimate device scaling: Intrinsic performance comparisons of carbon-based, InGaAs, and Si field-effect transistors for 5 nm gate length," *Proc. IEDM*, pp.11.2.1-4, 2011.

[7] H. Kam, T.-J. King-Liu, E. Alon, and M. Horowitz, "Circuit-level requirements for MOSFET-replacement devices," *Proc. IEDM*, pp.1.1. 15-17, 2008.

[8] C. Augustine, A. Raychowdhury, Y. Gao, M. Lundstrom, and K. Roy, "PETE: A device/circuit analysis framework for evaluation and comparison of charge based emerging devices," *Proc. ISQED*, pp.80-85, 2009.

[9] C. Pan and A. Naeemi, "System-Level Optimization and Benchmarking of Graphene PN Junction Logic System Based on Empirical CPI Model," *IEEE Proc. Intl. Conf. IC Design & Technology,* May. 2012.

[10] A. M. Ionescu and H. Reil, "Tunnel field-effect transistors as energy-efficient electronic switches," *Nature*, vol. 479, no. 7373 pp. 329-337, 2011.

[11] R. S. Ghaida and P. Gupta, "DRE: a Framework for Early Co-Evaluation of Design Rules, Technology Choices, and Layout Methodologies," *IEEE Trans. CAD*, vol. 31, no. 9, pp. 1379-1392, Sep. 2012.

[12] J. A. Davis, V. K. De, and J. D. Meindl, "A Stochastic Wire-Length Distribution for Gigascale Integration (GSI)—Part II: Applications to Clock Frequency, Power Dissipation, and Chip Size Estimation," *IEEE TED*, vol. 45, no. 3, pp. 590-597, Mar. 1998.

[13] J.-H. Ryu, S. Kim, and H. Wan, "Pareto front approximation with adaptive weighted sum method in multiobjective simulation optimization," *Proc. Winter Simulation Conference (WSC)*, pp.623-633, 2009.

[14] J. P. Benson *et al.*, "Car-park management using wireless sensor networks." *Proc. Conf. Local Computer Networks*, pp. 588-595, 2006.

[15] T. He *et al.*, "Achieving Real-time Target Tracking Using Wireless Sensor Networks," *Proc. RTAS Symp.*, pp. 37-48, 2006.

[16] S. P. Boyd and L. Vandenberghe, *Convex Optimization*. Cambridge, 2004.

[17] A. Pan and C. O. Chui, "A Quasi-Analytical Model for Double-Gate Tunneling Field-Effect Transistors," *IEEE EDL*, vol. 33, no. 10, pp. 1468-1470, Oct. 2012.

[18] A. Pan, S. Chen, and C. O. Chui, "Electrostatic Modeling and Insights Regarding Multigate Lateral Tunneling Transistors," *IEEE TED*, vol. 60, no. 9, pp. 2712-2720, Sept. 2013.

[19] K. Jeon *et al.*, "Si Tunnel Transistors with a Novel Silicided Source and 46 mV/dec Swing," *2010 VLSI Symp.*, p. 121-122, 2010.

[20] G. Leung and C. O. Chui, "Stochastic Variability in Silicon Double-Gate Lateral Tunnel Field-Effect Transistors," *IEEE Trans. Electron Devices*, vol. 60, no. 1, pp. 84-91, 2013.

[21] G. Leung and C. O. Chui, "Interactions between Line Edge Roughness and Random Dopant Fluctuation in Non-Planar Field-Effect Transistor Variability," *IEEE Trans. Electron Devices*, vol. 60, no. 10, pp. 3277-3284, 2013.

[2]PROCEED will be made publicly available as open-source software.

Modeling and Design Analysis of 3D Vertical Resistive Memory - A Low Cost Cross-Point Architecture

Cong Xu†, Dimin Niu†, Shimeng Yu‡, Yuan Xie†,§

†Pennsylvania State University, {czx102,dun118,yuanxie}@cse.psu.edu

‡Arizona State University, shimeng.yu@asu.edu

§AMD Research, yuanxie@amd.com

Abstract—Resistive Random Access Memory (ReRAM) is one of the most promising emerging non-volatile memory (NVM) candidates due to its fast read/write speed, excellent scalability and low-power operation. Recently proposed 3D vertical cross-point ReRAM (3D-VRAM) architecture attracts a lot of attention because it offers a cost-competitive solution as NAND Flash replacement. In this work, we first develop an array-level model which includes the geometries and properties of all the components in the 3D structure. The model is capable of analyzing the read/write noise margin of a 3D-VRAM array in the presence of the sneak leakage current and voltage drop. Then we build a system-level design tool that is able to explore the design space with specified constraints and find the optimal design points with different targets. We also study the impact of different design parameters on the array size, bit density, and overall cost-per-bit. Compared to the state-of-the-art 3D horizontal ReRAM (3D-HRAM), the 3D-VRAM shows great cost advantage when stacking more than 16 layers.

I. INTRODUCTION

ReRAM is one of the most promising candidates for next-generation memory subsystems. Compared to conventional NAND Flash, ReRAM has superior read/write access latency, orders of magnitude higher endurance, better scalability, much lower operating voltage and byte addressability. The key challenge for ReRAM to place NAND Flash is to improve the integration density in terms of cost-per-bit, given that multi-level-cell (MLC) NAND Flash continues to scale beyond 20nm technology node and 3D NAND Flash is emerging [1]. To realize low cost design, cross-point ReRAM architecture have been widely studied, featuring a cell size of $4F^2$. By simply stacking the cross-point structure layer by layer [2]–[4], the bit density of ReRAM is further improved. For example, a 32Gb 2-stack cross-point ReRAM prototype with a NAND Flash-compatible interface was demonstrated [4]. However, this approach, referred as 3D horizontal ReRAM (3D-HRAM), requires critical lithography and other process for every stacked layer, and this fabrication cost overhead increases linearly with the number of stacks. Recently proposed 3D vertical ReRAM (3D-VRAM) architecture that tilts the horizontal ReRAM by 90 degrees [5], [6] attracts a lot of attention because it offers an alternative low-cost solution. Significant cost saving is achieved by the elimination of the cost-consuming process during the fabrication of the intermediate layers [6].

Most research on 3D-VRAM still focus on device level. A full 3D circuit model with sufficient accuracy is not well established. There is still a big gap between the device optimization and system-level design analysis. For example, the impact of the design parameters in a 3D-VRAM on the array-level and system-level metrics is not clear yet. Moreover, a detailed comparison between the 3D-VRAM and 3D-HRAM cannot be done without a comprehensive model. To facilitate these studies, we present a 3D-VRAM model from device, array to macro. Our device model captures the nonlinear I-V characteristics in an ReRAM element. The array model accounts for most of the important components in a 3D-VRAM structure including the plane and pillar electrodes, the access select transistor and so on. We carefully determine the abstraction level of these modules to maintain a good balance between accuracy and simulation speed. As

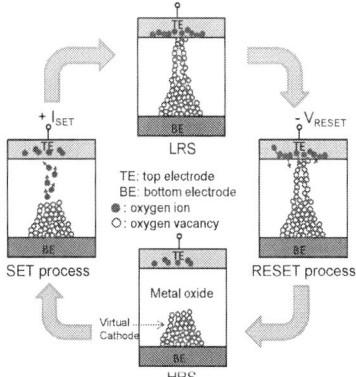

Fig. 1.   Demonstration of a ReRAM cell and its SET/RESET operations

cost-per-bit is the single most factor when adopting a new memory technology, we also develop a macro-level cost model that takes the detailed 3D ReRAM fabrication process into considerations. Then we combine these models in a design flow to enable the design space exploration under various specified constraints. Our tool is able to find the optimal design point(s).

We use the model to evaluate the voltage drop, array capacity, bit density, and cost-per-bit with different settings. The results suggest that optimizing etching aspect ratio and metal layer thickness are critical when designing a 3D-VRAM. We find that there is a tradeoff between the array capacity and bit density when tuning some of the design parameters such as the metal layer thickness. Our analysis also provides key insights on how to optimize 3D ReRAM design with fewer stacks and more stacks. The methodology we proposed in this work should be valuable for device-circuit-architecture co-design of 3D-VRAM.

II. PRELIMINARY

A. ReRAM Basics

The basic structure of an ReRAM cell is illustrated in Figure 1. One metal oxide layer sandwiched by two metal electrodes - the top electrode (TE) and the bottom electrode (BE). A low resistance state (LRS) represents digital '1' while a high resistance state (HRS) represents digital '0'. The switching from HRS to LRS is defined as a SET operation while the opposite switching is defined as a RESET operation. Here we focus on bipolar switching, which means that a SET and RESET occurs at opposite voltage polarities. When a positive voltage is applied, a SET process leads to the formation of conductive filaments (CFs) made of oxygen vacancies [7]. Once the CFs are formed, the ReRAM cell is in LRS. In contrast, when a negative voltage is applied across the cell, a RESET process leads to the rupture of the CFs, switching the cell into HRS.

Compared to the NAND Flash, ReRAM has much faster speed ($< 10ns$), orders of magnitude higher endurance (up to 10^{12}), better scalability ($< 10nm$) and much lower operating voltage ($< 3.3V$). These advantages not only make ReRAM a NAND Flash replacement candidate with high performance and low power, but also ease the

---

Xu, Niu and Xie are supported in part by SRC grants, NSF 1218867, 1213052. This material is based upon work supported by the Department of Energy under Award Number DE - SC0005026.

978-1-4799-2817-0/14 $31.00 © 2014 IEEE

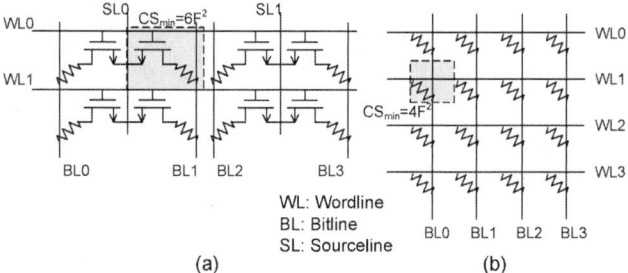

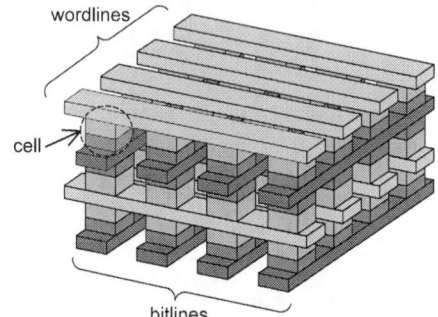

Fig. 2. Schematic view of ReRAM array structures: (a) 1T1R; (b) Cross-point

design from dealing with wear-out problem and multi-stage on-chip charge pumps required to provide high operating voltage.

B. Planar ReRAM Structure

As shown in Figure 2, there are two typical structures of a planar ReRAM array. The 1T1R structure illustrated in Figure 2(a) uses one dedicated MOSFET transistor as the access transistor to provide the write current required for cell switching. The transistor is able to isolate the selected cell from other unselected cells. In this form, the minimum cell size is $6F^2$, which is the same as the current DRAM technology. Figure 2(b) shows the cross-point structure where ReRAM cells are sandwiched between wordlines and bitlines. The minimum cell size is $4F^2$.

The biggest challenge of a cross-point design is the existence of multiple sneak leakage paths in the array, even if the V/2 voltage biasing scheme is applied [8]. When one wordline and one bitline is activated, we expect the current to completely pass through the selected cell at their intersection. However, the current will also flow through the half-selected cells (the cells in the activated wordline or bitline other than the selected cell) and the unselected cells (the cells in the half-biased wordlines and bitlines), referred as sneak current. The sneak current increases the total current flow on the activated wordline and bitline, and thus incurs significant area overhead of on-pitch write drivers and multiplexers. It also worsens the voltage drop problem on the wordline and bitline resistance.

To suppress the sneak current of the half-selected cells, the nonlinearity in the I-V curve of a ReRAM cell is introduced by either connecting a diode in serial with the cell [3], [4] or engineering a self-rectifying property of the cell [2], [9]. Nonlinearity means that the equivalent resistance of the cell increases when the applied voltage on it decreases.

C. 3D ReRAM Structure

To further improve the bit density of ReRAM, many 3D structures have been proposed and demonstrated [2]–[6], [10]. One straightforward approach is to stack planar cross-point structure layer by layer, namely 3D-HRAM, as shown in Figure 3. The adjacent layers share their wordlines and bitlines alternatively. Chen et al. [10] discussed the addressing scheme of 3D-HRAM. To maximize the density of 3D-HRAM, stacking more layers is desired. However, every additional layer introduces extra fabrication process steps, including lithography, etching and chemical-mechanical planarization (CMP). These steps may eventually prevent the cost reduction.

As an alternative 3D ReRAM solution, 3D-VRAM was proposed to reduce the fabrication steps for high density ReRAM design. The schematize view of the 3D-VRAM architecture is illustrated in Figure 4. Each ReRAM array consists of L wordline planes, N_b bitlines and N_s sourcelines. Two adjacent wordline plane electrodes are separated by a dielectric isolation layer. The cell is now located at every cross point of a vertical pillar electrode and a wordline plane. The key cost saver of 3D-VRAM is the elimination of the critical

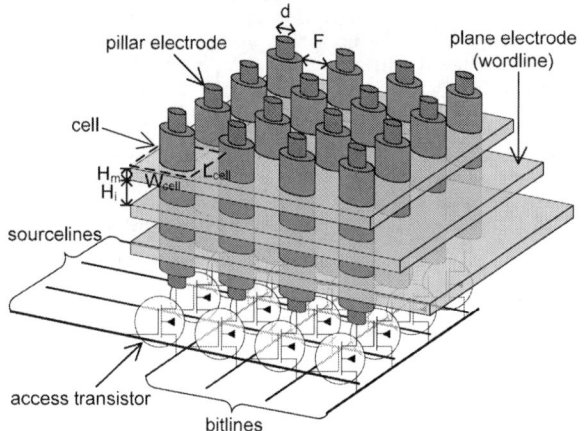

Fig. 3. Schematic view of 3D Horizontal ReRAM

Fig. 4. Schematic view of 3D Vertical ReRAM

lithography and etching steps of the intermediate layers. The wordline planes and isolation layers are deposited consecutively. The process of defining the pillar electrodes and cells is involved only after the top most layer is deposited, and only two critical lithography and etching steps are required (one for patterning the pillar electrode, one for opening the contact for wordline planes). Chen et al. [6] have demonstrated the detailed fabrication process.

To address such an array, one access transistor is introduced at the bottom of each vertical pillar electrode. During a write operation, V_g is applied on one selected sourceline to turn on the N_b access transistors alone the selected sourceline while all the other transistors remain off by grounding the unselected sourcelines. This operation basically activates a vertical plane, which is a de facto cross-point structure. Therefore normal voltage biasing schemes for writing and reading a cross-point structure can be applied on the activated plane.

III. MODELING

A. Modeling of a ReRAM Element

For the sake of simplicity, most prior work [8] use a linear resistor (either HRS or LRS) to represent the ReRAM element in a cross-point structure. Such an approach results in an unacceptable simulation error of the sneak current and voltage drop when the cell has a large nonlinearity. To consider the effect of a nonlinear ReRAM cell, Niu et al. [11] multiplies the half-selected cells by a nonlinearity constant. However, the error can still vary depending on the shape of I-V curve of the cell. Another approach is to build a SPICE-compatible model [12] for ReRAM with full dynamics by incorporating the differential equations for the state variable of a ReRAM cell, which produces accurate results but the run time could go unbounded when simulating an array with $> 10^5$ cells. To maintain both good accuracy and simulation speed, we implement the representative I-V relationship of a typical ReRAM cell, based on the experimental results in Yu et al's work [13], as an HSPICE subcircuit. We take

978-1-4799-2817-0/14 $31.00 © 2014 IEEE

TABLE I
HSPICE SUBCIRCUIT OF THE RERAM MODEL

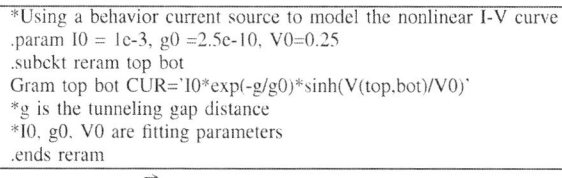

*Using a behavior current source to model the nonlinear I-V curve
.param I0 = 1e-3, g0 =2.5e-10, V0=0.25
.subckt reram top bot
Gram top bot CUR='I0*exp(-g/g0)*sinh(V(top,bot)/V0)'
*g is the tunneling gap distance
*I0, g0, V0 are fitting parameters
.ends reram

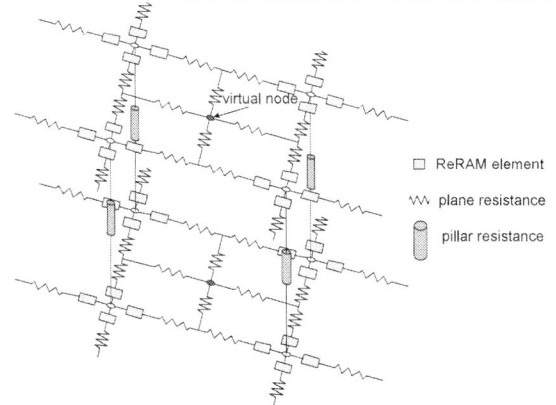

□ ReRAM element

∧∧∧ plane resistance

pillar resistance

Fig. 5. Circuit model: abstraction of a 2-layer 3D-VRAM array

out the equations of the switching dynamics which is the most time-consuming part of the simulation, as the focus of this work is to perform DC analysis of 3D-VRAM. The description of the ReRAM element in HSPICE is shown in Table I.

B. Modeling of a 3D-VRAM Array

We develop a circuit model of the 3D-VRAM array by approximating the ReRAM cells and plane resistance with segmented elements. As Figure 5 shows, each ReRAM cell is represented by four ReRAM elements, defined in Section III-A. One advantage of 3D-VRAM over 3D-HRAM is that the wordline is a metal plane rather than multiple metal wires, making the effective resistance between adjacent cells smaller than the wire resistance in the 3D-HRAM counterpart. To model such effect, one virtual node is added to emulate the two-dimensional current flow through the wordline plane using discrete resistors. We also tried to add four and nine virtual nodes and our results will show that the error of adding one virtual node is already very small. Not shown in Figure 5, access transistors are implemented below the bottom layer using 22nm PTM model [14]. In order to minimize the voltage drop on the access transistor, we assume that the gate voltage is boosted when the NMOS is on.

C. Geometry Parameters and Design Constraints

To enable the exploration of the large design space, we parameterize some of the important geometries in the 3D-VRAM array and summarize them in Table II. Under these definitions, the height of a vertical stack (one wordline layer plus one isolation layer) is $H_s = H_m + H_i$. The pitch is defined as the minimum distance from the center of a cell to the center of its neighboring cell,

$$P = D + 2T_{ox} + F \qquad (1)$$

When modeling the cell size, the width of a cell is bounded by either the pitch or the width of the underlying access transistor assuming standalone memory design rule,

$$W_{cell} = max(P, W_{tran} + F) \qquad (2)$$

and the length of a cell is also bounded in the similar manner,

$$L_{cell} = max(P, L_{tran} + 2F) \qquad (3)$$

where L_{tran} is typically assumed to be a fixed value (F in this work).

TABLE II
GEOMETRY PARAMETERS OF THE 3D-VRAM ARRAY

Metric	Description	Explored Values
H_m	Height of a wordline plane	20, 30, 0nm
H_i	Height of an isolation layer	20nm
H_s	Height of a vertical stack	-
T_{ox}	Thickness of the switching layer	5nm
D	Diameter of a pillar electrode	-
F	Feature size of the design	22nm
P	Minimum distance from cell to cell	-
W_{tran}	Gate width of an access transistor	22, , nm
L_{tran}	Gate length of an access transistor	22nm
W_{cell}	Cell width along bitline direction	-
L_{cell}	Cell length along sourceline direction	-
AR	Etching aspect ratio	10, 20, 30
N_s	Number of sourcelines per array	1 ∼ 25
N_b	Number of bitlines per array	1 ∼ 25
L	Number of vertical stacks	2 ∼

The etching aspect ratio defines the maximum ratio of the total height of vertical stacks to the diameter of a pillar electrode,

$$AR = \frac{H_s \times L}{D + 2T_{ox}} \qquad (4)$$

From equation (1) to equation (4) we get,

$$A_{cell} = \begin{cases} P^2 & \text{if Case 1} \\ P \times (W_{tran} + F) & \text{if Case 2} \\ 3F \times (W_{tran} + F) & \text{if Case 3} \end{cases} \qquad (5)$$

and the conditions for the 3 cases are,

$$Cond \begin{cases} Case1: & \frac{H_sL}{AR} \geq max(W_{tran}, 2F) \\ Case2: & (\frac{H_sL}{AR} - 2F)(\frac{H_sL}{AR} - W_{tran}) < 0 \\ Case3: & \frac{H_sL}{AR} \leq min(W_{tran}, 2F) \end{cases} \qquad (6)$$

Equation (5) indicates that the cell size is bounded by the etching aspect ratio when building a 3D-VRAM array with many stacks and/or a thick metal layer (case 1). In case 1, we can afford to increase the width of the underlying transistor without increasing the cell size. The up-sizing in turn relaxes the design constraints from the perspective of sneak current and voltage drop due to the stronger driving capability of the wider access transistors, and potentially increases the maximum vertical stacks L_{max}.

Another important metric is the bit density, defined as L/A_{cell}. Combining equation (5) with equation (1) the bit density can be calculated as following,

$$D_{bit} = \begin{cases} \frac{1}{\frac{H_s^2L}{AR^2} + \frac{F^2}{L} + \frac{2H_sF^2}{AR}} & \text{if Case 1} \\ \frac{1}{(\frac{H_s}{AR} + \frac{F}{L}) \times (W_{tran} + F)} & \text{if Case 2} \\ \frac{1}{\frac{3F}{L} \times (W_{tran} + F)} & \text{if Case 3} \end{cases} \qquad (7)$$

As can be seen from Equation (7), in case 2 and 3 the bit density is improved when adding more stacks in the array (increasing L). One interesting observation for case 1 is that the bit density is actually a decreasing function of L given the boundary condition of case 1. That means adding more stacks will reduce the bit density when the cell size is bounded by the etching aspect ratio.

D. Read and Write Operations

As mentioned in Section II-C, during a read or write access, only one sourceline is selected to activate a vertical plane of cross-point structure. Within the cross-point structure, the voltage biasing for write and read operation are similar to that of a 2D cross-point structure, as shown in Figure 6.

Sneak current and voltage drop are two major issues in a cross-point design, as explained in Section II-B. In 3D-VRAM, the driving capability of the access transistor specifies another constraint in the array design: the total sneak current through a selected bitline should remain under the turn-on current of the NMOS so that there won't be

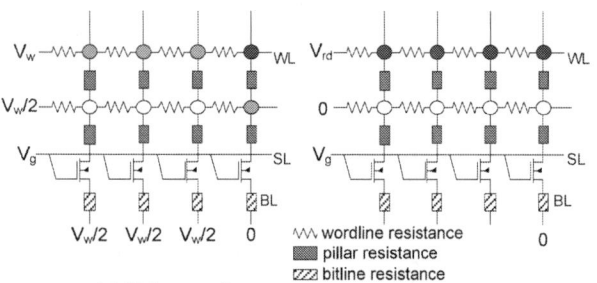

(a) Write operation   (b) Read operation

Fig. 6. Voltage biasing in 3D-VRAM for (a) write and (b) read operations

TABLE III
MAJOR PROCESS ADDERS OF 3D ReRAM DESIGNS WITH L STACKS

Item	3D-HRAM	3D-VRAM
Metal deposition	$L+1$	L (wordline) + 1 (pillar)
Interlevel dielectric deposition	$L+1$	$L-1$
Switching layer deposition	L	1
Critical lithography	$L+1$	2
Etching (metal/oxide)	$L+1$	2
CMP	$L+1$	2

a large voltage drop on it. This constraint is the key limiting factor on maximum vertical stacks L_{max}.

E. Area and Cost Modeling

Cost-per-bit is the single most important factor when adopting a new memory technology. The ultimate goal of technologies scaling, cell structure innovation, as well as chip yield improvement is to reduce the cost-per-bit of a memory chip. Since there is a direct relationship between the die area and the die cost, we need to model the total area of a 3D-VRAM die first.

Our modeling framework is based on an open source software NVSim [15], which is a circuit-level area, timing, and power model for various non-volatile memories including NAND Flash, ReRAM, Phase-Change Memory etc. To estimate the area of a ReRAM chip, we first use NVSim to break down the area of a NAND Flash die into cell arrays, local/global decoders, sense amplifiers/latches, and charge pump circuits etc. The results are then calibrated with industrial NAND Flash chips. We assume the ReRAM and NAND Flash share the same interface design and keep their original design with the same silicon footprint. We calculate the area of the components with different circuit designs or transistor sizings for ReRAM in the heavily modified NVSim and replace the values in NAND Flash die area breakdowns. A key difference is that our ReRAM requires much lower operating voltage ($< 3.3V$) than NAND Flash ($> 15V$) and therefore charge pumps with much smaller overhead are needed. We use the model presented by Palumbo et al. [16] to calibrate the area of the charge pump circuits,

$$A_{charge\_pump} = k \cdot \frac{N^2}{(N+1) \times V_{DD} - V_{Out}} \frac{I_L}{f}, \quad (8)$$

where k is a technology-dependent constant, N is the number of stages in the charge pump, V_{Out} is the output voltage, I_L is the write current and f is the working frequency.

After the die area is obtained, the total cost of a ReRAM die can be calculated as,

$$C_{die} = C_{Wafer} \times Y_{Wafer}/N_{gd} \quad (9)$$

where C_{Wafer} is the cost of a wafer, Y_{Wafer} is the wafer yield, and N_{gd} is number of good dies in the wafer. N_{gd} depends on the die area A_{Die}, the diameter of the wafer d_{Wafer} and defect density D_0,

$$N_{gd} = (\frac{\pi d_{Wafer}^2}{4A_{Die}} - \frac{\pi d_{Wafer}}{\sqrt{2A_{Die}}})/(1 + \frac{D_0 A_{Die}}{\alpha})^{-\alpha} \quad (10)$$

TABLE IV
OTHER PARAMETERS OF A ReRAM DESIGN

Metric	Description	Value
V_{core}	Core voltage	1. V
V_w	Write voltage of a selected cell	$\pm$ V
V_g	Boosted gate voltage on selected sourceline	. V
V_{rd}	Read voltage of a selected cell	0. V
-	Technology node of peripheral circuitry	nm
-	Copper resistivity	$\mu\Omega \cdot cm$
-	Aspect Ratio of bitline metal	1.

The next critical step is to include the fabrication process in the wafer cost,

$$C_{Wafer} = C_{Wafer_0} + C_{Wafer+} - C_{Wafer-} \quad (11)$$

where C_{Wafer_0} is the cost of a baseline NAND Flash wafer. C_{Wafer+} and C_{Wafer-} represent the cost of extra process steps associated with 3D ReRAM fabrication (i.e. additional lithography, etching, deposition etc.) and the redundant process cost of NAND Flash compared to ReRAM (i.e. floating gate fabrication), respectively. Most of the cost parameters in the model are collected from the IC Knowledge LLC [17], which has data for industrial 20nm-class NAND Flash. To estimate C_{Wafer+} we carefully break down the fabrication steps in a typical 3D ReRAM process flow [6] and summarize the major process adders compared to a planar NAND Flash process in Table III. The cost modeling tool allows the user to customize these process adders as an optional input and calculates the wafer cost overhead automatically. As Table III shows, 3D-HRAM does not save fabrication steps or masks because in each stack we need lithography, etching and CMP steps to pattern the features and therefore its cost-per-bit is expected to remain high. In contrast, 3D-VRAM requires only 2 critical lithography steps (1 for patterning cells and 1 for exposing the wordline electrodes) thus it is expected as a promising approach for low cost-per-bit.

IV. EXPERIMENTAL RESULTS AND DISCUSSIONS

A. Simulation Methodology

As one of the goals for this work is to identify the optimal design corners in a huge design space, we explore a set of design parameters in Table II. Our simulation methodology works in the following way: given the design target (i.e. array capacity, integration density etc.) and constraints (i.e. maximum vertical stacks, allowed voltage drop, minimum sensing margin etc.), our model tries all the design choices by exploring, if necessary, the parameters with multiple values or a range in Table II (parameters with single value means little flexibility). Then we set other parameters of a memory design according to Table IV. For each design point, our tool automatically generates a HSPICE netlist file and performs the simulation. Then we get the worst-case voltage drop, sneak current, sensing margin etc. and check if they are under the constraints: store the solution if so or ignore it otherwise. After all possible solutions have been obtained, the tool chooses one or a subset of them to meet the design target. For example, assuming we want to find a 3D-VRAM design with the maximum bit density given the following constraints: (a) 2kb array capacity (that is, $N_b \times N_s \times L = 2048$); (b) voltage drop $> 2/3V_w$; (c) sensing current difference $> 0.1\mu A$. Our model is going to search the optimal design within: (a) $H_m = 10nm$ and $AR = 30$ (no exploration needed); (b) $W_{tran} = 22$ or $44nm$ (limited exploration needed); (c) $(N_s, N_b, L) = (16, 16, 4)$ or $(32, 32, 2)$. And it may find the design with $W_{tran} = 22nm$ and $(N_s, N_b, L) = (16, 16, 4)$ satisfies all the constraints and has the maximum bit density ($0.667b/F^2$). Note that we limit the search range by setting $N_b = N_s$ and later in this section the format of "128 * 16" is used to denote a 3D ReRAM array size of $128 \times 128 \times 16$. We can get rid of this constraint if symmetric array design is not required or simulation time is not a concern.

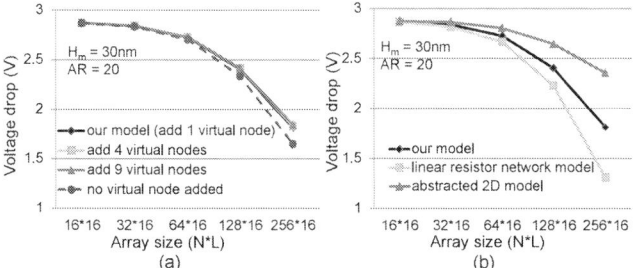

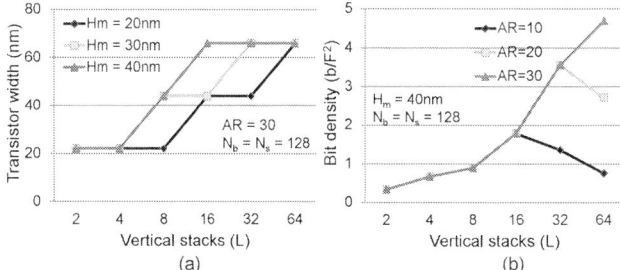

Fig. 7. Simulations errors of voltage drop by (a) adding different number of virtual nodes and (b) using different models

Fig. 9. (a) Required access transistor sizing with increasing vertical stacks and (b) Bit density versus vertical stacks with different AR

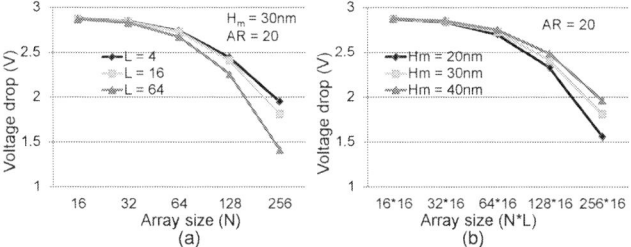

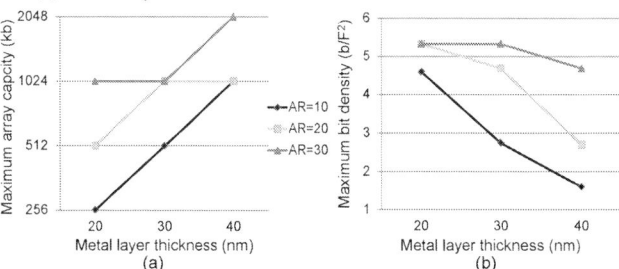

Fig. 8. Voltage drop versus array size when varying (a) number of vertical stacks and (b) thickness of metal layer

Fig. 10. Impact of H_m and AR on (a) maximum array capacity and (b) maximum bit density

B. Model Accuracy

First we show our model maintains a reasonable accuracy. In our model, we add one virtual node in the sub-circuit shown in Figure 5 to emulate the current flow through the two-dimensional wordline plane electrode. This accuracy can be improved if more virtual nodes are added, making the current flow path more closer to the reality. However, adding more virtual nodes complicates the model significantly and results in increasing the simulation time by one or two orders of magnitude. Figure 7(a) illustrates the relation between the worst-case voltage drop and 3D array size. Note that we use N to represent both N_b and N_s since we assume they are identical. As can be seen in Figure 7(a), adding 1, 4 and 9 virtual nodes results in very similar trend (and also absolute values) in voltage drop. The error between our model and the one with adding 9 virtual nodes remains within 3% for an array size of $256 \times 256 \times 16$. We can also tell from the figure that the model without adding virtual node overestimates the voltage drop by more than 15% for a large 3D array.

Next we will show that most of the other simplified models are not suitable for large 3D ReRAM array simulation. For comparison purpose, we implemented a 3D model by building a pure-resistor based network, where the resistance have only several discrete values. In another abstracted 2D model, we focus on the analysis within the activated vertical plane, which is a de facto cross-point structure. As can be seen in Figure 7(b), the abstracted 2D model underestimates the voltage drop problem because it ignores the sneak path from the unactivated vertical planes while the linear resistor network model overestimates the voltage drop problem because it can not model the changing nonlinearity of the half-selected cells along the selected wordline/bitline. And the errors of both approaches go beyond 30% when simulating a large 3D array.

To summarize, our model turns out to be a good balance between accuracy and simulation time.

C. Impact of Geometry Parameters

The number of vertical stacks L that affects the pillar resistance and the thickness of the metal layer H_m that affects the plane resistance are two important design parameters in determining the array size, density, and write/read noise margin. Figure 8 illustrates their impact on the voltage drop. As can be seen in Figure 8(a), the voltage drop

gets worse as we add more stacks because the voltage loss on the pillar electrodes increases, given the same H_m and AR. In this case, the array size of $256 \times 256 \times 16$ is not workable because the voltage drop is below $V_w/2$ [11]. On the other hand, increasing H_m will reduce the resistivity of the wordline plane and thus alleviate the voltage drop along that direction, as illustrated in Figure 8(b). We will show the negative effect of increasing H_m later.

Then we fix the two-dimensional array geometry (128×128 in this case) and add more stacks. In order to maintain a workable 3D array with voltage drop and sensing current under constraints, we have to size up the access transistor to provide enough driving current. Figure 9(a) presents the trend that width of transistor increases as we want to build more 3D stacks. Then the increasing rate of the transistor sizing becomes superlinear with L and eventually begins to dominate the cell size. That is, adding more stacks will hurt the bit density adversely, and Figure 9(b) demonstrates the effect that the turn point occurs in the bit density curve. One important observation is that a large AR may help shift the turn point to the right or diminish it. That can be explained by the condition of case 1 in equation 6: increasing AR means a larger threshold of L to enter case 1 for the access transistor to dominate the cell size.

Given the target capacity of a 3D-VRAM chip, its cost-per-bit depends on both the array capacity (or the number of arrays in the chip) that affects the peripheral circuitry overhead and the bit density that determines the total cell area. We try to identify the impact of H_m and AR on maximum array capacity and maximum bit density that can be achieved, as shown in Figure 10. Interestingly, H_m plays an opposite role in affecting the two metrics: increasing H_m improves array capacity but hurts the bit density. On one hand, a larger H_m reduces the voltage drop as observed in Figure 8(b). On the other hand, a larger H_m also means a higher vertical stack, which increases the cell size when the etching aspect ratio limits the pitch. This is also why a larger AR slows down the bit density degradation with increasing L in Figure 10(b).

D. 3D-VRAM VS 3D-HRAM

In this Section we perform a comprehensive comparison between 3D-VRAM and 3D-HRAM in terms of voltage drop, array capacity, bit density, die area and cost-per-bit.

978-1-4799-2817-0/14 $31.00 © 2014 IEEE

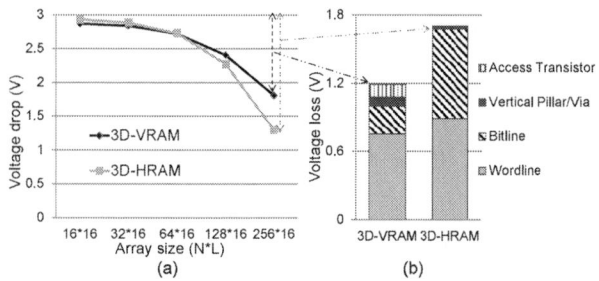

Fig. 11. (a) Voltage drop of 3D-VRAM and 3D-HRAM and (b) Voltage loss breakdown of 3D-VRAM and 3D-HRAM arrays of × × 1 cells

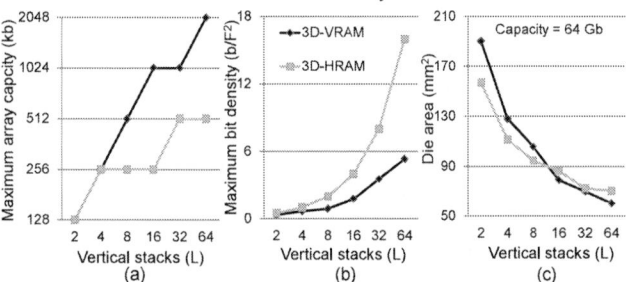

Fig. 12. 3D-VRAM VS 3D-HRAM on (a) array capacity, (b) bit density, (c) area of a 64Gb chip

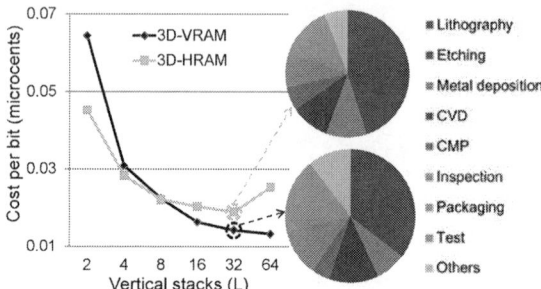

Fig. 13. Cost per bit projections with an example of cost breakdowns

optimized when targeting certain goals. The comparisons between 3D-VRAM and 3D-HRAM indicate that the cost-per-bit of 3D-VRAM is only half that of 3D-HRAM at 32 stacks. A comprehensive timing and power model remains to be developed to fully evaluate the performance and power metrics of a 3D-VRAM system.

As for voltage drop, the results in Figure 11(a) shows that 3D-VRAM has worse voltage drop at smaller array size because the voltage drop on the access transistor dominates the voltage loss. However, 3D-VRAM demonstrated significant better voltage drop at larger size and the reason is two fold, as examined in Figure 11(b). First, the wordline plane in 3D-VRAM has lower effective resistivity than the wordline wire in 3D-HRAM. Second, the total current on the selected bitline in 3D-VRAM (1 full selected cell plus L half selected cells) is way smaller than that in 3D-HRAM (1 full selected cell plus N half-selected cells) since $L < N$ at large array size. Due to this reason, 3D-VRAM allows larger capacity than 3D-HRAM given the same L, as illustrated in Figure 12(a). But the bit density of 3D-VRAM is not as high as that of 3D-HRAM, as shown in Figure 12(b), because the underlying access transistor and the etching aspect ratio both limit the cell size of 3D-VRAM. However, the overall effective density depends on both the array capacity and bit density. Particularly, when the bit density is large enough, the die area will be dominated by the peripheral circuitry. In that case, the array capacity has a larger impact on the total die area. Figure 12(c) shows that the die area of 3D-VRAM is significantly larger than that of 3D-HRAM for 2-stack and 4-stack counterpart but it shows great area advantage at 64-stack counterpart.

Figure 13 projects the cost-per-bit for 3D-VRAM and 3D-HRAM. As expected, the cost-per-bit of 3D-VRAM continues to go down when building more stacks is feasible since there is not any significant process adder to that. But the trend is different for 3D-HRAM: the reduction rate in its cost-per-bit slows down significantly when adding more stacks and the cost-per-bit may even increase beyond 32 stacks. The cost breakdowns for 32-stack 3D ReRAM designs are also demonstrated in Figure 13.

V. CONCLUSION AND FUTURE WORK

ReRAM has a great potential to replace NAND Flash if its cost-per-bit can be optimized. 3D-VRAM provides such an opportunity. In this paper, we build a full circuit model for 3D-VRAM with sufficient accuracy and reasonable speed. The design analysis shows a high etching aspect ratio improves both the array capacity and bit density, while a thick metal layer helps the former but hurts the latter. The results also suggest that some design parameters has to be co-

REFERENCES

[1] A. Nitayama and H. Aochi, "Bit cost scalable (BiCS) technology for future ultra high density storage memories," in *Proceedings of IEEE Symposium on VLSI Technology*, 2013.

[2] C. Chevallier *et al.*, "A 0.13 um 64Mb multi-layered conductive metal-oxide memory," in *Proceedings of the IEEE Solid-State Circuits Conference Digest of Technical Papers (ISSCC)*, Feb. 2010, pp. 260–261.

[3] A. Kawahara *et al.*, "An 8Mb multi-layered cross-point ReRAM macro with 443MB/s write throughput," in *Proccedings of the IEEE International Solid-State Circuits Conference (ISSCC)*, Feb. 2012, pp. 432–434.

[4] T.-Y. Liu *et al.*, "A 130.7mm2 2-layer 32Gb ReRAM memory device in 24nm technology," in *Proccedings of the IEEE International Solid-State Circuits Conference*, Feb. 2013, pp. 432–434.

[5] I. Baek *et al.*, "Realization of vertical resistive memory (VRRAM) using cost effective 3D process," in *Proceedings of the IEEE International Electron Devices Meeting (IEDM)*, 2011, pp. 31.8.1–31.8.4.

[6] H.-Y. Chen *et al.*, "HfOx based vertical resistive random access memory for cost-effective 3d cross-point architecture without cell selector," in *Proceedings of the IEEE International Electron Devices Meeting (IEDM)*, 2012, pp. 20.7.1–20.7.4.

[7] H.-S. Wong *et al.*, "Metal oxide RRAM," *Proceedings of the IEEE*, vol. 100, no. 6, pp. 1951 –1970, June 2012.

[8] J. Liang, S. Yeh, S. S. Wong, and H.-S. P. Wong, "Effect of wordline/bitline scaling on the performance, energy consumption, and reliability of cross-point memory array," *J. Emerg. Technol. Comput. Syst.*, vol. 9, no. 1, pp. 9:1–9:14, Feb. 2013.

[9] J. J. Yang *et al.*, "Engineering nonlinearity into memristors for passive crossbar applications," *Applied Physics Letters*, vol. 100, no. 11, p. 113501, 2012.

[10] Y.-C. Chen *et al.*, "3D-HIM: A 3D High-density Interleaved Memory for bipolar RRAM design," in *Proceedings of the IEEE/ACM International Symposium on Nanoscale Architectures*, 2011, pp. 59–64.

[11] D. Niu *et al.*, "Design trade-offs for high density cross-point resistive memory," in *Proceedings of the ACM/IEEE international symposium on Low power electronics and design (ISLPED)*, 2012, pp. 209–214.

[12] W. Fei *et al.*, "Design exploration of hybrid cmos and memristor circuit by new modified nodal analysis," *IEEE Transactions on Very Large Scale Integration (VLSI) Systems*, vol. 20, no. 6, pp. 1012–1025, 2012.

[13] S. Yu *et al.*, "A neuromorphic visual system using rram synaptic devices with sub-pj energy and tolerance to variability: Experimental characterization and large-scale modeling," in *Proceedings of the IEEE Electron Devices Meeting (IEDM)*, 2012, pp. 10.4.1–10.4.4.

[14] W. Zhao and Y. Cao, "New generation of predictive technology model for sub-45 nm early design exploration," *IEEE Transactions on Electron Devices*, vol. 53, no. 11, pp. 2816 –2823, Nov. 2006.

[15] X. Dong *et al.*, "NVSim: A Circuit-Level Performance, Energy, and Area Model for Emerging Nonvolatile Memory," *IEEE Transactions on Computer-Aided Design of Integrated Circuits and Systems*, vol. 31, no. 7, pp. 994–1007, 2012. [Online]. Available: http://nvsim.org

[16] G. Palumbo and D. Pappalardo, "Charge pump circuits: An overview on design strategies and topologies," *IEEE Circuits and Systems Magazine*, vol. 10, no. 1, pp. 31–45, Quarter 2010.

[17] IC Knowledge LLC., "IC cost model revision 1202a." [Online]. Available: http://www.icknowledge.com

978-1-4799-2817-0/14 $31.00 © 2014 IEEE

The Stochastic Modeling of TiO$_2$ Memristor and Its Usage in Neuromorphic System Design

Miao Hu[1], Yu Wang[2], Qinru Qiu[3], Yiran Chen[1], and Hai Li[1]

[1]Dept. of Electrical & Computer Engineering, University of Pittsburgh, Pittsburgh, PA, 15261, USA
[2] Dept. of Electrical Engineering, Tsinghua University, Beijing, 100084, China
[3]Dept. of Electrical Engineering & Computer Science, Syracuse University, Syracuse, NY, 13210, USA
mih73@pitt.edu,   yu-wang@tsinghua.edu.cn,   qiqiu@syr.edu,   yic52@pitt.edu,   hal66@pitt.edu

Abstract—**Memristor, the fourth basic circuit element, has shown great potential in neuromorphic circuit design for its unique synapse-like feature. However, though the continuous resistance state of memristor has been expected, obtaining and maintaining an arbitrary intermediate state cannot be well controlled in nowadays memristive system. In addition, the stochastic switching behaviors have been widely observed. To facilitate the investigation on memristor-based hardware implementation, we built a stochastic behavior model of TiO$_2$ memristive devices based on the real experimental results. By leveraging the stochastic behavior of memristors, a macro cell design composed of multiple parallel connecting memristors can be successfully used in implementing the weight storage unit and the stochastic neuron – the two fundamental components in neural network (NN)s, providing a feasible solution in memristor-based hardware implementation.**

1. Introduction

As traditional von Neumann computing systems based on CMOS technologies gains less performance increment and energy efficiency from device scaling, neuromorphic hardware systems that potentially provide the capabilities of biological perception and information processing within a compact and energy-efficient platform have gained great attentions [1][2]. However, the hardware development of NNs in traditional VLSI circuits still falls behind from the following perspectives. First, the weight matrix storage by digital-analog convertors, capacitors, or floating gates, has low precision, high power consumption, and high area overhead. Second, the voltage-based matrix computation induces many design issues including voltage offset, noise generation and voltage saturation. Last but not the least, the architecture and connection of such neuromorphic systems are hard to scale up, limiting the size and function of hardware implementations [3].

Theoretically, an idea memristor exhibits similarly as a synapse in bio-tissues [4]: it "remembers" the total electric flux through the device as its memristance M, which can be leveraged as the weight between an input voltage and an output current such as $I = V/M$. Such device feature potentially provides a complementary solution in neuromorphic design.

However, at current stage, a large gap exists between the theoretical memristor characteristics and the experimental data obtained from real devices, raising severe concerns in feasibility of memristor-based hardware design. For instance, the memristor theory expresses a continuous and stable memristance change. Though an arbitrary intermediate state can be obtained by carefully setting current compliance and period in a single metal oxide memristor, the corresponding realization at large scale, *e.g.*, crossbar array, is very difficult after including intrinsic design constrains, process variations, *etc.* Keeping a memristor in its ON or OFF state (R_{on} or R_{off}), on the con-

trary, is much more controllable. Thus, memristors nowadays are utilized as "memristive switches" [9].

Moreover, metal oxide based memristor behaves stochastically and hence even a single memristive device demonstrates large variations in performance. More specific, the *static states* of a single memristive switch, *i.e.*, R_{on} and R_{off}, are not fixed, but have large variations with skewed distributions and heavy tails [10]. The switching mechanism of a memristive switch, that is, its *dynamic behavior*, performs as a stochastic process [11], which has been widely demonstrated in various materials [19][20]. Previous statistical analyses on memristors were limited to the binary switching as data storage. However, as an analog device in NN application, it is necessary to understand and model memristor's analog stochastic characteristic. Here, we refer memristor to TiO$_2$ thin-film device.

In this work, we built a stochastic behavior model of TiO$_2$ memristive devices based on the real measurement results [9][10] to better facilitate the exploration of memristive switches in hardware implementation. The model bypasses material-related parameters while directly linking the device analog behavior to stochastic functions. Simulations show that the proposed stochastic device model fits well to the existing device measurement results.

To overcome the gap between the theoretical and real characteristics of memristive devices, we propose a macro cell design composed of a group of parallel connected memristive switches. It utilizes multiple memristors to represent an analog value by leveraging the stochastic behavior. Though the design sacrifices the design density, it is still more efficient than the CMOS implementations in floating gates or capacitors [18]. The usage of macro cells in weight storage unit and stochastic neuron, the two fundamental elements of neuromorphic system [12], is then demonstrated. The macro cells can be naturally integrated into memristor crossbars that previously were proposed as weight storage in neuromorphic computation [21].

The remainder of the paper is organized as follows. Section 2 provides the preliminary knowledge. Section 3 describes the stochastic model and calibrates it with experimental data. Section 4 presents the macro cell design and shows its usages. At last, we conclude the paper in Section 5.

2. Preliminary
2.1 Fundamental Components in Neural Network

Inspired by biological system, NNs mimic neuron-synapse networks, in which synapses transmit weighted signals and neurons process these signals based on activation functions. Many NN functionalities can be obtained through different network topologies, training methods, and activation functions. However, two fundamental components are always essential:

- **The weight carrier** for weight storage and signal modulation. The weights shall be represented by continuous analog state (or at least highly accurate digital states).
- **The stochastic neuron** can be taken as neuron with a probabilistic activation function. It has been widely used in modern NNs, *e.g.*, *Restricted Boltzmann Machine* (RBM).

2.2 Memristor and Its Potential in Neural Network

As illustrated in Figure 1(a), a memristor describes the relationship between flux (φ) and charge (q). The first physical demonstration of memristor was announced in 2008 through a TiO_2 thin-film material [14] depicted in Figure 1(b). The basic theoretical model contains of the static state represented by memristance M as $V = I \cdot M(w)$ and the dynamic behavior described by the movement of an internal state w under electrical excitation as $\frac{dw}{dt} = f(V, w)$. Note that the internal state w is physically meaningful. For example, w in a TiO_2 thin-film material is the width of its barrier. Memristor is considered as a potential candidate for efficient neuromorphic circuit realization. Many researches on theoretical analysis [5][6] and hardware implementation [7][8] were demonstrated.

2.3 Characteristics of Real Memristive Devices

Compared to theoretical characteristics of ideal devices, many non-ideal features have been revealed in real memristive devices. For example, a geometrical variation aware TiO_2 device model illustrated in Figure 1(c) was developed [15]. More importantly, although a single memristor can be tuned to arbitrary analog state, it is difficult to generalize this approach to large-scale designs because of the sneak paths. We face the unfortunate reality that only "memristive switches" presenting binary states are practical for designs with nano-devices [4].

Moreover, the stochastic behavior in dynamic switching process and large variations in static states have been widely observed in experimental results of metal oxide materials. In brief, the time to successfully change the state of a single memristive switch is not deterministic but follows a long tail distribution [9]. And R_{on} and R_{off} follow skewed distributions [10]. These non-ideal characteristics shall be considered in hardware implementations built with memristive switches.

Though many physical memristor models were built based on insight mechanisms [11][16][17], they cannot reflect the large variation induced by stochastic switching behavior. Stochastic models can better link the statistical measurement data to probability functions. But the existing stochastic models are limited to only the binary switching behaviors [9][10] and hence cannot capture the intermediate analog state.

3. Stochastic Model for Memristive Switch

We proposed a stochastic model for TiO_2 memristive switch

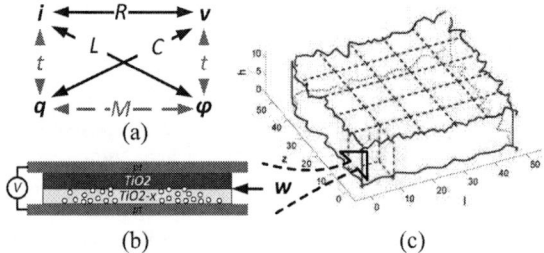

(a)

(b)   (c)

Figure 1. (a) Memristor is the fourth basic circuit element. (b) The ideal TiO_2 memristor model. (c) The geometric variation model.

based on both the inspection of the physical mechanisms [11][17] and the statistical analysis of experimental data [9][10]. Our model describes the stochastic memristive switching based on the behavior analysis in both static states and dynamic switching process.

3.1 On and OFF Static States

The static stochastic behavior can be described by the distributions of R_{on} and R_{off}. In TiO_2 memristor, the initial barrier width w follows a normal distribution and the device resistance exponentially depends on w. Therefore the distribution of state resistance follows the lognormal *probability density function* (pdf) function, which is [11]:

$$f_x(x; \mu, \sigma) = \frac{1}{x\sigma\sqrt{2\pi}} \cdot \exp(-\frac{(\ln x/\mu)^2}{2\sigma^2}), \quad x > 0. \quad (1)$$

Here, μ is the normal mean and σ is the standard deviation. Note that R_{on} or R_{off} does not change within a given static state because w remains constant. Therefore, we can use lognormal function (Lognorm) to generate the sampling data, such as

$$R_{on} = \text{Lognorm}(\mu_{Ron}, \sigma_{Ron}), \text{ and} \quad (2a)$$
$$R_{off} = \text{Lognorm}(\mu_{Roff}, \sigma_{Roff}). \quad (2b)$$

3.2 Dynamic Switching Process

The dynamics in TiO_2 memristor is a complex oxide electroforming process. It can be explained as an electro-reduction and vacancy creation process caused by high electric fields and enhanced by electrical Joule heating. Usually the barrier width w is used to model the vacancy channeling mechanism. Although the vacancy channeling mechanism has been evidenced by experiments [11], it is difficult to match it to a pure physical model. Instead, our model is based on the analysis of three major behaviors; we start with a mathematical analysis of the analog stochastic switching behavior from the statistical aspect, and then bridge the parameters in mathematical expression with the physical excitation. At last, the impact of over tune is integrated into the stochastic model.

3.2.1 Analog Stochastic Switching Behavior

The stochastic resistance changing has been observed in high frequency measurement at low voltage [17]. The time dependency of switching probability can be approximated by the *cumulative probability function* (CDF) of lognormal distribution, such as [9]:

$$P(\text{Success switch}) = F(t_{switch}; \mu_t, \sigma_t) = \frac{1}{2}\text{erfc}\left[-\frac{(\ln t_{switch}/\mu_t)^2}{\sqrt{2}\sigma_t^2}\right] \quad (3)$$

Here, t_{switch} represents the pulse width of activation time. And μ_t and σ_t are related to the external voltage V.

Instead of studying the complicated physical mechanism and its impact, we use mathematical method to analyze the ON-OFF switching probability. According to Eq. (3), the ON-OFF switching probability can be approximated by a CDF of lognormal distribution, differentiation of $P(\text{Success switch})$ at t_{switch}, then is a pdf of the lognormal distribution, such as

$$\frac{dP(\text{Success switch})}{dt_{switch}} = f_{t_{switch}}(t_{switch}; \mu_t, \sigma_t). \quad (4)$$

Eq. (4) describes the distribution of the increment of switching probability $dP(\text{Success switch})$ at time t_{switch} when applying a signal with a short pulse width dt_{switch}.

The switching mechanism of a memristive device is intrinsic. Hence, the characteristic of the stochastic behavior remains unchanged and follows the same probability function during its switching process. From its physical meaning perspective, Eq. (4) reflects the increment of switching probability at time t_{switch}, which can be associated to the resistance change ΔR. Physically, a successful switching event with a pulse of t_{switch} indicates that the device resistance changes from R_{on} to R_{off}, or vice versa, that is, $\Delta R = |R_{\text{off}} - R_{\text{on}}|$.

Considering that both ON and OFF switching are the cumulative results of the analog resistance changing and the increment of switching probability is directly reflected by the change of resistance, the change of analog resistance at time t_{switch} can be generated by mapping to the distribution of the increment of switching probability, leading to

$$\frac{dR}{dt} = (R_{\text{off}} - R_{\text{on}}) \cdot f_{t_{\text{switch}}}(t_{\text{switch}}; \mu_t, \sigma_t). \quad (5)$$

3.2.2 Time & Voltage Dependency of Switching Probability

Time and voltage dependency of switching probability describes the switching probability of memristive switch under applied voltage V and activation time t_{switch}. The switching process resulted from the cumulative impact of input signals can be modeled with CDF function. The lognormal switching time distribution comes from the nonlinear switching dynamics of the devices. Considering that the median switch time (μ_t) is exponentially dependent on the applied voltage amplitude V, we approximate μ_t as an exponential function, such as:

$$\mu_t = \exp(aV + b), \quad (6)$$

where a and b are fitting parameters.

Since σ_t has only a weak dependence on V, we can approximate the relationship between σ_t and V by a hard threshold squashing function, such as

$$\sigma_t = \begin{cases} \sigma_{\text{thres\_H}} & (\sigma_t \geq \sigma_{\text{thres\_H}}) \\ cV + d & (\sigma_{\text{thres\_L}} < \sigma_t < \sigma_{\text{thres\_H}}) \\ \sigma_{\text{thres\_L}} & (\sigma_t \leq \sigma_{\text{thres\_L}}) \end{cases}. \quad (7)$$

Where, c and d are fitting parameters. $\sigma_{\text{thres\_H}}$ and $\sigma_{\text{thres\_L}}$ are the upper and lower boundaries, respectively. Our model applies two individual sets of fitting parameters to ON and OFF switching processes.

3.2.3 The Resistance Shifting Due To Over Tune

Over tune stands for the behavior when one or more external voltage pulses continue being applied in the switching direction after the state switching of memristor already succeeds. For example, apply an ON switching signal to a device already in ON state. Based on the vacancy channeling mechanism, the over tune in OFF state continues eliminating the oxygen vacancy until all the oxygen vacancies disappear and the device becomes an insulator. In ON state, the over tune creates more oxygen vacancies to form more conducting channels. The device mechanism becomes less appropriate to be modeled with barrier width w since the channel frontier no longer exists. The resistance shifting in real devices is even more complex after including thermal, electron kinetic energy, and other physical issues. During over tune, a memristor device remains in the same static state and the resistance shifting follows the static resistance distribution. However, a systematic impact on μ_{Ron} and μ_{Roff} has been observed [10].

Here, we use a statistical method to analyze the impact of over tune on the resistance shifting. The charge q flowing through the device is used as the input variable, which has a direct impact on the number of oxygen vacancies and the device resistance. To exhibit the trend of resistance shifting, a linear approximation can be assumed between the passing charge q and the mean shifting μ_{shift} as [14]:

$$\mu_{\text{shift}} = e \cdot q = e \cdot \left(\frac{V}{M}\right) \cdot t \quad . \quad (8)$$

Here, e is the fitting parameter that describes the shift speed of mean, M is the current memristor resistance. The new μ_{Ron} and μ_{Roff} can be calculated from Eq. (7):

$$\mu'_{\text{Ron}} = \mu_{\text{Ron}} - \mu_{\text{on-shift}} = \mu_{\text{Ron}} - e_{\text{on}} \cdot q, \ \mu'_{\text{Ron}} \geq 0. \quad (9a)$$
$$\mu'_{\text{Roff}} = \mu_{\text{Roff}} + \mu_{\text{off-shift}} = \mu_{\text{Roff}} + e_{\text{off}} \cdot q. \quad (9b)$$

Though more complicated fitting equations can be established, such an approach is impractical and unnecessary at current stage considering of insufficient experimental data available. The resistance shifting caused by over tune is constrained within the target resistance state, demonstrating less impact on the overall memristor characteristic compared to the ON-OFF switching.

3.3 Stochastic Model Verification

We verified the proposed stochastic model from perspectives of static states and dynamic switching process.

Static States: Figure 2 shows the resistance distributions of a memristive switch in ON and OFF states. The blue bars in the figure are real measurement data of a TiO$_2$ memristive switch [10]. The results show that the lognormal distribution fits well to the real device data in ON state. However, in OFF state, the heavy tail is captured but the median value is slightly skewed. Though the distribution of R_{off} is not perfectly fitted, the error in distribution fitting of R_{off} has ignorable impact in the circuit simulation since R_{off} is more than two orders of magnitude higher than R_{on}.

Dynamic Switching Process: Figure 3 shows the time dependencies of ON and OFF switching probability at different applied voltages. The results have high approximation to the experimental results [9]. The error mainly comes from the approximation of the relationship between σ_t and V. As aforementioned, establishing a more reliable estimation of σ_t requires more experimental data.

Figure 4 shows the simulated analog resistance changing process of a TiO$_2$ memristor to better demonstrate the time and voltage dependency of switching probability and the resistance shifting due to over tune. The external voltage is set as $-3.0V$ to switch the memristor from R_{off} to R_{on}. The 100 curves in the figure represent the resistance changings by repeating 100 times of the ON switching procedure for the same device. The distribution of 100 tests agrees well with the switching probability curve at -3.0V in Figure 3(a): about 40% of the curves reach R_{on} before 0.1 S.

Considering the obvious stochastic behavior of memristive device at nanometer regime, traditional device modeling based on curve fitting is not enough. In this work, we built a stochastic model for TiO$_2$ memristor by bridging the key physical

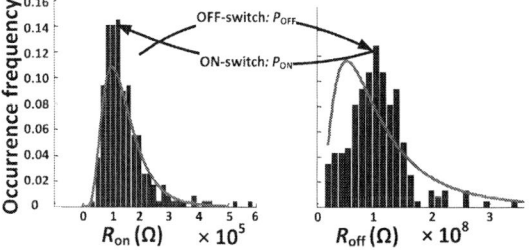

Figure 2. The static state distributions of a memristive switch.

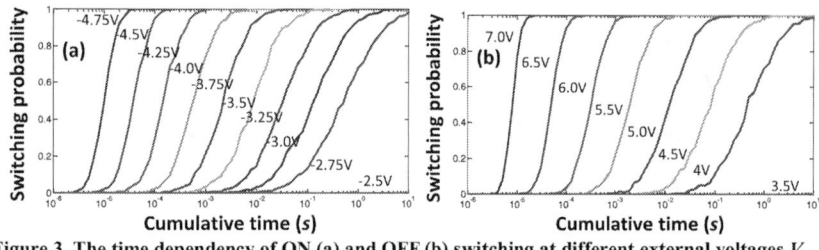
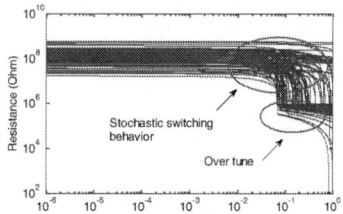

Figure 3. The time dependency of ON (a) and OFF (b) switching at different external voltages *V*.

Figure 4. The analog switching process of a TiO₂ memristor.

mechanisms and the experimental data fitting. *The model combines the stochastic characteristics in static states and dynamic switching process together, and extends the stochastic study to the analog state while still holding high approximation to the existing data.* The accurate and fast estimation on the distribution of device's analog states makes the proposed model more meaningful for higher level circuit and system designs. This model can be generalized to other metal oxide memristors [19][20] for the same stochastic nature, that is, the percolation property of the thin dielectric soft breakdown [22]. The proposed model can be further enhanced by integrating with reliable physical model that precisely describes the stochastic switching mechanism. The complex and slow physical model generates the required distribution data to develop the proposed fast stochastic model.

4. Memristive Switches in Neural Network

Our primary interest is to effectively utilize memristive switches and provide feasible designs for NN hardware. Rather than digging into specific NNs, we realized two fundamental NN components with memristive switches: the weight storage unit and the stochastic neuron for binary/continuous value generation. To ease the following discussion, we change the expression of memristive switches from resistance R (in unit Ω) to conductance G (in unit S), where $G = 1/R$.

4.1 Weight Storage Units

Storing high-precision continuous weight is beyond the capability of a single memristive switch. We proposed a ***macro cell*** design composed of a group of parallel connected memristive switches for weight storage.

4.1.1 Characterization of Multiple Memristive Switches

Multiple memristive switches connected in serial or in parallel can provide multi-level conductance (resistance) values by simply combining the ON and OFF states of these devices. Comparing the two connection topologies, the design of parallel connection can be easily adapted on crossbar arrays. Also, it can provide a linear function of the read-out current, mitigating the pressure on sensing circuit. Thus, a group of parallel

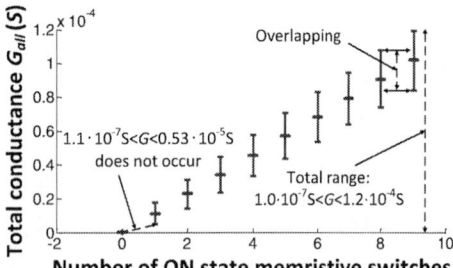

Figure 5. The conductance distribution of parallel connected memristive switches.

connected memristive switches is adopted in our design. The programming/detecting on the different ON and OFF combinations is realized through the peripheral circuit.

Here we take 9 parallel connected switches as an example. Figure 5 shows the distribution of its overall conductance G_{all}. To evaluate the impact of resistance in OFF state, we gradually increase the device number in ON state and remain the others in OFF state. The simulation result shows that the mean and deviation of G_{all} linearly grows as the number of ON memristive switches increases. Moreover, when all the memristive switches are in OFF state, the variation is negligible, indicating the variation in OFF State has little impact on the total conductance. In other words, the ON-state variation dominates the distribution of G_{all}. Thus, with more memristors in a macro cell, it can achieve larger conductance range, roughly proportional to G_{ON} times number of memristors.

4.1.2 Macro cell – A Continuous Weight Storage Unit

The parallel connection of memristive switches can be easily adopted in crossbar arrays. Let's use a 3×3 memristive switch crossbar in Figure 6(a) as an example. By combining the three inputs wires together and connecting the three output wires, the 9 memristive switches in this structure are parallel connected. We name such a structure as a ***macro cell***.

The given example has 10 possible ON-OFF device combinations, corresponding to 10 different conductance levels. Ideally, the 10 conductance levels can be differentiated by tuning the number of memristive switches in ON state. Unfortunately, as the simulation result in Figure 5 shows that the large resistance variation of ON state causes overlapping of conductance distribution, which is problematic in realizing traditional digitalized data storage for lacking of noise margin between adjacent levels. However, it also indicates continuous analog weight storage since a macro cell can achieve any arbitrary conductance within the overlapping range. For instance, the total conductance G_{all} of the macro cell in Figure 6(a) ranges from $0.53 \cdot 10^{-5}$S to $1.2 \cdot 10^{-4}$S. The unreachable conductance ranges from $1.1 \cdot 10^{-7}$S to $0.53 \cdot 10^{-5}$S, corresponding to the region from the upper bound of 9 switches in OFF state to the lower bound when only one switch is in ON state.

4.1.3 Feedback Attempt Scheme

A feedback attempt scheme can be used to achieve target conductance in macro cells. Figure 6(a) illustrates the conceptual diagram of the programming scheme. First, the number of memristive switches in ON state is determined to tune the overall conductance roughly. The output current is detected to check if the target G has been reached. A feedback control then is given to finely tune the macro cell memristor conductance. If the detected current is not within the absolute error threshold E, an ON state memristor is randomly selected to reset and then set again. Under a given operation condition, the

978-1-4799-2817-0/14 $31.00 © 2014 IEEE

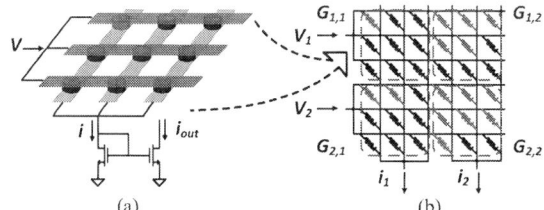

Figure 6. (a) A macro cell containing of 9 memristive switches on a 3x3 crossbar. (b) Partitioning a 6x6 memristive switch crossbar to obtain a 2x2 macro cell crossbar for continuous weight storage.

target conductance may not be obtained within a certain number of tryouts, indicating that either one or more memristors are too conductive or too resistive. We need to gradually reduce or increase the memristor number on ON state until the total conductance falls into tunable range. Then, restart the programming through the random attempt scheme.

In weight storage unit design, voltage pulses are used to control the switching of memristive switches. The pulse width t_{switch} is fixed, which is determined by the speed requirement. The amplitude V with ~100% switching probability is required to ensure the deterministic switching.

Figure 7 summarizes the average and the worst-case reconfiguration cycles to approach the different target conductance. The target conductance G_{Tar} can be generated by comparing to a reference current signal. Each data in the figures represents the statistical results of 1 million samples. The simulation results show that reconfiguration cycles increases linearly as the target conductance rises, while dramatically increases as the threshold error E decreases. More importantly, it shows that the proposed feedback attempt scheme can already achieve high precision programming within affordable attempts: any target conductance can be approached within on average 25 attempts with the error of $E = 0.1 \cdot 10^{-5}$S, corresponding to only 1% of the achievable conductance range. In the worst-case study, when $E = 0.3 \cdot 10^{-5}$S, the macro cell reaches the target conductance within 50 attempts in most cases. A rough calculation of $(G_{max} - G_{min})/2E$ implies that the macro cell can achieve at least 17 non-overlapping conductance levels rather than 10 levels obtained from ON/OFF combinations. If more attempts are affordable, we can increase to 50 non-overlapping conductance levels by reducing E to $0.1 \cdot 10^{-5}$S.

4.1.4 Macro Cells in Crossbar

The proposed macro cell can be easily adopted on larger crossbar structure. Figure 6(b) shows an example in which a 6×6 memristive switch crossbar is partitioned into 4 macro cells to implement a 2×2 weight matrix. The design sacrifices density while offering a practical and reliable way to realize continuous resistance state for analog storage and computation via binary switching memristors. The biggest advantage of

such a design is the dramatical decrement of programming complexity: a complex and slow feedback scheme is necessary when tuning a memristor to a specific analog state. In contrast, binary switching of memristors can adopt the existing memory programming scheme that is simple and reliable.

Compared with the crossbar array implementation by using floating gates or capacitors [18], memristors enable simpler structure. Moreover, the charge-based CMOS devices require certain dimension to guarantee data accuracy, while the memristor technology can easily shrink. Thus, though a macro cell employs multiple memristors, it still provides better area efficiency (1~2 orders) over CMOS technologies.

4.2 Stochastic Neurons

The stochastic switching process is a severe issue for non-volatile memories with memristive switches. However, with careful design, it can be leveraged in designing stochastic neurons. Generally, stochastic neurons can be categorized into the binary neuron and the continuous value stochastic neuron.

4.2.1 Binary Stochastic Neuron

Binary stochastic neuron generates random binary pulse signals, which uses external voltage signals to control the probability of 0 (OFF) or 1 (ON) generation. Figure 8 illustrates the design of a binary stochastic neuron with a memristive switch. The operation timing diagram is given in the inner set of the figure. Figure 9 shows the voltage dependency of ON and OFF switching of a TiO_2 memristive switch. Each curve has a fixed pulse width. The voltage dependency shows a normal dependency between the applied voltage and the switching probability, where t_{switch} has a log impact on the means and deviations of switching distributions.

Accordingly, the binary stochastic neuron can control the probability of random numbers by applying a fixed pulse width t_{switch} and adjusting voltage amplitude V. Figure 9 also demonstrates the tradeoff between t_{switch} and the tunable range of V. The longer pulse width results in the lower applied voltage and the wider tunable range, which alleviates the hardware design complexity but the speed of circuit operation exponentially reduces. The shorter pulse width makes the circuit run much faster, at the cost of smaller tunable range and the increased risk of device damage. A related work partially verified our design by using contact-resistive random-access-memory to build random number generator [13].

4.2.2 Continuous Value Stochastic Neuron

Continuous value stochastic neuron generates random pulse signal. The voltage amplitude of the pulse signal is an analog value, which falls into a given distribution with controllable mean and noise. As illustrated in Figure 8, a continuous value stochastic neuron can be constructed by replacing the single memristive switch in a binary stochastic neuron with a macro

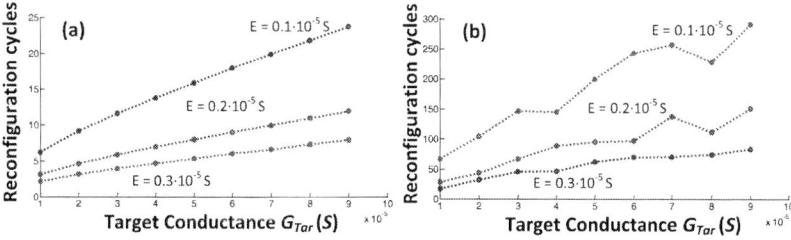

Figure 7. Average (a) and Worst-case (b) reconfiguration cycles to reach target conductance with different absolute error E.

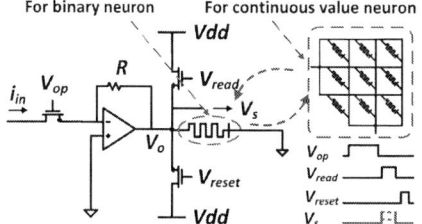

Figure 8. Binary/continuous value stochastic neuron design.

978-1-4799-2817-0/14 $31.00 © 2014 IEEE   835

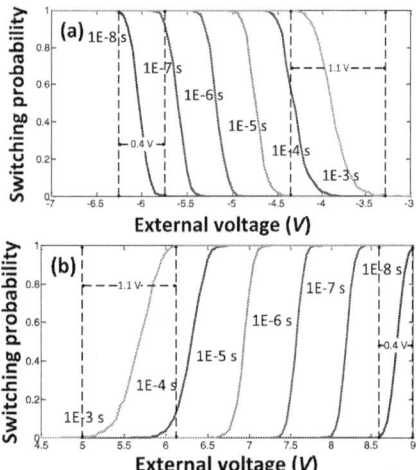

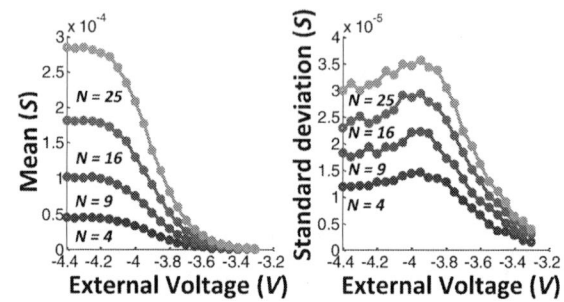

Figure 10. Voltage dependency of macro cell conductance. N is the number of memristive switches.

opinions, findings and conclusions or recommendations expressed in this material are those of the authors and do not necessarily reflect the views of DARPA, NSF, or their contractors.

Figure 9. Voltage dependency of ON (a) and OFF (b) switching at different pulse widths t_{switch}.

cell. The noise and mean are controlled by the external voltage signal and the number of memristive switches in a macro cell.

Figure 10 shows the means and standard deviations of the noise generated by the proposed continuous value stochastic neuron. The designs with different macro cells containing $N = 4, 9, 16, 25$ memristive switches are compared. The means and the deviations of total conductance are controllable through the applied voltage. When a zero-mean noise signal is required, an offset current/voltage source can be added at the output V_s to cancel out the mean shifting considering that the voltage amplitude dependency of mean follows a normal CDF.

The variation comes from both the stochastic ON switching process and the randomness of ON state resistance. When $V > -3.95\,V$, the major contribution to variation comes from stochastic switching. Hence, the standard deviation decreases as the voltage amplitude drops down. When $V < -3.95\,V$, a memristive switch has >80% probability to successfully change to ON state, as shown in Figure 9. Thus, the randomness of ON state dominates and the deviation is saturated. After all, using memristive switches, it is possible to replace the traditional continuous stochastic neuron [12] with memristive switch based circuit to obtain higher area/ power efficiency.

5. Conclusion

In this paper, we proposed a stochastic memristor model from the macro perspective of stochastic characteristics in memristive switches. With the help of the model, we evaluated the performance of practicable memristive switches on two fundamental NN components. Weight storage unit for continues value is realized by using parallel connected memristive switches, or macro cell. A programming scheme is provided to tune the macro cell to any desired approachable conductance with high precision. For stochastic neurons, we made use of the stochastic behavior of memristive switches to benefit the natural generation of binary/continuous random values with controllable mean and variation. The controllability of noise has been also analyzed and demonstrated.

6. Acknowledgment

This work was supported in part by DARPA D13AP00042, NSF EECS-1311747, ECCS-1202225, CNS-1253424, CNS-1342566, and HP Labs Innovation Research Program. Any

7. References

[1] K. Hornik, et al., "Multilayer feedforward networks are universal approximators," *Nerual Networks*, vol.2, pp.359-366, 1989.

[2] E. M. H. Hassoun, "Associative neural memories: Theory and implementation," in *Oxford University Press*, 1993.

[3] S. P. Eberhardt, et al., "Analog VLSI neural networks: Implementation issues and examples in optimization and supervised learning," *IEEE Trans. on Industrial Electronics*, vol. 39, pp. 552-564, 1992.

[4] S. D. Ha and S. Ramanathan, "Adaptive oxide electronics," *J. Appl. Phys.*, vol.110, pp.071101, 2011.

[5] K. Cantley, et al., "Hebbian learning in spiking neural networks with nano-crystalline silicon TFTs and memristive synapses," *IEEE Trans. on Nanotechnology*, vol.10, pp.1066-1073, 2011

[6] H. Kim, et al., "Neural synaptic weighting with a pulse-based memristor circuit," *TCAS-I*, vol.59, pp.148-158, 2012.

[7] K.-H. Kim, et al., "A functional hybrid memristor crossbar-array/CMOS system for data storage and neuromorphic applications," *Nano Lett.*, vol.12, no.1, 389-395, 2012.

[8] J. J. Yang, et al., "Engineering nonlinearity into memristors for passive crossbar applications," *APL*, vol.100, pp.113501, 2012.

[9] G. Medeiros-Ribeiro, et al., "Lognormal switching times for titanium dioxide bipolar memristors: Origin and resolution," *Nanotechnology*, vol.22, no.9, pp.095702, 2011.

[10] W. Yi, et al., "Feedback write scheme for memristive switching devices," *Appl. Phys. A*, vol.102, pp.973-982, 2011.

[11] J. J. Yang, et al., "The mechanism of electroforming of metal oxide memristive switches," *Nanotechnology*, vol.20, pp.215201, 2009.

[12] H. Chen, et al., "Continuous-valued probabilistic behavior in a VLSI generative model," *IEEE Trans. on Neural Networks*, vol.17, pp.755-770, 2006.

[13] C.-Y. Huang, et al., "A contact-resistive random-access-memory-based true random number generator," *IEEE Electron Device Lett.*, vol.33, pp.1108-1110, 2012.

[14] D. B. Strukov, et al., "The missing memristor found," *Nature*, vol. 453, pp.80-83, 2008.

[15] M. Hu, et al., "Geometry variations analysis of TiO$_2$ thin film and spintronic memristors," in *ASPDAC*, pp.25-30, 2011.

[16] F. Miao, et al., "Force modulation of tunnel gaps in metal oxide memristive nanoswitches," *Appl. Phys. Lett.* vol.95, pp.113503, 2009.

[17] Pickett, M.D. et al., "Switching dynamics in titanium dioxide memristive devices," *J. Appl. Phys.*, vol.106, pp.074508, 2009.

[18] V. Srinivasan, et al., "Floating-gates transistors for precision analog circuit design: an overview," *MSCAS*, pp.71-74, 2005.

[19] S. Yu, et al, "An electronic synapse device based on metal oxide resistive switching memory for neuromorphic computation," *IEEE Transactions on Electron Devices*, vol. 58, pp. 2729-2737, 2011.

[20] W. Lu, et al, "Stochastic memristive devices for computing and neuromorphic applications," *Nanoscale*, 2013.

[21] R. E. Pino, et al, "Statistical memristor modeling and case study in neuromorphic computing," in *DAC* 2012, pp. 585-590.

[22] S. Blonkowski, "Filamentary model of dielectric breakdown," *J. Appl.Phys.*, vol. 107, no. 8, p. 084109, 2010.

Through-Silicon-Via Inductor: Is it Real or Just A Fantasy?

Umamaheswara Rao Tida

Missouri S & T
Rolla, MO 65409
utxvc@mst.edu

Cheng Zhuo

Intel Research
Hillsboro, OR 97124
cheng.zhuo@intel.com

Yiyu Shi

Missouri S & T
Rolla, MO 65409
yshi@mst.edu

Abstract - Through-silicon-vias (TSVs) can potentially be used to implement inductors in three-dimensional (3D) integrated systems for minimal footprint and large inductance. However, different from conventional 2D spiral inductors, TSV inductors are fully buried in the lossy substrate, thus suffering from low quality factor. In this paper, we propose a novel shield mechanism utilizing the micro-channel, a technique conventionally used for heat removal, to reduce the substrate loss. This technique increases the quality factor and the inductance of the TSV inductor by up to 21x and 17x respectively. It enables us to implement TSV inductors of up to 38x smaller area and 33% higher quality factor, compared with spiral inductors of the same inductance. To the best of the authors' knowledge, this is the first proposal on improving quality factor of TSV inductors. We hope our study shall point out a new and exciting research direction for 3D IC designers.

I Introduction

Three-dimensional integrated circuits (3D ICs) are generally considered to be most promising alternative that offers a path beyond Moore's Law. Instead of making transistors smaller, it makes use of the vertical dimension for higher integration density, shorter wire length, smaller footprint, higher speed and lower power consumption, and is fully compatible with current technology [1].

The through-silicon-vias (TSV) is a critical enabling technique for 3D ICs, which forms vertical signal, power and thermal paths. While many challenges still exist in 3D ICs, a big one is related to TSVs: they are large in size, typically 5-10x larger than the standard cells in 32nm process [8]. Yet their diameters do not scale with the devices due to imposed limitations of wafer handling and aspect ratios. International Technology Roadmap for Semiconductors (ITRS) suggests that the TSV diameter will remain almost constant in 2012-2015 [2]. On the other hand, a large number of TSVs are needed to deliver signal and power, to dissipate heat and to provide redundancy. Moreover, to guarantee high yield rate, foundries typically impose a minimum TSV density rule. For example, Tezzaron requires that at least one TSV must exist in every 250 μm x 250 μm area [3]. To satisfy this rule, lots of dummy TSVs need to be inserted, which further increase the area overhead.

To alleviate the problem, there have been efforts in the literature to make use of those dummy TSVs for alternative purposes. In this paper, we are particularly interested in the application of TSVs towards on-chip inductors, which are the critical component in various microelectronic applications, e.g. on-chip voltage regulators, voltage control oscillators, power amplifiers and radio frequency (RF) circuits.

Conventional implementation of on-chip inductors uses multi-turn planar spiral structure. This structure occupies a significant area and requires special RF process for higher quality factor. For example, [19] reported an inductor which occupies 78,400 μm^2 routing area, equivalent to the area of 62K gates in 45 nm technology. In 3D ICs, however, it is possible to utilize through-silicon-vias (TSVs) to build vertical inductors [4-7]. One example of a toroidal TSV inductor in a two-tier 3D IC is shown in Fig. 1. An apparent advantage of such TSV inductors is the minimal footprint on routing layers and accordingly high inductance density. However, since it is completely buried in the lossy substrate, its quality factor is inferior compared with that of the 2D spiral inductor. Accordingly, as pointed out in [7], TSV inductor can be used when area is the only concern. This essentially declares that such a TSV inductor is only a "fantasy", useless in practice.

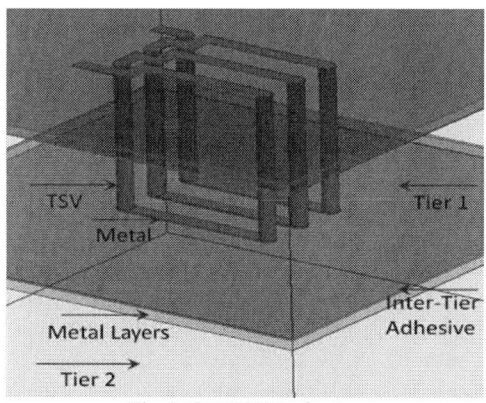

Fig 1. 3D TSV Inductor

To make TSV inductors "real", fundamental question need to be understood: is there any shield mechanism that can be used to reduce substrate loss? This paper provides the answer. Specifically, the main contributions of our work are as follows.

We put forward a novel shield technique using the micro-channel, which has been used in 3D IC industry including IBM and Nanonexus as a low-cost cooling technique [16], to reduce the substrate loss. Experimental results indicate that it can boost the quality factor and the inductance of the TSV inductor by up to 21x and 17x respectively. It changes the TSV inductor concept from just a fantasy to something practical - with the technique, TSV inductors can achieve up to 38x smaller area and 33% higher quality factor, compared with spiral inductors of the same inductance. This suggests that TSV inductors with micro-channel shields are a much better option than spiral inductors in 3D ICs.

The remainder of the paper is organized as follows. Section II reviews the existing efforts on the TSV inductor and the motivation of our work. Section III proposes a novel micro-channel shield mechanism to increase the inductance and the quality factor of the TSV inductor. Concluding remarks are given in Section IV.

II. Background and Motivation

Due to the limited space on an integrated circuit chip and highly competitive chip market, on-chip inductors must fit within a limited space and be inexpensive to fabricate. In that regard, it is desirable for an on-chip inductor to have a high inductance L per unit area. The spiral structure in a square shape can typically achieve lower than 100 nH/mm^2 density [10, 11].

Another important inductor metric is the quality factor Q, which is the ratio of its inductive reactance to its resistance and is used to measure energy efficiency [10]. To achieve high quality factor, on-chip inductors are typically implemented using thick metal on top metal layers in RF process. To reduce the EM coupling between the inductor and any metal wires beneath it, Patterned Ground Shield (PGS) technique is typically used, further occupying valuable routing resources.

The general structure of existing 3D TSV inductors [4-7] is shown in Fig. 1(previous page), which is composed of front/back metals and TSVs in a toroidal structure. The most attractive advantage of such a TSV inductor is its minimal footprint on the silicon surface. In addition, no PGS is necessary as the majority of the magnetic flux run in parallel with metal wires (in the horizontal plane).

Moreover, according to the literature, the quality factor of the TSV inductor is significantly less than its 2D spiral counterpart, mainly due to the loss from the substrate. Unlike the 2D spiral inductor, the entire TSV inductors is buried in the silicon substrate, which is lossy at high frequencies. As a result, Bontzios et al [7] suggested that for 50 μm substrate thickness and below, TSV inductor should be used when area is the only concern, which means it is of little practical value. One question that rises here is: Is there any way that we can reduce substrate losses for TSV inductors, so that their quality factor can be at par or even better compared with spiral inductors for practical use?

The practical range of interest for various parameters with their type is listed in the Table I.

There are three things worthwhile to note here: 1) While existing works only use two tiers (T=2) to implement the inductor, in this paper we extend the study to designs of up to four tiers (according to [17], 3D ICs of up to five tiers have already been fabricated). Since the bottom tier does not

need any TSV, the actual inductor is formed in the top T-1 tiers. 2) To achieve maximum quality factor, the cross-sectional area should be square. In other words, once we fix the number of tiers T, the TSV pitch should be (T-1)H, where H is the height of a single tier. 3) The 150 MHz operating frequency represents applications such as on-chip voltage regulator applications, while 1/5/10 GHz represents resonant clocking or RF applications. The nominal settings to study the quality factor and inductance variation with frequency are shown in fig. 2.

TABLE I. LIST OF PARAMETERS, THE RESPECTIVE DEFAULT UNIT AND RANGES OF INTEREST.

Notation	Meaning	Range
N	Number of turns	1-6
T	Number of tiers	2-4
P(μm)	Loop pitch[1]	13-23
W(μm)	Width of metal strip	3-12
f(GHz)	Operating frequency	0.15, 1,5, 10

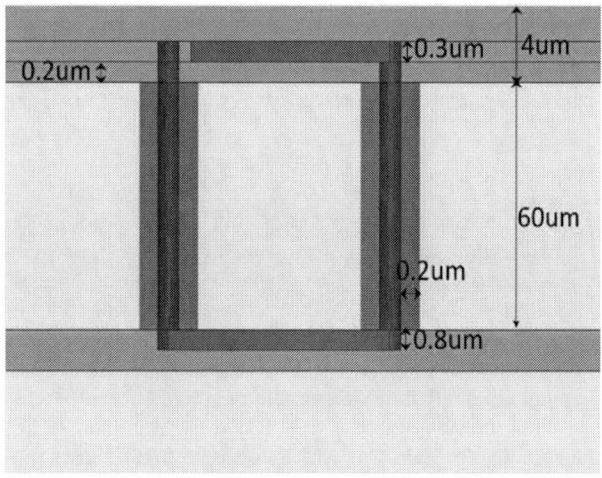

Fig. 2. Nominal settings (not to scale)

The corresponding inductance and quality factor vs. frequency plot for the above nominal settings are shown in Fig. 3.

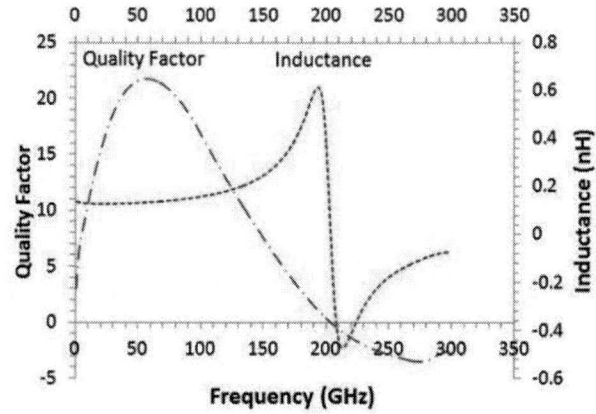

---

[1] Loop pitch is defined as the separation between adjacent turns.

Fig. 3. Quality factor and inductance vs. frequency for the TSV inductor with nominal settings

III. Loss Reduction Via Micro-Channel Shields

The substrate loss dominates when the frequency is over 1 GHz and when the substrate conductivity is over 10 S/m (which is normal for digital applications). In other words, the TSV inductor is subject to severe efficiency loss over 1 GHz. To tackle the issue, we are interested in devising effective shield mechanism to reduce substrate loss.

To help understand the distribution of eddy current in the TSV inductor, we simulate it with the nominal setting shown in Section III. The resulting E and H fields are plotted in Fig. 4 (a) and (b), respectively. From the figure, we can clearly see that the E field decreases as we get farther from the TSVs, while the H field completely penetrates through the area between the TSVs. As such, we can expect that most of the eddy current loss comes from the silicon substrate near the TSVs, which inspires us to think: Why not remove the silicon substrate in that area?

This reminds us about a seemingly irrelevant technique, micro-channel, which has been widely used as a low-cost heat-removal technique in 3D ICs (e.g.[12, 13]). Simply speaking, the technique etches a channel from the bottom surface of the substrate for liquid cooling and only requires extra two lithography steps, which are relatively cheap to implement. The fabrication process of micro-channel is already mature—an example of such process from IBM and Nanonexus is shown in Fig. 5 [16].

In our situation, we can place such channels adjacent to the TSVs to remove part of the substrate. Specifically, we etch four identical channels, one on each side of the two TSVs. An illustration of such a structure is shown in Fig. 5 for a two-tier design. For multiple tiers, we need to place four channels at each tier, adjacent to the TSVs. The micro-channels can either be filled with coolant like conventional micro-channels, or just open with air. Note that these channels are etched on the backside of the silicon substrate, and will not affect any devices. An extra benefit of such technique is the reduced temperature at the inductor. When inductors are used to form antennas, they typically bear high temperature [14]. Accordingly, with the micro-channels the heat will be able to dissipate faster.

To verify the effectiveness of this approach, we vary the height (H_c) and the width (W_c) of the four micro-channels and compare the improvement of Q and L at 10 GHz based on a structure with two tiers (T=2) and six turns (N = 6). All the other parameters conform to the nominal settings discussed in Section II. The micro-channels are placed 5 μm from TSV center to the nearer edge of the channel. The resulting Q and L are reported in Table II. The improvements over the case without micro-channel shields are also reported in parentheses. From the table, we can easily see that both channel height and width have profound impact on Q. For the maximum height of 60 μm and width of 25 μm, a 71.0% improvement of Q over the TSV inductor without micro-channel can be observed. Considering reliability and

manufacturability, the aspect ratio of the channel is limited [15]. Accordingly, designers should carefully consider the tradeoff between the micro-channel dimension and Q. On the other hand, L remains almost constant with various W_c and H_c.

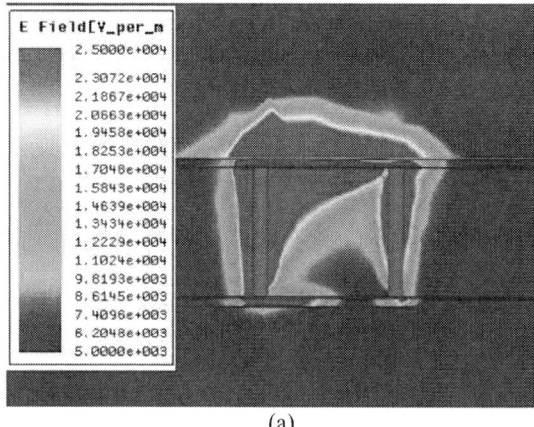

(a)

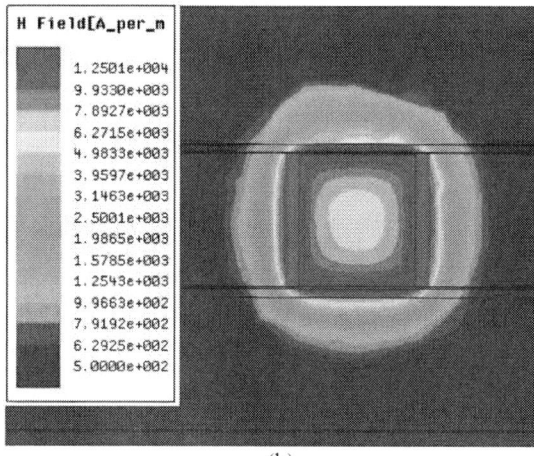

(b)

Fig. 4 (a) E field and (b) H field distributions

We further study how Q and L change when using maximum micro-channel dimensions for different number of turns N, number of tiers T and frequency f. The results on Q and L at 10 GHz are reported in Tables III and IV, respectively. To show the effect at different frequencies, the results on Q at 1 GHz are reported in Table V. Note that we omit the table for L at 1 GHz as it remains constant with or without the micro-channel. In all the tables, improvement over the case of the same N and T but without the micro-channel is also reported in parentheses.

From the tables, we can draw the conclusion that micro-channel technique is more important at larger N and T, and at higher frequencies – both Q and L improves significantly. This is in accordance with the intuition that substrate losses become larger with larger N, T, or higher f. At 10 GHz, up to 21x increase in Q and 17x increase in L are observed (Tables III and IV, N=3, T=5), while at 1 GHz only Q is improved by up to 3x (Table V, N=6, T=5).

978-1-4799-2817-0/14 $31.00 © 2014 IEEE

(a) Chip after FEOL and BEOL processing

(b) Electrical through via fabrication

(c) Silicon etch from micro-channels (one lithography step)

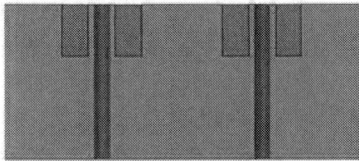

(d) Spin coat and polish a sacrificial polymer material

(e) Avatrael cover for micro-channels spin coated, patterned and cured (one lithography step)

(f) Assemble chip-on-chip

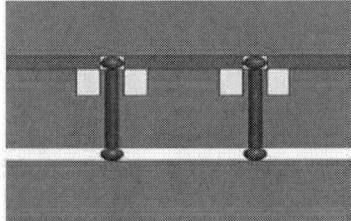

(g) Assemble chip on substrate

Fig. 5. Micro-channel fabrication steps (only two extra lithography steps (c) and (e) are required) [16]

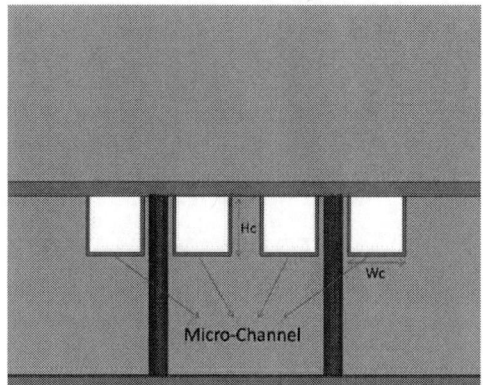

Fig. 6. Micro-channel shields for substrate loss reduction

One more thing worth mentioning here is that the increased self-resonant frequency brought by the micro-channel. For example, in Table VI, when N = 5 and T = 4, the TSV inductor without micro-channels ceases to work as an inductor (Q<0), while the TSV inductor with micro-channels still provides positive quality factor. For that reason, no improvement is reported.

TABLE II. Q AND L VS. MICRO-CHANNEL DIMENSION (10 GHZ, N = 6, T = 2) THE RELATIVE IMPROVEMENT OVER THE CASE WITHOUT MICHROCHANNEL SHIELDS IS RPEORTED IN PARENTHESES.

		W_c (μm)					
		Quality Factor			Inductance (nH)		
		10	20	25	10	20	25
H_c (μm)	10	6.71 (4.5%)	6.92 (8.1%)	7.03 (9.6%)	1.396 0.0%	1.395 0.0%	1.394 0.0%
	20	7.02 (9.6%)	7.29 (13.8%)	7.46 (16.3%)	1.393 0.0%	1.390 0.0%	1.390 0.0%
	30	7.34 (14.6%)	7.76 (21.1%)	7.94 (23.9%)	1.391 0.0%	1.386 0.0%	1.384 0.0%
	40	7.73 (20.5%)	8.28 (29.2%)	8.59 (34%)	1.388 0.0%	1.386 0.0%	1.382 0.0%
	50	8.25 (28.8%)	8.98 (40.1%)	9.41 (46.1%)	1.388 0.0%	1.380 0.0%	1.376 0.0%
	60	9.12 (42.2%)	10.34 (61.4%)	10.96 (71.0%)	1.386 0.0%	1.374 0.0%	1.377 0.0%

Finally, we set up three different sets of target inductance and operating frequency, and compare the resulting 2D spiral inductor, conventional TSV inductor without micro-channel shields, and our TSV inductor with micro-channel shields in terms of quality factor and area. The results are reported in Table VI. The 2D spiral inductors are implemented through a special RF process, which includes a total of 9 metal layers of 8 μm thick in total. The spiral inductor is implemented on M9 of 4 μm thick (to improve Q). The PGS as shown in [10] is also embedded. T, D, d and W denote the number of turns, outer diameter, loop pitch, and metal width for the spiral inductor respectively. For the TSV inductors with and without micro-channel shields under the same design spec, we use the same geometries for comparison. Their notations are shown in Table I and other process details are listed in Section III. The area for all inductors is measured by the total routing resource occupied. For TSV inductors, the area also includes the substrate occupied by the TSVs.

TABLE III. QUALITY FACTOR (Q) AND AREA (A) COMPARISON BETWEEN 2D SPIRAL INDCUTORS (W/ PGS) AND 3D TSV INDUCTORS (BOTH W/O AND W/ MICRO-CHANNEL SHIELDS) UNDER SAME DESIGN SPECS (L AND F).

Design Specs			Spiral Inductor						TSV Inductor						
#	f (GHz)	L (nH)	Geometry				Q	A (μm^2)	Geometry				Q		A (μm^2)
			T (μm)	D (μm)	d (μm)	W (μm)			N	T	W (μm)	P (μm)	w/o shield	w/ Shield	
1	1	6.5	2	560	535	10	5.7	313,600(1)	6	2	7	20	4.6	7.6	8,255(1/37.9x)
2	5	2.50	2	400	355	20	6.9	160,000(1)	4	3	6	18	4.3	10.3	9,358(1/17.1x)
3	10	0.95	1	330	320	10	10.0	108,900 (1)	2	3	6	18	5.8	10.1	4,679(1/23.3x)

From the table, we can easily see that the TSV inductors without micro-channel shields has much inferior quality factor compared with the spiral inductor of the same inductance, while the TSV inductors with the micro-channel shields can achieve higher quality factor compared with the spiral inductor. For example, at 1 GHz, while 6.5 nH is achieved by all the three inductors, our TSV inductor with micro-channel shields can achieve a 38x area reduction with 33% quality factor improvement compared with the spiral inductor.

TABLE IV. Q VS. NUMBER OF TURNS(N) AND NUMBER OF TIERS(T) FOR MAXIMUM MICRO-CHANNEL DIMENIONS (MEASURED AT 10 GHZ). THE RELATIVE IMPROVEMENT OVER THE CASE WITHOUT MICHROCHANNEL SHIELDS IS RPEORTED IN PARENTHESES.

Q		T			
		2	3	4	5
N	1	10.96 (5.88%)	13.12 (14.5%)	14.53 (38.9%)	14.90 (70.3%)
	2	11.36 (11.7%)	11.31 (78.52%)	7.59 (168%)	6.14 (359%)
	3	11.89 (26.3%)	9.15 (167%)	4.65 (406%)	2.00 (2034%)
	4	11.89 (42.4%)	7.46 (269%)	2.93 (1007%)	1.03 N/A
	5	11.37 (55.3%)	5.97 (371%)	1.87 N/A	0.19 N/A
	6	10.98 (71%)	4.74 (483%)	1.01 N/A	-2.08 N/A

TABLE V. L VS. NUMBER OF TURNS (N) AND NUMBER OF TIERS (T) FOR MAXIMUM MICRO- CHANNEL DIMENIONS (MEASURED AT 10 GHZ). THE RELATIVE IMPROVEMENT OVER THE CASE WITHOUT MICHROCHANNEL SHIELDS IS RPEORTED IN PARENTHESES.

L (nH)		T			
		2	3	4	5
N	1	0.135 (0.0%)	0.344 (0.0%)	0.577 (0.0%)	0.828 (1.2%)
	2	0.344 (0.0%)	0.958 (0.0%)	1.729 (0.0%)	2.708 (11.9%)
	3	0.594 (0.0%)	1.700 (0.0%)	3.523 (39.9%)	5.882 (1615%)
	4	0.843 (0.0%)	2.741 (1.0%)	5.959 (424%)	8.855 N/A
	5	1.093 (0.0%)	3.390 (2.2%)	8.577 N/A	3.055 N/A
	6	1.376 (0.0%)	5.206 (58.5%)	10.634 N/A	-3.518 N/A

TABLE VI. Q VS. NUMBER OF TURNS (N) AND NUMBER OF TIERS (T) FOR MAXIMUM MICRO- CHANNEL DIMENSIONS (MEASURED AT 1 GHZ). THE RELATIVE IMPROVEMENT OVER THE CASE WITHOUT MICHROCHANNEL SHIELDS IS RPEORTED IN PARENTHESES.

Q		T			
		2	3	4	5
N	1	3.03 (0.0%)	3.74 (0.5%)	4.15 (1.0%)	4.44 (2.0%)
	2	3.36 (0.0%)	4.72 (3.2%)	5.37 (4.8%)	5.90 (9.9%)
	3	3.76 (0.2%)	5.28 (1.7%)	6.39 (15.8%)	6.83 (36.4%)
	4	4.02 (0.8%)	6.04 (11.1%)	7.11 (37.4%)	7.57 (88.3%)
	5	4.13 (0.1%)	6.49 (17.8%)	7.65 (67.0%)	7.78 (153.0%)
	6	4.29 (2.7%)	6.81 (26.5%)	7.92 (98.5%)	8.01 (235.1%)

IV. Conclusions and Future Work

In this paper, we have systematically examined how various parameters affect their performance. In addition, we have proposed a novel shield mechanism utilizing the micro-channel technique to drastically improve the quality factor and the inductance. To the best of the authors' knowledge, this is the very first in-depth study on TSV inductors along with a technique to make them practical. In the future, we will try to implement benchmark applications such as on-chip voltage regulators and transceivers using the proposed TSV inductors.

References

[1] Franzon, P.D.; Davis, W.R.; Thorolffson, T.; , "Creating 3D specific systems: Architecture, design and CAD," in Proc.Design, Automation & Test in Europe Conference & Exhibition (DATE), pp.1684-1688, 8-12 March 2010.

[2] 2011 ITRS roadmap, http://www.itrs.net/Links/2011ITRS /2011Chapters/2011Metrology.pdf

[3] Dae Hyun Kim et al; , "3D-MAPS: 3D massively parallel processor with stacked memory," Solid-State Circuits Conference Digest of Technical Papers (ISSCC), 2012 IEEE International , vol., no., pp.188-190, 19-23 Feb. 2012.

[4] Gary VanAckern, "Design Guide for CMOS Process On-Chip 3D Inductor Using Thru-Wafer Vias," Master Thesis, 2011

[5] Zhang Bo et al., "3D TSV Transformer Design for DC-DC/ACDC Converter", 60th Electronic Components and

Technology Conference (ECTC), pp. 1653 - 1656, Jun 2010.

[6] Huang et al, "Interleaved Three-Dimensional On-Chip Differential Inductors and Transformers," US Patent 2008/0272875, Nov. 2008.

[7] Bontzios, Y.I.; Dimopoulos, M.G.; Hatzopoulos, A.A.; , "Prospects of 3D inductors on through silicon vias processes for 3D ICs," VLSI and System-on-Chip (VLSI-SoC), 2011, IEEE/IFIP 19th International Conference on , vol., no., pp.90-93, 3-5 Oct. 2011

[8] Loi, I.; Angiolini, F.; Fujita, S.; Mitra, S.; Benini, L.; , "Characterization and Implementation of Fault-Tolerant Vertical Links for 3-D Networks-on-Chip," Computer-Aided Design of Integrated Circuits and Systems, IEEE Transactions on , vol.30, no.1, pp.124-134, Jan. 2011

[9] HFSS websit:"http://www.ansys.com/

[10] C. Pattrick Yue and S. Simon Wong, "On-Chip Spiral Inductors with Patterned Ground Shields for Si-Based RF IC's," IEEE Journal of Solid-State Circuits, vol. 33, no. 5, pp. 743-752, May 1998.

[11] J. Zhang, "Indcutor with Patterned Ground Plane," US Patent 2009/0250262, 2008

[12] Yoon Jo Kim et al, "Thermal Characterization of Interlayer Microfluidic Cooling of Three-Dimensional Integrated Circuits With Nonuniform Heat Flux," Journal of Heat Transfer-transactions of The Asme - J HEAT TRANSFER , vol. 132, no. 4, 2010

[13] Bing Shi and Ankur Srivastava, "TSV-constrained micro-channel infrastructure design for cooling stacked 3D-ICs," ACM International Symposium on Physical Design, pp. 113-118, 2012

[14] http://gradient-da.com

[15] K. Gantz and M. Agah, "Predictable three-dimensional microfluidic channel fabrication in a single-mask process," Technical Digest of the 14th International Conference on Solid-State Sensors, Actuators, and Microsystems (Transducers07), pp. 755-758, June 10-14, 2007

[16] D. Sekar et al, "A 3D IC Technology with Integrated Microchannel Cooling," Interconnect Technology Conference, IITC 2008.

[17] Z. Tao, W. Kui, F. Yi, C. Yan, L. Qun, S. Bing, X. Jing, S. Xiaodi, D. Lian, X. Yuan, C. Xu and L. Youn-Long," A 3D SoC design for H.264 application with on-chip DRAM stacking," 3D Systems Integration Conference (3DIC), 2010 IEEE International, 2010.

[18] D. S. Gardner, G. Schrom, F. Paillet, T. Karnik and S. Borkar, "Review of on-chip inductor structures with magnetic films," IEEE Trans. Magnetics, 45 pp. 4760-4766, 2009

[19] Xinhai Bian et al," Simulation and modeling of wafer level silicon-base spiral inductor," Electronic Packaging Technology and High Density Packaging (ICEPT-HDP), 13th International Conference, 13-16 Aug, 2012

Design and Control Methodology for Fine Grain Power Gating based on Energy Characterization and Code Profiling of Microprocessors

Kimiyoshi Usami[1], Masaru Kudo[1], Kensaku Matsunaga[1], Tsubasa Kosaka[1], Yoshihiro Tsurui[1],
Weihan Wang[2], Hideharu Amano[2], Hiroaki Kobayashi[3], Ryuichi Sakamoto[3],
Mitaro Namiki[3], Masaaki Kondo[4] and Hiroshi Nakamura[5]

[1]Shibaura Institute of Technology, 3-7-5 Toyosu, Koto-ku, Tokyo 135-8548, Japan
[2]Keio University, 3-14-1 Hiyoshi, Kouhoku-ku, Yokohama, Kanagawa 223-8522, Japan
[3]Tokyo University of Agriculture and Technology, 2-24-16 Naka-cho, Koganei-shi, Tokyo 184-8588, Japan
[4]The University of Electro-Communications, 1-5-1 Chofugaoka, Chofu-shi, Tokyo 182-8585, Japan
[5]The University of Tokyo, 7-3-1, Hongo, Bunkyo-ku, Tokyo 113-8656, Japan

This paper presents a design and control scheme of a microprocessor whose internal function units are power gated at instruction-by-instruction basis. Enabling/disabling the power gating is adaptively controlled under the support of on-chip leakage monitors and the operating system to minimize energy overhead due to sleep-in and wakeup. Measured results of the fabricated chip in the 65nm CMOS technology demonstrated that our approach reduces energy to 21-35% in the range of 25-85°C as compared to the non power-gated case. Energy dissipation was reduced by up to 15% as compared to the conventional fine-grain power gating technique in the same temperature range.

I. Introduction

Leakage power is a primary concern in modern devices from high-end microprocessors down to ultra low power sensors. In high-performance processors [1, 2], it is reported that leakage power occupies 20-30% of the total power. It is also reported that leakage power becomes more dominant in microprocessors in near-threshold voltage operation because dynamic power reduces drastically [3]. As a technique to reduce leakage, power gating is a promising approach and hence employed in a commercial high-end microprocessor [4] and an SoC for portable applications [5]. So far, power gating has been adopted in various spatial and temporal granularities, as summarized in Fig. 1. Most of today's microprocessors or system LSIs employ core-level or IP block-level power gating with the temporal granularity of tens of microsecond to one millisecond of powered-off period. Thus, power gating is typically applied for standby period of the cores.

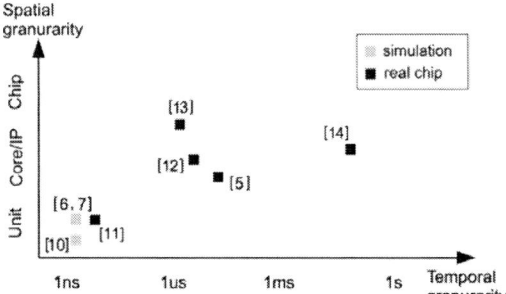

Fig. 1. Temporal and spatial granularity of PG.

As leakage power becomes a major contributor, leakage saving is required not only for the standby period but also for the active period of the core. Fine-grain power gating (FGPG) has emerged as a technique to power gate the internal logic blocks of a processor in a fine-grained manner while running the entire processor [6, 7]. However, since the powered-off period is short in FGPG, energy dissipated at the power-off and power-on becomes a critical overhead in energy saving. Unless leakage energy saved during the powered-off period exceeds the overhead energy, the total energy is not reduced or even increased by powering off. The minimum powered-off time to gain in energy savings is referred to as the break-even time (BET). In [6], the authors proposed a Time-Based (TB) sleep policy to power gate function units based on BET. In this policy, when an idle event is detected, a counter starts to count idle cycles. When the counted value reaches BET, the power switch of the function unit is turned off. This approach suppresses the power gating for the idleness shorter than BET to avoid wasting energy. To perform this, however, a counter circuit to count idle cycles, a comparator to compare the counted number with BET, and a control circuit are required. Since these circuits operate at every clock cycle after detecting the idleness, they consume dynamic power, resulting in energy overhead. Papers [8, 9] proposed approaches to improve efficiency of leakage reduction in the TB policy. However, they still use the idle-cycle counters and hence are accompanied by the energy overhead.

In contrast to the TB policy, another sleep policy named WIPS (Whenever Idle Put to Sleep) is employed in [7, 11]. In this policy, function units such as ALU, shifter, multiplier and divider in a microprocessor are powered off as default. Only when the function unit is used, it is powered on. After finishing the execution, it is powered off again. Compared to the TB policy, the WIPS approach does not require an idle-cycle counter, the comparator or the control circuit, resulting in not incurring energy overhead due to those circuits. Meanwhile, WIPS has a drawback that it powers off the function unit at the short idleness that does not lead to energy saving (or even leads to energy increase). How to suppress this without using idle-cycle counters has not been solved yet.

Contributions of this paper can be summarized as follows:
(1) We demonstrated that the advantage of the TB policy is dismissed due to energy overhead of the control circuit, as compared to the WIPS policy.

(2) We proposed an Adaptive WIPS (A-WIPS) policy to suppress ineffective power gating of the WIPS policy.

(3) We demonstrated effectiveness of the A-WIPS policy in energy savings at the operating test chip.

The rest of this paper is organized as follows. Section II describes the conventional TB and WIPS policies and discusses inherent problems. Section III describes the proposed approach. Section IV presents the results.

II. Conventional FGPG techniques and problems

When improving the FGPG control for FUs, it is essential to know the characteristics of idleness. We analyzed the length of idleness and its occurrence at FUs in a 32-bit MIPS based microprocessor by simulation. We investigated five different application programs [15] such as quick sort algorithm (qsort), Discrete-Cosine-Transform (DCT), JPEG encoding (JPEG), security algorithm (blowfish) and Cyclic Redundancy Check for data transmission (CRC). Results are depicted in Fig.2. Here, we define the long sleep as the idleness whose idle time is longer than BET and the short sleep vice versa. It should be noted that the long sleep does not exist in ALU irrespectively of the temperature or application programs. ALU is in the active state for 80-90% of the entire execution time. The remaining 10-20% is in the idle state but idleness is shorter than BET. On the contrary, the divider is inactive for almost all the time because it is rarely used. In the JPEG program, the divider is used but the time is only 1%. Thus, in the divider, all idleness is the long sleep. In contrast, in the shifter and multiplier, the long and short sleeps are mixed and their share changes with the temperature and applications.

From this observation, the best way to get gain in energy savings is to always disable PG for ALU and to enable PG for the divider. The remaining issue is how to control the shifter and the multiplier. To study this, we actually designed control circuits for the TB policy and compared energy dissipation between the TB and WIPS policies. The control circuit for the TB policy contains a counter and a comparator. The counter starts counting when the idleness is detected and counts up at every clock cycle. The counter's output is compared with BET of FU and if they are equal we stop

counting. The control circuit is provided for each FU. We created two designs for the control circuit: one is the original design and the other is with low power techniques such as clock gating and frequency division. We stop the clock of the counter by clock gating. Additionally, we lower the clock frequency of the counter to half of the CPU's by using a frequency divider. Although dynamic power can be reduced more by further lowering the frequency of the counter, it sacrifices the accuracy of BET detection. Control circuits were designed in RTL, synthesized using a commercial 65nm cell library and simulated for evaluation. Fig.3(a) and (b) show the results.

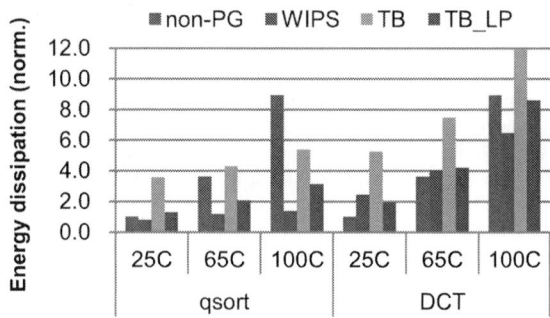

Fig. 3(a). Comparisons of energy dissipation for the shifter. (Energy is normalized to that of non-PG at 25°C.)

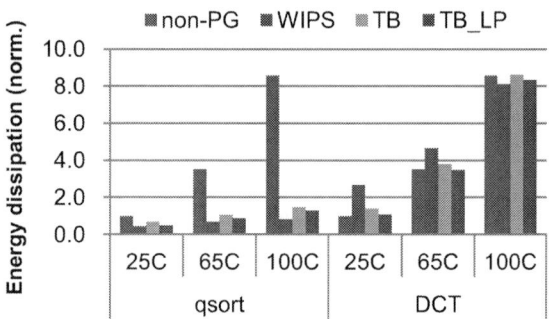

Fig. 3(b). Comparisons of energy dissipation for the multiplier. (Energy is normalized to that of non-PG at 25°C.)

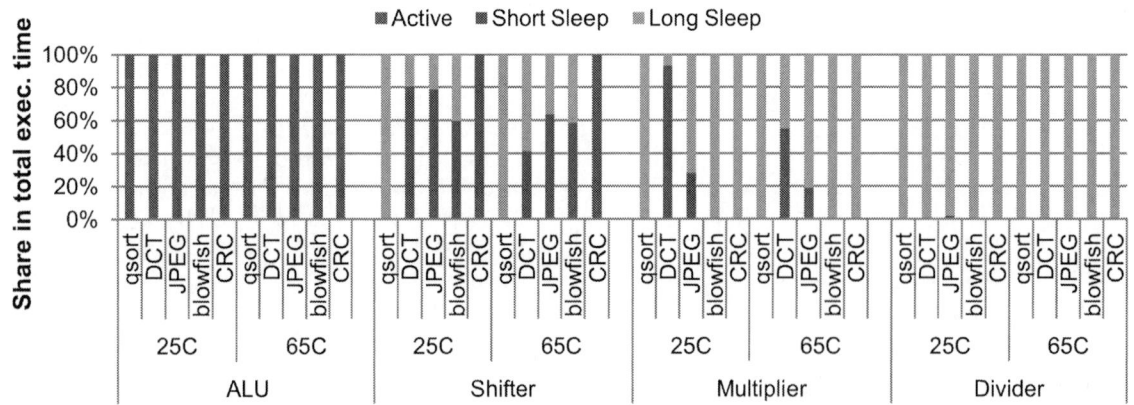

Fig. 2. Share of the active state, the short sleep, and the long sleep in the total execution time for different application programs.

978-1-4799-2817-0/14 $31.00 © 2014 IEEE

For the shifter, it is demonstrated that the TB policy does not reduce energy but instead increases it as compared to the non-PG at 25°C and 65°C for qsort, and at all temperatures for DCT. It was found that the TB control circuit consumes almost comparable energy to the function unit itself at the maximum. Low power techniques to the control circuit reduced energy by 42-63%, and thereby the TB with low power technique (TB_LP) achieves lower energy than the non-PG at 65°C and 100°C for qsort, and 100°C for DCT. However, energy is still larger at 25°C for qsort, and 25°C and 65°C for DCT as compared to the non-PG. In contrast, the WIPS policy successfully reduces energy over the non-PG for qsort, while it fails for DCT at 25°C and 65°C.

For the multiplier, both TB and TB_LP reduce energy against the non-PG but dissipate more energy than the WIPS policy at all temperatures for qsort. For DCT, the WIPS policy consumes the largest energy at 25°C and 65°C. This is because the short sleeps occur very frequently as shown in Fig.2. In contrast, at 100°C the WIPS policy achieves the minimum energy. Since sub-threshold leakage drastically increases with the temperature, the WIPS policy to power-off instantly becomes more effective at higher temperature. These observations motivated us to develop a new approach to control the WIPS policy that suppresses the wasteful power-off without adding a power-consuming controller.

III. Adaptive Fine Grain Power Gating

To get gain in energy savings in FGPG considering the energy overhead, we propose Adaptive Fine Grain Power Gating (A-FGPG) that does not use an idle-cycle counter or comparator. In A-FGPG, we switch the mode between PG-enabled and PG-disabled for each FU depending on application programs and the temperature. We provided with control registers in the processor to disable PG for each FU and made them changeable by the operating system (OS). The key components of A-FGPG are the following:
- A microprocessor with function units for which FGPG can be enabled/disabled by on-chip control registers
- An on-chip leakage monitor to measure temperature-dependent leakage current
- Operating system to activate the leakage monitor, reads the result and change the value in the control registers
- Adaptive WIPS (A-WIPS) sleep policy based on energy characterization and code profiling, under which the operating system switches enabling/disabling FGPG for each function unit.

The novelty of this paper is a proposal of the A-WIPS sleep policy and the demonstration of energy reduction at the fabricated real chip.

A. Design of a microprocessor with FGPG

We designed a microprocessor, *Geyser-3*, based on the 32-bit MIPS architecture with a 5-stage pipeline and 8KB caches for instruction and data. Implementation was performed in the Fujitsu 65nm CMOS technology. FUs such as ALU, shifter, multiplier and divider are independently power gated at instruction-by-instruction basis. The power supply for FUs is separated from others for precise measurement. The FU to be woken up is detected by pre-decoding an instruction at the instruction-fetch (IF) stage, as presented in [7]. The unit is fully powered on during the instruction-decode (ID) stage before the operation at the execution (EX) stage without causing pipeline stalls. To power on the FU at one cycle (i.e. 5ns at 200MHz) while suppressing the ground bounce, we divided the power switch into small pieces and turned them on with small time differences. An unbalanced buffer tree structure to drive the power switches achieved 3ns wakeup time at 45mV ground bounce.

The power switches are nMOS footer sleep transistors. We used the triple-well process to separate the p-well of the power switches and that of nMOS transistors of FU. The source nodes of nMOS transistors of FU are connected to the drain nodes of the power switches by the virtual ground line. The p-well of FU is connected to the virtual ground line as well. The layout of the power gated multiplier is shown in Fig.4. The body of the multiplier is located in the center and is surrounded by power switches and buffers to drive them.

Fig.4. Layout of the power-gated multiplier.

B. On-chip leakage monitor

We provided with an on-chip leakage monitor in the microprocessor. Our leakage monitor outputs the number which changes with the temperature. The circuits to monitor pMOS leakage and nMOS leakage are depicted in Fig.5(a) and (b), respectively. The basic structure is identical to that in [16] but we designed the circuit such that the leakage monitors can communicate with OS. When OS activates both monitors, the enable signal EN_{SA} becomes '1', whereas EN_{LS} once becomes '1' and changes to '0'. In the pMOS leakage monitor, the VGND node is gradually charged up due to leakage current in the leakage sensor. At higher temperature, charging up is faster because of larger leakage. The number of clock cycles it takes until the VGND voltage reaches V_{REF} is measured using a counter and the result is reported as the monitor output. In the nMOS monitor, discharging time of the VVDD node is measured. By averaging the reported results of the pMOS/nMOS monitors, OS identifies the temperature. Area of the leakage monitor is $4057\mu m^2$, which is only 4% of the total area of FUs.

978-1-4799-2817-0/14 $31.00 © 2014 IEEE

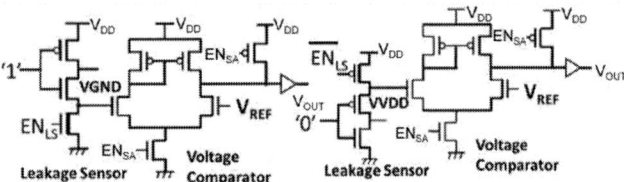

Fig. 5(a). pMOS leakage monitor, (b). nMOS leakage monitor.

OS activates the leakage monitor, reads the result, and determines if it should change the control bit to disable PG. Since the leakage at lower temperature is not so large, FGPG could lead to energy wasting. To suppress this, OS disables PG if the temperature detected by the monitor is below the pre-characterized number. Notice that this procedure should be done only at every few milliseconds because the chip temperature varies in time on that scale [17]. Hence, energy overhead due to this procedure is negligible.

C. Energy characterization and code profiling

The OS disables FGPG if the temperature detected by the leakage monitor is below the pre-determined value. To obtain this number, we performed energy characterization for each FU and code profiling for application programs. In energy characterization, we conducted circuit simulations for extracted parasitics from the layout. Energy dissipation of each FU at enabling/disabling PG was analyzed for various idle times and temperatures. We define the saved energy as

$$E_{SAVED} = E_{PG\_DISABLED} - E_{PG\_ENABLED}. \qquad (1)$$

Fig.6 shows the results for the multiplier.

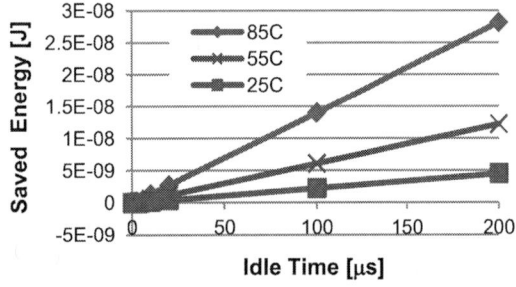

Fig. 6. Results from energy characterization for multiplier.

The saved energy increases almost linearly with the idle time. However, for shorter idle times than 10μs the saved energy changes nonlinearly. This is because the transient-glitch energy [18] at the wakeup becomes visible.

Code profiling was performed to analyze how often idle events occurred and how long each idle event continued at each FU for a target program. Combining the results from energy characterization and code profiling, we compute the total saved energy for the target program at various temperatures using the following equation:

$$E_{tot\_saved} = \sum_i E_{SAVED,i} \times n_i \qquad (2)$$

where $E_{SAVED,i}$ is the saved energy for idle cycles i and n_i is the number of occurrences of idle cycles i. For experiment, we conducted code profiling for five application programs summarized in Table 1 and computed the total saved energy at 25°C, 55°C and 85°C.

Table 1. Application programs used at the experiment.

Program	Description
matrix	Computes inner products for two 40x40 matrices
qsort	Executes quick sort for 500 elements for 200 times
Dhrystone	Executes Dhrystone 1.1 for 100K times
bitcount	Counts numb. of bits of "1" in 32b data, 100K times
blowfish	Executes blowfish encryption for 40byte rand. data

Fig.7 shows the result of the computed total saved energy for the program "matrix". This indicates that we should disable PG of the multiplier under 63°C for this program due to the negative gain. In contrast, the results for other programs showed that the total saved energy is the positive number across all temperatures. Based on this, we enable PG for the multiplier at all temperatures for them. We perform the same procedure for other FUs as well.

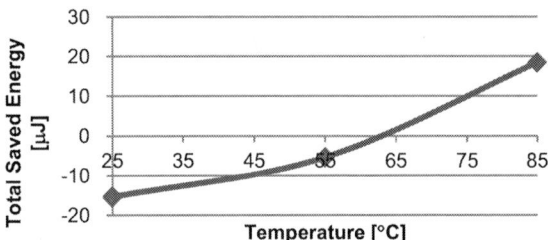

Fig. 7. Computed total saved energy for multiplier at "matrix".

Based on the computed total saved energy, we make a decision whether we should enable or disable PG for each FU and each temperature. For above application programs, the decision made for ALU was to disable PG for all temperatures. In contrast, the decisions made for the divider was just the opposite. Decisions made for the shifter and the multiplier are summarized in Table 2. In the table, "D" and "E" indicate the disable-PG and enable-PG, respectively.

Table 2. Decisions based on the computed total saved energy. "mat", "qs", "Dh", "bit" and "bf" stand for matrix, qsort, Dhrystone, bitcount and blowfish, respectively. D and E stand for disable-PG and enable-PG, respectively.

	Shifter					Multiplier				
	mat	qs	Dh	bit	bf	mat	qs	Dh	bit	bf
25°C	D	D	D	D	D	D	E	E	E	E
55°C	D	D	E	D	D	D	E	E	E	E
85°C	D	D	E	D	E	E	E	E	E	E

IV. Results

The photo of the fabricated chip is shown in Fig.8. Table3 summarizes the feature of the chip. We also implemented OS and confirmed that it successfully starts the pMOS/nMOS leakage monitors, reads the results and switches the control bit to enable/disable PG accordingly.

978-1-4799-2817-0/14 $31.00 © 2014 IEEE

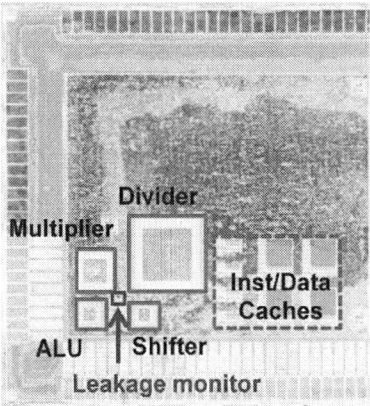

Fig.8. Photo of the fabricated microprocessor chip, *Geyser-3*.

Table 3. Feature of the fabricated chip

Process	Fujitsu 65nm CMOS, 12metals
Area of function units [um]	ALU: 121.4 x 113.4 Shifter: 116.4 x 114.8 Multiplier: 199.4 x 199.4 Divider: 369.0 x 368.6 Total: 1610 x 1443
V_{DD}	1.2 [V]
V_{REF} of leak. monitor	0.6 [V]
Clock frequency	200MHz (max.)
L1 cache	Inst. : 8KB, 64B-line, 2way Data : 8KB, 64B-line, 2way
Synthesis	Synopsys Design Compiler
Layout	Synopsys ICC

We conducted measurements using the chip shown above. Fig.9 shows a photo of our evaluation environment.

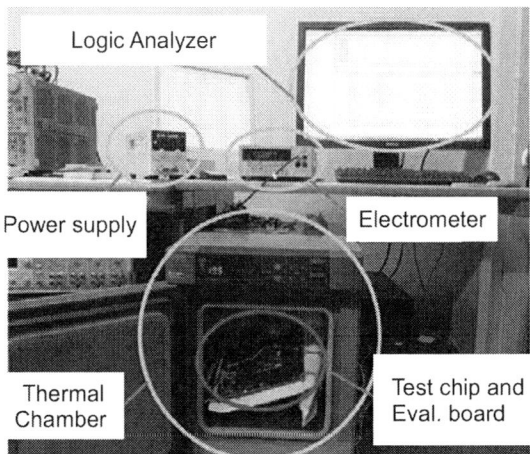

Fig. 9. Photo of our evaluation environment.

To evaluate the chip at various temperatures, we put the chip into a thermal chamber and measured current dissipation of the chip. There are two independent power supply domains in the chip; one is for four FUs and the other is for the entire chip except the FUs. At the evaluation of current dissipation, we only measured the current flowing into FUs. Although the FGPG core itself can operate at the clock frequency up to 200MHz, we used 50MHz core clock frequency at the evaluation because the I/O bus connecting to the evaluation board cannot be operated at high temperature at high clock frequency. Fig.10 shows the measured results of the pMOS/nMOS leakage monitors. The average of both monitor outputs is stored in a look-up table (LUT) and OS can identify the temperature by referring to LUT.

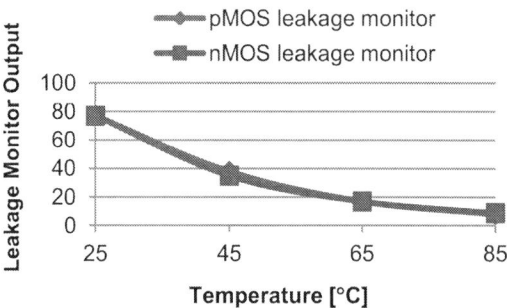

Fig. 10. Measured results for leakage monitors.

Fig.11(a), (b) and (c) show the measured total current dissipation of the FUs for the application programs summarized in Table 1 at 25°C, 55°C and 85°C, respectively. The bars marked as "proposed" are the results from our A-WIPS policy. As compared to the WIPS policy [7], the proposed scheme reduced energy dissipation of the FUs by 5-15% at 25°C, by 6-11% at 55°C and by 2-6% at 85°C. As compared to non-PG, energy is reduced to 21-35% by our approach.

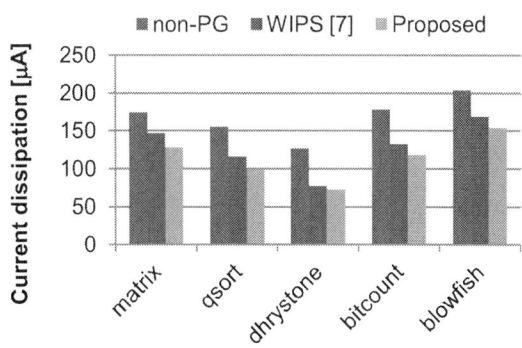

Fig. 11(a). Measured current dissipations for function units at 25°C.

978-1-4799-2817-0/14 $31.00 © 2014 IEEE   847

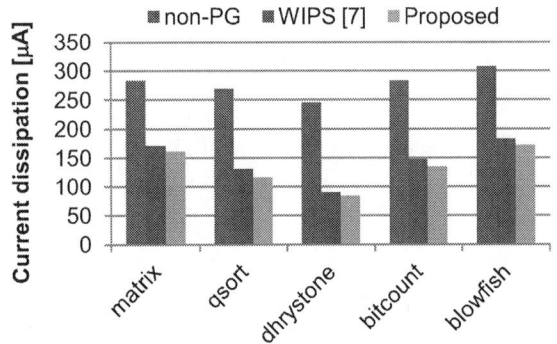

Fig.11(b). Measured current dissipations for function units at 55°C.

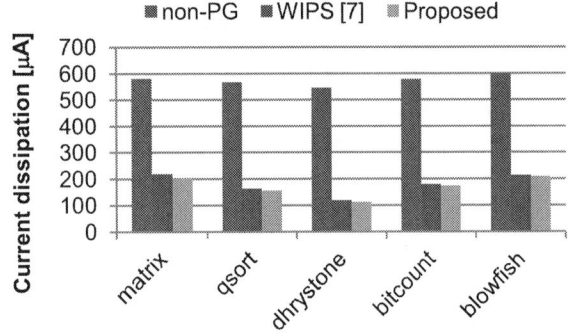

Fig. 11(c). Measured current dissipations for function units at 85°C.

V. Conclusions and future work

We designed and implemented an embedded processor in which internal function units are power gated. We proposed an Adaptive Fine Grain Power Gating technique to control enabling/disabling PG based on energy characterization and code profiling under the support of OS and leakage monitors. Measured results demonstrated effectiveness of the proposed approach under the temperature variation. As a future work, we will study the influence of the process variation.

Acknowledgments

This work was partially supported by JSPS KAKENHI S Grant Number 25220002. In addition, this work was supported by VLSI Design and Education Center (VDEC), the University of Tokyo in collaboration with Synopsys, Inc, Cadence Design Systems, Inc. and Mentor Graphics, Inc.

References

[1] J. Hart et al., "3.6GHz 16-core SPARC SoC processor in 28nm," Proc. 2013 IEEE International Solid-State Circuits Conference (in presentation slide), pp. 48-49, 2013.

[2] W. Hu et al., "Godson-3b1500: A 32nm 1.35GHz 40W 172.8Gflops 8-core processor," Proc. 2013 IEEE International Solid-State Circuits Conference, pp. 54-55, 2013.

[3] S. Jain et al., "A 280mV-to-1.2V wide-operating-range IA-32 processor in 32nm CMOS," Proc. 2012 IEEE International Solid-State Circuits Conference, pp. 66-68, 2012.

[4] R. Kumar and G. Hinton, "A family of 45nm IA processors," Proc. 2009 IEEE International Solid-State Circuits Conference, pp. 58-59, 2009.

[5] Y. Kanno et al., "Hierarchical power distribution with 20 power domains in 90-nm low-power multi-CPU processor,"" Proc. 2006 IEEE International Solid-State Circuits Conference, pp. 540-541, 2006.

[6] Z. Hu et al., "Microarchitectural techniques for power gating of execution units," Proc. the 2004 International Symposium on Low Power Electronics and Design, pp. 32-37, 2004.

[7] N. Seki et al., "A fine grain dynamic sleep control scheme in MIPS r3000," Proc. 26th IEEE International Conference on Computer Design, pp. 612-617, 2008.

[8] A. Youssef, et al, "Dynamic standby prediction for leakage tolerant microprocessor functional units," Proc. IEEE/ACM International Symposium on Microarchitecture, pp. 371-384, 2006.

[9] A. Lungu, et al, "Dynamic power gating with quality guarantees," Proc. the 14th International Symposium on Low Power Electronics and Design, pp. 377-382, 2009.

[10] S. Dropsho et al., "Managing static leakage energy in microprocessor functional units," Proc. 35th Annual IEEE/ACM International Symposium on Microarchitecture, pp. 321-332, 2002.

[11] D. Ikebuchi et al., "Geyser-1: A MIPS R3000 CPU core with fine grain runtime power gating," Proc. the 2009 IEEE Asian Solid-State Circuits Conference, pp. 281-284, 2009.

[12] J. Koppanalil et al., "A 1.6 GHz dual-core ARM Cortex A9 implementation on a low power high-k metal gate 32nm process," Proc. 2011 International Symposium on VLSI Design, Automation and Test, pp. 1-4, 2011.

[13] Atmel SAM4L Series Summary, Atmel, 2010.

[14] S. Sawant et al., "A 32nm Westmere-EX Xeon enterprise processor," Proc. 2011 IEEE International Solid-State Circuits Conference, pp. 74-75, 2011.

[15] M. Guthaus et al., "Mibench: A free, commercially representative embedded benchmark suite," Proc. 2001 International Workshop on Workload Characterization, pp. 3-14, Dec. 2001.

[16] S. Koyama, et al, "Design and Analysis of On-chip Leakage Monitor using an MTCMOS circuit," The 23rd International Technical Conference on Circuits/Systems, Computers and Communications, pp. 205-208, Jul. 2008.

[17] N. Weste and D. Harris, *CMOS VLSI Design*, 4th ed., Addison-Wesley, p.243, 2011.

[18] K. Usami, et al, "Adaptive Power Gating for Function Units in a Microprocessor," IEEE International Symposium on Quality Electronic Design, pp. 29-37, March 2010.

978-1-4799-2817-0/14 $31.00 © 2014 IEEE

A Hybrid Random Walk Algorithm for 3-D Thermal Analysis of Integrated Circuits

Yuan Liang

Department of Computer Science
and Technology,
Tsinghua University,
Beijing 100084, China
e-mail: liangyuan10@mails.tsinghua.edu.cn

Wenjian Yu[*]

Department of Computer Science
and Technology,
Tsinghua University,
Beijing 100084, China
e-mail: yu-wj@tsinghua.edu.cn

Haifeng Qian

IBM T. J. Watson Research
Center,
Yorktown Heights,
NY 10598, USA
e-mail: qianhaifeng@us.ibm.com

Abstract − In this work, a hybrid random walk method is proposed for the thermal analysis of integrated circuits. Preserving the advantage of generic random walk method (GRW), i.e. the suitability for simulating local hot-spots, the proposed techniques largely reduce its runtime for accurate high-resolution simulation, and is suitable for the realistic pyramid-shape IC model. This is achieved by combining the GRW and the floating random walk techniques, and a novel usage of rectangular cuboid transition domain. The techniques to handle the Neumann boundary and convective boundary in thermal simulation are also discussed. Numerical experiments on several IC test cases validate the efficiency and accuracy of the proposed techniques, and demonstrate more than 100X speedup over the GRW method.

I. Introduction

The continuous scaling trend of the CMOS technology has led to the drastic increase of the number of devices in integrated circuit (IC) and the ratio of interconnect delay to device delay. The heat dissipation has become a problem that threatens circuit reliability and performance [1]. Specifically, accurate and efficient chip-level thermal analysis is also indispensable for many design-time circuit optimizations.

Some thermal analysis algorithms have been proposed for chip-level analysis, such as the geometric multigrid solver [2], Green's function based fast algorithm [3], and the preconditioned conjugate gradient (PCG) algorithm [4]. The volume discretization has been employed to transform the problem into a linear equation system, and the temperature profile of the whole simulated domain is solved in most existing works. However, the entire temperature profile is not really required, because what we want is often the temperature of hot-spots at IC device layer.

Different from the deterministic methods solving linear equations, another kind of method for thermal analysis is the random walk method. Generally speaking, the random walk method is most efficient when point values or linear functionals of the solution are needed. Therefore, it should be suitable for the thermal analysis where the temperatures of some target hot-spots are needed. The random walk algorithms have been proposed for power grid analysis [5-7]. Based on the similarity of thermal analysis and power grid

analysis, the random walk techniques in [5] was applied to the problems of thermal via planning [8] and thermal analysis [17]. In [6], the importance sampling technique was proposed to accelerate the convergence rate of the random walk based power grid analysis. Besides, the random walk algorithms have also been successfully applied to capacitance extraction [9, 10, 18].

A major limitation of many existing works on thermal analysis [2, 3, 8] is that they consider a rectangular simulation domain, with a simplified boundary assumption (often the Dirichlet condition) accounting for the effect of heat dissipation. In practice, the heat spreader and heat sink attached to the die are much wider than IC die (see Fig. 1). By approximating the whole thermal system with a single rectangular domain, substantial error (up to tens of degrees in temperature) may be introduced [4]. In a recent work [4], a preconditioned conjugate gradient (PCG) algorithm was proposed for the realistic pyramid IC model, which takes the solution of a larger and approximate rectangular domain by fast Poisson solver as the preconditioner.

In this work, we propose a fast random walk algorithm for thermal analysis. The pyramid-shape IC model is considered. We focus on the application where only the temperatures of target hot-spots are needed, such as that in the floorplanning stage. The proposed method takes advantages of generic random walk (GRW) and floating random walk (FRW), to reduce the length of each walk. Employing pre-characterization techniques to handle the Neumann boundary and convective boundary, we further accelerate the hybrid random walk method. Compared with the generic random walk algorithm [5, 8], the proposed method achieves several orders of magnitude speedup while preserving high accuracy.

II. Background

A. 3-D Thermal Model

In the pyramid-shape IC model (see Fig. 1), the IC region mainly includes two parts: silicon substrate and the interconnect layer. The former is made of silicon, while the latter is filled with metal and dielectrics. To simplify the

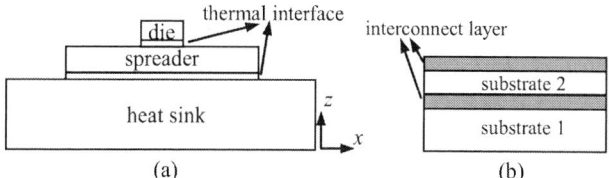

Fig. 1. The pyramid-shape IC geometry for thermal simulation: (a) the side view, (b) the details of a IC with two tiers of dies.

This work was supported in part by NSFC under Grant 61076034, Tsinghua University Initiative Scientific Research Program, and Fundamental Research Funds for the Central Universities under Grant 2011JBZ002.
*W. Yu is the corresponding author.

978-1-4799-2817-0/14 $31.00 © 2014 IEEE

discussion, we consider the interconnect layer as a homogeneous layer with an effective thermal conductivity calculated using the percentage of metal volume. This is typical in most previous works. Note that there is no problem for the random walk method to handle the model with lateral variation of thermal conductivity as considered in [8]. Similarly, we ignore the heat dissipation through packaging and board in this work, since that is a minor dissipation channel.

The steady-state thermal analysis involves solving the temperature profile $T(x,y,z)$ from the 3-D Poisson equation:

$$k\left[\frac{\partial^2 T(x,y,z)}{\partial x^2}+\frac{\partial^2 T(x,y,z)}{\partial y^2}+\frac{\partial^2 T(x,y,z)}{\partial z^2}\right]=-p(x,y,z), \quad (1)$$

where k is thermal conductivity and $p(x,y,z)$ is the internal heat generation density at point (x, y, z). The heat generation is due to the device modules, or function blocks located around the top surface of the silicon die. Eq. (1) holds for a homogeneous region. For a problem with multiple homogeneous regions, the equation of continuous heat flux should be applied at the interface between two regions.

The finite volume method (FVM) is conventionally used for 3-D thermal simulation, where the domain is discretized into cells and each cell is associated with a temperature [1-4]. Similar to simulating the steady-state electric current field with electric resistors, we can define and calculate thermal resistor to model the heat flow through the interface between any two adjacent cells. Fig. 2 illustrates the finite volume discretization of a 3-D rectangular domain. For two cells with different thermal conductivities or different length along the aligned direction, the thermal resistance connecting them can be calculated as the series of resistors. For example, at the interface of two homogeneous layers [see Fig. 2(c)], the across-interface thermal resistance R_{lz} is:

$$R_{lz}=\frac{1}{2k_1}\cdot\frac{h_{z1}}{h_x h_y}+\frac{1}{2k_2}\cdot\frac{h_{z2}}{h_x h_y}, \quad (2)$$

where h_x and h_y are edge sizes of cell along x-axis and y-axis, respectively. h_{z1} and h_{z2} are the heights of the two adjacent cells; k_1 and k_2 are the thermal conductivities.

The heat source resembles the current source in electric circuit. Thus, an equivalent circuit with resistors and current source is generated. With the nodal analysis approach, a linear equation system:

$$AT=f \quad (3)$$

is formed, where A is a sparse symmetric positive definite matrix, f is the vector of current sources, and T is the temperature vector. The temperature profile can be obtained by solving (3) with linear equation solvers.

There are different boundary conditions for the simulation domain. At the bottom surface of heat sink, a convective condition should be set, which models the heat transfer mechanism at the interface of heat sink and air:

$$k\frac{\partial T}{\partial \vec{n}}+h(T-T_{amb})=0, \quad (4)$$

where $\vec{n}$ is the out normal direction of the boundary, T_{amb} is the ambient temperature, and h is the convective coefficient. The partial derivative in (4) can be approximated with finite difference formula. By defining

$$R_{amb}=1/(h\cdot h_x\cdot h_y), \quad (5)$$

where h_x and h_y are the edge sizes of cell along x-axis and y-axis respectively, we can model the effect of convective boundary with the thermal resistors of value R_{amb}. They connect the nodes of boundary cells to a virtual node with temperature T_{amb}. For other boundaries of the domain, Neumann boundary (adiabatic condition) is usually assumed. It is naturally modeled by the equivalent circuit.

B. The Random Walk Method

There are mainly two kinds of random walk method: the discrete random walk (DRW) method (also called "fixed" random walk method) [5-8], and the floating random walk (FRW) method (also called "walk inside the domain" method) [9-11]. The DRW method relies on an existing mesh grid, where walk jumps from a grid node to its adjacent nodes. In the FRW method, the cubic [9, 10] or spherical [11] transition domains with variable size are employed, and each step of walk is from the center to the boundary of a transition domain. As shown in Fig. 3, a walk in DRW method usually involves larger number of steps than a walk in FRW method.

The DRW has been applied to the problems of power grid analysis and thermal analysis [5-8], which is called *generic random walk* (GRW) method. In a game of GRW, a walker starts at a node in the grid, for which the voltage/temperature is calculated. The walker then randomly visits a neighbor node. The probability of each neighbor j being chosen from node i is

$$P(i,j)=\frac{G_{ij}}{\sum_{j}^{d(i)}G_{ij}}, \quad (6)$$

where $d(i)$ is the edge degree of node i, and G_{ij} is the electric/thermal conductance between node i and its neighbor j. At each node, the walker will receive a reward of

$$T_r(j)=\frac{p_j}{\sum_{j}^{d(i)}G_{ij}}, \quad (7)$$

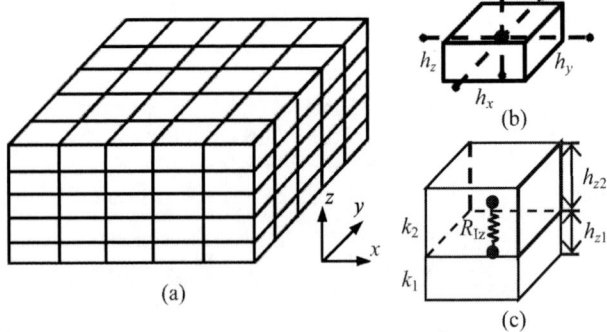

(a)

Fig. 2. A rectangular domain is discretized into grid cells, and each cell is assumed with constant temperature. (a) The discretization of a homogeneous domain. (b) A cell and six thermal resistors connecting it to its neighboring cells. (c) Two cells adjacent to the interface of two subdomains with different thermal conductivities.

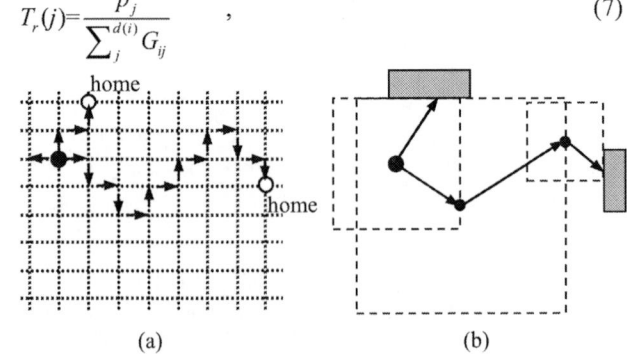

(a)   (b)

Fig. 3. Discrete random walks on a mesh grid (a), and floating random walks using cubic transition domains (b).

where p_j is the current/power injected into current node j. The walk ends if the walker hits a node with known voltage/temperatue (home node), where the walker receives the final reward of the value of voltage or temperature. The total amount of money collected by the walker is an estimation of the voltage/temperature of the starting node. Due to the central limit theorem, the average money of many walks obeys a normal probability distribution, whose variance is inversely proportional to the number of walks. This gives a tradeoff between runtime and accuracy.

Several techniques have been employed to accelerate the GRW method. One is to set the limit on path length of a random walk. If suitably set, the time of performing random walk can be reduced with negligible error [5, 8]. Another technique is setting the cell whose temperature has been calculated as a new home cell. This cuts down the average length of walks significantly, and is very useful for solving the voltage/temperature for many nodes.

Note that, R_{amb} is larger than the thermal resistance of cells for several orders of magnitude, in the thermal model with convective boundary. Therefore, the random walk needs more steps to terminate, if compared with the thermal model with the Dirichlet boundary. This brings difficulty to the computational efficiency.

To improve the efficiency of GRW, a hierarchical random walk method was proposed in [5], whose basic idea is to divide the grid into a global grid and multiple local grids, and build transition matrix for each local grid. The transition matrix contains the transition probabilities among the external nodes of a local grid. Special technique is also proposed to sparsify the transition matrix [5], for reducing the time of performing random walk.

The FRW method is suitable for the Laplace equation or Poisson equation (1) without preset mesh grid in the simulation domain. For capacitance extraction, where the Laplace equation is solved, the FRW method has large advantages over the deterministic methods for problems involving a large number of conductors [9]. To fit the Manhattan interconnect geometries, cubic transition domain is used [see Fig. 3(b)]. And, to make the walks across dielectric interface easily, finite difference techniques have been presented in [9] to pre-characterize the transition cube with two dielectric layers. For the thermal problem governed by Poisson equation (1), the power item $p(x, y, z)$ brings difficulty to FRW. Only for some special cases, such as in [11] where the power density within the whole domain is uniformed, the FRW method can be efficiently used.

In the FRW method, the probabilities from the center point to the boundary panels of transition cube are stored. This costs much less memory than the hierarchical random walk method in [5], for which the transition probabilities among any two boundary nodes are needed. Because the transition domain in FRW method is usually scalable, the characterization of it can usually be performed offline. This means shorter runtime of FRW method than the hierarchical method. Because most portion of simulation domain of a thermal analysis problem is governed by the Laplace equation (i.e. no power item), the FRW method would be applied there to enable high performance.

III. A Hybrid Random Walk Algorithm for the Thermal Analysis

In this section, a hybrid random walk method for thermal analysis is proposed. It combines the generic random walk method and the idea of floating random walk. Techniques handling Neumann boundary and convective boundary are also discussed for the reduction of runtime.

A. The Basic Idea

The advantage of FRW over GRW is due to the hop from the center of a transition domain to its boundary, which would reduce the length of walk (see Fig. 3). Our idea to improve the GRW for thermal analysis is to perform FRW in the region without power dissipation. This is illustrated in Fig. 4, where we show a walk starting from the GRW region and then behaving in the FRW manner. Therefore, we get a hybrid random walk scheme, and avoid the difficulty of FRW for handling the power item in general heat problem.

The choice of transition domain affects the efficiency of FRW. In the problem of thermal analysis, the random walk is performed in the pyramid-shape IC geometry, where the silicon die has much larger width than its height. The transition domain of cuboid shape, rather than the cubic domain for capacitance extraction [9], should be used to achieve better efficiency. This enables jumping far in lateral directions, and reduces the length of walk. And, to make the walk crossing the interface of different thermal materials, we also need the cuboid transition domain with two halves of different materials. These two kinds of transition domains are labeled with "I" and "II" in Fig. 4. Note that the FRW transition domain is bounded by the boundaries of the geometric model, and should not intersect any power dissipation region. If the walk location is approaching to these boundaries, the FRW will be degraded to the GRW.

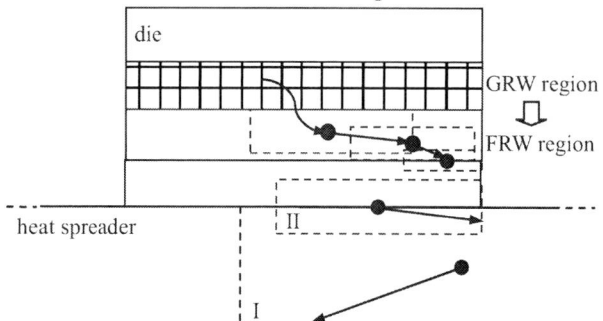

Fig. 4. A portion of the pyramid-shape IC model for the illustration of the hybrid random walk method. Several transition domains with cuboid shape are shown.

The aspect ratio of the cuboid transition domain may affect the performance of the hybrid random walk algorithm. We simply set the lateral size of the domain (length and width) to be 10× of its height. In Section IV, we will present the experimental results on test cases with different sizes, which reveal how the aspect ratio of the cuboid affects the performance.

B. Scalable Transition Domain and Its Characterization

For an arbitrary transition domain where the Laplace equation of temperature $T(\mathbf{r})$ holds, we have:

$$T_c = \oint_S G_s(r) T(r) dr \quad \cdot \qquad (8)$$

Here, T_c is the temperature at the domain's center point, and S is the boundary of the domain. $G_s(r)$ is the surface Green's function, which has non-negative value and can be regarded as the probability density function (PDF) for selecting a random point on S. If we discretize S into small panels, with $G_s(r)$ we can calculate the transition probabilities to these panels, which makes the FRW procedure feasible [9]. More importantly, these transition probabilities can be pre-calculated and recalled during the FRW procedure. This largely reduces the runtime of the FRW algorithm.

The surface Green's function $G_s(r)$ has the analytical expression only for the cubic or sphere transition domain with homogeneous material [9]. For the transition domain with other geometry or including inhomogeneous materials, we have to use numerical techniques to characterize its transition probabilities. Rather than using the finite difference techniques as in [9], we use the GRW method in this work to accomplish the characterization of transition domain. We make a dense discretization mesh on a cuboid transition domain, which corresponds to a resistance network. Then, random walks are started from the geometric center of this domain, and terminate at the domain's boundary. In the characterization, we perform M random walks, and store the numbers of their destination boundary cells in an array of length M. Because M can be view as the resolution of the probability estimation, we set it to be a large number in our experiment. While performing the FRW, for each hop we randomly pick an integer k from $[0, M)$. Then, the k-th element of the array indicates this hop's target. This approach enables much faster transitions in the FRW procedure than the conventional approach using the transition probabilities directly. The tradeoff may be the memory usage of the array for destinations, which is affordable in our problem of thermal analysis.

It should be pointed out that the transition probabilities are only relevant to the shape of the transition domain. In other words, the transition probabilities do not change if the domain is proportionally scaled. So, for an aspect ratio of the cuboid, we only need to pre-characterize a single transition domain. When performing a hop in FRW, we should firstly traverse all boundaries and power regions, and then choose the largest "safe" transition domain. By scaling the pre-characterized transition domain to this largest one, we can therefore execute the hop.

Different from the scalable FRW transition domain, another approach may be employing multiple fixed-size transition domains [5, 12]. This kind of transition domain is more generalized, and useful for handling complex medium and conductor topology. For the thermal analysis, the scalable transition domain is more efficient and costs much less memory storage than the approach using multiple fixed-size transition domains.

C. Handling the Neumann and Convective Boundaries

In the thermal model, Neumann boundary is assumed at the sidewalls. If a walk stops at the Neumann boundary, no FRW transition domain is available to continue the walk. In this case, it has to behave in the GRW manner, which definitely increases the length of walk. A remedy to this situation is the path reflection technique [13], which can be illustrated as Fig. 5(a). Neumann boundary condition is equivalent to the situation where the original domain is mirrored with respect to the boundary. In such a way, the Neumann boundary is no more considered as an obstacle to the random walk. Note that it is not necessary to enlarge the original simulated domain in the mirroring operation. In fact, if the walk overcomes the boundary jumping to an external point, its position is transformed to the corresponding mirrored point inside the domain [see Fig. 5(a)].

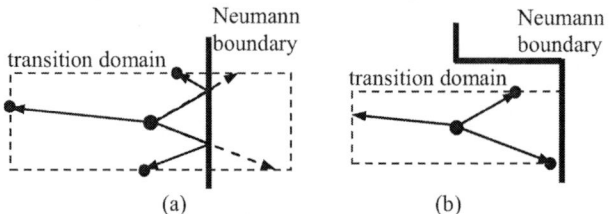

Fig. 5. Handle the Neumann boundary with the path reflection technique (a), or a special transition domain (b).

For some situations, the path reflection technique does not work. An example is shown in Fig. 5(b), where the inflated transition domain will intersect three Neumann boundary faces. In such cases, the equivalent mirrored domain for the original domain is not available. To settle this problem, we employ a special transition domain, which is restricted by Neumann boundary but avoid the walk stopping on it [see Fig. 5(b)]. This transition domain is defined as the one with a sidewall face being Neumann boundary. Using the similar GRW procedure described in last subsection, we can characterize the transition probabilities from the domain's center to the left five boundary faces. Combining the path refection technique and the Neumann-specific transition domain, the efficiency of FRW procedure is preserved.

Another special boundary in thermal analysis is the convective boundary. If a walk stops at the convective boundary, a following GRW hop can terminate the walk in certain probability. However, this probability is very small due to the large value of R_{amb}. To overcome this difficulty, we can define a convective-specific transition domain. It has the same size as the normal transition domain touching the convective boundary, but forbids the walk stopping at the convective boundary. The walk jumps to the left five boundary faces, or terminates at the virtual node with temperature T_{amb}. The transition probabilities are pre-characterized also with the GRW procedure. It is obvious that this convective-specific transition domain helps to reduce the length of walk. However, from (5) we see that R_{amb} depends on the edge sizes of discretization cell, which means the variable convective-specific transition domain cannot fit this discretization sizes. This may introduce error, and we will evaluate it through the simulation experiments.

IV. Numerical Results

To validate the efficiency of the proposed random walk techniques, several chip structures with different power profiles are tested. The random walk algorithms have been implemented in C++. While characterizing each transition domain, $M=10^6$ random walks are performed. The efficient Mersenne twister random number generator [16] is used. Unless otherwise stated, we use the cuboid transition domain with aspect ratio of 10. For the basic hybrid random walk

978-1-4799-2817-0/14 $31.00 © 2014 IEEE

implementation, two transition domains (described in Section III.A and III.B) are pre-characterized. For the treatment of Neumann boundary and convective boundary, two additional transition domains are needed.

The termination criterion of random walk algorithms is always that the 1-σ error becomes smaller than 1% of the mean value, equivalent to a relative error within 3% (in 99.7% confidence level). All experiments are carried out with serial computing on a PC with 2.70GHz dual-core Pentium CPU, 6GB memory.

A. Test Cases

The following chip structures are used in experiments.

- **Structure 1**: A four-core 2-D chip artificially generated, which is the Testcase no. 2 in [4]. The transverse dimensions of the die, spreader, and sink are 1cm×1cm, 3cm×3cm and 7cm×7cm. The thickness and material parameters are listed in Table I. The power map (totally 176W) models a scenario with one core idle, one core with peak load and two others with median loads (see Fig. 6).
- **Structure 2:** A 2-D chip imitating the POWER6 microprocessor [14]. Except that the die is of 1.6cm×2cm, the transverse dimensions are the same as Structure 1. The thickness and material parameters are listed in Table I. The power map is similar to Testcase no. 1 in [4] (shown in Fig. 6), with total power of 175W.

The heat convective coefficient at the bottom surface of heat sink is 8700 W/(K·m²) [3]. For each case, the power sources are assumed to present in a 5μm-thickness layer at the top of substrate. The ambient temperature is set to 20°C.

TABLE I
The Thickness and Material Parameters

Thickness of	Structure 1 (μm)	Structure 2 (μm)	Thermal conductivity (W/(K·cm))
Interconnect layer	50	10	0.3
Silicon substrate	450	480	1.25
Thermal interface	200	200	0.3
Heat spreader	1500	1500	3.95
Heat sink	5000	5000	3.95

For each test structure, with different resolution of discretization we obtained three test cases. Case 1-1, Case 1-2, Case 1-3 are obtained from Structure 1, while Case 2-1, Case 2-2 and Case 2-3 are from Structure 2.

(a)   (b)

Fig. 6. The power maps used for (a) Structure 1 and (b) Structure 2.

B. Efficiency and Accuracy Validation

To evaluate the efficiency of proposed techniques, three versions of random walk methods are defined.

- GRW: The GRW method for thermal analysis [8].
- Hybrid0: The hybrid method combining GRW and FRW,

as described in Section III.A and III.B.
- Hybrid1: The hybrid random walk method using the techniques handling Nuemann boundary.
- Hybrid2: The hybrid random walk method using the techniques for handling both Nuemann and convective boundaries.

They are used to calculate the node temperatures on the power-source layer for each test case. The average runtime for a node is listed in Table II. In the second column of the table, we list the number of discretization nodes in each case. The average number of walks (#walk) performed for simulating a node and average number of hop in a walk (#hop) are also given.

TABLE II
The Average Runtime of Random Walk Methods for Calculating the Temperature of a Node (time in unit of second)

Test case	#node	GRW			Hybrid random walk				
		time	#walk	#hop	hybrid0	hybrid1	hybrid2	#hop	Sp*
1-1	5.24e5	49.8	5471	2.34e5	28.65	23.4	2.61	7.52e3	19
1-2	4.19e6	199	5522	9.28e5	48.2	33.68	3.66	8.11e3	54
1-3	6.55e7	949	6409	5.64e6	57.37	41.33	4.32	1.21e4	220
2-1	5.33e5	35.5	3576	2.52e5	32.86	29.46	1.46	1.08e4	24
2-2	4.26e6	143	3709	9.90e5	43.71	38.38	2.73	1.57e4	52
2-3	6.66e7	762	3281	5.98e6	72.7	48.41	5.04	2.71e4	151

*Speedup ratio of the hybrid2 to GRW.

From Table II, we can see that for the cases with coarse grid discretization there is no remarkable advantage brought by the hybrid random walk over GRW. This is because a hop in GRW is executed much faster than a hop in FRW. With the increase of discretization resolution, the runtime of GRW increases rapidly, since the length of walk in GRW depends on the number of grid nodes. Because the runtime of the FRW and proposed hybrid method hardly depends on the discretization of the domain, the advantage of proposed method over GRW becomes remarkable. For the largest cases (Case 1-3 and 2-3), the speedup ratios of Hybrid1 and Hybrid2 reach **23** and **220** respectively, while comparing with GRW. Actually, the number of hops per walk in Hybrid0 can be one order of magnitude fewer than that in GRW. After using the techniques handling Neumann and convective boundaries, the number of hops per walk is reduced further. The experiments also reveal that the hybrid random walk method hardly affects the number of walks for convergence.

While running Hybrid2, the pre-calculated transition probabilities should be loaded, which costs about 81MB in our experiments. The pre-characterization is executed with the approach given in Section III.B. In our implementation, we partition each side of the cuboid transition domain to 20 segments. Under this setting, the pre-characterization costs about 33 minutes which is affordable.

To validate the accuracy of the random walk methods, we use the Matlab "\" operator [15] to solve Case 1-1 and Case 2-1. The results are regarded as the standard value. For the other cases, the direct equation solver in Matlab "\" cannot complete the computation due to memory limitation. For Case 1-1, Matlab "\" costs 2198 seconds and 5.5GB memory, and the results show the highest temperature is 83.0°C. The highest temperature obtained from Hybrid1 is 86.0°C, whose error is consistent to the set accuracy criterion. For Case 2-1, Matlab "\" produces the result in 1579 seconds, with the highest temperature of 61.0°C. For this cases, the highest temperature obtained from Hybrid1 is 63.1°C. For the temperatures of nodes in power region, the errors of Hybrid1 are also calculated. The results show that the average error is less than 0.7°C. The comparisons validate the accuracy of the hybrid random walk method.

The accuracy of Hybrid2 is also evaluated, which shows that its error can be as much as 20°C. After careful analysis, we find out that this is due to the mismatch of R_{amb} used in GRW (i.e. the FVM discretization) and R_{amb} essentially used with the convective-specific transition domains. Therefore, the treatment of convective boundary in Hybrid2 is not valid due to loss of accuracy.

Finally, to evaluate how the aspect ratio of the cuboid transition domain in FRW affects the efficiency of the proposed hybrid method, we change it from 10 and rerun the experiments for Case 1-3. With different kind of cuboid transition domains, the average runtime of Hybrid1 for a node is listed in Table III. From this Table, we can see that using the cuboid with aspect ratio of 10 brings **4.1X** speedup as compared with the cubic transition domain. And for the cuboids with similar aspect ratio, the performance of the hybrid random walk method is similar.

TABLE III
The Runtime of the Hybrid Random Walk Method with Different Aspect Ratios of Transition Cuboids (in unit of second)

Aspect ratio	1	5	8	10	12	15	20
Time/node	170	43.3	41.9	41.3	40.9	41.8	42.5

V. Conclusions

In this paper, a hybrid random walk method is presented for the thermal simulation with chip, heat spreader and heat sink regions. By combining the generic random walk on discretization grids and the floating random walk in medium domain, the hybrid method achieves large efficiency improvement for the thermal analysis of hot-spots in IC. Efficient techniques are proposed to pre-characterize and utilize a rectangular cuboid transition domain, and handle the Neumann boundary in the floating random walk procedure. Experimental results validate the proposed techniques and demonstrate its speedup of up to 23X over the generic random walk method, while keeping high accuracy.

In the future work, we will further investigate efficient and accurate technique to handle the convective boundary in thermal analysis. The fast random walk based thermal solver may also be applied to the floorplanning stage of IC design.

References

[1] S.-C. Lin and K. Banerjee, "Thermal challenges of 3D ICs," in *Wafer- Level 3D ICs Process Technology*, Springer Inc. 2008, pp. 307-332.

[2] P. Li, L. T. Pileggi, M. Asheghi, and R. Chandra, "IC thermal simulation and modeling via efficient multigrid-based approaches," *IEEE Trans. Computer-Aided Design*, Vol. 25, no. 9, pp. 1763-1776, 2006.

[3] Y. Zhan and S. S. Sapatnekar, "High efficiency Green function-based thermal simulation algorithms," *IEEE Trans. Computer-Aided Design*, Vol. 26, no. 9, pp. 1661-1675, 2007.

[4] H. Qian, S. S. Sapatnekar, and E. Kursun, "Fast Poisson solvers for thermal analysis," *ACM Trans. Design Automation*

of Electronic Systems, Vol. 17, no. 3, article 32, June 2012.

[5] H. Qian, S. R. Nassif, S. S. Sapatnekar, "Power grid analysis using random walks," *IEEE Trans. Computer-Aided Design*, Vol. 24, no. 8, pp. 1204-1224, 2005.

[6] T. Miyakawa, K. Yamanaga, H. Tsutsui, H. Ochi, and T. Sato, "Acceleration of random-walk-based linear circuit analysis using importance sampling," in *Proc. GLSVLSI*, 2011, pp. 211-216.

[7] B. Boghrati and S. S. Sapatnekar, "Incremental power network analysis using backward random walks," in *Proc. ASP-DAC*, 2012, pp. 41-46.

[8] E. Wong and S. K. Lim, "3D floorplanning with thermal vias," in *Proc. DATE*, 2006, pp. 878-883

[9] W. Yu, H. Zhuang, C. Zhang, G. Hu, Z. Liu, "RWCap: A floating random walk solver for 3-D capacitance extraction of VLSI interconnects" *IEEE Trans. Computer-Aided Design*, Vol. 32, no. 3, pp. 353-366, 2013.

[10] T. A. El-Moselhy, I. M. Elfadel, and L. Daniel, "A hierarchical floating random walk algorithm for fabric-aware 3D capacitance extraction," in *Proc. ICCAD*, pp. 752-758, 2009.

[11] A. Haji-Sheikh, E. M. Sparrow, "The floating random walk and its application to Monte Carlo solution of heat equation," *Journal SIAM on Applied Mathematics,* Vol. 14, pp. 370-389, 1966.

[12] T. A. El-Moselhy, I. M. Elfadel, and L. Daniel, "A capacitance solver for incremental variation-aware extraction," in *Proc. ICCAD*, pp. 662-669, 2008.

[13] P. Maffezzoni and A. Brambilla, "Analysis of substrate coupling by means of a stochastic method," *IEEE Electron Devices Letters*, Vol. 23, pp. 351-353, 2002.

[14] H. Q. Le, W. J. Starke, J. S. Fields, F. P. O'Connell, D. Q. Nguyen, B. J. Ronchetti, W. M. Sauer, E. M. Schwarz, M. T. Vaden, "IBM POWER6 microarchitecture," *IBM J. Res. Devel.*, Vol. 51, no. 6, pp. 639–662, 2007.

[15] Y. Chen, T. A. Davis, W. W. Hager, and S. Rajamanickam, "Algorithm 887: CHOLMOD, supernodal sparse Cholesky factorization and update/downdate," *ACM Transactions on Mathematical Software*, Vol. 35, no. 3, article 22, 2008.

[16] M. Matsumoto and T. Nishimura, "Mersenne twister: A 623-dimensionally equidistributed unifrom pseudo-random number generator," *ACM Trans. Modeling and Computer Simulation*, vol. 8, no. 1, pp. 3-30, Jan. 1998.

[17] J. Guo, S. Dong, and S. Goto, "Random walk algorithm for large thermal RC network analysis," in *Proc. ASICON*, Oct. 2009, pp. 771-774.

[18] C. Zhang and W. Yu, "Efficient space management techniques for large-scale interconnect capacitance extraction with floating random walks" *IEEE Trans. Computer-Aided Design*, Vol. 32, no. 10, pp. 1633-1637, 2013.

LightSim : A Leakage Aware Ultrafast Temperature Simulator

Smruti R. Sarangi

Computer Science and Engg.
Indian Institute of Technology
Hauz Khas, New Delhi, 110016
Tel: +91-11-2659 7065
e-mail: srsarangi@cse.iitd.ac.in

Gayathri Ananthanarayanan

School of Information Technology
Indian Institute of Technology
Hauz Khas, New Delhi, 110016
Tel: +91-11-2659 6041
e-mail: gayathri@cse.iitd.ac.in

M. Balakrishnan

Computer Science and Engg.
Indian Institute of Technology
Hauz Khas, New Delhi, 110016
Tel: +91-11-2659 1285
e-mail: mbala@cse.iitd.ac.in

Abstract— In this paper, we propose the design of an ultra-fast temperature simulator (LightSim) that can perform both steady state and transient thermal analysis, and also take the effect of leakage power into account. We use a novel Hankel transform based technique to derive a transient version of the Green's function for a chip, which takes into account the feedback loop between temperature and leakage. Subsequently, we calculate the temperature map of a chip by convolving the derived Green's function with the power map. Our simulator is at least 3500 times faster than HotSpot, and at least 2.3 times faster than competing research prototypes [4, 12]. The total error is limited to 0.18 °C .

I. INTRODUCTION

Since the last ten years, on-chip temperature is increasingly being regarded as a first class design constraint. Temperature has a direct effect on the amount of leakage power, and lifetime reliability. Hence, it is necessary for designers to get estimates of chip temperature at an early stage of the design process. In response to this requirement researchers in both industry and academia have proposed a wide variety of simulation tools for estimating on-chip temperature. The existing simulation tools primarily solve two variants of thermal problems – estimating steady state temperatures [1. `steady state` problem], and estimating transient temperatures [2. `transient` problem]. If the feedback loop between temperature and leakage is considered, then we have two more problems namely the [3. `steady state leakage` problem], and the [4. `transient leakage` problem].

Researchers have mostly built on generic methodologies to model temperature such as the finite difference and finite element methods. To take the leakage feedback loop into account most tools run the temperature simulation several times till the temperature and leakage power values converge. Unfortunately, most of the finite element and finite difference based methods are fairly slow. Some of the fastest tools developed over the last decade such as HotSpot 6 [14], and Sesc-Therm [19] take several minutes to solve the `transient` and `transient leakage` problems. Hence, researchers have proposed to use novel Green's function based methods [18, 17] that have been shown to be 2-3X faster [19] than finite element/difference methods. Their main drawback is that they are not suitable for transient thermal analysis.

In this paper, we propose the design of LightSim, an ultra-fast thermal simulator that solves all the four thermal problems

in less than 13 ms on a standard Intel i7 desktop processor. It is primarily designed for estimating the on-chip temperature of multicore processors. As pointed out in prior work [14, 4], the aim of designing an architecture level temperature estimator is to broadly estimate the high level thermal profile, and thus an extremely high level of accuracy is not necessary. Hence, we make a set of simplistic assumptions to speed up our simulation time (similar to [4, 12, 7]). LightSim is at least 3500 times faster than the most popular open source simulator, HotSpot, and is about 2-4 times faster than the fastest research prototypes – CONTILTS [4] and PowerBlur [12].

The crux of our technique is to create a transient version of the Green's function (impulse response of a point power source) that takes the leakage feedback loop into account. We use a novel Hankel transform based technique to arrive at this approximate version of the Green's function. Subsequently, we derive the thermal profile using a standard method [18, 17], which convolves the derived Green's function with the power profile. Our approach effectively combines the positive aspects of Green's function based methods (speed) with that of traditional finite difference based methods (incorporation of transient, and leakage effects). Our approach does have some deficiencies though. At the moment, the leakage values at the rim of the chip are not derived rigorously, and use empirical factors. Even though the error in estimation is small, we intend to propose a more rigorous framework in the future.

II. BACKGROUND AND RELATED WORK

A. Source of Power Dissipation

There are two sources of power dissipation in a processor – dynamic and static(leakage) power. Dynamic power is dissipated because of the switching activity in circuits. It is independent of temperature. It is dependent on the workload, and the micro-architecture.

However, the static or leakage power is a function of temperature, and is traditionally described by Equation 1 in the simplified BSIM 4 [8] model.

$$P_{leak} \propto v_T^2 * e^{\frac{V_{GS}-V_{th}-V_{off}}{\eta * v_T}} \left(1 - e^{\frac{-V_{DS}}{v_T}}\right) \quad (1)$$

v_T is the thermal voltage (kT/q), V_{th} is the threshold voltage, V_{off} is the offset voltage in the sub-threshold region and η is a constant.

978-1-4799-2817-0/14 $31.00 © 2014 IEEE

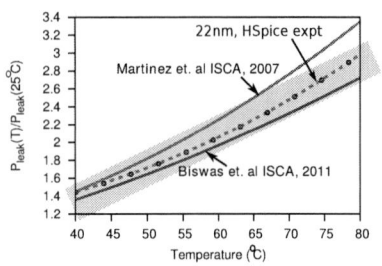

Fig. 1. Leakage models

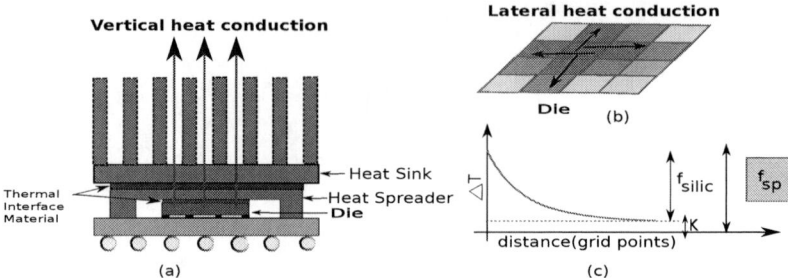

Fig. 2. (a) Processor package (b) Lateral heat conduction (c) f_{sp} and f_{silic}

P_{leak} has a complex dependence on temperature, and is exponentially dependent on temperature for high values($>$ $500mV$) of the threshold voltage (V_{th}). In modern processors, the threshold voltages are about 150 mV, and the relationship between leakage power and temperature becomes approximately linear (based on Equation 1). Two highly cited recent papers ([10] and [1]) have also arrived at the same conclusion using accurate physical measurements and HSpice simulations. We also conducted HSpice simulations for finding the dependence of leakage on temperature for the 22nm technology node, and the results are shown in Figure 1, along with the results of [10, 1]. The operating range of a processor is typically between $45°C$ and $70°C$. In this range, leakage is approximately linear as we can see in Figure 1. This fact has been used to speed up thermal simulators [9], and dynamic thermal management algorithms [5, 6]. All the authors have reported less than 5% error with a linear model of leakage. Note that to estimate temperature, we need to repeatedly estimate leakage, and then the resultant temperature, till convergence.

B. Lateral Heat Conduction

Figure 2(a) shows a diagram of the processor's package. The processor die generates heat because of dynamic and leakage power dissipation. There is some amount of lateral heat conduction as shown in Figure 2(b); however, a larger fraction of the heat escapes through the heat spreader and sink to the ambient (environment). The heat spreader is a copper plate that helps to homogenize the temperature distribution of the silicon die.

Now, if we apply a point heat source at the center of the die, then the temperature distribution is mostly isotropic (independent of direction), and the relation between ΔT, and the radial distance (measured in grid points) is conceptually shown in Figure 2(c). We observe that the heat spread function f_{sp} can be divided into two parts. The first part is a rapidly decaying function (f_{silic}) that captures the temperature rise in the adjoining grid points. It represents the fact that lateral heat conduction is limited to a small region because of the conductivity of silicon (140 W/m-K) is lower than the conductivity of copper (400 W/m-K). The second part is a constant κ that captures the effect of the entire die heating up because of a point power source. This happens because some of the heat is transferred to the spreader, which in turn transfers some energy to all the points in the die. This heat spread function (f_{sp}) is also known

as the Green's function of the system at the center of the die. We thus have:

$$f_{sp} = f_{silic} + \kappa \qquad (2)$$

By conducting empirical measurements using the Hotspot simulator, we concluded that the radially symmetric f_{sp} function accurately describes the temperature rise for most of the points on the die other than the rim (around 10% of the total area). This is because f_{silic} decays very quickly.

C. Related Work on Temperature Simulation

All the temperature simulation algorithms essentially solve the Fourier's heat equation:

$$\frac{\partial \mathcal{U}}{\partial t} = \alpha_1 \nabla^2 \mathcal{U} + \alpha_2 Q \qquad (3)$$

Here, $\mathcal{U}(x, y, z, t)$ is the temperature field, ∇ is the Laplacian operator, Q is the power profile, α_1, and α_2 are constants. The boundary conditions typically specify adiabatic (no heat flow) side/bottom boundaries, and a constant temperature above the die.

The most commonly used method to solve Fourier's equation is the finite difference method. The finite difference method models the points in the package as a grid, and then for each grid point it transforms the Fourier equation into a linear recurrence relation. For example it transforms $\partial\mathcal{U}(x, y, z, t)/\partial t$ to $(\mathcal{U}(x, y, z, t + h) - \mathcal{U}(x, y, z, t))/h$ for grid point (x, y, z). We can solve the resulting set of recurrence relations using standard techniques in linear algebra. The accuracy of this method is a function of the number of grid points, n. Since matrix multiplication, and inversion are essentially $O(n^3)$ operations, this method is very slow. It can be accelerated by the alternating direction implicit method [15], model order reduction using Krylov subspaces [16], and multigrid methods [16]. The popular tool, Hotspot [14], creates an electrical circuit based on the recurrence equations, and processes it using existing circuit simulation methods. Similarly, the CONTILTS [4] simulator significantly speeds up the simulation by using a piece wise constant approximation for the variation of power against time, and by using results from the theory of linear systems. Researchers have experimented with the slower finite element method [3]. However, architectural thermal simulations do not need such high level of accuracy, at the cost of simulation speed.

978-1-4799-2817-0/14 $31.00 © 2014 IEEE   856

A recent set of approaches [18, 17, 12] compute the 2D impulse response (Green's function, G) of an unit power source in the center of the die, and compute the change in the temperature field ($\mathcal{U}$) as:

$$\mathcal{U} = G \star Q \quad (4)$$

Here, $\star$ is the 2 dimensional convolution operator, and Q is the power map. The Green's function is essentially f_{sp} in 2D Cartesian co-ordinates for most of the die (see Section B). We wish to create a time varying version of the Green's function that takes leakage power into account, and uses Equation 4 for the `transient` and `transient leakage` problems.

III. DERIVATION OF THE GREEN'S FUNCTIONS

Symbol	Full Form	Meaning
$\mathcal{U}$	$\mathcal{U}(r,t)$	Temperature field
P_{dyn}	$P_{dyn}(x,y)$	Dynamic power
P_{leak}	$P_{leak}(x,y)$	Leakage power
β		dP_{leak}/dT
f_{sp}	$f_{sp}(r,t)$	Steady state Green's function (polar co-ordinates)
f_{silic}	$f_{silic}(r,t)$	Local heat spread function (polar co-ordinates)
κ		Global heat spread ($f_{sp} - f_{silic}$)
κ_1	$\beta\kappa \int\int_A f_{sp} dx.dy$	
κ_2	$2\pi\beta(\kappa+\kappa_1)(\mathcal{H}(f_{silic})\,|_{s=0})$	
ϕ	$\kappa + \kappa_1 + \kappa_2$	
f_{leaksp}	$f_{leaksp}(r,t)$	Leakage aware Green's function (polar co-ordinates)
f_α	$f_\alpha(s)$	$2\pi C(1 + 2\pi\beta\mathcal{H}(f_{silic}))$
f_{inv}	$f_{inv}(r,t)$	$\mathcal{H}^{-1}(\mathcal{H}(f_{leaksp})\times$ $e^{-\frac{t}{f_\alpha(s)(\kappa\delta(s)/s+\mathcal{H}(f_{silic}))}})$
		$f_{inv} = f_{inv_0}^\epsilon + f_{inv_\epsilon}^\infty$
f_{trans}	$f_{trans}(r,t)$	Transient Green's function with the effect of leakage
Functions: $\mathcal{U}, f_{sp}, f_{silic}$ are also used in Cartesian co-ordinates		

TABLE I
GLOSSARY

A. *Steady State Leakage* Problem

Let us start from Equation 4, and use $f_{sp}(x,y)$ as the Green's function without considering the effects of leakage. Secondly, let us split the power into two parts – P_{dyn} (dynamic), and P_{leak} (leakage). Let $\mathcal{U}_0$ be the temperature field with no dynamic power, and $\mathcal{U}_P$ denote the final temperature field. Let P_{leak0} be the leakage field at $\mathcal{U}_0$. We thus have:

$$\mathcal{U} = \mathcal{U}_P - \mathcal{U}_0 = f_{sp} \star (P_{dyn} + P_{leak}) - f_{sp} \star P_{leak0}$$
$$= f_{sp} \star (P_{dyn} + \Delta P_{leak}) \quad (5)$$

Let us now assume P_{dyn} to be the Dirac delta function, δ (a point source with total power of 1 Watt). Hence, $f_{sp} \star P_{dyn} =$

f_{sp}. Secondly, since we approximate leakage with a linear model. We have $\Delta P_{leak} = \beta\mathcal{U}$ (β is a constant of proportionality). Thus:

$$\mathcal{U} = f_{sp} + \beta f_{sp} \star \mathcal{U} \quad (6)$$

Using Equation 2 ($f_{sp} = f_{silic} + \kappa$) we get:

$$\mathcal{U} = f_{silic} + \kappa + \beta f_{silic} \star \mathcal{U} + \beta(\kappa \star \mathcal{U})$$
$$= f_{silic} + \kappa + \beta f_{silic} \star \mathcal{U} + \beta\kappa \int_{-\infty}^{\infty}\int_{-\infty}^{\infty}\mathcal{U}dx.dy$$
$$\approx f_{silic} + \kappa + \beta f_{silic} \star \mathcal{U} + \underbrace{\beta\kappa \int\int_{\mathcal{A}} f_{sp}dx.dy}_{\kappa_1} \quad (7)$$
$$= f_{silic} + \kappa + \beta f_{silic} \star \mathcal{U} + \kappa_1$$

We make a simplifying assumption here by assuming that the integral of $\mathcal{U}$ over the area of the chip ($\mathcal{A}$) is approximately equal to the integral of f_{sp} over $\mathcal{A}$. Let us now compute 2D Fourier transforms of the LHS and RHS. The 2D Fourier transform is defined as:

$$\mathcal{F}(s,t) = \frac{1}{2\pi}\int_{-\infty}^{\infty}\int_{-\infty}^{\infty} e^{-j(xs+yt)}dxdy \quad (8)$$

Note that the Fourier transform (as defined by Equation 8) of a convolution is equal to the product of Fourier transforms multiplied by 2π. We thus have from Equation 7:

$$\mathcal{F}(\mathcal{U}) = \mathcal{F}(f_{silic}) + (\kappa+\kappa_1)\mathcal{F}(1) + 2\pi\beta\mathcal{F}(f_{silic})\mathcal{F}(\mathcal{U}) \quad (9)$$

Now, f_{silic} is radially symmetric because we are assuming a large (theoretically infinite) die size. As mentioned in Section B, this assumption holds for about 90% of the area of a typical 2x2 cm^2 die. Let us now use zero order Hankel transforms($\mathcal{H}$) [11] to reduce the 2D problem to a 1D problem. A zero order Hankel transform is equivalent to 2D Fourier transforms of radially symmetric functions in polar co-ordinates [11]. A Hankel transform is defined as:

$$\mathcal{H}(k) = \int_0^{\infty} f(r)\mathcal{J}_0(sr)rdr \quad (10)$$

Here $\mathcal{J}_0$ is a Bessel function of the first kind of order 0. The inverse Hankel transform has exactly the same form with r replaced by the transform variable s. The Hankel transform of a 2D convolution is the product of the transforms multiplied by 2π. From Equation 9, we have after converting to polar co-ordinates.

$$\mathcal{H}(\mathcal{U}) = \mathcal{H}(f_{silic}) + (\kappa+\kappa_1)\mathcal{H}(1) + 2\pi\beta\mathcal{H}(f_{silic})\mathcal{H}(\mathcal{U})$$
$$\Rightarrow \mathcal{H}(\mathcal{U}) = \frac{\mathcal{H}(f_{silic}) + (\kappa+\kappa_1)\mathcal{H}(1)}{1 - 2\pi\beta\mathcal{H}(f_{silic})} \quad (11)$$

β is typically a small value (0.09 to 0.006), and the peak value of $\mathcal{H}(f_{silic})$ is limited to small values(≈ 1) as observed in our simulations. Thus, we can simplify Equation 11 by considering the fact that $(1-x)^{-1} \approx 1+x$, when $x \ll 1$. We use the result: $\mathcal{H}(f(r) \star g(r)) = 2\pi\mathcal{H}(f(r)\mathcal{H}(g(r))$.

$$\mathcal{H}(\mathcal{U}) = (\mathcal{H}(f_{silic}) + (\kappa + \kappa_1)\mathcal{H}(1)) \times (1 + 2\pi\beta\mathcal{H}(f_{silic}))$$
$$= \mathcal{H}(f_{silic}) + 2\pi\beta\mathcal{H}(f_{silic})^2 + \mathcal{H}(\kappa + \kappa_1)$$
$$+ 2\pi\beta(\kappa + \kappa_1)\mathcal{H}(1)\mathcal{H}(fsilic) \tag{12}$$

$\mathcal{H}(1)$ is $\delta(s)/s$, and is thus 0 everywhere other than 0. Hence, $\mathcal{H}(1)\mathcal{H}(f_{silic}) = \mathcal{H}(1)\,(\mathcal{H}(f_{silic})\,|_{s=0})$. Thus:

$$\mathcal{U} = f_{silic} + 2\pi\beta\mathcal{H}^{-1}(\mathcal{H}(f_{silic})^2) + \kappa + \kappa_1$$
$$+ \underbrace{2\pi\beta(\kappa + \kappa_1)(\mathcal{H}(f_{silic})\,|_{s=0})}_{\kappa_2}$$
$$= f_{silic} + 2\pi\beta\mathcal{H}^{-1}(\mathcal{H}(f_{silic})^2) + \underbrace{(\kappa + \kappa_1 + \kappa_2)}_{\phi} \tag{13}$$

$$\boxed{f_{leaksp} = f_{silic} + 2\pi\beta\mathcal{H}^{-1}(\mathcal{H}(f_{silic})^2) + \phi}$$

Here $\mathcal{U}$ (referred to as f_{leaksp}) is the modified Green's function that takes into account the effect of leakage.

B. *Transient Leakage Problem*

If we consider the transient case, the resultant temperature field is defined by: (refer to [14]).

$$\mathcal{U} = f_{sp} + \beta f_{sp} \star \mathcal{U} - C f_{sp} \star \frac{\partial \mathcal{U}}{\partial t} \tag{14}$$

Here, C is a constant (thermal capacitance). By following the same set of steps that we followed to derive Equation 12, and the Leibnitz rule, we get:

$$\mathcal{H}(\mathcal{U}) = \mathcal{H}(f_{leaksp}) - $$
$$\underbrace{2\pi C(1 + 2\pi\beta\mathcal{H}(f_{silic}))}_{f_\alpha} \times \mathcal{H}(f_{sp})\mathcal{H}\left(\frac{\partial\mathcal{U}}{\partial t}\right)$$
$$= \mathcal{H}(f_{leaksp}) - f_\alpha\mathcal{H}(f_{sp})\mathcal{H}\left(\frac{\partial\mathcal{U}}{\partial t}\right)$$
$$= \mathcal{H}(f_{leaksp}) - f_\alpha\mathcal{H}(\kappa + f_{silic})\frac{\partial\mathcal{H}(\mathcal{U})}{\partial t} \tag{15}$$

Solving for t, and applying the boundary conditions: $\mathcal{H}(\mathcal{U}) = 0\,|_{t=0}$, and $\mathcal{H}(\mathcal{U}) = \mathcal{H}(f_{leaksp})\,|_{t=\infty}$.

$$\mathcal{H}(\mathcal{U}) = \mathcal{H}(f_{leaksp}) \times \left(1 - e^{-\frac{t}{f_\alpha\mathcal{H}(\kappa + f_{silic})}}\right)$$
$$\Rightarrow \mathcal{H}(\mathcal{U}) = \mathcal{H}(f_{leaksp}) \times (1 - e^{-\frac{t}{f_\alpha(s)(\kappa\delta(s)/s + \mathcal{H}(f_{silic}))}}) \tag{16}$$

The Hankel transform of κ is $\kappa\delta(s)/s$. It is ∞ for $s = 0$, and is 0 for all other values of s. Let us now evaluate the inverse Hankel transform (f_{inv}) of $\mathcal{H}(f_{leaksp}) \times e^{-\frac{t}{f_\alpha(s)(\kappa\delta(s)/s + \mathcal{H}(f_{silic}))}}$.

Let us approximate $\delta(s)$ as a function that is equal to $1/\epsilon$ from 0 to ϵ, and is 0 everywhere else. Here, $\epsilon \to 0$. We can thus break the inverse Hankel transform f_{inv} into two parts – f_{inv0}^ϵ,

and $f_{inv_\epsilon}^\infty$. $f_{inv_\epsilon}^\infty$ can be calculated by using the formula for the inverse Hankel transform, and numerical integration.

$$f_{inv_\epsilon}^\infty(r,t) = \int_\epsilon^\infty \left(\mathcal{H}(f_{silic}) + 2\pi\beta\mathcal{H}(f_{silic})^2\right) \times$$
$$e^{-\frac{t}{2\pi C(1+2\pi\beta\mathcal{H}(f_{silic}))(\mathcal{H}(f_{silic}))}} \mathcal{J}_0(sr)sds \tag{17}$$

Let us now compute f_{inv0}^ϵ. Since $\delta(s)/s \gg \mathcal{H}(f_{silic})$ when $s < \epsilon$, we can ignore all the terms with $\mathcal{H}(f_{silic})$. We can thus approximate $\mathcal{H}(f_{leaksp}) = \phi\delta(s)/s$ (Equation 13). We thus have:

$$f_{inv0}^\epsilon(r,t) = \int_0^\epsilon (\phi\delta(s)/s) \times e^{-\frac{t}{f_\alpha(s)(\kappa\delta(s)/s)}} s\mathcal{J}(sr)ds$$
$$= \int_0^\epsilon (\phi/\epsilon) \times e^{-\frac{ts\epsilon}{f_\alpha(s)\kappa}} \mathcal{J}(sr)ds \qquad (\delta(s) = 1/\epsilon) \tag{18}$$

Since $\epsilon \to 0$, the product $sr \to 0$, and thus $\mathcal{J}_0(sr) \to 1$. Secondly, f_α tends to ($f_{\alpha 0} = 2\pi C(1 + 2\pi\beta\mathcal{H}(f_{silic})(0))$, because s tends to 0. By making these substitutions, we get:

$$f_{inv0}^\epsilon(r,t) = \int_0^\epsilon (\phi/\epsilon) \times e^{-\frac{ts\epsilon}{f_{\alpha 0}\kappa}} ds$$
$$= \frac{\phi f_{\alpha 0}\kappa}{t\epsilon^2}\left(1 - e^{-\frac{t\epsilon^2}{f_{\alpha 0}\kappa}}\right) \tag{19}$$

Note that when t is small, f_{inv0}^ϵ tends to ϕ. It becomes zero as $t \to \infty$. To conclude, the Green's function, f_{trans}, for the `transient leakage` problem is:

$$\boxed{f_{trans}(r,t) = f_{leaksp}(r) - f_{inv0}^\epsilon(r,t) - f_{inv_\epsilon}^\infty(r,t)} \tag{20}$$

f_{inv0}^ϵ incorporates the effect of heating up of the entire package. This is a slowly increasing function. $f_{inv_\epsilon}^\infty$ is a much faster growing function and models the transient temperature rise in the neighborhood of the power source. Lastly, note that we assume that the leakage power increases instantaneously (quasi-static assumption). Thermal time scales are at least of the order of microseconds, and this is too large for non quasi static effects to set in [8, 2].

C. Corrections for the Edges and Corners

We use the technique proposed by Park et. al. [12] to make corrections for the rim of the chip. Park et. al. use three different Green's functions corresponding to the center, edge, and corner of the chip. We do the same and use the appropriate version of f_{sp} in Equations 6, and 14. We need to also make a correction to the term $\beta f_{sp} \star \mathcal{U}$ in Equations 6 and 14. This term represents the feedback component in the temperature leakage loop. A primary power source increases the temperature in the neighborhood, and each point in the neighborhood starts acting as a secondary power source dissipating leakage power. Since the size of the neighborhood is reduced to a quarter for the corners, and reduced to half for the edges, we divide the term $\beta f_{sp} \star \mathcal{U}$ by 4 for the corners, and 2 for the edges. These two corrections help us significantly reduce the error in temperature estimation for the rim of the chip.

978-1-4799-2817-0/14 $31.00 © 2014 IEEE

D. Using the Green's Functions

This part of the algorithm is the same as prior approaches that use Green's functions [18, 17]. They compute the 2D convolution of the Green's function and the power map. This is typically done by computing the Fourier transform of both the functions, multiplying them, and then computing the inverse transform. We use Equations 12, and 16, to get the 2D Fourier transform of the Green's function, and then proceed to obtain the temperature map. We accelerate the process of temperature estimation by pre-computing the Green's function, and using a fast lookup table based approach similar to [17].

IV. RESULTS

A. Setup

We implemented LightSim in two parts. The offline part is written in R [13], and the online part in C. The offline part first invokes the popular Hotspot [14] tool to compute three different Green's functions (center, edge, corner) for a chip with the parameters shown in Table II.

Parameter	Value
Die size	400 mm^2
Silicon conductivity	130 W/m-K
Spreader conductivity	370 W/m-K
Heatsink conductivity	237 W/m-K
Convection resistance	0.1 K/W
Spreader thickness	3.5 mm
Heatsink thickness	24.9 mm

TABLE II
HOTSPOT CONFIGURATION

We conducted experiments with 4 to 1024 mesh points, and concluded that using 256 mesh points is appropriate for architectural simulation (error was limited to 2%, also see [19]). The offline component of LightSim processes the data from Hotspot, and finds the Hankel transforms of the Green's functions for the center, rim, and corner. Subsequently, the online part computes the Hankel transform of the final temperature field for a given time interval (only for `transient` and `transient leakage` problems). It then converts the Hankel transform into a 2D FFT by converting to Cartesian coordinates. We divide the power map into three portions – center (90% of area), corners (2%), and edges (8%). We compute the 2D FFT of each power map, and convolve it with the 2D FFT of the corresponding Green's function. The final temperature field is a superposition of the three individual temperature fields. We assume 20% leakage at ambient temperature (30°C). We collect all our results on a Quad core, Intel i7 (3.1 GHz) desktop processor (memory 4GB) running Ubuntu Linux 12.1.

B. Accuracy

We compare LightSim with the popular Hotspot simulator. Figure 3 shows a plot of the Green's function for the `steady state leakage` problem. For Hotspot, we get the Green's

function with leakage by applying a point heat source to the 136th grid point (die center), and then running the simulation multiple times till convergence. For LightSim, we plot *leakspread*. Note that the x-axis shows the number of the grid point (printed row wise). The error is less than 2%. Figure 4 compares the obtained Green's functions using both the approaches for the `transient leakage` problem (for the origin). We observe a maximum divergence of 0.18°C (4.5%) at 1 ms. For 100 random power maps, we show the maximum error in Table 5 computed at 2ms. For Hotspot, we compute the leakage using the BSIM4 model. For a sample power map(Fig. 6), we show the results of transient simulation at 0.01 ms (Fig. 7), 0.1ms (Fig. 8), 1 ms (Fig 9), and 5 ms (Fig. 10). The y-axis shows the increase in temperature, ΔT.

C. Speed

Simulator	Steady state leakage	Transient leakage (5ms)
Hotspot 5	30.3ms	45s
CONTILTS	25.2 ms	206.4 ms
Liu et. al. [7]	39.3 ms	264.8 ms
PowerBlur	20.1ms	58.2 ms
LightSim	8.7 ms	12.8 ms

TABLE III
COMPARISON OF EXECUTION TIMES

We compare LightSim with Hotspot, CONTILTS, Liu et. al.[7], and PowerBlur in Table III. We simulate all the algorithms (other than Hotspot) in C. The open source C code for Hotspot 5 is freely available for download. Secondly, for the purpose of fair comparison, we precompute all the Green's functions and steady state matrices before hand. These parameters are specific to a chip, and do not change with the power map, or transient simulation interval. For the steady state case, LightSim is at least 2.3 times faster than PowerBlur. For transient simulation with leakage, LightSim is 3500 times faster than Hotspot for a 5ms interval, and is 4.5 times faster than the nearest competitor (PowerBlur). Fundamentally, LightSim is faster than Green's function based methods (PowerBlur) because it incorporates the effect of leakage, and does not require multiple iterations. Moreover, it is significantly faster than finite difference based methods because they rely on costly $O(n^2)$ or $O(n^3)$ time matrix operations, whereas LightSim uses faster FFT based methods ($O(nlog(n))$ time), and using Hankel transforms makes the computation of the Green's function a 1D problem.

V. CONCLUSION

We conclude that our Hankel transform based approach to quickly compute the Green's function for steady state and transient analysis with leakage, is useful for fast architecture level temperature estimation.

978-1-4799-2817-0/14 $31.00 © 2014 IEEE

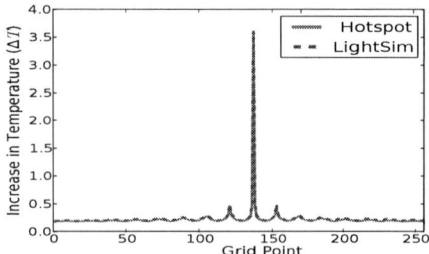

Fig. 3. Steady state Green's function

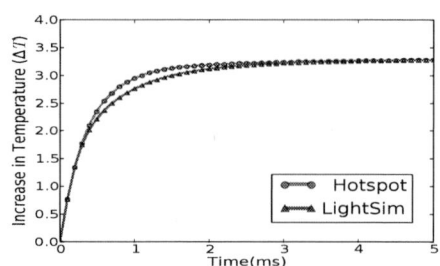

Fig. 4. Transient Green's function at the origin

Location	Error (%)	
	steady state	transient(at 2ms)
Center	0.6%	1%
Edge	1%	1.8%
Corner	1.8%	2.4%

Fig. 5. Errors of different configurations

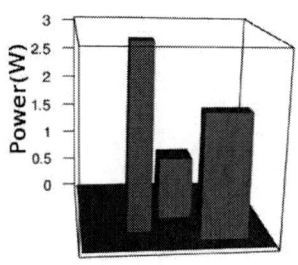

Fig. 6. Power Map

Fig. 7. ΔT at 0.01 ms

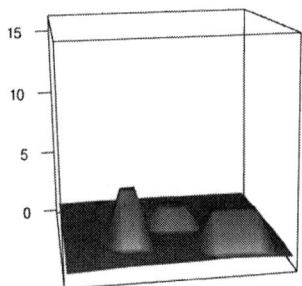

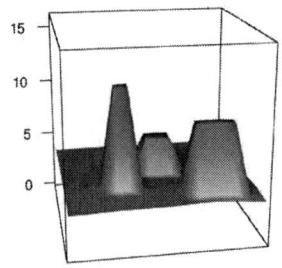

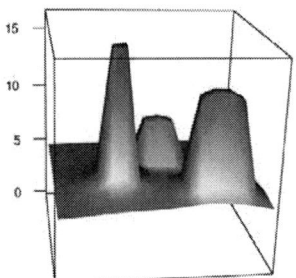

Fig. 8. ΔT at 0.1 ms

Fig. 9. ΔT at 1 ms

Fig. 10. ΔT at 5 ms

REFERENCES

[1] S. Biswas, M. Tiwari, T. Sherwood, L. Theogarajan, and F. T. Chong. Fighting Fire with Fire: Modeling the Datacenter-Scale Effects of Targeted Superlattice Thermal Management. In *ISCA*, 2011.

[2] Elmar Gondro, Oskar Kowarik, Gerhard Knoblinger, and Peter Klein. When do we need non-quasistatic cmos rf-models? In *Custom Integrated Circuits, 2001, IEEE Conference on.*, pages 377–380. IEEE, 2001.

[3] B. Goplen and S. Sapatnekar. Efficient thermal placement of standard cells in 3d ics using a force directed approach. In *ICCAD*, 2003.

[4] Y. Han, I. Koren, and C. M. Krishna. Tilts: A fast architectural-level transient thermal simulation method. *J. Low Power Electronics*, 3(1):13–21, 2007.

[5] H. Huang, V. Chaturvedi, G. Liu, and G. Quan. Leakage aware scheduling on maximum temperature minimization for periodic hard real-time systems. *J. Low Power Electronics*, 8(4), 2012.

[6] H. Huang, G. Quan, and J. Fan. Leakage temperature dependency modeling in system level analysis. In *ISQED*, 2010.

[7] P. Liu, Z. Qi, H. Li, L. Jin, W. Wu, S. X. D. Tan, and J. Yang. Fast thermal simulation for architecture level dynamic thermal management. In *ICCAD*, pages 639–644, 2005.

[8] W. Liu, K.M. Cao, X. Jin, and Chenming Hu. Bsim 4.0.0 technical notes. Technical Report UCB/ERL M00/39, EECS Department, University of California, Berkeley, 2000.

[9] Y. Liu, R. P. Dick, L. Shang, and Huazhong Y. Accurate Temperature-dependent Integrated Circuit Leakage Power Estimation is Easy . In *DATE*, 2007.

[10] F. J. Mesa-Martinez, J. Nayfach, and J. Renau. Power Model Validation through Thermal Measurements . In *ISCA*, 2007.

[11] P. K. Mittal. *Integral Transforms for Engineers and Physicists*. Har Anand Publishers, 2007.

[12] J. Park, S. Shin, J. Christofferson, A. Shakouri, and S. Kang. Experimental validation of the power blurring method. In *SemiTherm*, 2010.

[13] R Core Team. *R: A Language and Environment for Statistical Computing*. R Foundation for Statistical Computing, Vienna, Austria, 2013.

[14] K. Skadron, M. R. Stan, W. Huang, S. Velusamy, K. Sankaranarayanan, and D. Tarjan. Temperature-Aware Microarchitecture. In *ISCA*, 2003.

[15] T. Wang and C.C.P. Chen. Thermal-adi - a linear-time chip-level dynamic thermal-simulation algorithm based on alternating-direction-implicit (adi) method. *Very Large Scale Integration (VLSI) Systems, IEEE Transactions on*, 11(4):691–700, 2003.

[16] T. Wang and C.C.P. Chen. Spice-compatible thermal simulation with lumped circuit modeling for thermal reliability analysis based on modeling order reduction. In *ISQED*, 2004.

[17] Y. Zhan and S. S. Sapatnekar. Fast computation of the temperature distribution in vlsi chips using the discrete cosine transform and table look-up. In *ASPDAC*, 2005.

[18] Y. Zhan and S.S. Sapatnekar. A high efficiency full-chip thermal simulation algorithm. In *ICCAD*, 2005.

[19] A. Ziabari, E. K. Ardestani, J. Renau, and A. Shakouri. Fast thermal simulators for architecture level integrated circuit design. In *SemiTherm*, 2011.

978-1-4799-2817-0/14 $31.00 © 2014 IEEE

Fast Vectorless Power Grid Verification using Maximum Voltage Drop Location Estimation[*]

Wei Zhao , Yici Cai and Jianlei Yang

Tsinghua National Laboratory for Information Science and Technology
Dept. of Computer Science and Technology, Tsinghua University, Beijing, P. R. China
caiyc@mail.tsinghua.edu.cn   yjl09@mails.tsinghua.edu.cn

Abstract - Power grid integrity verification is critical for reliable chip design. Vectorless power grid verification provides a promising approach to evaluate the worst-case voltage fluctuations without the detailed information of circuit activities. Vectorless verification is usually required to solve numerous linear programming problems to obtain the worst-case voltage fluctuation throughout the grid, which is extremely time-consuming for large-scale verification. In this paper, a maximum voltage drop location estimation approach is proposed for efficient vectorless verification. The power grid nodes are grouped into disjoint subsets, and an estimation strategy is utilized to roughly locate the nodes which have the worst-case voltage drop in each group. Consequently, the verification problem size can be significantly reduced compared with accurate verification. Experimental results show that the proposed approach can achieve remarkable speedups with acceptable accuracy loss.

I INTRODUCTION

The design quality of the on-die power grid has a direct impact on chip performance and reliability. A well-designed power grid should provide sufficient power supply to all transistors on the chip and guarantee a certain noise margin during all possible circuit operations. As the supply and threshold voltages are decreased for lower power consumption and better performance in high performance circuits design, the functionality and performance of modern integrated circuits are becoming more and more vulnerable to power supply noises. Hence, power grid verification has become a crucial procedure to ensure a reliable chip design.

Traditional simulation-based power grid verification techniques suffer from the prohibitively high computation cost in enumerating an extremely large number of possible input current patterns. Another disadvantage of these methods is that early-stage verification cannot be performed since detailed information about the current waveforms is still unavailable. To overcome these issues and enable early stage power grid verification, a series of vectorless power grid verification techniques has been proposed in [1], and further studied in [2-10]. These approaches adopt the notion of current constraints to capture the circuit uncertainty at an early design stage, and evaluate the worst-case voltage fluctuations on each grid node by solving linear programming problems under the feasible space specified by these current constraints. In recent years, great efforts have been made to reduce the computation cost of vectorless power grid verification. Hierarchical matrix inversion algorithm [5] and $\mathcal{H}$-matrix-based approximate matrix inversion

algorithm [7] is proposed to speed up the sub-problem of linear system solution. A convex dual algorithm [4] is proposed to speed up the sub-problem of linear programming solution. The sparse approximate inverse (SPAI) technique is proposed in [3] to accelerate the matrix inversion problem as well as reduce the number of variables in linear programming problems. However, the computation cost is still too high to be practical in the prototype vectorless power grid verification framework due to the extremely large number of linear programming problems that need to be solved. So there are also several research works to improve the verification efficiency. A geometric approach is introduced in [6], a selected inversion approach is proposed in [8], and a partial differential equation (PDE) constrained optimization based multilevel verification framework is proposed in [9].

In this paper, we propose a new efficient vectorless power grid verification framework. Major contributions of this paper are listed as follows:

1) We present a maximum voltage drop location estimation algorithm. For a group of power grid nodes which are properly selected, the proposed algorithm can find out the possible nodes which may have the maximum worst-case voltage drop without the accurate calculation of the exact worst-case voltage drops at each node in this group.

2) Based on the location estimation technique, we propose a modified framework for vectorless power grid verification. In this framework, power grid nodes are checked in a group-by-group manner. We first try to pick out the possible "worst" power grid nodes in a certain group, and then perform accurate verification on these nodes. By following this strategy, the number of power grid nodes which need accurate verification is effectively reduced, and thus the overall vectorless power grid verification becomes much more efficient.

3) Since the accuracy of the maximum voltage drop location estimation algorithm depends to some extent on the grouping method for selecting the nodes which will be analyzed in one group, we propose a modified algebraic circuit partitioning technique which has a better ability to deal with the impact of VDD pads than methods that simply adopt existing graph partitioning algorithms to decompose the power grid.

The rest of this paper is organized as follows. Power grid modeling and prior vectorless power grid verification

---

[*]This work was supported by National Natural Science Foundation of China (NSFC) under Grant No.61274031.

978-1-4799-2817-0/14 $31.00 © 2014 IEEE

methods are summarized in Section II. Details of the proposed approach are presented in Section III. Experimental results are provided and analyzed in Section IV. Concluding remarks are given in Section V.

II. BACKGROUND

A. Power Grid Model

The modeling methods have a strong influence on the power grid verification methodology. For RC model, since voltage levels can only be inferior to VDD, we just need to check that the voltage on every node of the power grid does not drop by more than some critical threshold. However, for RLC model, the presence of inductances can cause the voltage on a given node to fluctuate in both directions, either increasing or decreasing, which makes the power grid verification problem more complicated.

With RC model in [1], each branch of the power grid is represented as a resistor and there is a capacitor from every grid node to ground. Let v(t) be the vector of voltage drops, and i(t) be the vector of current excitations. Then according to [1], the following system equation which describe the functional relationship between voltage drops and current excitations must be satisfied,

$$G\mathbf{v}(t) + C\mathbf{v}'(t) = \mathbf{i}(t) \qquad (1)$$

where G is an $n \times n$ conductance matrix and C is an $n \times n$ diagonal matrix of node capacitances.

An RLC model is introduced in [2]. In this model, each branch of the power grid is represented either by a resistor, referred to as an r-branch, or by a resistor in series with an inductor, referred to as an rl-branch. The system equations can be written as:

$$G\mathbf{v}(t) + C\mathbf{v}'(t) - M\mathbf{i}(t) = \mathbf{i}_S(t) \qquad (2)$$
$$M^T\mathbf{v}(t) + L\mathbf{i}'(t) = 0 \qquad (3)$$

where M is an $n \times m$ incidence matrix whose elements are either ± 1 or 0, and L is an $m \times m$ diagonal matrix of inductance values.

B. Current Constraints

To address the requirement of early-stage power grid verification, the framework of current constraints which defines a feasible space of current excitations is adopted by vectorless power grid verification algorithms to capture the circuit uncertainty at the early design stage.

There are mainly two kinds of current constraints used in the existing vectorless power grid verification algorithms: local constraints and global constraints. Local constraints define upper bounds on individual current sources,

$$0 \le \mathbf{i}(t) \le I_L \quad \forall t \ge 0 \qquad (4)$$

where $I_L \ge 0$ is an $n \times 1$ upper bound vector. Global constraints define upper bounds for certain groups of current sources,

$$U\mathbf{i}(t) \le I_G \qquad (5)$$

where U is an $m \times n$ matrix which contains only 0s and 1s and $I_G \ge 0$ is an $m \times 1$ upper bound vector.

These current constraints give DC upper bounds on tran-

sient waveforms. If the upper bounds are also defined by a set of transient waveforms, the result of the vectorless power grid verification may be more realistic while the involved problems will be more difficult to solve. In [10], hierarchical current and power constraints are used to represent the transient constraints approximately.

C. Vectorless Power Grid Verification

Vectorless power grid verification aims to check the safety of the power grid by estimating the worst-case voltage fluctuations under given current constraints. By establishing a computable explicit expression for the vector of maximum node voltage fluctuations, vectorless power grid verification boils down to solving linear programs which depend on a certain matrix inverse. In [1], a DC model of the power grid is used to give an upper bound of the voltage drops in the corresponding RC model. So the upper bound of the worst-case voltage drop vector can be evaluated as follows:

$$\mathbf{v}_{max}(t) = emax_{\forall \mathbf{i} \in \mathcal{F}}(G^{-1}\mathbf{i}) \qquad (6)$$

where the notation $emax(\cdot)$ means an element-wise maximization. For the same RC model, a more careful analysis is given in [6], which results in a tighter upper bound:

$$\mathbf{v}_{max}(t) = (I + G^{-1}\frac{C}{\Delta t})emax_{\forall \mathbf{i} \in \mathcal{F}}(A^{-1}\mathbf{i}) \qquad (7)$$

where $A = G + C/\Delta t$. Upper and lower bounds of the voltage fluctuations in the RLC model are analyzed in [2].

Most of the existing approaches aim to compute the worst-case voltage fluctuation for each node by solving linear programming problems which is time-consuming for large-scale power grids. A geometric approach is introduced in [6]. By exploiting particularities of the power grid at design time and the geometry of current feasibility space, the solution of the vectorless verification problem is reduced to a user-limited number of linear system solutions. In [9], a scalable multilevel vectorless power grid verification framework based on the key ideas of multilevel PDE-constrained optimization methods is proposed. By taking advantage of a series of coarsest to coarser grid verifications, the finest power grid verification can be accomplished in a more efficient way. The selected inversion in [8] selectively constructs the matrix inverse by exploiting the grid locality and constraint locality to improve the inversion efficiency and also speedup the linear programming which can provide a possible way for the large-scale power grid verification.

III. PROPOSED APPROACH

A. General Framework

In this paper, we mainly discuss the RC power grid model in which only the IR drops need to be considered. We will first use the upper bound defined by (6) to show how the proposed algorithm works since it is relatively simpler. And then we will also give some explanation about how to handle the more complicated upper bound defined by (7).

To guarantee the safety of the power grid, we should check that the worst-case voltage drop of every power grid node does not exceed some critical threshold. Since the up-

978-1-4799-2817-0/14 $31.00 © 2014 IEEE

per bound of the worst-case voltage drop vector defined by (6) includes an element-wise maximization notation, most vectorless power grid verification algorithms also follow the element-wise checking strategy, i.e., these methods compute the worst-case voltage drops for every node on the grid. However, if we can filter out the nodes which have relatively smaller worst-case voltage drops in advance, the computation cost of the whole vectorless verification will be reduced. For example, it is easy to see that all the power grid nodes with no current source attached cannot have the maximum worst-case voltage drop in the DC model since given an arbitrary waveform combination, only the nodes that have current sources connected to ground can become the "worst" power grid node, i.e., the node which has the maximum voltage drop.

The main idea of this paper is to give a generalized method to pick out the power grid nodes which may have the maximum worst-case voltage drop. The vectorless power grid verification can be viewed as a maximization problem:

$$Maximize \quad v_{max} = max\{v_{max_1}, v_{max_2}, \cdots, v_{max_n}\}$$

$$s.t. \quad v_{max_k} = max_{i \in \mathcal{L}} (G^{-1}e_k)^T i, \mathcal{L} = \{i | Li \leq I_m, i \geq 0\} \quad (8)$$

And the prototype vectorless power grid verification framework can be described as follows:

Algorithm 1. Prototype Vectorless Power Grid Verification Framework
---

1. For $k = 1$ to n
2.   Maximize $v_k = (G^{-1}e_k)^T i$ s.t. $i \in \mathcal{L}$
3.   Let $v_{max_k} = max \, v_k$
4. End For
5. Find $v_{max} = max\{v_{max_1}, v_{max_2}, \cdots, v_{max_n}\}$

The general framework of the proposed approach which is based on the maximum voltage drop location estimation technique is shown in algorithm 2. In this framework, power grid nodes are checked in a group-by-group manner. We first try to pick out the possible "worst" power grid nodes in a certain group, and then perform accurate verification on these nodes.

In fact, the proposed verification framework is an approximate approach since the maximum voltage drop location estimation algorithm may make a wrong prediction. However, both of the theoretical analysis and experimental results have shown that if we can choose a good estimating function and select the nodes which will be analyzed in one group properly, the accuracy of the maximum voltage drop location estimation algorithm will be quite good. We can further improve the estimation accuracy by checking the power grid nodes which may have the largest k_0 voltage drops ($k_0 < k$) in one group based on the prediction made by the estimation algorithm.

If we use the upper bound defined by (7), the problem will become more difficult to solve. It seems that we must solve the element-wise maximization problem completely in order to get the upper bound of worst-case voltage drop vector. But we can combine the proposed approach with the sparse approximate inverse technique to overcome this difficulty. Let

$$V_a = emax_{\forall i \in \mathcal{F}}(A^{-1}i) \quad (9)$$

The framework described in Algorithm 2 can be used to make an estimation of the largest q elements in V_a with minor modifications. Sparse approximate inverse methods such as SPAI [3] can be used to find an approximation to $I + G^{-1}C/\Delta t$ which basically ignore the extremely small values in this matrix. So we can find out the node that may have the maximum voltage drop based on these results and then compute the exact value of the maximum voltage drop.

Algorithm 2. Vectorless Power Grid Verification Using Maximum Voltage Drop Location Estimation
---

Node grouping based on circuit partitioning

1. Divide the set of power grid nodes into k subsets: $N = N_1 \cup N_2 \cup \cdots \cup N_k$
2. Let $num = max\{|N_1|, |N_2|, \cdots, |N_k|\}$
3. For $j = 1$ to num
4.   Select one node from each subset unless it is empty: $G_j = \{P_{j1} \in N_1, P_{j2} \in N_2, \cdots, P_{jk} \in N_k\}$
5. $N_1 = N_1 \setminus \{P_{j1}\}, \cdots, N_k = N_k \setminus \{P_{jk}\}$
6. End For

Verification procedure

7. For $j = 1$ to num
        Maximum voltage drop location estimation
8.   Maximize $f(i) = f(v_{P_{j1}}, v_{P_{j2}}, \cdots, v_{P_{jk}})$ s.t. $i \in \tilde{\mathcal{L}}$
9.   Let $i_j^* = argmax_{i \in \tilde{\mathcal{L}}} f(i)$
10.   Find
    $$v_{P_j}(i_j^*) = max\{v_{P_{j1}}(i_j^*), v_{P_{j2}}(i_j^*), \cdots, v_{P_{jk}}(i_j^*)\}$$
        Accurate verification
11.   Maximize $v_{P_{j*}} = \left(G^{-1}e_{P_{j*}}\right)^T i$ s.t. $i \in \mathcal{L}$
12.   Let $v_{max_j} = max \, v_{P_{j*}}$
13. End For
14. Find $v_{max} = max\{v_{max_1}, v_{max_2}, \cdots, v_{max_{num}}\}$

B. Estimating Function

The construction of the objective function (line 8 in algorithm 2) used in the maximum voltage drop location estimation procedure is the most critical aspect in the proposed verification framework. The estimating function should be easy to optimize while the estimation accuracy should also be guaranteed.

First, consider the maximum function

$$g(i) = max\{v_{P_{j1}}, v_{P_{j2}}, \cdots, v_{P_{jk}}\} \quad (10)$$

If we use $g(i)$ as the objective function for maximum voltage drop location estimation, the resulting maximization problem:

$$Maximize \quad g(i) = max\{v_1, v_2, \cdots, v_n\} \quad s.t. \quad i \in \mathcal{L} \quad (11)$$

is exactly equal to the problem defined by (8). The function

$$h(i) = ln(e^{v_{P_{j1}}} + e^{v_{P_{j2}}} + \cdots + e^{v_{P_{jk}}}) \quad (12)$$

is a differentiable approximation to $g(i)$. In fact, $g(i)$ can be also interpreted as the infinity norm of the upper bound vector v_{max}. So every function which is constructed based on the general p-norm

$$r(\mathbf{i}) = \left(v_{P_{j1}}^p + v_{P_{j2}}^p + \cdots + v_{P_{jk}}^p \right)^{1/p} \ (p \geq 1) \quad (13)$$

can be used as an approximation to $g(i)$.

However, since $g(\mathbf{i})$ is a convex function, when it is used in the nonlinear programming which is usually formulated as a minimization problem, the objective function $-g(\mathbf{i})$ becomes a concave function. So the resulting optimization problem is very difficult to solve. Even if we just want to obtain an ε-optimal solution, the computation speed is still not fast enough. Other functions such as $h(\mathbf{i})$ and $r(\mathbf{i})$ $(p > 1)$ also suffer the similar disadvantage. Hence, we choose the function which can be interpreted as the 1-norm of the upper bound vector $\mathbf{v}_{max}$ as the estimating function:

$$f(\mathbf{i}) = v_{P_{j1}} + v_{P_{j2}} + \cdots + v_{P_{jk}} \quad (14)$$

The optimization of this linear function is quite easy. But $f(\mathbf{i})$ is not a very good approximation to $g(\mathbf{i})$ since the difference between the 1-norm and the infinity norm is relatively large. To ensure the estimation accuracy, we must take advantage of the locality effect of the power grid [3, 11].

A possible better choice is to use the weighted 1-norm:

$$f(\mathbf{i}) = w_1 v_{P_{j1}} + w_2 v_{P_{j2}} + \cdots + w_k v_{P_{jk}} \ (w_l > 0) \quad (15)$$

to handle the influence of current constraints in $\tilde{\mathcal{L}}$. For example, we can compute the number of conflicts between a certain node P_l and the other nodes in the same group and denote it by c_l. We can also compute the maximum distance d_l between P_l and the other nodes approximately by using the distance between the general parts they belong to (This will be introduced in the next section). So it appears to be a reasonable choice to set

$$w_l = 1 + 0.5 \frac{c_l}{c_{max}} + 0.5 \frac{d_l}{d_{max}} \quad (16)$$

where $c_{max} = max_{l \in G_j}\{c_l\}$ and $d_{max} = max_{l \in G_j}\{d_l\}$.

The analysis of the estimation accuracy can be established as follows. For an arbitrary node P_a on the power grid, let

$$\mathbf{i}_{P_a}^* = argmax_{\mathbf{i} \in \mathcal{L}} \, v_{P_a}(\mathbf{i}) \quad (17)$$

and decompose the vector into two non-overlapping parts:

$$\mathbf{i}_{P_a}^* = \mathbf{i}_{P_a}^1 + \mathbf{i}_{P_a}^2 \quad (18)$$

Then according to the locality effect of the power grid, we can assume that the voltage drop on this node is mainly caused by $\mathbf{i}_{P_a}^1$, i.e.,

$$v_{P_a}(\mathbf{i}_{P_a}^1) \geq (1 - \delta_1) v_{P_a}(\mathbf{i}_{P_a}^*) \ (\delta_1 > 0) \quad (19)$$

while $\mathbf{i}_{P_a}^1$ contains only a few nonzero elements. If the power grid nodes which are selected to be analyzed in the same group are well separated, we can ensure that all of the nonzero elements in $\mathbf{i}_j^1 = \mathbf{i}_{P_{j1}}^1 + \mathbf{i}_{P_{j2}}^1 + \cdots + \mathbf{i}_{P_{jk}}^1$ will also be contained in $\mathbf{i}_j^* = argmax_{\mathbf{i} \in \tilde{\mathcal{L}}} f(\mathbf{i})$ by making minor modifications to the original feasible region $\mathcal{L}$. Since the difference between $\tilde{\mathcal{L}}$ and $\mathcal{L}$ is not very large, we can make the assumption that for any node P_a in this group, the following inequality holds:

$$v_{P_a}(\mathbf{i}_j^*) \leq (1 + \delta_2) v_{P_a}(\mathbf{i}_{P_a}^*) \ (\delta_2 > 0) \quad (20)$$

Combing (19) and (20), we can get that

$$\mu v_{P_a}(\mathbf{i}_j^*) \leq v_{P_a}(\mathbf{i}_{P_a}^*) \leq \lambda v_{P_a}(\mathbf{i}_j^*) \quad (21)$$

where

$$\lambda = \frac{1}{1 - \delta_1} \quad \text{and} \quad \mu = \frac{1}{1 + \delta_2} \quad (22)$$

Now we can analyze the accuracy of the proposed algorithm under the assumption that the exact worst-case voltage drop $v_{P_a}^* = v_{P_a}(\mathbf{i}_{P_a}^*)$ at P_a can be treated as a uniformly distributed random variable. Suppose that we have already calculated the approximate values of the worst-case voltage drops at two grid nodes P_a and P_b by optimizing the objective function $f(\mathbf{i})$, and the result is $v_{P_a} > v_{P_b}$. So the maximum voltage drop location estimation algorithm will predict that $v_{P_a}^* > v_{P_b}^*$. The probability that we have made a right prediction is:

$$P = \begin{cases} 1 & , \ \mu v_{P_a} \geq \lambda v_{P_b} \\ 1 - \frac{\left(\lambda v_{P_b} - \mu v_{P_a} \right)^2}{\left(\lambda v_{P_a} - \mu v_{P_a} \right)\left(\lambda v_{P_b} - \mu v_{P_b} \right)} & , \ \mu v_{P_a} < \lambda v_{P_b} \end{cases} \quad (23)$$

The inequality (21) also provides a way to compute a conservative upper bound of the maximum voltage drop.

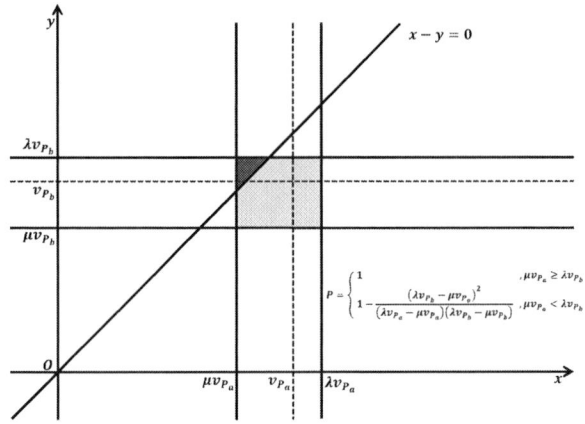

Fig. 1. Estimation accuracy analysis.

C. Node Grouping Based on Circuit Partitioning

From the analysis above, we can see that the main purpose of the node grouping procedure is to guarantee that all nodes in the same group are well separated. In order to achieve this goal, we first partition the power grid into several non-overlapping parts. Then each time we select one node from each part to generate a new group. Therefore, a good circuit partitioning algorithm is needed in the proposed verification framework.

There are many existing power grid partitioning algorithms. The partitioning approach introduced in [11] is a "shell"-based partitioning technique that specifically takes advantage of the locality effect of the power grid under C4 packaging. The disadvantage of this method is that it needs the detailed geometric information of the power grid. As a result, the partition strategy will become much more complicated when dealing with a 3-D irregular power grid. Alge-

978-1-4799-2817-0/14 $31.00 © 2014 IEEE

braic circuit partitioning techniques are also been studied in many works [5, 12]. These methods try to decompose the power grid by adopting existing graph partitioning algorithms. The method used in [12] is in fact an algebraic $\mathcal{H}$-matrix splitting technique. In this paper, we use a similar technique which is introduced in [13] and make some simplification since the hierarchical property is not needed.

However, although these algebraic circuit partitioning techniques can handle the irregular structure of the power grid very well, they cannot deal with the impact of VDD pads. The basic idea of these methods can be interpreted as estimating the influence between any two nodes by computing the shortest path length connecting them in the resistance network. Hence, only the off-diagonal elements in the matrix are used while the information about the location of voltage sources is actually contained in the values of diagonal elements. This can be seen in Fig. 2. The lengths of shortest paths from P_0 to P_1, P_2 and P_3 can reflect the influence between each pair of nodes correctly. However, if there is a VDD pad nearby, this strategy will fail to give a right estimate.

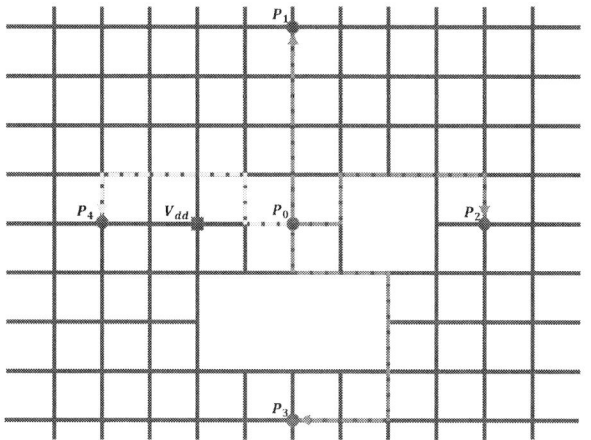

Fig. 2. The impact of VDD pads.

To overcome this difficulty, we propose a technique that takes the impact of VDD pads into consideration by changing the original values of the off-diagonal elements in the matrix. The modification is done in a level-by-level manner. The level of each node is determined by the breadth first search (BFS) algorithm. First, we pick out all the nodes which are connected to voltage sources and denote the set of these nodes by S_1. This can be done purely algebraically since the following inequality only holds at these nodes.

$$g_{ii} > \sum_{j \neq i} |g_{ij}| = g_{sum} \qquad (24)$$

We change the value of g_{ij} for all these nodes synchronously.

$$\eta = e^{1 - \frac{g_{ii}}{g_{sum}}}, \quad g_{ij}^* = \eta g_{ij}, \quad g_{ji}^* = \eta g_{ji} \qquad (25)$$

Then the inequality (24) will also be satisfied at the nodes that are connected to the nodes in S_1. So the modification procedure can be performed at the next level. If an off-diagonal element has been modified, it will be marked and

no longer changed a second time. The stopping criteria can be controlled by

$$\eta < \varepsilon_0 \quad or \quad Level > Level_{max} \qquad (26)$$

D. Modification of Feasible Region

In order to ensure that the nonzero elements in the current vector $\mathbf{i}_j^1 = \mathbf{i}_{P_{j1}}^1 + \mathbf{i}_{P_{j2}}^1 + \cdots + \mathbf{i}_{P_{jk}}^1$ will also be contained in $\mathbf{i}_j^* = argmax_{\mathbf{i} \in \tilde{\mathcal{L}}} f(\mathbf{i})$, we may need to make some small modifications to the original feasible region $\mathcal{L}$. So it is need to find the local support region $supp(P_a)$ of a certain power grid node P_a, which is in fact an approximation to the set of nonzero elements in the vector $\mathbf{i}_{P_a}^1$.

If we are dealing with the upper bound defined by (7), since a sparse approximation to $I + G^{-1}C/\Delta t$ will be computed, we can use this information to estimate the major elements in the inverse of $A = G + C/\Delta t$ and find the local support region for each node. If we only consider the upper bound defined by (6), a method which requires much less computational cost can be used. We can adopt the Dijkstra's algorithm to find the nearest n_0 grid nodes to P_a in the modified resistance network which has been used in the power grid partitioning procedure and get the local support region of P_a. Notice that although a larger local support region can guarantee a smaller δ_1, it may also cause conflict to the global current constraints, which will result in a larger δ_2. Hence, we restrict the size of each local support region by setting the value of n_0 to be about 25~41 and checking that there is no singular local support region that can cause conflict to any global current constraint.

After all the local support regions of the power grid nodes in one group have been found, we begin to modify the original feasible region $\mathcal{L}$. First, we add equality constraints to set every current source in these local support regions to its maximum allowable value. And then we check the original global current constraints and relax the ones which cannot be satisfied.

E. Complexity Analysis

In terms of complexity, since the node grouping procedure based on circuit partitioning can be performed in almost linear complexity, i.e., $O(n\log^\alpha n)$ with moderate parameter α [12] and needs to be done only once, the main factors that affect the total runtime are still the linear system solution and the linear programming solution in the for-loop. Verifying a single node on the power grid needs one linear system solution and one linear programming solution in the original verification framework. And we denote the runtime by T_e. The maximum voltage drop location estimation procedure require to solve one linear system to obtain the objective function $f(\mathbf{i})$, to solve one linear programming, and to solver another linear system to get the estimation result. So it is known that $2T_e < T_g < 3T_e$, where T_g is the runtime for verifying a group of nodes in the proposed framework. Hence, we can achieve about $\frac{n}{3num} \approx \frac{k}{3}$ times of speedups over the original method, in which num is the number of node groups and k is the number of partitions of the power grid.

IV. EXPERIMENTAL RESULTS

The proposed maximum voltage drop location estimation based vectorless power grid verification framework is implemented in C++. We use Cholmod [14] to solve the linear equations and lp_solve 5.5 [15] to solve the involved linear programming problems. For comparison, we also implement the original vectorless power grid verification framework based on Algorithm 1. All these experiments are performed on a 64-bit Linux machine with 2.33GHz Intel Xeon E5345 processor and 8GB RAM. The test cases are generated based on the power grids used in [5]. These power grids are mainly small and medium sized 3-D regular grids. In order to verify the accuracy of the maximum voltage drop location estimation method under different circumstances, we also generate several different types of test cases. The parameters of the test cases are shown in Table I.

TABLE I.   TESTCASE PARAMETERS

Power Grid	Type	#Nodes	#VDD Pads	#Global Constraints
PG1	2-D irregular	4082	9	6
PG2	3-D regular	5875	9	10
PG3	2-D irregular	9964	16	10
PG4	3-D irregular	11706	18	10
PG5	3-D regular	22939	25	10
PG6	3-D irregular	35568	36	12

Experimental results of the entire framework are shown in Table II and Fig. 3. The function $\tilde{f}(\mathbf{i})$ in (15) is used as the estimating function for maximum voltage drop location estimation. In terms of verification accuracy, since accurate verifications are performed for all the nodes selected from each group, we can get a precise result of the worst voltage drop across the entire grid if the estimation algorithm can pick out the real "worst" node from a certain group. Even if the estimation algorithm makes several wrong predictions, the whole framework is still able to find a power grid node which has a relatively large worst-case voltage drop. In terms of runtime, since we do not implement any acceleration technique for the linear programming problem, the runtime cannot be compared with the results in [5] directly. From the experimental results we can see that the modified vectorless power grid verification framework can achieve about 15X speedups over the original method, which may be further improved on lager power grids according to the complexity analysis made above.

TABLE II.   EXPERIMENTAL RESULTS

Test Case	#Partitions	Runtime			Error (mV)
		Original	Modified	Speedup	
PG1	16	122.46 s	21.16 s	5.79	0.17
PG2	16	293.75 s	48.63 s	6.04	0
PG3	25	996.41 s	105.43 s	9.45	7.53
PG4	32	25.36 m	122.76 s	12.40	0
PG5	36	1.72 h	424.35 s	14.60	0
PG6	36	5.14 h	20.66 m	14.92	2.81

V. CONCLUSIONS

In this paper, we present a modified vectorless power grid verification framework using a maximum voltage drop location estimation technique. By picking out the nodes which may have the maximum worst-case voltage drop in advance, the whole verification cost is significantly reduced. Experimental results have confirmed that the proposed estimation approach is accuracy and efficient for practical use. The modified resistance network which is introduced for the node grouping procedure of the estimation algorithm can be also used on other occasions such as the preconditioner construction for iterative linear solvers.

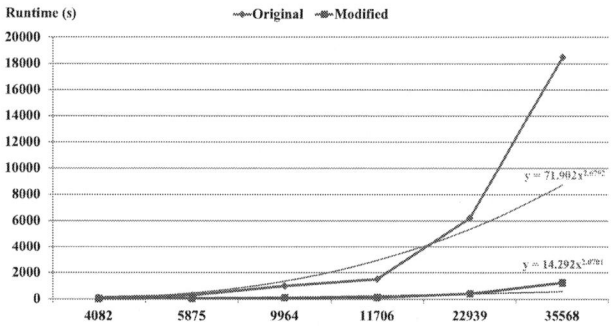

Fig. 3. Runtime comparison.

REFERENCES

[1] D. Kouroussis and F. N. Najm, "A static pattern-independent technique for power grid voltage integrity verification," In proceedings of DAC, pages 99-104, 2003.

[2] N. H. Abdul Ghani and F. N. Najm, "Fast vectorless power grid verification under an RLC model," IEEE Trans. Computer-Aided Design, vol. 30, no. 5, pages 691-703, May 2011.

[3] N. H. Abdul Ghani and F. N. Najm, "Fast vectorless power grid verification using an approximate inverse technique," In proceedings of DAC, pages 184-189, 2009.

[4] X. Xiong and J. Wang, "An efficient dual algorithm for vectorless power grid verification under linear current constraints," In proceedings of DAC, pages 837-842, 2010.

[5] X. Xiong and J. Wang, "A hierarchical matrix inversion algorithm for vectorless power grid verification," In proceedings of ICCAD, pages 543-550, 2010.

[6] I. A. Ferzli, F. N. Najm and L. Kruse, "A geometric approach for early power grid verification using current constraints," In proceedings of ICCAD, pages 40-47, 2007.

[7] Wei Zhao, Yici Cai and Jianlei Yang, A multilevel $\mathcal{H}$-matrix-based approximate matrix inversion algorithm for vectorless power grid verification, In proceedings of ASP-DAC, pages 163-168, 2013.

[8] Jianlei Yang, Yici Cai and Wei Zhao, "Selected inversion for vectorless power grid verification by exploiting locality," In proceedings of ICCD, pages 257-263, 2013.

[9] Zhuo Feng, "Scalable multilevel vectorless power grid voltage integrity verification," IEEE Trans. Very Large Scale Integration Systems, 2012.

[10] Chung-Kuan Cheng, Peng Du, Andrew B. Kahng and etl., "More realistic power grid verification based on hierarchical current and power constraints," In proceedings of ISPD, pages 159-166, 2011.

[11] E. Chiprout, "Fast flip-chip power grid analysis via locality and grid shells," In proceedings of ICCAD, pages 485–488, 2004.

[12] J. M. S. Silva, Joel R. Phillips and L. Miguel Silveira, "Efficient simulation of power grids," ACM Trans. on CAD, 29(10):1523-1532, October 2010.

[13] Grasedyck L, Kriemann R and Le Borne S, "Parallel black box $\mathcal{H}$-LU preconditioning for elliptic boundary value problems," [J]. Computing and Visualization in Science, 11(4-6): 273-291, 2008.

[14] http://www.cise.ufl.edu/research/sparse/cholmod

[15] http://lpsolve.sourceforge.net/5.5

The 19th Asia and South Pacific Design Automation Conference
Author Index

Table of Contents:

A

Abe, Keiko (Toshiba)	p. 6 (1S-2)
Abraham, Jacob (Univ. of Texas)	p. 390 (5S-3)
Adam, Kostas (Mentor Graphics)	p. 155 (2B-4)
Aguilera, Paula (Univ. of Wisconsin - Madison)	p. 726 (8A-4)
Ahourai, Fereidoun (Univ. of California, Irvine)	p. 97 (2S-2)
Al Faruque, Mohammad Abdullah (Univ. of California, Irvine)	p. 97 (2S-2)
Amano, Hideharu (Keio Univ.)	p. 843 (9C-1)
Amaru, Luca (EPFL-LSI)	p. 628 (7B-3)
Ambrose, Jude Angelo (Univ. of New South Wales)	p. 267 (4S-1)
Ananthanarayanan, Gayathri (IIT Delhi)	p. 855 (9C-3)
Angiolini, Federico (iNoCs)	p. 337 (4B-3)
Arasu, M. Annamalai (A*STAR)	p. 27 (1A-4)
Asai, Hideki (Shizuoka Univ.)	p. 774 (8C-4)
Atienza, David (École Polytechnique Fédérale de Lausanne)	p. 714 (8A-2)
Aungskunsiri, Kanin (Univ. of Bristol)	p. 795 (9S-3)
Axelos, Nicholas (National Technical Univ. of Athens)	p. 652 (7C-2)

B

Badr, Yasmine (Univ. of California, Los Angeles)	p. 61 (1B-4)
Baharani, Mohammadreza (Univ. of Tehran)	p. 292 (4A-1)
Bai, Shang-Ya (National Cheng Kung Univ.)	p. 507 (6B-1)
Balakrishnan, M. (IIT Delhi)	p. 855 (9C-3)
Banno, Koji (Fujitsu Semiconductor)	p. 670 (7C-5)
Batterywala, Shabbir H (Synopsys India Pvt.)	p. 149 (2B-3)
Bertels, Koen (Delft Technical Univ.)	p. 213 (3A-3)

Bertozzi, Davide (Univ. of Ferrara) — p. 337 (4B-3)

Beyeler, Michael (UC Irvine) — p. 570 (7S-3)

Bhattacharya, Sambuddha (Synopsys India Pvt.) — p. 149 (2B-3)

Bi, Xiaojun (National Univ. of Singapore/A*STAR) — p. 27 (1A-4)

Bichler, Olivier (CEA-LIST) — p. 563 (7S-2)

Billoint, Olivier (CEA-LETI) — p. 79 (1C-3)

Bishnoi, Rajendra (Karlsruhe Inst. of Tech.) — p. 700 (8S-4)

Bjerregaard, Tobias (Teklatech) — p. 337 (4B-3)

Blehadj, Bilel (INRIA) — p. 563 (7S-2)

Bogdan, Paul (Univ. of Southern California) — p. 323 (4B-1)

Bonneau, Damien (Univ. of Bristol) — p. 795 (9S-3)

Borkar, Nitin (Intel) — p. 388 (5S-2)

Borkar, Shekhar (Intel) — p. 388 (5S-2)

Bosio, Alberto (Univ. of Montpellier/LIRMM) — p. 544 (6C-3)

Braun, Andreas (FZI Research Center for Information Technology) — p. 800 (9A-1)

Brayton, Robert (Univ. of California, Berkeley) — p. 250 (3C-1)

Bringmann, Oliver (FZI Research Center for Information Technology/Univ. of Tübingen) — p. 800 (9A-1)

Brisk, Philip (Univ. of California, Riverside) — p. 231 (3B-2)

Bruschini, Claudio (EPFL) — p. 788 (9S-2)

Burg, Andreas (EPFL-TCL) — p. 628 (7B-3)

Burger, Andreas (FZI Research Center for Information Technology) — p. 800 (9A-1)

Burri, Samuel (EPFL) — p. 788 (9S-2)

C

Cai, Yici (Tsinghua Univ.) — p. 455 (5C-3)

Cai, Yici (Tsinghua Univ.) — p. 525 (6B-4)

Cai, Yici (Tsinghua Univ.) — p. 861 (9C-4)

Cao, Yan (Pennsylvania State Univ.) — p. 762 (8C-2)

Cao, Yu (Arizona State Univ.) — p. 161 (2C-1)

Caramanis, Michael C. (Boston Univ.) — p. 105 (2S-3)

Carlson, Kristofor D. (UC Irvine) — p. 570 (7S-3)

Carrion Schafer, Benjamin (Hong Kong Polytechnic Univ.) — p. 616 (7B-1)

Chabi, Djaafar (Univ. Paris-Sud) — p. 676 (8S-1)

Chakrabarty, Krishnendu (Duke Univ.) — p. 244 (3B-4)

Chakraborty, Samarjit (TU Munich) — p. 119 (2A-2)

Chakraborty, Samarjit (TU Munich) — p. 646 (7C-1)

Chakraborty, Samarjit (TU Munich) — p. 812 (9A-3)

Chandra, Vikas (ARM) — p. 474 (6S-2)

Chang, Naehyuck (Seoul National Univ.) — p. 379 (4C-5)

Chappert, Claude (Univ. Paris-Sud) — p. 676 (8S-1)

Charbon, Edoardo (Delft Univ. of Tech.) — p. 788 (9S-2)

Chen, Deming (Univ. of Illinois, Urbana-Champaign)   p. 73 (1C-2)

Chen, Deming (Univ. of Illinois, Urbana-Champaign)   p. 373 (4C-4)

Chen, Fu-Wei (National Tsing Hua Univ.)   p. 658 (7C-3)

Chen, Gengsheng (Fudan Univ.)   p. 424 (5B-2)

Chen, Hao (Boston Univ.)   p. 105 (2S-3)

Chen, Hung-Ming (National Chiao Tung Univ.)   p. 519 (6B-3)

Chen, Meng-Ling (National Chiao Tung Univ.)   p. 519 (6B-3)

Chen, Mike (Nimbus Automation Technologies)   p. 525 (6B-4)

Chen, Quan (Univ. of Hong Kong)   p. 262 (3C-3)

Chen, Shi-Hao (Global Unichip)   p. 519 (6B-3)

Chen, Shuai (Nanyang Technological Univ.)   p. 191 (3S-2)

Chen, Xiaoming (Tsinghua Univ.)   p. 161 (2C-1)

Chen, Yiran (Univ. of Pittsburgh)   p. 67 (1C-1)

Chen, Yiran (Univ. of Pittsburgh)   p. 355 (4C-1)

Chen, Yiran (Univ. of Pittsburgh)   p. 361 (4C-2)

Chen, Yiran (Univ. of Pittsburgh)   p. 592 (7A-2)

Chen, Yiran (Univ. of Pittsburgh)   p. 831 (9B-3)

Chen, Yunji (Chinese Academy of Sciences)   p. 201 (3A-1)

Chen, Zhijie (Univ. of Pittsburgh)   p. 592 (7A-2)

Cheng, Jiandong (Shanghai Jiao Tong Univ.)   p. 443 (5C-1)

Cheng, Yuanqing (LIRMM)   p. 544 (6C-3)

Cher, Chen-Yong (IBM)   p. 385 (5S-1)

Chi, Baoyong (Tsinghua Univ.)   p. 29 (1A-5)

Chien, Hsi-An (National Tsing Hua Univ.)   p. 418 (5B-1)

Cho, Hyungmin (Stanford Univ.)   p. 390 (5S-3)

Choi, Kiyoung (Seoul National Univ.)   p. 285 (4S-4)

Chowdhury, Salim (Oracle)   p. 513 (6B-2)

Chu, Chris (Iowa State Univ.)   p. 137 (2B-1)

Chui, Chi On (Univ. of California, Los Angeles)   p. 818 (9B-1)

Clermidy, Fabian (CEA-LETI)   p. 578 (7S-4)

Clermidy, Fabien (CEA-LETI)   p. 79 (1C-3)

Clermidy, Fabien (CEA-LETI)   p. 563 (7S-2)

Coskun, Ayse K. (Boston Univ.)   p. 105 (2S-3)

Cui, Tiansong (Univ. of Southern California)   p. 167 (2C-2)

Cui, Xiaotong (Chongqing Univ.)   p. 131 (2A-4)

D

Daneshtalab, Masoud (Univ. of Turku)   p. 298 (4A-2)

Davis, Kristan D. (IBM)   p. 385 (5S-1)

Davoodi, Azadeh (Univ. of Wisconsin - Madison)   p. 640 (7B-5)

De Micheli, Giovanni (EPFL-LSI)   p. 628 (7B-3)

De, Vivek (Intel) — p. 388 (5S-2)
Deng, Wei (Tokyo Inst. of Tech.) — p. 21 (1A-1)
Deng, Wei (Tokyo Inst. of Tech.) — p. 25 (1A-3)
Dilillo, Luigi (CNRS/LIRMM) — p. 544 (6C-3)
Dinh, Trung Anh (Ritsumeikan Univ.) — p. 225 (3B-1)
Dombrowa, Marc B. (IBM) — p. 385 (5S-1)
Drechsler, Rolf (Univ. of Bremen/Cyber Physical Systems DFKI GmbH) — p. 483 (6A-1)
Drechsler, Rolf (Univ. of Bremen/Cyber Physical Systems DFKI GmbH) — p. 489 (6A-2)
Du, Zidong (Chinese Academy of Sciences) — p. 201 (3A-1)
Dubrova, Elena (Royal Inst. of Tech.) — p. 634 (7B-4)
Duranton, Marc (CEA-LIST) — p. 563 (7S-2)
Dutt, Nikil (UC Irvine) — p. 570 (7S-3)

E

Ebrahimi, Mojtaba (Karlsruhe Inst. of Tech.) — p. 700 (8S-4)
Ecker, Wolfgang (Infineon Technologies AG) — p. 806 (9A-2)
Eftaxiopoulos, Nikolaos (National Technical Univ. of Athens) — p. 652 (7C-2)
Eles, Petru (Linköping Univ.) — p. 436 (5B-4)
Esen, Volkan (Infineon Technologies AG) — p. 806 (9A-2)

F

Fakhouri, Sani (Univ. of California, Irvine) — p. 91 (2S-1)
Faldamis, Georgios (Cavium) — p. 329 (4B-2)
Farahini, Nasim (KTH) — p. 556 (7S-1)
Farahini, Nasim (KTH) — p. 578 (7S-4)
Fattah, Mohammad (Univ. of Turku) — p. 349 (4B-5)
Fei, Wei (Nanyang Technological Univ.) — p. 191 (3S-2)
Fiske, Johnathan (Univ. of California, Riverside) — p. 231 (3B-2)
Fujimori, Yoshikazu (Rohm) — p. 17 (1S-4)
Fujita, Shinobu (Toshiba) — p. 6 (1S-2)
Fukami, Shunsuke (Tohoku Univ.) — p. 684 (8S-2)

G

Gaillardon, Pierre-Emmanuel (EPFL-LSI) — p. 628 (7B-3)
Gamrat, Christian (CEA-LIST) — p. 563 (7S-2)
Gao, Jhih-Rong (Univ. of Texas, Austin)

Gao, Ping (Aries Design Automation) — p. 143 (2B-2)

Gao, Ping (Aries Design Automation) — p. 750 (8B-4)

Ghaida, Rani (GLOBALFOUNDRIES) — p. 61 (1B-4)

Ghiribaldi, Alberto (Univ. of Ferrara) — p. 337 (4B-3)

Giannopoulou, Georgia (ETH Zurich) — p. 125 (2A-3)

Gill, Gennette (D.E. Shaw Research) — p. 329 (4B-2)

Girard, Patrick (CNRS/LIRMM) — p. 544 (6C-3)

Gong, Fang (Univ. of California, Los Angeles) — p. 424 (5B-2)

Gooding, Thomas M. (IBM) — p. 385 (5S-1)

Goswami, Dip (TU Munich) — p. 119 (2A-2)

Grissom, Daniel (Univ. of California, Riverside) — p. 231 (3B-2)

Große, Daniel (solvertec) — p. 800 (9A-1)

Guo, Jie (Univ. of Pittsburgh) — p. 592 (7A-2)

Guo, Yongxin (National Univ. of Singapore, Singapore/National Univ. of Singapore (Suzhou) Research Institute) — p. 27 (1A-4)

Gupta, Mukul (Qualcomm) — p. 61 (1B-4)

Gupta, Puneet (Univ. of California, Los Angeles) — p. 61 (1B-4)

Gupta, Puneet (Univ. of California, Los Angeles) — p. 155 (2B-4)

Gupta, Puneet (UCLA) — p. 467 (6S-1)

Gupta, Puneet (Univ. of California, Los Angeles) — p. 818 (9B-1)

H

Haedicke, Finn (solvertec/Univ. of Bremen)   p. 800 (9A-1)
Han, Kyuseung (Seoul National Univ.)   p. 285 (4S-4)
Han, Qingrui (Huawei Technologies)   p. 367 (4C-3)
Han, Yinhe (Chinese Academy of Sciences)   p. 394 (5A-1)
Han, Yinhe (Chinese Academy of Sciences)   p. 720 (8A-3)

Hara-Azumi, Yuko (Nara Inst. of Science and Tech./JST, PRESTO) p. 85 (1C-4)

Haring, Ruud A. (IBM) p. 385 (5S-1)

Hayashikoshi, Masanori (Renesas Electronics) p. 12 (1S-3)

Hayes, Jerry (IBM) p. 91 (2S-1)

He, Lei (Univ. of California, Los Angeles) p. 424 (5B-2)

Heliot, Rodolphe (CEA-LETI) p. 563 (7S-2)

Hemani, Ahmed (KTH) p. 556 (7S-1)

Hemani, Ahmed (KTH) p. 578 (7S-4)

Henkel, Jörg (Karlsruhe Inst. of Tech.) p. 274 (4S-2)

Ho, Tsung-Yi (National Cheng Kung Univ.) p. 225 (3B-1)

Ho, Tsung-Yi (National Cheng Kung Univ.) p. 244 (3B-4)

Ho, Tsung-Yi (National Cheng Kung Univ.) p. 507 (6B-1)

Hoskote, Yatin (Intel Labs) p. 219 (3A-4)

Howard, Jason (Intel) p. 388 (5S-2)

Hsu, Po-Yi (National Tsing Hua Univ.) p. 501 (6A-4)

Hu, Kai (Duke Univ.) p. 244 (3B-4)

Hu, Miao (Univ. of Pittsburgh) p. 831 (9B-3)

Hu, Xing (Univ. of Chinese Academy of Sciences) p. 550 (6C-4)

Hu, Yu (Univ. of Chinese Academy of Sciences) p. 550 (6C-4)

Huang, Chung-Yan (Ric) (National Taiwan Univ.) p. 744 (8B-3)

Huang, Hui (Univ. of California, Los Angeles) p. 219 (3A-4)

Huang, Kai (Zhejiang Univ.) p. 406 (5A-3)

Huang, Kai (Technical Univ. Munich) p. 708 (8A-1)

Huang, Pengcheng (ETH Zurich) p. 125 (2A-3)

Huang, Ryan H.-M. (National Chiao Tung Univ.) p. 664 (7C-4)

Hwang, Leslie K. (Univ. of Illinois, Urbana-Champaign) p. 73 (1C-2)

Hwang, TingTing (National Tsing Hua Univ.) p. 658 (7C-3)

I

Igarashi, Makoto (Tohoku Univ.) p. 185 (3S-1)

Ikeda, Sho (Tokyo Inst. of Tech.) p. 23 (1A-2)

Ikeda, Shoji (Tohoku Univ.) p. 684 (8S-2)

Inoue, Hiroaki (NEC) p. 282 (4S-3)

Ishida, Tsutomu (Fujitsu Labs.) p. 670 (7C-5)

Ishihara, Noboru (Tokyo Inst. of Tech.) p. 23 (1A-2)

Ishikuro, Hiroki (Keio Univ.) p. 31 (1A-6)

Ito, Hiroyuki (Tokyo Inst. of Tech.) p. 23 (1A-2)

Izumi, Shintaro (Kobe Univ.) p. 17 (1S-4)

J

Jabeur, Kotb (Spintec Laboratory, CEA-INAC/CNRS/UJF/G-INP)	p. 692 (8S-3)
Jagannathan, Ashok (Intel Technology India Pvt.)	p. 622 (7B-2)
Je, Minkyu (A*STAR)	p. 27 (1A-4)
Jeong, Hui-Sung (Korea Aerospace Univ.)	p. 39 (1A-10)
Jiang, Ke (Linköping Univ.)	p. 317 (4A-5)
Jiang, Pisu (Univ. of Bristol)	p. 795 (9S-3)
Jiang, Wei (Univ. of Electronic Science and Tech. of China)	p. 317 (4A-5)
Jiang, Weiwei (Columbia Univ.)	p. 329 (4B-2)
Jiang, Yingtao (Univ. of Nevada, Las Vegas)	p. 298 (4A-2)
Jin, Ning (GLOBALFOUNDRIES)	p. 61 (1B-4)
Jin, Song (North China Electric Power Univ.)	p. 720 (8A-3)
Jones, Alex K. (Univ. of Pittsburgh)	p. 67 (1C-1)
Jones, Alex K. (Univ. of Pittsburgh)	p. 355 (4C-1)
Jou, Jing-Yang (National Central Univ./National Chiao Tung Univ.)	p. 604 (7A-4)
Juan, Da-cheng (Carnegie Mellon Univ.)	p. 323 (4B-1)
Jun, Minhee (Carnegie Mellon Univ.)	p. 256 (3C-2)
Jung, Junwon (Korea Univ.)	p. 35 (1A-8)

K

Kagalwalla, Abde Ali (Univ. of California, Los Angeles)	p. 155 (2B-4)
Kamimura, Tatsuya (Tokyo Inst. of Tech.)	p. 23 (1A-2)
Kanazawa, Yuzi (Fujitsu Labs.)	p. 670 (7C-5)
Kang, Wang (Univ. Beihang, China/Univ. Paris-Sud)	p. 676 (8S-1)
Kanoun, Karim (École Polytechnique Fédérale de Lausanne)	p. 714 (8A-2)
Karnik, Tanay (Intel)	p. 388 (5S-2)
Karthik, Aadithya V. (Univ. of California, Berkeley)	p. 250 (3C-1)
Kasha, Ravi (Indian Inst. of Tech. - Madras)	p. 598 (7A-3)
Kashif, Hany (Univ. of Waterloo)	p. 113 (2A-1)
Kasper, Michael (Fraunhofer Institute for Secure Information Technology)	p. 780 (9S-1)
Kauer, Matthias (TUM CREATE)	p. 646 (7C-1)
Kauer, Matthias (TUM CREATE)	p. 812 (9A-3)
Kawaguchi, Hiroshi (Kobe Univ.)	p. 17 (1S-4)
Kawai, Hiroyuki (Renesas Electronics)	p. 12 (1S-3)
Keng, Brian (Univ. of Toronto)	p. 732 (8B-1)
Kim, Chulwoo (Korea Univ.)	p. 33 (1A-7)
Kim, Chulwoo (Korea Univ.)	p. 35 (1A-8)
Kim, Heejun (Korea Univ.)	p. 35 (1A-8)
Kim, Jae-Joon (POSTECH)	p. 179 (2C-4)
Kim, Jungmoon (Korea Univ.)	p. 33 (1A-7)

Kim, Jungmoon (Korea Univ.) p. 35 (1A-8)

Kim, Tae-Hwan (Korea Aerospace Univ.) p. 39 (1A-10)

Kim, Taemin (Intel Labs) p. 219 (3A-4)

Kim, Won-Tae (Korea Aerospace Univ.) p. 39 (1A-10)

Kim, Younghyun (Seoul National Univ.) p. 379 (4C-5)

Klein, Jacques-Olivier (Univ. Paris-Sud) p. 676 (8S-1)

Knechtel, Johann (Dresden Univ. of Tech.) p. 53 (1B-3)

Knoll, Alois (Technical Univ. Munich) p. 708 (8A-1)

Kobayashi, Hiroaki (Tokyo Univ. of Agri. and Tech.) p. 843 (9C-1)

Koh, Cheng-Kok (Purdue Univ.) p. 41 (1B-1)

Kohira, Yukihide (Univ. of Aizu) p. 173 (2C-3)

Kondo, Masaaki (Univ. of Electro-Communications) p. 843 (9C-1)

Kondo, Toshio (Mie Univ.) p. 400 (5A-2)

Konigsmark, Sven Tenzing Choden (Univ. of Illinois, Urbana-Champaign) p. 73 (1C-2)

Kopcsay, Gerard V. (IBM) p. 385 (5S-1)

Kosaka, Tsubasa (Shibaura Inst. of Tech.) p. 843 (9C-1)

Krämer, Juliane (Univ. Berlin) p. 780 (9S-1)

Krichmar, Jeffrey L. (UC Irvine) p. 570 (7S-3)

Krishnaiah, Gummidipudi (Intel Technology India Pvt.) p. 622 (7B-2)

Ku, Jerry C. Y. (National Chiao Tung Univ.) p. 664 (7C-4)

Kudo, Masaru (Shibaura Inst. of Tech.) p. 843 (9C-1)

Kuehlmann, Andreas (Coverity/Univ. of California, Berkeley) p. 738 (8B-2)

Kunimoto, Masaya (NAIST) p. 85 (1C-4)

Kuo, Hsien-Kai (National Chiao Tung Univ.) p. 604 (7A-4)

Kuroda, Tadahiro (Keio Univ.) p. 31 (1A-6)

L

Labios, Alvin (Canon Information Systems Research Australia (CiSRA)) p. 267 (4S-1)

Lai, Chien-Yu (National Taiwan Univ.) p. 744 (8B-3)

Lai, Chih-Yen (National Chiao Tung Univ.) p. 604 (7A-4)

Lai, Liangzhen (UCLA) p. 467 (6S-1)

Laing, Anthony (Univ. of Bristol) p. 795 (9S-3)

Lam, Michael (Mentor Graphics) p. 155 (2B-4)

Lansner, Anders (KTH) p. 556 (7S-1)

Lansner, Anders (KTH) p. 578 (7S-4)

Le, Bao (Univ. of Toronto) p. 732 (8B-1)

Lee, Gwang-Ho (Korea Aerospace Univ.) p. 39 (1A-10)

Lee, Hyung Gyu (Daegu Univ.) p. 379 (4C-5)

Lee, Sangyeop (Tokyo Inst. of Tech.) p. 23 (1A-2)

Lee, Yoojong (KAIST) p. 47 (1B-2)

Lee, Youngjoo (KAIST) p. 37 (1A-9)

Li, Boxun (Tsinghua Univ.) — p. 361 (4C-2)

Li, Hai (Univ. of Pittsburgh) — p. 831 (9B-3)

Li, Hai (Helen) (Univ. of Pittsburgh) — p. 610 (7A-5)

Li, Hong Wei (Nokia Research Centre) — p. 795 (9S-3)

Li, Min (Univ. of Wisconsin - Madison) — p. 640 (7B-5)

Li, Nan (Royal Inst. of Tech.) — p. 634 (7B-4)

Li, Shuai (Purdue Univ.) — p. 41 (1B-1)

Li, Xiaowei (Chinese Academy of Sciences) — p. 394 (5A-1)

Li, Xin (Carnegie Mellon Univ.) — p. 256 (3C-2)

Li, Yong (Univ. of Pittsburgh) — p. 67 (1C-1)

Li, Yong (Univ. of Pittsburgh) — p. 355 (4C-1)

Li, Zhiming (Guangzhou Institute of Advanced Technology, CAS) — p. 298 (4A-2)

Li, Zhuoyuan (Nimbus Automation Technologies) — p. 525 (6B-4)

Liang, Haichao (Kyushu Inst. of Tech.) — p. 185 (3S-1)

Liang, Yi (Univ. of Illinois, Urbana-Champaign) — p. 373 (4C-4)

Liang, Yuan (Tsinghua Univ.) — p. 849 (9C-2)

Lienig, Jens (Dresden Univ. of Tech.) — p. 53 (1B-3)

Liljeberg, Pasi (Univ. of Turku) — p. 349 (4B-5)

Lin, David (Stanford Univ.) — p. 478 (6S-3)

Lin, Louis Y. -Z. (National Chiao Tung Univ.) — p. 664 (7C-4)

Lin, Xue (Univ. of Southern California) — p. 167 (2C-2)

Lingamneni, Avinash (Rice Univ.) — p. 201 (3A-1)

Liu, Xiaoxiao (Univ. of Pittsburgh) — p. 355 (4C-1)

Liu, Yongpan (Tsinghua Univ.) — p. 379 (4C-5)

Liu, Zhenyu (Tsinghua Univ.) — p. 367 (4C-3)

Lobino, Mirko (Griffith Univ.) — p. 795 (9S-3)

Lu, Zhonghai (Royal Inst. of Tech.) — p. 343 (4B-4)

Lukasiewycz, Martin (TUM CREATE) — p. 646 (7C-1)

Lukasiewycz, Martin (TUM CREATE) — p. 812 (9A-3)

Luo, Rong (Tsinghua Univ.) — p. 379 (4C-5)

Lye, Aaron (Univ. of Bremen) — p. 489 (6A-2)

M

Ma, De (Hangzhou Dianzi Univ.) — p. 406 (5A-3)

Ma, Jun (Chinese Academy of Sciences) — p. 394 (5A-1)

Ma, Yue (Univ. of Notre Dame) — p. 317 (4A-5)

Mak, Terrence (Chinese Univ. of Hong Kong) — p. 298 (4A-2)

Mak, Wai-Kei (National Tsing Hua Univ.) — p. 137 (2B-1)

Mak, Wai-Kei (National Tsing Hua Univ.) — p. 501 (6A-4)

Mao, Mengjie (Univ. of Pittsburgh) — p. 67 (1C-1)

Marculescu, Diana (Carnegie Mellon Univ.) — p. 323 (4B-1)

Marculescu, Radu (Carnegie Mellon Univ.)   p. 323 (4B-1)

Mariani, Giovanni (Univ. della Svizzera Italiana - ALaRI, Switzerland/Politecnico di Milano)   p. 213 (3A-3)

Martin-Lopez, Enrique (Univ. of Bristol)   p. 795 (9S-3)

Maruyama, Yuki (Delft Univ. of Tech.)   p. 788 (9S-2)

Masahiko, Yoshimoto (Kobe Univ.)   p. 17 (1S-4)

Mastronarde, Nicholas (State Univ. of New York at Buffalo)   p. 714 (8A-2)

Masu, Kazuya (Tokyo Inst. of Tech.)   p. 23 (1A-2)

Matsukura, Fumihiro (Tohoku Univ.)   p. 684 (8S-2)

Matsunaga, Kensaku (Shibaura Inst. of Tech.)   p. 843 (9C-1)

Matsuzawa, Akira (Tokyo Inst. of Tech.)   p. 21 (1A-1)

Matsuzawa, Akira (Tokyo Inst. of Tech.)   p. 25 (1A-3)

Meeuws, Roel (Delft Technical Univ.)   p. 213 (3A-3)

Mehdipour, Farhad (Kyushu Univ.)   p. 292 (4A-1)

Meyer, Brett (McGill Univ.)   p. 537 (6C-2)

Mirkhani, Shahrzad (Univ. of Texas)   p. 390 (5S-3)

Mishchenko, Alan (Univ. of California, Berkeley)   p. 250 (3C-1)

Mitra, Subhasish (Stanford Univ.)   p. 390 (5S-3)

Mitra, Subhasish (Stanford Univ.)   p. 478 (6S-3)

Miwa, Shinobu (Univ. of Tokyo)   p. 1 (1S-1)

Miyahara, Masaya (Tokyo Inst. of Tech.)   p. 21 (1A-1)

Miyamori, Takashi (Toshiba)   p. 311 (4A-4)

Morie, Takashi (Kyushu Inst. of Tech.)   p. 185 (3S-1)

Morrow, Katherine (Univ. of Wisconsin - Madison)   p. 726 (8A-4)

Mukherjee, Tamal (Carnegie Mellon Univ.)   p. 256 (3C-2)

Muller, K. Paul (IBM)   p. 385 (5S-1)

Munns, Jack (Univ. of Bristol)   p. 795 (9S-3)

Musa, Ahmed (Tokyo Inst. of Tech.)   p. 21 (1A-1)

Musta, Thomas E. (IBM)   p. 385 (5S-1)

Mutyam, Madhu (Indian Inst. of Tech. - Madras)   p. 598 (7A-3)

N

Nakabayashi, Tomoyuki (Mie Univ.)   p. 400 (5A-2)

Nakada, Takashi (Univ. of Tokyo)   p. 1 (1S-1)

Nakamura, Hiroshi (Univ. of Tokyo)   p. 1 (1S-1)

Nakamura, Hiroshi (Univ. of Tokyo)   p. 843 (9C-1)

Nakashima, Yasuhiko (NAIST)   p. 85 (1C-4)

Nam, Gi-Joon (IBM)   p. 91 (2S-1)

Namiki, Mitaro (Tokyo Univ. of Agri. and Tech.)   p. 843 (9C-1)

Nassif, Sani (IBM)   p. 91 (2S-1)

Nazarian, Shahin (Univ. of Southern California)   p. 167 (2C-2)

Negi, Rohit (Carnegie Mellon Univ.)   p. 256 (3C-2)

Niemann, Philipp (Univ. of Bremen) p. 483 (6A-1)
Niskanen, Antti O. (Nokia Research Centre) p. 795 (9S-3)
Nitta, Izumi (Fujitsu Labs.) p. 670 (7C-5)
Niu, Dimin (Pennsylvania State Univ.) p. 762 (8C-2)
Niu, Dimin (Pennsylvania State Univ.) p. 825 (9B-2)
Nock, Richard W. (Univ. of Bristol) p. 795 (9S-3)
Noguchi, Hiroki (Toshiba) p. 6 (1S-2)
Nomura, Kumiko (Toshiba) p. 6 (1S-2)
Noori, Hamid (Ferdowsi Univ. of Mashhad) p. 292 (4A-1)
Nowick, Steven M. (Columbia Univ.) p. 329 (4B-2)
Nuzzo, Pierluigi (Univ. of California, Berkeley) p. 250 (3C-1)

O

O'Brien, Jeremy L. (Univ. of Bristol) p. 795 (9S-3)
Oboril, Fabian (KIT) p. 207 (3A-2)
Oboril, Fabian (Karlsruhe Inst. of Tech.) p. 700 (8S-4)
Ohno, Hideo (Tohoku Univ.) p. 684 (8S-2)
Okada, Kenichi (Tokyo Inst. of Tech.) p. 21 (1A-1)
Okada, Kenichi (Tokyo Inst. of Tech.) p. 25 (1A-3)
Ong, Zhong-Liang (National Univ. of Singapore) p. 610 (7A-5)

P

P. D., Sai Manoj (Nanyang Technological Univ.) p. 461 (5C-4)
Palem, Krishna (Rice Univ.) p. 201 (3A-1)
Palermo, Gianluca (Politecnico di Milano - DEIB) p. 213 (3A-3)
Palermo, Gianluca (Politecnico di Milano) p. 304 (4A-3)
Pan, Andrew (Univ. of California, Los Angeles) p. 818 (9B-1)
Pan, David Z. (Univ. of Texas, Austin) p. 143 (2B-2)
Pan, David Z. (Univ. of Texas, Austin) p. 513 (6B-2)
Pan, Gung-Yu (National Chiao Tung Univ.) p. 604 (7A-4)
Panda, Preeti Ranjan (Indian Inst. of Tech. Delhi) p. 622 (7B-2)
Parameswaran, Sri (Univ. of New South Wales) p. 267 (4S-1)
Parameswaran, Sri (Univ. of New South Wales) p. 412 (5A-4)
Park, In-Cheol (KAIST) p. 37 (1A-9)
Park, Sangyoung (Seoul National Univ.) p. 379 (4C-5)
Patel, Hiren D. (Univ. of Waterloo) p. 113 (2A-1)
Peddersen, Jorgen (Univ. of New South Wales) p. 267 (4S-1)
Peddersen, Jorgen (Univ. of New South Wales) p. 412 (5A-4)
Pedram, Massoud (Univ. of Southern California) p. 167 (2C-2)

Pedram, Massoud (Univ. of Southern California) — p. 495 (6A-3)

Pei, Songwei (Beijing Univ. of Chemical Tech.) — p. 720 (8A-3)

Pekmestzi, Kiamal (National Technical Univ. of Athens) — p. 652 (7C-2)

Pendina, Gregory Di (Spintec Laboratory, CEA-INAC/CNRS/UJF/G-INP) — p. 692 (8S-3)

Peng, Zebo (Linköping Univ.) — p. 436 (5B-4)

Pilania, Arun Kumar (Intel Technology India Pvt.) — p. 622 (7B-2)

Pileggi, Lawrence (Carnegie Mellon Univ.) — p. 256 (3C-2)

Plosila, Juha (Univ. of Turku) — p. 349 (4B-5)

Poremba, Matt (Pennsylvania State Univ.) — p. 586 (7A-1)

Prenat, Guillaume (Spintec Laboratory, CEA-INAC/CNRS/UJF/G-INP) — p. 692 (8S-3)

Q

Qi, Nan (Tsinghua Univ.) — p. 29 (1A-5)

Qi, Zhongdong (Tsinghua Univ.) — p. 525 (6B-4)

Qian, Haifeng (IBM) — p. 849 (9C-2)

Qian, Zhiliang (Hong Kong Univ. of Science and Tech.) — p. 323 (4B-1)

Qin, Evean (Vennsa Technologies) — p. 732 (8B-1)

Qiu, Qinru (Syracuse Univ.) — p. 831 (9B-3)

R

Rajagopalan, Subramanian (Synopsys India Pvt.) — p. 149 (2B-3)

Rarity, John G. (Univ. of Bristol) — p. 795 (9S-3)

Rath, Alexander Wolfgang (Infineon Technologies AG) — p. 806 (9A-2)

Ravelosona, Dafiné (Univ. Paris-Sud) — p. 676 (8S-1)

Ray, Sayak (Univ. of California, Berkeley) — p. 250 (3C-1)

Regazzoni, Francesco (ALaRI - USI) — p. 788 (9S-2)

Rosenstiel, Wolfgang (FZI Research Center for Information Technology/Univ. of Tübingen) — p. 800 (9A-1)

Rotenberg, Eric (North Carolina State Univ.) — p. 400 (5A-2)

Roychowdhury, Jaijeet (Univ. of California, Berkeley) — p. 250 (3C-1)

Ryoo, Jihyun (Seoul National Univ.) — p. 285 (4S-4)

S

Saeedi, Mehdi (Univ. of Southern California) — p. 495 (6A-3)

Sakamoto, Ryuichi (Tokyo Univ. of Agri. and Tech.) — p. 843 (9C-1)

Salami, Bagher (Ferdowsi Univ. of Mashhad) — p. 292 (4A-1)

Samukawa, Seiji (Tohoku Univ.) — p. 185 (3S-1)

Sankaranarayanan, Sriram (Univ. of Colorado, Boulder) — p. 449 (5C-2)

Sano, Toru (Toshiba)	p. 311 (4A-4)
Sapatnekar, Sachin (Univ. of Minnesota)	p. 430 (5B-3)
Sarangi, Smruti R. (IIT Delhi)	p. 855 (9C-3)
Sarhan, Hossam (CEA-LETI)	p. 79 (1C-3)
Sasaki, Takahiro (Mie Univ.)	p. 400 (5A-2)
Sato, Hideo (Tohoku Univ.)	p. 684 (8S-2)
Sato, Yohei (Renesas Electronics)	p. 12 (1S-3)
Satterfield, David L. (IBM)	p. 385 (5S-1)
Scheffer, Louis (Howard Hughes Medical Institute)	p. 197 (3S-3)
Schneider, Josef (Univ. of New South Wales)	p. 412 (5A-4)
Schneider, Reinhard (TU Munich)	p. 119 (2A-2)
Schneider, Reinhard (TU Munich)	p. 812 (9A-3)
Seifert, Jean-Pierre (Univ. Berlin)	p. 780 (9S-1)
Sekimoto, Ryota (Keio Univ.)	p. 31 (1A-6)
Sekine, Tadatoshi (Shizuoka Univ.)	p. 774 (8C-4)
Senger, Robert M. (IBM)	p. 385 (5S-1)
Sengupta, Deepashree (Univ. of Minnesota)	p. 430 (5B-3)
Sha, Edwin (Chongqing Univ.)	p. 131 (2A-4)
Shafaei, Alireza (Univ. of Southern California)	p. 495 (6A-3)
Shafique, Muhammad (Karlsruhe Inst. of Tech.)	p. 274 (4S-2)
Shah, Hardik (Technical Univ. Munich)	p. 708 (8A-1)
Shao, Zili (Hong Kong Polytechnic Univ.)	p. 592 (7A-2)
Sharma, Namita (Indian Inst. of Tech. Delhi)	p. 622 (7B-2)
Shi, Guoyong (Shanghai Jiao Tong Univ.)	p. 443 (5C-1)
Shi, Yiyu (Missouri Univ. of Science and Tech.)	p. 837 (9B-4)
Shikata, Akira (Keio Univ.)	p. 31 (1A-6)
Shim, Minseob (Korea Univ.)	p. 35 (1A-8)
Shim, Seongbo (KAIST)	p. 47 (1B-2)
Shimizu, Toru (Renesas Electronics)	p. 12 (1S-3)
Shin, Insup (KAIST)	p. 179 (2C-4)
Shin, Youngsoo (KAIST)	p. 47 (1B-2)
Shin, Youngsoo (KAIST)	p. 179 (2C-4)
Silvano, Cristina (Politecnico di Milano - DEIB)	p. 213 (3A-3)
Silvano, Cristina (Politecnico di Milano)	p. 304 (4A-3)
Sima, Vlad-Mihai (Delft Technical Univ.)	p. 213 (3A-3)
Siriburanon, Teerachot (Tokyo Inst. of Tech.)	p. 21 (1A-1)
Siriburanon, Teerachot (Tokyo Inst. of Tech.)	p. 25 (1A-3)
Skadron, Kevin (Univ. of Virginia)	p. 537 (6C-2)
Somenzi, Fabio (Univ. of Colorado, Boulder)	p. 449 (5C-2)
Song, Yang (Huawei Technologies)	p. 367 (4C-3)
Song, Yang (Nanyang Technological Univ.)	p. 461 (5C-4)
Song, Zheng (Tsinghua Univ.)	p. 29 (1A-5)
Stamelakos, Ioannis (Politecnico di Milano)	p. 304 (4A-3)

Stan, Mircea (Univ. of Virginia) — p. 537 (6C-2)
Steinhorst, Sebastian (TUM CREATE) — p. 812 (9A-3)
Stensgaard, Mikkel (Teklatech) — p. 337 (4B-3)
Stoimenov, Nikolay (ETH Zurich) — p. 125 (2A-3)
Stucki, Damien (ID Quantique) — p. 788 (9S-2)
Subramoney, Sreenivas (Intel Technology India Pvt.) — p. 622 (7B-2)
Sugavanam, Krishnan (IBM) — p. 385 (5S-1)
Sugawara, Yutaka (IBM) — p. 385 (5S-1)
Sugiyama, Tomoyuki (Mie Univ.) — p. 400 (5A-2)
Sun, Guangyu (Peking Univ.) — p. 67 (1C-1)
Sun, Yilai (Kyushu Inst. of Tech.) — p. 185 (3S-1)
Sun, Zhenyu (Univ. of Pittsburgh) — p. 610 (7A-5)
Sung Kim, Nam (Univ. of Wisconsin - Madison) — p. 726 (8A-4)
Svensson, Christer (Linköping Univ.) — p. 578 (7S-4)

T

Tahoori, Mehdi (KIT) — p. 207 (3A-2)
Tahoori, Mehdi (Karlsruhe Inst. of Tech.) — p. 700 (8S-4)
Takahashi, Atsushi (Tokyo Inst. of Tech.) — p. 173 (2C-3)
Takasaki, Takahiro (Shizuoka Univ.) — p. 774 (8C-4)
Takeda, Susumu (Toshiba) — p. 6 (1S-2)
Takenaka, Takashi (NEC) — p. 282 (4S-3)
Tan, Sheldon (Univ. of California, Riverside) — p. 455 (5C-3)
Tanabe, Jun (Toshiba) — p. 311 (4A-4)
Tang, Puying (Univ. of Electronic Science and Tech. of China) — p. 455 (5C-3)
Tao, Jun (Carnegie Mellon Univ.) — p. 256 (3C-2)
Tatenguem Fankem, Herve (Univ. of Ferrara) — p. 337 (4B-3)
Temam, Olivier (INRIA) — p. 201 (3A-1)
Temam, Olivier (INRIA) — p. 563 (7S-2)
Tenhunen, Hannu (Univ. of Turku) — p. 349 (4B-5)
Thiele, Lothar (ETH Zurich) — p. 125 (2A-3)
Thompson, Mark G. (Univ. of Bristol) — p. 795 (9S-3)
Thuries, Sebastien (CEA-LETI) — p. 79 (1C-3)
Tida, Umamaheswara Rao (Missouri Univ. of Science and Tech.) — p. 837 (9B-4)
Tim, Yenni (National Univ. of Singapore) — p. 610 (7A-5)
Ting, Hui-Ling (National Tsing Hua Univ.) — p. 658 (7C-3)
Todri-Sanial, Aida (CNRS/LIRMM) — p. 544 (6C-3)
Tohara, Takashi (Kyushu Inst. of Tech.) — p. 185 (3S-1)
Tsai, Tu-Hsiung (National Chiao Tung Univ.) — p. 519 (6B-3)
Tschanz, James (Intel) — p. 388 (5S-2)
Tsoumanis, Kostas (National Technical Univ. of Athens) — p. 652 (7C-2)

Tsui, Chi-Ying (Hong Kong Univ. of Science and Tech.)   p. 323 (4B-1)

Tsurui, Yoshihiro (Shibaura Inst. of Tech.)   p. 843 (9C-1)

U

Ueki, Hiroshi (Renesas Electronics)   p. 12 (1S-3)

Ukhov, Ivan (Linköping Univ.)   p. 436 (5B-4)

Usami, Kimiyoshi (Shibaura Inst. of Tech.)   p. 843 (9C-1)

Usui, Hiroyuki (Toshiba)   p. 311 (4A-4)

V

Valentian, Alexandre (CEA-LETI)   p. 563 (7S-2)

van der Schaar, Mihaela (Univ. of California, Los Angeles)   p. 714 (8A-2)

Vangal, Sriram (Intel)   p. 388 (5S-2)

Velev, Miroslav (Aries Design Automation)   p. 750 (8B-4)

Veneris, Andreas (Univ. of Toronto)   p. 732 (8B-1)

Venkata Kalyan, T. (Indian Inst. of Tech. - Madras)   p. 598 (7A-3)

Viehl, Alexander (FZI Research Center for Information Technology)   p. 800 (9A-1)

Villani, Mattias (Linköping Univ.)   p. 436 (5B-4)

Virazel, Arnaud (Univ. of Montpellier/LIRMM)   p. 544 (6C-3)

W

Wabnig, Joachim (Nokia Research Centre)   p. 795 (9S-3)

Wakabayashi, Kazutoshi (NEC)   p. 282 (4S-3)

Wang, Cong (Tsinghua Univ.)   p. 379 (4C-5)

Wang, Danghui (Northwestern Polytechnical Univ.)   p. 592 (7A-2)

Wang, Dongsheng (Tsinghua Univ.)   p. 367 (4C-3)

Wang, Jianxing (National Univ. of Singapore)   p. 610 (7A-5)

Wang, Ke (Univ. of Virginia)   p. 537 (6C-2)

Wang, Shaodi (Univ. of California, Los Angeles)   p. 818 (9B-1)

Wang, Ting-Chi (National Tsing Hua Univ.)   p. 418 (5B-1)

Wang, Weihan (Keio Univ.)   p. 843 (9C-1)

Wang, Xiaohang (Guangzhou Institute of Advanced Technology, CAS)   p. 298 (4A-2)

Wang, Yanzhi (Univ. of Southern California)   p. 167 (2C-2)

Wang, Ying-Chih (Carnegie Mellon Univ.)   p. 256 (3C-2)

Wang, Yu (Tsinghua Univ.)   p. 161 (2C-1)

Wang, Yu (Tsinghua Univ.) — p. 361 (4C-2)

Wang, Yu (Tsinghua Univ.) — p. 831 (9B-3)

Wang, Yuhao (Nanyang Technological Univ.) — p. 191 (3S-2)

Wang, Yuzhi (Tsinghua Univ.) — p. 361 (4C-2)

Wang, Zhaohao (Univ. Paris-Sud) — p. 676 (8S-1)

Wang, Zhihua (Tsinghua Univ.) — p. 29 (1A-5)

Waszecki, Peter (TUM CREATE) — p. 646 (7C-1)

Wei, Zhulin (Huawei Shannon Laboratory) — p. 191 (3S-2)

Welp, Tobias (Univ. of California, Berkeley) — p. 738 (8B-2)

Wen, Charles H.-P. (National Chiao Tung Univ.) — p. 664 (7C-4)

Weng, Chuliang (Huawei Shannon Laboratory) — p. 191 (3S-2)

Wille, Robert (Univ. of Bremen/Cyber Physical Systems DFKI GmbH/Technical Univ. Dresden) — p. 483 (6A-1)

Wille, Robert (Univ. of Bremen/Cyber Physical Systems DFKI GmbH/Technical Univ. Dresden) — p. 489 (6A-2)

Wong, Martin D. F. (Univ. of Illinois, Urbana-Champaign) — p. 73 (1C-2)

Wong, Martin D.F. (Univ. of Illinois, Urbana-Champaign) — p. 531 (6C-1)

Wong, Ngai (Univ. of Hong Kong) — p. 262 (3C-3)

Wong, Weng-Fai (National Univ. of Singapore) — p. 610 (7A-5)

Wu, Cheng-Yin (National Taiwan Univ.) — p. 744 (8B-3)

Wu, Chengyong (Chinese Academy of Sciences) — p. 201 (3A-1)

Wu, Kaijie (Chongqing Univ.) — p. 131 (2A-4)

Wu, Po-Hsun (National Cheng Kung Univ.) — p. 507 (6B-1)

Wu, Sheng-Kai (National Tsing Hua Univ.) — p. 501 (6A-4)

Wu, Wei (Univ. of California, Los Angeles) — p. 424 (5B-2)

X

Xie, Yuan (AMD, China/Pennsylvania State Univ.) — p. 550 (6C-4)

Xie, Yuan (AMD, China/Pennsylvania State Univ.) — p. 586 (7A-1)

Xie, Yuan (AMD, China/Pennsylvania State Univ.) — p. 762 (8C-2)

Xie, Yuan (AMD, China/Pennsylvania State Univ.) — p. 825 (9B-2)

Xiong, Yong Zhong (A*STAR) — p. 27 (1A-4)

Xiu, Siwen (Zhejiang Univ.) — p. 406 (5A-3)

Xu, Cong (Pennsylvania State Univ.) — p. 825 (9B-2)

Xu, Hui (Toshiba) — p. 311 (4A-4)

Xu, Yi (Macau Univ. of Science and Tech., Macau/AMD) — p. 550 (6C-4)

Xu, Yi (AMD Research) — p. 586 (7A-1)

Xydis, Sotirios (Politecnico di Milano) — p. 304 (4A-3)

Y

Yachide, Yusuke (Canon Information Systems Research Australia (CiSRA)) p. 267 (4S-1)

Yamanouchi, Michihiko (Tohoku Univ.) p. 684 (8S-2)

Yamashita, Shigeru (Ritsumeikan Univ.) p. 225 (3B-1)

Yan, Guihai (Chinese Academy of Sciences) p. 394 (5A-1)

Yan, Rongjie (Chinese Academy of Sciences) p. 406 (5A-3)

Yang, Huazhong (Tsinghua Univ.) p. 161 (2C-1)

Yang, Huazhong (Tsinghua Univ.) p. 361 (4C-2)

Yang, Huazhong (Tsinghua Univ.) p. 379 (4C-5)

Yang, Jianlei (Tsinghua Univ.) p. 861 (9C-4)

Yang, Mei (Univ. of Nevada, Las Vegas) p. 298 (4A-2)

Yao, Yuan (Royal Inst. of Tech.) p. 343 (4B-4)

Ye, Zuochang (Tsinghua Univ.) p. 768 (8C-3)

Yeung, Jackson Ho Chuen (Chinese Univ. of Hong Kong) p. 238 (3B-3)

Yoo, Hoyoung (KAIST) p. 37 (1A-9)

Yoshioka, Kentaro (Keio Univ.) p. 31 (1A-6)

Young, Evangeline F. Y. (Chinese Univ. of Hong Kong) p. 53 (1B-3)

Young, Evangeline F.Y. (Chinese Univ. of Hong Kong) p. 238 (3B-3)

Yu, Bei (Univ. of Texas, Austin) p. 143 (2B-2)

Yu, Hao (Nanyang Technological Univ.) p. 191 (3S-2)

Yu, Hao (Nanyang Technological Univ.) p. 461 (5C-4)

Yu, Shimeng (Arizona State Univ.) p. 825 (9B-2)

Yu, Tan (Univ. of California, Riverside) p. 455 (5C-3)

Yu, Ting (Univ. of Illinois, Urbana-Champaign) p. 531 (6C-1)

Yu, Wenjian (Tsinghua Univ.) p. 756 (8C-1)

Yu, Wenjian (Tsinghua Univ.) p. 849 (9C-2)

Z

Zervakis, Georgios (National Technical Univ. of Athens) p. 652 (7C-2)

Zhan, Jia (Pennsylvania State Univ.) p. 586 (7A-1)

Zhang, Chao (Tsinghua Univ.) p. 756 (8C-1)

Zhang, Jun (Chongqing Univ.) p. 131 (2A-4)

Zhang, Licong (TU Munich) p. 119 (2A-2)

Zhang, Moning (Tsinghua Univ.) p. 768 (8C-3)

Zhang, Pei (Xi'an Jiaotong Univ.) p. 795 (9S-3)

Zhang, Runjie (Univ. of Virginia) p. 537 (6C-2)

Zhang, Xia (Univ. of Electronic Science and Tech. of China) p. 317 (4A-5)

Zhang, Xiaoxu (Zhejiang Univ.) p. 406 (5A-3)

Zhang, Yan (Univ. of Colorado, Boulder) p. 449 (5C-2)

Zhang, Yaojun (Univ. of Pittsburgh) p. 355 (4C-1)

Zhang, Yilin (Univ. of Texas, Austin)   p. 513 (6B-2)

Zhang, Youguang (Univ. Beihang)   p. 676 (8S-1)

Zhang, Yue (Univ. Paris-Sud)   p. 676 (8S-1)

Zhao, Junfeng (Huawei Shannon Laboratory)   p. 191 (3S-2)

Zhao, Wei (Tsinghua Univ.)   p. 861 (9C-4)

Zhao, Weisheng (Univ. Paris-Sud)   p. 676 (8S-1)

Zhao, Wenhui (Univ. of Hong Kong)   p. 262 (3C-3)

Zhnag, M. S. (National Univ. of Singapore)   p. 27 (1A-4)

Zhou, Qiang (Tsinghua Univ.)   p. 525 (6B-4)

Zhu, Jia (Tsinghua Univ.)   p. 367 (4C-3)

Zhuo, Cheng (Intel Research)   p. 837 (9B-4)

Zou, Qiaosha (Pennsylvania State Univ.)   p. 762 (8C-2)